Franz Josef Land
Spitsbergen
Novaya Zemlya
KARA SEA
Bear I.
BARENTS SEA
Jan Mayen
WHITE SEA
Pechora
Ob'
RUSSIAN FEDERATION (RUSSIA)
Irtysh
Faeroes
NORWAY
SWEDEN
FINLAND
Lake Onega
Lake Ladoga
Volga
ESTONIA
LATVIA
LITHUANIA
Kaliningrad
DENMARK
BRITAIN
NETHERLANDS
BELGIUM
LUX.
GERMANY
Elbe
Vistula
POLAND
Odra
BELARUS'
UKRAINE
Dnepr
Dnestr
CZECH REPUBLIC
SLOVAKIA
Danube
Rhine
Seine
FRANCE
SWITZ.
AUSTRIA
HUNGARY
MOLDOVA
SLOVENIA
CROATIA
Sava
BOSNIA HERZEGOVINA
SERBIA
RUMANIA
Po
Rhône
ITALY
Corsica
Sardinia
Sicily
MALTA
BULGARIA
MAKEDONIJA
ALBANIA
GREECE
Crete
BLACK SEA
Don
Volga
Ural
KAZAKHSTAN
ARAL SEA
UZBEKISTAN
KYRGYZSTAN
TAJIKISTAN
TURKMENISTAN
CASPIAN SEA
GEORGIA
AZERBAIJAN
ARMENIA
TURKEY
AFGHANISTAN
IRAN
SYRIA
Tigris
IRAQ
Euphrates
KUWAIT
CYPRUS
LEBANON
ISRAEL
JORDAN
BAHRAIN
QATAR
U.A.E
SAUDI ARABIA
OMAN
YEMEN REPUBLIC
RED SEA
MEDITERRANEAN
TUNISIA
LIBYA
EGYPT
Nile
CHAD
TIBESTI
SUDAN
NIGER

The Birds of the
Western Palearctic

Concise Edition Volume 2

The Birds of the Western Palearctic

Concise Edition
based on *The Handbook of the Birds of Europe, the Middle East, and North Africa*

Volume 2
Passerines

D. W. Snow and C. M. Perrins

Robert Gillmor
Brian Hillcoat
C. S. Roselaar
Dorothy Vincent
D. I. M. Wallace
M. G. Wilson

Oxford   New York
OXFORD UNIVERSITY PRESS
1998

Oxford University Press, Great Clarendon Street, Oxford OX2 6DP
Oxford New York
Athens Auckland Bangkok Bogota Bombay
Buenos Aires Calcutta Cape Town Dar es Salaam
Delhi Florence Hong Kong Istanbul Karachi
Kuala Lumpur Madras Madrid Melbourne
Mexico City Nairobi Paris Singapore
Taipei Tokyo Toronto Warsaw
and associated companies in
Berlin Ibadan

Oxford is a trade mark of Oxford University Press

Published in the United States
by Oxford University Press Inc., New York

© Oxford University Press, 1998

All rights reserved. No part of this publication may be reproduced, stored in a retrieval system, or transmitted, in any form or by any means, without the prior permission in writing of Oxford University Press. Within the UK, exceptions are allowed in respect of any fair dealing for the purpose of research or private study, or criticism or review, as permitted under the Copyright, Designs and Patents Act, 1988, or in the case of reprographic reproduction in accordance with the terms of licences issued by the Copyright Licensing Agency. Enquiries concerning reproduction outside those terms and in other countries should be sent to the Rights Department, Oxford University Press, at the address above.

This book is sold subject to the condition that it shall not, by way of trade or otherwise, be lent, re-sold, hired out, or otherwise circulated without the publisher's prior consent in any form of binding or cover other than that in which it is published and without a similar condition including this condition being imposed on the subsequent purchaser.

A catalogue record for this book is available from the British Library

Library of Congress Cataloging in Publication Data
(Data available)

ISBN 0 19 850187 0 (Volume 1)
ISBN 0 19 850188 9 (Volume 2)
ISBN 0 19 854099 X (Set)
(Available only as two volume set)

Typeset by Latimer Trend & Company Ltd., Plymouth
Printed in China

Dedicated to the memory of

Stanley Cramp (1913–87)
Editor of *Birds of the Western Palearctic* 1977–87

and

William Wilkinson (1932–96)
Chairman of West Palearctic Birds Ltd 1987–96

CONTENTS

Order **PASSERIFORMES**

Tyrant Flycatchers Family Tyrannidae — 1009
  Acadian Flycatcher *Empidonax virescens* — 1009
  Eastern Phoebe *Sayornis phoebe* — 1010

Larks Family Alaudidae — 1011
  Kordofan Bush-lark *Mirafra cordofanica* — 1011
  Chestnut-headed Finch Lark *Eremopterix signata* — 1012
  Black-crowned Finch Lark *Eremopterix nigriceps* — 1013
  Dunn's Lark *Eremalauda dunni* — 1014
  Bar-tailed Desert Lark *Ammomanes cincturus* — 1015
  Desert Lark *Ammomanes deserti* — 1017
  Hoopoe Lark *Alaemon alaudipes* — 1019
  Dupont's Lark *Chersophilus duponti* — 1020
  Thick-billed Lark *Rhamphocoris clotbey* — 1022
  Calandra Lark *Melanocorypha calandra* — 1024
  Bimaculated Lark *Melanocorypha bimaculata* — 1026
  White-winged Lark *Melanocorypha leucoptera* — 1027
  Black Lark *Melanocorypha yeltoniensis* — 1029
  Short-toed Lark *Calandrella brachydactyla* — 1032
  Hume's Lark *Calandrella acutirostris* — 1034
  Lesser Short-toed Lark *Calandrella rufescens* — 1034
  Asian Short-toed Lark *Calandrella cheleënsis* — 1036
  Crested Lark *Galerida cristata* — 1037
  Thekla Lark *Galerida theklae* — 1040
  Woodlark *Lullula arborea* — 1041
  Skylark *Alauda arvensis* — 1043
  Small Skylark *Alauda gulgula* — 1046
  Razo Lark *Alauda razae* — 1047
  Shore Lark *Eremophila alpestris* — 1047
  Temminck's Horned Lark *Eremophila bilopha* — 1052

Swallows Family Hirundinidae — 1053
  Tree Swallow *Tachycineta bicolor* — 1053
  Brown-throated Sand Martin *Riparia paludicola* — 1053
  Sand Martin *Riparia riparia* — 1055
  Banded Martin *Riparia cincta* — 1058
  African Rock Martin (includes Pale Crag Martin) *Hirundo fuligula* — 1058
  Crag Martin *Hirundo rupestris* — 1059
  Swallow *Hirundo rustica* — 1061
  Ethiopian Swallow *Hirundo aethiopica* — 1064
  Red-rumped Swallow *Hirundo daurica* — 1064
  Cliff Swallow *Hirundo pyrrhonota* — 1066
  House Martin *Delichon urbica* — 1066

Pipits, Wagtails Family Motacillidae — 1070
  Richard's Pipit *Anthus novaeseelandiae* — 1070
  Blyth's Pipit *Anthus godlewskii* — 1072
  Tawny Pipit *Anthus campestris* — 1072
  Berthelot's Pipit *Anthus berthelotii* — 1075

Long-billed Pipit *Anthus similis* — 1076
Olive-backed Pipit *Anthus hodgsoni* — 1077
Tree Pipit *Anthus trivialis* — 1079
Pechora Pipit *Anthus gustavi* — 1081
Meadow Pipit *Anthus pratensis* — 1082
Red-throated Pipit *Anthus cervinus* — 1084
Water Pipit *Anthus spinoletta* — 1086
Rock Pipit *Anthus petrosus* — 1089
Buff-bellied Pipit *Anthus rubescens* — 1092
Yellow Wagtail *Motacilla flava* — 1094
Citrine Wagtail *Motacilla citreola* — 1098
Grey Wagtail *Motacilla cinerea* — 1100
Pied Wagtail/White Wagtail *Motacilla alba* — 1103
African Pied Wagtail *Motacilla aguimp* — 1106

Bulbuls Family Pycnonotidae — 1109

White-cheeked Bulbul *Pycnonotus leucogenys* — 1109
Yellow-vented Bulbul *Pycnonotus xanthopygos* — 1110
Common Bulbul *Pycnonotus barbatus* — 1111
Red-vented Bulbul *Pycnonotus cafer* — 1112

Waxwings, Hypocolius Family Bombycillidae — 1113

Waxwings Subfamily Bombycillinae — 1113

Waxwing *Bombycilla garrulus* — 1113
Cedar Waxwing *Bombycilla cedrorum* — 1115

Hypocolius Subfamily Hypocoliinae — 1116

Gray Hypocolius *Hypocolius ampelinus* — 1116

Dippers Family Cinclidae — 1118

Dipper *Cinclus cinclus* — 1118

Wrens Family Troglodytidae — 1122

Wren *Troglodytes troglodytes* — 1122

Mockingbirds Family Mimidae — 1125

Northern Mockingbird *Mimus polyglottos* — 1125
Brown Thrasher *Toxostoma rufum* — 1126
Gray Catbird *Dumetella carolinensis* — 1127

Accentors Family Prunellidae — 1128

Dunnock *Prunella modularis* — 1128
Siberian Accentor *Prunella montanella* — 1131
Radde's Accentor *Prunella ocularis* — 1132
Black-throated Accentor *Prunella atrogularis* — 1133
Alpine Accentor *Prunella collaris* — 1134

Chats, Thrushes Family Turdidae — 1137

Rufous Bush Robin *Cercotrichas galactotes* — 1137
Black Bush Robin *Cercotrichas podobe* — 1139
Robin *Erithacus rubecula* — 1140
Thrush Nightingale *Luscinia luscinia* — 1143
Nightingale *Luscinia megarhynchos* — 1145

Siberian Rubythroat *Luscinia calliope*	1147
Bluethroat *Luscinia svecica*	1149
Siberian Blue Robin *Luscinia cyane*	1152
Red-flanked Bluetail *Tarsiger cyanurus*	1153
White-throated Robin *Irania gutturalis*	1154
Eversmann's Redstart *Phoenicurus erythronotus*	1155
Black Redstart *Phoenicurus ochruros*	1157
Redstart *Phoenicurus phoenicurus*	1161
Moussier's Redstart *Phoenicurus moussieri*	1163
Güldenstädt's Redstart *Phoenicurus erythrogaster*	1164
Blackstart *Cercomela melanura*	1166
Whinchat *Saxicola rubetra*	1167
Canary Islands Stonechat *Saxicola dacotiae*	1169
Stonechat *Saxicola torquata*	1170
Pied Stonechat *Saxicola caprata*	1173
Ant Chat *Myrmecocichla aethiops*	1174
Isabelline Wheatear *Oenanthe isabellina*	1175
Wheatear *Oenanthe oenanthe*	1178
Pied Wheatear *Oenanthe pleschanka*	1180
Cyprus Wheatear *Oenanthe cypriaca*	1183
Black-eared Wheatear *Oenanthe hispanica*	1183
Desert Wheatear *Oenanthe deserti*	1185
Finsch's Wheatear *Oenanthe finschii*	1187
Red-rumped Wheatear *Oenanthe moesta*	1188
Red-tailed Wheatear *Oenanthe xanthoprymna*	1190
Eastern Pied Wheatear *Oenanthe picata*	1191
Mourning Wheatear *Oenanthe lugens*	1193
Hooded Wheatear *Oenanthe monacha*	1194
Hume's Wheatear *Oenanthe alboniger*	1195
White-crowned Black Wheatear *Oenanthe leucopyga*	1196
Black Wheatear *Oenanthe leucura*	1198
Rock Thrush *Monticola saxatilis*	1201
Blue Rock Thrush *Monticola solitarius*	1204
White's Thrush *Zoothera dauma*	1206
Siberian Thrush *Zoothera sibirica*	1207
Varied Thrush *Zoothera naevia*	1208
Wood Thrush *Hylocichla mustelina*	1209
Hermit Thrush *Catharus guttatus*	1210
Swainson's Thrush *Catharus ustulatus*	1210
Gray-cheeked Thrush *Catharus minimus*	1211
Veery *Catharus fuscescens*	1211
Tickell's Thrush *Turdus unicolor*	1212
Ring Ouzel *Turdus torquatus*	1212
Blackbird *Turdus merula*	1215
Eye-browed Thrush *Turdus obscurus*	1218
Naumann's/Dusky Thrush *Turdus naumanni*	1219
Red-throated/Black-throated Thrush *Turdus ruficollis*	1220
Fieldfare *Turdus pilaris*	1222
Song Thrush *Turdus philomelos*	1225
Redwing *Turdus iliacus*	1228
Mistle Thrush *Turdus viscivorus*	1230
American Robin *Turdus migratorius*	1234

Old World Warblers and Allies Family Sylviidae ... 1235

- Cetti's Warbler *Cettia cetti* ... 1235
- Fan-tailed Warbler *Cisticola juncidis* ... 1237
- Graceful Warbler *Prinia gracilis* ... 1239
- Cricket Warbler *Spiloptila clamans* ... 1241
- Scrub Warbler *Scotocerca inquieta* ... 1241
- Pallas's Grasshopper Warbler *Locustella certhiola* ... 1243
- Lanceolated Warbler *Locustella lanceolata* ... 1244
- Grasshopper Warbler *Locustella naevia* ... 1245
- River Warbler *Locustella fluviatilis* ... 1247
- Savi's Warbler *Locustella luscinioides* ... 1249
- Gray's Grasshopper Warbler *Locustella fasciolata* ... 1251
- Moustached Warbler *Acrocephalus melanopogon* ... 1251
- Aquatic Warbler *Acrocephalus paludicola* ... 1254
- Sedge Warbler *Acrocephalus schoenobaenus* ... 1255
- Paddyfield Warbler *Acrocephalus agricola* ... 1258
- Blyth's Reed Warbler *Acrocephalus dumetorum* ... 1260
- Cape Verde Cane Warbler *Acrocephalus brevipennis* ... 1262
- Marsh Warbler *Acrocephalus palustris* ... 1263
- Reed Warbler *Acrocephalus scirpaceus* ... 1265
- Clamorous Reed Warbler *Acrocephalus stentoreus* ... 1267
- Great Reed Warbler *Acrocephalus arundinaceus* ... 1269
- Basra Reed Warbler *Acrocephalus griseldis* ... 1271
- Oriental Reed Warbler *Acrocephalus orientalis* ... 1272
- Thick-billed Warbler *Acrocephalus aedon* ... 1273
- Olivaceous Warbler *Hippolais pallida* ... 1273
- Booted Warbler *Hippolais caligata* ... 1277
- Upcher's Warbler *Hippolais languida* ... 1278
- Olive-tree Warbler *Hippolais olivetorum* ... 1280
- Icterine Warbler *Hippolais icterina* ... 1282
- Melodious Warbler *Hippolais polyglotta* ... 1284
- Marmora's Warbler *Sylvia sarda* ... 1286
- Dartford Warbler *Sylvia undata* ... 1288
- Tristram's Warbler *Sylvia deserticola* ... 1290
- Spectacled Warbler *Sylvia conspicillata* ... 1291
- Subalpine Warbler *Sylvia cantillans* ... 1293
- Ménétries's Warbler *Sylvia mystacea* ... 1295
- Sardinian Warbler *Sylvia melanocephala* ... 1296
- Cyprus Warbler *Sylvia melanothorax* ... 1299
- Rüppell's Warbler *Sylvia rueppelli* ... 1300
- Desert Warbler *Sylvia nana* ... 1301
- Arabian Warbler *Sylvia leucomelaena* ... 1303
- Orphean Warbler *Sylvia hortensis* ... 1305
- Barred Warbler *Sylvia nisoria* ... 1306
- Lesser Whitethroat *Sylvia curruca* ... 1308
- Whitethroat *Sylvia communis* ... 1310
- Garden Warbler *Sylvia borin* ... 1314
- Blackcap *Sylvia atricapilla* ... 1316
- Eastern Crowned Leaf Warbler *Phylloscopus coronatus* ... 1319
- Greenish Warbler *Phylloscopus trochiloides* ... 1320
- Arctic Warbler *Phylloscopus borealis* ... 1322

Pallas's Warbler *Phylloscopus proregulus*	1324
Brooks's Leaf Warbler *Phylloscopus subviridis*	1325
Yellow-browed Warbler *Phylloscopus inornatus*	1325
Hume's Leaf Warbler *Phylloscopus humei*	1327
Radde's Warbler *Phylloscopus schwarzi*	1327
Dusky Warbler *Phylloscopus fuscatus*	1328
Bonelli's Warbler *Phylloscopus bonelli*	1329
Eastern Bonelli's Warbler *Phylloscopus orientalis*	1331
Wood Warbler *Phylloscopus sibilatrix*	1333
Plain Willow Warbler *Phylloscopus neglectus*	1335
Caucasian Chiffchaff *Phylloscopus lorenzii*	1336
Chiffchaff *Phylloscopus collybita*	1337
Willow Warbler *Phylloscopus trochilus*	1340
Ruby-crowned Kinglet *Regulus calendula*	1342
Goldcrest *Regulus regulus*	1342
Canary Islands Goldcrest *Regulus teneriffae*	1345
Firecrest *Regulus ignicapillus*	1346
Golden-crowned Kinglet *Regulus satrapa*	1348

Old World Flycatchers Family Muscicapidae — 1349

Brown Flycatcher *Muscicapa dauurica*	1349
Spotted Flycatcher *Muscicapa striata*	1349
Red-breasted Flycatcher *Ficedula parva*	1353
Semi-collared Flycatcher *Ficedula semitorquata*	1355
Collared Flycatcher *Ficedula albicollis*	1357
Pied Flycatcher *Ficedula hypoleuca*	1358

Babblers Family Timaliidae — 1362

Bearded Tit *Panurus biarmicus*	1362
Iraq Babbler *Turdoides altirostris*	1365
Common Babbler *Turdoides caudatus*	1367
Arabian Babbler *Turdoides squamiceps*	1368
Fulvous Babbler *Turdoides fulvus*	1370

Long-tailed Tits and Allies Family Aegithalidae — 1372

Long-tailed Tit *Aegithalos caudatus*	1372

Tits Family Paridae — 1375

Marsh Tit *Parus palustris*	1375
Sombre Tit *Parus lugubris*	1378
Willow Tit *Parus montanus*	1379
Siberian Tit *Parus cinctus*	1382
Crested Tit *Parus cristatus*	1383
Coal Tit *Parus ater*	1385
Blue Tit *Parus caeruleus*	1388
Azure Tit *Parus cyanus*	1392
Great Tit *Parus major*	1393

Nuthatches Family Sittidae — 1398

Krüper's Nuthatch *Sitta krueperi*	1398
Corsican Nuthatch *Sitta whiteheadi*	1399
Algerian Nuthatch *Sitta ledanti*	1400
Red-breasted Nuthatch *Sitta canadensis*	1401
Nuthatch *Sitta europaea*	1402

Eastern Rock Nuthatch *Sitta tephronota*	1404
Rock Nuthatch *Sitta neumayer*	1406

Wallcreepers Family Tichodromadidae — 1408

Wallcreeper *Tichodroma muraria*	1408

Treecreepers Family Certhiidae — 1411

Treecreeper *Certhia familiaris*	1411
Short-toed Treecreeper *Certhia brachydactyla*	1414

Penduline Tits and Allies Family Remizidae — 1416

Penduline Tit *Remiz pendulinus*	1416

Sunbirds and Allies Family Nectariniidae — 1419

Pygmy Sunbird *Anthreptes platurus*	1419
Nile Valley Sunbird *Anthreptes metallicus*	1421
Palestine Sunbird *Nectarinia osea*	1422

Old World Orioles and Allies Family Oriolidae — 1424

Golden Oriole *Oriolus oriolus*	1424

Shrikes Family Laniidae — 1428

Bush-shrikes and Allies Subfamily Malaconotinae — 1428

Black-crowned Tchagra *Tchagra senegala*	1428

Typical Shrikes Subfamily Laniinae — 1430

Brown Shrike *Lanius cristatus*	1430
Isabelline Shrike *Lanius isabellinus*	1431
Red-backed Shrike *Lanius collurio*	1433
Long-tailed Shrike *Lanius schach*	1436
Lesser Grey Shrike *Lanius minor*	1436
Great Grey Shrike *Lanius excubitor*	1440
Southern Grey Shrike *Lanius meridionalis*	1442
Grey-backed Fiscal Shrike *Lanius excubitorius*	1444
Woodchat Shrike *Lanius senator*	1445
Masked Shrike *Lanius nubicus*	1447

Crows and Allies Family Corvidae — 1450

Jay *Garrulus glandarius*	1450
Siberian Jay *Perisoreus infaustus*	1454
Azure-winged Magpie *Cyanopica cyanus*	1456
Magpie *Pica pica*	1457
Nutcracker *Nucifraga caryocatactes*	1460
Alpine Chough *Pyrrhocorax graculus*	1464
Chough *Pyrrhocorax pyrrhocorax*	1466
Jackdaw *Corvus monedula*	1468
Daurian Jackdaw *Corvus dauuricus*	1471
House Crow *Corvus splendens*	1472
Rook *Corvus frugilegus*	1475
Carrion Crow/Hooded Crow *Corvus corone*	1478
Pied Crow *Corvus albus*	1481
Brown-necked Raven *Corvus ruficollis*	1481
Raven *Corvus corax*	1483
Fan-tailed Raven *Corvus rhipidurus*	1486

Starlings Family Sturnidae — 1489
- Tristram's Grackle *Onychognathus tristramii* — 1489
- Daurian Starling *Sturnus sturninus* — 1491
- Grey-backed Starling *Sturnus sinensis* — 1491
- Starling *Sturnus vulgaris* — 1492
- Spotless Starling *Sturnus unicolor* — 1496
- Rose-coloured Starling *Sturnus roseus* — 1498
- Common Myna *Acridotheres tristis* — 1500

Sparrows, Rock Sparrows, Snow Finches Family Passeridae — 1503
- House Sparrow *Passer domesticus* — 1503
- Spanish Sparrow *Passer hispaniolensis* — 1506
- Dead Sea Sparrow *Passer moabiticus* — 1509
- Iago Sparrow *Passer iagoensis* — 1511
- Desert Sparrow *Passer simplex* — 1511
- Tree Sparrow *Passer montanus* — 1513
- Sudan Golden Sparrow *Passer luteus* — 1515
- Pale Rock Sparrow *Carpospiza brachydactyla* — 1517
- Yellow-throated Sparrow *Petronia xanthocollis* — 1518
- Rock Sparrow *Petronia petronia* — 1519
- Snow Finch *Montifringilla nivalis* — 1522

Weavers and Allies Family Ploceidae — 1526
- Village Weaver *Ploceus cucullatus* — 1526
- Streaked Weaver *Ploceus manyar* — 1526
- Red-billed Quelea *Quelea quelea* — 1528

Waxbills, Grassfinches, Mannikins Family Estrildidae — 1529
- Red-billed Firefinch *Lagonosticta senegala* — 1529
- Red-cheeked Cordon-bleu *Uraeginthus bengalus* — 1530
- Orange-cheeked Waxbill *Estrilda melpoda* — 1530
- Common Waxbill *Estrilda astrild* — 1531
- Red Avadavat *Amandava amandava* — 1533
- Indian Silverbill *Euodice malabarica* — 1534
- African Silverbill *Euodice cantans* — 1534

Vireos Family Vireonidae — 1536
- Yellow-throated Vireo *Vireo flavifrons* — 1536
- Philadelphia Vireo *Vireo philadelphicus* — 1537
- Red-eyed Vireo *Vireo olivaceus* — 1537

Finches Family Fringillidae — 1539
Chaffinches Subfamily Fringillinae — 1539
- Chaffinch *Fringilla coelebs* — 1539
- Blue Chaffinch *Fringilla teydea* — 1544
- Brambling *Fringilla montifringilla* — 1545

Typical Finches Subfamily Carduelinae — 1548
- Red-fronted Serin *Serinus pusillus* — 1548
- Serin *Serinus serinus* — 1550
- Syrian Serin *Serinus syriacus* — 1552
- Canary *Serinus canaria* — 1554
- Citril Finch *Serinus citrinella* — 1556
- Greenfinch *Carduelis chloris* — 1557

Goldfinch *Carduelis carduelis*	1561
Siskin *Carduelis spinus*	1564
Pine Siskin *Carduelis pinus*	1568
Linnet *Carduelis cannabina*	1568
Twite *Carduelis flavirostris*	1571
Redpoll *Carduelis flammea*	1574
Arctic Redpoll *Carduelis hornemanni*	1578
Two-barred Crossbill *Loxia leucoptera*	1580
Crossbill *Loxia curvirostra*	1582
Scottish Crossbill *Loxia scotica*	1586
Parrot Crossbill *Loxia pytyopsittacus*	1587
Crimson-winged Finch *Rhodopechys sanguinea*	1589
Desert Finch *Rhodospiza obsoleta*	1591
Mongolian Trumpeter Finch *Bucanetes mongolicus*	1592
Trumpeter Finch *Bucanetes githagineus*	1594
Scarlet Rosefinch *Carpodacus erythrinus*	1596
Sinai Rosefinch *Carpodacus synoicus*	1599
Pallas's Rosefinch *Carpodacus roseus*	1600
Great Rosefinch *Carpodacus rubicilla*	1600
Pine Grosbeak *Pinicola enucleator*	1602
Long-tailed Rosefinch *Uragus sibiricus*	1606
Bullfinch *Pyrrhula pyrrhula*	1606
Azores Bullfinch *Pyrrhula murina*	1609
Yellow-billed Grosbeak *Eophona migratoria*	1610
Japanese Grosbeak *Eophona personata*	1610
Hawfinch *Coccothraustes coccothraustes*	1610
Evening Grosbeak *Hesperiphona vespertina*	1613

New World Wood-warblers Family Parulidae — 1615

Black-and-white Warbler *Mniotilta varia*	1615
Golden-winged Warbler *Vermivora chrysoptera*	1615
Tennessee Warbler *Vermivora peregrina*	1616
Northern Parula *Parula americana*	1617
Yellow Warbler *Dendroica petechia*	1618
Chestnut-sided Warbler *Dendroica pensylvanica*	1619
Black-throated Blue Warbler *Dendroica caerulescens*	1620
Black-throated Green Warbler *Dendroica virens*	1620
Blackburnian Warbler *Dendroica fusca*	1620
Cape May Warbler *Dendroica tigrina*	1621
Magnolia Warbler *Dendroica magnolia*	1622
Yellow-rumped Warbler *Dendroica coronata*	1623
Palm Warbler *Dendroica palmarum*	1623
Blackpoll Warbler *Dendroica striata*	1623
American Redstart *Setophaga ruticilla*	1625
Ovenbird *Seiurus aurocapillus*	1625
Northern Waterthrush *Seiurus noveboracensis*	1626
Common Yellowthroat *Geothlypis trichas*	1627
Hooded Warbler *Wilsonia citrina*	1627
Wilson's Warbler *Wilsonia pusilla*	1628
Canada Warbler *Wilsonia canadensis*	1629

Tanagers Family Thraupidae — 1631

Summer Tanager *Piranga rubra*	1631
Scarlet Tanager *Piranga olivacea*	1632

Buntings and Allies Family Emberizidae — 1633

Buntings, New World Sparrows, and Allies Subfamily Emberizinae — 1633

- Rufous-sided Towhee *Pipilo erythrophthalmus* — 1633
- Field Sparrow *Spizella pusilla* — 1635
- Lark Sparrow *Chondestes grammacus* — 1635
- Savannah Sparrow *Ammodramus sandwichensis* — 1635
- Fox Sparrow *Passerella iliaca* — 1636
- Song Sparrow *Melospiza melodia* — 1637
- Swamp Sparrow *Zonotrichia georgiana* — 1637
- White-crowned Sparrow *Zonotrichia leucophrys* — 1637
- White-throated Sparrow *Zonotrichia albicollis* — 1639
- Dark-eyed Junco *Junco hyemalis* — 1639
- Lapland Bunting *Calcarius lapponicus* — 1640
- Snow Bunting *Plectrophenax nivalis* — 1642
- Black-faced Bunting *Emberiza spodocephala* — 1645
- Pine Bunting *Emberiza leucocephalos* — 1646
- Yellowhammer *Emberiza citrinella* — 1648
- Cirl Bunting *Emberiza cirlus* — 1651
- White-capped Bunting *Emberiza stewarti* — 1653
- Rock Bunting *Emberiza cia* — 1653
- Meadow Bunting *Emberiza cioides* — 1655
- House Bunting *Emberiza striolata* — 1655
- Cinereous Bunting *Emberiza cineracea* — 1658
- Cinnamon-breasted Rock Bunting *Emberiza tahapisi* — 1659
- Ortolan Bunting *Emberiza hortulana* — 1659
- Grey-necked Bunting *Emberiza buchanani* — 1662
- Cretzschmar's Bunting *Emberiza caesia* — 1663
- Yellow-browed Bunting *Emberiza chrysophrys* — 1664
- Rustic Bunting *Emberiza rustica* — 1665
- Little Bunting *Emberiza pusilla* — 1667
- Chestnut Bunting *Emberiza rutila* — 1669
- Yellow-breasted Bunting *Emberiza aureola* — 1670
- Reed Bunting *Emberiza schoeniclus* — 1672
- Pallas's Reed Bunting *Emberiza pallasi* — 1676
- Red-headed Bunting *Emberiza bruniceps* — 1677
- Black-headed Bunting *Emberiza melanocephala* — 1679
- Corn Bunting *Miliaria calandra* — 1681

Cardinal-grosbeaks and Allies Subfamily Cardinalinae — 1686

- Dickcissel *Spiza americana* — 1686
- Rose-breasted Grosbeak *Pheucticus ludovicianus* — 1686
- Blue Grosbeak *Guiraca caerulea* — 1688
- Indigo Bunting *Passerina cyanea* — 1689
- Lazuli Bunting *Passerina amoena* — 1689
- Painted Bunting *Passerina ciris* — 1689

New World Blackbirds, Orioles, and Allies Family Icteridae — 1690

- Bobolink *Dolichonyx oryzivorus* — 1690
- Brown-headed Cowbird *Molothrus ater* — 1691
- Rusty Blackbird *Euphagus carolinus* — 1692
- Common Grackle *Quiscalus quiscula* — 1692
- Eastern Meadowlark *Sturnella magna* — 1692
- Red-winged Blackbird *Agelaius phoeniceus* — 1693

	Yellow-headed Blackbird *Xanthocephalus xanthocephalus*	1693
	Black-vented Oriole *Icterus wagleri*	1693
	Baltimore Oriole *Icterus galbula*	1694
APPENDIX	English names of the birds of the western Palearctic	1695
INDEXES		
	Scientific names	[1]
	English names	[10]
	Deutsche Namen	[15]
	Nombres españoles	[19]
	Noms français	[23]
	Nomi in Italiano	[27]
	Nederlandse namen	[31]
	Svenska namn	[35]
	Русские названия	[39]

Order PASSERIFORMES

The perching birds, known colloquially as passerines. Largest order of class Aves, comprising well over half (c. 5300) of the total of living species. Diverse group of tiny to fairly large landbirds of many adaptive types, mainly arboreal but also terrestrial and aerial. Well characterized by possession of a syrinx (resonating chamber at lower end of trachea consisting of bony rings supplied with complex system of muscles and vibrating membranes) and of perching feet suitable for gripping slender branches (equipped with set of 4 toes joined at same level, hind toe of which is often stronger than others and non-reversible). Cosmopolitan, except Antarctica and some oceanic islands; represented in all terrestrial habitats throughout the world, from high Arctic to low Antarctic and from almost waterless desert to tropical rain-forest. Some species are long-distance migrants.

Two main subdivisions generally recognized: suboscines, with simpler syrinx and vocal repertoires, mainly in tropics and New World; and oscines, with complex syrinx and voice (true 'songbirds'). All west Palearctic passerines oscine, except two New World vagrants (Acadian Flycatcher and Eastern Phoebe).

Tyrant Flycatchers Family Tyrannidae

Small to medium-sized suboscine passerines; mainly arboreal, often feeding on insects caught by sallying out from perch, though some species terrestrial. About 360 species in numerous genera. Occur over whole of New World but main diversity in Neotropics. Northern forms migratory. Two representatives in west Palearctic, recorded only as stragglers.

Acadian Flycatcher *Empidonax virescens*

PLATE: page 1126

Du. Groene Elftiran   Fr. Moucherolle vert   Ge. Buchentyrann   It. Tiranno acadico
Ru. Восточная белоглазая мухоловка   Sp. Papamoscas verde   Sw. Akadtyrann

Field characters. 11.5 cm; wing-span 21–24 cm. Slightly larger (especially about head) than Red-breasted Flycatcher, with proportionately bigger bill. Rather small, high-crowned bird with flycatching habits. Upperparts essentially olive-green with striking double pale yellow wing-bar and similarly coloured edges to tertials and inner secondaries. Underparts dusky-olive, fading below breast to yellow-white. 1st autumn less colourful, browner above and whiter below, with buffier wing-bars and tertial edges. Flight like *Ficedula* flycatcher; wags tail but does not hold it cocked.

Above description will distinguish Acadian Flycatcher from west Palearctic flycatchers but quite insufficient to separate it from 4 congeners which are equally or even more likely to cross North Atlantic. Field identification of this group of flycatchers the most difficult of all North American birds.

Only calls likely from vagrant are characteristic thin 'peet' and sharp emphatic 'weece'. Call of next most likely congener, Yellow-bellied Flycatcher *E. flaviventris*, is distinctive loud sneezy 'chew'.

Habitat. Breeds in Nearctic in lower middle latitudes, in moist or swampy lowland woods, especially deciduous floodplain forests. Prefers deep shade of fairly mature stands with ample open areas for flycatching.

Distribution. Breeds in extreme southern Ontario (Canada), and in USA from south-east Minnesota east to Connecticut and south to eastern Texas, Gulf coast, and central Florida.

Accidental. Iceland: specimen, perhaps ♂, November 1967.

Movements. Migratory. Occurs on southward migration mainly in south-east USA, regularly on Caribbean coast of Mexico, and in Belize, Guatemala, and Honduras. Winters in eastern Nicaragua and on Caribbean and Pacific slopes of Costa Rica and Panama, also in Colombia, Ecuador, and western Venezuela. Spring migration apparently follows similar route.

Wing-length: ♂ 73–80, ♀ 68–75 mm.

Eastern Phoebe *Sayornis phoebe*

PLATE: page 1125

Du. Phoebetiran   Fr. Moucherolle phébi   Ge. Phoebe   It. Febe orientale
Ru. Восточный феб   Sp. Fibi primordial   Sw. Grå fibi

Field characters. 17–18 cm; wing-span 24–27 cm. Larger than Spotted Flycatcher, but with similar form. Dark brownish-grey above, with sooty head, wings, and tail, wings only faintly marked with dull bars across coverts in adults but these marks and pale fringes to tertials obvious on 1st winter; lacks eye-ring and has wholly dark bill and legs. Dull white below, with brownish grey patch by shoulder and furrows down flanks and in 1st winter and autumn adult yellow suffusion, particularly on lower flanks. Flight and behaviour recall Spotted Flycatcher; frequently flicks tail.

Easily distinguished from Spotted Flycatcher by lack of streaks on crown and breast. Much larger than any other Palearctic flycatcher, but other potentially vagrant Nearctic flycatchers are even bulkier. Call a rather loud, tinny 'tsy'.

Habitat. Throughout year in woodland and edge habitats near water, farmyards, gardens, etc., always with trees near by, from boreal to southern temperate North America; in non-breeding season south to tropical Central America. Feeds aerially on flying insects, and takes some fruits in winter. Nest usually built on rock, under bridge, or on building.

Distribution. Breeds from north-west to south-east Canada and throughout eastern half of USA except extreme south-east, though continues to expand in that direction.

Accidental. Britain: Lundy (Devon), April 1987.

Movements. Partial migrant, wintering from southern parts of breeding range to south-east Mexico; rare but regular winter visitor in southern California. Leaves north of range September, remainder October–November, returning March to early May.

Wing-length: ♂ ♀ 77–90 mm.

Larks Family Alaudidae

Rather small oscine passerines (suborder Passeres); terrestrial, feeding on insects, seeds, and plants. About 80 species in 15 genera; represented by 25 species in west Palearctic, 22 breeding.

Differ from all other oscines in structure of tarsus and syrinx: back of tarsus rounded, covered with small scutes (not sharply edged and smooth). Body rather robust; neck short. Sexes often differ in size, with ♂ larger. Bill shape highly variable, even within species (as in *Melanocorypha*) or between sexes (as in Razo Lark *Alauda razae*); adapted to special feeding method and/or diet (e.g. long and curved for digging or short and stout for crushing hard seeds). Wing fairly long and often pointed. Flight typically strong and undulating, with periodic closure of wings, but rather fluttering and wavering over short distances; steeply climbing or circling song-flights are characteristic. Tails short or of medium length. Legs short to fairly long; usual gait a walk or run. Hind claw straight and often long, especially in species living on soft soil with short close vegetation, but short in those on bare and hard ground; toes and claws relatively shortest in expert runners, such as Hoopoe Lark *Alaemon alaudipes*. Bathing in standing water does not occur, but rain-bathing and dusting typical.

Plumage usually cryptic, unmarked or streaked; often adapted to colour of local soil, species occurring in large range of habitats tending to have several races differing in colour, particularly when non-migratory. Some species show conspicuous marks on wing or outer tail-feathers, especially in flight. Sexes usually similar. Monogamous mating system the rule. Nest-building and incubation normally by ♀; both sexes feed young. Nestlings have rather scanty down, confined to upperparts; often spotted. Mouth with some contrasting dark spots. Young usually leave ground nest before able to fly.

Kordofan Bush-lark *Mirafra cordofanica*

PLATES: pages 1012, 1051

Du. Kordofanleeuwerik   Fr. Alouette du Kordofan   Ge. Kordofanlerche
Ru. Кордофанский кустарниковый жаворонок

Field characters. 15 cm; wing-span 23–25 cm. Similar in size to sympatric Dunn's Lark but with slightly smaller, less deep bill, squarer head (with slight crest) and slightly shorter wings. Quite small but robust lark with noticeably warm, golden-buff and rufous appearance; best distinguished from Dunn's Lark by pale open face (lacking blackish marks around eye), obviously but sparsely streaked breast, more rufous upperparts and striking pattern of tail, which has rufous centre, then black panels and bright white outer tail feathers (recalling Short-toed Larks). Bill pale flesh-clay with browner culmen and tip; legs pale flesh-pink. Immature shows darker mottles on crown and wing coverts.

Habitat. Arid, open plains in desert and semi-desert with scattered bushes and grasses (especially *Aristida* and *Stipagrostis*) in northern tropical Africa.

Distribution and population. In west Palearctic sections of Mauritania and Mali, widespread but uncommon.

Beyond west Palearctic, breeds Mauritania and Mali (chiefly north of 15°N), also south-west Niger and Sudan (Darfur and Kordofan); recorded Sénégal without proof of breeding.

Movements. Local wanderings depending on rains and food availability; in Mali generally moves south of 18°N from May to August.

Food. Invertebrates and seeds.

Social pattern and behaviour. ♂ seen singing from top of low bush. No other information.

Voice. Song of ♂ described as musical and sweet.

Breeding. Season. In Mali and Mauritania lays eggs at beginning of rainy season May–August; young bird collected in November; eggs recorded in Sudan in May. No further information on nest, eggs, or young.

Wing-length: ♂♀ 78–87 mm.

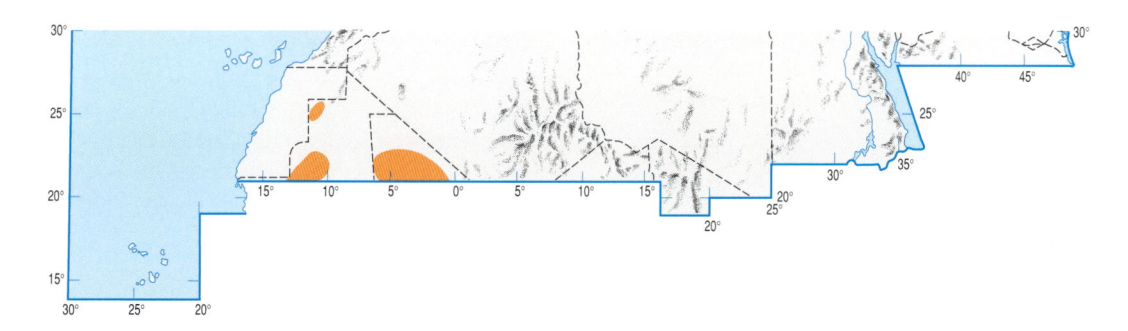

Black-crowned Finch Lark *Eremopterix nigriceps*. *E. n. albifrons*: **1** ad ♂ fresh (autumn), **2** ad ♂ worn (spring), **3** ad ♀ fresh (autumn), **4** ad ♀ worn (spring), **5** juv. *E. n. nigriceps*: **6** ad ♂ fresh (autumn), **7** ad ♀ fresh (autumn). *E. n. melanauchen*: **8** ad ♂ fresh (autumn), **9** ad ♀ (autumn). Chestnut-headed Finch Lark *Eremopterix signata*: **10** ad ♂, **11** ad ♀. Kordofan Bush-lark *Mirafra cordofanica* (p. 1011): **12** ad.

Chestnut-headed Finch Lark *Eremopterix signata*

PLATES: pages 1012, 1023

Du. Somalische Vinkleeuwerik   Fr. Moinelette d'Oustalet   Ge. Harlekinlerche   It. Allodola testacastana
Ru. Пятнистый воробьиный жаворонок   Sp. Gorrialondra tuticastaña   Sw. Brunhuvad finklärka

Field characters. 11–12 cm; wing-span 21–22 cm. About 15% smaller than Short-toed Lark but with much dumpier form, due particularly to stubby conical bill on proportionately large head; close in size and form to Black-crowned Finch Lark. Closely related to 3 other congeners in sub-Saharan and north-east Africa, with all ♂♂ showing strongly decorated plumage and particularly pale bill, dark head with white cheek patch and wholly or partly black throat, centre to breast and underbody. ♂ of Chestnut-headed Finch Lark distinguished by chestnut crown with white central patch, chestnut invasion of black areas on head and breast and narrow white collar around nape becoming white band from rear cheeks down sides of breast and along flanks. Females much duller than male, lacking continuous black marks except from lower belly to undertail, and grey rather than brown, also with pale, orange-buff supercilium. Immature like ♀ but underparts wholly white from breast to undertail. At all ages, black underwing-coverts of flying bird catch eye and pattern of short tail suggests Short-toed Lark.

Adult ♂ unmistakable but separation of ♀ and immature from congeners not fully studied. Calls include sharp 'chip-up' and sibilant 'tsssp'.

Habitat. Dry grassy plains, stony, scrubby semi-desert and desert in tropical north-east Africa below *c.* 2000 m; feeds on seeds and insects on ground, occasionally perching in low bushes; often seen drinking at waterholes. Outside breeding season, sometimes occurs in flocks of hundreds.

Distribution. South-east Sudan, central and south-east Ethiopia, Somalia, and northern and eastern Kenya.
Accidental. Israel: Eilat, adult ♂, May 1983.

Movements. Sedentary and dispersive, making irregular and extensive movements probably depending on climatic factors, e.g. in south-east Kenya, rather erratic increase in numbers over 2 years in April–May, September–October, February, and May–September. Eggs laid in Somalia and Sudan May–June.

Wing-length: ♂♀ 75–82 mm.

Black-crowned Finch Lark *Eremopterix nigriceps*

PLATES: pages 1012, 1023

Du. Zwartkruinvinkleeuwerik   Fr. Moinelette à front blanc   Ge. Weißstirnlerche   It. Allodola capinera
Ru. Белолобый воробьиный жаворонок   Sp. Alondra cabecinegra   Sw. Svartkronad finklärka

Field characters. 10–11 cm; wing-span 20–22 cm. 20–25% smaller than any other west Palearctic lark; hardly larger but broader-winged than Serin. Tiny, rather dumpy lark, with stubby, conical bill, short, broad wings, rather short legs, and (on ground) compact form suggesting small finch. ♂ more boldly patterned than any other west Palearctic lark, with totally black underparts and sandy upperparts; grey-white head decorated with black on crown and loral band. ♀ and juvenile strikingly different from ♂, with sandy-brown upperparts and no obvious characters except for dull dark median-covert bar and black under wing-coverts.

In west Palearctic, small size, black under wing-coverts, and mainly black tail diagnostic of genus (but see also Chestnut-headed Finch Lark). Flight noticeably light, with rapid flutter on take-off, easy bursts of fast wingbeats, and then undulating descent. Walks and runs, crouching on stopping; also stands upright.

Habitat. A tropical species based on saharo-sahelian savannas, extending into west Palearctic only along lowland coastal belts with moisture regime favouring vigorous seasonal growth of herbage, especially grasses and succulents. Associated soils tend to be fine and sandy, often in depressions or beds of wadis, or on flats subject to occasional brief flooding.

Distribution and population. Iraq. In 1920s only one breeding site known; no recent information. Kuwait. Breeding confirmed in north-east May 1996; first evidence of breeding since claims of 1979. Egypt. Breeds on south-east border with Sudan, where fairly common. Western Sahara. Breeds only in area rarely if ever visited by ornithologists, thus no information on current status. Latest records early 1970s. Mauritania. Locally fairly common. Chad. Tibesti: possibly breeds. Cape Verde Islands. Locally common to abundant Santiago, Fogo, Boavista, and Maio, rare Brava and São Nicolau.

Accidental. Israel, Algeria.

Beyond west Palearctic, breeds south of Sahara from Sénégal to Somalia, thence north-east through Arabia to north-west India.

Movements. Resident and partial short-distance migrant. Present all year in Sahel breeding zone, though in central Chad at least some birds move further north, to breed, in wet season August–October. Nomadic tendency indicated by extensions to breeding range when conditions suitable. Extent of movement obscured by uncertainty over northern limits of breeding range, which probably vary from year to year according to rainfall.

Food. Seeds and insects. Feeds either by picking up food from ground or by taking it directly from vegetation—usually by reaching up from ground, but also by fluttering up to grass heads, presumably to snatch seeds or, possibly, insects. May run erratically and occasionally chases flying insects, showing considerable speed and agility.

Social pattern and behaviour. Gregarious outside breeding season; flocks relatively small and loose—up to 60 birds. Breeding territory defended by both members of pair. Song-flight: ♂ takes off and legs dangling, ascends fairly steeply but linearly with rapidly beating wings, usually singing as he climbs. At peak of ascent, usually c. 6–10(−20) m, flight becomes undulating: rapid fluttering on outstretched wings interspersed with slight drops with wings briefly and partially folded. Bird typically describes roughly circular path of relatively small diameter. Duration of song-flight varies, but usually not more than 1 min. At end of flight, bird initiates descent sometimes by a dipping action of body and outspread wings, or by briefly closing wings. Descends in stepped series of swooping glides, wings held stiffly out and slightly above horizontal; at end of each glide, bird rises slightly before gliding further.

Voice. Freely used in breeding season, especially by ♂, but birds mostly silent at other times. Song a regularly repeated single phrase, typically given by ♂, rarely by ♀. Final unit usually prolonged and descending; overall effect sweet and plaintive. Contact-alarm call an oft-repeated 'jip', 'tchip', 'jeep', or 'tchup', given by both sexes, in the air or on ground. Warning-call similar, but a more reedy and lower-pitched 'dzeep' or 'jreep', or 'zree'.

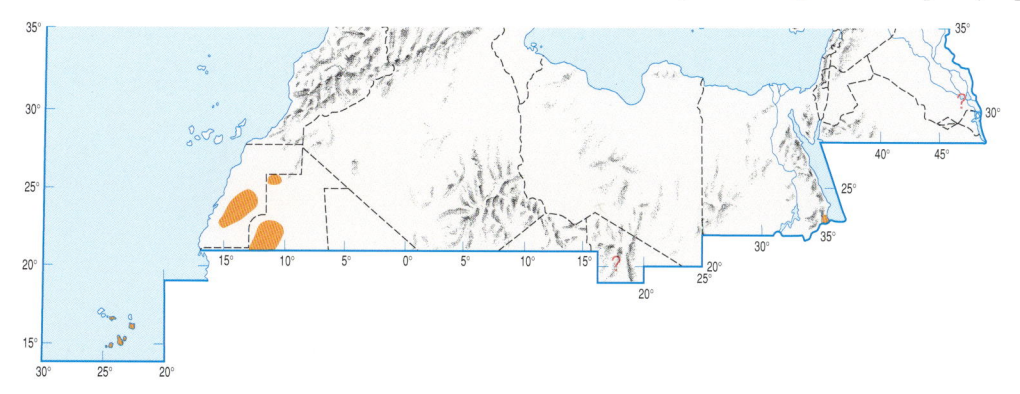

Breeding. Season. Cape Verde Islands: eggs laid September to January or February. Southern Morocco: breeding begins late April. Eastern Saudi Arabia: north of *c.* 25°N (where migratory), eggs laid by end of April. Site. On ground in shelter of tussock or stone. Nest: scrape, saucer-like depression (hoof-print can be used), or quite deep cup-shaped hole (apparently dug specially) with top of nest flush with ground; lined with grass, hair, or feathers on base of fine twigs; may have rim of pebbles surrounded by twigs, with centre of nest bare so that eggs laid directly on sand. Eggs. Sub-elliptical, smooth and fairly glossy; greyish or dirty white, evenly speckled and blotched light brown, with underlying purplish-grey spots. Clutch: 2–3. Incubation. 11(–12) days. Fledging Period. *c.* 12–14 days.

Wing-length: *E. n. nigriceps*: ♂ 74–80, ♀ 73–77 mm.

Geographical variation. Slight in size, more pronounced in colour. Nominate *nigriceps* from Cape Verde Islands smallest; Saharan race *albifrons* has slightly longer wing and tail; Arabian *melanauchen* larger still. ♂ *albifrons* like ♂ nominate *nigriceps* in colour, but upperparts more uniform sandy-grey with less dark grey-brown of feather-bases showing, appearing less spotted dull grey. ♂ *melanauchen* differs from both nominate *nigriceps* and *albifrons* in having more extensively black crown and black upper mantle, latter largely separated from black of hindcrown by white band across hindneck.

Dunn's Lark *Eremalauda dunni*

PLATES: pages 1015, 1023

Du. Dunns Leeuwerik   Fr. Alouette de Dunn   Ge. Einödlerche   It. Allodola di Dunn
Ru. Малый вьюрковый жаворонок   Sp. Alondra de Dunn   Sw. Streckad ökenlärka

Field characters. 14–15 cm; wing-span 25–30 cm. 10% smaller than Desert Lark; in spite of overlapping measurements, noticeably bulkier than Bar-tailed Desert Lark. Rather small but bulky lark, with strikingly heavy bill contributing to large-headed appearance, broad, well-rounded wings, and (at times) long-looking legs. Plumage essentially pale sandy with dark-moustached and bold-eyed face, and black panels on sides of tail recalling *Calandrella* larks. Streaking of upperparts and chest variably distinct but important to separation from plain *Ammomanes* larks.

All too easy to overlook among populations of *Ammomanes* larks; distribution in west Palearctic incompletely known. Previously thought to be indistinguishable from Bar-tailed Desert Lark but now known to possess several invariable, diagnostic characters, of which massive bill, head-marks, and tail pattern are the most visible. In addition, structure differs from *Ammomanes* larks in larger head, distinctly broad and round wings (with short points when folded), and broader tail. Allows close approach and often content to run or creep away from observer, using rocks and shrubs as cover. Wing-shape yields characteristically flapping or even floppy flight-action.

Habitat. In very warm lower latitudes, tropical and subtropical, on flat arid lowlands, not rocky, stony, or of broken ground, and apparently overlapping little into full desert. In eastern Saudi Arabia breeds in small lightly vegetated sandy wadis and shallow depressions within undulating gravelly desert, apparently avoiding sand desert. Winters sparsely on firm stony desert, apparently living on plant seeds left over from spring growth of grass; also on old middens where nomad bedouin have corralled sheep.

Distribution and population. Regular in northern Saudi Arabia, but elsewhere in west Palearctic breeding apparently erratic. Israel. Usually occasional nomadic visitor (e.g. 6 records 1978 to spring 1988). In November 1988, several hundred birds invaded Arava valley (including Jordanian side), apparently linked to high rainfall previous winter; 8 pairs stayed to breed in Israel in 1989. Jordan. Has occurred and bred, although possibly only erratically, at localities mapped (e.g. bred Azraq 1965–6); see also Israel. Egypt. Various recent records in Sinai suggest rare and irregular breeding. Mauritania. Bred Zemmour 1970 (4–5 nests found). One collected near Chegga 1930.

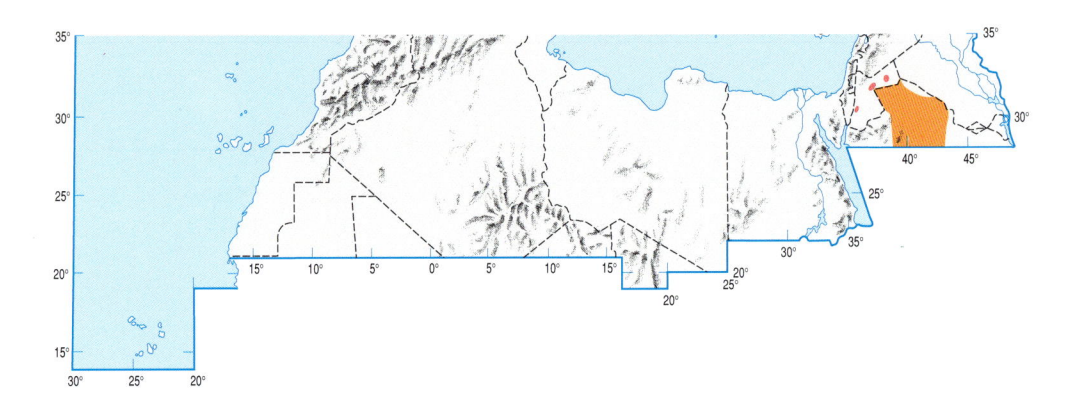

Dunn's Lark *Eremalauda dunni*. *E. d. eremodites*: **1** ad fresh (autumn), **2** ad worn (spring), **3** juv. Nominate *dunni*: **4** ad fresh (autumn), **5** ad worn (spring). Bar-tailed Desert Lark *Ammomanes cincturus*. *A. c. arenicolor*: **6** ad fresh (autumn), **7** ad worn (spring), **8** juv. *A. c. cincturus*: **9** ad fresh (autumn).

Accidental. Lebanon, Kuwait.

Beyond west Palearctic, breeds discontinuously along southern edge of Sahara from Mauritania to Sudan and in Arabia.

Movements. Resident; also some dispersal of uncertain, but probably small, extent. As in other desert larks, likely that main component of movement is nomadism during drought conditions.

Food. Seeds and insects. Picks food from surface of ground, and also digs with bill: uses hammering or rapid sideways flicking movements, throwing up spurts of sand and creating a pit. Also examines tussocks and undersides of shrubs, sometimes jumping up to snatch at items.

Social pattern and behaviour. Gregarious outside breeding season, occurring in groups of 2–20, often with other larks. Territorial in breeding season. Territory advertised by song-flight, in which ♂ typically rises into wind to 30 m, sometimes 50 m or more. While singing, stays more or less stationary, but swinging from side to side, with slow, lazy, almost owl-like wing-beats.

Voice. Song of ♂ a scratchy warbling, interspersed with a melodious but melancholy whistling of short phrases; longer and more developed than Bar-tailed Desert Lark. Contact-call a single or repeated 'ziup' or 'chiup chiup' given in flight. A louder, more emphatic version of this call, 'chee-oop', given in alarm.

Breeding. SEASON. Israel: eggs laid mid-April. Mauritania: breeding once recorded in January, following autumn rains. SITE. On ground in shelter of tussock. Nest: scrape lined with fresh vegetation. EGGS. Similar in shape and form to those of *Mirafra* larks; white, with blackish and lavender spots and blotches. Clutch: 2–3. INCUBATION. 13–16 days. (Fledging period not recorded.)

Wing-length: *E. d. dunni* (Sahara): ♂ 95–101, ♀ 85–92 mm.

Geographical variation. Rather marked. Nominate *dunni*, from Sahara, smaller than *E. d. eremodites* from Arabia, and upperparts more rufous-coloured, less blackish on crown.

Bar-tailed Desert Lark *Ammomanes cincturus*

PLATES: pages 1015, 1023

Du. Rosse Woestijnleeuwerik Fr. Ammomane élégante Ge. Sandlerche It. Allodola del deserto minore
Ru. Чернохвостый вьюрковый жаворонок Sp. Terrera colinegra Sw. Sandökenlärka

Field characters. 15 cm; wing-span 25–29 cm. 5–10% smaller and distinctly slimmer than Desert Lark; noticeably smaller-headed and less bulky than Dunn's Lark. Rather small, neatly made lark, with rather bunting-like small bill and round head. Plumage essentially pale grey-buff, more uniform than in other desert-dwelling larks, but with obvious orange glow on upper- and underwing and on tail-base. Primaries and tail-feathers variably tipped black, with inverted T pattern on tail usually strong enough to recall wheatear.

Although congeneric with Desert Lark, bunting-like character and more active behaviour distinctive; has noticeably freer gait and faster flight. At first sight confusable with Desert Lark, Dunn's Lark, and juvenile Temminck's Horned Lark, but combination of relatively small bill, rounded head, and dark transverse, terminal marks on outer wing and rather short tail distinctive. Does not allow close approach, escaping with fast, long-legged run or lengthy flight. Song-flight either a series of steep undulations or a quite high and circular sweep, with action more flapping than in normal flight. Stance frequently upright, emphasizing spindly legs and, again, recalling bunting.

Habitat. In subtropical and tropical lower latitudes, in contrast to Desert Lark on wide flat or gently sloping bare stony or sandy deserts, rarely with more than sparse vegetation, and without access to water or relief from intense heat. In such terms ranks as most essentially desert-dwelling of west Palearctic larks, and penetrates furthest into interior of Sahara, although highest density found near Atlantic coast. In north-west Africa also frequents low dunes and hollows between them, or sandy wadis amidst mountains. In some parts of range, replaces Crested Lark as common roadside bird, especially at lower elevations.

Distribution and population. ISRAEL. 400–500 pairs in 1980s; declining. JORDAN. Widely distributed but uncommon. IRAQ. Bred Al Hadr in north in 1949; no further information. KUWAIT. First confirmed breeding May 1996, in north-east. SAUDI ARABIA. Widespread and often common. EGYPT. Fairly common. LIBYA. Tripolitania: widespread and probably commoner than Desert Lark. Cyrenaica: common. TUNISIA. Common in south. MOROCCO. Common in south. Discovered 1982–5 north-west of main breeding range, but breeding not proved. MAURITANIA. Widespread and fairly common. CAPE VERDE ISLANDS. Abundant Sal, Boavista, and Maio; locally common Santiago and São Nicolau; rare Fogo.

Accidental. Balearic Islands, Italy, Malta, Canary Islands.

Records in Syria (May 1972, August 1976, April 1994) perhaps result from dispersion rather than vagrancy.

Beyond west Palearctic, breeds discontinuously along southern edge of Sahara from Mauritania to Sudan, and from Arabia to Pakistan.

Movements. Resident, also dispersive and nomadic to an uncertain extent.

Food. Seeds and other plant material, and insects. Will feed by digging into ground, but less often than Dunn's Lark (at least when accompanying that species) and picks from surface more. Feeding flock will move rapidly forward, alternately running and pausing.

Social pattern and behaviour. Gregarious outside breeding season, though less so than some other larks, generally occurring in small flocks. Flying flocks constantly twist and turn in perfect unison, exposing reddish wing colour not visible when perched. Breeding territories tend to be in neighbourhood group. From start of breeding season, ♂ performs song-flight similar to Desert Lark and Black-crowned Finch Lark. Flight-path deeply undulating and markedly meandering but roughly circular, ending with steep descent to ground.

Voice. Song of ♂ a pure, almost ethereal series of quiet trilling and fluting phrases, interrupted by louder, characteristic, mournful 'see-oo-lee', like creaking door. Contact-alarm call a thin, high 'peeyu'; also a rather chirrupy 'chweet' and a dry, rattling chirrup, 'rreep'.

Breeding. SEASON. Cape Verde Islands; eggs laid September–April, probably affected by occurrence of rains. North Africa: February–April(–May). Eastern Saudi Arabia: from March. SITE. On ground usually in shelter of tussock or small stone. Nest: shallow depression lined with vegetation, often with rim of small stones all round or on exposed side. EGGS. Sub-elliptical, smooth and glossy; white, lightly spotted with black, grey, and purple. Clutch: 2–4. (Incubation and fledging periods not recorded.)

Wing-length: *A. c. cincturus*: ♂ 91–97, ♀ 84–91 mm.

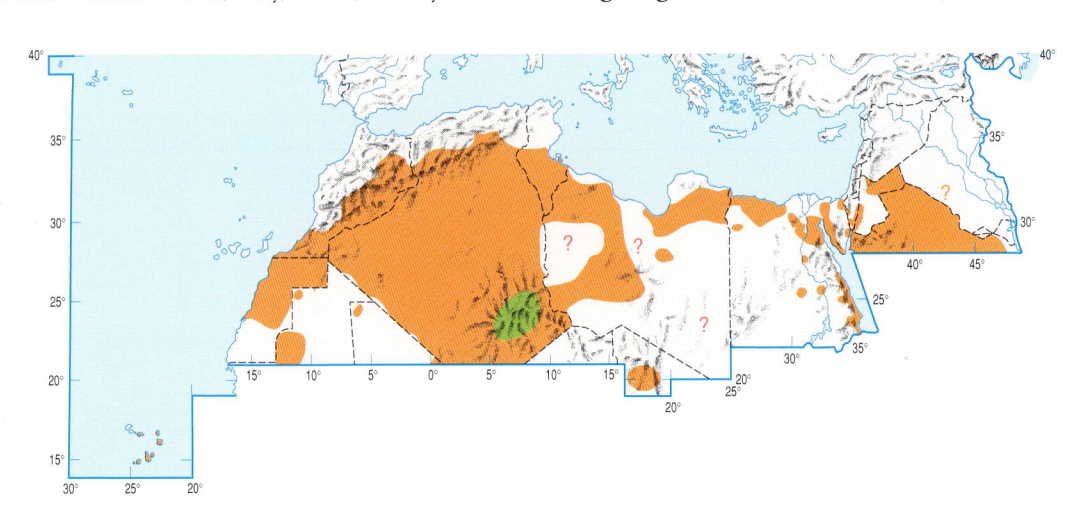

Geographical variation. Slight, in colour only. Nominate *cincturus* from Cape Verde Islands deeper rufous-cinnamon on upperparts, flight-feathers, and tail than *arenicolor* from Sahara, Arabia, and Middle East. Sides of neck and all chest distinctly streaked grey on vinous-cinnamon ground (almost uniform pink-cinnamon in *arenicolor*), throat almost completely mottled grey on cream-buff ground (uniform cream-white in *arenicolor*), remainder of underparts deeper pink-cream.

Desert Lark *Ammomanes deserti*

PLATES: pages 1018, 1023

Du. Woestijnleeuwerik   Fr. Ammomane isabelline   Ge. Steinlerche   It. Allodola del deserto
Ru. Пустынный жаворонок   Sp. Terrera sahariana   Sw. Stenökenlärka

Field characters. 16–17 cm; wing-span 27–30 cm. 10% larger than Bar-tailed Desert Lark and Dunn's Lark. Medium-sized, robust lark with usually long, evenly pointed bill, rather large head, full body, and quite long wings, but relatively short tail. Plumage always dull but subject to tonal variation within habitats, being sandy on sand, grey on rocks, and dusky on dark basalt. At close range, straw-yellow base to bill, pale fore-supercilium and eye-ring, crown-streaks, dark mottling on throat and chest, and particularly dark distal panel on rusty tail show.

Extremely variable plumage tone, amorphous pattern, and variation in bill length often provoke misidentification. Important therefore to remember that (1) it is essentially a lark of broken, rocky, and, above all, usually sloping habitats, not overlapping those of Bar-tailed Desert Lark and Dunn's Lark, and (2) its general character is less reminiscent of other desert-dwelling larks, being more like Skylark (or even a large pipit). Flight action recalls Skylark but progress lacks uneven quality of that species and is slower than Bar-tailed Desert Lark.

Habitat. Despite misleading name, not found in flat open desert, being closely attached to low rock-faces, flanking escarpments, boulder-strewn or stony slopes, and scree; habitats without some nearly vertical element are generally avoided. At least in parts of range finds rock faces surrounding springs especially attractive; also readily attaches itself to such structures as an isolated police post or even a desert monastery. Sedentary and usually reluctant to move far either on foot or in flight. Will, however, range alongside tracks and roads and over bare ground especially after breeding season.

Distribution and population. TURKEY. Small population discovered 1983 on barren limestone plateau above Birecik in south-east, and confirmed annually since; maximum number of birds seen 15. SYRIA. Breeds many places in interior. Marked annual fluctuations. ISRAEL. At least 10 000 pairs in 1980s. JORDAN. Widespread and common. IRAQ. Widespread. KUWAIT. Breeds in small numbers in west, but only in years of good winter rainfall. SAUDI ARABIA. Common in almost all rocky areas. EGYPT. Common. TUNISIA. Common in rocky hills. MOROCCO. Common in some areas in south. Eastern population has recently expanded north into eastern Rif. MAURITANIA. Rather uncommon.

Accidental. Cyprus, Lebanon.

Beyond west Palearctic, breeds discontinuously along southern edge of Sahara from Mauritania to northern Somalia, and from Arabia to Turkmenistan and north-west India.

Movements. Highly sedentary; lack of movement reflected in high degree of subspeciation and adaptation of plumage colour to local soil or rock type. Dispersal, where this occurs, is very local.

Food. Seeds and insects. Seeds taken from ground and by hammering at goat droppings. Insects taken from ground and among stones, also from small bushes, by reaching up from ground below or by clambering about in them.

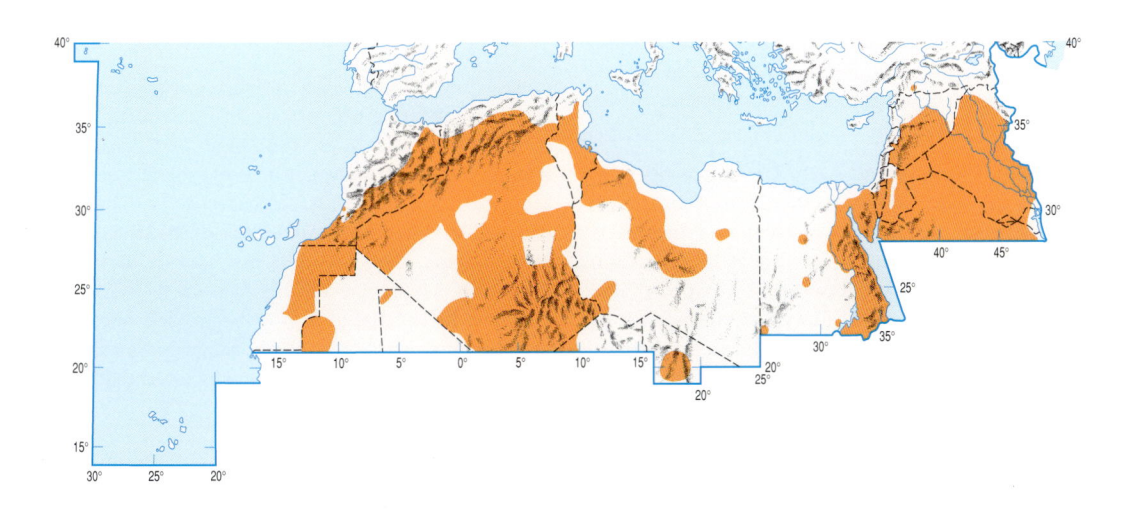

Desert Lark *Ammomanes deserti* (p. 1017). *A. d. payni* (southern Morocco): **1** ad fresh (autumn), **2** ad worn (spring), **3** juv. *A. d. algeriensis* (Algerian Sahara): **4** ad fresh (autumn), **5** ad worn (spring). *A. d. deserti* (Nile valley to Saudi Arabia): **6** ad fresh (autumn). *A. d. isabellinus* (southern Nile valley to Saudi Arabia): **7** ad fresh (autumn), **8** ad worn (spring). *A. d. annae* (Azraq area, Jordan): **9** ad fresh (autumn).

Social pattern and behaviour. Only mildly gregarious outside breeding season, never forming large dense flocks. Often encountered alone or in twos and threes. Breeding territories usually well-dispersed. ♂ has brief, undulating song-flight. Consists either of horizontal flight between 2 eminences, or of steep ascent from and return to ground, with or without horizontal flight in between; descent may be vertical; horizontal phase has undulations *c.* 2 m deep due to alternation of bouts of deep floppy wing-beats with equal periods of closed wings.

Voice. Song of ♂ melodious, rather sad, and (unlike Bar-tailed Desert Lark and Dunn's Lark) far-carrying: a repeated phrase, typically of 2–3 syllables: e.g. 'chur-rer-ee', 'trreeooee'. Given during song-flight; often also on ground, when typically incomplete. Contact-call a soft quiet undertone, 'chu'; sometimes a slurred 'chee-lu' becoming rapid and twittering when excited.

Breeding. SEASON. North Africa: start of season varies from January–February in south and near west coast to March–April in north. SITE. On ground, usually in shelter of tussock or stone. Nest: shallow scrape lined with available vegetation, with rim or ramp of stones, all round or on exposed side. EGGS. Sub-elliptical, smooth and glossy; greenish-white to pink, finely spotted dark or reddish-brown, and some purplish-grey, spots sometimes concentrated at broad end. Clutch: 1–5; 1–4 in desert areas of North Africa, and 3–5 further north. (Incubation and fledging periods not recorded.)

Wing-length: *A. d. deserti*: ♂ 98–103, ♀ 92–98 mm.
Weight: Varies racially: ♂♂ mostly 24–28, ♀♀ 20–25 g.

Geographical variation. Marked and complex. 17 races have been described for west Palearctic alone, with 10 more just outside its limits which may straggle into our area or influence populations living just within its borders. Not all these now considered worthy of racial status, however. Colour often directly related to colour of local soil and intensity of sunlight. Birds from sandy habitats mostly buff-coloured, those of stony ground various shades of grey, rufous, or brown. In some areas, pale and dark birds live side-by-side (notably in Ahaggar in southern Algeria, in Nile valley, and along shores of Gulf of Aqaba), though no real dimorphism, as intermediates usually occur; in adjacent areas, pale or dark birds may occur alone, similar to one of nearby variants. Size also varies, and length and shape of bill. Long-winged populations (*mya*, *whitakeri*) inhabit central Sahara from Mauritania to central Libya, characterized also by long and rather heavy bill. Remaining populations of Mauritania, northern Algeria, Tunisia, Egypt, Middle East, and Arabia are small in size. As marked sexual size-dimorphism occurs, ♂♂ of small-sized races can be as large as ♀♀ of large-sized ones. Geographical division between long- and short-winged birds not sharp; intermediate areas occupied by birds of intermediate size.

Hoopoe Lark *Alaemon alaudipes*. *A. a. alaudipes*: **1** ad rufous morph fresh (autumn), **2** ad rufous morph worn (spring), **3** juv rufous morph. *A. a. doriae* (Iran east to India): **4** ad fresh (autumn). *A. a. boavistae*: **5** ad rufous morph fresh (autumn). Thick-billed Lark *Rhamphocoris clotbey* (p. 1022): **6** ad ♂ fresh (autumn), **7** ad ♂ worn (spring), **8** ad ♀ partly worn, **9** juv.

Hoopoe Lark *Alaemon alaudipes*

PLATES: pages 1019, 1023

Du. Witbandleeuwerik   Fr. Sirli du désert   Ge. Wüstenläuferlerche   It. Allodola beccocurvo
Ru. Пестрокрылый пустынный жаворонок   Sp. Alondra ibis   Sw. Ökenlöplärka

Field characters. 18–20 cm, of which bill up to 4 cm; wing-span 33–41 cm. Body size close to Skylark but bill, wings, and legs much longer. ♂ up to 20% larger than ♀. Quite large, attenuated lark, with character on ground recalling large pipit until long, decurved bill (often held above horizontal) seen. Plumage sandy to grey above and stone-white below, with (on ground) little relief except for narrow black eye-stripe, moustache, and lower cheek border, dark spots on breast, and transverse dull black and white bands on folded wing. Appearance in flight dramatically different, with black and white bands along both wing surfaces (recalling Hoopoe) and mainly black, white-edged tail. ♀ and juvenile have fainter facial pattern.

Adult unmistakable, though at distance in haze confusion not only with large pipit but also courser possible. Juvenile unmistakable in flight but on ground could suggest rufous morph of Dupont's Lark (but that species always spotted on throat and chest, with typical lark wing pattern). Flight action loose and free but with illusion of weak flutter caused by wing pattern. Song-flight comprises short vertical climb, distinctive tip-over, and sudden descent. Walks and runs, often for several hundred metres. Stands erect in alarm with bill and head held up often recalling gaunt, attenuated Mistle Thrush.

Habitat. From arid tropics and subtropics to fringe of Mediterranean zone, mainly in flat or gently undulating lowlands. Attracted to coastlines, even down to seashore; in north-west Africa, highest density recorded in sublittoral zone. Requires ready access to sand or soft soil for probing, and thus favours wadis with sandy beds, small or even large dunes, abandoned cultivation, vehicle tracks, roadsides, sandy steppe with *Stipa* grass and saltmarsh, with either dense or sparse vegetation, especially where plants pile up heaps of wind-blown sand.

Distribution and population. SYRIA. Local in southern interior; commoner towards east. ISRAEL. 100–150 pairs in 1980s; declining. JORDAN. Fairly common. IRAQ. Common. KUWAIT. Breeds in all desert areas, including islands, in years of good winter rainfall. SAUDI ARABIA. In Harrat Al Harrah reserve, breeds in small numbers. TUNISIA. Common. ALGERIA. Most common between 28 and 35°N. May breed Lake Boughzoul in north. MOROCCO. Evidence of recent spread north into Moulouya valley spring 1983, but perhaps temporary and dependent on climatic conditions. CAPE VERDE ISLANDS. Only occurs Boavista and Maio, where common.

Accidental. Malta, Greece, Lebanon.

Beyond west Palearctic, breeds discontinuously along

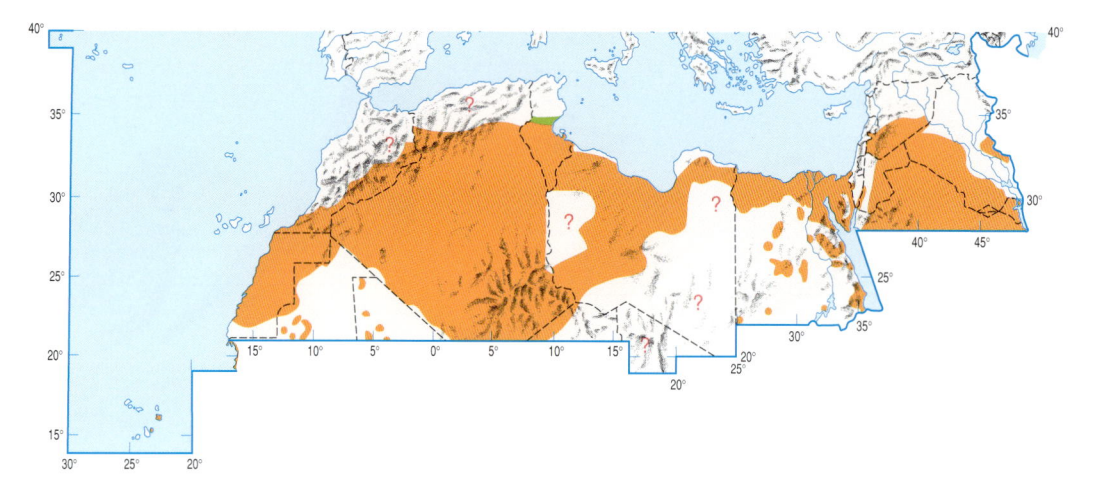

southern edge of Sahara from Mauritania to Somalia, and from Arabia to north-west India.

Movements. A true desert bird, generally very sedentary. Movements mainly apparent from extralimital occurrences; e.g. several November–February records from coastal plains of western Morocco and from northern Algeria, small parties not infrequent in winter on coastal dunes of Tripoli, Libya, and at least 26 records from Malta, most August–December.

Food. Largely insects and, where available, snails; also a little plant material. Much animal food obtained by digging with bill.

Social pattern and behaviour. Usually solitary outside breeding season. Breeding territories typically large. ♂ advertises territory by song-flights, often repeated at short intervals, taking off from top of bush or other low prominence; springs up, singing, on fluttering wings and with tail spread, rising more or less vertically for c. 1–4(–10) m, flips over, even somersaulting (thus displaying striking plumage pattern), then, holding wings close to body, nose-dives back to same perch; sings on descent and opens wings at last moment before landing.

Voice. Freely used in breeding season, when song of ♂ is characteristic feature of desert. Song typically of 3 parts: a series of pure fluting sounds. Beginning and ending of song may be given when perched, but in typical song-flights 1st series of fluting sounds given on ascent, trill as bird flips over at peak of ascent, and final part on descent. Contact-call a buzzing 'zeee' becoming a harsh 'shweee' in excitement.

Breeding. SEASON. Cape Verde Islands: from October, perhaps to February or March. North Africa: begins from end of February in southern desert areas, to May in north. Iraq: laying from mid- to end of May. SITE. In or on top of tussock or low shrub, 30–60 cm above ground; also, perhaps less often, on ground in shelter of rock or bush. Nest: when above ground, constructed of twigs woven into bush, and lined with softer material; on ground, in shallow scrape lined with twigs and with soft inner lining of wool and plant down. EGGS. Sub-elliptical, smooth and glossy; white to pale buff, variably spotted and blotched reddish-brown, with underlying lavender and grey markings, particularly at broad end. Clutch: 2–4. INCUBATION. About 2 weeks. FLEDGING PERIOD. Young stay in nest 12–13 days, fledging several days later.

Wing-length: *A. a. alaudipes*: ♂ 123–134, ♀ 111–120 mm.
Weight: *A. a. alaudipes*: ♂ 39–47, ♀ 30–39 g.

Geographical variation. Two colour morphs occur, one more rufous, the other greyer. Geographical variation rather slight, both in colour and size. Birds of Cape Verde Islands (*A. a. boavistae*) and North Africa (nominate *alaudipes*) rather small, size clinally larger through Middle East towards Iran. Rufous morph of *boavistae* has broader and deeper rufous pink-cinnamon fringes along feathers of upperparts than nominate *alaudipes* of North Africa, not sandy-pink; grey morph darker and purer grey than buffish-grey North African birds.

Dupont's Lark *Chersophilus duponti*

PLATES: pages 1021, 1023

DU. Duponts Leeuwerik   FR. Sirli de Dupont   GE. Dupontlerche   IT. Allodola del Dupont
RU. Жаворонок Дюпона   SP. Alondra de Dupont   SW. Dupontlärka

Field characters. 18 cm, of which bill 2 cm; wing-span 26–31 cm. Apart from bill, 15% shorter than Skylark but noticeably more robust than *Calandrella* larks. Small to medium-sized uncrested lark, with quite long, decurved bill, rather long neck, bulky body, and spindly legs; folded wing-points of rather short, broad wings, almost cloaked by long tertials, do not extend down tail as in most larks. Plumage pattern much as Skylark but noticeably scaled or evenly fringed on scapulars

Dupont's Lark *Chersophilus duponti*. *C. d. duponti*: **1** ad fresh (autumn), **2** ad worn (spring), **3** juv. *C. d. margaritae*: **4** ad fresh (autumn). *C. d. duponti-margaritae* intergrade: **5** ad fresh (autumn). Woodlark *Lullula arborea* (p. 1041). Nominate *arborea*: **6** ad fresh (autumn), **7** ad worn (spring), **8** juv. *L. a. pallida*: **9** ad fresh (autumn).

and folded wing; copious streaks on head interrupted by pale crown-stripe, supercilium, eye-ring, and half-collar below cheeks. Ground-colour of plumage varies from dark brown to chestnut.

Unmistakable when seen well, but rufous race *margaritae* subject to risk of confusion with juvenile Hoopoe Lark when size, wing pattern, and plain chest of Hoopoe Lark obscured. When bill not visible, build may provoke confusion with crested larks *Galerida* but Dupont's Lark lacks crest. Flight fast, with wing shape producing more flapping action than *Alauda*; normal flight and song-flight both end in sudden rapid perpendicular plunge, with flutter 2–3 m above ground. Prefers to escape by persistent running among ground cover. Inconspicuous except when singing and calling in first and last hours of daylight.

Habitat. Ecologically as well as geographically restricted; confined to dry and usually warm Mediterranean lowlands and plateaux. In Spain, on northern fringe of range, occurs both near coasts at heights of 50–120 m and far inland at 340–1200 m, at temperatures ranging from $-10°$ to $40°C$ and with very low rainfall. Always in open country, breeding on flat areas or on slopes not exceeding 25% gradient, with vegetation cover not more than 30–50 cm tall. Outside breeding season, such areas abandoned in favour of cereal fields (especially barley and oats), where birds assemble in mixed flocks with Skylark and Calandra Lark.

Distribution. Breeds only in west Palearctic. SPAIN. Range has declined in recent decades, at least regionally. Vulnerable to increasing cultivation of shrub-steppe habitat. PORTUGAL.

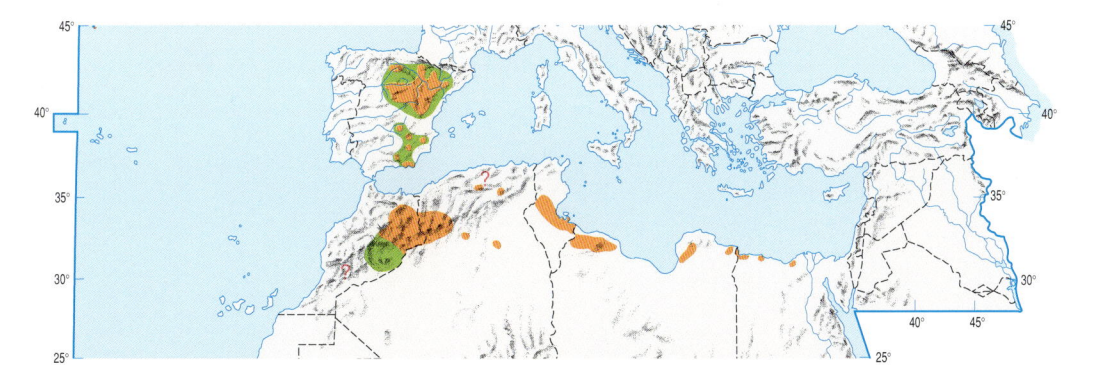

Recorded near Lisbon in 19th century and presumed to breed there then. LIBYA. Thinly distributed in Tripolitania, and probably rare and local in Cyrenaica. TUNISIA. Marked decline in north of range during 20th century. MOROCCO. Considered common on Hauts Plateaux (eastern Morocco) in 1950s–60s; still present there, but status inadequately known.

Accidental. France, Italy, Malta, Greece, Cyprus, Canary Islands.

Population. SPAIN. 13 000–15 000 pairs; slight decrease. No population figures for North Africa, but data suggest total numbers approximate to those in Spain. Regarded as scarce in Egypt and Tunisia, and nowhere very common in Algeria.

Movements. Resident. Some dispersal occurs, but very little known of this and it may be irregular. Has been found occasionally in Algerian Sahara, south of breeding range and outside normal habitat. Further, vagrants known from Mediterranean basin mainly September–December.

Food. Mainly insects and small seeds. Feeds by using bill like woodpecker to dig into ground or into sandy bases of plant tufts; also by splitting open balls of horse dung.

Social pattern and behaviour. Poorly known, partly because very inconspicuous. Not very sociable, and usually occurs singly or in pairs, though small parties sometimes recorded. Song of ♂ given in flight, on ground, or from stone or low plant. In song-flight, song delivered during ascent, in which bird may rise to *c.* 100–150 m, thus sometimes higher than Skylark and lost to view. May remain aloft for *c.* 30 min or even nearly 60 min. Descent sudden, vertical, and very rapid, with closed wings which are opened 2–3 m from ground.

Voice. Song of ♂ unlike that of any other lark; in timbre and overall quality, closer to twittering of Linnet. Commonest call, given in flight and from ground, a very human, fluting 2-syllable whistle, 2nd syllable rising sharply in pitch and rather creaky and nasal: 'hoo hee', 'coo-chic', or 'pu-chee'. Alarm-call a soft, rather quiet 'tsii' or whistling call.

Breeding. SEASON. Northern Algeria and Tunisia: eggs laid 1st week March to early June; perhaps 2 broods. SITE. On ground, set into tussock, hidden under bush, or sheltered by stone. Nest: scrape lined with available vegetation, including rootlets, fibres, small twigs, and hair, sometimes little or no softer lining. EGGS. Sub-elliptical, smooth and glossy; white to pink, densely spotted reddish-brown, with purplish-grey undermarkings. Clutch: 3–4 (2–5). (Incubation and fledging periods not recorded.)

Wing-length: *C. d. duponti*: ♂ 99–106, ♀ 88–95 mm.
Weight: *C. d. duponti*: ♂♀ 32–47 g.

Geographical variation. Marked. *C. d. margaritae* from southern Atlas mountains to north-west Egypt very distinctly rufous, not black and rufous like nominate *duponti* from Iberia; tail and hind claw distinctly shorter, bill distinctly longer. Differences in colour in part bridged by populations from Morocco to north-west Tunisia, which are locally rather variable.

Thick-billed Lark *Rhamphocoris clotbey*

PLATES: pages 1019, 1030

Du. Diksnavelleeuwerik   FR. Alouette de Clotbey   GE. Knackerlerche   IT. Allodola beccoforte
RU. Толстоклювый жаворонок   SP. Alondra piquigruesa   SW. Tjocknäbbad lärka

Field characters. 17 cm; wing-span 36–40 cm. At least 10% shorter than Calandra Lark but with much deeper bill. Medium-sized to large lark with remarkably deep, heavy bill and broad head, upright stance and long legs, and relatively

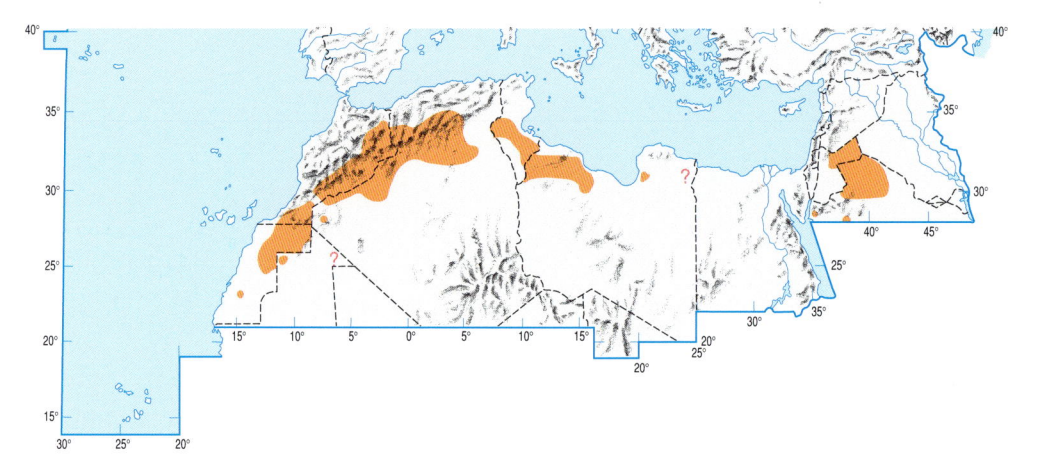

Black-crowned Finch Lark *Eremopterix nigriceps* (p. 1013): **1–2** ad ♂, **3** ad ♀. Chestnut-headed Finch Lark *Eremopterix signata* (p. 1012): **4** ad. Dunn's Lark *Eremalauda dunni* (p. 1014): **5** ad. Bar-tailed Desert Lark *Ammomanes cincturus* (p. 1015): **6–7** ad. Desert Lark *Ammomanes deserti* (p. 1017). *A. d. annae*: **8** ad. *A. d. algeriensis*: **9–10** ad. Hoopoe Lark *Alaemon alaudipes* (p. 1019): **11–12** ad. Dupont's Lark *Chersophilus duponti* (p. 1020). *C. d. duponti*: **13** ad. *C. d. margaritae*: **14** ad.

long-winged and short-tailed flight silhouette. Adult has uniform pink grey-brown upperparts, black-splashed face, bold black spotting from upper chest to vent, and broad white trailing edge to wings. Appears strikingly dark head-on, with pale bill, white throat, white eye-ring, and white vent obvious in contrast. Juvenile has only faint facial marks and weak spotting below.

Adult unmistakable; given clear sight of bill, juvenile also. Confusion with *Melanocorypha* larks possible at distance, particularly when only dark underwing catches eye, but unlikely to persist when bill and plumage details show. Flight powerful, with stronger action than in Skylark and recalling *Melanocorypha* larks. Song-flight apparently involves slow descent. Walks and hops; will run from danger. Stance noticeably upright, bird reaching up to tug at plants and standing noticeably erect in alarm.

Habitat. In Mediterranean and arid subtropical zones bordering or overlapping edges of desert, especially on open stony hammada where some sparse vegetation may be present. Accordingly ranked as a typical desert lark, but also found on wadi beds with grass and other plants or on stony-clay terrain with relatively rich vegetation, including succulents such as *Aizoon*, on clay beds with hillocks, and on enclosed areas between villages.

Distribution and population. Breeds almost exclusively in west Palearctic; probably extends further south in Arabia. SYRIA. Bred Syrian desert 1930–31, but uncertain whether within present-day borders. JORDAN. Scarce. May also breed in Sharra Highlands. SAUDI ARABIA. In Harrat Al Harrah reserve, breeds in small numbers on ground slopes without vegetation. EGYPT. 1st breeding record south of Marsa Matruh, Western Desert, 1995. LIBYA. May extend south to Hammada el Homra in west. TUNISIA. Scarce. MOROCCO. Uncommon to abundant. MAURITANIA. Occurs in north, but no breeding recorded.

Sporadic records in non-breeding areas are probably of nomadic dispersal rather than accidental status. Thus in Israel, 17 records 1960–89 in Arava valley and southern Negev, of which 13 (of 1–8 individuals) between 1980 and 1989. In Egypt, most records from Cairo–Suez road area (including flock of 50 birds in 1946); exceptional west of Nile valley. Also recorded occasionally from western Kuwait.

Movements. Basically resident, though subject to nomadic dispersal outside breeding season. Few precise data, but indication that some birds move towards coastal plains for winter.

Food. Seeds, green plant material, and insects. Reaches up to plant heads to pull them down to ground, and uses bill as clippers to cut green shoots. Will dig for food.

Social pattern and behaviour. Little studied. Outside breeding season, usually occurs in small scattered parties. Breeding pairs widely scattered. Song-flight of ♂ seldom witnessed and no detailed description available. ♂ apparently rises to considerable height, calls, and throws itself right and left in zigzag flight; descends to ground 'parachute' fashion.

Voice. Song of ♂ consists of medley of notes, tinkling and warbling, sweet and rather quiet. Contact-call, commonly associated with flight, a sharp 'prit'; other calls include a low 'coo-ee' and 'sree'.

Breeding. SEASON. North-west Africa: late March to late May, exceptionally February or even January. SITE. On ground, usually under bush or stone; less often in the open. Nest: shallow depression filled with lining of vegetation, often with rim of stones on open side. EGGS. Sub-elliptical, smooth and glossy; creamy-white to pink, finely speckled with red-brown or chestnut, with underlying greyish mottling; when fresh, characteristic pink or bright pink tint. Clutch: 3–5 (2–6). (Incubation and fledging periods not recorded.)

Wing-length: ♂ 125–134, ♀ 119–125 mm.

Calandra Lark *Melanocorypha calandra*

PLATES: pages 1025, 1030

Du. Kalanderleeuwerik   Fr. Alouette calandre   Ge. Kalanderlerche   It. Calandra
Ru. Степной жаворонок   Sp. Calandria común   Sw. Kalanderlärka

Field characters. 18–19 cm; wing-span 34–42 cm. About 10% longer and 25% bulkier than Skylark, with proportionately broader wings, shorter tail, and much heavier bill; about 10% larger than Bimaculated Lark. Large, robust, heavy-billed lark, with large wing area but no crest. Upperparts essentially brown and well streaked; underparts off-white, little streaked but with large black patches on sides of upper breast. Face dominated by heavy, rather conical bill and quite prominent buff-white supercilium and eye-ring. White trailing edge emphasizes almost black underwing; tail has white outer feathers.

Distinctive character, with weight of bill exceeding all other larks except Thick-billed Lark, and body bulk and leg length always obvious. Of the 4 *Melanocorypha* in west Palearctic, Calandra Lark the most widespread but subject to confusion with smaller, yet closely similar Bimaculated Lark, particularly in Near East where they may intermingle on passage. Hence neither of them instantly identifiable; see Bimaculated Lark. Flight free, with long and broad-based wings affording easy, even powerful flight action which lacks hesitancy of Skylark but has typical lark undulations and sweeps. Walks and runs more than hops and perches freely on shrubs. Stance generally fairly upright.

Habitat. In lower, middle, and marginally upper middle latitudes, subtropical, Mediterranean, steppe, and temperate, on open lowland plains and upland plateaux. Avoids rocky, gravelly, saline, and other infertile or degraded soils, and semi-deserts, but tolerates low and uncertain rainfall and regular summer heat up to 32°C. Essentially a steppe bird, at opposite end of spectrum from desert-living *Ammomanes* larks. Occurs primarily on grasslands, ranging from virgin steppe to cultivated crops, areas of profuse mixed herbage, and even water-meadows. Sometimes occurs among shrubs, bushes, or even well scattered low trees.

Distribution. Decline in range and numbers, attributed chiefly to changing farming practices. Range decrease reported from France, Iberia, Ukraine, and Moldova, and probably Albania, Greece, and Italy. FRANCE. Now absent in former breeding

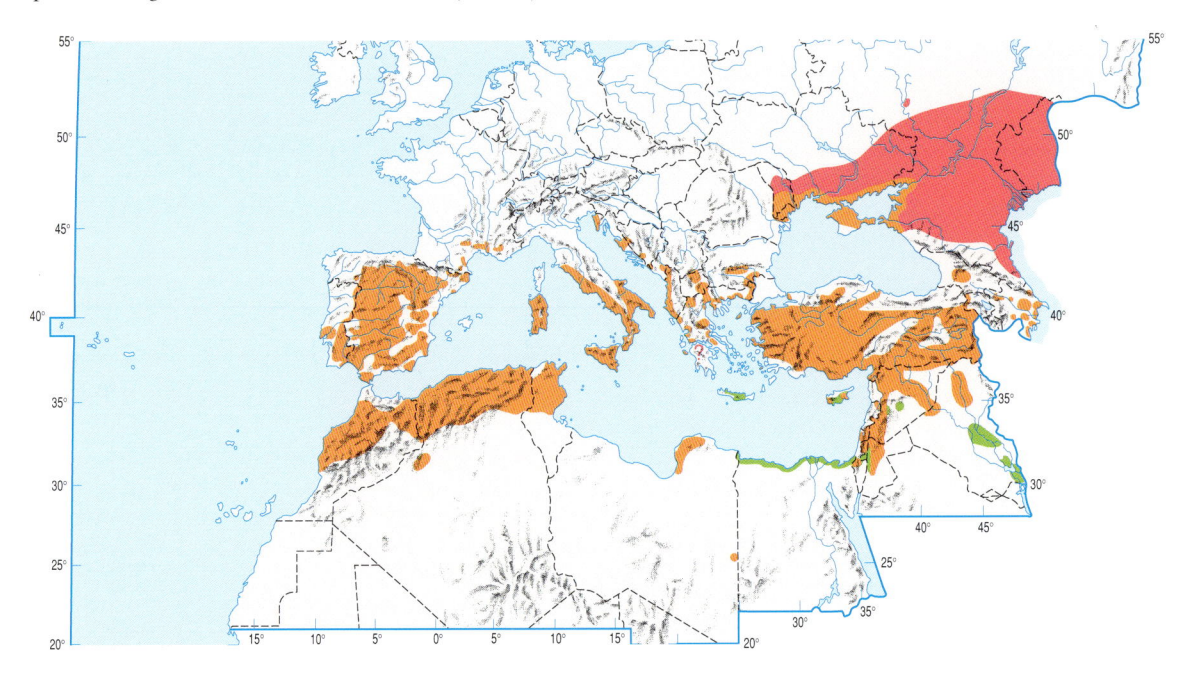

Calandra Lark *Melanocorypha calandra*. *M. c. calandra*: **1** ad fresh (autumn), **2** ad worn (spring), **3** juv. *M. c. hebraica*: **4** ad fresh (autumn). Bimaculated Lark *Melanocorypha bimaculata* (p. 1026). *M. b. rufescens*: **5** ad fresh (autumn), **6** ad worn (spring), **7** juv. Nominate *bimaculata*: **8** ad fresh (autumn), **9** ad fresh (autumn) with smaller chest-crescents.

areas, e.g. Camargue. Corsica: bred rarely 19th century. RUSSIA. Has spread north in Voronezh region. TURKEY. Apparently absent only from parts of Black Sea and Mediterranean coastlands, and scarce or absent at higher altitudes in east. CYPRUS. Range has increased.

Accidental. Britain, Luxembourg, Netherlands, Germany, Norway, Sweden, Finland, Poland, Czech Republic, Austria, Switzerland, Balearic Islands, Madeira, Canary Islands.

Beyond west Palearctic, extends east to central Asia, south to Iran.

Population. Some eastern populations stable, but many countries report decline. FRANCE. 50–150 pairs; decreasing. SPAIN. 1.03–3.4 million pairs; slight decrease. PORTUGAL. 10 000–100 000 pairs 1978–84; slight decrease. ITALY. 5000–15 000 pairs 1983–93; slight decrease. GREECE. 2000–5000 pairs. Probably slow decline since 1950s. ALBANIA. 300–800 pairs 1981; slight decrease. YUGOSLAVIA: CROATIA. 100–150 pairs; stable. BULGARIA. 1000–5000 pairs. RUMANIA. 25 000–40 000 pairs 1986–92; stable. RUSSIA. 1–10 million pairs; stable. UKRAINE. 400–1500 pairs in 1984; marked decrease. MOLDOVA. 3500–5000 pairs in 1990; slight decrease. ARMENIA. Very rare. AZERBAIJAN. Common. In south-east Shirvan, less common than Lesser Short-toed Lark and Crested Lark. TURKEY. 100 000–1 million pairs. CYPRUS. Probably 10 000–25 000 pairs; trend uncertain, but probably no decrease. ISRAEL. 4000–6000 pairs in 1980s. JORDAN. Rather scarce. TUNISIA. Common. MOROCCO. Common, especially north of Marrakech and in east.

Movements. Resident in southern Europe, Near East, and North Africa, though migratory to partially migratory in FSU. Though resident over much of west Palearctic range, birds not necessarily sedentary. Forms large flocks in autumn and winter, often mixing with Corn Bunting, and these wander to unknown extent. A rare and irregular passage migrant to Malta, March to early April and mid-September to early November; such birds probably too far west to be FSU emigrants. Main passages in FSU appear to be in October and March, though in autumn cold-weather exodus continues into November; spring return movement often obscured by nomadic movements of locally-wintering flocks.

Food. In summer, largely insects; in winter, seeds and grass shoots. Runs during feeding, taking items from ground. Digs for pupae with bill and can use it to crack frozen snow crust. Occasionally flies up to inspect tops of bushes from the air.

Social pattern and behaviour. Often markedly gregarious outside breeding season. Solitary and territorial when breeding, though neighbourhood groups may be formed in areas of high density. ♂ will sing on ground or from bush, etc., but more often initially when flying in regular circles at height of *c.* 10 m, then ascending and continuing to sing, sometimes very high. During circling flight wings fully extended and slow, deep wingbeats recall wader; dark undersurfaces prominent. Singing birds apparently stimulate one another to perform song-display; thus, song-flight commonly performed in small groups.

Voice. Song of ♂ not unlike Skylark but louder, more complex, and richer. Normal song a continuous flow of short phrases or single notes (often imitative of other species) and characteristic rapid chirping or vibrating sounds. Flight-calls all more or less sharp and trilling and characteristic 'kleetra' note may precede or interrupt song.

Breeding. SEASON. Spain: laying begins early April. Algeria: laying from early April to early June. Greece: laying from first half of April to early June. Southern FSU: 1st eggs late March in southern Ukraine, early April in Aral-Caspian area, mid-April in north-west Kazakhstan. 2 broods. SITE. On ground, under tussock. Nest: shallow depression, lined with grass stems and leaves, with inner lining of softer vegetation. EGGS. Sub-elliptical to oval, smooth and slightly glossy; whitish, sometimes greenish or yellowish, heavily spotted and sometimes blotched dark brown or red-brown, and pale purple; blotching sometimes concentrated at broad end. Clutch: 4–5 (3–6).

INCUBATION. 16 days. FLEDGING PERIOD. Unknown but young fed in nest for *c.* 10 days.

Wing-length: *M. c. calandra:* ♂ 126–141, ♀ 115–122 mm.
Weight: *M. c. calandra:* ♂ 54–73, ♀ 44–66 g.

Geographical variation. Slight, both in colour and size. 3 races recognized in west Palearctic. Slight clinal increase in size towards east, irrespective of races recognized, but birds of Algerian Hauts Plateaux large. Nominate *calandra* from southern Europe and North Africa, east to Ural steppes, is darkest race, feather-centres of upperparts black, sides dull grey-brown, brighter olive-brown or rufous-brown only when newly moulted; chest greyish, distinctly spotted black; sides of breast and flanks dull olive-grey, hardly rufous. *M. c. hebraica*, from Levant, paler; and *M. c. psammochroa*, from northern Iraq, paler and more sandy coloured.

Bimaculated Lark *Melanocorypha bimaculata*

PLATES: pages 1025, 1030

Du. Bergleeuwerik   Fr. Alouette monticole   Ge. Bergkalanderlerche   It. Calandra asiatica
Ru. Двупятнистый жаворонок   Sp. Calandria bimaculada   Sw. Asiatisk kalanderlärka

Field characters. 16–17 cm; wing-span 33–41 cm. About 10% smaller and relatively longer-winged than Calandra Lark. Medium-sized to large lark, of closely similar appearance to Calandra Lark but differing in whiter supercilium (in adult), darker lores, narrower but usually longer black crescents on upper breast, and absence of obvious white trailing edge to wings and sides to tail, which instead shows white terminal spots. In autumn, cheeks rusty. Underwing less dark than Calandra Lark.

Wing and tail pattern allow ready separation from Calandra Lark in close flight view but care always needed in observations of bird on ground or flying at distance. General character differs most in more pointed bill, shorter tail, and lesser bulk; proportionately longer wings noticeable on take-off.

Habitat. In lower-middle warm continental latitudes, in higher and rougher parts of steppe zone, replacing Calandra Lark up to at least 1650 m in Afghanistan, 2000 m in Armenia. Ascends mountains to limit of crops, and occurs on cultivated plateaux at high altitudes; also on dry heath, shrubland, stony tracts, and almost bare steppes. During migration uses lake shores, and fields of rice and stubble. Winters in India in barren semi-desert, sparse cultivation, harvested and fallow fields, margins of jheels, and on dry tidal mudflats.

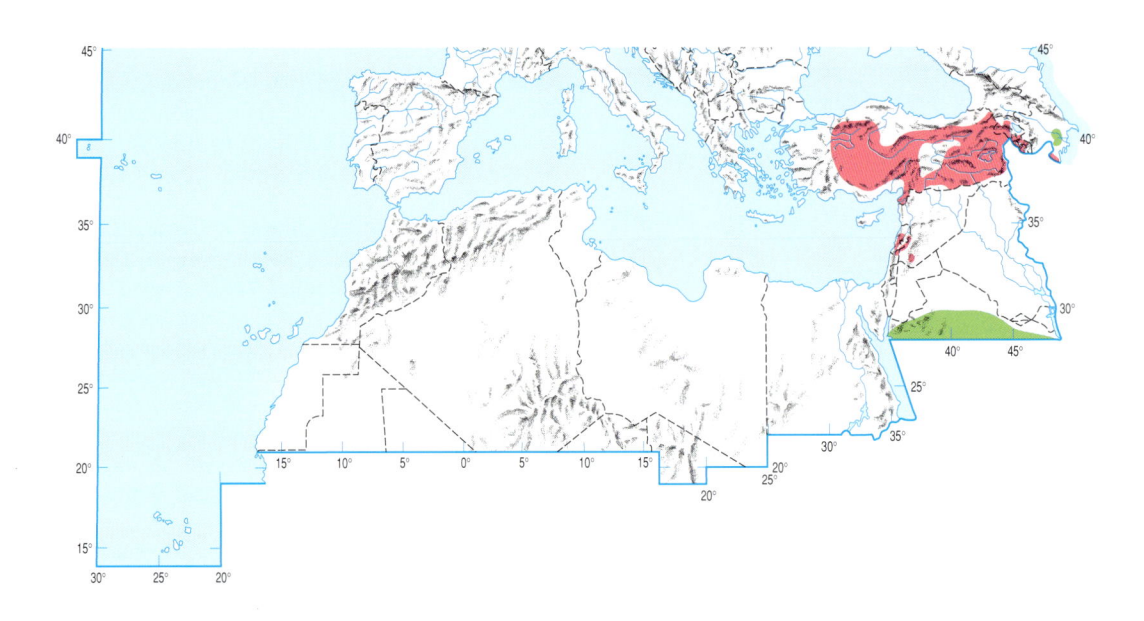

Distribution and population. ARMENIA. Widespread and locally very common. AZERBAIJAN. Uncommon to common in south. Very common in Nakhichevan region. TURKEY. Breeds central, east, and south-east Anatolia, also Taurus mountains; in some places extends into Mediterranean coastlands. 5000–50 000 pairs. SYRIA. Range concentrated on slopes of Anti-Lebanon. May also occur on spurs of Jabal ad Durūz in south. ISRAEL. Occurs only at 2 sites on Mt Hermon foothills. *c.* 40 pairs. IRAQ. In early 1900s several pairs found breeding along Samarra-Tekrit railway (Tigris area, 34–35°N); no further reports. KUWAIT. Breeding recorded only 1978.

Accidental. Britain, Sweden, Finland, Italy, Greece.

Beyond west Palearctic, breeds east to central Asia, south to Iran.

Movements. Migratory. Some winter on southern edge of breeding range, but majority migrate south to Pakistan and north-west India, Middle East, and north-east Africa. Timing of passages poorly known but, as in Calandra Lark, main periods appear to be October–November and March–April.

Food. Insects and seeds. When taking cultivated *Sorghum*, feeds either by nipping out grain from low seed heads (especially when seed still soft), by perching on large seed heads (more often as seed ripens), or by collecting fallen grain. In Sudan, recorded doing considerable damage by digging up freshly sown grain.

Social pattern and behaviour. Generally similar to Calandra Lark, including song-flight.

Voice. Song of ♂ substantially similar to Calandra Lark. Flight-call a 'prrp' or 'chirp', like Skylark but more mellow.

Breeding. SEASON. Armenia: laying begins first few days of May; latest fresh clutches mid-July. Possibly 2 broods. SITE. On ground in shelter of tussock or low bush, usually shaded from sun at hottest time of day. Nest: shallow scrape, lined with grasses and rootlets, walls of coarser material than floor; also clad in bits of old dung, rag, paper, etc. EGGS. Sub-elliptical to oval, smooth and glossy; white, greyish, or brownish, occasionally with olive tinge, heavily or lightly spotted and speckled, sometimes in patches, with light or dark brown and some pale purple. Clutch: 3–5, rarely 6. INCUBATION. 12–13 days. FLEDGING PERIOD. Leave nest at *c.* 9 days, before able to fly.

Wing-length: *M. b. bimaculata*: ♂ 118–128, ♀ 110–119 mm.
Weight: All races: ♂♀ 47–62 g.

Geographical variation. Slight in colour, virtually none in size. 2 races in west Palearctic. *M. b. rufescens* from central southern Turkey, northern Syria, Iraq, and Lebanon is more rufous than nominate *bimaculata* from northern Turkey and southern Transcaucasia.

White-winged Lark *Melanocorypha leucoptera*

PLATES: pages 1028, 1030

DU. Witvleugelleeuwerik   FR. Alouette leucoptère   GE. Weißflügellerche   IT. Calandra siberiana
RU. Белокрылый жаворонок   SP. Calandria aliblanca   SW. Vitvingad lärka

Field characters. 18 cm; wing-span 33–37 cm. Close in size to Calandra Lark but with more attenuated silhouette, due to relatively longer, even slightly forked tail; 15% longer and much bulkier than Snow Bunting (with which confusion possible). Upperparts chestnut on crown (paler on ♀ and juvenile) and wing-coverts; underparts essentially white with streaked chest and flanks. Flight feathers black, contrasting with thick white trailing edge above and also with white underwing coverts below.

In flight, shape and action recall Skylark as much as Calandra Lark.

White-winged Lark *Melanocorypha leucoptera* (p. 1027): **1** ad ♂ worn, **2** ad ♀ worn (spring), **3** juv. Black Lark *Melanocorypha yeltoniensis* (p. 1029): **4** ad ♂ fresh (autumn), **5** ad ♂ worn (spring), **6** ad ♀ fresh (autumn), **7** ad ♀ worn (spring), **8** ad ♀ very worn, **9** juv.

Habitat. In continental warm arid middle latitudes, lowland and to some extent upland, avoiding rocky and mountainous, forested, and wetland areas, and those humanly settled or much disturbed, although sometimes overlapping cultivation, especially on migration or in winter. Tolerates saline soils, but also lives on sand, clay, gravel, and sometimes on dark earth. Prefers vegetation neither tall nor dense.

Distribution and population. RUSSIA and KAZAKHSTAN. Range has decreased in last 100 years with ploughing up of steppes and creation of shelter-belts. Less numerous than most other larks. In Russia, perhaps 10 000–100 000 pairs.

Accidental. Britain, Germany, Norway, Finland, Poland, Slovakia, Austria, Switzerland, Italy, Malta, Greece, Yugoslavia, Turkey (regular winter visitor until about start of 20th century).

Beyond west Palearctic, extends east between *c.* 47° and 55°N to western foothills of Altai and Zaysan depression.

Movements. Short-distance migrant, mainly within FSU. Birds flock in midsummer, when breeding completed, and some movement begins then. Exodus protracted, and some birds still present in southern Urals in December; however, main passage occurs mid-August to October. Return movement proceeding in March; southern areas reoccupied by mid-April, though in northern parts main arrivals in late April or even early May.

Food. Insects and seeds in summer, seeds in winter.

Social pattern and behaviour. Normally gregarious outside breeding season; flocks (often large) are formed for migration, local movements, and winter feeding. Breeding territories small, closely grouped so as almost to constitute colonies. Song-flight usually performed at height of *c.* 10–20 m and not very long; bird ascends with fairly rapid and shallow wing-beats; wings then beaten slowly and deeply so that tips meet above and below body. Also, often sings on ground, from some eminence.

Voice. Song of ♂ of fine quality and similar to Skylark. Contact- and flight-calls include a slightly tinny, clear 'wed' or 'wäd', given at intervals; a drawn-out, quiet and rather hoarse squeal, rather like angry cat; and a call similar to Skylark's.

Breeding. SEASON. Southern FSU: laying starts late April, continuing to mid-July; most eggs laid in May. 1–2 broods. SITE. On ground, usually well-concealed and sheltered by tussock or other plant. Nest: shallow scrape, lined with carelessly woven basket of vegetation. EGGS. Sub-elliptical, smooth and glossy; very variable—whitish, pale green, or pale yellow, variably speckled and spotted olive, grey-brown, and grey, usually with ring of spots at broad end. Clutch: 5–6 (4–7). INCUBATION. 12 days. FLEDGING PERIOD. Unknown, but young leave nest still unable to fly.

Wing-length: ♂ 119–127, ♀ 111–117 mm.
Weight: ♂ 40–52, ♀ 36–48 g.

Black Lark *Melanocorypha yeltoniensis*

PLATES: pages 1028, 1030

Du. Zwarte Leeuwerik   Fr. Alouette nègre   Ge. Mohrenlerche   It. Calandra nera
Ru. Чёрный жаворонок   Sp. Calandria negra   Sw. Svartlärka

Field characters. 19–21 cm; wing-span 34–41 cm; ♂ 10% larger than ♀. Largest of genus in west Palearctic, ♂ exceeding Calandra Lark in average wing and tail length. Large, robust lark, of similar form to Calandra Lark except for slightly less deep-based bill. ♂ essentially black, except for yellow-horn bill and wide off-white or buff tips to head- and body-feathers in autumn and winter. ♀ much less distinctive, lacking uniformity of ♂ but extensively black-brown above and on flanks, with pale rump. Juvenile even less distinctive but displays adult's diagnostic dark under wing-coverts.

♂ almost unmistakable at any season; ♀ and juvenile subject to some risk of confusion with Thick-billed Lark (smaller in body but with larger bill and uniform pinkish crown and back), juvenile White-winged Lark (thick white trailing edge to wing), and Bimaculated Lark (white spots on end of tail). Underwing pattern differs from all other large larks with dark underwing in lacking white trailing edges; furthermore, in Black Lark, under wing-coverts distinctly darker than dusky undersurfaces of flight-feathers. Tail also almost uniformly dark, lacking obvious white edges of many larks. Flight, gait, and habits much as Calandra Lark.

Habitat. In middle continental latitudes, concentrated in steppe zone of warm dry summers and snowy winters, on broad open plains and rolling country, where wormwood *Artemisia* (especially) or feather-grass *Stipa* are dominant, with plenty of short grass; sometimes on saline or alkaline soils, and on clay, usually in neighbourhood of water, including freshwater or saline lakes. In winter, spends most time in areas where snow thinnest, either through wind or artificial action; thus will forage on roads, follow herds of horses, or even straggle into human settlements.

Distribution and population. RUSSIA and KAZAKHSTAN. Only breeding areas in west Palearctic. Decline in range (including some contraction eastwards) and numbers due chiefly to ploughing of dry steppe grasslands; remaining grassland vulnerable to intensification of cattle and sheep farming. Some populations perhaps stable in favourable habitat. In Russia, perhaps 6000–10 000 pairs.

Accidental. Germany, Norway, Sweden, Finland, Poland, Czech Republic, Austria, Italy, Malta, Greece, Rumania, Turkey, Lebanon. Rare and irregular in Azerbaijan, probably occurring only in cold winters.

Beyond west Palearctic, extends east between *c.* 47° and 55°N to Zaysan depression.

Movements. Dispersive and perhaps nomadic, mainly within FSU. Throughout breeding range, present all winter in roaming flocks (which may include birds from more distant breeding areas). Birds wander beyond breeding range in winter (sometimes in large numbers), extent probably varying from year to year according to severity of weather. Adult ♂♂ predominate in winter in breeding areas, almost exclusively so in colder seasons, which suggests that ♀♀ and young birds tend to disperse furthest.

Food. Insects and seeds. Digs through snow to reach food. In loose snow, makes distinctive channels 15–20 cm long, probably by moving snow with head and breast. If necessary, digs down up to 8 cm, then makes side tunnels of *c.* 10–12 cm to reach seeds. In late winter, especially, gathers by roads or around animal herds to feed on disturbed ground. Will drink saline water.

Social pattern and behaviour. Gregarious outside breeding season, often markedly so. Said to be generally in pairs on breeding grounds but far more ♂♂ present than can be

Thick-billed Lark *Rhamphocoris clotbey* (p. 1022): **1** ad ♂, **2** ad ♀. Calandra Lark *Melanocorypha calandra* (p. 1024): **3–4** ad. Bimaculated Lark *Melanocorypha bimaculata* (p. 1026): **5–6** ad. White-winged Lark *Melanocorypha leucoptera* (p. 1027): **7–8** ad. Black Lark *Melanocorypha yeltoniensis* (p. 1029): **9** ad ♂ fresh (autumn), **10** ad ♂ worn (spring), **11** ad ♀ fresh.

explained by relatively unobtrusive nature of ♀♀; mating system possibly thus not wholly monogamous. Dispersion in breeding season very uneven, both locally and over large area; compact neighbourhood groups occur and similar habitat near by may hold none or only a few scattered pairs. Song-display peculiar and striking, owing to variety of postures adopted, types of movement, and beauty of flight accompanied by rather melodious song. ♂ sings from ground or some eminence—hummock, clump of earth or snow, pile of straw—with wings drooped and tail cocked. At higher intensity, gradually changes to posture shown in Fig. 1: wings extended more to side, then forwards and flapped. Song perhaps given more often from ground than in flight, but, at least in interaction with ♀, ♂ regularly interrupts bouts of courtship to perform song-flight: ascends while singing, beating wings evenly and deeply so that they meet over back in soft wing-clap; whole effect like displaying pigeon. At height of *c.* 20–30 m, ♂ briefly stalls and initially glides down with wings raised or held below body, then plummets almost to ground, brakes, and glides horizontally for a few metres before landing gently.

Voice. Song of ♂ of astounding vocal virtuosity, richness of repertoire, and capacity for mimicry; contains rich purring trills and twitters, and perfect imitations of many other species. Other calls apparently mainly trilling or 'fizzing', but repertoire not well studied.

Fig. 1.

(FACING PAGE) Short-toed Lark *Calandrella brachydactyla* (p. 1032). *C. b. brachydactyla*: **1** ad fresh (autumn), **2** ad worn (spring), **3** juv. *C. b. rubiginosa*: **4** ad fresh (autumn). *C. b. longipennis*: **5** 1st autumn. *C. b. artemisiana*: **6** ad fresh (autumn).
Lesser Short-toed Lark *Calandrella rufescens* (p. 1034). *C. r. minor*: **7** ad fresh (autumn), **8** ad worn (spring), **9** juv. *C. r. nicolli*: **10** ad fresh (autumn). *C. r. apetzii*: **11** ad fresh (autumn). *C. r. heinei*: **12** ad fresh (autumn), **13** ad worn (spring). *C. r. aharonii*: **14** ad fresh (autumn). *C. r. persica*: **15** ad fresh (autumn). *C. r. rufescens*: **16** ad fresh (autumn). *C. r. polatzeki*: **17** ad fresh (autumn).
Asian Short-toed Lark *Calandrella cheleënsis* (p. 1036): **18** ad fresh, **19** ad worn.
Hume's Lark *Calandrella acutirostris* (p. 1034): **20** ad fresh, **21** ad worn.

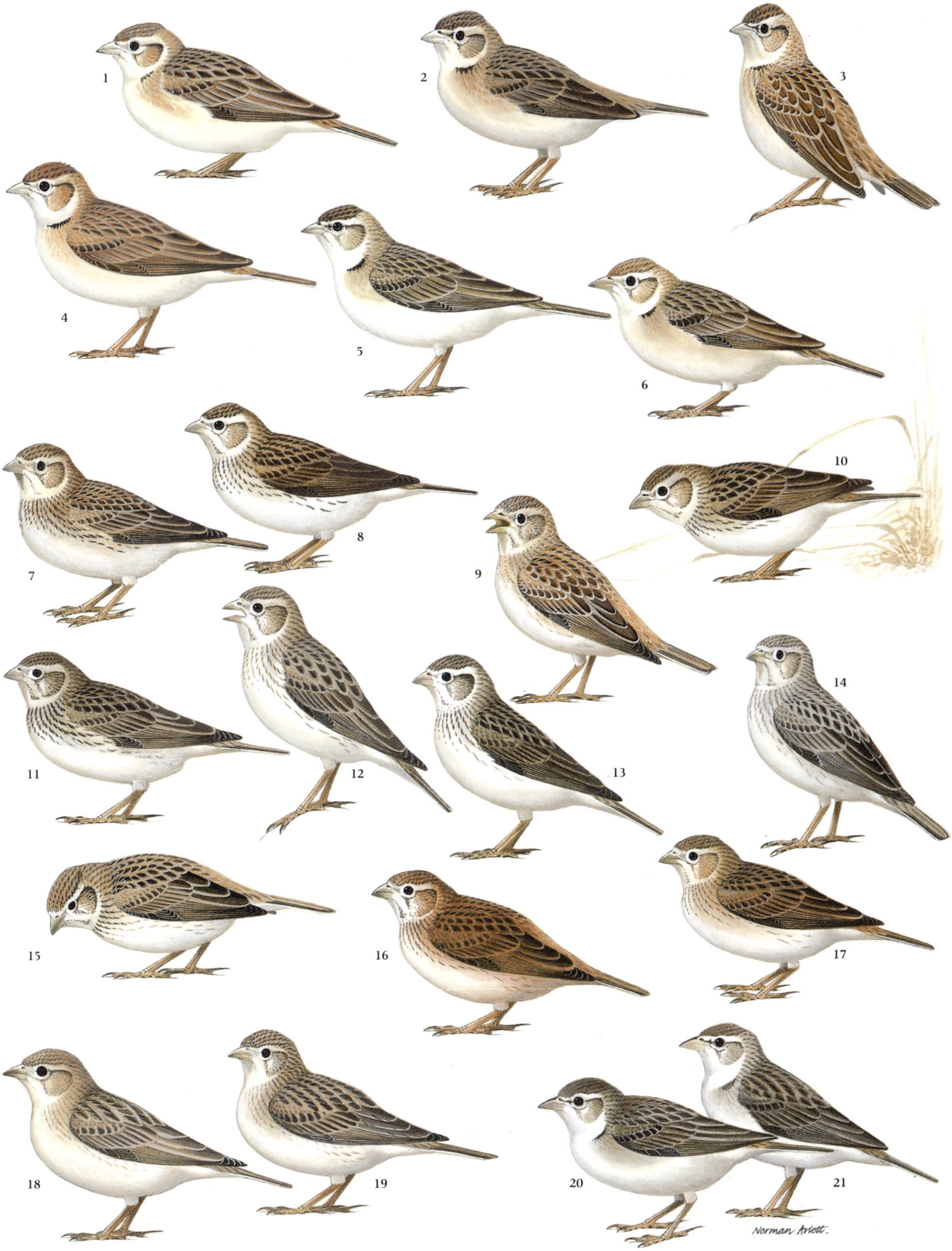
1
2
3
4
5
6
7
8
9
10
11
12
13
14
15
16
17
18
19
20
21
Norman Arlott.

Breeding. SEASON. Transvolga: first eggs laid late March. Orenburg region: first eggs laid late April. Kazakhstan: first eggs late April, latest clutch beginning of August. SITE. On ground in shelter of tussock. Nest: shallow depression lined with grass stems and other vegetation. EGGS. Sub-elliptical, smooth and glossy. Pale blue or olive-green with light brown, often olive, mottling and blotching, coalescing at broad end. Clutch: 4–7. INCUBATION. 15–16 days. (Fledging period not recorded.)

Wing-length: ♂ 132–142, ♀ 117–125 mm.
Weight: ♂ 56–76, ♀ 51–68 g.

Short-toed Lark *Calandrella brachydactyla*

PLATES: pages 1031, 1051

DU. Kortteenleeuwerik FR. Alouette calandrelle GE. Kurzzehenlerche IT. Calandrella
RU. Малый жаворонок SP. Terrera común SW. Korttålärka

Field characters. 13–14 cm; wing-span 25–30 cm. About 30% smaller than Skylark, with shorter, more finch-like bill, no crest, and rather more compact form. Small, rather flat- and square-headed lark, usually of pale, cryptic coloration and (when adult) lacking streaks on chest. Within west Palearctic, colour variable with western birds essentially warm sandy-buff above and eastern ones pale grey-ochre. Upperparts have typical lark pattern; underparts usually little-marked except for buff breast and sometimes-prominent small dark patch at shoulder. On a few (probably immature) spring vagrants and breeding birds, chest may show more streaks in scattered gorget. Tertials almost overlap tips of primaries, unlike Lesser Short-toed Lark. Tail pattern distinctive, with buff centre, then black panels and bright white outer feathers.

Commonest member of genus in west Palearctic, but unobtrusive and sometimes difficult to observe closely. Size and appearance overlap with Lesser Short-toed Lark (see that species) and with 8 other species of juvenile or adult lark: Dunn's Lark (similar size and long tertials but with relatively more massive bill and no chest-marks); ♀ and juvenile Black-crowned Finch Lark (20% smaller, with relatively deeper bill and black under wing-coverts); Desert Lark (15% larger, with longer, more pointed bill, no obvious streaking, no chest-marks, and dull tail pattern); Bar-tailed Desert Lark (same size but with bunting-like, not finch-like, character, no obvious streaking, no chest-marks, and dark terminal tail-band); juvenile Temminck's Horned Lark (10% larger, with no obvious streaking, no chest-marks, and longer tail); Skylark (25% larger, with short crest, and heavy streaks above and over all chest); Woodlark (15% larger, with short crest, heavy streaks above and over chest, and bold supercilium) and Hume's Lark (see that species).

Flight light and fast, slightly undulating, ending in sudden descent. Flying flocks keep in tight formation, like small finches. Song-flight begins with fluttering ascent, continues in undulating, high but small circles and ends in plummet.

Habitat. Ecologically intermediate between larks of desert or semi-desert and those adapted to more vegetation cover, breeding in middle and lower middle latitudes in steppe, Mediterranean, and fringing temperate zones. Basically a steppe bird, favouring dry open plains and uplands, terraces, slopes, and undulating foothills, which can be of sand or clay, sometimes stony or gravelly, with a variety of vegetational cover, from low bushy *garrigue* with mosaic of small patches of bare soil to neglected farmland, weedy fallows, stubble, harvest fields, and cereal crops growing up to *c.* 70 cm.

Distribution. FRANCE. Marked range reduction in 20th century. SLOVAKIA. First bred 1992. AUSTRIA. Bred Neusiedler See area 1966. SWITZERLAND. Only breeding record 1989 in Valais. SPAIN. Range has decreased. MALTA. Range decrease through loss of habitat. RUSSIA. Northward expansion in Voronezh region. UKRAINE. Slight decrease. KAZAKHSTAN. Ural delta: numbers severely reduced by rising water levels in recent years. JORDAN. No recent evidence of breeding Azraq. IRAQ. Distribution inadequately known. In early 1960s, bred alongside Lesser Short-toed Lark in Jazia area near Baghdad. EGYPT. Recorded nesting Wadi el Natrun (in north) in early 1900s. Breeding at various sites northern Sinai discovered 1989–90.

Accidental. Iceland, Britain (annual), Ireland, Belgium, Netherlands, Germany, Denmark, Norway, Sweden (now annual), Finland (annual), Poland, Austria, Switzerland, Madeira.

Beyond west Palearctic, extends east through central Asia to *c.* 120°E, south to Iran.

Population. Decline in Spain probably due mainly to fragmentation and loss of habitat through agricultural intensification; this likely to be cause of decreases elsewhere also. FRANCE. 1000–10 000 pairs in 1970s; slight decrease. SLOVAKIA. 1–10 pairs. HUNGARY. 10–15 pairs 1990; stable. SPAIN. 2.2–2.6 million pairs; slight decrease. PORTUGAL. 100 000–1 million pairs 1978–84; stable. ITALY. 15 000–30 000 pairs 1983–93. MALTA. 2000–3000 birds; numbers fluctuate. GREECE. 20 000–40 000 pairs. ALBANIA. 1000–3000 pairs 1981; slight decrease. YUGOSLAVIA: CROATIA. 1000–1500 pairs; stable. BULGARIA. 5000–10 000 pairs; stable. RUMANIA. 6000–8000 pairs 1986–92; stable. RUSSIA. 100 000–1 million pairs; slight increase. UKRAINE. 7000–11 000 pairs in 1984; slight decrease. AZERBAIJAN. Common, but less numerous than Lesser Short-toed Lark. TURKEY. 10 000–100 000 pairs. Locally common in west, scarce in east. CYPRUS. Perhaps 5000–20 000 pairs; probably stable. ISRAEL. A few thousand pairs in 1980s; declining. JORDAN. Breeds in small numbers. TUNISIA. Very common. MOROCCO. Common.

Movements. Migratory in Palearctic, except perhaps locally in southern parts of range where may be only partially migratory. Passage occurs on broad front across Mediterranean, Sahara, and Middle East to winter quarters in Africa south to Sahel and Red Sea, mainly within arid zone 14–17°N.

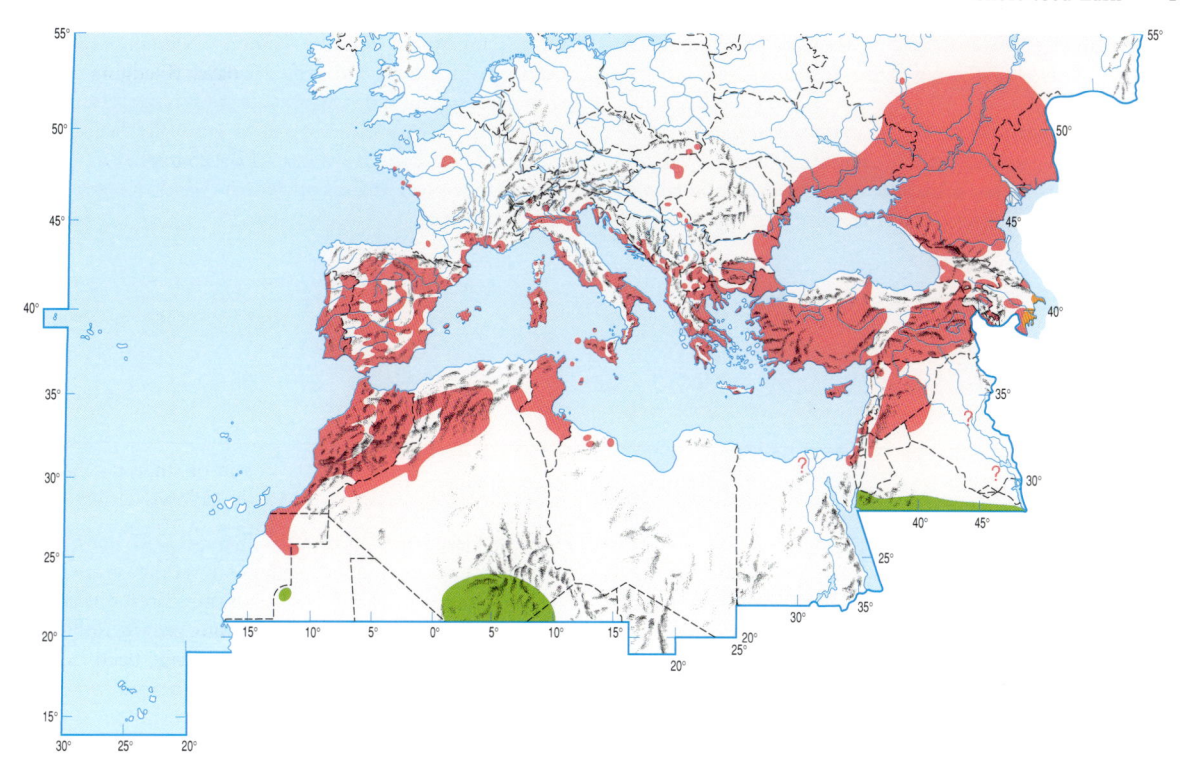

After flocking during July, autumn movement begins mid-August, at peak late August and early September in Europe and Turkey though continuing into October. Spring passage may begin early in west, in late January, but main passage of eastern breeders, through Malta and Cyprus, not until March.

Food. Insects and seeds in summer, mostly seeds only at other times. Takes food from ground or low plants, and often digs with bill.

Social pattern and behaviour. Highly gregarious outside breeding season, typically occurring in flocks of a few birds or up to several thousand. Flocks compact and highly coordinated. Breeding territories typically clustered, forming neighbourhood groups of 10–20 pairs; usually 10–20(–30) km between groups. ♂ sings mostly in flight but also not uncommonly on ground. Elaborate song-flight performed over nest-territory. Bird ascends steeply with rapidly beating wings to $c.$ 8–15 m; flies, frequently in spiral path, drifting from side to side, and giving introductory part of song; thereafter, gives main part of song, this continuing up to 30–50 m, whereupon bird extends wings, motionless, and, on final note of song-phrase, closes wings and descends, or may open wings to effect slower gliding descent. Before reaching ground, beats wings again a few times to achieve a much lesser and usually silent ascent, then drops down again and initiates a new major ascent, repeating sequence of song as for 1st ascent, and so on. Final descent to ground, usually a silent headlong dive, arrested $c. \frac{1}{2}$ m above ground by spreading wings and performing slightly undulating flight like pipit for 10–20 m before landing.

Voice. Used throughout the year, but especially in breeding season when song of ♂ dominant feature. In ascent phase of song-flight, often begins with single accelerating 'dip-dip...', not infrequently interspersed with mimicked calls, of a variety of species. After ascending to a certain height, introductory notes give way to main song: series of 10–20(–60) phrases, each phrase $c.$ 8–10 units, and phrases repeated persistently at short intervals, but with characteristic falling cadence at end. Commonest contact-calls, frequently given in flight, at all times of year, a soft 'chup' or a chirrup like sparrow. Excitement and alarm expressed by slight variants of contact-calls, e.g. 'girrtititt', 'tschrrt', or soft low 'kirk kirk kirk'.

Breeding. SEASON. North Africa: laying from beginning of April to early June. Spain: laying from early May to July. South-eastern Europe, including Cyprus: laying mid-April to June. Hungary: first eggs early May. 2 broods. SITE. On ground, usually in shelter of tuft of vegetation; sometimes in the open. Nest: shallow depression, lined grass leaves and stems, rootlets, etc., with inner lining of softer vegetation, feathers, thistle down, wool, etc. EGGS. Sub-elliptical, smooth and glossy. Whitish, creamy-white, or sometimes greenish or greyish, variably marked in 2 types: (1) heavily but evenly mottled pale brown and lavender grey; (2) spotted and blotched darker brown, with some pale purplish-grey, marks larger and denser at broad end, forming zone or cap. Clutch: 3–5(–6). INCUBATION. 13 days. FLEDGING PERIOD. 12–13 days; young leave nest at 9–10 days.

Wing-length: *C. b. brachydactyla*: ♂ 91–96, ♀ 88–93 mm.
Weight: *C. b. brachydactyla*: ♂ ♀ 20–26 g.

Geographical variation. Slight and predominantly clinal; boundaries between most races arbitrarily drawn. Variation especially difficult to assess because of marked individual variation and marked influence of abrasion and bleaching. In general, upperparts and ear-coverts become greyer and less rufous towards east, and supercilium and underparts become whiter, less pale buff or cream. *C. b. brachydactyla* from southern Europe east to western Yugoslavia generally heavily streaked black or black-brown on upperparts. 6 other races, slightly paler and/or more rufous, in North Africa, Levant, and other areas east of range of *brachydactyla*.

Hume's Lark *Calandrella acutirostris*

PLATES: pages 1031, 1051

Du. Tibetaanse Leeuwerik   Fr. Alouette de Hume   Ge. Tibetlerche   It. Calandrella di Hume
Ru. Тонкоклювый жаворонок   Sp. Terrera de Hume   Sw. Höglandslärka

Field characters. 12–14 cm; wing-span 27–30 cm. Close in size to Short-toed Lark but typically with longer, more pointed bill and squarer wing tip. South Asian, montane counterpart of Short-toed Lark, with similar character and behaviour. Appearance of nominate western subspecies very close to grey, eastern race *longipennis* of Short-toed Lark and separation from that form impractical except at close range or in hand. On average, crown and back even paler, less brown (making pale supercilium less conspicuous) but conversely rump and upper tail-coverts slightly more rufous in tone; pattern of tail similar but white outer edge noticeably narrower (due to white being confined to outer web of outermost feather and not covering both webs as in Short-toed Lark); underparts whiter, with typically any dark mark by shoulder duller or more broken up.

Beware marked variation of Short-toed Lark, of which not only *longipennis* but also Caucasian race *artemisiana* has convergent appearance; in latter, bill as long as Hume's Lark. In hand, distinguished with certainty by equal length of 4 longest primaries (4th distinctly shorter in Short-toed Lark). Calls include sharp 'trree' and disyllabic 'tre-lit'.

Habitat. Breeds at high altitude (up to 5000 m) in rocky valleys, montane steppe, and alpine meadows in south-west and southern central Asia, frequenting low stony foothills, semi-desert, and fallow land at lower levels in non-breeding season. Feeds, at times in largish flocks, on ground on seeds, buds, and insects, and nests on ground semi-colonially.

Distribution. Breeds in scattered populations from eastern Iran through Afghanistan, northern Pakistan, north-west India, extreme southern Kazakhstan, Tajikistan, and Kyrgyzstan to Nepal, Tibet, and central China.

Accidental. Israel: Eilat, adult trapped February 1986.

Movements. Migratory, some populations possibly only partially, wintering mainly in lowlands on northern and central Indian subcontinent from Pakistan to Bangladesh; some in (e.g.) Iran and Nepal move to lower altitude. In Kazakhstan, arrives end of April or beginning of May, and departs September–October.

Wing-length: ♂ ♀ 81–102 mm.

Lesser Short-toed Lark *Calandrella rufescens*

PLATES: pages 1031, 1051

Du. Kleine Kortteenleeuwerik   Fr. Alouette pispolette   Ge. Stummellerche   It. Pispoletta
Ru. Серый жаворонок   Sp. Terrera marismeña   Sw. Dvärglärka

Field characters. 13–14 cm; wing-span 24–32 cm. Most races are of similar bulk to Short-toed Lark but all show longer wing-point (since tertials fall noticeably short of primary-tips) and even shorter, less pointed bill. Small lark with very similar character to Short-toed Lark but differing from adult of that species in more heavily streaked and browner upperparts and chest. Lacks discrete dark marks at shoulder; pale supercilia fully join over bill, forming pale forehead.

Eminently confusable with juvenile congeners but less likely than Short-toed Lark to be mistaken for other small or medium-sized larks, since plumage pattern far less variable and closely matches only that of much larger Skylark. Best separated from Short-toed Lark by voice (see below). Normal flight and gait as Short-toed Lark but relatively rounder wing sometimes visible in flight silhouette.

Habitat. Breeds in middle latitudes, in continental steppe, Mediterranean, and semi-desert zones, overlapping widely with Short-toed Lark, and sometimes sharing breeding territories, as on *Salicornia* patches in marismas of Coto Doñana (Spain), although not on grassier lowlands used exclusively by Short-toed Lark. Another area of overlap, around Azraq (Jordan), finds Short-toed Lark in small minority, while Lesser Short-toed Lark flourishes so much on sandy or silty ground with low to medium shrub cover as to be ranked the commonest of all local passerines. Seems to subsist on barer, poorer, drier, more saline, or more clayey or gravelly sites than Short-toed Lark, thus reducing real degree of competition even where they exist side by side, filling subtly complementary roles.

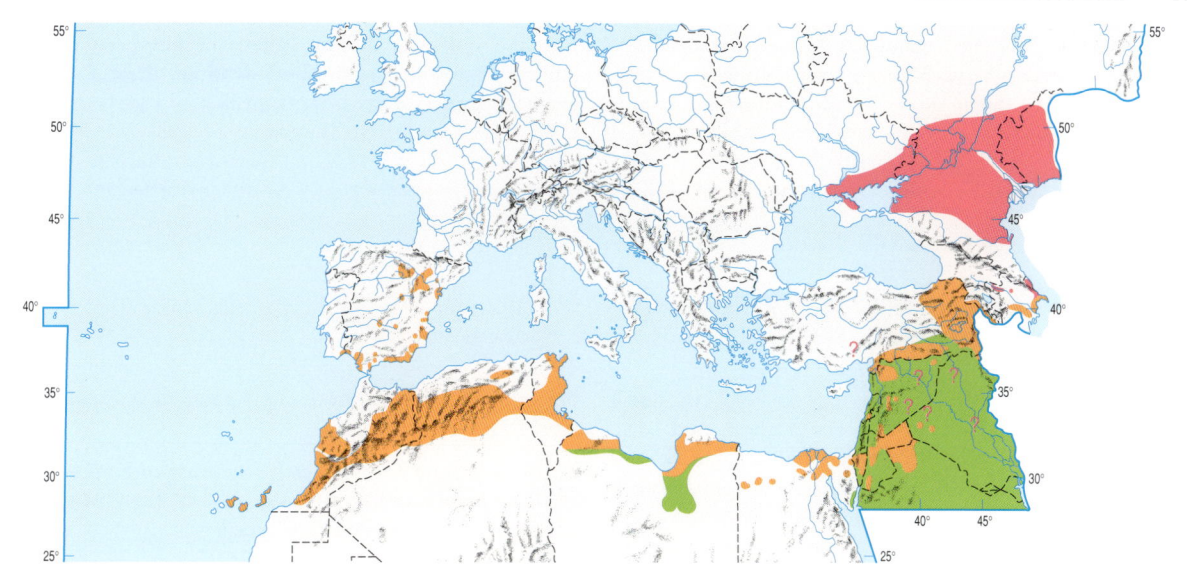

Distribution. TURKEY. Records of breeding coastal western Anatolia and southern coastlands require confirmation. SYRIA, IRAQ. Range incompletely known. JORDAN. Very common in eastern desert, especially around Azraq. SAUDI ARABIA. In Harrat Al Harrah reserve, probably the most numerous breeding bird, occurring wherever there is vegetation. MOROCCO. Widespread and plentiful; locally common in east. CANARY ISLANDS. Common Lanzarote and Fuerteventura, decreasing Gran Canaria and Tenerife. Perhaps also breeds La Gomera.

Accidental. Britain, Ireland, France (Corsica), Germany, Norway, Sweden, Finland, Austria, Switzerland, Balearic Islands, Italy, Malta, Greece, Bulgaria, Rumania. Some records, especially in south-east, may involve Asian Short-toed Lark.

Beyond west Palearctic, breeds Iran north-east to northwest Altai mountains.

Population. SPAIN. 230 000–260 000 pairs; slight decrease. PORTUGAL. 10–100 pairs 1990. RUSSIA. 100 000–1 million pairs; stable. UKRAINE. 10 000–17 000 pairs in 1984; slight decrease. AZERBAIJAN. Common to very common. TURKEY. 10 000–100 000 pairs (includes Asian Short-toed Lark). ISRAEL. A few thousand pairs. EGYPT. Fairly common. TUNISIA. Common. ALGERIA. Less common than Short-toed Lark. CANARY ISLANDS. One estimate of 17 000–19 000 pairs.

Movements. Resident to dispersive (nomadic) in western parts of west Palearctic range, dispersive to migratory in north-east. Map shows general area of wintering.

Food. Largely insects in summer, more seeds in spring and autumn, and presumably largely seeds in winter.

Social pattern and behaviour. Generally similar to Short-toed Lark, but song-flight lower and in wider circles; lacks obvious undulations but involves changes of speed.

Voice. Song quite unlike Short-toed Lark: jaunty, jerky jangle of continuous phrases, with only brief interjections of calls and rare plaintive notes; may include excellent mimicry. Commonest call less dry, more protracted, and sharper than Short-toed Lark: loud, rattled or buzzed 'prrt' or 'prrirrick', often 'chirrit'; clipped double 'r' and terminal 't' or 'ck' much more audible than in Short-toed Lark.

Breeding. SEASON. Canary Islands: eggs laid from 2nd week of March. Algeria and Tunisia: eggs laid from 2nd week of April to early June. Spain: eggs laid from April. Southern FSU: eggs laid April–June. Probably 2 broods. SITE. On ground in shelter of tussock. Nest: shallow scrape, lined with vegetation. EGGS. Sub-elliptical, smooth and glossy; very variable, usually whitish, yellowish, or buff, more or less spotted and blotched dark brown. Clutch: (2–)3–5. INCUBATION. Not recorded. FLEDGING PERIOD. Young leave nest at 9 days.

Wing-length: *C. r. apetzii* (Spain): ♂ 87–92, ♀ 81–87 mm. *C. r. heinei* (south European FSU): ♂ 95–102, ♀ 89–98 mm.
Weight: *C. r. heinei*: ♂ 23–29, ♀ 21–30 g.

Geographical variation. Marked, complex. Involves colour (mainly ground-colour and width of shaft-streaks of upperparts and chest, and amount of white in tail), size (as expressed in length of wing and tail) and shape of bill (heavy and conical or small and fine). 9 races in west Palearctic. Nominate *rufescens* (Tenerife), *apetzii* (Spain), *nicolli* (Nile delta), and *pseudobaetica* (eastern Turkey) are darkest races, with upperparts with broad and heavy black streaks, chest also heavily marked. Ground-colour of upperparts of nominate *rufescens* rufous-brown or cinnamon-brown, on chest rufous-cinnamon. Iberian *apetzii* has even more extensive black marks than nominate *rufescens*, below extending to throat and even chin, but ground-colour paler and more strongly contrasting, pale grey-brown or cream-buff on upperparts, cream or off-white on chest. Eastern races generally largest, western races smallest.

Asian Short-toed Lark *Calandrella cheleënsis*

PLATES: pages 1031, 1051

Du. Mongoolse Kortteenleeuwerik   Fr. Alouette de Swinhoe   Ge. Salzlerche   It. Calandrella della Mongolia
Ru. Солончаковый жаворонок   Sp. Terrera mongólica   Sw. Asiatisk korttålärka

Field characters. 14–15 cm; wing-span 25–33.5 cm. Small lark of similar size and structure to Lesser Short-toed Lark. Only recently separated from Lesser Short-toed Lark and incompletely studied in field. In race inhabiting central plateau of Turkey, upperparts pale stone-grey, well-streaked with black; underparts white, faintly washed cream beneath gorget of narrow, short black streaks; outer tail feathers show more white than in Lesser Short-toed Lark. Within west Palearctic range, overlaps with rather large race of Short-toed Lark (*artemisiana*) which has noticeably longer bill, usually rufous crown, buffier grey upperparts, unstreaked breast, and different voice.

Habitat. Breeds in middle latitudes in dry continental steppe and semi-desert, but particularly on barren, flat salt and soda plains with scattered scrubby vegetation (especially wormwood), and saline clay depressions; also at edges of spring floodwaters and after breeding often by lakes and rivers in wet meadows, etc. and in fields.

Distribution and population. TURKEY. Confined mainly to barren fringes of salt and soda lakes of Central Plateau; locally very common. SYRIA. Collected in El Qaryatayn in central Syria, but no proof of breeding. IRAQ. Breeds central and eastern areas. Map shows general area of probable wintering.
  Accidental. See Lesser Short-toed Lark.
  Beyond west Palearctic, breeds Iran and Turkmenistan east to north-east China and Baykal region.

Movements. Apparently mainly sedentary but some populations at least partially migratory; in Turkey, winter numbers on central plateau much lower than between April and late September. In Kazakhstan, winters within breeding range, only a small number of birds moving south but still remaining close to breeding grounds. In Mongolia, probably some overwinter but others head south from late August (mainly September), returning from April.

Food. Adult and larval insects, particularly caterpillars, beetles, and Orthoptera; in spring also seeds, which are main food in autumn and winter, including cereals.

Social pattern and behaviour. Solitary and territorial, forming loose groups of 2–3 broods after fledging, and some records of flocks of 10–20 birds, perhaps more on migration. Forms pairs from end of February, and post-breeding flocks from late July. Behaviour as Lesser Short-toed Lark.

Voice. Has much in common with Lesser Short-toed Lark; varied and includes much mimicry, e.g. of Desert Wheatear. Song is simple and flowing, not very melodious, delivered from ground, low shrub, or in generally low song-flight; Skylark-like but not so 'jubilant', also long series of 'tije-tije-tije . . .'. Described as considerably longer, more flowing, and richer in notes than Short-toed Lark. Calls (described for race of south-east Iran) include a distinctive percussive rattle, strongly recalling Lapland Bunting, and a sonorous, ringing 'üt üt üt'.

Breeding. SEASON. In Turkey, young recorded in nest mid-June; eggs laid in Kazakhstan and Mongolia April–July; 2 broods seem to be usual. SITE. In scrape in ground usually under small shrub. Nest: solid and carefully-woven structure of fine herb and grass stems lined with plant down, feathers, etc. EGGS. Sub-elliptical, smooth and glossy; yellowish-white with olive-brown spots concentrated at broad end. Clutch: (3–)4–5. INCUBATION. No information. FLEDGING PERIOD. *c.* 9 days.

Wing-length: ♂♀ 87–107 mm.
Weight: ♂♀ 19–27 g.

Crested Lark *Galerida cristata*

Du. Kuifleeuwerik   Fr. Cochevis huppé   Ge. Haubenlerche   It. Cappellaccia
Ru. Хохлатый жаворонок   Sp. Cogujada común   Sw. Tofslärka

PLATES: pages 1038, 1051

Field characters. 17 cm; wing-span 29–38 cm. Slightly shorter overall than Skylark but distinctly bulkier about head and body, with rather long, strong bill and rather short, broad tail. Medium-sized lark, with long spiky crest on rear crown, portly character on ground due to deep belly and rather short tail, and compact silhouette in flight due to broad wings and short tail. Plumage pattern and colours recall dull Skylark but differ distinctly in stronger facial marks, heavy moustaches, more open chest-streaks on paler ground, more uniform upperparts and buff outer tail-feathers. Wing lacks white trailing edge but underwing of European races glows orange-buff.

In Europe (north of Iberia) and throughout Middle East, only representative of its genus and only liable to confusion (on ground) with dull Skylark showing more crest than usual; confusion quickly dispelled when differences in wing and tail pattern visible. In Iberia, southern France, and North Africa (east to Libya), often occurs close to Thekla Lark and all too easy to confuse with it, since both specific and racial characters subject to overlap and even reversal. For distinctions, see Thekla Lark. Flight over short distances noticeably flapping and floating, with low glides and heavy flutter on landing. Song-flight irregular but will ascend to heights of up to 150 m and sing for 3 min or more while circling; wing-beats far less rapid than in Skylark. Shares with Thekla Lark characteristic shuffle when feeding, this action being exaggerated by deep, often loose belly feathering.

Habitat. Widely spread across and beyond continental west Palearctic, from fringe of boreal zone through temperate, steppe, Mediterranean, arid semi-desert, and desert zones, including oases. Prefers lowland plains and levels, although ascending in Atlas mountains of north-west Africa to 1260 m. Basic habitat is open, dry, often warm and dusty, flat or gently sloping, with very low or sparse vegetation not covering more than c. 50% of territory, any trees or shrubs present being widely spaced and not hemming it in. Attracted by various man-modified areas simulating semi-desert features, such as railway yards, parade grounds, airfields, harbour surrounds,

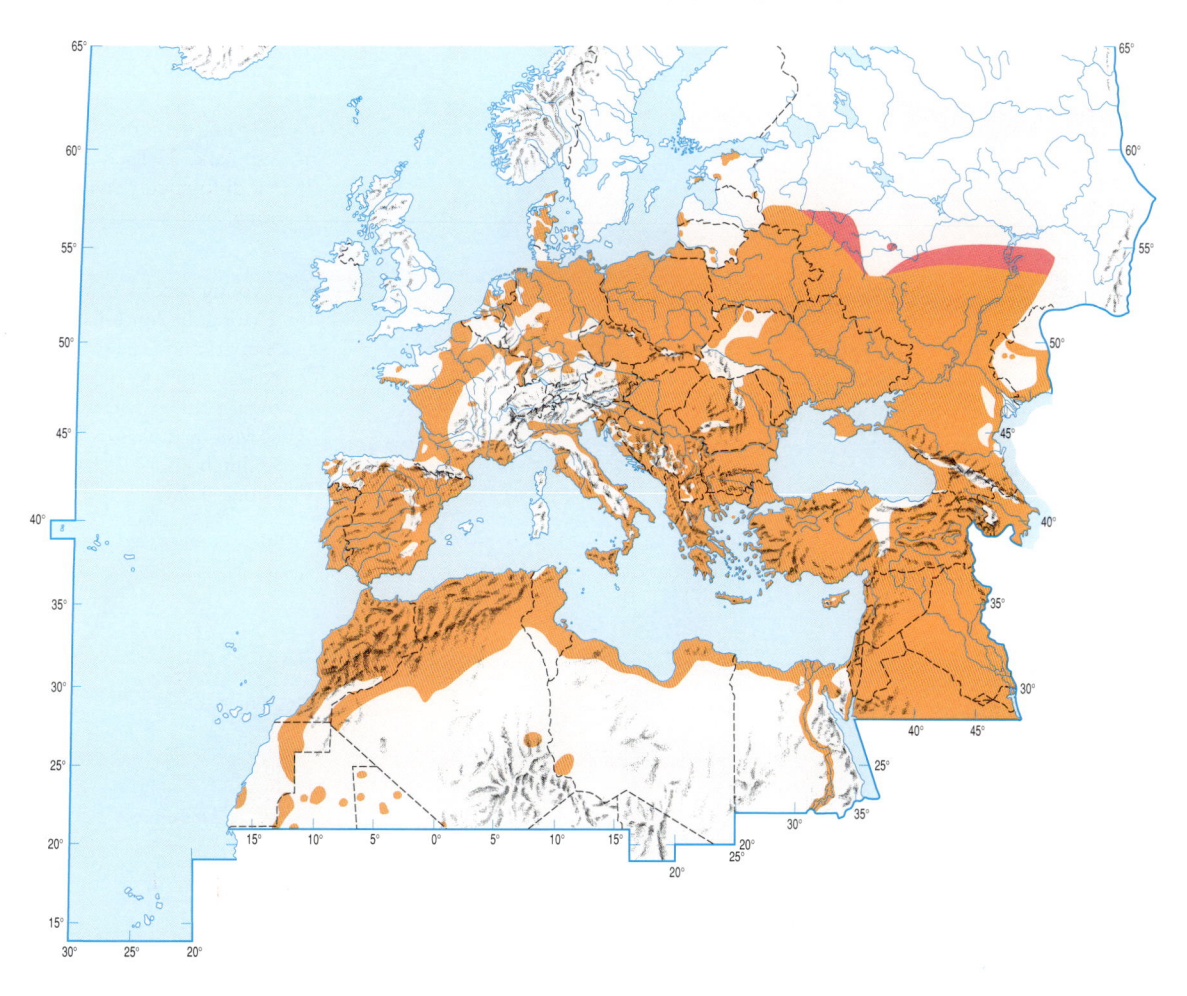

Norman Arlott

gravel pits, refuse dumps, and urban and industrial wastelands.

Distribution. Widespread decline in range and numbers in western and central Europe over several decades, probably chiefly due to loss of habitat through changing farming practices, also to reduction in number of horses (seeds in animal-dung being frequent food source). Now extinct in several countries, with last breeding records in Luxembourg 1973, Norway 1972, Sweden 1989, and Switzerland 1980s. JORDAN. Very common in north; far fewer in south. IRAQ. Widespread and common. KUWAIT. Breeds very commonly all desert areas including islands, only in years of good rainfall. SAUDI ARABIA. In Harrat Al Harrah reserve, numerous in sandier areas. ALGERIA. Very common in north.

Accidental. Britain, Finland, Balearic Islands, Malta, Chad, Canary Islands.

Beyond west Palearctic, extends east through central and southern Asia to north-east China and eastern India, south to Arabia; in Africa, breeds from Mauritania and Sénégambia east to northern Somalia, locally further south.

Population. Many European countries report decreases. FRANCE. 1000–10 000 pairs in 1970s. BELGIUM. In 1990, 70 pairs at coast, c. 35 pairs inland; c. 400 pairs 1971. NETHERLANDS. 400–450 pairs 1989–91; 3000–5000 pairs 1973–7. GERMANY. 12 000–18 000 pairs. DENMARK. Probably only 50–100 pairs; 300–500 pairs 1987–8. ESTONIA. c. 50–100 pairs 1991. LATVIA. 10–50 pairs in 1980s. LITHUANIA. Extremely rare. POLAND. 3000–5000 pairs. CZECH REPUBLIC. 1100–2200 pairs 1985–9. SLOVAKIA. 1000–2000 pairs 1973–94. HUNGARY. 50 000–60 000 pairs 1979–93; stable. AUSTRIA. Continuing decline; one estimate 150–200 pairs. SPAIN. 400 000–1 million pairs. PORTUGAL. 10 000–100 000 pairs 1978–84; stable. ITALY. 200 000–400 000 pairs 1983–93. GREECE. 40 000–70 000 pairs; probably stable. ALBANIA. 10 000–20 000 pairs 1981. YUGOSLAVIA: CROATIA. 20 000–25 000 pairs; stable. SLOVENIA. 800–1000 pairs. BULGARIA. 100 000–1 million pairs; stable. RUMANIA. 200 000–400 000 pairs 1986–92; stable. RUSSIA. 100 000–1 million pairs; stable. BELARUS'. 3500–4000 pairs in 1990; fluctuating, but some evidence of decline. UKRAINE. 15 000–18 000 pairs in 1988. MOLDOVA. 18 000–25 000 pairs in 1990; stable. AZERBAIJAN. Commonest lark; probably not less than tens of thousands of pairs. TURKEY. 1–10 million pairs. CYPRUS. Probably 100 000–150 000 pairs; probably stable. ISRAEL. A few hundred thousand pairs in 1980s; declining. EGYPT.

(FACING PAGE) Crested Lark *Galerida cristata*. *G. c. cristata* (central Europe): **1** ad fresh (autumn), **2** ad worn (spring), **3** juv. *G. c. subtaurica* (central Turkey): **4** ad fresh (autumn). *G. c. riggenbachi* (western Morocco): **5** ad fresh (autumn). *G. c. macrorhyncha* (northern Sahara): **6** ad fresh (autumn). *G. c. arenicola* (northern Sahara): **7** ad fresh (autumn). *G. c. festae* (Cyrenaica): **8** ad fresh (autumn). *G. c. nigricans* (south-west Libya): **9** ad fresh (autumn). *G. c. maculata* (Nile Valley): **10** ad fresh (autumn). *G. c. brachyura* (southern Cyrenaica): **11** ad fresh (autumn).

Thekla Lark *Galerida theklae* (p. 1040). *G. t. theklae*: **12** ad fresh (autumn), **13** ad worn (spring), **14** juv. *G. t. ruficolor* (central Morocco, coastal Algeria and Tunisia): **15** ad fresh (autumn). *G. t. erlangeri*: **16** ad fresh (autumn). *G. t. carolinae*: **17** ad fresh (autumn). **18** ad worn (spring).

Abundant. TUNISIA. Very common. MOROCCO. Uncommon, only locally abundant in sandy habitats. MAURITANIA. Common.

Movements. Largely migratory in north of FSU breeding range. Mainly resident elsewhere; some dispersal occurs, but scale uncertain. Apparently sedentary in North Africa and Middle East, where birds show much subspeciation and adaptation of plumage colour to that of local soils.

Food. Plant material (seeds, also leaves) and invertebrates (especially beetles); fewer invertebrates in winter. Most food taken from on or below ground surface. Digs with blows of bill to left and right, making funnel-shaped hole c. 2 cm across and c. 2 cm deep; towards the end, both hacks and pushes bill into ground. Often extracts seeds from seed heads on ground and sometimes turns over leaves. From low plants (up to 10–12 cm high) will take seeds directly, also tips of shoots (etc.) and invertebrates. Will take insects by aerial-pursuit. Most seeds and fruit swallowed whole but will de-husk cereal grain and grass seed by knocking it against ground.

Social pattern and behaviour. Not markedly gregarious. Outside breeding season, sometimes occurs singly, though generally in pairs or small parties of 3–4 or up to 10–15. May roost communally, each bird hollowing out a small pit for itself when roosting in sand or loose soil. Typically solitary and territorial in breeding season. ♂'s song given from ground, perch, or in flight. In song-flight, ♂ usually takes off from elevated point, ascends at angle into wind and starts loud song only at c. 30–70 m; may continue ascent to c. 100–200 m. Singing bird uses slow fluttering wing-beats, frequently hovering in wind, then continues silently on undulating path, sings again, etc. Overall, flight less hovering than Skylark and flight-path involves more circles, arcs, and undulations.

Voice. Song of ♂, given in flight or from perch or ground, comprises long, soft fluting sounds, short whistles, pulsating elements, loud double notes and tremolos, and odd twittering sounds. Characteristically includes many imitations of other bird species. Courtship-song quiet, rambling and chattering, not unlike twittering song of Swallow. Contact-alarm calls, commonly given by both sexes from ground, perch, or in flight, throughout the year, characteristically pure, musical, liquid, and lilting; rather thin and ethereal. Renderings include 'peeleevee', 'di-dji-djii', 'twee-tee-too'. Shrill and loud when excited, soft and sweet when few birds feeding together. Flight-call a gentle fluting 'djui' or 'too-hea'.

Breeding. SEASON. France and Germany: March–July. Spain: April–June. North Africa: March–June. Southern FSU: from mid-April. 2–3 broods. SITE. On ground in the open, or in shelter of low shrub or tussock, also under low bank. Nest: shallow depression with untidy lining of grass or other vegetation; can be domed, giving shelter from sun, dome incorporating lowest branches of shrub. EGGS. Sub-elliptical, smooth and glossy; off-white to grey-white, finely spotted and speckled buff-brown and grey, markings sometimes gathered into zone or cap at broad end. Clutch: 3–5(–7). INCUBATION. 11–13 days. FLEDGING PERIOD. Young usually leave nest at 9 days, either returning to nest over next 3 days or abandoning it altogether. Fledge at 15–16 days.

Wing-length: *G. C. cristata* (most of central and western Europe): ♂ 105–111, ♀ 97–106 mm.

Weight: *G. C. cristata*: ♂ 40–52, ♀ 37–55 g.

Geographical variation. Marked and complex, involving mainly ground-colour and intensity of streaking, less so size and bill-shape. In resident populations, colour variation mainly related to aridity or humidity of environment and amount of sunshine, but also to local variations in colour of soil; in migrant populations (mainly those of FSU), colour rather constant over large areas despite distinct local variations in soil colour. 24 races in west Palearctic. Differences between races slight, often visible only in series of skins.

Thekla Lark *Galerida theklae*

PLATES: pages 1038, 1051

Du. Thekla Leeuwerik   Fr. Cochevis de Thékla   Ge. Theklalerche   It. Cappellaccia di Theklea
Ru. Короткоклювый хохлатый жаворонок   Sp. Cogujada montesina   Sw. Teklalärka

Field characters. 17 cm; wing-span 28–32 cm. Same size as Crested Lark but with more fan-like crest, rather slighter form, longer legs, and (in Europe) shorter bill. Medium-sized lark, closely similar to Crested Lark and differing constantly only in structure, heavier black-brown streaks on chest, on and around cheeks, and on hindneck and upper mantle, and (in Europe) grey underwing.

In all parts of range in west Palearctic also inhabited by Crested Lark, differences between the two species neither striking nor easy to see and much care needed to allow for geographical variation in plumage. Since adult's chest pattern is a much used character, important to realise that forms of Crested Larks with heavy chest-streaks are restricted to Turkey and highlands of Israel, Jordan, and Syria where Thekla Lark does not occur; conversely, forms of Thekla Lark with less heavy streaks occur widely in east, central, and southern Morocco and in coastal districts of Algeria, and these, with their more rufous or paler, sandier plumage tones, invite serious confusion with sympatric races of Crested Lark. Identification in North Africa, particularly in region of Tanger (where darkest forms of both species occur within sight of each other), must therefore rely heavily on differences in habitat, structure, behaviour, and voice. Escape-flight shorter, with bird tending to break back behind observer (perhaps due to more pronounced territoriality). Perhaps a little nimbler on ground than Crested Lark, while most postures definitely less hunched in appearance (due to less deep belly). Perches on plants more than Crested Lark and is less easy to flush from ground cover. Prefers rocky and grassy interfaces between open ground and adjacent scrub or heath.

Habitat. In lower middle latitudes, mainly Mediterranean, but overlapping warm temperate and sub-tropical. Overlap and competition with Crested Lark limited to restricted parts of range, where problems of field identification have sometimes aggravated difficulties in distinguishing habitat differences. Characteristic of Thekla Lark is requirement for mixture of contrasting elements within compact territory. Thus occupies (e.g.) ample dry bare open soil, side by side with rocks and boulders or dense shrub cover, even with trees; roadsides alongside walled cornfields or quiet heaths; or dry stream beds with clumps of oleander or flanked by bushes on steep slopes. Such edge-effect mosaics often ecologically impermanent. Where they have succeeded clearance of earlier forest cover, they imply historical expansion of range over land previously unsuitable, as in south-east France where birds live on shrub-heath of kermes oak with rosemary, *Cistus*, and other second growth plants.

Distribution. SPAIN. Range has decreased. Vulnerable to agricultural intensification and afforestation. Abundant in Mediterranean coastal mountains, especially in south-east. EGYPT. Perhaps breeds in extreme north-west: present Salum 1920, 1931, also 1991, 1994 (area rarely visited by ornithologists in intervening years).

Accidental. Occasional records of dispersing or vagrant birds in northern Mauritania.

Beyond west Palearctic, breeds Ethiopia, Somalia, and northern Kenya.

Population. FRANCE. 10–100 pairs in 1970s. SPAIN. 1.4–1.6 million pairs; slight decrease. PORTUGAL. 10 000–100 000 pairs 1978–84; stable. LIBYA. Locally common. TUNISIA. Common. ALGERIA, MOROCCO. Locally more plentiful than Crested Lark.

Movements. Resident, and generally very sedentary.

Food. Insects and seeds. Most food taken from ground, often by searching under stones: pushes bill under stone then flips it over with quick sideways jerk. Unlike Crested Lark, not known to dig for food, though this perhaps a consequence of more stony habitat on which usually seen.

Social pattern and behaviour. Much less well known than Crested Lark, but generally similar. Song-flight resembles that of Crested Lark; bird describes fairly big circles, flying with weak, slow, fluttering wing-beats, or moves horizontally back and forth, tail and wings fully spread, at times quite high up.

Voice. Repertoire more restricted than in Crested Lark. In loud song, long fluting sounds alternate with short whistles, tremolos, and double notes; numerous good imitations of other bird species interpolated. Contact-alarm call, given by both sexes, is distinctive, and one of the best means of separating Thekla Lark from Crested Lark: a soft, melodiously fluting whistle, of 2–4 notes, 'doo-dee-doo-deeee', lower pitched than equivalent in Crested Lark. Most importantly, last note is longest of series and first rises in pitch, then falls; marked emphasis on final note due to its length and frequency pattern.

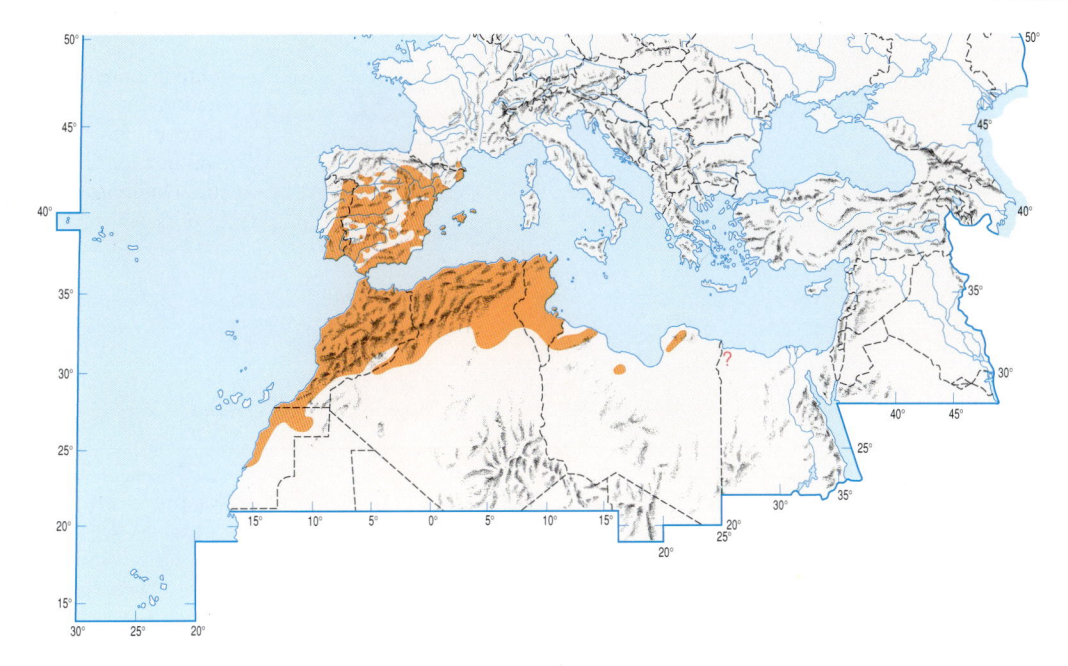

Breeding. SEASON. Spain and Portugal: eggs laid February–June. North Africa: laying period in central, northern, and eastern Algeria and Tunisia early April to early June; in western Morocco, mid-February to late May. 2 broods. SITE. On ground in the open or in shelter of tussock. Nest: very similar to Crested Lark, though generally smaller. EGGS. Sub-elliptical, smooth and glossy; very similar to Crested Lark, perhaps indistinguishable. Clutch: 3–4(2–7). INCUBATION. No information. FLEDGING PERIOD. *c.* 15 days, young leaving nest at 9 days.

Wing-length: *G. t. theklae*: ♂ 102–108, ♀ 92–104 mm.
Weight: *G. t. theklae*: ♂♀ mostly 31–42 g.

Geographical variation. Marked in colour, rather slight in size; mainly clinal and some populations rather variable, but variation not as complex as in Crested Lark. 6 races recognized in west Palearctic. *G. t. erlangeri* from northern Morocco is darkest race, ground-colour of upperparts browner than in nominate *theklae* from Iberia, streaks on upperparts and chest broader and deeper black. Small-sized *carolinae*, extending through northern Sahara from Figuig (Morocco) to north-west Egypt, the palest and most rufous.

Woodlark *Lullula arborea*

PLATES: pages 1021, 1051

DU. Boomleeuwerik   FR. Alouette lulu   GE. Heidelerche   IT. Tottavilla
RU. Юла   SP. Totovía   SW. Trädlärka

Field characters. 15 cm; wing-span 27–30 cm. 20% shorter than Skylark and more delicately built. Smallest of medium-sized larks with erectile crest, short square tail, broad rounded wings, and fine bill, all combining into subtly but distinctly different form from Skylark. Plumage basically buff, well streaked above and on chest, with bold white supercilium which reaches nape, dark-edged, rufous cheeks, and conspicuous black and white marks on carpal feathers. Tail tipped (not edged) white. Flight buoyant but strikingly hesitant, even recalling small bat, as does blunt-winged and short-tailed silhouette.

May be confused with short-tailed juvenile Skylark and Small Skylark but neither shows diagnostic carpal 'headlights' of Woodlark. Song-flight lower but wider than Skylark.

Habitat. Breeds in milder upper and lower middle latitudes of west Palearctic, scattered over the more benign situations between July isotherms of *c.* 17°C and 31°C. Avoids severe cold and windy wet conditions and also hot arid areas. Mainly a temperate and Mediterranean species, but overlaps boreal and steppe zones. On edge of range, in Britain, occurs mainly on lowland (but often sloping) well-drained sites on sand, gravel, or chalk, with short grass for feeding, longer grass or heather for nesting, and scattered trees, bushes, or other suitable fixed song-posts as alternative to delivery of song in flight. Avoids intensive agriculture, but often favours neglected or abandoned farmland. Habitat requirements on continent differ somewhat. In west, most characteristic are heathlands, low moors, rundown farmlands with ragwort or similar weeds, scrubby hillsides, thinly timbered parkland, outskirts of woods, and felled woodland. Burnt-over areas, open woods of birch or oak, golf courses, and sand-dunes with trees or shrubs also favoured. In central Europe, optimal habitat is dry warm open pine-heath, often with heather. In North Africa, occurs on stony hillsides with forests of cork oak and holm oak, and maquis or open woods at 600–1800 m, but in High Atlas up to 3000 m.

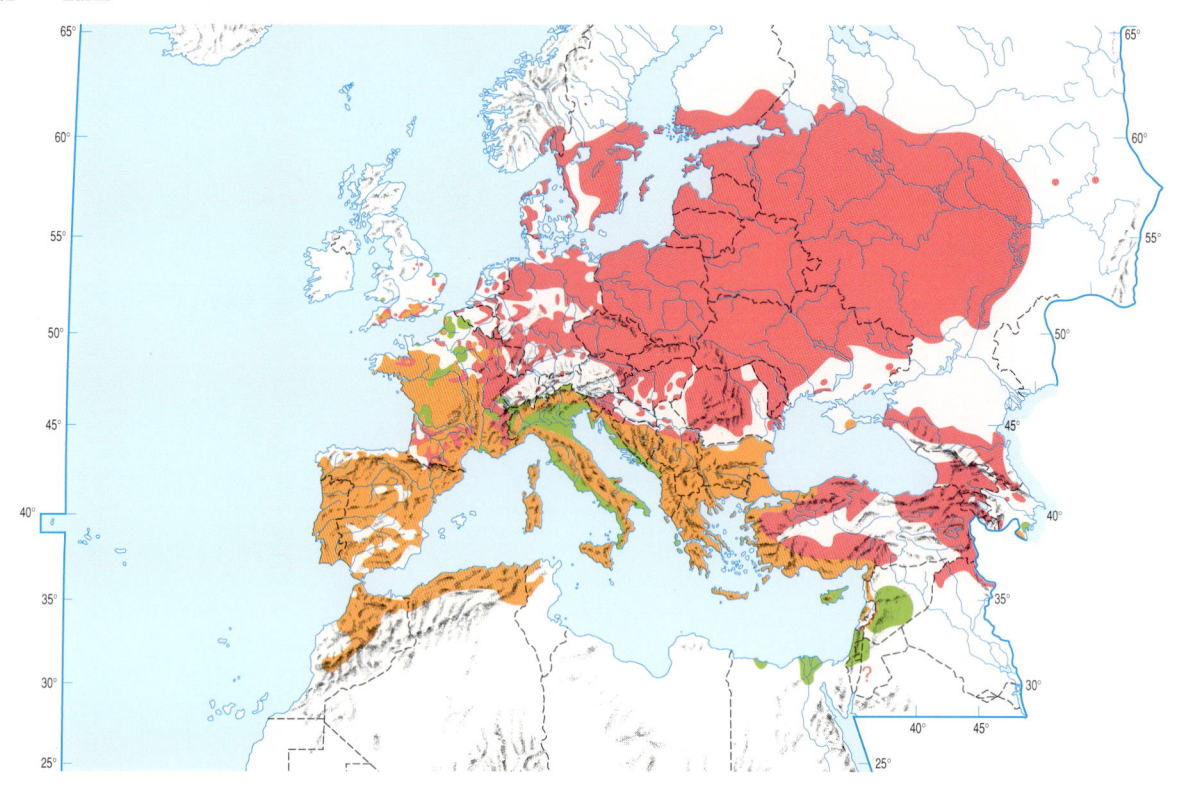

Distribution. Widespread decrease in range and population, due chiefly to loss of habitat (heath and dry grassland). Has withdrawn southward at northern limit. Contraction reported Britain (e.g. no longer breeds Wales), France, Belgium, Luxembourg, Denmark, Norway, Finland, Czech Republic, Austria (perhaps), Switzerland, Spain, Slovenia, and Ukraine. IRELAND. Bred 19th century, also 1905, 1954. AZERBAIJAN. Range poorly known.

Accidental. Faeroes, Ireland, Kuwait, Libya.

Beyond west Palearctic, extends east to Iran and southern Turkmenistan.

Population. Most countries report decreases. BRITAIN. 350 pairs 1988–91; fluctuates. FRANCE. Under 100 000 pairs in 1970s. BELGIUM. 450–550 pairs 1989–91; decline from 1400 pairs late 1960s, now fluctuating. LUXEMBOURG. 50–100 pairs. NETHERLANDS. 2700–3500 pairs 1989–91; locally rather common. Marked fluctuations. GERMANY. 29 000 pairs in mid-1980s. DENMARK. 300–400 pairs 1987–8. NORWAY. 50–200 pairs 1970–90; stable. SWEDEN. 1000–3000 pairs in late 1980s. FINLAND. 800–1000 pairs in late 1980s. ESTONIA. 2000 pairs in 1991. LATVIA. 1000–6000 pairs in 1980s. LITHUANIA. More numerous in south and south-east. POLAND. 15 000–30 000 pairs; stable. CZECH REPUBLIC. 600–1100 pairs 1985–9. SLOVAKIA. 1500–3000 pairs 1973–94; stable. HUNGARY. 5000–8000 pairs 1979–93; stable. AUSTRIA. Sparse and local; one estimate of 170–210 pairs. SWITZERLAND. 250–300 pairs 1985–93. SPAIN. 560 000–1.3 million pairs. PORTUGAL. 100 000–1 million pairs 1978–84; stable. Decreasing in Algarve in recent years. ITALY. 20 000–40 000 pairs 1983–93; stable. GREECE. 4000–10 000 pairs; probably stable. ALBANIA. 2000–5000 pairs 1981. YUGOSLAVIA: CROATIA. 10 000–12 000 pairs; stable. SLOVENIA. 800–1000 pairs. BULGARIA. 10 000–100 000 pairs; stable. RUMANIA. 30 000–50 000 pairs 1986–92; stable. RUSSIA. 10 000–100 000 pairs; stable. BELARUS'. 12 000–17 000 pairs in 1990. UKRAINE. 4500–5200 pairs in 1986. MOLDOVA. 1500–3000 pairs in 1990; stable. AZERBAIJAN. Uncommon. TURKEY. 10 000–100 000 pairs. CYPRUS. Probably 100–1000 pairs. ISRAEL. 25–30 pairs in 1980s. TUNISIA. Uncommon.

Movements. Migratory in northern half of breeding range, partially so in central Europe and southern FSU, but mainly resident in maritime climate of western Europe and in Mediterranean basin (including North Africa). Winters within southern half of breeding range, or a little beyond it into Egypt and northern Middle East. Less conspicuous as a migrant than many other larks, since seldom forms large and compact flocks for diurnal passage.

Food. In breeding season, largely medium-sized insects and spiders; at other times, apparently mostly seeds. Direct observation, and analysis of food brought to young, indicate food collected from ground surface, uppermost layer of soil, and from low parts of plants. Feeding bird bustles about swiftly, pecking among plants and in earth. Rubs caterpillars repeatedly against ground. Most seeds swallowed complete with husks; only coarser seeds with awns de-husked by beating against ground.

Social pattern and behaviour. Overall, less gregarious than some other larks, notably Skylark. Loose-knit parties of *c.* 15–20 (usually less) occur outside breeding season. Solitary and territorial when breeding, ♂♂ tending to remain same territory from year to year, mating with different ♀♀. Several pairs often nest fairly close together while apparently suitable areas alongside not occupied. Song-flight important for self-advertisement and territory proclamation; performed mainly by single ♂♂ looking for mate, also in period soon after pairing. ♂ makes angled ascent, normally from tree-top; spirals up, then circles at fairly constant height—usual maximum *c.* 100 m if unpaired, *c.* 50 m if paired. Sings all the while, moving in irregular loops and spirals, dipping and rising in sweeping curves. Descent may be gradual and spiralling with continuous song, or (less frequently), song broken off at 50–30 m, bird then plummeting with closed wings to same or neighbouring perch or ground. Song-flight often preceded or followed by song from perch or ground.

Voice. Most calls characteristically melodious. Song of ♂ unlikely to be confused with other west Palearctic larks. Series of remarkably rich and full-sounding phrases separated by pauses. Clear, liquid, mellow, pleasantly fluting sounds given in descending pattern. Phrases tend to start quietly followed by crescendo or by crescendo, diminuendo, and accelerando. Single-segment phrases in particular start quietly; this followed by crescendo, and always by accelerando. Contact-alarm call (harder or softer in tone depending on mood) best-known and most frequently uttered call. Given throughout the year in a variety of situations (perched or in flight): when feeding, resting, on nest, or tending young; also during courtship and to express anxiety. Consists of 1–3 syllables, characteristically sweet and musical: 'lit-loo-eet', 'tew leet', 'd'lui', etc., individually variable.

Breeding. SEASON. Britain and western Europe: egg-laying from mid-March. FSU: laying starts up to 1 month later. North Africa: main laying period end of March to end of May. SITE. On ground, in shelter of scrub, bracken, or grass, sometimes at base of tree stump or sapling; rarely, on bare ground in the open. Nest: deep depression in ground, excavated by the building bird, lined with vegetation. Base layer of thicker stems, leaves, pine-needles, and some moss; main layer of grass leaves and stems; inner lining of finer grasses and some hair. EGGS. Sub-elliptical, smooth and fairly glossy; whitish, sometimes tinged olive, with fine spots and sometimes larger blotches of brown or grey-brown, often formed into band round broad end. Clutch: 3–5 (2–6). INCUBATION. 12–15 days. FLEDGING PERIOD. 10–13 days.

Wing-length: *L. a. arborea*: ♂ 94–101, ♀ 91–97 mm.
Weight: *L. a. arborea*: ♂♀ mostly 25–35 g.

Geographical variation. Slight, predominantly clinal, in colour only. Two races recognized: nominate *arborea* in north of range, *pallida* in south. Typical *pallida* from Iran as well as populations from Turkey, Levant, Greece, southern Italy, and north-west Africa have fringes of crown, scapulars, and back greyish pink-cinnamon or grey-buff in fresh plumage, not cinnamon-brown as in nominate *arborea*; rump and upper tail-coverts and ground-colour of hindneck and upper mantle distinctly grey or olive-grey, less olive- or rufous-brown; ground-colour of sides of breast and chest pale cream or off-white, similar to remainder of underparts, not buff and rufous as in nominate *arborea*.

Skylark *Alauda arvensis*

PLATES: pages 1044, 1051

DU. Veldleeuwerik   FR. Alouette des champs   GE. Feldlerche   IT. Allodola
RU. Полевой жаворонок   SP. Alondra común   SW. Sånglärka

Field characters. 18–19 cm; wing-span 30–36 cm. 30% larger than short-toed larks; 20% larger than Woodlark; as long but less bulky than Crested Lark; shorter and less bulky than *Melanocorypha* larks. Commonest and most widespread lark of west Palearctic, with character of family most exemplified by strong bill, fairly stout and long body, quite long legs, wings, and tail, streaked brown plumage, and terrestrial habits. Chief marks of Skylark are short crest, open-faced appearance (due to pale eye-ring and supercilium), fully streaked, quite sharply demarcated chest, white trailing edge to wings, and white tail-sides.

Often only lark in many habitats throughout west Palearctic, but can suggest many other smaller or larger larks, being also confused with Lapland Bunting, Snow Bunting, and Corn Bunting. Prominent, long, white outer tail-feathers shared as single most obvious mark only by much smaller *Calandrella* larks, slimmer Shore Lark, and much bigger Calandra Lark but white trailing edges to wings only shared by last. For distinctions from Small Skylark, see that species. Flight over short distances more weaving and fluttering than most larks, with characteristic loose hover before landing; over long distances and on migration, more purposeful but with characteristically erratic rhythm, involving bursts of loose, rather floppy or noticeably flapped wing-beats alternated with short glides or almost complete wing-closures. Loose, slightly rising and falling and above all hesitant progress over open country displayed by no other common passerine and diagnostic even at long range. Song-flight much more confident: bird rises with rapid wing-beats, and after prolonged hovering falls by 'parachute'. Gait essentially a noticeably even-paced walk, but becomes a shuffle when legs are flexed and body lowered.

Habitat. Breeds in upper and lower middle latitudes across and beyond west Palearctic, spreading from continental to

Skylark *Alauda arvensis*. *A. a. arvensis*: **1** ad fresh (autumn), **2** ad worn (spring), **3** juv. *A. a. scotica*: **4** ad worn (spring). *A. a. sierra* (central and southern Iberia): **5** ad worn (spring). *A. a. dulcivox* (steppes of lower Volga eastwards): **6** ad fresh (autumn). Small Skylark *Alauda gulgula* (p. 1046): **7** ad fresh (as recorded in Israel, October—presumably *inconspicua*, from west of range). Razo Lark *Alauda razae* (p. 1047): **8** ad fresh (autumn), **9** juv.

oceanic climates and from temperate into boreal zone. Markedly less Mediterranean than Woodlark and Crested Lark, which need more warmth and are more patchily distributed owing to more specific habitat requirements. Inhabits open surfaces of firm, level or unobstructed soils, neither arid nor muddy, although often moist, and preferably well clothed with grasses (including cereals) or low green herbage. Presumed to have spread from natural steppe grasslands with deforestation and expansion of crops and pastures encouraging massive habitat changes, especially through 19th century.

Distribution. Decreases reported Ireland, Netherlands, Germany, Iberia, Slovenia, and slight increase Ukraine. ALGERIA. Widespread in north-west of range; more local further east and south. MOROCCO. May have spread northward since early 1980s.

Accidental. Bear Island, Iceland, Mauritania, Azores, Madeira.

Beyond west Palearctic, widespread northern Asia east to north-east Russia, south to Iran and northern China. Introduced to Vancouver Island (Canada), Hawaii, and Australasia.

Population. Almost all countries of northern and western Europe—east to Czech Republic, Slovenia, and Italy—report recent declines, sometimes marked. Attributed to changing farming practices, notably reduction in crop diversity, loss of winter stubbles, and use of insecticides and herbicides. Apparently chiefly stable further east, with slight increase in Ukraine. FAEROES. 5–10 pairs; probably declining, but never common. BRITAIN. 2 million territories 1988–91; dramatic decline since late 1970s. IRELAND. 570 000 territories 1988–91. FRANCE. 300 000–1.3 million pairs. BELGIUM. 115 000 pairs 1973–7. LUXEMBOURG. 15 000–20 000 pairs. NETHERLANDS. 150 000–175 000 pairs. GERMANY. 4.9 million pairs in mid-1980s. DENMARK. 1.3 million pairs 1983–5. NORWAY. 100 000–500 000 pairs 1970–90. SWEDEN. 700 000–1 million pairs late 1980s. FINLAND. 300 000–400 000 pairs late 1980s. ESTONIA. 100 000–200 000 pairs 1991. LATVIA. 1.1–1.8 million pairs 1980s. LITHUANIA. More numerous in centre and north-west. POLAND. 4–9 million pairs. CZECH REPUBLIC. 800 000–1.6 million pairs 1985–9. SLOVAKIA. 200 000–400 000 pairs 1973–94. HUNGARY. 150 000–300 000 pairs 1979–93. AUSTRIA. Common; one estimate of 40 000–50 000 pairs. SWITZERLAND. 40 000–50 000 pairs 1985–93. SPAIN. 2–6 million pairs. PORTUGAL. 100 000–1 million pairs 1978–84. ITALY. 500 000–1 million pairs 1983–93. GREECE. 2000–5000 pairs. ALBANIA. 500–1000 pairs 1981. YUGOSLAVIA: CROATIA. 200 000–250 000 pairs. SLOVENIA. 8000–12 000 pairs. BULGARIA. 1–10 million pairs. RUMANIA. 600 000–1 million pairs 1986–92. RUSSIA. 1–10 million pairs. BELARUS'. 1.2–1.7 million pairs. UKRAINE. 900 000–1 million pairs in 1988. MOLDOVA. 50 000–70 000 pairs in 1989. AZERBAIJAN. Locally common, e.g. in Shemakha upland. TURKEY. 50 000–500 000 pairs. TUNISIA. Common.

Movements. Shows gradation from being wholly migratory in north and east of breeding range to making no more than local movements in south. Northern and central Europe largely vacated in winter, with movement south-west on broad front, reaching western seaboard and Mediterranean area. Autumn passage begins September, peaking in northern Europe in first half of October; lasts into first half of November in southern Europe. Large-scale cold-weather movements can occur at any time during winter. Return passage begins January in south, becoming heavy in February and early March. Arrivals in Scandinavia mid-February to late March, in Russia in March and early April.

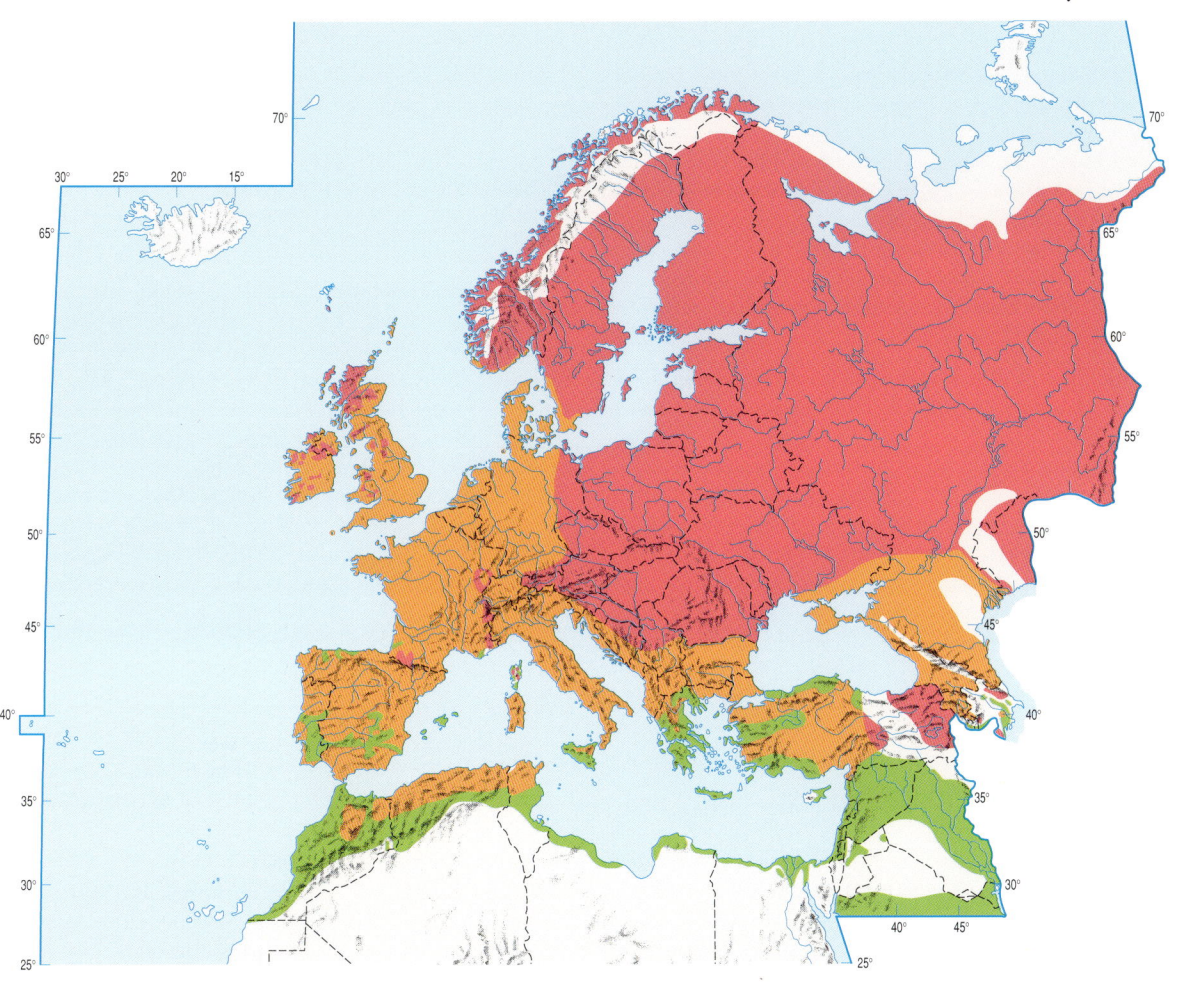

Food. Plant and animal material taken at all times of year, but insects especially important in summer, cereal grain and weed seeds in autumn, leaves and weed seeds in winter, and cereal grain in spring. Walks over ground taking items from soil surface or pecking at leaves, flowers, or seed heads. Seems to locate all food visually and to peck at it directly, though sometimes digs with bill in loose soil for newly sown, partly exposed grain, and uproots cereal seedlings up to c. 4 cm tall to peck off attached grain.

Social pattern and behaviour. Often in flocks outside breeding season. Monogamous mating system, with rare cases of bigamy; mate-fidelity regular from year to year when both survive. Breeding pairs territorial; in favourable habitat, territorial boundaries contiguous. Song of ♂ given from ground, open perch (e.g. post, tree) or in flight. For song-flight, usually takes off silently and ascends at c. 20–70°, always into wind. From c. 10–20 m, steep spiralling ascent accompanied by song, or starts singing immediately after take-off. Vigorous song during ascent typical of period with pair-formation and territorial disputes. Ascends, with tail spread and characteristically fluttering wings, to c. (20–)50–100 m. Next (normally longest) phase not preceded by any noticeable change in wing action or song, involves hovering, often alternating with slow horizontal circling over territory or, not infrequently, beyond. Song-flight usually lasts 1–5 min (average 2–2.5), rarely up to 10 min, 20 min exceptional. Descends in slow spirals, wings fully spread and motionless, and continues to sing while gliding down to land, or (more often) suddenly stops singing at c. 10–20 m, then plummets with nearly closed wings.

Voice. Song of ♂ loud and melodious, but with fairly restricted range of sounds, although the many different combinations and long, apparently unbroken flow give it attractive quality. Usually gives 1–2 'trii' or 'trli' sounds soon after take-off, then continuous song containing frequent series of varyingly modulated 'trli' or 'dji' whistles in variety of pitch patterns. Frequent repetition typical, e.g. of a high-pitched 'tee-ee tee-ee', then lower-pitched 'tyu-yu tyu-yu'. Will mimic other birds and animals, e.g. songs and/or calls of other larks, waders. Song given on ground basically similar to song in flight but extremely variable; usually shorter, quieter, more warbling and melodious and with more pauses. Calls variable and far less distinct: commonest a liquid, rippling, disyllabic 'chirrup', but also varied confusingly into huskier, drier, less protracted, slurred and whistled notes.

Breeding. SEASON. Britain and north-west Europe: egg-laying from late March or early April. Little variation over rest of range, except for delay in onset of breeding related to increasing latitude or altitude. Up to 4 broods per season, fewer in northern latitudes. SITE. On ground in the open or among short vegetation such as grass or growing crops. Nest: shallow depression lined with grass leaves and stems, with inner lining of finer material; rarely with rampart of small stones. EGGS. Sub-elliptical, smooth and fairly glossy; grey-white, often tinged greenish, thickly spotted brown or olive. Clutch: 3–5(–7). INCUBATION. 11 days. FLEDGING PERIOD. c. 18–20 days, but young usually leave nest at 8–10 days.

Wing-length: *A. a. arvensis*: ♂ 108–123, ♀ 99–112 mm.
Weight: *A. a. arvensis*: ♂ mostly 34–50, ♀ 26–43 g.

Geographical variation. Slight, involving size and colour. Size increases slightly towards east, and in western Europe cline of decreasing wing length runs towards south. Differences in colour slight and obscured by strong individual variation; most races separable only by comparing series of birds at same stage of plumage wear. Nominate *arvensis* in western part of range usually olive-brown above with broad black streaks, while paler and more greyish birds predominate towards east; variation strong everywhere, however. *A. a. scotica* from Ireland, north-west England, Scotland, and Faeroes similar to dark and brown variants of nominate *arvensis*. 8 races recognized in west Palearctic.

Small Skylark *Alauda gulgula*

PLATES: pages 1044, 1051

DU. Kleine Veldleeuwerik   FR. Alouette gulgule   GE. Kleine Feldlerche   IT. Allodola orientale
RU. Малый полевой жаворонок   SP. Alondra india   SW. Mindre sånglärka

Field characters. 15.5–16.5 cm; wing-span 26–30 cm. About 15% smaller than Skylark, with structure recalling Woodlark and differing distinctly from Skylark in longer bill, shorter, broader-based, and rounder wings (lacking obvious extension of primaries when folded), and shorter and rather narrower tail; legs proportionately longer than Skylark. Plumage closely similar to Skylark but tends to look greyer than western forms of that species, showing at close range paler fore-face and supercilium (forming ⊤-shaped mark), rusty rear cheeks and wing-panel, more narrowly streaked breast-band on buffier underparts, slightly darker lining to wing (in flight), and buff (not white) trailing edge to wings and sides to tail in flight.

Race of Israeli birds uncertain but these readily distinguished from Skylark on characters given above. In flight, beats of short, rounded wings produce slightly more flapping action than Skylark; action and more compact silhouette again suggesting Woodlark. Gait and behaviour much as Skylark but birds easier to approach, sitting tight and in flock last to rise.

Song similar to Skylark but including buzzing notes. Flight-calls distinctive, transcribed as harsh, buzzing 'bzeep' or 'pzeebz' (final part with twanging quality) and 'baz baz' or 'baz-terrr', most reminiscent of Yellow Wagtail or Richard's Pipit; also a soft 'pyup', similar to Ortolan Bunting.

Habitat. Breeds in east Palearctic and Oriental regions where it largely replaces Skylark in equivalent habitats, showing similar adaptation from moist natural grasslands to field cultivation at fairly wide range of altitudes, and similar tendency to avoid closed, wooded, or broken terrain.

Distribution. Breeding widespread in southern Asia east to Philippines and north to Kazakhstan and Tibet.

Accidental. Azerbaijan: specimen collected at seashore near Lenkoran pre-1911. Kuwait: one, November 1986. Egypt: Sinai, October 1990 (2), December 1993 (1), October 1995 (2). In Israel, has wintered regularly (up to 16 birds) at Eilat since at least 1984, with a few records elsewhere.

Movements. Resident or migratory. North-west populations (which breed nearest to west Palearctic) present in FSU only during breeding season. Depart mainly during early October though some remain until mid-December; return February to mid-March. Wintering range of these north-west populations uncertain; probably plains of northern India and Pakistan.

Wing-length: ♂ 94–102, ♀ 85–96 mm.

Razo Lark *Alauda razae*

PLATES: pages 1044, 1051

Du. Razo-leeuwerik   Fr. Alouette de Razo   Ge. Razolerche   It. Allodola di Razo
Ru. Жаворонок острова Разо   Sp. Terrera de Cabo Verde   Sw. Razolärka

Field characters. 12–13 cm; wing-span 22–26 cm. Less than $\frac{3}{4}$ size of Skylark, with 30–40% shorter and rounder wings (primaries entirely hidden beneath tertials of closed wing), and shorter tail, but with rather heavier and longer bill and rather long sturdy legs; appears curiously 'front-heavy'. Rather small, compact, short-crested lark. Plumage mealy-grey, streaked dull black above; pale cream below, with dark-streaked buff chest. Short tail noticeably dark, with prominent white outer edge, but short rounded wings of adult lack obvious pale trailing edges of Skylark.

Habitat. Entire species occurs only on small windswept waterless island of Razo (Cape Verde Islands). Most live on a flat area of decomposing lava and tufa carrying meagre growth of herbs and low scrub, but a few spread to wide ravines with fine sandy bottoms, or feed on stretch of black rocks close to the ocean, involving an altitude range of *c*. 150 m.

Distribution and population. Razo. Restricted to vegetated parts of plains and valleys, which represent less than half of total area (7 km^2) of island. *c*. 250 birds; stable.

Movements. Sedentary. No records from anywhere other than Razo.

Food. Seeds and insects. Digs, using bill to prise pebbles free of soil, presumably to expose food items. Bill length difference of 20.7% exists between sexes, and feeding ecologies thus likely to differ.

Social pattern and behaviour. Occurs mainly in groups of up to 25 birds. Large gathering of adults and immatures recorded in early January; perhaps most of the population. Very little information on breeding behaviour, but apparently similar to Skylark. ♂ sings while ascending with gentle wing-beats; ascent vertical, unlike spirals of Skylark; continues to sing while hovering at *c*. 10 m.

Voice. Similar to Skylark, but song much less varied.

Breeding. SEASON. Probably erratic, governed by rains; well-grown young seen in nest in April; birds courting and eggs found in October of same year. SITE. On ground among short grass, under boulder or creeping plants. Nest: frail structure of grass, placed in small depression. EGGS. Whitish with fine brownish or greyish spots, increasing towards broad end. Clutch: one of 3 reported. (Incubation and fledging periods uncertain.)

Wing-length: ♂ 83–89, ♀ 76–80 mm.

Shore Lark *Eremophila alpestris*

PLATES: pages 1048, 1051

Du. Strandleeuwerik   Fr. Alouette hausse col   Ge. Ohrenlerche   It. Allodola golagialla
Ru. Рогатый жаворонок   Sp. Alondra cornuda   Sw. Berglärka   N. Am. Horned Lark

Field characters. 14–17 cm; wing-span 30–35 cm. About 10% smaller and slighter than Skylark; 15% bigger than Temminck's Horned Lark. Medium-sized lark with form like Skylark except for shorter bill, proportionately longer tail and (in adult ♂) thin 'horns' above eyes. Breeding adult essentially pink-brown above and white below, with yellow face strikingly decorated with black horns, thick 'comma' from lores to mid-cheeks, and gorget. In winter, facial pattern less contrasting. Juvenile shows similar ground-colours, but is heavily spotted except on pale yellow face and white belly. In flight, contrast of long white vent and broad black centre to undertail often catches eye. Southern FSU race *brandti* greyest form, with white, not yellow ground to face.

North of Mediterranean, adult unmistakable, but in Levant and across North Africa liable to confusion with Temminck's Horned Lark (see that species). Almost complete geographical separation of arctic communities of Shore Lark presents few chances to confuse juvenile with young Skylark, but close or sympatric breeding with Woodlark and Skylark in alpine habitats from south-east Europe east to Turkey and Iraq could lead to more frequent risk of mistake. Short bill and yellowish face of Shore Lark are, however, diagnostic. Flight action light, recalling (with length of tail) large pipit as much as other medium-sized larks. Wing-beats noticeably rapid, with almost complete closures, causing shooting accelerations and first cruciform, then slim silhouette, as bird or flock flickers over ground. Gait includes both shuffling and high-stepping walk, crouching and more-normal run, and (particularly on shingle) hopping. Feeding bird constantly on the move, with quiet and unobtrusive progress hiding considerable speed.

Habitat. Breeds in west Palearctic in high latitudes or at high altitudes, in regions widely separated by forested, cultivated, wetland, and other unsuitable terrain. Mainly in subarctic or arctic lowland tundra, or in barren steppes and arctic-alpine zones of middle and lower-middle latitudes, much of intermediate area being occupied by Skylark. In winter, occurs on coastal dunes, salt-marshes, and beaches, and some arable

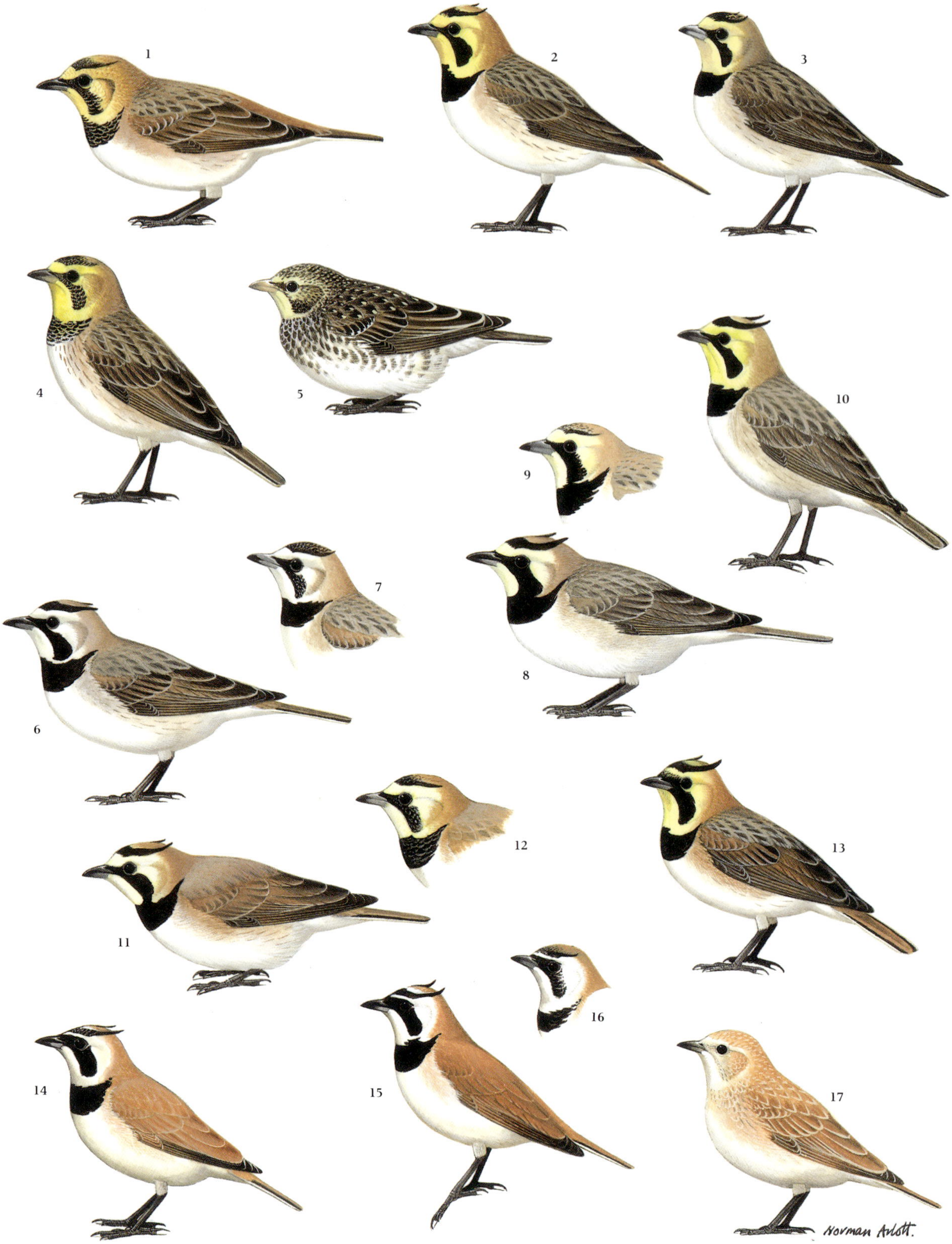
1
2
3
4
5
10
9
7
8
6
12
13
11
16
14
15
17
Norman Arlott.

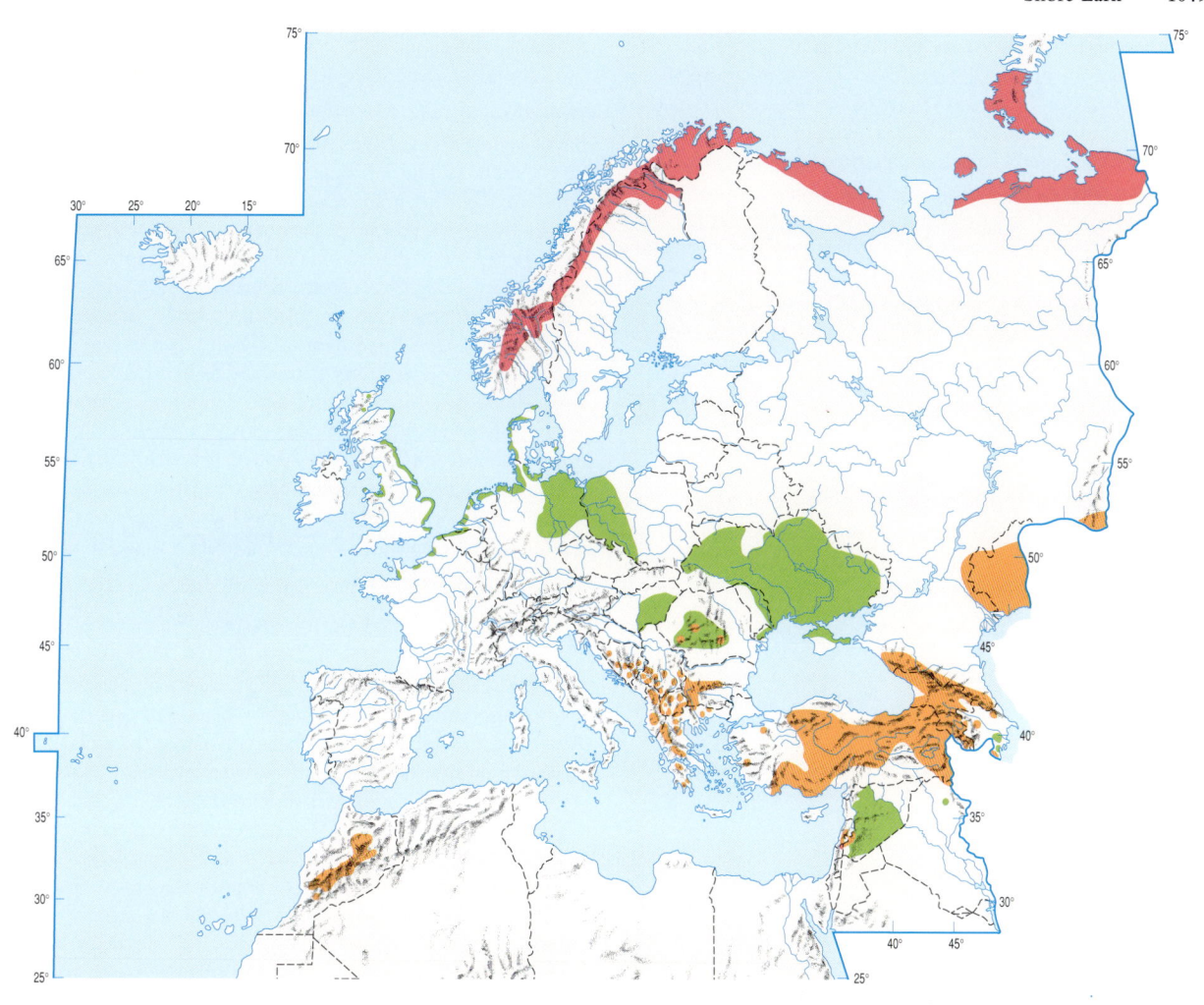

lands in lowlands which are free of snow, or have plants emerging above snow cover. Does not usually occur on same terrain as other larks.

Distribution. Fenno-Scandian range and population have declined markedly; reasons largely unknown, but perhaps involve changes in main wintering area, as breeding habitat stable. BRITAIN. Scotland: summered 1972–3, possibly breeding 1973; bred 1977. RUMANIA. Breeds at a few isolated sites. AZERBAIJAN. Distribution poorly known. KAZAKHSTAN. Ural delta: numbers severely reduced by rising water-levels in recent years. TURKEY. Widespread breeder in practically all mountainous regions. MOROCCO. Locally common in Moyen Atlas and Haut Atlas.

(FACING PAGE) Shore Lark *Eremophila alpestris*. *E. a. flava*: **1** ad ♂ fresh (autumn), **2** ad ♂ worn (spring), **3** ad ♀ worn (spring), **4** 1st winter, **5** juv. *E. a. brandti*: **6** ad ♂ worn, **7** ad ♀ worn. *E. a. penicillata*: **8** ad ♂ worn, **9** ad ♀ worn. *E. a. atlas*: **10** ad ♂ worn. *E. a. bicornis*: **11** ad ♂ worn, **12** ad ♀ worn. *E. a. alpestris*: **13** ad ♂ worn.
Temminck's Horned Lark *Eremophila bilopha* (p. 1052): **14** ad ♂ fresh (autumn), **15** ad ♂ worn (spring), **16** ad ♀ worn, **17** juv.

Accidental. Spitsbergen, Bear Island, Iceland, Faeroes, Ireland, Luxembourg, Switzerland, Spain, Malta, Cyprus.

Beyond west Palearctic, highly discontinuous distribution continues eastwards: breeds along north Siberian coast, and from Iran and east Caspian east to *c.* 120°E. Also widespread in North America; in South America, isolated population in Colombian Andes.

Population. NORWAY. 2000–10 000 pairs 1970–90; decreasing, notably in north. SWEDEN. 500–2000 pairs in late 1980s. FINLAND. Dramatic decline since 1960s (10 000 pairs in 1950s). Population almost extinct; some pairs may still breed in Finnish Lapland. GREECE. 500–1000 pairs; stable. ALBANIA. 100–500 pairs 1981. YUGOSLAVIA: CROATIA. 100–120 pairs; stable. BULGARIA. 1000–10 000 pairs; stable. RUMANIA. Probably 20–30 pairs 1986–92; slight increase. RUSSIA. 100 000–1 million pairs; stable. AZERBAIJAN. Uncommon. TURKEY. 10 000–100 000 pairs. ISRAEL. 60 pairs in 1980s.

Movements. Eurasian arctic and subarctic race wholly migratory, wintering coasts of southern North Sea and western Baltic, in fluctuating numbers inland in north-central and eastern Europe, and in large numbers across southern FSU (e.g. reaching Sea of Azov, northern Caucasus, Kazakhstan steppes, and further east). Regular wintering in eastern Britain dates from about 1870, and comparable 19th century increase occurred on German side of North Sea; not clear whether this was due to changed breeding range or density in Scandinavia, or altered migration pattern.

Autumn departures from Finnmark (Norway) in late September and first half October, but earlier from Russian breeding grounds. Passage through Baltic area and arrivals in North Sea mid-October to mid-November. Onward movement can occur in severe weather, becoming more numerous in southern wintering areas and stragglers penetrating beyond normal range. Return movement begins in March, with breeding areas reoccupied from late April in Finnmark to second half May in Novaya Zemlya and Yamal peninsula, often while ground still snow-covered.

All populations of Balkans, Morocco, Turkey, Middle East, and south-central Asia basically resident. These breed in mountainous regions, and many birds descend in autumn to foothills, plains, and cultivations, therefore becoming more widespread during winter.

Food. In summer, insects and some seeds; in winter, seeds. Takes food from ground, walking about or running in short spurts. Takes seeds from plant stems, pulling vigorously if necessary; 2–3(–4) birds may work one plant, stripping seeds from bottom to top. To reach food under snow, scratches with feet, tossing it aside with bill.

Social pattern and behaviour. Usually gregarious outside breeding season, often associating with other species, especially buntings and finches, but not usually integrated with them. ♂♂ of migratory populations tend to arrive on breeding grounds before ♀♀. Song of ♂ given from ground or slight elevation (rarely from tree or building), and in flight. Ascent usually silent and fairly steep (c. 60°) into strong wind, in wide spirals if wind light. Climbs to c. 80–250 m, alternating bouts of rapid fluttering wing-beats with brief periods of coasting on closed wings, so that ascent undulating. Continuous song given while beating wings slowly and deeply; may intermittently close wings and tail. While singing, usually heads into wind and may remain almost stationary. At conclusion, closes wings and drops head-first, more or less vertically or at steep angle, with audible whizzing sound; close to ground, opens wings to fly horizontally for some distance. Sometimes lands on starting point.

Voice. Song and calls lack volume of most larks and have characteristic sibilant, piping, even rippling quality. Song twittering or tinkling, with frequently repeated, drawn-out note recalling that of Corn Bunting. Calls include 'tseep', like long note of Meadow Pipit (but less anxious in tone and more incisive), and rippling 'tsee-sirrp', often uttered by members of escaping or excited flock.

Breeding. SEASON. Northern Scandinavia: for double-brooded birds, laying of 1st clutch usually starts in last week of May, 2nd clutch at end of June; single clutches usually laid in 2nd–3rd week of June; season up to 3 weeks later in arctic Russia. SITE. On ground in the open; often in short vegetation, or sheltered by small tussock; occasionally on slight hummock; nest-entrance usually protected from prevailing wind or direct sun. Nest: depression in ground, usually excavated but sometimes natural, lined with small twigs, rootlets, grass stems, and leaves, with finer lining of grass and plant down. Small stones and other items placed round nest, sometimes forming path. EGGS. Sub-elliptical, smooth and glossy; variable in colour but most greenish-white, heavily spotted yellowish-brown, often with brown hair-streaks; may also have larger spots or dark zones. Clutch: 2–4(–5). INCUBATION. 10–11 days. FLEDGING PERIOD. Young leave nest at 9–12 days but do not fly until 16–18 days.

Wing-length: *E. a. flava* (arctic Eurasia): ♂ 108–116, ♀ 100–107 mm.
Weight: *E. a. flava*: ♂ 31–46, ♀ 26–44 g.

Geographical variation. Marked and complex. 2 main groups discernible: (1) *alpestris* group (northern Eurasia and the Americas) in which black of cheeks separated from black chest by white or yellow streak from throat up to sides of neck; (2) *penicillata* group (central Asia west to Balkans) in which broad black cheek-patch broadly connected with extensively black chest, isolating small pale throat-patch (though black connection at lower sides of neck sometimes narrow and indistinct in ♀). Marked variation occurs within both groups, mainly involving colour of pale areas of face and throat (yellow or white), general colour and streaking of upperparts, colour of lesser upper wing-coverts, and size; in *penicillata* group, also variations in amount of white on forehead and in bill shape. In west Palearctic, *alpestris* group comprises *flava* (arctic areas), *brandti* (Caspian, eastwards), *atlas* (North Africa), and (as a straggler from eastern Canada) nominate *alpestris*; *penicillata* group comprises *balcanica* (Balkans), *penicillata* (Asia Minor, eastwards), and *bicornis* (Lebanon).

(FACING PAGE) Short-toed Lark *Calandrella brachydactyla* (p. 1032). *C. b. brachydactyla*: **1–2** ad. Lesser Short-toed Lark *Calandrella rufescens minor* (North Africa and Middle East) (p. 1034): **3–4** ad. Asian Short-toed Lark *Calandrella cheleënsis* (p. 1036): **5** ad. Hume's Lark *Calandrella acutirostris* (p. 1034): **6** ad. Kordofan Bush-lark *Mirafra cordofanica* (p. 1011): **7** ad. Crested Lark *Galerida cristata* (p. 1037): **8–9** ad. Thekla Lark *Galerida theklae* (p. 1040): **10–11** ad. Woodlark *Lullula arborea* (p. 1041): **12–13** ad. Skylark *Alauda arvensis* (p. 1043). **14–15** ad. Small Skylark *Alauda gulgula* (p. 1046): **16** ad. Razo Lark *Alauda razae* (p. 1047): **17** ad. Shore Lark *Eremophila alpestris* (p. 1047). *E. a. flava*: **18** ad *E. a. bicornis*: **19** ad. Temminck's Horned Lark *Eremophila bilopha* (p. 1052): **20–21** ad.

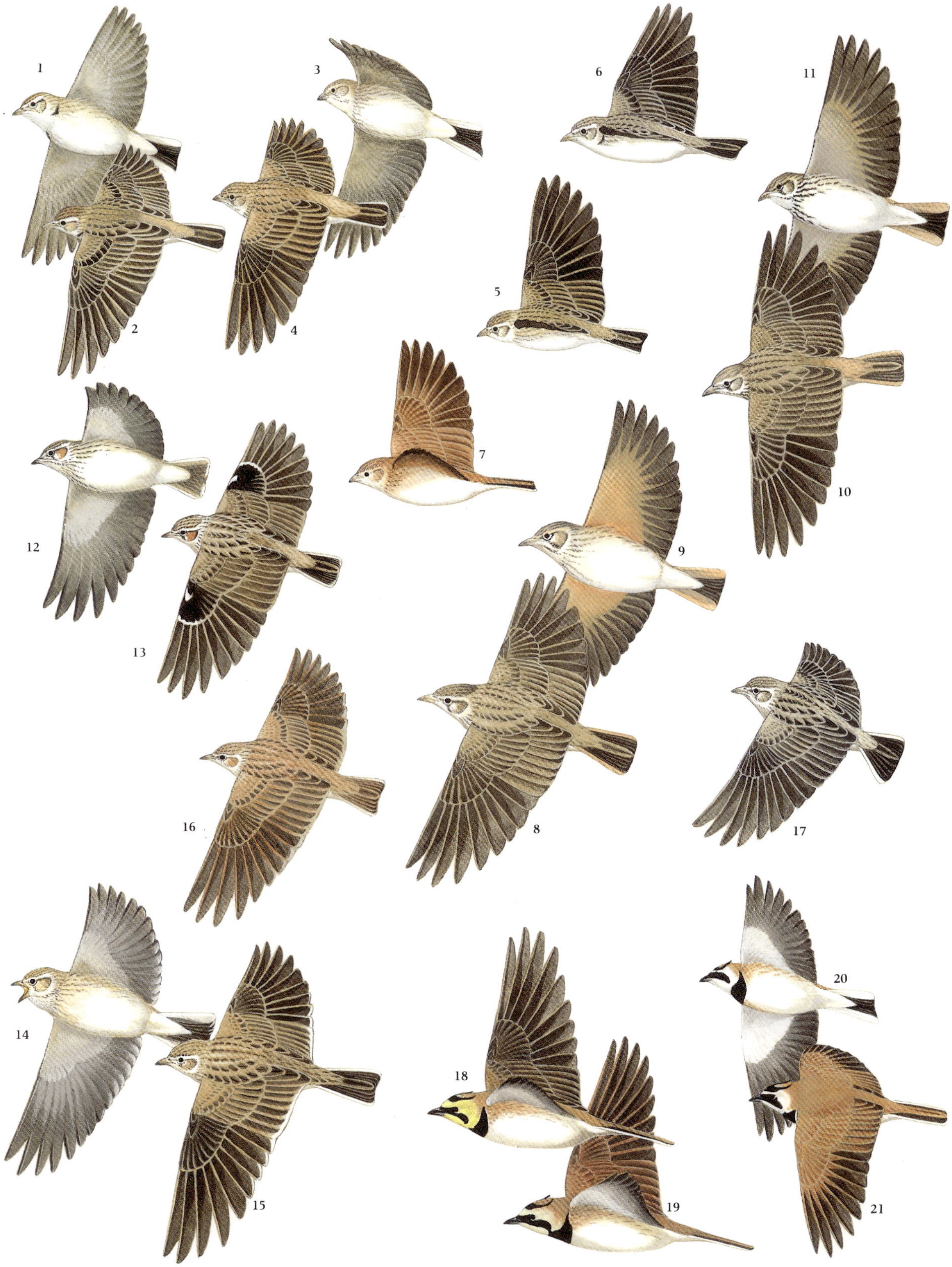
1
2
3
4
5
6
7
8
9
10
11
12
13
14
15
16
17
18
19
20
21

Temminck's Horned Lark *Eremophila bilopha*

Du. Temmincks Strandleeuwerik   Fr. Alouette bilophe   Ge. Saharaohrenlerche   It. Allodola cornuta africana
Ru. Малый рогатый жаворонок   Sp. Alondra cariblanca   Sw. Ökenberglärka

Field characters. 13–14 cm; wing-span 26–31 cm. 20% smaller and shorter-tailed than Shore Lark; close in size to Bar-tailed Desert Lark, but less upstanding. Rather small, delicate lark, sharing 'horns' and basic plumage pattern of Shore Lark, but has pale areas of face white, upperparts sandy-pink, and underparts buff-white. Juvenile lacks all facial pattern and shows faint pale spots above.

In Levant and across North Africa, may overlap with wandering Shore Lark but most birds of both species usually well divided by altitude and habitat. Temminck's Horned Lark essentially a smaller, paler bird adapted to level steppe or desert regions. Adults unmistakable, with slightness, white face, and plumage uniformity obvious. Juvenile much paler than Shore Lark and best distinguished from similarly-coloured desert larks by black-panelled, white-edged tail.

Habitat. In lower middle latitudes in arid, warm, usually level lowlands, bare or sparsely vegetated, extending from Mediterranean to oceanic climate in Morocco. In north-west Africa, only below 1000 m, whereas Shore Lark lives above 2000 m and is completely separated in habitat. Although occurring on stony plains and deserts, prefers stony plateaux and steppes with solid soil; in sandy areas stays on patches where soil compact, avoiding pure sand.

Distribution and population. SYRIA. Characteristic bird of Syrian desert. ISRAEL. Not more than a few tens of pairs following sharp decline in 1980s due mainly to loss of habitat. JORDAN. Most widespread and common lark of non-mountainous hammada deserts; scarcer further south. IRAQ. Common at least formerly in western deserts. KUWAIT. Fairly common in most years, but scarce after dry winters. SAUDI ARABIA. In Harrat Al Harrah reserve, numerous in all undulating areas. EGYPT. Records from Eastern Desert March–May 1990 suggest possible (temporary?) range extension. Fairly common. LIBYA. Range in south uncertain. TUNISIA. Fairly common. MOROCCO. Common.

Accidental. Malta.

Beyond west Palearctic, breeds only in Arabia.

Movements. Resident. Some dispersal occurs; this probably at individual rather than population level in North Africa, but may be a more regular feature in Arabia, where winter visitors occur in large flocks south of breeding range.

Food. Seeds, and occasional insects and fruits. Feeding flocks in Morocco keep tightly together, and birds seen turning over stones many times their own weight.

Social pattern and behaviour. Poorly known, but most aspects probably similar to Shore Lark. Song of ♂ given from ground and in the air. Song-flight similar to Shore Lark, but ascent rather feeble and never to any great height.

Voice. Overall, similar to Shore Lark, though quieter. Song of ♂ consists of disconnected bursts of soft, melodious twittering or quiet, very fine warbling; mostly a somewhat monotonous repetition (with some variation) of short phrases: 'dee dee-eeee', 'chep seee-eee', or 'chep-ep seeee'.

Breeding. SEASON. Algeria: eggs laid April–May. Western Morocco: eggs laid mid-February to April. Jordan: eggs and young found late April and early May. SITE. On ground in the open, or in shelter of tussock. Nest: shallow depression lined with grass, twigs, and rootlets, with inner lining of soft grass-heads; mud lining with rag and wool also recorded; usually a rampart of small stones. EGGS. Sub-elliptical, smooth and glossy; virtually indistinguishable from Shore Lark. Clutch: 2–4. (Incubation and fledging periods not recorded.)

Wing-length: ♂ 96–106, ♀ 88–96 mm.

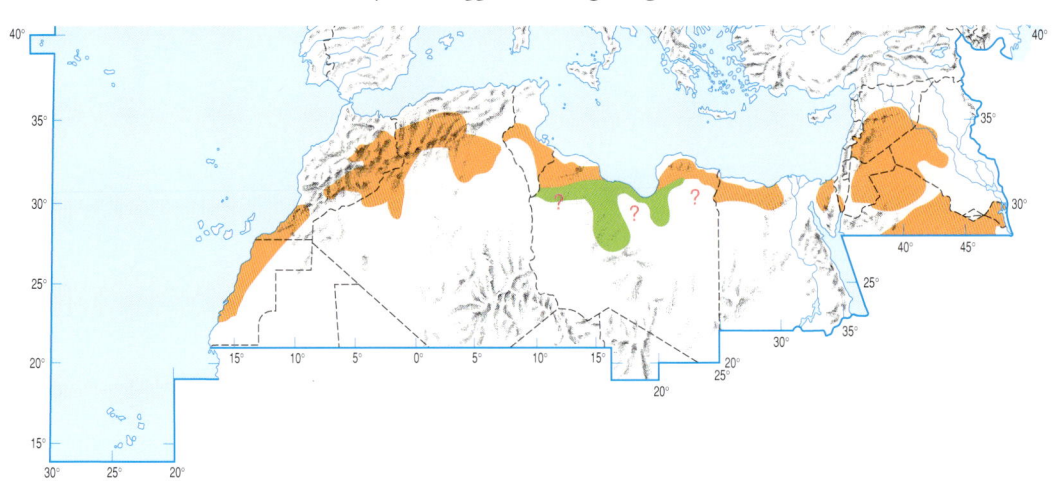

Swallows Family Hirundinidae

Highly specialized group of small oscine passerines (suborder Passeres); flight-feeders on flying and wind-borne invertebrates, spending much time on the wing. 70–80 species in 16–19 genera. Cosmopolitan but absent from polar regions and most oceanic islands. Many highly migratory. Family represented by 7 species breeding in west Palearctic, and 4 vagrants.

Differ from other Passeres in distinctive structure of syrinx. Body slender, neck short. Sexes similar in size. Bill very short and flattened and mouth wide; a few weak rictal bristles present in some species. Wing very long and pointed. Flight light and graceful, with high manoeuvrability. Tail often forked, outer feathers greatly elongated in some species. Legs and toes short and slender, but claws strong; front toes more or less united at base. Bathe by dipping in flight in series of brief splashes without landing on surface. Drink from flight and, less often, by sipping from puddles (etc.). Sunning common, sometimes with one wing lifted.

Plumage compact. Usually dark (black, blue, or brown), often with metallic sheen and contrasting underparts (white, chestnut, or greyish) and/or rump (white, buff, or rufous); underparts sometimes with chest-band, tail sometimes with contrasting spots or patches. Tarsus and toes sometimes partly or fully feathered. Sexes usually alike or nearly so. Nestlings have rather scanty 1st down soon after hatching, but denser 2nd down develops after *c.* 1 week.

Tree Swallow *Tachycineta bicolor*

PLATE: page 1054

Du. Boomzwaluw   Fr. Hirondelle bicolore   Ge. Sumpfschwalbe   It. Rondine arboricola
Ru. Американская древесная ласточка   Sp. Golondrina canadiense   Sw. Trädsvala

Field characters. 12–13 cm; wing-span 34–37 cm. Size between Sand Martin and Swallow; form recalls former, lacking streamers to tail fork. Adult ♂ iridescent emerald-blue on head cap, back, rump and inner wing-coverts, duller, more oily brown on wings and tail, pure white below; ♀ less iridescent above; immature uniform oily greyish-brown above, with suggestion of clouded breast band. At all ages, white lateral tail-coverts obvious on sides of narrow rump. Flight light as Sand Martin.

Adult unmistakable; immature may suggest Sand Martin and also potentially vagrant Rough-winged Swallow *Stelgidopteryx ruficollis* but breast marks never form discrete band of former or broad suffusion on flanks of latter.

Habitat. Open areas over lakes, marshes, coasts, fields, and meadows in vicinity of trees, in airspace below *c.* 50 m (but much higher on migration) from boreal to southern temperate North America and tropical South America. Feeds on flying insects and can subsist on seeds and berries in winter; nests in cavity of (usually dead) tree or in nest-box.

Distribution. Breeds from Alaska and across Canada, where northern limit of range is around treeline, south to southern California and east to northern parts of southern states of USA; range expanding southwards.

Accidental. Britain: St Mary's (Isles of Scilly), adult ♂, June 1990.

Movements. Migratory, wintering around Gulf of Mexico, occasionally as far north as New England, south to most of Mexico, Central America, Caribbean islands, and northern South America; perhaps resident in southern California. Eastern populations migrate south along east coast of USA, others probably follow Mississippi and Rocky Mountains. Leaves breeding areas July–August, returning mid-March to early April. Vagrant to Greenland.

Wing-length: ♂♀ 109–125 mm.

Brown-throated Sand Martin *Riparia paludicola*

PLATES: pages 1055, 1067

Du. Vale Oeverzwaluw   Fr. Hirondelle paludicole   Ge. Braunkehl-Uferschwalbe   It. Topino africano
Ru. Малая береговушка   Sp. Avión zapador africano   Sw. Brunstrupig backsvala

Field characters. 12 cm; wing-span 26–27 cm. Fractionally smaller than Sand Martin. Smallest hirundine of west Palearctic with character and behaviour of Sand Martin, but rather more compact form. Upperparts and underparts contrast less than on Sand Martin; lacks pale mouse to dull white chest-band of that species.

Instantly separable from Sand Martin when plain underparts visible, but liable to confusion with Crag or Rock Martins when small size and uniformly dark tail not apparent. Flight less fast and agile than Sand Martin, at times recalling small bat; lacks obvious momentum of Crag or Rock Martins. Behaviour as Sand Martin.

Tree Swallow *Tachycineta bicolor*: **1** ad, **2** juv. Banded Martin *Riparia cincta* (p. 1058): **3** ad, **4** juv. Ethiopian Swallow *Hirundo aethiopica* (p. 1064): **5** ad, **6** juv. Cliff Swallow *Hirundo pyrrhonota* (p. 1066): **7** ad, **8** juv.

Habitat. Replaces Sand Martin in low latitudes near southern limits of west Palearctic. Consequently tolerates much warmer climates, but exposed to oceanic influences in Moroccan breeding area which reaches up to Mediterranean climatic zone.

Distribution and population. Confined in west Palearctic to Morocco, where extension of breeding range near Rabat in 1980s. Uncommon. Population fluctuating, with many breeding colonies unstable.

Accidental. Israel, Saudi Arabia, Egypt.

Beyond west Palearctic, breeds discontinuously from Uzbekistan and Afghanistan east to Taiwan and Philippines, from Sénégambia east to Ethiopia and south to Cape Province, and in Madagascar.

Movements. Moroccan population sedentary, breeding during winter. Vagrants of unknown origin have occurred within west Palearctic (see Distribution) February–May, and November, and outside it at Abqaiq (eastern Saudi Arabia) in May and in Oman September–October.

Food. No information from west Palearctic. Elsewhere, diet

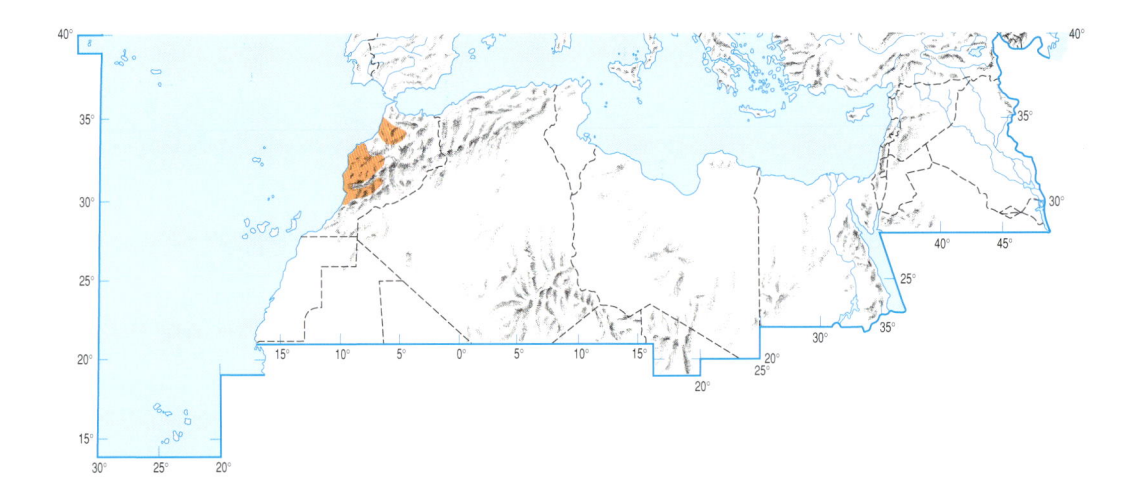

Brown-throated Sand Martin *Riparia paludicola mauritanica* (Morocco): **1** ad, **2** juv. Sand Martin *Riparia riparia*. *R. r. riparia*: **3** ad, **4** juv. *R. r. diluta*: **5** ad. *R. r. shelleyi*: **6** ad. House Martin *Delichon urbica* (p. 1066): **7** ad breeding, **8** ad non-breeding, **9** juv.

consists of small flying insects. Takes prey in flight, often over water.

Social pattern and behaviour. Little known, but apparently similar to Sand Martin.

Voice. Song twittering. Call disyllabic, recalling Sand Martin but clearer toned—'sree-sree'.

Breeding. SEASON. Morocco: laying begins November–December and lasts to February, exceptionally April. SITE. In excavated tunnel in river or gorge bank, or side of road bank or pit; typically 0.8–3 m above ground. Nest: in chamber at end of tunnel 30–80 cm long, often rising slightly; nest cup of feathers and grass. EGGS. Sub-elliptical, smooth and glossy; white. Clutch: 3–4. INCUBATION. About 12 days. FLEDGING PERIOD. *c.* 20 days.

Wing-length: ♂♀ 97–104 mm.

Sand Martin *Riparia riparia*

PLATES: pages 1055, 1067

DU. Oeverzwaluw   FR. Hirondelle de rivage   GE. Uferschwalbe   IT. Topino
RU. Береговушка   SP. Avión zapador   SW. Backsvala   N. AM. Bank Swallow

Field characters. 12 cm; wing-span 26.5–29 cm. 20–25% smaller than Swallow, with distinctly slighter structure and slightly forked tail. Rather small, slightly built hirundine, smallest in Europe and 2nd smallest of west Palearctic. Dark brown above, white below; only obvious character dark brown chest-band, but dark dusky brown underwing and undertail also noticeable.

Unmistakable in clear, close view. From above, may suggest Brown-throated Sand, Crag, or Rock Martins but none of these shows dark chest-band above white breast and belly. Flight fast and agile but without grace of Swallow or steadiness of House Martin; involves less gliding, more fluttering, fewer changes in height, and more sudden changes in direction.

Habitat. Mainly aerial, usually in low airspace. Breeds from Mediterranean through steppe, temperate, and boreal to upper subarctic zones, in continental and also oceanic climates, with summer temperatures ranging down almost to 10°C, and varying regimes of wind and rainfall. Rarely on ground, spending much of day in flight, resting on overhead wires and other suitable perches; roosts gregariously, often in reedbeds. As an aerial feeder, largely independent of nature of underlying terrain, provided it supports adequate flying insects. Avoids densely wooded or built-up areas, mountains and broken terrain, deserts, and narrow valleys, preferring neighbourhood of sandy, loamy, or other workable banks, excavations, cliffs, and earth-mounds, suitable for tunnelling to form nest-chambers. Such sites often occur naturally along rivers and streams, or by lakes and on sea coasts, or artificially where sand or other materials are extracted in conditions leaving some faces undisturbed.

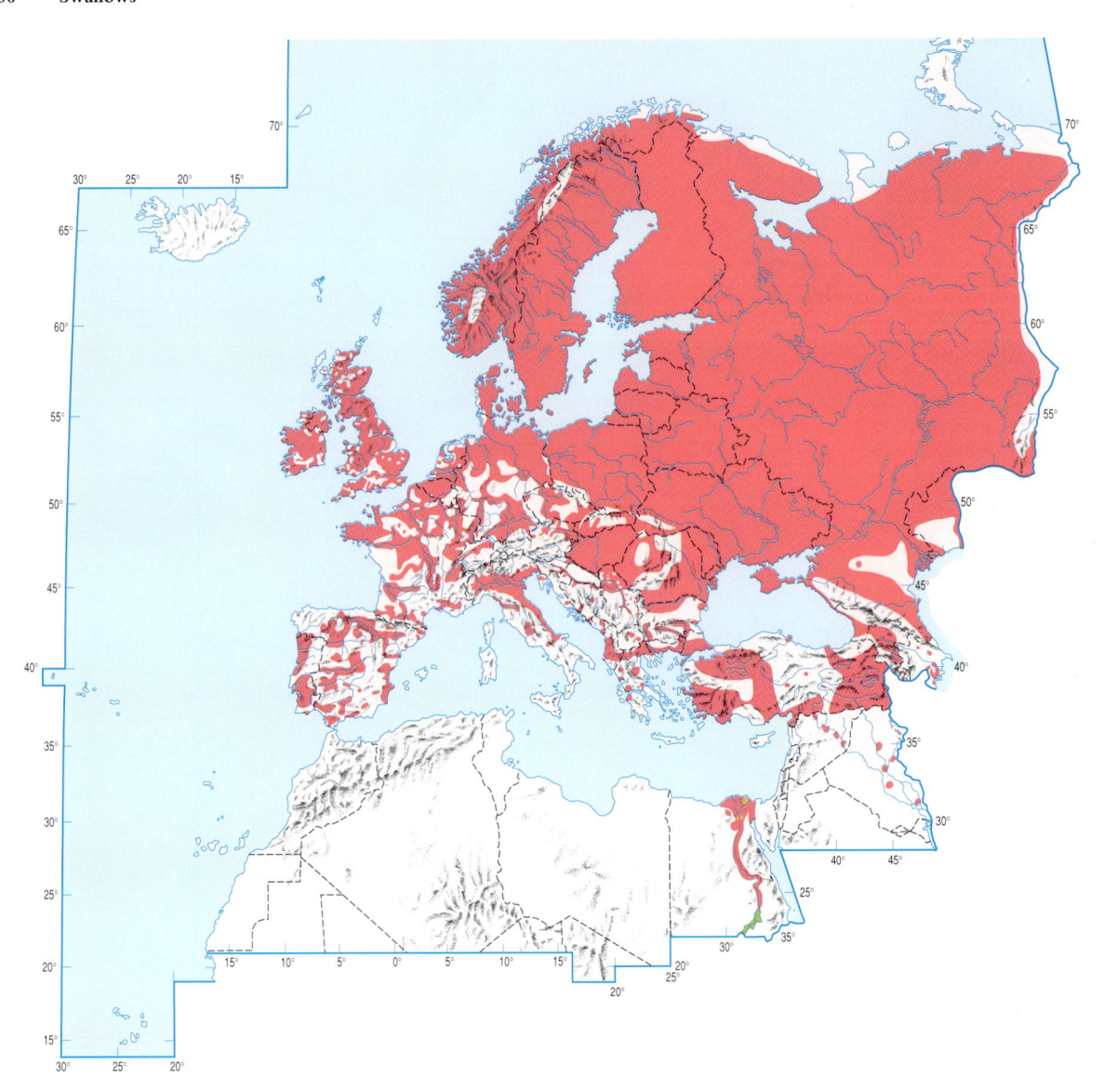

Distribution. Spread with creation of new sand and gravel pits in several countries, but marked recent decrease in range Netherlands, less so Britain, Ireland (in 1970s), Belgium, Luxembourg, Denmark (probably 25% reduction), Austria (perhaps), Spain, Italy, Ukraine, and Moldova. Mostly stable elsewhere, though fluctuating Switzerland and Slovenia, and increased Portugal. GERMANY. Largest colonies (1000+ holes) in east are on coastal cliffs. BALEARIC ISLANDS. Has never bred on any island. RUMANIA. Common in lowlands, with colonies in banks of all large rivers. ARMENIA. Local. AZERBAIJAN. Distribution inadequately known; no information from northeast. TURKEY. Likely to occur everywhere in suitable habitat, but probably severely under-recorded. Distribution also obscured by migration up to early June and from July onwards. SYRIA. Breeding and breeding-season records from Euphrates region; probably breeds at least locally in north-east. Not yet clear whether range continuous or fragmented as shown. ISRAEL. Reported breeding in past, but no recent evidence, though summering occurs. TUNISIA. Last bred in 1960s. ALGERIA. Bred 19th century and perhaps 1978. MOROCCO. Occasional breeder, perhaps formerly more regular; latest record 1969.

Accidental. Iceland, Faeroes, Madeira, Cape Verde Islands.

Beyond west Palearctic, breeds throughout much of Asia south to southern Pakistan and central China; also widespread in North America. Breeding grounds of *eilata* unknown.

Population. Declined over much of north-west Europe, also in some countries in south, but generally not in east. Reduction in numbers attributed primarily to drought in winter quarters. Population apparently stable Norway, Sweden, Latvia, Lithuania, Poland, Czech Republic, Hungary, Croatia, Bulgaria, Rumania, Russia, and Belarus'; fluctuating France, Switzerland, Portugal, Albania, Slovenia, and slight increase reported Estonia and Slovakia. BRITAIN. 77 500–250 000 pairs in 1988–91. Marked decline, notably 1968–9 (nearly 1 million pairs before this) and 1983–4; recent partial recovery. IRELAND. 16 000–24 000 pairs; decrease since 1969. FRANCE. 10 000–100 000 pairs in 1970s. BELGIUM. Marked decrease: now under 5000 pairs, with 2930 pairs Flanders in 1991, and under 2000 pairs Wallonia in 1994. LUXEMBOURG. 60–80 pairs; decreasing since 1980s. NETHERLANDS. Strong decline (attributed to habitat changes) from 25 000 pairs 1963–4 to 9000–12 000 pairs 1989–91. Some recovery in east and south since mid-1980s. GERMANY. Estimated 130 000 pairs in mid-1980s. Revised estimate (also taking slight decline into account) 65 000–90 000 pairs in early 1990s. DENMARK. 20 000–40 000 pairs 1987–8; stable or perhaps decreased 1981–94. NORWAY. 100 000–250 000 pairs 1970–90. SWEDEN. 100 000–250 000 pairs in late 1980s. FINLAND. 80 000–120 000 pairs in late 1980s; slight decrease. ESTONIA. 20 000–50 000 pairs in 1991. LATVIA. 30 000–100 000 pairs in 1980s. LITHUANIA. Common; population apparently stable. One estimate of 15 000–20 000 pairs 1985–8. POLAND. 150 000–200 000 pairs. CZECH REPUBLIC. 18 000–36 000 pairs 1985–9. SLOVAKIA. 15 000–30 000 pairs 1973–94. HUNGARY. 60 000–80 000 pairs 1979–93. AUSTRIA. Fluctuates with availability of man-made sites; no apparent changes. One estimate of 9000–15 000 pairs. SWITZERLAND. 3000–5000 pairs 1985–93; stable after decline since 1960 of 37%. SPAIN. 540 000–750 000 pairs; slight decrease. PORTUGAL. 10 000–100 000 pairs 1978–84. ITALY. 8000–9000 pairs 1983–93; slight decrease. GREECE. 10 000–20 000 pairs. ALBANIA. 2000–5000 pairs in 1981. YUGOSLAVIA: CROATIA. 25 000–30 000 pairs. SLOVENIA. 150–250 pairs. BULGARIA. 10 000–100 000 pairs. RUMANIA. 30 000–50 000 pairs 1986–92. RUSSIA. 1–10 million pairs. BELARUS'. 200 000–250 000 pairs in 1990. UKRAINE. 140 000–150 000 pairs in 1988; slight decrease. MOLDOVA. 6500–7000 pairs in 1985; slight decrease. AZERBAIJAN. Common. TURKEY. 10 000–100 000 pairs. EGYPT. Abundant (probably over 100 000 pairs).

Movements. Migratory; most of breeding range vacated in winter. Main west Palearctic populations plus all those of Siberia winter in African Sahel zone (from Sénégal eastwards) and in East Africa south to Mozambique.

Extensive ringing, particularly in Britain and Ireland, has revealed much detail of movements within western Europe. During late summer and autumn, early-brood juveniles undertake local movements, which, by beginning of August, become oriented southwards; those leaving Britain choose short sea crossing. Passage of British and Irish birds continues SSW to Biscay coast of France and skirts north-west end of Pyrénées. Some birds then move down Ebro valley to Mediterranean coast of Spain, and others appear to move overland to Coto Doñana area. Birds cross Mediterranean into Morocco, reaching Sahel region by October or early November. In wintering areas, birds congregate where food available, which is often (but not always) associated with water. Appear to be nomadic, and ringing evidence from British birds suggests that many move east through Sahel region during winter to make their return northwards via Niger inundation zone (Mali) or further east. In a normal year, first arrivals in southern Britain occur well before end of March.

Food. Small airborne invertebrates. Feeds almost exclusively in flight, exceptionally by running about on ground.

Social pattern and behaviour. Often highly gregarious, at all times of year. Flocks of up to several hundreds, even thousands, common during migration periods and in winter quarters, sometimes associated with other hirundines. Monogamous mating system of seasonal duration probably the rule, but ♂♂ commonly attempt to copulate with ♀♀ other than mate, and some ♀♀ obtain new mate for 2nd brood. Small nest-area territory defended around mouth of tunnel. Pairing linked to nest-site selection; completed before end of excavation. In Britain, song given late April to early July, occasionally up to early September. ♂ takes initiative in selecting site. c. 1 week before young fledge, some ♀♀ leave mates to raise young while they pair with new ♂♂ to raise 2nd broods.

Voice. Song a harsh twittering, little more than a sequence of contact-calls. Contact-call, given perched or in flight, a harsh grating 'tschrd' or 'tschr', not as trilling as House Martin.

Breeding. SEASON. Britain and western Europe: egg-laying from end of April or early May. Little variation across range, though up to 2 weeks earlier in south. 2 broods except in north and east of range. SITE. Hole in river bank, sand quarry, or sea-cliff, occasionally in drain pipe or other artificial hole. Nest: excavated hole, mean length 65 cm (35–119, but most 46–90), ending in chamber 4–6 cm in diameter. Nest-cup made of feathers, grass, leaves, etc. EGGS. Sub-elliptical, smooth and fairly glossy; white. Clutch: 4–6 (2–7). INCUBATION. 14–15 days. FLEDGING PERIOD. Average 22.3 days.

Wing-length: *R. r. riparia*: ♂♀ 103–111 mm.
Weight: *R. r. riparia*: ♂♀ mostly 11–16 g.

Geographical variation. Slight, mainly clinal; involves size, depth of ground-colour of upperparts and chest-band, and width and contrast of chest-band. 4 races in west Palearctic, 3 of them marginal; nominate *riparia* occupies most of range. Size probably clinally smaller towards south. Nominate *riparia* from northern and central Europe east to Siberia largest, birds from southern Europe, north-west Africa, Turkey, and Levant perhaps slightly smaller. *R. r. diluta* from Ural river area distinctly paler and chest-band less clear-cut; *R. r. shelleyi* from Egypt slightly paler, chest-band clear-cut. *R. r. eilata*, a small dark-plumaged race, occurs on passage in Israel; perhaps breeds in northern Iraq.

Banded Martin Riparia cincta

PLATES: pages 1054

Du. Witbrauwzwaluw   Fr. Hirondelle à collier   Ge. Weißbrauen-Uferschwalbe   It. Topino dai sopraccigli bianchi
Ru. Белобровая береговушка   Sp. Avión cinchado   Sw.

Field characters. 16–17 cm; wing-span 39–41 cm. Size close to Crag Martin; 25% larger than Sand Martin, with hardly forked tail. Plumage pattern suggests dark Sand Martin, but at close range white fore-supercilium contrasting with blackish tail and broad brown breast band bulging downwards towards fore-belly distinctive. Flight less agile than Sand Martin, appearing sluggish at times.

When size not established, could be confused with Sand Martin and even Alpine Swift. Infrequent call a quiet chatter.

Habitat. Low airspace over ponds, marshes, coasts, savanna, pasture, etc. in sub-Saharan Africa below $c.$ 3000 m; forages for flying insects in slow deliberate flight, singly, in pairs, or in small flocks, and not gregarious on the whole; often around large herbivores and above grass fires. Nests, but not colonially, in burrow usually near water.

Distribution. Rather complex and not very well known; breeds in scattered populations throughout Africa south of Sahara; apparently mainly non-breeding visitor to areas north and west of Zaïre, though breeds in extreme southern Sudan, Ethiopia, Cameroon, and possibly parts of Nigeria.

Accidental. Egypt: Elephantine Island (Lake Nasser), November 1988.

Movements. Complicated and poorly understood; basically moves north and south with rains, breeding generally in rainy season. Migratory in Zimbabwe and South Africa; probably resident with local movements in Sudan, Ethiopia, Zaïre, Angola, and East Africa, though situation unclear because of influxes of non-breeders. Eggs laid in Ethiopia May–August. Vagrant to northern Yemen.

Wing-length: ♂ ♀ 114–140 mm.

African Rock Martin (includes Pale Crag Martin)
Hirundo fuligula

PLATES: pages 1059, 1067

Du. Vale Rotszwaluw   Fr. Hirondelle isabelline   Ge. Steinschwalbe   It. Rondine rupestre africana
Ru. Африканская скалистая ласточка   Sp. Avión roquero africano   Sw. Blek klippsvala

Field characters. 12.5 cm; wing-span 27.5–30 cm. Only marginally larger but more robust than Sand Martin; $c.$ 15% smaller than Crag Martin. Similar to Crag Martin in structure and flight, and in rather uniform plumage, showing white spots on vent and inner webs of all but central tail feathers but, in comparison, looking paler and greyer with white (unstreaked) throat and less contrasting dark under wing-coverts.

Paler races can be mistaken for Sand Martin and Brown-throated Sand Martin; separation best based on vent and tail patterns.

Habitat. In warm dry lower latitudes, from fringe of Mediterranean through subtropical to tropical, forming counterpart of Crag Martin in tropical and arid mountainous and rocky areas. In North Africa frequents gorges and ravines in desert areas, but in Egypt found near monuments (e.g. Abu Simbil) and in certain desert towns (e.g. Aswan).

Distribution and population. Desert distribution, also numbers and trends, little known. ISRAEL. Breeding on houses in Arava and Negev since mid- or late 1970s. Total population

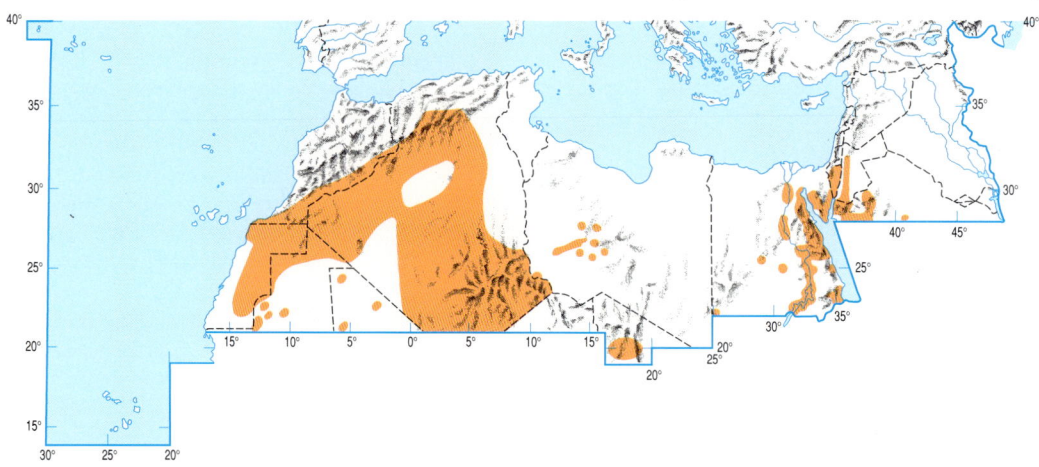

African Rock Martin *Hirundo fuligula*. *H. f. obsoleta*: **1** ad, **2** juv. *H. f. spatzi*: **3** ad. *H. f. perpallida*: **4** ad. Crag Martin *Hirundo rupestris*: **5** ad fresh (autumn), **6** ad worn (spring), **7** juv.

increased to a few thousand pairs in 1980s. JORDAN. Very common resident where mapped. IRAQ. Status unclear: perhaps present in gorges in north, but only Crag Martin identified there in 1950s. EGYPT. Common (in range 10 000–100 000 pairs). ALGERIA. Locally common in Sahara, Hoggar, and Tassili; scarce in north of range. MOROCCO. Scarce. MAURITANIA. Local and uncommon.

Accidental. Kuwait.

Beyond west Palearctic, breeds Arabia east to Pakistan, and widely distributed in sub-Saharan Africa.

Movements. Mainly resident, but also some seasonal movement of unclarified extent.

Food. Insects, taken in flight. Like Crag Martin, flight involves much steady gliding. Apparently hunts usually at no great height.

Social pattern and behaviour. Little information for west Palearctic. Saharan birds usually nest in isolated pairs.

Voice. A quiet bird; song and calls apparently resemble Crag Martin.

Breeding. SEASON. North-west Africa: eggs laid February–April, perhaps some January. SITE. On vertical rock surfaces, usually under overhang, mainly on cliffs, but also under bridges, on houses and other buildings, particularly under ledges and verandahs, etc. Nest: deep half-cup of mud without incorporated vegetable matter, lined with fine plant material and feathers. EGGS. Elongated, sub-elliptical, smooth and slightly glossy; white, spotted and speckled black and purple-grey, often concentrated at broad end. Clutch: 2–3. INCUBATION. About 17 days. FLEDGING PERIOD. 25–30 days.

Wing-length: ♂♀ 110–125 mm.

Geographical variation. 4 races recognized in west Palearctic, based mainly on depth of colour, which varies clinally, boundaries hard to define: *H. f. obsoleta* (Egypt to northern Arabia, considered separate species, Pale Crag Martin, by some), *perpallida* (north-east Arabia and southern Iraq), *presaharica* (most of north-west Africa), *spatzi* (southern Algeria).

Crag Martin *Hirundo rupestris*

PLATES: pages 1059, 1067

Du. Rotszwaluw   Fr. Hirondelle de rochers   Ge. Felsenschwalbe   It. Rondine montana
Ru. Скалистая ласточка   Sp. Avión roquero   Sw. Klippsvala

Field characters. 14.5 cm; wing-span 32–34.5 cm. Head, body, wings, and tail all broader or bulkier than those of any other west Palearctic hirundine; 15% larger than African Rock Martin. Chunky hirundine with almost unforked tail. Essentially uniform dusky brown, with almost black under wing-coverts; at close range shows dark-mottled throat, paler and warmer buff-brown forebody, dull white chevrons on sides of vent, and white spots on underside of tail. Flight least energetic of west Palearctic hirundines, involving much steady gliding.

In Europe, field identification of Crag Martin easy since African Rock Martin restricted to desert areas of North Africa and Middle East. Much more localized than other hirundines, spending much time around breeding areas.

Habitat. Breeds in west Palearctic in low middle latitudes, from temperate to Mediterranean zone, mainly in warm dry continental climates, but attachment to mountainous regions involves exposure to sharply differing temperatures. Deep shadow, winds, and snow are generally avoided. In Switzerland, occurs from lowlands to above treeline, most frequently in warm, dry, and sheltered situations.

Distribution. FRANCE. Apparently no longer breeding Seine-Maritime since 1970s. GERMANY. Tiny population is northernmost outpost of extensive Alpine range; typical of peripheral population in that has probably become extinct several times, yet returned to breed. AUSTRIA. Range increase; motorway viaducts colonized. SWITZERLAND. Range increase: now in whole Jura. SPAIN. Winter range perhaps more extensive than shown. PORTUGAL. Algarve: range expansion. YUGOSLAVIA. Range reported to have expanded north in 1980s. BULGARIA. Slight increase. RUMANIA. Spread north since 1968. Recently discovered breeding in some gorges of southern Carpathians. TURKEY. Some gaps perhaps due to under-recording. Nests exceptionally almost at sea-level in mountains near coast. SYRIA. Local breeder several places in Anti-Lebanon; a few winter (perhaps breeds) Euphrates area. ISRAEL. First found breeding Hermon massif 1978. JORDAN. Possible winter range mapped. IRAQ. Probably breeding in north. EGYPT. Rare and irregular winter visitor; some sight records presumably misidentified Pale Crag Martin. TUNISIA. No recent evidence of breeding.

Accidental. Britain, Belgium, Denmark, Finland, Ukraine, Kuwait (or perhaps scarce passage-migrant), Madeira, Canary Islands.

Beyond west Palearctic, breeds south-west Arabia, and from Iran east to central Asia and northern China.

Population. FRANCE. 1000–10 000 pairs in 1970s. GERMANY. *c.* 10 pairs in mid-1980s; marked fluctuations. AUSTRIA. Perhaps over 500 pairs, and slightly increasing. SWITZERLAND. 2000–2500 pairs in 1985–93; increase. SPAIN. 84 000–100 000 pairs; stable. PORTUGAL. 10 000–100 000 pairs 1978–84; stable or (in Algarve) increasing. ITALY. 5000–10 000 pairs 1983–93; stable. GREECE. 10 000–20 000 pairs. ALBANIA. 2000–5000 pairs in 1981. YUGOSLAVIA: CROATIA. 1000–1500 pairs; stable. SLOVENIA. 50–100 pairs; slight increase. BULGARIA. 1000–10 000 pairs; slight increase. RUMANIA. 50–60 pairs in 1986–92; marked increase. AZERBAIJAN. Uncommon; more common in Nakhichevan region. TURKEY. 10 000–100 000 pairs. CYPRUS. 2000–5000 pairs; probably stable. ISRAEL. 2–5 pairs Mt Hermon. MOROCCO. Uncommon.

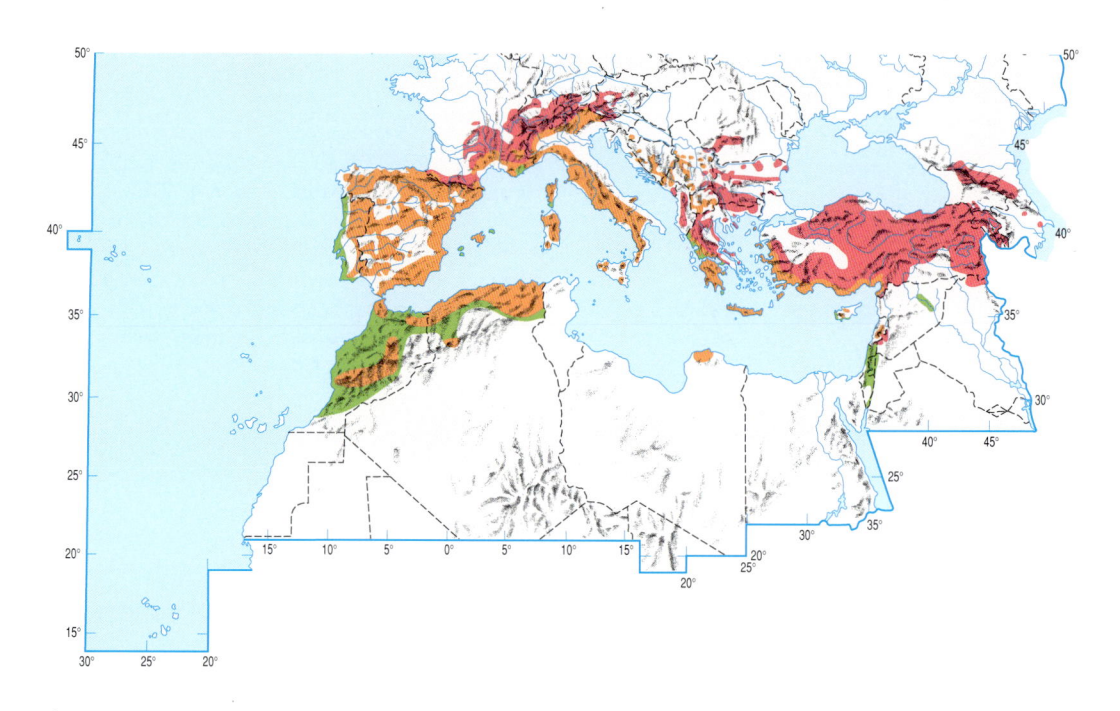

Movements. Northern populations partially migratory; mainly resident elsewhere though making altitudinal movements. Though largely migratory across southern Europe, from Greece to Spain, some winter regularly on northern side of Mediterranean basin and especially at western end. However, large numbers cross Straits of Gibraltar into Morocco, with peak passages October and March. Winter range in Africa mainly northern parts of Morocco, Algeria, and Tunisia, though small numbers also occur in Sénégal, along Nile valley as far south as northern Sudan, and on Red Sea coast and in western highlands of Ethiopia. Resident in Cyprus and Turkey, though leaves upland areas for winter, when found mostly in coastal and other lowland places.

Food. Small insects. Prey mostly taken in flight, which typically involves much more steady gliding than other west Palearctic hirundines except African Rock Martin. Often hunts close to cliff-faces; will take insects disturbed from rock surfaces by the bird's own flight, and will pick insects from rocks as it passes.

Social pattern and behaviour. Gregarious outside breeding season, sometimes markedly so where food locally abundant. Breeds solitarily or in small loose colonies, rarely more than 10 pairs. Highly aggressive towards conspecifics and many other species encroaching on nest-territory.

Voice. Freely used throughout the year, and birds possess wide repertoire, but most calls quiet. Song a throaty and persistent rapid twittering sound, given in flight; not very conspicuous. Commonest call consists of variations on 'prrit' or 'chwit'.

Breeding. SEASON. Main laying period from mid-May, with 2nd clutches laid in July. SITE. On vertical rock wall, in crevice, small hollow, or occasionally in shallow tunnel, and usually under an overhang. Frequently on building. Height above ground up to 40 m. Nest: half-cup of mud, lined with feathers and plant material. EGGS. Long sub-elliptical, smooth and slightly glossy; white, sparsely spotted red and grey, concentrated at broad end. Clutch: (1–)3–5. INCUBATION. 13–17 days. FLEDGING PERIOD. 24–27 days.

Wing-length: ♂♀ 126–136 mm.
Weight: ♂♀ mostly 17–30 g.

Swallow *Hirundo rustica*

PLATES: pages 1062, 1067

DU. Boerenzwaluw   FR. Hirondelle rustique   GE. Rauchschwalbe   IT. Rondine
RU. Деревенская ласточка   SP. Golondrina común   SW. Ladusvala   N. AM. Barn Swallow

Field characters. 17–19 cm, of which tail-streamers 2–7 cm; wing-span 32–34.5 cm. Most attenuated hirundine of west Palearctic; less bulky than Red-rumped Swallow. Medium-sized hirundine with classic form of small bill, wide gape, streamlined body, long wings, and long forked tail with outer feathers forming streamers in adult. Upperparts shiny blue-black; underparts buff-white to red-buff, relieved by red-chestnut face and throat and blue-black chest-band and by noticeably dark under-surface of flight-feathers and white-spotted tail. Flight light and graceful, with characteristic sweeps and swoops after insects.

Habitat. Breeds across west Palearctic from subarctic through boreal, temperate, steppe, and Mediterranean zones in both continental and oceanic climates; missing only from arctic tundra and desert belts. Entirely dependent on constant supply of small flying insects taken in flight in lower airspace over surface, either of shallow water or clothed with low moist green vegetation. Usually avoids densely wooded, precipitous, arid, and densely built-up areas, preferring pasture grazed by large animals, meadows, and farm crops, especially where accessible open structures such as barns, sheds, stables, outhouses, and porches provide suitable nest-sites near by, with roof-ridges, overhead wires, and bare branches or twigs as perches and places to sun and preen, and ready access to water. In Swiss Alps, breeds freely up broader valleys to 1775 m, but avoids narrower and more shady side-valleys and exposed tops. In Caucasus, follows human settlement to heights of 2400–3000 m.

Distribution. Breeding range mostly stable in recent decades, though fluctuating Faeroes, slight decrease reported Austria, Spain, Albania, Rumania, Ukraine, and Moldova, slight increase Switzerland. ICELAND. Occasional breeder (up to 5 pairs): since 1979, bred 1986, 1988–90. FAEROES. Regular visitor, breeding some years (0–1 pair in recent decades). GERMANY. Density generally lower in east. MALTA. Bred 1974, 1995. TURKEY. Widespread and common throughout. CYPRUS. Very common around towns and villages. SYRIA. Widespread breeder in cultivated landscape of west, also in north. LEBANON. Distribution poorly known; perhaps more widespread than shown. JORDAN. Common resident in Jordan Valley. ALGERIA. Common in north of range; scarcer further south.

Accidental. Spitsbergen, Bear Island, Jan Mayen, Iceland (annual), Azores, Madeira.

Beyond west Palearctic, breeds virtually throughout Asia except extreme north, India south of Himalayas, and South-East Asia; also throughout much of North America.

Population. Known to fluctuate, but widespread recent decline affecting north-west Europe, Fenno-Scandia and Baltic States, also some central European and Mediterranean countries, and east to Rumania, Moldova, and Ukraine. Likely causes include loss of feeding habitat and reduced supply of insect food through changes in farming practices; also climate change on breeding and wintering grounds. BRITAIN. 570 000 territories 1988–91; steady decline in 1980s, some recovery in 1988. IRELAND. 250 000 pairs 1988–91. FRANCE. 1–5 million pairs; stable. BELGIUM. 125 000 pairs 1973–7; continuing decrease. LUXEMBOURG. 15 000–20 000 pairs. NETHERLANDS. 170 000–300 000 pairs 1979–85; marked decline. GERMANY. 1.51 million pairs in mid-1980s; marked decline. Reports of general decline in east refer mainly to 1960s; detailed information lacking for more recent period. DENMARK. 200 000–300 000 pairs 1983–5. NORWAY. 100 000–400 000 pairs 1970–90. SWEDEN. 200 000–400 000 pairs in late 1980s. FINLAND. 150 000–200 000 pairs in late 1980s. ESTONIA. 100 000–200 000 pairs in 1991; decrease, especially in 1960s–70s. LATVIA. 117 000–475 000 pairs in 1980s; stable. LITHUANIA. Common, but decreasing. One estimate of

Swallow *Hirundo rustica*. *H. r. rustica*: **1** ad ♂ fresh (autumn), **2** ad ♂ worn (spring), **3** ad ♀ fresh (autumn), **4** juv. *H. r. savignii*: **5** ad fresh (autumn). *H. r. transitiva*: **6** ad. Red-rumped Swallow *Hirundo daurica* (p. 1064). *H. d. rufula*: **7** ad fresh (autumn), **8** ad worn (spring), **9** juv. *H. d. daurica*: **10** ad fresh (autumn).

20 000–40 000 pairs 1985–8. POLAND. 1.5–2.9 million pairs. Stable, but strong decrease in large towns in recent decades. CZECH REPUBLIC. 400 000–800 000 pairs 1985–9; slow decrease. SLOVAKIA. 200 000–400 000 pairs 1973–94; slight decrease. HUNGARY. 150 000–200 000 pairs 1979–93; stable. AUSTRIA. Common; apparently stable. One estimate 250 000–300 000 pairs. SWITZERLAND. 100 000–150 000 pairs 1985–93; slight decrease. SPAIN. 783 000–812 000 pairs; slight decrease. PORTUGAL. 1 million pairs 1978–84; stable. ITALY. 500 000–1 million pairs 1983–93; slight decrease. GREECE. 20 000–50 000 pairs; slight decline. ALBANIA. 20 000–50 000 pairs in 1991; slight decline. YUGOSLAVIA: CROATIA. 500 000–600 000 pairs; slight decrease. SLOVENIA. 200 000–300 000 pairs; stable. BULGARIA. 500 000–5 million pairs; stable. RUMANIA. 300 000–500 000 pairs 1986–92; slight decrease. RUSSIA. 1–10 million pairs; stable. BELARUS'. 800 000–830 000 pairs in 1990; stable. UKRAINE. 800 000–850 000 pairs in 1988. MOLDOVA. 25 000–35 000 pairs in 1988. AZERBAIJAN. One of commonest breeding species. Probably more than several tens of thousands of pairs. TURKEY. 100 000–1 million pairs. CYPRUS. Probably 50 000–100 000 pairs; stable. ISRAEL. 10 000 pairs in 1980s. Marked decrease in 1950s attributed to pesticides; some recovery since early 1980s. IRAQ. Locally common. EGYPT. Abundant (over 100 000 pairs). TUNISIA. Common. MOROCCO. Locally common.

Movements. Migratory. A few aberrant individuals winter every year in southern and western Europe as far north as Britain and Ireland, and recorded annually in winter in southern Spain; small numbers winter regularly in North Africa; also small resident or partly resident populations in east Mediterranean countries. Otherwise, west Palearctic birds are long-distance migrants.

European and north-west Asian birds winter largely in Africa, mainly south of equator, though also locally numerous in West Africa. Passage broad-front, including large transdesert movements into and out of Africa across Sahara and Middle East. Juvenile dispersals begin July and become oriented southwards by early August as migration begins. Autumn passage protracted, with peak exodus from north-west Europe in September and first half of October. Mediterranean passage and arrivals in Africa north of equator are at height mid-September to late October, and birds become numerous in wintering regions south of equator in November. Return movement begins February. In North Africa, Mediterranean basin, and Middle East, peak spring movement occurs mid-March to late April. Early birds return to north-west Europe in second half of March, though main arrivals mid-April to mid-May.

Food. Almost wholly flying insects; in breeding season, especially flies (Diptera), but apparently usually fewer aphids than taken by House Martin, and average prey size much greater than for House Martin and Sand Martin. Prey taken almost entirely by aerial-pursuit, but many records of other methods, used especially when non-flying prey temporarily abundant or conditions unsuitable for aerial-pursuit.

Social pattern and behaviour. Highly gregarious outside breeding season. Large numbers typically occur in pre-migratory assemblies, sometimes on migration, and at roost-sites. Monogamous mating system the rule, but polygyny (♂

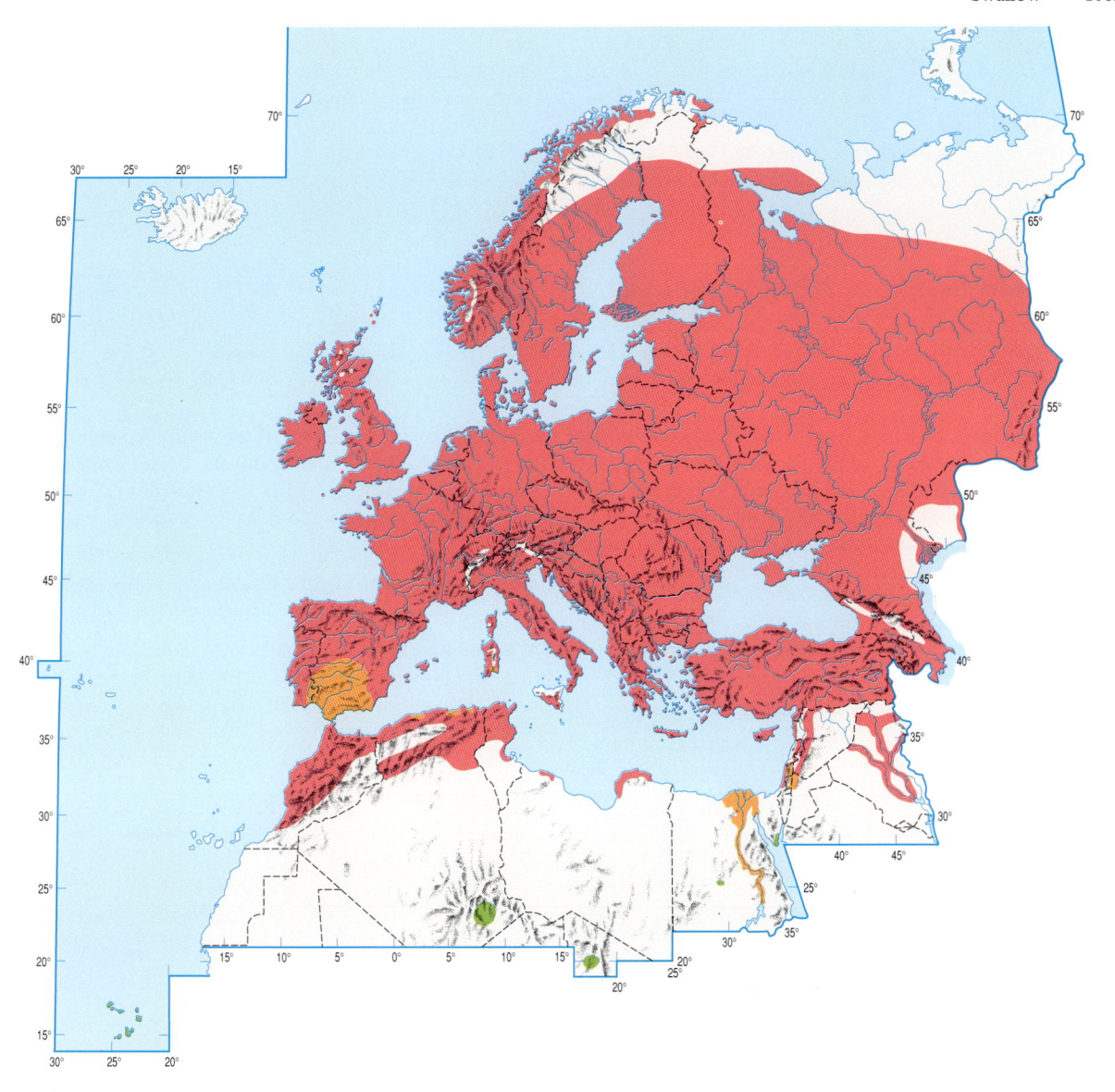

and ♂ 2♀♀) sometimes occurs, and birds sometimes copulate with non-mates if they are nearest neighbours in a colony. Pair-bond typically maintained for 2nd clutch and not infrequently for life. Nests mostly solitary, occasionally colonial. Colonies usually less than 5 pairs, occasionally many more. Both members of pair defend territory around nest. ♂ takes initiative in nest-site selection. Soon after he arrives, ♂ of established pair may build nest or refurbish an old one before ♀ returns; or ♂ may wait until she arrives. Once ♀ accepts site, she typically does most building while ♂ sings near by, in the air, or perched on vantage point near nest-site. Breeding success of ♂♂ depends significantly on length of tail-streamers; birds with long streamers most attractive to ♀♀; acquire mates earliest and breed most successfully.

Voice. Song a slightly spluttering but cheerful warble; often has distinctive terminal 'su-seer', last note higher pitched but falling; loud at close range but hardly carrying far. Calls include clear 'witt' or 'witt-witt' and loud distinctive 'splee-plink' in alarm or excitement.

Breeding. SEASON. Britain and western Europe: egg-laying from end of April or early May. Up to 2 weeks later in Scandinavia, and 1 week earlier in southern Europe. Southern Spain and north-west Africa: first eggs March. 2 broods usual, except in far north. SITE. On small ledge against vertical surface, e.g. beam or window-ledge in building, less often without support beneath; rarely in cave or tree. Nest: shallow half-cup or cup of mud pellets, usually mixed with plant material, with inner lining of feathers; normal dimensions 20 cm across and 10 cm deep. EGGS. Elongated elliptical or oval, smooth and glossy; white, lightly marked with red-brown spots, plus some lilac and grey. Clutch: 4–5 (2–7). INCUBATION. 11–19 days. FLEDGING PERIOD. 18–23 days.

Wing-length: *H. r. rustica*: ♂ 120–129, ♀ 118–125 mm.
Weight: *H. r. rustica*: ♂ ♀ mostly 16–22 g.

Geographical variation. Slight over most of west Palearctic. Nominate *rustica* throughout, except in extreme south-east. *H. r. transitiva* (Lebanon, southern Syria, Israel) has rusty-pink or rufous-buff underparts; *H. r. savignii* (Egypt) has rufous-chestnut underparts and is smaller. Birds from transition zones from nominate *rustica* into *transitiva* in south-east Europe and Turkey are strongly variable individually.

Ethiopian Swallow *Hirundo aethiopica*

PLATE: page 1054

Du. Ethiopische Zwaluw   Fr. Hirondelle d'Ethiopie   Ge. Fahlkehlschwalbe   It. Rondine etiopica
Ru. Эфиопская ласточка   Sp. Golondrina etiópica

Field characters. 13–15 cm. Somewhat smaller and slighter than Swallow, with forked tail having much shorter streamers on outer feathers. Upperparts, wings, and tail glossy, blue-black relieved by rufous-chestnut fore-crown and subterminal white spots on tail feathers; underparts pure white, interrupted by small but noticeable black patch on chest-sides, near shoulder.

Unmistakable. Throat may show rufous tinge but always lacks dark bib of Swallow. Call a weak twitter.

Habitat. Grasslands, savanna, open woodland, and bush up to almost 3000 m between Sahara and equator in tropical Africa; also in towns and villages, over inland waters and marshes as well as coasts. Feeds on flying insects often in lowest airspace, sometimes in large flocks. Loosely colonial; nest is mud cup attached to building (usually inside), bridge, rock, in large hollow tree or cave; man-made structures responsible for continuing range expansion to south and west.

Distribution. Breeds in scattered populations from Sénégal east to Ethiopia and Somalia and south to north-east Tanzania.
Accidental. Israel: Bet She'an, adult trapped March 1991.

Movements. Probably resident, or wanders locally, in parts of (e.g.) Kenya, Nigeria, Sudan, and Ethiopia; otherwise nomadic, following rains since breeds in rainy season, so present in West Africa March–October, Central Africa August–December. Lays eggs in Sudan March–July, in Ethiopia April–October.

Wing-length: ♂ ♀ 97–112 mm.

Red-rumped Swallow *Hirundo daurica*

PLATES: pages 1062, 1067

Du. Roodstuitzwaluw   Fr. Hirondelle rousseline   Ge. Rötelschwalbe   It. Rondine rossiccia
Ru. Рыжепоясничная ласточка   Sp. Golondrina dáurica   Sw. Rostgumpsvala

Field characters. 16–17 cm, of which tail-streamers 5–6 cm; wing-span 32–34 cm. Medium-sized hirundine similar in general form to Swallow but bulkier and with even stubbier bill, marginally shorter and slightly more rounded wings, and slightly shorter, blunter, and less wire-like tail-streamers. Plumage differs from Swallow in lack of dark chest-band and white spots in tail, and in presence of pale rufous rump area, black (not pale) under tail-coverts, and chestnut nape. Flight rather deliberate; usually slower and with more gliding than Swallow.

In west Palearctic, unmistakable when closely seen but requires care at long range. Beware also occurrence of House Martin with rump stained pink, Swallow with rump made pale by wear, and hybrid between these two species with intermediate appearance; any vagrant Red-rumped Swallow must be closely observed and all characters carefully checked.

Habitat. In west Palearctic, breeds in lower middle latitudes in warm temperate, steppe, and especially Mediterranean zones, extending into oceanic climates, from sea-level in south-west Spain to over 1000 m in Cyprus, where sea-cliffs and caves are occupied as well as mountains. Compared with Swallow, has retained closer links with primitive breeding and hunting habitats and has not yet become so committed to dependence on human structures and land uses, except in certain regions. Much more dependent on reliable warm climate, and less adaptable to sudden or drastic change. Winters in dry African grasslands, or in cultivation and forest clearings.

Distribution. Northward expansion reported in 1970s Spain, France, Italy, Rumania, and Bulgaria. More recent range increase in France and (marked) in Portugal. BRITAIN. Single bird in House Martin colony Dorset July–September 1988, sang frequently and helped to feed young at nest. FRANCE. First bred 1965. Corsica: has bred occasionally since 1962. SPAIN. Before 1929 only in extreme south, but reached central Spain 1951–3 and Gerona 1960. BALEARIC ISLANDS. Has never bred on any of the islands. ITALY. First bred Apulia 1963. Bred Sardinia 1965. Sicily: birds seen collecting mud 1970, but breeding not confirmed. YUGOSLAVIA. Marked and continuing expansion from 1950s. BULGARIA. Marked increase. RUMANIA. First bred 1975. TURKEY. Widespread, locally common only in areas within influence of Mediterranean and Sea of Marmara. Apparently spreading recently into western Black Sea coastlands. CYPRUS. Commonest hirundine in hills and mountains. SYRIA. Apparently rather irregular breeder with restricted range; probably breeding regularly only Kassab and Ra's al Basīt in north-west.

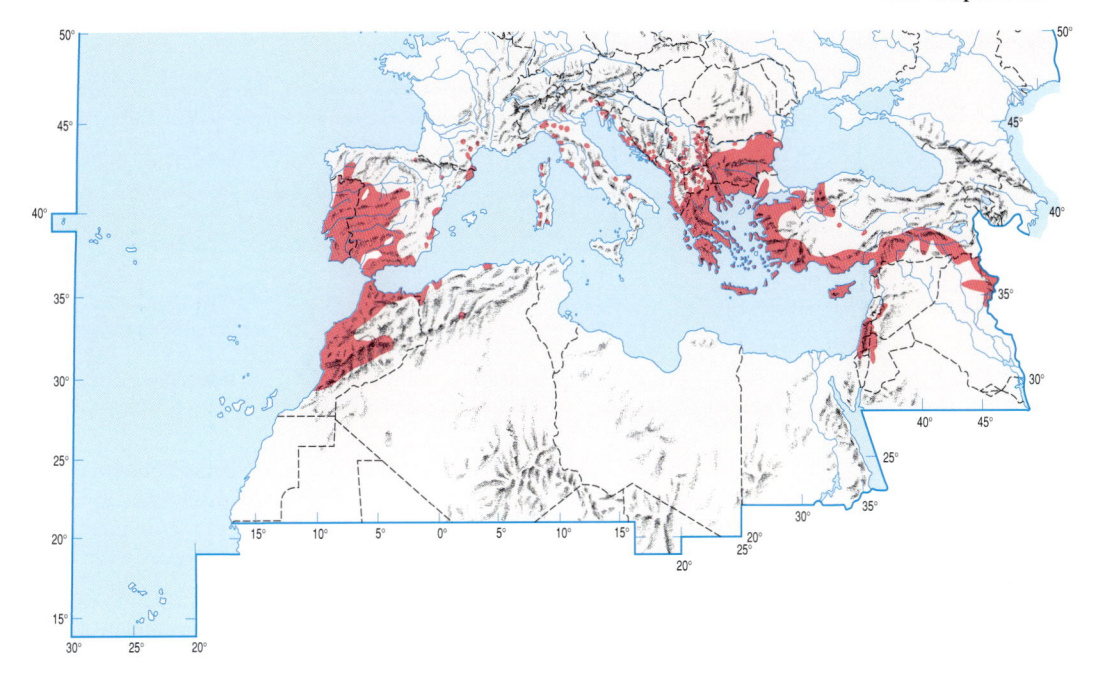

Accidental. Iceland, Britain (annual), Ireland, Belgium, Luxembourg, Netherlands (annual in recent years), Germany, Denmark, Norway, Sweden (now annual), Finland, Poland, Austria (almost annual), Switzerland, Ukraine, Armenia, Azores, Madeira.

Beyond west Palearctic, breeds in much of Asia (except south-east) south of c. 50–54°N, in south-west Arabia, and patchily in sub-Saharan Africa south to Malawi.

Population. FRANCE. 10–20 pairs; increasing. SPAIN. 9600–13 200 pairs; slight increase. PORTUGAL. 10 000–100 000 pairs 1978–84; marked increase. ITALY. 15–25 pairs 1983–93; slight decrease. GREECE. 5000–20 000 pairs; stable. ALBANIA. 2000–5000 pairs 1981. YUGOSLAVIA: CROATIA. 100–150 pairs; slight increase. BULGARIA. 5000–10 000 pairs; slight increase. RUMANIA. c. 10–40 pairs 1986–92. TURKEY. 10 000–100 000 pairs. CYPRUS. Perhaps 50 000–100 000 pairs; probably stable. SYRIA. Probably considerable annual fluctuations. ISRAEL. 10 000–15 000 pairs in 1980s. JORDAN. Locally fairly common. MOROCCO. Uncommon.

Movements. Migratory, but winter quarters still unconfirmed; presumed to lie in savanna zone of northern Afrotropics, where birds inseparable in the field from resident African populations. Pronounced spring and autumn passage across Straits of Gibraltar. Migrants occur across whole of North Africa though are rare in Tunisia and Libya; from this, assumed that Spanish and Moroccan birds migrate towards West Africa, those from south-east Europe and south-west Asia to north-east Africa.

Autumn passage at Straits of Gibraltar mainly September, continuing into October. Evidently rather earlier from eastern breeding areas, since gone from Turkey by end of September and main Cyprus passage late August to mid-September; many fewer autumn observations than in spring, presumably due to more long-range overflying in autumn. Spring passage begins February, continuing into April.

Food. Invertebrates. Takes airborne prey by aerial-pursuit, at up to 100 m or more. Hunting flight involves more steady gliding and less rapid wing-beats than Swallow.

Social pattern and behaviour. Gregarious outside breeding season. Typically breeds solitarily or in colonies of up to 3–4 pairs, occasionally much larger; nests may be touching. Near prospective nest-site, ♂ performs display-flight in which he circles ♀, giving soft contact-calls which ♀ sometimes reciprocates. ♂ and ♀ may fly alongside each other with synchronized wing-beats.

Voice. Less vociferous than Swallow. Song a short, sweet, twittering warble, lacking vehement bursts and interspersed calls of Swallow. Commonest call recalls Swallow but more like sparrow—'djuit'.

Breeding. SEASON. Spain: 1st clutches laid second half of April, last in September. Greece and Bulgaria: laying from mid-May to end of July. North Africa: laying from end of April to July. 2–3 clutches. SITE. Overhanging rock ledge, inside cave, under bridge or culvert, or in ruined building, etc. Nest: rounded bowl attached to overhanging, usually horizontal surface, with extended entrance tunnel also along surface. Constructed of mud pellets reinforced with plant and grass stems; lined softer vegetation, wool, and feathers. EGGS.

Long, sub-elliptical, smooth and slightly glossy; white, occasionally with fine red-brown speckling. Clutch: 4–5 (2–7). INCUBATION. 13–16 days. FLEDGING PERIOD. 22–27 days.

Wing-length: *H. d. rufula*: ♂ 120–128, ♀ 118–127 mm.

Weight: *H. d. rufula*: ♂ ♀ 19–28 g.

Geographical variation. *H. d. rufula* is only race breeding in west Palearctic, but nominate *daurica* (larger, plumage slightly darker; breeding eastern Asia) has occurred as vagrant.

Cliff Swallow *Hirundo pyrrhonota*

PLATE: page 1054

DU. Amerikaanse Klifzwaluw   FR. Hirondelle à front blanc   GE. Fahlstirnschwalbe   IT. Rondine rupestre americana
RU. Белолобая ласточка   SP. Avión roquero americano   SW. Rödkindad stensvala

Field characters. 12.5–14.5 cm. Head and body size as Swallow but form differs strikingly in virtually square tail. Plumage pattern and colours recall Red-rumped Swallow; dark, glossy, blue-black upperparts relieved by whitish forehead, pale rufous neck-collar and rump, and white streaks down mantle; dark cream underparts show rufous throat, blackish patch on centre of breast and black-mottled undertail-coverts. Immature lacks gloss on upperparts, has paler rufous areas, duller forehead and breast patch. Flight slower than native west Palearctic hirundines, with characteristic soaring and planing. Call a single liquid note.

Unmistakable in adult plumage, when compared with other west Palearctic swallows; basic pattern of plumage and tail shape diagnostic in combination.

Habitat. Variety of open landscapes (often near water) and including towns, villages, and farms, from boreal to subtropical North America, in non-breeding season in tropical and subtropical South America. Feeds aerially on flying insects in small flocks; nests (sometimes large colonies) in roofed mud cup with entrance tunnel on cliff, building, or bridge, and occasionally in sand burrow.

Distribution. Breeds from central Alaska east across Canada to Nova Scotia, and south throughout USA (except in Florida and surrounding regions of south-east where scattered and local) to southern Mexico.

Accidental. Iceland: Kvísker, juvenile ♀, October 1992. Britain (all juveniles): Isles of Scilly, October 1983; December 1995; Cleveland, October 1988; Humberside, October 1995. Canary Islands: Tenerife, juvenile, September 1991, this species or Cave Swallow *H. fulva*.

Movements. Migratory, wintering in Brazil, Paraguay, Argentina, and sometimes Chile; occasionally in Central America. Breeds April–August in southern part of range, from May in north. Vagrants recorded in Arctic (Wrangel Island and St Lawrence Island) and southern Greenland.

Wing-length: ♂ ♀ 105–113 mm.

House Martin *Delichon urbica*

PLATES: pages 1055, 1067

DU. Huiszwaluw   FR. Hirondelle de fenêtre   GE. Mehlschwalbe   IT. Balestruccio
RU. Городская ласточка   SP. Avión común   SW. Hussvala

Field characters. 12.5 cm; wing-span 26–29 cm. 10% smaller than Swallow (ignoring latter's tail-streamers), but head and body appear hardly less bulky due to proportionately shorter wings and tail. Medium-sized, bull-headed hirundine, with distinctly forked tail. Upperparts blue-black with broad white rump; underbody white. Juvenile much duller, dusky above with distinctive white tips to tertials.

Flight less rapid and often at greater height than Swallow. Perches and clings; waddles on ground more easily than Swallow, often raising wings and tail to do so.

Habitat. Breeds sparsely from subarctic and boreal, and more abundantly through temperate to steppe and Mediterranean zones of west Palearctic, in oceanic as well as continental climates, tolerating wet, windy, and chilly conditions but usually avoiding extremes of temperature, and vulnerable to their effects on insect prey. In suitable weather, tends to forage in airspace above lower levels favoured by Swallow. Although transition from primitive rock-nesting to general use of buildings, bridges, and other artefacts is virtually complete over much of Europe, in some regions nesting on natural rock-faces with suitable surfaces and pitches remains locally common, e.g. in Switzerland, although one of highest sites is on a building at nearly 2200 m. In Cyprus, majority prefer cliffs and rocks at all elevations. In western Europe, much more ready than Swallow to live in large cities, not only in suburbs but, where air clean, even in centre. Often mixes with other swallows, both at breeding places and in foraging, but tends

Brown-throated Sand Martin *Riparia paludicola mauritanica* (Morocco) (p. 1053): **1–2** ad. Sand Martin *Riparia riparia riparia* (p. 1055): **3–4** ad. African Rock Martin *Hirundo fuligula obsoleta* (p. 1058): **5–6** ad. Crag Martin *Hirundo rupestris* (p. 1059): **7–8** ad. Swallow *Hirundo rustica rustica* (p. 1061): **9–10** ad. Red-rumped Swallow *Hirundo daurica rufula* (p. 1064): **11–12** ad. House Martin *Delichon urbica*: **13–14** ad breeding.

to concentrate round sites often fixed by long usage, leaving wide areas unvisited. Presence of water often an attraction, although not essential.

Distribution. Decrease reported Netherlands and Ukraine, increase Azerbaijan, Cyprus, and Israel. SPITSBERGEN. May have bred 1924. ICELAND. Bred 1966, 1976, 1990, and 1992. FAEROES. Regular visitor; bred 1956, 1966. BRITAIN. Some peripheral changes related to population fluctuations. MALTA. Bred 1981–2. AZERBAIJAN. Considerable extension of range in 20th century; now widespread in lowlands as well as foothills and mountains. TURKEY. Largely absent Central Plateau, also much of east and south-east, though probably underrecorded. CYPRUS. Range expansion since 1970 still continuing. SYRIA. Range difficult to delineate exactly. Regular at higher altitudes in Anti-Lebanon, also further north at An Nabk and in coastal belt; probably only local and exceptional breeder Damascus. ALGERIA. Common in northern towns. MOROCCO. Rarely nests south of Atlas mountains.

Accidental. Spitsbergen, Bear Island, Iceland (annual), Azores, Madeira.

Beyond west Palearctic, widely distributed in central and northern Asia, south to Iran, Himalayas, and southern China; breeds sporadically in South Africa and Namibia.

Population. Fluctuating. Decline reported Belgium, Netherlands, Germany, Fenno-Scandia (stable Norway), and Ukraine, increase Lithuania, Switzerland, Moldova, and Cyprus. BRITAIN. 250 000–500 000 pairs 1988–91; no major change known. IRELAND. 70 000–140 000 pairs 1988–91. FRANCE. Under 1 million pairs in 1970s. BELGIUM. 70 000–130 000 pairs 1989–91; probably declining, at least locally. LUXEMBOURG. 10 000–12 000 pairs. NETHERLANDS. Estimated 450 000–500 000 pairs in *c*. 1970, 131 000–164 000 pairs 1989–91. GERMANY. 1.45 million pairs in mid-1980s. Long-term decline many areas, but considerable local increases also reported. In east, 600 000 pairs in early 1980s, and considerable fluctuations some areas. DENMARK. 17 000–170 000 pairs 1987–8. NORWAY. 200 000–500 000 pairs 1970–90. SWEDEN. 100 000–250 000 pairs in late 1980s; probable decrease. FINLAND. 100 000–150 000 pairs in late 1980s; marked decrease. ESTONIA. 100 000–200 000 pairs in 1991. Stable, following apparent increase in 1960s(–70s). LATVIA. 90 000–320 000 pairs in 1980s; stable. POLAND. 400 000–600 000 pairs. Stable, but slight decrease in urban areas. CZECH REPUBLIC. 600 000–1.2 million pairs 1985–9; stable. SLOVAKIA. 500 000–1 million pairs 1973–94; stable. HUNGARY. 100 000–150 000 pairs 1979–93; stable. AUSTRIA. Common; probably stable. SWITZERLAND. 200 000–300 000 pairs 1985–93. SPAIN. 2.14–2.16 million pairs; stable. PORTUGAL. 100 000–1 million pairs 1978–84; stable. ITALY. 500 000–1 million pairs 1983–93. GREECE. 100 000–200 000 pairs; stable. ALBANIA. 90 000–150 000 pairs in 1991. YUGOSLAVIA: CROATIA. 200 000–300 000 pairs; stable. SLOVENIA. 50 000–100 000 pairs; stable. BULGARIA. 500 000–5 million pairs; stable. RUMANIA. 150 000–200 000 pairs in 1986–92; stable. RUSSIA. 1–10 million pairs; stable. BELARUS'. 250 000–350 000 pairs in 1990; stable. UKRAINE. 140 000–150 000 pairs in 1988. MOLDOVA. 30 000–40 000 pairs in 1988. AZERBAIJAN. Common. TURKEY. 100 000–1 million pairs. CYPRUS. Increase since 1970, still continuing. Perhaps now 40 000–60 000 pairs. ISRAEL. 2000 pairs in late 1980s. TUNISIA. Common. MOROCCO. Uncommon.

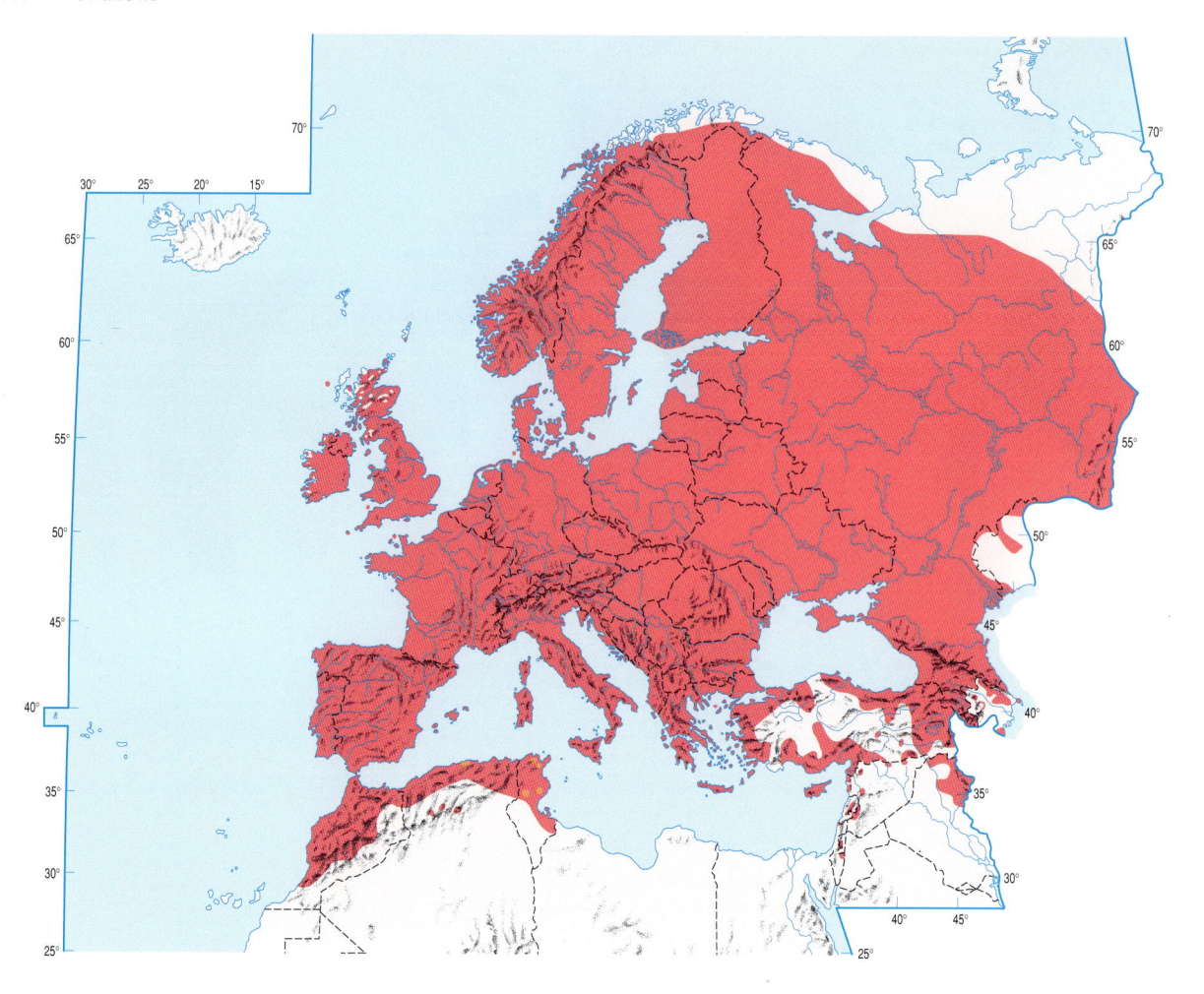

Movements. Migratory. Winter records exist from Mediterranean basin and western Europe, north to Britain and Ireland, but west Palearctic and west Asian birds otherwise winter wholly in Afrotropics. Within Afrotropics, very wide scatter of observations indicates extensive size of winter range there, including humid zone north of equator, but such sightings few in relation to huge numbers of birds which must be involved. This presumably due to highly aerial lifestyle, with high-altitude foraging like swifts.

A relatively late migrant in autumn, since late broods often still in nest in late August or September. Main southerly movement through Europe September–October, with significant minority still transient in November. Return to breeding colonies as early as January in early years in north-west Africa and southern Iberia; progressively later further north, with main return over most of Europe in second half of April and first half of May.

Food. Almost wholly flying insects; in breeding season, especially flies (Diptera) and aphids. Prey taken almost entirely by aerial-pursuit, though many reports of birds feeding while perched on ground or trees; also (presumably while perched) from walls, rock faces, and reeds. Insects normally taken from below, bird shooting steeply up with rapid wing-beats to catch it; then usually glides down to previous height.

Social pattern and behaviour. Highly gregarious throughout the year. Flocks of several hundred common on migration and in winter quarters. Pair-bond monogamous, typically of seasonal duration or shorter. Typically breeds in colonies, divided into sub-colonies. Established breeders show marked year-to-year fidelity to nest-site, or its vicinity. Markedly aggressive in defence of small nest-area territory. ♂ takes initiative in nest-site slection and in early stages of building; thereafter both sexes share nest-building. Before laying, pair usually spend some time in nest together. In colonies, often lay in half-completed nests; add nest-lining during laying.

Voice. Freely used throughout the year. Song soft, sweet, long twitter of melodious chirps. Contact-call a hard but merry 'chirrrp'; in alarm, a shrill, plaintive 'tseep'.

Breeding. SEASON. North-west and central Europe: first eggs at beginning of May, rarely late April, main laying period begins mid-May; last young in nest until mid-October in mild year. Northern and north-east Europe: eggs laid from end of May or June; last young in nest in September. North Africa: main laying period May–June, rarely from late March, uncommonly after July. 2 broods in central part of range but rare in northern and southern parts; 3 broods very occasionally recorded. SITE. Most frequent on outer walls of buildings, under eaves or other overhang; also under bridges, culverts, etc.; natural sites are on cliffs and outcrops, coastal and inland, also under overhangs; makes free use of specially designed nest-boxes. Nest: half-cup of mud pellets, down, feathers, and other light material; cup formed against vertical wall and overhanging 'roof', with small oval entrance at top; may be contiguous with other nests, so that overall shape very variable from part-spherical to very irregular; variant nests include open-topped ones like Swallow, and half-spheres (against vertical walls) with side-entrances. EGGS. Sub-elliptical, smooth and slightly glossy; white, very occasionally with fine, light red spotting. Clutch: 3–5 (1–7). INCUBATION. 14–16 days. FLEDGING PERIOD. 22–32(–40) days.

Wing-length: ♂♀ 105–116 mm.
Weight: ♂♀ mostly 15–23 g.

Pipits, Wagtails Family Motacillidae

Well-defined family of small oscine passerines (suborder Passeres); mainly terrestrial and insectivorous. 54–58 species in 5 genera in 2 closely related groups—pipits (3 genera) and wagtails (2 genera). Cosmopolitan but most species in Eurasia and Africa; occur in open country, often near water. Many migratory. Family represented in west Palearctic by 18 species, with 10 *Anthus* and 5 *Motacilla* breeding.

Body slender, elongated (shape often exaggerated by long tail); neck short. Sexes similar in size. Bill thin and pointed; notched. Wing medium to long. Flight strong, undulating (especially in wagtails); song-flights, launched from ground or perch, are characteristic of both pipits and wagtails. Tail medium to long; longest in *Motacilla* which move it up and down persistently, especially during feeding. Legs medium to long, slender; hind claw often elongated—least so in forms which regularly perch in trees or live on hard, bare ground. Gait typically a walk or run. Plumage soft; streaked and cryptic in *Anthus*, brightly and contrastingly coloured (with black, yellow, white, or orange) in other pipits and in wagtails. Many species have paler, often white, outer tail-feathers. Except in most *Anthus*, sexes often slightly different.

Richard's Pipit *Anthus novaeseelandiae*

PLATES: pages 1071, 1091

Du. Grote Pieper   Fr. Pipit de Richard   Ge. Australspornpieper   It. Calandro maggiore
Ru. Степной конек   Sp. Bisbita de Richard   Sw. Större piplärka

Field characters. 18 cm; wing-span 29–33 cm. Differs structurally from Tawny Pipit in stouter bill, more rounded crown, broader back, up to 10% longer wings and fuller tail, 20% longer and much stouter legs, and 80% longer hind claw, and from Blyth's Pipit in 10–15% longer and stouter bill, marginally longer wings (with folded primary-tips showing beyond longest tertial), 15% longer tail, 10% longer and stouter legs, and 30–40% longer hind claw. 2nd largest pipit in west Palearctic, with rather dark, brown and buff, heavily streaked plumage. Dark streaks most obvious on crown, back, and chest; wing shows pale buff to white double wing-bar and tertial-fringes; tail edged white.

Identification of Richard's Pipit and other large pipits difficult. Characters crucial to separation of this species are structure (see above), pale lores (shared by Blyth's Pipit), pointed dark centres to median coverts (in adult) and particularly heavy streaks on crown and mantle. Flight powerful, matched only by Long-billed Pipit and composed of obvious bursts of wing-beats with 'shooting' bounds of bird on closed-up wings; often flutters with dropped legs before landing. Gait high-stepping and strong; large feet noticeable when lifted off ground.

Song a monotonous, simple phrase, written 'chi-chi-chee-chee-chee' (last 3 notes falling in pitch). Commonest call in flight explosive, rather nasal, 'shreep' or 'sh-rout', recalling

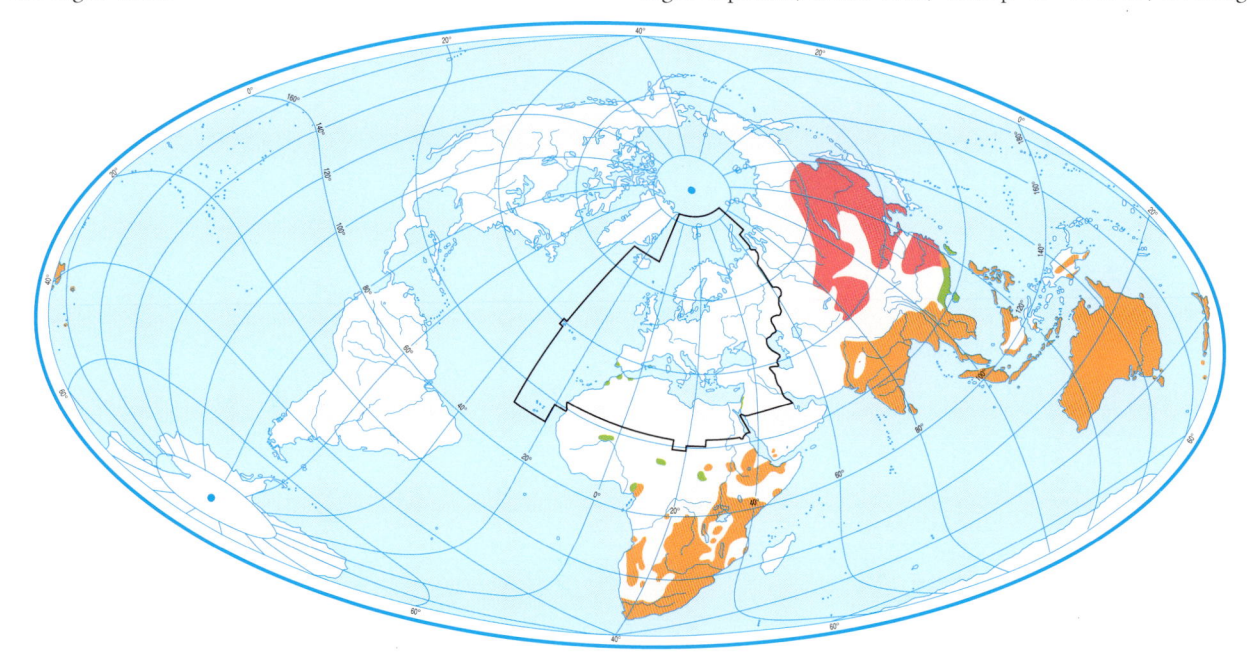

Richard's Pipit *Anthus novaeseelandiae*. *A. n. richardi*: **1** ad fresh, **2** ad worn, **3** 1st winter, **4** juv. *A. n. dauricus* (from eastern Palearctic): **5** ad fresh, **6** 1st winter. Blyth's Pipit *Anthus godlewskii* (p. 1072): **7** ad fresh, **8** 1st winter.

House Sparrow; other notes written as soft 'chup' and 'chirp'.

Habitat. Breeds extralimitally in continental Asian middle latitudes to tropics, on mainly open lowland level or gently sloping ground of steppe, grassland, or cultivated type, warm and sunny but not arid, below 1800 m in FSU but somewhat higher and occasionally to 3000 m in Himalayan foothills. Ecologically a counterpart in east Palearctic of Meadow Pipit, geographically overlapping with Tawny Pipit from which it differs in choosing more fertile, moister grassland. Prefers short not-too-dense grass, often in river valleys and sometimes in forest-steppe; avoids taiga and other tree-grown, wetland, or broken and precipitous areas, although tolerating tussocky patches and perching freely on grass tufts and bushes. Also occurs on fallows, edges of cultivation, low dry crops, roadsides, and even stony terrain.

Distribution. Accidental or passage-migrant, in a few cases wintering (see Movements). Spitsbergen, Britain (annual), Channel Islands, Ireland, France, Belgium (annual), Luxembourg, Netherlands, Germany (annual), Denmark, Norway, Sweden (annual), Finland (regular), Latvia, Poland, Czech Republic, Slovakia, Austria (almost annual), Switzerland, Spain, Portugal (1–2 each autumn in Algarve, once 7 together; small numbers also winter), Balearic Islands (1–3 wintering in recent years), Italy (rare, but perhaps regular on passage, Sicily), Malta (almost annual), Greece (rare and irregular), Yugoslavia, Bulgaria, Georgia (many collected April and September late 19th century), Turkey (very rare, irregular), Cyprus, Syria, Lebanon, Israel, Jordan, Kuwait, Egypt (rare), Algeria, Morocco (scarce in winter, records more frequent in recent years), Canary Islands.

Movements. Northern races are long-distance migrants, wintering among local races from Pakistan to Indo-China and south to Malaysia. Other races of Afrotropics, central and south-east Asia, and Australasia are resident or largely so, some making altitudinal or short-distance movements.

A feature of the movements of the south-west and central Siberian race *richardi* is the high frequency with which birds appear well to the west of normal breeding and wintering ranges, even wintering regularly in southern Iberia. Birds occur mostly as individuals but also in small parties, in western Europe from Scandinavia and Shetland to Ireland and Portugal, while some reach North Africa from Morocco to Egypt. On Malta, rare but almost annual in recent years, late September to late February, some birds probably overwintering. Scale of this westward drift demonstrated by British and Irish records: over 1100 to 1982, and in every year recently; most records September to mid-November with a few in December; spring records far fewer but with April peak.

Food. Diet mainly invertebrates, taken from among ground vegetation and from crevices in logs and rocks; sometimes takes flying insects by jumping up or in short flights.

Wing-length: *A. n. richardi* (Europe): ♂ 97–102, ♀ 92–94 mm.
Weight: *A. n. richardi* (Europe): ♂♀ 25–36 g.

Geographical variation. Rather slight in Asia, mainly involving size; more marked in Australasia and (especially) Afrotropics. Migratory races of north-central Asia, of which only *richardi* definitely known to occur in west Palearctic, much larger than non-migratory races of southern Asia, and hence sometimes considered separate species *A. richardi*.

Blyth's Pipit *Anthus godlewskii*

PLATES: pages 1071, 1091

Du. Mongoolse Pieper Fr. Pipit de Godlewski Ge. Steppenpieper It. Prispolone di Blyth
Ru. Забайкальский конек Sp. Bisbita de Blyth Sw. Mongolpiplärka

Field characters. 17 cm; wing-span 28–30 cm. Close in size to Richard's Pipit but less heavily built, with 10% shorter and finer-tipped bill, marginally shorter wings (with longest tertial cloaking primary-tips), 15% shorter tail, 10% shorter and finer legs, and 25–30% shorter hind claw. Large, rather bright, buff, streaked pipit, with general character closer to Tawny Pipit than Richard's Pipit but shorter-tailed than both. In all plumages, pattern similar to Richard's Pipit and immature Tawny Pipit but general tone more ochre or orange than Richard's Pipit, and less ochre or grey than Tawny Pipit. In all plumages, rather softer streaks, more indistinct cheek, rather paler hind-neck pattern, and (particularly) more orange-buff tone of whole underbody including under tail-coverts provide subtle distinctions from Richard's Pipit.

Adult has noticeably demarcated, oblong dark centres to median coverts (unlike softly pointed internal marks of adult Richard's Pipit). Blyth's Pipit increasingly reported in western Europe; in addition to structural and plumage characters, flight action may be helpful, being lighter than Richard's Pipit and lacking fluttering stall before landing. Nimbler on the ground than Richard's Pipit.

Commonest note recalls Tawny Pipit, not Richard's Pipit, being short, quite stony 'chup' or 'chep'; also utters more drawn-out notes, recalling Richard's Pipit but softer and higher pitched, written 'psheeoo'.

Habitat. Breeds extralimitally in middle, lower middle, and low latitudes in continental steppe zone and in uplands and mountains, inhabiting dry rocky mountain slopes with scant vegetation. On migration prefers swampy land, and in winter resorts to dry ricefields, fallow, edges of cultivation, and grassland.

Distribution. Breeds from eastern Altai (Russia) east to Manchuria and south to Tibet.

Accidental. Britain: one collected Sussex, October 1882. 4 records October–December 1990–94. Belgium: one, November 1986. Netherlands: one collected, November 1983. Norway: one Nord-Trøndelag, November 1995. Finland: 6 individuals, October–November, 1974–90 (probably now very rare but regular visitor). Israel: one Eilat, November 1987.

Movements. Migratory. Widespread and locally common in main winter quarters—Pakistan, Bangladesh, and peninsular India.

Wing-length: ♂ 89–98, ♀ 84–93 mm.

Tawny Pipit *Anthus campestris*

PLATES: pages 1073, 1091

Du. Duinpieper Fr. Pipit rousseline Ge. Brachpieper It. Calandro
Ru. Полевой конек Sp. Bisbita campestre Sw. Fältpiplärka

Field characters. 16.5 cm; wing-span 25–28 cm. Noticeably less heavily built than Richard's Pipit, with narrower bill, shorter wings, tail, and legs, less broad back, and much shorter hind claw. Noticeably long, slim pipit, with size, form, and appearance of adult suggesting pale wagtail. Adult shows diagnostic combination of pale supercilium, dark lores, virtually unstreaked chest, bold wing-covert bars contrasting with indistinct markings on rest of upperparts, and tail broadly edged white. Juvenile heavily streaked well into 1st winter, suggesting Richard's Pipit and Blyth's Pipit before adult of own species.

Important to note that while Tawny Pipit shows palest head of all large pipits at distance, contrast of pale supercilium and dark <-mark formed by front of eye-stripe and moustachial stripe across lores forms obvious diagnostic feature at close range. Most difficulty of identification arises from overlap of more streaked immature plumage with those of other large pipits, but dark loral and moustachial marks present from fledging.

Flight markedly undulating but lacks power of Richard's Pipit; strongly reminiscent of wagtail, as is slim, long-winged, and long-tailed silhouette. Carriage most horizontal and gait most free of all large pipits, with walk, run, and alert pauses all recalling wagtail. Wags tail frequently, with much greater freedom than only occasional similar movement of other large pipits.

Habitat. In lower middle and middle continental latitudes, from Mediterranean and steppe through temperate zones, preferring dry but not arid ground. Avoids steep or rocky terrain, and tall or dense vegetation. Favoured habitats tend to

Tawny Pipit *Anthus campestris*. *A. c. campestris*: **1** ad breeding (fresh), **2** ad breeding (worn), **3** 1st winter, **4** juv. *A. c. griseus*: **5** ad breeding (fresh). Long-billed Pipit *Anthus similis captus* (p. 1076): **6** ad breeding (fresh), **7** ad breeding (worn), **8** juv–1st winter.

be more frequent in sunny continental lowlands, but locally occur at 2600 m in Armenia. In Germany, breeds on dry warm wastelands and heaths, often beside Woodlark and Stone Curlew, and on sandy arable fields and sandy banks of rivers or lakes; similar habitats occupied in other parts of western Europe. In north-west Africa, occupies dry mountain slopes and plateaux up to 2400 m, and is abundant in Atlas above treeline, some up to 3000 m. Other situations recorded include coastal sand-dunes (even close to high-water mark), dry watercourses, stony flats, roadsides, vineyards, and dry hillsides. In winter in Africa, preference for arid ground accentuated: common in coastal zone, in steppe, *Acacia* bush, and in barest parts of thorn-savanna, and even on edge of desert; associates with grazing cattle.

Distribution. Range decrease reported France, Benelux countries, Germany, Denmark, Sweden, Estonia, Czech Republic, Slovakia, Austria, Iberia, Italy, and Ukraine. Attributed (like simultaneous population decline) primarily to loss of habitat brought about by changes in farming practices. FRANCE. Apparently now stable after retreat towards south noted in 1970s. BELGIUM. Population estimated at 150 pairs in early 1970s, but now extinct; last bred 1986. LUXEMBOURG. Small population (*c.* 5 pairs) from 1950s until 1975–80; extinct since. NETHERLANDS. Marked decrease; last bred on coast *c.* 1967. GERMANY. Became extinct several states in west 1966–70. More widespread in east (where main stronghold in south-east and virtually disappeared from coastal dunes) because of habitat requirements. SWEDEN. Bred further north in 19th century; following decline, became remnant population within forest areas, and range now stable. FINLAND. Breeding confirmed 7 times on south-west coast, 1966–81. SWITZERLAND. Has bred regularly at one site since 1981, and more recently (perhaps only temporarily) at 2 others. MALTA. Bred 1993 (2 pairs). YUGOSLAVIA: SLOVENIA. Fluctuating range. TURKEY. Widespread, but sometimes locally absent from apparently suitable habitat. SYRIA. Fairly widespread, locally not at all rare, in mountains of west.

Accidental. Iceland, Britain (annual), Ireland, Norway, Finland, Madeira, Cape Verde Islands.

Beyond west Palearctic, extends east across Asia to Mongolia, south to Iran and Tibet.

Population. Has suffered marked decline Netherlands, Germany, Czech Republic, and Austria; slight decline Denmark, Estonia, Slovakia, Spain, Portugal, Italy, Ukraine, and Israel; mostly stable otherwise. FRANCE. 20 000–30 000 pairs. NETHERLANDS. Decreased in 20th century. 60–80 pairs 1989–91. GERMANY. Estimates vary widely: 650 pairs in mid-1980s; 2200 ± 800 pairs in east alone in early 1980s. Total in early 1990s perhaps 1000–1500(–3000) pairs pending results of mapping schemes in eastern strongholds. DENMARK. 29–31 pairs in 1989. SWEDEN. 150–300 pairs in late 1980s; now stable after earlier decline. ESTONIA. *c.* 50 pairs in 1991. LATVIA. 200–500 pairs in 1980s. LITHUANIA. Rare; more common in east. POLAND. 2000–4000 pairs. Decreased Silesia. CZECH REPUBLIC. 40–80 pairs 1985–9. SLOVAKIA. 70–150 pairs 1973–94. HUNGARY. 10 000–20 000 pairs 1979–93; increase. AUSTRIA. *c.* 10 pairs. Marked decline since late 1950s. SWITZERLAND. 1–2 pairs regularly 1985–93. SPAIN. 400 000–640 000 pairs. PORTUGAL. 1000–10 000 pairs 1978–84. ITALY. 15 000–40 000 pairs 1983–93. GREECE. 10 000–20 000 pairs. ALBANIA. 2000–5000 pairs in 1981. YUGOSLAVIA: CROATIA. 3000–6000 pairs. SLOVENIA. 50–100 pairs; fluctuating. BULGARIA. 350–1000 pairs. Rumania. 15 000–20 000 pairs in 1986–92. Russia. 10 000–100 000 pairs. Belarus'. 3000–8000 pairs in 1990. Fluctuating, but apparently stable overall. Ukraine. 1500–2000 pairs in 1988. Moldova. 4000–5500 pairs in 1988. Azerbaijan. Common. Turkey. Not rare, though rather inconspicuous. 50 000–500 000 pairs. Israel. Only a few tens of pairs in 1980s. Tunisia. Uncommon. Algeria. Local and uncommon. Morocco. Scarce to locally common.

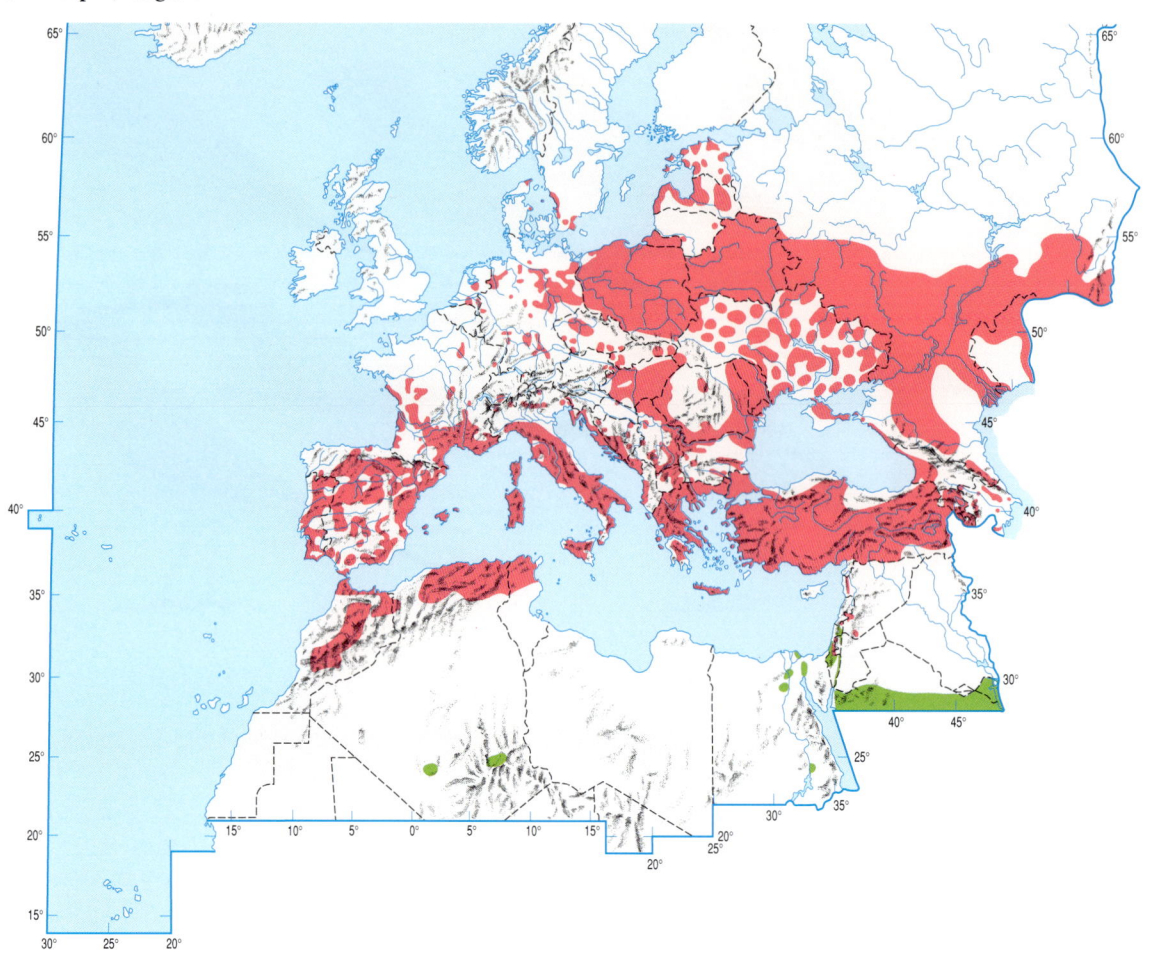

Movements. Essentially migratory. Within west Palearctic breeding range, wintering occurs only in Turkey and Levant. West Palearctic birds winter in Africa, mainly in Sahel zone south of Sahara, and Arabia.

Movement occurs across entire length of Mediterranean, mainly mid-August to mid-October. Evacuation of African winter quarters probably starts as early as February, with main passage in April in Mediterranean area, mid-April to mid-May further north. Vagrants to Britain and Ireland occur April to mid-June, but chiefly in autumn, mainly late August to mid-October.

Food. Chiefly insects; also some seeds, mainly in winter. Feeds on ground and amongst low herbage, taking insects in stop-run-peck manner like small plover, occasionally leaping up, or rarely after brief aerial pursuit.

Social pattern and behaviour. Solitary or in loose flocks outside breeding season. Mating system essentially monogamous, but polygyny not infrequently recorded: ♂ paired with 2 ♀♀ in same territory. Breeding usually solitary and strictly territorial. Song of ♂ given mainly in flight, also from tree, hummock, etc.; less commonly from flat ground. For song-flight, takes off from perch or ground and ascends silently with fluttering wing-beats to c. 20–30(–150) m; ascent may be almost vertical. Sings while flying with series of deep undulations across territory, with 1 song-unit given during descent of each undulation. Gliding descent resembles other pipits, with wings and tail raised.

Voice. Song short and undeveloped, a ringing repetition of loud, tinny disyllable 'chir-ree chir-ree ...'. Flight-calls include common, loud 'tzeep', recalling Yellow Wagtail but slightly less plaintive, and less common, hoarse 'chup', recalling House Sparrow; none has harsh, explosive quality of Richard's Pipit.

Breeding. Season. Western Europe: egg-laying from mid-May. Southern Sweden; main laying period from mid-June. Central and southern Europe: laying from mid-May to July. North Africa: on low ground, laying begins second half of April, continuing to June; c. 2 weeks later on high ground.

1–2 broods. SITE. On ground in shallow hollow, often under plant tuft; bird sometimes makes scrape. Nest: cup of grass stems and leaves, and roots, lined with finer plant material and hair. EGGS. Sub-elliptical, smooth and glossy; whitish, heavily marked with brown and purplish-grey spots and blotches.

Clutch: 4–5(3–6). INCUBATION. 11.5–13 days. FLEDGING PERIOD. 13–14 days.

Wing-length: ♂ 87–101, ♀ 83–94 mm.
Weight: ♂♀ mostly 24–32 g.

Berthelot's Pipit *Anthus berthelotii*

PLATES: pages 1076, 1091

DU. Berthelots Pieper   FR. Pipit de Berthelot   GE. Kanarenpieper   IT. Pispola di Berthelot
RU. Конек Бертелота   SP. Bisbita caminero   SW. Kanariepiplärka

Field characters. 14 cm; wing-span 21–23.5 cm. Marginally smaller than Meadow Pipit; form and actions recall Rock Pipit and Water Pipit as much as Tawny Pipit. Smallest pipit of west Palearctic. Brownish-grey above and buff-white below, with obvious pattern on pale face, pale double wing-bar and tertial fringes, and streaked crown and chest.

The only common pipit on Canary Islands and Madeira; confusion may arise with vagrant congeners, of which most likely is Meadow Pipit (see that species). Appearance in flight intermediate between small and large pipits, with light and rapid action lacking exaggerated 'shooting' bounds of Tawny Pipit. Stance and gait closer to Rock Pipit and Water Pipit than to Tawny Pipit, with noticeable head movement when walking; runs quickly over flat ground, and jumps nimbly over boulders; has distinctive habit of clambering over small plants when feeding.

Habitat. Only on oceanic subtropical islands of Canaries group and Madeira. Habitat varies somewhat between islands but generally in every zone from sea-level to highest ground at *c.* 1600–2100 m or even higher. Particularly numerous on rocky or maritime plains and plateaux with scant vegetation, or on sun-baked rock-strewn hillsides covered with bushes of *Euphorbia* affording shade and nest-sites. Also common on vine-clad slopes and in coastal tomato patches.

Distribution and population. Restricted to Madeira and Canary Islands, where populations apparently stable in recent decades. MADEIRA. 1000–1500 pairs. CANARY ISLANDS. Common on all islands. Estimated 15 000–20 000 pairs.

Movements. Resident, with little or no evidence of even local movement.

Food. Insects and seeds. Feeds almost exclusively on ground, picking invertebrates while walking or (also while perched) snatching them from the air if close; sometimes makes short aerial pursuit.

Social pattern and behaviour. Usually solitary or in pairs. Generally reported to be extremely tame. Song-flight similar to that of Tawny Pipit: ♂ sings while circling in deep undulations, then plunges silently to ground.

Voice. Song a plaintive, 'tsiree' or 'tchelee', monotonously repeated about every second. Commonest call a low-pitched, husky 'tsik'.

Breeding. SEASON. Canary Islands: eggs found first half of January to late May; on Tenerife, main season January–February at low altitudes, March–April higher up. Madeira: eggs found early February to August, rarely from late January. SITE. On ground, under bushes or low plants, or sheltered by stone. Nest: neat cup of plant stems and fibres, lined with hair, wool, and feathers. EGGS. Sub-elliptical, smooth and glossy; pale grey, with pinkish or yellowish-brown (rarely dark grey) speckles and flecks; occasionally with dark hair-streaks at broad end. Clutch: 2–5 on Canary Islands. (Incubation and fledging periods not recorded.)

Wing-length: ♂ 75–81, ♀ 73–77 mm.

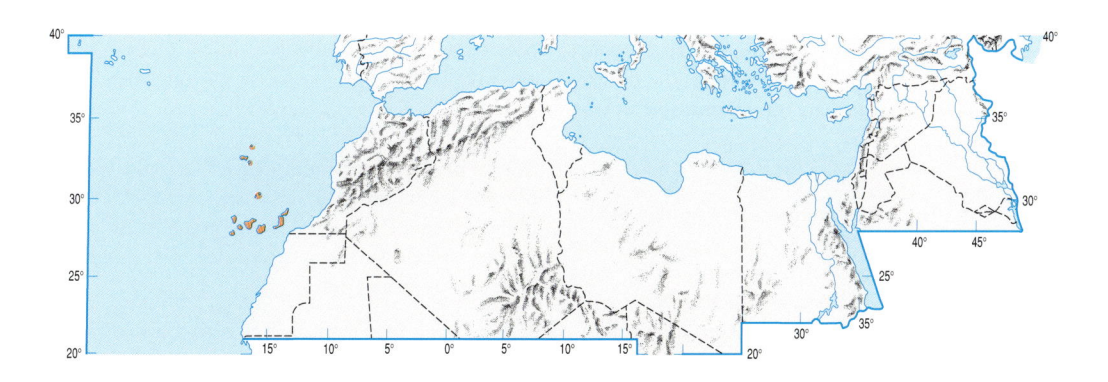

Berthelot's Pipit *Anthus berthelotii*. *A. b. berthelotii* (Canary Is.): **1** ad fresh (autumn), **2** ad worn (spring), **3** juv. Meadow Pipit *Anthus pratensis* (p. 1082). *A. p. pratensis*: **4** ad fresh (autumn, grey-olive type), **5** ad fresh (autumn, brown type), **6** ad worn (summer), **7** juv. *A. p. whistleri*: **8** ad fresh (autumn).

Long-billed Pipit *Anthus similis*

PLATES: pages 1073, 1091

Du. Langsnavelpieper Fr. Pipit à long bec Ge. Langschnabelpieper It. Pispola beccolungo
Ru. Длинноклювый конек Sp. Bisbita piquilargo Sw. Långnäbbad piplärka

Field characters. 19 cm; wing-span 27–34 cm. Has longest bill and tail but shortest hind claw of genus in west Palearctic; differs structurally from Tawny Pipit in *c.* 15% longer bill, longer wings, and tail. Noticeably long, heavy-chested pipit, with large wings, long and rather full, dark tail, and otherwise rather featureless grey-brown and ochre plumage; long, heavy bill and short hind claw diagnostic in combination. In all plumages, rather diffuse streaks of upperparts and chest, and rather dull face and usually pale buff outer tail-feathers form useful characters.

Flight powerful, with similar bulky silhouette and actions to Richard's Pipit including surging take-off and habit of hovering over ground cover.

Identification not difficult when compared with complex problems posed by other large pipits. The pale race occurring in Middle East, *captus*, is confined to mountains and nearby areas of Levant.

Habitat. Breeds in lower middle and lower latitudes, warm and largely semi-arid, in hilly and marginally in mountainous country; typically at low and medium elevations, on dry and sometimes steep, grassy and stony slopes with boulders, shale, or rock outcrops. In winter in India, descends to lower valleys and foothills, frequenting grassy plains, open low scrub jungle, dry watercourses, grassy canal banks, fallow land, wheatfields, and sand-dunes.

Distribution and population. SYRIA. Probably breeds extreme south: 3 observations 1976–7, including singing ♂ in March. ISRAEL. 1500–2000 pairs in 1980s; decrease. JORDAN. Localized; more common in north than south of range.

Accidental. Cyprus, Iraq, Kuwait.

Beyond west Palearctic, distribution patchy: from Arabia to Himalayas, in western India and Burma. Widespread in southern and eastern Africa, more local further west.

Movements. Most populations upland or montane residents with at most local vertical movements, but west Palearctic populations are total or partial short-distance migrants, present all year but wintering at lower altitude than breeding.

Food. Mainly insects, taken on the ground.

Social pattern and behaviour. Generally solitary or in pairs, but parties of up to 6 or more occur in winter. For song-flight, ♂ usually ascends from perch with quivering wings; ascends (while singing) to considerable height and may stay aloft for several minutes.

Voice. Song rather simple but with distinctly longer and more disjointed phrases than Tawny Pipit: 2–4 notes, loud and deliberate, but variable in tone and sequence. Commonest call when flushed a loud, ringing 'che-vlee', unlike commonest calls of Tawny Pipit.

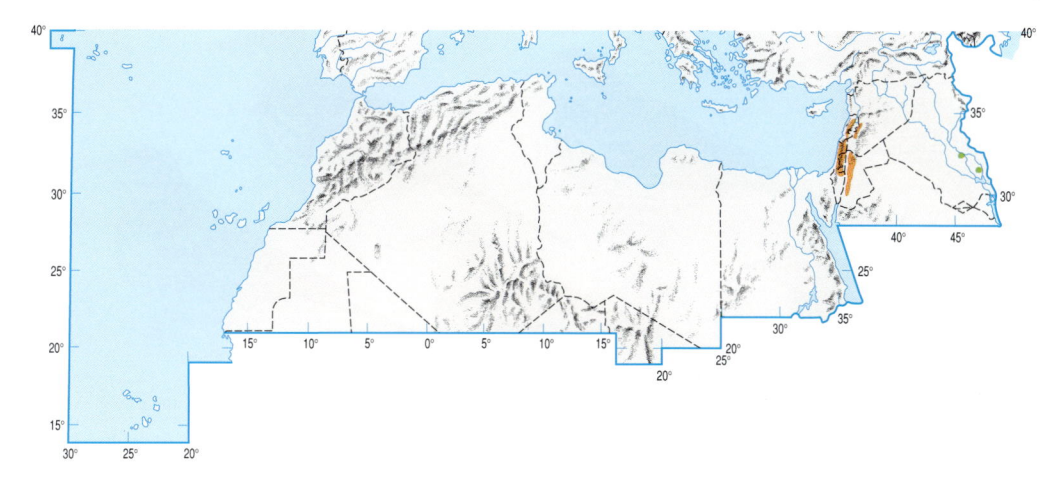

Breeding. SEASON. Middle East: April–July. Israel: April–May. SITE. On ground, in shelter of rock or tuft of vegetation, often on slope. Nest: shallow depression with cup of vegetation, lined with finer material. EGGS. Sub-elliptical, smooth and glossy; whitish or grey-white, heavily marked with brown spots and freckles. Clutch: 3 (2–5). INCUBATION. 13–14 days. FLEDGING PERIOD. c. 2 weeks.

Wing-length: ♂ 94–98, ♀ 89–92 mm.

Olive-backed Pipit *Anthus hodgsoni*

PLATES: pages 1078, 1091

DU. Groene Boompieper   FR. Pipit à dos olive   GE. Waldpieper   IT. Prispolone indiano
RU. Пятнистый конек   SP. Bisbita de Hodgson   SW. Sibirisk piplärka   N.AM. Olive Tree-Pipit

Field characters. 14.5 cm; wing-span 24–27 cm. Size between Tree Pipit and Red-throated Pipit, with similar structure except for rather long forehead, receding chin, and full chest. Small but sleek pipit, with behaviour most recalling Tree Pipit. Upperparts noticeably pale green-olive, with only faintly streaked back and plain rump; underparts noticeably clean, with ground-colour mainly white and beautifully decorated with evenly spread lines of large black spots. Face also well marked, with broad pale buff fore and white rear supercilium and (on most) white and black rear cheek spots. Bill rather small; black-horn above, flesh- or buff-horn below. Legs flesh-pink; feet with short hind claw like Tree Pipit.

Subject to confusion with both Tree Pipit and Red-throated Pipit at distance and in flight, since all 3 species have similar silhouette and wing action, commonly utter one not dissimilar call, and Olive-backed Pipit and Tree Pipit both enter trees freely. Flight free and buoyant; lacks hesitancy of Meadow Pipit. Escape-flight also like Tree Pipit, bird taking cover in canopy or going into dense ground cover. Will even walk confidently along branches—a skill apparently lacking in other pipits. Tail movement most obvious of all small pipits: often strongly pumped rather than wagged.

Habitat. For breeding occupies large slot in upper middle and middle latitudes broadly below range of Pechora Pipit and complementing (with extensive overlap) that of Tree Pipit, which covers nearly all corresponding part of west Palearctic, from boreal through temperate zones to subtropics. Northern population spreads through coniferous taiga forest, mainly in its sparser sections and at its edges along river banks and on fringes of bogs and marshes, but also in birchwoods, alder thickets, and larch groves. In winter, in south-east Asia, resorts to coffee plantations, mango groves, and other suitable wooded terrain; also found on ground under trees at forest edge, on forest footpaths, and along shady highways or on outskirts of villages.

Distribution and population. Range now known to extend further west than previously thought (perhaps even to White Sea at Arkhangel'sk), but detailed information still lacking. RUSSIA. Fairly rare: 100–1000 pairs; apparently stable.

Accidental. Faeroes, Britain (annual), Ireland, France, Netherlands, Germany, Denmark, Norway, Sweden, Finland, Estonia, Poland, Switzerland, Balearic Islands, Portugal, Malta, Turkey, Cyprus, Israel (regular in autumn), Kuwait.

Beyond west Palearctic, breeds throughout much of northern Asia, also from Himalayas north-east to Japan.

Movements. Essentially a long-distance migrant, though birds breeding in Himalayas may winter in adjacent areas of northern India. Northern populations move to southern Japan, Philippines, and parts of peninsular India.

Olive-backed Pipit *Anthus hodgsoni yunnanensis* (northern Eurasia): **1** ad breeding, **2** 1st winter (brown type), **3** 1st winter (olive type). *A. h. hodgsoni*: **4** 1st winter. Tree Pipit *Anthus trivialis*: **5** ad fresh (autumn, olive-type), **6** ad breeding (spring), **7** 1st winter (brown type), **8** juv.

Some records of westward vagrants to Britain regarded as linked to strong anticyclonic activity in Siberia. Up to 1983, 27 records in Britain and Ireland: 25 occurred September–November (mostly late September and October) and 2 in spring (April–May); also, one present well inland February–March 1984 had presumably overwintered.

Food. Chiefly insects in summer and seeds in winter. Feeds on ground amongst low herbage. In Philippines in winter, frequently feeds in pines, walking along branches probing for insects among needles and cones.

Social pattern and behaviour. Little information, but apparently similar to Tree Pipit, including song flight.

Voice. Song recalls Tree Pipit but harsher and more sibilant. Calls include quiet 'tseep' or 'tsee' and distinctive 'teaze', loud and strident in full alarm and recalling Redwing.

Breeding. SEASON. Western Siberia: June–August. Normally 2 broods. SITE. On ground, in shelter of rock or tuft of vegetation. Nest: shallow depression containing cup of moss and grass, lined with fine grass and hair. EGGS. Sub-elliptical, smooth and glossy; very variable in ground-colour and pattern; dark brown, brown, or grey, with more or less darker brown spotting and streaking. Clutch: 4–5. INCUBATION. 12–13 days. FLEDGING PERIOD. Leave nest at 11–12 days and fed for at least 1–2 days further, by which time flying.

Wing-length: ♂ 84–90, ♀ 80–86 mm.
Weight: ♂♀ mostly 17–25 g.

Tree Pipit *Anthus trivialis*

Du. Boompieper   Fr. Pipit des arbres   Ge. Baumpieper   It. Prispolone
Ru. Лесной конек   Sp. Bisbita arbóreo   Sw. Trädpiplärka

PLATES: pages 1078, 1091

Field characters. 15 cm; wing-span 25–27 cm. Slightly bulkier than Meadow Pipit, with slimmer rear body making tail length more obvious, and slightly longer wings with narrower point. Rather small, sleek, and elegant pipit, with somewhat more attenuated form than Meadow Pipit and Red-throated Pipit. Plumage pattern typical of small pipits but differs subtly from typical Meadow Pipits and all Red-throated Pipits in combination of noticeably pale eye-ring, warm but not rufous, streaked upperparts, striking wing-bars, yellow-buff throat and boldly spotted breast obvious above little-streaked flanks. At close range, quite large bill, noticeably pink legs, plain rump, and short hind claw are useful characters.

Most observers too ready to rely on supposedly distinctive call for identification, but 4 other pipits give similar calls and/or have similar appearance: Olive-backed Pipit, Pechora Pipit, buff race of Meadow Pipit, and Red-throated Pipit. Because of overlaps between small pipits, most plumage characters are suspect, though streaked rumps of Pechora Pipit and Red-throated Pipit separate them from plain-rumped Meadow Pipit and this species. Important to use calls and bare part structure to reduce confusion.

Habitat. Breeds in middle and upper middle latitudes, and in Scandinavia up through subarctic to borders of Arctic. Mainly in continental but spreading marginally into oceanic climates, between July isotherms of 10–26°C but avoiding more exposed windy and wet as well as torrid and arid conditions. Like congeners, basically a ground-feeder and ground-nester, but unique among them in west Palearctic in attachment to trees and bushes as look-outs and song-posts, no less essential in breeding territory than suitable foraging terrain and nest-sites. Accordingly shuns both open treeless and shrubless habitats and those where density of woody vegetation leaves insufficient open low herbage accessible. The remaining acceptable blends include parkland-savanna types, heathland or grassland in earlier stages of tree colonization, mature hedgerows, birch scrub, young conifer plantations 1 m or more high, open woods of oak, ash, and pine, and even closed high forest, provided there is no thick shrub layer. Ascends mountains as high as trees grow, breeding in Swiss Alps even higher, in zone of dwarf conifers up to 2300 m. Winters in Africa among good-sized well-spaced trees, as a rule at least 10 m high with ground surface accessible.

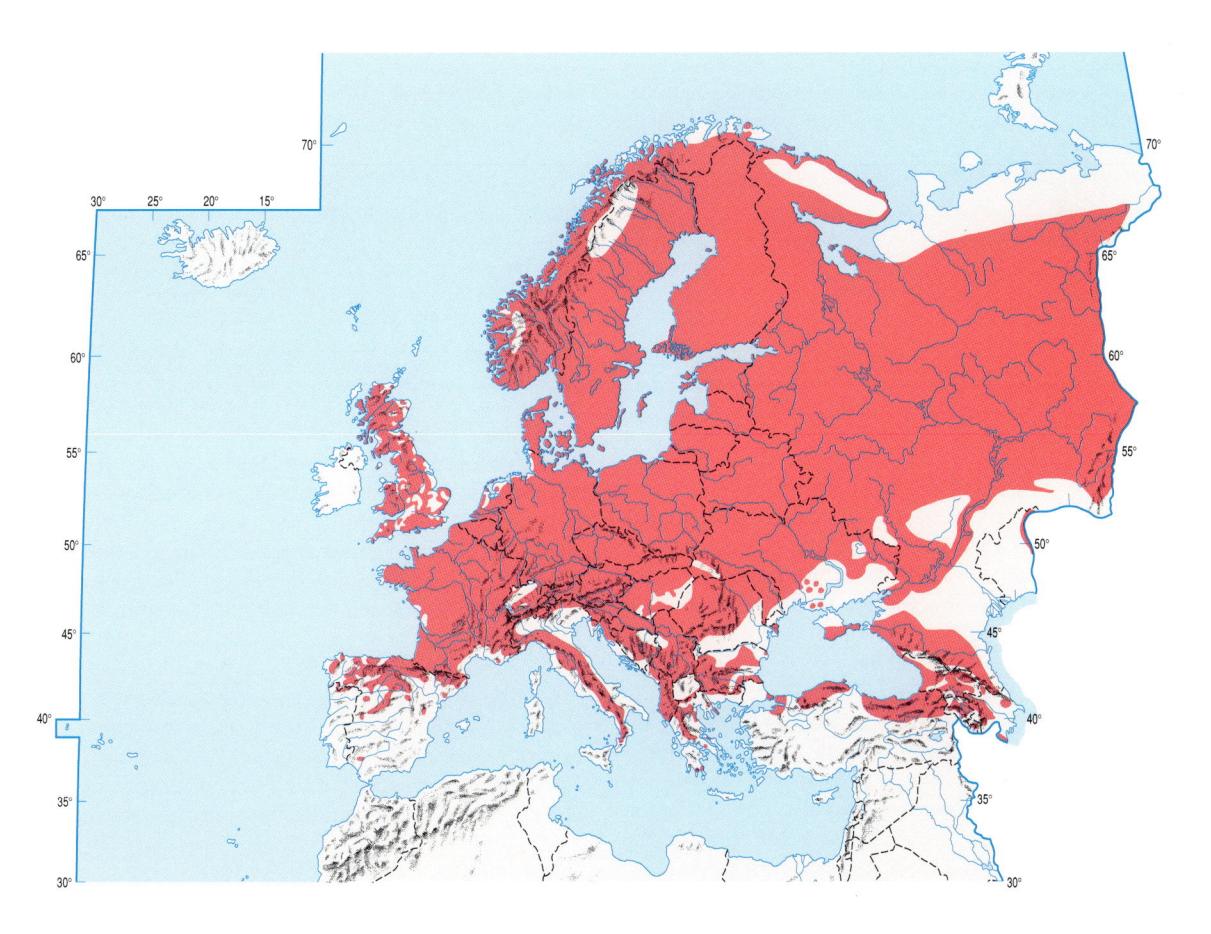

Distribution. Breeding range has decreased Britain, Denmark, and Switzerland. BRITAIN. Spread northward in Scotland from early 20th century. Striking decline more recently, especially in central and southern England. IRELAND. Perhaps a few pairs breed. GERMANY. Scarce in north-west and on Lower Rhine. PORTUGAL. Breeding confirmed in north in late 1980s/early 1990s, after earlier sight records in same area. ITALY. Bred Sicily 1973. ALGERIA. Perhaps nested 1978.

Accidental. Spitsbergen, Jan Mayen, Iceland, Faeroes (probably overlooked regular visitor), Madeira, Cape Verde Islands.

Beyond west Palearctic, extends east to c. 140°E, south to northern Iran and north-west Himalayas.

Population. Decline reported Britain, Belgium, Denmark, Switzerland, and Spain; mostly stable elsewhere, though increase Estonia and Czech Republic. BRITAIN. 120 000 territories 1988–91. Fluctuating, but slight downward trend since 1970. FRANCE. Under 1 million pairs in 1970s. BELGIUM. 27 000 pairs 1973–7. Has declined since, nearly to extinction on coast. LUXEMBOURG. 10 000 pairs. NETHERLANDS. Probable marked decrease reported since 1900, but probably now stable after population increased slightly to 40 000–55 000 pairs 1979–85. GERMANY. 1.28 million pairs in mid-1980s. DENMARK. 11 000–110 000 pairs 1987–8. Decreased 1988–94 following increase 1976–88. NORWAY. 1–2 million pairs 1970–90. SWEDEN. 3.5–7 million pairs in late 1980s. FINLAND. 3–4 million pairs in late 1980s. Now stable, though slight decrease (in north) 1974–7, and slight increase 1978–83. ESTONIA. 500 000–1 million pairs in 1991; increase 1970s–80s. LATVIA. 500 000–900 000 pairs in 1980s. LITHUANIA. Common. POLAND. Fairly numerous. CZECH REPUBLIC. 500 000–1 million pairs 1985–9. SLOVAKIA. 200 000–400 000 pairs 1973–94. HUNGARY. 200 000–400 000 pairs in 1979–93. AUSTRIA. Common. SWITZERLAND. 25 000–40 000 pairs 1985–93. SPAIN. 300 000–400 000 pairs. PORTUGAL. 50–100 pairs in 1992. ITALY. 40 000–80 000 pairs 1983–93. GREECE. Probably over 100 pairs. ALBANIA. 1000–2000 pairs in 1981. YUGOSLAVIA: CROATIA. 40 000–60 000 pairs. SLOVENIA. 20 000–30 000 pairs. BULGARIA. 2500–10 000 pairs. RUMANIA. 150 000–200 000 pairs 1986–92. RUSSIA. 1–10 million pairs. BELARUS'. 1.8–2 million pairs in 1990. UKRAINE. 150 000–160 000 pairs in 1988. MOLDOVA. 15 000–25 000 pairs in 1988. AZERBAIJAN. Uncommon; locally very common in Great Caucasus. TURKEY. 10 000–100 000 pairs.

Movements. Long-distance total migrant. Winters irregularly in Israel, in Persian Gulf states, where scarce, and possibly in northern Iran, but otherwise wholly in Afrotropics and Indian subcontinent. Main wintering area in Africa extends across from Guinea coast at 10°N to Ethiopia—in the west, south only to northern edge of equatorial rain forest, but in east extending south to Natal and Transvaal.

In autumn, all populations breeding west of c. 15°E, plus those from Finland and even from north-west Russia east to c. 40°E, move between south-west and just west of south into western Mediterranean basin and to Portugal, and thence into Africa. Northward passage in spring on a broad front. Higher proportion of birds overfly Mediterranean area in autumn than in spring.

Leaves north European Russia September to early October. Autumn passage through Switzerland occurs late July to late October; at Gibraltar, autumn exodus from end of August to late September with continuance on minor scale until 3rd week of October. In southern African winter quarters, present late October–March. Northward passage begins March, with main movement through Mediterranean area in April and arrival on breeding grounds from early April to late May in north.

Food. Chiefly insects with some plant material taken in autumn and winter. Food taken mostly from ground, low herbage, and leaf litter, more rarely from twigs, branches, and tree trunks. Occasionally takes insects after short aerial pursuit from ground.

Social pattern and behaviour. Often solitary outside breeding season, though small flocks and sometimes larger concentrations occur for migration, feeding, and roosting. Territorial when breeding, with little tendency to form neighbourhood groups. Song of ♂ given from perch, typically top of isolated tree in clearing or protruding branch of tree at edge of wood; also from ground, and in characteristic song-flight. Song-flight consists of c. 60° ascent (from perch or ground) to c. 15 m, occasionally up to c. 30–35 m, above take-off point, followed by gliding descent with wings spread to form parachute, tail raised and spread, and legs dangling; lands on same or adjacent perch, sometimes on ground. Usually starts singing near peak of ascent and continues during descent. Song occurs from arrival to mid- or end of July.

Voice. Song excels Meadow Pipit, Rock Pipit, and Water Pipit in both quality and power, carrying much further than in Meadow Pipit: single, long, rising and falling stanza of (1) repeated monosyllables, (2) slight stuttering phrase, and (3) increasingly forceful disyllables, sequence sounding like 'chik-chik ... chiachia-wich-wich-tsee-a-tsee-a-tsee-a ... '; hence a continuous, increasingly confident but decelerating trill, with last part strongly recalling Canary. Commonest call much stronger than the three other common pipits and audible at longer range: 'teez', 'tseep', or 'skeeze', sounding quite high-pitched but buzzing in tone.

Breeding. SEASON. North-west Europe: first eggs laid late April, with main laying season May–June; last young in nest August. 1–2 broods, rarely 3. SITE. On ground flat or sloping; in low cover, or more or less in the open. Nest: shallow depression holding substantial cup of dry grass leaves and stems, often with moss foundation, lined with finer grasses and hair. EGGS. Sub-elliptical, smooth and glossy; extremely variable in colour—often brown, grey, or reddish, but also pale blue, pink, green, or dark brown; may be evenly speckled, have dark zone at broad end, or be blotched, streaked, or spotted; some have black hair-streaks; a few are blue with no markings, or dark brown all over. Clutch: 2–6. INCUBATION. 12–14 days. FLEDGING PERIOD. 12–14 days, but young normally leave nest 2–3 days before, earlier if disturbed, and do not return.

Wing-length: ♂ 83–96, ♀ 79–90 mm.
Weight: ♂♀ mostly 18–29 g; migrants may be heavier.

Pechora Pipit *Anthus gustavi*: **1** ad breeding, **2** ad breeding, **3** 1st winter. Red-throated Pipit *Anthus cervinus* (p. 1084): **4** ad ♂ breeding (typical), **5** ad ♂ breeding (dark), **6** ad ♀ breeding, **7** ad ♀ non-breeding, **8** 1st winter.

Pechora Pipit *Anthus gustavi*

PLATES: pages 1081, 1091

Du. Petsjorapieper   Fr. Pipit de la Petchora   Ge. Petschorapieper   It. Pispola della Pechora
Ru. Сибирский конек   Sp. Bisbita del Pechora   Sw. Tundrapiplärka

Field characters. 14 cm; wing-span 23–25 cm. Slightly less attenuated and less bulky than Tree Pipit, with 15% shorter tail; close in size and form to fledgling Meadow Pipit. 2nd smallest pipit of west Palearctic, with rather long and fine bill and rather short tail. Warm buff plumage sharply and copiously streaked, with no other markings showing at distance except for white belly and buff-white in outer tail-feathers; last lack cold and clean tone of tail-sides of other small pipits. Diagnostic marks apparent at close range include buff-white mantle-stripes, obvious pale double wing-bar, and fully streaked rump.

Long thought difficult to identify, but well-marked bird actually the most decorated of small pipits occurring in west Palearctic, with warm, heavily streaked, and bright appearance. Flight consists of erratic bursts of wing-beats interspersed with short bounds, floats, and glides; somewhat hesitant progress thus recalls Meadow Pipit. Flight silhouette also recalls Meadow Pipit, with rather short, straight-sided tail suggesting juvenile of that species. Shy, skulking in dense cover such as long grass and centre of bush. Difficult to flush, escaping first in low flits, then in high flight.

Habitat. Breeds from fringe of west Palearctic eastward along a mainly subarctic band, apparently sandwiched between Olive-backed Pipit to south and Red-throated Pipit to north. Inhabits bushy tundra and remote taiga swamps, but not pure tundra, apparently preferring overgrown areas with tall dense sedge, reed-grass, and plentiful shrubs or even trees, mainly in lowlands, along rivers and coasts.

Distribution and population. Russia. Still little known; no recent confirmed breeding.

Accidental. Iceland, Britain, France, Norway, Sweden, Finland, Lithuania, Poland.

Beyond west Palearctic, extends east in narrow band across northern Siberia; also breeds Komandorskie islands, and isolated population at Lake Khanka (Ussuriland, south-east Russia).

Movements. Migratory, wintering in East Indies. Because of thinly spread population, dates of movement remain obscure. In Britain (where 28 recorded up to 1985, almost all on Fair Isle, Scotland), occurs late August to mid-November (mostly late September and early October) and once (in Suffolk) in late April.

Food. Chiefly insects. Forages mainly on ground picking food from lower parts of plant stems and leaves.

Social pattern and behaviour. Mostly solitary or in twos outside breeding season. Breeding pairs tend to occur in neighbourhood groups, dictated at least to some extent by habitat availability. ♂ sings in flight, also (often as a continuation) from bush, etc., or ground. Ascends silently and fairly rapidly but not steeply to c. 30–70 m then flies in wide circles or smaller loops; at intervals hovers for a few seconds with tail spread and sings. Reminiscent of lark in wheeling and hovering flight.

Voice. Song of west Palearctic race consists of trilling sounds followed by guttural warbling. Commonest call a stony 'pwit' or 'p(r)it'.

Breeding. SEASON. Siberia: eggs laid late June and July. Probably one brood. SITE. On ground in low cover or in shelter of tuft of vegetation or low scrub. Nest: substantial cup of grass and other leaves, lined with finer vegetation. EGGS. Sub-elliptical to oval, smooth and glossy; pale grey, sometimes pink-tinted or even dark red-brown, with grey speckling overall, and occasional black hair-streaks or blotches at broad end. Clutch: 4–5(–6). INCUBATION. About 12–13 days. FLEDGING PERIOD. 12–14 days.

Wing-length: ♂ 83–87, ♀ 78–84 mm.

Meadow Pipit *Anthus pratensis*

PLATES: pages 1076, 1091

DU. Graspieper FR. Pipit farlouse GE. Wiesenpieper IT. Pispola
RU. Луговой конек SP. Bisbita común SW. Ängspiplärka

Field characters. 14.5 cm; wing-span 22–25 cm. Somewhat smaller than all other small pipits except Berthelot's Pipit and Pechora Pipit, with more rounded wings; much shorter and less bulky than large pipits, with proportionately shorter tail. Rather small, sleek but dumpy, active pipit; epitome of genus but lacking striking diagnostic field characters. Much more readily identified by call than plumage. Typically olive or brown above, heavily streaked except on plain rump, and ochre- or grey-white below, heavily spotted and streaked on chest and flanks; only cold white tail-sides catch eye. At close range, rather slender bill, rather indistinct face pattern, distinct but usually dull wing-bars, pale brown legs, and long hind claw form useful characters.

Commonest pipit in western Europe; general character, rather anonymous appearance and all-important diagnostic call soon learnt but risk of confusing buffer or silent individuals with all other small congeners high. Most frequent confusions are with Tree Pipit and Red-throated Pipit (see those species). Flight has characteristic hesitant, jerky action, shared only by Pechora Pipit. Escape flight usually high and wandering, accompanied by hysterical calls. Gait most creeping of genus, exaggerated by usually low stance.

Habitat. Breeds in middle, upper middle, and upper latitudes of west Palearctic, from temperate through boreal to fringe of arctic climatic zones, and from continental to oceanic regimes, accepting rainy, windy, and chilly conditions, but avoiding ice and prolonged snow cover as well as torrid and arid areas, within rather narrow temperature range of 10–20°C on Eurasian mainland. Chooses, as a ground-dweller, open areas of rather low fairly complete vegetation cover. Avoids extensive bare rock, stones, sand, soil, and close-cropped grass or herbage, and on the other hand tall dense vegetation, including woods, forests, and reedbeds. Flourishes, however, in plots of young planted trees, and perches freely on fence-posts, telegraph wires, stone walls, and other points of vantage. Winters opportunistically in terrain often similar to that used for breeding, but sometimes very dissimilar; noted in Tunisia in glades of cork oak forests near 700 m, and also on tussocky plateaux on fringe of pine forest.

Distribution. Range generally stable, though some decrease Britain, Ireland (perhaps), and Denmark (probably by 10–15%), slight expansion Czech Republic and Switzerland. BRITAIN. Decline in central and southern areas. GERMANY. Much commoner from Hunsrück, Spessart, Rhön, and Erzgebirge north, especially in large river valleys and on coast. AUSTRIA. Very local; main breeding area in north. SPAIN. Irregular breeding in Asturias and Santander. ITALY. Breeds irregularly in Central

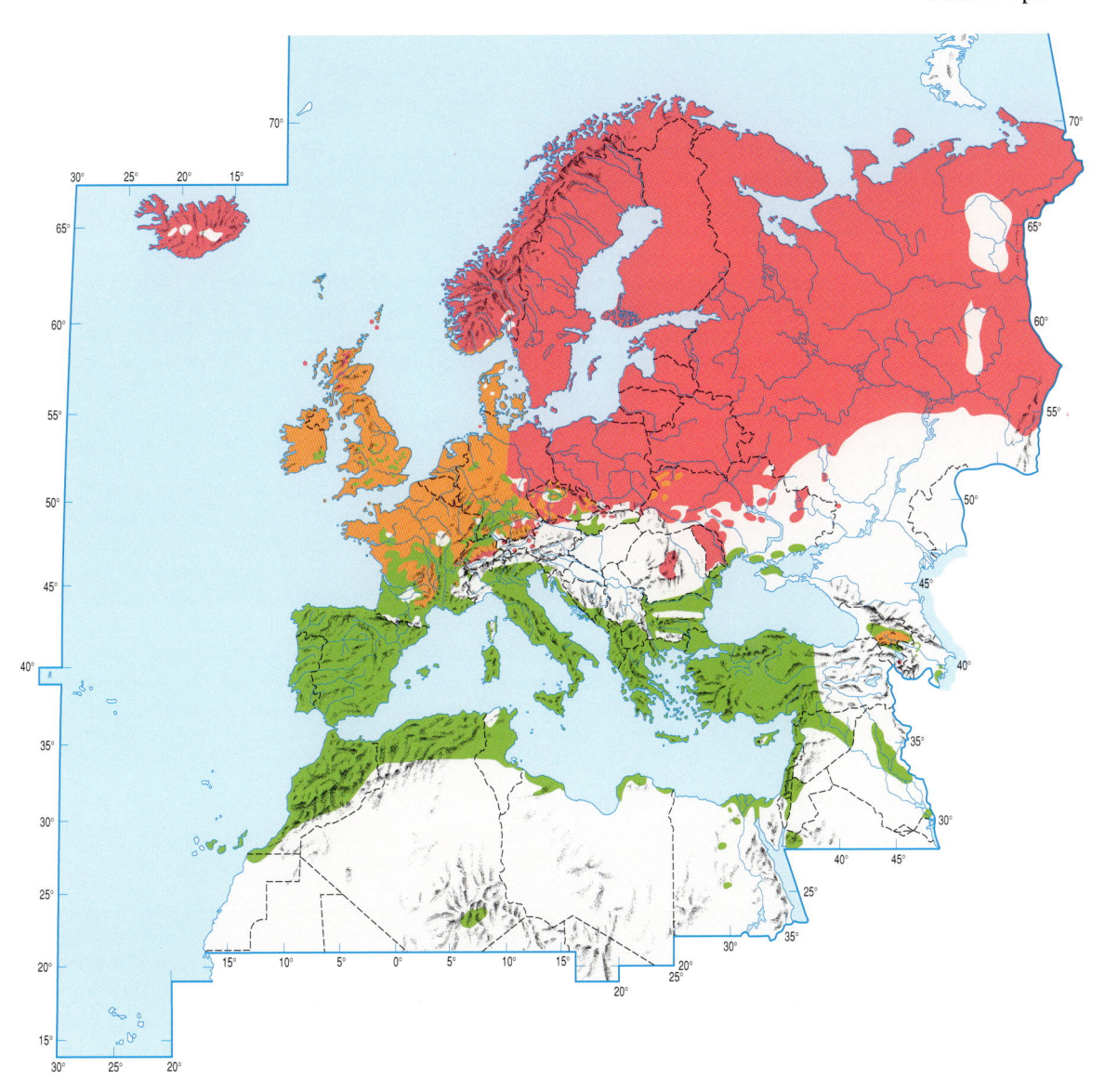

Appennines. YUGOSLAVIA. Bred 1923, 1932, and 1947; no records in recent years. RUMANIA. Rare; confined to a few sites in Transylvania.

Accidental. Spitsbergen, Bear Island, Jan Mayen, Kuwait (perhaps scarce passage-migrant), Azores, Madeira.

Beyond west Palearctic, extends east only to Ob' river; also breeds south-east Greenland.

Population. Decline reported Britain, Netherlands (probable), Denmark, Estonia, Lithuania, and Poland. Habitat changes (Netherlands, Poland) only possible cause suggested. Marked increase in Czech Republic, slight increase Finland, Austria, and Switzerland. Mostly stable elsewhere. ICELAND. 500 000–1 million pairs in late 1980s. FAEROES. 200–500 pairs. BRITAIN. 1.9 million territories 1988–91. Decline since early 1980s. IRELAND. 900 000 territories 1988–91. FRANCE. Under 100 000 pairs in 1970s. BELGIUM. 35 000 pairs 1973–7; no clear trend. LUXEMBOURG. 1000–1500 pairs. NETHERLANDS. 70 000–100 000 pairs 1979–85. Decrease at least in part due to habitat changes. GERMANY. 176 000 pairs in mid-1980s. 80 000 pairs in east in early 1980s, and revised estimate for whole country of 80 000–100 000 in early 1990s. DENMARK. 3000–40 000 pairs 1987–8; decreased 1987–94. NORWAY. 1–5 million pairs 1970–90. SWEDEN. 500 000–1 million pairs in late 1980s. FINLAND. 1–1.5 million pairs in late 1980s. ESTONIA. 50 000–100 000 pairs in 1991; decreased 1960s–80s. LATVIA. 50 000–100 000 pairs in 1980s. POLAND. 20 000–30 000 pairs; slow and continuing decline. CZECH REPUBLIC. 30 000–60 000 pairs 1985–9. SLOVAKIA. 250–500 pairs 1973–94; trend unknown. AUSTRIA. Increasing since 1970s. SWITZERLAND. 300–500 pairs 1985–93. RUMANIA. 100–500 pairs 1986–92. RUSSIA. 1–10 million pairs. BELARUS'. 180 000–230 000 pairs in 1990.

UKRAINE. 35 000–38 000 pairs in 1988. MOLDOVA. 650–800 pairs in 1988.

Movements. Resident or partial migrant in western Europe, but northern and eastern populations are medium-distance total migrants, though in some milder winters only extreme north completely vacated. Movement almost entirely diurnal. Northern limits of wintering areas in western Europe not easy to define and vary with severity of season: normally, wintering occurs throughout Britain and Ireland (except uplands) and in central Europe north to Denmark and east to western Germany. Otherwise, wintering centred around Mediterranean basin. Autumn passage may begin as early as mid-August and extend to late October. Movement in autumn predominantly south-west in western Europe, passage at Gibraltar peaking mid-October to early November. Northward passage of western populations in spring follows essentially same route as autumn, including changes in direction involved in passage via Iberia, Britain, and Iceland. Within north-west mainland Europe, including Scandinavia, movement close to north-east in almost all cases. Takes place mainly in March–April, with northernmost breeding areas occupied mid-May.

Food. Mainly invertebrates, with some plant seeds in autumn and winter. Feeds almost exclusively on ground, walking at steady rate picking invertebrates from leaves and plant stems. Occasionally takes insects in flight which it has disturbed, but never flies after them.

Social pattern and behaviour. Often in flocks outside breeding season, but these rather loose-knit. Monogamous mating system, though polygyny recorded occasionally. Pair-bond may be maintained over successive years. Breeding territories may be concentrated in small neighbourhood groups, but more usually nests well-spaced, averaging *c.* 100 m apart. ♂ sings mainly in flight, though also from high or low perch or from ground. For song-flight, ascends with fluttering wing-action and tail widely spread, rather like Skylark, moderately fast and not quite vertically or at *c.* 40°, to *c.* 5–35 m. Then either descends immediately or flies with markedly shallower wing-beats, sometimes circling, more or less level, for some distance. Descends in parachute style like Tree Pipit: tail first slightly spread and horizontal, then raised and usually closed; wings bent stiffly back and raised, legs dangling. 1st part of song given from take-off or only after ascending a few metres, remainder during descent, part sometimes after landing.

Voice. Song increasingly fast, then slowing sequence of repeated, tinkling, feeble notes, rising in pitch to become more musical trill; written 'tsee-tsee tseek tseek tsee-er tsee-er'; lacks loud terminal flourish of Tree Pipit. Commonest call, rather thin, sibilant squeak, written 'weesk' or 'sreep'; often repeated or yelled in alarm, conveying sense of hysteria. Other calls include short, tinny 'chip' and notes suggestive of monosyllable of Red-throated Pipit (but not hard note of Pechora Pipit).

Breeding. SEASON. Central and western Europe: first eggs from first half of April, last eggs laid beginning of August. Britain: from second half of April with peak 2nd week of May in south and 3rd week in north. Swedish Lapland: laying begins mid-June to early July. 2 broods normal in central and western Europe, but rare in north. SITE. On ground, usually concealed in vegetation. Nest: cup of grasses and other plant material, lined finer vegetation and hair. EGGS. Sub-elliptical, smooth and glossy; quite variable, usually brown, grey, or reddish, spotted or mottled, sometimes finely streaked brown, black, or grey. Clutch: 3–5(2–6). INCUBATION. 11–15 days. FLEDGING PERIOD. 10–14 days.

Wing-length: ♂ 78–86, ♀ 73–81 mm.
Weight: ♂♀ 15–22 g.

Geographical variation. Slight and clinal. 2 subspecies recognized. Nominate *pratensis* occupies most of range. *A. p. whistleri* from Ireland and western Scotland on average deeper and redder olive-brown above, and with slightly heavier black streaks; ground-colour of chest, sides of breast, and flanks markedly deeper, cinnamon-buff.

Red-throated Pipit *Anthus cervinus*

PLATES: pages 1081, 1091

DU. Roodkeelpieper   FR. Pipit à gorge rousse   GE. Rotkehlpieper   IT. Pispola golarossa
RU. Краснозобый конек   SP. Bisbita gorgirrojo   SW. Rödstrupig piplärka

Field characters. 15 cm; wing-span 25–27 cm. Noticeably plumper than Meadow Pipit in foreparts. Rather small, sleek, but quite robust pipit, with colourful, changing plumage patterns. Combination of little-marked face, broadly streaked upperparts and rump, and (in winter) buff underparts with large-spotted chest and flanks creates darker, richer plumage than Tree Pipit and most Meadow Pipits. In summer, ♂ less streaked below with variable pink or red-buff suffusion on face, throat, and even chest, producing diagnostic appearance; summer ♀ usually has less pink and more streaks below. At close range, rather short bill, usually brown legs, and long hind claw form useful characters.

Typical breeding adult virtually unmistakable; winter adult and immature more difficult to distinguish from (1) Olive-backed Pipit (similar call, but markedly distinct plumage colours and pattern), (2) Tree Pipit (sharing similar call and buff ground to plumage, but with unstreaked rump and short hind claw, (3) Pechora Pipit (sharing rich plumage tone to upperparts and streaked rump, but with strikingly white belly and distinctive call, and (4) Meadow Pipit (sharing similar plumage tones in

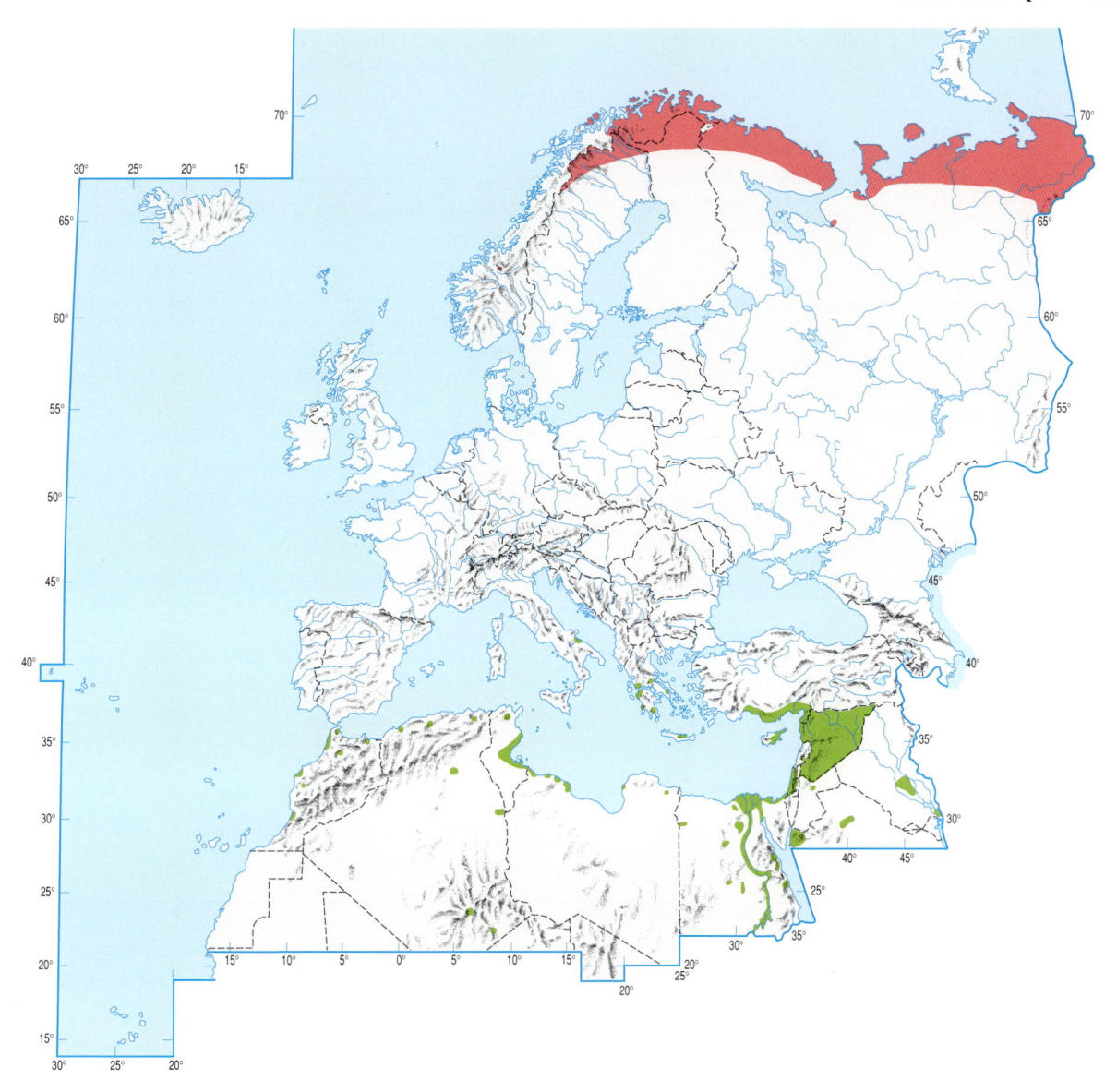

Irish and Scottish race and having same long hind claw, but with unstreaked rump and distinctive call). Flight as Tree Pipit but silhouette somewhat bulkier, with rather broader wing-tip. Escape-flight as Meadow Pipit; song-flight well developed, higher than Meadow Pipit and often sustained longer. Gait as Meadow Pipit.

Habitat. Arctic and subarctic, between July isotherms of 2–15°C, north of forest limits and mainly on shrubby or mossy tundra, although locally in Scandinavia up to 1000 m, and near water in swamps of willow and birch. Near settlements will adapt to drained and cultivated land, as well as damp grassy flats. Overlaps in northern Europe with Meadow Pipit and Tree Pipit, both of which are less adapted to arctic conditions. In Russia, favours hummocky humid tundra and boggy levels overgrown with sedge and other herbage, including osier thickets. In winter associated with short grass produced by grazing animals and areas that have patches of very shallow water, as in cattle-tramped mud.

Distribution. No major range changes reported. NORWAY. Bred in south (62°20′N, Hedmark) 1975.

Accidental. Bear Island, Faeroes, Britain (near annual), Ireland, Latvia (only 2 records, but more likely scarce passage-migrant, overlooked before), Lithuania, Albania, Madeira.

Beyond west Palearctic, extends east in narrow band across northern Siberia; also breeds Bering Strait area of Alaska.

Population. NORWAY. 5000–20 000 pairs 1970–90; stable. SWEDEN. 100–1000 pairs in late 1980s; stable. FINLAND. 2000–5000 pairs in late 1980s; slight decrease. RUSSIA. 10 000–100 000 pairs; stable.

Movements. Migratory, wintering largely in tropics of Africa and south-east Asia. Migration thus longer than that of any other pipit except Pechora Pipit. In Europe, winters regularly only in Italy and Greece. Also regular in scattered areas of southern Turkey, Middle East, and North Africa, and abundant in Nile valley of Egypt. Winters all across Sahel zone, mainly in the east, south in the east to Tanzania. In accordance with eastern Africa being more important wintering area than western, the most westerly breeding populations move mainly east of south in autumn to pass east of Baltic into central Europe. Records scanty through west-central Europe; accidental or uncommon in Iberia.

Leaves southern Russian tundra usually between late August and early October. Passes southern Sweden and Denmark early September to mid-October. Occurs on Malta and Cyprus mid-October to November. Present in East Africa from late October. Northward passage in spring protracted: in Israel begins early February and may continue until June. Birds arrive on Russian tundra in late May.

Westward vagrants move with Meadow Pipit in both spring and autumn, perhaps reflecting shared migration route between north-west Europe and north-west Africa, and often found in mixed flocks or parties. Pattern of records in Britain and Ireland fits well enough with that for migration in western Europe, birds occurring mainly late August to early November and (fewer) in May.

Food. Chiefly insects, also small water snails and a few seeds. Feeds on ground by pecking and probing amongst vegetation; on seashore, probes amongst washed-up seaweed. After capture, largest prey items are vigorously pounded on ground before swallowing.

Social pattern and behaviour. Gregarious outside breeding season, with mostly loose-knit flocks of varying size occurring for migration, feeding, and roosting. Breeding behaviour generally similar to Meadow Pipit; song-flight often higher and longer, bird gliding in wide arc on stiffly extended wings.

Voice. Song has form between those of Tree Pipit and Meadow Pipit. 2 common calls distinctive: (1) in migration flight and in alarm, a quite loud call of 1–2 syllables—hissing and rasping at a distance but piercing at close range—written 'skeeze' or 'sii(s)', and 'skee-eaz', 'sssii', or 'pee-ez', with extended form 's(k)ee-eze'; distinguished (at close range) from similar notes of Tree Pipit and Olive backed Pipit by quite marked pulse in volume; (2) mainly from wintering birds in flock, full, rather abrupt 'teu', 'chup', or 'chit', uttered both singly and in rapid series; not unlike shorter, quieter notes of Meadow Pipit but more musical than monosyllable of Pechora Pipit.

Breeding. SEASON. At southern extent of range, first eggs laid end of May; further north, laying from early or mid-June to early July; last young, from late or repeat layings, fledge up to mid-August. One brood. SITE. On ground, in side of hummock or bank, or sheltered by low scrub; sometimes at end of short 'tunnel' in mossy hummock. Nest: hollow in moss or ground, filled with cup of grass leaves and stems, with some moss and dead leaves in base; minimal lining of finer grass, hair, and some feathers. EGGS. Sub-elliptical, smooth and glossy; variable, mainly grey, buff, olive, or pinkish, with fine spots, speckles, or blotches of brown, grey, or red-brown; sometimes with fine black hair-streaks. Clutch: 5–6 (2–7). INCUBATION. 11–14 days. FLEDGING PERIOD. 11–15 days.

Wing-length: ♂ 85–90, ♀ 82–86 mm.
Weight: ♂♀ mostly 17–24 g.

Water Pipit *Anthus spinoletta*

PLATES: pages 1087, 1091

DU. Waterpieper FR. Pipit spioncelle GE. Bergpieper IT. Spioncello
RU. Горный конек SP. Bisbita ribereño SW. Vattenpiplärka

Field characters. 17–17.5 cm; wing-span 24–29 cm. Structure as Rock Pipit except for slightly longer tail and legs. Plumage differs consistently from Rock Pipit in white outer tail-feathers and almost white underwing and (in spring and summer) much more contrasting, colourful pattern, with noticeably greyer upperparts and paler, hardly streaked white underparts. Long pink-white supercilium, greyish double wing-bar, and strong (♂) and faint (♀) buff or pink wash from throat to mid-belly catch eye. In winter and 1st autumn, more like Rock Pipit, particularly Fenno-Scandian race *littoralis*, but most remain noticeably cleaner and less diffusely and widely streaked below, inviting confusion with Buff-bellied Pipit (see that species). Often has rather paler brown legs than Rock Pipit. Flight and gait similar to Rock Pipit.

Habitat. Montane. Breeds in west Palearctic in middle and lower-middle latitudes at considerable elevations, in Switzerland infrequently below 1400–1800 m and thence up to 2600 m or even higher. Prefers zone of stunted trees with sparse ground cover or moist meadows, often close to glaciers and on steep nearly bare crags, even above snowline. Habitat in FSU varies, especially extralimitally where arctic birch thickets and stone screes with rhododendron thickets and rocky tundra may be occupied. Descends in winter to lower ground, or banks of mountain streams, occurring in spring on boggy lowlands with shrubs, sandy lowlands, and arable land. In western Europe, descends to flooded or damp meadows, watercress beds, estuaries and seashores, including mudflats.

Distribution. GERMANY. Mainly in Alps with small, isolated populations in Bayerischer Wald and Schwarzwald, occasionally breeding other uplands: e.g. Harz (first bred 1964), also Hoher Meißner, Hessen (1958, 1983).

Winter mapping tentative. Status may need revision in (southern) Scandinavia: hitherto considered rare visitor to Denmark, but recent (unconfirmed) report of influx (c. 100)

Water Pipit *Anthus spinoletta*. *A. s. spinoletta*: **1** ad breeding fresh (spring), **2** ad non-breeding fresh (autumn). *A. s. Coutellii*: **3** ad breeding fresh (spring), **4** ad non-breeding fresh (autumn). *A. s. blakistoni*: **5** ad breeding fresh (spring). Buff-bellied Pipit *Anthus rubescens* (p. 1092). *A. r. rubescens*: **6** ad breeding fresh (spring), **7** ad non-breeding fresh (autumn). *A. r. japonicus*: **8** ad breeding fresh (spring), **9** ad non-breeding fresh (autumn).

in winter 1995–6; first confirmed record in Sweden October 1993, 19 birds present from December 1994 and further reports (unconfirmed) 1995–6. ISRAEL. Buff-bellied Pipit now known to be regular winter visitor, occurring alongside Water Pipit, at least in north. TUNISIA. Probably scarce in winter, but only Rock Pipit positively identified.

Accidental. Spitsbergen, Bear Island, Jan Mayen, Ireland, Mauritania (this species and/or Rock Pipit), Canary Islands.

Beyond west Palearctic, breeds from northern Iran to Lake Baykal and north-west China.

Population. Apparently stable throughout range. FRANCE. Under 100 000 pairs in 1970s. Presumably includes both this species and Rock Pipit. GERMANY. 4000 pairs in mid-1980s. POLAND. 2500–3500 pairs; mostly stable, but recent slow decrease in Sudety mountains. CZECH REPUBLIC. 260–380 pairs 1985–9. SLOVAKIA. 700–1100 pairs 1973–94. AUSTRIA. Common in alpine habitats above 1000 m. SWITZERLAND. 50 000–100 000 pairs in 1985–93. SPAIN. 16 000–32 000 pairs. PORTUGAL. 1–5 pairs 1978–84. ITALY. 30 000–70 000 pairs 1983–93. GREECE. 500–2000 pairs. ALBANIA. 500–1000 pairs in 1981. YUGOSLAVIA: CROATIA. 1000–2000 pairs. SLOVENIA. 500–800 pairs. BULGARIA. 1000–10 000 pairs. RUMANIA. 50 000–70 000 pairs 1986–92. RUSSIA. See Rock Pipit. UKRAINE. 4000–4500 pairs in 1988. AZERBAIJAN. Common. TURKEY. 10 000–100 000 pairs.

Movements. Populations from mountains of central and southern Europe recorded in winter almost throughout inland central Europe north-west to Belgium and western Netherlands, rather rarely on coast. Small numbers regularly winter in southern England. Many thus move north in autumn, though movement in Alps often purely altitudinal, birds descending from mountains to adjacent lakes and rivers. Considerable numbers make longer movements, mostly south-west to Iberia and North Africa. No evidence that populations of Spanish mountains make other than local altitudinal movements. Birds breeding in Turkey, Caucasus, and northern Iran make altitudinal or longer movements, presumably accounting for birds wintering in Middle East, including Egypt.

Food. Mainly invertebrates; also some plant material. Feeds mainly on ground, but occasionally catches insects in flight by making short leaps or flying from perch. In cold spells in high mountains during breeding season, feeds around burrow-entrances of marmots.

Social pattern and behaviour. Solitary or gregarious outside breeding season. In Britain, birds reported to use same area (even same few square metres) for feeding over several weeks. Flocks occur for roosting and (usually loose-knit) for feeding: often in groups of 2–5, sometimes 20–60 or up to 200 or more. Territorial in breeding season, and normally monogamous. ♂ song-flight basically similar to Meadow Pipit. Bird uses rapid, shallow wing-beats, ascending at *c.* 30–45°, rarely spiralling, to *c.* 10–30 m. From peak of ascent, circles or flies in long arc, often crossing territory, to perform return flight to starting point. Slow, gliding descent similar to that of Tree Pipit: wings widely spread and well forward, tail raised and partly spread.

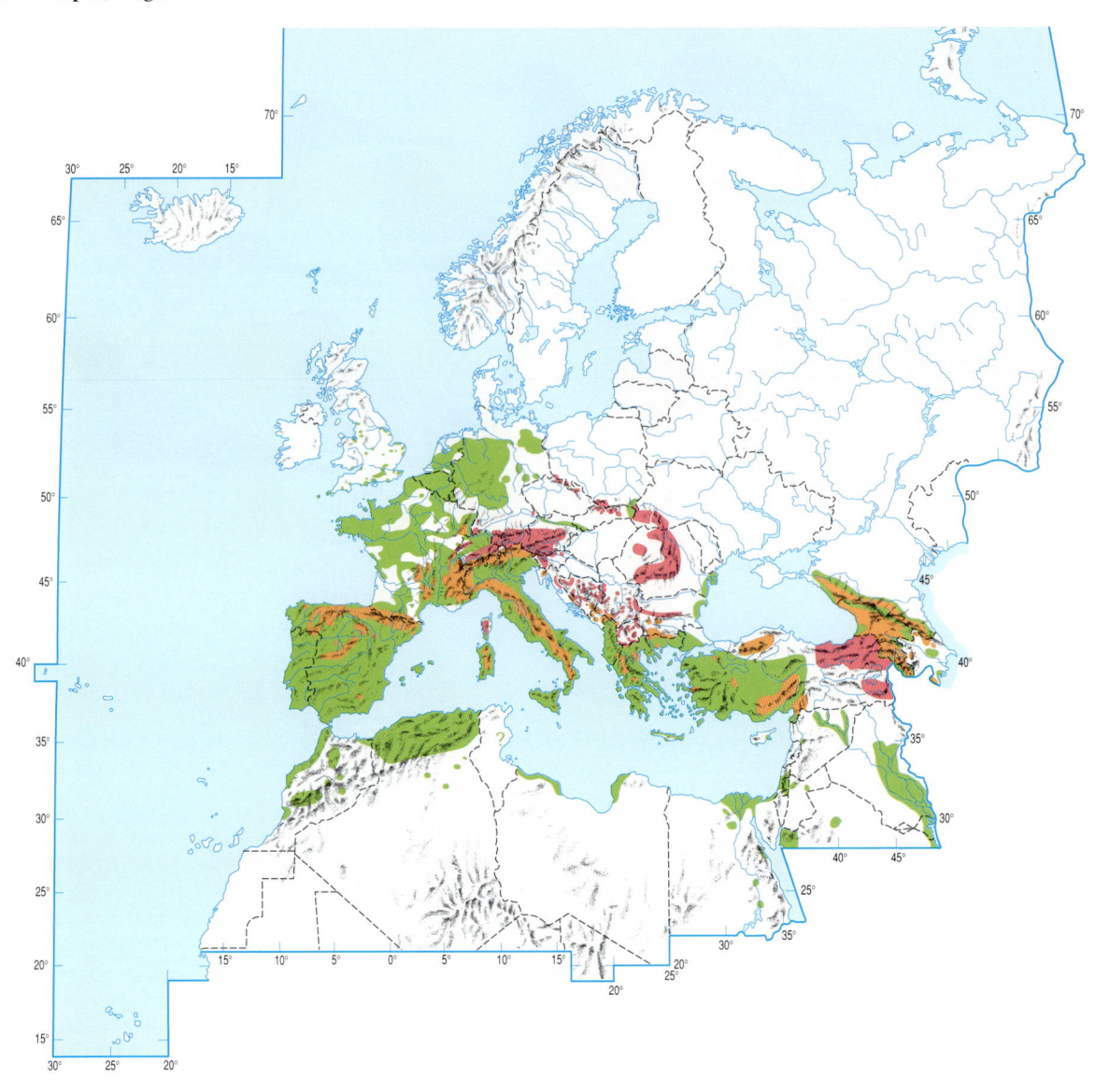

Voice. Repertoire much as in Meadow Pipit; song similar, and confusable with Meadow Pipit's but generally more melodious, approaching Tree Pipit in quality. Contact-alarm call a 'tsiip', 'peeht' or 'weeeez', appreciably fuller, less feeble and squeaky, rather more grating than Meadow Pipit, and sharper and more drawn-out than Rock Pipit.

Breeding. SEASON. Southern Europe: eggs laid end of April to early July. 2 broods. SITE. In side of steep bank or hollow, well concealed by overhanging vegetation; sometimes at end of short tunnel. Nest: cup of grass stems and leaves, and moss, with slight lining of finer leaves and a few hairs. EGGS. Sub-elliptical, smooth and glossy; grey-white, heavily mottled brown and grey, sometimes with dark zone or cap at broad end, and occasionally with black hair-streaks. Clutch: 4–6(–7). INCUBATION. 14–15 days. FLEDGING PERIOD. 14–15 days.

Wing-length: ♂ 88–96, ♀ 82–90 mm.
Weight: ♂♀ mostly 19–27 g.

Geographical variation. Most of west Palearctic range occupied by nominate *spinoletta*. *A. s. coutelli* of eastern Turkey and Caucasus similar, but upperparts in non-breeding plumage paler brown, marked with slightly more pronounced dark olive-brown feather-centres; streaks on underparts narrower and more restricted; in breeding plumage, forehead, crown, and hindneck paler grey, mantle and scapulars down to upper tail-coverts paler sandy olive-brown with more pronounced dark streaks on mantle and scapulars; underparts paler, pink-buff or sandy-buff, much less vinous.

Rock Pipit *Anthus petrosus*. *A. p. petrosus*: **1** ad breeding fresh (spring), **2** ad breeding worn (summer), **3** juv. *A. s. kleinschmidti*: **4** ad fresh (autumn). *A. p. littoralis*: **5** ad breeding fresh (spring), **6** ad breeding worn (summer). *A. p. petrosus/littoralis*: **7** ad fresh (autumn).

Rock Pipit *Anthus petrosus*

PLATES: pages 1089, 1091

Du. Oeverpieper   Fr. Pipit maritime   Ge. Strandpieper   It. Spioncello marino
Ru. Скальный конек   Sp. Bisbita costero   Sw. Skärpiplärka

Field characters. 16.5–17 cm; wing-span 22.5–28 cm. 10–15% larger than Meadow Pipit, with longer and stronger bill and legs, and fuller tail. Quite large pipit with more robust form than smaller pipits. Shares Water Pipit's mottled (not streaked) upperparts, and dark bill and legs, but differs in grey and white or smoky outer tail-feathers, dusky underwing, and (in spring and summer) retention of dusky, heavily-streaked underparts.

Dark bare parts and characteristic call (see below) are distinctive but confusion arises between Rock Pipit and Meadow Pipit or Red-throated Pipit which can breed in areas abutting or even overlapping with Rock Pipit (see those species). Separation from Water Pipit often difficult and still incompletely studied; Fenno-Scandian race *littoralis* may suggest either species. Flight easy and confident, lacking hesitant flutter of Meadow Pipit; does not have, however, urgent bounds of larger pipits. Flight silhouette fuller-bodied and broader-tailed than smaller pipits but far less long-tailed than large ones. Song-flight like Meadow Pipit, but usually from and to rock. Gait lacks persistent long runs of large pipits.

Habitat. Coastal. Ranges from middle to upper latitudes in temperate, boreal, and arctic zones, rarely penetrating more than a short distance inland and almost entirely attached to rocky sea-cliffs and crags, rarely much higher than 100 m and often down to shore level. Avoids totally exposed situations, preferring sheltered gullies or inlets, and islands, even far offshore; not troubled by high and gusty winds or prolonged heavy rain and salt spray.

Distribution. Confined to west Palearctic. ICELAND. Has bred: small population (1–2 pairs) 1987–92. DENMARK. Spread in islands of Kattegat. NORWAY. Some range extension. FINLAND. First bred Valassaaret islands 1962. ESTONIA. Nesting first confirmed 1961.

Winter mapping tentative. *A. p. littoralis* perhaps regular at some inland sites (e.g. Mecklenburg lakes, Germany) but these not plotted. TUNISIA. See Water Pipit.

Accidental. Iceland, Austria.

Population. FAEROES. 3000–6000 pairs (estimate for early 1980s too low). BRITAIN. 34 000 pairs 1988–91. Probably stable, perhaps local declines. IRELAND. 12 500 pairs 1988–91. FRANCE. See Water Pipit. DENMARK. 100 pairs in 1989; stable. NORWAY. 50 000–200 000 pairs 1970–90; stable. SWEDEN. 4000–10 000 pairs in late 1980s; slight decrease. FINLAND. 1500–2000 pairs in late 1980s; probable increase. ESTONIA.

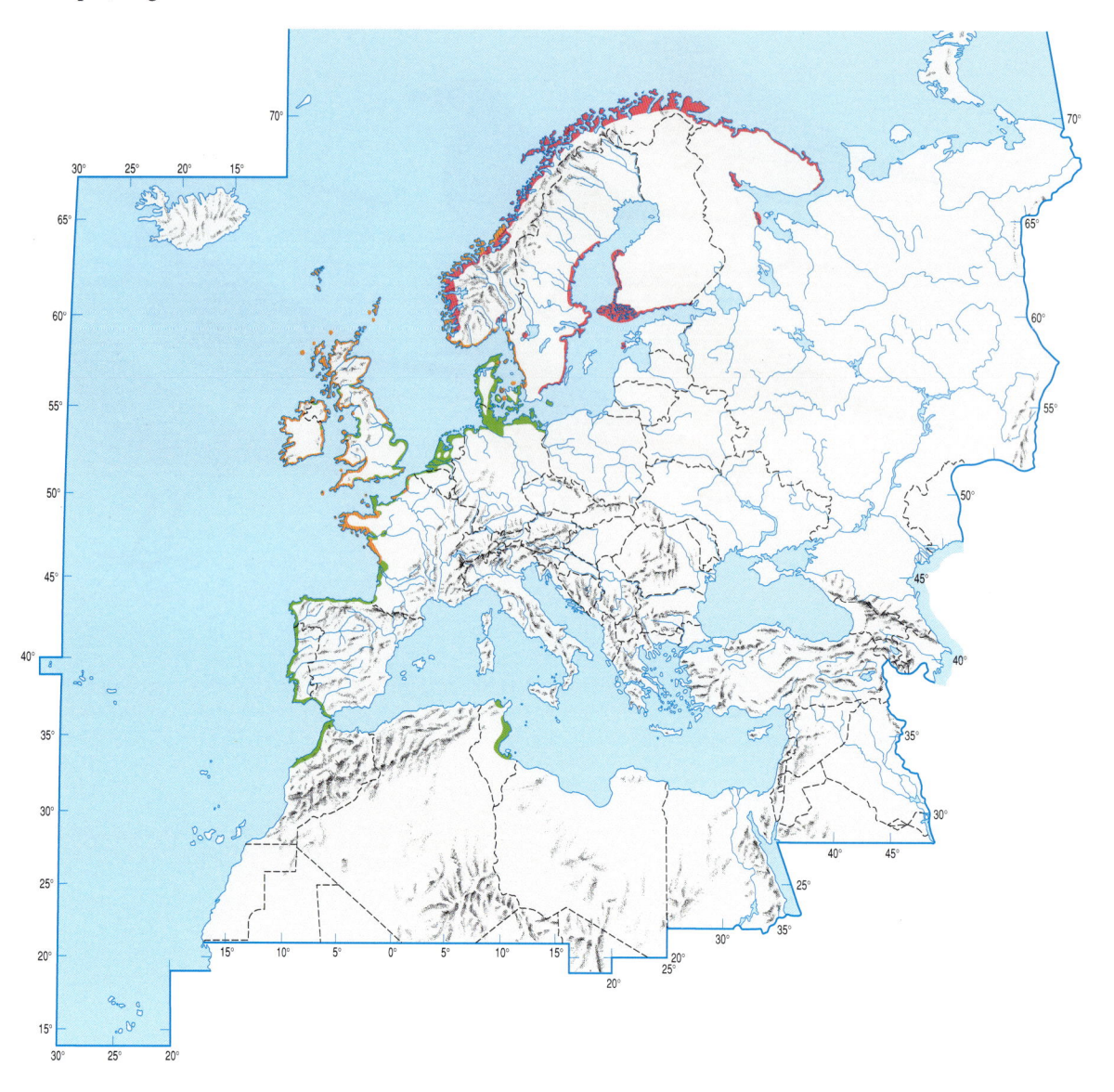

3–10 pairs in 1991; stable. RUSSIA. 100–1000 pairs. Presumably refers only to this species.

Movements. Faeroes population winters almost exclusively within breeding range. Most 1st-year birds from Fair Isle leave in autumn, and move south to Scotland or even Netherlands. Populations of mainland Britain and Ireland are basically resident with local dispersive movements, birds appearing away from breeding areas from September. Baltic and northern populations vacate breeding areas in winter, moving between WSW and south to Britain, Iberia, and western Mediterranean. Some birds reach North Africa.

Food. Mainly invertebrates. Feeds on ground among tide wrack and rocks, especially on isopods and small marine molluscs, rarely making short pursuits to catch insects in flight. Frequently wades in sea water, following waves as they retreat.

(FACING PAGE) Red-throated Pipit *Anthus cervinus* (p. 1084): **1** 1st winter.
Pechora Pipit *Anthus gustavi* (p. 1081): **2** 1st winter.
Meadow Pipit *Anthus pratensis pratensis* (p. 1082): **3** 1st winter.
Tree Pipit *Anthus trivialia* (p. 1079): **4** 1st winter.
Olive-backed Pipit *Anthus hodgsoni* (p. 1077). *A. h. yunnanensis*: **5** 1st winter.
Rock Pipit *Anthus petrosus* (p. 1089). *A. p. petrosus*: **6** ad non-breeding. *A. p. littoralis*: **7** ad breeding.
Berthelot's Pipit *Anthus berthelotii* (p. 1075): **8** ad fresh.
Buff-bellied Pipit *Anthus rubescens* (p. 1092). *A. r. rubescens*: **9** ad non-breeding.
Water Pipit *Anthus spinoletta* (p. 1086): **10** ad non-breeding.
Richard's Pipit *Anthus novaeseelandiae* (p. 1070): **11** 1st winter.
Blyth's Pipit *Anthus godlewskii* (p. 1072): **12** 1st winter.
Long-billed Pipit *Anthus similis* (p. 1076): **13** ad fresh.
Tawny Pipit *Anthus campestris* (p. 1072): **14** ad breeding, **15** 1st winter.

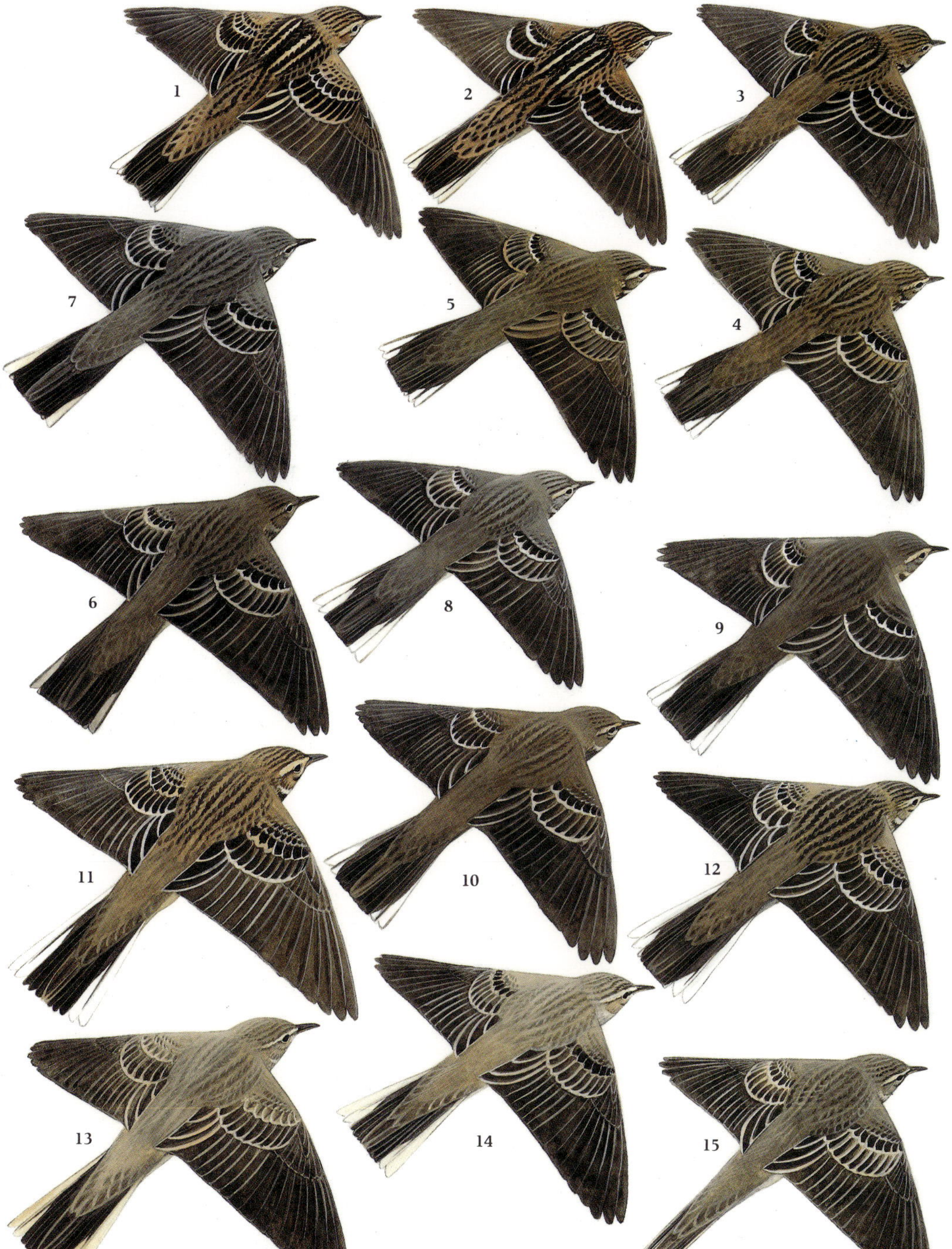
1
2
3
7
5
4
6
8
9
11
10
12
13
14
15

Social pattern and behaviour. Normally territorial throughout the year, at least in parts of range. Rarely in regular flocks except on migration. Highly territorial in breeding season, when normally monogamous, but evidence of regular polygyny in Sweden. Breeding behaviour and displays, including song-flight, very similar to Meadow Pipit.

Voice. Song recalls Meadow Pipit but much louder, more musical and more tinkling, with stronger terminal trill. Commonest call 'phist', 'feest', or 'weesp', less of a cheep than Meadow Pipit with less squeaky, fuller, yet still sibilant tone.

Breeding. SEASON. Britain and Ireland: egg-laying from early or mid-April. Scandinavia: from mid-May in south but not before June in north. 2 broods in south of range, 1 in north. SITE. In hole or hollow in cliff, from near base to top, or in bank or under thick vegetation, never far from shore. Nest: as Water Pipit but with inclusion of seaweed as material. EGGS. Sub-elliptical, smooth and glossy; grey-white, finely but heavily spotted olive-brown and grey, sometimes with cap at broad end. Clutch: 4–6. INCUBATION. 14–15 days. FLEDGING PERIOD. 16 days.

Wing-length: ♂ 87–96, ♀ 80–90 mm.
Weight: ♂♀ mostly 20–27 g.

Geographical variation. Two main races, nominate *petrosus* (most of Britain, Ireland, France) and *littoralis* (Fenno-Scandia, including Denmark, and north-west Russia); and two races with limited ranges, *meinertzhageni* (Outer Hebrides) and *kleinschmidti* (Faeroes, Shetland, Fair Isle). *A. s. petrosus* closely similar to *littoralis* in non-breeding plumage, though fringes of mantle and scapulars perhaps slightly more pure olive. Rather different in breeding plumage, however: though pre-breeding moult almost as extensive as in *littoralis*, new breeding feathers rather similar to non-breeding, showing hardly any reduction of spotting on chest. *A. s. meinertzhageni* similar to *petrosus* in non-breeding plumage, but underparts slightly yellower (like *kleinschmidti*, but upperparts darker). *A. s. kleinschmidti* has virtually no supercilium and pale wedges of outer tail-feathers brown-white; in non-breeding more yellow-olive on upperparts than other races.

Buff-bellied Pipit *Anthus rubescens*

PLATES: pages 1087, 1091

DU. Pacifische Waterpieper FR. Pipit farlousane GE. Pazifikpieper IT. Spioncello del Pacifico
RU. Американский конек SP. Bisbita acuático americano

Field characters. 16–17 cm; wing-span 21.5–26 cm. Nearctic race slightly smaller than Water Pipit, approaching size of Tree Pipit and Meadow Pipit; Siberian race overlaps with Water Pipit. Medium-sized to quite large, robust pipit, with appearance convergent with Fenno-Scandian race of Rock Pipit, Water Pipit, and buffier morphs of Meadow Pipit. Differs from first two in combination of white outer tail-feathers and sharp, blackish spots and streaks on always buff underparts, present in all plumages but most marked in 1st autumn and winter (when Rock Pipit has dullest, most heavily marked, almost blotched underparts and Water Pipit shows much paler, almost white ground from throat to vent). Breeding adult also lacks bluish tone to head and mantle (typical of Water Pipit). Siberian race has consistently rather brown legs.

Identification of Nearctic race (in Britain and Ireland) and of Siberian race (in Middle East) being made with increasing confidence in autumn and winter, since pattern of underparts diagnostic then (but in spring appearance of underparts of some Rock Pipits of Fenno-Scandian race closely similar, differing only in greater weight of streaking below). Resemblance to Meadow Pipit heightened by similarity of call and, in Siberian race, paler legs.

Habitat. Arctic, subarctic, and alpine zones throughout breeding range; in north in rocky tundra, birch and dwarf shrub thickets, scree, stream courses, bogs, etc., and further south in alpine meadows, fell fields, steppe, etc. up to *c.* 4000 m. On migration and in winter quarters in low-altitude open places often without vegetation, such as shores, sand dunes, paddyfields, mudflats, or fields. Feeds on ground, often in flocks in winter, and also nests on ground usually near or under rock or vegetation.

Distribution. Breeds from Lake Baykal in southern Siberia east to Sakhalin island, Kamchatka, Chukotskiy peninsula, Aleutians, Alaska, northern Canada (south to Nova Scotia in east), to western Greenland, and south to Mexican border in USA; absent roughly east of Rocky Mountains and south of Hudson Bay. Race *japonicus* possibly breeds further west in Asia than at present known since regular though scarce passage migrant and winter visitor in Israel.

Accidental. Iceland, Britain, Ireland, Germany (Helgoland), Italy, Egypt. One Italian bird was race *japonicus*; all other European records most likely refer to nominate *rubescens*.

Movements. Migratory and partially migratory; nominate *rubescens* moves to lower altitude and also south to southern

(FACING PAGE) Yellow Wagtail *Motacilla flava* (p. 1094). *M. f. flavissima*: **1** ad ♂ breeding, **2** ad ♀ breeding, **3** ad ♂ non-breeding, **4** juv. *M. f. flava*: **5** ad ♂ breeding, **6** ad ♀ breeding, **7** ad ♂ non-breeding, **8** juv. *M. f. thunbergi*: **9** ad ♂ breeding, **10** ad ♀ breeding, **11** ad ♂ non-breeding. *M. f. cinereocapilla*: **12** ad ♂ breeding, **13** ad f breeding, **14** ad ♂ non-breeding, **15** juv. *M. f. feldegg*: **16** ad ♂ breeding, **17** ad ♀ breeding, **18** ad ♂ non-breeding, **19** juv.

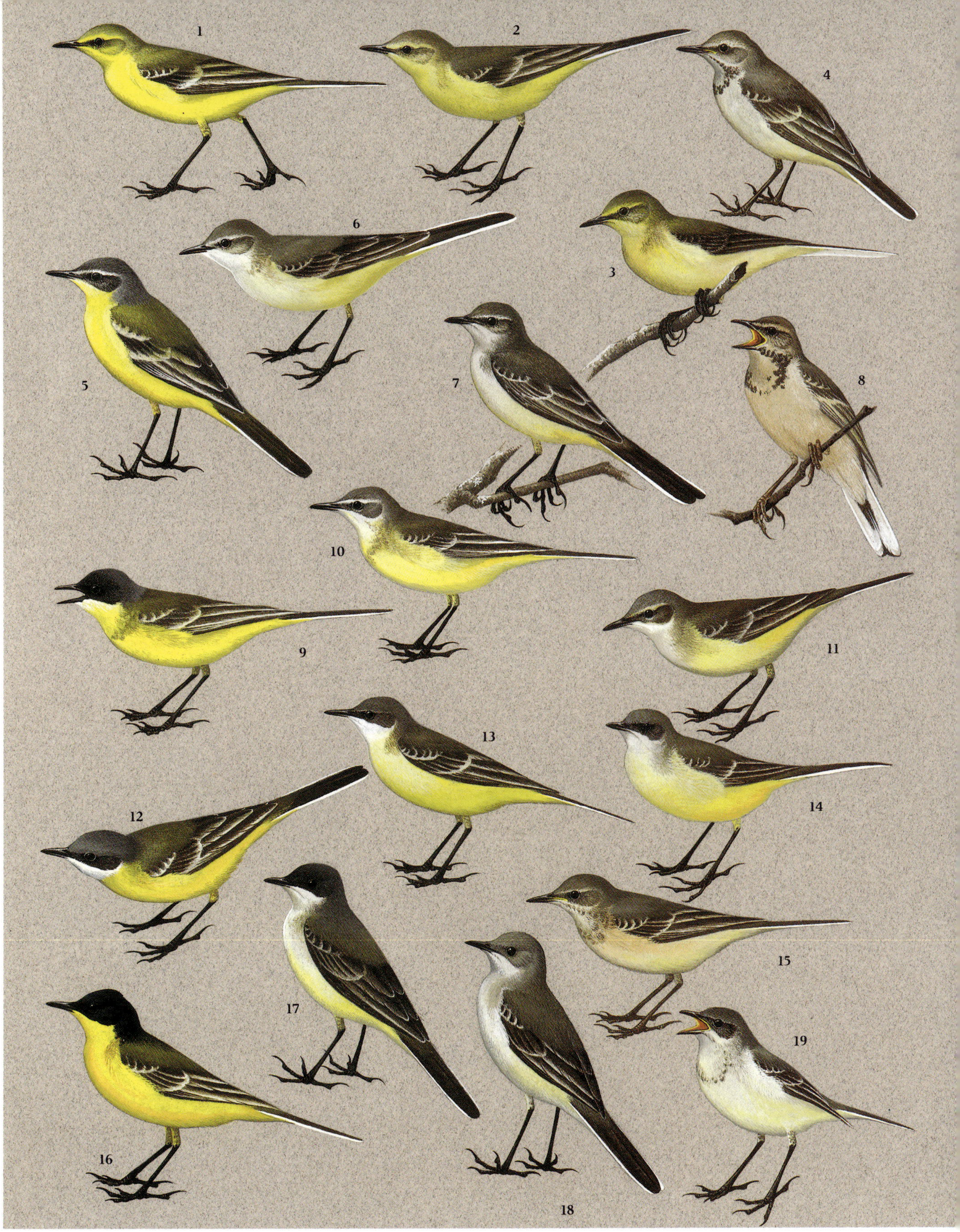
1
2
4
6
3
5
7
8
10
9
11
13
14
12
15
17
19
16
18

USA (furthest north on coasts, i.e. to Washington state in west, Virginia in east), and throughout Mexico to parts of Central America. Race *pacificus* migrates to parts of Rocky Mountains in south-west USA, race *alticola* probably to southern USA approximately west of Texas, and Mexico. In Newfoundland, nominate *rubescens* departs mid-September to October, and arrives from late April, wintering in Mexico from late September to mid-May; in Alaska, *pacificus* leaves towards end of August, returning May–June. Race *japonicus* winters in Pakistan and northern India, south-east Asia, and Japan, but any further destination of small numbers moving through Israel unknown. Arrives Japan late October to November, departing late March to May, and recorded in southern Israel mainly in November (a few, perhaps most, remaining all winter) and on return passage up to April.

Wing-length: ♂♀ 76–92 mm.

Yellow Wagtail *Motacilla flava*

PLATES: pages 1093, 1095, 1107

Du. Gele Kwikstaart   Fr. Bergeronnette printanière   Ge. Schafstelze   It. Cutrettola
Ru. Жёлтая трясогузка   Sp. Lavandera boyera   Sw. Gulärla

Field characters. 17 cm; wing-span 23–27 cm. Somewhat smaller than Pied Wagtail, with sleeker form and 20% shorter tail; noticeably less attenuated than Grey Wagtail, with 25% shorter tail. Smallest, most compact of west Palearctic wagtails, with form and silhouette more like large pipit than any of the others. Plumage of both adult and (most) 1st-winter birds basically yellow below and on patterned edges of wing-feathers. Adult breeding ♂♂ of the many races differ in pattern of green, bluish grey, and black crowns, variable yellow or white supercilia, and variable white or yellow chins and throats. ♀♀ and immatures may show hints of ♂♂'s pattern but many racially inseparable. In winter even ♂♂ may lose racial characters. Identification also complicated by frequent occurrence of racial hybrids and unusually pale birds.

Wagtail with complex systematics and morphology, subject to both confusion with other species and doubts in racial identifications. British and extreme west European race *flavissima* most distinctive, with green and yellow head, but blue- and grey-headed forms less easy to separate. Unlikely to be mistaken for Pied Wagtail, except as juvenile when the 2 species are closest in appearance. Also unlikely to be mistaken for Grey Wagtail, since that species is yellow-rumped and has long white wing-band in addition to very long tail. Much more serious risk of confusion is with Citrine Wagtail in both adult and sub-adult plumage (see that species). Flight bounding but not as 'shooting' as in White Wagtail and Grey Wagtail. Flight silhouette recalls longer-tailed pipits. Gait as Pied Wagtail but stance on ground less horizontal; much addicted to following cattle.

Habitat. Breeds in west Palearctic from lower middle to high latitudes, near July isotherm of 10°C, in arctic tundra and subarctic, boreal, temperate, steppe, and Mediterranean zones, mainly continental but marginally oceanic, largely on level or gently sloping lowlands, but in Caucasus and elsewhere in damp meadows on river banks and lake shores up to 2000–2500 m. In west of range (e.g. Germany, Switzerland) confined however to lowlands, avoiding mountains and broken, arid, sandy, stony, or bare ground, forests, and enclosed landscapes. In breeding season, occupies fringes of wetlands, such as riversides, lakesides, upper levels of salt-marshes, floodlands, moist pastures, water-meadows, excavations, subsidences, grazed fens with scattered small trees, sedge marshes, slacks within sand-dunes, and similar habitats both natural and artificially managed, especially where herbage low, dense, luxuriant, moist, and preferably near shallow surface water. In some parts of range will also breed on relatively dry farmland, heath, moor, and upland areas. In winter in Africa, as to some extent in breeding season, exploits a niche left open by native species, in close association with large herbivores, both wild and domestic.

Distribution. No longer breeds regularly in Ireland, and range decreased (presumably at least in part because of habitat changes, though not clear whether drought in winter quarters also implicated) Britain, France, Germany, Denmark, Norway, Czech Republic, Slovakia, and Ukraine. Slight expansion reported Netherlands and Switzerland. BRITAIN. *M. f. flavissima* is main race. Marked range contraction since 1930s affecting Cornwall and much of Scotland. Nominate *flava* breeds occasionally. IRELAND. *M. f. flavissima* is main race. Formerly 2 large colonies, but extinct by 1941. Occasional breeding records later, but none since 1983. FRANCE. Apparently some contraction since 1936. Nominate *flava* is main race, but *flavissima* breeds

(FACING PAGE) Yellow Wagtail *Motacilla flava*. Ad ♂ breeding: **1** *flavissima*, **2** *lutea*, **3** *leucocephala* (Mongolia, Sinkiang), **4** *beema*, **5** *M. f. flava*, **6** *simillima* (far eastern Asia), **7** *iberiae*, **8** *pygmaea*, **9** *thunbergi*, **10** *cinereocapilla*, **11** nominate *flava-feldegg* intergrade ('*dombrowskii*'), **12** *feldegg*. *M. f. iberiae*: **7** ad ♂ breeding, **13** ad ♀ breeding, **14** ad ♂ non-breeding, **15** juv, **16** 1st winter ♂. *M. f. pygmaea*: **8** ad ♂ breeding, **17** ad ♀ breeding, **18** ad ♂ non-breeding, **19** juv, **20** 1st winter ♂. Variants, ad ♂ breeding: **21** 'old' *flavissima*, **22** 'brown' *flavissima*, **23** 'blue' *flavissima*, **24** '*superciliaris*' (variant of *thunbergi*). 1st ad ♂ non-breeding: **16** *iberiae*, **20** *pygmaea*, **25** *thunbergi*, **26** *cinereocapilla*, **27** nominate *flava-feldegg* intergrade ('*dombrowskii*'), **28** *flavissima*, **29** *lutea*, **30** *beema*, **31** *simillima*, **32** *leucocephala*, **33** nominate *flava*, **34** *feldegg*.
Citrine Wagtail *Motacilla citreola* (p. 1098), 1st winter ♂: **35** *M. c. citreola*, **36** *werae*.
Pied Wagtail and White Wagtail *Motacilla alba* (p. 1103), 1st winter ♂: **37** *M. a. alba* (White Wagtail), **38** *yarrellii* (Pied Wagtail).

1
3
5
2
4
6
7
11
9
8
10
12
13
17
21
22
14
18
15
19
23
24
20
26
16
25
27
28
29
30
32
34
31
33
35
36
37
38

coast from Somme to Finistère, and *iberiae* (with tendency towards *cinereocapilla* in appearance) in Pays Basque and on Mediterranean coast. NETHERLANDS. Nominate *flava* is main race and range has increased locally; overlaps with *M. f. flavissima* on coast, but no interbreeding (separated by habitat). GERMANY. Some range contraction in west. Main race is nominate *flava*, but *flavissima* breeds sporadically North Sea islands, and *cinereocapilla* occurs, probably with hybridization in south-west. DENMARK. Range decreased. Nominate *flava* is main race, with *flavissima* breeding (mostly west coast of Jylland) since 1970. NORWAY. Range of main race *thunbergi* stable, while that of nominate *flava* and *flavissima* has decreased. SWEDEN. *M. f. thunbergi* breeds north of 60°30′N, nominate *flava* south of there. FINLAND. *M. f. thunbergi* predominates over nominate *flava*, interbreeding frequently. CZECH REPUBLIC and SLOVAKIA. Nominate *flava* is main race, but birds resembling *cinereocapilla* breed in small numbers central Slovakia. AUSTRIA. Nominate *flava* (main race) intergrades with *cinereocapilla* in south. *M. f. feldegg* almost annual breeder Neusiedler See area and elsewhere since 1969. ITALY. *M. f. cinereocapilla* is main race, *feldegg* local, nominate *flava* rare. RUMANIA. Both nominate *flava* and *feldegg* breed, latter more widespread in south. TURKEY. *M. f. feldegg* widespread, though apparently uncommon Black Sea Coastlands (except some deltas) and in plains of south-east. SYRIA. Range difficult to determine exactly. JORDAN. Breeds (small population) only Azraq, where may be restricted to wet years. LIBYA. Possibly breeding Benghazi 1993 where 3 juveniles seen 25 July. MAURITANIA. Banc d'Arguin: breeds several islands.

Accidental. Spitsbergen, Bear Island, Iceland, Azores, Madeira, Cape Verde Islands.

Beyond west Palearctic, extends throughout northern Asia south to Iran, north-west Himalayas, and northern China; also breeds Alaska and northern Yukon.

Population. Widespread decrease affecting Britain, Benelux countries, Germany, Fenno-Scandia, Lithuania, Czech Republic, Slovakia, Austria, Spain, Greece, Croatia, and Ukraine. Mostly stable elsewhere, though increase reported Sweden, Switzerland, and Slovenia. BRITAIN. 50 000 territories 1988–91. Some decline in 1980s affecting mainly southern England and Wales; marked decrease earlier in Scotland. FRANCE. 10 000–100 000 pairs in 1970s. BELGIUM. 8500–11 000 pairs 1989–92; decrease some areas, increase in others. LUXEMBOURG. 40–60 pairs; decrease since 1960s. NETHERLANDS. Nominate *flava*: 40 000–70 000 pairs 1979–85; decrease, with some local increases. *M. f. flavissima*: 200–350 pairs 1979–85; decrease. GERMANY. 90 000 pairs in mid-1980s. In east, 56 000 pairs in early 1980s. DENMARK. 2000–21 000 pairs 1987–8. NORWAY. 100 000–500 000 pairs of *thunbergi* 1970–90; stable. Marked decrease reported for nominate *flava* (50–100 pairs) and *flavissima* (10–20 pairs). SWEDEN. 100 000–200 000 pairs in late 1980s. *M. f. thunbergi* by far the more numerous and has probably increased; nominate *flava* 10 000 pairs at most. FINLAND. 500 000–800 000 pairs in late 1980s. ESTONIA. 10 000–20 000 pairs 1991; now stable following decline in 1950s–60s. LATVIA. 10 000–25 000 pairs in 1980s. POLAND. 100 000–200 000 pairs. CZECH REPUBLIC. 600–1200 pairs 1985–9. SLOVAKIA. 2500–4000 pairs 1973–94. HUNGARY. 30 000–50 000 pairs 1979–93. AUSTRIA. Marked decrease in east, but increasing Bodensee area. SWITZERLAND. 100–150 pairs 1985–93. SPAIN. 70 000–240 000 pairs. PORTUGAL. 10 000–100 000 pairs 1978–84. ITALY. 20 000–40 000 pairs 1983–93. GREECE. 10 000–20 000 pairs. ALBANIA. 1000–3000 pairs in 1981. YUGOSLAVIA: CROATIA. Perhaps 14 000–18 000 pairs. SLOVENIA. 200–300 pairs. BULGARIA. Perhaps 500 000–1 million pairs. RUMANIA. Nominate *flava*: 1–1.5 million pairs 1986–92. *M. f. feldegg*: perhaps 10 000–12 000 pairs, though difficult to estimate. RUSSIA. 1–10 million pairs. BELARUS'. 470 000–530 000 pairs in 1990. UKRAINE. 47 000–50 000 pairs in 1988. MOLDOVA. 8000–10 000 pairs in 1988. AZERBAIJAN. Common. TURKEY. 100 000–1 million pairs; trend unknown. CYPRUS. Scarce; probably stable. ISRAEL. *c.* 15 pairs in Hula valley. EGYPT. Common (in range 10 000–100 000 pairs). TUNISIA. Locally common. MOROCCO. Common. MAURITANIA. 12–15 pairs in 1995.

Movements. Most populations migratory, wintering Afrotropics, India, and south-east Asia. Egyptian race largely resident, and some parts of breeding range in north-west Africa and southern Spain occupied through the winter, with possibility that some individuals are resident.

Several factors make this a particularly well documented migrant: large populations; conspicuous (mostly diurnal) movement; use of huge communal roosts, both on migration and in winter, facilitating ringing; assumption by ♂♂ of racially distinct breeding plumage shortly before spring migration. On the other hand, confusion can arise through racial intermediates and disjunct pattern of geographical variation. Precise wintering areas of the various races are not well established but in the main lie between south-east and south-west of respective breeding areas.

Movement broad-front in both spring and autumn, with numerous sightings of migrants at sea in all areas. Autumn passage in Switzerland has been noted as early as late July but main passage begins second half of August and peaks through September usually to end abruptly in early October, though individuals have been noted still passing in first third of November. At Straits of Gibraltar, passage extends from early August to early November peaking mid-September. Arrives in Afrotropics in late September, further south in October. Movement north in spring, after build-up of fat just south of Sahara, is also on broad front, starting in March and extending to early May. ♂♂ reach breeding grounds before ♀♀; arrivals are from late March in south, west, and much of central Europe, from mid-April in Moscow area, and from early May or early June in Lapland.

Many records occur of birds resembling a particular race well outside that race's normal range, but some (at least) of these are part of the species' normal variability and do not necessarily indicate vagrancy. Birds showing the characters of several races have been recorded in Britain, for example, mainly in spring and sometimes well outside their normal range: continental nominate *flava* occurs regularly and has bred occasionally.

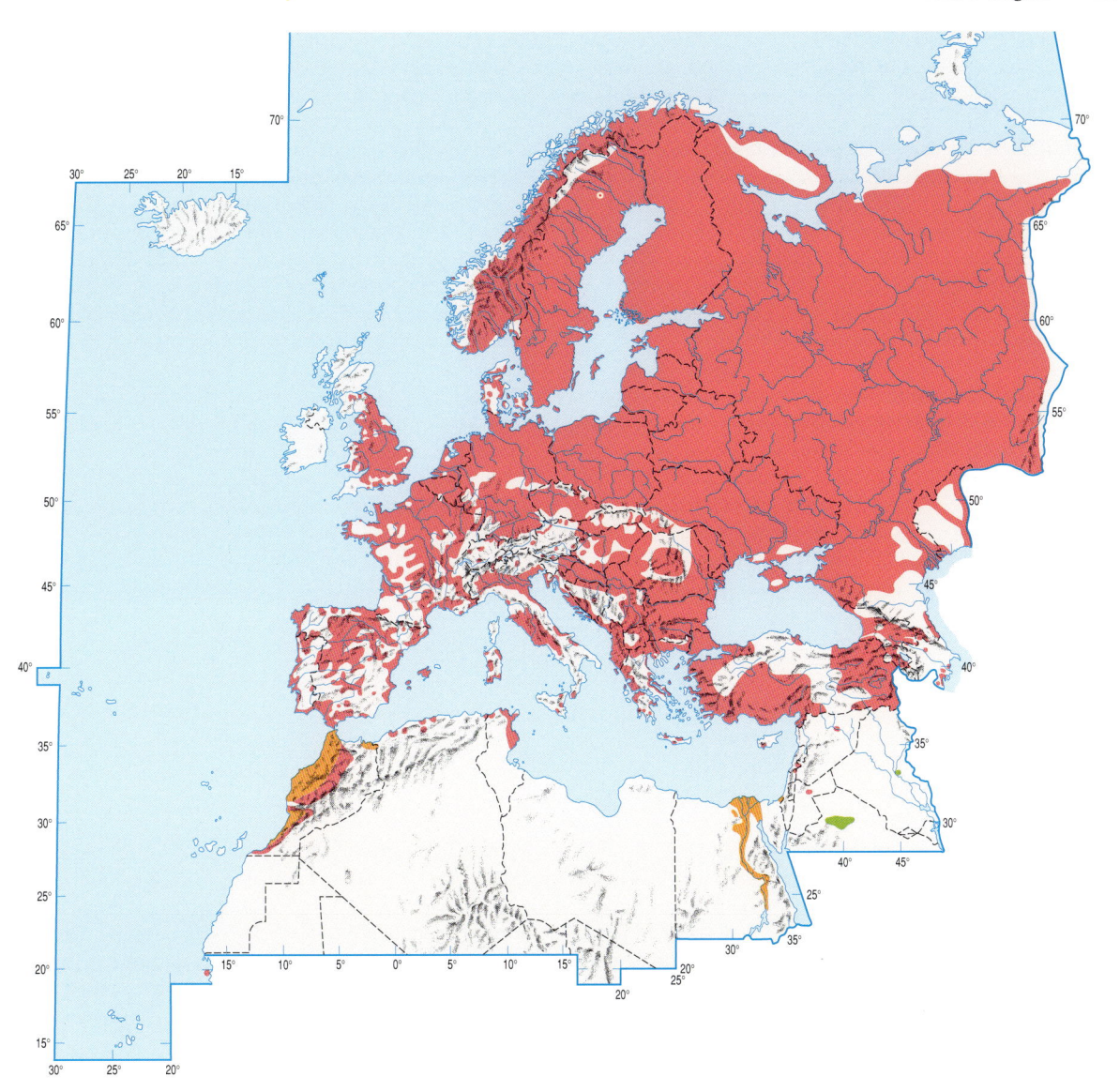

Food. Small invertebrates. 3 main foraging techniques. (1) Picking. Picks items from ground or water surface while walking. (2) Run-picking. Makes quick darting run at prey, picking it up either from surface or as it takes off. (3) Fly-catching. Makes short flight from ground or perch, catching prey in mid-air—either in bill or by knocking it down with wings. Occasionally takes insects from plants in hovering flight, or flies low over water snatching insects from surface. Often feeds in association with grazing cattle and sheep, taking insects disturbed by animals or blood-sucking species from animals themselves.

Social pattern and behaviour. Gregarious outside breeding season. In winter quarters, dispersion varies with food supply, but majority form small flocks. Monogamous mating system the rule; pair-bond of seasonal duration. Breeding territories tend to form neighbourhood groups and overlap, especially when feeding young. ♂ sings from perch and often in display-flight. Song-flight varies in length and duration; essentially a series of long wavering undulations.

Voice. Song of ♂ not very loud, but often prolonged; a rapid sequence of twittering sounds, comprising loosely connected syllables of contact-call-type, given from ground or elevated perch, or in flight. Contact-alarm call, most commonly heard, differs slightly between races: a shrill, drawn-out, quite musical but plaintive 'tsweep', with slight terminal accent.

Breeding. SEASON. Northern Scandinavia: most eggs laid June. Southern Scandinavia: laying from last week of May. Britain and Ireland: first eggs mid- to late April, main laying period May, last eggs found early August. Southern and south-east Europe: from end of April to early June. North Africa: from end of April to end of May. 1–2 broods, depending on latitude.

Nest: cup of grass leaves and stems placed in shallow scrape, lined with hair, wool, or fur. EGGS. Sub-elliptical, smooth and glossy; grey-white to buff, densely spotted various shades of brown, often with dark hair-streaks. Clutch: 4–6 (3–8). INCUBATION. 11–13 days. FLEDGING PERIOD. Fledge at *c.* 16 days, but leave nest at 10–13 days.

Wing-length: *M. f. flava:* ♂ 77–86, ♀ 74–82 mm. Other races not markedly different except *pygmaea* (Egypt): ♂ 74–79, ♀ 70–75 mm.

Weight: *M. f. flava:* ♂♀ mostly 14–21 g; spring migrants in Africa, before departure north, to 28 g.

Geographical variation. Marked and complex; mainly involves colour of ♂ breeding plumage, less so other plumages; also (but scarcely) size. 2 complexes recognized, often considered separate species: (1) *lutea* complex (*lutea, flavissima, extralimital taivana*); (2) *flava* complex (all other races). Every member of *lutea* complex overlaps partly or fully with *flava* complex, apparently with limited interbreeding, though some gene-flow between the complexes occurs, and members of *lutea* complex are perhaps not closely related to each other as measurements and structure of each are closer to the neighbouring member of *flava* complex than to other members of *lutea* complex. *Flava* complex subdivided into 3 groups: grey-headed *thunbergi* group in north, blue-headed nominate *flava* group in mainly temperate latitudes, and black-headed *feldegg* group in south, from Balkan countries to eastern Kazakhstan. Each of these groups sometimes considered a separate species also, but as they are connected by hybridization zones of variable width, better combined into a single highly polytypic species. Breeding ♂♂ of all races readily separable in colour, apart from birds of unstable local populations in hybridization zones.

Following races breed in west Palearctic: *lutea* (basin of lower Volga north to Kazan' and Perm', eastwards); *flavissima* (Britain and locally on continental coast of north-west Europe); nominate *flava* (most of Europe); *cinereocapilla* (Italy and north-west Yugoslavia); *iberiae* (south-west France, Iberia, and north-west Africa); *pygmaea* (Egypt); *thunbergi* (Norway east to northern Russia); *feldegg* (Balkans east to Caspian); *beema* (lower Volga).

Citrine Wagtail Motacilla citreola

PLATES: pages 1095, 1099, 1107

DU. Citroenkwikstaart FR. Bergeronnette citrine GE. Zitronenstelze IT. Cutrettola testagialla orientale
RU. Желтоголовая трясогузка SP. Lavandera cetrina SW. Citronärla

Field characters. 17 cm; wing-span 24–27 cm. Slightly larger than Yellow Wagtail, with 10% longer tail; less deep-chested than White Wagtail. Most constant marks are slate-grey upperparts, and striking double white wing-bar and white fringes to tertials. Breeding ♂ shows fully yellow head and underparts; contrast of yellow head with black neck-shawl diagnostic (but this lost in winter). Breeding ♀ suggests dusky-backed Yellow Wagtail, and juvenile and 1st-winter birds recall pale Pied Wagtail; ♀ may lack yellow tint below breast and immature always does so but both show noticeably grey flanks. Distinction of 1st-winter bird from paler, particularly mutant grey Yellow Wagtails difficult; requires close observation of head pattern. In Citrine Wagtail, supercilium deep and white (often tinged buff in front and onto forehead) and continues around ear coverts as pale surround (if this obvious and complete, diagnostic). Beware, however, hybrids.

Thus important to study suspect Citrine Wagtail fully for form, voice, and behaviour; all offer useful additional clues. Flight silhouette more robust than Yellow Wagtail, with slightly longer tail. Flight action slightly more powerful, without hint of stalling so obvious in Yellow Wagtail. Gait and stance, perching, and feeding behaviour closer to Yellow Wagtail than to White Wagtail.

Habitat. Breeds from arctic and subarctic through boreal and temperate dry continental zones, and extralimitally to subtropics, ranging from sea-level to upper level of meadow vegetation at above 4500 m in Pamirs, where densest population occurs at 3500 m. In tundra belt, inhabits osier thickets on coast and on islands in large river deltas, perching often on bushes. Also by lakes on marshy-shrubby tundra, on wet sections of mountain tundra, and among willow bushes on tussocky mountain meadows. In south in upper part of forest belt and in mountains, often in very damp places; also in river valleys in peaty hummocky bogs or marshy meadows covered with sparse low shrubs. In winter frequents marshes, squelchy grassy margins of ponds, and irrigated ricefields; also occurs by rivers, canals, drains, swamps, pools, flooded areas, and reed-beds, never away from water.

Distribution. Notable and apparently still continuing westward range expansion in FSU and beyond. BRITAIN. Adult ♂ feeding 4 young 1976; ♀ not seen. GERMANY. First (unsuccessful) breeding attempt near Greifswald (Mecklenburg-Vorpommern) in 1996; sightings during passage periods include ringed bird probably from Polish population. SWEDEN. Adult ♂ feeding 3 young 1977; ♀ not seen and young not identified. FINLAND. Mixed breeding with Yellow Wagtail confirmed 6 times in south since 1983. First breeding by Citrine Wagtail pair 1991. ESTONIA. First recorded (single ♂) in 1990, same bird returning in 1991 and breeding successfully in mixed pair with Yellow Wagtail; another bird in 1993. LATVIA. First bred Jelgava 1993. At least 3 pairs bred 1994. LITHUANIA. First bred Žuvintas reserve in south in 1986, and 2 pairs there 1988–9. 3 pairs Čepkeliai in south in 1992. POLAND. First bred (at least 4 pairs) at 2 sites on Gulf of Gdańsk in 1994; total of 12–13 pairs bred at 3 sites 1995–6. CZECH REPUBLIC. Bred northern Moravia 1977. FSU. Has spread west in both southern and

Citrine Wagtail *Motacilia citreola*. *M. c. citreola*: **1** ad ♂ breeding, **2** ad ♀ breeding, **3** ad ♂ non-breeding, **4** 1st winter. *M. c. werae*: **5** ad ♂ breeding. Grey Wagtail *Motacilla cinerea* (p. 1100). *M. c. cinerea*: **6** ad ♂ breeding, **7** ad ♀ breeding, **8** ad ♂ non-breeding, **9** juv. *M. c. canariensis*: **10** ad ♂ breeding. *M. c. schmitzi*: **11** ad ♂ breeding.

(less conspicuously) northern parts of range. BELARUS'. First bred 1982; distribution still patchy. UKRAINE. Breeding at least from 1970s. TURKEY. First recorded 1964 and first bred 1981. Perhaps previously overlooked. 5 known breeding localities in eastern Anatolia since 1980s and breeding likely at a further 5. Pairs observed Sultan Marshes and Tuz Gölü perhaps indicate has spread further west. ISRAEL. Pair attempted to breed Eilat 1986.

Accidental. Iceland, Britain (annual), Channel Islands, France, Belgium, Netherlands, Germany, Denmark, Norway, Sweden (annual), Finland (annual), Estonia, Latvia, Lithuania, Poland, Czech Republic, Slovakia, Hungary, Austria (almost annual since mid-1980s), Switzerland, Spain, Balearic Islands, Italy, Greece (regular), Yugoslavia, Bulgaria, Rumania (probably regular on coast), Kuwait, Morocco.

Beyond west Palearctic, northern population extends east to *c.* 115°E, southern population east to *c.* 125°E, south to Afghanistan and Himalayas.

Population. RUSSIA. Perhaps 100 000–1 million pairs; stable. BELARUS'. 50–150 pairs in 1990. UKRAINE. Perhaps *c.* 200–400+ pairs in 1991; slight increase. TURKEY. 500–5000 pairs; trend unknown.

Movements. Migratory, wintering mainly in India and southeast Asia. Rather little information on timing of movements. Autumn departure from Russia starts mainly in early September. Northern birds arrive India in September and depart mostly March–April. Occurrence in Middle East evidently peripheral to main movement: scarce winter visitor and passage migrant, October–March.

Most vagrants to Britain occur September–November with highest numbers on Fair Isle (Scotland). Spring records in Europe have increased in recent years, in line with range expansion.

Food. Invertebrates, often aquatic. Usually forages in or near wet habitat. 3 foraging techniques. (1) Picks items from ground or water surface while standing or walking on ground or wading in shallow water—even up to belly, long legs being well adapted to this. Hunts prey flushed by grazing animals, bird walking around and between legs of cattle, etc. (2) Bird plunges head into water to catch insect larvae. (3) Snatches insects flying past with brief upward flutter.

Social pattern and behaviour. Quite gregarious during breeding season, markedly so at other times. Most aspects of behaviour similar to Yellow Wagtail, with which it regularly associates and has been occasionally recorded interbreeding.

Voice. Song of ♂ a sequence of phrases composed of units resembling contact-calls, variously combined together. Contact-calls variable; usually monosyllabic, sometimes disyllabic. Monosyllabic call widely agreed to be harsher, and perhaps shriller and shorter than Yellow Wagtail; to the practised ear, readily distinguishable. Rendered 'sreep', 'drreep', or 'sweeip', constant in pitch and with rasping quality reminiscent of Tree Pipit.

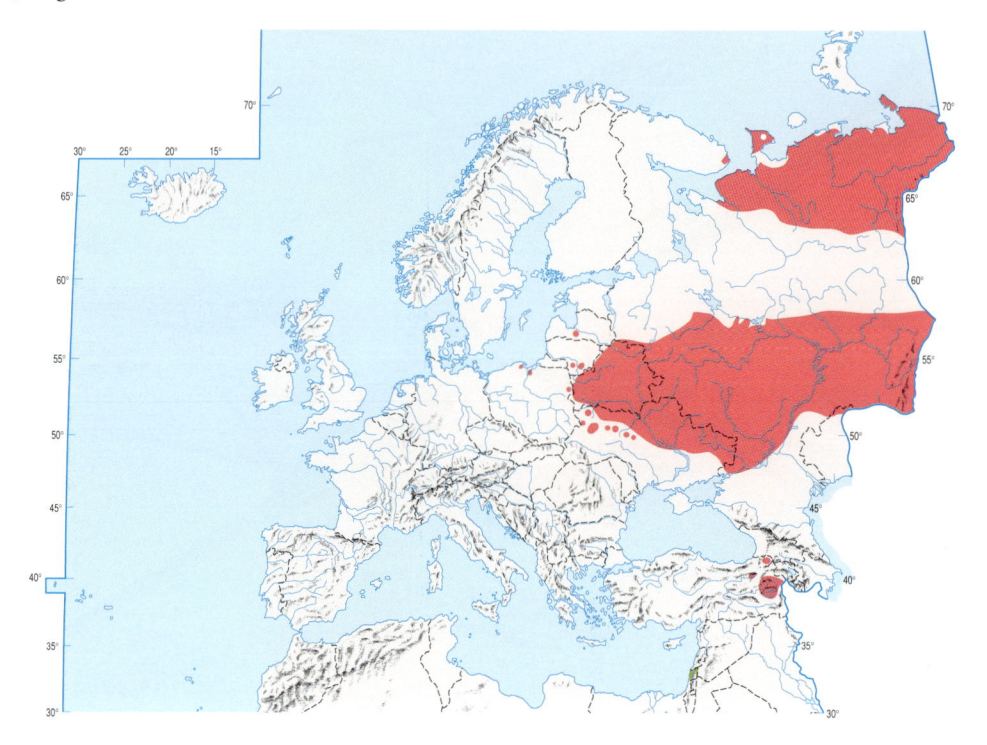

Breeding. SEASON. Northern Russia: eggs laid mid-June. Southern FSU: eggs laid late April to June. 1–2 broods. SITE. On ground in hollow in bank, or under thick vegetation or stone. Nest: cup of moss and plant leaves and stems, lined with hair, wool, and feathers, thicker in north of range. EGGS. Sub-elliptical, smooth and glossy; buff or pale grey, finely speckled grey or grey-brown; sometimes mottled light brown. Clutch: 4–6(–7). INCUBATION. 14–15 days. FLEDGING PERIOD. 13–15 days.

Wing-length: *M. c. citreola*: ♂ 85–90, ♀ 80–85 mm.

Weight: *M. c. citreola*: ♂♀ mostly 18–25 g.

Geographical variation. 2 races breeding in west Palearctic. Breeding ♂ of nominate *citreola* (northern Russia) has upperparts dark grey with olive tinge, broad black band across upper mantle and down to upper sides of breast, and olive-grey flanks. ♂ *werae* (southern plains) similar but upperparts slightly paler grey, black band on average narrower, flanks and sides of breast less extensively washed with paler olive-grey, and yellow of head and underparts slightly paler; a little smaller.

Grey Wagtail *Motacilla cinerea*

PLATES: pages 1099, 1107

DU. Grote Gele Kwikstaart   FR. Bergeronnette des ruisseaux   GE. Gebirgsstelze   IT. Ballerina gialla
RU. Горная трясогузка   SP. Lavandera cascadeña   SW. Forsärla

Field characters. 18–19 cm; wing-span 25–27 cm. More attenuated than any other west Palearctic wagtail, with exceptionally long tail (up to 35% longer than in Yellow Wagtail). Very graceful, lithe, slim wagtail, with almost constantly moving tail. Plumage essentially grey above, with olive-yellow rump, and yellow below; black wings show obvious white central band in flight. ♂ has black bib in summer. ♀ and immature have paler, more buff-white underbody but always yellow vent.

Bird in atypical surroundings may suggest Yellow Wagtail momentarily but no other west Palearctic wagtail is as long-tailed, or shows combination of yellow rump, single white bar on both upper- and underwing, and pale brown-flesh legs. Flight markedly bounding, with translucent panel in midwing remarkably obvious at times, and 'shooting' curve or fall exaggerated by marked acceleration and attenuation and 'whipping' or 'streaming' tail. Flight silhouette much more attenuated than any other west Palearctic wagtail. Gait noticeably delicate. Stance most horizontal of all west Palearctic wagtails, with tail even held up as bird dashes about.

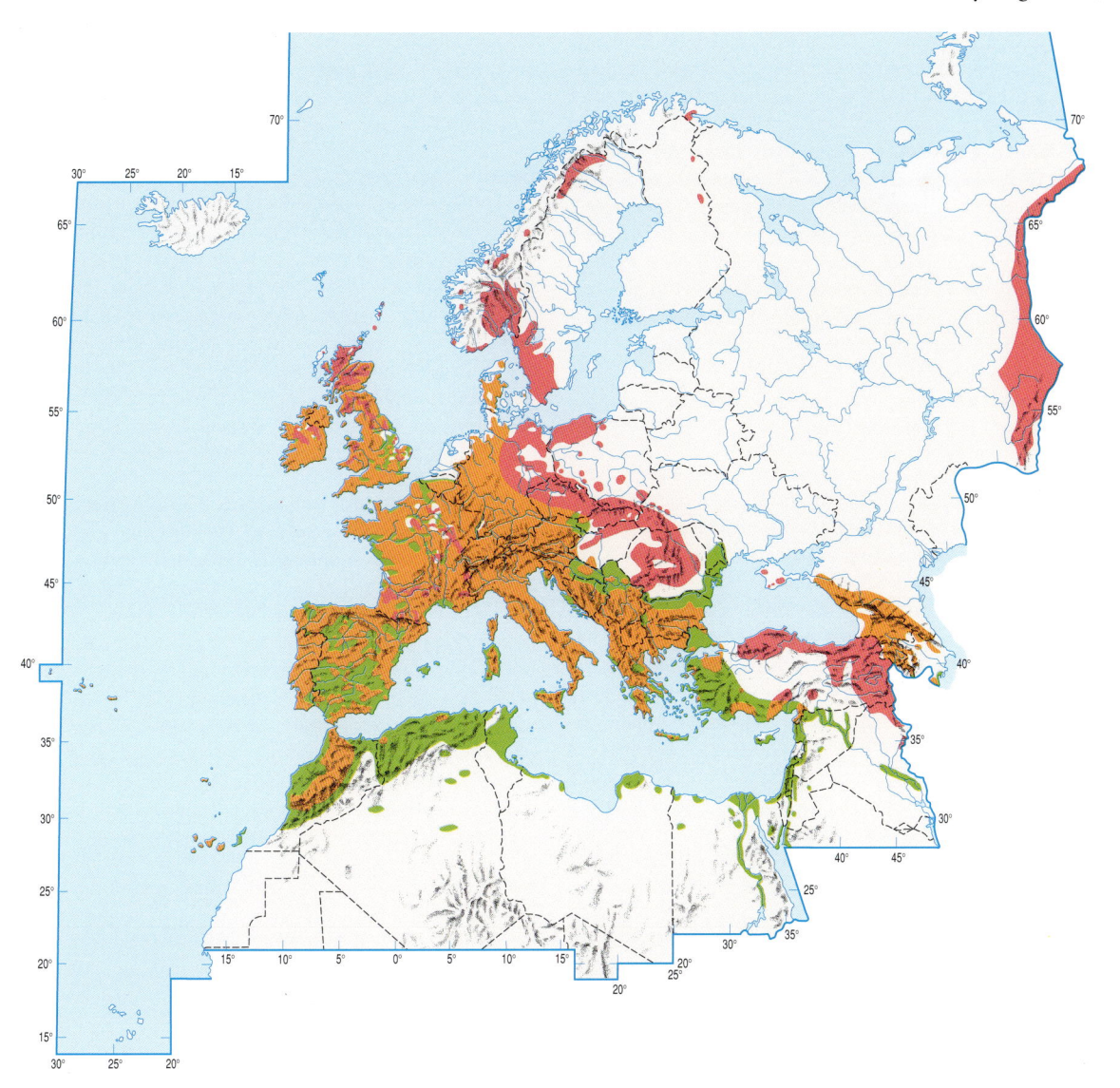

Habitat. In west Palearctic, occurs mainly in temperate middle and lower-middle latitudes, overlapping sparingly into boreal and Mediterranean. Typical breeding habitats include combination of: (1) fresh water, especially fast-running streams and rivers, but also canals, lowland streams, and margins of lakes, both oligotrophic and eutrophic; (2) rock slabs, boulders, vertical rock faces, shingle or gravel stream beds, or artefacts such as sluices, weirs, locks, culverts, walls, or roofs; (3) sheltering trees, shrubs, or dense herbage; (4) holes, ledges, or hollows for nesting. Upland and mountain streams often provide ample choice of such requirements; lowland streams, lakes, and reservoirs generally less; lowland canals, woodland pools, habitations with artificial tanks or drains, and other peripheral types of site are occupied locally or infrequently. In winter, shifts generally to lowlands, estuaries, coasts, and artificial situations such as sewage farms, retaining more marked attachment to water than congeners.

Distribution. Became established in central Europe shortly after 1850. Colonized several countries and spread in 20th century. BRITAIN. Some spread in eastern and southern England in 1950s, continuing more recently in east; irregular breeder Orkney and Outer Hebrides. NETHERLANDS. Probably colonized 1850–1900; first bred c. 1915. GERMANY. In east, most in uplands, but some in lowlands perhaps overlooked. DENMARK. First bred 1923 and spread (probably by 7% in recent years). NORWAY. First bred Oslo 1919. Marked and continuing increase. SWEDEN. First bred 1916. Following range expansion, now fairly common in south-west and breeding at low density in mountains of north-west (north to Lapland). FINLAND. First bred 1967 in south. May now breed regularly in north-east. ESTONIA. Probably irregular breeder, but proved only 1975. LATVIA. First bred Sigulda 1991, then near Cēsis 1993. POLAND. Confined to mountains mid-19th century, spreading to lowlands from 1865 and breeding there in 20th century. Recent small local decreases. TURKEY. Widespread, but nowhere common. Largely absent central and south-east Anatolia (where most watercourses run dry). CYPRUS. Perhaps breeds occasionally. CANARY ISLANDS. Widespread Gran Canaria, Tenerife, La Palma, and La Gomera; breeding yet to be confirmed other islands.

Accidental. Iceland, Faeroes, Lithuania.

Beyond west Palearctic, extends east to Pacific, south to Iran, Himalayas, and northern China.

Population. Apparently stable in most countries, though fluctuates with hard winters. BRITAIN. 34 000 pairs 1988–91. Decline since 1970s; fluctuating in eastern and southern England 1950s–60s. IRELAND. 22 000 pairs 1988–91. FRANCE. 10 000–100 000 pairs in 1970s. BELGIUM. 2300–3650 pairs 1989–92. LUXEMBOURG. 300–400 pairs. NETHERLANDS. 110–175 pairs mid- to late 1970s; 170–210 pairs 1991. Fluctuating. Decline 1940s–70s due to hard winters and water pollution. GERMANY. 82 000 pairs in mid-1980s. Another estimate (for early 1990s) of 35 000–60 000 pairs. DENMARK. 200 pairs in 1988; increase. NORWAY. 500–2500 pairs in 1994; marked increase. SWEDEN. 500–1000 pairs in late 1980s; slight decrease reported, but range expansion suggests tendency for numbers to increase in recent years. FINLAND. 5–20 pairs in late 1980s. POLAND. 2000–4000 pairs. Decrease in lowlands, except Pomerania. CZECH REPUBLIC. 20 000–40 000 pairs 1985–9. SLOVAKIA. 10 000–20 000 pairs 1973–94. HUNGARY. 400–600 pairs 1979–93. AUSTRIA. Common along smaller rivers. SWITZERLAND. 8000–11 000 pairs 1985–93. SPAIN. 13 500–17 000 pairs. PORTUGAL. 10 000–100 000 pairs 1978–84. ITALY. 20 000–50 000 pairs 1983–93. GREECE. 10 000–15 000 pairs. ALBANIA. 2000–5000 pairs in 1991. YUGOSLAVIA: CROATIA. 12 000–16 000 pairs. SLOVENIA. 5000–10 000 pairs. BULGARIA. 10 000–100 000 pairs. RUMANIA. 200 000–350 000 pairs 1986–92. RUSSIA. 1000–5000 pairs. UKRAINE. 3500–3800 pairs in 1988. AZERBAIJAN. Common. More numerous than White Wagtail in Zakataly district (Great Caucasus). TURKEY. 10 000–100 000 pairs. MOROCCO. Scarce to uncommon. AZORES, MADEIRA. Common.

Movements. Mainly a partial migrant, but wholly migratory or resident in some parts of range. Found in winter throughout most of European breeding range but some move to Africa as far south as southern Malawi. Winter range also includes Middle East, and birds from central and eastern Asia move to India and south-east Asia as far as New Guinea, with single record from northern Australia. Populations of Azores, Madeira, and Canary Islands are resident.

In autumn, movements within Europe of over 200 km mainly between north-west, south-west, and south-east in central and western Europe, and mainly around south-west in Britain. Coastal autumn passage occurs in Britain and Ireland, most marked at southern headlands and inshore islands and almost synchronous throughout, peaking mid-September. Northward passage in spring generally inconspicuous.

Food. Largely insects. 2 main foraging techniques. (1) Picking. Bird walks or runs, repeatedly picking up small items or chasing more mobile prey, with tail wagging and snapping up, down, or to one side, apparently to flush insects; may also wade in shallow water picking up tadpoles or lunging for small fish. (2) Flycatching. Flies from perch or ground; if from ground, flight a steep or near-vertical fluttering leap to maximum c. 6 m. In addition, may hover (intermittently or continuously) to obtain flying insects or prey from leaves or tree crevices, and may take prey in aerial-pursuit using zigzag flight, bird tumbling and circling, tail apparently acting as rudder.

Social pattern and behaviour. Outside breeding season, dispersion varies from solitary to relatively gregarious, depending on food supply. At some rich feeding sites, large numbers congregate but each defends individual-distance, birds thus appearing scattered. Breeding pairs territorial; territories usually linear, along streams, often not conspicuous. ♂ sings both perched and in song-flight. Sings repeatedly from elevated perch (e.g. tree, rock), sometimes quivers wings and ruffles rump feathers. Song-flight often compared to Tree Pipit: typically, ♂ descends parachute-fashion from high perch, wings outspread and held steady or fluttering, tail raised slightly, and rump exposed. Gives trilling song during descent, then contact-calls before landing on ground or low perch.

Voice. Song a staccato, then more melodious trill, with opening notes effectively an extension of commonest call—a high-pitched, metallic disyllable, 'tzitzi', staccato or stuttered, and shorter or more clipped than Pied Wagtail. Warning- and alarm-call a plaintive, drawn-out, rising '(t)weee'.

Breeding. SEASON. North-west Europe: first eggs laid in last few days of March; main laying period April–May, last eggs laid early August. Central and eastern Europe; first eggs second half of April. North Africa: laying from late March to May. Canary Islands: early March to June. 2 broods, occasionally 3. SITE. In hole or crevice in wall or bank, under bridge, or among tree roots. Nest: cup of grass, roots and small twigs, often with moss, lined hair; size variable, shaped to fit crevice. EGGS. Sub-elliptical, smooth and glossy; whitish, cream or grey-buff, faintly marked grey or grey-buff. Clutch: 4–6(3–7). INCUBATION. 11–14 days. FLEDGING PERIOD. 13–14 days.

Wing-length: *M. c. cinerea* ♂ 82–89, ♀ 80–86 mm.
Weight: *M. c. cinerea*: ♂♀ mostly 14–22 g.

Geographical variation. Slight; mainly involves tail length, depth of colour of body, extent of supercilium, and amount of black on tail. 4 races in west Palearctic; 3 in Atlantic islands, nominate *cinerea* elsewhere. *M. c. patriciae* (Azores), *canariensis* (Canary Islands), and *schmitzi* (Madeira) all slightly smaller than nominate *cinerea*, but tarsus and hind claw proportionately short in *canariensis* and bill long in *schmitzi* and (in particular) *patriciae*.

M. c. schmitzi is distinctly darker slate-grey on upperparts and ear-coverts than nominate *cinerea*; white supercilium reduced to narrow stripe behind eye, white stripe on lower cheeks narrow and indistinct; black of throat deeper; underparts often deep yellow; tail with more extensive black. *M. c. patriciae* similar to *schmitzi*, but differs in extent of black on tail.

Pied Wagtail/White Wagtail *Motacilla alba*
(Pied Wagtail refers to *M. a. yarrellii*, White Wagtail to all other races)

PLATES: pages 1095, 1105, 1107

Du. Witte Kwikstaart Fr. Bergeronnette grise Ge. Bachstelze It. Ballerina bianca
Ru. Белая трясогузка Sp. Lavandera blanca Sw. Sädesärla

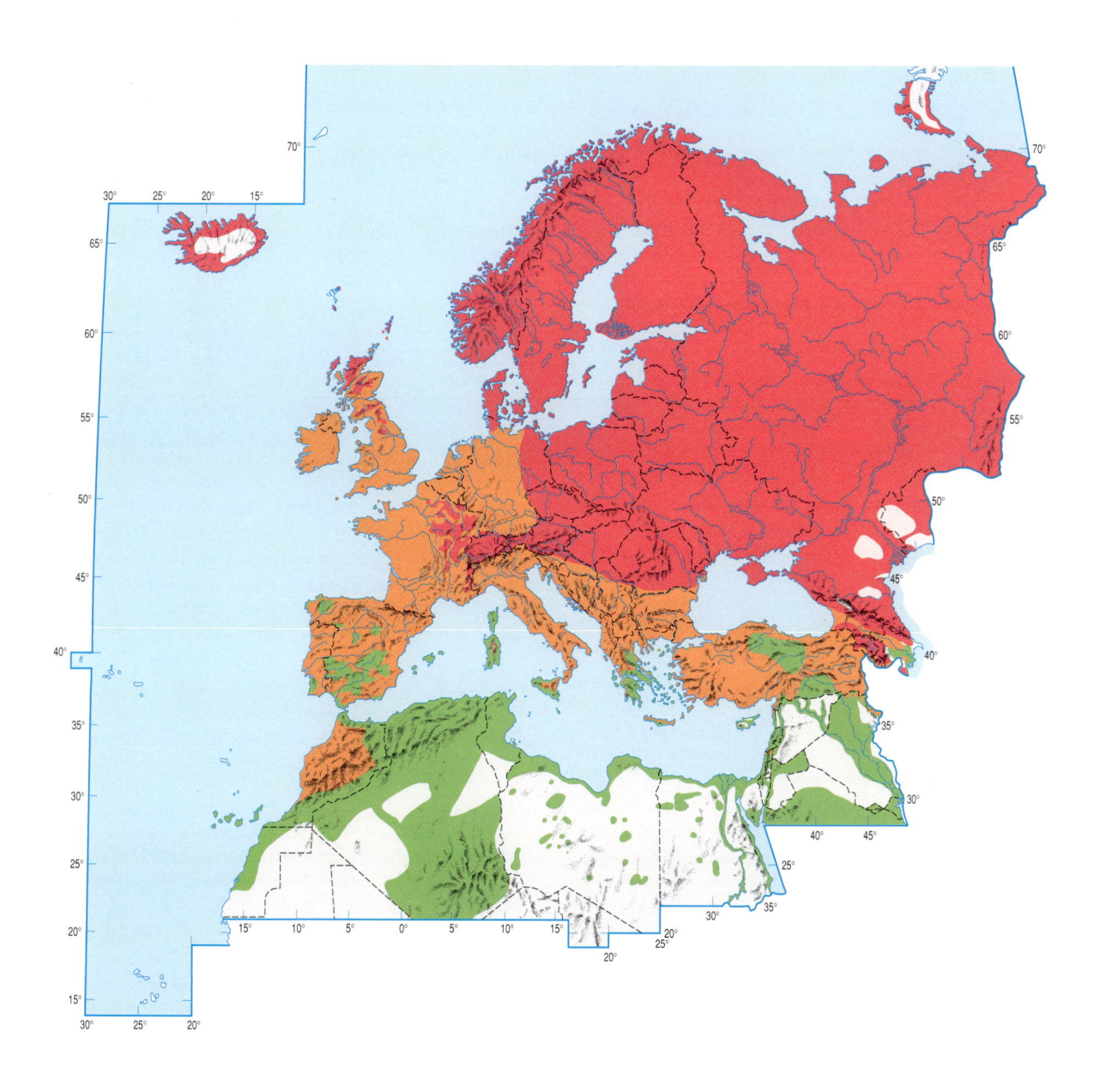

Field characters. 18 cm; wing-span 25–30 cm. Somewhat bulkier and markedly less attenuated than Grey Wagtail, with 25% shorter tail; slightly larger than Yellow Wagtail, with 20% longer tail. Ground-haunting, active insectivore, sharing form and actions of pipit but differing markedly in mainly pied plumage and long tail, often wagged. Adult essentially black (or grey) and white, ♂♂ of the various races differing mostly in back colour (black in Pied, grey in White) and more or less patched face. In winter, even ♂♂ may lose racial characters.

Specific identification not difficult, with only African Pied Wagtail (marginal in west Palearctic) providing closely matching appearance and yet easily distinguished by striking differences in forepart and wing patterns. In western Europe, separation of nominate *alba* (with grey back) from British *yarrellii* (black back) usually dependable but occasional intermediates not assignable. Beware convergent appearance of juvenile with some juvenile Yellow Wagtails and juvenile Citrine Wagtails, best separated by call (see below). Flight noticeably free and bounding, consisting of alternating short bursts of rapid wing-beats and long, 'shooting' curves or falls with wings closed. Flight silhouette essentially diamond-shaped, with thin tail which does not 'whip' like Grey Wagtail. Gait remarkably free, with nimble walk, short or long run, hops, and leaps. Especially when walking, head moves backwards and forwards more than other wagtails.

Habitat. From highest mainland latitudes to middle oceanic as well as continental zones, arctic and subarctic, boreal, temperate, steppe, Mediterranean, and even desert fringes, from July isotherm of 4°C to subtropics, penetrating into cold windy rainy regions and into arid and hot climates. Correspondingly adaptable as between wide variety of waterside habitats: lakes, rivers, streams, canals, estuaries, and sea coasts; also others distant from water, especially where agriculture, pastoralism, human settlements, roads, tracks, airfields, parks, gardens, gravel-pits, or other human intervention have provided essential bare spaces or very low vegetation cover. Differs from Yellow Wagtail in avoidance of tall or dense vegetation, except for roosting, then using reedbeds, bushes, and palm-groves and also artefacts such as large urban buildings, sugar-cane plantations, tree-lined urban streets, and horticultural glasshouses. To some extent shares attachment to grazing animals, and extends this to livestock in farmyards and small pens, especially where water provided. Differs also in readiness to breed in enclosed situations, including gardens with lawns or urban open spaces, or contrastingly where there are extensive bare tracts either of arable land or waste, even semi-desert.

Distribution. JAN MAYEN. Bred 1908. BRITAIN. Bred Isles of Scilly early 20th century. Some expansion in northern Scotland. A few nominate *alba* breed occasionally, mainly Shetland (Scotland), but also Channel Islands (spread from continent), where *yarrellii* bred 1952 and 1966. IRELAND. Spread in west, e.g. Galway and Mayo. FRANCE. Breeds irregularly Corsica. *M. a. yarrellii* breeds occasionally, mainly coastal Pas de Calais and Somme; mixed pairs frequent. NETHERLANDS. 4–20 pairs of *yarrellii* (some mixed) bred 1973–85, mainly along west coast. GERMANY. *M. a. yarrellii* rare breeder in north-west. UKRAINE. Slight increase. TURKEY. Common and widespread, but only local Central Plateau, south-east, and Aegean Coastlands. CYPRUS. Occasional breeder. SYRIA. Bred Rás al-Basit 1981–3, 1991, but may breed elsewhere in north. ALGERIA. May breed occasionally; no proof. MOROCCO. Since 1960s has extended range north and east, spreading chiefly along river valleys and wetlands.

Accidental. Spitsbergen, Bear Island, Jan Mayen, Azores, Madeira, Cape Verde Islands.

Beyond west Palearctic, extends throughout Asia except peninsular India and south-east; also breeds Alaska and south-east Greenland.

Population. Apparently stable in most of range. ICELAND. 10 000–50 000 pairs in late 1980s. FAEROES. 2–10 pairs. BRITAIN. 300 000 territories 1988–91. Fluctuates according to winter weather. IRELAND. 130 000 territories 1988–91; slight increase. FRANCE. Under 1 million pairs in 1970s. BELGIUM. Estimate (probably too low) 25 000 pairs 1973–7. In last 10 years, probably stable following decline. LUXEMBOURG. 12 000–15 000 pairs. NETHERLANDS. 60 000–120 000 pairs of nominate *alba* 1979–85. GERMANY. 1.46 million pairs in mid-1980s. In east, 130 000 pairs in early 1980s. DENMARK. 50 000–300 000 pairs 1987–8; increase 1978–94. NORWAY. 100 000–500 000 pairs (including 20–100 pairs of *yarrellii*) 1970–90. SWEDEN. 500 000–1 million pairs in late 1980s. FINLAND. 1–1.5 million pairs in late 1980s; slight decrease. Increased in north 1941–77. ESTONIA. 50 000–100 000 pairs 1991; stable in last 50 years. LATVIA. 150 000–300 000 pairs in 1980s. LITHUANIA. Stable. POLAND. 150 000–250 000 pairs. CZECH REPUBLIC. 100 000–200 000 pairs 1985–9. SLOVAKIA. 50 000–100 000 pairs 1973–94. HUNGARY. 50 000–80 000 pairs 1979–93. AUSTRIA. Very common. SWITZERLAND. 100 000–150 000 pairs 1985–93. SPAIN. 112 000–370 000 pairs. PORTUGAL. 10 000–100 000 pairs 1978–84. ITALY. 60 000–120 000 pairs 1983–93; perhaps slight decrease. GREECE. 2000–5000 pairs. ALBANIA. 1000–2000 pairs 1981; slight decrease. YUGOSLAVIA: CROATIA. 40 000–50 000 pairs. SLOVENIA. 25 000–50 000 pairs. BULGARIA. 100 000–1 million pairs. RUMANIA. 1.5–2.5 million pairs 1986–92. RUSSIA. 1–10 million pairs. BELARUS'. 380 000–420 000 pairs in 1990. UKRAINE. 380 000–400 000 pairs in 1988; slight increase. MOLDOVA. 25 000–35 000 pairs in 1988. AZERBAIJAN. Very common. TURKEY. 50 000–500 000 pairs. ISRAEL. A few tens of pairs in 1980s. MOROCCO. Uncommon.

Movements. Varies from wholly migratory to more or less resident. Most northern populations in west Palearctic migrate south to Mediterranean area, tropics and subtropics of Africa;

(FACING PAGE) Pied Wagtail and White Wagtail *Motacilla alba*. *M. a. alba* (White Wagtail): **1** ad ♂ breeding, **2** ad ♀ breeding, **3–4** ad ♂ non-breeding, **5** ad ♀ non-breeding, **6** juv. *M. a. yarrellii* (Pied Wagtail): **7** ad ♂ breeding, **8** ad ♀ breeding, **9** ad ♂ non-breeding, **10** 1st summer ♀. *M. a. dukhunensis* (White Wagtail): **11** ad ♂ breeding. *M. a. subpersonata* (White Wagtail): **12** ad ♂ breeding, **13** ad ♀ breeding. *M. a. personata* (Iran eastwards) (White Wagtail): **14** ad ♂ breeding, **15** ad ♀ breeding.
African Pied Wagtail *Motacilla aguimp vidua* (Afrotropics except extreme south) (p. 1106): **16** ad ♂ breeding, **17** ad ♀ breeding, **18** 1st winter ♂, **19** 1st winter ♀, **20** juv.

Norman Arlott.

extralimital eastern populations to peninsular India and south-east Asia. Autumn passage occurs across entire length of Mediterranean.

Passage of Icelandic birds (nominate *alba*) through Britain and Ireland occurs mostly August–October. In southern Finland, passage begins late August and peaks mid-September with only stragglers in October. In Switzerland, autumn departure generally begins *c.* 10 September, peaks mid-October, but continues regularly well into December. Return movement in spring is early. Arrival of nominate *alba* over wide areas of central Europe may be as early as February but mainly March–April, while arrival in southern Scandinavia is late March and in northern Scandinavia around mid-April. On Fair Isle, passage may start mid-March, but is usually early April to early May.

Food. Small invertebrates. 3 main foraging techniques. (1) Picking. Picks items from ground or water surface while walking; will also walk on floating vegetation. (2) Run-picking. Makes quick darting run at prey, picking it up either from surface or as it takes off. (3) Flycatching. Makes short flight from ground, catching prey in mid-air. Will also take food from water while hovering, and may hover repeatedly to take small swarming insects.

Social pattern and behaviour. Dispersion outside breeding season varies from gregarious to solitary, even territorial, depending on food supply. Territorialism (for feeding) common in winter quarters. Migrates mostly in small flocks. Communal roosting, often in large numbers, a characteristic and sometimes conspicuous feature in winter; less usual in breeding season. No evidence for other than monogamous mating system. Pair-bond usually lasts only for duration of breeding season, though may occasionally form on winter territory. Territorial when breeding; territories typically used year after year. ♂ sings rather little, and then mostly from elevated perch, often a roof-top, or from ground; more often attracts ♀ with series of contact-calls. Song-flight little used compared with other wagtails.

Voice. Song simple but jaunty and twittering, consisting largely of repeated slurred contact-calls. At least 3 regular calls: commonest flight-call 'tschizzik' (tends to be softer in nominate *alba*), but one contact-call, 'tzeurp', lacks disyllabic form and can suggest commonest monosyllable of Yellow Wagtail.

Breeding. SEASON. North-west Europe: first eggs laid beginning of April, main period late April to mid-May, last eggs early August. Iceland and northern Scandinavia: laying begins early June. Central Finland: first eggs in 2nd week of May; main period mid-May to early June, last eggs 2nd week of July; up to 1 week earlier in southern Finland. South and south-east Europe: late April to mid-July. North Africa: eggs found from late May to June. 2 broods, rarely 3, in south of range, mainly one in north. SITE. Hole or crevice in wide variety of natural and artificial sites, including building, wall, bank, cliff, pile of debris, dense bush, old nest of other species. Nest: cup of twigs, grass stems and leaves, roots, and moss, lined with hair, wool, and feathers. EGGS. Sub-elliptical, smooth and glossy; whitish, blue-white or grey, with fine, even freckling of grey-brown or grey, occasionally with brown spots. Clutch: 5–6 (3–8). INCUBATION. 11–16 days. FLEDGING PERIOD. 11–16 days.

Wing-length: *M. a. alba*: ♂ 87–96, ♀ 85–92 mm.
Weight: *M. a. alba*: ♂♀ mostly 17–25 g.

Geographical variation. 4 races in west Palearctic. Nominate *alba* occupies most of range. *M. a. yarrellii* of Britain and Ireland differs mainly in much darker, blacker upperparts; white tips of wing-coverts and tertials slightly wider but this often not noticeable; difference in size negligible. *M. a. dukhunensis* (extreme south-east of west Palearctic range) similar to nominate *alba* in all plumages, but grey of upperparts on average slightly paler and white tips of median and greater upper wing-coverts wider. *M. a. subpersonata* of Morocco markedly different from nominate *alba* in breeding plumage: head, neck, and chest black, but forehead, supercilium, and eye-ring white; white fringes along median and greater upper wing-coverts and along tertials very wide, often completely hiding black feather-bases, especially in ♂.

(FACING PAGE) Yellow Wagtail *Motacilla flava* (p. 1094). *M. f. flavissima*: **1** ad ♂ breeding, **2** ad ♀ breeding, **3** 1st winter ♂. *M. f. flava*: **4** ad ♂ breeding, **5** ad ♀ breeding, **6** 1st winter ♂. *M. f. lutea*: **7** ad ♂ breeding. *M. f. feldegg*: **8** ad ♂ breeding, **9** ad ♀ breeding, **10** 1st winter ♂. *M. f. thunbergi*: **11** ad ♂ breeding.
Citrine Wagtail *Motacilla citreola citreola* (p. 1098): **12** ad ♂ breeding, **13** ad ♀ breeding, **14** 1st winter.
Grey Wagtail *Motacilla cinerea cinerea* (p. 1100): **15** ad ♂ breeding, **16** 1st winter.
Pied Wagtail and White Wagtail *Motacilla alba* (p. 1103). *M. a. alba* (White Wagtail): **17** ad ♂ breeding, **18** ad ♀ breeding, **19** 1st winter. *M. a. yarrellii* (Pied Wagtail): **20** ad ♂ breeding, **21** ad ♀ breeding. *M. a. personata* (White Wagtail): **22** ad ♂ breeding.
African Pied Wagtail *Motacilla aguimp vidua* (p. 1106): **23** ad ♂ breeding, **24** ad ♀ breeding.

African Pied Wagtail *Motacilla aguimp*

PLATES: pages 1105, 1107

DU. Afrikaanse Witte Kwikstaart   FR. Bergeronnette pie   GE. Witwenstelze   IT. Ballerina nera africana
RU. Африканская трясогузка   SP. Lavandera pía   SW. Afrikansk sädesärla

Field characters. 18.5–19 cm; wing-span 26–31 cm. Rather greater size than White Wagtail most evident in plumper body, enhanced by plumage pattern of head and forebody. Invariably black and white, differing from White Wagtail most in more linear face pattern, with long white supercilium, completely black lores and cheeks, and long white throat. Bold white transverse panel on folded wing becomes even more obvious in flight when large white bases of flight-feathers exposed.

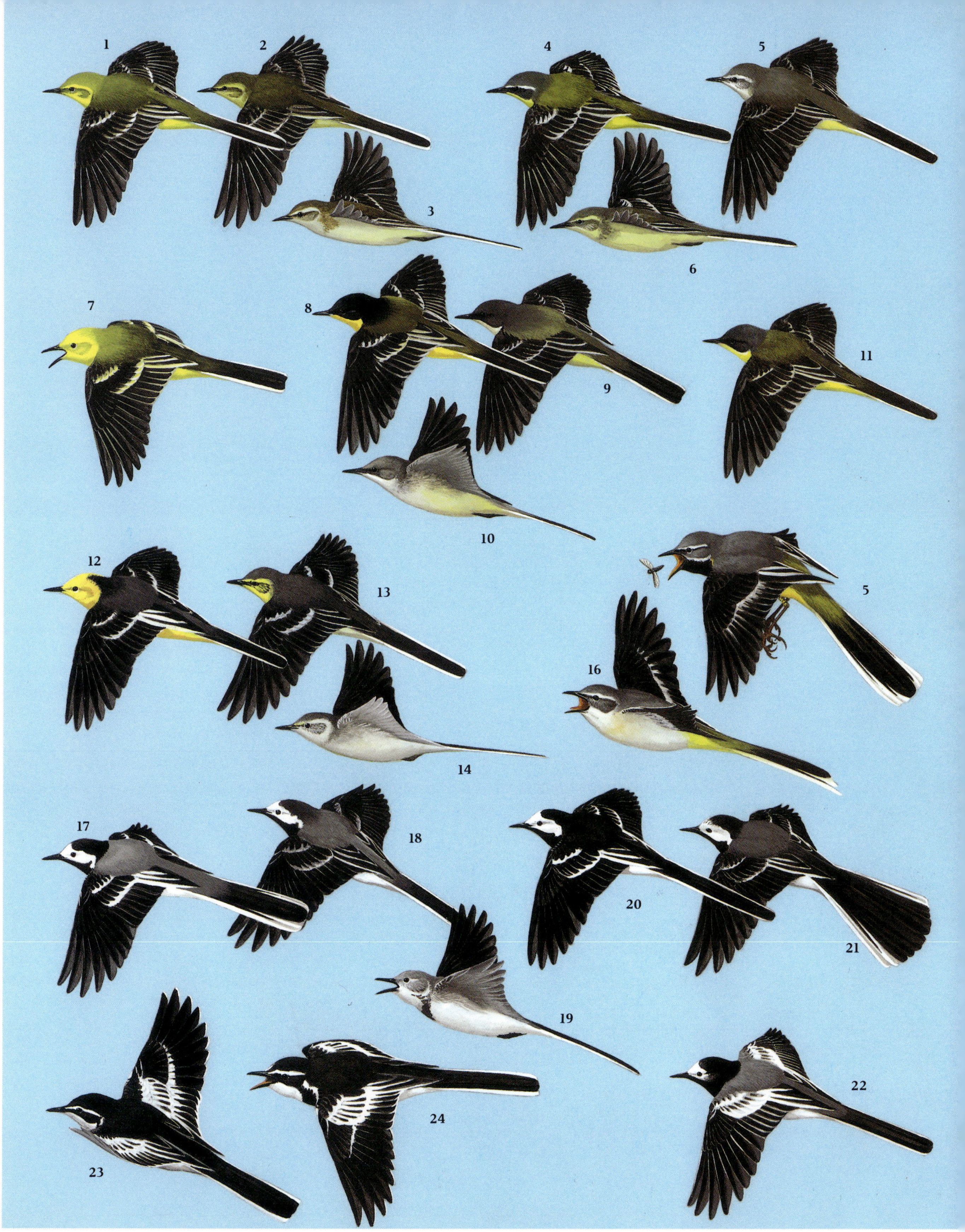
1
2
3
4
5
6
7
8
9
10
11
12
13
5
16
14
17
18
20
21
19
22
23
24

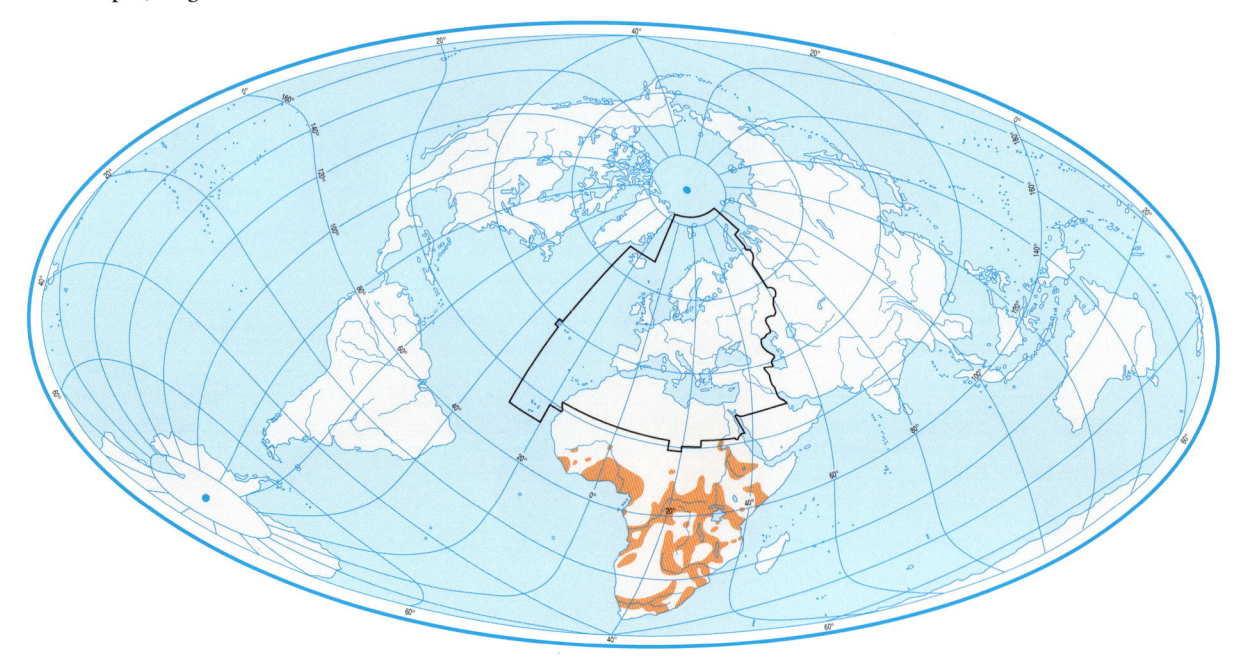

Juvenile confusable with darker juveniles of White Wagtail but easily distinguished by large area of white on flight-feathers and coverts.

Plumper form obvious to observer long familiar with White Wagtail and this visible even in flight, with silhouette less narrow-tailed and action rather less bounding. Gait and behaviour as White Wagtail.

Habitat. Largely lowlands, though recorded up to 3000 m, throughout tropical and subtropical Africa south of Sahara; often along rivers on sand-banks, shingle, etc., especially in dry areas, and on coastal lagoons; tame and commonly near man (even entering houses), particularly in gardens, fields, playing fields, roads, etc. in villages and cities.

Distribution and population. EGYPT. Rare breeder Lake Nasser area. 5 pairs (1 pair feeding young) 1994.

Movements. Generally sedentary, making only local wanderings usually dependent on changing water-levels, and for this reason unknown whether actually resident in Egypt. In South Africa, breeding visitor to eastern Cape September–March, and non-breeding migrant in Transvaal uplands, though resident at lower elevations. Occasional non-breeding visitor to Zanzibar and Pemba islands, Tanzania.

Food. Adult and larval insects of all kinds particularly by water, and opportunistically takes small fish; also domestic scraps near houses. Forages in typical wagtail fashion, walking or running on ground, occasionally jumping up to seize flying insect.

Social pattern and behaviour. Usually occurs singly, in pairs, or in family parties, but outside breeding season in loose flocks of up to c. 100, especially at feeding sites. Hundreds roost together in city trees or on buildings. Territorial and solitary breeder. Pair defends nesting territory c. 100 m in diameter, usually by river, plus adjacent feeding territory of 1–2 ha.

Voice. Song a medley of mellow and high-pitched notes; melodious, recalling Canary: 'tsip weet-weet, twip-twip-twip, weep-weep, tip-tip-tip, weet-woo-woo'. Mimics other species. Call, usually uttered on taking flight, loud disyllabic 'chizzit', more slurred and less divided than equivalent in White Wagtail.

Breeding. SEASON. On Lake Nasser, Egypt, eggs recorded April; eggs laid in Mali February–May; elsewhere breeds throughout year principally in local rainy season. SITE. On Lake Nasser, in rock crevice on tiny rocky islets close to shore; otherwise on ground in vegetation, in shallow cavity in river bank, in flood debris, tree cavity, etc., and frequently on or in houses and other structures. Nest: rough cup of small twigs, grass, stems, leaves, etc. lined with hair, feathers, fine grasses. EGGS. Sub-elliptical, smooth, and glossy; white speckled with brown and with pale grey blotches. Clutch: 3–4 (2–7). INCUBATION. 13–14 days. FLEDGING PERIOD. 15–16 days.

Wing-length: ♂♀ 85–102 mm.
Weight: ♂♀ 22–33 g.

White-cheeked Bulbul *Pycnonotus leucogenys mesopotamiae* (Iraq to southern Iran): **1** ad, **2** juv. Yellow-vented Bulbul *Pycnonotus xanthopygos* (p. 1110): **3** ad, **4** juv. Common Bulbul *Pycnonotus barbatus* (p. 1111). *P. b. barbatus*: **5** ad, **6** juv. *P. b. arsinoe*: **7** ad.

Bulbuls Family Pycnonotidae

Small to medium-sized oscine passerines (suborder Passeres); many arboreal but some terrestrial to greater or lesser extent; most frugivorous and insectivorous, some also taking nectar and pollen. About 120 species in *c.* 14 genera. Found in Old World only, mainly in forest and parkland: tropics of Africa and Asia, north to Middle East and Japan, east to Philippines and Indonesia. Family represented in west Palearctic by 4 species of *Pycnonotus*, three of which breed regularly.

Body moderately slender; neck short. Sexes of similar size in most species but ♂ larger in some. Bill short to medium in length and rather curved; often slender and pointed, sometimes hooked, with or without a notch. Wing rather short, broad and rounded. Flight relatively weak-looking in most species, but swift and agile in some (if only over short distances). Tail medium to long; tip usually round, square, or graduated. Legs short, toes weak. Gait a hop. Plumage soft and, especially on rump, dense; often a patch of hair-like feathers on nape, sometimes concealed. Colour mainly sombre—brown, olive, or green but often with bright or contrasting head markings and/or red, yellow, orange, or white under tail-coverts. Some species crested. Sexes alike.

White-cheeked Bulbul *Pycnonotus leucogenys*

PLATES: pages 1109, 1116

Du. Witoorbuulbuul   Fr. Bulbul à joues blanches   Ge. Weißohrbülbül   It. Bulbul guancebianche
Ru. Белощекий бюльбюль   Sp. Bulbul cariblanco   Sw. Vitkindad bulbyl

Field characters. 18 cm; wing-span 25.5–28 cm. 10% smaller than Common Bulbul and Yellow-vented Bulbul, but of similar form. Black, white-cheeked head, and white-tipped tail diagnostic. Juvenile's head browner than adult's, with white cheeks less clear-cut.

Unmistakable in west Palearctic. Behaviour, flight, and other actions much as Common Bulbul; even more commensal.

Habitat. Mainly subtropical, in warm dry areas from coastal mangroves to *c.* 1800 m in hills of Baluchistan (Pakistan) and up to *c.* 2400 m in Nepal. A bird of open country not of forest, and of bushes rather than trees, also frequenting palm-groves and gardens.

Distribution and population. SYRIA. Status unclear:

recorded only Palmyra oasis March 1977 and 1979. IRAQ. Reported to be widespread and common, with expansion up Euphrates valley, in 1950s, also in Baghdad and around Babylon in autumn 1983. KUWAIT. Regular in small numbers; fairly stable. No range changes.

Accidental. Records from Israel and Jordan (Aqaba, April 1990) perhaps refer to escapes.

Beyond west Palearctic, breeds eastern Arabia east to Assam.

Movements. Essentially resident.

Food. Mainly insects, fruit, and berries; also seeds, buds, and nectar. In Iraq, ripe dates a favourite item, also unripe figs.

Social pattern and behaviour. Outside breeding season, in pairs or small flocks of 5–6, sometimes more at favourable food source. No evidence for other than monogamous mating system. In Iraq, however, associations of 3 birds commonly occur at nests under construction, raising possibility of helpers, as in Common Bulbul. Pair-bond maintained all year.

Voice. Song a fruity, musical 4-syllable unit, frequently repeated, suggesting speeded-up Golden Oriole; may include bubbling, chuckling, and warbling sounds, also a protracted series of 'chip-chop' phrases. Contact-call a 1–3-syllable rather squeaky, 'rusty' sound, rendered 'k-zee zee kr-zer ze'; rather like high-pitched sparrows.

Breeding. SEASON. Iraq: eggs found April to mid-July, mainly mid-April to mid-May. Eastern Saudi Arabia: eggs laid early March to July. 2 (possibly 3) broods per season. SITE. In low bush or sometimes in branches of low tree. Nest: substantial cup of grass stems and leaves, roots, and thin twigs, lined with finer rootlets, lichens, and grass. EGGS. Sub-elliptical, smooth and glossy; pinkish-white, heavily marked with spots, blotches, and streaks of red, with underlying small purple spots. Clutch: 3(2–5). INCUBATION. About 12 days. FLEDGING PERIOD. 9–11 days.

Wing-length: ♂ 91–98, ♀ 87–92 mm.

Yellow-vented Bulbul *Pycnonotus xanthopygos*

PLATES: pages 1109, 1116

DU. Arabische Buulbuul FR. Bulbul d'Arabie GE. Gelbsteißbülbül IT. Bulbul capinero
RU. Желтопоясничный бюльбюль SP. Bulbul capirotado SW. Levantbulbyl

Field characters. 19 cm; wing-span 26.5–31 cm. Size, appearance, and behaviour close to north-west African race of Common Bulbul. Differs in blacker head, grey-white eye-ring, and yellow vent, with combination of last two characters diagnostic.

In brief view, general appearance so similar to Common Bulbul that instant identification quite impossible. Beware escapes in normally separate ranges.

Habitat. Mediterranean and subtropical, but habitat otherwise scarcely distinguishable from that of Common Bulbul, and broadly similar to White-cheeked Bulbul. In Arabia, a bird of gardens, palm groves, and fairly thick bush in wadi beds. In Lebanon, in moister valleys with trees by rivers at no great altitude, and in orange and banana plantations as well as in gardens, orchards, groves, and thickets, chiefly in coastal strip.

Distribution and population. TURKEY. Recent records further west than known limit (Kemer and Antalya area), but range extension requires confirmation. 5000–50 000 pairs. SYRIA. Local but regular breeder in west. Range poorly known; species easily overlooked, and also commonly kept in captivity. ISRAEL. Increased and expanding to all areas. A few hundred thousand pairs in 1980s. JORDAN. Very common in Jordan valley. IRAQ. First recorded Baghdad (several, September 1983) and Al Qaim (3 in October 1983). EGYPT. Fairly common (in range 1000–10 000 pairs).

Accidental. Kuwait.

Beyond west Palearctic, breeds only in Arabia.

Movements. Apparently sedentary.

Food. Mainly fruit, seeds, and insects; occasionally leaves, flowers, and nectar. Insects often taken in flight. In Lebanon, gathers in large flocks to feed on flying ants. Recorded chasing Hoopoes and stealing mole-crickets from them.

Social pattern and behaviour. Gregarious throughout the year when food locally abundant, otherwise mostly in pairs or 'duos'. Duo comprises 2 siblings of same or different sex, associating closely. Where density high, Israel, flocks comprise a few hundred birds or even thousands. Territorial when breeding, with territories small and usually contiguous in favourable habitats. Monogamous mating system, with pair-bond maintained all year. Song given by territorial ♂ from an elevated perch in territory. May also be given in snatches while bird feeds and preens in treetops, rarely in flight.

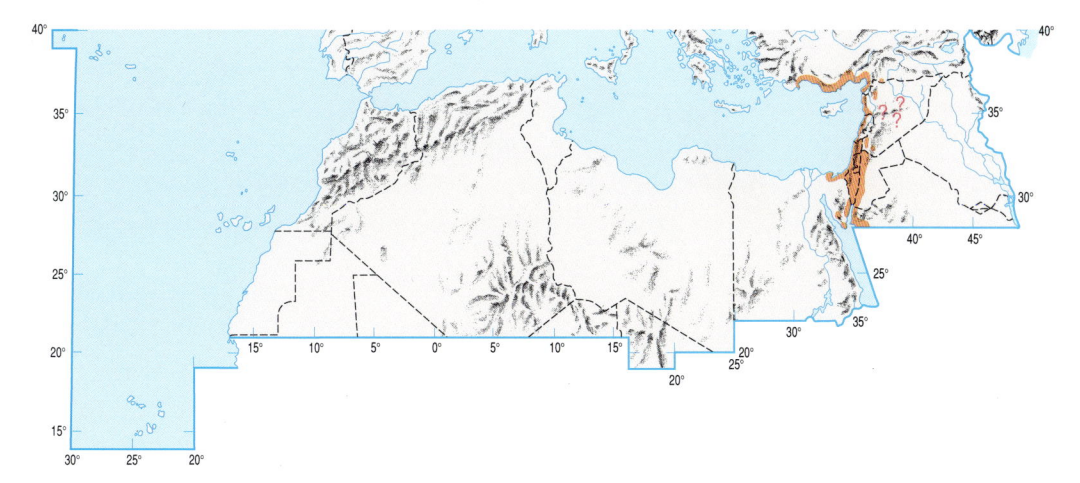

Voice. Song a monotonous repetition of a single phrase which, in Israel, varies somewhat between regions; comprises 2–8 syllables, variously stressed. Sometimes rich and flute-like, sometimes like contralto whistle (but lower) of Blackbird, sometimes husky and deep, nearly always in short disjointed snatches. Occasionally mimics other birds. Calls very varied: chirping, bubbling, whistling, or sharp and scolding.

Breeding. SEASON. Middle East: eggs laid end of May to beginning of July. Israel: April–August. 2–3 broods. SITE. In bush or low palm. Nest: small cup of thin twigs, grass stems, moss, and leaves, sometimes also string and wool; lined with hair, shredded bark and rootlets. EGGS. Sub-elliptical, smooth and glossy; light violet to pinkish-white, well marked with violet- or red-brown and grey spots and speckles. Clutch: 3 (2–4). INCUBATION. About 14 days. FLEDGING PERIOD. 13–15 days.

Wing-length: ♂ 96–103, ♀ 90–95 mm.
Weight: ♂♀ 35–46 g.

Common Bulbul *Pycnonotus barbatus*

PLATES: pages 1109, 1116

Du. Grauwe Buulbuul Fr. Bulbul des jardins Ge. Graubülbül It. Bulbul golanera
Ru. Обыкновенный бюльбюль Sp. Bulbul naranjero Sw. Trädgårdsbulbyl

Field characters. 19 cm; wing-span 26.5–31 cm. Slightly larger than Corn Bunting; 20% smaller than any *Turdoides* babbler. Medium-sized passerine, with high-crowned head, rather broad wings, and long, slim body and tail; general character most recalls small, long-tailed thrush or sober shrike. Plumage sombre dusky-brown, relieved by noticeably darker, umber-brown head and paler greyish underparts.

Liable to confusion only with Yellow-vented Bulbul, when much blacker head and yellow vent of that species not seen. Flight supposedly weak but no less strong than Jay; bird capable of rapid escape and dramatic plunge through cover; hint of instability comes from erratic wing-beats which give fluttering action to broad wings, exaggerated by apparent waving and spreading of long tail. Gait includes hopping, leaping, and clambering.

Habitat. Tropical and subtropical; in west Palearctic, also breeds marginally in warm Mediterranean zone. In north-west Africa, above all a plains bird, but ascends freely to 700–900 m and locally in Haut Atlas to 2300 m. Always in green and fertile places, including wooded streams, gorges, and oases, as well as gardens and orchards.

Distribution and population. EGYPT. Common. Marked extension of range in Nile delta and valley in 20th century. Parts of Suez Canal area colonized since 1945. Now known to breed Abu Simbil (southern Lake Nasser). TUNISIA. Common resident in suitable habitats (permanent wadis). ALGERIA. Very common. MOROCCO. Fairly common. CHAD. Tibesti: very common where vegetation plentiful May–June 1961.

Accidental. Spain.

Beyond west Palearctic, widespread throughout sub-Saharan Africa except south-west.

Movements. Resident and (often at least) sedentary.

Food. Mainly fruit and insects; also seeds, flowers, young leaves, nectar, and even crystallized gum. Insects commonly caught on ground by scouring vegetation, but aerial feeding is also common.

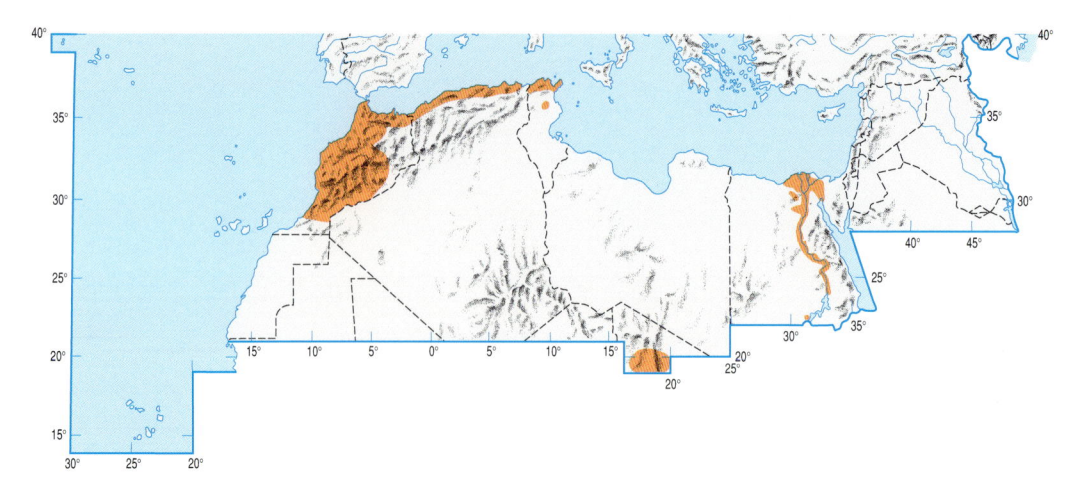

Social pattern and behaviour. Little studied; apparently very similar to Yellow-vented Bulbul.

Voice. Rich repertoire freely used. Readily mimics great variety of other species. Given to sudden, staccato outbursts of noise, with both calls and song rich and fluting.

Breeding. SEASON. North-west Africa: eggs from mid-May to August. SITE. In tree or bush, in variety of positions. Nest: cup of grass stems and leaves, lined hair. EGGS. Sub-elliptical, smooth and glossy; pinkish-white to white, heavily spotted and speckled with red-brown and lilac. Clutch: 2–3(–4). INCUBATION. 12–14 days. FLEDGING PERIOD. 12–14 days.

Wing-length: Nominate *barbatus*: ♂ 100–107, ♀ 93–100 mm.

Geographical variation. Rather slight in west Palearctic. *P. b. arsinoe* from Egypt and northern Sahel zone smaller than nominate *barbatus* from north-west Africa; face more extensively and deeper glossy black, reaching crown, ear-coverts, cheeks, and chin, merging into dark brown on hindcrown, sides of neck, and throat.

Red-vented Bulbul *Pycnonotus cafer*

DU. Roodbuikbuulbuul FR. Bulbul à ventre rouge GE. Rußbülbül IT. Bulbul dal sottocoda rosso
RU. Розовобрюхий бюльбюль SP. Bulbul ventrirrojo SW. Rödgumpad bulbyl

Widespread resident in Pakistan, India, Sri Lanka, Burma, and western Yunnan (China). Successfully introduced into Fiji, Tonga, and Samoan and Hawaiian Islands, unsuccessfully to Australia and New Zealand. In Kuwait, has bred in different areas of Kuwait city and Ahmadi to the south; 1 pair bred Shuweikh 1994, and recorded in pairs elsewhere in city; not clear if permanently established.

Waxwings, Hypocolius Family Bombycillidae

Medium-sized oscine passerines (suborder Passeres) in 2 sub-families: Bombycillinae (waxwings) and Hypocoliinae (single species, Grey Hypocolius *Hypocolius ampelinus*), both represented in west Palearctic.

Waxwings Subfamily Bombycillinae

Comprises a single genus *Bombycilla* of 3 species confined to temperate and subarctic Northern Hemisphere. Migratory and nomadic, with periodic irruptions.

Body plump-looking; neck short. Sexes of similar size. Bill short, thick, and broad at base; slightly hooked and notched. Wing long and pointed. Tail short and square. Leg short but toes strong, middle and outer united at base; claws long.

Plumage soft, dense, and silky; a delicate vinous-brown with paler yellow or whitish belly and contrasting velvet-black face-mask and throat-patch, partly bordered by white lines. Lores and narrow frontal band, including nostrils, covered with short, dense, plush-like feathers. Crest of short dense feathers present in all species.

Waxwing *Bombycilla garrulus*

PLATES: pages 1114, 1116

Du. Pestvogel   Fr. Jaseur boréal   Ge. Seidenschwanz   It. Beccofrusone
Ru. Свиристель   Sp. Ampelis europeo   Sw. Sidensvans   N. Am. Bohemian Waxwing

Field characters. 18 cm; wing-span 32–35.5 cm. Size and general form similar to Starling. Medium-sized, vinaceous-brown passerine, with bold crest on head and (usually) bright 'waxy' appendages on secondaries. At close range, black bib (of adult), contrasting grey rump and rich brown vent, yellow or white tips and fringes to primaries, and yellow terminal band on black tail obvious. Juvenile shows dull white supercilium and upper cheeks.

Unmistakable at close range. For distinctions of Cedar Waxwing, see that species. Important to note that plumage tones of Waxwing vary widely according to light intensity and background colour—can look as dark in silhouette as Starling. Flight silhouettes and actions strongly recall Starling, but shorter bill, fuller head (due to laid-back crest), and bulkier body produce less angular form, while actions somewhat slower, with less rapid wing-beats and more floating glides and turns; capable of hovering. Wing-beats make rattling sound. Flocks fly in compact formations, again recalling Starling. Agile in tree and bush foliage, with feeding actions and gait reminiscent of large tit or crossbill. When feeding on ground, gait restricted to hops and shuffles. Often astonishingly tame and frequently 'lazy'.

Habitat. Breeds in west Palearctic in upper middle latitudes in subarctic and boreal zones up to 10°C July isotherm, stopping short of treeline, in belt of dense tall taiga, especially of spruce and pine, sometimes mixed with broad-leaved species such as birch. Occurs largely in lowlands and valley forests, but also in uplands, although apparently not in mountains. Prefers for breeding old stunted conifers festooned with hanging witch-hair lichen *Usnea*; dense forest interiors and fringes by peat swamps or dwarf heath are both acceptable sites. On switching in autumn to diet of berries, often confronts choice between finding adequate supplies for winter in native forests or launching eruptive movements to alternative supply sources in temperate lands, seeking profuse crops of fruits of rowan, rose, or other trees and shrubs, including introduced garden varieties. At this season occurs on roadsides, in parks and gardens, along hedgerows, and wherever berries can be found, regardless of human presence, and abandoning any special attachment to conifers.

Distribution. Little information on range changes. Moves further south than mapped in irruption years. Norway. Slight decrease in breeding range. Sweden. Breeds irregularly further south. Estonia. Occasional breeder, verified only in 1968.

Accidental, or occasional irruptive migrant and winter visitor. Bear Island, Jan Mayen, Iceland (annual in very variable numbers since 1984), Faeroes, Ireland (almost annual), France, Spain, Portugal, Malta, Greece, Albania, Armenia, Turkey, Cyprus, Israel, Algeria.

Beyond west Palearctic, extends east across Siberia to Pacific, south to *c.* 51°N. Also breeds north-west North America (south to north-west USA).

Population. Norway. 500–2000 pairs 1970–90; stable. Sweden. 5000–50 000 pairs in late 1980s; trend unknown.

Waxwing *Bombycilla garrulus*: **1** ad ♂, **2** ad ♀, **3** 1st winter ♂, **4** 1st winter ♀, **5** juv. Grey Hypocolius *Hypocolius ampelinus* (p. 1116): **6** ad ♂, **7** ad ♀, **8** juv ♀.

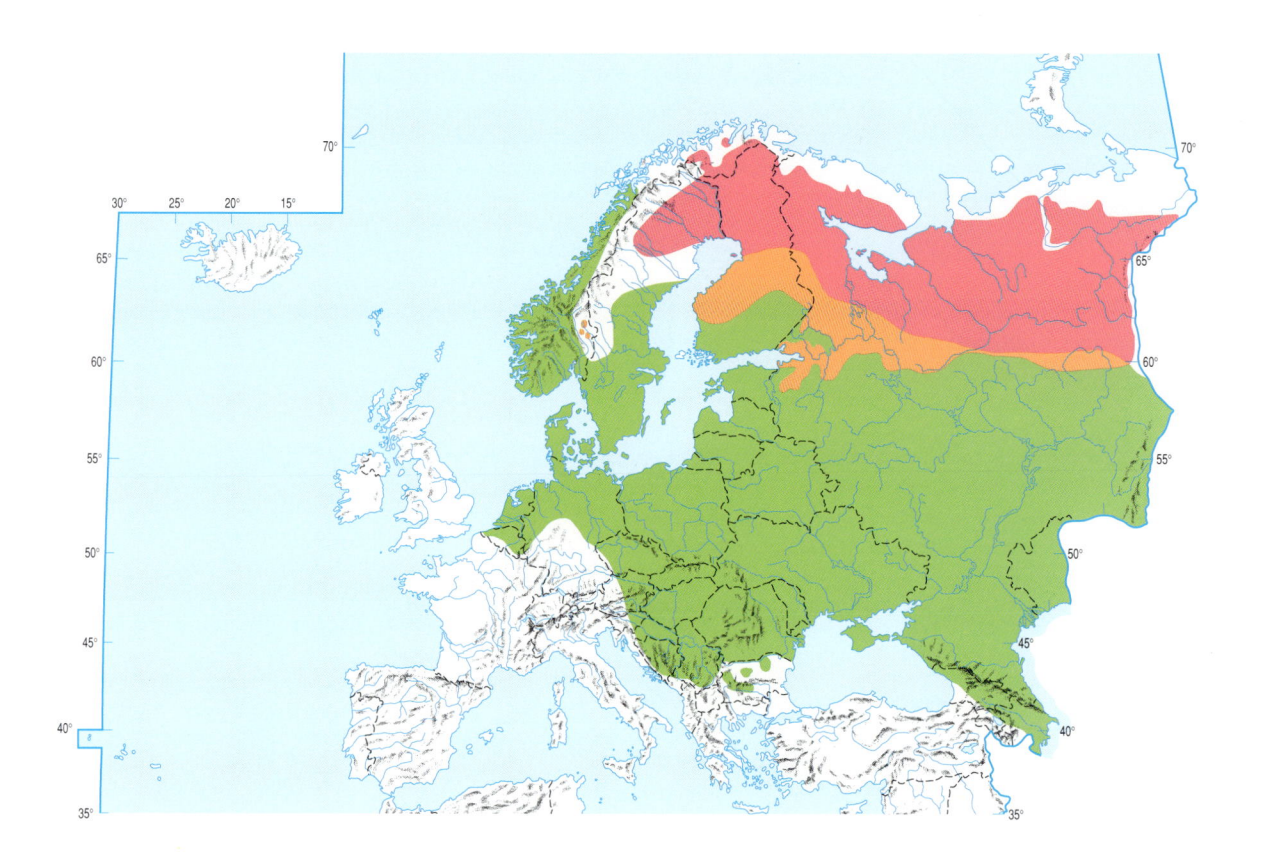

FINLAND. 20 000–50 000 pairs in late 1980s; slight increase. RUSSIA. 100 000–1 million pairs; stable.

Movements. Partial migrant, often making eruptive movements. In northern Europe regularly overwinters within southern part of breeding area and also makes annual limited movements to southern Sweden and Denmark with recent extension to north-central Europe. Breeding populations of northern Fenno-Scandia, and probably from further east, move both south-west to Britain and western Europe and also south to south-east to central and eastern Europe, and there are occasional records of more westerly movement to Iceland, the Faeroes, and eastern Greenland.

Invasions of Britain occur October–March, with maximum numbers in mid-winter, and occasionally extend to April, rarely to May, and 2 invasions in July. In most years only a few birds occur, but irregularly there are large numbers, mostly in eastern areas from Shetland to Kent. Best-documented eruption is that of 1965–6. In autumn 1965, acute imbalance between population size and food supply in Fenno-Scandia led to large-scale eruption, with large flocks reaching Britain and central Europe, and small numbers to Mediterranean islands.

Food. In summer, mainly insects, especially mosquitoes and midges; in winter, chiefly fruit (of many kinds), also buds and flowers. Change from insects to fruit (and back again) occurs gradually according to weather and abundance of insects. Insects caught mainly by flycatching from tops of trees, shrubs, telegraph poles, etc. Fruit taken mainly from tree but sometimes from ground below it. Berries picked with slight stooping motion, held briefly in bill and swallowed with quick toss of head. Sometimes bird clings to underside of branch and eats berry from below; occasionally takes one in hovering flight. Usually makes short visits to food, feeding intensively. May consume 2–3 times own body weight of berries per day.

Social pattern and behaviour. Highly gregarious, especially outside breeding season. Compact flocks of varying size occur for migration, feeding, and roosting. Nothing to suggest mating system other than monogamous. Only ♀ incubates; both sexes feed young. Breeding pairs, where numerous, almost colonial, with little evidence of territoriality. On breeding grounds, ♂ sings from tree-top near nest. Birds of both sexes in winter flocks will sing from perch in chorus.

Voice. Commonest call feeble though distinctive high, sibilant trill, 'sirrrrr'. Song essentially a variation of this. Wing-rattling at take-off and landing audible up to *c*. 30 m away and as characteristic of the bird as its call.

Breeding. SEASON. Northern Scandinavia: laying normally begins mid-June, but from 1st week in early seasons, or even last few days of May. One brood. SITE. In tree, 3–15 m above ground; in low pine or scrub, usually close to stem, but in taller tree, often out on branch. Nest: cup with base of thin twigs 2–15 cm long, then grass and reindeer moss, with lining of dry grass and sometimes fine lichens. EGGS. Sub-elliptical to oval, smooth and glossy; grey-blue to pale blue, sometimes buffish, lightly spotted black and grey. Clutch: 5–6 (4–7). INCUBATION. 14–15 days. FLEDGING PERIOD. 14–15 days.

Wing-length: ♂♀ 114–125 mm.
Weight: ♂♀ mostly 50–75 g.

Cedar Waxwing *Bombycilla cedrorum*

PLATE: page 1431

FR. Jaseur d'Amérique GE. Zedernseidenschwanz
RU. Американский кедровый свиристель

Field characters. 14.5 cm; wing-span 30–31 cm. 10% smaller and noticeably slimmer than Waxwing. Small species with similar form to Waxwing but differing distinctly in less decorated wings, which are virtually unmarked in immature, and in adult have only white inner edges to tertials and thin wax-red spines on innermost secondaries (lacking the conspicuous white or yellow tips and outer webs to primaries of Waxwing). Plumage also differs from Waxwing in browner tone, white, not rufous-cinnamon vent and under tail-coverts (at all ages) and in narrow white frontal band and supercilium above black eye-panel and yellowish, not buff-white belly (in adult).

Distinctive at close range when rather smaller, slimmer form, and less decorated or colourful wings and vent obvious. Flight and behaviour as Waxwing. Call like Waxwing but more quavering.

Habitat. Open, occasionally dense woodland, trees and bushes bordering fields and orchards or by marshes, lakes, and rivers, and also in large gardens and parks in boreal to southern temperate North America; in winter to tropical Central America, in drier places by streams and in canyons. Feeds on fruits and insects, in large flocks in non-breeding season; nests on branch of tall shrub or tree.

Distribution. Breeds from south-east Alaska east across central Canada to Newfoundland, south to northern California, Colorado, and western North Carolina.

Accidental. Iceland: Gerðar, April–July 1989. Britain: Nottingham, February–March 1996 (under review).

Movements. Migratory and partially migratory; can be irregular and nomadic depending on food supply. West Palearctic records followed occurrence of unusually high numbers in north-east North America in winter 1988–9 and spring–summer 1989, and in winter 1995–6, when coincided with exceptionally large Waxwing influx. Moves south within breeding range (some wintering in southern Canada) and to Mexico, Central America, and parts of northern South America.

Wing-length: ♂♀ 89–99 mm.

White-cheeked Bulbul *Pycnonotus leucogenys mesopotamiae* (p. 1109): **1–2** ad. Yellow-vented Bulbul *Pycnonotus xanthopygos* (p. 1110): **3–4** ad. Common Bulbul *Pycnonotus barbatus* (p. 1111): **5–6** ad. Waxwing *Bombycilla garrulus* (p. 1113): **7–8** ad ♂. Grey Hypocolius *Hypocolius ampelinus*: **9–10** ad ♂, **11** ad ♀.

Hypocolius Subfamily Hypocoliinae

Comprises a single species: Grey Hypocolius *Hypocolius ampelinus*, breeding Turkmenistan to Iraq. Marginal breeder in west Palearctic.

Closely resembles Bombycillinae (waxwings) in many aspects of anatomy and habits, with rather similar short and stubby bill, black face-mask, thick crest, contrastingly patterned flight-feathers and tail, and soft plumage, though colour more subdued—grey and buff predominating and red absent. Tail much longer and wing shorter, with more rounded tip. ♀ duller than ♂, with no face-mask and less obvious wing and tail pattern.

Grey Hypocolius *Hypocolius ampelinus*

PLATES: pages 1114, 1116

Du. Zijdestaart   Fr. Hypocolius gris   Ge. Seidenwürger   It. Ipocolio
Ru. Свиристелевый сорокопут   Sp. Ampelis gris   Sw. Grå palmfågel

Field characters. 23 cm; wing-span 28–30 cm. Almost as long as but slighter than Southern Grey Shrike, with tail proportionately longer and not graduated, and bill without hook. Unique, sleek, long-tailed passerine, with somewhat shrike-like appearance but with behaviour recalling both Waxwing and babbler. Plumage essentially pale grey above and pale isabelline-buff below, with black primaries tipped white and tail tipped black. ♂ has black face mask extending to nape. Juvenile entirely pale sandy-brown, though ♂ has dull black tips to tail-feathers.

In brief view may suggest shrike, particularly when flicking tail, or babbler, with direct flight into dense cover particularly recalling latter's escape behaviour.

Flight strong, direct or circling, without undulations of shrike; action whirring when climbing, fast-flapping over distance, and gliding in descent and before landing. Usually unobtrusive, keeping to thick palm cover, but apparently becomes tame in close association with human habitat.

Habitat. Subtropical and tropical, in more vegetated belts within arid lowlands, level or gently undulating; often in river valleys fringing desert or semi-desert with patchy or thin scrub, open broad-leaf scrub, groups of trees, irrigated areas, gardens, or palm groves.

Distribution and population. Iraq. Described as breeding widely but locally in 1st half of 20th century. More recent information lacking.

Accidental. Turkey (one, August 1986), Israel (4 November–April records, 1987–90). Egypt: 2, Abu Simbil, November–December 1995; record in 1938 extralimital (Gebel Elba). In Kuwait, scarce passage-migrant, recorded annually in both seasons since 1982.

Beyond west Palearctic, breeds in southern Iran and Turkmenistan, and probably Afghanistan.

Movements. Most birds apparently short-distance migrants. Recorded on breeding grounds in winter only in Karun district of western Iran. Data available so far suggest it winters largely in Saudi Arabia, being generally uncommon, but locally numerous in western areas from Hejaz mountains northwards, in central Arabia, and in oases of the Eastern Province.

Food. Fruit and some insects. Searches for food among trees, rarely descending to ground. Noticeably deliberate in feeding movements when perched on bush, stretching and balancing to reach berries with, at times, tail angled well downwards. Chews fruit, rejecting skin and stones, though small stones and pips swallowed and later excreted. Will also fly down from perch like shrike apparently to take insects on ground, and recorded flying up to *c.* 3–4 m to catch insect, then returning to perch.

Social pattern and behaviour. Gregarious, especially outside breeding season when forms mostly small flocks of 5–10 birds. Mating system evidently monogamous. Pair-formation apparently takes place after arrival on breeding grounds and bond strong once established. All nest-duties shared by both sexes. Breeds in small loose colonies, each pair (both sexes) defending small nest-site territory (nest and immediate vicinity). No territorial song.

Voice. Includes a variety of mewing or whistling sounds, and a fairly low-pitched and harsh monosyllable, 'chirr' or 'kirr'. Loud, continuous 'kirrr' calls a feature of courtship display.

Breeding. SEASON. Iraq: eggs from early May to June or July. 2 broods. SITE. In bush or low tree, 1–4 m above ground, often well hidden in densest part of bush but may be more exposed in tree. Nest: base of small twigs, with cup of grass and tufts of vegetable down, lined with more down, sometimes wool and hair. EGGS. Sub-elliptical to oval, smooth and rather glossy; white to very pale grey, sometimes tinged green when fresh, usually with zone of lead-grey to grey-brown blotches round broad end, sometimes forming band, occasionally scattered or sometimes unmarked. Clutch: 3–4(–5). INCUBATION. *c.* 14 days. FLEDGING PERIOD. Leave nest at 13–14 days.

Wing-length: ♂♀ 97–106 mm.
Weight: ♂♀ 48–57 g.

Dippers Family Cinclidae

Quite small to medium-sized oscine passerines (suborder Passeres); mainly terrestrial and aquatic—being unique among Passeres in living in close contact with water, foraging largely below surface of flowing streams and rivers for invertebrates and, to lesser extent, fish. 5 closely similar species in single genus *Cinclus*. 1 species in west Palearctic, breeding widely but locally.

Though expert divers, no special morphological adaptations for life under water apart from dense plumage and broad membrane above nostrils which can be closed when head submerged; oil-gland, however, larger than that of most other Passeres. Body short and rotund, robust-looking; neck short.

♂ larger than ♀ in most species. Bill slender, appearing slightly upcurved; compressed laterally and slightly hooked and notched. Wing short and broad, tip rounded; concave beneath, fitting sides of body closely. Tail short, square or nearly so. Legs long and stout, with strong toes and claws (middle claw sometimes pectinated). Feet not webbed but used for swimming on surface as well as for foraging under water (moving against current) and wading in shallow water. Gait a walk. Plumage soft, long, and dense with thick layer of underlying down. Sombre in colour, largely brown or black, with or without white on head and chest. Sexes alike.

Dipper *Cinclus cinclus*

PLATE: page 1119

Du. Waterspreeuw   Fr. Cincle plongeur   Ge. Wasseramsel   It. Merlo acquaiolo
Ru. Оляпка   Sp. Mirlo acuático   Sw. Strömstare

Field characters. 18 cm; wing-span 25.5–30 cm. Size between wheatear and small thrush; bulk shows in deep chest and belly, made more obvious by short wings and half-cocked tail. Medium-sized, rotund, aquatic passerine, with shape suggesting huge Wren. Plumage generally black-brown, relieved (in west Palearctic races) by broad white throat and chest in adult and white-mottled underparts in juvenile.

Unmistakable (no other congener in west Palearctic), but racial identifications between 'chestnut-bellied' and 'black-bellied' forms not always safe (see Geographical Variation). In Britain, occurrences of 'black-bellied' birds traditionally assumed to stem from movements of northernmost nominate *cinclus*; in rest of Eurasia, geographical variation highly complex, some populations differing within same mountain range. Flight direct and rapid along straight runs of water but more jinking among boulders and round bends. Gait involves walking, running, and occasional hopping, but most obvious action is characteristic bobbing of body, accompanied by downward flick of tail and blinking of white eyelid. Enters and submerges in water freely, walking over stream bed or swimming upstream.

Habitat. Unique in west Palearctic: beside, on, and under swift-running streams and rivers of mountainous and hilly regions and dispersed over middle, higher, and lower middle latitudes, continental and oceanic, between July isotherms of 10–22°C in boreal, temperate, Mediterranean and steppe climatic zones. Will descend locally and infrequently to lowlands where fast-flowing water created, even artificially, and sometimes occurs by lakes or on seashore. Typical habitat contains plenty of rock faces and boulders but mainstay of foraging habitat is shallow water, often with gravel bottom and aquatic or bankside vegetation.

Distribution. BRITAIN. Bred Isle of Man (now recolonized) and Orkneys before 1950. Slight extension of range Midlands and south-central England late 1960s–early 1970s. More recent decrease west Wales, south-west and north-east England, also parts of Scotland may be due in part to stream acidification. FRANCE. Formerly bred Normandy and (to 1930s at least) Brittany. NETHERLANDS. 7 breeding records, including 1993 (1 pair) and 1994 (2 pairs). GERMANY. In lowlands of east, no confirmed breeding 1978–82 (atlas survey), though bred there 1966 (Eberswalde) and 1972 (Altentreptow). ESTONIA. Breeding confirmed pre-1875 and in 1915. Since 1971, frequent cases of breeding, in 1980s perhaps even annual. LATVIA. Breeds more or less regularly in 3 areas shown. BULGARIA. Slight decrease. BELARUS'. Regular all year on one river, but breeding still not proved. UKRAINE. Slight decrease. MOLDOVA. Status uncertain: only July–August records, but almost annual. CYPRUS. Extinct, last recorded 1945. IRAQ. Winter and spring records; probably breeds. MOROCCO. Unknown in Middle Atlas until mid-1960s, and discovered in eastern Middle Atlas only in early 1980s.

Accidental. Spitsbergen, Faeroes, Malta, Tunisia.

Beyond west Palearctic, breeds Iran, and Himalayas east to Baykal region and central China.

Population. Mostly stable, though decreases reported Iberia, Greece, Bulgaria, and Ukraine, and slight increase Denmark. BRITAIN. 7000–21 000 pairs 1988–91; local declines and increases. IRELAND. 6500–8000 pairs. Increase in north-west 1972–82. FRANCE. 10 000–50 000 pairs. BELGIUM. 740 pairs 1973–7. LUXEMBOURG. 250–300 pairs. GERMANY. Estimated 14 000 pairs in mid-1980s perhaps too high: put at 8000–11 000 in early 1990s. In east, *c.* 560 pairs in early 1980s. DENMARK.

Dipper *Cinclus cinclus*. *C. c. gularis*: **1–3** ad fresh (autumn), **4** ad worn (spring), **5** juv. *C. c. cinclus*: **6** ad fresh (autumn), **7** NW Iberian form ('atroventer') ad fresh (autumn). *C. c. aquaticus*: **8** ad fresh (autumn). *C. c. caucasicus*: **9** ad fresh (autumn). *C. c. rufiventris*: **10** ad fresh (autumn).

Up to 2 pairs 1976–81, 4–5 pairs 1992. NORWAY. 5000–25 000 pairs 1970–90. SWEDEN. 5000–50 000 pairs in late 1980s. FINLAND. 250–300 pairs in late 1980s (of which southern population 20–30 pairs). ESTONIA. Up to 5 pairs 1991. LATVIA. 1–5 pairs. POLAND. 400–800 pairs. Now stable, after recent decline. CZECH REPUBLIC. 1000–2000 pairs 1985–9. SLOVAKIA. 3000–6000 pairs 1973–94. HUNGARY. 10 pairs 1979–93. AUSTRIA. Common, and apparently stable (very locally negative influences due to river regulation). SWITZERLAND. 2000–3000 pairs 1985–93. SPAIN. 6200–8000 pairs. PORTUGAL. 1000–10 000 pairs 1978–84. ITALY. 4000–8000 pairs 1983–93. GREECE. 100–300 pairs. ALBANIA. 100–500 pairs in 1981. YUGOSLAVIA: CROATIA. 2000–3000 pairs. SLOVENIA. 1000–3000 pairs. BULGARIA. 1000–5000 pairs. RUMANIA. 35 000–50 000 pairs 1986–92. RUSSIA. 1000–10 000 pairs. UKRAINE. 400–500 pairs 1988. AZERBAIJAN. Uncommon to locally common. TURKEY. 500–5000 pairs. MOROCCO. Scarce to uncommon resident.

Movements. Most populations resident but undertake local post-breeding dispersal movements, often involving altitudinal change. However, north European populations are subject to medium- or long-distance partial migration, some birds staying in breeding areas as long as water continues to flow, even beneath ice perforated by air-holes. In Alps in winter (September or October to February or March) occurs at lower altitudes away from breeding areas, notably at lake-sides, but also found above treeline at up to 2600 m. In winter in Britain, mountain streams regularly vacated in favour of lower reaches; some birds cross watersheds and in severe weather move to coasts and estuaries. Some winter immigration to Britain from Europe.

Food. Large invertebrates of stream beds, especially larvae of caddis flies. Feeds predominantly while submerged, walking on stream bed. In calm water, wings held to side; in rougher water, tail spread and bird progresses by use of wings. Leg movements continue as if on land. To surface, spreads tail, holds wings open and accelerates rapidly up. Actions underwater thus a combination of swimming, walking, and 'flying'. Will submerge like grebe while swimming or 'belly floating' on surface. In calm conditions will walk directly into water and submerge without recourse to swimming; also jumps into water from rocks or directly from flight, and submerges immediately. Main feeding technique under water is to move stones and feed on items exposed underneath. Swallows smaller items under water, others (e.g. fish, larvae of caddis flies) brought to surface.

Social pattern and behaviour. Territorial throughout the year. Dispersion outside breeding season varies. Usually solitary, established breeders of either sex defending individual territories. Monogamous mating system the rule; exceptionally, ♂ bigamous. ♀ alone incubates and broods but both sexes feed young until independent. Breeding territories often separated by unoccupied length of waterway, but sometimes abut. Song given by both sexes, throughout year except when moulting (July–August); resurgence from September associated with establishment of winter territory. Both sexes sing perched or in flight, in almost any part of territory.

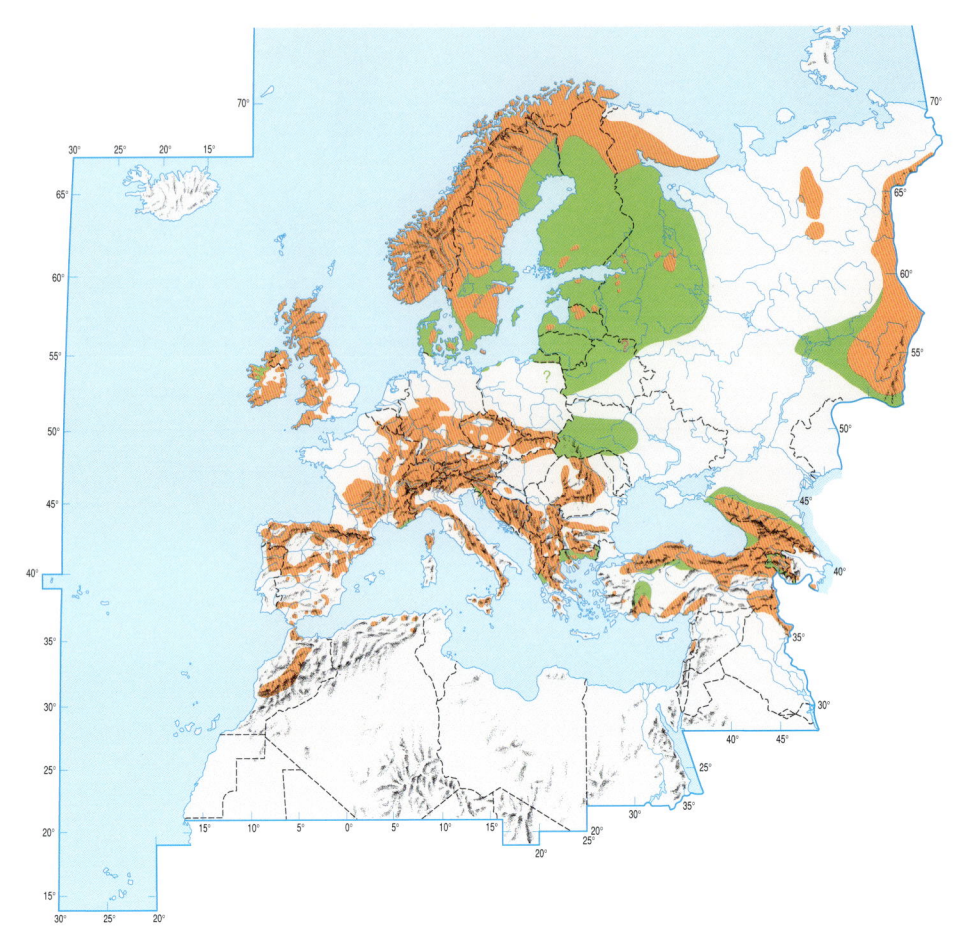

Voice. Song a very sweet rippling warble. Song of ♂ comprises a variety of notes in apparently any order and repeats short phrases and units; song of ♀ a less sweet series of whistles and disconnected units, usually easily distinguishable from ♂ by being more scratchy and less melodious. Commonest call a loud high-pitched 'zit', given usually 2(–4) times in succession, when excited or alarmed, also commonly on approaching nest, perched or more often in flight.

Breeding. SEASON. British Isles: eggs laid last week of February to mid-June, with peak late March to late April; little regional variation. North-west and central Europe: laying begins mid-March, but delayed by cold springs, and becomes later with increasing altitude. North Africa: laying from mid-March to May. Scandinavia: from early May in southern Norway and southern Finland, but from mid-May to early June in northern Finland. 1–2 broods, rarely 3. SITE. In cavities or on ledges above, often overhanging water; much use made of artificial sites where available; also uses nest-boxes. Nest: domed structure of moss and grass stems and leaves, with wide entrance, usually pointing down towards water, with inner cup of stems, rootlets, leaves, and hair. EGGS. Sub-elliptical, smooth and glossy; white. Clutch: 4–5 (1–8). INCUBATION. Mean 16 days (12–18). FLEDGING PERIOD. 20–24 days.

Wing-length: *C. c. gularis*: ♂ 92–99, ♀ 82–91 mm. *C. c. cinclus*: ♂ 95–101, ♀ 87–91 mm.
Weight: *C. c. gularis*: ♂ 60–76, ♀ 50–67 g. *C. c. cinclus*: ♂ 58–84, ♀ 49–72 g.

Geographical variation. Marked and complex; involves colour of head and nape and width of dark feather-fringes on remainder of upperparts (both strongly affected by bleaching and wear), colour of breast and belly (often with marked individual variation), and (slightly) size. 7 races in west Palearctic, but boundaries not clear-cut and some variation within races. *C. c. gularis* from Britain is rather dark above, crown and nape rather dark greyish-brown, drab-brown, or chocolate-brown; underparts rather dark, but with pronounced chestnut-brown breast and central belly, merging into brownish-black vent. *C. c. hibernicus* from Ireland and western Scotland darker above than any other west Palearctic race; crown and nape dark chocolate-brown; rufous of breast duller, darker, and more restricted, belly more extensively brownish-black. Nominate *cinclus* from northern Europe has upperparts similar to *gularis*, but breast and belly completely blackish-brown; some rufous-brown tinge often visible along border with white chest, but not as much as in *gularis*. *C. c. aquaticus* from central Europe (south to north-west France, western Germany, and Czech Republic rather pale above and bright rufous-chestnut or deep chestnut on breast and belly; crown and nape rather pale grey-brown. *C. c. minor* from north-west Africa near *aquaticus* in colour and size of wing, but bill longer. *C. c. caucasicus* from Caucasus and Transcaucasia differs from other west Palearctic races in colour of upperparts; dull grey-brown of crown and nape extends to mantle and scapulars; breast and belly uniform dull grey-brown, rather like crown and nape. *C. c. rufiventris*, from Lebanon, known for only 2 specimens.

Wrens Family Troglodytidae

Tiny to fairly small oscine passerines (suborder Passeres); insectivorous, living mainly in scrub close to ground or on ground close to scrub—though some found in marshes or in rocky areas. 52–60 species in up to 14 genera. Found solely in the Americas, except for a single species—Wren *Troglodytes troglodytes*—in Eurasia (including many islands of North Atlantic and North Pacific). Most species sedentary.

Body short and rotund; neck short. ♂ slightly larger than ♀ in many species. Bill slender and somewhat curved, with sharp tip; usually rather short but fairly long in some terrestrial species. Wing short and rounded. Flight direct and rapid with fast, whirring wing-beats. Tail short to long, often cocked up; feathers soft and often narrow. Leg and toes strong, with long claws; front toes partly fused at bases. Plumage thick and soft; generally closely barred or streaked in various shades of brown though underparts often partly uniform buff to white; often with some contrasting spots on wing and tail; some species have distinct pale supercilium. Sexes alike.

Wren *Troglodytes troglodytes*

PLATE: page 1123

Du. Winterkoning   Fr. Troglodyte mignon   Ge. Zaunkönig   It. Scricciolo
Ru. Крапивник   Sp. Chochín   Sw. Gärdsmyg   N. Am. Winter Wren

Field characters. 9–10 cm; wing-span 13–17 cm. Not smallest but shortest bird in west Palearctic, due to habit of holding tail erect; appears about half size of Dunnock. Tiny, restless, and pugnacious passerine, usually seen at or near ground level and at any distance appearing warm brown overall. At close range, rather long, thin bill, pale buff supercilium, barred wings and flanks, and often cocked tail catch eye.

Unmistakable in west Palearctic. In flight, no characters show and bird becomes just a compact, round-winged, almost bee-like creature. Gait mouse-like, creeping, hopping, and climbing over ground (and less usually along branches).

Habitat. Breeds in west Palearctic throughout middle latitudes and in parts into higher latitudes, from Mediterranean through temperate to boreal zones, with oceanic rather than continental tendency, between July isotherms of *c.* 10–22°C, thus avoiding extremes of cold and heat. In suitable situations will nest above treeline (in Switzerland, to 2000 m or even 2400 m), but mainly in lowlands from sea-level; also a successful colonist of both inshore and offshore oceanic islands. Within predominantly moist mild climatic range, suitable habitat offered by wide variety of low cover and foraging opportunities, including herb and field layers of plant growth (within or outside woodland), crops and aquatic vegetation, fallen trees and branches or heaps of brash, hedgerows, gardens, parks, and shrubberies. Is attracted to earthen banks, stone walls, outhouses and other free-standing structures, and natural crags, fissures, sea-cliffs, and other faces or slopes providing cavities, crevices, and interstices which can be profitably explored or used for roosting or nesting.

Distribution. No evidence of any marked range changes. Denmark. Increase. Belarus'. Mapped as resident breeder throughout, but only small proportion of total population winters. Moldova. Density higher in north than in south. Turkey. Breeding proved only Thrace, Black Sea coastlands and Western Anatolia, likely also in Taurus mountains. Syria. Resident breeder in western mountains. No proof of breeding in lowlands. Jordan. Small resident population in North-West Highlands.

Accidental. Kuwait, Egypt, Azores, Madeira.

Beyond west Palearctic, breeds from northern Iran and south-central Asian mountains east and north to south-east Russia, also in North America south to California and Appalachians.

Population. Generally stable, though marked fluctuations reported many countries after hard winters. Iceland. 1000–3000 pairs in late 1980s; perhaps increasing. Faeroes. 500–1200 pairs in early 1990s; estimate of 250–500 pairs in early 1980s too low. Britain. 7.1 million territories 1988–91. 10-fold increase 1964–74 Britain and Ireland after 2 hard winters. Ireland. 2.8 million territories. France. Over 1 million pairs in 1970s. Belgium. 240 000 pairs 1973–7. Luxembourg. 12 000–15 000 pairs. Netherlands. 300 000–400 000 pairs 1979–85. Germany. 1.52 million pairs in mid-1980s. Population can sink to only 10% of normal years following severe winter. Denmark. 290 000 pairs in 1985; increase. Norway. 100 000–500 000 pairs 1970–90. Sweden. 100 000–500 000 pairs in late 1980s. Now stable, though 10-fold decrease 1975–9. Finland. 20 000–60 000 pairs in late 1980s; slight decrease. Estonia. 50 000–100 000 pairs 1991; stable, following increase in 1950s–60s. Latvia. 200 000–300 000 pairs in 1980s. Lithuania. Stable. Poland. Fairly numerous. Czech Republic. 100 000–200 000 pairs 1985–9. Slovakia. 100 000–200 000 pairs 1973–94. Hungary. 20 000–25 000 pairs 1979–83. Austria. Very common. Switzerland. 250 000–350 000 pairs 1985–93. Spain. 2.1–4 million pairs. Portugal. 100 000–1 million pairs 1978–84. Italy. 1–2.5 million pairs 1983–93. Greece. 50 000–100 000 pairs. Albania. 5000–20 000 pairs 1981. Yugoslavia: Croatia. 160 000–200 000 pairs. Slovenia. 50 000–70 000 pairs. Bulgaria. 500 000–1 million pairs. Rumania. 100 000–200 000 pairs in 1986–92. Russia. 1–10 million pairs. Belarus'.

Wren *Troglodytes troglodytes*. *T. t. troglodytes*: **1–3** ad fresh (autumn), **4** ad worn (spring), **5** juv. *T. t. islandicus*: ad fresh (autumn). *T. t. zetlandicus*: **7** ad fresh (autumn). *T. t. hirtensis*: **8** ad fresh (autumn). *T. t. hyrcanus*: **9** ad fresh (autumn). *T. t. kabylorum*: **10** ad fresh (autumn).

450 000–500 000 pairs in 1990. UKRAINE. 180 000–200 000 pairs in 1988. MOLDOVA. 10 000–14 000 pairs in 1988. AZERBAIJAN. Common. TURKEY. 50 000–500 000 pairs. CYPRUS. Common. ISRAEL. At least a few thousand pairs in 1980s. TUNISIA. Common. ALGERIA. Locally common. MOROCCO. Uncommon to locally common.

Movements. Migratory, partially migratory, and resident. Continental populations exhibit long- and short-distance movements, either on basically north-south axis or altitudinal, with many in southern parts of range sedentary. In Europe and probably in Asia, apparently unable to endure winter further north than *c.* −7°C January isotherm. Some individual birds make substantial journeys, e.g. British birds 50–250 km, Swedish birds up to 2500 km or longer. Movement essentially nocturnal; may begin as early as August.

Food. Chiefly insects, especially beetles; also spiders. Mostly taken from surface of leaves (upper- and underside), twigs, and bark or from crevices in bark, rocks, walls, and leaf litter. Usually feeds within 2 m of ground, lower in winter than in summer.

Social pattern and behaviour. In sedentary populations, ♂♂ territorial for most of year, less so when chick-rearing, moulting, and in winter. Migrants show marked fidelity to winter quarters. Outside breeding season, birds roost communally in hard weather, clustered together in sheltered sites; seldom more than 10 together, but up to 61 recorded. Mating system varies with race and with quality of habitat. Island races thought to be typically monogamous. In western Europe, *c.* ½ ♂♂ monogamous, rest successively polygynous with 2–3(–4) ♀♀ such that broods overlap. In primeval forest, Białowieża (Poland), on average less than 20% of ♂♂ polygynous, never breeding with more than 2 ♀♀ per season, and then only in optimal (ash and alder) habitat. Pair-bond, even in cases of monogamy, never close, ♂ and ♀ associating only for courtship and mating. ♂ may take significant share in care of nestlings in monogamous pairs; polygynous ♂♂ often play little or no part. Song given mainly by ♂, occasionally by ♀, throughout the year except when moulting. Rate of territorial-song shows strong diurnal rhythm; at maximum on leaving roost, then declines both in rate and length until noon, staying sporadic thereafter, though bird may sing briefly before roosting.

Voice. Territorial-song of ♂ a well-structured rattling warble of clear shrill notes, delivered with remarkable vehemence. Varies in length; typically *c.* 4–6 s. Courtship-song usually much abbreviated, softer, sweeter, more warbling, and rapidly repeated. Whisper-song given mainly by ♀, rarely ♂, resembles distant song of Swallow, varying from occasional 'whit' sounds to sustained twittering. Commonest call, a single, double or repeated sharp note, like large pebbles knocked together. Given to communicate presence, mainly by ♂ throughout the year, in almost any context, including presence of predators. Variable in pitch and rate of delivery: often a higher pitched 'tick', like beads knocked together, given by both sexes throughout the year. Alarm-call a rapid chittering or reeling, given rapidly in short series.

Breeding. SEASON. Britain and north-west Europe: egg-laying from late March or early April. Central Europe: up to 1 week later. European Russia: main laying period second half of May. Cyprus: eggs found early April to May. SITE. Very variable, but essentially a hollow, crevice, or hole at ground level or at up to c. 10 m high, average height increasing during season; in side of wall, tree, or steep bank, often inside building, and recorded in numerous artefacts. Nest: domed structure of leaves, grass, moss, and other vegetation, lined with feathers and hair. Building: main structure by ♂, lined by ♀, occasionally by ♂; ♀ very occasionally helps build main nest. EGGS. Sub-elliptical, smooth and glossy; white, sometimes with speckling of black or brown at broad end. Clutch: 5–8 (3–9); ♀♀ breeding with more polygynous ♂♂ lay bigger clutches than those with less polygynous ♂♂. INCUBATION. Average 16 days (12–20). FLEDGING PERIOD. 14–19 days.

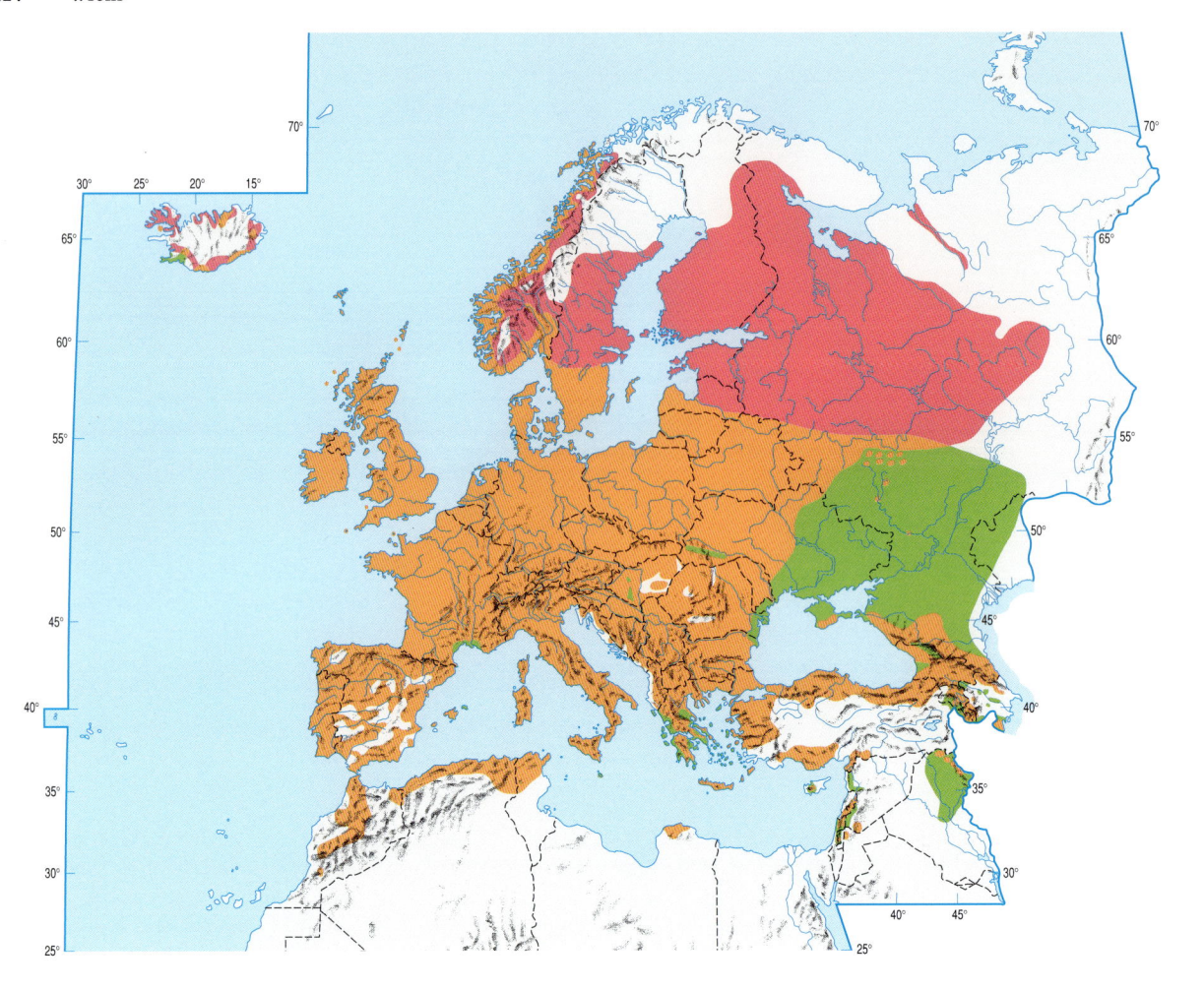

Wing-length: *T. t. troglodytes*: ♂ 47–52, ♀ 45–50 mm.
Weight: *T. t. troglodytes*: ♂ mostly 8–12, ♀ 7–11 g.

Geographical variation. Marked and complex; 13 races in west Palearctic. Variation mainly clinal in continental Europe, but obscured by marked individual variation; some isolated island races more strongly differentiated. Variation involves depth of ground-colour of body, extent and width of barring, size, and relative size of bill, tarsus, and foot. A number of geographical trends apparent. Nominate *troglodytes* occupies most of continental Europe. Wing, tail, bill, tarsus, and foot gradually longer towards north in sequence *fridariensis* (Fair Isle), *hirtensis* (St Kilda), *zetlandicus* (Shetlands), and *borealis* (Faeroes), ending in large *islandicus* from Iceland (in particular, bill of *zetlandicus* and *borealis* proportionately long). Towards south, populations of arid Mediterranean France, central Spain, southern Italy, Sicily, and Greece are more diluted rufous-brown or greyish-brown above, purer white below with more contrasting dark bars, tending towards *kabylorum* (north-west Africa, Balearics, and southern Spain), *koenigi* (Corsica and Sardinia), and *cypriotes* (Crete, Rhodes, Cyprus, and Levant). *T. t. juniperi* from Cyrenaica (Libya) similar to *kabylorum*, but upperparts slightly greyer and bill longer. *T. t. hyrcanus* from Caucasus to Iran close in colour to nominate *troglodytes*, but barring slightly more extensive and bill longer. *T. t. indigenus* from Ireland, northern England, and Scotland slightly darker above than nominate *troglodytes*, less bright and rufous.

Northern Mockingbird *Mimus polyglottos*: **1** ad. Varied Thrush *Zoothera naevia* (p. 1208); **2** ad ♂, **3** ad ♀. Eastern Phoebe *Sayornis phoebe* (p. 1010); **4** ad breeding, **5** 1st winter.

Mockingbirds Family Mimidae

Medium-sized oscine passerines (suborder Passeres); most species terrestrial or living in scrub or low trees close to ground, feeding on ground-living invertebrates, fruits, and seeds. 29–31 species in *c.* 11–13 genera. Most species non-migratory, except those in higher latitudes. 3 species accidental in west Palearctic.

Body rather elongated and thrush-like in most species. Sexes similar in size. Bill medium to long, strong; straight to sharply decurved. Used for flicking over leaves and other ground debris when feeding, also for digging in some species. Wing short in most species, broad, and rounded. Flight rather slow and laboured, with deliberate wing-beats; typically of short duration. Tail long. Legs and toes strong. Plumage variable in colour but generally rather uniform grey or brown, sometimes more rufous; underparts either similar or paler grey-brown to white, often with dark spots; under tail-coverts or tip of tail sometimes contrastingly coloured. Wing frequently has contrasting bars. Sexes virtually alike.

Northern Mockingbird *Mimus polyglottos*

PLATE: page 1125

Du. Spotlijster   Fr. Moqueur polyglotte   Ge. Spottdrossel   It. Mimo poliglotta
Ru. Североамериканский певчий пересмешник   Sp. Sinsonte norteño   Sw. Nordlig härmtrast

Field characters. 22–23 cm; wing-span 31–36 cm. Size close to Blackbird. Somewhat larger than Great Grey Shrike, with slender, pointed bill, relatively short wings, and only rounded, not graduated tail, and longer legs. Plumage suggests grey shrike, since grey inner and black outer wing broadly banded white across bases of primaries and barred white on median and greater coverts, and black tail broadly edged white. Upperparts dusky-grey but head shows pale, staring yellowish eye (not black mask), ruling out shrike; face and underparts greyish-white.

Affinity with thrashers, thrushes, and wrens evident in terrestrial behaviour; thus likely to suggest shrike only momentarily, whilst on perch or in flight. In flight, curiously slow wingbeats make wingmarks even more obvious; on ground, frequently flicks tail and swings it from side to side. Commonest call a loud 'tchack', recalling thrasher; loud song containing much mimicry unlikely to be heard from transatlantic vagrant.

Habitat. Forest edges, parkland, and second-growth habitats

Acadian Flycatcher *Empidonax virescens* (p. 1009): **1** ad breeding, **2–3** autumn (fresh). Brown Thrasher *Toxostoma rufum*: **4–6** ad fresh (autumn). Gray Catbird *Dumetella carolinensis*: **7–9** ad fresh (autumn).

at low altitude in northern temperate to subtropical America, also common in suburbs and in agricultural country. Forages for invertebrates on ground, especially in short grass and on lawns, and also takes fruits; nests in bush or dense tree.

Distribution. Breeds in extreme southern Canada from coast to coast, all of USA, Mexico almost to southern border, and some Caribbean islands.

Accidental. Britain: Saltash (Cornwall), August 1982; Hamford Water (Essex), May 1988. Netherlands: Schiermonnikoog (Friesland), October 1988.

Movements. Poorly understood; probably partial migrant in northern part of range, with distances of up to 800 km recorded, moving south within breeding range though some winter at northern edge of distribution; movements in south-west USA and elsewhere local and complex.

Wing-length: ♂ 111–113, ♀ 103–108 mm (averages).

Brown Thrasher *Toxostoma rufum*

PLATE: page 1126

Du. Rosse Spotlijster Fr. Moqueur roux Ge. Rote Spottdrossel It. Mimo rossiccio
Ru. Коричневый пересмешник Sp. Sinsonte castaño Sw. Rödbrun härmtrast

Field characters. 24 cm; wing-span 29–31 cm. Close to Mistle Thrush in size, but form more babbler-like since bill decurved at tip and rounded tail almost $\frac{1}{2}$ length of bird. Plumage thrush-like, with red-brown upperparts, relieved by double white wing-bar, and black-spotted, white underparts. Pale yellow eye striking.

Unmistakable in west Palearctic. Flight and gait suggest small nimble Magpie. Cocky, robust, but skulking. Calls of vagrant loud, harsh 'tschek' or 'chip'.

Habitat. Breeds in Nearctic middle temperate latitudes in lowlands and sparsely to *c.* 1700 m on elevated plains. Frequents open brushy woods, scattered patches of brush, small trees in open areas, and also shelterbelts, copses, and planted shrubberies in suburbs, where dense thickets offer suitable nest-sites.

Distribution. Breeds in North America from south-east Alberta east to New England and south to Gulf Coast.

Accidental. England: 1, Dorset, November 1966 to February 1967.

Movements. Migratory over northern two-thirds of range, with breeding and wintering ranges overlapping in southern USA, north on east coast to Massachusetts.

Wing-length: ♂ 100–112, ♀ 99–106 mm.

Gray Catbird *Dumetella carolinensis*

PLATE: page 1126

Du. Katvogel   Fr. Moqueur chat   Ge. Katzenvogel   It. Uccello gatto
Ru. Кошачья птица   Sp. Pájaro gato   Sw. Kattfågel

Field characters. 18.5 cm; wing-span 24–26 cm. Size close to Redwing but has different structure, with rounded tail almost as long as body. Plumage essentially dark grey, with black cap and rusty under tail-coverts.

Unmistakable. Flight chat-like, with bursts of strong wing-beats allowing dashes into and out of cover, and flicking and spreading of tail obvious in tight manoeuvres. Hops, using balancing tail movements on ground.

Calls include cat-like mew and harsh 'kak kak kak'.

Habitat. In Nearctic middle temperate latitudes, in continental lowlands. Breeds in thickets, woodland edges, shrubby marsh borders, orchards, parks, and similar habitats with dense vegetation. Ventures into the open only where there is immediate access to dense cover.

Distribution. Breeds in North America from southern British Columbia to Nova Scotia and south to Texas and Georgia.

Accidental. Britain: 1, Jersey (Channel Islands), mid-October to December 1975. Ireland: 1, Cape Clear (Cork), November 1986. Germany: 1 collected Helgoland, October 1840; 1, Leopoldshagen, May 1908.

Movements. Migratory over most of range, with some overlap of breeding and wintering ranges on east coast and in deep south of USA. Winters south to Panama and Caribbean islands.

Wing-length: ♂ 88–96, ♀ 86–92 mm.

Accentors Family Prunellidae

Small oscine passerines (suborder Passeres); mainly terrestrial, often feeding close to scrub or boulders; food mainly insects in summer, but seeds and fruits also important in winter. 13 species in single genus *Prunella*, widespread in Eurasia, with greatest density of species in mountains of central Asia. Some species sedentary, others migratory or move altitudinally in winter. 5 species in west Palearctic, all breeding.

Body compact; neck short. Sexes of similar size. Bill peculiar: of medium length, tapering rather evenly to point; hard, wide at base with laterally swollen appearance; culmen rounded in cross-section. Used for flicking over leaves and other debris when feeding. Wing short to moderately long, rounded or rather pointed. Flight strong and rapid in most species, straight or undulating; usually sustained over short distances only and made at low height. Tail moderately long in most species, short in others; tip square or notched. Leg fairly short with strong toes; hind claw the longest. Usual gait a hop or shuffling walk, typically while moving close to ground with legs well bent; some species run. Plumage thick and rather coarse. Generally stone-coloured, tinged with grey, brown, or rufous; often streaked above; uniform dull grey or buff below, with contrasting black or black-and-white spotted or barred throat-patch and with bright rufous or deep buff chest or flanks; many species have uniform grey patch at side of neck; often a pronounced supercilium. Sexual differences slight.

Dunnock *Prunella modularis*

PLATES: pages 1129, 1136

Du. Heggemus   Fr. Accenteur mouchet   Ge. Heckenbraunelle   It. Passera scopaiola
Ru. Лесная завирушка   Sp. Acentor común   Sw. Järnsparv

Field characters. 14.5 cm; wing-span 19–21 cm. Same length as House Sparrow but less bulky, with fine bill, evenly domed head, and slimmer rear body and tail. Sleek but plump, ground-creeping passerine, with warbler-like bill, rather round head, and constant, seemingly nervous wing-twitching. Plumage essentially grey on head and chest, brown elsewhere (except for whitish belly), copiously streaked on upperparts and flanks.

Though without striking markings, has distinctive character. Little risk of confusion with congeners but in brief view can also be mistaken for other small ground-hugging passerines (e.g. juvenile Robin, small bunting, dark warbler). Flight normally low, quite fast and whirring, with rather round wings and quite long, full tail obvious as bird ducks into cover. Gait includes characteristic mouse-like shuffle or creep (with legs almost hidden), as well as more active hopping, with simultaneous flicking of wings and occasional jerking of tail.

Habitat. In upper and middle latitudes, mainly in temperate but marginally in subarctic, boreal, and Mediterranean zones (and in south especially in montane zone), between July isotherms 13–26°C. Apparently, like congeners, Dunnock evolved in scrub and stunted coniferous arctic-alpine and wooded tundra habitats which are still occupied in south and north-east of range. In southern parts occurs mainly near tree-line in mountains. In north of range, still mainly in spruce but also in mixed and broad-leaved woodlands, especially along rivers and streams. In Belgium, departure from montane and coniferous habitats much more marked, although young conifer plantations and conifers in parks and gardens still used, but also thickets, brambles, hedges, wooded marshes, and edges of large forests, main population being in lowlands. In Britain shift away from montane and coniferous habitat is almost complete, having occurred apparently *c.* 200 or more years ago. Now a pioneer species, commonly invading a wide variety of scrub-grown situations, it has adapted to field hedgerows, farms, railway embankments and cuttings, churchyards, parks, gardens, and vacant urban land, as well as many semi-natural bushy and shrubby areas, including both inshore and offshore islands where high winds and exposure inhibit forest growth.

Distribution. Few changes reported. Britain. Colonized Outer Hebrides and Orkneys (Scotland) in 2nd half of 19th century, and bred Fair Isle (Shetland) 1974. Germany. Expanded range in southern Bayern in early 1980s. Now stable overall. Sweden. Expanded range to south-east 1900–1950s. Hungary. Range increase. Azerbaijan. No information from Little Caucasus. Range probably more extensive than shown. Algeria. May have bred recently, but no proof.

Accidental. Bear Island, Jan Mayen, Iceland, Faeroes (probably overlooked almost annual visitor), Kuwait, Libya, Algeria, Mauritania.

Beyond west Palearctic, breeds only in northern Iran. Introduced to New Zealand.

Population. Increased markedly Finland and Estonia, less so Netherlands, Poland, Czech Republic, Hungary, and Austria; decreased Britain, Denmark, mostly stable elsewhere. Britain. 2 million territories 1988–91; shallow decline since mid-1970s. Ireland. 810 000 territories 1988–91. France. Under 1 million pairs in 1970s. Belgium. 210 000 pairs 1973–7. Luxembourg. 12 000–15 000 pairs. Netherlands. 125 000–200 000 pairs 1979–85. Increase since 1960s. Germany. 1.49 million pairs in mid-1980s. Denmark. 20 000–200 000 pairs 1987–8; decreased 1976–94. Norway. 500 000–1.5 million pairs 1970–90. Sweden. 1–2.5 million pairs in late 1980s. Finland.

Dunnock *Prunella modularis*. *P. m. modularis*: **1** ad ♂ fresh (autumn), **2** ad worn (spring), **3** ad ♀ fresh (autumn), **4** juv. *P. m. occidentalis*: **5** ad ♂ fresh (autumn). *P. m. obscura*: **6** ad ♂ fresh (autumn). *P. m. hebridium*: **7** ad ♂ fresh (autumn).

300 000–500 000 pairs in late 1980s. ESTONIA. 50 000–100 000 pairs in 1991; marked increase since 1950s, still continuing. LATVIA. 150 000–300 000 pairs in 1980s. LITHUANIA. Stable. POLAND. 150 000–300 000 pairs; still slowly increasing. CZECH REPUBLIC. 200 000–400 000 pairs 1985–9. SLOVAKIA. 300 000–500 000 pairs 1973–94. HUNGARY. 1200–1500 pairs 1979–93. AUSTRIA. Common; increasing since 1960s. SWITZERLAND. 110 000–220 000 pairs 1985–93. SPAIN. 400 000–1 million pairs. PORTUGAL. 10 000–100 000 pairs 1978–84. ITALY. 100 000–200 000 pairs 1983–93. GREECE. 1000–2000 pairs. ALBANIA. 500–1000 pairs in 1981. YUGOSLAVIA: CROATIA. 5000–6000 pairs. SLOVENIA. 20 000–30 000 pairs. BULGARIA. 1000–10 000 pairs. RUMANIA. 150 000–250 000 pairs 1986–92. RUSSIA. 1–10 million pairs. BELARUS'. 120 000–200 000 pairs in 1990. UKRAINE. 150 000–160 000 pairs in 1988. AZERBAIJAN. Uncommon. TURKEY. 10 000–50 000 pairs.

Movements. Resident, partial migrant, and, in northern and central Europe, total migrant. Main continental populations, especially those breeding in northern areas (Fenno-Scandia, northern Germany, Poland, and northern FSU east to Urals) and to lesser extent those from southern areas (central France to Corsica, Sardinia, and central Italy), move to winter in south-west Iberia, Mediterranean area, southern FSU, and Turkey. Considerable passage to Mediterranean islands, but apparently rather rare in North Africa, and only at all regular in northern Tunisia. Movement on wide front and close to a north-east/south-west axis. Main migration periods September–November and March–April.

Food. Largely insects, plus significant proportion of small seeds in winter. Predominantly a ground feeder, spending much of time in cover. Feeds under bushes, hedges, young conifers, and piles of twigs, and amongst roots and leaf litter. In summer feeds more often in vegetation, up to 8 m above ground. Major foraging technique is a steady hop along ground with body horizontal, accompanied by ceaseless pecking movements; always moves forward and does not retrace steps. Will also pick seeds from seed-heads, and other food directly from vegetation.

Social pattern and behaviour. Essentially solitary outside breeding season, occupying individual home-ranges, though local feeding aggregations may give appearance of gregarious behaviour. Home-ranges of ♂ and ♀ are independent, and occupants of home-ranges are dominant over intruders. Ranges of 2–6 birds (of either sex) may overlap locally and all may congregate temporarily at rich feeding patches within areas of overlap. Typically, mating system essentially monogamous or involves polyandrous trio of birds (♀ plus 2♂♂); in addition, polygyny regular (♂ plus 2, occasionally 3, ♀♀), and 'polygynandry' occasional (2–3 ♂♂ plus 2–4 ♀♀). Trios arise when ♀ frequents territories of 2 ♂♂, and these territories then tend to coalesce; conflict between the 2 ♂♂ gradually declines and one (usually the older) becomes dominant over the other, though both use same perches for singing. Less often, trio formed by ♂ persistently intruding on territory of monogamous pair until he is accepted. Generally, ♂♂ of a trio that have copulated with a ♀ feed her young, and those that have not do not. Territory and nest-site fidelity of ♀♀ well marked; ♂♂ more likely to move from year to year, depending on opportunities to occupy more favourable habitat and feeding areas. Song (normally by ♂ only, occasionally by ♀) usually delivered from exposed perch with wide range of heights depending on vegetation structure; occasionally given in flight. Main song-period from early spring to end of breeding season. Aggressive behaviour over territories involves song duels and, at close quarters, flicking of wings, either rapid, or slower ('wing-waving') with wings held up for appreciable fraction of second, either both wings together or alternately. Mating behaviour peculiar, unique to genus. Pre-copulation display elaborate. ♀ crouches, ruffles body feathers, shivers wings, and quivers tail, raising it to expose cloaca. ♂ hops from side to side behind her and pecks cloaca for up to 2 min before mating. Copulation itself extraordinarily brief: ♂ appears to jump over ♀ (and action has been so interpreted by some observers), cloacal contact lasting for fraction of second. During pecking by ♂, ♀'s cloaca becomes pink and distended and from time to time makes strong pumping movements, sometimes ejecting a droplet which contains a mass of sperm; ♂ looks intently at such droplets and copulation follows immediately after. Pecking of ♀'s cloaca more prolonged when another ♂ has spent more time with her. This process interpreted, in view of complex mating system, as part of ♂'s strategy of attempting to increase his share of paternity.

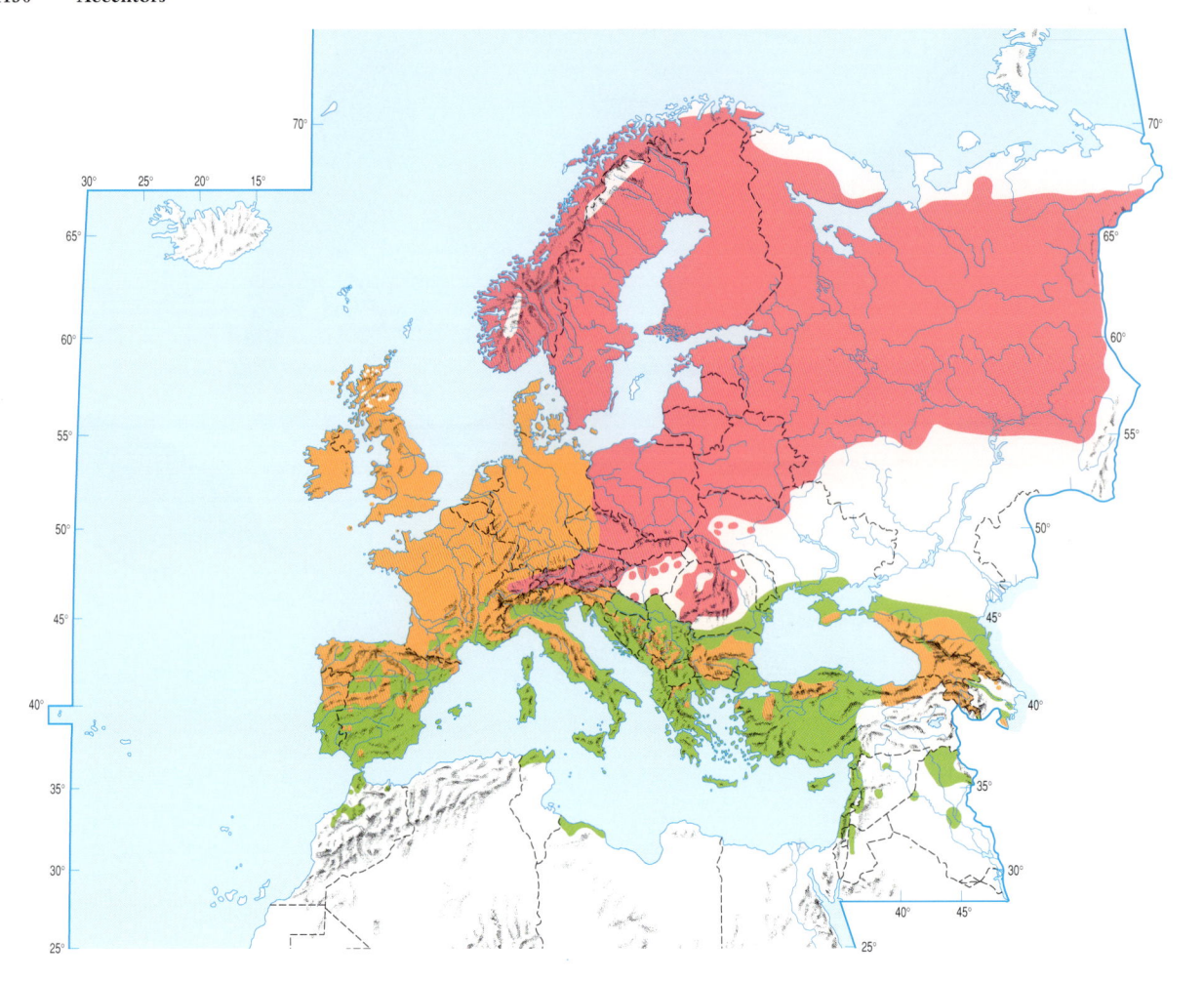

Voice. Song of ♂ a short, undistinguished, rather high-pitched warble, somewhat recalling short section of song of Wren. Song mostly 2–3.5 s long, occasionally up to 4.5 s; delivered in bouts of usually 2–8. Song of ♀ infrequent but probably regular in some individuals; shorter and less elaborate than ♂'s. Most familiar call a loud single 'seep'; sometimes 2(–3) given in quick succession, intervals only a little longer than calls themselves. This call used almost exclusively outside breeding season (September–February) by both sexes. Trill-call, rapid 'ti-ti-ti' or 'he-he-he' of 3–4(-5) notes, fundamentally a contact-call, uttered by both sexes, often alternately by paired ♂ and ♀. Often immediately precedes flight, and may be continued in flight and after landing. Alarm-call a loud single note, similar to seep-call but somewhat higher pitched. Varies in volume and rate of repetition: at most intense, loud and piercing, uttered with bill wide open.

Breeding. SEASON. Britain and north-west Europe: egg-laying from late March or early April. Leningrad region: laying begins mid-May. Sweden: eggs laid late April to June. 2 broods, occasionally 3. SITE. In bush, hedge, or low tree, 0.5–3.5 m above ground, sometimes in side of bank; normally well concealed. Occasionally uses old nest of another bird. Nest: quite substantial cup of twigs, leaves, stems, roots, and other plant parts, lined with wool, hair, moss, and sometimes feathers. EGGS. Sub-elliptical, smooth and glossy; bright blue, rarely with a few reddish spots. Clutch: 4–6 (3–7). INCUBATION. 12–13 days. FLEDGING PERIOD. 11–12 days.

Wing-length: Nominate *modularis*: ♂ 68–74, ♀ 65–72 mm.
Weight: Nominate *modularis*: ♂ mostly 17–25, ♀ 16–24 g.

Geographical variation. 8 races in west Palearctic, forming cline across Europe. Variation mainly involves general colour; size varies little. *P. m. occidentalis* from England, Wales, and eastern Scotland differs from nominate *modularis* from central and northern Europe in more extensively dusky olive-brown crown and hindneck, marked with slightly longer and broader black-brown streaks; mantle and scapulars slightly duller rufous-brown; underparts on average slightly darker grey. *P. m. hebridium* from Inner and Outer Hebrides heavily mottled black-brown on crown and hindneck; mantle, scapulars, and tertials have broad and poorly defined black feather-centres and rather narrow dull brown (not rufous-brown) feather-sides; grey of underparts slightly darker than in *occidentalis*, extending further down belly. *P. m. mabbotti* from south-west and south-central France, and *P. m. meinertzhageni* from Yugoslavia and Bulgaria, poorly differentiated from nominate *modularis*. *P. m. euxina* from Asia Minor and *P. m. fuscata* from Crimea a little paler than nominate *modularis*; *P. m. obscura* from Caucasus area and eastern Turkey most distinct, browner than other races, underparts more extensively washed olive-brown on paler grey ground than any other race; ground-colour of mantle and scapulars duller rufous-brown than nominate *modularis*, streaks on upperparts browner, less black; throat to breast extensively marked with off-white or pale buff feather-tips, forming scaly pattern.

Siberian Accentor *Prunella montanella*

PLATES: pages 1132, 1136

Du. Bergheggemus   Fr. Accenteur montanelle   Ge. Bergbraunelle   It. Passera scopaiola asiatica
Ru. Сибирская завирушка   Sp. Acentor siberiano   Sw. Sibirisk järnsparv

Field characters. 14.5 cm; wing-span 21–22.5 cm. Close in size to Dunnock; marginally smaller than Black-throated Accentor. Similar in form to Dunnock but plumage much more distinctive with bold buff-cream supercilium, black cheek-patch, little-streaked rufous-brown upperparts, and ochre-buff chest splashed there and along flanks with black and rufous-brown.

Easily distinguished from typical Black-throated Accentor by lack of black or largely black throat, more uniform upperparts (streaked rufous rather than blackish), rufous-brown fringes to larger wing-feathers, and rufous-brown (not grey-brown) flank-streaks. Flight, gait, and behaviour apparently as congeners.

Habitat. In boreal and subarctic zones of continental upper latitudes. Breeds along river banks, in tangles of shrubbery, chiefly willows, and also on mountains to treeline, where it occurs in stunted spruces and birch crowns in sparse woodlands.

Distribution and population. RUSSIA. Record of singing ♂ near Nar'yan-Mar on Pechora river (67°37′N) July 1992 may indicate range extends considerably further west than shown. Population perhaps 100–1000 pairs; stable.

Accidental. Denmark, Sweden, Finland, Poland, Czech Republic, Slovakia, Austria, Italy, Lebanon. Record in Norway October 1992 either this or Black-throated Accentor.

Beyond west Palearctic, breeds in 2 narrow (and discontinuous) bands converging in eastern Siberia: across northern Siberia and in southern Siberia from north-east Altai eastwards.

Movements. Migratory, whole population wintering in Korea

Siberian Accentor *Prunella montanella*: **1** ad fresh (autumn), **2** ad worn (spring), **3** juv. Radde's Accentor *Prunella ocularis* **4** ad fresh (autumn), **5** ad worn (spring), **6** juv. Black-throated Accentor *Prunella atrogularis*: **7** ad fresh (autumn), **8** ad worn (spring), **9** juv.

and eastern China. In north of range, vacates northern Urals early September, returning early June.

Food. Diet largely insects; in winter, also seeds. Forages on ground, pecking at soil, dead leaves, and grass; also regularly in trees and bushes.

Social pattern and behaviour. Very little known.

Voice. Song of ♂ close to Dunnock, more melodious than Black-throated Accentor. Contact-call a quiet trisyllable 'dididi' or 'tsee-ree-see'.

Breeding. SEASON. Siberia: June–July. SITE. In thick shrub or fork of low tree, close to main stem. Nest: compact cup of twigs, plant stems and leaves, and moss, lined finer material and hair. EGGS. Sub-elliptical, smooth and glossy; deep blue-green. Clutch: 4–6. INCUBATION. 10 days at one nest. (Fledging period not recorded.)

Wing-length: ♂ 72–78, ♀ 70–73 mm.
Weight: ♂♀ mostly 17–20 g.

Radde's Accentor *Prunella ocularis*

PLATES: pages 1132, 1136

DU. Steenheggemus   FR. Accenteur de Radde   GE. Steinbraunelle   IT. Sordone di Radde
RU. Пестрая завирушка   SP. Acentor alpino perso   SW. Kaukasisk järnsparv

Field characters. 15.5 cm; wing-span 22–23 cm. Noticeably larger than Dunnock and Siberian Accentor. Rather bulky accentor, with basic plumage pattern most like Siberian Accentor but colours differing in bold white supercilium, black (not rufous) streaks on upperparts, and greyer rump and wings; usually shows broken malar stripe.

Flight typical of genus but hops on straight legs (not bent like Dunnock); also walks with twitching and hopping movements.

Habitat. In lower middle and lower continental latitudes, in warm arid zone. In Transcaucasia, breeds at altitudes between 2000 m and 3000 m on rocky or stony mountain slopes carrying scrub vegetation, chiefly junipers. In the Ala Dagh (Turkey), with dry Mediterranean climate, found breeding in area of sheep and goat pastures and gullies with scattered bushes of barberry *Barbarea* between 2400 m and 2600 m. Winters at lower altitudes in shrubby growth bordering mountain streams.

Distribution and population. Poorly known. Apparently rare and with fragmented distribution; main threat habitat degradation. ARMENIA and AZERBAIJAN. Dispersing birds (including ♂♂ still in song) in July–August main source of erroneous ideas about extent of breeding range. Total population probably not exceeding several hundred pairs. TURKEY.

May be more widely distributed than shown. Population perhaps 500–5000 pairs, but extrapolations from transect counts in south-east suggest considerably higher.

Winter visitor (probably regular) Israel and Syria.

Beyond west Palearctic, breeds only in Iran (no proof of breeding in Kopet-Dag, Turkmenistan).

Movements. Apparently resident in Turkey and Caucasus. Populations in Iran winter south of breeding ranges, at lower altitude.

Food. Small seeds and insects. Forages mainly on ground among snow, rocks, scrub, and low vegetation, and alongside streams.

Social pattern and behaviour. Closely similar to Dunnock, though mating system not yet elucidated.

Voice. Song of ♂ quiet, sweet, and gentle, with twittering or clear bubbling quality and closely resembling Dunnock rather than Alpine Accentor. Contact-alarm calls include a 'trill' of 3–4 syllables, and a 'seep', like Dunnock but quieter, thinner, and less penetrating.

Breeding. SEASON. Southern FSU: nests found June–August. SITE. In low bush, in one case nearly 1 m above ground. Nest: cup of twigs, leaves, and stems, with lining of finer material, including hair, wool, grass, and moss. EGGS. Sub-elliptical, smooth and glossy; blue. Clutch: 3–4. INCUBATION. 11–12 days. FLEDGING PERIOD. Leave nest at 12 days, before able to fly.

Wing-length: ♂ 77–79, ♀ 73–74 mm.

Black-throated Accentor *Prunella atrogularis*

PLATES: pages 1132, 1136

Du. Zwartkeelheggemus   Fr. Accenteur à gorge noire   Ge. Schwarzkehlbraunelle   It. Sordone golanera
Ru. Черногорлая завирушка   Sp. Acentor gorginegro   Sw. Svartstrupig järnsparv

Field characters. 15 cm; wing-span 21–22.5 cm. Slightly larger than Dunnock and Siberian Accentor. Of similar form to Siberian Accentor but with more heavily marked head, black on chin and throat, black-streaked brown back, and rather pale brown rump. Juvenile has more diffuse head pattern and sometimes only mottled throat.

Birds with full black bib of breeding season unmistakable, but some show only speckled throat and are less distinctive, suggesting both Siberian Accentor and Radde's Accentor. Observer faced with dark-cheeked accentor must concentrate on colour of and heaviness of streaking on both upper- and underparts and on throat pattern. Flight, gait, and behaviour much as Dunnock, but more robust and less shy, perching in open even when disturbed.

Habitat. Breeds in upper and middle continental latitudes, in north in subalpine belt in clumps of stunted spruce shrubs, while in central Asia it inhabits tall conifer forests; also nests in broad-leaved forests and in scrub with plenty of juniper, or in impassable thickets. Avoids open areas, living in low dense and often thorny bushes or on ground. In India, winters in hills up to *c.* 2500 m, but mostly below 1800 m, in scrub jungle, tea gardens, orchards, and bushes near cultivation.

Distribution and population. RUSSIA. Still poorly known, but not confined to northern Urals as previously supposed: small isolated 'colonies' discovered eastern coast of Mezen' bay (White Sea, near Arctic Circle) in 1950s, and breeding proved (1 nest found, up to 5 singing ♂♂) near Nar'yan-Mar (on Pechora) 1993, single bird also seen there August 1994. Total population tentatively estimated at 1000–2000 pairs.

Accidental. Germany, Sweden, Finland, Israel, Kuwait. For Norway, see Siberian Accentor.

Beyond west Palearctic, breeds in south-central Asian mountains.

Movements. Urals population (nominate *atrogularis*) migratory, central Asian population (*huttoni*) partially so, also moving altitudinally. Combined winter quarters lie in Afghanistan, Pakistan, Kashmir, and mountains of central Asia.

Autumn departures from Urals begin August, birds subsequently appearing scattered widely to south and south-east. Reaches Orenburg (southern Urals) in October.

Food. Diet largely insects, supplemented by seeds (mainly in winter). Will also take other small arthropods. Feeds on ground, favouring woods, dense shrubs, grassy clearings, and stream banks; in winter, normally in patches of weeds, reeds, bushes, and alongside ditches.

Social pattern and behaviour. Little information; no indication of major differences from congeners.

Voice. Song closely resembles that of Dunnock and Siberian Accentor. Calls include quiet 'teeteetee'.

Breeding. SEASON. Urals: eggs found June. SITE. On branch of tree or shrub. Nest: cup of twigs and moss, with some grass leaves and stems, lined with finer material and hair. EGGS. Sub-elliptical, smooth and glossy; deep blue-green. Clutch: 3–5 (1–6). INCUBATION. 11–14 days. FLEDGING PERIOD. 11–14 days.

Wing-length: ♂ 72–76, ♀ 70–74 mm.

Geographical variation. *P. a. huttoni* from mountains of central Asia has longer bill and more rounded wing-tip than nominate *atrogularis* from north-west Russia, and black of throat directly bordered by buff of chest, without white gorget.

Alpine Accentor *Prunella collaris*

PLATES: pages 1135, 1136

Du. Alpenheggemus   Fr. Accenteur alpin   Ge. Alpenbraunelle   It. Sordone
Ru. Альпийская завирушка   Sp. Acentor alpino   Sw. Alpjärnsparv

Field characters. 18 cm; wing-span 30–32.5 cm. At least 25% larger and much more robust than any other accentor. Quite strong-billed, bulky accentor, with form somewhat recalling lark or pipit. Plumage patterned as Dunnock but much more colourful, being basically dull blue-grey, with mottled brown back, white-edged black median coverts forming striking panel across forewing, white-speckled throat, rufous-splashed flanks, and white-tipped dark brown tail.

Unmistakable in close view, but much less distinctive at long range, with pale-speckled throat and chestnut flanks difficult to see; overall pattern and colour then suggest dark Skylark or pipit, this appearance enhanced by undulating flight, winter habit of feeding in flocks, and rather lark- or pipit-like calls. Flight more free than any west Palearctic congener, with fluent action and undulating progress. Song-flight short, but sequence recalls lark. Gait a quick walk varied with little runs and hops. Carriage more erect than Dunnock, with head usually held higher. Flutters and jerks wings and tail like Dunnock.

Habitat. Breeds exclusively in mountain ranges of middle latitudes, from *c.* 1800–2000 m up to snowline; normally up to *c.* 2600–3000 m in Switzerland and in Caucasus, but extralimitally to 4000 m in central Asia and even higher in Himalayas. Prefers patches of alpine grassland strewn copiously with boulders or large stones, often on flat plateaux but sometimes on slopes or scree, and almost always well above treeline and relatively free of shrub growth. A ground-feeder and ground-nester, occasionally perching on low bush or plant-stem, and singing from rock or ground or in low song-flight. In winter, sometimes stays in breeding area but usually shifts to lower slopes, remaining commonly above 1800 m, sometimes entering villages or visiting chalets, hotels, etc. Rarely and locally shifts to rocky and scrubby ground in lowlands, or migrates further afield.

Distribution. Slight decrease in breeding range in Ukraine only change reported. GERMANY. Part of Alpine population, breeding only above treeline, so German population can only be small. PORTUGAL. May breed, but no proof. TURKEY. Local. ALGERIA. First breeding record 1978 in Djurdjura mountains in north; also present there 1979.

Accidental. Britain, Belgium (perhaps escapes), Luxembourg, Netherlands, Denmark, Norway, Sweden, Finland, Hungary, Jordan.

Beyond west Palearctic, breeds from Iran east across central Asia to south-east Russia and Japan.

Population. Apparently mostly stable. FRANCE. Under 10 000

Alpine Accentor *Prunella collaris*. *P. c. collaris*: **1** ad ♂ fresh (autumn), **2** ad worn (spring), **3** ad ♀ fresh (autumn), **4** juv. *P. c. subalpina*: **5** ad ♂ fresh (autumn). *P. c. montana*: **6** ad ♂ fresh (autumn).

pairs in 1970s. GERMANY. *c.* 950 pairs in mid-1980s. POLAND. 230–380 pairs. CZECH REPUBLIC. 15–20 pairs 1985–9. SLOVAKIA. 300–400 pairs 1973–94. AUSTRIA. Common. SWITZERLAND. 4000–7000 pairs 1985–93. SPAIN. 17 000–20 000 pairs. ITALY. 10 000–20 000 pairs 1983–93. GREECE. 1000–2000 pairs. ALBANIA. 500–1000 pairs 1981. YUGOSLAVIA: CROATIA. 800–1000 pairs. SLOVENIA. 200–300 pairs. BULGARIA. 1000–10 000 pairs. RUMANIA. Common at high altitudes, especially on scree slopes. 4000–6000 pairs 1986–92. UKRAINE. 40–120 pairs in 1988; slight decrease. AZERBAIJAN. Uncommon to locally common. TURKEY. 1000–10 000 pairs. MOROCCO. Scarce.

Movements. Resident or subject to local altitudinal or more distant movements. Most descend in winter below snowline or seek snow-free patches. Wintering in lowland southern Europe apparently widespread though birds presumably sparsely distributed and occurrences perhaps irregular. A weak but distinct southerly movement occurs through several Alpine passes in October. In France, birds from Alps reach

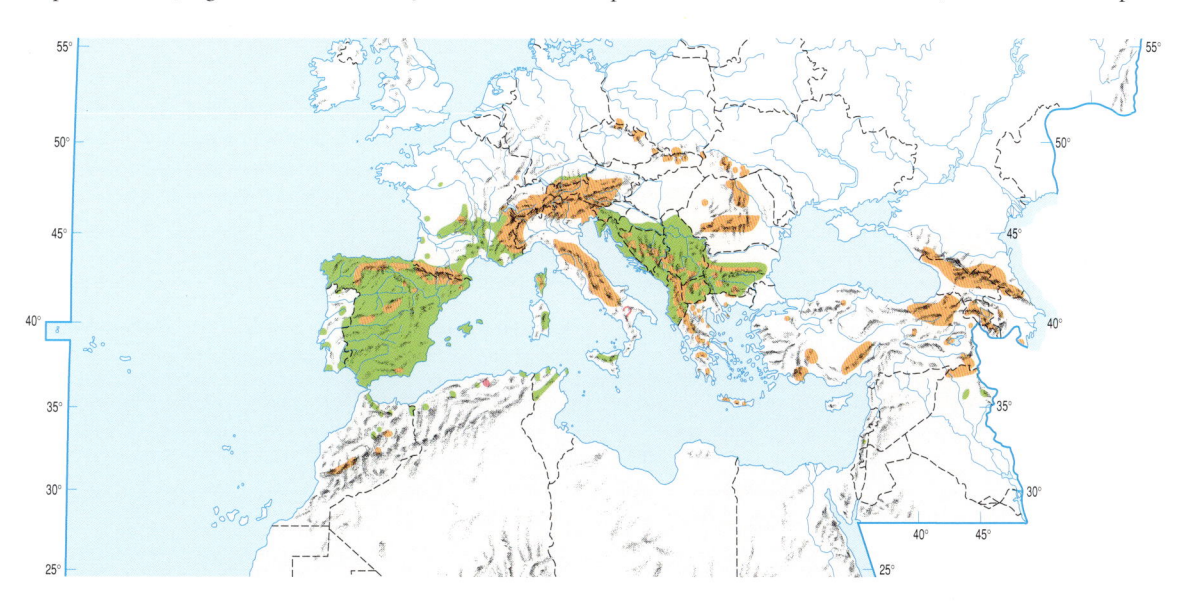

Dunnock *Prunella modularis* (p. 1128): **1–2** ad fresh (autumn). Siberian Accentor *Prunella montanella* (p. 1131): **3–4** ad fresh (autumn). Radde's Accentor *Prunella ocularis* (p. 1132): **5–6** ad fresh (autumn). Black-throated Accentor *Prunella atrogularis* (p. 1133): **7–8** ad fresh (autumn). Alpine Accentor *Prunella collaris*: **9–10** ad ♂ fresh (autumn).

Basse-Provence and Côte d'Azur. Upland breeding areas start to be fully reoccupied from mid-March, but mainly April.

Vagrants occur west and north to Britain and Fenno-Scandia. Less regular now in Britain than previously: only 10 records in period 1958–95, though total of 29 before that; mostly August–January, with a few March–June.

Food. Largely insects, plus significant proportion of plant seeds. Food taken principally on ground, bird moving with quick small hops or walking quietly. Sometimes makes aerial sallies and will chase active prey on foot. Forages among rocks, stones, lichens, moss, grassy vegetation, snow fields, and streams; particularly favours edge of melting snow.

Social pattern and behaviour. More gregarious than Dunnock outside breeding season. Mating system probably essentially monogamous and pair-bond of seasonal duration, but regular occurrence of extra helpers at nests, and of birds in threes and fours in breeding season, suggests more complex systems perhaps as in Dunnock. ♂ sings mainly from rock, stone, edge of rocky shelf, sometimes from low bush or other plant or flat ground, also in flight. Song-flights short and direct, or with high soaring and hovering and some gliding with wings held stiffly up in steep **V**. Mating behaviour has much in common with Dunnock. In soliciting-posture, ♀ flattens and crouches, wings half-spread and quivering; holds tail closed and steeply raised, moving it rapidly from side to side; ruffles vent feathers, exposing bright rosy-red cloaca which is constantly in motion; copulation extremely rapid and brief.

Voice. Noisy on breeding grounds where density high, also in autumn feeding flocks. Song of ♂ a phrase of varying length or several in succession giving a varied, well-sustained, continuous chattering warble at moderate speed; slightly slower than Dunnock and generally superior to that species, being more developed and musical; often likened to Skylark. Commonest and most characteristic call typically rolled or rippling. Often loud, clear and pleasant sounding, though sometimes quieter and, with oddly ventriloquial effect, can be difficult to locate. Number of syllables variable and renderings include 'tchirririRIP', 'truiririp', 'tritritri', etc. Other calls include a more tinkling variant of this and a shorter, huskier 'churrp' or 'teurrp'.

Breeding. SEASON. Pyrénées and Alps: egg-laying from late May. Little apparent variation across range. 2 broods. SITE. In crevice in cliff or rocks, or between boulders. Nest: loosely made of grass leaves and stems, with neat inner cup of moss lined with hair and feathers. EGGS. Sub-elliptical, smooth and glossy; uniform pale blue. Clutch: 3–4(–6). INCUBATION. (11–)14–15 days. FLEDGING PERIOD. *c.* 16 days or more.

Wing-length: *P. c. collaris*: ♂ 101–110, ♀ 95–100 mm.
Weight: *P. c. collaris*: ♂♀ mostly 37–43 g.

Geographical variation. Slight in west Palearctic; 3 races recognized. Nominate *collaris* occupies most of west Palearctic range; *P. c. subalpina* Balkans and western Turkey; *P. c. montana* northern and eastern Turkey, and Caucasus. Plumage becomes greyer towards east.

Chats, Thrushes Family Turdidae

Small to medium-sized oscine passerines (suborder Passeres); of several adaptive types, feeding on invertebrates (often spiders, worms, and snails as well as insects) and fruits (often berries). About 300 species, most occurring in woodland, parkland, and scrub, though many feed on ground and some are entirely terrestrial; a few closely attached to water. Almost cosmopolitan in distribution but with largest diversity in tropical Africa and Asia. Many species migratory. Well represented in west Palearctic.

Turdidae here treated as comprising 2 subfamilies: Turdinae (chats, thrushes) and Enicurinae (forktails). Latter wholly extralimital. Within Turdinae, chats sometimes considered to form separate tribe ('Erithacini' or 'Saxicolini') from true thrushes ('Turdini'). 38 species of chats in west Palearctic: 4 accidental, 1 migrant only (Eversmann's Redstart *P. erythronotus*), and remainder breeding. 19 species of thrushes: 11 accidental, rest breeding.

Body compact in most species; neck short. Sexes generally of similar size. Bill usually slender but fairly stout in some species; of medium length. Used by some species (e.g. Blackbird) for flicking aside debris when searching for food. Wing shape varied, from short and rounded to long and pointed. Flight varied, from weak to strong, straight to undulating. Tail of medium length to fairly long in many species but short in some and very long in others; mostly square or rounded. Leg and foot usually strong. Foot used in feeding by a few species at least (e.g. Blackbird, which scratches backwards with one foot to uncover food hidden in earth or ground litter). Gait a run and/or hop.

Plumage generally cryptic and often spotted below in larger thrushes; much brighter in some thrushes (e.g. *Monticola*) and in many chats, especially on side of head, underparts, and tail where often a contrasting pattern. Occurrence of sexual dimorphism highly variable: within many genera, sexes alike in some species, dissimilar in others, with ♀ the duller, more cryptic sex. Nest-building, incubation, and brooding by ♀ alone; care of young by both sexes. Juvenile plumage always cryptic and typically spotted.

Rufous Bush Robin *Cercotrichas galactotes*

PLATE: page 1138

Du. Rosse Waaierstaart   Fr. Agrobate roux   Ge. Heckensänger   It. Usignolo d'Africa
Ru. Тугайный соловей   Sp. Alzacola común   Sw. Trädnäktergal

Field characters. 15 cm; wing-span 22–27 cm. 10% smaller than Nightingale, with rather slimmer form ending in long, fan-shaped tail. Medium-sized, strong-billed, long-tailed, and sprightly chat, with posture frequently recalling Wren. Plumage essentially bright rufous- to grey-brown above and buff-white below, with obvious pale supercilium, double wing-bar, and diagnostic orange-rufous tail broadly tipped black and white.

Unmistakable in west Palearctic. Flight rather less flitting

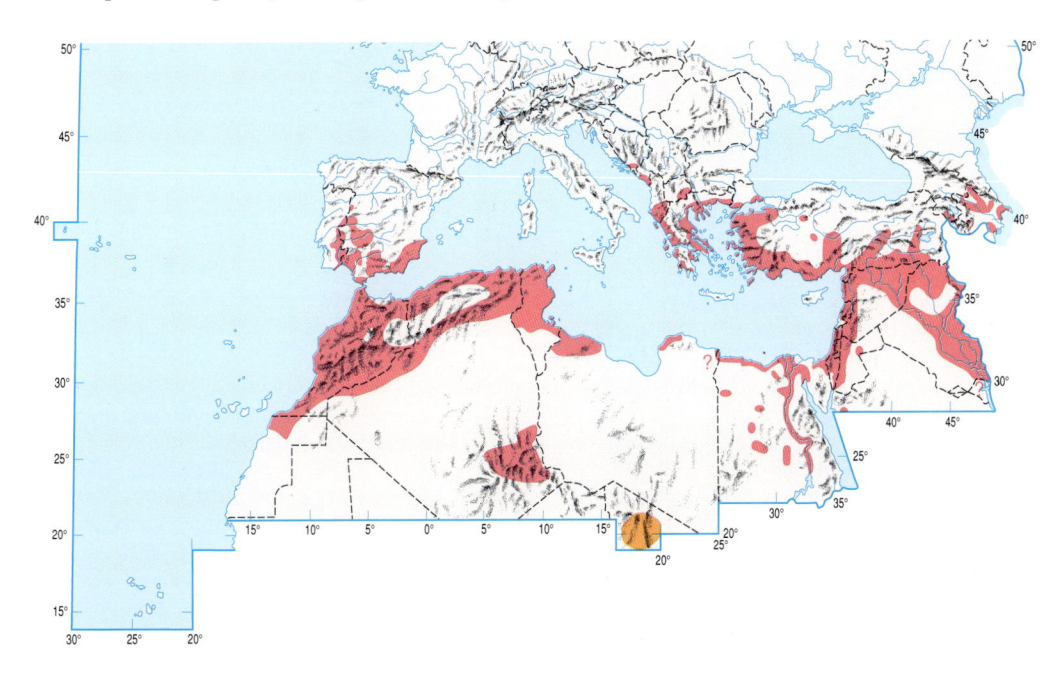

Rufous Bush Robin *Cercotrichas galactotes*. *C. g. galactotes*: **1** ad fresh (autumn), **2–3** ad worn (spring), **4** juv. *C. g. syriacus*: **5–7** ad worn (spring). *C. g. familiaris*: **8** ad worn (spring). Black Bush Robin *Cercotrichas podobe podobe* (Africa): **9** ad.

than more typical chats, with flutter and dash more reminiscent of large warbler. However, any confusion with warbler quickly dispelled by instant assumption when landed of postures and wing and tail movements strongly recalling Nightingale and even Wren. Display flight 'butterfly-like'. Most characteristically, cocks tail right up over back, producing almost U-shaped bird set upon rather long legs. Gait hopping, accompanied by tail-jerks or cock.

Habitat. Breeds in dry middle and lower middle latitudes, in Mediterranean, steppe, and desert fringe zones, above 25°C July isotherm, mainly in lowlands. In North Africa, not attracted to natural maquis and forest, and avoids both mountains and bare plains. More attracted by man-made habitats such as parks, orange groves, gardens, and groups of prickly pear; in steppes, favours areas planted with bushes and trees. Often associated with man, living near houses. In Lebanon, found in coastal strip, lower hills, and river gorges, in bushes, plantations, groves, and arid scrub. In FSU, considered aptly named in Russian as 'river forest nightingale' from preference for valleys with bottomland forests, shrubs, and bulrush beds. In Africa, winters in dry *Acacia* steppe, sometimes around human settlements on coastal plain but also in dry uplands below about 1000 m.

Distribution. AZERBAIJAN. Widespread in lowlands; has probably extended range since 19th century. SYRIA. Regular, though not very common, where mapped. ISRAEL. Expansion in desert agricultural zones. KUWAIT. Fairly common summer visitor, though no firm evidence of breeding. LIBYA. Range in east uncertain. Locally common Tripolitania, chiefly in coastal zone. ALGERIA. May breed Ahaggar.

Accidental. Britain, Ireland, France, Germany, Norway, Switzerland, Balearic Islands, Bulgaria, Rumania, Ukraine, Madeira, Canary Islands.

Beyond west Palearctic, breeds from eastern Arabia northeast to Kazakhstan and western Pakistan, and in sub-Saharan Africa from Mauritania and Sénégambia east to Somalia.

Population. SPAIN. 12 000–24 000 pairs; stable. PORTUGAL. 1000–10 000 pairs 1978–84; stable. GREECE. 1000–2000 pairs. ALBANIA. 100–500 pairs in 1981. AZERBAIJAN. Common. TURKEY. Locally common; 5000–50 000 pairs. ISRAEL. 20 000–40 000 pairs in 1980s. Major decline due to pesticides in 1950s; slow recovery from early 1970s. JORDAN. Common. IRAQ. Very common. EGYPT. Fairly common. TUNISIA. Common. MOROCCO. Uncommon.

Movements. Eurasian and North African populations are migratory, wintering in northern Afrotropics. Autumn passage everywhere is poorly documented, but exodus from breeding areas occurs September to early October. More North African records in spring, when birds arrive early April in southern Algeria and Tunisia and later in April or into early May further north; reaches southern Iberia in late April and early May, but not until late May in central Spain and apparently some not until early June in Portugal.

Food. Mostly insects and earthworms, often rather large; occasionally fruit. Feeding method varies with prey. Pursues ants, Orthoptera, etc., on ground. Takes small Diptera and Hymenoptera from flowers, sometimes hovering to do so. Locates earthworms by probing in soft ground, throwing earth aside with bill once worm found; also hunts worms like

Blackbird by using fast run followed by pause with head cocked to one side and quick jab with bill. Takes Lepidoptera in flight.

Social pattern and behaviour. Not gregarious at any time. Territorial in breeding season, with territories contiguous in suitable habitat. ♂♂ start singing almost immediately upon arrival on breeding grounds; sing from exposed perch, also not infrequently in song-flight. In song-flight, bird flies up from bushes and glides down or horizontally with wings raised and tail fanned. In aggressive display to trespasser, ♂ stands upright, head horizontal, bill sometimes gaping; fans tail and constantly moves it up and down; when raised, tail begins to close as it passes beyond the vertical, and is fully closed as it arcs forwards to almost touch the head. Meanwhile, partially spreads and lowers wings until they nearly touch ground, flicks them well forwards and, with undersides facing forwards, holds them there for 1–2 s. Similar display given to ♀ during pair-formation.

Voice. Song rich, varied, and beautiful, but phrases disjointed: combines series of clear ringing notes recalling lark, pulsing notes suggesting Nightingale, and murmuring warble. Commonest call a hard 'teck teck'. Contact-alarm call a sibilant 'sseeep' or 'tseeeet', sometimes shortened to 'zip', 'zetk', 'tsip' (etc.), given in winter, on spring passage, and among family members prior to autumn migration.

Breeding. SEASON. Iberia, Greece, and North Africa: main laying period second half of May and early June. Iraq: eggs laid early or mid-May to late June. Southern FSU: main laying season late May and June. 2 broods. SITE. In thick bush or low tree, often near trunk. Occasionally uses old nest of another species. Nest: loosely-constructed untidy structure of fine twigs, grasses, and rootlets, lined with vegetable down, wool, hair, and feathers, and (in southern parts of range) often a piece of snake skin. EGGS. Sub-elliptical, smooth and fairly glossy; white or very pale grey, sometimes tinged blue or green, heavily marked with brown, purplish-brown, or purplish-grey spots, speckles, and small streaks. Clutch: 4–5 (2–6). INCUBATION. 13 days. FLEDGING PERIOD. 12–13 days.

Wing-length: *C. g. galactotes*: ♂ 84–92, ♀ 81–89 mm.
Weight: *C. g. galactotes*: ♂ ♀ 21–26 g.

Geographical variation. Marked; 2 main groups separable, each with 2 races in west Palearctic, differing in general colour, wing formula, and relative tail length. In all races, rump to tail rufous-cinnamon; in western group (nominate *galactotes*, *minor*), colour of upperparts rufous-brown, rather like rump and tail, but in eastern group (*syriacus*, *familiaris*) upperparts grey-brown or drab-brown, contrasting markedly with rufous-cinnamon rump and tail; tail of eastern group slightly shorter relative to wing length; wing-tip slightly more pointed.

Black Bush Robin *Cercotrichas podobe*

PLATE: page 1138

DU. Zwarte Waaierstaart FR. Agrobate podobé GE. Rußheckensänger IT. Usignolo podobè
RU. Чёрный тугайный соловей SP. Alzacola negro SW. Svart trädnäktergal

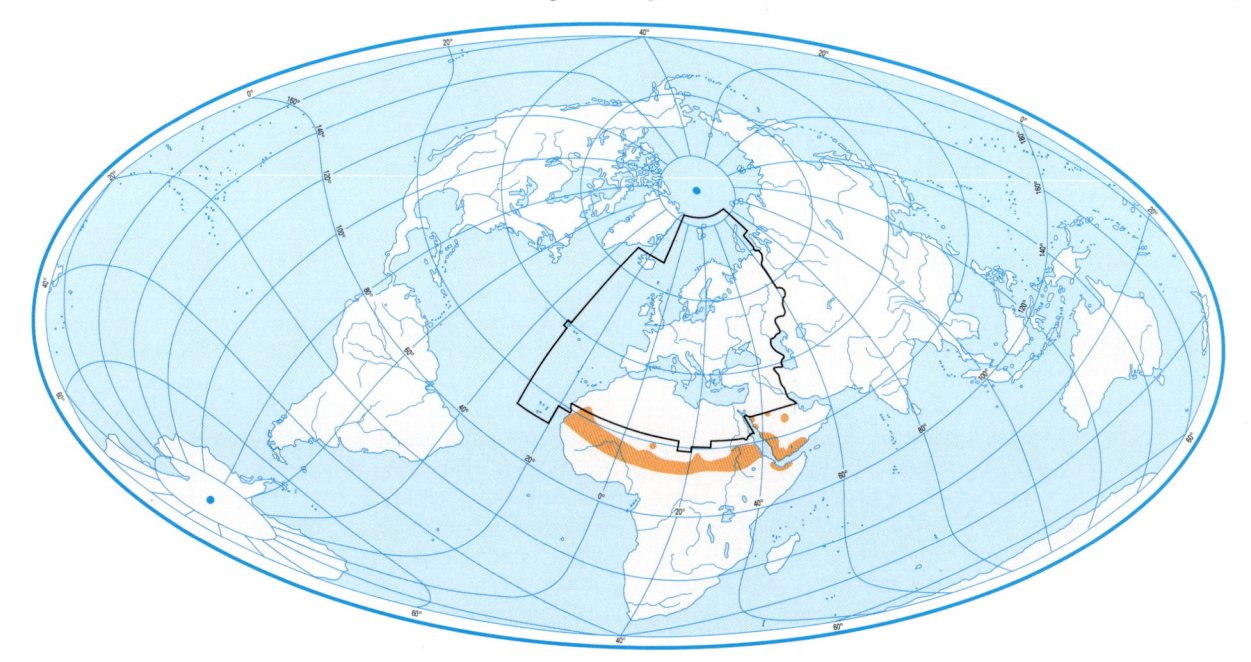

Field characters. 18 cm; wing-span 23–28 cm. Close in size and structure to Rufous Bush Robin but with 40% longer tail and 10% longer legs. Sprightly, bush-haunting chat with similar general character to Rufous Bush Robin but all-black except for white tips to under tail-coverts and bold white tips to graduated outer tail-feathers, last always conspicuous on cocked tail.

Unmistakable. Flight, gait, and behaviour strongly reminiscent of Rufous Bush Robin but spends more time on ground.

Habitat. Tropical and marginally subtropical, in arid mainly lowland regions from fringe of desert through scrub and acacia savanna, especially on sandy soils. In northern Africa in arid savanna and thorn-scrub of scattered acacia bushes or clumps of palms and tamarisk, and in gardens and hedges, up to at least 1500 m, also in hot arid semi-desert and wadis with dense thickets; avoids close stands of trees and banks of rivers, using shrubs rather than trees as song-posts; in Arabia, prefers dry bush vegetation and gardens for breeding, avoiding summits of mountains and juniper zone. In south-east Israel, in acacia, tamarisk, and similar trees near cultivated land and gardens.

Distribution and population. Breeds in band from Mauritania and Sénégal east to Red Sea coast, and in Arabia north to Ha'il region. Extends north to west Palearctic in Mauritania (e.g. breeds at Akhmakou 21°11′N 11°54′W); population size unknown, but likely to be small. In Arabia, sight records within west Palearctic.

Accidental. Israel: 1st record 1981 and 21 records 1985–90, all March–July in southern Arava; a few others 1991–5; possibly regular summer visitor and breeder. Algeria: 2 birds on 8 February 1968 at Tamanrasset (23°N) and another almost certain in nearby Ahaggar massif 12 February 1968.

Movements. Apparently largely or wholly resident in both Africa and Arabia, and possibly sedentary in Chad; some altitudinal, and perhaps some north–south, movement noted in winter since very uncommon non-breeding visitor to Somalia November–March. In Yemen, records up to 2400 m may refer to migrants. Spreading northwards in Arabia following increased irrigation and becoming regular in southern Israel.

Food. Unknown in detail; forages mainly on ground, probing in bare soil or sand or in leaf litter under bushes; also feeds in low shrubs, hopping frequently down to ground and back again.

Social pattern and behaviour. Little studied; in pairs all year round, territorial and pugnacious, though also confiding; ♂ and ♀ often well apart but in vocal contact. Non-breeding (presumably) population density sometimes high, e.g. 22 birds in small wadi in Yemen in November and up to 47 together in spring.

Voice. Song fairly short but pleasantly varied, babbling phrase, with predominantly sweet, fluted notes interspersed with more scratchy or throaty ones. Call thrush-like.

Breeding. SEASON. Egg-laying recorded in Lake Chad area April–June; in Mali, February–September; in Sudan and Ethiopia May and July; in Arabia, end of March to end of April. SITE. Crevice in tree trunk, between fronds of date palm, in roof of disused building, or in low bush. Nest: often untidy cup of dry leaves of grass, palm fibres, twigs, rootlets, sometimes cloth, lined with soft grass and hair. EGGS. Grey-white or pale greenish-white, with olive, blue-grey, or red-brown speckles often concentrated towards broad end. Clutch: 2–4. (Incubation and fledging periods not recorded.)

Wing-length: ♂ 90–102, ♂ 85–92 mm.
Weight: ♂♀ 24–27 g.

Robin *Erithacus rubecula*

PLATES: pages 1141, 1156

DU. Roodborst   FR. Rougegorge familier   GE. Rotkehlchen   IT. Pettirosso
RU. Зарянка   SP. Petirrojo común   SW. Rödhake

Field characters. 14 cm; wing-span 20–22 cm. 15–20% smaller than Nightingale. Small, robust chat, with rather large head and fat chest. Grey to brown above, with warm brown tail; face and chest diagnostically drenched orange-buff to rufous-chestnut above dull white body. Juvenile brown above and buff below, copiously spotted pale buff.

Adult unmistakable in good view but liable to suggest other small chats (e.g. nightingales, Red-flanked Bluetail) in brief glimpse or in spotted juvenile plumage. Flight usually low and flitting, with quick turns and dives amongst and into cover. Stance usually rather upright, with bold head carriage, prominent chest, and frequently flicked wings and tail. Uses both short and long hops in quick succession, movement characteristically interrupted by pauses and wing- and tail-flicks. Pugnacious. British race *melophilus* tame (seemingly addicted to human habitation and affection), but paler continental races noticeably wilder and more skulking, this behaviour persisting on migration and in winter quarters.

Habitat. Breeds in upper and especially middle latitudes of west Palearctic, in boreal, temperate and Mediterranean zones, in oceanic and continental mainly humid lowlands and wooded mountains to treeline (up to 2200 m in Switzerland) between July isotherms 13–23°C. Preferred habitat includes elements of cool shade, moisture, cover of at least medium height and not more than medium density, patches or fringes of open ground (free of tall or obstructive vegetation, stones, ridges, or marked irregularities) and song-posts giving adequate view without undue exposure. In some parts of range occupies coniferous forest, especially with moist mossy floor and ample dead wood, but in others is attached to broad-leaved or mixed woodland and, especially in Britain and parts of western Europe,

Robin *Erithacus rubecula*. *E. r. rubecula*: **1** ad fresh (autumn), **2** ad ♂ worn (spring), **3** juv. *E. r. tataricus*: **4** ad ♂ worn (spring). *E. r. melophilus*: **5** ad ♂ worn (spring), **6** 1st winter. *E. r. superbus*: **7** ad ♂ worn (spring). *E. r. caucasicus*: **8** ad ♂ worn (spring).

to parks, gardens with trees and shrubs, road verges, burial grounds, and other humanly managed and disturbed habitats.

Distribution. ICELAND. Breeding attempt (eggs infertile) 1961. FAEROES. Bred 1960, 1964, 1966. FINLAND. Has expanded north in 20th century. TURKEY. Widespread in damp forests of northern Anatolia; local and very rare elsewhere. IRAQ. May breed in Taurus and Zagros mountains in north. AZORES. Absent from Flores and Corvo. CANARY ISLANDS. Widespread and common in forests on Gran Canaria, Tenerife, La Palma, La Gomera, and El Hierro.

Accidental. Spitsbergen, Bear Island, Jan Mayen, Iceland (annual maximum total 170 in spring 1994), Mauritania.

Beyond west Palearctic, extends east from Urals to upper Ob'; also breeds northern Iran.

Population. Some fluctuations, but generally stable. BRITAIN. 4.2 million territories 1988–91. IRELAND. 1.9 million territories 1988–91. FRANCE. 3–6 million pairs. BELGIUM. 210 000 pairs 1973–7. LUXEMBOURG. 15 000–20 000 pairs. NETHERLANDS. 275 000–375 000 pairs 1979–85. GERMANY. 3.3 million pairs in mid-1980s. DENMARK. 40 000–700 000 pairs 1987–8. NORWAY. 500 000–1.5 million pairs 1970–90. SWEDEN. 3–6 million pairs late 1980s. FINLAND. 2–3 million pairs late 1980s; marked decrease, but fluctuates. ESTONIA. 200 000–500 000 pairs 1991. LATVIA. 700 000–1 million pairs in 1980s. CZECH REPUBLIC. 500 000–1 million pairs 1985–9. SLOVAKIA. 500 000–1 million pairs 1973–94. HUNGARY. 350 000–500 000 pairs 1979–93. SWITZERLAND. 450 000–600 000 pairs 1985–93. SPAIN. 1.2–3 million pairs; slight increase. PORTUGAL. 10 000–100 000 pairs 1978–84; slight increase. ITALY. 1–2.5 million pairs 1983–95. GREECE. 50 000–100 000 pairs. ALBANIA. 10 000–30 000 pairs in 1981. YUGOSLAVIA: CROATIA. 1–1.2 million pairs. SLOVENIA. 200 000–300 000 pairs. BULGARIA. 500 000–5 million pairs. RUMANIA. 700 000–1 million pairs 1986–92. RUSSIA. 10 million pairs. BELARUS'. 1.1–1.2 million pairs in 1990. UKRAINE. 400 000–450 000 pairs in 1986. MOLDOVA. 30 000–40 000 pairs in 1988. AZERBAIJAN. Common. TURKEY. 10 000–100 000 pairs. ALGERIA, MOROCCO. Uncommon. AZORES. Common. MADEIRA. Rare.

Movements. Most populations partially migratory, with ♂♂ more sedentary than ♀♀; totally migratory in north-east of range and probably largely sedentary in extreme south. Thus British and Irish populations largely resident, but some birds migrate SSW at least as far as southern Iberia. Winters south to Saharan oases and Middle East. A nocturnal migrant, though some local movements occur by day.

Main wintering area lies from Ireland and Britain south to Morocco and south-east through Europe with few birds north-east of line from southern Denmark to Bosporus (Turkey), though some December records as far north-east as southern Sweden and north-east Poland. High ground of central Europe largely vacated. Notable concentrations in Mediterranean basin, including areas where breeding does not occur, e.g. parts of Iberian and Yugoslavian coasts, southern Turkey, and most Mediterranean islands.

Autumn passage on broad front. Passage on Polish Baltic coast mid-August to early November, peaking late September and October; passage in south appears to occur slightly later than in north, and also slight tendency for adults to pass later than juveniles. Arrivals at wintering sites in Andalucía (Spain) occur from early September, peaking in October, and North Africa reached by late September (exceptionally late August), with largest influxes during October. Most records from Saharan oases in period late December to March. First signs of return passage in early February with increasing numbers along coasts of Algeria and Morocco. By late February, first birds arrive in Switzerland and southern Germany, exceptionally southern Norway. Peak passage throughout most of range in March. Although Baltic coast reached in mid-March, birds do not reach Urals until early May.

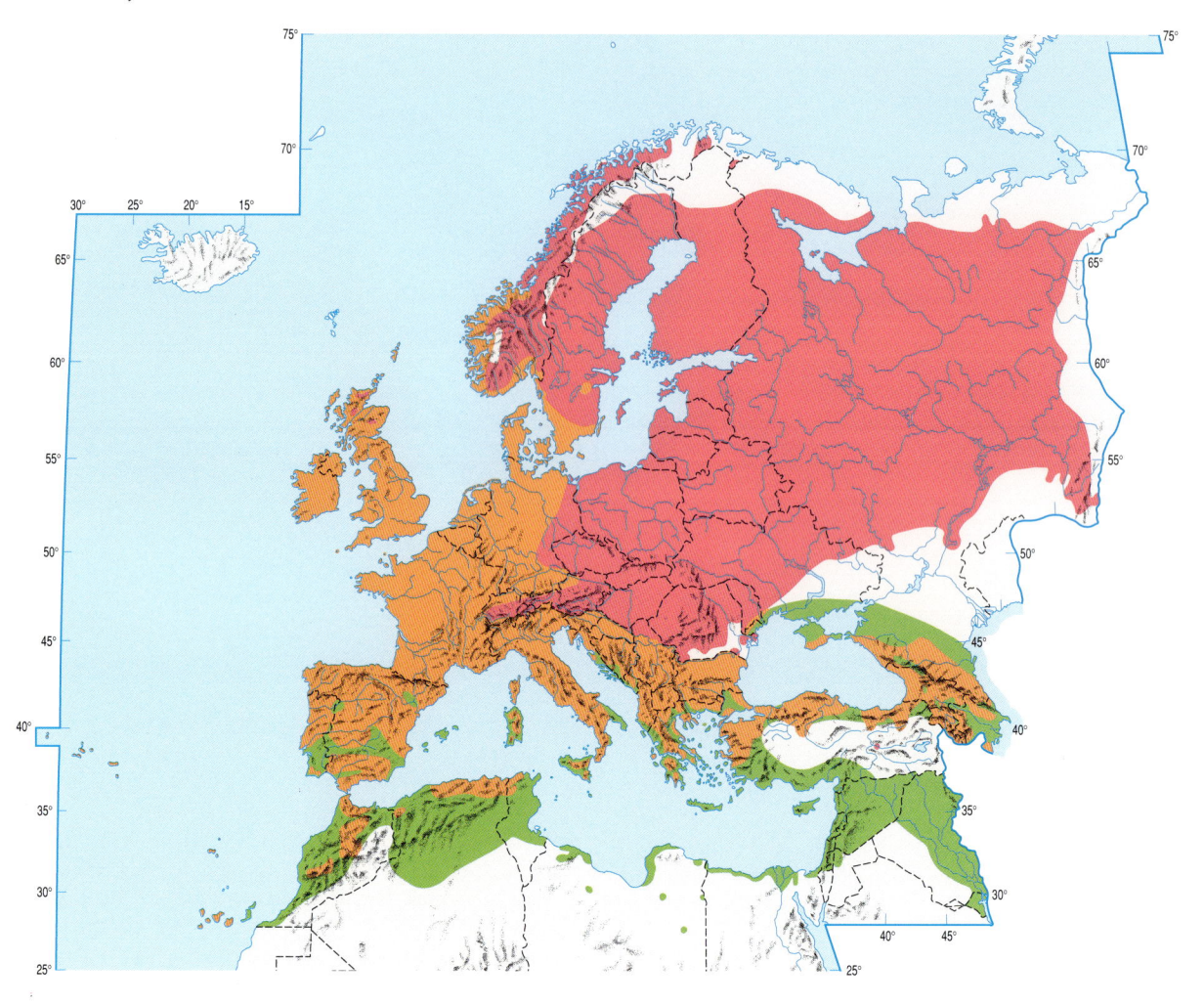

Food. Invertebrates, especially beetles and (in southern Spain at least) ants; also fruit and seeds in winter. Uses 2 main methods to locate prey: (1) perching on bush or low branch, flying down to eat prey, then returning to perch (may fly up to c. 20 m to pick up tiny insect); (2) hopping on ground, but usually not turning over leaves, etc. In woodland, thus depends primarily on animals moving on surface of litter layer, but also sometimes takes prey from tree-trunk, branch, or leaf, and occasionally catches items in the air. When ground frozen, will accompany Pheasant, mole, man, etc., to feed at areas where hard surface broken; also (when no frost) will feed on ground disturbed by mole tunnelling beneath; well-known habit of accompanying digging gardeners presumably an extension of this. Takes fruit either perched or, frequently, with rapid flight sally.

Social pattern and behaviour. Solitary and strongly territorial for most of year, defence of territory typically relaxed only during severe winter weather and moult. Outside breeding season, both sexes defend individual territories, and sing; exceptionally, pair-bond maintained. Fidelity to wintering sites high in sedentary populations, with most ♂♂ and some ♀♀ defending same site throughout life. Migrants often return to same wintering sites in successive years. Mating system typically monogamous, but small proportion of ♂♂ practise simultaneous bigamy. Strongly territorial in breeding season, with territories often more aggregated than expected from habitat. ♂ advertises territory by song; neighbours frequently sing against one another. Aggression most frequent and violent when territories being established or boundaries changing during pairing period. Serious injury to eyes and legs during intraspecific fighting sometimes common in high-density populations. Vast majority of encounters, however, resolved without physical contact. Threat display involves simultaneous singing and display of orange breast. ♂ starts regular courtship-feeding of ♀ within a few days of ♀'s completion of nest, continuing through egg-laying and incubation and providing most of her food in latter period.

Voice. Highly vocal throughout year, except during primary moult when virtually silent. Song a melodic varied warble, typically with desultory delivery and phrases of varying length, and with switching between high and low frequencies. Autumn song softer and more wistful than spring song and tends to contain longer phrases. Commonest call a short sharp 'tic', mainly used in territorial defence, but also frequently as alarm-call. Very variable rate of delivery and tonal quality; as urgency increases, call repeated more rapidly and sound more deep-throated. Contact-alarm call a high-pitched, thin 'tswee'; in more intense alarm, high-pitched, sharp, and generally thin 'tseep' notes; sometimes an elongated 'tseeeee . . . ' up to 0.8 s long which typically sounds plaintive and appears to fade away.

Breeding. SEASON. Britain and Ireland: egg-laying from early March to June; rarely, nests also found in all other months of year. Canary Islands: from mid- or late March to early June. Madeira: eggs laid April–June. North-west Africa: eggs laid mid-April to late May. Southern Europe: laying begins April. Central Europe: laying from end of April to late July. Scandinavia: from very end of April in south but not before mid- or late May in north. FSU: from mid-May to mid-June in north; from mid-April in south. 2 broods, rarely 3, except in north of range where normally 1. SITE. Natural hollow in tree-stump, bank, among tree-roots, rock-crevice, or hollow tree, from ground level to *c.* 5 m up, rarely to 10 m. Makes much use of artificial holes in wide variety of man-made objects, often in or attached to buildings; uses nest-boxes. Nest: bulky structure of dead leaves forming base, on which cup of moss, grass, and leaves is built, lined with finer material including hair, vegetable fibre, and occasionally feathers. EGGS. Sub-elliptical, smooth but not glossy; white or faintly bluish, variably marked with sandy-red freckles and small blotches, often sufficient to colour entire egg uniform reddish, but can be very sparse or (rarely) absent, or gathered towards broad end; much more heavily marked in Canary Islands race. Clutch: 4–6 (2–8). INCUBATION. Average 13.7 days; slightly longer in March–April (14.0 days) than in May–June (13.5 days). FLEDGING PERIOD. Average 13.4 days (10–18).

Wing-length: *E. r. rubecula.* ♂ 72–74, ♀ 68–73 mm.
Weight: *E. r. rubecula.* ♂ mostly 15–21, ♀ 14–19 g (to 25 g when egg-laying).

Geographical variation. Slight; 8 races recognized in west Palearctic but variation strongly clinal except for isolated *superbus* from Gran Canaria and Tenerife (Canary Islands). Involves colour of upperparts, depth of rufous on face and chest, presence of rufous on upper tail-coverts and tail-base, and (to slight extent) size. In addition to *superbus*, populations of nominate *rubecula* from Scandinavia, of *melophilus* from Britain and Ireland, and of *hyrcanus* from south-east Transcaucasia and northern Iran are the only well-characterized forms; all other races and populations either similar to one of these (differing only in measurements) or more or less intermediate. Nominate *rubecula* from Scandinavia characterized by brownish olive-green upperparts and orange face and chest. *E. r. melophilus* from Britain differs in fresh plumage by warmer dark olive-brown upperparts, more rufous-brown on upper tail-coverts and tail-base, deeper orange-rufous face and chest, and deeper buff-brown flanks. *E. r. hyrcanus* from northern Iran browner on upperparts than nominate *rubecula*, rather like *melophilus* but less warm olive; face and chest rufous-orange like *melophilus*, flanks rather pale, like nominate *rubecula*. *E. r. superbus* from Gran Canaria and Tenerife dark greyish-olive on upperparts, darker and greyer than nominate *rubecula*; wider band of ash-grey on forecrown and from side of crown down to side of breast, ash-grey tinge often extending to central crown and across upper mantle; face and chest even deeper rufous-chestnut than in *melophilus*; belly and vent whiter; small. Other races, intergrading with nominate, are *witherbyi* (Tunisia and Algeria), *tataricus* (Urals), *valens* (Crimea), and *caucasicus* (Caucasus and Transcaucasia).

Thrush Nightingale *Luscinia luscinia*

PLATES: pages 1144, 1156

DU. Noordse Nachtegaal   FR. Rossignol progné   GE. Sprosser   IT. Usignolo maggiore
RU. Обыкновенный соловей   SP. Ruiseñor ruso   SW. Näktergal

Field characters. 16.5 cm; wing-span 24–26.5 cm. Similar in size to Nightingale, but with more pointed wings and marginally longer tail than European population of that species. Differs from that species in drabber, more olive-grey tone to plumage, and even richer voice. From above and behind, best distinction is reduced contrast of dull rufous-brown tail with dull brown upperparts; from below and in front, shows pale throat and mottled chest. Juvenile darker than adult, with buff and dark brown spots.

Separation from Nightingale fairly easy when differences in plumage tone, tail contrast, and fore-underpart pattern fully visible, but more rufescent birds with indistinctly mottled chests perhaps not identifiable if silent. Flight and other behaviour apparently identical to Nightingale but usually shyer, often escaping by running and crouching.

Habitat. Close counterpart of Nightingale; habitat, like geographical range, approaches and sometimes overlaps with it, but maintains clear specific separation. Inhabits more continental, easterly and northerly temperate breeding grounds, overlapping

Thrush Nightingale *Luscinia luscinia*: **1** ad fresh (autumn), **2** ad worn (spring), **3** juv. Nightingale *Luscinia megarhynchos*. *L. m. megarhynchos*: **4** ad fresh (autumn), **5** ad worn (spring), **6** juv. *L. m. hafizi*: **7** ad fresh (autumn).

boreal and steppe zones in middle latitudes between July isotherms 17–25°C. Essentials of habitat—comprising deep soft humus with (usually) ground cover of dead leaves, tall and dense but patchy herbage, and plenty of tall bushes, shrubs, or low trees forming thicket or open woodland—are typically found along river banks or near standing water, normally in river valleys or on lowland plains. In African winter quarters, highly localized in dense thickets, sometimes shared with Nightingale, mostly below 1500 m.

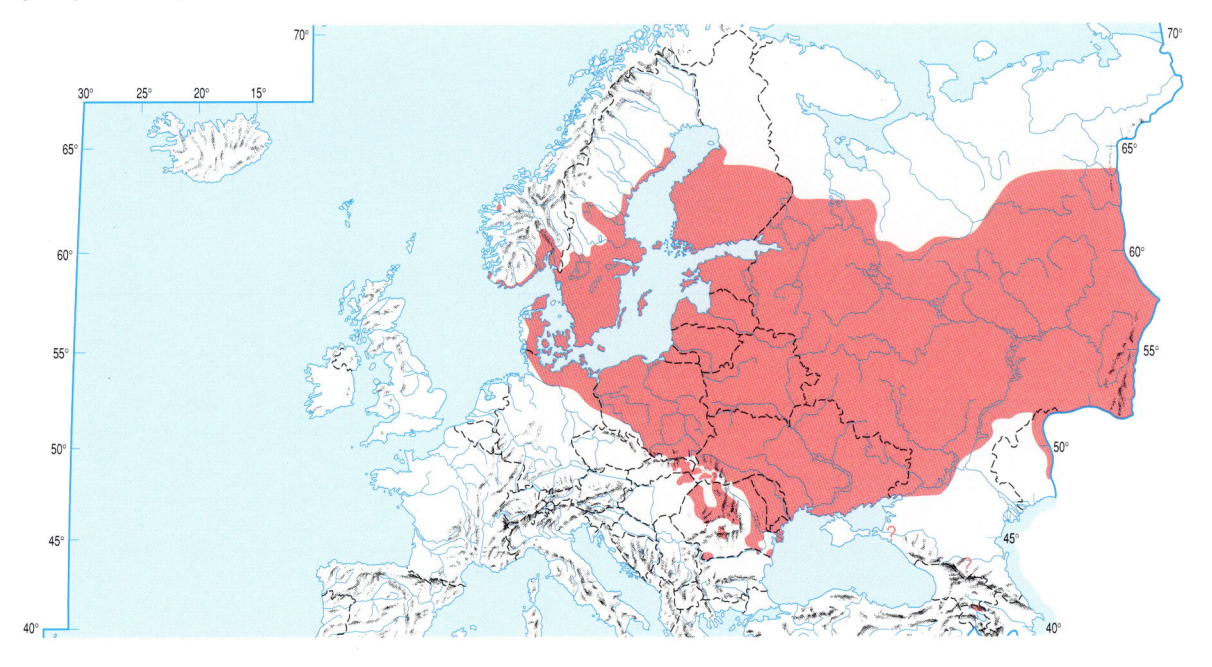

Distribution. Has extended range west and north in 20th century. BRITAIN. Singing ♂ Shetland 1993; 3 singing ♂♂ (2 south-east England, 1 Scotland) May–June 1994. NETHERLANDS. First breeding record 1995. GERMANY. Clear expansion south-west, birds reaching Frankfurt/Oder area (in increasing numbers) from early 1980s, with isolated breeding records further south-west, including Berlin. FENNO-SCANDIA. Recent increase. POLAND. Significant expansion south-west since 1950s, locally up to 100 km. CZECH REPUBLIC. First breeding record 1989. AUSTRIA. Bred in east until early 19th century. YUGOSLAVIA. Occasional breeder. RUSSIA. In Leningrad region has expanded north-east since 1900. AZERBAIJAN. Possibly nested Lenkoran lowland (in south-east) at end of 19th century.

Accidental. Britain, Ireland, France, Belgium, Netherlands, Switzerland, Spain, Balearic Islands, Malta, Libya, Canary Islands.

Beyond west Palearctic, extends east in narrow band to c. 90°E.

Population. Stable in most countries. GERMANY. 20 000 pairs in mid-1980s; vast majority in Mecklenburg-Vorpommern. DENMARK. 11 000–120 000 pairs 1987–8; fluctuating. NORWAY. 300–1000 pairs 1970–90; marked increase. SWEDEN. 20 000–50 000 pairs in late 1980s. FINLAND. 15 000–20 000 pairs in late 1980s; slight increase. ESTONIA. 20 000–50 000 pairs in 1991; gradual increase. LATVIA. 50 000–150 000 pairs in 1980s. LITHUANIA. Decreasing. POLAND. 45 000–65 000 pairs; probably stable now, following increase. SLOVAKIA. 1000–1500 pairs 1973–94. HUNGARY. 50–100 pairs. RUMANIA. 75 000–120 000 pairs in 1986–92. RUSSIA. 100 000–1 million pairs. BELARUS'. 160 000–175 000 pairs in 1990. UKRAINE. 250 000–300 000 pairs in 1986. MOLDOVA. 30 000–40 000 pairs in 1988.

Movements. Migratory, wintering entirely in eastern Africa—largely south of equator with some north to southern Ethiopia and some as far south as Natal. West Baltic population heads south-east before turning south into north-east Africa.

Leaves breeding areas mainly from early August, although first passage in Crimea by late July. Passage on Turkish Black Sea coast from mid-August to late September, up to late September on Bosporus. Passes through Middle East from late August to early October, peaking mid-September. Movement into Kenya begins end of October. Main arrival in southern Africa from late November.

Leaves winter quarters in March, and exodus complete by early April. Occurs mid-April to mid-May on European coasts of Black Sea, and reaches south-west of breeding range (Rumania) from mid-April. In European FSU, arrives from late April in south of range to early or mid-May in north. Arrives in Sweden in first half of May. Spring records in Britain far more frequent than autumn records, possibly linked to range expansion.

Food. Arthropods and some fruit. Feeds largely on ground, hopping around and disturbing leaves to search; particularly during nestling phase also feeds in herb and shrub layers and even recorded foraging in crown of tree. Will also take flying insects in brief aerial-pursuit.

Social pattern and behaviour. Generally very similar to Nightingale (with which it has hybridized in captivity), the two species being antagonistic when in contact, both in breeding and in winter quarters.

Voice. Song immediately recalls Nightingale, but louder, no less beautiful, and even more impressive; more often lacks magnificent crescendo, instead having solemn, somewhat religious quality, with staccato phrases and pure bell-like peals dominant, interspersed with rasping 'dserr' notes. Calls rather more clipped and higher pitched than Nightingale with monosyllabic 'whit' recalling *Ficedula* flycatcher and more grating croak.

Breeding. SEASON. Laying begins mid- to late May over most of range for which there is information. One brood. SITE. On ground, among dead branches, roots, or thick leaf litter, frequently in heavily shaded position. Nest: loose and bulky structure, with basal pad of leaves and cup of grass leaves and stems, lined with finer material and hair. EGGS. Sub-elliptical, smooth and slightly glossy; variable, buff, olive, greenish, or grey-blue, with reddish markings normally giving tinge to whole egg, occasionally just to broad end; often with irregular chalky-white marks. Clutch: 4–5. INCUBATION. c. 13 days. FLEDGING PERIOD. 9–10 days.

Wing-length: ♂ 87–94, ♀ 84–90 mm.
Weight: ♂ ♀ mostly 24–30 g.

Nightingale *Luscinia megarhynchos*

PLATES: pages 1144, 1156

DU. Nachtegaal   FR. Rossignol philomèle   GE. Nachtigall   IT. Usignolo
RU. Южный соловей   SP. Ruiseñor común   SW. Sydnäktergal

Field characters. 16.5 cm; wing-span 23–26 cm. 15% larger than Robin, with proportionately smaller head and longer tail giving well-balanced form suggesting small thrush. Medium-sized, graceful chat, with alert, rather upright carriage, noticeably uniform plumage, and skulking habits. Russet-brown above, warmest on tail; dull brown-grey below, with paler throat and vent; pale eye-ring emphasizes gentle expression. Juvenile liberally spotted buff and dark brown.

Only breeding nightingale in western Europe and around Mediterranean, and thus readily identifiable—only slight possibility of confusion with similarly-coloured species such as Rufous Bush Robin and juvenile Robin or White-throated Robin. Across central Europe (from Germany south-east to Black Sea) in breeding season and around North Sea in migration periods, closely similar Thrush Nightingale also occurs (see that species). Flight recalls that of Robin, but less flitting, consisting of bursts of wing-beats, leading into sweeping glides, sudden turns, and dives into cover. Moves on ground by short or long hops on rather long legs, usually with rather erect carriage. Movements accompanied by frequent flicks of wings and tail; cocks tail when excited.

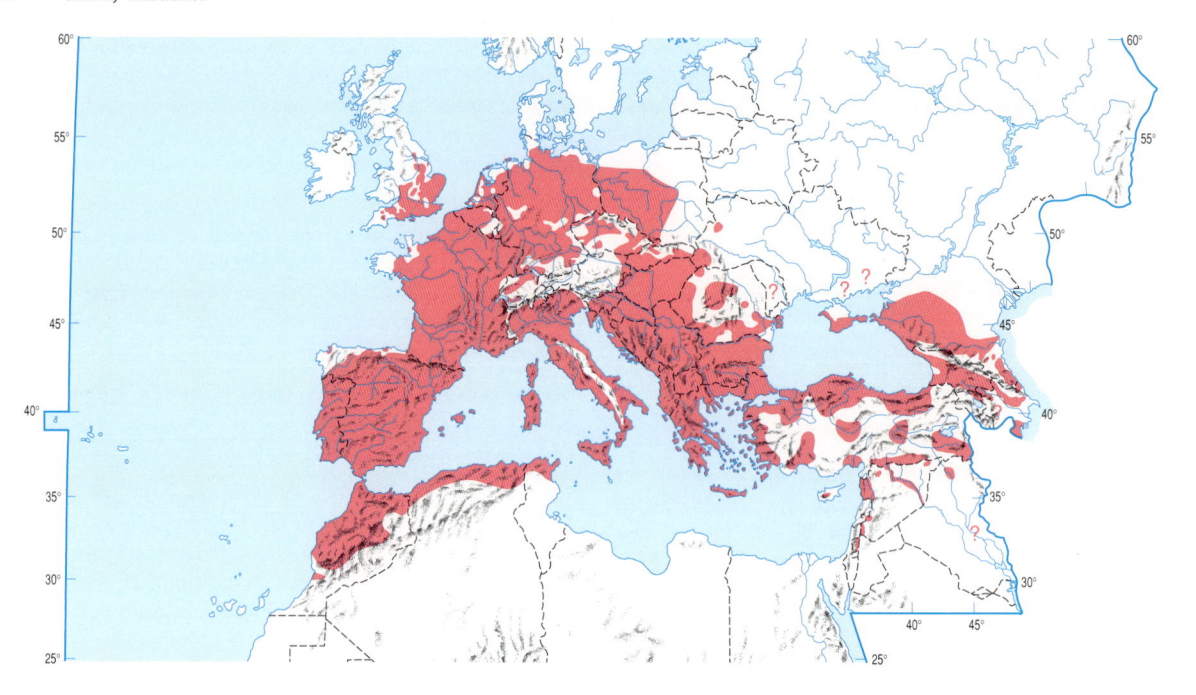

Habitat. Breeds in west Palearctic in middle and lower-middle latitudes, with some oceanic bias, in mild and warm temperate, Mediterranean, and steppe climatic zones between July isotherms 17–30°C. Differs from Thrush Nightingale in more southerly, westerly, and generally somewhat warmer breeding range, less restricted to lowlands, valleys, and neighbourhood of water in most regions, and more ready to inhabit drier sandy soils and sunny hillsides. In Switzerland, breeds exceptionally up to 1100 m although only locally above 600 m.

Occupies 3 distinguishable habitat types, of which 1st closely resembles that of Thrush Nightingale, in thickets or woods near water. 2nd habitat type is on drier soil, no surface water, with thicket, scrub, or managed open woodland including various open spaces, not necessarily flat, and offering unobstructed feeding ground with ample leaf-litter and sunny as well as shady conditions. Smaller patches of cover, often in quite open areas, comprise 3rd main habitat type, characteristic of Mediterranean region, on dry and warm hillsides or valleys, often near human settlements, and even involving use of telegraph poles and rooftops for singing.

In winter in tropical Africa, frequents savanna woodland, thorny scrub, river gallery edges in mountains, humid forest edges and clearings, low second growth, and tangles of small trees, bushes, and rank herbage fringing watercourses.

Distribution. BRITAIN. Contraction has continued, especially at northern and western fringes. GERMANY. Has expanded in east. CZECH REPUBLIC. Slow increase. MALTA. First breeding record 1995. UKRAINE. Has withdrawn southward. TURKEY. Widespread but very sporadic. SYRIA. Old reports of breeding in north and on coast; probably still breeds at least locally. JORDAN. May breed along Jordan river. IRAQ. Perhaps no longer breeding central Iraq.

Accidental. Iceland, Faeroes, Ireland, Denmark, Norway, Sweden, Finland, Estonia, Lithuania, Madeira, Cape Verde Islands.

Beyond west Palearctic, extends north-east from Iran to Altai and western Mongolia.

Population. BRITAIN. 5000–6000 pairs 1988–91; fluctuating. FRANCE. Over 1 million pairs in 1970s; more common in south. BELGIUM. 2400–3200 pairs 1989–91; fluctuates, tending to decrease in some areas. LUXEMBOURG. 150–200 pairs. NETHERLANDS. 7500–10 000 pairs 1979–85. Strong increase in west, strong decrease in south and east. GERMANY. 95 000 pairs in mid-1980s; increasing in east, where estimated 50 000 pairs in early 1980s. POLAND. 40 000–60 000 pairs; slow decrease. CZECH REPUBLIC. 6000–12 000 pairs 1985–9; slow increase. SLOVAKIA. 10 000–20 000 pairs 1973–94; stable. HUNGARY. 100 000–150 000 pairs 1979–93; stable. AUSTRIA. Common in east; scarce elsewhere. Apparently stable. SWITZERLAND. 2000–2500 pairs 1985–93; stable. SPAIN. 450 000–1.7 million pairs; stable. PORTUGAL. 10 000–100 000 pairs 1978–84; stable. ITALY. 500 000–1 million pairs 1983–95. GREECE. 100 000–150 000 pairs; slight decrease. ALBANIA. 20 000–50 000 pairs 1981; slight decrease. YUGOSLAVIA: CROATIA. 200 000–400 000 pairs; slight increase. SLOVENIA. 2000–4000 pairs; stable. BULGARIA. 100 000–500 000 pairs. RUMANIA. 80 000–150 000 pairs 1986–92; stable. RUSSIA. 10–100 pairs; stable. UKRAINE. 100–150 pairs in 1986; slight decrease. AZERBAIJAN. Common, e.g. in lowland along southern slope of Great Caucasus. TURKEY. 50 000–500 000 pairs. CYPRUS. Common, probably stable. ISRAEL. 500–1000 pairs in 1980s. TUNISIA, ALGERIA, MOROCCO. Common.

Movements. Migratory, wintering in Afrotropics. Western populations (nominate *megarhynchos*, breeding in Europe, western Turkey, and north-west Africa) winter between Sahara and rain forest from West Africa east to Uganda. European breeding birds leave in autumn between end of July and September. Movement through Europe broadly south-west, with birds occurring throughout Mediterranean region though commonest in west. The relative scarcity in much of North Africa and also Middle East in autumn suggests Mediterranean and Sahara normally crossed in one continuous flight. Present in winter quarters from early November to early April. Some present in Afrotropics until early May, but spring passage through Nigeria concentrated in late March and early April with arrivals in North Africa and southern Europe at this time. Unlike autumn, many records in spring along North African coast and on Mediterranean islands and even commonly inland in Algeria and Libya, so passage obviously on broad front.

Food. In breeding season, terrestrial invertebrates, especially beetles and ants; in late summer, also berries. Feeds on ground, taking food mostly from litter layer but also from bare ground and from leaves or twigs or while gripping bark. Moves on ground by long hops with frequent pauses but will also drop on to prey from perch and catch insects in flight.

Social pattern and behaviour. Mostly solitary. Local concentrations occur on spring migration. Early ♂♂ arrive on breeding grounds alone; later ♂♂ in small groups, sometimes with ♀♀. Territorial in winter quarters, where regularly sings. Density of breeding territories varies greatly. Territories tend to be traditional, and within a given area the same (best) ones are always occupied first. ♂♂ markedly site-faithful, occupying same territory from year to year, ♀♀ less so. ♂ sings chiefly from low undergrowth, also low branches of trees, sometimes fully exposed; occasionally in flight. By day, sings from several perches, often in the open, regularly changing perch with start of next song; nocturnal song given mainly from one perch, used several nights in succession. Diurnal song mainly an interaction with approaching rivals, nocturnal song more long-distance advertisement to ♀♀. Nocturnal song thus associated mainly with pair-formation. Song period rather short; song declines before mid-May, with little from end of May/beginning of June.

Voice. Song of ♂ remarkable for its richness, variety and vigour; a succession of phrases 2–4 s long, separated by pauses of about same length. Many units have clear, rich, liquid, bubbling, or piping quality, other sequences a kind of musical chuckle, and even rather toneless or unmusical notes and variants of alarm-call occur. Most striking are the rapid loud sequence of 'chooc' units, and the fluting, much higher-pitched 'pioo' repeated rather slowly in striking crescendo. Capable of mimicry, including song of Thrush Nightingale. Calls also varied. Most characteristic is alarm call, a croaking frog-like 'krrrrr'. Other calls include a soft whistling 'hweet', very like Chiffchaff or Willow Warbler, a low 'tuk tuk', like Blackbird, and a harsh 'raäk' or 'praäk', like Jay, given in extreme alarm, but also signalling anger and annoyance.

Breeding. SEASON. Egg-laying from late April or early May. Little variation over whole range, with similar dates in southern England, central Europe, south-west FSU and North Africa. 2 broods in south of range, 1 in north. SITE. On ground or slightly above, in twigs and undergrowth below scrub, or in dense herbage. Nest: bulky cup of dead leaves and grass, lined with finer material and feathers, occasionally domed. EGGS. Sub-elliptical, smooth and very slightly glossy; pale blue, green, or green-blue, finely speckled and mottled pale red-brown, often no more than a general reddish tinge; sometimes forming more marked zone at broad end. Clutch: 4–5 (2–6). INCUBATION. 13 days. FLEDGING PERIOD. 11 days.

Wing-length: *L. m. megarhynchos*: ♂ 81–87, ♀ 78–85 mm.
Weight: *L. m. megarhynchos*: ♂♀ mostly 17–24 g.

Geographical variation. Marked, but clinal, involving general colour, and length of wing and tail. Strongly saturated umber-brown, with chest clouded olive-buff, in populations from Britain, western France, north-west Spain, and Portugal; gradually paler eastward. Wing and tail shortest in populations from north-west Africa, Iberia, and Corsica, gradually longer eastwards. Two main races in west Palearctic, without clear boundaries: nominate *megarhynchos* (Europe and North Africa, east to central Turkey and Levant); *africana* (Caucasus and eastern Turkey); the latter grading into very pale *hafizi* of central Asia.

Siberian Rubythroat *Luscinia calliope*

PLATES: pages 1148, 1156

DU. Roodkeelnachtegaal   FR. Calliope sibérienne   GE. Rubinkehlchen   IT. Calliope
RU. Соловей-красношейка   SP. Ruiseñor calíope   SW. Rubinnäktergal

Field characters. 14 cm; wing-span 22.5–26 cm. Marginally larger than Robin; close in size and structure to Bluethroat but with 10% longer tail. Rather small but sturdy, long-legged, ground-haunting chat, with elegant character, flight, and behaviour most recalling Bluethroat. Essentially brown above and brown-grey below, with strikingly pale vent and under tail-coverts obvious when dark brown tail cocked; ♂ has striking face pattern, with white supercilium and sub-moustachial stripe and glossy pink or red throat; ♀ has white throat and buff-white supercilium.

Habitat. Approximately replaces Robin in breeding niche eastwards from Urals region across upper middle and middle continental latitudes in boreal and cool temperate zones, mainly in lowlands from flat sea coast up river valleys, but also occurring above treeline in stunted tree and subalpine shrub growth. Although widespread in taiga, tends to avoid dense coniferous stands, preferring thickets of bird cherry, birch, and willow, but also larch and pine with fallen trees, heaps of broken branches, tall grass, small bogs, and regrowth after burning; sometimes in open meadows or clearings, especially

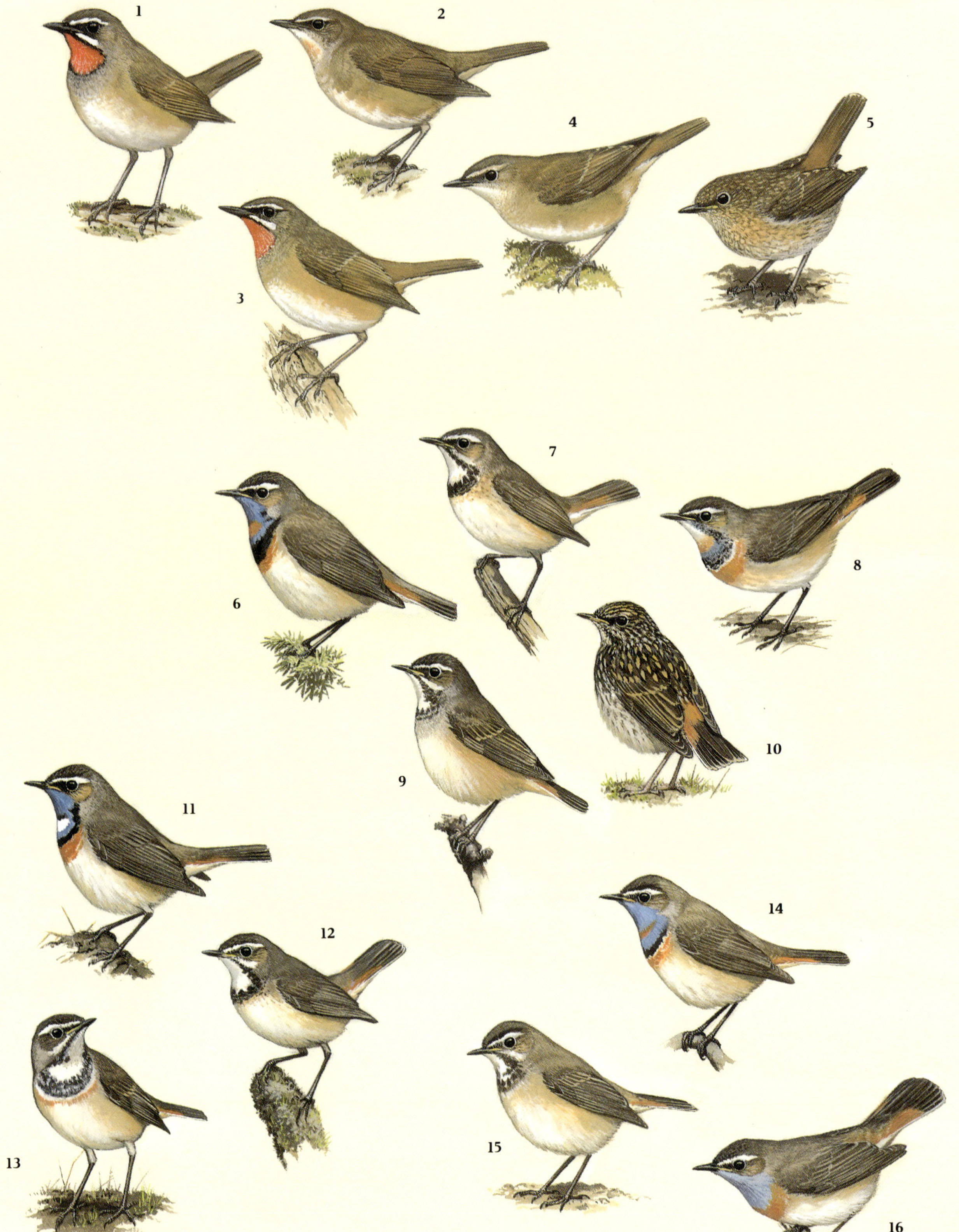
1
2
3
4
5
6
7
8
9
10
11
12
13
14
15
16
Hilary Burn

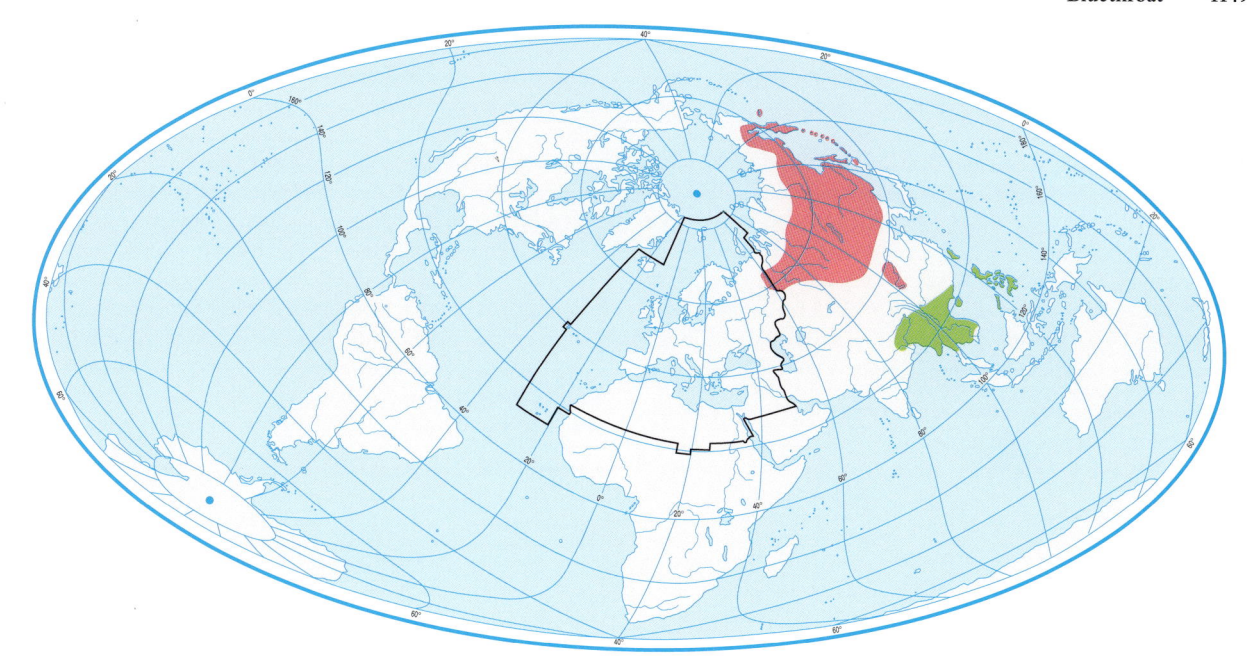

near rivers or coast. In winter in India resorts to dense scrub near water, hedges near villages, underbrush along sides of country roads, long grass, sugarcane, reeds, and sometimes tea plantations, up to 1500 m.

Distribution and population. RUSSIA. 10–100 pairs; stable.
Accidental. Iceland, Britain, Denmark, Finland, Estonia, Italy, Egypt.

Movements. Long-distance migrant. Passes through Mongolia, China, Korea, and Japan, to winter in south-east Asia from Philippines and southern China to Assam (eastern India) and Bangladesh.

Food. Mainly insects and spiders, and small molluscs. Feeds mostly on ground, but also among low bushes and reeds; runs rapidly in short spurts.

Social pattern and behaviour. Little known, but apparently similar to Robin and nightingales.

Voice. Song of ♂ strikingly beautiful, rich, varied, melodious, and sustained warbling with some hard 'squeezed' notes. Often loud and delivered with great vigour and intensity. Contact-alarm calls include a loud, melodious, drawn-out, and rather plaintive whistle, and a harsh and sometimes loud 'tacking' or 'chacking', like a thrush.

Breeding. SEASON. In Tomsk area (Siberia), laying begins early June, and main fledging period 10–20 July. One brood. SITE. On ground in thick tussock or under thick low bush, occasionally just above ground. Nest: loosely constructed of extremely fine stems and vegetable fibres, slightly lined with hair and plant fluff. EGGS. Sub-elliptical, smooth and glossy; pale blue, finely marked with red-brown speckles and mottles, giving overall greenish tinge, sometimes gathered at broad end. Clutch: 5 (4–6). (Incubation and fledging periods unknown.)

Wing-length: ♂ 73–85, ♀ 73–79 mm.
Weight: ♂♀ mostly 72–85 g.

Bluethroat *Luscinia svecica*

PLATES: pages 1148, 1156

DU. Blauwborst   FR. Gorgebleue à miroir   GE. Blaukehlchen   IT. Pettazzurro
RU. Варакушка   SP. Pechiazul común   SW. Blåhake

Field characters. 14 cm; wing-span 20–22.5 cm. Marginally smaller than Robin, with markedly slimmer form and proportionately longer legs. Small, graceful, elegant chat, with noticeably erect carriage and characteristic cocking and fanning of tail. Bright chestnut patches at bases of outer tail-feathers diagnostic; rest of plumage essentially dark brown above, silky buff-white below with dusky flanks. Breeding ♂ has blue throat, bordered below with black-white-chestnut bands, ♀ has white throat and black-splashed necklace. Marked racial variation in tone of upperparts and colour and shape of spot on ♂'s throat (white in west and south, red in north and east). Juvenile recalls juvenile Robin, but already has diagnostic tail pattern.

(FACING PAGE) Siberian Rubythroat *Luscinia calliope*: **1** ad ♂ worn (spring), **2** ad ♀, **3** 1st winter ♂, **4** 1st winter ♀, **5** juv. Bluethroat *Luscinia svecica*. *L. s. svecica*: **6** ad ♂ breeding, **7** ad ♀ breeding, **8** ad ♂ non-breeding, **9** 1st winter ♀, **10** juv. *L. s. cyanecula*: **11** ad ♂ breeding, **12** ad ♀ breeding, **13** 1st winter ♂. *L. s. pallidogularis*: **14** ad ♂ breeding, **15** 1st winter ♂. *L. s. magna*: **16** ad ♂ breeding.

Chats, Thrushes

Unmistakable; no other small west Palearctic chat has similar tail pattern, except for mountain- and desert-haunting Red-tailed Wheatear which also has rufous rump and lacks blue throat or necklace. Flight free but nearly always close to ground, with flitting action and characteristic terminal flat glide or sweep into base of cover; flight silhouette recalls Robin but with apparently broader rump and tail. In the open, gait like Robin, with bold, upright carriage and long (even bounding) hops, varied by fast, short runs; when in cover, creeps furtively like mouse and capable of moving through densest vegetation. Frequently holds tail up and then looks remarkably long-legged, especially when bobbing in anxiety or excitement.

Habitat. Breeds from arctic and boreal upper latitudes to temperate and steppe middle latitudes and montane regions, continental and mainly cool. Patchiness in south of range suggests approach to relict status. Best adapted to regions intermediate between forest and open plains or valleys, such as wooded tundra with marshy glades among spruce, dwarf willows, and junipers, woods of birch, and shrubby wetlands, ascending from sea-level to high Scandinavian fjells. Also on floodplains and banks of rivers and lakes in dense but low woody vegetation. South European race *cyanecula* differs ecologically in stronger preference for bushy sites by water, including flood levels with alders and reedbeds, while southern populations in Spain at up to 2000 m frequent dry stony slopes covered with Spanish broom, singing from topmost twigs. In tropical African winter quarters, occupies edges of watery places, as a rule swamps, even very small ones, but also moist marshes with clumps of shrub.

Distribution. Northern and eastern race (nominate *svecica*)

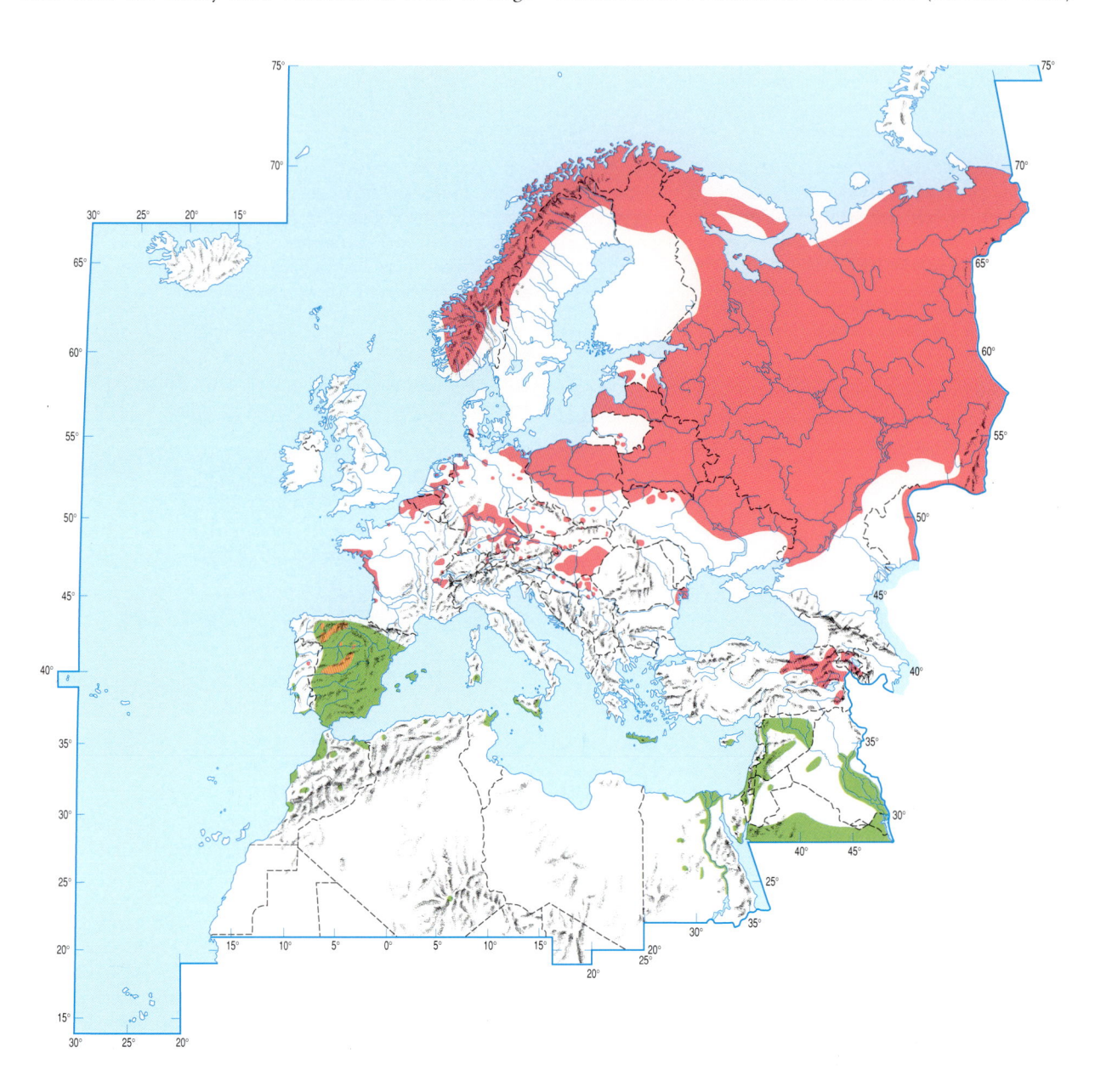

found to be also breeding central Europe in recent decades. BRITAIN. Scotland: in 1968, ♀ with clutch of eggs found, but no ♂ was seen; in 1985 pair reared 2 young, and in 1995 pair reared 3 young. FRANCE. Both *cyanecula* (in north and west) and *namnetum* (in west) increasing range. BELGIUM. Range increase. LUXEMBOURG. No breeding record since 1902. NETHERLANDS. Range has increased since 1970s, following marked decrease. GERMANY. Now only remnants of former almost continuous range. DENMARK. Bred 1858–73 and from 1992. ESTONIA. Range formerly more continuous. POLAND. Breeding in south (nominate *svecica*) only since early 1980s; otherwise mostly *cyanecula*. CZECH REPUBLIC. Breeding since 1978 (nominate *svecica* breeds northern mountains, *cyanecula* elsewhere). SLOVAKIA. Slight range increase. AUSTRIA. Breeding of nominate *svecica* confirmed at 13 sites 1975–90; *cyanecula* increasing, sites predominantly man-made, especially gravel-pits. SWITZERLAND. Breeds irregularly (1–4 pairs). ITALY. 2–3 pairs bred in Alps 1983–5, then apparently disappeared. RUMANIA. Bred Danube delta 1927, 1967–8, also more recently.

Accidental. Iceland, Faeroes, Britain (annual), Ireland.

Beyond west Palearctic, extends east across central and northern Asia to north-east Siberia and northern and western China; also breeds Alaska.

Population. FRANCE. Under 10 000 pairs in 1970s. BELGIUM. 1850–2150 pairs 1989–91; recent increase. NETHERLANDS. 5500–7500 pairs in 1989–91; increase from 1970s, following strong decrease. GERMANY. 3800 pairs in mid-1980s; decrease in east, where estimated *c.* 270 pairs in early 1980s. DENMARK. 1–3 pairs. NORWAY. 500 000–1 million pairs 1970–90; stable. SWEDEN. 100 000–300 000 pairs in late 1980s; stable. FINLAND. 100 000–200 000 pairs in late 1980s; slight increase. ESTONIA. 100–200 pairs in 1991; apparently stable. LATVIA. 50–200 pairs in 1980s; slight decrease. LITHUANIA. Rare. POLAND. 1000 pairs; recent slight increase following strong decrease. CZECH REPUBLIC. 130–190 pairs 1985–9; marked increase. SLOVAKIA. 15–30 pairs 1973–94; slight increase. HUNGARY. 400–500 pairs; stable. AUSTRIA. Very local. One estimate 300–400 pairs. SPAIN. 9000–12 800 pairs; stable. ALBANIA. 2000–5000 pairs in 1991. YUGOSLAVIA: CROATIA. 50–100 pairs; stable. RUMANIA. Probably 10–20 pairs; slight decrease. RUSSIA. 100 000–1 million pairs; stable. BELARUS'. 3000–6000 pairs in 1990; pronounced decline last 25–30 years. UKRAINE. 7000–8000 pairs in 1989; stable. TURKEY. 500–1000 pairs.

Movements. Mainly migratory, west Palearctic populations having extensive wintering area extending from Mediterranean basin south to northern Afrotropics, and east to Indian subcontinent.

Northern race, nominate *svecica* (breeding Scandinavia east across northern Russia), winters patchily right across Mediterranean and over entire African winter range of the species; probably also in Indian subcontinent.

South-west races, *cyanecula* (breeding Spain and central Europe) and *namnetum* (breeding western France), move between south and west, with concentration on autumn passage in southern Spain and western Portugal, majority then moving south into Africa. Return passage starts early, from late February or early March.

Food. Largely terrestrial invertebrates, mostly insects; in autumn, also some seeds and fruits. Feeds on ground, hopping, running briefly, and pausing; also takes items from low vegetation and will catch insects in the air. Searches for food by turning over leaves and soil.

Social pattern and behaviour. Normally solitary outside breeding season; apparently territorial in winter quarters. Mating system probably essentially monogamous, with occasional polygyny. Breeding territories may be clumped, with nests quite close together, probably primarily due to habitat constraints. During ♂'s territorial song, red spot of nominate *svecica* or white spot of *cyanecula* may briefly become twice normal size, thus probably an important accompanying signal in territory advertisement; throat and breast pattern as well as red marks at base of tail similarly have important signal function in other antagonistic and heterosexual displays. ♂ sings from within undergrowth or, when more excited, from exposed perch up to *c.* 6–10 m high, or in flight.

Voice. Most outstanding feature is remarkable mimetic song, snatches of which are also given in winter (by both sexes). Full song by ♂ only, but ♀ gives quiet and loud but normally short bursts. Song loud, sweet, and exceptionally varied; reminiscent of Nightingale but not so rich or full in tone, having merrier, more tinkling quality; characteristic phrases include throaty 'torr-torr-torr-torr' and metallic, ringing 'ting-ting-ting'. Mimicks insects and mechanical sounds as well as many other bird species. Calls include hard 'tacc tacc', lower croaked 'turrc turrc', and plaintive 'hweet' like Nightingale and *Phylloscopus* warbler.

Breeding. SEASON. Central Europe: laying begins late April. Scandinavia: eggs laid from late May. Finland: laying at 69°N normally takes place during 2 weeks beginning 9 June. One brood in north of range, 2 in south. SITE. On ground in dense vegetation, in tussock, under bush, or in hollow in low bank. Nest: cup of grass stems and leaves, with roots and moss, lined with hair and finer vegetation. EGGS. Sub-elliptical, smooth but only slightly glossy; pale blue, green, or blue-green, finely marked red-brown, often indistinct, giving rusty tinge to shell, occasionally more heavily mottled. Clutch: 5–6 (4–8). INCUBATION. 13–14 days. FLEDGING PERIOD. 14 days, but may leave nest 1–2 days earlier.

Wing-length: *L. s. svecica*: ♂ 74–81, ♀ 73–77 mm.
Weight: *L. s. svecica*: ♂ mostly 15–25, ♀ 15–22 g.

Geographical variation. Marked and complex; 6 races recognized in west Palearctic. Much individual variation locally, especially in apparent zones of secondary intergradation, such as central European FSU. *L. s. cyanecula* from Belgium and eastern France east to Carpathians and approximately to Smolensk, Novgorod, and St Petersburg is rather large, and ♂ has rounded silky-white spot or short bar on lower throat in breeding plumage, only rarely absent. *L. s. namnetum* from western France similar, but smaller in size. Birds from mountains of northern and central Spain usually included in *cyanecula*, but in fact intermediate between this race and similarly isolated southern mountain race *magna* from Caucasus area; large in

Siberian Blue Robin *Luscinia cyane*: **1** ad ♂, **2** ad ♀, **3** 1st winter. Red-flanked Bluetail *Tarsiger cyanurus*: **4** ad ♂ bright morph fresh (autumn), **5** ad ♂ bright morph worn (spring), **6** ad ♀ worn (spring), **7** 1st winter, **8** juv.

size and throat-spot often absent. Nominate *svecica*, with red throat-spot, occupies all northern areas of west Palearctic. Throat-spot variable in complex populations to south-east; *pallidigularis* (from Volga eastwards) the most well-marked, ♂ with lower cheeks, chin, and throat glossy pale cerulean-blue (much paler than in *cyanecula*) and spot on lower throat rufous-cinnamon, shaped as a large broad bar or broadly triangular spot.

Siberian Blue Robin *Luscinia cyane*

PLATES: pages 1152, 1156

Du. Blauwe Nachtegaal   Fr. Rossignol bleu   Ge. Blaunachtigall   It. Usignolo azzurro siberiano
Ru. Синий соловей   Sp. Ruiseñor coliazul   Sw. Blånäktergal

Field characters. 13.5 cm; wing-span 20–21 cm. Close in size to Robin and Red-flanked Bluetail but with 10–15% shorter wings, 25–30% shorter tail, and 10% longer legs. Rather small but compact, robust chat, with relatively large bill and head, rather short wings and tail, and long legs; shape, and habit of running on ground, recall small crake *Porzana*. ♂ intensely coloured, dark blue above, with deep black face-mask, and white below; ♀ and immature dull, olive-brown above, fulvous-brown and white below. Legs pale flesh.

Adult ♂ unmistakable in west Palearctic but ♀ and immature liable to confusion with Red-flanked Bluetail from which distinction best based on different form, pale legs, lack of orange flanks, and lack of strongly blue-toned tail. Flight light, but fast wing-beats yield rather fluttering action, recalling Robin before other chats. Gait remarkably free, incorporating short and long hops, and high-stepping run. Constantly quivers short tail (unlike Red-flanked Bluetail). Rarely leaves ground cover and then only briefly. Alarm-call a rapid 'chuck-chuck-chuck'.

Habitat. Breeds in middle and lower-middle latitudes of continental and oceanic east Palearctic, characteristically in deep taiga of spruce, fir, birch, aspen, and other (mainly coniferous) forest trees, with dense shady canopy and fallen trees but no undergrowth, often by riversides and near meadows with tall herbage. In Borneo in winter, occupies primary and secondary forest, especially along streams, from sea level to *c.* 1800 m.

Distribution. Breeds in southern Siberia from upper Ob' east to Sea of Okhotsk, south to north-east China and Japan.

Accidental. Channel Islands: 1 probably 1st-year ♀, Sark, October 1975.

Movements. Migratory. Winters in southern Asia from southern China, Indochina, Thailand, and southern Burma south to Philippines, Borneo, Malaya, and Sumatra, straggling to eastern India (Bengal, Manipur). Passage seems to be essentially through Mongolia and China, i.e. to east of Himalayas and associated mountain systems.

Wing-length: ♂ 74–81, ♀ 73–79 mm.

Red-flanked Bluetail *Tarsiger cyanurus*

PLATES: pages 1152, 1156

Du. Blauwstaart   Fr. Robin à flancs roux   Ge. Blauschwanz   It. Codazzurro
Ru. Синехвостка   Sp. Coliazul cejiblanco   Sw. Blåstjärt

Field characters. 14 cm; wing-span 21–24.5 cm. Close in size and structure to Robin, though slightly shorter-billed and longer-tailed. Small, fairly compact chat, with general character recalling both Robin and redstart; behaviour often suggests flycatcher. Typical ♂ intensely coloured, dark sheeny blue above, white and grey below, with variable whitish supercilium and long splash of orange along flanks. ♀ and dull ♂ olive-brown above, with pale eye-ring and darker, blue-washed rump and tail; dull white below, with clean throat emphasized by brown chest and orange flank-panel. Juvenile like adult ♀ but liberally spotted with cream and scalloped dark brown on head and back and mottled dark on chest and flanks.

Unmistakable given view of orange panel on flanks, but ♂ from behind, ♀ and immature may suggest Siberian Blue Robin. Combines shape of foreparts, gait, and perching behaviour of Robin with shape of rear parts and flight of Redstart, pouncing hunt of *Saxicola* chat, and tail movement and flycatching of *Ficedula*, being often high in trees and never as persistently on ground as Siberian Blue Robin. Flight free and light, with fluent wing-beats and rather loose-tailed appearance. Gait essentially hopping, whether along branches, in foliage, or over ground. Constantly flicks wings and tail open, like *Saxicola*.

Habitat. Breeds in upper-middle and marginally in upper continental latitudes, exclusively boreal and montane, in thick mossy conifer forest, especially taiga, on moist soil, generally with undergrowth, and with July temperatures of 15–24°C.

Distribution and population. FINLAND. First recorded 1949. May breed annually in north-east near Russian border (but only 2 nests found). Number of singing ♂♂ has increased in recent years (maximum 8). ESTONIA. Pair bred 1980. RUSSIA. 100–1000 pairs.

Accidental. Britain, Channel Islands (Sark), France, Netherlands, Germany, Denmark, Norway, Sweden, Estonia, Poland, Slovakia, Italy, Cyprus, Lebanon, Israel.

Beyond west Palearctic, two disjunct populations: in northern Asia south to Mongolia and Japan, and from Himalayas to central China.

Movements. West Palearctic populations are long-distance migrants (wintering from Burma east to southern China and Taiwan). (Southern race *rufilatus*, breeding Himalayan region and western China, mainly shows short-distance altitudinal movements.) West Palearctic birds therefore make long easterly movements (in autumn), passing north of major central Asian mountain systems, before turning south through Mongolia and China. Autumn migration begins early September; northern edge of range deserted by mid-September. Return passage begins April, vanguard reaching southern Siberia in second half of April. Spreads north and west during May, reaching Arkhangel'sk region around 20 May–4 June.

Food. Insects; also fruits and seeds outside breeding season. Feeds in low trees, shrubs, and on ground. Catches insects by hopping about on ground, by perching and flying down to take items located, and by brief aerial-pursuit like flycatcher.

Social pattern and behaviour. Little known; apparently

generally similar to robins. ♂ sings usually from tree-top or other prominent perch, sometimes lower down and half in cover. Song-period rather long; in Finland, 12 May–20 July.

Voice. Song of ♂ consists of short but distinctive and often loud phrases, audible up to several hundred metres and of several different types; suggesting a pure-voiced thrush as much as a chat. A bird will sing several different song-types but sometimes gives one repeatedly with short pauses over long period. Commonest call from breeding birds 'weep' or 'peep', often repeated. Call from vagrants a short 'teck-teck', recalling Robin.

Breeding. SEASON. Pechora basin (Russia): 1st broods on the wing late June, 2nd in mid-August. SITE. On ground in hollow among tree roots, or in hole in bank, or slightly above ground in stump or fallen log. Nest: cup of moss, grass, and roots, lined with softer grass, wool, hair, and sometimes conifer needles. EGGS. Sub-elliptical, smooth and slightly glossy; white, sometimes lightly marked with brownish blotches, usually at broad end. Clutch: 5–7. INCUBATION. No information. FLEDGING PERIOD. 15 days.

Wing-length: ♂ 77–84, ♀ 73–79 mm.
Weight: ♂♀ mostly 12–16 g.

White-throated Robin *Irania gutturalis*

PLATES: pages 1155, 1156

Du. Perzische Roodborst   Fr. Iranie à gorge blanche   Ge. Weißkehlsänger   It. Pettirosso golabianca
Ru. Соловей-белошейка   Sp. Petirrojo turco   Sw. Vitstrupig näktergal

Field characters. 16.5 cm; wing-span 27–30 cm. Nearly 20% larger than Robin; close in size to nightingales, with similar form and structure except for slightly longer wings and tail. Quite robust and bulky chat, more recalling robin-chats *Cossypha* of Africa than Palearctic relatives. Diagnostic combination of rather long black tail, white vent, and rufous-buff flanks. ♂ striking, with black face-mask contrasting with white throat, narrow white supercilium, dark blue-grey upperparts, and rich rufous-orange chest. ♀ much less colourful, with brown-grey head and back. Juvenile like adult ♀ but with buff spots above and dark mottling on chest.

♂ unmistakable in west Palearctic but not so in winter range of East Africa, where several similar robin-chats *Cossypha* may cause brief confusion when their rufous rumps and tails not seen. ♀ and immature puzzling, suggesting several other chats, e.g. Thrush Nightingale and ♀ Redstart until black tail visible (a character shared only with Blackstart but that species 15% smaller, with more uniform, paler grey or brown plumage). Flight free and light; due to relatively long wings and tail, flight silhouette can suggest a thrush as much as a chat. Gait and stance as nightingale. Inveterate skulker in ground cover and difficult to flush.

Habitat. Breeds in warm dry continental lower-middle latitudes, typically in uplands, on stony rocky hillsides with fairly dense to more open scrub. Occurs in similar habitats in winter quarters.

Distribution and population. AZERBAIJAN. Uncommon. Breeds mostly in Nakhichevan region. TURKEY. Widespread in interior. Very local in northern part of range. 5000–50 000 pairs. SYRIA. Distribution still very poorly known; assumed to be breeding higher altitudes and extensions of Anti-Lebanon and Allovit mountains. ISRAEL. *c.* 10 pairs on slopes of Mt Hermon in some years. IRAQ. Breeds in Ser Amadiya mountains.

Accidental. Britain, Netherlands, Norway, Sweden, Greece. Vagrant or rare passage-migrant to Cyprus and Egypt.

Beyond west Palearctic, extends east from Iran to western Tien Shan.

Movements. Migratory, wintering in rather restricted area of East Africa, mainly in Kenya in plateau country north and east of highlands and in Tanzania mainly in north-east and dry interior. Arrives in Turkey from mid-April, with continuing arrivals into May; arrival apparently synchronous across entire range. Most leave breeding grounds by end of August. Nowhere common on passage, probably because east Mediterranean and north-east Africa are usually overflown.

Food. In breeding season at least, mainly insects; in autumn, also fruits. Feeds mainly on ground, turning over leaves to search; also in trees and bushes.

Social pattern and behaviour. Not very sociable. Wintering birds territorial; sing regularly, usually from low perches. On breeding grounds ♂ sings from an exposed and elevated perch—top of bush or tree—sometimes from middle of tree,

White-throated Robin *Irania gutturalis*: **1** ad ♂ fresh (autumn), **2** ad ♂ worn (spring), **3** ad ♀ worn (spring), **4** 1st winter ♂, **5** juv.

or in flight. May ascend into song-flight after singing from perch. Detailed description of song-flight lacking, but a number of variants (or perhaps simply phases) occur. May ascend while singing, then, continuing to sing loudly, glide slowly or rapidly with wings and tail widely spread and rigid for several tens of metres, to land (tail widely spread) on bush or rock. Several song-flights sometimes performed in quick succession, bird thus moving from perch to perch across territory.

Voice. Song a short phrase of clear bell-like notes, of similar quality and volume to those of nightingale. Calls include 'tirric', 'churr', and 'chick', also recalling nightingale.

Breeding. SEASON. Armenia: eggs laid from first half of May. SITE. In lower part of shrub or small tree, on stump, or in tree crevice; 15–125 cm above ground. Nest: cup of dry grass leaves, twigs, and bark, lined with vegetable down and hair; often some feathers, bits of rag, paper, sheep's wool, etc. EGGS. Sub-elliptical, smooth and fairly glossy; greenish-blue, with yellowish or rusty-brown spots usually coalescing at broad end. Clutch: (3–)4–5. INCUBATION. 13 days. FLEDGING PERIOD. Young leave nest at 9–10 days old, not able to fly.

Wing-length: ♂ 92–101, ♀ 91–99 mm.
Weight: ♂ 19–27, ♀ 16–25 g.

Eversmann's Redstart *Phoenicurus erythronotus*

PLATES: pages 1158, 1165

DU. Eversmanns Roodstaart FR. Rougequeue d'Eversmann GE. Sprosserrotschwanz IT. Codirosso di Eversmann
RU. Красноспинная горихвостка SP. Colirrojo de Eversmann SW. Altairödstjärt

Field characters. 16 cm; wing-span 25.5–27 cm. 10–15% larger and bulkier than Redstart. Rather large redstart, with proportionately larger head, less slim body, and slightly shorter tail than smaller congeners. ♂ unusually patterned, having fully rufous fore-underparts and back, and white patches along scapulars and on primary coverts. ♀ fawn-coloured, with prominent eye-ring, pale fulvous tips to wing-coverts and tertials, and whitish vent.

♂ unmistakable but ♀ and immature far less distinctive, recalling Redstart but less orange-buff below, with paler buff rump, duller, less bright-sided tail, more conspicuous eye-ring, and paler tips to wing coverts. Flight and behaviour not well studied but apparently as Redstart, though tail movement consists of distinct upward flirt without nervous quiver.

Commonest call a soft, slurred croaking monosyllable; croaking 'gre-er' in alarm.

Habitat. Breeds in lower-middle latitudes, extralimitally, in temperate continental montane regions up to 5400 m at upper limit of spruce, but mainly much lower in sparse woodlands, often of stunted trees, broad-leaved or coniferous, or of tall shrubs, on dry stony rather than moist soil. In winter in Indian subcontinent, usually below 2100 m in arid country: waste land, scrub jungle, olive groves, orchards, dry river beds,

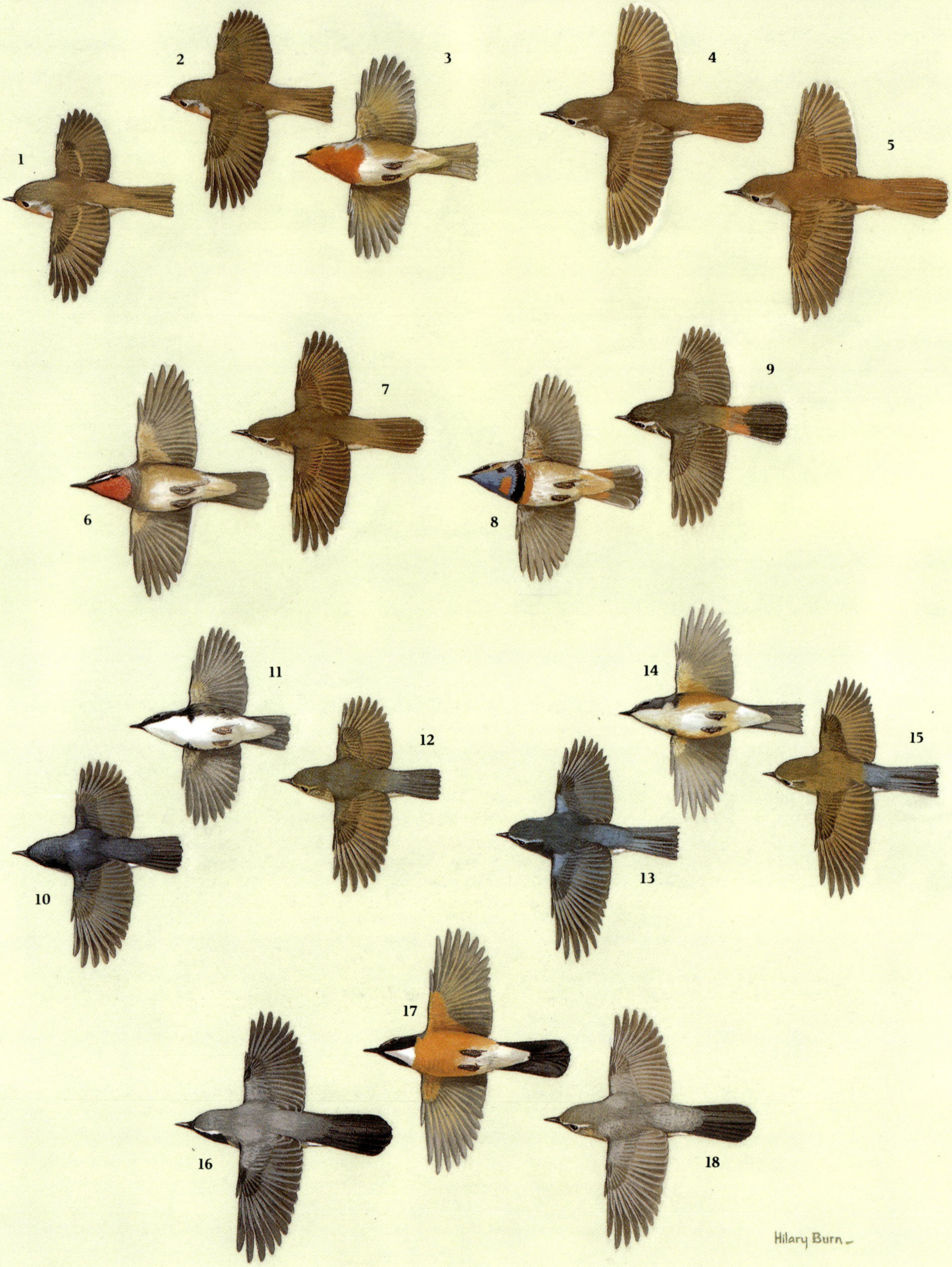

1
2
3
4
5
6
7
8
9
10
11
12
13
14
15
16
17
18
Hilary Burn

wooded compounds, and avenues and groves of *Acacia* and similar trees.

Distribution. Breeds central Asian mountains: Tien Shan, Tarbagatay, Altai, and from north-west Mongolia to west and south of Lake Baykal.

Accidental. Israel: ♂ near Merom Golan, November 1988. Kuwait: single birds December 1970, December 1987. Rare winter visitor to southern Iraq.

Movements. Vary from altitudinal displacement to short- or medium-distance migration, with a few birds penetrating south-west as far as Persian Gulf. Wintering areas vary from year to year, even month to month, as direct result of weather and food availability; birds often forced to move well away from breeding grounds to find suitable conditions. True migrant in north-east of breeding range (Altai region, *c.* 50–55°N).

Food. Largely fruit and seeds in mid-winter, largely insects at other times. Feeds by (1) picking items from ground (sometimes rummaging in leaf-litter or clearing away thin snow layer) or from vegetation (including bushes), (2) flying on to prey from low perch, or (3) taking aerial prey in brief flight.

Wing-length: ♂ 84–92, ♀ 83–89 mm.

Black Redstart *Phoenicurus ochruros*

PLATES: pages 1159, 1165

Du. Zwarte Roodstaart   Fr. Rougequeue noir   Ge. Hausrotschwanz   It. Codirosso spazzacamino
Ru. Горихвостка-чернушка   Sp. Colirrojo tizón   Sw. Svart rödstjärt

Field characters. 14.5 cm; wing-span 23–26 cm. Marginally larger than Redstart, with 10% longer wings and 10–15% longer tail. Of closely similar form to Redstart, shares same bright red tail but differs distinctly in otherwise less colourful plumage (in western races), less attenuated form, ground-hugging behaviour, and running (as well as hopping) gait. Dominant colour in all plumages dusky-black in ♂, dusky grey-brown in ♀ and immature; in all plumages, striking chestnut rump and tail and (in western adult ♂♂) white wing-panel. Eastern races have underparts below chest increasingly saturated deep chestnut, suggesting Redstart and Güldenstädt's Redstart. Juvenile like adult ♀ but more uniform dusky-brown and liberally flecked and barred dark brown (not pale-spotted like Redstart).

Western races *gibraltariensis* and *aterrimus* unmistakable in all plumages, with uniform dusky appearance quite distinct from any congeners. Adult ♂♂ of eastern races with chestnut underbody recall Redstart but quickly separable on blacker back and lack of large pale vent; ♀♀ and 1st-years of eastern races more likely to be confused with Redstart but differences in form and gait should prevent confusion, while lack of any warm rufous or buff on underbody above vent is apparently trustworthy. Flight like Redstart, but rather less flickering wing-beats and much less loose-tailed appearance make flight silhouette more compact. Gait a brisk hop like Redstart but, unlike that species, often also runs quickly like wagtail; pauses in alert, upright posture. Less shy than Redstart and markedly more terrestrial.

Habitat. Breeds in west Palearctic in middle latitudes, oceanic as well as continental, mainly in sunny warm or mild temperate, Mediterranean, and steppe or montane climates, avoiding persistently wet or humid situations and dense vegetation of any height. Favours rocky, stony, boulder-strewn broken or craggy terrain, including cliffs, right up to snowline. Frequency and convenience of nest-sites in walls or roofs of buildings, outbuildings, and wide variety of other structures have evidently led to evolution of close commensalism with man, spreading from montane to lowland regions and facilitating extensive northward spread across plains and valleys. This has extended even into large cities where absence or scarcity of water, trees, shrubs, and grasslands more than compensated for by presence of waste patches colonized by weed species, with many bare disturbed areas of soil, and a choice of commanding song-posts and cavities suitable for nesting.

Distribution. Long-term expansion of range to north and north-west in northern Europe, with Denmark colonized mid-19th century, and first breeding Sweden 1910, Latvia 1923, Lithuania 1939, Norway 1944, Belarus' 1956, Estonia (breeding regularly since 1960s), Finland (breeding regularly since 1970s). Range increase reported Slovenia, Russia, Ukraine, and Moldova. BRITAIN. Bred Durham 1845, Sussex 1909 (perhaps), 1923, and for some years in Cornwall, then annually since 1939. Has bred in most English counties, but predominantly in south-east. Colonization remains very slow. FRANCE. Corsica: first breeding record 1986. RUSSIA. In June 1993, pair recorded prospecting at 68° 30′ N near mouth of Pechora river, and another ♂ in song at 67° 37′ N.

Accidental. Iceland, Faeroes, Mauritania, Madeira.

(FACING PAGE) Robin *Erithacus rubecula* (p. 1140). *E. r. rubecula*: **1** 1st winter. *E. r. melophilus*: **2–3** ad.
Thrush Nightingale *Luscinia luscinia* (p. 1143): **4** ad.
Nightingale *Luscinia megarhynchos* (p. 1145): **5** ad.
Siberian Rubythroat *Luscinia calliope* (p. 1147): **6** ad ♂, **7** ad ♀.
Bluethroat *Luscinia svecica svecica* (p. 1149): **8** ad ♂ breeding, **9** ad ♀ breeding.
Siberian Blue Robin *Luscinia cyane* (p. 1152): **10–11** ad ♂, **12** ad ♀.
Red-flanked Bluetail *Tarsiger cyanurus* (p. 1153): **13–14** ad ♂, **15** ad ♀.
White-throated Robin *Irania gutturalis* (p. 1154): **16–17** ad ♂, **18** ad ♀.

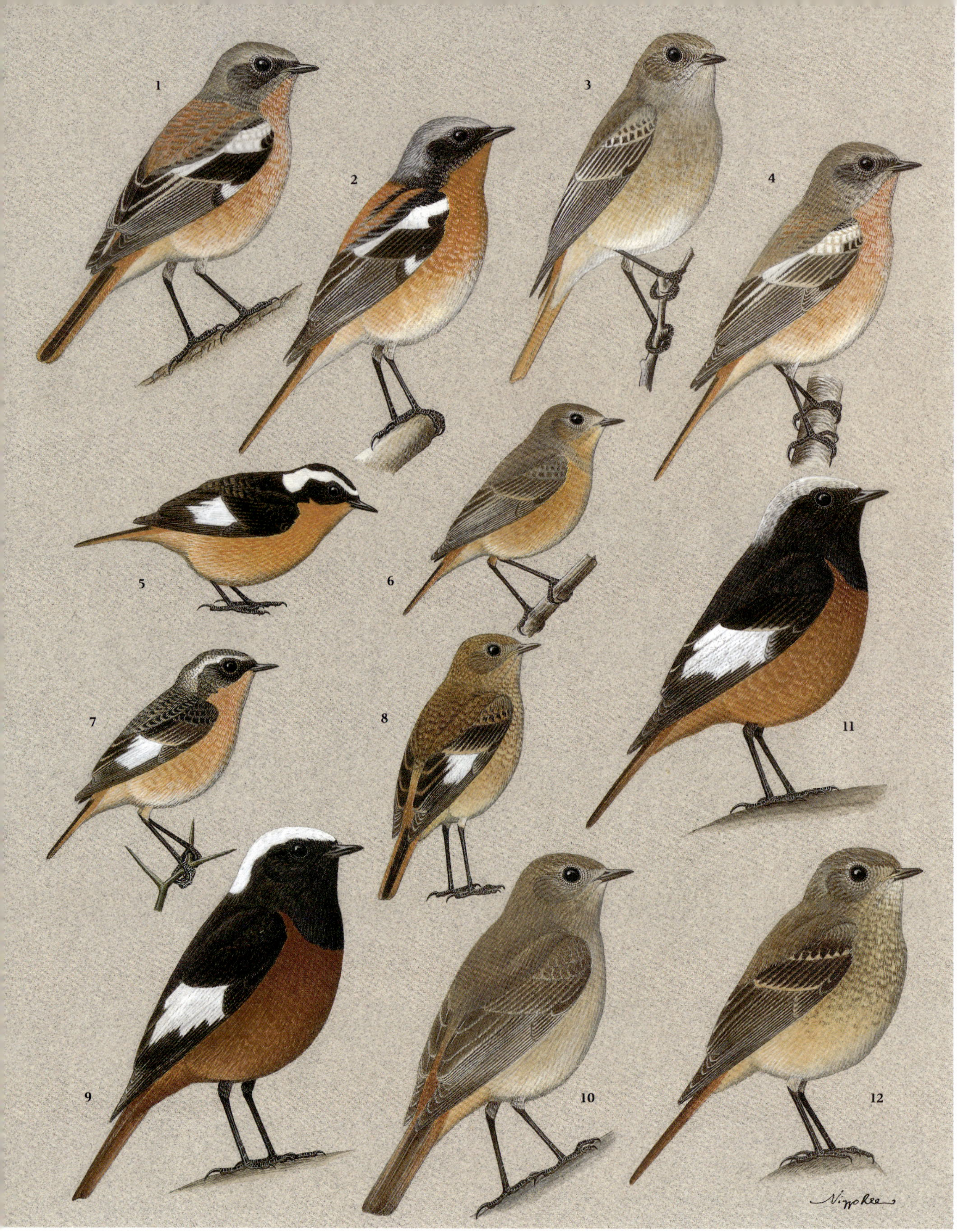
1
2
3
4
5
6
7
8
9
10
11
12

Black Redstart *Phoenicurus ochruros*. *P. o. gibraltariensis*: **1** ad ♂ fresh (autumn), **2** ad ♂ worn (spring), **3** ad ♀ fresh (autumn), **4** ad ♀ worn (spring), **5** 1st summer ♂, **6** juv. *P. o. aterrimus*: **7** ad ♂ worn (spring). *P. o. semirufus*: **8** ad ♂ worn (spring). *P. o. phoenicuroides*: **9** ad ♂ worn (spring), **10** ad ♀ worn (spring), **11** 1st winter ♂.

Beyond west Palearctic, breeds from Iran east to central Asia and central China.

Population. Increase north-west Europe, Norway, Finland, Estonia, Latvia, Czech Republic, Croatia, Slovenia, Rumania, Belarus', Ukraine. Otherwise stable; trend in Middle East and Morocco not known. BRITAIN. 80–120 pairs 1988–91. FRANCE. Under 1 million pairs in 1970s. BELGIUM. 15 000 pairs 1973–7. LUXEMBOURG. 12 000–15 000 pairs. NETHERLANDS. 20 000–28 000 pairs 1979–85. GERMANY. 1.15 million pairs in mid-1980s. DENMARK. 400–1000 pairs 1987–9. NORWAY. 10–100 pairs 1970–90. SWEDEN. 500–1000 pairs in late 1980s. FINLAND. 20–30 pairs in late 1980s. ESTONIA. 100–200 pairs 1991. LATVIA. 400–800 pairs in 1980s. LITHUANIA. Highest numbers in south and west. POLAND. 50 000–200 000 pairs. CZECH REPUBLIC. 200 000–400 000 pairs 1985–9. SLOVAKIA. 100 000–200 000 pairs 1973–94. HUNGARY. 50 000–60 000 pairs 1979–93. AUSTRIA. Very common. SWITZERLAND. 250 000–500 000 pairs 1985–93. SPAIN. 400 000–900 000 pairs. PORTUGAL. 10 000–100 000 pairs 1978–84. ITALY. 200 000–400 000 pairs 1983–95. GREECE. 10 000–15 000 pairs.

(FACING PAGE) Eversmann's Redstart *Phoenicurus erythronotus* (p. 1155): **1** ad ♂ fresh (autumn), **2** ad ♂ worn (spring), **3** ad ♀ fresh (autumn), **4** 1st winter ♂.
Moussier's Redstart *Phoenicurus moussieri* (p. 1163): **5** ad ♂ worn (spring), **6** ad ♀ worn (spring), **7** 1st winter ♂, **8** juv ♂.
Güldenstädt's Redstart *Phoenicurus erythrogaster* (p. 1164): **9** ad ♂ worn (spring), **10** ad ♀ worn (spring), **11** 1st winter ♂, **12** juv ♀.

YUGOSLAVIA: CROATIA. 30 000–40 000 pairs. SLOVENIA. 50 000–80 000 pairs. BULGARIA. 50 000–500 000 pairs. RUMANIA. 30 000–80 000 pairs 1986–92. RUSSIA. 1000–10 000 pairs. Increase in Voronezh region. BELARUS'. 20 000–35 000 pairs in 1990. UKRAINE. 70 000–75 000 pairs in 1986. MOLDOVA. 2500–4000 pairs in 1988. AZERBAIJAN. Common. Very common in Nakhichevan mountains. TURKEY. 50 000–500 000 pairs. ISRAEL. 100–200 pairs in 1970s–80s. IRAQ. Abundant in northern mountains. MOROCCO. Scarce and local.

Movements. Resident, partial migrant, or migrant in different parts of range, main wintering area for west Palearctic breeders being Mediterranean basin, with a few on west coast of Europe and some southward extension to north-west and (especially) north-east Africa.

P. o. gibraltariensis (breeding in most of European range east to Ukraine and Crimea) winters in small numbers in Britain and west and central France; regular in winter in Belgium, and occasional in Switzerland and Germany. Northern limit of wintering varies from year to year. Southern limit is northern edge of Sahara with occasional records further south. However, bulk winter in Mediterranean basin, being common on coasts and islands in western Mediterranean. Nominate *ochruros* (breeding eastern Turkey and Caucasus east to northern Iran) makes only short movements to winter in western Zagros mountains of Iran and in Iraq; present on breeding grounds from end of March or early April to end of September.

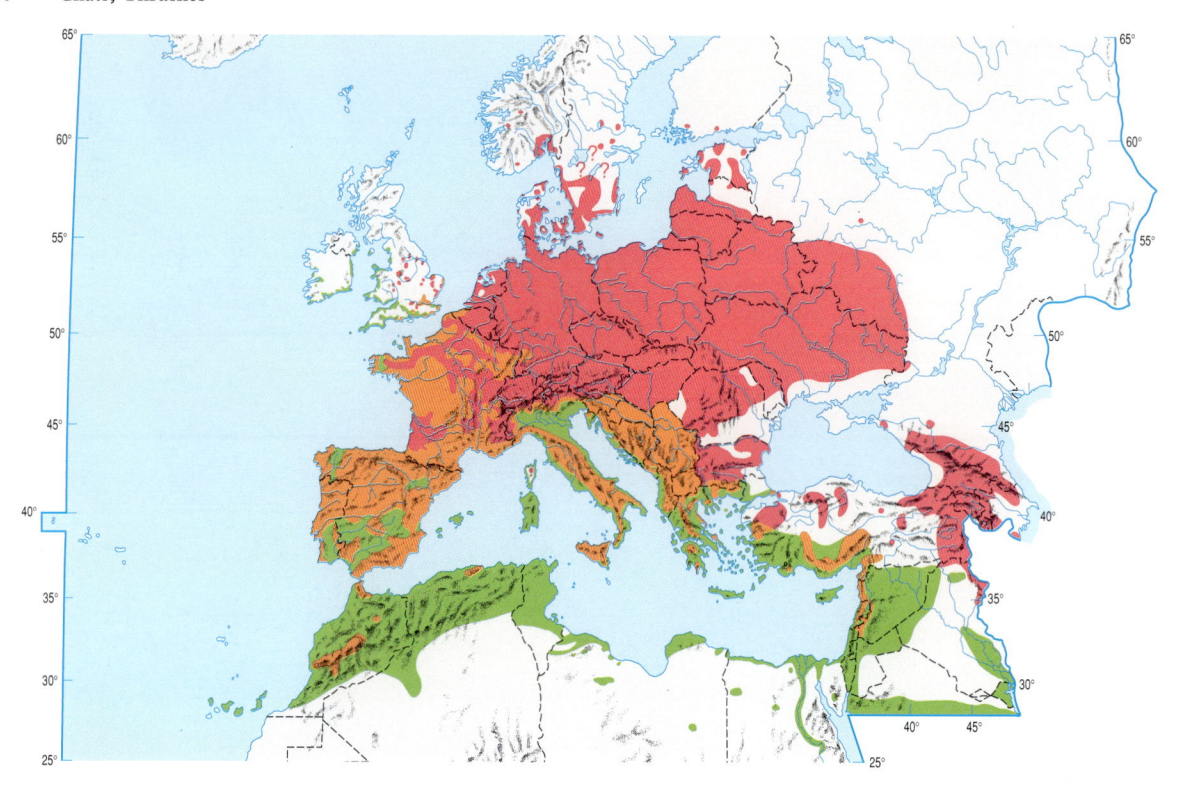

Food. Small or medium-sized invertebrates and fruit. Feeds on ground, hopping or running (and recorded digging 2–4 cm into hard surface to get at larvae), by flying from perch on to ground prey (and usually returning quickly to perch), and by taking aerial prey in short flight; also takes items from walls, foliage, etc., by hovering.

Social pattern and behaviour. Outside breeding season, solitary, in pairs, occasionally in small groups. Breeding pairs territorial, but neighbouring nests may be as little as 6 m apart, occasionally less. ♂ usually sings from high, exposed perch. Distinguished among small ground-feeding passerines for conspicuous high song-posts, often at 20 m or more, on rocks or buildings.

Voice. Song a rather rushed warble, quieter than Redstart; highly individual subterminal phrase suggests rattle of ball-bearings shaken together or poured into bottle; often ends in louder burst of rushed, ringing notes. Commonest call a short 'tsip' (which does not recall *Phylloscopus* warbler, unlike common call of Redstart); this note often introduces scolding 'tucc-tucc'; in alarm, rapid rattle 'tititicc'.

Breeding. SEASON. Western Europe, including England: egg-laying from 2nd half of April. Eastern Europe: up to 2 weeks later. Southern Europe and North Africa: similar to western Europe. 2 broods, occasionally 3. SITE. On ledge in cave or building, or in hole or crevice in rock or wall. Up to 45 m above ground, mostly 1–4 m. Nest: loose cup of grass leaves and stems, moss, and other plant material, lined with wool, hair, and feathers. EGGS. Sub-elliptical, smooth and glossy; in *gibraltariensis*, white, rarely tinged blue or with faint brownish spotting; in nominate *ochruros*, pale blue to blue-green. Clutch: 4–6 (2–8). INCUBATION. 13–17 days. FLEDGING PERIOD. 12–19 days.

Wing-length: *P. o. gibraltariensis*: ♂ 85–91, ♀ 83–90 mm.
Weight: *P. o. gibraltariensis*: ♂♀ mostly 13–19 g.

Geographical variation. Marked: in ♂, involves extent of black and grey on upperparts, colour of underparts and underwing, presence of white on forehead, and amount of white on wing; in ♀ and juvenile, mainly involves general colour; in both sexes, size. 2 main subspecies-groups: (1) *gibraltariensis* group with *gibraltariensis* and *aterrimus*, occurring Europe and North Africa east to Crimea and (probably) western Turkey; (2) *phoenicuroides* group in east of range, breeding mainly beyond west Palearctic; *semirufus* in Levant. Nominate *ochruros* from eastern Turkey, Caucasus, and northern Iran combines characters of both main groups, apparently as a result of secondary intergradation; much individual variation, but in general nearer to *phoenicuroides* group. ♂♂ of *gibraltariensis* group characterized by uniform grey forehead and crown, black breast merging into grey-white central belly and vent, and large white wing-patch formed by broad white outer fringes of secondaries and outer primaries; ♀ and juvenile generally dark brownish-grey. ♂ *gibraltariensis* usually has upperparts uniform medium grey, but lower mantle and inner scapulars frequently black; ♂ *aterrimus*, of south and west Iberia, generally blacker above and below. Typical ♂ of *phoenicuroides* from central Asia differs markedly from *gibraltariensis* by fully deep rufous-cinnamon sides of breast, belly, flanks, vent, axillaries, and under wing-coverts, sharply contrasting with black chest; forehead often white, contrasting with black rim along base of upper mandible (white sometimes concealed, especially in fresh plumage); white wing-patch absent; small. P. o. semirufus from Levant small, like *phoenicuroides*;

♂ similar to *phoenicuroides*, but upperparts usually with much black (sometimes up to crown and down to back); black of chest reaches slightly further down.

Redstart *Phoenicurus phoenicurus*

PLATES: pages 1162, 1165

Du. Gekraagde Roodstaart   Fr. Rougequeue à front blanc   Ge. Gartenrotschwanz   It. Codirosso
Ru. Обыкновенная горихвостка   Sp. Colirrojo real   Sw. Rödstjärt

Field characters. 14 cm; wing-span 20.5–24 cm. As long as Robin but much more attenuated, with flatter crown, longer wings, and slim rear body extending into rather long tail; slightly smaller and less robust than Black Redstart. Small, elegantly dressed, and graceful chat, with brilliant rufous-chestnut rump and tail always eye-catching whether loosely flirted in flight or characteristically quivered on perch. ♂ blue-grey above, with white forehead and supercilium, black face and throat, and rufous-orange underbody other than white vent. ♀ brown-grey above, buff to white below, with pale eye-ring and edges to inner wing feathers; has characteristic demure expression. Juvenile like ♀ but spotted pale buff above and barred dark brown below; may show pale wing-bars. In winter, ♂ loses immaculate appearance due to hoary tips to face and breast feathers.

Far from unmistakable, with (1) adult ♀ subject to confusion particularly with eastern races of Black Redstart, (2) immature ♂ on passage in Middle East confusable with Eversmann's Redstart, and (3) ♀ and juvenile confusable with 6 other species of *Phoenicurus* of similar sex or age. Best marks of Redstart are (1) slim, elegant form, (2) blue tone of ♂'s crown and back, (3) clean tone of ♀'s underparts, (4) almost constant

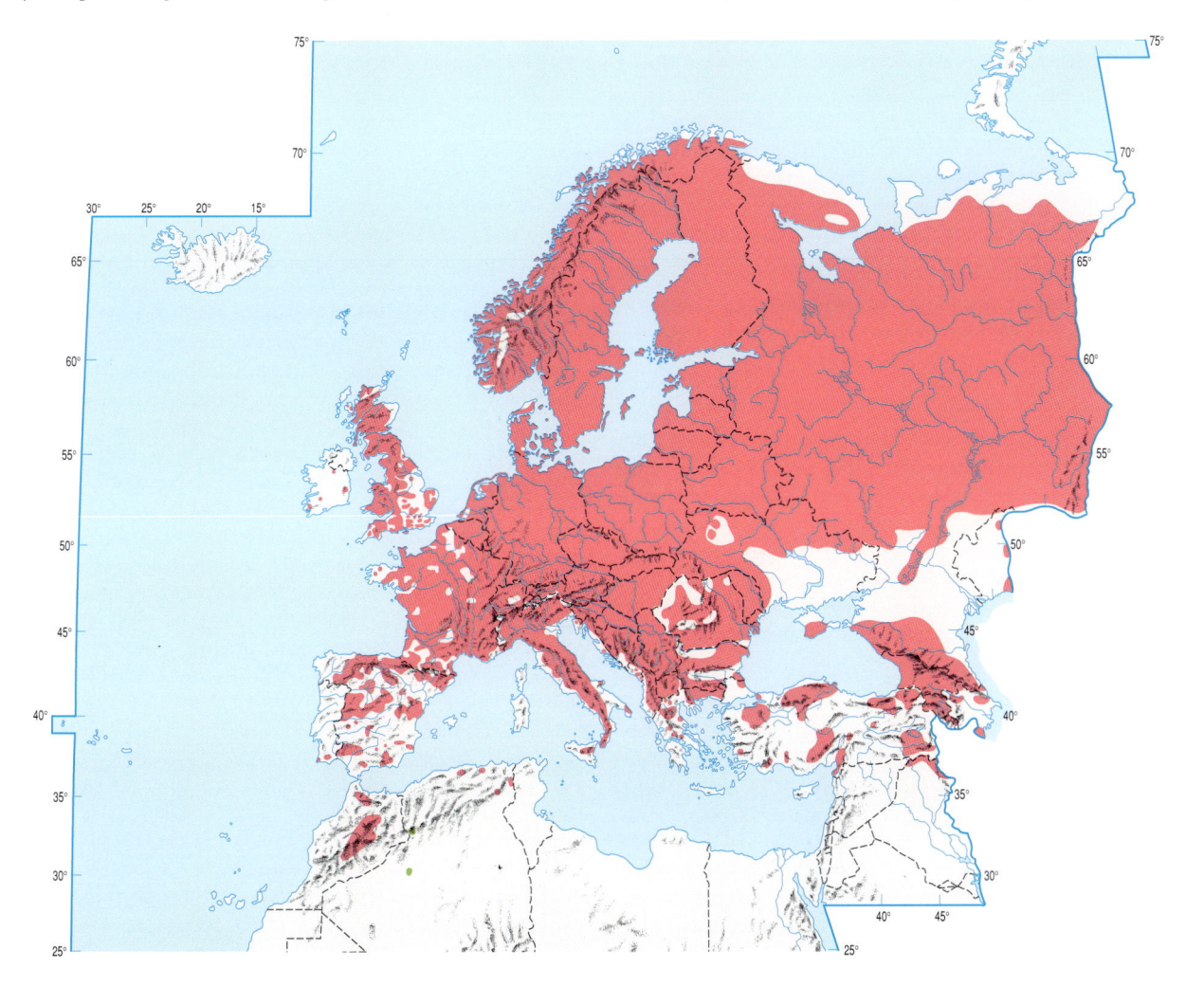

Redstart *Phoenicurus phoenicurus*. *P. p. phoenicurus*: **1** ad ♂ fresh (autumn), **2** ad ♂ worn (spring), **3** ad ♀ worn (spring), **4** 1st winter ♂, **5** 1st winter ♀, **6** juv. *P. p. samamisicus*: **7** ad ♂ worn (spring), **8** ad ♀ worn (spring), **9** 1st winter ♂.

tail-quivering (shared only by Black Redstart), (5) distinctive calls, and (6) arboreal behaviour. Flight fluent and agile, with slight undulations. Carriage usually half-upright on open perch, more upright on ground but much less so in foliage; length of tail enhanced by almost constant, neurotic quivering.

Habitat. Breeds in west Palearctic from upper to middle latitudes, mainly continental and lowland, in boreal, temperate, steppe, and Mediterranean zones between July isotherms 10–24°C. Requires sheltered but fairly open wooded or parkland areas with access to dry secure nest-holes in trees, rocks, walls, banks, or other places and without too dense or tall unbroken undergrowth or herbage. At least in west of range prefers broad-leaved or mixed trees, but in some parts occupies open pinewoods, and is adapted to woodland edges, streamside and roadside trees, orchards, and gardens in human settlements; also heaths and commons with scattered trees or copses, pollard willows along streams or ditches, open hilly country with loose stone walls, old ruins, quarries, and rocky places. In winter quarters, occurs in wide variety of woodland types as well as thickets and scrub.

Distribution. Contraction of range reported Britain, Netherlands, Slovakia, Hungary, Slovenia, and Ukraine. BRITAIN. Marked decline, notably in central, southern, and eastern England. NETHERLANDS. Decrease since 1970s; recent slight recovery. KAZAKHSTAN. Ural delta: singing ♂♂ June 1992, May 1993 suggest possible breeding. CYPRUS. May have bred. SYRIA. Breeding possible, not yet confirmed. IRAQ. Breeds in northern mountains. TUNISIA. No recent evidence of breeding.

Accidental. Bear Island, Iceland, Azores, Madeira, Cape Verde Islands.

Beyond west Palearctic, extends from Urals broadly south-east to Lake Baykal; also breeds Iran.

Population. Long-term decline probably mainly due to loss of old forests and mature trees; also, sharp decrease followed severe drought in winter quarters 1968–9. Most countries report continuing decline; some recent increase or recovery Britain, Provence region of France, Netherlands, southern Finland, and Croatia. BRITAIN. 90 000 pairs 1988–91, probably closer to true figure than estimated 330 000 territories. Sharp fall 1969–73; overall increase since, but decrease continues at edge of range. IRELAND. 5–25 pairs. FRANCE. Under 1 million pairs in 1970s. BELGIUM. Estimate of 4600 pairs 1973–7 following drastic decline probably too low. LUXEMBOURG. 5000–7000 pairs. NETHERLANDS. 33 000–45 000 pairs 1989–91. GERMANY. 190 000 pairs in mid-1980s; in east, marked decline since 1960s–70s, but apparently stabilized at much lower level more recently. DENMARK. 14 000–160 000 pairs 1987–8. NORWAY. 50 000–500 000 pairs 1970–90. SWEDEN. 100 000–300 000 pairs in late 1980s. FINLAND. 300 000–400 000 pairs in late 1980s. ESTONIA. 10 000–20 000 pairs in 1991. LATVIA. 60 000–100 000 pairs in 1980s; stable. POLAND. 20 000–60 000 pairs. CZECH REPUBLIC. 30 000–60 000 pairs 1985–9. SLOVAKIA. 10 000–15 000 pairs 1973–94. HUNGARY. 8000–12 000 pairs 1979–93. AUSTRIA. Fairly common; one estimate of 5000–8000 pairs. SWITZERLAND. 10 000–15 000 pairs 1985–93. SPAIN. 75 000–94 000 pairs. PORTUGAL. 100–1000 pairs 1978–84; stable. ITALY. 30 000–50 000 pairs 1983–95. GREECE. 2000–5000 pairs. ALBANIA. 500–1000 pairs in 1981. YUGOSLAVIA: CROATIA. 5000–7000 pairs. SLOVENIA. 3000–5000 pairs. BULGARIA. 500–5000 pairs. RUMANIA. 200 000–250 000 pairs 1986–92. RUSSIA. 100 000–1 million pairs; sharp decrease in Leningrad region since 1960s. BELARUS'. 50 000–60 000 pairs in 1990; stable. UKRAINE. 9000–10 000 pairs in 1986. MOLDOVA. 35 000–50 000 pairs in 1988; stable. AZERBAIJAN. Common. TURKEY. 10 000–100 000 pairs. IRAQ. Perhaps fairly common. TUNISIA. No recent evidence of breeding. MOROCCO. Scarce.

Movements. Migratory. Movement mainly nocturnal, with broad-front trans-desert passages across Africa and Middle East. 2 distinct populations: nominate *phoenicurus* breeds Europe, Siberia, and north-west Africa, and winters across Afrotropics north of equator; *samamisicus* breeds around Black and Caspian Seas and in northern Middle East, and winters in Arabia, Sudan, and Ethiopia.

Autumn movement through Europe mainly south-westward. Leaves breeding grounds in second half of August, with peak passage through north-west Europe in early September and numbers there diminishing gradually during October. Relatively rare in autumn in North Africa, suggesting tendency for Mediterranean and northern Sahara to be overflown. First arrivals south of Sahara in first half of September, but not common there until mid-October. Spring passage in Africa more conspicuous than in autumn. Vanguard arrives northern Europe in first half of April; main arrivals there mid-April to mid-May, with passage declining into June.

Food. On breeding grounds at least, largely insects (especially adult and larval Lepidoptera and Coleoptera) and spiders. 4 main feeding methods. (1) Picks items from ground; apparently does not probe for worms (etc.) and rarely searches in leaf-litter. (2) Feeds in trees and other vegetation, picking items from trunks, branches, and leaves, including by hovering near foliage, etc. (3) Flies from perch on to prey on ground, normally returning to perch to eat it. (4) Takes aerial prey in brief flight from perch. Small fruits also regularly eaten, especially in late summer and autumn.

Social pattern and behaviour. In winter, apparently usually solitary or in pairs. On migration, occurs singly or in parties of up to 50–60 or more. Territorial, in pairs, in breeding season. ♂ (occasionally ♀) sings from high and exposed perches, same ones used throughout breeding season. About 2 weeks after arrival, ♂ suddenly switches from full song to strangled, monotonous song given near one favoured nest-site, at same time performing nest-showing display, either to prospective mate or, often, when no ♀ is present. ♂ enters prospective nest-hole, looks out of entrance again, and occasionally sings with head out of hole, displaying white forehead; or may fly in and out (sometimes very frequently), showing rufous tail and rump, clinging just below hole, occasionally with brief song or ticking call—or may appear to enter but not do so, holding tail fanned. ♂ generally goes to a number of different sites and offers them to ♀.

Voice. Song sweet, rather melancholy in tone and weak in volume, with characteristic squeaky terminal jangle. Comprises 2-part phrases of *c.* 2 s duration, but highly variable. 1st part normally species-specific, relatively pure-toned, and consisting of a high-pitched, slightly drawn-out tone followed by 2–4 lower-pitched or also higher-pitched, slightly shorter units: 'ji-gjü gjü gjü . . .' or 'jü-jik jik jik . . .'. 2nd part contains different-sounding, partly clicking or rattling, partly pure-sounding passages which contain rich mimicry. Calls include loud, plaintive, *Phylloscopus*-like 'hweet'; rather liquid, slightly explosive 'tuick'; often run together as tremulous, scolding 'hwee-tucc-tucc'.

Breeding. SEASON. North-west Europe: egg-laying from end of April or early May. Southern Europe: up to 2 weeks earlier. Northern Finland: late May to late June. 2 broods over most of range, but only 1 in north. SITE. Hole in tree, rocks, or building, less often in bank, among tree-roots, or heap of stones; readily uses nest-box. Nest: loose cup of grass, moss, and other vegetation, lined with wool, hair, and feathers. EGGS. Sub-elliptical, smooth and glossy; pale blue. Clutch: 5–7 (3–10). INCUBATION. 12–14 days. FLEDGING PERIOD. 14–15 days (13–17).

Wing-length: *P. p. phoenicurus*: ♂ 77–84, ♀ 75–81 mm.
Weight: ♂♀ mostly 11–19 g.

Geographical variation. Slight in size, more marked in colour of flight-feather fringes. 2 races recognized. ♂ *samamisicus* from Crimea, eastern Turkey, and Levant eastwards differs from nominate *phoenicurus* in broad white fringes along outer webs of flight-feathers, forming conspicuous white area—rather variable in extent, sometimes consisting of whole outer webs of tertials and secondaries (except tips) and rather broad fringes along primaries, sometimes formed by rather narrow fringes on secondaries only; in worn plumage, fringes partly wear away and white area sometimes less obvious.

Moussier's Redstart *Phoenicurus moussieri*

PLATES: pages 1158, 1165

DU. Diadeemroodstaart   FR. Rougequeue de Moussier   GE. Diademrotschwanz   IT. Codirosso algerino
RU. Беловровая горихвостка   SP. Colirrojo diademado   SW. Diademrödstjärt

Field characters. 12 cm; wing-span 18.5–20.5 cm. 15% shorter than Redstart, with relatively shorter wings and tail and sturdy body. Smallest redstart of west Palearctic, with rather compact form recalling *Saxicola* chat (as does ♂'s plumage). ♂ black above with large white wing-patch and huge white circlet round crown and dark rufous throat and underbody. ♀ resembles ♀ Redstart, but underparts have much rufous. Juvenile suggests ♀ but dully spotted above and scaled on fore-underparts.

♂ unmistakable; ♀ and immature less distinctive, but small

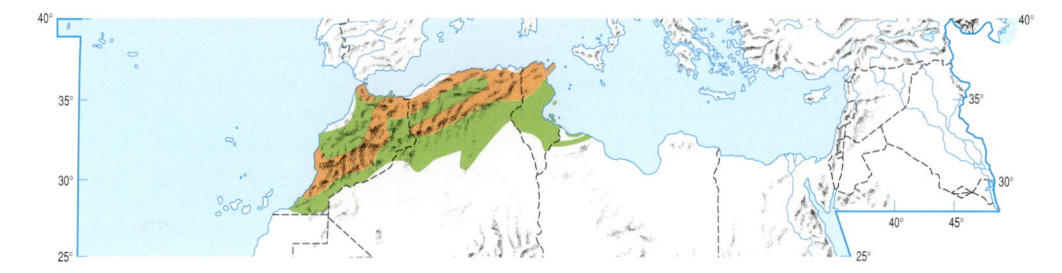

size, compact form, and warm rufous underparts soon learnt. Rather fluttering wing-beats and shorter-tailed silhouette recall Stonechat. Usually stays close to ground level. Hops. Shivers tail like Redstart.

Habitat. In lower-middle latitudes in warm dry Mediterranean climate, at all elevations from sea-level to 1900 m in Algeria and Tunisia and to 3000 m in Moroccan Haut Atlas. In east of range, in broken maquis, on dry grassy, stony, or rocky slopes, and in old or degraded forests on broken terrain. In Atlas region, occurs at forest base and on stony summits, forest summits, and denuded plateaux, and at higher elevations among bushes and xerophytic plants. In winter, shifts to lower ground along fringe of Sahara, in bushes in wadis, and in *Zizyphus* scrub on plains.

Distribution and population. Breeds only in west Palearctic. LIBYA. Breeding suspected Tripolitania. TUNISIA. Common. MOROCCO. Uncommon to common; probably stable.

Accidental. Britain, France, Spain, Italy, Malta, Greece.

Movements. Resident, dispersive, and perhaps migratory over relatively short distances, normally staying within North Africa.

Food. Takes insects on ground; frequently digs with bill. Makes brief flights from ground in pursuit of flying prey. Also, at least occasionally, takes fruit.

Social pattern and behaviour. Little information, but not known to differ significantly from congeners.

Voice. Song of ♂ a thin reedy warble, recalling Dunnock. Call a loud 'wheet' or 'beezp', usually followed by a rasping 'tr-rr-rr'.

Breeding. SEASON. Algeria and Tunisia: eggs laid 1st week of April to mid-June. Moroccan Sahara: laying from mid-March. One brood, possibly 2. SITE. On ground, under tussock or low bush, also in hole in tree or wall. Nest: loose cup of light vegetation, lined with feathers and hair. EGGS. Sub-elliptical, smooth and glossy; white or very pale blue, proportions of different coloured eggs varying with region. Clutch: 4–5 (3–6). (Incubation and fledging periods not recorded.)

Wing-length: ♂ 65–70, ♀ 62–71 mm.
Weight: ♂♀ 14–15 g.

Güldenstädt's Redstart *Phoenicurus erythrogaster*

PLATES: pages 1158, 1165

DU. Witkruinroodstaart   FR. Rougequeue de Güldenstädt   GE. Riesenrotschwanz   IT. Codirosso di Güldenstädt
RU. Краснобрюхая горихвостка   SP. Colirrojo coronado   SW. Bergrödstjärt

Field characters. 18 cm; wing-span 28–30 cm. Close in length to Rock Thrush; nearly 30% larger than Redstart, with proportionately longer wings and shorter tail. Biggest redstart of west Palearctic, with rather long-crowned head and more robust appearance than smaller congeners. ♂ has diagnostic combination of white crown and nape and large white wing-patch, but otherwise recalls ♂ Black Redstart of eastern races. ♀ suggests ♀ Redstart but has rather uniform dusky-buff underparts and duller tail. Juvenile like ♀ but obscurely mottled.

♂ and (when size apparent) ♀ unmistakable. Flight more floating than smaller congeners.

Habitat. Breeds mainly extralimitally in elevated rugged middle latitudes of central Asia up to perennial snow and glacier line, at *c.* 5000 m, in severe climates with summer snow, hail, and sleet on plateaux and mountain peaks, stonefields, mountain tundra, steep crags with scree, stony patches with scattered bushes, and alpine meadows. In Caucasus, inhabits uppermost belts of mountains and narrow defiles traversed by

Eversmann's Redstart *Phoenicurus erythronotus* (p. 1155): **1** ad ♂, **2** ad ♀. Black Redstart *Phoenicurus ochruros gibraltariensis* (p. 1157): **3** ad ♂, **4** ad ♀. Redstart *Phoenicurus phoenicurus* (p. 1161). *P. p. phoenicurus*: **5** ad ♂, **6** ad ♀. *P. p. samamisicus*: **7** ad ♂. Moussier's Redstart *Phoenicurus moussieri* (p. 1163): **8** ad ♂, **9** ad ♀. Güldenstädt's Redstart *Phoenicurus erythrogaster* (p. 1164): **10** ad ♂, **11** ad ♀.

rapid mountain streams; also scree and detritus of glacial moraines.

Distribution and population. Breeds in Caucasus area, where rare. Rough estimate only 2500–3000 birds; highest numbers on northern slopes of central Caucasus between Mt Elbrus and Mt Kazbek (Russia/Georgia).

Accidental. One record in Tabuk in northern Saudi Arabia, in early 1980s.

Beyond west Palearctic, breeds in central Asian mountains.

Movements. Mainly a short-distance altitudinal migrant, many descending to foothills, valleys, and plains for winter months, while some birds disperse further.

Food. In summer, insects, especially beetles; berries in winter. Feeds by picking items from ground, ice, water, or bank of stream, hopping or running about—even feeds inside animal carcasses, though presumably taking insects attracted to them; also flies on to prey from low perch, sometimes pursuing it along ground, or takes aerial prey in brief flight.

Social pattern and behaviour. Poorly known, but generally similar to congeners. In winter, birds that move away from breeding areas may occur in loose flocks. Breeding territories usually large. ♂ sings from prominent perch, with tail spread or slightly lowered; also in song-flight, gliding with quivering wings and tail widely spread. However, generally rather silent with song rarely heard even during courtship period.

Voice. Song consists of short, clear, melancholy whistles, somewhat reminiscent of Blackbird in tone, with twittering or ticking sounds interpolated. Contact-call a weak 'lik'. 'Tsee-tek tsee-tek-tek' given by alarmed adults in vicinity of nest or young, also in winter quarters.

Breeding. SEASON. Altai mountains: eggs laid from June. SITE. On ground, in hole or crack in rocks, on building (e.g. under roof), and also recorded using nest-box. Nest: bulky cup of grass stems and leaves, lined with wool, hair and feathers. EGGS. Sub-elliptical, smooth and slightly glossy; blue. Clutch: 4 (3–5). INCUBATION. 12–16 days. FLEDGING PERIOD. 14 days.

Wing-length: ♂ 100–108, ♀ 95–101 mm.
Weight: ♂♀ 22–32 g.

Blackstart *Cercomela melanura*. *C. m. melanura*: **1** ad ♂ worn (spring), **2** ad ♀ worn (spring), **3** juv. *C. m. lypura* (north-east Africa): **4** ad ♂ worn (spring). *C. m. airensis*: **5** ad ♂ worn (spring). Ant Chat *Myrmecocichla aethiops* (p. 1174): **6** ad ♂ worn (spring), **7** 1st autumn ♀, **8** juv.

Blackstart *Cercomela melanura*

PLATES: pages 1166, 1200, 1202

Du. Zwartstaart   Fr. Traquet à queue noire   Ge. Schwarzschwanz   It. Sassicola codanera
Ru. Чернохвостка   Sp. Colinegro real   Sw. Svartstjärt

Field characters. 14 cm; wing-span 23–27 cm. Close in size to Redstart. Small, rather long-billed and noticeably slim chat with rather uniform grey plumage uniquely relieved by black rump and tail. Juvenile browner than adult, lacking spotted appearance of other chats.

Unmistakable. Flight noticeably light and floating, recalling that of Redstart but with shorter tail evident. Gait and stance like Black Redstart. Fans tail open constantly, action accompanied by flicking of drooped wings.

Habitat. Breeds in warm arid lower latitudes, temperate, subtropical, and tropical. Shows preference for thorny bushes in rocky ravines. Normally found on steep terrain rather than on level areas.

Distribution and population. Syria. Recorded end October to early January in Yarmuk valley in south, probably involving dispersal from upper Jordan valley. Israel. A few tens of thousands of pairs in 1980s. Jordan. Common resident of wadis of Rift Margins. Egypt. Sinai: fairly common. Niger. Common in Aïr. Chad. Tibesti: common in all vegetated areas.

Beyond west Palearctic, breeds Arabia, and in sub-Saharan Africa from Mali east discontinuously to Somalia.

Movements. Resident or even sedentary throughout range.

Food. Insects, also berries as available. Feeds mainly by perching on rocks, trees, and bushes, dropping on to prey on ground; also by searching vegetation and occasionally in brief hover or aerial-pursuit.

Social pattern and behaviour. Social organization poorly known but apparently similar to redstarts. Characteristically fearless, with restless flitting between rocks or small trees, frequently fanning wings and conspicuous black tail, interspersed with occasional singing. Constant opening and closing of wings and tail occurs in a measured way—not like quick flick of Wheatear: tail expanded to c. 45° each side, primaries well fanned but inner wing only slightly extended. Song delivered often from tree-top or other perch, probably uttered during most of year.

Voice. Song of ♂ a series of loud, well-spaced notes 'chree chrew chitchoo chirri chiwi ...', etc.; occasional chirps recall House Sparrow. Contact-calls include a loud, liquid 'chura lit' or loud 'tyootrit', like fragment of song. Alarm-call a high-pitched whistling 'feefee' or 'whee'.

Breeding. Season. Southern Sahara: March–June. Probably similar in Egypt and Israel. Site. Crevice in rock, up to 0.5 m from entrance. Nest: cup of grass stems and leaves, lined with hair and finer vegetation. May have platform of small pebbles.

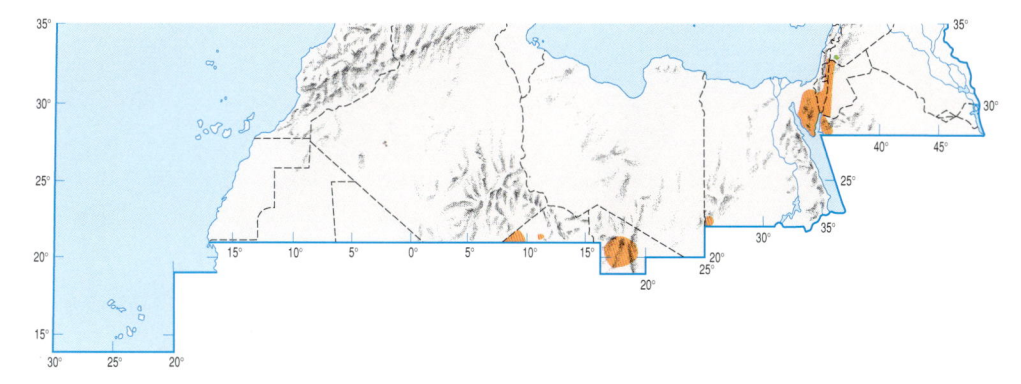

EGGS. Sub-elliptical, smooth and fairly glossy; very pale blue, finely speckled red to red-brown, speckles sometimes concentrated at broad end. Clutch: 3–4. INCUBATION. 13–14 days. FLEDGING PERIOD. 13–15 days.

Wing-length: *C. m. melanura*: ♂ 81–86, ♀ 76–82 mm.

Geographical variation. Rather slight, clinal, involving general colour of body only. Nominate *melanura* (Levant) rather pale and grey. Colour gradually darker and browner through Sahel zone towards west: *airensis* (northern Niger and Chad) intermediate, extralimital *ultima* darkest and strongly brown.

Whinchat *Saxicola rubetra*

PLATES: pages 1168, 1174

Du. Paapje   Fr. Tarier des prés   Ge. Braunkehlchen   It. Stiaccino
Ru. Луговой чекан   Sp. Tarabilla norteña   Sw. Buskskvätta

Field characters. 12.5 cm; wing-span 21–24 cm. Close in size to Stonechat but somewhat slimmer, with less rounded head (appearing so partly due to more linear face pattern) and noticeably longer wings (with long point when folded). Small but quite robust, rather heavy-billed chat, with compact outline when perched and broad cruciform silhouette when wings extended in flight. Adult shows diagnostic combination of striking white line across inner wing-coverts and dark oval or diamond-shaped cheek-panel between long pale supercilium and pale lower border; streaked brown and black above, with white bases to sides of tail, and warm buff below. ♂'s head vividly patterned, with black-brown crown, white supercilium, black-brown cheeks with white lower border. ♀ similarly patterned but colours subdued, with supercilium and cheek-border buffier. Immature duller than ♀, with even more subdued head pattern and no white line on wing-coverts.

Distinctive but not unmistakable. Dark pale-lined face, dark upperparts, and orange underparts of breeding bird present easy target. Much buffier, even sandy and rather uniform look of 1st-winter and non-breeding birds invites confusion with eastern races of Stonechat; juvenile can closely approach appearance of juvenile of any race of Stonechat. Identity of juvenile confirmed by well-marked supercilium (at least behind eye), wide rufous-buff fringes and white spots on back, and pale underparts (particularly throat); also, tail shows white triangles on sides of base. Flight usually low, level, and fast, with noticeably faster wing-beats than wheatears and rather long, broad wings dominating silhouette which ends in rather short, square tail. Looks strongly coloured in flight, with conspicuously warm glow above in non-breeding plumages, dark underwing usually obvious, and uniform underparts striking. Stance on ground typically half-upright but adopts more level posture when perched, unlike Stonechat. Gait a rapid hop, with occasional run. Less nervous than Stonechat.

Habitat. In west Palearctic, breeds largely to north of Stonechat, in boreal and temperate but only marginally in steppe and Mediterranean zones of middle and upper middle latitudes, with more continental bias than Stonechat, although favouring moister and less rough habitats. Accepts sparser and less robust perches than Stonechat, often using posts, fences, or tall weeds; accordingly less dependent on heaths and moors and on coastal situations (except open dunes) and more attracted to grassy areas, including some farmland types, young conifer plantations with grassy ground cover, railway embankments, verges of quiet roads, fringes of wetlands, and grassy uplands. In winter in Africa, widespread in vegetated areas wherever open places occur with suitable perches and access to ground.

Distribution. BRITAIN. Marked decrease in lowland areas. IRELAND. Marked recent decrease. SLOVAKIA. Slight decrease. AZERBAIJAN. Distribution poorly known.

Accidental. Spitsbergen, Bear Island, Iceland, Madeira.

Beyond west Palearctic, extends east from Urals in narrowing band to *c.* 98° E; also breeds northern Iran.

Population. Declining in western and central Europe (also Finland); stable Denmark (fluctuating), Norway, Sweden, Latvia, Poland (but some decrease in south-west), Hungary, Portugal, Balkans, and FSU; increasing Estonia and Czech Republic. BRITAIN. 14 000–28 000 pairs 1988–91. Decline, at

Whinchat *Saxicola rubetra*: **1** ad ♂ breeding, **2** ad ♀ breeding, **3** 1st winter ♂, **4** juv. Canary Islands Stonechat *Saxicola dacotiae*: **5** ad ♂ worn (spring), **6** ad ♀ worn (spring), **7** 1st winter ♂, **8** juv.

least in south and east, since 1950s, probably mainly due to severe loss of habitat. IRELAND. 1250–2500 pairs 1988–91. FRANCE. Under 100 000 pairs in 1970s; decreasing. BELGIUM. 350–500 pairs 1989–91; steep decline, continuing. LUXEMBOURG. 80–120 pairs; decreasing since 1960s. NETHERLANDS. 700–1100 pairs 1989–91; strongly declining, locally to over 75% compared with early 1960s. GERMANY. 40 000–60 000 pairs; strong decline. DENMARK. 1000–21 000 pairs 1987–8. NORWAY. 50 000–300 000 pairs 1970–90. SWEDEN. 200 000–500 000 pairs in late 1980s. FINLAND. 300 000–400 000 pairs in late 1980s. ESTONIA. 50 000–100 000 pairs in 1991. LATVIA. 300 000–500 000 pairs in 1980s. LITHUANIA. Decreasing. POLAND. 100 000–250 000 pairs. CZECH REPUBLIC. 10 000–20 000 pairs 1985–9. SLOVAKIA. 10 000–20 000 pairs 1973–94; slight decrease. HUNGARY. 30 000–50 000 pairs 1979–93. AUSTRIA. Only locally fairly common; declining since 1960s. SWITZERLAND. 5000–7000 pairs 1985–93; decreasing. SPAIN. 15 000–20 000 pairs; slight decrease. PORTUGAL. 10–100 pairs 1978–84. ITALY. 10 000–15 000 pairs 1983–95. GREECE. 2000–3000 pairs. ALBANIA. 500–1000 pairs in 1981. YUGOSLAVIA:CROATIA. 10 000–20 000 pairs. SLOVENIA. 8000–12 000 pairs. BULGARIA. 600–6000 pairs. RUMANIA. 200 000–250 000 pairs 1986–92. RUSSIA. 100 000–1 million pairs. BELARUS'. 550 000–650 000 pairs in 1990. UKRAINE. 27 000–32 000 pairs in 1986. MOLDOVA. 5000–7000 pairs in 1988. AZERBAIJAN. Uncommon. TURKEY. Local; 500–5000 pairs.

Movements. Essentially a trans-Saharan migrant, wintering in tropical Africa, though also regularly in Algeria and Iraq; other wintering records north of Sahara are exceptional but widely scattered through Mediterranean basin and western seaboard of Europe north to Britain. Wintering range extends from Sénégal through Nigeria and Zaïre to Uganda, and uncommonly in Kenya and Tanzania, south to Malawi and Zambia.

Birds leave north European breeding grounds in late August and September, with peak numbers on passage in western Europe in early September. First arrivals at wintering sites are in mid- or late September. Return passage begins February–March, continuing into early May.

Food. Invertebrates; occasionally berries. Hunts from perch, flying to and taking prey mainly from ground or in vegetation, sometimes in flight like flycatcher.

Social pattern and behaviour. Solitary in winter quarters. Flocks of up to *c.* 30 occur on migration, but usually fewer than 6–8. Territorial and mostly solitary in breeding season, though neighbourhood groups of pairs form around returning breeders. Song of ♂ usually given from elevated perch, occasionally in flight. Song-flight similar to Stonechat: bird flies up, singing, with body erect and tail spread, making tail pattern conspicuous; followed by drop back to perch.

Voice. Song variable in length, phrase, and quality, in part recalling Stonechat, Wheatear, Redstart, and even Bluethroat; always contains harsh introductory phrase, skirling notes, and silvery warble; notable for mimicry of wide range of bird species. Commonest call a disyllabic, somewhat scolding 'tick-tick', like Stonechat but softer and often preceded by more musical syllable—hence 'tuee-tick-tick'; short, hard, rattling 'churr' in alarm.

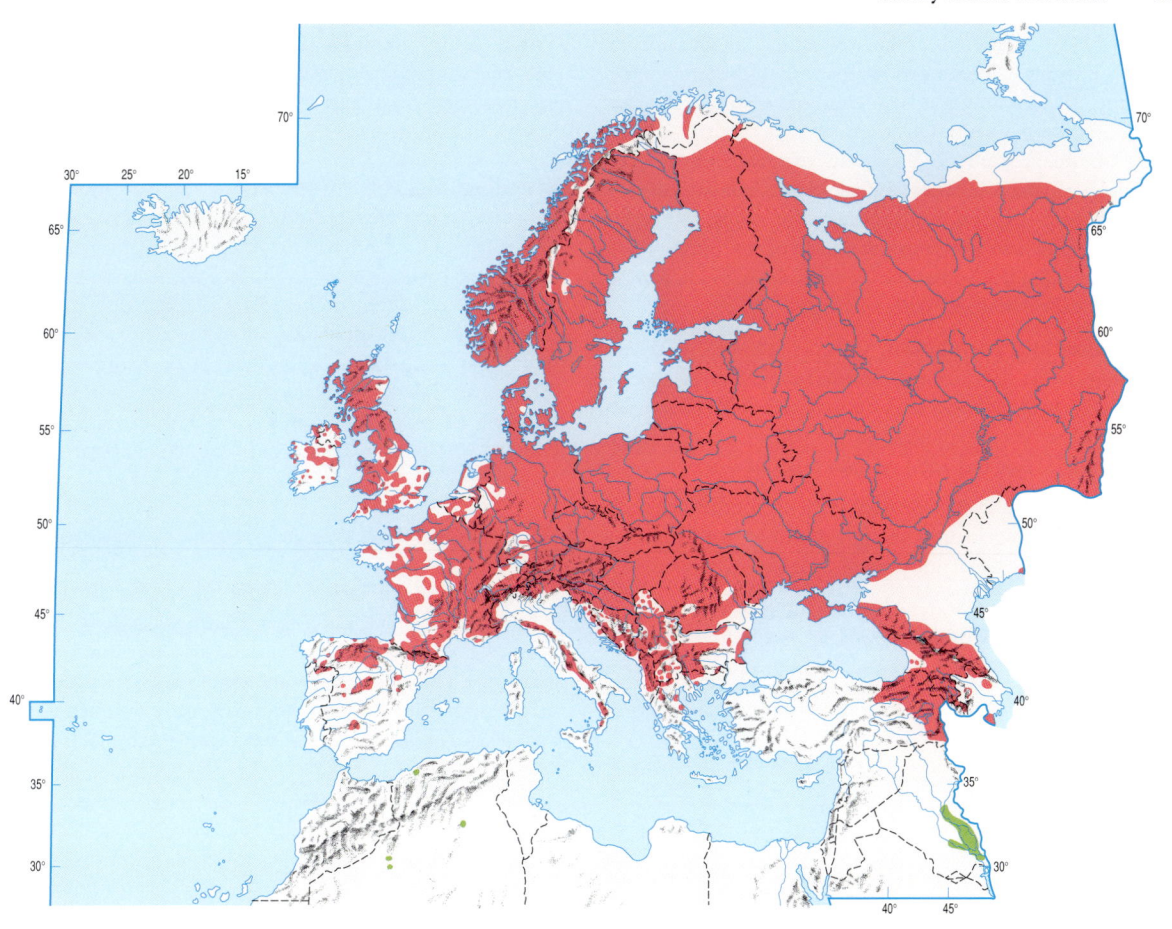

Breeding. SEASON. North-west Europe: egg-laying from late April or early May. Little variation across range. One brood, occasionally 2. SITE. On ground in vegetation, usually well hidden. Nest: cup of grass stems and leaves and moss, lined with finer material and hair. EGGS. Sub-elliptical, smooth and glossy; pale blue, with very fine speckling of red-brown (sometimes sparse, or concentrated towards broad end) giving rusty appearance. Clutch: (2–)4–7. INCUBATION. 12–13(–15) days. FLEDGING PERIOD. Leave nest at 12–13 days if undisturbed, starting to fly at 17–19 days.

Wing-length: ♂ 74–81, ♀ 73–78 mm.
Weight: ♂♀ mostly 14–19 g; migrants to 26 g.

Canary Islands Stonechat *Saxicola dacotiae*

PLATES: pages 1168, 1174

Du. Canarische Roodborsttapuit   Fr. Tarier des Canaries   Ge. Kanarenschmätzer   It. Saltimpalo delle Canarie
Ru. Канарский чекан   Sp. Tarabilla canaria   Sw. Kanariebuskskvätta

Field characters. 12.5 cm; wing-span 19–20.5 cm. Closely approaches size of Stonechat, with similar form and with appearance strongly recalling its paler, eastern races. Spring ♂ dark black-brown above, with narrow white supercilium, broad white collar extending from throat, and white patch on inner wing-coverts; buff-white to pink-white below, with orange-rufous bib on breast. ♀ drab brown above, greyer white below, with indistinct supercilium, narrow white patch on inner wing, and only faintly buff chest. Juvenile like ♀ but head and mantle spotted buff-white and chest speckled and barred brown.

Only small *Saxicola* chat breeding in Canary Islands (and never recorded elsewhere), but subject to confusion there with Stonechat (fairly regular on passage and in winter) and Whinchat (regular on passage). Best distinguished from Stonechat by white throat and narrow white supercilium, from Whinchat by pale or only orange-patched underparts and all-black tail.

Habitat. Now restricted to Fuerteventura, an oceanic and subtropical island. Climate very warm with night temperatures exceeding 32°C, but windy on west coast which is mainly avoided. Widely distributed from mountains to seashore, mainly on steep, stony, and sparsely vegetated ground, frequenting both open hillsides and secluded shallow valleys or ravines. Found also on a lava stream or slope, on barren volcanic terrain, and low hills and stony plains, although not generally a plains bird. Sometimes in cultivation, fields, gardens, or verdant valley bottoms with tamarisks.

Distribution and population. CANARY ISLANDS. Breeds on Fuerteventura; in 1985 census, estimated 750 ± 100 pairs. A few birds recently observed on Lanzarote. Bred Alegranza until at least 1913, when an apparent family party was seen also on Montaña Clara; not recorded on either island since.

Movements. Apparently sedentary, with no confirmed records away from breeding islands, though the only birds ever seen on Montaña Clara were not proved to have bred there.

Food. Invertebrates. For flying insects, flutters from perch on (e.g.) rock or tree, usually returning to new perch. Also flies down to catch prey on ground: usually hops but will also run.

Social pattern and behaviour. Poorly known; appears generally similar to Stonechat.

Voice. Song resembles that of Stonechat: a rather scratchy 'bic-bizee-bizeeu', etc.; also sings in flight with a repeated 'liu-liu-liu-screeiz', 1st part mellow, 2nd rasping. Call a sharp 'chep', like Stonechat.

Breeding. SEASON. Earliest eggs in January, but most mid-February to late March, with start perhaps governed by onset of winter rains. 2 broods recorded. SITE. On ground among stones and rocks or low down in wall, not more than 0.5 m above ground, often sheltered by overhanging stone or low bush. Nest: firmly built cup of plant stems and roots; lined with goat hair. EGGS. Sub-elliptical to short sub-elliptical, smooth and glossy; light green-blue, with fine speckling or mottling of red-brown, usually thicker towards broad end. Clutch: 4 (2–5). INCUBATION. 13–15 days. FLEDGING PERIOD. 16–18 days.

Wing-length: *S. d. dacotiae*. ♂ 60–66, ♀ 60–64 mm.

Geographical variation. Slight. *S. t. murielae* (apparently extinct) from small islet off northern Lanzarote differs from nominate *dacotiae* of Fuerteventura in more uniform rufous-cinnamon underparts without white on belly; crown slightly lighter, more red-brown, underparts more uniform vinaceous-buff in fresh plumage, chest-patch less sharply defined.

Stonechat *Saxicola torquata*

PLATES: pages 1171, 1174

Du. Roodborsttapuit   Fr. Tarier pâtre   Ge. Schwarzkehlchen   It. Saltimpalo
Ru. Черноголовый чекан   Sp. Tarabilla común   Sw. Svarthakad buskvätta

Field characters. 12.5 cm; wing-span 18–21 cm. Close in size to Whinchat but with 10% shorter, more rounded wings (with only short point when folded) and noticeably round head (due partly to plumage pattern). Smallest widespread chat of west Palearctic, with virtually diagnostic character: large, round, busby-like head, compact form, upright stance, often bold behaviour (of ♂), constant nervous twitching of wings and tail, and whirring flight. When perched, recalls guardsman on sentry post; in flight, huge bumble-bee. Adult ♂ has wholly black head (lacking pale throat of other *Saxicola* chats in breeding plumage) and chestnut breast. ♀ and (particularly) immatures far less distinctive, with most characters subject to complex geographical variation. Eastern races have noticeably pale rumps, black under wing-coverts, and (on immatures) pale supercilia and throat, recalling Whinchat. Juvenile resembles dull ♀, but has only speckled or pale throat and no obvious collar, being instead spotted and streaked buff on head and back, washed with rufous on rump, and finely spotted and barred dark brown on chest and flanks.

Adult ♂ unmistakable, no other common small ♂ chat being wholly black-headed. Adult ♀♀ of western and Caspian races also unmistakable, but adult ♀♀ of Siberian races and all juveniles subject to marked overlaps of characters. Beware particularly that ♀ and juvenile Siberian birds resemble ♀ and juvenile Whinchat in face and body pattern. Flight more whirring than flitting, with rapid, jerky bursts of wing-beats producing flat and direct flight; will hover; turns with great agility, ducking into cover or bouncing on to perch, immediately flicking wings and tail on landing. Gait a bouncing hop, but spends less time on ground than redstarts or wheatears, feeding mostly from perch.

Habitat. Breeds in west Palearctic in middle and lower middle latitudes, in temperate, steppe, and Mediterranean zones, except for disjunct range of eastern race *maura* which extends into boreal zone of north-west of the region. Absent from high-altitude mountainous regions in north of range, and from high forest, wetlands, and open expanses which are bare or have

(FACING PAGE) Stonechat *Saxicola torquata*. *S. t. rubicola*: **1** ad ♂ worn (spring), **2** ad ♀ worn (spring), **3** 1st winter ♂, **4** 1st winter ♀, **5** juv. *S. t. hibernans*: **6** ad ♂ worn (spring), **7** ad ♀ worn (spring), **8** 1st winter ♂, **9** 1st winter ♀. *S. t. variegata*: **10** ad ♂ worn (spring), **11** 1st winter ♂. *S. t. armenica*: **12** ad ♂ worn (spring), **13** ad ♀ worn (spring), **14** 1st winter ♂. *S. t. maura*: **15** ad ♂ worn (spring), **16** 1st winter ♂. *S. t. stejnegeri* (eastern Siberia): **17** ad ♂ worn (spring), **18** 1st winter ♂.
Pied Stonechat *Saxicola caprata rossorum* (Transcaspia to Pakistan) (p. 1173): **19** ad ♂ worn (spring), **20** ad ♀ worn (spring), **21** 1st winter ♂.

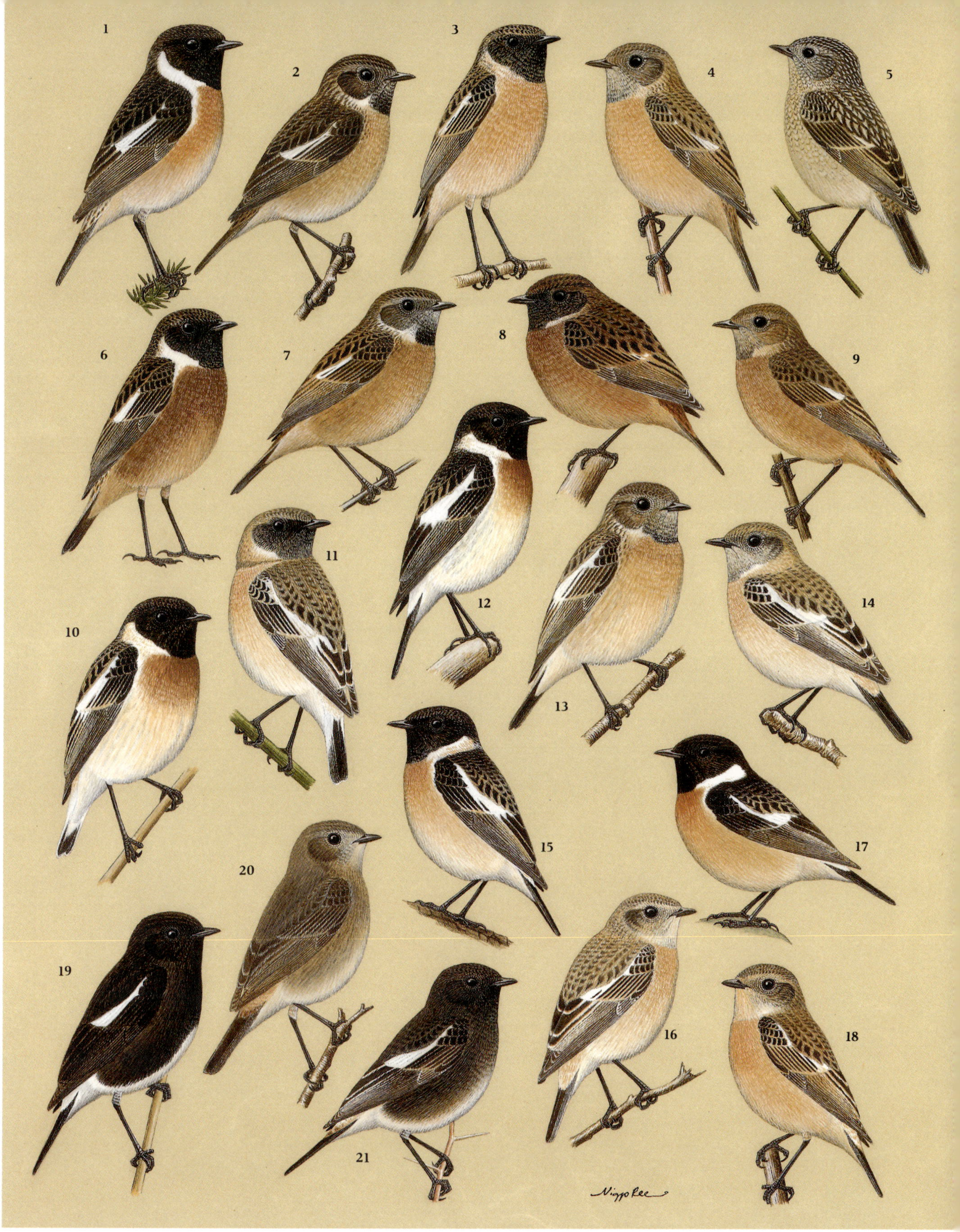
1
2
3
4
5
6
7
8
9
10
11
12
13
14
15
16
17
18
19
20
21

only sparse or low vegetation. Within these limitations, inhabits wide variety of dry plains and hillsides, often submarginal for agriculture, characterized by scattered bushes, shrubs, stones, walls, or fences 1 m or more high, used as look-outs or songposts commanding lower heathland, grassland, or bare patches.

Distribution. Range has increased Denmark, Poland, Slovakia, Switzerland, Slovenia, and Russia. BRITAIN. Has withdrawn from east of range, and become more fragmented in west. IRELAND. Serious decline. BELGIUM. Decrease. GERMANY. Main populations Niedersachsen and Pfalz. NORWAY. Sporadic breeder; first bred 1974 Hordaland. FINLAND. First breeding record 1992: 2 pairs at Kuusamo *c.* 66° N. TURKEY. Higher density in East Anatolia than elsewhere.

Accidental. Iceland, Faeroes, Norway, Sweden (annual), Finland (annual), Estonia, Latvia, Lithuania, Belarus', Madeira.

Beyond west Palearctic, breeds in much of Asia north of Himalayas, in Iran and Arabia; widespread in sub-Saharan Africa (local in west) and Madagascar.

Population. Widespread decline in western Europe—Britain and Ireland south to Spain (stable Portugal), east to Germany – due mainly to agricultural intensification, e.g. destruction of hedges, afforestation of fallow land. Mostly stable further east, with increases reported Slovakia, Italy, and Slovenia. BRITAIN. 8500–21 500 pairs 1988–91. IRELAND. 7500–18 750 pairs 1988–91. FRANCE. Under 1 million pairs in 1970s. BELGIUM. 1150–1400 pairs 1989–91. LUXEMBOURG. 120–180 pairs. NETHERLANDS. 1800–2300 pairs 1989–91. GERMANY. 3800 pairs in mid-1980s; in east, where incompletely known, provisional estimate of *c.* 50 pairs in early 1980s. DENMARK. 6–8 pairs. NORWAY. Up to 100 pairs. POLAND. 10 000–15 000 pairs. Increase in Silesia; stable elsewhere. CZECH REPUBLIC. 2500–5000 pairs 1985–9. SLOVAKIA. 20 000–40 000 pairs 1973–94. HUNGARY. 70 000–80 000 pairs 1979–93. AUSTRIA. Fairly common in east and south-east, scarce elsewhere. One estimate 3000–5000 pairs. SWITZERLAND. 200–250 pairs 1985–93. SPAIN. 300 000–700 000 pairs. PORTUGAL. 10 000–100 000 pairs 1978–84. ITALY. 200 000–300 000 pairs 1983–95. GREECE. 50 000–100 000 pairs. ALBANIA. 1500–3000 pairs in 1991. YUGOSLAVIA: CROATIA. 10 000–20 000 pairs. SLOVENIA. 8000–12 000 pairs. BULGARIA. 10 000–100 000 pairs. RUMANIA. 50 000–80 000 pairs 1986–92. RUSSIA. 10 000–100 000 pairs. UKRAINE. 2000–3500 pairs in 1986. MOLDOVA. 4500–6000 pairs in 1988. AZERBAIJAN. Uncommon to locally common. TURKEY. 10 000–100 000 pairs. TUNISIA. Locally common. ALGERIA. Breeds in small numbers. MOROCCO. Uncommon.

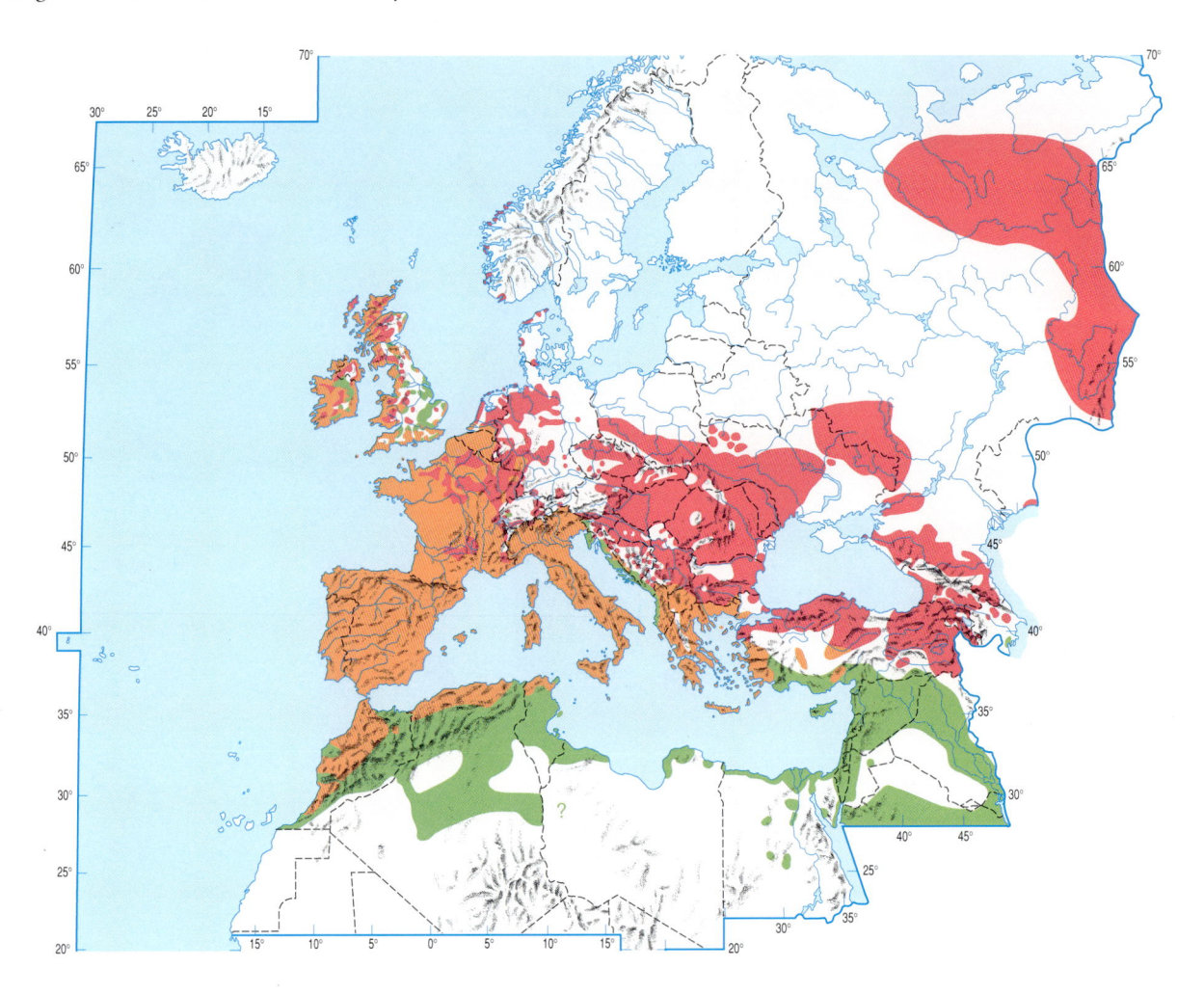

Movements. Varies from migratory to resident in different parts of range, being sensitive to cold winter weather. Migratory European populations winter in south of breeding range, with notable concentrations in southern and eastern Spain, Balearic islands, and Algerian coast. Mediterranean populations apparently resident, North African probably so. *S. t. hibernans* (breeding Britain, Ireland, north-west France, and western Iberia) partially migratory, at least in north, some (mainly older) birds resident, others (mainly young) migratory as far as Iberia and North Africa, or making shorter movements to south and west coasts. *S. t. maura* (breeding Ural area and north-west Russia) is long-distance migrant, majority wintering in northern India and north-west Burma.

Both *maura* (mostly) and *stejnegeri* (breeding eastern Siberia) are vagrants to western Europe, occurring mainly in period 9 September–7 November. *S. t. variegata* (breeding north Caspian area) has occurred in Norway in June, and in Britain September–October.

Food. Small and medium-sized insects and other invertebrates; also berries. Locates terrestrial prey from elevated perch (e.g. bush), then flies, glides, or hops to ground, picking prey up on landing or while standing on ground; may return to same perch or new one.

Social pattern and behaviour. Mostly solitary or in pairs outside breeding season, though occasional feeding aggregations occur. Pairs that do not disperse after breeding (i.e. especially those on coasts) defend same or similar territories through winter. Mating system essentially monogamous, though small proportion of ♂♂ polygynous. ♂ sings, perched on top of tree or bush, or occasionally in flight. Song-posts higher than perches used while feeding; therefore these activities rarely mixed. On perch, ♂ stands erect, head often raised and white wing-coverts exposed. Song-flights include phrases of typical song plus buzzes and whistles; flight slow and jerky, with shallow wing-beats and periodic brief hovers, bird sometimes rising and falling several times without forward movement; keeps forebody slightly raised, tail lowered, and feet dangling, white rump and neck- and wing-patches conspicuous. Most singing is in isolation or near to mate; rarely in territorial interactions, though song-duels with other ♂♂ occur occasionally.

Voice. Highly vocal during breeding season, notably in territorial advertisement (♂) and nest defence (♂ and ♀). Calls (mostly alarm) rarer outside breeding season, and birds almost silent in winter. Song of ♂ a variable, rather melancholy warbling sequence of phrases, sometimes delivered in 'scrappy' short bouts with unfinished phrases, but also unbroken sequences of several dozen phrases. Compared with Whinchat, shrill (fewer low-pitched sounds) and monotonous due to repetition of phrases and smaller pitch changes between successive units. Calls include a plaintive 'hweet' and a most characteristic, insistent, hard 'tsak tsak' (particularly from scolding ♂); often run together into 'hwee-tsak-tsak' or 'hwee-tsak hwee-tsak', harsher notes sounding like clash of 2 pebbles.

Breeding. SEASON. North-west Europe: egg-laying from March. From mid-April in south and central Europe. Northern Russia: from mid-May. 2–3 broods; 4 also recorded. SITE. On or close to ground in dense vegetation, at base of bush, in tussock, or low down in thick scrub. Nest: loose, unwoven cup of dry grass stems and leaves, lined with hair and feathers sometimes with wool. EGGS. Sub-elliptical, smooth and moderately glossy; pale blue to green-blue, variably marked with red-brown—often very finely but sometimes more heavily and with cap round broad end. Clutch: 4–6 (2–7). INCUBATION. 13–14 days. FLEDGING PERIOD. 12–16 days.

Wing-length: *S. t. rubicola* ♂ 63–70, ♀ 62–68 mm.
Weight: *S. t. rubicola* ♂♀ mostly 13–17 g.

Geographical variation. Marked; involves general proportions, size, general colour (especially head and upperparts), and amount of white on upper tail-coverts and tail-base. In Palearctic, 2 main groups: (1) *rubicola* group in west, with *rubicola* (most of west and south of Europe, North Africa, and most of Turkey and Caucasus) and *hibernans* (Britain, Ireland, western Brittany, and western Iberia), and (2) *maura* group in eastern Europe and Asia (in west Palearctic, *maura* in northeast, *armenica* in eastern Turkey and Transcaucasia, *variegata* in Volga steppes). Wing generally longer in *maura* group than in *rubicola* group, and bill shorter; plumage differences not striking but complex, with cline of decreasing saturation of colours from west to east.

Pied Stonechat *Saxicola caprata*

PLATES: pages 1171, 1174

DU. Zwarte Roodborsttapuit FR. Tarier pie GE. Mohrenschwarzkehlchen IT. Saltimpalo nero e bianco
RU. Чёрный чекан SP. Tarabilla pía SW. Svart buskskvätta

Field characters. 13.5 cm; wing-span 21–23 cm. 10–15% larger than Stonechat. Robust chat with typical *Saxicola* character but almost black plumage, relieved on ♂ only by white wing-patch, rump, and vent, and on ♀ by rusty or buff on rump and vent.

♂ unmistakable but ♀ and immature suggestive of darker races of Stonechat. Flight, gait and behaviour much as Stonechat.

Song brisk and whistling, 'chip-chepee-chewee chu'; calls include repeated, plaintive 'chep chep-hee' or 'chek chek trweet', and sharp, scolding 'chuh' in strong alarm.

Habitat. In continental lower middle latitudes, in plains and hills, but avoiding mountains and also forests and steppes, preferring low scrub, often on stony hillsides, moist places with thickets near reedbeds and coarse grass, especially beside

Whinchat *Saxicola rubetra* (p. 1167): **1** ad ♂ breeding, **2–3** 1st winter ♂. Canary Islands Stonechat *Saxicola dacotiae* (p. 1169): **4** ad ♂ worn (spring), **5–6** 1st winter ♂. Stonechat *Saxicola torquata* (p. 1170). *S. t. rubicola*: **7** ad ♂ worn (spring), **8–9** 1st winter ♂. *S. t. maura*: **10–11** 1st winter ♂. *S. t. stejnegeri*: **12–13** 1st winter ♂. Pied Stonechat *Saxicola caprata rossorum* (p. 1173): **14** ad ♂ worn (spring), **15–16** 1st winter ♂.

rivers, canals and ponds, and where tamarisk clumps, willows, and grass alternate with cultivation; also orchards, gardens, cultivation or semi-cultivation and damp meadows, especially near water.

Distribution. Breeds from Iran and Turkmenistan east through southern Asia and much of Indonesia to Philippines and New Guinea.

Accidental. Cyprus: ♂ Larnaca, November 1986. Israel: single birds January 1979, October 1994, November 1996. Iraq: reported November and March at Fao in south.

Movements. Largely migratory in west of breeding range, but eastern tropical races sedentary. Populations of Transcaspia, eastern Iran, and Afghanistan move south, with most individuals wintering from south-east Iran along Makran coast into plains of north-west Pakistan.

Wing-length: ♂ ♀ 73–79 mm.

Ant Chat *Myrmecocichla aethiops*

PLATES: pages 1166, 1200, 1202

Du. Bruine Miertapuit   Fr. Traquet brun   Ge. Ameisenschmätzer   It. Sassicola mangiaformiche
Ru. Африканский чекан   Sp. Tarabilla termitera   Sw. Svart termitbuskskvätta

Field characters. 17 cm; wing-span 28.5–32 cm. About 15% smaller than Rock Thrush but of similar form, being noticeably bulkier than any other west Palearctic chat. Large, rather short-tailed, stumpy, ground-loving chat, with almost wholly brown-black plumage relieved only by striking white panel on primaries visible in flight.

Flight weak, slower than wheatears and with noticeably more flapping wing-beats. Walks, runs, hops, and leaps, frequently cocking and spreading tail and drooping wings. Perches on plants and telegraph wires; much attracted to hollows.

Vocal, with pairs and family parties frequently calling, sometimes in unison. Song a whistled warble, based on piping 'tee-chu'.

Habitat. Tropical dry lowlands from border of desert through very open bush (intermediate between orchard-bush and thorn-country) to cultivation and seashore—will occupy a derelict village and even enter a town amidst a level dusty plain. A ground feeder, but less terrestrial than wheatears; often perches on walls or even telegraph wires. For breeding habitat, an essential appears to be presence of unlined wells or large artificial holes in ground, suited to excavation within them of horizontal shafts or hollows to serve for nesting and roosting.

Distribution. Breeds in narrow band from Sénégambia to central Sudan; also isolated population in Kenya and northern Tanzania.

Accidental. Chad: 1 collected Tibesti 1954.

Movements. Little information; probably largely sedentary.

Wing-length. Nominate *aethiops* (Sénégal to Lake Chad and northern Cameroon): ♂ 108–115, ♀ 110–115 mm.

Isabelline Wheatear *Oenanthe isabellina*

PLATES: pages 1176, 1200, 1202

Du. Isabeltapuit   Fr. Traquet isabelle   Ge. Isabellsteinschmätzer   It. Culbianco isabellino
Ru. Каменка-плясунья   Sp. Collalba isabel   Sw. Isabellastenskvätta

Field characters. 16.5 cm; wing-span 27–31 cm. On average, larger than Greenland race of Wheatear (of which pale individuals frequently mistaken for this species), with longer, slightly hooked bill; more bull-headed look; basally broader but still long, though slightly less pointed wings; rather shorter and broader, more rounded tail; more exposed thigh and tarsus. Largest and palest wheatear in west Palearctic, with less contrast between upperparts and underparts than even ♀♀ of most other species. Essentially isabelline-brown above and buff-white below, marked only by cream supercilium, black lores (in ♂), darker rear wings (especially when broad pale fringes worn off), broad white rump, and wide black-brown band on tail, not shaped in classic ⊥ of most wheatears. In flight, wings show diagnostic pattern of dark rim above and below, with wholly pale under wing-coverts. Juvenile somewhat buffier above, with even less obvious face pattern and faintly paler streaks and dark tips on upperparts and dark brown freckles on breast.

No absolute differences in behaviour from Wheatear, but the following are characteristic: (1) noticeably taller stance; (2) much more developed loping run; (3) jerkier wagging and flagging of tail; (4) more flapping, less flitting flight action, ending in 'parachute' landing (due to large area of wings and tail being strikingly displayed). Little chance of confusion with other wheatears except for ♀ and immature Desert Wheatear and Red-rumped Wheatear (see those species).

Habitat. Breeds in west Palearctic in lower middle and middle

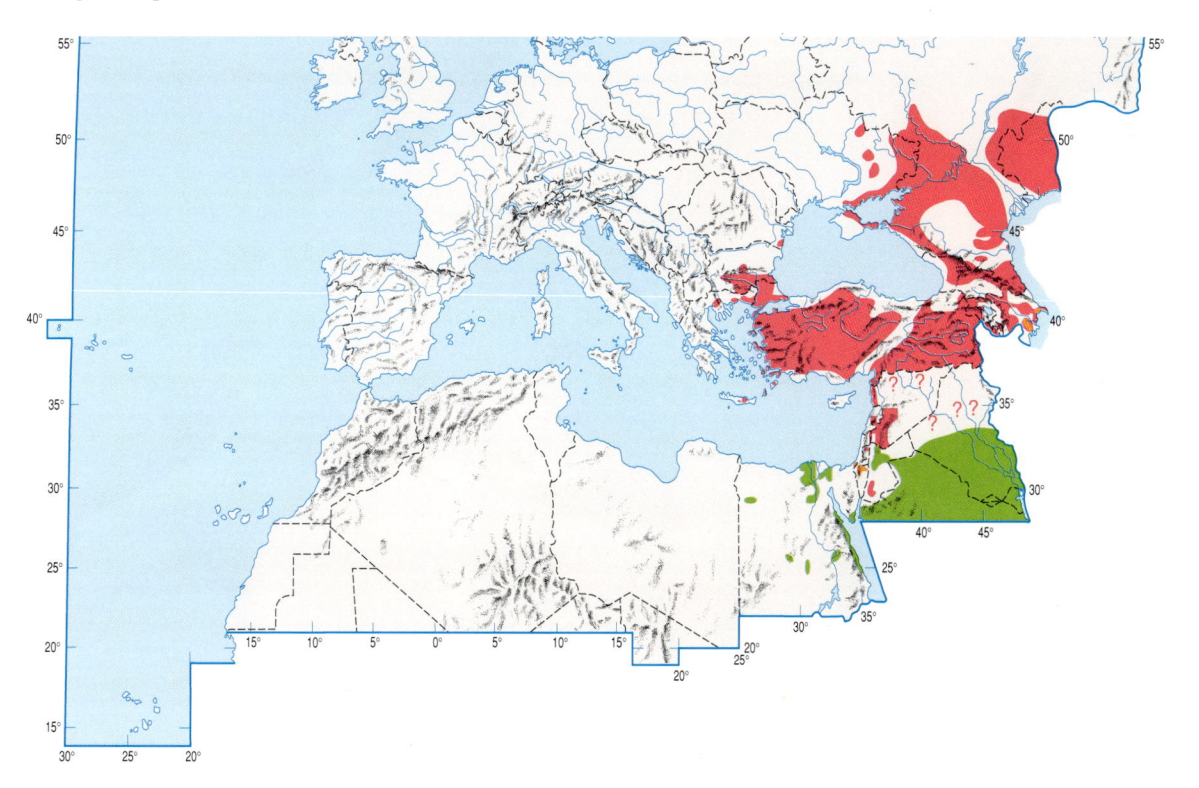

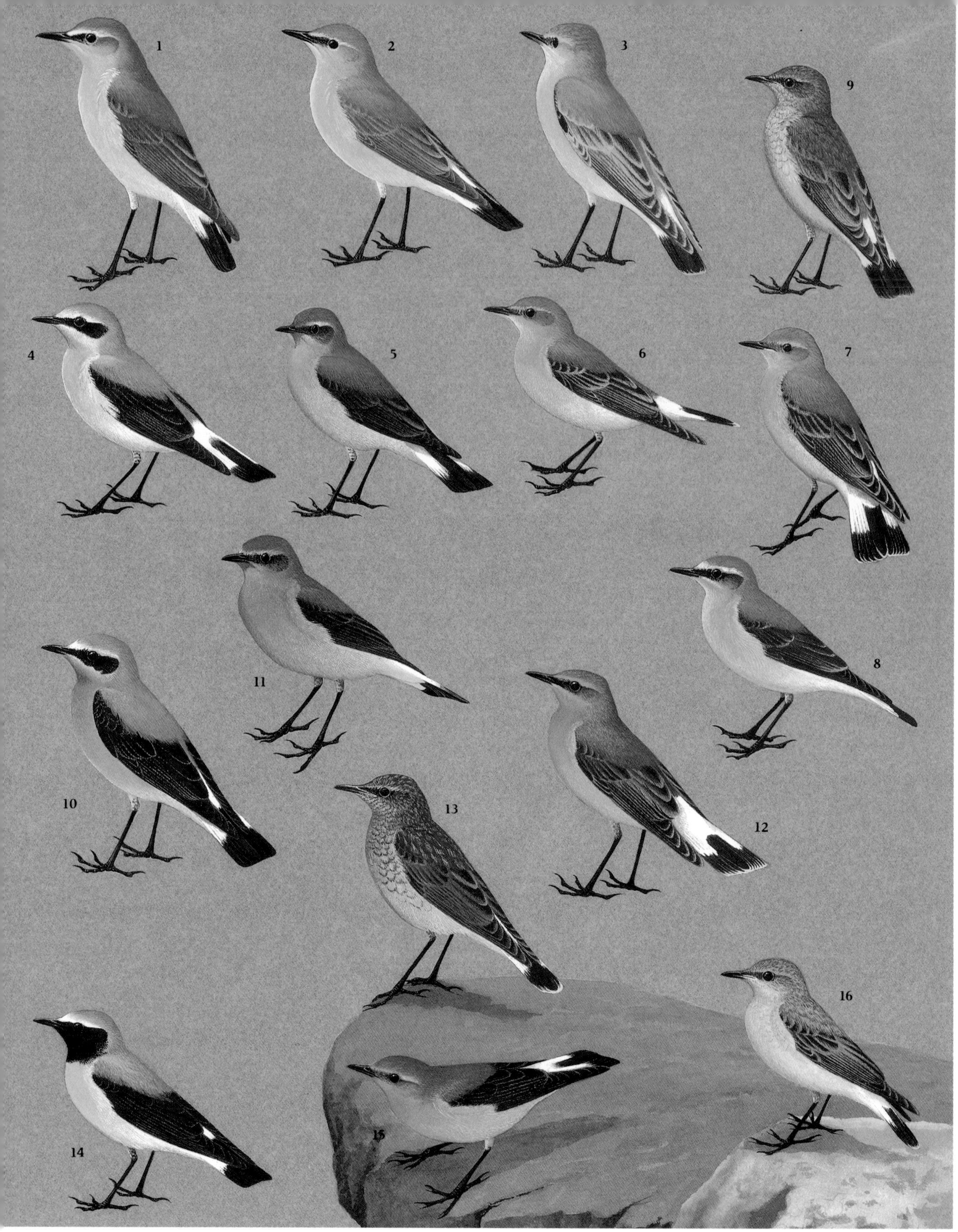
1
2
3
9
4
5
6
7
11
8
10
13
12
16
14
15

continental latitudes on plains and plateaux up to 3500 m in warm arid climate. Prefers level or gently sloping terrain, open but with sufficient isolated shrubs or large rocks, and with clay or sandy soil but not loose sand or surface gravel. Found locally on river banks with rich grass cover, and even, in passing, on mown lawns, but prefers very short, sparse vegetation with ample bare patches. Accordingly, largely a steppe and steppe-desert bird, dependent on opportunities for nesting in burrows. In winter in Africa, ranges from coastal flats to 2500 m, mainly in drier and more barren open places in East Africa, but also in short grass at over 1500 m; in West Africa, especially favours burnt ground.

Distribution. BULGARIA. First breeding record 1972. RUMANIA. Recorded only in south of Dobrogea. FSU. Some spread south in 20th century, also west of Don and north in Voronezh region. AZERBAIJAN. Most widespread of the wheatears. KAZAKHSTAN. Ural delta: numbers severely reduced by rising water-levels in recent years. TURKEY. Absent from mountain-tops, where replaced by Wheatear. IRAQ. No proof of breeding, but thought likely in salt deserts.

Accidental. Britain, Ireland, France, Denmark, Norway, Sweden, Finland, Poland, Morocco, Madeira, Canary Islands.

Beyond west Palearctic, extends through central Asia east to Baykal region and north-east China, south to Iran.

Population. GREECE. 200–500 pairs. BULGARIA. 100–1000 pairs 1980–90; increasing. RUMANIA. Probably 10–15 pairs 1986–92. RUSSIA. 100 000–1 million pairs; stable. UKRAINE. 10–50 pairs in 1986; stable. AZERBAIJAN. Very common. TURKEY. 100 000–1 million pairs. ISRAEL. At least a few hundred pairs in 1980s. JORDAN. Common.

Movements. Migratory. Winters up to northern edge of Sahel zone of Africa, from east to west, south in east to northern Tanzania; also in Egypt and Middle East, and in Pakistan and north-west India.

Migration protracted; signs of movement as early as late January and late July, but some birds remain in wintering areas furthest west until mid-March, and in breeding areas furthest east until mid-October. Passage more conspicuous in spring than in autumn in Mediterranean areas furthest west. Passage largely nocturnal and on broad front, probably mainly in a WSW or south-west direction in autumn.

(**FACING PAGE**) Isabelline Wheatear *Oenanthe isabellina*: **1** ad ♂ (spring), **2** ad ♀ (spring), **3** 1st winter ♂.
Wheatear *Oenanthe oenanthe* (p. 1178). *O. o. oenanthe*: **4** ad ♂ breeding, **5** ad ♀ breeding, **6** 1st winter ♂, **7** 1st winter ♀, **8** 1st winter ♂, **9** juv. *O. o. leucorhoa*: **10** ad ♂ breeding, **11** ad ♀ breeding, **12** 1st winter ♂, **13** juv. *O. o. seebohmi*: **14** ad ♂ breeding, **15** ad ♀ breeding, **16** juv.

Food. Mainly invertebrates; ants and beetles particularly important. Usually forages by making quick dashes along ground after prey. Sometimes uses perch (e.g. bush, stone) to watch for prey, drops down to ground, and eats item before flying up to same or new perch. Digs in soil with bill to extract invertebrates, especially in early spring when few on surface. Seeks prey in rodents' burrows, particularly in early spring or when ground covered by snow. Relatively long legs well suited to cursorial foraging, and relatively short tail to upright searching stance.

Social pattern and behaviour. Normally solitary and territorial in winter, though apparent pairs recorded towards end of winter and during migration when birds sometimes loosely gregarious, also associating with other wheatears. Mating system apparently varies between areas, from wholly monogamous to regularly bigamous. Breeding territories vary greatly in extent; may be small and contiguous, or widely scattered with little contact between pairs. ♂ sings from low perch (pile of stones, bush, etc.) or in flight. In full (rather variable) song-flight, ascends steeply with fluttering wing-beats, tail widely spread and lowered (colour pattern conspicuous), to *c.* 10–20 m. Tail rarely raised during ascent; feet may dangle. Hovers for a few seconds at peak, sometimes also briefly once or twice during ascent and 1st part of descent. Descent is otherwise gliding (sometimes in spirals) with wings and tail spread, and/or plummeting and still accompanied by loud song which continues after landing, sometimes near ♀, on perch. ♂ generally takes up territory first and selects nest-hole; shows hole to prospective mates by running or flying to burrow in song-flight; stands with tail cocked up before entering.

Voice. Song loud, rich, and varied, with croaks, pipes, and whistles flung out with abandon and interspersed with mimicry of other ground birds. Calls usually loud, with rich, piped 'weep' or 'dweet', rather high-pitched, disyllabic 'wheet-whit', and quiet 'tcheep' or 'cheep'.

Breeding. SEASON. In Transcaspia, laying begins end of March, though sometimes considerably earlier in south. In lowland Kazakhstan, most young fledge late May, in steppes between Volga and Ural rivers early June. 2(–3) broods in south of range, normally 1 in north. SITE. Normally in burrow of rodent or sometimes of bee-eater, occasionally in natural hole or crevice in ground or rock. Nest: bulky cup of dried grass, roots, and hair, lined with hair, wool, and feathers. EGGS. Sub-elliptical, smooth and glossy; pale blue, rarely with faint reddish specks. Clutch: 5–6 (4–7). INCUBATION. 12 days. FLEDGING PERIOD. Young leave nest at 13–15 days before able to fly.

Wing-length: ♂ 97–106, ♀ 93–100 mm.
Weight: ♂ mostly 28–34, ♀ 25–36 g.

Wheatear *Oenanthe oenanthe*

Du. Tapuit   Fr. Traquet motteux   Ge. Steinschmätzer   It. Culbianco
Ru. Обыкновенная каменка   Sp. Collalba gris   Sw. Stenskvätta

PLATES: pages 1176, 1200, 1202

Field characters. 14.5–15.5 cm; wing-span 26–32 cm. Eurasian race, nominate *oenanthe*, 10% longer than Robin. Structure and bulk noticeably variable, with Greenland race *leucorhoa* up to 15% larger and longer-winged than nominate *oenanthe*, approaching Isabelline Wheatear in size and even exceeding it in wing length; otherwise, usually larger and longer-winged but shorter-tailed than other wheatears. Bold, bouncy, ground-loving chat, epitome of *Oenanthe* and occurring more widely than any other. White rump and white tail with black ⊥ make striking flight character (shared by 8 other wheatears). Specific characters most obvious in spring and summer, with fully blue-grey crown, nape, and back of ♂ diagnostic, and always prominent, clean supercilium and pale or clean throat and breast of ♀ helpful. Juvenile has plumage colours of ♀, copiously spotted buff and/or scaled dark brown on fore- and upperparts. North-west African race *seebohmi* wears diagnostic black throat and Greenland race *leucorhoa* is noticeably bulky, with richer, darker plumage, particularly in autumn.

Most widespread and best-known wheatear, but only adult ♂ (and, in *libanotica*, adult ♀) easily identified. ♀ and immature liable to confusion with several other wheatears (particularly Black-eared Wheatear), which see for main distinguishing characters. Flight like that of all chats but silhouette fuller across wing-bases, rump, and tail; action essentially flitting but progress rapid and direct, usually just above ground. On ground, usually makes series of hops broken by pauses or brief perching on raised ground; also uses loping run on flat surfaces; always accompanied by frequent flicks of wings and tail, and tail also spread at times. Stance usually half-upright, with fully erect posture when inquisitive or alarmed.

Habitat. Breeds from high and low Arctic through boreal and temperate zones to steppe, Mediterranean, and subtropical arid zones, from July isotherms of 3°C to over 32°C, and from extreme continental to extreme oceanic climates. Much of this expansion must have occurred since the last glaciation, and far surpasses that of other wheatears with which, however, it shares constraints of requiring ready-made rock or burrow nest-site immediately neighbouring seasonally insect-rich bare patches or short swards for easy foraging. Breeding habitat highly diversified: flat, sandy, sparsely vegetated arctic tundra at sea-level, sand-dunes, coastal islands, shingle, cliff-tops, heaths and downland closely grazed by rabbits or sheep, roadsides with short grass, moors, steppe, bogs, clearings in forests, riverside bluffs, embankments, walled fields, rocky alpine meadows and defiles, and sparsely vegetated mountain-top plateaux up to 3000 m, even among snowfields. On migration, usually occurs in lowlands and on managed grasslands or arable. In African winter quarters found on bare soil from sea-level to over 3000 m, favouring hillsides and rocky outcrops.

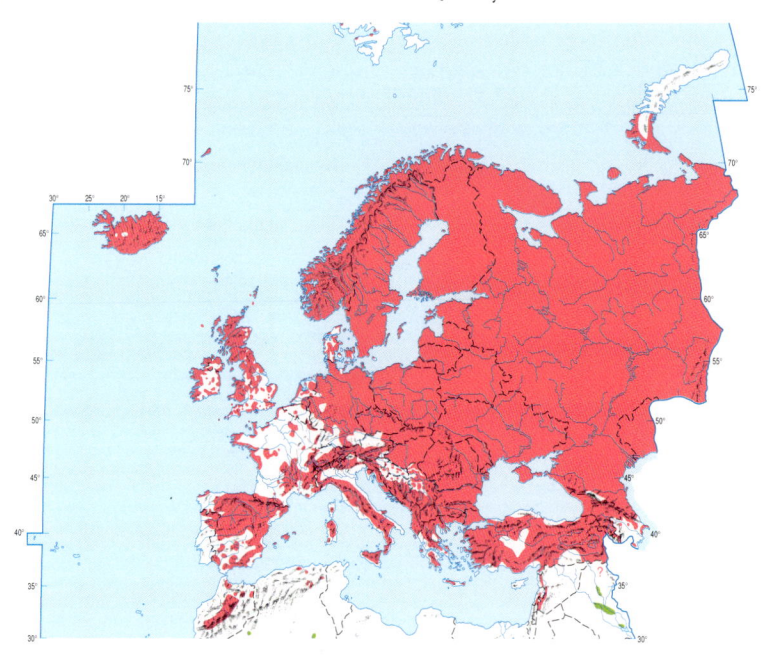

Distribution. Range decrease Britain (notably in central, southern, and eastern England), Ireland, France, Denmark, Czech Republic, Slovakia; increasing Switzerland, Ukraine. GERMANY. Highest populations along middle Elbe. AZERBAIJAN. Widespread at 2300–2800 m. No data for Little Caucasus or Nakhichevan region. TURKEY. Widespread; very local Inner Anatolia. SYRIA. Breeding proved mountains of Anti-Lebanon around Bloudan, and probable Burgush and Allovit mountains. IRAQ. May breed in north. MOROCCO. Scarce and local in Rif and eastern Morocco, common in Moyen Atlas and Haut Atlas.

Accidental. Azores, Madeira.

Beyond west Palearctic, extends east throughout much of central and northern Asia, and south to Iran. Also breeds Greenland and arctic North America.

Population. Decline western Europe (Britain and Ireland south to Iberia, east to Poland and Czech Republic, also Finland and Lithuania) probably chiefly due to habitat changes; Sahel drought in late 1960s perhaps also implicated. Increasing Ukraine; elsewhere stable. Spitsbergen. Under 10 pairs. Jan Mayen. Probably regular breeder. Iceland. 10 000–50 000 pairs in late 1980s. Faeroes. 2000–4000 pairs. Britain. Above 55 000 pairs 1988–91; long-term decline in southern half of England. Ireland. 12 000 pairs 1988–91. France. Under 100 000 pairs in 1970s. Belgium. 30–50 pairs 1989–91 (280 pairs 1973–7). Luxembourg. 20–30 pairs. Netherlands. Locally fairly common. 1500–1900 pairs 1989–91. Germany. 23 000 pairs in mid-1980s. Another estimate gives 7000–12 000 pairs in early 1990s. Denmark. 1000–3000 pairs 1987–8. Norway. 500 000–1 million pairs 1970–90. Sweden. 100 000–500 000 pairs in late 1980s. Finland. 200 000–300 000 pairs in late 1980s. Estonia. 10 000–20 000 pairs in 1991. Latvia. 10 000–30 000 pairs in 1980s. Poland. 4000–10 000 pairs. Czech Republic. 500–1000 pairs 1985–9. Slovakia. 6000–9000 pairs 1973–94. Hungary. 20 000–30 000 pairs 1979–93. Austria. Locally common. Declining locally in north and east; no apparent changes in Alpine population. Switzerland. 3000–5000 pairs 1985–93. Spain. 326 000–361 000 pairs. Portugal. 1000–10 000 pairs 1978–84. Italy. 100 000–200 000 pairs 1983–95. Greece. 20 000–50 000 pairs. Albania. 5000–10 000 pairs in 1991. Yugoslavia: Croatia. 10 000–15 000 pairs. Slovenia. 200–300 pairs. Bulgaria. 50 000–500 000 pairs. Rumania. 80 000–150 000 pairs 1986–92. Russia. 1–10 million pairs. Belarus'. 60 000–70 000 pairs in 1990. Ukraine. 17 000–20 000 pairs in 1988. Moldova. 6000–8000 pairs in 1989. Azerbaijan. Common. Turkey. 50 000–500 000 pairs. Israel. *c.* 70 pairs in 1980s.

Movements. Migratory, though North African race *seebohmi* probably only partially so. Winter quarters of entire world population, including birds breeding in Nearctic, in tropical Africa—in broad belt south of Sahara from West African coast to Indian Ocean, and south in eastern Africa to northern Zambia; records of wintering elsewhere few and probably exceptional. Passage occurs on broad front across southern Europe, Mediterranean, and full length of North African coast; recorded in about equal abundance in both seasons, in contrast to many passerines.

Migration seasons notably protracted. Birds leave breeding grounds chiefly from August; some movement southward noted from mid-July, with passage continuing until *c.* 3rd week of October, and stragglers into November. Departure from winter quarters protracted, probably especially in west, with passage noted from late January in southern Morocco, and records from mid-February to May in Algeria. Passage across North African coast and Mediterranean chiefly March–April, tailing off to mid-May. In north-west Europe, a notably early spring migrant. Thus, often the first passerine to reach Britain, where sometimes recorded early March (exceptionally late February), but more usually from mid-March with peak in early April. First arrivals in Netherlands mid-March. In Norway, arrives in south from mid-March but not present in arctic regions until mid-May.

Iceland, Greenland, and east Canadian population winters from Sénégal and Sierra Leone east to Mali. Autumn migration involves south-east crossing of North Atlantic, and frequency of records from ships south-east of Greenland is clear evidence that large numbers fly non-stop from Greenland to western Europe.

Food. Chiefly insects; also spiders, molluscs, and other small invertebrates, supplemented by berries. Normally locates prey visually, chiefly on ground or in low vegetation. 2 main foraging techniques, which may be used in same area. (1) Running: in flat areas of short turf, runs (or sometimes hops) short distance, stops to pick up item or to scan ground ahead, and then runs on. (2) Perching: in areas of scattered perches (e.g. stones, low bushes), uses these to scan ground near by, drops down for item, and then returns to perch or moves to new one; tends first to scan area closest to perch, then progressively further. Frequently flies (or from very low elevation runs or hops) to new perch, whether or not prey capture attempted.

Social pattern and behaviour. Mainly solitary outside breeding season, but birds of all ages often in parties on migration, sometimes forming larger assemblies particularly at island stop-overs and in bad weather. Short-term territories held by birds on passage, and more stable territories in winter quarters. Mating system essentially monogamous, but occasionally polygynous (♂ plus 2 ♀♀); paired ♂♂ sometimes promiscuous. Monogamous pair-bond lasts for breeding season only, but renewed annually through strong bond to territory; bird of either sex holds territory and takes new partner if former mate fails to return at start of breeding season. Territorial-song given by ♂ mostly from favoured song-posts—usually some low eminence (e.g. stone, fence), occasionally telegraph wire or tree-top; also in song-flight, less often in normal flight. Territorial-song rarely heard (except in intense territorial disputes) before ♂ is paired; thereafter given chiefly as self-advertisement, loudly and harshly in territorial defence, more softly in presence of mate. Song-flight resembles that of pipit: bird flutters jerkily upwards, typically singing, wings beating rapidly in short bursts, tail fanned and flirted (i.e. momentarily spread wider). At peak of flight, wings do not beat continuously so that bird tends to dance up and down. In final phase, may dive obliquely back to ground, still singing, or precede dive by straight-line flight of up to *c.* 50 m or by circling for up to 10 s with deep slow wing-beats and fanned tail ('butterfly flight') before diving down. In aggressive reaction to intruder on territory, ♂ commonly performs 'flashing-display', often from some eminence: turns towards or away from rival with head raised, exposing throat and breast, and giving loud subsong; tail is fanned and, every 1–2 s, flirted, while wings partially flicked out. Bird facing rival may expose fanned tail above back by bobbing or by crouching and lowering head so that bill almost touches ground.

Voice. Territorial-song of ♂ a short but vigorous, pleasantly modulated warble in which melodious, rather lark-like notes are mingled with harsh creaky and rattling sounds, including mimicry. Song of North African *seebohmi* is more measured, melodious and sonorous; compared with nominate *oenanthe*, units longer and lower pitched. Commonest call a harsh 'chack-chack' or (in mild alarm) 'weet-chack-chack'. Alarm-calls of breeding ♀ *libanotica* and far eastern *oenanthe* include loud 'weet', dangerously suggestive of Isabelline Wheatear.

Breeding. SEASON. Britain and north-west Europe: egg-laying from mid-April to June. South and central Europe: early May to June. Iceland: late May and June. Scandinavia: early to mid-May to early July. 1–2 broods. SITE. In hole in wall, among stones or rocks, in burrow, or in ruined building; will use nest-box; also in holes in wide variety of man-made objects. Nest: foundation (absent in nests in rock crevices) comprises large, untidy mass (up to 25 cm across) of dried stems of bracken, heather, and other plants, plus grass and occasional large feathers; cup more tightly woven of finer grass stems and leaves, with some moss and lichen. EGGS. Sub-elliptical, smooth and not glossy; very pale blue, unmarked or with a few red-brown flecks at broad end. Clutch: 4–7 (2–9). INCUBATION. Average 13.05 days (10–16). FLEDGING PERIOD. Average 15 days (10–21), though most young leave actual nest in burrow and move around in it at *c.* 10 days.

Wing-length: *O. o. oenanthe*: ♂ 90–104, ♀ 88–103 mm. *O. o. leucorhoa* (Greenland, Iceland): ♂ 99–110, ♀ 96–108 mm.
Weight: *O. o. oenanthe*: ♂ ♀ 18–29 g.

Geographical variation. Slight and clinal on Eurasian continent, but difference of isolated races *seebohmi* of north-west Africa and *leucorhoa* from Iceland, Greenland, and adjacent Canada more marked. Nominate *oenanthe* occupies most of west Palearctic range. Birds from Fenno-Scandia and north European Russia south to Spain and Turkey all closely similar in wing length, only birds of Crete slightly smaller. *O. o. libanotica* from southern Europe through central Asia east to Levant rather poorly differentiated from nominate *oenanthe*; bill longer, and plumage slightly paler; tail usually with less extensive black. North-west African *seebohmi* similar to *libanotica*, but cheeks, chin, throat, and sides of breast of ♂ black (throat occasionally mixed buff); tail-tip has even less black. *O. o. leucorhoa* considerably larger than other races, and plumage more deeply coloured.

Pied Wheatear *Oenanthe pleschanka*

PLATES: pages 1181, 1200, 1202

Du. Bonte Tapuit   Fr. Traquet pie   Ge. Nonnensteinschmätzer   It. Monachella dorsonero
Ru. Каменка-плешанка   Sp. Collalba pía   Sw. Nunnestenskvätta

Field characters. 14.5 cm; wing-span 25.5–27.5 cm. Build distinctly lighter than Wheatear, with slightly shorter bill and wings and 15% shorter legs, but with tail 5% longer and proportionately appearing more so; very similar in size, weight, and structure to Black-eared Wheatear except for slightly longer wings. Slight, round-headed and long-tailed wheatear; rather shy but with somewhat shrike-like perching and hunting behaviour. Rump and tail pattern basically as Wheatear but with longer black tips on outer tail-feathers. Black under wing-coverts obvious but not diagnostic; under tail-coverts white. Adult ♂ breeding has silver-white crown and nape contrasting with black face and throat which join black back and wings (making bird appear broad-shouldered from behind). ♀ rather dark and drab, with indistinct greyish supercilium and usually dark or mottled throat. In autumn, ♀ and immature show faint pale tips on mantle and usually lack pale throat.

♂ distinctive but in area of overlap needs to be separated from Mourning Wheatear (buff under tail-coverts and white webs to primaries showing in flight). ♀ and immature troublesome, resembling particularly eastern race of Black-eared Wheatear, but usually show more hoary tipping to feathers of head and back. In flight, length of tail contributes to more floating, less flitting action. Gait hopping and stance on ground less erect than most wheatears (due to rather short legs).

Habitat. Typically occupies desolate stony places with scattered boulders, fallow fields at margin of cultivated areas, earth cliffs, rocky valleys, grassy, stony, and wooded steppes, human settlements, and similar places with shrubs and small trees, up to 1800 m (3600 m in extralimital Afghanistan) in continental mid-latitudes of eastern sector of west Palearctic. In winter in similar habitats up to *c.* 2500 m in tropical and subtropical Africa and Arabia, though also in town gardens.

Distribution. GREECE. Isolated breeding records in north-east, e.g. 1–2 pairs 1989. YUGOSLAVIA. Bred 1966. BULGARIA. Slight increase. RUMANIA. Found only in south-east and in Dobrogea. FSU. Distribution at times sporadic. AZERBAIJAN. Range poorly known. ARMENIA. Breeds occasionally. TURKEY.

(FACING PAGE) Pied Wheatear *Oenanthe pleschanka*. *O. p. pleschanka*: **1** ad ♂ fresh (autumn), **2** ad ♂ worn (spring), **3** ad ♀ worn (spring), **4** 1st winter ♀, **5** juv.
Cyprus Wheatear *Oenanthe cypriaca* (p. 1183) **6** ad ♂ worn (spring).
Black-eared Wheatear *Oenanthe hispanica* (p. 1183). *O. h. hispanica*: **7** ad ♂ pale-throated morph worn (spring), **8** ad ♂ black-throated morph worn (spring), **9** ad ♀ worn (spring), **10** 1st winter ♂, **11** 1st summer ♂, **12** juv. *O. h. melanoleuca*: **13** ad ♂ black-throated morph worn (spring), **14** ad ♂ pale-throated morph worn (spring), **15** ad ♀ worn (spring), **16** 1st winter ♂.
Finsch's Wheatear *Oenanthe finschii* (p. 1187): **17** ad ♂ worn (spring), **18** ad ♀ worn (spring), **19** 1st winter ♂, **20** juv.

Norman Arlott.

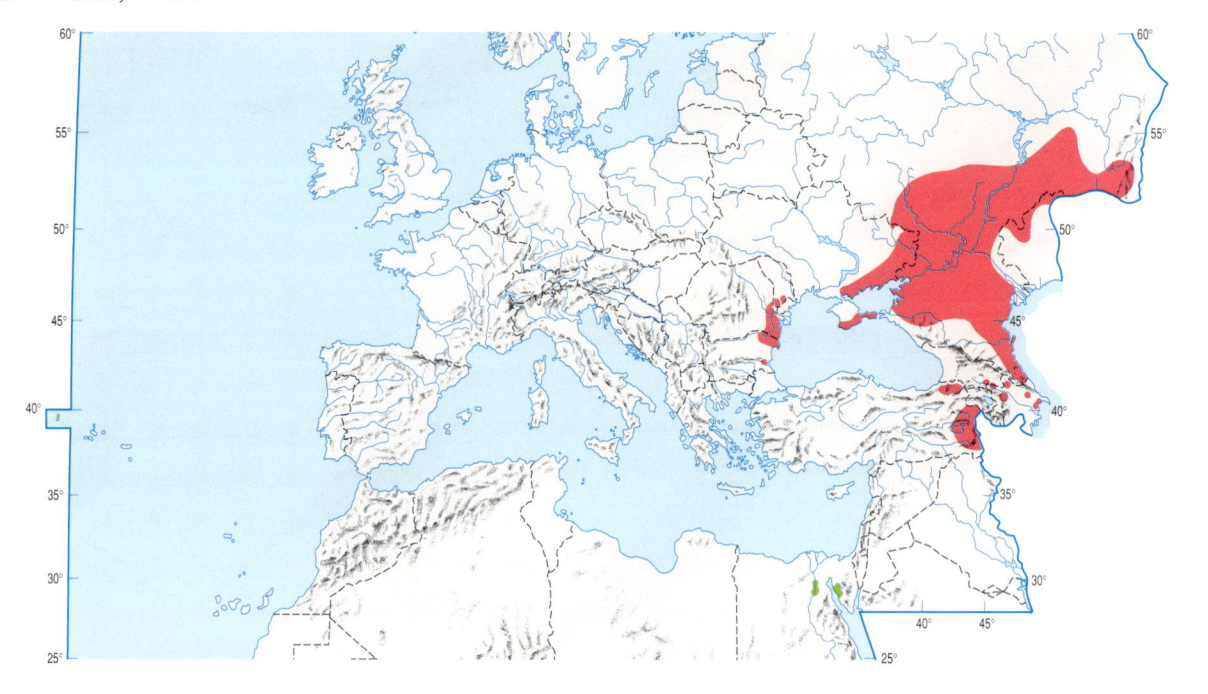

Regular only in extreme east; increasingly more sporadic and irregular further west. Syria, Lebanon. See Cyprus Pied Wheatear.

Accidental. Britain, Ireland, France, Netherlands, Germany, Denmark, Norway, Sweden, Finland, Poland, Hungary, Austria, Italy, Malta, Yugoslavia, Cyprus, Libya.

Beyond west Palearctic, extends east through central Asia to c. 120° E, south to Iran.

Population. Bulgaria. 100–500 pairs 1985–91; slight increase. Rumania. 1200–1500 pairs 1986–92; slight decrease. Russia. 100 000–1 million pairs; stable. Ukraine. 100–800 pairs in mid-1980s; stable. Moldova. 70–100 pairs in 1989; stable. Azerbaijan. Uncommon. Turkey. 50–1000 pairs.

Movements. Migratory; winters in eastern Africa (Sudan to Somalia south to north-east Tanzania) and south-west Arabia (Yemen). Moves to south and south-west on broad front via Turkey, Middle East, and Pakistan; leaves breeding grounds August–October, returning mid-March to early May (from early April in Rumania); recorded in Eritrea from early September to early May, in Kenya mid-October to early April.

Food. Almost entirely insects, taken mostly from bare ground but occasionally from low vegetation. Typically watches for prey from perch c. 1 m high, flies down to make capture and returns to perch immediately. Berries seasonally important.

Social pattern and behaviour. Solitary and territorial, but often in dense neighbourhood groups (perhaps due to habitat constraints) with small territories predominant. Behaviour, especially postures and displays, shares many features with Black-eared Wheatear, e.g. courtship-display flights by ♂ and ♀ together; raising of wings over back so that carpal joints almost touch; prolonged wing-shivering during whole period of pair-consolidation, nest-building, etc., and wider use of dancing-flight. ♂ sings from perch and in flight; in song-flight, ascends almost vertically, hovers briefly at c. 25–40 m, then glides down at angle with final plummet sometimes to take-off point; normally starts to sing c. 3–4 days after occupying territory. Pair-bond monogamous and apparently seasonal. Courtship-display much as in other wheatears, though probably less aerial than in e.g. Black-eared Wheatear. Hybridizes with that species, and near species contact zone in Iran 65% of population were hybrids, on west shore of Caspian 49%.

Voice. Much in common with Black-eared Wheatear; song also used in winter. Song of ♂ short (6–12 units), varied, typical wheatear phrases comprising whistles, smacking or chattering, trilling, and strangled sounds; frequency modulation produces buzzy quality. Mimicry of other bird species also typical. Song rendered 'geretschiretschö' or 'giretschiretschere', and can be reminiscent of Whitethroat. Calls, or other sounds, include clicking (which can merge into series, giving snoring sound), threat-rattle, whistle-, rasp-, and bugling-calls.

Breeding. Season. In Ukraine, laying starts early May. Site. Hole in rock or bank, under stone, sometimes in building. Nest: cup of dry grass and stems, lined with finer grasses, roots, wool, hair, etc. Eggs. Sub-elliptical, smooth and glossy; pale blue to green-blue, lightly speckled red-brown. Clutch: (3–) 4–6. Incubation. 13–14 days. Fledging Period. 13–14 days.

Wing-length: ♂ 88–100, ♀ 86–98 mm.
Weight: ♂♀ 15–25 g.

Cyprus Wheatear Oenanthe cypriaca

PLATE: page 1181

Du. Cyprus-tapuit   Fr. Traquet de Chypre   Ge. Zypernsteinschmätzer   It. Monachella di Cipro
Ru. Кипрская каменка   Sp. Collalba de Chipre   Sw. Cypernstenskvätta

Field characters. 13.5 cm; wing-span 24–25.5 cm. 5–10% smaller than Pied Wheatear in all measurements, with more rounded wings; close in size to Black-eared Wheatear but with slightly shorter tail. Counterpart of Pied Wheatear on Cyprus but with very different song. Breeding ♂ differs in duller crown, more extensive black back restricting white rump (to only 60% of Pied Wheatear's area), slightly lower border to black throat and buffier underparts, with pink tone on breast and vent usually retained into summer (when Pied Wheatear then worn to all white below). Breeding ♀ very different from Pied Wheatear, resembling unworn ♂ with white edge to dusky crown, mottled nape, and buff underparts.

Behaviour and flight as Pied Wheatear but song diagnostic.

Habitat. Rough, rocky open ground in hills and mountains up to *c.* 1800 m with scattered trees, but also thick woodland; common in mountain pine forest, and probably most arboreal wheatear; also agricultural land and commonly around houses, industrial areas, etc. Less common in central plain and wetlands, but is otherwise absent only from beaches, town centres, and perhaps dense scrub on level ground.

Distribution and population. Cyprus. Most numerous breeding bird, at very high densities in hills and mountains. Very common around human habitations and has probably benefited from tourist development. 150 000–300 000 pairs—probably towards upper end of range. Syria, Lebanon. Following reports, when Pied Wheatear and Cyprus Wheatear regarded as conspecific, perhaps more probably latter species: said to have bred near Beirut (Lebanon) and in Syria, but proof not known; 2 adult ♂♂ recorded June 1991 on Palmyra to Damascus road (Syria) may suggest local breeding.

Movements. Migratory, moving in broad front to winter quarters in southern Sudan and Ethiopia, a few in Egypt; regular on passage in very small numbers (mostly in spring) in Middle East from Egypt to Saudia Arabia; apparently occurs regularly in Turkey, especially on south coast in spring. Leaves breeding grounds August–September(–October), returning from late February to April; moves through Jordan and Syria mainly mid-March to early April.

Food. Almost wholly insects, particularly ants, beetles, and grasshoppers, also berries; recorded taking lizard. Forages mainly on bare ground and in vegetation; catches flying insects and pounces on ground prey from perch.

Social pattern and behaviour. Solitary, in pairs, or family parties; can form loose flocks on migration, e.g. 20 seen in spring, 30–40 mid-October. Breeding density can be very high; 50–100 pairs per km^2 in favoured habitat. Song almost always delivered from high (5–10 m, 25 m recorded) song-post and song-flight unusual (in contrast to most other wheatears); usually from tree top, but also from telephone pole or wire, rock, or building. ♂♂ very pugnacious, fighting when no ♀ present; arrive *c.* 1 week before ♀♀.

Voice. Song of ♂ is monotonous, harsh, purring series of cicada-like 'bizz-bizz-bizz' or 'zee-zee-zee' sounds, each unit uttered 3-6 times per s, each song lasting on average *c.* 8 s, exceptionally up to 1 min. Calls apparently similar to Pied Wheatear.

Breeding. Season. Eggs laid early April to early July, mainly May, *c.* 3 weeks later at 1500 m than in lowlands; 2 broods perhaps not unusual. Site. Usually in hole in wide range of places, such as earth bank, under boulder, among tree roots, in cave, rock crevice, tree cavity, and frequently in or on building. Nest: bulky, loose, untidy structure of twigs and other plant material, lined with hair, grass, and rootlets; sometimes with entrance 'path' of small twigs. Eggs. Sub-elliptical, smooth and glossy; bright blue, usually spotted reddish brown near broad end. Clutch: 4–5. Incubation. *c.* 14 days. Fledging Period. *c.* 14 days.

Wing-length: ♂♀ 78–92 mm.
Weight: ♂♀ 12–23 g.

Black-eared Wheatear Oenanthe hispanica

PLATES: pages 1181, 1200, 1202

Du. Blonde Tapuit   Fr. Traquet oreillard   Ge. Mittelmeer-Steinschmätzer   It. Monachella
Ru. Чернопегая каменка   Sp. Collalba rubia   Sw. Medelhavsstenskvätta

Field characters. 14.5 cm; wing-span 25–27 cm. Distinctly lighter in build than Wheatear, being similar in size and structure to Pied Wheatear except for slightly shorter wings. Rather slim elegant wheatear, with long, conspicuous tail giving slimmer, lengthier outline than most others of similar plumage. Rump and tail pattern basically as Wheatear but black terminal band less uniformly broad, though more black along outer edge than in any other wheatear. Both sexes have pale-throated and black-throated morphs, with former increasingly common towards east of range. Spring ♂ has wholly or partly black scapulars and wings more obviously divided by pale or whitish back and white rump and tail than any other wheatear (making bird appear narrow-shouldered from behind). ♀ has stronger pattern than many ♀ wheatears, having black wings contrasting boldly with sandy back and chest, but shows only short indistinct supercilium (unlike Wheatear). Black under wing-coverts striking, particularly in western race. Juvenile like ♀ but upperparts spotted pale buff and chest and fore-flanks faintly scaled dark brown.

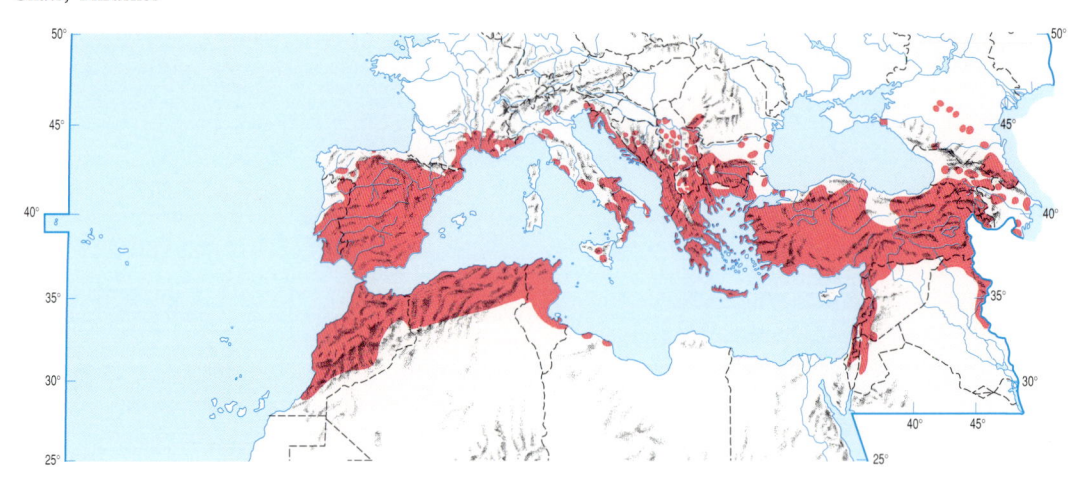

Adult ♂ and spring adult ♀ easy to distinguish, but winter ♀ and immature subject to confusion, with some nominate *hispanica* approaching appearance of pale individuals of Wheatear and most *melanoleuca* closely resembling Pied Wheatear. Dark under wing-coverts eliminate Wheatear, leaving separation from Pied Wheatear as main problem and one not easily resolved, though paler (sandier or greyer) tone of mantle, back, and scapulars, and absence of hoary fringes on them, considered diagnostic of Black-eared Wheatear. Flight like Wheatear but rather more agile, with lighter wing-beats and apparently looser tail. Gait, stance, and behaviour as Pied Wheatear, with similarly frequent perching on vegetation.

Habitat. Breeds at lower middle latitudes in warm mainly continental Mediterranean and steppe regions, above 23°C July isotherm. Within these limits, largely replaces Wheatear in habitats below *c.* 600 m, but in Caucasus ascends in mountains to 2000–2300 m, also inhabiting steppes with rocky outcrops or stony hillocks and slopes, and cliff-like river banks. More generally in open or lightly wooded arid country; also on warm rocky lowlands and stony ground, limestone hills, slopes with debris, dry river valleys, dry and stony fields, Mediterranean heaths with kermes oak, vineyards with stone banks, and dry cultivations. In winter resorts to *Acacia* and thornbush steppe and semi-desert country, rocky hills, and gardens.

Distribution. FRANCE. Apparently extended further north formerly. IBERIA. Slight decrease in range. MALTA. Bred June 1982. RUMANIA. Has spread, but only a few scattered breeding localities. AZERBAIJAN. Very common in Nakhichevan region. TURKEY. Widespread; commonest on lower slopes and hills of west and south. IRAQ. Very common in Taurus and Zagros mountains. EGYPT. May breed north-east Sinai. ALGERIA. More common along north-west coast than further east.

Accidental. Britain, Ireland, Belgium, Netherlands, Germany, Denmark, Norway, Sweden, Finland, Poland, Czech Republic, Hungary, Austria, Switzerland, Ukraine.

Beyond west Palearctic, breeds in western Iran and south-west Kazakhstan.

Population. Decline reported Iberia and Italy, probably mainly because of winter habitat changes in Sahel due to drought, but breeding habitat vulnerable to agricultural intensification and changes in forestry practices. FRANCE. 100–500 pairs; stable. SPAIN. 513 000–620 000 pairs. PORTUGAL. 10 000–100 000 pairs 1978–84. ITALY. 1000–2000 pairs 1983–95. GREECE. 30 000–60 000 pairs; stable. ALBANIA. 5000–10 000 pairs in 1991. YUGOSLAVIA: CROATIA. 40 000–60 000 pairs; stable. BULGARIA. Locally numerous; 1000–10 000 pairs. RUMANIA. *c.* 30–50 pairs 1986–92; slight increase. TURKEY. 50 000–500 000 pairs. ISRAEL. At least a few thousand pairs in 1980s. JORDAN. Fairly common. LIBYA. Breeds sparingly. TUNISIA. Fairly common. MOROCCO. Common.

Movements. Migratory. Winters in semi-desert and *Acacia* savanna belt across northern tropical Africa from Sénégal to Ethiopia. Nominate *hispanica* (breeding south-west Europe and North Africa) winters south of *c.* 18°N, mainly in northern Sénégal, south-west Mauritania, and Mali. Eastern race, *melanoleuca* (breeding east from south-east Italy), tends to replace nominate *hispanica* on wintering grounds east of 0–5°E, although there is overlap in Mali.

Departure from breeding grounds August–September. Arrival on wintering grounds from mid-September (*hispanica*) or September–October (*melanoleuca*); departure mainly March–April, arriving on breeding grounds mainly from late March to early May.

Food. Almost entirely insects; also fruits, especially in late summer. Prey taken mainly from bare ground or short vegetation up to 10 cm tall. Usually watches from perch up to 3 m above ground; flies down to make capture, returning to same perch. Light weight allows it to perch on flimsy vegetation unusable by other heavier wheatears.

Social pattern and behaviour. Solitary and apparently territorial in winter. In breeding season, less tolerant of high density than Pied Wheatear and territorial system stricter. Nevertheless, concentrations (neighbourhood groups) with small territories do occur if suitable habitat restricted or relief impedes visual contact. ♂ sings from perch (rock, overhead wire, etc.) spending much of day there and singing loudly; usually 1–3 favourite perches in open parts of territory. Also performs song-flights (probably to attract mate) over territory. With fluttering flight and tail widely fanned, ascends rather jerkily at *c.* 60° to *c.* 20–30(–50) m, singing loudly; makes slight deviations (1–2 m) to side. For up to $1\frac{1}{2}$(–2) min circles or flutters about over territory, making sharp turns. Descent usually rapid, with closed-wings plummet for last 5–10 m, bird typically landing on take-off point or another song-post.

Voice. Song of ♂ similar to Pied Wheatear. Typically comprises short, loud phrases of 7–11 units; lasts *c.* 1–1.5(–2) s. A cheerful, rapid warbling, rich and full-toned, rising and falling, and with characteristic dry, buzzy, or scratchy and raucous quality. Also includes mimicry, and shows much variation within and between individuals. Calls very similar to Pied Wheatear; include a variety of clicking sounds, and whistles; in alarm, clicks and whistles combined.

Breeding. SEASON. Algeria and Tunisia: eggs laid from end of April to early June. Spain: laying begins late April or early May; in Guadalajara and Soria (north-central Spain), most clutches laid 16–31 May. Greece: from early May. Armenia: laying begins late April or early May. Usually 1 brood. SITE. On ground in shallow hole, under stone, in thick vegetation, or at base of dense bush. Nest: cup of grass and moss, lined with finer material including hair. Bits of twig often placed by nest (entrance), forming platform. EGGS. Sub-elliptical, smooth and glossy; pale blue, with fine markings of red-brown, sometimes forming cap at broad end. Clutch: 4–5(–6). INCUBATION. 13–14 days. FLEDGING PERIOD. 11–12 days.

Wing-length: *O. h. hispanica*: ♂ 86–92, ♀ 86–90 mm. *O. h. melanoleuca*: ♂ 88–95, ♀ 86–91 mm.
Weight: *O. h. hispanica*: ♂♀ mostly 14–21 g. *O. h. melanoleuca*: ♂♀ 12–22 mm.

Geographical variation. Rather marked, with 2 distinct races differing mainly in general colour (both sexes) and in extent of black marks on face and throat (♂ only); also, some variation in tail pattern, in relative numbers of black- and white-throated morphs, and (very slightly) in size. Adult ♂ *melanoleuca* (southern Italy, and Yugoslavia eastwards) differs in particular from nominate *hispanica* (western parts of range) in narrow strip of black extending along base of forehead: in nominate *hispanica*, black of lores extends narrowly to nostrils, but generally does not meet at base of culmen; in *melanoleuca*, black meets narrowly and forms black strip *c.* 1–3 mm wide. Adult ♂ black-throated (*stapazina*) morph of *melanoleuca* more extensively black on throat than black-throated nominate *hispanica*. General colour of adult ♂ *melanoleuca* close to nominate *hispanica* in autumn, but *melanoleuca* with more distinct grey tinge on crown, nape, and mantle (nominate *hispanica* virtually uniform deep pinkish buff-cinnamon). In spring, difference more marked: by March–April, crown and hindneck of *melanoleuca* white with traces of brown-grey feather-tips, lower mantle tinged pale buff-cinnamon (in nominate *hispanica*, uniform darker buff-cinnamon); by May–June, crown, hindneck, and mantle of *melanoleuca* white except for slight cream tinge on outer mantle (in nominate *hispanica*, off-white with distinct cinnamon wash all over).

Desert Wheatear *Oenanthe deserti*

PLATES: pages 1186, 120 , 1202

DU. Woestijntapuit FR. Traquet du désert GE. Wüstensteinschmätzer IT. Monachella del deserto
RU. Пустынная каменка SP. Collalba desértica SW. Ökenstenskvätta

Field characters. 14–15 cm; wing-span 24.5–29 cm. Averages smaller and appears dumpier than Wheatear, with less pointed wings but proportionately longer tail; noticeably smaller and slighter than Red-rumped Wheatear. Rather small, round-headed, compact wheatear. All-black tail and pale webs of flight feathers diagnostic. ♂ distinguished by noticeably sandy plumage, relieved by black face and throat and whiter rump and inner wing-coverts, ♀ by more uniform sandy-pink appearance than congeners (with only black-brown primaries and tail obvious). Juvenile like autumn ♀, but upperparts faintly spotted pale, breast lightly scaled brown-buff, and rump whiter.

All-black tail quickly distinguishes it from Black-eared Wheatear and most other similarly coloured wheatears, but note that tail pattern and noticeably pale underwing of ♀ and immature closely matched by ♀ Red-rumped Wheatear and approached by Isabelline Wheatear (see those species). Flight essentially as Wheatear but more flitting and less fluent, with hint of *Saxicola* chat at times. Stance usually half-upright, with round-topped head held well up even when neck retracted and this attitude contributing much to rather compact silhouette.

Habitat. In lower middle latitudes, mainly continental, warm and arid, in steppe, Mediterranean, and desert zones; on wide variety of terrain from sea-level (especially in west of range) to high plateaus and even mountain summits extralimitally in Asia. In North Africa, occurs on Atlantic coast and on degraded steppe at edge of Sahara. Ecologically replaces Black-eared Wheatear towards desert. Fond of old cultivation which has reverted to desert. In FSU, inhabits primarily low-lying places with elongated barchan sand-dunes or clay-covered steppeland, with stony soil of moraines carved by run-off gullies and

Desert Wheatear *Oenanthe deserti*. *O. d. homochroa*: **1** ad ♂ worn (spring), **2** ad ♀ worn (spring), **3** 1st winter ♂, **4** 1st summer ♂, **5** juv. *O. d. atrogularis*: **6** ad ♂ worn (spring), **7** 1st winter ♂. *O. d. deserti*: **8** ad ♂ worn (spring).

ravines. Evidently adaptable to varying habitats over extensive range, but ecologically the name 'Desert' is misleading: penetrates little beyond desert fringe.

Distribution. GEORGIA. Recorded nesting in low mountains near Tbilisi in early 1900s. AZERBAIJAN. Considered to be nesting in eastern Transcaucasia in early 1900s; specimen collected *c.* 1884 near Baku without nesting evidence. No recent sightings. TURKEY. First recorded breeding Birecik on Euphrates 1985, but probably regular and previously overlooked. SYRIA. May also breed in Euphrates region. JORDAN. Very characteristic bird of Eastern Desert and Azraq. SAUDI ARABIA. Common in Harrat Al Harrah reserve, where breeding first proved in 1989. EGYPT. Breeding proved only in Sinai, but observations indicate probable breeding where mapped. LIBYA. Southern limits uncertain. In Tripolitania, common in most semi-desert areas and locally near coast. ALGERIA. Very common in west part of Hauts Plateaux; mostly rare in northern Sahara.

Accidental. Britain, Ireland, France, Belgium, Netherlands, Germany, Denmark, Norway, Sweden, Finland, Estonia, Hungary, Switzerland, Spain, Malta, Greece, Madeira, Canary Islands.

Beyond west Palearctic, breeds east through central Asia to north-central China, south to Himalayas and Iran; breeding also recorded south of west Palearctic border in Mauritania.

Population. ISRAEL. 1000–2000 pairs in 1980s. Nomadic, with annual variations. EGYPT. Scarce. TUNISIA, MOROCCO. Common.

Movements. Most populations migratory, some only partially. Winters in Africa south to Sahel zone, in south-west Asia and east to central India, and in eastern Himalayas. Frequent vagrant over large area north to Sweden, west to Canary Islands, and east to Kuril Islands (eastern Russia). Timing of movements and routes taken poorly known.

Food. Predominantly insects, particularly ants, beetles, and larvae; occasionally spiders, worms, small lizards, and seeds. Takes food mainly from bare ground; sometimes from low vegetation or in flight like flycatcher. Typically searches for food from low or elevated perch (up to *c.* 1.5 m high), launching flying attack usually against prey within 10 m. Light weight allows use of flimsier perches than most other wheatears. Occasionally hunts, in manner of Kestrel, by hovering into wind a few metres above ground and diving on to prey.

Social pattern and behaviour. Outside breeding season normally solitary and territorial. Breeding dispersion varies from clumped, in neighbourhood groups, to very sparse and scattered, with neighbouring pairs rarely coming into contact. Territorial-song usually delivered from 2–3 favoured song-posts such as top of bush or rock. During breeding season, song-flights often performed: ♂ usually ascends at moderate angle to height of 8–10 m and, singing most of the time, circles territory with measured, wide-amplitude wing-beats before closing wings and plunging back to earth. During song-flight, sometimes folds wings and, calling, plummets precipitously with tail slightly spread before pulling out of dive and resuming normal song-flight; such plummets during main part of song-flight unique in wheatears.

Voice. Territorial-song of ♂ a monotonous succession of brief

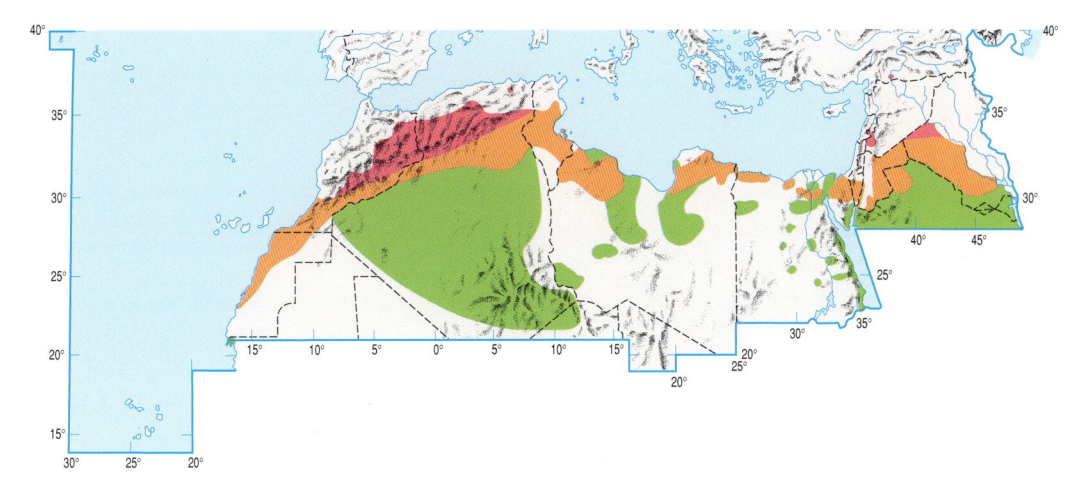

(2–4 notes), rather stereotyped, mournful phrases—roughly 'swee-you', likened to rusty hinge—delivered slowly whilst perched or in song-flight. Contact-alarm call a rather plaintive hoarse or fluting whistle. Commonest sign of agitation a harsh 'tuk', much sharper than Wheatear, turning into a rattle in more intense alarm.

Breeding. SEASON. Algeria and Tunisia: eggs found 10 March–26 May. Middle East: eggs found April–May. Kazakhstan: laying begins late April. SITE. In hole in ground, or among rocks, often in old rodent burrow. Nest: bulky cup of grass, dead leaves, and roots, lined with hair, feathers, and wool. EGGS. Sub-elliptical, smooth and slightly glossy; pale blue, with variable red-brown specks and spots, often mainly at broad end forming ring. Clutch: 4–5 (3–6). INCUBATION. 13–14 days. FLEDGING PERIOD. 13–14 days.

Wing-length: *O. d. homochroa*: ♂ 88–95, ♀ 84–90 mm.
Weight: *O. d. homochroa*: ♂♀ 15–23 g.

Geographical variation. 3 races in west Palearctic. *O. d. homochroa* from North Africa has mantle and scapulars pink-cinnamon or slightly vinous, chest pink-buff, white wedge on inner webs of flight-feathers narrow. Nominate *deserti*, from Levant, and *atrogularis*, from Transcaucasia, have more saturated greyish-cinnamon or sandy-grey mantle and scapulars and deeper buff or cinnamon-buff chest; differ from each other only in size (slightly).

Finsch's Wheatear *Oenanthe finschii*

PLATES: pages 1181, 1200, 1202

Du. Finsch' Tapuit   FR. Traquet de Finsch   GE. Felsensteinschmätzer   IT. Monachella di Finsch
Ru. Черношейная каменка   Sp. Collalba oriental   Sw. Finschstenskvätta

Field characters. 14 cm; wing-span 25–27 cm. Close in structure and size to Pied Wheatear and Black-eared Wheatear but with proportionately larger head and more rounded wings; smaller (but heavier) than Mourning Wheatear; thicker-legged than all. Small to medium-sized but robust, often shy wheatear, with plumage pattern overlapping those of at least 3 congeners. Tail pattern distinctive, with anchor, not ⊥ shape and narrow white terminal rim. ♂ most recalls black-throated Black-eared Wheatear but differs in black of throat joining black of wings and pale tips to flight feathers in both fresh and worn plumage. ♀ less distinctive, but has rufous cheeks and usually dark throat-band. Juvenile like ♀ but has even broader pale fringes to wing-feathers; lacks spots and scales of most similar juvenile wheatears.

Identification less studied than most other wheatears; most confused with Pied Wheatear and Mourning Wheatear (see those species). Flight often darting or banking, with sudden rises or falls and changes of direction.

Habitat. In continental lower-middle latitudes of south-east sector of west Palearctic, in dry warm temperate and steppe zones. Breeds in FSU on bare clay sands, rocky steppes, and ravines heaped with stones in low mountains, up to zone of pistachios at *c.* 1400–1600 m. In Lebanon, occurs on stony ground and outcrops of rock in mountain ravines or on hillsides, especially rock-rims in arid country. In Anatolia (Turkey), confined to rocky hillsides, preferring steep re-entrants.

Distribution and population. BULGARIA. Recorded breeding 1988 in south-west. AZERBAIJAN. Common. Very common in Nakhichevan region. TURKEY. Occasional west of main breeding area, and may be expanding range slowly. 1000–20 000 pairs. SYRIA. Quite widespread in western mountains; many move down into deserts and steppes of interior, where commonest wheatear in winter. ISRAEL. Pair bred Mt Hermon 1981. IRAQ. Probably breeds, but no proof.

Accidental. Greece, Bulgaria, Kuwait.

Beyond west Palearctic, breeds from Iran and east Caspian east to *c.* 68°E.

Movements. Partially migratory. Difficult to determine status in some parts of range, with wintering birds of unknown origin swelling local, probably resident, populations. Altitudinal differences in migratory status may occur. Wintering records from southern Turkmenistan, Transcaucasia, and Kazakhstan, but main west Palearctic wintering areas lie from southern Turkey and northern Iraq south to north-east Egypt and northern Saudi Arabia; locally common on Cyprus.

Food. Mainly insects (especially ants and beetles); also some seeds and other plant material.

Social pattern and behaviour. Migratory birds and at least some sedentary ones defend individual territories outside breeding season. Mating system essentially monogamous but occasionally polygynous; after ♀♀ start incubating, ♂♂ tend towards bigamy, sometimes pairing simultaneously with another ♀ in a 2nd territory. Breeding pairs in sedentary southern populations often isolated (e.g. in Tajikistan, pairs 1.5–2 km apart), but migratory populations further north form neighbourhood groups, locally in dense concentrations. ♂ sings regularly during pair-formation, from perch and in song-flight. Each ♂ has 2–3 preferred song-posts, usually on highest point of a hill. Singing ♂ runs back and forth in characteristic posture with tail widely spread and head raised, then launches into song-flight: ascends, singing, with shivering wing-beats and tail widely spread, then lands on another song-post. Sometimes hovers for a few seconds, and may describe 2–3 tight circles.

Voice. Song given by both sexes but mainly ♂, extremely varied, often melodious but interspersed with harsh sounds. Description of song as scratchy but including rich fluty whistles is typical. Mimicry rare, but includes calls of wide variety of bird species and of tree-frog. Calls include 'zik' and 'chek', combined in alarm with descending 'seep'.

Breeding. SEASON. Turkey and Middle East: eggs laid April. FSU: starts earlier than all other wheatears, in extreme south laying sometimes from mid-February, in north not before early April. 1–2 broods in north of range, up to 3 in south. SITE. In hole in rock outcrop, among stones, or in bank; less commonly in rodent burrow. Nest: flat saucer of grass and small twigs, usually lined with finer grass and hair, wool, feathers, etc. Usually an accumulation of stones (etc.) in nest-entrance and tunnel. EGGS. Sub-elliptical, smooth and glossy; pale blue, with variable red-brown speckling or spotting. Clutch: 5 (4–6). INCUBATION. 12–13 days. FLEDGING PERIOD. 15–16 days.

Wing-length: *O. f. finschii*: ♂ 85–92, ♀ 82–87 mm.
Weight: *O. f. finschii*: ♂ ♀ 21–32 g.

Geographical variation. Rather slight; involves general colour and size. Nominate *finschii* from central Turkey and Taurus mountains south to Dead Sea region is smaller than *barnesi* from eastern Turkey eastwards. In colour, ♂ *barnesi* very similar to nominate *finschii*, but under tail-coverts slightly deeper cream-buff, and forehead, crown, mantle, and central back tinged pinkish cream-buff (in nominate *finschii*, pale buff-grey); races inseparable in colour when plumage worn. ♀ *barnesi* more sandy-buff on upperparts (less sandy-grey).

Red-rumped Wheatear *Oenanthe moesta*

PLATES: pages 1189, 1200, 1202

DU. Roodstuittapuit FR. Traquet à tête grise GE. Fahlbürzel-Steinschmätzer IT. Monachella testagrigia
RU. Краснопоясничная каменка SP. Collalba de Tristram SW. Rödstjärtad stenskvätta

Field characters. 16 cm; wing-span 25–29 cm. Larger than Desert Wheatear, with bigger bill and head and slightly longer, broader wings and tail. Rather bull-headed, compact wheatear with dull rufous rump and nearly all-dark tail. ♂ lacks sharp contrasts of most ♂ wheatears, with grey crown and nape, black face, throat, shoulders, and back, always pale-fringed wings and dull underparts. ♀ much more distinctive than most ♀ wheatears, with noticeably rufous crown and cheeks diagnostic. Juvenile like ♀ but with less rufous head and faint spots and streaks.

Red-rumped Wheatear *Oenanthe moesta*: **1** ad ♂ worn (spring), **2** ad ♀ worn (spring), **3** 1st winter ♂, **4** juv ♂. Red-tailed Wheatear *Oenanthe xanthoprymna* (p. 1190). *O. x. chrysopygia*: **5** ad worn (spring), **6** ad heavily worn (summer), **7** 1st winter ♂, **8** juv. *O. x. xanthoprymna*: **9** ad ♂.

In brief glimpse or at distance, may be confused with Desert Wheatear (which has similar rump and tail pattern but is smaller, without black back in ♂ or chestnut head in ♀ and immature). Flight distinctive, bird tending to drop from perch and skim over ground, with rather loose wing-beats producing floating progress recalling Skylark. Gait a long hop.

Habitat. Across narrower belt of Afro-Asian lower-middle latitudes than Desert Wheatear, also ranging from warm arid continental to oceanic climates, especially under Mediterranean influences. Mainly frequents flat ground, often in vicinity of saline and barren areas, but not absolute desert. Habitat almost identical with Desert Wheatear. Intrudes into narrow belt of scrub between Mediterranean and western desert of Egypt. In southern Morocco, seems to use patches more densely vegetated than Desert Wheatear would; also occurs in rocky hills. Distribution seems to owe its patchiness to availability of rodent holes.

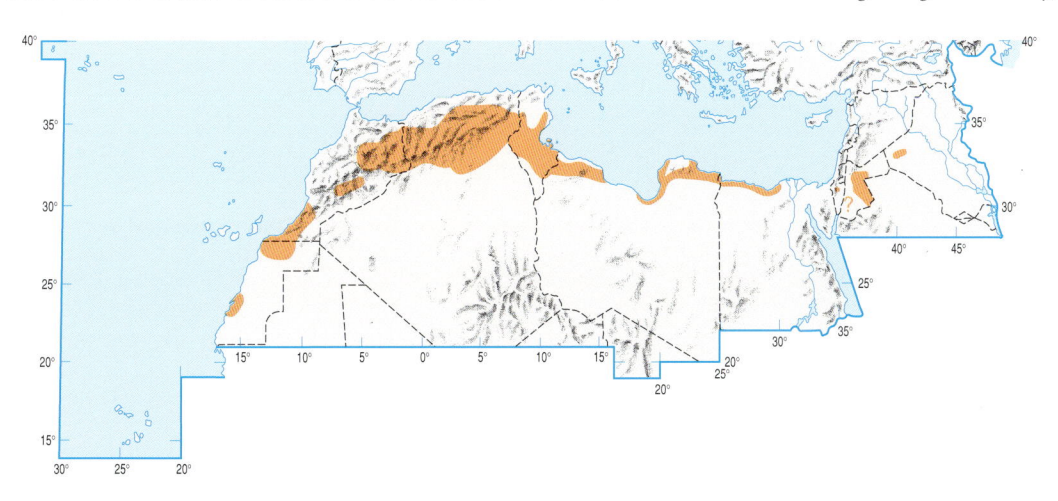

Distribution and population. Breeds only in west Palearctic. SYRIA. Bred formerly, but no recent records. ISRAEL. Very rare; perhaps 5–10+ pairs. JORDAN. Localized. SAUDI ARABIA. Perhaps bred formerly; no recent evidence. EGYPT. Fairly common. Bred northern Sinai 1928; present status there unknown. LIBYA. Limits uncertain, but widely distributed in north-east. Fairly common to common. TUNISIA. Possible extension northward since beginning of century, due to desertification. Common. ALGERIA. More common in west. MOROCCO. Uncommon.

Movements. Mainly sedentary, with some evidence for partial migration and local movements.

Food. Mainly insects. Prey usually caught by hopping along ground and jabbing; also by perching higher, e.g. on top of bush, and pouncing at items on ground.

Social pattern and behaviour. Not well known; no evidence of marked differences from congeners.

Voice. Territorial-song given apparently by both sexes (probably less often by ♀), evidently varies markedly in quality and timbre. In winter, Morocco, described as short and rather rattling in ♂, short and harsh in ♀. Other accounts report much more melodious songs: in Tunisia, song of ♂ described as short and rippling, singularly sweet and pathetic; in Cyrenaica (Libya), a pleasant musical 'twee-chirr-rur-rur-rur' from ♂, repeated slowly without much variation. Courtship-song most remarkable: a long wavering warbling whistle, rising progressively in pitch, aptly likened to whistling kettle coming to the boil. Usually given in antiphonal duet by pair (each bird starting at bottom of scale as its partner reaches the top, but sometimes overlapping). Contact-alarm calls brief, typically of hard or harsh clicking quality.

Breeding. SEASON. Algeria and Tunisia: prolonged, with eggs laid from 1st week of February and late nests in June. Jordan: newly-fledged young seen late April. 2 broods, possibly 3. SITE. Up to 2 m deep in hole in ground, usually of rodent or other small mammal, sometimes natural hole or hole in wall; entrance may be concealed under bush or root. Nest: cup of leaves, rootlets, and stems, lined with (e.g.) wool, hair, feathers, and not uncommonly snakeskin. EGGS. Sub-elliptical, smooth and slightly glossy; very pale blue, with variable red-brown speckling. Clutch: 4–5. (Incubation and fledging periods not recorded.)

Wing-length: *O. m. moesta*: ♂ 89–96, ♀ 86–90 mm.

Geographical variation. Very slight; birds from Middle East usually separated as *brooksbanki* on account of slightly larger bill, somewhat greyer (less black) mantle and scapulars, and whiter rump and upper tail-coverts of ♂, and greyer (less rufous) upperparts of ♀.

Red-tailed Wheatear *Oenanthe xanthoprymna*

PLATES: pages 1189, 1200, 1202

Du. Roodstaarttapuit   Fr. Traquet à queue rousse   Ge. Rostbürzel-Steinschmätzer   It. Monachella codarossa
Ru. Златогузая каменка   Sp. Collalba persa   Sw. Rödgumpad stenskvätta

Field characters. 14.5 cm; wing-span 26–27 cm. Close in size and structure to Wheatear. Mountain-haunting species, with general character and greyish appearance much recalling Wheatear but showing diagnostic combination of rufous-chestnut rump and black ⊥-mark on tail; basal sides of tail chestnut or (in some nominate *xanthoprymna*) white. Crown and back dull greyish brown, with face and throat fairly nondescript or solid black (♂ and most ♀ nominate *xanthoprymna*). Juvenile like pale ♀.

♂ and black-throated ♀ nominate *xanthoprymna*

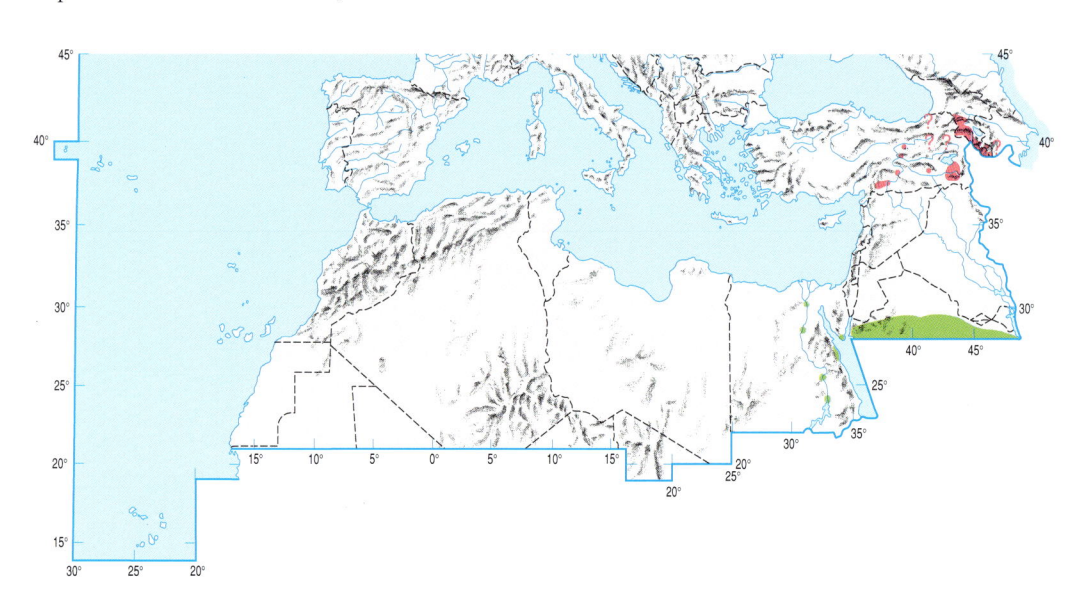

unmistakable. Pale-throated ♀ of that race and both sexes of eastern *chrysopygia* less distinctive: beware confusion with Red-rumped Wheatear or ♀ Wheatear (see those species). Flight and behaviour much as Wheatear but more prone to perch on ground or lowest plants, even hiding in holes.

Habitat. Breeds in continental lower middle latitudes, in arid and steppe climates, mainly in upland and mountain regions, favouring steep and bare terrain with ready access for nesting and refuge to burrows, fissures, scree, and other sites offering convenient holes. In winter quarters, occurs on rocky hills and arid bushy country; rarely in true desert.

Distribution and population. FSU. Numbers extremely low, though sometimes numerous in favoured habitat. Armenia. Very local. Azerbaijan. Rare. Distribution poorly known. Inhabits mountains of Nakhichevan region, and possibly Zangelan and Djabrail districts. Turkey. First recorded 1967. Breeding discovered from early 1980s; proved only in foothills and mountains bordering steppe plateau of south-east. May also breed further north. 100–1000 pairs.

Accidental. Cyprus, Libya. Occurs occasionally in Israel as passage-migrant or winter visitor. Birds recorded Syria and Jordan presumably passage-migrants, but very rare.

Beyond west Palearctic, breeds from Iran east to western Pamirs and extreme west of Pakistan.

Movements. Largely migratory, wintering in eastern Africa and south-west Asia. Present in winter quarters October to mid-February or March.

Food. Diet largely insects, especially ants, beetles, and larval Lepidoptera. Typically finds food on bare ground using dash-and-jab technique or launches flying attacks from firm, elevated perch (e.g. stone or boulder—rarely vegetation). Food also picked off vegetation and dug out of ground with bill (e.g. beetle larvae).

Social pattern and behaviour. Outside breeding season normally solitary and territorial. Except during short stops on migration, both sexes normally hold individual territories which are defended against conspecific birds, socially subordinate wheatear species and some other birds. Breeding territories vary greatly in size, according to habitat, with variable amount of overlap between neighbours. ♂ sings from perch (e.g. bush, high rock) or in flight. Each ♂ usually has several (normally up to 3) favoured song-posts affording good view over territory. Song-flight performed at height of *c.* 10–15 m. Characteristically smooth and slow, with exaggerated wing-beats and smooth changes of height. Unlike other wheatears, sometimes hangs almost motionless in the air, with only slight movement of fully extended wings, legs often dangling; or makes smooth turns.

Voice. Territorial song of ♂ consists of short phrases separated by short pauses, or pauses reduced until phrases merge to form more continuous variant. Song of simple structure, with constant repetition of a melodious whistling motif and only slight variations; mimicry virtually absent. Attractive tonal quality recalls song of Blue Rock Thrush. Contact-alarm calls include clicks, rasping sounds, and whistles.

Breeding. Season. Eastern Turkey: nests with young recorded early to mid-June and 8 August. Transcaucasia: April–June, sometimes from late March. 2 broods. Site. In hole in rocks, among stones, or in wall of building, occasionally in burrow of Bee-eater; nest 20–50(–90) cm from entrance. Nest: loosely constructed, fairly shallow cup of dry grass and coarser plant fibres, lined with finer fibres stripped from plant stems; sometimes on thick base of debris. Often with adjacent platform (less commonly also base) of small flat stones. Eggs. Sub-elliptical, smooth and slightly glossy; very pale bluish-white, or pure white with barely discernible bluish tinge, unmarked or with sparse red-brown spotting, sometimes concentrated at broad end. Clutch: 4–6. Incubation. 13 days. (Fledging period not recorded.)

Wing-length: ♂ 92–98, ♀ 86–93 mm.
Weight: ♂♀ 20–27 g.

Geographical variation. Marked, in colour only. Nominate *xanthoprymna* (south-east Turkey, east into Iran) markedly different from *chrysopygia* (Transcaucasia, east to Afghanistan): upperparts darker, sides of head, throat, and under wing-coverts black in ♂, rump and tail-coverts deeper rufous-cinnamon, and tail-base white (in *chrysopygia*, plumage almost uniform sandy-brown with rufous tail-base). The 2 races sometimes considered separate species, but some intergradation occurs and measurements and structure similar, hence treated as single polytypic species.

Eastern Pied Wheatear *Oenanthe picata*

PLATES: pages 1192, 1200, 1202

Du. Picata-tapuit Fr. Traquet variable Ge. Elstersteinschmätzer It. Monachella variabile
Ru. Чёрная каменка Sp. Collalba pía oriental Sw. Orientstenskvätta

Field characters. 15 cm; wing-span 25.5–27 cm. Three morphs, taxonomy complex. Size and structure close to Pied Wheatear and Mourning Wheatear; 10–15% smaller than Hume's Wheatear, Black Wheatear, and White-crowned Black Wheatear, lacking their relatively large heads. Medium-sized wheatear of variable appearance, subject to confusion with all five species named above. ♂ of westernmost morph pied; closely resembles Hume's Wheatear but distinguishable by deeper black breast bib (not ending just below throat but abutting shoulder), far less extension into back of white rump

Eastern Pied Wheatear *Oenanthe picata*. *opistholeuca* morph: **1** ad ♂ worn (spring), **2** ad ♀ worn (spring), **3** 1st winter ♂, **4** juv. *picata* morph: **5** ad ♂ worn (spring). *capistrata* morph: **6** ad ♂ worn (spring). Hume's Wheatear *Oenanthe alboniger* (p. 1195): **7** ad ♂ worn (spring), **8** ad ♀ worn (spring), **9** 1st winter ♂, **10** juv.

(the upper edge of which is straight and not rounded), and dull tone of black plumage (not glossed). ♂ of pied eastern morph recalls Pied Wheatear but its pale crown and nape always cleaner (and pure white when worn); also suggests Mourning Wheatear but lacks white inner webs to flight feathers and warm buff under tail-coverts of that species. ♂ of mainly black eastern morph recalls Black Wheatear, immature White-crowned Black Wheatear, and particularly black morph of Mourning Wheatear; best distinguished from Black Wheatear and White-crowned Black Wheatear by distinctly smaller size, lack of dull wings (of Black Wheatear) and lack of mainly white outer tail-feathers (of White-crowned Black Wheatear); separated at close range from black morph Mourning Wheatear by smaller bill and lack of noticeably pale inner webs to flight feathers. ♀♀ of all races have plumage patterns like ♂♂ but dark feathers are sooty or dusky-brown, with rufous tone to ear coverts. Juveniles unstudied in field.

Levant records from 1960s shown to be invalid (following discovery of black morph Mourning Wheatear in basalt desert) but recent occurrence in Israel has restored need to consider this species in pied wheatear identification. Flight and behaviour typical of genus. Song rather scratchy in tone, far less pleasing than Hume's Wheatear. Calls not distinctive.

Habitat. Arid, boulder-covered hilly country with scattered vegetation, fallow fields, ruins, and around villages in south-central and south-west Asia; generally above 500 m and up to 3300 m in Pakistan. Winters in desert, stony semi-desert, rocky hills, ravines, cultivated ground near settlements, etc. at lower elevations down to sea level. Feeds on invertebrates and berries; nests in hole in bank or sometimes tree, rock crevice, or on building.

Distribution. Breeds from central Iran east to Tajikistan and Pakistan.

Accidental. Israel: Eilat, dark morph, February 1986.

Movements. Migratory and partially migratory, movements often only over short distances and principally altitudinal; birds in Tajikistan and Turkmenistan migrate to north-west India, Pakistan, and perhaps southern Iran where common in winter; winters regularly in small numbers in United Arab Emirates and Oman. In Pakistan, migrates from north and west to lower elevations in south and east, particularly to Indus valley, as well as to north-west India, though some remain in valleys in north of country. Leaves Turkmenistan late September, returning mid-March; departs mountainous areas of Pakistan mainly September, present in winter quarters in south and in India early August to end of March, and returns to breeding grounds from late February to mid-April.

Wing-length: ♂ ♀ 85–98 mm.

Geographical variation. Marked, but only in distribution of 3 colour morphs (*picata*, *capistrata*, *opistholeuca*). Those breeding nearest to west Palearctic, in Iran, virtually all *picata* morph.

Mourning Wheatear *Oenanthe lugens*

PLATES: pages 1194, 1200, 1202

Du. Rouwtapuit Fr. Traquet deuil Ge. Schwarzrücken-Steinschmätzer It. Monachella lamentosa
Ru. Траурная каменка Sp. Collalba fúnebre Sw. Sorgstenskvätta

Field characters. 14.5 cm; wing-span 26–27.5 cm. Bulkier than Pied Wheatear, with 10% longer bill and legs (but less robust than Finsch's Wheatear). Medium-sized, rock-haunting, strongly pied wheatear, with complex plumage variations and no obvious diagnostic characters except for combination of white wing-panel (in flight) and (usually) pinkish or rusty under tail-coverts. Typical ♂ patterned much as Pied Wheatear except for wing-panel and longer white rump. ♀ like ♂ or much duller. Juvenile essentially sandy-buff, streaked dull brown above and cream-buff below; faintly scaled grey on chest. ♂ has striking black mask on face. In black basaltic shield of Syria and Jordan, black morph occurs but still shows specific pale wing panel.

♂ and ♀ of Middle East races unmistakable, with extent of white on inner webs of flight-feathers much greater than any other similar wheatear and obvious in flight, and buff under tail-coverts usually discernible on ground in close view, so that confusion with Pied Wheatear and Eastern Pied Wheatear (*capistrata* morph with white crown and belly) unlikely to persist. ♂ and ♀ of western race *halophila* far less distinctive, and ♀ particularly puzzling, requiring close study for separation from similar wheatears, though usually with more complete grey-black throat and rosy vent. Flight light and flitting, recalling Black-eared Wheatear. Less prone than Pied Wheatear to perch on vegetation.

Habitat. Breeds along narrow band of lower middle latitudes in warm arid Mediterranean and subtropical desert climates, not generally extending either to sea-coast or beyond desert fringe, but being tied to rocky exposures, clay hill-slopes, and banks of wadis. In Red Sea Province of Egypt, inhabits most desolate wadis and rocky gorges, often in neighbourhood of Hooded Wheatear.

Distribution and population. IRAQ. Possibly breeds in north at least occasionally. ISRAEL. A few thousand pairs in 1980s. JORDAN. Characteristic bird of Rift Margin avifauna. EGYPT. Common. LIBYA. Tripolitania: breeds sparingly; southern limit uncertain. TUNISIA. Fairly common. ALGERIA. Rather local. MOROCCO. Scarce.

Accidental. Turkey, Cyprus, Canary Islands.

Beyond west Palearctic, breeds Iran and Arabia, and in 3 isolated populations in eastern Africa: Ethiopia, Somalia, and Kenya–Tanzania border.

Movements. Partially migratory, with considerable variation between populations in proportion of birds migrating and in distances involved. North African populations (*halophila*), breeding from Morocco east to Libya, are partially migratory with migrants making short-distance movements. Some birds seen as far south as Aïr and Ahaggar massifs in January–February but probably most move less than 50 km. East Mediterranean populations (nominate *lugens*), breeding from eastern Egypt north through Israel to Syria and northern Iraq, apparently less migratory than *halophila*. In Israel, dispersal starts late July, but most marked from September to late February.

Food. Mainly insects; occasionally berries. Usually catches prey in sallies to ground from (preferably) firm perch, e.g. boulder or stone. Sometimes hops along ground in brief dash after prey, or jabs at ground or under stone from stationary position. Also creeps into crevices, and makes darting aerial-pursuits from low perch.

Social pattern and behaviour. Territorial during breeding season; at other times of year occurs in pairs in some areas, but more usually singly, ♂♂ and ♀♀ holding individual territories. In breeding season, ♂ sings either in song-flight or from exposed perch (e.g. rock, bush). In song-flight, ascends from perch in gentle, gliding curve, tail fanned, white underside of flight-feathers contrasting with black under wing-coverts and axillaries, and lands on same or different perch; may rise to *c.* 70 m, hovering with wings and tail outspread before gliding down.

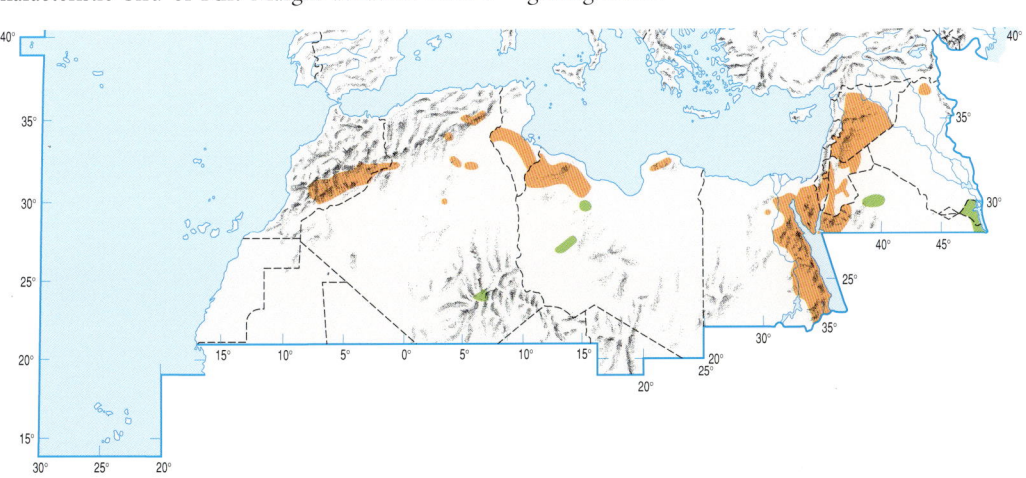

Mourning Wheatear Oenanthe lugens. O. l. halophila: **1** ad ♂ worn (spring), **2** ad ♀ worn (spring), **3** 1st winter ♂, **4** juv ♀. O. l. lugens: **5** ad ♂ worn (spring). O. l. persica: **6** ad ♂ worn (spring). Hooded Wheatear Oenanthe monacha (p. 1194): **7** ad ♂ worn (spring), **8** ad ♀ worn (spring), **9** 1st winter ♂, **10** juv.

Voice. Song varied; a pleasing warble, which may be loud or subdued; phrases, of varying length, often interspersed with pauses or with call-type units. In winter territories, song given by both sexes, with little or no differentiation. Commonest call a repeated 'chack-chack'. Other calls include a low, harsh 'zeeb' and, in alarm, a high-pitched squeak.

Breeding. SEASON. Israel: laying usually begins mid-March, continuing to early June. Egypt: eggs laid February–June. North-west Africa: eggs found early March to late April. 2 broods probably usual. SITE. Deep in crevice in rock, under boulder, in rodent burrow, or in hole in bank. Nest: shallow cup of vegetation, lined with (e.g.) hair, feathers, or wool. Platform of small stones placed in front of, beneath, and around nest, in varying quantity—or none at all where crevice too narrow. EGGS. Sub-elliptical, smooth and glossy; light green-blue, spotted red-brown, often with ring of larger blotches round broad end. Clutch: 4–5 (3–6). INCUBATION. Period not recorded. FLEDGING PERIOD. Young leave nest at 14–15 days, before able to fly.

Wing-length: O. l. lugens: ♂ 94–98, ♀ 89–94 mm. O. l. halophila: ♂ 89–97, ♀ 85–90 mm.
Weight: O. l. halophila: ♂ 22–25, ♀ 19–22 g.

Geographical variation. Marked, especially in ♀♀, as these are either similar to ♂ (nominate lugens, persica) or markedly different (halophila). Birds breeding in basalt deserts of southern Syria and northern Jordan, and formerly believed to be Eastern Pied Wheatear, are a largely black morph of O. lugens lugens which interbreeds with typical birds. Typical ♂ nominate lugens from eastern Egypt and southern part of Middle East closely similar to ♂ halophila, and inseparable in worn plumage (except for flight-feathers); in fresh plumage, nominate lugens on average slightly paler and greyer on central crown, underparts slightly paler cream-white. O. l. persica from southern Iran (some wintering just within west Palearctic) similar to nominate lugens, but larger, crown browner, under tail-coverts darker rufous-cinnamon, tail more extensively black, less extensive white on inner webs of flight-feathers.

Hooded Wheatear Oenanthe monacha

PLATES: pages 1194, 1200, 1202

Du. Monnikstapuit   Fr. Traquet à capuchon   Ge. Kappensteinschmätzer   It. Monachella dal cappuccio
Ru. Каменка-монашка   Sp. Collalba pechinegra   Sw. Munkstenskvätta

Field characters. 17.5 cm; wing-span 29.5–30.5 cm. Noticeably larger than Mourning Wheatear, with 15% longer bill, wings, and tail but 5–10% shorter legs, all contributing to more attenuated form, most obvious in length of rear body and tail. Long-billed, rather large-headed and lengthy wheatear. ♂ has distinctly long white crown, black foreparts, back, and wings, and white underbody, lower back, rump, and tail; tail has only black central line. ♀ has generally pale sandy-brown body and wings, and distinctive fawn-buff rump and tail-sides. Combination of pale underparts and tail pattern diagnostic in both sexes. Juvenile like adult ♀ but spotted cream-buff above and dully scaled or mottled black below.

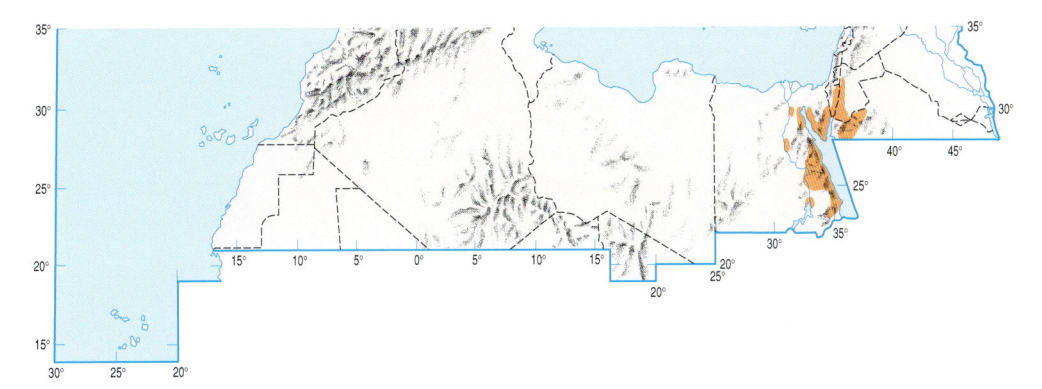

Unmistakable in good view; caution necessary only with suddenly retreating ♂, in which length of rump and tail suggests Hume's Wheatear and tail pattern recalls White-crowned Black Wheatear. As ♀ paler and more uniform than other ♀ wheatears, may suggest other dull passerines. Flight noticeably flowing and floating, sometimes likened to butterfly but actually more like large redstart. Gait hopping; stance low and rather level.

Habitat. In arid hot lower middle and lower latitudes, demanding rocky hills of great barrenness; not found breeding west of the Nile. In Arabia, occurs in most desolate wadis and ravines, avoiding vegetation, and defends large territories against other insectivores owing to low insect populations. Needs rock faces with fissures or cavities as places of refuge if alarmed.

Distribution and population. ISRAEL. 100–200 pairs in 1980s. JORDAN. Thinly distributed. SAUDI ARABIA. Scarce. EGYPT. Fairly common in central Red Sea mountains, and in southern Sinai.

Accidental. Cyprus, Iraq, Kuwait.

Beyond west Palearctic, breeds Arabia, Iran, and southern Pakistan.

Movements. Largely sedentary with a few individuals making short movements.

Food. Probably takes mainly insects and a few seeds. Searches for food from elevated perch (stone, post, etc.), launching flying attacks against prey on ground and (more so than other wheatears) in the air, sometimes pursuing prey to height of 50–100 m. Taking both ground and aerial prey permits feeding throughout the day, as ground prey relatively abundant during cooler periods of morning and evening while flying prey continue to be available during heat of day.

Social pattern and behaviour. Poorly known; apparently similar to other desert-living wheatears.

Voice. Song, given apparently only by ♂, a sweet medley of whistles and thrush-like notes. Calls include a harsh 'zack'.

Breeding. Very little known. SEASON. Israel: eggs laid from May. Jordan: pair seen courting in late April; ♂ with food in bill seen late May. Egypt: young seen early June. SITE. One nest in Iran deep in hole 2.5 m above base of north-facing bank of wadi. Nest: of straw and weeds, lined soft material. EGGS. Pale sky-blue with tiny rust-coloured spots. INCUBATION. 14–15 days. FLEDGING PERIOD. 14–15 days.

Wing-length: ♂ 102–111, ♀ 98–104 mm.
Weight: ♂♀ 18–23 g.

Hume's Wheatear *Oenanthe alboniger*

PLATES: pages 1192, 1200, 1202

DU. Hume's Tapuit   FR. Traquet de Hume   GE. Schwarzkopf-Steinschmätzer   IT. Monachella di Hume
RU. Белочёрная каменка   SP. Collalba de Hume   SW. Svartvit stenskvätta

Field characters. 17 cm; wing-span 29–30.5 cm. Distinctly larger than Eastern Pied Wheatear, with 25% longer and stouter bill, heavier and more rounded head, 10% longer wings, larger feet, and (due to different plumage pattern) apparently longer back and tail. Rather large, long-billed, bull-headed, and long-backed wheatear, with rather erect carriage and totally glossy black and pure white plumage. Roundness of head exaggerated by restriction of black on underparts to chin and throat. Tail pattern a classic ⊥, with white rump and tail-sides extended by white lower back into longer area of white than on any other wheatear. Combination of all-black head and long white rump and tail diagnostic. ♀ duller than ♂, juvenile even more so.

Lack of white crown sufficient to distinguish it from several other similar wheatears except nominate *picata* morph of Eastern Pied Wheatear. Flight free and floating, but rather less active than Eastern Pied Wheatear, sitting on perches for longer.

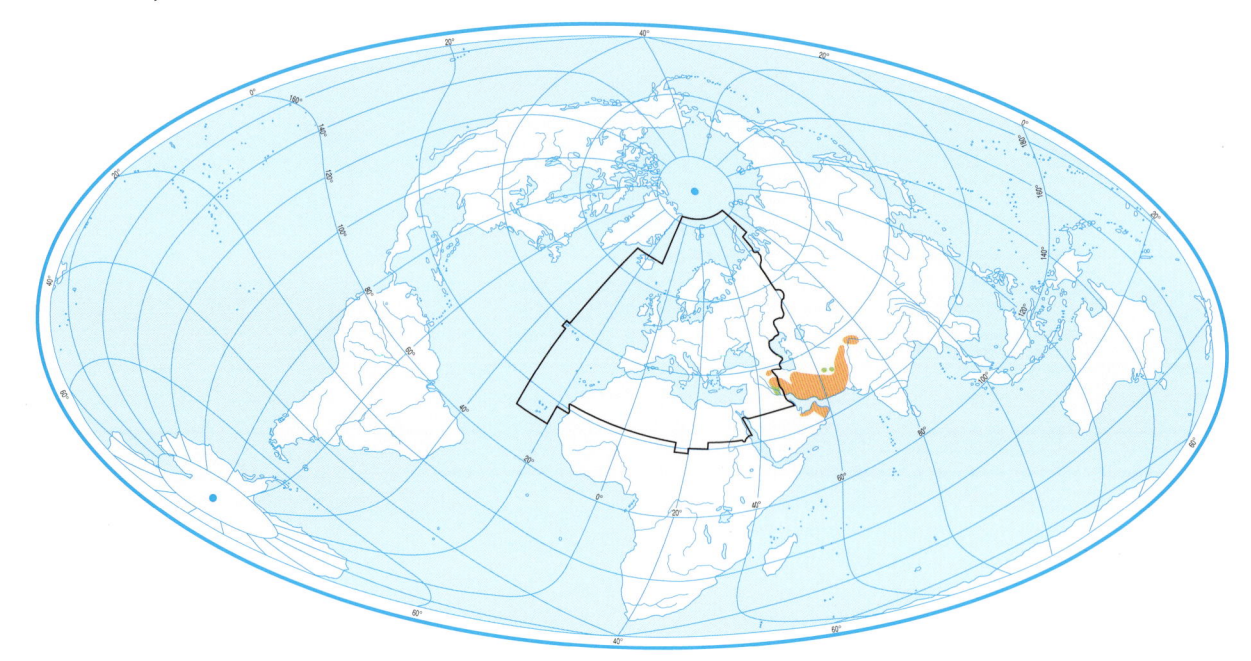

Habitat. Breeds commonly where hills outcrop from plains, and in valleys and ravines in warm steppe zone, from sea-level to c. 1900 m in steppe, sub-steppe, and deforested zones of lower middle to lower continental latitudes near fringe of west Palearctic. Requires mainly bare ground for feeding, usually gently sloping or flat, although high up on steep rocky hillsides less steep patches sometimes acceptable. In all cases, steep, rocky hillsides or broken cliffs essential, serving as song-posts, look-outs for prey, sources of shade, perches for loafing, backgrounds of cryptic value, refuges from predators, roosts, and nest-sites.

Distribution and population. IRAQ. In 1950s, not uncommon in Khanaqin area in east.
Accidental. Kuwait.

Movements. Information extremely limited; appears to be basically resident with (sometimes at least) some local movement outside breeding season.

Food. Mainly insects; some seeds. Taken mainly from open ground, usually by dash-and-jab technique or by aerial-pursuit, launching attack on prey from low perch.

Social pattern and behaviour. Little known. Territories defended throughout the year against all intruding wheatears and other similar birds; interspecific territorial boundaries noted with Finsch's Wheatear and Red-tailed Wheatear. Fidelity to territory recorded in successive winters.

Voice. Song loud and cheerful, lacking discordant notes of most wheatears: 'chew-de-dew-twit' on rising scale; 'chi-roochirichirrichiri', etc. Calls include short, sharp whistle given 3–4 times; in alarm, uses harsh, grating monosyllable and quiet 'chit-tit-tit'.

Breeding. SEASON. Iran: eggs laid April–May; in south-west, fledglings seen from second half of April to second half of May. SITE. In hole or crevice in rock face, sometimes high up in cliff, or in wall of old building. Nest: shallow cup of twigs and plant stems, plastered with mud containing limestone chips; poorly lined with grass, hair, and feathers; platform of small stones in front of nest, shelving towards entrance of hole. EGGS. Sub-elliptical, smooth and glossy; very pale blue-white, unmarked or with a few speckles of pale red. Clutch: 4–5. (Incubation and fledging periods not recorded.)

Wing-length: ♂ 102–111, ♀ 97–104 mm.
Weight: ♂ mostly 23–28, ♀ 22–27 g.

White-crowned Black Wheatear *Oenanthe leucopyga*

PLATES: pages 1197, 1200, 1202

DU. Witkruintapuit   FR. Traquet à tête blanche   GE. Saharasteinschmätzer   IT. Monachella nera testabianca
RU. Белогузая каменка   SP. Collalba negra de Brehm   SW. Vitkronad stenskvätta

Field characters. 17 cm; wing-span 26.5–32 cm. Somewhat less plump than Black Wheatear, with slightly longer, more pointed wings. Rather large, oval-headed and oval-bodied, glossy black wheatear, with bold white rump and tail often showing only black central line. Adults of both sexes have white crown; duller juvenile and 1st-year have all-black head.

White-crowned Black Wheatear *Oenanthe leucopyga*: **1** ad ♂ worn (spring), **2** ad ♀ worn (spring), **3** 1st winter ♂, **4** 1st ad ♂ moulting into ad (1st summer), **5** juv. Black Wheatear *Oenanthe leucura leucura* (p. 1198): **6** ad ♂ worn (spring), **7** ad ♀ worn (spring), **8** 1st winter ♂, **9** juv.

Differs distinctly in underwing (all-black) and tail pattern (black restricted almost wholly to centre) from Black Wheatear. Flight free and floating, even recalling Swallow in its easy tumbles and sweeps along hillsides and over adjacent desert. Gait and behaviour much as Black Wheatear.

Habitat. Across Afro-Arabian lower middle latitudes, Mediterranean to subtropical and tropical. A true Saharan species characteristic of desert with less than 100 mm annual precipitation. Frequents the most impoverished localities, at all altitudes up to 3000 m in Ahaggar (Algeria) and Tibesti (Chad), especially rocky and sometimes earthen banks of wadis, but also oases. Sole passerine seen in 160 000 km^2 of low desert east of Aïr (Niger) and sole resident passerine at Egyptian desert oases; at home in desert wherever there is broken ground, and frequents houses in oases and cemeteries on their edges.

Distribution and population. SAUDI ARABIA. Common in Harrat Al Harrah reserve. ISRAEL. At least a few thousand pairs in 1980s. JORDAN. Common resident of arid and rocky Rift Margins. EGYPT. Common. LIBYA. Scarce on Jefara and Jebel Nafusa in Tripolitania, common in more arid areas further south, and in Fezzan. Status in Libyan desert uncertain. TUNISIA. Uncommon. ALGERIA. Found everywhere except the ergs proper. MOROCCO. Common. MAURITANIA. Locally common. MALI. Uncommon, localized. NIGER. Common and widespread in Aïr. CHAD. Common south of Libyan border.

Accidental. Britain, Spain, Malta, Greece, Turkey, Cyprus, Kuwait.

Extends locally south of west Palearctic in Africa and Arabia.

Movements. Largely sedentary throughout range, though some individuals or populations may make short-distance movements in winter.

Food. Mainly insects, but diet notably diverse, including plant material and small reptiles. Catches prey in flight, on ground, or in bushes. Typically perches on low vantage points (shrub, stone, etc.) and drops down or sallies forth, up to 10(–50) m away, to take prey from ground in manner of a shrike; at moment of capture, frequently spreads wings, sometimes repeatedly, perhaps to confuse and entrap prey. Also sallies, sometimes steeply upwards, to catch aerial prey, or hops through foliage, gleaning insects or plucking berries.

Social pattern and behaviour. Not markedly gregarious, but, like Black Wheatear, young may remain with parents on parental territory during winter, members of family party feeding quite close together. Pair-bond life-long (lasting until one partner dies or is evicted by rival); maintained all year. In A-Rabba plain (Sinai), spacing of territories determined by dispersion of human dwellings, every cluster of houses forming nucleus for a territory; topographical features (e.g. undulations, wadis) usually form natural territorial boundaries. ♂ usually sings from top of a bush, cliff, etc., accompanied by spasmodic movements causing forebody to rock. Also sings while flying normally between perches, and during song-flight: bird launches itself from perch where it has been singing, and glides to another perch on stiff, sometimes quivering wings, with tail spread.

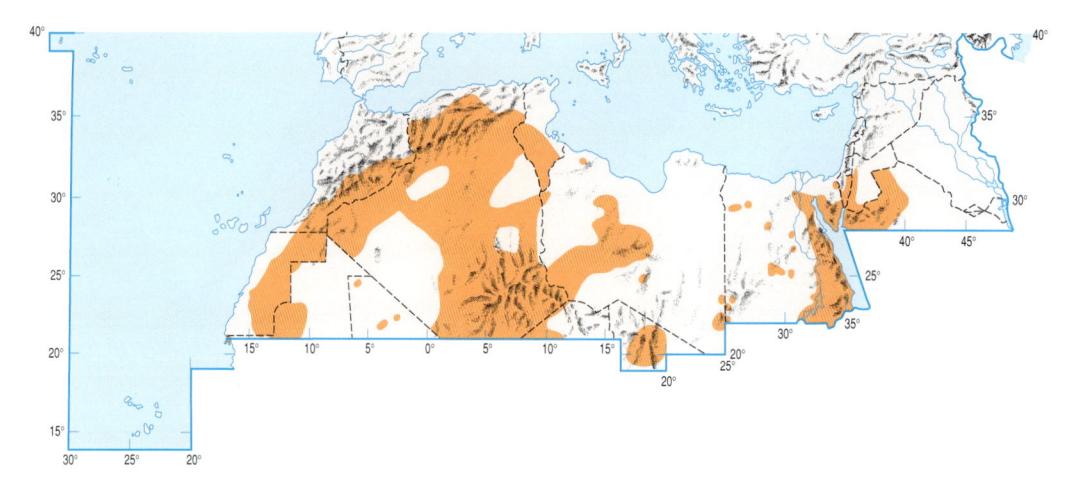

Voice. Song of ♂ very varied; loud, combining musical warbles with more discordant chanted notes like thrush, commonly also mimicry of other species including mammals. Commonest call a short, harsh, rather grating 'dzik', given by both sexes; in alarm, a repeated, strong, far-carrying 'hwee-weet', also scolding or rattling calls.

Breeding. SEASON. North-west African Sahara: eggs found mid-February to late May. Probably similar in Egypt and Sinai. Sometimes 2 broods. SITE. In hole (up to 25 cm deep) in rocks, under stones, in bank, or occasionally in wall of building. Nest: cup of dry grass, lined with wool and feathers, sometimes with base of twigs or bits of wood; approach often paved with platform of pebbles. EGGS. Sub-elliptical, smooth and glossy; very pale blue to whitish-blue, sparsely spotted red-brown at broad end. Clutch: 3–5 (2–6). INCUBATION. *c.* 2 weeks. FLEDGING PERIOD. *c.* 2 weeks.

Wing-length: *O. l. leucopyga*: ♂ 102–110, ♀ 98–101 mm.
Weight: *O. l. leucopyga*: ♂♀ 24–39 g.

Geographical variation. Slight; involves size, and depth of gloss on body. *O. l. ernesti* from east of Nile valley a little larger than nominate *leucopyga* from west, and plumage gloss stronger and more bluish, but distinctions not clear-cut.

Black Wheatear *Oenanthe leucura*

PLATES: pages 1197, 1200, 1202

Du. Zwarte Tapuit   Fr. Traquet rieur   Ge. Trauersteinschmätzer   It. Monachella nera
Ru. Белохвостая каменка   Sp. Collalba negra   Sw. Svart stenskvätta

Field characters. 18 cm; wing-span 26–29 cm. Somewhat rounder-headed and bulkier than White-crowned Black Wheatear but no larger, with wings slightly shorter and rounder. Rather large, big-headed, and deep-chested black wheatear with broad white rump and black ⊥ on white tail. ♀ duller than ♂; juvenile even more so.

Virtually free of confusion in Iberia, but easily confused at distance in north-west Africa with wholly black-headed immature White-crowned Black Wheatear. Separation best based on tail pattern which in White-crowned Black Wheatear lacks prominent terminal black band. Flight thrush-like: exceptionally buoyant, with fluent beats of rather broad, round wings; action floating in level flight, but also makes bold tumbles down slopes and apparently effortless ascents up slopes and crags. Carriage usually half-erect, but ♂ assumes upright stance on look-out perch. Behaviour includes exaggerated spreading of wings and tail.

Habitat. In contrast to other west Palearctic wheatears, confined to west Mediterranean lower middle latitudes, largely under coastal and even oceanic influences rather than arid or continental, but in warm band of July isotherms 24–32°C. In Spain, essentials for habitat are general aridity, denuded soil, and presence of a rock-wall or equivalent. Subject to these, occurs in variety of situations from Rock of Gibraltar and craggy sea-cliffs with boulders, to inland foothills and high sierras—in gorges and rocky or boulder-strewn places from sea-level to *c.* 2000 m. In northern Morocco, typical of Hauts Plateaux and ascends Atlas mountains from base to *c.* 3000 m, avoiding trees as much as true desert. Also avoids flat terrain, including wetlands, and infrequently in contact with man over most of range. Represents extreme manifestation of attachment of wheatears to rocky and perpendicular elements in habitat, and hints at ecological convergence with rock thrushes.

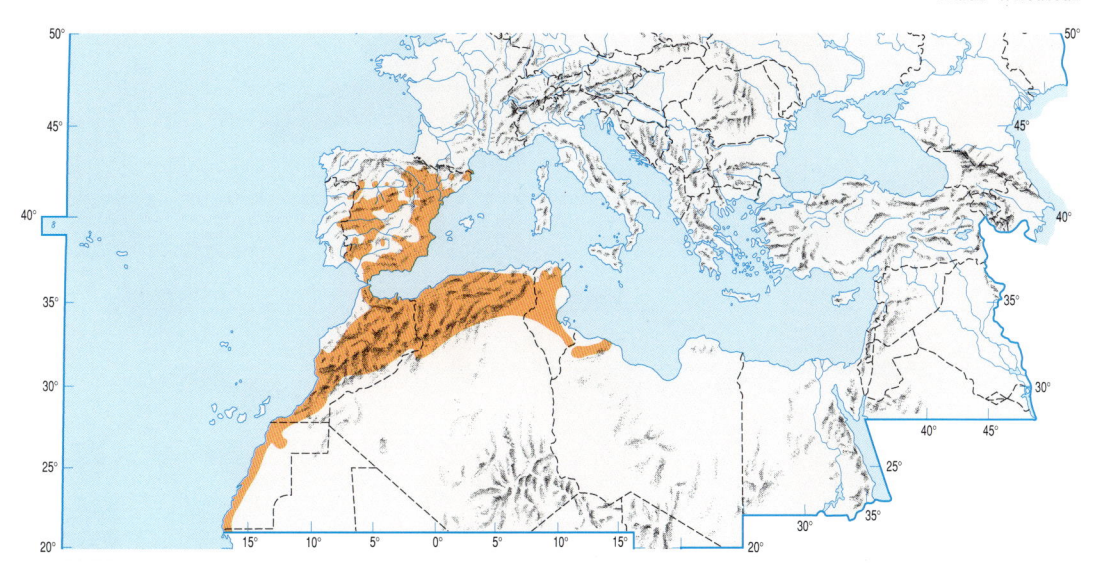

Distribution and population. Breeds only in west Palearctic. Range and numbers have declined in Europe. Vulnerable to predation (snakes, rats, and foxes) and to loss of preferred habitat such as ruined buildings and man-made caves. FRANCE. Now only 1–2 pairs. Probably bred Gard and Alpes Maritimes in 19th century, and until at least 1938 in Var. SPAIN. 4000–15 000 pairs. PORTUGAL. 100–1000 pairs 1978–84. ITALY. Perhaps formerly occasional breeder Liguria, Tuscany; listed as uncommon breeder Sicily in 1870s, but no details. LIBYA. Common in Jebel Nafusa, scarce elsewhere. TUNISIA. Locally common. MOROCCO. Uncommon to locally abundant. WESTERN SAHARA. Presence south to border with Mauritania recently confirmed. MAURITANIA. A few sightings; status uncertain.

Accidental. Britain, Norway, Austria, Balearic Islands, Malta, Greece, Yugoslavia, Bulgaria, Israel. No certain records from Egypt.

Movements. Generally sedentary, although some individuals disperse after breeding, and partial or total altitudinal migration occurs in some mountain regions, e.g. Atlas (Morocco) and Sierra Nevada (Spain).

In view of sedentary nature, has occurred as a vagrant over a surprisingly wide area, north to Shetland (Scotland) and Norway and east to Bulgaria and Israel. Origin of stragglers unknown; timing of records suggests that many may be dispersing juveniles.

Food. Mainly insects. Prey usually caught on ground by 'hop and search' technique; occasionally in flight. May also fly from perch (e.g. rock, bush) to catch prey on ground. Will search around large rocks or probe cracks and holes for prey, and scratch for food under bushes or other vegetation. Berries and other small fruits regularly taken in autumn and winter.

Social pattern and behaviour. Not markedly gregarious, but more so than most other wheatears; in south-east Spain, loose groups of up to 6 birds commonly feed together at any time of year. Pair-bond probably life-long and maintained throughout the year, though loosely outside breeding season. ♂ advertises with territorial-song, delivered in song-flight resembling that of pipit. Almost always starts singing from exposed perch, usually rock, bush, or tree, then often ascends with fluttering wing-beats, spreads wings, fans tail, and sings while gliding to another perch. After landing, ♂ usually performs dancing-display; lowers breast, raises tail slowly, simultaneously fanning it, and begins to ruffle belly feathers. All body plumage except crown is ruffled. Steeply raised, widely fanned tail is shivered up and down, while wing-tips, which lie over base of tail, are flicked. Display accompanied by courtship-song, also by dancing (tripping) on the spot or in a small arc.

Voice. Song not loud, sounding distant even when delivered by close bird: begins with chortle, then brief mellow and melodious warble, then chortle, then further warble, and usually ending in chortling chatter, more sibilant than earlier phrases. Commonest call tri-syllabic, a quiet but penetrating 'pee-pee-pee'. When disturbed, 'chack' sounds, typical of other wheatears.

Breeding. SEASON. Eastern Pyrénées (France): eggs laid from mid-April. Southern Spain: eggs laid from mid-March. Northeast Spain: latest eggs 1 July. Algeria and Tunisia: earliest eggs late February, most March or April to June. One brood, occasionally 2. SITE. Hole in rock wall, cliff, cave, or man-made wall. Nest: cup of dead grass and rootlets, incorporating feathers and wool; scant lining, or lining integral with main structure. Normally builds platform of small stones at sides of nest; can be 10–15 cm wide or even more, and incorporating several hundred stones, but in such cases undoubtedly built up over several years. EGGS. Sub-elliptical, smooth and glossy; very pale blue to bluish-white, variably speckled red-brown, markings usually concentrated at broad end. Clutch: 3–5(–6). INCUBATION. 14–18 days. FLEDGING PERIOD. 14–15 days.

Wing-length: *O. l. syenitica*: ♂ 96–105, ♀ 92–97 mm.

1
2
3
4
5
6
7
8
9
10
11
12
13
14
15
16
17
18
19
20
21
22
23
24
25
26
27
28
29
30
31
32

Geographical variation. Slight; mainly involves colour of body, perhaps slightly size, and amount of black on tail-tip. Adult ♂ nominate *leucura* from southern Europe distinctly blacker than adult ♂ *syenitica* from North Africa, without its brown tinge, sometimes slightly glossy above. ♀ nominate *leucura* darker than ♀ *syenitica*, above and below.

Rock Thrush *Monticola saxatilis*

PLATES: pages 1204, 1218

Du. Rode Rotslijster   Fr. Monticole de roche   Ge. Steinrötel   It. Codirossone
Ru. Пестрый каменный дрозд   Sp. Roquero rojo   Sw. Stentrast

Field characters. 18.5 cm; wing-span 33–37 cm. 10% smaller than Redwing and Blue Rock Thrush. Strong-billed, rather long-bodied, rock-haunting bird, with shape suggesting large short-tailed chat rather than thrush. ♂ essentially blue, black, and white above, with rufous body and tail; ♀ basically buff, pale-spotted above and barred below except on rufous tail. In winter and 1st autumn, plumage copiously scaled with pale tips which obscure above patterns. Juvenile like ♀ but paler due to even more copious mottling.

Almost unmistakable among wild birds of west Palearctic (♀ and immature always pale-throated, unlike Blue Rock Thrush), but some risk of confusion with escaped Asian and African congeners of similar appearance. Flight usually low and floating, with full beats of wings as slow as *Turdus* thrushes but with short tail giving appearance of large chat; capable of considerable acceleration when escaping. Stance strongly recalls large wheatear, with usually upright attitude; characteristically wags tail, with obvious upward jerk and loose flirt. Gait essentially hopping but long-paced, allowing even progress on flat ground and nimbleness over rocks.

Habitat. Breeds in west Palearctic in lower middle latitudes in continental warm temperate, steppe and Mediterranean montane zones, on sunny, dry often stony hollows or terraces, preferably dotted with stunted trees or shrubs serving as perches. In southern Switzerland, also on rocky heaths and in vineyards from 500 m, but mainly at 1500–2700 m. Forages over some distance from nest, down to hayfields and farmland, using rocks, walls, roofs of buildings, and bare branches or treetops as hunting look-outs. In Spain, nests on barren hillsides with boulders and crags, chiefly at *c.* 1250–2300 m. In Germany, as elsewhere towards north of European range, many sites occupied last century (e.g. ruined castles on the Rhine, heaps of debris) were deserted before or soon after its end. In winter in tropical west Africa, lives in savanna and in erosion areas with scattered low bushes and stony gullies, or better-wooded land, even gardens.

Distribution. Long-term decline in range (especially southward contraction) and numbers; causes unclear, but probably include loss of habitat (breeding and perhaps winter) and possibly climatic change. Recent range decrease reported from Slovakia, Hungary, Iberia, Ukraine, and Moldova. FRANCE. Range retraction since 19th century; apparently now stabilized. BELGIUM. Bred 19th century. LUXEMBOURG. Presumed to have bred in 19th century, e.g. 1865, but no proof; no sightings since 1900. GERMANY. Bred in 19th century north to Niedersachsen; extinct since. CZECH REPUBLIC. Extinct since 1979. AZERBAIJAN. Distribution poorly known. SYRIA. Breeding suspected above Bloudan (Anti-Lebanon).

Accidental. Britain, Ireland, Belgium, Luxembourg, Netherlands, Germany, Denmark, Norway, Sweden, Finland, Estonia, Madeira, Canary Islands.

Beyond west Palearctic, extends east through central Asia to Lake Baykal and north-east China, and south to Iran.

Population. Most countries report decreases, marked in Slovakia, Hungary, and Ukraine. Stable Switzerland, Greece, Rumania, and perhaps Slovenia and Bulgaria, with apparent increase Croatia; trend uncertain Albania and Turkey. FRANCE. Probably over 1000 pairs. POLAND. 5–10 pairs 1985–91. SLOVAKIA. 15–30 pairs 1973–94. HUNGARY. 20–50 pairs. AUSTRIA. Scarce and very local. SWITZERLAND. 500–700 pairs 1985–93. SPAIN. 3500–4800 pairs. PORTUGAL. 100–1000 pairs 1978–84.

(FACING PAGE) Blackstart *Cercomela melanura melanura* (p. 1166): **1–2** ad ♂ worn (spring).
Ant Chat *Myrmecocichla aethiops* (p. 1174): **3–4** ad ♂ worn (spring).
Isabelline Wheatear *Oenanthe isabellina* (p. 1175): **5–6** ad ♂ worn (spring).
Wheatear *Oenanthe oenanthe oenanthe* (p. 1178): **7–8** ad ♂ worn (spring).
Pied Wheatear *Oenanthe pleschanka pleschanka* (p. 1180): **9–10** ad ♂ worn (spring).
Black-eared Wheatear *Oenanthe hispanica hispanica* (p. 1183): **11–12** ad ♂ pale-throated morph worn (spring).
Desert Wheatear *Oenanthe deserti homochroa* (p. 1185): **13–14** ad ♂ worn (spring).
Finsch's Wheatear *Oenanthe finschii* (p. 1187): **15–16** ad ♂ worn (spring).
Red-rumped Wheatear *Oenanthe moesta* (p. 1188). **17–18** ad ♂ worn (spring).
Red-tailed Wheatear *Oenanthe xanthoprymna chrysopygia* (p. 1190): **19–20** ad ♂ worn (spring).
Eastern Pied Wheatear *Oenanthe picata opistholeuca* (p. 1191): **21–22** ad ♂ worn (spring).
Mourning Wheatear *Oenanthe lugens halophila* (p. 1193): **23–24** ad ♂ worn (spring).
Hooded Wheatear *Oenanthe monacha* (p. 1194): **25–26** ad ♂ worn (spring).
Hume's Wheatear *Oenanthe alboniger* (p. 1195): **27–28** ad ♂ worn (spring).
White-crowned Black Wheatear *Oenanthe leucopyga* (p. 1196): **29–30** ad ♂ worn (spring).
Black Wheatear *Oenanthe leucura leucura* (p. 1198): **31–32** ad ♂ worn (spring).

1
2
3
4
5
6
7
8
9
10
11
12
13
14
15
16
17
18
19
20
21
22
23
24
25
26
27
28
29
30
31
32

ITALY. 5000–10 000 pairs 1983–95. GREECE. 10 000–20 000 pairs. ALBANIA. 2000–5000 pairs in 1981. YUGOSLAVIA: CROATIA. 3000–5000 pairs. SLOVENIA. 200–500 pairs. BULGARIA. 1000–5000 pairs. RUMANIA. 250–400 pairs 1986–92. UKRAINE. 20–50 pairs in 1986. MOLDOVA. 400–600 pairs in 1989.

(FACING PAGE) Blackstart *Cercomela melanura melanura* (p. 1166): **1–2** ad ♀ worn (spring).
Ant Chat *Myrmecocichla aethiops* (p. 1174): **3–4** ad ♀ worn (spring).
Isabelline Wheatear *Oenanthe isabellina* (p. 1175): **5–6** ad ♀ worn (spring).
Wheatear *Oenanthe oenanthe oenanthe* (p. 1178): **7–8** ad ♀ worn (spring).
Pied Wheatear *Oenanthe pleschanka pleschanka* (p. 1180): **9–10** ad ♀ worn (spring).
Black-eared Wheatear *Oenanthe hispanica hispanica* (p. 1183): **11–12** ad ♀ worn (spring).
Desert Wheatear *Oenanthe deserti homochroa* (p. 1185): **13–14** ad ♀ worn (spring).
Finsch's Wheatear *Oenanthe finschii* (p. 1187): **15–16** ad ♀ worn (spring).
Red-rumped Wheatear *Oenanthe moesta* (p. 1188). **17–18** ad ♀ worn (spring).
Red-tailed Wheatear *Oenanthe xanthoprymna chrysopygia* (p. 1190): **19–20** ad ♀ heavily worn (summer).
Eastern Pied Wheatear *Oenanthe picata opistholeuca* (p. 1191): **21–22** ad ♀ worn (spring).
Mourning Wheatear *Oenanthe lugens halophila* (p. 1193): **23–24** ad ♀ worn (spring).
Hooded Wheatear *Oenanthe monacha* (p. 1194): **25–26** ad ♀ worn (spring).
Hume's Wheatear *Oenanthe alboniger* (p. 1195): **27–28** ad ♀ worn (spring).
White-crowned Black Wheatear *Oenanthe leucopyga* (p. 1196): **29–30** ad ♀ worn (spring).
Black Wheatear *Oenanthe leucura leucura* (p. 1198): **31–32** ad ♀ worn (spring).

AZERBAIJAN. Uncommon. TURKEY. 5000–50 000 pairs. ISRAEL. 7–12 pairs in 1980s. MOROCCO. Scarce; probably some decline.

Movements. Migratory. Most winter in Afrotropics, birds from eastern China travelling at least 7500 km from breeding to wintering grounds. A few birds appear to winter in Africa north of Sahara and in Arabian peninsula. Nocturnal migrant, usually travelling singly or in loose aggregations, often with Blue Rock Thrush. Main wintering area lies north and east of central African rain forests: from northern Nigeria and Cameroon (south to *c.* 8°30′N) east to Eritrea and from there south to at least 9°S in Tanzania.

Mediterranean populations of southern Europe and north-west Africa begin to disperse from breeding sites in August, most having left by late September. Appears to cross Sahara on broad front from Morocco to Sinai, but especially common in central section. Reaches Chad mid-October, Nigeria late November. Occasional November–January records in Morocco, Ahaggar massif, Libya, and Egypt may indicate wintering north of Sahel zone by very small number. Most sites south of Sahara vacated by mid-March with stragglers remaining until at least mid-April. Passage noted in Sahara and on North African coast March–May with peak in late March and early April. However, first arrivals at southern breeding sites are usually in February, demonstrating that early passage in Africa overlooked. Northernmost European breeding sites usually reached April.

Food. Mostly large insects (especially beetles, Lepidoptera larvae, and Orthoptera); also a variety of berries. Feeds mainly by flying from perch (rock, tree, etc.) on to prey on ground; may eat several items while on ground, sometimes running or hopping a few metres between each before returning to perch.

Social pattern and behaviour. Usually solitary. Small loose-knit flocks, notably of young birds, occur on spring migration.

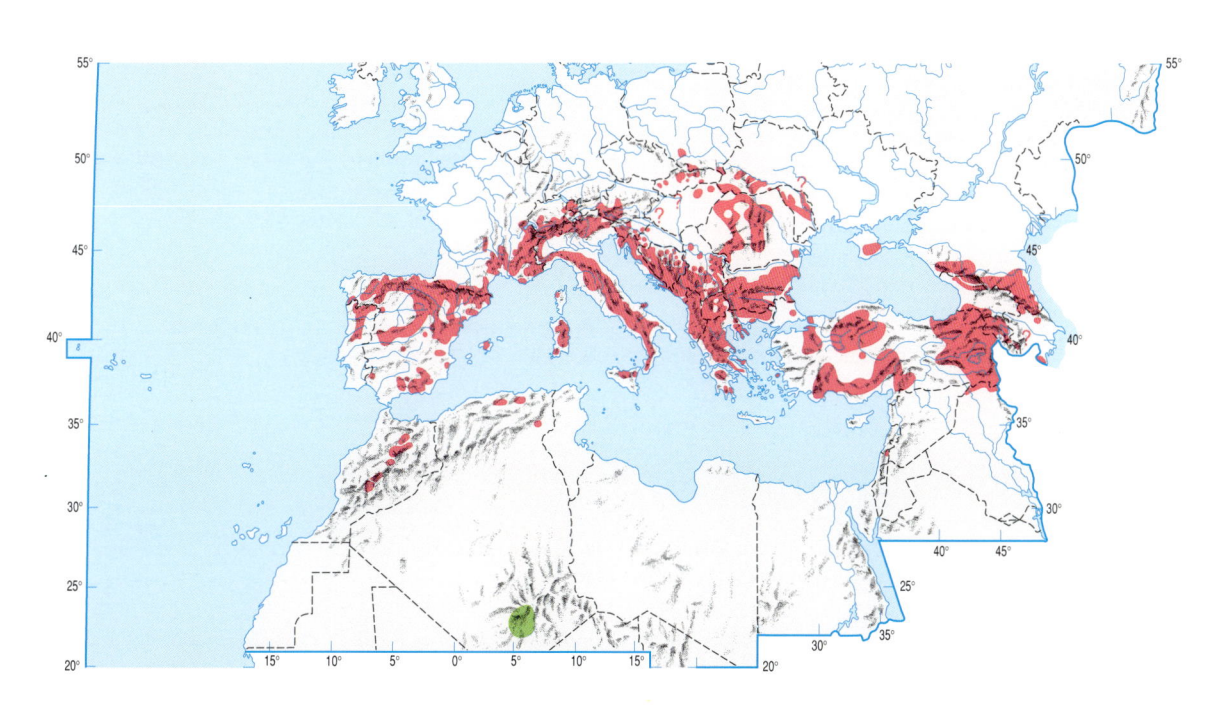

Rock Thrush *Monticola saxatilis*: **1** ad ♂ breeding, **2** ad ♀ breeding, **3** ad ♂ non-breeding, **4** 1st winter ♂, **5** juv. Blue Rock Thrush *Monticola solitarius*. *M. s. solitarius*: **6** ad ♂ worn (spring), **7** ad ♀ worn (spring), **8** 1st winter ♂, **9** juv. *M. s. longirostris*: **10** 1st winter ♂.

In winter, Tanzania, ♂ and ♀ probably defend separate territories. No evidence for other than monogamous mating system. Pair-bond presumably breaks down outside breeding season, but may be renewed for several years on breeding grounds. Song-display mainly by ♂, much less often ♀, given perched, in song-flight, or normal flight. Song from perch often a response to intrusion by nearby ♂, leading to song-duel. Song-flight begins from perch. ♂ takes off suddenly, initially staying low, then ascends steeply with slow powerful wing-beats. Bird begins singing during ascent, reaching maximum output at top of ascent where bird soars, flutters rapidly, and typically introduces mimicry into song; then suddenly plummets, not singing, for 15–20 m with wings and tail outspread. Usually, bird does not land after descent but, apparently using momentum of plummet, ascends for a 2nd song-flight, less high, however, than 1st. Depending on intensity, bird repeats song-flight 2–3 times or more. Song-flight always ends on perch, with low variant of song.

Voice. Song, given by both sexes but mainly by ♂, similar to Blue Rock Thrush, but softer and more flowing. Comprises melodious fluting phrases, often with obvious mimicry. Many species mimicked, Chaffinch song most regularly. Contact-alarm call single or short series of 'tak' sounds, often accompanied by tail-flicking. Warning- and alarm-calls include a plaintive mournful pipe, not unlike Bullfinch, and in greater alarm, an emphatic rapidly repeated 'schack-schack'.

Breeding. SEASON. Earliest eggs late April, main season May–June, apparently throughout range. 1–2 broods. SITE. Horizontal crevice in rock-face, wall, ruin, or crag, under boulder on steeply sloping ground, or occasionally in tree-hole. Nest: neat cup of grass, rootlets, and moss, lined with finer rootlets and moss. EGGS. Sub-elliptical, smooth and glossy; pale blue, often unmarked, or with some faint speckles of red-brown at broad end. Clutch: 4–5(–6). INCUBATION. 14–15 days. FLEDGING PERIOD. 14–16 days.

Wing-length: ♂ 118–131, ♀ 115–126 mm.
Weight: ♂♀ mostly 43–63 g.

Blue Rock Thrush *Monticola solitarius*

PLATES: pages 1204, 1218

DU. Blauwe Rotslijster FR. Monticole bleu GE. Blaumerle IT. Passero solitario
RU. Синий каменный дрозд SP. Roquero solitario SW. Blåtrast

Field characters. 20 cm; wing-span 33–37 cm. 20 % shorter than Blackbird; 10 % longer than Rock Thrush. Shape, stance, and habits much as Rock Thrush, but noticeably bulkier with longer tail. Plumage dark and relatively uniform at all ages and seasons: ♂ dusky blue, ♀ mottled dusky, juvenile mottled dark brown.

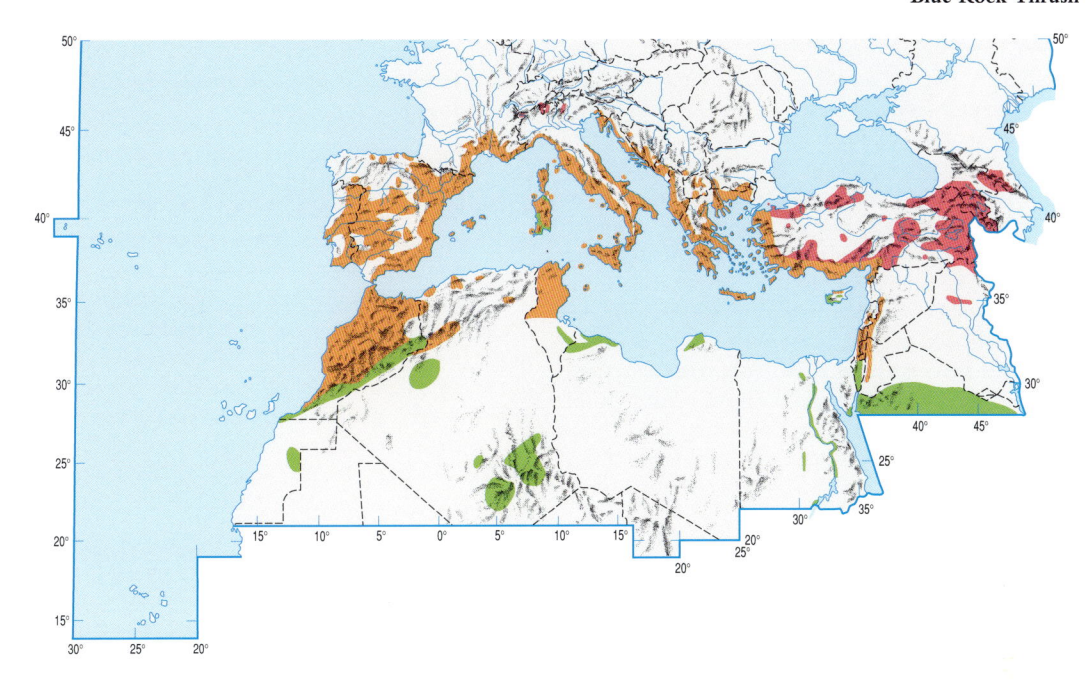

Almost unmistakable (♀ and immature always dark-throated, unlike Rock Thrush), but could be confused with Blackbird (or even Starling) in brief glimpse. More serious risk of confusion is with escaped captives of own Far Eastern races and escaped Himalayan whistling thrushes *Myiophoneus*. Flight, gait, and behaviour much as Rock Thrush but appearance more typically thrush-like, with longer tail obvious both in flight and on ground.

Habitat. Breeds in west Palearctic in middle and lower middle latitudes in warm dry temperate, Mediterranean and steppe climatic zones, montane and coastal, rocky and nearly always in part precipitous. In western Europe, frequents 3 main habitat types—sea-cliffs or other rocky coastlines, mountain valleys and faces (usually below 800 m), and major structures (e.g. castles and ruins) in cities or other settlements. In Switzerland, after recent abandonment of many former breeding sites, now confined almost entirely to quarries of suitable aspect with high broken faces and adequate height and breadth. In winter, appears widely in habitats rarely used for breeding, ranging from olive orchards in Spain to urban areas in Morocco.

Distribution. FRANCE. Bred further north formerly. POLAND. 1947 breeding record rejected. AUSTRIA. No indications of breeding since 1930. SWITZERLAND. Bred further north in 19th century. RUMANIA. Presumed to be irregular or only occasional breeder; no confirmed records. AZERBAIJAN. Common in Nakhichevan region; otherwise rare. TURKEY. Main stronghold in south and south-west. SYRIA. Breeding proved Kassab area and spurs of Anti-Lebanon in Wadi el-Karir; may breed elsewhere in mountain range. JORDAN. Nests locally in Rift Margin highlands and wadis. EGYPT. Perhaps breeds occasionally. ALGERIA. Breeding and wintering probably more widespread than shown.

Accidental. Britain, Sweden, Slovakia, Ukraine, Canary Islands.

Beyond west Palearctic, breeds from Iran and Kazakhstan east through Himalayas to China (widespread) and Japan, also in Philippines and Malay peninsula.

Population. Numbers and trends difficult to judge, but probably some decrease overall. Vulnerable to coastal development and (in Spain) to construction of reservoirs. In Malta, nestlings often taken for rearing in captivity. FRANCE. 1000–10 000 pairs. SWITZERLAND. 20–25 pairs 1985–93; stable. SPAIN. 12 500–16 800 pairs. PORTUGAL. 1000–10 000 pairs 1978–84. ITALY. 10 000–20 000 pairs 1983–95. MALTA. 300–500 birds; decrease in last 10 years. GREECE. 5000–20 000 pairs. ALBANIA. 1000–2000 pairs in 1981. YUGOSLAVIA: CROATIA. 3000–5000 pairs. SLOVENIA. 10–30 pairs. BULGARIA. 50–150 pairs. TURKEY. Locally common. 5000–50 000 pairs. CYPRUS. Probably 100–200 pairs; apparently stable. ISRAEL. At least a few thousand pairs in 1980s. TUNISIA. Common. ALGERIA. Locally common in north. MOROCCO. Locally common.

Movements. Partially migratory in west Palearctic (extralimitally in eastern Asia primarily migratory). Vertical displacements common. Main wintering areas of migrants lie in North Africa and Arabia. Migrates singly or in loose aggregations, primarily at night, often with Rock Thrush. Timing of movements poorly known; passage rarely observed, even in Mediterranean basin, and difficult to distinguish between long-distance migration and local dispersal. Vagrants occur north of breeding range as far as Sweden. Most records in August or early September, suggesting reverse migration.

Food. Mainly invertebrates, also lizards and plant material. Feeds on ground, by pouncing on prey from perch, and by making short chases after flying prey. Plant material mainly fruits, taken largely in autumn and winter.

Social pattern and behaviour. Very shy throughout the year. Mostly solitary outside breeding season, rarely in small groups. Monogamous mating system. Pair-bond breaks down outside breeding season; not known if renewed in subsequent breeding seasons. Breeding territories may have considerable overlap between neighbours where density high. Song-display mainly by ♂, but ♀ also sings. ♀ sings from perch, ♂ either perched, in song-flight, or normal flight. During flight, song may be given continuously or with pauses. Mostly starts from a rock, dead tree (etc.), often beginning with upward glide, bird spreading wings and tail like bee-eater for lift; less often uses beating ascent or downward glide. Following this, bird usually loses height by downward glide, almost vertical drop, or angled drop like pipit. At end of song-flight, bird glides upwards to a perch-site or brakes hard and lands softly. After landing, bird hops or runs a short way or stays still, then often makes wing movements similar to flicking of Pied Flycatcher or singing Starling. Seen from behind, wings appear to beat slowly and asymmetrically.

Voice. Perched song of ♂ deliberate, loud and melodious, recalling Blackbird or Mistle Thrush, but phrases simple, short, and repetitive; may include mimicry. Aerial song may be louder and phrases markedly longer. Song of ♀, apart from being given less often, seems not to differ consistently from ♂ song. Other calls include an abrupt hard 'tchuc-tchuc' like Blackbird, a very high-pitched plaintive 'peep' given repeatedly in presence of intruder, e.g. man near nest, and a disyllabic 'uit' rather like Nuthatch.

Breeding. SEASON. Iberia: egg-laying from mid-April. Northwest Africa: eggs found late April to late May. Malta: laying begins late March. 2 broods. SITE. In hole or crevice in cliff, under overhanging rock, in cave or quarry, in wall of old building, occasionally in horizontal drainage pipe or hole in tree. Nest: rather bulky but loosely built shallow cup of coarse dry grass, moss, and some roots, lined with softer and finer roots, and grasses, occasionally with feathers and plant down. EGGS. Sub-elliptical, smooth and fairly glossy; very pale blue to blue-green, unmarked or with fine reddish, reddish-brown, or brown speckling and mottling, particularly at broad end. Clutch: 4–5 (3–6). INCUBATION. 12–15 days. FLEDGING PERIOD. c. 18 days.

Wing-length: *M. s. solitarius*: ♂ 123–133, ♀ 116–128 mm.
Weight: *M. s. solitarius*: ♂♀ mostly 57–64 g.

Geographical variation. Slight. Most of west Palearctic occupied by nominate *solitarius*. *M. s. longirostris* from eastern Iraq (and eastwards) distinctly smaller; ♂ slightly paler than ♂ nominate *solitarius* in fresh plumage, but similar when worn; ♀ and juvenile distinctly paler and greyer on upperparts, less brown; underparts of ♀ paler and distinctly less heavily barred.

White's Thrush *Zoothera dauma*

PLATES: pages 1231, 1233

Du. Goudlijster   Fr. Grive dorée   Ge. Erddrossel   It. Tordo dorato
Ru. Пестрый дрозд   Sp. Zorzal dorado   Sw. Guldtrast

Field characters. 27 cm; wing-span 44–47.5 cm. Largest thrush of west Palearctic. As long as Mistle Thrush but heavier about bill, head, and body and proportionately shorter-tailed. Structure unusual: bill long and heavy, head large, and wings relatively long (though bluntly pointed) in comparison with tail; combined with undulating flight, gives woodpecker-like appearance. Golden- or olive-buff above and yellow-white below, copiously scaled with black crescents on head and body and softly banded dark across primaries and primary coverts and along ends of all flight feathers. Underwing striped white, black, white, and grey from front to rear. When spread, tail and upper tail-coverts show distinctive pattern: pale golden centre contrasting with blackish panels and those emphasizing largely white outermost feathers.

Can be confused with white-spotted juvenile Mistle Thrush, but that species never truly scaled (nor banded black and white under wing), and differs also in longer, plainer tail and narrower body. Main problems in identification are predilection for dense cover and difficulty in seeing diagnostic underwing pattern due to lack of prominent upstroke of wing. Flight recalls Mistle Thrush, but bounding action produced by alternating bouts of deep wing-beats and closed wings is also strongly reminiscent of woodpecker, particularly at long range. Gait apparently restricted to walk and loping run. Ground-loving, though at slightest disturbance flies up into dense foliage, swooping up to perch in characteristic glide.

Habitat. Breeds in upper middle latitudes, mainly in boreal continental zone of taiga coniferous forest, largely within range of Siberian Thrush, which apparently shows greater preference for neighbourhood of water. Habitat includes dense spruce along river valleys, adjoining mixed or broad-leaved stands on ridges or slopes, including open woods with larch, birch, and aspen, often at headwaters of streams. In Himalayas, breeds to at least 3300 m, inhabiting densely forested hillsides on broken ground. In winter occurs in tropical and subtropical woodlands.

Distribution and population. RUSSIA. Confined to northeast where range and numbers little known, partly owing to skulking habits. 100–1000 pairs; stable.

Accidental. Iceland, Faeroes, Britain, Ireland, France, Belgium, Netherlands, Germany, Denmark, Norway, Sweden, Finland, Poland, Austria, Balearic Islands, Italy, Greece, Yugoslavia, Rumania.

Movements. Varies from wholly migratory to sedentary in different parts of range. Northern race *aurea* wholly migratory.

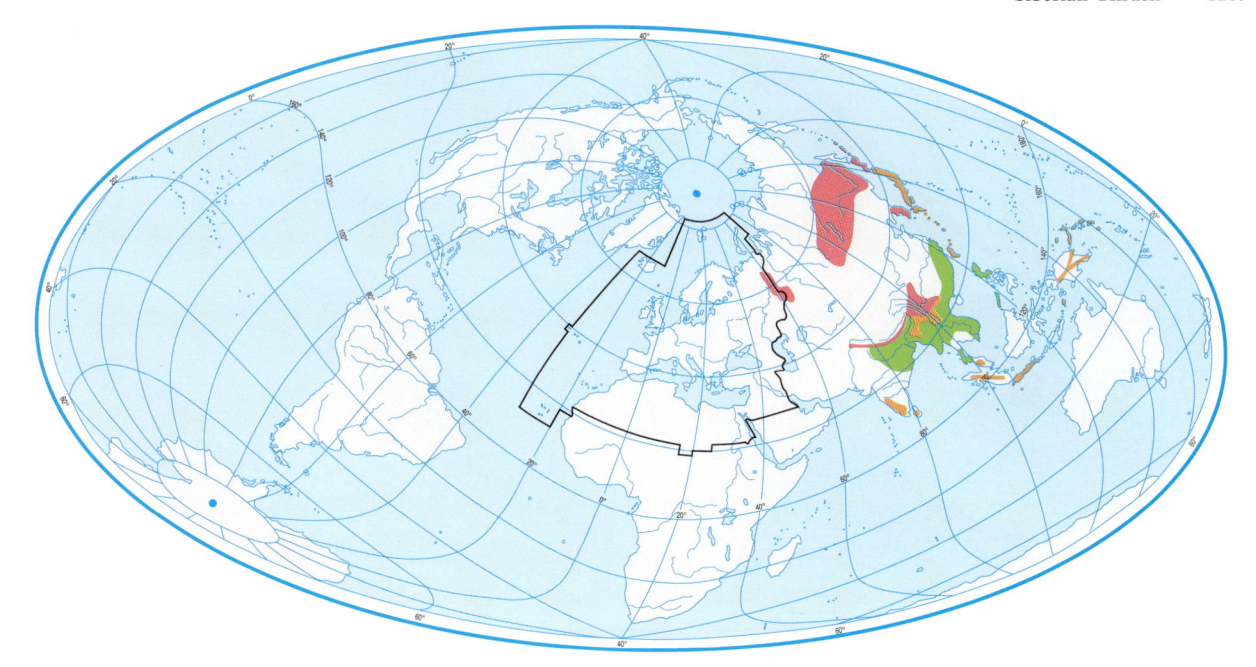

Winters in Philippines, China south of Yangtze to Kwangtung and Yunnan, Hong Kong (where scarce with numbers varying considerably between years), Taiwan, Assam, and Indo-China. Migrates south-east from west of breeding range with passage across Sinkiang and north-west Mongolia. European records chiefly October–January; spring records perhaps of over-wintering individuals.

Food. Insects, worms, and berries. Feeds on ground, turning over leaves with bill. Flushes insects by suddenly opening wings and tail, and apparently brings worms to surface by raising itself up on toes and rapidly vibrating whole body for several seconds.

Social pattern and behaviour. Poorly known. Normally solitary outside breeding season, including during migration, though small flocks reported in both spring and autumn. In breeding season presumed monogamous; not known if territorial. ♂ sings from tree, less commonly from ground. Sings mostly in evening after sunset, at night, and in morning before dawn, though will do so at other times of day, especially in dull weather.

Voice. Generally rather quiet. Song of ♂ completely unlike any other west Palearctic thrush: characteristically slow, haunting, melancholy, and rather monotonously repeated soft fluting whistles; loud, carrying up to 0.5–1 km or more. May continue for hours, with pauses between whistles remaining roughly of equal length. Contact-alarm calls include plaintive whistles, similar to song, and drawn-out 'zieh' like that of many thrushes, but higher pitched, longer and much sharper, even penetrating. High-intensity alarm-call given at nest a muffled snoring or growling 'rrra' or 'krrrua'.

Breeding. SEASON. Western Siberia: fresh eggs found 1 June, newly-fledged young in August. One brood, possibly 2. SITE. In fork of tree, 2–6 m up, occasionally on ground among stones and plants. Nest: base of dry ferns supporting cup of leaves, twigs, and moss, poorly plastered on inside with mud; thick lining of thin rootlets, grass, and leaves. EGGS. Oval, smooth and moderately glossy; greenish-blue, almost covered with fine reddish speckling, though often 1 per clutch with stronger ground-colour and fewer and bolder markings. Clutch: 4–5. (Incubation and fledging periods not recorded.)

Wing-length: ♂ 170–176, ♀ 164–172 mm.
Weight: ♂♀ mostly 110–170 g.

Siberian Thrush *Zoothera sibirica*

PLATES: pages 1208, 1233

Du. Siberische Lijster   Fr. Grive de Sibérie   Ge. Schieferdrossel   It. Tordo siberiano
Ru. Сибирский дрозд   Sp. Zorzal siberiano   Sw. Sibirisk trast

Field characters. 22 cm; wing-span 34–36 cm. Close in size to Song Thrush. Rather flat-crowned thrush; adult ♂ mainly dark slate, ♀ and immature buff-brown, mottled and barred below, but in all plumages identifiable by prominent white or pale supercilium and tail-corners, and black- and white-barred underwing.

Only much larger, paler, and scaled White's Thrush shows similar underwing pattern but note that smaller *Catharus*

Siberian Thrush *Zoothera sibirica*: **1** ad ♂, **2** ad ♀, **3** 1st winter ♂, **4** 1st winter ♀. Tickell's Thrush *Turdus unicolor* (p. 1212): **5** ad ♂, **6** 1st winter ♀. Eye-browed Thrush *Turdus obscurus* (p. 1218): **7** ad ♂ worn (spring), **8** 1st winter ♂.

thrushes are similarly but less contrastingly marked. On ground, adult and 1st-winter ♂ unmistakable, but ♀ and juvenile ♂ confusing, recalling both Song Thrush and Redwing and best distinguished by close observation of underpart markings. Flight, gait, and stance little studied, but brief references suggest *Turdus*-like carriage and actions, unlike congeneric White's Thrush. Calls include gruff squawk and soft 'zit', recalling Song Thrush.

Habitat. Breeds in east Palearctic, from upper to lower middle latitudes, mainly in boreal coniferous taiga zone, largely lowland but partly montane. The sparse data suggest preference for dense stands of trees or shrubs, especially spruce and broad-leaved species such as poplars on moist ground in flood plains of rivers or in neighbourhood of water. In winter in India, frequents hill forest up to at least 1800 m.

Distribution. Breeds in Siberia from *c.* 85°E to Sea of Okhotsk, south to north-east China and Japan.

Accidental. Britain, Ireland, France, Belgium, Netherlands, Germany, Norway, Sweden, Poland, Hungary, Austria, Switzerland, Italy, Malta.

Movements. Migratory. Siberian populations migrate south from breeding range in eastern Siberia (or south-west from east of range) across Mongolia and China, mainly east of 100°E, to winter quarters in south-east Asia. Leave breeding grounds from early September but some birds still present until mid-October; arrive India from October. Northward migration from late March, arriving on breeding grounds from late May.

European records August–March (mainly October–February) include flocks of 17–18 birds in Poland in January and March and *c.* 25 birds in Hungary in mid-February.

Wing-length: ♂ 117–127, ♀ 115–124 mm.

Varied Thrush *Zoothera naevia*

PLATE: page 1125

Du. Bonte Lijster   Fr. Grive à collier   Ge. Halsbanddrossel   It. Tordo vario
Ru. Ошейниковый дрозд   Sp. Zorzal abigarrada   Sw. Blåryggad trast

Field characters. 20–24 cm; wing-span 36–39 cm. Slightly smaller than Blackbird, with relatively shorter tail. Basic pattern and colours of plumage similar to American Robin but at close range ♂ lacks broken white spectacle, showing much more variegated appearance, with rufous rear-supercilium, black breast-band, conspicuous orange-buff tips and fringes to larger coverts, tertials, and flight feathers, barred lower flanks, and pale belly. ♀ and immature duller than ♂, with dull buff rather

Wood Thrush *Hylocichla mustelina*: **1** ad worn (spring), **2** 1st winter. Hermit Thrush *Catharus guttatus faxoni* (southern Canada and north-east USA) (p. 1210): **3** 1st winter. Swainson's Thrush *Catharus ustulatus swainsonii* (north-eastern North America) (p. 1210): **4** 1st winter. Gray-cheeked Thrush *Catharus minimus minimus* (northern North America) (p. 1211): **5** 1st winter. Veery *Catharus fuscescens* (p. 1211). *C. f. fuscescens* (north-eastern North America): **6** 1st winter. *C. f. fuliginosus* (southern Quebec and Newfoundland): **7** 1st winter.

than rufous ground to underparts and only dusky breast band. Rare pale morph, with whitish ground to underparts, occurs. In flight, spread wing shows white band across inner webs of flight feathers.

Separation from American Robin easy but beware convergence of head and upperpart pattern with those of ♀ and immature Siberian Thrush; see that species.

Habitat. Breeds in dense, moist, shady woodlands, especially coniferous forests, in boreal to northern temperate western North America; in winter also in woods, ravines, and thickets south to sub-tropics. Nests in tree; forages mostly on forest floor on invertebrates, feeding in flocks in autumn on berries.

Distribution. Breeds from north-west Alaska, central Yukon and north-west Mackenzie south to north-west California, northern Idaho, and north-west Montana.

Accidental. Britain: Nanquidno (Cornwall), 1st-year, November 1982.

Movements. Migrant and partial migrant, wintering from southern Alaska in mild winters, but mainly southern British Columbia south to central, western, and southern California to extreme northern Baja California (Mexico); many also apparently follow river valleys eastwards to southern Canada and northern USA, where regular but rare along eastern seaboard south to Virginia. Departs Alaska September, present as winter visitor in California October–April, and returns to breeding grounds in Alaska April to early May.

Wing-length: ♂♀ 122–129 mm.

Wood Thrush *Hylocichla mustelina*

PLATES: pages 1209, 1233

Du. Amerikaanse Boslijster   Fr. Grive des bois   Ge. Walddrossel   It. Tordo boschereccio
Ru. Американский лесной дрозд   Sp. Zorzal charlo americano   Sw. Fläckskogstrast

Field characters. 19 cm; wing-span 30–34 cm. Only 15% smaller than Song Thrush, being noticeably larger than *Catharus* thrushes with more *Turdus*-like form obvious in larger bill and head and plumper body. Small Nearctic thrush, immediately suggesting small, bright, and clean *Turdus*. Red cap contrasts with otherwise mainly tawny-brown upperparts; underparts pure white covered with round black spots from lower throat to fore-belly and rear flanks. Obvious white eye-ring emphasizes large, dark eye.

Unmistakable, being (1) larger and much more boldly and fully spotted below than any *Catharus* thrush, even rufous-crowned Veery, and (2) smaller, much redder above (especially on head), and much whiter and more boldly spotted below than Song Thrush. Flight like *Turdus* thrush but lighter and faster, with rapid turns and ducks into cover. Typically hops but also uses fast loping run.

Calls include low, medium-pitched 'quirt' in alarm and sharp 'pit pit', which may be extended into rapid 'pip-pip-pip-pip'.

Habitat. Breeds in temperate middle latitudes of Nearctic, mainly in lowland moist shady broad-leaved woodlands with ample undergrowth beneath tall trees, especially along streams and lake borders in swamps, keeping usually to ground and lower branches of trees.

Distribution. Breeds extreme south of Canada west to Manitoba, south through USA to Gulf coast.

Accidental. Iceland: ♂, October 1967. Britain: one, Isles of Scilly, October 1987. Azores: one in 19th century (undated specimen, identified only after 1962). Madeira: one, Porto Santo, January 1986.

Movements. Migratory. Moves south-west in autumn, and recorded as a migrant and wintering from south Texas south to Panama and north-west Colombia; winters mainly on Caribbean slopes. Dispersal and autumn migration occur July–August, with few records on breeding grounds after September. Spring migrants reach southern USA in late March and southern Canada by early May.

Wing-length: ♂ 106–116, ♀ 103–110 mm.

Hermit Thrush *Catharus guttatus*

PLATES: pages 1209, 1233

Du. Heremietlijster Fr. Grive solitaire Ge. Einsiedlerdrossel It. Tordo eremita
Ru. Дрозд-отшельник Sp. Zorzal común americano Sw. Eremitskogstrast

Field characters. 17 cm; wing-span 25–28.5 cm. Slightly smaller than other *Catharus* thrushes; 20% smaller than any Palearctic *Turdus* thrush. Smallest North American thrush with form and upperparts recalling Thrush Nightingale but underparts strongly spotted on breast and dappled on flanks; contrast of olive-brown back with chestnut tail, rump, and upper tail-coverts diagnostic within genus. Habit of cocking and then slowly lowering tail also distinctive.

Flight less flapping than *Turdus* thrush, with rather flitting wing-beats, reminiscent of nightingale. Carriage fairly upright, with raised tail also recalling nightingale—but appears neckless at times. Gait a hop. Not shy but secretive, escaping into low canopy or thick ground-cover. Commonest call a low 'chuck'.

Habitat. Breeds in pure or mixed coniferous woodlands from lowlands up to mountain forest zone. In winter, found in woodland, thickets and bushy areas, including city parks.

Distribution. Breeds from central Alaska east across much of forested Canada to Newfoundland and south to southern California, New Mexico, Wisconsin, and Maryland.

Accidental. Iceland, Britain, Germany, Sweden.

Movements. Almost wholly migratory, some south-western populations perhaps partly resident. Main wintering areas are central and southern USA and west coast from southern British Columbia south to Mexico, Guatemala, El Salvador, and northern Bahamas. Autumn migration starts at end of September, peaking in mid-October. Spring migration is from early April in southern part of breeding range to May in north.

Wing-length: ♂ 93–100, ♀ 89–94 mm.

Swainson's Thrush *Catharus ustulatus*

PLATES: pages 1209, 1233

Du. Dwerglijster Fr. Grive à dos olive Ge. Zwergdrossel It. Tordo di Swainson
Ru. Свенсонов дрозд Sp. Zorzal chico Sw. Beigekindad skogstrast

Field characters. 18 cm; wing-span 27–30 cm. About 10% larger than Thrush Nightingale; averages slightly smaller than Gray-cheeked Thrush. Small thrush with form recalling nightingale, olive upperparts and buff to dusky-white underparts indistinctly spotted on throat and breast; closely similar to Gray-cheeked Thrush in general appearance but with usually less grey, even faintly russet tinge to upperparts, bold buff eye-ring and buff ground-colour to cheeks, and buffier ground-colour to throat, breast, and flanks.

From above or behind, might suggest dull nightingale to Palearctic observer, but larger, and tail not obviously redder than rest of upperparts. Not difficult to distinguish from Hermit Thrush (which is slightly smaller, lacks buff tone to face, and shows markedly rufous rump and tail) or Veery (not strongly spotted on breast, lacks obvious eye-ring, and has different call); dullest birds do, however, invite confusion with Gray-cheeked Thrush, with close attention to face pattern and call essential to distinction (see that species). Flight, gait and behaviour much as Hermit Thrush but cocks and lowers tail more slowly and far less often. Commonest call an emphatic 'whit' or 'wick'; also a short, high-pitched 'heep' or 'queep' in flight.

Habitat. Breeds in coniferous, mixed, and broad-leaved woodlands preferring lower and damper areas, especially near streams. Winters in forest and woodlands of various kinds, also plantations and bushy areas.

Distribution. Breeds from central Alaska east across forested Canada to Newfoundland, and south to California, New Mexico, the Great Lakes area, and West Virginia.

Accidental. Iceland, Britain, Ireland, France, Germany, Norway, Finland, Austria, Ukraine.

Movements. Migratory. Moves through southern Canada and USA (scarce in south-west), less commonly through Bahamas and west Caribbean islands, to winter from Mexico to north-west Argentina. Autumn migration from late August to late October. Regular on Bermuda in early October. In spring, recorded (rarely) in western West Indies 19 March–10 May. Most birds reach breeding grounds by early May.

Wing-length: ♂ 94–107, ♀ 93–102 mm.

Gray-cheeked Thrush *Catharus minimus*

PLATES: pages 1209, 1243

Du. Grijswangdwerglijster   Fr. Grive à joues grises   Ge. Grauwangendrossel   It. Tordo di Baird
Ru. Малый дрозд   Sp. Zorzal carigrís   Sw. Gråkindad skogstrast

Field characters. 18 cm; wing-span 28.5–32 cm. Slightly larger than Swainson's Thrush, with up to 10% longer legs; largest of genus. Small thrush with form recalling nightingale, grey-olive upperparts, and grey-white underparts spotted black-brown on throat and breast; closely similar to Swainson's Thrush in general appearance but with grey- rather than green-toned upperparts (never looking as warm) in post-juvenile plumages, usually less obvious eye-ring, off-white mottling on grey cheeks, and only faintly buff-grey ground-colour to more distinctly spotted breast and dappled flanks.

Specific diagnosis must be careful, since (apart from face pattern) browner individuals come close in appearance to greyer individuals of Swainson's Thrush while some immatures show more prominent and whiter eye-rings than adults, leaving colour of lores and cheeks (always buff in Gray-cheeked Thrush, never with more than grey-brown or off-white ground-colour in post-juvenile Gray-cheeked Thrush) as most visible diagnostic character. Also important to recognize that throat and centre of breast whiter in Gray-cheeked Thrush than Swainson's Thrush. Flight, gait, and carriage as Swainson's Thrush but apparently never cocks tail. No acceptable west Palearctic records of newly separated Bicknell's Thrush *C. bicknelli*, and differences in field not fully studied.

Commonest call a long mono- or disyllable, 'wheu', 'quee-a', or 'vee-a', down-slurred and sounding higher pitched and more nasal than similar call of Veery and quite different from short monosyllable of Swainson's Thrush.

Habitat. Breeds in boreal forests, north to tree limit. Throughout northern fringe of stunted spruces, willows, and alders of arctic Canada and Alaska, finds suitable habitat beyond recognized treeline. On migration, occurs wherever there is sufficient cover, in or on edge of woodland, and in thickets along streams, roadside shrubbery, village gardens, and city parks. In winter in Venezuela, occurs at up to 3000 m in damp thickets, clearings, open woodland, and forest edge.

Distribution. Breeds north-east Siberia (east of *c.* 150°E), and Alaska through northern Canada to Labrador and New-foundland.

Accidental. Iceland, Britain, Ireland, France, Germany, Norway.

Movements. Migratory. Autumn migration primarily through eastern North America (from Bermuda in the east to Great Plains and eastern Texas), Bahamas, Greater Antilles, and southern Central America. Winters mainly in northern South America (Colombia, Venezuela, Guyana, Ecuador). Autumn migration begins early September, peaking late September and early October. Present in Venezuela September–May. Reaches central USA by 1 May and northern Alaska by 1 June.

Wing-length: ♂ 101–111, ♀ 98–107 mm.

Veery *Catharus fuscescens*

PLATES: pages 1209, 1243

Du. Veery   Fr. Grive fauve   Ge. Wilsondrossel   It. Tordo usignolo bruno
Ru. Вертлявый дрозд   Sp. Zorzal solitario   Sw. Rostskogstrast

Field characters. 17 cm; wing-span 28–31.5 cm. 5% larger than Nightingale, with 10% longer wings; close in size to Swainson's Thrush. Small thrush, with slightly longer tail than congeners; upperparts uniform rusty-brown recalling Nightingale, as does particularly chestnut rump and tail; underparts buff, with indistinctly spotted chest recalling Thrush

Nightingale. Within genus, lack of well-marked dark-spotted chest diagnostic.

Distinguished from Nightingale by bulkier, rather neck-less form, faintly indicated wing-bars (shared with other *Catharus* but most obvious on this species), and well-indicated malar stripe. Gait, carriage, and behaviour as Hermit Thrush but its tail movement is less pronounced. Commonest call a low 'phew' or 'view', down-slurred and often extended to 'whee-u'.

Habitat. Breeds in Nearctic middle latitudes, mainly in temperate lowlands but ascends suitably wooded mountain slopes. Having the most southerly range of North American thrushes except Wood Thrush, is correspondingly less attached to coniferous forest, occupying wholly broad-leaved as well as mixed woodland, especially where it is more open with broad-leaved undergrowth; also frequents second growth, willows, or alders along lakes and streams. Winters in Venezuela on edge of rain forest up to 950 m and in second growth.

Distribution. Breeds south-east British Columbia east to Newfoundland, south in Rockies to northern Arizona and in Alleghenies to Georgia.

Accidental. Britain: single birds Cornwall, October 1970, Devon, October–November 1987. Sweden: Svenska Högarna, September 1978.

Movements. Migratory. From most parts of breeding range, autumn migration is essentially south-east, migrants passing through eastern USA west to Rocky Mountains and central Texas and occurring on Caribbean slopes of Central America and on Bahamas and Cuba. Winters in South America in Colombia, Ecuador, Venezuela, Guyana, and north-west Brazil. Autumn migration across northern and eastern USA occurs mid-August to mid-October, peaking early September.

Wing-length: ♂ 99–105, ♀ 93–101 mm.

Tickell's Thrush *Turdus unicolor*

PLATES: pages 1208, 1233

Du. Tickells Lijster   Fr. Merle unicolore   Ge. Einfarbdrossel   It. Tordo di Tickell
Ru. Одноцветный дрозд   Sp. Zorzal unicolor   Sw. Gråtrast

Field characters. 21 cm; wing-span 33–38 cm. 10% smaller than Eye-browed Thrush but with similar proportions. Small and slight thrush, with plainer plumage than any other occurring in west Palearctic; ♂ mainly blue-grey, with white underbody, ♀ and immature olive-brown above and on breast, with pale face, brown-streaked white throat, tawny flanks and white underbody. Underwing orange-buff. Bill and orbital ring yellow.

♂ virtually unmistakable. ♀ far trickier, inviting confusion with dull ♀ and immature Eye-browed Thrush. Flight and behaviour typical of genus. Contact- and alarm-call 'juk-juk'.

Habitat. Breeds in alpine zone at 1500–1800 m and up to 2700 m, in broad-leaved open forest with grassy carpet or thin undergrowth, and in willow groves, orchards, and gardens. Although found in Kashmir in any wooded area, and even in heavy mixed forest or open scrub, has conspicuously adapted to settlements, foraging on lawns and flowerbeds, singing in tree-tops, and nesting lower in trees. Shifts in winter to similar habitats at lower level.

Distribution. Breeds in Himalayas from Chitral to eastern Nepal and perhaps Sikkim.

Accidental. Germany: adult ♂, specimen, Helgoland, October 1932.

Movements. Short-distance migrant. Arrives on breeding grounds at end of March and in April, leaving in September–October to move east along Himalayas to winter quarters lying along foothills from Kangra to Arunachal Pradesh, and in north-east peninsular India, Bangladesh, and northern Baluchistan. An unlikely species to reach western Europe.

Wing-length: ♂ 122–127, ♀ 115–121 mm.

Ring Ouzel *Turdus torquatus*

PLATES: pages 1213, 1218

Du. Beflijster   Fr. Merle à plastron   Ge. Ringdrossel   It. Merlo dal collare
Ru. Белозобый дрозд   Sp. Mirlo de collar   Sw. Ringtrast

Field characters. 23–24 cm; wing-span 38–42 cm. Slightly smaller and rather less stocky than Blackbird, but with 10% longer wings. Medium-sized, restless thrush; round-headed but otherwise rather attenuated, having noticeably long, sharp-cornered tail. Differs from all other west Palearctic thrushes in combination of sooty plumage, white (♂) or pale-barred (♀) chest-band (lacking in juvenile) and wing-panel, and more or less prominent greyish scaling of underparts, particularly in winter and 1st autumn. Pale wing-panel and underpart scaling more pronounced in Alpine and eastern races.

Ring Ouzel *Turdus torquatus*. *T. t. torquatus*: **1** ad ♂ worn (spring), **2** ad ♀ worn (spring), **3** 1st winter ♂, **4** 1st winter ♀, **5** juv.
T. t. amicorum: **6** ad ♂ worn (spring), **7** 1st winter ♂. *T. t. alpestris*: **8** ad ♂ worn (spring), **9** ad ♀ worn (spring), **10** 1st winter ♂.

Adult ♂ and ♀ can only be confused with partial albino Blackbird with white chest-band. Immatures more open to confusion with Blackbird, so important to recognize that Ring Ouzel has distinctive, rather rakish form, always paler wings than Blackbird, with pale fringes on wing coverts individually invisible in flight but still casting grey-brown appearance over whole area above and below, and distinctive voice (see below). Flight easy, rapid, and direct; lacks hesitant action of Blackbird.

Habitat. Breeds in upper and middle latitudes, largely oceanic upland in former and continental montane in latter, tolerating exposure to high winds and rainfall, but generally avoiding ice and persistent snow. Habitat of nominate *torquatus* in north-west Europe typically open moorland or fell with rarely more than sparse and stunted trees, occasionally at sea-level but normally at 250 m or higher, in Scotland up to 1200 m. Most nesting territories in Britain include small crags, gullies, screes, boulders, or broken ground. Alpine populations in south breed in open coniferous forest; in Switzerland normally above 1100–1300 m, in Carpathians from 250 m to treeline. Winters mainly in north-west Africa, especially in Atlas Saharien on dry and bare slopes or crests with juniper woodland.

Distribution. Some decrease Britain, Ireland, Sweden, Spain. No evidence of breeding in Baltic States. FAEROES. Bred 1981–2. BRITAIN. In 19th century scattered breeding in lowland England. Some more recent fragmentation of range. IRELAND. Decreased during 20th century. BELGIUM. Formerly irregular breeder (e.g. 1901, 1940, 1944). Since 1973 has bred in various years in Hautes Fagnes (Ardennes): 1990 (11 territories), 1992 (2 territories). NETHERLANDS. Perhaps bred mid-19th century, but no proof. GERMANY. Found most uplands with altitude 800–1000 m, also Alps. Preference for open stands means likely to benefit from tree deaths due to acid rain. DENMARK. Bred 1935. SWEDEN. Formerly bred north-west coast, but declined in 20th century and last bred there 1966. FINLAND. Bred Korppoo (south-west archipelago) in 1981. CZECH REPUBLIC, SLOVAKIA. Range increase and breeding at lower altitudes 1950–75. TURKEY. Range apparently more continuous in east. Probably underrecorded. ALGERIA. Breeds Djurdjura.

Accidental. Jan Mayen, Iceland, Latvia, Lithuania, Syria, Israel, Jordan, Kuwait, Mauritania, Madeira, Canary Islands.

Beyond west Palearctic, breeds in Iran and Kopet-Dag (Turkmenistan).

Population. Apparently stable in most countries, though decrease reported Britain, Ireland, and Spain, and increase Germany and (perhaps) Italy. BRITAIN. 5500–11 000 pairs 1988–91; now stable, though declined earlier Scotland (changes perhaps due to climate and/or competition with other thrushes). IRELAND. 180–360 pairs 1988–91. FRANCE. 1000–10 000 pairs in 1970s. GERMANY. 17 000 pairs in mid-1980s; increase in alpine areas. Much lower estimate (5000–10 000) for early 1990s. In Brocken area (Harz mountains) 35–40 pairs in early 1990s, habitat changes perhaps causing increase. NORWAY. 10 000–100 000 pairs 1970–90. SWEDEN. 3000–10 000 pairs in late 1980s; trend unknown. FINLAND. 100–200 pairs in late 1980s. POLAND. 1000–3000 pairs. Decrease in Silesia, probably stable elsewhere. CZECH REPUBLIC. 1500–2500 pairs 1985–9. SLOVAKIA. 3000–5000 pairs 1973–94. AUSTRIA. Common. No recent changes. SWITZERLAND. 15 000–25 000 pairs 1985–93. SPAIN. 6000–7000 pairs. ITALY. 10 000–20 000 pairs 1983–93. GREECE. 50–200 pairs. ALBANIA. 20–50 pairs 1981. YUGOSLAVIA: CROATIA. 800–1200 pairs. SLOVENIA. 2000–3000 pairs. BULGARIA. 1000–10 000 pairs. RUMANIA. 60 000–70 000 pairs 1986–92. RUSSIA. 10–100 pairs. UKRAINE. 16 000–19 000 pairs 1986. AZERBAIJAN. Common. TURKEY. 1000–10 000 pairs.

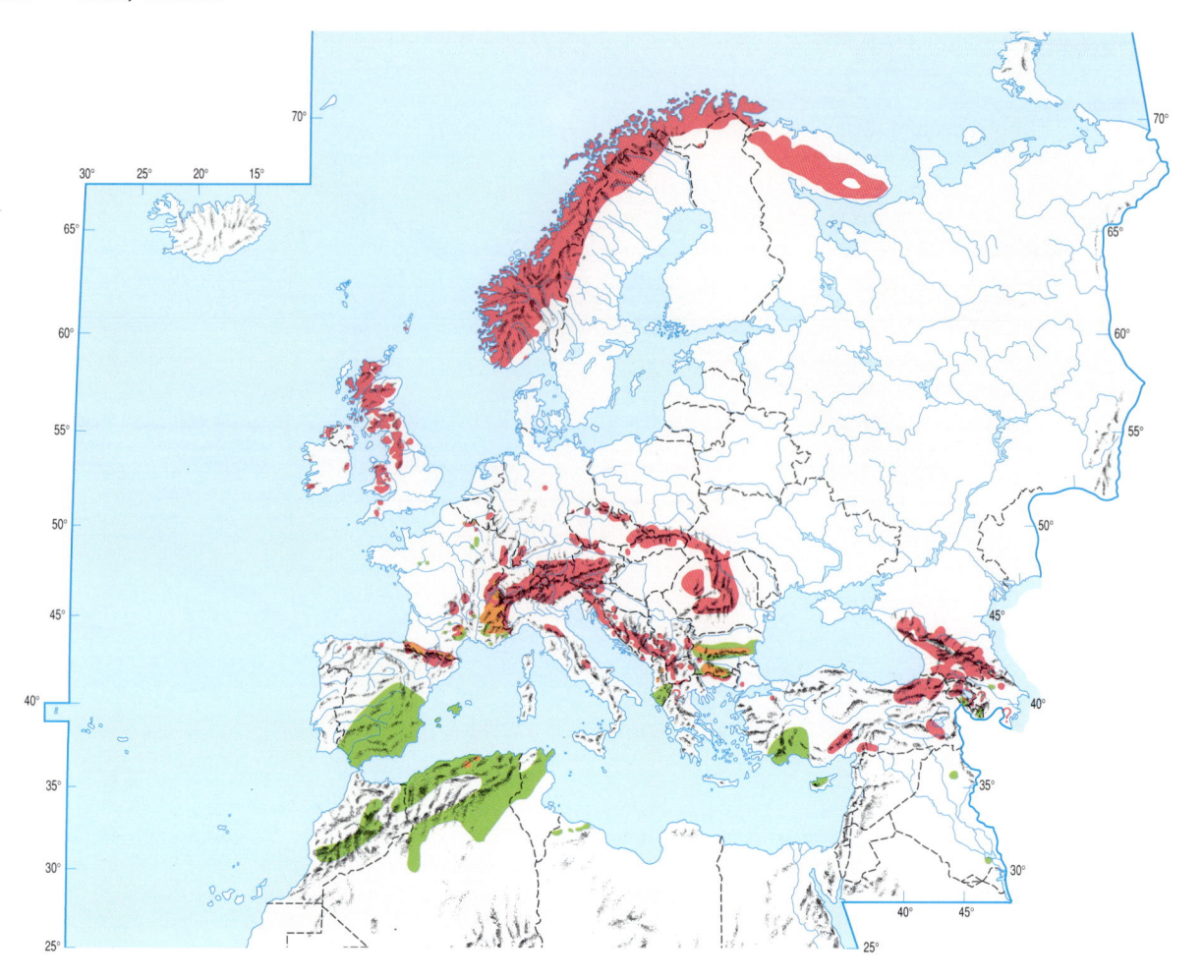

Movements. Migratory to locally resident in south. North European and British race, nominate *torquatus*, known to winter in southern Spain and north-west Africa, mainly in Atlas mountains, particularly on mountain tops of Atlas Saharien from Tunisia to Morocco. South European and west Turkish race *alpestris* winters in south of breeding range (some birds thus apparently only short-distance migrants or perhaps resident or making only altitudinal movements) as well as in north-west Africa (see above), Malta (scarce), and Cyprus. Caucasus, east Turkish, and Turkmenistan race *amicorum* presumably winters largely in Iran and southern Turkmenistan.

Southward migration starts in September. Birds of both European races arrive in north-west Africa from mid-October but main arrival is from mid-November. In spring, nominate *torquatus* starts leaving North Africa in March–April. Arrival in England begins from 2nd week of March and passage through Fair Isle extends from early April with bulk in May. Arrives on Norwegian breeding grounds April–May.

Food. In spring and early summer, adult and larval insects and earthworms; at other times, mainly fruit. Birds wintering in north-west Africa feed mainly on juniper berries in montane woodlands.

Social pattern and behaviour. Often much shyer than other thrushes. Usually gregarious outside breeding season. At end of breeding season, gathers into flocks for feeding, often associating loosely with other thrushes. No evidence for other than monogamous mating system. Breeding territories tend to form neighbourhood groups. ♂ sings from exposed elevated perch in territory, e.g. tree-top, rock, wall, heather clump, also often in flight. Song period extends from start of breeding to end of June/early July.

Voice. Song of ♂ a repetition of melancholy, plaintive phrases of 2–4 simple, monotonous, fluting piping notes; distinct pause between phrases. May also incorporate other short melodious or twittering phrases. Contact-calls include a chuckling rattle 'tchook-tchook-tchuc' or 'ti-ti-tjuck', and in flight, including migration, a rolling 'tjuirr'. Alarm-call a rapid 'tac-tac-tac', rather metallic and somewhat like Blackbird but deeper and mellower; in greater alarm, a loud rattling chatter.

Breeding. SEASON. Britain and Ireland: egg-laying from mid-April to late June. Scandinavia: from early May to end of June in south, and from late May to early August in north. Alps: similar to Britain and Ireland. 1–2 broods. SITE. In Britain (nominate *torquatus*), on or close to ground in low vegetation, or on rock-ledge or in crevice, rarely in tree. In Poland (*alpestris*), in trees (mainly coniferous), placed close to trunk and supported by twigs or 1–2 thicker branches. In Caucasus (*amicorum*), nests normally in crevices of rocks. Nest: composed of 3 parts—thick and compact external layer of twigs; thin, and sometimes incomplete, plastering of mud mixed with broken grass leaves and moss, covering bottom and lower part of walls; thick lining (normally concealing mud) of delicate grass blades or, occasionally, rootlets. Rim of wall thickened with strongly woven grass leaves, stalks, and twigs, to 3 cm in width. Humus sometimes used instead of earth for plastering. EGGS. Sub-elliptical, smooth and glossy; pale blue, with evenly distributed small red-brown, reddish-purple, and purplish-grey blotches; sometimes with reddish wash overall. Clutch: 4 (3–6). INCUBATION. 12–14 days. FLEDGING PERIOD. 14–16 days.

Wing-length: *T. t. torquatus*: ♂ 138–145, ♀ 136–143 mm.
Weight: *T. t. torquatus* (on migration): ♀♂ 92–138 g.

Geographical variation. Mainly involves width of pale feather-fringes and chest-band, the 3 races differing distinctly. Little variation in size. *T. t. alpestris*, breeding in central and southern Europe, differs from nominate *torquatus* in much broader pale edges to body feathers (not wearing off as in nominate *torquatus*) and more extensive whitish edges to wing-feathers; underparts of ♀ sometimes white with limited number of dark marks. *T. t. amicorum*, breeding from Anatolia and Caucasus to Transcaspia, has body similar to nominate *torquatus* except for broader white chest-band, but white fringes of upper wing-coverts and flight-feathers much broader than in both nominate *torquatus* and *alpestris*, especially on secondaries.

Blackbird *Turdus merula*

PLATES: pages 1216, 1218

DU. Merel   FR. Merle noir   GE. Amsel   IT. Merlo
RU. Чёрный дрозд   SP. Mirlo común   SW. Koltrast

Field characters. 24–25 cm; wing-span 34–38.5 cm. At least 50% bulkier and longer-tailed than Starling; 5% longer and 10% stockier than Ring Ouzel. Medium-sized, round-headed, rather long-tailed, noisy thrush, differing from all other west Palearctic *Turdus* in uniformly or mainly dark plumage—black in ♂, umber or rufous-brown in ♀, and rufous-brown in juvenile. Only obvious features are yellow bill and eye-ring in ♂, pale throat in ♀, and streaking and mottling in juvenile.

Adult ♂ in normal plumage unmistakable but high incidence of plumage aberrations and less uniform dress of ♀ and immature bring in whole range of pitfalls, especially confusion with Ring Ouzel and other rarer thrushes. Best answer to these pitfalls is full knowledge of Blackbird's plumage variation. Normal flight rapid and agile, with easy movement through dense cover. Migratory flight also distinctive, with characteristic bursts of fluttered wing-beats producing 'shooting' surges of speed along fairly level track or dramatic plummeting dive; lacks slow rise and fall of larger thrushes (e.g. Fieldfare). Runs and hops quickly, with typical start-stop-start progress when feeding; movements alert and sprightly.

Habitat. Exceptionally diverse, including dense woodland, varied types of farmland, heaths, moors, some wetlands, and settled sites including inner cities. Found in middle and overlapping to lower middle and upper latitudes of west Palearctic, including oceanic islands and coasts as well as milder boreal and temperate continental regions. Given shelter, will tolerate wet, windy, and cool situations better than very warm and dry ones; prefers moisture and shade, with ample access to bare ground, layers of dead leaves or short grass and herbage, even where overshadowed by low bushes and shrubs or tree canopy; avoids distances from cover exceeding *c.* 100–200 m.

Distribution. Colonized Faeroes, Cyprus, and Egypt, and spread (including increased breeding in towns in some cases) Scotland, Ireland, Sweden, Finland, Baltic States, Poland, Slovakia, Spain, Italy, Bulgaria, Ukraine, Israel, and Canary Islands. ICELAND. Occasional breeder, regular last few years: since 1979, bred 1985, 1988, 1991–4. FAEROES. First bred 1947. BRITAIN. Spread to Scotland, notably Shetland, in 20th century. IRELAND. Spread to islands and coastal regions in west. SWEDEN. Spread north of 60°30′N mostly in last 50 years. FINLAND. Marked spread north since 1913. AZERBAIJAN. Established in Baku since late 1970s. TURKEY. Very local in mountain valleys of east and south-east. CYPRUS. Probably breeding since 1979, mainly Mt Troodos. ISRAEL. Marked spread, chiefly after 1960s. EGYPT. Colonized north-east Sinai in mid-1970s, central and southern Nile delta in mid-1980s, range expansion continuing since. CANARY ISLANDS. Spread to dry south on all islands.

Accidental. Spitsbergen, Bear Island, Jan Mayen.

Beyond west Palearctic, breeds from Iran discontinuously east to eastern China, south to Sri Lanka. Introduced to Australia and New Zealand.

Population. Increase reported Faeroes, Ireland, Denmark, Estonia, Spain, Italy, Croatia, Slovenia, Bulgaria, and Ukraine; decrease Britain, Finland, Albania, apparently stable elsewhere. FAEROES. 40–75 pairs. BRITAIN. 4.4 million territories 1988–91. IRELAND. 1.8 million territories 1988–91. FRANCE. Over 1 million pairs in 1970s. BELGIUM. 540 000 pairs 1973–7. LUXEMBOURG. 40 000 pairs. NETHERLANDS. 600 000–900 000 pairs 1979–85; stable after increase. GERMANY. 8.5 million pairs in mid-1980s. DENMARK. 1.7 million pairs 1983–5. NORWAY. 100 000–1 million pairs 1970–90. SWEDEN. 1–2 million pairs

Blackbird *Turdus merula*. *T. m. merula*: **1** ad ♂, **2** ad ♀, **3** 1st winter ♂, **4** 1st winter 'stockamsel' variety, **5** juv. *T. m. syriacus*: **6** ad ♀, **7** juv. *T. m. azorensis*: **8** ad ♂, **9** ad ♀. *T. m. mauritanicus*: **10** ad ♀.

in late 1980s. FINLAND. 150 000–250 000 pairs in late 1980s; marked decrease after increase 1978–83. ESTONIA. 100 000–200 000 pairs 1991. Explosive increase during last 100 years. LATVIA. 150 000–200 000 pairs in 1980s. POLAND. Fairly numerous. CZECH REPUBLIC. 2–4 million pairs 1985–9. SLOVAKIA. 400 000–800 000 pairs 1973–94. HUNGARY. 300 000–600 000 pairs 1979–93. AUSTRIA. Very common. No recent changes. SWITZERLAND. 800 000–1 million pairs 1985–93; stable after increase. SPAIN. 2.3–5.9 million pairs. PORTUGAL. 100 000–1 million pairs 1978–84. ITALY. 2–5 million pairs 1983–93. GREECE. 50 000–100 000 pairs. Probably stable though reported to have decreased earlier because of hunting. ALBANIA. 10 000–40 000 pairs 1981. YUGOSLAVIA: CROATIA. 2–2.5 million pairs. SLOVENIA. 200 000–300 000 pairs; marked increase. BULGARIA. 1–10 million pairs. RUMANIA. 1.5–2 million pairs 1986–92. RUSSIA. 10 000–100 000 pairs. Stable overall, though increase Leningrad region. BELARUS'. 500 000–550 000 pairs in 1990. UKRAINE. 150 000–170 000 pairs 1986. MOLDOVA. 40 000–70 000 pairs 1989. AZERBAIJAN. Common to numerous. TURKEY. 100 000–1 million pairs. CYPRUS. Probably 5–20 pairs. ISRAEL. At least a few hundred thousand pairs in 1980s. JORDAN. Common resident in forests of northern highlands. TUNISIA. Common. MOROCCO. Abundant. AZORES. Common. MADEIRA. Very common. CANARY ISLANDS. Common on Gran Canaria, Tenerife, La Palma, La Gomera, and El Hierro.

Movements. Resident and migratory, with northern populations moving south or west to winter in southern or western Europe chiefly within boundaries of breeding range. Chiefly a nocturnal migrant with many killed on North Sea crossings at lighthouses. In contrast to other thrushes, little affected by severe winters and hard-weather movements are unusual.

Autumn movements begin late September with main passage in October and early November, but juveniles disperse from breeding areas during July and early August; directions usually random. Return movement to north or north-east begins late February, with main passage in March and early April and some birds still on passage in early May.

Food. Mainly insects and earthworms; also fruit throughout year, with peak in autumn and winter. Feeds largely on ground throughout year, also in trees and bushes. Often searches for food among leaf litter, etc., flicking loose material aside with bill, or seizing it momentarily to throw it aside, and at same time may bring foot forward to level of head and scratch backward; can dig thus through 5–7 cm of snow. Pulls earthworms out of ground, apparently usually locating them by seeing tip protruding from burrow, but can also detect and locate by ear invertebrates moving under soil. Recorded diet includes many unusual items, such as small fish, newts, and lizards; also wide variety of fruits, wild and cultivated.

Social pattern and behaviour. Generally shy in woodland and other habitats unfrequented by man, but often confiding in towns and suburbs. Tendency to gregarious behaviour outside breeding season strongly dependent on migratory/resident status of individuals concerned. In areas providing winter food supply (e.g. gardens) territory-holders generally remain in territories all year, ♂ and ♀ outside breeding season tending to occupy different areas within former breeding territory. Migratory birds moderately gregarious outside breeding season, typically migrating in loose flocks and feeding in groups of up to 10–20. Mating system monogamous, though a few exceptional cases of bigamy recorded. In areas with resident populations, established pairs usually remain together in successive breeding seasons, as long as both partners survive. Most new pairs formed in late winter and early spring. Breeding territories may be small and densely packed in optimal (garden) habitats, much sparser in woodland and open country. Song (by ♂ only) delivered from perch; occasionally sings in flight between perches. Song-period several months, from late winter to end of breeding season; time of onset much dependent on weather, stimulated by mild and damp conditions. In northern Europe, may start late December if weather mild, rarely even earlier. Courtship and aggressive displays of ♂ similar, most striking feature being humping-up of back and rump feathers and lowering and fanning of tail. Displaying bird may pick up and brandish a leaf, sometimes persistently. From autumn to spring, ritualized aggressive behaviour in groups of up to c. 20 birds may take place near roosting sites, suggesting 'communal' display.

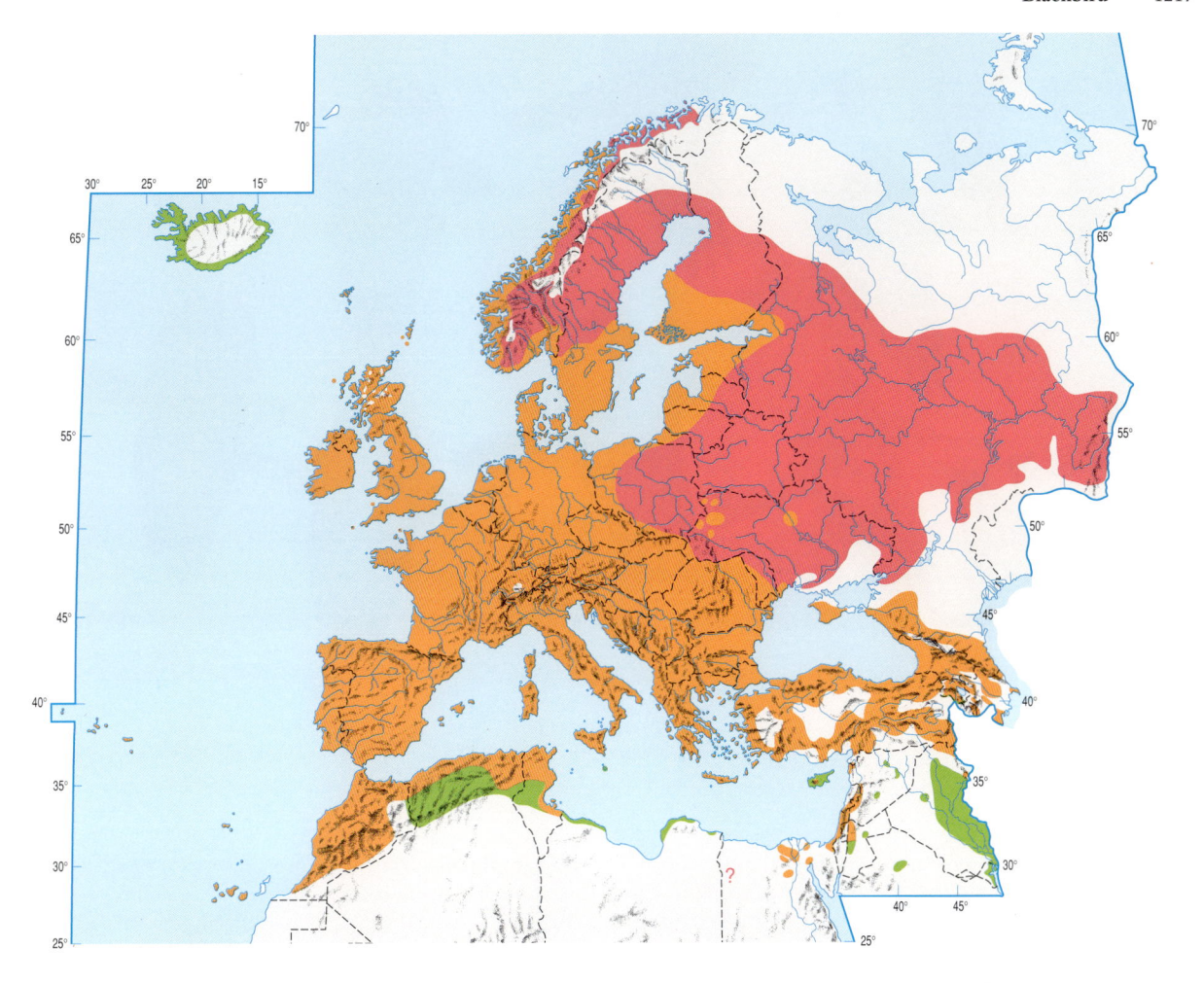

Voice. Song a rich, mellow, fairly low-pitched warble, with characteristically languid delivery and fluted quality; lacks 'shouted' phrases of other thrushes and often tails off into creaky, chuckling notes. Each adult ♂ has large repertoire of song-phrases exhibiting great individual variation. Calls loud and varied: most distinctive are low 'pook' and 'chook' sounds often given in warning and half-alarm situations (and accompanied by simultaneous tail-cocking and wing-flicking), and hysterical chatter and screaming rattle in full alarm. Groups join in distinctive 'chink'-ing chorus at dawn and dusk.

Breeding. SEASON. Rather little variation across range. Western Europe and Britain: egg-laying usually from March to late June; in Britain, timing significantly correlated with latitude, up to 2 weeks later in Scotland than in southern England. Finland: laying begins mid-April in advanced springs, but main period usually from last week of April; north of 62°N, first eggs early to mid-May. Czechoslovakia: laying begins last 10 days of April, annual variation significantly correlated with climatic factors. Canary Islands: eggs found March–June, mainly May–June. 2–3 broods regular (except in north of range, where 3 rare), with 4(–5) recorded in (e.g.) Britain. SITE. Typically against trunk of small tree or bush supported by small branches and twigs, or among branches; frequently in or on wall, outside or inside building, among pile of brushwood or other debris, or occasionally on ground. Nest: substantial cup of grass, straw, small twigs and other plant material, usually on foundation of

Rock Thrush *Monticola saxatilis* (p. 1201): **1–2** ad ♂, **3** ad ♀, **4** 1st winter ♂. Blue Rock Thrush *Monticola solitarius solitarius* (p. 1204): **5–6** ad ♂, **7** ad ♀. Ring Ouzel *Turdus torquatus torquatus* (p. 1212): **8–9** ad ♂, **10** ad ♀. Blackbird *Turdus merula merula* (p. 1215): **11–12** ad ♂, **13** ♀.

moss, occasionally incorporating decoration of paper, foil, etc.; plastered inside with mud (rarely not) and lined with fine grass. Eggs. Sub-elliptical, smooth and glossy; usually pale greenish-blue, mottled and speckled light red-brown, though varying from brownish tint all over to unmarked. Clutch: 3–5 (2–6); varies within season (smallest at beginning and end). Incubation. Usually 12–14 days. Fledging Period. Average 13.6 days (10–19).

Wing-length: *T. m. merula*: ♂ 126–140, ♀ 121–131 mm.
Weight: *T. m. merula*: ♂♀ mostly 80–125 g.

Geographical variation. Clinal, except for some island populations; mainly involves colour of ♀ and size. 7 races recognized in west Palearctic. In general, size smaller towards south, but mountain birds have longer wings than lowland ones. *T. m. cabrerae* (Madeira and Canary Islands) and *T. m. azorensis* (Azores) are small, ♂♂ glossier and deeper black than nominate *merula*, ♀♀ darker. Other southern races, *aterrimus* (southern Balkans eastwards), *syriacus* (Turkey and Levant), *mauritanicus* (north-west Africa) are all darker, especially ♀♀, than nominate *merula*.

Eye-browed Thrush *Turdus obscurus*

PLATES: pages 1208, 1232

Du. Vale Lijster   Fr. Grive obscure   Ge. Weißbrauendrossel   It. Tordo oscuro
Ru. Оливковый дрозд   Sp. Zorzal rojigrís   Sw. Gråhalsad trast

Field characters. 23 cm; wing-span 36–38 cm. Size between Song Thrush and Redwing. Rather small and rakish thrush, with general character most like Redwing. Grey- to olive-brown above, with obvious white supercilium, eye-ring, and throat; upper breast grey but chest and flanks orange or buff, contrasting with white belly and vent. Underwing pale grey. Sexes closely similar; ♀ and particularly 1st-autumn bird duller than ♂.

Unmistakable when seen well at close range, with lack of obvious spots (except in juvenile), grey underwing, and uniformity of basic plumage colours ruling out confusion with Naumann's Thrush. Less distinctive at distance, when size, appearance, flight, and behaviour strongly recall Redwing. Flight fast, with rakish silhouette; over long distance, direct and not undulating, recalling Starling. Song clear and rich, mostly a repetition of 2–3 ringing notes. Calls include: thin,

Dusky Thrush and Naumann's Thrush *Turdus naumanni*. *T. n. eunomus* (Dusky Thrush): **1** ad ♂ worn (spring), **2** 1st winter ♂. *T. n. naumanni* (Naumann's Thrush): **3** ad ♂ worn (spring), **4** 1st winter ♂. American Robin *Turdus migratorius migratorius* (northern North America) (p. 1234): **5** ad ♂ worn (spring), **6** 1st winter ♂.

quiet 'zip-zip', 'plip', or 'tlip', recalling pipit; soft 'tchuck' or 'tchick'; in full alarm, loud 'ke(w)k' recalling Redwing but less hoarse.

Habitat. Breeds in upper middle and higher latitudes of east Palearctic, in continental boreal lowland and montane habitats, from fringe of taiga in larch to dense forests of spruce and fir, especially in sheltered valleys or near water. In winter, occurs in woods, open country, and gardens.

Distribution. Breeds in Siberia from *c*. 75°E east to Sea of Okhotsk (sporadically Sakhalin island), south and south-east to Lake Baykal, Mongolia, Amur- and Ussuriland. Japan: reported in summer, but breeding not confirmed.

Accidental. Britain, France, Belgium, Netherlands, Germany, Norway, Sweden, Finland, Poland, Czech Republic, Portugal, Italy, Malta, Israel.

Movements. Migratory. Passes through Mongolia, Manchuria, Korea, and China to winter in eastern India, Burma, Vietnam, southern China, and Japan south to Malay peninsula, Philippines, and Indonesia. Stragglers have reached Arabia and Alaska, as well as west Palearctic.

Wing-length: ♂ 123–129, ♀ 119–127 mm.

Naumann's/Dusky Thrush *Turdus naumanni*

PLATES: pages 1219, 1232

Du. Bruine Lijster/Naumanns Lijster Fr. Grive de Naumann/à ailes rousses Ge. Naumanndrossel It. Cesena di Naumann
Ru. Дрозд Науманна/Бурый дрозд Sp. Zorzal de Naumann Sw. Bruntrast

Field characters. 23 cm; wing-span 36–39 cm. Larger than Redwing (wings and tail nearly 15% longer). Medium-sized, robust thrush, with rather stout bill and body. Diagnostic combination of broad pale supercilium and much rufous or chestnut under and on wings and on rump. Northern race *eunomus* (Dusky Thrush) strongly patterned on face and spotted and barred black below, southern race *naumanni* (Naumann's Thrush) less strongly patterned on face and spotted rufous below—all marks geometrically dappled, rather than lined as in most spotted thrushes.

Both races unmistakable at close range. Most telling characters of nominate *naumanni* are dull pink-red suffusion and dappling of underparts and red-chestnut tail, brighter than rump. Best features of *eunomus* are broad cream supercilium contrasting with dark rear cheek-patch, pale half-collar, bright chestnut wing-panel and rump (warmer than tail), and black-splashed underparts. In brief glimpse or at longer range, however, both races liable to be confused with other species. Nominate *naumanni* may suggest Red-throated Thrush (but lacks its markedly uniform plumage colours), Eye-browed

Thrush (but lacks its grey-brown upper breast and uniform upperparts), and American Robin (but is smaller and lacks its dusky upperparts, black-streaked white throat, and fully brick-red underparts). Dusky Thrush liable to be passed over as being Redwing since face patterns very similar, breast markings of adult Redwings form gorget and dark rufous flanks of Redwing can simulate wing-panel of *eunomus*, while dark rufous underwing common to both in all plumages. Appearance in flight most recalls Song Thrush, being plump and not slight and rakish like Redwing. Gait and behaviour on ground recall both Song Thrush and Redwing, with noticeably uptilted bill at times. Song very similar in phrasing and tone to Song Thrush but less chanted and strong. Calls varied, recalling Blackbird and Redwing.

Habitat. Breeds in higher latitudes of east Palearctic, mainly in lowlands from fringe of tundra through taiga and wooded steppe, in riverain woods of willow and poplar, with dense tall bushy undergrowth; also in birch and alder and more rarely in larch along terraces and low scrub. Occurs in thinly wooded regions and on outskirts of forest. In winter, occurs in open fields, grasslands, and thinly wooded country.

Distribution. Breeds in Siberia east of Yenisey river and Sayan mountains.

Accidental. Both nominate *naumanni* and *eunomus* recorded Britain, France, Belgium, Germany, Finland, Poland, Austria, Italy, and Cyprus. Nominate *naumanni*: Czech Republic, Hungary, Belarus'. *T. n. eunomus*: Faeroes, Netherlands, Denmark, Norway, Yugoslavia, Kuwait. Race not determined: Ukraine, Israel.

Movements. Migratory. North Siberian race, *eunomus*, migrates through Kuril Islands, Sakhalin, Hokkaido, Korea, Manchuria, and Mongolia to winter in southern Japan, Taiwan, southern China, north-west Thailand, northern Burma, Assam (India), and occasionally Nepal and Pakistan. South Siberian race, nominate *naumanni*, migrates through Mongolia and Korea to winter in China north to southern Manchuria and south to Yangtse river and in Korea, with a few in Japan and Taiwan.

Recorded chiefly autumn–winter (*eunomus* more frequently than nominate *naumanni*).

Wing-length: *T. n. eunomus*: ♂ 128–136, ♀ 122–129 mm. Nominate *naumanni*: similar to *eunomus* or wing up to *c.* 5 mm longer.

Red-throated/Black-throated Thrush *Turdus ruficollis*

PLATES: pages 1221, 1232

Du. Roodkeel-/Zwartkeellijster Fr. Grive à gorge rousse/Grive à gorge noire Ge. Bechsteindrossel It. Tordo golarossa/golanera
Ru. Краснозобый/чернозобый дрозд Sp. Zorzal papirrojo/papinegro Sw. Taigatrast

Field characters. 25 cm; wing-span 37–40 cm. Close in size to Blackbird, but tail 20% shorter. Medium-sized, bulky thrush, with well-balanced form most recalling Blackbird but behaviour and plumage pattern somewhat reminiscent of Fieldfare. Head and upperparts pale grey- to umber-brown; throat and chest dark, underbody dull white, and underwing

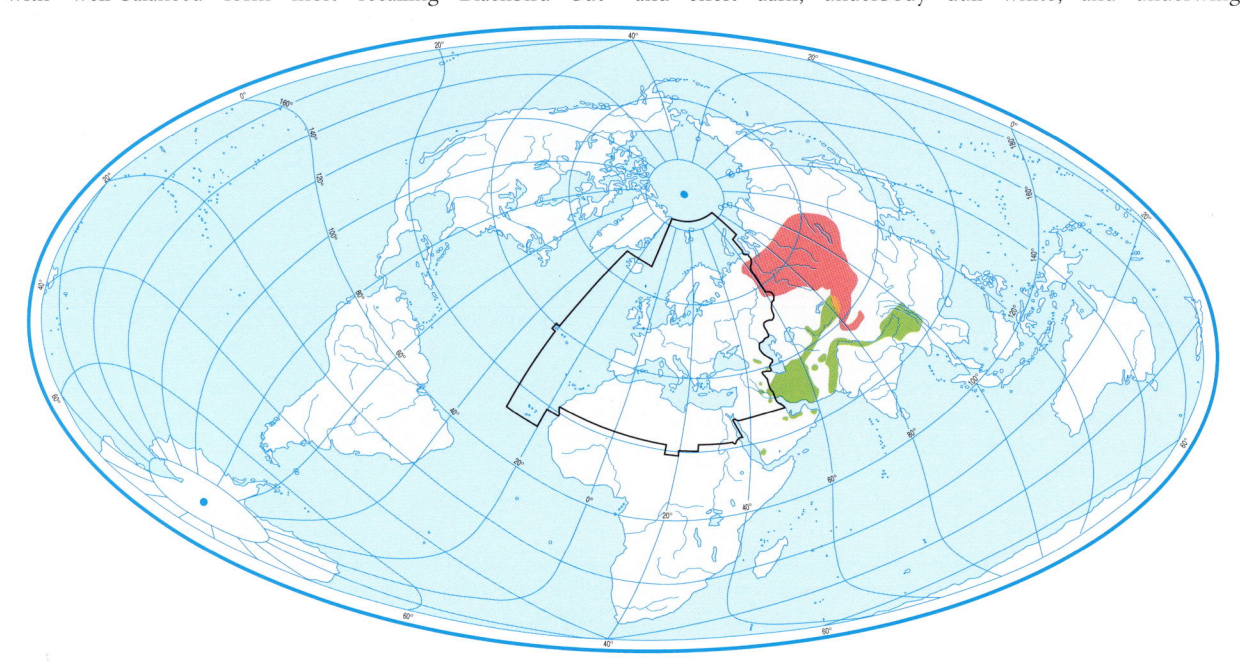

Black-throated Thrush and Red-throated Thrush *Turdus ruficollis*. *T. r. atrogularis* (Black-throated Thrush): **1** ad ♂ worn (spring), **2** ad ♀ worn (spring), **3** 1st winter ♂, **4** 1st winter ♀, **5** juv. *T. r. ruficollis* (Red-throated Thrush): **6** ad ♂ worn (spring), **7** ad ♀ worn (spring), **8** 1st winter ♂.

rufous-buff. ♂ of western race (Black-throated Thrush) has black chest and black-brown tail. ♂ of eastern race (Red-throated Thrush) has dull red chest and tail. ♀♀ and immatures of both races have less contrasting plumage patterns.

Adult of both races distinctive, with dark chest, rather uniform upperparts, and pale rear underbody producing pattern only reminiscent of Eye-browed Thrush (much smaller, with obvious white supercilium and orange-buff lower breast and flanks). However, less typically marked ♀♀ of nominate *ruficollis* and immatures of both races may invite confusion with American Robin and (particularly) Naumann's Thrush. Flight more like Fieldfare than Blackbird, with bursts of strong wing-beats and momentary closures of wings producing direct, little-undulating progress; within cover and when feeding, more like Blackbird. Gait as Blackbird but with markedly long hops and more upright carriage, sometimes suggesting wheatear.

Habitat. Breeds in central and marginally in west Palearctic in upper to middle latitudes from lowlands and boreal continental to montane temperate zones, with marked ecological differences between northern race *atrogularis* and south-eastern nominate *ruficollis*. Nominate *ruficollis* inhabits sparse mountain forests, mossy scrub tundra above them, taiga on plateaux, and bottomland forests by mountain rivers. *T. r. atrogularis* frequents borders and open parts of various types of forest including low swampy taiga; also riversides, sparse dry woods or clusters of (e.g.) larches scattered over subalpine steppe, and in south of range up mountains to *c*. 2000–2200 m. Most winter in Himalayas and adjoining mountains up to 3000–4200 m, descending to plains under stress of weather, and resorting to cultivation, stubble fields, pastures, grassy slopes, fallow land with sparse scrub, and edges of forest.

Distribution and population. RUSSIA. Poorly known, but no apparent changes in range or numbers. Perhaps 1000–10 000 pairs; stable.

Accidental. Both nominate *ruficollis* and *atrogularis* recorded Britain, France, Germany, Norway, Poland, Italy. *T. r. atrogularis*: Belgium, Netherlands, Denmark, Sweden, Finland, Estonia, Latvia, Czech Republic, Austria, Spain, Greece, Bulgaria, Rumania, Ukraine, Israel, Egypt.

Movements. Migratory, but some stay to winter in Siberia in years with good berry crop. West Siberian race *atrogularis* migrates through northern Iran, Afghanistan, and Mongolia to winter abundantly all across Indian subcontinent as well as from Arabia and Iraq to Burma and south-west China north to Turkmenistan and Himalayas. East Siberian race, nominate *ruficollis*, migrates through Mongolia and China west to Sinkiang and Tibet to winter in Afghanistan, northern Pakistan, northern Kashmir to Assam, northern Burma, and China mainly in north and west.

Vagrants in west Palearctic are mostly *atrogularis*. In Britain and Ireland, most occur late autumn or winter; may be birds which have not moved south of breeding grounds until onset of severe weather.

Social pattern and behaviour. Most aspects poorly known. Often in flocks outside breeding season. Some evidence that breeding may be loosely colonial.

Voice. Most characteristic feature of song is its hoarse timbre and low pitch; single phrase of 3–4 units separated by pauses of equal length. In alarm, gives throaty 'which-which-which', recalling chuckles of Blackbird but softer in tone. Flight-call a thin 'see', similar to Redwing.

Breeding. SEASON. North-west Russia: main laying period May. One brood. SITE. On low stump, or in tree, close to ground. Nest: external layer of grass stems and leaves, a thick rim of stalks, thickly plastered with mud, and inner lining of finer grass. EGGS. Sub-elliptical, smooth and glossy; light blue with fine reddish-brown speckling and mottling, sometimes thicker at broad end. Clutch: 5–6 (4–7). INCUBATION. 11–12 days. (Fledging period not recorded.)

Wing-length: *T. r. atrogularis*: ♂ 134–142, ♀ 130–136 mm. *T. r. ruficollis* similar.

Weight: *T. r. atrogularis*: ♂♀ mostly 80–100 g.

Geographical variation. 2 quite distinct races, which overlap and regularly hybridize in area of overlap. Compared with *atrogularis*, nominate *ruficollis* (breeding south-east of *atrogularis*, from Altai east to Transbaykalia) has black of head, throat, and chest replaced by russet-brown; upperparts rather paler, and tail extensively rufous. Intermediates locally more numerous than the typical forms.

Fieldfare *Turdus pilaris*

PLATES: pages 1223, 1233

DU. Kramsvogel FR. Grive litorne GE. Wacholderdrossel IT. Cesena
RU. Рябинник SP. Zorzal real SW. Björktrast

Field characters. 25.5 cm; wing-span 39–42 cm. 5–10% shorter than Mistle Thrush but proportionately longer-tailed; 20–25% longer than Redwing. Large, bold, long-tailed, often noisy thrush, with rather rakish form both on ground and in the air. Plumage more boldly variegated and richly coloured than any other west Palearctic thrush, with blue-grey head, vinous-chestnut back, grey rump, and almost black tail obvious on ground, and heavily speckled breast and flanks, white vent, and black undertail obvious from below. Juvenile duller and more evenly spotted below. Combination of grey rump, black tail, and white underwing diagnostic.

Unmistakable; no other west Palearctic thrush shows as much contrast between rump and tail, and white underwing shared only by Mistle Thrush. Flight noticeably loose and leisurely, with bursts of wing-beats alternating with brief glides on extended wings and short 'shooting' glides with wings closed. Escape-flight less panicky than other gregarious thrushes, ending not by ducking into cover but in birds festooning bare branches of 'safe tree', often in company with Redwings.

Habitat. Breeds in middle and higher latitudes of west Palearctic, in subarctic, boreal, and temperate zones, in woods of birch, pine, spruce, alder, and mixed species, usually in open growth or on fringes of moist areas with grass cover; often along rivers or in groups of trees in fens or bogs, in sheltered but cool and humid situations. In Scandinavia, nests above 1000 m where juniper and dwarf birch afford sufficient shelter; also on rocky outcrops in folds of exposed fells, or on stony slopes with a few straggling bushes, resembling habitat of Ring Ouzel, and even on open tundra beyond treeline. In winter, prefers fringes between open and wooded country, needing big rough fields (including cropfields) and avoiding forests, moorlands, wetlands, and (except in hard weather) human settlements. Attraction of opportunities to feast on berry crops is overriding at the right season.

Distribution. Marked expansion began after 1750. Has colonized and/or spread in many countries of north-west, central, south and south-east Europe (including south and west in west of FSU). ICELAND. Occasional breeder: in 1950s, several years in 1980s, and 1990. BRITAIN. First bred Orkney 1967, annually Shetland 1968–70, then on mainland south to Staffordshire, but breeding confirmed only in Scotland in recent years. FRANCE. First bred 1953. BELGIUM. First bred 1967. Rapid increase in range and numbers in 1970s, still continuing. LUXEMBOURG. First recorded breeding 1969. NETHERLANDS. Bred 1903, 1905, 1925, 1936, and regularly from 1972. GERMANY. Most important advances in 1960s after immigration from east beginning of 19th century; continuing expansion. DENMARK. First bred 1960. Range expanded by at least 17% last 2 decades. SWEDEN. Slow expansion into Skåne since *c.* 1972. POLAND. Confined to north-east until 19th century; spread thereafter until early 20th century and, after decline, in some areas more recently. CZECH REPUBLIC. Established mid-19th century, but some spread more recently. SLOVAKIA. Main colonization and spread after 1950. SWITZERLAND. First bred 1923. Spreading in 1980s especially in alpine regions, occasionally to 2000 m. ITALY. Range increasing in Alps. GREECE. First bred Krikellon (central Greece) 1981, then Voras mountain (Makedhonia) *c.* 1986, and Vardoussia mountains (centre) 1992. RUMANIA. Bred Moldavia 1966–71. Range expansion to west in Carpathians and Transylvania. RUSSIA. Recorded nesting at 68°30′N on Russkiy Zavorot peninsula 1992–4.

Accidental. Spitsbergen, Bear Island, Jan Mayen, Kuwait, Libya, Tunisia, Algeria, Morocco, Azores, Madeira.

Beyond west Palearctic, extends east through Siberia to *c.* 117°E, south to northern Kazakhstan and Lake Baykal.

Population. Increase (continuing in several countries) with colonization and range expansion France, Low Countries, Germany, Denmark, Estonia, Poland, Hungary, Austria, and Ukraine, marked in Italy, Slovenia, and Rumania. Marked decrease Finland, apparently stable elsewhere. BRITAIN.

Fieldfare *Turdus pilaris*: **1** ad ♂ worn (spring), **2** ad ♀ worn (spring), **3** 1st winter ♀, **4** juv. Redwing *Turdus iliacus* (p. 1228). *T. i. iliacus*: **5** ad worn (spring), **6** 1st winter, **7** juv. *T. i. coburni*: **8** ad worn (spring), **9** 1st winter.

Generally under 10 pairs breeding annually late 1960s to early 1980s. Under 25 pairs 1988–91. FRANCE. Under 10 000 pairs in 1970s. BELGIUM. 10 000–14 000 pairs 1989–91; rapid and continuing increase. LUXEMBOURG. 3000–4000 pairs. NETHERLANDS. 700–900 pairs 1986. GERMANY. 761 000 pairs in mid-1980s. Increase in Brandenburg and Mecklenburg-Vorpommern. DENMARK. 1000–5000 pairs 1993–4. NORWAY. 1–3 million pairs 1970–90. SWEDEN. 750 000–1.5 million pairs in late 1980s. FINLAND. 800 000–1.2 million pairs in late 1980s. ESTONIA. 100 000–200 000 pairs 1991. Increase probably continuing after setback in 1970s. LATVIA. 40 000–150 000 pairs in 1980s; stable or slightly decreasing. POLAND. 150 000–400 000 pairs. CZECH REPUBLIC. 70 000–140 000 pairs 1985–9. SLOVAKIA. 10 000–20 000 pairs 1973–94. HUNGARY. 50–150 pairs 1979–93. AUSTRIA. Common, especially between 400 and 1200 m. SWITZERLAND. 100 000–200 000 pairs 1985–93. ITALY. 5000–10 000 pairs 1983–93. GREECE. Under 10 pairs. YUGOSLAVIA: SLOVENIA. 1000–3000 pairs. RUMANIA. 7000–10 000 pairs 1986–92. RUSSIA. 1–10 million pairs. Apparently stable overall, though some local declines or increases. BELARUS'. 600 000–700 000 pairs 1990. UKRAINE. 110 000–130 000 pairs 1988.

Movements. Migratory, though in some years of winter abundance of food some resident or move only short distances. Winters mainly in western, central, and southern Europe, Turkey, and Iran, also south to Canary Islands and Persian Gulf states. Usually reaches southernmost parts of Europe only in bad winters and rarely occurs on North African coast in good numbers.

Birds flock prior to departure, becoming increasingly restless and making local movements. Spring passage generally more visible than in autumn with some impressive continual movements of birds often totalling several thousand passing along lines of hills or valleys in a matter of hours. Sudden movements of large numbers as a consequence of severe weather are commonplace across the entire wintering range. Individuals do not necessarily return to same area in successive winters with some subsequently recovered in winter up to 1600 km distant.

Southward migration begins late September or early October and continues into November. Return often begins early, birds wintering in south-central Europe making partial return movements in February. Main arrivals in Norway from mid-April and in Sweden and Finland from late April.

Food. Wide range of invertebrates; also fruits at almost all times of year, as available. Wintering birds may feed for considerable periods entirely on fruit. Feeds on ground and in trees and bushes. Will turn over clods of earth, etc. (even stones up to 10 cm across), and scratch through snow to take food beneath. To take flying insects will fly up high in the air like Starling. Recorded entering shallow water to take fish, also pecking at dead fish.

Social pattern and behaviour. Most gregarious of west Palearctic thrushes outside breeding season, with flocks of up to several thousand on migration. A few individuals, probably adult ♂♂, may temporarily defend concentrated fruit sources in hard weather. Monogamous mating system; no polygyny recorded. Pair-bond maintained only for breeding season, and breeding by same pair in 2 successive years apparently exceptional. Breeds solitarily or colonially; often no clear distinction and, though most birds in colonies of up to 40–50 pairs, dispersion varies regionally. In south of range, more often solitary, and colonies smaller, than in north. Commonly associates with other species for breeding, notably passerines evidently attracted by defensive advantages of proximity to Fieldfare; also associates amicably with shrikes and falcons. ♂ frequently sings in display-flight, especially at start of breeding season, mostly in early morning and evening. Song-flight distinctive, not undulating like normal flight but horizontal, with slower more deliberate wing-beats. Most notable feature of breeding behaviour is aggressive communal defence of nests against potential predators. Attack-flight direct, with shallow rapid wing-beats; often swoops low over predator, braking and veering off at last moment, defecating as it does so. Attacks persist until predator retreats, but by then its plumage may be so soiled and matted that it is grounded, and several reported dying as a result.

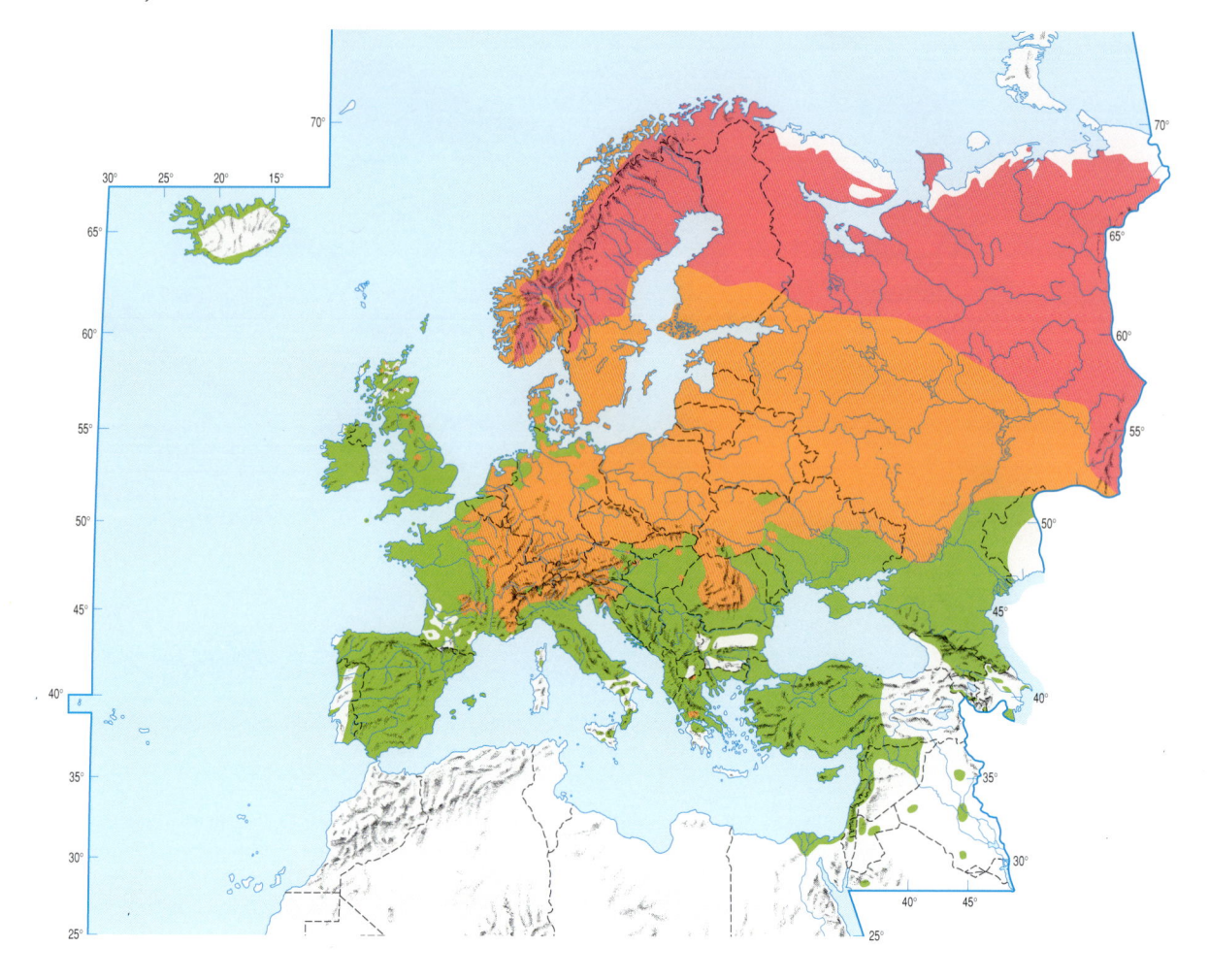

Voice. Song at best a feeble string of chuckles, whistles, squeaks, and normal calls. Commonest (diagnostic) call harsh, aggressive-sounding cackle 'chacker chack chack' or 'chac-chac-chac-chack'; also gives a soft, drawn-out 'seeh' and a squawk.

Breeding. SEASON. Scandinavia: egg-laying from early May. Central Europe: laying begins early April. Lapland: laying begins late May and early June. 1–2 broods. SITE. In tree, placed in crotch of branch against trunk, or on side branch; exceptionally on ground or in depression among rock. Nest: bulky though compact structure with outer parts of grass reinforced with twigs, roots, etc., lined with thick layer of mud, and inner lining of fine grasses and a few roots; woven ring of grass glued with mud, rarely with a little horse hair and fur. EGGS. Sub-elliptical, smooth and glossy; pale blue, very variably marked with red-brown speckles, often very heavily, less often blotched, sometimes forming cap at broad end. Clutch: 5–6 (3–7). INCUBATION. 10–13 days. FLEDGING PERIOD. 12–15 days.

Wing-length: ♂ 139–152, ♀ 136–147 mm.
Weight: ♂♀ mostly 80–120 g.

Song Thrush *Turdus philomelos*. *T. p. philomelos*: **1** ad worn (spring), **2** 1st winter, **3** juv. *T. p. clarkei*: **4** ad worn (spring), **5** 1st winter. *T. p. hebridensis*: **6** ad fresh (autumn), **7** juv.

Song Thrush *Turdus philomelos*

PLATES: pages 1225, 1233

Du. Zanglijster   Fr. Grive musicienne   Ge. Singdrossel   It. Tordo bottaccio
Ru. Певчий дрозд   Sp. Zorzal común   Sw. Taltrast

Field characters. 23 cm; wing-span 33–36 cm. Slightly larger than Redwing, with stockier build, noticeably blunter wing-point, and rump and tail slightly longer in proportion; 15% smaller than Mistle Thrush, with proportionately shorter tail. Medium-sized thrush, with well-balanced form, upright carriage, brown-toned upperparts, and boldly spotted underparts. Within west Palearctic *Turdus*, has diagnostic combination of faint face pattern and golden-buff underwing. Juvenile buff-spotted above.

Commonest and most widespread spotted thrush in Fenno-Scandia and temperate Europe, with little risk of confusion with Mistle Thrush (much larger and longer-tailed), Redwing (smaller and more patterned), Siberian Thrush (similarly sized but with barred underwing), Nearctic thrushes *Hylocichla* and *Catharus* (much smaller and less spotted). Flight typical of smaller *Turdus*, with clear affinity to that of wheatears and nightingales but more powerful and faster; wing-beats fast, with obvious bursts producing acceleration but not marked undulations of larger thrushes. Alternates short runs or series of hops with pauses and alert posture. Flicks wings and tail in excitement.

Habitat. In upper and middle latitudes of west and central Palearctic, both continental and oceanic, largely temperate but also boreal and marginally subarctic. Tolerates cool, humid, and windy but not arid, very warm, nor persistently frosty and snowy climate. Birds can exist almost anywhere where trees or bushes accompany open grassland, patches of dead leaves under trees, or moist ground supporting ample invertebrate food organisms. Basically a bird of primitive forests, both coniferous and broad-leaved, where beech, birch, and oak give some shade and humidity. Requires ample undergrowth, either of young spruce or fir or (increasingly in recent times) of equivalent deciduous cover. Modern conversion of lowlands to agricultural, industrial, and urban uses has stimulated, especially in western Europe, switch to small woodlands, parklands, hedgerows, railway embankments, roadsides, cemeteries, gardens, and even interiors of cities. Habitat in winter differs from that in breeding season only in so far as high or outlying areas and woodland are vacated.

Distribution. No significant range changes. BRITAIN. Formerly bred Shetland regularly. IRELAND. Some islands off west coast colonized in 20th century. FRANCE. Corsica: no proof of breeding. ITALY. Expanding in hills at 100–200m in 1980s.

Accidental. Spitsbergen, Bear Island, Jan Mayen, Iceland (almost annual), Mauritania, Azores, Madeira.

Beyond west Palearctic, extends east from Urals in broad

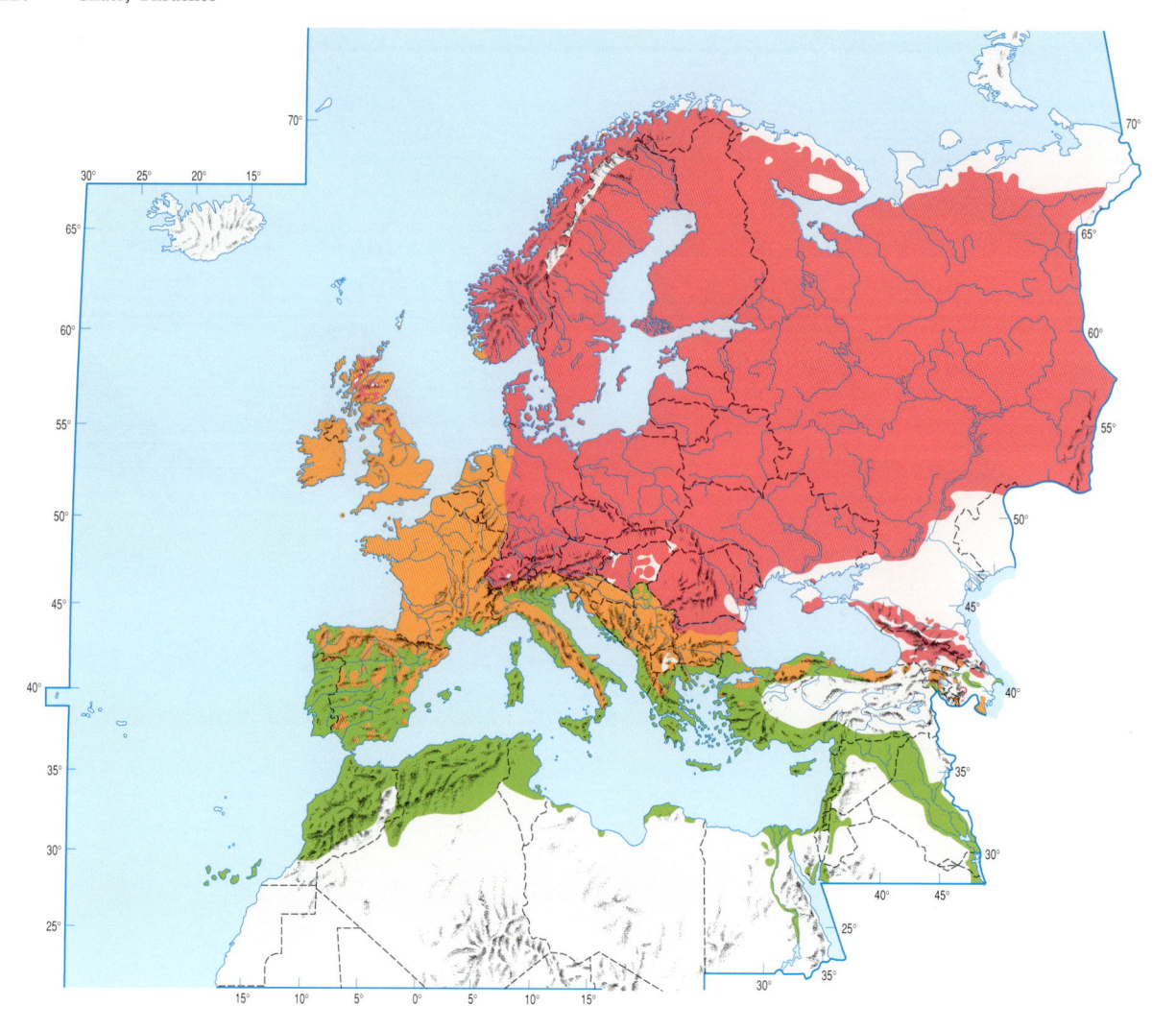

band to Lake Baykal; also breeds northern Iran. Introduced to Australia and New Zealand.

Population. Decline, notably in Britain and Netherlands, also Ireland, Belgium, Estonia. Mostly stable elsewhere, though increase reported Finland and Iberia. Decline attributed to hard winters, but other factors (perhaps including pesticides) probably also involved. BRITAIN. 990 000 territories 1988–91; steep decline since mid-1970s. IRELAND. 390 000 territories 1988–91. FRANCE. 400 000–2 million pairs; fluctuating. BELGIUM. 200 000 pairs 1973–7; probably some decline early 1980s. LUXEMBOURG. 20 000 pairs. NETHERLANDS. 100 000–125 000 pairs 1989–91. Marked decline since early 1980s, slight recovery since 1993. GERMANY. 2.4 million pairs in mid-1980s. DENMARK. 30 000–290 000 pairs 1987–8. NORWAY. 500 000–1 million pairs 1970–90. SWEDEN. 1.5–3 million pairs in late 1980s. FINLAND. 1–1.5 million pairs in late 1980s. Increase reported after 1930 (in north) and more recently. ESTONIA. 200 000–500 000 pairs 1991. Decline since 1960s, more marked since 1970s. LATVIA. 200 000–250 000 pairs in 1980s. POLAND. 500 000–1 million pairs. Marked decline in urban habitats, stable elsewhere. CZECH REPUBLIC. 400 000–800 000 pairs 1985–9. SLOVAKIA. 300 000–600 000 pairs 1973–94. HUNGARY. 150 000–200 000 pairs 1979–93. AUSTRIA. Common. No apparent changes. SWITZERLAND. 200 000–250 000 pairs 1985–93. SPAIN. 200 000–400 000 pairs. PORTUGAL. 10–100 pairs 1991. ITALY. 100 000–300 000 pairs 1983–93. GREECE. 2000–5000 pairs. ALBANIA. 200–1000 pairs 1981. YUGOSLAVIA: CROATIA. 200 000–400 000 pairs. SLOVENIA. 100 000–150 000 pairs. BULGARIA. 100 000–1 million pairs. RUMANIA. 400 000–500 000 pairs 1986–92. RUSSIA. 100 000–1 million pairs. BELARUS'. 700 000–800 000 pairs 1990. UKRAINE. 180 000–200 000 pairs 1986. MOLDOVA. 35 000–60 000 pairs 1989. AZERBAIJAN. Common. TURKEY. 10 000–100 000 pairs.

Movements. Mostly resident in south and west, but northern populations partially or entirely migratory; more birds move if weather severe. In contrast to (e.g.) Redwing and Fieldfare, populations show strong affinity to regular wintering areas. Most nominate *philomelos* from Fenno-Scandia, Germany, Switzerland, Poland, and FSU are migratory, moving south-west or south-east through Europe to winter in southern England, France (mainly towards south-west), Spain, and Portugal. Those from furthest north, especially 1st-year birds, winter furthest south to Canary Islands, Morocco, Algeria, Tunisia, Libya, and Cyprus. Birds from Denmark, Netherlands, Belgium, and north-east France are partially resident with most others moving only short distances south or south-west, though considerable numbers from Netherlands winter in Britain and Ireland. Birds from east-central Europe winter correspondingly east of birds from Fenno-Scandia and western Europe: mainly in Italy, Yugoslavia, Greece, Balkans, and Cyprus. Many birds breeding Britain and Ireland (*clarkei*) winter north-west France, northern Spain, and Portugal to Balearics. Birds from Outer Hebrides and Skye (*hebridensis*) are largely sedentary but some move to Ireland.

Southward departures in autumn begin in August but main passage September to early November; movement of birds into Ireland continues even into February. Birds wintering around Mediterranean arrive mid-October with frequent influxes until mid-April. During severe weather over Europe, large-scale mid-winter arrivals occur regularly in North Africa. Returning birds leave North Africa late March to mid-April. Northward movements from Portugal, Spain, western France, and through Britain and Ireland also at this time. Movement through Netherlands, Helgoland, and Denmark begins March and continues to mid-May. Finnish birds back on breeding grounds by mid-April and those in northern Sweden by early May.

Food. Wide range of invertebrates, especially earthworms; also fruit, as available, throughout year. Snails important in diet, especially during droughts and hard weather, when other food unavailable. Deals with snails by beating them against any hard surface, often a stone ('anvil'): holds snail in bill, either by lip of shell or by snail itself, through shell's opening; holds head to left or right, brings head and neck rapidly down, turning head at same time; usually only a part of shell removed thereby, and bird then flicks out snail's body, picks it up, and wipes it on ground before eating it; behaviour apparently largely innate.

Social pattern and behaviour. Outside breeding season more or less solitary or in small feeding or roosting aggregations, except during migration when regularly in large but loosely coordinated flocks. In areas where populations partially resident, many ♂♂ and smaller numbers of ♀♀ occupy individual winter territories, territories of ♂♂ being more or less the same as previous or subsequent breeding territories. Monogamous mating system. Pair-formation from early spring; process gradual and sometimes interrupted by cold weather, often not completed until shortly before nesting begins. Song usually delivered from perch, often tree-top or other high vantage point, occasionally from ground. In areas where populations totally migratory, song begins immediately after arrival of ♂♂ on breeding grounds, depending mainly on latitude; ends about mid-July in all areas. Where some ♂♂ resident (e.g. southern England), song begins when territories re-established in late autumn, usually late October or November; becomes intermittent in winter.

Voice. Song a loud, clear, vigorous succession of simple but mainly musical phrases distinguished by their repetitive character, great variety, and clear enunciation; more penetrating and less rich than Blackbird, lacking wild skirling quality of Mistle Thrush. Unmusical, harsh, or chattering sounds regularly intermixed with pure notes, also mimicry of other species. Commonest call a short 'tsip', given chiefly in flight and when flushed, often during nocturnal migration. Alarm-calls include a subdued 'djük' or 'dukduk' in low intensity alarm; in greater excitement, a rapid succession of higher-pitched and sharper units, producing a rattling 'kikikik. . .'.

Breeding. SEASON. Britain and western Europe: egg-laying usually from March to July; season prolonged and nests recorded in virtually all months of year. Central and eastern Europe: laying begins mid- to late April. Finland: laying from late April in south and mid-May in north. 2–3(–4) broods, not more than 2 in north of range. SITE. In trees and shrubs, often against trunk supported by twigs or branch, or among dense twigs; also in creepers on wall, on ledge, in bank, and on ground among thick vegetation. Nest: neat structure of twigs, grass, and some moss, loose towards outside, compacted towards inside; thickly lined with hard plaster material made from mud, dung, and (especially) rotten wood, often mixed with leaves. EGGS. Sub-elliptical, smooth and slightly glossy; bright pale blue, lightly spotted and speckled dark purple-brown or black; rarely unmarked. Clutch: 3–5 (2–6). INCUBATION. Average 13.4 days (10–17). FLEDGING PERIOD. Average 13.2 days (11–17).

Wing-length: *T. p. clarkei*: ♂♀ 111–119 mm.
Weight: *T. p. clarkei*: ♂♀ mostly 65–100 g.

Geographical variation. Clinal: main trend towards increasing paleness of plumage running from Hebrides through Scotland, England, and continental Europe to FSU, and minor trend towards increasing darkness of plumage from north to south in western continental Europe. 3 races recognized in Europe, each a more or less well-marked stage on east-west cline. *T. p. hebridensis* from Outer Hebrides and Isle of Skye is darker earth-brown above than *clarkei*, except for greyish rump and upper tail-coverts; underparts have more numerous, larger, and blacker spots contrasting more with paler buff ground-colour; flanks darker smoke-brown. *T. p. clarkei* from remainder of Britain and Ireland and adjacent parts of European continent characterized by warm brown upperparts; rump and upper tail-coverts olive-brown or slightly rufous, not as grey as *hebridensis* and nominate *philomelos*; chest and flanks rather deep yellow-buff; underparts profusely spotted. Typical nominate *philomelos* from Scandinavia, Poland, and eastern Rumania east to western Siberia and Caucasus area greyer olive-brown on upperparts, rump and upper tail-coverts olive-grey without rufous or brown tinge; buff ground-colour of underparts paler, more cream, less extensive. No appreciable variation in size in west Palearctic.

Redwing _Turdus iliacus_

Du. Koperwiek   Fr. Grive mauvis   Ge. Rotdrossel   It. Tordo sassello
Ru. Белобровик   Sp. Zorzal alirrojo   Sw. Rödvingetrast

PLATES: pages 1223, 1233

Field characters. 21 cm; wing-span 33–34.5 cm. Noticeably slighter and more rakish than Song Thrush, with slightly longer and more pointed wings but slightly shorter, sharp-cornered tail; slightly smaller than Dusky/Naumann's Thrush and Eyebrowed Thrush. Rather small, slight thrush, with noticeably pale-striped head and spots on underbody; red-chestnut underwing and flanks combine with rather dark upperparts to provide distinctive but not diagnostic appearance. Adult has chest, most of flanks, and sides of belly well marked, but rear flanks and vent noticeably white; immature has less obvious rufous flanks and fuller pattern of spots and streaks.

Adult easily distinguished from adult Song Thrush, with face pattern, chestnut underwing and flanks, and white rear body obvious in comparison, and not to be confused with any other common thrush. Immature may however suggest Eyebrowed Thrush (close in size, differing at distance only in lack of spotted gorget, and pale underwing), Dusky Thrush (larger, with greater rufous suffusion on underparts, wings, rump, and tail), and ♀ Siberian Thrush (larger and with barred underwing obvious in flight). Flight faster than Song Thrush, with sharp wing-point and tail-corners contributing to rakish silhouette which suggests Starling, and rather more undulating track, with periods of closed wings between bouts of wing-beats, also recalling Starling. Gait and general behaviour as Song Thrush but stance rather less upright. Normally much less in cover than Song Thrush, frequently sharing open feeding grounds with Fieldfare.

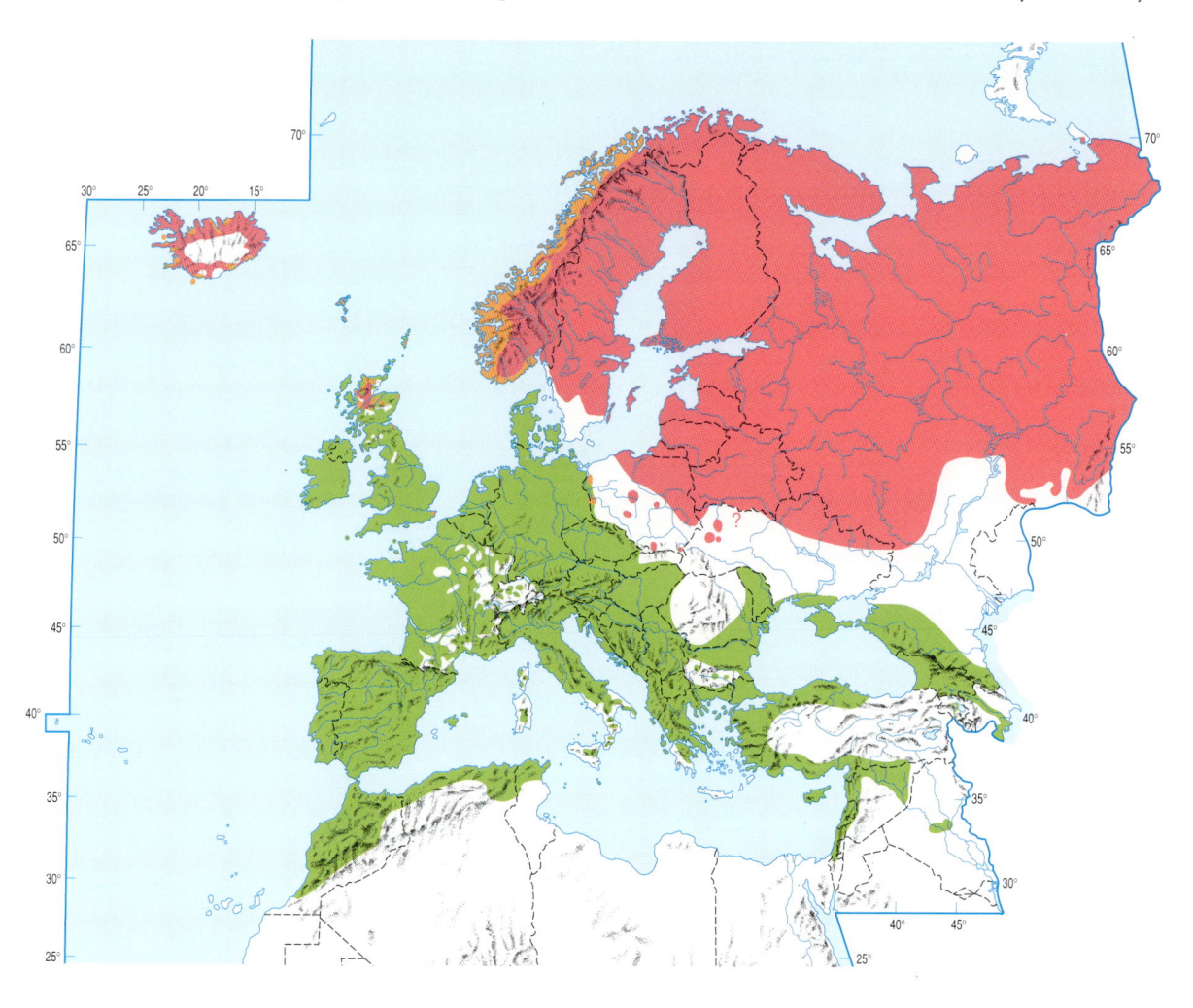

Habitat. Breeds in upper and upper middle latitudes of west Palearctic, mainly in subarctic and arctic lowlands and uplands, but avoiding snow and ice, and exposed chilly or stormy situations. Likes cover of birch or mixed woodland, often with many pines and spruces, especially along rivers and on floodlands, but also in low thickets of scrub birch, dwarf willow, and juniper, preferably on swampy ground. In Iceland breeds in broken, often rock-strewn country, commonly in birch scrub but also on ground sites among rocks even where scrub almost absent. In Scandinavia and Iceland, also nests locally in town parks and gardens. Recent colonists in Scotland have nested in grounds of large private houses, in hedgerows, on edges of oakwoods, in hillside birchwoods, and in variety of other sites combining open ground with some suitable cover. In winter occurs in a wide range of habitats, from open woodland to fields and, in smaller numbers, gardens; in autumn (especially) primarily related to availability of fruit.

Distribution. Colonized Britain, and bred occasionally (in some cases perhaps becoming established) Belgium, Germany, Denmark, Czech Republic, Slovakia, Austria, and Italy. Range expansion to south in Sweden and parts of FSU. SPITSBERGEN. Bred 1987. BRITAIN. First bred Scotland 1925; annually since 1967. Breeding confirmed fewer squares 1988–91 than during 1968–72 atlas survey. Recent breeding England, notably Kent (1975–91). IRELAND. Breeding attempt 1951. GERMANY. No regularly occupied sites, but breeding almost annual (c. 3 pairs in mid-1980s), especially pre-Alpine area and in east. Range has expanded west. DENMARK. First bred 1967. Occasional breeding since, latest 1–2 pairs 1983. SWEDEN. Expanded south in 1950s–60s and more recently. CZECH REPUBLIC, SLOVAKIA. First bred 1881 (Slovakia); of 20 subsequent reports, 17 since 1951. AUSTRIA. Bred exceptionally at least 1939, 1974, 1975, 1977. ITALY. Occasional breeding (perhaps by escaped birds); last 1991, Alps of Lombardy. FSU. Spread Lithuania, Belarus', Ukraine, Leningrad and Voronezh regions.

Accidental. Spitsbergen, Bear Island, Jan Mayen, Jordan, Kuwait, Libya, Azores, Madeira.

Beyond west Palearctic, extends east through Siberia to Kolyma river and Lake Baykal.

Population. Decrease Faeroes, Finland, and Estonia, increase Czech Republic, Belarus', and Ukraine. Stable (or fluctuating) most other countries. ICELAND. 100 000–300 000 pairs in late 1980s. Stable after increase (from c. 1930s). FAEROES. 5–15 pairs; small decrease in 1980s. BRITAIN. 40–80 pairs 1988–91, but seriously underrecorded and population could be well above this in western and northern Scotland alone. Stable or fluctuating, though apparently decline from peak of perhaps 300 pairs 1972. NORWAY. 1–1.5 million pairs 1970–90. SWEDEN. 1–2 million pairs in late 1980s. FINLAND. 2–3 million pairs. Marked decrease, though earlier reported to be fluctuating, or increasing considerably (in south). ESTONIA. 100 000–200 000 pairs 1991. Decline since late 1960s, more marked since 1970s. LATVIA. 60 000–100 000 pairs in 1980s. LITHUANIA. Population size unknown, though fewer in south. POLAND. 500–1500 pairs; stable after increase. CZECH REPUBLIC. 2–10 pairs 1985–9. SLOVAKIA. 1–5 pairs 1973–94; fluctuating. RUSSIA. 100 000–1 million pairs. BELARUS'. 140 000–180 000 pairs 1990. Some increase in 1960s–70s, especially in southern parts. UKRAINE. 40–70 pairs 1988.

Movements. Migratory or partially migratory. Winter range of whole population only just extends outside west Palearctic, so east Siberian birds must travel at least 6500 km WSW to reach winter quarters. Iceland and Faeroes population winters in Scotland, Ireland, western France, and Iberia. Since early 1930s, has also wintered increasingly in Iceland, particularly in larger towns. British and mainland Eurasian population winters in western Europe south from Scotland, coastal Norway, and south-east Baltic area, and around Mediterranean, Black, and south Caspian Seas.

Autumn movement out of Sweden and Norway occurs late September to mid-November, sporadically to December. Departure from Tomsk (south-central Siberia) starts at end of August and a few remain to late October. Arrives in France from late September, weather and feeding conditions there determining when birds move on to Iberia: may reach Spain and Portugal in November, but in some years not until January. Return passage begins in February, takes place mainly March–April, continuing into May in north. Lack of fixed wintering areas is a notable feature of Redwing's migration. Only a few ringing recoveries demonstrate year-to-year winter site-fidelity, and there are many to show that birds may winter in widely different localities in different winters, e.g. many birds ringed in Britain in winter have been recovered in subsequent winters in Italy, Greece, and localities even further to south and east.

Food. Wide variety of invertebrates; in autumn and winter, and to a lesser extent in spring, also berries. Feeds on ground and in trees and bushes. In foraging on open ground in winter, runs or hops in short bursts, usually 1–5 paces, halting between each run to scan ground in immediate vicinity. Also sweeps bill sideways to remove loose material; in woodland, will walk through leaf litter searching continually thus, and uses same technique while standing by cattle dung to search for larvae in it.

Social pattern and behaviour. Gregarious outside breeding season. Typically migrates in loose flocks. Pair-bond maintained for 2nd brood or re-nesting, but breaks down after breeding season. Breeding territories often well dispersed, but in optimum habitats may form loose colonies. ♂ sings from high exposed perch, e.g. rock, tree, overhead wire, occasionally from the ground; by night as well as day.

Voice. Song less developed than Song Thrush and variable in content; commonest phrase contains 3–4 fluted notes (sometimes faintly disyllabic or with others interspersed) and characteristic throaty chuckle at end recalling Blackbird but without richness of that species. Commonest call a thin, quite low-pitched but nevertheless slightly rasped, penetrating, and prolonged 'seez', 'seeih', or 'seeip' remarkably far-carrying, and often heard from night sky in late autumn and winter. In excitement or alarm, 'chittuck' or abrupt 'kewk'.

Breeding. SEASON. Scandinavia: egg-laying from early May; only slightly later in far north. Iceland: laying begins mid-May. 2 broods normally. SITE. On ground under bushes, or in thick vegetation, tree, bush, or stump. Nest: bulky cup with outer layer of twigs, grass, and moss, plastered inside with mud, plus some fragments of vegetation, with inner lining of fine grass stems and leaves; often very thin, but with thicker rim, up to 2 cm. EGGS. Sub-elliptical, smooth and glossy; pale blue to greenish-blue, profusely marked with fine red-brown speckling and mottling, though marks often small and indistinct. Clutch:

4–6 (3–7). INCUBATION. 12–13 days. FLEDGING PERIOD. 8–12 days.

Wing-length: *T. i. iliacus*: ♂♀ 110–121 mm.
Weight: *T. i. iliacus*: ♂♀ mostly 50–75 g.

Geographical variation. Two races recognized. *T. i. coburni* from Iceland and Faeroes differs from nominate *iliacus* in slightly larger size, slightly darker upperparts, and in particular in darker underparts: black streaks on throat heavier; breast, sides of belly, flanks, and under tail-coverts washed more extensively with olive-brown; brown marks on chest tend to merge together and extend further down to breast and sides of belly; sides of head, chest, and under tail-coverts strongly tinged buff.

Mistle Thrush *Turdus viscivorus*

PLATES: pages 1231, 1233

Du. Grote Lijster   Fr. Grive draine   Ge. Misteldrossel   It. Tordela
Ru. Деряба   Sp. Zorzal charlo   Sw. Dubbeltrast

Field characters. 27 cm; wing-span 42–47.5 cm. 25–30% larger-bodied than Song Thrush, with proportionately longer wing and tail contributing to marked attenuation of rear body and markedly heavy-chested and long silhouette; slightly longer and much bulkier than Fieldfare. Large thrush, with bold, upright carriage emphasized by length of tail, powerful bill, and sturdy legs; combination of olive-grey-brown upperparts, large discrete black spots on underparts, and white underwing diagnostic. Noticeable pale area on lore and around eye; corners of tail almost white, obvious from behind.

Large size, greyer upperparts, and white underwing immediately rule out smaller spotted thrushes. Juvenile traditionally mistaken for White's Thrush, but plumage pattern actually very different, lacking bold, heavy barring on underparts and black-and-white-banded underwing and strongly marked tail. Flight powerful and direct, with bursts of loose wing-beats giving noticeable momentum to heavily built and long-tailed bird; closes wings after each burst (producing long elliptical silhouette) but flight not undulating. Often flies rather high, even in breeding season reaching 30 m or more. Has

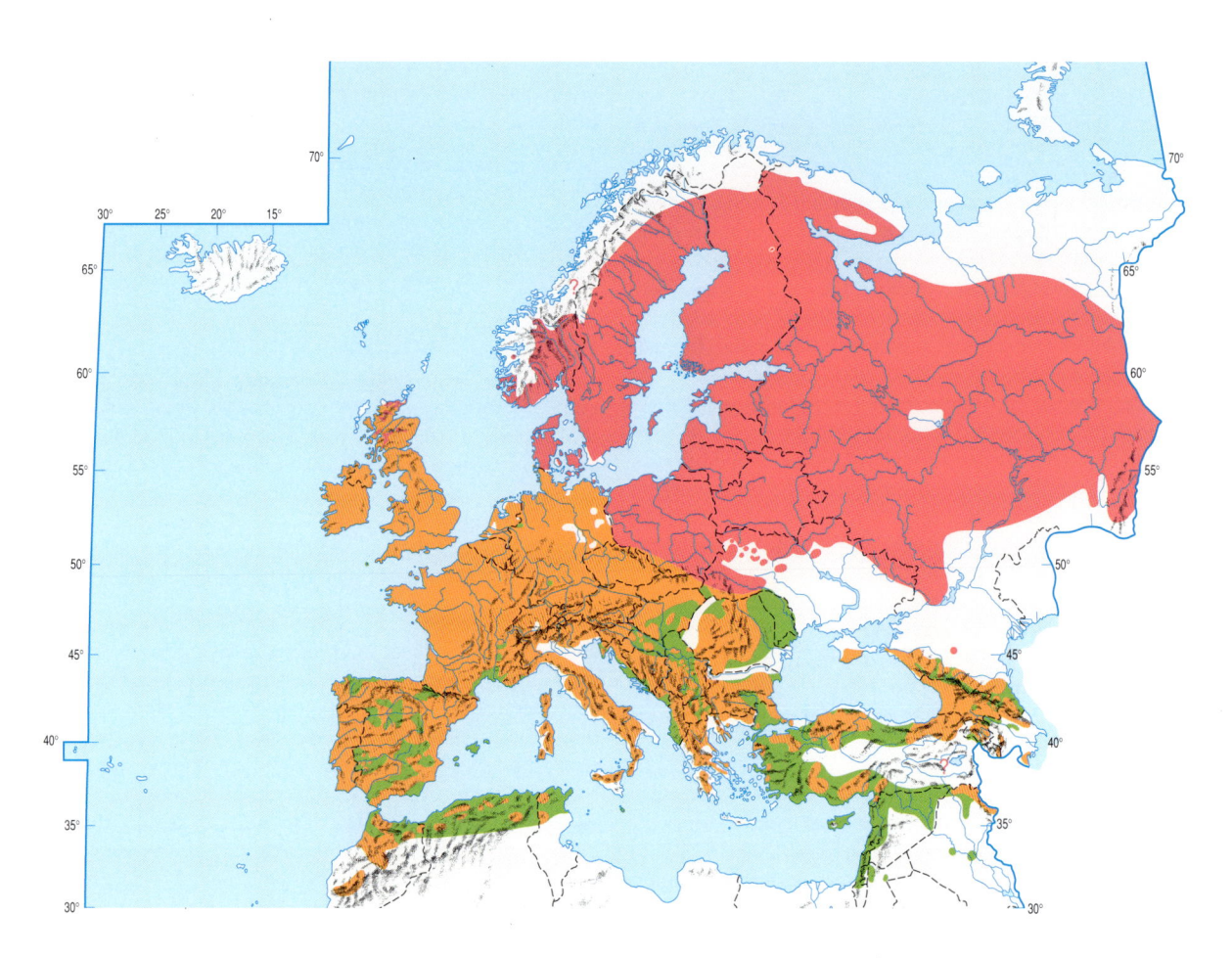

Mistle Thrush *Turdus viscivorus*. *T. v. viscivorus*: **1** ad worn (spring), **2** 1st winter, **3** juv. *T. v. deichleri*: **4** ad worn (spring). White's Thrush *Zoothera dauma aurea* (Russia) (p. 1206): **5** ad worn (spring), **6** 1st winter, **7** juv.

more bounding action when taking off or flying low, and may momentarily suggest other non-related species, e.g. Green Woodpecker or cuckoo. Gait as Song Thrush but behaviour bolder, being more aggressive. Wary in winter. Migrates in much smaller parties than other thrushes; forms loose associations of families in autumn but usually solitary in winter.

Habitat. Breeds in west Palearctic, historically in more continental upper and middle latitudes than other thrushes, but has during last 200 years extended into oceanic zone, especially in Britain and Ireland. Despite bold and robust appearance and conduct in wind and rain, is vulnerable to cold, especially snow and ice, and avoids arid and very warm areas. In mountain regions prefers middle altitudes, *c.* 800–1800 m in Europe. In boreal, temperate, and Mediterranean zones, widely but thinly distributed where there is conjunction of stands of tall trees with open grassland or short herbage, not too close to human settlements and with ready access to seasonal berry fruits. Avoids dense forests, but also treeless or sparsely wooded areas, broken or bare terrain, and wetlands. Has recently, in parts of range, overcome reluctance to inhabit urban parks and gardens, and has simultaneously expanded range as well as diversity of habitats, assisted in some areas by afforestation and planting for amenity.

Distribution. Spread Britain and Ireland and, more recently, Netherlands, Denmark, Norway, Hungary, and Austria. ICELAND. Infertile eggs laid 1988. BRITAIN. Rare northern England and Scotland end of 18th century, then spread in 19th (markedly in 1st half) and 20th centuries. Irregular breeding Orkney and Lewis. IRELAND. Colonized most of country by 1850 after first breeding 1807. NETHERLANDS. Rare and restricted to northeast before *c.* 1870. Spread to all provinces since 1900–1920, but expansion halted 1970–75, since when stable or declining. AUSTRIA. Spreading and colonizing lowlands in 1980s. IRAQ. Common in Ser Amadiya mountains and probably breeds there. MADEIRA. Probable breeder: regular observations (including adults with food and juveniles) Montado do Basreiro.

Accidental. Iceland, Faeroes, Israel, Kuwait, Azores, Canary Islands.

Beyond west Palearctic, extends east to *c.* 100°E, south to Iran and north-west Himalayas.

Population. Decline reported Britain, Belgium, Netherlands, Denmark, Estonia, Lithuania, Italy, and Ukraine; slight increase Ireland, Finland, Hungary, and Austria, apparently stable (or fluctuating) elsewhere. BRITAIN. 230 000 territories 1988–91. Recent decline after increase (this marked in 19th century, also in Ireland). IRELAND. 90 000 territories 1988–91. FRANCE. 50 000–200 000 pairs. BELGIUM. 35 000 pairs 1973–7. LUXEMBOURG. 2000–3000 pairs. NETHERLANDS. 25 000–35 000 pairs 1979–85. GERMANY. 462 000 pairs in mid-1980s. Another estimate (for early 1990s) of 240 000–350 000 pairs. DENMARK. 1400–14 000 pairs 1987–9. Decrease, but this fits badly with range expansion. NORWAY. 10 000–50 000 pairs 1970–90. SWEDEN. 75 000–200 000 pairs in late 1980s; trend unknown. FINLAND. 50 000–80 000 pairs in late 1980s. Probable increase following marked decrease. ESTONIA. 2000–5000 pairs 1991. Marked decline since at least 1960s. LATVIA. 30 000–60 000 pairs in 1980s. LITHUANIA. Rare and decreasing; more abundant in east. POLAND. 20 000–30 000 pairs. CZECH REPUBLIC. 35 000–70 000 pairs 1985–9. SLOVAKIA. 40 000–80 000 pairs

Eye-browed Thrush *Turdus obscurus* (p. 1218): **1–2** 1st winter. Dusky Thrush and Naumann's Thrush *Turdus naumanni* (p. 1219). *T. n. eunomus* (Dusky Thrush): **3–4** 1st winter. *T. n. naumanni* (Naumann's Thrush): **5–6** 1st winter. Black-throated Thrush and Red-throated Thrush *Turdus ruficollis* (p. 1220). *T. r. atrogularis* (Black-throated Thrush): **7** ad ♂, **8–9** 1st winter ♀. *T. r. ruficollis* (Red-throated Thrush): **10** ♂, **11** 1st winter ♀. American Robin *Turdus migratorius migratorius* (p. 1234): **12** ad ♂, **13** 1st winter.

1973–94. HUNGARY. 5000–10 000 pairs 1979–93. AUSTRIA. Slight increase in north-east. SWITZERLAND. 40 000–50 000 pairs 1985–93. SPAIN. 330 000–790 000 pairs. PORTUGAL. 10 000–100 000 pairs 1978–84. ITALY. 50 000–100 000 pairs 1983–93. GREECE. 10 000–30 000 pairs. ALBANIA. 2000–5000 pairs 1991. YUGOSLAVIA: CROATIA. 100 000–150 000 pairs. SLOVENIA. 20 000–25 000 pairs. BULGARIA. 10 000–100 000 pairs. RUMANIA. 70 000–100 000 pairs 1986–92. RUSSIA. 10 000–100 000 pairs. BELARUS'. 120 000–160 000 pairs 1990; fluctuating, but apparently stable overall. UKRAINE. 5000–6000 pairs 1988. AZERBAIJAN. Common in mountains; rare in lower part of forest zone. TURKEY. 5000–50 000 pairs. TUNISIA, MOROCCO. Uncommon.

Movements. Varies from migratory in north and east of range to sedentary or dispersive in west and south, though even in least migratory populations some birds make substantial movements. Some central and south European birds winter within breeding range of the relatively sedentary *deichleri* of north-west Africa, Corsica, and Sardinia. Main winter range of central European and Scandinavian birds extends from Belgium through western and southern France to north-east Spain. British birds tend to be sedentary or make movements of less than 50 km, though proportion travel considerably further, some to France.

Autumn migration in Britain mainly August–November although juveniles and adults may form flocks in July and start to move south. Return to breeding areas in spring is early. First birds appear on central European breeding grounds in February. In central Sweden, average arrival date 27 March.

Food. Wide variety of invertebrates, also fruits especially in autumn and winter. Feeds on ground (though not usually in undergrowth) and in trees and bushes. To take flying insects will fly up high in the air like Starling. Will fly up to *c.* 1.5 m from ground, without hovering, to take berries. For defence of feeding trees in winter, see next section.

Social pattern and behaviour. Moderately gregarious at times outside breeding season, especially in late summer and early autumn. From October, pairs or single birds defend concentrated fruit sources, in England most often holly trees, in France and Germany mistletoe. Spend most time near tree in mid-winter and defend it successfully through to start of

(FACING PAGE) White's Thrush *Zoothera dauma aurea* (p. 1206): **1–2** ad.
Siberian Thrush *Zoothera sibirica* (p. 1207): **3–4** ad ♂, **5** ad ♀.
Wood Thrush *Hylocichla mustelina* (p. 1209): **6–7** 1st winter.
Hermit Thrush *Catharus guttatus faxoni* (p. 1210): **8–9** 1st winter.
Swainson's Thrush *Catharus ustulatus swainsonii* (p. 1210): **10–11** 1st winter.
Gray-cheeked Thrush *Catharus minimus minimus* (p. 1211): **12–13** 1st winter.
Veery *Catharus fuscescens fuscescens* (p. 1211): **14–15** 1st winter.
Tickell's Thrush *Turdus unicolor* (p. 1212): **16–17** 1st winter.
Fieldfare *Turdus pilaris* (p. 1222): **18** ad ♂, **19–20** 1st winter.
Song Thrush *Turdus philomelos philomelos* (p. 1225): **21–22** 1st winter.
Redwing *Turdus iliacus iliacus* (p. 1228): **23–24** 1st winter.
Mistle Thrush *Turdus viscivorus viscivorus* (p. 1230): **25–26** 1st winter.

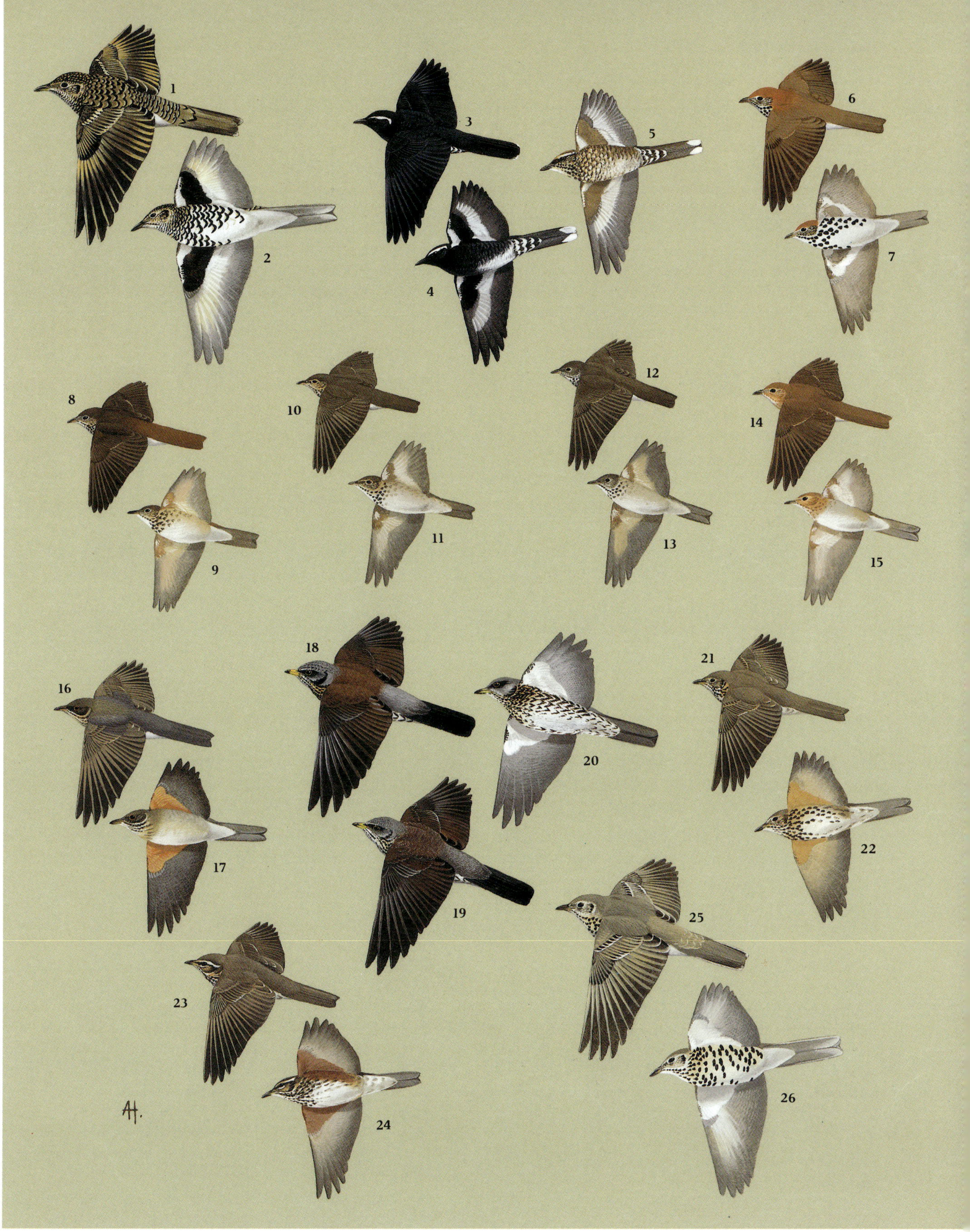
1
2
3
4
5
6
7
8
9
10
11
12
13
14
15
16
17
18
19
20
21
22
23
24
25
26
AH.

breeding season if conditions mild, though severe weather brings invading flocks of other thrushes which may overwhelm defenders. Much of defence takes place when Mistle Thrushes are mainly taking other food (from nearby trees or ground); tree-fruit thus conserved can, if slowly depleted, be food-source in spring and early summer. Breeding territories typically very large in comparison with Song Thrush and Blackbird in same areas. ♂ sings normally from high perch, typically tree-top; exceptionally from ground and not infrequently in flight. Song given in almost any weather, unlike other thrushes, which are inhibited by wind and rain; a little song in autumn, at time of establishment of fruit defence, but most from beginning of breeding season to early June.

Voice. Song loud and challenging, phrases typically of 3–6 notes with notable skirling quality; overall rather monotonous compared with Song Thrush and (especially) Blackbird but volume greater than either, carrying up to 2 km. Commonest calls essentially chattering: ticking 'churr' in flight, a dry, staccato 'tuck-tuck-tuck' on perch and loud excited rattle.

Breeding. SEASON. Western and central Europe: egg-laying from 2nd half of March. Britain: earliest nests late February. North Africa: laying from late March. Finland: earliest nests late April, most begun in first half of May. 2(–3) broods. SITE. On stout branch against trunk of tree, or on fork of horizontal branch; also on ledge of building (including inside ruin), cliff-face, or bank. Nest: large cup comprising 3 layers—outermost of sticks, grass, moss, and roots, loosely woven then compacted by middle layer of mud, often containing grass and leaves and some rotten wood which sometimes penetrates to outside of nest; thicker inner lining of fine grasses. EGGS. Sub-elliptical, smooth and glossy; pale blue, green-blue, or buff-tinted, spotted and blotched red-brown to purple, markings sometimes gathered at broad end. Clutch: 3–5(–6). INCUBATION. 12–15 days. FLEDGING PERIOD. 12–15 days.

Wing-length: *T. v. viscivorus*: ♂ 149–164, ♀ 142–162 mm.
Weight: *T. v. viscivorus*: ♂♀ mostly 100–150 g.

Geographical variation. Clinal throughout much of west Palearctic range; also considerable individual variation in plumage colour. 2 races recognized in west Palearctic: nominate *viscivorus* in most areas, *deichleri* (paler and greyer) in north-west Africa, Corsica, and Sardinia.

American Robin *Turdus migratorius*

PLATES: pages 1219, 1232

Du. Roodborstlijster   Fr. Merle d'Amérique   Ge. Wanderdrossel   It. Tordo migratore
Ru. Странствующий дрозд   Sp. Robín americano   Sw. Vandringstrast

Field characters. 25 cm; wing-span 35–39.5 cm. Length similar to that of Blackbird but head and body heavier. Quite large, robust thrush; mainly dark grey above and brick-red below, with bold but broken white eye-ring and white vent. Sexes somewhat dissimilar; ♀ browner above and duller red below.

Unmistakable, though in brief view ♀ and immature might be confused with Eye-browed Thrush or Naumann's Thrush (both noticeably smaller, with obvious supercilium and far less orange or red on underparts) or with Red-throated Thrush (which has red on face to chest and on tail, not from chest to vent). Flight powerful with action recalling both Blackbird (at ground level) and Fieldfare (on migration). Gait typical of genus but nervous on ground and movements noticeably restless, often accompanied by tail- and wing-flicking. Calls similar in tone and structure to those of Blackbird. Include challenging trisyllabic 'kwik kwik kwik' (given with jerk of tail); low-pitched 'tchook-tchook-tchook' and 'seech each-each-each' given in anxiety; quiet, disyllabic 'pit pit'; quiet, monosyllabic 'tseep'.

Habitat. Probably the most fully known example of modern transformation from a natural habitat of open forest and forest clearings to one associated closely with man, recorded over *c.* 300 years in middle and lower-middle Nearctic temperate latitudes. Settlements with gardens containing grassy lawns and shrubberies, including suburbs and even inner cities, have not only been colonized but have become concentrations of higher density than was attainable in earlier natural habitat.

Distribution. Breeds over much of North America to treeline or just beyond, and south to southern Mexico and Guatemala.
   Accidental. Iceland, Britain, Ireland, Belgium, Germany, Denmark, Norway, Sweden, Czech Republic, Austria.

Movements. Migratory over most of its range, scale and distance of migration varying with berry crops and severity of weather. Winters casually in much of breeding range, but primarily in USA (numbers increasing towards south) and in Bahamas, Cuba, Mexico, and Guatemala. Birds leave Alaskan breeding grounds in August, central Canada in September, and northern USA in October. Present on Bermuda late October or early November to March, and on Bahamas and western Cuba late October to April. Late spring migrants leave Florida early to mid-April, but breeding birds arrive earlier in much of range: e.g. late February in central USA, early March in New England, mid- to end of March in North Dakota and Utah, mid-May in Newfoundland.

Wing-length: *T. m. migratorius*: ♂ 127–136, ♀ 122–132 mm.

Old World Warblers and Allies Family Sylviidae

Tiny to small oscine passerines (suborder Passeres). A few semi-terrestrial but most cover-haunting and mainly arboreal, feeding chiefly on insects and other small invertebrates, obtained mostly by picking from vegetation; small fruits (often berries) and even seeds also taken by some species, also vegetable matter and liquids from flowers. Some 400 species, occurring mostly in woodland, parkland, scrub, and aquatic vegetation. Mainly of Palearctic, Afrotropical, and Oriental distribution, with largest diversity in tropical and southern Africa. Most species in northern Eurasia (i.e. majority in west Palearctic) highly migratory, travelling by night and feeding by day; winter mainly in tropics. 65 species in west Palearctic, all but 11 breeding.

Sexes generally of similar size. Bill usually thin but fairly stout in some genera; usually straight and of medium length. Wing variable: short and rounded in many non-migratory species, longer in migratory ones; 10 primaries, p10 generally short in non-migratory species with rounded wing-tip, tiny in migratory species with more pointed wing-tip. Aerial flight-displays by ♂♂ common, often accompanied by song. Tail highly variable: of short to medium length in most species but fairly long in some; mostly square or rounded but graduated or fan-shaped in some species. Monogamous mating system the rule, but some polygynous. Nests vary from simple open cups to more complex domed structures; built by both sexes, or largely by ♀. Building of 'cock' nests (unlined and not necessarily used for breeding) by ♂♂ a widespread feature of behaviour in *Sylvia*.

Plumage usually plain-coloured in leaf-inhabiting species, with upperparts various shades of brown, olive, or green, underparts paler, whitish, yellowish, or olive. Species inhabiting grass and reeds often profusely streaked, especially on upperparts. Some species show contrasting pattern on tail-tip or outer tail.

Cetti's Warbler *Cettia cetti*

PLATES: pages 1236, 1274

Du. Cetti's Zanger   Fr. Bouscarle de Cetti   Ge. Seidensänger   It. Usignolo di fiume
Ru. Широкохвостая камышовка   Sp. Ruiseñor bastardo   Sw. Cettisångare

Field characters. 13.5 cm; wing-span 15–19 cm. Head and body close in size to Reed Warbler but wings 10% shorter and tail 20% longer. Medium-sized, quite robust warbler, with short round wings, and relatively long, much rounded tail. Plumage rather dark and dull: essentially rufous-brown above and greyish below, with dull white supercilium and dull white spots on brown under tail-coverts. Sudden loud outburst of song diagnostic.

Only west Palearctic member of essentially east Asian genus (bush warblers); easily identified by voice and behaviour (see

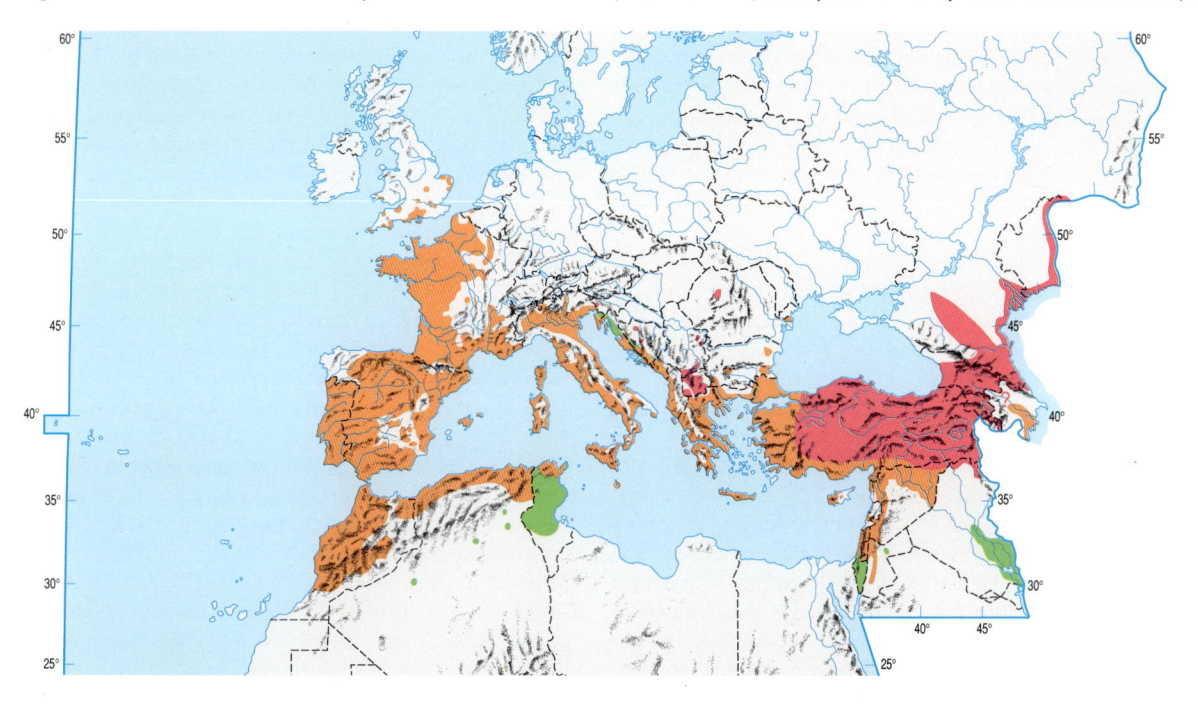

Cetti's Warbler *Cettia cetti*. *C. c. cetti*: **1–2** ad, **3** juv. *C. c. albiventris* (Iran eastwards): **4** ad. Fan-tailed Warbler *Cisticola juncidis*. *C. j. juncidis*: **5–6** ad, **7** juv. *C. j. neurotica*: **8** ad.

below), and marked preference for wet or damp areas with thickets. Although appearance becomes distinctive with experience, silent bird liable to initial confusion with brown species of *Locustella* and *Acrocephalus* warblers or even with nightingales *Luscinia*. Best structural character is prominent broad-feathered and rounded tail, often fanned, flicked, and cocked both on perch and in flight—but this feature sometimes displayed by *Locustella*. Best plumage characters at distance are markedly uniform upperparts and greyish underparts, less pale-centred than any confusion species. At close range, fine bill, rather domed, deep head, narrow, straight supercilium, and long, pale-tipped under tail-coverts form diagnostic combination. Flight rapid, bird dashing at low level between patches of dense cover with whirring beats of rounded wings and frequent tail-flirting.

Habitat. In lower middle and now in some middle latitudes of west Palearctic, in temperate, Mediterranean, steppe, and (marginally) some desert zones. Lives basically in warm situations where thick shelter normally enables survival throughout year, especially waterside and swamp vegetation, and thick bushy cover away from woodlands. Since *c.* 1920 has successfully expanded over cooler oceanic lowlands fronting Bay of Biscay, English Channel, and North Sea. Sites here usually below 100 m, but breeds up to much higher elevations (1400–2100 m) in southern Europe, Mediterranean islands, and North Africa. Habitats in FSU are more varied than elsewhere, including meadows with isolated bushes, hedgerows protecting orchards, mountain streamsides flanked by impenetrable vegetation, and reedbeds or dense stands of grass.

Distribution. In west of range, long-term expansion northwards, interrupted by hard winters. BRITAIN. First sighting 1961, and first confirmed breeding 1973 (near Canterbury, Kent, where strongly suspected 1972); dramatic colonization since (singing ♂♂ north to Shropshire), but some retreat by late 1980s after cold winters. Jersey (Channel Islands): first proved breeding 1973 (probable 1971–2). FRANCE. In mid-19th century almost confined to Provence; gradual expansion became conspicuous 1920s, reaching English Channel coast by 1961, and practically all low-lying parts of northern and western France by early 1970s. Local withdrawals or drastic reduction in numbers especially in recent decades, notably in north of range after severe 1978–9 winter, and more widespread after 1984–5 winter. Almost disappeared from Camargue in population crash after winters 1940–41 and 1986–7. BELGIUM. First bred 1964; recent retraction following spread. LUXEMBOURG. First recorded 1983, but not yet known to have bred. NETHERLANDS. First seen 1968. Breeding proved only in 1976, but suspected 1977–83, when 6–60 singing ♂♂. Only 0–3 present (probably under-reported) since 1984–5. GERMANY. Bred Niedersachsen 1975. SWITZERLAND. First seen 1965. Present at various sites 1973–83, and bred at least in 1975 and 1978. Since 1991, up to 4 singing ♂♂ suggest breeding, not yet proved. ITALY. Central and western Po valley colonized since *c.* 1970. MALTA. First proved breeding 1970, but probably established since at least early 1950s. RUMANIA. Breeds at 2 sites, recently reconfirmed. UKRAINE. Status uncertain; no confirmed observations, though perhaps overlooked. AZERBAIJAN. Distribution perhaps more widespread than known. CYPRUS. Increased greatly since mid-1970s.

JORDAN. Small population established late 1980s. IRAQ. Breeds very commonly along rivers in Ser Amadiya mountains. TUNISIA. Localized breeder. MOROCCO. Widespread and common in west, more local south and east of Atlas.

Accidental. Ireland, Sweden, Poland, Egypt. In Kuwait, reported as scarce winter visitor and passage-migrant.

Beyond west Palearctic, extends east to Zaysan depression, south to Iran.

Population. BRITAIN. 450 pairs 1988–91. Rapid growth, but decreases after cold winters, this effect less severe in milder south-west than in south-east and eastern England. Jersey: increased to *c.* 30 pairs by 1978; decrease since 1984; 20+ birds in early 1990s. FRANCE. Under 100 000 pairs in 1970s. For fluctuations, see Distribution. BELGIUM. *c.* 5 pairs in 1994. 240 pairs in mid-1970s, crashed due to cold winter 1978–9. SPAIN. 18 000–19 000 pairs; stable. PORTUGAL. 100 000–1 million pairs 1978–84; slight increase. ITALY. 200 000–300 000 pairs 1983–95; slight increase. MALTA. Up to 100 singing ♂♂. GREECE. 20 000–50 000 pairs. ALBANIA. 5000–10 000 pairs in 1981. YUGOSLAVIA: CROATIA. 2000–3000 pairs; slight increase. SLOVENIA. 80–120 pairs in 1978. BULGARIA. 100–1000 pairs; slight increase. RUMANIA. *c.* 5–10 pairs 1986–92. RUSSIA. 7000–15 000 pairs; stable. AZERBAIJAN. Varies from rare to common at various sites. TURKEY. 50 000–500 000 pairs. CYPRUS. Perhaps 5000–20 000 pairs; has increased greatly. SYRIA. Common, reaching very high densities locally. ISRAEL. At least a few thousand pairs 1970s–80s.

Movements. Varies from migratory to sedentary in different parts of range. West Palearctic populations mainly sedentary, with some northerly dispersal in autumn, and vertical displacement and southerly extension of range in winter.

Food. Chiefly insects, also other invertebrates, taken mostly from or near ground; diet not uncommonly includes aquatic invertebrates, perhaps more often in winter, presumably taken from water's edge or by reaching down to surface from raised stems, etc.

Social pattern and behaviour. Skulking, keeping low in vegetation for long spells. In spite of habitat preference, however, not really shy. At least in southern England, ♂♂ and ♀♀ usually sedentary throughout the year, but minority of ♂♂ and ♀♀ (mates) regularly vacate breeding territories for winter, departing September and returning April. Mating system varies from monogamy to polygamy, but more often the latter, ♂ practising successive polygyny with 2 or more ♀♀ such that broods overlap. Role of sexes in care of young very different. ♀ alone incubates, broods and does most, sometimes all feeding of young in nest, but ♂ usually very active in feeding fledged young. Solitary and territorial when breeding. ♂-territories mutually exclusive with fairly clear boundaries which appear to be respected, without dispute, from at least mid-April. Size and shape of territory vary, but most often linear (for up to several hundred metres) along watercourses.

Voice. Freely used throughout the year, especially song, given mainly by ♂, also occasionally by ♀. Highly distinctive: a sudden explosion of clear penetrating notes lasting typically 2.5–5 s; ceases as abruptly as it begins and may not be repeated for several minutes. By day, ♂ sings while patrolling territory, typically delivering each phrase from a different perch. Phrase includes an exclamatory opening unit, thus: 'TCHI tchitchirititchitchirititchi'. Calls include a hard, staccato 'chip', soft 'wheet', harsh clicking sounds, and prolonged metallic rattle.

Breeding. SEASON. West and south-west Europe (including Britain), also Greece and southern FSU: main laying period mid-April to end of June. 2 broods. SITE. In thick vegetation, e.g. supported on stems of reed or nettle, among twigs, or occasionally in stouter branches of dense tangled scrub; in zone where ground vegetation tangled with woody stems. Mainly at 30–45 cm above ground level, rarely to 2 m. Nest: bulky, untidy cup, with base of leaves and stems, finer stems and roots above; lined with feathers, hair, reed flowers, and other fine material. EGGS. Sub-elliptical, smooth and glossy; chestnut-red to deep red, sometimes paler; occasionally with band of darker colour at broad end. Clutch: (2–)4–5. INCUBATION. 16–17 days. FLEDGING PERIOD. 14–16 days.

Wing-length: West Palearctic populations, ♂ 58–66, ♀ 50–62 mm.
Weight: ♂ usually 12–17, ♀ 9–15 g.

Geographical variation. Slight. *C. c. orientalis*, from Caucasus eastwards, slightly paler and larger than nominate *cetti* from rest of west Palearctic.

Fan-tailed Warbler *Cisticola juncidis*

PLATES: pages 1236, 1274

Du. Graszanger   Fr. Cisticole des joncs   Ge. Cistensänger   It. Beccamoschino
Ru. Веерохвостая камышовка   Sp. Buitrón común   Sw. Grässångare

Field characters. 10 cm; wing-span 12–14.5 cm. 25% smaller than Sedge Warbler; similar in head and body size to Graceful Warbler and Scrub Warbler but much shorter-winged and (especially) shorter-tailed than both. Except for goldcrests, smallest of west Palearctic Sylviidae, with stumpy fan-tail. Plumage warm-buff, with heavy dark streaks above and paler throat and vent; pale area round eye, and black-and-white tips to tail-feathers. Song and song-flight distinctive.

In Europe, confusion with small streaked *Acrocephalus* warblers is unlikely to persist, as Fan-tailed Warbler is much smaller, less patterned on head, and much shorter-tailed; also separated from *Acrocephalus* by preference for long grass, low tussock marsh, or open margins of reedbeds. In Middle East, confusion with Graceful Warbler more likely (habitats overlap), but short, often-cocked tail instantly distinctive. Flight distinctive, normal action consisting of short flits or bobs on whirring wings, together with short spread tail producing compact, rather bee-like form. Song-flight lengthy and undulating, each bound synchronized with song-note. Gait normally a hop, varied by sidling and creeping movements, but may also stride like pipit; makes remarkable, easy progress through dense cover and rapid ascent of stems, often with feet on separate plants. Busy, seldom still, and often inquisitive and excitable.

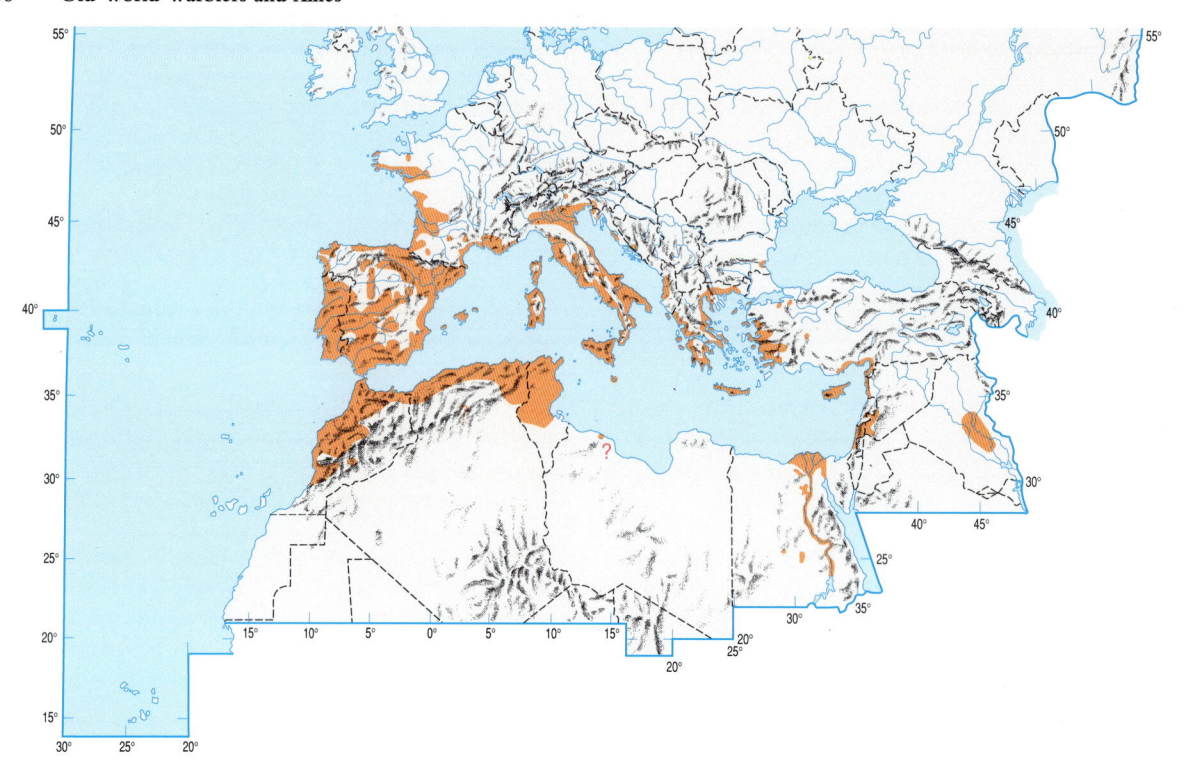

Habitat. In west Palearctic, habitat lies primarily in Mediterranean and neighbouring warm temperate lowlands, both continental and oceanic, but basically a bird of tropical grasslands, often those subject to destructive seasonal fires. Occupies wet as well as dry habitats, almost always, however, dominated by grasses or other plants, not stiff or woody, and below *c.* 1 m tall. These must permit access to more or less level soft unobstructed soil beneath, and for nesting afford access to such flexible plants as soft, narrow-bladed grasses. Dry crops, such as cereals, lucerne, and sugar-cane often colonized, as are wet ricefields and wetlands affording suitable vegetation. Generally avoids arid sandy, stony, and rocky areas. In Europe, most habitats are coastal and low-lying, often modified by man.

Distribution. In western Europe, periodic dramatic northward expansions reversed by severe winters. Has extended range in Balkan peninsula. FRANCE. In 19th century, confined to Mediterranean coast. Major spread from 1912–13, withdrawing to Mediterranean coast after 1939–40 winter. New wave of expansion from late 1960s, with first breeding records Brittany and Normandy in early 1970s (and vagrants recorded England 1976 and 1977); but retreat and population crash throughout France following severe winters 1984–5, 1986–7; renewed expansion began 1988. BELGIUM. First bred 1977 (probably 2 pairs); colonization stopped following cold winter 1978–9. NETHERLANDS. First bred 1974; occasional breeder since. AUSTRIA. First breeding record Rhine delta, 1995. SWITZERLAND. First recorded 1972; bred 1975. SPAIN. Has spread in coastal areas of north. ITALY. Dispersed or declined in newly colonized areas in Po valley after hard winters. GREECE. Map shows maximum range following series of mild winters. Crete: first recorded 1967, proved breeding 1975. MALTA. Breeding since early 1970s. YUGOSLAVIA. Bred Istra valleys from 1974, spreading south to central Dalmatia late 1970s. BULGARIA. First recorded (5–6 singing ♂♂) Lake Atanasov saltpans near Burgas 1984. JORDAN. Azraq population became extinct in 1980s. A small number may breed elsewhere. MADEIRA. Possibly bred 1909, but evidence (based on clutch of eggs) unsatisfactory, identity of birds being unproven.

Accidental. Britain, Channel Islands, Ireland, Germany, Austria, Bulgaria, Kuwait, Canary Islands.

Beyond west Palearctic, breeds widely in sub-Saharan Africa, and from southern Asia south to northern Australia; local in Arabia and Iran.

Population. FRANCE. Over 1000 pairs in 1970s, but liable to strong fluctuations (see Distribution). SPAIN. 14 300–57 000 pairs; slight increase. PORTUGAL. 1–10 million pairs 1978–84; stable. ITALY. 100 000–300 000 singing ♂♂ 1983–95. MALTA. 500–1000 pairs. GREECE. 5000–10 000 pairs. ALBANIA. 2000–

5000 pairs in 1981. YUGOSLAVIA: CROATIA. 500–1000 pairs; slight increase. SLOVENIA. 8–12 pairs in 1978. TURKEY. 1000–10 000 pairs. CYPRUS. Common; probably stable. ISRAEL. Common. EGYPT. Abundant (over 100 000 pairs). TUNISIA. Common. Increasing with creation of irrigated cultivation. ALGERIA, MOROCCO. Common.

Movements. Chiefly sedentary but dispersive; also eruptive, with northern limits of range varying with temperature fluctuations; evidence of regular migration in western Mediterranean.

Eruptive movements result in sharp fluctuations in numbers: e.g. in Camargue, where suddenly abundant after drastic decline due to hard winters, and on Catalonian coast (north-east Spain), where very numerous in some years, scarce in others; similar fluctuations reported in Israel. Eruptions may also lead to rapid extension of range, notably in western Europe in early 1970s.

Food. Chiefly insects, taken on or near ground.

Social pattern and behaviour. Little information from west Palearctic, but studies elsewhere indicate mating system typically polygynous (♂ plus 2 or more ♀♀). ♀ alone broods and feeds young. Territory size very variable. Song-flight of ♂, essentially to attract ♀♀, highly conspicuous. ♂ performs song-flights for much of the day during breeding season: flies up to height of c. 30 m and, with tail fanned, follows lengthy undulating path in otherwise horizontal plane, singing so that each song-unit coincides with an upswing. Paths of neighbouring ♂♂ typically overlap and criss-cross, with no obvious antagonism. ♂ spends considerable part of his time on territory building so-called 'courtship-nests' for purpose of attracting ♀♀ throughout breeding season.

Voice. Rather a silent bird apart from song of ♂, which is freely used: sharp, high-pitched, slightly harsh or rasping 'tsip tsip tsip ...', units given at regular intervals of c. 0.5–1 s, in song-flight or from perch. Calls include persistent, far-carrying 'zip' or 'chip', used to maintain contact or in slight alarm, and other short, monosyllabic units uttered singly or in rapid series.

Breeding. SEASON. Prolonged in all areas. Eggs from late March, North Africa and southern Spain; from early May, northern France. 2–3 broods. SITE. Low down in marshy vegetation, typically clumps of grass, rushes, etc., from less than 10 cm to 50 cm above ground, occasionally more. Nest: elongated pear- or bottle-shaped structure with entrance at or towards top, made of grasses bound together with cobwebs, lined with more cobwebs, flowers, hair, and down. EGGS. Short oval to sub-elliptical, smooth and glossy; very variable, with white or light blue ground-colour, unspotted or with specks or larger blotches of red, purple, or black, from sparse to dense, sometimes concentrated into zone at large end. Clutch: 4–6(–7). INCUBATION. (12–)13 days. By ♀ only. FLEDGING PERIOD. 14–15 days.

Wing-length: ♂ 49–56, ♀ 45–50 mm.
Weight: ♂♀ mostly 8–12 g.

Geographical variation. Slight and clinal. Nominate *juncidis* from Corsica east to Turkey more contrastingly streaked than *C. j. cisticola* from western Europe and north-west Africa; *C. j. neurotica* from Middle East similar to nominate *juncidis*, but ground-colour paler.

Graceful Warbler *Prinia gracilis*

PLATES: pages 1240, 1274

DU. Gestreepte Prinia   FR. Prinia gracile   GE. Streifenprinie   IT. Prinia gracile
RU. Изящная славка   SP. Buitrón elegante   SW. Streckad prinia

Field characters. 10 cm (including tail 4.5 cm); wing-span 12.5–13.5 cm. Small, short-winged, strikingly attenuated warbler, brown streaked blackish above and dull white below; long graduated tail with black-and-white feather-tips, often drooped or cocked.

Sole west Palearctic member of Afrotropical and south Asian

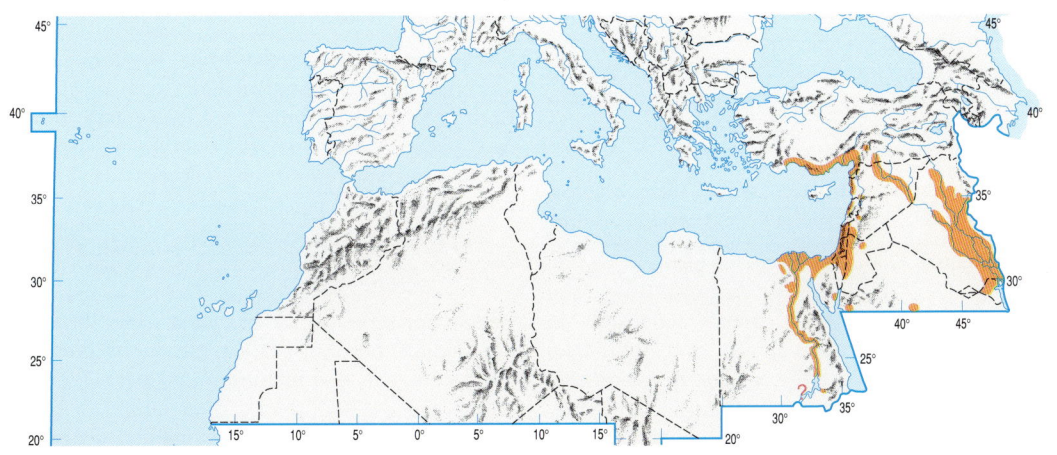

Scrub Warbler *Scotecerca inquieta*. *S. i. saharae*: **1** ad, **2** juv. *S. i. theresae*: **3** ad. *S. i. inquieta*: **4** ad. Graceful Warbler *Prinia gracilis*. *P. g. palaestinae* (Near and Middle East): **5** ad ♂, **6** ad ♀, **7** juv. *P. g. deltae* (Nile delta to western Israel): **8** ad ♂. *P. g. gracilis* (Nile valley from Cairo south): **9** ad ♂. Cricket Warbler *Spiloptila clamans*: **10** ad.

genus. Rather scruffy and *un*graceful, looking like mechanical toy bird at times. General slimness gives quite different character from Fan-tailed Warbler. Some caution needed with short-tailed newly fledged juvenile which might suggest latter species, but lack of strong warm tones in plumage, and yellow legs, quickly remove confusion. Flight distinctive, with almost pencil-slim silhouette of head and body, short, whirring wings, and long tail, often jerked or even side-switched; progress through tops of plants noticeably direct and level, ending in sudden plunge; at greater height, more bouncing and erratic. Gait essentially a hop, varied by sidling; adept at slipping through ground cover, escaping by this method as much as by flight. Skulking rather than shy.

Habitat. In Mediterranean, dry subtropical, and tropical zones of lower middle latitudes in south-east sector of west Palearctic; in lowlands and uplands. In Egypt, characteristic of low vegetation, natural or cultivated, from reeds along waterways to public gardens and suburbs, avoiding higher trees which inhibit ground vegetation. Generally intermediate in habitat preference between arid and moist vegetation, with tendency to favour moist and to be a ground bird mainly where there is ample plant cover. Avoids both desert and tall dense woodland.

Distribution. TURKEY. Common in coastal plains of eastern Mediterranean; local and rare south-east Anatolia. Range and numbers apparently fluctuate. IRAQ. Breeds abundantly in Mesopotamian plain. KUWAIT. Has spread south from Iraqi border. First recorded breeding 1990; apparently 5 territories 1995. EGYPT. Some local extensions of range; apparently spreading along Lake Nasser.

Accidental. Cyprus.

Beyond west Palearctic, breeds from Sudan to Arabia and in narrow band from Iran east to Assam.

Population. TURKEY. 500–5000 pairs. SYRIA. Locally common. ISRAEL. Abundant. JORDAN. Common. EGYPT. Abundant (over 100 000 pairs).

Movements. Generally sedentary. In southern Turkey, some dispersal apparently occurs outside breeding season. Very few records outside breeding range.

Food. Chiefly insects, taken from foliage or on ground. Hops about energetically and will make long leap if necessary to pounce on prey; occasionally catches an insect in flight or hovers to pick caterpillar off leaf. Regularly hangs upside down to feed from foliage.

Social pattern and behaviour. Territorial all year. Monogamous mating system. Pair-bond maintained closely all year. Both sexes incubate, brood, and feed young. Song given only by ♂, typically from one of several favoured song-posts, usually near boundary of territory, e.g. telegraph wire, tree-top, fence-post; may change perch frequently. Delivers song with bill wide open, exposing conspicuous black mouth, head turning from side to side, plumage ruffled, and tail vibrating. ♂ also has conspicuous threat display. Common response to rival is wing-snapping—a quick crackling triple 'brrrp-brrrp-brrrp', not audible beyond *c*. 10 m, produced possibly by wings brushing rapidly past the upward flicked tail. Bird may wing-snap when perched, during short jumps, or (most often) in flight. At highest intensity, ♂ combines wing-snapping with erratic aerial dancing-display: shoots almost vertically upwards 'for a few feet' with very rapid wing movements and repeated tail-cocking, giving whole display a jerky appearance; at peak of ascent, may either flit about erratically before perching again, or dive straight down to original perch (or one nearby) before repeating performance.

Voice. Freely used throughout the year. Song of ♂ a far-carrying, simple repetition of a hard, thin, somewhat slurred and disyllabic 'ze(r)wit' which sometimes has slight tinkling quality. Contact-alarm call a drawn-out 'breep' or 'zeet', of variable quality and rate of delivery, given by both sexes; also various ticking or twittering calls.

Breeding. SEASON. Eggs from March to July or August. 2–3(–4) broods. SITE. In grass or other low vegetation, bush, or occasionally creepers, sometimes in crop but more often along overgrown ditch, at edge of fields or in tamarisk thicket. Nest: occasionally touches ground but normally above, to 1.5 m, occasionally higher: untidy domed oval of dry grass mixed with vegetable down, leaves, bark fibre, rootlets, etc., lined with softer, downy material, often from Compositae seeds; entrance-hole over $\frac{2}{3}$ of the way up side. EGGS. Sub-elliptical, smooth and glossy, very variable in colour and markings. North Africa and Israel: whitish to very pale pink, blotched and speckled reddish-brown particularly at broad end. Iraq: green to pale blue-green, variably speckled and spotted red-brown, usually with band at broad end. Clutch: (2–)4–5. INCUBATION. 11–13 days. FLEDGING PERIOD. 12–13 (11–16) days.

Wing-length: ♂ 43–47, ♀ 41–45 mm.
Weight: ♂♀ mostly 6–7 g.

Cricket Warbler *Spiloptila clamans*

PLATE: page 1240

DU. Krekelprinia FR. Prinia à front écailleux GE. Schuppenkopfprinie
RU. Чешуйчатая славка SW. Sahelsångare

Field characters. 9–10 cm; wing-span 11–12 cm. Close in size and structure to Graceful Warbler. Rather small, perky, long-tailed warbler with variegated plumage. Plumage basically vinaceous-tawny or -buff above and pale cream on face and below, but these tones strikingly interrupted by black and white mottling on top of head, even bolder black centres and white edges of wing coverts and tertials, and white terminal and black subterminal bands to grey tail feathers; rump pale yellow. ♂ has pale grey shawl. Immature has browner upperparts and dark brown streaks on crown.

Unmistakable. Highly social; usually seen in small parties of 5 or 6 birds, moving restlessly between patches of low scrub. Has habit of jerking tail up and down and side to side, while uttering monotonous, tinkling and cricket-like call 'du-du-du-du-du-du'. Sweet song given from high perch.

Habitat. Breeds in thornscrub belt (especially in acacia) in Sahel zone of northern Africa, including true desert with sufficient vegetation; also in savanna with broad-leaved trees, avoiding grass plains or dense bush. Generally remains below *c.* 1000 m. Feeds on insects; nests low down in thorny shrub or in tussocky grass.

Distribution. Resident in narrow band from northern Sénégal and southern Mauritania east through northern Nigeria and Chad to Sudan and Eritrea.

In Mauritania, wanders just inside southern boundary of west Palearctic, but no proof of breeding there.

Movements. Only local and seasonal; perhaps partial migrant at northern edge of range since some move south in dry season (April–May in Mali), returning north in rains (July–September). Eggs recorded in Mali and Mauritania mainly July (June–September); almost all year round, though principally September–October, in Sénégal, and January–April and August in Sudan.

Wing-length: ♂♀ 43–50 mm.

Scrub Warbler *Scotocerca inquieta*

PLATES: pages 1240, 1274

DU. Maquiszanger FR. Dromoïque du désert GE. Wüstenprinie IT. Codanera
RU. Скотоцерка SP. Buitrón desertícola SW. Ökensnårsångare

Field characters. 10 cm (including tail 3.5 cm); wing-span 13.5–14 cm. Close in size to Fan-tailed Warbler but with longer tail, and head and body shaped more like Wren; more robust and less attenuated than Graceful Warbler, with tail *c.* 20% shorter. Small, quite robust, rather wren-like, and usually terrestrial warbler, with generally pale, sandy-grey plumage lacking obvious streaks (except on long domed crown and across face) and graduated tail with white feather-tips.

Puzzling rather than difficult to identify. When first seen, likely to suggest Fan-tailed Warbler or Graceful Warbler, but these unlikely to occur in Scrub Warbler's usual habitats; in any case, distinctive in wren-like character, less excitable and essentially terrestrial behaviour and blackish eye-stripe; bill never all-black as in some Graceful Warblers; loose tail movement and liquid voice add final diagnosis. Flight whirring and lacking in power, usually short and at low level. Can be skulking but often feeds openly around shrublets and lines of stones.

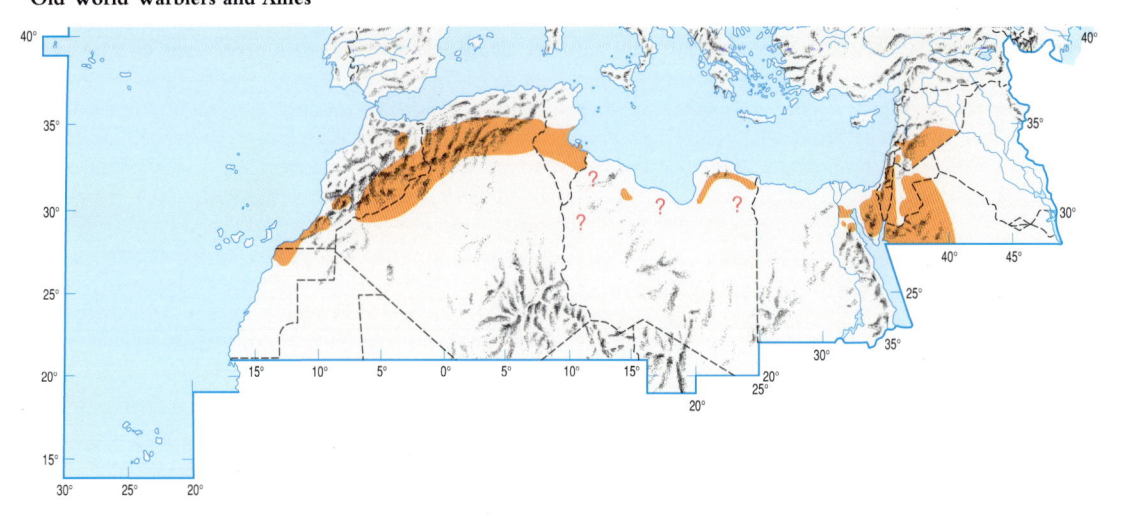

Habitat. In warm largely arid lower middle latitudes. In northwest Africa, extends from edge of sandy Saharan desert dunes across narrow strip of semi-desert and steppe towards Atlas range, indiscriminately occupying tracts of sand or firm soil, wadi beds, and rocky slopes, where there is a sprinkling of low shrubs and tufts offering shelter. Similar habitats occupied further east in west Palearctic range.

Distribution. SYRIA. Widely dispersed and evidently only local. Long overlooked; range extends far further inland than previously supposed. ISRAEL. Spread north in 1950s–60s, but some contraction in 1980s, due to development, afforestation, and also apparently locally to nest parasitism by Cuckoo. MOROCCO. Has extended range towards north-east. MAURITANIA. Recorded in Adrar area February–April. Status unknown; suggests considerable southward extension of known range.

Beyond west Palearctic, breeds Arabia north-east to west-central Asia.

Population. ISRAEL. Locally common; some decline in 1980s. JORDAN. Common. EGYPT. Scarce. TUNISIA. Fairly common. ALGERIA. Locally common.

Movements. Almost entirely sedentary in west Palearctic.

Food. Chiefly insects, but seeds may be important (or even predominate) in winter. Forages mainly on or near ground, typically rummaging in plant debris under bushes or even disappearing into cavities among stones for several seconds, but will also feed in canopy of bushes and small trees.

Social pattern and behaviour. Little known about west Palearctic races. Territorial, apparently with monogamous mating system. ♂ typically sings close to ground either during brief exposure on top of bush or in flight.

Voice. Freely used throughout the year. Descriptions of ♂'s song rather diverse, and not known to what extent this reflects geographical or individual differences. Call-notes typically interspersed between phrases. Quality described as jingling and cheery; a clear and pleasant melody. In Jordan, a descending 'tsi-tsi-tsi-tsi-hue'; also rendered 'sti-sti-tu-tu-tu' and 'tu-tu-tu-tu-tu-tu'. In same region, a quiet opening phrase recalling tit, followed by a descending 'tee-tee-tee-lu-lu-lu' recalling Woodlark. Commonest call a clear, high-pitched 'che-wee' or 'too-weet'; also a ringing trill, 'trrrrr', rasping or scolding alarm-calls, and rapid 'zit-zit zit', exchanged between group members.

Breeding. SEASON. Algeria and Tunisia: eggs laid late February to early June. SITE. In low scrub from near ground-level to 1.5 m. Nest: domed structure of twigs, grass stems, and other plant material, lined with feathers, fur, and plant down; side entrance (sometimes 2, one used as exit). EGGS. Sub-elliptical, smooth and glossy; white, with very fine red or purple-red spots and speckles, usually forming band at large end. Clutch: 3–4 (2–5). INCUBATION. 13–15 days. FLEDGING PERIOD. 13–15 days.

Wing-length: ♂ 45–51, ♀ 44–50 mm.
Weight: ♂♀ 6–9 g.

Geographical variation. 3 races in west Palearctic, differing mainly in ground-colour of body plumage: *S. i. saharae* (Algeria to Libya) palest; *S. i. theresae* (Morocco) darker than both *saharae* and nominate *inquieta* (Egypt eastwards).

Pallas's Grasshopper Warbler *Locustella certhiola sparsimstriata* (southern Siberia): **1** 1st winter. Lanceolated Warbler *Locustella lanceolata* (p. 1244): **2** 1st winter. Grasshopper Warbler *Locustella naevia* (p. 1245). *L. n. naevia*: **3** ad, **4** worn 1st winter, **5** juv fresh. *L. n. straminea*: **6** ad. *L. n. obscurior*: **7** ad.

Pallas's Grasshopper Warbler *Locustella certhiola*

PLATES: pages 1243, 1274

Du. Siberische Sprinkhaanzanger   Fr. Locustelle de Pallas   Ge. Streifenschwirl   It. Locustella di Pallas
Ru. Певчий сверчок   Sp. Buscarla de Pallas   Sw. Starrsångare

Field characters. 13.5 cm; wing-span 16–19.5 cm. Up to 10% longer and bulkier than Grasshopper Warbler, though wings and tail relatively shorter. Small to medium-sized, ground-cover-haunting warbler. Essentially brown and buff, softly streaked, paler birds showing pattern of marks recalling Sedge Warbler and even Moustached Warbler. Tail diagnostically marked, with rounded feathers showing subterminal dark bands and pale tips.

Up to 1980, difficulties of close observation and of plumage variation in Grasshopper Warbler led to doubts on practicality of field identification. Since then, however, observers visiting Siberia and Mongolia have been able easily to identify Pallas's Grasshopper Warbler. Compared to Grasshopper Warbler, noticeably plumper and less uniformly coloured, with usually paler supercilium and hindneck, noticeable white tips to inner webs of tertials, brighter and warmer rump and upper tail-coverts (apparent even side-on), and terminal tail-marks. From behind, colour and contrast of rump recall Sedge Warbler, but supercilium never as striking as in that species. Flight low; flitting and whirring, with wings less obvious in silhouette than broad rump and full, graduated, often fanned and flirted tail. Flight interspersed with jinking and ducking, usually ending in sudden dive into dense cover. Markedly reluctant to leave ground cover, hiding even in boulder piles, and disturbed bird thus difficult to follow.

Quiet, but migrants and wintering birds utter short, rolled 'chirr' or 'cher(k)'; also disyllabic variant 'chi-chirr' or 'chirr-chirr', occasionally followed by another 'chirr'.

Habitat. Breeds in east Palearctic, extending to higher middle latitudes than western counterpart Grasshopper Warbler and through warmer continental summer climates. Also ascends higher, to sparse woodlands and alpine meadows up to 1900 m, and to margins of lakes and small bogs in mountains. Mainly, however, in valleys and lowlands, generally requiring stands of tall grass, tussocks, and hummocky meadows or river banks.

Distribution. Breeds from Irtysh river (Siberia) and eastern Tien Shan east to Sea of Okhotsk and north-east China.

Accidental. Britain, Ireland, France, Belgium, Netherlands, Germany, Norway, Latvia, Poland, Israel.

Movements. Migratory. Winters north-east India east to Indochina, south to Greater Sunda Islands. Occasional autumn records in Europe presumably due to reversed migration; British records (mostly on Fair Isle, Scotland) all September–October. Records in Afghanistan and Israel (1 record, February) presumably vagrants.

Wing-length: ♂ ♀ 59–74 mm (data from different races combined).

Lanceolated Warbler *Locustella lanceolata*

PLATES: pages 1243, 1274

Du. Kleine Sprinkhaanzanger   Fr. Locustelle lancéolée   Ge. Strichelschwirl   It. Locustella lanceolata
Ru. Пятнистый сверчок   Sp. Buscarla lanceolada   Sw. Träsksångare

Field characters. 12–12.5 cm; wing-span 13–16.5 cm. 10% smaller than Grasshopper Warbler with proportionately stubbier bill and shorter tail. Smallest, most skulking, and most streaked *Locustella*, sharing with Grasshopper Warbler rather variable ground-colour to upper- and underparts but lacking any greenish tone above. Best distinguished from Grasshopper Warbler by streaked (not spotted) crown and back, narrower, less diffuse pale edges to tertials and deeper gorget of black streaks on breast.

If seen closely and well, heavily streaked, pipit-like plumage is distinctive, and small size also helpful in ruling out other *Locustella*. However, unusually pale birds are liable to confusion with unusually heavily marked Grasshopper Warblers. Main problem can be finding bird and achieving more than a few glimpses or flight views. Flight and other behaviour as Grasshopper Warbler but even more skulking, hiding in grass tufts, sidling through densest ground cover, and breaking from it for only short flights or runs.

Habitat. Breeds in upper middle latitudes of (chiefly) east Palearctic, broadly overlapping Pallas's Grasshopper Warbler but ranging up mountains only to *c.* 800 m and stopping short of main steppe zone. Appears more attracted to wet or moist situations and to presence of bushes and shrubs, although similarly resorting to tall herbage, reedbeds, and tussocky meadows of grasses and sedges, as well as grass and shrubby openings in pine-wood taiga or sparser parts of tall forests, and thickets of willow or other shrubbery and sedge surrounding marshes.

Distribution and population. RUSSIA. Breeding range not well known, but extends further north than formerly supposed. Westernmost extension of range probably contracting east towards Perm'. Total population tentatively estimated at 10 000–100 000 pairs in early 1990s. In Kirov region (northeast of Moscow), 2 sites (at *c.* 48–49°E) each contained 5 singing ♂♂ in June 1986.

Accidental. Bear Island, Britain, France, Belgium, Netherlands, Germany, Denmark, Norway, Sweden, Finland (including singing ♂♂, June–August—4 in 1993), Yugoslavia.

Movements. Migratory. Winters south-east Asia south to Greater Sunda Islands, west to northern India and Andaman Islands, east to Philippines. Records in northern Europe almost entirely in autumn (with marked increase since 1973), and extend from Ouessant (north-west France) north to 75°N (125 km north of Bear Island, Arctic Ocean); most are on Fair Isle (Scotland). Autumn occurrences include considerable proportion of juveniles, and presumably result from reversed migration.

Food. Mainly adult and larval insects, especially beetles and Hymenoptera; also spiders.

Social pattern and behaviour. Little information; apparently similar to congeners.

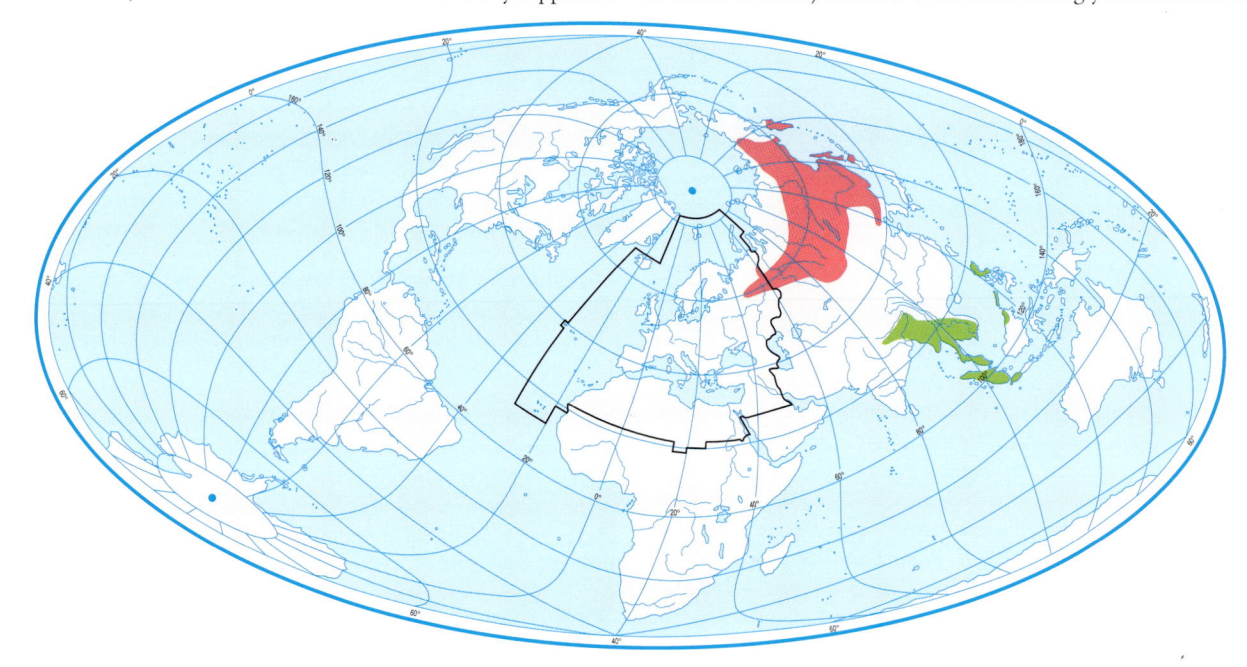

Voice. Song similar to chirring or 'reeling' of some other *Locustella*, and also confusingly like certain crickets, etc. Vibrating, thin, and high-pitched reeling or ticking trill suggests Grasshopper Warbler, but slightly slower and sharper, like stridulation of locust (Acrididae); pulsating quality (typical also of River Warbler) makes it even more strongly reminiscent of bush-cricket (Tettigoniidae). Calls include low-pitched 'chk' and a variety of metallic chacking sounds.

Breeding. Little information from west Palearctic. SEASON. Russia: eggs from mid-June. SITE. On or near ground in thick vegetation, often well concealed in tussock. Nest: deep, thick-walled cup of dry grass stems and leaves, moss, and other leaves, lined with finer grass. EGGS. Sub-elliptical, smooth and glossy; very pale pink to white, finely spotted and speckled red or pink, often profusely and more towards broad end. Clutch: 3–5. INCUBATION. 13–14 days.

Wing-length: ♂ ♀ 52–62 mm.
Weight: ♂ ♀ mostly 10–13 g.

Grasshopper Warbler *Locustella naevia*

PLATES: pages 1243, 1274

DU. Sprinkhaanzanger FR. Locustelle tachetée GE. Feldschwirl IT. Forapaglie macchiettato
RU. Обыкновенный сверчок SP. Buscarla pintoja SW. Gräshoppsångare

Field characters. 12.5–13.5 cm; wing-span 15–19 cm. About 10% shorter than Reed Warbler with relatively smaller head and shorter wings but longer tail. Rather small, dull, uniformly coloured and softly streaked warbler; epitome of *Locustella*, differing distinctly from *Acrocephalus* warblers in finer bill, shorter, rounder wings, usually rather broad rump, long tail-coverts, and full, graduated, and (when fanned) round tail, with even more skulking behaviour and distinctive reeling song. Few plumage features (even at close range) and variable olive to yellow-buff ground-colour combined with variable spotting and streaking provokes confusion with other *Locustella*. Upperparts softly to quite intensely spotted dark brown from crown to back. Typically, diffuse buff or reddish outer fringes to tertials, flight-feathers, and larger coverts just visible. Sides of throat and chest and upper flank warm buff, flecked or streaked dark brown, particularly on juvenile.

Commonest and most widespread *Locustella* in west Palearctic, but difficult to observe due to skulking behaviour and addiction to low tangled ground cover; hence potential confusion with all other *Locustella* except large Gray's Grasshopper Warbler. Singing male on open perch can be closely approached, but most views otherwise restricted to glimpses of small, flitting, short-winged, fan-tailed, olive-brown bird or small, often very slim, dull 'mouse' slipping through ground cover. Flight flitting (even whirring like Wren) and low; darting over short distance between plants, or rapid and more wavering over open spaces; ends in dive into cover. May appear to jink or 'buck' in flight, with lengthy-looking, broad rump and tail obvious; this impression strongest when tail fanned and flirted, particularly on entry to cover and in sudden twist or dive. Gait includes shuffle and hop up plant stem, horizontal creep and walk, remarkable high-stepping and loping run, and hop or jump on low plants or ground.

Habitat. In middle, mainly temperate, latitudes of west Palearctic, mostly continental but marginally oceanic, from July isotherms of 30°C to below 17°C. Generally in lowlands or on low hills. For breeding, favours various types of spaced-out dense low growth, including the following: overgrown bottomlands beside small streams; dry meadows with nettle-beds or scattered shrubs; patches of young trees, coniferous or broad-leaved, either spontaneous or planted, in forest clearings or along margins; osiers or coppices recently cut low, especially among tussock sedge; heaths, commons, and cliff-tops with plenty of brambles or gorse; moist or swampy ground such as slacks among sand-dunes with plenty of bushes or hedgerows; marshy or boggy rough grasslands with scattered or grouped willows, birches, or alders; thickets along river banks; fens and reedbeds with some bushes; occasionally, also standing crops of cereals or legumes, especially where bordered by drainage ditches. Such preferences lead to a need to shift from sites which have grown up too tall or too thick to be any longer suitable and thus to instability in local distribution. A special case of this occurs in woodlands, where regenerating low growth or new plantations provide suitable conditions for a few years only.

Distribution. Range increased markedly in Czech Republic and Hungary, less so in Fenno-Scandia, Estonia, and Slovakia; contracted in Britain, Ireland, Belgium, Rumania, and Moldova; fluctuating in France (though some expansion overall), and probably some other countries (see Population). BRITAIN. General thinning throughout range. IRELAND. Spread west before more recent decline. GERMANY. In east, density lower in south-east than elsewhere. DENMARK. First bred 1950s. NORWAY. First bred (Rogaland) 1967; spread thereafter. SWEDEN. First recorded breeding 1929; spread sparsely up to 1955 with subsequent infilling. FINLAND. First bred 1889; rapid spread from south-east in 1950s–60s. ITALY. Possibly breeds. GREECE. May breed; no proof. AZERBAIJAN. Perhaps bred late 19th century. TURKEY. Small population (8–10 singing ♂♂) upper Murat valley (in east; close to breeding grounds in Georgia and Armenia) in 1965, but not confirmed subsequently.

Accidental. Iceland, Faeroes, Malta (perhaps scarce passage-migrant), Cyprus, Lebanon, Israel, Jordan, Iraq, Kuwait, Egypt, Madeira, Cape Verde Islands.

Beyond west Palearctic, extends east to *c.* 100°E.

Population. Marked decline in Britain and Rumania, less so Benelux countries, Denmark, and Moldova; increased markedly Hungary, slightly in Norway, Finland, Estonia, locally in Germany and Austria; apparently mostly stable (or fluctuating) elsewhere. BRITAIN. 10 500 pairs 1988–91. Marked decline from 1971, following increase in 1960s; local fluctuations. IRELAND. 5500 pairs 1988–91. FRANCE. 10 000–100 000 pairs in 1970s; fluctuating. BELGIUM. 880–1580 pairs 1989–91. LUXEMBOURG. 200–300 pairs; decrease since 1980s. NETHERLANDS.

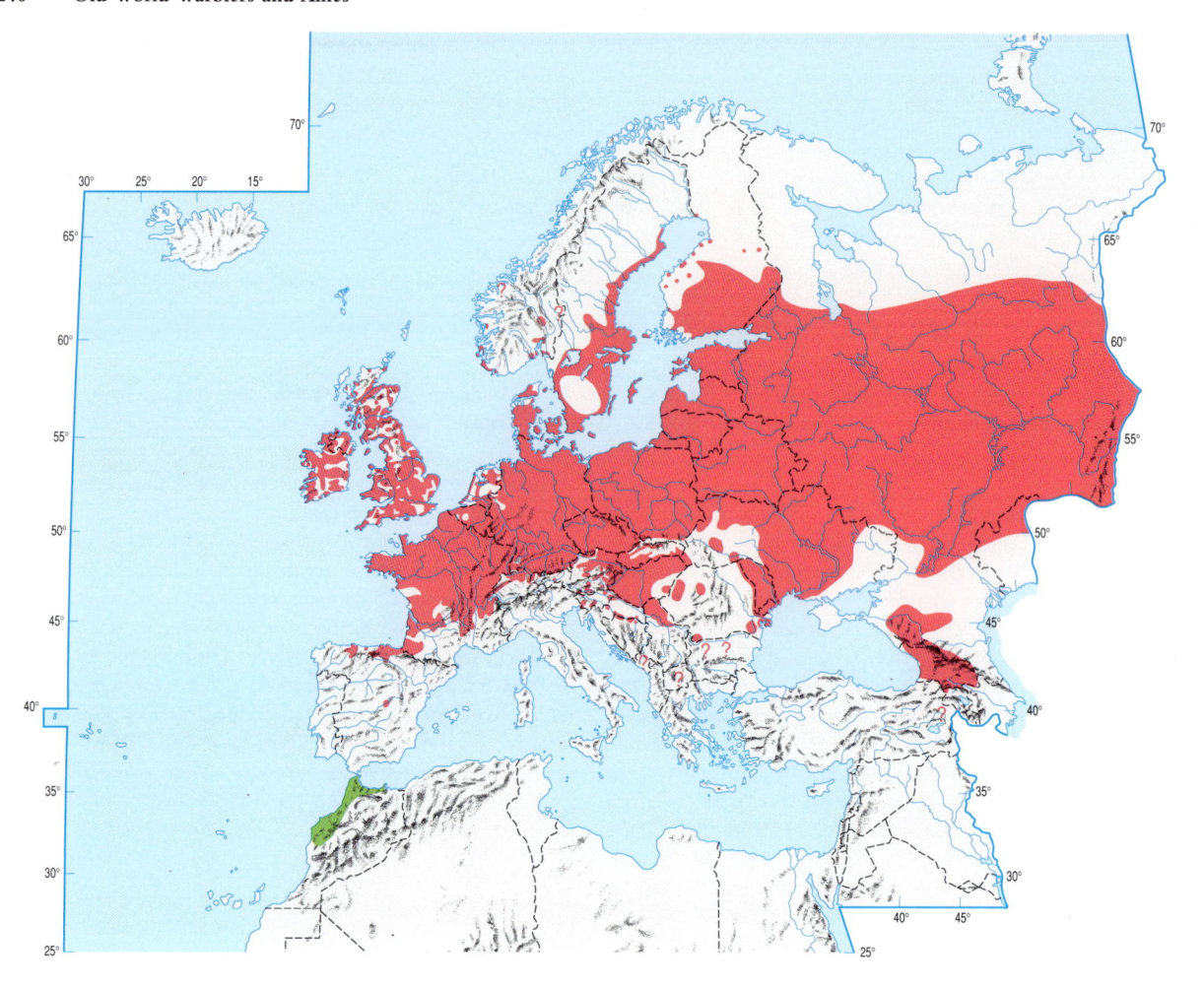

3000–5000 pairs 1979–85. Probable decline associated at least in part with loss of habitat. GERMANY. 80 000–160 000 pairs. Positive trend in some areas of east where 40 000 ± 20 000 pairs in early 1980s. DENMARK. 500–700 pairs 1987–8. Decrease 1989–94 following increase 1970–89. NORWAY. 30–100 pairs 1970–90. SWEDEN. 2000–10 000 pairs in late 1980s. Stable after sevenfold increase 1957–68. FINLAND. 4000–6000 pairs in late 1980s. Increase in 1950s–60s, probably slowing down in 1980s, but marked annual fluctuations. ESTONIA. 10 000–20 000 pairs 1991. Long-term increase, still continuing. LATVIA. 30 000–80 000 pairs in 1980s. LITHUANIA. Not abundant. Believed stable. POLAND. 4000–10 000 pairs. CZECH REPUBLIC. 15 000–30 000 pairs 1985–9. SLOVAKIA. 300–500 pairs 1973–94. HUNGARY. 2000–3000 pairs 1979–93. AUSTRIA. Not uncommon. SWITZERLAND. 200–250 pairs 1985–93. SPAIN. 250–300 pairs. YUGOSLAVIA: CROATIA. 200–300 pairs. SLOVENIA. 200–400 pairs. RUMANIA. 1000–2000 pairs 1986–92. Formerly common, now persistent decline. RUSSIA. 100 000–1 million pairs. BELARUS'. 12 000–17 000 pairs in 1990. UKRAINE. 1200–2000 pairs in 1988. MOLDOVA. 1600–2000 pairs in 1986.

Movements. Migratory. Winter quarters of west Palearctic birds not well known; apparently mainly in West Africa south of Sahara, with migration route concentrated through Iberian peninsula and along African west coast. Autumn passage chiefly August–September; prolonged spring passage February–May.

Food. Mainly insects. Food obtained while moving restlessly through vegetation and on ground. Rummages among dead leaves and stays long in one place dealing with prey once found. In reeds *Phragmites*, carefully examines each stem from top to bottom, working through all leaf axillae and from time to time descending to prostrate stems to dig for insects among them. Also hovers to pick prey from leaves and sometimes takes insects in flight.

Social pattern and behaviour. Little studied. Solitary in winter; territorial and apparently monogamous in breeding season. ♂ sings from low perch, usually below 1 m, either exposed or in cover, at all times of day and night.

Voice. Song of ♂ a rapid, monotonous, high-pitched (inaudible to some ears) and mechanical trill, not unlike sound of line running off angler's reel, hence popularly known as 'reeling'. Quiet start followed by crescendo; characteristically ventriloquial with peculiar apparent waxing and waning of volume and impression of changing pitch or tonal quality, especially noticeable on windy days, and presumably due to head-turning. Carrying power variable: up to 250 or 550 m, even *c.* 1 km in still conditions. Song may continue for many minutes, with brief, barely discernible pauses about every half-minute. Calls include soft, jingling variants of song, and variety of sharp ticking or more subdued clucking and rasping units.

Breeding. SEASON. Britain and western Europe: eggs from late April or May to July. 2 broods typical in southern England and probably elsewhere. SITE. On or just above ground in thick vegetation, often in tussock. Nest: thick cup of grass and plant stems and leaves, on base of dead leaves, lined with finer material, sometimes including feathers, horse hair, and plant down. EGGS. Sub-elliptical, smooth and glossy; white, finely but densely spotted and speckled lilac and purple or purple-brown, often tinting whole egg and usually with darker cap at broad end. Clutch: 5–6 (3–7). INCUBATION. 12–15 days. FLEDGING PERIOD. 10–12(–15) days.

Wing-length: Western Europe, ♂ 60–68, ♀ 62–66 mm.
Weight: ♂♀ 11–16 g; up to 20 g on migration.

Geographical variation. Slight in west Palearctic. *L. n. obscurior* from eastern Turkey and Caucasus greyer and more heavily spotted and streaked than nominate *naevia*. Latter intergrades with eastern race *straminea* (smaller and paler) on eastern fringes of west Palearctic.

River Warbler *Locustella fluviatilis*

PLATES: pages 1248, 1274

DU. Krekelzanger   FR. Locustelle fluviatile   GE. Schlagschwirl   IT. Locustella fluviatile
RU. Речной сверчок   SP. Buscarla fluvial   SW. Flodsångare

Field characters. 13 cm; wing-span 19–22 cm. 10% larger than Grasshopper Warbler; slightly shorter than Savi's Warbler overall but wing and tail average longer. Small to medium-sized, rather long-tailed, lithe, dull dark brown warbler, unmarked but for dull white mottling on throat, chest, and under tail-coverts and warmer tone on upper tail-coverts and tail; plumage may have grey, olive, or umber tone.

Quite distinctive, given its marked breeding habitat preferences, diagnostic song, mottled underparts (when visible), and broad and unusually substantial vent and tail in side view; also, tail much rounded. In flight, action and appearance as Savi's Warbler, but silhouette even fuller tailed. Gait and other behaviour typical of *Locustella* but, like Savi's Warbler, uses more exposed song-posts than Grasshopper Warbler.

Habitat. Breeds in upper middle and middle latitudes of warm continental boreal, temperate, and steppe zones of west-central

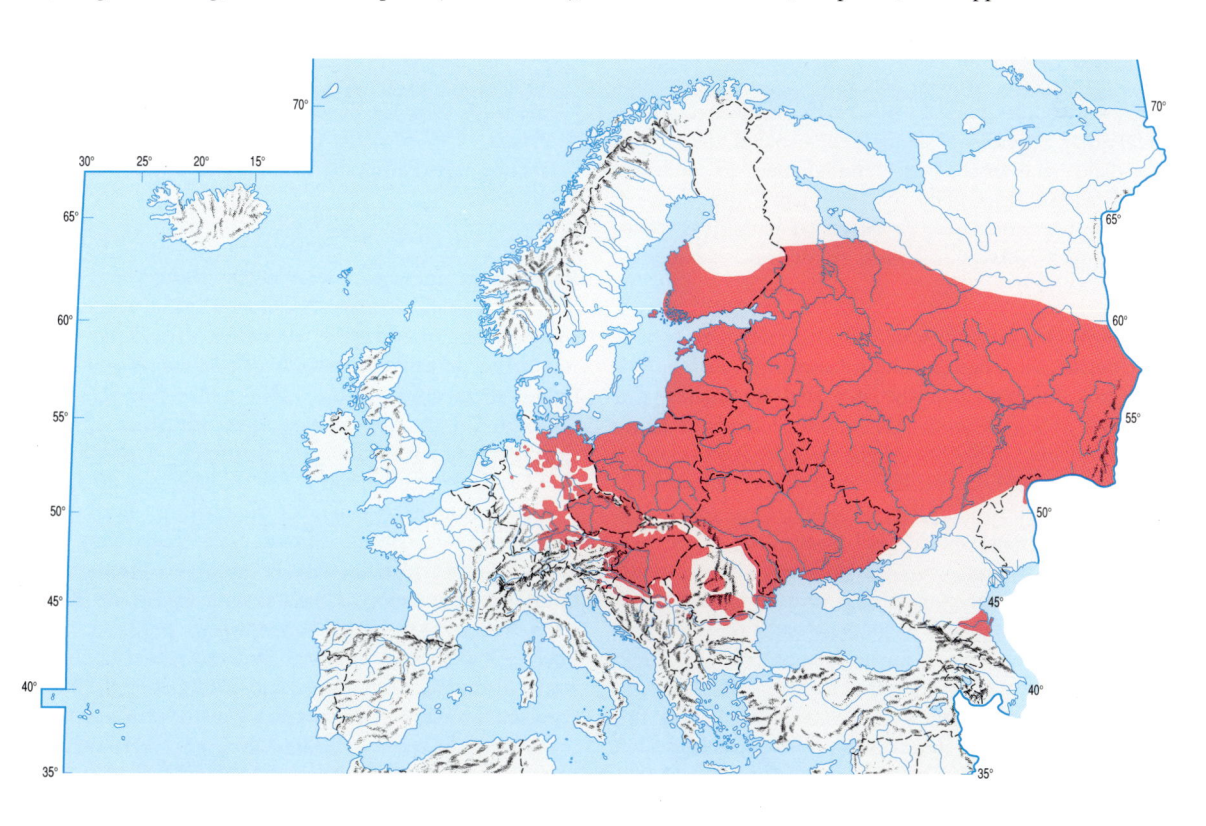

River Warbler *Locustella fluviatilis*: **1** ad, **2** juv. Savi's Warbler *Locustella luscinioides*. *L. l. luscinioides*: **3** ad, **4** juv. *L. l. fusca* (southern Asia): **5** juv worn (1st winter). Gray's Grasshopper Warbler *Locustella fasciolata fasciolata* (p. 1251): **6** worn 1st winter.

Palearctic, between isotherms of *c.* 17 and 23°C, forming western counterpart of Gray's Grasshopper Warbler. Requires ample stands of very dense but rarely tall vegetation, growing on shady bare soils, accessible to foraging and easy, concealed movement. Such cover may include thickets of grass and nettles among young growth of hazel, dogwood, alder, birch, hornbeam, ash, willow, and other trees characteristic of moist carr woodland and of floodlands, backwaters, damp forest clearings, bottomlands, bogs, sedge marshes, depressions in steppeland, and even parks and abandoned orchards, sometimes within cities. The required dense, low, tangled, often woody vegetation may be associated with river banks, meadows, edges of lakes, marshes, or swamps, woodlands, and forests.

Distribution. Has spread west in central and northern Europe since 1950s. Recent range expansion Germany, Sweden, Finland, Austria, and Bulgaria; also indicated by singing ♂♂ Britain (since 1984), Belgium (1962), and Netherlands (most of the 15 records 1976–92). Slight decline Rumania and Moldova, fluctuating Ukraine. Earlier expansion (at beginning of 20th century) followed by retreat in following years. GERMANY. Slow westward spread, still continuing. DENMARK. Several probable breeding records (singing ♂♂ apparently holding territory), but no confirmation to date. Of 25 records 1992, most were of singing ♂♂, and *c.* 31 such birds in 1995. SWEDEN. First recorded 1937, annually since 1950; no doubt breeds regularly, but no proof yet. FINLAND. First recorded 1869 and 1879; annual since early 1960s and breeding first confirmed 1974. CZECH REPUBLIC, SLOVAKIA. Regular since *c.* 1900; spread after 1945. GREECE. Perhaps breeds in north. RUMANIA. Probably more widespread than map suggests. RUSSIA. Records on Lower Don (south of Veshenskaya) suggest only sporadic breeding. Southern limit east of Don not well known. KAZAKHSTAN. No confirmed breeding south of Chapaevo in Ural valley. TURKEY. Singing birds frequent in spring, but no evidence of breeding.

Accidental. Spitsbergen, Iceland, Britain, France, Belgium, Netherlands, Norway, Switzerland, Spain, Italy, Malta, Algeria, Morocco.

Beyond west Palearctic, extends east in narrowing band to Irtysh river.

Population. Increase at least locally (though fluctuations apparently typical) Germany, Sweden, Finland, Estonia, Austria, and Bulgaria; mostly stable elsewhere. GERMANY. 1300–2500 pairs. Also, in east (where bulk of population and increasing, though marked fluctuations), 2600 ± 1100 pairs in early 1980s. SWEDEN. Continuous increase in records since 1937; to 131 singing ♂♂ in 1983, when increase stopped. FINLAND. 200–500 pairs in late 1980s (many ♂♂ remain unpaired); marked increase, though considerable fluctuations. ESTONIA. 2000–5000 pairs in 1991; increase, still continuing. LATVIA. 50 000–100 000 pairs in 1980s. LITHUANIA. Common. POLAND. Locally fairly numerous. CZECH REPUBLIC. 10 000–20 000 pairs 1985–9. SLOVAKIA. 10 000–15 000 pairs 1973–94. HUNGARY. 20 000–30 000 pairs 1979–93. AUSTRIA. Not uncommon in east. Locally increasing since mid-1960s, though decline also reported through loss of habitat. YUGOSLAVIA: CROATIA. 7000–9000 pairs. SLOVENIA. 1000–2000 pairs. BULGARIA. 20–100 pairs. RUMANIA. 800–1200 pairs 1986–92. RUSSIA. 100 000–

1 million pairs; stable, though some local increases and decreases reported. BELARUS'. 100 000–140 000 pairs in 1990. UKRAINE. 9000–10 000 pairs in 1991; fluctuating. MOLDOVA. 1700–2300 pairs in 1986; slight decrease.

Movements. Migratory. Winter quarters not well known, but lie from northern South Africa north to Zambia, Mozambique, and perhaps extreme south of Malawi. Migration very inconspicuous, perhaps due (in autumn) as much to skulking behaviour as to long-distance flights, as span of dates suggests passage protracted; no staging posts known within Europe. General direction is via east Mediterranean and Levant, requiring south-east heading for western breeders, and progressively more westerly heading for eastern breeders.

Food. Flying and non-flying arthropods, especially Homoptera, Diptera, small beetles, and spiders. Food obtained in dense herbaceous and bushy vegetation, by running about in grassy vegetation, and among fallen leaves of alder and nettles or in woods along rivers and streams.

Social pattern and behaviour. Essentially similar to Grasshopper Warbler; sings from higher perches, and predominantly (up to 90% of output) at night.

Voice. Song of ♂ a curious, mechanical, pulsating or rhythmic series of 'chuffing' sounds not unlike distant steam engine running at high speed or sewing machine. Similarity to certain insects also frequently emphasized; perhaps more like bush-cricket (*Decticus*, *Tettigonia*) than any other *Locustella*, and risk of confusion thus arises. Some birds have harder, more metallic quality and sound more like cicada (Cicadidae) than bush-cricket; mimicry of these insects perhaps involved, or changes may occur according to time of day. At irregular intervals, high-pitched, shrill, squeaky, or metallic sounds are interjected (perhaps connected with intake of breath); not audible with equal clarity from all distances and directions. At start of bout, bird usually gives only single bursts of song, but at full intensity song almost continuous with only brief pauses—ignoring breaks of less than 1 s, continuous song recorded for up to 72 min. Commonest call 'zick-zick', confusingly similar to alarm-call of Savi's Warbler.

Breeding. SEASON. Eastern Europe: from end of May to mid-July. One brood. SITE. On or within 30 cm of ground in thick vegetation or at base of bush. Nest: loose cup of grass stems and leaves, lined with finer grass and hair. EGGS. Sub-elliptical to long elliptical, smooth and glossy; white, finely but densely speckled and spotted brown and red-purple, often more at broad end; markings more contrasting than Grasshopper Warbler. Clutch: 5–6 (3–7). INCUBATION. 14–15 (13–16) days. FLEDGING PERIOD. 14–16 days.

Wing-length: ♂ 73–80, ♀ 69–77 mm.
Weight: ♂♀ mostly 15–19 g; up to 26.5 g on migration.

Savi's Warbler *Locustella luscinioides*

PLATES: pages 1248, 1274

DU. Snor   FR. Locustelle luscinioïde   GE. Rohrschwirl   IT. Salciaiola
RU. Соловьиный сверчок   SP. Buscarla unicolor   SW. Vassångare

Field characters. 14 cm; wing-span 18–21 cm. Fully 10% larger than Grasshopper Warbler, with proportionately longer wings; 10% longer than Reed Warbler, with relatively smaller head and tail graduated over half visible length. Small to medium-sized warbler, not as shy as smaller *Locustella* and with uniform brown plumage strongly suggesting some *Acrocephalus* warblers.

Commonest and most widespread of the 3 unstreaked *Locustella*. Song diagnostic to the experienced, but, in poor view, bird can be confused with (1) poorly marked, buffier form of Grasshopper Warbler (best distinguished by its voice, smaller size, shorter bill, and usually yellower legs), (2) uncommon, but similarly uniform, dark brown form of River Warbler (distinguishable by its voice, softly streaked and mottled throat and chest, and obvious white tips on under tail-coverts which reach tip of tail), (3) red-brown-toned species of *Acrocephalus*, particularly eastern race of Reed Warbler (best distinguished by its voice, tail shape, relatively short under tail-coverts, duller blue- or grey-toned legs, and bolder behaviour), and (4) Cetti's Warbler with its similar plumage tone and pattern (best distinguished by its voice, relatively obvious supercilium, colder, greyer underparts, short wings and under tail-coverts, and persistent cocking rather than fanning of tail). Flight less darting than Grasshopper Warbler, with less whirring but still typically flitting action and associated movement of much rounded, often broadly fanned tail. Generally less skulking than Grasshopper Warbler, ♂ often singing in full view from reed-tops and birds can easily be seen in chasing flights; restless, and likely to be seen in higher levels of ground cover. Essentially a reedbed bird, much more likely to be seen in company with Reed Warbler than Grasshopper Warbler.

Habitat. Breeds in west Palearctic lowlands across middle and lower middle latitudes between July isotherms of 18° and 32°C—in ample and unbroken cover of swamps, wetlands, floodlands, and inundated fringes of fresh or brackish surface waters, including sometimes fairly deep lakes and rivers, nest-site being usually approachable only by wading. In Europe, generally in fens and swamps with shallow water, where reeds are not too dense and have a varied underlayer of bur-reed, sedges, and rushes, sometimes with scattered bushes and small trees. Occasionally occurs in sedgy marshes without reeds.

Distribution. Moderate expansion in western and northern parts of range, especially from 1960s. Recent decrease reported Belgium, Netherlands, Portugal, Italy, and Greece. BRITAIN. Small numbers bred in south-east England until mid-19th century, when became extinct. Bred in Kent from 1960 and has since spread. FRANCE. Recent expansion to north, but details uncertain. In 19th century known only from south and west. Apparently flourishing colony on Atlantic coast 1860–80, but no record thereafter until 1920. Major part of country apparently colonized after 1920. BELGIUM. First recorded breeding 1937. GERMANY. In main breeding area of north-east, first recorded breeding Mecklenburg 1926. Schleswig-Holstein and Hamburg: regularly present in breeding season from 1960 after first recorded 1949–50. Bayern: spread to new breeding areas from 1950s. DENMARK. Probably breeds regularly; singing ♂♂ 1990–92 mapped. SWEDEN. First recorded 1947; now annual (10–16 records a year since 1978). No breeding records, but heard annually at one site in Skåne. FINLAND. First bred 1984 (also date of first record), then 1992. ESTONIA. Slowly expanding range. LATVIA. First recorded late 1960s; spreading north-east. LITHUANIA. First recorded 1926. POLAND. More numerous in northern half of country. CZECH REPUBLIC, SLOVAKIA. Bred 19th century in southern Slovakia, since 1933 in western Slovakia, since 1946–7 in southern Moravia, and since 1947–51 in Bohemia, Silesia, and eastern Slovakia; further expansion during 1969–70. SWITZERLAND. First bred *c.* 1950; spread during 1970s. BALEARIC ISLANDS. Several pairs said to have bred Ibiza 1963 and 1967, not recorded 1968. UKRAINE. Slight increase. KAZAKHSTAN. May be more widespread along northern Caspian than map suggests. Singing ♂♂ noted Ural delta; presumably breeds there. JORDAN. Probably bred Azraq 1960s to early 1980s. TUNISIA. No longer breeds: disappeared from Ichkeul (only known site) 1985.

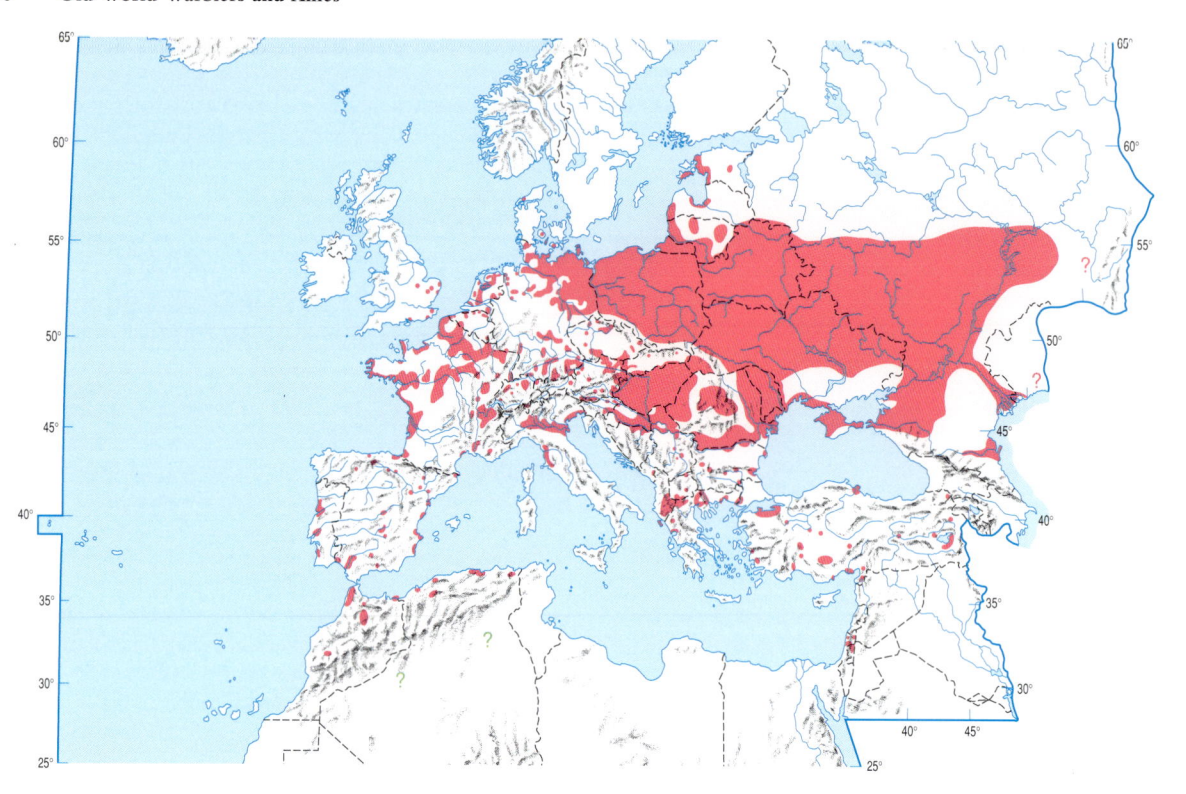

Accidental. Ireland, Norway, Sweden (annual), Finland (annual).

Beyond west Palearctic extends east patchily to north-west Mongolia.

Population. Decrease Britain, Belgium, Netherlands, Portugal, Italy, Croatia, Moldova, and Israel. Increased markedly Latvia, slightly Estonia, Poland (stable in recent years), and Ukraine; apparently stable elsewhere. BRITAIN. Perhaps 13 singing ♂♂ 1973–4, then peak of up to 6 confirmed pairs, possible maximum 28–30, in 1978–80, with decrease 1981–5. Estimated 8–22 pairs 1988–94. FRANCE. 1000–10 000 pairs in 1970s. BELGIUM. Continuing decline: *c.* 400 pairs 1950, 200–250 in 1969, 50–100 in 1982, probably only 10 pairs in early 1990s. NETHERLANDS. 1000–1600 pairs 1989–91, over 70% decline at many sites compared with 1965–75. GERMANY. 2700–3700 pairs. Bulk of population in east where probably generally increasing, despite considerable fluctuations some areas. ESTONIA. 50–100 pairs in 1991. LATVIA. 400–800 pairs in 1980s. LITHUANIA. Common. POLAND. 5000–15 000 pairs. CZECH REPUBLIC. 400–750 pairs 1985–9. SLOVAKIA. 1000–2000 pairs 1973–94. HUNGARY. 10 000–20 000 pairs 1979–93. AUSTRIA. 3000–5000 pairs; no clear trend. SWITZERLAND. 250–300 pairs 1985–93. Stable (after increase in 1970s). SPAIN. 1450–1900 pairs. PORTUGAL. 100–1000 pairs 1978–84. ITALY. 1000–2000 pairs 1983–93. GREECE. 500–1000 pairs. ALBANIA. 50–200 pairs 1981. YUGOSLAVIA: CROATIA. 2000–2500 pairs. SLOVENIA. 100–150 pairs. BULGARIA. 100–1000 pairs. RUMANIA. 100 000–200 000 pairs 1986–92. RUSSIA. 10 000–100 000 pairs. BELARUS'. 400–1500 pairs in 1990. UKRAINE. 3500–4000 pairs in 1986. MOLDOVA. 3500–5000 pairs in 1986. TURKEY. 1000–10 000 pairs. ISRAEL. Local and uncommon. MOROCCO. Scarce.

Movements. All populations migratory. Wintering grounds not well known, but apparently lie in sub-Saharan Africa north of forest zone, extending (perhaps discontinuously) from Sénégal to Eritrea. West European birds apparently head south or south-west in autumn (reverse in spring), central and east European birds south-east towards Levant, and Asian birds south-west.

Food. Mainly adult and larval arthropods, also snails. Feeds mainly low in dense vegetation, taking food also from water surface and from ground. When feeding on ground either walks relatively slowly and deliberately or hops, picking items from ground and stems.

Social pattern and behaviour. Essentially similar to Grasshopper Warbler.

Voice. Song of ♂. Introduction, audible only at close range, is series of ticking units (recalling Robin), typically lasting 1–3 or up to 5 s and gradually accelerating. Introductory units gradually merge into continuous vibrant and buzzing (reeling) trill, rich full 'zzrrreeee...', which shows rapid crescendo, then continues level in pitch and intensity for up to 30 s or more. As in Grasshopper Warbler, crescendo-diminuendo or ventriloquial effect also results from head-turning. Not especially loud or vigorous, but carries up to c. 300 m or even c. 1000 m in still conditions. Calls are variable, mainly monosyllabic 'pit', 'wett', 'zick', etc., given singly or in more or less rapid succession.

Breeding. SEASON. Eggs from late April, western and central Europe; from mid-April, southern Europe. 1–2 broods. SITE. In tall, aquatic or semi-aquatic vegetation over swampy ground or water. Nest: well-concealed cup with outer layers loosely constructed of dead water-plant leaves and inner section often more tightly woven from grass stems; inner cup of finer leaves and fibres. EGGS. Sub-elliptical, smooth and glossy; white, very finely and densely speckled brown, purplish, or greyish-brown, more densely at broad end, often giving colour to whole shell. Clutch: 3–6. INCUBATION. 10–12(–14) days. FLEDGING PERIOD. 11–15(–16) days.

Wing-length: ♂ 69–73, ♀ 66–71 mm.
Weight: Very variable; averages (♂♀) 14.5–16.5 g, extremes c. 11 and 20 g.

Geographical variation. Slight. *L. l. sarmatica* from c. 35°E eastwards slightly paler than nominate *luscinioides* to the west.

Gray's Grasshopper Warbler *Locustella fasciolata*

PLATES: pages 1248, 1274

Du. Grote Krekelzanger   Fr. Locustelle fasciée   Ge. Riesenschwirl   It. Locustella di Gray
Ru. Таежный сверчок   Sp. Buscarla de Gray   Sw. Större flodsångare

Field characters. 15 cm; wing-span 21–24 cm. Largest of all *Locustella*, approaching size of Great Reed Warbler but retaining rather small head and full rump and tail so characteristic of *Locustella*; 10–15% larger than River Warbler. Large, unstreaked *Locustella* differing distinctly from smaller River Warbler in size, even more rounded tail, dull, uniformly grey breast, and buff under-parts with unmottled dull orange under tail-coverts. In brief glimpse, size and colour may recall smaller, duller races of Great Reed Warbler, or even small thrush. Flight and behaviour much as River Warbler but size and weight usually apparent in all actions. Stance on perch, with full body and slightly raised tail, enhances resemblance to chat or thrush. Calls include loud 'tschrrok tschrrok' and sharp 'tchirp'.

Habitat. Breeds in warm dry continental middle latitudes of east Palearctic, mainly in lowland and coastal areas, from fringes of taiga to wooded steppeland, riverain bottomlands, meadows, glades, forest clearings, grassy mountain slopes, and thickets, including bamboo thickets.

Distribution. Breeds from upper Ob' and north-east Altai in western Siberia east to Sakhalin island, northern Japan, and Korea.

Accidental. France: two 1st-years, both on Ouessant, September 1913 and September 1933. Denmark: juvenile, Lodbjerg, September 1955.

Movements. Migratory, travelling east (from Siberian breeding grounds) and then south through eastern Asia, and south-west from Japan, to winter quarters. Reverse migration probably accounts for the 3 occurrences in western Europe.

Wing-length: *L. f. fasciolata* (eastern Asia): ♂ 78–86, ♀ 74–82 mm. *L. f. amnicola* (Sakhalin, northern Japan): ♂ 72–80, ♀ 75, 76 mm.

Moustached Warbler *Acrocephalus melanopogon*

PLATES: pages 1252, 1274

Du. Zwartkoprietzanger   Fr. Lusciniole à moustaches   Ge. Mariskensänger   It. Forapaglie castagnolo
Ru. Тонкоклювая камышовка   Sp. Carricerín real   Sw. Kaveldunsångare

Field characters. 12–13 cm; wing-span 15–16.5 cm. Western race *melanopogon* 10% smaller than Sedge Warbler (eastern races c. 15% larger than nominate *melanopogon*). Small, neat, perky warbler, sometimes looking long-legged, with size and plumage resembling Sedge Warbler but with head patterned even more strongly and upperparts and flanks darker and more rufous in

Moustached Warbler *Acrocephalus melanopogon*. *A. m. melanopogon*: **1** ad worn (late spring), **2** juv. *A. m. mimica*: **3** ad fresh (autumn). Aquatic Warbler *Acrocephalus paludicola* (p. 1254): **4** ad worn (late spring), **5** juv (August). Sedge Warbler *Acrocephalus schoenobaenus* (p. 1255): **6** ad worn (late spring), **7** (August).

western birds, of which some even show faintly streaked, rusty band over breast. Strong head and face pattern, with broad black crown, square-ended white supercilium, dusky eye-stripe, and dark-rimmed cheeks, reminiscent of Firecrest from side; hindneck and rump almost concolorous with mantle and tail when fresh. 3 races in west Palearctic, 1 differing distinctly from others; eastern *albiventris* and *mimica* much more difficult to separate from Sedge Warbler than is western nominate *melanopogon*.

Distinction from Sedge Warbler never easy, particularly with single bird and when comparison of colour tones or other characters is difficult. Separation by plumage characters should, if possible, be supported by some or all of the following: narrower, more compressed bill; more rounded wing and rather longer and more smoothly rounded tail, giving impression of Wren at times, especially when often cocked; less fluent and more flitting flight, not showing broad or depressed tail; less extrovert behaviour; voice (see below).

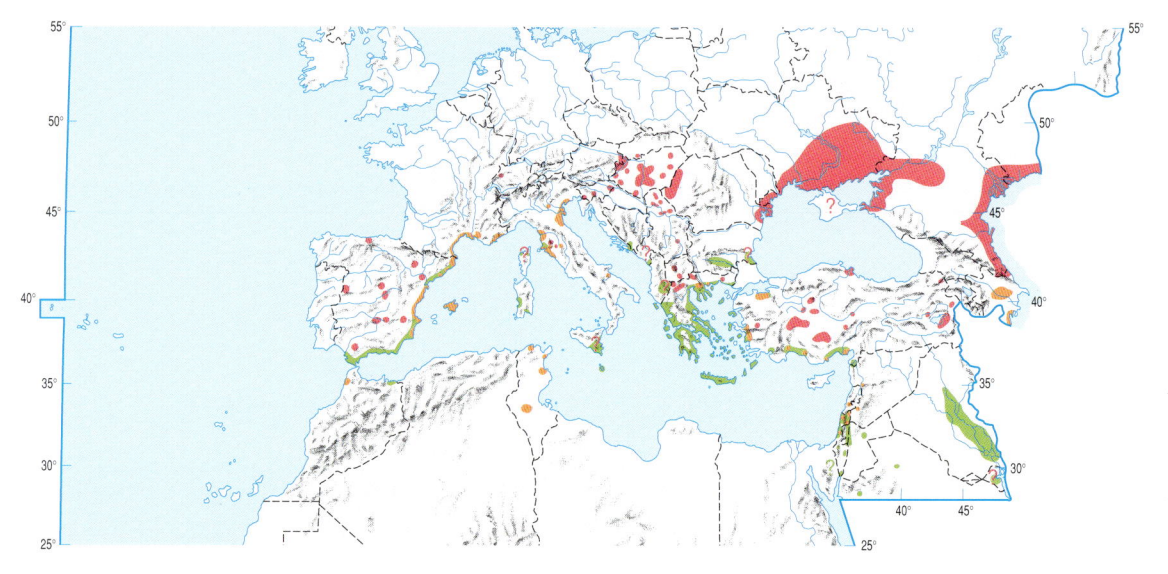

Habitat. In west Palearctic, largely a southerly resident counterpart to the more northerly migratory Sedge Warbler in warmer temperate lower mid-latitudes, and in Mediterranean and steppe zone, between July isotherms of 22°C and more than 32°C. Unlike most congeners not confined to lowlands, ascending in Caucasus to 1950 m. Habitat varies regionally, but generally prefers low aquatic vegetation, especially reeds, reedmace, sedges, etc., often with admixture of bushes. Structurally adapted to foraging in vertical-stemmed emergent vegetation.

Distribution. BRITAIN. Bred near Cambridge 1946. FRANCE. Corsica: a few pairs probably breed. GERMANY. First recorded 1957; small population in southern Bayern 1981–4, with breeding proved 1984, possible 1981. CZECH REPUBLIC. Regular at 1 site in Moravia since 1971, but no proof of breeding. SLOVAKIA. Regular breeder at one locality only; known since 1959. SWITZERLAND. First proved breeding 1981, Lac de Neuchâtel; not yet elsewhere. ITALY. Formerly bred Sicily; now only a possible breeder. GREECE. Inadequately known. BULGARIA. Recorded in May coastal lagoons; status unclear. AZERBAIJAN. Inadequately known. Rare in Lenkoran lowland and Kizil-Agach reserve, common mid-Kura and Samur delta. TURKEY. Situation obscured by widespread occurrence of migrants and wintering birds, also liable to confusion with Sedge Warbler and hence possibly underrecorded. ISRAEL. Habitat and population reduced by draining of swamps. No longer breeding Jezreel valley or northern plain. JORDAN. No longer breeding: Azraq population became extinct in 1980s. IRAQ. May breed in reedbeds of Mesopotamian plain. MOROCCO. Probably breeds occasionally Oued Massa estuary, and perhaps also in marshes of Lower Moulouya in north-east.

Accidental. Britain, Ireland, Germany, Denmark, Poland, Portugal (status not clearly understood), Algeria (occasionally winters).

Beyond west Palearctic, extends east to *c.* 75°E, south very locally to Arabia.

Population. Stable in much of range. FRANCE. 1000–2000 pairs. SLOVAKIA. 10–20 pairs 1973–94; fluctuating. HUNGARY. 1000–1200 pairs 1979–93. AUSTRIA. Estimated 9000 pairs in 1993, all at Neusiedler See. Increasing slightly. SWITZERLAND. Up to 5 pairs (or singing ♂♂) 1985–93. SPAIN. 2400–3200 pairs. Estimate (in early 1980s) of 5000–7000 pairs Albufera marsh, Mallorca (Balearic Islands) alone certainly too high. ITALY. 1000–3000 pairs 1983–93; perhaps slight decrease. GREECE. 100–500 pairs; trend unclear, perhaps stable. YUGOSLAVIA: CROATIA. 500–1000 pairs. SLOVENIA. 5–10 pairs. RUMANIA. 10 000–20 000 pairs 1986–92. RUSSIA. 50 000–500 000 pairs 1980–90. UKRAINE. 10–50 pairs in 1984. TURKEY. 5000–15 000 pairs. 1000–1500 pairs Kizilirmak delta. ISRAEL. Locally rare to scarce. 10–20 pairs in Bet Shean valley, Tabigha and Hula valley in 1970s–80s. Decline from 1970s. TUNISIA. Scarce. MOROCCO. Uncommon.

Movements. Sedentary or partially migratory to migratory, birds from north of breeding range moving south to winter within and south of breeding range. Eastern populations most migratory, with extensive wintering area in Pakistan and north-west India.

Food. Almost exclusively arthropods, especially small beetles, but water snails regularly taken. Food obtained by picking and probing from vegetation at or near water surface; also, especially when collecting items for nestlings, from water; only to lesser extent taken in the air.

Social pattern and behaviour. No comprehensive study. Apparently monogamous mating system, with both sexes sharing equally in incubation and care of young. Song given from exposed perch, typically a reed-top. ♂♂ sing throughout much of the year: best information on seasonal variation for southern France, especially Camargue, where it is the only warbler of wetland habitat to be heard in mid-winter.

Voice. Song of ♂ relatively continuous, as in Marsh Warbler and Reed Warbler, and richly varied, containing frequent double or treble units as in Reed Warbler. Diagnostic feature, notably different from Sedge Warbler, is the introductory series of relatively low-pitched pure-sounding notes recalling, but quieter than, crescendo of Nightingale, and also recalls richly fluted notes of Woodlark. Rest of song a scratchy warble, recalling Sedge Warbler but softer, thinner and sweeter, mostly lacking the frequent harsh rattling, jarring, and chattering sounds of that species. Calls include short, rolled 'trk' or 't-trrt', hard 'tak', quieter 'tuc tuc tuc' and similar monosyllabic units, given singly or in rapid succession; also grating 'churr'.

Breeding. SEASON. Eggs from mid-April. Little apparent variation across range. One brood. SITE. Over water, usually 30–60 cm above surface, in dense stands of reed, reedmace, rush, and low shrubs. Nest: deep untidy cup of loosely woven leaves and stems of aquatic plants, with denser inner lining of flowers of reed and some feathers; suspended from several vertical plant stems by loops of material, often with at least partial roof. EGGS. Sub-elliptical, smooth and glossy; white to grey-white, finely speckled and lightly mottled light olive overall; whole egg sometimes with olive-green tint. Clutch: 3–5(–6). INCUBATION. 14–15 days. FLEDGING PERIOD. *c.* 12 days.

Wing-length: ♂ 58–62, ♀ 55–59 mm.
Weight: ♂♀ mostly 10–14 g.

Geographical variation. Rather slight and clinal, with trend to greater size and paler coloration from west to east. 3 races: nominate *melanopogon* in west, *albiventris* from Ukraine east to lower Don river, *mimica* east from lower Don, also Turkey and Levant.

Aquatic Warbler *Acrocephalus paludicola*

PLATES: pages 1252, 1274

Du. Waterrietzanger   Fr. Phragmite aquatique   Ge. Seggenrohrsänger   It. Pagliarolo
Ru. Вертлявая камышовка   Sp. Carricerín cejudo   Sw. Vattensångare

Field characters. 13 cm; wing-span 16.5–19.5 cm. Slightly less bulky than Sedge Warbler, with appearance subtly altered by shorter bill on long-looking head (due to plumage pattern) and less broad tail with rather pointed feathers. Small, seemingly large-headed but slim, unobtrusive warbler, almost invariably haunting lowest levels of marsh cover; as secretive as small *Locustella* warblers. Has most striped head of all west Palearctic warblers, with long pale golden supercilium and central crown-stripe and dark lateral crown-stripes. Lacks discrete warm rump of Sedge Warbler and has much blacker streaks on pale olive- or golden-buff mantle and upper rump, forming long stripes. Side of underbody of adult finely streaked.

Distinctive in good view, but in brief glimpse easily confused with Sedge Warbler and vice versa. Necessary to check diagnostic (not just eye-catching) characters, these being not central crown-stripe or pale legs (these also shown by juvenile and 1st-winter Sedge Warbler), but (1) long, dark, lateral crown-stripe and pale supercilium, (2) longer and sharper stripes on back, continuing as streaks on rump and upper tail coverts, and (3) tail structure. Flight lighter and more darting than Sedge Warbler but with similar jinks and dives into cover; rump and particularly tail appear less broad, but tail can be depressed in flight as in Sedge Warbler.

Habitat. Breeds in mid-latitudes of west Palearctic, in mainly continental lowlands, within July isotherms 18–26°C, in circumscribed and fragmented range partly attributable to shrinkage of specialized habitat through climatic and human pressures. In particular, decline is associated with drainage and fragmentation of habitat disrupting typical group dispersion and promoting competition with Sedge Warbler.

Nests in marshes, favouring clumps of sedge and iris rather than reedbeds and willows, preferring low open tracts to taller growth, either over water or dry ground. Overlaps in places with Moustached Warbler, which favours wetter nest-sites over shallow water, and locally with Marsh Warbler and Sedge Warbler, which occur more often on drier sites.

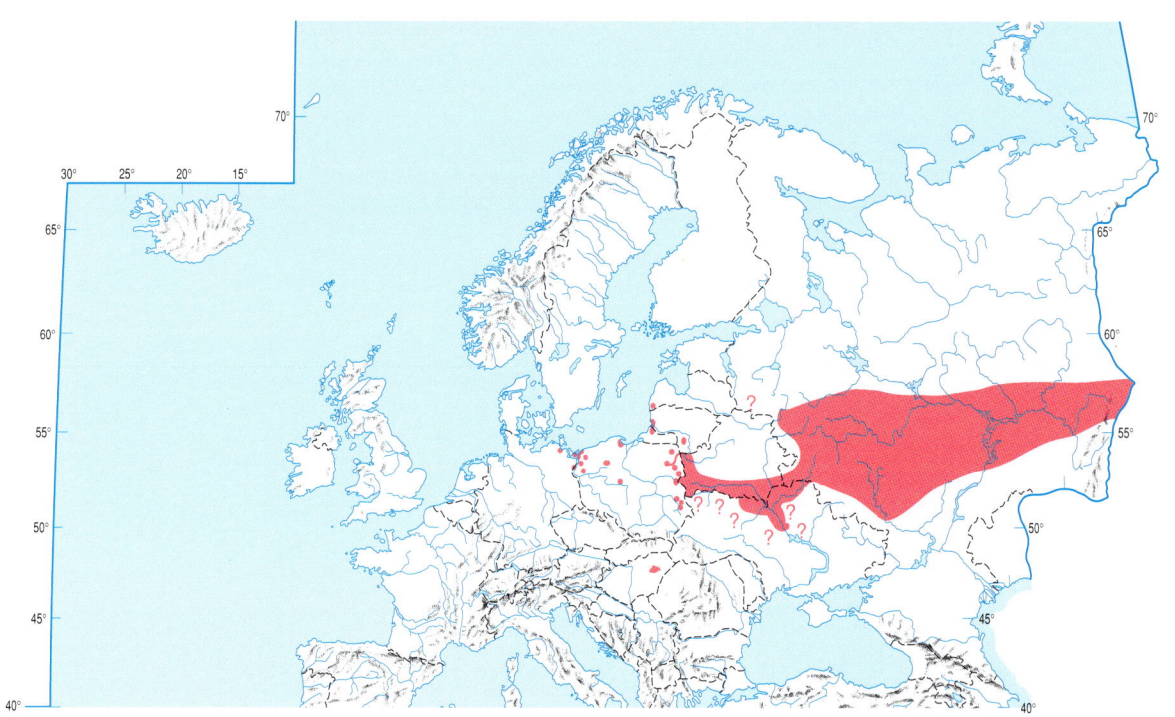

Distribution. Breeding range poorly known, especially in FSU, but probably always fragmented owing to habitat constraints. Now, both range and population further reduced through extensive drainage. Some older records dubious; has probably never bred Sweden, Finland, or Denmark. FRANCE. Bred Marne 1961. BELGIUM. Bred 1872, 1875, but no breeding records in 20th century. NETHERLANDS. Occasional breeder in 19th and early 20th centuries (last 1941). GERMANY. Now apparently extinct as breeding bird in west. Last proved breeding Niedersachsen 1972, suspected west of Hamburg 1980. In east, many formerly occupied sites now drained and abandoned. Formerly large population of Brandenburg extinct since 1977. Unstable population still on lower Oder (known since early 1970s) and new population established on coast. LATVIA. Breeding most likely Lake Pape; perhaps more widespread in south. POLAND. In 19th century, bred in most of lowlands, but range much decreased since. CZECH REPUBLIC. Irregular, sporadic breeder in Bohemia and Moravia. SLOVAKIA. Last bred 1974. AUSTRIA. Formerly bred irregularly Neusiedler See area. Last confirmed breeding 1940, but present in breeding season in early 1950s. ITALY. Extinct as breeding species, perhaps before 1950. YUGOSLAVIA, RUMANIA. Apparently extinct. RUSSIA. Extremely poorly known. Mapping tentative. BELARUS'. Significant population discovered in south in 1995. UKRAINE. Range fluctuating, but generally not well known.

Accidental. Ireland, Denmark, Norway, Sweden, Finland, Estonia, Balearic Islands, Malta, Greece, Turkey (exceptionally during breeding season), Jordan, Egypt, Azores, Canary Islands.

Beyond west Palearctic, extends east (not known whether continuously) to Ob' river.

Population. Estimates mostly given in pairs, but complex mating system means difficult to equate counts of singing ♂♂ with number of breeding ♀♀. GERMANY. 20–50 pairs; marked decline. Lower Oder: c. 50 pairs declining to 1–2 ♂♂, then rising again to 15 ♂♂ 1986–8. Greifswald: increased from 0 to 29 ♂♂ 1973–88, and total at 4 sites 42–46 ♂♂ in 1988. LATVIA. Perhaps 10–50 pairs. LITHUANIA. Rare and decreasing. Total unknown, but 25–30 pairs bred Žuvintas reserve 1986. POLAND. 1500–4000 pairs. Marked decline last 30 years, still continuing. HUNGARY. 400–420 pairs in 1994. Only population to have shown a marked recent increase. RUSSIA. Poorly known: perhaps 1000–10 000 pairs 1970–90; stable. BELARUS'. Estimated 3000 pairs in 1995. Previously, little known and often confused with Sedge Warbler. UKRAINE. 1–10 pairs in 1990; fluctuating.

Movements. Migratory. Winter quarters poorly known, but apparently lie in part (but presumably not for easternmost populations) in West Africa south of Sahara. Autumn passage through western Europe (north to southern England) and extreme western Africa; spring passage (at least in part) apparently more direct to breeding grounds. Passages prolonged.

Food. Predominantly insects. Taken almost exclusively while climbing or scurrying about in dense, low vegetation, though also recorded on warm autumn days feeding agilely higher up in (e.g.) bushes of willow.

Social pattern and behaviour. Unlike in other west Palearctic *Acrocephalus*, ♀ does all incubation and rears young alone, ♂ maintaining no bond with ♀ and rarely coming near nest. On breeding grounds, ♀♀ especially very secretive: if flushed, normally fly away low and silently, soon disappearing. ♂ sings from perch or in flight, moving about quite widely during day. Normally uses elevated perch (e.g. post, reed, low bush), sitting 0.5–2 m up; usually lower down in evening, but not hidden. For song-flight, takes off from elevated perch (normally less than 1 m) or, less commonly, from lower down in cover. Makes rapid, fluttering, fairly steep, and silent ascent to c. 3–30 m, starting to sing in last phase of ascent or at peak of ascent, when fanned tail raised vertically or even held slightly forwards, while head thrown back; this posture produces braking effect followed by even steeper descent and disappearance into vegetation—or descent can be shallower than ascent and gliding. Continues to sing while descending, but usually ceases shortly before or at latest upon landing near take-off point or up to 20–50 m away. Song-period April–August; peak song-activity in early morning towards sunrise and in late evening, sometimes well after sunset.

Voice. Song of ♂ relatively short and simple, as typical of polygynous species. Thus normally easily distinguishable from other *Acrocephalus*, even Sedge Warbler: monotonous and almost rigid compared with fluent irregularity and musical invention of that species, though similar units present in songs of both. Songs sometimes comprise only 1–2 motifs lasting c. 1 s. Typical song one or more introductory rattles followed by short rapid sequence of more tonal units: 'trrtrtr-jü-jü', 'tärrr dü dü dü', or 'trrt-di-di'; several such phrases often strung together to make a run. Mimicry of other birds probably a regular feature. Complete song lasts no more than 7 s. Calls include monosyllabic 'tack', characteristically soft 'tucc tucc' and similar units, singly or in series; also rasping 'trrr'.

Breeding. SEASON. Central and eastern Europe: eggs from early May to late July. 1–2 broods. SITE. In dense vegetation (usually old sedge clump) over swampy ground or water. Mean height above surface: 17.0 cm (3–30). Nest: neatly but rather loosely constructed cup of grass, plant stems and leaves, spiders' webs, and plant down, lined finer material; usually with roof, and more like nest of Reed Bunting than typical *Acrocephalus*. EGGS. Sub-elliptical, smooth and glossy; white or pale buff-olive, finely spotted and mottled olive-buff, often completely obscuring ground-colour; occasional thin and broken black hair-streaks. Often darker than eggs of Sedge Warbler, more like Tree Sparrow. Clutch: 4–6 (3–8). INCUBATION. 12–15 days. FLEDGING PERIOD. 13–14 days.

Wing-length: ♂♀ (little difference between sexes) 60–66 mm.
Weight: ♂♀ mostly 10–14 g.

Sedge Warbler *Acrocephalus schoenobaenus*

PLATES: pages 1252, 1274

DU. Rietzanger   FR. Phragmite des joncs   GE. Schilfrohrsänger   IT. Forapaglie
RU. Камышовка-барсучок   SP. Carricerín común   SW. Sävsångare

Field characters. 13 cm; wing-span 17–21 cm. Close in measurements to Willow Warbler but noticeably bulkier, with longer bill and broader rump and tail; slightly shorter and less bulky than Reed Warbler. Rather small, quite robust warbler, epitome of streaked *Acrocephalus*. Well marked above, with striking cream supercilium offset by dark crown, and tawny rump glowing between dull olive-brown, rather softly streaked back and brown tail.

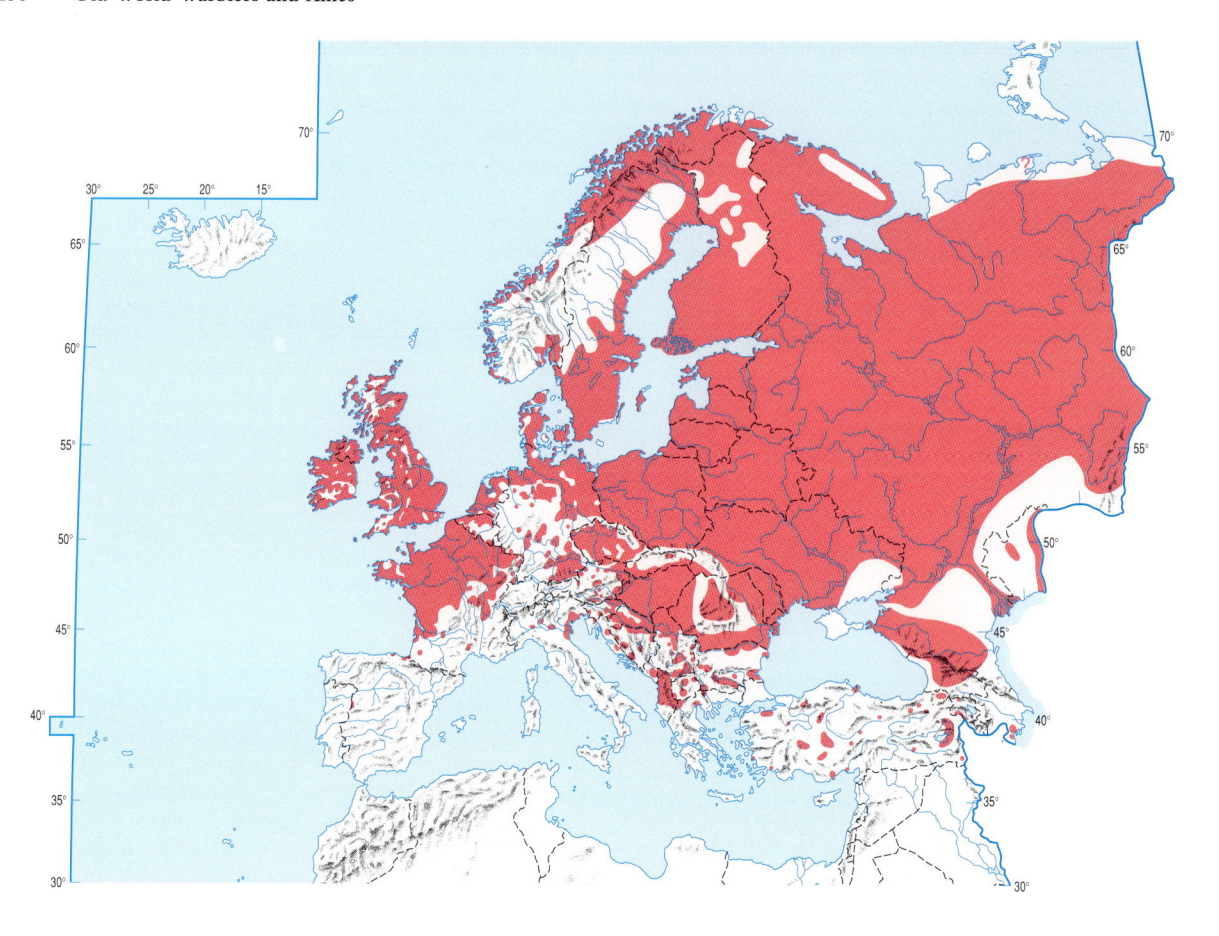

Most widespread and easiest to observe of *Acrocephalus*, but subject to confusion with 3 other small to medium-sized warblers in *Locustella* and *Acrocephalus*; risk of mistake highest in late summer or autumn when both faded adult and fresh, warm-coloured juvenile with paler-mottled crown centre present atypical appearance with latter in particular suggesting Aquatic Warbler. For differences, see Pallas's Grasshopper Warbler, Moustached Warbler, and Aquatic Warbler. Flight light and fluent, usually low and short with bird jinking among cover and suddenly turning or diving into it; during flight, tail (and apparently rump) frequently spread and even depressed below level of body, contributing to rather short-tailed shape. Gait typical of genus. Regularly flicks tail but only rarely cocks it. Not shy, frequently feeding in open foliage and on ground and singing from exposed song-post. Irascible and inquisitive, coming quickly to imitation of its call.

Habitat. In contrast to other *Acrocephalus*, extends from high arctic down to mid-latitudes, from boreal through temperate but only marginally to Mediterranean zone, ranging east not far beyond west Palearctic; predominantly in lowlands. Accordingly adapted to cool, often cloudy and moist climates, between 12 and 30°C July isotherms. Breeds in wide variety of low dense vegetation (by no means exclusively near water) or in moist depressions. Attracted to clay- and gravel-pits. Less closely linked to lakesides, river banks, and wetlands than most congeners, and usually avoids wetter reedbeds in standing water. Nearest water may be 500 m or more distant. Active in cover, ranging *c.* 10–150 cm above ground. Avoids trees and tall bushes, hard artefacts, rocks, and open surfaces generally, and comes into open view only where an immediate retreat into concealment is at hand. Will however breed in dry situations offering suitable cover, including neglected orchards, farm hedgerows, nettlebeds, fields under rice, barley, beans, oilseed rape, clover, and other crops, and even in dry young conifer plantations with trees up to *c.* 2 m high or in dense scrub of sea buckthorn.

Distribution. Range has contracted in parts of north-west Europe (Britain and Ireland, Benelux countries, Germany, Denmark), also Italy, Greece, Rumania, and Moldova. Increased slightly Ukraine. BRITAIN. Colonized Orkney in mid-19th century and Outer Hebrides in 20th century. Disappeared from many areas more recently. IRELAND. Decrease, mainly in south. BELGIUM. Population inland close to extinction, that on coast stable. LUXEMBOURG. Probably became extinct at end of 1980s. GERMANY. In east, retreat from south-west associated with population decline. NORWAY. Northern and southern populations separate and ecologically distinct. Northern population may have spread south (or perhaps overlooked in previous years). In south, first bred Rogaland 1923 (from Denmark) and Vestfold 1947 (from southern Sweden). FINLAND. Rapid expansion last 100 years due to man-made habitat changes; range now stable. SWITZERLAND. Surprisingly rare as breeding species. Only 2 confirmed breeding records, near Geneva 1903 and in Tessin 1972, but indications of breeding elsewhere in recent years. SPAIN. Breeding confirmed Basque country and Salamanca; no proof Marismas (mouth of Guadalquivir). GREECE. May breed at a few other sites in north-west. RUSSIA. Observations August 1992, 1994 and especially June and August 1993 suggest may breed at 68°30′N on Russkiy Zavorot peninsula. AZERBAIJAN. Poorly known. TURKEY. Widespread occurrence of singing ♂♂ on passage up to late May makes assessment of breeding distribution difficult. ALGERIA. Bred Lake Fetzara in north-east in early 1900s.

Accidental. Spitsbergen, Iceland, Faeroes, Madeira, Canary Islands.

Beyond west Palearctic, extends east to Yenisey river.

Population. Strong fluctuations, and long-term decline, probably due to variety of factors on breeding and wintering grounds. Decrease most marked Britain, Benelux countries, Germany, but also evident France, Denmark, Finland, Poland, Italy, Greece, Croatia, and Moldova. Slight increase Norway, Austria, and Ukraine; apparently mostly stable elsewhere. BRITAIN. Estimated 250 000 territories 1988–91. Some local increases (e.g. Orkney, Inner Hebrides), but marked decrease overall from 1969, partial recovery since 1985. IRELAND. 110 000 territories 1988–91. FRANCE. 10 000–100 000 pairs in 1970s. BELGIUM. 500–600 pairs 1989–91. Continuing decline (1160 pairs 1973–7). NETHERLANDS. 12 000–18 000 pairs 1989–91. Decline of at least 50% since c. 1970. GERMANY. 12 000–19 000 pairs. In east, 9000 ± 3500 pairs in early 1980s; perhaps only lower end of range by early 1990s. DENMARK. 900–11 000 pairs 1987–8. NORWAY. 10 000–100 000 pairs 1970–90. Some decrease in south-east from 1970, but overall slight increase. SWEDEN. 50 000–200 000 pairs in late 1980s. Apparently stabilized after 50% decrease 1965–74, but indications of continuing decrease southern and central areas, increase Baltic coastal areas of extreme north. FINLAND. 300 000–500 000 pairs late 1980s. Rapid increase from late 19th century, but slight decrease more recently. ESTONIA. 50 000–100 000 pairs 1991. Now stable following increase in 1960s–70s. LATVIA. 80 000–200 000 pairs in 1980s. LITHUANIA. Abundant. POLAND. 6000–20 000 pairs. Tendency to decrease in south-west. CZECH REPUBLIC. 40 000–80 000 pairs 1985–9. SLOVAKIA. 10 000–16 000 pairs 1973–94. HUNGARY. 150 000–200 000 pairs 1979–93. AUSTRIA. 6000–12 000 pairs. Locally increasing; not uncommon in east. SPAIN. 120–200 pairs. ITALY. 30–100 pairs 1983–93. GREECE. 500–1000 pairs. ALBANIA. 50–200 pairs in 1981. YUGOSLAVIA: CROATIA. 15 000–20 000 pairs. SLOVENIA. 500–1000 pairs. BULGARIA. 100–1000 pairs. RUMANIA. 300 000–500 000 pairs 1986–92. RUSSIA. 1–10 million pairs 1970–90. BELARUS'. 100 000–160 000 pairs in 1990. UKRAINE. 55 000–60 000 pairs in 1986. MOLDOVA. 5000–7000 pairs in 1986. AZERBAIJAN. Uncommon. KAZAKHSTAN. Ural delta: numbers severely reduced by rising water-level. TURKEY. 1000–10 000 pairs.

Movements. Migratory, entire breeding population wintering in Africa south of Sahara, from Sénégal east to Ethiopia and south to eastern Cape Province (South Africa) and northern Namibia. Widespread and locally numerous in much of range. Initial heading in autumn is generally south to south-west, sometimes more south-easterly. Movement begins in July, with dispersal to premigratory feeding-grounds, adults preceding juveniles. Emigration from Britain probably under way by early August, and passage virtually ended by early October. At Ottenby (southern Sweden), passage recorded late July–early October, chiefly August–mid-September. Transit through Mediterranean area chiefly August and September.

Departure from winter quarters mainly in early April. First arrivals on breeding grounds in western France late March, most in early April; in Britain, usually early April, but some records in March, main arrival mid- or late April to early May. Reaches breeding grounds in Poland 2nd half of April, and southern Finland in early or mid-May.

Food. Chiefly insects, also some plant material outside breeding season. Feeds predominantly low down in dense vegetation, notably reeds and rushes, also in cereal fields and bushes. Feeding techniques adapted to slow-moving or stationary prey: mainly 'picking' (perched bird picks insects from leaves and twigs, occasionally hovering to do so) and 'leap-catching' (bird takes flying insect in the air as it flits between perches); feeds mostly in the hours just after dawn and just before dusk, capitalizing on lower mobility of prey at these times.

Social pattern and behaviour. Solitary outside breeding season, individuals defending territories as long as food available, then moving to do likewise elsewhere. Mating system typically monogamous; occasionally, ♂♂ bigamous (simultaneously or successively), and promiscuous to unknown extent. Breeding territory relatively small (but much larger than that of Reed Warbler); typically 0.1–0.2 ha. In breeding season, ♂ bold and conspicuous before pairing, but more discreet and elusive once paired. Sings from exposed perch, e.g. top of reed-stem, bush, or low tree, interspersed with song-flights. ♂♂ start singing vigorously within hours of arrival on breeding grounds (including regular singing at night). Arrival of ♀♀ and pairing have instantly suppressive effect on song output with subsequent and perhaps compensatory increase in calling. Thereafter, song continues at low level for rest of breeding season due to late-arriving ♂♂, failure to pair, or mate-desertion.

Voice. Song of ♂ a hurried, vigorous, and varied medley. Sweet musical passages freely interspersed with harsh, strident, chattering ones, often including mimicry. Appears to serve almost entirely for mate-attraction rather than territorial defence. Renderings of song fail to convey vigour, rhythm, complexity (etc.), but, with these qualifications, one described as 'chit-chit-tuk-tuk-tuk-chitterwee-terwee-tit-tit-tit-twee-twee-tit-it-it-it-cherwee...', another a more hurried 'chit-it-it-it-it ter-ter-ter-ter-ter-richee-terrichee-see-see-see...'. For differences from Moustached Warbler, see that species. Compared with Reed Warbler, faster, lacking the almost metronomical repetition of units characteristic of that species, also more imitative, though not nearly as accomplished a mimic as Marsh Warbler. Calls include a variety of monosyllabic, scolding or grating notes 'tucc', 'chirr', etc., often run rapidly together in alarm, producing stuttering rattle.

Breeding. SEASON. Britain and north-west Europe: eggs from early May. Finland: laying begins early June, exceptionally end of May. Usually 1 brood, sometimes 2. SITE. In variety of tall vegetation or low bushes. On ground or up to 50 cm above, rarely more. Nest: deep cup, rounded to cylindrical, with loosely woven outer structure of grass, plant stems and leaves, moss, and sedges, often with spiders' webs, with thick inner layer of finer leaves and stems, lined with reed flowers, hair, and plant down. Supported on vertical stems of plants, with outer part of nest woven round them. EGGS. Sub-elliptical, smooth and glossy; very pale green or olive-buff, finely and liberally speckled and mottled olive overall; some eggs uniformly coloured, obscuring ground-colour. Clutch: 5–6 (3–8). INCUBATION. 13–15 days (12–16). FLEDGING PERIOD. 13–14 days (10–16).

Wing-length: ♂ 65–71, ♀ 62–67 mm.
Weight: ♂♀ mostly 9–15 g; migrants with large fat deposits to 19 g.

Paddyfield Warbler *Acrocephalus agricola*

PLATES: pages 1259, 1274

DU. Veldrietzanger   FR. Rousserolle isabelle   GE. Feldrohrsänger   IT. Cannaiola di Jerdon
RU. Индийская камышовка   SP. Carricero agrícola   SW. Fältsångare

Field characters. 13 cm; wing-span 15–17.5 cm. Close in size to Reed Warbler but with shorter bill, noticeably shorter wings, and slightly longer tail. Medium-sized, quite bold but skulking warbler, resembling Reed Warbler. Plumage ground-colour varies from dusky olive-brown through reddish- to pale sandy-brown; shows rather short, but quite broad, cream or buff supercilium under dusky edge to crown, dark-centred tertials, and long rufous rump. Behaviour includes pronounced raising of tail.

Fully rufous spring migrants most striking, with orange-foxy glow exceeding that of any Reed Warbler; pale sandy-orange autumn migrants also catch eye. Conversely, dull brown breeding bird in reedbed is a typical unstreaked *Acrocephalus* and differences in bill and head pattern have to be looked for. Happily, voice quite distinctive; calls certainly more useful than plumage or structure. Flight as Reed Warbler, though short wing sometimes evident in slightly more whirring, less flitting action, and length and shape of tail show when turning. Gait and other behaviour as other unstreaked *Acrocephalus*, though tail-flicking and tail-cocking, and raising of crown feathers, are particularly characteristic. Not as dependent on reedbeds as Reed Warbler; migrants occur in low cover and will feed on open ground nearby.

Habitat. Breeds in continental, dry, warm mid-latitudes, in lowlands. In Russia, breeds by reed-fringed lakes (often quite small) in deserts, plains, wooded steppe, river valleys, etc.; chiefly in low, sparse reed or reedmace, also mixed with (e.g.) willow and tamarisk. On large lakes will breed far out from shore in unbroken reedbeds, but on Black Sea coast generally avoids them, and on Black Sea islands breeds in dense *Artemisia*. On migration, occurs in tall grass of marshy meadows, in thickets, and even in small trees such as apple.

Distribution. Several breeding records well north of main range. FINLAND. First recorded 1980; bred 1991. LATVIA. Lake Pape: probably bred 1987–8, 1990. RUMANIA. Regular and

Paddyfield Warbler *Acrocephalus agricola capistrata* (Siberia to China): **1** ad fresh (winter), **2** ad worn (late spring), **3** 1st winter. Reed Warbler *Acrocephalus scirpaceus* (p. 1265). *A. s. scirpaceus*: **4** ad (late spring), **5** juv. *A. s. fuscus* (Asia): **6** ad, **7** 1st winter.

common breeder. RUSSIA. Fairly common breeder at fishfarm in central Kirov region (*c.* 58°N) in 1995. UKRAINE. Crimea: sporadic in reedbeds up to 1960s; later found in most districts where rice grown. North-west Black Sea area: up to 1976, recorded only Danube delta and Black Sea Nature Reserve; after 1976, proved breeding at 3 other sites, including Dnestr estuary. ARMENIA. First recorded and first breeding in 1995. TURKEY. First recorded Van marshes 1986; singing birds there 1987–8 suggest breeding. Karabulak (foot of Mt Ararat): 3 birds July 1989.

Accidental. Faeroes, Britain, Ireland, France, Belgium, Netherlands, Germany, Denmark, Norway, Sweden, Finland, Estonia, Latvia, Poland, Hungary, Spain, Portugal, Italy (Sardinia: 3 ringed winter 1993–4, one of which retrapped winter 1994–5), Malta, Greece (Crete), Israel.

Beyond west Palearctic, extends east through southern Siberia and central Asia to Yenisey river and western China; also breeds Manchuria and Ussuriland (south-east Russia). Range incompletely known.

Population. BULGARIA. Perhaps 10–50 pairs; stable. Estimate in 1980s of 80–130 pairs Durankulak lake. RUMANIA. Probably over 500–1500 pairs 1986–92. RUSSIA. 10 000–100 000 pairs 1970–90; stable. UKRAINE. 4000–5000 pairs in 1986; stable. Crimea: very numerous on Lebyazh'i islands up to end of 1960s, followed by major decline.

Movements. All populations migratory, wintering in Indian subcontinent. Leave breeding areas August–September; return to lower Ural river and Crimea late April or early May; main arrival in Rumania end of May.

Food. Chiefly insects, taken from surface of emergent and waterside vegetation. Will also cling sideways on plant stem and lunge to snap up insects from water surface.

Social pattern and behaviour. No detailed studies. Nothing to indicate other than monogamous mating system. Incubation apparently by ♀ alone, but nestlings fed (equally) by both parents. ♂ sings usually from upper part of reed stem and clearly visible like Great Reed Warbler. On breeding grounds in Russia, sings from arrival (or only after period of silence), with peak May–June, continuing to July.

Voice. Song of ♂ most like Marsh Warbler, but quieter and has quicker and more even tempo and lacks harsh 'zi-chay' units typical of that species. Differs from Reed Warbler in being more or less continuous and fluently chattering, with no obvious division into phrases; more melodious and lacking harsh churring. As in Marsh Warbler, but perhaps less conspicuously, song characterized by very varied rhythm, rapid passages alternating with slower ones, and flow frequently interrupted briefly by single or series of (mimicked) calls, 'trills', and short pauses. Imitative powers remarkable though inferior to Marsh Warbler. Contact-alarm calls apparently of at least 2 kinds, as combined 'chr...chuck' (constantly uttered) reported for wintering birds in India. Very soft but affirmative ticking 'check', 'chik-chik', or 'chac', quieter than Sedge Warbler and unlike any call of Reed Warbler, given while moving about and feeding.

Breeding. SEASON. Black Sea: nesting begins mid-May; peak laying May or early June and extends (apparently 2nd broods) into July. SITE. In aquatic and semi-aquatic vegetation, especially sparse reeds and reedmace, also sedges and low shrubs. Nest built over land or water. Nest: tightly constructed cylindrical cup of reed and grass leaves and stems, woven round 2–8 vertical stems of water plants. Lined with finer grasses and reed flowers, occasionally with plant down. EGGS. Sub-elliptical, smooth and glossy; pale green, variably marked with fine speckles and heavier spots and blotches of darker green, olive, and grey, often forming cap at broad end. Occasionally pinkish or buff with darker markings. Clutch: 3–6. INCUBATION. Probably 12 days.

Wing-length: ♂ 56–61, ♀ 55–60 mm.
Weight: ♂♀ mostly 8–13 g.

Blyth's Reed Warbler *Acrocephalus dumetorum*

PLATES: pages 1261, 1274

DU. Struikrietzanger FR. Rousserolle des buissons GE. Buschrohrsänger IT. Cannaiola di Blyth
RU. Садовая камышовка SP. Carricero de Blyth SW. Busksångare

Field characters. 13 cm; wing-span 17–19 cm. Close in size to Reed Warbler but slimmer, with finer-looking, noticeably tapering bill and slightly shorter wings; much less plump than Marsh Warbler, with *c.* 10% shorter wings. Medium-sized, rather slim, unobtrusive but restless and active warbler, with rather cold and uniform, grey to olive-brown upperparts and rather pale underparts (noticeably so in adult). Has dark, tapering bill, and whitish fore-supercilium and eye-ring, often combining to form 'spectacle', and noticeably dull-centred tertials. Concave posture (head and tail raised), frequent tail movement, and dull grey legs are also useful clues. At least one call distinctive.

Subject to serious confusion with Reed Warbler (particularly greyer eastern race) and Marsh Warbler in worn plumage (when bright, full olive and cream or yellow tones lost). Unhappily, also approaches size, structure, and appearance of dull Paddyfield Warblers and 3 (typically or occasionally) grey *Hippolais* species: Olivaceous Warbler, Upcher's Warbler and Icterine Warbler. However, Blyth's Reed Warbler does have individual aspects to its structure, behaviour, postures, movements, and voice. Following characters partly tentative but worthy of close observation: (1) length of fine, tapering bill emphasized by long forehead and rather low crown; (2) short, rounded wings, giving shortest folded wing-point of unstreaked *Acrocephalus*; (3) rather slim build, with body less bulky than most Reed Warblers and all Marsh Warblers though this partly illusory, being enhanced by (4) holding of bill, head, and usually tail above body line in somewhat *Sylvia*-like manner, unlike typically 'horizontal' posture of Reed Warbler and normally rather upright, tail-depressed stance of Marsh Warbler; (5) rather dainty feet, with shorter toes and claws than Reed Warbler; (6) frequent tail movement, including obvious lifting, flicking, cocking, and fanning, with fanning particularly eye-catching and contributing to prominence of tail and apparent

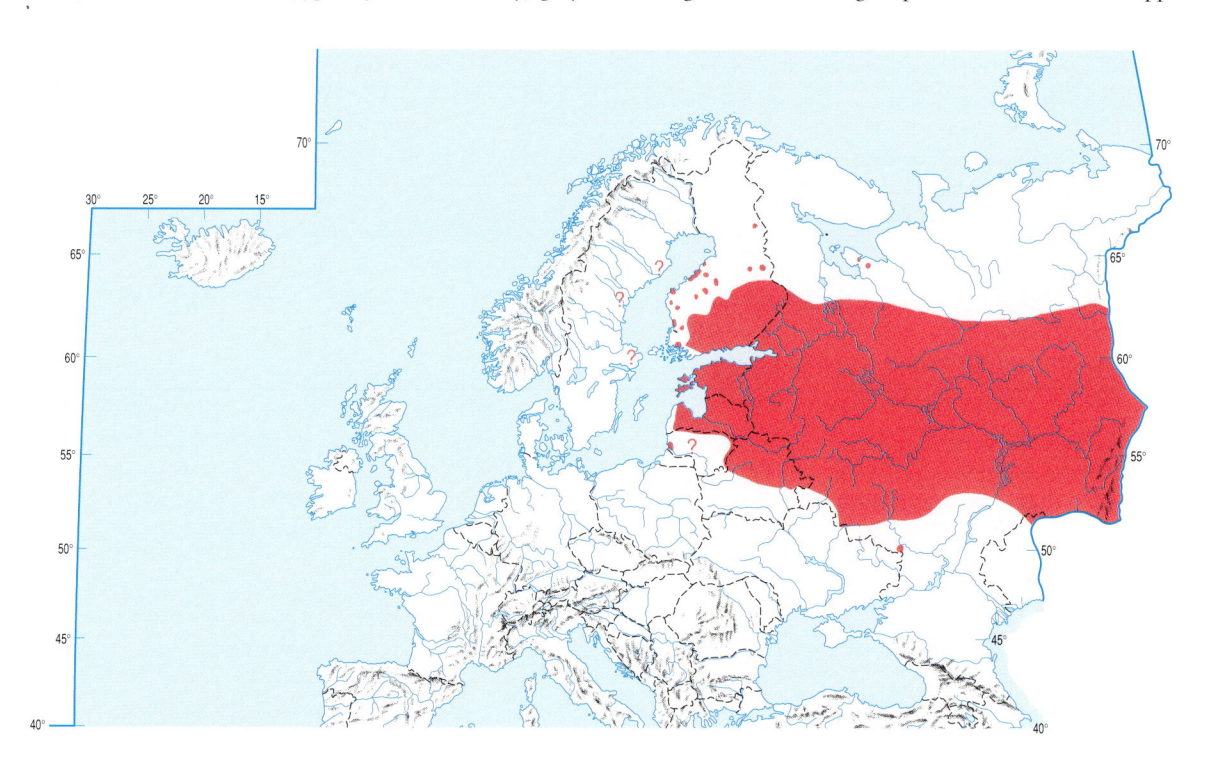

Blyth's Reed Warbler *Acrocephalus dumetorum*: **1** ad (late spring), **2** 1st winter. Cape Verde Cane Warbler *Acrocephalus brevipennis* (p. 1262): **3** ad, **4** juv. Marsh Warbler *Acrocephalus palustris* (p. 1263): **5** ad (late spring), **6** 1st winter.

shortness of under tail-coverts. In addition, feeds in side or top foliage of trees and bushes, moving freely to ground cover and open floor in way far more rarely shown by other *Acrocephalus*. Flight typically flitting, with round wings and broad tail creating quite different silhouette from Paddyfield Warbler and recalling Cetti's Warbler as much as *Acrocephalus*.

Habitat. Has recently spread west across west Palearctic in upper middle continental latitudes through temperate zone to margin of boreal zone, overlapping range and habitat of Marsh Warbler to south and west. Although found in willow and alder beds along brooks and rivulets, along canals and irrigation ditches overgrown with reeds and tall grass, and at edges of marshes, such aquatic habitats play much less important role than drier and more wooded ones. These include riverain deciduous forests; forest gullies grown with bushes, nettles, etc.; tangles of wild raspberries in glades among damp spruce stands; scattered bushes or piles of wind-thrown trees in open pinewoods; birch clumps in wooded steppe; tangles of tamarisk on dry riverbeds; thickets of dog rose and wild cherry on slopes of hills and valleys; abandoned or tended orchards, parks, and vegetable gardens. Some of these varied types are located at elevations unusual for *Acrocephalus* along mountain streams and on plateaux up to 1000 m or even 1200 m. Prime requisite for nesting is combination of dense bushy or herbaceous vegetation with plenty of neighbouring well-lit more open areas around. Thus contrasts with most other *Acrocephalus* in breadth of tolerance and adaptation to varied habitats.

Distribution. Has spread west in Fenno-Scandia and Baltic area, north and west in FSU, in last 50 years. NORWAY. First breeding attempt Møre og Romsdal (and 7 other singing ♂♂ elsewhere) 1995. SWEDEN. First recorded 1958, annual since 1969. Breeding confirmed only 1984, but perhaps breeds (almost) annually. FINLAND. First recorded 1934, first breeding confirmed 1947; rapid expansion to north-west then stabilized. ESTONIA. First recorded 1890–93; spread in 1930s and 1950s. Widespread records at 59 sites in 1960s, 128 in 1970s, but density at present still higher in east. LATVIA. First recorded breeding 1944. LITHUANIA. After 2 singing ♂♂ in north in 1976 and 1 in west in 1988, observed annually. Bred in west in 1983. POLAND. Several records of singing ♂♂ 1967–86, and pair seen 1983 in extreme north, suggesting incipient colonization, but no proof of breeding. RUMANIA. Perhaps rare summer visitor, commoner on passage. No definite proof of breeding. FSU. Scarce and scattered on southern fringe of range. RUSSIA. First recorded Leningrad region 1869 and northern limit probably ran through region then; subsequently moved north, with breeding recorded Kareliya in 1970s. Kaliningrad region: recorded from 1976.

Accidental. Britain, Ireland, France, Belgium, Netherlands, Germany, Denmark, Norway, Poland, Switzerland, Spain, Italy, Malta, Bulgaria, Turkey, Syria, Israel, Kuwait.

Beyond west Palearctic, extends east to upper Lena river and south-east Altai (Russia), south to northern Afghanistan.

Population. SWEDEN. Slow increase in records since 1970s, c. 20♂♂ now recorded annually. FINLAND. 5000–8000 pairs in late 1980s; marked increase. ESTONIA. 2000–3000 pairs in 1991. Continuous upward trend since 1961. LATVIA. 5000–10 000 pairs; increase. LITHUANIA. Rare. RUSSIA. 100 000–1 million pairs 1970–90; stable. BELARUS'. 100–300 pairs in 1990; fluctuating, but apparently stable overall.

Movements. Migratory, entire breeding population moving south or south-east to winter in Pakistan, India, Nepal, Sri Lanka, and east to Burma. Birds leave breeding grounds in July and August. In Finland, departs from mid-July and all have left by late August. Spring departures from winter quarters March to May. Birds appear in Moscow region from mid-May onwards. Reaches Finland from late May, and most birds have arrived by mid-June. Records (very infrequent) in Britain and Ireland almost entirely in autumn, presumably resulting from reversed migration.

Food. Principally insects, also spiders and snails. Feeds by moving about actively in tree canopy (especially in winter) and in shrub and herb layer; quite often also on ground, and takes insects in flight.

Social pattern and behaviour. Mating system apparently varies from monogamy to polygyny, though further study required. Territories often clustered, giving impression of 'semi-coloniality' even in optimal habitat, intervening suitable habitat remaining unoccupied. ♂ sings from perch and in flight. Perches used include top of bush or prominent side-branch, small trees, and telegraph wires. Until paired, sings for long periods, mainly at night. Once paired, no longer sings at night and gives only short subdued bursts in day.

Voice. Rich and mimetic song, also given outside breeding season, is most conspicuous feature and can be useful aid to identification. Distinctive combination of constantly and energetically repeated, often attractive pure whistled, clearly articulated units or motifs, sometimes harsher chirping or chattering, mimicked sounds (of many species), and frequently interpolated calls in relatively short phrases. Phrases can be separated by pauses or more continuous, with individual ones not easily discernible. Delivered at slow rate with characteristic time patterning; leisurely delivery and repetition recall song of Song Thrush. Loud song of mellifluous, full and rounded, unhurried phrases with regular pauses typically given at night, while day-time songs can be more hurried and without pauses, though frequently interrupted. Calls include distinctive soft, but penetrating 'thik', and harsh churring 'trrk'.

Breeding. SEASON. Finland: laying begins early June with some first clutches as late as mid-July. One brood. SITE. In dense vegetation including reeds, nettles, shrubs, and small trees. Usually 10–40 cm above ground, occasionally up to 1 m. Nest: neat and compact cup, sometimes conical, of leaves, stems, and plant down, often with spiders' webs; lined finer stems and hair. EGGS. Sub-elliptical, smooth and glossy; very pale greenish, creamy, greyish, or pinkish with spots and small blotches of olive-green, buffish-, greyish-, or reddish-brown, pale grey, or purple, often concentrated at broad end. Clutch: 3–6. INCUBATION. 12–14 days. FLEDGING PERIOD. 11–13 days.

Wing-length: ♂ 61–66, ♀ 61–64 mm.
Weight: ♂ 9–15, ♀ 10–14 g; migrants much heavier, to 24 g.

Cape Verde Cane Warbler *Acrocephalus brevipennis*   PLATE: page 1261

DU. Kaapverdische Rietzanger   FR. Rousserolle du Cap-Vert   GE. Kapverden-Rohrsänger   IT. Cannaiola di Capo Verde
RU. Короткоперая камышовка   SP. Carricero de Cabo Verde   SW. Kap Verde-sångare

Field characters. 13.5 cm; wing-span 17–19 cm. Distinctly larger than Reed Warbler, with 15% longer tail and 10% longer legs and feet. Medium-sized, long-billed, and rather long-tailed warbler resembling dull Reed Warbler. Grey-brown above, brownish- to greyish-white below, with no obvious markings. Actions recall *Hippolais* more than *Acrocephalus*.

Geographical isolation on Cape Verde Islands prevents confusion (though Olivaceous Warbler has been recorded there). Flight action fluttering, with frequent fanning of tail when moving between perches; interestingly, escape-flight invariably directed to tree canopy (where it usually feeds), not ground cover.

Habitat. Resident exclusively on Cape Verde Islands, with dry windy climate, and virtually stripped of main natural vegetation during c. 500 years of human colonization and exploitation. Favours well-vegetated valleys, avoiding drier areas, but occurs up to c. 1400 m above sea-level. Has adapted well to artificial habitats, especially plantations of sugar-cane and banana, but optimal nesting habitat appears to be undisturbed patches of reed in fields on valley bottoms or on slopes up to c. 800 m. Also nests in manioc, orange, and coffee bushes. Secretive in reed thickets, but elsewhere lives openly in gardens and plantations.

Distribution and population. Now breeds only on Santiago (Cape Verde Islands). Formerly occurred, but now apparently extinct, on 2 other islands. Brava: scarce 1897, increased by 1951, last recorded 1969. São Nicolau: not uncommon 1865, numerous in 1897, and collected 1924, but not recorded since. Extinctions probably due to drought and subsequent disappearance of habitat. Santiago: sparse but well distributed in 1897, scarce in 1966. Estimated 500 pairs 1988–93; probably more or less stable, at least during recent decades, apart from 'normal' fluctuations with drought or rain.

Movements. None reported; apparently sedentary.

Food. Mainly small to medium-sized insects, and several times seen eating adult ant-lion. Importance of fruit not known, but recorded feeding on figs. Feeds usually in canopy of small trees, sometimes in low vegetation.

Social pattern and behaviour. Little known; apparently similar to other *Acrocephalus*.

Voice. Song a distinctive liquid bubbling, like bulbul, with no harsh notes but containing characteristic trills. Commonest call a throaty 'ker(r)-chow' or 'grug', like Great Reed Warbler.

Breeding. SEASON. Prolonged; probably 2 main periods, spring (February–March) and during and after summer rains (June–November). SITE. Mainly in reeds; also from 0.6 to 5 m up in bushes and trees and at up to 2 m in sugar-cane. Nest: deep cup of dry grass suspended in outer sprays of bush canopy with twigs passing through rim; nests in sugar-cane or reed are suspended between stems, resembling those of Great Reed Warbler. EGGS. Sub-elliptical, smooth and glossy; bluish-grey-white, streaked darker grey and black. Clutch: 2 nests had 2 and 3 eggs. (Incubation and fledging periods not known.)

Wing-length: ♂ 65–70, ♀ 63–69 mm.
Weight: ♂ ♀ 15–17 g.

Marsh Warbler *Acrocephalus palustris*

PLATES: pages 1261, 1274

DU. Bosrietzanger   FR. Rousserolle verderolle   GE. Sumpfrohrsänger   IT. Cannaiola verdognola
RU. Болотная камышовка   SP. Carricero políglota   SW. Kärrsångare

Field characters. 13 cm; wing-span 18–21 cm. Close in size to Reed Warbler and Blyth's Reed Warbler but with slightly shorter bill, 10% longer wings, and much plumper body in most attitudes. Medium-sized, quite heavy, rather pear-shaped warbler, with relatively long wings. Plumage varies with wear from bright olive-green to dull grey-brown; upperparts little-marked except by fairly distinct pale fore-supercilium and eye-ring and bright fringes to tertials and primary-tips (last absent in Blyth's Reed Warbler); underparts yellowish or cream, with pale fore-face and wide bright pale throat often contrasting with olive suffusion on side of breast and flank. Fresh-plumaged adult thus the brightest, most greenish, even most yellowish *Acrocephalus* but worn adult and immature much less distinctive. Legs pale and bright.

Separation from Reed Warbler long-debated, and, in recent years, evidence of hybridization has compounded difficulties. Thus important to recognize that Marsh Warbler has distinctive character which once learnt allows observer to depend less on bewildering details of colour tones, degrees of wear, etc. Within unstreaked *Acrocephalus*, Marsh Warbler is plumpest, heaviest, and longest-winged, with (1) typically rather short, wide-based bill, (2) rather domed crown, typically lacking as low a forehead as Reed Warbler and Blyth's Reed Warbler and thus tending to enhance short-billed appearance, (3) seemingly fuller chin and throat, enhancing round-headed appearance, (4) rather long wings with folded wing-point at least $\frac{3}{4}$ length of exposed tertials, displaying fully 8–9 bright primary tips (with noticeably even spacing indicative of least round shape within unstreaked *Acrocephalus*) and extending to end of, or even beyond, upper tail-coverts, (5) pear-shaped or pot-bellied body which gives impression of bird carrying most weight lower than other *Acrocephalus*, and (6) long, full under tail-coverts, cloaking $\frac{3}{4}$ of tail. In addition, less agile or energetic than Reed Warbler with rather upright carriage both at rest and on the move, with tail usually held down (and rarely above wing-points), and rather slower, even at times almost reluctant or clumsy movements, recalling *Hippolais*. Flight flitting, with longer wings than Reed Warbler hardly visible but clearly contributing to slightly more fluent action; lack of contrasting rump in adult tends to make bird look less broad across rump and tail. Like Blyth's Reed Warbler (but only a few Reed Warblers), breeding habitat essentially comprises dense tree, bush, or tall weed stands along wetland edges, frequently ascending to greater height in such habitat than Reed Warbler.

Habitat. Breeds in west Palearctic mainly in cool temperate middle latitudes, continental except in England and northern France, and largely lowland, although breeding up to 2100 m in Alps and to 3000 m in Georgia. Prefers somewhat damper and ranker sites than Blyth's Reed Warbler, but, in contrast to Moustached Warbler, Aquatic Warbler, and some other *Acrocephalus*, is rarely attracted to vegetation growing out of or alongside water or swamps. Prefers rank, tufty, and fairly tall herbage, especially nettles, meadowsweet, willowherb, loosestrife, wild rose, young osiers, bird cherry, alder, and other woody or erect herbaceous vegetation, which may neighbour or even be overshadowed by taller trees or bushes. Such vegetation, which may be interspersed with reeds, is characteristic of moist or seasonally flooded soils and of neglected edges or depressions. Readily exploits suitable areas created through human intervention such as gravel extraction, drainage works, and spread of waste patches near canals or industrial sites, sometimes enabling penetration within built-up areas; e.g. standing crops, especially where their edges are mingled or verged with swamp plants and rough hedgerows or ditches. In England, margin of wet swamp grading into drier vegetation, classed as high marsh, is the most typical breeding site.

Distribution. Significant northward range expansion in Fenno-Scandia and north-west Russia, and range increase also France and Ukraine. Declined Britain, Rumania, Moldova. BRITAIN. Range never extensive, but contracted from c. 1930s, with loss of small isolated breeding populations. Ceased breeding Somerset 1961, Gloucestershire 1984, but some signs of recovery and tendency to expand in early 1990s. DENMARK. Expansion of c. 10% 1974–94. NORWAY. First bred 1970, Vestfold. Spread north and west slightly during 1970s–80s. SWEDEN. Probably bred 1920, but first confirmed 1934. Still spreading north, and since 1977 recorded annually to 64°N along Gulf of Bothnia. FINLAND. First recorded 1944; first confirmed breeding 1950. Increase and spread northwards since 1980s. SPAIN. Single nests found Ebro delta 1961 and 1962. No subsequent breeding evidence. RUMANIA. Widely distributed; no data from Moldavia and Walachia. RUSSIA. Leningrad region: in 1950s still rare, even in south; now common near St Petersburg and breeding annually in extreme north of region. AZERBAIJAN. Breeds mostly in mountains. TURKEY. Locally common in upper Kara and Aras valley in east. Systematic observations and breeding records lacking for (probable) western population. Widespread singing birds mid-May to early June (in north, south-west and south) likely to refer to migrants rather than breeders. SYRIA. Small *Acrocephalus* singing and once (July 1980) evidently breeding perhaps Marsh Warbler, but other species not to be excluded.

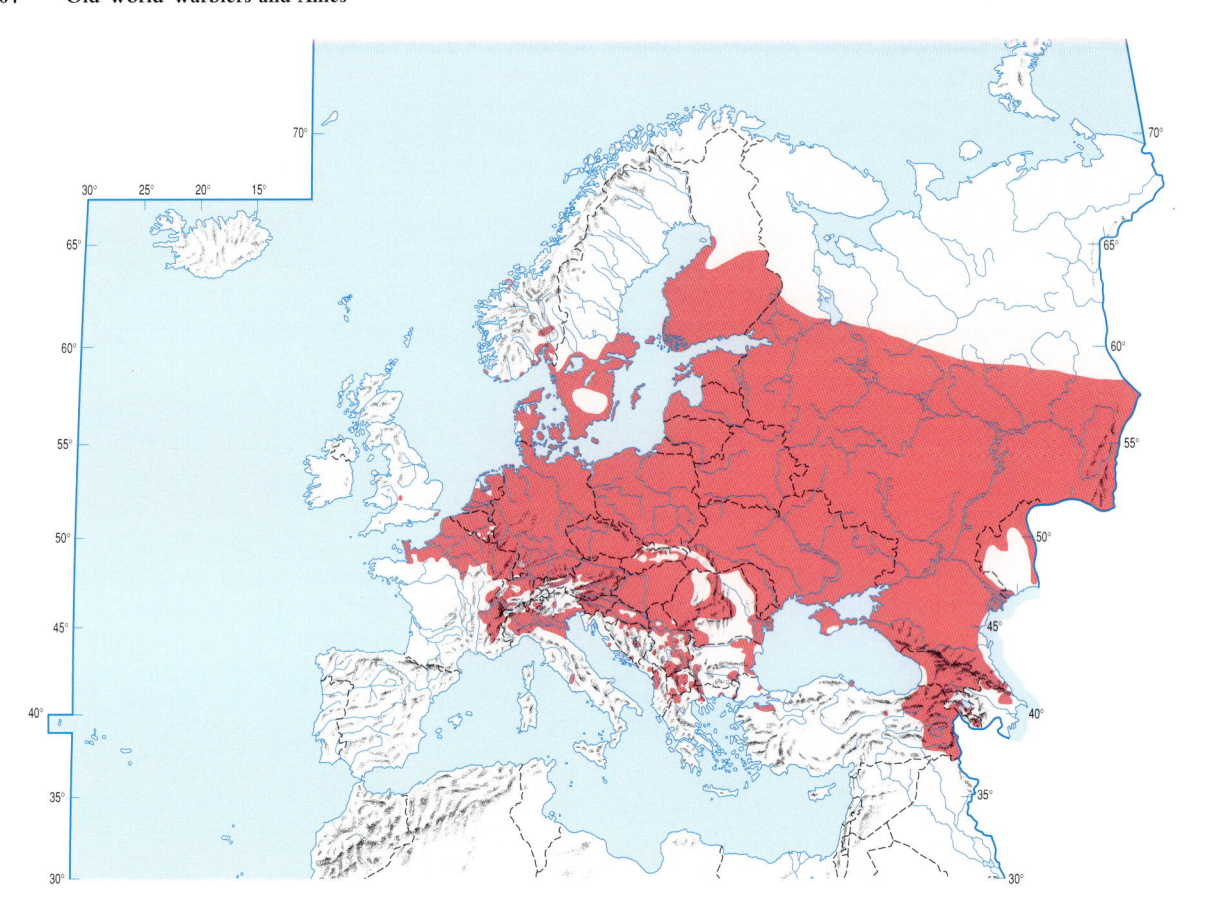

Accidental. Iceland, Faeroes, Portugal, Balearic Islands, Malta, Jordan, Tunisia, Morocco, Madeira.

Beyond west Palearctic, extends from north Caspian east to *c.* 70°E.

Population. Marked increase Sweden and Finland, less so France, Netherlands, Norway, Estonia, and Ukraine. Decrease Britain, Luxembourg, Denmark, Czech Republic, and Moldova, mostly stable elsewhere. BRITAIN. Decreased since *c.* 1950. Estimated 50–80 pairs 1968–72. Marked fluctuations 1975–86; 10–21 pairs 1987, 12 pairs 1988–9. Continuing to fluctuate (or perhaps slowly recovering and increasing) in 1990s: 12–58 pairs breeding 1993. FRANCE. 1000–10 000 pairs in 1970s. BELGIUM. 7400–19 000 pairs 1973–7; stable, though with local increases and decreases. LUXEMBOURG. 1400 pairs in 1960s; 500–800 pairs in early 1990s. NETHERLANDS. 15 000–22 500 pairs 1973–7, 40 000–70 000 pairs 1979–85. GERMANY. 463 000 pairs in mid-1980s; fluctuating. In east, where 110 000 ± 50 000 pairs in early 1980s, stable (unlike other *Acrocephalus*), locally even increasing. DENMARK. 16 000–17 000 pairs 1987–8. Slight decrease reported, though this does not accord well with range expansion. NORWAY. 50–200 pairs in 1980s. SWEDEN. Under 500 pairs in 1943, *c.* 5000 by early 1970s, 10 000–30 000 in late 1980s. FINLAND. Rapid increase since 1960, reaching 4000–6000 pairs by late 1980s. ESTONIA. 50 000–100 000 pairs in late 1980s. Increase, especially in 1960s and 1980s. LATVIA. 70 000–120 000 pairs in 1980s. LITHUANIA. Common. POLAND. 100 000–400 000 pairs. CZECH REPUBLIC. 80 000–160 000 pairs 1985–9. SLOVAKIA. 40 000–50 000 pairs 1973–94. HUNGARY. 50 000–80 000 pairs 1979–93. AUSTRIA. Common. SWITZERLAND. 2000–2500 pairs 1985–93. ITALY. 10 000–30 000 pairs 1983–93. GREECE. 2000–5000 pairs. ALBANIA. 500–1000 pairs in 1981. YUGOSLAVIA: CROATIA. 60 000–90 000 pairs. SLOVENIA. 5000–10 000 pairs. BULGARIA. 1000–10 000 pairs. RUMANIA. 300 000–500 000 pairs 1986–92. RUSSIA. 100 000–1 million pairs 1970–90. BELARUS'. 80 000–130 000 pairs in 1990. UKRAINE. 45 000–50 000 pairs in 1988. MOLDOVA. 4500–6000 pairs in 1986. AZERBAIJAN. Locally common. KAZAKHSTAN. Ural delta: numbers severely reduced by rising water-level. TURKEY. 1000–10 000 pairs.

Movements. Migratory. Winters mainly in south-east Africa from Zambia and Malawi south to Natal and Cape Province, South Africa. Migrates from breeding grounds into Middle East, birds from west of range taking south-east or ESE heading, thence southwards across Arabia. Adults leave breeding grounds soon after young independent, juveniles depart *c.* 2 weeks after independence, some showing local dispersal first, mainly from August but some not until late September. Departure from winter quarters from March until late April. Return passage rapid. Arrival on breeding grounds from late April on Georgian Black Sea coast, but not until May further north and west. Arrival on northern and western limits of range not until late May or early June.

Food. Chiefly insects and arachnids, with some snails and rarely berries in late summer and autumn. Feeds mainly by gleaning from vegetation in grass and shrub layer, also sometimes from lower branches of trees.

Social pattern and behaviour. Spends 3 times as long in Africa as on breeding grounds, where birds (apparently of both sexes) sing and establish individual home-ranges. Mating system apparently varies between monogamy and opportunistic polygyny. Majority of birds monogamous. Nest built by ♀ alone; incubation (by ♀ only at night) and feeding of young by both sexes equally. Breeding territories tend to be clumped, often patchily distributed; absent from some apparently suitable sites, but where birds are present all available habitat is fully occupied. ♂ sings in upright posture, crown and throat feathers ruffled and bill wide open, white throat and orange mouth prominent. Normally sings from perch (exposed or concealed), rarely in flight between song-posts. Song sustained throughout much of day and night by unpaired ♂♂; after pairing, as in Reed Warbler, song output decreases markedly (sometimes stops completely, certainly always nocturnal song).

Voice. Song of ♂ a continuous liquid warbling chatter, blending remarkable variety of mimicked sounds into fast-flowing sequences of striking beauty and vivacity; rate of delivery and musical association of motifs give it pleasant fluidity and sweet silvery quality. At times low, rolling, blurred, and gurgling, babbling or gabbling like Sedge Warbler, with panting, sighing, wheezing, and nasal sounds; harsh 'zi-chay' is characteristic, as also high-pitched, clear liquid trills and tremolos like Canary. Mimetic ability quite outstanding and song may be almost entirely imitative: at least 99 European species copied and 113 African species (80 passerines) identified. Calls varied, including typical *Acrocephalus* 'chirr' in alarm, also sharp, loud 'tchak', quiet 'tuc' or 'tchuck', sharp 'tweek' and extended 'wheet-wheet-wheet'.

Breeding. SEASON. Eggs from late May in western and central Europe. One brood. SITE. In tall, often dense vegetation, also low to medium scrub; from 20 cm to over 200 cm above ground. Nest: cylindrical, sometimes tapering, cup of leaves and stems of dry grass and other plants, with more compact lining of finer material, plus hair and some plant down. Distinct handles woven from rim round supporting plant stems. EGGS. Sub-elliptical, smooth and glossy; very pale green or blue, sometimes grey, boldly and irregularly spotted and blotched olive-green and grey, plus some very faint speckles; markings often concentrated at broad end. Clutch: 3–5(–6). INCUBATION. 12–14 days. FLEDGING PERIOD. 10–11 days.

Wing-length: ♂ 67–73, ♀ 65–71 mm.
Weight: ♂♀ mostly 10–15 g; migrants to 20 g.

Reed Warbler *Acrocephalus scirpaceus*

PLATES: pages 1259, 1274, 1286

DU. Kleine Karekiet   FR. Rousserolle effarvatte   GE. Teichrohrsänger   IT. Cannaiola
RU. Тростниковая камышовка   SP. Carricero común   SW. Rörsångare

Field characters. 13 cm; wing-span 17–21 cm. Close in measurements to Sedge Warbler but bulkier, with peaked crown in most attitudes, larger wings, and broader rump and tail; also close in size to Marsh Warbler but with longer bill, 10% shorter wings, and slimmer body. Medium-sized, compact, robust, and skulking but inquisitive warbler. Plumage varies from brown-olive to -grey in adult and to rufous-brown in juvenile; shows few features at any age except for rufous rump, with supercilium and eye-ring less distinct than any other *Acrocephalus*.

Common and widespread—epitome of the most confusing group of small to medium-sized *Acrocephalus* (and of all west Palearctic Sylviidae), which includes the largely sympatric Marsh Warbler and partly overlapping Blyth's Reed Warbler and Paddyfield Warbler. For main differences, see under those species.

Habitat. Breeds in middle latitudes of west Palearctic, mainly in lowlands with continental climate between July isotherms 10–32°C and with July rainfall of up to 75 mm, nests then being vulnerable to heavy downpours. Spreads into oceanic climatic zone in western France, and also thinly in England and Wales. Attachment to mature beds of reed *Phragmites* with strong stems taller than 1 m produces patchy distribution. Stands of reeds used for nesting may be quite small, often by margins of sluggish rivers, ponds, or shallow lakes, or in narrow lines along ditches. Broader reedbeds in fresh or brackish waters tend to be less favoured, especially if dense and exposed to waves. Regularly, and sometimes predominantly, feeds in other vegetation adjacent to reedbeds, such as cultivated crops or woodland.

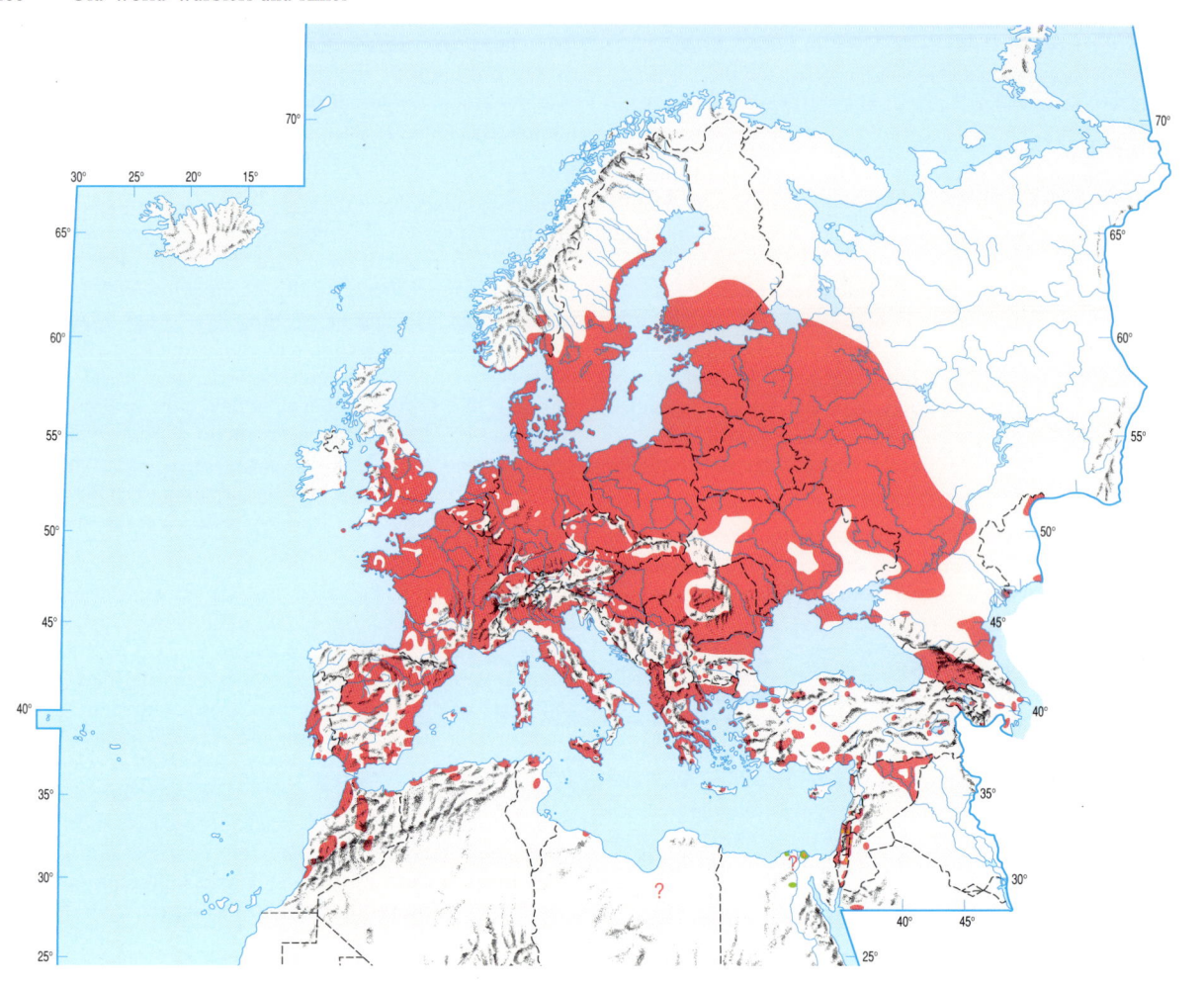

Distribution. Range expansion in Britain, Ireland, Netherlands, Fenno-Scandia, Baltic States, north-west Russia; slight decrease Belgium, Luxembourg, Greece, Albania, Ukraine, and Moldova. BRITAIN. Spread northward and westward. IRELAND. Bred 1935, but colonized since 1980, with breeding first proved 1981. GERMANY. Absent or very sparsely distributed in some uplands owing to lack of habitat. In east, c. 80% of population in Mecklenburg-Vorpommern and Brandenburg. NORWAY. Established in Oslofjord area during 1940s. SWEDEN. Major and continuing expansion: in mid-19th century found only in extreme south; by 1920s spread to central lowlands, and by 1940 found north of Dalälven river (c. 60°30′N). FINLAND. First bred 1920s; spread inland (favoured by habitat changes) from 1960s. ESTONIA. First recorded 1871, but breeding confirmed only 1934. Marked expansion in 1950s. LATVIA. Few records until 1930s; marked spread in recent decades. MALTA. Bred 1977, 1995–6. RUSSIA. Leningrad region: first recorded breeding 1964; whole of southern shore of Gulf of Finland, also of Lake Ladoga, colonized by 1980. AZERBAIJAN. Poorly known, especially in central areas. TURKEY. Probably occurs all suitable reedbeds throughout country. KUWAIT. Very common passage-migrant, probably breeding since 1995. EGYPT. First recorded breeding 1983.

Accidental. Iceland, Faeroes (probably overlooked almost annual visitor), Madeira.

Beyond west Palearctic, extends east to Altai, south to Iran.

Population. Decrease noted Belgium, Luxembourg, Germany, Denmark, Portugal, Italy, Croatia, Ukraine, and Moldova; increase Ireland, Netherlands, Fenno-Scandia, Estonia, Latvia, and Switzerland, mostly stable elsewhere. BRITAIN. 40 000–80 000 pairs 1988–91. Now stable or perhaps slight increase, though abrupt decrease between 1968 and 1969. IRELAND. 50–150 pairs. FRANCE. 10 000–100 000 pairs. BELGIUM. 4700 pairs 1973–7 following decline (c. 9000 in 1960s), but stable since. LUXEMBOURG. 120–150 pairs. NETHERLANDS. 35 000–50 000 pairs 1976–7; 70 000–110 000 in 1979–85. GERMANY. 80 000–130 000 pairs. 46 000 ± 23 000 pairs in east in early 1980s. DENMARK. 2300–34 000 pairs 1987–8. NORWAY. 1000–10 000 pairs 1970–90. SWEDEN. 250 000–500 000 pairs in late 1980s. FINLAND. 15 000–20 000 pairs in late 1980s. Huge increase following eutrophication of waters and expansion of reedbeds, but probably decrease in 1980s. ESTONIA. 20 000–50 000 pairs 1991. LATVIA. 20 000–40 000 pairs in 1980s. LITHUANIA. Common, more numerous in west. POLAND. 30 000–150 000 pairs. CZECH REPUBLIC.

50 000–100 000 pairs 1985–9. Slovakia. 10 000–20 000 pairs 1973–94. Hungary. 100 000–150 000 pairs 1979–93. Austria. At least 30 000–60 000 pairs at Neusiedler See; local elsewhere. No clear trend. Switzerland. 7000–9000 pairs 1985–93. Spain. 7100–13 600 pairs. Portugal. 1000–10 000 pairs 1978–84. Apparently stable overall, though report of marked decrease Algarve during past decade. Italy. 30 000–60 000 pairs 1983–93. Greece. 10 000–20 000 pairs. Earlier report of decline due to habitat destruction, but apparently now stable. Albania. 2000–5000 pairs in 1981. Yugoslavia: Croatia. 8000–12 000 pairs. Slovenia. 200–300 pairs. Bulgaria. 1000–10 000 pairs. Rumania. 2–2.5 million pairs 1986–92. Russia. 10 000–100 000 pairs 1970–90. Belarus'. 5000–10 000 pairs in 1990; fluctuating, but apparently stable overall. Ukraine. 3000–5000 pairs in 1988. Moldova. 5000–7000 pairs in 1986. Azerbaijan. Common. Turkey. 5000–50 000 pairs. Kizilirmak delta: 1500–2000 pairs; more numerous (unusually) than Great Reed Warbler. Cyprus. Locally common; probably stable. Israel. Common, locally abundant: e.g. 30 000 pairs Jezreel valley in 1980s. Egypt. Rare (*c.* 1–100 pairs). Tunisia. Scarce and localized. Algeria. Locally common. Morocco. Uncommon.

Movements. All populations migratory, wintering in Africa south of Sahara and south at least to Zambia. From much of Europe, birds head between WSW and SSW (reverse in spring) to Iberia. West and north European populations start to leave breeding grounds mid- or late July, adults preceding juveniles. South-central European and eastern populations take SE, S or SW headings, entering Africa via eastern Mediterranean area or Middle East. Vacates winter quarters chiefly March–April, reaching west European breeding grounds (e.g. north-east Spain, Camargue) in 2nd half of April, exceptionally late March. In Britain, main immigration not until May, but recorded from early April. Arrives in Rheinland (Germany) in 1st half of May, Leningrad region (Russia) in late May.

Food. Chiefly insects and spiders, some small snails, occasionally some plant material. An opportunist, able to take advantage of local, variable, and short-lived sources of abundant food supply. Feeds mostly at middle height in reeds and rushes and in centres of bushes, and occasionally on ground, thus in more open situations than Sedge Warbler.

Social pattern and behaviour. Solitary outside breeding season. Markedly faithful, within and between years, to small, sometimes overlapping, feeding territories in winter quarters. Mating system typically monogamous, but ♂♂ rarely bigamous. Both sexes incubate and brood young, but ♀ more than ♂; both sexes feed young. In breeding season, territorial ♂ typically has 2–3 main song-posts and 1–2 subsidiary ones; song-post often a denser clump of reeds, occasionally an unusually tall clump. In Britain, song-period late April to August, late singers perhaps typically unpaired ♂♂. Unpaired ♂♂ sing with striking persistence and intensity, especially around dawn; after pairing, song continues, but much reduced.

Voice. Song a series of rather low, mostly guttural churring phrases, with markedly metrical (almost metronomical) quality resulting from repetition of units; units usually given 2–3 times, but occasionally up to 8 times in available recordings: e.g. 'kerr-kerr-kerr chirruc chirruc chirruc kek-kek-kek chirr-chirr', delivered almost continuously in erratic bursts or sequences lasting usually 5–20 s, occasionally 3 min or more; at night, in shorter bursts of up to 15 s. Song starts slowly and is followed by main section (more or less constant in intensity of sound and speed) which ends abruptly. Less of a mimic than congeners and, like Sedge Warbler, apparently only a minority incorporate mimicry. In territorial defence, both ♂ and ♀ (the latter occasionally) give short bursts of song; sustained song (♂ only) used to advertise for mate. Commonest call, a conversational 'churr-churr', usually rather low and soft. Alarm call a harsh, grating 'churrrr'.

Breeding. Season. Western and central Europe: eggs from mid-May. 1–2 broods. Site. In vegetation over water, especially reeds; also in other tall vegetation and low shrubs over dry ground. Nest: deep, cylindrical cup of grass and reed stems and leaves, plus plant down and spiders' webs, woven round plant stems; lined with finer material including hair. Eggs. Sub-elliptical, smooth and glossy; very pale green to greenish-white, more or less heavily spotted, blotched, and speckled olive, green, and grey, often forming cap at broad end; sometimes more evenly distributed finer markings. Clutch: 3–5 (2–7). Incubation. 9–12 days. Fledging Period. 10–12 (9–13) days.

Wing-length: ♂♀ (sexes similar) 63–69 mm.
Weight: ♂♀ mostly 10–16 g.

Clamorous Reed Warbler *Acrocephalus stentoreus*

PLATES: pages 1268, 1274

Du. Indische Karekiet   Fr. Rousserolle stentor   Ge. Stentorrohrsänger   It. Cannareccione stentoreo
Ru. Туркестанская камышовка   Sp. Carricero ruidoso   Sw. Papyrussångare

Field characters. 18–20 cm; wing-span 21–24 cm. Close in size to Great Reed Warbler but slimmer, with bill, tail, and legs averaging fractionally longer but wing 10–20% shorter and blunter, while tail also rounder and bill thinner and more decurved. Large, but rather slender warbler, similar to Great Reed Warbler but with different structure (as above), also flatter crown, usually less distinct face pattern and somewhat paler underparts. Supercilium usually narrower and duller,

Clamorous Reed Warbler *Acrocephalus stentoreus stentoreus* (west Palearctic race): **1** ad (spring), **2** 1st winter. Great Reed Warbler *Acrocephalus arundinaceus. A. a. arundinaceus* (most of west Palearctic): **3** ad (spring), **4** juv. Oriental Reed Warbler *Acrocephalus orientalis* (p. 1272): **5** ad. Basra Reed Warbler *Acrocephalus griseldis* (p. 1271): **6** ad. Thick-billed Warbler *Acrocephalus aedon* (p. 1273): **7** 1st winter.

obvious only before and above eye; eye-stripe only indistinct, and cheeks duller; thus has less fierce expression, enhanced by narrower bill.

Extremely unlikely to occur in Europe, but from Egypt eastwards (and in winter quarters to south) distinction from Great Reed Warbler is difficult except by song. Flight and behaviour much as Great Reed Warbler but movements of tail apparently more pronounced, certainly in display.

Habitat. Breeds from lower middle to subtropical latitudes in warm arid continental climate replacing more northern Great Reed Warbler as specialized large warbler of emergent aquatic vegetation, but preferring papyrus to reeds where choice exists.

Distribution and population. SYRIA. Few records (in at least one case Great Reed Warbler not excluded) and no confirmation of breeding. ISRAEL. Locally quite common. Disappeared from some areas and population declined 1960s–80s. JORDAN. Most likely resident along Yarmuk and Jordan rivers and at As Safi. No longer breeds Azraq. IRAQ, KUWAIT. Arabian and south-central Asian race (*brunnescens*) apparently scarce winter visitor. EGYPT. Spreading in Nile valley and Wadi el Rayan following completion of Aswan dam and creation of lakes. Abundant (over 100 000 pairs).

Beyond west Palearctic, breeds discontinuously from African Red Sea coast north-east to Kazakhstan, and east through Indian sub-continent to south-west China; also from Philippines and southern Indonesia east to Solomon Islands, south to Australia and Tasmania.

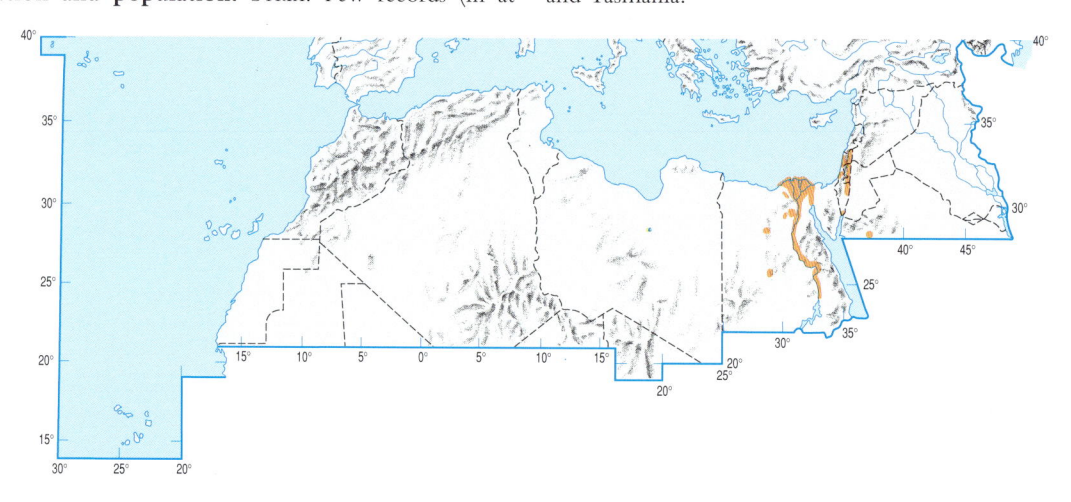

Movements. Varies from sedentary to migratory in different parts of range. West Palearctic populations chiefly resident, with local dispersal outside breeding season. In Egypt, recorded in areas where it does not breed, including middle of Cairo.

Food. Largely insects and other invertebrates.

Social pattern and behaviour. Mainly studied beyond west Palearctic. In Western Australia, mating system varies between monogamy and opportunistic polygyny. Territorial when breeding, but often forms close-packed neighbourhood groups of many pairs within comparatively small area; nests sometimes only a few metres apart in Australia. ♂ usually sings from well down in reeds, intermittently from reed top or tree; both by day and by night. In west Palearctic, song recorded February–May and September.

Voice. Song shows some similarity to Great Reed Warbler in combining harsh, frog-like and chattering noises with variety of sweeter sounds or loud peculiar squeaks. Tempo similar to Great Reed Warbler, but song overall less raucous, more melodious; phrases given 3–4 times. A 'karra-karra-kareet-kareet-kareet' or 'prit-prit-pritik' are typical, but many variations. Contact-alarm calls include loud, harsh 'ke', 'chur-r chur-r' and 'tak'.

Breeding. SEASON. Israel: April to end of July. Egypt: March to late June. SITE. In reeds and other aquatic vegetation, normally over water. 10–15(–65) cm above water or ground. Nest: deep, cylindrical cup of reed stems and leaves, lined with finer material. EGGS. Sub-elliptical, smooth and glossy; apparently identical in colouring and markings to those of Great Reed Warbler. Clutch: 3–6. INCUBATION. 13–14 days. FLEDGING PERIOD. 11–13 days.

Wing-length: ♂ 80–85, ♀ 75–80 mm.
Weight: ♂♀ mostly 21–27 g.

Great Reed Warbler *Acrocephalus arundinaceus*

PLATES: pages 1268, 1274

DU. Grote Karekiet FR. Rousserolle turdoïde GE. Drosselrohrsänger IT. Cannareccione
RU. Дроздовидная камышовка SP. Carricero tordal SW. Trastsångare

Field characters. 19–20 cm; wing-span 24–29 cm. Over 50% larger than Reed Warbler, with proportionately larger bill, more peaked head, more robust body, longer wing-points, and lengthier rump and tail combining into bold, heavy form; heavier (by 20%) and stockier than Clamorous Reed Warbler, with proportionately shorter and seemingly deeper bill, more peaked crown, longer wing-point equal in length to exposed tertials, and rather shorter tail with less graduated end. Large, robust warbler, warm olive-brown above and cream below; rufous-buff tones show on inner wing feathers and particularly on flanks and under tail-coverts. At most ranges, face pattern distinctive, with cream-buff supercilium, whitish eye-ring, and dusky eyestripe combining to give stern expression; in fresh plumage, tail feathers tipped dull white. Eastern race colder in plumage tone, greyer above and whiter below. Legs mostly pale brown, unlike bluish-grey of most Clamorous Reed Warblers.

Easily distinguished from smaller *Acrocephalus* by size and song but, particularly in Middle East, subject to confusion with Clamorous Reed Warbler and Basra Reed Warbler, which see. Also similar to vagrant Oriental Great Reed Warbler and Thick-billed Warbler, which see. Flight lacks flitting action of smaller *Acrocephalus*, having instead rather laboured or thrashing bursts of wing-beats producing heavy, even jerky progress over short distance and with heaviness enhanced by length and spread of long rump and tail; tail often held loosely below level of body; tends to crash into cover without sudden jink or turn. Gait much as Reed Warbler but much less nimble, with weight of bird evident not only in slower, even clumsier movements but also in greater disturbance of plant stems or other foliage; on ground, hops powerfully, recalling thrush.

Habitat. Breeds in middle latitudes of west Palearctic, in both cool and warm and arid and moist climates, between July isotherms 17–32°C, mainly in temperate, steppe, and Mediterranean zones. In west, occurs mainly in lowland, even in Switzerland nesting only up to *c.* 650 m. Mostly concentrated in aquatic vegetation emerging from shallow standing water, fresh or brackish, especially in strong, tall, and dense reeds fringing banks or swamps, or islanded above the shallow bottoms of lakes or sluggish rivers. Early in season, before reeds fully grown, will frequent treetops around ponds, and often feeds in them.

Distribution. Expansion in northern and eastern Europe probably due to climatic amelioration and concurrent development of suitable habitats. Range has decreased more recently, notably in north-west. BRITAIN. Spring vagrants frequently sing, and single birds seen with nest-material at same site 1993, 1994. LUXEMBOURG. Ceased regular breeding by 1963. Now irregular (1–3 pairs). NETHERLANDS. Strong range contraction, partly through loss of habitat, but now much apparently suitable habitat unoccupied. GERMANY. Reduced to small remnant populations in west. Many sites also abandoned in east (especially south-western part) in last 30 years. DENMARK. Decreased 1970–92; latest confirmed breeding 1988. Singing ♂♂ 1990–92 mapped. SWEDEN. First recorded breeding 1917. FINLAND. Colonized south coast in 1930s; slight spread inland from 1960s. BALTIC STATES. Marked spread inland from coastal sites in 20th century. CZECH REPUBLIC. Range decrease. RUMANIA. Widespread, but local: breeds all large reedbeds, especially in lowlands. RUSSIA. Colonized whole of Moscow region, where breeding confirmed 1977. AZERBAIJAN. Range apparently more extensive than all other *Acrocephalus*. Cyprus. Bred 1982, 1985. Israel. Common breeder Hula and Bet Shean valleys, and northern coastal plain before draining of swamps in 1950s. Reduced to a few pairs (not all breeding) by late 1980s. Jordan. Sole site (Azraq) now abandoned. Iraq. Range probably more extensive than shown.

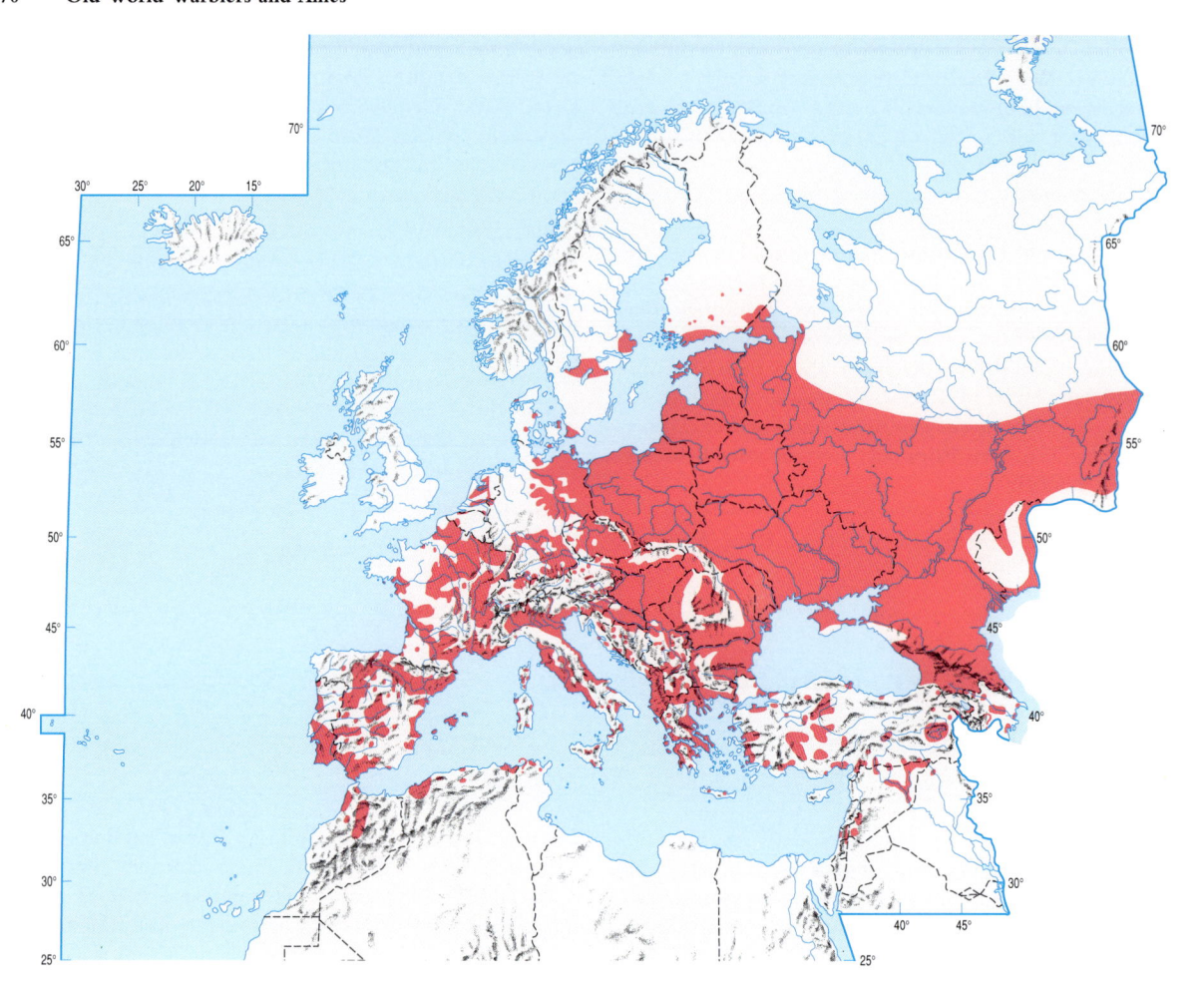

Accidental. Faeroes, Britain (near-annual), Ireland, Norway, Azores, Madeira, Canary Islands.

Beyond west Palearctic, extends across Siberia and central Asia to Altai, Zaysan depression, western Sinkiang (China), and Tajikistan. Bred Eastern Province (Saudi Arabia) 1995.

Population. Marked decrease France, Benelux countries, Germany (where habitat destruction one cause), Czech Republic, Slovenia, and Israel; less marked or only local decline Denmark, Poland, Slovakia, Austria, Switzerland, Portugal, Italy, Greece, and Israel. Increase Sweden, Estonia, and Latvia; apparently stable elsewhere. France. 1000–5000 pairs. Belgium. 1000 pairs in 1960s, 130 in 1973–7, 4–5 in 1989–91. Netherlands. 400–550 pairs 1989–91; decreased by *c.* 80% since 1970. Germany. 7400–12 000 pairs. In east, 4700 ± 1400 pairs in early 1980s. Decline from 1950s or 1960s. Denmark. 8–15 pairs 1987–90. Sweden. 100–200 pairs in late 1980s. Slight increase since 1960s. Finland. Average annual number recorded (marked fluctuations) 3.6 in 1950s, 7.5 in 1960s, 30.2 in 1970s, mostly singing ♂♂. 70–100 pairs (many unpaired ♂♂) in late 1980s. Estonia. 5000–10 000 pairs 1991. Remarkable long-term increase. Latvia. 10 000–20 000 pairs in 1980s. Lithuania. Common. Poland. 8000–30 000 pairs. More or less stable, though perhaps decrease in south-west. Czech Republic. 1500–3000 pairs 1985–9. Slovakia. 1000–2000 pairs 1973–94. Hungary. 30 000–50 000 pairs 1979–93. Austria. Over 8000 pairs Neusiedler See; relatively scarce elsewhere. Locally declining; otherwise no clear trend. Switzerland. 200–250 pairs 1985–93; stable, following decline. Spain. 9500–12 600 pairs. Portugal. 1000–10 000 pairs 1978–84. Stable, though decreased somewhat in Algarve over last decades. Italy. 20 000–40 000 singing ♂♂ 1983–93. Decline: e.g. under 50 pairs Sicily in 1980s where formerly much more abundant. Greece. 15 000–30 000 pairs. Albania. 5000–10 000 pairs 1981. Yugoslavia: Croatia. 20 000–25 000 pairs. Slovenia. 300–400 pairs. Bulgaria. 10 000–100 000 pairs. Rumania. 400 000–500 000 pairs 1986–92. Russia. 500 000–5 million pairs 1970–90. Increase Moscow region in 1980s. Belarus'. 60 000–90 000 pairs in 1990.

Ukraine. 30 000–35 000 pairs in 1990. Moldova. 20 000–30 000 pairs in 1986. Azerbaijan. Common to very common. Kazakhstan. Most numerous species in reedbeds of Ural delta. Turkey. 10 000–100 000 pairs. Generally the most numerous *Acrocephalus*. 275–325 pairs Kizilirmak delta. Tunisia. Scarce and localized. Morocco. Uncommon.

Movements. All populations migratory. West Palearctic populations winter in Africa south of Sahara, eastern populations in south-east Asia. In autumn, European birds head initially between south-west and south-east, chiefly between SSW and SSE. Movement south begins chiefly in August, passage within breeding range continuing mostly to end of September. Reaches winter quarters very late, and most birds evidently pause in northern tropics, where many moult. Spring migration begins in March; present in wintering areas mostly until early April. Arrives in southern Europe from mid-April, progressively later further north. Reaches northern Denmark and southern shore of Gulf of Finland only from mid-May. In Britain, majority of records are in south-east, chiefly in spring.

Food. Mainly insects, with some spiders, snails, and small vertebrates; some fruits and berries outside breeding season. Feeding in trees and bushes (e.g. willows, oaks) is regular at many sites and may provide main source of food. Many of the more frequent prey items (dragonfly nymphs, aquatic beetles and their larvae, aquatic Hemiptera, fish fry) are presumably taken from water surface or just beneath.

Social pattern and behaviour. Mating system usually monogamous, but simultaneous or successive polygyny frequent. Nest duties largely undertaken by ♀, ♂ guarding nest but otherwise sharing only in feeding of young. Territories tend to be clustered in neighbourhood groups near good feeding sites. Unpaired ♂♂ usually sing from exposed perch near top of reed, sometimes from solitary bush, occasionally from tree (up to 6–8 m) or telegraph wire. After pairing, song infrequent and normally given from within dense vegetation. Sings more or less throughout day from arrival on breeding grounds or at least once territory established, with pre-dawn chorus of remarkable volume; persistent song even around midday characteristic of unpaired ♂; snatches also given throughout night, especially if moonlit.

Voice. Song well known for remarkable volume (carrying up to *c.* 1 km) and low-pitched, guttural (frog-like) croaking or harsh grating sounds, including churrs and rattling components, interspersed with various tones, some piping or creaky. General impression is of harsh, low-frequency song (lowest of European *Acrocephalus*), with noises predominant; great interval leaps are typical, low-pitched and grating units being combined with high-pitched but often still harsh tones. Given in phrases of *c.* 3–4 s or more, separated by pauses of 1–3 s and in slow or moderato tempo: 'kar-kar keet-keet-keet karra-karra orre-orre tchu-tchu-tchu karra-kee' etc. Calls include harsh 'chack', 'tack', or 'ak', given when disturbed; harsh, staccato 'kaa-kaa-kaaa' like chattering of Magpie; and, low-pitched croaking churr, given as alarm at nest or when accompanying fledged young.

Breeding. Season. Central and western Europe: eggs from mid-May. Southern Europe: begins early to mid-May. Usually 1 brood, occasionally 2. Site. Dense stands, usually of reeds, in water. Preference for areas with thick reed stems. Nest: deep, cylindrical cup of reed stems and leaves, plus plant down, spiders' webs, and reed flowers, lined with finer material including hair and sometimes feathers; 10–130 cm above water. Eggs. Sub-elliptical, smooth and glossy; pale green, blue, or blue-green, occasionally white, spotted, sometimes speckled and blotched dark brown, olive-green, and pale blue-grey. Clutch: 3–6. Incubation. About 14 days. Fledging Period. 12–14 days.

Wing-length: ♂ 94–101, ♀ 89–95 mm.
Weight: ♂ mostly 29–40, ♀ 24–36 g; migrants up to 51 g.

Basra Reed Warbler *Acrocephalus griseldis*

plates: pages 1268, 1274

Du. Basra-karekiet   Fr. Rousserolle d'Irak   Ge. Basrarohrsänger
Ru. Иракская камышовка

Field characters. 17–18 cm; wing-span 21–24.5 cm. 15–20% smaller (and almost 40% lighter) than Great Reed Warbler, with proportionately longer and more slender, finely pointed bill, shorter legs and rather shorter, less graduated tail; wing point long, at least equalling length of exposed tertials (unlike Clamorous Reed Warbler). Rather large but slim warbler, intermediate in size between Great Reed Warbler (and Clamorous Reed Warbler) and Reed Warbler, and with less rufous plumage than Great Reed Warbler; differs most distinctly in darker brown flight feathers and tail (in fresh plumage), whiter underparts which virtually lack breast streaks and have creamy yellow wash on flanks, and strongest head pattern of all large *Acrocephalus*, with always whitish supercilium emphasized by usually dark eyestripe. Long wing-point emphasized by noticeably pale fringes and tips to primaries. Only recently separated from Great Reed Warbler and still subject to confusion with smaller individuals of its greyer eastern race; observations on general character lacking.

Habitat. Breeds among reeds, reedmace, and similar thick marshy waterside vegetation in Middle East. In east African winter quarters remains in vicinity of water, sharing bushy thickets and rank undergrowth in wet places, along ditches, etc. with Marsh Warbler, Reed Warbler, and Great Reed Warbler.

Distribution and population. Breeds only in west Palearctic. IRAQ. Little recent information. Range extends from Baghdad area to Al Faw. KUWAIT. Uncommon passage-migrant; almost certainly breeding Jahra Pool since 1995.

Accidental. Cyprus, Israel.

Movements. Migratory; winters in eastern Africa from southern Somalia south through eastern Kenya (where apparently commonest wintering *Acrocephalus* on lower Tana floodplain in south-east), and eastern Tanzania to Malawi and Mozambique, to at least 17°S. Winter range thus far more extensive than known breeding range. Leaves breeding grounds August–September, passes through Arabia and crosses Red Sea into north-east Africa, arriving in Sudan and Ethiopia from late August; present in East Africa from late October to early April. In spring apparently makes long unbroken flights on return migration since few records from Africa; common in north-east Saudi Arabia April and May and arrives in Iraq from April.

Food. No details, but presumably as Great Reed Warbler.

Social pattern and behaviour. Not studied on breeding grounds but probably as Great Reed Warbler. Density of c. 10–21 per ha recorded in Kenya, and winter song from January to March.

Voice. Known in detail only from winter quarters. Song in Kenya subdued and rhythmic sequence of low-pitched sounds like 'chuc-chuc-churruc-churruc-chuc', quieter and less strident than Reed Warbler and without the guttural grating and croaking quality of Great Reed Warbler song. Rich song quality of birds in Somalia thought to be this species suggested nightingale. Main call a harsh 'chaarr', slightly louder than equivalent in Reed Warbler.

Breeding. No information; probably as Great Reed Warbler in most, if not all, aspects.

Wing-length: ♂♀ 72–88 mm.
Weight: ♂♀ 12–29 g.

Oriental Reed Warbler *Acrocephalus orientalis*

PLATES: pages 1268, 1274

DU. Chinese Karekiet FR. Rousserolle d'Orient GE. Chinarohrsänger
RU. Восточная камышовка

Field characters. 18.5–19.5 cm; wing-span 23–26 cm. Most slighter (and 20% lighter) than Great Reed Warbler and some noticeably smaller, all showing proportionately 15% shorter wings, with point equalling or only $\frac{2}{3}$ length of exposed tertials; all measurements overlap with Clamorous Reed Warbler but tail is somewhat shorter and less rounded. Large warbler with appearance intermediate between Great Reed Warbler and Clamorous Reed Warbler; converges closely with Red Sea race of latter, being separable only by somewhat shorter, stubbier bill, distinct striations on lower throat and chest, and broad whitish tips to tail feathers which (in fresh plumage) provide best distinction from both Clamorous Reed Warbler and eastern race of Great Reed Warbler.

Only recently separated from Great Reed Warbler and observations on general character lacking. Song recalls other large reed warblers but less rhythmic, written 'kawa-kawa-kawa-gurk-gurk-eek-eek-kawa-gurk'. Commonest call not differentiated from Great Reed Warbler.

Habitat. Principally reedbeds, both coastal and freshwater, but also meadows with long grass, flooded willow beds, cultivated land with bordering reeds, etc. in cool temperate to subtropical east Palearctic and Oriental regions. Overwinters in similar habitats; in Malaysia, ♀♀ largely confined to reedbeds, ♂♂ more at edges or in scrub, so attachment to water seems to lose force outside breeding season; on migration, frequents also parks and gardens.

Distribution. From north-west Mongolia and southern Transbaykalia eastwards to Amurland, Ussuriland, Sakhalin island, eastern China, Korea, and Japan.

Accidental. Israel: Eilat, February–April 1988, May 1990.

Movements. Migratory, wintering from north-east India, throughout Indochina, Malaysia, Philippines, and Indonesia. Departs eastern Siberia chiefly late August to early September and very common on passage through Hong Kong mainly September–October; leaves Japan August to mid-September, a few remaining in south into October. Present in Indonesia and Philippines September–May, spring migration beginning in April with peak passage in Hong Kong mid-April to mid-May. Arrives in Ussuriland from mid-May, Amurland end of May, and Japan from mid-April into June.

Wing-length: ♂ 81–95, ♀ 72–86 mm.

Thick-billed Warbler *Acrocephalus aedon*

PLATES: pages 1268, 1274

Du. Diksnavelrietzanger   Fr. Rousserolle à gros bec   Ge. Dickschnabel-Rohrsänger   It. Cannareccione beccoforte
Ru. Толстоклювая камышовка   Sp. Carricero picogordo   Sw. Tjocknäbbad sångare

Field characters. 18–19 cm; wing-span 21–24 cm. Less bulky than Great Reed Warbler, with noticeably shorter bill, more domed (less angular) head, much shorter wings, and longer and distinctly more graduated tail; structure distinctive, with weight apparently set forward and rump and tail looking relatively narrow or slight. Large, stubby-billed warbler, with less full rear body and tail than Great Reed Warbler; plumage similarly uniform warm olive-brown over fore-parts and wings but can appear markedly rufous over rump and tail. Lacks contrasting supercilium. Cocks tail.

Recent observations have shown Thick-billed Warbler to have rather shrike-like shape, more telling than its few plumage characters. Trailing and (when not spread) narrow tail always noticeable in flight and especially when cocked by perched bird. At close range, round head with short bill and dark eye can recall Garden Warbler. Carriage less upright than Great Reed Warbler, with stance on perch or ground again recalling *Sylvia* warbler or shrike.

Song unlike Great Reed Warbler: essentially more warbling and chortling, in longer melodious phrases that lack guttural repetitions and include much mimicry. Commonest call an abrupt, deep, harsh 'chok-chok', 'tschok-tschok', 'tschuk', 'tack', or 'tschak-tschak', harder notes recalling Stonechat.

Habitat. Breeds extralimitally in middle latitudes of east Palearctic, mainly in dry warm continental lowlands. Prefers thickets of bushes to reedbeds, independently of presence of water.

Distribution. Breeds in Asia from Ob' river and northern Mongolia east to Ussuriland and north-east China.

Accidental. Britain: single birds, Shetland, October 1955 and September 1971. Finland: one, Norrskär (Mustasaari), October 1994. Egypt: one, monastery of St. Katherine, Sinai, November 1991.

Movements. All populations migratory, moving south through Mongolia and central and eastern China to winter in foothills of Himalayas and from southern and eastern India east to southern China and Vietnam. Vagrants to west Palearctic are presumably reverse migrants and most likely to be Siberian race.

Wing-length: ♂ 78–86, ♀ 76–82 mm.

Olivaceous Warbler *Hippolais pallida*

PLATES: pages 1276, 1286

Du. Vale Spotvogel   Fr. Hypolaïs pâle   Ge. Blaßspötter   It. Canapino pallido
Ru. Бледная пересмешка   Sp. Zarcero pálido   Sw. Eksångare

Field characters. 12–13.5 cm, wing-span 18–21 cm. Western race (Iberia and north-west Africa) slightly larger than Reed Warbler, with similar structure except for slightly longer bill and wings, 10% longer and less rounded tail, deeper belly, and shorter under tail-coverts; other races up to 25% smaller, closer in size to Reed Warbler though with 10% shorter bill and 5–10% shorter wings and tail. Medium-sized to quite large warbler, with rather long to long bill, (typically) flat crown, rather short wings, and sturdy legs. Closed wing-point forms $\frac{1}{4}$ of total wing length (or equal to $c. \frac{1}{2}$ length of exposed tertials), with primary-tips only just reaching end of upper tail-coverts. Plumage recalls Garden Warbler: dull grey or brown above and dull white below, with dull supercilium, brighter eye-ring, pale edges to inner flight-feathers, and pale edges and corners to tail. Never shows clear yellow tones. Habitually lowers and raises tail.

Large Iberian and north-west African race (*opaca*) not difficult to identify, but care necessary to eliminate pale grey or brown

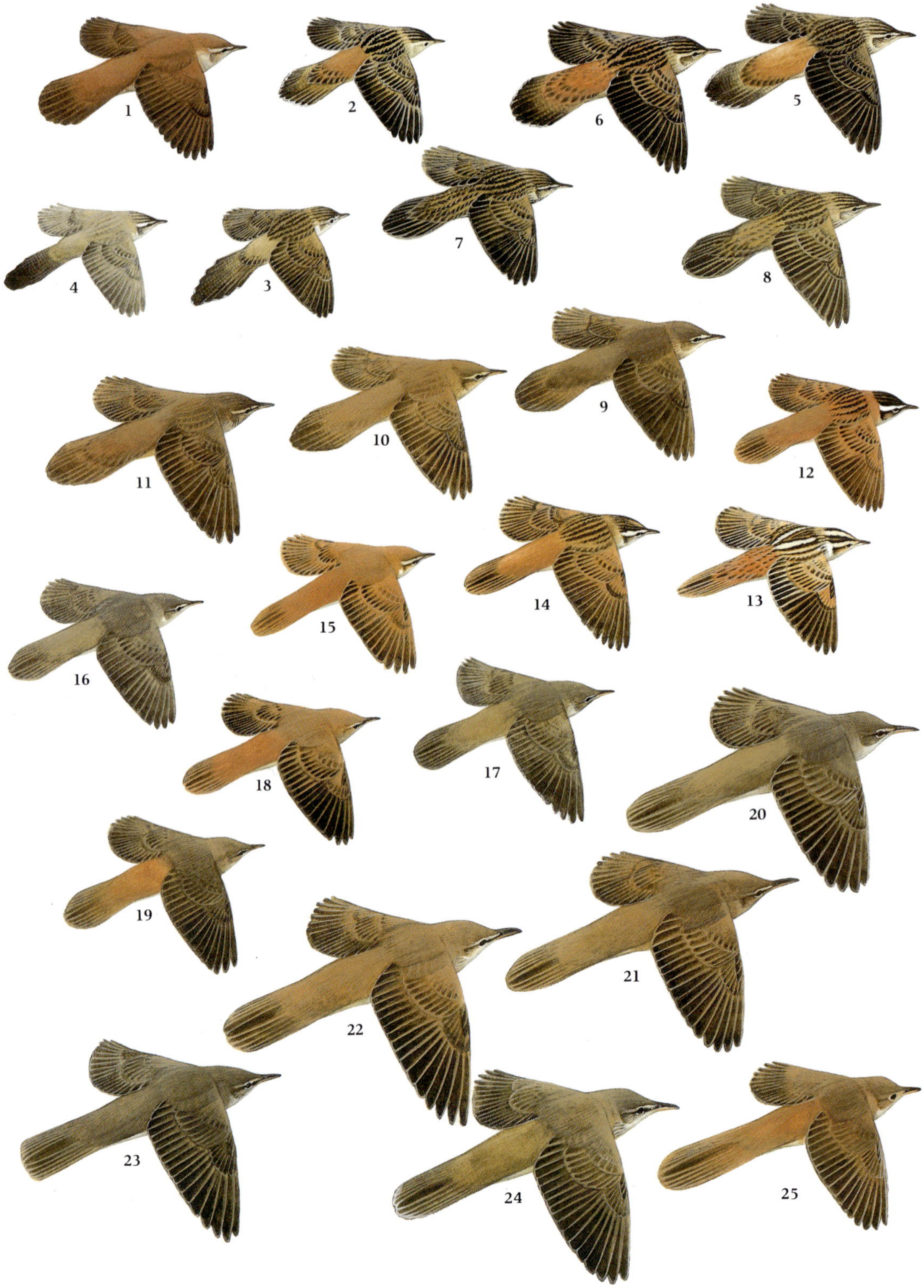
1
2
6
5
4
3
7
8
9
11
10
12
14
13
16
15
18
17
20
19
22
21
23
24
25

variants of Icterine Warbler and Melodious Warbler. Small south-east European race (*elaeica*) liable to confusion with Booted Warbler. Also confused with Garden Warbler, though differences in bill length and shape (short, even stubby in Garden Warbler) and head pattern and shape (virtually unmarked and with domed crown in Garden Warbler) are usually striking. Much more distinctive in Iberia and North Africa than elsewhere. In flight, combination of fluttering action, roundness of wings, and weight of bird gives impression of rather heavy or unsteady and laboured progress, particularly in turns or on entry to cover; lacks fluent wing-beats of long-winged *Hippolais*. Carriage distinctly level and gait rather heavy or clumsy; uses distinctive powerful, heavy foraging movements through foliage, with frequent upwards stretching of head and neck, forward or sideways leaning towards food, and purposeful tug of bill to capture it. Frequent tail-pumping (more constant than in other *Hippolais*) consists of shallow, loose, up-and-down movement below level of back.

Habitat. Breeds in lower middle latitudes of west Palearctic, geographically complementing Icterine Warbler, and accordingly experiencing warmer mainly Mediterranean and steppe climate, ranging to subtropics, from July isotherms of 22°C to more than 32°C. Seems to breed in lower shrubs and in drier surroundings than other *Hippolais*, including shrub growth in steppe and semi-desert, and among scattered broad-leaved or coniferous trees on dry river banks or in warm river valleys; also in tamarisk clumps, palms in oases, orchards, parks, and gardens.

Distribution. HUNGARY. Long-term range increase. BALKANS. Marked extension of range since mid-20th century, but with fluctuations. RUSSIA. Expansion in recent decades in southern Dagestan. UKRAINE. May nest in south-west, but no proof. TURKEY. Common in west and centre, more scarce and local in higher valleys of east. ISRAEL. Very common in centre and north; fewer in south. EGYPT. Considerable southward

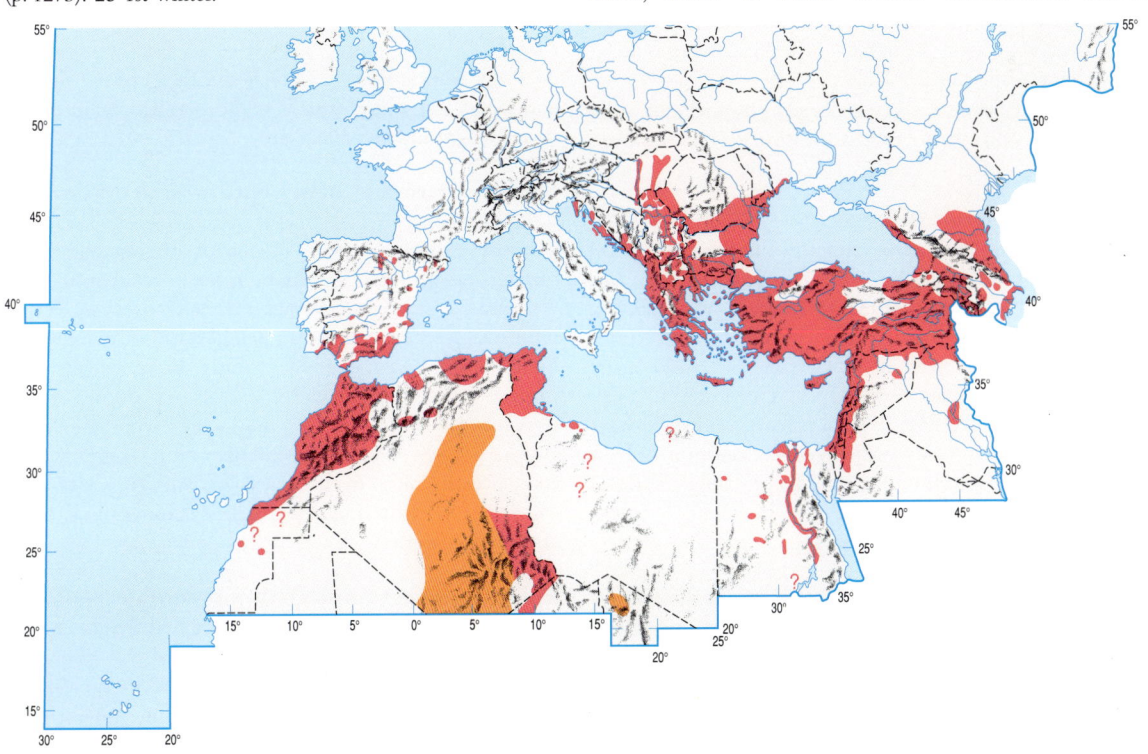

(FACING PAGE) Cetti's Warbler *Cettia cetti* (p. 1235): **1** ad. Fan-tailed Warbler *Cisticola juncidis* (p. 1237): **2** ad. Graceful Warbler *Prinia gracilis palaestinae* (p. 1239): **3** ad. Scrub Warbler *Scotocerca inquieta saharae* (p. 1241): **4** ad. Pallas's Grasshopper Warbler *Locustella certhiola* (p. 1243). *L. c. sparsimstriata*: **5** worn 1st winter. *L. c. rubescens*: **6** worn 1st winter. Lanceolated Warbler *Locustella lanceolata* (p. 1244): **7** worn 1st winter. Grasshopper Warbler *Locustella naevia naevia* (p. 1245): **8** ad. River Warbler *Locustella fluviatilis* (p. 1247): **9** ad. Savi's Warbler *Locustella luscinioides luscinioides* (p. 1249): **10** ad. Gray's Grasshopper Warbler *Locustella fasciolata fasciolata* (p. 1251): **11** worn 1st winter. Moustached Warbler *Acrocephalus melanopogon melanopogon* (p. 1251): **12** ad. Aquatic Warbler *Acrocephalus paludicola* (p. 1254): **13** juv (autumn). Sedge Warbler *Acrocephalus schoenobaenus* (p. 1255): **14** ad. Paddyfield Warbler *Acrocephalus agricola* (p. 1258): **15** ad. Blyth's Reed Warbler *Acrocephalus dumetorum* (p. 1260): **16** ad. Marsh Warbler *Acrocephalus palustris* (p. 1263): **17** ad. Reed Warbler *Acrocephalus scirpaceus* (p. 1265). *A. s. scirpaceus*: **18** ad. *A. s. fuscus*: **19** ad. Clamorous Reed Warbler *Acrocephalus stentoreus brunnescens* (p. 1267): **20** ad fresh: **21** *A. s. stentoreus* ad. Great Reed Warbler *Acrocephalus arundinaceus* (p. 1269): **22** ad. Basra Reed Warbler *Acrocephalus griseldis* (p. 1271): **23** ad. Oriental Reed Warbler *Acrocephalus orientalis* (p. 1272): **24** ad fresh. Thick-billed Warbler *Acrocephalus aedon* (p. 1273): **25** 1st winter.

Olivaceous Warbler *Hippolais pallida*. *H. p. opaca*: **1** ad worn (summer), **2** ad fresh (autumn), **3** ad 1st winter. *H. p. elaeica*: **4** ad fresh (autumn), **5** ad 1st winter. *H. p. pallida*: **6** ad. *H. p. reiseri*: **7** ad. Booted Warbler *Hippolais caligata*. *H. c. caligata* (most of west Palearctic range): **8** ad fresh (autumn), **9** 1st winter. *H. c. rama* (lower Volga eastwards): **10** ad, **11** 1st winter.

extension along Nile in 20th century. ALGERIA. Less common in north than in south.

Accidental. Britain, Ireland, France, Belgium, Germany, Sweden, Finland, Czech Republic, Slovakia, Austria, Italy, Malta, Madeira, Cape Verde Islands.

Beyond west Palearctic, breeds locally in north of sub-Saharan Africa, and from Arabia north-east to southern Kazakhstan.

Population. Trend varies. HUNGARY. 200–250 pairs 1979–93; increase. SPAIN. 500–1300 pairs; stable. PORTUGAL. 100–1000 pairs 1978–84; stable. GREECE. 100 000–150 000 pairs. ALBANIA. 10 000–20 000 pairs in 1981; fluctuating. YUGOSLAVIA: CROATIA. 6000–9000 pairs; slight increase. BULGARIA. 1000–10 000 pairs; slight increase. RUMANIA. 2500–4000 pairs 1986–92; possible decline. AZERBAIJAN. Common to very common. TURKEY. 500 000–5 million pairs. CYPRUS. Perhaps 20 000–40 000 pairs; probably stable. SYRIA. Common, at least in west. JORDAN. Fairly common. IRAQ. Has probably decreased since 1960s. EGYPT. Abundant (over 100 000 pairs). TUNISIA. Fairly common. MOROCCO. Common.

Movements. European and Asian populations migratory; some African populations sedentary at south of range. Migratory populations winter mainly in dry areas of Africa north of the equator, from Sénégal east to Eritrea and Somalia, extending to c. 7°S in Tanzania. From west of range, passage (both seasons) through north-west Africa on broad front; from south-east Europe, passage mostly through Levant. Spring migration rather late; main arrival north-west Africa in April, passage through Strait of Gibraltar continuing to early June. Reaches southern Balkans late April, northern Balkans in May.

Food. Chiefly insects; also fruit in late summer. Forages restlessly among foliage, generally within upper half of bushes and trees, and taking insects while perched or in flight. Also drops to ground to collect prey.

Social pattern and behaviour. In winter quarters, mostly solitary and territorial, with evidence of fidelity to wintering site over successive years. No evidence for other than monogamous mating system. Nest built by both sexes; incubation usually by ♀; both parents feed nestlings and fledged young. Territorial, with strong tendency to form neighbourhood groups, reaching high densities in favourable habitat. Song usually delivered from perch within cover of foliage, sometimes from more exposed spot; also sings on the wing, while moving about among foliage, and intermittently while foraging. Seasonal song-period long, from arrival in breeding area (also occasionally in wintering area) to late summer, and daily song also long-sustained.

Voice. Song shows marked regional variation. Song of Iberian and north-west African birds high-pitched, bright and pleasing in tonality, and brisk in delivery; in quality a rather scratchy warble, with some fluty notes, which occur typically at start of phrases as well as intermittently; song usually comprises repeated phrases of fluctuating pitch, sometimes separated by pauses, sometimes delivered more continuously, though not shaped to produce cyclic pattern. Song of Saharan populations apparently similar, but weaker and thinner. Song of Egyptian birds vigorous and uneven, a stuttering sequence of notes with sweeter ones interspersed; duration of songs usually *c.* 6 s, occasionally up to 30 s. Songs of populations in south-east Europe and Near East consists of complex phrases 2–3 s long, with characteristic progression of pitch, repeated without pause in cyclic pattern. Calls include a sharp 'tack' or 'tchack' (commonest call); continuous 'tset-tset . . .', when disturbed; and series of sharp ticking sounds (like small stones being struck together), e.g. when alarmed at nest.

Breeding. Season. North-west Africa: eggs laid early May (exceptionally April) to early July. Cyprus and Levant: eggs laid May–June. Greece: laying begins late May. 1–2 broods. Site. In tree, bush, creeper, etc. 0.4–9 m above ground. Nest: well-built cup of small twigs, grass, plant stems, lined with finer material including hair, rootlets, and some plant down. Eggs. Sub-elliptical, smooth and glossy; pale grey-white, occasionally tinged pink, sparingly spotted and speckled black, rarely with black lines. Clutch: 2–5. Incubation. 11–13 days. Fledging Period. 11–15 days.

Wing-length: *H. p. opaca*: ♂ 68–74, ♀ 67–73 mm. *H. p. reiseri* (Sahara): ♂ 64–68, ♀ 62–68 mm. *H. p. elaeica* (Balkans): ♂ 66–71, ♀ 62–68 mm.
Weight: *H. p. opaca* (Iberia, north-west Africa): ♂ ♀ mostly 8–13 g.

Geographical variation. Marked, complex. Perhaps better considered to comprise 3 incipient species, *opaca*, nominate *pallida* (with races *reiseri* and *laeneni*) and *elaeica*; relations of these to Booted Warbler not yet elucidated and species limits as presented here perhaps not satisfactory.

Booted Warbler *Hippolais caligata*

PLATES: pages 1276, 1286

Du. Kleine Spotvogel Fr. Hypolaïs bottée Ge. Buschspötter It. Canapino asiatico
Ru. Бормотушка Sp. Zarcero escita Sw. Gråsångare

Field characters. 11.5–12 cm; wing-span 18–21 cm. Smaller than Olivaceous Warbler, with noticeably shorter bill, more domed head, and even shorter wings and tail. Close in size to Bonelli's Warbler but with 5–10% longer bill and 10% longer legs; outer wing most rounded of *Hippolais*. Rather small and fairly slim warbler, with (for a *Hippolais*) unusually short (but still wide) bill, quite round head, short and round wings, and sturdy legs; smallest *Hippolais* and subject to as much confusion with plain-plumaged *Phylloscopus* as with smaller individuals of Olivaceous Warbler and pale variant of Melodious Warbler. Closed wing-point forms $\frac{1}{4}$ of total wing length (or equal to $\frac{1}{3}-\frac{1}{2}$ exposed length of tertials), with primary-tips not reaching end of upper tail-coverts. Essentially greyish-brown above and buff-white below, with dusky linear smudge above quite long pale supercilium, whitish eye-ring, and off-white outer tail-feathers.

Perhaps the most confusing of all small nondescript west Palearctic warblers, with small, sometimes dainty, sometimes dumpy, apparently long-tailed form which, combined with short, fine bill, suggests particularly Bonelli's Warbler rather than *Hippolais*. Prolonged observation necessary to establish breadth of bill-base, pot-belly, short under tail-coverts (real cause of long-tailed appearance), and square end to tail which together indicate *Hippolais*. Important to recognize in Europe that small pale variant Melodious Warbler is most likely bird to be mistaken for Booted Warbler; when worn, upperparts of Melodious Warbler may even become milky tea colour (a tone wrongly considered unique to Booted Warbler) and its yellow-stained throat may only be visible in close frontal view. Flight lighter than Olivaceous Warbler and Melodious Warbler, with less impression of labouring but still with most rounded wings of *Hippolais*; flight looks heavier than all *Phylloscopus* but able to catch flies. Often more constantly active than other *Hippolais*.

Habitat. Breeds mainly in middle latitudes of central Palearctic, extending within west Palearctic to lower boreal zone and extralimitally in Asia into subtropics, but predominantly in warm continental, temperate, and steppe zones from July isotherms of 18°C to over 32°C. In southern Russia, ascends to 1600 m, ranging from rich shrubbery and grasses on plains, along lower parts of rivers and dense reedbeds, to orchards, riverside plantations, vegetable gardens, melon patches, trees, copses, forests, bottomland woods by rivers, clumps of herbage or shrubs in semi-desert, thorn thickets, and pistachio groves in mountain river valleys. More northerly and westerly populations breed both in taiga and broad-leaved forest in bushy glades, regrown patches of burnt or felled timber, meadows with shrubs, weedy borders of cornfields, woods of low birch in steppe zone, coastal maquis, and sometimes reeds.

Distribution and population. Russia. 10 000–100 000 pairs; stable. Kazakhstan. Ural delta: numbers severely reduced by rising water-levels in recent years.

Accidental. Britain, France, Belgium (including October 1993 bird ringed Britain in September), Netherlands, Germany, Denmark, Norway, Sweden, Finland (annual since 1986), Estonia, Greece, Ukraine, Turkey, Israel, Egypt.

Beyond west Palearctic, extends east to Yenisey valley and extreme western China, south to northern Oman.

Movements. Migratory, all populations moving south or south-east (initially south-west from east of range) to winter chiefly in India. Autumn migration chiefly August–September. In north-west of range (Moscow region), departures and passage during August; further east, breeding grounds in Urals deserted by mid-August. Spring migration chiefly April–May. In Volga-Kama region, vanguard from early May, main arrival 2nd half of May, and reaches Moscow region from mid-May.

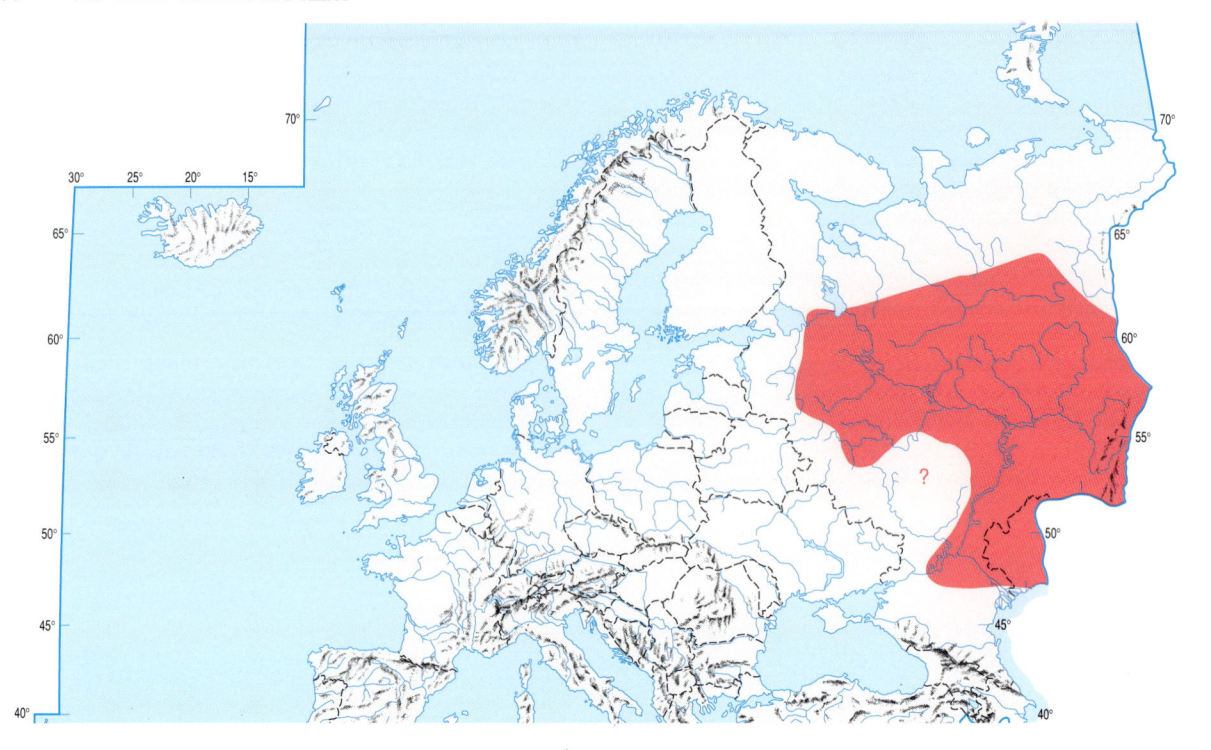

Records of vagrancy in Finland (and 1 in Estonia) are late May to July, suggesting overshooting, but elsewhere in northern Europe recorded almost entirely in autumn, probably reverse migration.

Food. Chiefly insects. Forages very actively at various levels—in tree canopy, in bushes, among undergrowth, and on ground. Moves restlessly among foliage, with action recalling *Phylloscopus* warbler, and often flies out from extremity of branch to take flying insects. Will stretch upwards towards leaf or bark-crevice, or hang almost head downwards.

Social pattern and behaviour. No comprehensive study. In winter quarters, usually solitary and territorial. Mating system probably usually monogamous. Incubation and brooding mainly by ♀, though ♂ participates. Young fed by both parents. Territorial in north of range. Further south, where locally abundant, dense breeding concentrations occur, suggesting modification of usual territorial behaviour. ♂♂ begin singing on arrival on breeding grounds, intensively at first, especially during courtship, including at night; then falling off, virtually ceasing during incubation.

Voice. Song a fast ecstatic chatter, rather suggesting Mediterranean *Sylvia*. Quality variable; may be sweet and slightly liquid warble, spirited though not loud, or may consist of guttural, harsh, and nasal sounds. Commonest calls are abrupt, monosyllabic 'chick', 'tschak', 'tsik', etc., very variable; also, in alarm, prolonged rasping or churring sounds.

Breeding. SEASON. Russia: eggs laid late May to early July. Usually 1 brood in north of range, 1–2 broods in south. SITE. From close to ground to 180 cm in tall herbs or low shrubs. Nest: strong cup of twigs, roots, stems, and leaves, lined with finer material including feathers, hair, and plant down. EGGS. Sub-elliptical, smooth and glossy. Very pale pink, sparsely spotted and blotched black or black-brown. Clutch: 4–6 (2–7). INCUBATION. (11–)12–14 days. FLEDGING PERIOD. 13 (11–14) days.

Wing-length: ♂ 59–65, ♀ 57–62 mm.
Weight: ♂♀ mostly 8–11 g.

Upcher's Warbler *Hippolais languida*

PLATES: pages 1279, 1286

Du. Grote Vale Spotvogel   Fr. Hypolaïs d'Upcher   Ge. Dornspötter   It. Canapino di Upcher
Ru. Пустынная пересмешка   Sp. Zarcero Upcher   Sw. Orientsångare

Field characters. 14 cm; wing-span 20–23 cm. Size close to Icterine Warbler but with 5–10% longer bill, slightly shorter wings, and 15–20% longer tail. Medium-sized warbler, with long bill, quite round crown, rather long wings, and noticeably

Upcher's Warbler *Hippolais languida*: **1** ad worn (summer), **2** ad fresh (autumn), **3** 1st winter. Olive-tree Warbler *Hippolais olivetorum* (p. 1280): **4** ad worn (summer), **5** ad fresh (autumn), **6** 1st winter.

long tail; frequent tail movement reminiscent of *Sylvia* warbler or chat. Closed wing-point forms $\frac{1}{3}$ of total wing length (or equal to $\frac{1}{2}$ exposed length of tertials), with primary-tips reaching to or beyond end of upper tail-coverts. Essentially brownish-grey, with dusky or even bluish tone on fore-upperparts set off by rather clean but dull white underparts, with narrow pale supercilium, quite bright eye-ring (contrasting with mottled face), fairly conspicuous wing-panel, pale rump, and palest-edged tail in *Hippolais*.

After Olive-tree Warbler, 2nd most distinctive *Hippolais*. Rather grey plumage may suggest Olivaceous Warbler but exaggerated movements of tail and call (see below) diagnostic. Flight confident, with fluent beats of noticeably pointed wings and 'rigid' body-tail line. Gait and behaviour apparently typical of *Hippolais* except for highly individual habit of exaggerated and apparently uncoordinated tail movement which displays full white outer edges: includes loose downward flicking typical of *Hippolais*, cocking, fanning, and circular waving, last recalling shrike.

Habitat. Breeds in lower middle latitudes of central Palearctic in warm continental semi-arid climates of steppe zone, from lowlands to 1000–1600 m. Characteristic *Hippolais* of semi-desert habitats and of cultivation interspersed with them, frequenting sand dunes with bushy scrub, thickets of tamarisk and scrub around crusted clay sands, and stony desert slopes overgrown with buckthorn. Also in orchards forming enclaves in desert, in pistachio groves, and in scrub of wild almond, maple, and ash.

Distribution and population. Range not fully known.

Armenia. Local. Azerbaijan. Breeds in south-west, also Araks valley in Nakhichevan region; range otherwise unknown. Uncommon. Turkey. Occurs mainly on fringe of steppe country in south-east. Isolated colony east of Ankara since 1974. 1000–10 000 pairs. Syria. Breeding first recorded 1991, when virtually ubiquitous in gardens above Halbun on slopes of Anti-Lebanon; may also occur elsewhere in same area. Israel. A few tens or hundreds of pairs in each of several localities. Jordan. Small population recently discovered near Amman National Park; breeding first confirmed 1990, but probably long overlooked. May prove to be widespread in similar habitats.

Accidental. No reports of vagrancy. In Egypt, 4 recent records, all in Sinai, but regular occurrence there is likely.

Beyond west Palearctic, breeds southern Iran and western Pakistan north to southern Kazakhstan.

Movements. Migratory, entire population moving south or south-west (reverse in spring) to winter in East Africa, from Eritrea and Somalia south to Tanzania. Route lies through Middle East, and no records from Cyprus or west of Sinai in Egypt. Autumn migration begins early but is protracted. Spring migration begins March. Vanguard reaches breeding grounds at Mt Hermon (northern Israel) from mid-April and reported in Turkey from early May.

Food. Apparently chiefly invertebrates for much of year, but not known whether fruit important in late summer as in other *Hippolais*. Forages chiefly within trees and bushes, and fairly low, but also in more exposed parts of canopy; in breeding season, feeds not infrequently on shaded ground between bushes.

Social pattern and behaviour. Little studied. In winter quarters, usually well dispersed. No indication of other than monogamous mating system. ♂♂ begin to sing on arrival on breeding grounds, and sing very actively while setting up territories and during pair-formation; after egg-laying, song becomes infrequent. Song delivered from exposed perch, also from within foliage, and in song-flight with slow wing-beats.

Voice. Song shares with other *Hippolais* bright and pleasing tonality, overall high pitch, brisk delivery, and tendency to include (a) intermittent brief series of call-type units and (b) repetition (often without pause) of musical phrases of varying length and complex tonal quality. Regarded as distinctive and unmistakable; entirely different from song of sympatric Olivaceous Warbler, and more melodious and flowing than *Acrocephalus* warblers. Various ticking and thin piping units usually precede and may follow song, being sometimes widely and irregularly spaced and pitched (some very high)—sometimes stuttering, sometimes more rapid and regular. Most common call a dry, guttural, rather quiet 'tjug', softer and less forceful than Olivaceous Warbler; given frequently on passage and on wintering grounds.

Breeding. Season. Armenia: eggs from late April to June. Site. In bushes and low trees 50–200 cm above ground; may be supported on branching twigs or suspended from them. Nest: very neat, rounded cup, or cylindrical with rounded base, of fine grasses, stems, fur, and plant down, plus spiders' webs (particularly at twig supports); lined finer material, including plant down, feather grass, and fur; outside of nest frequently incorporates insect and spider cocoons. Eggs. Sub-elliptical, smooth and glossy; pale lilac-pink, purplish, or grey, very sparsely spotted or speckled grey or blackish, occasionally with blackish scrawls. Clutch: (3–)4–5. Incubation. 13–14 days. Fledging Period. 14–15 days.

Wing-length: ♂ 73–80, ♀ 72–77 mm.
Weight: ♂♀ mostly 10–14 g.

Olive-tree Warbler *Hippolais olivetorum*

PLATES: pages 1279, 1286

Du. Griekse Spotvogel   Fr. Hypolaïs des oliviers   Ge. Olivenspötter   It. Canapino levantino
Ru. Средиземноморская пересмешка   Sp. Zarcero grande   Sw. Olivsångare

Field characters. 15 cm; wing-span 24–26 cm. Close in size to Barred Warbler but with 15% longer bill and slightly shorter tail; 15% larger than western Olivaceous Warbler, with stronger bill, 10% longer wings, rather longer tail, and much stouter legs and feet. Large pear-shaped warbler, with heavy bill, rather flat crown, noticeably long wings, and heavy legs; largest and deepest-billed *Hippolais*, with bulk recalling larger *Acrocephalus* warbler at times. Closed wing-point forms $\frac{1}{3}$ of total wing length (or equal to over $\frac{1}{2}$ exposed length of tertials), with primary-tips reaching beyond end of upper tail-coverts. Dusty- or brownish-grey above, dusty-white below, with quite broad fore-supercilium, fairly distinct eye-ring, indistinct pale wing-bar, pale edges to inner flight-feathers forming wing-panel, and greyish wash from side of neck to rear flank.

Most distinctive *Hippolais*, with dull plumage, long, prominent bill, large body, long wings, and heavy legs. Likely to be confused only briefly with pale variant Icterine Warbler, Upcher's Warbler, and large Olivaceous Warbler. When bill obscured, may also suggest immature Barred Warbler. Flight confident, with fluent wing-beats; looks noticeably heavy but not laboured. More skulking and less demonstrative than other *Hippolais*.

Habitat. Breeds in lower middle latitudes of west Palearctic in east Mediterranean zone between July isotherms of 26 and 32°C and thus in warmest climate of any *Hippolais*. Also more coastal, insular, and arboreal, not ranging deep into continental hinterland but inhabiting marine islands and frequenting open-canopy oakwoods, olive groves, orchards, almond plantations, and areas of other well-spaced trees with ample crowns. In Bulgaria, usually in areas of bushes rather than trees. In Greece, occurs in vineyards, brushwood, and scrubby vegetation, often on hillsides. In African winter quarters, favours acacia savanna and dry bush country, and also occurs in dry woodland, usually below 1200 m.

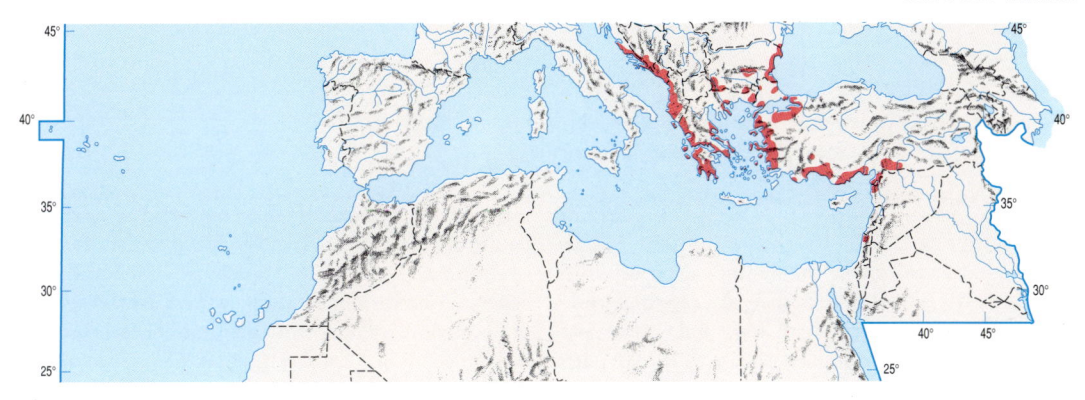

Distribution. Breeds only in west Palearctic. BULGARIA. Marked range extension. TURKEY. Very local and generally rare, breeding chiefly Mediterranean coast and Sea of Marmara. Isolated sites perhaps not regularly occupied. SYRIA. Several dozen discovered May 1982 in thickets at Kassab in north-west, probably continuation of Turkish range. ISRAEL. Breeds in extreme north-west, at least in some years, and possibly elsewhere.

Accidental. Italy, Kuwait, Algeria.

Population. Little studied. Apparently stable, with increase Bulgaria. GREECE. 5000–10 000 pairs. ALBANIA. 1000–3000 pairs in 1981. YUGOSLAVIA: CROATIA. 500–1000 pairs. BULGARIA. 100–1000 pairs. TURKEY. 1000–10 000 pairs.

Movements. Migratory, all birds wintering in eastern and southern Africa, from Kenya south to Natal (South Africa). Reported only from scattered localities, usually in acacia country. Recorded infrequently on passage in both seasons, chiefly single birds. Departs south chiefly end of July to early September. Arrives in breeding range early May, occasionally in April.

Food. Presumably chiefly invertebrates; figs recorded in late summer. Forages mainly within canopy of trees, also in bushes, and recorded feeding methodically on ground.

Social pattern and behaviour. Apparently solitary and territorial outside breeding season. Sings regularly in winter quarters, indicative of territorial defence. No information on mating system. Both ♂ and ♀ brood and feed nestlings. Territories tend to be clumped in neighbourhood groups, as in other *Hippolais*. Song given chiefly within crown of trees hidden from view, also from shrubs, frequently while moving through foliage. Sings persistently during courtship and continues during incubation; may be heard throughout night as well as by day, especially on moonlit nights; persists in June heat when other birds are silent.

Voice. Song has very distinctive deep, rich, and throaty timbre; basically fairly uniform in pitch, lower than other *Hippolais* and slower in delivery. Comprises succession of discrete, rather loud 'chroik' or 'chuck' units, highly complex, and giving impression of combining deep tone with creaks and squawks. Delivered at slow and sustained pace, and may recall Great Reed Warbler. No evidence of mimicry. Most common call a deep 'tuk', 'chuk', 'tak', or 'chack', recalling Great Reed Warbler in pitch, given singly or in unhurried series, and evidently varying in intensity from soft to harsh according to context. Alarm-call a 'trrr', sometimes very prolonged.

Breeding. SEASON. Eggs laid 2nd half of May and June. One brood. SITE. In low tree, especially olive or oak, or in bush. Height above ground 0.5–3 m. Nest: deep, rounded cup of grass, plant stems, strips of bark, and rootlets, including plant down, lined with fine pieces of grass, root fibres, sparingly with horse-hair, sometimes with plant down; outside covered with thick layer of spiders' webs. EGGS. Sub-elliptical, smooth and glossy; very pale pink, lightly spotted and speckled blackish, with some black scrawls. Clutch: 3–4. (Incubation and fledging periods not accurately known.)

Wing-length: ♂♀ 82–92 mm.
Weight: migrants in Africa: ♂♀ 13.7–23.1 g.

Icterine Warbler *Hippolais icterina*

Du. Spotvogel   Fr. Hypolaïs ictérine   Ge. Gelbspötter   It. Canapino maggiore
Ru. Пересмешка   Sp. Zarcero icterino   Sw. Härmsångare

PLATES: pages 1283, 1286

Field characters. 13.5 cm; wing-span 20.5–24 cm. Close in size to Blackcap, but with 15% longer bill, slightly longer wings, and 15% shorter tail giving differently balanced, more front-heavy form. Medium-sized warbler (occupying mid-point in size range of *Hippolais*), with long bill, rather flat crown, long wings, and relatively slim, square tail. Like Melodious Warbler, basically green above and yellow below but with confusing pale variants. Closed wing-point forms $\frac{1}{3}$ of total wing length, with primary-tips at least reaching end of upper tail-coverts. Markings include on adult pale fore-face, distinct yellow supercilium and eye-ring, conspicuous yellow wing-panel, and blue-grey legs; on immature, yellow marks may appear whitish.

Typical bright green and yellow adult in fresh plumage distinctive, with combination of wing pattern and length and bluish legs sufficient to eliminate typical Melodious Warbler. Atypical dull or worn adult and juvenile far less easy to separate from Melodious Warbler and in most dilute plumage may also suggest Upcher's Warbler and Olivaceous Warbler; important therefore to identify on structure as well as plumage. Distinct wing-panel and/or long wing-point noticeably extending beyond bunch of inner secondaries and tertials also eliminates Olivaceous Warbler (whose wing-point forms only $\frac{1}{4}$ of total visible wing length when folded and looks rounded in flight), leaving Upcher's Warbler to be separated on tone of upperparts, nature of tail movement, and commonest call. General character of Icterine Warbler also somewhat individual: long bill exaggerated by prominent head with flat fore-crown but often quite obvious peak above eye; quite deep chest but less deep belly than most *Hippolais* (giving slim appearance); nervous, flighty behaviour giving impression of alertness. Flight confident and fluent, even dashing, with pointed wings showing well and no impression of labouring; recalls other long-winged warblers and even Spotted Flycatcher. Gait as Olivaceous Warbler but carriage more graceful and upright at times, contributing to impression of alertness.

Habitat. Breeds in west Palearctic in middle and upper latitudes, boreal and temperate, mainly continental, between July isotherms 15–25°C; areas occupied are thus cooler than in other *Hippolais*, even reaching into low arctic. Mainly in lowland and river valleys, but in Urals and Carpathians climbs far into hills, and in Switzerland breeds up to 1500 m. An arboreal rather than a forest bird, liking sunny but fairly moist places, preferably with glades or along woodland edges. Favours crowns of well-spaced trees with plenty of tall undergrowth. In north of range occurs in forests of pure birch and in south also in pure oak. Frequents conifers mainly in mixed open

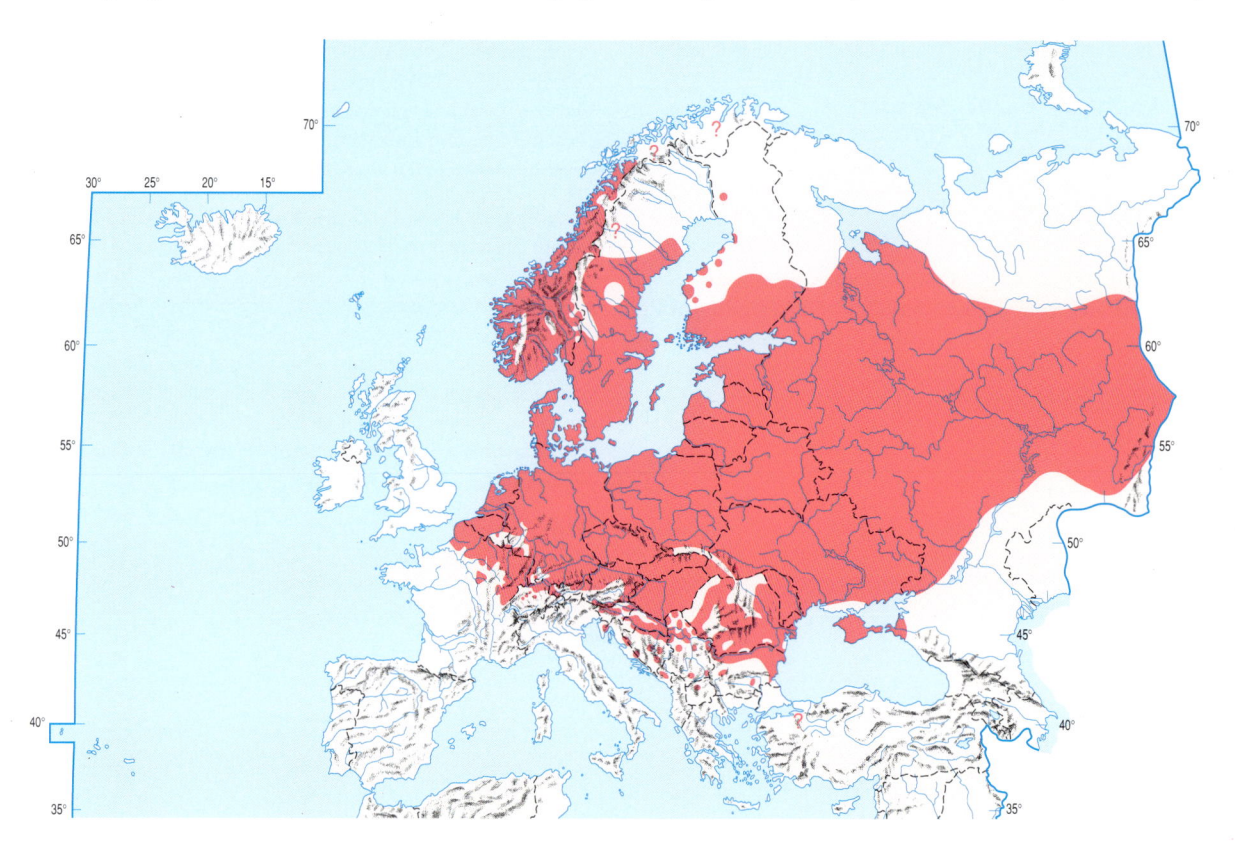

Icterine Warbler *Hippolais icterina*: **1** typical ad fresh (spring), **2** ad worn (summer), **3** 1st winter, **4** juv. Melodious Warbler *Hippolais polyglotta* (p. 1284): **5** ad fresh (spring), **6** ad worn (summer), **7** 1st winter, **8** juv.

stands, but prefers broad-leaved copses, spinneys, strips of trees and bushes (e.g. along railway lines), orchards, parks, groves, gardens, shelterbelts, and tall hedges with trees.

Distribution. BRITAIN. First confirmed breeding Scotland 1992; possibly bred southern England late 19th century and 1907, and Yorkshire 1970. DENMARK, SWITZERLAND. Range has decreased. GREECE. May breed in north-east, but no proof. AZERBAIJAN. May have nested in early 1900s; no recent records. TURKEY. Perhaps breeds occasionally, though confirmation required.

Accidental. Iceland, Faeroes, Britain (annual), Ireland (annual), Kuwait, Morocco (or probably very scarce spring passage-migrant), Madeira.

Beyond west Palearctic, extends in narrowing band east to upper Ob' river.

Population. Recent decline western Europe (Denmark, Low Countries, France, and Switzerland); stable Baltic States south to Austria and Hungary; trend varies elsewhere. FRANCE. 1000–10 000 pairs in 1970s. BELGIUM. 8400–10 000 pairs 1973–7. LUXEMBOURG. 20–40 pairs. NETHERLANDS. 35 000–55 000 pairs 1979–85. GERMANY. 165 000 pairs in mid-1980s. Other estimates 300 000–400 000 pairs in early 1990s, and 350 000 ± 161 000 pairs in east alone in early 1980s. DENMARK. 21 000–225 000 pairs 1987–8; decrease 1983–94, following increase 1976–83. NORWAY. 50 000–300 000 pairs 1970–90; stable. SWEDEN. 50 000–100 000 pairs in late 1980s; slight increase. FINLAND. 20 000–40 000 pairs in late 1980s; slight increase. ESTONIA. 20 000–50 000 pairs in 1991; now stable following decrease in 1970s. LATVIA. 50 000–110 000 pairs in 1980s. POLAND. 60 000–300 000 pairs. CZECH REPUBLIC. 50 000–100 000 pairs 1985–9. SLOVAKIA. 10 000–20 000 pairs 1973–94. HUNGARY. 3000–4000 pairs 1979–93. AUSTRIA. Regionally common. SWITZERLAND. 150–200 pairs 1985–93. YUGOSLAVIA: CROATIA. 200–400 pairs; slight increase. SLOVENIA. 1–5 pairs; marked decrease. BULGARIA. 50–500 pairs; slight increase. RUMANIA. 10 000–20 000 pairs 1986–92. RUSSIA. 1–10 million pairs; stable. BELARUS'. 570 000–600 000 pairs in 1990; fluctuating, but apparently stable overall. UKRAINE. 45 000–49 000 pairs in 1986; slight increase. MOLDOVA. 3000–5000 pairs in 1988; stable.

Movements. Migratory, entire population wintering in sub-Saharan Africa, chiefly south of equator. Arrives on breeding grounds late, and departs early. Autumn departure late July to early September, with peak passage early August in southern Sweden. Vanguard in spring reaches Malta mid-April; northward passage through Europe mainly late April to early June, arriving northern breeding areas late May.

Food. Chiefly insects, also fruit in late summer. Forages restlessly among foliage of trees and bushes, taking insects while perched or while fluttering. Will fly out from ends of branches to capture aerial prey. Foraging action typically slightly clumsy in comparison with *Phylloscopus* warblers.

Social pattern and behaviour. Mostly solitary outside breeding season. Territorial in winter quarters, where song regularly heard. No evidence for other than monogamous mating system. ♀ alone broods; both parents feed nestlings and fledged young. In some areas territories form neighbourhood groups, and may reach quite high densities. ♂ usually sings within cover of foliage in tree-tops. Song intensive and persistent; begins soon after arrival on breeding grounds. Much reduced after pair-formation, only snatches of song given during nest-building and thereafter. Occasionally sings at night.

Voice. Song strikingly vigorous, varied, and far-carrying; a fluent and strongly rhythmic outpouring of rich, musical, harsh, and strident sounds, often juxtaposed in sharp contrast. Noted for great diversity of fine mimicry. Song a blend of 3 components: (i) repeated brief figures; (ii) mimicked calls (usually repeated); (iii) more prolonged musical passages with considerable range in pitch. Recalls *Acrocephalus*, especially Marsh Warbler, but more rapid and powerful. Energetic repetition of brief melodious figures may recall Song Thrush. Most commonly heard and distinctive call a 3-unit 'chi chi-vooi' (2nd unit lighter than 1st, 3rd higher pitched and more musical, a questioning squeak); given by both sexes in a variety of circumstances, probably solely during breeding season. Other calls include a drawn-out 'dedewiih', indicating intense alarm, various monosyllabic calls, and harsh churring, in defence of nest.

Breeding. SEASON. Central Europe: eggs from early May. Finland: laying early June to early July, with peak mid-June. One brood. SITE. In fork of tree or bush, often ornamental or fruit. Usually 1–4 m above ground, exceptionally to 14 m. Nest: neat, substantial cup of grasses, roots, moss, and plant down, lined with hair, fine grasses, and roots, often attached to supporting twigs with fine grass loops. May be decorated on outside with bark, wool, paper, rags, etc. EGGS. Sub-elliptical, smooth and glossy; pale purple-pink, often paler towards large end, sparsely spotted and speckled black, with occasional streaks. Clutch: 4–5 (2–7). INCUBATION. 13–15 (12–16) days. FLEDGING PERIOD. 13–14 (12–16) days.

Wing-length: ♂ 76–83, ♀ 75–80 mm.
Weight: ♂ mostly 10–15, ♀ 11–15 g.

Melodious Warbler *Hippolais polyglotta*

PLATES: pages 1283, 1286, 1313

DU. Orpheusspotvogel   FR. Hypolaïs polyglotte   GE. Orpheusspötter   IT. Canapino
RU. Многоголосая пересмешка   SP. Zarcero común   SW. Polyglottsångare

Field characters. 13 cm; wing-span 17.5–20 cm. Noticeably smaller than Icterine Warbler, with slightly shorter bill, more rounded head, 15% shorter and more rounded wings, and slightly shorter tail; size overlaps with small individuals of Olivaceous Warbler and approaches large individuals of Booted Warbler, both of which also have short wings. Rather small to medium-sized and rather compact warbler, with long bill, (typically) round crown, and short wings; 2nd smallest *Hippolais*. Closed wing-point forms $\frac{1}{4}$ of total wing length, with primary tips not reaching end of upper tail-coverts. Plumage as Icterine Warbler (including pale variants) but typically less bright, browner-green above and richer yellow below, with less striking head pattern, only indistinct pale edges to inner flight-feathers, and dull brown legs.

Typical adult or juvenile not difficult to distinguish from Icterine Warbler, though practice required to distinguish different bill and head shape, shorter wings, and less dashing behaviour. Bright adult with conspicuous wing-panel all too easy to pass over as Icterine Warbler; separation has to be based on structure and behaviour. Atypical pale birds also liable to confusion with small Olivaceous Warblers and large Booted Warblers, since all 3 species have short wings and similar plumage without striking markings. Separation best based on bill shape (strong but not appearing long in Melodious Warbler, strong, long, and extended by flat crown in Olivaceous Warbler, short but still wide-based in Booted Warbler) and face pattern (most bare-faced in Melodious Warbler, less developed but often with noticeably dusky lores and isolated eye-ring in Olivaceous Warbler, again less developed but often with dark linear smudge over short supercilium and hint of eye-stripe in Booted Warbler) and body size (usually noticeably small and less pot-bellied in Booted Warbler). Important also with indeterminate bird to obtain front view of throat and chest, as even palest Melodious Warbler shows some dashes of full yellow there. Fluttering flight often suggests fledgling; otherwise gait and behaviour as congeners.

Habitat. Breeds in milder and warmer lower middle latitudes of west Palearctic, stopping short of main continental climate; tolerates higher precipitation than other *Hippolais* but settles in markedly warmer regions than Icterine Warbler, between July isotherms 19–30°C. Although breeding in southern Switzerland up to 800 m or higher, and to 1700 m in north-west Africa, is predominantly a bird of wooded lowlands.

In Camargue (southern France), breeds mainly in riverain forest, in formerly open areas becoming overgrown with pioneering tamarisk, elm, and white poplar. Advantage is also taken of trees and bushes occurring around cultivation. Further north, in eastern France, favours acacia, willow, and other light-foliaged smallish trees near roads, and also rather low luxuriant coppice of alder, oak, and acacia, often near water. In Coto Doñana (southern Spain), breeds in thickets of bramble and tree-heath, especially where scattered cork oaks are present, usually with ground cover of bracken. Cannot tolerate small gardens or fragmented habitats that Icterine Warbler may occupy.

Distribution. Breeds only in west Palearctic. Expanding along northern and north-eastern borders of range since mid-1930s. FRANCE. Has spread east, also north-west into Brittany. Corsica: breeding probable 1988, proved 1989; probably recent colonization. BELGIUM. Records increased from 1967, with first confirmed breeding 1981 at southernmost tip; has reached border of Flanders. LUXEMBOURG. First recorded 1975; first confirmed breeding 1986. NETHERLANDS. Pair bred Flevoland 1990. GERMANY. First confirmed breeding Baden-Württemberg, 1984, when also suspected Saarland; rapid expansion since. POLAND. ♂ with nest material recorded May 1987. SWITZERLAND. First breeding discovered Tessin 1960, but probably nested earlier; continuing increase. BALEARIC ISLANDS. Pair summered Mallorca 1994 (breeding not confirmed). ITALY. Sicily: bred Palermo 1973. YUGOSLAVIA. Breeding first proved 1947, Istra peninsula.

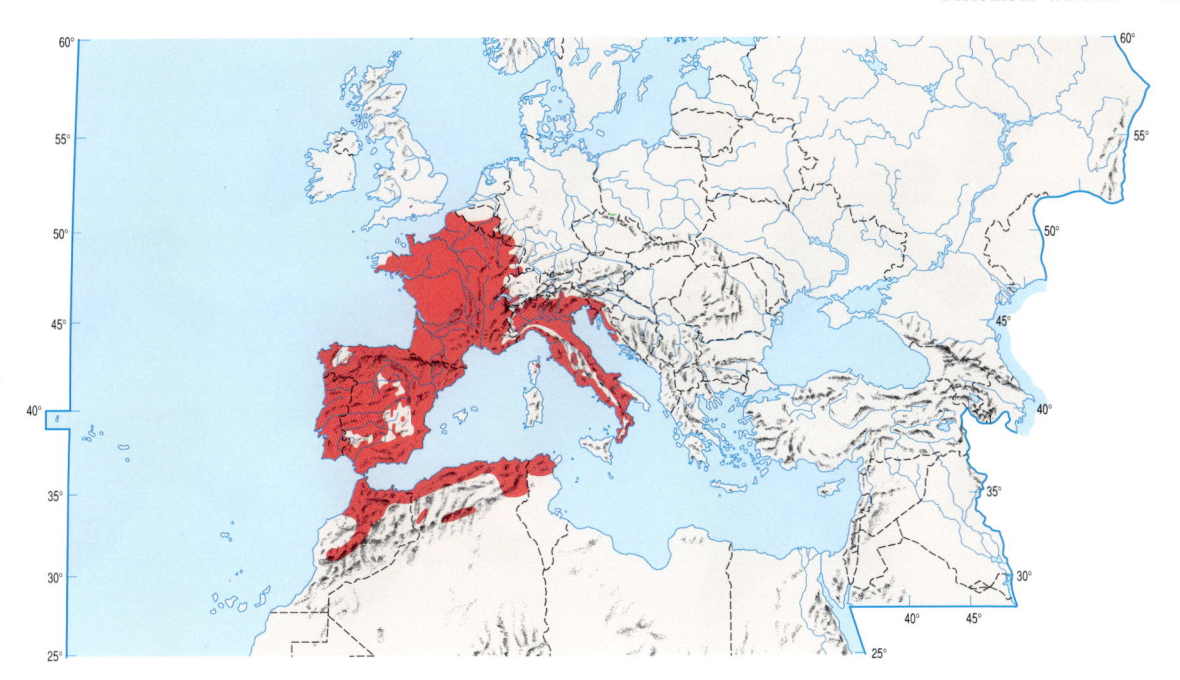

Accidental. Iceland, Britain (annual), Ireland (annual), Netherlands, Denmark, Norway, Sweden, Poland, Czech Republic, Austria, Malta, Greece, Libya, Madeira.

Population. Has increased rapidly in recently colonized areas; stable in south of range. FRANCE. At least 100 000 pairs; increasing. BELGIUM. 50 pairs in 1984, and has increased since. LUXEMBOURG. 10–20 pairs. GERMANY. *c*. 100 pairs. SWITZERLAND. 300–350 pairs 1985–93. SPAIN. 700 000–1.5 million pairs. PORTUGAL. 100 000–1 million pairs 1978–84. ITALY. 50 000–150 000 pairs 1983–95. YUGOSLAVIA: CROATIA. 3000–5000 pairs. SLOVENIA. 1000–2000 pairs. TUNISIA. Scarce. MOROCCO. Uncommon.

Movements. Migratory, entire population wintering in West Africa. Widespread throughout woodland savanna north of rain forest, from Gambia and Sierra Leone east to Nigeria and Cameroon.

Birds head mostly south-west initially (reverse in spring) to enter Africa via southern Iberia, where common and widespread at both seasons. Autumn migration begins late July. Passage in Strait of Gibraltar area from late July, mostly August to early September with stragglers continuing well into October. Spring migration begins early but is protracted, with late arrival on breeding grounds. Passage at Gibraltar late March to early June, peaking in May.

In Britain and Ireland, 1958–85, 871 records, of which 94% in autumn (mainly August to mid-October), mostly in south and south-west, and 6% in spring (mainly mid-May to mid-June), mostly in south-east.

Food. Chiefly adult and larval insects; fruit taken prior to migration. No detailed information. Feeds restlessly within shrubs and trees, recalling *Phylloscopus* warblers, carefully examining foliage to pick off insects before moving on abruptly, and darting through branches to snap up passing aerial prey. Like other *Hippolais*, has almost clumsy foraging action.

Social pattern and behaviour. Mostly solitary outside breeding season. Territorial in winter quarters, where song regularly heard. Monogamous mating system probably the rule. Incubation, and probably brooding, by ♀ alone; both parents feed young. Territorial, with strong tendency to form neighbourhood groups, leaving apparently equally favourable habitat unoccupied. Song usually given from exposed perch—treetop or protruding branch, but also from deep cover. ♂♂ start singing immediately upon arrival on breeding grounds; song continuous and tireless, given especially in early morning and at end of day; in calm weather, one of the few passerines to sing at midday. When nesting starts, song much reduced.

Voice. Song typically in 2 parts or phrases, the 1st consisting of simple motifs or single units, each repeated several times and sometimes imitative, leading without pause into 2nd, a sustained, fast, chattering warble with fairly limited range of pitch and volume but complex time patterning and tonal quality. Considerable variation between performers: poor singers almost unrecognisable as same species; some individuals scarcely progress beyond 1st phrase, but good singers continue with fluent and well-sustained 2nd phrase; either phrase may be more prominent, depending on individual performer. Despite differences, resemblance of 2nd phrase of song to Icterine

Reed Warbler *Acrocephalus scirpaceus* (p. 1265): **1** 1st winter. Olivaceous Warbler *Hippolais pallida opaca* (p. 1273): **2** 1st winter. *H. p. elaeica*: **3** 1st autumn. Booted Warbler *Hippolais caligata caligata* (p. 1277): **4** 1st winter. *H. c. rama*: **5** 1st winter. Upcher's Warbler *Hippolais languida* (p. 1278): **6** 1st winter. Olive-tree Warbler *Hippolais olivetorum* (p. 1280): **7** 1st winter. Icterine Warbler *Hippolais icterina* (p. 1282): **8** ad fresh, **9** 1st winter. Melodious Warbler *Hippolais polyglotta*: **10** ad fresh, **11** 1st winter.

Warbler may be noticeable. Repertoire of mimicry far less extensive than in Icterine Warbler. Most frequent and characteristic call strongly recalls House Sparrow: a harsh chatter of varying intensity, either in single units or in series.

Breeding. SEASON. France and Iberia: eggs from early May. Usually one brood. SITE. In fork of branch or twigs of low tree or shrub. From 0.3–5.5 m above ground. Nest: deep cup, often tapering down into fork, of plant stems and leaves, with plant down and spiders' webs, lined with hair, rootlets, and plant down, and sometimes feathers. EGGS. Sub-elliptical, smooth and glossy; pale pink or purple-pink, usually more purple towards small end, more pink towards large end; sparingly spotted and speckled black, with occasional black hair streaks. Clutch: (3–)4–5. INCUBATION. 12–13(–14) days. FLEDGING PERIOD. 11–13 days.

Wing-length: ♂ 65–71, ♀ 62–67 mm.
Weight: ♂♀ mostly 11–14 g; migrants to 17.3 g.

Marmora's Warbler *Sylvia sarda*

PLATES: pages 1287, 1317

Du. Sardijnse Grasmus   Fr. Fauvette sarde   Ge. Sardengrasmücke   It. Magnanina sarda
Ru. Сардинская славка   Sp. Curruca sarda   Sw. Sardinsk sångare

Field characters. 12 cm; wing-span 13–17.5 cm; tail 4–5 cm. Close in size to Dartford Warbler but with 5–10% shorter tail (and thus less attenuated form). Small warbler, with spiky bill, high crown, short wings, and long tail; often has perky stance with weight forward. Plumage essentially dull blue-grey in ♂, browner below in ♀ and browner below and above in juvenile. Eye of adult ochre to red; eye-ring orange to red; bill base strikingly pink to orange. One call distinctive.

Small size, long slim tail (lacking strong white edges), and leaden plumage make adult distinctive, but confusion possible with Dartford Warbler in all but close view. Juvenile much less distinctive and identification requires close observation of grey-brown underparts (buffier in Dartford Warbler) and more compact structure with shorter tail. Flight, gait, and behaviour much as Dartford Warbler, though appears less skulking (due partly to more open habitat). Feeds at low level in cover, even foraging on ground like Dunnock. Quite tame and curious, investigating disturbance and openly scolding intruder.

Habitat. Breeds strictly within Mediterranean coastal and island areas, in warm situations with average July temperatures up to 24–26°C and relatively frost-free winters. Ascends to

Marmora's Warbler *Sylvia sarda*. *S. s. sarda*: **1** ad ♂, **2** ad ♀, **3** juv. *S. s. balearica*: **4** ad ♂. Dartford Warbler *Sylvia undata* (p. 1288). *S. s. undata*: **5** ad ♂, **6** ad ♀, **7** juv. *S. u. toni*: **8** ad ♂. *S. u. dartfordiensis*: **9** ad ♂.

400–500 m on Mallorca, on hillsides and mountains up to nearly 1000 m on Corsica. Prefers fairly uniform low cover, below height attractive to competitors, especially Dartford Warbler and Sardinian Warbler, as also Stonechat, all of which at times overlap with it. Mainly concentrated on parts of heathland (garigue) with much heath, palmetto or dwarf fan palm, and some grass (usually maintained as a result of poor soil, exposure, or fire) on islets and coastal slopes or hillsides, such areas occasionally being already invaded by scattered trees.

Distribution and population. Breeds only on Mediterranean islands of west Palearctic; no evidence for other than very rare, accidental occurrences on mainland of Europe. FRANCE. Corsica: widespread both along coast and inland. 10 000–20 000 pairs. SPAIN. Balearic Islands: 14 000–25 000 pairs. On Menorca, last reported breeding 1974, last sighting 1979; replaced by Dartford Warbler. ITALY. Breeds Sardinia and Pantelleria island. 5000–10 000 pairs 1983–95. In Sicily, formerly winter visitor, now rare and irregular. GREECE. Crete, Cyclades: status uncertain; several recent records (including apparent breeding); further study required.

Accidental. Britain (north to south-east Scotland), France (mainland), Spain (mainland), Italy (mainland), Malta, Egypt (Salum in north-west), Morocco (or rare winter visitor).

Movements. Corsican and Sardinian populations partially migratory, many birds remaining within breeding range all year; populations of Balearic Islands mostly sedentary. Passage to and wintering in north-west Africa reported mostly November–March. Occasional reports north of breeding areas are presumably vagrants or overshooting spring migrants.

Food. Chiefly small arthropods. Feeds mainly in low vegetation and on ground; also occasionally in trees and higher shrubs, not uncommonly sallying for insects like flycatcher.

Social pattern and behaviour. Mostly territorial (either singly or in pairs) outside breeding season. No evidence for other than monogamous mating system, pairs probably remaining together throughout year where sedentary. Both members of pair incubate and tend young. Song usually given from conspicuous perch, often topmost spray of bush. Song-flight a steady, fluttering or dancing ascent to 4–7 m (recalling Whitethroat), ending with plunge into cover. Song recorded January, but mainly in breeding season; also (at least in Corsica) well-marked second song-period early autumn.

Voice. Song a rather weak high-pitched sweet warble of limited variety; comprises phrases characterized by brief, high-pitched, tonal opening leading into sequence of stuttered, slurred and trilled notes. Probably shows differences in detail on different islands; in comparison with Dartford Warbler, usually far more repetitive and less varied, though confusion could arise, depending on range of individual repertoire of either species; lacks grating component typical of Dartford Warbler. Most commonly heard call abrupt, almost explosive, but differs markedly between islands: in Corsica, a very hard, guttural 'crrep' or 'crrip'; in Balearic Islands, a nasal, tersely disyllabic 'tschrerk', close juxtaposition of 2(–3) syllables giving slightly jerky effect. Alarm- and warning-call a hard, guttural, rather hollow-sounding 'g'k g'k g'k' or 'ch' ch' ch'', given as rapid series and accelerating frequently to rattled bursts; in intense agitation, a drawn-out 'zerr'.

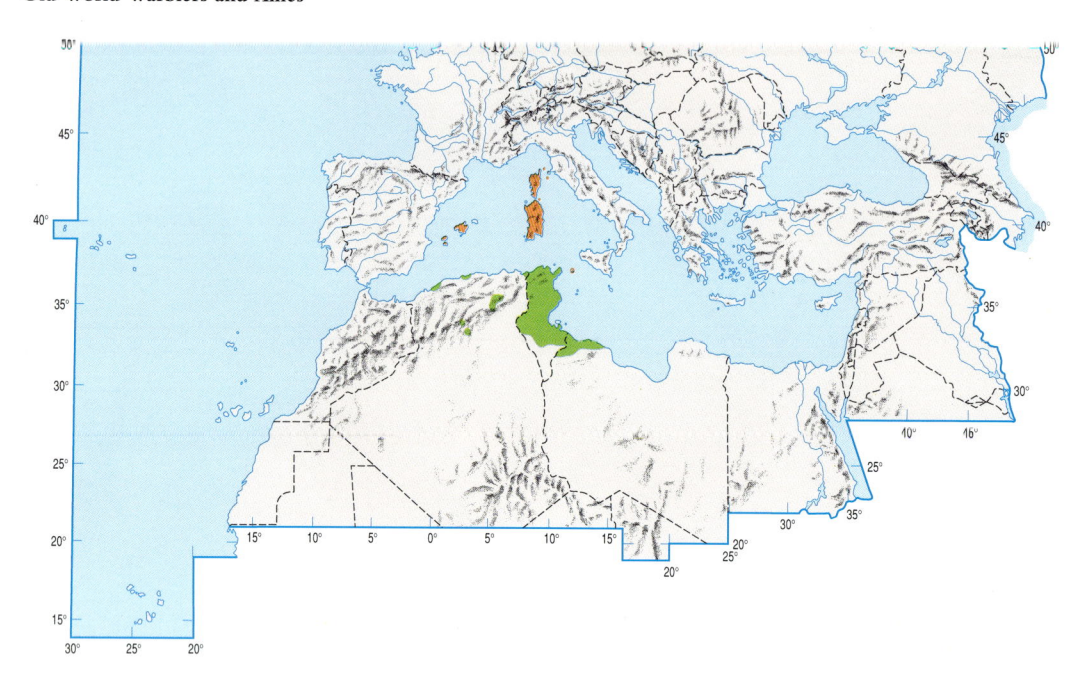

Breeding. SEASON. Balearic Islands: eggs laid late March to early June. Corsica: eggs laid (mid-)late April to early July. 2 broods, perhaps occasionally 3. SITE. In low scrub; 90–120 cm above ground, exceptionally to 190 cm. Nest: well-constructed and substantial cup, often with thickened rim, of dry grasses, stems, and leaves, often with vegetable down and sometimes wool, bits of bark, and spiders' webs and cocoons; lined with finer material including grass, roots, hair, plant down, and occasionally a few feathers. EGGS. Sub-elliptical to short sub-elliptical, smooth and slightly glossy; white or grey-white, with buff to red-brown and grey spots often concentrated at broad end, or sometimes with heavy red-brown blotching and lighter grey and brown speckles. Clutch: 3–4 (2–5). INCUBATION. 12–15 days. FLEDGING PERIOD. c. 12 days.

Wing-length: *S. s. sarda* (Corsica, Sardinia, Elba) ♂ 53–59, ♀ 51–56 mm. *S. s. balearica* (Balearic Islands) ♂ 48–54, ♀ 48–51 mm.
Weight: ♂♀ mostly 8–12 g.

Geographical variation. *S. s. balearica* smaller than nominate *sarda* (see above), and with paler plumage.

Dartford Warbler *Sylvia undata*

PLATES: pages 1287, 1317

Du. Provence-grasmus   Fr. Fauvette pitchou   Ge. Provencegrasmücke   It. Magnanina
Ru. Провансальская славка   Sp. Curruca rabilarga   Sw. Provencesångare

Field characters. 12.5 cm; wing-span 13–18.5 cm. Smaller than Lesser Whitethroat with much shorter and rounder wings and long, slim tail; close in size to Subalpine Warbler (but tail 15–20% longer) and Marmora's Warbler (but tail 5–10% longer). Small warbler, with spiky bill, rather high crown, short wings, and strikingly long tail; has attitude rather like Wren on perch, but looks long and slim in flight. Essentially dark slaty-brown above, brown-pink below in ♂, somewhat paler in ♀ and juvenile; white tail-edges not prominent. Eye and eye-ring of adult dirty orange to red; bill base yellowish.

Smallness, darkness, long- and slim-tailed form, and busy, skulking behaviour all make Dartford Warbler easy to distinguish from most other *Sylvia*, but care needed to distinguish ♀♀ and immatures of Dartford Warbler and Marmora's Warbler. Flight whirring, looking weak, with characteristic undulations even over usual short distance and with pumping or wagging action of tail; straight and usually low, just above vegetation. Makes sudden break from and entry into cover.

Habitat. Breeds in Mediterranean and mild temperate lower middle and middle latitudes of west Palearctic in areas with average July temperatures of 30°C or above, down to January level of 4°C, barely surviving not-uncommon falls in temperature to below 0°C with accompanying ice or snow cover. Shows some preference for maritime regions and islands, but in south of range ascends to uplands, and even to *c.* 1500 m in Spain. In southern Europe, mainly frequents open garigue; in north-west Spain, also in low pine woods; in North Africa, confined to largely hilly coastal scrub with kermes oak. In England, breeding habitat is almost entirely lowland heath dominated either by gorse or by heaths; in some places, bramble and, to less extent, bracken or grasses may serve as substitutes.

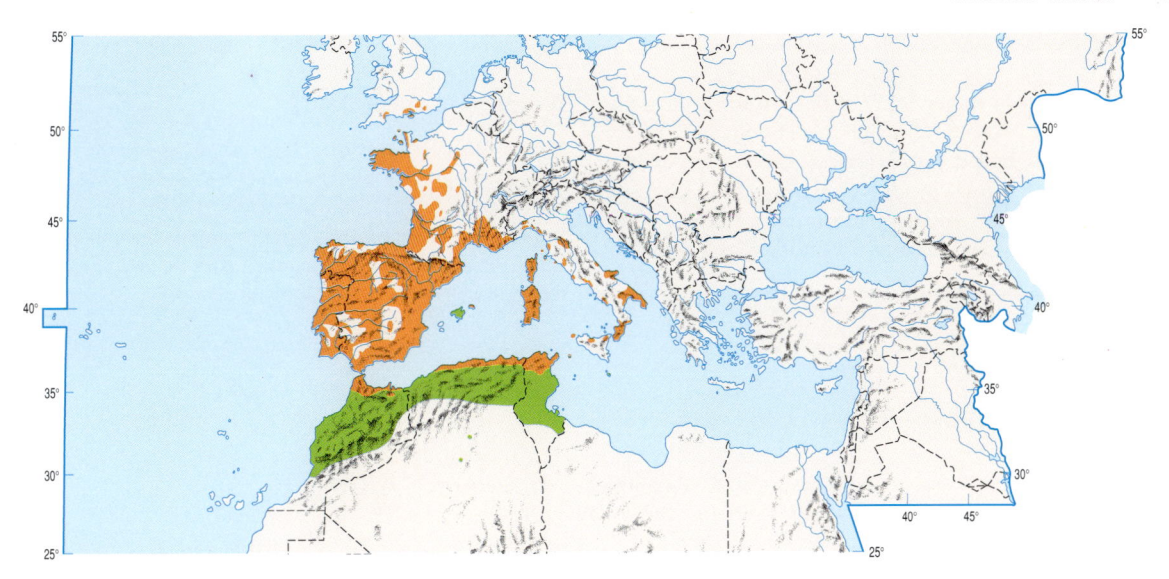

Distribution. Breeds only in west Palearctic. BRITAIN. Marked contraction since 19th century, when bred north to Oxfordshire (also Shropshire), and from Cornwall to Kent. Present-day more restricted range fluctuates, due to high vulnerability to severe weather; recent spread, including first breeding record Somerset 1993. FRANCE. Some evidence for expansion north-east since 19th century. SPAIN. Range has decreased in recent decades through loss of suitable habitat. Balearic Islands: first recorded breeding Menorca 1975.

Accidental. Ireland, Belgium, Netherlands, Germany, Sweden, Finland, Czech Republic, Switzerland, Greece, Yugoslavia, Libya.

Population. BRITAIN. Up to 950 pairs 1988–90; 1600–1670 pairs in 1994. Recent increase, but strong fluctuations (e.g. reduced to 11 pairs after harsh 1962–3 winter). Jersey (Channel Islands): *c.* 50 birds. FRANCE. Under 1 million pairs in 1970s; fluctuating. SPAIN. 1.7–3 million pairs; slight decrease. PORTUGAL. 10 000–100 000 pairs 1978–84; stable. ITALY. 10 000–30 000 pairs 1983–95; stable. TUNISIA, ALGERIA. Locally common. MOROCCO. Uncommon.

Movements. Partially migratory and dispersive. Many birds remain all year on breeding grounds, but autumn and winter records are frequent in non-breeding areas, probably chiefly juveniles. In Mediterranean, presence in winter on islands where none breed shows that some birds undertake sea crossings: scarce but regular visitor to Malta and Balearic Islands (and Menorca now colonized: see Distribution). Some European birds reach north-west Africa, vagrants occur far afield.

Food. In Britain, exclusively arthropods, but occasionally also fruit in autumn and winter in continental Europe. Feeds mainly in low scrub, occasionally in trees.

Social pattern and behaviour. Secretive and generally shy. Mostly territorial, singly or in pairs, outside breeding season. Mating system essentially monogamous. Where sedentary, pair-bond life-long, and closely maintained throughout year. Both sexes incubate, and tend young. Song delivered most frequently from topmost spray of gorse or other bush, also in song-flight and from within cover. Not a persistent singer, and may remain silent for hours; heard most frequently early in morning and at dusk. Sings chiefly on still, sunny days, and typically very quiet in inclement weather. Song declines very little after young hatch, and continues while feeding young and in anticipation of 2nd brood. After summer silence, song renewed (with re-establishment of territory) in autumn and sporadic in winter.

Voice. Song not unlike other *Sylvia*. Comprises phrases of *c.* 1–2 s, separated by pauses of the same length or longer; medley of sweet piping notes and softly metallic rattling sequences. Spirited in delivery; phrase openings and endings often high-pitched and tonal, contributing to characteristic cheerful effect. Rather weak in volume, though may carry *c.* 200 m in favourable conditions. Most common and characteristic call a rather grating 'tchirr', more plaintive and less scolding than Whitethroat, more subdued and soft; begins and ends abruptly, and suggests small stifled sneeze. Given in variety of contexts, typically when disturbance causes bird to emerge from concealment on to vantage-point. Alarm-calls include a hard, incisive 'tucc', becoming rapid 'tututututucc' when excited; and quickly repeated 'tzi-tzi-tzi', indicating intense agitation.

Breeding. SEASON. Southern England: eggs from mid-April. Jersey (Channel Islands): eggs end of March to July. Provence (southern France) and southern Spain: eggs from early April; on Menorca, from late March. Algeria: eggs early April to mid-June. 2 broods, exceptionally 3. SITE. In dense, often evergreen bushes. Nest: compact cup of grass leaves and stems, and bits of heather, usually with vegetable down, cobwebs, and occasionally feathers, sometimes with distinct middle layer of plant down; lining (inner layer) of finer material including rootlets and hair. EGGS. Sub-elliptical, smooth and glossy; white, sometimes tinged grey or green; finely but variably marked with dark brown to grey-brown speckles, spots and mottles, sometimes concentrated at broad end. Clutch: 3–5(–6). INCUBATION. 12–14 days. FLEDGING PERIOD. 12 ± 2 days.

Wing-length: Continental European and north-west African population, ♂ 52–58, ♀ 51–55 mm. English birds, ♂ 50–54, ♀ 51–53 mm.

Weight: ♂ ♀ mostly 9–10 g.

Geographical variation. Rather slight, involving colour only. 3 races recognized, but many populations intermediate: nominate *undata* (continental populations) with slate-grey mantle and scapulars; *dartfordiensis* (Atlantic populations, including England) browner, less grey above, dark below; *toni* (southern Iberia, north-west Africa) grey-brown above, paler below.

Tristram's Warbler *Sylvia deserticola*

PLATE: page 1291

Du. Atlasgrasmus FR. Fauvette de l'Atlas GE. Atlasgrasmücke IT. Sterpazzola di Tristram
RU. Атласская славка SP. Curruca de Tristram SW. Atlassångare

Field characters. 12 cm; wing-span 13–17 cm. Close in size to Subalpine Warbler (but with 15% shorter wings) and Spectacled Warbler (but with 10% longer tail); less attenuated than Dartford Warbler, with 15% shorter tail. Small warbler, with small bill, rather high crown, short wings, and slim tail. ♂ mainly grey above, with bright orange-brown edges to black-centred inner wing-feathers; pink-buff to vinous-brown below. ♀ and juvenile grey-brown above and paler buff below, with whitish chin and moustachial stripe. Eye yellow to brown, set in whitish eye-ring.

Not easy to identify, as ♂ intermediate in appearance between Subalpine Warbler and Dartford Warbler and ♀ and juvenile approach Subalpine Warbler, Spectacled Warbler, and Whitethroat. Separation of ♂ from Dartford Warbler best founded on tail (medium length and white-edged in Tristram's Warbler, not long and dull-edged), wing colour (splashed orange-rufous, not dull brown), and eye (yellowish with white eye-ring, not red with red eye-ring). Separation from ♂ Spectacled Warbler not difficult: in any prolonged view, that species' white throat, warmer, paler upperparts, only pink-washed breast and flank, and white underbody make underparts overall much paler than ♂ Tristram's Warbler.

Habitat. Breeds in lower middle latitudes in Mediterranean and steppe-desert zones, in hilly areas covered with scrub and in poorest forests of holm oak where more bushy conditions occur. In Atlas mountains of Morocco, inhabits bushes between large well-spaced cedars which occur above 1600 m; also in relatively dense shrub cover of juniper, *Cistus*, *Buxus*, *Pistacia*, etc. Winters in Morocco in lower and more open bushy habitats, even to desert fringes, including oases, saline areas, and river beds.

Distribution and population. Breeds only in west Palearctic. TUNISIA. Uncommon. MOROCCO. Extension of previously known range to eastern Moyen Atlas and Anti-Atlas. Uncommon to locally common.

Accidental. Spain (Gibraltar), Malta, Canary Islands.

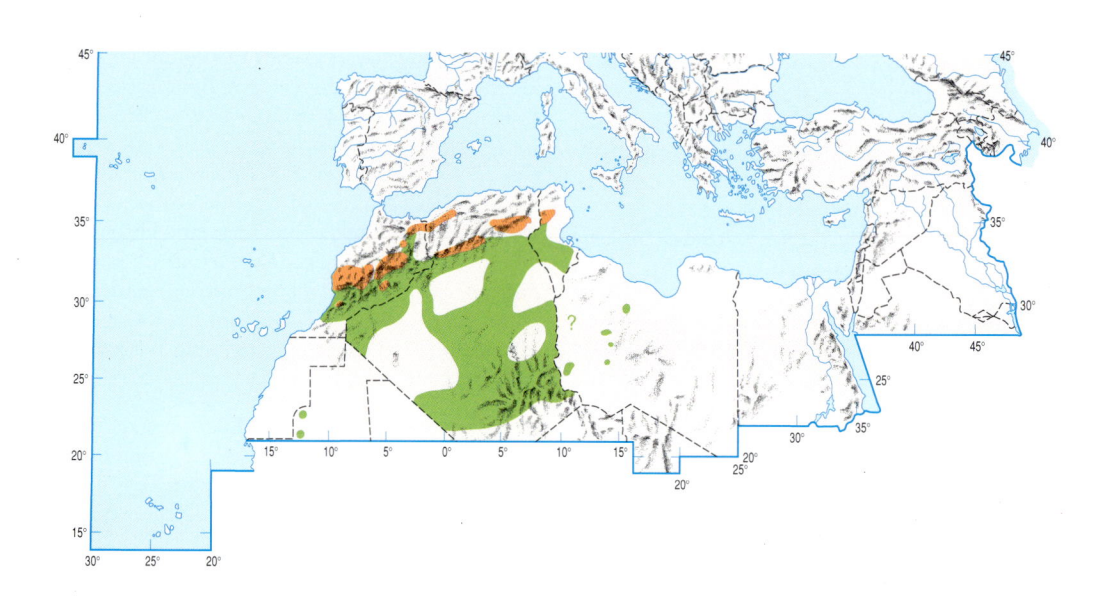

Tristram's Warbler *Sylvia deserticola*. *S. d. deserticola*: **1** ad ♂ breeding, **2** ad ♀ breeding, **3** juv. *S. d. maroccana*: **4** ad ♂ breeding.
Spectacled Warbler *Sylvia conspicillata*. *S.c. conspicillata*: **5** ad ♂ breeding, **6** ad ♀ breeding, **7** juv. *S. c. orbitalis*: **8** ad ♂ breeding.

Movements. Partial migrant, making altitudinal and mainly short-distance movements, occasionally reaching southern edge of Sahara. Some birds present mid-winter in breeding range.

Food. Chiefly insects. Habits similar to Dartford Warbler: explores bushes, daintily hopping from twig to twig, and also forages among clumps of grass.

Social pattern and behaviour. Little information. General habits similar to Dartford Warbler. ♂ sings from bush or occasionally in song-flight; both in wintering areas and on breeding grounds.

Voice. Song a chattering sequence interspersed with rattles, often with a more tuneful middle section. Shows strong resemblance to Subalpine Warbler, but not as soft. Calls include a sharp 'chit' or 'chit-it', a 'tacking' rattle, reminiscent of Wren, and a 'tscherr' or 'zerr'.

Breeding. SEASON. Algeria and Tunisia: eggs from last week of April to end of May. Morocco: March–July. SITE. In bush, 1–1.5 m above ground. Nest: deep cup of coarse grasses lined with finer grasses and plant down and sometimes horse hair. EGGS. Sub-elliptical, smooth and glossy; pale green to greenish-white, well marked with dark brown and some grey spots and blotches, often concentrated at broad end. Clutch: 3–5. (No information on incubation and fledging periods.)

Wing-length: ♂ ♀ 52–58 mm.
Weight: ♂ ♀ mostly 8–10 g.

Geographical variation. Slight, involving general colour only. *S. d. maroccana* from Atlas Mountains generally darker than nominate *deserticola* from areas to the east.

Spectacled Warbler *Sylvia conspicillata*

PLATES: pages 1291, 1317

DU. Brilgrasmus   FR. Fauvette à lunettes   GE. Brillengrasmücke   IT. Sterpazzola di Sardegna
RU. Очковая славка   SP. Curruca tomillera   SW. Glasögonsångare

Field characters. 12.5 cm; wing-span 13.5–17 cm. Averages slightly smaller than Subalpine Warbler, with slightly longer bill, 10–15% shorter wings, and 10% shorter tail; slightly bulkier than Tristram's Warbler but with 10% shorter tail. Small warbler, with short bill, quite high crown, seemingly large head, and short wings. Breeding plumage essentially grey- to sandy-brown above, with strikingly orange or cinnamon fringes to wing-feathers; pink below, with white chin and grey throat.

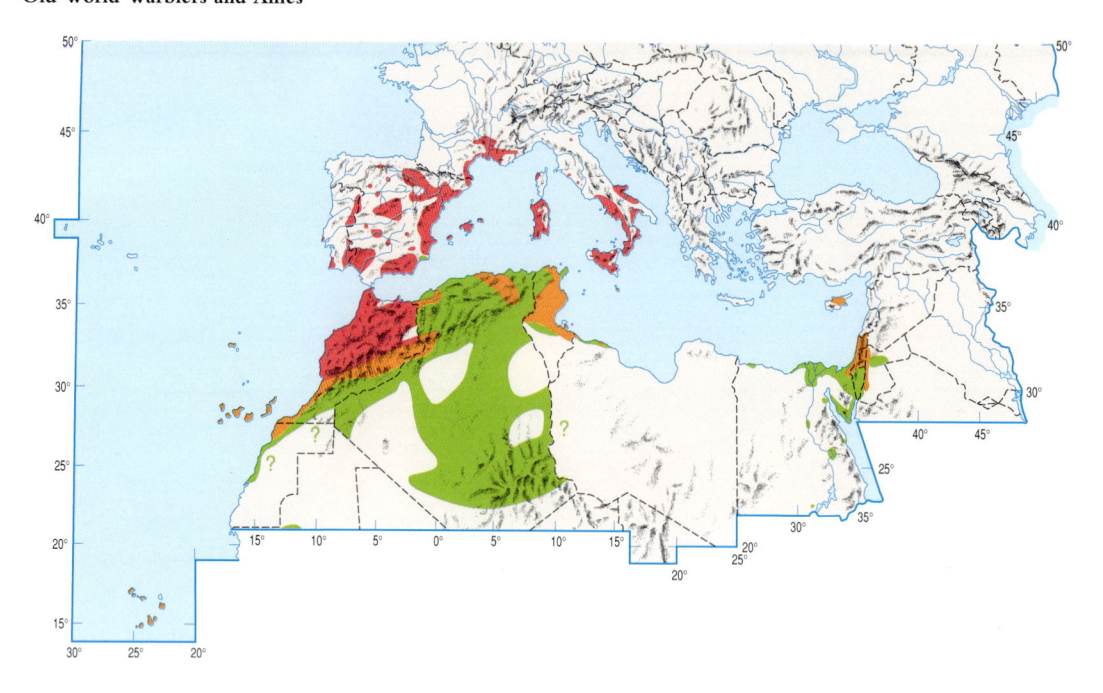

In adult, black tail white-edged and eye ochre to orange, set in red bare ring and white eye-ring. Non-breeding adult and juvenile recall Whitethroat more than any other small *Sylvia*.

Except for adult ♂ in spring and summer, not easy to identify, being more or less similar (according to state of plumage) to several other medium-sized to small *Sylvia*. Mostly confusable with Whitethroat, but that species 15–20% larger, with fuller tail that trails in freer flight and with plumage pattern that lacks black or dusky fore-face, fully pink underparts, and noticeably dark tail, while its well-known chestnut wing-marks are in fact narrower and less orange than those of Spectacled Warbler. Flight free and bouncy, with whirring wing-beats and tail often spread and flicked. Gait essentially a bouncy hop but with hint of clumsiness (unlike Subalpine Warbler), varied by creeping or sidling.

Habitat. Breeds in lower middle latitudes of west Palearctic, principally in Mediterranean zone, in warm dry lowlands and hilly country, between July isotherms 23–32°C. Generally breeds in low scrub of garigue, salt flats, arid cultivated fields, or semi-desert; on Atlantic islands inhabits denser growth such as tree-heath or gorse.

Distribution. Breeds only in west Palearctic. SWITZERLAND. First recorded 1989, when pair bred Valais. SYRIA. Clarification of status needed; perhaps breeds, and may have bred formerly. EGYPT. Status unclear. Probably breeds north-east Sinai; no recent records of breeding elsewhere. LIBYA. Breeds on north-west coast. CAPE VERDE ISLANDS. Local and uncommon São Vicente, widespread and locally common elsewhere.

Accidental. Britain, Netherlands, Germany, Switzerland, Greece, Yugoslavia, Turkey, Iraq.

Population. FRANCE. Probably fewer than 10 000 pairs. SPAIN. 140 000–300 000 pairs; stable. PORTUGAL. 1000–10 000 pairs 1978–84; stable. Algarve: slight increase over past decade. ITALY. 10 000–20 000 pairs 1983–95; stable. MALTA. Drastic decline in 1981. No sign of recovery, although recent slight increase. Fewer than 500 birds. CYPRUS. Perhaps 20 000–50 000 pairs; probably stable. ISRAEL. Fairly common. Negev population *c.* 300 pairs in 1980s. JORDAN. Breeds at low densities. TUNISIA. Uncommon. MOROCCO. Scarce to common. MADEIRA. Rare. CANARY ISLANDS. Common and widespread.

Movements. Varies from migratory to sedentary in different parts of range. Chiefly migratory in southern France, with a few remaining to winter in Camargue; some winter in Murcia (south-east Spain), but migratory elsewhere in Iberia and apparently exclusively a summer visitor to Corsica, Sardinia, and mainland Italy. Some winter sporadically on islands off Italy. Resident on Malta and Cyprus. In north-west Africa, present all year, but some breeding areas entirely vacated. Probably only a minority go beyond northern edge of Sahara. Passage movements inconspicuous within Europe and across African coast, presumably due to relatively low numbers as well as to overflying; no evidence of concentrations. Destination and distance of movement of individual populations little known.

Food. Mainly invertebrates but recorded taking fruit in spring and autumn.

Social pattern and behaviour. Birds live in pairs, and territories maintained in winter in some areas; but in small flocks in Cape Verde Islands. No evidence for other than monogamous mating system. Both sexes incubate and feed the young. ♂ sings from prominent but minor vantage points (typically top of bush) around his territory, also in song-flight. On Malta (where most complete study made), ♂ sings regularly from end of January, also to some extent earlier in warm weather; song continues unabated until hatching, thereafter diminishing gradually, almost ceasing by June; some resurgence in autumn, starting in September, but song then of relatively low intensity.

Voice. Song a short sweet exhilarating warble, often mixed with rattles; not stereotyped, and quickly repeated, resembling Sardinian Warbler but less raucous. Most common call a high-pitched 'tseet' used continually at all times, especially between mates; also, in alarm, variable churring calls, becoming metallic rasping churr in extreme alarm.

Breeding. SEASON. Malta: eggs laid from 1st week of March, continuing to June. North Africa: eggs mid-March to late June. South-west Morocco: eggs from mid-February. Canary Islands: eggs early March to early May. Cape Verde Islands: all months except June–August, peaking October–November and April–May. SITE. In low, dense vegetation, e.g. matted grass, tussocks, small shrubs, thistles, etc., usually from ground level to 0.7 m, rarely to 2 m. Nest: neat but loosely constructed deep cup of dried grass stems, rootlets, and leaves, often including rag, wool, cobwebs, and paper; edge usually forms distinct rim; lined with layer of soft plant down and then final lining of fine roots and a little hair. EGGS. Sub-elliptical, smooth and glossy; very pale green to buff-white, finely marked with light buff to olive speckles and spots, sometimes concentrated at large end. Clutch: 3–5. INCUBATION. Usually 12–13 days. FLEDGING PERIOD. 11–12 days.

Wing-length: ♂♀ 53–59 mm; little geographical or sex variation.
Weight: ♂♀ mostly 9–11 g.

Geographical variation. Slight. Birds of Atlantic islands (*S. c. orbitalis*) darker and more richly coloured than nominate *conspicillata* (rest of range).

Subalpine Warbler *Sylvia cantillans*

PLATES: pages 1294, 1317

DU. Baardgrasmus   FR. Fauvette passerinette   GE. Weißbart-Grasmücke   IT. Sterpazzolina
RU. Рыжегрудая славка   SP. Curruca carrasqueña   SW. Rödstrupig sångare

Field characters. 12 cm; wing-span 15–19 cm. Noticeably larger than Tristram's Warbler and Spectacled Warbler, but still 10% smaller than Lesser Whitethroat, with rather short and fine bill and rather higher crown; size between Ménétries's Warbler and Sardinian Warbler but proportionately shorter-tailed. Rather small, quite slim, and elegant warbler, with rather short bill. ♂ blue-grey above, dark pink-chestnut below, with white moustachial stripe. ♀ grey-brown above, buff-white below. Juvenile has pale sandy-brown fringes to inner wing-feathers recalling Whitethroat. Eye gold to red; bare ring usually red in ♂ but pale in some ♀♀ and all immatures, provoking confusion particularly with Spectacled Warbler.

Traditionally considered distinctive, but appearance actually

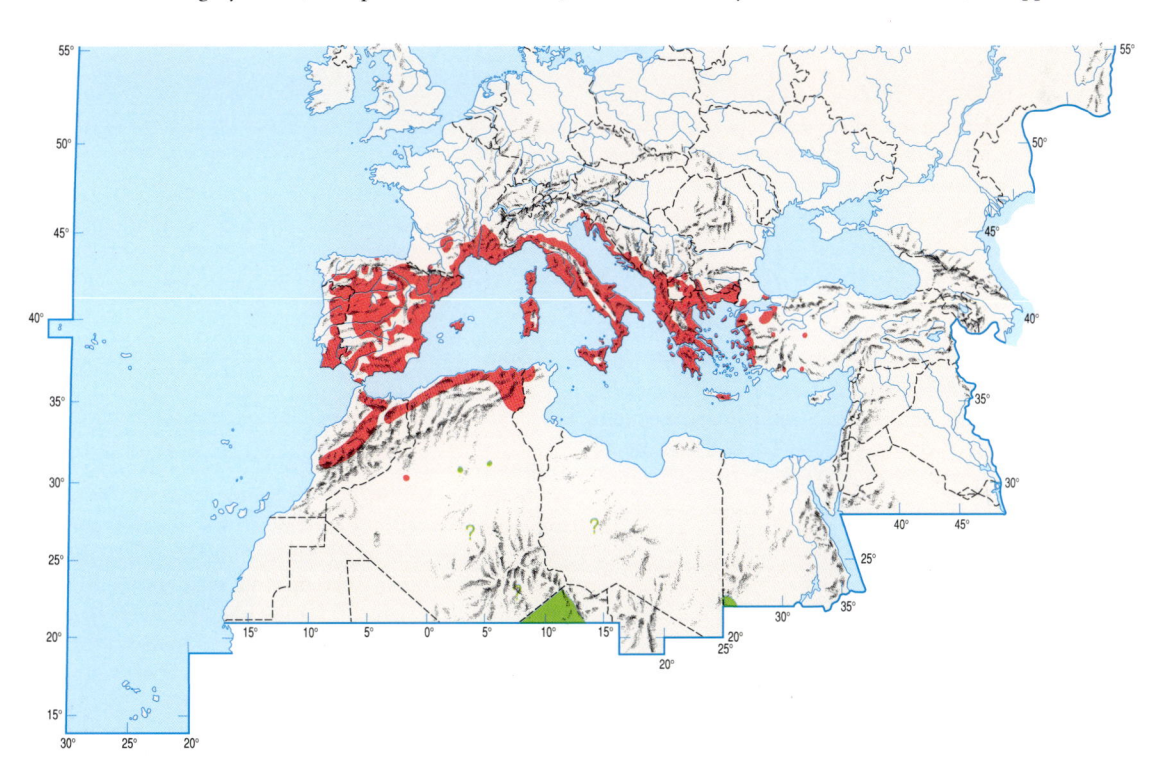

Subalpine Warbler *Sylvia cantillans*. *S. c. cantillans*: **1** ad ♂ breeding, **2** ad ♀ breeding, **3** 1st winter, ♂, **4** 1st winter ♀, **5** juv. *S. c. albistriata*: **6** ad ♂. Ménétries's Warbler *Sylvia mystacea*: **7** ad ♀, **8** ad ♀, **9** juv.

similar in various ways (depending on age and sex) to 8 other *Sylvia*, so considerable care required with identification. Within its genus, Subalpine Warbler occupies mid-point in size/structure, and differences in plumage tones of its races, compounded by much greater warmth of juvenile plumage, promote wide range of pitfalls only partly elucidated. Subalpine Warbler has however 2nd shortest bill of genus, relatively long wings, rather sharp-cornered tail, proportionately rather short legs, and marked white moustachial stripe (weak only in juvenile). Apart from juvenile with its bright tertial-edges, has more uniformly-coloured head and upperparts than all confusion species. ♀ and juvenile have off-white outer eye-ring shared only with Spectacled Warbler and Tristram's Warbler. Among confusion species, shares only with Spectacled Warbler flight silhouette and actions strongly recalling small, slight Whitethroat. Flight light and free, with action recalling Whitethroat but wing-beats faster, while tail held less loosely and never appearing to trail (though occasionally flicked and even cocked on entry to cover). Active, often feeding with constant movement more typical of *Phylloscopus* than larger *Sylvia*; given to flicking wings and (particularly) cocking tail, though never holding tail up as do Dartford Warbler and Marmora's Warbler. Noted as secretive but this due at least partly to preferred habitat of dense scrub.

Habitat. Breeds in Mediterranean region, in dry, warm summer climate between July isotherms 23–30°C, from sea-level to at least 1000 m in Spain, to 1100 m in Greece, and to *c.* 2000 m in Haut Atlas of Morocco. In habitat spectrum of Mediterranean *Sylvia* warblers, differs most from Marmora's Warbler and Dartford Warbler which normally avoid trees, and least from Sardinian Warbler which is even more arboreal. Lives in dense, xerophytic, often prickly scrub on sunny hillsides, often clad mainly in holm oak and kermes oak, but even down to lowest and poorest thorn scrub. May occur in maquis or in scrubby ravines, but more frequently inhabits lower garigue consisting mostly of *Cistus*, broom, and other small shrubs or low heath scrub. Also breeds not infrequently in less arid situations, such as thick hedgerows with oak, bramble, and bracken, and alongside streams with oleander.

Distribution. Breeds only in west Palearctic. BULGARIA. Slight range increase. TURKEY. Common in southern Thrace and north-west Anatolia; fewer further south. LIBYA. Probably bred 1966. MOROCCO. Formerly bred also further south.

Accidental. Iceland, Britain (now annual), Ireland, Belgium, Netherlands, Germany, Denmark, Norway, Sweden (perhaps annual), Finland, Poland, Austria, Switzerland, Ukraine, Madeira, Cape Verde Islands.

Population. FRANCE. In 1970s, *c.* 24 000 pairs in Languedoc, probably up to 100 000 in whole country. SPAIN. 1.1–2.3 million pairs; stable. PORTUGAL. 10 000–100 000 pairs 1978–84; stable. Algarve: marked fluctuations from year to year, but probably declining over past decade. ITALY. 30 000–50 000 pairs 1983–95; stable. GREECE. 100 000–300 000 pairs; stable. ALBANIA. 10 000–30 000 pairs 1981. YUGOSLAVIA: CROATIA. 60 000–90 000 pairs; stable. SLOVENIA. *c.* 5–10 pairs. BULGARIA. 500–5000 pairs; slight increase. TURKEY. 1000–10 000 pairs. TUNISIA, MOROCCO. Uncommon. ALGERIA. Locally fairly common.

Movements. European populations wholly and north-west African population mostly migratory, all populations wintering in Africa, chiefly along southern edge of Sahara, from Mauritania and Sénégal to Sudan. In south-west Europe, migration more conspicuous in autumn than spring. Southward movement begins in August, and is leisurely, with long stopovers. The earliest migrant to reach Malta, where very numerous in autumn (common in spring); recorded from mid-July (mostly 1st-year birds for 1st month of passage), chiefly mid-August to 3rd week of September; many remain (up to 19 days), leaving with sufficient fat for trans-Saharan crossing. Autumn passage of south-east European population far less conspicuous than spring, evidently due to long-stage overflying; for many birds, heading must have substantial westerly component. Northward movement in spring extends further east than autumn movement; thus no satisfactory autumn records in Cyprus, but usually fairly common in spring.

Numbers of vagrants in Britain and Ireland have increased steadily, especially in spring (76% of 143 records, 1958–85); records scattered, almost all on coast.

Food. Chiefly adult and larval insects, also fruit in late summer and autumn. Feeds in scrub, also in foliage of olive, oaks, and other trees; occasionally on ground.

Social pattern and behaviour. Social organization outside breeding season little known; may occur in flocks on migration; not known to hold territories in winter quarters, but sings not uncommonly. No evidence for other than monogamous mating system. Both sexes incubate, ♀ probably doing most, and feed the young. ♂ sings usually from the cover of scrub or lower branches of tree, but sometimes from an exposed spray, or in song-flight from one bush or tree to another. Song-flight dancing and fluttering, like Whitethroat: bird may simply fly up, then descend, singing the while, or sing while fluttering from one vantage point to another. Song period mainly April–June.

Voice. Song lively and clear, reminiscent of Whitethroat but more sustained and more musical, with some rather prolonged sweet notes, sometimes also some quite harsh rattling sounds though often without these; lacks the peculiarly harsh strident noises usually incorporated by Sardinian Warbler. Calls include a hard but not very loud 'tec' given singly or in series, accelerating in excitement or alarm, thus 'tec tec tec-tec...'; also a low, rolled 'krrrrrr'.

Breeding. SEASON. Southern Europe: early April to late June. Algeria: early May to mid-June. Usually 2 broods. SITE. In low bushes, or occasionally low trees, 30–130 cm above ground. Nest: deep, well-built cup of grass stems and leaves, also leaves and stalks of other herbs, with some roots and plant down, almost always with cobwebs in outer layers; lined with finer grasses, herbs, rootlets, and hair. EGGS. Sub-elliptical, smooth and glossy; whitish, sometimes tinted green or pink, finely speckled, blotched, and spotted red-brown, olive, or buff, markings often gathered at large end; thus several distinct colour types occur. Clutch: 3–4(–5). INCUBATION. 11–12 days. FLEDGING PERIOD. 11–12 days.

Wing-length: Spain, France, and Italy (*S. c. cantillans*): ♂ 58–63, ♀ 57–61 mm. Balkans and Asia Minor (*S. c. albistriata*): ♂ 62–64, ♀ 61–64 mm.

Weight: ♂♀ mostly 8–13 g; migrants to 19 g.

Geographical variation. Rather slight; most pronounced in colour, slight in size (see Wing-length). Breeding ♂ *albistriata* (Balkans and Asia Minor) has darker brick-red chin to breast and slightly broader moustachial stripe than nominate *cantillans* from rest of Europe. *S. c. inornata* (North Africa) also more brick-red below, with deep rufous flanks (pink in nominate *cantillans*, greyer in *albistriata*).

Ménétries's Warbler *Sylvia mystacea*

PLATE: page 1294

Du. Ménétries' Zwartkop   Fr. Fauvette de Ménétries   Ge. Tamariskengrasmücke   It. Occhiocotto di Ménétries
Ru. Белоусая славка   Sp. Curruca de Menetries   Sw. Östlig sammetshätta

Field characters. 13.5 cm; wing-span 15–19 cm. Close in size to Sardinian Warbler. Rather small to medium-sized warbler, with quite heavy bill, usually flat crown, and quite long tail. Plumages suggest Sardinian Warbler and several other small *Sylvia*. ♂ dull black on head, dark grey-brown above, pink-white to brown-pink below, usually darkest from throat to fore-belly. ♀ and juvenile grey- to sandy-brown above, rusty on forehead and brightest on fringes of inner wing-feathers, brown- to buff-white below. Bright yellow to red-brown eye set in pinkish to yellow or brown bare ring.

In normal range, most likely to be confused with Sardinian Warbler, but also risk of mistake with ♂ Subalpine Warbler and ♀ Rüppell's Warbler. However, ♂ Sardinian Warbler has longer dark bill, darker orange-red bare ring round eye, no or only vestigial pink on underparts, always some grey on flank, and rather more rounded tail; ♂ Subalpine Warbler is distinctly smaller and has bright red eye-ring, uniform, greyer head and upperparts, paler tail, and typically much darker pink or brick-red underparts; ♀ Rüppell's Warbler is distinctly larger and has only dull flesh base to bill, darker orange-red bare ring round eye, darker ashy or brown upperparts with noticeably pale tertial fringes, (on some) black flecks on throat, and only buff tinge on underparts. Perched bird habitually raises tail above body angle and waves it from side to side and up and down, as if awkwardly hinged.

Habitat. Breeds in lower middle latitudes in arid continental climates, in a variety of habitats, e.g. scrub along river banks, wooded lower slopes of mountains, transition vegetation between steppe and desert zones, palm groves and other cultivation associated with irrigation.

Distribution and population. RUSSIA. 100–1000 pairs; stable. AZERBAIJAN. Perhaps the most widespread and common of the *Sylvia* warblers. TURKEY. Local in east and south-east; probably seriously underrecorded in mountains. 1000–10 000 pairs. SYRIA. Quite widespread locally. LEBANON. May breed locally.

Accidental. Portugal, Canary Islands. Probably status of passage-migrant in Israel and Jordan, though very rarely recorded.

Beyond west Palearctic, breeds Iran north-east to Syr-Dar'ya valley.

Movements. All populations migratory, moving south or south-west to winter in north-east Africa (Sudan to Somalia), Arabia, and southern Iran. Entry into Africa apparently via Arabia; very rare in Israel. Leaves breeding grounds late and returns early, especially in south of range. In central and southern Iraq, passage early September to early November, and local birds absent only in coldest part of winter. Local birds return to central Iraq mid- or late February to early March. Returns to breeding grounds in Syria and south-east Turkey from early April.

Food. Chiefly invertebrates, but fruit readily taken in autumn and seeds in winter. Feeds by taking insects from branches, stems, and foliage, flitting to top of bush and working quickly downwards with much waving of tail, and later flying out and up to top of next bush.

Social pattern and behaviour. Nothing to suggest other than monogamous mating system. Both parents incubate, brood and feed young. ♂ sings from a bush, often on exposed spray; also in song-flight. ♂♂ start singing almost upon arrival on breeding grounds, and period of settlement associated with intense display, notably song-flights.

Voice. Song of ♂ a melodious, mostly quiet conversational chatter, interspersed with harsher notes; more variable and longer than Whitethroat. Commonest call a low, harsh, buzzing 'chrrr' or harsh 'cheee'; also a rattling chatter.

Breeding. SEASON. Iraq: eggs early April to early June. SITE. In low scrub or grassy vegetation, 10–90 cm above ground. Nest: deep cup, often thin-walled, of twigs, plant stems, grass leaves, and stems, with some rootlets, plant fibres, down, and spiders' webs, lined with finer grasses and rootlets, plus small feathers and hair. EGGS. Sub-elliptical, smooth and glossy; white or faintly tinged green, profusely speckled, spotted, and blotched all over with brown, olive-brown, and grey, sometimes more heavily at broad end. Clutch: 4–5. INCUBATION. 11–13 days. FLEDGING PERIOD. 10–11 days.

Wing-length: ♂ 57–63, ♀ 55–62 mm.
Weight: ♂♀ mostly 9–11 g.

Sardinian Warbler *Sylvia melanocephala*

PLATES: pages 1297, 1317

DU. Kleine Zwartkop FR. Fauvette mélanocéphale GE. Samtkopf-Grasmücke IT. Occhiocotto
RU. Масличная славка SP. Curruca cabecinegra SW. Sammetshätta

Field characters. 13.5 cm; wing-span 15–18 cm. Close in size to Lesser Whitethroat but with slightly longer bill, 15% shorter wings, and 10% longer tail. Rather small to medium-sized, slim warbler, with spiky bill, short wings, long tail, and alert angry expression due to frequently steep forehead and red eye and bare ring set in dark face. ♂ black on head and

Sardinian Warbler *Sylvia melanocephala*. *S. m. melanocephala*: **1** ad ♂, **2** ad ♀, **3** 1st winter ♀, **4** juv. *S. m. momus*: **5** ad ♂. *S. m. norrisae*: **6** ad ♂. Cyprus Warbler *Sylvia melanothorax* (p. 1299): **7** ad ♂, **8** ♀, **9** juv.

dusky elsewhere above, off-white with dusky-washed flank below. ♀ and juvenile dusky-brown above, dirty brown and white below. Both ♂ and ♀ have dark tail with white edges and corners.

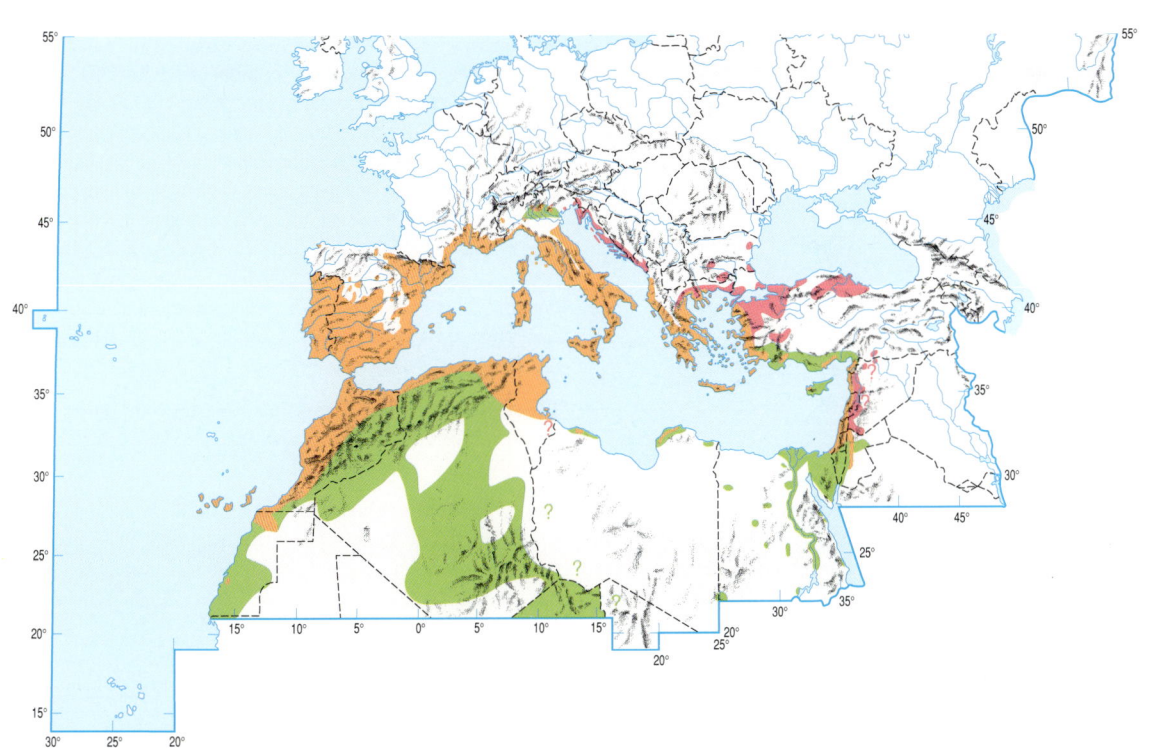

Has distinctive character arising from relatively long and slim shape (*contra* many rather plump portrayals in field guides), dark-hooded head, dusky upperparts, strongly suffused body sides, rather long, often narrow-looking and white-edged tail, skulking behaviour, and frequent scolding calls. Identification not difficult in west of range, but in east Mediterranean area confusion with several other species possible, especially Ménétries's Warbler and Rüppell's Warbler (which see for main distinguishing features). Flight whirring and fast, bird making characteristic dart and dive between patches of cover, often spreading, even fanning tail to show white edges clearly. Bold, curious, and irascible, with noisy calls frequently announcing presence when no movement visible; tends to inspect disturbance at just above ground level but will also pop briefly out of bush-top, frequently cocking tail and raising crown feathers.

Habitat. Breeds in Mediterranean zone, within July isotherms 23–30°C, mainly in dry coastal regions and on islands, generally at low elevations. Equally at home in crowns of close-growing trees (especially pine and oak), tall undergrowth or maquis away from trees, low shrubs and garigue, or even in herb layer or on ground, not excluding bare rocks and clifftops. Indifferent to nearness of dwellings and human activities.

Distribution. Breeds only in west Palearctic. ITALY. Has spread north in recent decades. MALTA. First bred 1874. BULGARIA. Marked range increase. RUMANIA. First recorded 1970 in Dobrogea, where probably breeds. CYPRUS. Breeds north-west coastal plain and inland hills, where now more numerous than Cyprus Warbler; first recorded breeding 1988. SYRIA. Generally scattered and in small numbers. ISRAEL. Has apparently expanded south in 2nd half of 20th century. EGYPT. Breeds north-east Sinai. Formerly bred Lake Qarun, Faiyum (endemic race *norrisae*), but due to salination and loss of vegetation this population now apparently extinct; last recorded 1939. ALGERIA. Very common in north-west. CANARY ISLANDS. Local on Fuerteventura and Lanzarote; common on other islands.

Accidental. Britain, Ireland, Belgium, Netherlands, Germany, Denmark, Norway, Sweden, Finland, Czech Republic, Hungary, Austria, Switzerland, Ukraine, Madeira.

Population. FRANCE. 10 000–100 000 pairs in 1970s. SPAIN. 990 000–1.9 million pairs; slight increase. PORTUGAL. 100 000–1 million pairs 1978–84; stable. ITALY. 300 000–600 000 pairs 1983–95; stable. MALTA. Very common; increase. GREECE. 500 000–1 million pairs. ALBANIA. 20 000–50 000 pairs in 1981. YUGOSLAVIA: CROATIA. 30 000–50 000 pairs; stable. SLOVENIA. 30–50 pairs; slight increase. BULGARIA. 100–1000 pairs; slight increase. TURKEY. 10 000–100 000 pairs. CYPRUS. At least 50 pairs in 1995; increasing rapidly. ISRAEL, JORDAN, TUNISIA. Common. MOROCCO. Common to abundant.

Movements. Varies from partially migratory to sedentary in different parts of range. In west of range, winter quarters include most of breeding range, extending south in Africa to c. 17°N. In east of range, breeding grounds furthest north (Yugoslavia, Bulgaria and northern Turkey) apparently entirely vacated. Autumn movements span late August to December. Spring migration mostly late February or early March to April.

Food. Chiefly insects, but also fruit in autumn and winter; in south of breeding range, fruit predominates in diet for much of the year. Feeds mainly in low scrub but also on ground and in canopy.

Social pattern and behaviour. Mainly solitary or in pairs outside breeding season. Where resident, territories occupied throughout the year. Birds wintering in non-breeding areas may establish temporary feeding territories, but rarely sing. No evidence for other than monogamous mating system. Both sexes share in incubating, brooding, and feeding the young. ♂ sings mostly from elevated perch, typically top of a bush, also in song-flight. Where resident, some song may be given in most or all months of the year; but main song-period from shortly before breeding season to June or July.

Voice. Song like Whitethroat but more sustained and of better quality; quite musical and pleasing warbling freely interspersed with hard rattling alarm-type units. Harsh rattles make song highly locatory. Most striking call a loud rattling sound likened to machine-gun fire or to twirling small wooden rattle, e.g. loud harsh 'kre kre krekrekre' or 'tratratratra'—commonly 2–6 bursts of 2–5 units each. Given mainly by ♂, throughout the year, often 'spontaneously' in general excitement, but notably outside breeding season when it serves (as substitute for song) in territorial defence, eliciting counter-calling from neighbours. Other calls include repeated scolding notes and, in alarm, a harsh stuttering 'stititititiicc'.

Breeding. SEASON. South-west Europe: from mid-March, with peak late April and early May, though late young in nest to early July. Malta: mid-February to mid-July, with peak March–May. Greece: eggs laid late April to mid-June. Northwest Africa: main laying period early April to early June. 2 broods, but, on Malta, early breeders may raise 3. SITE. In low scrub, tall grasses, brambles, etc. 0.20–1.80 m above ground, usually 0.75–1.35 m. Nest: compact cup of grass leaves and stems, plant stalks, vegetable down, and cobwebs, lined with finer grasses and some rootlets, usually with distinct thicker rim. EGGS. Sub-elliptical, smooth and glossy; very variable in colour and markings; white often tinged green, pink, or buff; profusely (sometimes sparsely) spotted, speckled, and blotched brown, olive, grey, buff, or red-brown, markings sometimes concentrated at broad end. Clutch: 3–5(–6). INCUBATION. 13 (12–15) days. FLEDGING PERIOD. 12–13 days.

Wing-length: Western populations (east to Italy), ♂ 57–64, ♀ 57–62 mm. Levant populations, ♂ 55–60, ♀ 55–58 mm. **Weight:** ♂♀ mostly 10–15 g.

Geographical variation. Rather slight. *S. m. momus* (Levant) a little smaller and paler than other populations. *S. m. leucogastra* (Canary Islands) variable between islands, not clearly definable. All intermediate areas occupied by nominate *melanocephala*.

Cyprus Warbler *Sylvia melanothorax*

PLATE: page 1297

Du. Cyprusgrasmus   Fr. Fauvette de Chypre   Ge. Schuppengrasmücke   It. Occhio di Cipro
Ru. Кипрская славка   Sp. Curruca ustulada   Sw. Cypernsångare

Field characters. 13.5 cm; wing-span 15–18 cm. Close in size to Sardinian Warbler and Rüppell's Warbler but tail up to 10% shorter than former's. Rather small to medium-sized warbler, with general character of Rüppell's Warbler and similar plumage in ♀ to Sardinian Warbler except for faint throat and breast mottling. ♂ has distinctive black and dusky upperparts, paler tertial and covert edges, white moustachial stripe, and mottled black and white underparts. Juvenile usually lacks black mottling of adult but already shows pale tertial edges. Eye yellow or brown, surrounded by yellowish bare orbital ring, and white eye-ring.

♂ unmistakable; ♀ less distinctive, as ♀ Rüppell's Warbler also shows black mottling on throat; Rüppell's Warbler differs only in lack of white outer eye-ring (though narrow and also sometimes absent on Cyprus Warbler), redder legs, and cleaner underparts.

Habitat. Occupies a region with warm dry sunny climate, breeding on scrub-covered coastal plains in maquis vegetation, nesting low down under dense cover; also in mountains. Apparently avoids open ground, including cultivation, as well as dense forests, orange plantations, and wetlands. Besides maquis, frequents forest edges with scattered pines and *Cistus*, stands of oak, and undergrowth along valley bottoms, or open places in forest.

Distribution and population. Breeds only in Cyprus, where very widespread and common in suitable habitat (*c.* 20% of island). *c.* 100 000 pairs. Probably stable.

Accidental. Turkey.

Movements. Partial migrant. Wintering grounds imperfectly known. In Cyprus, vacates breeding grounds above *c.* 1000 m; on low ground, less common in winter than summer but more widespread, and found especially along coast; many migrate, departing mainly late September to October, apparently heading SSE towards Israel initially. Exceptional in Lebanon, but regular on passage in Israel, October and February–March, a few remaining to winter. Very few records eastern Egypt, but may winter there; November records Red Sea hills of north-east Sudan. Returns to Cyprus late February to March.

Food. Mainly invertebrates. Both sexes feed typically by moving along branches of trees and inside bushes to pick off small insects; also often dart from bush to take butterflies from low vegetation.

Social pattern and behaviour. Few details of dispersion outside breeding season. Prolonged song-period consistent with possibility of year-round defence of territory by sedentary ♂♂. Nothing to suggest other than monogamous mating system. Most, if not all, incubation by ♀. Both sexes feed young. Territorial ♂ sings mainly from favoured elevated perches and sometimes, in situations of high arousal, in song-flight. Sings in all months except July–September.

Voice. Song a rapid sequence of rattling sounds interspersed with more tonal ones. Similar to Sardinian Warbler but timbre more wooden due to generally lower frequencies. Despite dry rattling character, quality of gaiety derives from vigorous delivery, variation in loudness between successive units (or subunits), almost continuous change of emphasis from rather high (squeaky) to rather low pitch, and variability in length of phrases and in gaps between phrases. Calls include a very brief, sharp staccato sound, 'tchek' or 'dzak'; a dry rasping 'zrik' or 'zrirk'; and, in alarm, a prolonged rattle.

Breeding. SEASON. First eggs laid end of March on low ground, continuing to June; first eggs later at altitude. SITE. In low scrub; 30–110 cm above ground; rarely on ground. Nest: cup of grass stems and leaves, bound together with cobwebs, lined with finer grasses and a few hairs, also including cobwebs; foundation often contains strips of juniper bark even if none growing nearby. EGGS. Sub-elliptical, smooth and glossy; very pale green, occasionally whitish, with olive-brown and violet-grey spots, speckles, and fine blotches, sometimes concentrated at broad end. Clutch: (3–)4–5. (Incubation and fledging periods not recorded.)

Wing-length: ♂ 59–63, ♀ 60–63 mm.
Weight: ♂♀ mostly 10–15 g.

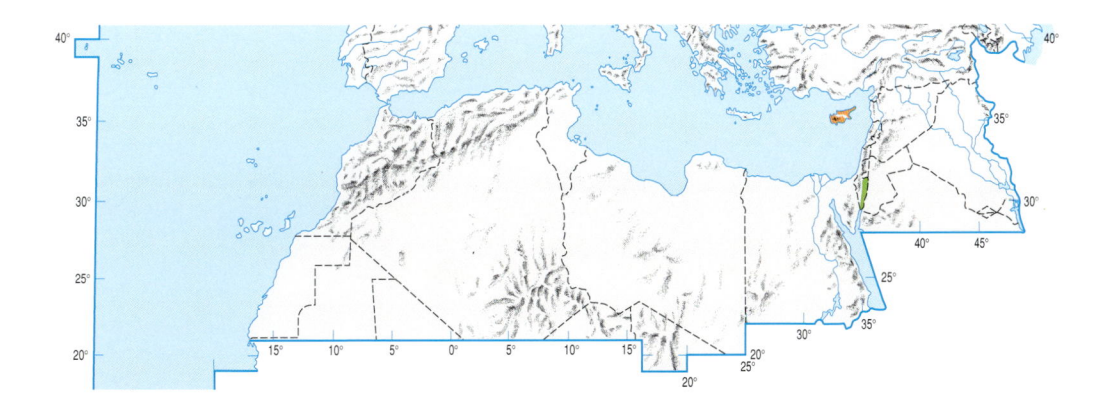

Rüppell's Warbler *Sylvia rueppelli*

Du. Rüppells Grasmus   Fr. Fauvette de Rüppell   Ge. Maskengrasmücke   It. Silvia del Rüppel
Ru. Славка Рюппеля   Sp. Curruca de Rüppell   Sw. Svarthakad sångare

PLATE: page 1301

Field characters. 14 cm; wing-span 18–21 cm. Close in size and shape to Whitethroat but with 10% longer bill and proportionately slightly shorter wings and tail; largest of scrub-haunting warblers in Mediterranean region. Medium-sized, robust, and evenly balanced warbler, with quite long bill, quite long and square tail, and general character more like Whitethroat than Sardinian Warbler. ♂ distinctive, with black forehead and throat divided by white moustachial stripe; rest of plumage grey above with conspicuous pale tertial-fringes and -tips and white-edged black tail; basically cream below with grey flank. ♀ grey-brown above, buff-cream below, sometimes showing black mottling on throat and pale moustachial stripe but others lack distinctive markings though all and juvenile show conspicuous pale tertial fringes. Eye orange to chestnut, bare ring orange-brown to -red.

♂ and well-marked ♀ have diagnostic combination of white moustachial stripe and black front half of head, throat, and upper breast. Poorly marked ♀, juvenile, and 1st-year far less distinctive and confusion possible with other warblers, especially ♀ Sardinian Warbler, which is smaller, with proportionately longer and more rounded tail, swarthier appearance with browner flank and no striking moustachial stripe, and flesh-brown (not red-brown) legs. Skulking but not shy; generally less lively than most *Sylvia*.

Habitat. Within limited east Mediterranean breeding range, occurs in dry warm climates between July isotherms 25–30°C, minimum average July temperature experienced thus being the highest of any European *Sylvia*. Habitat consists of dry thorny scrub on rocky slopes or in narrow rock fissures and ravines, or undergrowth of old, open woods of oak or cypress predominantly in mountains, up to 1600 m in Crete.

Distribution and population. Breeds only in west Palearctic. Greece. 3000–10 000 pairs; apparently stable. Turkey. Common in south of western Anatolia and in south coastlands, local near Sea of Marmara. 5000–50 000 pairs. Syria. Confined to north-west, where breeding first confirmed Saladin's Castle 1993.

Accidental. Faeroes, Britain, France, Denmark, Finland, Rumania, Ukraine, Tunisia, Algeria.

Movements. Migratory, entire population moving south to winter mainly in Chad and Sudan. Leaves breeding grounds chiefly late August to September. Comparatively late arrival (from October) in winter quarters suggests birds use intermediate staging-posts. Vacates winter quarters mid-February to late March, with peak passage on North African coast mid- or late March to early April. First arrivals on southern coastlands of Turkey in early or mid-March, but main arrival in south-east Europe late March to early April.

Food. Adult and larval insects, with some fruit in autumn. Feeds mainly under cover, hopping about inside canopy of low trees and bushes.

Social pattern and behaviour. Dispersion outside breeding season little known. Birds sing in winter quarters in few weeks before departure. On migration, sometimes assemble in large numbers at stopovers. Nothing to suggest other than monogamous mating system. Both sexes share incubation, brooding, and feeding of young. ♂ sings from conspicuous elevated perch in middle of territory, or in song-flight with slow-flapping 'butterfly'-flight and parachute-descent. Song heard from arrival on breeding grounds, but duration of song-period not recorded.

Voice. Song composed of hard, rapidly chattering, usually brief phrases, comprising mainly short harsh units (including 'tak' and

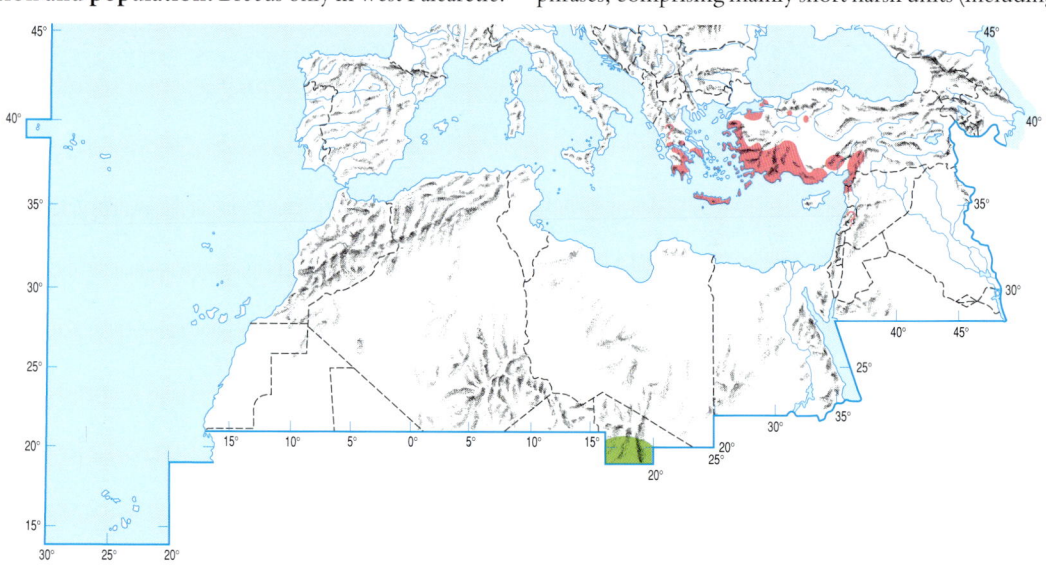

Rüppell's Warbler *Sylvia rueppelli*: **1** ad ♂, **2** ad ♀, **3** 1st winter ♀, **4** juv. Barred Warbler *Sylvia nisoria* (p. 1306): **5** ad ♂ breeding, **6** ad ♀, **7** 1st winter, **8** juv.

'trrr' call-type units) with single pure tones. Likened to Sardinian Warbler, notably for rattling sounds, and, at least in Greece, may be difficult to distinguish from that species. In Turkey, more like Cyprus Warbler than Sardinian Warbler. Phrases typically not more than about 2 s long, given at rate of up to 16 per min; but song becomes prolonged at times when bird is excited, as in song-flights; also much louder and more varied in song-flight. Calls include a hard 'tak', indistinguishable from Sardinian Warbler, and a rattling 'tictictictictic...', compared to winding of a clock.

Breeding. SEASON. Crete: main laying period 2nd half of April extending into 1st half of May; similar period elsewhere in Greece. SITE. In thick, often thorny, scrub; 45–75 cm above ground. Nest: well-built cup of grass leaves and stems and some vegetable down, lined with finer material. EGGS. Sub-elliptical, smooth and glossy; white, tinged green or buff, with profuse fine speckling, spotting, and mottling of green, olive, brown, and grey, sometimes with dark cap at broad end, and occasionally with fewer, larger blotches. Clutch: 4–5(–6). INCUBATION. About 13 days. (Fledging period not recorded.)

Wing-length: ♂ 67–74, ♀ 65–72 mm.
Weight: ♂♀ mostly 12–15 g.

Desert Warbler *Sylvia nana*

PLATES: pages 1302, 1317

Du. Woestijngrasmus   Fr. Fauvette naine   Ge. Wüstengrasmücke   It. Sterpazzola nana
Ru. Пустынная славка   Sp. Curruca sahariana   Sw. Ökensångare

Field characters. 11.5 cm; wing-span 14.5–18 cm. Noticeably smaller than Whitethroat, with 15% shorter wings and 20% shorter tail (though with not dissimilar general structure and shape); smallest *Sylvia* in west Palearctic, hardly exceeding larger *Phylloscopus* warblers in bulk and length. Small, seemingly nervous, rather slim but quite robust warbler, with rather peaked head and obvious tail. Plumage palest and least marked of *Sylvia*, with pale grey-brown or sandy upperparts and almost white underparts relieved mainly by white-edged rufous tail. At close range, dark tip and yellow base to bill and yellow eye give rather fierce expression. Has diagnostic habit of spreading and then flicking tail.

Unmistakable; paler and drabber, as well as smaller, than any other *Sylvia*, with diagnostic rufous tone to upper tailcoverts and all but outer fringes of tail. Quite different in structure from Scrub Warbler, which in any case has softly streaked upperparts

Desert Warbler *Sylvia nana*. *S. n. deserti*: **1** ad, **2** juv. *S. n. nana*: **3** 1st winter. Lesser Whitethroat *Sylvia curruca* (p. 1308). *S. c. curruca*: **4** ad worn (spring), **5** ad fresh (autumn), **6** juv. *S. c. blythi*: **7** 1st winter. Whitethroat *Sylvia communis* (p. 1310). *S. c. communis*: **8** ad ♂ breeding, **9** ad ♀ breeding, **10** juv. *S. c. volgensis*: **11** ad ♂ breeding. *S. c. icterops*: **12** ad ♂ breeding.

and spindly tail. Flight free and fast, with whirring wing-beats; silhouette quite compact, like small Lesser Whitethroat, and with rather sharp corners to tail sometimes noticeable. Escape-flight low and short, and direct to cover, from which bird hard to flush. Spends much time on ground; carriage reminiscent of alert Dunnock.

Habitat. Breeds in west Palearctic mainly in semi-deserts of western Sahara, in fairly level open tracts of sand or clay, sometimes stony, bearing sparse low patches of shrubby vegetation or herbage; sometimes also on sand dunes, in patches of thorn scrub or tamarisk, and on stony hillsides and foothills of mountains.

Distribution and population. GERMANY. Unpaired ♂ in song and built 2 nests, Schleswig-Holstein 1981. RUSSIA. 1000–10 000 pairs; stable. TUNISIA. Not scarce. MOROCCO. Restricted to sandy deserts south of Haut Atlas.

Accidental. Britain, Netherlands, Germany, Denmark, Sweden, Finland, Italy, Malta, Ukraine, Azerbaijan, Turkey, Cyprus, Canary Islands, Cape Verde Islands.

Beyond west Palearctic, extends from Caspian east to central China. Range incompletely known.

Movements. Resident to migratory. North African population chiefly resident, though some dispersal possible in autumn and in extremely dry conditions. Eastern populations migrate through Turkestan, Iran, and Afghanistan to winter from north-east Africa east to deserts of north-west India. In autumn, vacation of north Caspian breeding grounds is gradual, from September to end of October. Arrival in spring is from end of March.

Vagrancy records in north-west Europe all eastern race *S. n. nana*, and chiefly mid-October to early November.

Food. Chiefly insects; also some seeds and berries. Feeds mainly in low scrub and commonly on ground where it is well camouflaged.

Social pattern and behaviour. Little known about dispersion outside breeding season, but apparently occurs singly or in pairs. No evidence for other than monogamous mating system. Apparently both parents incubate and feed young. ♂ sings mostly from top of, or within, bush, also while hopping about (often while feeding) in bush, and sometimes on ground. Singing bird often spreads and partly erects tail, exposing white outer rectrices and tips. Less often performs short song-flight from one bush to another, similar to Whitethroat or Spectacled Warbler. In Algeria, sings December–April.

Voice. Song of North African population recalls Whitethroat, notably in overall duration and somewhat erratic pitch intervals, but timbre more like Blackcap; most phrases start with a harsh, emphatic and descending 'krrrrr', followed immediately by a jaunty tuneful warble (difficult to render but with 'tui' or similar sounds prevailing) ending with a rising whistle. Song of Asian population very different, consisting of a stereotyped phrase, central portion a repetition of identical short notes, e.g. 'turrr-ti-ti-ti-ti-ti-ti-teu'. Calls include churring or purring 'drrrrrr', 'tirr', etc.; in alarm, rattling, sparrow-like 'ch-ch-ch-ch...'.

Breeding. SEASON. Western Sahara, and probably southern Algeria: eggs laid January to early March. Central and north-east

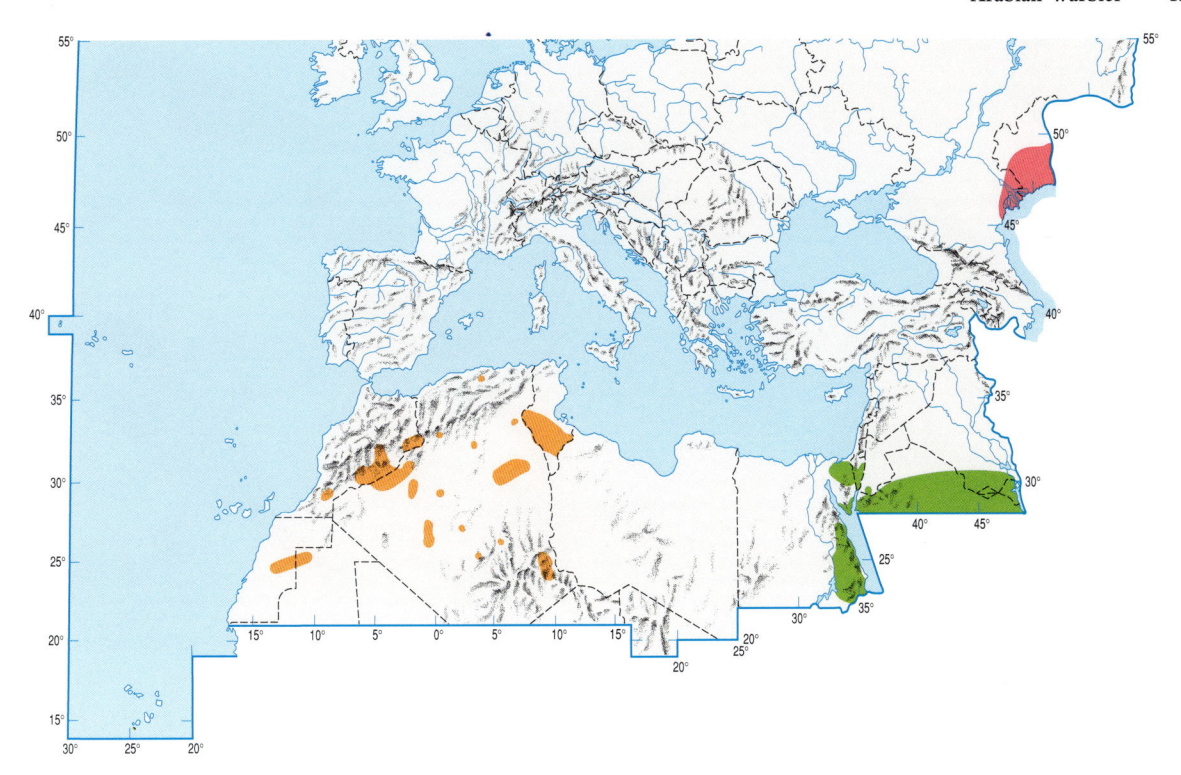

Algeria: eggs laid mid-March to May. 1 brood in northern Sahara, perhaps normally 2 in Western Sahara. Turkmenistan: eggs laid late April to early June. 2 broods. SITE. In low scrub, up to 110 cm above ground. Nest: substantial, thick-walled cup of twigs, grass stems, and leaves, with plant down and cobwebs, lined with finer grasses and fibres and, usually, more down. EGGS. Sub-elliptical, smooth and glossy; white, or faintly tinged blue, with fine brown and grey speckles and scrawls, heavier at broad end. Clutch: Asia: 4–5(–6). North Africa: 2–5. (Incubation and fledging periods not recorded.)

Wing-length: Central Asian population (nominate *nana*), ♂ 58–62, ♀ 54–58 mm. North African population (*S. n. deserti*), ♂ 56–59, ♀ 54–60 mm.
Weight: Central Asian population, ♂ ♀ mostly 8–10 g. North African population, ♂ ♀ 9–10.6 g.

Geographical variation. Slight in size. *S. n. deserti* of North Africa markedly paler on upperparts than nominate *nana* of Asia, and whole plumage with cream-pink suffusion.

Arabian Warbler *Sylvia leucomelaena*

PLATE: page 1304

DU. Arabische Zwartkop   FR. Fauvette d'Arabie   GE. Akaziengrasmücke   IT. Bigia del mar Rosso
RU. Пегая славка   SP. Curruca arabe   SW. Rödahavssångare

Field characters. 14.5 cm; wing-span 18–20.5 cm. Approaches Orphean Warbler in size but structure differs in slightly shorter bill, 10–15% shorter and more rounded wings, and (relative to wings) noticeably longer and more rounded tail. Medium-sized to rather large, quite bulky, rather upright and seemingly long-tailed warbler, with plumage recalling both Orphean Warbler and Sardinian Warbler. Diagnostic characters of adult ♂ of west Palearctic race (compared with Orphean Warbler) apparently restricted to noticeably black head, narrow, white-flecked eye-ring enclosing black bare ring and dark eye, cleaner grey upperparts, paler underparts, and lack of obvious white edges to tail, though white under-tips to outermost tail feathers may show when tail spread. ♀, immature, and juvenile more like Orphean Warbler, but eye always dark and white in tail inconspicuous. Frequently drops tail in circular movement below body-line.

Needs to be carefully distinguished from Orphean Warbler. Important to note that wing is blunter, with primary projection only $\frac{1}{3}$ or less of exposed tertials (c. $\frac{1}{2}$ in Orphean Warbler).

Habitat. Almost entirely warm and arid, but in Hejaz and Asir mountains (Saudi Arabia) including temperate summits above 3000 m. In Israel and Jordan, generally confined to relatively dense acacia groves; elsewhere breeds also in arid scrub.

Distribution and population. ISRAEL. First recorded breeding

Arabian Warbler *Sylvia leucomelaena*. **1** ad ♂, **2** ad ♀, **3** juv. Orphean Warbler *Sylvia hortensis*. *S. h. hortensis*: **4** ad ♂, **5** ad ♀, **6** 1st winter ♂, **7** juv. *S. h. jerdoni* ('*balchanica*', Iran): **8** ad ♂.

1972. Uncommon in Arava valley; at 2 main concentrations, 50 and 16 pairs in 1980s. JORDAN. Localized resident of Wadi Araba (alongside Arava valley); first identified 1963.

Beyond west Palearctic, breeds on Red Sea and adjacent coasts south to Somalia, east to Oman.

Movements. Chiefly resident or presumed resident. In Israel and Jordan, present all year on breeding grounds; some 1st-winter birds move south or south-west to Eilat and eastern Sinai.

Food. Chiefly insects but readily takes fruit when available. Most commonly feeds by moving along branches and trunks of acacias like woodpecker, sometimes even hanging upside down, and drawing out insects (especially Lepidoptera larvae) from crevices. Also forages for small insects on ground beneath trees by probing and moving bill from side to side through sand and litter.

Social pattern and behaviour. Pairs probably maintain territories throughout the year. Mating system evidently monogamous. Both parents incubate, brood, and feed young. ♂♂ sing openly from perches in tree-tops; also occasionally in flight. ♀ also sometimes sings. ♂♂ sing almost throughout the year.

Voice. Song of ♂ a series of short warbled phrases, rather like thrush; far-carrying and audible at hundreds of metres. Alarm-calls include a low, thrush-like 'chuck' or 'chack', and a soft churring rattle.

Breeding. SEASON. Israel: mid- or late February to mid-July. Usually 2 broods, not uncommonly 3. SITE. In periphery of acacia trees, 80–300 cm from ground. Nest: cup of plant stems and leaves, mainly of annuals, woven with fibres, lined with finer material; sometimes robust, but sometimes very thin-walled. EGGS. Sub-elliptical, smooth and glossy; speckled with grey and brown markings. Clutch: 2–3(–4). INCUBATION. 15–16 days. FLEDGING PERIOD. 14–17 days.

Wing-length: ♂ 71–75, ♀ 67–72 mm.
Weight: ♂♀ mostly 11–15 g.

Orphean Warbler *Sylvia hortensis*

PLATES: pages 1304, 1317

Du. Orpheusgrasmus   Fr. Fauvette orphée   Ge. Orpheusgrasmücke   It. Bigia grossa
Ru. Певчая славка   Sp. Curruca mirlona   Sw. Mästersångare

Field characters. 15 cm; wing-span 20–25 cm. Noticeably larger than Whitethroat, with almost twice its bulk and 25% longer wings; close in size to Barred Warbler. Medium-sized to large and robust warbler, with strong bill and heavy head. Plumage dark dusky-brown above, mostly grey or buff below, with noticeable features restricted in adult to pale eye set in dull black (♂) or dusky (♀) face, white throat, and clear white edges to tail. Most ♂♂ of eastern populations, immature, and some ♀♀ have dark eye.

Distinctive, due to large size and basically 2-toned, rather featureless plumage. Confusion restricted in Europe and around Mediterranean to Sardinian Warbler and Ménétries's Warbler (both much smaller, with contrasting upperpart pattern, red eye and bare ring, and different behaviour) and ♀ Rüppell's Warbler (25% smaller, with reddish eye and bare ring, prominent pale tertial-fringes, and, in adult ♀, black throat and white moustachial stripe). In south of Dead Sea depression, migrants (and possible breeding birds of Jordan uplands) need to be carefully distinguished from resident Arabian Warbler. Flight confident, with fluent action typical of large warbler and smaller chat; intersperses glides with bursts of wing-beats and makes heavy entry into cover, spreading tail noticeably. Escape-flight often quite long, with bird heading for canopy like Blackcap and not merely plunging into nearest thicket.

Habitat. Breeds in lower middle latitudes of west Palearctic, mainly in warm, dry Mediterranean climate but secondarily in steppe and warm temperate zones, not only in lowland but on hillsides and mountain foothills. In Switzerland, occurs between 400 and 880 m, and in French Alps also to *c.* 800 m; in northwest Africa and Taurus (Turkey) to *c.* 1600 m and in Armenia often up to *c.* 1900 m and occasionally to 2580 m. Generally inhabits areas with mixture of open woodland and tall bushy growth, including olive groves, orchards, and gardens; is inclined to favour broad-leaved species, but will accept pines and juniper. In north-west Africa, concentrates in open oakwoods especially of sparse evergreen oak with little undergrowth.

Distribution. Range and population have decreased Spain and Italy, probably increased Bulgaria; loss of habitat through changed farming practices likely cause of declines. FRANCE. Distribution outside Mediterranean area unstable, with overall reduction since 19th century. BELGIUM. Bred 1915. LUXEMBOURG. Bred 1865, 1898, and 1900. SWITZERLAND. Breeding discovered Valais 1966, but probably bred earlier. AZERBAIJAN. Range very poorly known. SYRIA. Perhaps also breeds Euphrates area. JORDAN. Perhaps breeds locally in southern Rift Margins. TUNISIA. Locally common in north.

Accidental. Britain, Belgium, Luxembourg, Germany, Slovakia, Austria, Malta, Ukraine, Kuwait, Madeira, Canary Islands.

Beyond west Palearctic, breeds Iran east to western Tien Shan and western Pakistan.

Population. FRANCE. Under 10 000 pairs in 1970s. SWITZERLAND. 10–15 pairs 1985–93; stable. SPAIN. 170 000–440 000 pairs. PORTUGAL. 100–1000 pairs 1978–84; stable. ITALY. 1000–2000 pairs 1983–95. GREECE. 3000–10 000 pairs. ALBANIA.

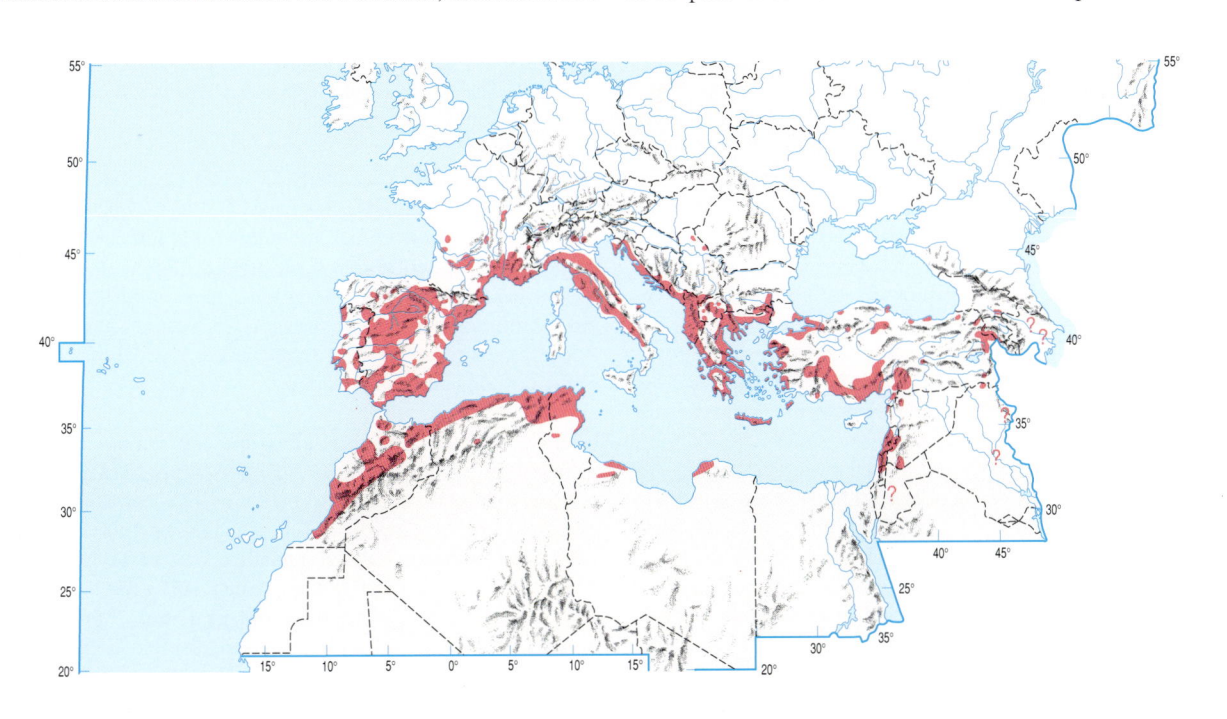

1000–2000 pairs 1981. YUGOSLAVIA: CROATIA. 10 000–15 000 pairs; stable. BULGARIA. 100–1000 pairs. AZERBAIJAN. Probably rare. TURKEY. Very local; nowhere common. 5000–50 000 pairs. ISRAEL. Local, fairly common. ALGERIA. Mostly uncommon, but locally common. MOROCCO. Scarce.

Movements. All populations migratory. West European and north-west African populations winter in sub-Saharan Africa, chiefly 14–17°N, from southern Mauritania and northern Sénégal east to western Sudan, with most records from Mali, Niger, and Chad. Departure south begins late July; reaches winter quarters September–October.

Departure north begins late February or early March, but continues into May; reaches European breeding grounds mainly from late April to May. East European populations winter in north-east Africa, with bulk of passage across eastern end of Mediterranean and through Levant. Extralimital Asian populations winter in Arabia, Iran, and Indian peninsula.

Food. Chiefly invertebrates, also berries. Prefers to feed in larger bushes and trees where it flits about canopy picking small insects from branches.

Social pattern and behaviour. Little studied outside breeding season. No evidence for other than monogamous mating system. Both sexes incubate, feed and brood young. Territory large for *Sylvia*; song-posts extend up to 200 m from nest, while parents may fly 200 m to collect food for young. ♂ sings from cover of tall bushes, rarely in horizontal song-flight. Song-period May–July; most song early in morning, and from late afternoon until dusk.

Voice. Song differs markedly between western birds (subspecies *hortensis*) and eastern birds (*crassirostris*). Song of western birds a vigorous far-carrying warble, combining 2–4 sorts of units into short clearly separated phrases, with strong tendency to repetition of phrases. Song of eastern birds fuller, more accomplished and varied, recalling Nightingale, or midway between that species and Blackcap, with beautiful rich fluting phrases recalling Blue Rock Thrush; phrases longer and almost continuous, much more diverse, incorporating excellent mimicry often repeated several times. Calls include a single 'tak' or a loose series of short 'tek-tek' units; in alarm, a rattling or churring 'trrrr'.

Breeding. SEASON. Southern Europe: end of April to late June, with main laying period 1st half of May. South-east Europe: eggs laid mainly 2nd week May to early June. North-west Africa: eggs mid-April to early June. One brood. SITE. In branches of small trees and shrubs, 0.5–3.5 m above ground. Nest: well-constructed cup of grass and plant stems, with some twigs, bound together with cobwebs, moss, fibres, and plant down, lined with finer grasses, fibres, and sometimes spider cocoons; usually with distinct rim. EGGS. Sub-elliptical, smooth and glossy; white or very faintly tinged bluish, sparsely spotted, speckled, and blotched brown, black or olive, and grey, markings heavier at broad end. Clutch: 3–5(–6). INCUBATION. 12–13 days. FLEDGING PERIOD. 12–13 days.

Wing-length: ♂ 77–85, ♀ 74–86 mm.
Weight: ♂♀ mostly 16–25 g; migrants to 31 g.

Geographical variation. 2 races recognized in west Palearctic. Populations east of Adriatic Sea and Gulf of Sirte (*S. h. crassirostris*) whiter below and with longer bill than nominate *hortensis* to the west.

Barred Warbler *Sylvia nisoria*

PLATES: pages 1301, 1317

DU. Sperwergrasmus FR. Fauvette épervière GE. Sperbergrasmücke IT. Bigia padovana
RU. Ястребиная славка SP. Curruca gavilana SW. Höksångare

Field characters. 15.5 cm; wing-span 23–27 cm. Marginally largest *Sylvia*; 10–15% longer and bulkier than Garden Warbler, with fuller-ended and 25% longer tail. Quite large, robust warbler, with strong bill, rather big head, large feet, and long and full tail. Adult basically grey-toned above and cream- to brown-white below, with best-marked ♂♂ showing grey upperparts, staring pale eye, and copious but well-spaced barring particularly on underparts; dullest ♀♀ have browner upperparts and less barring. Juvenile more buff in ground-colour above and virtually unbarred below. At all times has diagnostic combination of large size, whitish tips to median and greater coverts (forming dull bars), and pale edges and quite bright tips to tertials. Looks heavy in flight, with pale corners or tip of tail often showing. Often erects crown feathers above steep forehead.

At distance, barring of even adult ♂ far from striking, and invisible at all ranges on juvenile and many 1st-winter birds. Immature's slight pattern of dull wing-bars and tertials may also become indistinct, thus often giving rather featureless appearance recalling Garden Warbler. Barred Warbler has, however, characteristic combination of pale-based bill, rather big-headed and robust form, lengthy wings, and rather long, full tail (with pale corners or tip often visible from behind). Flight strong, with weight nearly always evident; at times rapid and free, at others more fluttering and erratic; over longer distances, tends to develop undulations, recalling small shrike. Gait laboured, with heavy hopping and clumsy sidling.

Habitat. Breeds in upper middle latitudes of warm continental west and central Palearctic, in temperate, steppe, and marginally in boreal zones, from July isotherms of 17°C to above 32°C; mainly in lowlands but ascends in Caucasus to 1500 m, and to 1400 m in northern Italy. Essentially a bird of bushy terrain; avoids both arid and wetland areas and is not a forest bird, but will inhabit narrow shelterbelts and plantations, clearings in broad-leaved and mixed woodlands with plenty of undergrowth (especially of thorny bushes), and early stages of regrowth of felled or burnt timber. Also frequents bushy hillsides, rough growth on woodland margins or by pasture or meadowland, hedgerows, roadside verges, parks, orchards, riverain thickets, and (in Asia) steppe fringes.

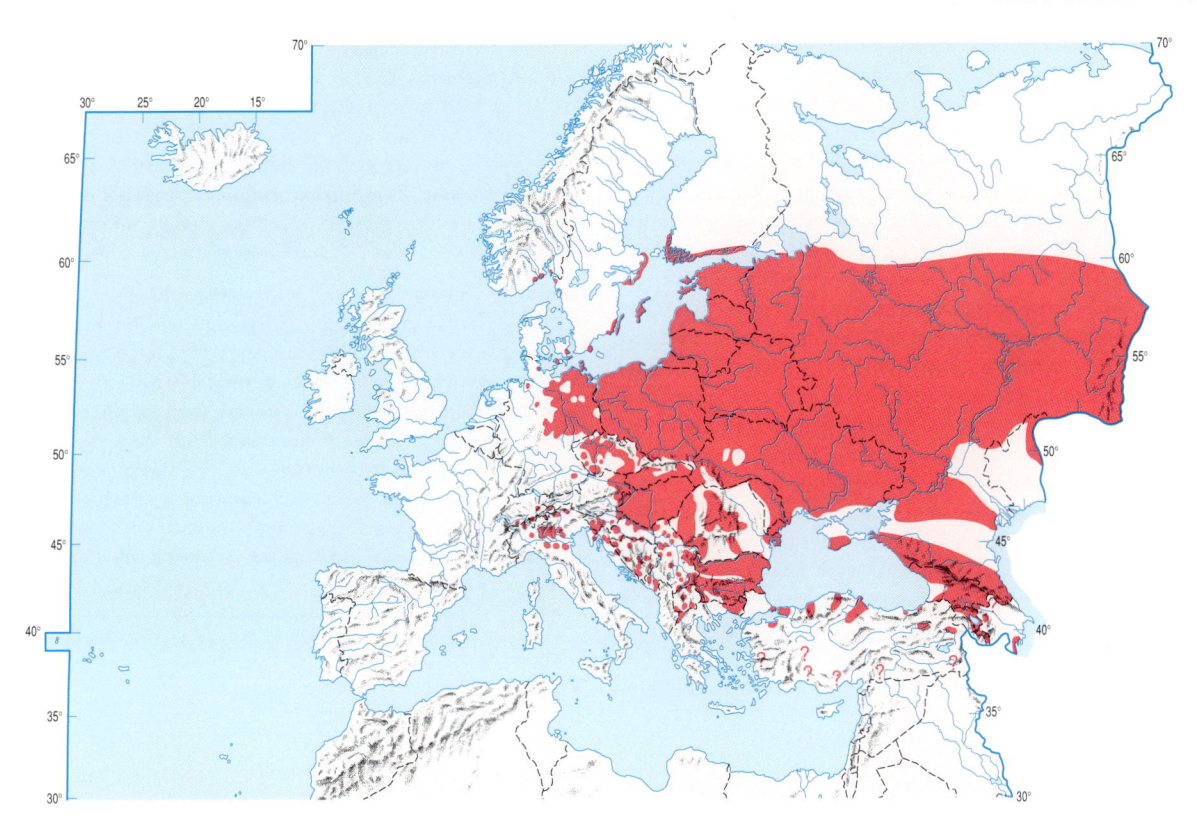

Distribution. FRANCE. Bred Lorraine 1908. GERMANY. Range continuous only in east. DENMARK. Map shows breeding 1990–92 (3–8 pairs); probably none bred 1994. NORWAY. First bred 1972, Vestfold. SWEDEN. Has spread since 1960; some sites with few pairs and may be temporary. FINLAND. Established 1920s; has spread eastwards along southern coast since 1970s; probably now slowing down. AZERBAIJAN. Range very patchy. TURKEY. Locally common in north and east. SWITZERLAND. Breeding first recorded 1952, Grisons.

Accidental. Jan Mayen, Iceland, Spain, Balearic Islands, Malta. Regular autumn vagrancy north-west to Britain and Ireland, and probably almost annual in Faeroes.

Beyond west Palearctic, extends east to Mongolia, south perhaps to northern Iran.

Population. GERMANY. In east (where great majority breeds), 4400 ± 1300 pairs in early 1980s. Another estimate (for entire country) of 9000 pairs in mid-1980s. NORWAY. 10–20 pairs in late 1980s. SWEDEN. 250–1000 pairs in late 1980s; stable. FINLAND. 2000–3000 pairs in late 1980s; probable increase. ESTONIA. 10 000–20 000 pairs in 1991; increasing. LATVIA. 1000–6000 pairs in 1980s; stable. LITHUANIA. Rare; stable. More numerous in central and southern areas. POLAND. 5000–15 000 pairs; probably stable. CZECH REPUBLIC. 1500–3000 pairs 1985–9; slight decrease. SLOVAKIA. 3000–6000 pairs 1973–94; stable. HUNGARY. 20 000–40 000 pairs 1979–93; stable. AUSTRIA. Locally not uncommon in east. SWITZERLAND. 10–20 pairs 1985–93; stable. ITALY. 1000–2000 pairs 1983–95; slight increase. GREECE. 100–1000 pairs. YUGOSLAVIA: CROATIA. 1000–2000 pairs; stable. SLOVENIA. 400–600 pairs; slight increase. BULGARIA. 1000–10 000 pairs; stable. RUMANIA. 20 000–35 000 pairs 1986–92. RUSSIA. 100 000–1 million pairs; stable. BELARUS'. 10 000–15 000 pairs in 1990; fluctuating, but apparently stable overall. UKRAINE. 2700–3300 pairs in 1988; slight decrease. MOLDOVA. 22 000–30 000 pairs in 1986; slight decrease. AZERBAIJAN. Uncommon. TURKEY. 500–5000 pairs.

Movements. Migratory, all populations wintering in East Africa, from Sudan to Tanzania. Arrives on breeding grounds late (not until mid-May in north), and departs early (mainly August). Autumn migration very inconspicuous, presumably due mainly to skulking habits and long-stage overflying; north of initial concentrations, spring movement also often unobtrusive. Passage of all populations both seasons is chiefly funnelled through Levant and Middle East. In view of standard direction, overshooting spring migrants surprisingly rare in Britain: only 7 records during 1958–85. In autumn, however, vagrants (chiefly juveniles) appear regularly west and north-west of breeding range, presumably due to reversed migration.

Food. Chiefly invertebrates but in late summer and autumn mainly berries. Forages mostly in low bushes for insects. Rarely feeds on ground, hopping briefly near cover in search of earthworms, or in open fields.

Social pattern and behaviour. Among most skulking of *Sylvia* and, especially in breeding season, easily overlooked; but conspicuous concentrations may occur on migration. Mating system includes monogamy and polygamy, many ♂♂ obtaining second mate after first ♀ starts laying. Most striking aspect of behaviour is regular association with Red-backed Shrike. Territories of the two species often overlap and their nests are frequently close together, not uncommonly in same bush. Advantage of nesting association may be mutual warning of danger (but similarity of plumage to ♀ shrike suggests possibility of protective mimicry). ♂ sings from perch in tree or bush (unlike some other Sylviidae, not usually while hopping about), in short flights, or in ritualized song-flight. In song-flight, bird climbs steeply, singing from outset, in curved path, returning after 5–10 s to original perch, or, more often, one nearby; ascent rather irregular, even jerky, to 10–15 m above ground. On ascent, wing-beats slow and deep; after peak of ascent, bird raises wings over back or half-spreads them and, still singing, makes gliding descent while rocking from side to side. ♂ starts singing a few days after arrival, coinciding with appearance of ♀♀. Song-activity especially high mid-May but declines markedly in first 10 days of June.

Voice. Relatively complex repertoire in breeding season, but birds largely silent at other times. Song may include mimicry of Red-backed Shrike, and many calls similar to that species. Song loud and vigorous, rich in warbled and chattering sounds, interspersed with harsher ones; may include mimicry, incidence and accuracy of which seem to vary with habitat and individual. Most typical contact-call, by both sexes: hard coarse rattling 'trrrrt', 1–2 s long, duration and emphasis varying with context. Other calls include a 'tsek', recalling similar call of Blackcap, repeated 'cha-cha-cha. . .', and buzzing, nasal 'czow'.

Breeding. SEASON. Central and western Europe: eggs from mid-May. Finland: main laying period June. One brood. SITE. In young trees, shrubs, and brambles, with preference for thick or thorny growth. Height above ground 30–200 cm. Nest: substantial, well-built cup of stalks and grass stems, with some twigs, rootlets, and spider cocoons and cobwebs, lined with finer material and hair; supported on twigs, shoots, and suckers, though not attached to them. EGGS. Sub-elliptical, smooth and glossy; whitish or variably tinted green, pink, or buff, with irregular and sometimes sparse spots and blotches of brown, olive, buff, red-brown, and grey. Clutch: 4–5 (2–6). INCUBATION. 12–13 days. FLEDGING PERIOD. 10–12 days.

Wing-length: ♂ 88–94, ♀ 87–90 mm.
Weight: ♂♀ mostly 22–28 g; migrants to 36 g.

Lesser Whitethroat *Sylvia curruca*

PLATES: pages 1302, 1317

Du. Braamsluiper   Fr. Fauvette babillarde   Ge. Klappergrasmücke   It. Bigiarella
Ru. Славка-завирушка   Sp. Curruca zarcerilla   Sw. Ärtsångare

Field characters. 12.5–13.5 cm; wing-span 16.5–20.5 cm. Western races noticeably smaller and more compact than Whitethroat, with less peaked crown and 10% shorter wings and tail. Small to medium-sized, slim but not strikingly long-tailed warbler, with rather demure appearance and often more secretive behaviour than Whitethroat. Dull grey-brown above and dull white below, with dusky head (showing obvious blackish mask) and white-edged dusky tail; usually no contrasting wing-panel. Sexes closely similar.

Not difficult to identify in north of breeding range, since not only rather dull plumage but also behaviour and song strikingly different from Whitethroat; from southern part of breeding range to winter quarters, open to confusion with at least 5 other *Sylvia*. Within *Sylvia*, occupies mid-points in size and structure: smaller than ♂ or immature Rüppell's Warbler, lacking its reddish eye and legs; similar in size to Ménétries's Warbler and Sardinian Warbler but not totally dark-headed and less dusky above, also lacking their red eye or eye-ring and flesh-brown legs; larger than ♀ Subalpine Warbler, lacking its dull throat, uniformly-coloured head, pale base to bill, and pale legs. Note also that racial variation within Lesser Whitethroat marked in both size and strength of face mask; smallest, least marked birds are particularly confusing. Flight less jerky than Whitethroat. Usually shy, staying within cover for long periods; far less curious than Whitethroat, generally ignoring imitations of its call.

Habitat. Breeds mainly in middle and upper middle latitudes of west Palearctic, in continental warm temperate, steppe, and boreal zones, largely in lowlands but in Switzerland mainly at 1300–2000 m. Flourishes in habitats intermediate between extensive closed forest and open country, resorting to well-spaced often tall bushes, shrubs, hedgerows, plantations, well-grown gardens, parks, cemeteries, and similar situations where dense cover well broken with pronounced vertical structure, often facing glades, clearings, or grasslands. In England, favours tall hedgerows on farmland or along grass-verged country roads, small thickets, or patches of scrub, taller than those preferred by Whitethroat and often including tree song-posts and regrowth in clearings or young plantations, sometimes of conifers. In central Europe young coniferous trees often favoured, and in Switzerland broken ground such as margins of torrents, edges of scree, and avalanche corridors. Increasingly in some areas, breeds within human settlements.

Distribution. Evidence of expansion in north and west of range. FAEROES. Bred 1964, 1981. BRITAIN. Expanding north and west; regular breeding Cornwall since 1977. In Scotland, first breeding 1974; now recorded north to Orkney. Formerly bred Isle of Man. IRELAND. First breeding record 1990. FRANCE. Gradual spread westward; first bred Brittany 1980s. FINLAND. Expansion to north in 20th century, probably due to fragmentation of forests. AZERBAIJAN. Distribution very patchy.

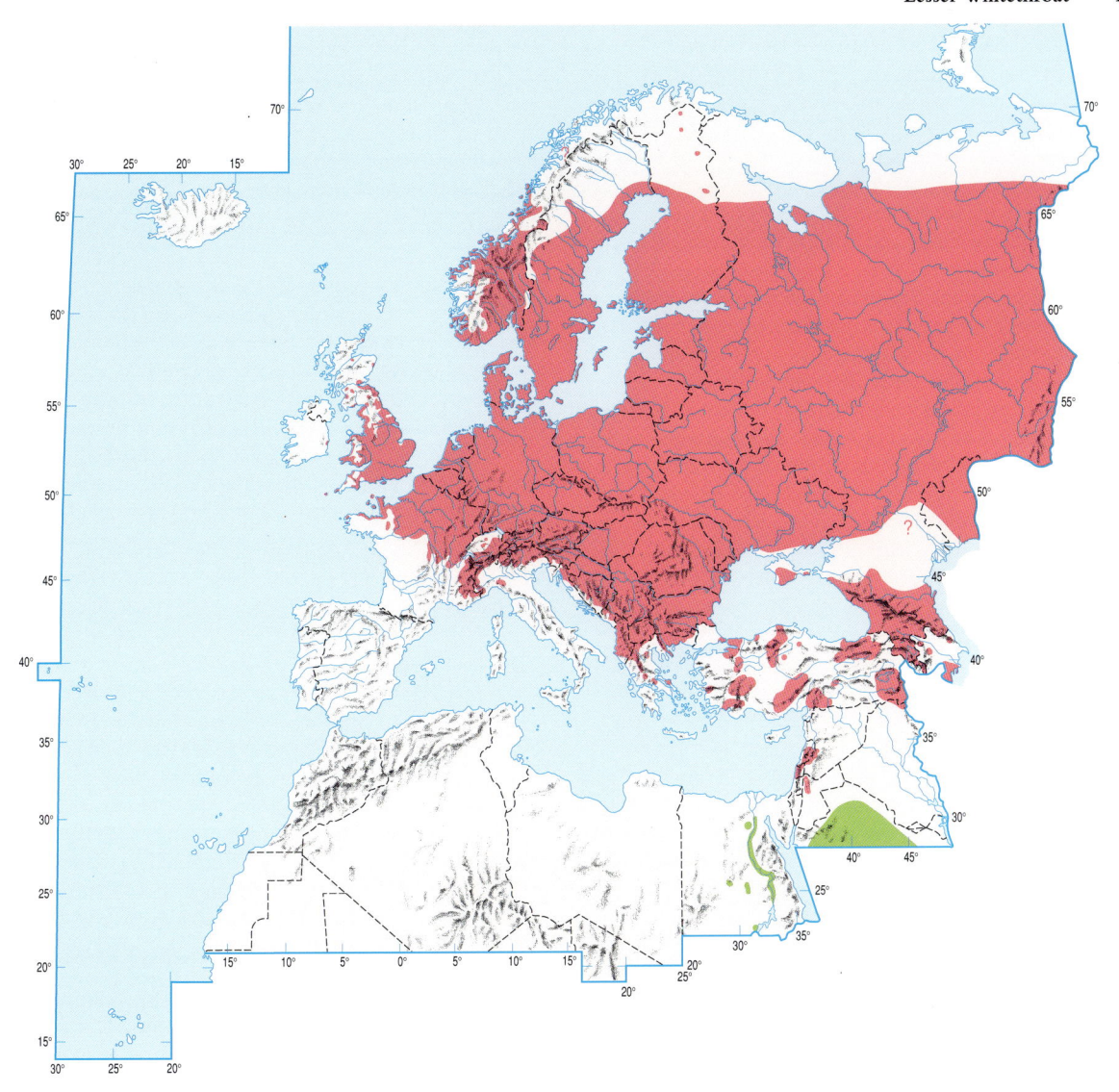

TURKEY. Widespread in hills and mountain valleys, but often local. SYRIA. Probably breeds in south-west, but not confirmed. JORDAN. Breeding first proved 1990. IRAQ. Breeds Ser Amadiya mountains.

Accidental. Iceland, Madeira. Vagrant (or perhaps rare migrant) Iberia, Tunisia, Algeria, Morocco.

Beyond west Palearctic, extends east to Lena river and north-central China, south to Iran and Pakistan.

Population. BRITAIN. 80 000 territories 1988–91; marked fluctuations, but no clear trend. FRANCE. Probably not more than 10 000 pairs; slight increase. BELGIUM. 12 000 pairs 1973–7; fluctuates. LUXEMBOURG. 3000–5000 pairs. NETHERLANDS. 15 000–30 000 pairs 1979–85; apparently decreasing. GERMANY. 516 000 pairs in mid-1980s. DENMARK. 26 000–275 000 pairs 1987–8; recent decrease. NORWAY. 10 000–100 000 pairs 1970–90; stable. SWEDEN. 150 000–400 000 pairs in late 1980s; stable. FINLAND. 200 000–400 000 pairs in late 1980s; slight decrease. ESTONIA. 20 000–50 000 pairs in 1991; stable or slight decrease. LATVIA. 40 000–80 000 pairs in 1980s; stable. LITHUANIA. Rare in north-east, abundant in central part, common in south. Decreasing. POLAND. 50 000–200 000 pairs; stable. CZECH REPUBLIC. 50 000–100 000 pairs 1985–9; slight decrease. SLOVAKIA. 40 000–80 000 pairs 1973–94; stable. HUNGARY. 60 000–80 000 pairs 1979–93; stable. AUSTRIA. Uncommon but widespread. SWITZERLAND. 2000–4000 pairs 1985–93; stable. ITALY. 10 000–40 000 pairs 1983–95; stable. GREECE. 2000–5000 pairs. ALBANIA. 1000–2000 pairs in 1981. YUGOSLAVIA: CROATIA. 3000–5000 pairs; stable. SLOVENIA. 5000–8000 pairs; slight increase. BULGARIA. 1000–10 000 pairs; stable. RUMANIA. 400 000–500 000 pairs 1986–92. RUSSIA. 100 000–1 million pairs; stable. BELARUS'. 140 000–150 000 pairs in 1990; stable. UKRAINE. 85 000–100 000 pairs 1988; slight decrease.

Moldova. 12 000–15 000 pairs 1986; slight decrease. Azerbaijan. Locally common. Turkey. 1000–10 000 pairs. Israel. Locally quite common in north, with a few in central areas. Jordan. Scarce.

Movements. All populations migratory, wintering in arid country south of the Sahara from the upper Niger east to Sudan and Eritrea, also in Egypt and Arabia. In autumn, European populations west of *c.* 30°E take heading between ESE and SSE (reverse in spring), reaching winter quarters via eastern Mediterranean. Thus, in France, not generally encountered west of breeding range. In Britain, departures and arrivals are concentrated on eastern and (especially) southern coasts.

Southward movement begins mid-July in Britain, with peak departures from coast at end of August and beginning of September. Switzerland traversed mostly late August to late September. In transit through Israel and Sinai mainly August–October. Reaches western Sudan and Chad from mid-October.

Northward movement starts early and is prolonged. Passage through Ethiopia and Eilat (southern Israel) begins late January, continuing to late April or early May. Recorded in southern Turkey from early March, and passage through central Europe late March to May. Peak arrival on south coast of Britain late April to early May, and in Fenno-Scandia and northern Russia not until mid- or late May.

Food. Chiefly invertebrates; also berries in late summer and autumn. Forages mostly in bushes and trees, taking insects from leaves, twigs, and bark. Records of drinking nectar and eating anthers of flowers in spring indicate that this is regular source of energy, especially for birds on migration.

Social pattern and behaviour. Somewhat gregarious outside breeding season; apparently not territorial in winter quarters. No evidence for other than monogamous mating system. Sexes share incubation, brooding, and feeding of young. ♂ sings usually from thick cover, often while moving about, or at intervals while moving considerable distance, often under cover, along hedgerows, exceptionally in flight. Song heard from arrival on breeding grounds until mid- (or late) July, though a few sporadic phrases may be heard until mid-September following moult, exceptionally early October. Song intensity higher when ♀♀ arrive on breeding grounds; highest during nest-building; rare after laying. About 1 week after arrival of ♂, and typically before arrival of ♀ in territory, ♂ builds cock nests (basic structure minus lining) which ♀ subsequently inspects; if she accepts nest, both birds then share in its completion.

Voice. Song of ♂ distinctive, and in 2 parts, though these not invariably in close sequence. A loud wooden rattling repetition of double unit (i.e. a trill), e.g. 'chikka-chikka-chikka-chikka-chikk', often, but not always, preceded (and rarely followed) by short, quiet, varied, musical warbling or chattering somewhat reminiscent of Whitethroat; warbling introduction audible only at close quarters and sometimes combined with very high-pitched sounds. Trill carries over 200 m and at a distance is the only audible part of song. Commonest call a hard, low 'tuk', usually in loose irregular series; also a hoarse, grating churr, and in extreme alarm a nasal rattling 'trrrrrrrrr', very similar to Wren.

Breeding. Season. North-west Europe: egg laying from early May. Central Europe: up to 2 weeks earlier. Finland: eggs laid 2nd half of May to end of June, peaking early June. Normally 1 brood, but 2 recorded. Site. In bushes and small trees, and occasionally perennial herbs, especially those with thorns, fine leaves (e.g. evergreens), or suckers on trunk, facilitating attachment of nest. From close to ground to 2.5 m above it. Nest: deep cup of grass and herb stems and leaves, usually with small twigs and rootlets, plus moss and spiders' webs and cocoons, lined with finer grass and rootlets, hair, and some plant down. Eggs. Sub-elliptical, smooth and glossy; white to creamy-white, sparsely marked with olive or olive-buff and grey speckles, spots, and blotches, often denser at broad end. Clutch: 4–6 (2–7). Incubation. 11–14 days. Fledging Period. 10–13 days.

Wing-length: nominate *curruca*, ♂ 64–69, ♀ 63–67 mm.
Weight: ♂♀ mostly 10–14 g; migrants to 19.5 g.

Geographical variation. Complex, and species limits in east not definitely established. Within west Palearctic, nominate *curruca* occupies most of range; *S. c. caucasica*, darker and greyer and slightly smaller, breeds in Balkans and Asia Minor. Birds belonging to extralimital eastern races are likely to occcur on migration in south-east of west Palearctic.

Whitethroat *Sylvia communis*

PLATES: pages 1302, 1317

Du. Grasmus   Fr. Fauvette grisette   Ge. Dorngrasmücke   It. Sterpazzola
Ru. Серая славка   Sp. Curruca zarcera   Sw. Törnsångare

Field characters. 14 cm; wing-span 18.5–23 cm. Close in size to Blackcap but with more peaked crown and proportionately longer tail giving distinctively more attenuated shape. Medium-sized, quite slim but rather large-headed and long-tailed, perky warbler. Epitome of west Palearctic *Sylvia*, nearly all of which have distinctive ♂♂ but rather similar ♀ and immature plumages. Crown and cheeks of ♂ grey, of ♀ and immature brown; both sexes and immature share white throat, rufous fringes to inner wing feathers (forming distinctive panel) and white outer tail-feathers. At close range, shows whitish eye-ring and pale brown eye, giving irascible expression.

Commonest *Sylvia* in north European farmland. Throughout north European range, separation from typically smaller, shorter-tailed, and less colourful Lesser Whitethroat not difficult. In southern parts of range, confusion possible with ♂ Spectacled Warbler and ♀♀ and immatures of Spectacled Warbler,

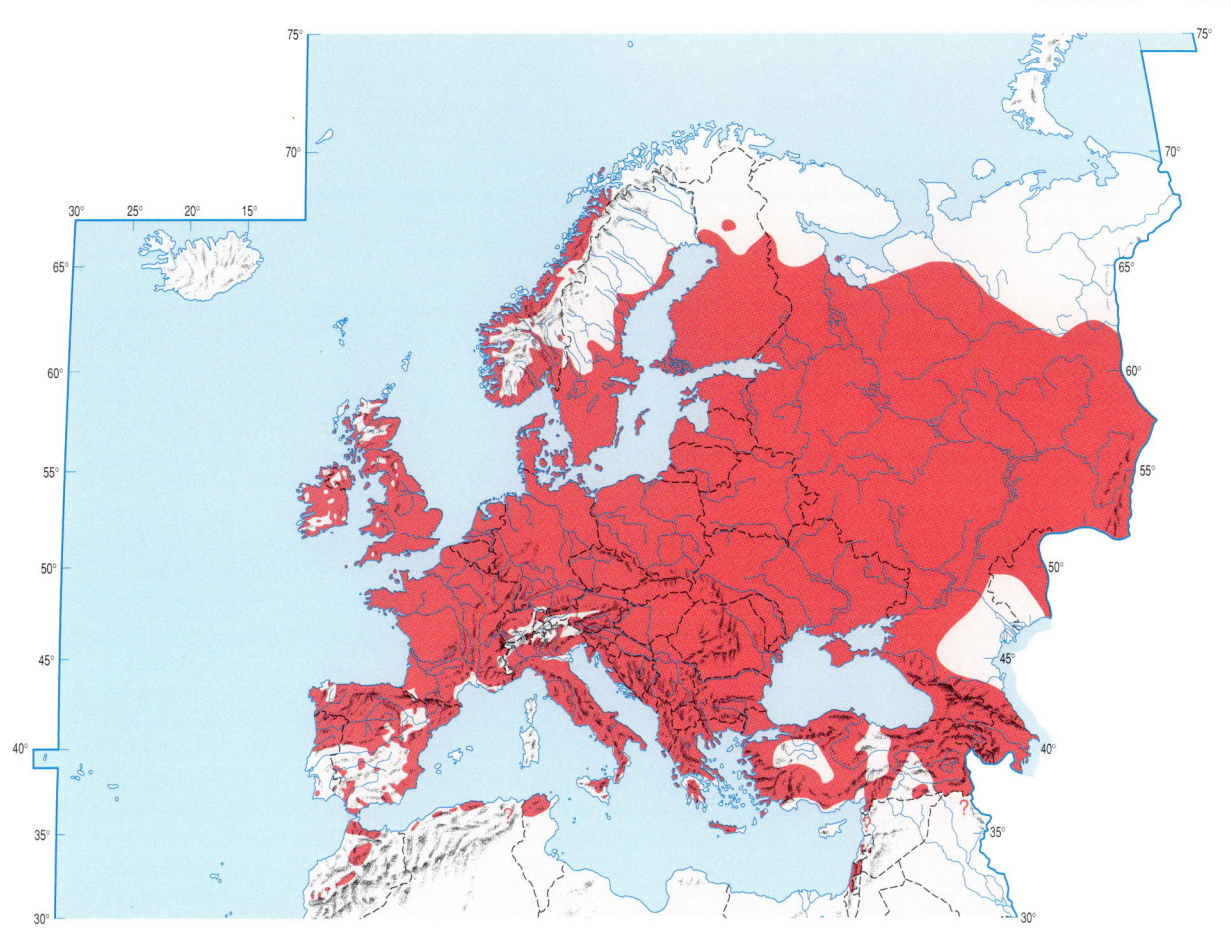

Tristram's Warbler, and Subalpine Warbler; all are noticeably smaller and slighter than Whitethroat but have similarly-marked wing and tail. Flight action not constant, strong impression of flitting and undulating, then darting, jerky, and jinking progress often accompanied by flirts or fans of long tail. Gait essentially hopping or sidling, allowing easy and fast progress through even tangled vegetation. Carriage noticeably horizontal, with head usually held up and tail only occasionally below horizontal and often held up (though full cocking or flirting normally restricted to interactive and sexual behaviour). Quite tame and curious, investigating slight disturbances and imitations of its call.

Habitat. Breeds over continental and oceanic west Palearctic in upper middle to lower middle latitudes, from boreal through temperate to steppe and Mediterranean zones, within July isotherms $c.$ 14–32°C; mainly in lowlands, but in Switzerland up to 1300 m or even 1500 m and in Caucasus to subalpine zone at 1500–2000 m. Avoids tall closed forest and densely vegetated wetland, requiring ample but discontinuous well-mixed and open cover of tall herbage, low bushes, and shrubs, usually on more or less dry, level or gently sloping, and fairly sunny terrain, sometimes by water or in marshy or fen areas, or in open woodland glades and edges, more often in woods of broad-leaved than of coniferous trees. Often occurs amid cultivation and sometimes in subalpine scrub or on moors or cliff slopes.

Distribution. No evidence of overall change, though drastic reduction in population has resulted in local disappearances. BRITAIN. Marked retreat from many northern areas. First bred Isles of Scilly 1965; sporadic in Outer Hebrides, Orkney, and Shetland. Highest numbers in eastern England. IRELAND. Strong decrease. SYRIA. Occasional records of breeding and probable breeding. Reported to be common throughout summer in Allovit mountains. MOROCCO. Range apparently decreasing in Rif and Moyen Atlas.

Accidental. Iceland, Madeira.

Beyond west Palearctic, extends east to Lake Baykal, south to northern Iran.

Population. Decline of 50–100% in different parts of western and central Europe from mid- or late 1960s to 1970s (especially 1969), due to drought in Sahel zone of Afrotropics. Some countries report continuing declines. BRITAIN. 660 000 territories 1988–91. Major crash of $c.$ 80% from 1969. Now fluctuating around new lower level. IRELAND. 20 000–40 000 pairs; much reduced. FRANCE. Under 1 million pairs in 1970s. BELGIUM. 29 000–50 000 pairs 1973–7. Slight increase in 1980s, but now tending to decline. LUXEMBOURG. 7000–9000 pairs. NETHERLANDS. 80 000–95 000 pairs 1989–91. Some recovery following strong decline. GERMANY. 326 000 pairs in mid-1980s; only a fraction of former population. In east, apparently stable at new lower level in recent years. DENMARK. 140 000–1.6 million pairs 1987–8; stable 1976–94. NORWAY. 50 000–300 000 pairs 1970–90; stable. SWEDEN. 500 000–1 million pairs in late 1980s; slight increase. FINLAND. 250 000–400 000 pairs in late 1980s; slight increase, but fluctuates. ESTONIA. 100 000–200 000 pairs in 1991; probable recent increase. LATVIA. 300 000–500 000 pairs in 1980s; stable. LITHUANIA. Common; decreasing. POLAND. 400 000–1 million pairs; stable. CZECH REPUBLIC. 90 000–180 000 pairs 1985–9; slight decrease. SLOVAKIA. 60 000–120 000 pairs 1973–94; stable. HUNGARY. 50 000–70 000 pairs 1979–93; stable. AUSTRIA. Still fairly common in extra-alpine area; locally marked decline. SWITZERLAND. 800–1000 pairs 1985–93; decrease. SPAIN. 450 000–600 000 pairs; slight decrease. PORTUGAL. 1000–10 000 pairs 1978–84; stable. ITALY. 50 000–200 000 pairs 1983–95; stable. GREECE. 100 000–200 000 pairs. ALBANIA. 5000–20 000 pairs in 1981; slight decrease. YUGOSLAVIA: CROATIA. 30 000–50 000 pairs; stable. SLOVENIA. 2000–3000 pairs; marked decrease. BULGARIA. 10 000–100 000 pairs; stable. RUMANIA. 600 000–1 million pairs 1986–92. RUSSIA. 1–10 million pairs; stable. BELARUS'. 400 000–450 000 pairs in 1990; stable. UKRAINE. 350 000–370 000 pairs in 1988; slight increase. MOLDOVA. 20 000–30 000 pairs in 1986; stable. AZERBAIJAN. Common. TURKEY. 50 000–500 000 pairs. ISRAEL. Fairly common. TUNISIA, ALGERIA. Uncommon. MOROCCO. Scarce, apparently declining.

Movements. All populations migratory, wintering in sub-Saharan Africa, from Sénégal east to Ethiopia and south to South Africa.

Initial autumn heading from continental Europe shows weakly expressed migratory divide: from west of 10°E, birds head west of south to Iberia; east of this, heading mainly south or east of south. Autumn migration begins late July, with main movement through central Europe August to mid-September, and stragglers to early October. Earliest records in Sénégal end of August, with main arrival a month later. Reaches Zimbabwe usually after mid-November, with main arrival December.

Spring migration from Africa begins March. Reaches Gibraltar from mid-March, Camargue and Sicily from early April. Arrivals tend to be earlier in western than central Europe. Reaches Britain from early or mid-April, apparently on broad front, peaking early May in south-west, mid-May elsewhere; central Sweden from mid-May; Leningrad region from 2nd week of May, passage continuing to mid-June.

Food. During breeding season mainly insects, especially beetles (Coleoptera), larvae, and bugs (Hemiptera); in late summer, proportion of fruit taken increases, and on autumn migration and in winter quarters feeds predominantly on berries. Food obtained in bushes and herb layer by searching foliage and small branches.

Social pattern and behaviour. Somewhat gregarious in immediate post-breeding period and on migration; in winter quarters essentially solitary. Pair-bond essentially monogamous but some ♂♂ polyterritorial, practising simultaneous bigamy. Both sexes incubate, brood, and feed young. In cases of 2nd broods, ♀ alone incubates 2nd clutch (started immediately after 1st brood fledges) while ♂ rears 1st brood to independence, thereafter helping mate to rear 2nd brood. ♂ sings mostly from dense cover, also, when excited and especially when unpaired, from exposed perch, occasionally in normal flight between song-posts. Especially in excitement, ♂ often performs song-flight: rises from song-post to height of 1–10 m, raising crown and spreading tail, before making, with exaggerated wing-beats, graduated descent in a series of jerky swoops, finally often with headlong plummet, to starting point or another perch some distance away. ♂♂ silent on first arrival on breeding grounds, and subsequently also if they fail to win a territory; start singing as soon as territories established. Once paired, ♂ ceases or reduces song; unpaired ♂♂ keep singing until July or August. Before or after ♀ arrives, ♂ may build several cock nests. Cock nests are flimsy, unlined structures, the outer framework often adorned up to rim with, variously, plant down and fibres, petals, and spider cocoons. ♂ attracts ♀ to nest by constantly flying ahead of her, turning back, and flying on again, singing persistently. If ♀ accepts a cock nest, she strips it of decoration, then completes it; or ♀ may also choose her own site and both then help to build.

Voice. Song a short, brisk, lively, rapidly-uttered warble of rather poor quality, though some more musical than others, e.g. roughly 'cheechiwee-cheechiweechoo-chiwichoo'. Usually begins with a drawn-out unit, followed by alternately high and low-pitched units, becoming more variable towards end and not uncommonly finishing with a short mimicked sound; each ♂ has several different phrases, but often repeats same phrase. In song-flight, song especially loud and protracted, up to 10 s long; also more flowing, and enriched with mimicry. Calls include a sharp 'tack tack', indicating alarm; a conversational 'wheet'; and, characteristic and most often heard when scolding or warning, a croaking 'churr' or 'charr'.

Breeding. SEASON. North-west Europe: eggs from 2nd half of April. Finland: eggs laid 3rd week of May to 2nd week of July, peaking 1st half of June. Southern Germany: eggs laid late April to mid-July. North Africa: eggs found mid-April to July. Usually 1 brood, sometimes 2; 2 perhaps normal in south. SITE. In low bush or shrub, or tall grass, or herbs. Nest: cup-shaped, usually quite deep, sometimes hemispherical, occasionally distorted probably by nestlings to give elliptical rim, constructed of grass and herb stems and leaves with some roots, plant down, and cobwebs, lined with finer grasses and rootlets and long hair. EGGS. Sub-elliptical, smooth and glossy; variable, usually very pale blue or green, rarely white, with fine green to olive and dark grey spots, speckles, and blotches, sometimes very sparse. Clutch: 4–5 (1–7). INCUBATION. Usually 11–12 days. FLEDGING PERIOD. 10–12 days.

(FACING PAGE) Melodious Warbler *Hippolais polyglotta* (p. 1284): **1** pale juv.
Garden Warbler *Sylvia borin* (p. 1314). *S. b. borin*: **2** ad, **3** 1st winter, **4** juv. *S. b. woodwardi*: **5** 1st winter.
Blackcap *Sylvia atricapilla* (p. 1316). *S. a. atricapilla*: **6** ad ♂, **7** ad ♀, **8** 1st winter ♂, **9** juv ♂, **10** juv ♀. *S. a. dammholzi*: **11** ad ♂. *S. a. pauluccii*: **12** ad ♀. *S. a. heineken*: **13** ad ♂, **14** ad ♀, **15** melanistic ad ♂.

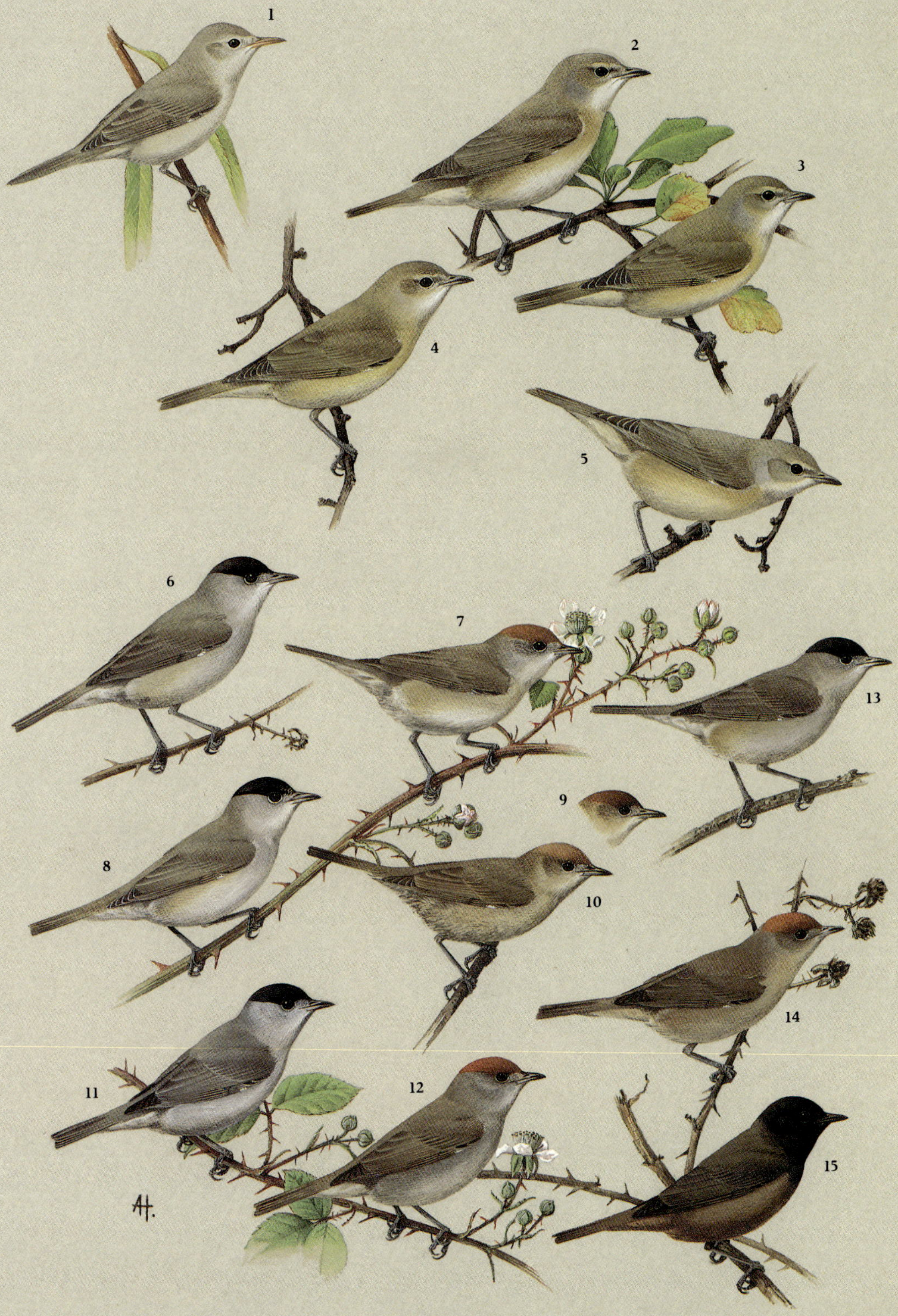
1
2
3
4
5
6
7
13
9
8
10
14
11
12
15
AH.

Wing-length: ♂ 70–77, ♀ 68–75 mm.
Weight: ♂♀ mostly 13–18 g; migrants to 25 g.

Geographical variation. Slight, but complex and clinal. 3 races recognized in west Palearctic: nominate *communis* in greater part of range; *S. c. volgensis* (paler and greyer) from eastern Poland, Hungary and Bulgaria eastwards; *S. c. icterops* (with darker upperparts) in Turkey, Caucasus, and Levant.

Garden Warbler *Sylvia borin*

PLATES: pages 1313, 1317

Du. Tuinfluiter   Fr. Fauvette des jardins   Ge. Gartengrasmücke   It. Beccafico
Ru. Садовая славка   Sp. Curruca mosquitera   Sw. Trädgårdssångare

Field characters. 14 cm; wing-span 20–24.5 cm. Slightly larger and plumper than Blackcap. Medium-sized to large warbler, with rather stubby bill, somewhat domed head, plump body, rather long wings, and full but not long tail. Plumage plain dull brown above and pale buff below; lacks any prominent character but dark eye emphasized by short pale buff to greyish fore-supercilium and eye-ring. Sexes similar.

A plague to all inexperienced observers, since its appearance (though not its general character, once this is learnt) can suggest not only other *Sylvia* but also medium-sized *Acrocephalus* (particularly less rufous, more olive species) and *Hippolais* (particularly brown-grey species and brown variants of green and yellow species). With practice, rather plump, compact, usually demure character, with gentle facial expression unlike any other medium-sized warbler, is as distinctive as near-uniform coloration. Flight confident, fast, and generally straight. Lacks nervousness of most other *Sylvia*; no wing- or tail-movements except in excitement.

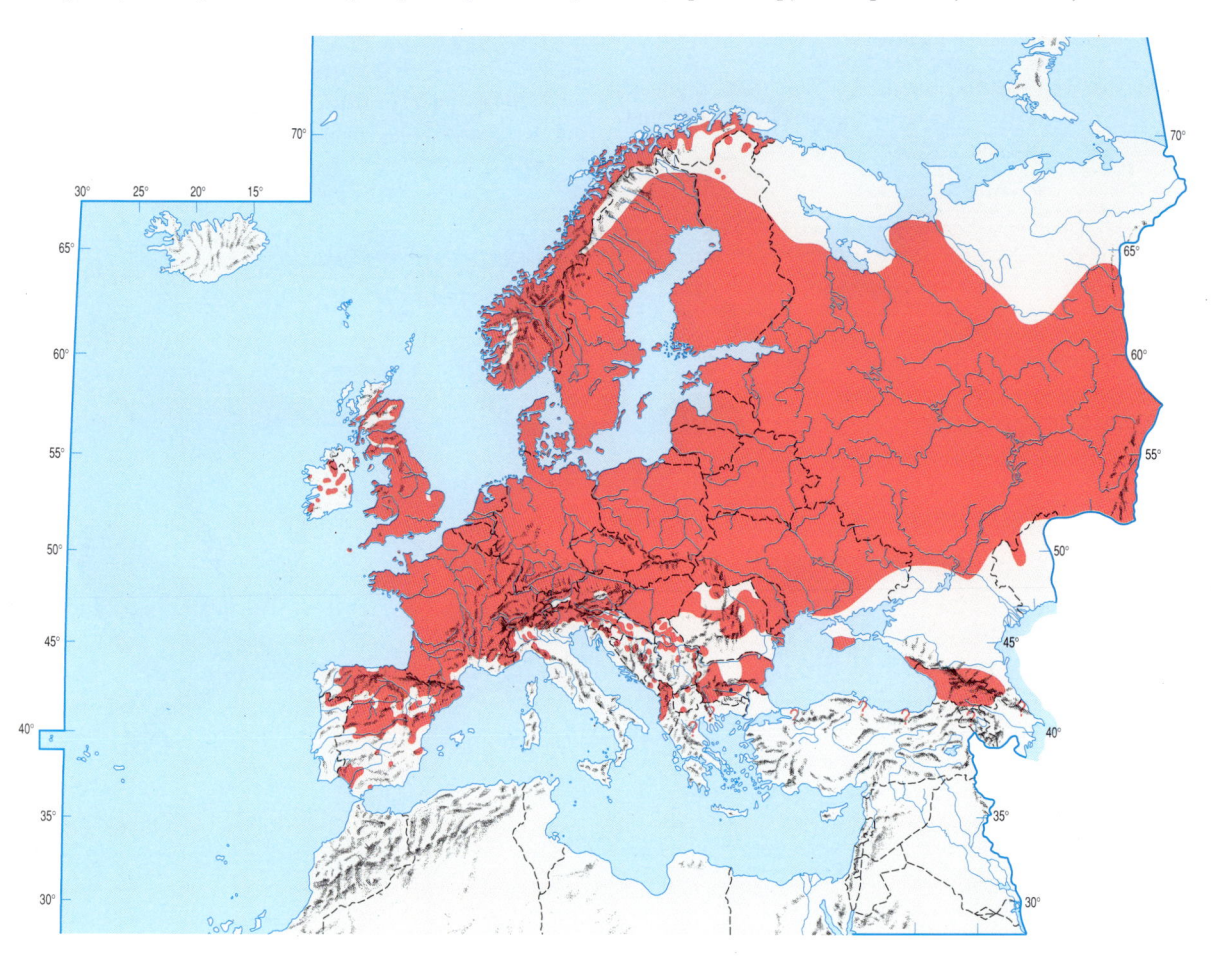

Habitat. Breeds in middle and upper middle latitudes of west Palearctic in continental and marginally oceanic, mainly temperate climate within July isotherms 12–28°C, thus tolerating cooler conditions than any other *Sylvia* in the region. In contrast to Lesser Whitethroat and Whitethroat, primarily a woodland bird, but in comparison with Blackcap generally prefers a more even, open canopy accompanied by much fairly dense and tall scrub or a shrub layer. Thus occurs mainly on woodland fringes, in glades and regrowth in clearings, or in shrubby growth on hillsides and along streams. In Britain, coppice woods, young conifer plantations, osier beds, and rhododendron thickets are sometimes occupied. In Switzerland, preference for shrubby clusters of green alder enables breeding locally above 2000 m, in common with Lesser Whitethroat. Unlike Blackcap, put off by disturbance; English name thus misleading.

Distribution. FAEROES. Bred 1948 and 1979. BRITAIN. Recent expansion northwards; now breeding sporadically in most north-east and highland counties of Scotland. Bred Orkney 1964. Highest numbers in Wales and southern England. GREECE. Perhaps only occasional breeder. TURKEY. Status inadequately known; evidently breeds rarely in Black Sea coastlands and perhaps in east, but confirmation required.

Accidental. Spitsbergen, Bear Island, Iceland (annual), Madeira.

Beyond west Palearctic, extends east through western Siberia to Yenisey river.

Population. Stable in much of range. BRITAIN. 200 000 territories 1988–91; continuing recovery since decline in early 1970s. IRELAND. 180–300 pairs 1988–91. FRANCE. 2–3 million pairs. BELGIUM. 75 000 pairs in 1980s; stable. LUXEMBOURG. 15 000–18 000 pairs. NETHERLANDS. 100 000–200 000 pairs 1979–85. GERMANY. 1.6 million pairs in mid-1980s. DENMARK. 27 000–380 000 pairs 1987–8. NORWAY. 200 000–700 000 pairs 1970–90. SWEDEN. 1–3 million pairs in late 1980s; slight increase. FINLAND. 1–2 million pairs in late 1980s. ESTONIA. 200 000–500 000 pairs in 1991; increasing throughout 20th century. LATVIA. 500 000–700 000 pairs in 1980s. LITHUANIA. Decreasing; rarer in south. POLAND. 200 000–400 000 pairs. CZECH REPUBLIC. 200 000–400 000 pairs 1985–9; slight increase. SLOVAKIA. 200 000–400 000 pairs 1973–94. HUNGARY. 5000–10 000 pairs 1979–93. AUSTRIA. Fairly common. SWITZERLAND. 140 000–170 000 pairs 1985–93. SPAIN. 400 000–700 000 pairs. PORTUGAL. 100–1000 pairs 1978–84. ITALY. 10 000–50 000 pairs 1983–95. ALBANIA. 500–2000 pairs in 1981. YUGOSLAVIA: CROATIA. 1000–2000 pairs. SLOVENIA. 3000–5000 pairs. BULGARIA. 200–2000 pairs. RUMANIA. 30 000–50 000 pairs 1986–92. RUSSIA. 500 000–5 million pairs. BELARUS'. 410 000–430 000 pairs in 1990. UKRAINE. 55 000–65 000 pairs in 1988. MOLDOVA. 5500–7000 pairs in 1986; slight decrease.

Movements. All populations migratory, wintering extensively in Africa south of c. 10°N in west and 3°N in east, south to South Africa. Within Europe, autumn heading chiefly south to southwest (and reverse in spring): to Iberia from western and central Europe (including Britain and western Scandinavia), to Italy from Finland and north-east Europe. Prolonged migration at both seasons. In Britain, movement south begins mid-July, larger number leaving from mid-August, with peak (including drifted continental birds) early September at eastern observatories, but later in west. Passage through Switzerland chiefly August, adults continuing to mid-September, juveniles markedly later, to mid-October. Prolonged passage early August to early November (mostly late August to early October) on Malta. Arrives in northern parts of winter quarters from mid-September, in southern Africa not until December. Leaves winter quarters from February. Passage through Mediterranean area mainly in early May; arrival at breeding grounds also mainly early May in south and central Europe, late May or early June in north.

Food. Chiefly insects in breeding season, mainly fruit at other times. In breeding season no significant differences from diet of Blackcap. Searches vegetation to pick insects from foliage and twigs, sometimes hovering or (less often) sallying out to catch flying insects. Like Blackcap, forages mainly in shrub layer below 6 m, but also makes sorties to canopy (up to 20 m), e.g. when aphids or larval Lepidoptera abundant there; seldom uses herb layer.

Social pattern and behaviour. Usually solitary outside breeding season, but sometimes (including on autumn migration) gathers in small parties to feed on fruit. Shows regular site fidelity in winter quarters, but no evidence of territorial defence. Mating system mainly monogamous, but ♂♂ occasionally practise successive polygyny. Both members of pair incubate (♂ less than ♀), brood, and feed young. Territories spaced out in suboptimal habitat, contiguous in preferred habitat; in best habitats may be very small, c. 0.2 ha. In areas where both occur, competes for space with Blackcap, and territories of the two species at least partly mutually exclusive. ♂ sings typically from low cover of trees or bushes, rarely from prominent perch, and often while moving about and feeding; occasionally sings in flight, especially when following ♀, or when approaching intruder on territory, but no ritualized song-flight. Song begins soon after arrival on breeding grounds; most frequent during pair-formation and nest-site selection, declines during nest-building, and little during incubation and nestling periods. In winter quarters, song (especially low-intensity type) given in all months but especially from start of spring passage when may be full or nearly so. ♂ builds rudimentary cock nests and displays them to ♀, who usually selects one and completes it in c. 3 days.

Voice. Song a sweet, even-flowing warble, more subdued and more sustained than Blackcap, though more subdued song of that species may be difficult or almost impossible to distinguish. May be sustained, with short pauses, for long periods. Occasionally includes accomplished mimicry of other species. Commonest call a hard 'vik', often abruptly repeated as disyllable and given by both sexes in various contexts ranging up to full alarm and distinct from sharper 'tak' call of Blackcap. Other calls include a low grating 'churrr', and a soft ascending 'duij' given in high-intensity alarm.

Breeding. SEASON. North-west Europe: eggs from early May. Southern Germany: eggs laid end of April to 2nd half of July. Southern Finland: eggs laid end of May to 2nd half of July peaking 1st half of June. Usually 1 brood; occasionally 2. SITE. In low tree or bush (especially bramble) or tall herbs (e.g. nettles, willowherb). 0–2 m above ground. Nest: substantial cup, loosely constructed on outside and underneath, of grass and herb stems and leaves, with some twigs and roots, up to 30 cm long, plus a little plant down and cobwebs, lined with finer grasses and rootlets and long hair. Eggs. Sub-elliptical, smooth and glossy; very variable, from white to buff, pink or green tinted, spotted and blotched (profusely or sparsely) buff, brown, olive, red-brown, grey, and purple, larger pale blotches often with dark centre, other spots with blurred edges, markings occasionally concentrated at broad end. Clutch: 4–5 (2–6). Incubation. Typically 11–12 days. Fledging Period. 10 days (9–12).

Wing-length: ♂ ♀ (no difference) 76–82 mm.
Weight: ♂ ♀ mostly 16–22 g; migrants to 35.5 g.

Geographical variation. Slight and clinal. 2 races recognized in west Palearctic: nominate *borin* in most of range; *S. b. woodwardi* (paler and somewhat larger) from White Sea (northern Russia) eastwards.

Blackcap *Sylvia atricapilla*

PLATES: pages 1313, 1317

Du. Zwartkop   Fr. Fauvette à tête noire   Ge. Mönchsgrasmücke   It. Capinera
Ru. Славка-черноголовка   Sp. Curruca capirotada   Sw. Svarthätta

Field characters. 13 cm; wing-span 20–23 cm. Size and structure close to Garden Warbler, but with flatter crown (appearing so partly due to plumage pattern) and somewhat slimmer vent and tail. Quite large and robust warbler, with rather long wings and legs but rather short tail. Essentially dusky-brown above, pale grey below; lacks white on tail-edges, and marked only by diagnostic short cap—black on ♂, brownish on ♀ and immature.

Unmistakable when small cap seen, but otherwise can be confused with similarly-sized Garden Warbler and other smaller and larger dark-capped *Sylvia*; however, Garden Warbler lacks any such marking and all other dark-capped warblers have white outer tail-feathers (and larger cap or wholly dark crown and face). Flight as Garden Warbler, with virtually identical silhouette. Stance rather more upright (especially singing ♂) and altogether less skulking than Garden Warbler.

Habitat. Breeds throughout middle latitudes of west Palearctic in temperate, boreal, and Mediterranean climates, oceanic as well as continental, within July isotherms 14–30°C. Highly arboreal, preferring to forage and sing in crowns of trees, often in more or less mature forest, although requiring also tall, not too dense shrubby undergrowth, especially for nesting. Coniferous woodland largely avoided. Compared with Lesser Whitethroat, Whitethroat, and Garden Warbler, Blackcap is found markedly more often in parkland, suburbs, and even in towns where enough tall trees with undergrowth occur.

Distribution. Faeroes. Bred 1985. Britain. Has spread northward, though still scarce and probably irregular in most of Scottish Highlands. Strong and continuing increase from at least 1950s. Ireland. Has spread west and south. Netherlands, Denmark. Recent expansion. Israel. Breeds irregularly and very rarely in north. Syria. Summer records suggest possible breeding. Algeria. May also breed in northern oases. Canary Islands. Breeds Gran Canaria, Tenerife, La Gomera, La Palma, and El Hierro; perhaps also Fuerteventura and Lanzarote. Cape Verde Islands. Breeds on all main islands except Sal, Boavista, and Maio; not reported on São Vicente since late 1960s.

Accidental. Bear Island, Jan Mayen, Iceland (annual).

Beyond west Palearctic, breeds northern Iran, and in west Siberia east to Ob' valley.

Population. Increasing Britain, Ireland (probably), France, Low Countries, Germany, Denmark, Sweden, Estonia, Czech Republic, Spain, Croatia, Ukraine; decrease Moldova and probably Finland. Elsewhere stable. Britain. 580 000 territories 1988–91. Ireland. 40 000 territories 1988–91. France. Over 1 million pairs in 1970s. Belgium. 63 000–130 000 pairs 1973–7. Luxembourg. 15 000–18 000 pairs. Netherlands. 70 000–120 000 pairs 1979–85. Germany. 3.3 million pairs in mid-1980s. In east, *c.* 450 000 pairs in early 1980s. Denmark. 27 000–370 000 pairs 1987–8. Norway. 200 000–700 000 pairs 1970–90. Sweden. 300 000–700 000 pairs in late 1980s. Finland. 50 000–80 000 pairs in late 1980s. Estonia. 100 000–200 000 pairs in 1991. Latvia. 350 000–500 000 pairs in 1980s. Lithuania. Abundant; rarer in south-east. Poland. 500 000–1.5 million pairs. Czech Republic. 600 000–1.2 million pairs 1985–9. Slovakia. 800 000–1 million pairs 1973–94. Hungary. 500 000–800 000 pairs 1979–93. Austria. Very common. Switzerland. 200 000–300 000 pairs 1985–93. Spain. 850 000–1.5 million pairs. Portugal. 100 000–1 million pairs 1978–84. Italy. 2–5 million pairs 1983–95. Greece. 100 000–200 000 pairs. Albania. 20 000–50 000 pairs in 1981. Yugoslavia: Croatia. 800 000–1 million pairs. Slovenia. 300 000–400 000 pairs. Bulgaria. 500 000–5 million pairs. Rumania. 1.2–2 million pairs 1986–92. Russia. 1–10 million pairs. Belarus'. 900 000–950 000 pairs in 1990. Ukraine.

(FACING PAGE) Marmora's Warbler *Sylvia sarda* (p. 1286): **1** ad ♂, **2** ad ♀.
Dartford Warbler *Sylvia undata* (p. 1288): **3** ad ♂, **4** ad ♀.
Spectacled Warbler *Sylvia conspicillata* (p. 1291): **5** ad ♀ breeding, **6** ad ♀ breeding.
Subalpine Warbler *Sylvia cantillans* (p. 1293): **7** ad ♂ breeding, **8** ♀ 1st winter.
Sardinian Warbler *Sylvia melanocephala* (p. 1296): **9** ad ♂, **10** ad ♀.
Desert Warbler *Sylvia nana nana* (p. 1301): **11** 1st winter.
Orphean Warbler *Sylvia hortensis* (p. 1305): **12** ad ♂.
Barred Warbler *Sylvia nisoria* (p. 1306): **13** ♀ 1st winter.
Lesser Whitethroat *Sylvia curruca curruca* (p. 1308): **15** ad worn (spring).
Whitethroat *Sylvia communis* (p. 1310): **16** ad ♀ breeding.
Garden Warbler *Sylvia borin* (p. 1314): **17** ad.
Blackcap *Sylvia atricapilla*: **18** ad ♂.

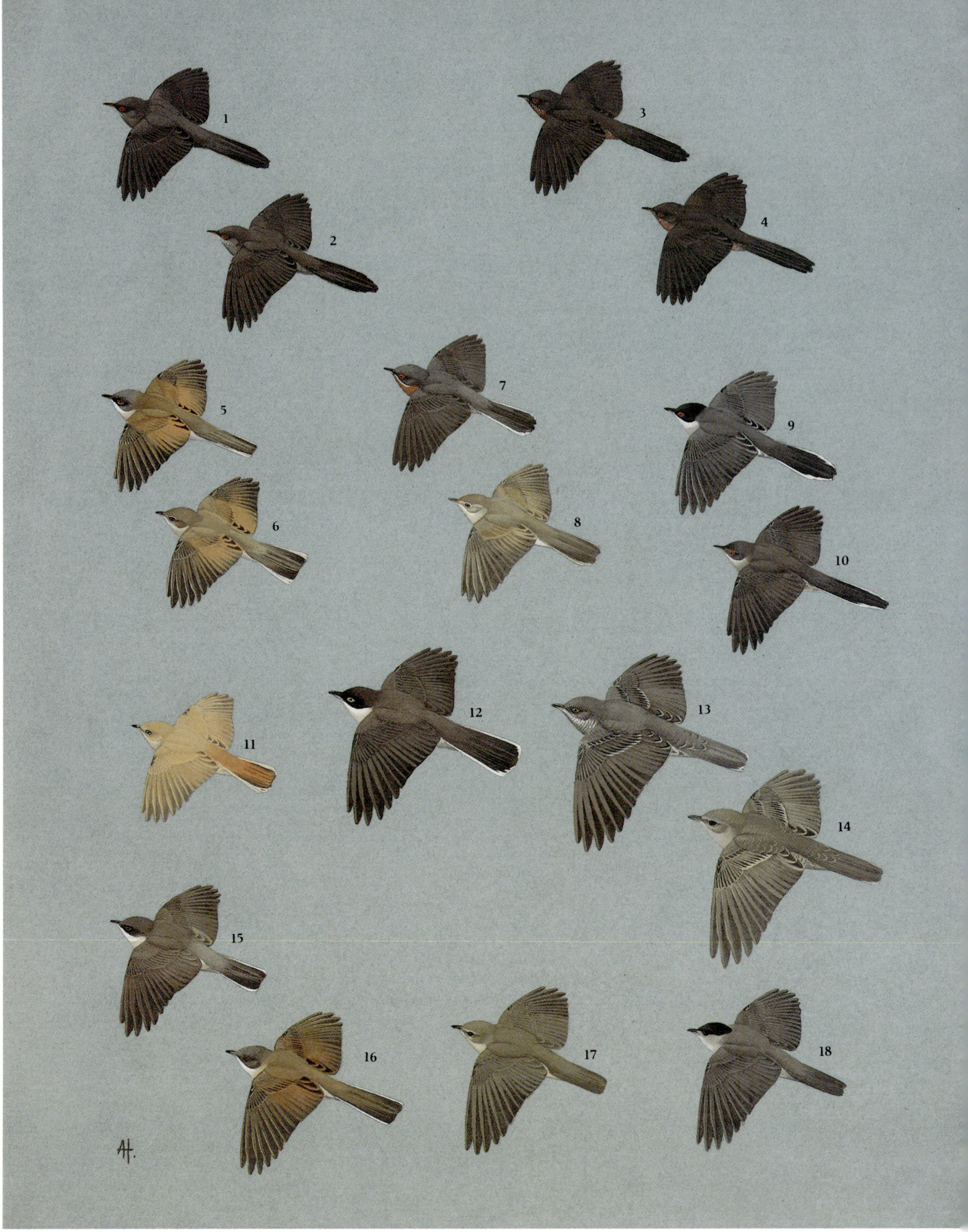
1
2
3
4
5
6
7
8
9
10
11
12
13
14
15
16
17
18
AH.

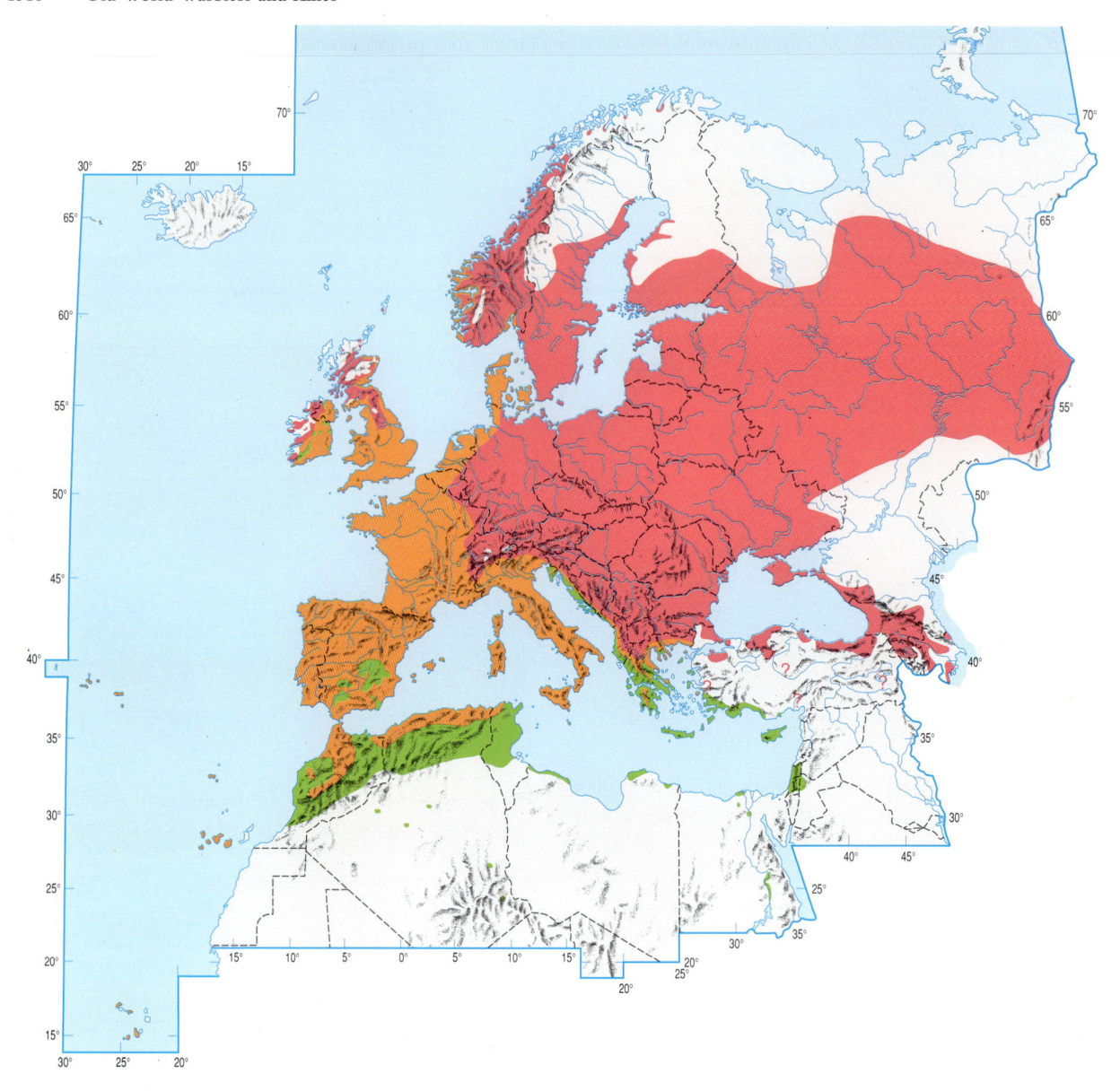

120 000–140 000 pairs in 1988. MOLDOVA. 60 000–80 000 pairs in 1986. AZERBAIJAN. Common. TURKEY. 50 000–100 000 pairs. TUNISIA. Common in north-west. ALGERIA. Locally common. MOROCCO. Scarce. AZORES. Common. MADEIRA. Very common. CAPE VERDE ISLANDS. Locally abundant.

Movements. Wide variety of strategies, populations from different parts of range varying from resident to migratory. Populations of Mediterranean and Atlantic islands chiefly resident or presumed resident. Northern and eastern birds wholly migratory, southern birds partially migratory, with most birds north of Mediterranean region leaving breeding area. 'Leap-frog' migrant: northern populations move longest distance, reaching south of winter range, and populations further south apparently move progressively less far. Winters within and south of breeding range, south to sub-Saharan Africa, north to Britain and south-west Norway. Main wintering areas in sub-Saharan Africa: West Africa west of Greenwich meridian; southern Sudan, Ethiopia, and Eritrea; equatorial East Africa south to Lake Nyasa. European populations show migratory divide, those west of 12°E heading chiefly south-west to southern France and Iberia, those east of 12°E chiefly south-east, funnelled from wide area towards Cyprus and Levant; also broad mixed area in central Europe and Scandinavia from which either south-west, south, or south-east heading is possible. Birds wintering in Britain are apparently all from continental Europe. Autumn migration begins chiefly in August. Northern birds leave earlier and migrate faster than southern ones; southern birds tend to leave after passage migrants have passed through. Main movement through northern and central Europe in September, diminishing through October. At British bird observatories, main passage starts late August, and most birds leave by end of September; arrival of winter visitors chiefly October. Spring migration begins early, with prolonged movement late February to May in Egypt and Levant, mid-February to mid-May in Strait of Gibraltar area. Earliest birds reach Britain late March, main arrival late April and early May. Earliest records in Helsinki region (Finland) and Leningrad region (north-west Russia) early May, usually from mid-May.

Food. Chiefly insects in breeding season, mainly fruit at other times. In breeding season, mainly picks insects from leaves and twigs, at heights of up to 20 m (canopy), not significantly different from feeding height of Garden Warbler. Feeding methods in winter more diverse, including ground foraging; also shows versatility in exploiting food from domestic sources, including bird-tables, in winter.

Social pattern and behaviour. Solitary or in small flocks outside breeding season. In East Africa, many birds forage (mainly for berries) in small mixed flocks, but later in winter some become territorial and establish regular song-posts. Birds wintering in Europe may remain sedentary for long periods where local food supply is suitable. Mating system essentially monogamous. Both sexes share incubation (more by ♀), brooding (more by ♀) and care of young. ♂ sings from cover of variable height, mostly 4.5–13 m. Typically has preferred song-post, used less after pairing; excited ♂ may also sing in flight, notably when approaching intruder on territory, but no ritualized song-flight. Migrants silent on first arrival on breeding grounds, but thereafter a lengthy song-period until early July or end of July, depending on region.

Voice. Song a sequence of clear rich warbling sounds, little slurred or discordant; fullest song, at start of breeding season, is especially intense, varied, and sustained (bouts of up to $2\frac{1}{2}$ min with gaps of only 1–2 s) and, at its best, has striking richness and beauty. Usually begins with chattering segment (jumble of harsh sounds) of varying length, followed by louder segment of pure fluting tones. Song of given individual tends to end in 1–2 regularly recurring patterns, such as 'rooty tooty rooty too'. Introductory segment commonly includes mimicry of various passerines, notably other warblers, thrushes, and finches. In localized, often isolated parts of western Europe (e.g. islands, peninsulas), full song regularly includes fluting variant consisting of 2 (1–3) notes differing in pitch and form, typically repeated 2 or more times, e.g. 'düdüdidüdidüdidü'; such songs often have little or no introduction. Commonest call throughout the year a hard scolding 'tacc' like 2 pebbles struck together, rapidly repeated when excited or alarmed. Other calls include a squeaky 'sweerr', given in alarm or as warning, and a churr similar to that given by Garden Warbler.

Breeding. SEASON. Western and north-west Europe: eggs from late April to early July. Finland: eggs laid 3rd week of May to early July. Southern Germany: eggs laid mid-April to late July. Southern France: eggs laid from mid-April. North-west Africa: eggs mid-April to early June. Cape Verde Islands: major breeding season from end of August to end of November, less marked 2nd season mid-January to end of March; presumably adaptation to major autumn and minor spring rainy seasons. 1–2 broods. SITE. In low brambles, shrubs, and trees, on branch or among trunk suckers; less often in creepers, tall herbs, or ferns. From close to the ground to 4.5 m above. Nest: finely-constructed cup, often with 'transparent' walls and/or bottom, of grass and herb stems and leaves, plus rootlets and small twigs, bound together with spiders' webs and cocoons, lined with finer grasses, rootlets, and hair. EGGS. Sub-elliptical, smooth and glossy; very variable in ground-colour and markings, mainly white or very pale buff, pink, or olive, with buff, brown, olive, red-brown, or purplish spots, speckles, and blotches, often quite sparse with larger blotches having darker centres, but occasionally blotches obscure ground-colour; rarely, few or no markings. Clutch: 4–6 (2–7). INCUBATION. Usually 11–12 days. FLEDGING PERIOD. 10–14 days.

Wing-length: North and west European populations, ♂ 71–80, ♀ 70–78 mm. Southern European, Mediterranean North African, and Atlantic island populations smaller, ♂ ♀ 67–76 mm. **Weight:** ♂ ♀ mostly 16–25 g; migrants to 31 g.

Geographical variation. Complex, but clinal and slight. 5 races recognized. Nominate *atricapilla* occupies most of range. *S. a. dammholzi* (Caucasus) greyer and paler; *S. a. pauluccii* (Mediterranean islands and parts of adjacent mainland) greyer and darker; *S. a. heineken* (Iberia, Morocco, Algeria, Madeira, Canary Islands) markedly darker and smaller); *S. a. gularis* (Cape Verde Islands, Azores) poorly differentiated from nominate *atricapilla* but wing usually shorter and bill longer.

Eastern Crowned Leaf Warbler *Phylloscopus coronatus*

DU. Kroonboszanger   FR. Pouillot de Temminck   GE. Kronenlaubsänger   IT. Luì coronato di Temminck
RU. Светлоголовая пеночка   SP. Mosquitero coronado   SW. Kronsångare

Breeds in eastern Siberia, from Argun' river along Amur river to its mouth, and south to western Manchuria, central and south-east Szechwan (China), Korea, and Honshu (Japan). Winters in Assam, Bangladesh, Burma, Thailand, Indochina, Malaya, Sumatra, and Java. A bird collected on Helgoland (Germany), 4 October 1843, was identified by Gätke as *P. coronatus*. The specimen is apparently no longer extant, but Gätke's description leaves little doubt that his identification was correct.

Greenish Warbler *Phylloscopus trochiloides*. *P. t. nitidus*: **1** ad breeding worn, **2** 1st winter. *P. t. viridanus*: **3** ad breeding worn, **4–5** 1st winter. *P. t. plumbeitarsus* (Two-barred Greenish Warbler): **6** 1st winter. Arctic Warbler *Phylloscopus borealis* (p. 1322). *P. b. talovka*: **7** ad breeding worn, **8–9** 1st winter.

Greenish Warbler *Phylloscopus trochiloides*

PLATES: pages 1320, 1347

Du. Grauwe Fitis   Fr. Pouillot verdâtre   Ge. Grünlaubsänger   It. Luì verdastro
Ru. Зеленая пеночка   Sp. Mosquitero troquiloide   Sw. Lundsångare

Field characters. 10 cm; wing-span 15–21 cm. Close in size to Chiffchaff, lacking relative bulk of Willow Warbler and (particularly) Arctic Warbler. Small, slim, graceful, and rather rakish *Phylloscopus*, light on the wing but neither as active nor as bold as Arctic Warbler. Plumage of western and northern form pale greyish-olive above, dull white below, with pale lower mandible, long yellowish-white supercilium (often turning up on nape), dark eye-stripe, usually short pale wing-bar on tips of outer greater coverts (in fresh plumage), and dusky legs. Birds of extralimital eastern population (*P. t. plumbeitarsus*, the so-called Two-barred Greenish Warbler) show 2nd less distinct bar on median coverts and are vagrant to west Palearctic. Call distinctive.

Hedged in by similarities to congeners, Greenish Warbler can rarely be identified instantly. Important to recognize that typical Greenish Warbler has (a) distinctive call shared only with Hume's Yellow-browed Warbler, (b) shape and behaviour most like Chiffchaff, (c) head and face pattern shared fully only with Arctic Warbler, (d) sharply-etched wing-bar(s), (e) almost wholly bright lower mandible, lacking in Chiffchaff and Arctic Warbler, and (f) dusky legs, not almost black ones as in eastern races of Chiffchaff with wing-bars. Thus in prolonged observation at close range, not difficult to eliminate Chiffchaff.

Eastern *P. t. plumbeitarsus* differs from northern west Palearctic populations in (a) darker olive-green, less grey upperparts, (b) longer supercilium, (c) yellowish-white bar across median coverts (breaking up with wear and frequently lost), (d) longer and much broader yellowish-white bar across greater coverts (not wearing off), and (e) cleaner, colder, virtually white underparts; can thus recall strange, lengthy Yellow-browed Warbler.

Southern *P. t. nitidus* (the so-called Green Warbler) differs from northern *P. t. viridanus* in (a) purer, olive-green upperparts, (b) usually yellow supercilium, (c) yellower cheeks, and (d) usually yellow tone to throat and breast. Actions include persistent fast flicking of wing-feathers, but tail movements not pronounced (unlike Chiffchaff). Active and agile, both within foliage and in flight; able to hover and catch flies. Feeds at all heights of vegetation.

Habitat. Breeds in west Palearctic in middle latitudes, in continental warm, boreal, and temperate climates, mainly in lowlands in the north but in Caucasus from foothills to 3000 m. Although arboreal and tolerant of coniferous, broad-leaved, and mixed forest, prefers fringes, clearings, or open stands to denser growth, and favours crowns of such trees as birch and aspen scattered among spruces. Also occurs outside forest in copses, parks, thickets, and subalpine meadows.

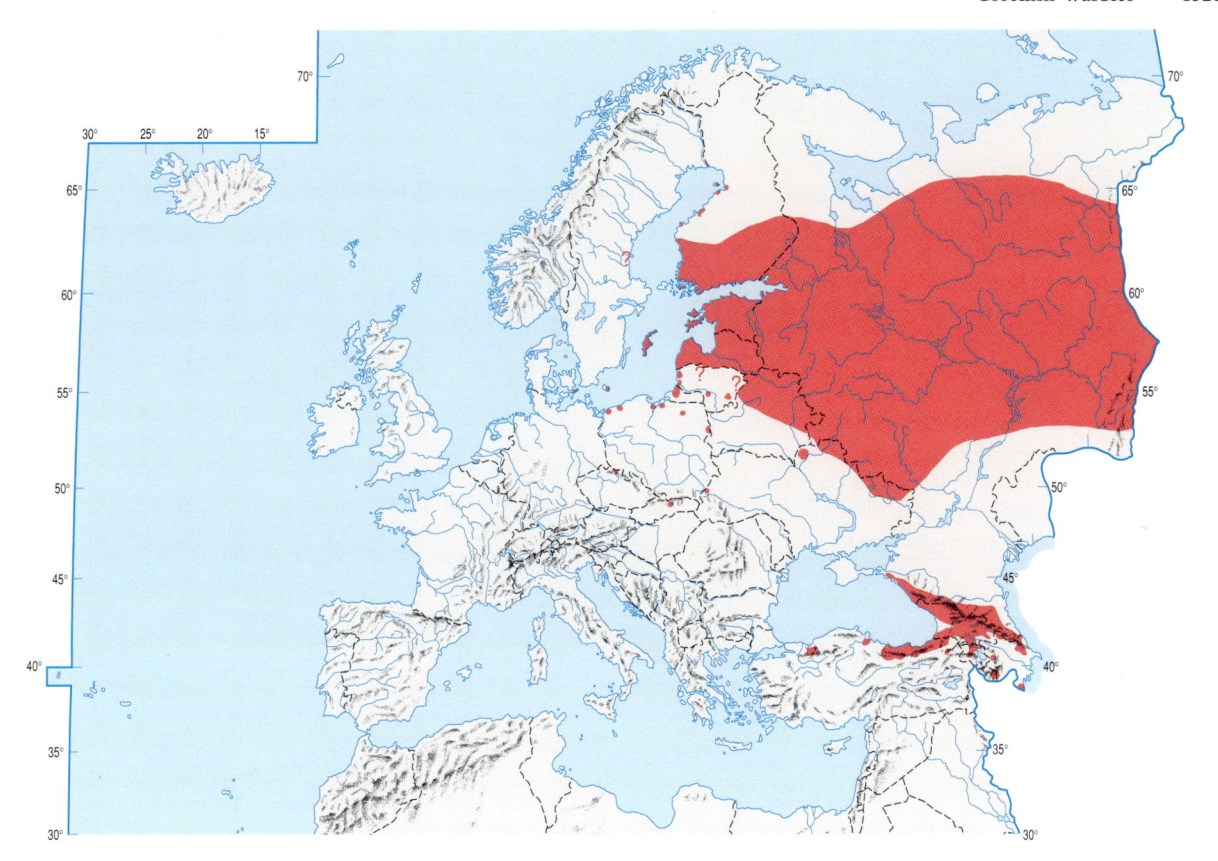

Distribution. Westward spread first noted in 2nd half of 19th century, sometimes in waves, but western limits of range constantly fluctuating and no proof of breeding in many areas where occurs more or less regularly. Invasion years (high May–June temperatures) alternate with years when absent or very scarce. Main waves of westward advance 1933–5, 1954–5, and 1961. BRITAIN. Recent increase in spring records and of singing birds. GERMANY. Nest-building recorded Mecklenburg 1935, breeding attempt Westerwald 1962, and first confirmed breeding Helgoland 1990. Single ♂♂ almost annual in east, especially Usedom and Rügen islands; c. 15 on Rügen in 1978 after warm May. DENMARK. Probably breeding from 1976; confirmed 1980, 1985, 1990, 1992–3. NORWAY. Single confirmed breeding record: Telemark 1991. SWEDEN. Breeds regularly Gotland at least since early 1980s (first nest found 1953); also confirmed at various sites on mainland, and perhaps regular along Baltic coast. FINLAND. First recorded breeding 1937, then rapid early expansion; reached Gulf of Bothnia 1949, coast at Oulu 1950. ESTONIA. Recorded from early 1900s, with spread in 1930s–50s and more recently; now patchily distributed throughout country. LATVIA. Probable increase. LITHUANIA. Single breeding pairs in centre 1961, 1963, east and west 1987. POLAND. Breeding confirmed in only few localities (first in 1958), but records of singing ♂♂ widespread, including southern mountains. CZECH REPUBLIC. Breeding recorded from 1992, though probably bred earlier. SLOVAKIA. First recorded breeding 1994: 2 breeding pairs (1 nest) and 2 singing ♂♂ in one locality; probably established earlier. FSU. Spread north and west in 19th century; further expansion around turn of century, leading to narrow wedge to Baltic. Absent or rare in following years; first recorded north of Lake Ladoga 1928. Southern limit of range evidently very fragmented; has not penetrated far into wooded steppe. *P. t. nitidus* spread since c. 1930 from Caucasus at least 180 km north-west into lowlands. UKRAINE. Probably breeding in north-east by 1980. TURKEY. *P. t. nitidus* local, not rare in montane forests of Black Sea coastlands west to about Bolu. Probably more widely distributed between Bolu and Ordu than known at present.

Accidental. Britain, Channel Islands, Ireland, France, Belgium, Netherlands, Norway, Spain, Cyprus. *P. t. nitidus*: Britain, Germany, Ukraine, Israel, Kuwait. *P. t. plumbeitarsus*: Britain, Netherlands, Sweden.

Beyond west Palearctic, extends east across Siberia to Kolyma mountains and Sea of Okhotsk, south through central Asian mountains (skirting much of Kazakhstan) to northern Iran, northern Pakistan, and south-west China.

Population. Marked fluctuations especially in west of range. SWEDEN. Overall increase, at least until 1978, when maximum

110 ♂♂. At least 1–10 pairs in early 1990s. Finland. 2000–5000 pairs in late 1980s; slight decrease. Estonia. 1000–5000 pairs in 1991; increase apparently continuing. Latvia. 500–2000 pairs in 1980s; slight increase. Lithuania. Rare. Poland. 50–300 pairs. Probably slight increase in 2nd half of 20th century. Czech Republic. 1–5 pairs. Russia. 100 000–1 million pairs; stable. Belarus'. 17 000–25 000 pairs in 1990. Fluctuating, but apparently stable overall. Ukraine. 1–15 pairs in 1988. Azerbaijan. *P. t. nitidus* common. No information on population or trends of this race elsewhere in FSU. Turkey. *P. t. nitidus*: 1000–10 000 pairs.

Movements. Migratory; in east of range, some birds probably altitudinal migrants. West Palearctic population winters in Indian sub-continent from Himalayan foothills of Nepal and Sikkim south to southern India, east to Bangladesh and west to Uttar Pradesh and Madhya Pradesh (India). Autumn movement begins July. Western populations depart early: at Lake Ladoga, last records (all juveniles) early August, and reported in Moscow region until mid-August. First arrivals on Indian plains mid-August, with main passage to end of September. Spring migration begins early (from March) but is protracted. Reaches Estonia usually at end of May, but earlier (c. 19 May) in invasion years, and Sweden from late May. Regular west of breeding range in autumn, with late August to early September peak in Britain and Ireland.

Food. Mostly small arthropods. Forages at all levels, from ground to canopy, picking items off vegetation; also captures insects in flight. Characteristically highly mobile, using hops and frequent short flights (also on ground), sometimes snatching insect from twig or leaf without pausing; flicks wings, perhaps to disturb prey.

Social pattern and behaviour. Solitary and territorial in winter, both sexes probably defending individual territories against conspecific birds. Birds regularly return to identical territories each winter. No evidence for other than monogamous mating system. ♂♂ arrive on breeding grounds a few days ahead of ♀♀. Incubation by ♀ alone; young fed by both sexes. Occasional song given by ♀ during breeding season, and both sexes probably sing when defending winter territories. ♂ sings almost always from upper habitat layer, occasionally lower down. No ritualized song-flight, but full song given not uncommonly in flight. Song-period extends generally from 2nd half of May to end of July.

Voice. Song of ♂ a rapidly delivered, often loud, lively, and slightly jerky phrase lasting c. 2–3(–5)s, composed of high-pitched liquid twittering warble or chattering jangle exhibiting wide pitch variation; 'frothy' quality reminiscent of ecstatic song of White Wagtail. Usually very steep rise and fall in pitch of individual units which, combined with speed of delivery, produces shrill effect and makes sounds difficult to differentiate. Trilling or rattle-type motifs sometimes incorporated. Song of ♀ a rather weak and simple version of ♂'s song. Commonest call, expressing excitement in various contexts (including territorial aggression), and apparently also used for contact, most strongly suggests call of Pied Wagtail, though chirruped quality sometimes also reminiscent of cheerful, squeaky sparrow. Variable; rendered as 'tiss-yip', 'chi-vee', 'wizip', etc. Other calls include a low-pitched, soft and quiet 'trrr' apparently serving as warning-call, and a sharp 'srrrt' given during fight.

Breeding. Season. Russia, Caucasus area: laying begins 1st or 2nd third June. Finland: 2 nests with eggs found 3rd week of June. One brood. Site. On ground in tall vegetation or low scrub, under stones, windfall, or tree roots, sometimes in crevices in banks, old walls, or tree stumps. Most nests well concealed and camouflaged. Nest: quite large and domed, with side entrance, constructed from grasses, plant fibres, moss, rootlets, and leaves, lined with finer material often including small amount of hair, fur, down, and feathers. Eggs. Sub-elliptical, smooth and matt; white, unmarked. Clutch: 3–7. Incubation. 12–13 days. Fledging Period. 12–16 days.

Wing-length: *P. t. viridanus*, ♂ 55–68, ♀ 54–64 mm.
Weight: ♂♀ mostly 6.5–9 g.

Geographical variation. Two races breed in west Palearctic: *P. t. viridanus* in north, *P. t. nitidus* in Caucasus area and northern Turkey. *P. t. plumbeitarsus* recorded as a vagrant.

Arctic Warbler *Phylloscopus borealis*

PLATES: pages 1320, 1347

Du. Noordse Boszanger   Fr. Pouillot boréal   Ge. Wanderlaubsänger   It. Luì boreale
Ru. Пеночка-таловка   Sp. Mosquitero boreal   Sw. Nordsångare

Field characters. 10.5–11.5 cm; wing-span 16.5–22 cm. Size between largest Willow Warbler and Wood Warbler. Structure, apart from somewhat short tail, recalls Willow Warbler; compared with Greenish Warbler, usually noticeably bulkier about bill, head, and body, with longer wings and under tail-coverts. Strong bill, angled crown, elliptical body, rather long wing-point, and relatively short tail may recall warblers of other genera as much as *Phylloscopus*. Medium-sized, rather strong-billed, fairly slim yet robust, and highly active *Phylloscopus*; looks as bulky in flight as Wood Warbler. Plumage bright olive-green above and grey-white below, with orange-yellow base to strong bill, long, often upturned white supercilium, prominent white eye-crescents interrupting dark eye-stripe, 1 distinct and 1 indistinct white wing-bar (in fresh plumage), and bright straw legs and feet. Call distinctive.

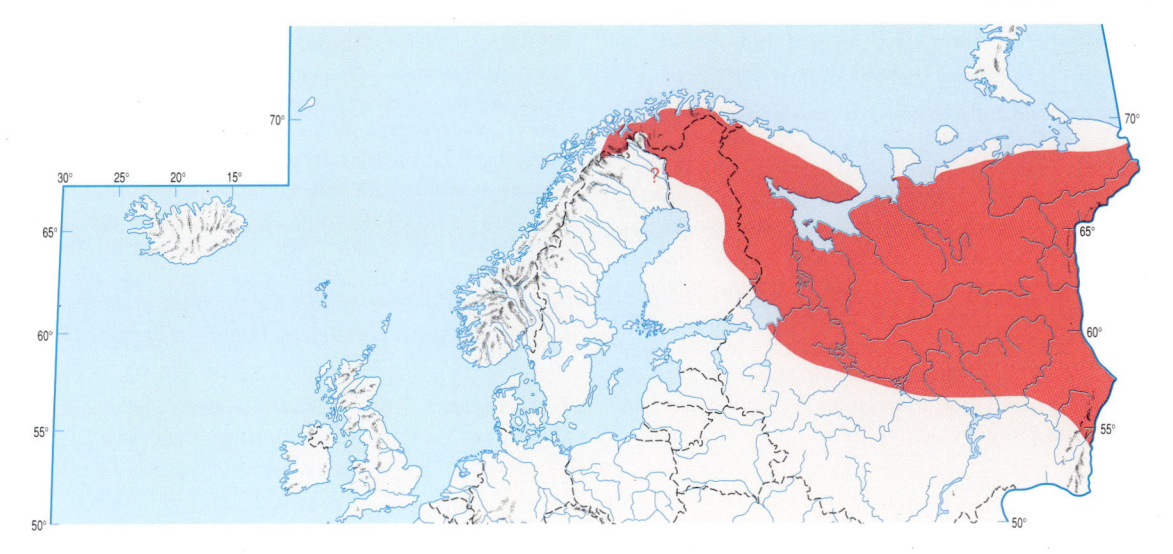

Distinctive if seen well at close range, with rather sturdy form, green and white appearance, long supercilium and eye-stripe, and bright bare parts; voice also unusual within *Phylloscopus*. However, care necessary to eliminate Greenish Warbler, and even much smaller Yellow-browed Warbler. Flight flitting rather than fluttering, and noticeably more dashing, confident, and faster than most *Phylloscopus*. Gait hopping, jumping, and sidling, again with strong hint of greater strength and weight than most *Phylloscopus*. Carriage noticeably horizontal.

Habitat. Might more accurately be named Subarctic Warbler, since breeding range falls entirely within that region, and as a tree-dweller is precluded from living beyond July isotherm of 10°C, beyond which tree growth is inhibited. Remarkable also for combining successful occupancy of so much of Holarctic taiga zone with absence from forests south of boreal climates except in east Asian mountain regions. Despite inhabiting largely coniferous forests, shows marked preference for birch, poplar, willow, and other broad-leaved trees or scrub wherever available, especially along river banks and near water, but in places also frequents dry, sparsely wooded slopes and even rhododendron thickets.

Distribution and population. NORWAY. Scarce breeder with most records restricted to Sør-Varanger area. 10–100 pairs 1970–90. SWEDEN. Decreasing. Possibly breeds annually, but if so only single pairs; total of only 6 breeding records. In 1950s, up to 10 ♂♂ recorded at one site. Perhaps up to 10 pairs in early 1990s. FINLAND. 3000–5000 pairs in late 1980s; probable increase. RUSSIA. 1–10 million pairs; stable.

Accidental. Faeroes, Britain (annual), Ireland, France, Netherlands, Germany, Denmark, Estonia, Lithuania, Poland, Spain (Gibraltar and Balearic Islands), Italy, Malta, Greece.

Beyond west Palearctic, breeds extensively across northern Asia reaching Alaska, also south to Japan and perhaps Korea.

Movements. All populations migratory, wintering in southern south-east Asia: widespread and generally common from Tenasserim (southern Burma) and (chiefly southern) Thailand east to Philippines and Moluccas, south to Greater and Lesser Sundas, and in Taiwan. Western populations migrate east across Russia, gradually turning southward east of Yenisey through eastern Mongolia and Manchuria to reach winter quarters via eastern China; route thus exceptionally long, at least 13 000 km for Fenno-Scandian birds.

From Fenno-Scandia and north-west Russia, autumn departures chiefly in August. Birds arrive Malay peninsula from mid-September. Leaves winter quarters late, mostly April to early May, arriving in west Palearctic breeding grounds mostly in 2nd half of June. Regular autumn vagrant to Britain in small numbers, with distinct peak mid-September; probably resulting mostly from reversed migration. Elsewhere in Europe, vagrancy rare but apparently increasing.

Food. Insects and a few other invertebrates. Forages actively in foliage of trees and bushes, sometimes in herbaceous vegetation and on ground; also catches insects in flight.

Social pattern and behaviour. Occurs in small flocks during migration, and in winter quarters sometimes associates with flocks of other small birds. Mating system probably essentially monogamous, but bigamy recorded. Incubation and brooding by ♀; young fed by both sexes. ♂ sings usually while perched on or near top of tree, sometimes hidden in foliage. When moving between song-posts, may perform remarkable wing-rattling display: sings from tree-top perch then, during flight to another tree, or short flit between branches, quivers or shuffles wings while braking (back hunched), thus producing rattling, whirring or buzzing noise. In winter quarters, some birds start to sing in March. On Fenno-Scandian breeding grounds, song period from arrival June to mid- or late July; most intensive song during 1st week after arrival.

Voice. Song a loud and vigorous, fast and musical, but rather monotonous trill, comprising repetition of *c.* 15–30 complex units and lasting up to *c.* 4 s; usually, but not invariably, preceded by one or more sharp, buzzy, rather metallic, but usually quiet 'tzick', 'zrik' or similar sounds. Trill shows some affinity with trilling songs of Wood Warbler and Bonelli's Warbler, but faster and usually lower pitched. Contact-alarm call quite unlike calls of other *Phylloscopus*: a short, hard, and sharply metallic 'zrik', 'dzik', 'tsirk', 'sirt', or 'tset', as at start of song but usually louder; reminiscent of call of Yellowhammer or Dipper. Other calls include a 'zit', given in short series, and a low, wooden rattling, churring, or chattering 'tr-tr-tr' or 'trr-trr'.

Breeding. SEASON. Northern Russia: laying begins late June. Northern Sweden: 1st eggs laid in last week of June or 1st week of July. Northern Norway: laying recorded 25 June to 1 July. SITE. On ground in vegetation, or in natural crevice in mossy bank or among tree roots, often well concealed. Nest: domed structure with side entrance, of dry thin grass stems and leaves, often with moss and dry leaves, lined with finer grasses, but only rarely with feathers or hair. EGGS. Sub-elliptical, smooth and glossy; white, finely and often sparsely speckled and spotted red-brown or dark brown. Clutch: (5–)6–7. INCUBATION. 11–13 days. FLEDGING PERIOD. 13–14 days.

Wing-length: ♂ 65–70, ♀ 61–68 mm.
Weight: ♂♀ mostly 8.5–12 g; migrants to 15 g.

Pallas's Warbler *Phylloscopus proregulus*

PLATES: pages 1325, 1347

DU. Pallas' Boszanger   FR. Pouillot de Pallas   GE. Goldhähnchen-Laubsänger   IT. Luì del Pallas
RU. Корольковая пеночка   SP. Mosquitero de Pallas   SW. Kungsfågelsångare

Field characters. 9 cm; wing-span 12–16.5 cm. Slightly smaller than Yellow-browed Warbler, with 10% shorter wings; close in size and rather large-headed appearance to Firecrest. Tiny, compact, and extremely active *Phylloscopus*, with general character like Yellow-browed Warbler but behaviour more recalling goldcrest (*Regulus*). Plumage rich but fairly pale green above and mainly white below, with pale yellow supercilium, and central crown-stripe, 2 yellowish-white wing-bars, fringes to tertials, and yellowish-white panel over upper rump.

Smallest and most beautiful *Phylloscopus* in west Palearctic, with combination of head pattern, pale band across upper rump, and clean underparts diagnostic. Most problems arise from difficulty of observation and, when view obscured, care necessary to avoid confusion with Yellow-browed Warbler (slightly larger, more attenuated silhouette, duller olive above, yellowish below, usually uniform crown concolorous with mantle, almost white wing-bars, no pale rump) and ♀ or immature Firecrest (similar in size but with almost black and white lateral head pattern, grey ear-coverts, almost white and black wing-bars, no pale rump). Flight rather direct and level; flitting, with fast wing-beats; hovers expertly in and around foliage, but less given to flycatching than Yellow-browed Warbler.

Song a loud cadence of rich notes, recalling both Willow Warbler (in phrasing and terminal fall) and Canary (in sweetness and repetitiveness). At least 2 calls freely given by vagrants: commonest a rather soft 'weesp', 'swee', 'tweet', or 'zit', less piercing than monosyllable of Yellow-browed Warbler, but also a louder disyllabic 'wee-esp', or 'ch-weet', more disyllabic than similar note of Yellow-browed Warbler.

Habitat. Breeds in coniferous forests in middle and lower middle latitudes of east Palearctic, in taiga towards north of range but mainly in mountains, and in south of range at high altitudes.

Distribution. Breeds in Asia north to *c.* 60°N, from Ob' valley east to Sea of Okhotsk and Sakhalin island, south to central China and Himalayas.

Accidental. Annual Britain, Sweden, and Finland, regular Denmark. Also recorded Faeroes, Channel Islands, Ireland, France, Belgium, Luxembourg, Netherlands, Germany, Norway, Estonia, Latvia, Lithuania, Poland, Czech Republic, Spain, Balearic Islands, Portugal, Italy, Malta, Ukraine, Israel, Morocco.

Movements. Northern populations are long-distance migrants, southern populations make shorter, mainly altitudinal movements. Main movements are through eastern Asia.

Regular autumn vagrant to north-west Europe; increasingly frequent since 1960s, previously rare. Occurrences involve mostly 1st-years and probably result chiefly from westward displacement in anticyclonic conditions, with reverse migration as possible additional factor. A few birds may occasionally overwinter.

Wing-length: ♂ 48–56, ♀ 47–53 mm.

Pallas's Warbler *Phylloscopus proregulus*. *P. p. proregulus*: **1** ad breeding worn, **2** 1st winter. Yellow-browed Warbler *Phylloscopus inornatus*. *P. i. inornatus* (Siberia): **3** ad breeding worn, **4** 1st winter. Hume's Leaf Warbler *Phylloscopus humei* (central Asia) (p. 1327): **5** ad breeding worn, **6** 1st winter. Radde's Warbler *Phylloscopus schwarzi* (p. 1327): **7** ad breeding worn, **8** 1st winter. Dusky Warbler *Phylloscopus fuscatus* (p. 1328). *P. f. fuscatus*: **9** breeding worn, **10** 1st winter.

Brooks's Leaf Warbler *Phylloscopus subviridis*

Du. Brooks' Bladkoning   Fr. Pouillot de Brooks   Ge. Brookslaubsänger   It. Luì di Brooks
Ru. Гималайская пеночка   Sp. Mosquitero de Brooks

Breeds in eastern Afghanistan (Safed Koh) and north-west Himalayas east to Murree hills and Gilgit. In winter descends to hills and plains of northern Pakistan and India from Punjab and Himachal Pradesh to western Uttar Pradesh. A record from Orenburg (Russia), at extreme eastern limit of west Palearctic, 17 September 1882, considered unproven.

Yellow-browed Warbler *Phylloscopus inornatus*

PLATES: pages 1325, 1347

Du. Bladkoning   Fr. Pouillot à grands sourcils   Ge. Gelbbrauen-Laubsänger   It. Luì forestiero
Ru. Пеночка-зарничка   Sp. Mosquitero bilistado   Sw. Taigasångare

Field characters. 10 cm; wing-span 14.5–20 cm. Up to 10% smaller than Chiffchaff, with finer bill and slightly shorter tail; 10% larger than Pallas's Warbler; noticeably longer than goldcrest (*Regulus*). Small, light, and strikingly active *Phylloscopus*, with general character and behaviour at times recalling goldcrest. Plumage essentially pale olive above and yellowish-white below, with long pale yellowish supercilium, 2 yellowish-white wing-bars, and bright white tertial edges—all with adjacent dark areas which create distinctive striped pattern on face and upperparts.

Commonest Siberian passerine vagrant to western Europe; key species in field identification of similarly marked *Phylloscopus*. Liable to confusion with same species as is Pallas's Warbler and Hume's Leaf Warbler (see those species), but can also suggest larger *Phylloscopus* with wing-bars. Inexperienced observer must beware mistake with Goldcrest (much smaller, more compact, and duller overall, with no supercilium nor eye-stripe but with similar wing pattern). Flight light and darting, recalling Willow Warbler in action (and silhouette) more than Pallas's Warbler, and over any distance never suggesting goldcrest; capable of fluttering and flycatching but less given to hovering than Pallas's Warbler.

Habitat. Breeds in high and central latitudes of east Palearctic (just into west Palearctic at western extreme of range), from treeline in north to *c.* 50°N in south; inhabits open forests with dense undergrowth up to altitude of *c.* 2000 m, often by rivers, preferring broad-leaved to coniferous; principally birch, poplar, willow, and larch, in far north in spruce/fir taiga and krummholz zone, and also in subalpine rhododendron. In European Urals, in upper slopes of thick alder and larch woodland. In winter in tropical and subtropical gardens, parks, street trees, open woodland, and shrubby places also up to *c.* 2000 m.

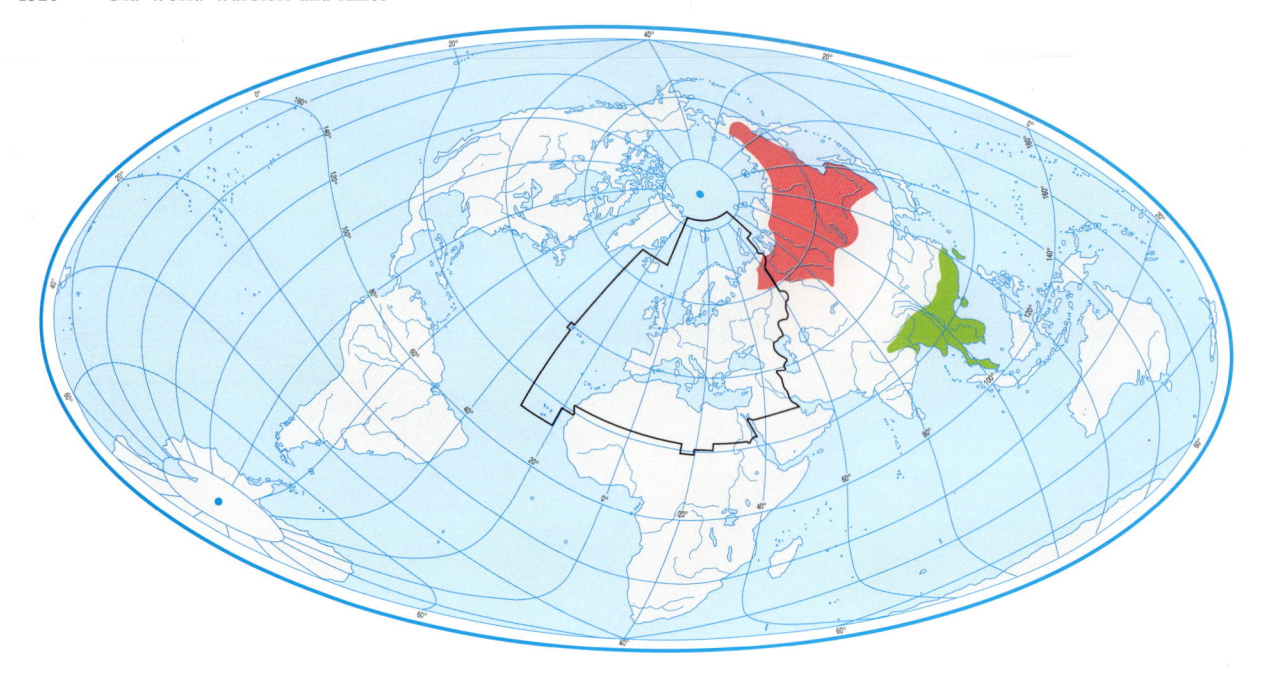

Distribution and population. RUSSIA. No range changes reported. Locally abundant; fairly scarce on slopes of Urals in early 1950s. Estimated 45 000–46 000 pairs in early 1990s.

Accidental. Status of rare but regular visitor, sometimes more numerous and generally increasing markedly in recent years, especially in countries of north-west. Annual or nearly so in countries marked with asterisk. Iceland, Faeroes*, Britain*, Channel Islands, Ireland*, France*, Belgium, Luxembourg, Netherlands*, Germany*, Denmark*, Norway*, Sweden*, Finland*, Estonia, Latvia, Lithuania, Poland*, Czech Republic, Hungary, Austria, Switzerland, Spain, Balearic Islands, Portugal*, Italy, Malta, Greece, Turkey, Cyprus, Israel, Kuwait, Egypt, Libya, Algeria, Morocco, Madeira, Canary Islands.

Movements. Migratory; winters from central Nepal (where uncommon), Sikkim, and Bangladesh east to eastern China, Hainan, and Taiwan, and south to Malay peninsula; by far commonest *Phylloscopus* species in Thailand and most common wintering warbler in Hong Kong. Leaves breeding grounds August–September, in north and west of range marked movement from mid-August; recorded on passage in Nepal mid-October to early November. Present in south-east Asia mainly mid-October to mid-April, and spring migration begins late March or early April; arrives from late May in western Siberia, at extremes of range in early June. Widespread autumn vagrant to central and western Europe (mostly mid-September to October in Britain and Ireland), with marked increase in recent years; now regular as far afield as southern Portugal.

Food. Insects and a few other invertebrates; highly active when foraging, impression being enhanced by constant fluttering and wing-flicking; picks or snatches items from twigs and leaves and regularly hovers to examine leaves; often pursues insects in agile flight. Generally forages in tree crowns, though also in herbs, and feeding on ground not uncommon.

Social pattern and behaviour. Little studied, and in areas of overlapping distribution often not distinguished from Hume's Leaf Warbler. On breeding grounds solitary, territorial, and most probably monogamous. In winter quarters usually solitary but sometimes in mixed flocks with other *Phylloscopus* species, tits, flycatchers, etc.; on migration in loose troops of 10–15 and more gregarious than most leaf warblers. In many places in Siberia commonest *Phylloscopus* species. In larch taiga of Kolyma mountains, north-east Siberia, 70 pairs per km^2 noted. Pair-formation rapid and egg-laying can begin 12–15 days after arrival on breeding grounds. In north-central Siberia, song noted from arrival mid- to late June almost up to mid-July.

Voice. Song of ♂ among simplest of all *Phylloscopus* songs and barely recognizable as such; completely different from Hume's Leaf Warbler, with characteristic short phrase comprising notes similar to main call, variously rendered 'tii-tyuiss', 'tyui-ti-tyu-tyuiss', or 'tssss-siu siu-tsss', lasting *c.* 1.5 s. These notes rather weak, thin, drawn-out and very high pitched, especially at beginning and end. Timbre can suggest calls of Coal Tit, treecreeper, or Penduline Tit, while timbre and structure like whistle-song of Hazel Grouse. Main call reminiscent of Coal Tit alarm-call: drawn-out, fine, shrill, penetrating and quite loud 'weest', 'weep', or 'swiist'; sometimes disyllabic (rising in pitch) and rendered 'weeist', 'siu-wiist', or 'tsie-wiet'. This also quite distinct from Hume's Leaf Warbler.

Breeding. SEASON. Eggs laid mainly from mid-June. SITE. On ground, generally sheltered by grass tussock, brushwood, tree roots, etc. Nest: domed structure of dry grass, moss, wood fibre, rootlets, etc. with side entrance; lined with similar but softer material and hair. EGGS. Sub-elliptical, smooth, and glossy; white, finely marked with red-brown, purple-brown, or purple-grey speckles sometimes concentrated at broad end. Clutch: 4–6 (2–7). INCUBATION. *c.* 11–14 days. FLEDGING PERIOD. 12–13 days.

Wing-length: ♂ ♀ 50–61 mm.
Weight: ♂ ♀ 4–9 g.

Hume's Leaf Warbler *Phylloscopus humei*

PLATES: pages 1325, 1347

DU. Hume's Bladkoning   FR. Pouillot de Hume   GE. Tienschan–Laubsänger
RU. Тусклая зарничка

Field characters. 10 cm; wing-span 16–20 cm. Marginally larger than Yellow-browed Warbler but without structural difference. Plumage differs from Yellow-browed Warbler in less green, greyer or duller upperparts and less striking pattern of stripes, particularly on wings where bar on median coverts and tertial fringes often indistinct or lacking. At close range, further distinguished from Yellow-browed Warbler by darker bill, buffy chest, and blackish legs.

Easily confused with dull Yellow-browed Warbler which may overlap in plumage tones (and which may lose stripes with wear), and other *Phylloscopus* with single wing-bars, particularly Greenish Warbler (see that species). Flight and behaviour as Yellow-browed Warbler. Commonest call usually disyllabic 'tiss-yip' or 'tze-weet', usually loud and ringing, recalling sparrow (whereas typical call of Yellow-browed Warbler suggests Coal Tit). Song a buzzing wheeze, with or without such calls.

Habitat. Breeds in montane and subalpine zones in middle latitudes of east Palearctic, generally at altitude of *c.* 2500–3500 m (recorded breeding at 4000 m), in open coniferous or mixed forest, at highest altitudes in birch, rhododendron, or krummholz zones, especially on sunny slopes, and also in bushes along mountain rivers. Winters in lower altitude broad-leaved woodland, gardens, and orchards. Feeds on insects; nests on ground under bush, grass tussock, fallen tree, etc.

Distribution. Breeds in central Asia in fairly narrow band from Sayan and Altai mountains of northern Mongolia and southern Siberia south-west through northern Sinkiang (western China), southern Kazakhstan, Kyrgyzstan, and Tajikistan to Himalayas. Overlaps with Yellow-browed Warbler to west and south-west of Lake Baykal, perhaps also in some places to south-east. Possibly isolated population of race *mandellii* in central China, mainly Szechwan, could extend further towards south-west than known at present.

Accidental. France, Belgium, Netherlands, Germany, Denmark, Norway, Sweden, Finland, Poland, Austria, Italy, Turkey, Israel, Iraq; in Britain, *c.* 20 claimed but yet to be confirmed.

Movements. Migratory; winters in Pakistan and Himalayan foothills east to Nepal (up to 2500 m), Sikkim, Bangladesh, northern Burma and Thailand, and widespread in Indian plains south to *c.* 16°N; in Delhi area, by far commonest *Phylloscopus* in winter apart from Chiffchaff. Also winters in southern Afghanistan, Iran, and a few in eastern Arabia. Leaves breeding grounds from August (mid-July at 3300 m in Kashmir), with main movement in western Siberia September to mid-October; reaches winter quarters from late September, and present in India mostly October–April. Northward movement begins late March, arriving in western Mongolia early May, and peak movement in western Siberia is 2nd half of May. Most vagrants to Europe and Levant recorded in late autumn (later than Yellow-browed Warbler) and overwintering regularly noted; commonly arrives together with vagrant Radde's Warbler, Dusky Warbler, and Pallas's Warbler.

Wing-length: ♂ ♀ 53–62 mm.

Radde's Warbler *Phylloscopus schwarzi*

PLATES: pages 1325, 1347

DU. Radde's Boszanger   FR. Pouillot de Schwarz   GE. Bartlaubsänger   IT. Luì di Radde
RU. Толстоклювая пеночка   SP. Mosquitero de Schwartz   SW. Videsångare

Field characters. 12.5 cm; wing-span 15.5–20.5 cm. 10–15% larger than Chiffchaff, with much stronger bill, larger head, broader back, and sturdiest, longest legs of west Palearctic *Phylloscopus*; at least as bulky as largest Wood Warbler and Arctic Warbler and broader-tailed than both. Large, rather large-headed and fairly plump *Phylloscopus*, with distinctive tit-like bill, short rounded wings, rather large and slightly rounded tail, and strikingly heavy, straw-coloured legs and feet. Plumage subtly but richly coloured: green- to brown-olive above, buff and yellow below, with long, deep cream supercilium, dark eye and eye-stripe, bright sheen over wings, and orange-buff under tail-coverts.

At first sight, may suggest Dusky Warbler or dark southern and eastern races of Chiffchaff, but unmistakable in good view at close range, showing heavy build, strong head pattern, and subtle but rich tones in rest of plumage creating vividness somehow unmatched even in other, brighter *Phylloscopus*. Flight can suggest small *Hippolais*: direct or sweeping and usually low, often ending in sudden plunge into cover; short, round wings sometimes give fluttering action. Has noticeable preference for ground cover on migration, and feeds on ground more than any west Palearctic *Phylloscopus* except Dusky Warbler.

Contact-alarm calls given regularly on migration (including by vagrants) and in winter. Several monosyllabic calls reported, with renderings suggesting considerable variety in pitch and timbre: short, clipped, nervous-sounding 'tchwit'; soft 'quip' or 'twit'; soft 'sok' or 'trock'; also multisyllabic calls. All calls may be repeated frequently, suggesting continuous state of nervous alarm.

Habitat. Breeds in central and east Palearctic in warm continental middle and lower middle latitudes in open forests, glades, clearings, regrowth in burnt and felled patches, strips of trees along riversides, and willow beds on lake shores and islands. Locally also in dense thickets of rhododendron in mountain ranges, where it ascends river valleys up to 1000 m. Spends most time in undergrowth and surrounding grasses and herbage, ascending to treetops only in breeding season; trees essential as song-posts, but must be well spaced, dense forest being avoided.

Distribution. Breeds from Novosibirsk and eastern Altai in southern Siberia east to Sakhalin island and Sea of Japan, south to north-east China and North Korea.

Accidental. Britain, Channel Islands, Ireland, France, Belgium, Netherlands, Germany, Denmark, Norway, Sweden (possibly annual), Finland, Poland, Czech Republic, Spain, Italy, Malta, Ukraine, Israel, Morocco.

Movements. Migratory, wintering in south-east Asia. Leaves breeding grounds mostly late August to mid-September. Spring migration through China mostly May to early June, arriving on Siberian breeding grounds about 2nd week of June.

Autumn vagrancy to western Europe, resulting from reverse migration or westward displacement in anti-cyclonic conditions, is less frequent than in some Asiatic *Phylloscopus*; spring records exceptional. In Britain and Ireland, 49 records 1958–85, all 26 September to 1 November.

Wing-length: ♂ 58–67, ♀ 56–65 mm.

Dusky Warbler *Phylloscopus fuscatus*

PLATES: pages 1325, 1347

Du. Bruine Boszanger   Fr. Pouillot brun   Ge. Dunkellaubsänger   It. Luì scuro
Ru. Бурая пеночка   Sp. Mosquitero sombrío   Sw. Brunsångare

Field characters. 11 cm; wing-span 14.5–20 cm. Size close to Chiffchaff but with slightly longer, more pointed bill, slightly shorter wings, rather longer and slightly rounded tail, and rather long legs; distinctly smaller (particularly about head) and more compact than Radde's Warbler. Small, fairly slight but energetic *Phylloscopus*, with spiky bill, rather short, rounded wings, and rather slender legs. Plumage essentially brown above and buff- and grey-white below, with rusty-white supercilium; adult lacks any fully green or yellow tones. Constantly flicks wings and tail.

Like Radde's Warbler, subject to confusion with dark southern and eastern races of Chiffchaff (particularly in poor light of late autumn), but in good view at close range, bright billbase, head pattern, and lack of green or (easily visible) yellow tones in plumage create distinctly different appearance, while persistent wing-flicking, frequent slight elevation of tail, and hard call are all distinctive.

Commonest calls of Siberian adults in spring and of European vagrants are short and hard: 'chip', 'tchick', 'chek', 'tack'; usually given twice, sometimes 4–6 times. Other calls include a staccato chattering, and a 'hweet' resembling Chiffchaff.

Habitat. Breeds in continental east Palearctic from boreal to warm temperate zone, from lowland plains and marshy river valleys in north to uplands and mountains further south, at altitudes of *c.* 400–3900 m. Unlike other *Phylloscopus* except Radde's Warbler is only marginally arboreal, being mainly found in shrub layers of open forest, in regrowth after fires, in willow beds along river and stream banks, and in upland thickets of dwarf birch.

Distribution. Breeds from upper Ob' and western Altai in Siberia east to Anadyr' region and Sakhalin island, south to eastern Himalayas and central and north-east China.

Accidental. Britain, Channel Islands, Ireland, France, Belgium, Netherlands, Germany, Denmark, Norway, Sweden (possibly annual), Finland (annual), Estonia, Latvia, Lithuania, Poland, Austria, Switzerland, Spain, Portugal, Italy, Malta, Greece, Russia (Leningrad region), Cyprus, Israel, Egypt, Morocco, Madeira.

Movements. Altitudinal and long-distance migrant. Northern populations winter from northern India and Nepal, east to southern China, Taiwan, Indochina, and Thailand.

Autumn vagrancy to western Europe, resulting from reverse migration or westward displacement in anti-cyclonic conditions, is widespread in small numbers. In Britain and Ireland, 47 records 1958–85; 45 between late September and November (one 18 August); bird recorded Isle of Man 14 May 1970 (recovered in Limerick, south-west Ireland, 5 December 1970), had probably wintered in western Europe. Record influx 1987: e.g. 17 Britain and Ireland, 10 Denmark, 6 Netherlands.

Wing-length: ♂ 57–67, ♀ 54–61 mm.

Wood Warbler *Phylloscopus sibilatrix* (p. 1333): **1** ad breeding, **2–3** 1st winter. Bonelli's Warbler *Phylloscopus bonelli*. *P. b. bonelli*: **4** ad breeding worn, **5** 1st winter. Eastern Bonelli's Warbler *Phylloscopus orientalis* (p. 1331): **6** ad breeding worn, **7** 1st winter.

Bonelli's Warbler *Phylloscopus bonelli*

PLATES: pages 1329, 1347

Du. Bergfluiter   Fr. Pouillot de Bonelli   Ge. Berglaubsänger   It. Luì bianco
Ru. Светлобрюхая пеночка   Sp. Mosquitero papialbo   Sw. Bergsångare

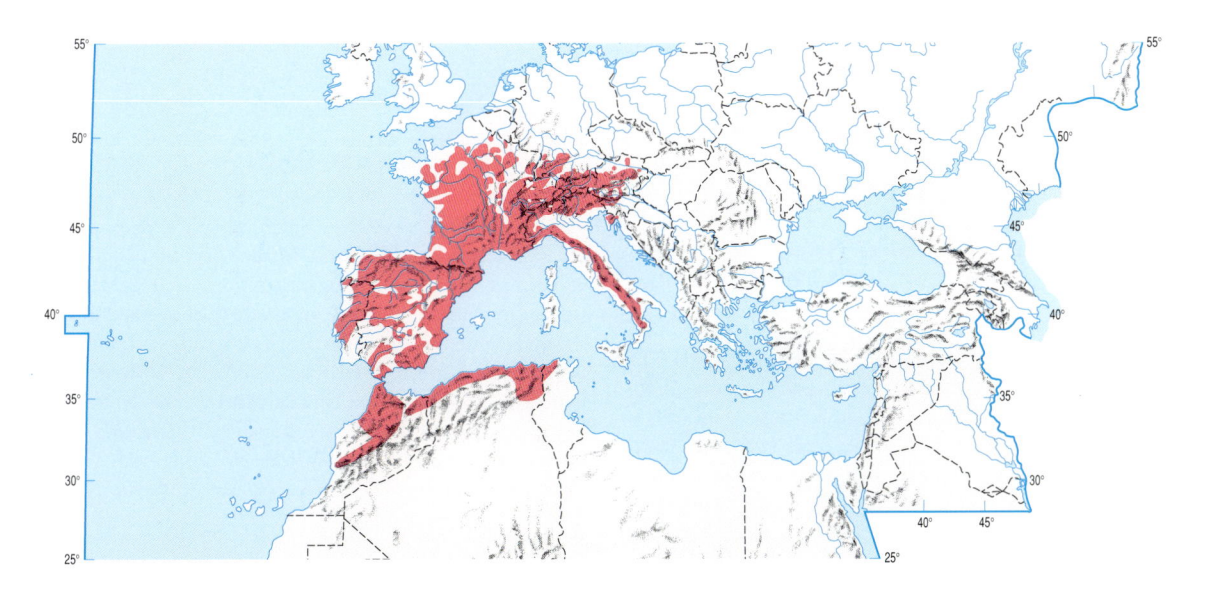

Field characters. 11.5 cm; wing-span 16–18.5 cm. Slightly larger than Chiffchaff, with somewhat longer, rather stubby bill and noticeably bulkier head and body. Medium-sized, quite bulky, rather round-headed, and full-tailed *Phylloscopus*. Plumage generally pale, dun-coloured, and relatively little-marked, but bright yellow-green area on forewing and folded flight-feathers is distinctive at close range; often a striking yellow patch on lower back and upper rump. Also shows restricted pink bill-base, large dark eye, rather grey head with diffuse supercilium, dark, pale-fringed tertials, and silky-white underparts.

Unusual pattern of dull head, bright yellow-green wings, and silky-white underparts makes Bonelli's Warbler in fresh plumage fairly easy to identify (clear view of rump not essential), but, with atypical or worn plumage, serious pitfalls exist in (1) worn adult Chiffchaff which can show mottled, grey-flecked head and back, broken supercilium, and dusty underparts (but has fine black bill, dull wings, characteristic tail-wagging, and different call), (2) worn or northern Willow Warbler (slighter billed, usually quite well-marked on head, and dull-winged), and (3) Booted Warbler which is of closely similar size and measurements and has plumage colours approached by dullest Bonelli's Warbler (but shows stronger face pattern, far less patterned tertials, and different structure, behaviour, and calls). Flight recalls Willow Warbler, being heavier and less fluttering than Chiffchaff, and paleness of underparts makes bird appear bulkier than both.

Habitat. Breeds only in middle and lower middle latitudes of west Palearctic in warm temperate and Mediterranean continental and oceanic climates, between July isotherms 19–31°C; mainly in lowlands towards the north of its range and mainly in mountains in south. At lower levels (from c. 100 m), often favours open deciduous woodland of oak, birch, beech, sweet chestnut, and other trees, provided canopy is thin and undergrowth suitable. Equally at home, however, in mixed or pure coniferous woods composed of pine, spruce, or larch. Shares some habitats with Chiffchaff, Willow Warbler, and Wood Warbler. In North Africa, breeds commonly in all oak and cedar woods except those completely lacking undergrowth.

Distribution. Breeds only in west Palearctic. Spread north in central Europe from mid-20th century; now apparently more or less halted. FRANCE. Distribution similar to that in 19th century, but spread to Sundgau area (Alsace) about 1940, and to Ardennes and Aisne in 1960s. BELGIUM. Occasional breeding: proved 1967, 1971. In 1980s, several singing ♂♂ most years Wallonia; no published records in early 1990s. NETHERLANDS. Occasional breeder: 1–5 pairs south Veluwe area 1973–7; last proved 1980. Otherwise vagrant. GERMANY. Bred Thüringen 1927, 1945 (2 pairs same locality 1944), 1963, several suspected breeding records Sachsen since 1980. POLAND. Bred Carpathians 1976. CZECH REPUBLIC. May have bred Krkonoše mountains 1977–8, but no proof.

Accidental. Britain (annual), Ireland, Netherlands, Denmark, Norway, Sweden, Finland, Poland, Slovakia, Hungary, Madeira.

Population. Apparently mostly stable. FRANCE. 100 000–1 million pairs in 1970s. GERMANY. 2300 pairs in mid-1980s. Decline Baden-Württemberg. AUSTRIA. Common in parts of eastern Alps. SWITZERLAND. 5000–9000 pairs 1985–93. SPAIN. 1.1–2.7 million pairs. PORTUGAL. 10 000–100 000 pairs 1978–84. ITALY. 50 000–100 000 pairs 1983–93. YUGOSLAVIA: SLOVENIA. 100–200 pairs. TUNISIA. Common in oak forests. MOROCCO. Uncommon to locally common.

Movements. Migratory, wintering in narrow belt along southern edge of Sahara, mostly 10–17°N, from Sénégal and southern Mauritania east to Lake Chad basin. Probably heads west of south initially in autumn; high concentrations occur within Europe, but passage inconspicuous in North Africa, presumably due to overflying. In spring, however, locally abundant in North Africa; migration is apparently on broader front. Autumn migration begins mid-July, with main movement in August, continuing to early or mid-September. Arrives early in Sénégal, from first days of September, and northern Nigeria early October. Spring migration begins late February or early March. Arrives on European breeding grounds from early April in south, mid-April in north.

Vagrant in north-west Europe. In Britain and Ireland, spring records are chiefly in south, suggesting overshooting; autumn records more widespread.

Food. Insects and a few other invertebrates. Feeding methods much as in other *Phylloscopus*. Most foraging done in tree crown, frequently on outermost branches and twigs.

Social pattern and behaviour. Solitary and apparently territorial in winter; sings frequently. Mating system apparently essentially monogamous. Incubation and brooding by ♀; young fed by both parents about equally. ♂ sings from perch in tree crown, sometimes while moving about, feeding and preening intermittently. Typically sings indefatigably at first, at virtually any time of day, stopping only if hot around midday or in afternoon or during bad weather. Song much reduced once paired and nest-site chosen (unpaired ♂♂ continue).

Voice. Song a short (up to c. 1 s) loose shivering trill of c. 7–13 units, usually all of same type within a phrase. Recalls at a distance song of Cirl Bunting and (especially when notes clearly separated) trilling part of song of Lesser Whitethroat. Compared with Wood Warbler, song-units are of more complex structure and lower pitched; inter-unit pauses also longer, so that song sounds slower; lacks descending scale and initial accelerando typical of Wood Warbler. Typical contact-alarm call differs strikingly from that of Eastern Bonelli's Warbler; commonest call a distinctly disyllabic 'chweet', 'hoo-eet', or 'clo-eee'.

Breeding. SEASON. North-west Africa: eggs laid mid-May to early July. France: laying begins end of April. Southern Germany: laying usually in 2nd week of May. Switzerland: occasionally from end of April or early May, but normally not before mid-May. Southern Spain: laying apparently 2nd and 3rd week of May. Usually 1 brood, occasionally 2. SITE. On ground, under overhanging vegetation, often in slight hollow (sometimes made by bird) in ground or bank. Nest: domed structure with side entrance, largely of grass, with small amounts of leaves and moss, lined with finer material including some hair. EGGS. Sub-elliptical, smooth and glossy; white, profusely marked with fine speckles and spots of dark red- or purple-brown, usually with heavier zone at broad end. Clutch: 5–6 (4–7). INCUBATION. 12–13 days. FLEDGING PERIOD. 12–13 days.

Wing-length: ♂ 62–67, ♀ 58–63 mm.
Weight: ♂♀ mostly 7–11 g.

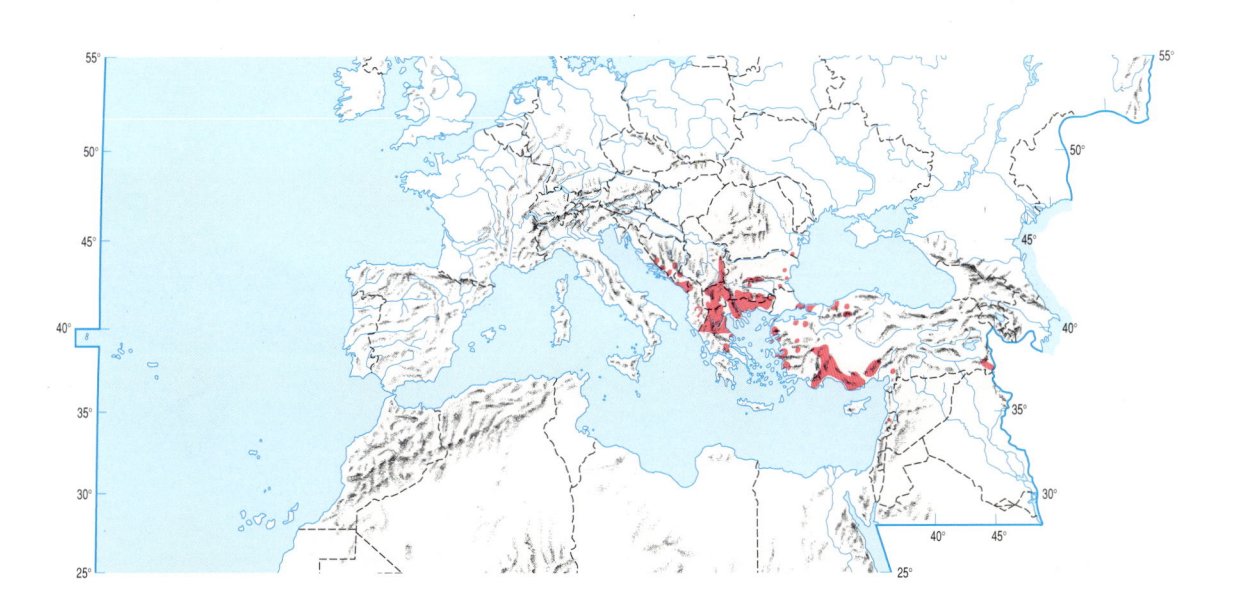

Eastern Bonelli's Warbler *Phylloscopus orientalis*

PLATES: pages 1329, 1347

Du. Balkanbergfluiter FR. Pouillot oriental GE. Balkanlaubsänger
RU. Восточная светлобрюхая пеночка

Field characters. 11.5 cm; wing-span 17–20 cm. Somewhat longer-winged than Bonelli's Warbler but otherwise size and structure similar. Plumage differs from Bonelli's Warbler in greyer or dirtier tone to upperparts, more contrasting yellowish-white edges to tertials (when fresh) but conversely less markedly green edges to wing-coverts and flight feathers (not forming often striking, broad panel of Bonelli's Warbler).

Best distinction from Bonelli's Warbler is distinctive, abrupt monosyllabic call (see Voice).

Habitat. Apparently no different from Mediterranean populations of Bonelli's Warbler; slopes and hills of maquis, pine and oak scrub, and open woodland with some undergrowth, from just above sea-level up to *c*. 1800 m in Greece and Turkey.

Distribution. Breeds only in west Palearctic. BULGARIA. Slight increase in range. RUMANIA. Recently recorded breeding in extreme south-east; probably expanding north. TURKEY. Apparently scarce, and confirmation of breeding needed Thrace and Amanus mountains. JORDAN. Bred 1991.

Accidental. Britain, Netherlands. Ukraine: one, L'vov region, April 1855; this or Bonelli's Warbler.

Population. GREECE. 10 000–30 000 pairs; stable. ALBANIA. 500–2000 pairs in 1981. BULGARIA. 50–500 pairs; slight increase. TURKEY. 1000–10 000 pairs. ISRAEL. Very rare breeder in north. 2–3 pairs bred Mt Hermon in 1990.

Movements. Migratory; winters separately from Bonelli's Warbler in north-east Africa (Sudan and Ethiopia), some perhaps further west. On migration, much less common autumn than spring on Malta and Cyprus, and between Libya and Syria (though apparently difference less pronounced in Egypt). In autumn, presumed this species conspicuous at Bosporus, western Turkey, mainly late July, and further south in Turkey last records early October. Scarce and irregular in Israel and Cyprus mainly August to early October (July to end of October); late migrants in Egypt towards end of November. Present in Sudan from end of August. Leaves winter quarters from late February, passing through Egypt and Israel principally mid-March to mid-May, and first birds arrive back on breeding grounds in southern Greece and western Turkey around beginning of April. Rare vagrant to western Europe.

Food. As Bonelli's Warbler.

Social pattern and behaviour. Not studied, but most likely very similar to Bonelli's Warbler.

Voice. Song a trill basically similar to Bonelli's Warbler, being (e.g.) also likened to song of Cirl Bunting. Differences reported (primarily from sonagraphic analysis) probably difficult to detect in the field. Main call completely unlike Bonelli's Warbler: a short, harsh, dry, rather wooden-sounding bunting-like 'pschitt', 'tüp', 'zjäpp', or Crossbill-like 'djip-djip', in series, but individual calls separated by quite long pauses. When feeding, quiet, metallic 'chirrip' or loud 'cheep' resembling sparrow *Passer* or Tawny Pipit, or incisive, abrupt 'tsioup' recalling Yellow Wagtail.

Breeding. SEASON. Eggs laid in Greece from about mid-May. Other aspects not studied but probably as Bonelli's Warbler.

Wing-length: ♂ 63–73, ♀ 60–69 mm.
Weight: ♂♀ 6–12 g.

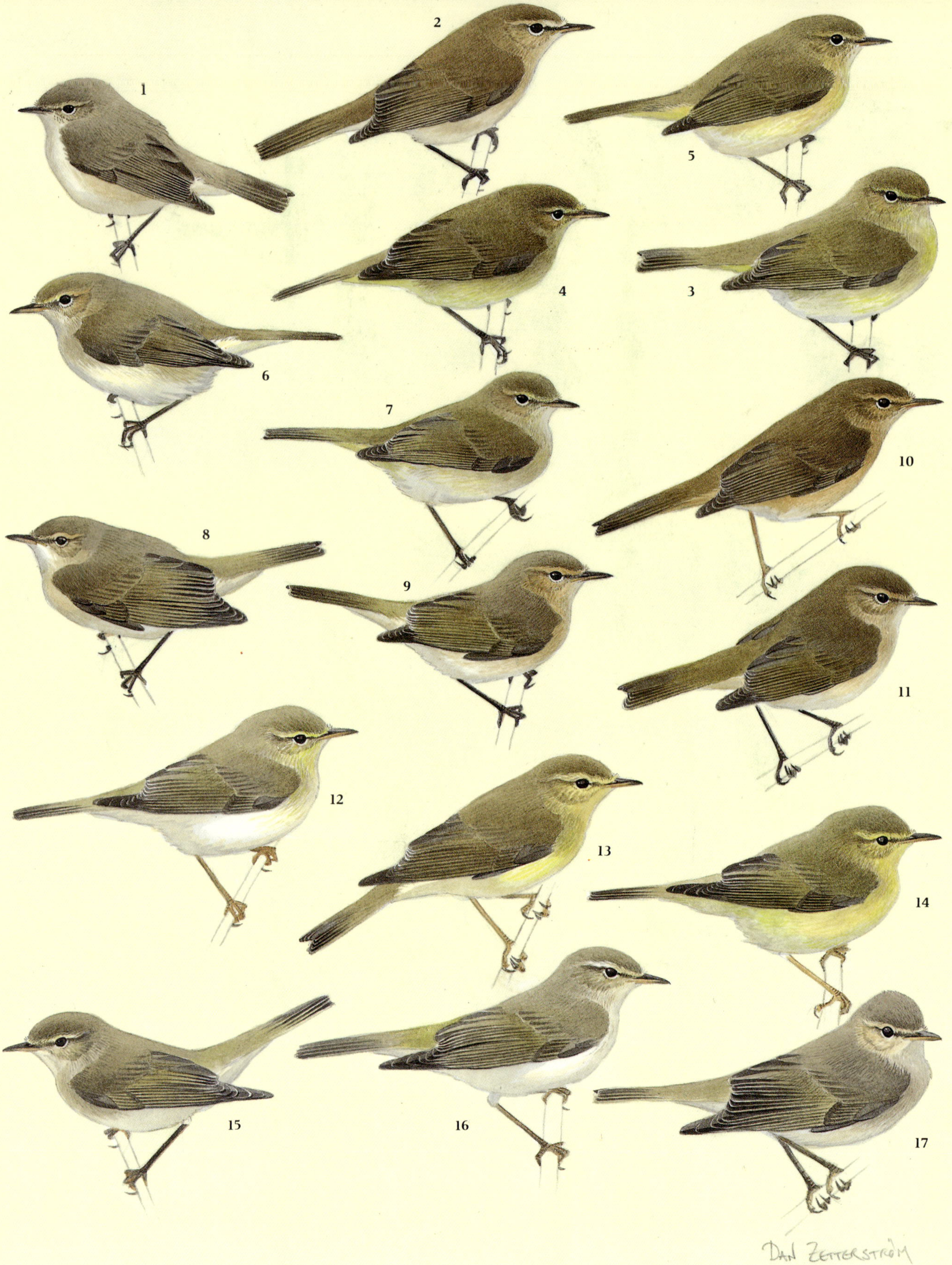
1
2
5
4
3
6
7
10
8
9
11
12
13
14
15
16
17
DAN ZETTERSTRÖM

Wood Warbler *Phylloscopus sibilatrix*

PLATES: pages 1329, 1347

Du. Fluiter   Fr. Pouillot siffleur   Ge. Waldlaubsänger   It. Luì verde
Ru. Пеночка-трещотка   Sp. Mosquitero silbador   Sw. Grönsångare

Field characters. 12 cm; wing-span 19.5–24 cm. Largest *Phylloscopus* in west Palearctic. Wings 10% longer in proportion than Willow Warbler and Arctic Warbler, with wing-point reaching beyond base of relatively short tail; even bulkier than Arctic Warbler. Large *Phylloscopus*, with strong bill, green upperparts, yellow supercilium and breast contrasting with white underbody, yellowish-white fringes to tertials, and quite strong, yellowish legs. Easiest *Phylloscopus* to identify, with plumage and voice equally distinctive.

Unmistakable in close view, when obvious bulk, long wing-points, plumage pattern and clean colours, and rather quiet behaviour combine in distinctive general character. A few adult Willow Warblers appear to have throats yellower than rest of underparts but they never show sharp demarcation on upper breast nor other markings of Wood Warbler. Confusion with Caucasus populations of Greenish Warbler possible at middle distance but not when its smaller size, longer supercilium, single wing-bar, and uniformly-toned underparts show. Flight fluent, including slow butterfly-like and whirring dragonfly-like actions in display, and expert hovering and agile flycatching while feeding. Carriage markedly horizontal, with tail held level and wings often relaxed so that their long points droop loosely below level of tertials and beyond vent.

Habitat. Breeds in temperate and boreal west Palearctic in middle and upper middle latitudes between July isotherms 15–24°C; largely in continental but marginally in oceanic climates, preferring hilly terrain to flat plains. Breeds regularly in Alps up to 1200–1300 m and exceptionally rather higher. Essentially a woodland bird, but type of woodland most favoured varies in different areas. In Switzerland, requires moist and shady woods with closed canopy and no or sparse undergrowth, being a typical bird of woods of beech, but often found also in mixed oak and hornbeam, sweet chestnut, spruce, and mixed woodlands. In Germany, also occurs in damp alder woods, birches, and in pine or spruce forest mixed with at least a few broad-leaved trees. In Belgium, prefers tall, closed beech forest, but with trees well spaced and sparse ground cover or bare soil; sometimes in coniferous or mixed woodland. In Britain, characteristic of woods of sessile oak and, while usually preferring mature canopy, will inhabit woods with trees only *c.* 7 m high, especially hillside birches. Selects woodland which has good canopy, little secondary growth, and sparse ground cover.

Distribution. Has spread in some northern parts of range, trend continuing more recently in Ireland and Fenno-Scandia, though recent slight decrease Denmark, also Moldova. BRITAIN. Local increases and decreases. Spread north in Scotland from mid-19th century, but little reliable evidence of spread this century. SPAIN. Bred Asturias 1986; may have bred Huesca province. RUSSIA. Range expansion north-east in Perm' region. TURKEY. Recorded northern Thrace late July and perhaps breeding; passage-migrant otherwise. LEBANON. Apparently bred 1887.

Accidental. Iceland, Portugal (annual in Algarve), Kuwait, Madeira.

Beyond west Palearctic, extends east to north-east Altai.

Population. Largely stable, though (marked) fluctuations reported in Belgium and Netherlands; increased Ireland, Norway, Sweden, Ukraine, markedly in Finland (where also strong fluctuations), decrease Denmark, Poland, and Moldova. BRITAIN. 17 200 territorial ♂♂ 1984–5. No recent change known, though reported to have decreased earlier in England. IRELAND. 10–50 pairs. FRANCE. 10 000–100 000 pairs in 1970s. BELGIUM. 31 000 pairs 1973–7, 20 000–60 000 in 1989–91. No signs of long-term increase or decline. LUXEMBOURG. 5000–8000 pairs. NETHERLANDS. 4000–7000 pairs 1979–85. GERMANY. 450 000–960 000 pairs. In east, 310 000 ± 149 000 pairs in early 1980s. DENMARK. 2200–23 000 pairs 1987–8. Fluctuating 1976–89, decrease 1989–94. NORWAY. 1000–10 000 pairs 1970–90. SWEDEN. 100 000–250 000 pairs in late 1980s. FINLAND. 150 000–300 000 pairs in late 1980s. ESTONIA. 500 000–1 million pairs in 1991; stable for decades. LATVIA. 1–1.3 million pairs in 1980s. LITHUANIA. Common. POLAND. Numerous and widespread. Probable decrease in recent years, especially in north-east. CZECH REPUBLIC. 80 000–160 000 pairs 1985–9. SLOVAKIA. 100 000–200 000 pairs 1973–94. HUNGARY. 100 000–150 000 pairs 1979–93. SWITZERLAND. 20 000–50 000 pairs 1985–93. ITALY. 10 000–50 000 pairs 1983–93. GREECE. 500–2000 pairs. YUGOSLAVIA: CROATIA. 1000–1500 pairs. SLOVENIA. 3000–5000 pairs. BULGARIA. 2000–20 000 pairs. RUMANIA. 120 000–180 000 pairs 1986–92. RUSSIA. Above 10 million pairs. BELARUS'. 2.4–2.6 million pairs in 1990. UKRAINE. 150 000–180 000 pairs in 1988. MOLDOVA. 35 000–50 000 pairs in 1986.

Movements. All populations migratory, wintering in sub-Saharan Africa from Sierra Leone and southern Guinea east to

(FACING PAGE) Plain Willow Warbler *Phylloscopus neglectus* (p. 1335): **1** ad.
Caucasian Chiffchaff *Phylloscopus lorenzii* (p. 1336): **2** ad worn (summer).
Chiffchaff *Phlylloscopus collybita* (p. 1337). *P. c. collybita*: **3** ad breeding (spring), **4** ad/1st winter, **5** juv. *P. c. abietinus*: **6** ad breeding (spring), **7** ad/1st winter. *P. c. 'fulvescens'* (brownish form, central Siberia): **8** 1st winter. *P. c. tristis* (Siberia and central Asia): **9** 1st winter. *P. c. canariensis*: **10** ad fresh (autumn). *P. c. exsul*: **11** ad fresh (autumn).
Willow Warbler *Phylloscopus trochilus* (p. 1340). *P. t. trochilus*: **12** ad breeding fresh (spring), **13** 1st winter, **14** juv. *P. t. acredula*: **15** ad breeding fresh (spring), **16** *'eversmanni'* (palest form of *acredula*) 1st winter. *P. t. yakutensis*: **17** 1st winter.

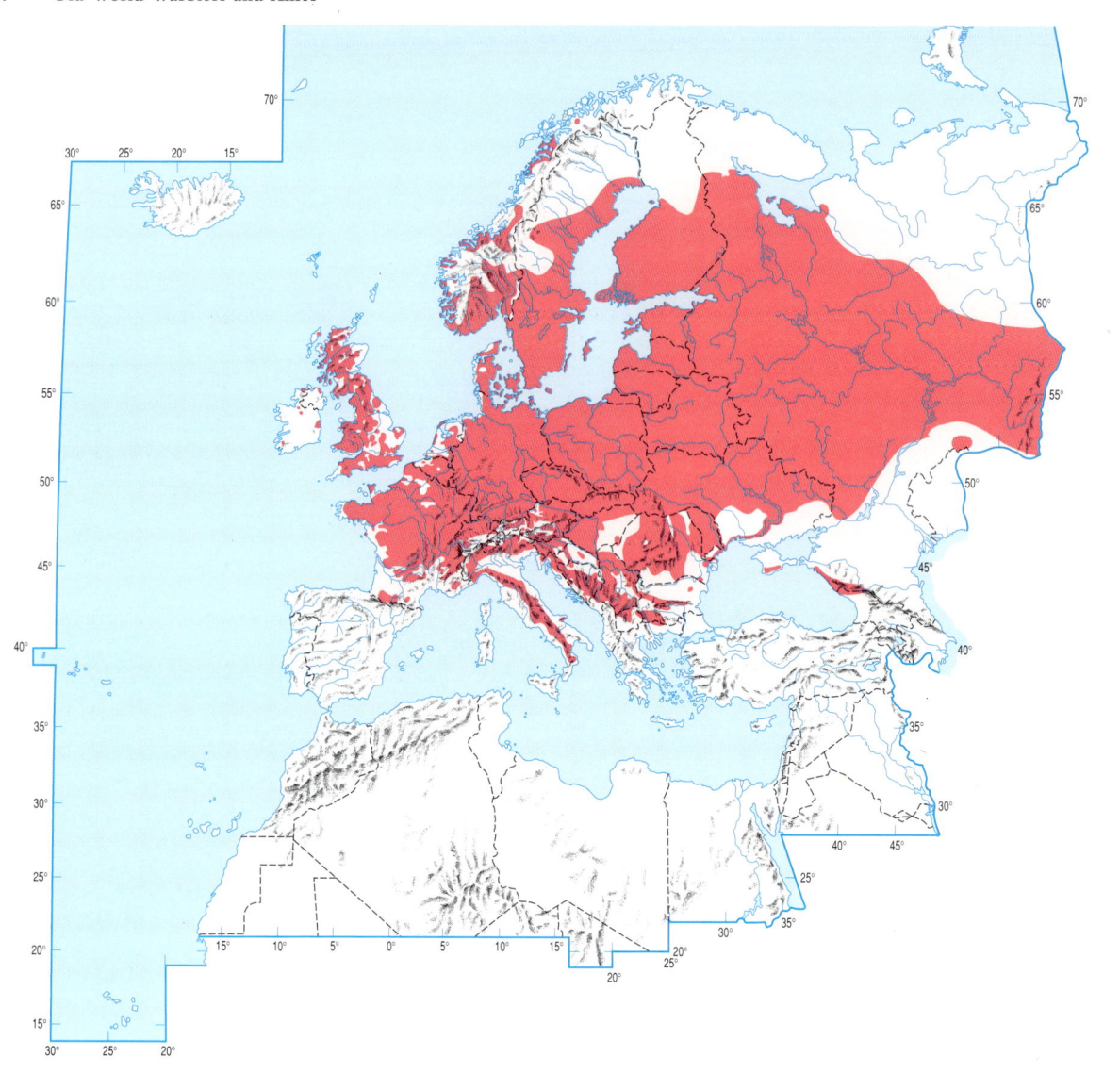

extreme south of Sudan and western Uganda, south to c. 6°S in Zaïre. In autumn, large numbers pass through central Mediterranean (Italy to Aegean), and highest concentrations winter immediately south of this region; passage regular in smaller numbers in east Mediterranean and Levant, and infrequent in west Mediterranean. Northward movement in spring follows similar course, but relatively more birds cross west Mediterranean.

Movement is rapid in both seasons. Autumn migration begins mid-July, with main departure of European breeders in August; last records (including passage birds) in western and central Europe mostly early or mid-September. Spring migration begins in March, chiefly April to mid-May. Recorded until April throughout winter range. Passage across Mediterranean late March to late May, chiefly in 2nd half of April. Movement through Europe is apparently in successive waves, with southern breeding grounds re-occupied first. Reaches western and central Europe mid- to late April, further north and east mostly in May.

Food. Insects and other invertebrates, with some fruit in autumn. Picks items off leaves (often from underside) and other parts of trees and bushes while moving through foliage (generally more slowly than Willow Warbler), sometimes fluttering, frequently hovering, or by making longer flycatching sallies.

Social pattern and behaviour. Usually solitary in winter; sometimes joins mixed-species feeding flocks. Occasional song, but no territorial behaviour. Mating system essentially monogamous but polygyny not unusual, most ♂♂ apparently attempting to attract 2nd ♀ after 1st successful pair-formation, i.e. during laying or incubation when ♂, having only territorial function up to hatching, is independent of ♀. Incubation and brooding by ♀, but young fed by both sexes; most feeding by ♂ for first 5 days (when ♀ brooding offspring), by both thereafter, share being about equal from day 7–8. Song serves to attract ♀: given strongly and persistently by ♂ (after few days' silence following arrival) before ♀ arrives, then declines sharply once pair-formation completed and breeding attempt started. Thereafter, song used for advertising occupancy and defence of territory. ♂ sings mainly in special display-area, from exposed branch below tree crown; also frequently in song-flight. In song-flight, flies with rapid, shallow, vibrating wing-beats on horizontal path between perches; repeatedly flies to certain favourite perches, and may move in criss-cross pattern about territory. Introductory part of song given from perch or once bird airborne and invariably protracted until next landing when trill normally follows.

Voice. ♂ has 2 common and distinctly different songs, not always given together. Trill-song most frequently given: a delicate series of almost identical detached units ('sip', 'tip', or 'vit', etc.) forming staccato preamble and fusing, via accelerando, crescendo, and pitch descent, into a normally shorter, fast, sibilant and ecstatically shivering trill in which units almost toneless, and sometimes metallic; the whole phrase normally lasting 2–3 s. Less frequently given piping-song is a gradually slowing series of 4–17 soft, drawn-out, clear piping 'pew' or 'tiu' notes, with musical but melancholy quality, which effect is increased by overall descent in pitch; some individual variation in quality of notes and tempo. Piping-song given independently or interpolated between trill-songs. Contact-alarm call similar to 'pew' of piping-song, but not so clear and sweet, more plaintive. Other calls include shrill, penetrating 'see-see-see' (in excitement), soft 'witwitwitwit', and other quiet high-pitched sounds.

Breeding. SEASON. Eggs mainly from mid-May, late May in north; laying may continue to early July. 1 brood usual, sometimes 2. SITE. On ground in vegetation, sometimes wedged under fallen tree or branch; often in slight depression usually made by bird. Nest: domed structure of dry grass leaves and stems and other plant material, including bark, lined with finer grasses and hair. EGGS. Sub-elliptical, smooth and glossy; white, heavily marked with dark red-brown, purple-brown, or dark brown spots and speckles, usually concentrated at broad end. Clutch: usually 5–7, 2nd clutches 3–6. INCUBATION. 12–14 days. FLEDGING PERIOD. 12–15 days.

Wing-length: ♂ 74–81, ♀ 70–78 mm.
Weight: ♂♀ mostly 8–12 g; migrants to 15 g.

Plain Willow Warbler *Phylloscopus neglectus*

PLATE: page 1332

DU. Dwergtjiftjaf FR. Pouillot modeste GE. Eichenlaubsänger IT. Luì grosso orientale
RU. Иранская пеночка SP. Mosquitero iraní SW. Dvärggransångare

Field characters. 9–10 cm; wing-span 12.5–16 cm. Smallest of plain *Phylloscopus*, hardly larger than Goldcrest and distinctly shorter-tailed than even smallest Chiffchaff or Caucasian Chiffchaff. Tiny, slight, noticeably short-tailed *Phylloscopus*, with shape, general behaviour, and 1 call all as reminiscent of goldcrest as of congener. Plumage dark and dingy, essentially grey-brown above and brown- or buff-white below; short cream supercilium and dusky eye-stripe; bare parts dark.

With its brownish plumage, goldcrest-like size and appearance, unlikely to be mistaken by experienced observer. Beware, however, that in south-east of west Palearctic it could occur with Caucasian Chiffchaff and brown form of Chiffchaff.

Song a brief twittering, recalling Goldfinch or Goldcrest. Contact-alarm calls unlike other *Phylloscopus*: rather harsh, low-pitched, resonant, and brief 'churr' or 'trrr trrr', likened to Lesser Whitethroat; also, distinctive hard, nasal, rather grating and sharp 'chit', not unlike Whitethroat.

Habitat. Breeds in lower middle latitudes of central Palearctic in dry, warm uplands, plateaux, and mountains, up to and above tree-line, in thickets, open country with sparse shrubs, or degraded woodland.

Distribution. Breeds from Iran and southern Turkmenistan east to south-central Asian mountains and western Pakistan.

Accidental. Sweden: one, Landsort island, south-east of Stockholm, October 1991. Jordan: one, Azraq, April 1963.

Movements. Partial short-distance migrant: some birds winter within breeding range, some make altitudinal movements, others move further afield.

Wing-length: ♂ 48–55, ♀ 47.5–51.5 mm.

Caucasian Chiffchaff *Phylloscopus lorenzii*

Du. Kaukasus-tjiftjaf   Fr. Pouillot de Lorenz   Ge. Kaukasuslaubsänger
Ru. Кавказская пеночка   Sw. Kaukasisk gransångare

Field characters. 10.5–11 cm; wing-span 18–21 cm. Close in size to Chiffchaff, but with up to 10% longer tail. Small leaf warbler, with structure, plumage, and behaviour recalling Siberian race of Chiffchaff. Upperparts warm grey-brown, lacking olive-green tones (even on fringes of tertials); underparts dull buff-white, wearing paler; best distinction from Chiffchaff is head pattern which shows usually pale, buff-white supercilia joining over bill and emphasized by dusky edge to front $\frac{2}{3}$ of crown and dark eye-stripe.

Only recently separated as monotypic species and subject to great risk of confusion with Chiffchaff, especially Siberian race (*tristis*); field study incomplete and identification of silent, untrapped vagrant untrustworthy.

Habitat. Restricted to upper part of forest belt and subalpine and alpine meadows of montane areas of Caucasus and neighbouring areas to south and west. In Caucasus, habitat very varied though breeds within narrow band *c.* 1.6 km wide between 1100 and 2700 m (mainly 1800–2300 m), generally higher than Chiffchaff. Highest densities noted in subalpine and krummholz broad-leaved and coniferous forest with well-developed undergrowth, as well as on wooded cliffs. Distribution correlates closely with that of *Rhododendron caucasicum*, especially where birches and rowans emerge above this shrub. Also breeds in damp places along rivers or streams, above tree-line on steep slopes with shrubs and tall herbs, in dark coniferous forest belt with some deciduous trees below subalpine zone, or in park-type oak forest. In north-east Turkey, in pine and juniper forest above *c.* 1700 m, mainly above range of Chiffchaff. In winter, sometimes in very different habitats such as subtropical Persian Gulf region.

Distribution. Uneven, with local concentrations typical. AZERBAIJAN. Occurs in Great and Little Caucasus mountains, but range poorly known. TURKEY. Confusion with Chiffchaff means distribution not well known.

Accidental. Israel.

Beyond west Palearctic, extends only to northern Iran.

Population. One of the most numerous species in many of its typical habitats in subalpine zone. GEORGIA. On Tviberi river, in Upper Svanetia, sometimes up to 10 singing ♂♂ audible from one spot. AZERBAIJAN. Uncommon to common. TURKEY. 1000–10 000 pairs.

Movements. Migratory, though many birds make only short-distance altitudinal movements within (e.g.) Azerbaijan, Armenia, and Georgia; more distant wintering grounds poorly known but thought to be mainly in valleys of Tigris and Euphrates in Iraq, and adjacent parts of Turkey, Syria, and Iran; recorded in Basra in southern Iraq, and in December north of Caucasus. Leaves breeding grounds in north-west Caucasus September–October, and passage or local birds recorded late October to November in Armenia and Iranian Azerbaijan. In spring recorded from mid-March in Armenia, but main movement apparently late April to mid-May, and first arrivals in north-west Caucasus probably end of April or early May. In Georgia in one year passage still conspicuous in high mountains at end of May.

Food. Mainly insects, which it frequently catches in flight, also spiders and small snails; young fed mainly caterpillars. Forages principally in crowns of trees and on tall herbs.

Social pattern and behaviour. On upper reaches of Tviberi river (Georgia) 5–7(–10) singing ♂♂ can be heard from one spot. Neighbourhood groups of 3–5 pairs nest close together, with 50–70 m between ♂ song-territories, though 100–120 m is more usual elsewhere. In zone of overlap at 1100–1900 m generally no interbreeding with Chiffchaff, and hybrids recorded only very rarely. Nests of the two species can be 150–200 m apart in Caucasus, but territories apparently mutually exclusive. Apart from at nest, remainder of behavioural repertoire hardly studied.

Voice. Song very like Chiffchaff, but with tendency to larger repertoire of different units and more variable phrase structure, with constant change of order of units. Commonest call distinctive 'psee' or 'pseeu', recalling Dunnock and eastern races of Chiffchaff.

Breeding. SEASON. Can be protracted and timing depends on weather, altitude, etc. Eggs laid from last third of May; most birds nesting at 1800–2100 m lay in first half of June, birds above $c.$ 2200 m sometimes only around mid-July; overlaps with Chiffchaff only rarely and in general $c.$ 1 month later. Normally 1 brood; rarely 2 at low altitudes. SITE. Usually on ground but sometimes in shrub or young tree (mainly rhododendron or small conifer) up to 1 m above ground. Nest: spherical structure of leaves and stems of grass, tree leaves, bark fibre, etc. with side entrance hole, lined with fine grasses, rootlets, hair, wool, plant down, and feathers. EGGS. Sub-elliptical to long oval, smooth and slightly glossy; white with reddish-brown streaks and short squiggles concentrated towards broad end. Clutch: 4–5(–6). INCUBATION. 14–15 days. FLEDGING PERIOD. 14–16 days.

Wing-length: ♂♀ 55–64 mm.
Weight: ♂♀ 8–9 g.

Chiffchaff *Phylloscopus collybita*

PLATES: pages 1332, 1347

DU. Tjiftjaf   FR. Pouillot véloce   GE. Zilpzalp   IT. Luì piccolo
RU. Пеночка-теньковка   SP. Mosquitero común   SW. Gransångare

Field characters. 10–11 cm; wing-span 15–21 cm. Slightly smaller and more compact than Willow Warbler, with slightly stubbier bill, more rounded head, and noticeably shorter wings but proportionately longer tail. Small, slight but often rather chesty warbler, less graceful and active than Willow Warbler and not as slim as Greenish Warbler. Tail movement distinctive. West-central and south-western races brownish-olive above and dull yellowish below, more uniform than adult Willow Warbler and lacking contrasting features except for dark bill, pale eye-ring within dull yellow supercilium, and usually dark legs. Northern and eastern races less warm, more olive, cooler brown or even grey above, and less yellow, even strikingly white below; many suggesting Greenish Warbler due to more distinct and paler supercilium, single pale wing-bar (extending over all tips of greater coverts at least in 1st autumn), and greater contrast between upper- and underparts. Dark bastard wing often visible as discrete mark below 'shoulder'.

Except for birds singing typical onomatopoeic song, one of the most confusing and difficult warblers but all races show rounded wings in flight and have diagnostic tail movement—a loose drop then sideways wag, much more obvious than short upwards flick of congeners. Commonest call of north-western birds a soft but quite emphatic 'hweet'; of eastern races a shriller and more plaintive 'sweeoo' or 'pseet', recalling young chicken in distress.

Habitat. Breeds in west Palearctic in upper and lower middle latitudes, between July isotherms 10–26°C in continental and oceanic boreal, temperate, and Mediterranean climatic zones. Basically a bird of mature lowland woodland with not too dense canopy and fairly copious variety of medium or tallish undergrowth, but may even extend (often thinly) to treeline in mountain woodland or to upper zone of closed forest. In west of range, prefers old deciduous or mixed woodland, extending into parks, large gardens with shrubberies, cemeteries, and coppiced woods and plantations, and even along hedgerows with tall trees. In comparison with Willow Warbler, however, much less attracted to younger, lower, and scrub or thicket types of vegetation; essentially a woodland bird, although less so than Wood Warbler.

Distribution. FAEROES. Bred 1981. BRITAIN. Has spread in Scotland since $c.$ 1950. IRELAND. In 1850, known in only 7 of 32 counties; by 1900, spread throughout, extending to western Kerry in 1950s. FRANCE. Corsica: first proved breeding 1986. NETHERLANDS. Recent spread; South Flevoland colonized. DENMARK. Increase, probably by 25%, 1974–94. NORWAY. Some northward spread since $c.$ 1970. SWEDEN. Small isolated colony established in Skåne since 1970, probably nominate *collybita*. TURKEY. Map includes some unidentified populations, this species or Caucasian Chiffchaff. SYRIA. Singing birds in June and gap of $c.$ 2 months between such records and last spring passage-migrants strongly suggest breeding in north-west. TUNISIA. Scarce breeder in oak forests of north-west.

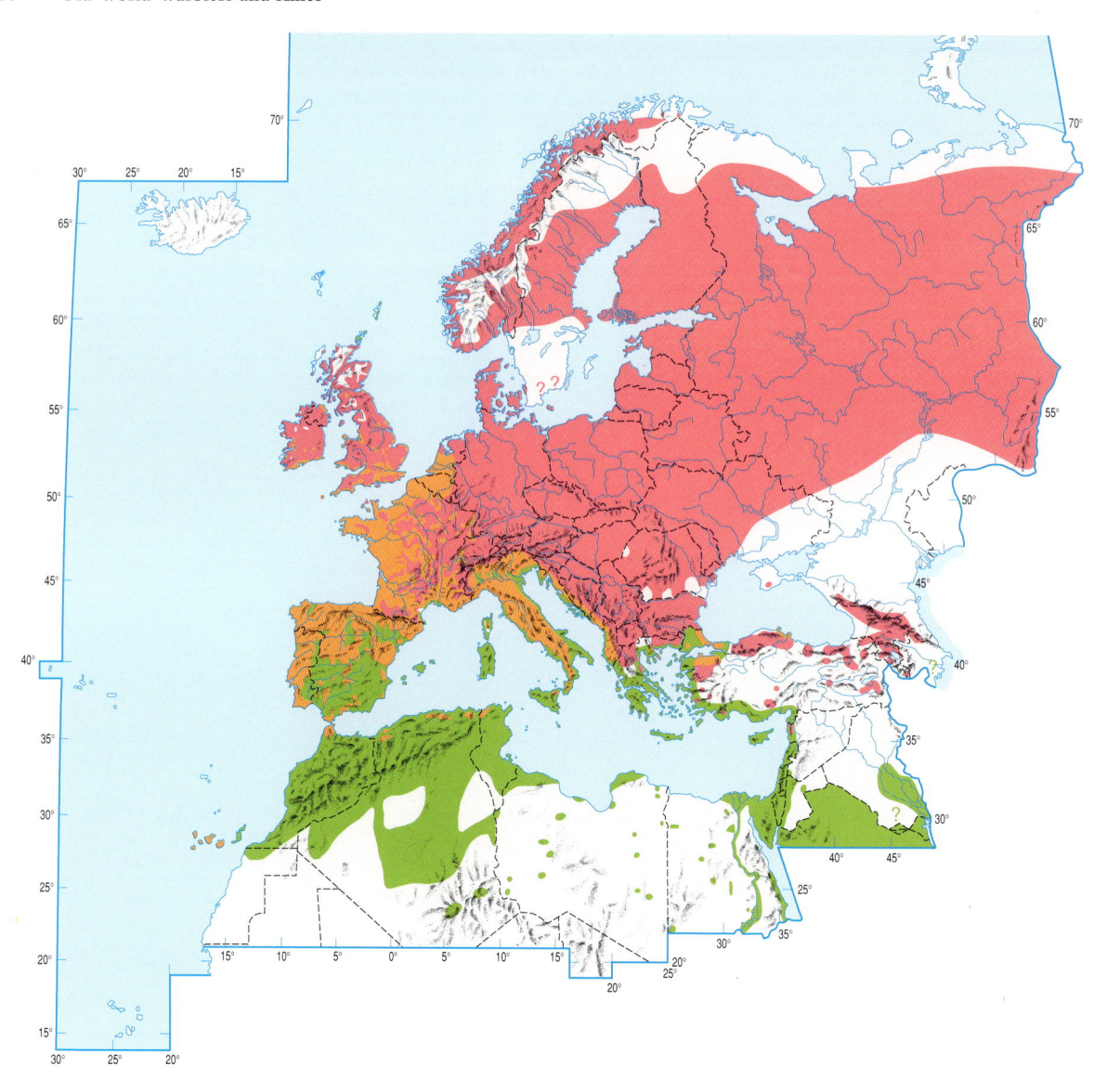

MOROCCO. Restricted to Tangier peninsula; scarce and very local. CANARY ISLANDS. Widespread and common on Gran Canaria, Tenerife, La Gomera, La Palma, and El Hierro. *P. c. exsul* (breeding Lanzarote, perhaps also Fuerteventura) very poorly known.

Accidental. Bear Island, Jan Mayen, Iceland (annual), Madeira, Cape Verde Islands.

Beyond west Palearctic, extends across Siberia to Yakutiya and Lake Baykal; also breeds northern Iran and Pamir-Alay mountains to north-west Himalayas.

Population. Stable in most of range, but increase Britain, Denmark (marked), and Ukraine, slight decline Finland and Moldova. BRITAIN. 640 000 territories 1988–91. Decline in 1970s; has fluctuated since, but currently increasing (in Scotland, since 1950s). IRELAND. 290 000 territories 1988–91. Increase since 1950s. FRANCE. Over 1 million pairs in 1970s. BELGIUM. 150 000–170 000 pairs 1973–7. LUXEMBOURG. 25 000–30 000 pairs. NETHERLANDS. 125 000–225 000 pairs 1979–85. Perhaps slight increase due to new parks and afforestation. GERMANY. 3.7 million pairs in mid-1980s. In east, 700 000 ± 301 000 pairs in early 1980s. DENMARK. 7000–87 000 pairs 1987–8. NORWAY. 100 000–500 000 pairs 1970–90. SWEDEN. 100 000–400 000 pairs in late 1980s. FINLAND. 150 000–300 000 pairs in late 1980s. Slight decrease after rather stable for decades. ESTONIA. 200 000–500 000 pairs in 1991. LATVIA. 500 000–600 000 pairs in 1980s. LITHUANIA. Common. POLAND. Widespread and numerous. CZECH REPUBLIC. 800 000–1.6 million pairs in 1985–9. SLOVAKIA. 600 000–1 million pairs 1973–94. HUNGARY. 400 000–600 000 pairs 1979–93. AUSTRIA. Very common. SWITZERLAND. 350 000–400 000 pairs 1985–93. SPAIN. 340 000–750 000 pairs; fluctuating. PORTUGAL. 10 000–100 000 pairs 1978–84. ITALY. 300 000–800 000 pairs 1983–93. GREECE. 10 000–40 000 pairs. ALBANIA.

2000–5000 pairs in 1981. Yugoslavia: Croatia. 200 000–250 000 pairs. Slovenia. 300 000–500 000 pairs. Bulgaria. 500 000–5 million pairs. Rumania. 1.2–1.6 million pairs 1986–92. Russia. 10 million pairs. Belarus'. 850 000–950 000 pairs in 1990. Ukraine. 200 000–230 000 pairs in 1988. Moldova. 65 000–80 000 pairs in 1986. Turkey. 10 000–100 000 pairs.

Movements. Most populations migratory; Canary Islands populations sedentary. Western populations winter within and south of breeding areas, eastern populations vacate breeding areas entirely. Winters south to northern Afrotropics, Arabia, and northern India. Within Britain, recoveries provide evidence that some breeding birds remain for winter, and that individuals return to same area in successive winters; also evidence of continental birds wintering in Britain. Autumn movement begins August and is protracted, though termination of passage often masked by overwintering. Spring departure from wintering areas begins in February, with main movement March–April. Migrants reach all south coast observatories in Britain by beginning of March; peak early April in west (where passage heavier and more sustained than in east), not until mid-April in east. Main arrival in Switzerland from mid-March, in mountains from early April. In southern Sweden, passage early April to mid-June, peaking in early May. Further east, movement is later: not before mid-April in Moscow region.

Food. Almost wholly insects; also some fruit in autumn and winter; occasionally nectar. Forages mainly high in tree canopy, also in bushes and lower down in dense thickets.

Social pattern and behaviour. Often solitary in winter, and may be territorial, with regular song; sometimes in flocks. Mating system apparently varies between monogamy and irregular polygyny. Incubation, brooding, and (almost all) nest-sanitation by ♀. Young also fed mainly by ♀, who is capable of rearing young alone. ♂'s participation varies considerably: generally little, especially for 2nd brood. Song of ♂ used for advertising territory and for contact with ♀; much reduced after pair-formation, especially if density low. ♂ sings normally from perch or while fluttering about in foliage and feeding, also in flight between perches (though no ritualized song-flight). Sings on breeding grounds with varying intensity from arrival up to July, with pause (for moult) July–August, and resurgence of high-intensity song in September(–October); some autumn song given by (territorial) adults, some by juveniles.

Voice. Song of ♂ in most of Europe a rather monotonous series of single clear units traditionally rendered 'chiff' and 'chaff' in English, 'zilp' and 'zalp' in German, i.e. one note higher pitched than the other; frequently a 3-note pattern. Given in phrases of variable length separated by pauses. Delivery measured and metronome-like, though some variation. Monotony countered to some extent by variation in number of unit-types per song (usually at least 3, but up to 10 per individual repertoire); also by irregularity of sequence. Song of birds breeding in south-west France and Iberia is distinctly different; normally of 3 segments (sequences of identical units), producing characteristic pattern: (i) 1–4(–7) metronome-like 'djep' units (closest to typical Chiffchaff), with relatively constant inter-unit pauses; (ii) 0–2(–3) emphasized 'swüid' units similar to hweet-call (see below) of nominate *collybita*; (iii) rapid trilling 'tetettettettett' of 1–4(–7) units, having accelerando and crescendo and some resemblance to song of Bonelli's Warbler. Song apparently similar in north-west Africa, but with halting rhythm, slowing up at end. Song of Canary Islands birds immediately recognizable as Chiffchaff, but units more complex; also generally faster, with shorter inter-unit pauses. Most frequently used and well-known call a soft and plaintive 'hweet', 'hüit', or 'fuit', resembling hooeet-call of Willow Warbler but almost monosyllabic, slightly more emphatic, higher pitched, and harsher. Some variation in different regions: e.g. calls of eastern and Iberian races typically shrill and descending in pitch. Repertoire includes a number of other monosyllabic or rattling calls.

Breeding. Season. Western Europe (including Britain): eggs mainly from late April. Central Europe: laying begins early May. Finland: laying begins 2nd half of May. Spain: in south, full clutches recorded mid-April to May. North Africa: eggs from early May. Canary Islands: nests with eggs recorded late January to mid-June; earliest at lower altitudes. 2 broods except in north of range. Site. On or close to ground concealed in tall vegetation, low bushes, tree branches, or creepers. Nest: domed structure with side entrance; made of dry grass stems and leaves, moss, and other plant leaves, lined with finer grasses and feathers. Eggs. Short sub-elliptical, smooth and glossy; white, sparingly spotted and speckled with dark purple, purple-brown, or black, rarely red-brown; markings often mainly at broad end. Clutch: 4–7 (3–9); 2nd clutches and replacement clutches smaller than 1st clutches. Incubation. 13–15 days. Fledging Period. 14–16 days.

Wing-length: *P. c. collybita* (most of western Europe): ♂ 57–64, ♀ 53–60 mm. *P. c. brehmii* (Iberia): ♂ 59–64, ♀ 54–59 mm. *P. c. exsul* (eastern Canary Islands): ♂ 50–52, ♀ 47–48.5 mm. *P. c. canariensis* (Tenerife): ♂ 52–57, ♀ 48–50.5 mm.

Weight: European populations: ♂♀ mostly 6–10 g.

Geographical variation. Slight in west Palearctic, with some intergradation. Nominate *collybita* occupies most of west and centre; *P. c. abietinus* (paler and greyer) north and eastern areas from northern Scandinavia and eastern Europe eastwards, and probably Turkey. *P. c. brehmii* of south-west France, Iberia and north-west Africa similar in size and plumage to nominate *collybita*, but distinct in voice (which see) and should perhaps be treated as separate species. *P. c. canariensis* and *P. c. exsul* of Canary Islands smaller, the former also with darker plumage.

Willow Warbler *Phylloscopus trochilus*

PLATES: pages 1332, 1347

Du. Fitis   Fr. Pouillot fitis   Ge. Fitis   It. Luì grosso
Ru. Пеночка-весничка   Sp. Mosquitero musical   Sw. Lövsångare

Field characters. 10.5–11.5 cm; wing-span 16.5–22 cm. Length similar to Blue Tit but slim and elliptical in shape; slightly larger and usually noticeably longer winged than Chiffchaff. Small, slight, graceful warbler, light on the wing and active in cover. Epitome of *Phylloscopus*, particularly of species which (typically) lack wing-bars and striking supercilium. Plumage usually quite pale and bright, with essentially greenish-olive upperparts and yellow-white underparts, yellow-white supercilium (not markedly contrasting), and (typically) yellow-brown legs. Song distinctive.

Phylloscopus warblers share rather small size, slim, elliptical shape, light flight, and rather subtle plumage colours, but identification of Willow Warbler and almost all other species requires careful study and frequent practice. Willow Warbler is commonest and most widespread species. Most singing ♂♂ are no problem, as song of Willow Warbler is one of the most easily learnt and remembered bird sounds. If not singing, Willow Warbler is traditionally confused mainly with Chiffchaff, being distinguished from it by relatively clean plumage tones, fairly pale legs, and longer, nearly disyllabic call. Important to understand that Willow Warbler arrives after and departs before Chiffchaff, is rather larger than average for *Phylloscopus*, and has shape characterized by spiky bill and lengthy silhouette, with quite marked extension of wing-points, tail, and slim rear body. Also shows rather calm behaviour except for sudden aggressive aerial pursuit of other passerines. Flight light and fast, with flitting action; flight-silhouette evenly balanced, with wings and tail appearing quite long and wings never showing rounded fan-shape of Chiffchaff. Regularly flicks tail but without sideways twitch of Chiffchaff.

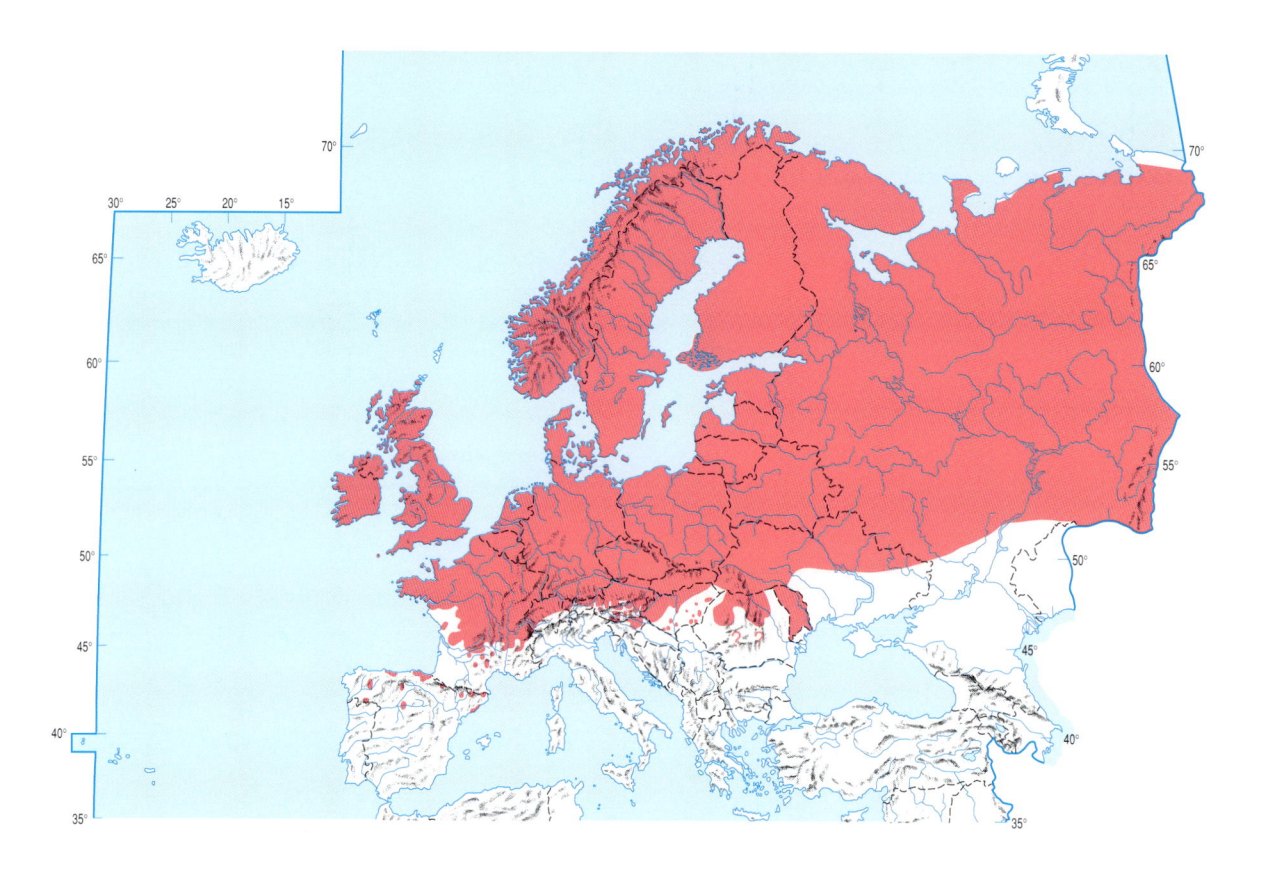

Habitat. Breeds in west Palearctic in middle and upper latitudes within temperate and boreal climatic zones, continental and oceanic, between July isotherms of 10 and 22°C. Unlike similar Chiffchaff, extends widely into arctic but only marginally into Mediterranean region; elsewhere, most of the breeding ranges and habitats of the 2 species overlap. Differs in being a bird less of mature woodland than of scrub, second growth, and transitions to more open habitats. Unlike Chiffchaff, will readily settle on shrubby, bushy, or cleared ground starting regrowth, even in absence of trees. Thus attracted to fringe areas such as birches beyond arctic taiga, heathlands, or forest clearings. Although resorting to woods of many different tree species, both deciduous and coniferous, prefers those which bear most accessible insect food, among which birch is outstanding. In far north, breeds on tundra in glades among willows. As an opportunist species can fit readily into small and shifting ecological niches and thus maintain large population fairly free from competition.

Distribution. FAEROES. Bred 1985. BRITAIN. Scotland: colonized Outer Hebrides and Orkney in 2nd half of 19th century. Bred sporadically Shetland (1906, 1949). Some spread in parts of Scotland and Wales. IRELAND. Perhaps slight spread west. HUNGARY. Slight increase. PORTUGAL. At least 1 pair bred Algarve 1989. ITALY. May breed in north. GREECE. Bred (probably exceptionally) near Thyrea in Evros valley (Thrace); date unknown. SLOVENIA. First recorded breeding for former Yugoslavia, Ptuj, 1979. TURKEY. Apparently only passage-migrant; no evidence of breeding.

Accidental. Spitsbergen, Bear Island, Iceland (annual), Madeira, Canary Islands.

Beyond west Palearctic, extends east across Siberia in narrowing band to c. 180°E.

Population. Largely stable, though increase Sweden (probably), Finland, Hungary, and Slovenia, decrease Britain, Belgium, Netherlands, Switzerland, and Moldova. BRITAIN. 2.3 million territories 1988–91. Stable (minor fluctuations) late 1960s to mid-1980s, but marked decline (47%) in south 1986-93. IRELAND. 830 000 territories 1988–91. FRANCE. 2.5–4.5 million pairs. BELGIUM. 130 000 pairs 1973–7. Apparently stable, but slight decline in some regions. LUXEMBOURG. 15 000–20 000 pairs. NETHERLANDS. 250 000–400 000 pairs 1979–85. Decline reported similar to that in southern Britain. GERMANY. 2.6 million pairs in mid-1980s. In east, 860 000 $\pm$ 327 000 pairs in early 1980s. DENMARK. 130 000–1.5 million pairs 1987–8. NORWAY. 2–10 million pairs 1970–90. SWEDEN. 8–12 million pairs in late 1980s. FINLAND. 8–13 million pairs in late 1980s. Long-term increase due to opening up of forests. ESTONIA. 1–2 million pairs in 1991. Recently stable; probable increase in 1950s–60s. LATVIA. 500 000–600 000 pairs in 1980s. LITHUANIA. Common. POLAND. Widespread and numerous. CZECH REPUBLIC. 500 000–1 million pairs in 1985–9. SLOVAKIA. 400 000–600 000 pairs 1973–94. HUNGARY. 8000–10 000 pairs 1979–93. AUSTRIA. Fairly common. SWITZERLAND. 6000–9000 pairs 1985–93. SPAIN. 20–100 pairs; fluctuating. ALBANIA. Up to 50 pairs in 1981. YUGOSLAVIA: SLOVENIA. 300–500 pairs. RUMANIA. 6000–10 000 pairs 1986–92. RUSSIA. 10 million pairs. BELARUS'. 950 000–1.1 million pairs in 1990. UKRAINE. 70 000–80 000 pairs in 1988. MOLDOVA. 5000–10 000 pairs in 1986.

Movements. All populations migratory, wintering extensively in sub-Saharan Africa from southern Sénégal east to Ethiopia, south to South Africa. From extreme east of breeding range, route thus exceptionally long—at least 12 000 km. Within Europe, birds from west of range head south to south-west in autumn; north Scandinavian and Finnish birds head south to south-east, indicating migratory divide in border area. Spring passage apparently on broader front, with recoveries of ringed birds averaging further east than in autumn. Afrotropical recoveries show that birds from west and central Europe, including Norway and southern Sweden, winter mainly in West Africa, birds from northern Scandinavia and Finland winter in central, East, and southern Africa. Migration earlier in autumn and later in spring than in Chiffchaff. Southward movement begins late July, and passage through Europe and across Mediterranean mostly completed by end of September. Reaches northern parts of African winter quarters in September, South Africa mostly from mid-October.

Spring migration begins late February to March. Passage through Mediterranean mid-March to mid- or late May, chiefly April, averaging later in east than west. First arrivals in Britain late March, main arrival in April. Central European breeding grounds also re-occupied at end of March and in April. Main arrival in southern and central Sweden 21 April to 15 May, northern Sweden 11–25 May. At Lake Ladoga (north-west Russia), most local birds arrive by 15–20 May, passage continuing to early or mid-June.

Food. Insects and spiders; in autumn also berries. Food obtained mostly by picking from leaves, twigs, and branches; also by flycatching.

Social pattern and behaviour. Solitary or in small parties in winter, sometimes with other species; some birds apparently territorial, at least for limited periods, and song not uncommon. Mating system varies between monogamy and polygyny (♂ plus 2–3 ♀♀). Factors favouring polygyny include long breeding season, ability quickly to replace lost clutch or brood, (locally) high density, less share taken by ♂ in care of offspring, mutual tolerance of ♀♀, and ability of ♀ to rear young alone. Polygyny sometimes associated with acquisition of secondary territory, but this not a prerequisite. Pair-formation takes place on breeding grounds where ♂♂ arrive ahead of ♀♀. All incubation and brooding by ♀. Both sexes feed young; ♂ often does less than ♀. ♂ usually sings from regular perch on top or side-branch of tree or bush. Unlike Chiffchaff, usually sings while still (not moving about), though typically turns head or body and gently wags tail. Apparently no ritualized song-flight, but will sing during chases. ♂♂ regularly sing on spring migration, and on breeding grounds from arrival. In Britain, song wanes from mid- or late June or early July; irregular but fairly frequent song given up to early August.

Voice. Song of ♂ a lyrical drooping melody of gentle, pure notes in silvery, rippling phrase, comprising several different segments (sequences of usually 2–5 similar notes) and lasting normally c. 3 s. Typically high pitched and faint at start, followed by crescendo to near middle, then dying away to quiet ending, also slowing down. Pitch may be constant within segments, but falls overall, especially at start. Each ♂ has several different song-types, and songs of neighbours usually distinctly different. Songs typically given at rate of 4–8 per min. Most commonly used and familiar call a plaintive, soft, and relatively quiet 'hooeet', similar to equivalent call of Chiffchaff, but rather lower pitched, less emphatic and, importantly, more distinctly disyllabic. Given by both sexes, primarily when disturbed; usually the only call from June, once song has faded. Other calls include a variety of chittering, rattling, or trilling sounds, given in extreme alarm, excitement, or distress.

Breeding. SEASON. Central and western Europe, including Britain: eggs from end of April to June. Southern Finland: eggs laid late May to early July. North-west European Russia: laying begins 2nd third of May and lasts to 2nd third of July, but great majority of birds lay last 5 days of May and first 5 days of June. Usually 1 brood; rarely 2. SITE. On ground, well concealed in vegetation, including herbs, bases of low shrubs and trees, and grass tussocks; can be up to 4.8 m above ground in tree, bank, crevice, or creeper. Nest: domed structure of dry grass, leaves, stems, moss, lichen, twigs, conifer needles, bark fibre, often with rootlets and pieces of rotten wood, lined with finer grasses, rootlets, animal hair, and feathers. EGGS. Short sub-elliptical, smooth and glossy; white with fine speckles of red and red-brown, sometimes with sparse larger spots of red, red-brown, or buff, sometimes concentrated at broad end. Clutch: 4–8 (3–9). INCUBATION. 12–14 days. FLEDGING PERIOD. 11–15 days.

Wing-length: ♂ 62–72, ♀ 60–68 mm.
Weight: ♂♀ mostly 7–12 g; migrants to 15.7 g.

Geographical variation. *P. t. trochilus* of western Europe replaced in Scandinavia and eastern Europe by variable populations with paler, browner plumage (*P. t. acredula*), merging into distinctly different and larger form of eastern Siberia (*P. t. yakutensis*).

Ruby-crowned Kinglet *Regulus calendula*

PLATE: page 1343

DU. Roodkroonhaantje   FR. Roitelet à couronne rubis   GE. Rubingoldhähnchen   IT. Regolo americano
RU. Рубиноголовый королек   SP. Reyezuelo crestado   SW. Rödhuvad kungsfågel

Field characters. 10 cm; wing-span 14.5–16–5 cm. Slightly but distinctly larger than west Palearctic *Regulus* but with similar structure. Instantly recalls other *Regulus* in general character and plumage, but lacks obvious crown- and eye-stripes. Bold whitish eye-ring diagnostic, far more striking than on Goldcrest.

Confusion with all other adult *Regulus* ruled out by lack of long, contrasting crown- and eye-stripes. Confusion of ♀ and juvenile with juvenile Goldcrest quite possible in late summer and early autumn: in close view, rather larger size, apparently rounder head, and particularly bolder, less diffuse eye-ring should show; stronger contrast between outer edges and centres of wing- and tail-feathers may also be of use. Flight and behaviour much as Goldcrest. Call a low-pitched, husky 'je-dit'.

Distribution. Breeds in North America from Alaska east to Labrador and Newfoundland, south to southern California, Guadalupe island (Mexico), New Mexico, northern Michigan, and Nova Scotia.

Accidental. Iceland: 1st-year found exhausted (specimen) Vestmannaeyjar, November 1987. Britain: claimed 19th century records Loch Lomond (Scotland) and Gloucester not accepted due to doubtful origin.

Movements. Chiefly migratory, wintering within and south of breeding range. Autumn movement chiefly September–October. In addition to vagrant in Iceland, several have been recorded in mid-Atlantic, late September and October.

Wing-length: ♂ 55–62, ♀ 52–59 mm.

Goldcrest *Regulus regulus*

PLATE: page 1343

DU. Goudhaantje   FR. Roitelet huppé   GE. Wintergoldhähnchen   IT. Regolo
RU. Королек   SP. Reyezuelo sencillo   SW. Kungsfågel

Field characters. 9 cm; wing-span 13.5–15.5 cm. 20–30% smaller than Chiffchaff, with proportionately even shorter wings and tail; slightly smaller and less bulky than Firecrest. Smallest bird of west Palearctic, with compact but essentially warbler-like form and busy tit-like behaviour. Plumage essentially olive above and buff-white below, with 2 white wing-bars (lower contrasting with black bases to flight-feathers). Dark eye set in pale face is more noticeable than (in adult) black-edged yellow/orange-centred crown. Flight lighter than even smallest *Phylloscopus* warbler; hovers frequently. Moves restlessly among foliage.

Tiny size, plump short-tailed form, and most features of plumage and voice shared fully by Firecrest, but, in brief distant view (or with wishful thinking), confusion also possible with small, striped *Phylloscopus*. Flight distinctive: over distance, a mix of whirring wing-beats and occasional jinking, giving impression of more effort than resultant progress; short flights while feeding are a mix of constant flitting, fast fluttering, and accomplished hovering—integrated with brief perching and creeping on branches and up and down trunks, generally giving impression of nervous agility second to no other small passerine.

Habitat. Breeds in middle and upper temperate and boreal latitudes of west Palearctic, between July isotherms 13–24°C and thus predominantly in cooler climates than Firecrest. In breeding season, strictly arboreal, attached to more or less dense stands of well-grown conifers, whether in lowlands or on mountains up to treeline (in Switzerland to 2200 m). Prefers spruce, silver fir, and mountain pine, and in artificial situations also Douglas fir and some other introduced conifers. Larch and

Goldcrest *Regulus regulus*. *R. r. regulus*: **1** ad ♂, **2** ad ♀, **3** juv. *R. r. azoricus*: **4** ad ♂. *R. r. inermis*: **5** ad ♂. Canary Islands Goldcrest *Regulus teneriffae* (p. 1345): **6** ad ♂. Firecrest *Regulus ignicapillus* (p. 1346). *R. i. ignicapillus*: **7** ad ♂, **8** ad ♀, **9** juv. *R. i. madeirensis*: **10** ad ♂. Ruby-crowned Kinglet *Regulus calendula* (p. 1342). **11** ad ♂.

Scots pine less attractive. Character and height of undergrowth irrelevant. Will inhabit broad-leaved woods only when at least a few spruce or firs are mixed in, and will colonize parks, cemeteries, and similar artificial areas only when they offer suitable conifers which are not otherwise locally available. After breeding season, ranges freely into deciduous trees and shrubs and out onto open heathlands or even into reedbeds.

Distribution. Some northward spread Scotland, Belgium, Norway, and Finland during 20th century. Benefits from spread of conifer plantations. FAEROES. A few pairs breeding in some years from 1979.

Accidental. Iceland (annual), Jordan, Morocco.

Beyond west Palearctic, breeds discontinuously east through southern Siberia to Sakhalin island and Japan, in Tien Shan mountains, northern Iran, and from Himalayas east to central China.

Population. Mostly stable in long term, but numbers fluctuate with winter conditions. BRITAIN. 560 000 territories 1988–91. IRELAND. 300 000 territories 1988–91. FRANCE. Under 1 million pairs in 1970s. BELGIUM. 100 000 pairs 1973–7. LUXEMBOURG. 12 000–15 000 pairs. NETHERLANDS. 40 000–60 000 pairs 1980–83. GERMANY. 1–1.5 million pairs; in east, 300 000 ± 141 000 pairs in early 1980s. DENMARK. 33 000–335 000 pairs 1987–8. NORWAY. 500 000–1 million pairs 1970–90. SWEDEN. 1–3 million pairs in late 1980s. FINLAND. 700 000–1.7 million pairs in late 1980s. ESTONIA. 100 000–200 000 pairs in 1991. LATVIA. 500 000–700 000 pairs in 1980s. LITHUANIA. Common. POLAND. 300 000–1 million pairs. CZECH REPUBLIC. 200 000–400 000 pairs 1985–9. SLOVAKIA. 150 000–300 000 pairs 1973–94. HUNGARY. 800–900 pairs 1979–93. AUSTRIA. Common. SWITZERLAND. 450 000–590 000 pairs 1985–93. SPAIN. 170 000–580 000 pairs. ITALY. 200 000–400 000 pairs 1983–95. GREECE. 1000–5000 pairs. ALBANIA. 500–2000 pairs in 1981. YUGOSLAVIA: CROATIA. 800–1200 pairs. SLOVENIA. 100 000–150 000 pairs. BULGARIA. 100 000–500 000 pairs. RUMANIA. At least 200 000–600 000 pairs 1986–92. RUSSIA. 10 million pairs. BELARUS'. 300 000–350 000 pairs in 1990. UKRAINE. 3500–4000 pairs in 1988. TURKEY. 10 000–100 000 pairs. AZORES. Common.

Movements. Resident to migratory. Movement both nocturnal and diurnal. European populations winter within and south of breeding range, and vacate entirely only extreme north of range in Fenno-Scandia and Russia. Autumn movement protracted. Fenno-Scandian birds depart late August to early November, with peak late September to mid-October coinciding with 1st cold spell. Immigrants reach east coast of Britain late August to early November, chiefly October. Spring movement (much lighter than autumn) begins February. Last records usually late March on Mediterranean islands, with weak passage March to mid-April in Camargue. Main movement in northern Europe late March to late April or early May, peaking early to mid-April. Vulnerability to extreme weather conditions may lead to large-scale disorientation or loss. Large numbers often gather on ships when night visibility poor, suggesting weak ability to maintain flight direction. In north-east Britain, unprecedented influx in October 1982 included 15 000 birds on Isle of May and 2000 on Fair Isle. Many records of stragglers to Iceland, chiefly in autumn. Weak spring passage in comparison with autumn also indicates heavy mortality among migrants.

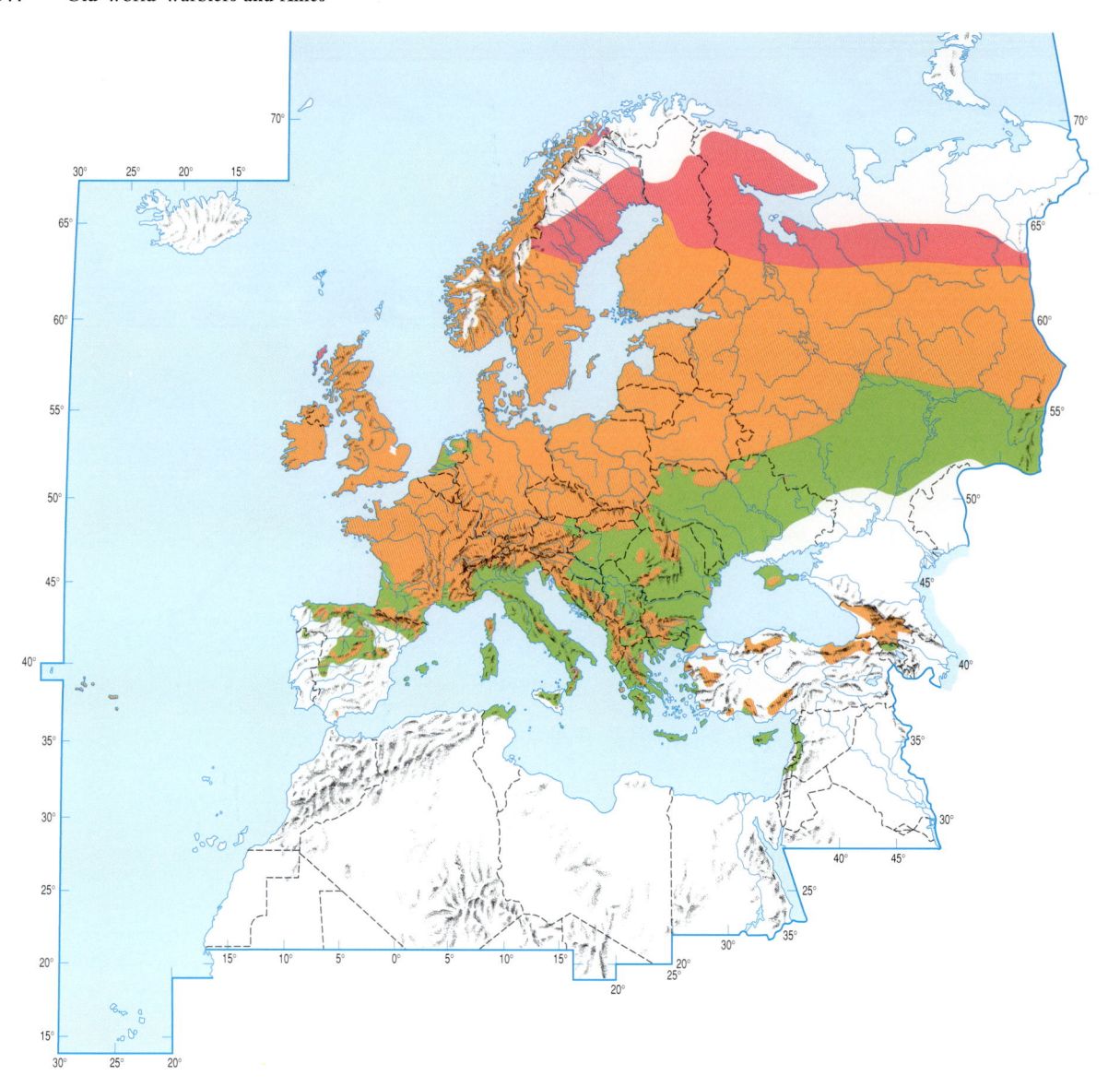

Food. Insects (especially Hemiptera, Collembola, and larval Lepidoptera) and spiders. Food obtained mainly from twigs in tree-crowns; less often from herb layer or ground. Feeds more in coniferous trees than Firecrest, and prefers denser branches. Flying insects taken in hovering flight but not pursued, and will hover in front of spider's web to take trapped insects.

Social pattern and behaviour. Outside breeding season, small groups maintain exclusive winter territories, which they apparently defend against neighbouring groups; regularly consort with foraging flocks of other species, especially tits. Mating system monogamous. Nest-building by both sexes; incubation and brooding of young by ♀ alone; feeding of young and care of fledged young by both sexes. Only ♂ gives full song, ♀ occasionally end section of song only. Exposed song-perches not used; bird usually sings while foraging. Song of ♂ serves for claiming and defence of territory, also attracting mate and maintenance of pair-bond. Main song-period same as breeding season, also some song February–March, before breeding begins, and in early autumn. Displays involve conspicuous raising of crest (inconspicuous at other times), presented to bird at which display is directed. Breeding pairs usually (if successful) rear 2 overlapping broods, made possible by strict division of labour at each nest between ♂ and ♀. Degree of overlap varies; probably in most cases, laying of 2nd clutch starts before young of 1st brood have left nest; overlap never so great that pair have young in both nests at same time. When ♀ begins to incubate 2nd clutch, ♂ is responsible for feeding 1st brood practically single-handed, first as nestlings and then as fledglings until they reach independence. As soon as 1st brood independent and has left territory, ♂ takes full share in feeding young of 2nd brood.

Voice. Song a very high-pitched but sweet-sounding phrase of set pattern, usually lasting 2.5–4 s. To human ear (which cannot distinguish fine details of structure) sounds like rhythmic repetition of 2- to 4-unit figure, e.g. 'pitEETily-pitEETily-pitEETily...' or 'eedle-eedle-eedle...', followed by terminal flourish. Terminal flourish very variable, and may include imitations of other species, such as 'pink' of Chaffinch. Other calls mostly very thin and high-pitched; similar to, and some hardly distinguishable from, equivalent calls of Firecrest, but tending to be perceptibly higher pitched. Commonest call a high-pitched 'siii'; as flight-call longer drawn-out and usually given in groups of 4, audible for *c.* 100 m. Serves to keep flock- and pair-members together. Trilling and clicking calls used in other contexts.

Breeding. SEASON. Eggs from late April or early May. 2 broods. SITE. Typically suspended in twigs near end of conifer branch, especially spruce and fir (exceptionally pine); rarely or locally (where typical sites not available) in fork of branches, or on side branch at junction with trunk of small tree; other evergreen plants used where conifers absent. Nest: almost spherical cup of moss, lichens, cobwebs, feathers, and hair, in 3 distinct layers—outer layer of cobwebs, moss, and lichens, with cobwebs used first to form link between twigs, middle layer of moss and lichens, and lining of feathers and hair. Small entrance at top often restricted by close proximity to twigs or branch above, and by rim of inward pointing feathers. EGGS. Sub-elliptical, smooth but not glossy; white to very pale buff, with very fine buff-brown, purplish, or grey-brown speckles, mainly at broad end forming ring or cap. Clutch: 9–11 (6–13). INCUBATION. 15–17 days. FLEDGING PERIOD. 17–22 days.

Wing-length: ♂ 51–59, ♀ 49–56 mm.
Weight: ♂♀ mostly 4.5–7 g.

Geographical variation. Slight and clinal, except on Azores where 3 races recognized: *inermis* (western Azores), *azoricus* (São Miguel) and *sanctae-mariae* (Santa Maria). All longer-billed than nominate *regulus*; first 2 darker-plumaged.

Canary Islands Goldcrest *Regulus teneriffae*

PLATE: page 1343

DU. Tenerife-goudhaantje FR. Roitelet de Ténérife GE. Teneriffagoldhähnchen IT. Regolo di Tenerife
RU. Канарский королек SP. Reyezuelo tinerfeño SW. Madeirakungsfågel

Field characters. 8.5 cm; wing-span 12.5–15 cm. Slightly smaller than Goldcrest. Strong crown pattern recalls Firecrest as much as Goldcrest; differs distinctly from latter in having black lateral crown stripes joined over forehead.

Behaviour and flight as Goldcrest.

Habitat. Upland (to *c.* 1600 m) pine or laurel forest, almost always with understorey of tree-heath, or in stands of pure tree-heath; rarely in pine forest where tree-heath absent.

Distribution and population. Breeds only in Canary Islands. Widespread and common in forests on Tenerife, La Gomera, La Palma, and El Hierro. 23 000–24 000 pairs.

Movements. Sedentary, probably moving to lower altitude in winter.

Food. Very small adult and larval insects, particularly, and in contrast to most Goldcrest and Firecrest populations, bush crickets, which are also fed to young; also spiders and harvestmen; captive bird showed only limited interest in springtails, unlike Goldcrest and Firecrest.

Social pattern and behaviour. Hardly studied in wild apart from some aspects of behaviour at nest; some details of repertoire (e.g. bathing, anti-ground predator response, nest building, feeding of incubating ♀ by mate, care of young, as well as feeding and aggressive behaviour) known only from one captive ♀ paired with ♂ Goldcrest, and appeared to be closer to that species than to Firecrest. Social interaction with Firecrest in captivity virtually non-existent; more aggressive than both those species in aviary.

Voice. Song and calls very similar to Goldcrest though harsher and slightly lower pitched; song briefer, without terminal flourish, described as ending instead in single note: 'see-see-see-see-charr', which is reminiscent of Firecrest song. Some ♂♂ give song ending in sound ('weia') resembling excitement-call of Firecrest. Song perhaps more variable than Goldcrest, especially end units; one ♂ sang 21 different end units in 29 songs. Excitement-calls more like those of Goldcrest than Firecrest.

Breeding. SEASON. Eggs laid from early March (nest building at 700 m can begin in mid-February); eggs in June probably replacements or 2nd broods. SITE. Usually in tree-heath *c.* 5 (2–14) m above ground, woven on to outer twigs, though these often horizontal unlike vertically hanging twigs used by Goldcrest. Nest: spherical, with entrance hole on top, perhaps occasionally at side, made of twigs, moss, and lichen and lined with soft grasses, rootlets, moss, feathers, spiders' webs, and plant down. EGGS. As Goldcrest, not pinkish like those of Firecrest. Clutch: 4–5 (3–6). INCUBATION. 16–17 days. FLEDGING PERIOD. *c.* 22 days.

Wing-length: ♂♀ 48–53 mm.

Firecrest *Regulus ignicapillus*

PLATES: pages 1343, 1347

Du. Vuurgoudhaantje   Fr. Roitelet à triple bandeau   Ge. Sommergoldhähnchen   It. Fiorrancino
Ru. Красноголовый королек   Sp. Reyezuelo listado   Sw. Brandkronad kungsfågel

Field characters. 9 cm; wing-span 13–16 cm. Marginally bulkier and seemingly larger-headed than Goldcrest. Close in character and plumage to Goldcrest but more beautiful, with head strongly striped with complete white supercilium and dusky eye-stripe and brighter plumage, greener above and often appearing frosty-white below. Patch of greenish-gold or -orange forms bright bronze shoulder-patch, lacking in Goldcrest.

In close, well-lit view, strength of head pattern and bright, clean colours of plumage allow easy identification. In poor view, glimpse of face- and wing-stripes may lead to confusion with small *Phylloscopus* (particularly Pallas's Warbler) and, when upper head hidden, Goldcrest. Warm bronze shoulder-patch not shown by any confusion species.

Habitat. Breeds in warm temperate and Mediterranean zones of mid-latitudes of west Palearctic between July isotherms of 16–18 and 24°C. Mainly in lowlands except in Mediterranean climates. Less closely associated with conifers than Goldcrest, breeding also in mixed woodland and, in southern Spain and north-west Africa, in broad-leaved woodland, especially oak; in parks, gardens, and small copses, as well as more extensive woodland. Recently colonized breeding sites in mature mixed woodland in Britain are dominated by conifers, particularly spruce; some sites, however, mainly deciduous, dominated by oak and beech with some holly.

Distribution. Breeds only in west Palearctic. Perhaps more widespread than known. BRITAIN. First breeding record 1962 New Forest (Hampshire) and has spread considerably though sparsely. Some sites probably yet to be discovered. FRANCE. Spread north-west since 1930s. BELGIUM. Has bred since at least 1916 (Brabant), and range has expanded with increase in conifer plantations. NETHERLANDS. First bred 1928. DENMARK. First bred 1961; range has probably decreased 1974–94. SWEDEN. First reported 1959; few records 1959–66, becoming more regular since; not known to have bred. LATVIA. Bred 1893. Singing ♂ May 1990 only breeding season record in 20th century; possibly rare breeder, overlooked. HUNGARY. Breeding first confirmed 1978. PORTUGAL. May be spreading in mountains. RUMANIA. Probably more widespread than shown. BELARUS'. No evidence of breeding, but possibly overlooked in west. Several old autumn and winter records. UKRAINE. Restricted and fragmented range. TURKEY. Perhaps also breeds other localities. MOROCCO. Small isolated population discovered north of Meknes since 1986.

Accidental. Norway, Finland, Estonia, Lithuania, Cyprus, Egypt.

Population. Apparently stable in most countries. BRITAIN. Rough estimate 80–250 singing ♂♂ 1988–91. After initial increase, numbers fluctuate markedly, probably due to differential winter survival. FRANCE. Under 100 000 pairs in

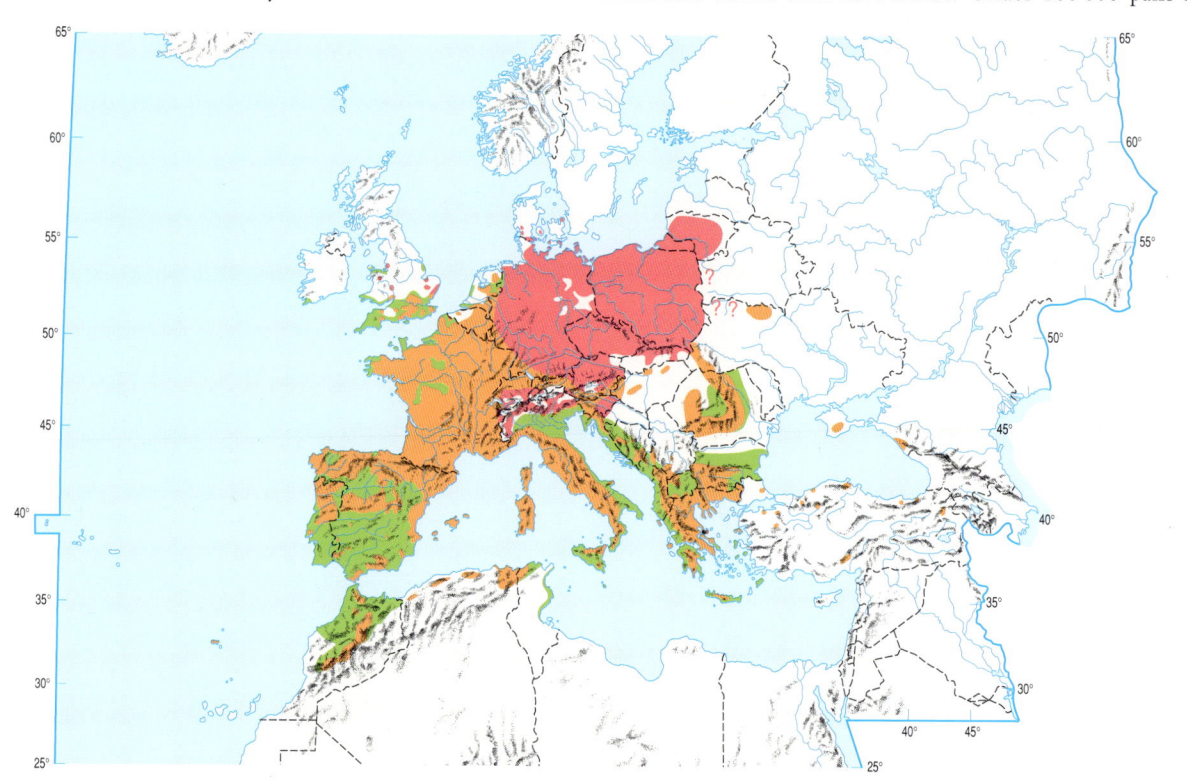

Arctic Warbler *Phylloscopus borealis* (p. 1322): **1** 1st winter. Greenish Warbler *Plylloscopus trochiloides* (p. 1320): **2** 1st winter. Pallas's Warbler *Phylloscopus proregulus* (p. 1324); **3** 1st winter. Yellow-browed Warbler *Phylloscopus inornatus* (p. 1325): **4** 1st winter. Hume's Leaf Warbler *Phylloscopus humei* (p. 1327): **5** 1st winter. Radde's Warbler *Phylloscopus schwarzi* (p. 1327): **6** 1st winter. Dusky Warbler *Phylloscopus fuscatus* (p. 1328): **7** 1st winter. Wood Warbler *Phylloscopus sibilatrix* (p. 1333): **8** ad. Bonelli's Warbler *Phylloscopus bonelli* (p. 1329): **9** 1st winter. Chiffchaff *Phylloscopus collybita* (p. 1337). *P. c. collybita*: **10** 1st winter. *P. c. tristis*: **11** 1st winter. Willow Warbler *Phylloscopus trochilus* (p. 1340): **12** ad. Firecrest *Regulus ignicapillus* (p. 1346): **13** ad. Eastern Bonelli's Warbler *Phylloscopus orientalis* (p. 1331): **14** ad.

1970s. BELGIUM. 100 000–150 000 pairs 1989–91; continuing increase (20 000 in 1968). LUXEMBOURG. 8000–12 000 pairs. NETHERLANDS. 5000–8000 pairs 1979–85. GERMANY. 1–1.2 million pairs; increasing. In east, 100 000 ± 50 000 pairs in early 1980s. DENMARK. Up to 15 pairs 1976–94. POLAND. 30 000–200 000 pairs. CZECH REPUBLIC. 50 000–100 000 pairs 1985–9. SLOVAKIA. 30 000–60 000 pairs 1973–94. HUNGARY. Probably fewer than 200 pairs. AUSTRIA. Regionally fairly common. Increasing since mid-1960s. SWITZERLAND. 350 000–400 000 pairs 1985–93. SPAIN. 910 000–2 million pairs. PORTUGAL. 10 000–100 000 pairs 1978–84; slight increase. ITALY. 100 000–300 000 pairs 1983–95. GREECE. 5000–20 000 pairs. ALBANIA. 500–2000 pairs in 1981. YUGOSLAVIA: CROATIA. 10 000–20 000 pairs. SLOVENIA. 50 000–100 000 pairs. BULGARIA. 10 000–50 000 pairs. RUMANIA. 100 000–220 000 pairs 1986–92. UKRAINE. 50–150 pairs in 1988. TURKEY. 1000–10 000 pairs. TUNISIA. Common. ALGERIA. Locally common. MOROCCO. Uncommon. MADEIRA. Very common.

Movements. Resident to partial migrant. Southern populations chiefly resident, northern and eastern populations chiefly migratory, heading between south and west (reverse in spring) to winter mainly in Mediterranean area and extreme west of Europe from Portugal north to Britain and (less regularly) Ireland. Nocturnal migrant. Limited information, due mostly to overall low numbers and inconspicuous behaviour. Apparently avoids lengthy sea-crossings: entry into Britain is chiefly via English Channel, and Mediterranean crossed at narrowest points. Autumn movement chiefly September–November; first arrivals at British observatories early September. Spring movement begins February, continuing to April or (in north) May. In Britain, passage (chiefly in south and south-east) begins in early March, peaking late March to early April. May–June coastal records suggest non-directional movements after main migration period.

Food. Arthropods, especially springtails, spiders, and aphids. Food obtained among twigs and branches in canopy, occasionally from ground.

Social pattern and behaviour. Behaviour in winter quarters little studied. Joins mixed-species foraging flocks of tits and other species. Mating system monogamous. Nest-building, incubation, and brooding of young by ♀ alone; feeding of young and care of fledged young by both sexes. ♂ usually sings while foraging, but stops still while singing, with bill open and head somewhat raised; crest may be raised. Song-period as breeding season, with in addition brief period in September–October, between end of moult and autumn migration. Striking face pattern, as well as brightly coloured crest, exhibited in displays by pointing bill at bird to which display is directed, thus differing from Goldcrest which bows head, making exhibition of crest main feature of displays.

Voice. Song given by ♂ only, a series of high 'si' notes, increasing in volume and speed; lacks rhythmic drive of song of Goldcrest; simpler, with less variation in pitch, and terminal flourish characteristic of Goldcrest is lacking or replaced by vibrant 'sirr'. Other calls nearly all high-pitched, commonly rendered 'zit zit zit', 'sisisi', etc., sometimes in long series or trills, perceptibly lower pitched and more incisive than equivalent calls of Goldcrest; also a harsh 'zerr', commonly given by aggressive birds in autumn and winter flocks.

Breeding. SEASON. Laying begins in western Europe from end of April, in central Europe from early May; 2nd clutches June–July. Alps: 1st clutches begun 2nd half of May, 2nd clutches July. North-west Africa, lowlands: laying begins mid-April. SITE. Typically, suspended in twigs near end of conifer branch; in habitats without conifers, suspended in twigs or leaves of variety of other plants, especially oak, ivy, and other climbing plants. Nest: almost spherical, elastic cup of moss, lichens, and cobwebs, of 3 main layers; outer layer of cobwebs, moss, and lichens, with cobwebs used first to fasten twigs together, middle layer of moss, and lining mainly of feathers (up to 3000) and hair; small entrance at top, restricted in size by close proximity of branch, twigs, and leaves. EGGS. Sub-elliptical, smooth without gloss or slightly glossy; pinkish-buff, with very fine red-brown spots (often so fine that eggs appear uniformly coloured), nearly always concentrated at broad end. Clutch: 7–12 in Europe; probably smaller in north-west Africa. INCUBATION. 14.5–16.5 days. FLEDGING PERIOD. Not recorded in the wild; in aviary, 22–24 days.

Wing-length: ♂ 51–56, ♀ 48–53 mm; Madeiran birds slightly larger.
Weight: ♂♀ 4–6.5 g.

Geographical variation. Slight. *R. i. balearicus* from Balearic Islands and North Africa is greyer on underparts than nominate *ignicapillus*. *R. i. madeirensis* from Madeira has shorter supercilium, duller orange crown, and longer bill.

Golden-crowned Kinglet *Regulus satrapa*

DU. Amerikaans Goudhaantje   FR. Roitelet à couronne dorée   GE. Indianergoldhähnchen   IT. Fiorrancino americano
RU. Золотоголовый королек   SP. Reyezuelo de oro   SW. Guldkronad kungsfågel.

Breeds in North America from southern Alaska east to southern Quebec and Newfoundland, south through highlands of Mexico to western Guatemala, and through Appalachians to North Carolina. Accidental in Bermuda.

Bird collected Lancashire (England), October 1897, not accepted due to doubt that it could cross Atlantic unaided. Several birds recorded on assisted passage up to almost half way across Atlantic.

Old World Flycatchers Family Muscicapidae

Small oscine passerines (suborder Passeres). Typically arboreal, feeding chiefly on insects caught during aerial sallies from branch or similar perch. 107–109 species, occurring mainly in woodland, parkland, orchards, and gardens. Mostly European, Asiatic, and African in distribution, with largest diversity in tropical Asia. All species in northern Eurasia migratory, travelling by night and feeding by day; winter mainly in tropics. 6 species in west Palearctic, 5 breeding.

Sexes generally of similar size. Bill rather short and flat; broad at base, upper mandible with small hook; mandibles audibly snapped at times in aggression. Rictal bristles well developed, covering nostrils. Wing rather short, bluntly pointed in most; 10 primaries, p10 reduced (25–40% of length of longest). Flight strong, swift, dashing, and agile; hovering reported as well as usual flycatching sorties and active pursuit of flying insects. Both wings characteristically flicked upwards in certain situations. Tail of short to medium length, typically square or slightly graduated; often jerked upwards in chat-like manner, sometimes when wings also elevated. Monogamous mating system in most species but some polygynous. Courtship-feeding occurs in all west Palearctic species. Open cup nest placed in hole, niche, or vegetation; built by ♀.

Plumage soft and full. Colour variable: dull in some species (e.g. many *Muscicapa*), bright in others—in which sexes usually differ. Juvenile plumage spotted (as in Turdidae).

Brown Flycatcher *Muscicapa dauurica*

PLATES: pages 1350, 1361

Du. Bruine Vliegenvanger Fr. Gobemouche brun Ge. Braunschnäpper It. Pigliamosche bruno asiatico
Ru. Ширококлювая мухоловка Sp. Papamoscas pardo Sw. Glasögonflugsnappare

Field characters. 11.5 cm; wing-span 19.5–21.5 cm. Noticeably smaller and more compact than Spotted Flycatcher, with proportionately larger bill, 20% shorter wings (extending little beyond end of upper tail-coverts), and 25% shorter tail; close in size to Red-breasted Flycatcher but with flatter head, slightly longer wings, and slightly shorter tail and legs evident in rather dumpier, less sprite-like form. 2nd smallest flycatcher of west Palearctic, with form and appearance intermediate between Spotted Flycatcher and ♀ or immature *Ficedula* flycatchers. No striking characters except for rather large dark eye offset by off-white loral stripe and eye-ring; at close range narrow dark malar stripe and pale yellowish base to lower mandible may show. Upperparts generally grey rather than brown in tone; underparts dull white, clouded or mottled on breast and flank; at close range, wings show pale fringes in pattern like Spotted Flycatcher, with pale tips to greater coverts forming obvious wing-bar in first winter plumage.

As a vagrant to west Palearctic, more likely to be confused with *Ficedula* flycatcher than with Spotted Flycatcher; distinction best based on uniform tail pattern (usually markedly edged white in *Ficedula*), wing pattern (much bolder in *Ficedula*, except Red-breasted Flycatcher), and head shape (rather round in *Ficedula* but flat or angled above eye in Brown Flycatcher). Behaviour rather quiet and less demonstrative than other flycatchers. Sits upright, like Spotted Flycatcher.

Commonest calls are high-pitched 'tzi' or 'see' similar to Spotted Flycatcher and trilling 't-r-rrr'.

Habitat. Breeds in east Palearctic middle latitudes in continental climate, largely in lowlands. Generally prefers mature deciduous trees, sometimes mixed with occasional conifers, and everywhere chooses least dense stands, with good undergrowth, near forest edge or by glades, clearings, or road verges.

Distribution. Breeds from the upper Ob' (Siberia) east to Sakhalin island and Japan, south very discontinuously through eastern China and Philippines to Lesser Sunda Islands; very local in India.

Accidental. Denmark: one at Blåvands Huk, September 1959. Sweden: 1st-winter bird, Svenska Högarna, September 1986. Greece: one at 40°48′N 25°50′E, east of Nestos river, September 1993.

Movements. Varies between migratory and resident across range. Northern populations long-distance migrants, arriving on breeding grounds May to early June, departing south from mid- or late August.

Wing-length: *M. d. dauurica* (breeding southern Siberia to Japan): ♂ 66–74, ♀ 64–73 mm.

Spotted Flycatcher *Muscicapa striata*

PLATES: pages 1350, 1361

Du. Grauwe Vliegenvanger Fr. Gobemouche gris Ge. Grauschnäpper It. Pigliamosche
Ru. Серая мухоловка Sp. Papamoscas gris Sw. Grå flugsnappare

Field characters. 14.5 cm; wing-span 23–25.5 cm. Overall 20% larger than Pied Flycatcher, with longer bill and head and at least 10% longer wings and tail. Medium-sized, flat- and broad-billed, lengthy flycatcher, biggest of family in west

Brown Flycatcher *Muscicapa dauurica* (p. 1349): **1** 1st winter. Spotted Flycatcher *Muscicapa striata. M. s. striata*: **2** ad breeding, **3** 1st winter, **4** juv. *M. s. neumanni* (Siberia): **5** 1st winter. *M. s. balearica*: **6** ad breeding. *M. s. tyrrhenica*: **7** ad breeding.

Palearctic and only species with well-streaked crown, throat, and breast. Plumage otherwise basically dun grey above and dull white below, with only thin pale margins on wing and tail feathers. Juvenile distinctive in appearance, being truly spotted buff-white on head, back, and fore-wing.

Commonest and most widespread flycatcher in west Palearctic. Generally considered unmistakable but nevertheless subject to confusion with smaller Brown Flycatcher (in adult plumage unstreaked on crown and breast, with pale loral streak and eye-ring) and 2 other potential vagrants from Asia, Sooty Flycatcher and Grey-streaked Flycatcher (Fig. 2). Feeding flight purposeful and accomplished, with flurry of wing-beats in diving take-off from look-out perch, long, remarkably flat glide towards prey, then instant turn, burst of wing-beats, and seemingly gliding ascent back to perch. Carriage distinctive, usually at least half-upright, with head usually held behind breast and seemingly sunk onto shoulders; long tail often appears as half overall length, and this, combined with long wing-points reaching almost half-way down it, produce remarkably attenuated rear to silhouette.

Habitat. Almost throughout west Palearctic, from Mediterranean and steppe through temperate and boreal to edge of Arctic zone, up to July isotherm of 11°C. Absent, however, from open areas devoid of trees and bushes and from densely forested, arid or exposed mountainous areas, although breeding freely in Switzerland up to 900 m and sporadically to *c.* 1500 m. Dependent on availability of raised perches, especially in such deciduous trees as beech, oak, and chestnut, and to lesser extent conifers such as pine and larch, growing in open order or by clearings, burnt patches, glades, or along streams, rivers, or edges of standing water. Requires ample accessible space for catching flying insects present at adequate density, and accordingly has adapted readily to avenues, parks, cemeteries, gardens, orchards, and other man-made habitats, being fairly tolerant of disturbance, and accepting nest-sites on inhabited houses, especially where overgrown by climbing plants.

Distribution. Little recent change, though declines reported Spain and Ukraine. BRITAIN. Breeding in Outer Hebrides and Orkney irregular. IRELAND. Range contracting in west. MALTA. In recent years 2–3 pairs have bred regularly. TURKEY. Breeds mainly north of 40°N. SYRIA. June records suggest possible breeding, but no proof. JORDAN. Breeding first proved 1990. IRAQ. Bred Ser Amadiya mountains in 1971. Reported as possible breeder in early 1950s. EGYPT. First breeding record 1989, Rafa (north-east Sinai); bred at same site 1990 (probably several pairs).

Accidental. Spitsbergen, Iceland, Madeira, Cape Verde Islands.

Beyond west Palearctic, extends east to Transbaykalia, south (skirting much of Kazakhstan) to Iran and western Himalayas.

Population. Decrease Britain, Ireland, Netherlands, France (probably), Germany, Finland, Czech Republic, Spain, and Ukraine; also in Belgium and Denmark, where apparently halted since mid-1980s. Causes uncertain, but those suggested for Britain perhaps apply more generally. Stable in other countries (increase reported Croatia). BRITAIN. 120 000 territories 1988–91. Long-term decline. Recent cool and wet early summers, and failure of rains 1983 and 1984 in Sahel (area traversed on passage) perhaps involved. Decrease more evident in south, suggesting habitat degradation and especially use of insecticides as other factors. IRELAND. 35 000 territories 1988–91. FRANCE. 100 000–1 million pairs in 1970s. BELGIUM.

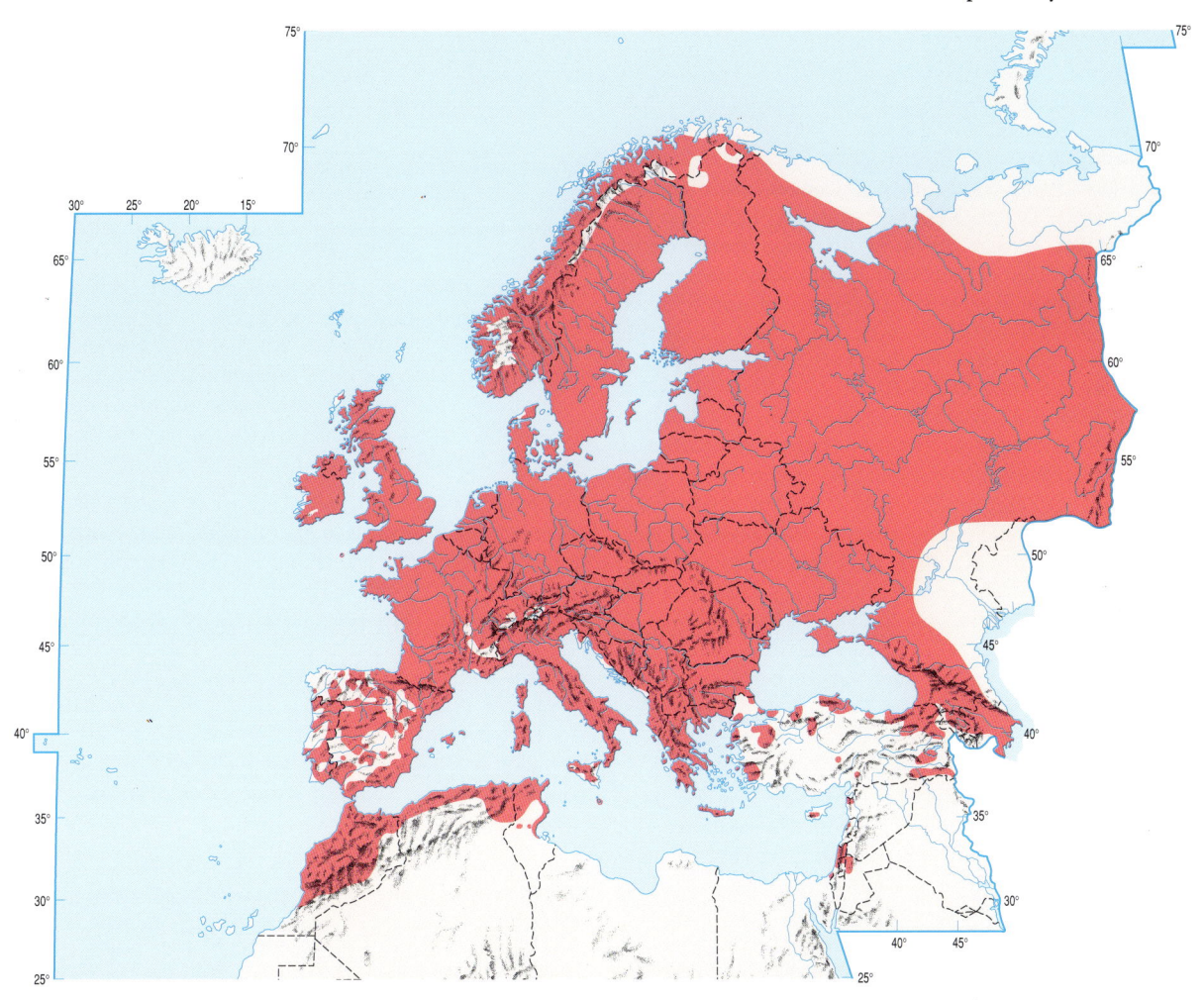

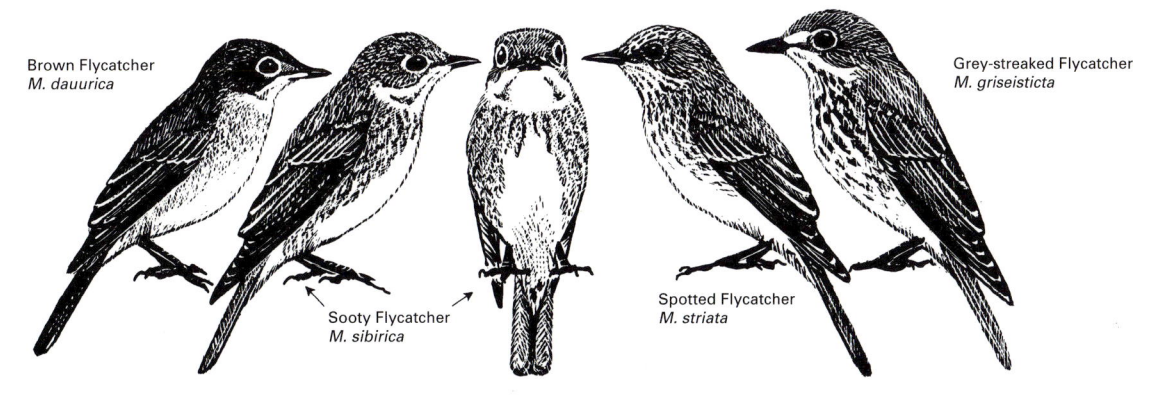

Fig. 2. Comparative drawings of 4 species of *Muscicapa* flycatcher, all in 1st-autumn/winter plumage. Brown Flycatcher is smallest, with pale loral patch and eye-ring and unstreaked head and body. Sooty Flycatcher is close in size to Spotted Flycatcher but has pale half-collar and obvious pale divide between indistinctly streaked flanks. Spotted Flycatcher is fully streaked on crown and fore-underparts. Grey-streaked Flycatcher is largest, with pale loral patch and eye-ring and strongly streaked underparts. All have similar wing patterns.

10 000–14 000 pairs 1989–91. LUXEMBOURG. 800–1000 pairs. NETHERLANDS. 50 000–100 000 pairs 1979–85, but 15–20% decline by 1989–91. GERMANY. 358 000 pairs in mid-1980s. DENMARK. 7000–92 000 pairs 1987–8. NORWAY. 100 000–500 000 pairs 1970–90. SWEDEN. 500 000–1.2 million pairs in late 1980s. FINLAND. 2–3 million pairs in late 1980s. ESTONIA. 100 000–200 000 pairs in 1991. LATVIA. 200 000–400 000 pairs in 1980s. LITHUANIA. Common. POLAND. 150 000–300 000 pairs. Stable with sometimes strong fluctuations. CZECH REPUBLIC. 30 000–60 000 pairs 1985–9. SLOVAKIA. 65 000–150 000 pairs 1973–94. HUNGARY. 80 000–120 000 pairs 1979–93. AUSTRIA. 25 000–40 000 pairs in 1992. SWITZERLAND. 30 000–60 000 pairs 1985–93. SPAIN. 640 000–690 000 pairs. PORTUGAL. 100–1000 pairs 1978–84. ITALY. 50 000–200 000 pairs 1983–95. GREECE. Above 10 000 pairs. ALBANIA. 2000–5000 pairs in 1981. YUGOSLAVIA: CROATIA. 20 000–25 000 pairs. SLOVENIA. 15 000–20 000 pairs. BULGARIA. 1000–10 000 pairs. RUMANIA. 50 000–80 000 pairs 1986–92. RUSSIA. 1–10 million pairs. BELARUS'. 1.35–1.45 million pairs in 1990. UKRAINE. 95 000–105 000 pairs in 1988. MOLDOVA. 17 000–25 000 pairs in 1988. AZERBAIJAN. Common. TURKEY. 5000–50 000 pairs. CYPRUS. Perhaps 100–200 pairs. ISRAEL. At least a few thousand pairs in 1980s; local increases and decreases. JORDAN. A few pairs nest in Northern Highlands. TUNISIA. Common breeder. ALGERIA. Fairly common. MOROCCO. Uncommon to locally common.

Movements. Long-distance migrant with all races moving to sub-Saharan Africa, majority wintering south of equator. Many birds are still on passage in central Africa, including areas south of equator, in October–November, spending only December–February in southernmost wintering areas. In spring this is one of the latest migrants to return to northern breeding grounds—early June in extreme cases.

Autumn recoveries in Europe and North Africa indicate migratory divide at c. 12°E: birds from Britain, Ireland, and western Europe initially head south-west to SSW through western France and Iberia; birds from central Europe and Fenno-Scandia head SSW to south-east through Italy, Greece, and Aegean. Despite migratory divide, passage through Mediterranean, North Africa, and Sahara on broad front (both seasons), and south European birds presumably head more directly south, with little or no division. Autumn departure begins late July, chiefly from August. Movement through Mediterranean region mid-August to mid-November, chiefly September–October. Reaches East Africa October–November, and South Africa late October to December.

Passage north starts late February or March. Passage through North Africa and Mediterranean region chiefly from mid-April, continuing to end of May. First arrivals in northern Europe in mid- or late April, but mainly in May, continuing to end of June.

Food. Mainly flying insects, especially Diptera and Hymenoptera; berries taken occasionally during breeding season, more regularly in autumn. Takes more prey from the air than other west Palearctic flycatchers, mainly by sallying out from perch, catching flying insect and returning to perch to swallow it.

Social pattern and behaviour. Typically solitary in winter quarters. Individuals probably hold succession of temporary feeding territories over course of winter. Mating system typically monogamous; rarely bigamous. ♀ typically does all incubation, though ♂ sometimes participates. Both sexes care for young. Song of ♂ soft and seldom heard except at close quarters. Given usually while perched unobtrusively in tree, sometimes in flight. ♂♂ sing, often almost all day, from arrival on territory. Song continues after pair-formation, during nest-site selection, and to lesser extent during nest-building by ♀; rare after egg-laying.

Voice. Apart from calls, relatively silent. Loud bill-snapping prominent in threat; at high intensity, rapid with rattling quality. Song of ♂ relatively quiet; an unobtrusive sequence of short, very high-pitched single or disyllabic units, variable in tonal quality, delivered at c. 1 unit per s, also short trills, e.g. 'sip sip sree sreeti sree sip'; no formation of phrases as such, and simple succession of call-type units resembles Hawfinch; thus different from songs of *Ficedula* flycatchers. Contact-alarm call a simple slightly drawn out 'zee' or 'zit'. At higher intensities of alarm, when disturbed at nest or when young threatened, a louder, rather variable 'see-tic', 'tzucc', 'tzee tsutzucc', etc.

Breeding. SEASON. Western and central Europe: eggs from mid-May, mainly late May or June. 2 broods usual except in north. SITE. On natural or artificial ledge, or in niche, requiring firm support below for loosely built nest, good view for incubating bird, and, often, overhang above nest for shelter: on tree-trunk supported by twigs; in creeper or in shallow crevice of tree or wall; on top of flat branch; on top of stacked wood, etc., piled against wall; in old nest of other bird; will use open-fronted nest-box. Nest: loosely built cup of fine twigs, rootlets, dry grass, moss, and lichens, lined with hair, feathers, and fine fibres. EGGS. Sub-elliptical, smooth but not glossy; very pale blue, green-blue, buff, or off-white, variably marked with red-brown and purple-grey mottling and blotching, often concentrated at large end. Clutch: 4–6 (2–7); some decline in clutch size through season. INCUBATION. 12–14 days. FLEDGING PERIOD. 12–16 days.

Wing-length: ♂ ♀ 83–90 mm (sexes not significantly different).
Weight: ♂ ♀ mostly 14–20 g.

Geographical variation. Slight. Nominate *striata* occupies most of west Palearctic. *M. s. inexpectata*, from Crimea, is darker above and more heavily streaked below; *M. s. balearica*, from Balearic Islands, paler and less streaked, with shorter wing; *M. s. tyrrhenica*, from Corsica and Sardinia, warmer brown on upperparts, with streaks on underparts replaced by coalescing spots.

Red-breasted Flycatcher *Ficedula parva*. *F. p. parva*: **1** ad ♂ breeding, **2** ad ♀ breeding, **3** 1st winter, **4** juv. *F. p. albicilla* (Siberia): **5** ad ♂ breeding, **6** 1st winter.

Red-breasted Flycatcher *Ficedula parva*

PLATES: pages 1353, 1361

Du. Kleine Vliegenvanger   Fr. Gobemouche nain   Ge. Zwergschnäpper   It. Pigliamosche pettirosso
Ru. Малая мухоловка   Sp. Papamoscas papirrojo   Sw. Mindre flugsnappare

Field characters. 11.5 cm; wing-span 18.5–21 cm. Up to 10% smaller than Pied Flycatcher but with proportionately longer tail and legs. Smallest, most nimble flycatcher in west Palearctic, with relatively tiny bill, neat, round head, and chesty body but slim vent and quite long tail. Tail often flirted to display unique pattern of dark centre and end, with bold white patches on sides. ♂ has red central throat and chest and grey neck and chest-sides reminiscent of Robin; ♀ and immature have buff or ochre chest.

Tail pattern (and chest pattern of ♂) unique in flycatchers of west Palearctic but some risk of confusion with potentially vagrant Mugimaki Flycatcher *F. mugimaki* (and other Himalayan and east Asian *Ficedula* escaping from cage-bird trade). Flight extremely agile, with more rapid wing-beats (and apparently rather long tail) used to perform tightest, most looping sallies of all west Palearctic flycatchers; usually hunts from perch in foliage of bushes and trees. Flicking of wings and tail (even cocking latter over back) more pronounced than in any other west Palearctic flycatcher.

Habitat. In west Palearctic in continental middle latitudes, mainly temperate but also boreal and montane, south of July isotherm of 15°C. A forest bird, in west of range mainly in mixed and deciduous stands (especially beech); in Germany from lowlands to *c.* 1200 m; in Armenia to 2350 m, higher than any other Muscicapidae. Prefers tall trees with much undergrowth, and likes neighbourhood of water and of glades or clearings. Also occurs in orchards and vineyards, and in north of range in forests (sometimes very dark) of spruce.

Distribution. NETHERLANDS. Several unconfirmed reports of breeding in recent years. GERMANY. Breeds mainly in north-east, otherwise local or sporadic; in east of range has recently expanded south and south-west. DENMARK. At least 17 breeding records, chiefly in east; no regular sites. NORWAY. 2 confirmed breeding records (Hedmark 1982, Østfold 1989), and thought to breed in small numbers. SWEDEN. First breeding record 1944, annual breeding records on mainland since 1959. POLAND. In west (except Pomerania), extremely scarce and local. CZECH REPUBLIC. Marked increase in range. GREECE. Breeding first recorded 1985, Rhodope mountains. YUGOSLAVIA: SLOVENIA. Slight increase. RUMANIA. Characteristic bird of alpine beechwoods. UKRAINE. Slight decrease. TURKEY. Restricted to scattered localities in Black Sea coastlands.

Accidental. Bear Island, Iceland, Faeroes, Britain (annual), Ireland (near annual), France (probably regular), Belgium, Luxembourg, Switzerland, Spain, Balearic Islands, Portugal, Tunisia, Algeria, Morocco, Madeira, Canary Islands. Status of 'vagrants' in south-west Palearctic uncertain; increase in records has led to speculation that small proportion of western population migrates south-west, perhaps to unknown West African winter quarters.

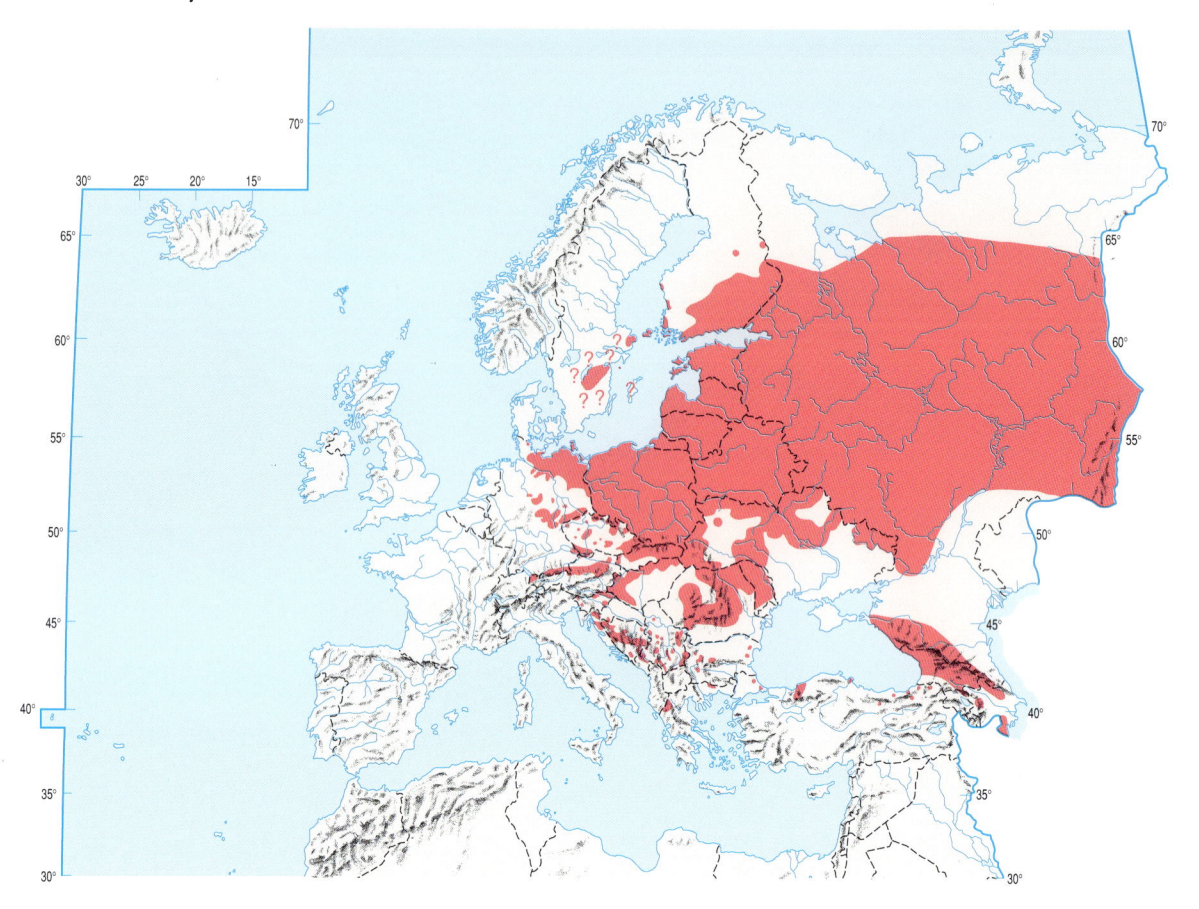

Beyond west Palearctic, extends east across Siberia to Kamchatka, Amur region, and Sakhalin island; also breeds northern Iran.

Population. Marked decrease Ukraine; some increase Germany, Czech Republic, and Slovenia; otherwise stable. GERMANY. In east, 1800 ± 590 pairs in early 1980s. Another estimate (for whole country) 3000–4000 pairs in early 1990s. NORWAY. Probably 1–20 pairs. SWEDEN. 100–1000 pairs in late 1980s. FINLAND. 1000–2000 pairs in late 1980s. ESTONIA. 10 000–20 000 pairs in 1991; more or less stable since increase 1950s–60s. LATVIA. 50 000–80 000 pairs in 1980s. LITHUANIA. Not abundant. POLAND. 2000–8000 pairs. CZECH REPUBLIC. 800–1400 pairs 1985–9. SLOVAKIA. 5000–10 000 pairs 1973–94. HUNGARY. 100–300 pairs 1979–93. AUSTRIA. Localized. GREECE. Probably not more than 10 pairs. YUGOSLAVIA: CROATIA. 1000–1500 pairs. SLOVENIA. 50–100 pairs. BULGARIA. 100–1000 pairs. RUMANIA. 30 000–40 000 pairs 1986–92. RUSSIA. 1–10 million pairs. BELARUS'. 130 000–140 000 pairs in 1990. UKRAINE. 2500–3000 pairs in 1988. MOLDOVA. 3500–5000 pairs in 1988. AZERBAIJAN. Common, more so in south-east. TURKEY. 1000–10 000 pairs.

Movements. All populations migratory, wintering in southern Asia, from Pakistan and India east to southern China, Indo-China, and Malay peninsula. Autumn movement protracted. European birds typically adopt south-east heading, but also regularly pass in small numbers south or south-east through central and eastern Mediterranean and north-east Africa. Some arrive in India mid-September, while others still in Europe late October or early November.

Departure from north-west India and Pakistan begins in mid-March, peaking April, straggling well into May. Main passage through Black Sea area late April to late May; arrival on European breeding grounds mainly mid-May to early June.

Increasing autumn vagrancy (especially of juveniles) to north-west Europe in recent years, typically during anticyclones, apparently due to reverse migration.

Food. Mainly insects and other invertebrates; occasionally fruit. In breeding season food taken mostly from trees (especially middle layer, less often lower parts of crown), though some is caught in the air (sallying from trees and bushes) and on the ground.

Social pattern and behaviour. Usually solitary outside breeding season; may defend feeding territory. No evidence for other than monogamous mating system. Only ♀ incubates and broods, but both sexes feed young. ♂ sings from dead branch near trunk, or in crown of tree, also in song-flight, fluttering with shivering wing-beats from branch to branch. Perched singer raises head, showing off throat. Song peaks in pair-formation, maintained during egg-laying, ends with onset of incubation; birds singing thereafter are unpaired, these continuing, often at high intensity, throughout summer. Song also occurs in autumn: starts in last 10 days of August, and most often heard from then until early September.

Voice. Song of ♂ resonant and loud; undoubtedly the most gifted of west Palearctic flycatchers. Has silvery melodious quality, often likened to *Phylloscopus* warbler, i.e. combination of 1st part like Wood Warbler and last part like Willow Warbler. Phrases usually last 3–4 s, often preceded by 'zit' calls. Rest of song comprises 3 sections: (i) usually a series of up to 6 similar 'tui' units; (ii) descending series of alternating units, variously rendered 'diü-tvi-diü-tvi' or 'didle-didle' or 'eida-eida'; (iii) final section (especially like Willow Warbler) continues to descend in pitch—series of full-sounding 'dlü-dlü' units.

Contact-alarm and excitement-calls include high-pitched 'zit' or 'zri', frequently given in general excitement, and short rattling 'zirrt'.

Breeding. SEASON. Central and eastern Europe: mid-May to end of June. Normally one brood. SITE. In hole in tree or wall, among side shoots against tree trunk, in nest-box, occasionally in bush. Nest: cup of moss, dry grass stalks and leaves, root fibres, and hair, lined with hair; nest in bush may be domed. EGGS. Sub-elliptical, smooth and glossy; whitish, sometimes faint buff or blue-green, with faint red-brown speckling, occasionally denser at large end. Clutch: 5–6 (4–7). INCUBATION. 12–13(–15) days. FLEDGING PERIOD. 12–13 (11–15) days.

Wing-length: ♂ 65–72, ♀ 63.5–71 mm.
Weight: ♂♀ mostly 8.5–11 g; migrants to 14 g.

Semi-collared Flycatcher *Ficedula semitorquata*

PLATE: page 1356

DU. Balkanvliegenvanger   FR. Gobemouche à demi-collier   GE. Halbringschnäpper   IT. Balia caucasica
RU. Полуошейниковая мухоловка   SP. Papamoscas semicollarino   SW. Balkanflugsnappare

Field characters. 13 cm; wing-span 23.5–24 cm. Virtually equal in size to Collared Flycatcher. Appearance intermediate between Collared Flycatcher and Pied Flycatcher, with diagnostic combination in breeding adult ♂ of white half-collar and upper wing-bar, grey rump-patches, and mostly white outer tail-feathers. Adult ♀ also shows white upper wing-bar, but non-breeding ♂, ♀, and immature dubiously separable from Collared Flycatcher and Pied Flycatcher on plumage characters. No discerned differences from congeners in behaviour, but has distinctive calls (see Voice).

Habitat. In south-east warm temperate lower middle latitudes of west Palearctic, finds forest requirements (which it shares with Collared Flycatcher) fulfilled mainly on slopes of mountains in belt occupied by mature deciduous trees, notably oak and hornbeam. Sometimes ascends higher, among firs; also in gorges and at lower level in orchards. Found up to *c.* 2000 m on mountain slopes. On plains, uses deciduous riverine forest; also occurs in old groves and gardens.

Distribution. Range and numbers poorly known. Has spread west in Balkan peninsula. RUSSIA. Teberda reserve, north-west Caucasus: very rare breeder in 1960s, regular from 1977, but total population still very small. AZERBAIJAN. Range includes Zakataly district (Great Caucasus), forests of Samur delta, and Bargushad range (Little Caucasus); perhaps also Talysh

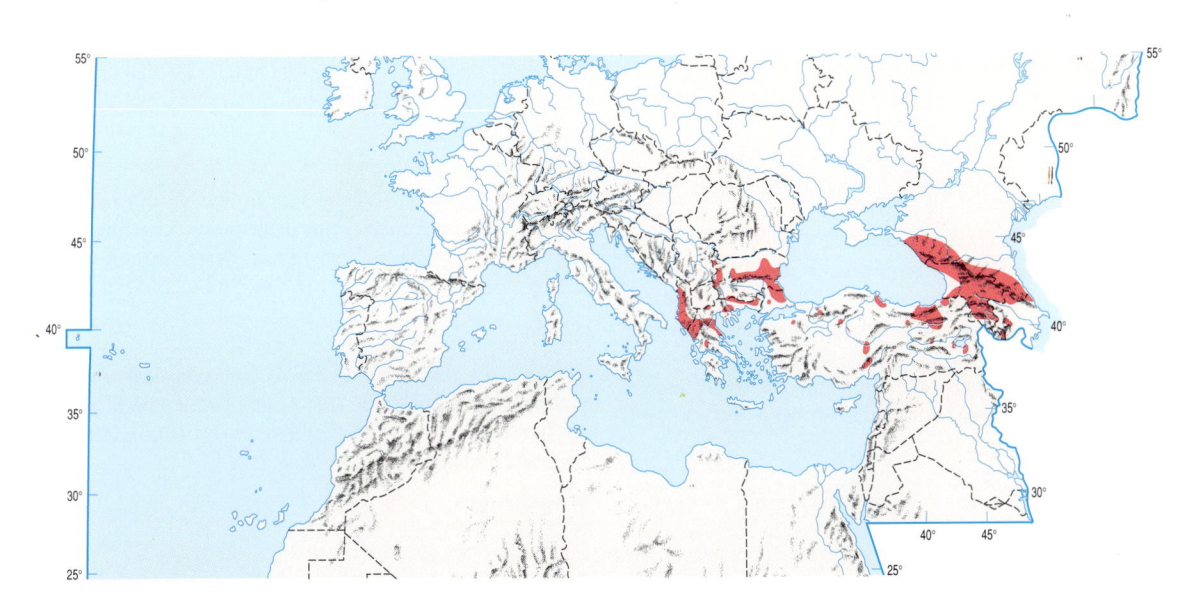

Semi-collared Flycatcher *Ficedula semitorquata*: **1–2** ad ♂ breeding, **3** ad ♀ breeding. Collared Flycatcher *Ficedula albicollis* (p. 1357): **4** ad ♂ breeding, **5** ad ♀ breeding. Pied Flycatcher *Ficedula hypoleuca* (p. 1358). *F. h. hypoleuca*: **6** ad ♂ breeding black type, **7** ad ♂ non-breeding fresh, **8** ad ♀ breeding, **9** 1st summer ♂ partly brown type, **10** 1st winter, **11** juv. *F. h. speculigera*: **12** ad ♂ breeding.

mountains and Nakhichevan region. TURKEY. Range includes North and East Anatolia, and (locally) central and southern Turkey. ALGERIA, MOROCCO. Observations widely separated from known range; further study required. In Algeria, 1991, this species or an unidentified *Ficedula* found breeding alongside Pied Flycatcher on southern slope of Djurdjura mountains (130 km south-east of Algiers); early research suggests breeding extends from Tunisia frontier west to Ouarsenis. In Morocco, several of this species or an unidentified *Ficedula* reported active at 3 sites in Moyen Atlas, May 1964; also several other records in passage periods (though confusion with Collared Flycatcher possible).

Accidental. Italy, Malta, Ukraine, Tunisia.

Beyond west Palearctic, breeding extends to northern Iran.

Population. GREECE. 1000–5000 pairs. BULGARIA. 500–5000 pairs. AZERBAIJAN. Uncommon. TURKEY. 1000–10 000 pairs.

Movements. Trans-Saharan migrant. Relatively small and circumscribed population, combined with identification difficulties and its common designation as race of Collared Flycatcher, result in incomplete knowledge of routes and winter quarters. Thus probably under-recorded almost everywhere. Wintering apparently limited to comparatively small area in East Africa from southern Sudan (south of 5°N) through western Uganda to Tanzania.

Food. Mainly flying insects. Feeding habits similar to Spotted Flycatcher, most prey being obtained by sallying out from perch after flying insect; rarely returns to same perch. Prey less often taken directly from leaves or branches (sometimes while hovering) or from ground.

Social pattern and behaviour. Little studied; essentially similar to Collared Flycatcher.

Voice. Song of ♂ similar in structure, timbre, and phrase length to Collared Flycatcher but tempo somewhat faster and, in this respect, more like Pied Flycatcher. Phrase lasts 1–7 s and is longer in morning, evening, and before pairing than at midday and later stages of breeding; however, length variable and even successive phrases in individual repertoire vary markedly from 3 to 7 s. Calls mostly similar to those of Collared Flycatcher and Pied Flycatcher, but rattling 'drrkt', given by ♂ when showing nest-hole to ♀, and penetrating whistling 'eep' or 'tseep', given in alarm, apparently quite distinct.

Breeding. SEASON. Caucasus: from mid-April (start of laying) to mid-July (presumably end of fledging), hatching mostly in first 3 weeks of May. One brood. SITE. Natural or artificial hole in tree; takes readily to nest-boxes. Nest: cup of dead leaves, dead plant stems, lichens, and moss, lined variously with fine rootlets, grasses, or bark fibre, less commonly hair, feathers, or plant down. EGGS. Sub-elliptical, smooth and slightly glossy; pale blue, unmarked, similar to Collared Flycatcher but darker. Clutch: 5–6 (4–7). INCUBATION. 13–14 days. FLEDGING PERIOD. 15 days (14–17).

Wing-length: ♂ 78–85.5, ♀ 75–83 mm.
Weight: ♂♀ mostly 13–15 g.

Collared Flycatcher *Ficedula albicollis*

PLATES: pages 1356, 1361

Du. Withalsvliegenvanger   Fr. Gobemouche à collier   Ge. Halsbandschnäpper   It. Balia dal collare
Ru. Мухоловка-белошейка   Sp. Papamoscas collarino   Sw. Halsbandsflugsnappare

Field characters. 13 cm; wing-span 22·5–24·5 cm. Size as Pied Flycatcher except for marginally longer wings. Small to medium-sized flycatcher, with similar general character to Pied Flycatcher. Breeding adult ♂ even more boldly pied than Pied Flycatcher and Semi-collared Flycatcher, with larger white forehead, striking full collar, complete and broad wing-bar (extending right across wing), and rump-patches (diagnostic in combination). Non-breeding ♂, ♀, and immature all greyer above (especially on nape and rump), cleaner white below than Pied Flycatcher, with more complete wing-bar and wider tertial-fringes, but similar to Semi-collared Flycatcher; all have almost completely dark tail, usually lacking white edges of Pied Flycatcher and Semi-collared Flycatcher. When wing folded, white bases to primaries form obvious pear-shaped mark (usually more obvious than smaller white patch of Semi-collared Flycatcher and vestigial spot of most Pied Flycatchers).

Breeding adult ♂ unmistakable, but in other plumages differs from Semi-collared Flycatcher only in ♀'s uniform median wing coverts. Within breeding range, separation from Pied Flycatcher on characters given above and call is practical, but on migration and in winter quarters distinction from Pied Flycatcher and particularly Semi-collared Flycatcher is more difficult since plumage characters overlap and only call gives clue.

Habitat. In continental middle latitudes of west Palearctic, in temperate and warm temperate climates, above July isotherm of 15°C. Appears to complement Pied Flycatcher as a bird of warmer more continental regions, more attached to crowns of trees than to their lower branches, and less frequently on ground. Breeds mainly in open broad-leaved woodland, also well-timbered parks and avenues, orchards and gardens with fruit trees.

Distribution. Breeds only in west Palearctic. LUXEMBOURG. May have bred in 19th century. GERMANY. Breeding areas isolated from main European range, and smaller ones very vulnerable. In east, irregular and sporadic breeding by one or a few pairs. DENMARK. ♂ paired with ♀ Pied Flycatcher bred 1986, Zealand; hybrid pair probably also bred 1994. SWEDEN. Formerly bred only in Gotland; established Öland since 1956. FINLAND. Annual spring visitor since 1967, occasionally breeding. Mixed breeding with Pied Flycatcher reported 1972–3, and 2 reports of presumed hybrids, 1975 and 1980. POLAND. Some recent northward spread; regular but infrequent hybridization with Pied Flycatcher. RUMANIA. Probably breeds in all broad-leaved forests of plateau and mountain zone. RUSSIA. Has colonized Kaliningrad region since late 19th century.

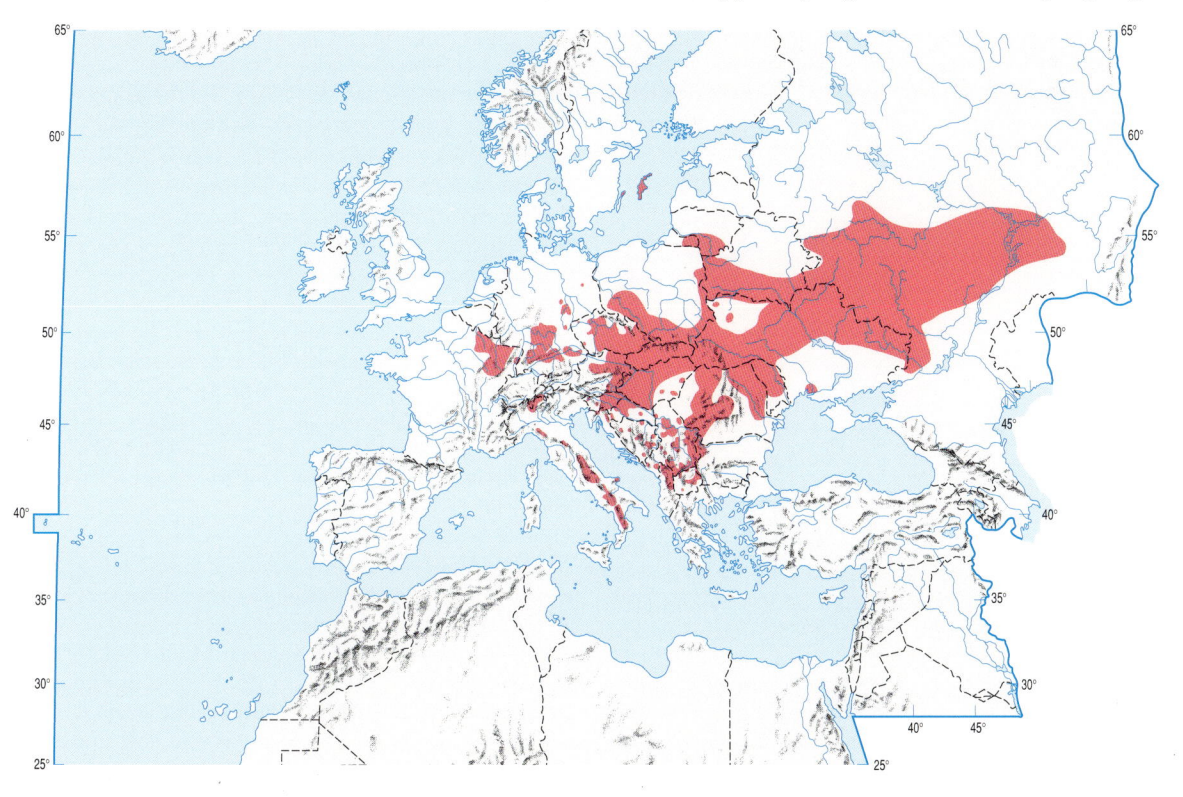

Accidental. Britain, Belgium, Luxembourg, Netherlands, Denmark, Norway, Estonia, Lithuania, Spain, Portugal. In Morocco, apparently vagrant or occasional spring migrant in south-east; some records predate separation of Semi-collared Flycatcher (which see) from this species.

Population. Stable in most countries, but decrease in Germany and perhaps Croatia, and slight increase Moldova. FRANCE. 1000–10 000 pairs. GERMANY. 13 000 pairs in mid-1980s. SWEDEN. 1000–5000 pairs in late 1980s. POLAND. 1000–6000 pairs. Recently fairly stable, after slow increase in 1960s–70s. CZECH REPUBLIC. 25 000–50 000 pairs 1985–9. SLOVAKIA. 70 000–150 000 pairs 1973–94. HUNGARY. 70 000–80 000 pairs 1979–93. AUSTRIA. 2000–3000 pairs in 1992. SWITZERLAND. 15–25 pairs 1985–93. ITALY. 1000–3000 pairs 1983–95. ALBANIA. 10–1000 pairs. YUGOSLAVIA: CROATIA. 10 000–20 000 pairs. SLOVENIA. 2000–3000 pairs. RUMANIA. 150 000–200 000 pairs 1986–92. RUSSIA. 5000–50 000 pairs. BELARUS'. 7000–10 000 pairs in 1990. UKRAINE. 20 000–25 000 pairs in 1990. MOLDOVA. 20 000–30 000 pairs in 1988.

Movements. Trans-Saharan migrant, wintering in Africa, mainly south of equator to *c.* 20°S.

Autumn heading from breeding grounds in southern Germany is narrowly confined, SSE to Italy, apparently followed by non-stop flight from Italy over Mediterranean and Sahara, as few records from Mediterranean islands and North Africa. In contrast to autumn, numerous spring records from Sahara, Mediterranean, and North Africa; heaviest passage appears to be through east Mediterranean. Autumn departure from breeding grounds mainly August; return mainly from mid-April to late May.

Records in Britain mostly on east coast: of 15 records, 1947–85, 14 between 4 May and 6 June, 1 in September; seasonal bias likely to arise, at least in part, from more distinctive spring than autumn plumage. For similar reason, nearly all records in Britain, and elsewhere away from breeding grounds, are of ♂♂.

Food. Arthropods, flying and non-flying; during breeding season, larval Lepidoptera important (for nestlings). Food obtained by sallying out from perch after flying prey, by picking directly from leaves and twigs, and from ground.

Social pattern and behaviour. Little information from outside breeding season; occurs in mixed feeding parties, apparently sometimes also territorial. Mating system essentially monogamous, but successive polygyny regular. Polygyny (typically bigamy) is associated with polyterritoriality of ♂: at times of laying at nest of 1st (primary) mate, ♂ starts displaying at other empty holes (up to 5) in his territory, and some ♂♂ thus attract new (secondary) ♀ and sometimes even a 3rd. Role of sexes in care of young varies where ♂♂ polygynous; in monogamous pairs, ♂ and ♀ share care of young—but only ♀ incubates and broods, irrespective of mating system. ♂ gives advertising-song mainly from nest-tree or neighbouring one; serves mainly to attract a mate, so declines sharply after pairing. Starts again, at lower intensity, when ♀ begins nest-building and stops immediately after start of laying unless ♂ moves to another territory. Where the two species occur together, Collared Flycatcher is dominant over Pied Flycatcher, forcing it to use suboptimal nest-sites in coniferous woods.

Voice. Advertising-song of ♂ a series of drawn-out units, mostly harsh whistles interspersed in some phrases with purer lower-pitched whistles. Described as wistful, recalling Robin, but delivery much slower. Compared with Pied Flycatcher, song tends to be longer, with slower tempo, and is less far-carrying. Ascending figure of 3 moderately loud whistling notes given in association with song, or in isolation. Other calls include a soft 'tsrr' contact-call, and a thin, clear, relatively drawn-out, far-carrying, easily locatable 'eep', used in alarm and also as advertising-call, in association with or as substitute for song.

Breeding. SEASON. Central Europe: eggs laid from late April. Sweden: in Gotland, eggs from 2nd half of May. One brood. SITE. Natural or artificial hole in tree, wall, or building; takes readily to nest-boxes. Preferred height above ground *c.* 15 m, rarely close to ground. Nest: cup of dry grass, leaves, and stalks, lined with fine grass. EGGS. Sub-elliptical, smooth and slightly glossy; very pale blue, unmarked. Clutch: usually 4–7; declines through season. INCUBATION. 12–14 days. FLEDGING PERIOD. 15–18 days.

Wing-length: ♂ 79–88, ♀ 77–84 mm.
Weight: ♂♀ mostly 12–16 g.

Pied Flycatcher *Ficedula hypoleuca*

PLATES: pages 1356, 1361

DU. Bonte Vliegenvanger FR. Gobemouche noir GE. Trauerschnäpper IT. Balia nera
RU. Мухоловка-пеструшка SP. Papamoscas cerrojillo SW. Svartvit flugsnappare

Field characters. 13 cm; wing-span 21.5–24 cm. Slightly smaller and noticeably shorter than Spotted Flycatcher, with stubbier bill and 15% shorter tail; close in size to Collared Flycatcher and Semi-collared Flycatcher, though with marginally shorter wings. Small to medium-sized, rather compact flycatcher. Adult ♂ breeding pied, other plumages essentially dun-brown above and dun-white below; all have white-edged tail and dark wings with bold but narrow white lines on tertial-edges and across at least tips of greater coverts.

♂ in full breeding plumage unmistakable: bold wing markings and lack of spots or streaks on foreparts in adult soon rule out Spotted Flycatcher, bold wing markings and lack of startling white basal patches on tail and warm buff breast quickly eliminate smaller Red-breasted Flycatcher, and bold wing markings on otherwise poorly marked ♀ and immature prevent confusion with smaller Brown Flycatcher. However, in all other plumages confusion is all too easy with Semi-collared Flycatcher and Collared Flycatcher. Effectively, atypical ♂, ♀, and non-breeding plumages in all 3 species show trend towards greyer upperparts and broader and longer wing-bar in sequence Pied—Semi-collared—Collared, with additional faint to quite distinct indication of at least half-collar and pale rump marks noticeably increasing in Semi-collared Flycatcher and Collared Flycatcher. Thus while separation of north-western Pied Flycatcher from other 2 species is often possible, separation of eastern Pied Flycatcher and distinction between Semi-collared Flycatcher and Collared Flycatcher is sometimes simply impossible. Flight intermediate between Spotted Flycatcher and Red-breasted Flycatcher, with similar basic action but shorter sallies and less tight or circular manoeuvres. Frequently moves tail up and down and (remarkably) flicks whole of folded wing from shoulder, particularly when disturbed and after settling from flight.

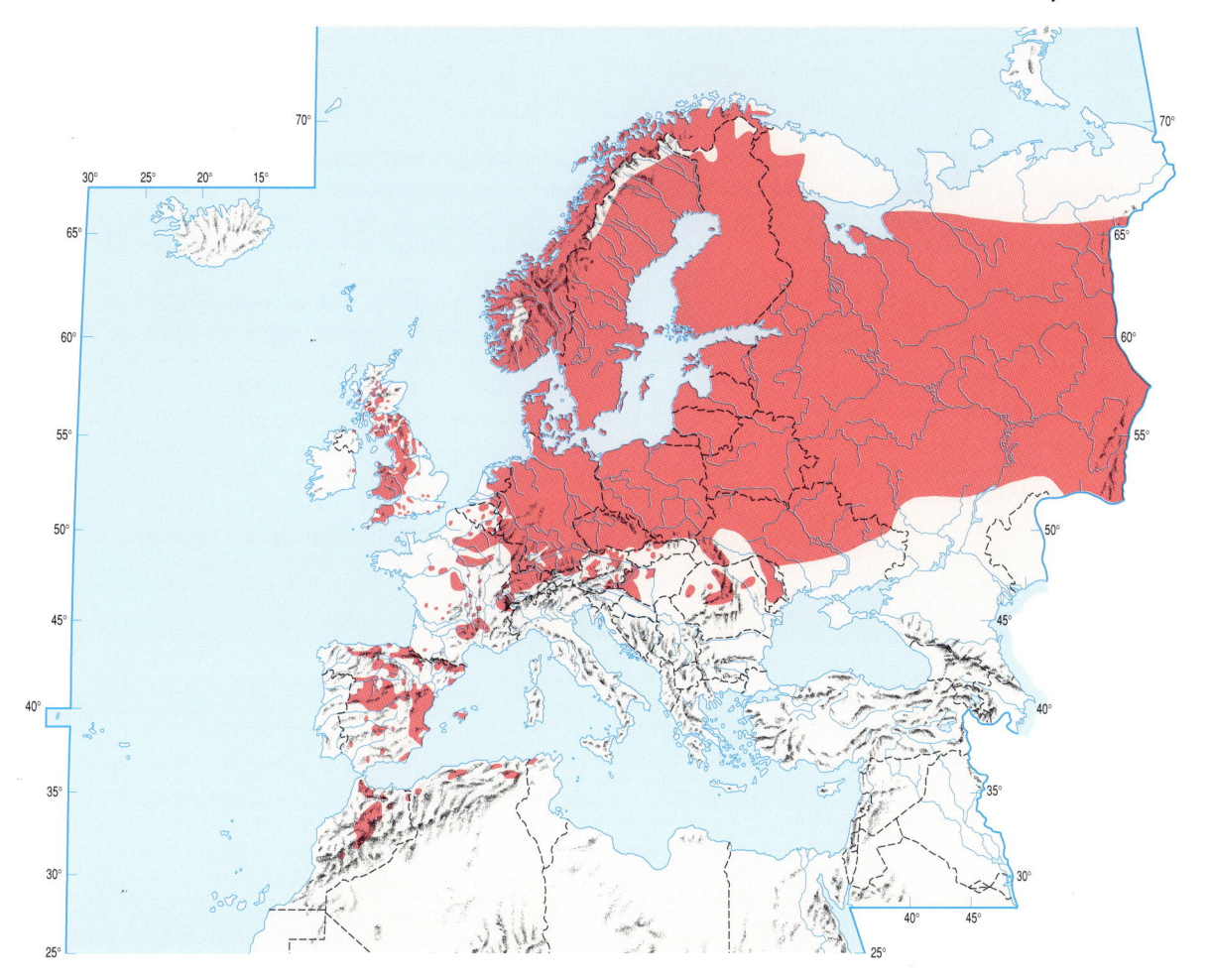

Best distinction of all 3 species lies in voice.

Habitat. In west Palearctic, breeds in higher latitudes than other *Ficedula*, in temperate but also in boreal and Mediterranean zones, mainly in lowlands or hilly country. In Algeria and Morocco, breeds at *c.* 1200–1800 m in forests of cedar, oak, and Aleppo pine. In Switzerland, has recently spread up alpine valleys, nesting at up to 1000 m or even to 1500–1600 m, on forest edges and by meadows and buildings; has also expanded into fringes and glades of conifer forests, but, as elsewhere in continental Europe, deciduous and mixed sunny open mature woodland with plenty of natural holes represents more basic element of habitat, supplemented by orchards, avenues, parks, and even gardens in low-density human settlements. In Britain, peculiar distribution related to local climate and available habitat. Range broadly coincides with occurrence of more than 1000 mm of annual rainfall, and with hilly terrain, and with area in which sessile oak is dominant tree, with little undergrowth. Conifers almost entirely avoided except where provision of nest-boxes makes an overriding attraction, which appears exceptionally strong for this species compared with other hole-nesters, and has even led to local extensions of range.

Distribution. Has spread north and west in some northern parts of range, south in Balkan peninsula. BRITAIN. Slight recent expansion of range, especially south-west Wales and north-west England. Breeds only very irregularly in most parts of Scotland. IRELAND. First recorded breeding 1985. BELGIUM. Range now more continuous. NETHERLANDS. Isolated breeding records in 19th century. By 1910 established in east; spread west since c. 1940; now practically stabilized, but some recent spread in north and west. FINLAND. Population established in far north-west Lapland in 1960s, following erection of nest-boxes in alpine birch forest. CZECH REPUBLIC. Marked increase. BALEARIC ISLANDS. Several pairs breeding southern Mallorca in recent years. ITALY. Occasional breeding; most recent Veneto 1986. RUMANIA. Recorded mainly in Transylvania. MOROCCO. Recently discovered breeding in Haut Atlas.

Accidental. Spitsbergen, Iceland, Madeira, Cape Verde Islands.

Beyond west Palearctic, extends east to upper Yenisey river (Siberia).

Population. Most populations stable. Increases in part due to provision of nest-boxes. BRITAIN. 35 000–40 000 pairs 1988–91. Probably stable since 1960s, following increase; nest-box schemes promote local increases. IRELAND. Under 5 pairs. FRANCE. 20 000–30 000 pairs. BELGIUM. 850–1650 pairs 1989–92; increasing, especially Flanders and Limburg. In 1967, at most 200 pairs. LUXEMBOURG. 800–1000 pairs. NETHERLANDS. 15 000–25 000 pairs 1979–85. Now nearly stable following increase (at least 6000 pairs 1972). GERMANY. 318 000 pairs in mid-1980s. DENMARK. 2000–32 000 pairs 1987–8; decrease 1980–94. NORWAY. 200 000–1 million pairs 1970–90. SWEDEN. 1–1.5 million pairs in late 1980s; slight increase. FINLAND. 1–1.5 million pairs in late 1980s; slight increase. ESTONIA. 200 000–500 000 pairs 1991. More or less stable now; increase in 1980s following decrease in early 1970s. LATVIA. 300 000–400 000 pairs in 1980s. LITHUANIA. Common. POLAND. 50 000–150 000 pairs. Mostly scarce; only locally fairly numerous. CZECH REPUBLIC. 10 000–20 000 pairs 1985–9. SLOVAKIA. 100–200 pairs 1973–94. HUNGARY. 20–50 pairs 1979–93. AUSTRIA. 300–400 pairs 1992. SWITZERLAND. 10 000–20 000 pairs 1985–93. SPAIN. 130 000–350 000 pairs; slight decrease. RUMANIA. 1000–1500 pairs 1986–92. RUSSIA. 1–10 million pairs. BELARUS'. 750 000–800 000 pairs in 1990. UKRAINE. 27 000–30 000 pairs in 1988. MOLDOVA. 400–700 pairs in 1988; slight decrease. TUNISIA. Uncommon breeder in oak forests. MOROCCO. Scarce.

Movements. Long-distance migrant, wintering in West Africa, south of the Sahara and mainly north of Gulf of Guinea, where various woodland habitats are utilized. No satisfactory records anywhere east of northern Zaïre and Central African Republic. On autumn migration, major stopover and fattening area in Iberia (mainly north-west) and probably northern Italy. Thus many northern migrants move far to west before turning south. ♀ failed breeders may migrate early, in June or early July. Normal autumn movements start late July or early August. Passage through Gibraltar from mid-August to end of October, chiefly September. Spring migration begins in March. Passage through Mediterranean peaks mid-April to early May. Return to breeding areas mid- to late April in southern sites, but up to mid-May in north.

Food. Arthropods, flying and non-flying, especially Hymenoptera, Diptera, and beetles (Coleoptera); during breeding season, larval Lepidoptera important (for nestlings); fruit taken regularly in small amounts in late summer and on migration, and when feeding conditions poor. Obtains food directly from trees or ground or by sallying out from perch after flying prey, usually for short distance.

Social pattern and behaviour. Among best studied of all west Palearctic passerines. Little information, however, outside breeding season. In winter quarters, apparently solitary and not uncommonly (perhaps typically) sedentary. Mating system essentially monogamous, but successive polygyny (usually with 2 ♀♀, occasionally 3) is regular. As in Collared Flycatcher, polygyny achieved through polyterritoriality of ♂. Degree of polyterritoriality and polygyny varies regionally, probably depending on differing likelihood of 'cheating'. ♂ may need to acquire distant (secondary) territory, so enabling him to deceive secondary ♀ about his already-mated status; low frequencies of polygyny may arise from high breeding densities preventing acquisition of secondary territory. When secondary ♀ starts laying, ♂ typically leaves her and returns to help feed primary brood; extent to which secondary broods are fed by ♂ seems to vary. In addition to polygyny, ♂♂ are highly promiscuous, and cuckoldry is common (notably when would-be polygamous ♂♂ leave their primary ♀ unguarded); such matings commonest between neighbours. ♀ alone incubates and broods. In monogamous pairs, feeding shared almost equally between sexes after 1 week. ♂ gives advertising-song from perch, usually high in tree, and accompanies song with slight drooping of wings and mild tail-flicking. Prior to arrival of ♀♀, ♂ uses song-posts in 'pre-territory', but in presence of ♀ focus of song switches to nest-hole, ♂ then singing from its immediate vicinity. In case of polygamous ♂♂, once primary ♀ begins laying, ♂ sings less frequently at primary site and stops visiting there, but sings to attract ♀ to secondary hole (and any subsequent ones). Song mostly restricted to May. ♂ attracts ♀ to nest-hole with 'nest-showing' display, combining conspicuous flight to hole with song.

Voice. Advertising-song of ♂ a short phrase of sweet tuneful notes, with markedly changing pitch, used mainly for mate-attraction. Clearly separated phrases, each c. 2 s long, with gaps of 5–30 s. Quite distinct 'strangled song'—lengthy, vivacious, whispering song, ending in harsh 'z z z z' sound—given most intensely immediately after arrival of ♀, when used in nest-showing display. Calls very varied. Commonest are a loud, penetrating 'whit', recalling Chaffinch; in alarm, a quite full, slightly explosive 'hweet', somewhat recalling common call of Chiffchaff; also a short, often repeated 'tic', not distinctive but often combined with former to produce disyllabic 'whit-tic' or 'whee-tic'.

Breeding. SEASON. Central Europe: eggs laid late April to early June. North-west Africa: eggs present 1st week of May to 1st week of June. England: laying starts between end of April and mid-May. Sweden: eggs laid late May to end of June. Moscow region: laying starts early May to mid-June. Finnish Lapland: laying starts mid-June. Generally 1 brood; 2 broods recorded only in England and Moscow region. SITE. Hole, especially of woodpecker, rot-hole caused by lost branch, and particularly in Scandinavia in old hole of Willow Tit.

Brown Flycatcher *Muscicapa dauurica* (p. 1349): **1–2** ad breeding. Spotted Flycatcher *Muscicapa striata* (p. 1349): **3–4** ad breeding. Red-breasted Flycatcher *Ficedula parva* (p. 1353): **5** ad breeding, **6** 1st winter. Collared Flycatcher *Ficedula albicollis* (p. 1357): **7** ad ♂ breeding, **8** ad ♀ breeding. Pied Flycatcher *Ficedula hypoleuca hypoleuca* (p. 1358): **9** ad ♂ breeding black type, **10** ad ♀ breeding.

Increasingly in nest-boxes where available, and these preferred to natural cavities. Nest: rough loose foundation of leaves, roots, fragments of bark, and grass; cup lined with feathers, wool, thin flakes of bark, hair, grass, fine roots, etc. Eggs. Sub-elliptical, smooth and slightly glossy. Pale blue, rarely with fine reddish-brown speckling. Clutch: 6–7 (5–8), decreasing as season progresses. Incubation. 13–15 days. Fledging Period. *c.* 14–17 days.

Wing-length: ♂ 77–84, ♀ 75–81 mm.

Weight: ♂ ♀ mostly 10–15 g; migrants to 20.5 g.

Geographical variation. Slight, mainly involving colour. Upperparts of ♂ vary from black to brown, proportions of different types varying regionally, black predominant in west and north-west of European range, browner types increasingly towards east. Nominate *hypoleuca* occupies most of west Palearctic; *F. h. iberiae*, with more extensive white on forehead and wings, Iberia; *F. h. speculigera*, with most extensive white, north-west Africa.

Babblers Family Timaliidae

Small to medium-large, mostly thrush-like oscine passerines (suborder Passeres). Most found in wide variety of deciduous and evergreen woodland, often dense and wet, though some species (including those in west Palearctic) inhabit dry scrubland and reedbeds. Some arboreal but most live close to or on ground, though usually within or near cover. Largely insectivorous and frugivorous but some omnivorous, even eating small vertebrates. 263–279 species, occurring mainly in tropical regions of Asia, with some in Africa, a few in northern Eurasia, and 1 in North America. Mainly sedentary. 5 species in west Palearctic, all breeding.

Sexes generally of similar size. Bill highly variable (short or long, thin or thick, straight or decurved, etc.). Wing usually rather short with rounded tip; 10 primaries, p10 somewhat reduced (40–70% of length of longest). Flight often weak—short and jerky between ground and elevated perch. Tail long (and often graduated) to short; 12 feathers. Leg and toes usually sturdy; leg quite long in the more terrestrial species. Foot employed by many (if not all) species to hold food objects. Often noisily vocal, with much reciprocal calling. ♂ song well developed in many species; may include imitations. Many species highly sociable, some being group-territorial and nesting communally. Unlike Turdidae, Muscicapidae, Sylviidae, and many other passerines, 'clump' in bodily contact with others in social group when loafing or roosting.

Plumage soft and lax. Colour variable: dull and sombre in some species (e.g. *Turdoides*) but highly colourful in others, with contrasting patterns on head, throat, breast, wing, or tail. Usually cryptic, however, at least to some extent, often with disruptive bars, spots, or streaks and sometimes with pale (often white) underparts to provide countershading. Sexes may differ (with ♀ duller) but usually similar. Some species crested. Juvenile plumage unspotted (unlike in Turdidae).

Bearded Tit *Panurus biarmicus*

PLATES: pages 1363, 1395

Du. Baardman   Fr. Panure à moustaches   Ge. Bartmeise   It. Basettino
Ru. Усатая синица   Sp. Bigotudo   Sw. Skäggmes

Field characters. 12.5 cm, of which tail 7 cm; wing-span 16–18 cm. Head and body close in size to Great Tit but with long and markedly graduated tail (15–20% longer than that of Long-tailed Tit). Stubby-billed, short-winged, and very long-tailed, somewhat sparrow- or tit-like bird, with essentially sandy- or tawny-russet plumage strikingly marked in ♂ by grey and white head with bold black moustaches and bold black under tail-coverts, and in all plumages by black, cream, and white marks near or on wings. Restless and vocal, with whirring, tail-trailing flight suggesting tiny pheasant, and commonest call, a ringing 'ping' or 'pzing', diagnostic.

Unmistakable whether discovered by sight or, more commonly, by call. Flight action includes 'winding' movement of tail in which it is first fanned then suddenly twisted. Usually low, but in pre-eruptive behaviour often flies up with excited dancing or bobbing action and then, if departing, noticeably less hesitant flight. Gait an agile hop, allowing jumps up reed, straddle between stems, and other acrobatic movements and postures. Also jumps and creeps over debris and mud, often waving or raising and fanning tail, then resembling small babbler. Often restless. Difficult to spot though easy to hear and not disturbed by close human presence; will come to imitations of its call.

Habitat. Breeds across middle latitudes of west Palearctic in cool and warm temperate climates, predominantly continental and lowland, concentrated in small often isolated fragments of suitable wetland. Range lies within July isotherms 17–32°C, but excludes uplands, all bare, rocky, and open areas, forests, coasts, farmland, open waters of all kinds, and (with rare exceptions) human settlements and artefacts. Concentrated in usually large tracts of reed and associated dense tall non-woody vegetation growing by or often in fresh or brackish water, or immediately adjoining marshes and swamps.

Distribution. Has spread (in some cases after earlier decline) in much of northern and eastern Europe and Iberia. Breeding areas much reduced after severe winters, and some populations extinguished, especially in north of range. Map therefore shows temporary breeding range. BRITAIN. Bred over much of central England in 1st half of 19th century, but by end of century, due to reclamation of coastal marshlands, persecution, and hard winters, largely restricted to East Anglia. Notable expansion began mid-1960s (in part due to immigration of Dutch birds), reaching other (mostly coastal) areas in eastern and southern England, and a few areas further west. Channel Islands: occasionally breeds Jersey. IRELAND. Bred 1976, 1982–5. FRANCE. Bred in several areas in north in 19th century, then declined; during 1st half of 20th century, breeding more or less confined to Camargue. Spread since 1947–51, especially after 1965, with settling of wintering birds from populations in Netherlands. BELGIUM. Probably bred 19th century; bred 1942, 1966–75

Bearded Tit *Panurus biarmicus*. *P. b. biarmicus*: **1** ad ♂, **2** typical ad ♀, **3** streaked ad, **4** juv ♂, **5** juv ♀. *P. b. russicus*: **6** ad ♂, **7** ad ♀.

(peak *c.* 50 pairs in 1969), 1981, 1988. In 1990s, increased numbers autumn and winter suggest possibility of renewed breeding. NETHERLANDS. Local and scarce until 1960s; new favourable habitat in Flevoland caused explosion in numbers in 1970s, with spread to suitable sites elsewhere in Netherlands and abroad. Only relatively small parts of Flevoland now suitable. GERMANY. A number of largely isolated breeding sites. Regular breeder in east since 1967. DENMARK. Breeding records 1967–9 (and possibly 1966), 1973–8, and 1986 onwards, with continuing increase. NORWAY. First sight record 1980, Østfold, and may have bred at same site 1981. In 1991–5, breeding confirmed Vest-Agder and Østfold; has also bred Rogaland. SWEDEN. First bred 1966. Range fluctuates: new sites colonized temporarily after good breeding seasons; Lake Tåkern the only permanent site. FINLAND. First bred 1986; regular since 1991. ESTONIA. First recorded 1978; early fluctuations, then since 1986 explosive increase. LATVIA. First recorded breeding 1972. LITHUANIA. First recorded breeding after 1964. POLAND. First recorded breeding 1910. CZECH REPUBLIC. Range increase. SLOVAKIA. First bred 1907. Marked expansion from 1945. Slight recent increase. SWITZERLAND. Breeding first confirmed Lac de Neuchâtel 1976 (possible from 1973). SPAIN. Slight increase. ITALY. Recently reoccupied former breeding areas in Lombardia. Formerly bred Sicily. RUMANIA. Common in reedbeds of Danube delta and coastal lagoons of Razelm-Sinoie. BELARUS'. First proved breeding 1993. UKRAINE. Slight decrease. TURKEY. Restricted to a few wetlands throughout country, mainly on Central Plateau.

Accidental. Kuwait, Egypt, Algeria, Morocco.

Beyond west Palearctic, extends from Caspian east across central Asia to north-east China.

Population. Strong fluctuations, especially in north of range. BRITAIN. Probably *c.* 400 pairs 1988–91; probably stable. FRANCE. 1900 pairs in 1970s, chiefly in Mediterranean. NETHERLANDS. 1300–2000 pairs 1989–91. Probably now more or less stable. Decreased following population explosion of 1970s (in 1975, minimum 10 000 pairs). GERMANY. *c.* 1500 pairs in mid-1980s. In east, 160±50 pairs in early 1980s. DENMARK. 500–10 000 pairs 1991–4; increase 1985–94. SWEDEN. 500–5000 pairs in late 1980s; slight increase. FINLAND. 200–300 pairs; increase. ESTONIA. 200–300 pairs 1991; rapid increase. LATVIA. 600–1000 pairs in 1980s; strong fluctuations. LITHUANIA. 30–40 pairs; highly variable dependent on winter conditions. POLAND. 100–300 pairs. Marked increase 1960s–70s, perhaps recent decline; strong fluctuations. CZECH REPUBLIC. 100–300 pairs 1985–9; marked increase. SLOVAKIA. 150–400 pairs 1973–94; fluctuating. HUNGARY. 5000–7000 pairs 1979–93; stable. AUSTRIA. At least 10 000 pairs (probably far more) in 1993; marked fluctuations. SWITZERLAND. 60–100 pairs 1985–93; increase. SPAIN. 2300–4500 pairs; slight increase. ITALY. 4000–10 000 pairs 1983–95; stable. GREECE. 3000–5000 pairs; stable. ALBANIA. 1000–3000 pairs in 1981. YUGOSLAVIA: CROATIA. 5–10 pairs; stable. BULGARIA. 50–200 pairs; stable. RUMANIA. 200 000–350 000 pairs 1986–92. RUSSIA. 50 000–500 000 pairs; stable. BELARUS'. Up to 10 pairs; probably increasing. UKRAINE. 1800–2300 pairs in 1990; slight decrease. MOLDOVA. 1200–1500 pairs in 1986; slight decrease. AZERBAIJAN. Common. KAZAKHSTAN. Uncommon in Ural delta. TURKEY. 1000–50 000 pairs; apparently fluctuating.

Movements. Sedentary to partial migrant. European populations basically fairly sedentary, but subject to eruptive (diurnal) movements; heading is dependent primarily on occurrence of reedbeds, irrespective of compass direction; birds travel typically in pairs and groups; some establish new colonies in wintering area, others return to area of provenance; extent of movement varies greatly and results in considerable fluctuation of range. Birds leave British breeding grounds from early to mid-September, chiefly in October, continuing to November; return movement late March to early May.

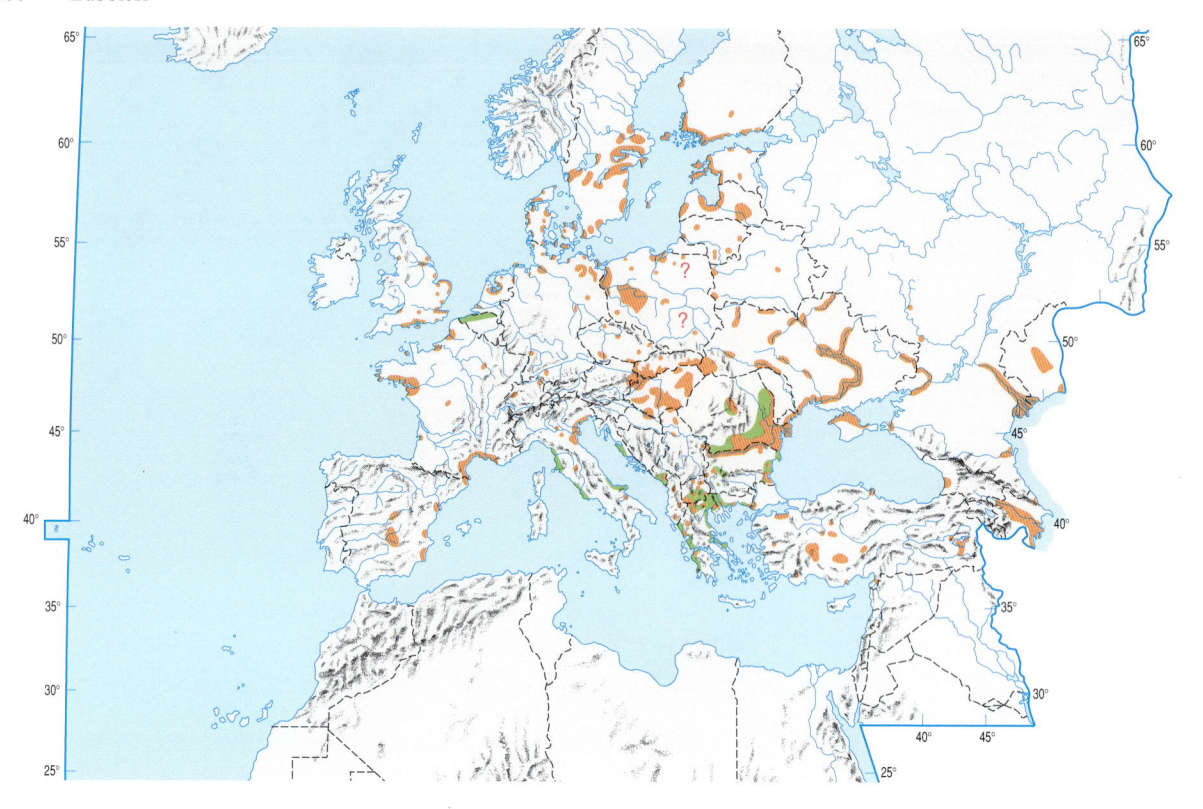

Food. Chiefly invertebrates in summer, mainly seeds in late autumn and winter. Forages almost exclusively in wetlands on reeds (etc.), muddy ground, and margins of open water. Moves across stems lying in water, feeding on insects from water surface, and from below surface, but not deeper than bill length. Very agile on reeds, mostly on lower parts. Often grasps a separate stem with each foot. When perched on stem, sometimes uses bill or foot to pull another seed-head closer, holding it pressed against stem (where perched) while feeding from bottom to top of seed-head. Will hang upside-down from seed-heads of reed or rush. Unlike tits *Parus*, not capable of breaking reeds open, but, especially outside breeding season, pecks at broken stems, cracks, or surface of stem for spiders, larvae, and eggs. While perched on reeds makes rapid darting movements to catch passing insects.

Social pattern and behaviour. Highly gregarious outside breeding season. Juveniles form flocks soon after independence, from mid-May in central Europe, from mid- or late June in eastern England; flocks increase in size and may contain 200 birds. Adults join flocks after breeding, in late summer or early autumn. Birds remain in groups of varying size and composition in winter; break up into pairs in February or March. Pairs loosely colonial, not territorial. Mating system normally monogamous. Nest-building, incubation, and care of young by both parents. After breeding, eruptive movements are preceded by conspicuous high-flying above reed beds, especially in early morning. Such flights begin in June; die down during moult, and then are renewed in more highly developed form in September and reach peak in October. Take place especially in still, sunny weather, from dawn to late afternoon. Restlessness develops in flocks: single calls given, then increasingly rapid and loud calling. With first loud call flocks (up to 25 birds) fly up, and loud calling continues throughout flight. Flocks rise steeply to several hundred feet, fly for some distance over reeds, then soon plunge down almost vertically and re-enter reeds. From 3rd week of September (mainly October) in mornings, small units of high-flying birds may break away from larger groups and fly out of sight, still calling.

Voice. Song not loud or striking, having no territorial function and probably serving to attract mate and maintain pair-bond; a 3-unit phrase lasting 2 s or a little longer: 'tschin-dschik-tschrää'; usually preceded by 3–5 introductory 'tschin' units, which may be omitted if song quickly repeated. Most characteristic and best-known call a ringing 'ping', 'pzing', or 'tschin'; used throughout year as contact-call between flock members, perched and in flight, especially September–October when groups of birds are on the move, least in winter. Excitement-call a harder 'tjick' or 'tschick', becoming sharper and more rapidly repeated as excitement increases. Other calls include a plaintive 'tūū', soft 'djüpp', reminiscent of Bullfinch, and harsh calls given by birds in danger.

Breeding. SEASON. Egg-laying from late March in south

(southern France, Austria, Hungary, southern Germany, Czechoslovakia, Azerbaijan) and from April in northern Europe, extending to July. 2–4 broods regular. SITE. Built among (not woven into) close-growing stems of reeds, sedges, and other marsh vegetation; typically among more or less vertical stems, but also among broken or flattened vegetation, in clusters of basal leaves, or (exceptionally) built into rim of nest of heron; nearly always protected above by 'roof' of sheltering vegetation. Nest: deep cup of dead leaves of reeds and other marsh plants, lined with reed flower-heads, usually also feathers, occasionally mammal hair. EGGS. Sub-elliptical, smooth and glossy; white to creamy-white, lightly and finely streaked, spotted, and speckled dark brown. Clutch: 4–8 (3–11); averages highest in May, decreasing in June. INCUBATION. 11–14 days, mostly 11–12. FLEDGING PERIOD. Usually 12–13(–16) days.

Wing-length: ♂ 58–64, ♀ 57–62 mm.
Weight: ♂♀ mostly 12–18 g.

Geographical variation. Rather marked in colour, slight in size. Nominate *biarmicus* occupies north and west of west Palearctic range. *P. b. russicus*, with generally paler plumage, occurs east from Austria and Balkans through southern FSU and Asia Minor, except extreme south. Isolated *P. b. kosswigi* from Amik Gölü (southern Turkey) is darker and more rufous- or pinkish-brown, above and below, than nominate *biarmicus*.

Iraq Babbler *Turdoides altirostris*

PLATES: pages 1366, 1369

Du. Irakese Babbelaar   FR. Cratérope d'Irak   GE. Rieddrossling   IT. Garrulo iracheno
RU. Камышевая дроздовая тимелия   SP. Tordalino iraquí   Sw. Irakskriktrast

Field characters. 22 cm; wing-span 21–24 cm. Size between bulbul and smaller thrush but with structure differing distinctly in decurved bill, shorter and more rounded wings, and much longer, fuller, and graduated tail; *c.* 10% smaller than Common Babbler, with slightly shorter bill and wings but rather longer tail. Dull brown babbler, indistinctly streaked above but unstreaked below, with slightly darker tail. Call distinctive.

Form, behaviour, and noise of *Turdoides* is distinctive, but this species subject to confusion not only with partly sympatric Common Babbler (slightly larger with paler greyer upperparts, streaked chest and fore-flanks, and usually yellow-toned legs) but also with larger *Acrocephalus* warblers which breed in or pass through shared habitat of reedbeds (particularly Great Reed Warbler, which is 15% smaller, with straight bill and much warmer and completely unstreaked plumage). Flight weak and brief, with trailing tail.

Habitat. Narrowly concentrated in south-east of west Palearctic, within arid region but linked with river margins, especially with reedbeds and moisture-loving trees such as poplar. Spreads onto neighbouring cultivated fields and thickets, and into palm groves. Outside breeding season, groups of birds are nearly always found in bushes and under trees, seldom in the open. Often shares habitat with Common Babbler, but differs in attachment to neighbourhood of water.

Distribution and population. IRAQ. Only breeding area in west Palearctic. Locally common, e.g. Baghdad province, Basra province.

Beyond west Palearctic, extends over Iraqi border to neighbouring parts of western Iran.

Movements. Resident, with only local dispersal of pairs for breeding.

Food. Chiefly insects and spiders. Seeks prey both on ground and in trees, rarely in the open. Group forages assiduously almost throughout day; moves as unit over ground, turning over fallen leaves and twigs and seizing prey thus exposed.

Social pattern and behaviour. Highly gregarious outside breeding season, always dispersed in groups which move over relatively wide home-range. Group size almost constant, typically 5–7 (3–15), comprising adults and juveniles. In spring, pair-formation occurs within group, pairs leaving to establish breeding territories and group thus disbanded by beginning of April; re-formed again after breeding season ends (August). Song occurs in group in spring prior to dispersal of pairs, and plays major role in mate attraction. Mating system monogamous. Both parents care for (incubate, brood, feed, and

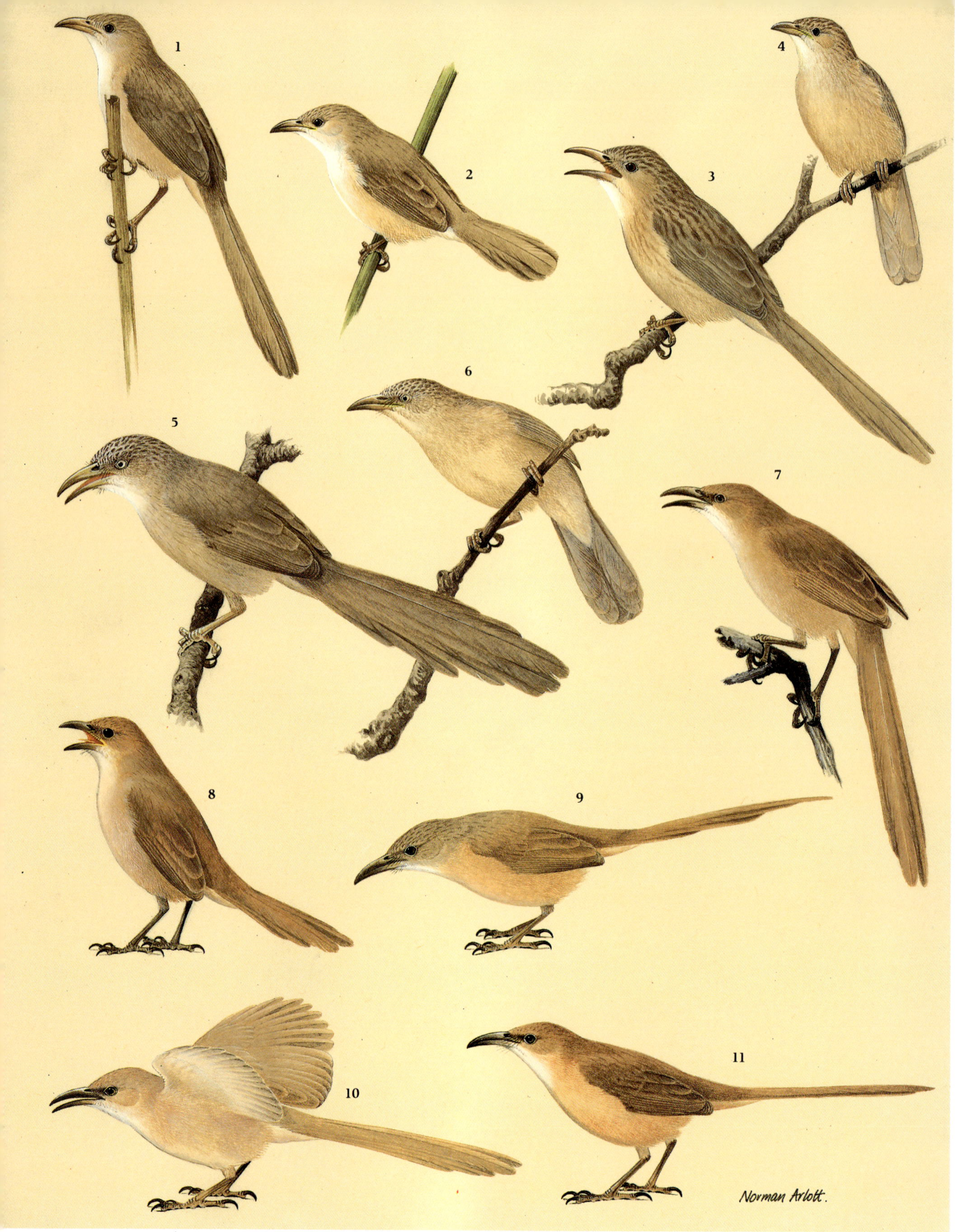

1
2
3
4
5
6
7
8
9
10
11
Norman Arlott.

defend) eggs and young, though much the greater share by ♀ (especially of incubation, brooding, and defence). Other adults may help in feeding nestlings.

Voice. Commonest and most characteristic call, especially outside breeding season, a whistling 'pherrrrrreree...' with descending pitch and volume throughout; sometimes, before fading out completely, ascends in pitch again, though not reaching initial pitch. Heard from any bird separated from rest of group, and answered by group; thus heard in chorus when disturbance disperses flock, and at roosting time (summons-call). Other calls include a noisy chattering 'pherrr pherrr pherrr', and higher-pitched 'phist' and 'phic' sounds. Song not distinctive: a somewhat more musical sequence of last mentioned calls.

Breeding. SEASON. Central Iraq: laying occurs March–July, mainly April–June. 2–3 broods. SITE. Mainly in trees (poplar, tamarisk), less often in reeds. Poplar site nearly always a fork of 3–5 branches adjoining main stem, well hidden by foliage. Nests in reeds built half-way up (at *c.* 1.5 m) and supported by (though not woven to) 4–5 stems. Nest: rather untidy deep cup of stems, with compact lining of grass, dead leaves, rootlets, fibres, reeds, occasionally some feathers. EGGS. Sub-elliptical, smooth and glossy; turquoise, more greenish than blue, with little variation, though one egg in clutch sometimes pale blue without usual gloss. Clutch: 3–4. INCUBATION. 13 days. FLEDGING PERIOD. 10 days (9–12).

Wing-length: ♂ 76–83, ♀ 72–78 mm.
Weight: ♂♀ average 33 g.

Common Babbler *Turdoides caudatus*

PLATES: pages 1366, 1369

DU. Gewone Babbelaar   FR. Cratérope de l'Inde   GE. Langschwanzdrossling   IT. Garrulo comune
RU. Длиннохвостая дроздовая тимелия   SP. Tordalino colilargo   SW. Orientskriktrast

Field characters. 23 cm; wing-span 22.5–25 cm. Iraq race (*salvadorii*) slightly larger than Iraq Babbler, with 10–15% longer wings, longer and less gradually decurved bill, and stronger legs. Quite large babbler, mainly dun-coloured but with buff face and distinct streaks on breast and fore-flank.

Overlaps with Iraq Babbler at head of Persian Gulf but widely separated from larger, darker Arabian Babbler and Fulvous Babbler; for distinction from Iraq Babbler, see that species. Flight weak, with oddly mechanical fluttering action leading into brief glide and untidy landing, long, loose tail trailing noticeably; birds often fly in 'follow-my-leader' sequence. Markedly terrestrial, progressing by strong leaps and hops. Lives all year in small groups.

Habitat. Breeds in tropical and lower subtropical latitudes, mainly extralimital. Inhabits part of arid south-east of west Palearctic, overlapping there with Iraq Babbler on cultivated land, but not sharing its attachment to water. Resident in arid and semi-desert areas and on stony lower hill slopes up to 2100 m, in dry thorn scrub or on sandy flood-plains dotted with clumps of tamarisk, shrubs, and sparse herbage, or on rocky terrain with sparse shrubs. Also in groves, shrubs, and trees of cultivation, orchards, and gardens.

Distribution and population. IRAQ. Only breeding area in west Palearctic. Common in Baghdad area in 1950s–60s.

Beyond west Palearctic, extends east to reach most of India except north-east.

(FACING PAGE) Iraq Babbler *Turdoides altirostris* (p. 1365): **1** ad, **2** juv.
Common Babbler *Turdoides caudatus*: **3** ad, **4** juv.
Arabian Babbler *Turdoides squamiceps* (p. 1368): **5** ad, **6** juv.
Fulvous Babbler *Turdoides fulvus* (p. 1370). *T. f. fulvus*: **7** ad, **8** juv. *T. f. acaciae*: **9** ad. *T. f. buchanani*: **10** ad. *T. f. maroccanus*: **11** ad.

Movements. Resident and apparently sedentary.

Food. Omnivorous, taking insects, fruit and seeds. Feeds mostly on ground among leaf litter, roots, etc., sweeping aside vegetation and digging with bill in soft earth; also a regular predator of eggs of small birds.

Social pattern and behaviour. Dispersed in small groups throughout year, each group occupying a territory. Group-territory has core-area, with less use made of periphery; very little overlap between groups. Typically 6–7 birds per group. Each group evidently comprises ♂-clan of common lineage with variable number of unrelated immigrant ♀♀. Apparently almost all juvenile ♀♀ leave natal group during 1st year. Within groups usually only one pair attempts to breed in a season, less often two. Social cohesion of group is marked. Members forage together, staying within *c.* 30-m radius of centre of group, and moving together (typically in single file) between different parts of territory. Groups apparently less cohesive during summer than winter, and spread over larger area while feeding. Dominance hierarchy evident in group throughout year, the dominant ♂ and ♀ forming the breeding pair.

Voice. Subtly varied repertoire, used throughout the year. Vocalizations difficult to schematize, most being variations (in pitch, intensity, rate of repetition, etc.) on a whistling sequence. Main contact-call a series of 15–25 whistling notes, heard in various situations, e.g. when assembling at and dispersing from roost, and during interactions between neighbouring groups. Song occurs, but undescribed.

Breeding. SEASON. Iraq: at least partly synchronous with Iraq Babbler, i.e. March–July, but no details. India: laying varies locally to cover almost whole year, but mostly March–July, extending into October. SITE. In Iraq uses rather different sites from Iraq Babbler, avoiding poplar and preferring low tamarisks and dense thorny bushes; not in reeds. NEST: foundation of thorny twigs, roots, and grass, with compact inner cup of finer grass stems and rootlets, often lined with hair, mosses, and leaves. EGGS. Sub-elliptical, smooth and glossy; turquoise, more bluish than green. Clutch: in Iraq 3–5. INCUBATION. In India, c. 13 days. FLEDGING PERIOD. In India, c. 11–12 days.

Wing-length: ♂ 87–93, ♀ 85–90 mm.
Weight: No data for Iraq.

Arabian Babbler *Turdoides squamiceps*

PLATES: pages 1366, 1369

DU. Arabische Babbelaar   FR. Cratérope écaillé   GE. Graudrossling   IT. Garrulo arabo
RU. Арабская дроздовая тимелия   SP. Tordalino arábigo   SW. Arabskriktrast

Field characters. 26–29 cm (of which tail 14–15 cm); wing-span 31–33.5 cm. Largest *Turdoides* in west Palearctic, close in size to Mistle Thrush. Large babbler, rather cold brown but with noticeably white throat and distinct scaling over crown, hindneck, and breast; rather long decurved bill, pale eye (of adult), long, graduated and loose tail, and strong legs and feet contribute to somewhat unattractive, scrawny, but arresting appearance.

Geographically isolated from other *Turdoides* and general character and behaviour unmistakable within its range, though appearance has been confused with Great Spotted Cuckoo. Flight consists of bursts of fluttering and whirring wing-beats, followed by brief glides or side-slips on outstretched short and very round wings and long, graduated and spread tail; ends in 'closed-up' bird diving into cover. Gait hopping and jumping, with strength of legs sometimes evident in amazing length and height of movements—all accompanied by balancing, cocking and lateral movements of tail which add to dramatic, theatrical performance unlike that of any other ground-dwelling passerine except for other *Turdoides*. Skulking, but betrays presence by bursts of chattering.

Habitat. In lower warm arid latitudes of south-eastern fringe of west Palearctic, not only in lowlands but locally above 2000 m and in northern Yemen even to 2800 m. In Arabia typically in bush country, with preference for acacia. In Israel does not penetrate to high elevations or to more humid northern regions, but will inhabit tamarisk clumps, salty swamps with reedbeds, gardens, vineyards, and even trees (e.g. palms) among houses; prefers open terrain, shunning dense vegetation.

Distribution and population. ISRAEL. Common wherever habitat suitable. Has apparently expanded in recent decades, adapting to developing agricultural settlements. At Hatzeva in Rift Valley, 65–220 birds (in 15–23 groups) over 17 years,

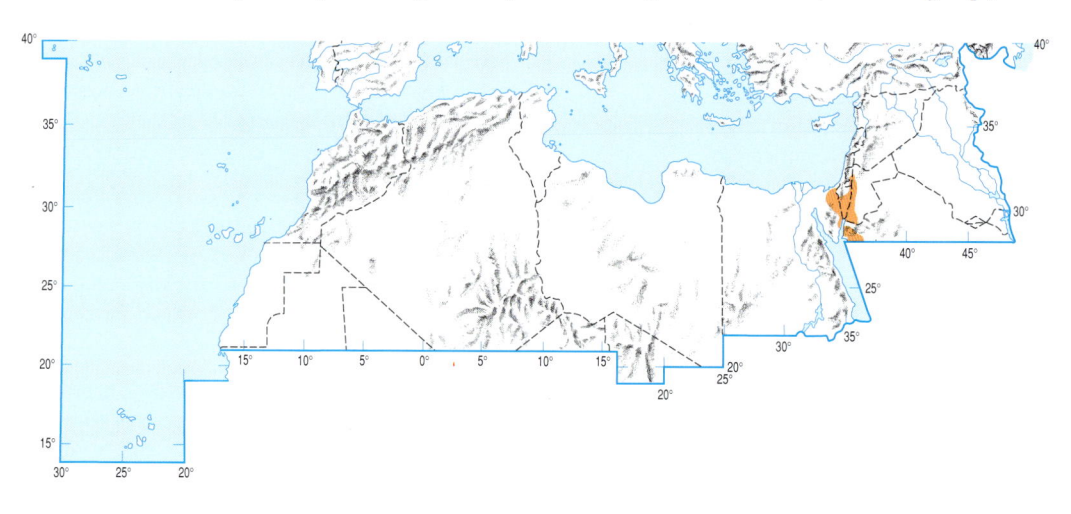

Iraq Babbler *Turdoides altirostris* (p. 1365). **1–2** ad. Common Babbler *Turdoides caudatus* (p. 1367): **3–4** ad. Arabian Babbler *Turdoides squamiceps* (p. 1368): **5–6** ad. Fulvous Babbler *Turdoides fulvus fulvus* (p. 1370): **7–8** ad.

decreasing after dry years. JORDAN. Characteristic bird of Rift Valley fauna. EGYPT. Local and rare in eastern Sinai.

Beyond west Palearctic, widespread in Arabia except coastlands of northern Arabian gulf.

Movements. Highly sedentary, with some local dispersal.

Food. Wide variety of invertebrates, small vertebrates, plant material (nectar, flowers, berries, leaves, seeds), and household scraps (etc.) from garbage dumps. All these food types recorded in one study in eastern Israel, demonstrating opportunism and versatility. Inquisitively inspects every novelty in search for food. Birds usually forage in groups, on ground, bushes, and trees, variously digging in soil, turning over stones and leaves, stripping bark, pecking at and prizing open branches, seed pods, and faeces.

Social pattern and behaviour. Highly social, with cooperative breeding system. Dispersed in territorial groups all year; group is essentially a coalition of birds united in defence of territory against other groups. 5–10% of birds are nomadic, living alone or in small groups. Non-territorial birds try to join territorial groups or, if possible, replace them; non-territorial birds do not breed but some may survive as outcasts for more than a year, then succeed in achieving breeding status within a territorial group, or join with other outcasts to form a new group on the occasional vacant territory. Size of territorial groups varies from 2 to 22. Only one ♂ and one ♀ within a group breed at any one time, usually the two dominant individuals of either sex. At night, only laying ♀ incubates; otherwise, incubation shared by all group members, especially the more dominant ones. All group-members share in brooding (almost until last day young in nest), feeding and guarding young (both in and out of nest), and nest-hygiene. Group members roost communally on tree branch. Site typically traditional, group using same tree and often same branch for several months or even years.

Voice. Rich repertoire, used throughout year. Song of ♂ a high-pitched but quiet warble, barely audible from a few metres, most often heard from young ♂♂ and serving to attract mates. Weak subsong given most notably by 1st-years of both sexes while feeding or while perched as sentinels. Communication (contact)-calls are sequences of discrete whistles (e.g. 'pee' or 'piu'), variable in tempo and other parameters according to context. Other calls include flight-call, a rapid repetition of 2–3 high-pitched units, apparently inviting rest of group to follow caller on relatively long flight; throaty trill, audible only over 20–30 m, given as advertisement of status when approaching another bird.

Breeding. SEASON. Israel: laying February-July, peaking May-June; rarely also October-December when rainfall and food supply suitable. In favourable seasons, up to 3 broods. SITE. Typically 1–7 m above ground inside cover, e.g. fork or crown of tree or bush. Nest: large, untidy, fairly deep cup of coarse grasses, rootlets, bark, twigs, and other dead plant material, with no distinction between outer structure and lining, though may be lined with a few feathers and animal hair; sometimes slightly domed. EGGS. Sub-elliptical, smooth and glossy; uniform turquoise. Clutch: 3–5 per ♀; early and late clutches usually 3, at peak of season often 5. Although only 1 nest per territory, sometimes more than 1 ♀ lays in nest, each laying 2–3 eggs, collectively thus up to 13 eggs per nest. INCUBATION. 13–14 days. FLEDGING PERIOD. *c.* 14 days.

Wing-length: ♂ 108–119, ♀ 107–115 mm.
Weight: ♂♀ 64–83 g.

Fulvous Babbler *Turdoides fulvus*

PLATES: pages 1366, 1369

DU. Bruingele Babbelaar   FR. Cratérope fauve   GE. Akaziendrossling   IT. Garrulo fulvo
RU. Сахарская дроздовая тимелия   SP. Tordalino rojizo   SW. Saharaskriktrast

Field characters. 25 cm; wing-span 27–30.5 cm. 2nd largest *Turdoides* in west Palearctic, close in size to Fieldfare. Rather large babbler, warm sandy-brown with faint scaling and streaking on head and mantle, underparts rather uniform buff.

Unmistakable, being isolated from other *Turdoides*. Distinctive within its range, though glimpse may briefly suggest other similarly-sized passerines (e.g. ♀ Blackbird, Black-crowned Tchagra). Flight fluttering, with oddly spasmodic bursts of wing-beats, glides on spread wings and tail, and untidy landing. Gait combines leaping, hopping, and running; usually escapes along ground. Usually in pairs or small parties.

Habitat. Occupies warm arid lower Mediterranean and subtropical latitudes of west Palearctic, mainly in lowlands; adapted to desert forms of *Acacia*, and also to *Zizyphus*, *Capparis*, and other arid-country species. In northern and western Sahara, optimal habitat is in desert savanna with such vegetation, but birds also occur in palm groves, and every type of tree and bush present seems to be occasionally used.

Distribution and population. EGYPT. Formerly common breeder in Nile valley from Aswan south to Wadi Halfa; now apparently extinct. TUNISIA. Localized. Breeding recorded north to Lac du Kelbia. ALGERIA. Not uncommon. MOROCCO. Previously recorded north of Atlas in Marrakech and Essaouira areas, but few recent sightings and uncertain whether still breeds there. Uncommon but widely distributed.

Beyond west Palearctic in Africa, extends from Mauritania and Sénégal east to Ethiopia.

Movements. Resident and mostly sedentary.

Food. Chiefly invertebrates and berries. Forages in groups, birds moving systematically from one patch of cover to next, threading their way through branches; from time to time, also forages on ground. Group covers extensive area in course of day, and some evidence that same route followed each day. In palm groves in Morocco, forages especially in dense secondary palm growth at bases of trees, probing in matting on trunks and at bases of leaflets, also by digging in soft ground with bill.

Social pattern and behaviour. Little known. Like most other *Turdoides*, lives in cohesive, highly social groups, with evidence of co-operative breeding.

Voice. Like other *Turdoides*, complex repertoire used all year. Main contact-call a series of descending whistles, 'peeoo-peeoo-peeoo...', heard in various circumstances, e.g. from roosting birds, and from group seeking refuge. Alarm-call a hollow metallic rattle, or soft whirring trill.

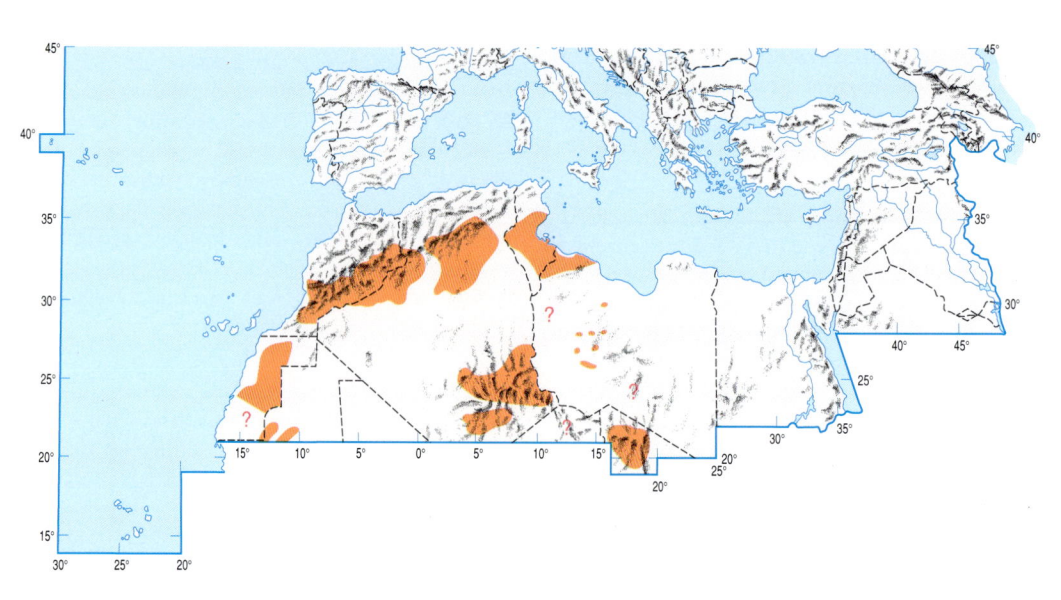

Breeding. SEASON. Laying period extends (across range) from (at least) beginning of November to July, but most records March–April. Western Sahara: January–May. Algeria: end of October or beginning of November to end of April, earlier in north than south. Tunisia: March–April(–July). SITE. Typically inside thorn bush, especially *Zizyphus,* less often in crown of date palm or pile of brushwood. Nest: deep cup of coarse grass and thin twigs, only thinly lined. EGGS. Sub-elliptical, smooth and glossy; uniform turquoise (intense blue-green). Clutch: 4–5 (3–6). (No information on incubation or fledging periods.)

Wing-length: *T. f. fulvus* (Tunisia and Algeria): ♂ 97–106, ♀ 95–102 mm. *T. f. maroccanus* (Morocco): ♂ ♀ 93–99 mm.
Weight: *T. f. fulvus*: ♂ 64–70, ♀ 62–63 g.

Geographical variation. Rather slight. 4 races recognized in west Palearctic, but poorly differentiated. *T. f. maroccanus,* in north-west, darkest in colour; *T. f. fulvus,* to east of *maroccanus,* largest, not so dark; *buchanani* and *acaciae,* further east and south, palest.

Long-tailed Tits and Allies Family Aegithalidae

Tiny to small oscine passerines (suborder Passeres). Found chiefly in woodland and scrub; mainly arboreal. 7 species, occurring in Eurasia and North and Central America. Mainly sedentary. One species in west Palearctic, breeding.

Sexes similar. Bill tiny—very short and somewhat laterally compressed; relatively thick, both mandibles strongly curved. Wing short and rounded; 10 primaries, p10 reduced (*c.* 40% of length of longest). Flight weak and laboured-looking, with rapid wing-beats; undulating over distances. Tail extremely long, graduated and emarginated; 12 rather narrow feathers, central pair shorter than 2nd pair. Leg somewhat more slender than in Paridae. Highly active, agile, and tit-like when searching for food and feeding amongst foliage; will hang upside down acrobatically by one or both feet. Like babblers (Timaliidae), highly sociable; mates and members of same social group clump together in body contact when loafing and roosting. Nest solitarily, however, and mating system monogamous, though some pairs assisted in rearing young by helpers. Nest a well-woven, purse-like, domed structure with side entrance, constructed of moss, lichen, and cobwebs and copiously lined with feathers; unlike in Paridae, built by both sexes.

Plumage soft, thick, and lax—more like Timaliidae than Paridae. Pattern striking: mainly black, grey, brown, and white. Narrow bare ring round eye.

Long-tailed Tit *Aegithalos caudatus*

PLATES: pages 1373, 1395

Du. Staartmees   Fr. Mésange à longue queue   Ge. Schwanzmeise   It. Codibugnolo
Ru. Длиннохвостая синица   Sp. Mito   Sw. Stjärtmes

Field characters. 14 cm, of which tail 9 cm; wing-span 16–19 cm. Head and body close in size to Coal Tit but their length doubled by extremely long and narrow tail. Tiny-billed, short-winged, almost pin-tailed tit, with essentially pied plumage washed with dusky and rosaceous tones. In northern race, head wholly white, but in other races head banded black through eye. Active and noisy, with whirring, bouncy flight.

Adult unmistakable, though juvenile may suggest *Parus* tit in brief glimpse. Flight silhouette and action as diagnostic as plumage pattern: impression of 'small blob' followed by 'flicking line' conveyed by no other west Palearctic bird occurring within its range. Extremely restless, with parties roving through woodland and along hedgerows in characteristic 'follow-my-leader' sequence, with occasional longer group flights.

Habitat. Breeds in west Palearctic in both continental and oceanic middle latitudes, and in Scandinavia beyond temperate into boreal climate; also partly in Mediterranean zone, within July isotherms *c.* 14–27°C. Year-round diet of insects, social foraging, and resident status bias habitat choice in winter towards deciduous woodland, typically of oak, ash, and locally sycamore. For nesting, strong preference shown towards scrub areas and even towards sites outside woodland in bushes and hedges. Occurs not uncommonly in orchards, parks, and gardens with suitable vegetation.

Distribution. BRITAIN. No evidence of long-term changes in status, though has bred Outer Hebrides. IRELAND. Has become less well-distributed. NORWAY. Has spread north since *c.* 1970. UKRAINE. Slight decrease in range. TURKEY. Local and scarce in east. IRAQ. Common in northern mountains.

Accidental. Balearic Islands, Tunisia, Morocco.

Beyond west Palearctic, breeds east across southern Siberia to Kamchatka and Japan, extending from east of range to south-west China; also in western Iran.

Population. Apparently stable over much of range, though fluctuating (with winter weather) Britain, Ireland, Netherlands, Estonia, and Belarus'. Marked decrease Sweden, Finland, less so Ukraine and Moldova. BRITAIN. 210 000 territories 1988–91. No long-term trend apparent, but marked decreases (as in Ireland) after hard winters. IRELAND. 40 000 territories 1988–91. FRANCE. 100 000–1 million pairs. BELGIUM. 11 000 pairs 1973–7. Probably stable in long term, some local increases. LUXEMBOURG. 3000–4000 pairs. NETHERLANDS. 20 000–40 000 pairs 1979–85. Slight increase up to 1980, slight decrease since. GERMANY. 160 000–200 000 pairs. No clear trend. In east, 55 000 ± 25 000 pairs in early 1980s. DENMARK. 700–9000 pairs 1987–8. NORWAY. 5000–20 000 pairs 1970–90. SWEDEN. 10 000–50 000 pairs in late 1980s. Marked decrease since 19th century, but trend not clear in recent decades. FINLAND. 4000–10 000 pairs in late 1980s. Probably long-term decrease due to modern forestry practices. ESTONIA. 5000–15 000 pairs in 1991. Decrease in 1950s–60s, more or less stable since 1970s. LATVIA. 20 000–40 000 pairs in 1980s. LITHUANIA. Common. POLAND. 20 000–40 000 pairs. CZECH REPUBLIC. 55 000–110 000 pairs 1985–9. SLOVAKIA. 60 000–120 000 pairs 1973–94. HUNGARY. 30 000–40 000 pairs 1979–93. AUSTRIA. 50 000–60 000 pairs in 1992. No apparent change. SWITZERLAND. 4000–6000 pairs 1985–93. SPAIN. 700 000–1.65 million pairs. PORTUGAL. 100 000–1 million pairs 1978–84. ITALY. 50 000–150 000 pairs 1983–95. GREECE. 2000–5000 pairs. ALBANIA. 1000–3000 pairs in 1981. YUGOSLAVIA: CROATIA. 15 000–20 000 pairs. SLOVENIA. 10 000–20 000 pairs. BULGARIA. 100 000–500 000 pairs. RUMANIA. 100 000–200 000

Long-tailed Tit *Aegithalos caudatus*. *A. c. caudatus*: **1** ad, **2** juv. *A. c. rosaceus*: **3** ad, **4** juv. *A. c. sibiricus* (Siberia): **5** ad. *A. c. macedonicus*: **6** ad. *A. c. irbii*: **7** ad. *A. c. tephronotus*: **8** ad.

pairs 1986–92. RUSSIA. 100 000–1 million pairs. BELARUS'. 35 000–50 000 pairs in 1990. UKRAINE. 6000–6500 pairs in 1988. MOLDOVA. 850–1000 pairs in 1988. AZERBAIJAN. Common. TURKEY. 10 000–100 000 pairs.

Movements. Mainly sedentary in most years over much of range, but irregular and sometimes massive irruptive movements, associated with high population levels, reported from central and northern areas. Birds generally move over short distances during day in cohesive flocks which may be maintained for many months. Autumn movements are late, often peaking mid-October or even later. Spring passage little observed, even after strong irruptions. Individuals of northern (white-headed) race very rarely reach eastern England and Scotland.

Food. Mainly arthropods, especially bugs, and eggs and larvae of Lepidoptera; plant matter only infrequently taken during autumn and winter. Food obtained mainly in top of shrub layer and especially in canopy of trees, but also in bushes, herb layer, and on thick tree trunks; only occasionally on ground. Plant matter includes seeds, flesh stripped from small fruits, and sap exuded from broken twigs or branches of birch or maple.

Social pattern and behaviour. Gregarious outside breeding season. Winter flocks (6–17 birds) consist of family parties (parents and offspring) from previous breeding season, together with any extra adults that helped to raise brood, remaining fairly constant in membership, except for losses due to death, from end of one breeding season to beginning of next. Flocks may amalgamate temporarily. Flocks occupy territories whose boundaries are vigorously defended against neighbouring flocks, but contact between neighbouring flocks infrequent. Flocks begin to break up into pairs in February or March, pairs setting up loosely defined territories within area occupied by winter flock. No territorial song. Twittering song, given by both sexes, juvenile and adult, apparently serves to maintain hierarchy within flock and avoid overt conflict. Monogamous mating system. Both sexes build nest and care for young; ♀ alone incubates. Helpers at nest regular during feeding of young. Helpers at nest are mainly failed breeders, siblings of ♂ parent at nest helped. Hover-display a regular feature of behaviour of parents and helpers at nests with young: bird flies steeply upwards 30–60 cm from perch and hovers briefly over nest before descending with flutter to same or another perch. Occurs before and after feeding young. Function problematical: very conspicuous, and being performed beside nest serves as accurate guide to position of nest with young; possibly serves to attract potential helpers; helpers markedly improve survival of young.

Voice. Both sexes give quiet twittering, trilling song, apparently of no fixed structure, especially in aggressive encounters with conspecifics. Distinctly different 'bubbling' song also given: comprises mainly the ordinary 'zee' and 'tsirr' notes, repeated very rapidly with a more melodious bubbly intonation than usual; regularly given during nest-building, apparently by both sexes simultaneously, and recorded from birds attending young in nest. Main call given by birds in flocks a soft, slightly metallic 'pit' or 'pt', sometimes preceded by high, thin 'see'. For longer-range communication, high-pitched, rather nasal 'zee' or 'srrip', given singly or repeated; usually given during flight of more than 3 m, especially when flock in rapid movement; also given a great deal by isolated individuals, and used for mutual contact between ♀ on nest and mate. In excitement or alarm, a rapidly trilled rippling 'tsirrrrrup' or 'tserr'; often used also during long flights of more than 20 m, when interspersed with 'eez' calls.

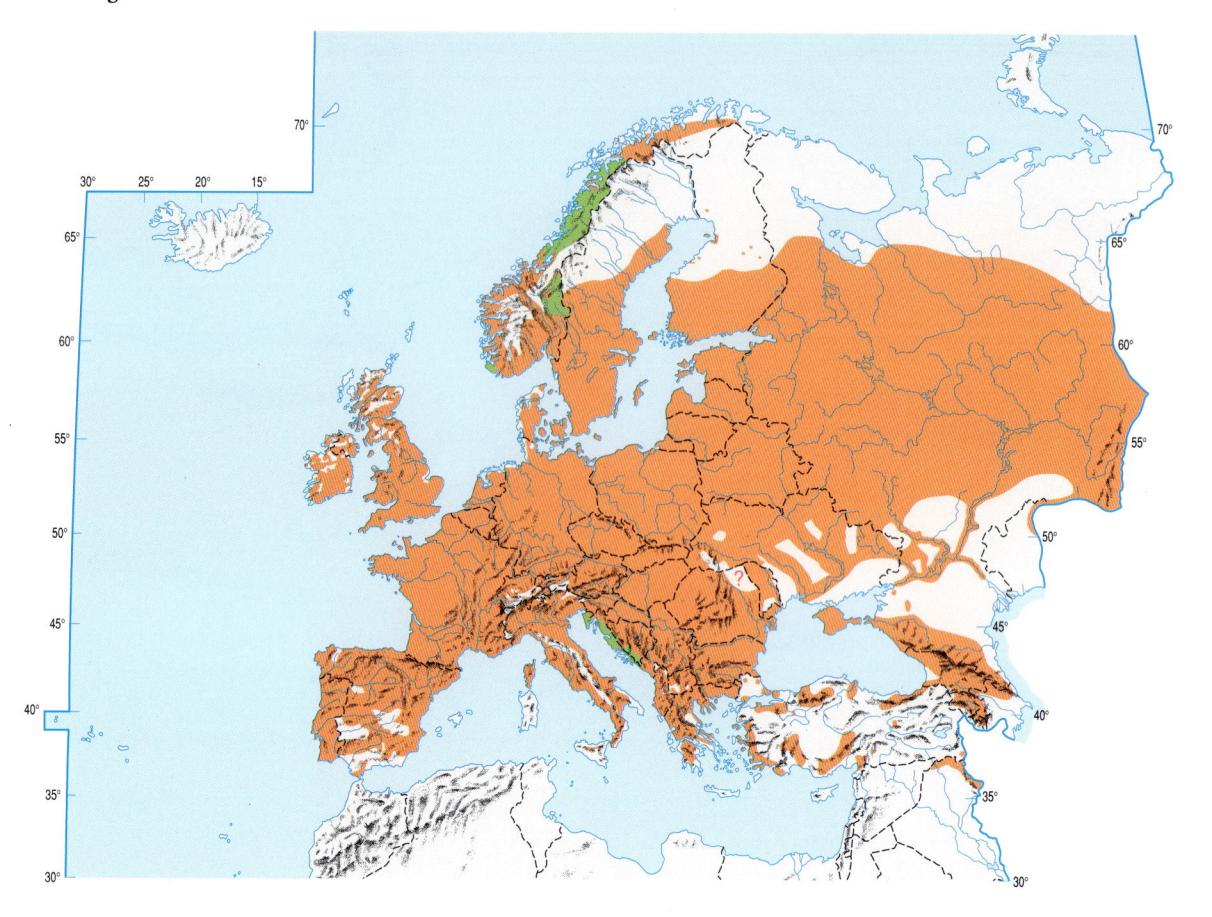

Breeding. SEASON. North-west and central Europe: eggs from late March, mainly early April. Finland: laying from late April to end of June. One brood. SITE. Extremely variable, from ground to tops of trees. In woodland, southern England, 2 distinct types: in low, usually thorny, bushes and brambles, mainly less than 3 m above ground; or in fork of tree or against trunk at 6–21 m. In mixed woodland, southern Germany, 2 types of site, according to height and supporting structure: 'standing' nests, based on tree fork, and 'hanging' nests among terminal twigs of conifers, rather evenly distributed by height, at 5–35 m. Nest: compact and domed structure of moss woven with cobweb and hair, covered on outside with camouflaging greyish lichen. Lined with abundant small feathers. Nest usually vertically elongated, with entrance hole in side near top; occasionally undomed or with 2nd hole near top. Hole and, to some extent, nest elastic, stretching with use. EGGS. Sub-elliptical; smooth and glossy; white, unmarked or with minute reddish speckling. Clutch: 8–12 (6–15). Increases with latitude. INCUBATION. Average 15.5 days for clutches begun before median laying date, 13.8 days after median date. FLEDGING PERIOD. 14–18 days.

Wing-length: *A. c. europaeus* (Netherlands): ♂ 60–67, ♀ 56–65 mm. *A. c. caudatus* (Russia, Scandinavia) slightly larger; British (*rosaceus*) and southern European populations (*taiti*, *irbii*) smaller.
Weight: ♂ ♀ mostly 7–10 g.

Geographical variation. Complex; 13 races recognized in west Palearctic. Most distinctive is nominate *caudatus* of Fenno-Scandia, Poland, Baltic States and Russia, with completely white head. *A. c. europaeus* of western and central Europe has black lateral crown-stripes, as have similar *rosaceus* (Britain and Ireland), *aremoricus* (western France), *taiti* (southern France and most of Iberia), and *macedonicus* (Balkans). Other European races have grey rather than black with variable pink backs: *irbii* (southern Iberia, Corsica), *italiae* (Italy), *siculus* (Sicily), *tephronotus* (Asia Minor) and *major* (Caucasus).

Tits Family Paridae

Small oscine passerines (suborder Passeres). Most found in deciduous, evergreen, and mixed woodland, including parks and gardens near human habitation. Arboreal but also descend freely to ground to feed; insects and similar prey form bulk of diet as well as seeds, nuts, fruit, and berries, especially in winter. Many species (but certainly not all) specialize in storing food, particularly seeds and nuts, hiding them singly with bill. Come readily to bird-tables and feeders provided by man. 42–50 species, occurring mainly in Eurasia with offshoots in North and Central America and in Africa. Mainly sedentary but some northern populations of some species migratory and others irregularly eruptive. 9 species in west Palearctic, all breeding.

Sexes close in size (♀ usually slightly smaller). Bill short but stout, pointed and strong. Wing rather short with rounded tip; 10 primaries, p10 reduced (c. 40% of length of longest). Sustained flight appears rather weak and undulating. Tail rather short and usually almost square; 12 feathers. Leg and toes short but strong. Highly active and agile when searching for food and feeding among foliage. Will hang upside down acrobatically by one or (usually) both feet. Foot employed to hold larger food items. Strongly vocal, but territorial song of ♂ unimpressive. Not highly sociable. Do not clump or allopreen. Some species territorial throughout year, solitary or in pairs in winter; others in loose flocks when not breeding. Courtship-feeding of ♀ by ♂ regular; plays essential role in ♀'s nutrition during egg-laying and incubation. Nest usually substantial, in hole; built (and hole excavated in some species) by ♀ only, and differing from nests of other hole-nesting passerines in that nest-cup is formed in mass of basal material (usually moss) which itself has no structure but is adjusted in its quantity to size of cavity. Before incubation, ♀ covers incomplete clutch with nest-material on leaving site; deters intruders with unique hissing display.

Plumage soft and thick. Typically brightly coloured, often with black, brown, or blue cap contrasting with white cheeks; remainder of upperparts uniform (mainly brown, grey, green, or bluish). Often a dark bib on chin and throat, remainder of underparts usually pale; wing in some species with contrasting bars on coverts. Some species crested. Sexes similar or nearly so. Juvenile similar to adult but duller, suffused with yellow in some species.

Marsh Tit *Parus palustris*

PLATES: pages 1376, 1395

Du. Glanskop   Fr. Mésange nonnette   Ge. Sumpfmeise   It. Cincia bigia
Ru. Черноголовая гаичка   Sp. Carbonero palustre   Sw. Entita

Field characters. 11.5 cm; wing-span 18–19.5 cm. No shorter but less robustly built than Willow Tit, with visibly shorter and smaller head, sleeker plumage, and usually more forked tail. Rather small, sharp-billed, evenly-balanced tit, particularly adept at holding food with feet. One of 4 tits with similar basic plumage pattern, consisting in Marsh Tit of glossy black cap and chin, dull brown upperparts, and off-white underparts. Slightly paler tertial-fringes do not form conspicuous pale shade or panel as in Willow Tit. One call ('pitchoo') diagnostic.

In areas where the two species coexist and are closely similar in plumage (Britain, western Europe), silent bird at middle or long range indistinguishable from Willow Tit; calling bird at close range unmistakable.

Habitat. Breeds in west Palearctic in middle and upper latitudes in temperate, and marginally in boreal, continental, and oceanic climates, within July isotherms 15–24°C, predominantly in lowlands but locally up to c. 1300 m in mountains. Shows strong preference for deciduous woodland and forest, typically oak or beech, moist rather than dry, and in relatively large rather than fragmented stands. Despite attraction towards extensive woodland, preference may be shown where choice exists for occupancy of alder carr, belts of riverain trees, orchards, and gardens or parks with suitable broad-leaf cover. Use often made of entire vertical spectrum from canopy to ground, although detailed studies suggest concentration on middle and lower layers. Requirement for availability of suitable ready-made nest-holes influences preference for at least a fair proportion of old or decaying trees, and avoidance of actively managed woodland, especially plantations.

Distribution. Britain. Now more sparsely distributed. Scotland: present in south-east 1921, but breeding not proved until 1945; spread since. Belgium. Range increased in 1980s–90s. Netherlands. Recent spread in north-east and south-west. Finland. Bred 1950. Spain. Spread slowly west. Turkey. Probably at very low density throughout northern Anatolia and perhaps more widespread in western Anatolia.

Accidental. Ireland, Finland, Balearic Islands, Portugal.

Beyond west Palearctic, breeds from western Altai in central Asia east to Sakhalin island and northern Japan, extending from east of range to south-west China and Burma.

Population. Decrease reported Britain, Germany, Denmark, Lithuania, and probable in Sweden, Estonia; probably increasing Netherlands, mostly stable elsewhere. Britain. 60 000 territories 1988–91. Shallow long-term decline. Estimates of 70 000–150 000 pairs in 1970s–80s. France. 100 000–1 million pairs in 1970s. Belgium. 30 000 pairs 1973–7. Luxembourg. 10 000–12 000 pairs. Netherlands. 13 000–20 000 pairs 1979–85. Germany. 700 000–900 000 pairs. In east, 65 000 ± 30 000

Marsh Tit *Parus palustris*. *P. p. palustris*: **1** ad, **2** juv. *P. p. dresseri*: **3** ad. Willow Tit *Parus montanus* (p. 1379). *P. m. rhenanus*: **4** ad, **5** juv. *P. m. borealis*: **6** ad. *P. m. salicarius*: **7** ad. *P. m. kleinschmidti*: **8** ad. *P. m. montanus*: **9** ad.

pairs in early 1980s, and some decrease; slightly more numerous than Willow Tit, but lower density. DENMARK. 8000–95 000 pairs 1987–8. NORWAY. 20 000–100 000 pairs 1970–90. Stable overall, but decrease due to modern forestry practices reported in south-east. SWEDEN. 100 000–250 000 pairs in late 1980s. ESTONIA. 50 000–100 000 pairs in 1991. LATVIA. 100 000–200 000 pairs in 1980s. LITHUANIA. Not abundant. POLAND. 150 000–300 000 pairs. CZECH REPUBLIC. 60 000–120 000 pairs 1985–9. SLOVAKIA. 80 000–160 000 pairs 1973–94. HUNGARY. 25 000–30 000 pairs 1979–93. AUSTRIA. 30 000–50 000 pairs in 1992. SWITZERLAND. 40 000–80 000 pairs 1985–93. SPAIN. 82 000–96 000 pairs. ITALY. 30 000–80 000 pairs 1983–95. GREECE. 1000–3000 pairs. ALBANIA. 1000–3000 pairs in 1981. YUGOSLAVIA: CROATIA. 200 000–300 000 pairs. SLOVENIA. 20 000–40 000 pairs. BULGARIA. 100 000–500 000 pairs. RUMANIA. 600 000–800 000 pairs 1986–92. RUSSIA. 10 000–100 000 pairs. BELARUS'. 70 000–80 000 pairs in 1990. UKRAINE. 10 000–12 000 pairs in 1986. MOLDOVA. 8000–12 000 pairs in 1988. TURKEY. 1000–10 000 pairs.

Movements. Sedentary, undergoing short-distance post-breeding dispersal over much of range but in northern areas part of population nomadic or southward-moving during winter. Seems not to participate in irregular eruptive movements of some other *Parus*.

Food. Mostly insects and spiders in spring and summer, also seeds, berries, and nuts at other times of year. Plant material more important than for other west Palearctic tits, and in some areas preferred to animal food outside breeding season; beechmast probably preferred plant food when available. In trees, pecks at bark on trunk or branch, vigorously tearing away moss or lichen with powerful body movements, or hammers at crevices with bill. Fruit and seeds dealt with by being held under one or both feet; fruit pulp then removed and eaten, or seed husk torn open by jabbing with upper mandible only; either removed or contents eaten through hole. Great amount of food material stored, from late summer to winter. Food stored among leaf litter, on dead stumps, under moss or lichen on tree branches, etc.; mostly removed and eaten within a few days of storing.

Social pattern and behaviour. Majority of population territorial, in pairs, throughout non-breeding season, maintaining same territories all year. Young birds without territories accompany mixed-species foraging flocks, subservient to and tolerated by territory-owners when in occupied territories. Territorial pairs accompany mixed-species flocks while passing through their territory, dropping out when border reached. Mating system monogamous. Nest-building, incubation, and brooding of young by ♀ alone; feeding of young and care of fledged young by both sexes. ♂ sings from canopy down to shrub layer, but not usually below *c*. 1 m. Song principally given by territory-holding ♂♂, but unestablished ♂♂ also sing. Song-duels common, often between birds considerable distance apart. Song-period varies little between different areas: in southern England begins mid-January, peaks February–March, and continues until late May; song very occasional in autumn and early winter.

Voice. Used freely throughout year. Song of ♂ very variable. Commonest type a simple repetition of notes with peculiar, liquid, bubbling quality or loud, ringing, bell-like rattle. Each ♂ has up to 5 different songs; usually gives series of songs of one kind, then switches to another. ♀ song occasional; as ♂'s, but less loud and units more variable. Most characteristic call (especially for distinguishing from Willow Tit) a distinctive, typically 2-unit 'pitchoo', with variants 'pitch-itchoo', etc.; also often single 'pitch' or 'pits' and 'choo'. With few exceptions, only given by birds holding or claiming territory; functions as deterrent to trespassers. 'Pitchoo' often followed by harsh, nasal 'tchee-tchee...' or 'tchaa-tchaa-tchaa-tchaa...'; may be speeded-up to a series of rapid scolding notes, commonly introduced by 1–2 purer notes, 'chickabeebeebeebee'. Other calls include a monosyllabic 'sip' used for contact at close quarters, and high-pitched thin 'tsip-sip' or 'sip-ip' notes, common in sexual and aggressive excitement and typical of all territorial disputes.

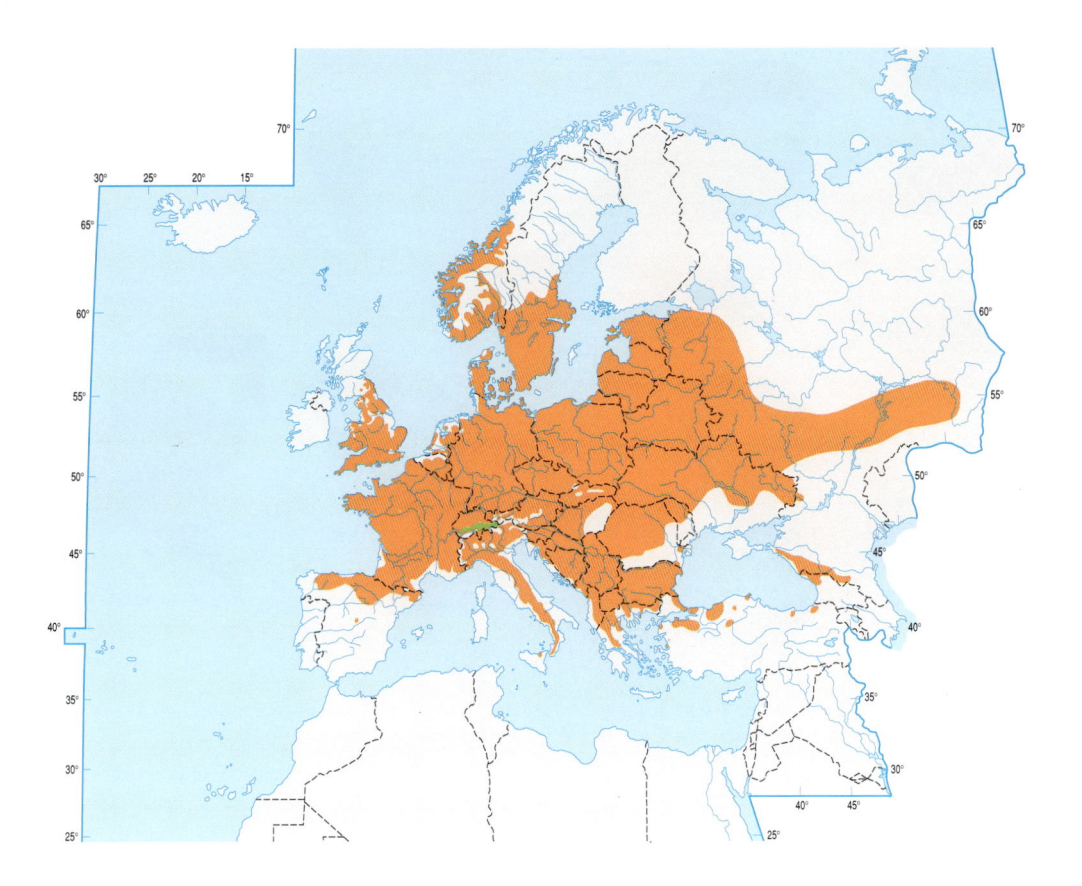

Breeding. SEASON. Rather little variation in laying date over at least most of Europe, but slightly earlier in north. Norway: laying 25 April to 15 May. Southern Sweden: median laying date 30 April. Southern Germany: average laying date 29 April. Southern England: average laying date 24 April. Laying continues until mid- or late May, later in rare cases of 2nd broods. SITE. In hole in tree or stump, or among tree roots; occasionally in wall or ground. Also occasionally uses nest-boxes. Nest: hole, which may be widened and deepened; does not normally excavate hole completely. Nest-cup a thick basal pad of moss, occasionally with mixture of other plant material, lined with hair and occasionally a few small feathers. EGGS. Sub-elliptical, smooth, not glossy; white, with usually very fine and sparse red-brown spots, often concentrated at broad end, and a few underlying violet-grey spots; occasionally unmarked. Clutch: usually 7–10. Average size of 1st clutch *c.* 8 in central and northern Europe. No geographical variation established, but perhaps slightly lower in peripheral areas than in central Europe; clutches larger in large than in small holes. INCUBATION. 13–17 days. FLEDGING PERIOD. 17–20 days.

Wing-length: Continental Europe: ♂ 64–70, ♀ 60–67 mm. *P. p. dresseri* (England): ♂ 62–66, ♀ 59–63 mm.
Weight: ♂ ♀ mostly 10–13 g.

Geographical variation. Slight and mainly clinal. Cline of decreasing size and increasing colour saturation runs from east to west. *P. p. dresseri* of England and western France smallest and darkest; *P. p. stagnatilis* of areas east of *c.* 25°E largest and palest. Nominate *palustris* occupies intervening areas.

Sombre Tit *Parus lugubris*. *P. l. lugubris*: **1** ad, **2** juv, **3** '*lugens*' (southern Greece). *P. l. dubius* (wintering north-east Iraq): **4** ad. *P. l. anatoliae*: **5** ad. *P. l. hyrcanus*: **6** ad. Siberian Tit *Parus cinctus* (p. 1382): **7** ad, **8** juv.

Sombre Tit *Parus lugubris*

PLATES: pages 1378, 1395

Du. Rouwmees Fr. Mésange lugubre Ge. Trauermeise It. Cincia dalmatina
Ru. Средиземноморская гаичка Sp. Carbonero lúgubre Sw. Balkanmes

Field characters. 14 cm; wing-span 21.5–23 cm. As long as but noticeably bulkier than Great Tit; *c.* 20% larger than Marsh Tit and Willow Tit. Medium-sized to large, bulky tit, with strong bill, broad head, and rather dull plumage similar in pattern to northern race of Willow Tit, though black bib larger and white cheeks less deep.

Unmistakable, being much larger than Marsh Tit and Willow Tit and having much bolder behaviour and distinctive voice; not as acrobatic as smaller tits, visibly using power rather than agility in some feeding actions and able to tear seed-heads apart like finch. Less restless than small tits, content to stay in discrete area until food source exhausted.

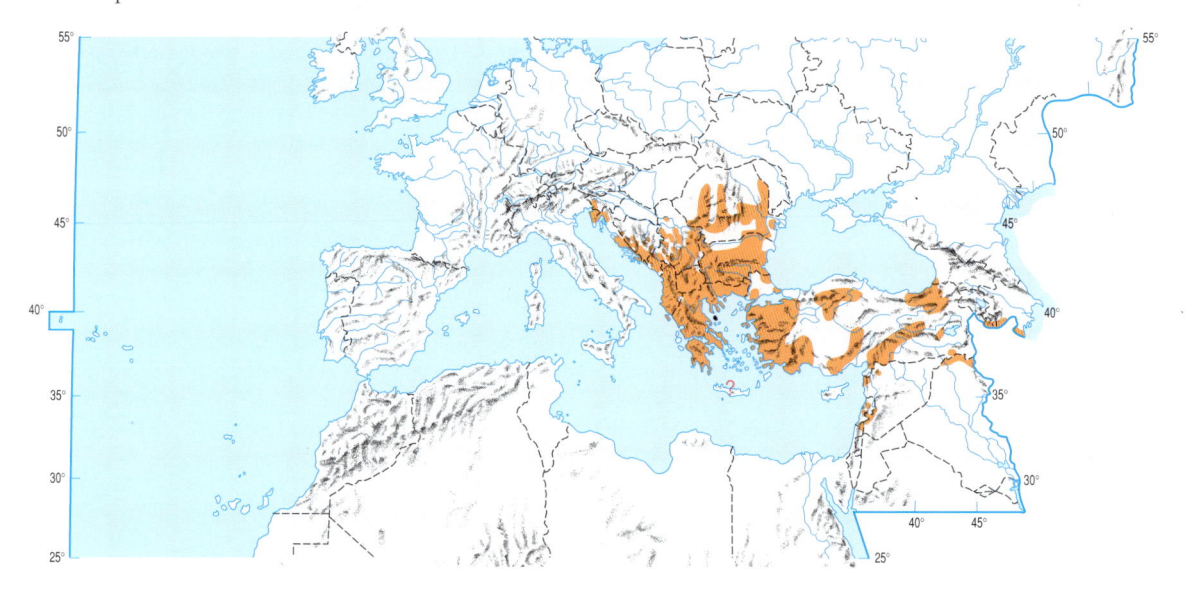

Habitat. Most southerly *Parus* breeding in west Palearctic, in warm continental middle and lower latitudes, mainly Mediterranean but also montane, within July isotherms 23–30°C. Inhabits open broad-leaved forests, especially of oak, and even dark conifers, riverside willow and poplar, orchards, and vineyards; regularly in scrubby terrain.

Distribution. RUMANIA. Probably more widespread than known at present. Probable recent spread north in Transylvania. AZERBAIJAN. Very local. TURKEY. Widespread, but only locally common; rare in mountain valleys of south-east. IRAQ. Very probably breeds in north.
Accidental. Italy.
Beyond west Palearctic, breeds only in Iran.

Population. Apparently stable throughout range. GREECE. 10 000–30 000 pairs. ALBANIA. 2000–5000 pairs in 1981. YUGOSLAVIA: CROATIA. 4000–6000 pairs. SLOVENIA. 10–20 pairs. BULGARIA. 100 000–500 000 pairs. RUMANIA. 1500–2000 pairs 1986–92. Nowhere numerous. ARMENIA. Moderate numbers in very restricted range. AZERBAIJAN. Rare. TURKEY. 10 000–100 000 pairs. ISRAEL. 50–80 pairs 1970s–80s.

Movements. Resident and, over most of range, sedentary, though moves about in mixed-species flocks in winter, 1st-years especially. A few autumn and winter records in central Turkey where it does not breed, and recorded in autumn in foothills of Zagros mountains of north-east Iraq, just beyond known breeding range.

Food. Mainly small invertebrates, especially caterpillars and other larvae; also seeds. Forages in trees and shrubs, also readily on ground, among rocks or leaves, but always flies up to nearest perch with food item. Often feeds on seed-heads of herbs.

Social pattern and behaviour. Pairs apparently stay together outside breeding season, at times joining flocks of other tits, finches, and buntings. No indication of other than monogamous mating system. Even in areas of apparently suitable habitat, population density tends to be low, but little exact information. Song rather infrequent, even in early stages of breeding.

Voice. ♂'s song of typical *Parus* type: a phrase consisting of same unit or motif repeated; very variable, but typically has buzzy quality due to very rapid modulation within units. Most characteristic call, apparently indicating general excitement or alarm, a deep, chattering 'chrrrrt' or 'chaerrrrr', quite unlike any other west Palearctic tit, not unlike alarm-call of House Sparrow; usually preceded by 2 high-pitched notes. Flight-call a rather hoarse 'snipp', like call of Yellowhammer.

Breeding. SEASON. Eggs laid from March in most of European range, in extreme north-west in 1st half of April; 2nd clutches in 2nd half of May and June, late broods still in nests early August. Much later in Turkish mountains at 1280–1521 m, average laying date of 1st clutches 1–6 May in 3 years. 2 broods apparently regular. SITE. Hole in tree or, apparently less often, in rocks. Nest: cup of plant material and wool, lined with feathers; moss apparently not used. EGGS. Sub-elliptical; white, with fine reddish speckles. Clutch: size not well known except near Ankara (Turkey), where average is *c.* 7 (4–9); in Europe apparently usually 5–7. INCUBATION. Average *c.* 13 days. FLEDGING PERIOD. *c.* 22 days.

Wing-length: *P. l. lugubris* (Balkans): ♂ 72–78, ♀ 71–75 mm. *P. l. anatoliae* (Asia Minor): ♂ 70–75, ♀ 68–73 mm.
Weight: ♂♀ 15–19 g.

Geographical variation. Slight. *P. l. anatoliae* of Asia Minor slightly darker above and whiter below, and smaller than nominate *lugubris* of Balkans. *P. l. hyrcanus* from south-east Transcaucasia, bordering Iran, smaller and more deeply coloured than other races of *lugubris*; sometimes treated as separate species.

Willow Tit *Parus montanus*

PLATES: pages 1376, 1395

DU. Matkop   FR. Mésange boréale   GE. Weidenmeise   IT. Cincia bigia alpestre
RU. Буроголовая гаичка   SP. Carbonero sibilino   SW. Talltita

Field characters. 11.5 cm; wing-span 17–20.5 cm. Slightly but distinctly bulkier than Coal Tit, with larger head and 10% longer tail; also bigger-headed and generally more robust-looking than Marsh Tit, but with slightly shorter wings and tail (tail usually with slightly rounded end, not square or slightly forked as in Marsh Tit). Rather small but bold and robust tit, with large head and sharp bill. Long, dull black cap (reaching mantle), long white cheeks, and pale fringes to tertials and innermost secondaries, creating wing-panel, distinctive at close range. Voice distinctive.
Except for larger and distinctly paler, grey-backed northernmost (*borealis*) and Alpine (*montanus*) races, difficult to distinguish from Marsh Tit on plumage, but differences in general character, behaviour, and voice (once learnt) allow certain identification of bird seen and heard well.

Habitat. Breeds across middle and upper middle latitudes of west Palearctic, mainly in temperate continental climates but extending through boreal and montane regions, between *c.* 12°C and 21–23°C July isotherms. In comparison with Marsh Tit, favours cooler and usually more elevated areas, up to treeline in mountains and on borders of tundra. 3 distinct patterns distinguishable. (1) In south of range, montane population inhabits mainly conifer forest provided it contains enough decaying tree-stumps for excavation of breeding holes and has access to clearings, rock-bands, scree, slopes under springs, or boggy patches with willows or alder. (2) At lower levels, mainly linked with open bushy damp stands or lines of low trees such as willow, alder, and birch, and lower woody shrubs such as elder, as well as copious growth of stiff plants such as hemp-nettle. In Britain examples of 1st pattern do not occur, all being related to 2nd: river valleys, borders of reservoirs and overgrown flooded gravel-pits, and farmland hedgerows and copses are favoured in preference even to broad-leaved woodland. (3) Northern race, *borealis*, occupies birch, conifer, and mixed woodland, damp places in birch and alder coppice, and stands of spruce or pine.

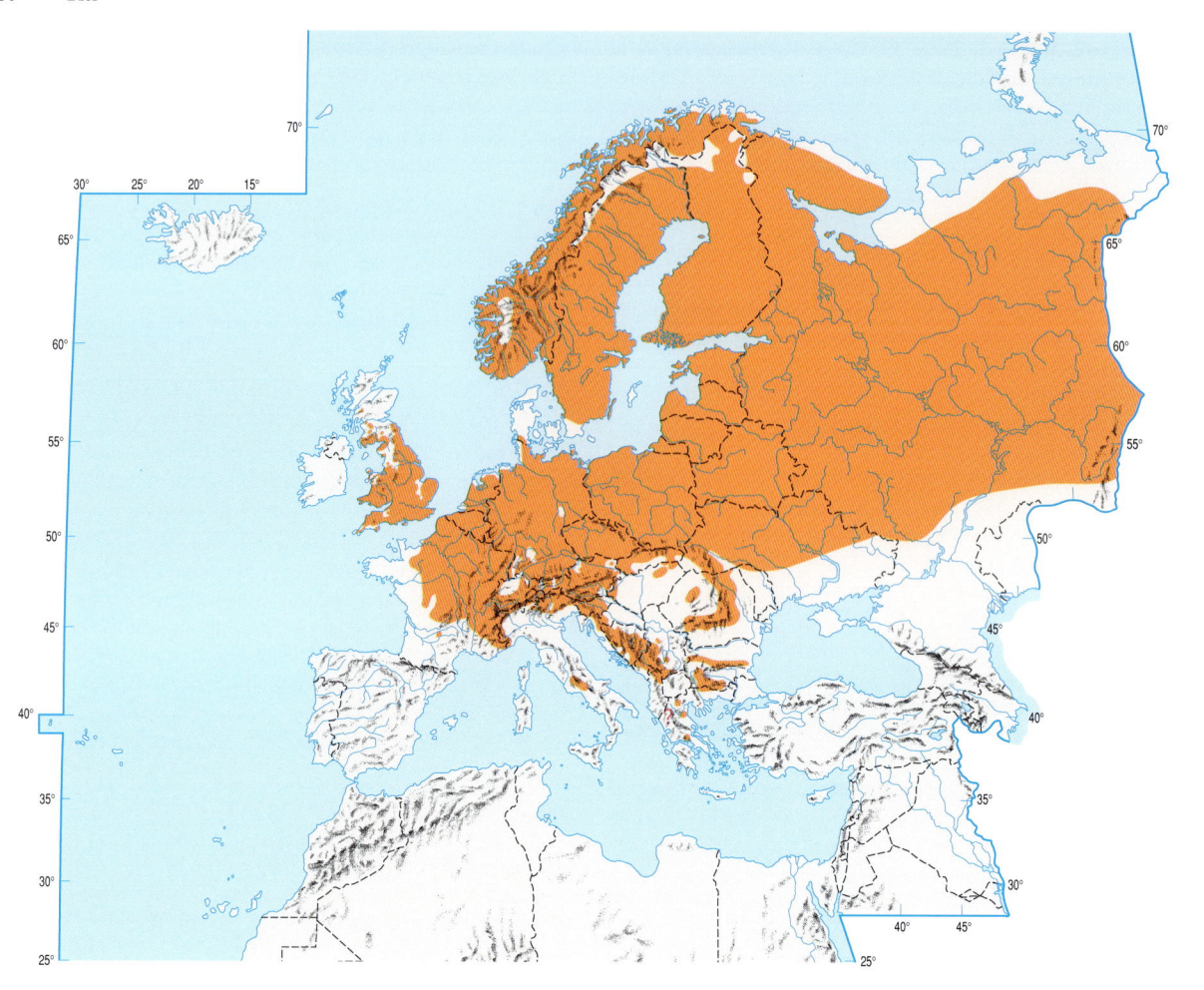

Distribution. Has spread in parts of northern and central Europe. BRITAIN. Marked contraction of range in Scotland since *c*. 1950: formerly occurred sparsely north to Inverness and Ross; now apparently confined to south, especially south-west. FRANCE. Spread recently south-west. NETHERLANDS. Range contraction in dunes in west, but gradual spread elsewhere. GERMANY. Marked expansion in sparsely occupied or unoccupied areas of Sachsen-Anhalt and north-west Sachsen. Colonized Harz mountains and northern foothills from the north 1957 to 1973–4 (north-east fully by mid- to late 1980s). DENMARK. Formerly absent. First recorded in south (spread from Germany) 1928, first recorded breeding attempt 1984, now regular. Single breeding record northern Sjaelland. CZECH REPUBLIC. Range expansion in Bohemia and Moravia since *c*. 1955. RUSSIA. Northern limit of range on Pechora river apparently runs through Nar'yan-Mar (67°37′N), where a few pairs and single birds seen August 1992 and 1994.

Beyond west Palearctic, extends broadly across northern Asia to Anadyr' region, Sakhalin island, and Japan.

Population. Stable over most of range, though increase Germany, Denmark, and Ukraine, and decrease Finland. BRITAIN. 25 000 territories in 1988–91. No clear trend: increased mid-1960s to mid-1970s, then shallow decline to former level, probably related to winter temperatures. FRANCE. 100 000–1 million pairs in 1970s. BELGIUM. 19 000 pairs 1973–7, following decrease from 40 000 pairs in 1968. Limburg province: decrease 1985–92, but seems stable in woodland. LUXEMBOURG. 10 000–12 000 pairs. NETHERLANDS. 40 000–60 000 pairs 1979–85. Has increased through afforestation, but some decrease in west. GERMANY. 200 000–300 000 pairs. Some increase in east where estimated 50 000 ± 25 000 pairs in early 1980s. DENMARK. 20–50 pairs 1987–92. NORWAY. 200 000–1 million pairs 1970–90. SWEDEN. 1–2 million pairs in late 1980s. FINLAND. 1–1.5 million pairs in late 1980s. Decrease since 1940s due to modern forestry practices. ESTONIA. 50 000–100 000 pairs in 1991. LATVIA. 180 000–250 000 pairs in 1980s. LITHUANIA. Not abundant; more numerous in west and north. POLAND. 100 000–300 000 pairs. CZECH REPUBLIC. 40 000–80 000 pairs 1985–9. Increase from *c*. 1955, now stable. SLOVAKIA. 60 000–150 000 pairs 1973–94. HUNGARY. 300–500 pairs 1979–93. AUSTRIA. 70 000–80 000 pairs in 1992. SWITZERLAND. 30 000–70 000 pairs 1985–93. ITALY. 30 000–50 000 pairs 1983–95. GREECE. 500–1500 pairs. ALBANIA. Up to 1000 pairs in 1981. YUGOSLAVIA: CROATIA. 7000–10 000 pairs. SLOVENIA. 10 000–20 000 pairs. BULGARIA. 100 000–500 000 pairs. RUMANIA. 30 000–60 000 pairs 1986–92. RUSSIA. At least 10 million pairs. BELARUS'. 150 000–170 000 pairs in 1990. UKRAINE. 60 000–70 000 pairs in 1986.

Movements. Sedentary over much of range, but some northern populations irruptive with large-scale movements recorded in some years.

Food. Invertebrates and seeds: mainly invertebrates in breeding season, with seeds, from autumn onwards, dominating in winter. An infrequent visitor to bird-tables in winter. Regularly stores food, especially in late summer and autumn; at least in north of range, in winter largely dependent on stored food.

Social pattern and behaviour. Northern populations gregarious outside breeding season, forming social groups which occupy fixed territories and defend them against neighbouring groups. Size and composition of winter social groups remarkably constant in any one area: thus in high-altitude birch woodland in central Norway, nearly all groups consist of 6 birds, adult pair and 2 pairs of young birds; at lower altitudes in central Norway, southern Sweden, and extreme northern Finland, groups consist mostly of 4 birds, typically old pair and 1 pair of young birds. In nearly all cases, old pair are birds which have bred together in same territory in previous breeding season. Social groups, with dominance hierarchies according to age and sex, established in late summer and autumn, and usually remain unchanged through winter, except for deaths. After break-up of groups at end of winter, territories occupied by original adult pair or, if widowed, adult and young bird from its group. In southern parts of range formation of such social groups does not occur: adult pairs maintain territories throughout winter; join mixed foraging flocks passing through territory, but much of time forage away from other species; apparently no regular association with juvenile pairs. Mating system monogamous; in all areas where detailed studies made, pairs remain together year after year in same territory. Territorial-song given mainly in late winter and spring, with slight resurgence after end of breeding season. In areas where Willow Tit coexists with Marsh Tit, latter is dominant; Marsh Tit regularly takes nest-holes excavated by Willow Tit, but at most times of year little conflict between the two species.

Voice. Used freely throughout year, but song, at least in some areas, relatively infrequent compared with other tits. Striking difference in song between Alpine and lowland populations, perhaps even suggesting incipient speciation. Territorial-song a simple repetition of pure-sounding notes, either more or less even in pitch or with marked downward inflection, these types having largely mutually exclusive distributions. Song of most European populations consists of usually 3–5, rarely 6 or more, fairly loud, downwardly inflected, drawn-out whistles, 'piū–piū–piū. . .', much like piping-song of Wood Warbler. Song of Alpine populations of central Europe very distinct; consists usually of 5–7(–10) much shorter notes, not downwardly inflected, even and soft in tone— 'düh düh düh düh. . .'. Alpine song-type mainly occurs in Alps (mostly above 1000 m), with isolated record from Jura; also in Slovenian–Croatian mountains of northern Yugoslavia and Tatra mountains on Czechoslovakia–Poland border. In Alps, songs are *only* of constant pitch, but songs without fall in pitch, i.e. presumably of Alpine type, also occur occasionally in other areas. In Alps, distribution of Alpine song-type corresponds closely with distribution of Alpine race, nominate *montanus*. Round northern edge of Alpine region, at altitudes mainly of 1000–1500 m, both song-types may occur together in same localities. In these areas, individuals generally sing one song-type only and do not react to songs of different type from their own more strongly than they do to songs of other species. Quite different kind of song occasionally recorded: rich, Nightingale-like notes, mingled with a soft musical warbling; uttered in brief snatches and not carrying far; apparently infrequent (perhaps commoner in Britain than elsewhere), and function unknown. Most characteristic call, apparently not showing significant geographical variation, a rather nasal, drawn-out 'tchay tchay. . .' or 'chaa chaa', often preceded by 1 or more high-pitched introductory units, e.g. 'zi-zi-tchaa-tchaa-tchaa'. Used mainly as contact-call; also, in alarm, given in louder, more penetrating form, usually without high-pitched introductory notes or with these replaced by harsher 'pett' or 'kett'. Very high-pitched, short 'sit' or 'zit' notes, similar to equivalent notes of other tits, frequently given, especially when foraging.

Breeding. SEASON. Laying begins about mid-April in south and central parts of European range, late April or early May in southern Fenno-Scandia, and mid-May in extreme north; continues into early June, or exceptionally (in north) into July. One brood. SITE. Usually in rotten tree trunk or stump, often as little as 7–8 cm in diameter so that nest-cavity separated from outside world only by bark; birch most favoured where available; alder, willow, elder, poplar, hazel, ash, and conifers also regularly used; also rotten posts. Nest-hole usually excavated entirely (by both sexes), but existing hole sometimes enlarged, e.g. rotted beginning of hole of woodpecker. Nest generally much slighter and less well formed than nests of other tits, composed mainly of plant fibres, including strips of bast and bark, dry grass, and plant down; lined with animal hair and plant fibres, occasionally a few small feathers; moss fragments sometimes used, not in quantity except rarely when cavity at bottom of hole needs filling, absence of moss and predominance of vegetable fibres making nest distinctively different from that of Marsh Tit. EGGS. Sub-elliptical, smooth, not glossy; white, variably marked with red-brown speckles and spots, often concentrated at broad end; underlying spots pale reddish-violet. Clutch: 4–11, mostly 6–9; strong correlation between clutch size and size of nest-cavity. INCUBATION. 13–15 days. FLEDGING PERIOD. 17–20 days.

Wing-length: *P. m. borealis* (Fenno-Scandia) and *P. m. montanus* (Alps): ♂ 64–70, ♀ 62–67 mm. Lowland west European populations smaller, ♂ mostly 57–62 mm.
Weight: ♂ mostly 10–14, ♀ 8–13 g.

Geographical variation. Complex, largely clinal. Size decreases and coloration becomes deeper from east to west in northern Europe; three races recognised, *borealis* in north and east, *salicarius* in centre, *rhenanus* in west, with extensive intergrading. *P. m. kleinschmidti*, from Britain, small and deeply coloured, represents detached end of cline. Nominate *montanus*, from Alps east to Carpathians and Greece, larger than other races, with more rufous-pink flanks. Many other races, of intermediate populations, have been named.

Siberian Tit *Parus cinctus*

PLATES: pages 1378, 1395

Du. Bruinkopmees   Fr. Mésange lapone   Ge. Lapplandmeise   It. Cincia siberiana
Ru. Сероголовая гаичка   Sp. Carbonero lapón   Sw. Lappmes

Field characters. 13.5 cm; wing-span 19.–21 cm. Slightly shorter overall than Great Tit, noticeably larger and much longer-tailed than Willow Tit, though with similar proportions of head and body. Rather large, long, bulky, and often fluffy tit, with plumage pattern basically like black-capped species but with dusky-brown crown and nape and much warmer buff-brown back and flank.

Unmistakable in west Palearctic: of black-capped tits, distribution overlaps substantially only with distinctive black-grey-white race of Willow Tit; instantly distinguishable from it in good view. Flight more like Great Tit than Willow Tit, with fluttering (not whirring) action giving less bouncing, more floating progress. Allows extremely close approach by observer.

Habitat. In west Palearctic sector of extensive range, inhabits high latitudes with cool boreal climate up to July isotherm of 10°C in high coniferous taiga, but also occurs in broad-leaved trees along banks of rivers running through coniferous forest. In winter in arctic Lapland, wanders over wide area of forest of pine and spruce at lower altitudes, contrasting with Willow Tit in not being a bird of birch forest. Will range beyond taiga into stunted forest, but becomes rare near northern forest limits, and remains at all times arboreal. Independent of man except locally where making use of refuse, and in very severe weather north of the arctic circle coming into small towns where food is supplied.

Distribution. FINLAND. Retreat from southern parts by *c*. 100–200 km in 20th century, probably due to climatic amelioration and, more recently, changing forestry practices. RUSSIA. Kirov region: recorded (single bird, pair, small flocks) March–April 1988 at *c*. 58–59°N. Probably example of nomadism south of breeding range, but possibility of range expansion not to be discounted.

Accidental. Estonia.

Beyond west Palearctic, extends across Siberia to Anadyr' region, Kamchatka, and Sea of Okhotsk, and south to Lake Baykal. In North America, breeds Alaska and adjoining areas of extreme north-west Canada.

Population. NORWAY. 10 000–50 000 pairs 1970–90. Stable after decrease reported in late 1950s. SWEDEN. 25 000–100 000 pairs in late 1980s. FINLAND. 100 000–150 000 pairs in late 1980s. Most recent trend is slight increase, but earlier reports of decrease, especially near southern range limit, so that increasing proportion of population in far north. RUSSIA. 100 000–1 million pairs; stable.

Movements. Chiefly sedentary and to some extent nomadic outside breeding season. More extensive movements southwards, especially of juveniles, recorded for some populations when numbers high.

Food. Small invertebrates and seeds; in winter also feeds at refuse tips and bird tables. Food stored throughout year at tips of twigs in needles, in crevices on branch or trunk, in clumps of lichen. Items stored all over territory, usually 3–4 items at each site, and each site used once only. Most activity takes place April–May (mainly caterpillars plus some seeds from opening cones) and

August–November (insects, caterpillars, and spiders, which are killed before storage). Stores hardly exploited outside winter. Apparently each bird finds own store.

Social pattern and behaviour. Gregarious outside breeding season, forming social groups which occupy fixed territories. Adults strictly sedentary, remaining in pairs throughout life; territories occupied by 1 adult pair, or 2 pairs may share a territory. Juveniles, after dispersal from natal area, settle in adult territories; groups thus formed consist of 2–9 birds, containing up to 7 juveniles. Within social groups, adults always dominant over younger birds; when in company with Willow Tit, Siberian Tit dominant. No evidence of other than monogamous mating system. Song little developed; not loud or far-carrying, and not given from established song-posts. Function not clear: may serve to strengthen pair-bond, and may be short-distance warning to intruders on territory.

Voice. Song a series of 3–4 (2–6) units, unmusical and sometimes of vibrant or rasping quality, variable in structure and tending to be irregular in tempo; 'cheep-cheep-cheep. . .' or 'prrrrree-prrrrree. . .'. Commonest call is a nasal 'tchay' or 'chaeer', like equivalent call of Willow Tit; often immediately preceded by high-pitched 'sip' or 'zit' units. Other calls include an aggressive 'tjillup', like wagtail, and soft, bubbling 'pip-pip-pip. . .', given during alarm at nest.

Breeding. SEASON. Finland: laying may begin early May, but usually late May, extending to early June. Season in Norway and Russia similar. One brood. SITE. Hole in tree (conifer, birch, aspen), often decaying stump; natural cavity or old hole of woodpecker, 0.3–5 m above ground; cavity often cleaned out and rotten wood removed. Also uses nest-boxes. Nest: typically of 3 layers, with base of decayed wood (including nests in nest-boxes), below variable amount of moss or grass (amount adjusted to size of cavity) and thick cup of hair. EGGS. Sub-elliptical or more or less oval, without gloss; milk-white with sparse rust-brown spots and underlying spots of pale olive-brown (distinct from other west Palearctic tits) and very pale violet-grey. Clutch: 6–10 (4–11). INCUBATION. 15–18 days. FLEDGING PERIOD. 19–20 days.

Wing-length: ♂ 65–71, ♀ 64–69 mm.
Weight: ♂ mostly 12.5–14, ♀ 11–12.5 g.

Crested Tit *Parus cristatus*

PLATES: pages 1384, 1395

Du. Kuifmees   Fr. Mésange huppée   Ge. Haubenmeise   It. Cincia dal ciuffo
Ru. Гренадерка   Sp. Herrerillo capuchino   Sw. Tofsmes

Field characters. 11.5 cm; wing-span 17–20 cm. Close in size to Marsh Tit but with head shape more like Blue Tit and backward-pointing crest unique in small arboreal passerines of west Palearctic. Small, rather compact tit, with large head. Rather dusty-brown, with black-and-white-banded face and black-and-white-mottled crown and crest. Unmistakable, even if only glimpsed in silhouette.

Flight action and form recall Blue Tit but not as confiding or inquisitive as most other tits. Flicks wings and tail and often changes angle of crest.

Habitat. From upper to lower middle latitudes of west Palearctic, mainly in dry cool or warm continental temperate climates within July isotherms *c.* 12–22°C, extending at fringe into Mediterranean montane zone, and ascending to treeline; also in low-lying coastal woods in western and south-western Europe. In northern Europe, prefers pine forest, mixed woods being occupied further south, and beech forest in Pyrénées; in southern Spain, common in cork oak.

Distribution. Has spread north in Denmark and Finland; minor range expansion in other western parts of range. BRITAIN. Scotland: formerly more widespread, before destruction of most of Caledonian pine forest. Some recent spread. FRANCE. Some spread this century in central France and Mediterranean lowlands. BELGIUM. Considerable expansion since 19th century, following spread of conifer plantations. NETHERLANDS. Colonized west 1940–50. FINLAND. Has spread during 20th century up to 1950s. Colonized Åland Islands 1926. SPAIN. Slight increase. UKRAINE. Occurrence in Crimea needs confirmation.

Accidental. Balearic Islands, Morocco.

Breeding range extends just beyond west Palearctic in Ural mountains (east to Sverdlovsk).

Population. Slight increase Belgium, Netherlands, Denmark, Spain, and Italy; decrease Sweden, Finland, and Czech Republic, apparently stable elsewhere. BRITAIN. 900 pairs in 1980. Stable or increasing locally. FRANCE. 100 000–1 million pairs. BELGIUM. 18 000 pairs 1973–7. Continuing increase since 1950s, due to spread and maturation of conifer plantations. LUXEMBOURG. 2500–3000 pairs. NETHERLANDS. 30 000–45 000 pairs 1979–85. Probably increased since 19th century through afforestation. GERMANY. 400 000–500 000 pairs. In east, 100 000 ± 50 000 pairs in early 1980s. DENMARK. 2000–22 000 pairs 1987–8. NORWAY. 50 000–200 000 pairs 1970–90. SWEDEN. 150 000–500 000 pairs in late 1980s. FINLAND. 150 000–300 000 pairs in late 1980s. Considerable increase up to 1950s, but decline since due to modern forestry practices. ESTONIA. 20 000–30 000 pairs in 1991. LATVIA. 300 000–370 000 pairs in 1980s. LITHUANIA. Common. POLAND. 50 000–200 000 pairs. CZECH REPUBLIC. 80 000–160 000 pairs 1985–9. SLOVAKIA. 25 000–50 000 pairs 1973–94. HUNGARY. 300–400 pairs 1979–93. AUSTRIA. 50 000–60 000 pairs in 1992. SWITZERLAND. 120 000–160 000 pairs 1985–93. SPAIN. 860 000–1.5 million pairs. PORTUGAL. 100 000–1 million pairs 1978–84. ITALY. 20 000–40 000 pairs 1983–95. GREECE. 2000–5000 pairs.

Crested Tit *Parus cristatus*. *P. c. mitratus*: **1** ad, **2** juv. *P. c. scoticus*: **3** ad. *P. c. cristatus*: **4** ad. *P. c. abadiei*: **5** ad. *P. c. weigoldi*: **6** ad.

ALBANIA. 1000–3000 pairs in 1981. YUGOSLAVIA: CROATIA. 1000–2000 pairs. SLOVENIA. 30 000–40 000 pairs. BULGARIA. 1000–10 000 pairs. RUMANIA. 200 000–300 000 pairs 1986–92. RUSSIA. 1–10 million pairs. BELARUS'. 640 000–660 000 pairs in 1990. UKRAINE. 6000–7000 pairs in 1986.

Movements. Extremely sedentary over most of range, movements generally confined to local summer dispersal of newly independent young and some redistribution at start of breeding season. Exceptional wandering in winter up to 50–100 km. In east of range reported to occur, rarely, well away from known breeding areas in winter.

Food. Mainly insects, spiders, and, especially outside breeding season, plant material (mainly conifer seeds). Both sexes commonly store food of all kinds, though mostly vegetable, mainly seeds of pine, spruce, and dead-nettle; also larval (and less often adult) Lepidoptera, spiders, and other invertebrates.

Social pattern and behaviour. Gregarious outside breeding season, forming small social groups which join mixed feeding flocks of other tits, treecreepers, Goldcrest, etc. Groups typically consist of an adult pair and a 1st-year ♂ and ♀; maintain a common territory which is defended against neighbouring groups. Monogamous mating system. Adult pairs remain together year after year, as long as both survive, forming core of social groups, as described above. When new pairs formed, individuals from same group highly preferred as partners; new pairs nest within group territory. Song very poorly developed.

Voice. Commonest call, apparently used for contact, alarm, and in other contexts, a rather soft, very distinctive, purring tremoloor trill, commonly but not always preceded by introductory 'sri', 'zi', or 'zizi' notes which are weaker and often not audible at a distance. Song little differentiated from other calls, consisting of persistently repeated, loud version of tremolo preceded by one or more high-pitched introductory notes, in variable pattern. Contact-call a rather urgent 's-si', 'si si si si', or 'si si si chup', with descent in pitch between successive units.

Breeding. SEASON. Laying may begin in March, but April is main laying period. Season of very variable length, depending mainly on incidence of 2nd broods, which do not occur in north of range but occur with increasing frequency towards south. SITE. Hole in tree or tree-stump, including old hole of woodpecker; occasionally hole in ground, exceptionally in nest of other bird. Nest-boxes regularly used. Nest: hole usually excavated (by ♀ alone), or existing hole much enlarged; nest-cup mainly of moss, lined hair, wool, and sometimes feathers. EGGS. Sub-elliptical, smooth, slightly glossy; white, spotted and blotched rust-red or violet-red, spots usually larger and more concentrated at broad end than in other tits. Clutch: 3–9; locally variable, but no clear geographical trend apparent except for decrease in north of range. Over much of central Europe, clutches of 6–7 predominate. INCUBATION. 13–16 days. FLEDGING PERIOD. 18–22 days.

Wing-length: *P. c. mitratus* (central Europe): ♂ 64–70, ♀ 61–65 mm. Scandinavian populations similar; Scottish and western European populations somewhat smaller.
Weight: ♂ ♀ mostly 10–13 g.

Geographical variation. Marked in colour, less so in size;

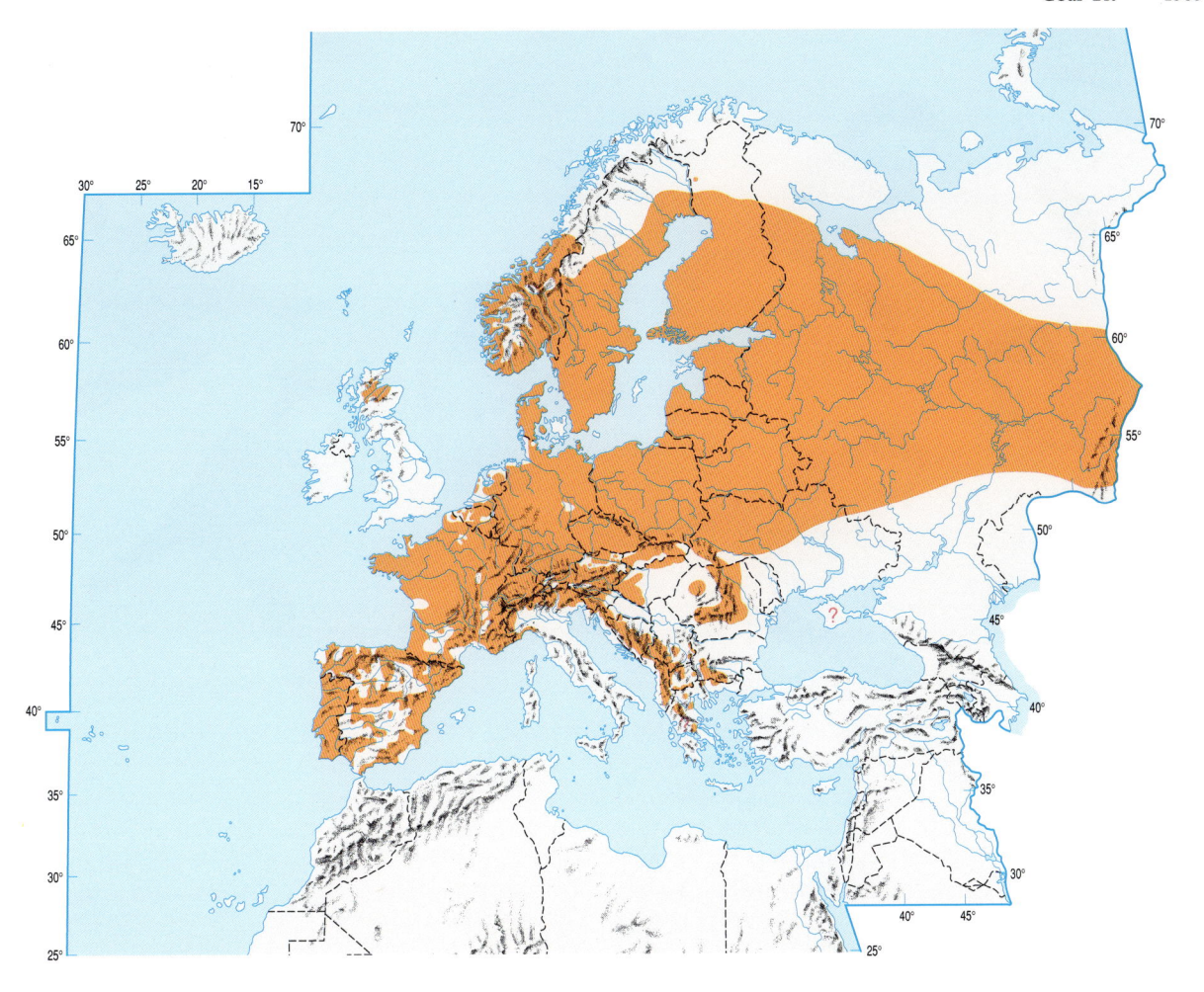

mainly clinal. Upperparts vary from rufous-brown at western and south-western extreme of range (*scoticus*, Scotland; *abadiei*, Brittany; *weigoldi*, southern Iberia) to olivaceous grey-brown in north-east and east (nominate *cristatus*, Fenno-Scandia to northern Russia; *bashkirikus*, southern Urals). Intermediate race *mitratus* occupies wide intermediate area.

Coal Tit *Parus ater*

PLATES: pages 1386, 1395

Du. Zwarte Mees   Fr. Mésange noire   Ge. Tannenmeise   It. Cincia mora
Ru. Московка   Sp. Carbonero garrapinos   Sw. Svartmes

Field characters. 11.5 cm; wing-span 17–21 cm. Slightly smaller, especially about head, and shorter-tailed than Blue Tit and all black-capped tits. Small, sprightly tit, with short tail, rather fine bill, and apparently small head. Basically olive- to slate-grey above and buff below; has diagnostic combination of black head with white nape and cheek, and 2 white wing-bars. Shows marked geographical variation; most strikingly different races in west Palearctic are north African, especially *P. a. ledouci* of Algeria and Tunisia, with strong yellow wash over plumage, and *P. a. cypriotes* of Cyprus, with brown back and dark flanks.

Unmistakable when wing markings and full head pattern visible, but subject to confusion with small black-capped Marsh Tit and Willow Tit from below, when unique characters hidden and then important to note voice (see below), size, structure, and extent of throat-patch. Coal Tit appears relatively smaller-headed, less full-chested, and much shorter- and narrower-tailed than all other west Palearctic *Parus* and has proportionately the largest black 'bib'. Flight closely resembles Blue Tit but even more rapid, with whirring wing-beats; darts suddenly between tree crowns and hovers around foliage.

Habitat. Breeds in west Palearctic from boreal through temperate to Mediterranean zones, in continental and oceanic upper and lower middle latitudes within July isotherms *c.* 12–22°C

Coal Tit *Parus ater*. *P. a. ater*: **1** ad, **2** juv. *P. a. hibernicus*: **3** ad. *P. a. britannicus*: **4** ad. *P. a. sardus*: **5** ad. *P. a. atlas*: **6** ad. *P. a. ledouci*: **7** ad. *P. a. michalowskii* (Caucasus): **8** ad. *P. a. cypriotes*: **9** ad.

and at all elevations from sea-level to treeline. Habitat selection closely related to size and morphology; in particular, foot morphology (long toes and claws, not opposable, unlike, e.g., Blue Tit) probably adapted to life in conifers, and fine bill for foraging for small food items, especially in conifers. In large parts of range conifers form almost the entire habitat, usually with preference for spruce. Where this is lacking, almost any other conifers will be accepted. Also occurs, but generally at lower densities, in pure broad-leaved woodland, e.g. in Britain and Ireland; and in some regions, such as Caucasus, parts of southern Europe, and North Africa, forests of beech or oak are main habitat.

Distribution. Spread France, Belgium, Germany, and Spain; expansion (and associated population increase) attributed to afforestation (in Belgium, also to provision of nest-boxes). BRITAIN. Bred Outer Hebrides early this century, and again 1965 and 1969. FRANCE. Rare or absent in western France before 1940s. Spread in 1940s, breeding Brittany 1943, Maine-et-Loire 1968, Vendée not until 1975. SYRIA. Apparently first definite record: *c*. 10 in forest with cedars, slopes of Allovit mountains at Slenfe in north-west, April 1992. MOROCCO. Small isolated population discovered and breeding suspected, Jbel Zerhoun north of Meknès in 1989.

Beyond west Palearctic, breeds east to Kamchatka, Sakhalin island, and Japan, extending south from central Siberia to Tien Shan mountains, and from eastern Siberia to eastern Himalayas; also in Iran and Kopet-Dag in Turkmenistan.

Population. Increase Britain, France, Belgium, Netherlands, Finland, Hungary, and Spain, slight decrease Germany (locally), Czech Republic, apparently mostly stable elsewhere. BRITAIN. 610 000 territories 1988–91. Relatively stable since late 1970s at new level after increase. IRELAND. 270 000 territories 1988–91. FRANCE. 10 000–100 000 pairs. BELGIUM. 57 000 pairs 1973–7; great increase since 1960s, some local increases more recently. LUXEMBOURG. 10 000–12 000 pairs. NETHERLANDS. 60 000–85 000 pairs 1979–85. Fluctuating, but probably increased since 19th century. GERMANY. 1.4–2.4 million pairs. In east, 360 000 ± 169 000 pairs in early 1980s. Harz mountains: decline since 1985 in spruce forest of at least 60 years and above 800 m mainly due to large-scale population crash of spiders and aphids caused by industrial emissions and acid rain. DENMARK. 8000–97 000 pairs 1987–8; stable 1976–94, but winter population increased by over 1000%. NORWAY. 100 000–500 000 pairs 1970–90. SWEDEN. 400 000–1 million pairs in late 1980s. FINLAND. 50 000–100 000 pairs in late 1980s. ESTONIA. 2000–20 000 pairs in 1991. Marked annual fluctuations; long-term trend not known, perhaps some decline. LATVIA. 60 000–100 000 pairs in 1980s; fluctuating. LITHUANIA. Rare. POLAND. 200 000–800 000 pairs. CZECH REPUBLIC. 450 000–900 000 pairs 1985–9. SLOVAKIA. 250 000–500 000 pairs 1973–94. HUNGARY. 15 000–20 000 pairs 1979–93. AUSTRIA. 500 000–1 million pairs in 1992. SWITZERLAND. 500 000–800 000 pairs 1985–93. SPAIN. 1.66–5.2 million pairs. PORTUGAL. 100 000–1 million pairs 1978–84. ITALY. 800 000–1.5 million pairs 1983–95. GREECE. 20 000–50 000 pairs. ALBANIA. 1000–3000 pairs in 1981. YUGOSLAVIA: CROATIA. 400 000–600 000 pairs. SLOVENIA. 100 000–200 000 pairs. BULGARIA. 500 000–1 million pairs. RUMANIA. 1.2–1.5 million pairs (probably underestimate) 1986–92. RUSSIA. 1–10 million pairs; fluctuating. BELARUS'. 50 000–60 000 pairs in 1990; fluctuating, but apparently stable overall. UKRAINE. 5000–6000 pairs in 1986. AZERBAIJAN. Uncommon to locally common. TURKEY.

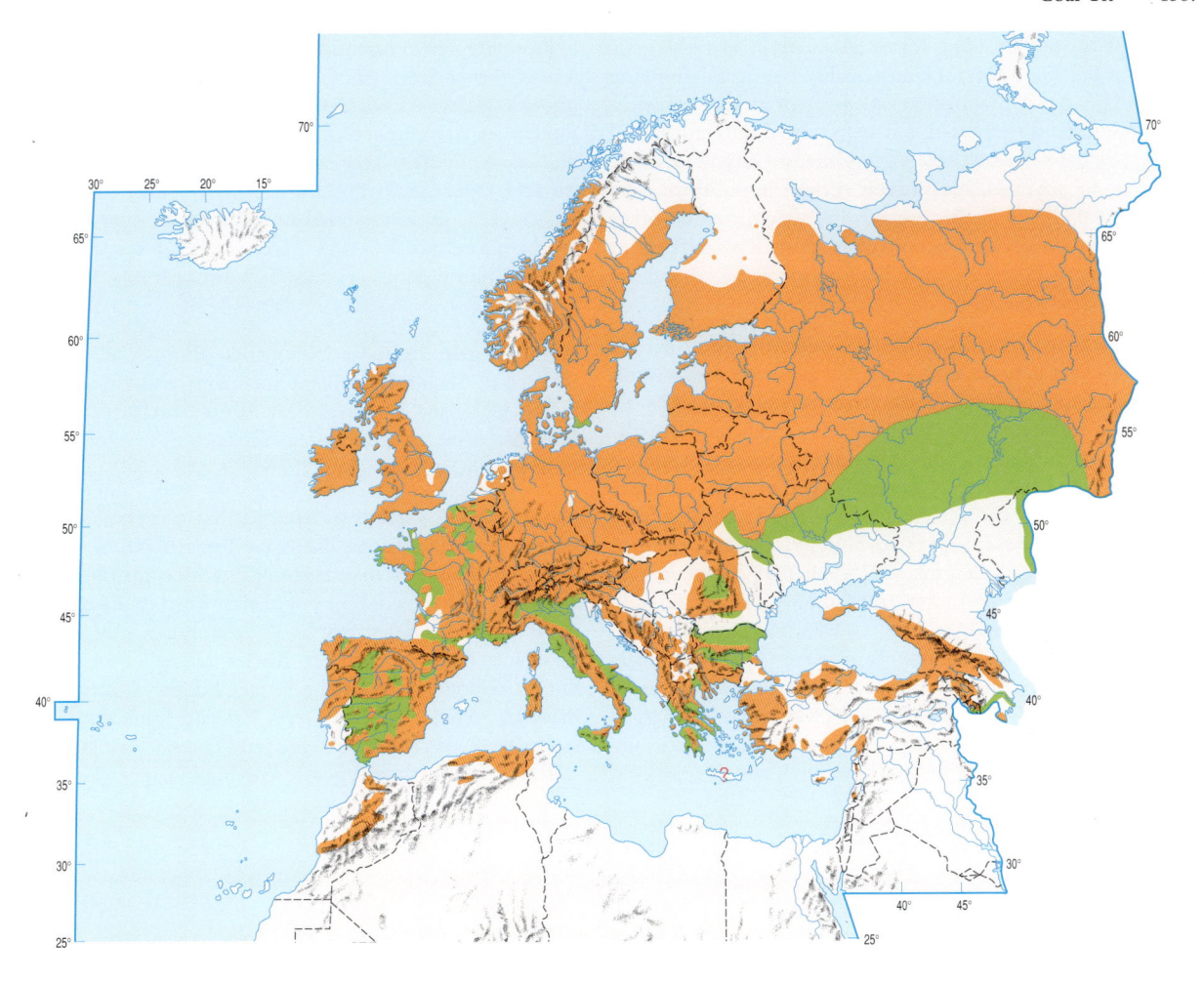

100 000–1 million pairs. CYPRUS. Common. TUNISIA. Common in oak forests of north-west. MOROCCO. Locally common.

Movements. Sedentary in south and west of range, but eruptive, sometimes in very large numbers, over much of remainder, and in some northern and eastern areas a fairly regular relatively short-distance migrant. Eruptions may involve large part of breeding area, or only isolated populations. Heading chiefly west or south-west, with tendency to avoid long sea crossings. During eruptions, continental birds regularly reach Britain. Timing of eruptive movements fairly similar throughout Europe, from end of August or beginning of September, usually peaking in 2nd half of September and continuing to end of October or beginning of November. Return passage March–May.

Food. Adult and larval insects and spiders, plus seeds in autumn and winter. Favourite plant material, seeds of spruce, often unavailable for years at a time when trees bear no cones. Moves along thickly needled branches, hopping, fluttering, and hovering. More so than other west Palearctic tits, shows great agility when foraging, frequently turning upside-down and hanging from cones or needles. Forages mostly on leaves, needles, and cones in upper parts of large conifers, seldom on thick branches or trunks, though this more common in Britain where more often in deciduous woodland. Forages by minute examination and gentle probing, searching even finest twigs, including vertically-hanging spruce branches.

Storing of food, animal and vegetable, occurs widely; seasonally variable according to availability of food; conifer seeds stored in large numbers at times of opening of cones.

Social pattern and behaviour. The smallest tit in all areas where it occurs, and socially subordinate to other species. Gregarious outside breeding season, joining mixed-species foraging flocks. Paired birds probably associate together in winter flocks and may be dominant over other conspecifics when within own territory. Monogamous mating system; partners usually remain together in successive years if both survive. ♂ sings, usually from high perch, throughout most or all of year, with peak in spring. ♀♀ sing regularly; usually less loud than ♂'s song; common during incubation and nestling periods, especially when disturbed.

Voice. Used freely throughout year. Of distinctive quality; almost all calls sweeter and clearer in tone than those of other west Palearctic tits, with almost complete absence of churring sounds. Song a repetition of 2–3 sweet notes, with notably alternated stress, e.g. 'teechü-teechü...', 'tchüee-tchüee...', 'chirriwee-chirriwee...', etc. Individual ♂♂ can use at least 16 different song-types in one breeding season. Alarm-calls variable; generally consist of pure notes of similar quality to song, but showing less rapid frequency modulation, and given in varying sequence (i.e. same note not repeated, but sequence of different notes) with pauses between each note. At low intensity (anxiety), single note may be repeated. North-west African populations have churring or dull trilling call as usual alarm-call, quite unlike other races and not unlike churring units of other tits. This call present in repertoire of European birds, but very rarely heard, only in situations of extreme alarm.

Breeding. SEASON. First clutches usually started in April. Season tends to be later in north than south, but geographical variation dependent more on local climate, especially as affected by altitude; also later on Corsica than would be expected from climate. Season of variable length, depending mainly on incidence of 2nd broods; exceptionally extends to end of July. Incidence of 2nd broods very variable geographically and annually, from none in some areas or some years to *c.* 50%, 3rd broods occasional; overall percentage of 2nd broods greater than in other west Palearctic tits. SITE. Hole in tree, tree-stump, rock crevice, wall, or ground (e.g. among tree roots, under stones, in old mouse holes). In many habitats commonly or mainly in ground; holes in ground almost certainly used because of competition with larger species for higher holes. Nest: hole, which may be enlarged if in rotten wood or ground; nest-cup of moss, generally distinguishable from other tits as whole nest made of same kind of moss; lined with hair and wool, often also a few feathers, but feathers never form main lining material. EGGS. Sub-elliptical, smooth, glossy; white, finely spotted or speckled red-brown, sometimes concentrated into band at broad end; occasionally unmarked. Clutch: geographically, annually, and seasonably variable. Over much of central Europe 1st clutch averages 8–9; full range 5–13, but extremes rare. Average slightly larger in southern England: 9.1–10.5. Much smaller in Corsica (average 5.9) and North Africa (average 5.7). INCUBATION. 14–16 days. FLEDGING PERIOD. Usually 19 days (18–20).

Wing-length: Most of Continental Europe: ♂ 60–67, ♀ 58–66 mm. Slightly larger in mountains in south of range, smaller in Britain, Ireland, Mediterranean islands, and Iberian lowlands. **Weight:** Most of Continental Europe: ♂♀ mostly 8–10 g; somewhat heavier or lighter elsewhere, as wing-length.

Geographical variation. Marked, but not complex. Clinal over continental Europe and Asia Minor, depth of coloration decreasing from west to east; most of area occupied by nominate *ater*, only extreme western population from Iberia recognized as distinct race, *vieirae*. Insular populations in west and south (*hibernicus*, Ireland; *britannicus*, Britain; *sardus*, Corsica and Sardinia) all recognized as racially distinct, but only Cypriot *cypriotes* markedly so (see Field characters). *P. a. atlas* from Morocco and, especially, *P. a. ledouci* from Algeria and Tunisia, also very distinct; *atlas* lacks yellow that is conspicuous in plumage of *ledouci*.

Blue Tit *Parus caeruleus*

PLATES: pages 1389, 1395

DU. Pimpelmees   FR. Mésange bleue   GE. Blaumeise   IT. Cinciarella
RU. Лазоревка   SP. Herrerillo común   SW. Blåmes

Field characters. 11.5 cm; wing-span 17.5–20 cm. Distinctly smaller than Great Tit, being 20% shorter overall, with proportionately larger head and shorter tail. Small tit, with short bill, short and round wings, short tail, and distinctive, seemingly square head usually carried above line of body in perky attitude. Essentially blue-green (or, in North Africa and Canary Islands, slate-blue) above and yellow below, with blue, black, and white markings on head, enhancing its squareness. Only tit of west Palearctic with both pale forehead and supercilium. Geographical variation marked.

Unmistakable. Flight whirring and fast, with noticeable bounce in action and square head visible in silhouette; behaviour as other tits, although generally boldest of family.

Habitat. Breeds in west Palearctic in continental and oceanic lower middle to upper middle latitudes, from Mediterranean and steppe through temperate to boreal zones, and from arid to humid climates, within July isotherms 14–30°C. Mainly inhabits lowlands, even in Swiss Alps not breeding above *c.* 1250 m, though higher in Caucasus. Essentially a bird of broad-leaved woodland, over much of Europe also extending to coppices, parks, gardens, etc., provided that suitable nest-holes are available, and even inner cities. Occurs in a wide variety of habitats in North Africa, including palm groves, and in montane conifer woodland in Canary Islands.

Distribution. BRITAIN. Bred Outer Hebrides from 1963 and Isles of Scilly from late 1940s. NORWAY. Has spread north since *c.* 1970. FINLAND. Very rare breeder in extreme south-west up to beginning of 20th century, subsequently spreading; present boundary reached in 1930s, but some extension to north probably still occurring. TURKEY. Widespread, but common only in north-west; otherwise, mainly at very low density. SYRIA. Breeds in Kassab area and Allovit mountains of north-west, but extent of range still not fully known. Damascus area: no proof of breeding, and no recent records. JORDAN. First recorded Wadi as Sir 1893; no further sightings until rediscovered 1984. Small population breeding Northern Highlands; first proved 1990.

(FACING PAGE) Blue Tit *Parus caeruleus*. *P. c. caeruleus* (most of continental Europe): **1** ad ♂, **2** ad ♀, **3** 1st ad (1st year), **4** juv. *P. c. obscurus* (Britain and Ireland): **5** ad ♂. *P. c. orientalis* (Volga river eastwards): **6** ad ♂. *P. c. teneriffae*: **7** ad ♂, **8** juv. *P. c. palmensis*: **9** ad ♂. *P. c. ombriosus*: **10** ad ♂. *P. c. degener*. **11** ad ♂. *P. c. ultramarinus*: **12** ad ♂. *P. c. cyrenaicae*: **13** ad ♂. Azure Tit *Parus cyanus* (p. 1392). *P. c. cyanus*: **14** ad, **15** juv. *P. c. hyperrhiphaeus* (south-west Siberia): **16** ad.

Chris Rose

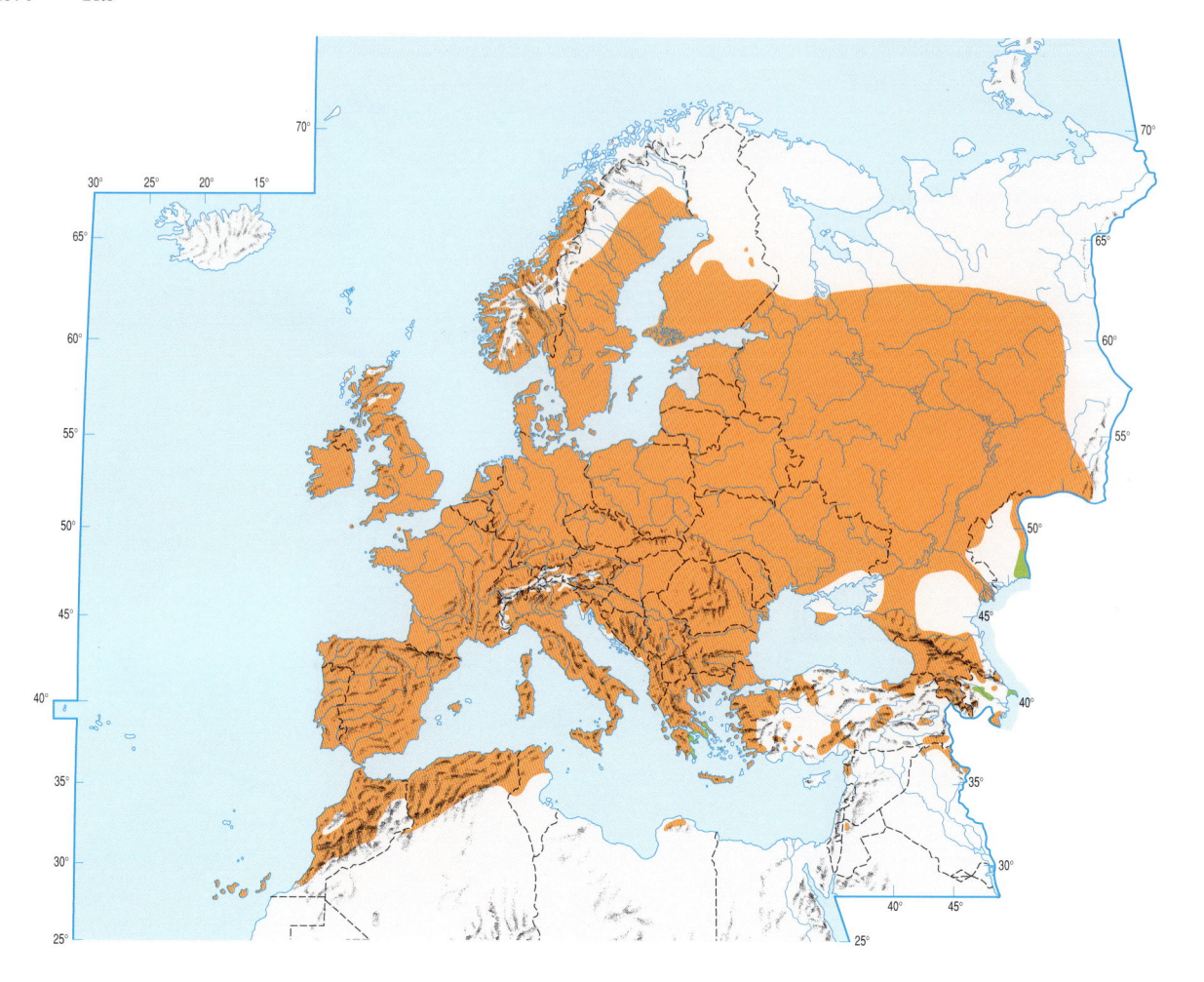

Accidental. Malta, Israel.

Beyond west Palearctic, breeds only in Iran and south-west Turkmenistan.

Population. Increase Britain, Netherlands, Finland, Estonia, and Ukraine; slight decrease Czech Republic, apparently mostly stable elsewhere. BRITAIN. 3.3 million territories 1988–91. Fluctuating around higher levels reached in 1970s; perhaps now increasing again. IRELAND. 1.1 million territories 1988–91. FRANCE. 100 000–1 million pairs in 1970s. BELGIUM. 130 000–200 000 pairs 1973–7. LUXEMBOURG. 30 000–40 000 pairs. NETHERLANDS. 125 000–200 000 pairs 1979–85. Marked increase since c. 1950. GERMANY. 2.7 million pairs in mid-1980s. In east, 430 000 ± 189 000 pairs in early 1980s. DENMARK. 290 000–2.9 million pairs 1987–8. NORWAY. 50 000–100 000 pairs 1970–90. SWEDEN. 400 000–1 million pairs in late 1980s. FINLAND. 100 000–150 000 pairs in late 1980s. Marked increase in recent decades mainly due to increased provision of nestboxes and winter feeding. ESTONIA. 20 000–50 000 pairs in 1991. Marked annual fluctuations, but considerable increase in recent decades. LATVIA. 100 000–140 000 pairs in 1980s. LITHUANIA. Common. POLAND. Numerous. CZECH REPUBLIC. 800 000–1.6 million pairs 1985–9. SLOVAKIA. 700 000–1.4 million pairs 1973–94. HUNGARY. 250 000–300 000 pairs 1979–93. AUSTRIA. 200 000–500 000 pairs in 1992. SWITZERLAND. 150 000–250 000 pairs 1985–93. SPAIN. 930 000–3.6 million pairs. PORTUGAL. 1 million pairs 1978–84. ITALY. 100 000–500 000 pairs 1983–95. GREECE. 100 000–300 000 pairs. ALBANIA. 20 000–50 000 pairs in 1981. YUGOSLAVIA: CROATIA. 100 000–200 000 pairs. SLOVENIA. 20 000–30 000 pairs. BULGARIA. 500 000–1 million pairs. RUMANIA. 500 000–800 000 pairs 1986–92. RUSSIA. 100 000–1 million pairs. BELARUS'. 330 000–370 000 pairs in 1990. UKRAINE. 85 000–100 000 pairs in 1986. MOLDOVA. 30 000–40 000 pairs in 1988. AZERBAIJAN. Common; in north-east, almost as numerous as Great Tit. TURKEY. 100 000–1 million pairs. TUNISIA, ALGERIA. Common. MOROCCO. Abundant. CANARY ISLANDS. Widespread and common all islands except Fuerteventura and Lanzarote where local and uncommon.

Movements. Basically resident, though over much of centre and north of range makes irregular eruptive movements, mainly to west and south. In southern areas mostly sedentary but makes altitudinal movements from highest breeding areas. Continental birds sometimes reach Britain in autumn, particularly but not exclusively in years of eruption.

Food. Chiefly insects and spiders, also fruits and seeds outside breeding season, nectar and pollen, especially in spring, and sap of trees. Diet reflects seasonal and other changes in food abundance. In trees, etc., actively examines twigs and leaves, advancing in small rapid hops to tips of branches; takes items while perched upright or hanging upside-down; can switch quickly between these postures, more readily than Great Tit. Examines all sides of branch, twig, or leaf, using bill or else bill and then foot to reach, examine, and turn or handle nearby ones. Does not store food.

Social pattern and behaviour. Gregarious outside breeding season, forming flocks which form part of larger, mixed-species flocks. Winter flocks consist of small nucleus of residents, and constantly drifting 'nomads'. Residents dominant over nomads, and ♂♂ of previous season's pairs over ♀♀; social dominance of residents related to territory, persistence of which (in potential form) is shown by dominance of individuals when within their own previous territories. Mating system basically monogamous, but low incidence of simultaneous bigamy probably regular, at least in optimal habitats. Pairs usually persist from year to year when both partners survive. New pairs mostly formed when flocks break up at end of winter. ♂ sings from various heights in trees and shrubs; not especially from conspicuous high perches, and song bouts not long sustained. Main song-period starts and finishes with period of territorial and reproductive behaviour, but also some song in late summer and autumn, especially from young birds. ♀♀ sing fairly often, especially when engaged in reproductive fighting. Song mainly used to proclaim territory; also important in maintaining contact between ♂ and ♀. Courtship or pair-bonding behaviour includes conspicuous slow flight with rapidly beating wings, or gliding flight, usually directed at nest-hole.

Voice. Used freely throughout year. Song of ♂ (European population) a clear, silvery, high-pitched tremolo, usually introduced by variable but smaller number of higher-pitched units of variable form: 'tsee tsee t t t t t t t t', 'tsee-tsee-tsee-tsü-tsühühühühühü'; lasts up to *c*. 2 s. Other forms of song consist of repetition of (usually) 2 unit types without tremolo, e.g. 'tee tee see see'; 'see-saw'. Less sharp and emphatic than 2-unit type songs of Great Tit, not as liquid and sweet as those of Coal Tit. Individual ♂♂ have repertoires of 3–8 different songs. Songs of isolated southern populations distinctly different. Main difference is lack of phrased song, i.e. songs are repetitions of same units or groups of units, thus similar to songs of other tits. Song very variable in Canary Islands; typical song in North Africa a repetition of phrase of 2–4 rather metallic units, e.g 'tizee tizee tizee...'. Most characteristic call is a rapid churring or scolding call, either given alone or preceded or followed by one or more 'tsee' units; like equivalent call of Great Tit, but softer and with faster tempo. Scolding and alarm-calls of this type used throughout year by both sexes, elicited by any (even mildly) alarming circumstance. Other calls include short, high-pitched 'tsee', used for contact between birds at close range, and drawn-out 'seeee', used as alarm-call against flying predators.

Breeding. SEASON. Average laying date of first clutches early April to mid-May, varying with latitude and altitude. Over much of central and northern Europe laying begins mostly in last week of April and first few days of May; at 50–55°N, retardation of 2.5 days per degree of latitude. Usually 1 brood; 2nd clutches uncommon or very rare in some areas, regular but generally undertaken by minority of breeding pairs in other areas. SITE. In hole in tree; also in wall or, where natural sites lacking, in artificial hole of any kind; rarely in ground, or in covered nest of another species. Nest-box readily used. Nest: pad of moss, often mixed with pieces of dead grass, straw, or with other plant material; cup lined with fine dry grass, hair, wool, some feathers, and often fine shavings of bark. EGGS. Sub-elliptical, smooth without gloss; white, with fine (sometimes larger) red-brown spots, usually concentrated at broad end, and some paler, greyer underlying spots; occasionally unmarked. Clutch: 2–18, varying with latitude, altitude, year, size of nest-cavity, and, especially, quality of habitat. Average size of 1st clutch mostly 10–12 (usual range 6–16) over much of continental Europe, showing slight increase from south to north; considerably lower in Mediterranean area and North Africa, and extremely low (average 3.5) in Canary Islands. INCUBATION. Average 14.2 days (13–16). FLEDGING PERIOD. 16–22 days, southern England; 19–23 days, Norway.

Wing-length: Most of Europe: ♂ 61–71, ♀ 61–69 mm. Smaller in Britain, Ireland, Mediterranean area, North Africa, and Canary Islands.

Weight: ♂♀ mostly 9.5–12.5 g; less in western and southern areas, as wing-length, and somewhat greater in cold northern and eastern areas.

Geographical variation. Marked, with 2 distinct groups: *caeruleus* group in Europe and Middle East, with blue crown and head stripes, greenish upperparts, and short, thick bill, and in which variation is mainly clinal and not pronounced; *teneriffae* group in North Africa and Canary Islands, with mainly blue-black crown and head stripes, slate-blue upperparts and rather long, thinner bill. 7–8 races recognized in *caeruleus* group, mainly reflecting darker plumage in more humid west and paler in north and east. 6 races of *teneriffae* group (*ultramarinus*, north-west Africa; *cyrenaicae*, Libya; *palmensis*, La Palma; *teneriffae*, central Canary Islands; *ombriosus*, Hierro; *degener*, Fuerteventura and Lanzarote) all rather distinct.

Azure Tit *Parus cyanus*

Du. Azuurmees   Fr. Mésange azurée   Ge. Lasurmeise   It. Cinciarella azzurra
Ru. Князек   Sp. Herrerillo cianeo   Sw. Azurmes

Field characters. 13 cm; wing-span 19–21 cm. Head and forebody only slightly larger than Blue Tit but rear body and tail distinctly longer. Medium-sized tit with quite long tail, having much of general character of Blue Tit but with rather ghostly blue-white-black plumage, pattern somewhat suggesting Long-tailed Tit: white head, with narrow black eye-stripe joined to black band across nape, 2 strikingly broad white panels on slate-blue wing, dusky-blue back, and white-edged bluish-black tail. Hybridizes with Blue Tit where breeding ranges overlap.

Unmistakable, though glimpse may suggest Blue Tit or even white-headed race of Long-tailed Tit. Flight action and silhouette recall both of these species, with strong bursts of wing-beats producing bounding, undulating progress said to be more powerful than Great Tit and even reminiscent of Bearded Tit. Restless and active but often rather difficult to find and see well, keeping to dense bushes or reeds and showing as irregularly as Bearded Tit. Flicks wings and tail; most recalls Blue Tit when tail cocked above line of body.

Habitat. Breeds in continental middle temperate latitudes of Palearctic. Inhabits light broad-leaved woods, especially along banks of rivers, streams, and lakes, preferably of willow, including tree and shrub growth. Also occupies floodlands covered with reeds and rushes, osier beds, and various kinds of moist or boggy thickets; extralimitally also in stands of willow and larch and woods of juniper or spruce up mountains to *c.* 2500 m, and in winter in semi-deserts. Such habitat types poorly represented in west Palearctic, and inability to adapt from them to closest available equivalents such as gardens and broad-leaved woods may at least partly explain failure of attempted westward expansion.

Distribution. Several waves of expansion to west into range of Blue Tit (with which it sporadically hybridizes), followed by retreat to usual limits; largest recorded wave in 1870s and 1880s, most recent 1973–9. Finland. Invasion in 1973–7; bred 1973, and mixed pair with Blue Tit 1975. FSU. Western range limit poorly known. Not clear to what extent regular west of Volga basin, though apparently well established in southern Belarus'. Russia. Moscow region: bred Yakhroma-Dubna lowlands 1981–2, where present in 1930s and 1950s; perhaps also breeds Moscow Meshchera region. Belarus'. First recorded (Vitebsk region, in north) 1843, first nest found 1904, with 4 further breeding records (in south) 1952–89. More systematic fieldwork in southern Poles'e region in 1980s suggested species occurs along Pripyat' river (and lower reaches of tributaries) east from Pinsk.

Accidental. France, Germany, Denmark, Sweden, Finland, Estonia, Latvia, Lithuania, Poland, Slovakia, Hungary, Austria, Yugoslavia, Rumania, Ukraine. Hybrids (with Blue Tit): Netherlands, Sweden, Finland, Latvia, Poland, Austria.

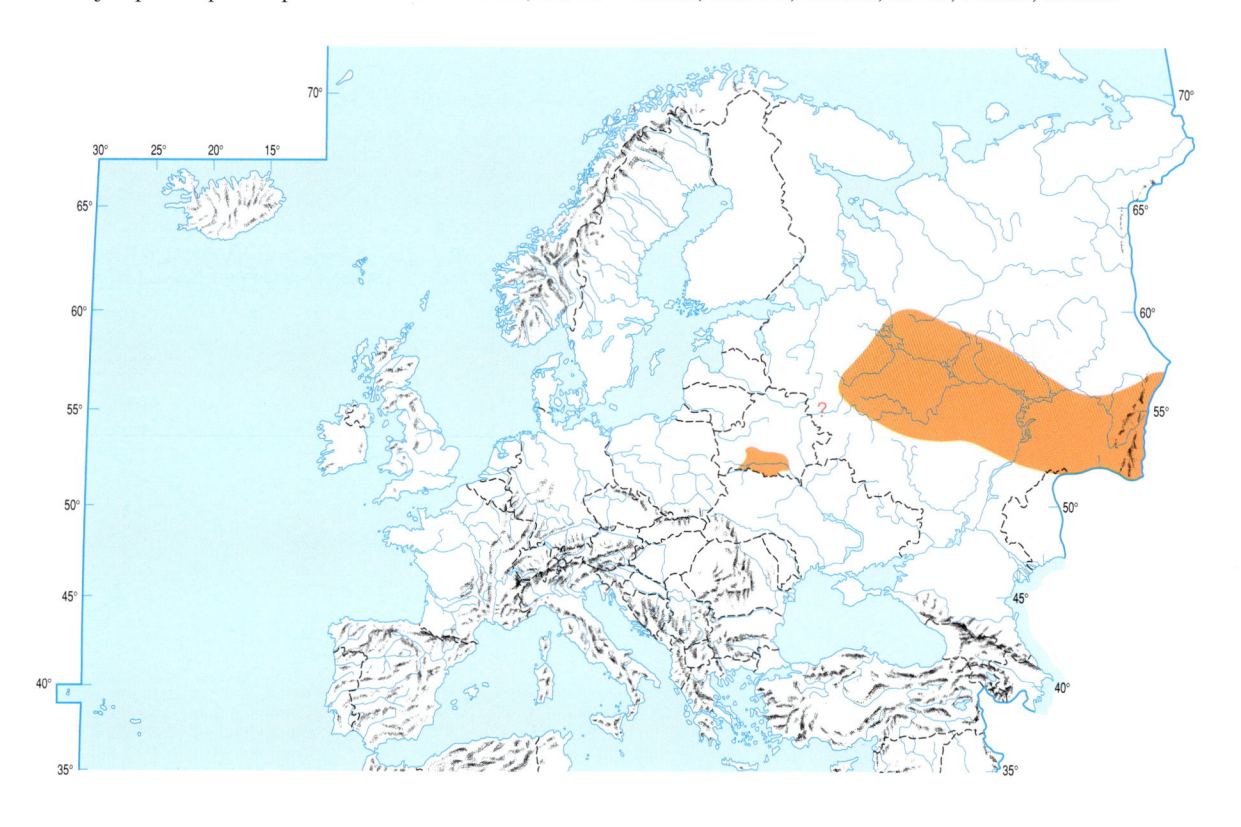

Beyond west Palearctic, extends east across central Asia to Ussuriland, south to Tien Shan mountains and northern China.

Population. Russia. 10 000–100 000 pairs; stable. Belarus'. 400–1000 pairs in 1990; slight increase in 1980s.

Movements. Resident over much of range, but also nomadic, with occasional records far from normal range; altitudinal migrant in some areas.

Food. Diet predominantly invertebrates, mainly adults, pupae, larvae, and eggs of Lepidoptera, and spiders; in autumn and winter more seeds and fruits. In autumn and winter works reed stems from top to bottom, pecking at nodes for larval Diptera: breaks open stems of dock and Umbelliferae for hibernating larvae.

Social pattern and behaviour. Scanty information indicates general similarity to Blue Tit. In small flocks or pairs outside breeding season. No evidence of other than monogamous mating system. During periodic invasions to west of usual breeding range, frequently forms mixed pairs with Blue Tit.

Voice. Generally rather similar to Blue Tit. Song a distinctive, brief trill. Calls include a rather quiet, disyllabic 'tirr tirr' and, when excited, a repeated 'tscherpink'.

Breeding. Season. European Russia: eggs laid from mid-May, occasionally earlier, young fledging on average in 2nd half June. Site. Usually hole in tree. Other kinds of hole, including nest-boxes and holes in ground or buildings, regularly used by extralimital races. Nest: cup of moss, dry grass, and wool or fur, lined with finer hair. Eggs. Sub-elliptical, smooth, not glossy; white, sparsely marked with red-brown speckles, spots, and blotches, often forming band at broad end. Clutch: 9–11. Incubation. 13–14 days. Fledging Period. 16 days.

Wing-length: ♂ 68–72, ♀ 67–72 mm.
Weight: ♂♀ 10.6–16 g Kazakhstan.

Great Tit *Parus major*

PLATES: pages 1394, 1395

Du. Koolmees Fr. Mésange charbonnière Ge. Kohlmeise It. Cinciallegra
Ru. Большая синица Sp. Carbonero común Sw. Talgoxe

Field characters. 14 cm; wing-span 22.5–25.5 cm. Over 20% bulkier and longer than Blue Tit. Large tit, with quite long and broad tail, quite heavy and spiky bill, and rather large domed head. Plumage basically green above and yellow below, with white-cheeked black head, black central stripe on underbody, and white wing-bar and tail-edges.

Unmistakable; large size and black and white head allow instant separation from Blue Tit. Flight most fluent of *Parus*, inviting comparison with medium-sized *Sylvia* warbler in its bursts of quite loose wing-beats, momentary closures of wings, side-slips, darts, and at times tail-heavy form.

Habitat. Breeds in west Palearctic from higher to lower middle latitudes, continental and oceanic, within July isotherms *c.* 12–32°C, in coolest and warmest forest zones, from subarctic to Mediterranean, and marginally in steppe and semi-desert. Extralimitally in Asia extends deep into tropics. Able to ascend mountains to treeline—in Switzerland exceptionally to 1900 m, in north-west Africa to 1850 m—but is much more a lowland species, disliking pure coniferous forest, and preferring mixed types and more open or even fragmented and scattered trees to dense pure deciduous forest. Nature of higher tree cover may be less important than structure and density of under-growth, and opportunities for ground feeding which is particularly important. Thus well able to live in gardens, city parks, cemeteries, orchards, hedgerows, and spinneys, as well as on farms where trees or shrubs present. However, height at which most feeding occurs varies with season. Like Blue Tit, ready and well adapted to take full advantage of human structures and other artefacts and of food and nest-sites made available incidentally or deliberately, and this is of special importance in successful overwintering, especially near limits of suitable range.

Distribution. Northward spread in Norway, Finland, and Scotland, probably due to climatic factors and (in Fenno-Scandia) ability to winter in areas of human settlement. Britain. Scotland: spread in north in 1st half of 20th century, continuing to 1980s; bred Outer Hebrides from 1962. Bred Isles of Scilly from 1920s. Finland. Marked expansion north in 20th century, but northernmost breeding population probably spread from northern Norway. Syria. Extensive range in west; probably only sporadic inland. Israel. Spread south in 1950s–70s, with spread of human settlement. Egypt. Established in northern Sinai by early 1970s. Morocco. Very local south of Haut Atlas, except for Sous valley, where (as in rest of range) common.

Accidental. Iceland, Malta.

Beyond west Palearctic, breeds extensively in Asia, north to southern Siberia (also Magadan and Kamchatka in north-east), east to Sakhalin island and Japan, south to Sri Lanka and Sunda Islands.

Population. Stable over most of range, but increase reported Netherlands, Estonia, Ukraine (marked), and Israel. Britain. 1.6 million territories 1988–91. No reason to suppose significant change compared with 1968–72. Ireland. 420 000 territories 1988–91. France. Over 1 million pairs in 1970s. Belgium. 400 000 pairs in 1960s. Estimate of 280 000 pairs 1973–7 probably too low. Luxembourg. 40 000–50 000 pairs.

Great Tit *Parus major*. *P. m. major*: **1** ad ♂, **2** 1st ad ♂ (1st winter), **3** juv. *P. m. newtoni*: **4** ad ♂. *P. m. excelsus*: **5** ad ♂. *P. m. corsus*: **6** ad ♂. *P. m. blanfordi*: **7** ad ♂.

NETHERLANDS. 250 000–500 000 pairs 1979–85. Marked increase since c. 1950. GERMANY. 10.4 million pairs in mid-1980s. In east, 1.5 ± 0.5 million pairs in early 1980s. DENMARK. 300 000–3.1 million pairs 1987–8. NORWAY. 500 000–1 million pairs 1970–90. SWEDEN. 1–2.5 million pairs in late 1980s. FINLAND. 1–1.5 million pairs in late 1980s. Now stable after rapid increase due to provision of nest-boxes and intensified winter feeding. ESTONIA. 150 000–200 000 pairs in 1991. Increase in recent decades and probably since mid-20th century. LATVIA. 400 000–500 000 pairs in 1980s. LITHUANIA. Common. POLAND. Numerous. CZECH REPUBLIC. 3–6 million pairs 1985–9. SLOVAKIA. 1.5–3 million pairs 1973–94. HUNGARY. 450 000–600 000 pairs 1979–93. AUSTRIA. 500 000–1 million pairs in 1992. SWITZERLAND. 500 000–700 000 pairs 1985–93. SPAIN. 1.6–4.3 million pairs. PORTUGAL. 1 million pairs 1978–84. ITALY. 1–2 million pairs 1983–95. GREECE. 200 000–500 000 pairs. ALBANIA. 30 000–50 000 pairs in 1981. YUGOSLAVIA: CROATIA. 2–3 million pairs. SLOVENIA. 200 000–300 000 pairs. BULGARIA. 1–10 million pairs. RUMANIA. 2.5–3 million pairs 1986–92. RUSSIA. At least 10 million pairs. BELARUS'. 1.5–1.7 million pairs in 1990. UKRAINE. 900 000–1.2 million pairs in 1986. MOLDOVA. 100 000–150 000 pairs in 1988. AZERBAIJAN. Common; locally very common. TURKEY. 1–10 million pairs. CYPRUS. Common. LEBANON. Common central areas in mid-1970s. ISRAEL. Abundant. JORDAN. Fairly common. TUNISIA. Common breeder in oak forests. ALGERIA. Common.

Movements. Resident over much of southern and central part of range and irregular eruptive migrant from northern areas, sometimes moving in huge numbers. Altitudinal migrant from some of highest breeding areas.

Food. Wide variety of insects, especially Lepidoptera and Coleoptera; also spiders; significant amount of seeds and fruit in winter. In winter, forages in wide variety of sites but mainly below 7 m (frequently in mixed-species flocks). In spring, feeding height generally rises suddenly to above 9 m when feeding on caterpillars. In winter, takes insects from bark, twigs, walls, and leaf litter, and may move nearer to human habitation to feed at bird-tables, etc. In northern Europe, ground foraging increases through winter due to increased dependence on fallen tree seeds, especially of beech and hazel, but varies between years due to variation in seed crops.

Social pattern and behaviour. Woodland populations mainly gregarious from late summer to spring, forming mixed foraging flocks with other species. Some resurgence of territorial behaviour by adult ♂♂ in autumn. In Israel, territorial throughout autumn and winter; flocks not formed; and in suburban areas elsewhere, where food regularly supplied by man, social hierarchies established at feeding sites and little flocking occurs. Mating system monogamous, with rare cases of bigamy. ♂

(FACING PAGE) Bearded Tit *Panurus biarmicus biarmicus* (p. 1362): **1** ad ♂, **2** ad ♀.
Long-tailed Tit *Aegithalos caudatus caudatus* (p. 1372): **3** ad.
Marsh Tit *Parus palustris palustris* (p. 1375): **4** ad.
Sombre Tit *Parus lugubris lugubris* (p. 1378): **5** ad.
Willow Tit *Parus montanus rhenanus* (p. 1379): **6** ad.
Siberian Tit *Parus cinctus* (p. 1382): **7** ad.
Crested Tit *Parus cristatus mitratus* (p. 1383): **8** ad.
Coal Tit *Parus ater ater* (p. 1385): **9** ad.
Blue Tit *Parus caeruleus caeruleus* (p. 1388): **10–11** ad ♂.
Azure Tit *Parus cyanus cyanus* (p. 1392): **12** ad.
Great Tit *Parus major major*: **13** ad ♂, **14** ad ♀.

Chris Rose

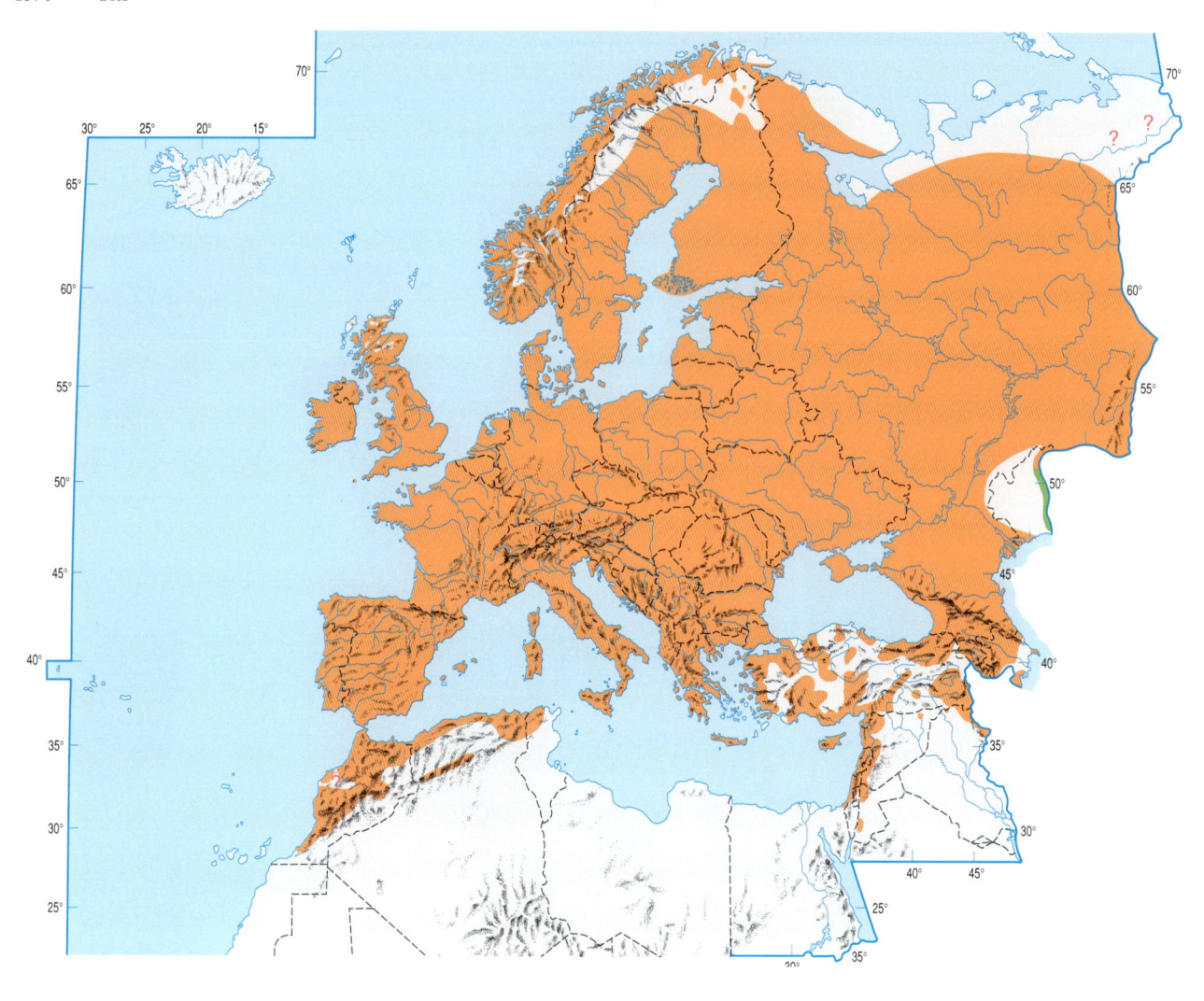

sings from various heights in trees and shrubs, often from highest perch available, occasionally on ground or in flight. Song-period coterminous with period of territorial and reproductive behaviour, mainly from late winter (when flocks break up) to end of nesting season, with some resumption during resurgence of territorial and sexual behaviour in autumn. Aggressive behaviour includes striking head-up posture, in which black chest-stripe (broader in ♂ than ♀) is displayed to opponent. Pair-bonding behaviour includes flight with shallow wing-beats, usually ending in short glide, often associated with nest-hole inspection.

Voice. ♂ notably vocal in almost all situations throughout year, ♀ considerably less so. Territorial-song of ♂ a repetition of usually 2- or 3-unit motifs in sequences of not usually more than 10 motifs; characteristically loud, sharp, and somewhat metallic, often including buzzy rather than tonal sounds, thus contrasting with rather similar songs of some other tits, e.g. sweeter, more liquid song of Coal Tit, thinner and higher-pitched song of Blue Tit. Variously rendered, e.g. 'teacher teacher...', 'teechuwee teechuwee...'. Other calls extremely varied, mainly consisting of churring sounds often associated with 'pee', 'tink' (very similar to common Chaffinch call) and other non-churring sounds. The 'tink' note used almost exclusively by ♂♂, in variety of situations, especially related to territory.

Breeding. SEASON. Laying begins April over most of west Palearctic; March–April in lowland areas in south, May in north. In extreme south (Israel), usually begins mid-February, early clutches 2nd half of January. Usually 1 brood; incidence of 2nd broods low and very variable. SITE. In tree-hole or, if not available, in wall or other man-made structure of any kind. Nest-box highly preferred, provision of adequate supply often resulting in their use by whole or almost whole population. Exceptionally, in old drey of squirrel, old bird's nest, or among dense twigs of tree or accumulated debris in hedge. Nest: foundation mainly of moss, often with some dry grass or other vegetable matter, thickly lined with hair, wool, and often feathers. EGGS. Sub-elliptical, smooth, not glossy. White, very variably speckled or spotted reddish-brown, often with concentration at broad end, underlying spots violet or violet-grey;

rarely almost or completely unmarked. Clutch: 3–18; usually 6–11 over much of Europe, clutches of over 13 rare, very small clutches most common in southern areas. INCUBATION. 12–15 days. FLEDGING PERIOD. 16–22 days.

Wing-length: Most of Europe: ♂ 73–81, ♀ 69–79 mm. Somewhat larger in mountains of North Africa; smaller on Mediterranean islands and in Israel.
Weight: ♂ mostly 16–22, ♀ 14–20 g.

Geographical variation. Mainly clinal, except insular populations; 10 races recognized. Nominate *major* occupies most of continental west Palearctic. Mediterranean and south-eastern races generally paler and greyer above and paler below, also smaller: *corsus* (southern Iberia, Corsica); *mallorcae* (Balearic Islands); *ecki* (Sardinia); *aphrodite* (southern Italy east to Cyprus); *niethammeri* (Crete); *terraesanctae* (Levant); *blanfordi* (northern Iraq). *P. m. newtoni*, of British Isles, similar to nominate *major* except for larger bill. *P. m. excelsus*, of north-west Africa, similar to *major* but bill larger and white on outer tail reduced.

Nuthatches Family Sittidae

Small to medium-sized oscine passerines (suborder Passeres). Found mostly in woodland, including parks and gardens, though 2 species haunt rocks and cliffs. Chiefly arboreal (or rock-haunting), feeding on insects, other small invertebrates (including snails), seeds, and nuts. All species store single items of food, especially seeds and nuts. Will come to bird-tables and feeders provided by man. 19–22 species, in Holarctic and tropical Asia. Mainly sedentary but some northern species prone to wander in certain years. 7 species in west Palearctic, 6 breeding.

Sexes of similar size (♀ averages smaller). Bill rather long (almost as long as head); straight and strong, sharply pointed. Though not as acrobatic as tits (Paridae), very agile in climbing trees (or rocks) when searching for food, placing one foot higher than other and progressing by a series of jerky hops; unlike treecreepers (Certhiidae), however, do not use tail as a prop. Able to move along branch thus while hanging upside-down and, unlike tits and treecreepers, able freely to descend head-first vertically as well as climbing upwards (though this habit not shared by all species in family). Unlike tits, do not use foot to hold larger food objects; instead, wedge hard food items (such as nuts and seeds) singly in crevices and hammer them with strong blows from bill. Noisily vocal; territorial song of ♂ loud but simple (often not obviously distinct from other calls). Palearctic species not sociable or gregarious; territorial in pairs for all or most of year, sometimes joining passing flocks of other species in winter. Monogamous mating system the rule. In most species, nest typically a simple cup in natural or excavated hole in tree; in west Palearctic species, both sexes involved in building. As anti-predator measure, some species reduce size of larger entrance holes with use of mud. Rock nuthatches atypical in enclosing rock crevice with elaborate, bottle-shaped outer nest of mud in which much other material (including insects) embedded; large cup-nest inside.

Plumage typically bluish-grey above, usually with contrasting black eye-stripe; some species with contrasting dark cap. Underparts wholly or partly white, rufous, or blue. Tail often with contrasting pale spots. Sexes often slightly different.

Krüper's Nuthatch *Sitta krueperi*

PLATE: page 1399

Du. Turkse Boomklever   Fr. Sittelle de Krüper   Ge. Türkenkleiber   It. Picchio muratore di Krüper
Ru. Черноголовый поползень   Sp. Trepador de medalla   Sw. Krüpers nötväcka

Field characters. 12.5 cm; wing-span 21–23 cm. 10% smaller and even more compact than Nuthatch but with proportionately longer bill. Small nuthatch, essentially slaty-blue above and white below, with sharply-patterned head and diagnostic broad reddish-brown chest-band.

Unmistakable; no other west Palearctic *Sitta* has similar discrete chest band. Flight and behaviour resemble Nuthatch, but more active and agile, reaching outermost twigs and inviting comparison to tit.

Habitat. Confined to coniferous woodlands. Breeds in lower middle latitudes of west Palearctic in warm dry Mediterranean and montane regions, ranging in Turkey up to 2500 m but locally down to sea-level.

Distribution and population. Confined to west Palearctic. GREECE. Occurs only on Lesbos. A few tens of pairs. FSU: RUSSIA, GEORGIA. Range and numbers in Caucasus poorly known. Common, locally numerous, in Teberda reserve

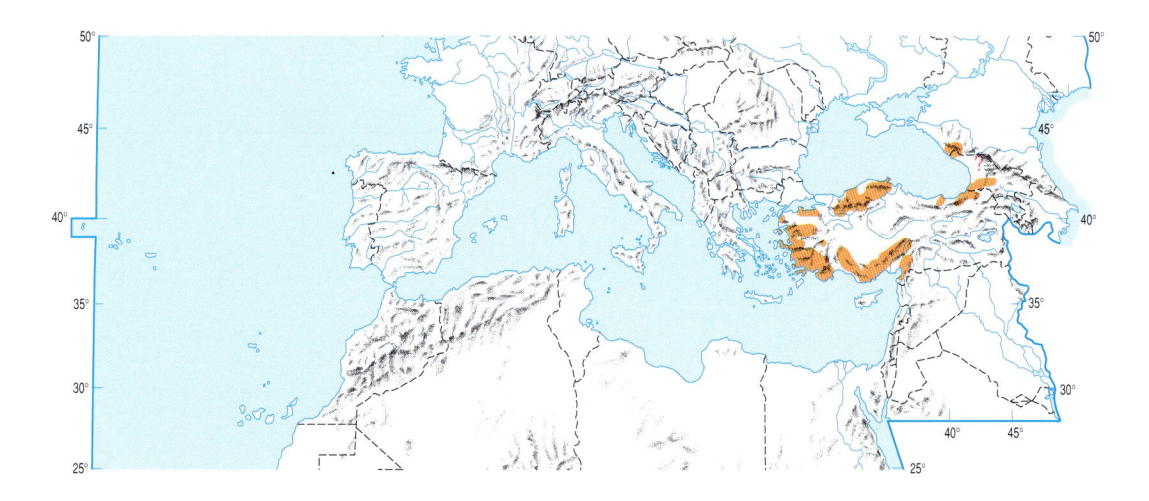

Krüper's Nuthatch *Sitta krueperi*: **1** ad ♂, **2** ad ♀, **3** juv. Corsican Nuthatch *Sitta whiteheadi*: **4** ad ♂, **5** ad ♀, **6** juv. Algerian Nuthatch *Sitta ledanti* (p. 1400): **7** ad ♂, **8** ad ♀, **9** juv. Red-breasted Nuthatch *Sitta canadensis* (p. 1401): **10** ad ♂.

(north-west Caucasus, Russia). TURKEY. Largely confined to areas of pine *Pinus brutia*. 10 000–100 000 pairs.

Movements. Chiefly sedentary; some birds disperse after breeding; some altitudinal movement in winter.

Food. Mainly invertebrates in summer; seeds taken mostly autumn and winter. Forages mainly in crowns of trees, especially on branches and twigs, regularly visiting cones to feed on insects and seeds. Also recorded taking flying prey on the wing.

Social pattern and behaviour. Not well known, the little available information indicating general similarity to other *Sitta* species, especially closely related Corsican Nuthatch.

Voice. Markedly more vocal than Corsican Nuthatch. Song of ♂ a trill usually lasting 2–5 s, sometimes up to 8 s; consisting of repetition of simple but very variable units. Tempo of song may gradually slow down, from rapid trills to slowly repeated, drawn-out units. Main contact call a rich 'doid', reminiscent of Canary or Greenfinch. Other calls include a 'hick', like soft version of call of Great Spotted Woodpecker, and, in excitement or alarm, a harsh 'schrä'.

Breeding. SEASON. Eggs from early April to mid-May, up to 4 weeks earlier in coastal areas than at high altitudes. Probably one brood. SITE. Hole, usually in conifer but can be in deciduous tree if conifers nearby. Excavates own hole or cleans out previous cavity. Nest comprises foundation of fragments of bark, rotten wood, large cone scales, lined with fibrous bark, moss, wool, feathers, hair, and fur. EGGS. Sub-elliptical, smooth; yellowish-white with rust-coloured or purplish speckles evenly distributed but often many at broad end. Clutch: 5–6. INCUBATION. Probably 14–17 days. FLEDGING PERIOD. 16–19 days.

Wing-length: ♂ 72–78, ♀ 69–74 mm.
Weight: ♂ 10–14.3 g.

Corsican Nuthatch *Sitta whiteheadi*

PLATE: page 1399

DU. Corsicaanse Boomklever FR. Sittelle corse GE. Korsenkleiber IT. Picchio muratore corso
RU. Корсиканский поползень SP. Trepador corso SW. Korsikansk nötväcka

Field characters. 12 cm; wing-span 21–22 cm. 10–15% smaller and noticeably slighter than Nuthatch, with proportionately finest bill, longest-looking head (due to linear plumage pattern), and shortest tail of west Palearctic *Sitta*. Noticeably small, rather slight nuthatch. Essentially grey and white, adult ♂ having sharply demarcated black crown and eye-stripe contrasting dramatically with otherwise white face; outer tail-feathers black tipped white. ♀ has dusky-blue crown and eye-stripe.

Confined to Corsica, where no other nuthatch. General

behaviour typical of *Sitta*, but actions on Corsican pine can be like tit, and movement along underside of branches recalls treecreeper. Flight and feeding actions more nimble than Nuthatch.

Habitat. Mountain forests of Corsican pine *Pinus nigra laricio*, mainly at 1000–1500 mm.

Distribution and population. Confined to west Palearctic, where occurs only in mountains of Corsica (France). Population estimated at 2000–3000 pairs. Some local decreases due to forest exploitation or burning (main threats to optimal habitat), but breeding range and population apparently stable overall.

Movements. Chiefly sedentary. Breeding pairs highly territorial throughout year.

Food. In winter, seems to feed mainly on seeds (especially Corsican pine) until May, then on insects and spiders until August. Caches food, almost exclusively seeds of Corsican pine, which are hidden behind bark or laid on thick branches and covered with small pieces of bark or lichen. Food-caching probably crucial for survival, when pine cones are closed in snow of early spring preventing access to seeds.

Social pattern and behaviour. Highly territorial behaviour typical of the family (see family Introduction), territory being maintained throughout year and used for all feeding and breeding activities. Territory size usually 7–10 ha, most activities taking place within core area of 4–6 ha.

Voice. Song of ♂ a trill of short notes delivered at rate of 14–15 per s, 'hididi. . .'; also slower sequence of drawn-out, pure, somewhat ascending whistled notes 'dühdühdüh. . .' or 'düidüidüi. . .'. Contact-call a short whistled note, which may be repeated rapidly as a trill: Alarm or excitement calls include a harsh, repeated 'wäd, wäd', suggesting distant Jay, and other sharp or harsh sounds.

Breeding. SEASON. Eggs laid end of April to 2nd week in May. SITE. Usually large- or medium-sized Corsican pine, typically with only partial or no bark. No plastering of nest-hole. Foundation of pine needles, wood chips, or bark fragments lined with hair, plant fibres, feathers, moss, or lichen. EGGS. Sub-elliptical, smooth, not glossy. Milky-white with pale to dark red speckling concentrated at broad end; occasional light brown and dark violet-grey speckles. Clutch: (5–)6. INCUBATION. Period not known, but hatching mid- to late May. FLEDGING PERIOD. 22–24 days.

Wing-length: ♂ 71–76, ♀ 70–73 mm.
Weight: ♂ ♀ mainly 12–14 g.

Algerian Nuthatch *Sitta ledanti*

PLATE: page 1399

DU. Algerijnse Boomklever FR. Sittelle kabyle GE. Kabylenkleiber IT. Picchio muratore algerino
RU. Алжирский поползень SP. Trepador argelino SW. Kabylnötväcka

Field characters. 12 cm; wing-span 21–22 cm. Close in size to Corsican Nuthatch; 20% smaller than Nuthatch. Plumage pattern close to Corsican Nuthatch and Red-breasted Nuthatch, differing only in less completely black crown of ♂, poorly marked dusky eye-stripe of ♀ and immature, and intermediate pale but warm buff tone on underparts.

Unmistakable. Range does not overlap with any other *Sitta*.

Habitat. High-altitude forest of fir (*Abies numidica*, a relict endemic species) with cedar, aspen, and oak, mainly above 1900 m; also lower forest (350–1121 m) dominated by oaks.

Distribution and population. Confined to west Palearctic, where known so far only from 4 sites in Petite Kabylie region of northern Algeria (not discovered until 1975), all within 40 km: Djebel Babor, Taza National Park (probably extending some distance beyond its boundary to south-west), and Tamentout and Djimla forests. Djebel Babor: population probably underestimated at 12–20 pairs when first discovered; 70 pairs in 1978, 80 pairs in 1982. Taza National Park: *c.* 360 individuals in 1989. No information on other sites.

Movements. Apparently highly sedentary, but little studied outside breeding season.

Food. In summer mainly invertebrates; in winter mostly nuts and seeds. Seeds are hoarded in thick moss cushions growing on trees and retrieved throughout winter.

Social pattern and behaviour. Little known, field studies since discovery in 1975 being concerned primarily with population size, essential ecological requirements, and conservation needs.

Voice. Song of ♂ a repetition of usually 7–12 sonorous, sometimes fluty, sometimes more nasal units, somewhat reminiscent of Wryneck: 'klieu-klieu-. . .' or 'klien-klien-. . .'. Call associated with aggression and territorial defence a toneless unmusical 'sh sh. . .', rapidly repeated.

Breeding. SEASON. Considerable variation caused by prevailing weather at breeding site, where deep snow often on ground in May: in one year, fledging started before 18 June, completed 25 June; next year, 5 broods fledged 6–8 July. SITE. Hole in dead or dying tree, usually in dead branch. Prefers huge old isolated tree. Nesting material includes wood chips, bristles of wild boar, dead leaves, and feathers. EGGS. No information. Clutch: record of 4 young fledging from 1 clutch. INCUBATION. Period not known. FLEDGING PERIOD. Probably 22–25 days.

Wing-length: ♂ 80–83, ♀ 77–81 mm.
Weight: ♂ 18.0, ♀ 16.5 g.

Red-breasted Nuthatch *Sitta canadensis*

PLATE: page 1399

DU. Canadese Boomklever FR. Sittelle à poitrine rousse GE. Kanadakleiber IT. Picchio muratore pettofulvo
RU. Канадский поползень SP. Trepador canadiense SW. Rödbröstad nötväcka

Field characters. 11.5 cm; wing-span 19–21 cm. 15% smaller than Nuthatch, with noticeably fine bill and slight, relatively slim form. Head pattern with black-and-white stripes and without discrete chestnut dappling on under tail-coverts.

Rather similar to Corsican Nuthatch and Algerian Nuthatch, but extremely unlikely to occur within their restricted ranges. Instantly distinguished from Nuthatch by smaller size and head pattern. Flight and general behaviour as other small *Sitta*.

Commonest call a high-pitched nasal 'nyak nyak nyak', like toy tin horn.

Habitat. Breeds in middle and upper middle Nearctic latitudes, mainly in coniferous forest especially of spruce and pine, and towards south of range mainly in mountains. On migration may be seen in beds of beach grass, or climbing over lichen-covered boulders and cliffs, or even inside farm buildings. Climbs freely about rocks and buildings in winter, and occurs in tree-bordered streets, orchards, and gardens.

Distribution. Breeds in southern Canada and USA south to California and Texas in west, and Appalachian mountains in east.

Accidental. Iceland: adult ♂, Vestmannaeyjar, May 1970. Britain: 1st-winter ♂, Norfolk, October 1989 to May 1990.

Movements. Irruptive migrant. Winters throughout most of breeding range, and south in some years to extreme south of USA (except southern Florida and arid regions of south-west) and northern Mexico. Winter distribution and abundance patterns vary considerably from year to year.

Wing-length: ♂ 68–71, ♀ 65–70 mm.

Nuthatch Sitta europaea

Du. Boomklever   Fr. Sittelle torchepot   Ge. Kleiber   It. Picchio muratore
Ru. Поползень   Sp. Trepador azul   Sw. Nötväcka

Field characters. 14 cm; wing-span 22·5–27 cm. Similar in length to largest tits but differs markedly in long and deep bill, long head, and short tail. Medium-sized, perky, and noisy passerine, with compact, spear-headed and short-tailed shape balanced on strong legs and feet apparently set well back on body. Essentially blue-grey above and buff and/or white below, with quite broad black eye-stripe, and orange-brown rear flank and dappling on under tail-coverts. West, central and south-east European race has lower throat buff, and breast and belly orange-buff, merging into flank marked with orange and chestnut; Fenno-Scandian and European Russian race, nominate *europaea* has mainly white underparts.

In most of its range, the only *Sitta* present, and thus unmistakable. In south-east Europe and Turkey, overlaps with Rock Nuthatch (widely) and Eastern Rock Nuthatch (narrowly); distinguished from these by smaller size, more slender bill, darker upperparts, and more buff and chestnut underparts, all south-eastern races having dappled under tail-coverts. Flight fast but fluttering, rarely over long distance. Flight silhouette distinctive, with spear-like head, round wings, and short, square tail; looks broad-ended when wings open, dart-shaped when wings closed. Gait essentially a short, jerky leap; moves thus in all directions on trunks and branches, without using tail as support. Behaviour excitable, with loud calls and jerky movements attracting attention.

Habitat. Breeds in west Palearctic across and marginally above middle continental latitudes, from warm temperate to cooler boreal climates within July isotherms 16–27°C. Mainly occupies lowlands except towards south of range, ascending in Switzerland to treeline, and to 1800 m in Caucasus. In North Africa, found in Moyen Atlas only at c. 1750–1850 m in cedar and mixed forest. In European Russia, breeds mainly in broad-leaved and mixed forest except in north, where nests readily in pure coniferous stands. In west European lowlands, however, closely attached to mature and well-grown or over-mature

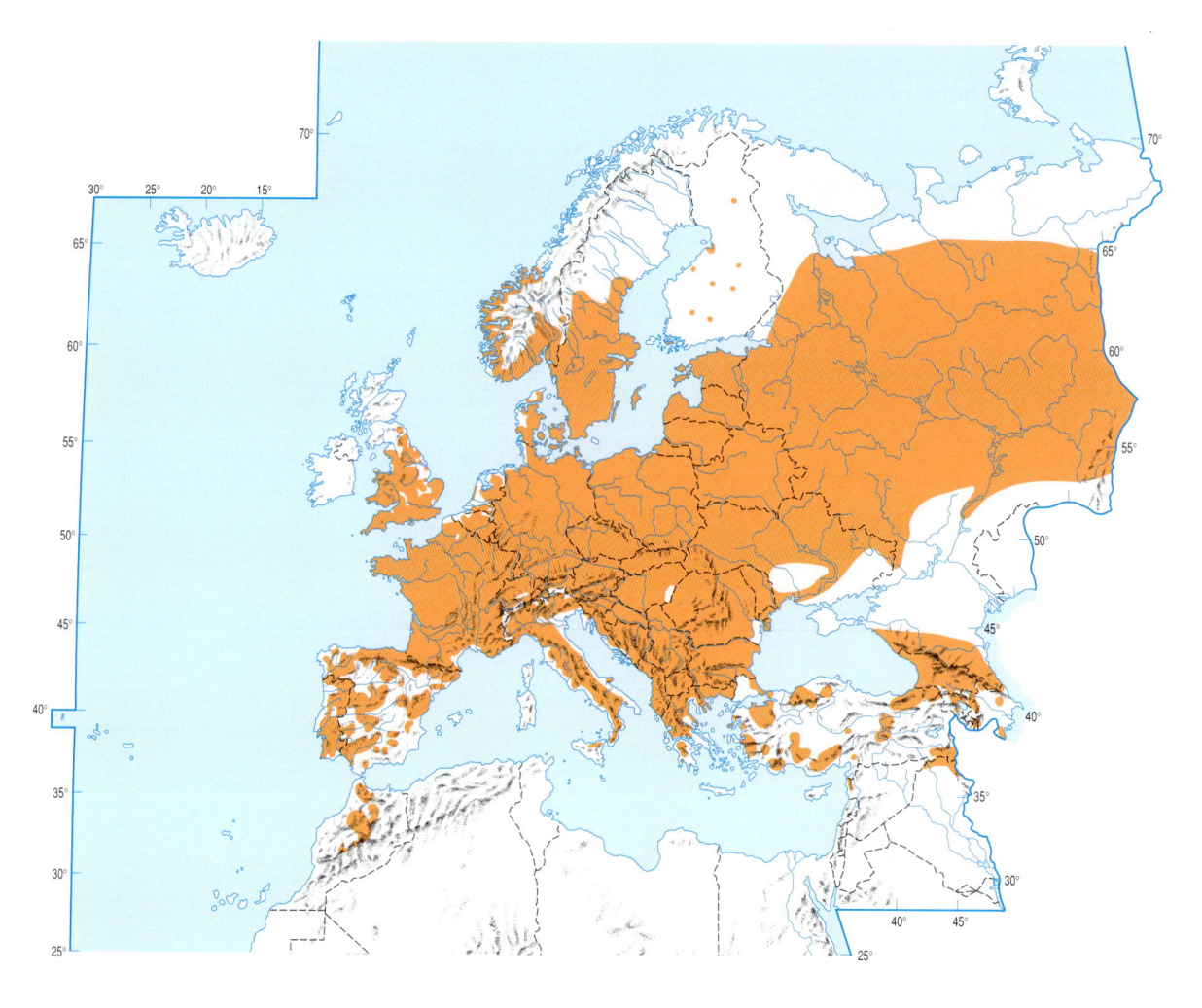

Nuthatch *Sitta europaea*: *S. e. caesia*: **1–2** ad ♂, **3** juv ♂. *S. e. europaea*: **4** ad ♂. *S. e. asiatica*: **5** ad ♂. *S. e. cisalpina*: **6** ad ♂. *S. e. levantina*: **7** ad ♂. *S. e. rubiginosa* (south-east Transcaucasia to Turkmenistan): **8** ad ♂.

and even decaying deciduous trees with large trunks and of at least medium height, carrying well spread crowns with many branches and clusters of twigs, which are used for foraging no less than main trunks and boughs. Old open parkland also favoured, provided trees not too widely spaced (as birds normally make only short flights). In Britain and some other parts of range, prefers oak, but also favours beech and sweet chestnut. Readily tolerates human neighbourhoods and freely visits bird-tables in winter.

Distribution. Has spread in Britain, Belgium, Netherlands, Denmark, Norway, and Morocco. BRITAIN. Has spread throughout Wales and over much of northern England; first recorded breeding Scotland 1989. BELGIUM. New regions colonized since 1980. NETHERLANDS. Colonized north since c. 1970 and more recent spread in south, reaching long-available suitable habitat. NORWAY. Has spread to inner parts of south since c. 1970. SWEDEN. *S. e. asiatica* breeds occasionally in north coastal area following autumn irruptions. FINLAND. *S. e. asiatica* breeds occasionally, following irruptions; in several localities (mostly central Finland) in 1977, following 1976 irruption. TURKEY. Very local and restricted to a few lower valleys in east and south-east. SYRIA. First breeding: Allovit mountains above Slenfe in north-west April 1992 (carrying nest material), 1993. MOROCCO. Extension of previously known range south to Haut Atlas.

Accidental. Channel Islands, Finland (nominate *europaea*).

Beyond west Palearctic, breeds widely in northern Asia east to Sakhalin island and Japan, continuing through north-east China south and south-west to south-east Asian mainland and Indian sub-continent; also breeds Iran.

Population. Apparently stable over much of range, but increase Britain, Belgium, Netherlands (probably), and Sweden, slight decrease Lithuania, fluctuating Denmark. BRITAIN. 130 000 territories 1988–91. Long-term gradual increase. FRANCE. 100 000–1 million pairs in 1970s. BELGIUM. 23 000 pairs 1973–7. No clear trend 1950–80, then increase: 17 000–32 000 pairs 1989–91. LUXEMBOURG. 18 000–20 000 pairs. NETHERLANDS. 10 000–17 500 pairs 1979–85. GERMANY. 1–1.3 million pairs. In east, where stable apart from losses attributed to hard winters, 190 000 ± 89 000 pairs in early 1980s. DENMARK. 3000–44 000 pairs 1987–8. NORWAY. 10 000–50 000 pairs 1970–90. SWEDEN. 100 000–250 000 pairs in late 1980s. FINLAND. Perhaps 10–30 pairs of *asiatica* after irruptions. ESTONIA. 20 000–50 000 pairs in 1991. LATVIA. 40 000–100 000 pairs in 1980s. LITHUANIA. Common. POLAND. Fairly numerous. CZECH REPUBLIC. 600 000–1.2 million pairs 1985–9. SLOVAKIA. 700 000–1 million pairs 1973–94. HUNGARY. 150 000–200 000 pairs 1979–93. AUSTRIA. 300 000–500 000 pairs in 1992. SWITZERLAND. 70 000–120 000 pairs 1985–93. SPAIN. 550 000–1.2 million pairs. PORTUGAL. 10 000–100 000 pairs 1978–84. ITALY. 50 000–200 000 pairs 1983–95. GREECE. 10 000–30 000 pairs. ALBANIA. 2000–5000 pairs in 1981. YUGOSLAVIA: CROATIA. 200 000–300 000 pairs. SLOVENIA. 60 000 pairs. BULGARIA. 500 000–1 million pairs. RUMANIA. 400 000–500 000 pairs 1986–92. RUSSIA. 500 000–5 million pairs. BELARUS'. 280 000–320 000 pairs in 1990. UKRAINE. 65 000–70 000 pairs in 1986. MOLDOVA. 30 000–50 000 pairs in 1988. AZERBAIJAN. Common. TURKEY. 10 000–100 000 pairs. MOROCCO. Not uncommon in mountains of north, but scarcer in Haut Atlas.

Movements. Chiefly sedentary; limited dispersal, perhaps only of 1st-year birds. In some years (e.g. 1976, 1995), when populations are high, autumn dispersals can take on the characteristics of eruptions and invasions, with movements of thousands of birds in all directions from WNW through S to SE; ringed birds recovered at distances of up to 460 km.

Food. Invertebrates and (mainly in autumn and winter) seeds. Most food obtained on trees, from both trunk and branches; feeds also on ground, especially spring and autumn. Will take winged prey in the air. Seeds and large and hard insects are wedged into cracks or crevices and smashed. Caches seeds, especially in autumn.

Social pattern and behaviour. Most notable aspect of social behaviour is markedly *un*social behaviour. Even newly fledged young keep separate (in contrast to other *Sitta*); threaten one another if they come close. Also aggressive to other species at feeding places. Territorial, mainly in pairs, throughout non-breeding season, with same pattern of territories maintained as during breeding season. Foraging confined to territory, which contains stored food; hence maintenance of territory important for winter survival. Territories sometimes, mainly in autumn, contain third bird, probably young bird unable to establish own territory; this bird just tolerated, frequently attacked, by owners. Monogamous mating system. ♂ sings to advertise occupation of territory, mainly December–May. No special song-posts used. ♂ sings in upright stance, apparently necessary for full volume of song; usually perches across branch. Courtship behaviour includes floating flight, with quivering wing-beats, from one tree to another; performed by both sexes; ♂ also performs circling flight, with head and bill raised and tail spread. Courtship-feeding occurs from just before egg-laying to end of incubation.

Voice. Freely used throughout year, associated with year-round maintenance of territory; many calls loud and far-carrying. Song of ♂ a uniform series of loud notes of simple structure; very variable in tempo, from drawn-out whistling notes slowly repeated to very short notes rapidly repeated; the two extremes, whistled song and trilled song, very distinct but linked by a range of song-types of intermediate tempo. Excitement-call, given by both sexes, a short, metallic 'twit'

or 'chwit', quite loud and of very distinctive quality, aptly described as like sound of pebble hitting ice on the slant; commonest, best-known call, given all year. Tempo variable, depending on degree of excitement. Contact-call a high-pitched 'sit', resembling equivalent calls of tits but louder. Other calls include toneless or harsh sounds, given especially during aggression or fighting.

Breeding. SEASON. Northern and western Europe: eggs laid early or mid-April to late May, occasionally end of March in western Switzerland. Generally single-brooded but very rare 2nd broods recorded if season early. SITE. Hole, natural or old hole of woodpecker most often of Great Spotted Woodpecker and Lesser Spotted Woodpecker; also in nest-box, rarely in wall cavity or on rock face. Entrance hole narrowed by plastering with mud if necessary; prefers old holes where tree growth has created short entrance tunnel. Nest: loose layered heap of flakes of bark, generally coniferous, especially Scots pine, rarely of deciduous leaf litter; this on foundation of wood chippings, fragments of soft and rotten wood, or rarely leaf litter; foundation large in larger cavities but small or lacking in nest-boxes of circular cross-section. Cavity often heavily plastered inside to protect against rain and wind. EGGS. Sub-elliptical, smooth and slightly glossy. White, sparsely speckled and blotched with bright red, reddish-brown, or reddish-purple, usually concentrated at broad end. Clutch: 6–11 (5–13). INCUBATION. 13–18 days. FLEDGING PERIOD. 23–24(–26) days.

Wing-length: Central Europe: ♂ 84–90, ♀ 81–89 mm. Slightly larger in northern Europe, smaller in Britain and southern Europe.
Weight: ♂♀ mostly 21–26 g.

Geographical variation. 7 races recognized in west Palearctic. Main division is between nominate *europaea*, with mainly white underparts (see Field characters), and *S. e. caesia*, with buff underparts. Nominate *europaea* occurs from Fenno-Scandia eastwards; *caesia* occupies most of Europe to south, with zone of intergradation from Denmark and eastern Poland eastwards. In southern areas from Iberia and Morocco east to Levant, populations generally smaller and more deeply coloured below than *caesia* (*hispaniensis*, Iberia south of *caesia*, Morocco; *cisalpina*, south of Alps, Sicily, Yugoslavian coast; *levantina*, Turkey and Levant; *caucasica*, Caucasus area). *S. e. asiatica*, a very small race with plumage similar to nominate *europaea*, occurs in extreme east (Urals area).

Eastern Rock Nuthatch *Sitta tephronota*

PLATE: page 1405

Du. Grote Rotsklever   Fr. Sittelle des rochers   Ge. Klippenkleiber   It. Picchio muratore di roccia orientale
Ru. Большой скалистый поползень   Sp. Trepador armenio   Sw. Östlig klippnötväcka

Field characters. 15.5 cm; wing-span 25–30 cm. Up to 20% larger than sympatric races of Rock Nuthatch, with even larger head and bill and showing more of its neck. Largest *Sitta* in west Palearctic, with seemingly huge head and bill emphasized by often upright stance. Plumage has basic pattern of Nuthatch

but differs in paler grey upperparts, longer and (at rear) broader eye-stripe, almost uniformly grey tail, and lack of discrete chestnut marks on flank and under tail-coverts.

Easily separated from Nuthatch by habitat, much larger size (especially of head and bill), and uniformly-coloured tail and

Eastern Rock Nuthatch *Sitta tephronota*: **1** ad, **2** juv. Rock Nuthatch *Sitta neumayer* (p. 1406). *S. n. neumayer*: **3–4** ad, **5** juv. *S. n. syriaca*: **6** ad. *S. n. rupicola*: **7** ad. *S. n. tschitscherini* (Iran): **8** ad.

under tail-coverts. Difficult to distinguish from Rock Nuthatch, but at close range shows rather longer, stronger bill, even bolder eye-stripe cum neck-blaze, slightly paler upperparts, brighter orange flank and vent, and (sometimes) paler legs. Flight fast and confident in spite of seemingly ridiculous, tail-heavy form which recalls Kingfisher; makes rapid ascents as well as free manoeuvres on cliff faces; action nevertheless not as speedy as Rock Nuthatch. Gait a powerful, jerky hop or leap, often developed into an apparent lope over flat surfaces. Perches freely on trees and bushes, unlike Rock Nuthatch.

Habitat. Breeds in dry montane lower middle latitudes of Asia, in Caucasus up to *c.* 2600 m on steep cliffs and in ravines. In Afghanistan, occurs at *c.* 1070–3000 m, inhabiting boulder-strewn mountain slopes as well as rock walls in canyons; less strictly rock-living than Rock Nuthatch. After breeding season in Caucasus, descends to mountain forests and orchards, then occurring also in trees, which it readily climbs.

Distribution and population. ARMENIA. Scarce. AZERBAIJAN. Status unclear. Probably occurs in extreme south: mountains of Zangelan and Kubatly districts, also Nakhichevan region (i.e. close to Megri district of Armenia where known to occur). TURKEY. First recorded 1881, but rediscovered only in 1972. Several areas where birds not specifically identified (possible confusion with Rock Nuthatch). Population 500–5000 pairs.

Beyond west Palearctic, breeds from Iran east to south-central Asian mountains and western Pakistan.

Movements. Chiefly sedentary. Some post-breeding and seasonal altitudinal movement.

Food. In summer, mainly insects and snails; from autumn to early spring, mainly seeds. Snails cached earlier in year are said to be important winter food.

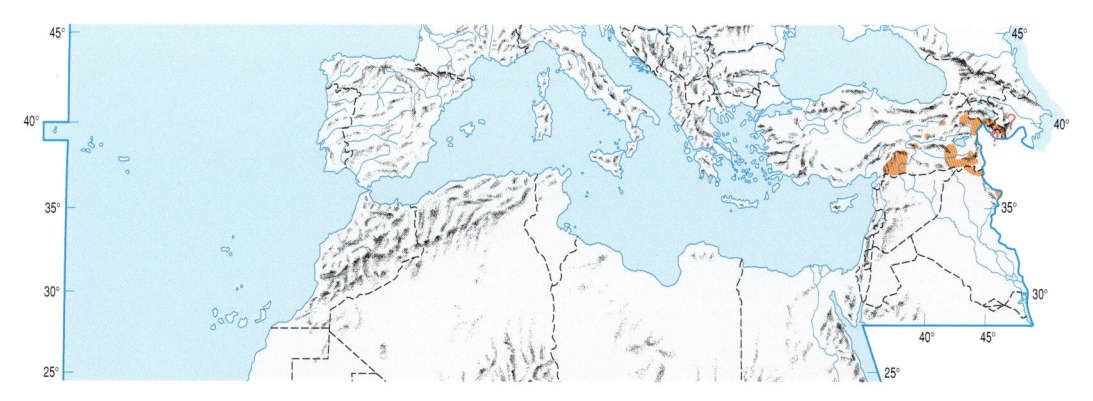

Social pattern and behaviour. Social organization outside breeding season not well known, but many pairs (perhaps all established pairs) evidently remain together in territory throughout year. Less commonly occurs solitarily or in parties of 3–10. No evidence of other than monogamous mating system, with long-term pair-bond and re-use of nest over several years. ♂ sings from rock, cliff-top, etc., usually near nest. Song given in March (from start of breeding) and continues through incubation.

Voice. Similar to Rock Nuthatch but generally deeper and slower in tempo. Freely used throughout year, perhaps less in autumn and winter. Main calls loud and penetrating. Song of ♂ loud, sonorous, and far-carrying; a trill or succession of less rapidly repeated units, very variable in unit structure, tempo, and duration. ♀ also sings a more stereotyped song, antiphonally with ♂. Other calls include a repeated 'pit pit...', used for contact between pair members, and harsh calls given in antagonistic encounters.

Breeding. Most information extralimital. SEASON. Transcaucasia; eggs laid early April. Two broods not uncommon. SITE. Wide crack, cavity, or hole in rock, tree, steep river bank, or building; hole may be excavated or pre-existing. Nest: cavity entrance heavily plastered over with mud forming wall up to 80 cm square. Less often a projecting entrance tunnel than in Rock Nuthatch and, if present, shorter. Cavity built of mud, animal excrement, and (if in tree) pistachio resin, mixed with hair, feathers, cloth, beetle wing-cases, small bones, fragments of vegetation, fibres, etc.; feathers may be cemented to outside wall and also placed in crevices 1 m from nest. Nest-cup in depression of dried mud, in one case lined with plant fibre, mouse hair, wool, and string. EGGS. Short sub-elliptical, glossy. Milky-white with sparse yellowish-brown to black specks and irregular blotches, sometimes concentrated at broad end; rarely purplish-grey markings, sometimes none at all. Clutch: 5–8 (4–9). INCUBATION. 12–14 days. FLEDGING PERIOD. 24–26 days.

Wing-length: ♂ 92–98, ♀ 88–95 mm.
Weight: ♂ 46–55, ♀ 43–45 g.

Rock Nuthatch *Sitta neumayer*

PLATE: page 1405

DU. Rotsklever   FR. Sittelle de Neumayer   GE. Felsenkleiber   IT. Picchio muratore di roccia
RU. Малый скалистый поползень   SP. Trepador rupestre   SW. Klippnötväcka

Field characters. 13.5–14.5 cm; wing-span 23–25 cm. European birds overlap in size with larger Nuthatches and smaller Eastern Rock Nuthatches, but birds from Asia Minor eastwards are smaller and more slender-billed so that in area of overlap with Eastern Rock Nuthatch in Turkey and Iraq, difference in bulk quite marked. 2nd largest nuthatch in west Palearctic with upright stance enhancing size, particularly in comparison with Nuthatch. Plumage essentially blue-grey above and dirty-white below, with striking black eye-stripe, pale buff rear underbody, and dark grey legs.

Loud voice, paler plumage, and (in west of range) larger size allow instant distinction from Nuthatch but not from Eastern Rock Nuthatch. At close range, where sympatric with Eastern Rock Nuthatch, distinguished by smaller size (especially of bill and head), much less obvious eye-stripe, slightly bluer upperparts, and duller underparts.

Voice loud and ringing compared with Nuthatch, but less loud and deep than Eastern Rock Nuthatch, with more rapidly uttered phrases.

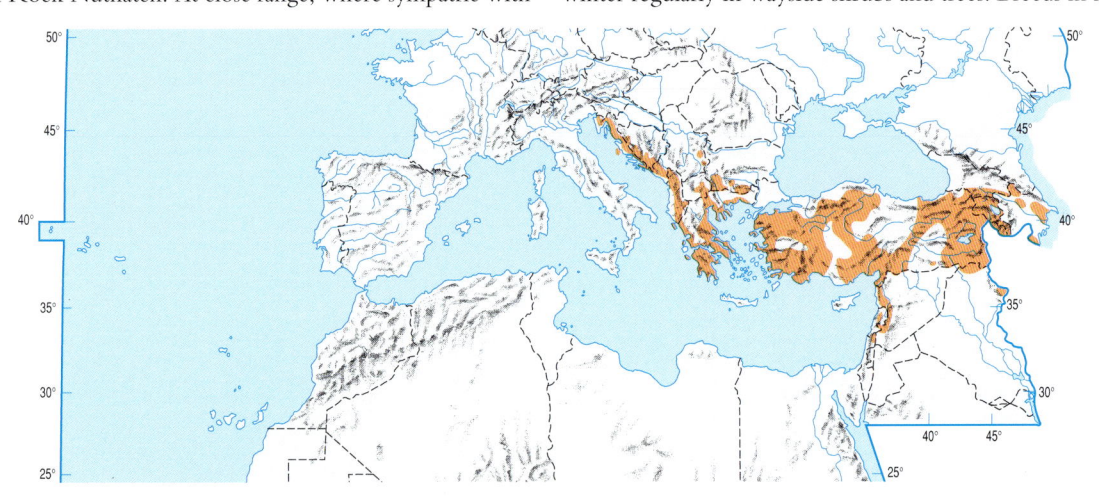

Habitat. Breeds in and near west Palearctic in warm Mediterranean lower middle latitudes of dry continental climate, within July isotherms *c.* 24–32°C. Occupies sunny, generally calcareous rocky slopes and steep faces with at most some dry shrub vegetation. Pronounced ground and rock bird, but in winter regularly in wayside shrubs and trees. Breeds in suitable isolated rocks at low levels but mainly on uplands and mountains to 3300 m; rare above 2650 m.

Distribution. BULGARIA. Recent extension of breeding range. TURKEY. Scarce and local in south-east.

Beyond west Palearctic, breeds only in Iran.

Population. GREECE. 10 000–30 000 pairs; stable. ALBANIA. 2000–5000 pairs in 1981. YUGOSLAVIA: CROATIA. 3000–5000 pairs; stable. BULGARIA. 100–500 pairs; slight increase. ARMENIA. Very common around Yerevan in 1960s. AZERBAIJAN. Uncommon to locally common. TURKEY. 50 000–500 000 pairs. ISRAEL. Mt Hermon: 18–50 pairs 1974–90.

Movements. Chiefly sedentary, with limited post-breeding dispersal and some seasonal altitudinal movement.

Food. In summer, mainly insects; in autumn and winter, mostly seeds, but snails also important. Prey taken mainly on ground, especially in crevices in rocks, etc., sometimes from branches and trunks of small bushes or occasionally large trees; also reported to take insects in the air.

Social pattern and behaviour. Social organization outside breeding season not well known, but established pairs remain together and maintain territories all year, as indicated by fact that pairs regularly use same nest year after year, refurbishing it not only in spring but also in autumn, and roosting together in it in winter. No evidence of other than monogamous mating system. Pair-formation probably takes place in autumn. ♂ sings from prominent rock or top of bush or small tree; adopts upright posture similar to Nuthatch. Pair-members also engage in antiphonal duetting, in both spring and autumn. Feeding of ♀ by ♂ regular during incubation.

Voice. Very freely used, slightest excitement being accompanied by loud, mainly trilling, calls which resemble song. Voice in many respects similar to closely related Eastern Rock Nuthatch. Song of ♂ loud, audible for several hundred metres; trill or succession of less rapidly repeated units; very variable in unit structure, tempo, and duration. Long songs may last up to 8 s, and towards end these typically fall in frequency and intervals between units become longer. Contact-calls variable, 'tsik', 'chik', or 'pit'; given singly or run together to form trill. Other calls include harsh or screaming calls, given during aggressive encounters.

Breeding. SEASON. Greece: eggs laid late April to mid-May, exceptionally late March. Armenia: eggs laid from end of March. Israel: nesting from March at low altitude, April at 2000 m. 2 broods said to occur. SITE. In slight concavity in rock face, usually sheltered from rain, or under overhang; also on buildings, even over doorway of inhabited house. Nest: entire flask-shaped structure of mud, animal excrement, hair, plant resin (etc.), built against rock; tunnel-like entrance up to 10 cm long, almost always projecting out from rock rather than to side. Mud often mixed with beetle wing-cases, berries, feathers of other birds, raptor pellets (etc.), with crushed insects and berries apparently important as binding material. Feathers, wool, tin-foil (etc.) also stuck onto rock and wedged into crevices surrounding nest-site, possibly as store of lining material. Nest-cup depression in dried mud mixture, lined with tightly-packed hair, wool, feathers, broken-up raptor pellets, grass, cloth, etc. EGGS. Sub-elliptical, smooth and glossy. Milky-white with speckles and small irregular blotches of yellowish or reddish-brown, light red, and reddish-purple; sparse, but may be concentrated at broad end, and some completely unmarked. Clutch: 8–10 (6–13). INCUBATION. 15–18 days. FLEDGING PERIOD. 23–25 days reported; probably usually a little longer.

Wing-length: ♂ 75–85, ♀ 75–83 mm.
Weight: ♂♀ mostly 24–31 g.

Geographical variation. 4 races recognized in west Palearctic; differ slightly in size, rather slightly in colour: nominate *neumayer* (south-east Europe), *zarudnyi* (western Asia Minor), *syriaca* (Levant), *rupicola* (eastern Asia Minor, Transcaucasia, northern Iraq).

Wallcreepers Family Tichodromadidae

Small oscine passerines (suborder Passeres). One species only: Wallcreeper, a long-billed, rock-climbing, insectivorous bird inhabiting high mountains in Eurasia.

Sexes similar in size. Tongue long and thin, divided at tip. Apart from bill, resembles Sittidae more than Certhiidae in most characters, including behaviour. Tail not used as prop when climbing rock-faces. Foot not used to hold larger food items; food-storing not recorded. Gait usually a hop on horizontal or angled rock, and will hop to adjacent rock when climbing; sometimes walks over short distance and will run.

Wallcreeper *Tichodroma muraria*

PLATE: page 1409

Du. Rotskruiper   Fr. Tichodrome échelette   Ge. Mauerläufer   It. Picchio muraiolo
Ru. Стенолаз   Sp. Treparriscos   Sw. Murkrypare

Field characters. 16.5 cm; wing-span 27–32 cm. Head and body close in size to Nuthatch, but longer decurved bill, longer, butterfly-like wings, and fuller tail all make bird larger and more loosely configured overall (especially in flight). Medium-sized, colourful, rock-creeping passerine, with long bill and large feet, and general character combining those of nuthatch and treecreeper. Plumage mix of dusky grey, black, and white, with large 'flashing' carmine wing-patch, unique in west Palearctic. Breeding ♂ has black lower face and fore-underparts. Marked seasonal variation in ♂ in non-breeding plumage, resembling ♀.

Extremely vivid and elegant plumage instantly diagnostic. Most actions on ground and voice suggest affinity to nuthatch rather than treecreeper (in spite of bill form). Flight recalls butterfly and Hoopoe, having erratic, flitting, and skipping action, seemingly desultory but actually quite powerful (allowing quite rapid ascents and manoeuvres around rock faces). Gait mainly a short, jerky hop, recalling that of smaller nuthatches. Wings constantly flicked open to display red areas more fully.

Habitat. Breeds in west Palearctic in montane regions of lower middle latitudes, in rocky, broken, or precipitous terrain, subject to varied weather stresses but normally cool or chilly and often dull and moist. In Switzerland, breeds at 350–2700 m (normally 1000–2000 m), using bare moist sites with overhanging rock faces, either in the open or in gorges; limestone preferred to granite, and presence of nest-cavities a limiting factor. Often near torrents, tunnels, caves, moraines, or scree. Moves down to valleys and plains in winter, occupying not only rock faces and quarries but castles, bridges, ramparts, churches, and other buildings, even within large cities.

Distribution. FRANCE. Breeding (perhaps overlooked earlier) first proved Jura 1973, Corsica 1978, Massif Central 1986. GERMANY. Bred Schwäbische Alb 1989. POLAND. Formerly bred Sudety mountains. GREECE. Bred Pelopónnisos 1989. RUMANIA. Range contracted since *c.* 1920; now probably more restricted than shown on map. UKRAINE. Status uncertain. No breeding-season records in last 100 years. AZERBAIJAN. Poorly known. TURKEY. Records near Antakya and Birecik in May and July perhaps late migrants and early post–breeding dispersal, respectively. Occurs in winter, but no proof of breeding, Syria, Lebanon, Israel, northern Iraq.

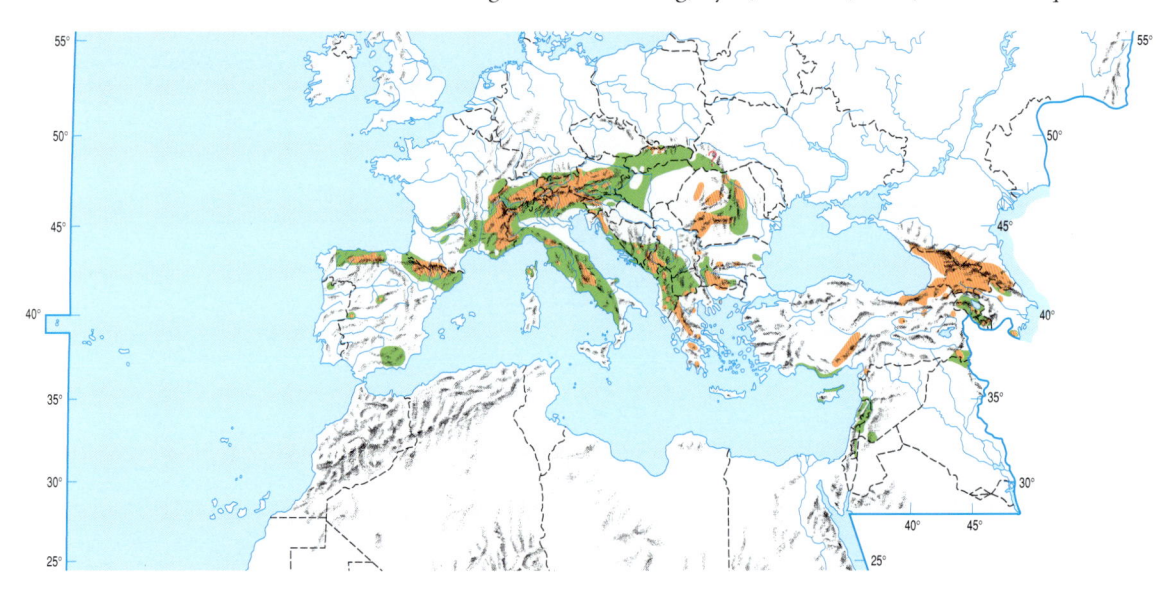

Wallcreeper *Tichodroma muraria*: **1–2** ad ♂ breeding, **3** ad non-breeding, **4** juv. Penduline Tit *Remiz pendulinus* (p. 1416). *R. p. pendulinus*: **5** ad ♂ fresh, **6** ad ♀, **7** juv. *R. p. macronyx*: **9** ad ♂.

Accidental. Britain, Channel Islands, Belgium, Luxembourg, Netherlands, Balearic Islands, Malta, Jordan, Algeria, Morocco.

Beyond west Palearctic, breeds northern Iran east through central Asian mountains to central China, north to Mongolia.

Population. Density low over much of range owing to typically large territory and strict habitat requirements. Apparently generally stable. FRANCE. 100–1000 pairs in 1970s. GERMANY. 75 pairs in mid-1980s. POLAND. 10–20 pairs. SLOVAKIA. 30–50 pairs 1973–94 (probably some overlap with Poland in Tatra mountains). AUSTRIA. 400–600 pairs in 1992. No trends discernible. SWITZERLAND. 500–800 pairs 1985–93. SPAIN. 9000–12 000 pairs. ITALY. 2000–6000 pairs 1983–95. GREECE. 100–500 pairs. ALBANIA. 10–100 pairs in 1981. YUGOSLAVIA: CROATIA. 2–5 pairs. SLOVENIA. 50–100 pairs. BULGARIA. 100–500 pairs. RUMANIA. 300–400 pairs 1986–92. AZERBAIJAN. Rare. TURKEY. 500–5000 pairs.

Movements. Altitudinal and short-distance migrant; some birds perhaps sedentary; others make longer movements which show regular pattern, though sporadic records furthest from breeding areas are probably attributable to drift; part of population remains within breeding range. Timing of autumn movement is regular, and apparently independent of weather conditions; basically similar throughout range, birds occurring outside breeding areas mostly October–April.

Food. Small insects and spiders. Prey obtained mainly on rock faces, also (especially in rainy weather) among pebbles on banks of streams and below cliffs, occasionally from trees. While foraging, frequently investigates crevices, often disappearing between rocks. Takes prey from small cracks in rock itself as well as from small grassy patches on it; also by flycatching; Diptera often picked off substrate while bird hovers.

Social pattern and behaviour. Normally solitary in winter, birds of both sexes defending individual feeding territories. Almost all birds descend to lower altitudes, though some (probably juveniles) move later or remain high up. Territories usually established by last third of October or mid-November. Some embrace single rock massif, others fragmented, taking in (e.g.) several quarries or parts of buildings. Exceptionally in quite large flocks outside breeding season. No evidence for other than monogamous mating system. Characteristic wing-flicking is slow movement performed constantly when alone or 2 together, feeding or resting briefly. Young wing-flick (slower than adults) as soon as they leave nest. White and red markings on large wings apparently have important signal effect, announcing presence and facilitating contact with conspecific birds, to some extent serving as substitute for contact-calls. ♂ sings probably mainly to attract ♀, also for territory advertisement and when disturbed at nest; in winter mainly to advertise territory. Sings mostly while perched or climbing. In early stages of breeding, ♂ performs conspicuous nest-showing display; makes circling flights starting and ending at nest-hole; flies on erratic course with sharp turns and steep descents and ascents.

Voice. Full song of ♂ an ascending, crescendo, and rallentando series of clear, musical, piping whistles (usually 4–5) followed by lower-pitched and exceptionally quiet tone lasting up to *c.* 1 s; last low-pitched note characteristic of highest-intensity song. Song of ♀ has same interval structure as ♂'s song, with

5 ascending tones and 1 final very quiet tone. Often sings incomplete songs (variation in intensity dependent on, e.g., season): may give only first 3–4 notes and more quietly than usual. Other calls include a short chirruping 'tui' or 'touiht', used for pair-contact; and varied 'rolling', relatively impure sounds, used as threat-calls.

Breeding. Season. Austria: at 2000 m eggs laid 1st half of June; at this altitude temperature rather constant so fledging over 10 years always in last third of July. Switzerland: at 1050–1100 m over 14 years fledging 5 July to 8 August, probably weather dependent; at 380 m, eggs laid end of April to beginning of May, fledging at end of June. One brood. Site. In cleft in rock face, behind rock, or in heap of boulders or scree; site must be inaccessible to mammalian predators, particularly stoat and marten. Height can range from base of rock wall to well over 100 m; rock faces with clumps of vegetation preferred to bare walls. Often in narrow gorge above mountain torrent, and wetness a common feature of nest sites. Nest: commonly 2 entrances to chamber; often one used as entrance and other as exit, or one by ♀ and other by ♂. Foundation of moss mixed with lichen, needles of pine, grass and roots, lined with hair, feathers, wool, rootlets, etc. Eggs. Sub-elliptical, slightly pyriform, smooth and moderately glossy. White, with sparse, fine, deep red to black speckling and occasional larger blotches, often concentrated at broad end. Clutch: (3–)4–5. Incubation. 18.5–20 days. Fledging Period. 29 days.

Wing-length: ♂ 94–104, ♀ 94–101 mm.
Weight: ♂♀ averages between 17.3 and 19.3 g; range (different sample) 15.0–19.6 g.

Treecreepers Family Certhiidae

Small oscine passerines (suborder Passeres). Found mostly in woodland, including parks and large gardens. Highly arboreal, feeding mainly on insects obtained on bark of trees. 5 species in Eurasia, with centre of distribution in central Asia. 2 largely sedentary species breeding in west Palearctic—one (Treecreeper) with extensive range in Eurasia, other (Short-toed Treecreeper) confined to west Palearctic.

Sexes similar. Bill rather long, sharp-pointed, and fine; laterally compressed and decurved, used as probe to take food from fissures in bark. Wing rather short and rounded; 10 primaries, p10 reduced (40% of length of longest). Flight tit-like, bird usually moving only short distances from tree to base of next tree. Tail quite long and graduated; well adapted to tree-climbing with 12 stiff feathers, shafts of which project at tips. Foot short, with rather long toes and strong claws (that of hind toe particularly long). Though can cling tit-like under foliage, not nearly so acrobatic as tits (Paridae) when feeding but adept at climbing trunks and branches of trees, using tail as prop like woodpeckers and with feet wide apart, typically starting at base of trunk and hopping upwards. Like nuthatches, able to move upside-down under branches but do not typically descend head first. Food-storing not recorded.

Plumage thick, long, and soft. Upperparts streaked and spotted cryptically with various shades of brown; underparts paler—whitish and/or brownish. Juvenile similar to adult or somewhat more spotted and barred.

Treecreeper *Certhia familiaris*

PLATE: page 1412

Du. Taigaboomkruiper   Fr. Grimpereau des bois   Ge. Waldbaumläufer   It. Rampichino alpestre
Ru. Пищуха   Sp. Agateador norteño   Sw. Trädkrypare

Field characters. 12.5 cm; wing-span 17.5–21 cm. Head and body close in size to Wren, but noticeably larger overall due to longer, distinctly decurved bill, longer, more paddle-shaped wings, and much longer, stiff, graduated and 'frayed' tail. Small, mouse-like, tree-trunk- and branch-climbing passerine, with rather long, decurved bill. Essentially brown above and white below, but at close range shows long white supercilium (from bill to nape), dull white mottled or lined back, white-barred and buff-banded wings, and rufous rump.

Except where geographically isolated from Short-toed Treecreeper (as in most of Britain, Ireland, Denmark, Fenno-Scandia, and Russia) distinction between the species varies from fairly easy (with calling bird) to impossible (with distant, silent bird) and can only be safely completed after exhaustive check of not just plumage but also structure, voice, habitat, and behaviour. Hence caution applied to all records of vagrant treecreepers. Of north-western populations of Treecreeper, north and central European race, nominate *familiaris*, has 'ghostly' appearance and (importantly) relatively short gently decurved bill not displayed by any race of Short-toed Treecreeper which, in case of northward- or eastward-straying vagrant, should look dingy and poorly marked by comparison and have relatively long, rather bent-tipped bill.

Flight extremely distinctive, with bursts of fluttered, rather butterfly-like wing-beats alternated with side-slips and tumbles on spread or closed wings just before landing on tree; produces both undulating and erratic progress, again giving rather ghostly impression over short distance and looking weak (as if buffeted by wind) in longer flight. Behaviour on tree or ground essentially mouse-like when creeping, with short hopping gait on widely splayed feet, aided by support from tail.

Habitat. Strictly arboreal, in west Palearctic occurring mainly in middle continental latitudes, but also in upper middle latitudes on oceanic fringe, and marginally into arctic coastlands, between July isotherms of 14–16°C and 23–24°C. Virtually confined to trees, of practically any species, having more or less perpendicular trunks, not spaced too far apart, and covered in bark providing crevices to contain invertebrate food organisms. Will also sometimes explore branches or twigs in crown, and occasionally walls, outsides of buildings, and bare or grassy ground. Where it overlaps with Short-toed Treecreeper, tends to be confined to higher ground and often to conifers. Where no such overlap (as in Britain), predominantly found in mature broad-leaved trees, either in woodland, parks, large gardens (even within cities), and hedgerow or other farmland timber. Managed woodlands generally unsuitable if felled before maturity and thus having few trees with the loose bark preferred for nesting and roosting (ornamentally planted *Sequoiadendron giganteum* has proved most attractive in this respect).

Distribution. Britain. First bred Outer Hebrides 1962; c. 10 pairs by 1966, only 1–2 pairs by 1983. Ireland. Significant decline since early 1970s. Netherlands. Breeding long suspected, but proved only in 1993–4 in extreme south-east. Denmark. Slight increase. Norway. Has spread north since c. 1970. Rumania. Probably more widespread than map suggests. Ukraine. Slight decrease. Turkey. Most records outside known range probably misidentified Short-toed Treecreeper.

Accidental. Faeroes, Channel Islands, Balearic Islands.

Beyond west Palearctic, extends east through southern Siberia and north-central Asia to Sakhalin island, Japan, and

Treecreeper *Certhia familiaris*. *C. f. familiaris*: **1–2** ad, **3** juv. *C. f. macrodactyla*: **4** ad. *C. f. britannica*: **5** ad. *C. f. persica*: **6** ad. Short-toed Treecreeper *Certhia brachydactyla* (p. 1414). *C. b. megarhyncha*: **7** ad, **8** juv. *C. b. brachydactyla*: **9** ad. *C. b. mauritanica*: **10** ad.

north-east China, thence south-west to Himalayas; also breeds northern Iran. Breeding population in North America now considered a separate species, Brown Creeper, *C. americana*.

Population. Stable (though often fluctuating, depending on severity of winter) over much of range, but increase Britain and perhaps Austria, slight decrease Finland, Czech Republic, Ukraine, and Moldova. BRITAIN. 200 000 territories 1988–91. Gradual increase, but fluctuating around cold winters. IRELAND. 45 000 territories 1988–91. FRANCE. 1000–10 000 pairs in 1970s. BELGIUM. 2300 pairs in 1960s, 1700 in 1973–7; probably stable. LUXEMBOURG. 2000–3000 pairs. NETHERLANDS. Zuid-Limburg: at least 13 pairs in 1993 (none in 1982). GERMANY. 437 000 pairs in mid-1980s. In east, 110 000 ± 55 000 pairs in early 1980s; less conspicuous than Short-toed Treecreeper and thus likely to be underestimated. DENMARK. 1500–16 000 pairs 1987–8; fluctuating. NORWAY. 20 000–100 000 pairs 1970–90. Report of marked fluctuations, but evidently stable overall. SWEDEN. 350 000–800 000 pairs in late 1980s. FINLAND. 50 000–100 000 pairs in late 1980s. Marked fluctuations, with peak in early 1970s; recent decrease due to changing forestry practices. ESTONIA. 20 000–50 000 pairs in 1991; marked annual fluctuations, but stable in long term. LATVIA. 110 000–160 000 pairs in 1980s. LITHUANIA. Common. POLAND. 60 000–150 000 pairs. CZECH REPUBLIC. 300 000–600 000 pairs 1985–9. SLOVAKIA. 100 000–150 000 pairs 1973–94. HUNGARY. 8000–10 000 pairs 1979–93. AUSTRIA. 50 000–80 000 territories in 1992. SWITZERLAND. 60 000–80 000 pairs 1985–95. SPAIN. 32 000–38 000 pairs. ITALY. 30 000–60 000 pairs 1983–95. GREECE. 2000–5000 pairs. ALBANIA. 500–2000 pairs 1981.

YUGOSLAVIA: CROATIA. 10 000–15 000 pairs. SLOVENIA. 15 000 pairs. BULGARIA. 1000–10 000 pairs. RUMANIA. 100 000–120 000 pairs 1986–92. RUSSIA. 100 000–1 million pairs. BELARUS'. 350 000–400 000 pairs in 1990. UKRAINE. 40 000–43 000 pairs in 1988. MOLDOVA. 4000–6000 pairs in 1988. AZERBAIJAN. Common. TURKEY. 1000–10 000 pairs.

Movements. Varies from sedentary to partial migrant across range. Movement both nocturnal and diurnal. Northern populations partially migratory, eruptive in some years, with heading chiefly south-west in autumn (reverse in spring). Extent of movement across Europe apparently slight, but not well known, as masked by resident populations. Spring passage weak in comparison with autumn, with small numbers even following irruptions. British, central and southern European populations sedentary.

Food. Insects and spiders throughout the year; some seeds in winter, mostly pine and spruce. Forages mainly on trunks of trees, starting low down and creeping spirally up with small hops, extracting food from crevices. Large branches are also investigated, often on their undersides.

Social pattern and behaviour. Normally solitary in winter, sedentary populations not moving far from nesting territory. Pair-members apparently sometimes stay together outside breeding season. Well known for association in winter with mixed flocks containing tits, Goldcrest, etc., which young join at independence; this association not close, Treecreeper not exploiting foraging successes of other flock members, but presumably benefiting from overall increased vigilance. Mating system apparently essentially monogamous, but some reports of polygyny. Only ♀ incubates and broods young, but young fed equally by both sexes, including after fledging. ♂ sings while stationary or climbing on tree and with bill wide open, often straight after landing. Exceptionally, sings while perched crossways on branch. Territory demarcation characterized by frequent song and flying to particular trees or branches. Song given mainly in breeding season; a less persistent singer than Short-toed Treecreeper. Roosts either singly or, in cold weather, communally, several birds huddled together in same niche, in hollows, cracks, or niches mainly in bark of trees; in Britain and Ireland, best known for using natural and excavated hollows in soft, fibrous bark of redwoods *Sequoiadendron* and *Sequoia*; presumably favoured for ease of excavation and insulating quality of bark.

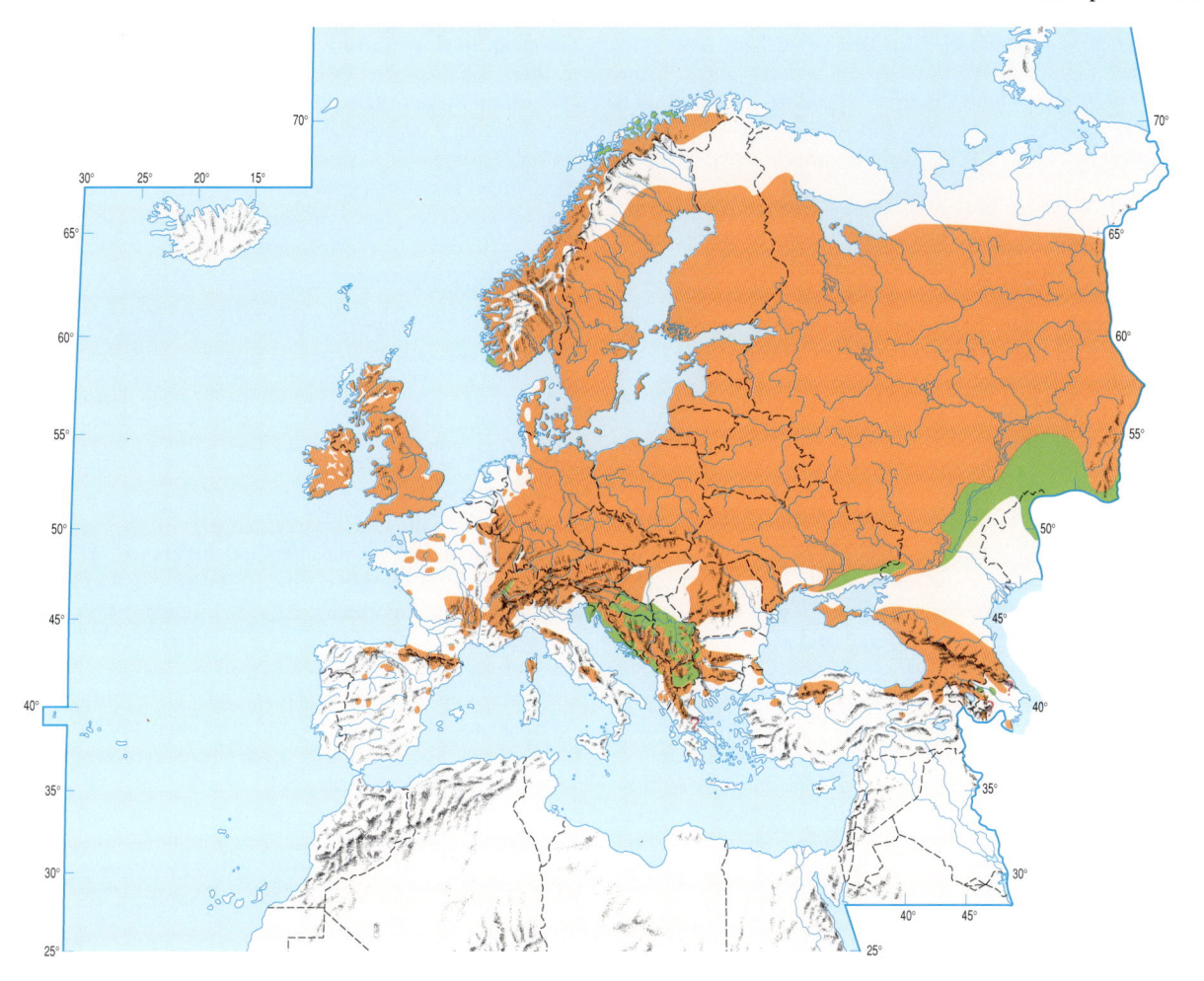

Voice. Song a thin, high-pitched and quite long (up to *c.* 3 s or more) phrase descending progressively in pitch and normally comprising 3 sub-phrases: (i) 2–3 or more 'tsree' units; (ii) 3–7 descending units, delicately thin and tremulous twittering quality being somewhat reminiscent of Goldcrest (but stronger, clearer, and louder), and (iii) 5–14 units in fine descending ripple which starts, after characteristic pause, higher than end of 2nd sub-phrase, but descends to lowest frequency; song usually ends with clear descending then rising whistle—'suih'. Silvery, thin, and tremulous start of song contrasts with increasingly fuller, confident quality, bird working up to flourish of more sibilant notes. In areas where Short-toed Treecreepers occur, 'mixed songs' (involving mimicry of the very different Short-toed Treecreeper) are common. Commonest call a characteristic high-pitched and shrill 'tsree', 'srieh', 'dsrrrieh', 'srihih', etc., with vibrato quality due to rapid frequency modulation. Given by both sexes, often in series, and serves as contact-call between pair members. Higher-intensity excitement call a very thin and sharp, almost hissing, 'tsee'; given by both sexes mostly in spring and summer. Threat-call a repeated screeching, almost harsh 'schräschrä...'. Flight-call a fine, short, high-pitched, tinkling 'si' or 'tsit'.

Breeding. SEASON. Eggs laid from late March to late June, somewhat later in north than in central and western Europe. 2 broods regular. Overlapping broods not unusual, where 2nd nest built and even eggs laid before 1st brood fledged. SITE. On tree trunk behind flap of loose bark or in crevice; also on or inside buildings; in wood stacks, and in special nest-boxes. Nest: foundation heap of small twigs up to 20 cm long, fibrous bark, dry grass, fragments of wood, and other debris dropped into fissure or behind bark flap, and often so narrow that sitting bird unable to turn round. Cup lined with hair, wool, seed heads, etc. EGGS. Sub-elliptical, smooth and not glossy. White, with very fine speckles or blotches of pink to reddish-brown, generally confined to broad end in ring or cap. Clutch: 5–6 (3–9). INCUBATION. 13–15 days. FLEDGING PERIOD. 13–18 days.

Wing-length: ♂ 60–69, ♀ 59–68 mm.
Weight: ♂♀ mostly 8–11 g.

Geographical variation. Slight; clinal in continental populations. 5 races in west Palearctic: nominate *familiaris* (northern and eastern Europe) paler above and below (pure white towards east) than *C. f. macrodactyla* of continental western and west-central Europe. *C. f. britannica* (Britain and Ireland) slightly darker than *macrodactyla*, more rufous below. *C. f. corsa* similar to *macrodactyla* but upperparts more sharply streaked white. *C. f. persica* (Crimea, Turkey, Caucasus) slightly darker and duller above than *macrodactyla*.

Short-toed Treecreeper *Certhia brachydactyla*

PLATE: page 1412

DU. Boomkruiper   FR. Grimpereau des jardins   GE. Gartenbaumläufer   IT. Rampichino
RU. Короткопалая пищуха   SP. Agateador común   SW. Trädgårdsträdkrypare

Field characters. 12 cm; wing-span 17–20.5 cm. Close in size to Treecreeper but structure differs subtly in rear body shape (due apparently to looser rear flank and rump feathers, their greater mass seeming to shorten tail) and slightly longer and more bent (rather than gently decurved) bill. General character and plumage closely resemble Treecreeper, particularly in maritime western Europe and Channel Islands, but over whole range upperparts generally duller (ground-colour less rufous, pale lines less obvious and less even), and underparts dirtier (from chest backwards). Supercilium often indistinct, broken or obscure in front of eye and rarely deep behind it.

No easy way of separating the two treecreepers in areas of sympatric occurrence or vagrancy. Length and tone of supercilium is best single character; difference in bill shape apparently not a reliable feature. Important also to recognize that Short-toed Treecreeper is (1) sole resident treecreeper in Channel Islands, most of Low Countries, much of France, Iberia (south of Pyrénées), most coasts of north Mediterranean countries, and north-west Africa and (2) in most of range primarily associated with oak forests or oak-scattered parkland and gardens. Voice generally louder than Treecreeper, much easier to hear and to locate.

Habitat. Occupies middle and lower middle latitudes of southwest Palearctic in continental and oceanic temperate and Mediterranean zones, between July isotherms of 17–18°C and

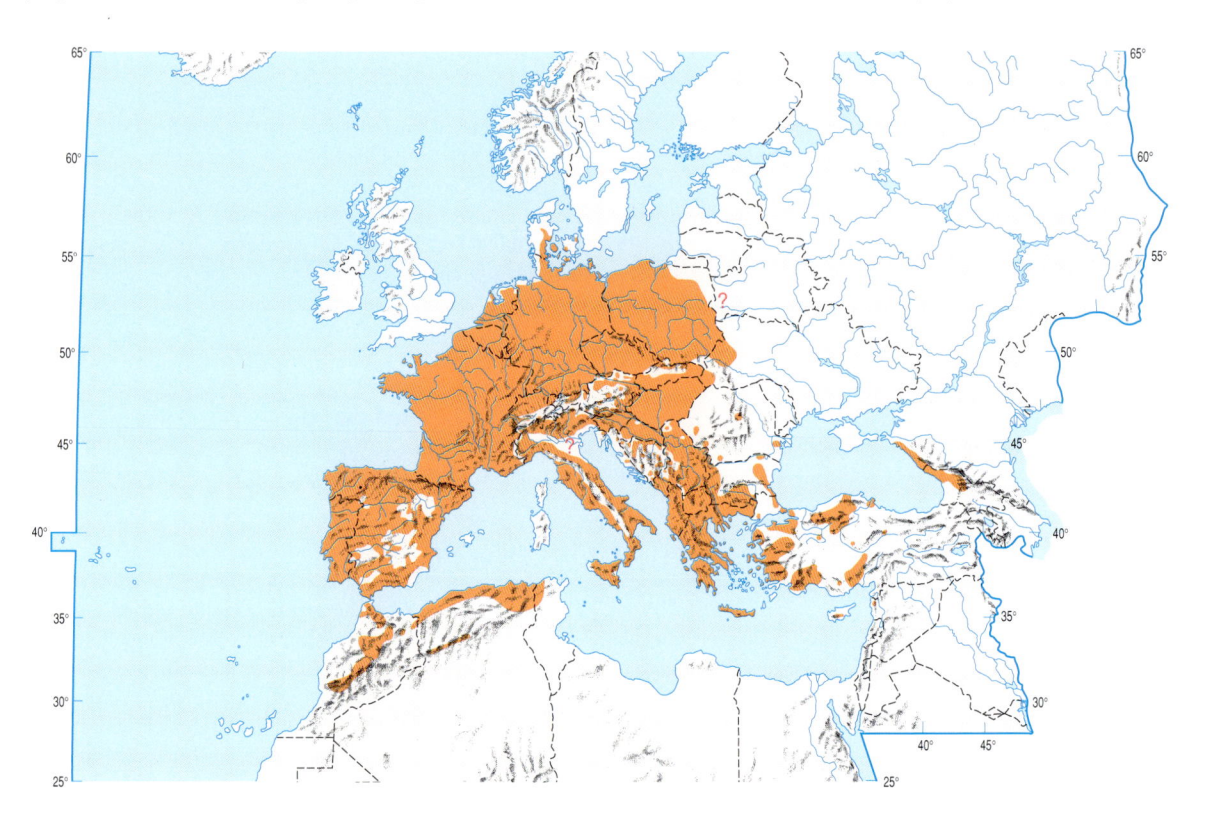

26°C. Occurs mainly in lowlands, but in Switzerland ascends slopes in south almost to 1400 m, in centre occasionally to 1000 m or more, and occurs in mountain woodlands in Turkey and north-west Africa. Typically inhabits groups of tall trees with rugged bark in avenues, parks, orchards, copses, and forest edges (especially broad-leaved or mixed), preferably with dense undergrowth; occurs freely in trees of cultivated areas and among houses.

Distribution. Breeds only in west Palearctic. Range contraction Ukraine, perhaps Austria and Italy; increase Denmark. BRITAIN. Channel Islands: common Jersey, Guernsey, uncommon Sark, rare Alderney. FRANCE. Corsica: 3 birds identified as this species, apparently giving song of North African race, tape-recorded April 1969. Perhaps indication of recently established population, but no subsequent records. GERMANY. In east, scarce or absent in mountains above 500 m. DENMARK. First bred 1946; continuing spread. POLAND. Little change in last 50 years. BELARUS'. Status uncertain: recorded in west in early 20th century including pair collected east of Slonim March 1917 and seen (not uncommonly) in north-east Grodno region May–October 1918. No proof of breeding and no recent records, but perhaps overlooked. RUSSIA, GEORGIA. Eastern limit of range in Caucasus poorly known. AZERBAIJAN. Status uncertain: reported Talysh mountains in early 1960s, but only Treecreeper confirmed there and in adjacent Lenkoran lowlands more recently. SYRIA. Presence long suspected; confirmed April 1994 when song heard at Saladin's Castle in north-west. ALGERIA. More common in east of range.

Accidental. Corsica, Sweden, Lithuania, Balearic Islands.

Population. Apparently stable over most of range, but increase Denmark and locally marked decrease Austria, slight decrease Ukraine. CHANNEL ISLANDS. Jersey: 200–500 pairs. FRANCE. 100 000–1 million pairs in 1970s. BELGIUM. 25 000 pairs in 1960s, 28 000 in 1973–7. LUXEMBOURG. 5000–8000 pairs. NETHERLANDS. 60 000–100 000 pairs 1979–85. GERMANY. 450 000–550 000 pairs. In east, 120 000 ± 60 000 pairs in early 1980s. DENMARK. 300–500 pairs. POLAND. 50 000–150 000 pairs. CZECH REPUBLIC. 75 000–150 000 pairs 1985–9. SLOVAKIA. 1000–3000 pairs 1973–94. HUNGARY. 30 000–40 000 pairs 1979–93. AUSTRIA. 10 000–15 000 pairs in 1992. SWITZERLAND. 30 000–60 000 pairs 1985–93. SPAIN. 1–3.3 million pairs. PORTUGAL. 100 000–1 million pairs 1978–84. ITALY. 100 000–500 000 pairs 1983–95. GREECE. 20 000–50 000 pairs. ALBANIA. 2000–5000 pairs in 1981. YUGOSLAVIA: CROATIA. 20 000–30 000 pairs. SLOVENIA. 30 000 pairs. BULGARIA. 1000–10 000 pairs. RUMANIA. Poorly known: tentative estimate 100–150 pairs 1986–92. UKRAINE. 200–300 pairs in 1988. TURKEY. 10 000–50 000 pairs. CYPRUS, TUNISIA. Common. MOROCCO. Uncommon.

Movements. Sedentary throughout range.

Food. Chiefly insect larvae and pupae and spiders, throughout the year. Main foraging method as Treecreeper, but tends to be slower, makes more spirals and hops.

Social pattern and behaviour. Less well studied than Treecreeper but most aspects apparently similar to that species; a more persistent singer. Apparently more sensitive to low temperatures than Treecreeper, and in cold weather forms communal roosts of up to 20 birds huddled together. Buildings commonly used, and tree sites similar to those used by Treecreeper.

Voice. Generally louder and more emphatic than Treecreeper. Song a relatively short (c. 1.1–1.5 s) stereotyped phrase of 6 (4–9) loud, clear, but rather plaintive whistling units, each separated by fairly even pause; rises in pitch at end. Delivery rapid and energetic, which together with duration, loudness, lower pitch of notes and distinct (more emphatic, slightly jolting) temporal pattern make it normally easily distinguishable from Treecreeper. Renderings include 'teet teet...teet-e-roi-i...titt', 'tü ti tü ti rüi sri'. Some songs abbreviated to 3–4 units; such songs apparently given by most birds in Denmark. Common, and most distinctive, call a shrill, loud, explosive, piping 'zeet', 'seek', 'peep', or 'tseep', recalling Dunnock and, particularly when repeated, Coal Tit; given singly or in slow to very fast series and sometimes combined with less common variant 'tür' or 'tüi'. Zeet-call used in alarm and in interaction with rivals, regularly in flight. Contact-call a clear, explosive, loud, penetrating 'tsree', 'sree', or 'srieh'; difficult to distinguish from homologous and similarly structured call of Treecreeper. Flight-call a thin, high-pitched, short, quiet 'si', 'sit', or 'pit' given almost constantly, except when resting, especially at take-off or in flight.

Breeding. SEASON. Eggs laid early April to mid-June; usually 2 broods. SITE. On tree trunk behind loose flap of bark or in deep crevice; on building or woodpile; in hole of other species especially woodpecker; in ivy; in specially designed nest-box; occasionally in foundation of raptor nest or squirrel drey. Nest: untidy foundation of twigs, conifer needles, grass, bark, plant fibres, cloth, paper, etc., lined with moss, lichen, cocoons, plant down, hair, rootlets, and many feathers; foundation can be very large and irregular, often filling available cavity. EGGS. Sub-elliptical, smooth, not glossy. White, with reddish-brown to reddish-purple spots and blotches concentrated towards broad end; markings sometimes so faint that egg appears white. Clutch: (4–)6–7. INCUBATION. 13–15 days. FLEDGING PERIOD. 15–18 days.

Wing-length: ♂ 59–67, ♀ 56–66 mm.
Weight: ♂♀ mostly 8–11 g.

Geographical variation. Slight and mainly clinal. 5 races recognized. Mainland Europe occupied by *megarhyncha* in north-west and nominate *brachydactyla* in south and east; latter generally darker and less rufous, but with much intergradation and minor differences within both races. *C. b. mauritanica* (north-west Africa) similar to nominate *brachydactyla* but a bit darker above and below; *C. b. dorotheae* (Cyprus) slightly greyer on upperparts than *brachydactyla*; *C. b. harterti* (Asia Minor) similar to *megarhyncha* but upperparts slightly duller rufous.

Penduline Tits and Allies Family Remizidae

Tiny to small oscine passerines (suborder Passeres). Found chiefly in open woodland. Mainly arboreal and insectivorous, though small seeds also taken. Construct remarkable pouch-like nests suspended from outer twigs of trees and shrubs, or from reeds. About 9 species, of which one in west Palearctic, breeding.

Penduline Tit *Remiz pendulinus*

PLATE: page 1409

Du. Buidelmees   Fr. Rémiz penduline   Ge. Beutelmeise   It. Pendolino
Ru. Ремез   Sp. Pájaro moscón   Sw. Pungmes

Field characters. 11 cm; wing-span 16–17.5 cm. 5–10% smaller than Blue Tit, with shorter wings but tail almost as long; 20% shorter than Bearded Tit. Small, compact tit-like passerine, with finely pointed bill and quite long tail. Plumage pattern and colours recall Red-backed Shrike, with typical adult showing grey-white head masked by long black eye-patches and contrasting with chestnut mantle. Juvenile has ochre-grey head and duller back.

Adult unmistakable, being smaller than any tit and having black mask and chestnut mantle. Juvenile more confusing and in reedbed glimpse could be mistaken for ♀ Bearded Tit until its lack of long, graduated, russet tail is confirmed. Flight has flutter and bounce of a tit but action and particularly quite long-tailed form at times also reminiscent of *Phylloscopus* warbler. Gait and behaviour resemble a tit, being just as lively and restless but rather less excitable and inquisitive.

Habitat. Occurs in west Palearctic middle latitudes, in warm continental temperate, steppe, and Mediterranean lowlands within July isotherms 18–32°C, in widely scattered and often small areas where its essential habitat needs are met. These are largely on deltas or by estuaries, lakes, rivers, canals, streams, or swamps, of fresh or brackish water, bearing luxuriant but usually not unbroken vegetation, especially reed mixed with tall herbage, tamarisk, willow, and poplar.

Distribution. Range has expanded to north, west, and south-west. In north-west, limit advanced *c.* 300 km west *c.* 1930–65, almost constant 1965–75, then advanced rapidly again *c.* 250 km west and 200 km north until 1985. BRITAIN. Records include ♂ building 2 nests Kent in 1990, and bird with large brood-patch trapped Cleveland 1992. FRANCE. No breeding in Camargue or Rhône valley since early 1980s, but number of breeding records increased west of Rhône delta since 1961. Expansion in north-east: first bred Alsace 1979, Lorraine in 1980s. BELGIUM. Recorded from 1966, annual since 1983. First nest-building (Hainaut) 1987, confirmed breeding 1988. LUXEMBOURG. First recorded breeding 1989. NETHERLANDS. First bred 1968; main colonization from 1981, but numbers low and territories varied up to 1984. Map shows range in 1987. GERMANY. A few breeding records as far west as Schleswig-Holstein and Bayern in 1930s. Continuing westward expansion (at least from 1950s), with most of country (except parts of west) lying within regular breeding range by 1985. DENMARK. First bred 1964; spread, and now breeding patchily all over country. NORWAY. Recorded since 1989; first bred (Vest-Agder) 1993. SWEDEN. First bred 1964. Lake Krankesjön (Skåne) only permanent breeding site, but explosive increase late 1980s, with breeding in 6 southern provinces, then major decline and disappearance from several strongholds. Northernmost populations (Östergotland, Närke) apparently doing best. FINLAND. Several breeding attempts from 1973; only successful confirmed breeding 1985. ESTONIA. First recorded breeding 1954 or 1955; spreading. LATVIA. Recorded breeding in 18th century; apparently recolonized in 20th century, with spread from late 1960s, over 20 sites occupied 1969–75, and *c.* 25 nests at one site in 1989. POLAND. Very rare until early 20th century. Marked spread since 1945; still spreading north, and now widespread (but usually scarce) throughout lowlands, including north-east, and locally fairly numerous in south-west. CZECH REPUBLIC, SLOVAKIA. Very rare in 19th century; spread western Slovakia, southern and northern Moravia 1914–30, throughout during 1945–60. More recent slight increase both countries. AUSTRIA. Has spread. Locally not uncommon in east, scarce elsewhere. SWITZERLAND. First recorded breeding 1952. Breeds every year, but sites vary. SPAIN. Marked expansion to west and south. PORTUGAL. Cartaxo: nest found 1982. Near Elvas: present in breeding season 1991, 1993, bred 1994. Algarve: first found breeding during atlas survey 1988–92. ITALY. Colonized plain of Lombardia and Lago Maggiore 1964. Recent expansion central Po valley. MALTA. First recorded 1972; wintering in increasing numbers from mid-1980s, then dropping in early 1990s. GREECE. Some decrease. YUGOSLAVIA: SLOVENIA. Range fluctuating. RUSSIA. Kaliningrad region: colonized in last 100 years. CYPRUS. Perhaps bred 1982. SYRIA. Reported breeding 1934, carrying nest material Homs 1978, feeding young Ghab 1979, probable breeding other sites. IRAQ. First recorded breeding 1966.

Accidental. Britain, Norway, Finland (annual), Tunisia, Morocco (or probably very scarce but regular winter visitor).

Beyond west Palearctic, extends broadly east across central Asia to *c.* 130°E, also south to Iran.

Population. Increase, marked in Netherlands and Spain, also reported Germany, Denmark (fluctuating), Sweden, Baltic States, Poland, Czech Republic, Slovakia, Austria, Italy, Belarus', and Moldova. Apparently stable Hungary, Switzerland,

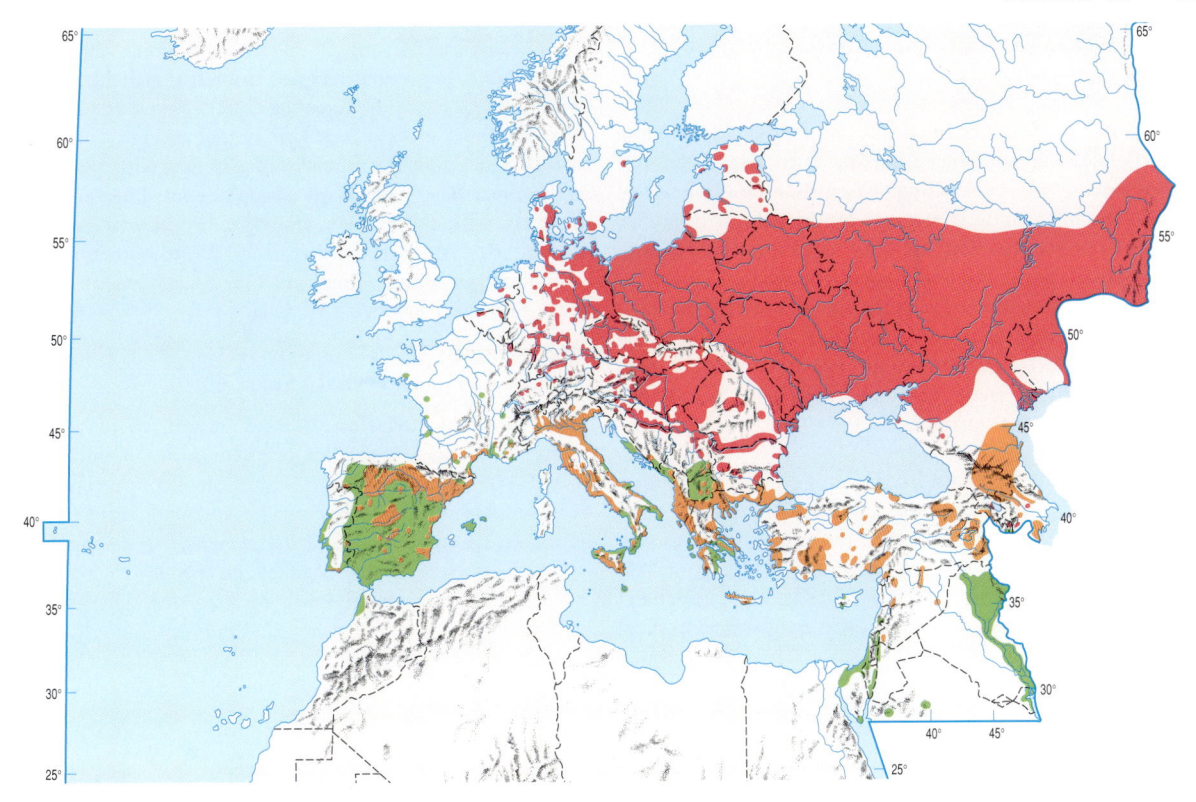

Croatia, Bulgaria, Russia, and Ukraine. Decrease southern France (since *c.* 1950), Sweden. FRANCE. 100–200 pairs. BELGIUM. 3–4 pairs 1989–90 and, after increase to at least 10 in 1991, also subsequently. LUXEMBOURG. Up to 5 pairs. NETHERLANDS. Remarkable recent increase: up to 7 pairs 1979–85, at least 30 in 1987, 225–250 territories in 1992. GERMANY. 3000 pairs in mid-1980s. Probably far higher by early 1990s: *c.* 5000 pairs in east alone 1990–91, only 1200 ± 360 there in early 1980s. DENMARK. 75–150 pairs. SWEDEN. 30–100 pairs in late 1980s. ESTONIA. 100–300 pairs in 1991. Long-term upward trend, though marked annual fluctuations. LATVIA. 300–800 pairs in 1980s. LITHUANIA. Common. POLAND. 6000–10 000 pairs. Considerable increase in 2nd half of 20th century; probably still continuing though less marked. CZECH REPUBLIC. 2500–5000 pairs 1985–9. SLOVAKIA. 5000–10 000 pairs 1973–94. HUNGARY. 2000–3000 pairs 1979–93. AUSTRIA. Continuing increase since 1950s. SWITZERLAND. 1–5 pairs 1985–93. SPAIN. 12 400–14 600 pairs. ITALY. 20 000–30 000 pairs 1983–95. GREECE. 1000–3000 pairs. ALBANIA. 500–2000 pairs in 1981. YUGOSLAVIA: CROATIA. 5000–10 000 pairs. SLOVENIA. 200–300 pairs; fluctuating. BULGARIA. 1000–10 000 pairs. RUMANIA. 5000–8000 pairs 1986–92. RUSSIA. 10 000–100 000 pairs. BELARUS'. 500–800 pairs in 1990. UKRAINE. 9000–12 000 pairs in 1988. MOLDOVA. 700–1000 pairs in 1988. AZERBAIJAN. Common. TURKEY. 50 000–500 000 pairs.

Movements. Migratory in north of range, resident in south, with intermediate populations partially migratory. West Palearctic birds migrate mainly between west and south to winter quarters in south-west and southern Europe; populations breeding east of 20°E probably mainly south. Migration patterns have changed markedly following great expansion of breeding population to west.

Main departure from northern breeding areas begins August or September, movement continuing to November. Departure from winter quarters begins February, some birds remaining until late March or April. Most birds reach northern breeding areas in April or May.

Food. Mainly larval insects (also adults and eggs), with spiders of considerable importance near beds of reed; plant material (mainly seeds of poplar, willow, and reed) mainly taken outside breeding season.

Social pattern and behaviour. Usually gregarious outside breeding season, but sometimes solitary. Rarely associates closely with other species. No suggestion of winter territoriality. Mating system complex, combining sequential polygyny and polyandry. Polygamy facilitated by constant nest-building by ♂♂ over long breeding season. Pair-bond typically loose, lasting 5–10 days (1–21), birds acting as pair usually only during nest-building. Incubation and care of young usually by ♀, less often by ♂. Often roost communally, chiefly in reed-beds, sometimes at considerable distance from breeding area. Song given by ♂ normally while perched on or near nest and apparently serves primarily to attract ♀. Singing ♂ stretches head up, ruffles crown, and makes sideways quivering movements of tail. Nest-building by both sexes. Both frequently work on nest simultaneously, one inside, one outside; ♂ will sing in nest while ♀ outside, or direct song into nest. ♀ may also work on nest eventually to be used for breeding while ♂ builds another nearby.

Voice. Song of ♂ rather finch-like, with trills and tinkling sounds. Typically given when attracting ♀ to nest, when may become intense and penetrating. Comprises short phrase of 1–4(–10) dissimilar units, sometimes longer trills or tremolos. Typically begins or ends with full-length or abbreviated 'seeoo' call. 'Seeoo' is most frequently used vocalization, given in variety of contexts and very variable: typically a high-pitched, soft, pure, plaintive, drawn-out, descending whistle. Calls given in winter (in flock) generally weaker than in breeding season. Alarm-call a 'srrii' or 'sreee'; some resemblance to 'seeoo', but sharp, trilled, and even-pitched.

Breeding. SEASON. Eggs laid from end of April to beginning of July, mostly from beginning of May to end of June. 2nd clutch after successful 1st brood not uncommon, usually overlapping, but often deserted; successful fledging of 2nd broods only in long favourable season. SITE. Nest suspended in fork of outermost hanging twigs, often over water; in small trees sometimes in crown; in reedbeds slung between 2–3 stems. Nest: large, free-hanging domed pouch-like structure with short downward-projecting entrance tube near top. Made of plant fibres, especially of hop, nettle, and grass, woven and compacted tightly to felt-like consistency with plant down, and animal hair, particularly sheep's wool; lined almost exclusively with plant down, more rarely feathers. EGGS. Long sub-elliptical, smooth and not glossy. White, sometimes faintly pink just after laying. Clutch: 6–8 (5–10). INCUBATION. 13–14 days. FLEDGING PERIOD. 18–26 days.

Wing-length: ♂♀ 54–59 mm.
Weight: ♂♀ mostly 8.5–10.5 g.

Geographical variation. 3 races recognized in west Palearctic. Nominate *pendulinus* occupies most of range; *R. p. menzbieri* (central Asia Minor, Levant and Transcaucasia) similar to nominate *pendulinus* but smaller and with thinner bill; *R. p. caspius* (southern Volga plains eastwards) with much more extensive chestnut in plumage. (*R. p. macronyx*, with all-black head, mainly extralimital, but hybridizes with *caspius* at mouth of Ural river.)

Sunbirds and Allies Family Nectariniidae

Tiny or small oscine passerines (suborder Passeres). Found in wide range of habitats, both open and dense, including forest, woodland, scrub, and gardens. Arboreal, feeding chiefly on insects, spiders, and nectar. About 118 species, mainly inhabiting tropical regions of Africa and Asia, with centre of distribution in Africa. Some species sedentary but many make seasonal movements in search of flowering plants. 3 species in west Palearctic, all breeding. A well-defined family of mostly specialized nectar-feeders, highly dependent on blossoms of flowering plants.

Sexes similar in size. Bill narrow, sharply pointed, and usually downcurved; medium length to long. Cutting edges of mandibles with fine serrations near tips for gripping large insects. Tongue long and modified for nectar feeding: tubular for most of length and divided at tip into 2 or 3 bristle-like prongs for extracting nectar. Wing short and rounded; 10 primaries, p10 reduced (small to tiny, 25–40% of length of longest primary). Flight strong and fast. Unlike hummingbirds, obtain most food while perched and do not typically hover before flowers when extracting nectar (though will do so briefly at times). Plumage of most ♂♂ brightly coloured and iridescent throughout year—often black with red, purple, or green lustre above and on throat, and red, orange, or yellow on rest of underparts; some also have coloured pectoral tufts. ♂♂ of some species, however, moult into ♀-like plumage after breeding; as this usually worn only for duration of wing-moult, can be considered a true eclipse plumage. ♀♀ usually much duller than ♂♂; unmarked grey, green, or brownish, without lustre.

Pygmy Sunbird *Anthreptes platurus*

PLATE: page 1420

Du. Kleine Honingzuiger Fr. Souimanga pygmée Ge. Grünbrust-Nektarvogel It. Nettarina pigmea
Ru. Карликовая нектарница Sp. Colibrí pigmeo Sw. Dvärgsolfågel

Field characters. ♂ with fully-grown tail 16.5 cm; non-breeding ♂, ♀, and juvenile 10 cm; wing-span 15–18 cm. Head and body close in size to Goldcrest but main part of tail wider and longer, with thin elongated central feathers of breeding ♂ adding 65% to bird's length; slightly smaller than Palestine Sunbird. Rather small, short-billed sunbird, constantly active and darting in flight. Breeding ♂ iridescent green on head, chest, and back, violet on rump, black on tail, and rich yellow on underbody; non-breeding ♂, ♀, and immature grey-brown above, yellow and white below, with pale supercilium (and in some ♂♂ green-black bib).

Among west Palearctic species, adult ♂ breeding can be confused only with Nile Valley Sunbird, but ranges do not overlap. Breeding plumage rather short-lived, however, so that ♂ in non-breeding plumage, ♀, and juvenile are for most of year small grey-brown and yellow-white birds which may suggest warbler to inexperienced observer. Distinctive sunbird characters are decurved bill, relatively large head, quite tubby

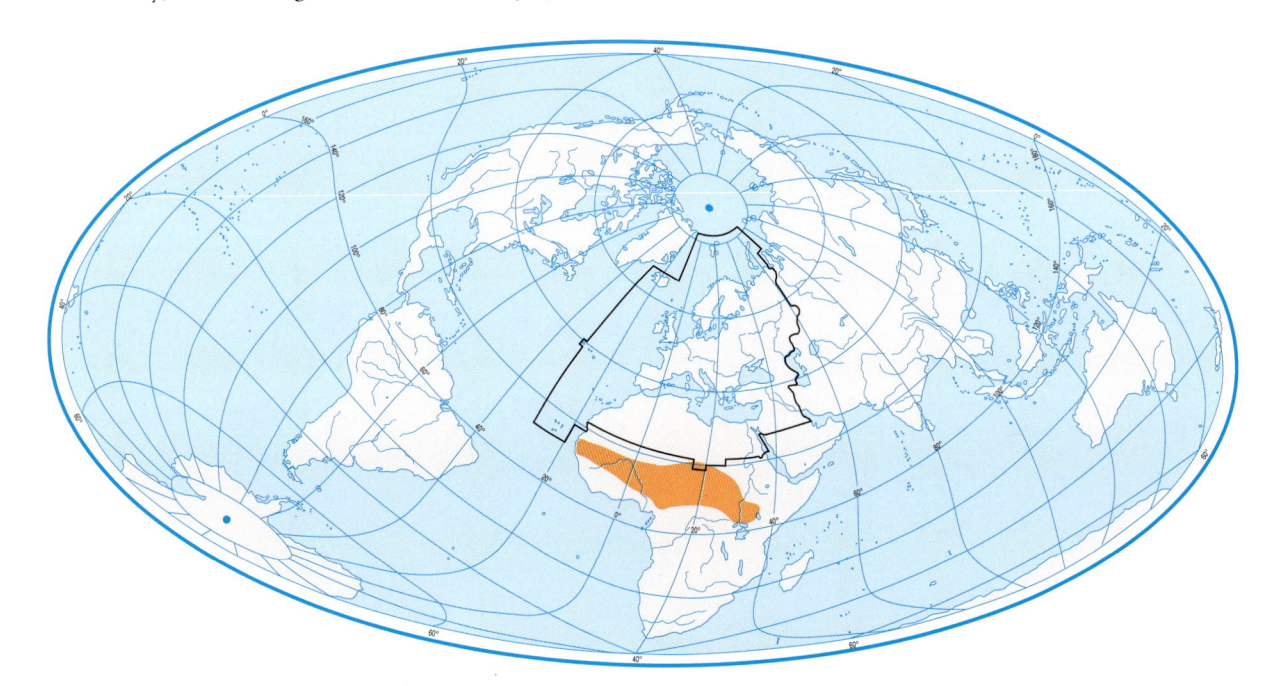

Pygmy Sunbird *Anthreptes platurus*: **1–2** ad ♂ breeding, **3** ad ♀, **4** juv. Nile Valley Sunbird *Anthreptes metallicus*: **5–6** ad ♂ breeding, **7** ad ♀, **8** juv. Palestine Sunbird *Nectarinia osea osea* (p. 1422): **9–10** ad ♂ breeding, **11** ad ♀, **12** juv.

body, well-cloaked, square-ended tail, and darting and hovering flight, as well as specialized feeding from flowers.

Habitat. As a tropical nectar-feeder, barely overlaps southern fringe of west Palearctic. Mainly inhabits savanna zone from Sahel southwards, especially in *Acacia* scrub, open woodland, and gardens.

Distribution and population. Map shows all areas of occurrence; status not known in many places. CHAD. Tibesti: small population, apparently breeding, found 1953.

Movements. Migratory (intra-Africa) or locally dispersive. Unlike sympatric *Nectarinia* sunbirds which breed in wet season, favours dry season for breeding over much of range, and shows movements (mainly north–south) accordingly. However, in Nigeria (as elsewhere) northern limits and movements unclear since birds occur only seasonally in some localities and throughout the year in others.

Food. Invertebrates and flower products (chiefly nectar).

Social pattern and behaviour. Little known, but presumably very similar to Nile Valley Sunbird, with which it is often considered conspecific.

Voice. Song of ♂ a beautiful, soft but vigorous silvery trilling, reminiscent of weak song by Skylark. Contact-call a 'cheek' or 'cheek-cheek'.

Breeding. SEASON. Nigeria: December–February(–April). 2 broods. SITE. Bush; c. 1.5–3 m above ground. Nest: roughly oval domed purse, upper half of rear side firmly attached by spiders' webs and fine fibre to suspending twig or main stem; small circular entrance in side, protected by porch c. 2.5 cm long. Outer structure of fine grass, plant fibre (notably cotton lint), spiders' webs, small leaves, and sometimes a few feathers, all skilfully and compactly interwoven; surface decorated variously with dead leaves, seeds, caterpillar droppings, and cocoons; chamber thickly lined with plant down, thinner on walls and roof. EGGS. Sub-elliptical, smooth and glossy; when fresh, pinkish, occasionally with small rufous speckling. Clutch: 1–2. INCUBATION. Period not precisely known, but in one case could not have exceeded c. 14 days. FLEDGING PERIOD. 12–15 days.

Wing-length: ♂ 54–60, ♀ 53–57 mm.
Weight: ♂♀ 5.2–7.3 g.

Nile Valley Sunbird *Anthreptes metallicus*

Du. Nijlhoningzuiger   Fr. Souimanga du Nil   Ge. Erznektarvogel   It. Nettarina metallica
Ru. Металлическая короткохвостая нектарница   Sp. Suimanga rabilarga   Sw. Nilsolfågel

Field characters. ♂ with fully grown tail 16 cm; non-breeding ♂, ♀ and juvenile 10 cm; wing-span 15–18 cm. Size, structure, plumage, voice, and behaviour similar to Pygmy Sunbird, but adult ♂ breeding has less fiery iridescence on foreparts and purple band across bottom of chest.

In good light, adult ♂ breeding not difficult to separate from ♂ Pygmy Sunbird as 'flash' from iridescent head and upperparts brilliant emerald, not bronzy or golden. Distinction of non-breeding ♂, ♀, and juvenile virtually impossible, but species' ranges do not overlap.

Habitat. Occurs in lower middle latitudes of west Palearctic, mainly in subtropical arid lowlands and river valleys. Dependence on nectar and on bushy nesting sites governs distribution in dry open scrub (especially *Acacia*), or on grassy areas with shrubs, or shrubby growth at edge of desert. Also occurs in gardens in irrigated areas. Normally avoids high altitudes, but recorded up to 2200 m in northern Yemen.

Distribution and population. EGYPT. Only breeding area in west Palearctic. Has extended range north along Nile valley in last 100 years; formerly bred only south of Aswan. Regular visitor to Cairo area, and breeding proved 1986. Singing ♂ at Hurghada, April 1990, far east of known range.

Beyond west Palearctic, breeds eastern Chad east to western and southern Arabia.

Movements. Short-range migrant or locally dispersive, including altitudinal movements. Pattern presumed to be influenced largely by sequence and abundance of flowering plants. In Egypt, moves down Nile valley in winter, then distributed throughout Nile valley, Faiyum and occasionally Suez Canal zone; in Cairo area, apparently mainly non-breeding visitor, early October to May, but see Distribution.

Food. Chiefly nectar, also invertebrates. Exploits nectar by perching (often hanging) on, or hovering in front of, flowers and inserting bill, often thus discolouring head with pollen.

Social pattern and behaviour. Gregarious outside breeding season, typically encountered in small feeding parties. Commonly associated with other species at feeding sites. No evidence of other than monogamous mating system. ♂ sings from perch, accompanied by vigorous wing-flicking, spreading and raising tail, body plumage shimmering with overall movement. Song common during breeding season. Both sexes help to build nest, but only ♀ incubates.

Voice. Song of ♂ a high-pitched warble, consisting of thin silvery trilling and hissing sounds, e.g. 'pruiit-prruiit-pruiit-tiririri-tiriri'. Contact-alarm call a grating, single or disyllabic, 'pee' or 'pee-ee', with penetrating nasal quality.

Breeding. SEASON. Egypt: March–September. SITE. Bush; typically 1.5–3 m above ground; nest rests in fork, or suspended from twig. Nest: oval flask-shaped purse with hole high on one side; outer fabric an intricate weave of (variously) plant fibres, rootlets, dead leaves, flower calyxes, plant down and seeds, bound together with spiders' webs and a few small feathers; lined with down and feathers. EGGS. Sub-elliptical, smooth and glossy. White, with pink flush when fresh, broad end faintly and finely speckled with rufous over larger grey underlying markings. Clutch: 2–3. (Incubation and fledging periods not recorded.)

Wing-length: ♂ 54–59, ♀ 51–56 mm.
Weight: ♂ ♀ 7–7.5 g.

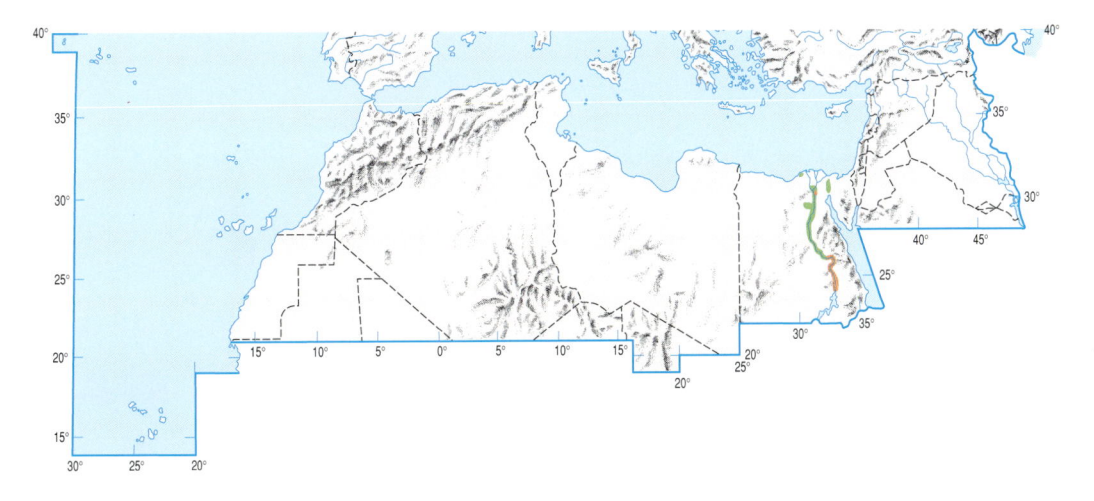

Palestine Sunbird *Nectarinia osea*

Du. Palestijnse Honingzuiger   Fr. Souimanga de Palestine   Ge. Jerichonektarvogel   It. Nettarina della Palestina
Ru. Палестинская нектарница   Sp. Suimanga palestina   Sw. Palestinasolfågel

Field characters. 10–11.5 cm; wing-span 14–16 cm. Close in size to Willow Warbler but with different structure. Small but fat-bodied passerine, with long decurved bill, rather broad-ended wings, straight tail, and rather long legs, adapted to nectar-feeding and frequently hovering to do so. Breeding adult ♂ looks all-black at distance but shows multi-coloured iridescence and red and yellow breast tufts at close range; non-breeding ♂, ♀, and juvenile olive-grey above, dusky white below. Restless, with fast, flitting flight between flower clumps. Unmistakable in west Palearctic.

Habitat. Occurs within small region in eastern lower middle and lower latitudes of west Palearctic, in Mediterranean and desert climates, with higher temperature than experienced by most Nectariniidae. Also uses wider range of habitats from lowlands near sea-level with gardens, orchards, bushy river banks, and rocky valleys to mountain summits clothed in juniper; in Jordan, in groves of cypress up to 1500 m.

Distribution and population. Syria. May breed north to southern Syria, though records are for winter. Lebanon. Bred 1947 or 1948. Israel. Rapid expansion and increase since 1930s–40s, favoured by spread of settlements. At least a few hundred thousand pairs. Jordan. Has expanded range along whole length of Rift Margin Highlands. Egypt. Sinai: not reported as present until 1979; first recorded breeding 1984, Rafa.

Beyond west Palearctic, breeds very patchily in north-central Afrotropics, also in western and southern Arabia.

Movements. Resident, locally dispersive, or short-distance migrant. In Israel, nomadic in winter, reaching Lebanon and Syria.

Food. Invertebrates, nectar, and (less commonly) other plant material. Nectar is taken with long, brush-tipped tongue from opening of tubular flowers and also from small flowers such as those of citrus trees. Capable of hovering near aperture of flower but more often sips nectar while perched alongside.

Social pattern and behaviour. Outside breeding season, sometimes in small feeding groups, otherwise solitary; not territorial. Monogamous mating system, pair-bond lasting through breeding season. Territorial during breeding season. ♂♂ promiscuous, sometimes seeking copulations with other ♀♀. Only ♀ incubates and broods, but both parents feed young. Once territory established, ♂ sings from regular high vantage points such as tree-tops or overhead wires. Rivals sometimes engage in song-duels, perching side by side and singing intensively at each other. In display to ♀, ♂ exposes red patches on chest, bobbing with head held erect on up-stretched neck, spreading tail and drooping wings.

Voice. Song of ♂ a sweet little jingle with a few introductory 'tew' notes followed by 'tyuh-tyuh-tyuh-tyuh-tyuh...'; may be interspersed with a few contact-alarm calls. Contact-alarm call a variable 'tsik', soft for communicating between pair-members, harder in alarm. Flight-call a repeated high 'tzik' given more or less singly or run together as quick trill.

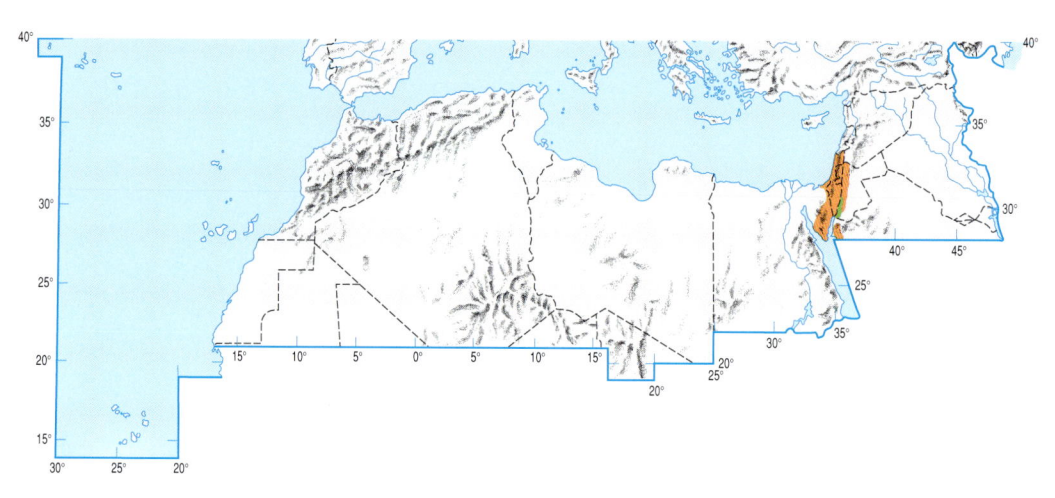

Breeding. Season. Israel: February–September in Mediterranean coastal plain, but sometimes in winter at Eilat. Site. At tip of hanging branch of tree or bush in sheltered place, e.g. close to wall of house, or vine covering ceiling of balcony. Nest: rather untidy pear-shaped purse, *c.* 18 cm long, 8 cm wide at base, with circular side-entrance near top and protected by small awning; trailing beard of leaves and twigs hangs from base. Outer fabric of thin stems, roots, leaves, plant down, and bark, bound with hair, wool, and cobwebs, and lined with feathers, wool, paper fragments, and leaves. Eggs. Sub-elliptical, smooth and glossy. Base colour pale grey or varying from yellowish to greenish to white, with small ill-defined violet-grey or brown-grey blotches and spots in loose circle at broad end; surface markings sometimes so faint as to barely darken base colour. Clutch: 1–3. Incubation. 13–14 days. Fledging Period. 14–21 days.

Wing-length: ♂ 52–57, ♀ 49–53 mm.
Weight: ♂ averages 7.6, ♀ 6.8 g.

Old World Orioles and Allies Family Oriolidae

Medium-sized, rather starling-like oscine passerines (suborder Passeres). Mostly inhabit tree-tops in forests, woodland, and parkland, descending to ground infrequently. Highly arboreal and secretive. Food mainly fruits, berries, and insects but nectar also taken by some species. Insects often large; include hairy caterpillars. 26–28 species, occurring chiefly in Africa and tropical southern Asia east to Indonesia and Philippines—some species extending into New Guinea and Australia, and one (Golden Oriole) into central Asia and Europe. Most species sedentary, others disperse in search of fruits, and a few truly migratory.

Sexes generally of similar size. Bill rather deep and pointed, straight but somewhat decurved, tip with fine hook; broad at base and about as long as head. Wing relatively large, long and pointed in most species; 10 primaries, p10 slightly reduced (*c.* 50% of length of longest). Flight strong and deeply undulating over longer distances, ending with woodpecker-like sweep into tree. Tail square and rather short; 12 feathers. Leg and foot short but strong. Song characteristically loud and melodious, liquid-sounding and fluting; some calls of similar nature, others harsher. Not sociable, and usually solitary outside breeding season—when, however, some species may form small groups and/or join mixed feeding parties; territorial when nesting, with monogamous mating system the rule. Nest a neat, deep, basket-like cup slung under fork of branch in tree, often high above ground; built mainly by ♀.

Golden Oriole *Oriolus oriolus*

PLATE: page 1426

Du. Wielewaal   Fr. Loriot d'Europe   Ge. Pirol   It. Rigogola Rigogolo
Ru. Иволга   Sp. Oropéndola   Sw. Sommargylling

Field characters. 24 cm; wing-span 44–47 cm. About 15% larger than Starling, with proportionately longer wings and tail producing shape rather like Mistle Thrush; $\frac{1}{3}$ smaller than Green Woodpecker. Rather large, colourful passerine, with strong bill, long wings, and quite long tail; shape on perch recalls Starling, shape and actions in flight suggest large thrush or woodpecker. ♂ bright yellow with black wings and black on tail; ♀ and immature sap-green above, cream with dull dusky streaks below, with blackish wings. A shy tree-dweller, but song loud and distinctive.

Adult ♂ unmistakable but dogged by possible confusion with escapes, especially Asian race *kundoo* (♂ at close range shows black streak extending behind eye, more yellow on tips of primary coverts and sides to tail). ♀ and immature much less distinctive; traditionally confused with *Picus* woodpeckers. Flight swift and powerful but sometimes looks unstable; action consists of bursts of rather loose, seemingly at times unequal wing-beats, sudden side-slips with wings spread, dives with wings closed, and striking upward-curving ascent into cover or on to perch; flight over long distance comprises long undulations. Active, but tends to stay within canopy and thus difficult to observe. Much more often heard than seen.

Habitat. In west Palearctic, breeds in middle latitudes, penetrating rather higher in continental interior and rather lower near warm ocean coasts. An arboreal but not a forest bird, and predominantly a lowland dweller, even in Switzerland not normally breeding above *c.* 600 m. Avoids large dense forests, especially of conifers, and also terrain which is treeless, or lacking in groups, lines, strips, or park-like open stands of mature deciduous trees with ample crowns well above ground. Nature and structure of undergrowth, sward, or herbage immaterial, as lower vegetation and ground surface are little visited. Range of habitats extends from parks, avenues, large gardens, spinneys, copses, riverain woodlands, and windbreaks, to moist open deciduous forest, always with more or less mature trees well enough apart individually or in groups to satisfy need for blend of good cover and easy mobility.

Distribution. Long-term increase in north of range. Britain. Probably bred regularly in Kent in mid- to late 19th century. Has probably bred regularly in East Anglia since *c.* 1967 (where nests almost exclusively in hybrid black poplars), with odd records in *c.* 10 other counties. Bred Scotland 1974. Netherlands. Some recent spread due to planting of poplars in formerly bare meadowland. Denmark. Reached south in 2nd half of 19th century, attaining present range early in 20th century; some recent decrease. Norway. Following increase in sight records, breeding first recorded 1972 (Vestfold), and has probably bred Jomfruland in Telemark. Sweden. First bred 1932; established as regular breeder after 1944. Finland. Range contracted from end of 19th century to 1930s, and since spread west. Turkey. Widespread and locally common wherever habitat suitable. Rumania. Widespread throughout country except in mountain regions. Syria. Presumed to breed in north-west, but not yet confirmed. Israel. First recorded breeding 1980. Iraq. Breeds in northern mountains.

Accidental. Iceland, Faeroes, Ireland (annual in spring, very rare in autumn), Azores, Madeira.

Beyond west Palearctic, extends east to Yenisey river, Mongolia, and eastern Tien Shan mountains, and from northern Iran to eastern Himalayas.

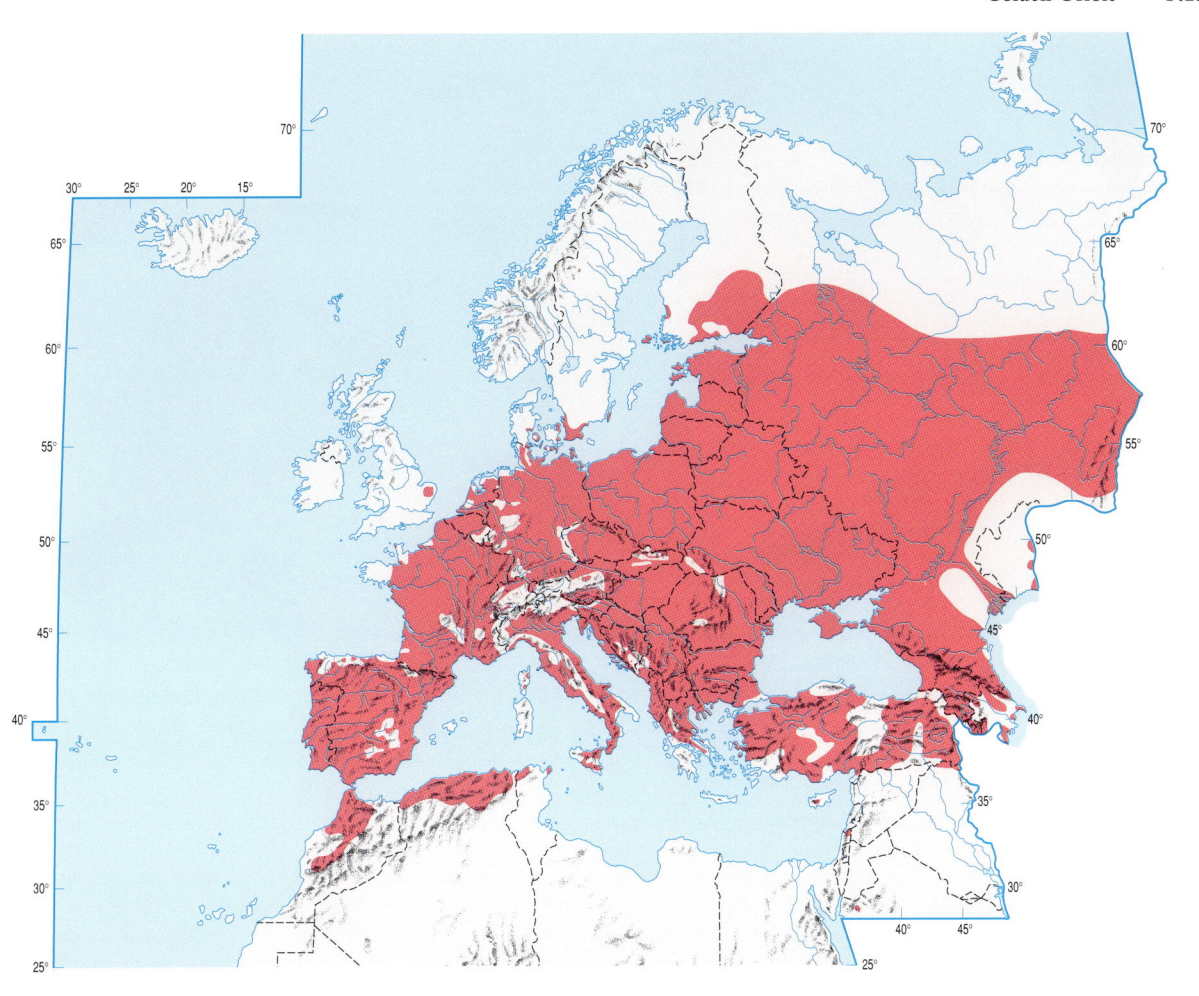

Population. Stable in most countries, though trend in north varies. BRITAIN. Always rare. Small population in 19th century, apparently declining 1st half 20th century; slight increase from 1950s, remaining vulnerable. In 1975–80, 2–7 pairs confirmed, 7–30 pairs maximum (including possible pairs); in 1988–94, 7–16 pairs confirmed, 28–42 pairs maximum. FRANCE. 10 000–100 000 pairs in 1970s. BELGIUM. 1300–2000 pairs 1988–91; steady decline continuing (2300 pairs 1973–7). LUXEMBOURG. 50–80 pairs; decreasing. NETHERLANDS. 7000–10 000 pairs 1989–91; local increases and decreases. GERMANY. 116 000 pairs in mid-1980s. Other estimates: 30 000–50 000 pairs in early 1990s; 45 000 ± 20 000 pairs in east in early 1980s. DENMARK. 20–100 pairs 1989–92; decrease 1974–94. SWEDEN. 10–100 pairs in late 1980s. FINLAND. 4000–6000 pairs in late 1980s; probable decrease. ESTONIA. 10 000 pairs 1991; decrease since 1970s. LATVIA. 30 000–50 000 pairs in 1980s. LITHUANIA. Not abundant. POLAND. 40 000–200 000 pairs. CZECH REPUBLIC. 8000–16 000 pairs 1985–9. SLOVAKIA. 7000–15 000 pairs 1973–94. HUNGARY. 80 000–100 000 pairs 1979–93. AUSTRIA. 2000–3000 pairs in 1992. SWITZERLAND. 500–1000 pairs 1985–93. SPAIN. 150 000–200 000 pairs. PORTUGAL. 10 000–100 000 pairs 1978–84. ITALY. 20 000–50 000 pairs 1983–95. GREECE. 20 000–50 000 pairs. ALBANIA. 10 000–30 000 pairs 1981. YUGOSLAVIA: CROATIA. 30 000–40 000 pairs. SLOVENIA. 4000–8000 pairs. BULGARIA. 100 000–1 million pairs. RUMANIA. 60 000–80 000 pairs 1986–92. RUSSIA. 100 000–1 million pairs. BELARUS'. 260 000–280 000 pairs in 1990. UKRAINE. 65 000–70 000 pairs in 1986; slight decrease. MOLDOVA. 18 000–25 000 pairs in 1988. AZERBAIJAN. Common. TURKEY. 50 000–500 000 pairs. CYPRUS. Scarce and local. ISRAEL. 2–5 pairs. TUNISIA. Uncommon and localized. ALGERIA. Scarce. MOROCCO. Scarce to locally common.

Movements. Western and northern race, nominate *oriolus*, migratory; central Asian race, *kundoo*, partially migratory. Nominate *oriolus* winters in sub-Saharan Africa, north to Cameroon, Central African Republic, Zaïre, and south-east Kenya, south to eastern Namibia and extreme south of South Africa; also regular on Zanzibar. Winter records very limited in comparison with observed passage, and mostly described as regular but uncommon; probably widespread in preferred habitat of densely foliaged trees, but inconspicuous, and liable to confusion with African Golden Oriole *O. auratus*.

Movement chiefly nocturnal; some diurnal movements noted, especially in spring; passage concentrated, with dates varying little from year to year; apparently migrates regularly through mountains, e.g. Carpathians and Swiss Alps.

Golden Oriole *Oriolus oriolus*. *O. o. oriolus* **1–3** ad ♂ breeding, **4–6** ad ♀, **7** ♂ 1st winter, **8** 1st winter ♀, **9** juv.

In autumn, heading within Europe ranges between south and east, with many recoveries in north-east and south-east Italy east to western Turkey of birds ringed western France east to Hungary; from Mediterranean, change to more southward direction required to reach winter quarters. Autumn passage regular throughout north-east Africa and as far west as Tunisia. Spring passage extends further west than autumn, indicating loop migration for many birds: widespread in central and northern Algerian Sahara, and conspicuous in North Africa west to eastern Morocco. In Mediterranean area south of 42°N, all autumn recoveries are east of 17°E, all spring recoveries west of 19°E. Probably a migratory divide between France and Iberia: evidence suggests Iberian and north-west African birds move west of south, presumably mostly in non-stop flight, to winter quarters as yet undiscovered, perhaps south-east of Sénégal in Guinea area. Autumn passage regular in small numbers at Strait of Gibraltar.

Autumn migration begins early, with most breeding areas vacated late July to August. Regular on passage throughout Switzerland mid-July to mid-September, and passage in Camargue ends mid-September. Many stop over in Mediterranean to build up fat reserves, feeding on fruit. First birds reach North Africa in August, with main passage September–October. Present in Cameroon and Central African Republic from October, and main arrival in East Africa October. Spring migration is late; vacates winter quarters March–April, and returns to breeding grounds late April to May, when trees in leaf.

Food. Insects (especially large caterpillars and, in spring, beetles) and berries (in late summer, autumn, and winter). Feeds mainly in tops of trees, picking items from foliage. Also catches insects in flight and sometimes feeds on ground among herbs. Before being swallowed, hairy larvae are skinned by being vigorously shaken and beaten against vegetation.

Social pattern and behaviour. Usually solitary, sometimes in small groups or larger feeding associations in winter. Mating system essentially monogamous. Pair-formation presumed to take place on breeding grounds where ♀♀ (anyway difficult to detect) arrive at same time as ♂♂, or ♂♂ precede ♀♀ by 4–8 (0–19) days. Breeding territories generally well dispersed. Many reports of nests visited by nomadic immature birds, often in parties (up to 7) and sometimes joined by non-breeding adults; probably 1–2-year-olds which return first to their natal territory. These supernumerary birds occasionally help at nest. Incubation mainly by ♀, ♂ taking over rarely and briefly; both sexes feed young more or less equally. ♂ advertises territory with loud whistling song; after pairing, song is given in courtship, when pair-members may sing in duet. Song-period in Europe and Turkey basically (late April–)May–July, with generally lower-intensity song noted late July and August. Song also sometimes given on migration and in winter quarters.

Voice. Large repertoire, but only whistling song and squalling call (both given by both sexes, though ♀ sings much less than ♂) are well known; several quiet calls given mainly at nest are audible only at close quarters. Whistling song of ♂ unmistakable: melodious, full-sounding, loud (initial units quieter) and clear fluted yodelling whistles in variable short (*c.* 1–2 s) phrases given in loose series and with characteristic jumps in pitch: 'weela-weeō', 'düdlio', 'dülioliu', etc. Whistling song of ♀ much as ♂'s, but higher pitched, quieter (though still audible at *c.* 300–400 m), and often shorter. Much quieter continuous twittering warbling song given when insufficiently stimulated to give whistling song, or expresses comfort and well-being. Squalling call a rather harsh 'wiächt', 'kyer', 'ääääh', 'räh', or similar; at times drawn-out, cat-like squalling; given by both sexes, mainly when disturbed, expressing excitement and (harder variants) discomfort. Loud, sharp, and shrill sounds given in series of 1–3 or more units and resembling calls of Kestrel, woodpecker, or Wryneck, given by both sexes, when attacking or pursuing predators (then especially shrill, almost screeching) and conspecific rivals.

Breeding. Season. Eggs laid early May to late June or early July, earlier in south than in north. One brood; replacement clutches only after early loss. Site. In fork of thin branch high in tree towards outer edge of crown, more rarely between parallel branches; occasionally hard against trunk. Nest: slung hammock-like below fork; foundation of plant fibres, grass, dry leaves, cloth, paper, string, wool, moss, bark, etc., held by grass or bark fibres 20–40 cm long looped, or stuck with saliva, and pulled more or less tight between support branches; lined with fine grass, wool, feathers, down, cocoons, small pieces of paper, etc. Eggs. Sub-elliptical to long sub-elliptical, silky and slightly glossy. White, cream, or very faint pink, with scattered well-defined black spots, occasionally with paler penumbra, sometimes concentrated towards broad end; seldom irregular blotches. Clutch: 3–4 (2–6). Incubation. 16–17 (15–18) days. Fledging Period. 16–17(–20) days.

Wing-length: ♂ 147–160, ♀ 142–156 mm.
Weight: ♂ ♀ mostly 56–79 g.

Shrikes Family Laniidae

Fairly small to medium-sized oscine passerines (suborder Passeres). Found in wide variety of habitats from dry, open bushland to woodland and forest. Bill strong, laterally compressed, and hooked with projection (tomial tooth) and notch in cutting edge of upper mandible. Many species bold and aggressive, using bill to kill and dismember relatively large prey (mainly insects and small vertebrates) typically caught by swoop to ground from exposed perch. Head relatively large, tail often rather long; leg and foot short and sturdy.

Bush-shrikes and Allies Subfamily Malaconotinae

39–44 species, almost wholly within Afrotropics. One species, Black-crowned Tchagra, extends into southern west Palearctic.

Generally similar to Laniinae as far as known, but bill in most genera relatively longer, less compressed laterally, less strongly hooked, and with less well-defined tomial tooth; wing much more rounded; tail more variable in length; leg and foot relatively heavier.

Black-crowned Tchagra *Tchagra senegala*

PLATES: pages 1429, 1449

Du. Zwartkruintjagra   Fr. Tchagra à tête noire   Ge. Senegaltschagra   It. Ciagra del Senegal
Ru. Черноголовая чагра   Sp. Chagra   Sw. Svartkronad busktörnskata

Field characters. 22 cm; wing-span 22–26 cm. 30% longer than Red-backed Shrike, with much deeper bill but wings no longer. Rather large, heavy-billed, long-tailed shrike, differing clearly from *Lanius* in skulking terrestrial behaviour, unusual plumage, and distinctive voice. General appearance recalls large bulbul, but shows diagnostic black-crowned and -lined head, rufous wings under scaled scapulars and tertials, and long, graduated black-ended tail with white terminal rim.

Unmistakable in west Palearctic. Flight flapping, with rather loose wing-beats; manoeuvres in restricted spaces executed with much spreading of wings and tail; normal action generally less confident than in *Lanius* shrikes. Gait well developed, with hopping and leaping in cover and hopping and loping run on ground; movements recall Magpie.

Habitat. Confined in west Palearctic to south-western warm arid lower latitudes, avoiding closed forest and mountains and preferring maquis, rough scrub of tamarisk, *Euphorbia*, or *Opuntia*, thorn bushes, and slopes of hills or ravines clad in wild olive, *Cistus*, and holm oak. Less frequently found in semi-arid or even sub-humid holm oak heathland or among cork oak with undergrowth of *Lentiscus*. Spends most of time on ground, commonly under dense thickets; occurs in ravines descending to sea on edge of city of Algiers.

Distribution and population. LIBYA. Bird making display-flights in suitable breeding habitat, Al 'Aziziyah (Tripolitania), 4 and 18 April 1969; no other records. TUNISIA. Locally common breeder. ALGERIA, MOROCCO. Uncommon.

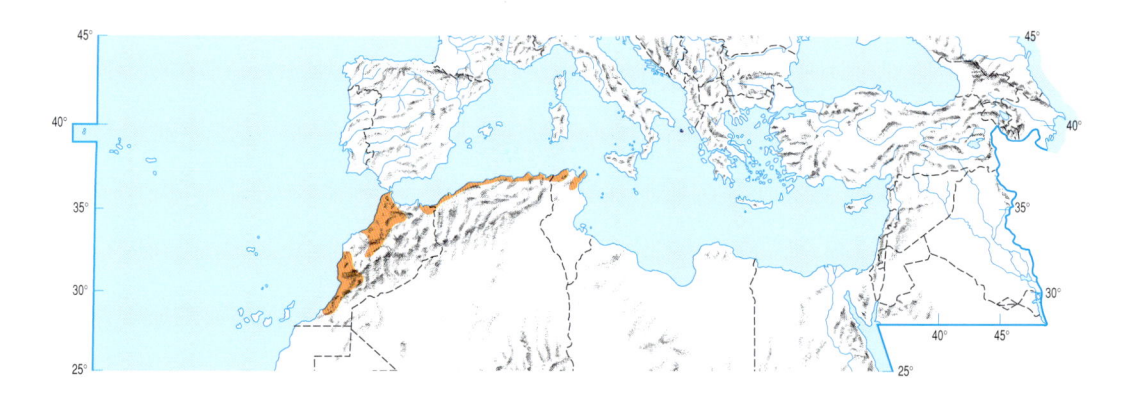

Black-crowned Tchagra *Tchagra senegala*: **1** ad worn (spring), **2** juv. Long-tailed Shrike *Lanius schach* (p. 1436): **3** ad ♂, **4** ad ♀, **5** juv moulting into 1st winter.

Beyond west Palearctic, widespread in sub-Saharan Africa, but absent from much of south; also breeds south-west and southern Arabia.

Movements. Resident and essentially sedentary, though family parties perhaps dispersive to some extent after breeding season.

Food. Invertebrates, notably beetles and grasshoppers, also amphibians, reptiles, and fruits. Will attack small mammals and birds but not recorded killing and eating them. Most prey seized on ground or in low bushes; runs (often swiftly), creeps in thick cover, or hops around bases of trees and bushes foraging in grass and flicking debris aside with bill like thrush; jumps up to pick invertebrates from low vegetation; pulls apart dung to get at beetles or termites, or scratches at soil.

Social pattern and behaviour. Little known. Although typically solitary or in pairs for most of year, family parties occur long after breeding. Sometimes joins mixed feeding parties, especially in winter. Mating system monogamous. Incubation by both sexes (mostly ♀), and both parents care for young. Established pairs apparently highly sedentary and site-faithful; same well-defined territories occupied year after year. ♂ sings from exposed perch or in song-flight, sometimes from ground.

Voice. Complex repertoire, little understood. ♂'s song heard all year. Song-flight accompanied by wing sounds, though not clear how these made. Song of ♂ a fluting, melodious, far-carrying phrase of *c.* 10 clearly separated notes, 2nd half descending; considerable individual variation. ♀ gives rattling, drawn-out call in duet with singing ♂, and other complex forms of duet include bubbling and tearing sounds. Alarm calls harsh, churring or grating.

Breeding. SEASON. Eggs recorded mid-April to early June. Possibly 2 broods. SITE. Horizontal branch or fork, generally low in dense cover in shrub or small tree. Nest: shallow cup of fine thorny twigs, plant stems, grass, rootlets, and vine tendrils, sometimes strengthened or attached to support with cobwebs, lined with grass and rootlets. EGGS. Sub-elliptical, smooth and slightly glossy. White to cream, scrawled, streaked, or smudged with claret to brown markings forming ring at broad end, or with spots and blotches towards broad end. Clutch: 2–3. INCUBATION. In East and southern Africa, 12–13 days. FLEDGING PERIOD. *c.* 16 days.

Wing-length: ♂ 88–97, ♀ 88–92 mm.
Weight: Not recorded in west Palearctic, but elsewhere, e.g. Kenya, 43.3–53.0 g.

Typical Shrikes Subfamily Laniinae

Birds mainly of open scrub, heath, and woodland, catching prey on ground, with swoop and pounce from perch or hover, and sometimes in flight. Some species (not all) will cache food (insect, vertebrate, and even fruit) or other items (e.g. own pellets and egg-shells), impaling each object on thorn, etc. Food sometimes concentrated thus in so-called larders, but chiefly scattered; also hidden singly in crevices.

Some 30 species in 3 genera, of which 2 occur only in Afrotropics. Main genus *Lanius*, with *c.* 26 species—widespread in Eurasia and Africa, extending into North America and New Guinea. Most Palearctic forms strongly migratory; 7 species breeding in west Palearctic (wintering mainly in Africa), plus 3 migrant or accidental.

Sexes of similar size in *Lanius* (to which rest of account confined). Bill particularly strong; heavily notched. Wing shape depends on extent of migratory journeys—more pointed in long-distance migrants, more rounded in others; 10 primaries, p10 reduced (*c.* 25–40% of length of longest primary, shortest in long-distance migrants). Flight strong and dashing; will also hover at times when feeding. Tail long, tip graduated; 12 feathers. In moments of excitement, tail frequently fanned and moved up and down or swung from side to side. Toes used to hold food objects by clamping and grasping. Calls loud and harsh; song of ♂ well developed in some species and duetting also occurs. Several west Palearctic species are accomplished mimics. Territorial when nesting; usually solitary and territorial outside breeding season.

Plumage soft. Upperparts typically grey or brown (black in Masked Shrike), sometimes with contrasting white, grey, or rufous forehead and/or crown. Most species have conspicuous black mask. Underparts usually white or cream, often with tawny flank and/or vent. Wing and tail black, wing with contrasting white bar or patches in many species. Sexes alike, or ♀ slightly duller than ♂, showing some reduction of black mask and some barring on flank, or (Red-backed Shrike) markedly different. Juvenile often brown or grey above, buff or white below, closely barred or scaled black.

Brown Shrike *Lanius cristatus*

PLATE: page 1431

Du. Bruine Klauwier Fr. Pie-grièche brune Ge. Braunwürger It. Averla bruna
Ru. Сибирский жулан Sp. Alcaudón colirrojo Sw. Bruntörnskata

Field characters. 18 cm; wing-span 26–28 cm. Slightly longer than Isabelline Shrike, with distinctly larger bill and head; even more so compared with Red-backed Shrike. Small to medium-sized, rather bull-headed, full-chested shrike, with proportionately heavy bill and long and slim-ended tail. Adult plumage superficially resembles western populations of Isabelline Shrike but differs in pale forehead patch, wide white supercilium, buff-brown mantle, back, and rump, rusty tail, and (sometimes) warm russet wash on underbody. Sexes usually similar.

Provided size and structure seen well, typical adult distinguishable from Isabelline Shrike by relatively large bill and head, longer, slimmer, and more rounded tail (with narrower feathers), and plumage differences listed above. Separation of immature from Isabelline Shrike and Red-backed Shrike not well studied in the field, though plumage should provoke

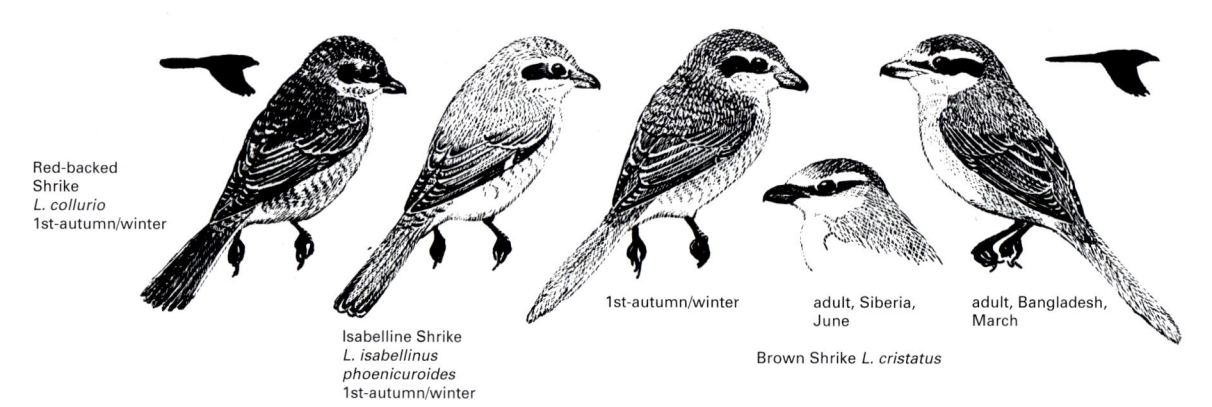

Fig. 3. Comparative drawings of 3 species of shrike *Lanius*. Red-backed Shrike is darkest and most barred, and has least patterned face and typically white-edged and -tipped tail). Isabelline Shrike is palest and least barred, especially on upperparts, with obvious face-patch, isolated contour lines on greater coverts and (before moult) on tertials, and usually pale patch on base of primaries. Brown Shrike has distinctly larger bill and head, slim tail, and always a discrete dark patch behind eye; on 1st-autumn/winter, barring is faint on upperparts and not as strong as Red-backed Shrike on underparts.

Brown Shrike *Lanius cristatus*: **1** ad, **2** juv. Grey-backed Fiscal Shrike *Lanius excubitorius* (p. 1444): **3** ad, **4** juv. Cedar Waxwing *Bombycilla cedrorum* (p. 1115): **5** ad, **6** juv.

confusion only with Red-backed Shrike. Brown Shrike differs in more striking facial pattern, less barring and paler tail; see Fig. 3. Flight as Red-backed Shrike but bird looks heavier about foreparts, while wings look somewhat shorter and side-on tail looks slim and tapering. Call a harsh, loud 'chr-r-r-r' or 'shark'.

Habitat. Breeds in east Palearctic, ranging uniquely among *Lanius* from semi-deserts and steppes through boreal to arctic latitudes. Also remarkable for ascending high mountain slopes. Breeds in a wide variety of semi-open habitats, generally with growth of bushes or low trees.

Distribution. Breeds from upper Ob' river and Tomsk region (Siberia) east to Anadyr' region, Kamchatka, and Sakhalin island, south to Mongolia, southern China, and Japan.

Accidental. Britain: adult, Shetland, 30 September to 2 October 1985. Denmark: 1st-year, trapped, Falster, October 1988.

Movements. All populations basically migratory, though slight overlap of breeding and winter range in southern China. Nocturnal migrant, capable of long unbroken flights, and rapid southward migration apparently typical. Most juveniles leave breeding grounds after adults. Winter site-fidelity recorded over several seasons.

Wing-length: ♂ 86–90, ♀ 84–89 mm.

Isabelline Shrike *Lanius isabellinus*

PLATES: pages 1432, 1449

Du. Isabelklauwier   Fr. Pie-grièche isabelle   Ge. Isabellwürger   It. Averla isabellina
Ru. Рыжехвостый жулан   Sp. Alcaudón isabel   Sw. Isabellatörnskata

Field characters. 17.5 cm; wing-span 25–28 cm. Slightly larger than Red-backed Shrike but with similar structure except for slightly longer tail. Rather small, compact but still long-tailed shrike, with behaviour of Red-backed Shrike but plainest, sandy to greyish upperparts of all west Palearctic shrikes. Only obvious features are rufous lower rump and red tail (on both surfaces) and, in adult ♂, black face-patch and white patch on primaries in western birds. ♀ slightly duller, with browner, less extensive eye-patch, reduced or absent patch at base of primaries, slightly scaled underparts, and pink bill-base. First winter shows remnants of juvenile barring above and below, with whitish-buff supercilium and only brown ear-patch, subterminal dark bars on tertials, upper tail-coverts, lower rump, and tail; white patch at base of primaries little developed. Not difficult to separate typical adult but far from simple to distinguish immature of western

Isabelline Shrike *Lanius isabellinus*. *L. i. phoenicuroides*: **1** ad ♂, **2** ad ♀, **3** 1st winter. *L. i. isabellinus* (eastern Asia): **4** ad ♂. Red-backed Shrike *Lanius collurio* (p. 1433). *L. c. collurio*: **5–6** ad ♂, **7** ad ♀, **8** 1st winter, **9** juv. *L. c. kobylini*: **10** ad ♂.

race *phoenicuroides* from Red-backed Shrike or to tell any plumage of either race from Brown Shrike; see those species. Hybridization with Red-backed Shrike produces birds not assignable to species.

Flight as Red-backed Shrike though action apparently lighter and silhouette less tail-heavy, but these differences due largely or wholly to differences (in adult) in plumage colours and patterns. Calls similar to Red-backed Shrike.

Habitat. Breeds in east Palearctic, in continental lower middle latitudes, in mountains up to 3500 m, but also on hills and barren plains. Occurs in tamarisk thickets in river valleys, patches of scrub in dry steppe, in mountains extending up to zone of prostrate juniper.

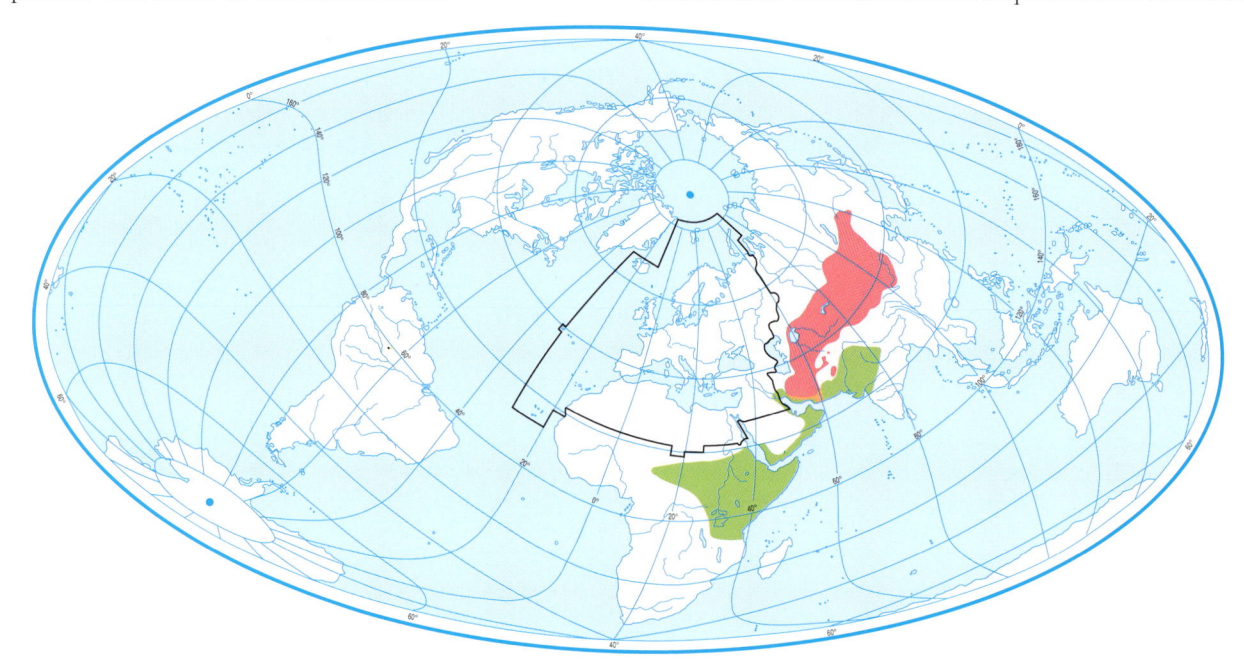

Distribution. Winters in southern Iraq and Kuwait. Otherwise (apart from vagrancy) occurs in west Palearctic as passage migrant.

Accidental. Britain, France, Belgium, Netherlands, Germany, Norway, Sweden, Finland, Latvia, Poland, Austria, Italy, Greece, Turkey, Cyprus, Canary Islands.

Movements. All populations migratory, wintering from north-west India and Afghanistan through southern Middle East to sub-Saharan Africa west to Nigeria and south to Tanzania. Heading in autumn varies between south-west (for most birds) to Africa, and south to south-east to India, while birds wintering in Nigeria must head almost due west within Africa. Return migration in spring apparently by same routes. A regular autumn passage migrant through Iraq; occasional in Israel and Sinai in spring.

In Britain and Ireland, *c.* 29 records 1958–89, 25 of them September–November and only 2 in spring; almost annual since 1975, but before then not treated as separate species so many earlier occurrences undocumented; up to 7 in one year (1988); recorded mainly Shetland and east coast, those in south-west having probably moved within Britain. Other northern and western European records all in autumn.

Wing-length: *L. i. phoenicuroides*: ♂ 91–97, ♀ 91–94 mm.
Weight: *L. i. phoenicuroides*: ♂♀ mostly 25–34 g.

Red-backed Shrike *Lanius collurio*

PLATES: pages 1432, 1449

Du. Grauwe Klauwier   Fr. Pie-grièche écorcheur   Ge. Neuntöter   It. Averla piccola
Ru. Сорокопут-жулан   Sp. Alcaudón dorsirrojo   Sw. Törnskata

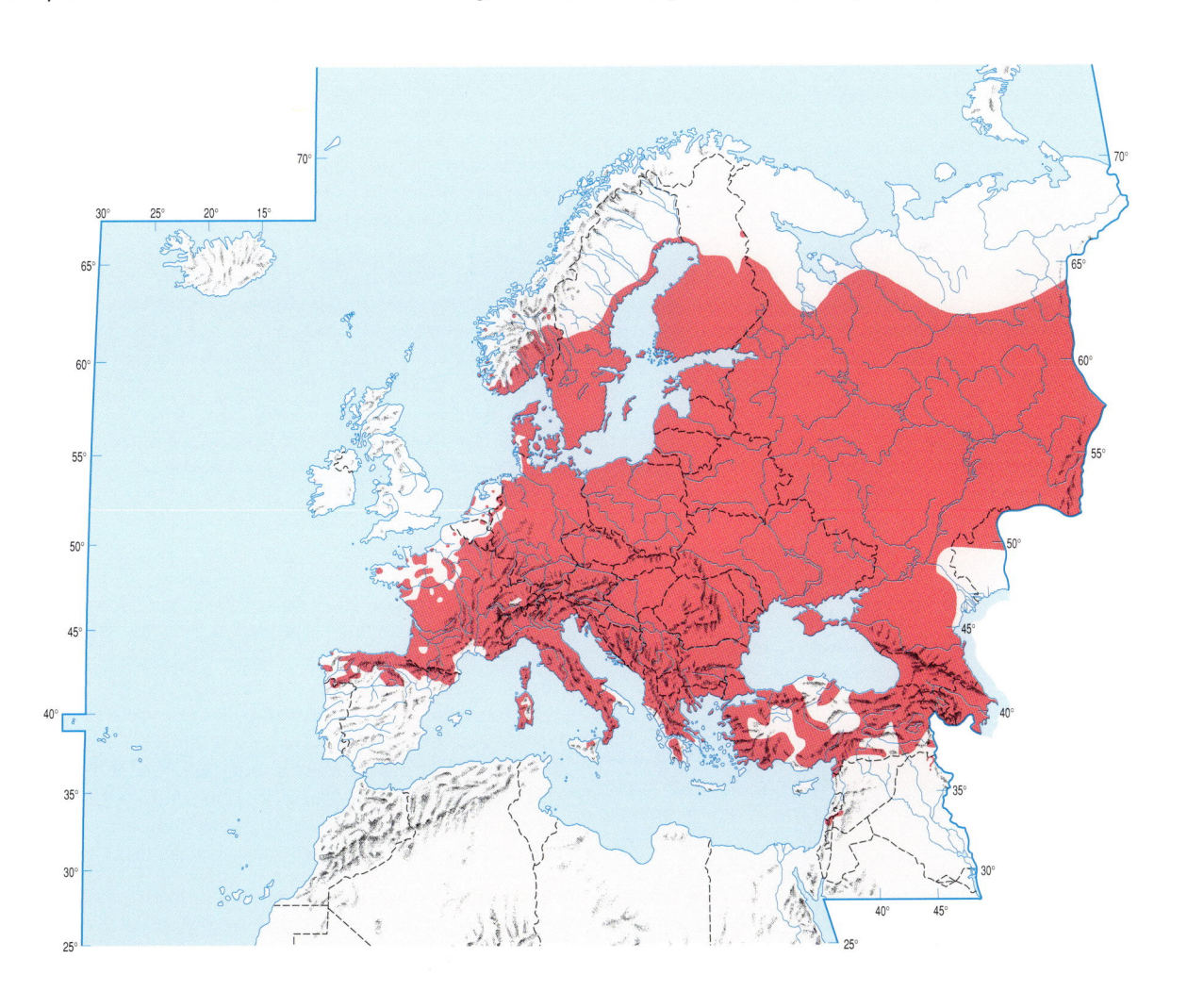

Field characters. 17 cm; wing-span 24–27 cm. Noticeably less bulky than Woodchat Shrike, with 5% shorter wings and tail; slightly smaller and shorter-tailed than Isabelline Shrike. Rather small, quite bold, raptorial passerine, with thick, hooked bill, fairly short wings, and relatively long tail; epitome of family in temperate Europe. Plumage of ♂ distinctive, with blue-grey and white head interrupted by black bill and mask, rufous back and inner wings, and white-edged black tail diagnostic; ♀ and immature essentially brown above, dull white below with much barring.

Adult ♂ unmistakable but paler, redder-tailed ♀ and immature subject to confusion with Brown Shrike and (particularly) Isabelline Shrike. Separation from Isabelline Shrike requires close, patient concentration on mantle, scapulars, and rump (always visibly barred in Red-backed Shrike but not so in Isabelline Shrike), tail (always white-sided and -tipped in fresh Red-backed Shrike but only white-tipped in fresh Isabelline Shrike, with diagnostic pale ginger sides), white mark at base of primaries (not recorded on ♀ and immature Red-backed Shrike but sometimes present on immature Isabelline Shrike), scaling on underparts (always visible or even strong on Red-backed Shrike but usually indistinct on Isabelline Shrike), and colour of bill-base (at most yellowish on Red-backed Shrike but pink on Isabelline Shrike). Flight between close perches is direct, swooping down from one and up to the other; fast and agile after prey, with characteristic 'untidy' downward pounce onto ground and usually immediate return to perch; flight over long distance markedly undulating, with shooting bounds. Habit of open perching on look-out for prey makes it conspicuous when feeding but at other times more secretive, sitting within outer foliage particularly when tired. When excited, indulges in much tail movement in form of either loose flick or curving swing, accompanied by partial spreading which causes 'flash' of white basal patches in ♂. Confiding if not pressed.

Habitat. Breeds in middle latitudes of west Palearctic in temperate, Mediterranean, and steppe climates, mainly continental and lowland, from July isotherm of 16°C upwards. Requires sunny, sheltered, warm, dry or even semi-arid, and level or gently sloping terrain, with scattered or open growth of bushes, shrubs, or low trees providing hunting look-out posts commanding areas of short grass, heath, or bare soil suitable for small prey. In Alps, ascends mostly to 1000 m, exceptionally 1850 m; in Caucasus usually to 2000 m, less often on subalpine meadows to 3200 m; in Sicily to 1400 m. In England, before recent virtual extinction, occupied neglected overgrown patches, heaths, open downs, overgrown orchards and gardens, hedgerows, and scrub along railways or roadsides, with tendency to nest near streams or pools attractive to insects.

Distribution. Range has contracted in Britain, Belgium, Netherlands, Denmark, Iberia, and Ukraine, but has expanded in Norway. Elsewhere, numbers greatly reduced in many areas, without apparent reduction of range on broad scale. Decline probably due mainly to loss and fragmentation of habitat resulting from afforestation and agricultural intensification, with increased use of pesticides causing loss of food resources. In northern and western periphery of range, breeding affected by cooler, wetter summers. BRITAIN. Major contraction south-eastward since late 19th century. Bred north to Yorkshire and Cumbria in 1850, but few north of Wash by 1950, and main stronghold in East Anglia by 1971. Now on verge of extinction.

Probable causes adverse climatic trend, fragmentation of habitat, and persecution. NETHERLANDS. Formerly bred over whole country except parts of west and north. NORWAY. Spread in south and west since 1970s, new breeding areas including small coastal islands and higher parts of inland valleys. SWEDEN. Has spread north 1970–90. LITHUANIA. Commonest in east and south. MALTA. May have bred 1972. RUSSIA. Locally common in Vologda region. BELARUS'. Widely distributed and common. UKRAINE. Widespread and evenly distributed in west. TURKEY. Widespread and locally very common in north-west, in Taurus range, and in eastern valleys; fewer elsewhere. SYRIA. June records at Kassab in 1981, and over several years at same site in Halbun (Anti-Lebanon), suggest probable breeding.

Accidental. Iceland, Faeroes, Ireland (near annual in autumn), Tunisia, Algeria, Morocco, Madeira, Canary Islands.

Beyond west Palearctic, extends east to upper Ob' and central Altai in West Siberia, also to north-west Iran.

Population. Stable Denmark (following decrease), Norway, Latvia, Hungary, Moldova, Belarus', Russia, Bulgaria. Most other countries report decreases, marked in Britain, Sweden, Finland, and Rumania. BRITAIN. Strong long-term decline, accelerating from mid-1900s. Over 300 pairs 1952, 80–90 pairs in 1971. In 1988–93, maximum total, including possible breeders, 5–13 pairs; but no confirmed breeding 1989 or 1993, only 1 pair 1988, 1990–92, 1994. FRANCE. Over 100 000 pairs. BELGIUM. 550–900 pairs 1989–91; only population in extreme south is stable. c. 570 pairs 1973–7, a tenth of 1930s population. LUXEMBOURG. 4000–5000 pairs. NETHERLANDS. 150–220 pairs 1992. In peat-moor reserve, Bargerveen, population grew from 2–4 pairs 1978 to 126 pairs 1994. GERMANY. 150 000 pairs in mid-1980s. In east, 60 000 $\pm$ 25 000 pairs in early 1980s. DENMARK. 1000–3000 pairs 1987–8. NORWAY. 5000–10 000 pairs 1970–90. SWEDEN. 20 000–100 000 pairs in late 1980s; over 50% loss 1970–90. FINLAND. 50 000–80 000 pairs in late 1980s. ESTONIA. 20 000 pairs in 1991. LATVIA. 20 000–40 000 pairs in 1980s. LITHUANIA. Common. POLAND. 50 000–300 000 pairs. CZECH REPUBLIC. 25 000–50 000 pairs 1985–9. SLOVAKIA. 65 000–130 000 pairs 1973–94. Decrease at lower elevations, stable or increasing at higher elevations. HUNGARY. 60 000–80 000 pairs 1979–93. AUSTRIA. 10 000–15 000 pairs in 1992. SWITZERLAND. 8000–12 000 pairs 1985–93. SPAIN. 240 000–500 000 pairs. PORTUGAL. 100–1000 pairs 1978–84. ITALY. 30 000–60 000 pairs 1983–95. GREECE. 20 000–50 000 pairs. ALBANIA. 10 000–30 000 pairs 1981. YUGOSLAVIA: CROATIA. 200 000–300 000 pairs. SLOVENIA. 20 000–30 000 pairs. BULGARIA. 100 000–1 million pairs. RUMANIA. At least 600 000–800 000 pairs 1986–92. RUSSIA. 100 000–1 million pairs. BELARUS'. 50 000–70 000 pairs in 1990. UKRAINE. 200 000–210 000 pairs in 1986; slight decrease overall, but stable with local fluctuations in west, where c. 80 000 pairs. MOLDOVA. 60 000–80 000 pairs in 1988. AZERBAIJAN. Common; locally very common on southern slope of Great Caucasus. TURKEY. 50 000–500 000 pairs. ISRAEL. Fairly common.

Movements. Migratory, wintering in eastern tropical and southern Africa; north to south-east and coastal Kenya, but main bulk of population from Zambia and Malawi southwards.

A classic case of loop migration, northward passage in spring following more easterly course than autumn passage, and also notable for concentration of migration routes across and round eastern end of Mediterranean, even by populations breeding in extreme west of Europe. Nocturnal migrant. Data suggest birds tend to feed on other passerine migrants on passage, rather than building up fat reserves prior to migration.

Birds leave northern, western and central European breeding grounds from late July, mostly in second half of August and early September. General direction of movement south-east or SSE towards eastern Mediterranean, but major changes in direction on passage through Europe regular at least in some populations. Thus breeders in extreme south-west of range (northern Spain and south-west France) migrate initially east, or even slightly north of east, to northern Italy and Greece, then alter course to south-east; and many Scandinavian and Finnish breeders migrate west of south to Italy before altering course to south-east. Birds crossing Mediterranean make landfall on North African coast almost entirely east of 20°E, with only isolated records to west, except Malta, where regular in small numbers. Passage through Egypt mainly mid-August to early November. First birds reach extreme southern wintering areas in late October.

Northward migration from winter quarters begins 2nd half of March; in extreme south, all birds gone by about middle of April. More easterly course of northward migration, compared with autumn, evident in East Africa. Thus spring passage mainly east of Lake Victoria (autumn passage especially marked west of 33°E). Further north, divergence between spring and autumn routes more pronounced: absent from Sudan in spring (common in north in autumn); in Somalia, spring passage outnumbers autumn passage 100:1, with corresponding copious arrivals on north shore of Gulf of Aden, and common in most of Arabia; spring migrants rare in Egypt, and mainly in east; in Israel also mainly in east. Birds arrive on breeding grounds in April in Israel, in Europe mostly May.

Food. Mainly insects, chiefly beetles; also other invertebrates, small mammals, birds, and reptiles. Most prey located from exposed, though usually low, perch using sit-and-wait strategy. Large moving insects spotted up to 30 m away, and caught in bill after shallow direct glide, sometimes with outstretched neck, which may terminate in brief hover before bird drops into vegetation; also drops straight onto prey below perch. Vehemence of this action may be shown by strikingly worn forehead plumage. Flying insects taken in rapid, sometimes lengthy, pursuit flight. Almost always carries even small prey back to perch for consumption or impaling. Many prey items impaled on thorns, broken twigs, barbed wire, etc., in caches (larders), though individual items may be widely scattered throughout territory. Invertebrate prey dealt with by having extremities, wing-cases, etc., removed by beating on substrate, or picked off by bill while held under, or in, foot or after impaling on thorn, etc. Vertebrate prey killed by blow to back of head or neck; brain often consumed on ground and animal decapitated before flying to perch or cache. Large vertebrate prey carried with difficulty back to perch, bird flying with body held at c. 45°. Unable to dismember vertebrate prey held only under foot so impaling necessary. Impaled vertebrates always have head and foreparts consumed first. Great majority of bird prey are nestlings and fledglings; adults taken are generally weak or injured, and any healthy ones probably taken at nest; probably only takes birds when insects unavailable.

Social pattern and behaviour. Mostly solitary outside breeding season, but occasionally several birds in loose association; ♀♀ apparently less solitary than ♂♂, often foraging closer together. Territoriality much reduced in winter quarters, except on passage, on first arrival and shortly before spring departure, but at all times each bird at least maintains small feeding territory. Mating system monogamous, but unpaired ♂♂ not uncommonly help to raise broods. Almost all incubation and brooding by ♀. Food for young initially brought exclusively by ♂, later by both sexes. Late broods often fed by only 1 parent, of either sex, other having left sometimes even when young still in nest. Also late in season (July onwards) unpaired ♂♂ (and less commonly ♀ failed breeders) increasingly tend to associate with families and help feed and defend nestlings and fledglings; unpaired ♂♂ often tend fledglings from several broods. ♂ establishing breeding territory takes circuitous route through area and sings from tree-tops and other high vantage points. Song and bouts of calling exchanged by neighbours. Very little song after nest-building and mating. ♂ arrives on breeding territory typically 1–3 days before ♀. Nest-building starts c. 2 days after pair-formation. Courtship-feeding a major constituent of courtship and of ♀'s nutrition during incubation and brooding.

Voice. Advertising song of ♂ a subdued jerky warbling, sometimes sustained for 10 min or more, incorporating remarkable mimetic ability, given typically from high exposed perch; rarely heard and almost only before pairing in spring, largely giving way to courtship song once ♀ arrives on territory. Courtship song consists of rapid excited twittering, associated exclusively with sexual display and thus never heard in ♀'s absence; largely replaces advertising song when ♀ arrives on territory, and heard especially until start of laying. Contact and excitement calls varied: include repeated loud, far-carrying, harsh chirp, 'chah' or 'kscha', given by ♂ at each vantage point, also occasionally in flight, when patrolling territory before ♀ arrives, thus proclaims ownership to rivals (who often reciprocate call) and prospective mates; also disyllabic 'chee-uck' or 'chu-ik', given by both sexes, especially during courtship. Commonest alarm-call, given by both sexes, short hard 'chack' or 'tek', often given in rapid series, accompanied by violent tail movements. Repeated drawn-out noisy 'tchraaa' also given in alarm and confrontations.

Breeding. SEASON. Eggs laid from early or mid-May (according to latitude) to July. Several replacement clutches may be laid, but 2 broods unusual. SITE. Generally low in dense, often thorny bush but sometimes high and easily visible in trees; in some areas not uncommonly in woodpiles. Nest: loose foundation of often green plant stems (some thick or woody), roots, grass, lichen, hair, etc., compactly lined with grass, hair, moss, fur, reed or reed-mace flower-heads, plant down, etc.; main structure often includes string, cloth, paper, etc. EGGS. Very variable in shape and colour; sub-elliptical to short sub-elliptical or oval to short oval, very slightly glossy, pale green, pinkish, buff, or creamy-white with band of light brown, olive, brownish-red, grey, or purple specks and small blotches near broad end; markings sometimes scattered over whole surface or even present only at narrow end. Clutch: 3–7. INCUBATION. 14 (12–16) days. FLEDGING PERIOD. 14–15 days.

Wing-length: ♂ 91–99, ♀ 89–98 mm.
Weight: ♂♀ mostly 25–35 g.

Long-tailed Shrike *Lanius schach*

PLATES: pages 1429, 1449

DU. Langstaartklauwier   FR. Pie-grièche schach   GE. Schachwürger   IT. Averla dal dorso rossiccio
RU. Длиннохвостый сорокопут   SP. Alcaudón cabecinegro   SW. Rostgumpad törnskata

Field characters. 20–23 cm, of which tail up to 10 cm; wing-span 25–28 cm. Up to 10% shorter overall than Great Grey Shrike, with slightly shorter bill, relatively longer and more graduated tail, and 15% shorter wings; 30% larger than Red-backed Shrike and allies. Plumage pattern oddly mixes those of grey and red-tailed *Lanius*: black face-mask and grey fore-upperparts recall Lesser Grey Shrike, but rufous lower back and rump and long black, buff-sided and -tipped tail are diagnostic.

Adult unmistakable in west Palearctic. Flight form and action recall Great Grey Shrike. Notably bold and fearless. Calls harsh, squealing, or yapping.

Habitat. Breeds extralimitally in low and tropical latitudes in warm arid climates, mainly in lowlands but also up to limits of deciduous tree cover in mountains, typically up to 1500–1700 m, but in valleys of small poplar-fringed rivers occurs in woodland almost up to glaciers at 2600 m, and in Himalayas even higher. In central Asia, attached to cultivation with frequent planted trees, such as shelter-belts along roads and railways, avenues and parks in towns, and orchards in villages.

Distribution. Breeds from Kazakhstan, Turkmenistan, and Afghanistan south-east through India, southern China, south-east Asia, and Indonesia to New Guinea.

Accidental. Hungary: Fehértó, April 1979. Turkey: Birecik, 1st-autumn (specimen), September 1987. Israel: adult ♂ at Sede Boqer, November 1982–February 1983.

Movements. Varies between migratory (including some altitudinal movements) and resident across range. Northern and western populations vacate northern parts of breeding range entirely. Information scanty on migration routes.

Wing-length: ♂ 94–102, ♀ 93–99 mm.

Lesser Grey Shrike *Lanius minor*

PLATES: pages 1437, 1449

DU. Kleine Klapekster   FR. Pie-grièche à poitrine rose   GE. Schwarzstirnwürger   IT. Averla cenerina
RU. Чернолобый сорокопут   SP. Alcaudón chico   SW. Svartpannad törnskata

Field characters. 20 cm; wing-span 32–34.5 cm. 15% smaller and more compact than Great Grey Shrike, with noticeably stubbier bill, longer wings, and shorter, less rounded tail; up to 15% larger than Woodchat Shrike, with (proportionately) wings 10% longer but tail rather shorter. Medium-sized to large but not strikingly long shrike, with rather stubby bill, apparently blunt head (due to plumage pattern). Adult essentially grey, black, and white above and pink-white below, with black face-mask (encompassing forehead), wide white bar across primaries, and wide white edges to tail. Juvenile and immature rather pale and not strongly barred, with quite long black-brown patch on ear-coverts and whitish tips to scapulars which form pale patch over folded wing.

Adult and 1st-winter bird distinctive if seen well: confusion with Great Grey Shrike and Southern Grey Shrike ruled out by Lesser Grey Shrike's combination of stubby bill, extensive black face-mask, much bolder white primary-patch, proportionately rather short, noticeably white-sided tail, and long, pointed wings (with wing-point equal to $c.\ 1\frac{1}{4}$ times length of

(FACING PAGE) Lesser Grey Shrike *Lanius minor*: **1** ad ♂ breeding, **2** ad ♀ breeding, **3** 1st winter, **4** juv, **5** '*turanicus*' (central Asia) 1st winter.
Great Grey Shrike *Lanius excubitor* (p. 1440). *L. e. excubitor*: **6** ad ♂, **7** ad ♀, **8** 1st winter, **9** juv. *L. e. homeyeri*: **10** ad ♂.
Southern Grey Shrike *Lanius meridionalis* (p. 1442). *L. m. meridionalis*: **11** ad ♂, **12** ad ♀, **13** juv. *L. m. algeriensis*: **14** ad ♂. *L. m. koenigi*: **15** ad ♂. *L. m. elegans*: **16** ad ♂. *L. m. aucheri*: **17** ad ♂. *L. m. pallidirostris*: **18** ad ♂, **19** 1st winter.

Norman Arlott 1987

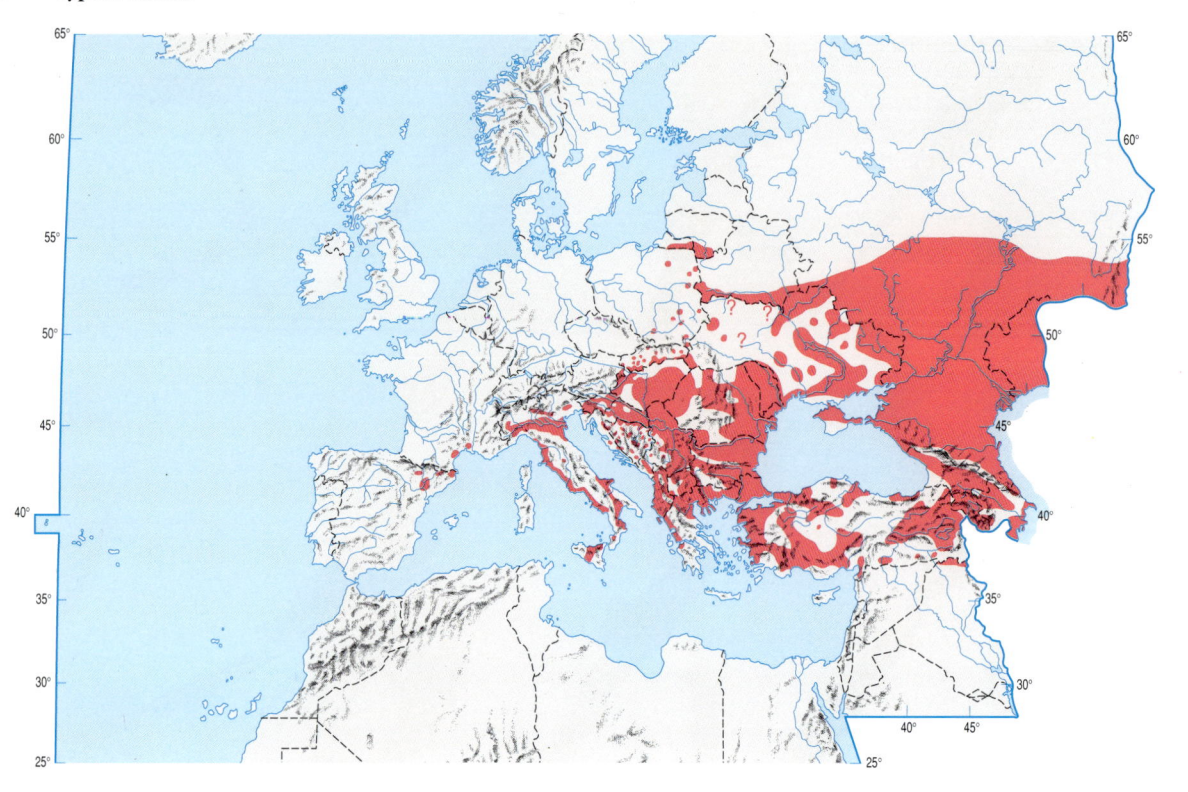

exposed tertials; c. $\frac{3}{4}$ of length of tertials in Great Grey Shrike). Even at some distance, rather erect stance and relatively compact form recognizable. Less distinctive in full or partial juvenile plumage, with pale edge to scapulars suggesting bolder mark of juvenile Woodchat Shrike, but that species is smaller and slighter, with more obvious barring above and below, more diffuse eye-patch, duller primary-patch, and narrower white edges to tail.

Habitat. Breeds in middle continental latitudes of west Palearctic, in temperate, Mediterranean, and steppe climates, in regions from July isotherm of 17°C upwards. Differs conspicuously from Great Grey Shrike in requirement for warmer and more benign climate, and open habitat with plenty of scattered or grouped trees and bushes. Need for drier and sunnier conditions than other European *Lanius* possibly connected with more specialized diet of large insects.

Distribution. After earlier northward expansion, range and numbers declined dramatically between mid-1800s and early 1900s in western, north-central, and northern areas; following regional recoveries in 1930s and 1960s, further sharp decline in recent decades. Causes of decline not fully known, but probably primarily climatic factors, low summer temperatures and heavy rainfall being very unfavourable for breeding. Monoculture practices in farming, reduction of prey due to pesticide use, and persecution presumably also involved, and perhaps adverse conditions in winter quarters. FRANCE. In 19th century, widespread except in north and south-west; dramatic decline since, and now very few sites. BELGIUM. 7 breeding records 1875–1930; none since. LUXEMBOURG. Rare breeder Moselle valley 19th century; no sightings since 1946. GERMANY. Once fairly widespread, but now apparently extinct. In west, restricted to Baden-Württemberg by 1975, where decreased from 120 pairs in 1950 to 7 pairs in 1970, 1 probable pair 1987. In east, latest breeding records 1965 and 1976. ESTONIA. Said to have bred in 1980s, but no proof. LATVIA. At end of 19th century common in west, sporadic in north-east; marked decrease beginning probably in 1920s. In 1980–84 one pair recorded, breeding not confirmed. POLAND. Perhaps no longer regular breeder. Widely distributed up to 1900. CZECH REPUBLIC. Now almost extinct. SLOVAKIA. Marked contraction southward, occurring chiefly on southern slopes of foothills. AUSTRIA. Now confined to small area near Neusiedler See. SWITZERLAND. Scarce and sporadic breeder in west up to 1950s; last recorded breeding 1972. SPAIN. Evidence of recent colonization. First observation 1947, first indications of breeding 1962. ITALY. Marked decrease in centre and north; now rare even in south. YUGOSLAVIA: SLOVENIA. Marked decrease. RUMANIA. Widespread, especially Dobrogea, Moldavia, and Walachia. BELARUS'. Now confined to extreme south. Fairly common in early 20th century. UKRAINE. Slight decrease. TURKEY. Apparently widespread.

Accidental. Faeroes, Britain, Ireland, Netherlands, Denmark, Norway, Sweden (annual), Finland (annual), Estonia, Portugal, Balearic Islands, Tunisia.

Beyond west Palearctic, extends east to upper Ob', northern Tien Shan mountains, and north-east Afghanistan.

Population. Most countries report declines, severe in France, Poland, Slovakia, Austria, Slovenia, and Belarus'. Apparently stable Croatia, Bulgaria, and Russia. In Rumania, variously reported as stable, or markedly declining. FRANCE. 25–30 pairs.

LITHUANIA. Very local and rare. SLOVAKIA. 400–600 pairs. HUNGARY. 5000–8000 pairs 1979–93. AUSTRIA. Under 10 pairs in 1993. SPAIN. 45–90 pairs. ITALY. 1000–2000 pairs 1983–95. GREECE. Not common, except very locally. 2000–3000 pairs. ALBANIA. 2000–5000 pairs in 1981. YUGOSLAVIA: CROATIA. 3000–4000 pairs. SLOVENIA. 20–30 pairs. BULGARIA. 1000–10 000 pairs. RUMANIA. 60 000–100 000 pairs 1986–92. RUSSIA. 10 000–100 000 pairs. BELARUS'. 50–200 pairs in 1990. UKRAINE. 3000–3500 pairs in 1986. MOLDOVA. 10 000–15 000 pairs in 1988. AZERBAIJAN. Common. TURKEY. 10 000–100 000 pairs. IRAQ. Rare.

Movements. Migratory, entire breeding population wintering in southern Africa, from extreme southern Angola and Namibia east to southern Mozambique and parts of South Africa.

Loop migrant, with spring passage further east than autumn. In autumn, birds from west of range head south or south-east over Greece and Aegean Sea, to enter Africa on narrow front principally through Egypt. On-going passage mainly between c. 20°E and Lake Victoria, requiring change of heading to west of south. Spring passage further east (in Zimbabwe chiefly east of c. 30°E, in autumn mainly west of this longitude) with higher numbers in East Africa, and main exodus apparently via Ethiopia and Somalia continuing through Middle East. Movement begins in central and western Europe late July to August, with peak in mid- to late August and stragglers into October. Crosses Aegean Sea mid-August to end of September, and passage in Cyprus late July to beginning of October. Arrives in Egypt from early August, with late birds early October to late November. Arrives in winter quarters from late October (Zimbabwe) to late November (Cape Province, South Africa).

Present in southern Africa until late March (exceptionally early April); main passage in Zambia first half of April. Widespread in Ethiopia and Somalia end of March to mid-May. Passage through Israel, Jordan, Iraq, and Syria mid-April to mid-May. Arrives on central European breeding grounds from early May.

In Britain and Ireland, 103 records 1958–89, mid-May to end of July (peaking end of May and beginning of June) and fewer mid-August to mid-November; roughly equal numbers in spring and autumn on Shetland and east coast. Annual in southern Sweden: 129 up to 1988, great majority May–August. Regular in Denmark, where at least 1 record almost every year 1965–88, generally May–August.

Food. Almost wholly insects, mainly beetles. Hunts principally from exposed look-out perch 1–6 m high (typically branch on side of bush or tree), flying down to take insects on ground below. Will use many perches throughout breeding territory, chief requirements being all-round view plus good sight of ground. Prey consumed whole if small, usually after return to perch, or if larger may be eaten held in foot. Impaling of food for storage or dismemberment very unusual.

Social pattern and behaviour. In African winter quarters mostly solitary and territorial, though sometimes apparently in pairs and not uncommonly in loose or more close-knit groups of up to 10 birds. Gatherings of spring migrants reported, and several birds typically arrive together on breeding grounds so have probably migrated and perhaps even wintered as a group. Little evidence for other than monogamous mating system. Incubation and brooding by ♀; young fed by both sexes, including after fledging. Breeding territories typically in neighbourhood groups of 3–7(2–10) pairs. Tendency to form groups more pronounced than in Red-backed Shrike and Woodchat Shrike. Nesting associations with falcons and Fieldfares (presumably for protection) regular in some areas. Loud song given from open perch almost exclusively by unpaired ♂♂; without any territorial function. Occupation of territory advertised mainly by slow display-flights in wide sweeps across territory, by both members of pair or (most persistently) by unpaired ♂♂. Courtship-feeding regular, from early in pair-formation to end of incubation.

Voice. Song of ♂ a varied and sustained babbling chatter, with some harsh, strangled, or grating sounds (includes own contact- and other calls), also whistles and trills; contains excellent mimicry of bird and mammal sounds. Loud song, given almost exclusively by unpaired ♂♂ without obvious stimulus (perhaps serves to attract ♀), is comparable in volume and delivery to Woodchat, much louder than Red-backed Shrike. Quieter song of roughly same composition given during courtship. Calls given in territorial advertisement or defence, for warning, or in alarm, very varied; mostly harsh, monosyllabic or disyllabic, uttered singly or, at high intensity, in rattling sequences.

Breeding. SEASON. Eggs laid May to early July. One brood. SITE. On lateral branch up to c. 4 m from trunk of tree (sometimes against trunk), in fork or in crown, generally at good height above ground. Nest: well-made structure with loose foundation of twigs, grass, rootlets, string, etc., often with high proportion of green plant stems, especially of aromatic species, with leaves and flowers attached; lined with rootlets, hair, feathers, etc., though often completely without lining. EGGS. Sub-elliptical to short oval and more rarely long oval, slightly glossy. Pale bluish-green, seldom cream or buff, with spots and small blotches of olive or olive-brown and lavender-grey concentrated towards broad end, remainder sparsely marked. Clutch: 5–6 (3–9). INCUBATION. (12–)15–16 days. FLEDGING PERIOD. 16–18 (13–19) days.

Wing-length: ♂ 114–126, ♀ 115–124 mm.
Weight: ♂♀ mostly 41–50 g.

Great Grey Shrike Lanius excubitor

Du. Klapekster   Fr. Pie-grièche grise   Ge. Raubwürger   It. Averla maggiore
Ru. Серый сорокопут   Sp. Alcaudón real   Sw. Varfågel

Field characters. 24–25 cm; wing-span 30–35 cm. Slightly larger and more attenuated than Southern Grey Shrike. 20% longer than Lesser Grey Shrike, with (proportionately) 10–15% longer bill, longer, more domed head, shorter wings, and 10% longer and more graduated tail. Largest shrike of west Palearctic, with notably aggressive look and beautiful grey, white, and black plumage. Most striking characters are quite long black bill and long black eye-patch (not extending over bill), white patch along scapulars, white bar across base of flight-feathers (varying with race), and long, graduated black, white-edged tail. Black in centre of tail is wide at base but narrow towards end (opposite of Lesser Grey Shrike). Recently separated from Southern Grey Shrike, to which species so-called Steppe Grey Shrike (*pallidirostris*) now assigned.

The only common shrike of northern taiga in summer and of temperate Europe in winter; range overlaps uncommonly with other shrikes except in late spring and autumn passage, while large size, attenuated form (but shorter wings), and head, wing, and tail pattern allow rapid distinction from Lesser Grey Shrike. Distinctions from Southern Grey Shrike covered under that species. Flight fastest and most powerful of *Lanius*; over short distance, characteristic sequence of dive to ground level, forward acceleration with short bounds, and then upward sweep (with wings and tail spread) to new perch; over longer distance, bursts of wing-beats produce markedly undulating progress, often at fair height. Gait hopping or leaping, looks clumsy and untidy on rare visits to ground. Perches more openly and higher than other *Lanius*, waving and spreading tail and occasionally bobbing body.

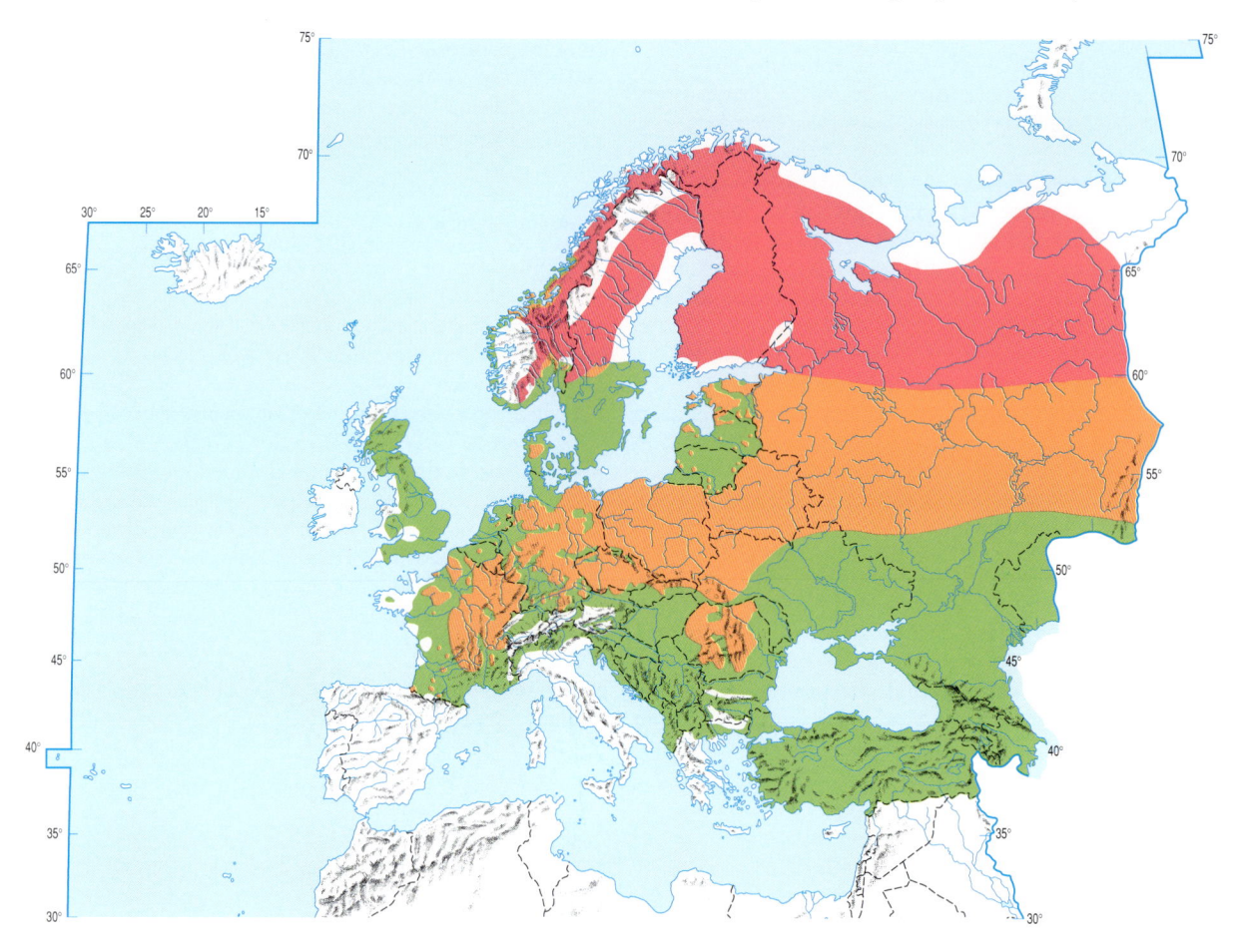

Habitat. Breeds in west Palearctic, from high to middle latitudes, from fringe of arctic and throughout subarctic and boreal to temperate climates, but avoiding oceanic and exposed areas, and not normally ascending above 1000 m. Avoids steep, rocky, bare, and densely forested areas, but in parts of range will occupy forest clearings, glades, and margins, and prefers taller trees than other west Palearctic shrikes, including coniferous as well as broad-leaved species. Over much of range, however, prefers open country with frequent bushes and trees, free standing or in small groups, which may be in marshes, heaths, parkland, or even cultivated areas and orchards, although infrequently near or within human settlements. Often finds overhead wires, poles, etc., serviceable as look-outs, especially where suitable trees or bushes are scarce. Where food plentiful, will accommodate itself to fairly sparse and stunted vegetation, and is characterized by unusual adaptability to any habitat offering minimum of look-out posts, nesting cover, and accessible prey of wide variety of animal species. Thus also able to winter in cold climates intolerable to other shrikes, using habitats equivalent although not identical to those used in summer.

Distribution. Widespread decline in range and population due chiefly to loss and degradation of habitat through agricultural intensification (e.g. fewer nest-sites and perches), with increased use of pesticides reducing insect prey. Some expansion in Fenno-Scandia. FRANCE, BELGIUM, NETHERLANDS. Range contraction. GERMANY. Patchily distributed following decline; southern breeding sites now so isolated that survival in doubt. Main strongholds in Thüringen and Mecklenburg-Vorpommern. DENMARK. First bred 1927. Recent increase. NORWAY. Has spread south since c. 1970. SWEDEN. Rapid expansion eastwards and southwards 1960–75, due to forestry changes producing large open (clear-felled) areas. FINLAND. Range contraction during early 1900s, by 1930 almost absent from southern half; since expanded in south. LATVIA. Has probably decreased. POLAND. Widespread but patchily distributed; in earlier years increased due to forest clearance. Probably commonest in Silesia. AUSTRIA. General decline since c. mid-1960s. Bred Rhine valley until 1978; now restricted to small area near northern border. SWITZERLAND. Widespread in 1950, but now extinct after rapid decrease since 1960s; last bred 1985. RUSSIA. May no longer breed in Moscow region (drainage of peat bogs); limit of breeding range may now be through Kalinin region. UKRAINE. Slight decrease. Breeds mostly in north-west.

Accidental. Faeroes, Ireland, Canary Islands. Vagrants in Spitsbergen, Bear Island, and Iceland probably also this species (rather than Southern Grey Shrike).

Beyond west Palearctic, widespread across northern Asia east to Chukotskiy peninsula, south to northern Kazakhstan, Tien Shan mountains, northern Mongolia, and Sakhalin island. In North America, breeds from Alaska across northern mainland of Canada to Labrador.

Population. Most populations decreasing, but stable Denmark (or perhaps increasing), Norway, Estonia (after marked decrease in 1960s), Russia, and Belarus', and fluctuating Sweden and Poland. FRANCE. In 1970s, 1000–10 000 pairs (including Southern Grey Shrike). BELGIUM. 130–160 pairs 1989–91. c. 350 pairs 1973–7. In Flanders still declining, but in Wallonia now apparently stable. LUXEMBOURG. 50–100 pairs. NETHERLANDS. 15–40 pairs 1989–91. GERMANY. 1200–1500 pairs 1990–91 (50% decline in 10 years in west). In east, 1900 ± 550 pairs in early 1980s. DENMARK. 10–25 pairs 1984–94. NORWAY. 5000–10 000 pairs 1970–90. SWEDEN. 1000–10 000 pairs in late 1980s. FINLAND. 5000–10 000 pairs in late 1980s. ESTONIA. 200–400 pairs 1991. LATVIA. 100–150 pairs in 1980s. LITHUANIA. Rare. POLAND. 2000–4000 pairs. CZECH REPUBLIC. 1000–2000 pairs 1985–9. SLOVAKIA. 500–1000 pairs 1973–94. AUSTRIA. Under 20 pairs in 1993. RUMANIA. 1000–3000 pairs 1986–92. RUSSIA. 100 000–1 million pairs. BELARUS'. 600–1200 pairs in 1990. UKRAINE. 900–1200 pairs in 1988.

Movements. Resident and migratory. Extreme northern populations vacate breeding areas completely, and some southern populations apparently sedentary; other populations consist of long-distance migrants, short-distance migrants, and sedentary birds. Except in France, migrations of nominate *excubitor* stop short of range of Southern Grey Shrike; winters within breeding range, west to Britain, and south to northern coast of Mediterranean (no evidence for regular crossing) east to Turkey and Caucasus. Race *homeyeri* partly resident, also migratory, moving to north-west China, Kazakhstan, northern Iran, northern Caucasus, Crimea, and parts of Balkans; very rare vagrant to western Europe. Marked annual variations in numbers of birds migrating in northern populations probably result from annual differences in breeding success and summer survival. Numbers also fluctuate in wintering areas, some individuals making hard-weather movements.

Peak of autumn passage in northern Europe mainly in October. Peak of spring passage in March or April; arrival on northern breeding grounds mainly in April.

Food. Large insects, chiefly beetles, small mammals, reptiles, and birds. Main hunting strategy waiting and watching from high vantage point; may turn round in complete circle while watching and frequently flies to new look-out. Flying insects, especially large Hymenoptera and beetles, taken after flight vertically upwards or pursued in direct flight; invertebrates on ground captured after swooping, undulating flight, often followed by hovering like Kestrel, or are pounced on if immediately below perch. May also forage for invertebrates on ground for long periods, especially in cold or wet weather, hopping like thrush. Invertebrates snapped up in bill and eaten at once or (if large) may be taken back to perch to be dismembered or impaled in cache. After swooping flight and vertical drop lands near vertebrate prey, covering any remaining distance in fluttering hops; strikes prey with bill at back of head without grasping with feet. Will attack mammals up to size of stoat, and pursue bats in the air. Predation on birds rather exceptional and generally only when weather prevents hunting for mammals or invertebrates; rarely outflies other birds and captures healthy adults only with difficulty.

Impales and wedges prey in caches ('larders') in bushes (with and without thorns), on barbed wire, etc., more than other west Palearctic shrikes, probably because of more northerly distribution, hence need to secure food supply in poor weather, and because unable otherwise to dismember vertebrate prey.

Social pattern and behaviour. Mostly solitary outside breeding season, though occasionally 2–3(–10) together on autumn passage. In north-central Europe, individuals of both sexes occupy extensive winter territories, from beginning of September at earliest to end of March at latest, mostly mid-October to mid-March; preliminary pair-bonding takes place in winter territory. Winter territories are near but usually displaced from breeding territories although one pair-member may occupy area roughly coincident with breeding territory. Unless severe conditions arise, individuals sedentary throughout winter and also faithful to same territories in successive years. Mating system typically monogamous. Almost all incubation and brooding by ♀. ♂ sings from conspicuous perch in spring and autumn, occasionally also on warm winter days; appparently serves for attracting ♀ rather than territorial demarcation. In central and northern parts of range, shows marked, and largely amicable, association with colonies of Fieldfare, not infrequently nesting less than 10 m from Fieldfare nest in same tree. The often high incidence of association relative to general rarity of the 2 species argues for positive attraction, based on mutual protection against predators, especially Corvidae.

Voice. Song (usually by ♂, less by ♀) a continuous, quiet, low chattering or melodious warble, varied by harsh and raucous noises, and interspersed with mimicked song and calls of numerous other birds. In ♂, serves mainly to attract ♀ and seldom heard in her absence or in another context. Given probably by both sexes at end of breeding season when can also be heard (most commonly of all) from juveniles. From October–November (end of moult) until nest-building, a more rudimentary kind of song consisting of a succession of single call-type units or motifs is more usual than full song. Contact and excitement-calls include a loud drawn-out whistle 'kwiet' or 'kwieht' (possibly long-distance contact-call); soft drawn-out 'trüh', by both sexes (possibly short-distance signal for attracting mate), and variable 'chliep', 'chli', 'up' calls, probably sexually motivated, directed mainly at mate. Threat and alarm calls harsh, rattling, or screeching.

Breeding. SEASON. Geographically variable; eggs laid from end of March in south (southern Germany), from mid-May in extreme north (Lapland); mostly from April or early May in central areas. SITE. In fork of tree or branch, in outer twigs, or on top of broad branch; in conifers often close to or against trunk; in absence of tall trees typically in low thorny bush. Nest: solid, bulky foundation (sometimes woven onto support) of twigs (which may be pulled from nest tree), plant stems, grasses, moss, string, plastic, etc., lined with rootlets, flowers, plant down, bark fibres, hair, feathers, etc. EGGS. Sub-elliptical, sometimes short sub-elliptical or oval, smooth and slightly glossy. Colour variable, also within clutch; buff to bluish or greenish-white, heavily marked over whole surface with small blotches and spots of olive, purplish-grey, buff, reddish-brown, or brown, sometimes with weak concentration at broad end. Clutch: 4–7 (3–9). INCUBATION. 15–17 (14–19) days. FLEDGING PERIOD. 15–18 (14–20) days.

Wing-length: ♂♀ 106–123 mm.
Weight: ♂♀ 48–81, mostly 60–75 g.

Geographical variation. Slight in west Palearctic. *L. e. homeyeri*, from Ukraine and Crimea eastwards through a zone at *c.* 49–53°N, paler than nominate *excubitor* (rest of west Palearctic), with more extensive white on wings.

Southern Grey Shrike *Lanius meridionalis*

PLATE: page 1437

DU. Steppeklapekster FR. Pie-grièche méridionale GE. Südlicher Raubwürger IT. Averla maggiore meridionale
RU. Иберийский серый сорокопут SP. Alcaudón real meridional

Field characters. 24–25 cm; wing-span 28–32 cm. Similar in length to Great Grey Shrike but 5–10% shorter-winged, with slightly longer and rather heavier bill and longer legs. Southern counterpart of Great Grey Shrike, differing in structure and more variable plumage, with white wing-bar usually confined to bases of primaries, and less obvious white rim to tail. Adult from southern France and Iberia has noticeably darker, lead-grey upperparts and vineous-pink tinged underbody, giving bird distinctly darker overall appearance than Great Grey Shrike; also shows narrow, white supercilium, deeper black mask behind eye, less broad but more contrasting white fringes to outer and lower scapulars, while whiter throat and vent contrast with rest of underparts. Juvenile also distinguishable from Great Grey Shrike; upperparts paler and browner than adult but underparts pale buffish-pink, faintly barred on sides of breast and flanks. Other North African and Middle East races less distinctive but underparts usually clouded with grey. Vagrants of Asian steppe race *pallidirostris* suggest washed-out Great Grey Shrike and show distinctive pale bill and lores and cream or faint pink wash to lower face and underparts.

Habitat. Warm and dry Mediterranean and steppe zones south of range of Great Grey Shrike, extending in west Palearctic south into Sahel zone in Africa; less arboreal than Great Grey Shrike, apparently avoiding high trees. In southern Europe, in open mosaics of fruit or olive plantations, garrigue and maquis landscapes with thorny shrubs and scattered low trees, grassy places, etc.; in Extremadura, Spain, in water meadows. In North Africa, Middle East, and south-central Asia, in similar landscapes but also in semi-desert and savanna-type communities, and in true desert with sparse dwarf shrubs and small trees in oases, wadis, etc.

Distribution. FRANCE. Apparently stable. Most northerly breeding-site in Lozère, 30 km south of range of Great Grey Shrike. SPAIN, PORTUGAL. Stable. SAUDI ARABIA. Scarce breeder at Harrat Al Harrah reserve. EGYPT. Now rare in Nile delta; commoner in 1930s. ALGERIA. In north of range, less widespread in east than west. Breeds in southern Saharan oases, but not in northern ones. CANARY ISLANDS. Breeds Tenerife, Gran Canaria, Fuerteventura, and Lanzarote.

Accidental. Britain, Netherlands, Germany, Norway, Cyprus.

Beyond west Palearctic, extends from Caspian east through

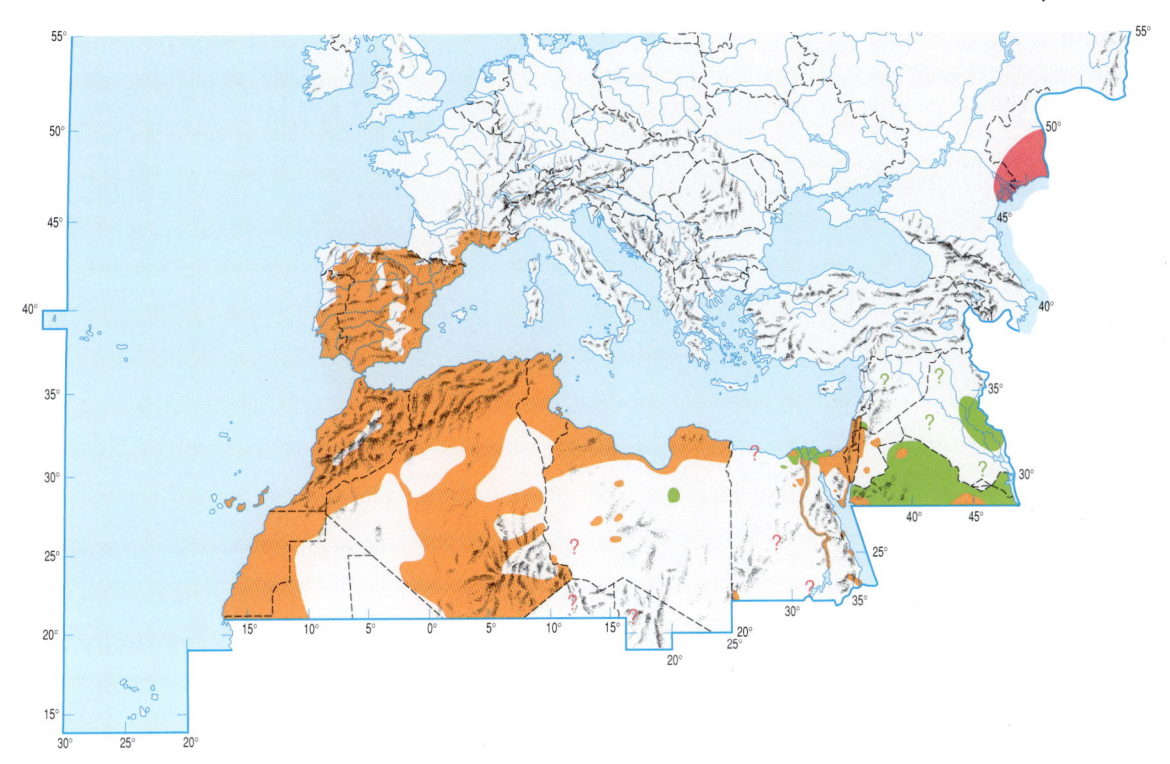

Uzbekistan, Kazakhstan (north to *c.* 48–49°N), and southern Mongolia to *c.* 110°E in northern China, and from Arabia and Iran to eastern India. Also breeds Africa in broad band from Mauritania east to Red Sea and northern Somalia.

Population. FRANCE. Perhaps 1000–2000 pairs. SPAIN. 200 000–250 000 pairs; slight decrease. PORTUGAL. 10 000–100 000 pairs 1978–84; stable. ISRAEL. Fairly common; some thousands of pairs. JORDAN, EGYPT. Fairly common. TUNISIA, MOROCCO. Common. CANARY ISLANDS. One tentative estimate of 1000–1500 pairs.

Movements. Many populations sedentary, some partially migratory; nominate *meridionalis* probably partially migratory, showing both erratic displacements (e.g. post-breeding wanderings in south-east France and Extremadura) and perhaps a regular migration of small numbers as far as North Africa via Straits of Gibraltar; however, most populations probably resident. Very rare vagrant to central France and north-west Europe. Birds in North Africa disperse in winter to unknown extent, but probably mainly sedentary; in Tunisia, some movement possible from inland areas to coast; numbers in south-east Morocco increase slightly outside breeding season and migratory movement reported in extreme west; Saharan race *elegans* common in winter in lower Sénégal, some reaching Gambia. Some Middle East populations regularly winter in north-east Africa, mainly Sudan. Race *pallidirostris* partially migratory, northern birds being long-distance migrants to north-east Africa perhaps as far south as Sudan–Kenya border, and to Middle East from Arabia and Iraq to Pakistan; rare vagrant to western Europe. Weak autumn movement at Straits of Gibraltar September–October; wintering birds present in north-east Africa and Arabia until March.

Food. As Great Grey Shrike but with higher proportion of reptiles; habitat characteristics mean that look-out perches are lower.

Social pattern and behaviour. In the main as Great Grey Shrike; generally monogamous, but in Negev, Israel, some ♂♂ simultaneously polygynous, holding much larger territories than monogamous birds (*c.* 750 compared with *c.* 60 ha). In Extremadura, in contrast to Great Grey Shrike, some evidence that pair may stay together, being sedentary all year, and in south-east Morocco breeding territories not always abandoned in winter. In Negev and parts of Spain, ♂♂ occupy same territory throughout year while ♀♀ disperse after breeding.

Voice. Apparently very similar to Great Grey Shrike, though in race *pallidirostris*, contact-call, adult begging-calls, and trills described as quite different.

Breeding. SEASON. In Extremadura, eggs laid March–June, mainly April; in Negev, from end of January. Up to 3 broods per year possible in Israel. SITE. In fork of tree or branch or in outer twigs; in arid places low (*c.* 1 m) in thorny bush, though sometimes high in palm in oasis; in Extremadura, average height 3 m in oak trees. Nest: bulky structure of sticks and twigs. EGGS. Very similar to Great Grey Shrike but some more strongly marked over whole surface and markings can be more warmly russet. Clutch: (3–)5–7. INCUBATION. 18–19 days. FLEDGING PERIOD. 14–15(–19) days.

Wing-length: ♂♀ 101–117 mm; 96–108 mm in Canary Islands.
Weight: ♂♀ 48–93, mostly 50–70 g.

Geographical variation. 7 races recognized in west Palearctic. In main part of range (North Africa) variation strongly clinal, with palest birds (*L. m. elegans*) in desert areas in south, darkest birds (*L. m. algeriensis*) in north. *L. m. pallidirostris*, from lower Volga, also pale, with extensive white on wings and pale bill. Nominate *meridionalis*, from Europe, much darker than North African races; *koenigi*, of Canary Islands, almost as dark. *L. m. aucheri* of Middle East, extending into north-east Africa, intermediate in colour between *algeriensis* and *elegans*.

Grey-backed Fiscal Shrike *Lanius excubitorius*

PLATE: page 1431

Du. Grijsrugklapekster   Fr. Pie-grièche à dos gris   Ge. Graumantelwürger
Ru. Сероплечий сорокопут

Field characters. Wing-span 32–36 cm. Over 10% larger in head and body and 30% longer- and also fuller-tailed than Southern Grey Shrike with rather untidy silhouette. Heavy black bill, deep black forehead and eye mask, black wings with bold white panel on bases of primaries recall Lesser Grey Shrike but easily distinguished from that species on size and structure, whitish edge to crown, and particularly different pattern to tail which lacks white sides and presents striking contrast between whitish base and fully black end. Distinction from Southern Grey Shrike also best achieved on size and structure, presence of black forehead, and absences of white edges to scapulars, inner wing feathers and tail. ♀ has dark chestnut patch on upper flanks (usually concealed by wing). Immature pale brown rather than grey above and buffish white below; dusky barring shows on chest, wing coverts, scapulars, and back and all black feathers are duller than adult's.

At distance recalls other large grey shrikes but unlike them, Grey-backed Fiscal Shrike gregarious, with babbler-like behaviour which includes dancing in groups over tree or bush canopy and flying in noisy parties between patches of cover. Flight action lazier than other grey shrikes, with looseness of wing beats somehow emphasized by unusual length of trailing tail.

Habitat. Wooded savanna, woodland, and open bush country with scattered thorny trees (especially acacia), shrubs, and grass in tropical Africa up to altitude of *c.* 2000 m; in Kenya, in rather humid areas with annual rainfall above 500 mm.

Distribution. Breeds in band from Mauritania and southern Mali east through Chad, northern Cameroon and southern Sudan to Ethiopia, south in eastern Africa to western Tanzania and eastern Zaïre. In Mauritania (where very uncommon to rare), breeding range just extends north to west Palearctic.

Movements. Poorly known but likely to be more or less resident in most parts of range. In Mali and Mauritania, most of population moves south in wet season (July–October) returning after rains to breed; in Kenya, probably sedentary in general; some movement into west of country during December–January.

Food. Insects, including caterpillars, and small vertebrates; spots prey on ground from look-out post and swoops down, returning to perch; parties chase flying ants. Young glean invertebrates from foliage.

Social pattern and behaviour. Little information apart from extralimital studies in Kenya of cooperative breeding system. Very gregarious at all times, commonly seen flying from tree to tree in noisy parties of *c.* 5–15, and general behaviour reminiscent of babblers. Strictly monogamous, and only one breeding pair per territory. In western Kenya, territorial displays ('rallies') take place between January and November, mostly around April–July at onset of major rains. Rallies include singing and elaborate displays by most group members, where birds gather in tall tree, rock back and forth, tail-spreading and wing-raising while singing. Single mated pair (pair-bond formed only at this time) appears to isolate itself from group and commences nest-building, then all other group members help at nest. Young interact with members of natal group for several years.

Voice. Very varied; flocks extremely noisy when flying between trees, screaming and chattering, which often ends in a squeak, but also utter low musical notes in harmony when perched. Apparently mostly silent at nest with no contact-calls reported.

Breeding. Season. In west Palearctic, eggs laid in Mauritania February–March. In Kenya, mainly June–July and November in association with rains, though attempts made all year round. Site. On branch or close to trunk of thorn tree or bush 1–12 m above ground. Nest: compact bowl of twigs (often thorny) and rootlets lined with soft grass, fibres, small feathers, etc. Eggs. Sub-elliptical, smooth, and slightly glossy; pale yellowish-grey or creamy pink with a few brown spots and grey undermarkings. Clutch: 3 (2–4). Incubation. 13–15 days. Fledging Period. 18–20 days.

Wing-length: ♂♀ 104–130 mm.
Weight: ♂♀ 46–64 g.

Woodchat Shrike Lanius senator

Du. Roodkopklauwier   Fr. Pie-grièche à tête rousse   Ge. Rotkopfwürger   It. Averla capirossa
Ru. Красноголовый сорокопут   Sp. Alcaudón común   Sw. Rödhuvad törnskata

PLATES: pages 1446, 1449

Field characters. 18 cm; wing-span 26–28 cm. Somewhat larger, particularly about head, than Red-backed Shrike; noticeably bulkier and larger-tailed than Masked Shrike. Medium-sized, somewhat skulking shrike. Adult ♂ pied except for chestnut rear crown and hindneck; has broad black frontal band and eye mask. ♀ duller than ♂, with white patch over bill reaching eye. Both sexes show diagnostic white rump. Immature intermediate in appearance between immature Lesser Grey Shrike and Red-backed Shrike. Crucial characters of immature Woodchat Shrike, compared with Lesser Grey Shrike, are (1) smaller size, (2) more heavily-barred crown and back, (3) less well-defined rear eye-patch, (4) barred sides to throat, (5) more contrasting pale scapular-patch, (6) smaller or no primary-patch, (7) less black wings, (8) dull, mottled but whitish rump, and (9) duller tail with narrower white edge. Compared with Red-backed Shrike has (1) bulkier form, (2) less rufous-brown ground-colour to upperparts (unless fully moulted), (3) more obvious eye-patch, (4) pale scapular-patch, (5) dull white patch at base of primaries, (6) brown wings, (7) mottled whitish rump, and (8) dark brown tail, with distinct isabelline margin and tip. Flight has actions typical of *Lanius* but noticeably direct and dashing, beginning with startling explosion—often from within cover—and not developing undulations as marked as in larger species, due perhaps to rather long wings and tail. Upward sweep to perch exaggerated by habit of perching on higher posts than smaller *Lanius*.

Habitat. Breeds in west Palearctic in middle and lower middle latitudes mainly in Mediterranean climatic zone, but extending in west into temperate zone within July isotherm of 19°C and in south-east to edge of steppe and desert. In some areas, shares habitat with Red-backed Shrike without apparent competition. Generally occurs in semi-open areas with bushes and well-spaced trees, such as open woodland, old orchards, gardens, and parks or hedges with large thorny bushes. In Spain, favours holm oak and cork oak up to at least 2000 m, though in Pyrénées only to 1500 m. In north-west Africa, also favours open oak forests, but occurs commonly in cultivated country with trees.

Distribution. Major long-term contraction southward and severe reduction in numbers, still continuing. Underlying cause perhaps climatic change, with rain in breeding season affecting breeding success by delaying egg-laying and reducing insect activity (and therefore availability). Additional factors include persecution, and (since mid-20th century) loss and degradation of habitat through afforestation and intensified farming techniques, also drought in Sahel winter quarters. FRANCE. Marked retreat southward in recent decades. BELGIUM. Now extinct. Regular in 19th century; reduced to a few pairs in extreme south in 1942, 2–3 pairs 1973–7. LUXEMBOURG. Rapid decrease since 1950s; now only occasional breeder (latest 1988). NETHERLANDS. Bred in extreme south up to 1963. GERMANY. Range much reduced; probably now extinct in east. POLAND. Now very scarce. CZECH REPUBLIC. Now only irregular breeder. SLOVAKIA. Marked decrease. HUNGARY. Formerly rare breeder, now exceptional; latest 1981. AUSTRIA. Last confirmed breeding 1982. SWITZERLAND. Rapid decrease since 1960s. MALTA.

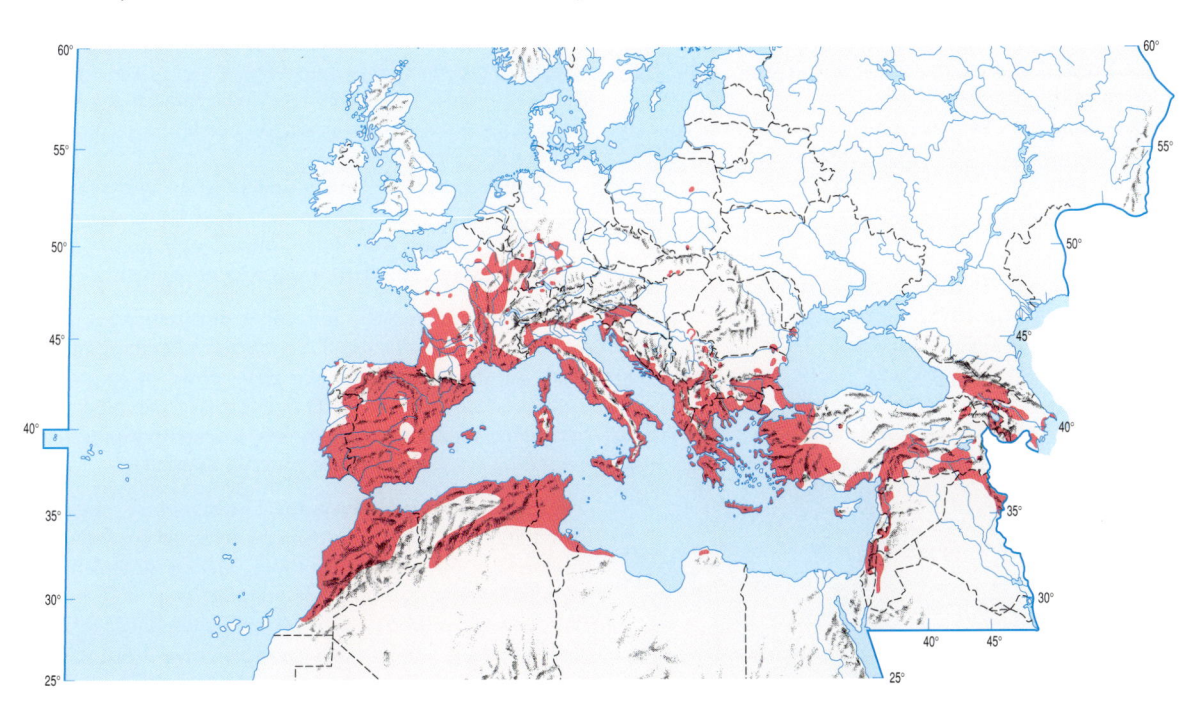

Woodchat Shrike *Lanius senator*. *L. s. senator*: **1** ad ♂ breeding, **2** ad ♀ breeding, **3** 1st winter, **4** juv. *L. s. badius*: **5** ad ♂ breeding. *L. s. niloticus*: **6** ad ♂ breeding. Masked Shrike *Lanius nubicus* (p. 1447): **7** ad ♂, **8** ad ♀, **9** juv.

Breeds occasionally. Until recently 2–3 pairs bred regularly, and formerly more common. BULGARIA. Marked range extension to north since 1950. RUMANIA. Probably sporadic breeder in Dobrogea. FSU: Recent reduction of breeding habitat in eastern Georgia, Armenia, and eastern Azerbaijan. RUSSIA. 2 sight records in extreme south-east (Dagestan), May 1990 and 1991. UKRAINE. Recorded breeding on Polish border in 19th century. CYPRUS. Bred 1980 and 1988, also several recent unconfirmed reports. Probably now regular breeder in small numbers. SYRIA. Locally quite numerous in west. IRAQ. Very common in northern mountains.

Accidental. Iceland, Britain (annual), Ireland (near annual), Denmark, Norway, Sweden (perhaps annual), Finland, Estonia, Belarus', Madeira.

Beyond west Palearctic, breeds only in Iran.

Population. Declines (often marked) reported from most of Europe except Portugal and south-east. FRANCE. 10 000–100 000 pairs in 1970s. GERMANY. *c.* 50 pairs in mid-1980s. POLAND. 20–100 pairs. SLOVAKIA. 1–25 pairs 1973–94. SWITZERLAND. 30–50 pairs 1985–93. SPAIN. 390 000–860 000 pairs. PORTUGAL. 10 000–100 000 pairs 1978–84. ITALY. 5000–10 000 pairs 1983–95. GREECE. 5000–20 000 pairs. ALBANIA. 2000–5000 pairs 1981. YUGOSLAVIA: CROATIA. 20 000–30 000 pairs; stable. BULGARIA. 300–3000 pairs; slight increase. RUMANIA. 10–15 pairs 1986–92. AZERBAIJAN. Mostly uncommon, but locally common. TURKEY. 5000–50 000 pairs. ISRAEL. Probably at least a few thousand pairs. JORDAN, TUNISIA, ALGERIA, MOROCCO. Common.

Movements. Migratory, wintering in sub-Saharan Africa north of Equator, and in small numbers (perhaps regular) in southern Arabia. Autumn movement probably broad-front in south to south-west direction; recorded in smaller numbers than in spring, presumably due chiefly to unbroken flights. Spring movement conspicuous throughout North Africa and Mediterranean, slower progression probably contributing to impression of seasonal difference in numbers.

Leaves European breeding grounds in second half of August or beginning of September (July to early October), adults before juveniles. Arrives in winter quarters August–October, mainly September. Departure from winter quarters protracted, February–May. Passage through Mediterranean area late March to early June, with main arrival on northern breeding grounds around mid-May.

Annual vagrant to Britain and Ireland, with average *c.* 14 birds (4–24) per year; peak period late May or early June and late August or early September. Annual vagrant in Netherlands, especially May–June with some August–September.

Food. Mainly insects and other invertebrates, principally beetles; occasionally small vertebrates. Perches on exposed look-out from which drops or glides down on to ground prey or makes sallying flights after flying insects. Method of dealing with prey similar to other shrikes. Impaling of prey recorded, including on passage, but apparently unusual.

Social pattern and behaviour. Usually solitary and territorial in winter. Some pair-formation probably takes place in winter quarters or on spring migration. Mating system apparently essentially monogamous, but in some areas ♂♂ outnumber ♀♀, and unpaired ♂♂ associate with pairs. Incubation usually by ♀; young fed by both sexes. ♂ sings from exposed or concealed perch in tree or bush; unpaired ♂♂ especially from high exposed perches and for many hours without interruption. Courtship-feeding occurs, from arrival on breeding grounds, but frequency varies greatly between pairs.

Voice. Song of ♂ superior to and much more regular than other European shrikes: very quiet or moderately loud, extremely varied and continuous fine, rich, musical warbling or (often predominantly) harsher, scratchy chattering with few high-pitched whistling tones and trills; contains much skilful mimicry, and usually preceded by characteristic calls which may also be incorporated in song. ♀ occasionally sings, sometimes in duet with ♂. Other calls very varied, mostly harsh or chattering; include distinctive harsh 'kiwick kiwick' or 'krähts krähts' recalling Partridge, and 3-syllable 'grack kjäck kack' in alarm.

Breeding. SEASON. Eggs laid early May to mid-July in central Europe; earlier in North Africa (from April) and Israel (from late March). 2 broods usual in south, but exceptional in Europe. SITE. In Europe in trees, especially fruit and olive trees; in Mediterranean region, also in dense or spiny bushes. Nest: strong cup of leafy plant material and roots lined with wool, hair, fine roots, cobwebs, moss, and lichen. Often incorporates flowers pulled from ground and sometimes composed only of green plant material. EGGS. Sub-elliptical, smooth and glossy. Pale to olive-green, also sandy or greyish-yellow and more rarely reddish-yellow or brown. Brown to pale olive speckles and blotches concentrated in circle at broad end, exceptionally also at narrow end; markings reddish-brown or purple on eggs with sandy or reddish ground-colour. Clutch: 5–6, rarely 7. INCUBATION. 14–15(–16) days. FLEDGING PERIOD. 15–18 days.

Wing-length: Central European birds: ♂ 98–102, ♀ 97–103 mm. Iberian and North African populations smaller, ♂♀ 91–98 mm; Balearic populations larger, ♂♀ 100–106 mm.
Weight: ♂♀ mostly 30–40 g; migrants to over 50 g.

Geographical variation. 4 races recognized. Nominate *senator* occupies most of mainland European range. *L. s. rutilans* (Iberia and north-west Africa) smaller; *L. s. badius* (Balearic Islands) larger, and white patch at base of primaries absent or largely hidden; *L. s. niloticus* (Cyprus and Levant eastwards) has extensive white on tail-base.

Masked Shrike *Lanius nubicus*

PLATES: pages 1446, 1449

DU. Maskerklauwier   FR. Pie-grièche masquée   GE. Maskenwürger   IT. Averla mascherata
RU. Маскированный сорокопут   SP. Alcaudón enmascarado   SW. Masktörnskata

Field characters. 17–18 cm; wing-span 24–26.5 cm. Close in size to Red-backed Shrike but with finer bill and generally slighter build, most evident in longer, more graduated tail. Small, rather slim shrike (not looking large-headed), with proportionately longest and slimmest tail of west Palearctic *Lanius* and much less bold behaviour than other species. ♂ essentially black above and white below, with white forehead, supercilium, and scapular panel, and pale rufous body-sides. ♀ duller than ♂, showing more pale fringes on wings. Immature less distinctive but has distinctive grey tone to barred upperparts and flanks, with indication of characteristic head and scapular markings.

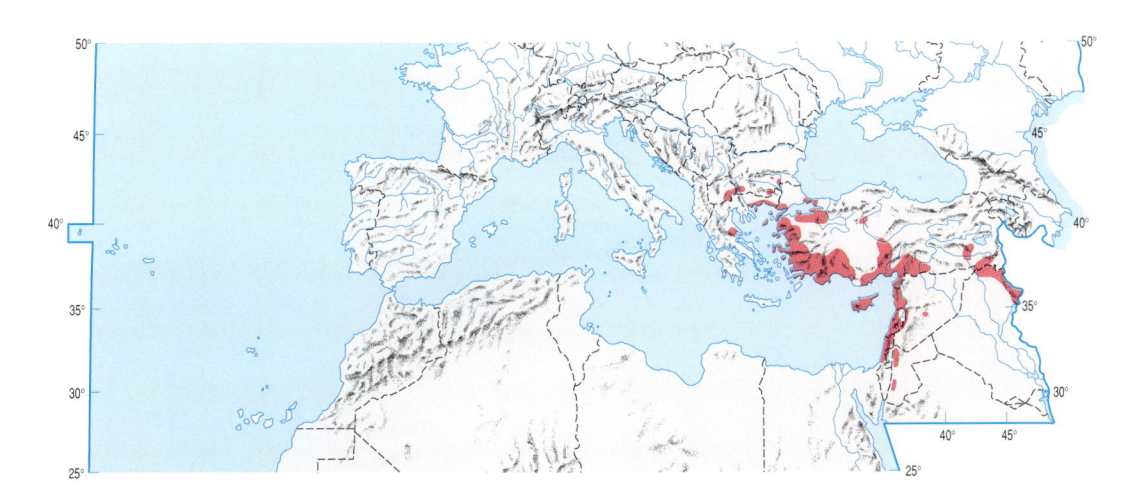

Adult ♂ unmistakable and ♀ unlikely to be confused with other *Lanius* due to slim build, pale head and scapular markings, dark rump, and generally less overt behaviour. Juvenile may, however, be confused with juvenile Woodchat Shrike: best distinguished by whitish forehead and supercilium accentuating dark eye-patch, greyer upperparts, and dark rump concolorous with back. Flight lightest of west Palearctic *Lanius*, recalling flycatcher in aerial chases after insects but still aggressive in pounces on ground prey. Perches erectly, often with long tail slightly cocked; also waves tail at times but generally sits quietly—on edge rather than on top of tree or bush.

Habitat. Breeds in west Palearctic in lower middle latitudes, mostly in Mediterranean warm zone and in hilly terrain, often less open and with higher tree cover than habitats favoured by most *Lanius*. Found in almost any kind of wooded country, in more open glades of forest and where there are isolated big trees in the open, as well as in high scrub, olive groves, and gardens.

Distribution. Has extended range north into Bulgaria. BULGARIA. First recorded breeding (3–6 pairs) 1976 in Kresna gorge in south-west. TURKEY. Widespread but only locally common Thrace, Western Anatolia, and Southern Coastlands; scarce and local in mountain valleys of south-east. CYPRUS. Very common in wooded areas of hills and mountains. Has recently extended range to Nicosia. SYRIA. Locally common in west, especially north-west coastal belt.

Accidental. France, Sweden, Finland, Spain, Balearic Islands, Malta, Libya, Algeria.

Beyond west Palearctic, breeding known only from western Iran, though spring migrants (presumably overshooting) have reached Turkmenistan.

Population. GREECE. Probably under 1000 pairs. YUGOSLAVIA: MAKEDONIJA. Rare and very localized. BULGARIA. 50–100 pairs; gradual increase. TURKEY. 5000–50 000 pairs. CYPRUS. Perhaps 10 000–20 000 pairs; no evidence of any decline. ISRAEL. Fairly common; probably a few thousand pairs. Decline since 1950s. JORDAN. Fairly common. IRAQ. Common, perhaps less so than Woodchat Shrike.

Movements. All populations migratory, wintering in sub-Saharan Africa, mainly south to *c.* 10°N in Sudan and Ethiopia, west to eastern Mali. Passage mainly through east Mediterranean at both seasons. Marked passage though Cyprus but exceptional in Crete.

Autumn passage in Mediterranean area mainly mid-August to mid-September, reaching winter quarters August–September. Leaves winter quarters usually from February onwards, passing through Mediterranean area mainly mid-March to April and arriving on breeding grounds mainly mid-April to mid-May.

Food. Insects, mainly grasshoppers and beetles, lizards, and small passerine birds, notably and perhaps exclusively exhausted migrants. Hunts from exposed perch on top of small tree or bush, or from cover on side of taller tree or in thick scrub; also perches on telephone wire, fence, etc. Drops or swoops on ground prey, often hovering briefly, either returning to perch, flying to another, or consuming prey on ground, or takes flying insects in rapid twisting aerial pursuit like flycatcher. Impales vertebrate and invertebrate prey.

Social pattern and behaviour. Not well known, but apparently similar to congeners. No evidence for other than monogamous mating system. Incubation by ♀, young tended by both parents. ♂ sings from exposed perch soon after arrival on breeding grounds. Courtship-feeding recorded during early stages of breeding and during incubation.

Voice. Song of ♂ usually a quiet pleasant warble but sometimes dry, toneless, and scratchy; not uncommonly sustained in 'bursts' of up to 1 min or more without breaks. Commonest calls a hard 'tsr' and a distinctive reedy, querulous 'keer keer keer', sounding more like squeal at distance; also a harsh 'krrr' in alarm.

Breeding. SEASON. Eggs laid early April to late June, peak period May. 2 broods usual. SITE. Nest slung below twigs at end of lateral branch, moulded on top of broad branch, in fork, or against trunk of tree, or in dense, often thorny bush. Nest: carefully constructed, compact, often inconspicuous structure of rootlets, bark strips, pine needles, plant down, plant stems (often with flowers attached), lichen, moss, cloth, string, cardboard, paper (some nests almost entirely of man-made materials), lined with sheep's wool, hair, rootlets, bark, fine stems, petals, paper, etc. EGGS. Sub-elliptical to short sub-elliptical, smooth and glossy or waxy; creamy to pale buff or yellowish, rarely white or greenish grey, with ring of pale to dark brown blotches towards broad end and large background blotches of pale to dark grey; remainder of egg sometimes hardly marked. Clutch: 4–7(–8). INCUBATION. About 15 days. FLEDGING PERIOD. 18–20 days.

Wing-length: ♂♀ 87–96 mm.
Weight: ♂♀ mostly 22–30 g.

(FACING PAGE) Black-crowned Tchagra *Tchagra senegala cucullata* (p. 1428): **1** ad worn (spring).
Isabelline Shrike *Lanius isabellinus phoenicuroides* (p. 1431): **2** ad ♂, **3** 1st winter.
Red-backed Shrike *Lanius collurio* (p. 1433): **4** ad ♂, **5** 1st winter.
Long-tailed Shrike *Lanius schach* (p. 1436): **6** ad ♂.
Lesser Grey Shrike *Lanius minor* (p. 1436): **7** ad ♂ breeding.
Great Grey Shrike *Lanius excubitor excubitor* (p. 1440): **8** ad ♂.
Woodchat Shrike *Lanius senator senator* (p. 1445): **9** ad ♂ breeding, **10** 1st winter.
Masked Shrike *Lanius nubicus*: **11** ad ♂, **12** ad ♀.

Norman Arlott.

Crows and Allies Family Corvidae

Medium to very large oscine passerines (suborder Passeres), including some of most adaptable and successful birds of entire order. Raven *Corvus corax* is largest of all passerines. Corvids found in many habitats, from forest, woodland, and steppe to tundra and desert. Many wholly or partially arboreal but some terrestrial. Most have generalized diet, taking wide variety of animal and vegetable food, often by scavenging, in trees and/or on ground. Some also specialize to greater or lesser extent in eating large seeds (e.g. those of oaks) and storing them for later use; concealment of surplus food, however, is characteristic of family as a whole. Of almost worldwide distribution but absent from Arctic, Antarctic, and most oceanic islands. 115 species; 16 in west Palearctic, 14 breeding (including introduced House Crow).

Sexes generally of similar size. Bill of most species stout, strong, and fairly long with slight hook, but (e.g.) rather longer and pointed in *Nucifraga* and long and decurved in *Podoces* and *Pyrrhocorax*. Nostrils usually rounded; open but closely covered by dense bristles. Wing shape variable but often broad; 10 primaries, p10 somewhat reduced (length 35–65% of longest). Flight typically strong and straight with rather deliberate wing-beats but more laboured-looking in shorter-winged, longer-tailed species. Tail variable, short to long, often graduated, central feathers much elongated in some species. Plumage colour variable but often wholly black, black and white, black and grey, or brown; some magpies and many jays, however, are highly colourful, often with contrasting blue, purple, or green patterns on head, throat, breast, wing, or tail. Sexes usually closely similar. Leg and toes sturdy; leg quite long in most species. Gait often a hop or bounding gallop (so-called 'polka step' in which legs leave ground one after the other), but most species also walk and run, some habitually.

Feeding methods varied. Foot employed to hold down food items by 'clamping', using one or both feet. Dropping of shellfish (etc.) from height on to hard surface in order to break them is typical example of often remarked 'intelligence' (especially in opportunistic feeders of genus *Corvus*), leading many systematists to consider them the 'highest' of all birds and place them last (or first) in sequence. This ability extends to other areas of foraging behaviour, including hiding and recovering of stored food, and indicates exceptionally good powers of memory and learning. Linked (e.g.) with skill and dexterity in use of bill and foot, curious and cautious nature, and (in larger species at least) prolonged period of dependence and semi-dependence when young, and relatively long life.

Voice unspecialized; calls mostly simple, loud, and harsh. Typical advertising song of most other passerine groups lacking, but quiet song (really subsong), usually uttered by lone birds, occurs in all or most species; mimicry (by both sexes) common. Often gregarious at times, mainly while feeding and roosting, but most species solitary and territorial when breeding. Nest an open cup or rough domed structure; built by both sexes. Incubation usually by ♀ only (fed by ♂); young fed by both sexes. Family bonds strong; post-fledging care, and association between adults and young and between siblings, prolonged in some species, especially larger *Corvus*.

Jay *Garrulus glandarius*

PLATES: pages 1452, 1463

Du. Gaai   Fr. Geai des chênes   Ge. Eichelhäher   It. Ghiandaia
Ru. Сойка   Sp. Arrendajo común   Sw. Nötskrika

Field characters. 34–35 cm; wing-span 52–58 cm. Somewhat larger than Jackdaw, with deeper body and broader wings. Rather small corvid, most colourful of family in west Palearctic, with short bill, domed head, broad wings, chesty body, and rather long tail. West and central European races pink- to grey-brown with dull white, streaked crown, black 'moustache', black, partly white-barred, and blue-splashed wings, bold white rump, and black tail; other races differ especially in head pattern, with distinctive dark, even black crown in North African, Mediterranean, and Crimean races. Flight jerky and weak-looking. Screeching call distinctive.

Unmistakable when seen well but subject to confusion with Nutcracker when only a dark silhouette flitting through trees and even Hoopoe when briefly glimpsed in sun and shadow. General character rather graceless, not improved by harsh voice. Flight lacks freedom and relative grace of crows *Corvus*, having characteristic action of rather quick, somewhat mechanical, uneven, half-depressed beats and full flaps of much extended and spread, rounded wings. Gait confident, with muscular hop and bounce along branches and ground; accompanied by frequent twitching and jerking of tail either up and down or side to side; also flicks out wings, exposing blue 'shoulder' more fully.

Habitat. Breeds across wooded middle and lower middle latitudes of west Palearctic to July isotherm of 14°C, mainly in continental temperate and Mediterranean climates, but marginally in oceanic, boreal, and wooded steppe zones. Predominantly lowland, but in Switzerland (infrequently) to 1200 m and in Carpathians to treeline at 1600 m. Strongly arboreal, and at home in fairly dense cover of trees, scrub, and woody undergrowth, especially in woodlands of oak, beech, and hornbeam, but also inhabits other broad-leaved and, in parts of range, coniferous forests. In some regions has spread into smaller outlying woodlands, spinneys, and copses, and even to urban and rural parkland, and to large gardens. Habit of burying food, especially acorns, indicates particularly strong and ancient association with oaks, and probable role in enabling them to spread uphill and over open country. After breeding season, tends to occur also in orchards, vineyards, and gardens, and in groups of trees scattered in steppes and semi-deserts, as well as around human settlements and along farm hedgerows.

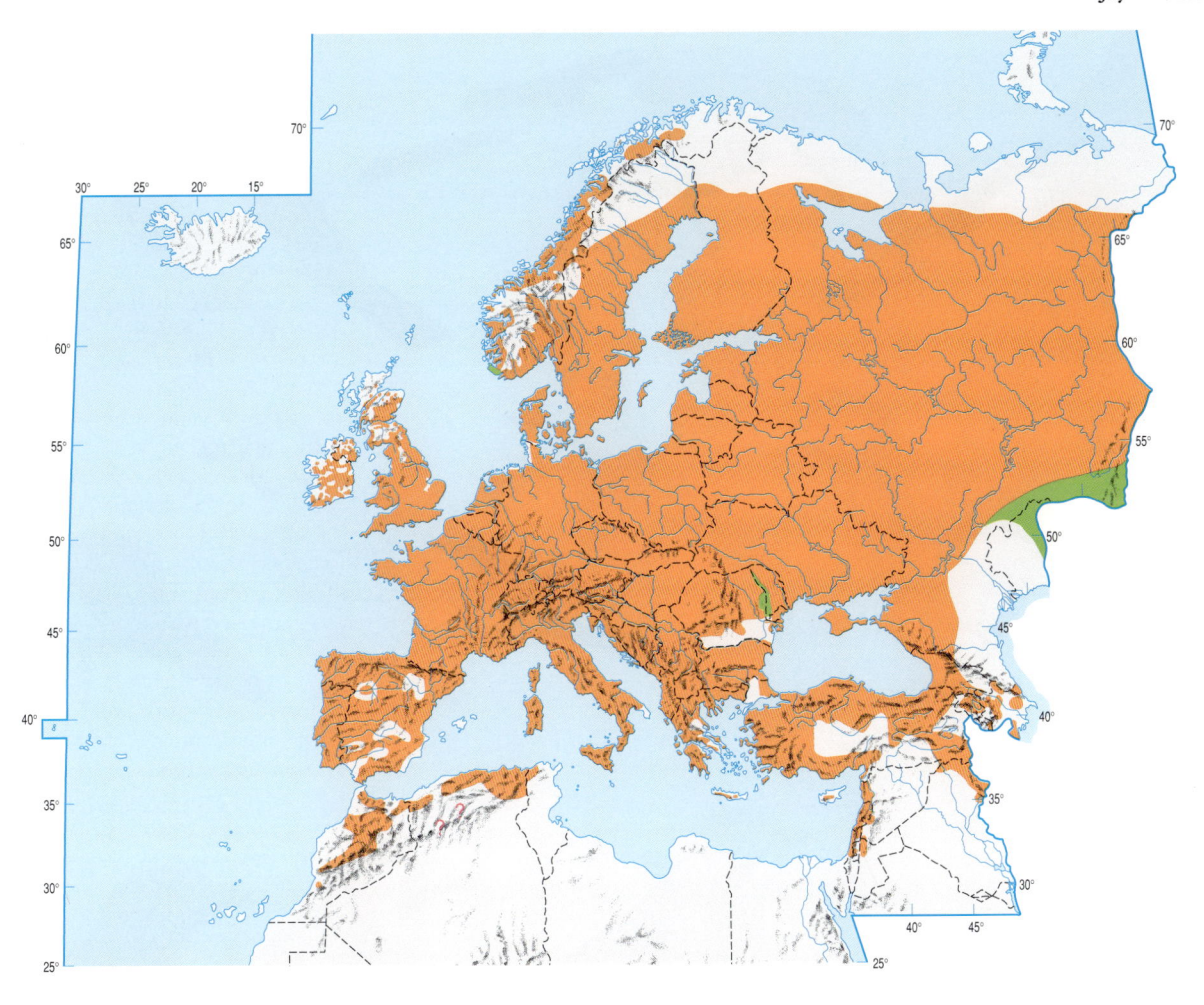

Distribution. No evidence of major changes in recent years, except for range expansion Scotland, Ireland, Denmark, Norway, Russia, Israel, and Morocco, and possible contraction in Algeria. Following decrease in human persecution, has spread into suburban and urban habitats in Britain, Belgium, Poland, Russia, and probably elsewhere. BRITAIN. Considerable northward expansion in Scotland, especially since early 1970s, associated with colonization of maturing conifer plantations. IRELAND. Expansion throughout 20th century, continuing in west. FRANCE. Recent spread to vineyards of Languedoc plains, only area where formerly absent. RUSSIA. Caucasus race, *krynicki*, has spread north in recent decades with increase and maturing of shelter belts. AZERBAIJAN. One of commonest birds of southern slope of Great Caucasus; also very common in Shemakha upland. TURKEY. Common and widespread; absent only from treeless steppe and semi-desert. SYRIA. Full extent of range still unknown. ISRAEL. Until 1940s confined to areas with natural woodland; expansion from 1950s, following development of settlements, agriculture, and afforestation. ALGERIA. Not recorded in Atlas Saharien in recent years. MOROCCO. Locally common in north, uncommon in Haut Atlas. Recent spread south-west.

Accidental. Balearic Islands (only 2 records, perhaps not of wild birds), Malta.

Beyond west Palearctic, extends from Urals east through Lake Baykal area to Sakhalin island and Japan; also breeds in northern and western Iran, and from Himalayas to Taiwan.

Population. Some evidence of increase Britain, Ireland, France, Norway, Estonia, Moldova, and (perhaps) Austria, largely stable otherwise. BRITAIN. 160 000 territories 1988–91; stable or, in Scotland and on farmland census plots in southern England, slightly increasing. IRELAND. 10 000 territories 1988–91. Increased throughout 20th century; more recently, increase in west, decline in centre and east. FRANCE. 100 000–1 million pairs; probably increased. Numbers in winter fluctuate according to irruptions of northern and eastern birds; last big invasion 1977. BELGIUM. 22 000 pairs 1973–7. Claims of increase (mainly by hunters) probably due to shift—birds now breeding more frequently near human habitation.

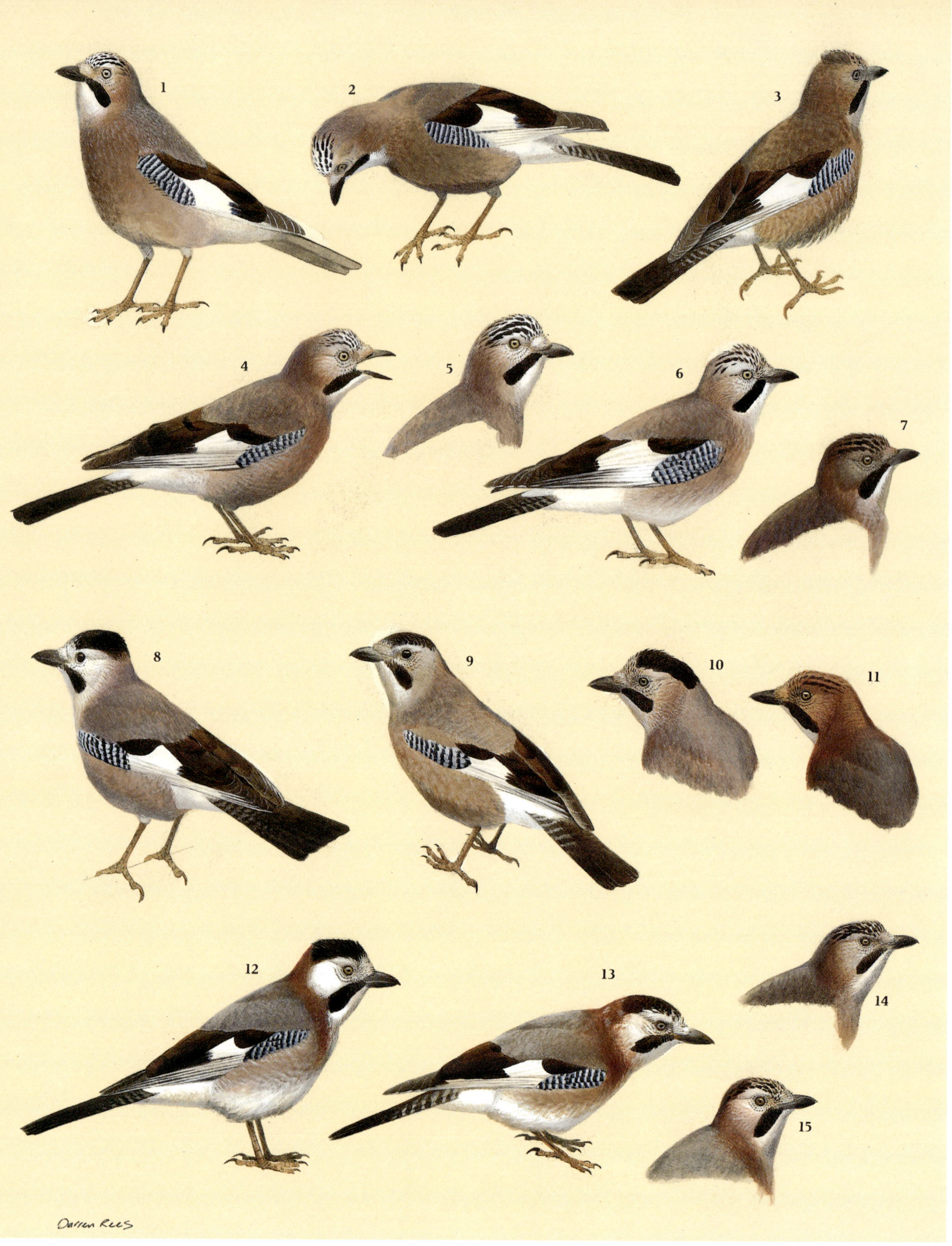
1
2
3
4
5
6
7
8
9
10
11
12
13
14
15
Darren Rees

Luxembourg. 15 000–20 000 pairs. Netherlands. 30 000–60 000 pairs 1979–85. Germany. 835 000 pairs in mid-1980s. In east, 100 000 ± 45 000 pairs in early 1980s. Denmark. 15 000–160 000 pairs 1987–8. Norway. 10 000–100 000 pairs 1970–90. Sweden. 200 000–500 000 pairs in late 1980s. Finland. 100 000–150 000 pairs in late 1980s. May have increased in north in last few decades, otherwise stable. Estonia. 30 000–50 000 pairs in 1991. Marked increase, at least since 1960s. Latvia. 20 000–40 000 pairs in 1980s. Lithuania. Common. Poland. Fairly numerous. Czech Republic. 150 000–300 000 pairs 1985–9. Slovakia. 15 000–30 000 pairs 1973–94. Hungary. 60 000–100 000 pairs 1979–93. Austria. 25 000–40 000 pairs in 1993. Stable or perhaps increasing slightly. Switzerland. 35 000–50 000 pairs 1985–93. Spain. 540 000–1.1 million pairs. Portugal. 10 000–100 000 pairs 1978–84. Italy. 50 000–200 000 pairs 1983–95. Greece. 20 000–50 000 pairs. Albania. 5000–10 000 pairs in 1981. Yugoslavia: Croatia. 150 000–200 000 pairs. Slovenia. 20 000–30 000 pairs. Bulgaria. 1–5 million pairs. Rumania. 350 000–500 000 pairs (probably more) 1986–92. Russia. 1–10 million pairs. Belarus'. 220 000–250 000 pairs in 1990. Ukraine. 55 000–60 000 pairs in 1988. Moldova. 35 000–50 000 pairs in 1988. Azerbaijan. Common; at least 1000–1500 pairs. Turkey. 100 000–1 million pairs. Cyprus. Common; probably stable. Israel. A few tens of thousands of pairs in 1980s. Jordan. Locally fairly common. Tunisia. Fairly common in oak forests. Algeria. Nowhere common.

Movements. Sedentary in west and south of range; mainly eruptive migrant in east and north, but small numbers move every year in extreme north. Eruptive (diurnal) migration of north and central European populations probably chiefly due to failure of acorn crop; notable years have included 1955, 1977, and 1983. Populations involved and extent of movement vary greatly. Heading usually ranges between west and SSW, with strong westerly component for eastern (Russian) birds. Most migrants are juveniles, and considerable proportion returns to area of origin in spring.

Autumn movement mid- or late September to early or mid-November, spring movement (in smaller numbers than autumn) March–June, thus continuing markedly late, when breeding season well under way. Birds are reluctant to cross sea; thus, passage migrants rare on Helgoland (Germany) and Ottenby (Öland island, Sweden); no records of sea crossings from Norway or Finland, and small number of Swedish records are mostly to Danish islands. However, British records and observations, especially in southern and eastern coastal areas, show that in some years continental birds reach Britain.

Food. Invertebrates, especially beetles and Lepidoptera larvae, fruits, and seeds, especially acorns; small vertebrates occasionally taken, most often in winter or when feeding young; also carrion and domestic scraps. During breeding season most food collected from leaves of trees, mainly caterpillars in oak, but otherwise forages principally on ground except when collecting acorns in autumn for storing. On ground, digs in leaf litter and soil surface with swinging movement of bill.

During September–October in central and western Europe intense collection of acorns takes place, which are then cached, generally in ground. Nuts of beech and hazel, and pine seeds, stored to much lesser extent. Birds arrive at oak stands from all directions and carry acorns away to own home-range, often covering considerable distances. Up to 9 acorns can be transported in gullet (90 large pine seeds, 15 beech nuts, or 10 hazel nuts), though 1–3 more usual with generally 1 in bill. Usual storage sites are under leaf litter, moss, in roots, etc. Usually 1, sometimes 2 or more, acorns per store; pushed into soil at $c.$ 45° and hammered a few times if still visible, hole then filled in with sideways movements of bill, and covered with leaves, stick, small stone, etc. Most acorns eaten in winter and spring come from caches; bird generally goes directly to site and digs up acorn, even in 40 cm of snow, thus exhibiting remarkable visual memory.

Social pattern and behaviour. Breeding pairs dispersed at low density on home-ranges throughout the year; in contrast to (e.g.) Magpie and Carrion Crow, non-breeders do not typically flock. Winter territoriality seems to be weak or non-existent, but in spring residents become territorial and no longer tolerate 1-year-olds which had coexisted on their home-ranges outside breeding season. 1-year-olds respond by dispersing to seek breeding vacancies elsewhere, becoming 'floaters' if they fail to settle. Mating system monogamous, pair-bond lifelong. Spring gatherings, of 3–20 birds or more, accompanied by display and great variety of calls, probably serve to bring together unpaired birds. Timid and wary of man where persecuted but readily becomes tame or indifferent in towns where not molested.

Voice. Innate repertoire supplemented by highly developed and widely used capacity of both sexes for accurate mimicry of various sounds, non-avian (including mechanical) as well as avian. Best-known call, typically given in alarm, is strident harsh rasping screech, given singly or more often twice in quick succession. Song, consisting mainly of medley of mimicked calls and phrases, given mainly by young unpaired birds and more often by ♂♂. Wide variety of low-volume mewing, clicking, chirruping (etc.) calls given in different social contexts.

Breeding. Season. Egg-laying begins April throughout most of west Palearctic range. One brood. Site. In fork or on branch of tree or bush, often thorny, usually close to or against trunk in middle of lower crown, or high in crown of young tree or of conifer; generally supported by 1–2 thick branches or many twigs; frequently in creeper (e.g. ivy, honeysuckle); perhaps increasingly on buildings, and rarely in tree-hole, large nest-box, or rock-face crevice. Nest: rough foundation of twigs $c.$ 0·3–1·5 cm in diameter, with inside layer of soft, thinner twigs, roots, stalks, etc., lined with rootlets, bast, grass, moss, leaves, hair, and rarely feathers; well-supported nests in ivy, etc., may have very little foundation. Eggs. Sub-elliptical, sometimes short sub-elliptical or oval, smooth and slightly

(FACING PAGE) Jay *Garrulus glandarius*. *G. g. glandarius*: **1** ad summer, **2** ad winter, **3** juv. *G. g. rufitergum*: **4** ad. *G. g. fasciatus*: **5** ad. *G. g. albipectus*: **6** ad. *G. g. glaszeri*: **7** ad. *G. g. atricapillus*: **8** ad, **9** juv. *G. g. krynicki*: **10** ad. *G. g. brandtii*: **11** ad. *G. g. cervicalis*: **12** ad, **13** juv. *G. g. whitakeri*: **14** ad. *G. g. minor*: **15** ad.

glossy or matt; pale green to blue-green, olive, or olive-buff, occasionally sand-coloured, finely speckled all over (sometimes very densely) with olive, light brown, or greyish-green; sometimes black hair-streaks at broad end where speckling may also be concentrated. Clutch: 5–7 (3–10). INCUBATION. 16–17(–19) days. FLEDGING PERIOD. 21–22 (19–23) days.

Wing-length: *G. g. glandarius*: ♂ 176–196, ♀ 170–188 mm. Most other west Palearctic races similar.

Weight: *G. g. glandarius*: ♂ mostly 150–190, ♀ 140–185 g.

Geographical variation. Marked and complex. 30–60 races have been recognized; many in west Palearctic, of which the most marked are shown in Plate (see also Field characters). *G. g. glandarius* occupies most of mainland Europe.

Siberian Jay *Perisoreus infaustus*

PLATES: pages 1455, 1463

DU. Taigagaai   FR. Mésangeai imitateur   GE. Unglückshäher   IT. Ghiandaia siberiana
RU. Кукша   SP. Arrendajo funesto   SW. Lavskrika

Field characters. 30–31 cm; wing-span 40–46 cm. Noticeably smaller, slighter, and proportionately longer-tailed than Jay, with much shorter, more pointed bill. Smallest and most delicate of west Palearctic Corvidae; rather drab brown-grey, with rufous-chestnut near wing-bend and on under wing-coverts, rump, and sides of tail setting bird 'on fire' in flight.

Unmistakable. Chief problem in identification is marked shyness during breeding season in distinct contrast to often tame behaviour in winter or time of food shortage. Flight differs from Jay: quite accomplished, but deeply undulating due to bird closing wings between burst of quick, easy, loose wing-beats. Gait a brisk, light hop, varied by sidling and jumping. Able to cling upside down like tit. Small groups often move in follow-my-leader line.

Habitat. Breeds across higher latitudes of west Palearctic, and is predominantly at all seasons a bird of coniferous forest, mainly of Norway spruce and Scots pine but also of larch and downy birch. Favours dense stands of forest, unmodified by man, rather than open growth. Contrasts with Jay in lack of fear of man, readily attaching itself to human travellers and their living quarters, but this has little effect on choice of habitat since normal range is largely uninhabited by people.

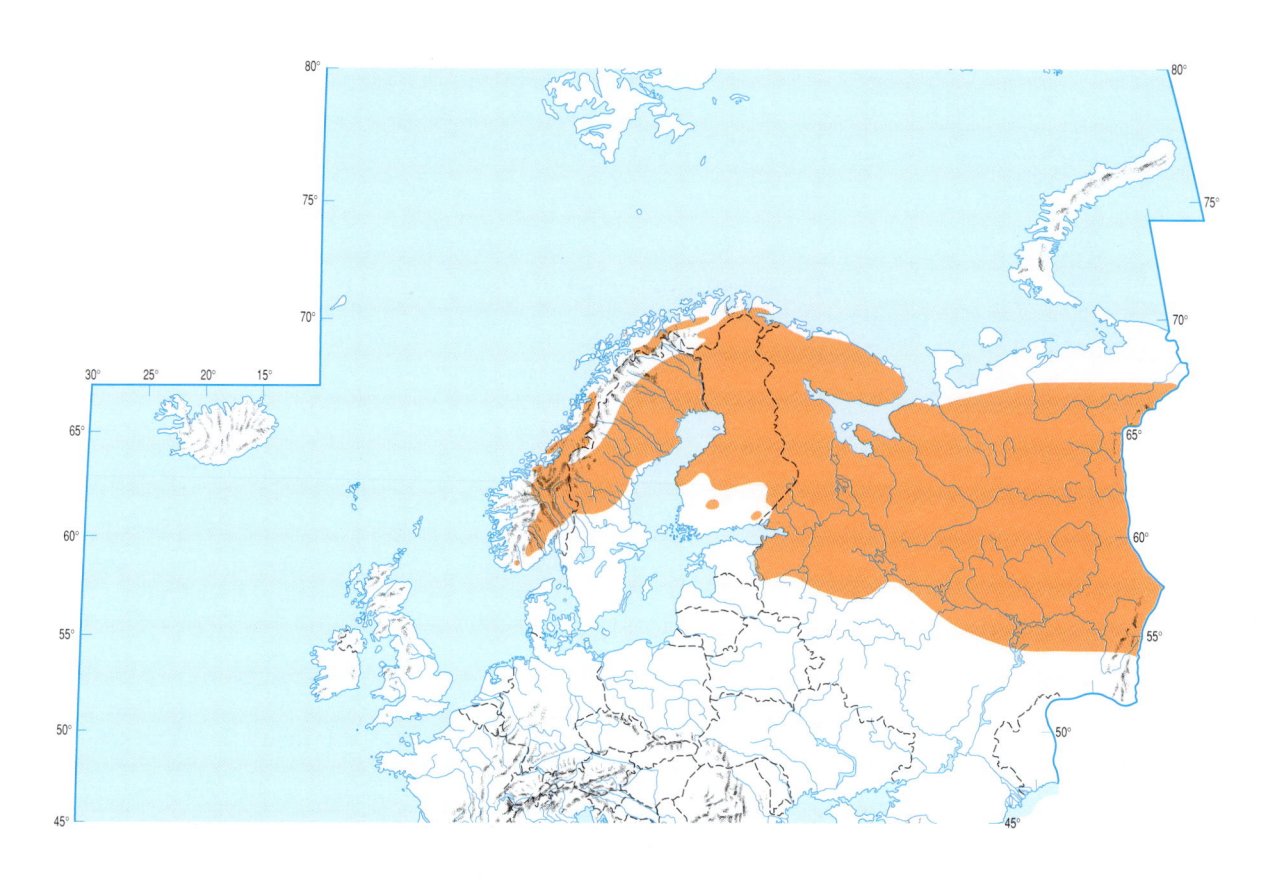

Siberian Jay *Perisoreus infaustus*. *P. i. infaustus*: **1** ad, **2** juv. *P. i. ruthenus*: **3** ad. Azure-winged Magpie *Cyanopica cyanus* (p. 1456): **4** ad, **5** juv in post-juvenile moult.

Distribution. SWEDEN. Probably bred as far south as 60°15′N in mid-19th century. Following earlier reports of withdrawal from southern parts of range, perhaps now spreading there slightly. January census in Gästrikland showed it to be more numerous than formerly believed. FINLAND. Striking range decrease in recent decades due primarily to destruction and fragmentation of large areas of mature coniferous forest with ensuing spread of cultivation and human settlements; scattered populations in south in danger of extinction. ESTONIA. Thought to have been regular breeder until 2nd half of 19th century, but no confirmed nesting cases known. Now irregular visitor; in last 50 years, only 5 breeding-season observations: 1956, 1957, 1974, 1979, 1986. LATVIA. Only 1 definite record in 20th century; recorded more often in past.

Accidental. Latvia, Poland, Slovakia, Belarus', Ukraine.

Beyond west Palearctic, extends broadly across northern Asia to Chukotskiy region, northern Sea of Okhotsk, and Sakhalin island.

Population. NORWAY. 10 000–50 000 pairs 1970–90; stable. SWEDEN. 50 000–200 000 pairs in late 1980s; slight decrease. FINLAND. 40 000–70 000 pairs in late 1980s; slight decrease. RUSSIA. Poorly known. 10 000–100 000 pairs; stable.

Movements. Resident, with all-year territory, but winter movements reported from outside Europe for northern population in eastern Siberia.

Food. Omnivorous all year. Captures and kills small mammals and small passerines up to size of tit; also plunders nests for eggs and nestlings. Feeds from all kinds of carrion. Takes variety of arthropods, largely insects. Plant material forms substantial portion of diet, especially various berries occurring in coniferous forest. Marked tendency to store food. Food-storing so intensive that it has been suggested that Siberian Jay overwinters successfully largely because of stores, and uses them to prepare for breeding and rearing young. Storing occurs in spring as well as in autumn and winter. Food stored in trees; typical hoarding sites are bark crevices, in lichens, or among conifer needles. Food carefully concealed and rendered almost invisible, small pieces of bark and lichens being used to cover hoarded items.

Social pattern and behaviour. Adults in resident territorial flocks all year. Movement of flock-members between trees is silent and unsynchronized. Unrelated flock-members are met with aggression (especially at food), including supplanting and chasing accompanied by calling. Dominance hierarchy exists whereby adult resident ♂♂ dominate all immigrants of both sexes, while adult resident ♀♀ dominate some but not all immigrants. Life-long pair-bonds make pair-formation difficult to observe; can take place at any time of year. Pairs may breed solitarily or in small groups. Not uncommon for extra birds to be associated with breeding pair, apparently as helpers.

Voice. Not highly vocal, but calls heard all year. Repertoire large; many calls flexible in pitch and speed of delivery.

Considerable mimetic ability. Song consists of a gentle flow of subdued phrases composed of low twittering and chattering mixed with melodious whistling and mewing, including mimicry. Common call, signalling excitement, fear, and anxiety, a harsh and usually loud 'eee', 'eeer', or 'kreee', given either singly or several in rapid sequence.

Breeding. SEASON. Regularly lays while terrain still snowbound, normally starts in first half of April. SITE. Almost always in conifer, close to trunk. Nest: rather loosely constructed platform of twigs supporting deep and well-insulated cup built mainly of lichens. Twigs interwoven with strips of birch bark and bound with spider cocoons. Cup lined mainly with feathers and reindeer hair. EGGS. Sub-elliptical, smooth and glossy. Very pale bluish-green, blue, or bluish-grey; spotted and finely blotched with olive-brown, grey, and lilac-grey, usually more heavily mottled at broad end. Clutch: 3–4 (2–5). INCUBATION. 19–20 days. FLEDGING PERIOD. *c.* 3 weeks.

Wing-length: *P. i. infaustus*: ♂ 139–152, ♀ 135–145 mm.
Weight: *P. i. infaustus*: ♂ 81–101, ♀ 73–93 g.

Geographical variation. Marked, but clinal throughout. Involves relative amount and depth of grey and rufous of body plumage, colour of cap, size and intensity of rufous of wing-patch, amount of grey on tips of tail-feathers, and size. Boundaries of subspecies impossible to define exactly; 3 recognized in west Palearctic and up to 13 extralimital. *P. i. infaustus* occupies Fenno-Scandia and Kola peninsula, grading into *ostjakorum* and *ruthenus* to east and south-east.

Azure-winged Magpie *Cyanopica cyanus*

PLATES: pages 1455, 1463

Du. Blauwe Ekster   Fr. Pie bleue   Ge. Blauelster   It. Gazza azzurra
Ru. Голубая сорока   Sp. Rabilargo   Sw. Blåskata

Field characters. 34–35 cm (of which nearly half tail); wingspan 38–40 cm. About 75% size of Magpie. Small, elegant corvid with rather attenuated form ending in long tail shaped like Magpie but looking narrower in flight. Body dove-brown, with black cap, white throat, and mainly azure-blue wings and tail. Flight recalls Magpie; other behaviour more like Jay.

Unmistakable in good view but may suggest Magpie if seen only as silhouette, and even Great Spotted Cuckoo in brief glimpse. Flight recalls Magpie but looser, with far less erratic and staccato bursts of wing-beats; progress thus more sustained, even dashing at times. Gait a light, bouncy hop; also clambers and jumps. Flicks wings and tail in excitement. Secretive rather than shy in breeding season; bold and confident at other times, roving widely in noisy groups.

Habitat. Breeds in west Palearctic only in warm lower middle latitudes in Iberia, locally in mountain gorges at *c.* 700 m but also at sea-level on sand-dunes overgrown with planted stone pine and introduced eucalyptus. In Coto Doñana (Spain) such modern woodlands are fully colonized, to virtual exclusion of the locally abundant Magpie, which nests there chiefly in low bushes outside woods. Arboreal, but forages freely on ground under canopy. In contrast to far separated Asian population, shows no marked preference for broad-leaved trees, or for river banks and islands. Also inhabits open cultivated or grass country with groups of trees, scrub, or hedgerows, and orchards or groves of olive and cork oak.

Distribution and population. FRANCE. Alleged breeding

attempt near Collioure (Pyrénées Orientales) in spring 1956 not sufficiently well documented to be acceptable. SPAIN. Both range and population stable. Occurrence in north-east occasional, but no evidence of breeding. Disperses in autumn into areas adjacent to breeding range. 240 000–260 000 pairs. PORTUGAL. Slight increase in range and numbers. 10 000–100 000 pairs 1978–84.

Beyond west Palearctic, breeds from Lake Baykal region east to Japan, south to central China.

Movements. Chiefly sedentary; after breeding, moves around in flocks in group-territory. Occasional records of presumed vagrants (mostly in spring) in north-east Spain and Pyrénées Orientales (southern France).

Food. Invertebrates, especially beetles, seeds, fruits, and more rarely small vertebrates; also carrion, scraps, and refuse. Forages generally in flocks, very often on ground. Stores food (acorns, olives, pine seeds) in caches in ground, presumably for later consumption.

Social pattern and behaviour. Highly gregarious throughout the year. Dispersed in stable flocks, each of which, after breeding season, comprises family parties originating from single breeding colony. Each flock defends extensive territory throughout year against other flocks. Mating system essentially monogamous but pair may receive help from other flock members. Dispersion of breeding pairs within colony loose, with each nest usually in separate tree or bush, but sometimes more than one in a single tree.

Voice. In keeping with highly gregarious lifestyle, repertoire complex. Calls often intergrade and may also be combined in rapid succession. Most frequently heard call, constantly heard from flocks on the move, a rather husky 'schrie' ascending in pitch. Various other harsh or more musical, piping calls given in different social contexts. Soft high-pitched chattering from ♂ displaying to ♀ may represent song.

Breeding. SEASON. Central Spain: at 300 m altitude, eggs laid early April to late May; at 1250 m, eggs laid early June to late July. Southern Spain and eastern Portugal: eggs laid end of April to early or mid-May. One brood. SITE. Generally in fork of branch towards edge of tree crown, top often covered by leaves of supporting branch. Preferred position is in mid-crown vertically but as far from trunk as possible without being visible in outer foliage; apparently best position for avoiding predators. Nest: rough, loose foundation of twigs, inside which layer of compacted pellets of earth, mud, or dung in form of bowl 2–5 cm thick, in which is embedded rough lower part of lining (roots, pine needles, small twigs) covered by rootlets, plant down, fibres, moss, lichen, animal hair, sheep's wool, feathers, etc., whole lining *c.* 1 cm thick. Fresh and aromatic stalks and flowers may be incorporated. EGGS. Sub-elliptical (short sub-elliptical to oval), smooth and faintly silky; from pale cream or olive-cream, rarely with bluish tinge, to pale yellowish-brown or brown-olive, with sparse small brown or olive-brown spots and greyish blotches, often concentrated at broad end, sometimes in circle. Clutch. 5–7 (4–9). INCUBATION. 15–16 (14–17) days. FLEDGING PERIOD. 14–16 days.

Wing-length: ♂ 131–143, ♀ 126–136 mm.
Weight: ♂ 70–78, ♀ 65–79 g.

Magpie *Pica pica*

PLATES: pages 1458, 1463

Du. Ekster   Fr. Pie bavarde   Ge. Elster   It. Gazza
Ru. Сорока   Sp. Urraca   Sw. Skata   N. Am. Black-billed Magpie

Field characters. 44–46 cm (of which over 50% is tail in adult); wing-span 52–60 cm. Bill, head, and body size close to Jackdaw but wings rather short, appearing fan-like, and tail very long and graduated. Medium-sized and markedly attenuated crow, with rather flat crown, apparently deep chest and short body, rather long legs, and long graduated tail. Boldly pied: black, with white belly and large white wing-panels. Diagnostic chattering call.

Unmistakable. Flight action alternates slightly 'desperate' bursts of rapid wing-beats with stalling glides, bird appearing to drag long straight tail. Over short distance and in and around cover, action more confident, dashing and sweeping with tail often spread; able to follow prey through thick foliage. Gait confident, with high-stepping walk and characteristic brisk, sidling hops and jumps, usually with tail held up; movements far less accomplished among branches, made seemingly awkward by trailing tail. Raises and fans tail when excited.

Habitat. Breeds in west Palearctic from upper to lower middle continental and oceanic latitudes, from boreal taiga through temperate to Mediterranean, steppe, and semi-desert zones, avoiding both densely forested and treeless regions, precipitous rocky terrain, and extensive wetlands. Occurs up to considerable altitudes, especially towards south of range, where ecological conditions are suitable, but is predominantly a lowland bird of open or lightly wooded country offering good opportunities for foraging on ground and for nesting, roosting, and taking cover in trees or shrubs. Inhabits woods of many different types, both broad-leaved and coniferous, wherever glades, clearings, or more open stands occur, and especially near margins of natural or cultivated grasslands and croplands. Spread into built-up areas and even major cities is increasingly apparent as result of overall population increase and as need for wariness diminishes; here will sometimes perch freely on artefacts including tall buildings. Adaptability and opportunism together with effects of human intervention in

Magpie *Pica pica*. *P. p. pica*: **1** ad, **2** 1st winter, **3** juv. *P. p. melanotos*: **4** ad. *P. p. mauritanica*: **5** ad.

spreading suitable mixed habitats have evidently contributed to historical advance of this species, in common with other Corvidae.

Distribution. Has colonized suburban habitats in Britain, Belgium, Netherlands, Finland, Estonia, Poland, and Russia, and probably elsewhere. Marked range expansion Slovenia and Ukraine, less so France, Czech Republic, and Slovakia (where spread to higher elevations in last 20 years), Iberia, Greece, Croatia, and Morocco. Spread (in some cases reoccupation of areas of former occurrence) and associated population increase follows decrease in human persecution. BRITAIN. Much of southern and eastern Scotland reoccupied since 1938. Upland forest plantations colonized more recently. FRANCE. Recent spread in south-east. Records east of Var river (Alpes-Maritimes) no longer uncommon; pair bred north of Menton, near Italian border, in 1987. Corsica: bred 1992. RUSSIA. Recent expansion east, to lesser extent north, Kola peninsula. Nar'yan-Mar (67°37′N on Pechora river): common in shrub tundra and breeding suspected, but not confirmed. LEBANON. Several reported 1967–8, one in 1986; apparently accidental. TUNISIA. Deemed nearly extinct in 1970s, but small population (some tens of pairs) discovered in 1989. ALGERIA. Widely dispersed, but very localized. MOROCCO. Slight extension in north. WESTERN SAHARA. Perhaps no longer breeds.

Beyond west Palearctic, breeds from Urals ESE to Ussuriland, extending broadly south to Iran, Himalayas, and southern China. Isolated populations in Kamchatka area and (very local) south-west Saudi Arabia. In North America, breeds southern Alaska and western Canada south to central California and Great Plains.

Population. Marked increase Britain, Estonia, Poland, Czech Republic, Slovenia, and Ukraine; same trend (less marked) Ireland, Denmark, Sweden, Finland, Slovakia, Austria, Switzerland, Spain, Greece, Croatia, Bulgaria, Belarus', and Moldova. Slight decrease Portugal, otherwise apparently stable. BRITAIN. 590 000 territories 1988–91; increase (noted in 1940s, and more recently with spread into suburban and urban habitats) still continues, though increase slowed on farmland and woodland census plots in southern England since 1970s. IRELAND. 320 000 territories 1988–91. FRANCE. 200 000–700 000 pairs; stable. BELGIUM. 19 000 pairs 1973–7. Increased in 1940s (wartime hunting ban), more recently only in towns, otherwise no indication of change. LUXEMBOURG. 8000–10 000 pairs. NETHERLANDS. 60 000–120 000 pairs 1979–85. GERMANY. Variously estimated at 500 000 pairs in mid-1980s, 210 000–280 000 pairs in early 1990s. In east, 70 000 ± 30 000 pairs in early 1980s. DENMARK. 31 000–320 000 pairs 1987–8. NORWAY. 200 000–500 000 pairs 1970–90. SWEDEN. 300 000–600 000 pairs in late 1980s. FINLAND. 150 000–200 000 pairs in late 1980s. Slight increase, especially in north, owing to disruption of forests. ESTONIA. 50 000–100 000 pairs in 1991. Marked increase, following decrease in 1940s–60s. LATVIA. 10 000–20 000 pairs in 1980s. LITHUANIA. Common. POLAND. 400 000–800 000 pairs. Marked increase since 1960s, still continuing. CZECH REPUBLIC. 40 000–80 000 pairs 1985–9. SLOVAKIA. 30 000–60 000 pairs 1973–94. HUNGARY. 100 000–140 000 pairs 1979–93. AUSTRIA. 4000–5000 pairs in 1992. Decreased in early 20th century; since 1950s, recovering and increase still continuing. SWITZERLAND. 20 000–40 000 pairs 1985–93. SPAIN. 220 000–1.2 million pairs. PORTUGAL. 10 000–100 000 pairs 1978–84. ITALY. 100 000–500 000 pairs 1983–95. GREECE. 30 000–80 000 pairs. ALBANIA.

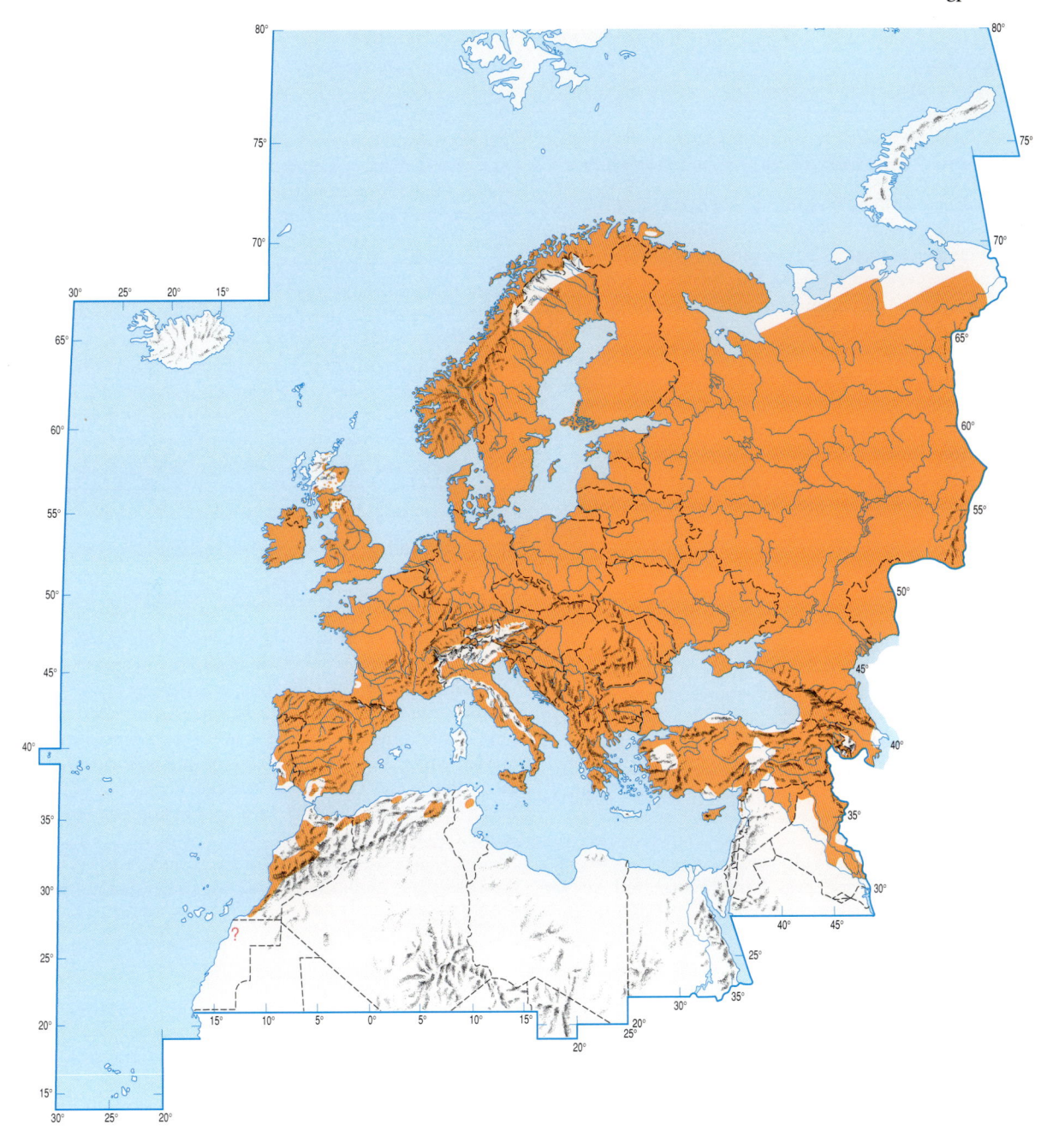

10 000–30 000 pairs in 1981. YUGOSLAVIA: CROATIA. 60 000–80 000 pairs. SLOVENIA. 8000–12 000 pairs. BULGARIA. 1–5 million pairs. RUMANIA. 100 000–200 000 pairs 1986–92. RUSSIA. 1–10 million pairs. BELARUS'. 480 000–500 000 pairs in 1990. UKRAINE. 1.3–1.5 million pairs in 1986. MOLDOVA. 60 000–70 000 pairs in 1988. AZERBAIJAN. Common, especially in foothills and lowlands. Above 2000 pairs. TURKEY. 1–10 million pairs. CYPRUS. Very common. SYRIA. Density extremely low. MOROCCO. Uncommon.

Movements. Sedentary, with limited dispersal, chiefly in north of range. Most ringing recoveries over 30–40 km involve birds from northern Europe, and show no preferred direction. Reluctant to cross sea.

Food. Invertebrates, especially beetles, fruits, and seeds; occasionally small vertebrates, including small birds (adults, sometimes seized in flight, and nestlings), and all kinds of carrion, refuse, and domestic scraps. Very opportunistic feeder, diet varying considerably according to habitat and local food sources: broadly, consumption of invertebrates highest in spring and summer, vertebrates and plant material in autumn and winter; if insects available all year then can comprise very high proportion of total diet.

Stores food like other Corvidae, but caches seem to be short-term, retrieved within 1–2 weeks at most and majority of stored items perishable. In central England, food stored all year except July, with marked peak September–December.

Social pattern and behaviour. Can be either solitary or gregarious throughout year. Outside breeding season, breeding birds may remain as pairs in their territories or abandon them, depending on quality of territory. Site-fidelity of residents is high: birds may remain on same territory for duration of their breeding life. Mating system essentially monogamous although ♂♂ commonly promiscuous. Pair-formation of non-breeders probably occurs in non-breeding flock. Within non-breeding flock, dominance hierarchy exists, ♂♂ generally dominant over ♀♀. High-ranking birds initiate ceremonial gatherings by flying into centre of established territories to provoke territorial response from owners. Noise and activity attracts all nearby breeders and non-breeders as spectators. Initiators usually evicted, but may return to and initiate gatherings at same location on several consecutive days, eventually establishing small territory there.

Voice. Complex repertoire with several calls intergrading. Best-known call, a harsh, rattling, far-carrying 'cha-cha-cha-cha'; signals alarm, annoyance, and attack–flee conflict, e.g. when nest threatened or when mobbing predators. At lower intensity, commonly a harsh 'shrak-ak', often repeated at brief intervals but never in continuous chatter. Song, associated with pair formation and courtship, relatively quiet, very variable, based on amalgamation of other calls. Other calls various; harsh, yelping, clicking, etc., depending on social context.

Breeding. SEASON. Varies little throughout west Palearctic range. Throughout Europe, peak laying period mid- to late April: in northern Finland, earliest laying c. 20 April; in Coto Doñana, south-west Spain, peak laying mid-April. One brood. SITE. Usually in tree, often thorny, but, where unavailable, in low bush or even on ground. In tree, nest placed in top of crown, fork of trunk, near trunk, or in outer twigs; supporting branches up to 4 cm thick may be built into nest, and weight often borne by live lateral twigs; in tall trees can be highly visible. Nest: loose, bulky outer layer of twigs and sticks usually extended to form roof (thus globular), often of thorny twigs, those with roof sometimes having 2 entrances; inside this, main part of nest is mud (occasionally dung) bowl incorporating twigs, roots, etc., this layer often forming distinct rim on top of nest; bound into this bowl is layer of more twigs, grass, etc., above which is lining of grass, hair, feathers, leaves, and other soft material. EGGS. Sub-elliptical, sometimes short or long sub-elliptical or even long oval, smooth and glossy; very variable in colour, also within clutch; pale or greenish-blue, or light to dark olive-brown, heavily speckled or blotched olive-brown and grey, only sometimes concentrated at broad end. Clutch: 5–7 (3–10). INCUBATION. 21–22 days. FLEDGING PERIOD. 24–30 days.

Wing-length: *P. p. pica* (Netherlands): ♂ 190–206, ♀ 177–195 mm.

Weight: *P. p. pica* (Netherlands): ♂ 210–272, ♀ 182–214 g.

Geographical variation. Marked, but mainly clinal. Involves size, relative tail length, colour of gloss of wing and tail, size of black tips on white primaries, and amount of white, grey, or black on rump; also, a bare spot behind eye in some populations. Plumage blackest in north-west African race *mauritanica*, which also has a relatively much longer tail than other races. Due to clinal character of variation in Eurasia, boundaries between races hard to define. Five intergrading races recognized in west Palearctic: nominate *pica* (most of Europe), *fennorum* (northern Fenno-Scandia, Baltic States, and northern Russia), *bactriana* (FSU, south of *fennorum*), *melanotos* (Iberia), and *mauritanica* (north-west Africa).

Nutcracker *Nucifraga caryocatactes*

PLATES: pages 1461, 1463

DU. Notenkraker   FR. Cassenoix moucheté   GE. Tannenhäher   IT. Nocciolaia
RU. Кедровка   SP. Cascanueces   SW. Nötkråka

Field characters. 32–33 cm; wing-span 52–58 cm. Close in size to Jay, with structure differing most in 10% longer, more pointed bill and 10–15% shorter tail, but with rather similar broad rounded wings. Rather small, long-billed, compact, and short-tailed corvid, with (uniquely in Corvidae) pale-spotted face and body and bold white vent and tail-rim all obvious against otherwise dark chocolate-brown plumage.

Unmistakable, with white-black-white bands under tail conspicuous from below or behind even at distances where pale spotting on body becomes invisible. Sharp-billed, short-tailed silhouette quickly apparent as distinct from that of Jay. Flight recalls Jay but rather more steady, with flapping, not so jerky wing-beats used during both level and undulating progress. Gait essentially a hop varied by sidling jumps and occasional bounces. Sociable, even gregarious during eruptions during which birds often become tame.

Habitat. Breeds in boreal or montane upper and middle latitudes of west Palearctic in cool continental forest lowlands and on mountains up to treeline, wherever essential requirements of coniferous forest and food are fulfilled. At home on ground as well as in trees. Within west Palearctic, resorts mainly to stands of Norway spruce in northern taiga zone or preferably Arolla pine in mountains, but sometimes

Nutcracker *Nucifraga caryocatactes*. *N. c. caryocatactes*: **1** ad winter, **2** summer, **3** juv. *N. c. macrorhynchos*: **4** ad.

larch or silver fir. Where Arolla pine absent, alternative reliance on storage for winter of hazel nuts renders access to these of vital importance, even if this involves repeated flights of several km; in Sweden at least, this factor is apparently indispensable.

Distribution. Recent range expansions in France, Belgium, Fenno-Scandia, and Czech Republic. Slight decrease Ukraine. FRANCE. Birds breeding in Ardennes probably of Siberian origin (spread from Belgium). Range increase also Alps and Vosges, and isolated populations now occur further west. Uncertain whether expansion involves both indigenous and Siberian birds. BELGIUM. Small population established in Ardennes 1968, following major invasion from Siberia; breeding suspected 1969, proved 1975. NETHERLANDS. Occasional breeder (*macrorhynchos*); proved only after 1968 invasion, at least 2 sites 1969–71. GERMANY. Higher densities almost always correlated with altitudes above 600 m (probably close correlation with spruce). Bred south-east Berlin in 1978. DENMARK. Occasional breeder after invasion years, both nominate *caryocatactes* and *macrorhynchos*. NORWAY. Recent spread to west, where now breeding in several localities. SWEDEN. Recent discovery of small populations north of limit of hazel at 60°15′N, apparently dependent on plantations of Arolla pine; appeared after influx of *macrorhynchos*. In 1980s, *c.* 30 birds at Umeå (*c.* 64°N), *c.* 10 at Skellefteå (*c.* 65°N), and fewer at 2 other places. FINLAND. Scattered records of breeding outside main range in south-west refer to *macrorhynchos*, which has established small isolated breeding poulations and is increasing. CZECH REPUBLIC, SLOVAKIA. Spread to lower elevations, noted after 1960. HUNGARY. No proof of breeding despite summer records in suitable habitat in north.

Accidental. Turkey (several records Bosporus 2nd half of 19th century, but only 1 since—November 1966). Also occurs irregularly in many other countries as occasional irruptive migrant, exceptionally reaching Spain.

Beyond west Palearctic, breeds broadly across northern Asia to Kamchatka, Sakhalin island, and Japan, extending south through eastern Kazakhstan to Tien Shan mountains; also along Himalayas, continuing from south-west to north-east China.

Population. Apparently mostly stable across range, though increases (generally local, after expansion to new areas) Finland, Estonia, Czech Republic, slight decrease in France, Ukraine. FRANCE. 3000–5000 pairs. BELGIUM. At least 180 pairs. No significant change since 1975. GERMANY. Estimates vary: 6000 pairs in mid-1980s; 11 000–15 000 pairs in early 1990s. In east, 800 ± 200 pairs, of which *c.* 550 in Thüringen, in early 1980s. DENMARK. Up to 10 pairs 1989–94. NORWAY. 100–1000 pairs 1970–90. SWEDEN. 5000–15 000 pairs in late 1980s. FINLAND. 1500–2000 pairs of nominate *caryocatactes* and 20–50 pairs of *macrorhynchos*. Marked increase. ESTONIA. 5000–10 000 pairs in 1991. Marked increase since 1970s, following earlier decline. LATVIA. 5000–20 000 pairs in 1980s. LITHUANIA. Not abundant. POLAND. 2000–5000 pairs. CZECH REPUBLIC. 2500–5000 pairs 1985–9. SLOVAKIA. 3000–6000 pairs 1973–94. AUSTRIA. Common in montane and subalpine regions. No apparent changes. SWITZERLAND. 20 000–30 000 pairs 1985–93. ITALY. 10 000–30 000 pairs 1983–95. GREECE. 100–600 pairs. ALBANIA. Up to 500 pairs in 1981. YUGOSLAVIA: CROATIA. 8000–12 000 pairs. SLOVENIA. 2000–3000 pairs. BULGARIA. 10 000–50 000 pairs. RUMANIA. 30 000–70 000 pairs 1986–92. RUSSIA. 10 000–100 000 pairs. BELARUS'. 4000–6000 pairs in

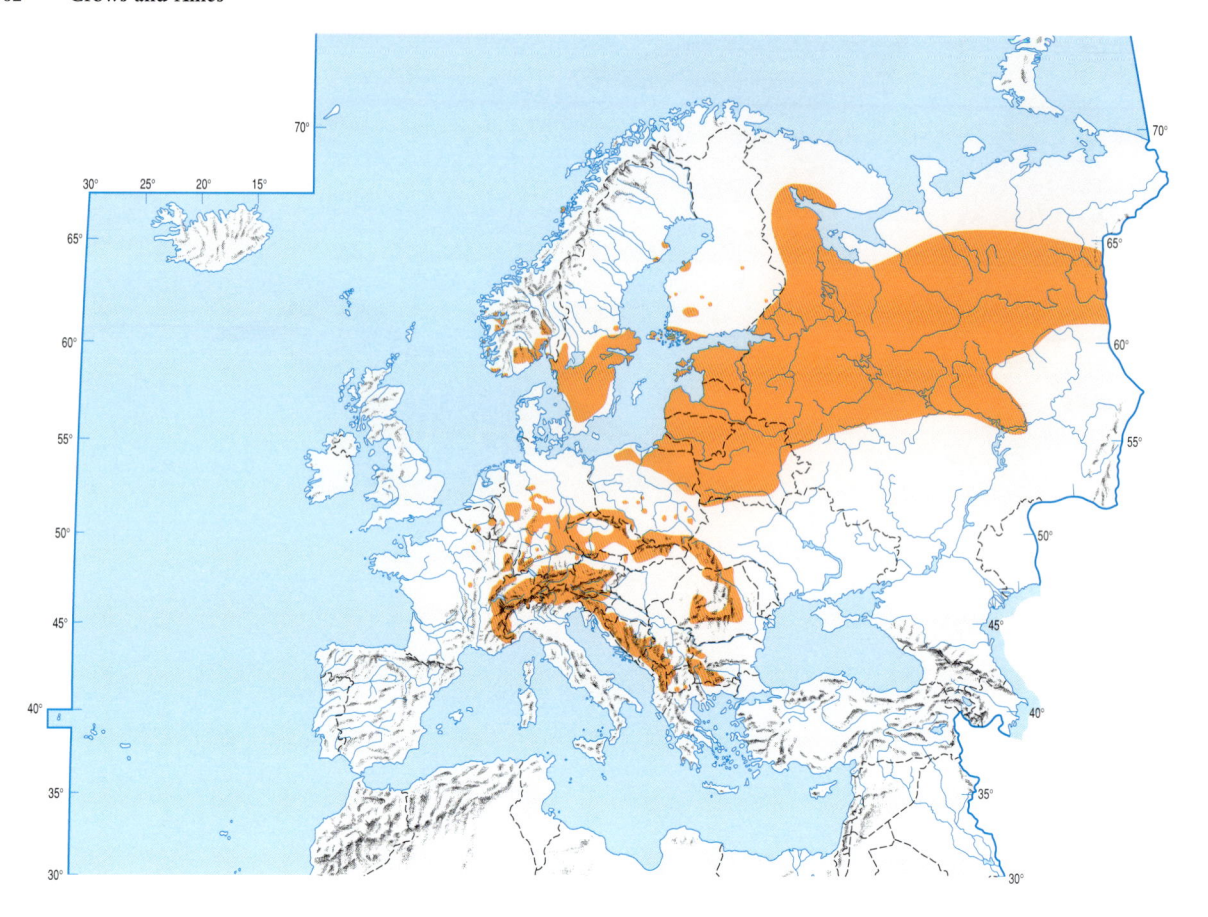

1990; fluctuating, but apparently stable overall. UKRAINE. 200–250 pairs in 1988.

Movements. Western race (nominate *caryocatactes*) and most Asian races chiefly resident, with weak invasions occasionally reported; Siberian race (*macrorhynchos*) eruptive migrant. Eruptive (diurnal) migration of Siberian race associated with failure of pine seed crop, especially Siberian stone pine. Migration exceptional in that little or no return movement occurs in following spring, most birds failing to survive, though a few breed in wintering areas in subsequent year(s); some return towards area of origin a few weeks after emigrating, however. Eruptions reach Britain, southern France, and northern Italy in exceptional years. Coastlines act as leading lines, and expanses of water as barriers. Recent invasions of western Europe in 1968 (after which breeding colonies established in Belgium), 1971, 1977 (after which birds bred in Finland and Sweden), 1985, and notably 1995.

Food. On breeding grounds, restricted to seeds, nuts, invertebrates (mainly insects, especially larvae of bees and wasps), and sometimes small vertebrates, but during irruptions birds virtually omnivorous; winter diet of sedentary birds almost wholly seeds and nuts from supplies stored in autumn. European nominate *caryocatactes* highly dependent on seeds of Arolla pine and nuts of hazel, while Siberian race *macrorhynchos* specializes on seeds of Siberian stone pine, other far-eastern pines, and spruce. Relationship with Arolla and Siberian stone pine so strong that many authors consider it an example of symbiosis or mutualism, since these seeds are heavy and unwinged, depending on agency other than wind for dispersal.

On breeding grounds in winter, completely dependent on supply of cached seeds and nuts. These transported in sublingual pouch, which expands to maximum size of *c.* 15 cm^3 in autumn when storing activity at peak. Caches are generally on ground in soft soil, under leaf litter, moss, etc., or on trunks of fallen trees or moss-covered rocks; always covered after deposition with leaf, stick, stone, etc. Often placed near object like tree trunk, stone, seedling tree, or tuft of grass, generally in more open areas in clearings, at forest edge, or just above treeline, and often under shelter of branches where snow cover thinnest in winter.

Social pattern and behaviour. Flocks occur during autumn seed-gathering period and for 'ceremonial gatherings' and migration, but basically solitary or in pairs and sedentary on territory throughout year; even birds dispersing in winter often in pairs. Territory contains food stores vital for winter survival and important also for rearing of young. Adults normally faithful to mate and territory throughout year and (probably) life. After attaining independence around early July, young have *c.* 6–8 weeks in which to establish territory (food-storing and wintering site). From July or August and through autumn, adults and (increasingly) juveniles spend most of day collecting nuts or seeds for storing. Birds may commute several km

Jay *Garrulus glandarius glandarius* (p. 1450): **1–2** ad summer. Siberian Jay *Perisoreus infaustus infaustus* (p. 1454): **3–4** ad. Azure-winged Magpie *Cyanopica cyanus* (p. 1456): **5–6** ad. Magpie *Pica pica pica* (p. 1457): **7–8** ad. Nutcracker *Nucifraga caryocatactes caryocatactes*: **9–10** ad summer.

(maximum 15 km, with 700 m altitudinal difference, in Switzerland) between roost or larders and collecting sites. When breeding, ♂ takes more equal share of nesting duties than other Corvidae. Unlike in other Corvidae, incubation is by both sexes. Both sexes also brood and feed young, ♂ brooding more than incubating. Period of dependence conspicuously long, up to 119 days. Most conspicuous feature of social behaviour is 'ceremonial gatherings' which perhaps facilitate meetings of unpaired birds: take place immediately before and during nest-building and also in late summer. Up to 10–12 birds assemble in a home-range, owners participating or watching and showing no aggression. Birds perform mutual bowing movements and chase one another, accompanying displays with wide variety of calls.

Voice. Generally silent during winter, but vocal with start of breeding (using varied repertoire of calls, especially in courtship), then silent again after laying; noisy in summer when adults with fledged young in territory and (stimulated by newcomers in area) at start of food-storing. Commonest call a repeated rasp or croak; also a drawn-out, low-pitched rattle. Song, apparently given by ♂ alone, a long, varied phrase of subdued, at times ventriloquial, harsh chattering, warbling, or babbling, with sweeter (even melodious) more tonal piping, whistling, whining, mewing, or bleating, interspersed with gurgling and clicking sounds, often containing mimicry of other birds.

Breeding. SEASON. Early, so in many areas throughout range whole breeding cycle (nest-building to fledging) can be in cold conditions with deep snow on ground. Sweden: egg-laying peaks last third of March to first third of April. Finland: laying peaks early to mid-April. Balkans: laying peaks end of March to beginning of April. One brood. SITE. Almost always in conifer, generally against trunk. Nest: compact, well-made structure containing 3–5 distinct layers providing good insulation; outer foundation of twigs generally of conifer (often green), though sometimes of other species such as birch, beech, or bramble, with roots, grass, and lichen, inside which layer of compressed beard-lichen *c.* 3 cm thick, then layer of fragments of decayed wood and moss *c.* 8 cm thick, followed by some earth mixed with various fibres, etc., and lastly lining of grass, rootlets, lichen, hair, etc., rarely feathers. EGGS. Sub-elliptical, sometimes short sub-elliptical, smooth and matt-glossy, shape variable even within clutch; ground-colour ranges from whitish to deep green but generally pale or greenish-blue, fine olive-brown and grey spots and speckles can be almost invisible or dense, sometimes concentrated at broad end, rarely in ring. Clutch: 3–4 (2–5). INCUBATION. *c.* 18 days. FLEDGING PERIOD. 24–25 days (23–28).

Wing-length: *N.'c. caryocatactes*: ♂ 188–198, ♀ 178–193 mm. *N. c. macrorhynchos* slightly smaller.
Weight: *N. c. caryocatactes*: ♂ mostly 150–190, ♀ 140–175 g.

Geographical variation. Rather slight in size, more marked in depth and width of bill, in size and extent of white spots, and in amount of white on tail. Bill longer and more slender, and white more extensive, in *N. c. macrorhynchos* (breeding Siberia, occasionally further west after irruptions) than in nominate *caryocatactes* (breeding Europe).

Alpine Chough *Pyrrhocorax graculus*

PLATES: pages 1465, 1471

Du. Alpenkauw   Fr. Chocard à bec jaune   Ge. Alpendohle   It. Gracchio alpino
Ru. Альпийская галка   Sp. Chova piquigualda   Sw. Alpkaja

Field characters. 38 cm (tail 12–14 cm); wing-span 75–85 cm. Close in size to Chough but with 40% shorter bill (decurved only towards tip), less broad and 'fingered' wings and 15–20% longer tail; 15% larger and less compact than Jackdaw. Medium-sized, rather small-headed, graceful crow. Black, with relatively small, short, decurved yellow bill and red legs.

Unless at close range, or calling, difficult to distinguish from Chough, though lacks full aerial grace and quite broad build of that species. Shortness of yellow bill diagnostic, but note that juvenile Chough has initially short, orange bill. Separation from Jackdaw in flight assisted by 2-toned (not uniform) underwing. Flight swift and skilful, with loose, deep wing-beats, folded wing attitudes, and fanning of full tail allowing high manoeuvrability at cliffs and easy sailing in upcurrents; even so, action slightly stilted compared with Chough. Flight silhouette differs from Chough due to rather narrower wings with only 4 separated primaries (5–6 in Chough) and longer tail. Gait combines shuffling walk and restricted hop. Often tame, scavenging near man. Sociable.

Habitat. Strictly montane, breeding in middle latitudes of west Palearctic in generally colder climates and at greater altitudes than any other bird species, in Switzerland up to 3000 m and only exceptionally below 1500 m, and in Morocco up to 3900 m. While demanding inaccessible nest-sites in steep rock-faces, often in caverns or fissures, is enabled by mastery of air to range over wide variety of foraging habitats, from snowline to treeline or lower, tending however to avoid snow cover and to favour alpine meadows, newly mown grasslands, and boulder slopes. Especially in winter, favours immediate neighbourhood of huts, hotels, settlements, ski lifts, and other tourist facilities, showing little aversion to human presence in small or large numbers.

Distribution. GERMANY. Breeding in small section of Alps with altitudes above 1000 m; northern limit of European range. POLAND. Former breeder in Tatra mountains (1850); since recorded only as accidental visitor. YUGOSLAVIA: SLOVENIA. Marked increase in range. TURKEY. Birds collected in winter near Izmir and Söke relate to winter dispersal or perhaps to small breeding population closer to Anatolian west coast. Also recorded at Yesilce (north-west of Gaziantep) in July and may breed near by. SYRIA. Mt Hermon: resident breeder at 2000–2800 m; 5–10 pairs apparently bred along Kedarim ridge in 1980s. Birds seen Anti-Lebanon mountains winter 1938–9 perhaps only visitors from Lebanon. IRAQ. Breeds northern mountains.

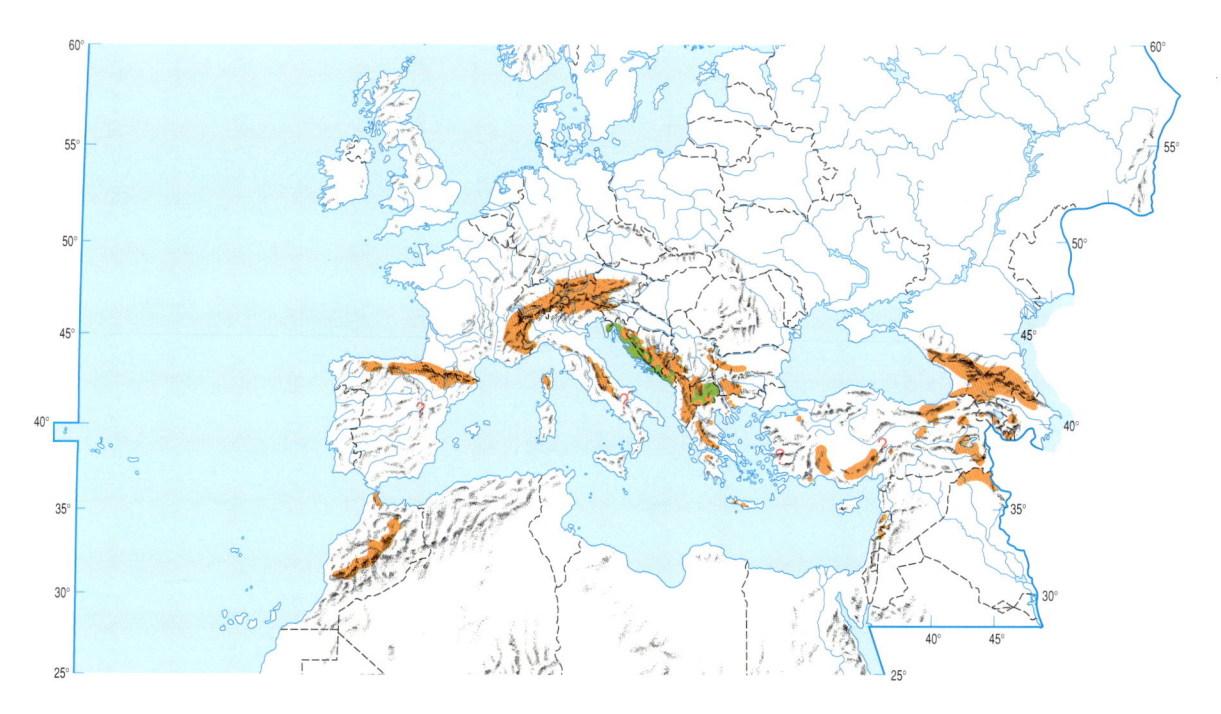

Alpine Chough *Pyrrhocorax graculus graculus*: **1** ad, **2** juv. Chough *Pyrrhocorax pyrrhocorax* (p. 1466). *P. p. pyrrhocorax*: **3** ad, **4** 1st winter, **5** juv. *P. p. docilis* (Balkans east to Transcaucasia): **6** ad. *P. p. barbarus* (north-west Africa and Canary Is.): **7** ad.

Accidental. Poland, Czech Republic, Slovakia, Hungary, Balearic Islands.

Beyond west Palearctic, breeds northern and western Iran, and from Afghanistan north-east to Sayan mountains, east to central China.

Population. Apparently stable in most countries. FRANCE. 10 000–30 000 pairs; some recent increase in Haute Savoie. GERMANY. 2000 pairs in mid-1980s. AUSTRIA. Widespread and locally common; no clear trend. SWITZERLAND. 5000–10 000 pairs 1985–93. SPAIN. 10 000–11 000 pairs. ITALY. 5000–10 000 pairs 1983–95. GREECE. 5000–10 000 pairs; slight decrease. ALBANIA. 100–500 pairs in 1981. YUGOSLAVIA: SLOVENIA. 800–1200 pairs; marked increase. BULGARIA. 1000–5000 pairs. AZERBAIJAN. Rare. TURKEY. 10 000–100 000 pairs. MOROCCO. Scarce to locally uncommon.

Movements. Mainly sedentary, except for altitudinal movements. Alpine populations often make diurnal altitudinal movements of up to several km between feeding and roosting sites. In Switzerland, daily movements may exceed 20 km in length and cover 1600 m altitude. More birds than formerly may remain high in Alps in winter due to development of skiing above 3000 m and consequent artificial food sources.

Food. From spring to autumn favours insects, particularly grasshoppers, crane-fly larvae, and beetles; in autumn and winter, berries; diet often includes refuse or scraps, particularly in winter. In summer, usually forages above treeline on short grass, rocks, scree, and cliffs; searches in crevices, under stones, among vegetation, in dung, and on buildings and walls; often feeds along receding snowline in early summer and occasionally picks items from snow surface. Frequently caches prey, especially in winter.

Social pattern and behaviour. Highly gregarious throughout year, commonly in small or large flocks, but sometimes in pairs or family parties. Many birds in flocks evidently paired throughout year. Pairs often breed solitarily but also in colonies of up to 20 or more. Flocks are highly vocal and habitually indulge in spectacular aerobatics, perhaps even more commonly than Chough and Jackdaw. Habitually glides and soars, flocks sometimes rising to great heights on strong updraughts, then often diving in apparent play or as prelude to landing.

Voice. Conspicuously vocal throughout year. Much the commonest call is high-pitched, penetrating 'chree' or 'tree', given singly or repeated, both with and without wing-flirting movements. Heard from flocks in flight or about to take off; apparently serves as advertising- and contact-call. Varies considerably, with extremes sounding like high, squeaky 'kee', 'squee', or 'skweea', usually given without wing movements, apparently as contact-call. The more aggressive the bird, the nearer the call is to rippling 'chree', and such calls always accompanied by wing-flirting. Succession of warbling, squeaky, chittering, and churring sounds reported from birds apparently at ease and in flocks or pairs feeding together.

Breeding. SEASON. Alps: eggs laid early May to mid-June, rarely April. Morocco: nests mid-May to July; young recorded early June at 2000 m and early July at 2550 m. Lebanon: probably June–July; fledged young mid-July. One brood. SITE. Ledge or crevice in cave, cliff, tunnel, shaft, or building. Access sometimes through small (e.g. 50 cm) entrance hole;

site sometimes more than 10 m from entrance and often in darkness. Nest: bulky mass of sticks (up to 50 cm long), roots, twigs, moss, and plant stems; lined with neat, compact cup of grass, fine twigs, rootlets, hair, and some feathers. EGGS. Sub-elliptical, smooth and glossy; whitish, tinged creamy or faintly buff, rarely greenish, profusely marked overall with small blotches, spots, and specks of dark brown, reddish-brown or olive-brown, with grey or lilac underlying markings. Clutch: 3–5(–6). INCUBATION. 18–21 days. FLEDGING PERIOD. 29–31 days.

Wing-length: *P. g. graculus*: ♂ 270–284, ♀ 251–267 mm.
Weight: *P. g. graculus*: ♂♀ 188–252 g.

Geographical variation. Slight, involving size (length of wing, tail, and bill) and relative length of tarsus and toes. Usually 2 races recognized, smaller nominate *graculus* in west (most of west Palearctic range), larger *digitatus* (southern Turkey and Levant eastwards) in east.

Chough *Pyrrhocorax pyrrhocorax*

PLATES: pages 1465, 1471

DU. Alpenkraai   FR. Crave à bec rouge   GE. Alpenkrähe   IT. Gracchio corallino
RU. Клушица   SP. Chova piquirroja   SW. Alpkråka

Field characters. 39–40 cm; wing-span 73–90 cm. Smaller-headed but broader-winged than Jackdaw; close in size to Alpine Chough but with 70% longer, more distinctly and evenly decurved bill, more oval head, somewhat longer, broader, and more 'fingered' wings, shorter tail, and slightly longer legs. Medium-sized, dashing, and graceful crow, with long, thin, decurved red bill (duller in juvenile). Plumage glossy velvet black. Flight most aerobatic of Corvidae. Call distinctive.

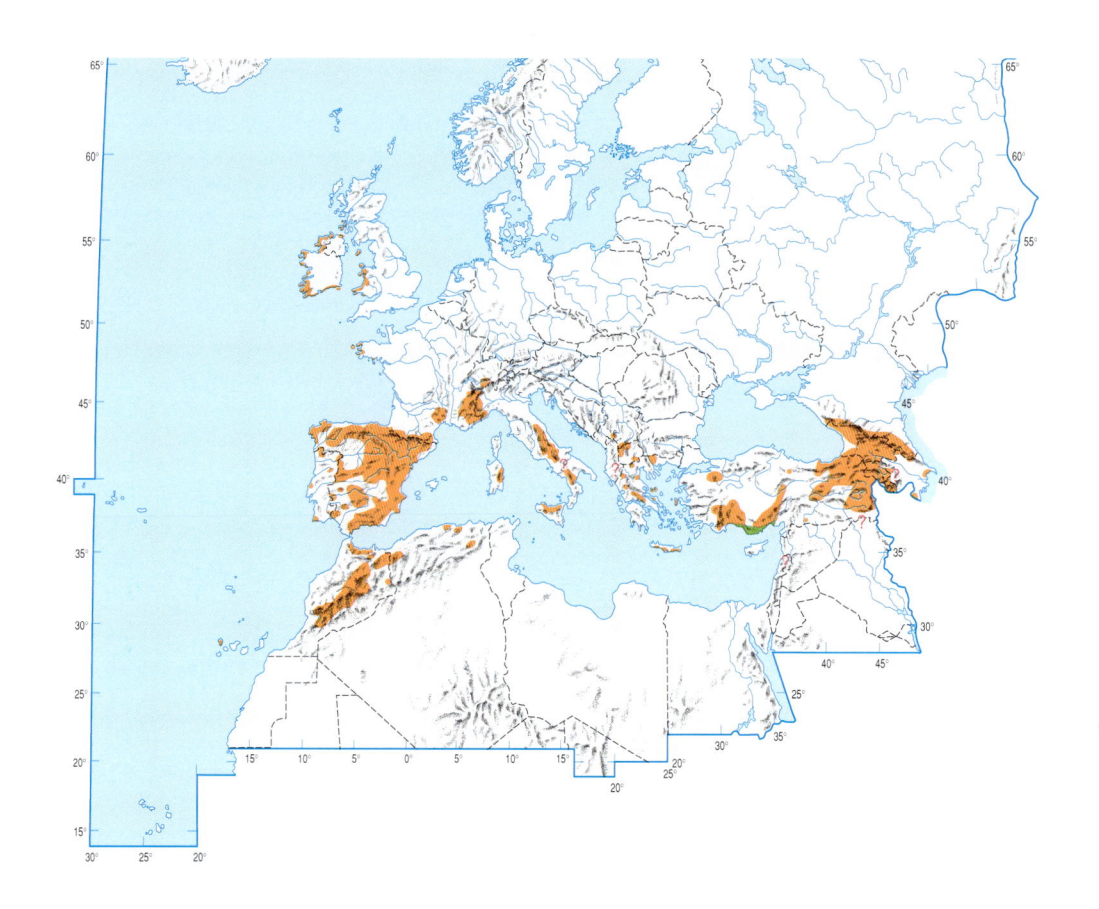

Not confined to rocky mountains and thus more widespread than Alpine Chough, and needing to be distinguished also from Jackdaw which frequently shares coastal habitats. Distinction from Jackdaw not difficult except at long range: Chough unmarked except for mainly blue-green iridescence, red bare parts, and silvery undersurface to flight-feathers, and is an altogether more aristocratic and graceful bird. Even dullest (British and Irish) Jackdaw shows distinctly grey rear of head, less sheen on black plumage, stubby black bill, and black legs; altogether more plebeian in general character. Distinction (especially of short-billed juvenile) from Alpine Chough much more difficult. In flight, shows quite broad and well-fingered wings and rather square tail; action buoyant and marvellously accomplished, including gliding and soaring, easy, almost leisurely acceleration into fast direct progress, sweeping dives, and even hurtling rolls and tumbling manoeuvres. No other crow, not even Alpine Chough, gives such impression of flight mastery and enjoyment, with flocks as well as individuals and pairs indulging in aerobatics up and down cliffs.

Habitat. In west Palearctic, breeds in temperate middle latitudes, either on coastal cliffs or inland crags, and locally on buildings or ruins, especially of stone. Elsewhere in continental mid-latitudes mainly in montane regions. In Switzerland, nests at 1200–1500 m; in Haut Atlas (Morocco) mainly 2000–2500 m, and in southern FSU inhabits mountains at 1200–3600 m, but sometimes occurs on crags at lower levels, nesting in crevices and caves usually near water. Normally breeds at lower altitudes than Alpine Chough. Feeds almost entirely on ground, in Britain and Ireland mainly along coasts where low-intensity farming combines patchy cereal cultivation with plenty of short grass grazed by sheep and cattle or wild rabbits, or kept down by high winds and salt spray, thus permitting essential access to invertebrate prey in soil. Will sometimes forage on stubbles, fallows, and ploughland, and in southern Europe may abound on plateaux with extensive cornfields, as on the Hoya de Guadix (Andalucía, Spain) at 900–1200 m. Sand-dunes, machairs, fields rich in dung deposits from livestock, and even beaches, are also used for foraging.

Distribution. Range has contracted markedly in north-west, in Alps, and Iberia. Following fragmentation of range, small, isolated populations in (e.g.) Brittany, Swiss Alps, Portugal, and Sardinia barely viable. Perhaps now stabilizing at least in Britain, Ireland, and France. Main cause of range and population decrease apparently loss of traditional pastoral farming, with persecution, disturbance, etc. probably also implicated. BRITAIN. Scotland: formerly occurred both inland and on east and west coasts; by early 19th century had vanished from inland areas and was declining on east coast. England: bred Devon until 1910, Cornwall until 1952. IRELAND. Formerly bred on east coast. FRANCE. Long-term contraction of breeding range. Extinct Normandy from 1910. AUSTRIA. Former breeder in Alps. Last known breeding site Wolaya (Kärnten) end of 19th century; suspected breeding in 20th century not confirmed. Now very rare visitor. SWITZERLAND. Disappeared as breeding bird in Grisons in 1967; only breeding area now Valais. ITALY. Has become extinct in central and eastern Alps; now only in western Alps. ALBANIA. Probably breeds in south (suggested population up to 50 pairs in 1981), but no proof. AZERBAIJAN. Distribution poorly known. TURKEY. Generally more common and widespread than Alpine Chough, occurring on all mountains which reach above 2000 m, also locally lower down in steppe country of south-east. Bred on cliffs near Bozüyük in 19th century. SYRIA. In first half of 20th century, still breeding in mountains around Al Qaryatayn and Palmyra. Now extinct. LEBANON. Status uncertain; flock near Faraya, 1969. IRAQ. Probably resident breeder in northern mountains. TUNISIA. Formerly bred; no records in 20th century. ALGERIA. Very localized. CANARY ISLANDS. Restricted to La Palma.

Accidental. Germany, Slovakia, Hungary, Balearic Islands, Israel.

Beyond west Palearctic, breeds from Iran to Himalayas, and north through central Asian mountains to Transbaykalia and northern China; local in Ethiopia.

Population. Decline, marked in Portugal and Italy, and (less severe) in France, Spain, and Greece. Apparently stable otherwise, though some increase Scotland and (locally) in Spain. BRITAIN. 315 pairs 1986–91. IRELAND. 906 pairs in 1992. FRANCE. 800–2000 pairs; has declined in north-west, but now apparently more or less stable elsewhere. Brittany: 35–45 pairs. Alps: over 30 pairs Verdon canyon and mountains north of it, Haute Provence; c. 20 pairs on Cheiron mountain, north of Grasse (Alpes-Maritimes). SWITZERLAND. 40–60 pairs 1985–93. SPAIN. 7000–9800 pairs; slight decrease. In limited area of central Spain, 324 pairs in 1990; increase 1975–90. PORTUGAL. 100–150 pairs in 1990. ITALY. 500–1000 pairs 1983–95. GREECE. 1000–5000 pairs. YUGOSLAVIA: CROATIA. 3000–5000 pairs. AZERBAIJAN. Uncommon, apart from Nakhichevan region where common. TURKEY. 5000–50 000 pairs. MOROCCO. Uncommon to locally abundant.

Movements. Mainly sedentary. Recorded far from breeding areas only exceptionally.

Food. Soil-living insects and other invertebrates, with grain and berries taken especially in winter or by upland populations. Forages generally in pairs or flocks on open ground with short vegetation (2–4 cm) or on bare, burnt, rocky, or disturbed ground. Feeding efforts often concentrated on interfaces, e.g. between vegetation and rock outcrops, between stones and earth, at the bases of shrubs and grass tussocks, and along edges of snow patches. Overturns stones to uncover prey; reaches over larger stones and pulls them towards body or pushes them to one side with partly open, down-pointed bill; flicks small stones aside. Searches dried dung of cow, sheep, or horse for invertebrates and grain fragments; breaks it open and turns it over, searching damp underside and ground beneath for prey. Takes berries and Lepidoptera larvae from trees and bushes with much fluttering and wing-flapping for balance.

Social pattern and behaviour. Throughout year, typically in small or (often temporary) large flocks, up to several hundred birds; often also in pairs, sometimes singly. Many flocking birds appear to be paired at all times of year and pairs commonly join and leave flocks. Pair-bond in adults almost certainly of long duration. Age at first breeding 2–4 years or later. Nesting solitary, occasionally in small loose colonies.

Voice. Conspicuously vocal throughout year. Commonest call a loud, yelping, drawn-out 'chwee-ow', subject to considerable variation; also rendered 'kyaa'. Often accompanied by wing-flirting when function is commonly self-advertising, but also given regularly without wing movements and in flight, when it probably functions as contact-call. Rarely gives succession of low warbling, chittering, and churring sounds, resembling 'songs' described for some other Corvidae; function obscure.

Breeding. SEASON. Britain and Ireland: first eggs laid early April, rarely late March, mainly mid- to late April or early May, upland pairs to mid-May. Alps: laying early to mid-April at 1400 m, mid-April to mid-May at 2300 m. Southern France: incubation early May. Usually one brood. SITE. Crevice in cliff or ledge in cave, shaft, or overhang; sometimes in or on building. Nest: bulky, untidy structure of dry twigs, roots, moss, and plant stems; base often solely heather stems, occasionally bound with mud, lined thickly with wool and occasionally other animal hair, man-made material, and thistle down. EGGS. Sub-elliptical, smooth and glossy; very pale, tinged greenish, creamy, or faint buff, marked overall with small blotches, spots, and streaks of olive-brown and grey; underlying markings, which often predominate, lilac-grey. Clutch: 3–5 (1–6). INCUBATION. 17–18(–21) days. FLEDGING PERIOD. 31–41 days.

Wing-length: *P. p. erythrorhamphus*: ♂ 282–315, ♀ 289–294 mm.
Weight: *P. p. erythrorhamphus*: ♂♀ mostly 260–350 g.

Geographical variation. Marked. Mainly involves size (as expressed in weight or wing length), relative measurements of tail, bill, tarsus, toes, and bill depth, colour of gloss on body, wing-coverts, flight-feathers, or tail, and (perhaps) wing shape. In general, size smaller in north and in birds of coastal cliffs or inland hills, larger in south and in high-mountain populations; gloss strongly purplish in populations living in humid climates, more greenish in arid regions. *P. p. pyrrhocorax* from Britain and Ireland is smallest race. Three other races recognized in west Palearctic, of which *erythrorhamphus* (Iberia, France, Alps, Italy, Sicily, Sardinia) is most widespread.

Jackdaw *Corvus monedula*

PLATES: pages 1469, 1471

DU. Kauw   FR. Choucas des tours   GE. Dohle   IT. Taccola
RU. Галка   SP. Grajilla común   SW. Kaja

Field characters. 33–34 cm; wing-span 67–74 cm. Less than 75% size of Rook; somewhat smaller and less broad-winged than Chough. Small, dapper, and bustling crow, with short bill and (on ground) quite high head carriage. Black, with grey rear to head (palest in east) and dull grey underparts from breast to vent. Eye pale in adult.

Unmistakable at ranges where short bill and grey rear of head are visible—or even just compact, bustling form—but sometimes difficult to separate on ground at distance when mixed with frequent companion Rook, while distant flight views may also provoke confusion with choughs. Head-on or from behind, silhouette and flight occasionally suggest small pigeon. Flight light and active, with quick, rather jerky beats of backward-directed wings which at distance may appear to 'twinkle'. Unlike larger crows, small size and relatively short legs produce more ground-hugging, bustling, even shuffling progress over ground. Highly sociable, often in hundreds with Rook.

Habitat. Breeds across middle and upper middle latitudes of west Palearctic, in boreal, temperate, steppe, and Mediterranean lowlands, continental and oceanic, up to July isotherm of 12°C. Tolerates wide ranges of precipitation and settled or unsettled weather, but avoids extremes of heat, ice, and snow. Needs sheltered nesting places, apparently adapting from main reliance on hollow or shady trees to rock crevices (inland or coastal), holes in buildings of various kinds, and even burrows of rabbit. Requirement for enough of these in sufficient proximity to satisfy gregarious instincts probably explains to some extent remarkably patchy and fluctuating distribution, and competition or commensalism with other birds subject to similar demands for breeding sites.

Distribution. Range decrease Germany, Czech Republic, Slovakia, Austria, Switzerland, Portugal, and Syria, with extinctions Malta and Tunisia. Range expansion France, Scandinavia, Spain, Italy, Croatia, Slovenia (marked), Ukraine, and Morocco. ICELAND. Almost annual accidental visitor. Unsuccessful breeding attempts 1977, 1987. FAEROES. Probably almost annual visitor, winter and spring. FRANCE. Spread from 1930s, especially in south-east; expansion continuing. Corsica: First observed 1982, first breeding recorded 1983. Sight records increasing; perhaps colonizing. NORWAY. Recent spread to coastal areas in south and west. SWEDEN. Spread north along Gulf of Bothnia; in 19th century did not breed north of Uppsala. ITALY. Recent range extensions in central and northern areas. MALTA. Formerly common resident; exterminated by human persecution in mid-1950s. AZERBAIJAN. Common in Kuba district in north-east. Probably absent from southern slope of Great Caucasus. SYRIA. Up to *c.* mid-20th century, bred around Aleppo and east of there, Euphrates (Ar Raqqah to Dayr az Zawr), also Damascus. More recently, only a few breeding-season observations of single birds or pairs. JORDAN. Bred 1993, and still present at same site 1994. IRAQ. Perhaps rare breeder in northern mountains. TUNISIA. Formerly bred in north-west, now extinct. MOROCCO. Locally distributed in mountains, but has recently colonized lowland town of Ouezzane.

Accidental. Iceland, Balearic Islands, Tunisia, Mauritania, Azores, Madeira, Canary Islands.

Beyond west Palearctic, extends east to upper Yenisey, eastern Tien Shan mountains, and north-west Himalayas.

Population. Decline reported Netherlands, Germany, Finland, Czech Republic (marked), Hungary, Austria, Switzerland, Portugal, and Rumania. Increase Britain, Ireland, France, Denmark, Lithuania, Spain, Italy, Croatia, Belarus', and Morocco, trend being marked in Slovenia and Ukraine. Apparently stable elsewhere. BRITAIN. 390 000 territories 1988–91. Increase

Jackdaw *Corvus monedula*. *C. m. spermologus*: **1** ad summer, **2** ad winter, **3** juv. *C. m. monedula*: **4** ad winter. *C. m. soemmerringii*: **5** ad winter. *C. m. cirtensis*: **6** ad winter.

since mid-1970s; perhaps now stabilized. On census plots in southern England, relatively stable, with moderate increase on farmland. IRELAND. 210 000 territories 1988–91; increase, apparently continuing. FRANCE. 75 000–400 000 pairs. BELGIUM. 21 000 pairs 1973–7. LUXEMBOURG. 1200–1500 pairs. NETHERLANDS. 60 000–120 000 pairs 1979–85. Locally declining. GERMANY. Estimates vary: 93 000 pairs in mid-1980s, 20 000–40 000, or 55 000–150 000, in early 1990s. In east, where decline in many regions, $10 000 \pm 4000$ pairs in early 1980s. DENMARK. 25 000–140 000 pairs 1987–8. Increase 1976–87, stable 1988–94. NORWAY. 1000–10 000 pairs 1970–90. SWEDEN. 150 000–500 000 pairs in late 1980s. Now stable following marked decrease during 20th century in central and southern areas. FINLAND. 40 000–60 000 pairs in late 1980s; decrease since 1956. ESTONIA. 30 000–40 000 pairs in 1991. Stable since 1970s, following increase. LATVIA. 10 000–50 000 pairs in 1980s; now stable after decrease 1950–68. LITHUANIA. Common. POLAND. Fairly numerous. CZECH REPUBLIC. 10 000–20 000 pairs 1985–9. SLOVAKIA. 3000–5000 pairs 1973–94. HUNGARY. 5000–10 000 pairs 1979–93. AUSTRIA. Tentative estimate of 2500–4000 pairs. Increase 1950s; decrease last decade attributed to intensification of forestry and agriculture; some local large colonies have disappeared altogether. SWITZERLAND. 950–1000 pairs 1985–93. SPAIN. 423 600–533 000 pairs. PORTUGAL. 1000–10 000 pairs 1978–84. ITALY. 50 000–100 000 pairs 1983–95. Increase in central and northern areas. GREECE. 100 000–200 000 pairs. ALBANIA. 20 000–50 000 pairs in 1981. YUGOSLAVIA: CROATIA. 60 000–80 000 pairs. SLOVENIA. 3000–5000 pairs. BULGARIA. 1–5 million pairs. RUMANIA. 40 000–60 000 pairs 1986–92. Decline in Transylvania over 60 years up to 1982. RUSSIA. 1–10 million pairs. BELARUS'. 350 000–400 000 pairs in 1990. UKRAINE. 80 000–85 000 pairs in 1986. MOLDOVA. 3500–5000 pairs in 1988. AZERBAIJAN. Uncommon to common. Probably under 1000 pairs. TURKEY. 1–10 million pairs. CYPRUS. Locally very common. ISRAEL. Fairly common. MOROCCO. Scarce to locally uncommon.

Movements. Resident to migratory, wintering almost entirely within breeding range; birds head mostly west or WSW, so some birds of northern race *C. m. monedula* and eastern race *soemmerringii* winter in range of western race *spermologus*. Arrivals and passage mask movements of local birds. Migrates by day in flocks, often in company of Rook. Juveniles migrate more than adults, and over longer distances. Chiefly resident in northern and western Europe, but less so further east. In many northern areas, e.g. Finland and Leningrad region (Russia), birds tend to concentrate near human habitations in winter. Autumn movement September–November. In Britain, continental birds arrive on east and south coasts October–November.

Spring movement is early, chiefly February–March, continuing to April or early May. Winter visitors leave Britain mid-February to 3rd week of April, and in northern Denmark, passage from mid-February, with sharp peak at end of March. Displacement westward across Atlantic not infrequent, underlining westerly component to heading in autumn. Many records from ocean weather stations in eastern Atlantic, sometimes involving parties of dozens; also recorded from Atlantic islands (Madeira and Azores). Some have reached eastern North America (*c.* 41–47°N) in recent years.

Food. Invertebrates, fruits, seeds, carrion, and scraps; sometimes small vertebrates or birds' eggs; food of nestlings predominantly invertebrates. Generally feeds in pairs or small flocks, almost wholly on ground, though will forage seasonally in tree-tops for defoliating caterpillars, beetles, or even acorns, though rarely seen on woodland floor. Feeds mostly in open areas of short or scattered vegetation in pasture, parks, etc., often in company with Rooks, and a common scavenger at rubbish tips, farmyards, and abattoirs; coastal birds readily forage between tidelines. Stores food to a far lesser extent than most other Corvidae.

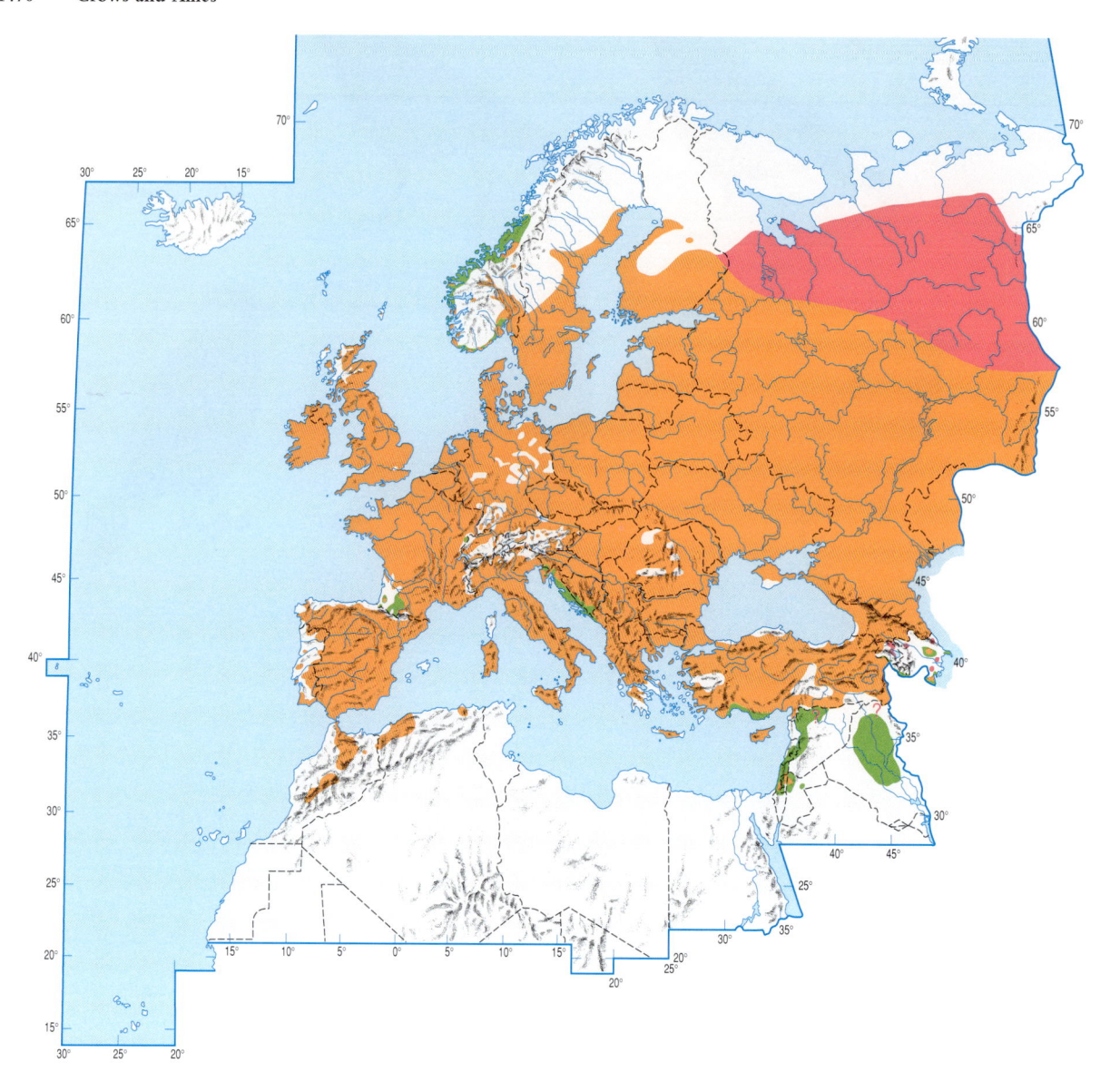

Social pattern and behaviour. Mostly gregarious outside breeding season, although life-long pair-bond means that pair is basic unit within flocks, e.g. pair continue to visit nest-sites almost throughout the year in parts of range where sedentary. Some pair-bonds known to persist at least 5 years. Even after several unsuccessful breeding seasons, pair-bond not usually dissolved, death being almost the only cause of severance. Pair-bonds form in 1st year but pair does not usually breed until 2 years old. Pairing seems to involve extended process of familiarization and is achieved with little obvious display.

Typically a colonial breeder, but dispersion largely dependent on availability of sites, and solitary pairs not uncommon.

Voice. Complex, especially at nest-site in breeding season. Many calls intergrade. Commonest call, given in flight or perched, 'kya' or 'KEak'; heard in variety of contexts, especially pair-contact, including invitation to fly or share company, often from bird announcing arrival at nest to mate and young. Loud, repeated, harsh, grating 'kaaarr' heard especially when warning about approach of, and when mobbing, predator. Song a quiet soliloquy which can last several minutes, given equally well by both sexes; a medley of call-type units with great variation in loudness and inflection; given when alone, perched, or flying. Other calls include a variety of mainly monosyllabic clucking, clicking, or hissing sounds, varying according to context.

Breeding. SEASON. Britain and Ireland: egg-laying from early April to mid-May; average date for laying of 1st egg 23–28 April. Little variation throughout Europe; e.g. average laying

Alpine Chough *Pyrrhocorax graculus graculus* (p. 1464): **1–2** ad. Chough *Pyrrhocorax pyrrhocorax pyrrhocorax* (p. 1466): **3–4** ad. Jackdaw *Corvus monedula spermologus* (p. 1468): **5–6** ad summer. Daurian Jackdaw *Corvus dauuricus*: **7–8** ad.

date in southern Finland 29 April (mid-April to end of May), in southern Spain also 29 April. One brood. SITE. Hole or cavity in tree, rock-face, man-made structure (especially chimney, bridge, or similar rather inaccessible place), and also in nest-boxes; very often in old tree-hole of Black Woodpecker; in some countries in disused burrow of rabbit, and in Finland commonly in nest-box erected for Goldeneye. Nest: very variable in size and structure depending on nature of cavity; in (e.g.) chimney or hollow tree sticks thrown in hole until they lodge, and nests often re-used, so foundation can be huge, or suspended just below cavity entrance, or can cover floor of nest-box or similar flat area. Foundation messy accumulation of largish twigs and sticks, often with irregular layer of lumps of mud or dung, lined with rootlets, rotten wood, stalks, moss, hair (sometimes taken from live animal), feathers, paper, etc. EGGS. Sub-elliptical, sometimes short sub-elliptical, smooth and glossy; pale light blue to greenish-blue with very variable specks and blotches of blackish-brown to light olive, pale grey, or greyish-violet, becoming larger towards broad end; some unmarked. Clutch: 4–6 (2–8). INCUBATION. 17–18 (16–20) days. FLEDGING PERIOD. 28–36 days.

Wing-length: *C. m. spermologus*: ♂ 228–252, ♀ 220–242 mm. *C. m. monedula* and *C. m. soemmerringii* very similar; *C. m. cirtensis* (Algeria) slightly smaller.
Weight: ♂ mostly 200–260, ♀ 180–250 g.

Geographical variation. Slight in colour, very slight in size. Variation in size notably small given wide distribution of species; bill of *spermologus* from western Europe slightly heavier and middle toe slightly longer than in other races. Variation in colour mainly involves depth of grey of rear of head and neck and of underparts, and presence and width of white crescent along side of neck at rear border of grey; also slight variation in colour of gloss on cap. Side of head and neck and hindneck of adult *C. m. monedula* (north and east Europe) on average paler grey than in adult *spermologus* (western Europe); some birds have short, narrow pale grey crescent on side of neck, but usually not a truly white, broad, and contrasting half-collar. *C. m. soemmerringii* (eastern Europe, where intergrades with *monedula*, east to central Asia) differs from other races in presence of distinct white crescent or half-collar on rear of neck. *C. m. cirtensis* (north-east Algeria) more or less intermediate between *C. m. monedula* and *C. m. spermologus*.

Daurian Jackdaw *Corvus dauuricus*

PLATES: pages 1471, 1472

DU. Daurische Kauw   FR. Choucas de Daourie   GE. Elsterdohle   IT. Taccola di Dauria
RU. Даурская галка   SP. Grajilla dáurica   SW. Klippkaja

Field characters. 33–34 cm; wing-span 67–74 cm. Size and structure as Jackdaw but with jowl under chin. Small, dapper, even elegant crow, with general character and behaviour much as Jackdaw. Typical adult mainly black with whitish collar and

Daurian Jackdaw *Corvus dauuricus*: **1** ad, **2** 1st winter. House Crow *Corvus splendens zugmayeri* (Iran to north-west India) (p. 1472): **3** ad, **4** juv.

underbody, suggesting pattern of 'Hooded Crow' races of Carrion Crow. Juvenile similar but duller, though subsequent 1st-year plumage mainly black.

Once size clearly established, pied adult and juvenile unmistakable. Note that some eastern Hooded Crows can, with wear, be as pale as duller individuals of Daurian Jackdaw but show pale back lacking in that species. Dark 1st-year potentially troublesome, closely approaching appearance of juvenile and dull adult Jackdaw of western race, but in close view Daurian Jackdaw shows diagnostic combination of dark eye, black bib, and only limited pale area on hindneck (not extending up to rear crown).

Voice said to resemble Jackdaw, but less cackling and lower pitched, recalling Carrion Crow, e.g. 'kaah', not 'kya'.

Habitat. Breeds extralimitally as eastern counterpart of Jackdaw in equivalent east Asian habitats, in hollow trees along river valleys, scattered in fields, or on river islands, and in mountainous Altai region on rock faces with access to meadows for foraging, up to 2000 m.

Distribution. Breeds from *c.* 96°E in southern Siberia east to Amurland and Ussuriland, south to central China.

Accidental. Netherlands: various sites in Noord- and Zuidholland, May 1995. France: Vendée, June 1995 (perhaps same individual). Sweden: Umeå, April 1985. Finland: Uusikaarlepyy (west coast), May 1883, shot.

Movements. Resident to migratory. Northern breeding areas mostly vacated; some birds remain at least in mild winters.

Wing-length: ♂ 220–244, ♀ 211–232 mm.

House Crow *Corvus splendens*

PLATES: pages 1472, 1487

Du. Huiskraai   Fr. Corbeau familier   Ge. Glanzkrähe   It. Cornacchia delle case
Ru. Блестящий ворон   Sp. Corneja india   Sw. Huskråka

Field characters. 41–43 cm; wing-span 76–85 cm. 10% smaller than Carrion Crow; has proportionately longer bill with deeper and more curved upper mandible, more domed crown, and (in some attitudes) longer neck and legs. Quite large attenuated crow, lacking bulk of common large European black crows and having distinctive bill and head profile. Plumage suggests hybrid between black and hooded forms of Carrion Crow: black on front of head and throat, abruptly grey on nape, neck, and chest, slate on underbody, and black on back, wings, and tail.

Typical bird unmistakable, but beware confusion with Hooded Crow where ranges overlap in northern Egypt. Plumage also recalls Jackdaw, but that species over 20% smaller, with short bill and legs. Flight powerful, with silhouette featuring long bill, narrow head, and long, round-cornered tail. Gait as Carrion Crow but appears more sprightly due to longer thighs.

Habitat. Original range lies in Indian subcontinent, in subtropical and tropical lowlands, also in hills up to somewhat more than 2000 m. Presence of some trees probably essential; roosts communally in mangroves, banyans, coconuts, and in tree plantations, often reached by long high-level flights. Has become intricately enmeshed with human activities in urban and even metropolitan areas, and to lesser extent in small hill-stations where opportunities for easy scavenging exist. The longest-established and most complete case of adaptation by Corvidae from natural to man-made habitats: now nearly always confined to human neighbourhoods, and generally successful in competition with other town scavengers. Can be a serious pest.

Distribution and population. In west Palearctic, confined to a few ports and areas along major shipping routes, where occurrence due to deliberate introductions by man and self-introductions by ship transportation. ISRAEL. First recorded 1976, at Eilat, where up to 12 birds 1976–87. At least 2–4 pairs apparently bred 1979–80, 8 pairs bred in 1988, and 32 birds in Eilat 1990. JORDAN. First recorded 1979, at Aqaba. At least 10–20 pairs in Aqaba in early 1990s, with concentrations in areas of mature palms along coast. KUWAIT. First recorded 1972, but 1957–8 records perhaps this species rather than hooded form of Carrion Crow. Scarce passage-migrant, most records April–May and October. Bred 1983–4, but not since. EGYPT. Established at Suez in or before 1922, thereafter spreading to other parts of Suez Canal area, Red Sea coast, and probably Sinai. 800–850 birds at Suez 1981, 1200–1500 in 1992.

Accidental. Certainly or presumably ship-assisted birds recorded Ireland, Netherlands, and Spain.

Movements. Almost entirely sedentary, but passage reported from Kuwait. In India, subject to altitudinal movements in cold northern areas.

Food. Very dependent on man's rubbish, scraps, offal, and sewage, otherwise any edible invertebrates, small vertebrates, plant material, or carrion. Feeds mostly on ground, but also in trees and on buildings. Forages mainly at rubbish tips, abattoirs, markets, farms, beaches near fisheries or tourist resorts, etc., travelling up to 16 km from roost to feeding place; carrion taken includes human corpses in India. Although cautious, will take any opportunity to scavenge or steal food at human habitations, even entering houses to take it from tables, making it a major pest species and health risk over much of its range.

Social pattern and behaviour. Gregarious throughout year. In Suez, flocks of 5–130, but within flocks commonly in pairs, also often trios. Mating system monogamous although promiscuity not uncommon. In India, pair-bond maintained all year and presumably for life. Territory occupied and defended during breeding season only. Communal roosting in traditional tree-sites a conspicuous element in daily routine; in India can involve thousands of birds. Start arriving before sunset and disperse again before sunrise.

Voice. Extensive repertoire to cater for contingencies such as suspicion, alarm, anger, invitation to copulate, announcement of food-finding, contentment, loss of mate, etc. Consists largely of variations on common contact-alarm call, harsh or rasping 'kwar kwar kwar' or 'waaa waaa waaa'.

Breeding. SEASON. Southern Israel: eggs laid April to late May. Kuwait: nest-building April to mid-May. SITE: Always close to human habitation; usually in fork near top of tree or in outermost branches. Has developed remarkable habit, particularly in India, of nesting in busy streets on buildings, street lamps, pylons, etc. In Kuwait, nested at least 20 m up on cranes although suitable trees nearby. Nest: untidy mass of twigs, often thorny, sticks, plastic, string, assorted pieces of metal, electrical cable, etc.; depression in centre lined with

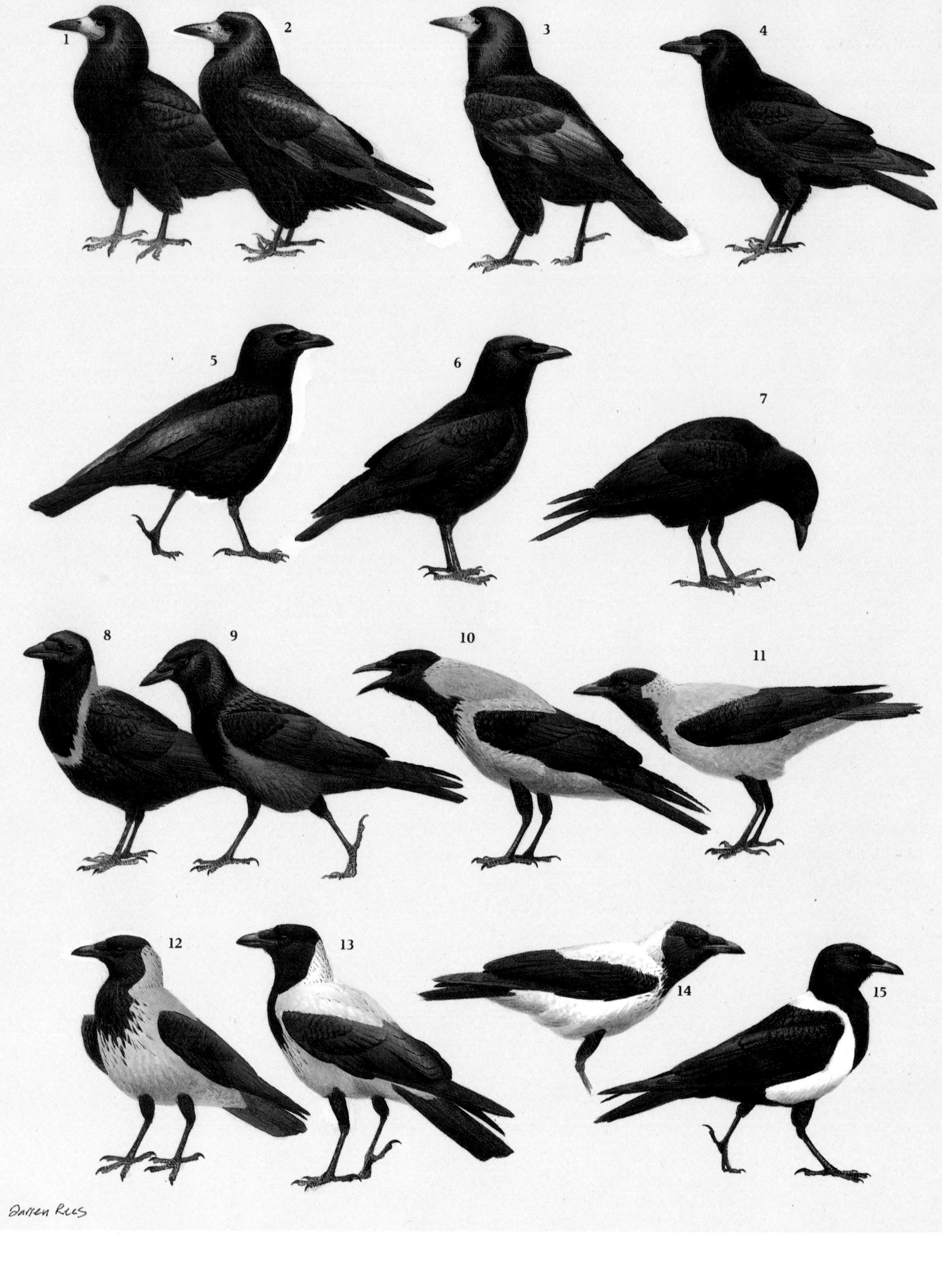
1
2
3
4
5
6
7
8
9
10
11
12
13
14
15
Darren Rees

fibres of wood or bark, grass, hair, cloth, and similar soft material, though can be unlined. In towns, often solely of wire and metal, including items such as spectacle frames, coat-hangers, bicycle pedals, metal sheeting, etc.; such nests may accumulate over years to weigh up to 25 kg and may apparently contain eggs of several pairs, though single nests can be over 8 kg and contain more than 250 m of wire; nests containing much metal last for years, often removing need to build new nest in subsequent years, and may be unlined. EGGS. Very variable in shape, size, and colour; generally short oval, some even pyriform; fairly glossy, pale bluish-green with brown or grey speckling, blotches, and streaks. Clutch: 3–5 (2–6). INCUBATION. 16–17 days. FLEDGING PERIOD. 21–28 days.

Wing-length: ♂ 263–285, ♀ 242–270 mm.
Weight: ♂ 300–362, ♀ 252–304 g.

Geographical variation. Marked, but largely clinal, involving only tone of grey on neck and chest. Widely introduced or self-established in coastal areas beyond native range, and such birds often difficult to assign to a race, either because they are derived from populations intermediate between 2 races, or are a mixture of various races.

Rook *Corvus frugilegus*

PLATES: pages 1474, 1487

DU. Roek   FR. Corbeau freux   GE. Saatkrähe   IT. Corvo
RU. Грач   SP. Graja   SW. Råka

Field characters. 44–46 cm; wing-span 81–99 cm. Slightly smaller than Carrion Crow, with more slender bill, proportionately smaller head with steeper forehead, and seemingly deeper body due to loose flank feathers cloaking thighs; in flight, more splayed wing-tips and rounder tail; over 30% larger than Jackdaw. Quite large and elegant crow, with slender bill, bare and pale face (in adult), and characteristic 'baggy trousers' above legs. Plumage black with heavy gloss. Unlike adult, juvenile has black nasal bristles and fully-feathered head, retained until December–April. Loose thigh feathers may be less evident. Rest of plumage duller and less glossy, with brown cast to hindneck and back. Easily confused with black races of Carrion Crow, but glossier plumage, steeper forehead, and less curved culmen of more slender bill usually evident.

Separation of adult from black races of Carrion Crow not difficult at closer ranges and in good light but may be impossible at distance, particularly in case of single bird when no clues of association or behaviour available. Important also to remember that Carrion Crow does regularly form flocks, particularly from late summer into winter, and exploits same food sources as Rook and Jackdaw at that time. Separation of young Rook, without bare face, tricky. Flight variable: around colony, remarkably agile; less accomplished when moving between feeding areas, with direct and deliberate progress along regular paths achieved by fast-flapping and slightly laborious action with more regular wing-beats and less gliding than in Carrion Crow, but birds erratically spaced, producing characteristic straggling flock (roosting flights are even more leisurely and disjunct). Flight silhouette shows noticeably narrow bill and face, well fingered ends to primaries, and rounded tail. Gait mainly a sedate and (probably due to 'baggy trousers') apparently slightly rolling walk; also includes heavy hopping and sidling movements.

Habitat. Breeds only in boreal and temperate middle latitudes of west Palearctic, in both continental and oceanic lowlands, to July isotherm of 12°C, but absent from warmer regions, except in winter. Range excludes most montane regions. Breeds in England up to 450 m, in Scotland usually well below 350 m and in Carpathians not above 600–700 m. Extralimitally, however, in Asia nests up to 2000 m or more. Requires for breeding fairly tall trees, either on edges of forests or woodlands or by preference in clumps, groves, or riverain or other linear forms fronting open grasslands or croplands. Avoids dense woodland, dry, hard, and rocky surfaces, wetlands and other tall dense vegetation. Breeds in towns and villages only where adjacent countryside is readily accessible. Dependence on agriculture, land improvement for pasture, and conservation of tall trees outside forests have expanded suitable habitats within modern times, favouring types of mixed arboreal and artificially short ground vegetation required.

Distribution. Range expansion France, Belgium, Scandinavia, Poland, Switzerland, Croatia, and Ukraine; decline Netherlands. FRANCE. Has spread south in 20th century, but this has slowed considerably in last 20 years. In 1930s, very rare breeder south of Loire. First breeding colonies near Lyon 1952, Vienne 1965; northern Gironde colonized 1974. NETHERLANDS. Virtually disappeared from north and west 1960–70; some recovery since. SWEDEN. Long-term spread to north. During first part of 19th century bred in only 3 provinces in south. New breeding sites established from 1860s, latest in Jämtland (*c.* 63°N) in 1981. FINLAND. Colonization began in 1880s. Large colonies established only in south-west; further north, only scattered pairs or occasional breeding. POLAND. Very local

(FACING PAGE) Rook *Corvus frugilegus frugilegus*: **1** ad summer, **2** ad winter, **3** 1st summer, **4** juv.
Carrion Crow *Corvus corone* (p. 1478). *C. c. corone*: **5** ad winter, **6** 1st summer, **7** juv. *C. c. cornix* × *corone*: **8–9** ad winter. *C. c. cornix*: **10** ad non-breeding, **11** juv. *C. c. sharpii* (Iraq): **12** ad winter. *C. c. capellanus* (Sardinia and southern Italy eastwards): **13** ad winter, **14** juv.
Pied Crow *Corvus albus* (p. 1481): **15** ad.

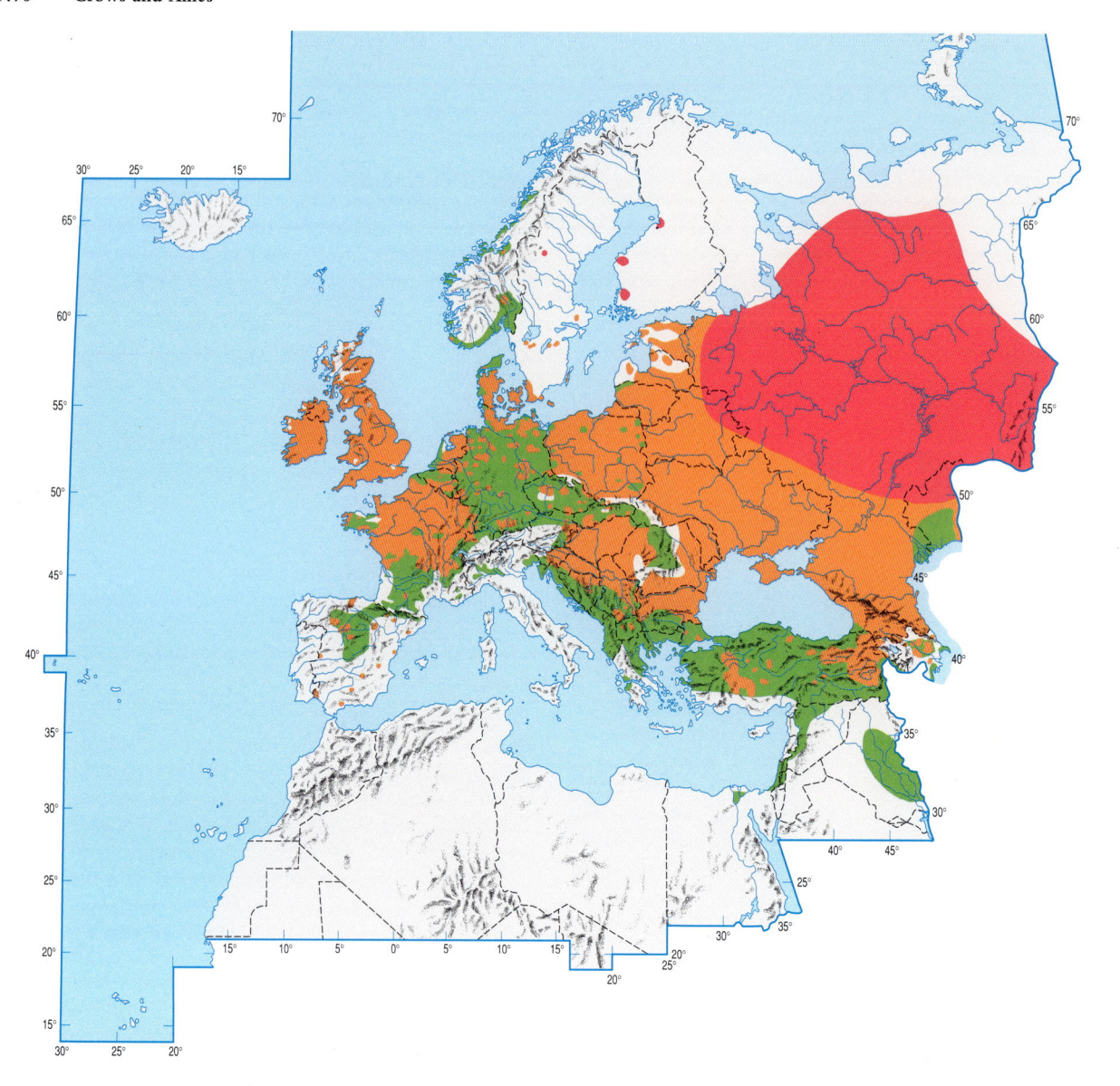

breeder until 2nd half of 19th century when becoming widespread. Further spread since 1945, colonizing human settlements and even city centres. AUSTRIA. Has bred in eastern Steiermark since 1986–7. SWITZERLAND. First bred 1963, subsequently establishing 2 main breeding centres. AZERBAIJAN. Mostly nests in Kura valley. KAZAKHSTAN. Colony of 70 nests Ural delta in 1993. TURKEY. Virtually restricted to steppe country and open cultivated valleys. SYRIA. Large deserted colony found on island in Euphrates 1965.

Accidental. Spitsbergen, Bear Island, Iceland (annual), Balearic Islands, Malta, Lebanon, Jordan, Kuwait, Tunisia, Algeria, Azores, Madeira.

Beyond west Palearctic, breeds across southern Siberia to Amurland and Ussuriland, extending south to Tien Shan mountains in central Asia, and to central China further east. Also breeds northern Iran. Introduced to New Zealand.

Population. Decline (due variously to persecution, changing land-use, and agricultural methods including use of pesticides, and probably other factors) in Britain, Ireland, Low Countries, Germany, Sweden, Finland, Estonia, Latvia, Poland, Bohemia (Czech Republic), Hungary, and Austria. Recovery reported in some of above countries, and recent increase also in Luxembourg, Norway, Lithuania, Slovakia, Switzerland, Croatia, and parts of FSU. BRITAIN. 853 000–857 000 pairs in 1980. General decrease (also Ireland) mid-1950s to mid-1970s, then slight recovery in some areas. IRELAND. 520 000 territories 1988–91. FRANCE. 100 000–300 000 pairs. Marked regional fluctuations. BELGIUM. All censuses since 1928 have given population in range 5000–9000 pairs; increase to 7000–10 000 pairs 1989–91. Decline until 1970s, increase from 1980s, due to better protection. LUXEMBOURG. 500–600 pairs in 1960, 1700 in early 1990s. Increase associated with urbanization; all colonies now in towns. NETHERLANDS. 40 000 pairs in 1944, 11 000 in early 1970s; some recovery to 28 000 pairs by 1985. Decline after 1950 mainly due to pesticides. GERMANY. 32 000–35 000 pairs. In east, 12 900 ± 2000 pairs in early 1980s; huge decline compared with beginning of 20th century. DENMARK. 12 000–140 000 pairs 1987–8; stable 1986–94. NORWAY. 500–700 pairs 1970–90. SWEDEN. 23 000–25 000 pairs in late 1980s. Increase, following marked decrease. FINLAND. 1100–1200 pairs in late 1980s. Increased to early 20th century, then decreased; stable during last decades. ESTONIA. 5000–10 000 pairs in 1991. Marked decline, at least in some areas, from early 1980s (following increase). LATVIA. 16 000 pairs in early 1970s, 7000–9000 in 1980s. LITHUANIA. Up to 100 000 pairs. POLAND. 300 000–900 000 pairs. Increased in 2nd half of 19th century (very scarce previously), then decreased; increased since mid-20th century, but more or less stable in recent years. CZECH REPUBLIC. 2600–3600 pairs 1985–9; stable. Bohemia: 1200 pairs in 1950, 400 in 1986. SLOVAKIA. 10 000–17 000 pairs 1973–94. HUNGARY. Decline in number of colonies, colony size, and mean density, especially in east; 254 361 pairs in 1980, 118 762 pairs in 1984. AUSTRIA. Increased 1950–75 to reach over 500 pairs in 1980; decreasing in recent years, with 450 pairs in 1992. SWITZERLAND. 300–500 pairs 1985–93. SPAIN. 1000–1500 pairs in 1979; stable. YUGOSLAVIA: CROATIA. 3470 pairs in 1993 census. SLOVENIA. 30–50 pairs; stable. BULGARIA. 10 000–100 000 pairs; stable. RUMANIA. 150 000–200 000 pairs (probably more) 1986–92. RUSSIA. 1–10 million pairs. BELARUS'. 700 000–1.2 million pairs in 1990. UKRAINE. 390 000–500 000 pairs in 1988. MOLDOVA. 60 000–70 000 pairs in 1988. AZERBAIJAN. Probably not more than 4000–5000 pairs. TURKEY. 10 000–50 000 pairs.

Movements. Resident to migratory, with more birds migrating in cold winters. Winters in Eurasia, within and south of breeding range. Migrates by day in flocks, often following leading-lines such as coastlines and river valleys and frequently accompanied by Jackdaws. Ringing data have revealed winter quarters of particular populations in unusual detail, and show that mountain ranges act as barriers, thus affecting winter distribution. Adults tend to move less far than juveniles.

British and Irish birds almost entirely resident; juveniles may disperse from natal area in 1st winter, but rarely move more than 100 km. Spanish birds also resident. Chiefly resident in France, though some birds from north migrate 100–400 km. Partial migrant in Low Countries, Germany, and Scandinavia. Chiefly migratory in Poland and Czechoslovakia. In FSU, present all year in southern areas, though some southern birds may move south in colder winters; in migratory areas further north, some birds stay irregularly in certain years. Within Europe, migrants head between west and south, so winter numbers far higher than summer in western Europe, and many migrate also to central and eastern Europe.

Following post-breeding dispersal, autumn departure begins September, with main movement October–November. Arrivals in Britain and France are late September to November, and wintering birds reach Rhône-Alpes (southern France) in last third of October. Hard-weather movements sometimes reported mid-winter. Return movement is early, February–March, exceptionally from January. Winter visitors leave Britain mid-February to 3rd week of April, and Rhône-Alpes by end of March or beginning of April. Swedish breeding birds return February–March; arrive on breeding grounds in Moscow region 2nd half of March.

Food. Invertebrates, mainly beetles and earthworms, plant material (principally cereal grain), small vertebrates, carrion, and scraps of all kinds. Primarily a bird of agricultural landscapes, foraging almost exclusively on ground; only rarely in trees, taking defoliating caterpillars, swarming beetles in spring, fruits, and locally walnuts, acorns, and pine cones. Forages on both pasture (taking invertebrates) and arable land (taking invertebrates and crops); in spring, feeds on newly-sown cereal or follows plough, etc., for exposed invertebrates, particularly larvae. Specializes more than other west Palearctic Corvidae in extracting invertebrates from below soil surface, digging with bill and deep-probing. Stores food in autumn, mainly in ground, for later consumption, mostly acorns, walnuts, and pine cones, though earthworms recorded as being stored throughout year—probably for more immediate consumption.

Social pattern and behaviour. Gregarious outside breeding season for feeding, roosting, and migration, though pair is the chief constituent unit. Mating system essentially monogamous, and pair-bond maintained throughout year, for several consecutive years, and, at least in established breeders, perhaps life-long. Within monogamy, pair share building and defence of nest, but usually only ♀ incubates (♂ feeding her on nest) and broods; ♂ alone brings food for young until nearly halfway through fledging, after which ♀ makes increasing contribution. Typically colonial breeder, with nests densely clustered in treetops ('rookery'). Nest density within colony varies and, where large numbers nest in uniform habitat, colony may be hard to define. Sexual cycle begins, and pair-bonds established, in autumn (regular colony attendance), enabling early spring nesting and fledging young before dry summer months when invertebrate food becomes scarce. Courtship-display in colony in autumn probably as intense as in spring. Formation of new pairs starts in autumn, and 1-year-olds visit colony from mid-March where show incipient courtship and nest-building activity, but usually do not breed until 2 years old.

Voice. Commonest call, used by both sexes all year for self-advertisement and contact, a cawing 'kaah', apparently harsher and 'flatter' than Carrion Crow; variable in timbre and pitch, both between individuals and in different contexts, and ♀'s call longer and higher pitched than ♂'s. Other calls, less loud but of similar quality, used in various social contexts. 'Song' of ♂ comprises medley of much or all of call repertoire, but components usually given more softly than in their respective contexts: various soft cawing, gurgling, rattling, and crackling calls, in sum resembling loud Starling. Given when perched, apparently to attract ♀, but no territorial function.

Breeding. SEASON. Britain: mean date of 1st egg varies from 7–23 March in southern England to mid-April in central Norway and Russia; later for first-time breeders than for older birds. One brood. SITE. In topmost crown of high tree, exceptionally on horizontal branch or against trunk; trees usually in rather isolated groups. No apparent preference for kind of tree, except for tall species. Nest: fairly regular hemisphere, sometimes slightly flattened; foundation of sticks and large twigs, inside which layer of thin pliable twigs very often of birch and willow, many with leaves, followed by compact mass of rootlets, moss, etc., mixed with clay to form small cup, which is lined with grass, moss, stalks, feathers, leaves, paper, etc. EGGS. Sub-elliptical, smooth and faintly glossy; light blue to dull green with olive-buff to blackish-olive specks, blotches, and hair-streaks, sometimes forming cap at broad end; some unmarked; often large variation within clutch. Clutch: 2–6 (1–7). INCUBATION. 16–18 days. FLEDGING PERIOD. 30–36 days.

Wing-length: ♂ 311–335, ♀ 297–320 mm.
Weight: ♂ mostly 300–340, ♀ 280–320 g.

Carrion Crow/Hooded Crow *Corvus corone*

PLATES: pages 1474, 1487

DU. Kraai   FR. Corneille noire   GE. Aaskrähe   IT. Cornacchia
RU. Чёрная ворона   SP. Corneja común   SW. Kråka

Field characters. 45–47 cm; wing-span 93–104 cm (85–110). Slightly larger than Rook, with deeper bill and curved upper mandible, heavier, flatter head, rather longer but less fingered wings, and slightly longer and squarer tail; 25% smaller than Raven, with markedly smaller bill, no shagginess to head, less full wings, and shorter, squarer tail. Large, quite powerful, heavy-billed crow, either all black (*corone* subspecies-group) or with contrasting grey back and underbody ('Hooded Crow', *cornix* subspecies-group). Wide range of intergradation between black and hooded birds in Britain, northern Europe, and Mediterranean region. Lacks any grace or attractive feature, unlike other similar common west Palearctic crows, and does not show loose thigh feathers of Rook.

Black birds need to be carefully distinguished from Rook and Raven, but note differences in structure (see above), flight, and behaviour. Hooded birds unmistakable in Europe and Levant, but racial hybrids with grey restricted to foreparts and just behind head may need care to prevent wishful thoughts of House Crow or even Pied Crow. Flight powerful but rather slow, with usually regular and emphatic but sometimes loose, untidy wing-beats producing steady and usually straight progress; capable of laborious hover; rarely soars; action varied much less than Rook. Gait a steady, direct walk, varied by clumsy hop or sidling jump, in more horizontal attitude than Rook.

Habitat. 'Hooded Crow' breeds in west Palearctic from subarctic and boreal through temperate to Mediterranean, steppe, and desert zones, up to 1000 m in Carpathians and Urals. In Scotland, predominates over nominate *corone* on higher ground, often moorland above *c.* 300 m, and is much more often found nesting on rocks, cliff ledges, and even on banks or islands on ground among heather, ranging up to *c.* 750 m; similar sites used in Norway. Apart, however, from these western populations subject to oceanic climates, habitats over greater part of continental range are no less arboreal than those of nominate *corone*, which in Swiss Alps breeds up to much higher levels, even up to 2000 m. In Russia, commonly found in breeding season in forest country, especially forest edges, groves, and river valleys.

Nominate *corone* in Continental Europe prefers open country with scattered trees, copses, and woodlands. In Britain, habitats are similar, but with marked attraction towards foraging on tidal estuaries, salt-marshes, and coasts. Recently, strong build-up has occurred within even largest towns, although even here arboreal nest-sites are commonly essential.

Distribution. Range increase France, Spain, Slovenia, Bulgaria, northern Russia, Ukraine, and Israel. BRITAIN. Scotland: hybrid zone nominate *corone*/*cornix* has shifted north-west in 20th century. FRANCE. Marked spread in Mediterranean area (Camargue to east). Breeding range of *cornix* stable (numbers unknown, but decreasing winter visitor in north). GERMANY. Overlap zone nominate *corone*/*cornix* 120–150 km wide; position and extent unchanged since 1920s. NORWAY. Nominate *corone* irregular breeder (up to 5 pairs) in coastal Rogaland. POLAND. Recently invading cities. AUSTRIA. Nominate *corone* slightly expanding east. RUSSIA. Kola peninsula: expanded from forest into forest tundra and to some extent tundra. Common breeder at Nar'yan-Mar (67°37′N) and recorded (no breeding evidence) further north on Russkiy Zavorot peninsula 1992–3. TURKEY. Scarce and local southern coastal zone, and steppe country of Central Plateau and south-east. ISRAEL. Recent range expansion follows development of new settlements in 1950s–60s. JORDAN. Formerly more widespread; now mainly in northern highlands and Jordan valley.

Accidental. Spitsbergen, Bear Island, Jan Mayen, Iceland, Malta, Libya, Tunisia, Azores.

Beyond west Palearctic, widespread in northern and central Asia east to Kolyma mountains, Sakhalin island, and Japan, south to Iran, north-west Himalayas, and central China.

Population. Recent dramatic decline in Nile Delta (Egypt), otherwise mainly increasing or stable. FAEROES. 500–1000 pairs. BRITAIN. Number of territories 1988–91: 790 000 (nominate *corone*), 160 000 (*cornix*), 20 000 (hybrids). Continuing increase, though has slowed on census plots in southern England since 1970s. IRELAND. 290 000 territories 1988–91. Increase (since 1924) continues. FRANCE. 200 000–1 million pairs. Spectacular increase in Mediterranean area; stable or increasing elsewhere. BELGIUM. 16 000 pairs 1973–7, but locally variable because of persecution. LUXEMBOURG. 8000–10 000 pairs. NETHERLANDS. 50 000–80 000 pairs 1979–85. GERMANY. 562 000 pairs in mid-1980s; another estimate of 400 000–500 000 pairs in early 1990s. In east, 125 000 ± 60 000 pairs (of which 48% *cornix*, 40% nominate *corone*, and 12% hybrids) in early 1980s. DENMARK. *C. c. cornix*: 21 000–220 000 pairs 1987–8; increase. *C. c. corone*: 300–4000 pairs. NORWAY. 200 000–600 000 pairs 1970–90. SWEDEN. 250 000–500 000 pairs in late 1980s; probably increasing. FINLAND. 200 000–300 000 pairs in late 1980s. Stable, though diminution of predators and increased availability of garbage probably caused some increase. ESTONIA. 50 000–100 000 pairs in 1991. Stable 1950s–70s, increased markedly in 1980s. LATVIA. 20 000–60 000 pairs in 1980s. LITHUANIA. Up to 70 000 pairs; increase. POLAND. 50 000–100 000 pairs. CZECH REPUBLIC. 12 000–24 000 pairs 1985–9. SLOVAKIA. 15 000–30 000 pairs 1973–94. HUNGARY. 70 000–80 000 pairs 1979–93. AUSTRIA. 30 000–50 000 pairs in 1992 (total for both races). SWITZERLAND. 80 000–150 000 pairs 1985–93; increase. SPAIN. 320 000–530 000 pairs; slight increase. PORTUGAL. 1000–10 000 pairs 1978–84. ITALY. 110 000–520 000 pairs 1983–95. GREECE. 150 000–200 000 pairs; slight increase. ALBANIA. 10 000–30 000 pairs in 1981. YUGOSLAVIA: CROATIA. 100 000–150 000 pairs. SLOVENIA. 8000–12 000 pairs; slight increase. BULGARIA. 500 000–1 million pairs; slight increase. RUMANIA. 30 000–50 000 pairs 1986–92. RUSSIA. 1–10 million pairs; slight increase. BELARUS'. 280 000–320 000 pairs in 1990; slight increase. UKRAINE. 450 000–500 000 pairs in 1986; marked increase. MOLDOVA. 7000–10 000 pairs in 1988; slight increase. AZERBAIJAN. At least 8000–10 000 pairs. TURKEY. 100 000–1 million pairs. Greatly outnumbered by Rook in east. CYPRUS. Common and widespread. ISRAEL. At least some tens of thousands of pairs in 1980s; increase. JORDAN. Common. EGYPT. Common (in range 10 000–100 000 pairs).

Movements. Varies from migratory in north of range to sedentary in south and west; many populations partially migratory. Winters almost entirely within breeding range. Migrants move mainly south-west or south. Movements diurnal, often in flocks.

C. c. corone (breeding western Europe) essentially sedentary. *C. c. cornix* (breeding northern and eastern Europe) includes essentially sedentary, partly migratory, and almost completely migratory populations. In Ireland and northern Scotland, ringing recoveries show no long-distance movements. Fenno-Scandian populations partially migratory, and more migratory in north than south. Migrants show strong tendency to follow coast in Baltic region. Autumn movement chiefly October–November; spring movement February–April.

Food. Principally invertebrates and cereal grain; also small vertebrates, birds' eggs, carrion, and scraps, proportions varying greatly according to local availability. In general, a ground-feeder and scavenger in agricultural landscapes, typically in pasture or rough grassland in spring and summer, arable fields in autumn and winter, when also nearer to towns, farms, woods, etc. Favourite sites include dung-rich pasture, hayfields, fields of cereal after harvest, areas by water (especially seashore), and rubbish tips, often exploiting rich food sources to exclusion of others; commonly follows plough.

Social pattern and behaviour. Single birds, pairs, and flocks occur at all times of year. Pairs and sometimes single birds hold territories, in all months in some regions, but principally during breeding season in others. Essentially monogamous, pair-bond of long duration. First breeding reported at 2 years old, but typically at 3 or later. Presence of pair in territory not always proof of breeding. Breeding dispersion essentially territorial; but nests sometimes so close together as to form loose colonies where suitable nest-sites are few and concentrated in extensive feeding areas. Interspecific territoriality may occur with Rook and Jackdaw; more common with Magpie, which it will attack and even kill. Where they nest near to each other there is mutual interference with nest-building and predation of nest contents.

Voice. Conspicuously vocal throughout the year. Commonest call a repeated harsh 'kraa' with vibrant or resonant quality but very variable; often delivered as sequence of bouts. Harsher and more vibrant than Rook. Each call generally drops perceptibly in pitch at the end but calls in same sequence are usually identical. Main variant is loud, harsh 'kraar', delivered in bouts of 2–6 calls with long pause between bouts; given with characteristic head and body movements; described as self-assertive, especially when given by ♂, and commonly given in territorial situations. Variants include more musical 'motor-horn' call lacking any terminal lowering of pitch within each call. ♀ sometimes gives mechanical-sounding, rattling 'klok klok klok', suggestive of machine-gun fire, in some territorial situations, at social gatherings, and sometimes in response to display or self-assertive calling of mate. 'Song', not commonly heard, a soliloquy of very variable calls that seem to represent low-intensity versions of most other calls and, possibly, some vocal mimicry.

Breeding. SEASON. Egg-laying from mid- to late March over most of west Palearctic range, continuing into May; from April in far north, from February in extreme south. One brood. SITE. High in tree at woodland edge, in small stand, or isolated; also on pylon or telephone pole, more rarely on cliff, rock, building, or ground; if no high trees or pylons available (e.g. in north of range) in small tree, bush, or dense low vegetation. Almost always in upper third of highest available local tree, in fork or on branch generally near trunk, or in topmost twigs of smaller species. Nest: rigid but elastic construction typically in 4 layers: foundation of stout, short twigs mostly snapped off trees and bushes, sometimes with leaves, held together by layer of turf and moss, which is followed by smaller twigs, stalks, roots, and commonly runners of couch grass, then lining of bast, bark strips, grass, wool, feathers, etc., and much soft man-made material. Animal bones and wire often incorporated, sometimes forming whole foundation. EGGS. Short sub-elliptical to long oval, smooth and slightly glossy; from light blue to green with very variable speckles, spots, blotches, and scrawls of olive-green to blackish-brown, sometimes very sparse, sometimes obscuring ground colour, often concentrated at broad end; great variation within clutch, and no difference between nominate *corone* and *cornix*. Clutch: 3–6 (2–7). INCUBATION. 18–19 days (17–20). FLEDGING PERIOD. 28–38 days.

Wing-length: *C. c. corone*: ♂ 318–340, ♀ 303–326 mm. *C. c. cornix* virtually the same.

Weight: *C. c. corone* and *C. c. cornix*: ♂ mostly 430–650, ♀ 370–570 g.

Geographical variation. 2 distinct groups. (1) All-black *corone* group, with 2 disjunct races: nominate *corone* in west and extralimital *orientalis* in east. (2) Grey-and-black *cornix* group ('Hooded Crow') with a number of races in northern and eastern Europe, from Corsica and Italy eastward, in Middle East, and in northern, western, and central Asia. Slight variation within nominate *corone*; birds from Spain average smaller than birds from England and other parts of western Europe, especially in bill depth. Within *cornix* group, variation slight and clinal, involving size, relative depth of bill, and tone of grey of body; populations become gradually smaller in size and in bill depth towards south and paler towards south and east. Geographical boundary between *corone* group and *cornix* group rather sharp, formed by relatively narrow zone in which extensive hybridization occurs. Due to secondary character of hybridization, the groups are sometimes considered separate species.

Pied Crow *Corvus albus*

PLATES: pages 1474, 1487

Du. Schildraaf   Fr. Corbeau pie   Ge. Schildrabe   It. Corvo bianco e nero
Ru. Пегий ворон   Sp. Corneja pía   Sw. Svartvit kråka

Field characters. 45 cm; wing-span 98–110 cm. Large, robust crow, slightly larger than Carrion Crow, with proportionately longer and deeper bill and longer legs. Black, with white chest and collar.

Unmistakable in Afrotropics, but vagrants to west Palearctic must be distinguished from bleached hooded forms of Carrion Crow (differing distinctly in greater extension of pale plumage to nape, back, rear underbody, and under wing-coverts). Flight powerful and (unlike Carrion Crow) bird often soars; wing shape recalls Raven, but body size and tail shape more like Carrion Crow.

Commonest call a deep, guttural croak, 'raark' or 'caw', recalling Rook or even Raven.

Habitat. Breeds in tropical low latitudes, usually avoiding arid regions and favouring more or less open country with trees, in which it normally nests; also occurs in forest clearings and in towns and villages, associating freely with man in cultivated or pastoral areas. Feeds largely on ground but is fond of aerial activities, soaring at considerable heights.

Distribution. Resident over almost all of sub-Saharan Africa, Madagascar, and Comores and Aldabra group in western Indian Ocean.

Accidental. Libya: Jalo, April 1931. Algeria: one reported in extreme south in 1961 (no further details), and another at In Azoua, December 1964.

Movements. Chiefly sedentary, especially in south; in some northern areas, many birds move north towards Sahara in rainy season.

Wing-length: ♂ 354–380, ♀ 341–367 mm.

Brown-necked Raven *Corvus ruficollis*

PLATES: pages 1482, 1487

Du. Bruinnekraaf   Fr. Corbeau brun   Ge. Wüstenrabe   It. Corvo del collobruno
Ru. Пустынный ворон   Sp. Cuervo desertícola   Sw. Ökenkorp

Field characters. 50 cm; wing-span 106–126 cm. 10% smaller than Raven, with proportionately more slender bill and longer head (both lacking depth of Raven), less shaggy throat, longer outer part of wing, less evenly-wedged end to tail, and longer-looking legs. Large crow, with general character most recalling Raven but with slighter build, brownish gloss on rear of head, and bill drooping downwards in flight.

At any distance, or with structure and colour of gloss uncertain, difficult to distinguish from Raven and both species thus subject to frequent confusion in narrow areas of overlap across North African desert boundary and from southern Levant eastwards. Even at closer ranges, brown neck may still not show, but differences in structure do allow certain identification. Flight much as Raven but less majestic, action being rather lighter in tight spaces and less powerful over long distances. Flight silhouette combines proportionately longer, narrower-based, and more pointed wings, rather thinner tail (from which central feathers extend to form slight blob), and drooping bill. On ground, extension of folded wing-tips to near or beyond tip of tail is also helpful, as wing-tips of Raven usually fall well short of tip of tail.

Habitat. Basically in deserts of lower middle latitudes, generally in very warm arid open plain country, but exceptionally resident on islands (Cape Verde), there also inhabiting cultivated land, which is usually avoided. In North Africa, closely linked to *Artemisia* steppe with groups of jujube trees *Zizyphus*, being more attached to desert than any other *Corvus*. Also resorts to date palms, tamarisks, and large shrubs, and (where available) to rock ledges and artefacts such as clay structures or telegraph poles. Prevailing confinement to lowland deserts and semi-deserts imposes difficulties in finding suitable breeding sites.

Distribution and population. TURKEY. Small flock reported at Cizre, July 1985; no evidence of breeding and record perhaps not entirely free of doubt (possible confusion with Raven). SYRIA. Pair seen Tall al-Qabli (south-east of Khan Abu Shamat) May 1976; perhaps breeding in area. ISRAEL. Has spread in recent decades, following development of agricultural settlements, army camps, and rubbish dumps. A few hundred pairs. JORDAN. Common. KUWAIT. Breeding probably decreasing despite several consecutive wet years, no sightings since May 1991. EGYPT. Common (in range 10 000–100 000 pairs). TUNISIA. Uncommon. ALGERIA. Very widespread, and apparently extending range to north. MOROCCO. Uncommon; apparently stable. MAURITANIA. Widespread and common. CANARY ISLANDS. Pair of uncertain origin perhaps bred Gran Canaria c. 1990. CAPE VERDE ISLANDS. Occurs all islands and islets, but scarce on Sal (no recent observations there). Common; stable.

Beyond west Palearctic, breeds from Arabia north-east to Kazakhstan; in Africa, extends south of west Palearctic to Sahel zone, and also to Kenya.

Movements. Essentially sedentary over west Palearctic range. In Israel, most of adult population resident, but non-breeders

Brown-necked Raven *Corvus ruficollis*: **1** ad winter, **2** 1st summer, **3** juv. Fan-tailed Raven *Corvus rhipidurus* (p. 1486): **4** ad, **5** juv.

(mainly immatures) make local movements to feeding sites and roosts, and sometimes longer seasonal or nomadic movements. Small numbers move beyond desert zone into Mediterranean zone, in autumn and winter and occasionally in summer.

Food. Ground-dwelling invertebrates, small vertebrates, and carrion; some fruit, grain, and other seeds. Forages generally in open ground, either cultivated or uncultivated, on soil, sand, stones, and short, often grassy, vegetation where sometimes takes contents of birds' nests; often feeds on rubbish-tips, dung-heaps, and animal carcasses. Perches and hangs head-down on camel and donkey to pick ectoparasites such as ticks from neck, back, sides, legs, and occasionally head; also feeds at cuts and sores.

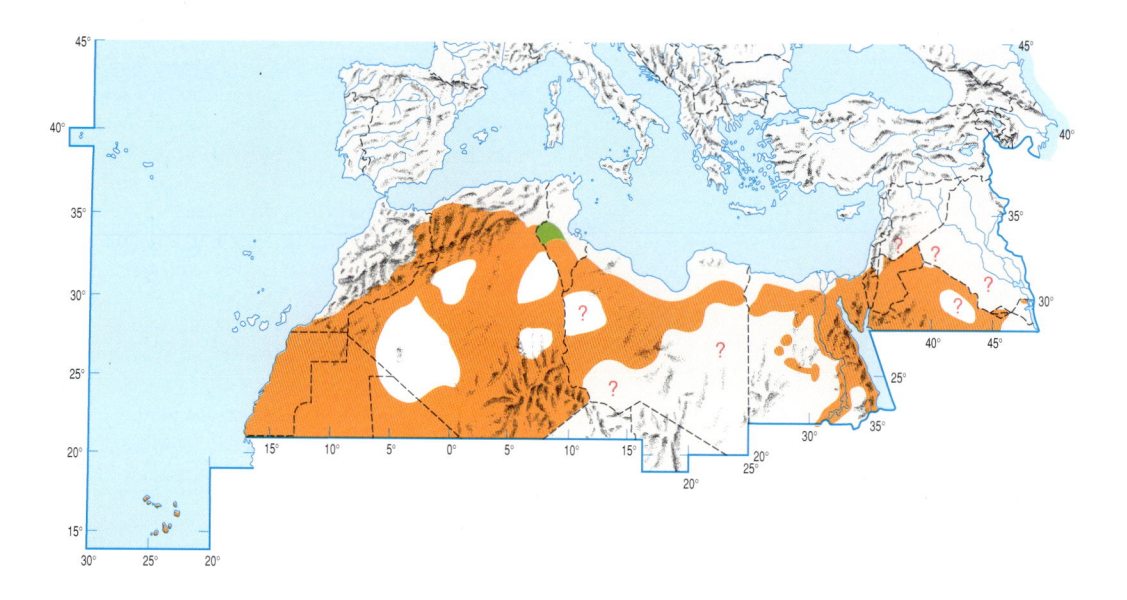

Social pattern and behaviour. Not well known, but generally similar to Raven. At all seasons occurs singly, in pairs, and (where numerous) in flocks. In Israel, adults mainly in territorial pairs all year, while remainder of population, especially immatures, lives in flocks of variable size which tend to congregate at food sources. Very shy and wary in many areas, but can be bold and fearless where not persecuted by man. In Israel, territorial birds will accompany military convoys and groups of hikers passing through their territories, to pick up scraps.

Voice. Quite freely used. Commonest call a harsh 'karr-karr-karr' or 'korr-korr', typically much less deep and croaking than most calls of Raven; evidently variable according to social context and circumstances.

Breeding. SEASON. Cape Verde Islands: eggs laid mid-November to mid-April, after rains. North-west Africa: eggs laid early January to late March. Egypt: eggs laid late February to early April. Israel: laying mainly mid-February to early March, some late January. SITE. Prefers crown of tree, but in arid areas, sometimes in low bushes; where trees scarce, on cliffs or rocky outcrops, either on open or concealed ledge, or in cracks in cliffs; or on ground among bushes. Uses man-made structures and frequently recorded on side-extensions of telegraph poles and power pylons. Nest: bulky structure of sticks similar to that of Raven, but smaller; lined thickly with plant fibre, grass, wool, hair, feathers, paper, etc. EGGS. Sub-elliptical, smooth and glossy; pale blue with spots, small longitudinal streaks or scribbles, and small blotches of olive-buff to olive-brown and blue-grey; markings often pale and sparse. Clutch: 2–5 (1–7). INCUBATION. 18–23 days. FLEDGING PERIOD. Leave nest at 35–38 days; flying at 42–45 days.

Wing-length: ♂ 378–439, ♀ 363–408 mm.
Weight: ♂ 580–795, ♀ 500–700 g.

Raven *Corvus corax*

PLATES: pages 1484, 1487

DU. Raaf   FR. Grand Corbeau   GE. Kolkrabe   IT. Corvo imperiale
RU. Ворон   SP. Cuervo común   SW. Korp

Field characters. 64 cm; wing-span 120–150 cm. As large as Buzzard, differing from all other west Palearctic crows in deep, massive bill, flat head, shaggy throat, long, deep body, long, broad wings, strong legs and feet, and rather long, wedge-ended tail; up to 25% larger than Brown-necked Raven and 30–35% larger than Carrion Crow and Rook. Huge, majestic, and powerful black crow, with superb flight but at times playful nature. Cruciform silhouette when soaring.

Unmistakable if seen well in north or west of range, with huge, powerful form unique among *Corvus*. Glimpsed or only seen at distance in North Africa and Levant, subject to confusion with Brown-necked Raven; separation best based on comparison of size and structure. Best characters of Raven are (1) deeper and longer bill and head (slighter and often drooped in Brown-necked Raven), (2) larger and broader-based and broader-tipped wings (narrower-based and more pointed in Brown-necked Raven), and (3) fuller, wedge-ended tail (narrower-based and with central feathers just protruding in Brown-necked Raven). Flight powerful and majestic; often performed at great height and includes more gliding (including hanging on the wind) and soaring than any other crow, so inviting confusion with broad-winged raptors. Active flight produced by series of powerful, rather ponderous, noticeably regular but not deep wing-beats, interspersed with glides and occasional falls and rises. Particularly in early breeding season, may indulge in aerial play, with rolls, tumbles, and dives all performed with evident enjoyment.

Habitat. So wide-ranging that concept of habitat is hardly applicable. Breeds almost throughout west Palearctic, except for certain densely settled and cultivated regions, from Arctic to tropics, even to July isotherm of 3°C; from sea-level to c. 2400 m in Alps. Overriding requirements are for nest-site of difficult access, normally on rock-face or tall tree, and wide, largely undisturbed foraging area with tracts of open surface of any kind on which long-range food-gathering, often involving high flights, can be practised. Thus avoids interior of large or dense forests, scrub woodland, thickets, shrubby terrain, wetlands with tall aquatic vegetation, orchards, plantations, field crops, and intensively farmed or grazed lands. Coasts with cliffs, even in windy and chilly climates, often satisfy, especially where they afford respite from human persecution.

Distribution. Recent expansion of range over much of northern, central, and parts of south-east Europe; trend most marked Czech Republic, Slovakia, Slovenia, Bulgaria, and Ukraine. Expansion due to active conservation, including reintroduction schemes, and to reduced persecution. Slight decline Portugal. BRITAIN, IRELAND. Formerly widespread, but range much reduced in 19th century. Considerable expansion in 20th century. FRANCE. Range and numbers declined in late 19th/early 20th century; expansion subsequently up to 1970s, trend-continuing mainly in Massif Central and inland Brittany since. BELGIUM. Becoming extinct from late 19th century; in 20th century, bred 1919, 1948; reintroduced in south in 1976. At least 2 (or up to 40) pairs in 1990s. LUXEMBOURG. Last bred 1946. NETHERLANDS. Sharp decline, with last breeding 1928, 1944. Reintroduced from 1966, first pairs breeding 1976, but numbers breeding and success increased only from 1987. GERMANY. Almost disappeared in east in 1940s, but recolonized from Schleswig-Holstein; expansion continuing in south (introduced Thüringen). NORWAY. Recent spread in south. FINLAND. Has recently spread over extensive areas in south where formerly absent. POLAND. Former extensive range much reduced, then recovery from 1930s with recolonization, from north-east, of all areas. AUSTRIA. Expansion in north-east;

Raven *Corvus corax*. *C. c. corax*: **1** ad winter, **2** 1st summer, **3** juv. *C. c. varius*: **4** ad. *C. c. laurencii*: **5** ad. *C. c. tingitanus*: **6** ad.

breeding Wienerwald (south-west of Vienna) since *c.* 1970. SWITZERLAND. Before 1950, breeding restricted to Alps; has since spread to Jura and lowlands. TURKEY. Very scarce or absent south-west Anatolia, south coast, heart of Central Plateau, and steppe of south-east. SYRIA. Widespread breeder, perhaps restricted by lack of secure nest-sites. Southern and north-eastern limits of range not clear. JORDAN. Apparently no longer breeds. Formerly common in Amman. IRAQ. Recent breeding status unknown. ALGERIA. Widespread and common. MOROCCO. Common north of Haut Atlas; small numbers breed locally further south.

Accidental. Spitsbergen, Bear Island, Malta, Azores, Madeira.

Beyond west Palearctic, widely distributed in northern and central Asia, south to Iran, Pakistan, and Himalayas. In North America, breeds from arctic Alaska across Canada south to Nicaragua in west, and to Appalachians in east. Also breeds Greenland.

Population. Increasing over much of range, markedly in Sweden, Finland, Estonia, Czech Republic, Slovakia, Austria, Slovenia, Bulgaria, Belarus', and Ukraine. Decline reported Britain, Portugal, Cyprus, and Canary Islands, apparently stable Iceland, Norway, Hungary, Spain, Italy, Moldova, and Israel. ICELAND. 1600–3000 pairs in late 1980s. FAEROES. 150–300 pairs. BRITAIN. 7000 pairs 1988–91. Small decline overall, more marked decline Northumberland and southern Scotland. IRELAND. 3500 pairs 1988–91. FRANCE. 1000–3000 pairs. NETHERLANDS. Increased from 17 pairs in 1988 to 50 in 1992, when also 31 territories held by non-breeders. GERMANY. 6000 pairs in mid-1980s, though another estimate (for early 1990s) of 3400–5000 pairs. In east, 3400±900 pairs in early 1980s. DENMARK. 350–400 pairs in 1990. NORWAY. 20 000–50 000 pairs 1970–90. SWEDEN. 10 000–20 000 pairs in late 1980s. Increase began in 1960s. FINLAND. 5000–7000 pairs in late 1980s. Increase (especially in south) due to decreased persecution and increased availability of garbage. ESTONIA. 500 pairs in late 1960s, 3000–5000 in 1991. LATVIA. 2000–4000 pairs in 1980s. Decrease in 2nd half of 19th century; recovery began 1940s. LITHUANIA. 15 000 pairs. POLAND. 3000–6000 pairs. Recovery began 1930s; still increasing. CZECH REPUBLIC. Only 5–10 pairs 1973–7, 250–400 pairs 1985–9. SLOVAKIA. 1500–2000 pairs 1973–94. HUNGARY. Major decline in 1970s, but recovered to 300–400 pairs by 1979–93. AUSTRIA. 2500–3500 pairs in 1992. After alarming decrease, recovery and marked increase since 1960s, still continuing. SWITZERLAND. 1500–2500 pairs 1985–93. Increase since 1950. SPAIN. 60 000–90 000 pairs. PORTUGAL. 1000–10 000 pairs 1978–84. ITALY. 3000–5000 pairs 1983–95. Probably stable overall, though increase reported in pre-Alpine region. GREECE. 5000–10 000 pairs. ALBANIA. 1000–3000 pairs in 1981. YUGOSLAVIA: CROATIA. 7000–10 000 pairs. SLOVENIA. 1000–1500 pairs. BULGARIA. 150–1000 pairs. Increase since 1950s; former decline attributed to poisoned bait against wolf. RUMANIA. 2000–3000 pairs 1986–92. RUSSIA. 100 000–1 million pairs. BELARUS'. 14 000–22 000 pairs in 1990. UKRAINE. 20 000–22 000 pairs in 1986. MOLDOVA. 200–300 pairs in 1988. AZERBAIJAN. Uncommon; probably several hundred pairs. TURKEY. 5000–50 000 pairs. CYPRUS. Scarce to fairly common. Less numerous than in 1950s. ISRAEL. Common up to 1960s (at least a few hundred pairs). Then drastically reduced by pesticides and perhaps other causes; recovered somewhat in 1980s, but still only *c.* 25 pairs. EGYPT. Rare. TUNISIA.

Common. CANARY ISLANDS. 622–667 pairs in 1987. Declining, possibly due to changes in agriculture, and persecution.

Movements. Populations south of *c.* 60°N essentially sedentary, but some immatures make extensive movements. Northern populations mainly sedentary and dispersive, but more prone to make southward movements in winter; longest recorded movements 350–510 km.

Food. Plant and animal material, taken opportunistically; animal food may be killed with powerful bill, or scavenged as carrion, refuse, etc.; also robs nests and takes invertebrates (especially molluscs on shore); plant material mainly cereals and fruits. Where carrion plentiful, usually takes food mostly by scavenging. Usually forages on ground away from cover, and commonly on rubbish-tips, near slaughterhouses, on tide-line, etc., and, where not persecuted, scavenges boldly around dwellings, particularly of nomadic herdsmen; will follow the plough, and frequently on fields where dung has been spread. Harries sick and injured animals, even species not normally preyed upon, making darting lunges with bill, often aimed at eyes; also takes advantage of tourist refuse, road casualties, remains of raptor kills, and infestations of defoliating caterpillars, and will hawk for flying ants. Often hides food, particularly when hungry; prefers to cache fat or fatty meat, but also recorded hiding whole eggs, bones, bread, dates, and dung; during breeding season, caches more than at other times and preferentially food suitable for young. Usually caches in holes or under stones; in snow, digs hole with sweeping motions of bill.

Social pattern and behaviour. Outside breeding season occurs solitarily, in pairs, or in flocks (sometimes large) which commonly include paired birds. Territory-holders commonly remain in territory all year, even in severe weather, but sometimes join flocks at food in and near to territory. Monogamous, and almost certainly pairs for life. Age of first breeding not established from marked birds, but probably 3 years or older. Behaviour appears more sophisticated than in other birds. Social and sexual displays mainly undifferentiated and highly variable, probably allowing communication of subtle detail of social information to mate or to well-known member of group. Members of pair readily recognize each other individually and transmit modified vocal information directed only at mate, even over long distances. Behaviour interpreted as play, often including aerobatics, apparently common. Play sequences prone to great individual variation and group-specific play combinations arise by mutual imitation.

Voice. Much used, throughout year. Most common calls are low-pitched with gruff, croaking tone, so, despite great variability, these are easily distinguishable from other European *Corvus*—though less easily from some calls of Brown-necked Raven and, especially, Fan-tailed Raven. Commonest call a short barking 'pruk'; varies considerably, between and within individuals. Given in many situations, in flight and perched. Often 3–4 calls in rapid succession in alarm, or, more slowly, as 'conversation' between pair members. Various other, mainly softer, calls are used in different social contexts. So-called song comprises long series of varied, mainly soft sounds, many musical or pleasing in tone. Not known whether one or both sexes sing.

Breeding. SEASON. Early. Egg-laying from February, or even late January, over much of European range; from March in far north. North Africa: apparently later than Europe; usually from beginning of April, sometimes end of March. One brood. SITE. High up on tree, isolated or generally at forest edge, inland or coastal cliff, or man-made structure; where undisturbed, can be much lower, also in tall shrubs or even on ground. Nest: basically in 4 fairly distinct layers: outer foundation of sticks, twigs, and woody stems up to *c*. 150 cm long and 2·5 cm thick, neatly woven with fresh twigs, which sometimes form distinct rim, then layer of earth, dung, grass, and roots, at times making bowl, but can be completely absent, particularly if ground still snow-covered; this is lined with moss, grass, rootlets, leaves, etc., then finally compacted layer of wool, hair, fur, grass, lichen, stems, etc.; wire and bones not infrequently found as material. EGGS. Sub-elliptical, smooth and slightly glossy; light blue to blue-green with very variable olive to blackish-brown blotches, spots, scrawls, or hair-streaks usually concentrated towards broad end, though at times completely absent, occasionally with greyish-violet undermarkings. Clutch: 4–6 (2–7). INCUBATION. 20–21 days. FLEDGING PERIOD. *c*. 45 days (35–49), young often leaving nest before able to fly.

Wing-length: *C. c. corax*: ♂ 407–452, ♀ 400–439 mm.
Weight: *C. c. corax*: ♂ 1080–1560, ♀ 800–1315 g.

Geographical variation. Slight in intensity of gloss, colour of feather-bases, and length of throat feathers, more pronounced in size (wing, tail, weight), relative tarsus length, and relative length and depth of bill. Within west Palearctic, North African race (*tingitanus*) and Canarian race (*canariensis*) smallest, nominate *corax* largest. Other races (*hispanus*, Iberia and Balearic Islands; *laurencei*, Greece, Turkey, Cyprus, eastwards) intermediate. *C. c. varius* (Faroes and Iceland) similar to *C. c. corax* but with tendency to white feather-bases.

Fan-tailed Raven *Corvus rhipidurus*

PLATES: pages 1482, 1487

DU. Waaierstaartraaf   FR. Corbeau à queue courte   GE. Borstenrabe   IT. Corvo coda a ventaglio
RU. Трубастый ворон   SP. Cuervo colicorto   SW. Kortstjärtad korp

Field characters. 47 cm; wing-span 102–121 cm. Head and body close in size to those of Brown-necked Raven but with shorter wings and much shorter, stump-like tail. Large, bat-winged black crow, with folded wing-tips extending well past tail and behaviour on ground recalling Rook.

Unmistakable. Due to unusual silhouette flight appears less powerful than other large crows, most recalling Rook; much given to soaring and group acrobatics. Gait a walk, varied by sidling jumps and occasional leaps; relative lack of tail makes legs appear long. Has distinctive habit of holding bill open, as if panting.

Habitat. Breeds in subtropics and tropics, mainly in semi-arid or arid regions, preferring presence of cliffs or crags suitable for nesting, and creating opportunities for habit of playing and soaring in thermals.

Distribution and population. SYRIA. Single record: Palmyra, August 1975. ISRAEL. Estimated 300 pairs; apparently more abundant formerly, perhaps adversely affected by competition with Brown-necked Raven which has spread. JORDAN. Common from Dead Sea south to Aqaba. EGYPT. Small numbers breed south Sinai.

(FACING PAGE) House Crow *Corvus splendens zugmayeri* (p. 1472): **1–2** ad.
Rook *Corvus frugilegus frugilegus* (p. 1475): **3–4** ad winter.
Carrion Crow *Corvus corone* (p. 1478). *C. c. corone*: **5–6** ad winter. *C. c. cornix*: **7–8** ad winter.
Pied Crow *Corvus albus* (p. 1481): **9–10** ad.
Brown-necked Raven *Corvus ruficollis ruficollis* (p. 1481): **11–12** ad winter.
Raven *Corvus corax corax* (p. 1483): **13–14** ad winter.
Fan-tailed Raven *Corvus rhipidurus*: **15–16** ad.

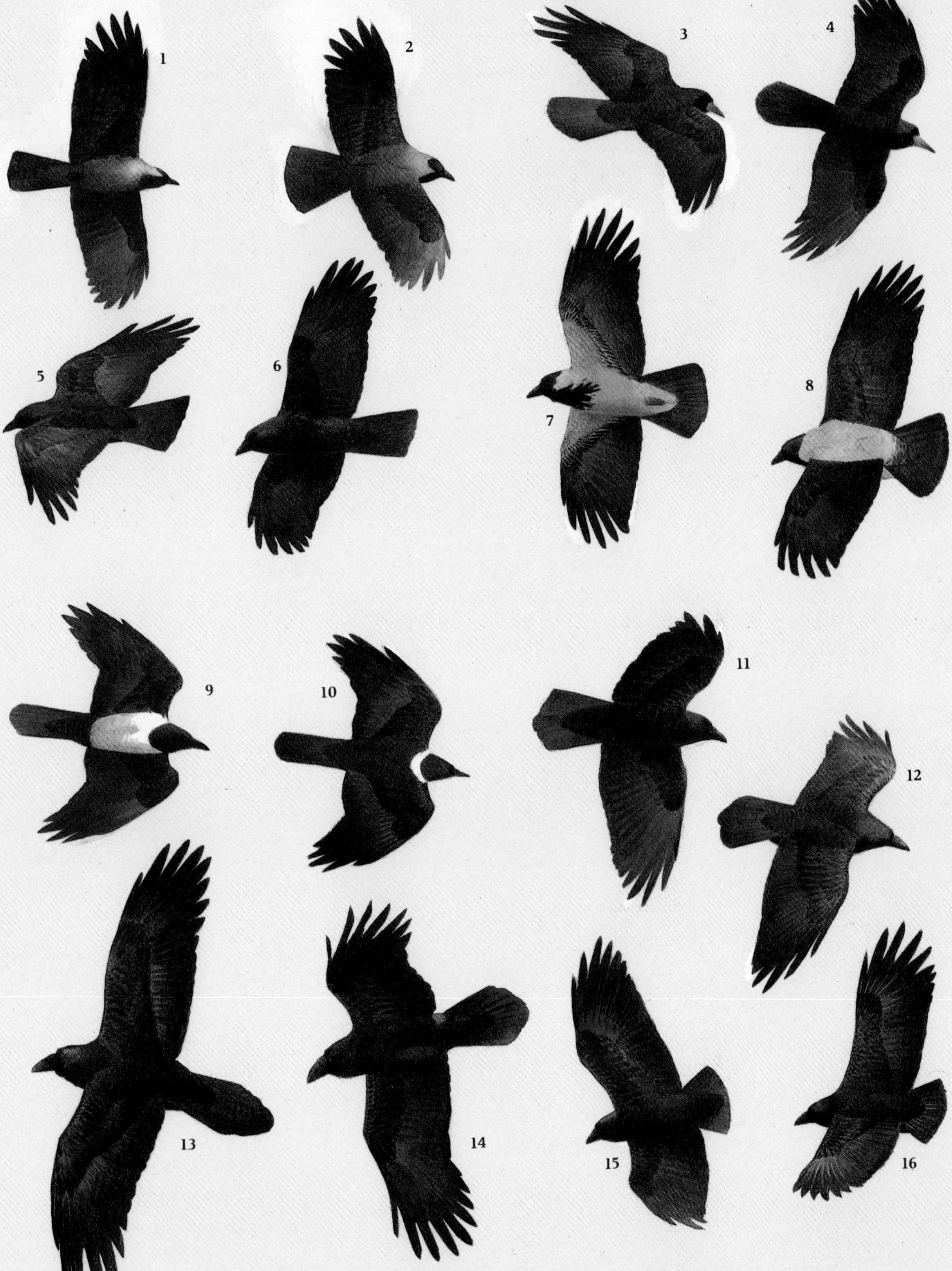
1
2
3
4
5
6
7
8
9
10
11
12
13
14
15
16
Darren Rees

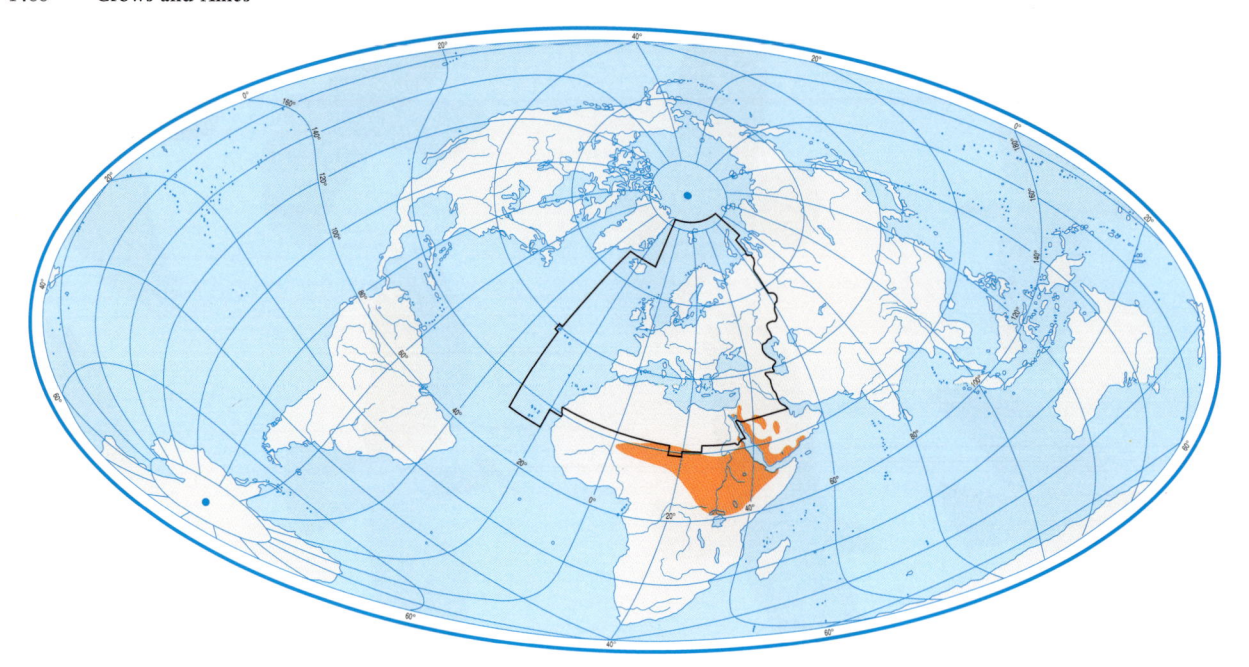

Movements. Essentially sedentary, with local movements. In Israel, mainly sedentary but recorded at Eilat in south chiefly in winter, and formerly wintered regularly in and near Jerusalem; wanderers occur singly or in groups along Arava valley.

Food. Mainly recorded scavenging near human settlements for offal, scraps, and rubbish; also takes insects and other invertebrates, grain pecked from animal droppings, and fruit. Frequently perches on backs of goats and camels, searching for ticks and other ectoparasites. Feeds on ripening dates from tree, and in Saudi Arabia farmers may put sacks over fruit to reduce losses.

Social pattern and behaviour. Very poorly known. Occurs singly, in pairs, and in groups (especially at good food sources) all year. Mating system and nature of pair-bond not known. No information on roles of sexes in breeding.

Voice. Best-known call, commonly given as series, a falsetto croaking very like White-necked Raven *C. albicollis* of Afrotropics, but perhaps even higher pitched. Usually shorter and somewhat higher pitched than Brown-necked Raven and much higher pitched than Raven. Show considerable variation in pitch, duration, intensity, and amount of vibrato. 'Song' a medley of soft clucks, high-pitched squeals, and loud tremolos suggestive of frogs.

Breeding. SEASON. Israel and Jordan: eggs laid early March to late April. SITE. Hole, crevice, or sheltered ledge, usually inaccessible in sheer cliff; rarely on building, exceptionally in tree. Nest: loosely constructed platform and cup of sticks, twigs, and roots, lined with wool, hair, cloth, freshly plucked twigs, and other soft material. EGGS. Sub-elliptical to long oval, smooth and glossy; pale greenish-blue, blotched and speckled or faintly streaked with olive-brown and dark brown, and less conspicuous violet-grey under-markings. Clutch: 3–4 (2–6). INCUBATION. 18–20 days. FLEDGING PERIOD. 35–40 days.

Wing-length: *C. r. stanleyi*: ♂ 359–370, ♀ 351–363 mm.
Weight: *C. r. stanleyi*: ♂ ♀ 330–550 g.

Geographical variation. Marked; involves size only. Birds from Sinai and from Jordan valley south through Saudi Arabia to Yemen and Oman markedly smaller than birds from Afrotropics; named *stanleyi* in honour of Stanley Cramp, instigator of *The Birds of the Western Palearctic*. Races do not differ in colour, but *stanleyi c.* 10% smaller in all measurements.

Starlings Family Sturnidae

Medium-sized oscine passerines (suborder Passeres). Habitat range wide: grassland, savanna, and steppes, woodland, forest, etc.; some species closely associated with human cultivation and habitation. Mainly arboreal to greater or lesser extent but some (especially in genus *Sturnus*) largely ground-feeders. Many species frugivorous and insectivorous but others more omnivorous, eating variety of animal and vegetable foods. Family of Old World origin: found mainly in Africa, Eurasia, Indonesia, and western Pacific islands, with main centres of distribution in Afrotropics and southern Asia. A few species commonly kept as cage or aviary birds. Some introduced elsewhere, 2 (Starling and Common Myna) widely, including North America, Australia, and New Zealand. Predominantly sedentary or nomadic but some species migratory. 113 species: 6 in west Palearctic, 5 breeding (including introduced Common Myna) and 1 vagrant.

Typically rather heavily built with sexes of similar size (♂ slightly larger). Head bristly and almost naked in a few species, some others having wattles, lappets, or bare skin round eye. Bill usually quite stout and pointed, sometimes with hook; often decurved to some extent; characteristically used by ground-feeding species to pry when feeding, i.e. by method known as open-billed probing in which closed mandibles inserted into hole and then pressed open in order to expose and then seize prey; linked with anterior narrowing of skull which enables feeding bird to look forward while prying. Wing broad and rounded or rather long and pointed; 10 primaries, p10 reduced (length 15–25% of longest) and often lanceolate. Flight strong, fast, and straight in longer-winged species, slower and more deliberate in others. Tail often fairly short and square but long and graduated in some; 12 feathers. Leg and toes sturdy. Gait a hop in some species but most walk and run, often with upright stance.

Voice relatively unspecialized; calls mostly simple, often loud, harsh, and grating, but uttered frequently and often delivered in garrulous manner. Advertising song also typically simple, usually consisting of musical whistles or warblings. Mimicry common, especially from ♂ in courtship song: some mynas can be trained to imitate human phrases with surprising accuracy. Most species highly gregarious at times, especially when feeding and roosting. Many loosely colonial when breeding, each pair defending small territory near nest, but some tropical species nest communally with helpers and maintain group territory. Monogamous mating system probably the general rule. Incubation usually by ♀ only, or by both sexes but with ♀ often taking larger share. Both sexes usually feed young.

Plumage variable: often black or black-and-white, sometimes with patches of rufous, brown, grey, pink, crimson, or yellow; silky and often with glossy, metallic iridescence. A few species have erectile or permanently raised crests. Sexes usually quite similar, with ♀ often duller and (e.g.) lacking bright facial colour of ♂; in a few species, sexes different. Seasonal variation quite marked in a few species, mainly due to changes caused by abrasion or by development of lappets, etc.

Tristram's Grackle *Onychognathus tristramii*

PLATES: pages 1490, 1501

Du. Tristrams Spreeuw   Fr. Rufipenne de Tristram   Ge. Tristramstar   It. Storno di Tristram
Ru. Тристрамов гракл   Sp. Estornino irisado   Sw. Sinaiglansstare

Field characters. 25 cm (tail 9 cm); wing-span 44–45 cm. Head and body close in size to Starling but wings and especially tail noticeably longer, with silhouette somewhat reminiscent of Ring Ouzel. Lengthy, strong-billed and strong-legged, starling-like bird, with long, square-cut tail. Plumage looks black at distance, with striking pale chestnut patches on primaries.

Unmistakable in west Palearctic, with other members of this African genus approaching no nearer than coast of lower Red Sea. Flight fast, action recalling both starling and thrush; often performs sweeping, diving, and climbing manoeuvres which, when indulged in by flocks, give impression of fearless and purposeful aerial play. Gait combines bouncing hop and striding walk.

Habitat. Restricted to Arabia and Levant, where basically confined to mountains or other rocky areas, inhabiting ravines, canyons, and cliffs, down to below sea-level. Roosts and nests in these, using holes or crevices in caves. Needs access to vegetation bearing food, from desert plants to dates, prickly pears, and grapes obtained on visits to villages and even towns, especially rubbish dumps. Association with man at Arad (Israel), 600 m above sea-level, has recently led to urban nesting in holes or crevices in uninhabited buildings. Recent increase in numbers and spread of range seem attributable to growing dependence on man, and greater tameness.

Distribution and population. ISRAEL. Up to 1950s, limited distribution in Judaean desert and Dead Sea depression; has since expanded range, following development of agriculture and settlements. 1000–2000 pairs in 1980s; increase. JORDAN. Locally common. 21–40 pairs Petra, 1983. EGYPT. Locally common, south Sinai.

Beyond west Palearctic, breeds western and southern Arabia.

Movements. Resident and dispersive, with flocks seen out of breeding season well away from breeding range.

Food. Mainly fruit and insects. Feeds in bushes and on ground, sometimes among cattle. Will crack snail shells on anvil like

Tristram's Grackle *Onychognathus tristramii*: **1** ad ♂ summer, **2** ad ♂ winter, **3** ad ♀, **4** juv. Daurian Starling *Sturnus sturninus* (p. 1491): **5** ad ♂, **6** ad ♀, **7** juv.

thrush. Picks ectoparasites from pelts of ibexes, donkeys, cattle, and camels.

Social pattern and behaviour. Usually in flocks outside breeding season. Monogamous, pair-bond retained all year and apparent in flocks. Both sexes participate in searching for nest-site, collecting material, and in building; ♀ incubates alone, both sexes feed young. During courtship, ♂ vibrates wings close to body during which wing-patches make 'blurring flash' in sun. During incubation, ♂ occasionally sings near nest.

Voice. A highly vocal species; differences between song and calls not always clear. Typical call/song a variety of loud musical whistles, e.g. 'dee-oo-ee-o', 'o-eeou', 'vu-ee-oo', 'sweee-to', 'tsoowheeo', 'wee-o-weee'. Other calls include a harsh rising 'weeeaagh' or 'kraaaah' rather reminiscent of alarm-call of Golden Oriole.

Breeding. SEASON. Israel: beginning of March to end of June, first clutches March–April. 2 broods. SITE. Crevices and holes in rocky ravines, on ledges, and in caves, high above ground;

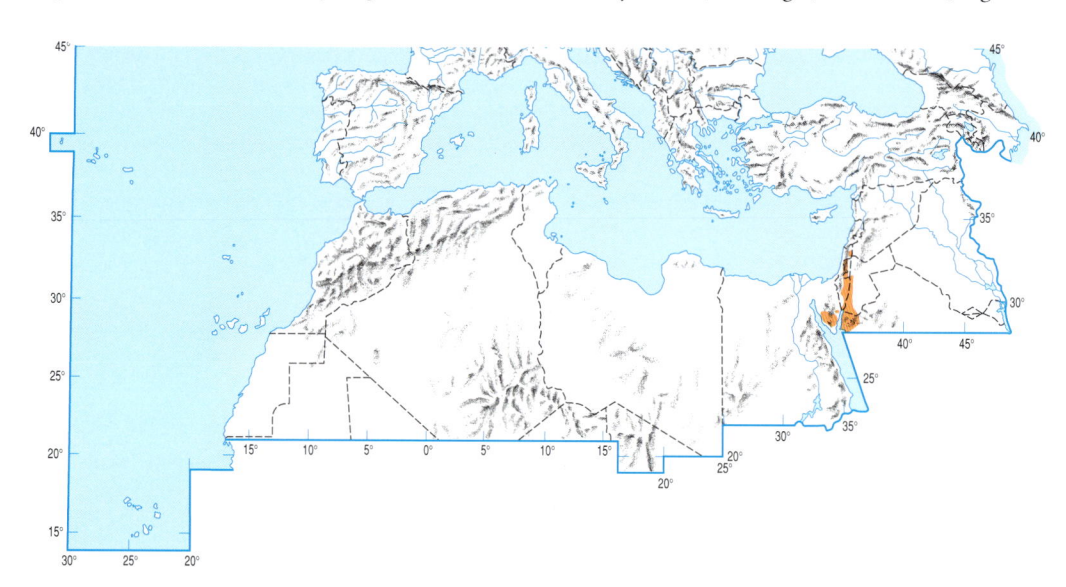

in Arad (Israel), has recently begun nesting on roofs and in shutter units of buildings, 6–21 m above ground. Nest: of twigs, tamarisk branches, rarely with green leaves, with lining of softer material including feathers, hair, and paper; nest size and shape adapted to fit cavity but usually a deep plate. EGGS. Sub-elliptical, smooth; turquoise-blue, sometimes mottled with sparse light brown markings or reddish speckles. Clutch: normally 3 but up to 5. INCUBATION. About 16 days. FLEDGING PERIOD. 28–31 days.

Wing-length: ♂ 149–158, ♀ 143–151 mm.
Weight: ♂♀ 98–140 g.

Daurian Starling *Sturnus sturninus*

PLATES: pages 1490, 1502

DU. Daurische Spreeuw   FR. Etourneau de Daourie   GE. Mongolenstar   IT. Storno di Dauria
RU. Малый скворец   SP. Estornino dáurico   SW. Amurstare

Field characters. 18 cm; wing-span 30–33 cm. About 15% smaller than Starling, with proportionately 10% shorter and much stouter bill, 20% shorter tail, and noticeably long toes, giving large-footed appearance. Small and short-billed starling, with remarkably variegated plumage: pale grey head and underbody contrast with dark back, tail, and wings; 2 whitish wing-bars.

To inexperienced observer, almost pied plumage might briefly suggest ♂ Snow Bunting, but once true character established, can only be confused with partially moulted immature Rose-coloured Starling (which lacks white wing-bars) or Grey-backed Starling (only one visible broad white band across wing-coverts; obvious white tips to outer tail-feathers; lacks dark back). Flight action varies: over short distance and when flycatching, relatively broad wings and changes in body angle produce loose, rocking progress reminiscent of roller, and steep ascents from ground and curving glides recall bee-eater; over long distance, action and progress much reminiscent of Starling.

Call on flushing a slow, soft, drawn-out 'chirrup', resembling similar call of Starling.

Habitat. Breeds in east Palearctic in upper middle latitudes in warm continental summer climate, avoiding mountainous regions and favouring river valleys with groves of trees scattered in fields and meadows, or villages and outskirts of lowland towns.

Distribution. Breeds in central and eastern Asia from Transbaykalia, Amurland, and Ussuriland south to northern China and Korea.

Accidental. Norway: juvenile shot Lillestrøm (Akershus) September 1985. Britain: record of adult ♂, Fair Isle (Shetland), May 1985 previously accepted as wild vagrant now considered to be of dubious origin, more likely relating to escape from captivity.

Movements. Migratory. Winters in large flocks in lowlands of peninsular Malaysia (south to Singapore) and Sumatra; apparently in smaller numbers in Tenasserim (southern Burma), Thailand (chiefly in south), Cochinchina and Tonkin (Vietnam), and northern Laos, with records also from southern China. In spring, reaches breeding grounds from late April, main arrival (fairly concentrated) not until mid- or late May.

Wing-length: ♂ 103–112, ♀ 101–110 mm.

Grey-backed Starling *Sturnus sinensis*

DU. Mandarijnspreeuw   FR. Etourneau mandarin   GE. Mandarinstar   IT. Storno della Cina
RU. Серый скворец   SP. Estornino chino   SW. Mandarinstare

A south-east Asian species, breeding in southern China and northern Indochina and extending south and south-west of breeding range in winter. 2 records at Lågskär (Finland): adult ♀, July–September 1975; May 1990; both may have been escapes.

Starling *Sturnus vulgaris*

Du. Spreeuw   Fr. Etourneau sansonnet   Ge. Star   It. Storno
Ru. Скворец   Sp. Estornino pinto   Sw. Stare

PLATES: pages 1493, 1502

Field characters. 21.5 cm; wing-length 37–42 cm. Head and body close in size to Blackbird but with rather long, pointed bill and sloping forehead (giving very slender appearance to front of head), distinctly triangular wings, and rather short, square tail; similar in size to Rose-coloured Starling and Spotless Starling. Medium-sized, rakish but full-bodied, bustling passerine, with cheeky mien but marked wariness; epitome of family. Adult looks blackish at any distance but actually intricately patterned and shot with iridescence, so quite beautiful at close range; white-spotted in winter. Juvenile dull brown at any distance, with pale throat. Bill strikingly pale yellow on breeding birds.

North of range of Spotless Starling, adult unmistakable at close range but paler juvenile may invite confusion with similarly-aged Rose-coloured Starling (see that species). At long distance and in poor light, confusable above all with waxwings *Bombycilla*, but also with small dark thrushes (especially Redwing), small *Calidris* waders, Little Auk, and even hirundines. Where range overlaps that of Spotless Starling, separation subject to considerable pitfalls which have attracted no real study (see that species). Flight swift and usually notably direct, bird even taking beeline from nest to food source; action combines rapid beats of triangular wings with occasional glides and momentary closed-wing attitudes producing level shooting progress; over short distance and in tight spaces, wheels and sweeps, but still retains shooting element, giving impression of almost too much speed at times and leading to apparently rushed tumbling landing (particularly of flock). When hawking for flying insects (particularly in autumn), flight slower and more graceful, recalling Swallow. Flight silhouette distinctive, with rather geometric outline in pointed bill and head, triangular wings and short, sharp-cornered tail, all combining into uncanny recall of V-winged aircraft. Gait most commonly a quick, confident walk. Gregarious and quarrelsome.

Habitat. In west Palearctic, breeds from upper to lower middle latitudes marginally to Arctic fringe and thence through boreal and temperate to steppe and Mediterranean climatic zones within July isotherms *c.* 10–30°C. Mainly in lowlands and uplands but in Alps breeds regularly up to 800 m and locally even to 1500 m; in Caucasus, up to 1850 m. During rather brief breeding season must concentrate where suitable holes are available, either naturally in hollow trees, rock or clay crevices, or previously excavated burrows or holes, or artificially in apertures in buildings or other structures. Readiness to fly frequently over considerable distances permits breeding in open forests or near woodland margins, along rocky coastlines and in not too dense human settlements, birds foraging on neighbouring grasslands, field crops, floodlands, vacant sites, and even airfields, refuse tips, and sewage disposal areas. After young fledge, parties move quickly to open country, including grazed hill pastures, upper parts of salt-marshes, heaths, rocky shorelines, and seasonally to orchards and thickets bearing soft fruits or berries. Avoids, however, lower dense cover such as bushes, undergrowth, or tall herbage, but will roost in reedbeds; on ground prefers sparse or low vegetation offering ease of movements and access to food organisms beneath soil surface. On buildings, prefers ledges or holes at no great height, except for roosting or where unusually attractive shelter is offered on towers or tall structures. Markedly ready to take advantage of human settlements, artefacts, and managed land, from which it has profited much.

Distribution. Long-term spread in north of range from 19th century to 1950s or 1960s, and more recent southward spread in France and Italy, reaching northern Spain. ICELAND. Long known as winter visitor; breeding first proved in 1935. BRITAIN, IRELAND. Marked range contraction in early 19th century followed by long-term expansion from *c.* 1830, continuing until northern and eastern areas colonized in 1950s or 1960s (*zetlandicus* survived from earlier in Shetland and Outer Hebrides). Recent decline highland areas of Scotland. FRANCE. Slow spread south (where formerly absent as breeder) since 1930s, almost all of southern France having been colonized in *c.* 50 years. Corsica: several pairs reported to have bred 1932, and breeding may still sometimes occur, but confirmation needed. SWEDEN. Expansion into interior of northern Sweden began *c.* 1850 (previously bred only south, centre, and along Gulf of Bothnia), completed *c.* 1950. FINLAND. Range expansion up to end of 1960s. ESTONIA. Regular wintering (mainly coastal west and north-west) since 1950s. AUSTRIA. Expanding along Alpine river valleys since 1960. SPAIN. Regular breeder only from 1960. Spread, from north-east along Bay of Biscay coast, slowed after contact (and apparently competition) with Spotless Starling. BALEARIC ISLANDS. Irregular breeder Mallorca, now apparently established in north-east where several pairs bred 1993–4. ITALY. Recent spread in centre and south, also up to maximum 2000 m in Alps. Sicily: small breeding population found 1979 probably present since at least 1974. MALTA. First breeding record (released birds Comino) 1993–4. YUGOSLAVIA: CROATIA. Slight spread. RUSSIA. Kola peninsula: vagrant early 20th century. Now breeds to *c.* 68°N. TURKEY. Scarce, local or absent parts of western Anatolia, Black Sea coastlands, Taurus, south coast and Hatay. SYRIA. Sporadic or local breeder. IRAQ. Rare breeder in north. First proved breeding 1988. AZORES. Common, especially in coastal areas, on all islands. CANARY ISLANDS. Recent colonist, breeding Gran Canaria and Tenerife.

Accidental. Spitsbergen, Bear Island, Jan Mayen, Mauritania, Madeira, Cape Verde Islands.

Beyond west Palearctic, extends from Urals east to Amga river and Baykal region and south through Altai to eastern

Starling *Sturnus vulgaris*. *S. v. vulgaris*: **1** ad ♂ summer, **2** ad ♀ summer, **3** ad ♂ winter, **4** ad ♀ winter, **5** 1st summer, **6** 1st ad winter, **7** juv. *S. v. zetlandicus*: **8** juv. *S. v. tauricus*: **9** ad winter. *S. v. purpurascens*: **10** ad winter.

Tien Shan; also from Iran to Pakistan. Introduced and now widespread in North America; also southern Africa, Australia, and New Zealand.

Population. Increased with range expansion, but more recent decline affecting Britain, Low Countries, Germany, Fenno-Scandia, Baltic States (stable after decrease in Lithuania), Poland, Czech Republic, and parts of Russia. Marked decrease also in numbers of wintering birds in Britain, France, Spain, and Israel. Cause of decline not clear, but probably includes persecution in winter and pesticides. ICELAND. 1500–2500 pairs in late 1980s; increasing slowly. FAEROES. 30 000–80 000 pairs. BRITAIN, IRELAND. Over 1.1 million territories in Britain, 360 000 in Ireland, 1988–91. Increase from *c.* 1830 after drastic decline early in 19th century. Marked decline in Britain 1960s–80s. FRANCE. 4.5–9.4 million pairs in 1988. Increase 1965–75, stable in 1980s, apparently declining since. BELGIUM. Increase to 430 000 pairs 1973–7, decrease 1983–93. LUXEMBOURG. 30 000–40 000 pairs. NETHERLANDS. Over 500 000 pairs 1979–85. Declining slowly, at least locally. GERMANY. 4 million pairs in mid-1980s. In east, where recent decrease, 1.5 million ± 600 000 pairs in early 1980s. DENMARK. 220 000–2.3 million pairs 1987–8. NORWAY. 200 000–500 000 pairs 1970–90. Slight decrease, mainly in north. SWEDEN. 750 000–1.5 million pairs; marked decrease. FINLAND. 50 000–80 000 pairs in late 1980s. Dramatic decrease (after increase up to 1960s) in 1970s–80s, still continuing. ESTONIA. 20 000–50 000 pairs in 1991. Slight decline in 1960s, crashed 1977–81. LATVIA. 50 000–250 000 pairs in 1980s. LITHUANIA. Common. POLAND. 3.5–5 million pairs. Marked increase from 2nd half of 19th century, probable decrease in recent decades. CZECH REPUBLIC. 800 000–1.6 million pairs 1985–9. SLOVAKIA. 400 000–800 000 pairs 1973–94; stable. HUNGARY. 300 000–500 000 pairs 1979–93; stable. AUSTRIA. 250 000–300 000 pairs in 1992; increase. SWITZERLAND. 150 000–220 000 pairs 1985–93; stable. SPAIN. 220 000–1 million pairs; marked increase. ITALY. 1–3 million pairs 1983–95; stable after increase. GREECE. 10 000–20 000 pairs. ALBANIA. 1000–3000 pairs in 1981. YUGOSLAVIA: CROATIA. 800 000–1.2 million pairs. SLOVENIA. 80 000–100 000 pairs; stable. BULGARIA. 1–10 million pairs; stable. RUMANIA. 1.5–2 million pairs 1986–92. RUSSIA. 5–50 million pairs. Apparently stable, though declined in Yaroslavl' region 1960–85, and decreased dramatically in Kareliya in 1970s–80s. BELARUS'. 1.5–1.7 million pairs in 1990; stable. UKRAINE. 1.2–1.5 million pairs in 1986; slight increase. MOLDOVA. 200 000–300 000 pairs in 1988; slight increase. AZERBAIJAN. Probably at least tens of thousands of pairs. TURKEY. 500 000–5 million pairs. AZORES. Common. CANARY ISLANDS. Apparently stable.

Movements. Generally migratory in north and east of breeding range, although increasing tendency to remain resident in urban areas; partial migrant or resident in south and west. Young disperse or, in some populations, undertake more extensive directional movements. Direction of autumn migration of adults predominantly to south-west, but more southerly in east of range and more westerly in west. Migrant populations winter western and southern Europe, Africa north of Sahara, Egypt, northern Arabia, northern Iran, and plains of northern India. Scottish and Faeroe island populations resident.

Northern and central Europe largely vacated in winter. While some birds cross North Sea to reach Britain, there is a major concentrated migration route along southern North Sea coast, through Low Countries. Passage September–November. Early migrants are predominantly Dutch breeders with destinations in Britain; German and Scandinavian birds pass mid-October, while November migrants originate mainly in Poland, Finland, and Russia, with destinations in Belgium and France.

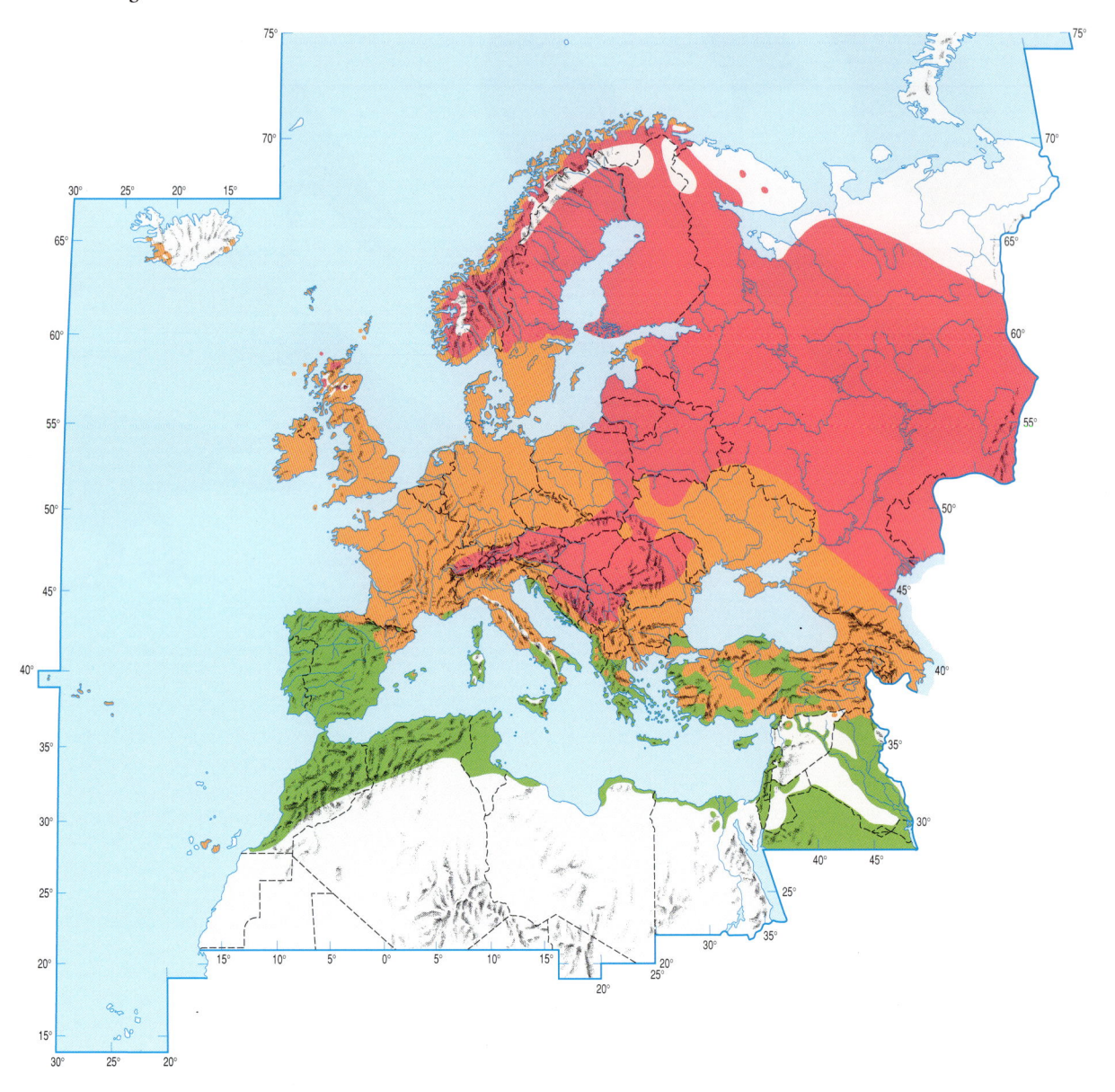

Where resident, as in Britain, juveniles disperse after attaining independence from parents. Juveniles of migratory populations undertake longer and more directional movements. Where a summer migrant, among earliest bird species to return to breeding area, e.g. from February in Belgium, March in much of northern Europe.

Food. Animal and plant material taken at all times of year but animal food predominates in spring and is fed almost exclusively to nestlings. Plant material forms high proportion of diet in autumn and winter. Seasonal changes in relative proportions of plant and animal foods paralleled by changes in intestine length: longer when eating plant than when taking animal food. Animal food mainly insects and their larvae. Soft fruits taken in summer and autumn, and seeds, including cereals, in autumn and winter. Forages mainly on ground in open areas of short grass (often in association with grazing ungulates) or other short or sparse vegetation, e.g. cereal stubbles, and sometimes follows plough; also feeds in intertidal zone, on sewage treatment beds, refuse tips, farmyards, feeding areas for domestic stock, including troughs and open faces of cereal silage clamps. Sometimes forages in taller vegetation but this common only in summer flocks of juveniles in coastal saltmarsh, upland rough pasture, or moorland. Sometimes arboreal, especially post-breeding flocks of adults and young where caterpillars abundant, and also when eating fruit. Sometimes forages among fur of ungulates, hawks insects, and drinks nectar.

Much food taken from surface or just below surface of soil or among grass roots. Walks or runs over ground and chases or pecks at items on surface, but specially adapted for probing into substrate (prying or open-bill probing): pushes closed bill into soil, litter, or grass roots, and opens bill to create hole; during bill-opening, eyes rotate forward to give binocular vision. These actions permitted by modifications of skull and musculature.

Social pattern and behaviour. Gregarious throughout year. After breeding, ranges of adults and young may separate (see Movements), when adults remain in small flocks but young can form flocks of hundreds, sometimes thousands, in suitable habitats. Migrates in flocks. Especially in winter, forms flocks with other species, often Lapwing, Rook, Jackdaw, gulls, and thrushes. Where resident, ♂ may defend nest-site for most or some of winter, taking in nest-material and roosting there, sometimes with ♀. Mating system both monogamous and polygynous; polygyny usually successive, occasionally simultaneous. Also regularly practises intra-brood parasitism; when laying parasitic eggs, ♀ sometimes removes one of host eggs. ♂ defends nest-site and builds most of nest but ♀ contributes to lining; ♀ incubates but ♂ sits on eggs, reducing heat loss, in absence of ♀; ♀ broods for first few days after hatching but both sexes feed young. Breeding usually colonial but sometimes solitary. Distance between nests usually determined by availability of nest-holes, and colonies appear loose, but high degree of synchrony of 1st clutches suggests social interactions important in colony maintenance. Territory restricted to small area around nest-site.

In breeding season, ♀ roosts in nest-site. Resident birds in winter may also roost in nest-site; migrants also roost in nests after arrival in breeding area. During breeding season small (hundreds) communal roosts comprise breeding ♂♂ and non-breeders; in summer, roosts larger with substantial segregation of adults and juveniles; size of roosts increases to maximum in winter, when numbers commonly exceed 100 000 and can reach 1 million. Roost sites include reedbeds, cane fields, plantations (especially young conifers), thickets (especially thorn), bridges, piers, and buildings in town centres.

In breeding season, only ♂ sings—from perch which may be branch, fence, television antenna, gutter, etc., usually close to nest-site. Adopts characteristic upright posture, with bill pointed slightly up, tail pointed down with rump feathers ruffled giving hunch-backed appearance, feathers of lower belly ruffled and long throat feathers puffed out. In more intense song, bill often pointed higher, exaggerating puffed-out throat feathers, and may flick half-open wings away from body (Fig. 4, left), or slowly wave wings (Fig. 4, right), the latter with main function of mate-attraction. ♂♂ sing in most months, except for a few weeks during post-breeding moult. Song most intense in spring, with smaller peak of singing in autumn after moult when nest building also occurs. In autumn and winter, when often in large flocks, birds periodically sing in day-time roosts. They also sing in night roost after evening arrival and before morning departure; some can be heard singing all night. Function of roost singing not known, though volume often striking.

Voice. Highly vocal all year (especially ♂) except during moult when virtually silent. Repertoire rich and varied. Song of male a lively rambling medley of throaty warbling, chirruping, clicking, and gurgling notes, often incorporating mimicry of other species, interspersed with musical whistles and pervaded by peculiar creaky quality. Each ♂ has unique repertoire of *c.* 20–70 song segments, sung in predictable sequence. An outstanding mimic, producing accurate copies of calls or songs of other bird species, frogs, and mammals (e.g. goat, cat, man) and of mechanical sounds. Other calls include a 'querrr', typically given in flight but also when perched; a short metallic 'chip', given in response to (usually flying) predator or indicating anxiety; and a snarling call given when mobbing predator at close quarters.

Breeding. SEASON. Breeding often in 3 phases: 1st clutches initiated synchronously throughout colony over period usually of 4–10 days; 2nd clutches less synchronous and started 40–50 days after 1st clutches; between 1st and 2nd clutches, 'intermediate' clutches include replacements of lost 1st clutches, clutches of ♀♀ of polygamous ♂♂, and clutches of late-arriving birds. 1st clutches initiated in April in most areas; late March to early April in south, late April to early May in north. In north and east of range, usually one brood. SITE. Hole in tree, cliff, building, pylon, etc.; also in nest-boxes, occasionally in holes in ground where alternatives scarce; uses old holes and will take over current ones of woodpeckers and Sand Martin. Nest: bulky base of dry grasses and fine twigs, sometimes pine needles, size of this base dependent largely on size of cavity, and in roof spaces can be *c.* 1 m across, 25 cm deep; cup constructed in part of nest cavity remote from entrance, with variable lining of finer materials, including grasses, rootlets, moss, feathers, wool, paper, etc.; lining may be thicker in cooler climates; in later stages of building, fresh leaves or flowers may be incorporated into lining; these perhaps contribute to mate attraction or to defence against nest parasites. EGGS. Sub-elliptical, sometimes ovoid, smooth with some gloss, various shades of pale blue and occasionally white; some eggs, especially in 2nd clutches, with reddish or blackish spots of dried blood from parasite bites on incubating ♀. Clutch: 3–8, generally averaging 4.5–5.4; later clutches somewhat smaller. INCUBATION. Average 12·2 days (11–15). FLEDGING PERIOD. *c.* 21 days.

Wing-length: *S. v. vulgaris*: ♂ 127–141, ♀ 123–137 mm.
Weight: *S. v. vulgaris*: ♂ mostly 70–90, ♀ mostly 60–90 g.

Fig. 4

Geographical variation. Marked variation in colour, mainly of gloss on head, body, upper wing-coverts, and streak on outer web of tertials and secondaries, and in general colour of juvenile. Less marked variation in size. Gloss of cap, chin, and throat of birds towards eastern limit of west Palearctic (*S. v. tauricus, purpurascens, caucasicus*) purple rather than green, but much variation. Populations of Atlantic islands largely similar to *S. v. vulgaris* in adult plumage, differing mainly in juvenile plumage and in structure. *S. v. faroensis* (Faeroes) large, leg and foot strong, bill longer, wider at base; juvenile sooty-black with restricted white on chin and belly, white with bold black spots on throat. *S. v. zetlandicus* from Shetland (except Fair Isle) intermediate between *faroensis* and *S. v. vulgaris*. *S. v. granti* from Azores similar to *S. v. vulgaris*, but slightly smaller, especially leg, and upperparts more often completely purple.

Spotless Starling *Sturnus unicolor*

PLATES: pages 1497, 1502

Du. Zwarte Spreeuw   Fr. Etourneau unicolore   Ge. Einfarbstar   It. Storno nero
Ru. Чёрный скворец   Sp. Estornino negro   Sw. Svartstare

Field characters. 21–23 cm; wing-span 38–42 cm. Slightly larger than Starling (particularly ♂), though with similar structure; adult has distinctly more elongated feathers on head and throat. Medium-sized starling of similar form and behaviour to Starling but adult differs noticeably in black and evenly glossed, virtually unspotted plumage (except in 1st-year ♀ and in fresh winter plumage). Juvenile differs from continental races of Starling in darker, browner plumage.

Restricted to Iberia (also marginally France, see Distribution), west Mediterranean islands, and north-west Africa. Breeding thus overlaps with Starling only in north of range. Separation in winter from Starling little studied, and clearly high risk of overlapping appearance between worn, relatively spotless adult Starling and fresh, relatively spotted immature Spotless Starling, as both species may moult early or late and show erratic wear. Winter flocks of Starling known to be joined by Spotless Starling in Iberia and north-west Africa. Best distinctions of full adult Spotless Starling appear to be (1) colour of gloss (essentially evenly purple in Spotless Starling, always partly brilliant green in Starling) and (2) lack of many pale feather-margins on wings and tail (particularly in ♂). Separation of immature depends on close observation of (1) size and shape of spots (small and arrow-shaped in Spotless Starling, small to quite large, and round in Starling) and again (2) wing markings (relatively faint in Spotless Starling, strong in complex linear pattern in Starling). Differences in leg colour less trustworthy, with much overlap, though extremes are helpful: pale flesh in Spotless Starling, bright but fully reddish-brown in Starling. Flight, gait, and behaviour much as Starling but said to fly even faster.

Common Myna *Acridotheres tristis* (p. 1500): **1** ad, **2** juv. Spotless Starling *Sturnus unicolor*: **3** ad ♂ winter, **4** ad ♂ summer, **5** juv.

Habitat. Within restricted warm west Mediterranean range, habitat corresponds closely to that of Starling elsewhere, nesting in buildings and in tree-holes. In Spain, prefers open woodland with access to short grass and herbage, and frequently found in association with cattle; in winter, prefers more open places such as irrigated and cereal fields. In Corsica, mainly on littoral plains of west coast, frequenting cultivation but also degraded maquis and outskirts of villages. In north-west Africa, much attracted to human habitations, but also nests colonially in holes in large cedars in Moyen Atlas, making long flights from forest to open plateaux to collect food, at altitudes around 1700–1800 m.

Distribution. Breeds only in west Palearctic. FRANCE. Has bred near Opoul (Pyrénées Orientales) since at least 1983, and in several villages near Leucate and Sigean (Aude). Still expanding. Corsica: apparently now more widespread. SPAIN. Began to spread east (from central and south-western parts) probably in 1950s; expansion probably due to changing agricultural practice, especially irrigation, and reached present limits in 1980s; now sympatric with Starling at or near north-eastern limit of range, competition with it probably restricting further range expansion. BALEARIC ISLANDS. Accidental visitor, but in 1994, bird paired with Starling raised at least 2 young. ITALY. Sicily: sparsely distributed. TUNISIA. Range expansion in north and west. MOROCCO. Locally common to 31°40′S.
Accidental. Malta, Greece, Libya, Madeira, Canary Islands.

Population. FRANCE. Under 300 birds on mainland; increase. SPAIN. 2–2.5 million pairs; slight increase. PORTUGAL. 100 000–1 million pairs 1978–84; stable. ITALY. 50 000–100 000 pairs 1983–95; stable. TUNISIA. Fairly common; increase. ALGERIA. Rare. MOROCCO. Uncommon to locally common; apparently stable.

Movements. Resident, or partial short-distance migrant, subject to nomadic dispersal.

Food. Principally invertebrates from early spring to summer, seeds and fruits during rest of year. Opportunistic feeder, taking food basically according to abundance; favourite foraging places are improved grassland and pasture, but also feeds in vineyards, olive groves, arable fields (especially stubble), and rubbish tips. Very strong association with cattle all year round, taking plant remains in dung, flushed insects, and parasites on body and head of animals.

Social pattern and behaviour. Gregarious all year. Feeding flocks average 90–110 birds in summer and winter, but only 8–12 in April–May. In immediate post-breeding period, juveniles form into large flocks, subsequently all ages together. Mating system both monogamous and (probably minority of ♂♂) polygamous, either serially or simultaneously. ♂ sings mostly from favoured song-post close to nest such as tree-branch, roof, overhead wire, television aerial, but also on feeding grounds when mate-guarding. Full song heard late February–June, but also on warm days in autumn (after moult) and winter. Displays generally similar to those of Starling.

Voice. Song resembles Starling in overall structure, duration, and variability, but noisier, and introductory whistles much louder. Whistles also given independently as calls, and are among the commonest sounds heard outside breeding season.

An accomplished mimic, incorporating imitations of many other bird species in song. Contact-call a sound roughly rendered as 'gaa-haa', somewhat like cackle of chicken. Threat- and alarm-calls include a chattering 'tretet', and sharp 'fiit...fiit', similar to 'chip' call of Starling.

Breeding. SEASON. Egg-laying begins March in North Africa, April in Spain and Sicily. 2 broods. SITE. Hole, usually in man-made situation such as under roof-tiles, in wall (often in large towns), agricultural structure, nest-box, etc.; also in tree or rockface. Nest: foundation of twigs, dry grass, and herb and cereal stalks lined thickly with rootlets, grass, leaves, flowers, and feathers. EGGS. Sub-elliptical, smooth, and slightly glossy; pale blue, c. 4% in 1st clutches with very small reddish spots, 37% in 2nd clutches. Clutch: 4–5 (2–9). Significant relationship between clutch size and size of nest-box; 5 commoner in larger boxes, 4 in smaller ones. INCUBATION. 10.5–11.6 days (10–15). FLEDGING PERIOD. 21–22 days (18–25).

Wing-length: ♂ 119–143, ♀ 117–138 mm.
Weight: ♂ average c. 91–96, ♀ average c. 86–90 g.

Rose-coloured Starling *Sturnus roseus*

PLATES: pages 1499, 1501

DU. Roze Spreeuw   FR. Etourneau roselin   GE. Rosenstar   IT. Storno roseo
RU. Розовый скворец   SP. Estornino rosado   SW. Rosenstare

Field characters. 21·5 cm; wing-span 37–40 cm. Size as Starling but with shorter bill and more domed crown softening outline of head. Medium-sized, rather short-billed, (in adult) crested starling with pale orange-pink legs. Adult in worn plumage basically blue- to purple-black, with striking pink jacket on back, flanks, and belly; spotted and streaked in fresh plumage. Juvenile pale dun-brown with whitish throat, eye-ring, and belly, prominent brownish-white margins to larger, distinctly darker wing-feathers, paler rump, and striking yellowish base to bill.

Adult unmistakable. Given reasonable view, so is juvenile, since plumage distinctions from juvenile Starling are further enhanced by dark under wing-coverts and rather less frenetic, more placid behaviour. Flight closely resembles Starling but wings look more leaf-like, less triangular in outline and appear to beat more loosely; certainly flocks sweep, circle, and land more slowly. Gait variably described, said to be more sprightly than Starling but that of vagrant juveniles often slow and methodical. Gregarious, with social behaviour like Starling; not adapted to urban roosts except where large trees available.

Habitat. In west Palearctic, ranges over lower middle latitudes, mainly in steppe, semi-desert, and Mediterranean lowland zones. Movements often governed by ephemeral localized abundance of gregarious invertebrate prey, concentrated in dry, open, often arid spaces, as well as grasslands and stony or rocky terrain. Requires ready access to water, but not dependent on wetlands or sea coasts. Resorts to trees and bushes, usually only in smallish groupings. Availability of natural or artificial rock piles, or cliff faces with plenty of holes or clefts, is normal requirement for nesting; generally below 400 m, although recorded much higher. Preference is for foothills and undulating terrain where suitable nest-sites available, which may include artefacts such as fortress walls, railway embankments, salt mines, and spaces between tombstones. Generally indifferent to human presence, often occurring in settled areas. Social roosting often occurs in groves, avenues, orchards, or tree plantations, or in thorn scrub or reedbeds, often at some distance from areas serving its normal ground-loving habits.

Distribution. Erratic, irruptive visitor to central and western Europe, with occasional breeding to west of usual limits. SLOVAKIA. Bred in 1918. HUNGARY. Small influx 1989; bred 1994–5. ITALY. Bred 1739, 1875. GREECE. Very unstable. Recorded breeding for first time in northern Makedhonia in 1985, at 2 sites in 1987, and 1 in eastern Makedhonia in 1988; sites not occupied in following years. YUGOSLAVIA. Breeds almost every year in Makedonija and southern Montenegro, but presumed extinct (recently only 1 pair) in Slovenia. BULGARIA. Many old breeding records. Now occurs almost annually on coast of Dobrogea, also some years further south. Bred near Balchik in 1971, at least 2 other breeding records in Dobrogea since; perhaps breeds annually. RUMANIA. Sporadic and irregular breeder, especially in south-east. RUSSIA. Range fluctuating. UKRAINE. Marked decrease. MOLDOVA. Apparently no confirmed breeding records. AZERBAIJAN. Common, but sporadic and irregular. TURKEY. Highly erratic nester; even huge colonies (several thousand birds) usually abandoned following year. Most sites Van Gölü area. SYRIA. Occasional breeder. ISRAEL. Known to have bred in 19th century, and probably did so after influx in 1945.

Accidental. Iceland, Faeroes, Britain, Channel Islands, Ireland, France, Belgium, Netherlands, Germany, Denmark, Norway, Sweden, Finland, Estonia, Latvia, Lithuania, Poland, Czech Republic, Slovakia, Austria, Switzerland, Spain, Balearic Islands, Portugal, Italy, Malta, Albania, Belarus', Libya, Tunisia, Algeria, Madeira, Canary Islands.

Beyond west Palearctic breeds Iran and extends east from Caspian to western and southern Altai and eastern Tien Shan.

Population. Wide fluctuations typical. HUNGARY. Hortobágy: 1600–1700 pairs (plus 1200 non-breeders in area) in 6 colonies in 1995. GREECE. Up to 1000 pairs. ALBANIA. Up to 500 pairs

Rose-coloured Starling *Sturnus roseus*: **1** ad ♂ summer, **2** ad ♂ winter, **3** ad ♀ summer, **4** 1st winter ♂, **5** 1st winter ♀, **6** juv.

in 1981. YUGOSLAVIA: CROATIA. 3–6 pairs. BULGARIA. Up to 1000 pairs. RUMANIA. Up to 500 pairs (or more) 1986–92. RUSSIA. 10 000–100 000 pairs. UKRAINE. 200–700 pairs in 1974; marked decrease. TURKEY. 5000–100 000 pairs.

Movements. Migratory, wintering south-east of breeding range in peninsular India and Sri Lanka; migrates in flocks (sometimes huge) by day. Populations from west of range migrate almost directly east before heading south-east into India.

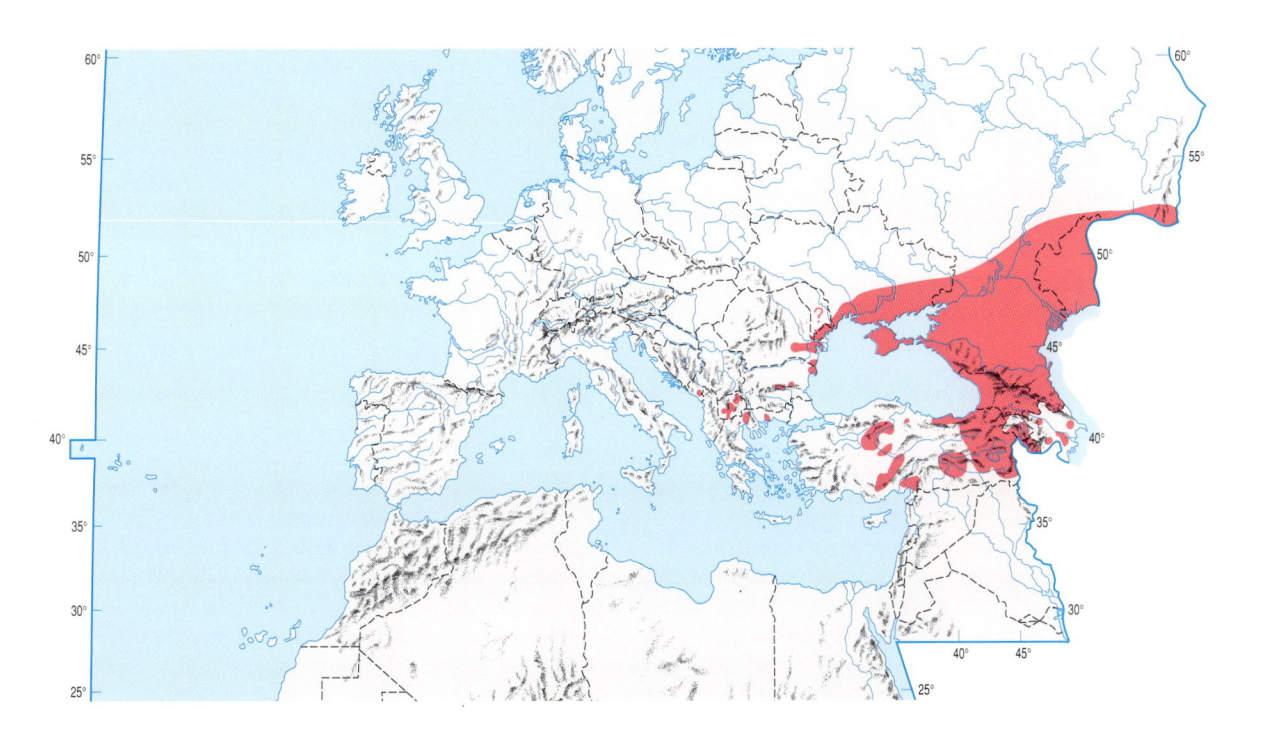

In line with irregular colonization of breeding areas, spring invasions occur west of normal range in south-east Europe (notably to Yugoslavia and Greece), apparently less frequent and smaller than formerly; similarly, at edge of breeding range, numbers may vary from thousands to few (e.g. in Hungary). At Villafranca in north-east Italy, 6000–7000 pairs bred in exceptional influx in 1875; but in 20th century only vagrant to Italy. Further north in Europe, vagrant both seasons. In Britain and Ireland, recorded chiefly mid-May to beginning of November; widely distributed, with many in Shetland (Scotland), chiefly May–July, and in Isles of Scilly, chiefly October; 4 records of successful overwintering. In Sweden, recorded mostly May–June. In France, 1900–89, 28 records involving 62 individuals, mainly in west and south; one overwintering record.

Food. In breeding season mainly insects, especially locusts and grasshoppers, and other swarming Orthoptera; after young fledge, major items are grapes and mulberries; takes fruit, nectar, and seeds autumn and winter. Acrididae remain flightless ('hoppers') for *c.* 40–50 days over summer, presenting ideal food resource. During breeding season, most food taken from surface of ground but some Orthoptera caught in the air. When taking Orthoptera from ground, large flocks move in one direction, with birds in front moving (usually running) faster than those behind, birds from rear flying to front in 'roller-feeding' fashion. Breeding birds fly 10 km or more from colony to feeding areas.

Social pattern and behaviour. Gregarious all year. Little information on size of feeding flocks but tens to hundreds reported in breeding season; in winter, forms small parties or large flocks, sometimes of 'swarm' proportions. Probably normally monogamous, pair-formation taking place when breeding colony is established, usually very rapidly. ♂♂ sing (Fig. 5, upper) usually near nest but sometimes sitting in tree; often chorus of (e.g.) 100 birds singing together in colony. In courtship, ♂ moves singing round ♀ in crouched posture but with head up, wings and tail continuously vibrated, crest raised, and throat ruffled (Fig. 5, lower).

Voice. Song of ♂ has same character as Starling but more harsh and unmusical, a jumble of discordant grating noises mixed with some melodious warbling notes; no mimicry of other bird species reported. Most frequently repeated sequences of song are short series of rather dry low-pitched squeaks 'kitch kitch kitch kitch'. Flocks at winter roosts keep up continuous excited chattering and bubbling, very like Starling, before settling down. Short flight-calls, similar to those of Starlings,

Fig. 5

but more yelp than scream in alarm, produced when taking off. Various harsh, rasping, or sharp repeated calls given in other contexts.

Breeding. SEASON. Throughout breeding range a late spring migrant. In invasions of Hungary and Greece, nest-building and egg-laying started late May to early June; colonies deserted mid-July to early August. Normally 1 brood but 2nd broods may occur in years of exceptional abundance of locusts. SITE. Hole, most commonly among stones; also in crack in rocks or cliff, under railway sleeper, in wall or bridge, under roof, in thatch, in nest-hole of Sand Martin, and sometimes in hole in tree. Nest: roughly made of thin twigs and grasses, lined with finer grass, often with fresh wormwood (*Artemisia*) and feathers, or of dry stems and leaves of annual herbs, especially giant fennel (*Ferula*) or grass. EGGS. Pale blue or pale azure without marks and with slight gloss. Clutch: 3–6 (2–10); 8 and above probably laid by 2 ♀♀. INCUBATION. *c.* 15 days. FLEDGING PERIOD. *c.* 24 days.

Wing-length: ♂ 127–139, ♀ 125–135 mm.
Weight: ♂♀ mostly 67–88 g.

Common Myna *Acridotheres tristis*

PLATES: pages 1497, 1501

Du. Treurmaina   Fr. Martin triste   Ge. Hirtenmaina   It. Maina comune
Ru. Обыкновенная майна   Sp. Miná común   Sw. Brun myna

Field characters. 23 cm; wing-span 33–36·5 cm. Only 10% longer but noticeably bulkier than Starling, with deeper bill, seemingly deeper chest and strong, lanky legs. Yellow bill and facial wattle obvious on black head, warm brown-grey body, brown-black wings with prominent white central panel and white under wing-coverts, white vent, and yellow legs.

Unmistakable in west Palearctic but beware always chance of captive origin. Flight has noticeably flapping action, lacking darting progress of *Sturnus* and reminiscent of Jay. Gait mainly a jaunty walk. Extremely self-possessed, being tame and bold around man and exploiting every possible food source.

Tristram's Grackle *Onychognathus tristramii* (p. 1489): **1–2** ad ♂ summer. Rose-coloured Starling *Sturnus roseus* (p. 1498): **3–4** ad ♂ summer, **5–6** juv. Common Myna *Acridotheres tristis tristis*: **7–8** ad.

Habitat. Natural breeding range lies within subtropical and tropical lower latitudes of Asia, mainly in lowlands. Breeds locally in high areas, and in central Asia irregularly into more northerly climates, much influenced by opportunities arising from human modification of habitats. Forages much in the open, from semi-desert to cultivation. Roosts in large trees, coconut groves, reedbeds, sugar cane plantations, railway stations, warehouses, and other safe sheltered situations.

Distribution and population. FRANCE. Several (undoubted escapes) in Dunkerque harbour since at least 1986. Up to 5 pairs breeding. RUSSIA, GEORGIA. Resident and breeding in small numbers on Black Sea coast in Sochi and Gagra areas (north-west Caucasus) at least from 1978; population probably originates from escaped cagebirds, though may be part of natural range expansion (striking feature in Kazakhstan and central Asia). TURKEY. Spread west along Black Sea coast to Trabzon (39°43′E). KUWAIT. Range increasing: breeding at Doha and Jahra Town (both west of Kuwait City). Fairly common; slight increase. CANARY ISLANDS. Feral population on Tenerife since 1993.

Movements. Mainly sedentary in native range in western Asia, and where introduced in west Palearctic.

Food. Omnivore and scavenger. Feeds predominantly on ground, sometimes in trees. Feeding sites include grassy areas, scrub, roads, pavements, and foliage of trees and shrubs; will forage around buildings and livestock or poultry farms; scavenges among rubbish and waste food and on foreshore at low tide; occasionally hawks for insects from perch.

Social pattern and behaviour. Essentially solitary or in pairs all year, occasionally forming small flocks and feeding aggregations. Monogamous pair-bond, life-long after a pair has bred successfully, and persisting all year. Pairs for first time within flocks during winter and spring. Territorial when breeding, but pairs may nest in close proximity. Both sexes sing throughout year, most often during breeding season and near nest. Both sexes participate in all phases of breeding, from nest-building to brooding and feeding of young.

Voice. Highly vocal during breeding season, and to lesser extent during rest of year. Song disjointed, rather noisy, and tuneless, comprising variety of different phrases, each usually repeated quite rapidly several times; may include mimicry of other species. Some units are gurgling, others disyllabic and whistling, others rather softer with almost strangled timbre. Flight-call a rather weak and querulous 'kwerrh'. Alarm-call a grating 'traaahh' somewhat like Nutcracker.

Breeding. SEASON. Very variable, according to geographical area; no information for small west Palearctic populations. SITE. Wide variety of holes and crevices. Natural sites include holes in trunk of trees, gaps in dense vegetation or clusters of fruit, and holes and crevices in earth banks and cliffs. Other types of hole include those made by other species of birds or

Daurian Starling *Sturnus sturninus* (p. 1491): **1–2** ad ♂. Starling *Sturnus vulgaris vulgaris* (p. 1492): **3** ad ♂ summer, **4–5** ad ♂ winter, **6** juv, **7** flock. Spotless Starling *Sturnus unicolor* (p. 1496): **8–9** ad ♂ summer.

animals and wide variety of cavities in man-made structures, even in machinery in use. Nest: untidy mass of natural and man-made material such as twigs, leaves, roots, straw, feathers, fur, paper, cloth, string, cigarette ends, etc., with or without shallow cup. Eggs. Elongated oval or often pear-shaped, with hard glossy texture; pale blue to sky-blue or greenish-blue, unmarked. Clutch: 4–5 (2–6). Incubation. 13–18 days. Fledging Period. 22–35 days; young may leave nest before able to fly.

Wing-length: ♂ 138–152, ♀ 134–147 mm.
Weight: ♂♀ mostly 110–130 g.

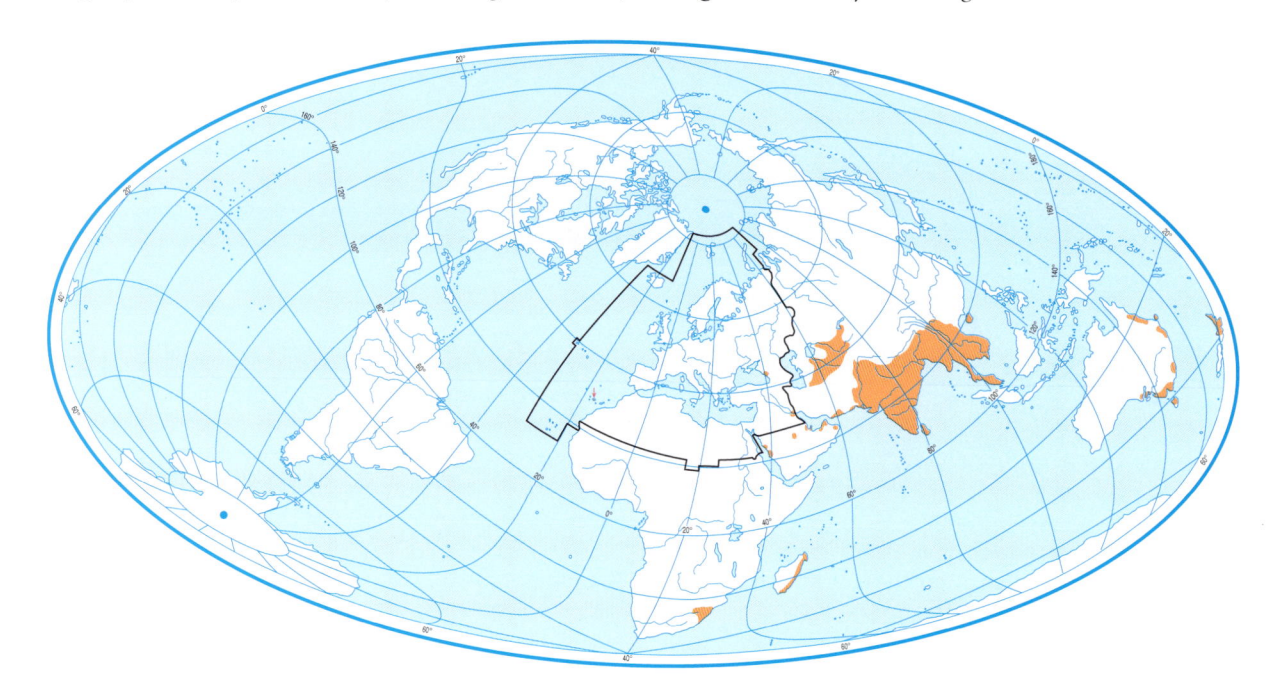

Sparrows, Rock Sparrows, Snow Finches Family Passeridae

Predominantly small, thick-billed oscine passerines. Most found in open, dry or semi-arid country: bush, savanna, and even desert; also (e.g.) in forest, woodland cultivation, high mountain country (*Montifringilla*), and human habitation—some species (most notably House Sparrow and, in Orient, Tree Sparrow) being highly urbanized and among most successful of all birds. Mostly feed on or near ground, seeds (especially of wild grasses and cultivated grain) forming bulk of diet in many species, but other vegetable food and insects also taken (insects especially for young); bread and other man-made foods predominate in towns and villages. Occur naturally in Old World only, with centre of distribution in Afrotropics, but 2 species of *Passer* introduced elsewhere (including New World and Australia). 34–37 species; 11 in west Palearctic, all breeding. Relationship to other deep-billed, seed-eating passerines much debated, but now generally considered to constitute separate family.

House Sparrow *Passer domesticus*

PLATES: pages 1504, 1513

Du. Huismus   Fr. Moineau domestique   Ge. Haussperling   It. Passera
Ru. Домовый воробей   Sp. Gorrión doméstico   Sw. Gråsparv

Field characters. 14–15 cm; wing-span 21–25.5 cm. Slightly larger than Tree Sparrow, with 10% longer wings but proportionately somewhat shorter tail; averages slightly smaller than Spanish Sparrow. Heavy-billed, rather large-headed, robust passerine, suggesting tubby finch but differing in broader wings and square-ended tail; epitome of genus. ♂ boldly patterned: warm brown above, with mainly grey crown and black eye-stripe and bib contrasting with dull white cheeks, dark streaks over back, 2 pale wing-bars, grey rump, and greyish underparts. ♀ rather featureless: dull brown with indistinct pale supercilium and 2 pale wing-bars. Flight often fast and direct, with more whirring action than finches. Populations hybridize with Spanish Sparrow, particularly in southern Switzerland, Italy, Malta, and Sicily.

♂ unmistakable in good view, but ♂ hybrids and all ♀♀ and juveniles subject to confusion with Spanish Sparrow (see that species). Distinction from Tree Sparrow not difficult, with clear differences in head pattern, structure, and voice soon learnt. ♀ and juvenile may also suggest ♀ and immature rosefinches and Lapland Bunting to inexperienced observer, but lack streaked underparts and distinctive calls of those species.

Habitat. Greatly affected by enormous spread of range within recent historical time, changing breeding habits and diet, and close and flexible association with man; thus liable only to limited interspecific competition, though in parts of range competes with Tree Sparrow and Spanish Sparrow which show similar but much more limited tendencies. In west Palearctic, has spread in recent times and is now established throughout except some small Atlantic islands, the more arid tracts of North Africa, and high Arctic fringe. Avoids closed or dense vegetation, from forests to plantations, large thickets, reedbeds, and some high-density built-up areas, especially where structures are tall and lacking in ledges and vegetation. Except for seasonal foraging in cornfields and on other crops, usually avoids open terrain lacking in shrub, tree, or other cover, and, unlike some congeners, shows little attraction to either fresh water or sea coasts. Wherever constant food supply is assured by human activities shows remarkable indifference to climatic constraints, extending north to 10°C July isotherm fringing tundra, and tolerating extremes of heat, aridity, and moisture.

Distribution. Has spread north to northernmost Europe in 2nd half of 19th and 20th centuries in wake of human settlements, and has recently expanded range in Middle East and Egypt. ICELAND. Breeding first recorded 1959. Annual breeding in west 1971–80; this population now extinct, but small population in south-east since 1985. FAEROES. Has bred since *c*. 1935. BRITAIN. Has declined in Scotland, and locally in England. IRELAND. Possible decline in west. NORWAY. Recent spread in north. AUSTRIA. Markedly expanding, and reaching higher altitudes in Alps. RUSSIA. Has spread north in 20th century; breeding on Murman coast by 1955–7. Common in extreme north of range at Nar'yan-Mar (67°37′N). UKRAINE. Marked increase in range. SYRIA, JORDAN. Breeds in all inhabited areas. ISRAEL. Up to beginning of 20th century bred mainly in Mediterranean areas. Has since expanded to whole country. IRAQ. Breeds in towns and villages. EGYPT. Dakhla and Kharga oases colonized by 1981. LIBYA. Some apparent range extension in last 30 years. WESTERN SAHARA. Spread south along coast continuing; nesting Layoune since 1980, Dakhla since 1988. MAURITANIA. At Nouadhibou (Cap Blanc), first sight record 1961, others in 1980s, becoming established by 1990. At Banc d'Arguin, 24 birds seen May 1988. Uncertain whether these records involve birds of ship-assisted origin spreading north from Nouakchott (extralimital coastal Mauritania) and Sénégal, or/and birds spreading south along coast from Western Sahara. AZORES. Introduced 1960; now common on all islands, except Flores (colonized 1983) and Corvo and Santa Maria (absent in 1985). CAPE VERDE ISLANDS. Occurs only on São Vicente (introduced, probably ship-assisted), where locally common.

Accidental. Bear Island, Madeira.

Beyond west Palearctic, extends east across Asia to northern Sea of Okhotsk, lower Amur valley, and western Mongolia,

House Sparrow *Passer domesticus*. *P. d. domesticus*: **1** ad ♂ summer, **2** ad ♂ winter, **3** ad ♀, **4** juv. × *italiae*: **5** ad ♂ summer. *P. d. tingitanus*: **6** ad ♂ summer. *P. d. niloticus*: **7** ad ♂ summer, **8** ad ♀ summer.

and south to Arabia, Sri Lanka, and Burma. In Russia, spread east from Urals from 19th century, reaching Pacific coast 1929. In addition, has become established and spread, with man's assistance, in the Americas, sub-Saharan Africa, Australasia, and many oceanic islands.

Population. Little change in much of range, except in consequence of expansion. Decreases in Britain, Fenno-Scandia, Slovakia, and probably Belarus'; increases reported from Ukraine and Moldova, and recently Faeroes, Belgium, and perhaps Portugal. ICELAND. 5–10 pairs. FAEROES. 5000–10 000 pairs. Increase to *c.* 1975, then decline, but recently new increase. BRITAIN. 2.6–4.6 million pairs 1988–91. Noticeable local decreases, perhaps mainly due to changed farming practices. IRELAND. 800 000–1.4 million pairs 1988–91. FRANCE. Over 1 million pairs in 1970s. BELGIUM. 2 million pairs in 1970s, decreasing to as low as 1 million pairs after hard winters of early 1980s; some signs of recovery in early 1990s. LUXEMBOURG. 35 000–40 000 pairs. NETHERLANDS. 1–2 million pairs 1973–7. GERMANY. 8–12 million pairs. DENMARK. 220 000–3.3 million pairs 1987–8. NORWAY. 300 000–800 000 pairs 1970–90. SWEDEN. 400 000–800 000 pairs in late 1980s. FINLAND. 300 000–500 000 pairs in late 1980s. ESTONIA. 100 000–200 000 pairs in 1991. LATVIA. 750 000–2 million pairs in 1980s; probably fewer than in 1950s. LITHUANIA. Common. POLAND. Abundant. CZECH REPUBLIC. 3–6 million pairs 1985–9. SLOVAKIA. 1.2–1.8 million pairs 1973–94. HUNGARY. 800 000–1 million pairs 1979–93. AUSTRIA. Abundant. SWITZERLAND. 800 000–1 million pairs 1985–93. SPAIN. 9.3–10 million pairs. Recently decreased markedly in north, with abandonment of cereal cultivation and loss of nest-sites due to new building techniques. PORTUGAL. 1 million pairs 1978–84. ITALY. 10 000–50 000 pairs 1983–95; *italiae* 5–10 million pairs 1983–95. GREECE. 500 000–1 million pairs. ALBANIA. 99 000–500 000 pairs in 1981. YUGOSLAVIA: CROATIA. 2.5–3 million pairs. SLOVENIA. 500 000–800 000 pairs. BULGARIA. 1–10 million pairs. RUMANIA. At least 4 million pairs 1986–92. RUSSIA. At least 10 million pairs. BELARUS'. 2.1–2.3 million pairs in 1990. UKRAINE. 3.5–4 million pairs in 1986. MOLDOVA. 250 000–400 000 pairs in 1988. AZERBAIJAN. Probably over 150 000–200 000 pairs. TURKEY. 1–10 million pairs. CYPRUS. Locally abundant. SYRIA. Especially high numbers around large farms. ISRAEL. At least a few million pairs. JORDAN. Widespread and very common. KUWAIT. Very common. EGYPT, MOROCCO. Abundant. MAURITANIA. *c.* 100 pairs 1990–91. AZORES. 50 000–60 000 pairs in 1984.

Movements. Most races sedentary, especially in west of range. Juveniles disperse locally from natal area, but once settled remain within a few km. A small proportion, mainly juveniles, makes more directed migration, mainly to south and south-west but usually limited in extent. Larger-scale movements occur sporadically, mainly involving northern populations.

Two other types of movement are of greater interest. Many populations undertake movement from colony area in late summer to ripening grain fields; this can be up to 2 km, birds remaining in area of grain fields and returning to breeding areas September–October. Second is dispersal: occurs October, April–May, and (juveniles wandering after becoming independent) June–August. Difficult to distinguish October and April–May movement from migration, but seems to involve birds that as result of overcrowding have failed to obtain breeding sites.

Food. Mainly plant material, though nestlings largely fed animal material during first half of nestling period, and some animal material taken by adults immediately prior to and during breeding season. Vegetable food principally seeds, but shoots, buds, and berries taken to lesser extent. Birds living in urban and suburban areas take wide range of household scraps, elsewhere exploit food put out for domestic animals. Forages mainly on low plants or seeds on ground, though commonly perches on ripening cereal heads, frequently breaking stems. Seeds also taken directly from trees. De-husks seeds by pressing them against palate with tongue. Elderberries eaten by squeezing out juice and pulp, dropping skin and seeds, but, with harder fruits like *Pyracantha* and *Cotoneaster*, eats seeds and discards pulp. Feeding methods highly variable and adaptable. Collects invertebrates by (e.g.) searching leaves and bark of trees for slow-moving prey, hovering close to plants and pouncing on prey, catching insects in flight, shaking leaf clusters by grasping with feet and fluttering wings to dislodge insects.

Social pattern and behaviour. Gregarious, with dispersion throughout year based mainly on small, loose, but discrete colonies. In areas of low breeding density and favourable feeding, colonies remain isolated from neighbouring ones except for minor contact, but at higher densities and where food more dispersed (e.g. urban areas) birds from different colonies associate at feeding areas within radius of *c.* 500 m. Once independent, young form small foraging flocks which later coalesce at suitable feeding places into larger aggregations of young from several neighbouring colonies. These aggregations grow as more young fledge and are joined by adults that have finished breeding. Once adults have completed annual moult they return to breeding colonies to repossess their old nest-sites. Mates remain together. Some polygamy occurs: both bigamy—♂ exceptionally maintaining 2 ♀♀ in nearby nests or 2 ♀♀ laying in one nest—and casual polyandry by ♀ soliciting copulation from ♂ other than mate; such promiscuity may be quite frequent. Both sexes build nest, incubate, brood young, and feed them for up to 2 weeks after fledging. No true song, but ♂ gives advertisement-call at nest-site or from nearby perch during renewed sexual activity September–November and again in spring, starting late January, to proclaim ownership. Unpaired ♂ switches to song-display at approach of possible mate, adopting advertising-posture with chest thrust forward, showing off black bib, wings held out slightly and lowered and partly rotated towards ♀, showing off white wing-bar, and tail raised and fanned, with grey rump feathers ruffled (Fig. 6). In this posture ♂ hops round ♀, bowing stiffly up and down; at high intensity, may shiver wings. Most pair-formation replacement of lost mate, either on return to nesting colony in autumn or later by ♂ advertisement-calling at nest-site. New pairs of young birds also formed in spring by 1st-year ♂♂ taking up nest-site and attracting mate by advertisement-calling. By far the greatest number of copulations are at nest-site. ♂ gives solicitation-display to ♀, similar to advertising-posture except that he crouches and shivers wings. ♀ not ready to accept mate performs threat display and flies off chased by ♂. This attracts neighbouring ♂♂ who follow, and group-display occurs when ♀ lands in bush or on ground: ♂♂ hop round ♀ in solicitation-display posture, attempting to peck her cloaca; ♀ threatens with forward-threat any ♂ approaching too closely and may fly off, again pursued by ♂♂, and process repeated; occasionally, forced copulation occurs, but usually unsuccessful.

Voice. Extensive range of calls used at nest and elsewhere. Song of ♂ a loose sequence of basic chirps and variants, given by unpaired ♂ at nest to proclaim ownership and attract ♀, speeding up and becoming more excited on her approach; also given by paired ♂ at end of nestling period to induce mate to start another clutch, and by ♂♂ in group-display. Flight-call may also be monosyllabic 'chirp', but more often disyllabic 'churrip', 'churrit', or 'turrip' with deeper 'u' sound in 1st syllable and 2nd syllable accentuated. Used in flight, presumably to maintain contact. Social-singing, consisting of quiet chirping sounds, regularly heard outside breeding season, particularly on winter afternoons, a number of birds collecting in tree and calling together. Threat-call a rattling 'churr-r-r-it-it-it-it' or 'chit-it-it-it-it', often used against both conspecific and other intruders at feeding sites, and at nest.

Breeding. SEASON. Start of egg-laying positively correlated with latitude. In Europe normally April, but March in Azores and not until early May in Finland. Up to 4 broods per year, but 2 or 3 more usual. SITE. Usually in hole: in buildings and other man-made structures; also in earth-bank (including nest-holes of Sand Martin) or cliff, foundations of occupied and unoccupied large nests (e.g. of Corvidae), free-standing in branches of tree, or in creepers on wall. Enclosed nests of swallows and martins frequently appropriated. Nest: free-standing nest is large, domed, roughly globular structure, with entrance at side; loosely woven of dried grass or straw; cup lined with feathers, hair, or other soft material, especially tree bast; in hole, available space normally filled with material, though nest may be reduced to cup if little space. EGGS. Sub-elliptical, smooth and only slightly glossy. White or faintly tinted greenish or greyish; very variably marked with spots, speckling, or small blotches of grey, blue-grey, greenish-grey, purplish-grey, black, or purplish-brown; rarely unmarked. Clutch: 3–5 (2–7). INCUBATION. Usually 11–14 days. FLEDGING PERIOD. 11–19 days.

Wing-length: *P. d. domesticus*, continental Europe: ♂ 75–85, ♀ 72–80 mm.
Weight: ♂♀ mostly 24–38 g.

Geographical variation. Marked; involves mainly depth of colour, to lesser extent width of streaking on upperparts, size, and relative bill depth. *P. d. domesticus* (most of Europe) is a dark race; populations from Britain and Ireland slightly smaller and darker than in northern and central Europe. *P. d. balearoibericus* from Mediterranean France, central and eastern Iberia, Balearic Islands, Balkans from Yugoslavia and southern and eastern Rumania south to Greece, and western and central Asia Minor, paler than *P. d. domesticus*, about intermediate between *P. d. domesticus* and very pale *biblicus* of Levant. *P. d. tingitanus* from North Africa (Morocco to Libya) similar in colour to *balearoibericus*, but ♂ has pronounced black streaking on cap, entire cap sometimes mottled dark grey and black. *P. d. niloticus* (Egypt) pale like *biblicus*, but much smaller.

Fig. 6

Spanish Sparrow *Passer hispaniolensis*

PLATES: pages 1507, 1513

DU. Spaanse Mus   FR. Moineau espagnol   GE. Weidensperling   IT. Passera sarda
RU. Черногрудый воробей   SP. Gorrión moruno   SW. Spansk sparv

Field characters. 15 cm; wing-span 23–26 cm. Averages slightly larger than House Sparrow, with slightly heavier bill and stronger plumage pattern contributing to bolder form. Rather large handsome sparrow differing distinctly from House

Spanish Sparrow *Passer hispaniolensis*. *P. h. hispaniolensis*: **1** ad ♂ summer, **2** ad ♂ winter, **3** ad ♀, **4** juv. *P. h. transcaspicus* (Iran eastwards): **5** ad ♂ winter, **6** ad ♀. Tree Sparrow *Passer montanus* (p. 1513): **7** ad, **8** juv.

Sparrow in ♂'s dark chestnut crown, whiter cheeks, and black-splashed chest, flanks, and back, and more streaked appearance, even on flanks of ♀ and juvenile. Voice richer and deeper than House Sparrow. Populations hybridize with House Sparrow, particularly in southern Switzerland, Italy, Malta, and Sicily (e.g. Italian Sparrow; see Geographical variation).

♂ easily separated from House Sparrow, given clear sight of more vivid head pattern and more extensive black tracts in plumage, but distinguished from hybrid, if at all, only by close inspection. Crucial characters of full-blooded Spanish Sparrow are voice (see below) and full black gorget, which joins black-splashed shoulders and back, extends into streaks and arrowheads along entire flanks, and confines pale underparts to belly, vent, and under tail-coverts.

Habitat. In contrast to House Sparrow, remains confined to narrow lower middle latitudes, largely Mediterranean but extending east into west-central Asia in steppe and semi-desert valleys, sometimes ascending foothills and locally breeding in mountains. Typically, however, a warm lowland moisture-loving species inhabiting trees, shrubs, thickets, and reedbeds along riversides or irrigation ditches, groves of olives, date palms, *Acacia*, and eucalyptus, and even glades in woods and forests, where nests are often built in foundations of nests of storks and eagles. In the course of recent evolution it seems that this species tended to diverge from House Sparrow partly by becoming adapted to less arid areas and even to moist habitats, and partly by preferring to nest in vegetation and less frequently occupying human cultivation and settlements.

Nevertheless, in countries on both sides of Mediterranean, has tended to converge ecologically with House Sparrow and has even formed stable hybrid populations.

Distribution. Has spread in Balkan region in recent decades. Colonization of Atlantic islands began in early 19th century and was completed in 2nd half of 20th century. YUGOSLAVIA. Expanding north and north-west, more slowly up coast than inland. Since 1950 has spread north to Vojvodina. BULGARIA. Spread north along coast to Balchik by 1960, continuing northward thereafter. RUMANIA. First recorded (in southern Dobrogea) 1964; subsequently spread north to southern edge of Danube delta, Bucharest, and up river Siret. MOLDOVA. Arrival from south since 1950. AZERBAIJAN. Occurs mostly in lowlands. TURKEY. Virtually absent from Black Sea coastlands (but common Kizilirmak delta) and from mountain valleys of east and south-east. CYPRUS. Colonizing new areas. SYRIA. Regular, often in large colonies, in north; elsewhere only sporadic and in small numbers. KUWAIT. More than 100 birds near Ahmadi throughout February 1955, with many showing breeding activity; but colony suddenly deserted and no eggs laid. Occasional nest-building activities in 1980s. EGYPT. First recorded breeding 1990, north-east Sinai (several pairs). ALGERIA. Density varies greatly. MADEIRA. Arrived May 1935 after persistent easterly winds and became established. CANARY ISLANDS. Said to have been introduced, but colonization probably a natural westward extension of range. First recorded in east 1828 (Lanzarote) reaching Hierro in south-west 1960. CAPE VERDE ISLANDS. First recorded 1832, Santiago. Now widespread and common Santiago, Fogo, Boavista, and Maio;

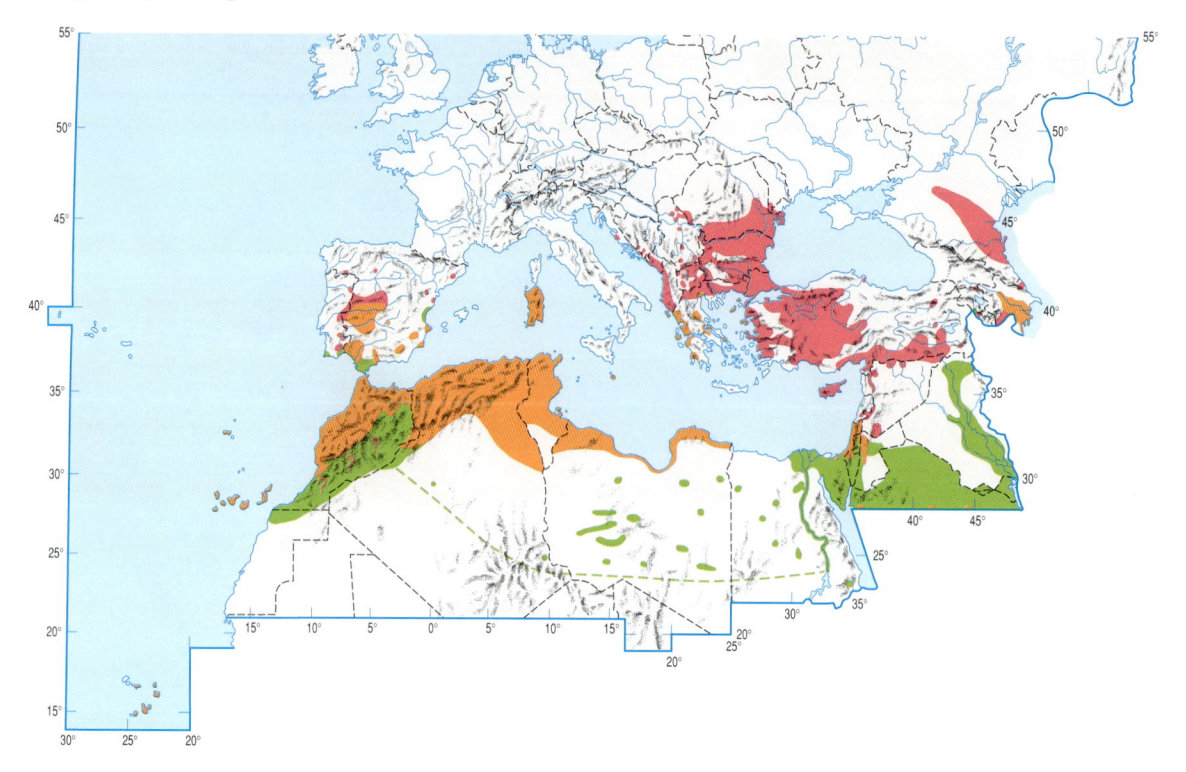

uncommon and irregular on São Nicolau and Sal. Formerly also reported from Brava and Santo Antão (but no recent observations there) and from São Vicente (where now only hybrids × *domesticus* remain in small numbers).

Accidental. Britain, Norway, Balearic Islands, Ukraine.

Beyond west Palearctic, extends from Iran north-east to eastern Kazakhstan.

Population. SPAIN. 12 000–16 000 pairs. PORTUGAL. 10 000–100 000 pairs 1978–84. ITALY. Sardinia: 300 000–500 000 pairs 1983–95. MALTA. Abundant. GREECE. 300 000–600 000 pairs. ALBANIA. 20 000–50 000 pairs in 1981. YUGOSLAVIA: CROATIA. 500–800 pairs; slight increase. BULGARIA. 100 000–1 million pairs; slight increase. RUMANIA. 500 000–600 000 pairs 1986–92. AZERBAIJAN. In mid-1960s, over 1 million birds at start of breeding season, but has decreased since. TURKEY. 100 000–1 million pairs. CYPRUS. Locally fairly common to common; increasing. ISRAEL. At least a few tens of thousands of pairs in 1970s–80s; some decrease in north, but increase east and south. JORDAN. Localized. TUNISIA. Common. MOROCCO. Common, but marked decline by end of 1980s following severe drought and chemical controls. MADEIRA. Rare. CAPE VERDE ISLANDS. Stable, local fluctuations, probably due to drought.

Movements. Pattern very complex. Some southern populations mainly sedentary, but others partially migratory. Populations in north-west Africa both migratory and nomadic. Eastern populations show more regular migratory behaviour, in some areas moving further north for successive breeding attempts. Winters in Spain, North Africa, Middle East, central Asia, northern Pakistan, and north-west India.

Food. Plant material (mainly seeds) and invertebrates. Plant material taken from low herbs or ground (shed seeds), but cereal seeds taken from ripening heads of grain at 'milk' stage; also buds and fruit from trees. Invertebrates taken mainly from ground, but also by searching leaves of bushes and trees, by fluttering in front of leaves and by catching insects in flight.

Social pattern and behaviour. Highly gregarious throughout year, more so than House Sparrow. Essentially monogamous, though not known if pair remains together for subsequent breeding attempts that are often in different locations. Breeding colonial, with some colonies very large and nests closely packed. Size of colony very dependent on local conditions, ranging from a few pairs to many thousands. Where population sedentary (e.g. Malta), same colony area used year after year and nests may be re-used; in migratory populations, same colony sites can be used for successive broods in one year and from year to year, but new sites frequently used and birds may move to new area for successive broods. ♂ sings at nest, to attract ♀ and to proclaim ownership. Sings with tail fanned, chest thrust forward showing off black throat, and wings held out, drooped, and violently shivered. Pair-formation occurs mainly during formation of colony. If unpaired ♀ approaches, ♂ hops round her in song-display. This period of intense activity and noise, as nest-sites and pairs are being established, lasts for only a few days, though in large colonies the cycle is not closely synchronized and adoption of nest-sites and pair-formation continue in newly established areas as colony grows from its starting point.

Voice. Generally very similar to House Sparrow, but advertising-calls typically fuller and louder with strident quality.

Song a rapidly repeated 'cheeli-cheeli-cheeli', more metallic and slightly higher pitched than House Sparrow. Basic 'chirp' and variants much as in House Sparrow. Threat-call a nasal 'churr-it-it-it', similar to House Sparrow, but slightly deeper and rattle shorter.

Breeding. SEASON. Laying begins March in North Africa and Canary Islands, early in April in south-west Europe. Usually 2, sometimes 3 broods. In Cape Verde Islands 2 separate breeding seasons: August–October, following rains, and February–March. SITE. Very large variety of sites used. Free-standing nests in trees, resting on branch forks, are most common. No particular preference shown for type of tree, pylons and telegraph poles providing suitable substitutes; more rarely low bushes, hedges, and reeds. Nests often built into foundations of nests of large birds, including birds of prey and White Stork, at times in company with House Sparrow. Also uses cavities. Nest: free-standing, large, untidy, roughly spherical, domed structure with entrance-hole on side. Normally constructed of dry grass or straw and thin twigs, but sometimes fresh green vegetation; leafy sprigs from trees and flower panicles sometimes incorporated. Nests in cavity of similar construction, filling most of available space. Nest-cup a more compact structure of fine grass, plant down, feathers, and animal hair. EGGS. Sub-elliptical, smooth and slightly glossy. White or faintly tinted blue or green, marked with specks, spots, or small blotches in various shades of grey, violet-grey, blackish-violet, or purplish, often with darker markings concentrated at large end. Clutch: 4–6 (2–8). INCUBATION. 11–13 days. FLEDGING PERIOD. 11–15 days.

Wing-length: ♂ 76–81, ♀ 73–78 mm.
Weight: ♂ mostly 26–36, ♀ 22–30 g.

Geographical variation. Very slight within populations of pure *P. hispaniolensis*, but situation complex in areas of hybridization with *P. domesticus*. Position of Italian Sparrow '*P. italiae*' and other populations of more or less similar appearance is problematical. Most likely that *italiae* as found in northern and central Italy is stabilized hybrid population between *P. domesticus* and *P. hispaniolensis*. Throughout mainland Italy and Sicily, populations gradually show more *P. hispaniolensis* characters towards south, with those of western Sicily and Malta almost pure *P. hispaniolensis*. In Algeria, Tunisia, and north-west Libya, where breaking up of ecological or spatial separation between *P. domesticus* and *P. hispaniolensis* is apparently more recent, populations show extreme individual variation, from pure *P. domesticus* to pure *P. hispaniolensis* and all forms between. Populations of Asia Minor and Levant verge towards Iranian race *transcaspicus*.

Dead Sea Sparrow *Passer moabiticus*

PLATES: pages 1510, 1513

DU. Moabmus   FR. Moineau de la Mer Morte   GE. Moabsperling   IT. Passera del Mar Morto
RU. Месопотамский воробей   SP. Gorrión del Mar Muerto   SW. Tamarisksparv

Field characters. 12 cm; wing-span 19–20 cm. Almost 20% smaller than frequent companion Spanish Sparrow, with proportionately smaller bill and more compact build. Distinctly trim sparrow, smallest and (♂) most colourful of genus. ♂ has grey head with pale supercilium, streak under eye, and submoustachial stripe turning up round cheek, and black bib; yellow-buff tone to rear supercilium and lower part of moustache unique within genus. ♀ buffier above and cleaner below than all other *Passer* except Desert Sparrow; also shows yellow in supercilium and on side of throat; as ♂, lacks bold wing-bars.

Unmistakable when small size, ♂'s unique head colours and pattern (lacking pale cheeks) and ♀'s head and upperparts well seen. Not known to overlap with other small *Passer*; streaked back of ♀ and juvenile quickly exclude Desert Sparrow.

Habitat. Patchily distributed in south-west Asia near watercourses or pools in arid usually lowland regions where shrubs such as tamarisk, thick scrub, or trees afford cover and nest-sites; in Israel, also in well-vegetated cultivated land.

Distribution. Has expanded in Israel, and spread north probably through Iraq to northern Syria and Turkey. Extension

Dead Sea Sparrow *Passer moabiticus moabiticus*: **1** ad ♂ summer, **2** ad ♂ winter, **3** ad ♀, **4** juv. Iago Sparrow *Passer iagoensis* (p. 1511): **5** ad ♂ summer, **6** ad ♀, **7** juv.

to Cyprus (also first record for Greece) apparently a continuation of westward spread in southern Turkey. TURKEY. Colony discovered at Birecik (upper Euphrates) 1964, and further colonies 1960s–70s; not certain whether this indicates range expansion or increased observer coverage. CYPRUS. Has bred on north shore of Akrotiri salt lake since at least 1976; first recorded 1973. SYRIA. Colony discovered 1968 extending south from Turkish border. Observations at various sites suggest regular breeding. ISRAEL. Up to 1930s limited to a few localities in Dead Sea area and southern Jordan valley. Has since spread, due to growth of agricultural settlements and consequent increased availability of water. IRAQ. Has probably spread north along Tigris: known only as far north as Baghdad area in 1915–22, and first found breeding on upper Tigris at Mosul in late 1940s.

Accidental. In Greece, *c.* 20 birds, Kalithea (Rhodes), 1st week of October 1972. In Egypt, small flocks recorded winter 1971–2 (*c.* 15 birds), and 30 October – 3 November 1987 (*c.* 10 birds).

Beyond west Palearctic, local in Iran and Afghanistan.

Population. TURKEY. 500–5000 pairs. CYPRUS. 35 nests in 3 colonies, 1985. ISRAEL. Some thousands of pairs in 1970s–80s; considerable increase in recent decades.

Movements. Many, or locally all, birds absent from breeding colonies October–March, but this appears to be more of a dispersal into feeding areas in cultivated land than a directed migration.

Food. Mainly seeds, especially of grasses. Most food sought on ground, but seen taking seeds from tamarisk trees and papyrus, and majority of insect food obtained by searching leaves of bushes and small trees such as tamarisk and willow.

Social pattern and behaviour. Largely gregarious. Normally in small flocks outside breeding season. Forms small loose breeding colonies, typically of 10–15 pairs, with nests not less than 8 m apart, typically 15–30 m. Breeding colony maintained from year to year. Breeding behaviour generally similar to House Sparrow.

Voice. Used mainly in breeding season. Song an excited, rhythmic, high-pitched 'chilling-chilling-chilling', shriller than Spanish Sparrow. Given by unpaired ♂ in song-display when ♀ approaches him at nest. Basic 'chirp' a regularly repeated 2-syllable 'chip-chew' or 3-syllable 'chip-chip-chew'.

Breeding. SEASON. Egg-laying in Israel from end of March, from late April in Cyprus. In Israel, commonly 2(–3) broods. SITE. Mostly 1–10 m above ground in branches of trees near water or standing in water; occasionally recorded at least 2 km from water. Nest: bulky, open globular or cone-shaped structure built of stiff dry twigs, 15–25 cm long, finely interwoven round branches of tree (resembling small nest of Magpie), lined with thick pad of plant down, seed panicles, fibres, and feathers. Domed, with entrance hole 40 mm wide spiralling down from top to cup. EGGS. Sub-elliptical, smooth and slightly glossy. Ground-colour white or buffish, but often completely obscured by purplish-brown or grey spots and speckling; 1–2 eggs in clutch much lighter with only sparse spotting at large end. Clutch: Israel, 4–5 (1–6). INCUBATION. 9–16 days. FLEDGING PERIOD. 11–13 days.

Wing-length: *P. m. moabiticus*: ♂ 60–64, ♀ 58–62 mm.

Weight: *P. m. moabiticus*: ♂♀ 11–17 g.

Geographical variation. Rather slight in west Palearctic. *P. m. moabiticus* (Israel and Jordan) smaller than *P. m. mesopotamicus* (southern Turkey, northern Syria, and Iraq).

Iago Sparrow *Passer iagoensis*

PLATES: pages 1510, 1513

DU. Kaapverdische Mus   FR. Moineau à dos roux   GE. Rostsperling   IT. Passera di Capo Verde
RU. Воробей Яго   SP. Gorrión grande   SW. Brunryggad sparv

Field characters. 13 cm; wing-span 17·5–20 cm. About 10% shorter than Spanish Sparrow, with noticeably slighter bill, 20% shorter wings, and 15% shorter tail. Rather small, compact sparrow, confined to Cape Verde Islands. Plumage colours and pattern recall House Sparrow but ♂ has black crown-centre, rufous supercilium, black rear eye-stripe extending round white cheek, and only narrow black bib. ♀ duller, with strong streaks on scapulars, contrasting pale supercilium, dark eye-stripe, and (on some) dusky bib.

On Cape Verde Islands, confusable only with Spanish Sparrow, but that species noticeably larger, with ♂ heavily marked black on chest, back, and flanks. Flight and behaviour typical of genus. Highly gregarious.

Habitat. In west Palearctic only in Cape Verde Islands, which, although tropical and oceanic and having main breeding season at end of the rains, have unusually cool climate, with tendency to increasing dryness. Climatic change appears to be modifying habitats, partly to the advantage of Iago Sparrow which is typically a bird of open desert, dry scrub, and rock faces, but also occurs in woodland and in cultivated areas and towns.

Distribution and population. Breeds on all islands and islets of Cape Verde Islands except Fogo. Stable, and mostly common and widespread, but only small, fluctuating numbers on Santa Luzia, Branco, and Sal.

Movements. Apparently sedentary.

Food. Limited observations suggest mainly plant material obtained from ground, especially seeds, also young leaves; insects taken by adults during breeding season and fed to young.

Voice. Song a loose series of calls 'cheep chirri chip cheep chirri chip chip'. Basic 'chirp', given in groups of 2–3 calls, lower pitched than call of House Sparrow. Threat-call a soft churring 'chur-it-it-it-it'.

Breeding. SEASON. Triggered by rains in August–September. SITE. Mainly in holes: lava cliffs, road cuttings, stone walls, buildings, wells, on ground under boulders, etc., but some free-standing nests in branches of trees. Nest: free-standing nest is globular, domed structure with side entrance; under boulder, compact open structure of thick plant stems. Outer structure coarse dry grass, cup lined with hair and feathers. EGGS. Similar to those of House Sparrow. Clutch: 3–5. (Incubation and fledging periods not recorded.)

Wing-length: ♂ 63–68, ♀ 57–62 mm.

Desert Sparrow *Passer simplex*

PLATES: pages 1512, 1513

DU. Woestijnmus   FR. Moineau blanc   GE. Wüstensperling   IT. Passera del deserto
RU. Пустынный воробей   SP. Gorrión sahariano   SW. Ökensparv

Field characters. 13.5 cm; wing-span 22–25 cm. About 10% smaller than House Sparrow, with stubbier bill, somewhat longer tail and 10% longer legs contributing to leggy stance. Rather small, relatively large-headed sparrow with pale plumage. ♂ essentially pale grey above and buff-white on rump and body, with bold black bill, short eye-stripe and bib, and pied wings. ♀ remarkably dissimilar, essentially pale sand-buff above, with horn bill and only vestigial wing markings.

Unmistakable if adult ♂♂ present, but single ♀ or juvenile could well puzzle observer inexperienced in desert passerines. Flight and behaviour typical of *Passer*.

Habitat. Confined to arid, subtropical, mainly lowland regions, in sandy areas, hollows among dunes, or dry wadis with shrubs, but locally ascends hills and even slopes of mountains. Nesting requirements lead to choice of locations near oases with trees, or sometimes isolated buildings such as forts, while patches of cultivation are also attractive.

Distribution and population. LIBYA. Locally common. TUNISIA. Localized and scarce. ALGERIA. Erratic in choice of breeding localities, seldom occupying one locality for more than 2 consecutive years. MOROCCO. Scarce and local; no recent changes reported.

Desert Sparrow *Passer simplex*. *P. s. saharae* (north and central Sahara): **1** ad ♂, **2** ad ♀, **3** juv. *P. s. simplex* (southern Sahara): **4** ad ♂, **5** ad ♀. Sudan Golden Sparrow *Passer luteus* (p. 1515): **6** ad ♂, **7** ad ♀, **8** juv.

Accidental. Egypt: Gebel Uweinat area in extreme south-west.

Beyond west Palearctic, breeds Turkmenistan and Uzbekistan, and extends in narrow belt from Mauritania to central Sudan in sub-Saharan Africa.

Movements. Some southward movement in winter, but erratic appearances suggest this may be more nomadic than directed.

Food. Casual records suggest chiefly plant material with some animal matter in breeding season. Seeds collected on ground and by flying up and pulling seed heads of larger plants to ground.

Voice. Song much more musical than House Sparrow: melodious series of trills which recalls Linnet and particularly Greenfinch. Calls chattering, recalling House Sparrow but can sound higher pitched: 'chip-chip' or subdued repeated 'chu'.

Breeding. SEASON. Northern Sahara: end of March to June; 2 broods. SITE. Variety of holes, from tree-cavities to cracks

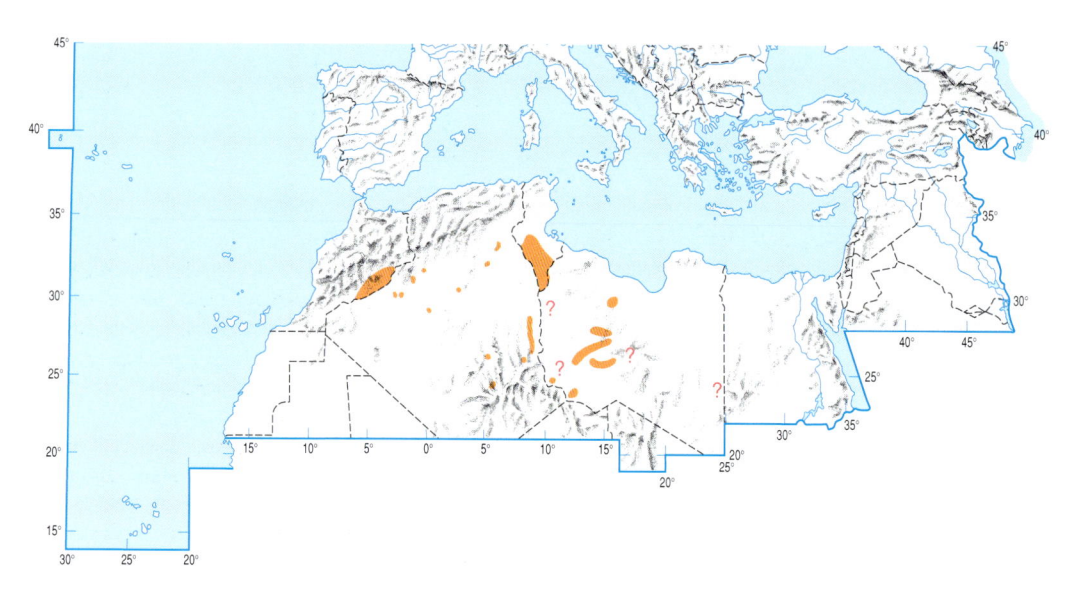

House Sparrow *Passer domesticus domesticus* (p. 1503): **1–2** ad ♂ summer, **3** ad ♀. Spanish Sparrow *Passer hispaniolensis hispaniolensis* (p. 1506): **4–5** ad ♂ summer. Dead Sea Sparrow *Passer moabiticus moabiticus* (p. 1509): **6** ad ♂. Iago Sparrow *Passer iagoensis* (p. 1511): **7** ad ♂. Desert Sparrow *Passer simplex saharae*: **8** ad ♂.

in rocks and human structures; occasionally among roof rafters. Also builds exposed nests closely interwoven with dense branches of trees, occasionally in foundations of nests of crows and vultures. Nest: free-standing nest built in 3 parts—loose, untidy outer structure of plant fibre and coarse grass, inner flask-shaped structure of twigs, grass, and tightly woven plant fibre with entrance on one side sloping up to nest-cup; cup made of fine plant fibre, grass, and feathers. Eggs. Sub-elliptical, smooth and slightly glossy. White, spotted and blotched with shades of brown and violet-grey. Clutch: 2–5. Incubation. 12–13 days. Fledging Period. 12–14 days.

Wing-length: ♂ 76–83, ♀ 73–81 mm.
Weight: ♂♀ 18–21 g.

Tree Sparrow *Passer montanus*

PLATES: pages 1507, 1522

Du. Ringmus   Fr. Moineau friquet   Ge. Feldsperling   It. Passera mattugia
Ru. Полевой воробей   Sp. Gorrión molinero   Sw. Pilfink

Field characters. 14 cm; wing-span 20–22 cm. Only slightly but distinctly smaller than House Sparrow and Spanish Sparrow, with proportionately less bulbous bill, trimmer head and body, and narrower and 10% shorter tail. Quite large but always tidy-looking sparrow, with dashing, direct flight even more pronounced than in House Sparrow. Sexes alike. Both adult and juvenile have diagnostic combination of black spot on white cheeks, long white collar emphasizing head, and 2 pale wing-bars.

Unmistakable at any range where head pattern visible. With experience, also separable on silhouette and voice (see below).

Habitat. In west Palearctic, breeds in middle and (locally) higher latitudes up to July isotherms 12–13°C, mainly in continental but also marginally in oceanic climates, preferring temperate to warm Mediterranean or high boreal regimes, in which House Sparrow appears to hold an advantage. Despite scientific name normally a lowland or low upland bird; in Switzerland infrequent above *c.* 700 m, although in northern Caucasus reaches *c.* 1700 m. Within above range, occupies suitable habitats only patchily and with prolonged fluctuations, involving inexplicable colonizations, and desertions of settled areas. Acceptable habitats, from oceanic to continental, appear to fit series of distinct types: coastal cliffs, especially with ivy; other coastlines, especially with empty or ruined buildings; pollarded willows and other trees with nest-holes along slow-flowing lowland watercourses; quarries and nest-holes of Sand Martin; free-standing trees along roadsides or in groups in parks, cemeteries, or farmland; woodlands, especially where they are small and isolated in open country with well-spaced mature broad-leaved trees; and spacious suburbs especially where nest-boxes provided. Such urban habitats are not occupied in west of range where they are monopolized by House Sparrow. In eastern Asia, the ultimate stage is reached of replacing House Sparrow, even in inner cities.

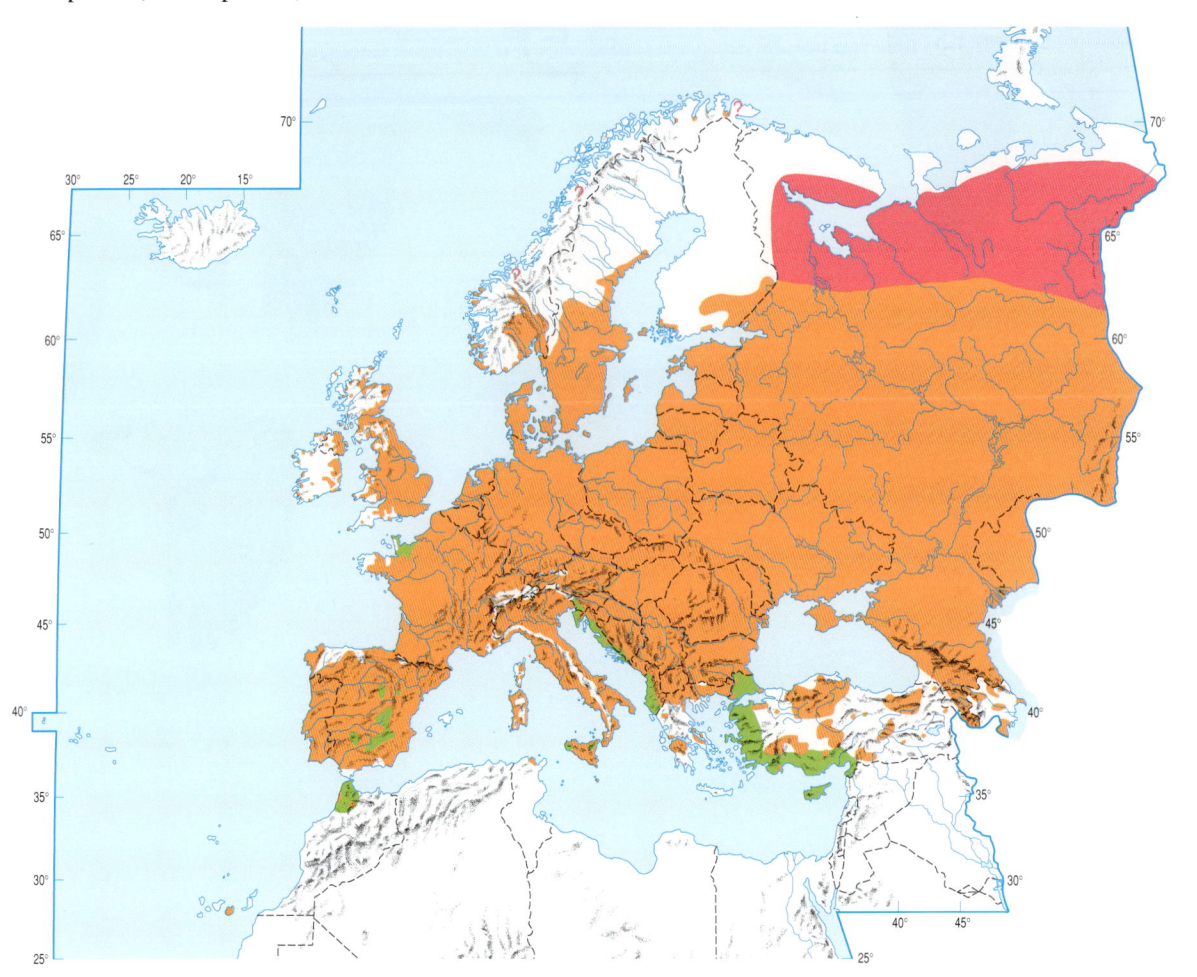

Distribution. Has recently spread in Fenno-Scandia, and extended range to north-west Africa. FAEROES. Bred in small numbers 1866 to before 1930, and perhaps again *c.* 1960. Only 6 sight records since. BRITAIN. Undergoes major fluctuations. Latest expansion from late 1950s, followed by decline and range contraction since 1976–7, especially in west and extreme south. IRELAND. Long-term decline and range contraction from 19th century to 1950s. No breeding recorded 1959–60, then recovery from 1961. FRANCE. Local decreases in north-west. NORWAY. Recent spread in parts of south-east, west, and north. SWEDEN. Bred on Finnish border in 19th and early 20th centuries. Began to breed in Gotland in 1920s, becoming common there in 1960s. Recently spread north into Medelpad (*c.* 62°30′N). FINLAND. Marked increase from east during 1980s. Many new local populations now in southern towns and villages. SPAIN, PORTUGAL. Slight increase. BALEARIC ISLANDS. Breeding suspected Mallorca 1993; in 1994, hybrid breeding attempts with House Sparrow at S'Albufera and S'Albufereta, and at least 1 pure pair at latter site. ITALY. Introduced to Sardinia at end of 19th century, and now widespread there. MALTA. First recorded breeding 1959; breeds in small scattered colonies or as isolated pairs on all 3 main islands. UKRAINE. Slight increase. TURKEY. Widespread but very local, mainly north of 40°N, on fringes of Central Plateau, and at foot of western Taurus. TUNISIA. Breeding discovered near Bizerte in 1990s. MOROCCO. Scarce and local breeder recently established in north; suspected 1985, subsequently proved. CANARY ISLANDS. Breeds Gran Canaria—first nests and birds found 1989. ♂♂ observed Tenerife, but breeding not confirmed.

Accidental. Iceland, Israel, Egypt, Algeria.

Beyond west Palearctic, widespread in Asia (except most of Indian subcontinent) east to Sea of Okhotsk, Sakhalin island, and Japan, south to Greater Sundas. Introduced in North America, Australia, eastern Indonesia, Philippines, and elsewhere.

Population. Increases reported in Norway, Finland, Estonia, Iberia, and Ukraine; decreases Britain, France, Low Countries, Germany, Denmark, Poland, and Switzerland; stable elsewhere. BRITAIN. 110 000 territories 1988–91; decline of 89% 1970–95. IRELAND. 1000–1500 pairs. FRANCE. 100 000–1 million pairs in 1970s. BELGIUM. 210 000 pairs 1973–7. Decrease since early 1980s, population at lowest 1985–90; since then slight recovery. LUXEMBOURG. 4000–8000 pairs. NETHERLANDS.

100 000–500 000 pairs 1979–85. Marked long-term decline, with short-term fluctuations. GERMANY. 2–3.5 million pairs. Decline reported Niedersachsen and several eastern regions. DENMARK. 26 000–375 000 pairs 1987–8. Decrease 1990–94 following increase 1976–90. NORWAY. 50 000–100 000 pairs 1970–90. Varies from very common in some areas to very rare in others. SWEDEN. 400 000–800 000 pairs in late 1980s. FINLAND. 7000–10 000 pairs in late 1980s. ESTONIA. 50 000–100 000 pairs in 1991. LATVIA. 150 000–300 000 pairs in 1980s. LITHUANIA. Common. POLAND. Numerous in lowland; locally decreasing in recent years. CZECH REPUBLIC. 500 000–1 million pairs 1985–9. SLOVAKIA. 300 000–600 000 pairs 1973–94. HUNGARY. 800 000–1 million pairs 1979–93. AUSTRIA. Common. SWITZERLAND. 70 000–100 000 pairs 1985–93. SPAIN. 2.5–4.1 million pairs. PORTUGAL. 100 000–1 million pairs 1978–84. ITALY. 500 000–1 million pairs 1983–95. GREECE. 20 000–50 000 pairs. ALBANIA. 5000–10 000 pairs in 1981. YUGOSLAVIA: CROATIA. 500 000–700 000 pairs. SLOVENIA. 100 000–200 000 pairs. BULGARIA. 100 000–1 million pairs. RUMANIA. At least 3 million pairs 1986–92. RUSSIA. 1–10 million pairs. BELARUS'. 900 000–950 000 pairs in 1990. UKRAINE. 800 000–950 000 pairs in 1986. MOLDOVA. 200 000–300 000 pairs in 1988. AZERBAIJAN. Numerous. TURKEY. 5000–50 000 pairs. TUNISIA. Common in sole locality.

Movements. Mainly sedentary, especially in west of range, with only small proportion undertaking relatively short-distance migration, mainly to south or south-west. Larger-scale autumn movements occur from time to time, particularly from more northerly parts of range. Numbers involved in these movements are subject to considerable fluctuations, forming irregular pattern that suggests eruptions rather than normal annual migration. Return movement in North Sea area occurs late March to May. Eruptive movements probably responsible for recent southward extensions of range that have occurred in 20th century in Corsica, Sardinia, and Malta, with some overwintering birds remaining to breed. Recolonization of Ireland after extinction in early 1950s probably arose in same way.

Food. Plant and animal material, proportions varying with both season and availability. Seeds comprise bulk of plant matter, with fewer buds and berries. Food predominantly sought on ground. Takes seeds from low growing plants, both by flying up and perching, e.g. on cereal stems (particularly at 'milk' stage), or by pulling seed-head to ground in case of weaker plants, and stripping off seed-head.

Social pattern and behaviour. Largely sociable all year. Outside breeding season, birds from adjacent loose breeding colonies tend to mix and wander over home-range of 10–100 km². After breeding season, adults join flocks of young birds that have already formed at suitable feeding places. Mating system essentially monogamous with pairs formed for life, though occasionally ♂♂ polygamous, taking over additional widowed ♀♀ in neighbouring nests. Breeding loosely colonial. Provided there is sufficient food, colony size depends on availability of suitable nest-sites. Nests in foundation of large nests of other bird species may be touching or within a few cm, but when nesting in individual holes in trees can be up to 1 km apart. Most pair-formation takes place at nest-hole. As substitute for song, ♂ also gives advertisement-calls from conspicuous perch at nest-site in enticement-posture during autumn and again in spring to proclaim ownership and attract a mate. Display also involves addition of nest-material to nest-hole and continues with increasing intensity until start of breeding.

Voice. No distinctive song as such, but excited series of 'tschirp' or 'tschilp' calls, typically with high and low calls rapidly alternating, given by unpaired ♂ at nest-site and directed by courting ♂ at ♀ away from nest. A variety of other, mainly monosyllabic chirping calls given in other social contexts. Flight-contact call a distinctive hoarse 'teck', producing chattering chorus from flock.

Breeding. SEASON. Egg-laying in Europe normally from mid-April to July. Usually 2–3 broods. SITE. Predominantly in hole: in tree, building, earth bank (including Sand Martin colonies), also in foundation of large nest (e.g. of crows, Grey Heron, birds of prey), and more rarely free-standing in branches of dense conifers and hawthorn. Nest: free-standing nest is flattened sphere, with entrance on side leading to nest-cup; in hole, available space normally filled with material, though roof can be omitted; built from plant stems, rootlets, and leaves, lined with moss, wool, hair, and feathers. EGGS. Sub-elliptical, smooth, slightly glossy; white to pale grey, heavily marked with spots, small blotches, or speckling, usually dark brown, sometimes purplish or greyish, often heavy enough to obscure ground; markings usually concentrated around broad end; great variation in size, shape, and colour. Clutch: typically 2–7 (1–8), 5 most frequent. Mean clutch size varies through season, with 2nd clutch largest and 3rd smallest. INCUBATION. 11–14 days. FLEDGING PERIOD. 15–20 days.

Wing-length: ♂ 68–74, ♀ 66–72 mm.
Weight: ♂ 19–29, ♀ 18–27 g.

Sudan Golden Sparrow *Passer luteus*

PLATES: pages 1512, 1522

DU. Bruinruggoudmus   FR. Moineau doré   GE. Braunrücken-Goldsperling   IT. Passera dorato
RU. Желтый воробей   SP. Gorrión aureo   SW. Guldsparv

Field characters. 13 cm; wing-span 18·5–20 cm. About 15% smaller and slighter than House Sparrow, with proportionately smaller bill and head. Small, highly gregarious sparrow, with yellowest plumage of west Palearctic *Passer*. ♂ has yellow head and body and chestnut back; ♀ plain yellow-buff on head and back; ♂ and ♀ share dark wing-coverts with 2 white to buff wing-bars.

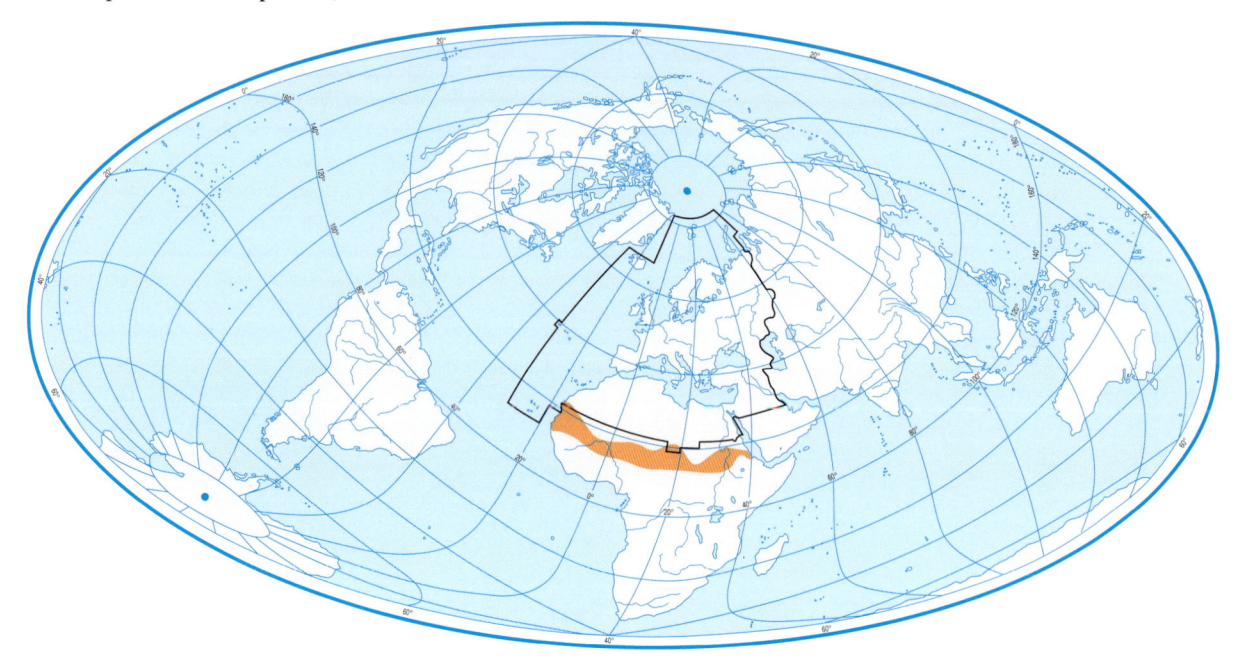

♂ unmistakable. ♀ and immature only confusable with other small sparrows when yellowish head or full strength of wing pattern obscured. Flight fast and agile, allowing both individual speed and group manoeuvrability; action as reminiscent of small African finches as of other sparrows.

Habitat. Mainly south of Palearctic in semi-arid tropical lowlands, distribution conforming closely to zone of dry woodland and steppe.

Distribution and population. A Sahel species, barely reaching southern fringe of western Palearctic. MAURITANIA. Within west Palearctic, breeds only at Cansado near Nouadhibou (Cap Blanc); first sight records 1978; 50–80 pairs 1990–91. ALGERIA. Occasionally breeds just south of west Palearctic limit: small population found In Guezzam (19°30′N 5°30′E) 1984. CHAD. Within west Palearctic, recorded breeding in Tibesti at Zouar (20°30′N 16°30′E) 1953.

Movements. Nomadic, following availability of food. Southward movement during breeding season in both Sénégal and Niger as birds seek new area with sufficient food to rear 2nd brood; ♂♂ move first, leaving ♀♀ to complete rearing of 1st brood.

Food. Mainly plant material, chiefly seeds, although nestlings fed largely insects. Seeds collected from seed heads of small plants and from ground, flock 'roller-feeding' with birds at rear overflying those at front.

Social pattern and behaviour. Highly gregarious. After breeding, occurs in flocks of 10–100 birds, frequently associating with weaver-birds and estrildine finches. Mating system monogamous. Forms large breeding colonies; nests may be within a few metres of each other.

Voice. Contains individual calls close to classic 'chirp' of House Sparrow but in chorus they run together into characteristically more sibilant twitter; when flushed, calls with fast rhythmic 'che-che-che-' (7–8 units), suggesting flight-call of Redpoll.

Breeding. SEASON. Breeding opportunistic, normally triggered by rains, more rarely in dry season, particularly in irrigated areas. SITE. Thorny trees and bushes. Nest: globular structure wedged between several branches with entrance hole in upper half. Built from 700–1200 thorny twigs, mostly less than 100 mm long, occasionally over 200 mm; more rarely of stiff grass stems; lower half compactly built, upper half rather looser; cup of grass, bark, leaves, flower heads, feathers. EGGS. Sub-elliptical, smooth and slightly glossy; white to light bluish- or greenish-grey, with brown to maroon spots and flecks. Clutch: in Niger, 3–5. INCUBATION. 10–12 days. FLEDGING PERIOD. 13–14 days.

Wing-length: ♂ 61–67, ♀ 61–65 mm.
Weight: ♂♀ mostly 12–16 g.

Pale Rock Sparrow *Carpospiza brachydactyla*

PLATES: pages 1518, 1522

Du. Bleke Rotsmus   Fr. Moineau pâle   Ge. Fahlsperling   It. Passera lagia pallida
Ru. Короткопалый воробей   Sp. Gorrión pálido   Sw. Blek stensparv

Field characters. 13·5–14·5 cm; wing-span 27–30 cm. As long, and as long-winged, as Rock Sparrow, but with slighter, sleeker form most noticeable in smaller bill and slimmer body; slightly smaller and noticeably shorter than Sinai Rosefinch, particularly in tail length. Fairly slim and rather anonymous sparrow-sized bird of uncertain affinity, recalling ♀ rosefinch or odd ♀ bunting as much as Rock Sparrow. Dull pale brown except for double buff wing-bar, pale panel on folded secondaries, and tail with white edges and tip; pale horn bill and bright orange-brown legs.

At first sight, a bird of remarkable anonymity. However, has diagnostic combination of rather short but strong and slightly bulbous bill, long, narrow outer wings with pale secondary-panel and double wing-bar, pale edges and white tip to tail (particularly obvious from below), and bright legs. Also differs distinctly from Rock Sparrow in slighter form and lack of streaks on head and mantle, from Yellow-throated Sparrow in somewhat larger size, pale bill, lack of warm brown lesser coverts and boldly white-tipped median coverts, from ♀ Desert Sparrow in larger size, greyer appearance, lack of pale, dark-rimmed wings, and from Dead Sea Sparrow in larger size and lack of mantle streaking. Voice also helpful in diagnosis (see below). Flight action lighter and even more fluent and bounding than Rock Sparrow, allowing flocks to swirl like finches; in silhouette, length of wings catches eye more than shortness of tail. Usually runs quite fast on exposed legs, suggesting (with sleek form) lark or even pipit.

Habitat. In west Palearctic, breeds in low middle latitudes in warm Mediterranean and subtropical zones, largely in arid areas, and often on hillsides and mountains, in Armenia predominantly at 700–2300 m; in Israel breeding at 1100–1800 m on Mt Hermon. Also frequents slopes of barren ravines, overgrown with grass and stunted pistachio bushes, or areas with scattered bushes and trees which occasionally form sparse thickets or groves. During hot part of day shelters in shade of cliffs, or resorts to water such as springs or streams. During cooler hours descends to ground in search of food, or perches on grass and shrubs such as wild almond, or on walls of abandoned clay dwellings. In winter, occurs in more open and cultivated areas further south, especially in cereal crops.

Distribution and population. ARMENIA. Limited range. Most frequently encountered in Azizbekov, Vedi, and Artashat areas. Also found nesting in vicinity of Yerevan and in Ashtarak region. Rare. AZERBAIJAN. Limited range. Nests in semi-arid mountains of Nakhichevan region, also in Zuvand upland (Talysh mountains). Common to very common. TURKEY. Breeds chiefly in hills of Mesopotamian plateau and along its border. Erratic, with not all sites occupied annually. Locally common: 100–1000 pairs. SYRIA. Occasional breeding season records suggest possible breeding. LEBANON. Recorded breeding only in Anti-Lebanon range. ISRAEL. Breeds only Mt Hermon area. Usually up to 10 pairs; sometimes 20–30 or none. JORDAN. Only breeding record Dana, 1994; possibly bred Azraq 1960s. KUWAIT. Breeding confirmed 1996. SAUDI ARABIA. Breeds in small numbers Harrat Al Harrah reserve.

Accidental. Cyprus.

Beyond west Palearctic, breeds Arabia, Iran, and Afghanistan.

Movements. Short-distance migrant, moving south or west of south to winter in Arabia and north-east Africa. Movements erratic, with great fluctuations in numbers; timing also variable.

Food. Mainly insects in breeding season, otherwise probably mostly plant material, especially seeds. Forages on ground and on low plants in sandy or rocky places. In winter quarters, feeds in large flocks of up to many thousands in fields of millet or sorghum; these cereal seeds taken on ground, or by hopping up to pull seed from head of grain, or by hanging onto seed head, weighing it down to ground.

Social pattern and behaviour. Study of behaviour has shown nothing in common with other Palearctic Passeridae (certainly not with Rock Sparrow) and it may be that Pale Rock Sparrow should be placed in the Fringillidae (see Geographical variation). Gregarious outside breeding season, occurring in flocks of up to *c.* 300 birds. Monogamous mating system apparently the norm. Pair-formation evidently takes place on breeding grounds or

Pale Rock Sparrow *Carpospiza brachydactyla*: **1** ad winter, **2** ad summer, **3** juv. Yellow-throated Sparrow *Petronia xanthocollis transfuga*: **4** ad ♂ summer, **5** ad ♀, **6** juv.

before arrival there. ♂ sings from ground, bush, or other low perch, sometimes higher up on cliff, less often in flight. Unpaired ♂ highly mobile in territory, flying from one song-post to another in display-flight, characteristic feature of which is that wings slightly retracted and vibrated (much faster wing-beats than normal) or held rigid, bird then gliding, always on straight trajectory.

Voice. Little or nothing in repertoire analogous to vocalizations of Rock Sparrow, and main calls are unlike those typical of Passeridae. Song of ♂ unlike sparrow, and more often likened to finch, bunting, or insect. Distinctive and persistent, sibilant or wheezing sound strongly recalling wheeze of Greenfinch; also reminiscent of 'chee-eese' ending of song of Yellowhammer and (superficially) like song of Corn Bunting. Other calls include a full liquid 'pluip', given by flying birds; a loud, clear, resonant tremolo reminiscent of excited calls of Rock Nuthatch; and low twittering or trilling sounds, heard from flocks.

Breeding. SEASON. Egg-laying from end of May; one brood. SITE. In low bush or tree, often thorny; often near water. Nest: open, bulky, untidy hemisphere of generally thorny twigs, stalks, roots, leaves, and grass, lined with smooth, felt-like mixture of plant down, flowerheads, soft leaves, bulb scales, and animal hair. EGGS. Sub-elliptical, smooth and glossy; white (rarely washed pink), with scattered black to reddish-brown spots or commas concentrated at broad end, and greyish undermarkings. Clutch: 4–5 (3–6). INCUBATION. 13–16 days. FLEDGING PERIOD. 10–11 days.

Wing-length: ♂ 96–102, ♀ 91–96 mm.
Weight: ♂ 21–25 g.

Geographical variation. None. This species is often included in *Petronia*; shows tail-spots similar to Rock Sparrow and general pale colour of (e.g.) Yellow-throated Sparrow. However, no yellow on chest, and bill not black in breeding ♂; nest different in location and structure (more like *Fringilla*); eggs quite different, showing sharp spots like *Fringilla*; nestling covered with dense down (almost naked in *Petronia*); voice and behaviour different. Morphology of jaw, tongue apparatus, and skull indicate it to be an early offshoot of finches (Fringillidae), sparrows (Passeridae), or even buntings (Emberizidae), and therefore removed from *Petronia* into monotypic genus *Carpospiza* pending further research.

Yellow-throated Sparrow *Petronia xanthocollis*

PLATES: pages 1518, 1522

DU. Indische Rotsmus   FR. Moineau à gorge jaune   GE. Gelbkehlsperling   IT. Passera lagia indiana
RU. Желтогорлый воробей   SP. Gorrión pintado   SW. Gulstrupig stensparv

Field characters. 12·5–13 cm; wing-span 23–27 cm. About 10% smaller than Pale Rock Sparrow with proportionately longer, finer bill. Rather small, quite slim sparrow, with noticeably long bill. ♂ most colourful west Palearctic member of rather nondescript genus: yellow patch on throat, chestnut patch on forewing, white tips to median coverts, and dark bill.

♂ unmistakable in good view, but ♀ and juvenile liable to confusion with Pale Rock Sparrow and other drab seed-eaters. Within west Palearctic *Petronia* and *Carpospiza*, combination of uniformly coloured tail, pale median covert wing-bar, and dull legs is diagnostic. Flight even lighter than Pale Rock Sparrow, with dipping undulations recalling small pipit.

Habitat. In west Palearctic, breeds in subtropical lower middle latitudes, mainly in lowlands; in Himalayan foothills, however, summers up to 750 m and at other seasons occurs locally up to *c.* 1200 m. In contrast to Rock Sparrow, favours open dry forest or forest scrub, oases, groves, gardens, and cultivated areas with scattered trees, shrubs, or hedgerows.

Distribution and population. TURKEY. Breeding recorded only since 1977, in a few localities in south-east. 50–1000 pairs. IRAQ. Localized breeder; easily overlooked.

Accidental. Israel, Kuwait (or rare migrant).

Beyond west Palearctic, extends east to eastern India; also breeds eastern Arabia.

Movements. Migratory in west Palearctic, wintering in southern Pakistan and India, with occasional winter records from Oman. Autumn migration begins early; leaves southern Iraq August–September. Spring migration March–April; reaches Iraq in April, south-east Turkey by early May.

Food. In breeding season, diet insects and plant material, at other times principally plant material, mainly seeds. In breeding season forages more often in trees than sparrows *Passer*; in remainder of year on ground, commonly in large flocks with *Passer*, finches, and buntings.

Social pattern and behaviour. Little known. Usually in loose-knit flocks outside breeding season, including on migration. Mating system probably monogamous. Breeds in loose colonies. ♂ sings persistently during breeding season, often more or less throughout day.

Voice. Song of ♂ a series (often long) of not-unattractive chirping sounds, 'chilp-chalp', 'chip-chip-chock', or 'chip' and 'chilup'; generally likened to *Passer*, but softer, more melodious, higher pitched, and slightly more liquid than House Sparrow. Main call similarly like *Passer*, but softer and more liquid and tuneful: 'cheep', 'chilp', or 'chirrup'.

Breeding. SEASON. Iraq: late April to late July; possibly 2 broods. SITE. Hole or crevice in tree, building, or other man-made structure; may use old hole of woodpecker or old, possibly usurped, nest of Red-rumped Swallow or of Dead Sea Sparrow. Nest: shapeless, untidy foundation, often fillng cavity, of dry grass, hair, wool, strips of bark, string, and other man-made material, thickly lined with feathers; cup often only a vague depression in thick pad of material. EGGS. Sub-elliptical (but great variation from elliptical to pyriform), smooth and very slightly glossy; very variable in colour, even within clutch, ranging from greenish to brownish-white, heavily speckled, blotched, and streaked with all shades of brown, often obscuring ground-colour. Clutch: 3–4 (2–6). (Incubation and fledging periods not recorded.)

Wing-length: ♂ 81–88, ♀ 77–82 mm.
Weight: ♂ 15–20, ♀ 14–20 g.

Rock Sparrow *Petronia petronia*

PLATES: pages 1520, 1522

DU. Rotsmus   FR. Moineau soulcie   GE. Steinsperling   IT. Passera lagia
RU. Каменный воробей   SP. Gorrión chillón   SW. Stensparv

Field characters. 14 cm; wing-span 28–32 cm. Close in size to House Sparrow, but with 30% longer, more conical bill, 25% longer wings, and almost 10% shorter tail. Bulky, long-winged, square-tailed sparrow, with heavy bill on rather large head (apparent size of head due partly to its linear patterning). Differs from other west Palearctic *Petronia* in larger size and strongly striped, streaked, and spotted greyish-brown plumage; combination of dark lateral crown stripes and white spots on tail-tip diagnostic.

To inexperienced observer, may suggest ♀ or immature of

Rock Sparrow *Petronia petronia*. *P. p. petronia*: **1, 6** ad ♂, **2** ad ♀, **3** juv. *P. p. barbara*: **4** ad ♂. *P. p. puteicola*: **5** ad ♂. House Sparrow *Passer domesticus domesticus* (for comparison): **7** ad ♀.

several *Passer*, especially Spanish Sparrow, but differs in structure, and obvious white tailspots of adult are not shared by any *Passer*. Flight much superior to *Passer* in power and speed, with pronounced bursts of fluent wing-beats giving swift, bounding progress and allowing it to make dramatic ascents of cliff faces. Flight silhouette differs from *Passer* in relatively heavier bill and head, much longer outer wings, and shorter, squarer tail, combining to suggest bulky projectile. Gait a strong hop; also jumps and shuffles. Generally more active on ground than *Passer*, with alert, rather upright stance often showing thighs above short legs.

Habitat. Breeds in middle latitudes of Palearctic in warm temperate Mediterranean, steppe, and desert climates, from sea-level up rocky slopes and hillsides to mountains at 2500 m in Armenia and to *c.* 4800 m in Himalayas. Generally frequents rather bare treeless terrain with scanty herbaceous vegetation, ranging from flat desert steppe to rocky slopes or outcrops, screes, stony patches, ravines, cliffs, crags, and clay or earth precipices. In some regions favours less severe environments, such as alpine meadows, grassy or shrubby riversides, vineyards, olive groves, stone walls, ruined castles and other structures on hilltops, and even human settlements, where it may come into competition with House Sparrow or Spanish Sparrow.

Distribution. Range contracted southwards in early part of 20th century. FRANCE. Disappeared from Alsace and Bourgogne early in 20th century. GERMANY. Formerly bred southern Baden, Nassau, Franken, and Thüringen; populations subject to rapid and wide fluctuations. Confined to Thüringen and adjoining areas by beginning of 20th century. Last records: pair nest-building in 1941, and sight-record 1944. POLAND. Breeding recorded in Sudety mountains 1897. AUSTRIA. A few observations from 1st half of 20th century, mainly from Salzburg Alps near border with Bayern; breeding suspected but no proof. No recent records. BALEARIC ISLANDS. No certain recent records on Mallorca, and now effectively extinct there. Still breeds Ibiza and Formentera. GREECE. Localized; distribution poorly known, especially in north. BULGARIA. Slight increase. AZERBAIJAN. In Great Caucasus, probably confined to Shemakha upland. Very common in Nakhichevan region. TURKEY. Widespread but local in all rocky areas near cultivation; apparently absent from coastal areas and from Thrace. Highest numbers in some mountain valleys of Taurus and in east. SYRIA. Breeds in north-west, also regular and locally even common in Anti-Lebanon. MOROCCO. Known for a long time in Moyen and Haut Atlas; recently discovered in Anti-Atlas and eastern Rif. CANARY ISLANDS. Breeds Gran Canaria, Tenerife (declining), La Gomera, El Hierro, and La Palma.

Accidental. Britain, Poland, Slovakia, Switzerland, Malta, Ukraine, Cyprus, Azores.

Beyond west Palearctic, extends east discontinuously to *c.* 120°E, north to upper Yenisey and Lake Baykal, south to Iran and south-west China.

Population. FRANCE. 1000–10 000 pairs in 1970s. SPAIN. 825 000–1.15 million pairs; stable. PORTUGAL. 10 000–100 000 pairs 1978–84; stable. ITALY. 10 000–20 000 pairs 1983–95; slight decrease. GREECE. 2000–5000 pairs. ALBANIA. 500–2000 pairs in 1981. BULGARIA. 10–100 pairs; slight increase. RUSSIA. 100–1000 pairs; stable. AZERBAIJAN. Locally common. TURKEY. 10 000–100 000 pairs. ISRAEL. A few thousand pairs in 1980s. JORDAN. Fairly common. At least 80 pairs in Petra area, 1983.

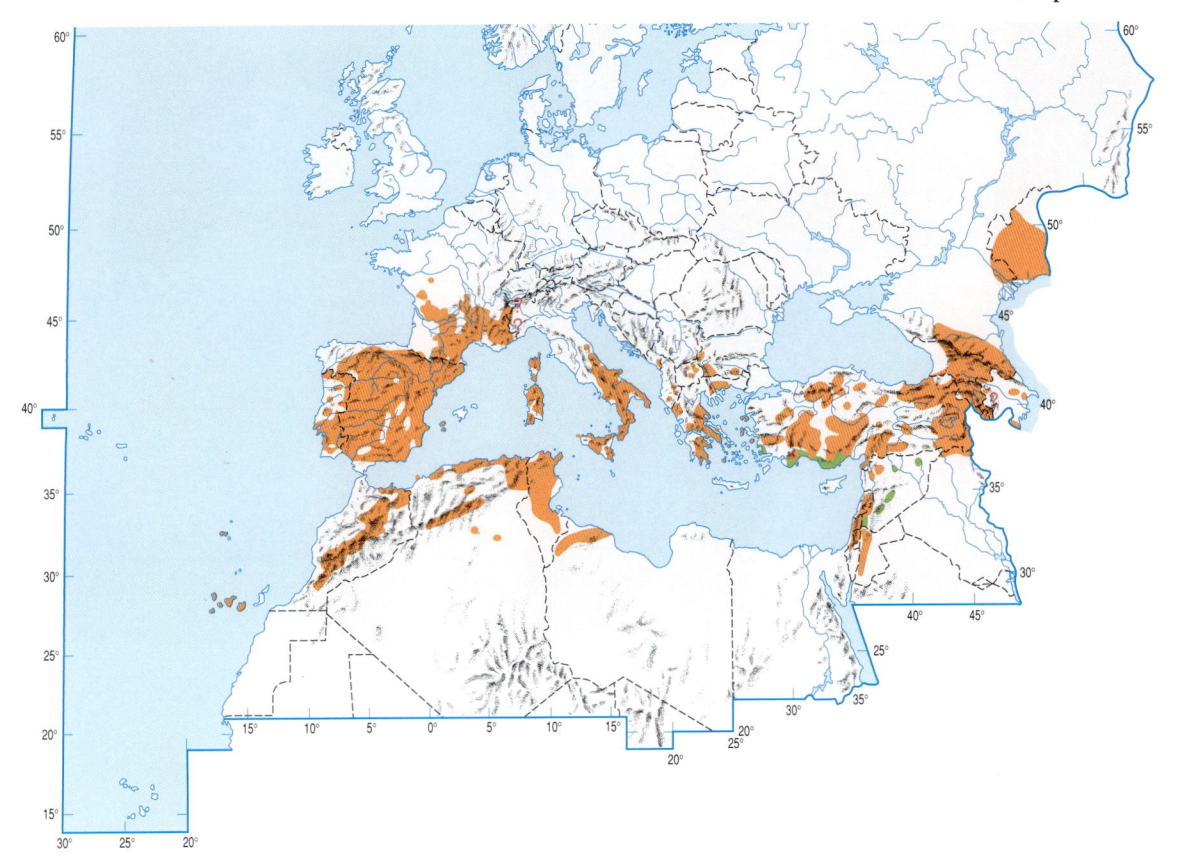

TUNISIA. Fairly common. MOROCCO. Uncommon to common. MADEIRA. Common, but much reduced since appearance of Spanish Sparrow.

Movements. Resident and to some extent dispersive; also altitudinal migrant. Resident in western Europe and northern Africa, with limited dispersal to cultivated areas in winter. Resident on Atlantic Islands, though on Madeira makes local visits to outlying islands in winter. In Turkey, distribution extends to lower altitudes in winter, birds occurring more widely in west, with fewer records from eastern breeding areas. In Israel, some populations wholly resident, others are altitudinal migrants.

Food. Mostly seeds throughout year, with berries in autumn, and in spring invertebrates, on which young almost exclusively fed, especially caterpillars and grasshoppers. Forages mainly on ground, running around like pipit, in low herbs and grass, among rocks, in fields of cereal or stubble, etc.; commonly in large flocks, particularly in winter, often with other species, especially finches.

Social pattern and behaviour. Typically in compact flocks, of up to many hundred, from late summer to spring. Mating system apparently regularly polygynous, ♂ advertising for and acquiring 2nd, sometimes even 3rd, mate after 1st has begun to nest. Breeds usually in small, loose colonies, less often in dense colonies. Successful polygyny dependent on ♂ acquiring new nest-hole(s); nest-building by ♀ alone.

Voice. More vocal than *Passer* sparrows, with large and varied repertoire. Main call a characteristic, oft-repeated, nasal, rather piercing 'pey-i' or 'peeyuee'; given from perch or in flight. Some calls clearly 2–3 syllables, others almost monosyllabic. Given singly or (especially in breeding season) as series, rapidly when excited, and serving to advertise territory or nest. Other calls include 'chi' or soft 'düj', given in flight, often in rapid sequence; and rapid rattling or chattering sound similar to *Passer*, given when disturbed or in threat.

Breeding. SEASON. France: end of April to August, peak June–July when 2nd clutches generally laid. Spain: eggs laid late April to early June. North Africa: 1st clutches laid mid-April to early May, 2nd probably late June. Canary Islands: eggs laid mostly in May, but eggs recorded late March and nestlings early August. Israel: eggs laid end of March to mid-June. SITE. Hole or cavity in rocks, earth bank, tree, building or other structure, disused well, etc.; commonly in old, or sometimes usurped hole of other species, particularly bee-eaters but also nuthatches, swallows and martins, or woodpeckers; also in old rodent burrow. Nest: largish untidy structure, very like that of House Sparrow; sometimes domed. Foundation principally of grass or straw, occasionally reduced to small pad, lined with feathers, hair, wool, string, cloth, paper, stalks of herbs, rootlets, etc. EGGS. Sub-elliptical, smooth and glossy; white to brownish-white, with grey or reddish- to blackish-brown speckles and blotches, concentrated at broad end. Clutch: 4–7(–8). Incubation: 11–14 days. FLEDGING PERIOD. 16–21 days.

Wing-length: *P. p. petronia*: ♂ 93–102, ♀ 89–100 mm.

Tree Sparrow *Passer montanus montanus* (p. 1513): **1** ad. Sudan Golden Sparrow *Passer luteus* (p. 1515): **2** ad ♂, **3** ad ♀. Pale Rock Sparrow *Carpospiza brachydactyla* (p. 1517): **4** ad winter. Yellow-throated Sparrow *Petronia xanthocollis transfuga* (p. 1518): **5** ad ♂. Rock Sparrow *Petronia petronia petronia*: **6–7** ad ♂.

Weight: *P. p. petronia*: ♂ ♀ 26–35 g.

Geographical variation. Rather slight; involves size, bill length, and colour (mainly tone of ground-colour, and colour, width, and contrast of dark streaks). Bleaching and abrasion have marked influence, freshly moulted birds tinged tawny or buff, turning to grey after a few months. Difference in colour between races often discernible only when series of specimens with same degree of abrasion compared. *P. p. petronia* (Europe and Atlantic islands) a dark race, ground-colour of head and upperparts greyish-brown when plumage fresh, dull cold grey when worn; dark streaks on cap, lower mantle, and scapulars distinct, black or black-brown. *P. p. barbara* from Algeria, Tunisia, and Libya paler and greyer than *P. p. petronia*; dark stripes on cap dark sepia-brown, less blackish; dark streaks on lower mantle and scapulars slightly narrower, shorter, and more restricted in extent. *P. p. puteicola* from Levant a well-marked race, large with giant bill; plumage distinctly paler than in previous races, ground-colour of upperparts grey-brown with sandy-buff tinge, less grey than in *barbara*; stripes on cap and streaks on lower mantle and scapulars greyish-olive-brown, contrasting only slightly with remainder of upperparts; underparts extensively whitish; throat spot paler yellow than in previous races.

Snow Finch *Montifringilla nivalis*

PLATE: page 1523

Du. Sneeuwvink   Fr. Niverolle alpine   Ge. Schneefink   It. Fringuello alpino
Ru. Снежный вьюрок   Sp. Gorrión alpino   Sw. Snöfink

Field characters. 17 cm; wing-span 34–38 cm. Size close to Snow Bunting but recalls particularly sparrow *Passer* in stronger bill, more angular head, slightly longer tail, and somewhat shorter wings. Quite large, montane passerine, mixing sparrow-like structure and behaviour with highly variegated plumage pattern much like Snow Bunting except for grey head and (in breeding season) solid black bib. Juvenile shows browner head and lacks visible bib.

(FACING PAGE) Snow Finch *Montifringilla nivalis*. *M. n. nivalis*: **1–3** ad ♂ summer, **4** ad ♂ winter, **5–6** ad ♀, **7** juv. *M. n. alpicola*: **8** ad ♂ summer.
Snow Bunting *Plectrophenax nivalis* (p. 1642). *P. n. nivalis*: **9–10** ad ♂ summer, **11–12** ad ♂ winter, **13** ad ♀ summer, **14–15** ad ♂ winter, **16–17** 1st winter ♂, **18** juv. *P. n. insulae*: **19** ad ♂ summer, **20–21** ad ♂ winter, **22** ad ♀ winter. *P. n. vlasowae*: **23** ad ♂ summer, **24** ad ♂ winter.

1
2
3
5
6
8
4
7
9
11
12
13
10
17
15
16
14
18
20
23
19
24
21
22
Ian Lewington '93

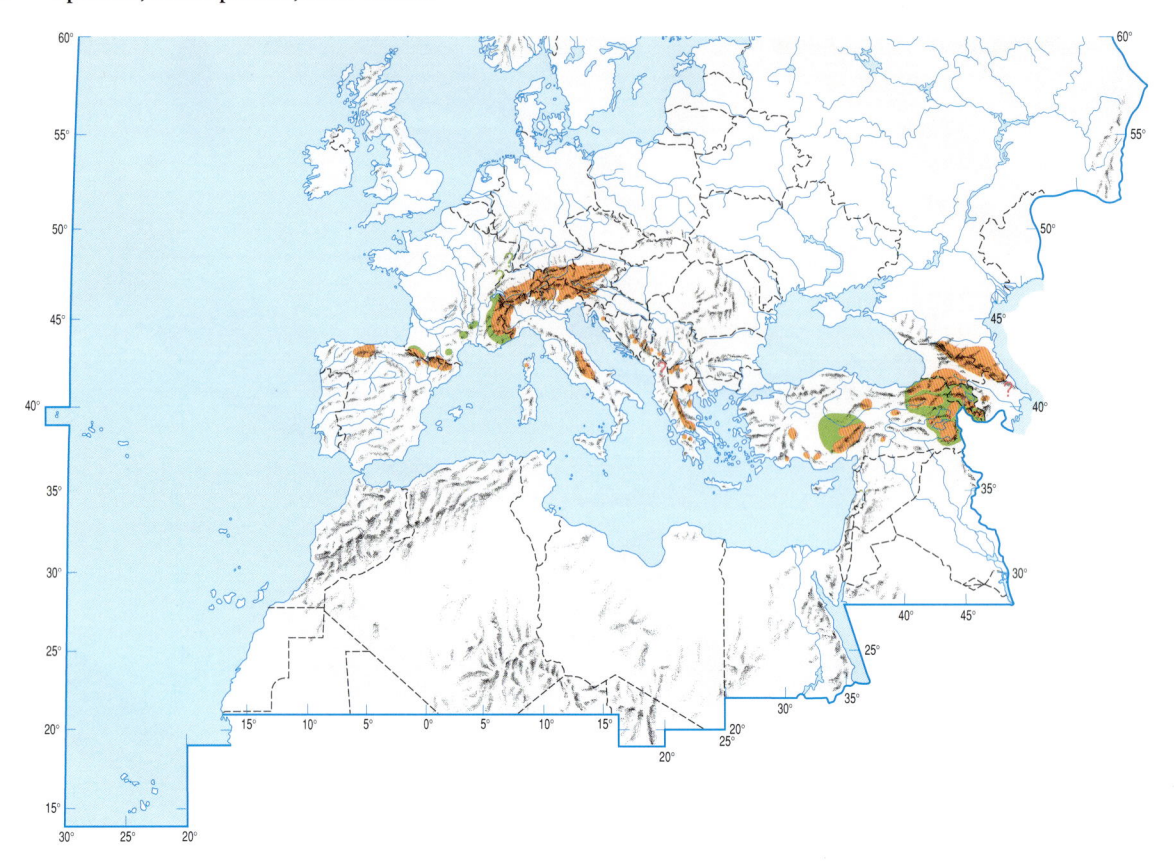

Comparisons with Snow Bunting mainly necessary to prevent wishful identification as vagrant Snow Finch. Given close clear view of grey head and black bib of adult, distinction from Snow Bunting easy, but at longer range both adult and much browner and duller-winged juvenile may require longer inspection. Flight fast; pied wing pattern produces apparently flickering action as in Snow Bunting but wing-beats actually rather stiffer, while wings lack loose attitudes so typical of Snow Bunting. General behaviour on ground recalls both finch and sparrow but does not perch on trees; often stands up, flicking tail upwards.

Habitat. Breeds in alpine and subalpine elevations of temperate and warm temperate zones across middle latitudes of Palearctic. Breeds in Abruzzi mountains (Italy) to above 2300 m, and in Swiss Alps only exceptionally below 1900 m or above 3000 m. In Caucasus, normally at 2750–3160 m, but in Tibet up to 5300 m. In winter commonly remains at similar altitudes, and only occasionally below 1000–1500 m. In Switzerland, inhabits grassy patches above treeline (avoiding even dwarf woody growth), scree slopes, and boulder patches on ridges and peaks or in passes; habitat usually includes suitable choice of crevices either in some kind of rock face (often side by side with Swift or Alpine Swift) or commonly on buildings such as cowsheds, structures protecting against avalanches, timber stacks, and even busy hotels, whose refuse may contribute significantly to diet. In Caucasus, where such artefacts less common at appropriate elevations, habitat consists more largely of boulders, rocks, precipices, screes, and heaps of stones, alternating with patches of alpine meadow watered by rivulets and streams.

Ecologically may be regarded as an alpine counterpart of House Sparrow and has taken similar advantage of man's now widespread exploitation of mountain habitats on which it formerly supported itself unaided in respect both of food and shelter.

Distribution. FRANCE. Corsica: small population discovered 1980. GERMANY. Restricted to southern Bayern. AUSTRIA. Widely distributed. ALBANIA. Perhaps breeds in high mountains in east. BULGARIA. May have bred Pirin mountains 1962, and central Rila mountains 1964, but records insufficiently documented. TURKEY. Widely distributed; perhaps less common in west of range.

Accidental. Luxembourg, Poland, Slovakia, Hungary, Portugal, Malta, Croatia, Rumania, Canary Islands.

Beyond west Palearctic, breeds very locally from Iran through central Asian mountains to south-east Altai (Russia) and Tibet.

Population. Apparently stable; no changes reported. FRANCE. 100–1000 pairs in 1970s. GERMANY. *c.* 200 pairs in mid-1980s. AUSTRIA. No apparent changes; perhaps slight increase due to tourism. SWITZERLAND. 2000–5000 pairs 1985–93. SPAIN. 5000–10 000 pairs. ITALY. 3000–6000 pairs 1983–95. GREECE. 1000–2000 pairs. YUGOSLAVIA: SLOVENIA. 100–200 pairs. AZERBAIJAN. Rare. TURKEY. 10 000–100 000 pairs.

Movements. Chiefly resident; some birds make altitudinal movements, especially in east of range; in west of range, some birds short-distance migrants. Most records of longer movements are from France, between north-west and south of Alpine breeding area, or north-east from Pyrénées. To west and north-west of Alps, reported from Jura and probably regular as far north as Vosges mountains. At southern edge of Alps in Haute-Provence (outside breeding area), winters regularly on Mont Ventoux and some other mountains. Further west, Cévennes perhaps also a regular wintering area.

Food. Invertebrates and seeds; in spring and summer insects and spiders, in winter almost wholly seeds, either in wild or provided by man along with scraps. Feeds on ground; in autumn, also on seed-heads of tall herbs, like *Carduelis* finch; often at hotels, tourist resorts (etc.) in winter and sometimes in summer.

Social pattern and behaviour. Typically gregarious outside breeding season, families joining together in flocks of up to 150 or more from July; some post-breeding flocks of juveniles only. Mating system monogamous. Solitary when breeding or in small compact or more dispersed neighbourhood groups. Territory established and defended (only during pair-formation, nest-building, and copulation) by ♂. ♂ sings for territorial advertisement, either perched or in undulating and gliding song-flight.

Voice. Song of ♂ consists of phrases of varying length: sometimes only one or a few motifs, or several units and motifs linked together in more continuous series of mainly chirping sounds of varying timbre like sparrow *Passer*; often preceded by 'szi' or 'pink', and these as well as other call-types are incorporated within song. Main call a highly variable impure or hoarse 'szi', 'tseeh', or 'zjih'; given singly or in loose series, in flight (solitary birds or flock) or on ground, and in wide variety of contexts: for contact, when disturbed near or far from nest, during antagonistic interactions at nest, and in encounters on feeding grounds. Alarm-call a 'pititit prrt' recalling (especially when preceded by high-pitched 'tsi') Crested Tit.

Breeding. SEASON. Egg-laying in 2nd half of May in Alps, mid-June in central Caucasus. Usually 2 broods. SITE. Crevice or cavity in rock face, boulder scree, earth bank, etc., in darkness at end of entrance passage or fully exposed; on building (even in centre of settlement), pylon, etc., or inside roof, also in nest-box or old mammal burrow. Nest: bulky but fairly neat construction with foundation of dry grass with roots, plant stalks, moss, and lichen, lined with fine plant material, feathers, hair, wool, etc.; very dry materials preferred, to eliminate dampness. EGGS. Sub-elliptical, smooth and matt or very faintly glossy; white. Clutch: 4–5 (3–6). INCUBATION. 13–14 days. FLEDGING PERIOD. 20–21 days.

Wing-length: Nominate *nivalis*: ♂ 117–127, ♀ 113–122 mm. *M. n. leucura* (southern and eastern Turkey): ♂ 111–119, ♀ 106–116 mm.

Weight: Nominate *nivalis*: ♂♀ average *c.* 35–45 g, seasonally variable.

Geographical variation. *M. n. nivalis* occurs throughout European range. In *M. n. leucura* from Asia Minor, wing shorter and bill longer; top of head and neck brown-grey, hardly contrasting with greyish-brown mantle and scapulars, which are lighter than in nominate *nivalis*, underparts whiter; lesser coverts sometimes black on basal half, outermost secondary often with some black visible on base of outer web. *M. n. alpicola* from Caucasus similar to *leucura* but longer winged.

Weavers and Allies Family Ploceidae

Small to medium-sized, often thick-billed oscine passerines. Found in great variety of habitats from semi-arid country (e.g. bush and savanna) to woodland, forest, and cultivation. Mostly feed above ground, seeds forming bulk of diet in many species, but other vegetable food and insects also taken (some species being partly or wholly insectivorous) and even nectar in a few cases. Some species are serious pests of cereal crops, most notably Red-billed Quelea. Of mainly Afrotropical distribution. Mainly sedentary, with some minor local or seasonal movement, but some species migratory. About 124 species; 1 introduced species in west Palearctic, Streaked Weaver.

Many species gregarious, feeding and roosting in flocks, sometimes (e.g. in Red-billed Quelea) of enormous size. Monogamous mating system in minority of solitary, territorial species (mainly insectivores of forest, savanna, and secondary bush), but majority polygynous and nest colonially to greater or lesser extent. Nests (of *Ploceus* especially) of remarkable complexity. In most species, typically a woven domed structure suspended from bush or tree and with elongated, downward pointing entrance tube; built by both sexes in monogamous species, by ♂ alone in polygamous ones.

Village Weaver *Ploceus cucullatus*

Du. Grote Textorwever   Fr. Tisserin gendarme   Ge. Dorfweber   It. Gendarme
Ru. Большой масковый ткач   Sp. Tejedor cabecinegro   Sw. Byvävare

Widespread resident in almost the whole of sub-Saharan Africa. On Cape Verde Islands, 2 records (of presumed introduced birds) in 20th century. In May 1924, 7 birds collected on Santiago. In June–July 1993, under 10 seen displaying and nest-building at Mindelo, São Vicente; probably bred, not confirmed. Recently reported breeding Algarve (Portugal).

Streaked Weaver *Ploceus manyar*

PLATE: page 1527

Du. Manyarwever   Fr. Tisserin manyar   Ge. Manyarweber   It. Tessitore striato
Ru. Маньярский ткач   Sp. Tejedor listado   Sw. Guldkronad rörvävare

Field characters. 15 cm; wing-span 20–22 cm. Size close to House Sparrow but with proportionately even heavier bill and head and shorter wings and tail. Suggests large-billed, chubby sparrow *Passer*. Adult ♂ has distinctive breeding plumage:

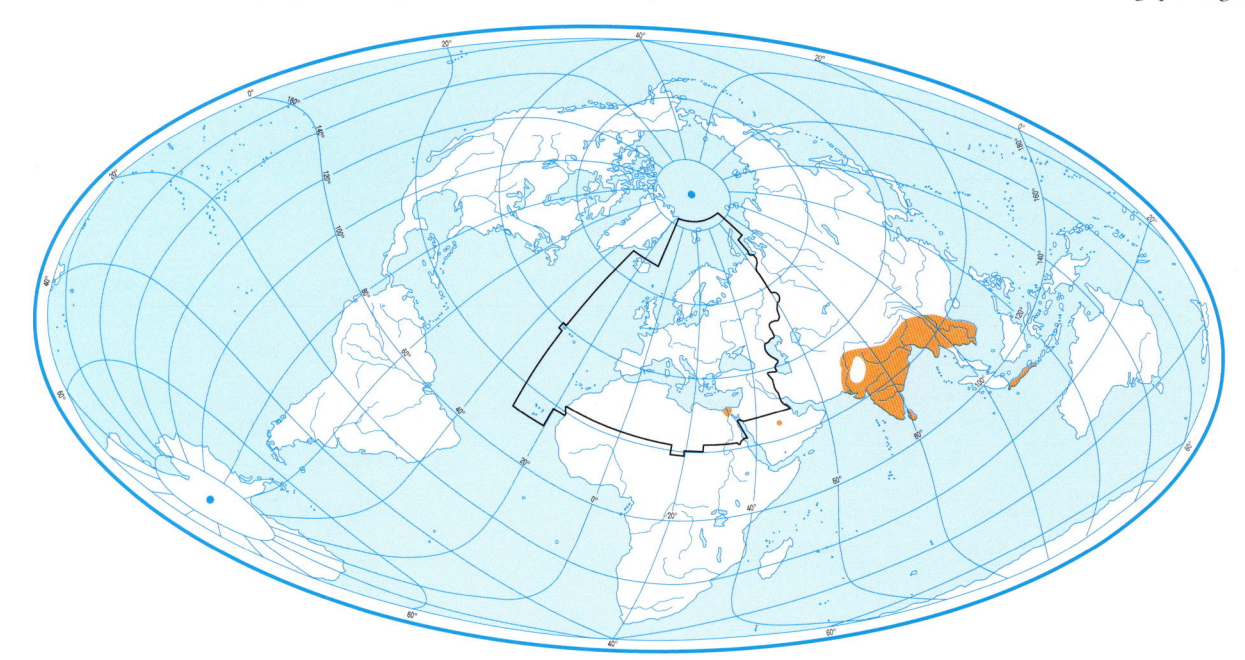

Streaked Weaver *Ploceus manyar*: **1–2** ad ♂ breeding, **3** ad ♂ non-breeding, **4** ad ♀ breeding, **5** juv. Red-billed Firefinch *Lagonosticta senegala* (p. 1529): **6–7** ad ♂, **8–9** ad ♀, **10** juv.

black head with bright yellow crown, conspicuously streaked back, breast, and flanks, blackish wings, and cream underparts. Non-breeding ♂ and ♀ show long, broad, decurved yellow supercilium and border to ear-coverts.

Adult ♂ in breeding plumage unmistakable but non-breeding ♂, ♀, and immature can be confused with similarly-sized *Ploceus* of Africa, which like Streaked Weaver may occur as escapes in west Palearctic. Flight, general behaviour, and colonial breeding recall sparrows. Like other *Ploceus*, looks front-heavy on perch or ground and in flight. Flock chorus recalls House Sparrow.

Habitat. In Nile delta, breeds in reeds up to 3 m high growing in shallow water, in one case in a band *c.* 1 km wide; habitat shared with Clamorous Reed Warbler and Fan-tailed Warbler. In natural range in India and Burma inhabits flat, swampy, and flooded land and riverbeds, especially reedmace and reeds standing in water, but is also found in tall grass.

Distribution. EGYPT. Only breeding area in west Palearctic. Introduced; fairly common in parts of Nile delta. Probably originated from birds escaped from Alexandria Zoo in 1971. First recorded breeding 1978; has spread rapidly across Nile delta and up the Nile; breeding confirmed Lake Manzala area 1990.

Movements. Sedentary. Moves about in flocks outside breeding season, though no information for introduced Egyptian population.

Food. Almost wholly granivorous; young fed mainly insects. As well as foraging on ground, often feeds on flower- and seed-heads of grasses, etc., clinging to stems or heads.

Voice. Song an attractive series of *c.* 6 high-pitched 'tsi' or similar short whistling notes, culminating in long, drawn-out wheeze. Commonest call 'chack', with stony quality suggesting Wheatear.

Breeding. SEASON. Egypt: eggs at end of 1st week of May, though nests still being built mid-May. India and Pakistan: breeding peaks during monsoon, June–September. SITE. In Egypt, in stands of reed in shallow water, nests attached to tops of 1–2 stems 2–3 m above surface. Nest: coarsely-woven retort-shaped brood pouch with short, downward-pointing entrance tube, made of long strips of grass or leaves of various widths; brood pouch often has clumps of mud and dung plastered on outside, and sometimes inside, commonly with embedded flower petals. EGGS. Sub-elliptical, white, smooth, and not glossy. Clutch: 2–3 (1–4). INCUBATION. Average 13.2 days. FLEDGING PERIOD. 15–20 days.

Wing-length: ♂ 67–73, ♀ 65–70 mm.
Weight: 2♂♂ (Thailand) 16.4, 18.4 g.

Red-billed Quelea *Quelea quelea*

Du. Roodbekwever   Fr. Travailleur à bec rouge   Ge. Blutschnabelweber   It. Quelea
Ru. Красноклювый ткач   Sp. Quelea común   Sw. Blodnäbbsvävare

Widespread and locally migratory in Afrotropics, often forming immense flocks and breeding in huge colonies. Flock of several hundred recorded on Tenerife (Canary Islands) 23 November 1965, and several dozen still present on 29 November. Regarded as vagrants, but birds poorly described, and escape possibility not considered.

Waxbills, Grassfinches, Mannikins Family Estrildidae

Tiny to small, often thick-billed oscine passerines; typically birds of open country—grassy savanna and thorn scrub—but also found in forest, reedbeds, etc. Feed chiefly on ground or in low vegetation, mostly eating grass seeds but insects also taken, and fruit by some. Of mainly tropical African and Asiatic distribution, with by far largest diversity in Afrotropics, extending to Australasia, where a further but less diverse adaptive radiation has occurred, and to many islands of Pacific Ocean. Some species introduced to areas outside natural ranges (including New World); several also long established as popular cagebirds. Largely sedentary though some dry-country species make local movements or are nomadic. c. 140 species; 7 recorded in west Palearctic, at least 3 now with established breeding populations.

Sexes generally of similar size. Bill often of typical seed-eater type, deep, short, and well adapted for de-husking seeds. Wing short and broad with rounded or bluntly pointed tip. Plumage colours variable: many waxbills are brown with contrasting red or blue rump, face, or underparts; grassfinches and allies are quite bright and contrastingly coloured green or brown above with red or black rump or tail, with red, black, or blue marks on face, and with white spots, streaks, or bands, or black bars, on underparts; most mannikins and allies more sombre, with contrasting patches of brown, black, and white. Bill often brightly coloured.

Voice often harsh and discordant. Song loud and musical in some species but mostly rather quiet and intimate without aggressive or territorial significance. Many species, especially those of open country, often highly gregarious outside breeding season, and some loosely colonial species continue to flock for feeding (etc.) even when nesting. Most, however, breed solitarily. Monogamous mating system the general rule, with strong pair-bond, often maintained throughout year. In most species, mates (and, in some, flock members too) clump together in close body contact when loafing and roosting.

Red-billed Firefinch *Lagonosticta senegala*

PLATE: page 1527

Du. Vuurvinkje   Fr. Amarante du Sénégal   Ge. Senegalamarant   It. Amaranto beccorosso
Ru. Обыкновенный амарант   Sp. Bengalí senegalés   Sw. Amarant

Field characters. 9 cm; wing-span 15–16 cm. Slightly smaller and dumpier than Common Waxbill, with proportionately shorter tail. Diminutive and nervous. ♂ fiery plum-red, with browner back and wings, black tail, and patch of tiny white spots or bars by shoulder; ♀ dull brown above, paler below, with contrasting red lore and rump and black tail. Bill pinkish-red, with black culmen emphasizing sharply pointed shape.

In Afrotropics, subject to confusion with at least 10 other Estrildidae; within west Palearctic, may be confused with escapes of imported congeners and Red Avadavat (see that species). Flight light and rapid, with fast whirring wing-beats suggesting tiny sparrow and allowing sudden manoeuvrability. Hops, but given bird's small size can seem like creeping. Forms small, lively, roving flocks.

Habitat. In Tamanrasset (southern Algeria), occurs near houses, feeding in streets, gardens, and orchards. In natural range in Afrotropics, inhabits dry areas with abundant *Acacia*, scrub, or other cover. Has largely colonized cultivated areas, villages, and old towns; avoids close-built modern towns where suitable food gathered on ground is lacking. Readily becomes tame, entering huts and houses.

Distribution. ALGERIA. Introduced at Tamanrasset c. 1940. A further colony since established c. 40 km to north-east in atypical rocky habitat. Has occurred, at least since 1972, at El Golea. At In Guezzam, just south of west Palearctic border, bred in 1954 but disappeared by 1960. Accidental. Morocco: one record involving 2 birds, Moyen Atlas, May 1964.

Movements. Chiefly sedentary, but local movements reported from some areas.

Food. Small grass seeds, mainly from ground, and some green plant material; some small insects. Forages in pairs or flocks on open ground and in low vegetation, rarely far from cover; often close to human habitation, where sometimes very tame. Picks most food from ground, or from low-growing seed-heads, though occasionally perches close to elevated seed-head; sometimes seizes and shakes stems or seed-heads to dislodge seeds. Small fragments of green vegetation taken by gripping leaf with bill and making backward jump.

Social pattern and behaviour. Less social than many members of family. Usually in pairs or small parties (typically up to 5 birds) when not breeding, but at any time will congregate at localized food source. Solitary and territorial when breeding. Territory limited to nest and its immediate surroundings. Most striking element in ♂'s courtship is feather-display; holds feather by tip of rachis and circles round ♀. Sight of detached feather exerts strong stimulus on ♂ from independence onwards; adult ♂ often tries to steal feather held by juvenile, and if feather drops from bill of one bird it is immediately retrieved by another before it reaches ground.

Voice. Song a short, rather feeble twitter. Commonest call a weak, piping 'tweet tweet' or 'teep teep', often uttered almost incessantly by flocks.

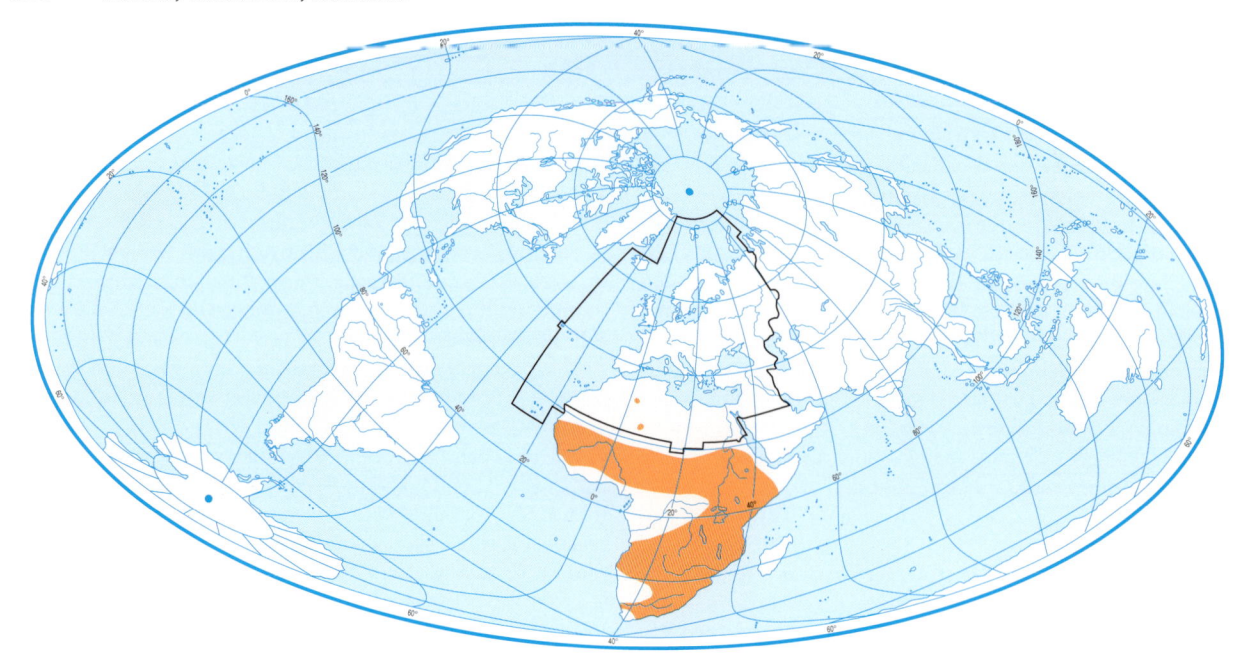

Breeding. SEASON. Ahaggar mountains (southern Algeria): eggs recorded mid- to late April and late August. Sénégal: main breeding season August–March (July–May) after rains, during which time may raise 5 broods. SITE. Well concealed, shaded cavity, depression, or platform among thick vegetation, on or in building, or on ground. Nests in vegetation placed on forked branches in leafy hedges and bushes, in leaf bases of bananas and palms, among tangled roots or in piles of brushwood; those in buildings are in walls, loose thatch, between or among stacked objects, etc.; those on ground are in thick vegetation, recess such as hoofprint, or hole in bank. Nest: well-camouflaged structure, mainly of grass stems c. 20 cm long, also straw, leaves, paper, etc., thickly lined with feathers and other fine material; depending on site, ranges from closed, domed structure to (less often) open cup; all types rather loosely constructed. EGGS. Sub-elliptical or oval, with little gloss; white. Clutch: 3–4 (1–6). INCUBATION. 11–14 days. FLEDGING PERIOD. 17–19 (14–20) days.

Wing-length: ♂♀ 48–51 mm.
Weight: ♂♀ 6–11 g.

Red-cheeked Cordon-bleu *Uraeginthus bengalus*

DU. Blauwfazantje   FR. Cordonbleu à joues rouges   GE. Schmetterlingsastrild   IT. Cordon blu guancerosse
RU. Красноухий астрильд   SP. Coliazul oidorrojo   SW. Rödkindad fjärilsfink

Breeds in semi-arid habitats in sub-Saharan Africa from Sénégal east to Ethiopia, south to Zambia and Tanzania; mostly resident. In Egypt, birds of captive origin occurred in El Maadi area near Cairo in mid-1960s, but no records since, and presumed not to have become established. Collected on São Vicente (Cape Verde Islands) 1924 (♂ on 20 January, ♀ on 13 October); not known whether escapes or introduced deliberately.

Orange-cheeked Waxbill *Estrilda melpoda*

PLATE: page 1531

FR. Astrild à joues orange   GE. Orangebäckchen
RU. Оранжевощекий астрильд

Breeds in sub-Saharan Africa from Sénégambia east to northern Chad, south and south-east to northern Angola, eastern Zaïre, and northern Zambia. Mostly resident, with local movements. Birds of captive origin breed in several localities in eastern Spain (Castellón province); first regular sightings 1990. Numbers and range have also increased, but uncertain whether fully established. Recent breeding also reported France and Algarve (Portugal).

Orange-cheeked Waxbill *Estrilda melpoda*: **1** ad, **2** juv. Lark Sparrow *Chondestes grammacus* (p. 1635): **3** ad, **4** juv.

Common Waxbill *Estrilda astrild*

PLATE: page 1532

Du. Sint Helenafazantje   Fr. Astrild ondulé   Ge. Wellenastrild   It. Astrilde
Ru. Волнистый астрильд   Sp. Estrilda común   Sw. Helenaastrild

Field characters. 11 cm (tail 4 cm); wing-span 12–14 cm. Length similar to Serin but structure distinctly different from finch: rather small head, short rounded wings, rather slim body, and long graduated tail combine to give slight but lengthy form. Epitome of essentially African genus. Grey-brown above with rufous rump, pale brown to rose below with almost black vent and tail; red bill and red face-patch.

Beware risk of escape of 2 common cagebirds: Black-rumped Waxbill *E. troglodytes* (with distinctive black upper tail-coverts) and Crimson-rumped Waxbill *E. rhodopyga* (black bill and red fringes to wing-coverts, tertials, and tail); both recently reported breeding Algarve (Portugal); hybrids also occur. Flight remarkably light, rapid, and direct, with fast whirring wing-beats; has sudden take-off and equally sudden landing. Hops, but often so close to ground that bird appears to shuffle or creep forward; climbs stems. Strongly gregarious, flock members keeping remarkably close together when feeding and in flight.

Habitat. Population on banks of Guadiana (Extremadura, Spain) occupies reedmace, reed, and *Arundo* growth, nesting in willow; others in tamarisk. In Portugal, recent spread linked to intensive use of stands of reed and *Arundo* for feeding and roosting. In Afrotropics, inhabits open country with long grass, marshes, marginal aquatic vegetation such as reedbeds, active and abandoned cultivation, grassy clearings and paths in forest or woodland, gardens, and surroundings of farms and dwellings where cover and seeds from tall plants are available.

Distribution and population. Breeding populations in several areas in south all certainly or probably originating from escapes or deliberate introductions. SPAIN. Established in several places; breeding from 1977. PORTUGAL. Established in several places since 1967; increase and spread continuing. Probably several thousand birds. AZORES. Flock of *c.* 100 birds present 1983–4; bred 1984, but subsequently disappeared. CAPE VERDE ISLANDS. Formerly reported from 6 islands, but now survives only Santiago, where widespread, locally abundant.

Beyond west Palearctic, breeds in much of Africa north to Cameroon and southern Sudan, also isolated population in Sierra Leone and Liberia. Introduced to Brazil and many islands.

Movements. Chiefly sedentary; local movements in drier parts of range. Introduced population in Iberia sedentary, with evidence of local movements; on Atlantic islands, apparently sedentary.

Food. Grass seeds, taken mainly from flower-heads; rarely some small insects. Forages generally in pairs or flocks, in low herbaceous vegetation, reeds, tall grasses, and on agricultural

Common Waxbill *Estrilda astrild*: **1–2** ad ♂, **3** ad ♀, **4** juv. Red Avadavat *Amandava amandava* (p. 1533): **5** ad ♂, **6–7** ad ♀, **8** juv. Indian Silverbill *Euodice malabarica* (p. 1534): **9** ad. African Silverbill *Euodice cantans* (p. 1534): **10–11** ad, **12** juv.

land; usually in grass-tops, on fallen grass flower stalks, or on ground.

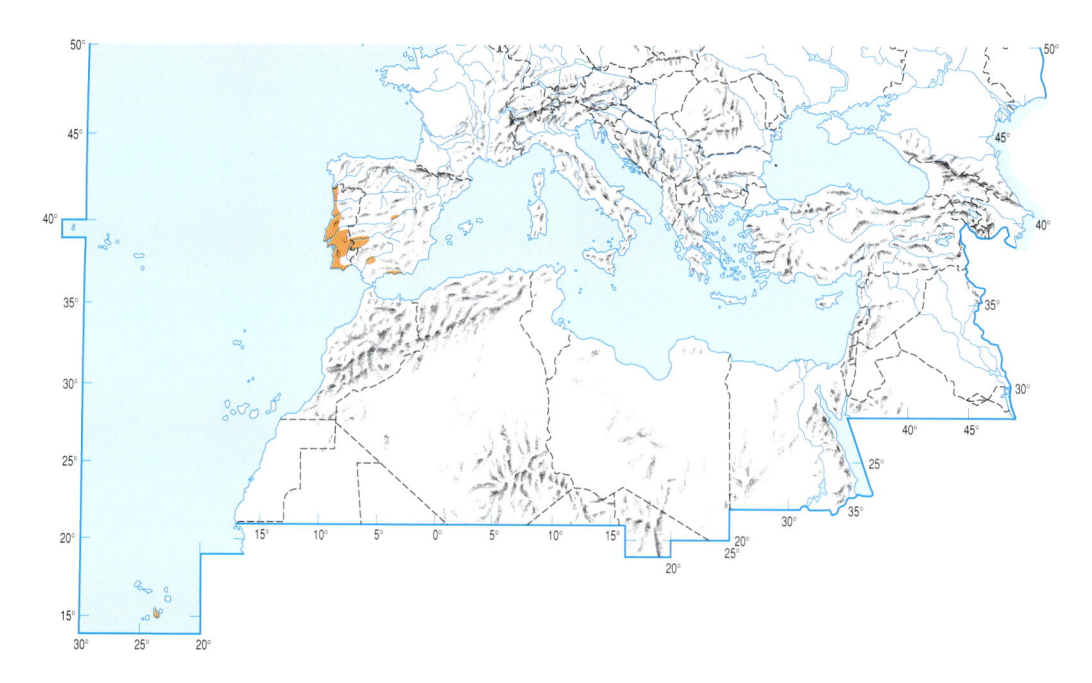

Social pattern and behaviour. Gregarious outside breeding season, usually in small flocks. In Portugal, flocks of up to 300 or more, but more often 5–50; in south-west Spain, flocks of up to 40. Solitary and territorial when breeding; territory limited to nest and its immediate surroundings. Though colonial breeding not known, several pairs may nest in relatively small areas of good food supply. During pair-formation and courtship, ♂ holds feather or piece of nest-material in bill, adopts upright posture (bill pointing upwards) with ruffled plumage on ventral side and flanks, often tilting himself away from ♀ to expose underside colours; tail simultaneously twisted towards ♀. In this posture, ♂ jerks stiffly up and down, but without jumping off perch, and starts singing loudly after a few upward movements.

Voice. Song 'tcher-tcher-preee' ('pree' a bubbling sound), rising in pitch. Calls include short but slightly buzzing monosyllable, 'tzep', also 'chip' and 'pit', twittering from flock.

Breeding. SEASON. Cape Verde Islands: nest-building starts early August, after first rains; first eggs laid late August, with subsequent nest-building from late September for 2nd brood. Portugal and Spain: breeds February–November. SITE. Cavity in thick vegetation; often low down among grass clumps or in bush, shrub, tree, or creeper. Nests in grass often rest on ground, or less than 50 cm up. Nest: large, untidy, inverted-pear-shaped, domed structure of grass stems, with downward-pointing entrance tube (occasionally absent) from side; often a 'false-nest' on dome; main structure of fresh grass stems, with or without flower-heads, criss-crossed, not woven together, to form hollow globe; not woven into surrounding vegetation. EGGS. Sub-elliptical, smooth and non-glossy; white, with faint pink flush when first laid. Clutch: 4–6 (3–9). INCUBATION. 11–12 days. FLEDGING PERIOD. 17–21 days.

Wing-length: ♂ 46–49, ♀ 44–48 mm.
Weight: ♂♀ (South Africa) 7.0–10.0 g.

Red Avadavat *Amandava amandava*

PLATE: page 1532

DU. Tijgervinkje FR. Bengali rouge GE. Tigerfink IT. Bengalino
RU. Тигровый астрильд SP. Bengalí rojo SW. Tigerfink

Field characters. 9.5 cm; wing-span 13–14.5 cm. Slightly longer than Red-billed Firefinch, but with similar form. Breeding ♂ red with brown wings, black tail, and variable white spots and speckling, most visible on side of breast, flank, tips of larger coverts, and rump (in fresh plumage); other plumages brown above and grey-buff and dusky-yellow below, with red rump (except juvenile) and black tail; juvenile shows 2 pale wing-bars.

Given their frequency as cagebirds, this and similar species could escape anywhere. ♂ Red Avadavat easily separated from Red-billed Firefinch by much more uniformly dark red appearance, black vent, and more extensive white spotting on breast, flanks, and wings. ♀ best distinguished from Red-billed Firefinch by somewhat darker, more olive-toned head and upperparts, blackish (not crimson) lore, and much darker, blackish-brown, faintly speckled wings.

Habitat. In Guadiana basin (Extremadura, Spain), introduced birds occupy wetlands dominated by reedmace and reed, as well as meadows and irrigated crops such as lucerne, maize, and tomatoes. In Granada (Spain), found in sugar-cane plantations and reedbeds. In Treviso (Italy), occupies reedbeds and stands of rush and sedge, often descending to feed on ground; often favours clay pits, especially those filled with water. In Nile valley (Egypt), lives mainly in reeds. In natural range in Asia, occurs in tall grass, reedbeds, bushes, and other rank growth, especially by lakes, rivers, and marshes; also sugar-cane, grassy clearings in jungle or open woodland, and cultivated areas and gardens.

Distribution and population. Breeding populations established in Spain, Italy, Egypt, and Portugal, originating from escaped cagebirds. SPAIN. Seen near Madrid from 1974, apparently breeding. In Extremadura seen since 1978, increasing year by year; major concentration along 110 km of river Guadiana, from south of Badajoz to village of Villanueva de la Serena. ITALY. 100–500 pairs 1995. First records mid-1970s, with range expansion after mid-1980s. At Sile river (Veneto), estimated 80–90 pairs 1983–4, now few individuals; at Venice Lagoon (Veneto), tens of pairs; in Tuscany, over 300 pairs; also breeding Latium and Molise, and occasionally elsewhere. KUWAIT. Sight records 1996. EGYPT. First recorded 1861. Now locally common, and continuing to spread. Recorded south to Luxor.

Beyond west Palearctic, breeds from Pakistan east to southwest China, south to Lesser Sundas. A few breeding records Arabia. Introduced to Japan and various islands.

Movements. Chiefly sedentary, with local movements.

Food. Grass seeds, from ground and vegetation; some small insects. Forages generally in pairs or flocks on ground or amongst herbaceous vegetation, particularly tall grasses and reeds, often close to water.

Social pattern and behaviour. Gregarious outside breeding season. In Spain, family parties form small wandering flocks from late November which congregate for roosting. In Treviso (Italy), in flocks in late winter and spring. In Extremadura, breeding is not colonial and nests are well separated. ♂ sings from favoured perch within nest-territory, evidently to attract ♀ and not to defend territory. In courtship, ♂ displays to ♀ with straw or similar material (not feather) in bill.

Voice. Song a high-pitched, mainly descending, soft, liquid twitter. Calls include high-pitched chirps and squeaks, one recalling commonest call of Penduline Tit.

Breeding. SEASON. Guadiana basin (Spain): eggs laid from late July, mainly August; latest eggs late November, before onset of autumn rains. Treviso (Italy): main nesting period November–December. Northern Egypt: at Suez, nest-building recorded early April, fledged young seen May–June; at Lake Qarun, breeding recorded November. SITE. Well hidden in bush, which is often overgrown with grass or other vegetation, or in reeds, often near, sometimes over, water. Always less than 1 m above ground or water, sometimes touching ground. Nest: hollow sphere of grass, slightly flattened above and below, often resting on flat platform of grass or twigs; central side entrance; exterior usually coarsest and sometimes held together with spiders' webs; lining usually of finer material, often grass flower-heads, also plant down, feathers, etc. EGGS. Sub-elliptical to short sub-elliptical, without gloss; white, though may appear pinkish due to translucency of shell. Clutch: in Guadiana 4–7, Treviso 5–6. INCUBATION. 13–14 days. FLEDGING PERIOD. 17–21 days.

Wing-length: ♂ 47–51, ♀ 46–50 mm.
Weight: Spain: ♂ average 10.0, ♀ 9.8 g.

Indian Silverbill *Euodice malabarica*

PLATE: page 1532

DU. Loodbekje   FR. Capucin bec-de-plomb   GE. Indischer Silberschnabel   IT. Becco d'argento indiano
RU. Малабарская амадина   SP. Picoplata de Malabar   SW. Indisk silvernäbb

A resident Asian species, distributed from southern Arabia east through Indian subcontinent to central Bangladesh, south to Sri Lanka. In Israel, has bred since 1988–9, evidently originating from escapes; recorded from various eastern localities; at least 10 pairs bred at Eilat in 1990. Birds of captive origin also recorded Cyprus, Jordan and Kuwait.

African Silverbill *Euodice cantans*

PLATE: page 1532

DU. Zilverbekje   FR. Capucin bec d'argent   GE. Afrikanischer Silberschnabel   IT. Becco d'argento africano
RU. Серебряноклювая амадина   SP. Monjita pico-de-plata   SW. Afrikansk silvernäbb

Field characters. 11 cm; wing-span 15–16 cm. About 20% larger than Common Waxbill, with proportionately huge, swollen conical bill and narrower, more graduated tail. Medium-sized, strong-billed, beady-eyed waxbill, with sharply pointed tail. Head and upperparts dun-brown with fine barring and vermiculations, underparts dull white, with black rump and tail. Bill proportionately much larger than in *Estrilda*, filling face; pale blue-grey.

Combination of bill shape and colour, quite large size, and pale plumage with black rump and tail separate it from all other species except Indian Silverbill, not yet established in west Palearctic (diagnostic off-white rump). Black-rumped Waxbill *E. troglodytes* may occur as escape (smaller, with bill red in adult or brown in immature, crimson streak through eye in adult, and whitish fringes and round end to tail). Flight action rather looser than in *Estrilda*, with short undulations.

Song a high-pitched 'trill', with falling then rising phrases; whispering quality at distance. Fast 'cheet-cheet-cheet' in flight, recalling Linnet but more tinkling.

Habitat. Across Afrotropical and Arabian lowlands, locally up to *c.* 2000 m. Inhabits dry savanna, thornscrub, grassy areas with *Acacia*, cultivated and settled areas, and neighbourhood of water in semi-desert country.

Distribution. Resident in dry savanna and thornscrub in Afrotropics from Sénégal to Sudan, Ethiopia, and Somalia, south in the east to northern Tanzania, and in southern Arabia.

Accidental. Algeria: one collected Tamanrasset, May 1952; 2 seen Amsel, 22 km south of Tamanrasset, April 1970.

Movements. Chiefly sedentary, with local movements.

Wing-length: ♂ 52–58, ♀ 51–56 mm.

Vireos Family Vireonidae

Small to medium-sized oscine passerines. Mostly arboreal; feed on insects (mainly by gleaning in outer foliage), also taking fruit (berries) and seeds, especially in autumn and winter. Solitary, sluggish, and rather tame. Except as vagrants, occur only in New World—with more species in Central and South America than in North America; many of more northerly species migratory. 50 species in 3 subfamilies (2 confined to Central and South America). 29 species in genus *Vireo* (typical vireos, mainly North and Central America), with 3 species accidental in west Palearctic.

Sexes almost similar in size (♀ smaller on average). Bill usually rather short; quite thick and slightly hooked. Wing long and pointed in migratory species. Plumages usually olive or grey above, white or yellow below, sometimes with contrastingly coloured cap, crown-stripes, eye-stripe, and/or eye-ring; wing with bars or plain. Iris of adults red or white in some species. Sexes alike or nearly so.

Yellow-throated Vireo *Vireo flavifrons*

PLATE: page 1536

Du. Geelborstvireo   Fr. Viréo à gorge jaune   Ge. Gelbkehlvireo   It. Vireo fronte gialla
Ru. Желтозобый вирeон   Sp. Vireo de garganta amarilla   Sw. Gulstrupig vireo

Field characters. 12 cm; wing-span 22–24.5 cm. Close in size to Red-eyed Vireo but with heavier, blunter-tipped bill, somewhat shorter tail, and even stronger legs. Plumage highly decorated, with diagnostic combination of yellow spectacle, throat, and breast, and bold double white wing-bar; bright green upperparts, bluish-grey rump, upper tail-coverts, inner forewing; black wings and tail; and mainly white rear underbody.

Plumage pattern and colours not matched in any similar passerine occurring in west Palearctic (including other *Vireo*).

Habitat. Breeds in temperate Nearctic lowland forests and kindred environments, feeding mainly in leafy crowns of tall mature deciduous trees, especially along woodland borders, streams, and roads, and in orchards, avoiding conifers and dense secondary growth. Has become attached to modern

Yellow-throated Vireo *Vireo flavifrons*: **1** ad. Philadelphia Vireo *Vireo philadelphicus* (p. 1537): **2** ad dark, **3** ad light. Red-eyed Vireo *Vireo olivaceus* (p. 1537): **4** ad autumn, **5** 1st winter.

residential areas, and has grown remarkably trustful of human presence.

Distribution. Breeds in eastern half of North America from southern Manitoba and southern Quebec south to Gulf of Mexico coast and northern Florida.

Accidental. Britain: Cornwall, September 1990.

Movements. Migratory. Winters in West Indies and from southern Mexico to extreme north of South America (east to Guianas). Autumn migration begins (chiefly late) August, with few remaining in northern states after end of September. First records mid-September in Central America, with main arrival in October, and present in South America only from November. Spring migration begins March, with last reports in Central America in late April.

Wing-length: ♂ 76–82, ♀ 73–79 mm.

Philadelphia Vireo *Vireo philadelphicus*

PLATE: page 1536

Du. Philadelphia-vireo   Fr. Viréo de Philadelphie   Ge. Philadelphiavireo   It. Vireo di Filadelfia
Ru. Тонкоклювый виреон   Sp. Vireo de Filadelfia   Sw. Canadavireo

Field characters. 10–11.5 cm; wing-span 20–22.5 cm. About 15% smaller and noticeably slighter than Red-eyed Vireo, with proportionately shorter bill and more slender legs; slightly longer than Chiffchaff but noticeably more stocky, with short tail. Greenish to dull brown above and yellowish below, with greyish crown, dull white supercilium, and dark brown eye.

At first glance, could be mistaken for (1) green and yellow Old World warblers, e.g. juvenile Willow Warbler, (2) one of the less patterned *Vermivora* New World warblers, or (3) other plain *Vireo*. Once short stubby bill clearly seen, confusion with warblers unlikely (much bigger Garden Warbler shows no supercilium, or green or yellow tones in plumage). Only necessary distinction is from other *Vireo*, of which only Red-eyed Vireo regularly crosses North Atlantic and is larger, with distinctly longer bill, much stronger head markings including longer, grey, more dark-edged crown and sharper rear supercilium, red eye (in adult), and no obvious yellow tone below except on under tail-coverts.

Habitat. Breeds in temperate Nearctic lowlands, in often moist broad-leaved woodlands, favouring edges, or second growth in old clearings and burnt areas, and willow or alder thickets by streams, ponds, or lakes, feeding both in treetops and lower growth.

Distribution. Breeds in North America, from north-east British Columbia east to central Quebec and Newfoundland, south to North Dakota, northern Michigan, and central Maine.

Accidental. Britain: 1st-winter, Isles of Scilly, October 1987. Ireland: Co. Cork, October 1985.

Movements. Migratory. Winters from Yucatán peninsula (Mexico) south to Panama and northern Colombia. Southward movement begins in August, when migrants appear across southern Canada and northern USA. Crosses Gulf of Mexico to Central American wintering areas chiefly in October. Northward movement begins late March, and by late April most birds have left winter range. Peak migration in May, and reaches main breeding areas only in late May or early June.

Wing-length: ♂ 66–70, ♀ 65–69 mm.

Red-eyed Vireo *Vireo olivaceus*

PLATE: page 1536

Du. Roodoogvireo   Fr. Viréo à oeil rouge   Ge. Rotaugenvireo   It. Vireo occhirossi
Ru. Красноглазый виреон   Sp. Vireo ojirrojo   Sw. Rödögd vireo

Field characters. 12.5 cm; wing-span 23–25 cm. Somewhat shorter than Garden Warbler, with 25% longer bill, less rounded head, up to 10% longer wings, and (sometimes when perched) apparently longer legs (due to exposure of thighs and seemingly shorter tail). Rather small but quite robust, weighty, warbler-like bird, with rather long and deep, stubby-ended bill, long head (exaggerated by strong linear pattern through eye), rather peaked crown, full body, pointed wings, and almost square tail. Bold, white, dark-bordered supercilium; eye red in adult, dark brown in 1st-winter. Upperparts mainly dull olive, underparts dull whitish.

In brief glimpse, can be confused with smaller, plain Old World warblers, but, when seen well, bill and head pattern separate Red-eyed Vireo from rather smaller, plain congeners occurring in west Palearctic. Juvenile more confusing but distinctive call always helpful. Stance rather level, often carrying tail at similar angle to back, with wing-points slightly drooped. Gait hopping but varied in rhythm so that bird moves both slowly (feeding sluggishly from foliage) and quickly (snatching at insect prey). Generally secretive, occasionally excitable, interacting with food competitors and then giving burst of calls.

Commonest call (in excitement or alarm) a nasal 'quee' or 'chway', with distinctly complaining timbre.

Habitat. Breeds in boreal, temperate, and subtropical zones of Nearctic, where before forest clearance it was held to be the most abundant bird species, as it still is wherever stands of trees exist within its range; inhabits mainly canopy, although nesting lower. Since forest clearance, has adapted to successor habitats on farms, and even in cities if trees are present.

Distribution. Breeds over much of North America (except northernmost areas and south-west), and in South America south to northern Argentina.

Accidental. Iceland, Britain and Ireland, France, Netherlands, Germany, Spain, Malta, Morocco.

Movements. North American populations migratory, South American populations apparently resident, except southernmost which withdraw northward. All populations winter in South America, south chiefly to Amazonia. Migrants begin to leave breeding areas before mid-August, and earliest birds appear in Central America and West Indies before end of August. Main movement September through northern and mid-latitude states. Northward migration starts early, with movement through Central America and from Texas to Georgia during March. Main passage into north-east breeding areas early to mid-May. Vagrant to Greenland. Rare autumn vagrant to west Palearctic, appearing with some regularity in Britain and Ireland, where 89 records up to 1995, especially south-west England and southern Ireland, mainly late September to 3rd week of October.

Wing-length: ♂ 80–84, ♀ 77–81 mm.

Finches Family Fringillidae

Small to medium-small, stout-billed, 9-primaried oscine passerines (suborder Passeres). Principally seed-eaters, found in many habitats in both Old World and New, including savanna, steppe, scrub, and even desert, but mostly in woodland, forest, parkland, cultivation, and the like. Bill strong and typically conical, but shows much adaptive variation; in all species, however, structurally designed internally for shelling seeds (with aid of tongue and strong jaw muscles). Unlike buntings (Emberizidae), finches much prefer dicotyledonous seeds, slicing them open using sharp edges of lower mandible. Now usually classified in 2 subfamilies: Fringillinae (chaffinches) and Carduelinae (typical finches).

Chaffinches Subfamily Fringillinae

Fairly small finches, inhabiting forest, woodland, parkland, and farmland. Pick seeds from ground and insects from foliage or, to lesser extent, by aerial fly-catching; insects and other small invertebrates figure prominently in summer diet, forming sole food of nestlings (unlike in cardueline finches). 3 closely related species in single genus *Fringilla*; all found in west Palearctic.

Sexes differ somewhat in size, ♂ slightly larger than ♀. Bill relatively long; pointed with straight culmen. Food brought to young in bill (not regurgitated as in cardueline finches). Wing fairly long, bluntly pointed. Flight typically undulating. Tail fairly long, slightly forked. Foot not used for holding or uncovering food (unlike some carduelines), or for clinging to vegetation when feeding. Lacks twittering calls of most cardueline finches. Song of ♂ of typical advertising type: relatively simple but loud, clear, and more stereotyped than longer, more generalized song of cardueline finches. Monogamous mating system with strong pair-bond. Nest a neat cup; built by ♀ only. Incubation by ♀ only; young fed by both sexes.

Chaffinch *Fringilla coelebs*

PLATES: pages 1542, 1553

Du. Vink   Fr. Pinson des arbres   Ge. Buchfink   It. Fringuello
Ru. Зяблик   Sp. Pinzón común   Sw. Bofink

Field characters. 14.5 cm; wing-span 24.5–28.5 cm. Similar in size to House Sparrow but with smaller, sharper bill, smaller, more peaked head (especially in ♂), and particularly longer wings and tail, giving classic form of *Fringilla*. Noticeably long, medium-sized, and rather elegant passerine. At all ages and in both sexes, plumage pattern dominated by contrasting white panels on leading wing-coverts, buff-white wing-bar on secondaries and inner primaries, and striking white outer tail-feathers which show well on perched bird and twinkle on flying one. Eurasian ♂ shows more colours than any other west Palearctic finch, with blue-grey crown, pale red to pink face and underparts, and mainly black ground to white wing and tail markings. ♀ and juvenile much more sober, with dusky olive-brown head and upperparts and dusky white underparts showing little pattern; wings and tail duller than ♂'s but show similarly striking marks.

Unmistakable at close range; no other finch or bunting of similar form has such bold white marks on wing and tail. Brambling shows similar wing pattern but has long white rump, all-black and more forked tail, and much whiter underbody. Flight light and free, with action and silhouette more reminiscent of pipit and bunting than smaller, shorter-tailed finches. Gait a hop, varied by distinctive tripping walk, accompanied by short quick steps and just visible nod of head. Stance on perch rather upright with tail often drooped and in singing ♂ head held up; on ground much less so, with tail length creating illusion of long body line.

Habitat. Breeds in west Palearctic in temperate wooded areas, from Mediterranean and marginally steppe zones up to boreal, and in places to edge of tundra; between July isotherms 12–30°C, usually occurs within shifting climatic boundary, beyond which Brambling replaces it to the north. Basically arboreal, and in breeding season occupies deciduous, mixed, and coniferous woods and forests, with preference for mixed deciduous woodland. In modern times, parklands, gardens, and farm hedgerows have also been extensively settled, even in suburbs and locally interiors of large cities. In winter, open areas of farmland up to some distance from tree cover are frequented where they offer sufficient food. In mountains, ascends generally to tree-line; in Caucasus, to 2200–2500 m.

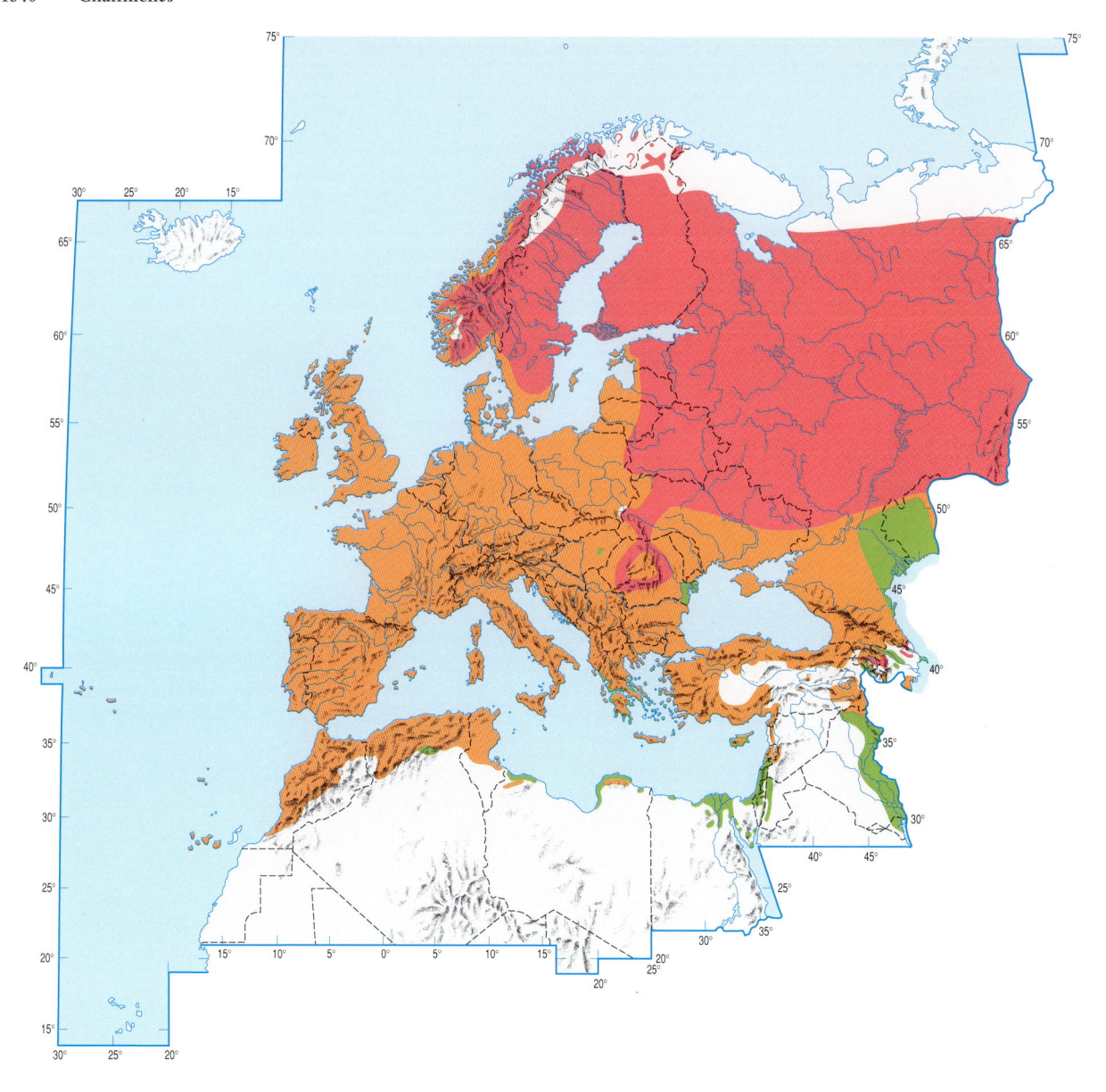

Distribution. Few changes reported. Slight range increase Spain, Croatia, and Ukraine. ICELAND. Bred 1986–9. FAEROES. Bred 1972 (2 records) and 1981. ITALY. Colonized Aeolian Islands in 1980s. MALTA. 2–4 pairs breed. TURKEY. More local in forest relicts on fringes of Central Plateau and in mountain valleys of east and south-east. Widespread and common in other wooded areas. ISRAEL. No evidence of breeding since reported in late 19th century as breeding in lower parts of Mt Hermon. IRAQ. Apparently breeds in north. CANARY ISLANDS. Breeds Gran Canaria, Tenerife, La Gomera, La Palma, and El Hierro.

Accidental. Bear Island, Iceland (annual).

Beyond west Palearctic, breeds northern Iran, and east in Russia to Irkutsk region. Introduced in New Zealand and Cape Town area of South Africa.

Population. Apparently stable over most of range, though slight increase Britain, Denmark, Spain, Croatia, and Ukraine, decrease Finland and Lithuania. BRITAIN. 5.4 million territories 1988–91. Shallow increase since early to mid-1970s. IRELAND. 2.1 million territories 1988–91. FRANCE. Over 1 million pairs in 1970s. BELGIUM. 250 000 pairs 1973–7. Stable, with local decline in western parts. LUXEMBOURG. 40 000–50 000 pairs. NETHERLANDS. 250 000–400 000 pairs 1979–85. GERMANY. 10.9 million pairs in mid-1980s. In east, 2.5 ± 0.9 million pairs in early 1980s. DENMARK. 310 000–3.2 million pairs 1987–8. NORWAY. 1–1.5 million pairs 1970–90. SWEDEN. 7.5–15 million pairs in late 1980s. FINLAND. 6–9 million pairs in late 1980s. Has increased in 20th century, perhaps owing to fragmentation of forests; most recent trend is slight decline. ESTONIA. 2–3 million pairs in 1991. LATVIA. 2.6–3.2 million pairs in 1980s. LITHUANIA, POLAND. Abundant. CZECH REPUBLIC. 4–8 million pairs 1985–9. SLOVAKIA. 3–5 million pairs 1973–94. HUNGARY. 650 000–800 000 pairs 1979–93. AUSTRIA. Very common. SWITZERLAND. 1.5–2 million pairs 1985–93. SPAIN. 2.6–6.4 million pairs. PORTUGAL. 1 million pairs 1978–84. ITALY. 1–2 million pairs 1983–95. GREECE. 200 000–500 000 pairs. ALBANIA. 50 000–90 000 pairs in 1981. YUGOSLAVIA: CROATIA. 3–4 million pairs. SLOVENIA. 500 000–1 million pairs. BULGARIA. 1–10 million pairs. RUMANIA. 3–4 million pairs 1986–92. RUSSIA. 10 million pairs. Declined Yaroslavl' region 1960–85. BELARUS'. 5.4–5.6 million pairs in 1990. UKRAINE. 2.5–2.8 million pairs in 1986. MOLDOVA. 150 000–250 000 pairs in 1988. AZERBAIJAN. Common. TURKEY. 1–10 million pairs. CYPRUS, TUNISIA, ALGERIA, AZORES. Common. MOROCCO, MADEIRA. Very common.

Movements. Sedentary to migratory, wintering chiefly within breeding range in Europe, but further south in Asia. In Europe, migrates south-west on fairly narrow front, with western populations wintering furthest west, and more eastern populations progressively further east. Winter visitors greatly augment populations of western and southern Europe (including Britain), and regularly reach North Africa. Migrates by day in flocks, most actively in morning hours, sometimes in company with Brambling. In many areas the most numerous visible autumn migrant.

British birds very sedentary; 90% move no further than 5 km from natal site, and the rest (almost entirely 1st-year birds) less than 50 km. Immigrants winter mainly in southern and central Britain (south of 54°N) and Ireland, and are chiefly of Scandinavian origin. French birds mostly sedentary; a few juveniles move west or south-west in autumn, and very occasionally cross Pyrénées to Spain. Birds from Low Countries and Scandinavia east to Russia and Czechoslovakia winter in France. In countries further east and north, proportion of migrants increases. In Sweden, overwinters in small numbers in south, and occasionally north to extreme north. In Norway, overwinters especially on coast. In southern Finland, small flocks commonly seen near human habitations in early winter. Winters only in very small numbers in Leningrad region; irregular in Moscow region.

Autumn migration August–December, chiefly September–November. In south of Mediterranean region, passage continues to early December, and even to January. Spring migration begins early and is prolonged, February–May, chiefly March to mid-April in central Europe. In Strait of Gibraltar, passage earlier than other finches, from start of February, peaking 1st half of March and continuing to early April. Further north, main passage in March and April, with arrival in northern breeding areas in 2nd half of April.

♀♀ more migratory than ♂♂, at least in some areas, tending to move further, and differential wintering occurs. Most birds remaining to winter in Sweden are ♂♂ (hence 'coelebs' = bachelor), and ♂♂ predominate in winter in Low Countries and Britain, but ♀♀ predominate in Ireland. In central Europe in winter, proportion of ♂♂ to ♀♀ c. 3:1 from Grenoble (south-east France) north-east to Poland, but further south in France proportion of ♀♀ increases, and they predominate in Rhône delta.

Food. Mainly seeds and other plant material; in breeding season mainly invertebrates. In Oxford (southern England), had widest-ranging and most varied diet of all Fringillidae in major study. Seeds taken generally on ground, notably freshly-turned soil, not direct from plant, except in shrubs and trees; feeds with rapid pecking action unsuited to removing seeds from herbs or grasses. Feeds most often in trees in spring and summer when taking invertebrates (especially defoliating caterpillars), and more on ground during rest of year; commonly forages in large flocks in open country outside breeding season. Larger seeds de-husked; small or long seeds crushed and swallowed, not de-husked.

Social pattern and behaviour. Mainly gregarious outside breeding season, forming flocks for feeding and migration. Flock composition and timing of flock-formation vary regionally according to local migratory status. Mating system essentially monogamous; exceptionally, successive bigamy by ♂ occurs. Monogamous ♂♂ readily promiscuous, and some evidence that ♀♀ also protagonists in promiscuity. ♀ alone builds nest, incubates, broods, and does most feeding of young until fledging. Territorial when breeding, but boundaries fluid and much feeding done outside territory. Song of ♂ serves to demarcate territory and attract ♀; typically delivered from high conspicuous perch in upright posture with head back, plumage relaxed, wing-bars partly concealed. In most pairs, song almost ceases for 1–2 weeks after pair-formation and then gradually returns, but never regains intensity of unpaired ♂♂. In migrant populations ♂♂ arrive on breeding grounds and occupy territories ahead of ♀♀. Pair-bonding behaviour includes 'moth-flight' by ♂, who flies at lower level than ♀ so that his upper side is visible to her, and crouching-lopsided display in which ♂ tilts body sideways towards ♀, at same time raising still-closed wing nearest ♀, exposing red flank and belly to her. Wing nearest ♀ is kept raised for several seconds while ♂ periodically glances at her. Then he relaxes and performs moth-flight to another perch to repeat display. In contrast to cardueline finches, courtship-feeding does not occur.

Voice. Song of ♂ over most of range a short accelerating and descending series of cheerful, quite musical notes, usually ending in trisyllabic flourish: 'chip-chip-chip-tell-tell-cherry-erry-erry-tissi-cheweeo' or 'chink-chink-chink-tee-tee-tee-terree-erree-erree-chissee-CHU-EE-OO'; acceleration of notes aptly compared to rhythm of bowler's run-up (in cricket). In wide areas of Europe, though only rarely in Britain, song may be followed by a 'kit' unit, similar to that of Great Spotted Woodpecker. ♂♂ have 1–6 song-types; typically give a series of one type before switching to another, often singing each type in turn before returning to 1st. In Azores, *moreletti* has 'trill' which is sometimes not clearly split into separate sections, with flourish shorter or absent. In Canary Islands, *canariensis* tends to have larger song repertoire, with longer units but without clear terminal flourish. Most frequent call a distinctive 'chink', given by both sexes during breeding season in varied

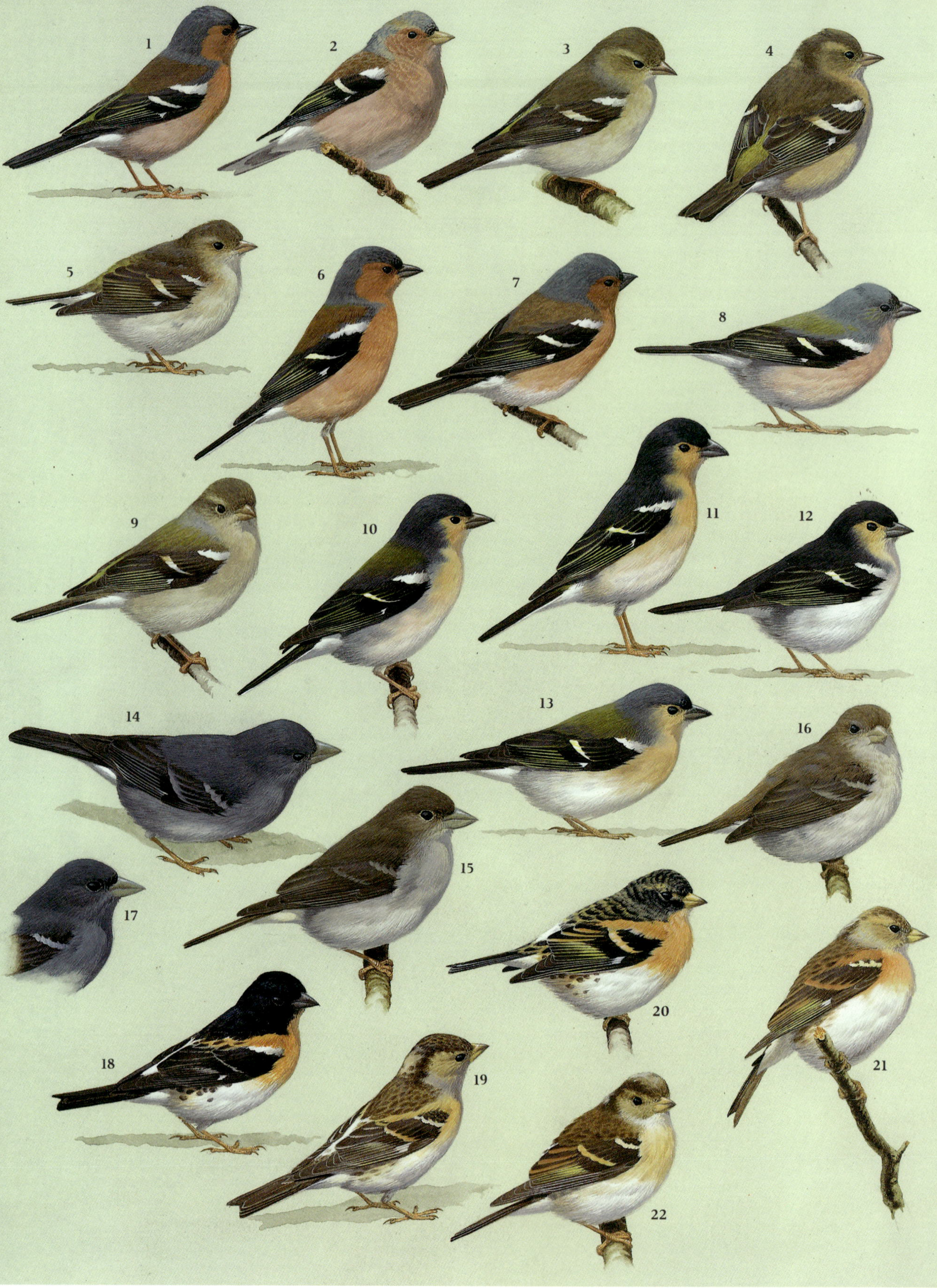
1
2
3
4
5
6
7
8
9
10
11
12
13
14
15
16
17
18
19
20
21
22

situations: mild alarm, mobbing, territory establishment, separation from mate. Flight-call a 'tupe' or 'tsüp', given by both sexes in flight and when about to take off. High-pitched thin 'seee', as in many passerines, is produced in strong alarm, especially in response to overhead hawk. So-called rain-call a regionally variable call given in many different contexts by breeding ♂, sometimes when alarmed but often apparently spontaneously. Occurs in long series. May take form often described as 'huit' but also, in England, as 'hreet' or 'breeze'; several other variants described, and boundaries between local dialects may be sharp.

Breeding. SEASON. Start and pattern of laying related to spring temperature, becoming later from south-west to north-east in Europe. Britain: most clutches started late April to mid-June, range mid-March to mid-July. Kaliningrad region (western Russia): eggs laid early May to early July, peak late May to early June. Germany: eggs laid beginning of April to mid-July. 2 broods apparently unusual. SITE. In fork of tree or bush, on branch or on several thin twigs. Nest: compact and neat, with firm walls and deep cup, clad with lichen and moss thus looking green or greyish; pliable and yielding to touch; outer layer of lichen, moss, bark, and fibres bound with spider silk, then grass and stalks lined with rootlets, hair, and feathers. EGGS. Sub-elliptical, smooth and slightly glossy; very variable, pale bluish-green to reddish-grey with purple-brown blotches (at times edged with pink), scrawls, and hair-streaks, concentrated at broad end. Clutch: 4–5 (3–6). INCUBATION. Average 12.6 days (10–16). FLEDGING PERIOD. Average 13.9 days (11–18).

Wing-length: Nominate *coelebs*: ♂ 86–95, ♀ 81–86 mm.
Weight: Nominate *coelebs*: ♂ mostly 20–29, ♀ 18–27 g.

Geographical variation. Marked and complex. 3 distinct groups of races, differing especially in pattern of head and colour of upperparts of ♂: *coelebs* group in western Eurasia and Middle East, *spodiogenys* group in North Africa, and *canariensis* group on Atlantic islands. Each group perhaps better considered a separate species.

(FACING PAGE) Chaffinch *Fringilla coelebs*. *F. c. coelebs* (Crimea eastwards): **1** ad ♂ summer, **2** ad ♂ winter, **3** ad ♀ summer, **4** ad ♀ winter, **5** juv. *F. c. gengleri* (Britain and Ireland): **6** ad ♂ summer. *F. c. solomkoi* (most of continental Europe): **7** ad ♂ summer. *F. c. africana*: **8** ad ♂ summer, **9** ad ♀ summer. *F. c. maderensis*: **10** ad ♂ summer. *F. c. canariensis*: **11** ad ♂ summer. *F. c. palmae*: **12** ad ♂ summer. *F. c. moreletti*: **13** ad ♂ summer.
Blue Chaffinch *Fringilla teydea* (p. 1544). *F. t. teydea*: **14** ad ♂ summer, **15** ad ♀ summer, **16** juv. *F. t. polatzeki*: **17** ad ♂ summer.
Brambling *Fringilla montifringilla* (p. 1545): **18** ad ♂ summer, **19** ad ♀ summer, **20** ad ♂ winter, **21** ad ♀ winter, **22** juv.

♂♂ of *coelebs* group characterized by vinous, rufous, or cinnamon ear-coverts and cheeks, contrasting with black lore and with blue-grey crown and side of neck, similar in colour to or slightly darker than underparts; also, lower mantle and inner scapulars various shades of brown, back to upper tail-coverts green. 8 races recognized within this group in west Palearctic.

In *spodiogenys* group of North Africa, crown and nape of ♂ medium blue-grey (in *africana*, occurring Morocco to north-west Tunisia) or light blue-grey (in *spodiogenys* of eastern and central Tunisia); nape with white spot; band on forehead as well as lore contrastingly black; side of head down to upper cheek and side of neck bluish ash-grey with contrasting white eye-ring (latter broken by black in front and behind); mantle and inner scapulars bright green, forming contrasting saddle, back and rump green, mixed green and blue-grey, or (occasionally) almost fully blue-grey, outer scapulars and upper tail-coverts blue-grey; lower cheek and chin down to upper belly and side of belly pale vinous-pink (*africana*) or pink-white (*spodiogenys*), side of breast and flank with restricted ash-grey, mid-belly to under tail-coverts white; tail more extensively white than in *coelebs* group.

In ♂ *canariensis* of *canariensis* group, occurring Gran Canaria, Tenerife, and Gomera (Canary Islands), upperparts largely deep slate-blue, black bar on forehead scarcely contrasting, but rump contrastingly bright green; in fresh plumage, feather-tips of upperparts broadly olive-brown, largely concealing plumbeous. Pattern on side of head differs strongly from *coelebs* and *spodiogenys* groups: lore, broad ring round eye, and front part of cheek ochre or orangey-buff, contrasting with dark blue-grey remainder of side of head and neck. Chin to upper belly and side of belly ochre, grading into cream-white or white mid-belly, vent, and under tail-coverts; belly tinged vinous; side of breast and lower flank rather contrastingly grey. Tail with less white than *coelebs* group. *F. c. palmae* of La Palma (western Canary Islands) similar to *canariensis*, but rump of ♂ dark blue-grey, not green; belly and upper flank white, contrasting sharply with ochre chest; larger, especially bill. *F. c. ombriosa* of Hierro (western Canary Islands) like *palmae*, but rump with traces of green and underparts slightly less extensively white, somewhat tending towards *canariensis*. *F. c. maderensis* from Madeira like *canariensis*, but nape and top of head of ♂ dark blue-grey, less deep plumbeous than in races of Canary Islands, black band on forehead contrasting more sharply; lower mantle, scapulars, back, and rump bright green, outer scapulars and upper tail-coverts blue-grey with green wash. *F. c. moreletti* of Azores like *maderensis*, but bill markedly longer and heavier at base, upperparts more extensively green, chest browner, belly less vinous; amount of white in tail markedly reduced, largely replaced by light grey.

Blue Chaffinch *Fringilla teydea*

Du. Blauwe Vink   Fr. Pinson bleu   Ge. Teydefink   It. Fringuello delle Canarie
Ru. Голубой зяблик   Sp. Pinzón azul   Sw. Kanariebofink

PLATES: pages 1542, 1553

Field characters. 16.5 cm; wing-span 26.5–31.5 cm. Up to 10% larger than Chaffinch, with similar silhouette but noticeably heavier and more robust, with distinctly longer bill. Plumage pattern resembles Chaffinch but much more uniform: ♂ bluish-grey, showing dull wing marks matching those of Chaffinch at close range. ♀ and juvenile dusky-olive, with brighter wing-bars.

Unmistakable, since even noticeably blue-backed Canarian races of Chaffinch have full white wing-marks, pink breast (in ♂), and fully white belly. Flight much as Chaffinch but larger size evident in rather slower or more gentle undulations.

Habitat. Restricted to small forest areas on Canary Islands in subtropical eastern Atlantic, mainly above 1200 m, and as high as 1830 m in the most favourable areas on southern slopes. Virtually confined to forests of pine of varying types, lowest with tangle of undergrowth of tree-heath, holly, and other shrubs, while at higher levels undergrowth is lacking, forest floor being strewn with pine needles on which much time is spent foraging for seeds and insects.

Distribution and population. Restricted to Tenerife and Gran Canaria in Canary Islands. Tenerife: common; distribution dependent on pinewoods, formerly in continuous belt round island, now reduced to isolated stands; reafforestation has resulted in slow but steady recovery locally. Gran Canaria: very scarce and declining, apparently since first half of 20th century, and primarily because of habitat destruction. Still present, but local, in Pinar de Tamadaba and Pinar de Pajonales, in 1984. Total population 1000–1500 pairs, with 180–260 birds on Gran Canaria.

Movements. Sedentary, making only very local movements in forests of pine. Occurs at high altitude even at times of deep snow cover, though probably makes limited altitudinal movements then. Recorded occasionally above treeline, and exceptionally at some distance from usual range.

Food. Seeds (principally Canarian pine *Pinus canariensis*) and invertebrates. On ground, seems to feed mainly in places where tall mature trees are closely spaced and understorey fairly low. In trees, searches crevices in pine bark for invertebrates, especially moths; starts at top of tree and examines each branch from base to tip before flying down to next one; also hangs from open cones to remove seeds. Commonly hawks after Lepidoptera.

Social pattern and behaviour. Relatively gregarious outside breeding season, forming small (up to 10 birds), loose, roving flocks, especially of younger birds, which break up at start of breeding season; flocks sometimes associate with other species, especially Chaffinch. Solitary and territorial in breeding season. Song given by ♂ from elevated perch. Sometimes heard on fine days in late winter, but song not full before April-May, at which time serves to demarcate territory.

Voice. Song not unlike Chaffinch but with simpler 'trill' and longer flourish. Most frequent call is 'whit-chooee', which appears equivalent to 'chink' of Chaffinch, but varies considerably between localities.

Breeding. SEASON. Tenerife: late breeder, with snow in early spring not uncommon on breeding grounds at *c.* 1000–2000 m; nest-building starts end of May, eggs laid 2nd half of June, in south-facing areas possibly May. Gran Canaria: perhaps a few weeks earlier because of lower altitude (*c.* 700–1200 m). Probably one brood. SITE. Almost always in pine, generally *c.* 10 m (1.5–20) above ground at end of thin branch, sometimes against trunk. Nest: foundation of twigs of pine, tree heath, or broom, herb stalks, pine needles, moss, lichen, and spiders' webs (last 2 materials giving many nests white appearance), lined with grass, plant down, hair, feathers, etc. EGGS. Sub-elliptical to long oval, smooth and slightly glossy; light blue or greenish-blue with chestnut or purplish speckles and streaks at broad end; faint undermarkings of violet or brownish-purple blotches. Larger and more brightly coloured than those of Chaffinch. Clutch: 2; apparently complete clutches of 1 sometimes recorded. INCUBATION. 13–14 days. FLEDGING PERIOD. 17–18 days.

Wing-length: *F. t. teydea*: ♂ 96–105, ♀ 88–98 mm.
F. t. polatzeki: ♂ 93–98, ♀ 87–93 mm.
Weight: *T. t. teydea* 2♂♂ 30.9, ♀ 29.2 g

Geographical variation. Slight in colour, more marked in size. *F. t. polatzeki* of Gran Canaria smaller than nominate *teydea* from Tenerife in all measurements; ♂ *polatzeki* differs from ♂ of *F. t. teydea* mainly in presence of velvet-black band on forehead (in adult), slightly duller and more restricted slate-grey on chin to chest, slightly paler greyish-white tips of median and greater coverts, and narrower slate-blue fringes of upper wing-coverts and tertials, upperwing appearing blacker.

Brambling *Fringilla montifringilla*

PLATES: pages 1542, 1553

Du. Keep   Fr. Pinson du Nord   Ge. Bergfink   It. Peppola
Ru. Юрок   Sp. Pinzón real   Sw. Bergfink

Field characters. 14 cm; wing-span 25–26 cm. Similar in body and wing size to Chaffinch but stubbier bill, seemingly larger head, and slightly shorter, more distinctly forked tail produce less attenuated silhouette on ground and in flight. Medium-sized, elegant finch, with general character and behaviour of Chaffinch and rather similar basic plumage pattern but different, less varied colours in ♂. Both sexes show diagnostic combination of long, oval white rump and almost completely black tail. ♂ distinguished in breeding season by glossy black head and mantle, bordered by orange blaze from breast across shoulder and below back; in winter by black-speckled face and crown and black-splashed mantle. ♀ and juvenile distinguished by mottled dark brown head, with broad buff supercilium and grey sides to neck.

Unmistakable when seen well but similar wing pattern may invite confusion with distant Chaffinch. Goldfinch and Bullfinch both show white rump and black tail but rest of plumage pattern and colours differ distinctly. Flight recalls Chaffinch but action differs in slightly quicker wing-beats, producing more bouncy progress, and apparently (perhaps due to lack of white in it) less spreading of tail in manoeuvres; flight silhouette less attenuated than Chaffinch.

Habitat. Breeds across boreal and subarctic zones of west Palearctic between 10° and 18–19°C July isotherms. Owing to northerly range and arboreal requirements does not extend much up mountains, but is common on uplands in the more open birch woods, and in mixed forests of birch and conifers. Sometimes ranges beyond into lower growth of juniper, willow, or alder. Tall and dense stands in forest appear to be less favoured than open growth with clearings. Change after breeding season from insectivorous to largely seed diet involves shift to ground feeding, sometimes in stackyards and on farmland, but attachment to woodland remains strong, especially in case of beech where crop of fallen mast is ample. Except in severe weather, remarkably independent of all kinds of human settlements, artefacts, and food resources.

Distribution. Marked annual fluctuations in breeding range. ICELAND. First recorded (as accidental) in 1930. Has bred nearly every year since 1978; up to 7 pairs per year. FAEROES. Bred 1967 and 1972, possibly also in 1982. BRITAIN. Breeding records widely scattered in Scotland (first recorded breeding in north in 1920, bred also, e.g. 1979, 1982) and eastern England. 1–2 pairs 1988–9. NETHERLANDS. Occasional breeder: scattered 10–150 pairs or single ♂♂ annually (♀♀ rare, seldom any young fledged). GERMANY. Repeated breeding records; perhaps descended from released decoy birds, but most sites Baltic area (similar to Scandinavian sites). 5 pairs in mid-1980s. DENMARK. Total of c. 10 breeding records. SWEDEN. Probable south-west spread since c. 1970, but obscured by changes from one year to another. Occasional breeding records south to c. 57°N. FINLAND. Range fluctuates markedly. ESTONIA. Range expansion over last 40 years so now patchily distributed in extensive forest areas throughout country. LATVIA. Probably bred 1892, confirmed 1960; observations of singing ♂♂ suggest very rare breeding. LITHUANIA. Bred 1959, 1978, 1983. CZECH REPUBLIC. Bred 1928, 1980. AUSTRIA. Bred 1952, 1976, 1988; further records of summering in Alps. ITALY. Occasional breeding (probably by escaped cagebirds), last in 1986. YUGOSLAVIA: SLOVENIA. Perhaps breeds. BELARUS'. Occasional breeder in northern half of country. Perhaps up to 10 pairs in 1990.

Accidental. Spitsbergen, Bear Island, Tunisia, Madeira, Canary Islands.

Beyond west Palearctic, extends broadly east across Russia to Kamchatka and Sea of Okhotsk.

Population. Largely stable, though marked fluctuations associated with changes in breeding distribution, and (in Norway and Sweden) with varying abundance of caterpillars. NORWAY. 1–2 million pairs 1970–90. SWEDEN. 500 000–2 million pairs in late 1980s. FINLAND. 2–3 million pairs in late 1980s. ESTONIA. 100–500 pairs in 1991. RUSSIA. 1–10 million pairs.

Movements. All populations migratory, wintering almost entirely south of breeding range. European birds head between west and south, chiefly south-west. Extent of movement is strongly dependent on food availability (chiefly seed of beech); local numbers wintering fluctuate greatly, and concentrations of millions of birds occur, especially in south-central Europe. Ringing data give evidence of winter site-fidelity, but also of individuals wintering in widely differing areas in different years. ♂♂ predominate in areas closer to breeding range. Migration mostly diurnal (chiefly in morning), especially inland, but nocturnal migration observed on coasts and at sea.

Autumn passage in north-central Europe begins mostly in 2nd half of September, and continues to early or mid-November; peaks 1st half of October in Finland, Poland, and eastern Germany, from mid-October further west. Present in winter quarters chiefly November–February. Spring migration February–May. ♂♂ leave winter quarters earlier than ♀♀ and arrive earlier on breeding grounds.

Food. Seeds, berries, and (in summer) invertebrates, especially Lepidoptera larvae and beetles; in winter quarters specializes in beechnuts. On breeding grounds feeds mainly in trees, but at other times mostly on ground, commonly in flocks. In winter and on migration feeds on agricultural land, especially cereal fields, and in beech woods. Takes beechnut on ground by pushing aside leaf litter with bill or by digging tunnels 30–40 cm deep through soft snow on slopes, using bill and 'swimming' action of wings. Sometimes eats beechnuts on trees, at times pulling them from cups while hovering; seeds of birch also taken *in situ*. In Switzerland, seeds of spruce important in winter diet at higher altitudes.

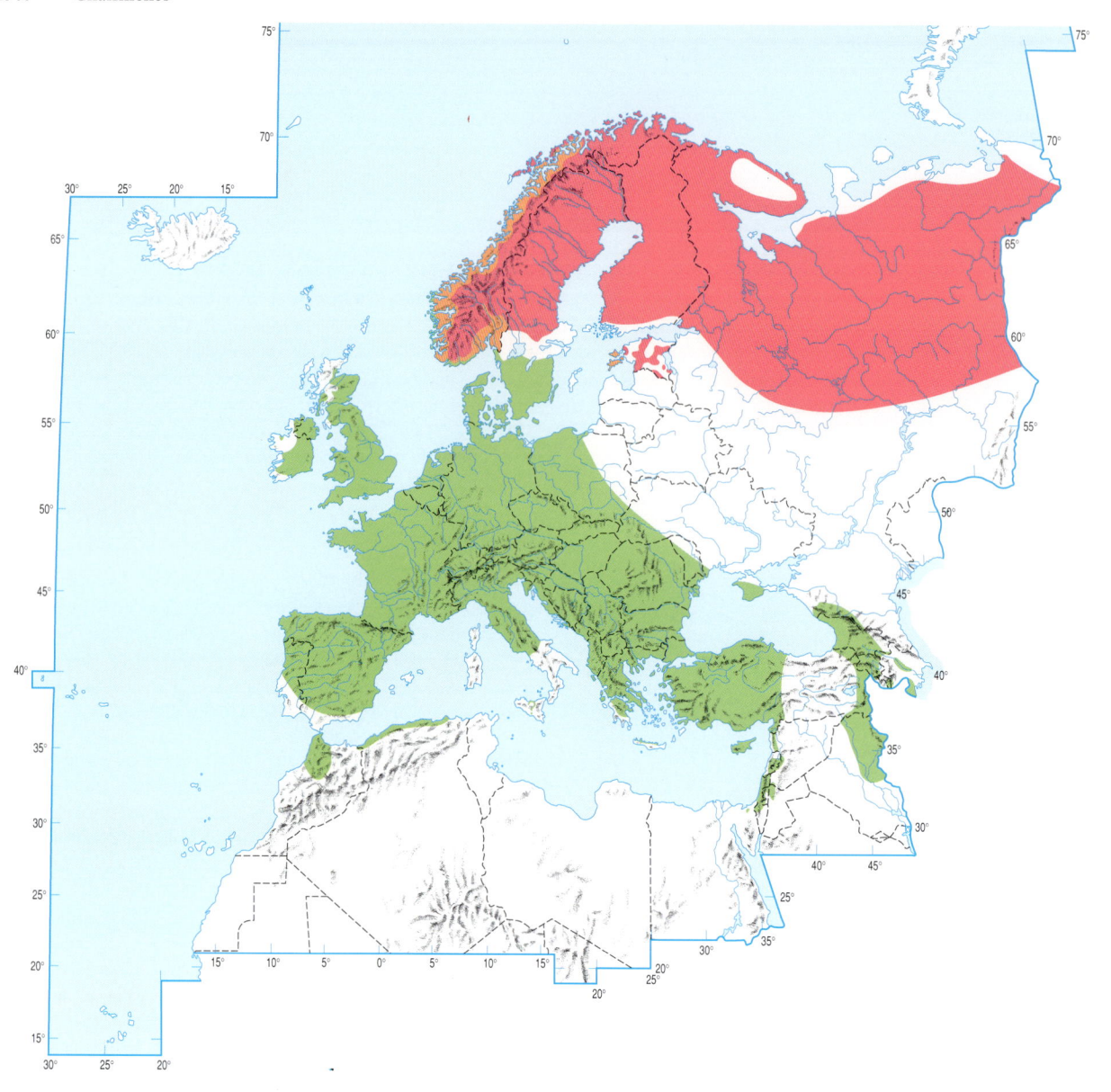

Social pattern and behaviour. Generally gregarious outside breeding season, including during migration, and sometimes forms enormous flocks for feeding, while those for roosting may be of many millions. Mating system monogamous and, unlike in Chaffinch, of ♀-defence rather than resource-defence type. Pair-formation often takes place in flocks on breeding grounds. Territorial when breeding, more commonly in loose neighbourhood groups. ♂ gives 'wheezing-song' from tree-top or other well-exposed perch. Each unit of song given with bill wide open and head thrown back (further than in Chaffinch) so that bill pointed up at c. 60° or almost vertical. Wings drooped and white rump exposed; wings sometimes also flicked outwards, and tail partly spread; will make small jumps between song-bouts and may turn 180°; also sometimes makes low and silent flights of c. 5–10 m, gliding or with slow, flicking wing-beats.

Voice. Song markedly different from Chaffinch, though occasionally mimics song and calls of that species. Deafening chorus of 'twittering' from enormous roosting flocks audible over several km, and considerable wing-noise created by such flocks leaving roost. 'Wheezing-song' of ♂ a drawn-out, somewhat dreary or melancholy wheezing or bleating sound with initial crescendo after quiet start: twanging, drawled 'dree-e-e'. Like wheeze of Greenfinch, but evenly pitched (not falling); given in persistent and monotonous series. Main call a characteristically nasal, hoarse, incisive and strangled, rather metallic sound—'wayeek' or 'wayk'. Given in flight or when perched, serving as contact-call in migrating flocks, also between pair-members on breeding grounds, and (especially loud and harsh) as main alarm-call. Flight-call a short, hard, slightly hoarse 'yeck', 'tjek', or 'chucc' given typically in series, rapid especially when taking off, and serving for close contact;

sometimes conversationally quiet and unhurried. Some similarity to flight-call of Chaffinch, but usually readily distinguishable through being more nasal, harder, and lower pitched.

Breeding. SEASON. Eggs laid from May to July. One or (at least in north-west Russia) 2 broods. SITE. Usually fairly high in tree, often against trunk of conifer or in fork of deciduous tree. Exceptionally in rocky scrub, even on ground. Nests can be fairly close together, and often within colonies of Fieldfare. Nest: similar to that of Chaffinch but larger and more loosely built; outer structure of moss, lichen, grass, heather, strips of birch or juniper bark, and cobwebs, lined with feathers (often many), moss, plant down, soft grass, hair, fur, and sometimes paper, string, etc. EGGS. Sub-elliptical, smooth and glossy; very like those of Chaffinch, but greener, and with same wide variation; from clear light blue to dark olive-brown, with sparse to dense pink to rusty spots and blotches, and sometimes fine hair-streaks. Clutch: 5–7 (3–8). INCUBATION. 11.5–12 days. FLEDGING PERIOD. 13–14 days.

Wing-length: ♂ 91–97, ♀ 84–91 mm.
Weight: ♂ mostly 19–30, ♀ 17–27 g.

Typical Finches Subfamily Carduelinae

Small to medium-sized, seed-eating finches inhabiting wide variety of habitats and specializing in extraction and breaking of edible seeds; these supplemented by buds in a few cases but insects (etc.) taken to much lesser extent than by fringilline finches, young being fed (by regurgitation) on seeds or mixture of seeds and invertebrates, never invertebrates exclusively. Found in both Old and New Worlds but mainly in Palearctic. About 126 species of which 27 breed in west Palearctic and 2 accidental (1 from North America).

Sexes almost the same size (♀ usually slightly smaller). Bill mainly of typical seed-eater type but very variable, even within genus; deepest (and as wide as high) in *Coccothraustes*, finest in some *Carduelis*; uniquely crossed at tip (for extracting pine seeds from cone) in *Loxia*. Both sexes of some species develop gular storage pouches in breeding season. Wing usually rather long and bluntly pointed. Flight fast and undulating, typically with periodic closure of wings; more erratic and flitting, with light, dancing action, in some smaller species. ♂♂ of several genera perform song-flights: butterfly-like with slow-beating wings or spectacularly undulating. Many species highly vocal especially when flocking, with a number of noisy often twittering calls. Song less specialized than in fringilline finches, being less stereotyped, longer, and quieter but sometimes highly melodious. Most species typically territorial when nesting; territories usually small and grouped in loose colonies. Monogamous mating system general, usually with seasonal pair-bond. After moult, many species have cryptically coloured feather-tips on head and body and typically attain brightly coloured breeding plumage by abrasion of feather-tips.

Red-fronted Serin *Serinus pusillus*

PLATES: pages 1549, 1553

Du. Roodvoorhoofdkanarie   Fr. Serin à front rouge   Ge. Rotstirngirlitz   It. Verzellino fronterossa
Ru. Королевский вьюрок   Sp. Verdecillo carinegro   Sw. Rödpannad gulhämpling

Field characters. 12 cm; wing-span 21–23 cm. Slightly larger than Serin but with similar stubby form on ground and in flight. Distinctive serin, with sooty head and breast, fiery orange-red forecrown, and heavily streaked back and flanks; rump orange in centre. Juvenile has rufous-buff face, cheeks, and throat.

Within *Serinus*, unmistakable at all ages, but beware confusion with (1) vagrant Redpoll which also has usually red but occasionally paler orange forehead in adult plumage and can appear quite swarthy and heavily streaked at distance, and (2) Turkish race of Twite whose black-splashed chest and flanks recall young Red-fronted Serin in 1st autumn transitional plumage.

Habitat. Breeds in south-east of west Palearctic, in middle and upper tree belts of mountains, subalpine meadows, and in wide and narrow ravines along rivers; in Caucasus, at 600–3000 m. Sings from upper branches of low birches or pines, or rock ledges, but is often on ground or on stony or rocky terrain, nesting in rock crevices but occasionally in lower branches of juniper, rose, or other shrub. Occurs in rhododendron zone and among juniper, descending in winter to valleys of lower mountain zone, and in snowy conditions even to foothill plains and town orchards, but rarely travels far. Very trusting, visiting courtyards and streets of mountain villages, and feeding in vegetable gardens.

Distribution and population. AZERBAIJAN. Range poorly known. Nests in mountains of Zakataly and Belokany districts of Great Caucasus, and in Murovdag (Little Caucasus), probably other areas in mountains. Uncommon. TURKEY. Occurs on most mountains which reach over 2000 m; westernmost breeding locality Ulu Dag. 10 000–100 000 pairs. IRAQ. Breeds Ser Amadiya mountains.

Accidental. Austria, Greece (mostly Chios), Cyprus, Syria, Egypt.

Beyond west Palearctic, extends from Iran east to Nepal and central Asian mountains.

Movements. Mainly altitudinal migrant. Chiefly resident in Turkey, dispersing to lower altitudes in winter (mostly November–March), and more widespread then in southern coastlands. Reaches Cyprus only exceptionally, but some Turkish birds move south inland, apparently especially in cold years, to winter locally in Syria, Lebanon, and northern and central Israel, chiefly above 800 m. Arrives in Israel from end of October to end of December, majority reaching winter sites December or beginning of January; movement inconspicuous, mostly 1st-year birds. Spring movement through Israel (also inconspicuous) from early February to mid-March, chiefly February. Winters irregularly highlands of northern Iraq.

Food. Seeds, fruits, and other plant material; sometimes small insects. Feeds on ground, in herbs, and in trees.

Social pattern and behaviour. Sociable all year, but largest flocks outside breeding season. Often associates for feeding with Serin and other finches. Pair-formation evidently takes place in flocks. Semi-colonial breeding apparently usual. Characteristic is communal singing and display (apparently linked to pair-formation) of up to 4 ♂♂ from perch or on ground when ♀(♀) nearby (similar when lone ♂ displaying to ♀):

Red-fronted Serin *Serinus pusillus*: **1** ad ♂ summer, **2** juv. Serin *Serinus serinus* (p. 1550); **3** ad ♂ summer, **4** ad ♂ winter, **5** ad ♀, **6** juv. Syrian Serin *Serinus syriacus* (p. 1552): **7** ad ♂ summer, **8** ad ♀ summer, **9** juv.

each ♂ sings, droops and partly spreads wings, fans and slightly raises tail, and prominently ruffles red crown; may crouch, also turn body or head to left and right in front of another ♂, or birds circle one another while singing. Usually ends with whole flock taking off suddenly. Singing in flight also frequent.

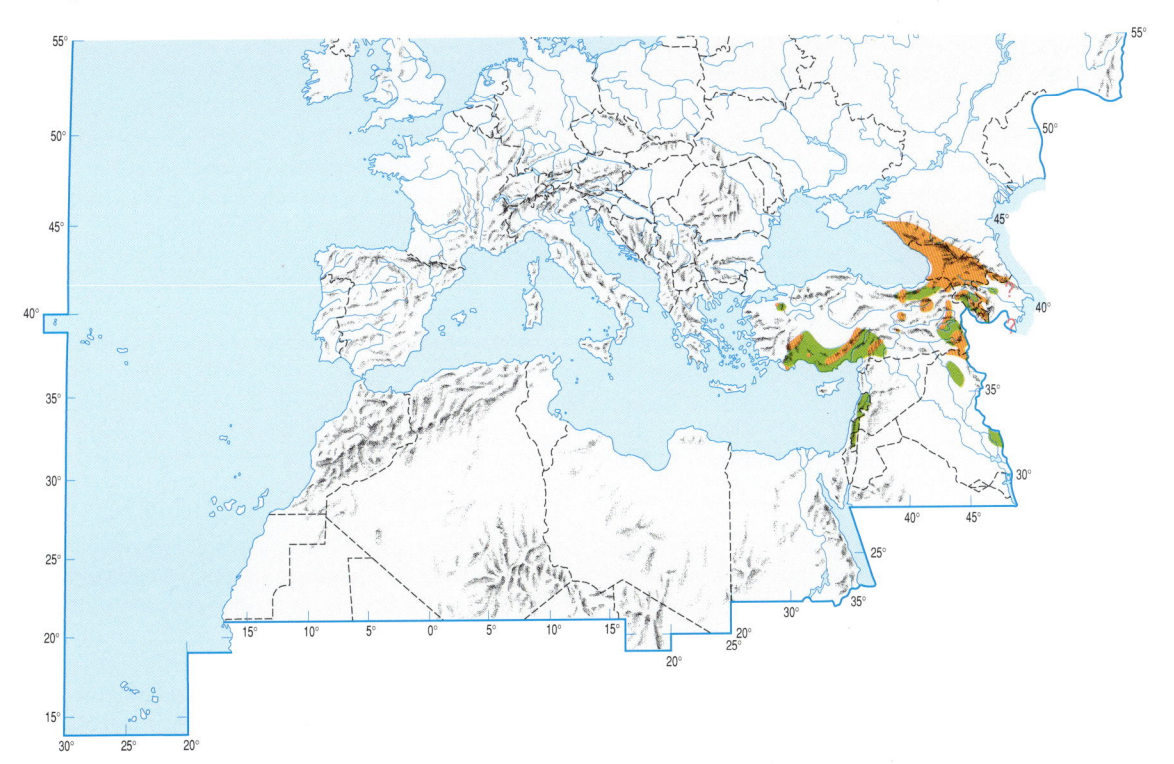

Voice. Song of ♂ an attractive and melodious, rapid and continuous, high-pitched bubbling twitter, with frequently interspersed ripples or (at times slightly hoarse) 'trills' of varying length, and hoarse 'kveeh' sounds; powerful outpouring delivery. Main call is 'ripple-call', a short, or sometimes longer, bubbling tremolo or high-pitched, soft, rippling twitter, resembling equivalent call of Serin but more delicate, weaker, and softer, with 'singing' and slightly nasal or metallic sound. Ripple-call given in flight and on ground or perch; serves for contact, perhaps also signals excitement or alarm. A 'dshUee' or 'djuee', closely resembling alarm-call of Serin, may have similar function.

Breeding. Season. Turkey: April–July; eggs laid mid-April at *c*. 1500 m. Caucasus: eggs laid 2nd half of May, 2nd clutch in July. Site. Low in dense bush or tree, generally growing at top or on ledge of inaccessible cliff, or high up in conifer, though still well-protected above by foliage; also in rock crevice, hole in scree, etc. Nest: neat and compact, appearing large and thick-walled for size of bird; foundation of dry grass, bark strips, stalks, moss, lichen, and sometimes twigs, lined thickly with plant down, feathers, etc., spiders' webs often incorporated. Eggs. Short sub-elliptical, smooth and faintly glossy; bluish-white, sparsely flecked, with pink or reddish-brown to purple-black scrawls, speckles, and blotches, mostly at broad end; sometimes unmarked. Clutch: 3–5. Incubation. *c*. 13 days. Fledging Period. 14–16 days.

Wing-length: ♂ 72–77, ♀ 68–75 mm.
Weight: ♂♀ mostly 10–13 g.

Serin *Serinus serinus*

PLATES: pages 1549, 1553

Du. Europese Kanarie   Fr. Serin cini   Ge. Girlitz   It. Verzellino
Ru. Канареечный вьюрок   Sp. Verdecillo común   Sw. Gulhämpling

Field characters. 11.5 cm; wing-span 20–23 cm. 15% smaller and noticeably more compact than Linnet; slightly smaller than other west Palearctic serins *Serinus* and Siskin but with proportionately long wings and deeply forked tail. Diminutive, stubby-billed, rather compact finch; epitome of genus. Adult has rather green, streaked upperparts with bright yellow rump; ♂ brilliantly yellow on forehead, face, throat, and breast; ♀ only dull yellow on face. Juvenile lacks yellow rump and is more heavily streaked below than either adult. Flight light, noticeably bouncing.

Below montane range of Citril Finch and north and west of Turkey, Serin overlaps widely only with Siskin; within and south of Turkey, overlaps only with distinctive Red-fronted Serin but approaches similar Syrian Serin. Anywhere in west Palearctic, further risk of confusion with escaped (wild caught or domestic) Canary and other (wild caught) African *Serinus*. Serin thus no easy target for inexperienced observer and worth long study when first found; important to remember that it has smallest size, heaviest and sharpest streaks, and stubbiest bill of west Palearctic *Serinus*. Song-flight flitting with almost bat-like wing-beats and glides; strongly recalls diminutive Greenfinch. Spends much time on ground, with rather fast hopping or seemingly creeping gait and low carriage when feeding but remarkably upright posture in excitement or alarm.

Habitat. Confined to west Palearctic, originally in Mediterranean zone, spreading north in 19th century into temperate drier and warmer regions of central Europe, and continuing in 20th century to fringe of boreal and steppe zones, and sparsely towards oceanic margins, not yet beyond July isotherm of 17°C. Vulnerable to cold wet weather and unable to cope with more northerly winters, or with higher altitudes except in south of range, where ascends to subalpine zone. Geographical spread has been accompanied by shift of habitat from mainly forest edges and clearings or scattered clumps and rows of trees on hills and mountain slopes to parkland, cemeteries, orchards, vineyards, garden suburbs, avenues, and other well-mixed, sunny, dry situations offering nest-sites, often in conifers, and song-perches, often on posts or cables.

Distribution. Major (mainly northward) spread began in 19th century; continued in 20th century. Recent expansion in Latvia, Lithuania, Hungary, Spain, Italy, Croatia, Russia, Belarus', Ukraine (marked), Moldova, and Cyprus. Decline reported Belgium, Denmark, Finland, and Estonia. Britain. Accidental until comparatively recently. Increase in records after 1960; first bred 1967, but breeding still sporadic (up to 2 pairs), not established. Jersey (Channel Islands): breeding suspected from 1972, first proved 1978. France. Spread north from 2nd half of 19th century, reaching Nord in 1950s, also more slowly west, reaching Finistère in 1970s. Spread has continued since, and general distribution more uniform, with gaps being filled up. Belgium. Distribution more patchy than formerly. Germany. Has bred in east since beginning of 20th century, spreading north to Baltic coast. Denmark. Sporadic breeding records 1948–65, then regular breeder for some years, no longer so by early 1990s. Sweden. First bred in 1940s; no obvious further spread in last decade. Finland. Probably not a regular breeder (up to 3 pairs). Marked increase in records in 1970s, then only 3–10 per year in 1980s. Bred 1967, 1976, 5 other probable records. Estonia. Breeding since 1950s. Latvia. First recorded 1935, first bred 1938. Lithuania. First breeding record 1957. Poland. First breeding recorded 1853; distributed throughout country, but still patchily. Czech Republic, Slovakia. Colonized 1840–60. Austria. Widespread and regionally common. Italy. Colonized Aeolian and Egadi islands in early 1980s. Malta. Occasional breeder. Greece. Not common in south, slightly commoner in north. Russia. Bred Voronezh region 1974. Reached south coast of Gulf of Finland in 1976. Belarus'. First bred in late 1950s. Turkey. Scarce and local eastern Black Sea coastlands and fringe of

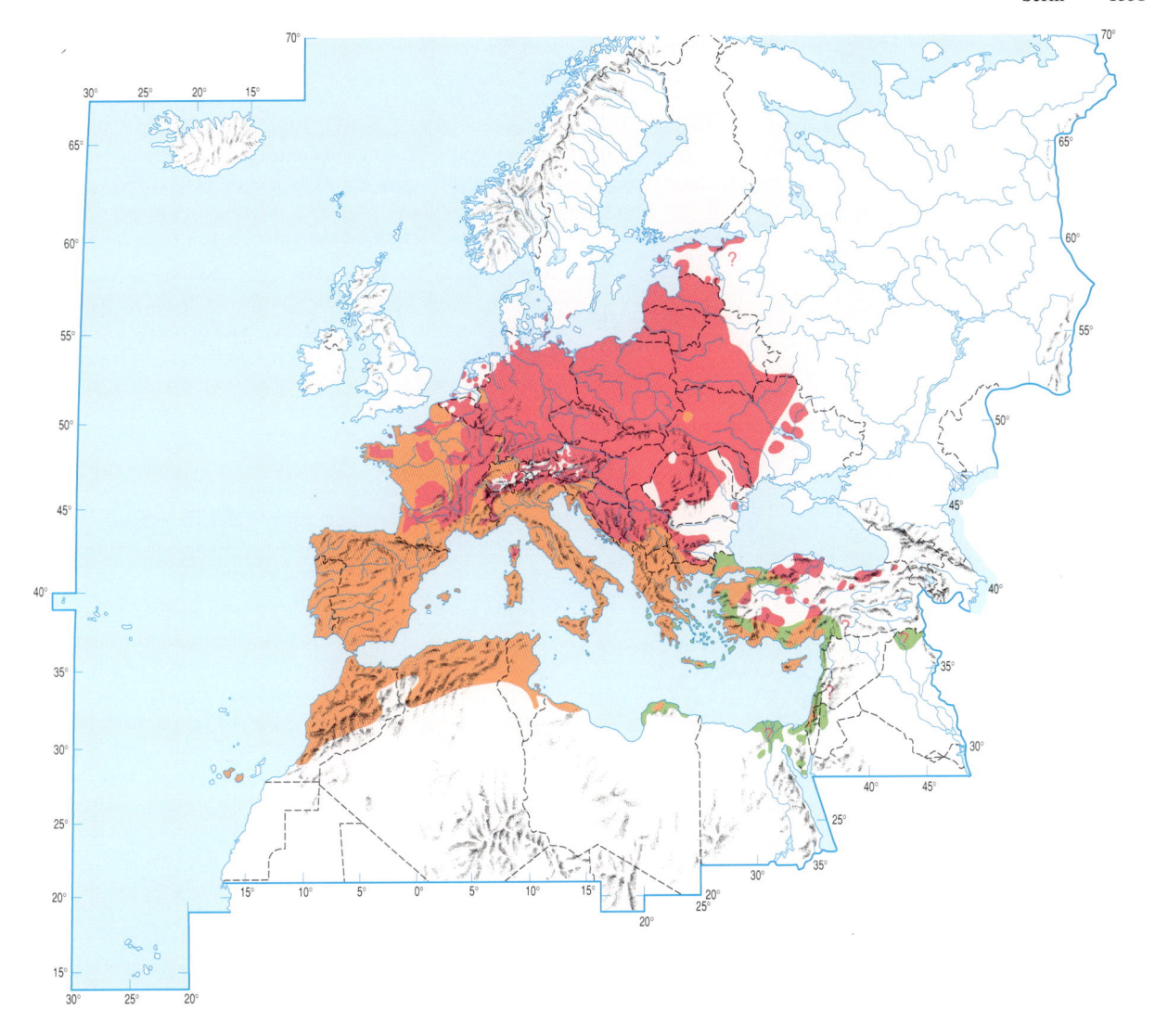

Central Plateau. CYPRUS. Extension of breeding range to lower altitudes since 1970. Locally common. SYRIA. Probably breeds near Damascus, but confirmation needed. ISRAEL. Breeding first recorded in 1977 on coastal plain; has since spread in coastal areas and western Negev. JORDAN. Bred Petra 1983. IRAQ. Probably rare breeder in north. EGYPT. Probably local breeder in Nile delta. CANARY ISLANDS. Breeding Gran Canaria and Tenerife since 1970s, perhaps originating from escapes.

Accidental. Ireland, Norway, Madeira.

Population. Increase Latvia, Lithuania, Spain, Italy, Croatia, Russia (Leningrad region), Ukraine, Moldova, and Israel; recent decline Low Countries, Germany, Fenno-Scandia, Estonia (marked), apparently stable elsewhere. BRITAIN. Channel Islands (Jersey): 20–30♂♂. FRANCE. 100 000–1 million pairs in 1970s. BELGIUM. 2500 pairs in 1970s, 650–1200 in 1989–91. LUXEMBOURG. 1000–1500 pairs. NETHERLANDS. Peak of 450–550 pairs 1978–9, falling to 100–150 in 1983–5. GERMANY. 563 000 pairs in mid-1980s; another estimate of 300 000–400 000 pairs in early 1990s. In east, 100 000 ± 50 000 pairs in early 1980s. DENMARK. 20–30 pairs 1976–80, up to 5 pairs in early 1990s. SWEDEN. 5–15 pairs in late 1980s. ESTONIA. Not more than 100 pairs in 1991. LATVIA. 50–300 pairs in 1980s. POLAND. 50 000–150 000 pairs; now apparently stable, following increase. CZECH REPUBLIC. 400 000–800 000 pairs 1985–9. SLOVAKIA. 50 000–100 000 pairs 1973–94. HUNGARY. 40 000–50 000 pairs 1979–93. SWITZERLAND. 20 000–40 000 pairs 1985–93. SPAIN. 4.1–6.6 million pairs. PORTUGAL. 1 million pairs 1978–84. ITALY. 200 000–600 000 pairs 1983–95. Recent increase includes Sicily. GREECE. 20 000–50 000 pairs. ALBANIA. 5000–20 000 pairs in 1981. YUGOSLAVIA: CROATIA. 120 000–180 000 pairs. SLOVENIA. 30 000–40 000 pairs. BULGARIA. 10 000–100 000 pairs. RUMANIA. 10 000–15 000 pairs 1986–92. RUSSIA. 100–1000 pairs. BELARUS'. 8000–15 000 pairs in 1990. UKRAINE. 600 000–650 000 pairs in 1986. MOLDOVA. 800–1000 pairs in 1988. TURKEY. 10 000–100 000 pairs. ISRAEL. A few hundred, perhaps a few thousand pairs in late 1980s; rapid increase. TUNISIA, ALGERIA, MOROCCO. Common.

Movements. Sedentary to migratory, wintering within and

south of breeding range. Most birds vacate northern parts of range, but winter records show that small numbers remain, at least in some years. In centre and south of range, amount of movement masked by passage and arrivals from further north, but even in Mediterranean countries a considerable number are migrants. Main autumn heading south-west for west European birds and south for east European birds (reverse in spring). Autumn movement (August–)September–November, chiefly October. Spring movement February–May, chiefly March–April.

In Britain, recorded in all months, with distinct peaks October–November and especially April–May. Long-term expansion of range across Europe shows marked north-east tendency (towards Leningrad region); expansion into Scandinavia slower, and still only sporadic in Britain, suggesting reluctance to cross open water.

Food. Seeds and other plant material; occasionally small invertebrates. Forages principally on herbs and on ground; tree-foraging probably mainly in spring. Especially in winter, forages in large flocks, often with other seed-eaters. Feeds energetically and with agility like Linnet (in tall herbs) or Siskin (in trees); extracts ripening seeds from heads of Compositae by carefully pulling down bracts surrounding inflorescence one by one, and removing petals or pappus using bill and tongue; pulls buds and catkins to pieces. Will use feet to hold items on ground while feeding.

Social pattern and behaviour. Gregarious outside breeding season, forming small flocks for feeding and migration, usually fewer than 100 birds but not uncommonly more on migration and in favoured wintering sites. In optimal conditions, breeds in neighbourhood groups, otherwise solitary. ♂ sings typically from tall exposed tree, overhead wire, etc., sometimes from much lower perch, even from ground; often followed by song-flight over and beyond territory. Launches almost vertically into the air, ruffles plumage, spreads wings and tail, and, with slow deep wing-beats (sometimes interspersed with glides) follows flitting erratic course c. 10–20 m above ground; path includes wide arcs and performer often throws himself from side to side in rolling movement; descent is in circles, with slow parachute drop at end to land on perch from which flight began, often beside ♀. There, continues singing or, after short pause, starts another song-flight.

Voice. Vocal all year, notably in continuing song outside breeding season. Song a fast jumble of high-pitched sizzling, tinkling (etc.) sounds given by ♂ from perch or in song-flight. Persistent succession of relatively long phrases of harsh jingling, chirping, twittering sounds, somewhat like splinters of glass being rubbed together. Reminiscent of Corn Bunting; utterly different from rich warbling of Canary; also recalls Siskin but much higher pitched. 'Ripple-call', a rather dry 'trillilit', given by both sexes as contact-call, typically (though not always) in flight. Other calls include a hard 'chit-chit-chit...', recalling Wren, and chirruping sparrow-like sounds. Common alarm-call a soft 'tsooeet', with upward inflection.

Breeding. Season. Egg-laying in February in North Africa, March in Spain, May in central and northern Europe. 2 broods usual. Site. Generally in conifer rather than broad-leaved tree, though also in bush; also commonly in fruit trees. Preferred position in outermost twigs, followed by top of conifer, then on branch against trunk. Nest: small and compact, of fine twigs, stalks, sometimes strips of bark, roots, grass, moss, or lichen, lined neatly and thickly with rootlets, hair, feathers, plant down, etc. Eggs. Sub-elliptical, smooth and slightly glossy; pale bluish-white, sometimes greenish-white, sparsely spotted and streaked rusty and purplish, mostly at broad end, sometimes forming circle. Clutch: 3–4 (2–5). Incubation. Average c. 12.7 days. Fledging Period. Average 15.2 days.

Wing-length: ♂ 68–78, ♀ 66–72 mm.
Weight: ♂♀ mostly 11–14 g.

Syrian Serin *Serinus syriacus*

PLATES: pages 1549, 1553

Du. Syrische Kanarie   Fr. Serin syriaque   Ge. Zederngirlitz   It. Verzellino di Siria
Ru. Сирийский вьюрок   Sp. Verdecillo sirio   Sw. Grönhämpling

Field characters. 12.5 cm; wing-span 21.5–24 cm. Nearly 10% larger than Serin, with less compact form most evident in relatively longer tail; matches Canary in size except for smaller bill. Small (not tiny), rather long-tailed serin, with rather subdued pale greyish-olive-yellow plumage which lacks strong streaks except on mantle. Breeding ♂ has remarkably open foreface, with almost orange forehead, pale yellow eye-ring, and conspicuous pale yellow greater coverts and edges to inner flight-feathers and tail. ♀ less colourful, greyer above and paler on face and below with indistinct streaks on rear flanks.

Not difficult to distinguish from Serin, due to larger size, more uniform, much less streaked plumage, and different voice, but beware confusion with potential vagrant Citril Finch which though greyer and less bright yellow on face and body shows similar wing and tail pattern.

Habitat. Breeds in restricted part of east Mediterranean sector of west Palearctic in dry warm sunny climate on high upland slopes and ridges, often rocky, carrying sparse open woodland or clumps of low bushes. Access to drinking water is essential. Early nests are at 900–1500 m, soon after snow melts, but 2nd

Chaffinch *Fringilla coelebs coelebs* (p. 1539): **1** ad ♂ summer, **2** ad ♀ summer. Blue Chaffinch *Fringilla teydea teydea* (p. 1544): **3** ad ♂ summer. Brambling *Fringilla montifringilla* (p. 1545): **4** ad ♂ summer, **5** ad ♀ winter. Red-fronted Serin *Serinus pusillus* (p. 1548): **6** ad ♂ summer. Serin *Serinus serinus* (p. 1550): **7** ad ♂ summer, **8** juv. Syrian Serin *Serinus syriacus*: **9** ad ♂ summer.

broods in July are reared at *c.* 1750 m. Sometimes breeding occurs in fruit orchards. In winter, migrates to lower ground, especially desert areas with some trees and water sources, or well-vegetated wadis or cultivated land. Regularly occurs on coast at Eilat (Israel), wintering mainly in wadis surrounding the town.

Distribution and population. Breeds only in west Palearctic. SYRIA. Breeds locally, in some cases in considerable numbers, slopes of Anti-Lebanon and Hermon massif. JORDAN. Small isolated population at Wadi Dana. 25–30 birds (including pairs and juveniles) in 1963, at least 20 in November 1991. ISRAEL. Breeds Mt Hermon and vicinity. 180 pairs 1982–3.

Movements. Resident, and altitudinal and short-distance migrant. Movements erratic and poorly known; more conspicuous in spring than autumn. Apparently mainly resident in Syria and Lebanon, though Israeli breeding area (Mt Hermon) vacated in autumn. Some birds descend to lower levels for winter; others move further, heading between north-east and south; some winter in Tigris valley in extreme north of Iraq, some in eastern and southern Syria, and Jordan; others reach Sinai (Egypt).

Food. No information in wild. In captivity prefers half-ripe or sprouting grass seeds.

Social pattern and behaviour. Little known, but apparently similar to congeners.

Voice. Song of ♂ a long phrase of trilling, chirping, harsh twittering, and high-pitched jingling sounds; includes 'siou' sounds and purring cardueline 'trrrr', and generally not loud. Calls include a variety of rippling, rattling, twittering, and grating sounds.

Breeding. SEASON. Israel: nest-building at 900–1500 m starts with snow-melt mid-April to May, and eggs laid May–June; most pairs ascend later to *c.* 1750 m July–August for 2nd brood. Syria and Lebanon: April–June, but very variable. SITE. Nests recorded 1–2 m above ground in oak, maple bush, cedar, juniper, hawthorn, and almond; can be rather conspicuous. Nest: rather like that of Goldfinch though less neat; cup shallow. EGGS. Sub-elliptical to oval, smooth and glossy; very pale blue, sparsely speckled reddish- or purplish-brown mostly at broad end; sometimes only spotted or scrawled in circle at broad end and rest of egg unmarked. Clutch: 4 (3–5). INCUBATION. 12–14 days. FLEDGING PERIOD. 14–16 days.

Wing-length: ♂ 75–80, ♀ 71–77 mm.
Weight: ♂♀ 10–14 g.

Canary *Serinus canaria*

PLATES: pages 1555, 1567

DU. Kanarie   FR. Serin des Canaries   GE. Kanarengirlitz   IT. Canarino
RU. Канарейка   SP. Canario   SW. Kanariefågel

Field characters. 12.5 cm; wing-span 20–23 cm. About 10% larger than Serin, with noticeably less stubby bill, proportionately shorter wings, and more attenuated rear body and tail. Small (not tiny), rather long-tailed finch; ancestor of larger domestic canary but resembling only its 'mule' variant, being far from wholly yellow and looking less green, more grey than Serin. Wing markings brightest of west Palearctic *Serinus*.

Unmistakable on Azores and western Canary Islands, where other *Serinus* not known. As an escaped bird elsewhere in west Palearctic (in wild form or as one of variety of domestic types) liable to confusion with all congeners except Red-fronted Serin. Short extension of folded primaries beyond tertials is always helpful. Flight fast, action resembling Serin but larger size and especially longer tail create silhouette recalling also Linnet and Redpoll.

Habitat. Resident on several west Palearctic Atlantic islands, at all altitudes from sea-level to 760 m or more in Madeira, to *c.* 1100 m in Azores, and even above 1500 m in Canary Islands. Sometimes in stands of pines, *Eucalyptus*, or in laurel forest and thickets of tamarisk, but more usually in open countryside with small trees, gardens, vineyards, orchards, and even on sand-dunes. Attracted, especially at nesting time, to banana trees bearing green clusters, camellias, and orange trees, and to shrubs such as heath and broom, as well as to cultivation of tomatoes and other crops, and to hedges. Vigorously aerial, especially in display, perching on highest treetops. Contrasts with other Fringillidae inhabiting the same islands in being highly adaptable and able to succeed over almost the entire range of available habitats.

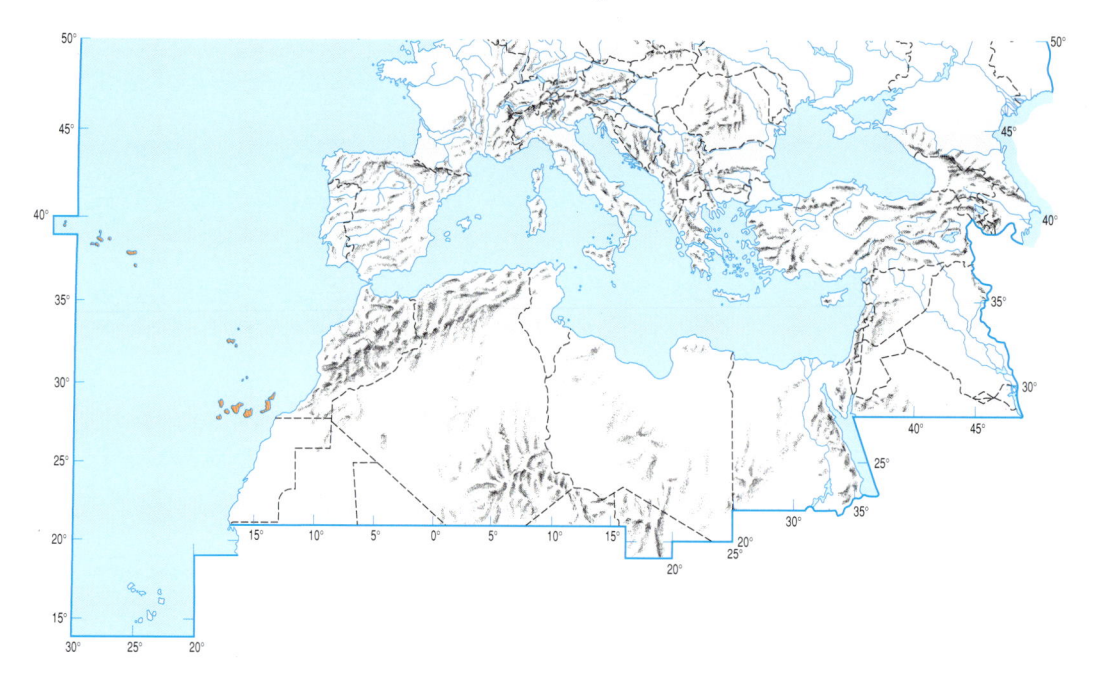

Canary *Serinus canaria* (p. 1554): **1** ad ♂ summer, **2** ad ♀ summer, **3** juv, **4** hybrid with Goldfinch *Carduelis carduelis*, **5** captive-bred variant. Citril Finch *Serinus citrinella* (p. 1556). *S. c. citrinella*: **6** ad ♂ summer, **7** ad ♀ summer, **8** 1st winter, **9** juv. *S. c. corsicana*: **10** ad ♂ summer.

Distribution and population. Breeds Madeira, Canary Islands, and Azores (all islands). Azores. Common. 30 000–60 000 pairs. Madeira. Very common. Canary Islands. Local Lanzarote (first breeding record 1995) and apparently Fuerteventura. Widespread and common on other main islands. 80 000–90 000 pairs.

Movements. Resident, with local movements. Tends to wander in flocks outside breeding season. Extent of inter-island movement apparently varies. In Canaries, not recorded from eastern islands. In Madeiran group, part of population leaves main island of Madeira in autumn; apparently rare in summer but common in winter on Porto Santo; birds arrive occasionally last week of August, but chiefly September–October, departing February–March. Inter-island movements also reported from Azores.

Food. Seeds and other plant material, occasionally small insects. Forages mainly on ground.

Social pattern and behaviour. Most detailed studies are of domesticated birds; no comprehensive studies of wild birds. Gregarious all year. During breeding season flocks persist for both feeding and roosting. Apparently breeds in neighbourhood groups, each pair defending not very large territory. ♂ sings from usually high perch; also performs song-flights in spring. Begins by ruffling plumage until he appears twice normal size, then (still ruffled) flies up from perch and, with slow bat-like wing-beats flies from tree to tree or in circle back to starting point. Final descent vertical or gliding with outstretched quivering wings (much like Tree Pipit), to alight on highest branch of tree, still singing.

Voice. Freely used throughout the year. Song of ♂ rich, varied, and musical. Consists of phrases or nearly continuous delivery of variously chirping, twittering, trilling, wheezing and piping sounds. Some organization into segments differing in duration, tempo, and timbre. Other calls include a loud twittering 'tjüdididi', typically given as flight-call; metallic 'dit-dit', similar to Citril Finch; and clear ascending whistle 'tweee'.

Breeding. SEASON. Canary Islands: eggs laid January to July. Madeira: March–June, mostly mid-April to end of May. Azores: eggs laid end of March to July, peak May–June. 2–3 broods. SITE. In tree or bush in woodland or hedge, commonly evergreen or species coming into leaf early; usually well hidden on fork or at end of branch, or in top of small tree. Nest: small, compact, often deep cup; foundation of twigs, stalks, rootlets, grass, moss, or lichen, lined with much plant down, also hair, feathers, or soft leaves. EGGS. Sub-elliptical, smooth, not glossy; pale light blue or bluish-green, darker than those of Serin and more heavily marked with violet, red, or rust spots and blotches, rather concentrated at broad end. Clutch: 3–4(–5). INCUBATION. 13–14 days. FLEDGING PERIOD. 15–17 days (14–21).

Wing-length: ♂ 71–76, ♀ 66–74 mm.
Weight: ♂ average 15.2, ♀ 15.3; latter including egg-laying ♀♀ of up to 18–20 g.

Citril Finch *Serinus citrinella*

Du. Citroenkanarie   Fr. Venturon montagnard   Ge. Zitronengirlitz   It. Venturone
Ru. Лимонный вьюрок   Sp. Verderón serrano   Sw. Citronsiska

Field characters. 12 cm; wing-span 22.5–24.5 cm. Slightly larger and less compact than Serin, with sharply tapered bill and proportionately longer wings; 20% smaller than Greenfinch. Small, elegant finch, form and behaviour recalling small *Carduelis* as much as any *Serinus*; plumage less contrasted and less streaked than any other west Palearctic *Serinus*. Mainland race basically grey-green, yellower on foreface, lower body, and rump; wings blackish with 2 greenish-yellow wing-bars.

Unmistakable within normal montane habitats in Europe but vagrant elsewhere could be confused with any west Palearctic *Serinus* except Red-fronted Serin. Appearance converges closely with Syrian Serin but that species slightly larger and paler, with much brighter yellow foreface in both sexes, fully yellow rump, prominent yellow fringes to tail-feathers, and white lower body. Distinction from Serin and from escaped adult Canary is simple, as both show distinct streaks, but separation of juveniles needs care.

Habitat. Largely restricted to montane south-west Palearctic, in cool alpine climates, upwards from 700 m in Schwarzwald (Germany) and from 1000 m in Switzerland up to treeline at *c.* 1300–1500 m, and after breeding to 3300 m or more. Heavy winter snowfalls usually render mainland breeding habitats untenable then, and emigrants to southern France and elsewhere near the Mediterranean often feed in birches and alders.

In breeding areas, prefers open light marginal woodland composed at least partially of spruce, often bordering on alpine meadows or clumps of spruce scattered on open terrain, and frequently having alpine huts as further attraction. While far from embracing man-made environments to extent of (e.g.) Serin, not reluctant to take advantage of their overlap with its natural habitat. Firmly tied to trees, but appears never to accept dense closed stands of forest.

Distribution and population. France. 1000–10 000 pairs in 1970s. Germany. 870 pairs in mid-1980s. Overall trends unclear, but probable decline Bayern and southern Schwarzwald, and considerable increase in northern Schwarzwald; conspicuous increase in number of records for Thüringer Wald since *c.* 1970. Austria. Locally common; no clear trend. Switzerland. 5000–30 000 pairs 1985–93; stable. Spain. Slight increase in range and population. 225 000–230 000 pairs. Italy. 5000–10 000 pairs 1983–95. Yugoslavia: Slovenia. 10–20 pairs; stable.

Accidental. Finland, Poland, Czech Republic, Slovakia, Balearic Islands, Turkey, Algeria, Morocco (Ceuta).

Movements. Short-distance and vertical migrant, wintering at middle altitudes, chiefly above 1000 m, though heavy snowfall causes birds to move lower temporarily; sedentary in some southern areas.

Most birds leave Alps in winter; in Switzerland, winters regularly (in varying numbers) only in Valais canton in south-west, less regularly on other south-facing slopes of Jura and Alps; numbers difficult to estimate as birds tend to move from place to place; more remain in milder winters, but probably never more than a quarter or third of population. Regularly remains in breeding areas of French/Italian maritime Alps; many French and Swiss birds winter in limited area in mountains of southern France—Cévennes, south-east Massif Central, and western edge of Alps from Vercors to Monts de Vaucluse. In Corsica and Sardinia, more widespread winter than summer, many birds moving to lower levels and to coasts.

Food. Small to medium-sized seeds (possibly more grass seeds than most other Fringillidae), and sometimes green material from wide variety of plants; and some insects. Seeds of spruce and pine important at times. Will hang from seed-head to extract seeds, though rarely completely upside-down, and in general appears less agile than (e.g.) Redpoll or Siskin, also when removing seeds from conifer cones in trees or feeding on catkins. Otherwise forages mainly on ground.

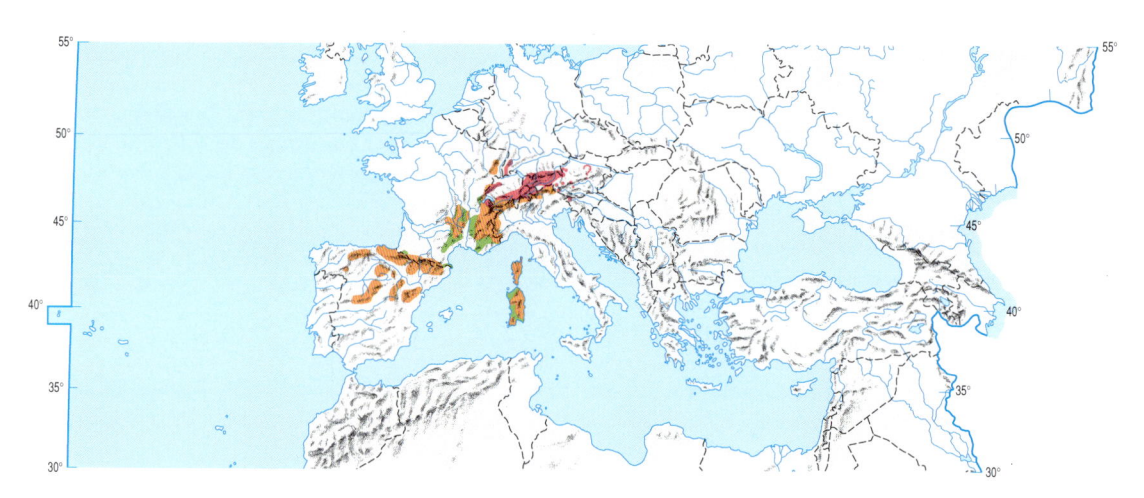

Social pattern and behaviour. Gregarious outside breeding season, with winter flocks of up to several hundred recorded in Switzerland and France, up to 20 in Corsica. In breeding season often in neighbourhood groups, with nests sometimes only a few metres apart. ♂ sings from top of tree or shrub; also in song-flight, in which bird circles, tail spread, beating wings slowly, giving bat-like impression; may sing while flying with shallow wingbeats from tree to tree or around tree. In alpine range, song noted in most months of year; irregular in winter, mainly when mild.

Voice. Song of ♂ nominate *citrinella* most often likened, in twittering and tinkling or rapid, at times strained, babbling quality to certain *Carduelis* finches and Serin. Often a short introduction of one to several well-spaced units followed by series of characteristically varied units or mostly short motifs: includes buzzes and rattles, trills, 'pi' or 'pink' sounds, and frequent flight-calls. Song of *corsicana* easily distinguished from nominate *citrinella* because of its segmented structure, with form closely akin to song of Wren. Corsican birds may also give slow song ('chant') comprising series of mostly long and very steady tones (6 types differing in pitch and timbre) of remarkable beauty delivered at very slow rate (*c.* 51 per min). Flight-call (in nominate *citrinella*) nasal, creaky and metallic 'di', 'dit', 'tweck', etc., sometimes given in rapid series; serves as contact-call in flight and at take-off, also frequently incorporated in song. Alarm calls mostly monosyllabic or disyllabic, clear and plaintive: 'tsüü', 'tsi-ew', 'hwee', etc.

Breeding. SEASON. Switzerland: eggs generally laid from 2nd half of April or early May, 2nd clutches June and 1st half of July; in exceptional years, eggs laid late February. Northern Italy: at 2500–3000 m, eggs laid late April to mid-June. Northeast Spain: at 500–1600 m, eggs laid early April to June. Commonly 2 broods. SITE. On mainland, almost always in conifer, usually close to trunk in upper part of tree and protected by dense twigs above, though sometimes out on branch. Nest-sites on Corsica very different; majority in tree-heath *c.* 1 m (*c.* 0.3–2) above ground, or in juniper. Nest: foundation of dry stalks, grass, roots, lichen, and spiders' webs, smoothly lined with hair, wool, feathers, rootlets, paper, etc.; on Corsica, flimsy and shallow, mostly of fine grass lined with hair, feathers, moss, or plant down. EGGS. Sub-elliptical, smooth and glossy; pale blue, sparsely marked towards broad end with small dark rust spots and scrawls and reddish-violet undermarkings. Clutch: (3–)4–5 in nominate *citrinella*; 3–4(–5) on Corsica. INCUBATION. 13–14 days. FLEDGING PERIOD. 16–17 (15–18) days.

Wing-length: *S. c. citrinella*: ♂ 75–83, ♀ 75–79 mm. *S. c. corsicana*: ♂ 70–77, ♀ 68–74 mm.
Weight: *S. c. citrinella*: ♂♀ average 12.2–12.9 g.

Geographical variation. 2 well-marked races, sometimes considered separate species. Isolated *corsicana* from Sardinia, Corsica, Capraia, and Elba differs from nominate *citrinella* from southern continental Europe in colour, size, habitat, and voice. Yellow of face of ♂ *corsicana* brighter yellow (less green) than in ♂ nominate *citrinella*; yellow-green of forecrown on average more restricted; mantle and scapulars conspicuously different, cinnamon-brown with pronounced dark grey streaks (in nominate *citrinella*, almost uniform yellow-green); rump duller yellow-grey (not bright green-yellow); black stripe on lore distinct (not faint and not mottled grey and yellow); chest purer yellow (less green); under tail-coverts mainly yellow-white or white (not mainly yellow). ♀ *corsicana* differs from ♂ in having brown-grey crown and hindneck, streaked darker brown, grey lore, and paler and more restricted yellow on underparts (in *S. c. citrinella*, ♀ differs from ♂ mainly in grey instead of yellow chin and throat; hindcrown, nape, mantle, and scapulars of ♀ somewhat browner than in ♂, belly less intensely yellow, but difference not as marked as in *corsicana*).

Greenfinch *Carduelis chloris*

PLATES: pages 1558, 1567

DU. Groenling   FR. Verdier d'Europe   GE. Grünling   IT. Verdone
RU. Зеленушка   SP. Verderón común   SW. Grönfink

Field characters. 15 cm; wing-span 24.5–27.5 cm. Close in size to House Sparrow, with strong bill and deep head and body also recalling that species but tail distinctly forked; 25% larger than Siskin and serins. Medium-sized, robust, plump and noticeably short-tailed finch, with stout conical bill; form most compact of west Palearctic finches except for even larger Hawfinch. ♂ olive-green and yellow, looking bright only in sunlight; ♀ dull olive-brown and yellowish-buff, faintly streaked on back; juvenile dirty buff-brown and pale buff, fully but not sharply streaked; all show striking yellow patches on primaries and side of tail, shining on ♂, duller on ♀, and dullest on juvenile. Bill pale flesh.

When size evident, adult unmistakable. All other yellow-green finches of west Palearctic smaller and much less uniformly patterned. When calling or in flight, juvenile not difficult to identify; if perched, its more strongly streaked plumage recalls ♀ and immature rosefinches, juvenile crossbills, and ♀ and immature sparrows; however, all lack yellow on primaries and sides of tail. Flight quite powerful, with long wings beaten in bursts which soon produce marked undulations; shows little agility and much flapping when feeding in trees, etc. Flight silhouette compact and noticeably short-tailed. Song-flight remarkably bat-like, with circular progress sustained in spite of seemingly too slow and erratic beats of splayed wings.

Habitat. Breeds almost throughout Europe, to south of Arctic Circle or of July isotherm of 14°C in boreal, temperate, Mediterranean, and steppe zones, extending also to North

Greenfinch *Carduelis chloris*. *C. c. chloris* (northern and central Europe): **1** ad ♂ summer, **2** ad ♂ winter, **3** ad ♀ summer, **4** juv ♂, **5** juv ♀. *C. c. aurantiiventris* (southern France, eastern Iberia, and Mediterranean Europe): **6** ad ♂ summer. *C. c. chlorotica* (Levant): **7** ad ♂ summer, **8** ad ♂ winter. *C. c. bilkevitchi* (Caucasus): **9** ad ♂ summer.

Africa and western Asia. Attached to tall densely leafed trees and to diet of seeds accessible under appropriate trees, on bushes, or on crop, weed, and other plants in fields. Has expanded from natural woodland edge, scrub, streambanks, and groups of trees on grassland to tall hedgerows, lines of planted trees, orchards, conifer plantations, cemeteries, churchyards, parks, large gardens, and other situations where tall trees, sunny aspects, and ready access to ground or other sources of seeds, fruits, and insect food are present together in breeding season. At other seasons, may utilize areas away from trees, on farm fields, salt-marshes, shingle banks, and other open sites. Mainly a lowland species, in Switzerland normally below 900 m, although found more sparsely up to *c.* 1400 m, especially in conifers, where, however, density is much lower than in lowland parks, etc.

Distribution. Has spread north in Fenno-Scandia, and also expanded in south of range. BRITAIN, IRELAND. Highest numbers in south and east. Due to changed farming practices, has become less abundant on farmland since 1950s, more numerous in towns and villages. NORWAY. Has spread in south; also northwards, especially in 1980s. SWEDEN. Expansion since 2nd half of 19th century, which has continued slowly. FINLAND. Recent northward spread; probably influenced by intensified winter feeding, also planting of rose bushes and abandonment of arable land. SPAIN. Slight increase. MALTA. Breeds occasionally. UKRAINE. Slight increase. CYPRUS. Colonized very rapidly in 1960s–70s; now widespread. ISRAEL. Has spread since 1970s, following development of agricultural settlements. IRAQ. Breeds in north. EGYPT. Common in north-east Sinai, and breeding recorded from 1985 in Nile delta. Apparently expanding range. AZORES. Introduced from Portugal *c.* 1890. MADEIRA. First proved to breed 1968; probably only accidental formerly. CANARY ISLANDS. Established Gran Canaria and Tenerife in late 1960s; now also breeding La Gomera, El Hierro, and Fuerteventura.

Accidental. Faeroes, Mauritania.

Beyond west Palearctic, breeds northern Iran; also isolated area in Uzbekistan, Kyrgyzstan, and southern Kazakhstan (east to *c.* 75°E). Introduced in New Zealand, south-east Australia, and very locally in south-east South America.

Population. Slight increase in Fenno-Scandia, and in other areas of range expansion. Decline reported Estonia (in 1980s, following increase 1970s), Czech Republic, and locally in Belgium. Elsewhere stable. BRITAIN. Estimated 530 000 territories 1988–91, but difficult to census; relatively stable since mid-1960s. IRELAND. 160 000 territories 1988–91. FRANCE. Over 1 million pairs in 1970s. BELGIUM. 70 000 pairs 1973–7. Apparently mostly stable, but recent study in Limburg showed decrease of 25% from early 1980s to early 1990s, mainly in agricultural areas. LUXEMBOURG. 18 000–25 000 pairs. NETHERLANDS. 40 000–80 000 pairs 1979–85. GERMANY. 2.9 million pairs in mid-1980s. DENMARK. 200 000–2.1 million pairs 1987–8. NORWAY. 100 000–500 000 pairs 1970–90. SWEDEN. 200 000–500 000 pairs in late 1980s. FINLAND. 100 000–150 000 pairs in late 1980s. ESTONIA. 50 000–100 000 pairs in 1991. LATVIA. 10 000–20 000 pairs in 1980s. LITHUANIA. Common. POLAND. Fairly common to common. CZECH REPUBLIC. 500 000–1 million pairs 1985–9. SLOVAKIA.

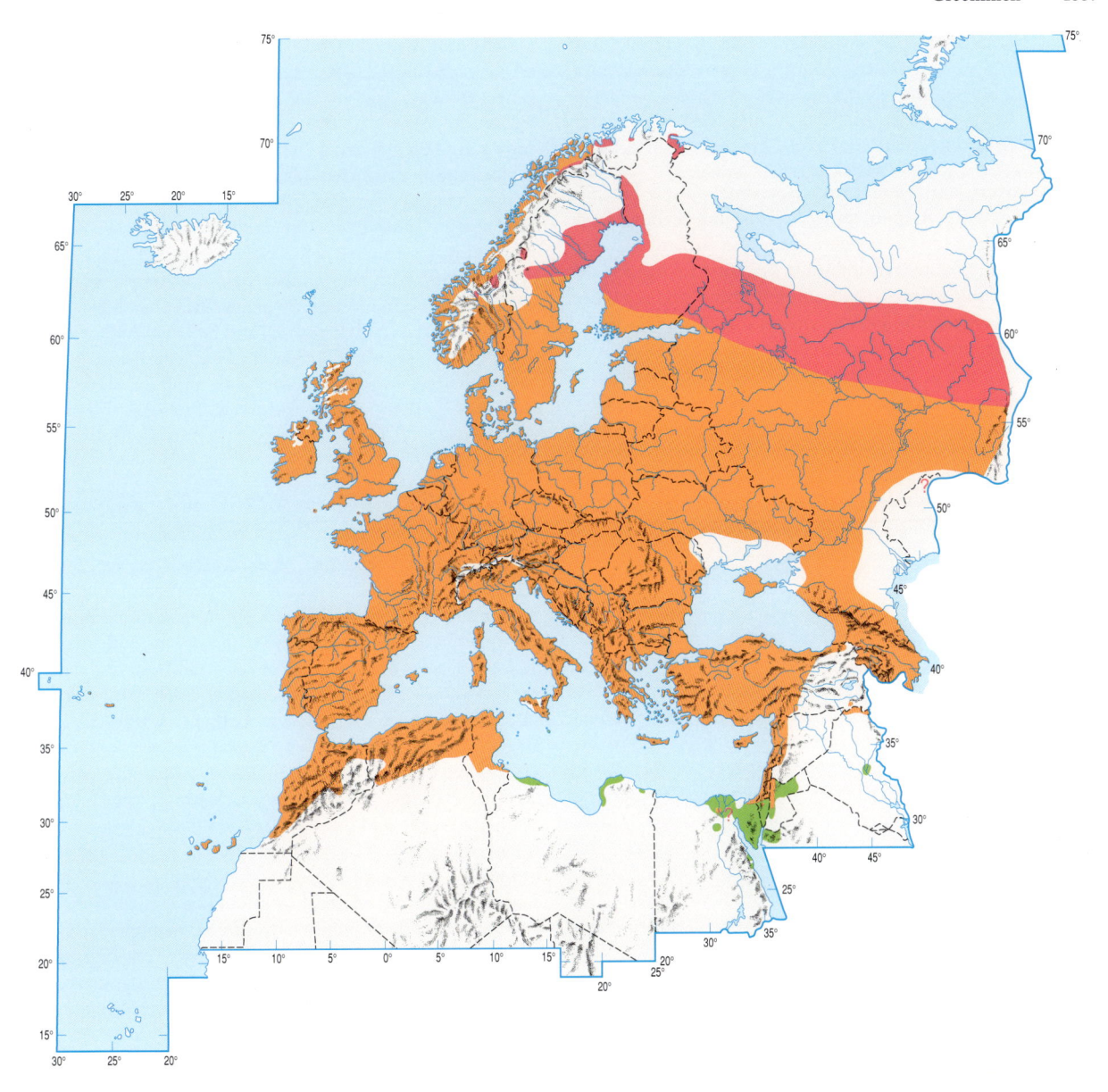

100 000–130 000 pairs 1973–94. HUNGARY. 200 000–250 000 pairs 1979–93. AUSTRIA. Common. SWITZERLAND. 150 000–250 000 pairs 1985–93. SPAIN. 106 000–3.6 million pairs. PORTUGAL. 1 million pairs 1978–84. ITALY. 200 000–600 000 pairs 1983–95. GREECE. 100 000–150 000 pairs. ALBANIA. 20 000–50 000 pairs in 1981. YUGOSLAVIA: CROATIA. 250 000–300 000 pairs. SLOVENIA. 50 000–80 000 pairs. BULGARIA. 100 000–1 million pairs. RUMANIA. 280 000–400 000 pairs 1986–92. RUSSIA. 100 000–1 million pairs. BELARUS'. 210 000–240 000 pairs in 1990. UKRAINE. 800 000–950 000 pairs in 1986. MOLDOVA. 35 000–50 000 pairs in 1988. AZERBAIJAN. Common. TURKEY. 50 000–500 000 pairs. CYPRUS. Common. ISRAEL. At least a few hundred thousand pairs in 1980s. JORDAN. Very common. TUNISIA. Common. MOROCCO. Uncommon to locally common. AZORES. Scarce. MADEIRA. Very rare.

Movements. Partially migratory in most of range; some southern populations apparently resident and dispersive. Birds head chiefly south-west to winter almost entirely within breeding range, with concentrations in Mediterranean region. Western populations winter furthest west, and more eastern populations progressively further east.

Autumn migration begins late; main movement October–November in most of range. Spring migration chiefly March–April.

Food. Fairly large (often hard) seeds, of herbs, trees, and shrubs, also of cereals; a few invertebrates taken in breeding season and also fed to young. Eats a wider range of seeds than probably any other Carduelinae in west Palearctic. Especially fond of rose hips. Perches in shrubs and trees to feed but more rarely on herbs, though will stand on stem to bend seed-head or flower onto ground; usually picks seeds from ground, jumps up to pull head down, or sometimes bites through stem so head falls; only occasionally clamps food under foot. Generally eats only seeds of fleshy fruits, discarding flesh. A frequent visitor to bird-tables, with preference for sunflower seeds.

Social pattern and behaviour. Gregarious outside breeding season from early autumn to spring, forming feeding flocks which are largest (up to several thousands in Britain) and most compact in late winter when food is well dispersed and highly localized. Mating system mainly monogamous but significant degree of polygyny also occurs. Breeds solitarily or in neighbourhood groups of 4–6 pairs. No territory as such, with defence confined to immediate nest-area; owner may forage at considerable distance from nest. ♂ sings from song-post in exposed tree-top or in song-flight (typically launched from song-post), sometimes also in normal flight. In song-flight, wing-beats deep and slow ('bat flight'), body rolling from side to side, path weaving erratically over and among tree-tops above breeding area, bird singing and calling the while. ♀ also sings but less often and less intensely than ♂.

Voice. Song of ♂ consists of groups of pleasant rolling tremolos, of which any particular bird has several variants differing subtly in pitch, timbre etc.; tremolos punctuated by more slowly delivered repetition of tonal and more noisy units, also by single longer and rather nasal 'chewlee' and the familiar nasal or buzzing wheeze. Simpler versions of song not uncommonly given, especially the nasal wheeze, which may be given repeatedly on its own. From any individual, wheeze can be quite variable, e.g. pitch sometimes even, at other times falling. Commonest call a rapid twittering 'chichichichichit' or 'chill ill ill ill'; also given as isolated units, thus repeated monosyllabic 'chüp' (recalls Crossbill but less hard and less explosive), 'cheu' or 'teu' and variants, commonly in flight but also when perched. Threat-call a repeated sharp rattle 'tsrr'. Alarm-call a slightly hoarse 'diUwee' or 'dshUee' like Canary.

Breeding. SEASON. Britain: clutches complete late April to mid-August, with peak mid-May and smaller peak in 2nd half of June. Fenno-Scandia: eggs laid late April to early July. Germany: eggs laid mid-March to late July, peak in May. Spain: late March to early August, with peaks in May and early July. 2 broods usual. SITE. Against trunk or in strong fork of dense bush, small tree (often in hedge), or creeper; conifers or other evergreens slightly preferred, especially early in season. Nest: stout, robust structure with foundation of dry twigs, grass, moss, and lichen lined with fine grasses, rootlets, plant down, hair, feathers, or man-made material. EGGS. Sub-elliptical, smooth and slightly glossy; greyish-white to bluish-white, or beige, sparsely spotted and blotched (rarely scrawled) reddish, purplish, or blackish, concentrated at broad end occasionally forming ring; some with pink or violet under-markings, others hardly marked. Clutch: 4–6 (2–7). INCUBATION. 11–15 days, average 12.9 days. FLEDGING PERIOD. 14–18 days.

Wing-length: *C. c. chloris*: ♂ 83–92, ♀ 81–90 mm.
Weight: *C. c. chloris*: ♂ 84–96, ♀ 81–94 g.

Geographical variation. Marked. Involves size (length of wing, tail, or tarsus), relative length, depth, and width of bill, colour and contrast of forehead, colour and intensity of green on upperparts, throat, and chest, and colour and intensity of yellow of belly to under tail-coverts, tail-base, and fringes of flight-feathers. Variation in size largely clinal, with larger birds in Fenno-Scandia and northern Russia, intermediate ones in Britain, France, mainland Italy, Balkans, Turkey, and Caucasus area, smaller ones in Iberia, north-west Morocco, north-east Tunisia, Balearic Islands, Corsica, Sardinia, Sicily, southern Greece, Cyprus, and Levant. Variation in colour more difficult to assess: depends greatly on age and sex of bird, and strongly influenced by bleaching and wear. Adult ♂ of *C. c. chloris* from northern and central Europe and of *harrisoni* from Britain and Ireland both brownish olive-green on upperparts, tinged grey on cap when fresh, fringed brown on remainder of upperparts then; when worn, olive-green of mantle and scapulars still distinctly tinged brown. Other races brighter green, lacking brown tinge. 9 races recognized in west Palearctic.

Goldfinch *Carduelis carduelis*. *C. c. carduelis*: **1** ad ♂ summer, **2** ad ♂ winter, **3** ad ♀ summer, **4** 1st summer ♂, **5** juv. *C. c. britannica*: **6** ad ♂ summer. *C. c. parva*: **7** ad ♂ summer. *C. c. niediecki*: **8** ad ♂ summer. *C. c. loudoni* (Iran, perhaps Turkey): **9** ad ♂ summer.

Goldfinch *Carduelis carduelis*

PLATES: pages 1561, 1567

Du. Putter   Fr. Chardonneret élégant   Ge. Stieglitz   It. Cardellino
Ru. Щегол   Sp. Jilguero   Sw. Steglits   N. Am. European Goldfinch

Field characters. 12 cm; wing-span 21–25.5 cm. Size close to Linnet but with 10% shorter tail. Small, delicate, beautifully marked finch, with noticeably pointed bill and light dancing flight. At all ages, displays diagnostic shining, golden-yellow panel along centre of black wing. Adult has unique head pattern of seemingly vertical bands of red-white-black and tawny-brown back contrasting with wings, bold whitish rump, and black tail. Juvenile brown and streaked, with pale 'unfinished' head, but shows diagnostic wing-panel.

Adult unmistakable but juvenile liable to confusion with Siskin and *Serinus*, best separated by obvious width of panel on black wings, black tail, amorphous head, and voice (see below). Flight light with noticeable acceleration and rise after burst of wing-beats and then glide or descent with wings closed up beside body, producing dancing, even skipping progress over short distances and more bounding undulations over long tracks than in larger finches.

Habitat. Breeds over west Palearctic north to July isotherm of 17°C in boreal, temperate, Mediterranean, and steppe zones, both Atlantic and continental. Predominantly in lowlands, but in Switzerland breeds generally up to 1000 m, quite commonly to 1300 m, and occurs in late summer and autumn up to 2400 m, resorting to alpine meadows and areas near chalets. In breeding season, shows preference for orchards, cemeteries, parks, gardens, avenues, and tree nurseries, often in or near human settlements, and especially where patches of tall weeds and other concentrated food sources are present. Also favours streamside or fen woodlands, open or fringe woodlands and heathlands, and commons with well-grown hawthorn, gorse, and other scrub or thicket species.

Outside breeding season, reliance on tall Compositae such as thistles, dandelions, ragwort, and burdocks dictates movements to rough grasslands, vacant sites, overgrown rubbish dumps, etc., although much use is made of woodlands in winter; alder and pine are also favoured food sources.

Distribution. Has spread in Britain, parts of Fenno-Scandia, Israel, and Egypt. BRITAIN. Has been spreading northwards over several decades, probably following cessation of exploitation. IRELAND. Some recent contraction of range, following long-term expansion. NORWAY. Recent expansion south-west and north. FINLAND. First bred 1860s, and expanded range up to 1950s. SPAIN. Slight increase. MALTA. Occasionally breeds. UKRAINE. Slight decrease. ISRAEL. Has expanded range in recent decades, following development of settlements and agriculture. IRAQ. Breeds in north. SAUDI ARABIA. Breeding near Tabuk since *c.* 1982. EGYPT. Has recently spread; evidence of breeding in Western Desert oases from 1981. AZORES. Introduced at end of 19th century; now breeds on all islands except Corvo.

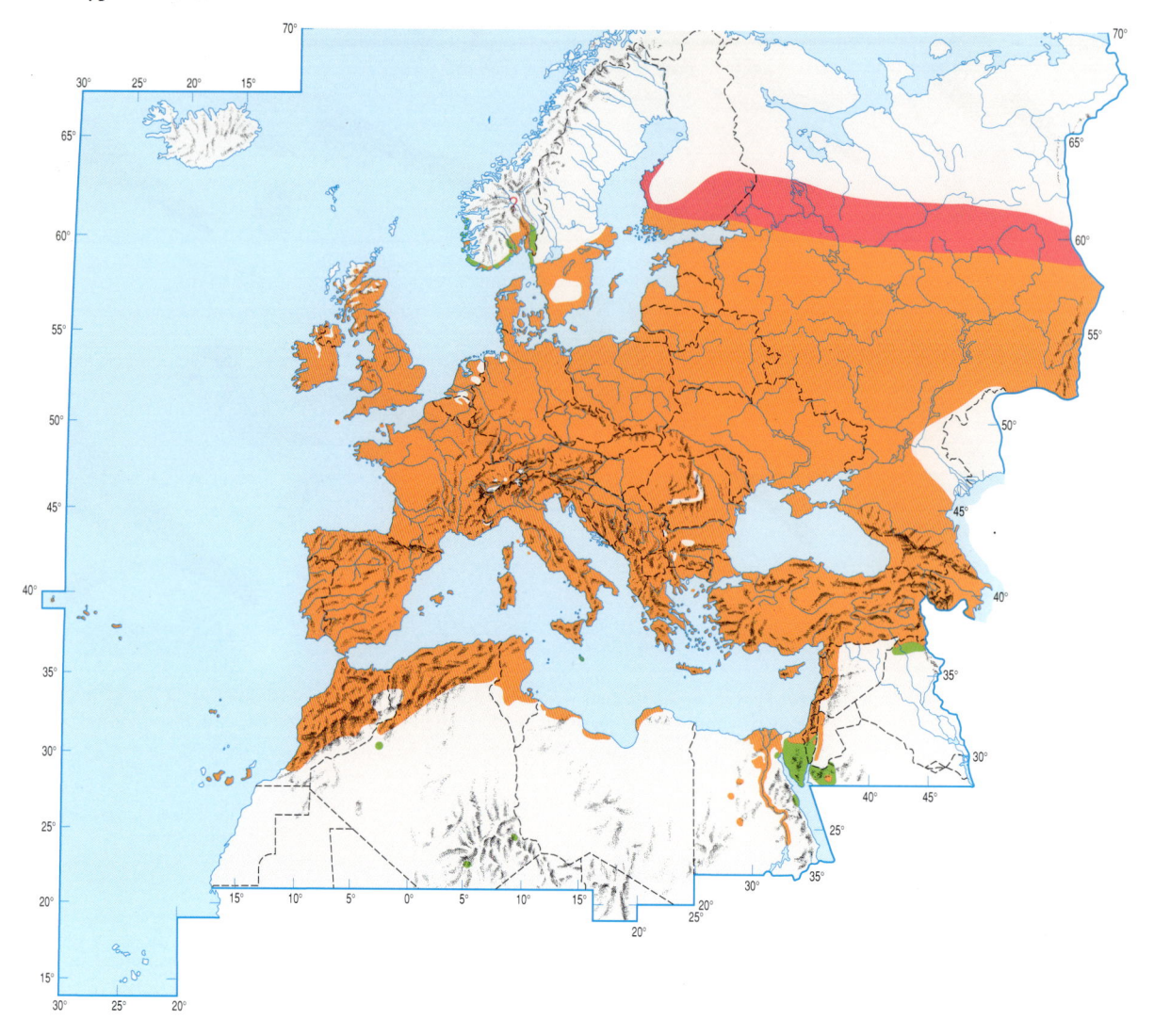

CANARY ISLANDS. Breeds Fuerteventura, Gran Canaria, Tenerife, and La Gomera. Birds observed on remaining islands, but breeding not confirmed. CAPE VERDE ISLANDS. Small introduced population present (and attempted to breed) Santiago 1963–5; not recorded since.

Accidental. Faeroes, Kuwait (or occasional winter visitor).

Beyond west Palearctic, extends east through south-west Siberia and northern Kazakhstan to Baykal region and north-west Mongolia, thence south-west to Iran and south to western Himalayas (including Nepal). Introduced in New Zealand, southern Australia, Bermuda, and very locally in south-east South America.

Population. Stable in most countries. BRITAIN. 220 000 territories 1988–91. Increase 1960s–70s, but decline since c. 1980, probably mainly due to increasing use of herbicides; small improvement on farmland 1987–94. IRELAND. 55 000 territories 1988–91. FRANCE. Over 1 million pairs in 1970s. BELGIUM. 3500–6500 pairs, mainly in Wallonia. Increase in early 1970s, probably due to control of bird trade; now stable. LUXEMBOURG. 6000–8000 pairs. NETHERLANDS. At least 4000–7000 pairs 1979–85; *britannica* (in west) increasing, nominate *carduelis* (in east) decreasing. GERMANY. 931 000 pairs in mid-1980s. Other estimates: 400 000–560 000 pairs in early 1990s; 190 000 ± 100 000 in east in early 1980s. DENMARK. 900–11 000 pairs 1987–8; increase 1988–94. NORWAY. 500–2000 pairs 1970–90. SWEDEN. 1000–5000 pairs in late 1980s; slight decrease. FINLAND. 3000–6000 pairs in late 1980s. Marked fluctuations: rapid recent increase following strong decline 1960s–80s. ESTONIA. 10 000–20 000 pairs in 1991; decrease. LATVIA. 15 000–50 000 pairs in 1980s. LITHUANIA. Common, but not abundant. POLAND. 200 000–400 000 pairs. CZECH REPUBLIC. 200 000–400 000 pairs 1985–9. SLOVAKIA. 100 000–150 000 pairs 1973–94. HUNGARY. 200 000–250 000 pairs 1979–93. AUSTRIA. Locally common. SWITZERLAND. 20 000–50 000 pairs 1985–93. SPAIN. 800 000–2.9 million pairs;

slight increase. PORTUGAL. 1 million pairs 1978–84. ITALY. 1–2 million pairs 1983–95. GREECE. 100 000–200 000 pairs. ALBANIA. 20 000–50 000 pairs in 1981. YUGOSLAVIA: CROATIA. 220 000–260 000 pairs. SLOVENIA. 50 000–60 000 pairs. BULGARIA. 100 000–1 million pairs. RUMANIA. 300 000–600 000 pairs 1986–92. RUSSIA. 100 000–1 million pairs. BELARUS'. 180 000–200 000 pairs in 1990. UKRAINE. 650 000–800 000 pairs in 1986; slight decrease. MOLDOVA. 40 000–60 000 pairs in 1988. AZERBAIJAN. Common. TURKEY. 1–10 million pairs. CYPRUS. Very common and widespread. ISRAEL. At least a few hundred thousand pairs in 1980s. JORDAN. Very common. SAUDI ARABIA. Increasing. EGYPT, TUNISIA, ALGERIA, MOROCCO, AZORES, MADEIRA. Common or very common. CANARY ISLANDS. Apparently declining.

Movements. Partially migratory, wintering almost entirely within breeding range, with concentrations in Mediterranean region. Areas of higher ground are vacated. Migrates by day in flocks. Western European migrants head mostly south-west or SSW on narrow front; east European birds range more widely, between west and south-east. Some Mediterranean populations perhaps sedentary.

Passage periods prolonged. Autumn migration August–December, chiefly September–November. Main departure from Britain mid-September to late October. In northern Denmark, movement from early August to late November, peaking in late October. On Polish coast, peaks late September and early October. At Strait of Gibraltar, main passage mid-October to mid-November; small numbers continue to move south in December, and some birds move north in January; movement thus virtually continuous. In Israel, winter visitors arrive October to mid-December, chiefly November. Leaves south of wintering range chiefly from mid-February; main passage in Gibraltar area March to early April. Passage to northern breeding areas continues to May.

Food. Small seeds, mainly Compositae; in breeding season, also small numbers of invertebrates. Prefers seeds in milky, half-ripe state, so changes food plants constantly over year, and continually on the move from one patch of suitable species to another, which can be several km away, sometimes following the same route every day. Generally takes seeds directly from flower or seed-head on plant, mostly on herbs, rarely grasses, in wasteland, open countryside, copses, etc., less often in parks or gardens; in winter regularly in trees, principally alder and pine. In breeding season, forages in pairs or groups of up to c. 30, at other times of year often in larger flocks. Among most animated and agile Fringillidae, making constant and skilful use of feet both to maintain balance when reaching for seeds and to hold seed-head (etc.) while feeding. Often lands at base of herb stem then moves up until flower-head bends over and can then be held under foot; on grass, several stems can be pulled together with bill then grasped in feet for support; in trees, frequently upside-down exploring cones or searching for insects in outermost leaves.

Social pattern and behaviour. Typically gregarious outside breeding season from late summer to spring, also often small flocks of off-duty birds and non-breeders in breeding season. In Europe, family parties begin to gather into loose nomadic flocks in mid-August, and as autumn approaches these in turn coalesce into more compact flocks of up to hundreds and sometimes even thousands. No evidence for other than monogamous mating system. Breeds mostly in neighbourhood groups of up to c. 9 pairs. Territory relatively small, serving for mating and nesting, but foraging almost entirely outside, up to 800 m from nest. Both sexes sing, but ♀ much less regularly than ♂. ♂ also has a song-flight: flies hesitantly with slow deep wing-beats, plumage ruffled, and tail spread, sometimes interspersed with gliding for 1–2 s; similar to song-flight of Greenfinch but not so erratic, less common. ♂ sings inside and outside territory, usually when accompanying mate; song-display by ♂ most intense when ♀ nest-building and laying. In western Europe, song reported throughout the year but mainly mid-March to mid-July, less often late January to late August, not uncommonly October. Arrival of birds already paired on breeding grounds indicates pair-formation occurs in flock.

Voice. Diverse repertoire used throughout the year. Song of ♂ a pleasing liquid twittering elaboration of call-type units with variations, recalling Canary; short snatches confusable with Linnet but less nasal and twangy. Mostly in phrases (separated by short pauses), sometimes almost continuous; phrases delivered at rapid tempo, often beginning with rapid sequence of diagnostic excitement-type calls (see below), followed by various 'trills' and, towards end, typically harsh drawn-out nasal units. Main contact-call consists typically of 2, less often 3 or more clear-sounding units of varying pitch 'teetut', 'titee', 'tütitee', etc., given by both sexes when perched or in flight, delivered in rapid succession; highly diverse, although pair-members tend to give same call-types. Other calls include a 'chlü', given while feeding in flock; mainly disyllabic calls showing definite slur between syllables, 'tuleep', 'weeyü', etc., given in excitement especially in territorial defence; and emphatic staccato 'titt wittit' when alarmed at nest.

Breeding. SEASON. Britain: eggs laid mainly mid-May to early August. Finland: eggs laid 1st half of May to early July. Poland: 1st clutch laid late April to June, 2nd clutch June to end of July. Germany: in south-west, peak laying end of May to mid-June. North Africa: eggs laid mostly April–May. Azores: eggs laid May–June. Usually 2, sometimes 3 broods. SITE. Well hidden in inaccessible outermost twigs of tree, and cover seems more important than support. Although nests vulnerable to wind, apparently no particular orientation preferred. Nest: very neat and compact cup of moss, roots, grass, and spider silk, which sometimes binds foundation to twigs, thickly lined with plant down, wool, hair, and occasionally feathers; sometimes 'decorated' on outside with aromatic flowers. EGGS. Sub-elliptical, smooth and slightly glossy; very pale bluish-white, sparsely spotted or scrawled reddish or purplish-brown, concentrated at broad end sometimes in faint ring; occasionally reddish-brown blotches or greyish-violet undermarkings. Clutch: 4–6 (3–7). INCUBATION. 11–14 days. FLEDGING PERIOD. 13–18 days.

Wing-length: *C. c. carduelis*: ♂ 78–84, ♀ 75–82 mm.

Weight: *C. c. carduelis*: ♂ ♀ mostly 14–19 g.

Geographical variation. Slight in west Palearctic, involving mainly wing length and general colour of upperparts (marked in Asia). 6 races recognized, but none well-marked. Colour of upperparts, considered to be important for identification of races, is highly influenced by bleaching and wear, grading from cinnamon-brown when fresh to dull brown-grey when worn in all races. *C. c. carduelis* from southern Scandinavia, eastern Netherlands and Belgium, and central France east to central Urals and south to mainland Italy, northernmost Yugoslavia, western Rumania, and northern Ukraine is rather large, upperparts and patches at side of belly saturated cinnamon-brown, ear-coverts with limited cinnamon (almost pure white when worn). *C. c. britannica* from Britain, Ireland, western and north-west France, and coastal Netherlands and Belgium slightly smaller, upperparts and belly-patches slightly duller brown when fresh, faintly tinged olive, less cinnamon, ear-coverts more extensively brown (greyish-cream-buff when worn), belly less pure white when worn than in *C. c. carduelis*. *C. c. parva* from Mediterranean France, Pyrénées, Iberia, Balearic Islands, North Africa, and Atlantic islands more distinctly demarcated from nominate *carduelis*, mainly due to small size, in particular wing and tail. *C. c. tschusii* from Corsica, Sardinia, Elba, and (apparently this race) Sicily is like *parva* in size, but bill slightly shorter and finer at base, and upperparts darker and duller earth-brown than in other races. *C. c. niediecki* from Asia Minor (except east), Levant, Egypt, Cyprus, northern Iraq, south-west Iran, Rhodes, Karpathos, and Transcaucasia similar in size to *C. c. carduelis*; in fresh plumage, upperparts and belly-patches slightly paler drab-brown than in nominate *carduelis*, less cinnamon. *C. c. balcanica* from Balkans and Greece (from Yugoslavia and southern Rumania south to Crete) and *colchicus* from northern and western slopes of Caucasus west to Crimea are poor races, similar in size to *niediecki* and *C. c. carduelis*; between them in colour, though nearer to *niediecki*.

Siskin *Carduelis spinus*

PLATES: pages 1565, 1567

Du. Sijs   Fr. Tarin des aulnes   Ge. Erlenzeisig   It. Lucherino
Ru. Чиж   Sp. Lúgano común   Sw. Grönsiska

Field characters. 12 cm; wing-span 20–23 cm. Slightly smaller than Goldfinch, with noticeably shorter tail and more compact form. Small, quite tubby but elegant finch with short but noticeably forked tail, sharing with Redpoll arboreal, tit-like behaviour and with serins yellow-green and streaked plumages. At all ages, shows fine bill (like Goldfinch and unlike all serins except Citril Finch) and striking combination of yellow band on wing and yellow basal patches on tail (unlike any confusion species). ♂ strikingly yellow and green, with diagnostic black crown, much black in wings and tail, emphasizing yellow marks, and greenish-yellow rump; ♀ more greenish, distinctly streaked above and below, especially from sides of breast to rear flanks. Juvenile buff-brown above, even more heavily streaked above and below; lacks pale rump but shows characteristic wing and tail pattern.

♂ unmistakable. ♀ and juvenile less distinctive, but fine bill distinguishes them from Serin, while streaked underparts separate them from Citril Finch. At distance, juvenile may also be difficult to isolate from frequent companion Redpoll, but at close ranges difference in wing pattern, fuller extent of streaks on underparts, and voice soon allow distinction. Flight light and quick, with bursts of wing-beats lifting bird into sudden ascent and soon developing rapidly dipping undulations in speedy direct progress.

Habitat. In west Palearctic, breeds in both lowland and mountain forest, coniferous or mixed, mainly in boreal and temperate zones, north to July isotherm of 13°C. Mainly occupies spruce but also fir and pine, especially where these are well-grown and well-spaced, and sometimes mixed with broad-leaved trees. Streamside locations are often preferred, especially outside breeding season where much foraging is in alders and birches along watercourses, often well away from conifers. Recently has begun to visit garden feeders in some areas. Has in recent years begun nesting more widely and frequently in fresh areas in England, apparently due to afforestation with conifers and to use of planted introduced conifers in parks and gardens, but in Switzerland in formerly neglected native stands, largely in montane regions at c. 1200–1800 m, but not infrequently in lowlands, with marked annual fluctuations.

Distribution. Has expanded range in Britain, Ireland, and other parts of western Europe, following spread of conifer plantations. Breeding range subject to local expansion and contraction, especially on southern fringe. ICELAND. First breeding record: at least 2 pairs bred 1994. FAEROES. Only 1 sight record before 1981, but sharp increase in numbers in 1980s. Bred 1985 and 1988. BRITAIN. Probably confined to Scots pine of Scottish highlands until mid-19th century. Has spread since, with major expansion from 1950, continuing as coniferous forests mature. IRELAND. Has expanded range in 20th century. FRANCE. Sporadic and fluctuating; e.g. in Jura, sometimes very rare, at other times in large numbers, depending on availability of spruce seeds. BELGIUM. Formerly irregular breeder. Apparently regular since at least 1960, but range contracted after 1977. LUXEMBOURG. Breeding long suspected; first proved 1982. GERMANY. Distribution very patchy. DENMARK. Increased probably by 13% 1974–94. NORWAY. Slight increase. POLAND. Breeds mainly in mountains and north-east forests. SPAIN. Small numbers breed in mountain woodlands, usually following high winter numbers. PORTUGAL. Probably bred in north (bird with incubation patch) 1984. In Algarve,

Siskin *Carduelis spinus*: **1** ad ♂ summer, **2** ad ♂ winter, **3** ad ♀ summer, **4** 1st winter ♀, **5** juv.

several pairs bred 1995 (and perhaps earlier). ITALY. Bred Sardinia 1949. Breeding proved Sicily (Mt Etna) 1984–5. YUGOSLAVIA. In some years breeds even in small lowland conifer plantations. RUMANIA. Breeds in small numbers along Carpathians. RUSSIA. At Nar'yan-Mar (67°37′N), well north of known range, at least 5 pairs present mid-June 1993, indicating potential breeding. TURKEY. Status still inadequately known. Locally common on mountain slopes in Black Sea coastlands, but numbers small in northern part of western Anatolia.

Accidental. Iceland, Azores, Madeira.

Beyond west Palearctic, breeds northern Iran and extends east across Russia to Sakhalin island and northern Japan.

Population. Has increased in line with range expansion. Marked fluctuations in numbers are characteristic of northern Europe; elsewhere, populations mostly stable overall, though with some fluctuations. BRITAIN. 300 000 pairs 1988–91. IRELAND. 60 000 pairs 1988–91. FRANCE. 100–1000 pairs. BELGIUM. 10–25 000 pairs. LUXEMBOURG. 2–10 pairs. NETHERLANDS. 300–700 pairs 1979–85. GERMANY. 10 000–20 000 pairs. DENMARK. 100–1000 pairs 1987–8. NORWAY. 100 000–1 million pairs 1970–90. SWEDEN. 400 000–1 million pairs in late 1980s. FINLAND. 500 000–1.5 million pairs in late 1980s. ESTONIA. 100 000–200 000 pairs in 1991. LATVIA. 100 000–200 000 pairs in 1980s. LITHUANIA. Common. POLAND. 10 000–30 000 pairs. CZECH REPUBLIC. 90 000–180 000 pairs 1985–9. SLOVAKIA. 20 000–40 000 pairs 1973–94. HUNGARY. 200–300 pairs 1979–93. SWITZERLAND. 5000–10 000 pairs 1985–93. SPAIN. 500–1100 pairs. ITALY. 500–1500 pairs 1983–95. YUGOSLAVIA: CROATIA. 10 000–20 000 pairs. SLOVENIA. 2000–3000 pairs. BULGARIA. 1000–5000 pairs. RUMANIA. 1500–4000 pairs 1986–92. RUSSIA. 1–10 million pairs. BELARUS'. 20 000–80 000 pairs in 1990. UKRAINE. 5000–6000 pairs in 1986. AZERBAIJAN. Uncommon. TURKEY. 500–5000 pairs.

Movements. Mostly migratory in northern breeding areas, but some southern populations may be resident. Many birds winter in different areas in different years but some are faithful to same area—even exactly the same site. Most are nomadic during winter, but minority becomes resident at same site for several months. Numbers migrating vary greatly from year to year, and more distant movements are recorded when large numbers of birds involved (eruption years). Availability of seed crops on favoured trees (alder, birch) seems to be major determinant of strength of movement away from breeding area. Recent habit of regular feeding in gardens noted particularly in March and early April when birds fattening for spring migration.

Autumn migration in northern areas may start as early as August, but generally peaks late September or October. Timing and strength vary from year to year, and significant numbers may move as late as December or even January. Spring departure from regular southern wintering areas is weak and may start in early February, but in most areas continues to mid-April. Passage in Switzerland from end of February to mid- or late April; peaks in April in northern Denmark. On south side of Gulf of Finland (where passage very conspicuous) and in Leningrad region, waves of migrants pass east or north-east until early or mid-May.

Food. Seeds, especially of conifers, alder, birch, and herbs; some invertebrates in breeding season. Very dependent on spruce (or pine) on breeding grounds. Feeds principally in trees, moving to tall herbs or ground when cones empty and seed has dropped; many invertebrates apparently taken from ground and at water's edge; also has habit, mainly in winter, of feeding on seeds washed up on shoreline, where favourite alder trees are common. Invertebrates picked from leaves and needles of trees, herbs, bracken, etc., and insects can be caught in flight. Moves restlessly through trees examining cones, acrobatically clinging to cones and twigs, often starting at crown and moving towards base; hangs and perches on stems and seed-heads of herbs with equal agility. Tweezer-like bill used to extract or pry out (bill inserted then opened) seeds (etc.) from tight spaces (e.g. cones, closed seed-heads, buds, catkins), very like Goldfinch, though unable to probe as deeply. In recent years (in southern England since early 1960s and Ireland since 1980s) frequents bird-tables and garden feeders in winter much more regularly than in past, particularly after alder seed crop exhausted; perhaps first entered gardens to feed on seeds of ornamental cypress, soon regularly taking peanuts. Another habit apparently acquired recently is eating of beech-mast, first recorded in central England in 1950s.

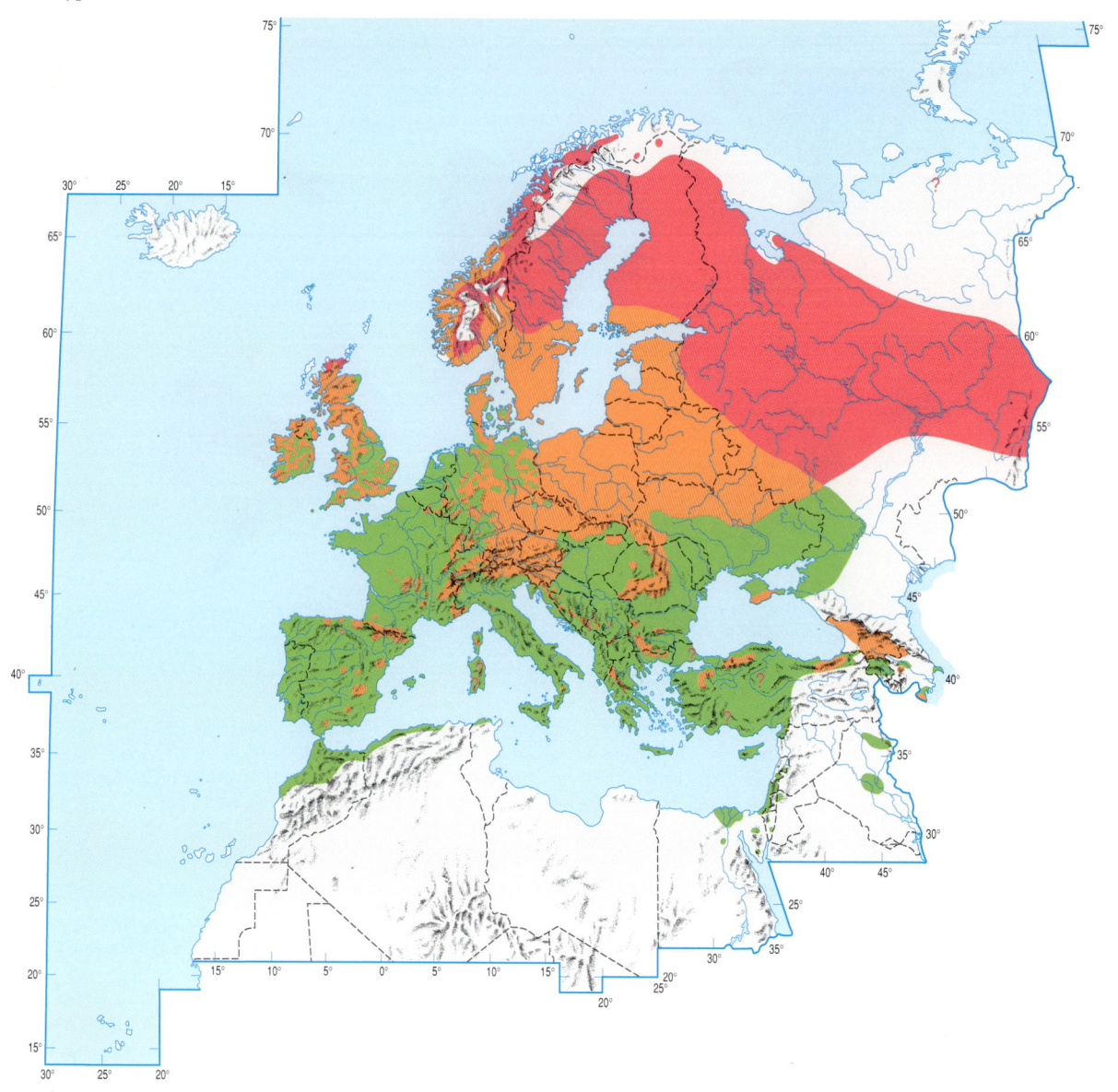

Social pattern and behaviour. Gregarious outside breeding season. Winter flocks of up to several hundred, sometimes several thousand. Feeding flocks sometimes associate with other finches, notably Redpoll and Goldfinch. Pair-formation evidently takes place in flocks; most birds paired when settle to breed. Territorial, but not conspicuously so, when breeding; usually in loose neighbourhood groups of 2–3 or up to c. 15 or more pairs, especially in areas of heavy cone crops. ♂ gives advertising-song from exposed perch on top of tree or concealed in canopy, also in flight. Several ♂♂ often sing close together, both on spring migration and when breeding, and song unlikely to have territorial function. Song-flights occur chiefly at start of breeding season (during courtship and

Canary *Serinus canaria* (p. 1554): **1** ad ♂ summer. Citril Finch *Serinus citrinella citrinella* (p. 1556): **2** ad ♂ summer. Greenfinch *Carduelis chloris chloris* (p. 1557): **3** ad ♂ summer, **4** ad ♀ summer. Goldfinch *Carduelis carduelis carduelis* (p. 1561): **5–6** ad ♂ summer. Siskin *Carduelis spinus*: **7** ad ♀ summer.

incubation), especially in sunny weather, several ♂♂ sometimes in the air together, so that their flight-paths cross and recross. May ascend (repeatedly) quite high, taking off in fluttering flight from perch where had previously sung. Bird has plumage ruffled and tail widely spread; flies in loops and circles at tree-top height or higher and over or close to nest-site, often with exaggeratedly deep wing-beats, so that wings appear almost to meet over back. Sings throughout performance, and also on perch between flights.

Voice. Several calls (and song) given all year, most familiar being attractive tonal calls such as 'dluee'. Advertising-song of ♂ consists of rapid, lively, persistent, and quite sweet twittering chatter, at times recalling Serin or Sedge Warbler; more or less continuous or sometimes in phrases. Often preceded by species-specific 'dluee' or 'diu-li' which, like several other calls, is also incorporated in song. Song contains repetitions of units or motifs in regular temporal pattern and, typically, numerous easily recognized imitations of other birds. Towards end, typically a feeble, drawn-out and nasal wheeze lasting *c.* 1 s. Contact-alarm calls varied; characteristically slightly melancholy, loud whistling sounds, some clear-toned, others with harsher, nasal component; much variation within and between individuals. Calls most commonly given in flight are a soft, whistling 'teeyu' descending in pitch, and an ascending 'tsoooee'. For short-range contact main call is 'twitter'-call:

short, dry 'tet', 'tut', or 'chek', often developing into longer twittering tremolo.

Breeding. SEASON. Southern Scotland: laying starts usually in April, but can be early to mid-March in years of good spruce cone crop, mid-May in poor years. Ireland: eggs can be laid in March; mostly from early April. Finland: eggs laid late April probably to beginning of August, with peak early May to early June; earlier in good spruce years. Switzerland: can lay late February in good spruce years. Russia: laying starts Leningrad region late April to May. Commonly 2 broods, at least in some areas. SITE. Generally inaccessible, at considerable height and in outer hanging twigs of conifer, usually spruce; also recorded against trunk. Nest: small hemispherical construction of (mostly) conifer twigs, heather, grass, moss, bark fibres, and spider's web lined with hair, fur, rootlets, plant down, and sometimes feathers, often with external camouflaging of moss and lichen; occasionally woven into hanging twigs. EGGS. Sub-elliptical, smooth, and slightly glossy; rather variable in colour, size, and shape; pale bluish-white or blue sparsely marked with purplish- to blackish-brown spots and scrawls, mostly at broad end; also rusty blotches and reddish-violet undermarkings. Clutch: 3–5(–6). INCUBATION. 12–13 days. FLEDGING PERIOD. 13–15(–17) days.

Wing-length: ♂ 71–76, ♀ 69–74 mm.
Weight: ♂ ♀ mostly 11–18 g.

Pine Siskin *Carduelis pinus*

Du. Dennensijs   Fr. Tarin des pins   Ge. Fichtenzeisig   It. Lucherino dei pini
Ru. Сосновый чиж   Sp. Lúgano pinariego   Sw. Tallsiska

A North American species, breeding mainly in coniferous forest from Alaska and Labrador south to southern Mexico, wintering throughout breeding range and in non-breeding areas in southern USA and Mexico. One occurred among large fall of migrants on eastbound ship in North Atlantic 8–11 October 1962, staying on board from *c.* 65°W to 32°W.

Linnet *Carduelis cannabina*

PLATES: pages 1569, 1577

Du. Kneu   Fr. Linotte mélodieuse   Ge. Bluthänfling   It. Fanello
Ru. Коноплянка   Sp. Pardillo común   Sw. Hämpling

Field characters. 13.5 cm; wing-span 21–25.5 cm. Similar in bulk to Twite but with shorter tail and thus looking slightly more compact; distinctly larger and longer than smallest Redpoll (*C. flammea cabaret*) but matched in size and structure by

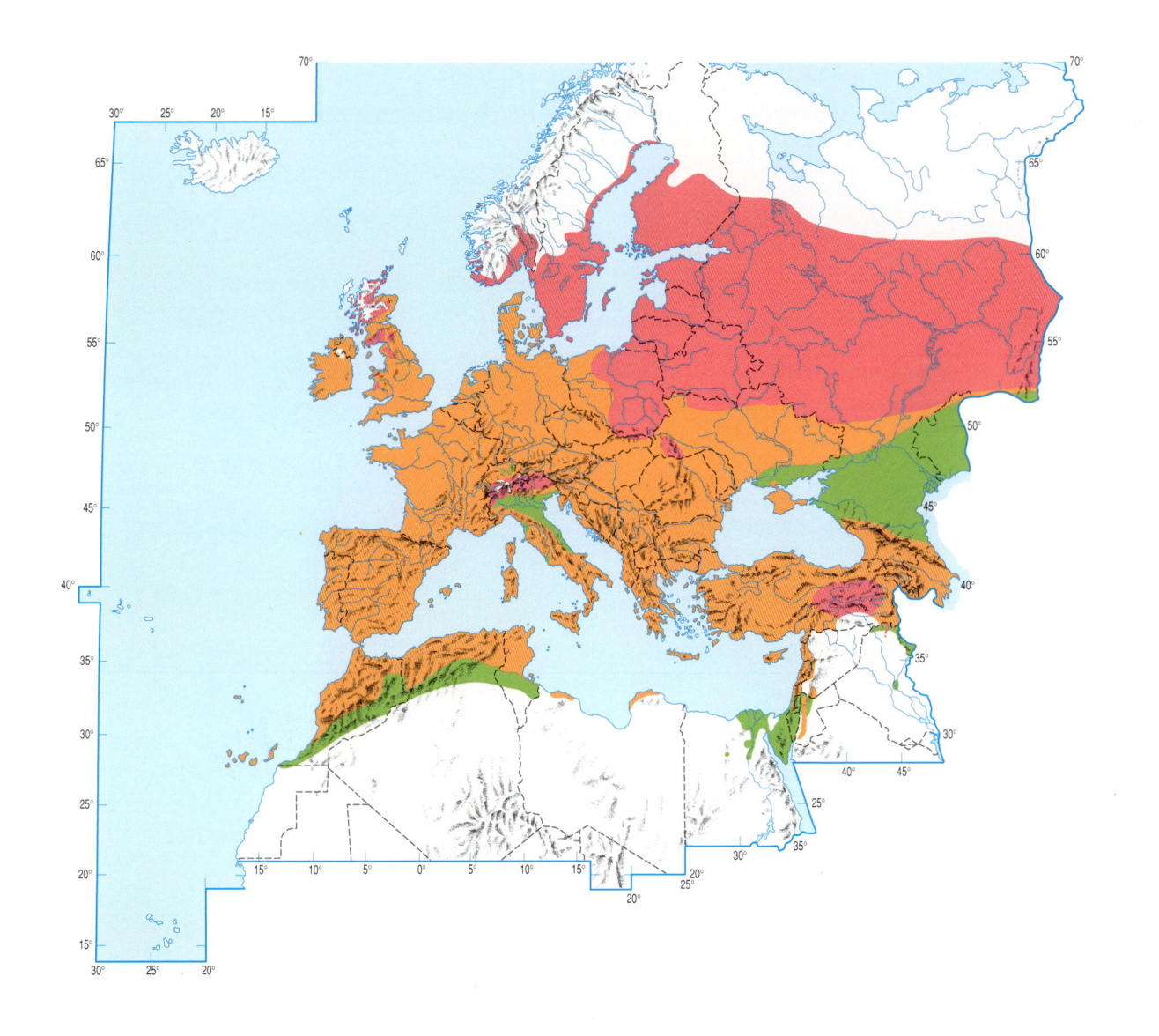

Linnet *Carduelis cannabina*. *C. c. cannabina*: **1** ad ♂ summer, **2** ad ♂ winter, **3** ad ♀, **4** juv. *C. c. autochthona*: **5** ad ♂ winter. *C. c. bella*: **6** ad ♂ summer. *C. c. guentheri*: **7** ad ♂ summer. *C. c. harterti*: **8** ad ♂ summer.

larger northern Redpolls. Small, quite compact, sociable but nervous finch; epitome of smaller, streaked members of *Carduelis*. Except in ♂, rather featureless plumage essentially brown above and pale buff-brown below, marked mainly by white edges to primaries and tail-feathers, quite heavy streaks on back and underparts, and paler eye-crescents and throat markings. ♂ in worn plumage colourful, with pale crimson forecrown on grey head, crimson chest and flanks, almost chestnut mantle, scapulars, and wing coverts (lacking obvious streaks). ♀ greyer headed than juvenile. Bill at all times grey or dull horn, never straw like Twite.

Adult ♂ in worn breeding plumage has diagnostic combination of grey head, white panel in wing, and white sides to tail. Non-breeding ♂ in fresh plumage, ♀, and juvenile liable to confusion particularly with Twite and also Redpoll, Siskin, and ♀ and juvenile serins *Serinus*. Thus important to study ♀ Linnet fully before attempting identification of other small streaked finches; voice (see below) as useful as rather unremarkable plumage. Flight light and rapid with easy take-off; progress wavering in short flights, but if longer shows short, steep undulations produced by bursts of wing-beats followed by wing-closure.

Habitat. In west Palearctic, breeds almost throughout European boreal, temperate, Mediterranean, and steppe climatic zones, extending to coastal North Africa and south-west Asia, in both continental and, to lesser extent, Atlantic climates. Mainly a lowland bird, but also widespread in suitable hilly regions, and in Swiss Alps nests up to 2200–2300 m on dry moors, alpine meadows, or scree where dwarf conifers or rough vegetation flourish. Generally avoids dense tall forests and woods except those of open type with clearings or glades and ready access to stands of seed-bearing wasteland or similar plants. Occupies fen woodlands but prefers scrub and heath vegetation with dry sunny aspect, farmland with hedges or low trees, vineyards, orchards, maquis, fields left uncultivated, young plantations, and untended forest edges; also, in some areas, rural or suburban gardens, and industrial wasteland. Outside breeding season will shift to more open habitats, such as salt-marshes, shingle banks, and sand-dunes, as well as farmlands.

Distribution. No major changes except in north, where range limit affected by varying severity of winters. BRITAIN. Highest numbers in east and south and along coast. IRELAND. Marked decline. BELGIUM. Local until *c.* 1930, since when has expanded over whole country; recent decrease in Limburg. NORWAY. Has disappeared from Nordland (decline from 1955; last observations 1970s); some spread in south-east and west. FINLAND. At beginning of 20th century, nearly disappeared from most parts of former range (south and centre). Slow recovery since late 1940s. During mild winters of early 1970s expanded range to north, but contraction followed hard winters of 1980s. AUSTRIA. Regionally common in east. SPAIN, UKRAINE. Slight increase. TURKEY. Widespread, in west generally in small numbers, in east often common. JORDAN. Common in north of range, less common further south. IRAQ. Probably breeds in north. ALGERIA. Common in Tell region, but rare in Saharan Atlas.

Accidental. Faeroes, Kuwait (or very rare winter visitor), Mauritania.

Beyond west Palearctic, extends east through West Siberia to *c.* 90°E, thence south-west through central Asian mountains (skirting much of Kazakhstan) to Afghanistan and Iran.

Population. Marked recent decrease Britain, Netherlands, Finland, and Estonia, and declines also reported Ireland, Belgium, Germany, Denmark, Scandinavia, Czech Republic, Slovakia, and Switzerland; probable cause intensification of agriculture. Slight increase Spain and Ukraine. Elsewhere apparently stable. BRITAIN. 520 000 territories 1988–91. Steep decline since at least 1977, probably due mainly to loss of food resources through increased use of herbicides; small improvement in farmland in early 1990s. IRELAND. 130 000 territories 1988–91. FRANCE. Over 1 million pairs. BELGIUM. 150 000 pairs 1973–7. Decrease since 1980s, particularly in farmland. LUXEMBOURG. 14 000–18 000 pairs. NETHERLANDS. 60 000–130 000 pairs 1989–91. Decline (still continuing) probably of 50–80% since c. 1960. GERMANY. 669 000 pairs in mid-1980s. In east, 180 000 ± 90 000 pairs in early 1980s. Considerable recent decline in Schleswig-Holstein and elsewhere. DENMARK. 160 000–1.8 million pairs 1987–8. NORWAY. 10 000–15 000 pairs 1970–90. SWEDEN. 100 000–250 000 pairs in late 1980s. FINLAND. 15 000–20 000 pairs in late 1980s. ESTONIA. 20 000–50 000 pairs in 1991. LATVIA. 10 000–25 000 pairs in 1980s. LITHUANIA. Common. POLAND. Locally fairly numerous. CZECH REPUBLIC. 60 000–120 000 pairs 1985–9. SLOVAKIA. 40 000–60 000 pairs 1973–94. HUNGARY. 60 000–100 000 pairs 1979–93. SWITZERLAND. 30 000–60 000 pairs 1985–93. SPAIN. 1.7–3.3 million pairs. PORTUGAL. 100 000–1 million pairs 1978–84. ITALY. 100 000–300 000 pairs 1983–95. MALTA. 5–10 pairs. GREECE. 50 000–80 000 pairs. ALBANIA. 1000–5000 pairs in 1981. YUGOSLAVIA: CROATIA. 35 000–40 000 pairs. SLOVENIA. 10 000–20 000 pairs. BULGARIA. 50 000–500 000 pairs. RUMANIA. 250 000–400 000 pairs 1986–92. RUSSIA. 10 000–100 000 pairs. BELARUS'. 130 000–180 000 pairs in 1990. UKRAINE. 900 000–1 million pairs in 1986. MOLDOVA. 25 000–40 000 pairs in 1988. AZERBAIJAN. Common. TURKEY. 1–10 million pairs. CYPRUS. Locally fairly common; probably stable. ISRAEL. A few thousand pairs in 1980s. JORDAN. More common in north than south of range. TUNISIA, MOROCCO. Common. MADEIRA. Rare. CANARY ISLANDS. Common and widespread.

Movements. Partially migratory, most birds moving south-west or SSW on narrow front to winter within and slightly south of breeding range, with concentrations in Mediterranean region. Areas of higher ground are vacated. Diurnal migrant. Atlantic island races sedentary, and probably also some Mediterranean populations.

Passage periods prolonged. Autumn migration August–November, peaking mid-September to mid-October in northern Europe, and October in south. Crosses Strait of Gibraltar earlier than other Fringillidae, beginning 2nd half of September, with mid-October peak sometimes extending into November. Passage later in central and east Mediterranean, mostly mid-October to November, to mid-December in Israel. Spring migration begins early, with movement from February in both south and centre of winter range. Arrives in Britain mid-March to early May, chiefly in April. Reaches Sweden end of March to April. Arrives in Leningrad region (north-west Russia) in March in early springs, but usually in April, with passage continuing throughout April–May.

Food. Small to medium-sized seeds; probably takes fewer invertebrates than any other west Palearctic finch apart from crossbills or Twite. In winter, forms large mixed flocks with other seed-eaters in open country, feeding much more on ground than in summer, though tends to remain together in groups within such flocks. Particularly dependent on weeds of open country and waste ground (especially Polygonaceae, Cruciferae, Caryophyllaceae, and Compositae), so habits determined to large extent by agricultural practices.

Social pattern and behaviour. Typically gregarious outside breeding season, forming flocks (sometimes of hundreds or even thousands) for feeding, roosting, and migration. Mating system apparently mainly monogamous, with occasional occurrence of polygyny (2 ♀♀ paired with same ♂). Territorial when breeding, though defended area apparently small, and loose neighbourhood groups of 2–12 or up to several dozen or more pairs often occur, with some nests only a few metres apart; many pairs solitary. As in other Carduelinae, song of little territorial significance and given almost throughout year, in early breeding season (also in late-summer and autumn flocks) often by several ♂♂ together. ♂ gives advertising-song within a few days of return to breeding grounds from elevated perch (occasionally on ground) and in flight. Sometimes sings in normal flight, but true song-flight also occurs. ♂ may take off from perch while singing and ascend c. 10 m, then sing while flying in erratic circles with rapid wing-beats, tail widely spread; descent fluttering or gliding in spiral, bird returning to same perch, or landing by ♀.

Voice. Advertising-song of ♂ a pleasing, musical and varied, rapid, lively twittering, incorporating many short units (some delicate), also trills or tremolos (some with metallic timbre), attractive and melodious drawn-out and flute-like whistles, and 'crowing' or twanging sounds; usually not very far-carrying. Typically begins with series of calls or modified call-units, 'gigigi' given in long accelerando sequence, sometimes rising and falling in pitch; certain other calls may precede or be included in song. Series of well-spaced delicate tinkling and cheerful twittering calls or motifs given by both sexes, sometimes as antiphonal duet; perhaps type of song. Main contact-call a slightly metallic twittering, e.g. 'kekeke-keke', 'gegege', 'tett-tett-terrett', 'chichichichit'; given singly, commonly 2 in quick succession, or longer series to integrate flock or family in flight or on ground. Alarm-call a plaintive, mewing, slightly strained 'hoooi' or 'tsooeet'.

Breeding. SEASON. Southern England: eggs laid mid-April to early August. Southern Finland: start of laying late April, peak period beginning of May to end of June. Brittany (western France): peak period for 1st clutch late April to early May; for 2nd clutch, 1st week of June. South-west Switzerland: at 2000 m, eggs laid mainly late May and early June; latest clutches 2nd half of July. Russia: in Leningrad region (north-west Russia), eggs laid April to August with peak in June. North Africa: laying from late March, mostly April. 2 broods, 3 in favourable conditions. SITE. Very low in dense, often thorny tree, bush, scrub, or hedge, or on ground; frequently in young conifer plantation. Evergreens commonly preferred early in season for 1st broods, later nests more often in deciduous trees and bushes when cover thick. Nest: foundation of small twigs, roots, stalks, and moss lined with hair, wool, plant down, sometimes feathers, paper, etc. EGGS. Sub-elliptical, smooth, and non-glossy, very variable in shape and colour; pale to whitish-blue, sometimes light grey, with distinct spots, speckles, and streaks of pink, purple, or purplish-brown, also larger blotches of various shades of red to blackish-purple; markings mostly at broad end. Clutch: 4–6 (3–7). INCUBATION. 12–14 days. FLEDGING PERIOD. 10–17 days.

Wing-length: *C. c. cannabina* ♂ 78–85, ♀ 76–82 mm.
Weight: *C. c. cannabina* ♂ mostly 17–22, ♀ 15–21 g.

Geographical variation. Rather slight; involves size (length of wing, tail, or tarsus), relative length, depth, and width of bill, and depth of colour, especially grey of hindneck of ♂ and brown of mantle and scapulars. Colours strongly affected by bleaching and wear, especially in southern populations; variation also confused by numerous *C. c. cannabina* wintering in breeding range of Mediterranean and Middle East races. 4 groups separable on size and bill depth: (1) *C. c. cannabina* (northern Eurasia from Ireland to western Siberia, south to Pyrénées, northern Italy, northern Balkans, and Ukraine) and *autochthona* (Scotland), with rather long wing and slender bill; (2) *bella* (Turkey, Cyprus, Levant, and Crimea through Caucasus area and Iran to south-west Turkmenistan) with wing length as *C. c. cannabina* or slightly longer but bill thicker at base; (3) Atlantic island races (*guentheri* on Madeira, *meadewaldoi* on western Canary Islands east to Gran Canaria, *harterti* on eastern Canary Islands), and *mediterranea* (Iberia, southern Italy, Dalmatia, Albania, Greece, and many islands in Mediterranean from Balearics to Crete and Karpathos in Greece) with rather short wing and slender bill (as *C. c. cannabina*); (4) populations of North Africa, as yet unnamed (occurring Morocco and mountains of Algeria and neighbouring Tunisia, perhaps also in lowlands of Algeria and Tunisia and in northern Libya) with short wing (as in 3rd group) and thick bill (as in 2nd). 2 distinct groups separable on colour, corresponding with grouping on bill depth in samples above: (1) darker, more slender-billed nominate *cannabina*, *mediterranea*, and Atlantic races; (2) paler, thicker-billed North African population and *bella*.

Twite *Carduelis flavirostris*

PLATES: pages 1572, 1577

DU. Frater   FR. Linotte à bec jaune   GE. Berghänfling   IT. Fanello nordico
RU. Горная чечетка   SP. Pardillo piquigualdo   SW. Vinterhämpling

Field characters. 14 cm; wing-span 22–24 cm. Close in size to Linnet but with over 10% longer tail and stubbier bill creating subtly different outline. Small, robust, but somewhat attenuated finch; counterpart of Linnet in mountains of south-west Asia and coastal periphery and hills of north-west Europe. Plumage pattern strongly recalls ♀ or juvenile Linnet but ground of upperparts more tawny in tone and of face, throat, chest, and flanks distinctly warmer buff, becoming even orange around eye and under bill, while belly clean white. Shares white wing-panel and tail-sides with Linnet but pale buff tips to median and greater coverts create more noticeable wing-bars. Bill grey in breeding adult but otherwise straw-yellow, recalling not Linnet but Redpoll. ♂ has long pink rump, recalling Redpoll.

Inseparable from Linnet at distance, but when closer shows distinctive pink rump (♂), darker, warm buff face and throat (adult), more distinct wing-bars (particularly lower), and 2-toned underparts. Flight, gait, and behaviour much as Linnet but flight silhouette differs in longer tail. Escape-flight often long. Associated with coastal barrens and open moorland but also at home in barren mountains.

Habitat. Occupies tundra, boreal, and marginally temperate zones, extending north to about July isotherm of 10°C. In contrast to Linnet, occupies terrain more or less free of trees and shrubs or bushy growth, in cool, windy, and often rainy climates without much sun or warmth, often on stony, rocky, or hilly ground, including sea-cliffs and inshore islands. In Britain and Ireland, largely a lowland bird, favouring heather moors, hill farms, and upland pastures, but not mountains or precipitous areas. In Scandinavia, breeds at high altitude on fjelds, and on barren slopes near crags or precipices, moving later to newly mown fields, and in coastal regions to gardens.

In winter, many shift to coastal lowlands, including salt-marshes. Other passerines share taste for these, but only Twite uses seaward fringe, feeding on seeds of *Salicornia* and *Aster*; sand-dunes, shingle banks, and cliffs are also well used. In Poland, has recently begun to make use of large beds of introduced goldenrod *Solidago*, while wartime bombing led to use of weeds on sites of destroyed buildings, and to roosting on high buildings in some German cities.

Distribution. FAEROES. Bred Nólsoy *c.* 1938–48; otherwise accidental. BRITAIN. Highest numbers in coastal areas of low-intensity farming, and South Pennine grouse moors. Range has contracted considerably in 20th century, but no evidence for overall change since 1968–72 survey. In Scotland, has recently spread south-east. In Wales, small population breeds regularly in north. IRELAND. Probable range reduction in west. NORWAY. One of commonest passerines along coast; fewer in uplands. More common in south than north. SWEDEN. Sporadic breeding (fewer than 10 records, but probably more frequent), chiefly along Norwegian border. FINLAND. Only one known breeding record, east of Lake Kilpisjärvi 1974, but a few pairs may breed regularly in extreme north-west. LITHUANIA. Bred in east 1978. AZERBAIJAN. Range poorly known. TURKEY.

Twite *Carduelis flavirostris*. *C. f. flavirostris*: **1** ad ♂ summer, **2** ad ♂ winter, **3** ad ♀ summer, **4** juv. *C. f. pipilans*: **5** ad ♂ winter. *C. f. brevirostris*: **6** ad ♂ winter.

Largely confined to mountain slopes in east; very local in eastern Taurus. IRAQ. May breed; flock of *c.* 30 birds seen end of April 1957 in north-east mountains.

Accidental. Switzerland, Spain, Portugal, Belarus'.

Beyond west Palearctic, extends discontinuously from northern Kazakhstan and northern Iran east to Mongolia, western China, and eastern Himalayas.

Population. BRITAIN. 65 000 pairs 1988–91. Widespread decline in Scotland in 20th century. IRELAND. 750–1000 pairs; declining. NETHERLANDS. Wintering population estimated 3000–6000, 1978–83. NORWAY. 100 000–500 000 pairs 1970–90; fluctuating. Temporary decline in 1960s due to agricultural pesticides. RUSSIA. 100–1000 pairs; stable. AZERBAIJAN. Rare. TURKEY. 1000–10 000 pairs.

Movements. Sedentary to migratory. British and Irish populations winter chiefly along coasts, with fewer records inland; in Scotland, upland areas are almost entirely vacated. Birds breeding in southern Pennines (north-central England) move south-east to winter chiefly on east coast from Lincolnshire to northern Kent; a few remain inland. Some birds cross North Sea, mainly to Low Countries. Populations breeding in Fenno-Scandia and Russia chiefly migratory, wintering in northern and eastern Europe; numbers wintering in east have increased in recent decades. Many winter around cities, roosting on buildings especially in Germany. Autumn migration gradual. Passage in southern Norway from last third of August, but peak not until 1st third of October. Main movement across continental Europe late October to November, continuing into December. Spring movement begins early, from mid- or late February; most birds leave central Europe by end of March. Passage in northern Denmark from late February, peaking early April, with stragglers in May.

Food. Small seeds; perhaps a few invertebrates in breeding season. Forages on ground or on low herbs, sometimes in trees; in breeding season, in open areas of pasture and cultivation, by roadsides, at tideline, and by fresh water, feeding mainly on seeds of Compositae, Polygonaceae, and Caryophyllaceae; in winter, in fields, waste ground, allotments, by rivers, etc., and very commonly on coastal salt-marshes, mostly on Chenopodiaceae and Compositae. Feeds in smallish groups in breeding season, but at other times in large mixed flocks of several thousand with other seed-eaters, notably Linnet.

Social pattern and behaviour. Outside breeding season typically in flocks (up to hundreds or thousands) for feeding, roosting, and migration. Mating system apparently mainly monogamous; pair-formation takes place in spring flocks. Pair-bond apparently strong, extending through and perhaps beyond one season and pair-members typically keep close company. Breeding territories widely dispersed where scarce or in extensive areas of suitable habitat, or often in neighbourhood groups, in which nests sometimes quite close together. ♂ sings from rock, post, tree, etc., or in flight. In song-flight (performed by birds in flocks, or over immediate nesting area once breeding), ♂ circles several times, sometimes with erratic manoeuvres; alternately flaps and glides, singing during gliding phase in which wings widely spread and lowered, tail also slightly lowered; resumes flapping after song, then descends like Tree Pipit.

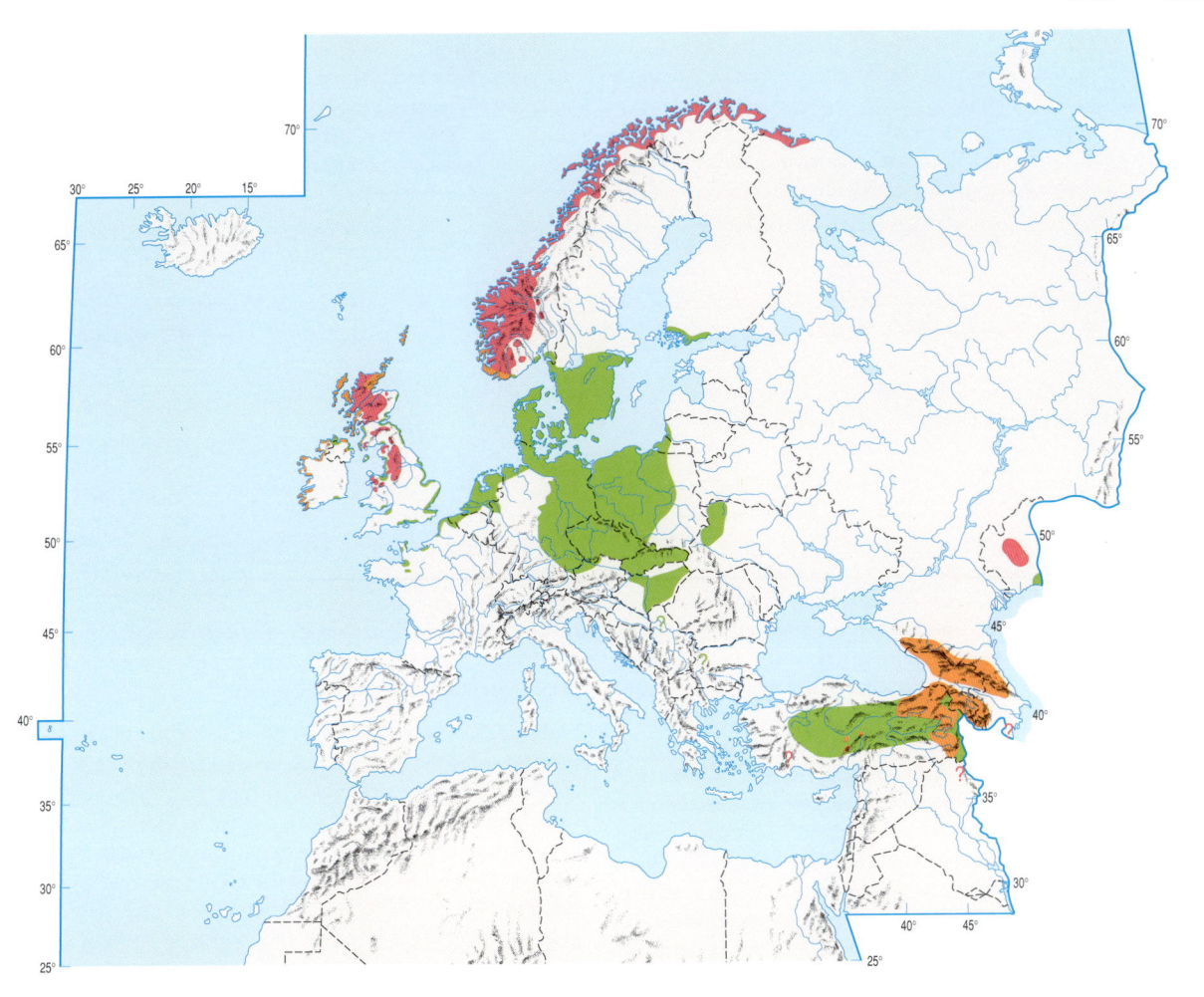

Voice. Song of ♂ has similarity to Linnet in general type, but not preceded by 'gigigi' typical of that species and overall much less musical and delicately twittering, more metallic and jangling with nasal timbre, harsher chatter, and pronounced resonant twangy quality; often includes buzzy sounds and hard rattles; units often smoothly linked but almost all sound strangled and closely resemble or are identical to calls; some quite musical, however, and song may comprise only these, e.g. 'teet-sweet teedle-eu twee-teedl-ee teedl-eu'. Song given rapidly and at uniform volume, in phrases of varying length, or continuous. Flight-call a bouncy chattering or twittering 'tup-up-up', 'chut chululutt'; close to equivalent call of Redpoll, perceptibly harder and more metallic than Linnet, often 3 units in succession, but sometimes gives longer series; speed of delivery also varies, with fastest series typically given in flight. Commonest call, serving for flock integration; given by both sexes throughout year, when feeding or resting, but mainly at take-off and in flight. Other calls include distinctive, hoarse, nasal, loud, and rather twangy rasping sound ascending in pitch and suggesting 'TWEit', 'tzeeip', 'tsooeek', etc., audible up to c. 400 m away and given, singly or in series, as main contact-call; and, in alarm, an ascending 'tooee' or 'tew'.

Breeding. SEASON. Britain: rather late breeder; in northern Scotland, eggs laid mid-May to mid-August, with peak around mid-June; in northern England, eggs laid from end of April, mostly late May and June. Norway: eggs laid from beginning of April to August; peak in central Norway mid-May to mid-June. Caucasus: eggs laid from about mid-May. 1–2 broods. SITE. On or very close to ground in heather, bilberry, bracken, grass tussocks, cotton-grass, rush, etc.; often under rock or in crevice and sometimes in dry-stone wall; also on cliff ledge with or without vegetation, and on young conifer in plantation. Nest: compact, well-built structure with thick, woven walls and deep cup; foundation of small twigs of heather, etc., roots, stalks, fronds of bracken, grass, moss, etc., lined thickly with felted mass of wool, hair, and sometimes feathers. EGGS. Sub-elliptical, smooth, slightly or non-glossy; very like Linnet but more often with scrawled pattern; pale to darkish-blue with varied specks, spots, small blotches, and scrawls of rust-red to purplish-brown towards broad end, and pink-violet undermarkings. Clutch: 4–6 (3–7). INCUBATION. 12–13 days. FLEDGING PERIOD. 11–12 days (10–15).

Wing-length: *C. f. flavirostris*: ♂ 74–83, ♀ 70–80 mm.
Weight: *C. f. flavirostris*: ♂ mostly 14–18, ♀ 13–16 g.

Geographical variation. Slight in western populations, mainly involving general colour; more pronounced in Caucasus area and central Asia, involving colour and size.

C. f. pipilans from Britain and Ireland (except Outer Hebrides) slightly darker in fresh plumage than nominate *flavirostris* from Scandinavia. *C. f. bensonorum* from Outer Hebrides more heavily marked with deeper black streaks than *pipilans*, ground-colour of upperparts duller brown. Ground-colour of most races of south-east Europe and Asia (in west Palearctic, *brevirostris* from Turkey and Caucasus area, *kirghizorum* from Volga–Ural steppe) markedly paler, contrasting distinctly with dark streaking; pink on rump generally paler, sometimes partly replaced by white.

Redpoll *Carduelis flammea*

PLATES: pages 1575, 1577

Du. Barmsijs   Fr. Sizerin flammé   Ge. Birkenzeisig   It. Organetto
Ru. Обыкновенная чечетка   Sp. Pardillo sizerín   Sw. Gråsiska

Field characters. 11.5–14.5 cm; wing-span 20–25 cm. Size varies noticeably: British and montane European race *cabaret* ('Lesser Redpoll') slightly smaller and more finely built than Linnet, with wing-span under 22.5 cm; Holarctic race, nominate *flammea* ('Mealy Redpoll'), is as large as Linnet, with wing-span over 21 cm (but smaller than Greenland race of Arctic Redpoll); Icelandic, Baffin Island, and Greenland race *rostrata* ('Greenland Redpoll') marginally larger than Linnet. Small, attractive finch, with form somewhat like tit and mainly arboreal behaviour in *cabaret* and nominate *flammea* (akin to Siskin), but more robust, with form more like sparrow and more terrestrial behaviour in *rostrata* (akin to Twite). At distance, recalls Linnet and Twite but at close range adult shows diagnostic combination of red forecrown, blackish frontal band and lores, black chin, double wing-bars, and pale tips to tertials. Between races, upperpart tone varies from most tawny in *cabaret* to darkest brown in *rostrata* and greyest in nominate *flammea*, with pale rump, particularly in nominate *flammea*, inviting confusion with Arctic Redpoll. When worn, ♂ has cheeks, lower throat, and sides of breast to fore-flanks rose-pink with buff or white feather-tips. Juvenile less easy to separate from ♀♀ and young of Linnet and Twite, but lores and chin greyish-black, while streaks on underparts much sharper and darker. In *cabaret*, specific identification quickly completed by sharply pointed bill, rather delicate form, great agility when feeding, dancing flight, and muttered flight-call; in nominate *flammea* and *rostrata*, structure more robust but flight action and call still distinctive.

When distinctive flight-call heard, can be confused only with Arctic Redpoll, but at distance or in poor view silent bird may need to be distinguished from Linnet, Twite, Siskin (frequent close companion on migration and in winter), and juvenile serins. Chief problem is difficulty of racial determination; often insurmountable due to extent of taxonomic variation and hybridization. Thus important to make racial claims in full context of either breeding distribution or pronounced irruptions (to which all northern races are particularly prone) and associated vagrancies of sympatric species; latter helpful in case of *rostrata*. Flight noticeably dancing, lighter and more steeply undulating or bounding than Linnet; in smaller races, recalls serins. Shy when breeding, but gregarious and often quite tame on migration and in winter, when tight-packed flocks—often in company with Siskin and rarely Arctic Redpoll—are relatively easy to observe in seed-bearing trees and plants. Behaviour and muttered calls convey impression of constant busyness. As aid to identification in Britain, most significant timings of movements are (a) post-breeding dispersal of *cabaret* from early September, (b) east coast passage or occasional marked irruptions of *C. f. flammea* in October–November, and (c) northern arrivals of vagrant *rostrata* from mid-September to early November.

Habitat. Varies widely across west Palearctic, being associated with climates ranging from arctic to relatively warm. In arctic zones, where Arctic Redpoll competes, will occupy treeless tundra, preferably with shrubby growth of creeping osier and dwarf-birch as well as stunted forests and open taiga. In southern Greenland, avoids foggy coastal zones, preferring warmer dry summer climate of interior below *c.* 200 m, especially sheltered places protected from wind, and slopes or hollows with luxuriant growth of willow, birch, juniper, alder, and rowan. Climatic amelioration has led to its replacement of Arctic Redpolls in certain areas. In Norway also, interior birchwoods are preferred to coastal regions, but birds breed from sea-level to *c.* 1000 m, and from dry to marshy ground, in birch and willow bushes but also in open birchwoods and even in stands of pine. Disjunct population in Britain has spread and increased remarkably, if erratically, during 20th century. Mainly distributed in open scrub woodland, often on hillsides and heaths and in field hedgerows, gardens, alder carrs along streams, and increasingly in young conifer plantations. In Switzerland, breeds principally in subalpine conifer woods, mainly above *c.* 1400 m, in sunny situations with trees at least 3 m tall (especially larch) adjoining alpine meadows and pastures.

An 'edge-species', evidently responsive to certain climatic influences and showing pronounced fluctuations and adaptations in habitat. Has taken advantage of some recent changes in landscape, e.g. in colonization of dunes in Netherlands and

(FACING PAGE) Redpoll *Carduelis flammea*. *C. f. flammea*: **1** ad ♂ summer, **2** ad ♂ winter, **3** ad ♂ summer, **4** ad ♀ winter, **5** 1st winter ♂, **6** juv. *C. f. cabaret*: **7** ad ♂ summer, **8** ad ♂ winter, **9** ad ♂ summer, **10** 1st winter ♂, **11** juv. *C. f. 'islandica'* (Iceland): **12** 1st winter ♂. *C. f. rostrata* (Greenland): **13** 1st winter ♂.
Arctic Redpoll *Carduelis hornemanni* (p. 1578). *C. h. exilipes*: **14** ad ♂ summer, **15** ad ♂ winter, **16** ad ♀ summer, **17** 1st winter ♂, **18** juv. *C. h. hornemanni*: **19** ad ♂ winter, **20** 1st winter ♂.

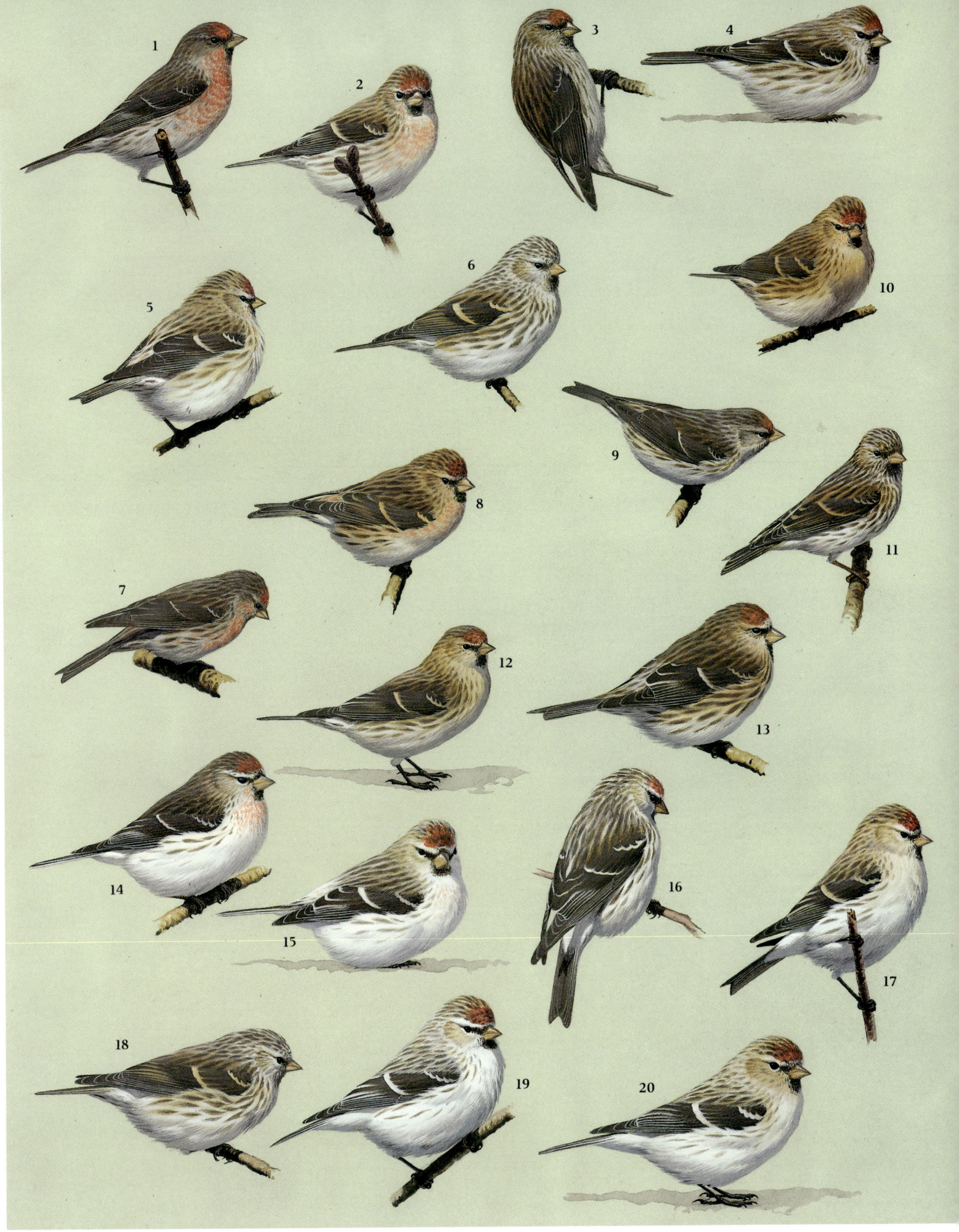
1
2
3
4
5
6
10
9
8
11
7
12
13
14
16
15
17
18
19
20

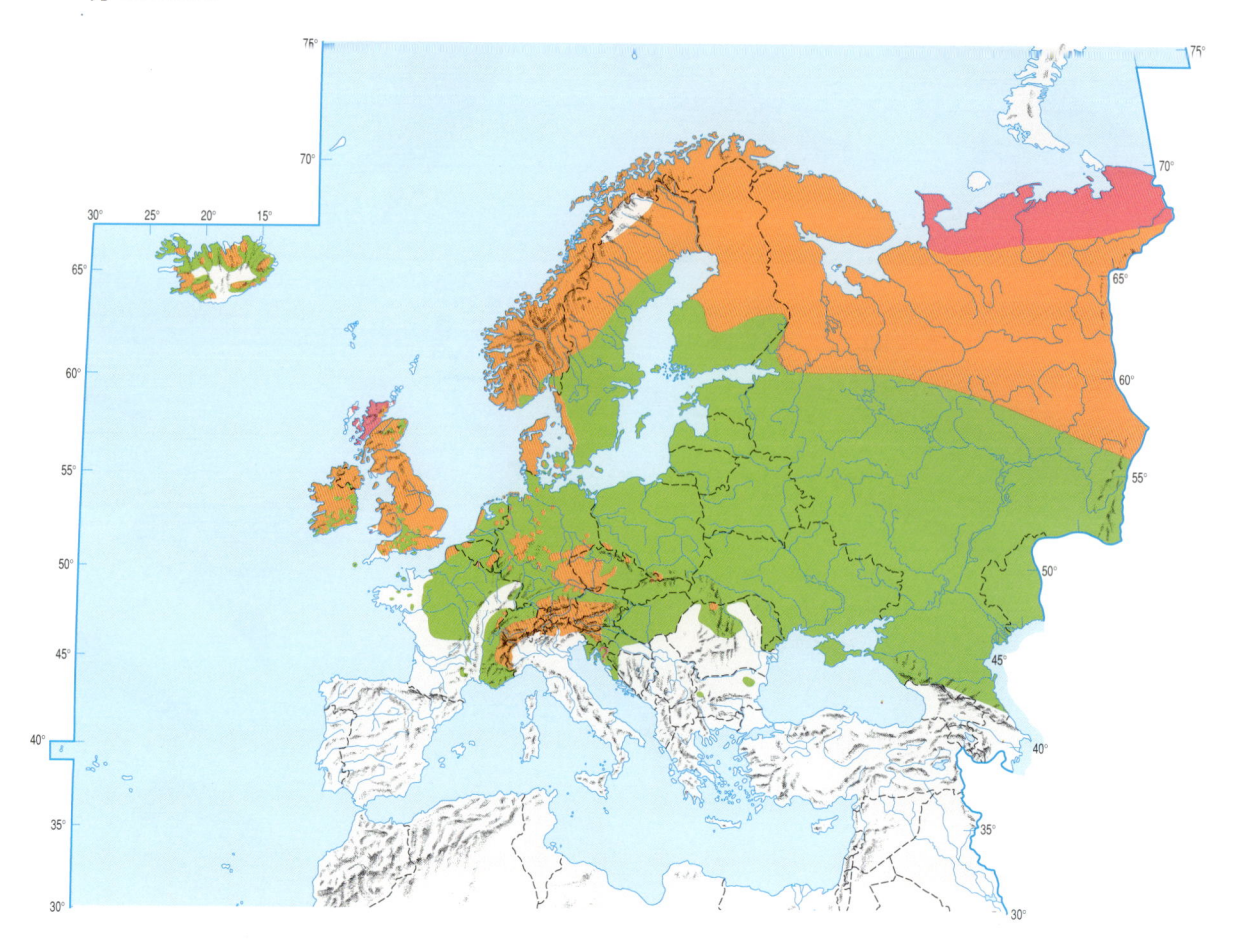

Denmark following spread of birch and alder since cessation of grazing.

Distribution. C. f. cabaret (until 1920 almost confined to Britain, Ireland, and Alps) has spread, apparently from Britain, to coastal areas of north-west Europe and inland to parts of central Europe. Expansion began in 2nd half of 19th century, stagnated in 1st half of 20th century, then resumed with new areas colonized especially in 1950s and 1970s. Main cause of expansion probably large-scale habitat changes due to planting of conifers for forestry, shelter-belts, and amenity. SPITSBERGEN. 1 breeding record, 1993, in west. FAEROES. Possibly bred Tórshavn 1960; otherwise scarce visitor. BRITAIN. Has spread since 1950, aided by afforestation; recently, marked local declines, probably due to habitat changes. IRELAND. Recent contraction in south and east. FRANCE. First bred 1966, near Boulogne; bred Jura from early 1970s. BELGIUM. Two separate populations. Has bred in Hautes-Fagnes in south-east since 1974, with recent increase, and on coast since 1975. NETHERLANDS. First bred (Wadden islands) 1942, spreading from 1960s; 80% breed mainly along coast, remainder inland in sandy country. GERMANY. Long-term expansion of range—since 1970s especially in lowlands. In Thüringen, probably first bred 1978; in Harz, first bred 1989. DENMARK. First bred 1954. Continuing spread; now very common in western Jylland. NORWAY. Has spread recently in lowland coastal areas south and west of previously known range; cabaret first bred 1987, Vest-Agder. SWEDEN. Small population (of cabaret) in south-west recorded from early 1970s. Dramatic expansion in early 1990s, and bred near Stockholm 1994. FINLAND. Very marked annual fluctuations. In peak years breeds well to south of range mapped. LATVIA. Breeding proved only once, 1931. First record of cabaret 1990. CZECH REPUBLIC. Breeding first recorded 1952; spread since, still continuing. YUGOSLAVIA. Breeds in restricted area in north-west, and possibly further south-east. RUMANIA. Very small population in northern mountains. RUSSIA. Range fluctuates.

Accidental. Spitsbergen, Bear Island, Jan Mayen, Spain, Balearic Islands, Portugal, Malta, Albania, Morocco. Rare winter visitor to Turkey and Cyprus.

Beyond west Palearctic, extends east throughout northern Asia south to c. 53°N; also breeds across Alaska and northern Canada and in Greenland. Introduced in New Zealand.

Population. Has increased in parts of western and central Europe, following range expansion. Northern populations (Fenno-Scandia, Russia) fluctuate markedly, according to abundance of birch and spruce seed crops. ICELAND. 2000–20 000 pairs in late 1980s (including pale birds, perhaps Arctic Redpoll—see Geographical variation of that species). Stable. BRITAIN. 160 000 pairs 1988–91; now declining following increase to mid-1970s. IRELAND. 70 000 pairs 1988–91;

Linnet *Carduelis cannabina cannabina* (p. 1568): **1** ad ♂ summer, **2** ad ♀, **3** juv. Twite *Carduelis flavirostris flavirostris* (p. 1571): **4** ad ♂ winter, **5** ad ♀ winter. Redpoll *Carduelis flammea* (p. 1574). *C. f. flammea*: **6** ad ♂ winter, **7** 1st winter ♂. *C. f. cabaret*: **8** ad ♂ winter, **9** 1st winter ♂.

declining. FRANCE. 1000–10 000 pairs in 1970s. BELGIUM. 60–260 pairs. Increase, though fluctuating between years. In 1988, some tens of pairs. NETHERLANDS. 600–1000 pairs 1979–85. Gradual increase since colonization in 1940s, decline since 1992. GERMANY. 7400 pairs in mid-1980s. In east, 1800±800 pairs in early 1980s; rapid increase. DENMARK. 1000–18 000 pairs 1987–8; increase 1990–94. NORWAY. 100 000–2 million pairs nominate *flammea* (northern race) 1970–90; fluctuating. 2000–20 000 pairs *cabaret* 1991–4; increasing. SWEDEN. 250 000–1 million pairs in late 1980s. Small population in south-west estimated at 125–150 pairs in 1st half 1980s. FINLAND. 300 000–600 000 pairs in late 1980s. POLAND. 30–80 pairs; slow increase since 1960s. CZECH REPUBLIC. 6000–12 000 pairs 1985–9; slight increase. SLOVAKIA. 300–600 pairs 1973–94; stable. AUSTRIA. Fairly common in Alps; increase since late 1950s. SWITZERLAND. 5000–15 000 pairs 1985–93. ITALY. 20 000–50 000 pairs 1983–95. YUGOSLAVIA: SLOVENIA. 500–1000 pairs. RUMANIA. At least 5–10 pairs 1986–92. RUSSIA. At least 10 million pairs.

Movements. Short-distance migrant in western Europe, but longer-distance from more northern and eastern breeding areas. Sometimes eruptive. Direction of movement chiefly southeast; distance varies from year to year, hence winter range also varies. Avoids long sea-crossings where possible (except *rostrata*).

British populations often winter within Britain but, in years when food scarce, also further south and east in Netherlands, Belgium, France, western Germany, and very occasionally Iberia. Isolated populations breeding central European Alps are largely resident, moving only to lower altitudes in winter. Fenno-Scandian populations move 1000–1500 km south-east to ESE in autumn to winter chiefly in European Russia, but variable numbers remain in Fenno-Scandia, more in years when seed crop of birch is large. *C. f. rostrata* migrate from southern Greenland to Iceland, and a few reach Britain and Ireland (chiefly north-west Scotland) each year. Icelandic population of *rostrata* probably resident.

Food. Very small seeds, especially birch; invertebrates in breeding season. Forages principally in trees, but moves to ground when seed in trees exhausted or has fallen; on ground, also eats seeds of herbs. Generally in trees in summer, shifting to ground increasingly throughout autumn and winter as birch crop diminishes, though may remain in birch or alder all winter; on ground, searches both below trees for fallen seed and in all kinds of open country, waste ground, etc.; in some places specializes in seeds of spruce or larch.

Social pattern and behaviour. Typically gregarious outside breeding season, and some flocking also when breeding. Especially large flocks occur on migration, particularly during irruptions. Breeds solitarily or (often) in neighbourhood groups; territorial, but not rigorously so, and defended area anyway small. Song given by ♂ without visual display when feeding, preening, or flying between perches. At high intensity serves to attract ♀ and has no territorial significance, being given before territory occupied.

Voice. Commonest call, in flight or perched, a chattering, twittering, or, at distance, muttering, staccato 'chuch-uch-uch-uch' or variant with metallic echo which gives ready differentiation from Linnet and Twite; chorus of flocks with buzzing timbre. Song combines common call with rattle, 'chuch uch uch crrrrrr'. Other common call when perched, a plaintive 'dsooee'.

Breeding. SEASON. Britain and Ireland: eggs laid from 2nd half of May or early June, occasionally April. Iceland: laying 1st half of June, from mid-May in mild springs. Finland: in north, eggs laid mid-May to late July, possibly beginning of August, with peak around 1st half of June; in central latitudes, from mid-April, with peak 1st half of May; early eggs can be laid with thick snow on ground and night temperature of −20°C. Russia: in far north, eggs laid from mid-June, mostly July. Germany: in Sachsen, laying from mid-April to beginning of August, with peak period in May. Switzerland: April–August, peak period late May and early June, mid-June at 1500 m. Northern Italy: eggs laid late April to late July at 2000 m. Generally 2 broods. SITE. In shrub or tree, at varying heights depending on habitat and species; occasionally on ground. Nest: foundation of twigs of spruce, juniper, birch, stalks, leaf stems, etc., with inner layer, often densely packed, of roots, grass, bark, moss, flower-heads, leaves, etc., thickly lined with plant down, hair, wool, and feathers, whole lining often very white in appearance. EGGS. Sub-elliptical, smooth, and slightly or non-glossy; bluish-white to pale blue-green with variable rust-red blotches, violet-pink undermarkings, and purplish-brown specks and scrawls towards broad end, often forming ring. Clutch: 4–6 (2–7). INCUBATION. 10–12 days. FLEDGING PERIOD. 9–14 days.

Wing-length: *C. f. flammea*: ♂ 72–78, ♀ 71–75 mm. *C. f. cabaret*: ♂ 68–74, ♀ 66–73 mm.

Weight: *C. f. flammea*: ♂ ♀ mostly 12–16 g. *C. f. cabaret*: ♂ ♀ mostly 9–12 g.

Geographical variation. Marked; involves size, relative depth and width of bill, and general colour. *C. f. cabaret* from Britain and north-west France to southern Sweden and south-west Norway as well as (isolated) in Alps and other mountain areas of central Europe is smallest race, wing mainly under 73 mm, tail mainly under 54 mm, but bill not much different from *C. f. flammea*; streaks on upperparts and sides of body slightly broader, blacker, and more sharply defined. *C. f. rostrata* from Iceland (sometimes named *islandica*, doubtfully distinct from Greenland race *rostrata*) distinctly larger than *C. f. flammea*, wing mainly over 79 mm in ♂, over 77 mm in ♀, tail mainly over 57 mm; bill deeper and wider at base, culmen and gonys clearly convex (not as straight as in *C. f. flammea*); darker than *C. f. flammea*, colour near *cabaret*, with width, extent, and colour of dark streaks as in *cabaret* but ground-colour paler buff-brown, less rusty or tawny.

Arctic Redpoll *Carduelis hornemanni*

PLATES: pages 1575, 1588

DU. Witstuitbarmsijs FR. Sizerin blanchâtre GE. Polarbirkenzeisig IT. Organetto artico
RU. Тундряная чечетка SP. Pardillo ártico SW. Snösiska N. AM. Hoary Redpoll

Field characters. 13–15 cm; wing-span 21–27.5 cm. Size varies considerably: main Holarctic race *exilipes* similar to north Eurasian and North American race of Redpoll but often looking plumper, particularly around neck; Baffin Island and

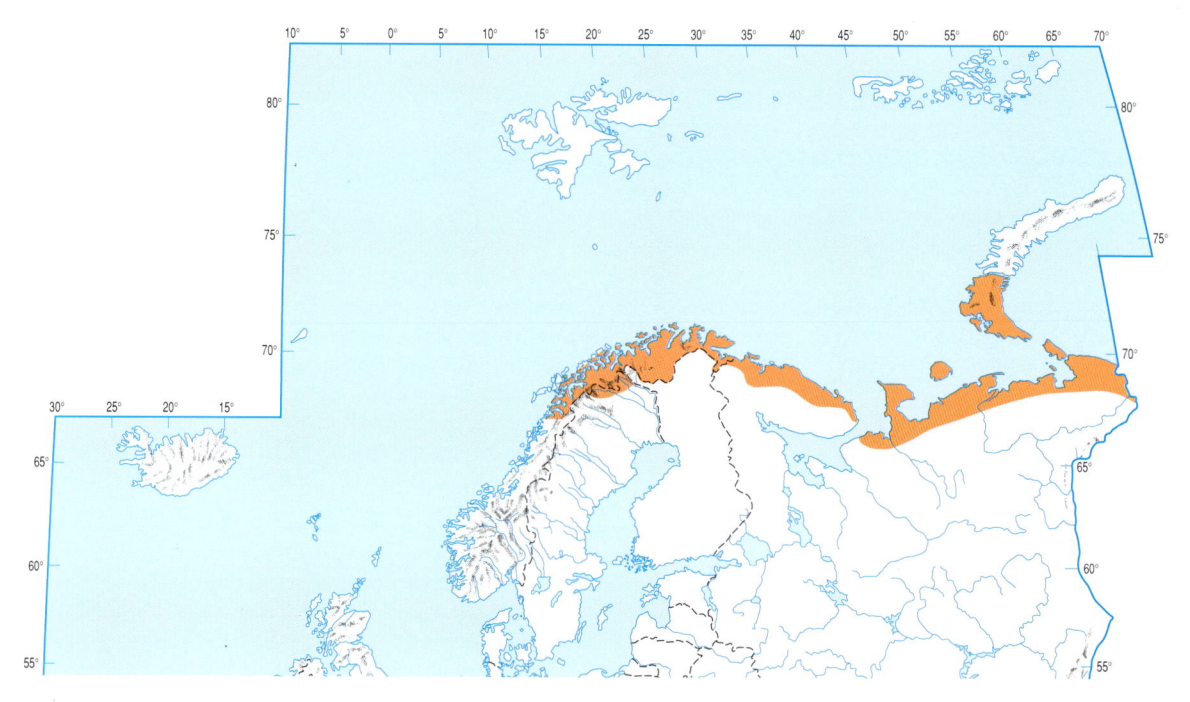

Greenland race, *C. h. hornemanni* ('Hornemann's Redpoll'), up to 10% larger, exceeding Greenland race of Redpoll and overlapping with Twite. Small but long, ghostly finch, with similar behaviour and flight to northern races of Redpoll, but rather loose plumage much greyer above and whiter below. Rump of adult and 1st-winter ♂ white and usually unmarked. Plumage pattern of adult and 1st winter ♂ like Redpoll but buff tones restricted to head, pale supercilium more marked, ear-coverts paler, wing marks white and more contrasting, long rump pure white, and underparts white and far less streaked. Juvenile and some 1st-winter ♀♀ much closer in appearance to northern race of Redpoll, showing streaked rump but less streaked paler underparts, especially below tail. Bill stubby, particularly in *exilipes*, and partly hidden by profuse feathering at base of bill.

Adult unmistakable when pure white rump seen but certain identification of Arctic Redpoll in juvenile and 1st-winter plumages dogged by its convergence, however caused, with Redpoll, particularly of sympatric *C. f. flammea*. Since some birds daunt museum taxonomists, they must defeat field observer; yet with extensive experience of Redpoll, many claims of Arctic Redpoll can be sustained given acute observation of characters noted above, particularly short bill, longer and denser frontal feathering (hiding bill and flattening face), thick-necked appearance, generally paler ground-colour to plumage, and little-streaked underparts. Flight action includes rather loose wing-beats (perhaps illusion due to white plumage).

Habitat. Distinguishable from that of Redpoll only by being confined to more northerly Arctic latitudes, where, however, the need for some kind of dwarf willow or other shrub growth remains indispensable in breeding season. In winter, main population remains in or near breeding latitudes, coping with night temperatures down to *c.* −60°C in central Alaska, and foraging for as long as possible in low light available. This degree of hardiness is all the more surprising since closely related forms of Redpoll inhabiting temperate climates quite commonly move south in winter, even to Mediterranean regions.

Distribution. No evidence of changes. Map shows breeding areas only, where largely resident; for other wintering areas, see Movements. SPITSBERGEN. One breeding record in west, 1948. FINLAND. Not mapped. In most years, a small percentage of Finnish redpoll population is *C. hornemanni*; confined to birch and willow zone of northern Lapland.

Accidental. Spitsbergen, Bear Island, Jan Mayen, Faeroes, Britain, France, Belgium, Netherlands, Germany, Lithuania, Poland, Czech Republic (up to 50 individuals), Slovakia, Hungary, Austria, Yugoslavia, Belarus', Ukraine. Records reflect distance of dispersive movement in some winters, rather than vagrancy.

Beyond west Palearctic, extends across extreme north of Asia and North America; also breeds Greenland.

Population. ICELAND. See Redpoll. NORWAY. 1000–10 000 pairs 1970–90; fluctuating. SWEDEN. 1000–10 000 pairs in late 1980s. FINLAND. 2000–10 000 pairs in late 1980s; slight increase. RUSSIA. 100 000–1 million pairs; stable.

Movements. Relatively short-distance, partial migrant from circumpolar breeding areas, usually to lower latitudes in winter, most birds probably moving south of Arctic circle. *C. h. exilipes* of European Arctic zone reaches as far west and south as 7°30′ E in Norway and southern Baltic (and sometimes beyond) in small numbers. In Britain, 109 records 1958–85 (chiefly autumn), almost entirely confined to east coast from Shetland to Kent.

Food. Diet comprises small seeds, particularly birch, alder, willow, various herbs, and grasses; some small invertebrates in summer; probably very similar to that of Redpoll. Forages in trees like Redpoll; in snow, above all in winter, when one of very few passerines to remain in Arctic, foraging restricted to scrub, tall herbs and catkins above snow, seeds on surface, snow-free patches at coasts or on windy slopes, roadsides, and rubbish tips.

Social pattern and behaviour. Apparently not significantly different from that of Redpoll.

Voice. Song similar to Redpoll. Flight-call slower and more spaced out than in Redpoll, written 'chut-chut'. Other calls include plaintive 'dyeeeu'.

Breeding. SEASON. Northern Sweden: eggs laid throughout June. Kola peninsula (north-west Russia): eggs laid from beginning of June to early July. SITE. Dwarf tree or shrub, almost always willow, poplar, birch, or alder; in crotch or on branch or twigs generally close to trunk, almost always less than *c.* 2 m above ground, though usually in upper part of shrub. Nest: robust structure with foundation of small twigs, bark, stems, roots, grass, catkins, etc., warmly lined with hair, fur, plant down, and many feathers, especially white ones of grouse; lining sometimes only of feathers, and often higher than outer wall. EGGS. Very like Redpoll, but perhaps larger and paler with heavier markings more often forming ring at broad end. Clutch: 4–5 (3–7). INCUBATION. 11–12 days. FLEDGING PERIOD. 10–12 days.

Wing-length: *C. h. exilipes*: ♂ 74–80, ♀ 70–76 mm.
Weight: *C. h. exilipes*: ♂ mostly 11–16, ♀ 10–15 g.

Geographical variation. Rather marked, involving size and general colour. *C. h. exilipes* from entire Holarctic except Iceland, Greenland, and arctic islands of eastern Canada rather small: wing mainly less than 80, tail less than 60, bill depth less than 6.3 mm. *C. h. hornemanni* from northern Greenland and neighbouring part of arctic Canada distinctly larger (wing mainly more than 80, tail more than 60, bill depth more than 6.3 mm), bill heavier, deeper and wider at base; plumage on average whiter. Pale birds on Iceland probably form separable unnamed population.

Relationships of redpolls complex and much debated. *C. hornemanni* considered by some authorities to be a race of *C. flammea*.

Two-barred Crossbill Loxia leucoptera

PLATES: pages 1581, 1588

Du. Witbandkruisbek   Fr. Bec-croisé bifascié   Ge. Bindenkreuzschnabel   It. Crociere fasciato
Ru. Белокрылый клест   Sp. Piquituerto franjeado   Sw. Bändelkorsnäbb   N. Am. White-winged Crossbill

Field characters. 15 cm; wing-span 26–29 cm. Slighter than Crossbill, with relatively weaker bill, smaller head, and longer tail creating rather slim form. Rather small, elegant crossbill, with somewhat more *Fringilla*-like outline compared with Crossbill and bold white wing-panels recalling Chaffinch. Adult ♂ has pink-red head, mantle, and underbody (lacking orange tone of other *Loxia*) and black scapulars, wings, and tail. ♀ also looks bright, with more streaked head and body than Crossbill and much brighter, purer yellow rump. Juvenile shows diagnostic wing-marks.

Unmistakable if full extent and shape of white wing markings seen well. Commonest calls also distinctive. Flight lighter than Crossbill, with slightly more pointed wings and proportionately longer tail combining with smaller head and slimmer body to give less robust, more *Fringilla*-like flight silhouette. Movements as acrobatic as Crossbill but noticeably less clumsy.

Habitat. After last glaciation, immense spread of coniferous forests across northern Palearctic presented an evolutionary challenge, owing to firm lock-up of great nutrient resources within billions of hard fir cones in crowns of spruces, pines, and larches. These resources, moreover, are in many cases retained on trees for much longer than most other fruits, and from time to time fail over large areas. To exploit them called for exceptional physical adaptations, and unusual flexibility in use of habitat. Challenge was successfully taken up by a group of cardueline finches evolving strong crossed mandibles backed by powerful musculature, including the legs, and by adaptability to different problems posed by the cones of various tree species, leading to evolution of distinct specialized forms of crossbill: Parrot Crossbill and Scottish Crossbill associated mainly with pine, Crossbill throughout its main boreal strongholds with spruce, and Two-barred Crossbill depending mainly on larch.

While all *Loxia* are constrained by dependence on conifers to conform to northerly distribution pattern, Two-barred Crossbill seems additionally inhibited by climatic factors, occupying the most northerly regions between July isotherms 13–20°C. (Extraordinary exception, however, has been establishment of small isolated population in Neotropics on Caribbean island of Hispaniola in stands of pine.) Can subsist on most conifers, and when necessary on berries and buds of other trees, but larch and, in some areas, spruce are main food trees. Although given to large-scale wanderings, on occasion has failed to settle in many areas where its food trees are abundant, or to establish colonies in more southern montane regions of Eurasia. While living commonly in dense coniferous forest, in some areas favours forest edges and open growth, or even detached groves.

(FACING PAGE) Two-barred Crossbill *Loxia leucoptera*: **1** ad ♂ summer, **2** ad ♂ winter, **3** ad ♀ summer, **4** 1st winter ♂, **5** juv. Crossbill *Loxia curvirostra* (p. 1582). *L. c. curvirostra*: **6** ad ♂ summer, **7** ad ♂ winter, **8** ad ♀ summer, **9–10** 1st winter ♂, **11** juv ♂, **12** juv ♀. *L. c. corsicana*: **13** ad ♀ summer, **14** ad ♀ summer. *L. c. balearica*: **15** ad ♂ summer, **16** ad ♀ summer. *L. c. poliogyna*: **17** ad ♂ summer. *L. c. guillemardi*: **18** ad ♂ summer. Scottish Crossbill. *Loxia scotica* (p. 1586): **19** ad ♂ summer. Parrot Crossbill *Loxia pytyopsittacus* (p. 1587): **20** ad ♂ summer, **21** ad ♀ summer.

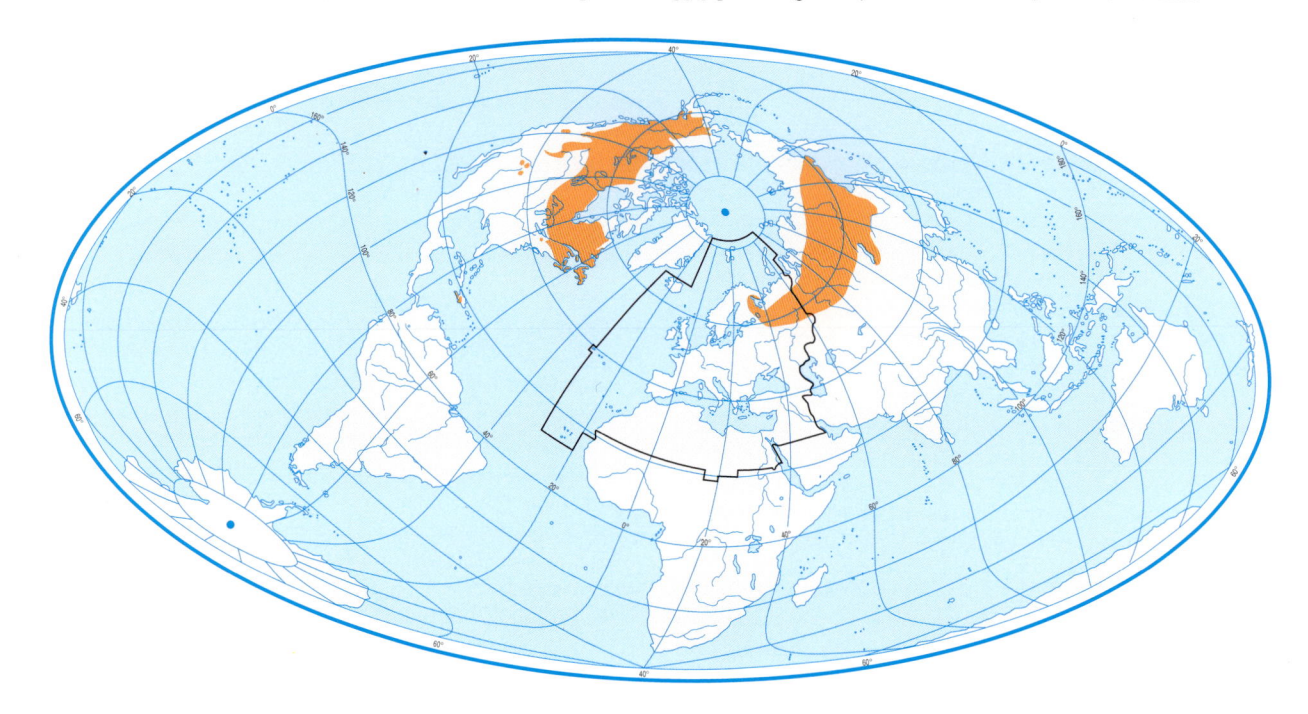

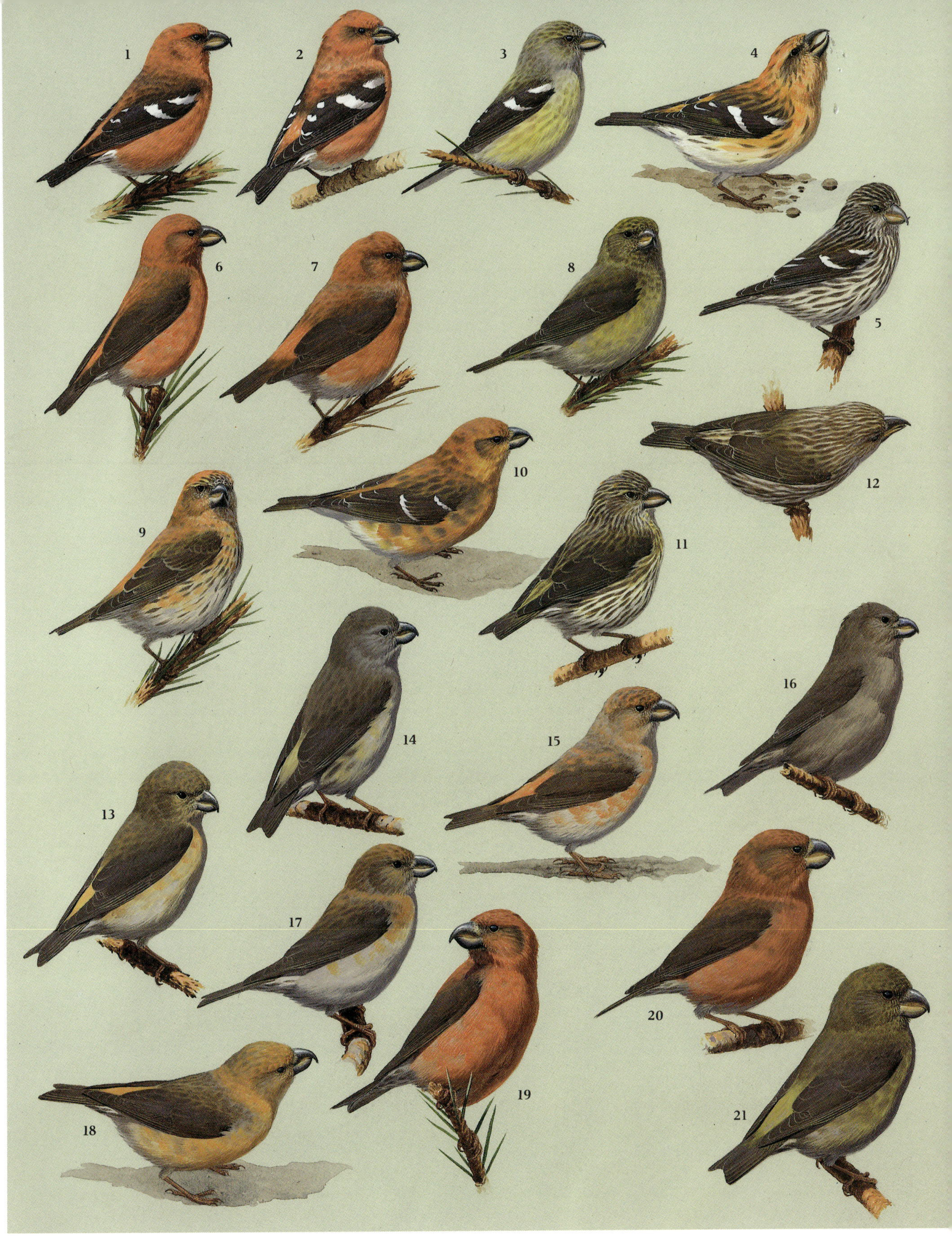
1
2
3
4
6
7
8
5
10
12
9
11
16
14
15
13
17
20
19
18
21

Distribution and population. West Palearctic is at or beyond western limit of regular breeding range. Breeding in Scandinavia (perhaps more frequent in last 100 years) and most of Finland irregular, usually after invasions, and cannot be precisely mapped. Population very variable. GERMANY. Bred in Berlin 1991. NORWAY. Breeds occasionally, after invasions: 1982 and (unsuccessful attempt) 1987. SWEDEN. Occasional breeder: several records in north after major influx 1985–7. FINLAND. Probably breeds in most years, but almost totally absent in years with poor seed crop. 50–50 000 pairs; marked annual fluctuations. RUSSIA. Apparently does not breed regularly in western part of European Russia. 10 000–100 000 pairs.

Accidental (status of rare and irregular visitor to much of Europe). Faeroes, Britain, Ireland, France, Belgium, Netherlands, Germany, Estonia, Latvia, Lithuania, Poland, Czech Republic, Slovakia, Hungary, Austria, Switzerland, Italy, Yugoslavia, Bulgaria, Belarus', Ukraine.

Movements. Resident and dispersive; also eruptive. In most years makes only limited movements in response to local food shortages. In occasional years, like Crossbill, makes eruptive movements, associated with shortage of preferred food (seed of larch *Larix*) and high population density. Birds move west or south-west, regularly reaching Finland and Sweden and occasionally various parts of eastern, central, and western Europe. In exceptional 1889 invasion, reported south to Switzerland, northern Italy, and Hungary, west to Britain and Netherlands. In more typical 1956 invasion, many observations in Finland, with small numbers continuing south-west to Sweden, Norway, Denmark, Germany, and Belgium. Invasions usually reach Scandinavia and Britain in July, continuing chiefly to September or October; recorded in central Europe later, mostly from September. Return movement February or March to early June.

In Sweden, occurs every year in strongly varying numbers, and increasingly since 1976; recent major invasions (600–800 birds recorded) in 1979 and in 3 consecutive years (probably for first time since records began in 1786) 1985, 1986, and 1987; largest ever in 1990 (more than 2300 birds), but only *c.* 170 in 1991. Sometimes breeds in spring following invasion (e.g. 1957, 1987), as also in other countries (see Distribution).

Food. Mainly conifer seeds, principally of larch and spruce, occasionally alder, birch and other plants; some invertebrates in breeding season. Larch is main food in some parts of range (e.g. eastern Russia), spruce in others (e.g. Fenno-Scandia, Murmansk region of north-west Russia); differences presumably due to distribution and abundance of these trees. Foraging methods identical to Crossbill; thinner bill more suitable for extracting seeds from between thin and relatively short cone scales.

Social pattern and behaviour. Not well known. Occurs in flocks all year, though pairs and single birds also recorded. Tends to nest in neighbourhood groups. Paired (perhaps also unpaired) ♂ sings from same area (radius *c.* 50 m) daily in early stages of nesting. Typical of *Loxia* in defending only ♀, and small area around nest before and during laying. Also regularly sings in flight, circling with slow wing-beats.

Voice. Distinctly different from other *Loxia*. Song comprises long series of different segments mostly delivered at very rapid rate; also some more well-separated units (calls). Segments include rattles, trills, and buzzy units; some harsh, others more musical; shows some similarity especially to Redpoll and domesticated Canary. Calls (perched and in flight) often disyllabic like Crossbill but distinctly higher pitched, drier, and less metallic. 3 main types: 'kip-kip' or 'tyip tyip'; 'chut-chut' suggesting Redpoll; remarkable sound like toy trumpet.

Breeding. SEASON. Murmansk region (north-west Russia): eggs laid February to mid-May in years of good seed crop of spruce and pine, with peak February–March; in years of poor larch or spruce crop, delayed until mid-June to end of August when pine seeds available. Finland: February–August, peak probably normally around June. SITE. In conifer (usually spruce), from low down to over 20 m above ground, almost always against trunk. Nest: foundation of dead conifer twigs, stalks, grass stems, lichen, bark, etc., lined with fine grass, rootlets, stems, and moss, then inside layer of hair, fur, rootlets, moss, feathers, and plant down. EGGS. Sub-elliptical, smooth and slightly glossy; very like those of Crossbill; pale whitish-blue to whitish-green, sparsely marked with specks, spots, and scrawls of purple or purplish-black, usually at broad end; sometimes undermarkings of pink or violet. Clutch: 3–5. INCUBATION. 14–15 days. FLEDGING PERIOD. 22–24 days.

Wing-length: ♂ 88–96, ♀ 87–94 mm.
Weight: ♂ mostly 30–38, ♀ 25–34 g.

Crossbill *Loxia curvirostra*

PLATES: pages 1581, 1588

DU. Kruisbek   FR. Bec-croisé des sapins   GE. Fichtenkreuzschnabel   IT. Crociere
RU. Клест-еловик   SP. Piquituerto común   SW. Mindre korsnäbb   N. AM. Red Crossbill

Field characters. 16.5 cm; wing-span 27–30.5 cm. Up to 15% larger than Greenfinch but with similar compact form; intermediate in size between Two-barred Crossbill and Scottish Crossbill; proportionately smaller-billed and smaller-headed than larger and heavier Parrot Crossbill. Beware sexual differences and racial variation in bulk; beware also adaptation of bill size to seed sources, south-east European race *guillemardi* approaching bulk of Parrot Crossbill. Large, powerful, somewhat clumsy and noisy finch, with heavy, crossed bill and sharply forked tail; epitome of genus. ♂ basically orange to red, with dusky wings and tail; ♀ and immatures grey or olive, juveniles heavily streaked. Plumage shows few features except for paler rump and vent in adult. Quite exceptionally, a few have narrow white tips on median and greater coverts and tertials; unlike Two-barred Crossbill, these marks extend into even narrower and fading white fringes on greater coverts and tertials. Flight strong and bounding, accompanied by distinctive bursts of loud disyllabic calls. Feeds and moves somewhat like small parrot.

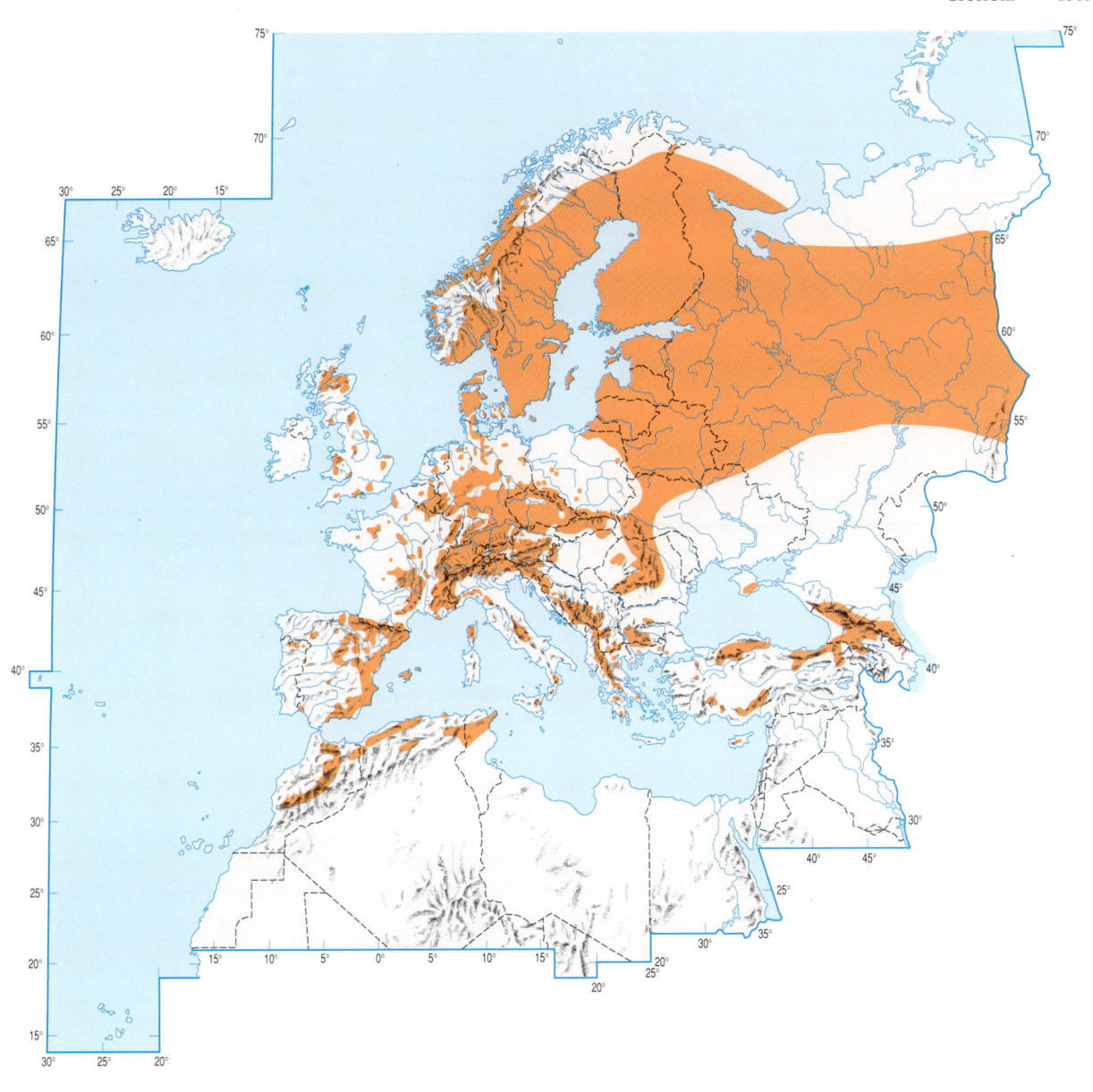

Unmistakable as a *Loxia* if heard or seen well; crossed bill and parrot-like form and behaviour distinctive. *L. curvirostra* constitutes, however, minefield for field observer, due to (1) rare type (1 in 1000) with thin white wing-bar suggesting Two-barred Crossbill and (2) racial variation in bill size and general bulk prompting confusion with Scottish Crossbill and Parrot Crossbill. 1st confusion well understood but 2nd largely ignored, with no tangible evidence on incidence of racial overlap and little or no practicality in field study. Accordingly, separation of Crossbill from its 2 larger congeners highly problematic, with only bill shape (not just size), food preference, and voice affording trustworthy clues even to expert observers, and subspecific identification a hopeless quest.

Habitat. In west Palearctic occurs mainly in boreal and subarctic coniferous forests, but also well represented, often by distinct races, in temperate and Mediterranean insular and montane areas. Ecologically divided largely into pine-dwelling populations in more southerly parts of range and spruce-based populations in more northerly, latter being more commonly subject to eruptive movements, during which they may remain to breed, temporarily or for longer, in hitherto unoccupied areas, especially where mature conifer plantations or shelterbelts have recently been developed.

Apparently equally at home in deep dense forest and on

edges or in open or detached stands, and readily occupies suitable mature conifers even where they grow in built-up areas, not excluding small towns, showing disregard of neighbouring human activities, and sometimes using artefacts such as overhead cables or drinking from rooftop water tanks. Resorts to ground only infrequently, being specialized in exploiting conifer seeds before these fall from tree; will switch to non-coniferous diet and accompanying habitat changes only when coniferous supplies fail, often forcing long-distance journeys to fresh habitat. Even where sedentary, given to fairly extensive flights, sometimes above lower airspace, apparently serving through frequent calling to make contact with conspecifics which have discovered other food sources. Frequently visits water, and this must influence choice of habitat in areas where such sources may be widely scattered.

Distribution. Breeding (map shows only this) sporadic in much of west and south of west Palearctic, following irruptions, though apparently rather less erratic in Mediterranean basin than in northern and central Europe. Slight increase Denmark. ICELAND. First (unsuccessful) breeding 1994. BRITAIN. Well established East Anglia, New Forest, and Kielder Forest, otherwise sporadic breeder. IRELAND. Scarce and irregular breeder. FRANCE. More widely distributed 1985–9 than 1970–75, especially in north, but perhaps only temporary effect. BELGIUM. First breeding after 1888–9 irruption; initially irregular, but probably annual breeder since 1950s. NETHERLANDS. Occasional breeder until mid-1970s, then regular. NORWAY. Breeding distribution not fully known (confusion with Parrot Crossbill); probably breeds in all areas with spruce woodland, predominating over Parrot Crossbill in south-east. POLAND. Scarce and irregular breeder in mountains and north-east. SPAIN. Irregular and scarce breeder in north, nearly always following invasions. PORTUGAL. Breeding range probably not permanent. ITALY. Sicily: breeding first confirmed 1981. GREECE. Crete: first recorded breeding 1984. YUGOSLAVIA. Regular breeder in native pine forest; occasional breeder in planted pinewoods. Range fluctuating in Slovenia. RUSSIA, UKRAINE. Fluctuating. AZERBAIJAN. Poorly known; no confirmed breeding. TURKEY. Locally fairly common Black Sea coastlands, Marmara, and Taurus mountains. Small flocks late April–July perhaps indicate local breeding: e.g. Beynam forest (Ankara), and Amanus mountains. ISRAEL. Occasionally breeds after irruptions, e.g. 1974, 1982–4.

Accidental. Bear Island, Iceland (but reached by several irruptions), Malta (occasional large influxes), Madeira, Canary Islands.

Beyond west Palearctic, extends east from Urals to southern Sea of Okhotsk and perhaps Sakhalin and Kuril Islands; isolated breeding areas in Tien Shan mountains, eastern Himalayas to south-west China, Vietnam, and Philippines. In North America, from south-east Alaska east to Newfoundland and north-east USA, and south in mountains to Nicaragua.

Population. Marked fluctuations in much of range, depending on conifer seed crop. Increase reported Britain, France, Belgium; stable Denmark (or fluctuating), Sweden, Finland, Czech Republic, Hungary, Switzerland, Greece, Croatia, Bulgaria, and Russia. BRITAIN. Probably fewer than 1000 birds in low years, several thousand in good years. Numbers unknown 1988–91. Increasing with afforestation. IRELAND. 20–150 pairs in good years. FRANCE. 1000–10 000 pairs in 1970s. BELGIUM. 3800 pairs 1975–6, 100–30 000 pairs 1989–91. Increase in 1970s; wide fluctuations thereafter, but general tendency to increase. LUXEMBOURG. Irregular: 5–200 pairs. NETHERLANDS. Highly variable: from fewer than 100 to over 5000 pairs. GERMANY. 16 000 pairs in mid-1980s, but more recent estimate of 20 000–80 000 pairs. In east, where numbers in lowlands not well known, 5000 ± 2500 pairs in early 1980s. DENMARK. 110–2200 pairs 1987–8. NORWAY. 100 000–500 000 pairs 1970–90. SWEDEN. 200 000–500 000 pairs in late 1980s. FINLAND. 50 000–400 000 pairs in late 1980s. Long-term changes poorly known. ESTONIA. Numerous in some years (up to 50 000 pairs), totally absent in others. LATVIA. 10 000–15 000 pairs in 1980s. LITHUANIA. Not rare. CZECH REPUBLIC. 30 000–100 000 pairs 1985–9. SLOVAKIA. 25 000–50 000 pairs 1973–94. HUNGARY. 100–300 pairs 1979–93. SWITZERLAND. 50 000–100 000 pairs 1985–93. SPAIN. 140 000–190 000 pairs. PORTUGAL. 100–1000 pairs 1992. ITALY. 30 000–60 000 pairs 1983–95. GREECE. 5000–10 000 pairs. ALBANIA. 500–2000 pairs in 1981. YUGOSLAVIA: CROATIA. 4000–8000 pairs. SLOVENIA. 10 000–20 000 pairs. BULGARIA. 2000–5000 pairs. RUMANIA. 45 000–60 000 pairs 1986–92. RUSSIA. 100 000–1 million pairs. BELARUS'. 2000–20 000 pairs in 1990. Large-scale fluctuations, no discernible trend. UKRAINE. 3000–5000 pairs in 1986. TURKEY. 1000–10 000 pairs. CYPRUS, TUNISIA. Fairly common. MOROCCO. Uncommon.

Movements. Resident and dispersive, also irruptive. In most years, birds disperse short distances in midsummer to find new feeding areas, moving in flocks in various directions but remaining within regular range. Local numbers may therefore fluctuate greatly from year to year, dependent on varying state of conifer seed-crops, especially spruce; timing of movement coincides with formation of new spruce cones. In irruption years (mostly involving *L. c. curvirostra*), birds move much further (up to 4000 km), mainly in one direction; such movements vary considerably in extent and duration, and tend to begin earlier and end later than in normal years. Irruptions probably result from high population levels coinciding with poor or moderate seed harvests; early departures suggest that crowding may sometimes alone stimulate movement. Birds frequently stay to breed in invasion areas, reinforcing local populations or colonising new sites; these settlements usually temporary, but occasionally permanent, e.g. colony in East Anglia (eastern England) dates from 1909 invasion.

Food. Conifer seeds, generally spruce, but in some parts of range (e.g. England and Mediterranean region) mostly pine. Very agile and acrobatic forager, easily fluttering from twig to twig, sidling along branches, hanging from cones, and clambering around, often using bill as help like parrot; either works at cones, usually riper ones, extracting seeds *in situ* while hanging on cone, or snips them off (sometimes taking whole sections of twig) to carry to perch, often in fork, where held under foot and seeds removed. Often flies to perch carrying cone as heavy as bird itself; quite able to hold loose cone against underside of branch and extract seeds while upside-down; legs and toes are adapted for grasping and securing cones, as bill is for extracting seeds. Inserts bill-tips between scales of cone from side, moves lower mandible (which is more angled than upper) sideways, flat against top scale, causing tip of upper mandible to push bottom scale downwards, then scoops out seed with tongue; on thin-scaled cones, upper mandible can be used to hook seeds out. If seed is still fast, can open scales further by inserting closed bill and turning. Away from conifers, readily feeds in broad-leaved trees, taking buds as well as fruits and insects, particularly caterpillars and aphids.

Social pattern and behaviour. Gregarious all year, though less so during breeding season when birds (breeders and non-breeders) nevertheless congregate to feed. Mating system apparently essentially monogamous; no evidence for bond being maintained beyond 1 breeding season, and this presumably unlikely to occur, at least in highly nomadic populations. Under optimal conditions, local populations may breed continuously for c. 9 months, but if cone crop poor, most or all do not breed. Often breeds in loose neighbourhood groups. Full song apparently given by ♂ only; ♀ also sings, but perhaps only subsong or similar, though both sexes participate in social-singing. ♂ sings from top of tall tree, sometimes lower down, usually close to nest or ♀; also regularly in flight, with slow wing-beats, sometimes accompanied by ♀.

Voice. Song variously described: loud but hesitant, even staccato phrase, interspersed by call, 'cheeree-cheree-cheuf-glipp-glipp-glipp-cheree . . .'; sweet, musical warble, somewhat like Greenfinch, 'chip-chip-chip-jee-jee-jee-jee', with first notes trilled but last loud and creaking and again interspersed with 3–4 calls; high-pitched twittering, interspersed with 'tiwee-tiwee' notes and more nasal sounds, altogether thinner, faster, and more ethereal than Parrot Crossbill; curious soft, at times almost inaudible phrases recalling Bullfinch or Starling. Commonest call in flight a disyllabic, explosive 'chip chip' or 'glipp glipp', emphasized in alarm. When feeding, a quieter 'chük chük'.

Breeding. SEASON. Stimulated to breed by abundance of food whenever it may occur. Britain and Ireland: in southern Scotland, in spruce, eggs laid August–April; in eastern England, in pine, December–June, peak February–April; in Ireland, c. 1 month later. Murmansk region (north-west Russia): in years of good spruce and pine seed crop, eggs laid February to mid-May; when only pine crop good, end of May to August, sometimes September. Finland: eggs found mid-January to mid-May, rarely late summer, peak mid-March to late April. South-west Germany: most clutches laid by mid-March; some pairs still nest-building early May. Switzerland: recorded breeding in every month with apparent exception of September, peak December–May. Pyrénées: end of February to July, peak mid-April to mid-May. Morocco: eggs laid November–June. SITE. High in conifer, usually standing isolated or at woodland edge, generally close to top of tree, covered from above by overhanging twigs. Nest: foundation of dead conifer twigs, strips of deciduous bark, moss, lichen, etc., lined with dry grass, decayed wood, plant down, hair, wool, and sometimes feathers. EGGS. Sub-elliptical, smooth and slightly glossy; creamy to bluish-white, very sparsely marked with dark purplish specks, spots, and short scrawls, concentrated at broad end, and violet-grey undermarkings. Clutch: 3–4 (2–5). INCUBATION. 14–15 (13–16) days. FLEDGING PERIOD. 20–25 (16–28) days.

Wing-length: *L. c. curvirostra*: ♂ 97–102, ♀ 94–99 mm. *L. c. balearica*: ♂ 91–98, ♀ 89–94 mm.

Weight: *L. c. curvirostra*: ♂ ♀ mostly 35–50 g.

Geographical variation. Rather slight and mainly clinal in west Palearctic, where 5 races recognized (more marked in central and eastern Asia and North and Central America). Involves depth of grey ground-colour, colour and extent of red (♂) or green (♀) on feather-tips of head and body, size, and relative length, depth, and width of bill. Colour of red in ♂ and green of ♀ strongly dependent on abrasion, becoming more glossy deep scarlet in ♂, bronzy- or orange-green in ♀ (but in some ♀♀, green of feather-tips largely worn off, plumage becoming mainly dull grey).

In Europe, grey ground-colour of body becomes gradually paler south from Cantabrian mountains, Pyrénées, and central Italy, bright colour of feather-tips paler and more reduced in extent, cline ending in *poliogyna* of North Africa. Ground-colour of *poliogyna* pale ash-grey, less dark and saturated than in nominate *curvirostra* from central and northern Europe eastward, feather-tips of head and body of adult ♂ pink-red, much grey of feather-bases visible; rump uniform rosy-pink; belly rosy-red with some white spots or streaks on feather-centres. Birds from central and southern Spain as well as Balearic Islands slightly paler and greyer than *L. c. curvirostra*, intermediate between *poliogyna* and *L. c. curvirostra*; *balearica* from Balearic Islands slightly smaller than both *L. c. curvirostra* and *poliogyna*, bill shorter and less deep. *L. c. corsicana* from Corsica has ground-colour of head and body slightly darker than in *L. c. curvirostra*, adult ♂ slightly darker scarlet-red, hardly distinguishable from ♂ *L. c. curvirostra*, but green of ♀ rather restricted, head and body mainly dark grey (less pale than *balearica*). *L. c. guillemardi* from Turkey, Crimea, Caucasus area, and Cyprus on average slightly larger than *L. c. curvirostra*, bill longer, thicker, and broader at base, like bill of Scottish Crossbill *L. scotica*.

Scottish Crossbill Loxia scotica

Du. Schotse Kruisbek   Fr. Bec-croisé d'Ecosse   Ge. Schottischer Kreuzschnabel   It. Crociere di Scozia
Ru. Шотландский сосновик   Sp. Piquituerto escocés   Sw. Skotsk korsnäbb

Field characters. 16.5 cm; wing-span 27.5–31.5 cm. Intermediate in size between Crossbill and Parrot Crossbill, with size and structure of bill and head closer to latter. Resident crossbill of relict forests of Scots pine in north-central Scotland, formerly regarded as race of one or other congener. All plumages as Crossbill but head larger than Crossbill and bill deeper (especially upper mandible) and hence blunter.

Crossbill known to have bred within breeding range of Scottish Crossbill and widespread occurrence there is considered possible, so assumption that all crossbills in Scotland from Perthshire northwards are Scottish Crossbill is not valid. No observed difference from Crossbill in flight, gait, or general behaviour, but Scottish Crossbill forms only small parties (up to 20 birds) after breeding, not erupting widely as Crossbill. Regrettably, the few identification clues that fit Scottish Crossbill also point to Parrot Crossbill, no longer a rare vagrant to Britain; thus, field identification of Scottish Crossbill is virtually impractical away from pines of breeding habitat, and may be unsafe even within them.

Habitat. Restricted within western boreal zone of west Palearctic to forests and smaller stands of Scots pine in north-east of Scottish Highlands, breeding in level lowlands and on gentle or steep slopes, up to sunny hilltops where trees are scattered. Mature plantations are no less attractive than primeval relict forest, and while old trees are preferred their height and size matter little. Breeding may occur in interior of well-spaced woodland, as well as in openings or clearings and on edges. Nests sometimes located in stunted pines on small islands or hillocks, and in forest bogs, also in clumps of pines undergrown with tall rank heather. Requires easy access to water, in waterholes, peat runnels, or other sources.

Distribution and population. Core areas north-west of Great Glen and in Strathspey and Deeside. Range limits uncertain in some peripheral areas (possible confusion with Crossbill). Distribution in valley of Dee and tributaries probably unaltered since *c.* 1800. Tentative estimates of 1500 birds in early 1970s and 1981–4, and 300–1300 pairs in 1988. Perhaps some decrease in distribution and population because of habitat loss, but nomadic habits, marked local variation in numbers, and identification difficulties mean numbers and trends very poorly known.

Movements. Resident and dispersive. In most years, birds disperse after breeding to seek better food supplies, settling in adjacent or fairly close woods or plantations in Scottish highlands. When large populations build up after several years of successful breeding, birds sometimes move further in general exodus; e.g. in summer of 1936 many birds in upper Strathspey dispersed in large flocks, and few remained there to breed in spring 1937.

Food. Conifer seeds, primarily of Scots pine; some small invertebrates in breeding season. Feeding methods as other *Loxia*. Moves acrobatically in groups through pines, usually high up in end branches, apparently selecting best cones; seeds extracted *in situ* or cone twisted off and taken to perch for handling, often a fork closer to trunk, where cone is steadied with feet on one part of fork while tail is braced against other part for increased leverage. Forages for invertebrates on forest floor; often reported eating mortar, putty, and similar mineral-rich substances, and visits drinking places frequently.

Social pattern and behaviour. Apparently no significant differences from Crossbill. Nests either solitarily or, especially in years of high numbers and good seed crop, in loose neighbourhood groups of 2–6 pairs; such clusters typically formed when pairs in small mobile flock simultaneously settle in same small wood or particular part of larger one. ♂ sings from perch (usually tree-top) or in flight; rarely sings for long or loudly near nest. Flock may give chorus of song, especially in sunny, calm, and frosty weather in late winter or early spring; loud phrases of individual ♂♂, in flight or when perched, stand out against background babble which includes various calls and subsong.

Voice. Apparently intermediate between Crossbill and Parrot Crossbill, but (in direct comparison) 'chup' generally distinguishable from equivalent 'chip' or 'dyip' of Crossbill.

Breeding. SEASON. North-east Scotland: eggs recorded in all months February–June; in general, synchronized with ripening of cones of Scots pine so that young can feed themselves before cones are empty of seed. SITE. Almost always in old pine, just in from woodland edge or clearing, high in fork at centre of crown or near end of spreading branch. Nest: bulky structure, with foundation of twigs of pine, larch, or birch, heather, moss, and grass, lined with lichen, fine grass, fragments of pine bark, dead leaves, fur, hair, and a few feathers. EGGS. Subelliptical, smooth and slightly glossy; creamy to bluish- or greenish-white, with sparse reddish to blackish blotches, specks, and sometimes short scrawls, concentrated at broad end; can be flushed pink. Clutch: 3–4 (2–6). INCUBATION. 12.5–14.5 days. FLEDGING PERIOD. c. 21 days (17–25).

Wing-length: ♂ 97–105, ♀ 92–103 mm.
Weight: ♂ 42–49, ♀ 36.5–46 g.

Parrot Crossbill *Loxia pytyopsittacus*

PLATES: pages 1581, 1588

DU. Grote Kruisbek   FR. Bec-croisé perroquet   GE. Kiefernkreuzschnabel   IT. Crociere delle pinete
RU. Клест-сосновик   SP. Piquituerto lorito   SW. Större korsnäbb

Field characters. 17.5 cm; wing-span 30.5–33 cm. ♂ noticeably larger and bulkier than Crossbill, with striking parrot-like bill (mandibles less crossed than in Crossbill), little or no forehead, flat crown on large head, and thick neck, pot-belly, and short-tailed appearance; some ♀♀ and juveniles 5% smaller, with less deep bill and more obvious forehead but still thick neck; size overlaps with Scottish Crossbill. Plumage similar to Crossbill but adult duller on wings. Flight as Crossbill but ♂ looks noticeably front-heavy.

Typical large, bull-necked, parrot-billed ♂ distinctive, but smaller individuals of either sex at any age converge with Scottish Crossbill and even with biggest Crossbill, so separation requires close observation of bill structure (see above) and hearing of apparently distinctive calls (see below). Flight and

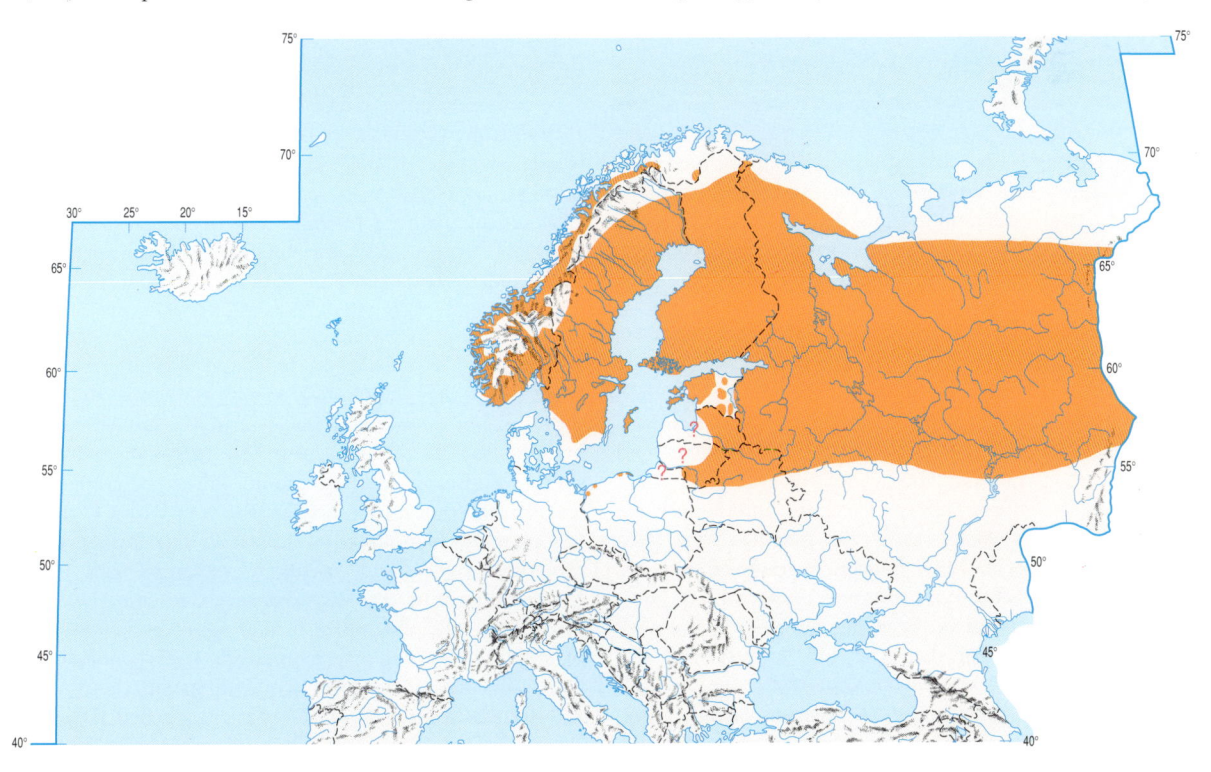

Arctic Redpoll *Carduelis hornemanni exilipes* (p. 1578): **1** ad ♂ winter, **2** 1st winter ♂. Two-barred Crossbill *Loxia leucoptera* (p. 1580): **3** ad ♂ summer, **4** ad ♀ summer. Crossbill *Loxia curvirostra* (p. 1582): **5–6** ad ♂ summer, **7** ad ♀ summer. Parrot Crossbill *Loxia pytyopsittacus* (p. 1587): **8** ad ♂ summer.

behaviour much as Scottish Crossbill and Crossbill, but shape in flight is most compact, heaviest (particularly at front), and most powerful of genus. Perched and (particularly) on ground, stance less upright than Crossbill and silhouette not only heavy at front but also pot-bellied and short- and thin-tailed.

Habitat. In boreal north-west Palearctic, between July isotherms *c.* 14–18°C. Distinguished from Crossbill by almost total specialization to pine forests, usually mature and open, but also on marshy land as well as more usual dry or mountain terrain. Sometimes in mixture of pine and other conifers. Despite successful adaptation to pines, has not spread to many pine forests occupied by Crossbill outside boreal zone, or even to some extensive pine forests within it.

Distribution. Not well known, owing to confusion with more widespread Crossbill. Breeding occurs west of main range following some influxes. BRITAIN. Irruptions occurred in 1962, 1982 and 1990. Following that in 1982, breeding attempted 1983 and proved 1984–5 (1–2 pairs); after 1990 influx, bred Scotland, northern England, and Norfolk in 1991 (1–6 pairs). BELGIUM. After 1989–90 influx, 1 certain and 3 probable breeding records in 1990; also bred 1995. NETHERLANDS. Bred successfully at 2 sites in 1983, perhaps in some other years. GERMANY. Bred (possibly more than once) before 1950. DENMARK. Bred 1994. NORWAY. Breeding distribution not well known; much commoner than Crossbill in west, and also occurs further north. LATVIA. No confirmed breeding 1980–84. POLAND. Very scarce breeder locally in north, mainly along coast. BELARUS'. Occasional observations of birds showing territorial behaviour over whole country, but no proof of breeding.

Accidental (or rare and irregular, irruptive visitor). Iceland, Britain, France, Belgium, Luxembourg, Germany, Czech Republic, Slovakia, Austria, Italy, Yugoslavia, Ukraine, Madeira.

Beyond west Palearctic, extends slightly east of Urals to Irtysh valley.

Population. NORWAY. 10 000–100 000 pairs 1970–90; fluctuates. SWEDEN. 10 000–50 000 pairs in late 1980s; stable. FINLAND. 10 000–100 000 pairs in late 1980s; slight increase. Never as abundant as Crossbill, seems to fluctuate less. ESTONIA. 2000–5000 pairs in 1991. LATVIA. 400–1000 pairs in late 1980s; fluctuating. LITHUANIA. Rare. RUSSIA. 10 000–100 000 pairs; stable.

Movements. Resident and dispersive; also eruptive. In most years, makes only limited movements in response to local food shortage, but occasionally makes eruptive movements, often in same year as Crossbill, but less extensive and reaching north-west Europe later in autumn. Pine-cone crops fluctuate less than those of spruce, and eruptions of Parrot Crossbill are less frequent than those of Crossbill; but former probably sometimes overlooked, owing to similarity with Crossbill. Migrating birds head chiefly south-west, reaching Denmark in most years, and extending further south and west in years of eruption. Recent major eruptions 1962 (coinciding with Crossbill), 1982, and 1990 (coinciding with both Crossbill and Two-barred Crossbill). Movement reached Belgium and Britain in 1962, but was more widespread in 1982, with records west to Britain and east to Mecklenburg (north-east Germany). Irruption in 1990 was also widespread, reaching Britain, Belgium, eastern France, and south-west Germany. Birds sometimes remain to breed in invasion areas, e.g. in Norfolk (probably also Suffolk), eastern England, 1983–85, Veluwe area (central Netherlands) 1983 and 1984, and at several sites in Denmark 1983.

Food. Conifer seeds, mainly pine, especially Scots pine, but also spruce; some invertebrates in breeding season. Foraging method identical to that of Crossbill.

Social pattern and behaviour. No comprehensive study, but apparently very similar to Crossbill.

Voice. Generally slightly deeper and louder than Crossbill, but individual calls may be indistinguishable given different degrees of emphasis and apparent regional variation. Song generally slower, better enunciated, and deeper. Calls considered specific to Parrot Crossbill: deep 'kop kop', 'choop choop', or 'chok', recalling ♂ Blackbird (given by feeding bird); very hard 'cherk cherk', given in alarm and deeper than equivalent call of Crossbill; 'püt püt püt', louder and more melodious than Crossbill.

Breeding. SEASON. Finland: eggs laid from beginning of February to late June, sometimes into August, peak mid-March to mid-May; influenced strongly by availability of pine seeds. Murmansk region (north-west Russia): February to mid-May in years of good pine and spruce crop, mid-May to August or September when only pine abundant. SITE. High in conifer at woodland edge, by clearing, track, etc., very rarely in dense forest; in spruce close to trunk, in pine in fork among dense twigs, usually a few metres from trunk; can be 20 m above ground, higher nests tending to be closer to trunk. Nest: foundation of dry conifer twigs, bark (often from deciduous trees), dead leaves, moss, lichen, etc., lined with dead grass, plant down and fibres, hair, sometimes feathers; near man, occasionally fragments of rope, etc.; very like that of Crossbill, perhaps larger, more robust. EGGS. Sub-elliptical, smooth, and slightly glossy; very like Crossbill, slightly larger with bolder markings; yellowish-white to pale blue-green with rust to purplish-brown spots, small blotches, and sometimes scrawls, mostly at broad end. Clutch: 3–4 (2–5). INCUBATION. 14–16(–17) days. FLEDGING PERIOD. 21–23 days (19–25).

Wing-length: ♂ 100–109, ♀ 100–105 mm.
Weight: ♂♀ mostly 48–61 g.

Crimson-winged Finch *Rhodopechys sanguinea*

PLATES: pages 1590, 1604

DU. Rode Woestijnvink   FR. Roselin à ailes roses   GE. Rotflügelgimpel   IT. Trombettiere alirosse
RU. Краснокрылый чечевичник   SP. Camachuelo ensangrentado   SW. Bergsökenfink

Field characters. 15 cm; wing-span 30–33.5 cm. 5–10% larger than Rock Sparrow but with rather similar bill, wing, and tail structure; 20% larger than Trumpeter Finch, with proportionately larger bill and head and shorter tail. Quite large, heavy-billed, robust, ground-haunting finch, with bounding flight and calls recalling Woodlark. Restricted to rocky mountainsides and summits above 1500 m in breeding season. At any distance, appears nondescript dark brown but at close range displays strikingly pink, dark-rimmed wings, dark crown, and intricate pattern of face markings and body spotting. ♂ shows pink basal patches and white tips to tail.

Unmistakable, with diagnostic combination of large size, dark cap, pale pink centres to wings, white under wing-coverts, and spotted or streaked underparts. Flight fast and powerful, with long wings obvious in silhouette; over distance, action produces weighty, deep undulations.

Habitat. Largely complementary to Trumpeter Finch, being situated somewhat further north, in warm temperate zone, and ranging over higher altitudes: from 2800 m upwards in Atlas of north-west Africa, above 1900 m in Israel. Generally found on stony slopes and ridges with sparse arid scrub and herbage, in scrub and juniper zone above trees, and on almost bare, dry eroded clay hills; wintering at lower altitudes on bare areas and arable cultivation.

Distribution and population. RUSSIA. Records in Terek valley in extreme south represent probable northward range expansion from Little Caucasus to Great Caucasus. GEORGIA. Spread from Armenia, now breeding Akhalkalaki district. AZERBAIJAN. Occurs in mountains of Nakhichevan region up to subalpine and alpine zones. Rare. TURKEY. Fairly common on mountain slopes west to Munzur Daglari (c. 39°E); further west, widespread but only locally common. 10 000–100 000 pairs. SYRIA. Probably breeds in Anti-Lebanon mountains. ISRAEL. Breeds only Mt Hermon; c. 30 pairs 1970s–80s. MOROCCO. Uncommon. ALGERIA. Recorded from Aurès mountains in 1840–42; then not again until rediscovered in July 1970.

Accidental. Iraq.

Beyond west Palearctic, extends discontinuously from Iran north-east to Tien Shan mountains and Zaysan depression.

Movements. Mainly altitudinal or short-distance migrant, descending to c. 2000 m in Atlas of Morocco in winter, and moving to below breeding range in Turkey. Birds breeding on Mt Hermon (Israel) depart after breeding, probably to adjacent Syria, returning to breeding sites in March and April.

Food. Diet comprises mainly seeds of low vegetation; a few invertebrates in breeding season. Forages mostly on bare rocky ground, scree, snowfields, etc., or in and around desert-type tussocky herbs and shrubs, in manner described as slow and heavy, taking seeds of limited range of plants, principally

Crimson-winged Finch *Rhodopechys sanguinea*. *R. s. sanguinea*: **1** ad ♂ summer, **2** ad ♂ winter, **3** ad ♀, **4** juv. *R. s. aliena*: **5** ad ♂ winter. Desert Finch *Rhodospiza obsoleta* (p. 1591): **6** ad ♂ summer, **7** ad ♀, **8** juv.

Chenopodiaceae, Boraginaceae, Cruciferae, and Compositae; also in overgrown gardens, sown fields, and in settlements, particularly in winter, where flocks feed (e.g.) on spilled seed of cultivated plants.

Social pattern and behaviour. Typically gregarious, including to some extent when breeding. Nothing to suggest other than monogamous mating system. Pair-formation takes place in flocks. Territorial when breeding, but apparently not markedly so; nests may be separated by a few tens of metres. ♂ sings from rock, tussock, or bush, also in flight. Song-flight at times spectacular, bird circling high and apparently singing in rhythm with deep undulations, alternating fluttering ascents and gliding descents.

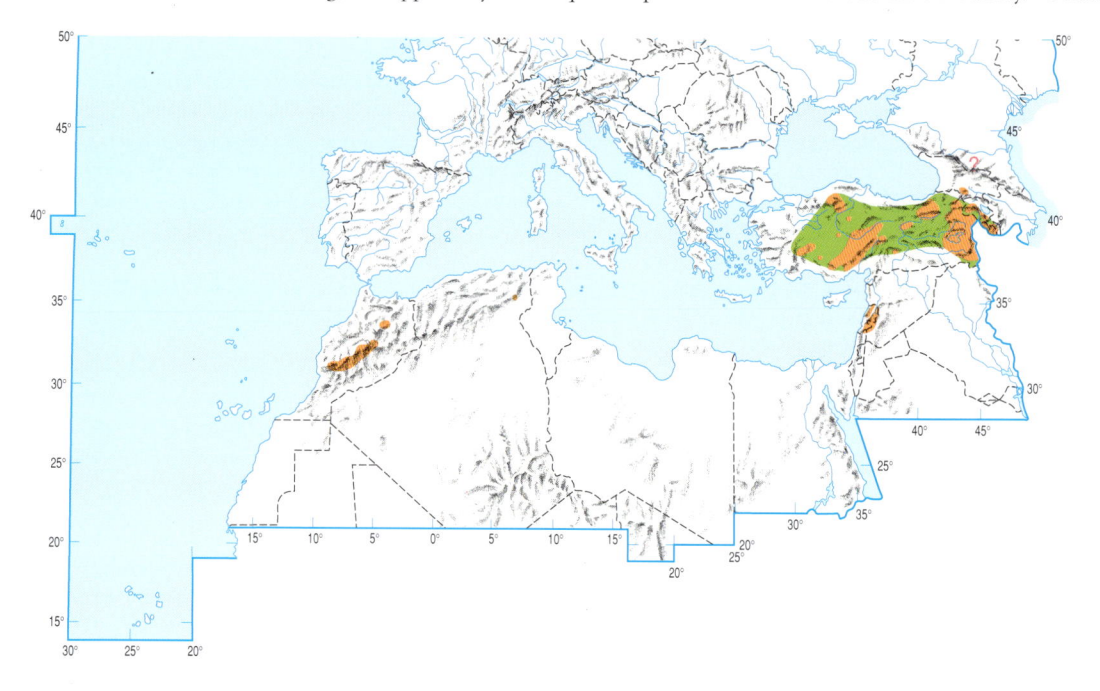

Voice. Song of ♂ a clear, melodious, but at times somewhat wheezy or grating short phrase, often given many times in quite rapid succession without much variety. Contact and alarm-calls include rich, musical, softly fluting whistles often reported as reminiscent of calls of Woodlark: 'dü-leet dü-leet', 'tureep tureep', or similar; also sparrow-like chirps.

Breeding. SEASON. Morocco: at 2200–3300 m, young recorded in nest mid-May and late June, and fledged young early July. Israel: at *c.* 2000 m, breeding begins when snows melt, and eggs laid from end of May or beginning of June. Southern Turkey: at 1100 m, laying starts from beginning of April to beginning of May. 1–2 broods. SITE. In stony places with little vegetation; on ground under overhanging rock, grass tussock, thorny cushion-type scrub, or in crevice between boulders; also in low bush and on cliff ledge. Nest: neat and loosely constructed, with foundation principally of tough dry grasses, including cereals, herb stalks and roots, lined with fine grass and plant fibres; rarely animal hair. EGGS. Sub-elliptical, smooth, and slightly glossy; light sky-blue with small violet-brown spots concentrated at broad end. Clutch: 4–5. INCUBATION. (12–)13–15 days. FLEDGING PERIOD. 13–15 (10–17) days.

Wing-length: *R. s. sanguinea*: ♂ 103–112, ♀ 100–106 mm.
Weight: *R. s. sanguinea*: ♂♀ 32–44 g.

Geographical variation. Fairly strong; involves colour only. North-west African race, *aliena*, markedly paler and duller in all plumages than *R. s. sanguinea* (Turkey and Levant eastwards), with fully ashy-grey nape, rose-white throat, wing-feathers only narrowly fringed rose, rump uniformly brown, and tail without obvious white. On at least some Moroccan birds, supercilium and surround to ear-coverts pale buff to cream.

Desert Finch *Rhodospiza obsoleta*

PLATES: pages 1590, 1604

DU. Vale Woestijnvink   FR. Roselin de Lichtenstein   GE. Weißflügelgimpel   IT. Trombettiere di Lichtenstein
RU. Буланый вьюрок   SP. Camachuelo desertícola   SW. Ökenfink

Field characters. 14.5 cm; wing-span 25–27.5 cm. Similar in size and in shape of bill and head to Greenfinch; slightly larger than Pale Rock Sparrow, with proportionately shorter wings but longer, forked tail. Quite large, dumpy finch, with dark stubby bill, well-forked tail, and diagnostic purring call. Head and body uniform pale buff, contrasting with dark-rimmed, pink and white wings, and dark, white-edged tail. ♂ has black loral stripe and (when breeding) bill. ♀ and juvenile duller than ♂.

Unmistakable if seen well, but in brief glimpse juvenile might suggest (smaller) Mongolian Trumpeter Finch, Trumpeter Finch, or Pale Rock Sparrow. Flight light and undulating; silhouette recalls small *Carduelis* finch. Perches upright in trees, dropping to ground to feed in manner of Greenfinch; note that *Bucanetes* finches do not normally use such perches.

Habitat. In west Palearctic, mainly in lowland arid and semi-arid areas like Trumpeter Finch but is less a desert bird and also ascends mountainous valleys. Occurs where some open tree or shrub growth present, in plantations, orchards, rows of trees, oases, areas of irrigation, and arid places with sparse herbage and scattered trees or shrub thickets, feeding in weedy or fallow cultivated areas. Like Trumpeter Finch, needs access to water. Feeds mostly on ground, but perches freely on bushes, trees, railings, and telegraph wires.

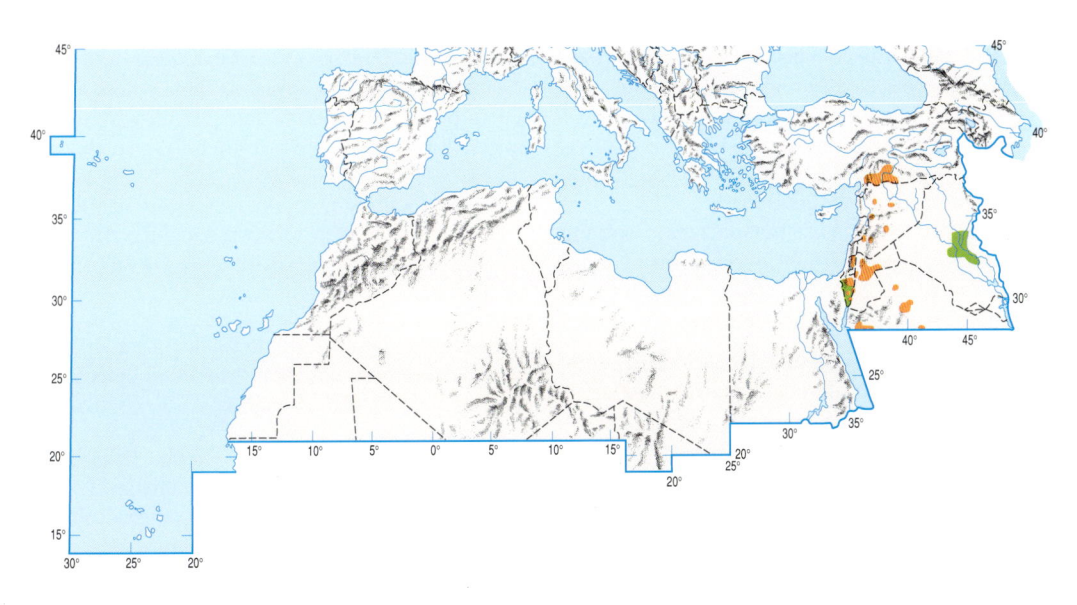

Distribution and population. TURKEY. Breeds plateau area of south-east Anatolia; recently recorded eastern Anatolia where evidently also breeding. 1000–10 000 pairs. SYRIA. Apparently only scattered single pairs, locally distributed in arid lowland wastes. ISRAEL. Up to 1950s an irregular winter visitor, some remaining to breed after large irruptions. Establishment of permanent breeding population began in late 1950s, following development of agricultural settlements. A few thousand pairs in 1980s. JORDAN. First sight record 1976; now fairly common. Has benefited from increase in agriculture and planting of trees in desert areas. EGYPT. First confirmed breeding: 11 birds (including young) at 3 localities northern Sinai, May 1994.

Beyond west Palearctic, breeds Arabia and from Iran east to north-east Kazakhstan, western Tibet, and Mongolia.

Movements. Chiefly sedentary, with small-scale movements, mainly within breeding range.

Food. Diet seeds and other parts of plants; a few insects in breeding season. Feeds mostly on ground, picking up seeds of desert plants in dry stony places, also in fields, orchards, etc.; sometimes in shrubs or trees taking buds and shoots.

Social pattern and behaviour. Gregarious all year, but especially outside breeding season. At end of breeding season, family parties congregate into flocks for feeding, drinking, roosting, and migration. Pair-bond long-lasting; in summer usually encountered feeding in pairs even when not actively nesting. Breeds solitarily or in small neighbourhood groups or colonies. After arrival on breeding grounds, ♂♂ sing from trees where nests will eventually be built. ♂ also sings when closely attending ♀ as she collects nest material and builds, also near nest-site during incubation.

Voice. Song of ♂ a quiet, introspective succession of mainly call-type units, notably tremolos, nasal, buzzing, and occasional harsh units, overall with a chattering or chortling quality reminiscent of Budgerigar; some songs distinctly phrased, others not. Main contact-call a quiet, soft, but melodious purring 'prrrrrrl', given when perched or in flight; varies in length, and often interspersed with short, slightly nasal 'dzhee'. An interrogative 'pink pink pink' is often given in breeding season.

Breeding. SEASON. Israel: eggs laid late March to mid-April; 2nd clutches May–June, probably mostly 2nd half of June. Southern Turkey: young recorded in nest mid-May. SITE. In horizontal or vertical fork of shrub or tree generally 1–5 m above ground, frequently in cultivated species in orchard, garden, etc. Nest: foundation of twigs and coarse herb stalks, lined with thick felt-like layer of plant down and other soft plant material, especially of cotton and poplar, sometimes fur, hair, cloth, etc. EGGS. Sub-elliptical, smooth and slightly glossy; white to pale greenish-blue, with small purplish-black specks and very fine hair-streaks at broad end. Clutch: 4–6 (3–7). INCUBATION. 12–15 days. FLEDGING PERIOD. 13–14 days (12–16).

Wing-length: ♂ 85–92, ♀ 83–89 mm.
Weight: ♂ 22–26, ♀ 17.5–28 g.

Mongolian Trumpeter Finch *Bucanetes mongolicus*

PLATES: pages 1593, 1604

Du. Mongoolse Woestijnvink   Fr. Roselin de Mongolie   Ge. Mongolengimpel   It. Trombettiere mongolo
Ru. Монгольский снегирь   Sp. Pirrula mongólica   Sw. Mongolfink

Field characters. 13 cm; wing-span 25.5–27.5 cm. Slightly larger than Trumpeter Finch; 10% smaller than Desert Finch, with proportionately longer wings. Medium-sized, stubby-billed, thick-headed, long-winged but rather short-tailed finch. Looks rather ghostly at distance but shows 2 strikingly pale panels across greater coverts and secondaries at close range. Unlike Trumpeter Finch, head sandy, and black-centred tail broadly fringed pink or off-white at all seasons; bill yellowish-horn, never red. Trumpeting call distinctive.

Closely related to Trumpeter Finch, but much more strongly patterned wing and tail immediately exclude that species. Desert Finch has forked tail and Crimson-winged Finch has dark cap and dappled underparts; both are larger.

Habitat. Breeds in dry sunny mountainous regions of south-central Palearctic, breeding at altitudes of 1000–4000 m or even higher. A ground bird, avoiding trees and even shrub growth when these are present in habitat. Favours steep and broken terrain with precipices and hollows, often including stony or clayey as well as rocky slopes and sparse grassy patches of desert or semi-desert type. Tends to occur at higher altitude than Trumpeter Finch.

Distribution and population. Early records (1911–15) from Transcaucasia and eastern Turkey–Armenia border area overlooked in most subsequent literature, and not generally recognized as occurring in west Palearctic until 1969. ARMENIA. Not known to occur at present. Site where recorded in 1915 lies in present-day eastern Turkey. AZERBAIJAN. Nakhichevan: majority of early records from Bulgan in this region, where found breeding (sympatrically with Trumpeter Finch) at Aza in 1969. Common, locally numerous in 1970. TURKEY. In 1989, rediscovered in east near Doğubayazit (where pair seen feeding fledged juveniles in 1990) and in 1992 near Igair. Count of 25 birds (including juveniles) in 1992 near Doğubayazit.

Movements. Some birds short-distance migrants, others make altitudinal movements; some remain in breeding areas all year, making only local movements.

Food. Seeds of grasses and low herbs. Forages almost wholly on ground on rocky slopes and mountainsides in sparse semi-desert type vegetation, very rarely on small shrubs or herbs.

Social pattern and behaviour. In flocks outside breeding

Mongolian Trumpeter Finch *Bucanetes mongolicus*: **1** ad ♂ summer, **2** ad ♂ winter, **3** ad ♀, **4** juv. Trumpeter Finch *Bucanetes githagineus* (p. 1594). *B. g. crassirostris*: **5** ad ♂ summer, **6** ad ♂ winter, **7** ad ♀, **8** juv. *B. g. amantum*: **9** ad ♂ summer.

season, and these generally larger than in Trumpeter Finch. In Azerbaijan, flocks usually of 4–20, sometimes associated with Spanish Sparrow and Rock Sparrow. Breeds semi-colonially in small neighbourhood groups, nests sometimes only 30–100 m apart; no obvious territoriality. Sings on ground; also regularly in flight. Song begins in spring, while still in flocks, several birds often singing together. Pair-formation apparently takes place in flocks, on feeding grounds, or during stopovers on migration to breeding areas.

Voice. A vocal species, calling in flight, when feeding, for contact with conspecifics flying past at some distance, etc. Song of ♂ a series of varied, more or less musical units; sometimes very similar to whistling song of Scarlet Rosefinch. Calls given by birds in flight or in feeding flocks variable: slightly nasal 'vzheen'; clear, somewhat melancholy 'vee-tyu'; melancholy 'piu'; or quite loud, short, attractive fluting sounds.

Breeding. SEASON. Nakhichevan (Azerbaijan): eggs laid

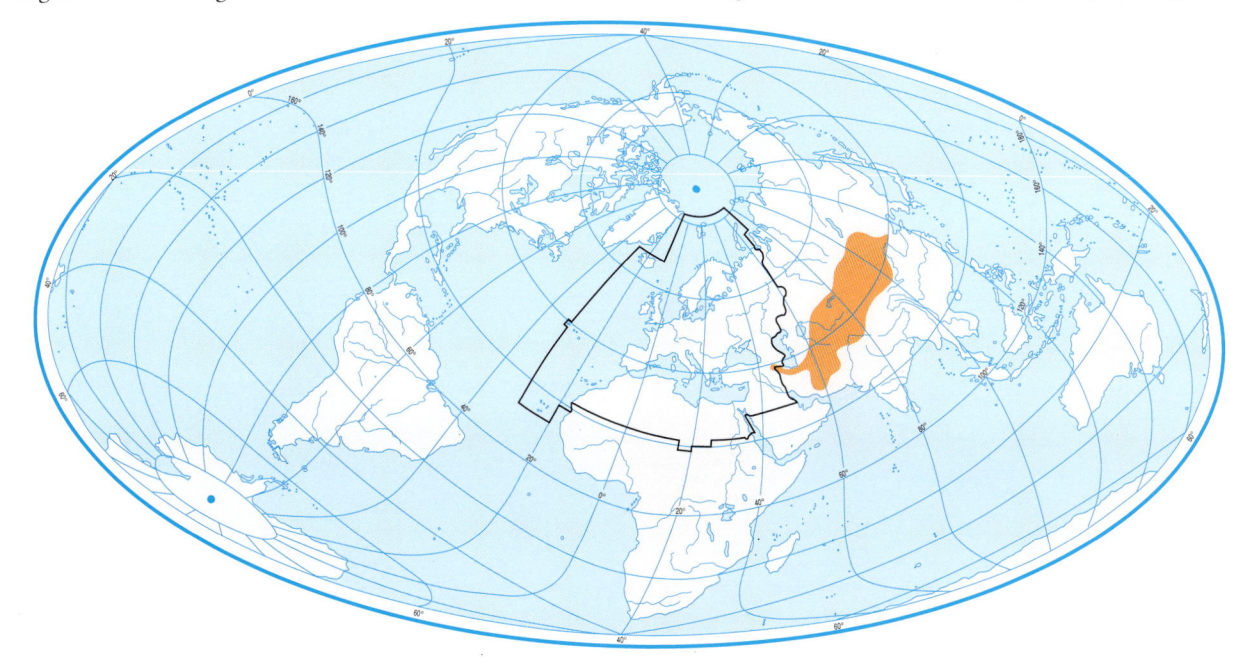

around 2nd half of April; fledged young recorded mid-May, and season very extended. Eastern Turkey: adults seen feeding young out of nest mid-July. SITE. On ground under bush, grass tussock, rock, etc., or in niche or crevice in rock face, between boulders in scree, or in clay wall of building or inside ruin; entrance tunnel of up to 40 cm recorded; some ground nests almost open. Nest: loose, bulky, flattish foundation of small twigs, rough stalks, stems, and leaves with inside layer of rootlets, grass, leaves, etc., lined with wool, hair, and sometimes plant material. EGGS. Sub-elliptical, smooth and slightly glossy; pale blue or slightly greenish-blue, sparsely marked with small brownish-black specks, rarely hairstreaks, at broad end. Clutch: 4–6 (3–8). (Incubation and fledging periods not known certainly.)

Wing-length: ♂ 87–92, ♀ 85–88 mm.
Weight: ♂♀ 18–26 g.

Trumpeter Finch *Bucanetes githagineus*

PLATES: pages 1593, 1604

DU. Woestijnvink FR. Roselin githagine GE. Wüstengimpel IT. Trombettiere
RU. Пустынный снегирь SP. Camachuelo trompetero SW. Ökentrumpetare

Field characters. 12.5 cm; wing-span 25–28 cm. About 10% larger than Serin, with similar dumpy form; slightly smaller than Mongolian Trumpeter Finch, with proportionately shorter wings. Rather small, stocky finch with bulbous bill, deep head, and rather short tail. Plumage rather uniform dusky- or sandy-pink, with orange- to wax-red bill (breeding ♂), large dark eye, darker flight- and tail-feathers, slightly paler rump, and orange-flesh legs. Song distinctive.

Remarkably featureless, requiring close approach for plumage details to be seen. Red bill of breeding ♂ diagnostic; uniformity of plumage virtually so, but shared in Levant by sympatric ♀ Sinai Rosefinch, which is 10% larger and longer-tailed, with ginger face and different voice. Beware also risk of confusion with Mongolian Trumpeter Finch (slightly larger, whitish wing-panel, different call), Desert Finch (15% larger, with usually dark bill, rosy-white wing-panel, black tertial-centres, primary-tips, and tail, and different call), and Pale Rock Sparrow (10% larger, with pale double wing-bar, white-tipped outer tail-feathers, and different call). Flight light and bounding, even skipping. Spends most of time on ground; gait a hop, varied by shuffle and creep.

Habitat. Patchily distributed across warm arid mainly lowland or hilly subtropical regions from western Sahara across North Africa and Middle East. Concentrates in deserts, semi-deserts, and steppes with minimum of vegetation and much stony or gravelly surface, preferably fronted by rocky crags or vertical exposures with plenty of crevices and sparse growth of small bushes and grasses. Lives mainly on ground but requires daily access to water and is ready to fly some distance to it, especially towards evening. Tolerates very high daytime temperatures. Sometimes enters deep wells with walls, and occasionally nests in holes or gutter-pipes on houses, although normally far from human habitation except for remote guard-posts, where its confiding disposition leads to its readily making itself a home.

Distribution. Recent spread Spain, Israel, and Morocco. SPAIN. Range expansion eastwards along coastal mountains to Manga del Mar Menor (Murcia) from arid zones of Almería; north-west spread to Guadix depression (Granada) only temporary. PORTUGAL. Apparently feral birds breeding between Portimão and Silves. ARMENIA. Pair recorded south of Vedi April 1962, 3 pairs in 1964, single ♂ near Yerevan 1963. Bred

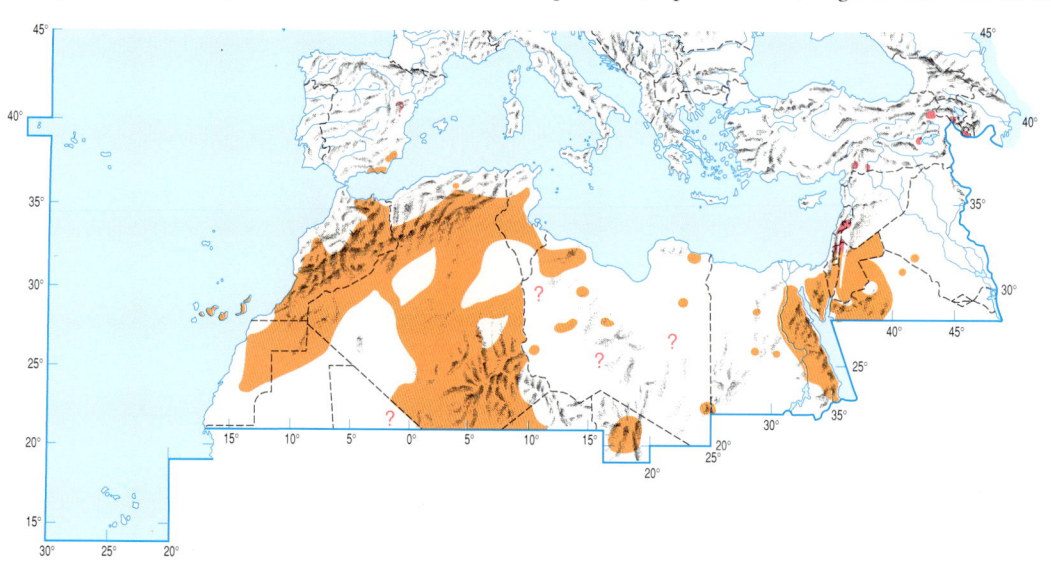

(3 pairs) near Vedi 1995. AZERBAIJAN. Nakhichevan region: up to 6 pairs and 2 single birds on slopes of Dary-Dag June 1962; ♀ and 2 fledglings collected, used nest found. Now known to breed at several sites in region, including sympatrically with Mongolian Trumpeter Finch at Aza (near Djulfa). TURKEY. First recorded 1974. Known to occur Yesilce area (north-west of Gaziantep), Nemrut Dag crater above Tatvan, and Aras valley. Seen in breeding season southern coastlands and near Birecik, but no proof of breeding. SYRIA. Breeding suspected, notably Anti-Lebanon, desert between Khān Abu Shamat and Sayqual (on Jordanian border), but still no proof. ISRAEL. Spread in 1970s–80s, following development of agricultural settlements and army camps providing constant water. SAUDI ARABIA. Fairly common Harrat Al Harrah reserve, but widespread and rather scarce in Arabia as a whole. MOROCCO. Has recently spread north to Moyen Atlas and Mediterranean coast; bred only temporarily (for a few years in 1980s) in Massif du Khatouate. Common in south, more local in north. CANARY ISLANDS. Breeds Lanzarote, Fuerteventura, Gran Canaria, Tenerife, and La Gomera.

Accidental. Britain, Channel Islands, France, Denmark, Sweden, Austria, Balearic Islands, Greece, Kuwait, Cape Verde Islands.

Beyond west Palearctic, breeds from Arabia north-east to central Kizil-Kum and northern Pakistan. Extends locally slightly south of west Palearctic border in Africa.

Population. SPAIN. 100–300 pairs. Slight increase overall, but slight recent decline of core population in Almería (perhaps only short-term fluctuation associated with rainfall). AZERBAIJAN. Nakhichevan: uncommon overall, though reported to be common, locally even numerous, in 1970. TURKEY. Poorly known; perhaps up to 50 pairs. ISRAEL. A few thousand pairs, but large annual variation. JORDAN. Widespread and fairly common. SAUDI ARABIA, EGYPT, TUNISIA. Fairly common. CANARY ISLANDS. Declining Tenerife; rare on La Gomera.

Movements. Resident and dispersive or nomadic. Need for daily water-supply leads to erratic movements, also to temporary colonization and frequent small-scale changes of range. Northward movements in north-west Africa in recent decades reflected in increasing records in southern Spain, with many wintering in Almería in 1969, and first recorded breeding in 1971; also 1st records elsewhere in western Europe (presumably chiefly this race); most such records mid-May to late July, coinciding with post-breeding dispersal.

Food. Seeds and other parts of grasses and low herbs; also a few insects. Forages almost wholly on ground, generally in rocky areas with scattered semi-desert vegetation; flits, creeps, and runs mouse-like around stones and shrubs searching for seeds. Digs fairly deeply into soil to find seeds, and perches on stems of grass to bend them over to get at seed-heads.

Social pattern and behaviour. More or less gregarious outside breeding season; flock sizes variable, but usually not large. Breeding dispersion varies from completely isolated pairs to groups of pairs nesting only short distances apart (but local concentrations of nests less characteristic than in Mongolian Trumpeter Finch); no obvious territoriality. ♂ sings on ground; also gives modified form of song in display-flight; flies fast in wide circles, with erratic changes of direction; regular alternation of active flight (4–5 vigorous wing-beats with gentle ascent) and gliding.

Voice. Remarkable for nasal/metallic sounds, comparable with sound made by child's tin or plastic trumpet. Calls freely, on ground and in the air. Song of ♂ variously described, and may vary geographically, but descriptions agree in giving drawn-out, nasal, trumpet-like unit as component, sometimes associated with shorter units (nasal, metallic or clicking sounds). Other calls also variable; consist principally of buzzing units (similar in quality to main component of song) and shorter units rendered 'chik', 'kek', 'tset', etc., without apparent buzzing or nasal quality.

Breeding. SEASON. Southern Spain: eggs laid about early May. Canary Islands: eggs laid 1st half of March (January to mid-May). North-west Africa: eggs laid February–June, mostly April. Israel: 1st clutch March–May, 2nd May–June. Nakhichevan (Azerbaijan): eggs laid from late April; 2nd clutches from mid-May. SITE. Depression on ground under rock, shrub, tussock, etc., in cleft between stones, or in cavity in rock face or wall of house. Nest: untidy foundation of small twigs, stalks, roots, rough grass, etc., neatly lined with dry grass, plant down, wool, hair, and (rarely) feathers; sometimes lined with grass only. EGGS. Sub-elliptical to short oval, smooth and slightly glossy; pale blue, sparsely marked with rusty to purplish-black spots and speckles, generally at broad end; sometimes small reddish-violet undermarkings. Clutch: 4–6. INCUBATION. 11–14 days. FLEDGING PERIOD. 13–14 days.

Wing-length: *B. g. zedlitzi* (Algeria and Tunisia): ♂ 87–93, ♀ 84–89 mm.
Weight: *B. g. zedlitzi*: ♂ 21–25, ♀ 19–22 g.

Geographical variation. Marked. Involves depth of ground-colour of (especially) upperparts (paler or darker drab-grey to drab-brown), extent and depth of rose-red on head, body, and wing, size, and depth and width of bill. 4 subspecies recognized: *zedlitzi* (North Africa and southern Spain), *crassirostris* (Sinai and Levant east to west-central Asia); and smaller *githagineus* (southern Egypt and neighbouring Sudan) and *amantum* (Canary Islands); *amantum* darkest in colour, with deepest pink.

Scarlet Rosefinch Carpodacus erythrinus

Du. Roodmus   Fr. Roselin cramoisi   Ge. Karmingimpel   It. Ciuffolotto scarlatto
Ru. Обыкновенная чечевица   Sp. Camachuelo carminoso   Sw. Rosenfink   N. Am. Common Rosefinch

PLATES: pages 1597, 1604

Field characters. 14.5–15 cm; wing-span 24–26.5 cm. Close in size to House Sparrow but with relatively smaller, more swollen bill and longer cleft tail; slightly smaller than Pallas's Rosefinch and Sinai Rosefinch. Medium-sized, bulbous-billed, quite stocky but long finch, with all plumages except adult ♂ reminiscent of ♀ House Sparrow and Corn Bunting. Adult ♂ drenched scarlet on head, rump, and fore-underparts; ♀ and immature dull olive-brown, softly streaked above and below. All ♀♀ and immatures show quite marked, narrow, whitish double wing-bar. Voice distinctive.

Traditionally regarded as the only rosefinch of Europe, confusion being most likely with escaped Purple Finch *C. purpureus* and House Finch *C. mexicanus* of North America. Purple Finch noticeably larger, with distinctive call—dull metallic 'tick' or 'pink'; ♂ noticeably more vinaceous than Scarlet Rosefinch (but not purple), ♀ and juvenile very heavily streaked on face and below. House Finch similar in size to Scarlet Rosefinch, call a distinctive, harsh, nasal 'che-urr', suggesting House Sparrow; ♂ bright red only on face, breast, and rump and heavily streaked on flanks, and ♀ and juvenile more continuously streaked below. Given potential vagrancy of Pallas's Rosefinch and other east Asian congeners, these species should be considered. Most serious problem appears to be confusion of young ♂ Scarlet Rosefinch with ♀ Pallas's Rosefinch (see that species). Voice important (see below). Flight slightly heavy, recalling both sparrow and Chaffinch; over distance, soon becomes undulating. Stance rather upright, with head seemingly sunk into shoulders so that bird can look bull-necked and dumpy, yet quite long-tailed, on perch.

Habitat. Extends into west Palearctic mainly in temperate continental climatic zone and in lowlands, but with disjunct population in foothills and mountains of south-east of region. Lowland form now extending much further west. In FSU, this form avoids desert zones and extends only moderately into forest steppe. Favoured habitats are thickets near forest edges, forest clearings, and patches of regrowth (coniferous or broad-leaf), groups of shrubs or isolated trees in humid meadows or river valleys, thickets of osier or bird-cherry, and sometimes orchards, graveyards, or thorn hedges. Further west in Europe,

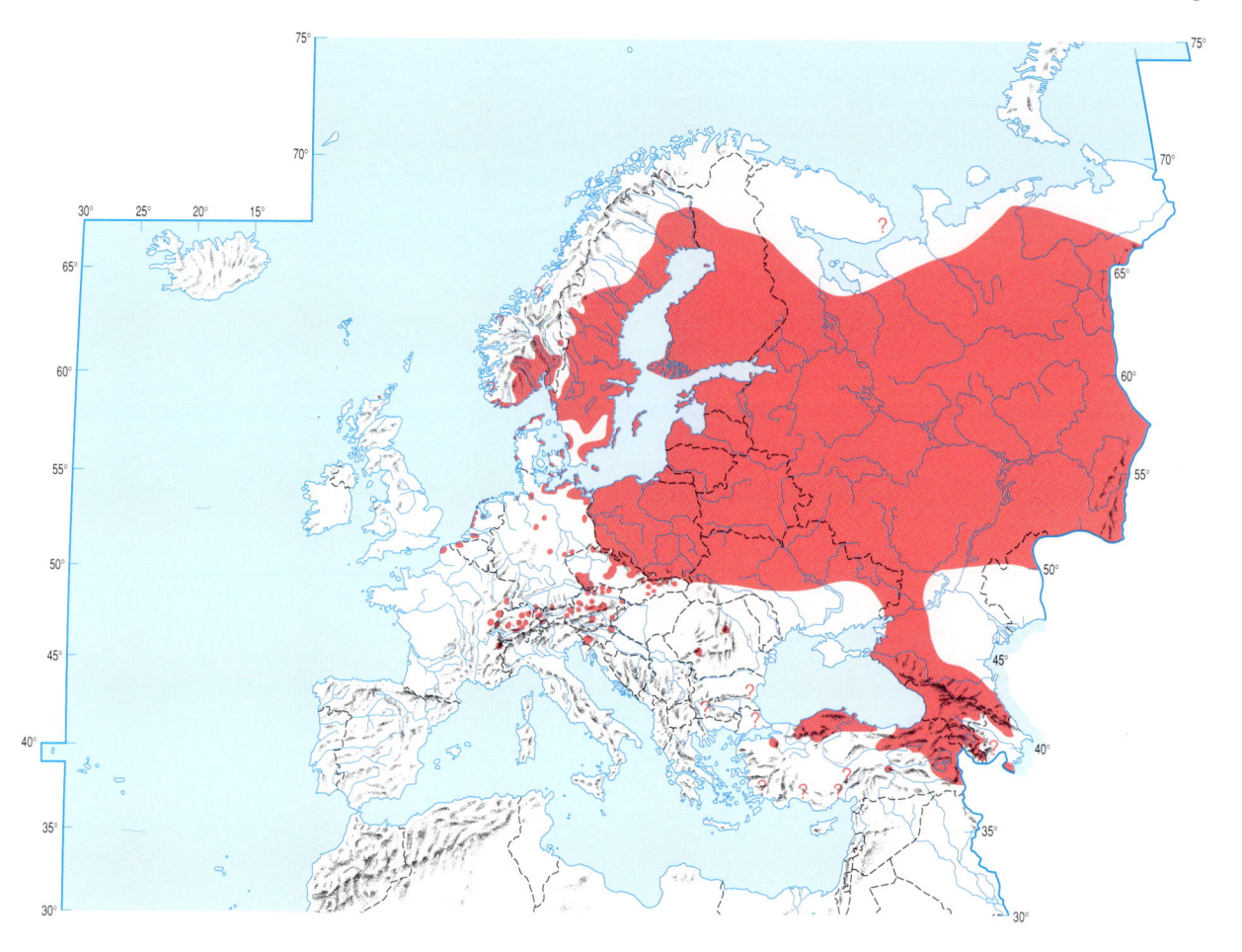

Scarlet Rosefinch *Carpodacus erythrinus erythrinus*: **1** ad ♂ summer, **2** ad ♂ winter, **3** ad ♀, **4** juv. Sinai Rosefinch *Carpodacus synoicus* (p. 1599): **5** ad ♂ summer, **6** ad ♂ winter, **7** ad ♀, **8** juv.

drier sites on farmland in briar and scrub patches or gardens are not favoured, choice of breeding places falling on fairly low thickets of alder, poplar, or willow, with a few taller trees as song-posts. In different parts of range, markedly different habitats are preferred, from moist (even swampy) bushy or scrub types in western lowlands to low open or marginal forests and drier fields with shrub or thicket patches in more eastern plains.

Distribution. Major westward expansion began in early, with regression in late, 19th century. 2nd stage (from 1930s) still in progress, and also taking place in east (few details). Spread attributed to long, warm autumns (favouring westward dispersal of 1st-year pioneers), also (in Finland) to increased breeding success in man-made habitats. BRITAIN. First bred (Highland, Scotland) 1982. Probable breeding at up to 5 localities (mostly Scotland) 1983–90, confirmed Shetland 1990. First bred England 1992 (5–20 pairs) following major influx. No breeding confirmed 1993–4. FRANCE. First recorded breeding (Doubs) 1985, but becoming established only from *c.* 1992. BELGIUM. First bred 1993. NETHERLANDS. Following increase in number of singing ♂♂, first confirmed breeding 1987 (suspected earlier). GERMANY. In east, regular breeder since 1967–8, initially on coast, gradually spreading inland (Oder valley 1974, Erzgebirge 1990, Thüringer Wald 1991). Spread west to Schleswig-Holstein, Niedersachsen and Helgoland, also Westfalen and Bayern in 1980s. DENMARK. First recorded 1943; breeding first recorded 1972. NORWAY. First bred 1970. Spread rapidly in 1970s–80s, reaching Rogaland in late 1980s; singing ♂♂ recorded north to Lofoten Islands. SWEDEN. Colonization began in 1930s. FINLAND. Sparse breeder in south-east until early 20th century. Major expansion since mid-1940s. POLAND. Scarce breeder with restricted range in early 19th century. Spread west along coast since *c.* 1900 (following retreat eastwards); now breeds throughout. CZECH REPUBLIC, SLOVAKIA. Continuing expansion (marked in Czech Republic) since first breeding in 1959. HUNGARY. No proof of breeding, despite increase in number of records since first in 1983. AUSTRIA. Rapid increase up to 1980s following first confirmed breeding in 1974. Very local. SWITZERLAND. Annual summer visitor in very small numbers since 1979. Now breeding regularly after first (unsuccessful) attempt in 1983. YUGOSLAVIA: SLOVENIA. Breeding since 1978. BULGARIA. At least 4 probable breeding localities by 1993. RUSSIA. Expansion to north and east since 1930s, reaching White Sea coast in 1966; observed north to Arkhangel'sk and Nar'yan-Mar (67°37′N on Pechora), and breeding recorded in south of Kanin peninsula. UKRAINE. Marked increase in breeding range. Crimea: first bred in 1991 (9 territorial ♂♂ in 1992). MOLDOVA. No breeding evidence. TURKEY. Range apparently expanding.

Accidental. Iceland, Faeroes, Ireland, Luxembourg, Spain, Balearic Islands, Malta, Greece, Cyprus, Jordan, Morocco.

Beyond west Palearctic, breeds widely across northern Asia (north of *c.* 45°N) to Kamchatka and Sea of Okhotsk, extending through central Asian mountains to northern Iran, Himalayas, and south-west China.

Population. Marked increase in virtually all areas affected by range expansion. At edge of range, prevalence of pioneering singing ♂♂ makes assessment of breeding population difficult. FRANCE. 3–37 pairs since 1992. BELGIUM. 5 territories in 1993, 9 in 1994. NETHERLANDS. 30 territorial ♂♂ 1991, 30–60 pairs 1994. GERMANY. 400–1000 pairs. In east, 280±80 pairs in early 1980s. DENMARK. 250–300 pairs. NORWAY. 1000–2500 pairs 1970–90. SWEDEN. 10 000–50 000 pairs in late 1980s. After colonization, population increased about threefold every 5th year. FINLAND. 200 000–300 000 pairs in late 1980s. At least 30-fold increase since mid-1940s. ESTONIA. 50 000–100 000 pairs in 1991. Marked increase in last 40–50 years. LATVIA. 20 000–50 000 pairs in 1980s. LITHUANIA. Not rare. POLAND. 20 000–50 000 pairs; increase since late 1960s. CZECH REPUBLIC. 30–50 pairs 1973–7, 350–450 in 1985–9. SLOVAKIA. 500–1000 pairs. SWITZERLAND. Up to 10 pairs. YUGOSLAVIA: SLOVENIA. 10–15 pairs. RUMANIA. 10–50 pairs 1986–92. RUSSIA. 1–10 million pairs; stable. BELARUS'. 140 000–160 000 pairs in 1990. Some increase in 1970s–80s, especially in south. UKRAINE. 250–600 pairs in 1990. AZERBAIJAN. Common. TURKEY. 5000–50 000 pairs.

Movements. Migratory. Most populations long-distance migrants; south-east populations in part altitudinal migrants. Winters south of breeding range from Pakistan east to eastern China, south to southern India (not Sri Lanka), northern Burma, Thailand (mostly in west), and northern Laos. In Pakistan, winters only in small numbers; in India, most abundant in central and western areas; winters up to c. 1500 m in Himalayan foothills. Very small numbers winter in eastern Israel, Oman, and Sinai, and new winter quarters in southern Europe may be beginning to be established. Western populations migrate east or south-east in autumn (central European birds perhaps initially heading north-east towards Leningrad and Moscow regions), passing north of Caspian Sea.

Autumn migration starts early, 2nd half of July and beginning of August, throughout boreal area of distribution; southern birds leave later. Arrives in Pakistan and India August–October. Spring migration late and rapid, April–May(–June). Leaves India and Pakistan mostly April to early May. Passage in Volga-Ural area almost entirely in May, and reaches European Russia in 2nd third of May. Arrives in Austria last third of May and 1st third of June.

In Europe, range has expanded in 2 directions, westward towards Britain and north-west Germany, and south-west towards Austria and Balkans (in Germany, most records are in north or south-east, and apparently few reach France); coasts of northern Europe, and rivers and mountain valleys of central Europe, act as leading lines. In Britain, increase in records has involved primarily 1st-year birds in autumn, apparently pioneers dispersing west; this probably linked with rapid increase in numbers breeding in Fenno-Scandia.

Food. Seeds, buds, and most other parts of plants; some invertebrates, mainly insects. Forages in grass and herbs, on arable land, in bushes, and in trees up to crown. Versatile in exploiting plants. Removes outer scales from bud, eating only soft nuclei; nibbles pieces from leaves and takes fresh conifer needles. Grass or cereal grains generally preferred unripe and milky; seeds of berries eaten and pulp usually discarded; skin sometimes left hanging *in situ*. When feeding on ground, pulls seed-heads down while standing on ground, or snips through stem; also perches on stem to bend it over, or reaches over from neighbouring twig, etc.; makes hole in side of Compositae heads and extracts seeds in bundle, biting off pappi.

Social pattern and behaviour. Occurs in flocks in winter. In spring, migrates singly (especially red-plumaged ♂♂) or in mostly small flocks. Arriving first on breeding grounds, ♂♂ form small flocks for feeding and are joined later by ♀♀. Generally monogamous, but some polygyny occurs. Territorial when breeding, with apparently some tendency to form neighbourhood groups. ♂ sings from song-post (tree-top, bush or overhead wire), occasionally in flight, from arrival on breeding grounds throughout breeding season; also occasionally in winter quarters and on migration.

Voice. Song of ♂ a far-carrying, sprightly whistling or piping phrase, characteristically clear, short, melodious, and attractive, comprising 3–5(–7) steeply ascending or descending notes given in stereotyped fashion; rendered 'WEEje-wü WEEja', 'tsitsewitsa', 'ste-weedye-vyu', etc. Contact-call a sprightly, clear-toned 'ueet', with same tonal quality as in song, or 'huit' like Chiffchaff, given when perched or in flight. Calls given by both sexes when disturbed (e.g. near nest) strangled, throaty and ascending in pitch, sounding di- or trisyllabic: 'dui', 'düei', or 'chräi'. Quiet, short 'zik', 'zit', or 'zlit' sounds given before take-off and in flight.

Breeding. SEASON. Late throughout region. Moscow region (central European Russia): eggs laid late May to mid-June; late-June clutches probably replacements. St Petersburg (north-west Russia): eggs laid from end of May to beginning of July. Southern Finland: end of May to early July; clutches started after mid-June considered replacements. Sweden: peak egg-laying 1st half of June. North-east England: eggs laid 1st half of June. Northern Turkey: May–July. SITE. Low in dense bush or young tree, generally well hidden close to trunk; sometimes in tangle of scrub, herbs (etc.), but only very rarely on ground. Nest: untidy, tangled foundation of twigs, stems, and grass, often including dried flowers and Umbelliferae stalks, with inner layer of finer grasses lined with rootlets, plant down, sometimes moss and lichen, and, where available, often large amounts of horsehair. EGGS. Sub-elliptical, smooth and glossy; light bluish-green with purplish or blackish-brown spots and hairstreaks at broad end, remainder of surface hardly marked; sometimes dark violet-grey undermarkings. Clutch: 4–6 (3–7). INCUBATION. 11–12(–14) days. FLEDGING PERIOD. 10–13 days.

Wing-length: *C. e. erythrinus*: ♂ 82–88, ♀ 80–85 mm.
Weight: *C. e. erythrinus*: ♂ mostly 20–25, ♀ 19–27 g.

Geographical variation. Very slight in size, more marked in general body colour. *C. e. kubanensis*, from Turkey and Caucasus area east, slightly larger than *C. e. erythrinus* (rest of west Palearctic) and ♂ darker rosy-red with more extensive red on body.

Sinai Rosefinch *Carpodacus synoicus*

PLATES: pages 1597, 1605

Du. Sinaï-roodmus Fr. Roselin du Sinaï Ge. Einödgimpel It. Ciuffolotto del Sinai
Ru. Бледная чечевица Sp. Camachuelo del Sinaí Sw. Sinairosenfink

Field characters. 14.5 cm; wing-span 25–27.5 cm. Slightly larger than Scarlet Rosefinch and Pale Rock Sparrow; bill proportionately a little finer than in Scarlet Rosefinch. Medium-sized, unobtrusive but nervous, ground- and cliff-haunting rosefinch, restricted to sand desert and hills. Adult ♂ almost wholly drenched carmine-red and pink; ♀ and juvenile pale fawn, showing only soft streaks. No other striking features except for pale legs.

In west Palearctic, occurs only in sandstone deserts of southern Levant and north-west Arabia where no possibility of confusion with other *Carpodacus* except rare vagrant Scarlet Rosefinch, but needs to be separated from similarly-sized Pale Rock Sparrow and smaller Trumpeter Finch which share its habitat and can show similar featureless plumage in brief view. Flight light and bounding, with fast ascent.

Habitat. In west Palaearctic, narrowly limited to highly arid, warm temperate regions with steep broken gorges, gullies, or crags, minimally vegetated. Extensive lowlands are rarely suitable; e.g. preferred sites in Jordan include ruins of Petra and precipices of Wadi Rum.

Distribution and population. ISRAEL. At least a few hundred individuals; large annual fluctuations. JORDAN. Characteristic resident of Southern Rift Margins and Rum desert. 41–80 pairs breeding at Petra, 1983. EGYPT. Southern Sinai: fairly common.

Beyond west Palearctic, very local in Afghanistan and western China (Sinkiang, Kansu, and Tsinghai).

Movements. Some populations sedentary, others dispersive, making chiefly altitudinal movements. In Israel, more widely dispersed winter than summer, as rains cause extension of suitable habitat. In Sinai, where breeds at 1000–2000 m in southern and central areas, many birds remain at 1000–1500 m in mild winters; in cold winters most birds move down to wadis and other low-lying districts, including Gulf of Suez area and northern Sinai.

Food. Mainly seeds; also leaves, buds, and fruit. Picks seeds from ground among rocks and plants; feeds also in fruit trees and wormwood (*Artemisia*) shrubs in flocks of up to 200. Often tamely forages amongst refuse left by tourists, and accustomed to artificial feeding, e.g. at St Katherine Monastery in Sinai. Feeds in early morning and late afternoon (except where food provided by man), and very frequently recorded drinking at water sources, up to 30 times per hr.

Social pattern and behaviour. Gregarious all year, though more so outside breeding season. Forms flocks of varying size for feeding, drinking, and roosting. Solitary and territorial when breeding, sometimes in loose neighbourhood groups with minimum 10 m between nests. In Israel, song-period mainly 2nd half of March and April. Pair-formation apparently takes place on breeding grounds when flocks break up between late March and mid-April.

Voice. Song varied and melodious, ♂ in display also giving buzzing note (beware confusion with Trumpeter Finch). Contact- and alarm-call a quiet, high-pitched 'tweet', 'tsweet', or 'ts-tsweet'. Flight-call a fairly rich 'trizp', suggesting Tree Pipit.

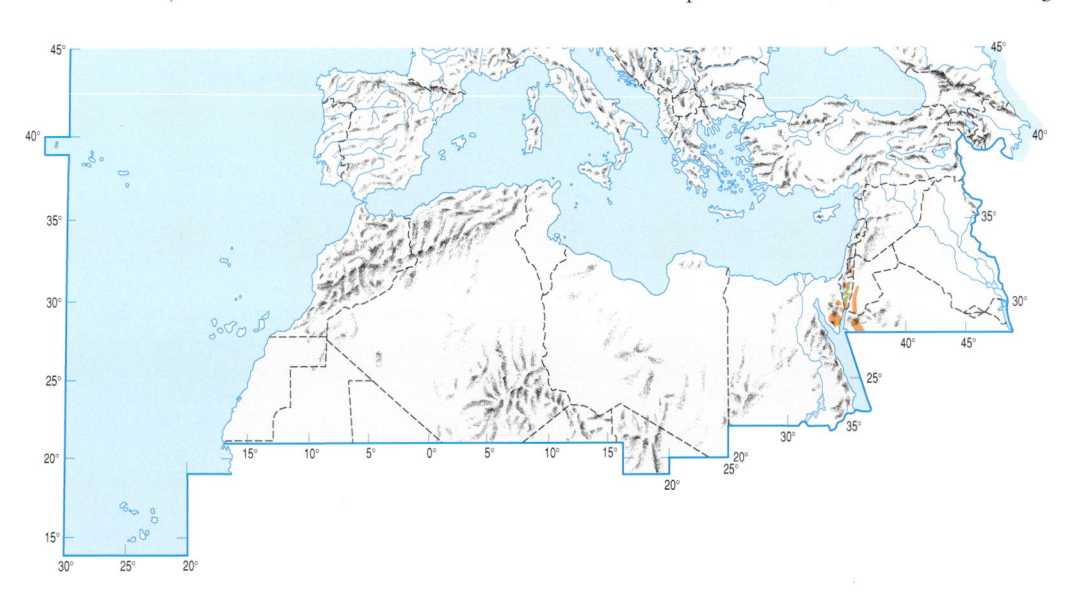

Breeding. SEASON. Southern Israel: eggs laid 1st half of April to May (late March to July). Sinai (Egypt): at 1500–2000 m, laying starts 2nd half of April; elsewhere, fledged young recorded 20 April. Jordan: eggs recorded from mid-April, and family parties at end of month. SITE. In crevice *c.* 50 cm deep in rock-face, at least 5 m above ground and often much higher in inaccessible cliffs; also recorded on ground. Nest: basket-shaped structure of delicate to rough stalks and stems, twigs, and long leaves lined with plant fibres, hair, and fur. Eggs. Sub-elliptical, smooth and glossy; light blue or blue-green with brown-black speckles at broad end, though these sometimes lacking. Clutch: 4–5. Incubation. 13–14 days. Fledging Period. 14–16 days.

Wing-length: ♂ 86–92, ♀ 82–85 mm.
Weight: ♂♀ 17–24 g.

Pallas's Rosefinch *Carpodacus roseus*

PLATES: pages 1601, 1605

Du. Pallas' Roodmus   Fr. Roselin rose   Ge. Rosengimpel   It. Ciuffolotto del Pallas
Ru. Сибирская чечевица   Sp. Camachuelo de Pallas   Sw. Sibirisk rosenfink

Field characters. 15.5–16 cm; wing-span 25.5–28 cm. Slightly larger than Scarlet Rosefinch, with less bulbous bill, proportionately slightly longer wings, and 10% longer tail. Medium-sized, quite robust but lengthy rosefinch, with rather dark, noticeably streaked mantle. Adult ♂ has rose-pink head, rump, and underbody, with silver feather-tips on crown and throat; brightest double wing-bar and tertial edges of similarly-sized west Palearctic *Carpodacus*. ♀ brown, suffused rose-red on face, chest, and rump and well streaked; darkest of similarly-sized *Carpodacus*. Even immature shows red tone on rump, ruling out confusion with Scarlet Rosefinch.

Adult ♂ unmistakable with its mantle streaks and bright pale wing-bars lacking in otherwise similar Sinai Rosefinch (a most unlikely vagrant in any case). Great care needed to distinguish ♀ and juvenile from Scarlet Rosefinch, potentially vagrant or escaped Purple Finch, and escaped House Finch.

Call a short, subdued whistle, repeated in song.

Habitat. Breeds in east-central Palearctic, in rather dry continental climatic zone, mainly in mountain forests.

Distribution. Breeds in central and eastern Siberia from Yenisey river and central Altai east to Kolyma river basin and Sea of Okhotsk, north to 67–68°N, south to northern Mongolia and Sakhalin island.

Accidental. Hungary: December 1850. Russia: Tatarstan (date unknown). Ukraine: ♂ and ♀ collected from small flock, Crimea, December 1902. More recent records in other countries (notably in Denmark, October 1987: previously accepted as wild vagrant) now regarded as escapes.

Movements. Partially migratory, also nomadic. Winter range extends far south of Siberian breeding range, reaching east-central China. Data suggest gradual movement southward in varying numbers. Some birds also move west and south-west of breeding range, reaching Tomsk (85°E) in most years, and sometimes Barnaul, Semipalatinsk, and Zaysan depression (80–84°E), probably only in years of heaviest snowfall. Also reported at Tobol'sk (68°12′E), and recorded exceptionally west of Urals. Records furthest west perhaps involve escaped cagebirds; above data show, however, that some westward vagrancy occurs.

Wing-length: ♂ 89–95, ♀ 86–91 mm.

Great Rosefinch *Carpodacus rubicilla*

PLATES: pages 1601, 1605

Du. Grote Roodmus   Fr. Roselin tacheté   Ge. Berggimpel   It. Ciuffolotto scarlatto maggiore
Ru. Большая чечевица   Sp. Camachuelo grande   Sw. Större rosenfink

Field characters. 20–21 cm; wing-span 34–36.5 cm. Largest finch reaching west Palearctic, 40% larger than Scarlet Rosefinch; approaches small thrush in length and bulk. Very large, long, but plump-bodied rosefinch, with stout bill and strong legs. Plumage rather dark and uniform, particularly at distance; ♂ drenched in crimson, with dull white spots below; ♀'s underbody dark spotted in strongly linear pattern.

Unmistakable. Large size unlikely to be missed, and no other rosefinch occurring in west Palearctic is so uniformly drenched rose-red and pale spotted below in ♂ nor so strongly streaked below in ♀. Flight slow and deeply undulating; flight silhouette long, recalling (with dark plumage) thrush.

Habitat. In west Palearctic, found only in Caucasus, breeding above 2500 m, sometimes at foot of glaciers where birds land on ice and peck snow. Mainly inhabit sunny alpine meadows, hopping about on grass or perching on jutting ledges of rocks or crags. Such habitats, above rhododendron zone, are characterized by piled-up boulders scattered amid stunted vegetation, with small clusters of birch, often flanked by large broken rock screes or steep rock faces, or isolated ledges overgrown with creeping rhododendron. In snowy winters descends to upper valleys, there occupying thickets of *Viburnum*, etc.

Distribution and population. Russia, Georgia. Breeds in highest parts of main Caucasus range, mainly on northern slopes of central Caucasus, between El'brus and Kazbek mountains, in Kabardino-Balkarskaya, northern Ossetia and Dagestan. Range may just extend into northern Azerbaijan. Population estimated at 500–1500 pairs. Has apparently declined (based on decrease

Pallas's Rosefinch *Carpodacus roseus*: **1** ad ♂ summer, **2** ad ♀, **3** 1st winter ♀. Great Rosefinch *Carpodacus rubicilla*: **4** ad ♂ summer, **5** ad ♂ winter, **6** ad ♀, **7** juv.

in size of winter flocks between late 19th century and 1960), perhaps because of over-exploitation by humans of sea buckthorn (main winter food plant) and other factors. TURKEY. 2–3 pairs at Kyrk Deirmen near Erzurum June–July 1910, and recorded in same area in 1940s. Birds singing, but no proof of breeding.

Beyond west Palearctic, breeds from Afghanistan and Tien Shan mountains east to Nepal and Eastern Tibet, north to Altai and Sayan mountains (Russia).

Movements. Altitudinal migrant; extent of movement depends on weather and food availability at high altitude.

In Caucasus, where breeds mainly at 3000–3500 m, birds usually remain above 2000 m in winter, feeding in alpine and subalpine zones, especially on steep and windswept slopes with little snow; they descend (predominantly young birds) to valleys at 900–1000 m only after heavy snowfall. Extent of such movements usually no more than 10–15 km, but up to 30–60 km in harsh winters with food shortage, even reaching foothills.

Food. Seeds, flowers, and other plant material; occasionally small insects. Forages on rocky slopes or scree among stunted alpine vegetation, generally at lower altitude than nesting or roosting sites, searching for seeds in cracks and crevices, or on herb-rich meadows with scattered boulders where birds perch to deal with flower- and seed-heads, or among plants on cliff ledges. In spring and summer, feeds almost wholly on ground, very rarely perching in lowest scrub and preferring to pick berries, etc., from ground, but in winter frequently enters bushes (e.g. sea buckthorn, rose, barberry, juniper) to take fruits; for most of year forages on, or at edges of, snow.

Social pattern and behaviour. Highly sociable, in generally small flocks most of year, rarely 100 or more. Mating system apparently essentially monogamous. Breeds solitarily, or sometimes a few pairs more clustered; few nests found and nothing to indicate territoriality, apart, perhaps, from song of ♂. ♂ gives loud advertising-song from ground or rock; body upright, head raised (even thrown back), crown and throat feathers ruffled. Song-period apparently extends through breeding season, at least to mid-August.

Voice. Advertising-song of ♂ a short flowing phrase comprising 1 or more segments, somewhat reminiscent of very rapid laugh; may carry up to 1 km or more. Contact-call similar to other *Carpodacus*: nasal 'kui' or 'kuii', recalling Bullfinch but sometimes shorter and more incisive, 'tvi' or 'tvit'. Flight-call a rapid twittering.

Breeding. SEASON. Caucasus: very late due to brief spring-summer period; earliest clutches in central Caucasus not before mid-July, many laid last third of month; young found still in nest at end of August. One brood. SITE. In cleft or crevice in rock face, or below boulder. Nest: foundation of thin twigs, stalks, grass, and moss, warmly lined with hair, wool, and in some cases many feathers. EGGS. Sub-elliptical to oval, smooth and glossy; intense sky-blue with slight greenish tinge, very sparsely marked at broad end with black specks, blotches, and scrawls. Clutch: 4–5(–6). INCUBATION. 16 days (in captivity). FLEDGING PERIOD. *c.* 17 days (in captivity).

Wing-length: ♂ 116–122, ♀ 112–118 mm.
Weight: Summer (East Siberia) ♂♂ 25.0–28.5(10), ♀ 27.5(1)

Pine Grosbeak *Pinicola enucleator*

PLATES: pages 1604, 1605

Du. Haakbek   Fr. Durbec des sapins   Ge. Hakengimpel   It. Ciuffolotto delle pinete
Ru. Щур   Sp. Camachuelo picogrueso   Sw. Tallbit

Field characters. 18.5 cm; wing-span 30.5–35 cm. Longer than any other finch in west Palearctic except Great Rosefinch; structure recalls Bullfinch but 15% larger than even northern race of that species. Very large, plump but attenuated finch, with stubby bill shaped like Bullfinch and secretive behaviour. In all plumages shows somewhat mottled, patchy colours with black wings and striking double white wing-bar, white tertial-fringes, and black tail; wing pattern recalls Two-barred Crossbill. Adult ♂ drenched in rose-red, ♀ and immature in orange, yellow, and grey. Juvenile resembles neither adult ♂ nor ♀, being rather dark sepia above, with only indistinct, narrow double wing-bar, and ashy-brown below, with only faint yellowish or buff wash on throat and breast.

Adult unmistakable, with diagnostic combination of large size, white wing markings, and grey belly. Juvenile puzzling, however, with dull plumage quite unlike adult's. Flight essentially finch-like, but large size, long silhouette, and strength of wing-beats also recall long-tailed thrush; over short distance, progress can appear slow and floating, but in full flight undulations become deep and powerful. Movements deliberate. Flicks up wings and tail when uneasy. Sociable and tame.

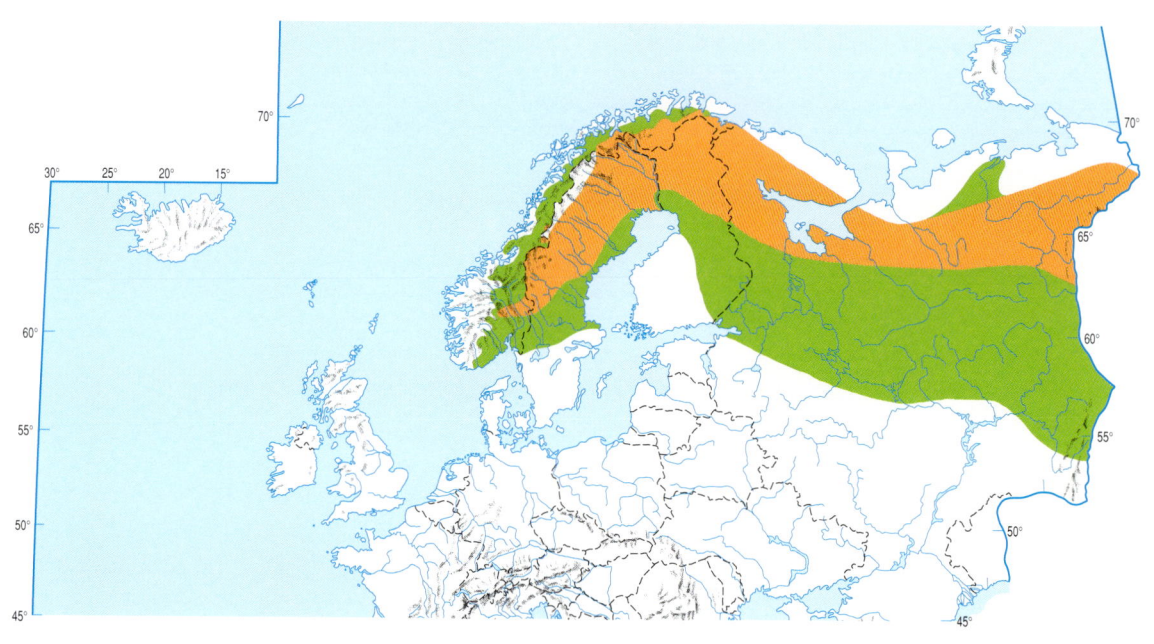

Habitat. In west Palearctic, breeds and largely winters in boreal forests north to treeline, within July isotherms 10–17°C. Uses all species of trees and shrubs, both coniferous and deciduous. Predominantly arboreal, descending in summer to take berries from shrubby plants, and sometimes to gather seeds on ground. In winter, migrants often attracted into towns where berried trees or shrubs present. Both geographically and in terms of acceptable habitat is among the most inflexible of west Palearctic birds.

Distribution. NORWAY. Recent spread in south. FINLAND. Considerable contraction of southern limit of breeding range to north in 20th century. CZECH REPUBLIC. Possibly bred in southern Bohemia, 1964 and 1966. RUSSIA. Nar'yan-Mar on Pechora river (67°37′N): pair, August 1992, and up to 5 ♂♂, August 1994; no breeding evidence, but may indicate range extends further north than known hitherto.

Accidental (or occasional irruptive migrant and winter visitor). Britain, France, Netherlands, Germany, Denmark, Lithuania, Poland, Slovakia, Hungary, Austria, Switzerland, Italy, Ukraine.

Beyond west Palearctic, extends east across northern Asia to Chukotskiy mountains, Kamchatka, Sakhalin island, and northern Japan, south to northern Mongolia. In North America, breeds in boreal forests of Alaska, much of Canada, and south through western mountains to Arizona.

Population. Apparently stable. NORWAY. 500–1000 pairs 1970–90. SWEDEN. 3000–15 000 pairs in late 1980s. FINLAND. 30 000–50 000 pairs in late 1980s. RUSSIA. 10 000–100 000 pairs.

Movements. Some populations migratory, most resident and eruptive. In northern Europe, some birds remain in breeding areas all year, but most move short distance south or southwest; data from northern Fenno-Scandia and adjoining Kola peninsula (Russia) show that areas north of Arctic Circle are vacated, though birds may overwinter only c. 100 km further south if sufficient food (often rowan) available; depart south August–October, returning February–April.

Irruptions into Europe apparently involve chiefly birds from Russia; numbers in central and southern Sweden (where irruptions most marked) too large to be accounted for by northern Scandinavian populations alone, and distribution of records suggests arrival from east. Birds reach central and western Europe only in exceptional years. Recent major irruptions have been in 1954, 1956, 1976, and 1989. 10 British records up to 1992, all 30 October to 15 May, in eastern Britain from Kent north to Shetland.

Food. Buds, shoots, and seeds, especially of spruce and other conifers, rowan, and berry-bearing shrubs; invertebrates in breeding season. Winter numbers, and to some extent movements, probably correlated with berry crop, particularly rowan and, where common, juniper. Diet rather like that of Bullfinch, but items larger on average; moves to herbs when preferred foods scarce. In spring and summer, feeds more often on or near ground taking fruits of understorey shrubs; in winter, when snow on ground, higher up in trees; winter flocks will remain in same tree for some time if undisturbed, methodically removing all fruits or buds. For a bulky bird, a fairly agile and skilful forager in trees, moving slowly and deliberately along thin twigs to reach buds and shoots, sometimes approaching head-down or stretching out from neighbouring perch; uses hooked upper mandible in climbing like crossbill and to grasp buds by tilting head 90° then straightening up to pull bud off; only eats soft kernel, discarding sticky outer scales, and peels soft green spruce cones to get at core; eats pulp and seeds of many berries but rejects skin; often sits quietly for long periods in same position between feeding bouts.

Social pattern and behaviour. Gregarious outside breeding season. Migrates south in large numbers in some years (see Movements): flocks during such irruptions often small, up to c. 50, sometimes several hundred or even several thousand. No evidence of other than monogamous mating system. In Russia, birds initially in flocks after arrival on breeding grounds; form pairs and take up territories from early May. Territorial and apparently solitary when breeding; no definite reports of neighbourhood groups such as occur in some other Carduelinae. Both sexes sing, though loud advertising-song perhaps given only by ♂, while ♀'s song may be of subsong type. Song given from exposed perch, sometimes while gliding between trees; no reports of a more elaborate, ritualized song-flight.

Voice. Loud advertising-song presumed to be given mainly or exclusively by ♂. Short yodelling phrases (c. 2 s) comprising series of clear, rich and melodious whistling or fluting motifs, including some resembling calls; song-phrases typically quieter at start and at end than in middle. Contact-calls various; include short series (usually 2–3) of attractive piping or fluting notes often descending in pitch overall and with tonal quality of Bullfinch; more-complex calls of same kind, including wholly tonal, ethereal, silken-soft, and quiet 'chliudwcee' of remarkable beauty; and series of ascending and descending whistles. Alarm-calls include loud excited 'tui-tui-tui'; subdued chirping sounds audible only at close range; and rather harsh, loud, explosive squeak or 'sneeze-whistle', rendered 'füd tschri-hüid' in which 'tschri' very loud.

Breeding. SEASON. Finland: eggs laid late May to mid–July, mainly 1st half of June with repeat clutches to end of June. Kola peninsula (north-west Russia): eggs laid late May. Probably one brood. SITE. Usually against, or close to trunk of pine or spruce, and in some areas juniper or birch; usually rather low. Nest: like that of Bullfinch, but larger and deeper. Rather untidy, loosely-built foundation of interwoven fine twigs of mainly spruce but also birch, juniper, or pine sometimes of considerable length; stiff dead twigs project from all sides of nest. Rootlets and grass blades, delicate shoots and soft fragments of lichen filaments sometimes woven into outer structure. Cup mainly dry grass, with moss, roots of bilberry, dry spruce twigs, and juniper shoots, lined with very thin roots and *Usnea* lichen, thin grass, moss, and sometimes animal hair. EGGS. Sub-elliptical, often long sub-elliptical; smooth, little or no gloss; variably green-blue, with blotches and spots of black or dark purple-brown and underlying pale violet-grey markings; sometimes freckled all over with small spots, others with a few very large blotches; markings often tend to be concentrated in zone around broad end. Clutch: 3–4 (2–5), mainly 4. INCUBATION. 13–14 days. FLEDGING PERIOD. 14 (13–18) days.

Wing-length: ♂ 106–113, ♀ 107–111 mm.
Weight: ♂♀ mostly 47–64 g.

Pine Grosbeak *Pinicola enucleator*: **1** ad ♂ summer, **2** ad ♂ winter, **3** ad ♀ summer, **4** ad ♀ winter, **5** 1st winter ♂, **6** juv.

Crimson-winged Finch *Rhodopechys sanguinea sanguinea* (p. 1589): **1** ad ♂ summer, **2** ad ♀. Desert Finch *Rhodospiza obsoleta* (p. 1591): **3** ad ♂ summer, **4** ad ♀. Mongolian Trumpeter Finch *Bucanetes mongolicus* (p. 1592): **5** ad ♂ summer, **6** ad ♀. Trumpeter Finch *Bucanetes githagineus crassirostris* (p. 1594): **7** ad ♂ summer, **8** ad ♀. Scarlet Rosefinch *Carpodacus erythrinus erythrinus* (p. 1596): **9** ad ♂ summer, **10** ad ♀.

Sinai Rosefinch *Carpodacus synoicus* (p. 1599): **1** ad ♂ summer, **2** ad ♀. Pallas's Rosefinch *Carpodacus roseus* (p. 1600): **3** ad ♂ summer, **4** ad ♀. Great Rosefinch *Carpodacus rubicilla* (p. 1600): **5** ad ♂ summer, **6** ad ♀. Pine Grosbeak *Pinicola enucleator* (p. 1602): **7** ad ♂, **8** ad ♀.

Long-tailed Rosefinch *Uragus sibiricus* (p. 1606): **1** ad ♂ summer, **2** ad ♀, **3** 1st winter ♀. Bullfinch *Pyrrhula pyrrhula pyrrhula* (p. 1606): **4** ad ♂, **5** ad ♀ summer, **6** ad ♀ winter, **7** juv.

Long-tailed Rosefinch *Uragus sibiricus*

PLATE: page 1605

Du. Langstaartroodmus   Fr. Roselin à longue queue   Ge. Meisengimpel   It. Ciffulotto siberiano
Ru. Длиннохвостый снегирь   Sp. Pirrula rabilargo   Sw. Långstjärtad rosenfink

Breeds in valleys and woodland of southern and eastern Siberia, northern Mongolia, northern and central China, south-east Tibet, and Japan. Winters in breeding range, and in west south to Tajikistan and western Sinkiang (China). Recorded as presumed escape in several European countries, e.g. Britain, Germany, Denmark, Sweden, Russia. Record from Finland, April 1989, formerly accepted as wild bird, but now presumed also to have been an escape.

Bullfinch *Pyrrhula pyrrhula*

PLATES: pages 1605, 1607, 1614

Du. Goudvink   Fr. Bouvreuil pivoine   Ge. Gimpel   It. Ciuffolotto
Ru. Снегирь   Sp. Camachuelo común   Sw. Domherre

Field characters. 14.5–16.5 cm; wing-span 22–29 cm. Size varies, being noticeably large in races or populations occupying highest latitudes and altitudes; typical bird of northern race over 15% larger than birds of temperate European races. Medium-sized to rather large, bull-headed, quite long-winged, and seemingly long-tailed finch, with short bulbous bill and distinctive plumage. ♂ and ♀ share black face and cap, black wings with white wing-bar, white rump, and black tail; ear-coverts and underbody pink in ♂, pale pink-brown in ♀. Juvenile like ♀ but lacks black cap.

Unmistakable. Plumage pattern and colours as unique as voice (see below). Separation of south-migrating bird of northern nominate race from smaller more sedentary birds of western and south-west Europe not difficult; larger size readily evident, particularly in flight when body bulk, wing-span, and tail length are eye-catching. Flight at times rather weak for finch, flitting within cover or along leading line—with white rump easy to follow; strong over distance, however, with action developing typical finch undulations and achieving considerable height between woods or on migration. Gait a clumsy hop, making bird look uncomfortable and untidy on ground. Feeds more slowly than other finches, visibly manipulating buds and seeds in bill. Has habit of flicking and twisting tail.

Habitat. Breeds across most of temperate west Palearctic in both oceanic and continental climates, and also summers over much of boreal zone, within July isotherms *c.* 12–21°C. Arboreal and mainly lowland, but in mountains in south of range ascends to breed up to treeline—in Caucasus up to 2500–2700 m. In FSU, largely in coniferous taiga forest, also in mixed and broad-leaved woods, but in Caucasus, in pine forest and tall beechwoods. Elsewhere, prefers mature mixed or coniferous forest with dense undergrowth. In winter often appears in orchards, towns, and suburbs. In Switzerland, only exceptionally in pure broad-leaved woodland, preferring spruce and usually keeping close to dense stands, but some nest in gardens and parks with thick undergrowth, even in large cities. In Britain, breeds mainly in broad-leaved woods, but commonly also in fringing or detached groups of trees, thickets, tall and dense hedgerows, large gardens, yews in churchyards, and even gorse on heaths; where conifers used, some preference for younger plantations or ornamental groups; orchards often visited in spring to pick buds but rarely favoured for nesting.

Due to its shyness is more readily observed outside woodland or along its edge, and this may lead to underestimation of use of interior of forests, especially where there are glades, clearings, or regenerating patches. Apparently, in parts especially of western Europe, habitats outside woodland have been increasingly colonized in recent decades, but questionable whether such habitats account for significant fraction of total population. Even within human settlements normally avoids contact with people, and does not usually show itself on ground, or even flying in the open.

Distribution. BRITAIN. Some recent local retractions, especially in Scotland. GERMANY. Distribution very uneven in east. DENMARK. Expansion by probably 7% 1974–94. NORWAY. Some recent spread in north and west. POLAND. In early 19th century known to breed only in mountains and in north-east; now widespread, expansion proceeding from Silesian mountains. PORTUGAL. First breeding record in Algarve 1988. RUSSIA. Common at Nar'yan Mar/Pechora delta, in extreme north of range. Further north, sight records June 1992, 1993, Russkiy Zavorot peninsula. UKRAINE. Slight increase.

Accidental. Iceland, Faeroes, Balearic Islands, Malta, Tunisia, Algeria, Morocco.

Beyond west Palearctic, extends east across northern Asia (north to 61–67°N) to Kamchatka, Sakhalin island, and Japan, south to northern Mongolia; also breeds northern Iran.

Population. Increases reported Denmark, Estonia (with marked fluctuations), Poland, and Ukraine, and decreases Britain, Belgium, Sweden (probably), and Lithuania. Stable elsewhere. BRITAIN. 190 000 territories 1988–91. Declining since mid-1970s, probably mainly due to hedgerow removal, perhaps also intensification of farming methods. IRELAND. 100 000 territories 1988–91; increasing throughout 20th century. FRANCE. 100 000–1 million pairs in 1970s. BELGIUM. 8000–16 000 pairs 1989–91. LUXEMBOURG. 10 000–12 000

Bullfinch *Pyrrhula pyrrhula*. *P. p. europoea*: **1** ad ♂, **2** ad ♀ summer. *P. p. pileata*: **3** ad ♂, **4** ad ♀ summer. *P. p. iberiae*: **5** ad ♂, **6** ad ♀ summer. Azores Bullfinch *Pyrrhula murina* (p. 1609): **7** ad ♂, **8** ad ♀.

pairs. NETHERLANDS. 15 000–20 000 pairs 1979–85. GERMANY. 438 000 pairs in mid-1980s. DENMARK. 500–6000 pairs 1987–8. NORWAY. 100 000–500 000 pairs 1970–90. SWEDEN. 200 000–500 000 pairs in late 1980s. FINLAND. 200 000–400 000 pairs in late 1980s. Has benefited from intensified winter feeding and increase of young spruce forests. ESTONIA. 50 000–100 000 pairs in 1991. LATVIA. 160 000–210 000 pairs in 1980s. LITHUANIA. Common. POLAND. 20 000–80 000 pairs. CZECH REPUBLIC. 100 000–200 000 pairs 1985–9. SLOVAKIA. 70 000–120 000 pairs 1973–94. HUNGARY. 200–300 pairs 1979–93. AUSTRIA. Common. SWITZERLAND. 50 000–80 000 pairs 1985–93. SPAIN. 118 000–170 000 pairs. PORTUGAL. 100–1000 pairs 1978–84. ITALY. 30 000–60 000 pairs 1983–95. GREECE. 1000–5000 pairs. YUGOSLAVIA: CROATIA. 20 000–30 000 pairs. SLOVENIA. 10 000–20 000 pairs. BULGARIA. 5000–50 000 pairs. RUMANIA. 200 000–250 000 pairs 1986–92. RUSSIA. 1–10 million pairs. BELARUS'. 80 000–100 000 pairs in 1990. UKRAINE. 1500–2100 pairs in 1986. AZERBAIJAN. Mostly uncommon, but common along southern slope of Great Caucasus. TURKEY. 1000–10 000 pairs.

Movements. Sedentary to migratory; probably most populations partially migratory. Winters chiefly within breeding range, those breeding at high levels tending to make altitudinal movements. Most migrants move short or medium distances, but some (apparently chiefly from Russia) move longer distances; in northern and central Europe, no evidence that northern populations move further than southern ones. North European birds move within wider compass than central European birds. Also eruptive migrant; numbers migrating show marked annual fluctuations; no link with particular food source established. Autumn migration begins late, and is fairly brief, mostly October–November; spring migration February–April.

Food. Seeds of fleshy fruits and herbs, buds, and shoots; invertebrates important in diet of young. Forages in woodland, thickets, hedgerows, etc., some moving to open country, large gardens, or orchards in late autumn to early spring; hardly ever feeds more than *c.* 10 m from cover. Generally extracts seeds from fruit on plant, since only large seeds can be picked up from ground because of bill shape; small seed-heads and sometimes small fruits may be bitten off and carried to ground for removal of seeds. In tree or bush, either extracts seeds from fruit *in situ*, leaving skin and pulp hanging, or mandibulates fruit by turning it in bill with tongue, removing pulp against lower mandible and swallowing seeds, sometimes without dehusking; bites into fruits and soft seed-heads much more than other Fringillidae of region and bill well-adapted to this.

Social pattern and behaviour. Differs markedly in many aspects of social behaviour from most other Fringillidae, showing some resemblance to Hawfinch. Most characteristic features include: persistence of sexual behaviour throughout year, with ♀ generally dominant over ♂; unusually intimate relationship between pair-members, with apparently persistent pair-bond; lack of display-flight; soft, quiet song, highly variable between individuals; lack of flocking and other social behaviour in breeding season. Occurs singly, in pairs, or in small groups (maximum *c.* 20) outside breeding season; groups form mostly at sites where food abundant. Migrates singly or in small scattered flocks. Monogamous mating system. Pair-bond lasts at least for duration of breeding season; pairs also seen together in winter, apparently faithful for successive breeding seasons. Pairs breed solitarily, but not obviously territorial. Song given by both sexes, but most commonly by ♂. Inconspicuous; not loud nor given from conspicuous vantage point, has no territorial function, but plays minor part in pair-formation and in ♂'s display to ♀, mainly in early stages of courtship. Pairing takes place in flocks (usually small groups), from which pairs then separate.

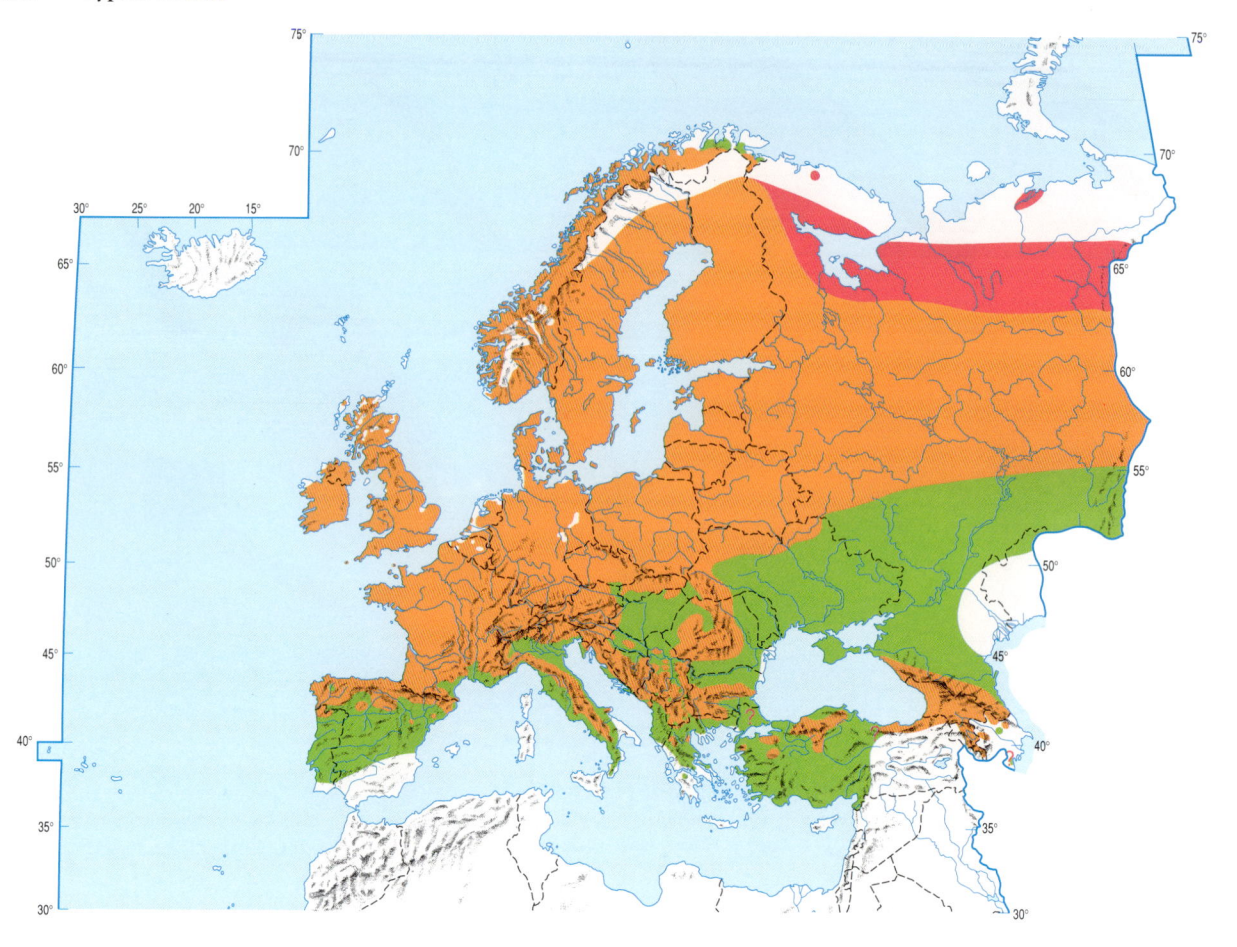

Voice. Vocal repertoire large, similar in ♂ and ♀; probably related to need for effective communication in dense vegetation in which vocal signals are usually given. In captivity has ability to learn and imitate human music. Song of distinctive timbre, but neither loud nor very different in quality from other calls; highly variable between individuals. Consists of varied units, many in pairs, often continuing (without marked pauses) for several minutes. Given either as 'directed song' as part of courtship-display, or as 'undirected song' usually at a distance from other individuals. Both categories of song given by both ♂ and ♀, most commonly by ♂. 'Directed song' audible only within a few metres, sometimes only evident from throat movements; 'undirected song' usually louder, audible up to c. 20 m. Most familiar call a plaintive, piping 'phew', given by both sexes all year. Birds separated from mates give repeated 'phew' calls which function as long-distance contact-calls, and variants of this call given in other contexts. Repertoire includes several other, mainly monosyllabic, calls, including hoarse, monosyllabic 'hhwhore', or 'phee-yore' (hiss preceded by whistled note similar to 'phew'), given with bill wide open, in head-forward threat posture; almost exclusively by ♀ when threatening ♂.

Breeding. SEASON. Britain: laying starts late April but mainly early May; exceptionally 1st half of April; repeat, 2nd, and possibly 3rd clutches into late August and even mid- to late September. Fenno-Scandia: late April or May to late July. Netherlands and Germany: laying starts April and continues to end of August. 2 broods usual, 3 possible in favourable conditions. SITE. Thick bushes and trees of many kinds, often evergreen, and in particular in dense branches of spruce. Nest: 2-layered structure similar to that of Hawfinch and Pine Grosbeak; loose base of dry twigs, which are often bitten from trees, with some moss and lichen; twigs may be up to 30 cm long so that they project from base. Nest-cup of dry grass and fine rootlets interwoven into foundation and lined with small quantity of fine rootlets and grasses, more rarely lichen, moss and leaves. EGGS. Sub-elliptical, though may be rather pyriform, smooth and slightly glossy. Ground-colour variable, but usually pale blue or clear greenish-blue with dark purple-brown spots, speckles and streaks which tend to be concentrated around broad end; sometimes also paler purple-violet blotches; varieties with reddish markings on white ground-colour also occur. Clutch: 4–5 (3–7). INCUBATION. 12–14(–15) days. FLEDGING PERIOD. 14–16 days (12–18).

Wing-length: See Geographical variation.
Weight: *P. p. pyrrhula* ♂♀ mostly 27–38 g. *P. p. europoea*: ♂♀ mostly 16–26 g.

Geographical variation. Marked in size, less so in colour (except in far east). 5 races recognized in west Palearctic. Races *pileata*, *europoea*, and *iberiae* from western Europe small, average wing of ♂ *c.* 81.5 mm (averages of various populations 80–83, measurements of individuals mainly 78–85); other measurements and bill size of these races also closely similar. Northern and eastern Europe inhabited by *P. p. pyrrhula*, with average wing length of ♂ *c.* 92.5 mm (averages of various populations 91.5–94.5, measurements of individuals mainly 89–98). For each sex, virtually no overlap in measurements between *P. p. pyrrhula* and smaller races. Large and small populations grade clinally into each other in wide zone through central Europe.

Differences between small western populations based on colour. *P. p. europoea* on continent from northern Germany and Netherlands to foothills of Pyrénées, Alps, and Vosges in France slightly darker grey on upperparts of ♂ than in *P. p. pyrrhula*, underparts slightly darker rosy-red, less rosy-pink; tips of outer greater coverts light grey, less whitish; ♀ distinctly darker and browner on upperparts and more vinous (less grey) on underparts. ♂ *pileata* from Britain and Ireland about as dark above as ♂ *europoea*, but more diluted dull pink-red below; ♀ slightly darker and browner above, distinctly so below, especially on flank. ♂ *iberiae* slightly paler grey on upperparts than ♂ *europoea*, underparts more fiery rosy-red. Birds from Caucasus area separated as *rossikowi*: rather like *P. p. pyrrhula* in colour and size, but upperparts of ♂ paler grey, underparts brighter red, less pinkish, ♀ darker and duller brown.

Azores Bullfinch *Pyrrhula murina*

PLATE: page 1607

DU. Azoren-goudvink   FR. Bouvreuil des Açores   GE. Azorengimpel   IT. Ciuffolotto delle Azzore
RU. Азорский снегирь   SP. Pirrula europea de las Azores

Field characters. 16–17 cm; wing-span 22–25 cm. Similar in size to northern race of Bullfinch but with slightly shorter wings and even longer bill. Differs distinctly from Bullfinch in lack of marked sexual dimorphism and white rump. Plumage otherwise similar to ♀ Bullfinch, with ♂ distinguished from ♀ and immature only by slightly ruddier tint to underparts.

Unmistakable. Restricted to São Miguel, Azores.

Habitat. Endemic mountain cloud forest and its margins on one island of Azores archipelago in mid-Atlantic; principally in laurel and associated understorey, and now probably restricted to one inaccessible area (perhaps only 5 km^2) of dense, 1–2 m high scrub on steep slopes, though also attracted to young plantations of introduced Japanese red cedar; apparently prefers thick undergrowth of ferns and fruit- or bud-producing plants which are sometimes absent in old laurel stands; rarely leaves cover.

Distribution and population. Breeds only on São Miguel in Azores, where limited to eastern part. 60–200 pairs.

Movements. Sedentary, though local movements, given small extent of habitat, can be relatively large at up to 3 km. Movements seemingly related to food availability (fruiting of plants); in summer appears to be more mobile than Bullfinch in Britain, with changes in position of 700–800 m not uncommon (up to twice as far as Bullfinch), and generally around 200 m in winter, when remains faithful to small areas for several months. Moves widely within home range like Bullfinch. Altitudinal movements also noted (e.g. from 700 to 300 m in May) but not known if this regular.

Food. Seeds and buds of fruits and herbs; principally herb seeds in summer (with some invertebrates) and seeds of fleshy fruits in autumn; also takes sporangia of ferns and tree seeds (especially introduced *Clethra*) in winter, and flower buds (particularly endemic *Ilex*), fern fronds, and moss tips in spring, when abundance of seeds and sporangia at lowest. Eating of fern fronds and spores is extremely unusual among birds and is perhaps result of scarcity of other foods. Preferences over year are complex and depend on availability and accessibility. Endemic plant species important in spring and autumn, but introduced species apparently preferred at other times, e.g. flowering heads of knotweed, a non-native herb of edge habitats, stripped 'voraciously' from May to July. In 19th century, when more widespread, was regarded as pest in orange orchards.

Social pattern and behaviour. Hardly studied, but most aspects apparently little different from Bullfinch. Population density high, and estimated at *c.* 20–30 pairs per km^2. Average home range in winter at 3.9 ha much like that of Bullfinch; perhaps around twice as large in summer. Clearly, restricted extent of habitat influences these figures. Density higher in native than in exotic vegetation. No evidence of territoriality (again as Bullfinch); birds seem to have multiple centres of activity within home range, and site fidelity in subsequent years very strong. Almost always singly or in pairs; one party of 8 seen in May, otherwise larger groups seen mainly after breeding and usually consist of juveniles. Observed courtship behaviour as Bullfinch and apparently more confiding than that species.

Voice. Not substantially different from Bullfinch.

Breeding. SEASON. Nest building seen May and July, eggs laid mid-June to August, mostly July; probably commonly double-brooded. SITE. Two nests were in Japanese red cedar *c.* 3 m above ground. Nest: outer layer of twigs, inner layer of dry grass, rootlets, and moss. EGGS. As Bullfinch. Clutch. probably smaller than in Bullfinch since most pairs apparently raise only 2 young. (No information on incubation or fledging periods.)

Wing-length: ♂♀ 85–93 mm.
Weight: ♂♀ c. 30 g.

Yellow-billed Grosbeak *Eophona migratoria*

DU. Witvleugeldikbek   FR. Grosbec migrateur   GE. Weißhand-Kernbeißer   IT. Frosone codanera
RU. Малый черноголовый дубонос   SP. Picogordo piquigualdo   SW. Mindre maskstenknäck

An east Palearctic species, breeding in south-east Russia (Ussuriland west to c. 117°E), north-east China, and Korea, and wintering mainly in eastern China; also an isolated population resident in southern China. Individuals recorded in Europe—e.g. Sweden (May 1981), Germany (January 1991), Faeroes (May 1992)—are thought to have been escapes.

Japanese Grosbeak *Eophona personata*

DU. Maskerdikbek   FR. Grosbec masqué   GE. Maskenkernbeißer   IT. Frosone mascherato
RU. Большой черноголовый дубонос   SP. Picogordo nipón   SW. Större maskstenknäck

An east Palearctic species, breeding in south-east Russia (Ussuriland and lower Amur region), north-east China, and Japan, and wintering within and to south and west of breeding range. Records in Europe—Sweden (May 1990), Norway (June 1989, April 1990), and Shetland (June 1992)—are probably of escapes.

Hawfinch *Coccothraustes coccothraustes*

PLATES: pages 1613, 1614

DU. Appelvink   FR. Grosbec casse-noyaux   GE. Kernbeißer   IT. Frosone
RU. Дубонос   SP. Picogordo común   SW. Stenknäck

Field characters. 18 cm; wing-span 29–33 cm. 20% larger than Greenfinch, with distinctive top-heavy, parrot-like proportions due to massive conical bill on large head, bull-neck, and short, square tail. Very large, huge-billed, big-headed, short-tailed, short-legged finch, bigger than all other common finches of temperate woodlands. Adult plumage warm buff, with bill blue-grey in breeding season, yellow in winter, emphasized by black lore and bib, grey nape, brown back, black flight-feathers boldly panelled white on larger coverts and across primaries, and white-tipped tail.

Unmistakable. Flight rapid with bursts of strong wing-beats propelling bird forward in shooting, then bounding, and soon deeply undulating progress; typically at considerable height (over 50 m) even when only moving between woods. Bird looks like whirring, front-heavy and short-tailed projectile. Shy, foraging quietly in treetops, but betrays presence by explosive call and remarkable territorial flight, with ♂ following 'roller-coaster' track over wood ending in dramatic plummet.

Habitat. Breeds in west Palearctic in lowland and hilly temperate zone, and parts of boreal, Mediterranean, and steppe zones, continental and to lesser extent oceanic, between July isotherms of c. 17–25°C. Most characteristically a specialist bird of natural open mixed oak and hornbeam forest, but extends freely to most other tall deciduous trees which carry large fruits within handling capacity of massive bill, especially beech, ash, wych elm, and sycamore or maple. Accordingly mainly found in crowns and forest canopy, liking to perch on topmost twigs. Significant secondary habitats are ribbons of mature trees along rivers or streams or fronting lakes and pools, and similar trees planted in avenues, parks, cemeteries, and large gardens. Occupies mixed broad-leaved/conifer woodlands and forests where broad-leaved predominate, but rarely breeds in conifers in west of range. Ascends freely in mountains to limits of deciduous forest.

Distribution. Some expansion in Scandinavia and western Russia. BRITAIN. Probably spread north and west in 2nd half of 19th and early 20th centuries. Elusive, and range probably incompletely known. Widespread in south-east, more local elsewhere; local decreases perhaps due to loss of broad-leaved woodland in 1970s–80s. IRELAND. Not known to have bred. Regular winter visitor in 19th century; then irregular, few records since 1910, and now apparently more or less accidental. FRANCE. Recent range increase. BELGIUM. Apparently increasing, but perhaps previously underrecorded. NORWAY.

Spreading. Very few breeding records before 1960; in 1960s–80s bred in several new localities in south and south-east. SWEDEN. Apparently spreading north; isolated breeding sites (e.g. Sundsvall, *c.* 62°30′N; Umeå, *c.* 64°N) recorded in 1980s. RUSSIA. First recorded breeding Kareliya in north-west in 1970s; breeding subsequently recorded up to 62°30′N. BELARUS'. Much rarer in north and east than elsewhere.

Accidental. Iceland, Faeroes, Libya, Madeira.

Beyond west Palearctic, extends east through West Siberia to upper Yenisey, thence south-east to Baykal region and northern Mongolia, and east to southern Sea of Okhotsk (also Kamchatka), Sakhalin island, northern Japan, and north-east China. Isolated population from eastern Afghanistan to Tien Shan and Dzhungarskiy Alatau mountains.

Population. Increases reported Belgium, Netherlands, Scandinavia, Estonia, Latvia, and Spain; slight decrease Ukraine. Elsewhere apparently mostly stable, though tending to fluctuate. BRITAIN. 3000–6500 pairs 1988–91. Local fluctuations; recent marked but uneven decline. FRANCE. 1000–10 000 pairs in 1970s. BELGIUM. 6500–12 000 pairs 1989–91. LUXEMBOURG. 5000–6000 pairs. NETHERLANDS. 9000–12 000 pairs 1985–6. GERMANY. 329 000 pairs in mid-1980s. In east, 85 000 ± 45 000 pairs in early 1980s. DENMARK. 1600–17 000 pairs 1987–8. NORWAY. 200–500 pairs 1970–90. SWEDEN. 5000–15 000 pairs in late 1980s. FINLAND. Rare and poorly known breeder. Perhaps 50–200 pairs in recent years. ESTONIA. 5000 pairs in 1991. LATVIA. 3000–10 000 pairs in 1980s. POLAND. 100 000–200 000 pairs. CZECH REPUBLIC. 150 000–300 000 pairs 1985–9. SLOVAKIA. 110 000–220 000 pairs 1973–94. HUNGARY. 80 000–120 000 pairs 1979–93. AUSTRIA. Fairly common; stable. SWITZERLAND. 4000–8000 pairs 1985–93. SPAIN. 4000–5000 pairs. PORTUGAL. 1000–10 000 pairs 1978–84. ITALY. 5000–15 000 pairs 1983–95. GREECE. 5000–15 000 pairs. YUGOSLAVIA: CROATIA. 15 000–20 000 pairs. SLOVENIA. 5000–10 000 pairs. BULGARIA. 10 000–100 000 pairs. RUMANIA. 150 000–180 000 pairs 1986–92. RUSSIA. 10 000–100 000 pairs. BELARUS'. 30 000–70 000 pairs in 1990. UKRAINE. 25 000–30 000 pairs in 1986. MOLDOVA. 12 000–15 000 pairs in 1988. AZERBAIJAN. Common. TURKEY. 1000–10 000 pairs. TUNISIA. Common.

Movements. Sedentary to migratory; northern populations migrate more than southern ones. Juveniles migrate more than adults, and ♀♀ more than ♂♂. European migrants head between west and south, wintering chiefly within breeding range; numbers fluctuate markedly from year to year. Makes local feeding movements in small flocks in wide variety of directions; longer movements are probably also associated primarily with food availability.

Many birds move south-west across Europe, and winter chiefly in northern Italy and southern France, also in north-east Spain, central Italy, and Balkans. Others cross Mediterranean, and some winter in Balearic Islands, Sardinia, and Corsica. Inconspicuous on passage, and fewer data than for most other finches. Autumn movement apparently protracted; reported from late August to early December, mostly mid- or late September to early November. Spring movement chiefly February–April.

Food. Large hard seeds, buds, and shoots of trees and shrubs; invertebrates, especially caterpillars, in breeding season. In spring and summer, forages mainly in woodland trees; in autumn and winter, in hedges and on ground. Seeds extracted from fleshy fruits by turning in bill, peeling off pulp (which is sometimes eaten) against lower mandible, then seeds cracked in bill; does not use feet. Massive bill adapted for splitting large hard seeds to get at kernel; 2 striated knobs at base of each mandible, which develop only in maturity, grip seed tightly so considerable pressure can be brought to bear. Suture of seed placed vertically lengthways along bill by combined movements of head and tongue, so 2 halves fall away to side when de-husked and kernel swallowed whole. Laboratory measurements have shown that force in excess of 50 kg can be employed, large muscles encasing skull providing power. Prefers some foods throughout region: seeds of hornbeam, beech, elm, maple, and Rosaceae, especially cherry and other *Prunus*; in Mediterranean area, seeds of olive, nettle-tree, and acorns important.

Social pattern and behaviour. Extremely wary compared with other Fringillidae, especially so when on ground, taking flight at slightest alarm and 'rocketing' silently up into tree-tops. Hence difficult to observe, and tends to be under-recorded. Avoids open areas, and when crossing open ground flies high and fast, plunging down into shelter of first trees encountered.

In flocks or smaller groups outside breeding season. Aggregations begin to form at end of June and become prominent in July, by coalescing of family parties at feeding sites; but no bonds between them, each family group keeping to itself. Monogamous mating system, perhaps with maintenance of pair-bond in winter. Breeds solitarily, or (especially in favourable habitats) in small groups. Solitary pairs defend definite boundaries of small territory round nest. In breeding colonies, only small area immediately round nest defended, allowing pairs to nest in close proximity. Pair-formation, involving displays and flight chases, takes place in early spring, while still in flocks. Song of ♂ never frequent, even where bird is abundant; not loud, not used for defining or defence of territory, and not delivered from special posts. Main function apparently related to pair-formation and courtship. During period of pair-formation in winter flocks, ♂♂ may withdraw from flock and sing in nearby trees. ♂♂ also sing from trees near nest, but do not defend them against other ♂♂, who may sing in same tree. Courtship displays of ♂ include waddling 'penguin-walk' with body almost vertical and wings drooped, and deep bowing with wings half-extended to show white spots on black primaries, sometimes with semicircular turn so that one wing is drooped in arc in front of ♀.

Voice. Song poor, a low halting 'deek-waree ree ree' with strained quality until more liquid and musical end to phrase, which suggests Goldfinch. Commonest call a clipped metallic 'tick', 'pix', or 'tzik', sometimes quickly repeated, used freely in flight (often only clue to bird's sudden passage overhead); not unlike Robin but much more powerful and explosive.

Breeding. SEASON. Britain: eggs laid 1st half of April to end of July, mainly late April to late June. Central Sweden: 2 weeks later than Britain. Netherlands: beginning of April to mid-August. Greece, Italy, and southern Spain: c. 3 weeks earlier than Britain. North Africa (Algeria and Tunisia): eggs laid end of March to end of May. Probably 1 brood. SITE. Prefers old, shrubby trees, especially oak and fruit trees. Position of nest varies, though generally well-lit and easy of access; clear preference for horizontal branches, especially in fruit trees; often attached to trunk or on supporting branch close by. Often in cover of ivy or honeysuckle. Nest: bulky foundation of dry twigs, distinct from 2nd layer of thin twiglets and blades of grass in which cup with soft plant matter is shaped. EGGS. Sub-elliptical, but varying in shape from short to long sub-elliptical, occasionally pyriform, smooth and slightly glossy or non-glossy; light blue or greyish-green, rarely pale buff or grey, sparsely, but usually fairly evenly marked with bold spots and scrawls and paler scribbling of blackish-brown, sometimes concentrated towards broad end. Usually faint underlying spots or streaks of pale ash-grey; considerable variation in colour includes unmarked slate-grey, white or greenish-blue; creamy-white with bold chocolate brown marks, and pure white with a few dark spots at broad end. Clutch: 4–5 (2–7). INCUBATION. 11–13 days. FLEDGING PERIOD. 12–13 days.

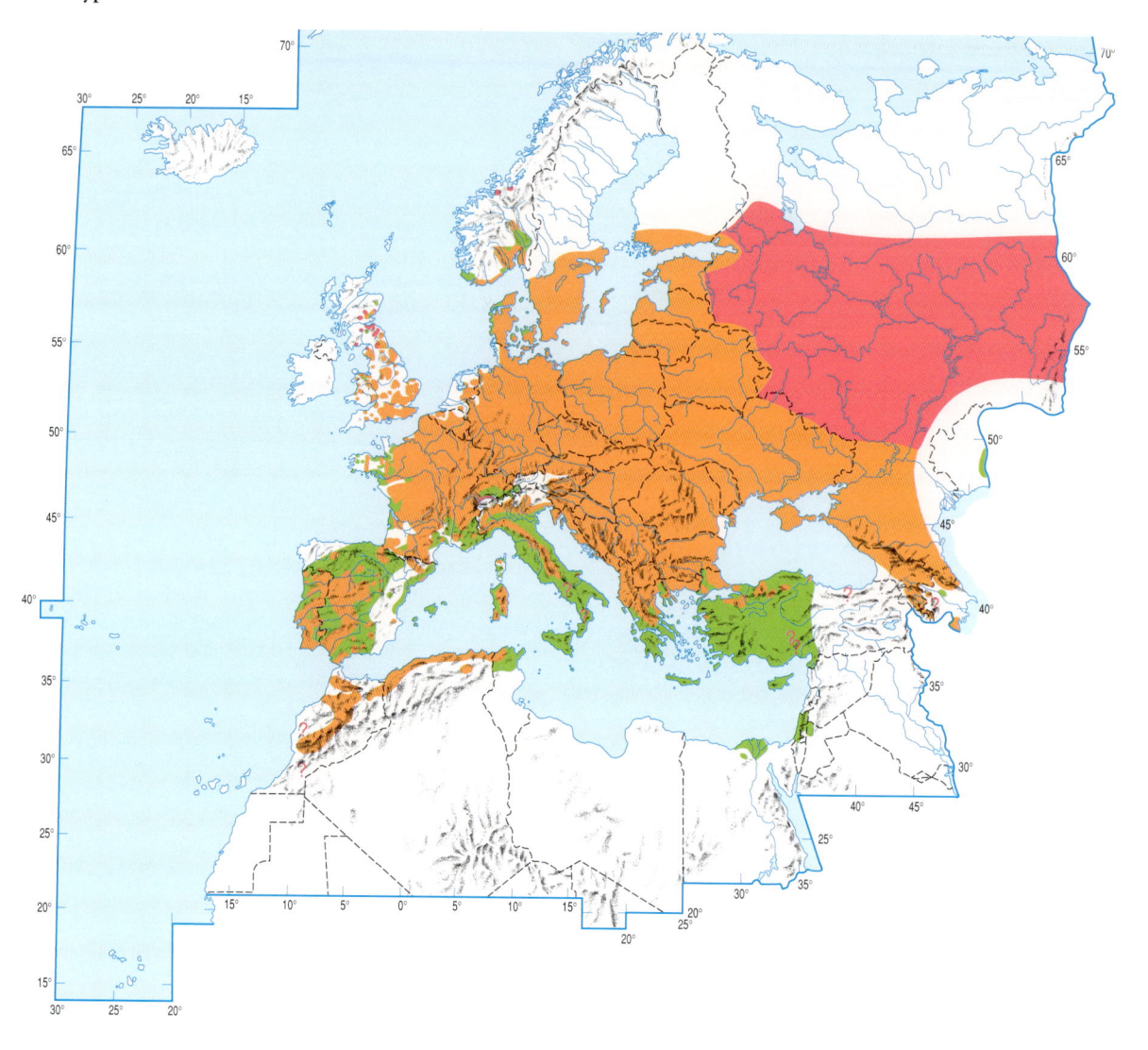

Wing-length: *C. c. coccothraustes*: ♂ 99–110, ♀ 97–105 mm.
Weight: *C. c. coccothraustes*: ♂ mostly 50–70, ♀ 46–65 g.

Geographical variation. Slight, largely clinal. 3 races recognized in west Palearctic, nominate *coccothraustes* occupying most of area. North African race *buvryi* distinguished by slightly paler and more diluted colours of head and body and less white on inner web of outer tail-feather. Variation in size more pronounced than in plumage, mainly involving wing-length and depth and width of bill at base. Size fairly large in *C. c. coccothraustes* from northern and central Europe south to Pyrénées, northern Italy, northern Yugoslavia, northern Rumania, and northern Ukraine; further south in Europe, birds slightly smaller in Iberia, Corsica, Sardinia, southern Italy, southern Balkan countries, Greece, Turkey, Crimea, and Caucasus area to northern Iran. Birds from Balkans eastward named *nigricans*, but differences from nominate race very slight.

Hawfinch *Coccothraustes coccothraustes*. *C. c. coccothraustes*: **1** ad ♂, **2** ad ♀, **3** juv. *C. c. nigricans*: **4** ad ♂. *C. c. buvryi*: **5** ad ♂. Evening Grosbeak *Hesperiphona vespertina* (p. 1613): **6** ad ♂, **7** ad ♀, **8** 1st winter ♂.

Evening Grosbeak *Hesperiphona vespertina*

PLATES: pages 1613, 1614

Du. Avonddikbek Fr. Grosbec errant Ge. Abendkernbeißer It. Frosono vespertino
Ru. Вечерний американский дубонос Sp. Picogordo vespertino Sw. Aftonstenknäck

Field characters. 17.5 cm; wing-span 32.5–34.5 cm. Close in size to Hawfinch but with proportionately rather smaller bill and head. Large, heavy, strong-billed, and short-tailed finch, with form somewhat recalling Hawfinch but in west Palearctic unique colours of pale bill, dusky and pale yellow head and body, and white-patched black wings and tail. ♂ has distinctive yellow forecrown, with white on flight-feathers restricted to bold patch on inner greater coverts, inner secondaries and tertials; ♀ lacks yellow on head and has greenish body with additional white panel over bases of primaries and bold white tips on central tail-feathers.

Unmistakable. In west Palearctic, only Hawfinch bears passing resemblance in shape, but differs distinctly in brown-buff body plumage and wholly dark tertials, with white panel on wing brightest towards carpal joint. Flight rather heavy, but swift and strong, with bursts of wing-beats producing bounding undulations; flies high above trees when moving between feeding areas. Commonest call a loud, ringing 'cleer', 'cleep', 'clee-ip', or 'chreep', recalling House Sparrow but louder and more strident.

Habitat. Breeds in boreal Nearctic coniferous forest, usually tall and mature, but also in mixed forest, including second growth, and at times in willows growing beside rivers, or in town gardens and shade trees. In winter, strongly attracted to bird-tables with sunflower seeds. Accessible salt is also an attraction. Recent changes have involved major eastward extension of range and more frequent wintering in and around human settlements, including parks and gardens.

Distribution. Breeds in North America, from south-west and north-central British Columbia east to Nova Scotia, south through Rocky Mountains to southern Mexico, and east of the mountains south to central Minnesota, southern Ontario, northern New York, and Massachusetts.

Accidental. Britain: ♂ St. Kilda (Outer Hebrides), March 1969; ♀ Nethybridge (Highland), March 1980. Norway: ♂ Østfold, May 1973; ♂ Sør-Trøndelag May 1975.

Movements. Erratic, nomadic, or inconsistently migratory. May arrive and depart regularly for several years, then not appear at all for a year or more. During range expansion to eastern North America, 1920–50, distribution and movements correlated well with widespread planting of box elder *Acer negundo*, on whose fruits it feeds in winter. Since 1950, summer occurrence correlates well with epidemics of spruce budworm, on whose larvae it feeds.

Wing-length: ♂ 111–118, ♀ 110–114 mm.

Bullfinch *Pyrrhula pyrrhula* (p. 1606). *P. p. pyrrhula*: **1** ad ♂ summer, **2** ad ♀ summer. *P. p. pileata*: **3** ad ♂. Hawfinch *Coccothraustes coccothraustes coccothraustes* (p. 1610): **4** ad ♂, **5** ad ♀. Evening Grosbeak *Hesperiphona vespertina* (p. 1613): **6** ad ♂, **7** ad ♀.

New World Wood-warblers Family Parulidae

Small 9-primaried oscine passerines (suborder Passeres); though unrelated to Old World warblers (Sylviidae), whose niche many of them occupy in New World, parulids as a group also known just as 'warblers' in North America—though a few species bear other names (e.g. 'chat', 'redstart'). Highly active; often arboreal though some (including Ovenbird and Northern Waterthrush) are terrestrial. Feed mainly on insects, gleaned from foliage and (in a number of species) by aerial flycatching; fruit (berries), seeds, and nectar also taken by some species. Usually solitary but some form flocks outside breeding season or on migration. Except as vagrants, occur only in New World (from Alaska to southern South America); northerly species migratory. 126 species in 29 genera of which 19 species of 8 genera accidental in west Palearctic.

Sexes virtually identical in size. Bill straight, slender, and pointed in many parulids (with inconspicuous rictal bristles) but wider and flatter (with well-developed rictal bristles) in more specialized flycatching ones. Wing often fairly long and pointed in migratory species, shorter and rounder in non-migratory; 9 primaries (p10 minute and hidden). Tail of medium size or fairly long; often slightly rounded. Leg and foot rather short and slender.

Plumages typically yellow, olive, or blue-grey, with patches or streaks of contrasting yellow, orange, red, black, or white, especially on head, rump, and chest; wing often with contrasting white, yellow, or red bars; tail often with white, yellow, or red spots. ♂ brighter than ♀ in most North American species; seasonal difference sometimes marked.

Black-and-white Warbler *Mniotilta varia*

PLATE: page 1616

Du. Bonte Zanger   Fr. Paruline noir et blanc   Ge. Kletterwaldsänger   It. Parula bianca e nera
Ru. Пегий певун   Sp. Reinita trepadora   Sw. Svartvit skogssångare

Field characters. 11.5–13 cm; wing-span 20.5–22.5 cm. Slightly smaller than Blackpoll Warbler, with subtly different structure: rather longer bill with noticeably sharp culmen, shorter square tail, and long toes. Rather small but lithe, bark-creeping Nearctic warbler, with black upperparts striped white, and white underparts streaked black; white central crown-stripe diagnostic.

Unmistakable among known transatlantic vagrants. Breeding ♂ Blackpoll Warbler somewhat similar but with wholly black crown lacking white central stripe, unstriped mantle, and bright brown legs. Black-and-white Warbler has unique habit among Parulidae of persistently creeping up and down tree trunks and along branches, with actions recalling both nuthatch and treecreeper. Often confiding. Calls include quite loud, hard 'chick' or 'tik', weak thin 'tsip' or 'tzit', and hiss in alarm.

Habitat. Breeds across cool to warm temperate Nearctic lowlands, on hillsides or ravines in all woodland types from mature deciduous or mixed stands to (more locally) northern conifer forests and also second growth. Forages on main branches or trunks of trees, rather than in foliage.

Distribution. Breeds in North America east of Rocky Mountains, from north-east British Columbia and south-west Mackenzie east to Newfoundland, south to eastern Texas and central Alabama.

Accidental. Iceland, Faeroes, Britain, Ireland.

Movements. Migratory. Winters from northern Mexico and extreme south and south-east of USA south through Central America and West Indies to north-west South America (Venezuela to northern Peru). Probably the most widely common parulid warbler across its winter range. Southward movement extends from Rockies to Atlantic, and even to Bermuda, continuing south through Mexico and Central America as well as across Gulf of Mexico and through Florida. Spring migration is reverse of that in autumn, including both trans- and circum-Gulf movements, on broad front.

Rare autumn vagrant to Atlantic seaboard of west Palearctic, mainly Britain and Ireland where 11 records up to 1995, majority September–October, but noted also March and December.

Wing-length: ♂ 68–79, ♀ 66–72 mm.

Golden-winged Warbler *Vermivora chrysoptera*

PLATE: page 1616

Du. Geelvleugelzanger   Fr. Paruline à ailes dorées   Ge. Goldflügel-Waldsänger   It. Vermivora alidorate
Ru. Златокрылый пеночковый певун   Sp. Reinita alidorada   Sw. Guldvingad skogssångare

Field characters. 11.5–12 cm; wing-span 19–20.5 cm. Noticeably smaller than Blackpoll Warbler, with wings and tail proportionately 5–10% shorter. Rather small, quite tit-like Nearctic warbler, with mainly blue-grey upperparts strikingly

Black-and-white Warbler *Mniotilta varia* (p. 1615): **1** ad ♂ breeding, **2** ad ♂ non-breeding, **3** ad ♀ non-breeding, **4** 1st winter ♂, **5** 1st winter ♀. Golden-winged Warbler *Vermivora chrysoptera* (p. 1615): **6** ad ♂, **7** ad ♀. Tennessee Warbler *Vermivora peregrina*: **8** ad ♂ summer, **9** ad ♂ winter, **10** 1st winter ♂.

marked by yellow forecrown and wing-blaze; bold black face-mask and bib; white panels on outer tail-feathers.

Unmistakable. Behaviour recalls tit; will forage in outer foliage, hanging downwards to look for invertebrates. Call a short 'chip', not noted as distinctive.

Habitat. Breeds in eastern Nearctic lowland woods of temperate zone. Prefers edges of mature deciduous woods of tall oaks, and maples shading meadows grown up with briars, tall herbage, bushes and grass, ideally by clear flowing stream. Sometimes breeds in overgrown clear-felled woodland patches or abandoned pastures in moist places. In South American winter quarters occupies equivalent habitats but favours higher altitudes in rain and cloud forests at 1000–3000 m.

Distribution. Breeds in eastern half of North America from south-east Manitoba and southern Ontario south to eastern Tennessee, northernmost Georgia, and south-east Pennsylvania.

Accidental. Britain: ♂, Kent, January–April 1989.

Movements. Migratory. Winters in Central and South America, from Guatemala and Honduras south to central Colombia and northern Venezuela. Autumn migration is mostly east of Great Plains and west of Appalachians, continuing across Gulf of Mexico; rare in Mexico and West Indies. Spring route is reverse of autumn, but includes more movement through eastern Mexico.

Wing-length: ♂ 61–68, ♀ 58–63 mm.

Tennessee Warbler *Vermivora peregrina*

PLATE: page 1616

Du. Tennessee-zanger   Fr. Paruline obscure   Ge. Brauenwaldsänger   It. Vermivora del Tennessee
Ru. Зеленый пеночковый певун   Sp. Reinita peregrina   Sw. Tennesseesångare

Field characters. 11–12 cm; wing-span 18.5–20.5 cm. Noticeably smaller than Blackpoll Warbler, with proportionately 5% shorter wings and 10% shorter tail; bill distinctive, rather short with sharply pointed tip. Rather small, usually dumpy but sharp-billed Nearctic wood warbler, with adult plumage suggesting small vireo and 1st-winter inviting confusion with other yellow Parulidae and *Phylloscopus* warblers. Yellowish lime-green above, with bright white vent and under tail-coverts. Adult ♂ has head dusky to bluish-grey, often sullied with green when fresh; rather short supercilium, lower eye-crescent, and throat almost white. ♀ as ♂, but grey of head less strong (more greenish) and not spreading to hindneck, while supercilium, rear face, and flank all more heavily marked yellow or buff. 1st-winter brighter and more yellowish above, and yellow below except for narrow whitish vent and under tail-coverts. At close range, shows yellow supercilium, bright whitish to yellow lower eye-crescent, obscure green eye-stripe, long diffuse yellowish bar across tips of greater coverts, indistinct dull yellowish bar on tips of median coverts, and (when fresh) narrow, almost white tips to primaries.

1st-winter traditionally confused with Arctic Warbler (much larger, with strong bill with bright flesh lower mandible, long rakish supercilium, dark eye-stripe, whitish wing-bars and underparts, and straw legs), but can also be confused with Philadelphia Vireo (stubby bill, grey always restricted to crown, no distinct wing-bars, more yellowish vent and under tail-coverts, blue-grey legs), Yellow Warbler (less pointed bill, no supercilium, wing-bar obvious only on inner greater coverts, deep yellow underparts), and other *Phylloscopus* warblers, particularly yellower, wing-barred forms of Chiffchaff and Greenish Warbler (neither, however, is as yellow below, or has similar bill form and colour, or bright rump). Has habit of moving body from side to side while keeping head still. Call a penetrating 'zit-zit' recalling Firecrest or 'zi' or 'zi-zi' suggesting tit.

Habitat. Breeds in cooler northern temperate Nearctic lowlands, in grassy or boggy woodland or forest openings and clearings, mixed with dense brush and clumps of young second-growth trees; needs *Sphagnum* moss for nesting; probably increased in modern times with spread of forest clearance. Migrants occur wherever there are trees, in gardens, orchards, cemeteries, parks, villages, and city suburbs. Wintering birds in Venezuela occur in deciduous, rain, and cloud forests and second growth up to 2200 m in mountains.

Distribution. Breeds throughout much of central and southern Canada, and in northern USA from northern Minnesota east to Maine.

Accidental. Iceland: specimen (Icelandic Museum of Natural History). Faeroes: September 1984. Britain: 2 different 1st-winter birds, Fair Isle, September 1975; 1st-winter, Orkney, September 1982; St Kilda (Western Isles) September 1995.

Movements. Migratory, wintering far south of breeding range, from southern Mexico south to central Colombia and northern Venezuela. Numbers at passage sites vary more than in other Parulidae, reflecting population changes following outbreaks of spruce budworm. Most birds move south between prairies and Appalachians, with few on east coast. Movement across Gulf of Mexico further west in spring than autumn, with more birds on western Gulf coast, and fewer in south-east states. Vagrant north to Hudson and Ungava Bays (northern Canada), and to southern Greenland.

Wing-length: ♂ 66–69, ♀ 61–65 mm.

Northern Parula *Parula americana*

PLATE: page 1618

Du. Blauwe Zanger   Fr. Paruline à collier   Ge. Meisenwaldsänger   It. Parula americana
Ru. Белоглазая парула   Sp. Reinita norteña   Sw. Messångare

Field characters. 10.5–11.5 cm; wing-span 17.5–18 cm. 20% smaller than Blackpoll Warbler with structure recalling goldcrest *Regulus*. Small, short-tailed, arboreal Nearctic warbler. Remarkably pretty: bluish head and upperparts with yellowish-green patch on mantle, double white wing-bar, incomplete white eye-ring, and bright yellow underparts.

Unmistakable. No other vagrant parulid is as small or as multicoloured. Tropical Parula *P. pitiayumi* (no eye-crescents or breast-bands) is very unlikely to stray; Canada Warbler much larger, with uniform blue-grey upperparts, no rufous breast-band, and entirely yellow underparts. Actions and behaviour include hovering and clinging upside-down to foliage, recalling both *Regulus* and small tit. Apparently silent as vagrant, but commonest call in Nearctic a sharp 'chick'.

Habitat. Breeds in temperate and warmer regions of eastern Nearctic, most commonly around woodland openings by swamps, ponds, and lakes, but sometimes elsewhere where trees carry either old man's beard lichen *Usnea* in north, or its moss-like counterpart 'Spanish moss' *Tillandsia* in south; infrequent where neither available. Moisture and tall mature trees are further associated requirements; most feeding done in treetops.

Distribution. Breeds in eastern half of North America from south-east Manitoba east to Nova Scotia south to Gulf of Mexico coast; scarce in most areas except in north-east (Maine to Nova Scotia) and south-east (Louisiana to South Carolina).

Accidental. Iceland, Britain, Ireland, France.

Movements. Migratory. Winters from southern Mexico to Costa Rica, and from extreme southern USA (mainly Florida) and Bahamas south throughout West Indies (rare in south). Migration is on broad front across eastern USA towards tropical America, but most birds pass through Florida to winter in West Indies. Rare autumn vagrant to Atlantic seaboard of west Palearctic, where 16 records from Britain and Ireland up to 1995, especially south-west England and southern Ireland, mainly from late September to 3rd week of October. Vagrant also in Greenland.

Wing-length: ♂ 57–65, ♀ 56–61 mm.

Yellow Warbler *Dendroica petechia*

PLATE: page 1618

Du. Gele Zanger   Fr. Paruline jaune   Ge. Goldwaldsänger   It. Dendroica gialla
Ru. Жёлтая древесница   Sp. Reinita amarilla   Sw. Gul skogssångare

Field characters. 11.5–13 cm; wing-span 17.5–20 cm. 15% smaller than Blackpoll Warbler, but with typical *Dendroica* structure. Quite small, sprightly, thicket-loving Nearctic warbler, always looking remarkably yellow and with diagnostic yellow panels on tail. At close range, shows pale eye-ring, double wing-bar, and rump.

Often looks all-yellow at distance, but until yellow marks on tail confirmed, subject to confusion with immature Tennessee Warbler (whitish vent and under tail-coverts) and immatures of *Wilsonia* warblers (clear yellow forehead and supercilium, no wing-bars). Call a rather full, soft 'tsep' or 'chip'.

Habitat. Widespread in Nearctic lowlands down to tropical regions. Around northern treeline, breeds in dense willow and alder thickets, and in Canada in similar thickets on edges of streams, lakes, bogs, and marshes as well as in ornamental garden shrubbery. Avoids both heavy forests and open grasslands lacking shrubs or trees, but is equally at home in both moist and dry habitats, including hedges, roadside thickets, orchards, and farms. In Caribbean breeds commonly in mangroves which are also frequented by wintering birds in South America.

Distribution. Breeds widely in North America from Alaska east to Newfoundland, mainly south to 35°N in central and eastern USA, but in west continuing through Mexico and Central America to coast of north-west South America (east to Venezuela, south to northern Peru) and the Galapagos; also southern Florida and West Indies.

Accidental. Britain: Bardsey Island (Wales), August 1964; Shetland, ♂, November 1990; Orkney, 1st-winter ♂, August 1992. Ireland: October 1995. Madeira: Selvagem Grande, September 1993.

Movements. Northern populations (breeding from Mexico northwards) migratory, wintering mostly from central Mexico south to central Peru and northern Brazil, with a few north to southern USA. Southern populations (breeding southern Florida, West Indies, and Mexico southwards) resident. Migration is broad-front, but eastern birds evidently bypass peninsular Florida and cross Gulf of Mexico further west, occurring east only to western Cuba (stragglers occur annually on east coast of North America, in October). Spring route is reverse of autumn, with birds that winter in South America taking circuitous path north-west into Central America before turning north across Gulf.

Wing-length: *D. p. aestiva* (eastern North America): ♂ 62–67, ♀ 59–65 mm.

Northern Parula *Parula americana* (p. 1617): **1** ad ♂ summer, **2** ad ♂ winter, **3** 1st winter ♂, **4** 1st winter ♀. Yellow Warbler *Dendroica petechia*: **5** ad ♂ breeding, **6** ad ♂ non-breeding, **7** 1st winter ♂, **8** 1st winter ♀.

Chestnut-sided Warbler *Dendroica pensylvanica*: **1** ad ♂ breeding, **2** ad ♀ breeding, **3** ad ♂ non-breeding, **4** 1st winter ♀. Black-throated Blue Warbler *Dendroica caerulescens* (p. 1620): **5** ad ♂, **6** 1st winter ♂, **7** 1st winter ♀. Blackburnian Warbler *Dendroica fusca* (p. 1620): **8** ad ♂ breeding, **9** ad ♂ non-breeding, **10** 1st winter ♀.

Chestnut-sided Warbler *Dendroica pensylvanica*

PLATE: page 1619

Du. Roestflankzanger   Fr. Paruline à flancs marron   Ge. Gelbscheitel-Waldsänger   It. Dendroica fianchicastani
Ru. Желтошапочный лесной певун   Sp. Reinita de costillas castañas   Sw. Brunsidig skogssångare

Field characters. 11.5–13 cm; wing-span 19–20.5 cm. 10% smaller than Blackpoll Warbler. Quite small, delicate, thicket-loving Nearctic warbler. Pale green upperparts with double yellow-white wing-bar; white underparts and white in outer tail. ♂ has back streaked with blackish lines behind almost completely black nape. Face shows black loral patch, eye-stripe, and surround to rear ear-coverts; loral patch also extends down from eye, narrowing and then joining thin, then broad, chestnut stripe from breast along flank. ♀ duller than ♂, with black lines and facial marks only dusky and chestnut stripe on breast and flank restricted or broken up. Appearance of adult much changed in winter, with rather pale green crown and upperparts and unmarked underparts except for chestnut stripe or spots on fore-flank on ♂ and a few ♀♀. Face shows only white eye-ring set off by grey lore and ear-coverts. 1st-winter resembles non-breeding ♀ but has yellower wing-bars.

Unmistakable, with face, breast, and flank pattern of breeding adult diagnostic, and green and white, almost ghostly, appearance of winter adult and 1st-winter equally distinctive. Beware slight chance of confusion with small, wing-barred *Phylloscopus* warblers (which always show prominent supercilium and eye-stripe and plain tail). Spends much time perched in the open, often hunting insects on ground or flycatching in the air. Cocks tail and frequently moves with it raised and wings drooped. Call a rich 'chip'.

Habitat. Breeds in eastern temperate Nearctic lowlands. Originally confined to limited ephemeral areas of early second growth in cleared or burnt woodland, and woodland edges created by streams, swamps, or other natural limits. Massive forest clearance, and widespread abandonment of subsequent farms and pastures, have greatly expanded suitable habitat, leading to much spread and increase.

Distribution. Breeds in North America from Saskatchewan east to Nova Scotia, south mainly to Minnesota, northern Ohio, and Maryland, and through Appalachians to northern Georgia and north-west South Carolina.

Accidental. Britain: single 1st-winter birds, Shetland, September 1985, and Devon, October 1995.

Movements. Long-distance migrant, wintering in Central America from Guatemala south to Panama. Movement through USA on relatively broad front, but scarce on east coast plain. Further south, data suggest birds mostly cross central Gulf of Mexico in autumn (regular east to western Florida), and western Gulf (including coast) in spring. Vagrant north to Middleton Island (southern Alaska) and southern Greenland.

Wing-length: ♂ 62–68, ♀ 58–65 mm.

Black-throated Blue Warbler *Dendroica caerulescens*

PLATE: page 1619

Du. Blauwe Zwartkeelzanger Fr. Paruline bleue Ge. Blaurücken-Waldsänger It. Dendroica blu gola nera
Ru. Синеспинный лесной певун Sp. Bijirita sombría Sw. Blåryggad skogssångare

Field characters. 12–14 cm; wing-span 17.5–19.5 cm. 15–20% smaller than Blackpoll Warbler, with rather more attenuated rear body and tail. Quite small, rather dark, bush-haunting Nearctic warbler; adult has diagnostic white patch at base of primaries. ♂ dark blue above, with black face, breast, and flank contrasting with otherwise white underparts; ♀ dusky-olive above and buff-yellow below, with narrow whitish supercilium and broken eye-ring. ♀ and immature have noticeable dark cheek.

Adult ♂ unmistakable—as are ♀ and immature with white primary patch, but confusing in other plumages and best distinguished by noticeably dark cheek and sharp, narrow whitish supercilium and eye-crescent. Commonest call soft, full 'tsep' or 'smack', recalling Dark-eyed Junco.

Habitat. Breeds in cool temperate eastern Nearctic lowlands and uplands, favouring hillsides and ridges carrying dense shady second-growth deciduous or mixed woodland, with dense undergrowth, especially laurel thickets. Often breeds near openings created by streams or roads and is active in treetops.

Distribution. Breeds in eastern North America from southern Ontario and north-east Minnesota east through Great Lakes region to New England and Nova Scotia, and south through Appalachians to northern Georgia.

Accidental. Iceland: adult ♂, Heimaey (Vestmannaeyjar), September 1988 to end of 1992.

Movements. Migrant, even southernmost breeders moving nearly 1500 km to winter chiefly in Bahamas and Greater Antilles, with small numbers on adjacent continental coasts. Migrates in vast numbers along Florida coasts. Vagrant (chiefly autumn) north to Newfoundland and nearby St-Pierre-et-Miquelon, and to Greenland.

Wing-length: ♂ 64–68, ♀ 61–65 mm.

Black-throated Green Warbler *Dendroica virens*

Du. Gekraagde Groene Zanger Fr. Paruline à gorge noire Ge. Grünwaldsänger It. Dendroica verdastra
Ru. Зеленый лесной певун Sp. Reinita papinegra Sw. Grön skogssångare

A North American species, breeding from east-central British Columbia east to Newfoundland, south to Minnesota, Great Lakes region, and Connecticut and along Appalachians to northern Alabama; also on coastal plain of Virginia and Carolinas. Winters mainly in eastern Mexico and Central America, with small numbers in West Indies; rare and irregular further north and south. Recorded in Germany (adult ♂, Helgoland, November 1858) and Iceland (immature ♀ found freshly dead on board ship in Reykjavík harbour, September 1984).

Blackburnian Warbler *Dendroica fusca*

PLATE: page 1619

Du. Sparrezanger Fr. Paruline à gorge orangée Ge. Fichtenwaldsänger It. Dendroica di Blackburn
Ru. Еловый лесной певун Sp. Silvia de Blackburn Sw. Orangestrupig skogssångare

Field characters. 11–13.5 cm; wing-span 19–21 cm. 10–15% smaller than Blackpoll Warbler. Quite small, notably arboreal Nearctic warbler, with dark ear-coverts, pale braces on back, yellow foreparts, and mainly white outer tail-feathers. Breeding ♂ has rich orange throat and breast and white panel on mid-wing; ♀ has foreparts yellow and 2 white wing-bars; immature like dull ♀.

Adult unmistakable, with unique pattern on head, back,

and foreparts, but immature troublesome, since most striking character of double white wing-bar shared by 4 known transatlantic vagrant Parulidae (including the commonest, Blackpoll Warbler) and 2 potential ones. Best distinguished from Blackpoll Warbler by much darker crown and cheek, pale braces on mantle, and fully yellow-buff fore-underparts; from Black-throated Green Warbler by streaked back, far less dusky wings and tail, and much yellower fore-underparts. Commonest call a rich 'chip.'

Habitat. Strictly arboreal, breeding in cooler temperate eastern Nearctic forest regions, mature stands of spruce, hemlock, and pine being the primary breeding habitat, although deciduous and mixed second growth are also used. Wintering birds in South American tropics mainly inhabit rain and cloud forests at 800–3100 m.

Distribution. Breeds in North America from central Alberta east to south-west Newfoundland and Nova Scotia, south to Great Lakes region and Connecticut, and in Appalachians to northern Georgia.

Accidental. Iceland: 1st-winter ♀ found exhausted on trawler, *c.* 65 km north-east of Horn (north-west Iceland), autumn 1987. Britain: Skomer (Dyfed), October 1961; 1st-winter ♂, Fair Isle, October 1988.

Movements. Long-distance migrant, wintering in Central and South America from Costa Rica south to Venezuela and central Peru. Ecologically restricted in summer and only locally common, but in main wintering area, Colombia, occurs in various forest strata and habitats and is most abundant of Parulidae. Migration mainly along or west of Appalachians, with fewer on Atlantic coast and in south-east states; most birds then cross (rather than fly round) Gulf of Mexico to Central America; spring route is further west than autumn, with yet fewer in south-east USA, and more on western Gulf coast and in Texas.

Wing-length: ♂ 66–73, ♀ 63–70 mm.

Cape May Warbler *Dendroica tigrina*

PLATE: page 1621

Du. Tijgerzanger   Fr. Paruline tigrée   Ge. Tigerwaldsänger   It. Dendroica di Capo May
Ru. Тигровый лесной певун   Sp. Reinita atigrada   Sw. Brunkindad skogssångare

Field characters. 11.5–13 cm; wing-span 19–21 cm. About 10% smaller than Blackpoll Warbler. Quite small, conifer-loving Nearctic wood warbler, with most variable plumage of all transatlantic vagrants and thus requiring close observation. Shares pale wing-bars, yellowish rump, and pale tail-spots with several congeners, but underparts more heavily and uniformly streaked than any. Only breeding ♂ distinctive, with chestnut ear-coverts, bold white wing-panel, and heavily streaked body.

Cape May Warbler *Dendroica tigrina*: **1** ad ♂ breeding, **2** ad ♂ non-breeding, **3** ad ♀ breeding, **4** 1st winter ♂, **5** 1st winter ♀.
Magnolia Warbler *Dendroica magnolia* (p. 1622): **6** ad ♂ breeding, **7** ad ♂ non-breeding, **8** ad ♀ breeding, **9** 1st winter ♀.

♀ and immature rather dull; combination of streaked body with (usually) pale spot behind ear-coverts provides best clue.

Breeding ♂ unmistakable. In all other plumages, important to note combination of pale half-collar (often reduced to yellowish to pale buff spot or patch behind ear-coverts) and uniformly streaked underparts. Immature confusing, suggesting Yellow-rumped Warbler (pure yellow rump and usually shoulder patch, blue fringes to tail and wing-feathers, and much browner upperparts), Blackpoll Warbler (dark rump, indistinct streaks on underparts, and bright yellowish to brown legs and feet), Palm Warbler (dark rump, yellow under tail-coverts, wags tail), and Pine Warbler (dark rump, unstreaked back, only light streaks below, and much bolder double wing-bar). Call rather hard, thin 'tsip'.

Habitat. Breeds in cool temperate forested lowlands of eastern Nearctic, especially where tall spruce and other conifers form open parklike stands, sometimes with patches of birch; also in mixed woods. Occurs on migration in various kinds of woods and thickets, also in trees and shrubbery near dwellings and along village streets; sometimes in orchards, thickets, and briar patches.

Distribution. Breeds in North America from southern Mackenzie and easternmost British Columbia east to Nova Scotia, south to North Dakota, northern Wisconsin, northern New York, and Maine.

Accidental. Britain: singing ♂, Paisley (Strathclyde), June 1977.

Movements. Migratory. Winter range much smaller than breeding range: in West Indies (chiefly Bahamas and Greater Antilles), with a few in southern Florida; casual in eastern Central America. Main movement between Appalachian Mountains and Mississippi River, despite occasional coastal concentrations. Many turn east farther south, to pass through Florida to West Indies. Autumn migration long drawn-out, with birds frequently lingering into November in most eastern states, and occasionally into December. Spring route is reverse of autumn, but some birds apparently fly from West Indies directly to Alabama and north-west Florida, passing by or over peninsular Florida.

Wing-length: ♂ 66–71, ♀ 63–68 mm.

Magnolia Warbler *Dendroica magnolia*

PLATE: page 1621

Du. Magnolia-zanger   Fr. Paruline à tête cendrée   Ge. Magnolienwaldsänger   It. Dendroica delle magnolie
Ru. Магнолиевый лесной певун   Sp. Reinita cejiblanca   Sw. Magnoliaskogssångare

Field characters. 11–13 cm; wing-span 17–19.5 cm. About 10% smaller than Blackpoll Warbler but with proportionately longer tail (further emphasized by rump and tail pattern). Rather attenuated, highly decorated Nearctic wood warbler, all plumages showing broad, centrally divided white band across tail (diagnostic) and yellow throat and rump. Breeding ♂ mainly black above, with white rear supercilium below grey crown and white panel across coverts, and yellow below, with strong black streaks from breast to flanks. Winter ♂ and ♀ duller, with ♀ showing only double white wing-bar and narrower body streaks. Immature shows striking pale spectacle and greyish band across breast.

No other similar passerine has white band midway along tail. Flight light and dancing, with tail appearing to trail at times. Calls include rather hard, high-pitched 'dzip' or 'tlep' and distinctive disyllabic 'chip chip' or 'tizic'.

Habitat. Breeds in cool temperate eastern Nearctic, mostly in young or low conifer woods or open mixed woods and edges. Migrants forage on trees in orchards and villages.

Distribution. Breeds in North America from south-west Mackenzie and central British Columbia east to Newfoundland, south to north-east Minnesota, central Michigan, Massachusetts, and in Appalachians to West Virginia.

Accidental. Iceland: Rangarvalla Sýsla, September–December 1995; Gullbringu Sýsla, October 1995. Britain: Isles of Scilly, September 1981.

Movements. Migratory. Winter range much smaller than breeding range: from eastern Mexico south to Panama, and (fewer) in Greater Antilles. Autumn migration is east of Rockies, mostly along or west of Appalachians; most birds then cross middle of Gulf of Mexico. Route in spring extends further west than autumn, following coast of Mexico into Texas as well as crossing Gulf further east.

Wing-length: ♂ 56–65, ♀ 56–61 mm.

Yellow-rumped Warbler *Dendroica coronata*

PLATE: page 1624

Du. Geelstuitzanger Fr. Paruline à croupion jaune Ge. Kronwaldsänger It. Dendroica coronata
Ru. Миртовый певун Sp. Reinita coronada Sw. Gulgumpad skogssångare

Field characters. 12.5–15 cm; wing-span 21–23.5 cm. Averages slightly larger than Blackpoll Warbler, with rather short bill but 10–15% longer tail giving lengthy silhouette. Medium-sized, relatively robust but graceful Nearctic wood-warbler. In all plumages, diagnostic combination of white eye-ring, white throat, double white wing-bar, bright yellow rump, usually yellow blaze by shoulder, and white spots on outer tail-feathers. Fringes of wing- and tail-feathers noticeably bluish; in winter adult and immature, back always noticeably brown and well streaked. Breeding ♂ has sharply etched forepart pattern of streaked grey crown, black face, and white marks around eye, with yellow crown-patch; breeding ♀ duller but shows similar crown-patch, usually lacking in immature.

Unmistakable, with length of tail producing less compact form than most relatives, and plumage patterns complex but distinctive. Flight noticeably light and jerky or undulating; bird often appears to dance in the air when flycatching; when hovering, tail trails noticeably. Often droops or flicks wings. Aggressive towards other passerines, often charging them in flight. Hardy, surviving on berries in winter. Calls include loud 'check' or 'chep', and metallic 'cheep' (all harder and more metallic than other *Dendroica*); also sharp thin 'tsi' and quiet 'prit'.

Habitat. Breeds in cooler northern latitudes of Nearctic, both in lowlands and mountains up to treeline, and even in willow scrub beyond it. Found mainly in coniferous and mixed woodland, especially if open. Migrants occur everywhere, especially in brushy areas, hedgerows, field borders, and weedy tangles, as well as at bird-tables.

Distribution. Breeds in North America from Alaska through much of Canada, extending southward in east to Great Lakes region, New England states, and (in Appalachians) West Virginia, and in west through mountains of western USA to Mexico and Guatemala.

Accidental. Iceland, Britain, Ireland.

Movements. Fully migratory (northern populations), partially migratory, including altitudinal movements, or resident (Mexican population). Winters in western, southern, and south-east USA, north to south-west British Columbia and southern Nova Scotia, south to Panama and West Indies. The only short-range migrant among northern paruline warblers, and by far the most hardy, wintering locally into snow zone where adequate food (especially bayberry) and cover coincide. Migration on broad front, narrowing southwards.

Rare autumn vagrant to Atlantic seaboard of west Palearctic. 22 records from Britain and Ireland up to 1995, especially south-west England and southern Ireland, mainly October, but noted also in September and November, twice in May, and a wintering individual was present in Devon, 4 January–10 February 1955. The records are usually comparatively late in the season, which reflects its rather late migration in North America.

Wing-length: ♂ 71–79, ♀ 70–76 mm.

Palm Warbler *Dendroica palmarum*

Du. Palmzanger Fr. Paruline à couronne rousse Ge. Palmenwaldsänger It. Dendroica delle palme
Ru. Пальмовый лесной певун Sp. Reinita palmera Sw. Brunhättad skogssångare

A North American species, breeding from central Mackenzie east to Newfoundland, south to central Alberta, north-east Minnesota, northern Wisconsin, and Maine. Winters from South Carolina (rare further north) south along Atlantic and Gulf of Mexico coasts to Louisiana, from Yucatán (eastern Mexico) south along coast to north-east Nicaragua, and in Greater Antilles and Bahamas. Adult ♂ found dead on tideline, Cumbria (England), May 1976.

Blackpoll Warbler *Dendroica striata*

PLATE: page 1624

Du. Zwartkopzanger Fr. Paruline rayée Ge. Streifenwaldsänger It. Dendroica di Blackpoll
Ru. Пестрогрудый лесной певун Sp. Reinita listada Sw. Vitkindad skogssångare

Field characters. 12.5–14.5 cm; wing-span 20.5–23 cm. Size between Willow Warbler and Wood Warbler, with proportionately slightly shorter wings but longer tail; somewhat larger than most other vagrant Parulidae but shorter than Yellow-rumped Warbler. Medium-sized, quite robust but graceful, usually arboreal Nearctic wood warbler; commonest

Yellow-rumped Warbler *Dendroica coronata* (p. 1623): **1** ad ♂ breeding, **2** ad ♂ non-breeding, **3** ad ♀ breeding, **4** 1st winter ♀. Blackpoll Warbler *Dendroica striata*: **5** ad ♂ breeding, **6** ad ♂ non-breeding, **7** ad ♀ breeding, **8** 1st winter ♀. Bay-breasted Warbler *Dendroica castanea* (for comparison): **9** 1st winter ♀.

passerine vagrant from Nearctic and hence epitome of genus in west Palearctic. In all plumages, shows double white wing-bar, white tail-spots, and diagnostic yellow feet. Breeding ♂ has striking black cap, white cheeks and black malar stripe and flank spots; breeding ♀ lacks black and white contrasts, showing streaked greenish crown and ear-coverts. Immature resembles ♀ but most buffier above face and below, where less strongly streaked.

Breeding ♂ unmistakable, recalling only wholly pied Black-and-white Warbler but easily separated by wholly black cap. ♀ and immature have less distinctive appearance, with double white wing-bar shared by 9 other known or potential vagrant relatives. Nevertheless, in autumn, plumage ground-colour of Blackpoll Warbler greener above and yellower below than most congeners, with brown to yellow legs and uniquely yellow feet, at least on soles. In Nearctic, Bay-breasted Warbler *D. castanea* presents pitfall: plumage pattern of non-breeding adult and immature similar but, unlike Blackpoll Warbler, shows little or no supercilium, bluish fringes to flight-feathers often unstreaked, wholly pale buff underbody (frequently washed chestnut on flanks), and dusky legs and feet. Another confusion species that has not yet crossed North Atlantic is Pine Warbler *D. pinus*, only differing constantly in more pronounced supercilium and eye-ring, unstreaked back, bolder wing-bars, and dusky legs and feet. Characteristically feeds in outer branches and foliage of trees, picking insects from underside of leaves, but transatlantic vagrants have fed in bush and ground cover. Feeding actions rather deliberate, sometimes suggesting *Hippolais* warbler. Calls include loud 'smack'; also rather hard 'tsip', similar to one call of Cape May Warbler.

Habitat. Breeds in northern and north-east Nearctic to near treeline, in mountains as well as lowlands, in coniferous woods, especially spruce, frequently stunted. Also inhabits mixed-wood edges, logged and burned areas, and alder thickets, favouring moist ground. While migrating may be found wherever trees grow, and often also along fences and stone walls in fields and pasture; in autumn even along weedy roadsides. On spring migration, common in orchards and shade trees. Winters in South America in deciduous, rain, and cloud forests up to 3000 m, and in lowlands in gallery forest, second growth, grassy fields with scattered vegetation, and coastal mangroves, foraging also on ground.

Distribution. Breeds in Alaska and across coniferous forests of northern and central Canada southward in east to eastern New York and Massachusetts.

Accidental. Iceland, Britain, Channel Islands, Ireland, France (Ouessant).

Movements. Migrant, with longest average migration among paruline warblers. Winters in north-west South America, south to Guianas, north-west Brazil, and western Bolivia (mostly east of Andes). There is a long-standing dispute whether autumn route continues mainly south across ocean from New England and eastern Canada, or mainly through south-east states and Lesser Antilles; both routes may be important. Route in spring is more westerly than in autumn, passing through southern states on broad front west to Louisiana and eastern Texas, with highest numbers in Florida; direct over-water movements between South America and New England (as in autumn) do not occur.

Rare autumn vagrant to Atlantic seaboard of west Palearctic. 34 records from Britain and Ireland up to 1995, especially south-west England and southern Ireland, mainly from late September to October.

Wing-length: ♂ 73–79, ♀ 70–76 mm.

American Redstart *Setophaga ruticilla*

PLATE: page 1626

Du. Amerikaanse Roodstaart   Fr. Paruline flamboyante   Ge. Schnäpperwaldsänger   It. Codirosso americano
Ru. Американская горихвостка   Sp. Candelita   Sw. Amerikansk rödstjärt

Field characters. 13–14.5 cm; wing-span 19.5–22 cm. About 10% smaller than Red-breasted Flycatcher which it recalls in plumage pattern and behaviour as much as any close relative. Quite small but long-tailed, elegant, extremely active Nearctic wood warbler, with bright orange (♂) to yellow (♀, immature) patches at shoulder, along bases of flight-feathers, and on bases of outer tail. ♂ otherwise black, with white belly; ♀ and immature otherwise greyish-green, with white spectacle and underparts. Flicks wings and spreads tail constantly.

Unmistakable; Red-breasted Flycatcher lacks wing-panel and has white tail-sides. Flight light and aerobatic, allowing bird to flit like butterfly through and round foliage and to dash out from cover in accomplished flycatching. Stance level with tail often raised above body line. Restless and excitable, constantly drooping and flicking wings and raising and fanning tail. Happy with trees of medium height, their understorey and scrub, when feeding behaviour may suggest *Sylvia* warbler. Tends to feed lower in trees and in more open areas than other Nearctic warblers. Calls include rather thin, clear 'tzit' or 'tseet' and rather hard, clicking 'tsip'.

Habitat. Breeds in cool and warm temperate zones of Nearctic, mainly in deciduous woodlands having openings or swampy places, but also in open second growth on moist lowlands. Also lives on borders of pastures, in orchards, and even among shade trees and garden shrubbery, sometimes near dwellings. Winters in South America in coastal mangroves, suburban areas, thorny thickets and forests to savannas, up to 3000 m.

Distribution. Breeds in North America from south-east Alaska east across western and southern Canada to Newfoundland, south to Utah in west (absent from much of Great Plains area) and to Gulf of Mexico coast further east.

Accidental. Iceland, Britain, Ireland, France, Azores, Madeira (Selvagem Grande).

Movements. Migrant; southernmost breeding areas only 600–800 km from northernmost wintering grounds, although most birds move further. Winters in West Indies and from Mexico south to northern Peru, northern Brazil, and Guianas. Migration in both seasons is on broad front, birds crossing Gulf of Mexico as well as skirting it to east and west. In Bermuda, where wintering is regular, very common on passage in autumn, but inconspicuous in spring.

Rare autumn vagrant to Atlantic seaboard of west Palearctic. In Britain and Ireland, 7 records up to 1990, mainly from south-west England and southern Ireland, October–November, but once in December.

Wing-length: ♂ 61–69, ♀ 59–65 mm.

Ovenbird *Seiurus aurocapillus*

PLATE: page 1626

Du. Ovenvogel   Fr. Paruline couronnée   Ge. Pieperwaldsänger   It. Seiuro coronato
Ru. Золотоголовый дроздовый певун   Sp. Reinita montana   Sw. Rödkronad piplärksångare

Field characters. 14–16 cm; wing-span 22–26 cm. Slightly longer than Northern Waterthrush; close in size to smaller pipits but dumpier, looking front-heavy. Quite large, terrestrial Nearctic wood warbler, with thrush-like plumage: crown striped orange in centre and black on sides, bold white eye-ring, olive upperparts, and heavily spotted and streaked underparts.

Unmistakable; lack of supercilium immediately excludes Northern Waterthrush. Flight more reminiscent of chat than smaller Parulidae, with bursts of wing-beats producing dashing progress through cover. Terrestrial, usually seen walking in dense undergrowth searching leaf litter; gait rather mincing, with tail raised above drooping wings and occasionally waved, further emphasizing frequent bobbing of rear body. Shy, hiding when disturbed. Calls include penetrating, loud, clicking 'tzick' and softer 'tseet'.

Habitat. Breeds in temperate Nearctic lowlands, in woodlands, foraging on forest floor. Favours well-drained bottomland, deciduous forest, not too thick with undergrowth. In winter in South America, inhabits deciduous forest, forest edge, and other wooded areas near sea-level.

Distribution. Breeds in North America from north-east British Columbia and southern Mackenzie east to Newfoundland, south to eastern Colorado, eastern Oklahoma, northern Alabama, and South Carolina.

Accidental. Britain: Shetland, October 1973; Devon, freshly

American Redstart *Setophaga ruticilla* (p. 1625): **1** ad ♂, **2** ad ♀, **3** 1st winter ♂, **4** 1st winter ♀. Ovenbird *Seiurus aurocapillus*: **5** 1st winter. Northern Waterthrush *Seiurus noveboracensis*: **6** 1st winter. Louisiana Waterthrush *Seiurus motacilla* (for comparison): **7** 1st winter.

dead, October 1985; Merseyside, tideline wing found, January 1969. Ireland: Lough Carra Forest (Mayo), freshly dead, December 1977; Dursey Island (Cork), September 1990.

Movements. Migrant. Breeding and winter ranges approach within 200 km in south-east USA, but most birds migrate over 1000 km. Winters mainly Florida, West Indies, and from northern Mexico south to Panama. Migration on broad front in both seasons, mainly east of Rockies, and from eastern Mexico to southern Florida.

Wing-length: ♂ 75–81, ♀ 71–77 mm.

Northern Waterthrush *Seiurus noveboracensis*

PLATE: page 1626

Du. Noordse Waterlijster   Fr. Paruline des ruisseaux   Ge. Drosselwaldsänger   It. Seiuro del nord
Ru. Речной певун   Sp. Reinita charquera   Sw. Nordlig piplärksångare

Field characters. 12.5–15.5 cm; wing-span 21–25 cm. Size close to Meadow Pipit but with plumper, proportionately shorter-tailed form. Medium-sized, plump but sleek Nearctic wood warbler adapted to ground-feeding on moist ground, where its horizontal posture, walking gait, and teetering of body and tail recall Common Sandpiper. Plumage markedly pipit-like but upperparts unstreaked; long narrow yellowish supercilium distinctive.

Genus unmistakable and separation from Ovenbird simple, as that species lacks supercilium and has much lighter, warmer olive upperparts and orange crown-stripe. Distinction from only slightly larger Louisiana Waterthrush *S. motacilla* (not yet recorded in west Palearctic) more difficult, and any vagrant waterthrush should be observed as closely as possible. Crucial to separation of the 2 species are bill size in relation to head (large in Louisiana; in proportion in Northern), shape of supercilium (bulging behind eye in Louisiana; uniformly narrow or slightly wider before eye in Northern), colour of rear supercilium (white in Louisiana; uniformly yellowish or buff when fresh in Northern), throat markings (diffuse in Louisiana; sharp, often forming gorget in Northern), and ground-colour of central flanks (strongly buff in Louisiana; lemon-yellow in Northern). Call distinctive: far-carrying, lengthy monosyllable with metallic, even explosive, quality, 'chwit', 'tsink', or 'peent'.

Habitat. Breeds in cooler temperate regions of Nearctic, up to northern treeline; in woodland, always near water, usually along streams, also by ponds and lakes. Never seen far from water except on migration, when it sometimes visits gardens, trees, and shrubbery around buildings and groves of trees. Wades in shallow waters like sandpiper, but is also at home among treetops. In winter in South America, stays always near water, by streams and lagoons or in mangroves and swamps, ascending to 2000 m.

Distribution. Breeds in North America from western Alaska and British Columbia east to Labrador and Newfoundland, south in western USA to northern Idaho and western Montana, and in east to Great Lakes region, West Virginia, and New England states.

Accidental. Britain, Channel Islands, Ireland, France, Canary Islands (this species or Louisiana Waterthrush).

Movements. Long-distance migrant, all birds moving over 1000 km (many much further) from breeding to wintering areas. Winters from Mexico south through Central America to Ecuador, northern Peru, northernmost Brazil, and the Guianas, and from southern Florida south throughout West Indies. Migration on broad front, mainly east of Rockies. In Bermuda, where winters regularly, common on passage in autumn, but inconspicuous in spring. Very widespread in winter range, and by far the most commonly reported of Parulidae in West Indies.

Rare autumn vagrant to Atlantic seaboard of west Palearctic. 6 records from Britain and Ireland up to 1995, especially south-west England and southern Ireland, from late August to 3rd week of October. Only spring record is from Jersey (Channel Islands), 17 April 1977.

Wing-length: ♂ 72–82, ♀ 72–78 mm.

Common Yellowthroat *Geothlypis trichas*

PLATE: page 1628

Du. Gewone Maskerzanger   Fr. Paruline masquée   Ge. Weidengelbkehlchen   It. Parula golagialla
Ru. Желтогорлый певун   Sp. Reinita gorgigualda   Sw. Gulhake

Field characters. 11–14 cm; wing-span 16–18 cm. Close in size to Chiffchaff but with plumper form enhanced by frequent raising of tail. Quite small Nearctic wood warbler; perky but skulks in ground cover. Uniform bright yellowish-olive upperparts and yellow and buff underparts. ♂ has diagnostic black mask; ♀ has short dull supercilium and eye-ring. Behaviour recalls both Wren and small *Sylvia* warbler.

♂ unmistakable but ♀ can be confused with ♀ or immature of other yellow parulids, particularly Yellow Warbler (more uniformly coloured, with indistinct wing-bars) and Wilson's Warbler (uniformly lemon-yellow below, with similarly coloured frontal band and rear supercilium). Flight light and flitting, recalling small *Sylvia*. Hops, also shuffles and clambers in dense cover. Stance level, often with tail raised well above body line in manner of Wren. Voice distinctive, with harsh quality recalling Wren, particularly when disturbed. Calls described as husky 'tchep' and 'chip', soft 'trep' and 'tep' as if clicking tongue, and quiet 'tic'.

Habitat. Breeds in temperate and subtropical regions of Nearctic in dense low cover in variety of sites, especially near water and in rank vegetation of marshes, such as cattails and bulrushes, and streamside thickets of willows. Despite preference for vicinity of water, occasionally occupies upland thickets of shrubs and small trees, poorly tended orchards, retired croplands, and weedy residential areas.

Distribution. Breeds from south-east Alaska across western and southern Canada to Newfoundland, and southward throughout USA to Oaxaca and Vera Cruz in Mexico.

Accidental. Britain. Single 1st-winter ♂♂: Lundy (Devon), November 1954; Isles of Scilly, October 1984; Kent, January–April 1989; ♂, Fair Isle, June 1984.

Movements. Varies from fully migratory to resident. In east, northern populations winter furthest south, overflying both short-range migrants and residents. Winters in southernmost states of USA, extending north to California in west and Virginia in east, also through Mexico to Panama, and in West Indies. Migration on broad front, including major movements through Florida and across Mexican Gulf.

Wing-length: Varies in different populations. *G. t. brachidactylus* (eastern USA and Canada; most likely vagrant to west Palearctic): ♂ 55–61, ♀ 52–56 mm.

Hooded Warbler *Wilsonia citrina*

PLATE: page 1628

Du. Kapzanger   Fr. Paruline à capuchon   Ge. Kapuzenwaldsänger   It. Parula del cappuccio
Ru. Капюшонная вильсония   Sp. Reinita encapuchada   Sw. Svarthakad citronsångare

Field characters. 12. 5–14 cm; wing-span 20–21.5 cm. Close in size to Blackpoll Warbler but with proportionately slightly shorter wings and noticeably longer tail; averages 10% larger than Yellow Warbler. Quite large, lengthy, Nearctic wood warbler fond of moist woodland. Bright olive to greenish-brown above and yellow below; ♂ has black hood around yellow face. Often spreads tail to show white panels on outer feathers.

♂ unmistakable. ♀ easily separated from ♀ Wilson's Warbler by yellow face and white tail-panels. Flight rapid, recalling

Common Yellowthroat *Geothlypis trichas brachidactylus*: **1** ad ♂, **2** ad ♀, **3** 1st winter ♂, **4** 1st winter ♀. Hooded Warbler *Wilsonia citrina* (p. 1627): **5** ad ♂, **6** ad ♀, **7** 1st winter ♂, **8** 1st winter ♀.

small *Acrocephalus* warbler. Stance level, with tail often raised and spread. In home range, prefers thickets and forest understorey. Call a loud, somewhat metallic, musical 'chink', 'chip', or 'tsyp'.

Habitat. Breeds in temperate and subtropical zones of eastern Nearctic, mainly in lowland woods or scrub, living in dense lower layers of vegetation, but not often seen on ground. Generally favours moist mature woodland.

Distribution. Breeds in south-east North America from south-east Nebraska and central Iowa east to southernmost Ontario and Rhode Island, south to south-east Texas and northern Florida.

Accidental. Britain: 1st-winter ♀, Isles of Scilly, September 1970; St Kilda (Western Isles), September 1992.

Movements. Short-distance migrant, with breeding and winter ranges separated by only 600–700 km in west (eastern Texas to northern Mexico). Winters from north-east Mexico south to Panama, with small numbers in Caribbean. Movement is across western Gulf of Mexico to eastern and southern Mexico, and only a few appear in peninsular Florida and Bahamas. Early migrants often overshoot or drift offshore, and in many years small numbers appear on islands around Nova Scotia, 600 km or more north-east and downwind from breeding range; many such in April or September.

Wing-length: ♂ 66–72, ♀ 62–68 mm.

Wilson's Warbler *Wilsonia pusilla*

PLATE: page 1629

Du. Wilsons Zanger   Fr. Paruline à calotte noire   Ge. Mönchswaldsänger   It. Parula di Wilson
Ru. Малая вильсония   Sp. Silvia de birrete del norte   Sw. Svartkronad skogssångare

Field characters. 11–12.5 cm; wing-span 15.5–17.5 cm. About 10% smaller than Hooded Warbler with proportionately shorter, finer bill; close in size to Yellow Warbler. Quite small, animated, willow-loving Nearctic wood warbler, with bright olive-green upperparts and lemon-yellow underparts; ♂ has black cap; no white in tail.

♂ unmistakable. ♀ easily distinguished from Hooded Warbler and Yellow Warbler by longer supercilium and more obvious eye-ring, due to increased contrast with more olive ear-coverts, and lack of white in tail, from Yellow Warbler also by lack of wing-bars. ♀ told from Tennessee Warbler by dark crown, less obvious supercilium, and yellow (not white) under tail-coverts. Remarkably active and restless, with light flight and expert flycatching (during which bill snaps audibly); droops and flicks wings, and twitches tail in almost rotary action. Stance level, often holding tail above

Wilson's Warbler *Wilsonia pusilla*: **1** ad ♂, **2** 1st winter ♂, **3** 1st winter ♀. Canada Warbler *Wilsonia canadensis*: **4** ad ♂ breeding, **5** ad ♀ breeding, **6** 1st winter ♂, **7** 1st winter ♀.

wings. Calls include sharp musical 'chip', harsher 'chut', short, lisped 'tsip', loud, rather liquid 'twick' (recalling Cetti's Warbler), and 3-syllable 'kick-kick-kick' (recalling Red-breasted Flycatcher).

Habitat. Breeds almost throughout Nearctic climatic zones, from Arctic tundra and montane valleys and slopes in Alaska, and up to nearly 500 m in southern California. Favours moist open shrubbery, including willow and dwarf birch, especially by streams, ponds, and bogs. On migration, more tolerant of drier situations.

Distribution. Breeds in North America from northern Alaska east to Newfoundland and south throughout most of Canada, extending in western USA to California and New Mexico, in eastern USA to northern Minnesota, northern Vermont, and Maine.

Accidental. Britain: ♂, Rame Head (Cornwall), October 1985.

Movements. Migratory, moving at least 800 km (most birds over 2500 km) between summer and winter ranges. Winter range (Mexico and southern Texas south to Panama) very limited in comparison with breeding range; highest numbers in Mexico. Migration on broad front in both seasons, mainly west of Appalachians, with most birds skirting Gulf of Mexico to west rather than overflying it; uncommon to rare in southeast states and West Indies. A few vagrant records July–September in arctic Canada and Alaska, on islands in Bering Sea, and in Greenland.

Wing-length: ♂ 53–59, ♀ 52–57 mm.

Canada Warbler *Wilsonia canadensis*

PLATE: page 1629

Du. Canadese Zanger   Fr. Paruline du Canada   Ge. Kanadawaldsänger   It. Parula del Canada
Ru. Канадская вильсония   Sp. Silvia canadiense   Sw. Kanadaskogssångare

Field characters. 12.5–14 cm; wing-span 20–22 cm. Close in size to Blackpoll Warbler but more robust, with noticeably heavier bill. Medium-sized, bold, undergrowth-haunting Nearctic wood warbler, with olive-grey to blue-grey upperparts, bright yellow underparts and yellowish-white spectacle in all plumages. ♂ has black forecrown, lore, forecheek, and broad necklace of black spots; ♀ shows shadows of similar marks. Legs pale.

Unmistakable. No other parulid has unmarked grey upperparts, wings, and tail. Calls include subdued 'chip' or 'tschip'; loud 'check' recalls House Sparrow.

Habitat. Breeds in temperate northern and eastern Nearctic deciduous and mixed forest zone, mainly in shrubby undergrowth of mature woodland, or in willows and alders along streams, in swamps, and in other moist places. Winters in South America in rain and cloud forests and second growth, in clearings, and at forest edges up to 2100 m.

Distribution. Breeds from north-central Alberta east through southern Canada to central Quebec and Nova Scotia, south in USA to central Minnesota and southern New England states, and in Appalachians south to Tennessee and north-west Georgia.

Accidental. Iceland: ♂ (specimen), September 1973.

Movements. Migratory throughout range, mostly breeding further north and wintering further south than Hooded Warbler. Winters in South America, from Venezuela and Colombia, mostly east of Andes, to central Peru, with a few in Central America. Migration is mainly in and west of Appalachians, avoiding south-east states, with ongoing route through eastern Mexico and across western Gulf. Vagrant to north Alaskan coast and Greenland.

Wing-length: ♂ 63–69, ♀ 60–66 mm.

Tanagers Family Thraupidae

Small to medium-sized 9-primaried oscine passerines (suborder Passeres). Some chat-like ground-feeders or specialized flycatchers but most species highly arboreal and frugivorous, diet of some also including (to greater or lesser extent) seeds, nectar, and insects; some tropical species follow ant columns. Except as vagrants, occur only in New World (from Canada to Argentina), mainly in tropics; northerly species migratory. About 261 species in *c.* 63 genera, of which 2 species of North American genus *Piranga* accidental in west Palearctic.

Sexes usually almost the same size. Bill variable, from stout seed-eater to insect-eater type, but often of short to medium length, rather conical in shape, and somewhat hooked; upper mandible noticeably decurved with notch near tip; notch well-developed in some species but small or vestigial in most. Wing variable, longish in some migratory species but quite short in most others; 9 primaries (p10 minute and concealed). Tail usually short or of medium length; shape variable but typically square-tipped or rounded. Leg usually quite short. Plumages of many species vividly and boldly coloured in almost rainbow-like manner; feathers sometimes with metallic sheen or opalescence. Sexes alike or ♂ brighter than ♀.

Summer Tanager *Piranga rubra*

PLATE: page 1632

Du. Zomertangare   Fr. Tangara vermillon   Ge. Sommertangare   It. Tanagra estiva
Ru. Алая пиранга   Sp. Candelo unicolor   Sw. Sommartangara

Field characters. 16–17 cm; wing-span 27–30 cm. Close in size to Corn Bunting but with form also recalling oriole *Oriolus*, particularly in long, pointed but swollen-looking bill; averages slightly longer-billed and longer-tailed than Scarlet Tanager. Quite large, tree-haunting passerine, with long, heavy bill, bulky, peaked head, plump oval body, and lengthy wings but relatively rather short tail. Adult ♂ red except for browner wings and tail, recalling ♂ crossbill; adult ♀ yellowish-olive above, strongly yellow below, like brightest ♂ Greenfinch. Juvenile and 1st winter much as ♀ but young ♂ in 1st summer partially red.

Tanagers are essentially insectivorous or frugivorous and lack bustling behaviour of finches. *Piranga* best distinguished by length and shape of pale bill and by sluggish behaviour. To experienced observer, tanager form unmistakable and identification of ♂♂ of the 2 west Palearctic species in full breeding plumage is simple, as Summer Tanager lacks solidly black wings and tail of Scarlet Tanager. ♀♀ and immatures far less easily distinguished, with larger green and yellow warblers (e.g. Icterine Warbler) presenting additional confusion species. Best separated from Scarlet Tanager on (1) larger, paler, untoothed bill, (2) more marked face, (3) more intense body colours, especially below, (4) yellow under wing-coverts (white in Scarlet Tanager), (5) lack of distinct contrast between body colours and those of wings and tail, and (6) sometimes multi-syllabic call. Flight action slower than finch, producing quite powerful but not flowing progress. Carriage usually rather level and front-heavy, with heavy bill and head contributing to somewhat neckless attitude. Calls include disyllabic 'pi-tuck' and distinctive longer phrase 'pik-i-tuck-i-tuck' or 'chicky-tuck-tuck', with rapid, staccato utterance and descending pitch.

Habitat. Breeds in warm temperate Nearctic lowlands, especially in tall open woods with scrubby oak undergrowth, such as drier pine and hickory, feeding and singing in treetops, but also tolerating fairly young second growth. Winters in tropical America in both woody and open situations including coastal mangroves, second growth, low open forest (including edges and clearings), coffee plantations, and scrubby grassland.

Distribution. Breeds in southern and south-central USA from California east to southern New Jersey, south to north-central Mexico.

Accidental. Britain: 1st-winter ♂, Bardsey Island (Wales), September 1957.

Movements. Short-range migrant. Winters from south-central Mexico south through Central America to eastern Peru, northern Bolivia, Amazonian Brazil, and Guianas. Eastern birds cross Gulf of Mexico to Central America, whereas western birds move overland through Mexico. Tends to overshoot eastern range in April, and especially in May, when recorded almost annually from New York to Nova Scotia; autumn vagrancy there is less frequent, but extends further, to St Pierre-et-Miquelon (off Newfoundland).

Wing-length: ♂ 93–99, ♀ 89–95 mm.

Summer Tanager *Piranga rubra*: **1** ad ♂, **2** ad ♀, **3** 1st winter ♂, **4** 1st winter ♀. Scarlet Tanager *Piranga olivacea*: **5** ad ♂ breeding, **6** ad ♂ non-breeding, **7** ad ♀ non-breeding, **8** 1st winter ♂. **9** 1st summer ♂.

Scarlet Tanager *Piranga olivacea*

PLATE: page 1632

Du. Zwartvleugeltangare   Fr. Tangara écarlate   Ge. Scharlachtangare   It. Tanagra scarlatta
Ru. Красно-черная пиранга   Sp. Candelo escarlata   Sw. Rödtangara

Field characters. 15·5–16 cm; wing-span 27–30·5 cm. Slightly smaller than Summer Tanager, with proportionately slighter and 10–15% shorter bill and 5–10% shorter tail. In adult ♂, bright red (breeding) or bright green-yellow (non-breeding) head and body contrast with wholly black wings and tail; adult ♀ and immature have greenish-olive upperparts merging with pale yellow underparts but mainly black or dusky wings and tail stand out; bright white under wing-coverts in all plumages.

Distinction from Summer Tanager easy in ♂♂ but less so in adult and immature ♀, particularly those with dullest wings and tail (see Summer Tanager). Call a low, toneless, hoarse or rasping disyllable, 'chip-burr', 'chip-kurr', or 'keep-back'.

Habitat. Breeds in temperate and warm temperate Nearctic in mature woodlands and groups of tall shade trees, even in suburbs. Prefers oak woods, especially in well-watered country, but will also occupy mixed woods, coppice, and orchards.

Distribution. Breeds in eastern North America from south-east Manitoba and North Dakota east to New Brunswick and Maine, south to eastern Kansas, central Arkansas, southern Appalachians, western North Carolina, and Maryland.

Accidental. Iceland, Britain, Ireland.

Movements. Long-distance migrant, breeding in temperate eastern North America and wintering within *c.* 10° of equator. Winters mainly in north-west South America, from western Colombia south to north-west Bolivia, rarely in Panama. Inconspicuous on migration, remaining in tree-tops except when driven to ground by scarcity of insects in cold weather. Probably most birds cross central Gulf of Mexico in both spring and autumn. Few birds overshoot to north; but occurs annually in Nova Scotia and Newfoundland in both seasons. Rare autumn vagrant to Atlantic seaboard of west Palearctic. In Britain and Ireland, 7 records up to 1995, especially south-west England and southern Ireland, from late September to 3rd week of October.

Wing-length: ♂ 92–102, ♀ 91–98 mm.

Buntings and Allies Family Emberizidae

Small to medium-sized, thick-billed 9-primaried oscine passerines (suborder Passeres). Occur in both New World and Old, with main diversity in former. Like finches (Fringillidae), emberizids are specialist seed-eaters but more conservative in choice of seeds and diet more varied, being often supplemented by insects and other small invertebrates and by fruit. Bill typically strong, hard, and deep; as in Fringillidae, structurally designed internally for shelling seeds (with aid of tongue and strong jaw muscles). Unlike finches, have distinct preference for monocotyledonous seeds (grasses) which they shell by crushing them in bill.

Buntings, New World Sparrows, and Allies
Subfamily Emberizinae

Small and medium-small emberizids variously known as buntings, longspurs, sparrows, juncos, towhees, etc.; occur in many open habitats, including savanna, steppe, alpine tundra, scrub, and desert, but mostly in parkland, fields, hedgerows, cultivation, etc. (especially in northern hemisphere). Often largely terrestrial, picking up seeds (etc.) from ground, though perching freely just above it. As in fringilline finches, insects and other small invertebrates eaten in summer, forming sole food of nestlings. Found in New World and Old but main radiation in former, with 58 species of 17 genera in North and Central America. Some 290 species; 30 in west Palearctic—19 breeding, 11 accidental (8 from North America).

In buntings (with exception of Corn Bunting), sexes almost the same size. Bill typically short, conical, and pointed—with cutting edges incurved. Corn Bunting has bony hump in roof of mouth against which seeds are crushed; this feature also found in many other buntings, but usually less well developed and absent in some. Wing rather long and bluntly pointed in migratory species, shorter and more rounded in sedentary ones; 9 primaries (p10 rudimentary and hidden). Tail rather short to long; slightly rounded, straight, or shallowly forked. Leg rather short or of medium length with strong toes and claws, hind claw long and straight in *Calcarius*. Foot not used for holding food.

Buntings in general not so vocal as cardueline finches, but have equally complex repertoire made up of simpler, less twittering, but quite loud calls. Song of ♂ of typical advertising type: varies from relatively short and simple to loud and musical. Although usually gregarious outside breeding season, feeding in small parties or larger flocks and roosting communally, buntings typically solitary and territorial when nesting. Monogamous mating system the general rule but polygyny occurs in some populations of Corn Bunting and Reed Bunting. Nest built by ♀ only, with ♂ in close attendance when she collects material. Incubation by ♀ only (fed by ♂). Young fed by both sexes, exceptionally by ♀ only. Young typically leave nest at early age, while still quite incapable of flight.

Except in a few species, sexes of buntings differ in plumage, ♂ the brighter bird to greater or lesser extent. Plumages of emberizines as a whole are varied, but many species are streaked with contrasting colours mostly on head, throat, and chest, forming bold pattern; many species (especially in Palearctic) have white spots on outer tail-feathers.

Rufous-sided Towhee *Pipilo erythrophthalmus*

PLATE: page 1634

Du. Roodflanktowie Fr. Tohi à flancs roux Ge. Rötelgrundammer It. Pipilo fianchirossi
Ru. Восточный тауи Sp. Chingolo punteado Sw. Brunsidad busksparv

Field characters. 17–18 cm; wing-span 25–30 cm. Close in size to Corn Bunting but with shorter wings, proportionately longer and more rounded tail, and longer legs. Rather large, robust passerine of dense cover, with striking tricoloured plumage and ample tail, often cocked. Adult ♂ has black hood, back, wings, and tail, contrasting with rufous flanks and vent and white belly, primary-patch, and outer tail; adult ♀ brown where ♂ black. Immature like adult, best distinguished by retained brownish juvenile flight feathers. Eye brown, not reddening until 1st autumn.

Unmistakable, with Spotted Towhee now considered separate species, distinguished by white spots on scapulars and wing coverts, lack of white primary patch and dull black tone of ♀'s and immature's upper parts. Bustling, noisy progress on ground as distinctive as vivid plumage contrasts. Gait a vigorous hop, becoming double-footed backwards kick when searching

Rufous-sided Towhee *Pipilo erythrophthalmus*: **1–2** ad ♂, **3** ad ♀. Lark Sparrow *Chondestes grammacus* (p. 1635): **4** ad. Savannah Sparrow *Ammodramus sandwichensis*. *A. s. princeps* (p. 1635): **5** ad. *A. s. labradorius*: **6** ad. Fox Sparrow *Zonotrichia iliaca iliaca* (p. 1636): **7** ad summer. Song Sparrow *Zonotrichia melodia melodia* (p. 1637): **8** ad summer.

through leaf litter and making audible scratching noise. Flight low and rushed, with beats of round wings producing fluttering sounds, and tail pumped and often spread in turns and landings. Calls include cloud 'towhee', also rendered as slurred 'chewink'; slightly metallic 'tyst' and soft 'heu'.

Habitat. Breeds in temperate Nearctic lowlands in scrub, and any sort of dense low woody vegetation, including abandoned or poorly kept pastureland, isolated forest openings, field and woodland edges, along streams and roadsides, where ground with plenty leaf litter is available for foraging. Not commonly found in tall mature trees, wetlands, or arid places and normally keeps close to cover. Will occupy city parks or suburbs with trees and tall shrubbery. Nests on ground or very low in bush; feeds on invertebrates and seeds.

Distribution. Breeds in eastern North America, from southern Manitoba, southern Ontario, south-west Quebec and Maine, south to Gulf of Mexico coast, west to Minnesota, Iowa, eastern Kansas, north-east Oklahoma, and Louisiana. Hybridizes with Spotted Towhee *P. maculatus* in narrow zone of central Great Plains.

Accidental. Britain: Lundy island (Devon), June 1966.

Movements. Partial migrant, wintering within breeding range. Southern populations mainly resident, northern populations withdraw southward from part of ranges in winter, though some remain in extreme southern Ontario and Maritime Provinces of Canada.

Movements in August bring birds to east coast from New York to Nova Scotia, but southward migration begins only in September, most birds having left Canadian prairies by end of month, when first migrants appear in Texas and Arizona and also along Atlantic coast. Recorded autumn movements south of 35°N are chiefly in October, as by November migrants are mingled with southern residents. Ringing data show broad-front movement.

Many spring reports represent start of singing rather than movement; most 'first records' in March are within winter range. Main movement extends from 2nd half of April to early May across north-east and to mid-May near northern range limits. Migrants regularly overshoot known range in eastern Canada.

Wing-length: *P. e. erythrophthalmus*: ♂ 85–95, ♀ 82–87 mm.

Field Sparrow *Spizella pusilla*

Du. Veldgors   Fr. Bruant des champs   Ge. Klapperammer   It. Passero dei campi
Ru. Малая воробьиная овсянка   Sp. Chingolito llanero   Sw. Åkersparv

A North American species breeding from Montana and Minnesota east to southern Quebec and Maine, south to Gulf of Mexico states. Northern populations winter from eastern Kansas east to Maryland and southern Massachusetts, south to north-east Mexico, Gulf of Mexico coast, and central Florida; southern populations resident. 5 came aboard eastbound ship off North America in 1962, 1 staying until at least 15°W on 12 October.

Lark Sparrow *Chondestes grammacus*

PLATES: pages 1531, 1634

Du. Roodoorgors   Fr. Bruant à joues marron   Ge. Rainammer   It. Passero calandra
Ru. Хондеста   Sp. Chingolo arlequín   Sw. Lärksparv

Field characters. 15.5–17 cm; wing-span 24.5–27 cm. Noticeably larger than Song Sparrow, with 25% longer wings and 10% longer and more rounded tail. Large bush-haunting Nearctic sparrow of dry open country. In all plumages, tail centrally brown, becoming black on outer feathers of which all show increasingly large white tips and (on outermost) complete white fringes; pattern of marks emphasizes fan shape of tail and forms diagnostic character. Adult also has vivid head pattern, with strong contrast between white or cream ground and black-chestnut lateral crown-stripes, black-edged, chestnut ear coverts (interrupted by white crescent under eye) and strong black malar stripe, and striking black spot in centre of breast. Immature has head colours less developed and breast clouded and streaked, lacking adult's spot. Rest of plumage essentially streaked above and pale, unmarked below; wings show double buff-white wing-bar and more prominently whitish patch at base of primaries which contrasts with blackish primary coverts.

At first glance, can be taken for Palearctic bunting but close observation of tail pattern prevents confusion. Flight strong and direct. Stance noticeably horizontal on ground but more upright on perches. Less persistently terrestrial than most Nearctic sparrows, happily entering bushes after breeding season. Call a weak 'chip'.

Habitat. Open country with trees and bushes, prairies, open woodland, forest edges, roadsides, farmland, etc. throughout temperate North America below *c.* 2000 m; often at farms and in villages. In winter, in oak savanna, open pine woods, weedy farmland, or cereal fields. Feeds on seeds and insects on ground; nests usually on ground under bush or in low tree or shrub in loose colonies, sometimes in tree.

Distribution. Breeds from south-west and south-central Canada south throughout most of USA (now irregular east of Mississippi) to northern Mexico.

Accidental. Britain: Suffolk, June–July 1981; Norfolk, May 1991.

Movements. Partial migrant, wintering from southern USA (some move to Atlantic coast) south to parts of Central America as far as El Salvador, occasionally to Cuba and Bahamas. Leaves breeding grounds from late July, returning from late March.

Wing-length: ♂♀ 79–94 mm.

Savannah Sparrow *Ammodramus sandwichensis*

PLATE: page 1634

Du. Savannah-gors   Fr. Bruant des prés   Ge. Grasammer   It. Passero delle praterie
Ru. Саванная овсянка   Sp. Chingolo sabanero   Sw. Gulbrynad grässparv

Field characters. 14–16 cm; wing-span 22.5–25 cm. Close in size to Reed Bunting but with proportionately longer, more pointed bill and less full and noticeably shorter, notched tail, with pointed feathers; relatively much shorter-tailed than Song Sparrow, with east Nearctic race *savanna* smaller than that species but Sable Island race *princeps* as large. Among vagrant Nearctic sparrows, relatively small to medium-sized and tubby; a pale, uniformly streaked, open-country species. Shows yellow fore-supercilium and eye-ring, narrow pale cream central crown-stripe, and buffish-white double wing-bar. Striking white submoustachial stripe contrasting with long dark malar stripe. *A. s. princeps* noticeably pallid, with cryptic sandy appearance.

In Nearctic, Savannah Sparrow shares streaked plumage with 12 other relatives, but confusion among transatlantic vagrants likely only with Song Sparrow, which differs in proportionately much longer and slightly rounded tail, greyish supercilium, chestnut streaks on head and back, large central dark spot on breast, more chestnut in wings and along tail-base, and pale brown (less pinkish) legs. Flight swift but rather erratic, even zigzagging over short distance, with bird soon landing with depressed tail; over longer distance, develops undulations even at low level. Calls include high, sharp, dry 'tsip' and sharp 'chirp' (in alarm).

Habitat. Breeds in Nearctic, from arctic tundra through boreal and temperate zones to subtropics and tropics, in various types of open unwooded habitat, especially with short herbage, either moist or dry, lowland (including coastal) or locally montane, as on higher mountain ranges of Alaska, where it also frequents shoreline driftwood and debris. On migration, found in pastures, weedy fields, orchards, and gardens.

Distribution. Breeds in North America, from western Alaska east to northern Labrador and Newfoundland, south to central California, northern New Mexico, Nebraska, Kentucky, Maryland, and New Jersey; also in highlands of Mexico, and perhaps west Guatemala. (In north-east Siberia, breeds in eastern part of Chukotskiy peninsula.)

Accidental. Britain: Portland, Dorset, April 1982 (*A. s. princeps*); 1st-winter, Fair Isle, September–October 1987.

Movements. Great variation, from long-distance migrants (northern inland-breeding races) through short-distance and partial migrants (northern coastal and mid-latitude races) to altitudinal migrants (breeding in southern alpine areas) and residents (southern coastal races). Frequents only open grass/herb layer habitat at all seasons, so birds retreating from winter snow (covering plant seeds on which they feed) may move considerable distance before finding suitable habitat; wintering is south of or below snow-line. Winters in North America north to Nevada, Missouri, and Tennessee (continuing north to British Columbia on west coast, and to Nova Scotia east of Appalachians), and south through Mexico to Honduras; also in western West Indies. Highest numbers winter in southern states (west to Texas) and southern California.

Wing-length: Considerable geographical variation. *A. s. savanna*, *mediogriseus*, and *oblitus* (breeding eastern Canada and eastern USA): ♂ 70–74, ♀ 67–70 mm. *A. s. princeps* (breeding Sable Island, off Nova Scotia) large; ♂ wing-length 73–83 mm.

Fox Sparrow *Passerella iliaca*

PLATE: page 1634

Du. Roodstaartgors   Fr. Bruant fauve   Ge. Fuchsammer   It. Passerella variabile
Ru. Пестрогрудая овсянка   Sp. Chingolo zorruno   Sw. Rävsparv

Field characters. 17–19 cm; wing-span 26–28 cm. Close in size to Corn Bunting but with proportionately smaller, conical bill and rather longer tail. Large robust thicket-loving Nearctic sparrow; habitually scratches through leaf litter. Within heavily-streaked appearance, rusty or fox-red tones on head, wing, rump and tail catch eye.

Unmistakable, being larger than House Sparrow and all Nearctic relatives; also oddly coloured, recalling Dunnock until rump and tail show. Flight strong and fast but with somewhat fluttering action and broad-rumped and broad-tailed silhouette. Stance quite erect, trailing tail. Commonest calls 'click' and 'chip'.

Habitat. Breeds in northern Nearctic lowlands and in some mountains further south, from Arctic to California. Lives in dense woodland, either coniferous or deciduous, favouring streamside growths of willow and alder, regrowth on burnt patches of forest, stunted conifers on coast, and woodland thickets and edges.

Distribution. Breeds in North America, from western Alaska east to northern Labrador and Newfoundland, extending south through western mountains to southern California, central Utah, and central Colorado, and east of Rockies south to central Alberta, central Ontario, central Quebec, and Nova Scotia.

Accidental. Iceland: Borg, November 1944. Ireland: Copeland Island (Down), June 1961. Italy: Liguria, 1936. Records in Germany (Mellum, May 1949, Scharhörn, April 1977) regarded as escapes.

Movements. Migrant, varying greatly in status. Boreal populations fully migratory; north-west populations show 'leapfrog migration' down west coast, northernmost wintering furthest south; western mountain populations vary from fully migratory to nearly sedentary with minor altitudinal movements. Winters from southern Alaska and south-west British Columbia south through Pacific states to Baja California (Mexico), and from New Mexico, Kansas, and southern Wisconsin across eastern and southern USA, north very locally to Canada. Records and ringing recoveries show marked passage along Atlantic coast, presumably chiefly to Newfoundland, where many breed. Vagrant to Alaskan and Canadian Arctic, and to Greenland.

Wing-length: *P. i. iliaca* (eastern Canada and eastern USA): ♂ 87–92, ♀ 81–89 mm.

Song Sparrow *Melospiza melodia*

PLATE: page 1634

Du. Zanggors   Fr. Bruant chanteur   Ge. Singammer   It. Passero cantore
Ru. Мелодичная овсянка   Sp. Chingolo melodioso   Sw. Sångsparv

Field characters. 15–16·5 cm; wing-span 20–21·5 cm. Close in size to Reed Bunting but shape and stance also reminiscent of Dunnock; most larger than mainland Nearctic races of Savannah Sparrow and has noticeably longer rounded tail. Medium-sized but quite long Nearctic sparrow; epitome of streaked members of that group. Plumage greyish-brown and sharply streaked, with broad rufous-brown lateral crown-stripe, rufous-brown rear eye-stripe and broken edge of ear-coverts obvious on greyish head, rather pale olive-grey nape, noticeably rufous and blackish wings, and black spot in centre of chest of adult formed by the coalescence of heavy rufous-black streaks on sides of throat and breast and which extend over flanks. No yellow on face and rest of underparts off-white. Tail has rufous sides at base and is characteristically 'pumped' in flight.

With size and appearance midway between Fox Sparrow and smaller races of Savannah Sparrow and not unlike Reed Bunting, Song Sparrow not easy to identify. Nevertheless, Reed Bunting shows white outer tail-feathers freely, and larger Fox Sparrow has fully chestnut rump and tail. Separation from Savannah Sparrow requires close observation of tail shape (noticeably short and notched in Savannah Sparrow), supercilium (not grey, but wholly or partly yellow or white in Savannah Sparrow) and wings (far less strongly rufous in Savannah Sparrow). Flight light and fast but made to look laboured and awkward by tail action. Stance rather level; often assumes slightly hunched posture recalling Dunnock. Calls include characteristic, slightly harsh 'chirup' or 'chepp', recalling House Sparrow but croakier; also fine, high-pitched 'tsii'.

Habitat. Breeds widely in temperate and adjoining climatic zones of Nearctic, mainly in lowland or upland, but locally to 1500 m or higher. A typical edge species, inhabiting thickets of shrubs and trees among grassland, brushy margins or openings of forest, brushy edges of ponds or lakes, shrub swamps, shelterbelts, farmsteads, and sometimes parks or suburbs.

Distribution. Breeds in North America, from southern Alaska (including Aleutian islands) east across southern half of Canada to Newfoundland, and south through USA to Puebla (central Mexico), northern New Mexico, north-central Arkansas, northern Georgia, and coastal South Carolina.

Accidental. Britain: 7 records up to 1995, 6 spring, 1 autumn. Norway: ♂, Østfold, May 1975.

Movements. Migratory status varies. In general, populations of northern coasts and of mid-latitudes inland are partly migratory, with some resident races in Aleutian Islands (Alaska); southern populations, especially in south-west states and Mexico, are sedentary. Winters from southern Alaska (Aleutian Islands) and coastal and southern British Columbia east through northern USA to south-east Canada, south throughout rest of breeding range and southern Texas, Gulf of Mexico coast, and southern Florida. No reports of vagrancy north to Arctic North America.

Wing-length: *M. m. melodia* and *M. m. euphonia* (eastern Canada and USA): ♂ 64–70, ♀ 62–67 mm.

Swamp Sparrow *Zonotrichia georgiana*

Du. Moerasgors   Fr. Bruant des marais   Ge. Sumpfammer   It. Passero delle paludi
Ru. Болотная овсянка   Sp. Chingolo pantanero   Sw. Träsksparv

A North American species, breeding east of Rocky Mountains from Mackenzie and north-east British Columbia east to Newfoundland, and south to Nebraska, Illinois, and Maryland. Winters in southern parts of breeding range (rarely north to Canada) south to southern USA and central Mexico. Up to 7 present on eastbound ship in North Atlantic in October 1962, at least 1 staying until *c*. 30°W on 11 October.

White-crowned Sparrow *Zonotrichia leucophrys*

PLATE: page 1638

Du. Witkruingors   Fr. Bruant à couronne blanche   Ge. Dachsammer   It. Passero corona bianca
Ru. Белобровая овсянка   Sp. Chingolo piquiblanco   Sw. Vitkronad sparv

Field characters. 16–18·5 cm; wing-span 23–25·5 cm. Similar in size to Reed Bunting; up to 10% longer than White-throated Sparrow with longer neck and often erect posture. Quite large, relatively elegant Nearctic sparrow, with

White-crowned Sparrow *Zonotrichia leucophrys*. *Z. l. leucophrys*: **1** ad summer, **2** 1st winter. *Z. l. gambelii*: **3** ad summer. White-throated Sparrow *Zonotrichia albicollis* (p. 1639): **4** ad white-striped morph, **5** ad tan-striped morph. Dark-eyed Junco *Junco hyemalis hyemalis* (p. 1639): **6** ad ♂, **7** ad ♀, **8** 1st winter ♂.

bunting-like form and stance. Plumage mainly unstreaked pearly-grey below and heavily streaked rufous above; adult has black and white crown-stripes, whitish throat, and double white wing-bar. 1st-winter differs distinctly in grey-buff central crown-stripe, dark brown head-stripes, cream throat, partly brown ear-coverts, and buff-brown (less grey) nape and mantle. Bill pale; supercilium restricted to behind eye.

In brief view, adult and immature easily confused with White-throated Sparrow, separation requiring clear view of head markings and (with immature) underparts: White-throated Sparrow always has sharp-etched white throat and at least yellowish fore-supercilium in adult, streaked chest and flanks in winter (particularly 1st winter), and mostly dark bill at all ages. Pine Bunting is another potential confusion species but differs in strongly rufous rump and white outer tail-feathers. Usually more extended neck and more erect stance give quite different posture from White-throated Sparrow. Calls include rather sharp, metallic 'pzit', suggesting alarm-call of Pied Flycatcher; also a thin, high 'tssiip' or 'seeet'.

Habitat. Breeds extensively in Nearctic from northern Alaska and Canada south through temperate zone. Throughout range, primarily a bird of woody shrubbery and thickets in more open situations; it also occupies bushy edges of woodlands, openings, old burns, and mountainside shrubbery. In New England, spring migrants frequent cultivated fields, pastures, roadsides, and bordering thickets; in autumn, feeds wherever weed seeds are abundant, in cornfields, potato fields, or by roadsides, preferring to be near cover.

Distribution. Breeds in North America, from northern Alaska eastward across much of northern mainland of Canada to Labrador and northern Newfoundland; from north-west Canada continues south through western USA to southern California, central Arizona, and northern New Mexico.

Accidental. Iceland: Heimaey (Vestmannaeyjar), October 1978. Britain: Fair Isle, and Humberside, May 1977. France: Barfleur (Manche), August 1965. Netherlands: December 1981 to February 1982.

Movements. Status varies: northern populations are long-distance migrants, some southern populations winter within breeding range. Winters in much of USA (except north-central and north-east areas) north to southern British Columbia, south to central Mexico; less regular in eastern coastal areas of USA from Massachusetts south to Florida, Bahamas, and Greater Antilles. Frequent vagrant north of breeding range (mainly spring), to Pribilof islands (Bering Strait), islands in Canadian Arctic, and Greenland.

Wing-length: *Z. l. leucophrys* (eastern Canada and eastern USA): ♂ 78–86, ♀ 73–80 mm.

White-throated Sparrow *Zonotrichia albicollis*

PLATE: page 1638

Du. Witkeelgors   Fr. Bruant à gorge blanche   Ge. Weißkehlammer   It. Passero golabianca
Ru. Белошейная воробьиная овсянка   Sp. Chingolo gorgiblanco   Sw. Vitstrupig sparv

Field characters. 15·5–18 cm; wing-span 22–25 cm. Similar in size to Reed Bunting; up to 10% shorter than White-crowned Sparrow, with neckless and less upright form. Quite large, long, and bunting-like Nearctic sparrow, with rather secretive behaviour. Adult dimorphic, with white or tan head-stripes; unstreaked below, with rest of plumage dominated by diagnostic black-edged white throat and yellow fore-supercilium. Almost all 1st-winter birds resemble tan-striped winter adult but distinguished by lack of clear grey on face and breast, these areas being mottled or more strongly but still diffusely streaked, and throat dull white, with broken malar stripe joining breast streaks. Dark ridge to bill.

Commonest vagrant Nearctic sparrow in west Palearctic, but difficult to separate in glimpse from White-crowned Sparrow; see that species. Paler immature may also be confused with Song Sparrow but is larger, has dark-striped head with buff (not grey) supercilium, and much brighter wing-bars. Flight active and fast, with flirting tail; usually at low level. Stance and posture recall *Passer* sparrow, with compact, neckless form. Secretive, usually on ground within cover. Commonest call a loud rather metallic 'chink'; also a thin, high, drawn-out 'tseet' or 'tseep'.

Habitat. Ranges across cool boreal and temperate forested regions of northern Nearctic. In Canada, mostly coniferous or mixed forest, especially in clearings cluttered with slashing, burntwoods, and open young woodlands and thickets. Further south in central USA, favours various semi-open wooded habitats such as coniferous forest with well-developed woody undergrowth, groves of aspen with shrubby understorey, marshes bordered with willow, and sometimes conifer plantations. On spring migration through New England, seems to prefer moist thickets, but in autumn many visit weedy gardens and cornfields, usually remaining on or near ground, rarely perching high in trees; also favours bush-bordered roads and edges of pinewoods.

Distribution. Breeds in North America, from south-east Yukon east to Newfoundland, south to central British Columbia, central Saskatchewan, North Dakota, and Great Lakes region east to northern New Jersey.

Accidental. Iceland, Britain, Ireland, Netherlands, Denmark, Sweden, Finland, Gibraltar.

Movements. Migratory. Winters in eastern half of USA from south-east Iowa, northern Ohio, Pennsylvania, and Massachusetts, south to Gulf of Mexico coast and Florida; also west from Texas to California and northern Mexico. Migration is on broad front across North America east of Rockies. Some birds overshoot breeding range in spring, giving records in arctic Alaska and Canada. Rare vagrant to west Palearctic, especially in spring. 19 records from Britain and Ireland up to 1995, of which 13 in April–June and 6 in October–December. Bird recorded 1 December 1984 remained until at least 7 April 1985.

Wing-length: ♂ 74–80, ♀ 70–75 mm.

Dark-eyed Junco *Junco hyemalis*

PLATE: page 1638

Du. Grijze Junco   Fr. Junco ardoisé   Ge. Junko   It. Zigolo ardesia
Ru. Темноглазый юнко   Sp. Chingolo pizarroso   Sw. Snöfågel

Field characters. 13·5–15 cm; wing-span 23–25 cm. Close in size to Tree Sparrow but with bunting-like bill and structure, including rather long tail; 30% smaller and more lightly-built than Eastern Towhee. Medium-sized, perky, rather tame Nearctic sparrow, differing from other vagrant Emberizidae in pink bill, and uniformly dusky-grey to brown hood, upperparts, and upper flanks contrasting with pure white underbody and tail sides.

Unmistakable. Flight light and fluttering, with quick wing-beats; tail noticeably straight in silhouette. Stance level or half-upright, with tail carried up and often flicked. Feeds on ground, usually in or near cover. Calls include characteristic smacking or clicking sound, run together into squeaky twitter, metallic 'clink', and slightly liquid 'chek' in alarm.

Habitat. Breeds in northern Nearctic, in Canada in coniferous and mixed woodland (especially openings and edges), on burntlands, and occasionally in gardens. A few breed south of conifer zone in mainly deciduous woodland, where felled areas with slash piles often attract them. In winter, frequents woodland and fields, but most occur along hedgerows and brushy field borders; feeds then on ground.

Distribution. Breeds in North America, from north-west Alaska east to Labrador and Newfoundland, south to northern Baja California (Mexico), southern New Mexico, north-west Nebraska, east-central Minnesota, central Michigan, Appalachians south to northern Georgia, and south-east New York.

Accidental. Iceland, Britain, Ireland, Netherlands, Denmark, Norway, Sweden, Poland, Gibraltar.

Movements. Status varies. Northern populations largely

migratory. Other populations partially migratory to sedentary according to latitude; some make altitudinal movements. Winters from southern Alaska and southern Canada east to Newfoundland, south throughout USA to northern Mexico. Migration (both seasons) on broad front east of Rockies, with channelling through valleys in west, but movement along coasts is not typical. Frequent vagrant to Canadian and Alaskan Arctic; also recorded north-west to Pribilof islands, and (both seasons) eastern Siberia.

Rare vagrant to west Palearctic, especially in spring. 18 records from Britain and Ireland up to 1995, of which 14 in April–May and 4 in December–February; 2 birds present December–March. Individual recorded at Gibraltar 18–25 May 1986 coincided there with arrival of White-throated Sparrow.

Wing-length: *J. l. hyemalis* (eastern Canada and north-east USA): ♂ 78–83, ♀ 70–80 mm.

Lapland Bunting *Calcarius lapponicus*

PLATE: page 1641

Du. IJsgors   Fr. Bruant lapon   Ge. Spornammer   It. Zigolo di Lapponia
Ru. Лапландский подорожник   Sp. Escribano lapón   Sw. Lappsparv   N. Am. Lapland Longspur

Field characters. 15–16 cm; wing-span 25·5–28 cm. Slightly smaller than Snow Bunting but with similar form, differing from typical *Emberiza* bunting in stubbier bill, larger head, bulkier build, noticeably oval, pointed wings, proportionately shorter forked tail, and long hind claw. Robust, slightly squat bunting, with flickering flight and plumage pattern like lark; often runs. In all adult plumages, shows at least partially rufous nape; breeding ♂ also has striking black head with white zigzag

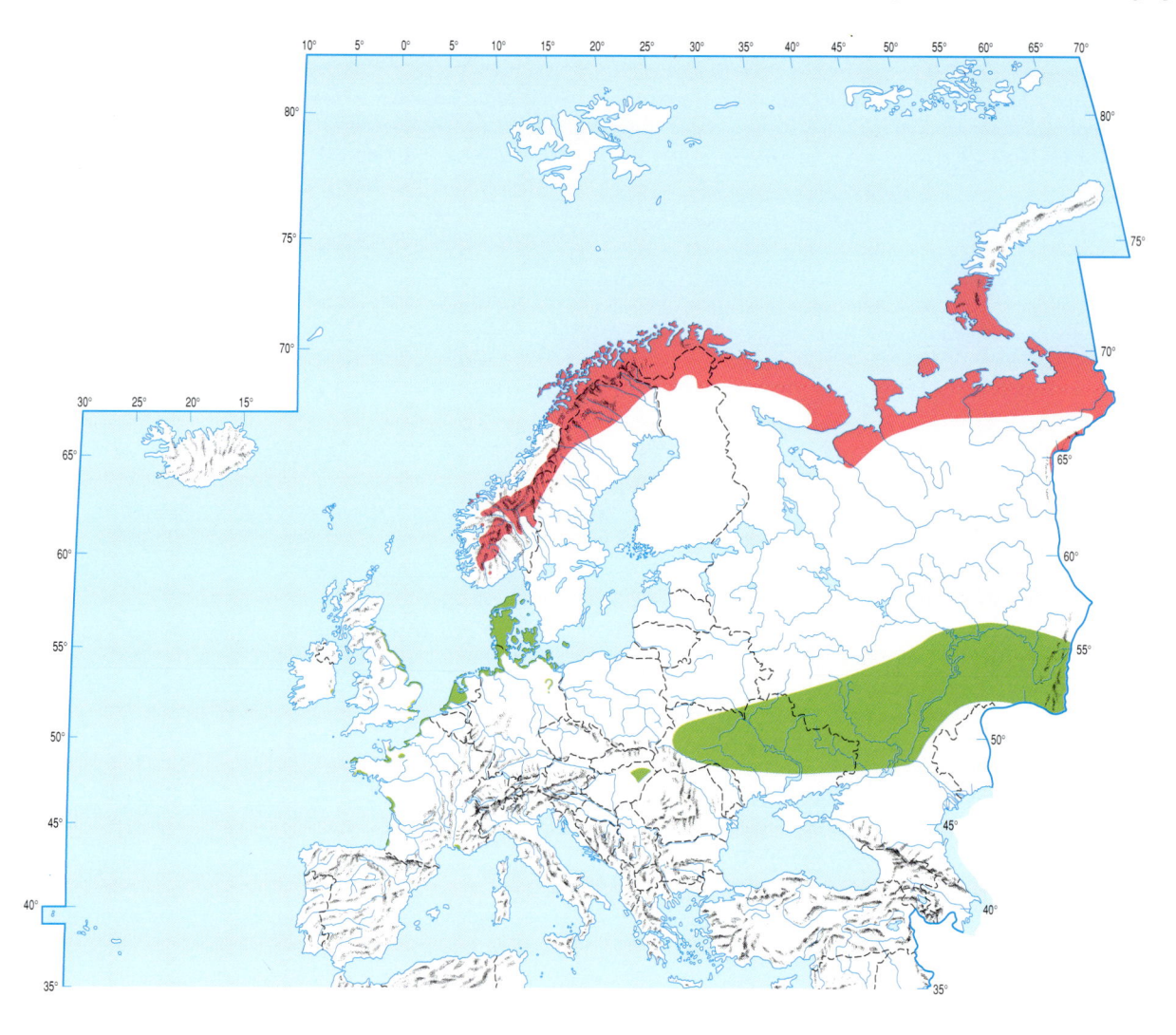

Lapland Bunting *Calcarius lapponicus*: **1–2** ad ♂ breeding, **3–4** ad ♂ non-breeding, **5** ad ♀ breeding, **6–7** ad ♀ non-breeding, **8** 1st summer ♂, **9** 1st winter ♂, **10** juv.

line from eye to nape and down neck. Underparts of adult noticeably white. Juvenile has bright reddish greater coverts, inviting confusion with Reed Bunting but distinguished by pale bill, more open face pattern, and characteristic dark mottling of central breast and foreflanks, also shown by ♀ and 1st winter ♂.

Breeding ♂ unmistakable if seen well, but in brief glimpse or when silent can be confused with Reed Bunting (less robust, with proportionately shorter wings and longer tail; lacks pale supercilium but has white submoustachial stripe, whitish-grey nuchal band, chestnut lesser wing-coverts, and often-flicked bright white outer tail-feathers) and Rustic Bunting (noticeably smaller and slighter, with silky-white underparts splashed chestnut across breast and along flanks). Until general character and voice learnt, winter ♂, ♀, and immature far less distinctive; can be confused with some larks, ♀ and young sparrows, and several other buntings. Most liable to cause confusion are House Sparrow, Pine Bunting, Reed Bunting, Snow Bunting (particularly juvenile and 1st-winter ♀ with little white in wing), and Corn Bunting; thus note especially structure (particularly extension of folded primaries which equals length of tertials, matched only by Snow Bunting), manner of flight (see below), habit of running (matched only by larks and Snow Bunting), and voice (see below). At close range, close inspection of nape for sign of rufous feathering usually provides diagnosis. Flight recalls both Snow Bunting and even more frequent companion Skylark, with closely similar action to both. Flight silhouette somewhat less bulky than Snow Bunting, again so closely recalling Skylark as to be indistinguishable from that species within distant flock but close to, showing longer, more oval outer wings and shorter, noticeably forked tail. Gait includes characteristic rapid run, loping walk, and hop.

Habitat. Extends across arctic tundra and boreal region of west Palearctic, between July isotherms of 2°C and 14–15°C. In contrast to equally arctic Snow Bunting, avoids rocky and precipitous or bare terrain, favouring low shrubby tundra and damp hummocky moss-tundra with dwarf birch, willow, and heath plants. Generally winters on flat and open grassy areas bordering coasts and estuaries, grass moorland, grass steppes, and bare open cultivated areas, in Britain favouring rough grassland or stubble at no great distance from sea.

Distribution. BRITAIN. 1–11 pairs bred Scotland 1977–80. FINLAND. Range has contracted; formerly extended south to northern end of Gulf of Bothnia. RUSSIA. Kola peninsula: expanding, from former range in tundra and forest tundra, into adjacent boreal taiga.

Accidental. Spitsbergen, Bear Island, Jan Mayen, Iceland (almost annual), Faeroes, Switzerland, Spain, Portugal, Malta, Yugoslavia, Bulgaria. Status in southern countries perhaps rare and irregular winter visitor, as reaches Italy regularly in small numbers.

Distribution circumpolar, extending from west Palearctic across northernmost Asia (south to Kamchatka in east), islands of Bering Sea, Alaska, northern Canada, and western and south-east Greenland.

Population. No evidence of long-term changes. NORWAY. 200 000–500 000 pairs 1970–90; stable. SWEDEN. 100 000–400 000 pairs in late 1980s; stable. FINLAND.

20 000–60 000 pairs in late 1980s; marked annual fluctuations. RUSSIA. 1–10 million pairs; stable.

Movements. Migratory. European birds head between south-west and (chiefly) south-east, Asian and North American birds chiefly south, and Greenland birds both south-west (to North America) and south-east (to north-west Europe).

European birds winter mainly in south European Russia and Ukraine, but no detailed information from this area. Most data relate to western and central Europe. Passage records are far more numerous than winter records, but fluctuate markedly from year to year. Winter populations in north-west Europe include birds from Greenland as well as Scandinavia. Winter distribution in Britain confined almost entirely to east coast from Kent to Firth of Forth, especially at estuaries and at various sites in Norfolk, with few inland. Number difficult to estimate, perhaps c. 200–500 birds, but many more in peak years. Many birds appear to move on in late winter. Passage records are chiefly in north and north-west, with some in south-west.

Autumn migration begins August. Most birds leave Greenland mid-August to end of September, and recorded chiefly mid-September to mid-October in Iceland. Vacates Sweden late August to October, with mid-September peak in south-west. Reported times of passage vary greatly. In France, earliest records in north-west at end of August or beginning of September, with gradual increase to October and November. First arrivals in Scotland and Ireland in late August also; long-term data at Fair Isle (Scotland) show earliest bird 23 August and build-up from early September, peaking 2nd half of September and gradually diminishing in October. Arrives in eastern Britain from 2nd week September, with numbers increasing October. Reaches Hungary from mid-September, with numbers building up October–November. Spring migration (February–)March–May. Main passage April in north-west Germany and Denmark, with latest reports early May. Reaches breeding grounds mostly in May.

Food. Invertebrates (especially flies Diptera) in peak breeding season, otherwise seeds of grasses and low herbs. On breeding grounds, forages busily on ground, running from tussock to tussock picking invertebrates from surface of vegetation, very rarely in bare places; sometimes in shrubs, even 2–3 m up in tree. Very often recorded plucking invertebrates from Rosaceae flowers (e.g. *Rubus*, *Dryas*). Jumps up to snatch flying insects and has been observed hawking c. 75 cm above ground.

Social pattern and behaviour. Gregarious at all times, much less so during breeding season, but even then small flocks of unpaired ♂♂ may occur. Regularly associates with other flocking species. Mainly monogamous pair-bond, but to some extent ♂♂ also polygamous and promiscuous. Territorial when breeding and in neighbourhood groups, probably due to clustering in favourable habitat. ♂ sings to attract mate and defend territory, perched on vantage point or in flight. Song-flight rather like pipit; ♂ ascends suddenly, and typically silently on steep diagonal; after reaching c. 6–15(–20) m, starts singing, swings from side to side, and turns to commence slow, widely spiral, gliding descent with outspread wings and tail fanned upwards; tends to land where he took off or, just before landing, glides to next eminence.

Voice. Song, usually given in display-flight, a short but lively and musical repeated phrase, suggesting Skylark, 'teeTOOree-teeTOOree-trree-oo' or 'kretle-KRLEE-trr-kritle-kretle-tru'. Calls on breeding grounds, very musical, piping 'teeleu' clearly enunciated in alarm, distinguished from rather metallic 'teeuu' or 'TEElu' and quiet tuneless but hard, clipped 'ticky-tick'. Characteristic (and diagnostic) call of migrants a dry, slightly rattled trill, usually ending in a more melodious fluting note: 'tick-tick-tick-teu', 'ticky-tick-teu' or 'prrrt...chu'.

Breeding. SEASON. In general, probably timed so that young leave nest during peak abundance of adult flies (Diptera). Swedish Lapland: eggs laid mid-June (end of May to mid-July); laying highly synchronized within population. Northern Russia: eggs laid 1st half of June, late June in cold springs; young from replacement clutches can leave nest as late as mid-August. One brood. SITE. On ground, commonly in slight depression or sheltered position in lee of hummock or tussock; in dry spot but often near water, very frequently protected by overhanging twigs of birch, willow, heather, or similar shrub. Nest: tightly built structure of dry grass, sedge, rootlets, leaves, lichen, etc., lined with soft grasses, plant down, hair, and many feathers, particularly of grouse; sometimes no lining. EGGS. Sub-elliptical, smooth and slightly glossy; very variable, pale greenish, greyish, or buffish, usually very heavily marked with olive-brown, rusty, or purplish-black blotches, spots, and scrawls. Clutch: 5–6 (3–7), very rarely 8. INCUBATION. 11–13 days. FLEDGING PERIOD. 9–10 days (8–11), young leaving nest c. 2–5 days before able to fly.

Wing-length: *C. l. lapponicus*: ♂ 93–99, ♀ 85–92 mm.
Weight: *C. l. lapponicus*: ♂♀ mostly 20–28 g.

Geographical variation. Rather slight, mainly involving colour of upperparts and size. *C. l. subcalcaratus* from Greenland and Canada (winter visitor to Europe) has bill slightly longer than nominate *lapponicus* (race breeding in west Palearctic), heavier and deeper at base, wing on average longer.

Snow Bunting *Plectrophenax nivalis*

PLATE: page 1523

DU. Sneeuwgors FR. Bruant des neiges GE. Schneeammer IT. Zigolo delle nevi
RU. Пуночка SP. Escribano nival SW. Snösparv

Field characters. 16–17 cm; wing-span 32–38 cm. Close in size to Corn Bunting, with even longer and more pointed wings but less chesty body; slightly larger than Lapland Bunting but with similar build, forming with it pair of highly terrestrial passerines. 2nd largest bunting of west Palearctic, with deep, stubby bill, rather round but usually flat-crowned head, rather long, quite deep body, proportionately long pointed wings, and noticeably forked tail; distinctive robust, low-slung silhouette on ground. Plumage predominantly white, strikingly pied on back and wings in breeding adult and softly variegated warm buff on head, back, and chest of winter adult and immature. In flight, shows bold white panels on wings and along sides of tail though former much restricted in juvenile and 1st-winter ♀. Flight powerful but action cum wing-pattern create partly illusory flickering or drifting progress. Runs freely.

Adult unmistakable in west Palearctic, there being no known overlap with usually montane Snowfinch (slightly larger, longer, and more upstanding, with grey head, brown upperparts, black bib in summer, and always fully white inner wings). Immature, particularly ♀♀ with least marked wings, subject to confusion with Lapland Bunting which shares both breeding and wintering habitats, and much behaviour, and has 2 similar calls. Mistake unlikely with bird at close range, however, as Snow Bunting always shows softly marked head and lacks (a) any discrete dark patches or streaks on underparts and (b) strong rufous on nape and wing-coverts; with birds in flight, frequent utterance by Snow Bunting of diagnostic lilting ripple allows instant identification. Rarely mixes with other species, and more likely to do so with Shore Lark than Lapland Bunting.

Habitat. Extends across arctic and higher boreal zones, beyond or above treeline, between July isotherms of 2°C and 14–15°C. Breeds in usually treeless, uncultivated, barren, rocky terrain, often near snow and ice, and even on isolated rocks deep within icecap. Often on seacliffs, including those with mass seabird colonies, but also on rocky terrain from lowland to plateau, bearing low and often sparse plant cover. In parts of range adapts to human settlements.

Migrating and wintering birds in FSU resort to open countryside, roads, threshing floors, and outskirts of settlements, and to forest edges. In Belgium, winters in littoral zone along beaches and treeless tracts. In northern Britain, some try to winter on upland moors, unless displaced by snowfall, while others favour stubble and turnip fields, remainder choosing marram grass behind sandy beaches. In England, favours shingle beaches and salt-marshes as well as sand-dunes and neighbouring stubble fields.

Distribution. FAEROES. Irregular breeder (up to 10 pairs). BRITAIN. Thinly but widely distributed throughout Scottish Highlands. Occasional breeding attempts at sea level in Northern and Western Isles. FINLAND. Up to early 19th century, bred regularly along northern Gulf of Bothnia; has occasionally bred on large inland lakes.

Accidental (or status of rare and irregular winter visitor). Balearic Islands, Malta, Turkey, Tunisia, Algeria, Morocco, Madeira, Canary Islands. Regular in small numbers in Azores, and probably Portugal.

Distribution circumpolar, extending from west Palearctic across northernmost Asia (south to Kamchatka in east), islands of Bering Sea, Alaska, northern Canada, and Greenland. Extends further north in Arctic than Lapland Bunting.

Population. SPITSBERGEN, BEAR ISLAND, JAN MAYEN. Regular breeder, but no estimates of population size. ICELAND. 50 000–100 000 pairs in late 1980s; stable. BRITAIN. 70–100 pairs 1988–91. Fluctuating, perhaps increasing. NORWAY. 100 000–500 000 pairs 1970–90; stable. SWEDEN. 25 000–100 000 pairs in late 1980s. FINLAND. 3000–6000 pairs in late 1980s. Fluctuating, but probably increasing. RUSSIA. 10 000–100 000 pairs; stable.

Movements. Partially migratory to migratory, many birds wintering far south of circumpolar breeding range; northernmost areas are vacated. In Europe, winters mostly in coastal areas and on inland plains. Numbers vary greatly from year to year, and also fluctuate over long periods. Present in Iceland all year, by far the commonest wintering passerine.

Autumn movement prolonged, September–December, with most passage records October–November. Spring movement northward begins early or mid-February. Leaves southern France February–March; latest record 28 February in Rumania, and rare by March in Hungary. Passage peaks end of February to early or mid-March in Denmark, north-east Germany, and Poland. Reaches southern Norway mid- or late March to April, and northern Norway at beginning of May. In north-east Scotland, spring departure rapid; most birds leave in March, a few still present in 1st half of April; ♂♂ depart c. 9 days before ♀♀ on average.

Food. Mainly seeds, with addition of insects in breeding season; young given only invertebrates. Feeds almost wholly on ground, sometimes perching on grasses or herbs to reach seeds. In winter quarters or on passage, feeds on arable land, especially winter crops, stubble, ploughed fields, etc.; pasture apparently often avoided; generally far from hedges, trees, and buildings; in eastern Germany, also in fields where dung spread, and on rough grassy edges and other weedy places.

Social pattern and behaviour. Usually gregarious outside breeding season, but quite often encountered singly. On return to breeding grounds, flock structure maintained until receding snow cover favours dispersal onto territories; flocks initially all ♂♂, but, with arrival of ♀♀, contain both sexes. Mating system typically monogamous (almost invariably so at high latitudes), exceptionally bigamous, i.e. ♂ with 2 ♀♀. Solitary and territorial when breeding; marked variation in territory size with region and habitat. Song serves for mate-attraction and territorial defence. ♂ sings mostly from ground (using high boulders, etc., in territory as preferred vantage points), occasionally in song-flight shortly before nesting. In song-flight, ♂ rises steeply and silently with rapid wing-beats for a few metres (often up to 10–15 m); at peak of ascent (and not until then) starts singing with outspread, slightly trembling wings, and glides down (in direction of rival if song-flight provoked by him) to vantage point; after landing, sometimes continues singing and may keep wings raised before closing them and crouching forwards. Until paired (and sometimes even after), resident ♂ performs advertising-display ('mannequin display') to every ♀ who lands in his territory: in upright posture, with wings spread back and down, and tail fanned and lowered, ♂ scuttles a short distance from ♀ (thus displaying bold piebald rear view to her), then shuts his wings and tail and runs back to her before turning and displaying anew.

Voice. Song musical, somewhat lilting and loud but curiously ventriloquial phrase, with fluted di- and trisyllabic notes that lack jingling quality of Lapland Bunting, 'turee-turee-turee-turiwee' or 'sweeto-swevee-weetuta-swee'. Local dialects striking. Flight-call diagnostic: soft, charming, musical, and distinctly rippled twitter, 'tirrirriripp', 'dirrirrt', or 'tirrirrillit'. Other calls: short, soft, plaintive, musical 'tuu' or 'piu', louder, more whistled or ringing, and less formed than similar note of Lapland Bunting, and usually commonest call of solitary bird.

Breeding. SEASON. Cairngorms (Scotland): eggs laid first 3 weeks of June (2nd half of May to 2nd half of July). Spitsbergen: eggs laid 2nd half of June (late May to July). Kola peninsula (north-west Russia): laying starts 1st half of June. Iceland: eggs laid from about mid-May into July. 1–2 broods. SITE. In cleft or crevice in scree, between rocks, in rock-face (including sea cliffs among auks), under boulder on grass, etc; more rarely, under turf or in hole in (e.g.) river-bank; nest up to c. 1 m inside cavity and entrance inaccessible to, or too narrow for, predators; seldom far from vegetation. Commonly in Arctic settlements inside buildings, under roofs, in nest-boxes, native cairns, graves, etc., and recorded inside any suitable object, (e.g.) tin can, box, human corpse. Nest: foundation of grass, stalks, leaf stems, moss, and lichen, lined with fine grass, hair, and many feathers. EGGS. Variable in size and markings; short to long sub-elliptical, smooth and slightly glossy; pale blue or greenish-blue, occasionally buffish, fairly evenly covered with reddish-brown to purplish-black spots and blotches, though can be concentrated at broad end; some have scrawls and violet-grey undermarkings. Clutch: 4–6 (3–8). INCUBATION. 12–13 days. FLEDGING PERIOD. *c.* 12–14 days (10–17).

Wing-length: *P. n. nivalis*: ♂ 104–118, ♀ 100–107 mm.

Weight: *P. n. vlasowae* (breeding) ♂ 33–50, ♀ 28–42 g. *P. n. nivalis* and *P. n. insulae* (winter) ♂ 28–44, ♀ 26–43 g.

Geographical variation. Involves colour of fringes of upperparts in fresh plumage, amount of black on rump, flight-feathers, and upper wing-coverts, and size (wing or tail). Little variation in size within west Palearctic. *P. n. nivalis* breeds North America, Greenland, and most of west Palearctic breeding range. *P. n. insulae* from Iceland is darkest race; black on wing and tail more extensive. *P. n. vlasowae* from Pechora basin (north European Russia) east to Wrangel Island and Anadyrland (eastern Siberia) paler than nominate *nivalis*.

Black-faced Bunting *Emberiza spodocephala*

PLATES: pages 1645, 1684

Du. Maskergors   Fr. Bruant masqué   Ge. Maskenammer   It. Zigolo mascherato
Ru. Седоголовая овсянка   Sp. Escribano enmascarado   Sw. Gråhuvad sparv

Field characters. 13·5–15 cm; wing-span 20–23 cm. Slightly smaller and proportionately a little shorter tailed than Reed Bunting but noticeably larger than Pallas's Reed Bunting. Rather small, quite slim bunting, with structure intermediate between Reed Bunting and Pallas's Reed Bunting but with long, stout, conical bill. Breeding adult ♂ has diagnostic bright pinkish to yellowish base to grey-black bill, black lores, greyish hood, and pale yellow underparts. Basic plumage colours and pattern of ♀ and immature recall Reed Bunting, but, with face and sides of neck clouded grey, also suggest Dunnock. Diagnosis therefore requires lengthy, close observation, with concentration on (a) grey to olive-brown lesser coverts (eliminating Reed Bunting), (b) grey-brown crown, (c) rather uniform pale brown to grey central ear-coverts (mottled rufous-brown in Reed Bunting), (d) pattern of median and greater coverts (wing-bars insignificant in Reed Bunting), and (e) inconspicuous supercilium (unlike Reed Bunting).

Breeding ♂ unmistakable but ♀ and immature lack distinctive features; ♀♀ have characteristic dull, rather cold and greyish, copiously streaked appearance unlike any other sympatric bunting, but this difficult to convey other than by likening to Dunnock. Differentiation from Reed Bunting and Pallas's Reed Bunting covered above; confusion with other *Emberiza* far less likely, common buntings of western Europe all being noticeably larger and other smaller vagrant relatives more boldly marked on head. Commonest call quiet but sharp

Black-faced Bunting *Emberiza spodocephala*: **1** ad ♂ breeding, **2** ad ♂ non-breeding, **3** ad ♀, **4** 1st winter ♀. Meadow Bunting *Emberiza cioides* (p. 1655): **5** ad ♂ winter, **6** ad ♀ winter.

'tzit', or quiet, slightly sibilant 'tsick' or 'tick', slightly thinner in tone than monosyllables of other buntings and often repeated.

Habitat. Breeding in east Palearctic in tall dense grass and shrubs, especially in river valley floodlands, in moist coniferous taiga forests and occasionally in mountain forest, which may be broadleaf, up to 600 m in Altai and 1500 m in Japan. In Indian winter quarters, feeds on ground in rice stubbles or on moist edges of pools, usually resorting to cover near water.

Distribution. Breeds in central and eastern Asia from upper Ob' valley and north-east Altai east to Sea of Okhotsk, Sakhalin island, southern Kuril Islands, and Japan, south to northern Tibet, south-west China, and eastern China.

Accidental. Britain: 1st-winter ♂, Greater Manchester, March–April 1994. Netherlands: 1st-winter ♂, Westenschouwen, November 1986; 1st-winter ♂, Friesland, October 1993. Germany: Helgoland, November 1910 and May 1980. Finland: ♂ Dragsfjärd, November 1981.

Movements. Chiefly migratory. Northern race, nominate *spodocephala*, migrates through Mongolia, south-east Russia, north-east China and Korea to winter in southern Korea, eastern and southern China from Hopeh (few) south to extreme south (including Hainan), west to Kwangsi and Hunan, also in Taiwan. Southern race *sordida* disperses widely between south-west and east, to winter from Bangladesh and eastern Nepal east through northern Burma to extreme north of Thailand (rare), northern Laos and northern Vietnam.

Wing-length: *E. s. spodocephala*: ♂ 70–75, ♀ 66–72 mm.

Pine Bunting *Emberiza leucocephalos*

PLATES: pages 1647, 1684

Du. Witkopgors   Fr. Bruant à calotte blanche   Ge. Fichtenammer   It. Zigolo golarossa
Ru. Белошапочная овсянка   Sp. Escribano cabeciblanco   Sw. Tallsparv

Field characters. 16·5 cm; wing-span 25–30 cm. Slightly larger than Yellowhammer, with tail usually 5–10% longer; slightly larger and more robust than Rock Bunting. Eastern counterpart and close relative of Yellowhammer. Unlike Yellowhammer, adult plumage of ♂ and ♀ strikingly different but both show white ground-colour to underparts, long rufous rump, and bright white outer tail-feathers. Breeding ♂ has striking white central crown and cheeks contrasting with bold black and chestnut stripes on face and chestnut throat; white underparts, interrupted by chestnut-spotted chest-band and flanks; similar but much duller, hoarier in fresh plumage. ♀ duller and patterned more like Yellowhammer but no trace of yellow, with dull white ground-colour to plumage most obvious in pale head, tips to median coverts, and belly. At close range, typical bird shows sharper, duller streaks on lateral crown-stripes and darker malar and chest streaking than Yellowhammer.

Certain identification bedevilled by (1) general similarity to Yellowhammer, (2) existence of dilute morphs of Yellowhammer, (3) hybridization and intergradation of all characters with Yellowhammer, and (4) convergent appearance of Rustic Bunting, Meadow Bunting, and Rock Bunting, though simultaneous occurrence of Rock Bunting with Pine Bunting unlikely. Adult ♂ unmistakable once precise head pattern confirmed but certain separation of ♀ and immature requires close, detailed observation. Separation most difficult with darker, more streaked immature; crucial to confirm pale whitish tips to median coverts and lack of any yellow below; if visible, lesser coverts show as rather greyish-brown patch, paler than on Yellowhammer. Important also to note that (1) outer fringes of all flight-feathers and greater coverts are paler in Pine Bunting than Yellowhammer, those on outer primaries particularly being white, not greenish-yellow, and (2) axillaries of Pine Bunting are white, not yellow. Elimination of dilute Yellowhammer and hybrids highly problematic. Flight and behaviour as Yellowhammer.

Habitat. Breeds mostly in Asia, overlapping in range and partly in ecological niche with Yellowhammer. Lacks subarctic element to match that of Yellowhammer, but is predominantly boreal and cool temperate in breeding distribution, and is accordingly a much more pronounced migrant. Within breeding range, favours well-lit forests of conifers, or in some regions birches and other deciduous trees, but avoids riverain deciduous woods, as well as mountain taiga. Will tolerate steppes if grassy, with clumps of trees. Winters commonly in flocks on foothills and plains of India and Pakistan, up to 1500 m, occasionally to nearly 2700 m; here it feeds on ground, perching in trees, on bush-covered grassy slopes, and on stubble and fallow fields.

Distribution. BRITAIN. Both pure and hybrid birds increasingly observed; 27 accepted up to 1995, indicating regular vagrancy and occasional wintering. POLAND. In May–June 1994, ♂ (only record) seen Biebrza valley; fed nestlings with ♀ Yellowhammer. ITALY. In winter, regular in north-east, and probably overlooked, so perhaps status of winter visitor. ISRAEL. Small numbers winter Mt Hermon, eastern Galilee, Jerusalem hills.

Accidental. Iceland, Britain, Ireland, France, Belgium, Netherlands, Germany, Denmark, Norway, Sweden, Finland, Poland, Czech Republic, Hungary, Austria, Switzerland, Spain, Malta, Greece, Yugoslavia, Bulgaria, Ukraine, Armenia, Turkey, Cyprus, Jordan, Morocco.

(FACING PAGE) Pine Bunting *Emberiza leucocephalos*: **1** ad ♂ spring, **2** ad ♂ winter, **3** ad ♀ winter, **4** juv, **5** ad ♀ intergrade with *E. citrinella*.
Yellowhammer *Emberiza citrinella* (p. 1648). *E. c. citrinella*: **6** ad ♂ spring, **7** ad ♂ winter, **8** ad ♀ winter, **9** 1st winter ♂, **10** juv. *E. c. caliginosa*: **11** ad ♂ spring. *E. c. erythrogenys*: **12** ad ♂ spring.
Cirl Bunting *Emberiza cirlus* (p. 1651): **13** ad ♂ spring, **14** ad ♂ winter, **15** ad ♀ spring, **16** ad ♀ winter, **17** juv, **18** ad ♂ spring *'nigrostriata'*.

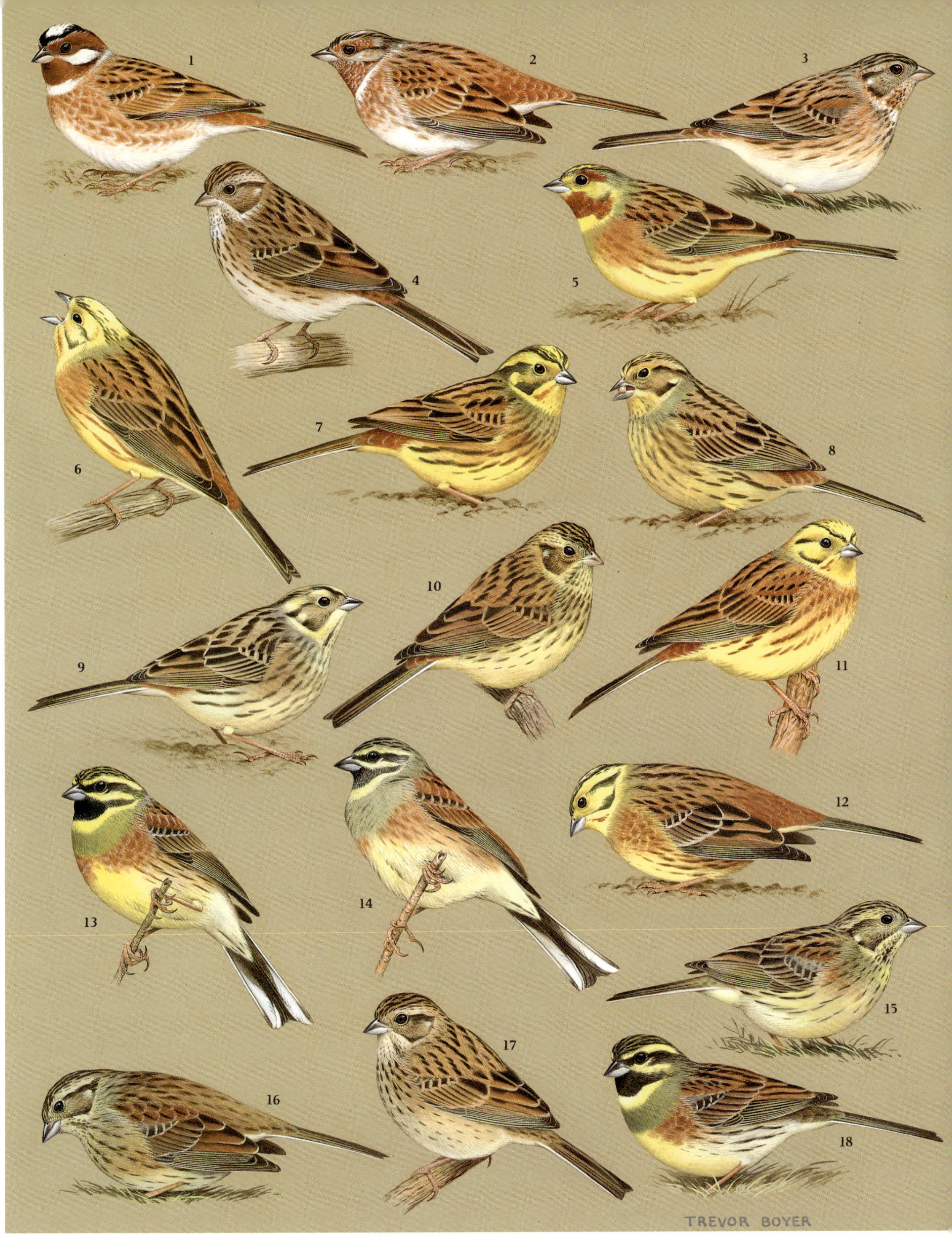
1
2
3
4
5
6
7
8
9
10
11
12
13
14
15
16
17
18
TREVOR BOYER

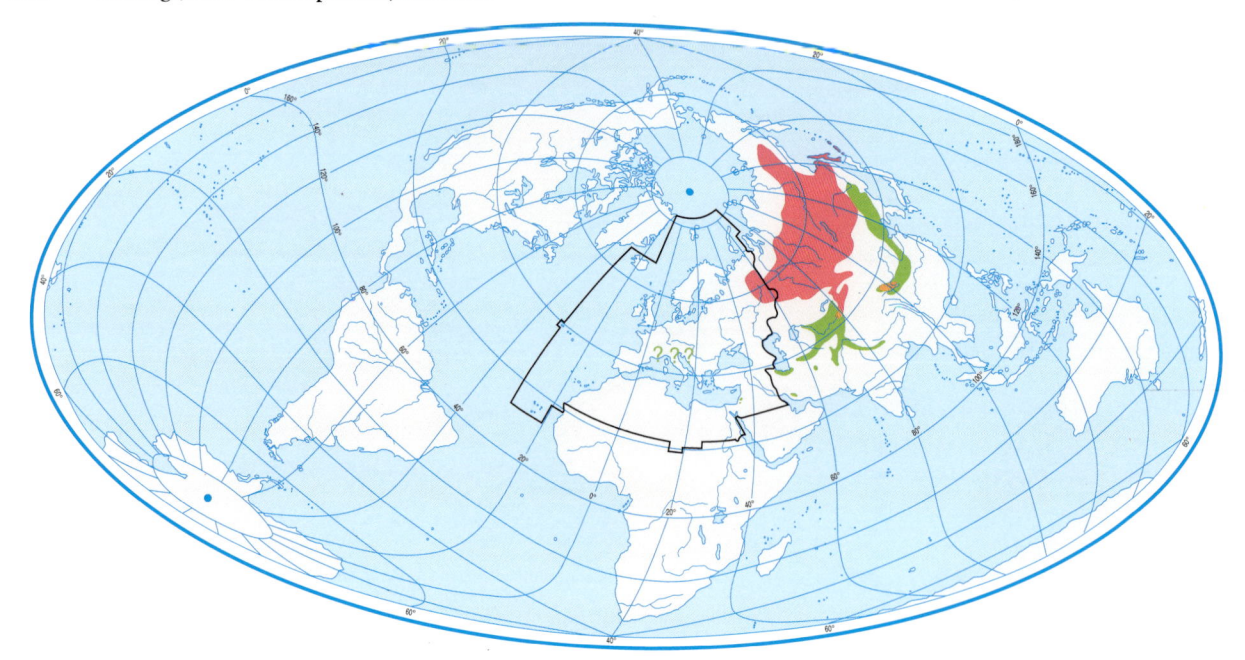

Movements. Migratory, birds moving chiefly south to winter in southern and central Asia. Winter range overlaps slightly with breeding range. In zone of sympatry with Yellowhammer (western Siberia), more migratory than that species and makes longer movements.

Food. Seeds and other plant material; insects in breeding season. Forages primarily on ground and in low bushes; on breeding grounds, feeds at forest edge or in large clearings; in winter quarters, where specializes on cereal grains, searches for food in flocks, often with other seed-eaters, on arable fields (bare soil or stubble), waste ground, in orchards, villages, parks, by roads and tracks, etc.; often near water and swampy places.

Social pattern and behaviour. Gregarious outside breeding season. No evidence for other than monogamous mating system. Hybridization with Yellowhammer common in sympatric zone. F1-hybrids are fertile, leading to varying degrees of departure from pure Pine Bunting or Yellowhammer. In sympatric zone, 2.5% of total population are F1-hybrids; *c.* 15% of all Pine Buntings and *c.* 20% of Yellowhammers show intermediate features indicative of varying degrees of mixed heredity. Song and other elements of behaviour very like Yellowhammer.

Voice. Song very like both Yellowhammer and Pine Bunting × Yellowhammer hybrids, and doubtfully separable from them. Similarly consists of rapid introductory series (with marked crescendo) of high-pitched tinkling buzzy units or motifs, typically followed by 1–2 units of differing structure which may be considerably drawn out. Other calls also similar to Yellowhammer, with, at most, subtle differences of tone or pitch.

Breeding. SEASON. Western Siberia: eggs laid from beginning of May; exceptional clutches at end of June are perhaps 2nd broods. Central Siberia: eggs laid about mid–May; 2nd broods fledge towards late July. SITE. In depression on ground, under bush, grass tussock, fallen branch or tree, etc. Nest: very like that of Yellowhammer; bulky foundation of tightly woven stalks, rootlets, and dry grass, lined with soft grasses and very often with horsehair. EGGS. Sub-elliptical, smooth and slightly glossy; very like Yellowhammer; very pale whitish-blue/green to pinkish or grey with faint purplish and lavender-grey spots and blotches plus a few brownish-black hairstreaks. Clutch: 4–5 (3–6). INCUBATION. 13 days. FLEDGING PERIOD. 9–10 (–14) days.

Wing-length: ♂ 88–100, ♀ 84–95 mm.
Weight: ♂ mostly 26–35, ♀ 24–34 g.

Yellowhammer *Emberiza citrinella* PLATES: pages 1647, 1684

DU. Geelgors   FR. Bruant jaune   GE. Goldammer   IT. Zigolo giallo
RU. Обыкновенная овсянка   SP. Escribano cerillo   SW. Gulsparv

Field characters. 16–16.5 cm; wing-span 23–29.5 cm. About 10% longer than Chaffinch, with distinctly longer and more forked tail; slightly longer and noticeably less compact than Cirl Bunting. Rather large bunting, with noticeably attenuated rear body and tail. Matched closely in size and form only by Pine Bunting. Adult plumage features basically lemon-yellow head (little marked and brilliant in ♂), streaked warm brown upperparts, long rufous-chestnut rump, yellow underparts with streaked chest, and bright white outer tail-feathers. Immature far less distinctive, with less yellow and more obvious streaks on underparts.

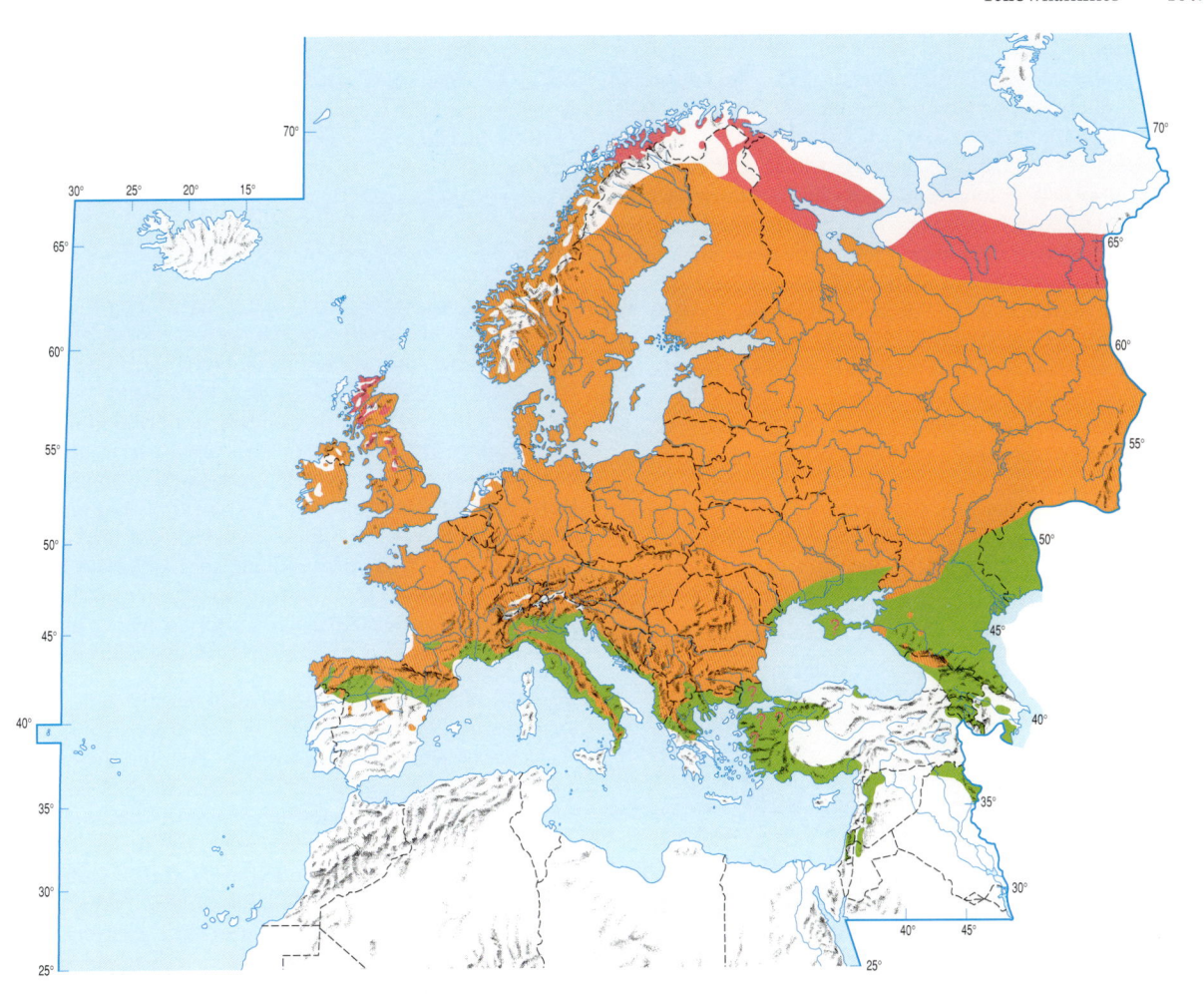

Commonest, most widespread bunting of west Palearctic. ♂ unmistakable. Adult and (particularly) juvenile ♀♀ constitute pitfalls for observers unaware of their marked plumage variation and eager to identify other congeners, particularly Cirl Bunting and closely related Pine Bunting; see those species for diagnosis, best based for Cirl Bunting on structure and voice and for Pine Bunting on lack of yellow in ground-colour of plumage. Flight recalls Chaffinch but action rather stronger, with often quicker bursts of wing-beats on take-off giving marked acceleration and rather shorter wing closures producing at times remarkably direct progress and always less marked and less regular undulations. Stance variable; often markedly upright on perch, with long forked tail below level of body, but noticeably horizontal on ground.

Habitat. Breeds across temperate and boreal zones of west Palearctic, within July isotherms 12–23°C, mainly in open lowlands or hilly country, in both continental and oceanic climates. Prefers dry sunny habitats with fairly rich and varied vegetation, avoiding dense forest, undrained wetlands, towns, or busy inhabited areas. Probably originally based on edges of open areas of forest (including coniferous taiga as well as broad-leaved woods) and fringing scrub of gorse, broom, hawthorn, and juniper, together with northern birch zone. Has profited by farming to extend widely across cultivated land with hedges, plantations, and paths or highways flanked by trees and bushes, but stops short of gardens, cemeteries, and ornamental parks.

Distribution. Decreases reported in north-west Europe, apparently due to changing land use and farming practices. BRITAIN. Highest numbers in eastern Britain and Midlands. Marked recent withdrawal from uplands and Scottish islands. IRELAND. Has disappeared from large areas of west; decline continuing. BELGIUM. Long-term contraction of range. NETHERLANDS. Retreating from many former breeding areas in west and north. NORWAY. Slight decrease. TURKEY. Reports of single singing ♂♂ late April to June in various years, especially in north-west; probably breeds at least in Thrace.

Accidental. Bear Island, Iceland, Faeroes, Balearic Islands (almost annual), Malta, Kuwait, Morocco.

Beyond west Palearctic, extends east across Siberia to c. 110°E. Introduced in New Zealand.

Population. Stable in most countries, but some declines, especially in north-west Europe. BRITAIN. 1.2 million territories 1988–91. Essentially stable on farmland, but decline in (sub-optimal) woodland habitat. IRELAND. 200 000 territories 1988–91. FRANCE. Over 1 million pairs. BELGIUM. 21 000–38 000 pairs 1989–91; long-term decline. LUXEMBOURG. 18 000–20 000 pairs. NETHERLANDS. 22 000–28 000 pairs 1989–91; decline of over 50% since 1960s. GERMANY. 1.7 million pairs in mid-1980s; in east, 280 000 ± 150 000 pairs in early 1980s, with decline in last 2–3 decades. DENMARK. 260 000–2.7 million pairs 1987–8. NORWAY. 200 000–500 000 pairs 1970–90; slight decrease. SWEDEN. 600 000–1.8 million pairs in late 1980s. FINLAND. 1–1.5 million pairs in late 1980s. ESTONIA. 100 000–200 000 pairs in 1991. LATVIA. 80 000–160 000 pairs in 1980s; slight decrease. LITHUANIA. Common; increasing. POLAND. Numerous. CZECH REPUBLIC. 2–4 million pairs 1985–9. SLOVAKIA. 800 000–1.5 million pairs 1973–94. HUNGARY. 350 000–400 000 pairs 1979–93. AUSTRIA. Very common. SWITZERLAND. 15 000–30 000 pairs 1985–93. SPAIN. 140 000–170 000 pairs. PORTUGAL. 100–1000 pairs 1978–84. ITALY. 20 000–50 000 pairs 1983–95; slight decrease. GREECE. 2000–5000 pairs. ALBANIA. 2000–5000 pairs in 1981. YUGOSLAVIA: CROATIA. 70 000–100 000 pairs. SLOVENIA. 30 000–50 000 pairs. BULGARIA. 10 000–100 000 pairs. RUMANIA. 450 000–600 000 pairs 1986–92. RUSSIA. At least 10 million pairs. BELARUS'. 160 000–200 000 pairs in 1990. UKRAINE. 300 000–350 000 pairs in 1986. MOLDOVA. 60 000–80 000 pairs in 1988.

Movements. Sedentary to migratory, with most populations partial migrants; also dispersive. Vacates entirely only extreme north of range, and winters chiefly within breeding range, especially in milder years. European migrants head chiefly south-west, usually moving only short or medium distances (up to c. 500 km in northern Europe, and c. 250 km in central Europe), so birds wintering south of range in Mediterranean region are mostly from central or southern Europe. Ringing data show that individuals winter in widely differing areas in different years: e.g. bird ringed winter in northern France recovered in south-west France in later winter, one ringed eastern France was recovered in northern Italy, and 2 ringed Germany recovered in Spain. Hard weather movements occur midwinter.

Autumn movement September–November(–December), peaking early October in northern Europe. In south of winter range, recorded chiefly December–February in Camargue (southern France) and Cyprus, November–February in Jordan and Israel. Spring movement February–May, mostly March–April. Reaches extreme north of Scandinavia late April to May.

Food. Seeds, chiefly of grasses; invertebrates in breeding season and casually throughout remainder of year. Some plant families (e.g. Cruciferae) completely ignored in wild although among commonest in habitat; seemingly avoids oily seeds, preferring those rich in starch. Feeds almost wholly on ground; in spring, forages near nest-sites, in woodland clearings and borders, by hedges and tracks, in newly-sown fields, etc.; in summer and autumn, on pasture and arable land, waste ground, stubble, and other harvested fields; in winter, also in agricultural areas, but in severe weather, particularly snow, comes to settlements, farmyards, animal feed, etc., though not often to gardens. Outside breeding season, feeds in flocks, often with other seed-eaters.

Social pattern and behaviour. Usually in flocks outside breeding season, but these usually loosely knit, primarily associations at good feeding sites, often with other species; during migration, flocks more closely integrated. Monogamous mating system; rarely polygynous. Solitary and territorial when breeding. Territories often more or less linear, e.g. along hedges or borders between woodland/scrub and open land. Song usually given from tree or bush; occasionally from clod of earth (etc.) on ground, rarely from building. Song-period long, from taking up of territories in early spring to July–August or later. Pair-formation takes place in ♂'s territory, beginning soon after break-up of winter flocks. ♀♀ seek out territorial ♂♂. Whole process, from first meeting to establishment of pair, may take several weeks. ♂'s courtship display typically consists of short runs on ground near ♀, during which plumage is ruffled and wings may be held up vertically, fully extended.

Voice. Song consists of a series of reiterated units, typically increasing in volume, followed by 1–2 drawn-out units; very characteristic (apart from similarity to Pine Bunting) and variously transcribed in different languages, in English as 'little-bit-of-bread-and-no-cheese'. In incomplete songs, which are common, final notes may be partly or entirely omitted; in latter case, song easily confused with Cirl Bunting. Well-marked but complex geographical dialects occur, affecting terminal part of song. Commonest call, used for contact throughout year, perched or in flight, and, with modifications, also in other contexts, 'zit' or 'tzit'. Flight-call, given also on taking off, a rapid burst of 'tit' units, producing a trilling 'tirr'. Usual alarm call a high-pitched 'see'.

Breeding. SEASON. Most of Europe: eggs laid beginning of April to July or August, rarely to beginning of September. Season begins later in far north and east. 2(–3) broods over west and south of range. SITE. Nearly always on or very close to ground, well hidden among grass or herbage. Typically against bank or base of hedge, small tree, bush, or well inside bramble. Nest: dry grass, plant stems, straw, leaves, and some moss lined with rootlets, fine grass, and horsehair. EGGS. Sub-elliptical to short sub-elliptical; smooth and slightly glossy, or non-glossy. Highly variable in colour; white, tinted bluish, greyish, or purplish, usually with faint, fine spotting or thin scribbles in pale violet-grey or reddish-purple, and with sparse, bold, irregular scrawls, small blotches, and fine hairlines in black or purplish-brown; hairstreaks may be distributed uniformly over whole surface or may form ring around broad end. Clutch: 3–5 (2–6). INCUBATION. 12–14 days. FLEDGING PERIOD. 11–13 days (9–18); young may leave nest before fully fledged.

Wing-length: *E. c. citrinella*: ♂ 80–96, ♀ 79–90 mm.
Weight: *E. c. citrinella*: ♂♀ mostly 25–36 g.

Geographical variation. Rather slight, mainly involving

colour. No differences in size except for birds of Britain, which are rather small. 3 races recognized. Nominate *citrinella* occupies most of west Palearctic range. *E. c. erythrogenys* in east of range (from *c.* 40°E in north and 25°E in south) generally paler, more sandy-brown above, and with extensive deep rufous wash on chest, side of breast, and flank. These races connected by broad zone in which highly variable populations occur. *E. c. caliginosa* from Scotland, Ireland, northern England, and Wales, forms end of cline of increasing colour saturation running from central Europe to north-west. Mantle and scapulars of *caliginosa* warm rufous-brown, with limited yellow-olive fringing along sides of feathers, limited olive in hindneck and across upper chest, faint greenish tinge to yellow of head and underparts, and extensive but dull rufous on chest and side of body; in fresh plumage, top and side of head rather dark brown-green.

Cirl Bunting *Emberiza cirlus*

PLATES: pages 1647, 1684

Du. Cirlgors   Fr. Bruant zizi   Ge. Zaunammer   It. Zigolo nero
Ru. Огородная овсянка   Sp. Escribano soteño   Sw. Häcksparv

Field characters. 15.5 cm; wing-span 22–25.5 cm. Slightly smaller than Yellowhammer, with 20% proportionately shorter bill and 5–10% shorter, rather more rounded wings, but almost as long tail. Medium-sized bunting, recalling Yellowhammer but with rather smaller, slighter, and more compact form most obvious in often flatter-headed and more hunched appearance. Plumage pattern also recalls Yellowhammer but shows at all times dull greyish-olive to greyish-brown rump. Adult ♂ colourful, but much less yellow than Yellowhammer, with striking grey-olive crown, black eye-stripe and bib, yellow supercilium and collar, and russet-sided olive chest. ♀ much less easy to distinguish from Yellowhammer, but ground-colour of plumage more buff than yellow with more linear face pattern, dark greyish-olive (not brown) lesser coverts, and finer streaks below.

Adult ♂ unmistakable, being separated from all other yellow-bellied *Emberiza* by its strongly linear face pattern and pale, dull rump. Immature ♂ and ♀ at all ages far less distinctive, easily confused with dull Yellowhammer and ♀♀ and immatures of Pine Bunting and Rock Bunting, as all these may hide their rufous rumps and share general buffy look of many Cirl Buntings. Secretive in tree and bush cover, even skulking among ground plants, though ♂♂ perch openly when singing.

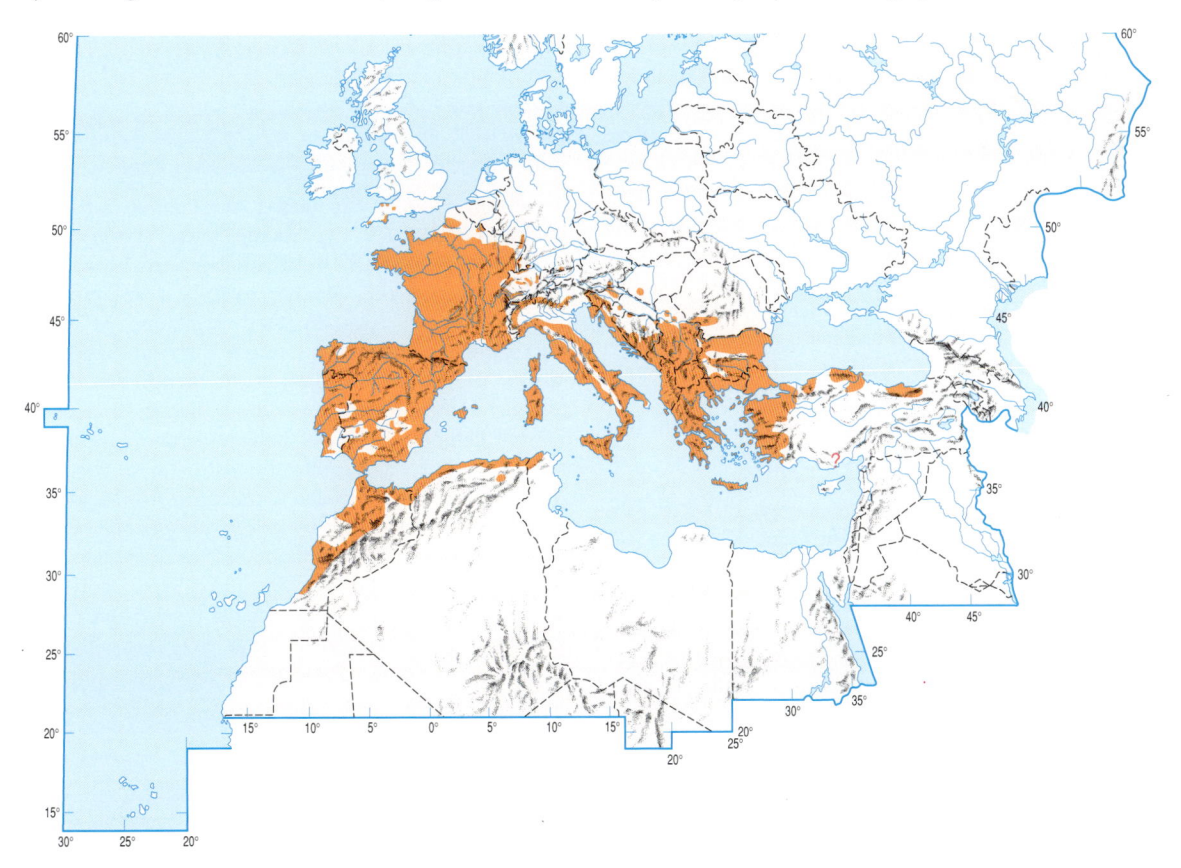

Habitat. Breeds within Mediterranean and adjoining oceanic temperate zones of south-west Palearctic; further limited by highly selective climatic, topographical, and ecological requirements. Except in England, is bounded by 20°C July isotherm. Extends to 17°C isotherm in southern Britain but perhaps more importantly limited as British resident to areas with mean January temperature above 6°C, and either at least 1500 hrs of sunshine or less than 105 cm of rainfall per year; other limiting factors are wind, night-frosts, altitude, slope, aspect, and exposure. In addition to such combination of sunshine, low rainfall, and mild winters, has equally exacting ecological requirements. These take somewhat different forms in north and south of range, although preference for benign, often sloping, and sunny terrain is general. In Britain, in contrast to Yellowhammer, avoids extensive open farmland, being confined mainly to small fields with plenty of hedgerow growth and tall trees, elms being favoured before their widespread demise through Dutch elm disease. Where such sites have been taken by human settlements their fringes are occupied, including large gardens and orchards.

Distribution. Range and numbers have decreased in northwest Europe, probably chiefly due to climatic change, but increased in parts of south-east Europe. BRITAIN. Breeding first recorded 1800, in south Devon; spread through southern counties throughout 19th century to peak in 1930s. Contraction noticeable from 1950s, and main range confined to coastal strip of south Devon by late 1980s. Decline attributed to climatic factors and especially to habitat and agricultural changes. Recent increase in last stronghold due to management measures, including habitat creation. FRANCE. Highest numbers in west and south. Some recent range contraction in north. BELGIUM. Former breeder, confined to warmest areas; last bred 1953, but singing ♂♂ present up to 1984. LUXEMBOURG. Bred until end of 1950s. GERMANY. At beginning of 20th century breeding extended north to Bonn and east to Unterfranken. Now virtually confined to south-west. HUNGARY. First recorded breeding 1975. AUSTRIA. Bred exceptionally in past. Observed regularly in Steiermark since 1980, and breeding first confirmed 1989. May also breed Inntal west of Innsbruck. YUGOSLAVIA. Spreading slowly northwards inland. RUMANIA. Expanding east along south Carpathians. ALGERIA. Common in Aurès region, and widespread in Tell region.

Accidental. Netherlands, Denmark, Poland, Czech Republic, Slovakia, Malta, Ukraine, Egypt, Canary Islands.

Population. BRITAIN. Long-term decline to 210–240 pairs 1973–6, maximum of 119 pairs in 1989. Increase since (see Distribution) to maximum of 241 pairs 1991, 412 pairs 1994. Channel Islands (Jersey): 20 singing ♂♂ in 1992. FRANCE. 100 000–1 million pairs in 1970s; decreasing locally. GERMANY. 250 pairs in mid-1980s; has declined since. HUNGARY. 1–10 pairs. AUSTRIA. 3–4 pairs 1990. SWITZERLAND. 400–800 pairs 1985–93; stable. SPAIN. 500 000–800 000 pairs; slight increase. PORTUGAL. 10 000–100 000 pairs 1978–84; stable. ITALY. 300 000–600 000 pairs 1983–95; stable. GREECE. 50 000–150 000 pairs. ALBANIA. 20 000–50 000 pairs in 1981. YUGOSLAVIA: CROATIA. 40 000–50 000 pairs; slight increase. SLOVENIA. 2000–3000 pairs; stable. BULGARIA. 10 000–100 000 pairs; stable. RUMANIA. At least 100–150 pairs 1986–92; marked increase. TURKEY. 100 000–1 million pairs. TUNISIA. Fairly common. MOROCCO. Uncommon; apparently stable.

Movements. Most populations essentially sedentary, but many leave colder parts of range in continental Europe in winter. Longest movements recorded up to *c.* 600–700 km, mainly in southerly and westerly directions.

Food. Seeds, mostly of grass or cereals; invertebrates in breeding season. Feeds almost wholly on ground, sometimes on stems of grasses or low herbs, most commonly on trampled or grazed grass in fields, by tracks, at vineyard edges, and similar weedy places, rarely on bare soil; in winter, very often on rough pasture and stubble.

Social pattern and behaviour. Outside breeding season some adult ♂♂ remain on territory, often accompanied by mate, otherwise occurs mainly in small parties and flocks. Generally monogamous; pair stays together in subsequent years, if both partners survive. Strongly territorial when breeding; ♂♂ vigorously exclude other ♂♂ from within *c.* 150 m of nest. Song typically given from high in tall tree, but also from telegraph wire, bush, wall, etc.; exceptionally on ground. Main songperiod (in Devon) late March to mid-August, but throughout range some song may occur in any month in fine weather.

Voice. Typical song a brief, rapid, rattling trill or tremolo, much recalling terminal rattle of Lesser Whitethroat: 'zezezeze. . .'. Commonest call, a single short, rather thin and quiet monosyllable, 'tzip', or, with melancholy timbre, 'tzepe'; noticeably less metallic than common call of Yellowhammer, recalling Song Thrush.

Breeding. SEASON. Extended throughout range. Devon (south-west England): 1st clutches started 1st week of May, exceptionally late April; latest clutches started towards end of August. North-central France: over 10 years, mean start of laying 23 April; eggs recorded beginning of September. Pfälzerwald and Rheinland (western Germany): eggs laid early April to late August. 2 broods usual, often 3. SITE. Low down and well hidden in dense tree, shrub, hedge, or creeper; often on wall behind vegetation; rather uncommonly on ground. Nest: rather bulky and untidy; foundation of rough stalks, roots, grass, leaves, and moss (which is sometimes main material), lined with fine stems and much hair but not feathers. EGGS. Subelliptical, smooth and very slightly glossy; greyish-white, often flushed pinkish or bluish, with many sepia spots, blotches, scrawls, and hairstreaks concentrated towards broad end, and pale grey undermarkings. Clutch 3–4 (2–5). INCUBATION. 12–13 days. FLEDGING PERIOD. 11–13 days.

Wing-length: ♂ 76–86, ♀ 74–81 mm.
Weight: ♂♀ mostly 21–29 g.

White-capped Bunting *Emberiza stewarti*

Du. Stewarts Gors   Fr. Bruant de Stewart   Ge. Silberkopfammer   It. Zigolo capobianco
Ru. Овсянка Стюарта   Sp. Escribano de Stewart   Sw. Svarthakad sparv

A central Asian species, breeding from southern Kazakhstan through Afghanistan to northern Pakistan, and wintering in Himalayan foothills and adjacent plains. ♀, showing no signs of captivity, was trapped in Herve (Belgium) on 9 August 1931, and kept in captivity until 10 April 1935. Not accepted for Belgian list, but considered by some authorities to have been a wild bird, taking into account the state of its plumage and the fact that the species was not known to have been imported.

Rock Bunting *Emberiza cia*

PLATES: pages 1654, 1684

Du. Grijze Gors   Fr. Bruant fou   Ge. Zippammer   It. Zigolo muciatto
Ru. Горная овсянка   Sp. Escribano montesino   Sw. Klippsparv

Field characters. 16 cm; wing-span 21.5–27 cm. Slighter than Yellowhammer with rather short, fine bill, somewhat shorter and rounder wings, and narrower tail, though tail averages longer; 20% larger than House Bunting. Medium-sized but relatively slim bunting, with long, thin tail contributing to more attenuated outline than any other congener. Shares rufous-buff ground-colour to plumage (particularly underparts) with 5 other *Emberiza*; best distinguished by rather small, lead-coloured bill, strong head pattern (particularly ♂) of blackish crown-stripes and complete black surround to ear-coverts on greyish ground, and strongly rufous rump. Terrestrial, rarely far from rocks.

♂ unmistakable if seen well, but in glimpse or in shadow may suggest Cirl Bunting (black throat, yellow and olive ground-colour to head and breast, more yellow underbody) or House Bunting (much smaller, and lacks bright wing-bars and deep rufous rump, but shares grey throat and breast and, in western race *sahari*, black stripes on head). ♀ and immature

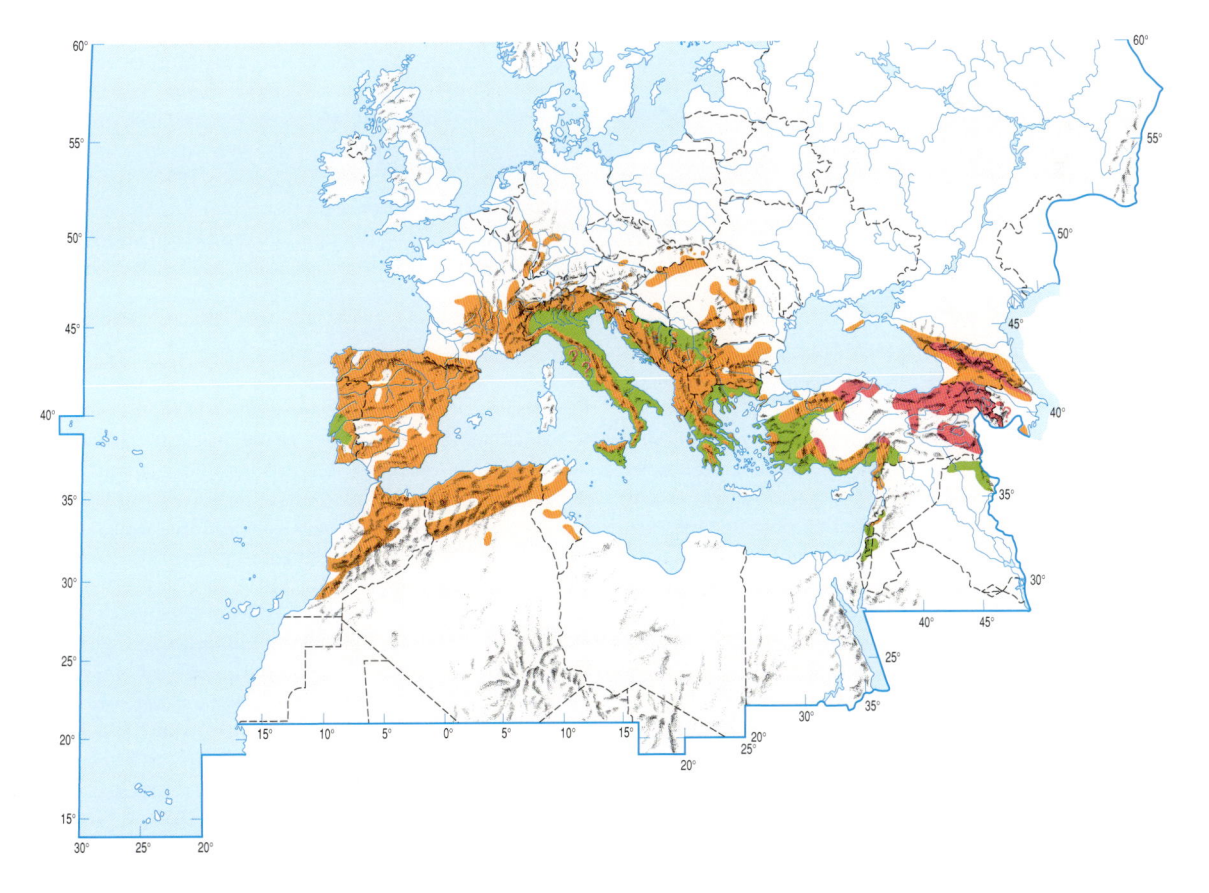

Rock Bunting *Emberiza cia*. *E. c. cia*: **1** ad ♂ spring, **2** ♂ winter, **3** ad ♀ spring, **4** ad ♀ winter, **5** 1st winter ♂, **6** juv, **7** ad ♂ spring. *E. c. prageri*: **8** ad ♂ spring. Cirl Bunting *Emberiza cirlus* (p. 1651): **13** ad ♂ spring, **14** ad ♂ winter, **15** ad ♀ spring, **16** ad ♀ winter, **17** juv, **18** ad ♂ spring 'nigrostriata' (heavily streaked form, Corsica and Sardinia).

troublesome, needing careful separation from Pine Bunting (slightly larger but less attenuated, with horn bill, buffish-white ground-colour to face, and dull white underparts, always well streaked on flanks), Yellowhammer (also larger and fuller-tailed, with at least yellowish hue to face and underparts), and House Bunting (much smaller, even more uniform rufous-buff in appearance, with less marked head and unstreaked underparts). Important to recognize that Rock Bunting is usually a bird of higher altitudes than any confusion species, mixing with House Bunting only during descent in winter to surroundings of arid mountains.

Habitat. Extends across lower middle latitudes of west Palearctic, from Mediterranean to Caucasus, from warm temperate to steppe climatic zones, within July isotherms of *c*. 20–30°C. Avoids most humid or wet situations, closed forest, and good agricultural land, preferring sunny semi-arid terrain, often stony or rocky, with more or less sparse shrub vegetation, and usually with no more than scattered trees. Often on slopes or hillsides, at up to 1900 m. Occupies open areas at upper forest limits, juniper scrub, subalpine meadows with shrubs and screes, stone-walled cultivated areas, and vineyards on hillsides. In northern part of European range seeks out warm dry southern or south-west slopes with oak scrub or other bushy cover, hedges, or patches of young conifers. Descends in winter to lower ground. European wintering birds feed on ground close to hedges, bushes, copses, and wayside trees.

Distribution. Rather localized. FRANCE. Some range retraction in north between 1930s and 1970s; local fluctuations. SLOVAKIA. Breeding first recorded 1954, followed by limited spread. HUNGARY. First recorded breeding in 1950s in north. SPAIN. Slight decrease. RUMANIA. Distribution very sporadic and fragmented. UKRAINE. Slight decrease. SYRIA. Presumed to breed in extreme north-west, but confirmation needed. ISRAEL. First sign of possible breeding Mt Hermon, juvenile trapped August 1967; a few pairs active 1970s. IRAQ. Probably breeds in northern mountains.

Accidental. Britain, Belgium, Luxembourg, Sweden, Poland, Czech Republic, Balearic Islands, Malta, Kuwait, Canary Islands.

Beyond west Palearctic, breeds in Iran and from western Himalayas north to Altai region.

Population. Stable in most countries. Declines probably due to habitat loss. FRANCE. 1000–10 000 pairs in 1970s. GERMANY. 600–700 pairs. Marked fluctuations and marked long-term decline. In southern Schwarzwald, declined by at least 35–40% between 1950s and 1988; main causes afforestation, agricultural changes, and disturbance; estimated *c*. 35 pairs 1986–8. SLOVAKIA. 150–250 pairs 1973–94. HUNGARY. 100–120 pairs 1979–93. AUSTRIA. Rare and very local. SWITZERLAND. 1000–2500 pairs 1985–93. SPAIN. 820 000–2 million pairs; slight decrease. PORTUGAL. 100 000–1 million pairs 1978–84. ITALY. 30 000–60 000 pairs 1983–95; slight decrease. GREECE. 10 000–20 000 pairs. ALBANIA. 5000–10 000 pairs in 1981. YUGOSLAVIA: CROATIA. 1000–2000 pairs. SLOVENIA. 1000–2000 pairs. BULGARIA. 1000–10 000 pairs. RUMANIA. 400–1000 pairs 1986–92. UKRAINE. 1–10 pairs in 1978; slight decrease. AZERBAIJAN. Locally common. TURKEY. 100 000–1 million pairs. Israel. *c.* 20 pairs in 1st half of 1980s; increase to 50–100 pairs in 1990. Tunisia, Morocco. Uncommon.

Movements. Most populations sedentary or dispersive, but partial migrant or migrant over short or medium distances from parts of range with coldest winters. Altitudinal movements in many high mountain areas.

Food. Seeds, mainly of grasses, and other parts of plants; invertebrates in breeding season. Feeds principally on ground among rocks and scrubby vegetation, or in short grass in fields, at woodland edges, etc., but not infrequently in bushes or tall herbs taking both seeds and insects. Mostly picks seeds from ground, but will also stand on stems, sometimes several at a time, to bend them over and reach seed-head, reaches over to seed-head from neighbouring perch, or pulls seed-head down while standing on ground. Catches flying insects in short sallies just above ground.

Social pattern and behaviour. Outside breeding season, occurs singly or in groups, of young and adults, sometimes in larger flocks. Territorial during breeding season. Song used by ♂ only, to mark and defend territory. ♂ typically sings from bare spike or other exposed perch at top of tree or bush, but also from vineyard stake or wire. In Switzerland, song-period early April to July or beginning of August, with some in October. Quiet song reported in mild still weather in winter from area outside breeding range.

Voice. Song of ♂ likened to song of Dunnock by many observers, although some variants suggest short song of Wren. Very variable, but typically fast, rather long (for *Emberiza*), with short, high-pitched units and musical phrases; often a squeaky timbre and somewhat jerky delivery. May incorporate mimicry of song and calls of other birds, including calls of Chaffinch and parts of song of Redstart. Calls include sharp 'tzit' like Cirl Bunting or thin, weak 'zeet', one or other occasionally repeated at least 3 times, and high, drawn-in 'seeee' or 'seea'.

Breeding. Season. Western Germany: eggs laid mainly about 1st half of May (mid-April to July, rarely August). Switzerland: eggs laid from late April or late May depending on altitude; 2nd clutches after mid-June, and eggs recorded mid-August. Hungary: early May to at least mid-July. Algeria: full clutches found early April to mid-June. 2–3 broods. Site. On or close to ground in cleft in rock or between boulders on slope, usually by bush, etc., generally hidden by vegetation though sometimes exposed; also in wall or earth bank, or low in dense tree or bush. Nest: foundation of dry grass, stalks, and roots, occasionally leaves and bits of bark, lined with fine grasses, rootlets, and some hair. Eggs. Sub-elliptical, smooth, and faintly glossy; very pale greyish- to purplish-white, heavily and intricately scrawled with long and meandering dark violet to black hairstreaks, sometimes forming ring at broad end. Clutch: 4–5 (3–6). Incubation. 12–14 days. Fledging Period. 10–13 days, young leaving nest before able to fly.

Wing-length: *E. c. cia*: ♂ 74–87, ♀ 72–83 mm.
Weight: *E. c. cia*: ♂♀ mostly 21–29 g.

Geographical variation. Slight, strongly clinal. 2–3 races recognized in west Palearctic, with extensive intergradation. Nominate *cia* in central Europe rather small with dark rufous-cinnamon upperparts; *par* in arid hills of west-central Asia (probably including Iraq) is large, with sandy-cinnamon upperparts with narrow black streaks and more uniform pinkish-rufous-cinnamon chest to under tail-coverts. *E. c. prageri*, of north-east Turkey, Caucasus area, and Crimea, distinctly paler and larger than any population of nominate *cia*, but hardly distinguishable from *par*; rufous-cinnamon of upperparts and belly brighter than in nominate *cia*, more pinkish, chin to chest paler grey; tips of median upper wing-coverts red-brown, not white as in nominate *cia*.

Meadow Bunting *Emberiza cioides*

PLATES: pages 1645, 1684

Du. Weidegors   Fr. Bruant à longue queue   Ge. Wiesenammer   It. Zigolo muciatto orientale
Ru. Длиннохвостая овсянка   Sp. Escribano cioide   Sw. Ängssparv

Breeds in open country and forest edges from central Asia (*c.* 80°E) east to eastern Siberia, Korea, Japan, and eastern China south to Kwangtung. Northern populations winter south of breeding range or at lower elevations within general breeding range; southern populations winter within breeding range. Record of singing ♂ in south-west Finland, May 1987, formerly accepted as wild bird but now presumed to be escape. Also unconfirmed record of 2 birds at Veneto (Italy), 1910.

House Bunting *Emberiza striolata*

PLATES: pages 1657, 1684

Du. Huisgors   Fr. Bruant striolé   Ge. Hausammer   It. Zigolo delle case
Ru. Домовая овсянка   Sp. Escribano sahariano   Sw. Hussparv

Field characters. 13–14 cm; wing-span 21.5–26 cm. 15–20% shorter and much slighter than Rock Bunting, Cretzschmar's Bunting, and Cinnamon-breasted Rock Bunting, with rather stubby bill, somewhat small head, and narrow tail. Rather small, delicate, cryptically plumaged bunting, with dark rufous-edged tail. Western race *sahari* tame and rather dull, with little-marked grey and rufous-buff plumage; larger eastern race nominate *striolata* shyer and more puzzling, with blackish stripes on head of typical ♂ suggesting Rock Bunting or (even more) Cinnamon-breasted Rock Bunting. Tail dull brownish-black with rufous (not white) edges to outer feathers.

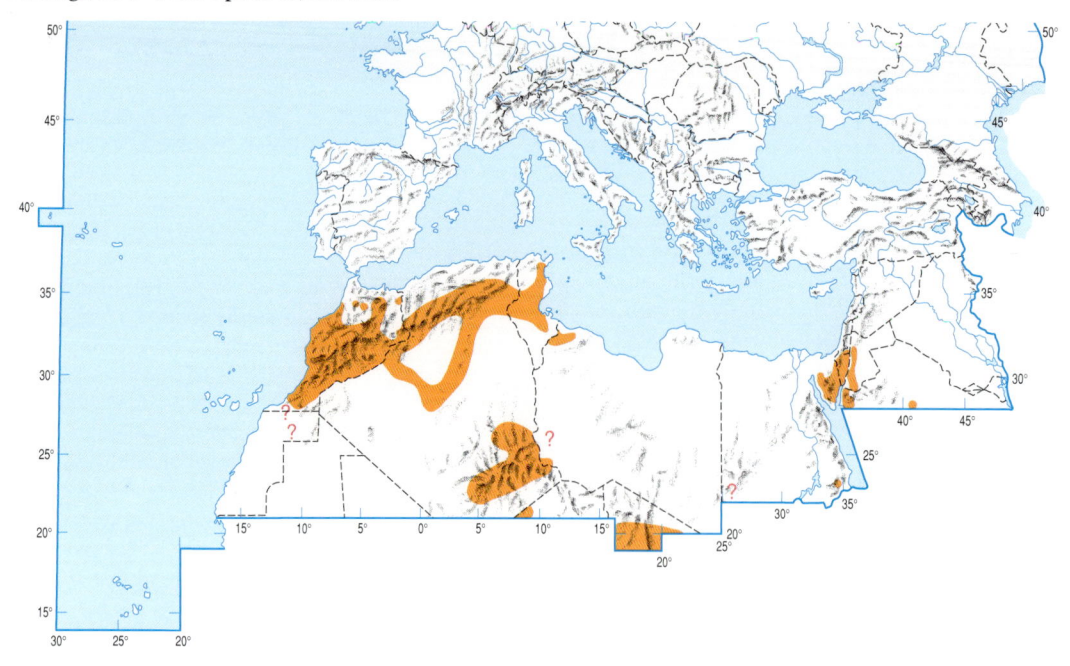

Confusing. Where not commensal with man, difficult to observe due to small size and cryptic plumage producing distinctive 'will of the wisp' character. When size not apparent, can be confused in Middle East with migrant Cretzschmar's Bunting and allies (pink bill, bright eye-ring, more streaked above and on wings, white outer tail-feathers); in north-west Africa, particularly in winter, with Rock Bunting (grey bill, white bar on median coverts, rufous-chestnut rump, white outer tail-feathers); and with potential vagrant from north-east Africa and Arabia, Cinnamon-breasted Rock Bunting (also short-winged but noticeably longer tailed, more strongly marked head with fully black or blackish throat and bib, much darker streaks on back and centres to wing-coverts, dark-spotted rump, and different calls).

Habitat. Situation confused by existence of ecologically distinct races, one of which (*sahari* of north-west Africa) has long been largely adapted to commensalism with man in inhabited settlements, while nominate *striolata* in Asia has remained attached to natural rocky habitat apart from having colonized some ruins of forts and other buildings. In north-west Africa, replaces House Sparrow in villages on edge of desert, even entering houses and shops. Also sometimes found in wild desolate places, at some altitude, but never in open desert. Nominate *striolata* in Asia lives in arid areas on rocky slopes with sparse vegetation; when found in desert and semi-desert is usually within reach of water. In winter, spreads to sandy plains, tamarisk scrub, and grass areas near canals, feeding on ground.

Distribution. ISRAEL. Widespread but scattered distribution. EGYPT. May also breed Gebel Uweinat in extreme south-west. TUNISIA, ALGERIA. Has spread north in 20th century. MOROCCO. Marked northward spread. Advance of more than 200 km along Atlantic coast recorded between 1867 and 1902. Has continued, reaching Casablanca 1965; breeding first recorded Oujda 1980, Rabat 1983, Fes 1985, Méknès 1989. WESTERN SAHARA. Extending southward from Morocco: recorded Layoune January 1995.

Accidental. Spain, Turkey, Cyprus, Kuwait, Canary Islands.

Beyond west Palearctic, breeds in Africa in narrow band from Mauritania and Sénégal east to Ethiopia and Somalia, thence south to northern Kenya and from Arabia north-east to north-west India.

Population. Has undoubtedly increased in north-west Africa with expansion of range, but no comparative data. ISRAEL. Rough estimate some thousands of pairs; sharp fluctuations, decreasing after dry winters. JORDAN. Uncommon. EGYPT. Fairly common. TUNISIA. Common in south; rarer in central areas. ALGERIA. Locally common. MOROCCO. Common.

Movements. Essentially sedentary, but with short-distance movements in some populations.

Food. Seeds, mostly of grasses; invertebrates in breeding season. In North Africa, very dependent on man, feeding in streets, inside houses, restaurants, etc., and on rubbish tips and dung heaps, though also on ground in rocky country. In east of range, however, generally forages far from human habitation.

House Bunting *Emberiza striolata*. *E. s. sahari*: **1** ad ♂ spring, **2** ad ♂ winter, **3** ad ♀, **4** juv. *E. s. striolata*: **5** ad ♂ spring, **6** ad ♀. Cinnamon-breasted Rock Bunting *Emberiza tahapisi* (p. 1659): **7** ad ♂, **8** ad ♀, **9** juv.

Social pattern and behaviour. In breeding season mainly in pairs; at other times often gathers in groups. During breeding season pairs seem to be monogamous and stable; pair-bond apparently year-long. Territoriality may be weak and rather fluid; perhaps undefended areas between territories. Territory apparently serves for feeding and nesting, and may include interior of occupied houses. Song used by ♂ to advertise territory. In Morocco, song-posts on conspicuous perches, commonly on walls of houses and gardens, including occupied buildings in city of Casablanca; in early spring, ♂ spends much time on song-post, with few movements away from it; some song throughout year. In Morocco, where regarded as sacred and receiving traditional protection, often very tame, feeding inside houses, shops, and mosques.

Voice. Song a short to medium-length, rather feeble phrase, with accelerated delivery recalling Chaffinch, 'wi-di-dji-du-wi-di-di' or 'witch witch a wee'. Calls include nasal 'tzswee', thin sharp 'tchiele', and 'sweee-doo'.

Breeding. SEASON. Morocco: eggs laid late January to October. Israel: eggs laid end of March to end of June. 1–2 broods in Israel, up to 3 in North Africa. SITE. In North Africa, in or on buildings but also in rocky country away from settlements; in east of range, generally avoids human habitation. Nest: small and cup-shaped; foundation of twigs, roots, grass stems, straw, etc., lined with hair, wool, plant down, and man-made material. EGGS. Sub-elliptical, smooth and faintly glossy; whitish, sometimes tinged faintly blue or green, speckled or spotted purplish or dark brown, markings increasing in intensity towards broad end. Clutch: 2–4(–5). INCUBATION. 13–14 (12–16) days. FLEDGING PERIOD. 17–19 days.

Wing-length: *E. s. sahari* (north-west Africa): ♂ 73–87, ♀ 72–78 mm.
Weight: *E. s. sahari*: ♂ 12–18, ♀ 13–16 g.

Geographical variation. Rather slight. Nominate *striolata* from Middle East slightly smaller than *sahari*, especially in bill and tail. Head pattern more prominent: cap grey with black streaks, but ground-colour of cap occasionally slightly pinkish-grey, less whitish, and black streaks sometimes less profuse; in worn plumage, prominent white stripe on central crown, supercilium and malar stripe whiter, eye-stripe and moustachial stripe blacker.

Cinereous Bunting *Emberiza cineracea*

Du. Smyrnagors   Fr. Bruant cendré   Ge. Türkenammer   It. Zigolo cenerino
Ru. Серая овсянка   Sp. Escribano cinéreo   Sw. Gulgrå sparv

PLATES: pages 1659, 1685

Field characters. 16–17 cm; wing-span 25–29 cm. Size between Yellowhammer and Black-headed Bunting, looking rather more attenuated and longer-winged than partly sympatric Cretzschmar's Bunting. Rather large, strangely featureless bunting, most resembling ♀ or immature Black-headed Bunting but showing white tail-feathers. At close range, shows faint plumage pattern converging with Cretzschmar's Bunting and allies; pale grey bill and at least faintly yellow throat crucial in identification. Sexes closely similar.

Lacks obvious characters but, compared with much more colourful migrant congeners, its featurelessness is immediately striking. At close range, head of ♂ may suggest Ortolan Bunting but grey bill and predominantly dull greyish plumage, lacking any rufous tones on underparts, allow ready distinction. ♀ and immature even more nondescript than ♂ but their white outer tail-feathers instantly exclude Black-headed Bunting and Red-headed Bunting. Note that Iranian race *semenowi* (occurring on migration in Middle East) is distinctly yellow on belly and under tail, initially suggesting last-named species more strongly. Immature may also be confused with palest Ortolan and particularly with Grey-necked Bunting but both these species show bright flesh- to reddish bill and at least buff-, if not rufous-, tinged underparts. Flight and behaviour as Ortolan Bunting.

Habitat. Imperfectly known, owing to scarcity of data from restricted and inaccessible areas of occurrence in south-east of west Palearctic and thinly inhabited winter quarters. Summer visitor to scrub-covered uplands in Turkey, in warm temperate or Mediterranean climate, occurring on high slopes, dry and rocky with sparse shrubby vegetation, and wintering often in dry coastal areas. Seems also to occur on passage in lowland deserts.

Distribution and population. Distribution patchy. GREECE. 100–200 pairs. In Aegean Sea, found to breed on Lesbos and Chios 1950s–60s, and Skyros 1994. In Ionian Sea, at least 3 singing ♂♂ Corfu, June 1991. TURKEY. 500–5000 pairs. Widely scattered in small numbers. SYRIA. Possibly breeds. IRAQ. Probably breeds in north (Kurdistan).

Beyond west Palearctic, breeds only in western Iran.

Movements. Migrant, moving south or south-west to winter (apparently) in eastern Sudan, Eritrea, Yemen, and south-west Saudi Arabia. Begins to leave breeding grounds in July, although juveniles sometimes stay as late as September. 2 main migration routes. Western route through southern Turkey via Syria, Lebanon, Jordan, Israel, and Egypt to Sudan and Eritrea used predominantly by *E. c. cineracea*, but also by smaller numbers of *semenowi*. Eastern route follows coast of Arabian Gulf and is used exclusively by *semenowi*; supposed wintering area for these is at south-west tip of Arabia. Arrives on breeding grounds early April. Uses same western and eastern migration routes as in autumn, with migrants recorded February–May.

Food. Diet seems to be principally seeds and small invertebrates, which are probably main food in breeding season. On passage in Israel, feeds on seeds in stubble in flocks with Cretzschmar's Bunting and Ortolan Bunting; also on rocky slopes and desert uplands with low vegetation and scrub.

Social pattern and behaviour. Little known; apparently similar to congeners.

Voice. Song a simple, ringing, tuneful phrase of 5–6 notes: 'drip-drip-drip-drip-drie-drieh'. Commonest call a short, metallic 'kjip', 'kup', 'kleup', or 'cluff'.

Breeding. SEASON. Turkey: in west, eggs recorded 2nd half of April and late May. SITE. On ground, on slope with sparse vegetation, though not usually arid, against rock partly hidden by overhanging grass, etc. Nest: one in eastern Turkey had foundation of stalks and stems, leaves of thistle, and grass-heads, lined with rootlets and hair; wall very thin where touching rock but thick and well-woven on opposite side. EGGS. Short sub-elliptical, smooth and slightly glossy; similar to other closely-related Emberizidae but perhaps more intensely blue, with fewer dark brown scrawls, and more spots and blotches rather evenly distributed over surface, with some concentration at broad end. Clutch: (3–)4–6. No further information.

Wing-length: *E. c. cineracea*: ♂ 86–96, ♀ 84–90 mm.
Weight: Iran, April: ♂ (6) 21.1–29.7, ♀ (2) 23.5, 24.8 g.

Cinereous Bunting *Emberiza cineracea*. *E. c. cineracea*: **1** ad ♂ spring, **2** ad ♂ winter, **3** ad ♀, **4** juv. *E. c. semenowi*: **5** ad ♂ spring, **6** ad ♀. Grey-necked Bunting *Emberiza buchanani* (p. 1662): **7** ad ♂ spring, **8** ad ♂ winter, **9** ad ♀, **10** juv.

Geographical variation. Marked, mainly involving colour and, to lesser extent, wing-length. Compared with nominate *cineracea* of western Turkey, *semenowi* in south-west Iran and probably neighbouring Iraq has cap of ♂ brighter yellow, less greenish yellow, green tinge extending faintly over hindneck, mantle, and rump; entire underparts bright sulphur-yellow, tinged olive-grey on chest, side of breast, and flank (chest less grey and belly less white than in nominate *cineracea*); under tail-coverts greyish-white with narrow dusky shaft-streaks and grey centre (more uniform off-white in nominate *cineracea*).

Cinnamon-breasted Rock Bunting *Emberiza tahapisi*

PLATE: page 1657

Du. Zevenstrepengors   Fr. Bruant cannelle   Ge. Bergammer   It. Zigolo pettocannella
Ru. Каштановая овсянка   Sp. Escribano tahapis   Sw. Afrikansk klippsparv

A resident, mainly Afrotropical species, widespread in open, rocky habitats south of Sahara, also in Socotra and southern Arabia. Record of this species from Sinai (Egypt) 1984, formerly accepted, now considered to relate to House Bunting.

Ortolan Bunting *Emberiza hortulana*

PLATES: pages 1660, 1685

Du. Ortolaan   Fr. Bruant ortolan   Ge. Ortolan   It. Ortolano
Ru. Садовая овсянка   Sp. Escribano hortelano   Sw. Ortolansparv

Field characters. 16–17 cm; wing-span 23–29 cm. Only marginally smaller but noticeably more compact than Yellowhammer; slightly larger on average than Grey-necked Bunting and Cretzschmar's Bunting. Relatively long-billed, rather round-headed, and rather plump bunting; epitome of trio which also includes Grey-necked Bunting and Cretzschmar's Bunting. Displays in all plumages common characters of bright eye-ring, pale sub-moustachial stripe contrasting with dark malar stripe, and rufous or at least warm buff underparts. ♂ and well-marked ♀ show diagnostic olive-toned head and breast isolating yellow throat; dull ♀ and immature less distinctive, requiring careful separation from allies.

Note that amongst trio of similar rufous-bodied species, only Ortolan Bunting shows extensive yellow on throat and

Ortolan Bunting *Emberiza hortulana*: **1** ad ♂ breeding, **2** ad ♂ non-breeding, **3** ad ♀, **4** 1st winter ♂, **5** juv. Cretzschmar's Bunting *Emberiza caesia* (p. 1663): **6** ad ♂ breeding, **7** ad ♂ non-breeding, **8** ad ♀, **9** 1st winter ♂.

underwing and green tone to head; also has much heavier streaks in all plumages than Grey-necked Bunting and dull brown, not rufous rump, unlike Cretzschmar's Bunting.

Habitat. Very varied, lying within July isotherms of 15–30°C, from high boreal through temperate, Mediterranean, and steppe zones, and to montane zones at *c.* 1500–2500 m in south of range. Attracted to trees, even breeding in forest glades and clearings, as well as pine forests, tree plantations, forest steppe with birch trees, slopes of low mountains overgrown with grass and small pistachio trees, and orchards. Contrastingly, occurs freely in steep ravines, on bare alluvial deposits, and on rocky ground scantily covered with prickly shrubs. Favours regions of high sunshine and low rainfall, regardless of latitude, and where food is readily available will spread widely over cultivated open land. Formerly ranged far north in Scandinavia, not only breeding at sea-level on islands in Gulf of Bothnia but up to *c.* 900 m on fjells of Norway, among junipers and other shrub cover, reflecting apparent preference for high sunlit slopes.

Distribution. Localized distribution, with many isolated populations. Decline in range and numbers (still continuing, and often considerable) widely reported in western and central Europe, mainly due to changes in farming methods, notably loss of hedgerows and lessening crop diversity. Stable Finland, Estonia, Poland, Bulgaria, and perhaps Russia. FRANCE. Major contraction of breeding range between 1960 and 1990, disappearing from 17 départements in north of country; now almost confined to southern half, with decrease continuing. Main cause agricultural intensification, but hunting in south-west probably contributory factor. NETHERLANDS. Formerly locally common in south and east. 130 pairs *c.* 1980, 32 pairs 1990, 13 pairs 1992; disappeared from last strongholds 1994. GERMANY. Close to extinction Niederrhein, Westfalen, and eastern Bayern, where formerly far more widespread. Commoner in east than west, but very unevenly distributed. DENMARK. A few probable breeding records 1980–94. NORWAY. Now confined to a few scattered localities in southeast. One pair bred Nordland 1987. SWEDEN. Disappeared from south in 1950s–60s, due to effect of pesticides; has since recovered but not re-occupied all former areas. In 1970s–80s began to breed in areas of intensive forestry, in clear-felled areas. FINLAND. Common in south and west, more patchily distributed in east (where almost absent in 19th century). LATVIA. Very rare in 19th century; from 1930s spread towards north-east, but expansion halted and now sporadic, patchily distributed. LITHUANIA. Commoner in south. POLAND. Localized. CZECH REPUBLIC. Breeding first recorded 1860; thereafter spread, halting 1950–60. HUNGARY. Only stable population now in Matra hills. RUMANIA. Common in lowlands, rarer in Transylvania. BELARUS'. Stable but fluctuating. TURKEY. Widespread and sometimes locally common, especially in mountains. SYRIA. May breed in north-west. ISRAEL. 6–8 pairs bred Mt Hermon 1979. IRAQ. Breeds northern mountains. ALGERIA. Formerly regarded only as passage migrant. Several tens of pairs have bred regularly in Djurdjura since at least 1977; has certainly bred in Petite Kabylie for about same period, probably also in Aurès mountains and perhaps south-east of Sétif.

Accidental. Iceland, Faeroes.

Beyond west Palearctic, breeds northern Iran and extends east from Urals to *c.* 100°E.

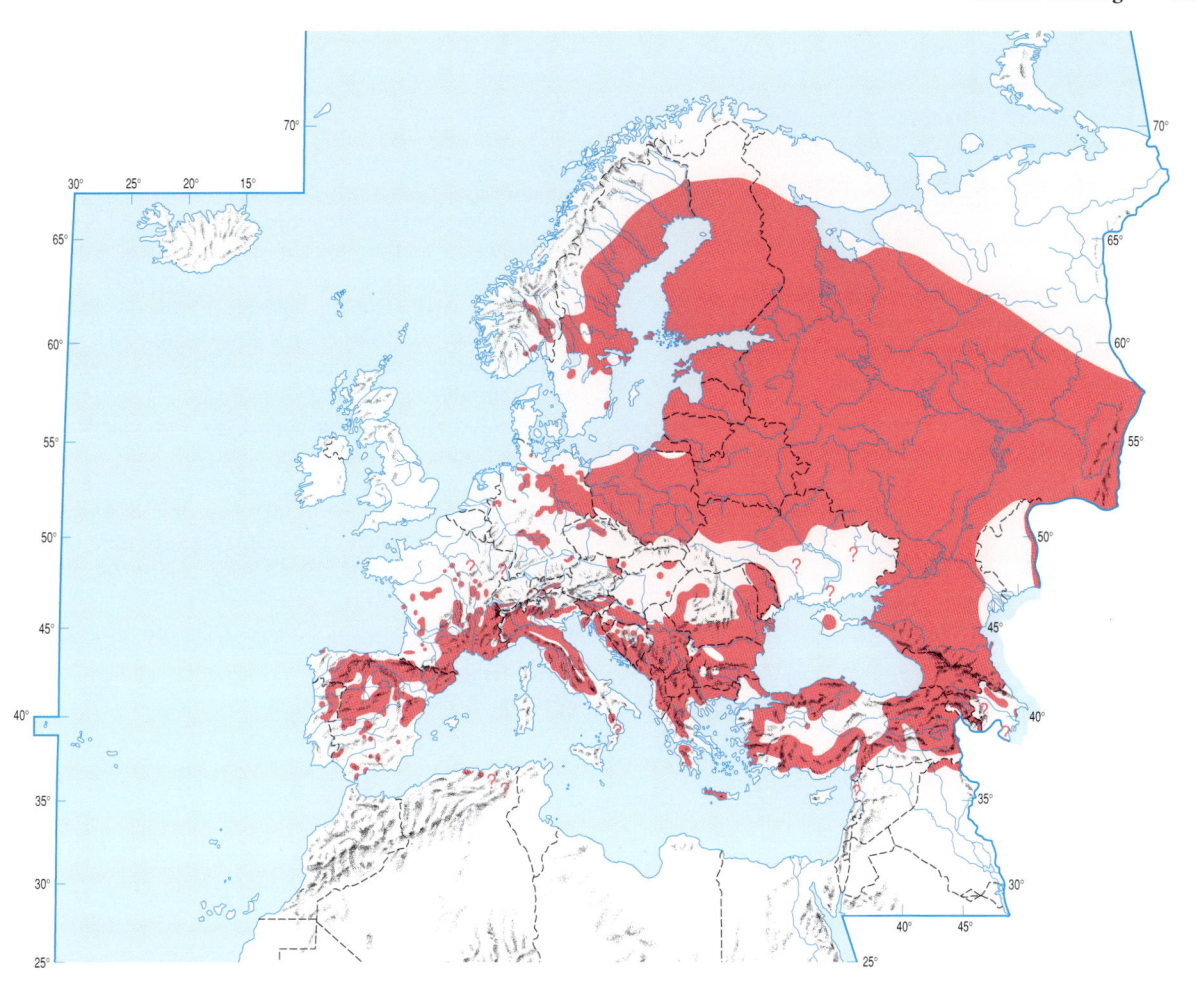

Population. For trend, see Distribution. FRANCE. 10 000–23 000 pairs. BELGIUM. 2–3 pairs 1990–94. 110 pairs 1973–7, when population already in decline. GERMANY. 3000 pairs in mid-1980s. Other estimates: 4000–5000 pairs in early 1990s; 7000 ± 3000 pairs in east in early 1980s. NORWAY. 100–500 pairs 1970–90. SWEDEN. 25 000–100 000 pairs in late 1980s. FINLAND. 150 000–200 000 pairs in late 1980s. ESTONIA. 5000–10 000 pairs in 1991. LATVIA. 500–2000 pairs in 1980s. POLAND. 10 000–100 000 pairs. CZECH REPUBLIC. 200–300 pairs 1985–9. SLOVAKIA. Up to 5 pairs 1973–94. HUNGARY. 10–30 pairs. AUSTRIA. Under 50 pairs 1990. SWITZERLAND. 200–250 pairs 1985–93. SPAIN. 200 000–225 000 pairs. PORTUGAL. 1000–10 000 pairs 1978–84. ITALY. 4000–8000 pairs 1983–95. GREECE. 20 000–30 000 pairs. ALBANIA. 1000–3000 pairs in 1981. YUGOSLAVIA: CROATIA. 1500–2500 pairs. SLOVENIA. 300–500 pairs. BULGARIA. 10 000–100 000 pairs. RUMANIA. 15 000–25 000 pairs 1986–92. RUSSIA. 10 000–100 000 pairs. BELARUS'. 1000–3000 pairs in 1990. UKRAINE. 800–2500 pairs in 1988. MOLDOVA. 6000–8000 pairs in 1988. AZERBAIJAN. Common. TURKEY. 1–10 million pairs.

Movements. Long-distance migrant, wintering in sub-Saharan Africa, north of 5°N. More reported from eastern than western areas of Africa; small numbers winter in southern Arabia. Wintering birds use open upland habitats at 1000–3000 m.

Autumn migration mostly inconspicuous; direction of movement from Fenno-Scandia and other parts of western Europe south-west (or SSW). Spring migration much more conspicuous in most areas. Passage through west and central Mediterranean late March to mid-May with most in 2nd half of April. Passage (or vagrant) birds in Britain and Ireland, mainly on North Sea coast, peak 1st half of May. Arrivals at breeding areas in Belgium and lower Rhine (Germany) from mid-April; main arrival in northern Sweden 2nd half of May. In Leningrad region, birds do not return until 2nd half of May and arrival prolonged into early June.

Food. Probably mainly invertebrates; also seeds, especially outside breeding season. In trees, forages up to topmost crown, and often hovers to pick off caterpillars dangling on silk or on leaves; commonly catches flying insects up to size of large beetles. Pointed, cone-shaped bill ideal for handling both seeds and insects; seeds picked up in tweezer-like bill-tip and manoeuvred with tongue under central ridge of upper mandible, long seeds positioned across bill; de-husking effected by rapid up-and-down movement so husk split on central ridge and expelled via bill-tip; kernel then similarly crushed and swallowed.

Social pattern and behaviour. Regularly occurs singly as well as in small flocks when on passage in northern and central Europe; further south, and in winter quarters, regularly in large flocks. Pair-bond essentially monogamous, apparently lasting for only one breeding attempt. ♂ sings from trees, rocks, bushes, telegraph wires or other elevated perches, sometimes from ground. Occasional song widely reported from migrants on spring passage in North Africa, Middle East, and Mediterranean region. ♂♂ begin singing in spring as soon as they reach breeding grounds, before ♀♀ arrive; regularly advertise territory by song-flights.

Voice. Frequently used in breeding season, less often at other times. Song varies individually and regionally, often with attractive ringing tone to 1st part and lower-pitched more melancholy 2nd part. Calls also vary: shrill 'tsee-up' and lower, fuller, more piping 'tseu' when breeding; shorter 'tsip' and incisive 'twick' in autumn; clear, metallic 'sleee', repeated loosely by overflying nocturnal migrants.

Breeding. SEASON. Sweden, southern Finland, and Leningrad region (north-west Russia): most clutches completed between 20 May and 20 June. Northern Bayern (central Germany): eggs laid early May to late June, mostly late May or early June. Catalonia (north-east Spain): start of laying mid-April, eggs recorded early July. Israel: at 1500–1900 m, from beginning of May to end of July. Normally 1 brood, perhaps occasionally 2. SITE. On ground; in north-west Europe, usually in cereals or other arable crop, often potatoes, frequently in depression in soil so top of nest-rim flush with ground; otherwise in vineyards, forest clearings, on rocky slopes, or in thick grass, heather (etc.), sheltered by overhanging rock or foliage. Nest: foundation of stalks, stems, roots, and leaves lined with fine grasses, rootlets, and hair. Sometimes when flush with soil, cup has no real foundation, and rough material arranged wreath-like on ground. EGGS. Sub-elliptical, smooth and faintly glossy; bluish, greyish, purplish, or pinkish, sparsely but evenly marked with brownish-black speckles, blotches, and scrawls, sometimes forming ring at broad end; greyish undermarkings. Clutch: 4–5 (3–6). INCUBATION. 11–12(–13) days. FLEDGING PERIOD. 12–13 days (9–14), usually leaving nest before able to fly.

Wing-length: ♂ 83–96, ♀ 77–92 mm.
Weight: ♂ 83–96, ♀ 81–93 g.

Grey-necked Bunting *Emberiza buchanani*

PLATES: pages 1659, 1685

DU. Steenortolaan FR. Bruant à cou gris GE. Steinortolan IT. Zigolo collogrigio
RU. Скалистая овсянка SP. Escribano cabecigrís SW. Bergortolan

Field characters. 15–16 cm; wing-span 24–27 cm. Slightly smaller than Ortolan Bunting, with somewhat shorter wings but proportionately longer tail. Eastern counterpart of Cretzschmar's Bunting but with slightly more pointed bill and rather shorter tail and characteristically faded, softly streaked plumage. Adult lacks grey breast-band of Cretzschmar's Bunting, having wholly vinous under-body. At all ages, pale brownish fringes to flight-feathers add to washed-out appearance. Sexes rather similar.

Adult instantly distinguished from Ortolan Bunting and Cretzschmar's Bunting by lack of dark breast-band. Juvenile more difficult to separate but has typical faded appearance of species; easily distinguished from Cretzschmar's Bunting but not from paler Ortolan Bunting, which however shows yellow on throat. Strictly montane in breeding season; always markedly terrestrial.

Habitat. In extreme south-east of west Palearctic; a counterpart to Cinereous Bunting further west. A more montane temperate to warm temperate species, occurring on dry rocky slopes of foothills and mountains, and on screes and rocky outcrops, as well as in ravines, favouring arid and barren terrain. Essential requirement seems to be patchy cover of grass and other xerophytic vegetation, including scattering of bushes, or stony desert. Wintering birds in India live on stony ground with sparse shrubs and on *Euphorbia*-covered broken hillsides: sometimes also on stubble.

Distribution and population. ARMENIA. Rare breeder, with restricted low-altitude range in north-west Araks valley region (Bagramian east to just east of Vedi). AZERBAIJAN. Rare; known from only a few sites. Nests in south of Nakhichevan region, and breeding also recorded in Zuvand upland (Talysh mountains), but distribution otherwise poorly known. TURKEY. Restricted to Van region and Kurdish Alps, where locally common or even abundant, west to Nemrut Dag above Tatvan. Perhaps also occurs in mountains bordering Aras valley. Population 1000–10 000 pairs. SYRIA. Status unclear: sole record above Bloudan (Anti-Lebanon) April 1993.

Accidental. Russia, Kuwait.

Beyond west Palearctic, breeds Iran north-east to Altai mountains and western Mongolia.

Movements. Migratory, wintering mainly in India. West Palearctic populations mainly move south-east to wintering grounds.

Food. Diet mainly seeds and other parts of plants, plus invertebrates in breeding season. Feeds on ground on rocky slopes with scattered scrubby vegetation, foraging slowly and methodically, often remaining motionless for long periods.

Social pattern and behaviour. Little studied, but as far as known closely similar to Ortolan Bunting.

Voice. Song a quite long, loud, rich 'trill', ascending towards higher-pitched and emphasized penultimate note, 'dze dze dze dzee-oo'; recalls Ortolan Bunting. Calls include 'tcheup', and 'choup' from perched bird and 'tsip' and 'tsik-tsik' in flight.

Breeding. Armenia: newly-fledged young and family parties seen 6–10 June. Iran: eggs recorded from end of April to July. Kazakhstan: eggs recorded late May to early July. SITE. On ground, usually well concealed in shelter of rock, shrub, or overhanging grass on stony slope. Nest: rather fragile foundation of coarse stalks and grass stems, occasionally some twigs, neatly and smoothly lined with fine grass, wool, and hair; on slope, rear wall of foundation sometimes lacking and front built up. EGGS. Sub-elliptical, smooth and slightly glossy; white, faintly tinged blue, green, or buff, sparsely speckled faint purplish-grey and with scattered spots and hairstreaks of purplish-black concentrated towards broad end; greyish-violet undermarkings. Clutch: 4–5 (3–6). (Incubation and fledging periods not recorded.)

Wing-length: ♂ 87–92, ♀ 80–87 mm.
Weight: ♂ mostly 19–26, ♀ 17–24 g.

Cretzschmar's Bunting *Emberiza caesia*

PLATES: pages 1660, 1685

DU. Bruinkeelortolaan   FR. Bruant cendrillard   GE. Grauortolan   IT. Ortolano grigio
RU. Красноклювая овсянка   SP. Escribano ceniciento   SW. Rostsparv

Field characters. 16 cm; wing-span 23–26.5 cm. Averages slightly smaller than Ortolan Bunting. Close counterpart of Ortolan Bunting, with similar structure and almost identical plumage pattern. Adult differs most in pure grey head and breast-band and orange-chestnut throat. Juvenile shows warmer, more rufous-orange ground-colour to plumage, with grey (not green) tinge to crown, white (not yellow) ground-colour to throat, and white underwing.

Adult unmistakable if true colours, including blue-grey of breast-band, are apparent. Important to remember that Cretzschmar's Bunting is darker-bodied than Ortolan Bunting and much more so than Grey-necked Bunting, with typically the darkest, most orange or rusty throat and vent of trio. Juvenile difficult to distinguish from Ortolan Bunting. Separation of juvenile from Grey-necked Bunting easier, since Cretzschmar's Bunting much more heavily streaked above and on breast, with darker flight-feathers, rufous rump and vent, and shorter tail. Head may appear bulbous, even peaked on rear crown.

Habitat. Breeds in east Mediterranean region in warm temperate climate, mainly not far from sea. Occurs on rocky hillsides and islands among sparse herbage, with some shrub or tree growth, usually below *c.* 1300 m. While overlapping with Ortolan Bunting, tends to spread more on to drier and more barren rocky slopes, being ecologically intermediate between those buntings inseparable from vegetation cover and those preferring bare open ground or rocks. Winters in dry savanna, steppe, cultivated areas, and gardens within arid regions.

Distribution and population. Breeds only in west Palearctic. Little information on trends, though apparently stable in Greece. GREECE. 5000–15 000 pairs. ALBANIA. Poorly known, but population apparently small: perhaps up to 500 pairs in 1981. TURKEY. Perhaps breeds occasionally outside normal range. 10 000–100 000 pairs. CYPRUS. 10 000–20 000 pairs. SYRIA. Large, continuous breeding area in north-west. Bred at Tall Tamir in north-east in 1945. Breeding also suspected in south, which would be link with populations of Mt Hermon and Golan. ISRAEL. A few thousand pairs. JORDAN. Uncommon.

Accidental. Britain, France, Netherlands, Germany, Sweden, Finland, Austria, Malta, Georgia, Kuwait, Libya, Algeria, Canary Islands.

Movements. Migrant, completely vacating breeding areas to move mainly southwards to winter in north-east Africa and perhaps west Arabia. Movement mainly nocturnal. Winters in Sudan, south to *c.* 11°N, and in Eritrea. Apparently migrates on broad front across and around east Mediterranean, in both autumn and spring. Autumn passage mainly August to October; spring passage mainly March and April, with earliest arrivals on breeding grounds (in Israel) end of February.

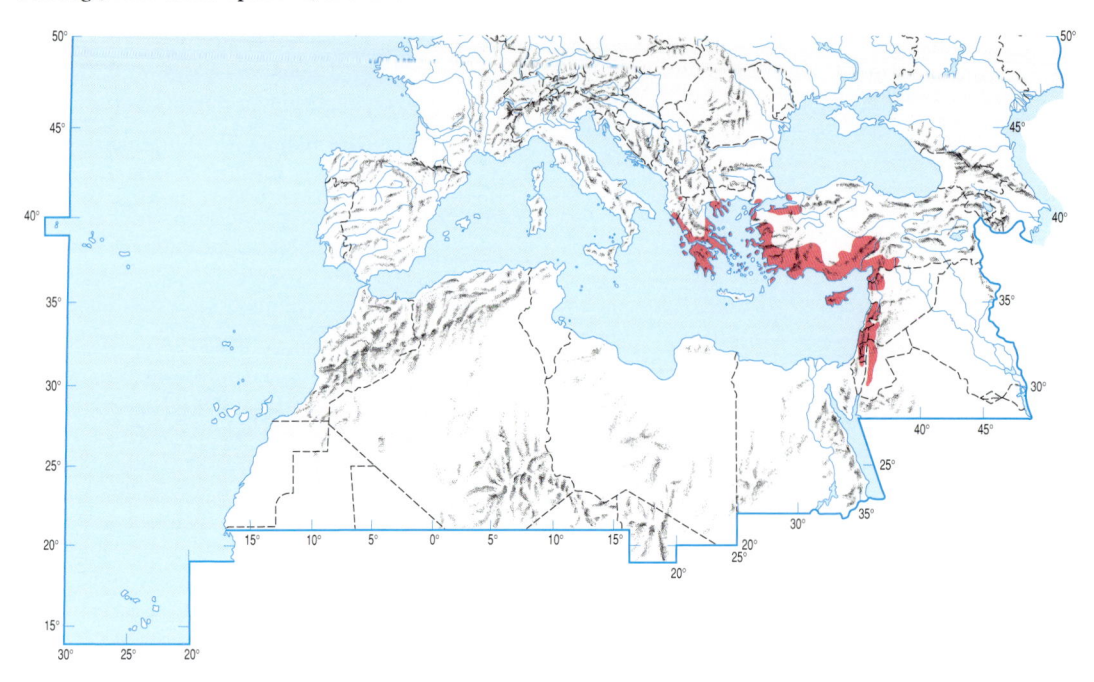

Food. Seeds and small invertebrates. Feeds almost exclusively on ground and said to be probably most terrestrial of Emberizidae of region.

Social pattern and behaviour. Very poorly known, but probably similar to close relatives.

Voice. Song similar to Ortolan Bunting but shorter and thinner-sounding, lacking pleasant ringing tone; usually 3–4 notes, last one longer, 'dzee-dzee-dzree'. Contact-call also like Ortolan Bunting but sharper, 'tchipp'.

Breeding. SEASON. Greece: eggs laid from April to mid-June. Cyprus: eggs laid from April at low altitude, from end of April at 700–1300 m. Israel: eggs laid late March to mid-July. 2 broods regular. SITE. On ground, often in depression, sheltered by rocks and vegetation, at times within roots of shrub, though frequently quite visible; usually on slope. Nest: foundation of stalks, roots, and grass, thickly lined with rootlets and hair; on slope, rear wall can be formed by rock or earth. EGGS. Sub-elliptical, smooth and slightly glossy; very pale yellowish-, bluish-, purplish-, or greyish-white with fine purplish-black speckling and hairstreaks over whole surface though mostly at broad end. Clutch: 4–5(–6). INCUBATION. 12–14 days. FLEDGING PERIOD. 12–13 days.

Wing-length: ♂ 79–88, ♀ 77–83 mm.
Weight: ♂♀ mostly 20–25 g.

Yellow-browed Bunting *Emberiza chrysophrys*

PLATES: pages 1665, 1685

Du. Geelbrauwgors   Fr. Bruant à sourcils jaunes   Ge. Gelbbrauenammer   It. Zigolo del sopracciglio giallo
Ru. Желтобровая овсянка   Sp. Escribano cejigualdo   Sw. Gulbrynad sparv

Field characters. 14–15 cm; wing-span 21.5–25 cm. Slightly but distinctly larger than Little Bunting, with longer and deeper bill, rather heavier head, and slightly plumper body; only slightly smaller than Rustic Bunting and Yellow-breasted Bunting and allies. Rather small bunting, with somewhat sparrow-like profile to bill and head; bill proportionately larger than any other *Emberiza* of similar size. Plumage of body, wings, and tail recalls Little Bunting, but head colours and pattern quite distinct, with at least partly yellow supercilium and white median crown-stripe contrasting strongly with blackish or black crown-sides and ear-coverts, latter distinctly spotted white at rear.

♂ unmistakable if seen well. Dull ♀ and immature more troublesome, inviting confusion with Rustic Bunting and Little Bunting and with ♀ and immature Black-faced Bunting which has however much duller head pattern, having no black on sides of crown, only pale buff supercilium, greyish cheeks and neck-sides, less discrete pale spot on rear ear-coverts, and less

Yellow-browed Bunting *Emberiza chrysophrys*: **1** ad ♂ spring, **2** ad ♂ winter, **3** ad ♀, **4** 1st winter ♂. Rustic Bunting *Emberiza rustica*: **5** ad ♂ breeding, **6** ad ♂ non-breeding, **7** ad ♀ non-breeding, **8** juv.

white, more diffusely streaked underparts. Commonest call a short, distinct 'zit'; likened to Little Bunting but softer and shriller; also resembles that of Rustic Bunting.

Habitat. Breeds in taiga forests of deep interior of south-east Siberia, inhabiting pines and larches, and also thickets in shrub layer, apparently in lowlands, shifting to similar habitats elsewhere for winter.

Distribution. Breeds in south-east Siberia, from valleys of Tunguska and Angara east to Stanovoy range.
    Accidental. Britain: Norfolk, October 1975; Shetland (Fair Isle), October 1980; Orkney, September 1992; Isles of Scilly, October 1994. Belgium: Tongeren, October 1966. Netherlands: Schiermonnikoog, October 1982. Ukraine: Lviv region, January 1983.

Movements. Migratory, but little information. Birds migrate via south-east Russia, north-east China and Korea to winter in eastern China from Kiangsu south to Kwangtung, and west along Yangtze valley to Red Basin. Vagrant as far south as Hong Kong. British records coincided with arrival of other Asian vagrants.

Wing-length: ♂ 78–84, ♀ 71–78 mm.

Rustic Bunting *Emberiza rustica*

PLATES: pages 1665, 1685

Du. Bosgors Fr. Bruant rustique Ge. Waldammer It. Zigolo boschereccio
Ru. Овсянка-ремез Sp. Escribano rústico Sw. Videsparv

Field characters. 14.5–15.5 cm; wing-span 21–25 cm. Slightly shorter than Reed Bunting due to shorter tail; 10% larger than Little Bunting. Medium-sized, rather upstanding, perky bunting, with rather square or peaked crown. All plumages show silky-white underparts with bold pattern of spots across breast and on flanks. ♂ beautiful: has black head with white central crown line, supercilium and spots on rear ear-coverts, strongly rufous breast, flanks, and rump, and bright white upper wing-bar and tail-edges. ♀ and immature less distinctive, lacking conspicuous head pattern but still showing wide rufous streaks on flanks and mottled warm rufous-buff rump.
    ♂ showing rufous breast and flanks unmistakable, but note that head and wing markings are similar in pattern to Yellow-browed Bunting (in close view, latter's yellow supercilium, grey lesser coverts, and striped breast dispel confusion). ♀ and immature much more liable to confusion with Yellow-browed Bunting, Reed Bunting, Little Bunting, and even Pine Bunting

and Meadow Bunting. In western Europe, Reed Bunting will present most frequent trap, but it lacks rufous on rump and along flanks and shows only dull wing-bars, while calls quite different. Flight light and fluttering, lacking uneven rhythm of Reed Bunting. Stance fairly level except when singing, but head usually held up, exaggerating depth of chest and displaying markings.

Habitat. Breeds in north Palearctic, in boreal zone between July isotherms of 12° and 22–23°C, having expanded in modern times from more easterly regions. Favours moist and wooded lowland situations, especially growth of willow, birch, and poplar on margins of coniferous taiga forest by fens or river banks, or moist mosses. Normal habitat requirements for breeding are scattered or marginal trees, thickets or dense undergrowth, moors or heaths and often streams or pools. Winters in cool temperate climates, on open as well as wooded terrain.

Distribution. Has expanded to west and south in 20th century. NORWAY. Recent spread south-west. SWEDEN. Range increase, apparently still continuing. First recorded breeding 1897. In 1950s, rather common north of 64°N along coast and in west; subsequently recorded with increasing frequency further south. FINLAND. Spread west and south. ESTONIA. First observed in 1978; breeding first recorded in 1979. Scattered pairs 1981–4, but observations much decreased since 1985, and cannot now be regarded as regular breeder. LATVIA. Sole breeding record 1985. RUSSIA. Slight range increase.

Accidental. Spitsbergen, Iceland, Britain (annual), Ireland, France, Belgium, Netherlands, Germany, Denmark, Latvia, Poland, Czech Republic, Austria, Switzerland, Spain, Portugal, Italy, Malta, Greece, Yugoslavia, Bulgaria, Ukraine, Azerbaijan, Turkey, Syria, Israel, Iraq, Kuwait, Egypt.

Beyond west Palearctic, extends east across Siberia to Bering Sea.

Population. NORWAY. 100–500 pairs 1970–90; slight increase. SWEDEN. 10 000–100 000 pairs in late 1980s. Previously reported to be increasing, but present trend unknown. FINLAND. 200 000–300 000 pairs in late 1980s. Increase since 19th century (reasons obscure) in connection with range expansion; slight decrease in recent decades. RUSSIA. 4.1–6.9 million pairs; stable.

Movements. All populations migratory. Western birds head east then south, and eastern birds head south or south-west, to reach winter quarters via south-east Russia and north-east China. Winters mainly in China and Japan.

Migration prolonged in both seasons. In autumn, begins in August in northern Europe. Birds leave Sweden end of August to September. In Leningrad region, movement probably begins in last third of August, and peaks early or mid-September; main migration ends in early October, with stragglers throughout month. Arrives in north-east China and Japan from mid-October. In spring, migrates through north-east China from late February, mostly to end of March, with small numbers to mid-April. In Leningrad region, passage usually begins in late April, but local birds arrive with main movement, in early May. First arrivals in southern Finland in early May, occasionally late April. Vagrant west and south of range both seasons in Europe. In Britain, now occurs annually; recorded in almost all months, suggesting some may overwinter.

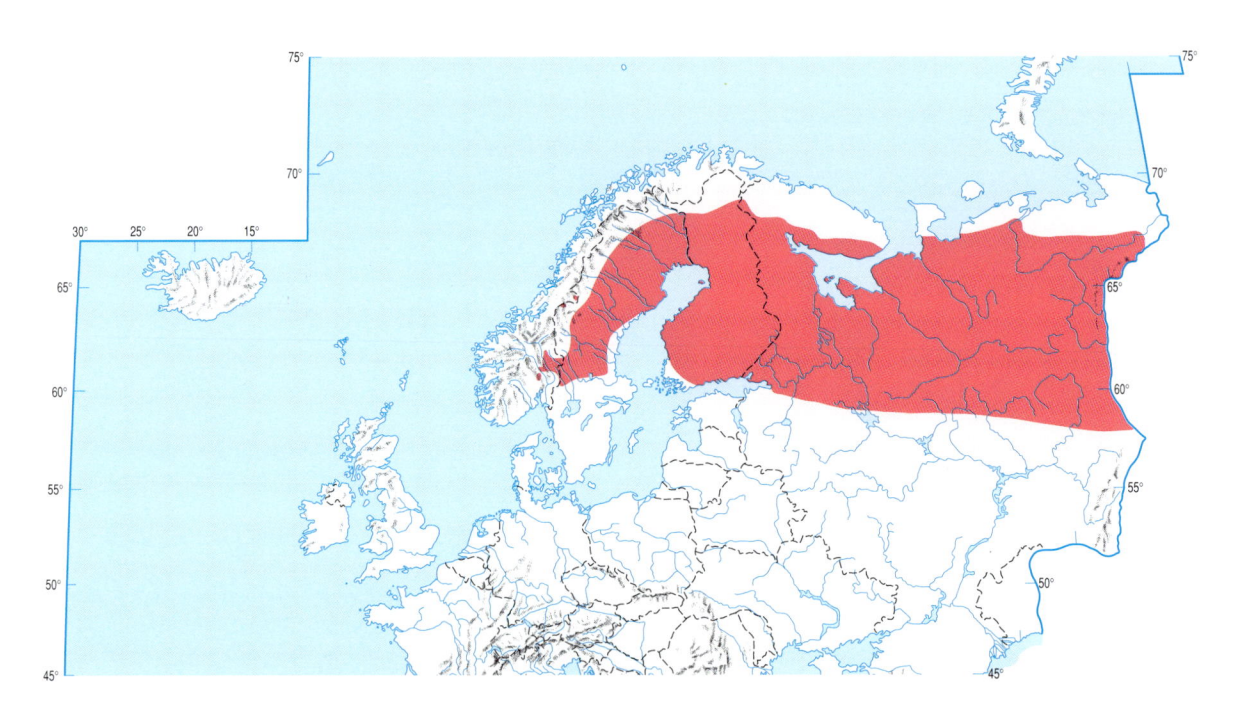

Food. Mainly seeds, plus insects and spiders in breeding season. Feeds on ground and in bushes, commonly in damp and marshy places, and much invertebrate prey associated with water. In forest, tosses leaves aside while foraging. In China and Japan in winter, most often seen feeding on ground and low vegetation, sometimes in flocks of hundreds; in woodlands, clearings, forest edges, open spaces including urban parks, etc., but especially in rice stubble, reedbeds, and on riverbanks.

Social pattern and behaviour. Gregarious outside breeding season. No evidence for other than monogamous mating system. Solitary and territorial when breeding. ♂ sings from elevated perch, with crown raised. Singing starts in flocks on spring migration, e.g. heard in March. In Fenno-Scandia, song starts beginning of May, continuing for only relatively short period, after which little heard till young fledge in June. In Leningrad region, resurgence of song in 2nd half of June coincides with dispersal of 1st broods and incubation of 2nd clutches.

Voice. Song a fairly short phrase, clear, mellow and melodious but varied, even irresolute in delivery (in last characteristic recalling Dunnock or broken song of Robin); can sound mournful, 'dudeleu-deluu-delee'. Commonest call much like Song Thrush but more distinct and higher pitched, 'zit'. Also has disyllabic 'tic tic', recalling Robin but with 2nd note sometimes lower pitched.

Breeding. SEASON. Sweden: start of laying mid-May to 1st half of June. Finnish Lapland: average date of 1st egg 6 June. Leningrad region (north-west Russia): eggs laid beginning of May to beginning of July. 1 or 2 broods. SITE. Generally on ground, often near water, in tussocky grass, moss cushions, among thick roots, etc., beside bush or tree or under overhanging grass, sedge, etc., or (rarely) in depression in level ground; sometimes above ground in tree or on stump. Nest: foundation of grass, sedge, horsetail, moss, leaves, needles, lichen (etc.), lined with fine plant material, hair, and sometimes feathers. EGGS. Sub-elliptical, smooth and faintly glossy; pale bluish-green or greenish-white, densely covered with fine olive-brown or greyish spots, blotches, and scrawls, though these can be absent; pale grey undermarkings. Clutch: 4–5 (3–6). INCUBATION. 11(–13) days. FLEDGING PERIOD. 7–10 days; young leave nest well before able to fly.

Wing-length: ♂ 78–83, ♀ 73–81 mm.
Weight: ♂♀ mostly 17–22 g.

Little Bunting *Emberiza pusilla*

PLATES: pages 1668, 1685

DU. Dwerggors   FR. Bruant nain   GE. Zwergammer   IT. Zigolo minore
RU. Овсянка-крошка   SP. Escribano pigmeo   SW. Dvärgsparv

Field characters. 13–14 cm; wing-span 20–22.5 cm. Distinctly less bulky than Reed Bunting; usually 10% smaller, with sharply pointed bill, flat sloping forehead, little or no neck, shorter, straight-edged tail, and shorter legs; close in size to but with less ample tail than Pallas's Reed Bunting; somewhat shorter and distinctly smaller-billed than Yellow-browed Bunting. Smallest bunting of west Palearctic, with delicate but compact form (lacking obviously long tail of larger *Emberiza*) and terrestrial behaviour recalling Linnet and Dunnock. At all ages, bright chestnut face with narrow but distinct cream eye-ring and dark stripes on crown-sides and around rear of ear-coverts (not reaching bill) distinctive. Pale double wing-bar most obvious feature of buff- to grey-brown upperparts; fine, discontinuous streaks on white underparts also characteristic. Flight suggests small finch rather than bunting.

Combination of small size, bill shape, pale eye-ring, and bright head pattern allow diagnosis, but Little Bunting nevertheless subject to persistent confusion with smallest Reed Bunting (particularly those with bright plumage), Rustic Bunting, Pallas's Reed Bunting, and Yellow-browed Bunting. Reed Bunting and Pallas's Reed Bunting easily separated on call and close observation of head and underpart patterns; Rustic Bunting and Yellow-browed Bunting have similar calls but Rustic Bunting shows much more obvious, dark rufous streaks on breast and flanks and intensely rufous-brown rump (dull, even greyish on Little Bunting), while Yellow-browed Bunting has distinctly black and yellow or white head markings, including much more distinct white spot on rear ear-coverts, and lacks rufous face. All confusion species also lack striking eye-ring. Twitches wings and occasionally flicks tail (but tail action lacks frequent slight spreading so characteristic of Reed Bunting).

Habitat. In west Palearctic, breeds only in boreal and arctic continental climatic zones, from 7°C July isotherm in north to 18–20°C in south, having more northerly range than any other Emberizidae except Lapland Bunting and Snow Bunting. Ecologically it comes between Lapland Bunting and Rustic Bunting, being a bird of moister and shrubbier tundra than Lapland Bunting but less wooded situations than Rustic Bunting. Favours willow zone along rivers through northern taiga, and open forest by river mouths. Towards west of range shows preference for undergrowth of dwarf birch or willow among taller trees, which may be birch, spruce, or other species. In forest country, selects open types with clearings close by. In winter quarters in eastern Asia, favours bracken and short grass on hillsides or along mule tracks; on plains, frequents stubble.

Distribution and population. Has spread west into Fenno-Scandia since 1930s, and north-east into Kola peninsula (north-west Russia). BRITAIN. So frequently recorded that no longer considered vagrant; pattern of occurrence shows annual late autumn arrivals, increasing wintering, and erratic spring appearances or withdrawals. NORWAY. Believed regular in Sør-Varanger, with up to 25 singing ♂♂ recorded. Has also bred at other sites in Finnmark. Up to 50 pairs, sporadically, 1970–90. SWEDEN. Few records, but may well breed annually, though

Little Bunting *Emberiza pusilla*: **1** ad ♂ breeding, **2** ad ♀ breeding, **3** ad non-breeding, **4** 1st winter ♂, **5** juv. Chestnut Bunting *Emberiza rutila* (p. 1669): **6** ad ♂ spring, **7** ad ♂ winter, **8** ad ♀, **9** 1st winter ♂.

no regular sites known. Up to 100 pairs in late 1980s, numbers apparently fluctuating between years. FINLAND. First recorded breeding 1935; then spread rapidly westward. In several years in 1980s, when marked increase, recorded breeding far beyond main area occupied. 5000–10 000 pairs in late 1980s. RUSSIA. No records for Kola peninsula in early 20th century. First recorded Khibin, and has now probably spread along whole coast to Ponoy estuary. 100 000–1 million pairs; stable.

Accidental. Iceland, Channel Islands, Ireland, France, Belgium, Netherlands (annual), Germany, Denmark, Estonia, Latvia, Poland, Czech Republic, Hungary, Austria, Switzerland, Spain, Portugal, Balearic Islands, Italy, Malta, Greece, Yugoslavia, Bulgaria, Ukraine, Turkey, Lebanon, Jordan, Kuwait, Egypt, Canary Islands.

Beyond west Palearctic, extends east across Siberia to western Chukotskiy peninsula and Sea of Okhotsk.

Movements. All populations migratory, wintering mainly from Nepal east to China and Indochina. Western birds head east from breeding grounds then south or south-east, and eastern birds head south, to reach winter quarters via Mongolia, south-east Russia, and north-east China. Autumn migration

August–November. Birds leave breeding grounds from mid-August to early or mid-September. Spring migration late March to June. One of latest migrants to reach north-east Finland; average earliest bird 6 June, and never earlier than 30–31 May.

Widespread records in west Palearctic, mostly in autumn. Annual in Britain, with 93 before 1958, and 522 in 1958–93 (including Ireland). Autumn records chiefly in Shetland (especially Fair Isle), on British east coast and in Isles of Scilly; spring records well scattered, and include a number of inland localities. Several reports midwinter, from Scotland south to Jersey (Channel Islands); in Merseyside (north-west England), bird remained from January to early April; other birds present at different sites in southern Britain may also have overwintered. On Finnish coast also, a few midwinter records. In Sweden, up to 1986, 209 records April–November, chiefly May to early July. In Netherlands, 68 records to 1994, chiefly September–November and a few February–May. In France, far more records than of Rustic Bunting; in 19th century, a few caught annually in autumn at Marseille in south-east; records in 20th century chiefly in west and south-west; recorded in all months September–April, chiefly October–November. In Israel, occurs in very small numbers; regular at Eilat, with 1–6 individuals each year, and 7 birds reported 1979–89 in Jerusalem hills; most records late October to mid-November.

Food. Seeds, also invertebrates in breeding season. On migration, most often feeds in crops, on turned soil, paths, and roads, almost wholly on ground. In winter quarters, in short grass, scattered woodland, marshy places, river banks, and (particularly) stubble and paddy fields.

Social pattern and behaviour. Gregarious outside breeding season. Birds wintering in Britain typically mix with (variously) Meadow Pipit, Tree Sparrow, Linnet, Yellowhammer, and Reed Bunting. No evidence for other than monogamous mating system. Solitary and territorial when breeding, possibly with tendency to form neighbourhood groups. ♂ sings typically from exposed tree-top, with bill raised and throat feathers ruffled, body vibrating with effort. Birds occurring Britain heard singing in April; in Russia, song heard during spring migration. In Timanskaya tundra (north European Russia) song-period 10 June to early July.

Voice. Song a fairly quiet, sweet, and varied phrase of unusual timbre, with tonal sounds, buzzy units, and clicks; phrase structure can suggest Rustic Bunting, Ortolan Bunting, and Reed Bunting. Calls given by migrants include 2 short, flat monosyllables: hard, sharp, clicking 'zik', 'tik', 'tzik', or 'pwick', recalling Hawfinch, Robin, and Rustic Bunting; quieter, lower, dry 'tick', 'tip', 'tsih', 'twit', or 'stip', suggesting Robin or Song Thrush; both often quickly repeated 2–3 times in alarm.

Breeding. SEASON. North-central Finland: eggs laid early June to mid-July. 1–2 broods. SITE. Usually on ground, on grass tussock or moss cushion sheltered by overhanging grass or twigs of (e.g.) alder, birch, willow, etc., also on tree stump. Nest: foundation of thin twigs, stalks of herbs, grass, sedge, horsetail, moss, lined with fine grass, lichen, and sometimes hair. EGGS. Sub-elliptical, smooth and slightly glossy; very pale green, olive, grey, or pink with sparse spots and blotches of purplish-black or dark brown, or scrawls and hairstreaks, sometimes forming vague circle at broad end; undermarkings of grey or violet. Clutch: 4–6 (3–7). INCUBATION. 11–12 days. FLEDGING PERIOD. 5–8(–11) days. In general, leaves nest unable to fly at 6–8 days, fully fledged c. 3–5 days later.

Wing-length: ♂ 69–76, ♀ 67–70 mm.
Weight: ♂ mostly 14–19, ♀ 13–18 g.

Chestnut Bunting *Emberiza rutila*

PLATES: pages 1668, 1685

DU. Rosse Gors   FR. Bruant roux   GE. Rötelammer   IT. Zigolo nocciola
RU. Рыжая овсянка   SP. Escribano herrumbroso   SW. Rödbrun sparv

Field characters. 14–15 cm; wing-span 21–23.5 cm. Slightly larger and noticeably less delicate than Little Bunting, with proportionately larger head, deeper chest, and longer wings; close in size to Yellow-breasted Bunting but with relatively shorter tail. In all plumages, shows yellow underparts, unstreaked chestnut rump, and little or no white on outer tail-feathers. ♂ bright chestnut on hood, back, and inner wing-feathers; ♀ and immature less distinctive, recalling Yellow-breasted Bunting but with less striped head and less sharply streaked underparts.

♂ unmistakable. ♀ and 1st-winter tricky when rufous rump hidden, closely resembling dull Yellow-breasted Bunting, which nevertheless shows distinct pale central crown-stripe, less distinct malar stripe, more distinctly streaked flanks, and narrow white edges to tail. Flight action and silhouette much as Yellow-breasted Bunting but general character on ground differs in distinctly dumpier form and unobtrusive behaviour. Commonest call similar to Little Bunting, a short monosyllabic 'zic'; also a thin high 'teseep'.

Habitat. Breeds in east Palearctic in temperate forest zone of Siberia, in open forests of larch and also broad-leaved trees such as alder and birch, apparently favouring rich ground-cover of herbaceous plants, and dense grass. Frequents mountain slopes and lake shores, and during spring migration also fields and gardens near villages. Wintering birds in India frequent rice stubbles and bushes in cultivation and forest clearings.

Distribution. Breeds eastern Siberia, from north-west Irkutsk region east to Sea of Okhotsk, south to Baykal region and probably northern Mongolia and northern Manchuria.

Accidental. 4 autumn records probably true vagrants: Netherlands, 1st-winter ♀, November 1937; single 1st-winter ♂♂, Norway, October 1974, Malta, November 1983, and Yugoslavia, October 1987. 5 records (June–July, September) for Britain, from Scottish and Welsh islands, of doubtful status; some or all may have been escapes from small numbers regularly imported as cagebirds.

Movements. Migratory, wintering mainly in southern China, Indochina and Burma. Migrates to winter quarters via Ussuriland (south-east Russia), eastern China (west to Shensi) and Korea. Rare records March–June of vagrancy west of winter range, in Sikkim, Nepal, Ladakh (north-west India), and Chitral (northern Pakistan). No reports in western Siberia.

Wing-length: ♂ 71–77, ♀ 69–75 mm.

Yellow-breasted Bunting *Emberiza aureola*

PLATES: pages 1671, 1685

Du. Wilgengors   Fr. Bruant auréole   Ge. Weidenammer   It. Zigolo dal collare
Ru. Дубровник   Sp. Escribano aureolado   Sw. Gyllensparv

Field characters. 14–15 cm; wing-span 21.5–24 cm. Slightly smaller and noticeably shorter-tailed than north-western race of Reed Bunting; slightly larger and longer-tailed than Chestnut Bunting; up to 10% larger than Little Bunting. Rather small but robust bunting, with spiky bill and rather short tail contributing to flight silhouette suggestive of *Fringilla*. In all plumages, ground-colour of little-streaked underparts yellowish; outer tail-feathers show white. Adult ♂ has narrow black-brown necklace, rich brown back, white lesser and median coverts (again recalling *Fringilla*), and whitish wing-bar; front of head and throat black when breeding. ♀ and immature have heavily striped head, broadly streaked back, double whitish wing-bar, and streaked flanks.

At close range, adult ♂ unmistakable but at distance or in poor view may briefly suggest ♂ Black-headed Bunting (10% larger, with proportionately larger bill and more attenuated build; lacks necklace, flank-streaks, and white wing-blaze). Female and immature distinctive compared with other large *Emberiza*, lacking fully streaked underparts and particularly lemon-yellow hue of Yellowhammer, but needing careful separation from Chestnut Bunting (same size but duller, more olive-toned above, with far less distinct head pattern and wing-bars, much more rufous rump and yellow under tail-coverts), Black-faced Bunting (same size but darker, more brown-toned above, with greyish shawl, cheeks, and collar, distinct dusky malar stripe, duller wing-bars, and greyish rump), and Red-headed Bunting (up to 10% larger, with more uniform, less streaked plumage and especially yellow rump and under tail-coverts). Flight light and fast, but, due to relatively compact silhouette, appearance and action may suggest *Fringilla* finch as much as typical bunting (particularly in adult ♂ showing white wing-blaze).

Habitat. In west Palearctic breeds in boreal zone between July isotherms of 12°C and 23–24°C, overlapping with Rustic Bunting and sharing with it a recent westward expansion over northern Europe. In European Russia, typically occupies relatively dry river valley meadows covered with dense tall grasses and scattered bushes of alder and willow. In taiga belt, inhabits also peat bogs, scorched areas with remains of birch and spruce trees, and alpine meadows with thick osier growths and solitary spruces. In Indian winter quarters, frequents grassland and farms, in or around small settlements, and also hedgerows and gardens in hills up to 1500 m.

Distribution and population. Westward spread began in 19th century. At least 2 recent waves of expansion, first in 1960–65, then 1974 onwards. Has also spread south to northern Ukraine. Population probably increasing slowly in west. NORWAY. Adults feeding young at Pasvik (near border with Finland) 1967, may have bred in Norway. FINLAND. First bred in 1920s; breeding areas in central Finland not occupied until 1940s and 1950s. 150–200 pairs in late 1980s. Rapid increase in middle of 20th century, but decrease since 1970s. RUSSIA. In 1930s, nested at Tsimlyansk on Lower Don. 10 000–100 000 pairs; stable. UKRAINE. Confirmed to middle Desna valley, where first found breeding 1930; c. 100 pairs.

Accidental. Iceland, Britain (annual), Channel Islands, Ireland, France, Belgium, Netherlands, Germany, Denmark, Norway, Sweden (perhaps annual), Estonia, Latvia, Poland, Czech Republic, Spain, Italy, Malta, Greece, Cyprus, Israel, Jordan, Egypt.

Beyond west Palearctic, extends east through Siberia and north-central Asia to Kamchatka, Sakhalin island and Japan.

Movements. All populations migratory. Western birds head east from breeding grounds then south, and eastern birds south or south-west, to reach winter quarters via Mongolia, south-east Russia, and China. Winters throughout south-east Asia.

Leaves breeding grounds early and returns late. Departure begins July, adults earlier than juveniles. Arrives in India, Nepal, and Burma in October; in south of range (Malay peninsula) not until November. Spring migration April–June; present in winter quarters until May, but most birds leave in April. In western Russia (west of Urals) arrivals late May to June. Most records of vagrancy in Europe are in Britain, Scandinavia, and Italy. Annual in Britain in recent years; almost all records in autumn, especially September (suggesting reverse migration). In Sweden, 23 records up to 1986, of which 19

Yellow-breasted Bunting *Emberiza aureola*: **1** ad ♂ breeding, **2** ad ♂ non-breeding, **3** ad ♀ breeding, **4** ad ♀ non-breeding, **5** 1st summer ♂, **6** 1st winter ♂, **7** 1st summer ♀, **8** juv.

in May–July (chiefly June) in north and south-east, suggesting overshooting, and 4 in August in south-east.

Food. Seeds, mainly of grasses, and other plant material; invertebrates in breeding season.

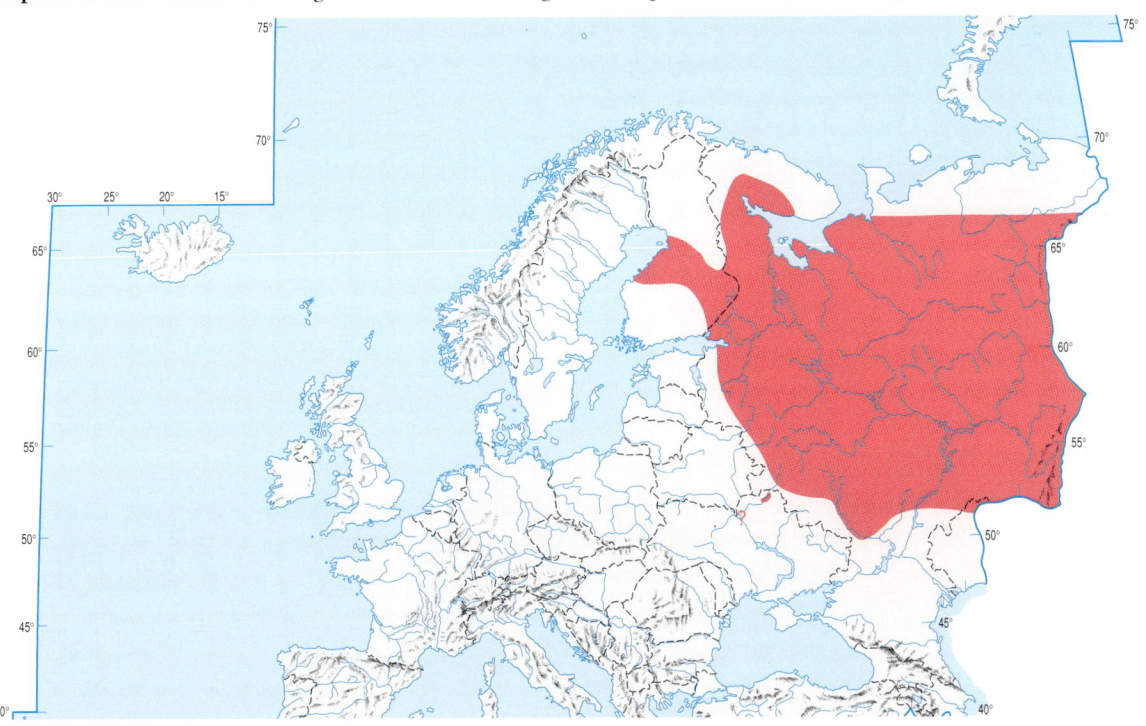

Social pattern and behaviour. Gregarious outside breeding season, in flocks of up to hundreds and sometimes thousands. Mating system mainly monogamous, but ♂♂ bigamous exceptionally. Typically breeds in neighbourhood groups, with relatively small territories leading to locally high densities; nests may be only 8–10 m apart in some well-vegetated sites, but 200 m or more if there is little cover. ♂ sings from exposed tree-tops and bushes, also from grass tussocks. Starts in winter quarters in March–April, shortly before spring migration, and heard on arrival (or soon after) on breeding grounds; continues to July.

Voice. Song recalls Ortolan Bunting but includes jingling or chiming also suggestive of Lapland Bunting and Reed Bunting; typical phrase terminally higher pitched and faster than Ortolan Bunting but with similar melodious timbre, 'tu tu-li tu-li-ti he-li li-lu-li'. Calls include short 'tik', 'zipp', or 'tzip', reminiscent of short call of Spotted Flycatcher and Little Bunting, and soft trilling 'trssit'.

Breeding. SEASON. Late and short throughout range. Finland: eggs laid around 2nd half of June to early July. Leningrad region (north-west Russia): eggs laid from mid-June. South of Moscow: most clutches laid 1st half of June. Volga-Kama region (east European Russia): eggs laid 2nd half of June. SITE. On ground, either on tussock or, where dry, in depression, nest-rim at times flush with ground, sheltered by scrub, commonly birch or willow, or in tree roots; where wet, often slightly above ground in bush or stout herb. Nest: foundation of dry grass and stalks lined with soft grass, rootlets, and sometimes hair. EGGS. Sub-elliptical, smooth and slightly glossy; greyish or greenish with olive or purplish-grey undermarkings, sparsely to heavily marked with brown to purplish-black blotches and scrawls. Clutch: 4–5 (3–7). INCUBATION. 13–14 days. FLEDGING PERIOD. (9–)11–14 days; leaves nest before able to fly.

Wing-length: ♂ 73–81, ♀ 72–76 mm.
Weight: ♂ mostly 20–26, ♀ 17–24 g.

Reed Bunting *Emberiza schoeniclus*

PLATES: pages 1673, 1685

DU. Rietgors   FR. Bruant des roseaux   GE. Rohrammer   IT. Migliarino di palude
RU. Камышовая овсянка   SP. Escribano palustre   SW. Sävsparv

Field characters. 15–16.5 cm; wing-span 21–28 cm. Size varies considerably between 2 main racial groups, being smallest in group containing nominate race of north-west Europe and allied northern races which average 10% shorter and noticeably slighter than Yellowhammer; intermediate in south European races, and largest in east European and Asian group in which birds overlap with Yellowhammer and have as long a tail; bulbous shape, and size, of bill also increase from north to south and east. Medium-sized to rather large bunting, seemingly large-headed and thick-necked (in ♂), with fairly lengthy form and distinctive voice. Breeding ♂ instantly recognized by black head and bib and white collar, shared only by Pallas's Reed Bunting; non-breeding ♂, ♀, and juvenile essentially brown above and buffish below, quite heavily streaked but lacking striking diagnostic character (since neither convex culmen nor diagnostic reddish tone of lesser wing-coverts easily seen). Paler eastern birds, rare small individuals, and (in ♀ and juvenile) occasional aberrant head plumage present serious pitfalls. Has distinctive habit of nervously spreading tail.

Breeding adult ♂ almost unmistakable, since only ♂ Pallas's Reed Bunting shows similar plumage but is instantly distinguishable by different call, typically smaller size, always whitish rump, and (when visible) grey lesser wing-coverts. In all other plumages (and remembering its quite marked geographical variation), adult ♂, ♀, and immature Reed Buntings are troublesome, particularly when silent, and, in these plumages, potential and known confusion species include, especially, Pallas's Reed Bunting (smaller, with usually pale rump, less marked head, usually more striped back, paler fringes to tertials and secondaries, finer streaks on underparts, and greyish or grey lesser coverts), Rustic Bunting (somewhat smaller and more compact, with bright rufous, not grey-brown rump), Little Bunting (usually noticeably smaller, with cream eye-ring, uniformly rufous, not mottled cheek), Black-faced Bunting (somewhat smaller and noticeably more compact, with similar contrast between pale submoustachial and dark malar stripes but greyish face and half-collar). Habitually flicks and spreads tail, exposing white on outer feathers. Territorial ♂ often as conspicuous as ♂ Yellowhammer, perching upright on top of reed or other tall plant, but ♀ (and both sexes in winter) unobtrusive, often remaining hidden in ground cover.

Habitat. Most widespread in range of west Palearctic breeding buntings, inhabiting oceanic islands and peninsulas, and continental plains from arctic through boreal, temperate, and Mediterranean to steppe and even desert climatic zones, between July isotherms as low as 10–11°C in north to above 32°C in south. Yet within this vast range, choice of occupied sites is ecologically restricted to particular types of dense and prolific fairly low vegetation, mainly associated with intense soil moisture. Avoids both closed forest and typical open country, as well as bare, rocky, or frozen surfaces, steep or broken ground, and areas of immediate human disturbance or settlement. Apparent attachment to marshes, fens, bogs, riversides, and inland waters occurs indirectly, rather through dependence on their associated vegetation types than being linked with any special need for water.

(FACING PAGE) Reed Bunting *Emberiza schoeniclus*. *E. s. schoeniclus*: **1** ad ♂ spring, **2** ad ♂ winter, **3** ad ♀ spring, **4** ad ♀ winter, **5** 1st summer ♀, **6** 1st winter ♂, **7** 1st summer ♀, **8** juv, **9** variant with pale crown. *E. s. pallidior*: **10** ad ♂ spring, **11** 1st ad ♂ winter. *E. s. incognita*: **12** ad ♂ spring, **13** 1st winter ♂. *E. s. reiseri*: **14** ad ♂ spring. *E. s. intermedia*: **15** ad ♂ spring. *E. s. pyrrhuloides*: **16** ad ♂ spring. *E. s. witherbyi*: **17** ad ♂ spring.
Pallas's Reed Bunting *Emberiza pallasi* (p. 1676): **18** ad ♂ spring, **19** ad ♂ winter, **20** ad ♂ spring, **21** ad ♀ winter, **22** 1st winter ♂, **23** juv.

TREVOR BOYER

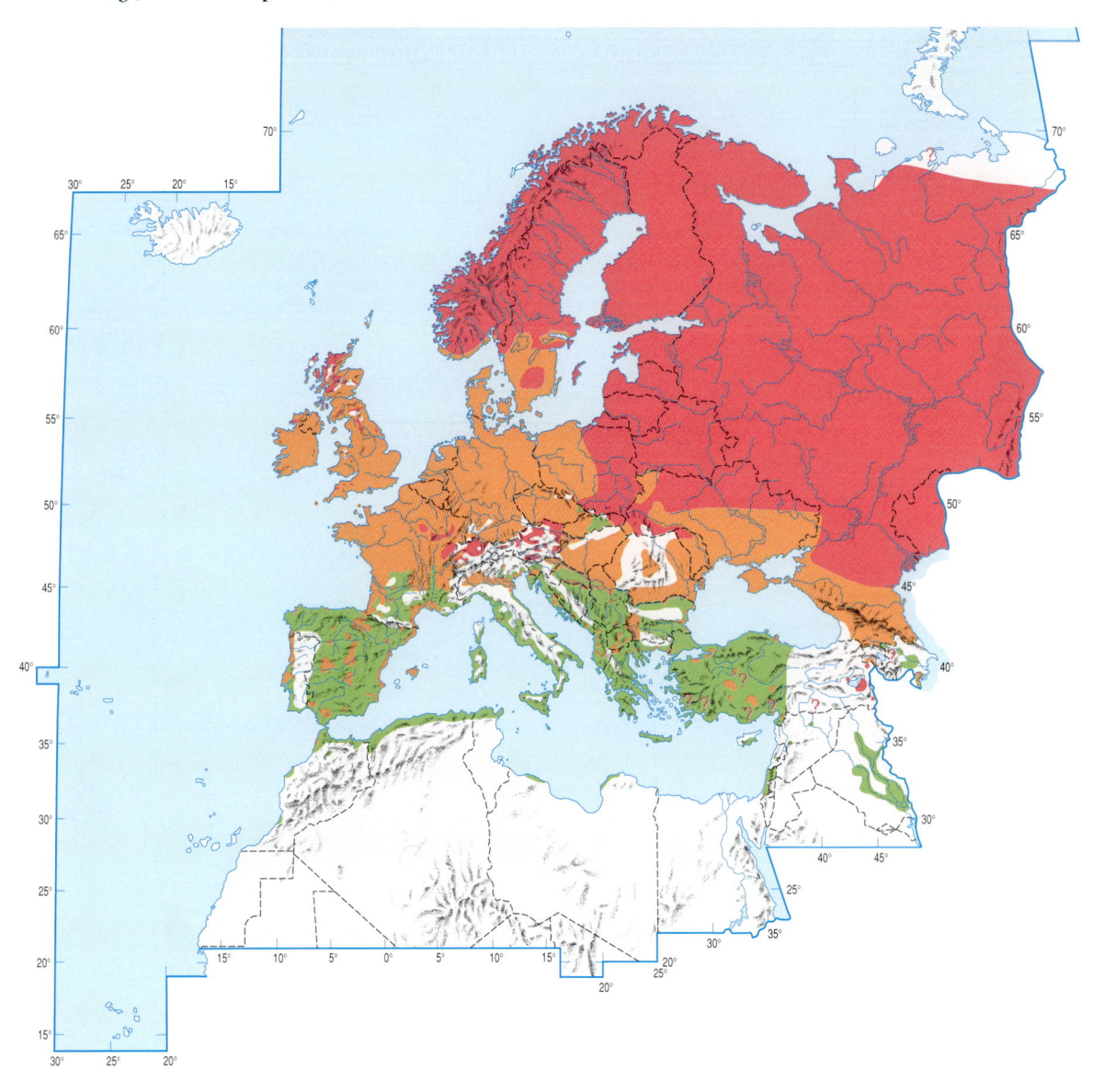

Occupies tall herbage and small shrubs found in marshy and swampy areas bordering fresh or brackish water of all kinds, normally in valleys and lowlands. Other common habitats include peat bogs, wet meadows with tall herbage, reedbeds, shrub tundra, swampy areas in grass steppes, and wet grassy clearings in forests. Abundance of suitable plant or animal food and unsuitability of terrain for competitors also influence habitat choice. In parts of range this has recently expanded to cover drier situations such as young conifer plantations, hawthorn scrub on waterless chalk downlands, and even middle of large fields of barley or other tall crops. Throughout range, whether resident or migratory, there is a shift after breeding season to drier and more open situations, such as fields, lake shores, and marram grass on sand-dunes, more similar to habitats of other buntings at this season.

Distribution. Range has contracted slightly Britain, Denmark, Portugal, and Greece, expanded markedly in Slovenia, less so France and Ukraine, probably also Russia. FAEROES. One breeding record, 1972; probably regular, overlooked summer visitor. BRITAIN. Decrease in north and west. FRANCE. Southward spread, still continuing. RUSSIA. First recorded breeding north of forest zone on Kola peninsula in late 1950s, spreading north along Voron'ya and Iokan'ga valleys to north coast. Potential breeder Russkiy Zavorot peninsula (68°30′N) where observations include pair in June 1993. AZERBAIJAN. Summer status not clear; possibly breeds, but no confirmed records. TURKEY. Common Meriç delta, Eber Gölü, marshes near Bulanik and Van, apparently rare most other areas. Does not breed annually at all sites or occurs in very small numbers and easily overlooked. SYRIA. Reported occurrence Tall Abyad (close to border with Turkey), May–July 1955 difficult to interpret. Morocco. No recent changes, but range now better known.

Accidental. Bear Island, Iceland, Kuwait.

Beyond west Palearctic, extends east across northern Asia to Lena river (with isolated population in Kamchatka), Sakhalin island and northern Japan, and south to Iran, Tien Shan mountains and northern China.

Population. Apparently stable in most countries, but decreases (probably temporary or restricted to particular habitats) reported Britain, France, Belgium, Finland, Lithuania, Italy, and Moldova. Britain. 220 000 territories 1988–91. Steep decline 1975–83, following high levels; now stable. Changes probably related to severity of winters. Ireland. 130 000 territories 1988–91. Increasing, and apparently expanding into drier habitats. France. 10 000–100 000 pairs in mid-1970s. Decrease through loss of marshland, but increase through adaptation to drier habitats. Belgium. 8000–11 000 pairs 1973–7. Probably decreasing again after severe winters of mid-1980s. Luxembourg. 500–800 pairs. Netherlands. 40 000–70 000 pairs 1979–85. Germany. 400 000 pairs in mid-1980s. In east, 150 000 ± 60 000 pairs in early 1980s. Denmark. 28 000–290 000 pairs 1987–8. Norway. 500 000–1 million pairs 1970–90. Sweden. 500 000–1 million pairs in late 1980s. Finland. 200 000–400 000 pairs in late 1980s. Marked increase in 20th century, due to eutrophication of waters and discontinuation of cattle grazing promoting reedy and shrubby growth along lake shores. Slight decrease in recent decades. Estonia. 50 000–100 000 pairs in 1991. Latvia. 50 000–70 000 pairs in 1980s. Lithuania. Common. Stable in wetlands, decreasing in agricultural areas. Poland. 20 000–100 000 pairs. Czech Republic. 40 000–80 000 pairs 1985–9. Slovakia. 20 000–40 000 pairs 1973–94. Hungary. 30 000–50 000 pairs 1979–93. Switzerland. 3000–5000 pairs 1985–93. Spain. 820–1560 pairs. Portugal. 1000–10 000 pairs 1978–84. Italy. 10 000–30 000 pairs 1983–95. Greece. Rare and very local; 2000–5000 pairs. Albania. 1000–3000 pairs in 1981. Yugoslavia: Croatia. 8000–12 000 pairs. Slovenia. 200–400 pairs; marked increase. Bulgaria. 100–1000 pairs. Rumania. 350 000–400 000 pairs 1986–92. Russia. 100 000–1 million pairs. Belarus'. 195 000–215 000 pairs in 1990. Ukraine. 50 000–75 000 pairs in 1986; slight increase. Moldova. 1000–2000 pairs in 1988. Turkey. 5000–15 000 pairs. Morocco. 10–25 pairs.

Movements. Northern group *schoeniclus* sedentary to migratory; southern *pyrrhuloides* group of races chiefly sedentary. Winters in areas with little or no snow cover (except *pyrrhuloides* group) making mid-winter flights if snowfall persists. In Europe, nominate *schoeniclus* migratory in north-east, increasingly sedentary towards south-west; migrants head between SSW and west. Some birds winter south of range in Mediterranean region. Thus, in Strait of Gibraltar (where few breed), locally common in winter, and regular on passage. In Corsica (where none breed), regular in small numbers. Widespread winter visitor in small numbers in north-west Africa, reported to edge of Sahara. Autumn movement chiefly mid-September to mid-November; spring movement chiefly mid-February to April.

Food. Seeds and other plant material; invertebrates in breeding season, and also opportunistically during remainder of year. Takes plant and animal material on ground among sedges, rushes, reeds, etc., in pasture and marshy grasslands, and also low in waterside bushes and trees, or on stems of reed; outside breeding season, more often on ground in open countryside and cultivated fields, weedy areas, woodland clearings, uplands, etc., well away from water, often in flocks with other seed-eaters.

Social pattern and behaviour. Regularly in flocks outside breeding season, often with other species (especially Fringillidae and other Emberizidae). ♂♂ usually leave flocks before ♀♀, visiting prospective territories in early morning. Monogamous mating system, but extra-pair paternity common; polygamy occasional. Solitary and territorial in breeding season. ♂♂ usually re-occupy territory held in previous year; ♀♀ tend also to do so, but less strongly. Territory used mainly for pair-formation, maintenance of pair-bond, and nesting. During pair-formation, some food obtained in territory, but generally obtained on neutral ground elsewhere. Song usually given from high point within territory, such as bush or reed stem; occasionally from ground. Song-period from taking up of territories in early spring to end of breeding: late January to early August, central England; beginning of March to beginning of August, Germany; early April to early July, Sweden at 60°N.

Voice. Song of ♂ a short series of rather unmusical, metallic units, of variable tempo; 'tweek tweek tweek tititick' gives good idea of typical song. Tempo varies greatly, depending on stage of breeding, with striking distinction between 'rapid' song (restricted to unpaired ♂♂) and 'slow' song (mainly restricted to paired ♂♂). In rapid song, units of song given in rapid succession, and intervals between phrases longer than length of phrase; in slow song, intervals between units longer, and intervals between phrases often not much longer than intervals between units (so that song series may be more or less continuous). Commonest call, used all year, when feeding or moving around in loose flocks with conspecifics, characteristic 'seeoo' or 'tseep', with soft, whistling timbre. Usually also given in slight alarm, and by solitary birds to make contact with others. In more intense alarm, call modified to higher-pitched 'see'. Note complete absence of ticking calls.

Breeding. Season. Egg-laying from May in most of European range, from late April in south; continues to July or August. 1–2 broods in most of range, perhaps 3 occasionally in south. Site. Usually well hidden on ground or on sedge tussocks, heaps of dead rushes, reeds, etc., by water; occasionally up to 4 m above ground in (e.g.) willow or alder. Nest: foundation of stems and blades of sedges, grasses, and other waterside plants, occasionally small twigs, lined with finer plant material, moss, rootlets, and sometimes hair or feathers. Eggs. Sub-elliptical, smooth and slightly glossy; very pale purplish, lilac-grey, or olive-brown, rarely buffish or greenish, with scrawls, spots, and blotches of purplish-brownish-black irregularly distributed over whole surface, or concentrated at either end; occasionally unmarked. Clutch: 4–5 (3–7). Incubation. 13

days (12–15). FLEDGING PERIOD. 10–12 days (9–13); leaves nest 3–5 days before able to fly.

Wing-length: *E. s. schoeniclus* (western and northern Europe): ♂ 76–87, ♀ 70–81 mm.
Weight: *E. s. schoeniclus*: ♂ mostly 17–25, ♀ 16–23 g.

Geographical variation. Marked, but largely clinal; involves depth of ground-colour and width of dark streaking on upperparts and flanks of both sexes and on chest of ♀, as well as size and shape of bill. 2 subspecies-groups recognized: thin-billed *schoeniclus* group in north, thick-billed *pyrrhuloides* group in south. Within each main group, 2 sub-groups recognizable, also based on bill size, but these less well-defined than main groups, apparently grading into each other. Sub-groups of thin-billed group in west Palearctic comprise: (1) nominate *schoeniclus* in western and northern Europe; (2) *stresemanni* in Carpathian basin, *ukrainae* from northern Moldova and northern Ukraine east to Saratov and Kuybyshev on middle Volga, *pallidior* in south-east European Russia. In thick-billed group, sub-groups in west Palearctic comprise: (1) *tschusii* from northeast Bulgaria and eastern Rumania through southern Moldova and Ukraine to Crimea and Sarpa area on lower Volga (*c.* 48–50°N), *incognita* in zone from Orenburg area eastward, *witherbyi* in Iberia, Balearic Islands, southern France, and (at least formerly) North Africa and Sardinia, *intermedia* on Corsica, Italy, and Dalmatian coast of Yugoslavia, *reiseri* from Albania to Turkey, *caspia* in eastern Transcaucasia and northern Iran and (perhaps this race) south-west Iran and Syria. In each sub-group, colour gradually paler towards east, and birds inhabiting same longitude are similar in colour, independent of bill size. Nominate *schoeniclus* a dark small-billed race.

Pallas's Reed Bunting *Emberiza pallasi*

PLATES: pages 1673, 1685

DU. Pallas' Rietgors   FR. Bruant de Pallas   GE. Pallasammer   IT. Migliarino del Pallas
RU. Полярная овсянка   SP. Escribano de Pallas   SW. Dvärgsävsparv

Field characters. 13–14 cm; wing-span 20.5–23 cm. Over 10% smaller than Reed Bunting but with similar form except for straight culmen, flatter crown, and slimmer body; only marginally larger than Little Bunting, with similarly pointed bill. 2nd smallest bunting of west Palearctic, closely resembling Reed Bunting in general character, plumage pattern and tail spreading, but having usually distinctive calls. At close range, compared with western Reed Bunting, both sexes show much more streaked upperparts (due to distinctively paler fringes to back and wing-feathers emphasizing black-brown centres), usually pale greyish rump, and only lightly marked underparts. Identification confirmed by dull (never rufous) lesser wing-coverts, bright double wing-bar, and pale panel on folded wing (but only 2nd of these obvious on juvenile). Adult ♂ further distinguished by yellowish to buffish tinge to rear collar, white rump, and virtually unstreaked underparts; adult ♀ and immature by much more uniform head (lacking obvious dark borders to crown and ear-coverts of Reed Bunting) and heavy black malar stripes turning into throat (and not breaking up to form obvious streaks on underparts).

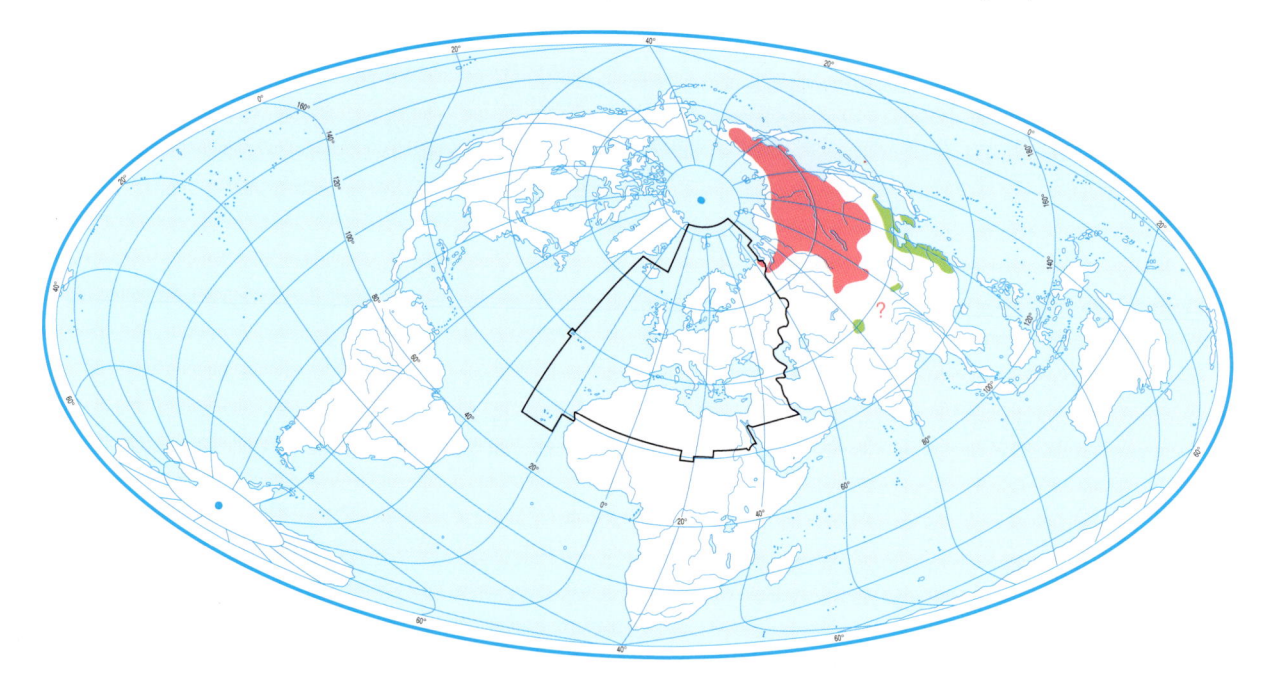

♂ showing pale rump unmistakable. ♂ hiding rump, ♀ and immature all require careful separation from Reed Bunting, and also from Little Bunting, ♀ and immature Black-faced Bunting, and even small Nearctic emberizid sparrows. Separation from other small buntings made easy by call and Little Bunting's rufous head with dark lateral crown-stripes and dark surround to cheeks not reaching bill, more striking cream eye-ring, duller ground-colour to upperparts and finely lined underparts; Black-faced Bunting's greyer face and upper mantle, duller browner upperparts, more rufous feather-margins on wings, buffier wing-bars, and fully streaked underparts; and small Nearctic sparrows' lack of white outer tail-feathers, yellowish or greyish supercilium, and usually fully streaked underparts. Flight lighter, more flitting than Reed Bunting; flying bird looks even smaller and more delicate than when perched.

Calls include 'tsleep', uttered when perched and recalling Tree Sparrow; 'chirrup'; 'pseeoo(p)' when flushed; 'tsee-see' in flight. Formerly considered diagnostic, but now known that Reed Bunting rarely utters 'chleep', similar to 1st call noted above.

Habitat. Breeds mainly in east Palearctic, in drier and cooler situations than overlapping Reed Bunting, occupying tundra with tall herbage and shrubs, but also shrubs and grass areas of steppes and semi-desert and, in south, mountain tundra. Northern populations inhabit river valleys with thickets of willow and alder in lowland tundra. Further south, breeds on high plateaux up to 2200–2500 m, in dwarf birch and other shrub growth. Winters in plains, preferring irrigated areas with shrubs and stands of reeds near rivers and lakes.

Distribution and population. RUSSIA. Has bred since at least 1981 in eastern Bol'shezemel'skaya tundra (north-east European Russia), range also extending onto western slopes of northern Urals. Western limit previously thought to be Taz basin. Occasionally recorded in southern Yamal peninsula, and sporadic breeding confirmed near Kharp. In Bol'shezemel'skaya tundra, breeds north to upper reaches of Kara river, west to source of Seyda river. Population estimated at 10 000–15 000 pairs.

Accidental. Britain: adult ♀, Fair Isle, September–October 1976; juvenile, Fair Isle, September 1981; 1st-winter ♂, Sussex, October 1990.

Movements. Northern populations long-distance migrants; southern populations short-distance and altitudinal migrants (perhaps dispersive rather than migratory in south-west). West Palearctic breeding birds winter mainly in China; in autumn, leave breeding areas August–September, return early or mid-June.

Food. Seeds and other plant material, invertebrates in breeding season.

Social pattern and behaviour. Little known; apparently similar to congeners such as Reed Bunting.

Voice. Song of ♂ a simple, short, somewhat monotonous phrase comprising series of repeated similar notes. For calls see Field characters.

Breeding. Almost all information extralimital. SEASON. North-east European Russia: eggs laid from late June or early July; young generally fledged by end of July. SITE. Well hidden on ground or tussock, or in depression in moss, lichen, etc., sheltered by shrub or grass; also less than *c.* 50 cm above ground in bush or small tree. Nest: rather flimsy foundation of dry stems and blades of grass and sedge, lined with similar but finer material, hair, and sometimes dry needles of larch. EGGS. Sub-elliptical, smooth and glossy; creamy-pink to reddish-brown, sometimes darker towards broad end, with scattered blackish-brown spots, small blotches, and hairstreaks and greyish-brown undermarkings and scrawls. Clutch: 4–5 (3–6). INCUBATION. About 11 days. FLEDGING PERIOD. In one case, young left nest at 10 days old.

Wing-length: ♂ 71–77, ♀ 68–73 mm.
Weight: ♂♀ mostly 12–16 g.

Red-headed Bunting *Emberiza bruniceps*

PLATES: pages 1678, 1685

DU. Bruinkopgors   FR. Bruant à tête rousse   GE. Braunkopfammer   IT. Zigolo testa aranciata
RU. Желчная овсянка   SP. Escribano carirrojo   SW. Stäppsparv

Field characters. 16 cm; wing-span 24.5–28 cm. Slightly smaller and more compact than Black-headed Bunting, with somewhat shorter, more rounded wings and slightly stubbier bill. Rather large bunting; structure and plumage of ♀ and immature much as Black-headed Bunting. ♂ has variable golden and chestnut head and bib, distinctly greenish mantle, bright yellow-green rump, and yellow underparts. ♀ resembles Black-headed Bunting but many show greenish-olive or grey tone on crown and back, while a few have buff-chestnut on forecrown, lower throat and upper breast. Juvenile plumage overlaps with Black-headed Bunting but typically less buff below. Hybridizes with Black-headed Bunting (♀ hybrids indistinguishable from that species).

♂ virtually unmistakable, but in glimpse on ground, beware confusion with Chestnut Bunting (15% smaller, with short tail and wholly rufous upperparts). ♀ and immature much more difficult, and in fresh plumage may be inseparable from Black-headed Bunting. Flight, actions, and behaviour much as Black-headed Bunting. Call distinctive.

Habitat. Adjoining and complementary to that of Black-headed Bunting in south Palearctic, but mainly in warmer,

Red-headed Bunting *Emberiza bruniceps*: **1** ad ♂ breeding, **2** ad ♂ non-breeding, **3** ad ♀, **4** 1st winter ♂, **5** juv. Black-headed Bunting *Emberiza melanocephala* (p. 1679): **6** ad ♂ breeding worn, **7** ad ♂ breeding fresh, **8** ad ♀ breeding, **9** ad ♀ non-breeding, **10** juv.

drier, and more open country, with less vigorous vegetation, in steppe, semi-desert, and desert oasis situations. Occupies all kinds of shrubby and herbaceous thickets, scattered in thin patches over relatively open countryside, but is highly typical of cultivated areas, seeking out water.

Distribution and population. RUSSIA, KAZAKHSTAN. Penetrated into west Palearctic only in 20th century, becoming widespread in Volga-Ural interfluve. Range limit advanced 800–900 km west, reaching Volgograd on Volga river by 1953. Subsequently, population decreased markedly and range contracted south-east to Caspian coast and Ural river. Range changes probably due to fluctuations in climate. Population size unknown, but widespread in suitable habitat.

Accidental. Many records regarded as dubious, relating to escaped cagebirds, but some genuine vagrancy probably occurs. Iceland, Faeroes, Britain, Ireland, France, Netherlands, Germany, Denmark, Norway, Sweden, Czech Republic, Switzerland, Spain, Italy, Turkey, Israel, Kuwait.

Movements. Migratory, all birds moving south-east to winter in India. Records of vagrancy to west Palearctic attributed mostly to escaped cagebirds, especially in northern Europe, where imported in large numbers until fairly recently. In Britain (and probably elsewhere), reports tend to be at migration seasons and in coastal areas, thus coinciding with vagrants of other species; this not surprising, however, since escaped individuals of migratory species are likely to show seasonal movements in appropriate directions, and to be observed especially at well-studied sites; also, observations inland not necessarily reported, owing to dubious origin. Some genuine vagrancy probably occurs, however; this supported by rapid colonization northwestwards in Kazakhstan, now reversed (see Distribution and population).

Food. Seeds (especially cereals) and other plant material, invertebrates in breeding season. Adults apparently eat much plant food throughout summer, though diet of young almost wholly invertebrates.

Social pattern and behaviour. Generally similar to closely related Black-headed Bunting, with which it hybridizes in contact zone in northern Iran. Available data (mostly extralimital) indicate that only ♀ builds nest, incubates, and rears (broods and feeds) young, at least to fledging. Hence likelihood of polygyny and/or promiscuity by ♂♂. Territorial in breeding season, with tendency to form neighbourhood groups. Once territory established, ♂ sings for long spells from favoured song-posts, also in flight between them, with dangling legs. In another, less common type of song-flight at peak of courtship period, ♂ flies high up and circles over territory; ascends rapidly, singing, followed by parachute-descent recalling song-flight of Tree Pipit.

Voice. Song very similar to Black-headed Bunting. Commonest call more penetrating than Black-headed Bunting, a brisk musical 'pwip', 'tweet', 'pweek', 'tliip', or 'tlyp', somewhat like House Sparrow but with liquid tone.

Breeding. All information extralimital. SEASON. Turkmenistan: eggs laid early May to last week of June. Kazakhstan:

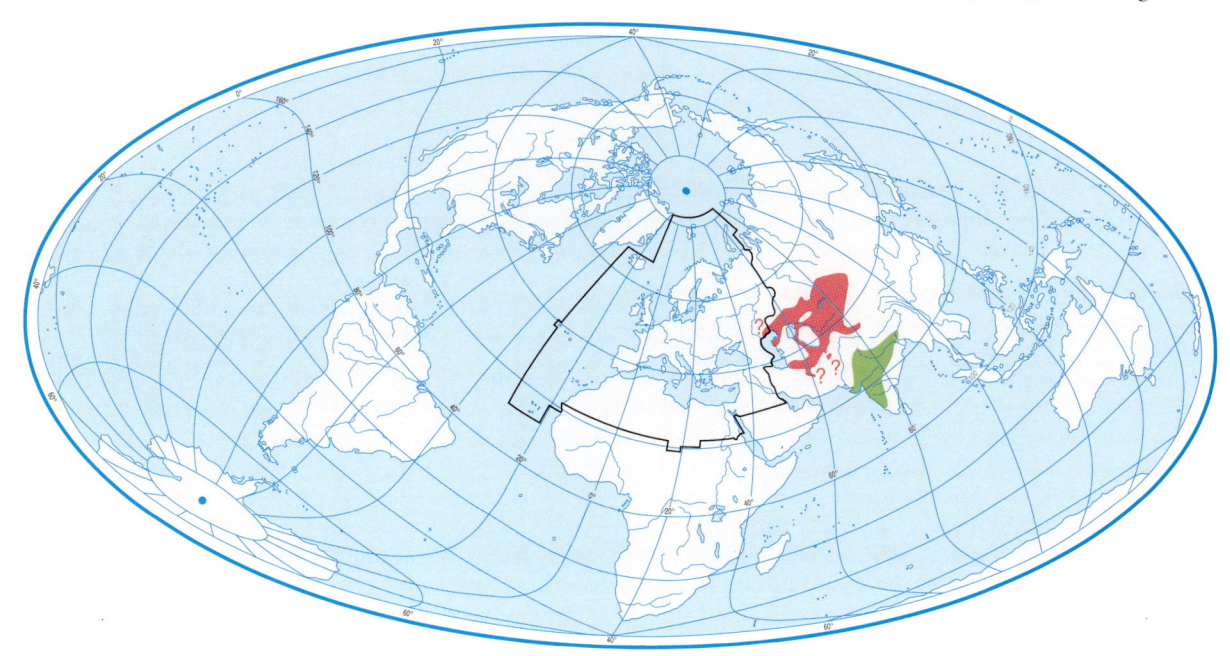

mid-May to late June. 2 broods. SITE. Low and well hidden in dense or thorny shrub, vine, fruit tree, etc., or very close to ground in thick grass. Nest: rather loose and untidy foundation of stems of cereals, rough grasses, Umbelliferae, Cruciferae, etc., often with flowers attached, sometimes pieces of bark or leaves; lined with fine grass, plant fibres, rootlets, and hair. EGGS. Sub-elliptical, smooth and slightly glossy; white or pale bluish-white, finely and sparsely spotted purplish-grey to brown, concentrated at broad end; no hairstreaks or scrawls. Clutch: 4–5 (3–6). INCUBATION. 12–13 days. FLEDGING PERIOD. (9–)12–13 days; often leaves nest before able to fly.

Wing-length: ♂ 84–92, ♀ 81–88 mm.
Weight: ♂ mostly 22–31, ♀ 18–30 g.

Black-headed Bunting *Emberiza melanocephala*

PLATES: pages 1678, 1685

DU. Zwartkopgors FR. Bruant mélanocéphale GE. Kappenammer IT. Zigolo capinero
RU. Черноголовая овсянка SP. Escribano cabecinegro SW. Svarthuvad sparv

Field characters. 16–17 cm; wing-span 26–30 cm. Looks noticeably larger than Yellowhammer with proportionately rather longer bill, more obvious neck, rather longer wings, and distinctly longer legs, but similar tail length; marginally larger than Red-headed Bunting, but no visible structural difference. 2nd largest bunting of west Palearctic, with rather long, tapering bill, rather long body, and noticeably long legs combining into characteristically heavy but sleek form shared only by Red-headed Bunting. Combination of uniformly pale, unstreaked underparts and lack of white outer tail-feathers excludes all other buntings except Red-headed Bunting. ♂ distinctive, with black head, chestnut back, and yellow underparts; ♀ and immature lack obvious characters and may not be separable from Red-headed Bunting.

♂ virtually unmistakable, but, in glimpse of bird on ground, beware confusion with similarly patterned ♂ Yellow-breasted Bunting (20% smaller, wholly chestnut and black head, chestnut necklace, darker upperparts, wing markings like Chaffinch, and streaked flanks) and ♂ of black-headed race of Yellow Wagtail (as long but much slimmer, with fine bill, greenish back, and white-edged tail). ♀ and immature always initially confusing and hard to separate not only from Red-headed Bunting but also from Cinereous Bunting. Distinction from similarly-sized Red-headed Bunting not always possible, but that species typically greyer or greener above, with brighter rump, and paler below, sometimes with more pronounced streaking on breast and flanks. Cinereous Bunting 15% smaller, with white outer tail-feathers. Flight as other *Emberiza* but noticeably strong, lacking rather erratic rhythm to wing-beats of smaller species; since flight silhouette lacks full end to tail of Yellowhammer, form and undulating progress may also suggest large pipit.

Habitat. Breeds in south-west Palearctic in warm temperate, Mediterranean, and steppe zones, between July isotherms of 23–32°C, generally in lowlands, avoiding both drier and wetter extremes. Favours fairly dense and tall bushy and scrub vegetation, including open maquis, wooded steppes, orchards, olive groves, and vineyards, and groves or thickets along streamsides, roadsides, or field borders. Also found in open forest with undergrowth, in open lowland grassland with scrub, especially thorn scrub, and on mountain slopes. Wintering birds in India feed in flocks and cultivated fields, sometimes causing serious damage to standing crops, also occupy scrub jungle, roosting in enormous concentrations with other species in thorn scrub and thickets.

Distribution. Now extinct Slovenia. FRANCE. Singing ♂♂ regular. ITALY. Locally common in scattered areas of south, east, and west coasts; has bred as far north as southern Lombardia. BULGARIA. Now breeding along whole Black Sea coast, following northward range expansion. Inland, Stara Planina forms northern limit. RUMANIA. Occurs at only a few localities, but slight range increase north to Danube delta area in 1960s. UKRAINE. Slight decrease. TURKEY. Sparse or absent in coastal zone of Black Sea coastlands. SYRIA. Breeding range less extensive than shown hitherto, possibly also changing between years. High density in extreme north-west. IRAQ. Rather abundant breeder in north.

Accidental. Iceland, Faeroes, Britain, Ireland, France, Netherlands, Germany, Denmark, Norway, Sweden, Finland, Latvia, Poland, Czech Republic, Slovakia, Austria, Switzerland, Spain, Malta, Kuwait(?), Tunisia, Algeria, Morocco.

Beyond west Palearctic, breeds only in Iran.

Population. Numbers and trends not well known, but perhaps slight increase Bulgaria and Rumania, slight decrease Ukraine, marked decrease Israel, apparently stable elsewhere. ITALY. 2000–4000 pairs 1983–93. GREECE. 100 000–200 000 pairs. ALBANIA. 10 000–30 000 pairs in 1981. YUGOSLAVIA: CROATIA. 15 000–20 000 pairs. BULGARIA. 5000–50 000 pairs. RUMANIA. Poorly known: probably 10–20 pairs 1986–92. RUSSIA. 1000–10 000 pairs. UKRAINE. 1–20 pairs in 1978. AZERBAIJAN. Common, locally very common. TURKEY. 1–10 million pairs. CYPRUS. 10 000–20 000 pairs; locally very common. ISRAEL. Some hundreds of pairs. Marked decline in 1960s–70s, mainly due to expansion of agriculture and afforestation. JORDAN. Small numbers.

Movements. Migratory, all birds moving south-east or ESE to winter in western and central India. Leaves breeding grounds early, and returns late. Departure (inconspicuous) late July to August, arriving India August–September. Occasional midwinter records from breeding range or intermediate areas, e.g. Israel. In spring, leaves winter quarters March–April; reaches Turkey mostly from late April; sometimes reported in Cyprus as early as March, but usually arrives in early or mid-April, with movement continuing to mid-May. Arrives on Aegean islands and Makedonija late April and early May. Vagrancy west of range is mostly in spring, suggesting overshooting.

Food. Seeds and other plant material; invertebrates in breeding season. In summer quarters, forages principally in cultivated areas: cereal or sunflower fields, vineyards, orange groves, etc., feeding both on ground and in shrubs or low in trees. Most foraging observations concern migrant birds or winter visitors,

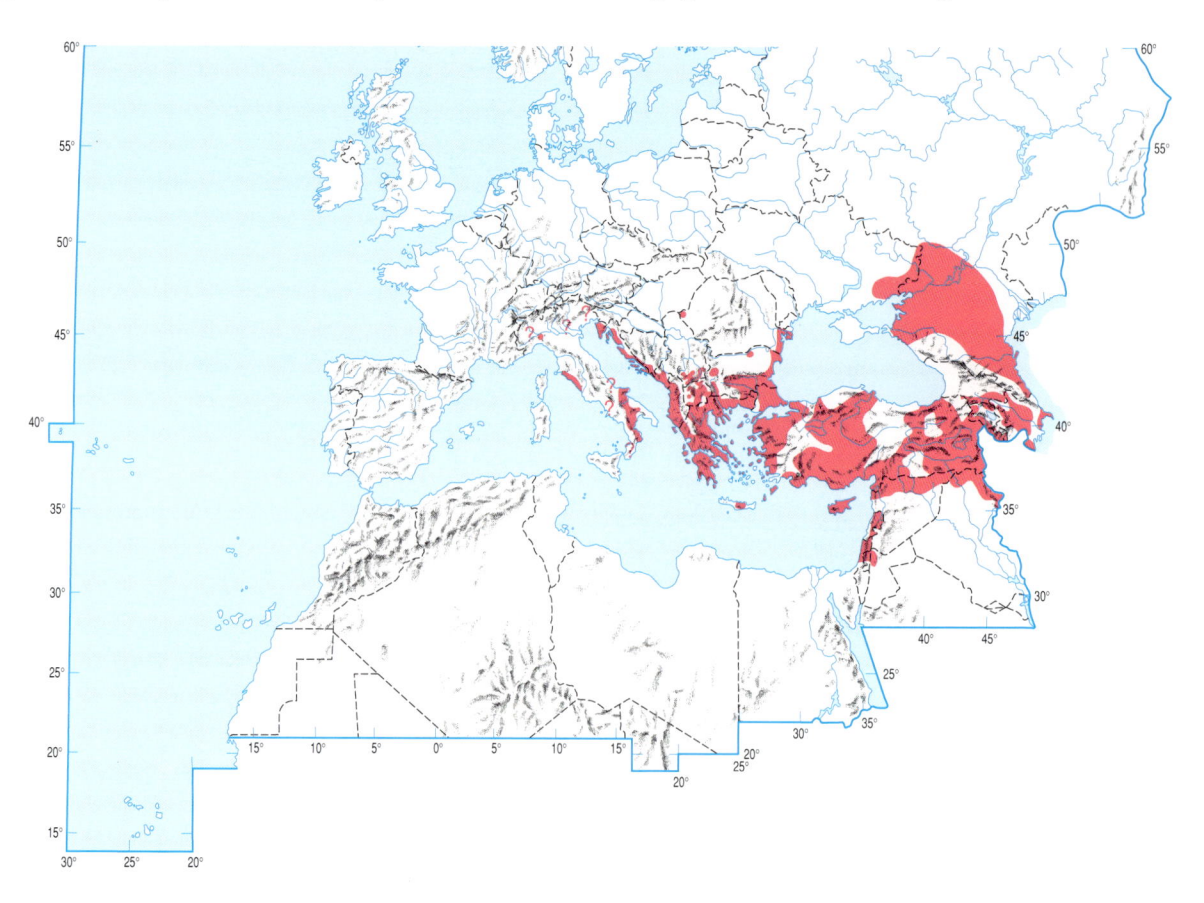

Corn Bunting *Miliaria calandra*. *M. c. calandra*: **1** ad spring, **2-3** ad winter, **4** juv. *M. c. buturlini*: **5** ad spring, **6** juv.

since species occurs then in huge numbers often causing considerable damage to millet and other cereals, maize, and rice.

Social pattern and behaviour. Gregarious outside breeding season, often in huge flocks in winter quarters. Mating system and breeding behaviour little studied, and not known to what extent, if at all, it differs from Red-headed Bunting (which see). In breeding season, ♂ sings indefatigably almost throughout day, with bill raised and crown slightly so, mostly from any convenient elevated song-post, typically tree, tall bush, vineyard post, telegraph pole or wire; also regularly (but less intensely) from ground while feeding, and in flight between song-posts, with shallow quivering wing-beats and dangling legs, similar to Corn Bunting. Less common, towering type of song-flight occurs, as in Red-headed Bunting.

Voice. Song quite short but more musical than usual in *Emberiza*; usually begins with rather grating twitter, then 2-3 accelerating, warbled phrases; suggests warbler and (partly) Lapland Bunting: 'sitt süt süt', then 'süterEE-süt-süte-ray'.

Calls include 'cheuh' or 'styu' recalling Yellowhammer, and clicking 'plüt' or 'chup' recalling Ortolan Bunting.

Breeding. SEASON. Short throughout region. Croatia: eggs laid from mid-May, mostly 1st half of June. Greece: eggs laid mid-May to end of June. Cyprus: laying starts mid-May. Levant: eggs laid early May to 1st half of July, peak May. Probably only one brood. SITE. Low down in dense, often thorny shrub, commonly on vine; usually 0.5-1.0 m above ground. Nest: loose, untidy foundation of stalks of herbs, grass (including reed), and leaves, lined with fine grasses, stems, rootlets, hair, and sheep's wool; fairly often with brightly-coloured flower-heads on outside. EGGS. Sub-elliptical, smooth and slightly glossy; very pale blue or greenish-blue, rarely buff, rather sparsely speckled purplish-grey to olive-brown, usually concentrated towards broad end; sometimes unmarked. Clutch: 4-5 (3-7). INCUBATION. About 14 days. FLEDGING PERIOD. 13-16 days.

Wing-length: ♂ 94-101, ♀ 86-94 mm.
Weight: ♂ 25-33, ♀ 24-32 g.

Corn Bunting *Miliaria calandra*

PLATE: page 1681

DU. Grauwe Gors   FR. Bruant proyer   GE. Grauammer   IT. Stillozzo
RU. Просянка   SP. Triguero   SW. Kornsparv

Field characters. 18 cm; wing-span 26-32 cm. Noticeably larger, bulkier, but proportionately shorter-tailed than Yellowhammer; close in size to Skylark. ♂ is largest bunting of west Palearctic with heavily streaked buff-brown plumage;

1682 Buntings, New World Sparrows, and Allies

♀ 10% smaller but still bulky, sharing ♂'s heavy bill and stout legs. Recalls ♀ sparrow or Skylark far more than other buntings. No white in tail. Size, flight, and voice all more important to identification than plumage details.

Within Emberizidae, large size, lack of obvious marks, flight, and voice form unmistakable combination. Inexperienced observer must, however, beware similarity of appearance to other streaked passerines which may share habitat of Corn Bunting or use it on migration. Commonest confusions are with larger streaked larks, ♀ and immature sparrows (*Passer* and *Petronia*), ♀ and immature rosefinches (*Carpodacus*), Rose-breasted Grosbeak, and Bobolink. Flight recalls rock sparrow (*Petronia*), differing from that of other buntings in loose, surging take-off (during which ♂ in breeding season often dangles and trails legs and bunched feet). Over longer distance, alternation of strong wing-beats and closed-wing attitudes produces powerful undulations, particularly noticeable in descent to feeding area or roost. Throughout flight, relatively large dark wings catch eye as much as dull, not proportionately long tail.

Habitat. Breeds in middle latitudes of south-west Palearctic, in cool and warm temperate, Mediterranean, and steppe climatic zones, within July isotherms of 17–32°C, including extremes of both oceanic and continental types. Mainly in lowlands, preferably undulating or sloping rather than level, and with pronounced liking for vicinity of sea coasts and, inland, cereals and other low-growing crops. Avoids forest, wetlands, rocky and broken terrain, and, in most regions, mountains or high plateaux, as well as built-up areas. Apart from need for perches to overlook territory and to serve as song-posts (low trees, bushes, overhead cables, fences, or walls), is at home in fully open country, and has minimal demands for cover, except to some extent for roosting, e.g. in reeds.

Distribution. Range has contracted in north-west, most

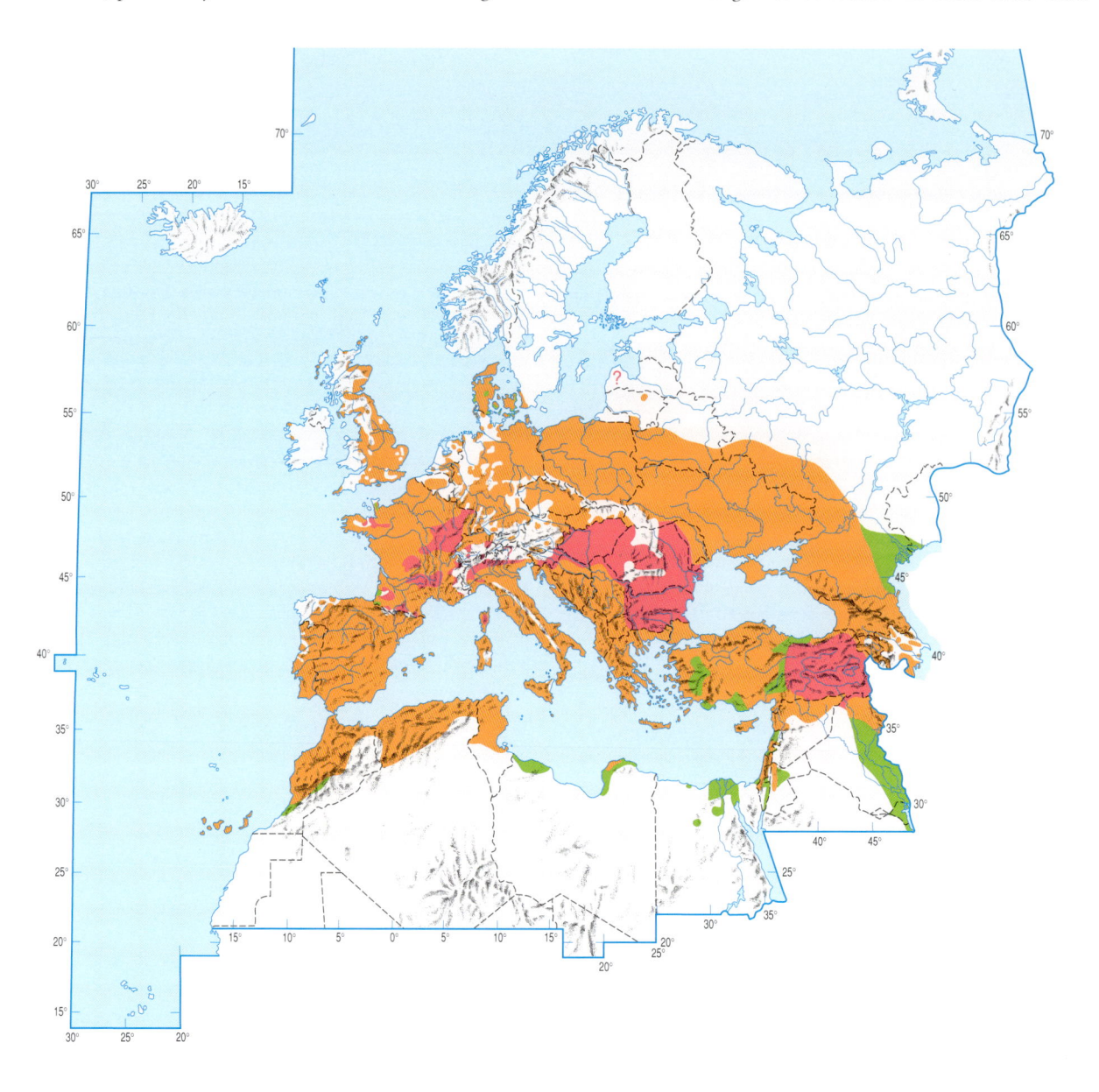

notably British Isles and Netherlands, in north (Sweden, Latvia, Lithuania), central Europe (particularly Switzerland, also Germany, Czech Republic, Slovakia, and Austria), and further south (Italy, Malta, Canary Islands) and south-east (Ukraine, Moldova). Probably due mainly to changing agricultural practice and land use. Slight increase Spain, fluctuating in Slovenia. Winter range unclear (notably Balkans). BRITAIN. Marked decrease especially in north and west, since *c.* 1930s. Has virtually disappeared from Wales and south-west England, and in Scotland now mainly in coastal areas. Decrease since early 1970s *c.* 45%. IRELAND. Formerly widespread; range contracted throughout 20th century. Now close to extinction. BELGIUM. Spread over large part of Lorraine since 1974. Expansion attributed to changing agricultural practice (very large fields), but most recent trend is decrease. GERMANY. Formerly occurred almost throughout country, now obvious gaps in all western Länder (states), and has similarly vanished from many areas in east as well. NORWAY. Formerly bred in south-west, but (except for possible breeding attempts in 1971 and 1977) no breeding records since 1928; now accidental visitor. ESTONIA. Breeding recorded only in 1947 and 1991 (perhaps up to 5 pairs); otherwise, few observations (e.g. 3 in 1991–2). LATVIA. Formerly very rare breeder; a few pairs may still breed. CZECH REPUBLIC, SLOVAKIA. Slow spread 1850–1960; marked contraction after 1965. AZERBAIJAN. In Great Caucasus, occurs only in lower part of forest zone. TURKEY. Widespread and often common or even abundant, especially in central and eastern areas. SYRIA. Quite common breeder, especially in north-west; at least local further east. JORDAN. A few linger after winter and breeding suspected, but not proven until 1990. TUNISIA. Common in north, rarer in central and southern areas. ALGERIA. Often common in cultivated areas. MOROCCO. Widespread and abundant in north, much scarcer and more local south of Haut Atlas. CANARY ISLANDS. Breeds on all islands.

Accidental. Faeroes, Finland.

Beyond west Palearctic, breeds Iran north-east to Dzhungarskiy Alatau (Kazakhstan) and adjoining north-west China.

Population. Widespread decline, marked in many countries. Slight increase only in Spain and Croatia, apparently stable France (decreasing locally), Luxembourg, Poland, Hungary, Portugal, Bulgaria, Russia, Cyprus, and Morocco. BRITAIN. As few as 20 000 territories in 1993. Decrease since *c.* 1930; massive decline (*c.* 75%) in last 20 years. Strong decrease north-east Scotland 1988–92 compared with 1944–8. IRELAND. Fewer than 30 territories 1988–91. Sharp decline. FRANCE. 100 000–1 million pairs in 1970s. BELGIUM. 3000–5500 pairs 1988–91. Decline since 1970s, still continuing. LUXEMBOURG. 20–200 pairs. NETHERLANDS. 1100 pairs 1975, 100–200 in 1989. Strong decline, probably from before 1970. GERMANY. 20 000–35 000 pairs. Marked decline, e.g. in Schleswig-Holstein: from 3000–4000 pairs in 1955 to only 40 in 1987, decline beginning 1960–65 and steepening rapidly after 1975, so that formerly considerable wintering population now practically extinct. Decline still continuing, at least in east, where 20 000 ± 8000 pairs in early 1980s. DENMARK. 11 000–120 000 pairs 1987–8; decrease 1981–94. SWEDEN. *c.* 100 birds in 1982, 5–10 pairs in late 1980s, and population perhaps no longer viable by early 1990s (6–7 singing ♂♂ 1992–3). Decrease from beginning of 20th century. LITHUANIA. Very rare; decreasing. POLAND. 50 000–200 000 pairs. CZECH REPUBLIC. 700–1400 pairs 1985–9. Marked decline since 1960s. SLOVAKIA. 4000–8000 pairs 1973–94; slight decrease. HUNGARY. 8000–12 000 pairs 1979–93. AUSTRIA. Population fluctuating, locally decreasing, in some parts markedly so, mainly because of habitat loss. SWITZERLAND. 200–250 pairs 1985–93. Marked decrease. SPAIN. 1.44–4.3 million pairs. PORTUGAL. 100 000–1 million pairs 1978–84. ITALY. 200 000–600 000 pairs 1983–95; slight decrease. MALTA. 400–600 birds; decreased in last 15 years. GREECE. 300 000–500 000 pairs; slight decrease. ALBANIA. 20 000–50 000 pairs in 1981. YUGOSLAVIA: CROATIA. 70 000–100 000 pairs. SLOVENIA. 1000–2000 pairs. BULGARIA. 100 000–1 million pairs. RUMANIA. 50 000–60 000 pairs 1986–92. RUSSIA. 1000–10 000 pairs. BELARUS'. 100–1000 pairs in 1990; fluctuating, but apparently stable overall. UKRAINE. 4000–7000 pairs in 1986; slight decrease. MOLDOVA. 1500–3000 pairs in 1988; slight decrease. AZERBAIJAN. Locally common. TURKEY. 1–10 million pairs. CYPRUS. 40 000–60 000 pairs. No evidence of any decline. ISRAEL. A few thousand pairs 1970s–80s. CANARY ISLANDS. Apparently declining.

Movements. Resident to partially migratory. Winters chiefly within breeding range, but also regularly south to North Africa and northern Arabia. Western migrants head mostly south-west or SSW, and some southern birds move west; at least some eastern birds head south or east of south. Data suggest central European birds migrate more than north European ones. Resident birds roam in flocks in winter, resulting in absence from some breeding localities.

Autumn migration protracted, August–December, and timing varies; formation of winter flocks (from August) masks departure and passage. Long-distance recoveries reported from October. Spring movement also protracted, (January–) February–May. In Strait of Gibraltar, passage late February to May. In southern England, reported returning to local breeding sites as early as January, more usually late February to March.

Food. Seeds (often of cereals), other plant material, and invertebrates, especially in breeding season. Feeds almost wholly on ground in arable fields, damp meadows, short rough grass, etc.; in autumn, commonly in stubble and fields where root crops have been harvested or dung spread; forages in farmyards, grain depots, etc., only in harsh winters, and much less so than, e.g., Yellowhammer.

Social pattern and behaviour. Generally gregarious outside breeding season; flocks typically very variable in size and composition, both geographically and in relation to weather. Mating system complex and typically variable within and between areas. ♂♂ regularly polygamous, with very variable number of ♀♀; up to 18 recorded. Bond between sexes not close; ♂ and ♀ hardly ever together on or off territory unless sexually or aggressively motivated. ♂♂ arrive on territories considerably before ♀♀ (seldom seen prior to settlement, may arrive in groups). ♀♀ construct nest and incubate alone. Nestlings provisioned mostly by ♀, contribution of ♂ varying from 0–50%. ♂ advertises territory by singing from exposed perch, less commonly in flight, and frequently on alighting. Perch-sites vary from ground-level, small thistles and grass clumps to tree-tops, but most commonly human artefacts (fence-wires and posts, hedges, walls, etc.). On leaving and approaching song-post, ♂, often while singing, commonly performs characteristic dangling-legs display: head lowered, feet dangled and legs pointed out slightly, accompanied by rapid, shallow wingbeats. ♀ does not sing. During breeding season, high level of song output maintained throughout day, even through midday.

Voice. Song diagnostic, a short, rushed jangle of ticking and discordant units: 'tük tük zik-zee-zrrississ', recalling rattle of bunch of small keys; often given repeatedly for long periods, forming characteristic sound of (particularly) barley fields in spring and summer. Calls varied but only 2 important for identification: on take-off and in flight, quite loud, initially slightly liquid but terminally clicked 'quit' or 'quick', often run together in accelerating series 'quit-it-it' and forming distinctive chorus when given by flock; rather harsh 'chip' as contact-call in breeding season.

Breeding. SEASON. Generally late. In northern and western Europe, egg-laying from late May or early June, continuing to July or August. Considerably earlier in extreme south. Israel: in lowlands, 2nd half of February to 2nd half of May; at higher altitude, mid-March to end of June. Canary Islands: eggs laid mid-March to mid-June. 2–3 broods in southern England, but 1–2 apparently more usual elsewhere. SITE. Generally on ground, in thick tangled grass or shrub, in depression in soil of arable field (sometimes perched 'awkwardly' on bare ground), or in pasture, often in clump of thick weeds. Nest: fairly large loose construction of stalks, grass stems (which can be green), and roots, lined with fine grass, rootlets, and sometimes hair. EGGS. Very variable in shape and colour; generally sub-elliptical, smooth and slightly glossy; whitish, often tinged with blue, purple, or buff, sparsely but boldly marked with blotches and meandering scrawls and hairstreaks of brownish-black or purple, though some with hardly any pattern or only large pale brown blotches; greyish-violet undermarkings often present. Clutch: 4–6 (1–7). INCUBATION. 12–14 days. FLEDGING PERIOD. 9–13 days, often leaving nest before able to fly.

Wing-length: *M. c. calandra*: ♂ 96–107, ♀ 87–96 mm.
Weight: *M. c. calandra*: ♂ mostly 43–63, ♀ 35–47 g.

Geographical variation. Slight and largely clinal. Size markedly constant throughout much of range, but birds from Britain, Canary Islands, and North Africa tend to be slightly smaller than those on continent, as are a number of populations on islands in western Mediterranean. Variation in colour clinal throughout Britain, continental Europe, and Asia, depth of ground-colour and width and extent of dark streaks on body decreasing from west to east; dark brownish western end of cline (*clanceyi* in western Scotland and western Ireland) and pale greyish eastern end (*buturlini* from inland Syria, Jordan, and south-east Turkey eastwards) fairly distinct, these two being only races considered validly separable from nominate *calandra* occupying rest of range.

Black-faced Bunting *Emberiza spodocephala* (p. 1645): **1** ad ♂ non-breeding. Pine Bunting *Emberiza leucocephalos* (p. 1646): **2** ad ♂ winter. Yellowhammer *Emberiza citrinella citrinella* (p. 1648): **3** ad ♂ winter, **4** ad ♀ winter. Cirl Bunting *Emberiza cirlus* (p. 1651): **5** ad ♂ winter, **6** ad ♀ winter. Rock Bunting *Emberiza cia cia* (p. 1653): **7** ad ♂ winter, **8** ad ♀ winter. Meadow Bunting *Emberiza cioides* (p. 1655): **9** ad ♂ winter. House Bunting *Emberiza striolata sahari* (p. 1655): **10** ad ♂ winter.

Cinereous Bunting *Emberiza cineracea* (p. 1658): **1** ad ♂ winter. Ortolan Bunting *Emberiza hortulana* (p. 1659): **2** ad ♂ non-breeding, **3** 1st ad ♂ non-breeding. Grey-necked Bunting *Emberiza buchanani* (p. 1662): **4** ad ♂ winter. Cretzschmar's Bunting *Emberiza caesia* (p. 1663): **5** ad ♂ non-breeding. Yellow-browed Bunting *Emberiza chrysophrys* (p. 1664): **6** ad ♂ winter. Rustic Bunting *Emberiza rustica* (p. 1665): **7** ad ♂ non-breeding. Little Bunting *Emberiza pusilla* (p. 1667): **8** ad ♂ non-breeding. Chestnut Bunting *Emberiza rutila* (p. 1669): **9** ad ♂ winter, **10** 1st ad ♂ winter.

Yellow-breasted Bunting *Emberiza aureola* (p. 1670): **1** ad ♂ non-breeding, **2** 1st winter ♂. Reed Bunting *Emberiza schoeniclus schoeniclus* (p. 1672): **3** ad ♂ winter, **4** 1st winter ♂. Pallas's Reed Bunting *Emberiza pallasi* (p. 1676): **5** ad ♂ winter, **6** 1st winter ♂. Red-headed Bunting *Emberiza bruniceps* (p. 1677): **7** ad ♂ non-breeding. Black-headed Bunting *Emberiza melanocephala* (p. 1679): **8** ad ♂ non-breeding, **9** ad ♀ non-breeding.

Cardinal-grosbeaks and Allies Subfamily Cardinalinae

Small to medium-sized emberizids, variously known as 'cardinals', 'grosbeaks', 'saltators', and even 'buntings', and collectively also as 'cardinal finches'. Mainly arboreal, feeding on seeds (which they crush in their strong bills), fruits, and insects. Except as stragglers, found in New World only from central Canada to northern Argentina. About 47 species in 15 genera, of which 4 species of 4 genera accidental in west Palearctic (all from North America).

Bill mainly large and stout, in some species recalling that of grosbeaks of subfamily Carduelinae (Fringillidae); differs internally in a number of features from bill of emberizines (and also from that of cardueline grosbeaks). Plumages markedly bright, with large and contrasting patches of different colours (red, yellow, blue, green, white, black, etc.), being especially multicoloured in *Passerina*.

Dickcissel *Spiza americana*

PLATE: page 1687

Du. Dickcissel   Fr. Dickcissel d'Amérique   Ge. Dickzissel   It. Spiza americana
Ru. Американская спиза   Sp. Gorrión cuadrillero   Sw. Dickcisselsparv

Field characters. 13–17·5 cm; wing-span 22·5–27·5 cm. Up to 10% larger than House Sparrow but with rather slimmer form except for square tail with pointed feathers and rather long legs; 5–10% smaller than Bobolink. Sparrow-like, with yellow on lores, supercilium, and submoustachial stripe, sharp blackish malar stripe, and bright bay-chestnut forewing-coverts in every plumage except 1st winter ♀. Adult ♂ has bluish bill (when breeding), grey crown and hind cheeks, and black bib contrasting with yellow breast. ♀ and immature less colourful but similarly patterned, lacking only bib.

♂ virtually unmistakable; only ♂ Dead Sea Sparrow has convergent plumage colours and pattern. ♀ and immature confusing, suggesting Bobolink, other Nearctic sparrows, and west Palearctic buntings. Important to realize that Dickcissel is ground-haunting and has distinctive tail shape and call; in addition, all but immature ♀ show restricted but vivid chestnut on forewing, lacking in all possible confusion species. Commonest call a low buzzing 'br-r-r-r-rt', rather like Long-tailed Tit; given in flight and during nocturnal migration.

Habitat. Breeds in Nearctic temperate open lowlands, either prairies and other grassland bearing tall grasses, herbs, or shrubs, or crops of grass, clover, or alfalfa; also in meadows, pastures, weed patches, and grain fields. A ground bird, but uses raised perches such as roadside fences or overhead wires for singing. Winters in South America in forest clearings, forest edges, llanos, open country, and cultivation such as rice-fields at low altitude.

Distribution. Breeds in North America from eastern Montana to Great Lakes area (mostly sporadic in extreme south of Canada), south to central Colorado, southern Texas, and Alabama; now only local or sporadic east of Appalachians (formerly reached Atlantic lowlands).

Accidental. Norway: adult, Måløy (Sogn og Fjordane), July 1981.

Movements. Migratory; erratic in distribution both winter and summer. Populations fluctuate greatly; breeding range east of Appalachians abandoned 1860–95, perhaps due to loss of winter habitat, but partially reoccupied 1928–54; has declined in most of range since then. Winters from Michoacán (central Mexico) south through Central America to northern Colombia, Venezuela, and the Guianas; locally also in small numbers in coastal lowlands of USA from southern New England to southern Texas. Autumn records in eastern North America, and attempted wintering where artificial food available, have increased greatly since 1949.

Wing-length: ♂ 81–88, ♀ 73–79 mm.

Rose-breasted Grosbeak *Pheucticus ludovicianus*

PLATE: page 1687

Du. Roodborstkardinaal   Fr. Cardinal à poitrine rose   Ge. Rosenbrust-Kernknacker   It. Beccogrosso pettorosa
Ru. Красногрудый дубоносовый кардинал   Sp. Candelo tricolor   Sw. Brokig kardinal

Field characters. 18–21·5 cm; wing-span 30–32·5 cm. Head and body similar in size to Corn Bunting but bill even larger and square-ended tail longer; at least 20% larger than medium-sized rosefinches (*Carpodacus*). Rather large, stout-billed, and quite long-tailed Nearctic passerine, with colourful underwing and double white wing-bar common to all plumages. Breeding ♂ black and white, with rose-pink breast and underwing. ♀ and immature recall *Carpodacus* finch but have more strongly

Dickcissel *Spiza americana*: **1** ad ♂ spring, **2** ad ♀ spring, **3** 1st winter ♂, **4** ad ♂ winter. Rose-breasted Grosbeak *Pheucticus ludovicianus*: **5–6** ad ♂ breeding, **7** 1st winter ♂, **8–9** ad ♀ spring, **10** ad ♂ non-breeding.

striped head including pale crown-centre, more heavily marked breast, pink (♂) and yellow (♀) underwing, and proportionately larger and paler bill.

♂ unmistakable. ♀ and immature recall *Carpodacus* finch, particularly heavily marked Purple Finch *C. purpureus* which might also cross Atlantic to western Europe, but Rose-breasted Grosbeak much larger, with size of bill, pale crown-centre, colourful underwing, strength of wing-bars, and pattern and unforked shape of tail adding to distinctive appearance. Important to recognize that Rose-breasted Grosbeak is no less bulky than Redwing or Pine Grosbeak. Flight strong, with beats of quite long wings obvious in bursts which produce both shooting acceleration, and marked undulations; in open flight, fore-silhouette and action can recall Hawfinch. Call a single, rather sharp, even squeaky, metallic 'peek', 'kick', or 'kink'; loud and far-carrying.

Habitat. Breeds in temperate Nearctic wooded lowlands, deciduous or mixed, strongly favouring edges between stands of large tall trees and thickets of tall shrubs, especially where an opening is made by a stream, pond, or marsh. Also in parks, on wooded farmland, and even in villages and large gardens which provide required edge effect. Winters in South America in open rain and cloud forest, rarely below 1000 m, and in secondary growth, brush, and cultivated land, especially near streams.

Distribution. Breeds in North America from north-east British Columbia and southern Mackenzie east across southern Canada to Nova Scotia, south to eastern North Dakota, eastern Kansas, southern Missouri, eastern Tennessee, and Maryland.

Accidental. Britain, Channel Islands, Ireland, France (Ouessant), Norway, Spain, Malta, Yugoslavia. Records (2) in Sweden regarded as escapes, and perhaps German record also; this possibility not to be discounted for records from at least some countries listed above.

Movements. Migrant. Winters from central Mexico south to Ecuador, Colombia, and Venezuela. Peak movement early to mid-September in northernmost states, mid- to late September in mid-latitudes, and October near Gulf of Mexico coast. Few birds reach winter range before late September, with most arrivals mid-October to December. In spring, peak migration starts mid- or late April in southern states, reaching southern Canada by mid-May.

Rare autumn vagrant to west Palearctic. 24 records from Britain and Ireland up to 1995, chiefly October; all concerned immature individuals.

Wing-length: ♂ 100–109, ♀ 98–105 mm.

Blue Grosbeak *Guiraca caerulea*: **1** ad ♂ summer, **2** ad ♂ winter, **3** 1st winter ♂, **4** ad ♀ summer. Indigo Bunting *Passerina cyanea* (p. 1689): **5** ad ♂ breeding, **6** ad ♂ non-breeding, **7** ad ♀ breeding, **8** 1st winter ♂. Lazuli Bunting *Passerina amoena* (p. 1689): **9** ad ♂ breeding, **10** ad non-breeding, **11** ad ♀ summer, **12** 1st winter ♀.

Blue Grosbeak *Guiraca caerulea*

PLATE: page 1688

Du. Blauwe Bisschop   Fr. Guiraca bleu   Ge. Azurbischof   It. Beccogrosso azzurro
Ru. Голубая гуирака   Sp. Cardenal azul   Sw. Blåtjocknäbb

Field characters. 14–17 cm; wing-span 26–28·5 cm. About 15% larger than Scarlet Rosefinch but with rather similar though stockier form; about 15% smaller than Rose-breasted Grosbeak, with proportionately shorter tail. Rather large, robust but shy Nearctic finch, with two rusty wing-bars in all plumages. ♂ deep blue, with blacker flight- and tail-feathers; ♀ and immature recall lightly marked *Carpodacus* finch but show some blue on wing and tail.

♂ unmistakable, being over 20% larger than main Nearctic confusion species, Indigo Bunting, which lacks heavy bill and wing-bars. ♀ and 1st-winter far less distinctive since appearance of bill, head, and wings converges with *Carpodacus* finches; essential to confirm presence of blue feathers and obviously deep, cream to rusty-buff upper wing-bar. Flight, gait, and behaviour all recall *Carpodacus* finch. Commonest call a sharp 'chink'.

Habitat. A Nearctic forest species, like Rose-breasted Grosbeak, firmly linked with edges between dense woodland and more open lower vegetation, as naturally occurs by streamsides, swamps, or where opened-up patches are undergoing second growth. Human intervention has partly replicated such terrain in roadside plantings, along farm hedgerows and ditches, and in weedy fields, into which range has recently expanded, although attachment to thick cover and to moist places persists and open ground is avoided.

Distribution. Breeds in southern North America from central California, Nevada, and the Dakotas east to New Jersey, south to central Florida and Gulf of Mexico coast, and through Mexico and Central America to central Costa Rica.

Accidental. Norway: ♂, Akershus, June 1970; ♂, Rogaland, November 1987. Records from Britain (1970, 1972, 1986) and Sweden (1980, 1983) thought to involve escaped birds.

Movements. Migrant (northern populations) to resident (southern populations). Migrants winter within and slightly south of residents' range: mainly from northern Mexico south to Panama, rarely in southern USA and western Caribbean.

Wing-length: ♂ 86–94, ♀ 81–87 mm.

Indigo Bunting *Passerina cyanea*

PLATES: pages 1688

Du. Indigogors   Fr. Passerin indigo   Ge. Indigofink   It. Ministro
Ru. Индиговый овсянковый кардинал   Sp. Pape azulejo   Sw. Indigofink

Field characters. 13–14·5 cm; wing-span 19·5–21·5 cm. Close in size and form to Linnet but with more slender tail; 25% smaller and much daintier than Blue Grosbeak. Small, active Nearctic bunting, with quite strong bill. Lacks obvious wing-bars except in juvenile. Breeding ♂ almost wholly deep blue. Winter ♂, ♀, and immature mainly buff-brown, often showing little blue and with faint plumage pattern suggesting *Carduelis* or *Carpodacus* finch.

♂ unmistakable in good view if size well judged. ♀ and immature nondescript but lack strong streaks typical of *Carduelis* and *Carpodacus* finches and also bulbous bill of *Carpodacus*. Flight and behaviour much as *Carduelis* finch. Commonest call a loud 'pwit', sometimes repeated.

Habitat. Breeds in temperate and warmer Nearctic lowlands, from coniferous forest zone southwards, including upland fringes but excluding most intensively cultivated and grazed areas, deserts, and closed-canopy forest, as well as human residential areas. Inhabits brushy and weedy fringes of cultivated lands, roads, railways, and rivers, as well as woods, where they are open, deciduous, and broken by clearings.

Distribution. Breeds in North America, from south-east Saskatchewan east to New Brunswick and Maine, south to southern New Mexico, Gulf of Mexico coast, and central Florida.

Accidental. Iceland (October 1951, October 1985), Ireland (October 1985). Other records (e.g. Britain 1973, Netherlands 1983, Denmark 1987) regarded as involving escaped birds.

Movements. Migratory, all birds wintering south of breeding range, mainly from central Mexico south to Panama, in Bahamas and Greater Antilles. Main autumn departure from north of range begins late August, with peak movement September in northern states, continuing to October further south, and to early November along Gulf of Mexico. Migration on broad front, including movements to West Indies. In spring, main movement northward April to early May, with passage and arrivals from early April in southern states. Most birds are in northernmost breeding areas soon after mid-May.

Wing-length: ♂ 67–72, ♀ 65–68 mm.

Lazuli Bunting *Passerina amoena*

PLATE: page 1688

Du. Lazuligors   Fr. Passerin azuré   Ge. Lazulifink   It. Papa lazuli
Ru. Лазурный овсянковый кардинал   Sp. Passerina lázuli   Sw. Lazulifink

A western North American species, breeding from southern British Columbia east to southern Saskatchewan and western South Dakota, south to northern Baja California (Mexico), central New Mexico, and western Texas. Records in west Palearctic (e.g. Faeroes 1981, Britain 1975, Norway 1991) presumed to involve escaped birds.

Painted Bunting *Passerina ciris*

Du. Purpergors   Fr. Passerin nonpareil   Ge. Papstfink   It. Papa della Luisiana
Ru. Расписной овсянковый кардинал   Sp. Pape arcoiris   Sw. Påvefink

A North American species, breeding in southern USA from south-east New Mexico, central Kansas, southern Missouri, and western Alabama south through Texas to extreme north of Mexico; also on Atlantic coast from North Carolina south to central Florida. Winters Mexico south to Panama, also Florida, Bahamas, and Greater Antilles. Records in west Palearctic (e.g. Britain 1978, Norway 1986, Finland 1982) thought to involve escaped birds.

New World Blackbirds, Orioles, and Allies Family Icteridae

Rather small to fairly large 9-primaried oscine passerines (suborder Passeres), known variously also as 'grackles', 'troupials', 'oropendolas, 'caciques', 'hangnests', etc. Most species arboreal but some terrestrial. Diet varied; includes insects and seeds (and even small vertebrates in case of certain larger species) but many frugivorous to greater or lesser extent, some also taking nectar. Except as vagrants, icterids occur only in New World, from Alaska to Tierra del Fuego, mainly in tropics; those breeding in North America mostly migratory. About 100 species in 23 genera of which 5 migrant North American species accidental in west Palearctic.

Sexes often markedly dissimilar in size, with ♂ the larger bird. Bill variable: conical and finch-like in some species, including cowbirds and Bobolink; quite slender and decurved in many others, including 'orioles' of genera *Quiscalus* and *Icterus*; straight and pointed in meadowlarks (*Sturnella*) and most 'blackbirds' (*Euphagus*, *Agelaius*, and *Xanthocephalus*); long, heavy, and pointed in oropendolas and caciques (*Cacicus*, etc.), with frontal casque. Plumages of many species black, often with metallic gloss, or partially so, combined with yellow, orange, green, red, etc.

Bobolink *Dolichonyx oryzivorus*

PLATE: page 1691

Du. Bobolink   Fr. Goglu des prés   Ge. Bobolink   It. Bobolink
Ru. Боболинк   Sp. Charlatán común   Sw. Bobolink

Field characters. 17–19 cm; wing-span 26–32 cm (♂ up to 20% larger than ♀). Size and wing length close to Corn Bunting, but with more pointed bill, proportionately smaller head, narrower neck when extended, plumper body, and shorter rounded tail with narrow pointed ends to feathers. Distinctive Nearctic bird of ground cover, with plumage of all but breeding ♂, and structure, suggesting strange weaver (Ploceidae), and also Quail. Head and mantle of winter ♂, ♀ and immature dramatically striped black-brown and golden-buff. Breeding ♂ black with yellow nape and white scapulars and rump.

Breeding ♂ unmistakable. Winter ♂, ♀, and 1st-winter distinctive but subject to confusion with other striped, buff passerines. Of these, greatest chance of mistake lies with escaped ♀ or immature weaver, since that family shares similar bill and tail shape and behaviour on ground. Of wild confusion species, needs to be separated from Rose-breasted Grosbeak (10% larger, with dark ear-coverts, fully streaked breast, pale patch at base of primaries, and long, square-ended tail) and Dickcissel (15% smaller, with no discrete black stripes on crown and face and diagnostic chestnut median and lesser wing-coverts). Hops, also jumps, and climbs quite nimbly on ground plants. Habitually raises head to look around, then recalling lark or quail. Commonest call distinctive, a soft but metallic 'pink', 'chink', or 'pint'.

Habitat. Breeds in temperate Nearctic open lowland country, originally in valley grasslands between forests and on coastal marshes, but, since widespread clearance for farming, has extensively colonized fields of grain, hayfields, and meadows. In autumn, shifts to river and coastal marshes and grain fields, and in south to ricefields.

Distribution. Breeds in North America from south-east British Columbia east to Nova Scotia, south to north-east California, northern Utah, north-east Kansas, central Illinois, West Virginia, and New Jersey.

Accidental. Britain, Ireland, France, Norway, Gibraltar, Italy.

Movements. Long-distance migrant, breeding in temperate North America and wintering beyond equator in South America, from south-east Peru south to northern Argentina. Main route in autumn for all populations is via south-east states (notably Florida), whence migration crosses Gulf of Mexico and West Indies (chiefly east of Mexico, west of Puerto Rico), some birds probably flying direct to South America. Spring route is reverse of autumn, main passage going from Colombia to Florida and radiating thence.

Vagrant northward to arctic Alaska, Hudson Bay and southern Labrador, also to Greenland. Rare autumn vagrant to Atlantic seaboard of west Palearctic. 18 records from Britain and Ireland up to 1995, September–October, of which 10 from Isles of Scilly. The only spring record is from Gibraltar: ♂, 11–16 May 1984.

Wing-length: ♂ 94–103, ♀ 86–95 mm.

Bobolink *Dolichonyx oryzivorus*: **1** ad ♂ breeding summer, **2** ad ♂ breeding spring, **3** ad ♀ breeding summer, **4** ad ♂ non-breeding, **5** 1st winter ♂. Brown-headed Cowbird *Molothrus ater ater*: **6** ad ♂, **7** ad ♀, **8** juv. Common Grackle *Quiscalus quiscula* (p. 1692). *Q. q. versicolor* (Canada and north-east and central USA): **9** ad ♂, **10** ad ♀. *Q. q. stonei* (east USA south of *versicolor*): **11** ad ♀.

Brown-headed Cowbird *Molothrus ater*

PLATE: page 1691

Du. Bruinkopkoevogel   Fr. Vacher à tête brune   Ge. Braunkopf-Kuhstärling   It. Molotro testabruna
Ru. Буроголовый коровий трупиал   Sp. Tordo cabecicafé   Sw. Brunhuvad kostare

Field characters. 17·5–21 cm; wing-span 29–35 cm. Nearly 10% larger than Starling, with deeper-based, rather short, noticeably conical bill, and often raised tail. Quite small but robust Nearctic icterid, with relatively dark and uniform plumage. ♂ dusky-black, with umber-brown hood extending to mantle and breast. ♀ and immature dusky mouse-brown, with paler throat.

Unmistakable, with finch-like bill, stocky, bustling character, and always raised tail when on ground. Flight rapid, with silhouette and action recalling Starling or thrush more than grackle (*Quiscalus*), but with more abrupt undulations. Commonest calls: 'chuck'; in flight, high whistled 'phee de de' or 'weee-titi' with accent on 1st note and then drop in pitch.

Habitat. In temperate to subtropical Nearctic, linked with prairies and pastures through ancient adaptation to feeding with bison, and later cattle and horses, which stir up insect food. Range enlarged by modern forest clearance. Habit of brood-parasitism has, however, led to attachment to woodland edges, thickets, and places where low or scattered trees are interspersed with grassland.

Distribution. Breeds in North America from British Columbia and southern Mackenzie east to central Quebec and southern Newfoundland, south throughout USA to central Mexico.

Accidental. Britain: ♂, Islay (Strathclyde), April 1988. Norway: ♀, Jomfru island (Telemark), June 1987.

Movements. Status varies from fully migratory (populations breeding north of 40°N in west, and north of 45°N in east), through partly migratory, to sedentary (most birds breeding south of 35°N). Winters in southern and eastern North America, north to north-central California, New Mexico, Kansas, southern Great Lakes region, and Nova Scotia, and south to southern Mexico. Wintering north of snow-line has increased greatly since 1950, especially in north-east.

Wing-length: ♂ 108–116, ♀ 97–102 mm.

Rusty Blackbird *Euphagus carolinus*

Du. Zwarte Troepiaal   Fr. Quiscale rouilleux   Ge. Roststärling   It. Merlo americano
Ru. Ржавчатый малый трупиал   Sp. Tordo canadiense   Sw. Myrtrupial

A North American species, breeding widely in Alaska and Canada (except south-west), and extending south in north-east USA to New York, Massachusetts, and Maine. Winters from Ontario (rarely north to Alaska) south (east of Rockies) to Gulf of Mexico coast. 2 recorded in Britain: one near Cardiff (Wales), October 1881, was perhaps a genuine drift-migrant; another in St. James's Park (London), July–August 1938, was presumably an escape.

Common Grackle *Quiscalus quiscula*

PLATE: page 1691

Du. Glanstroepiaal   Fr. Quiscale bronzé   Ge. Purpurgrackel   It. Gracchio americano
Ru. Обыкновенный гракл   Sp. Zánate común   Sw. Mindre båtstjärt

Field characters. ♂ 27.5–34, ♀ 25–30 cm; wing-span ♂ 39–45, ♀ 36–41 cm. Structure of bill, head, and body somewhat suggest large starling but with long wedge-shaped tail; Tristram's Grackle is the only west Palearctic species at all similar in form. Strange-looking, long-stepping, and long-tailed Nearctic icterid, with pale eye and all-dark plumage with astonishing multi-coloured iridescence.

Unmistakable. Occurrence within range of Tristram's Grackle virtually impossible; larger Boat-tailed Grackle *Q. major* and Great-tailed Grackle *Q. mexicanus* are not known to stray from North American ranges. Flight suggests strange thrush rather than starling: rapid and usually level, lacking rather abrupt undulations of other Nearctic relatives; ends in bouncing landing. References in field guides to tail having keel shape are unlikely to be relevant to observations of vagrants as this is only associated with ♂'s courtship display. Calls hoarse and grating, 'chuck', 'chack', and 'check'; also shrill 'cheer'.

Habitat. Breeds in Nearctic from cool to warm temperate regions, foraging on ground, especially where wet, in open places such as fields, pastures, lawns, golf courses, shores, marshes, or open wet woodlands. Nests in open woods, parks, groves, and shade trees and bushes. Invades cities to roost at night in shade trees, feeding on lawns and nesting in parks.

Distribution. Breeds throughout North America east of Rocky Mountains, north to north-east British Columbia, southern Mackenzie, James Bay, and south-west Newfoundland.

Accidental. Denmark: Gevninge (Zealand), March–May 1970.

Movements. Variable status; winters in southern half of breeding range, north to Kansas, Iowa, southern Great Lakes region, and New England, casually further north. Most populations breeding north of 45°N largely migratory, moving 1000 km or more to winter range, with only small proportion wintering north to Canada–USA border. Mid-latitude populations partly migratory, and south of 35°N most birds are sedentary.

Wing-length: ♂ 135–152, ♀ 123–137 mm.

Eastern Meadowlark *Sturnella magna*

Du. Witkaakweidespreeuw   Fr. Sturnelle des prés   Ge. Lerchenstärling   It. Sturnella orientale
Ru. Восточный луговой трупиал   Sp. Esturnela pechigualda   Sw. Östlig ängsstare

A North American species, breeding in south-east Canada, widely in eastern and central USA, in Cuba, Mexico, Central America and northern South America (Colombia east to Guianas and northernmost Brazil). Winters throughout breeding range except for northernmost parts. 4 records in England in 19th century have been dismissed as escapes, but the fact that 2 of the 3 dated records were in October, the most likely month for drift-migrants, suggests they may have been wild birds.

Red-winged Blackbird *Agelaius phoeniceus*

Du. Epauletspreeuw   Fr. Carouge à épaulettes   Ge. Rotschulterstärling   It. Ittero alirosse
Ru. Краноплечий луговой трупиал   Sp. Ameritordo común   Sw. Rödvingetrupial

A North American species, breeding in Canada north to southern Yukon, central Mackenzie, James Bay, and south-west Newfoundland, throughout USA and Mexico and south to Costa Rica; also in Bahamas, Cuba, and Isla de Pinos. Winters from southern British Columbia and southernmost Ontario south through breeding range. 16 dated European records from 19th century, 15 from Britain and 1 from Italy, and 2 undated British records. Some at least, and probably all, of the records must have been escapes, as the species has been a favourite cagebird; all records were of ♂♂, and concentration of records in south-east England in 1863–6 suggests escapes from consignment of birds imported at that time; also, bird recorded in Scotland was almost certainly one of several released not far away 20 days before.

Yellow-headed Blackbird *Xanthocephalus xanthocephalus*

PLATE: page 1694

Du. Geelkoptroepiaal   Fr. Carouge à tête jaune   Ge. Brillenstärling   It. Ittero testagialla
Ru. Желтоголовый трупиал   Sp. Tordo cabecidorado   Sw. Gulhuvad trupial

Field characters. ♂ 24–27·5, ♀ 20–25 cm; wing-span ♂ 38–45, ♀ 35–38 cm. ♂ noticeably larger than Blackbird but ♀ somewhat smaller; form differs in deep-based, long, pointed bill, stockier body, and looser tail. Large, lengthy, but also stocky Nearctic icterid; very dark plumage with yellow foreparts; adult ♂ black with yellow hood and mainly white carpal patches. ♀ dusky with yellow face and breast and mottled belly.

Unmistakable. Flight somewhat recalls starling: heavy and somewhat laborious, with marked abrupt undulations, but quite fast and direct over distance; flutters slowly with drooped tail before entering ground cover. Walks and runs. Stance usually half upright with head often sunk into shoulders and tail often hanging down when perched. Voice harsh and rasping, with guttural notes recalling suckling pigs and low monosyllabic 'krack' or 'kack'.

Habitat. Breeds in temperate Nearctic lowlands, favouring deep marshes fringing lakes and shallow river impoundments, where there are stands of cattail, bulrush, or reed. Forages in meadows and marshes, occasionally on grainfields and freshly ploughed land.

Distribution. Breeds in western half of North America from central British Columbia, northern Alberta, and south-west Ontario south to southern California, New Mexico, Kansas, and Wisconsin.

Accidental. Iceland: adult ♂, July 1983; Denmark: Kerteminde (Fyn), October 1918. Records from Britain, France, Norway, and Sweden regarded as probably referring to escapes from captivity.

Movements. Migrant, winter range (from southern USA south to southern Mexico) overlapping with extreme south of breeding range. Movements mainly through inland areas in both seasons; stragglers appear regularly in east coast areas August–October, with most in September. Vagrant to Greenland, and recorded at sea in Atlantic Ocean *c.* 480 km northeast of New York city.

Wing-length: ♂ 140–153, ♀ 115–122 mm.

Black-vented Oriole *Icterus wagleri*

Du. Waglers Troepiaal   Fr. Oriole cul-noir   Ge. Waglertrupial   It. Ittero di Wagler
Ru. Чернобрюхий цветной трупиал   Sp. Turpial de Wagler

A Mexican and Central American species, resident from northern Mexico south through highlands to north-central Nicaragua. One in Rogaland (Norway), July–November 1975, may have been an escape.

Yellow-headed Blackbird *Xanthocephalus xanthocephalus*: **1–2** ad ♂ summer, **3** ad ♂ winter, **4** ad ♀ summer, **5** 1st winter ♂.
Baltimore Oriole *Icterus galbula*: **6** ad ♀, **8–9** 1st winter ♂.

Baltimore Oriole *Icterus galbula*

PLATE: page 1694

Du. Baltimore-troepiaal   Fr. Oriole de Baltimore   Ge. Baltimoretrupial   It. Ittero di Baltimora
Ru. Балтиморский цветной трупиал   Sp. Bolsero de Baltimore   Sw. Baltimoretrupial

Field characters. 17–20 cm; wing-span 28–32 cm. Approaches size of Starling with rather similar bill and head shape but much longer body and rather long, slightly rounded tail. Colourful, arboreal icterid, with long pointed bill. Yellow-orange below at all ages, but tone varies. Adult ♂ striking: black hood, back, wings, and tail-centre, orange forewing, rump, underparts, and tail-edges, and white wing-bar. ♀ and immature duller, with double white wing-bar.

Unmistakable. Plumage colours and head and wing pattern may recall Brambling, but general character and behaviour distinctive. Note however that closely related Orchard Oriole *Icterus spurius* could conceivably cross Atlantic (slightly smaller than Baltimore Oriole, with slighter bill and relatively shorter wings and longer tail; underparts brick-red, not orange, in ♂, and greenish-yellow, not orange or warm yellow, in ♀ and immature). Flight free and rapid, with easy wing-beats; suggests large long-tailed warbler. Calls include rich, fluted whistle, 'pew-li', nasal 'ucht', and hard rattling 'cher-r-r-r-r'.

Habitat. Breeds in temperate Nearctic lowlands, favouring wooded river bottoms, upland forest, shelterbelts, and partially wooded residential areas and farmsteads. Absent from pure coniferous forest but after their clearance colonizes ensuing deciduous growth.

Distribution. Breeds in south-east Canada and eastern USA, west to a line from central Alberta to north-east Texas, hybridizing with Bullock's Oriole *I. bullockii* in south-east Alberta, North Dakota, eastern Colorado, western Nebraska, western Oklahoma, and north-central Texas.

Accidental. Iceland, Britain, Netherlands, Norway.

Movements. Chiefly migratory, wintering mainly from Mexico to north-west South America (Colombia and Venezuela). Since c. 1950, numbers wintering in temperate North America (especially on Atlantic seaboard) have increased greatly, notably where artificial food available. Autumn migration on broad front, but chiefly west of Florida, Cuba, and Yucatán (Mexico). Spring route mostly reverse of autumn, but birds more common in Florida in spring.

Vagrants have reached south-east Alaska, Churchill (Manitoba, 59°N), and south-west Greenland (65°N), and many others wander shorter distances beyond breeding range. Rare vagrant to Atlantic seaboard of west Palearctic, mainly in autumn. Of 18 records in Britain up to 1995, 13 in September–October, 2 in May, and 1 in December; also long-staying winter individuals Dyfed (1989), Essex (1991–2).

Wing-length: ♂ 94–101, ♀ 88–97 mm.

APPENDIX

English names of the birds of the western Palearctic

As explained in the Introduction, the English names used in the main work (*BWP*) have been retained in the Concise Edition. However, at present English names are in a state of flux, with many changes being proposed. To try to bring some order to this, an attempt is being made, under the auspices of the International Ornithological Congress, to create a single list of English names for all the birds of the world. The following list shows many of the new names proposed; it must be stressed that these are *provisional* in that the list of names has not been completed nor have the names yet found wide acceptance. Where *BWP* offered two English names, the species is not listed here if the first of those names is the one which is preferred.

In this list, the English name used in *BWP* is given in column 2, and the new name likely to be recommended is given in column 3. English names of species which are included in the Concise Edition, but were not in the full *BWP*, are as far as possible the same as the 'proposed' English names, so are not listed here.

The main principles underlying the new English names include: (1) avoiding changing well-established names wherever possible; (2) where the English name for a species is different in two or more regions of the world, precedence is given to the name used in the region where it primarily occurs (or breeds if it is a migrant); for example all North American Warblers would be called by their North American English name, even if other English names existed, say, in Europe and in South America; also, the European English name is used for a species which winters in Africa but has a different English name there; (3) adding an adjective to English names such as Heron, Kingfisher, Swift, Swallow, since on the world stage these are not sufficient to identify a species; (4) matching 'comparative' names such that there is not a White-fronted Goose and a Lesser White-fronted Goose, but a Greater and a Lesser White-fronted Goose; (5) in a few cases (diver to loon, skua to jaeger, guilllemot to murre) it has been necessary to opt for the name used on one side of the Atlantic or the other (in all these three examples, the apparently North American name is actually derived from an old European one). A further reason for recommending the use of jaeger and murre for some of the species previously known as skuas and guillemots is that both these names have been used for two genera of birds (*Catharacta* and *Stercorarius*, *Cepphus* and *Uria* respectively). The proposed names allow for these, in some ways quite different, birds to have different English names.

Although not using identical English names throughout, M. Beaman 1994 (*Palearctic Birds*, Harrier Publications, Stonyhurst, England) gives a reasoned argument for many of the proposed changes to English names. We are grateful to D. T. Parkin and F. B. Gill for help in compiling this list and for allowing it to be published here.

Scientific name	English name	Proposed new name
Gavia stellata	Red-throated Diver	Red-throated Loon
Gavia arctica	Black-throated Diver	Black-throated Loon
Gavia immer	Great Northern Diver	Great Northern Loon
Gavia adamsii	White-billed Diver	Yellow-billed Loon
Podiceps auritus	Slavonian Grebe	Horned Grebe
Fulmarus glacialis	Fulmar	Northern Fulmar
Hydrobates pelagicus	Storm Petrel	European Storm-petrel
Morus bassanus	Gannet	Northern Gannet
Phalacrocorax carbo	Cormorant	Great Cormorant
Phalacrocorax aristotelis	Shag	European Shag
Anhinga rufa	Darter	African Darter
Pelecanus onocrotalus	White Pelican	Great White Pelican
Botaurus stellaris	Bittern	Eurasian Bittern
Ixobrychus eurhythmus	Schrenck's Little Bittern	Schrenck's Bittern
Nycticorax nycticorax	Night Heron	Black-crowned Night Heron
Butorides striatus	Green-backed Heron	Striated Heron
Egretta alba	Great White Egret	Great Egret
Geronticus eremita	Bald Ibis	Northern Bald Ibis
Platalea leucorodia	Spoonbill	Eurasian Spoonbill
Cygnus columbianus	Bewick's (Tundra) Swan	Tundra Swan, but commonly split into *C. bewickii*: Bewick's Swan, *C. columbianus* Whistling Swan
Anser albifrons	White-fronted Goose	Greater White-fronted Goose
Tadorna tadorna	Shelduck	Common Shelduck
Anas penelope	Wigeon	Eurasian Wigeon
Anas crecca	Teal	Common Teal
Anas rubripes	Black Duck	American Black Duck
Anas acuta	Pintail	Common Pintail
Anas clypeata	Shoveler	Northern Shoveler
Marmaronetta angustirostris	Marbled Teal	Marbled Duck
Aythya ferina	Pochard	Common Pochard
Aythya marila	Scaup	Greater Scaup
Somateria mollissima	Eider	Common Eider
Histrionicus histrionicus	Harlequin	Harlequin Duck
Bucephala clangula	Goldeneye	Common Goldeneye
Pernis apivorus	Honey Buzzard	European Honey-buzzard
Gyps bengalensis	White-backed Vulture	Indian White-backed Vulture
Gyps fulvus	Griffon Vulture	Eurasian Griffon Vulture
Gyps rueppellii	Rüppell's Vulture	Rüppell's Griffon Vulture
Aegypius monachus	Black Vulture	Eurasian Black Vulture
Circus aeruginosus	Marsh Harrier	Western Marsh Harrier
Accipiter gentilis	Goshawk	Northern Goshawk
Accipiter nisus	Sparrowhawk	Eurasian Sparrowhawk
Buteo buteo	Buzzard	Common Buzzard
Aquila clanga	Spotted Eagle	Greater Spotted Eagle
Aquila heliaca	Imperial Eagle	Eastern Imperial Eagle
Falco tinnunculus	Kestrel	Common Kestrel
Falco subbuteo	Hobby	Eurasian Hobby
Falco rusticolus	Gyrfalcon	Gyr Falcon
Lagopus mutus	Ptarmigan	Rock Ptarmigan
Lagopus lagopus	Red Grouse	Willow Ptarmigan (Red Grouse acceptable for *L. l. scoticus*)

Appendix

Scientific name	English name	Proposed new name
Tetrao mlokosiewiczi	Caucasian Black Grouse	Caucasian Grouse
Tetrao urogallus	Capercaillie	Western Capercaillie
Colinus virginianus	Bobwhite	Northern Bobwhite
Alectoris chukar	Chukar	Chukar Partridge
Ammoperdix griseogularis	See-see	See-see Partridge
Perdix perdix	Partridge	Grey Partridge
Coturnix coturnix	Quail	Common Quail
Phasianus colchicus	Pheasant	Common Pheasant
Meleagris gallopavo	Turkey	Wild Turkey
Turnix sylvatica	Andalusian Hemipode	Small Button-quail
Crex crex	Corncrake	Corn Crake
Gallinula chloropus	Moorhen	Common Moorhen
Porphyrula martinica	American Purple Gallinule	Purple Gallinule
Porphyrio porphyrio	Purple Gallinule	Purple Swamp-hen
Fulica atra	Coot	Eurasian Coot
Fulica cristata	Crested Coot	Red-knobbed Coot
Grus grus	Crane	Common Crane
Grus leucogeranus	Siberian White Crane	Siberian Crane
Chlamydotis undulata	Houbara	Houbara Bustard
Rostratula benghalensis	Painted Snipe	Greater Painted-snipe
Haematopus ostralegus	Oystercatcher	Eurasian Oystercatcher
Haematopus meadewaldoi		Canary Islands Oystercatcher
Recurvirostra avosetta	Avocet	Pied Avocet
Charadrius hiaticula	Ringed Plover	Common Ringed Plover
Charadrius morinellus	Dotterel	Eurasian Dotterel
Pluvialis apricaria	Golden Plover	European Golden Plover
Vanellus spinosus	Spur-winged Plover	Spur-winged Lapwing
Vanellus tectus	Blackhead Plover	Black-headed Lapwing
Vanellus indicus	Red-wattled Plover	Red-wattled Lapwing
Vanellus gregarius	Sociable Plover	Sociable Lapwing
Vanellus leucurus	White-tailed Plover	White-tailed Lapwing
Vanellus vanellus	Lapwing	Northern Lapwing
Calidris canutus	Knot	Red Knot
Gallinago gallinago	Snipe	Common Snipe
Scolopax rusticola	Woodcock	Eurasian Woodcock
Numenius arquata	Curlew	Eurasian Curlew
Tringa totanus	Redshank	Common Redshank
Tringa nebularia	Greenshank	Common Greenshank
Arenaria interpres	Turnstone	Ruddy Turnstone
Phalaropus fulicarius	Grey Phalarope	Red Phalarope
Stercorarius pomarinus	Pomarine Skua	Pomarine Jaeger
Stercorarius parasiticus	Arctic Skua	Parasitic Jaeger
Stercorarius longicaudus	Long-tailed Skua	Long-tailed Jaeger
Larus canus	Common Gull	Mew Gull
Rissa tridactyla	Kittiwake	Black-legged Kittiwake
Sterna bergii	Swift Tern	Greater Crested Tern
Sterna antillarum	(considered here as subspecies of *S. albifrons* Little Tern)	Least Tern
Sterna saundersi	Saunders' Little Tern	Saunders' Tern
Chlidonias leucopterus	White-winged Black Tern	White-winged Tern
Uria aalge	Guillemot	Common Murre
Uria lomvia	Brünnich's Guillemot	Brünnich's Murre
Fratercula arctica	Puffin	Atlantic Puffin
Columba eversmanni	Yellow-eyed Stock Dove	Yellow-eyed Dove
Columba palumbus	Woodpigeon	Common Wood Pigeon
Columba trocaz	Long-toed Pigeon	Trocaz Pigeon
Columba bollii	Bolle's Laurel Pigeon	Bolle's Pigeon
Streptopelia decaocto	Collared Dove	Eurasian Collared Dove
Streptopelia turtur	Turtle Dove	European Turtle Dove
Streptopelia orientalis	Rufous Turtle Dove	Oriental Turtle Dove
Zenaida macroura	Mourning Dove	American Mourning Dove
Psittacula krameri	Ring-necked Parakeet	Rose-ringed Parakeet
Cuculus canorus	Cuckoo	Common Cuckoo
Otus brucei	Striated Scops Owl	Pallid Scops Owl
Otus scops	Scops Owl	European Scops Owl
Bubo bubo	Eagle Owl	European Eagle Owl
Surnia ulula	Hawk Owl	Northern Hawk Owl
Glaucidium passerinum	Pygmy Owl	Eurasian Pygmy Owl
Strix butleri	Hume's Tawny Owl	Hume's Owl
Caprimulgus europaeus	Nightjar	European Nightjar
Hirundapus caudacutus	Needle-tailed Swift	White-throated Needletail
Apus apus	Swift	Common Swift
Cypsiurus parvus	Palm Swift	African Palm Swift

Scientific name	English name	Proposed new name
Halcyon smyrnensis	White-breasted Kingfisher	White-throated Kingfisher
Alcedo atthis	Kingfisher	Common Kingfisher
Merops apiaster	Bee-eater	European Bee-eater
Coracias garrulus	Roller	European Roller
Upupa epops	Hoopoe	Eurasian Hoopoe
Jynx torquilla	Wryneck	Eurasian Wryneck
Picus viridis	Green Woodpecker	European Green Woodpecker
Eremopterix signata	Chestnut-headed Finch Lark	Chestnut-headed Sparrow-lark
Eremopterix nigriceps	Black-crowned Finch Lark	Black-crowned Sparrow-lark
Ammomanes cincturus	Bar-tailed Desert Lark	Bar-tailed Lark
Alaemon alaudipes	Hoopoe Lark	Greater Hoopoe Lark
Calandrella brachydactyla	Short-toed Lark	Greater Short-toed Lark
Calandrella acutirostris	Hume's Lark	Hume's Short-toed Lark
Alauda arvensis	Skylark	Eurasian Skylark
Alauda gulgula	Small Skylark	Oriental Skylark
Alauda razae	Razo Lark	Raso Lark
Eremophila alpestris	Shore Lark	Horned Lark
Eremophila bilopha	Temminck's Horned Lark	Temminck's Lark
Riparia paludicola	Brown-throated Sand Martin	Plain Martin
Hirundo fuligula	African Rock Martin (includes Pale Crag Martin)	Rock Martin
Hirundo rupestris	Crag Martin	Eurasian Crag Martin
Hirundo rustica	Swallow	Barn Swallow
Hirundo pyrrhonota	Cliff Swallow	American Cliff Swallow
Delichon urbica	House Martin	Common House Martin
Anthus spinoletta	Rock/Water Pipit	Water Pipit
Motacilla alba	Pied/White Wagtail	White Wagtail (Pied Wagtail acceptable for *M. a. yarrellii*)
Pycnonotus leucogenys	White-cheeked Bulbul	White-eared Bulbul
Pycnonotus xanthopygos	Yellow-vented Bulbul	White-spectacled Bulbul
Bombycilla garrulus	Waxwing	Bohemian Waxwing
Cinclus cinclus	Dipper	White-throated Dipper
Troglodytes troglodytes	Wren	Winter Wren
Cercotrichas galactotes	Rufous Bush Robin	Rufous-tailed Scrub Robin
Cercotrichas podobe	Black Bush Robin	Black Scrub Robin
Erithacus rubecula	Robin	European Robin
Luscinia megarhynchos	Nightingale	Common Nightingale
Phoenicurus phoenicurus	Redstart	Common Redstart
Saxicola dacotiae	Canary Islands Stonechat	Canary Islands Bushchat
Saxicola torquata	Stonechat	Common Stonechat
Saxicola caprata	Pied Stonechat	Pied Bushchat
Myrmecocichla aethiops	Ant Chat	Northern Anteater Chat
Oenanthe oenanthe	Wheatear	Northern Wheatear
Oenanthe chrysopygia	(considered here as subspecies of *O. xanthoprymna*: Red-tailed Wheatear)	Rufous-tailed Wheatear)
Oenanthe picata	Eastern Pied Wheatear	Variable Wheatear
Oenanthe leucopyga	White-crowned Black Wheatear	White-crowned Wheatear
Monticola saxatilis	Rock Thrush	Rufous-tailed Rock Thrush
Turdus merula	Blackbird	Common Blackbird
Turdus eunomus	(considered here as subspecies of *T. naumanni*: Naumann's Thrush)	Dusky Thrush
Turdus ruficollis	Red-throated/Black-throated Thrush	Dark-throated Thrush (if split *T. ruficollis* Red-throated Thrush, *T. atrogularis* Black-throated Thrush)
Cisticola juncidis	Fan-tailed Warbler	Zitting Cisticola
Prinia gracilis	Graceful Warbler	Graceful Prinia
Locustella naevia	Grasshopper Warbler	Common Grasshopper Warbler

Scientific name	English name	Proposed new name	Scientific name	English name	Proposed new name
Acrocephalus brevipennis	Cape Verde Cane Warbler	Cape Verde Warbler	*Sturnus roseus*	Rose-coloured Starling	Rosy Starling
Acrocephalus scirpaceus	Reed Warbler	European Reed Warbler	*Passer montanus*	Tree Sparrow	Eurasian Tree Sparrow
Sylvia communis	Whitethroat	Common Whitethroat	*Carpospiza brachydactyla*	Pale Rock Sparrow	Pale Rockfinch
Phylloscopus coronatus	Eastern Crowned Leaf Warbler	Eastern Crowned Warbler	*Petronia xanthocollis*	Yellow-throated Sparrow	Chestnut-shouldered Sparrow
Phylloscopus proregulus	Pallas's Warbler	Pallas's Leaf Warbler	*Montifringilla nivalis*	Snow Finch	White-winged Snowfinch
Phylloscopus bonelli	Bonelli's Warbler	Western Bonelli's Warbler	*Fringilla coelebs*	Chaffinch	Common Chaffinch
Phylloscopus collybita	Chiffchaff	Common Chiffchaff	*Serinus serinus*	Serin	European Serin
Regulus teneriffae	Canary Islands Goldcrest	Canary Islands Kinglet	*Serinus canaria*	Canary	Atlantic Canary
Muscicapa dauurica	Brown Flycatcher	Asian Brown Flycatcher	*Carduelis chloris*	Greenfinch	European Greenfinch
Ficedula hypoleuca	Pied Flycatcher	European Pied Flycatcher	*Carduelis carduelis*	Goldfinch	European Goldfinch
Panurus biarmicus	Bearded Tit	Bearded Reedling	*Carduelis spinus*	Siskin	Eurasian Siskin
Sitta europaea	Nuthatch	Eurasian Nuthatch	*Carduelis cannabina*	Linnet	European Linnet
Sitta neumayer	Rock Nuthatch	Western Rock Nuthatch	*Carduelis flammea*	Redpoll	Common Redpoll
Certhia familiaris	Treecreeper	Eurasian Treecreeper	*Loxia curvirostra*	Crossbill	Common Crossbill
Remiz pendulinus	Penduline Tit	Eurasian Penduline Tit	*Bucanetes mongolicus*	Mongolian Trumpeter Finch	Mongolian Finch
Oriolus oriolus	Golden Oriole	Eurasian Golden Oriole	*Carpodacus erythrinus*	Scarlet Rosefinch	Common Rosefinch
Lanius excubitor	Great Grey Shrike	Northern Grey Shrike	*Pyrrhula pyrrula*	Bullfinch	Eurasian Bullfinch
Garrulus glandarius	Jay	Eurasian Jay	*Pipilo erythrophthalmus*	Rufous-sided Towhee	Eastern Towhee
Pica pica	Magpie	Common Magpie	*Calcarius lapponicus*	Lapland Bunting	Lapland Longspur
Nucifraga caryocatactes	Nutcracker	Spotted Nutcracker	*Emberiza tahapisi*	Cinnamon-breasted Rock Bunting	Cinnamon-breasted Bunting
Pyrrhocorax pyrrhocorax	Chough	Red-billed Chough	*Emberiza schoeniclus*	Reed Bunting	Common Reed Bunting
Corvus monedula	Jackdaw	Western Jackdaw	*Icterus galbula*	Northern Oriole	Baltimore Oriole (now excludes *I. bullocki* Bullock's Oriole)
Corvus corax	Raven	Common Raven			
Onychognathus tristramii	Tristram's Grackle	Tristram's Starling			
Sturnus sturninus	Daurian Starling	Purple-backed Starling			
Sturnus sinensis	Grey-backed Starling	White-shouldered Starling			
Sturnus vulgaris	Starling	Common Starling			

INDEXES

SCIENTIFIC NAMES

aalge (Uria), 806
abyssinicus (Coracias), 972
Accipiter badius, 348
    brevipes, 349
    gentilis, 342
    nisus, 345
Accipitridae, 288
Accipitriformes, 288
Acridotheres tristis, 1500
Acrocephalus aedon, 1273
    agricola, 1258
    arundinaceus, 1269
    brevipennis, 1262
    dumetorum, 1260
    griseldis, 1271
    melanopogon, 1251
    orientalis, 1272
    paludicola, 1254
    palustris, 1263
    schoenobaenus, 1255
    scirpaceus, 1265
    stentoreus, 1267
Actitis hypoleucos, 680
    macularia, 683
acuminata (Calidris), 615
acuta (Anas), 223
acutirostris (Calandrella), 1034
adalberti (Aquila), 375
adamsii (Gavia), 10
aedon (Acrocephalus), 1273
Aegithalidae, 1372
Aegithalos caudatus, 1372
Aegolius funereus, 923
Aegypius monachus, 321
aegyptiacus (Alopochen), 195
aegyptius (Caprimulgus), 933
aegyptius (Pluvianus), 550
aequatorialis (Tachymarptis), 940
aeruginosus (Circus), 328
aethereus (Phaethon), 70
Aethia cristatella, 819
aethiopicus (Threskiornis), 148
aethiops (Myrmecocichla), 1174
affinis (Apus), 949
affinis (Aythya), 249
africanus (Gyps), 318
africanus (Phalacrocorax), 89
Agelaius phoeniceus, 1693
agricola (Acrocephalus), 1258
aguimp (Motacilla), 1106
Aix galericulata, 204
    sponsa, 204

ajaja (Platalea), 144
Alaemon alaudipes, 1019
Alauda arvensis, 1043
    gulgula, 1046
    razae, 1047
Alaudidae, 1011
alaudipes (Alaemon), 1019
alba (Calidris), 602
alba (Egretta), 126
alba (Motacilla), 1103
alba (Platalea), 152
alba (Tyto), 886
albellus (Mergellus), 274
albeola (Bucephala), 269
albicilla (Haliaeetus), 303
albicollis (Corvus), 1488
albicollis (Ficedula), 1357
albicollis (Zonotrichia), 1639
albifrons (Anser), 175
albifrons (Sterna), 790
alboniger (Oenanthe), 1195
albus (Corvus), 1481
Alca torda, 812
Alcedinidae, 952
Alcedininae, 956
Alcedo atthis, 956
alchata (Pterocles), 834
Alcidae, 806
alcinus (Machaerhamphus), 289
alcyon (Ceryle), 960
Alectoris barbara, 457
    chukar, 452
    graeca, 453
    rufa, 455
aleutica (Sterna), 784
alexandri (Apus), 941
alexandrinus (Charadrius), 569
Alle alle, 817
alle (Alle), 817
alleni (Porphyrula), 502
Alopochen aegyptiacus, 195
alpestris (Eremophila), 1047
alpina (Calidris), 620
altirostris (Turdoides), 1365
Amandava amandava, 1533
amandava (Amandava), 1533
americana (Anas), 210
americana (Aythya), 238
americana (Certhia), 1412
americana (Fulica), 509
americana (Parula), 1617
americana (Spiza), 1686

americanus (Coccyzus), 882
amherstiae (Chrysolophus), 476
Ammodramus sandwichensis, 1635
Ammomanes cincturus, 1015
    deserti, 1017
Ammoperdix griseogularis, 459
    heyi, 460
amoena (Passerina), 1689
ampelinus (Hypocolius), 1116
amurensis (Falco), 397 (pl.)
anaethetus (Sterna), 787
Anas acuta, 223
    americana, 210
    bahamensis, 225
    capensis, 217
    clypeata, 230
    crecca, 215
    cyanoptera, 228
    discors, 228
    erythrorhyncha, 225
    falcata, 210
    formosa, 214
    penelope, 207
    platyrhynchos, 218
    querquedula, 226
    rubripes, 222
    sibilatrix, 210
    smithii, 229
    strepera, 212
Anatidae, 158
Anatinae, 194
Anatini, 207
angustirostris (Marmaronetta), 232
Anhinga rufa, 91
Anhingidae, 91
Anous minutus, 803
    stolidus, 802
Anser albifrons, 175
    anser, 179
    brachyrhynchus, 173
    caerulescens, 182
    erythropus, 178
    fabalis, 170
    indicus, 182
    rossii, 184
anser (Anser), 179
Anseranas semipalmata, 159
Anseriformes, 158
Anserini, 162
Anthreptes metallicus, 1421
    platurus, 1419
Anthropoides paradisea, 516

    virgo, 516
Anthus berthelotii, 1075
    campestris, 1072
    cervinus, 1084
    godlewskii, 1072
    gustavi, 1081
    hodgsoni, 1077
    novaeseelandiae, 1070
    petrosus, 1089
    pratensis, 1082
    rubescens, 1092
    similis, 1076
    spinoletta, 1086
    trivialis, 1079
antigone (Grus), 511
antiquus (Synthliboramphus), 820
apiaster (Merops), 966
apivorus (Pernis), 290
Apodidae, 937
Apodiformes, 937
Apodinae, 940
apricaria (Pluvialis), 581
Apus affinis, 949
    alexandri, 941
    apus, 943
    caffer, 947
    pacificus, 947
    pallidus, 945
    unicolor, 942
apus (Apus), 943
aquaticus (Rallus), 484
Aquila adalberti, 375
    chrysaetos, 376
    clanga, 365
    heliaca, 371
    nipalensis, 370
    pomarina, 363
    rapax, 368
    verreauxii, 380
aquila (Fregata), 99
arabs (Ardeotis), 527
arborea (Lullula), 1041
arctica (Fratercula), 821
arctica (Gavia), 6
Ardea cinerea, 128
    goliath, 134
    herodias, 131
    melanocephala, 127
    purpurea, 132
Ardeidae, 101
Ardeinae, 109
Ardeirallus sturmii, 108

Scientific Names

Ardeola bacchus, 115
   grayii, 115
   ralloides, 113
ardeola (*Dromas*), 544
Ardeotis arabs, 527
ardesiaca (*Egretta*), 119
ardosiaceus (*Falco*), 409
Arenaria interpres, 686
   melanocephala, 686
Arenariinae, 686
argentatus (*Larus*), 741
ariel (*Fregata*), 99
aristotelis (*Phalacrocorax*), 84
armenicus (*Larus*), 748
arminjoniana (*Pterodroma*), 37, 40
arquata (*Numenius*), 658
arundinaceus (*Acrocephalus*), 1269
arvensis (*Alauda*), 1043
asiaticus (*Charadrius*), 575
Asio capensis, 921
   flammeus, 918
   otus, 915
assimilis (*Puffinus*), 55
astrild (*Estrilda*), 1531
ater (*Molothrus*), 1691
ater (*Parus*), 1385
Athene noctua, 903
atra (*Fulica*), 506
atratus (*Cygnus*), 165
atricapilla (*Sylvia*), 1316
atricilla (*Larus*), 716
atrogularis (*Prunella*), 1133
atthis (*Alcedo*), 956
audouinii (*Larus*), 732
auratus (*Colaptes*), 983
auratus (*Oriolus*), 1425
aureola (*Emberiza*), 1670
auritus (*Phalacrocorax*), 83
auritus (*Podiceps*), 22
avosetta (*Recurvirostra*), 542
Aythya affinis, 249
   americana, 238
   collaris, 241
   ferina, 238
   fuligula, 244
   marila, 247
   nyroca, 242
   valisineria, 240
Aythyini, 236

bacchus (*Ardeola*), 115
badius (*Accipiter*), 348
bahamensis (*Anas*), 225
bairdii (*Calidris*), 614
barbara (*Alectoris*), 457
barbatus (*Gypaetus*), 308
barbatus (*Pycnonotus*), 1111
Bartramia longicauda, 662
bassanus (*Morus*), 77
bengalensis (*Gyps*), 314
bengalensis (*Sterna*), 772
bengalus (*Uraeginthus*), 1530

benghalensis (*Coracias*), 973
benghalensis (*Rostratula*), 532
bergii (*Sterna*), 770
bernicla (*Branta*), 189
berthelotii (*Anthus*), 1075
biarmicus (*Falco*), 411
biarmicus (*Panurus*), 1362
bicalcaratus (*Francolinus*), 463
bicknelli (*Catharus*), 1211
bicolor (*Dendrocygna*), 160
bicolor (*Tachycineta*), 1053
bilopha (*Eremophila*), 1052
bimaculata (*Melanocorypha*), 1026
Biziura lobata, 159
bollii (*Columba*), 848
Bombycilla cedrorum, 1115
   garrulus, 1113
Bombycillidae, 1113
Bombycillinae, 1113
Bonasa bonasia, 428
bonasia (*Bonasa*), 428
bonelli (*Phylloscopus*), 1329
borealis (*Numenius*), 654
borealis (*Phylloscopus*), 1322
borin (*Sylvia*), 1314
Botaurinae, 101
Botaurus lentiginosus, 104
   stellaris, 101
brachydactyla (*Calandrella*), 1032
brachydactyla (*Carpospiza*), 1517
brachydactyla (*Certhia*), 1414
Brachyramphus marmoratus, 820
brachyrhynchus (*Anser*), 173
Branta bernicla, 189
   canadensis, 185
   leucopsis, 187
   ruficollis, 192
brevipennis (*Acrocephalus*), 1262
brevipes (*Accipiter*), 349
brevipes (*Heteroscelus*), 684
brevirostris (*Rissa*), 709
brucei (*Otus*), 890
bruniceps (*Emberiza*), 1677
brunnicephalus (*Larus*), 727
Bubo bubo, 893
bubo (*Bubo*), 893
Buboninae, 890
Bubulcus ibis, 116
Bucanetes githagineus, 1594
   mongolicus, 1592
Bucephala albeola, 269
   clangula, 272
   islandica, 269
buchanani (*Emberiza*), 1662
bullockii (*Icterus*), 1694
Bulweria bulwerii, 41
   fallax, 42
bulwerii (*Bulweria*), 41
Burhinidae, 546
Burhinus oedicnemus, 546
   senegalensis, 548
Buteo buteo, 353

hemilasius, 360
lagopus, 360
lineatus, 353
rufinus, 359
swainsoni, 353
butleri (*Strix*), 910
Butorides striatus, 111
   virescens, 113

cachinnans (*Larus*), 746
caerulea (*Guiraca*), 1688
caerulea (*Egretta*), 118
caerulescens (*Anser*), 182
caerulescens (*Dendroica*), 1620
caeruleus (*Elanus*), 294
caeruleus (*Parus*), 1388
caesia (*Emberiza*), 1663
cafer (*Pycnonotus*), 1112
caffer (*Apus*), 947
Cairinini, 203
calandra (*Melanocorypha*), 1024
calandra (*Miliaria*), 1681
Calandrella acutirostris, 1034
   brachydactyla, 1032
   cheleënsis, 1036
   rufescens, 1034
Calcarius lapponicus, 1640
calendula (*Regulus*), 1342
Calidridinae, 599
Calidris acuminata, 615
   alba, 602
   alpina, 620
   bairdii, 614
   canutus, 599
   ferruginea, 617
   fuscicollis, 613
   maritima, 618
   mauri, 606
   melanotos, 615
   minuta, 607
   minutilla, 612
   pusilla, 605
   ruficollis, 607
   subminuta, 612
   temminckii, 610
   tenuirostris, 599
californica (*Callipepla*), 447
caligata (*Hippolais*), 1277
calliope (*Luscinia*), 1147
Callipepla californica, 447
Calonectris diomedea, 42
   edwardsii, 45
   leucomelas, 46
camelus (*Struthio*), 1
campestris (*Anthus*), 1072
cana (*Tadorna*), 197
canadensis (*Branta*), 185
canadensis (*Grus*), 514
canadensis (*Sitta*), 1401
canadensis (*Wilsonia*), 1629
canaria (*Serinus*), 1554
cannabina (*Carduelis*), 1568

canorus (*Cuculus*), **875**
cantans (*Euodice*), 1534
cantillans (*Sylvia*), 1293
canus (*Larus*), 735
canus (*Picus*), 983
canutus (*Calidris*), 599
capense (*Daption*), 36
capensis (*Anas*), 217
capensis (*Asio*), 921
capensis (*Morus*), 79
capensis (*Oena*), 861
caprata (*Saxicola*), 1173
Caprimulgidae, 927
Caprimulgiformes, 927
Caprimulginae, 928
Caprimulgus aegyptius, 933
   europaeus, 929
   eximius, 933
   nubicus, 928
   ruficollis, 932
caprius (*Chrysococcyx*), 874
carbo (*Phalacrocorax*), 80
Cardinalinae, 1686
Carduelinae, 1548
Carduelis cannabina, 1568
   carduelis, 1561
   chloris, 1557
   flammea, 1574
   flavirostris, 1571
   hornemanni, 1578
   pinus, 1568
   spinus, 1564
carduelis (*Carduelis*), 1561
carneipes (*Puffinus*), 47
carolina (*Porzana*), 490
carolinensis (*Dumetella*), 1127
carolinus (*Euphagus*), 1692
Carpodacus erythrinus, 1596
   mexicanus, 1596
   purpureus, 1596, 1687
   roseus, 1600
   rubicilla, 1600
   synoicus, 1599
Carpospiza brachydactyla, 1517
caryocatactes (*Nucifraga*), 1460
caspia (*Sterna*), 766
caspius (*Tetraogallus*), 451
castanea (*Dendroica*), 1624
castro (*Oceanodroma*), 67
Cataracta maccormicki, 708
   skua, 704
Catharus bicknelli, 1211
   fuscescens, 1211
   guttatus, 1210
   minimus, 1211
   ustulatus, 1210
Catoptrophorus semipalmatus, 685
caucasicus (*Tetraogallus*), 449
caudacutus (*Hirundapus*), 939
caudatus (*Aegithalos*), 1372
caudatus (*Turdoides*), 1367
cauta (*Diomedea*), 29

Scientific Names [3]

cedrorum (Bombycilla), 1115
Centrocercus urophasianus, 427
Centropodinae, 884
Centropus senegalensis, 884
Cepphus grylle, 815
Cercomela melanura, 1166
Cercotrichas galactotes, 1137
    podobe, 1139
Certhia americana, 1412
    brachydactyla, 1414
    familiaris, 1411
Certhiidae, 1411
certhiola (Locustella), 1243
cervinus (Anthus), 1084
Ceryle alcyon, 960
    rudis, 959
Cerylinae, 959
cetti (Cettia), 1235
Cettia cetti, 1235
Chaetura pelagica, 939
Chaeturinae, 939
Charadriidae, 561
Charadriiformes, 532
Charadriinae, 561
Charadrius alexandrinus, 569
    asiaticus, 575
    dubius, 561
    hiaticula, 564
    leschenaultii, 574
    mongolus, 572
    morinellus, 577
    pecuarius, 567
    semipalmatus, 566
    tricollaris, 569
    vociferus, 567
cheleënsis (Calandrella), 1036
cherrug (Falco), 414
Chersophilus duponti, 1020
chilensis (Phoenicopterus), 156
Chlamydotis undulata, 525
Chlidonias hybridus, 794
    leucopterus, 799
    niger, 796
chloris (Carduelis), 1557
chloropus (Gallinula), 499
chlororhynchos (Diomedea), 29
Chondestes grammacus, 1635
Chordeiles minor, 936
Chordeilinae, 936
chrysaetos (Aquila), 376
Chrysococcyx caprius, 874
Chrysolophus amherstiae, 476
    pictus, 475
chrysophrys (Emberiza), 1664
chrysoptera (Vermivora), 1615
chrysostoma (Diomedea), 30
chukar (Alectoris), 452
cia (Emberiza), 1653
Ciconia ciconia, 140
    nigra, 138
ciconia (Ciconia), 140
Ciconiidae, 137

Ciconiiformes, 101
Cinclidae, 1118
Cinclus cinclus, 1118
cinclus (Cinclus), 1118
cincta (Riparia), 1058
cincturus (Ammomanes), 1015
cinctus (Parus), 1382
cineracea (Emberiza), 1658
cinerea (Ardea), 128
cinerea (Motacilla), 1100
cinereus (Xenus), 678
cioides (Emberiza), 1655
Circaetus gallicus, 324
Circus aeruginosus, 328
    cyaneus, 331
    macrourus, 335
    pygargus, 337
ciris (Passerina), 1689
cirlus (Emberiza), 1651
cirrhata (Lunda), 825
cirrocephalus (Larus), 727
Cisticola juncidis, 1237
citreola (Motacilla), 1098
citrina (Wilsonia), 1627
citrinella (Emberiza), 1648
citrinella (Serinus), 1556
Cladorhynchus leucocephalus, 539
clamans (Spiloptila), 1241
Clamator glandarius, 872
    jacobinus, 872
clanga (Aquila), 365
Clangula hyemalis, 261
clangula (Bucephala), 272
clotbey (Rhamphocoris), 1022
clypeata (Anas), 230
Coccothraustes coccothraustes, 1610
coccothraustes (Coccothraustes), 1610
Coccyzus americanus, 882
    erythrophthalmus, 882
coelebs (Fringilla), 1539
Colaptes auratus, 983
colchicus (Phasianus), 472
Coliidae, 869
Coliiformes, 869
Colinus virginianus, 448
collaris (Aythya), 241
collaris (Prunella), 1134
collurio (Lanius), 1433
collybita (Phylloscopus), 1337
Columba bollii, 848
    eversmanni, 845
    junoniae, 850
    livia, 839
    oenas, 842
    palumbus, 846
    trocaz, 848
columbarius (Falco), 400
columbianus (Cygnus), 166
Columbidae, 839
Columbiformes, 839
communis (Sylvia), 1310
concolor (Falco), 409

conspicillata (Sylvia), 1291
Coracias abyssinicus, 972
    benghalensis, 973
    garrulus, 970
Coraciidae, 970
Coraciiformes, 952
cordofanica (Mirafra), 1011
coromandelianus (Nettapus), 203
coronata (Dendroica), 1623
coronatus (Phylloscopus), 1319
coronatus (Pterocles), 828
corone (Corvus), 1478
Corvidae, 1450
Corvus albicollis, 1488
    albus, 1481
    corax, 1483
    corone, 1478
    dauuricus, 1471
    frugilegus, 1475
    monedula, 1468
    rhipidurus, 1486
    ruficollis, 1481
    splendens, 1472
Coturnix coturnix, 467
coturnix (Coturnix), 467
crecca (Anas), 215
Crex crex, 496
crex (Crex), 496
crispus (Pelecanus), 96
cristata (Fulica), 509
cristata (Galerida), 1037
cristatella (Aethia), 819
cristatus (Lanius), 1430
cristatus (Parus), 1383
cristatus (Podiceps), 17
crumeniferus (Leptoptilos), 143
Cuculidae, 871
Cuculiformes, 871
Cuculinae, 872
cucullatus (Lophodytes), 274
cucullatus (Ploceus), 1526
Cuculus canorus, 875
    gularis, 877
    saturatus, 878
cupido (Tympanuchus), 427
curruca (Sylvia), 1308
cursor (Cursorius), 552
Cursoriinae, 550
Cursorius cursor, 552
    temminckii, 554
curvirostra (Loxia), 1582
cyane (Luscinia), 1152
cyanea (Passerina), 1689
cyaneus (Circus), 331
Cyanopica cyanus, 1456
cyanoptera (Anas), 228
cyanurus (Tarsiger), 1153
cyanus (Cyanopica), 1456
cyanus (Parus), 1392
Cyclorrhynchus psittacula, 820
Cygnus atratus, 165

    columbianus, 166
    cygnus, 168
    olor, 162
cygnus (Cygnus), 168
cypriaca (Oenanthe), 1183
Cypsiurus parvus, 950

Daceloninae, 953
dacotiae (Saxicola), 1169
dactylatra (Sula), 73
Daption capense, 36
dauma (Zoothera), 1206
daurica (Hirundo), 1064
dauurica (Muscicapa), 1349
dauurica (Perdix), 467
dauuricus (Corvus), 1471
decaocto (Streptopelia), 853
delawarensis (Larus), 734
Delichon urbica, 1066
Dendrocopos leucotos, 1000
    major, 993
    medius, 998
    minor, 1002
    syriacus, 996
Dendrocygna bicolor, 160
    javanica, 161
    viduata, 160
Dendrocygnini, 160
Dendroica caerulescens, 1620
    castanea, 1623
    coronata, 1623
    fusca, 1620
    magnolia, 1622
    palmarum, 1623
    pensylvanica, 1619
    petechia, 1618
    pinus, 1624
    striata, 1623
    tigrina, 1621
    virens, 1620
denhami (Neotis), 522
deserti (Ammomanes), 1017
deserti (Oenanthe), 1185
deserticola (Sylvia), 1290
Diomedea cauta, 29
    chlororhynchos, 29
    chrysostoma, 30
    epomophora, 31
    exulans, 30
    melanophris, 27
    nigripes, 30
diomedea (Calonectris), 42
Diomedeidae, 27
discors (Anas), 228
Dolichonyx oryzivorus, 1690
domesticus (Passer), 1503
dominica (Pluvialis), 579
dougallii (Sterna), 777
Dromadidae, 544
Dromas ardeola, 544
Dryocopus martius, 989
dubius (Charadrius), 561

Pages 1–1008 appear in Volume 1, pages 1009–1694 in Volume 2

[4] Scientific Names

Dumetella carolinensis, 1127
dumetorum (*Acrocephalus*), 1260
dunni (*Eremalauda*), 1014
duponti (*Chersophilus*), 1020

eburnea (*Pagophila*), 761
ecaudatus (*Terathopius*), 326
Ectopistes migratorius, 865
edwardsii (*Calonectris*), 45
Egretta alba, 126
    ardesiaca, 119
    caerulea, 118
    garzetta, 122
    gularis, 121
    intermedia, 125
    thula, 120
    tricolor, 120
Elanus caeruleus, 294
elegans (*Sterna*), 776
eleonorae (*Falco*), 407
Emberiza aureola, 1670
    bruniceps, 1677
    buchanani, 1662
    caesia, 1663
    chrysophrys, 1664
    cia, 1653
    cineracea, 1658
    cioides, 1655
    cirlus, 1651
    citrinella, 1648
    hortulana, 1659
    leucocephalos, 1646
    melanocephala, 1679
    pallasi, 1676
    pusilla, 1667
    rustica, 1665
    rutila, 1669
    schoeniclus, 1672
    spodocephala, 1645
    stewarti, 1653
    striolata, 1655
    tahapisi, 1659
Emberizidae, 1633
Emberizinae, 1633
Empidonax flaviventris, 1009
    virescens, 1009
enucleator (*Pinicola*), 1602
Eophona migratoria, 1610
    personata, 1610
epomophora (*Diomedea*), 31
epops (*Upupa*), 976
erckelii (*Francolinus*), 464
Eremalauda dunni, 1014
eremita (*Geronticus*), 146
Eremophila alpestris, 1047
    bilopha, 1052
Eremopterix nigriceps, 1013
    signata, 1012
Erithacus rubecula, 1140
erythrinus (*Carpodacus*), 1596
erythrogaster (*Phoenicurus*), 1164
erythronotus (*Phoenicurus*), 1155

erythrophthalmus (*Coccyzus*), 822
erythrophthalmus (*Pipilo*), 1633
erythropus (*Anser*), 178
erythropus (*Tringa*), 662
erythrorhyncha (*Anas*), 225
Estrilda astrild, 1531
    melpoda, 1530
    rhodopyga, 1531
    troglodytes, 1531
Estrildidae, 1529
Eudocimus ruber, 144
Euodice cantans, 1534
    malabarica, 1534
Euphagus carolinus, 1692
eurhythmus (*Ixobrychus*), 107
europaea (*Sitta*), 1402
europaeus (*Caprimulgus*), 929
Eurystomus glaucurus, 974
eversmanni (*Columba*), 845
excubitor (*Lanius*), 1440
excubitorius (*Lanius*), 1444
exilis (*Ixobrychus*), 104
eximius (*Caprimulgus*), 933
exulans (*Diomedea*), 30
exustus (*Pterocles*), 831

fabalis (*Anser*), 170
falcata (*Anas*), 210
falcinellus (*Limicola*), 624
falcinellus (*Plegadis*), 144
Falco amurensis, 397 (pl.)
    ardosiaceus, 409
    biarmicus, 411
    cherrug, 414
    columbarius, 400
    concolor, 409
    eleonorae, 407
    naumanni, 391
    pelegrinoides, 424
    peregrinus, 419
    rusticolus, 417
    sparverius, 396
    subbuteo, 404
    tinnunculus, 393
    vespertinus, 397
Falconidae, 390
Falconiformes, 390
fallax (*Bulweria*), 42
familiaris (*Certhia*), 1411
fasciatus (*Hieraaetus*), 383
fasciolata (*Locustella*), 1251
feae (*Pterodroma*), 39
ferina (*Aythya*), 238
ferruginea (*Calidris*), 617
ferruginea (*Tadorna*), 197
Ficedula albicollis, 1357
    hypoleuca, 1358
    parva, 1353
    semitorquata, 1355
finschii (*Oenanthe*), 1187
fischeri (*Somateria*), 255
flammea (*Carduelis*), 1574

flammeus (*Asio*), 918
flava (*Motacilla*), 1094
flavifrons (*Vireo*), 1536
flavipes (*Tringa*), 672
flavirostra (*Limnocorax*), 496
flavirostris (*Carduelis*), 1571
flavirostris (*Rynchops*), 804
flaviventris (*Empidonax*), 1009
fluviatilis (*Locustella*), 1247
formosa (*Anas*), 214
forsteri (*Sterna*), 785
Francolinus bicalcaratus, 463
    erckelii, 464
    francolinus, 461
francolinus (*Francolinus*), 461
Fratercula arctica, 821
Fregata aquila, 99
    ariel, 99
    magnificens, 99
    minor, 99
Fregatidae, 99
Fregetta grallaria, 62
    tropica, 62
Fringilla coelebs, 1539
    montifringilla, 1545
    teydea, 1544
Fringillidae, 1539
Fringillinae, 1539
frugilegus (*Corvus*), 1475
Fulica americana, 509
    atra, 506
    cristata, 509
fulicarius (*Phalaropus*), 692
fuligula (*Aythya*), 244
fuligula (*Hirundo*), 1058
Fulmarus glacialis, 34
fulva (*Hirundo*), 1066
fulva (*Pluvialis*), 580
fulvus (*Gyps*), 315
fulvus (*Turdoides*), 1370
funereus (*Aegolius*), 923
fusca (*Dendroica*), 1620
fusca (*Melanitta*), 266
fusca (*Phoebetria*), 32
fuscata (*Sterna*), 789
fuscatus (*Phylloscopus*), 1328
fuscescens (*Catharus*), 1211
fuscicollis (*Calidris*), 613
fuscus (*Larus*), 737

gabar (*Micronisus*), 342
galactotes (*Cercotrichas*), 1137
galbula (*Icterus*), 1694
galericulata (*Aix*), 204
Galerida cristata, 1037
    theklae, 1040
gallicus (*Circaetus*), 324
Galliformes, 427
Gallinagininae, 633
Gallinago gallinago, 635
    media, 638
    megala, 642

    stenura, 640
gallinago (*Gallinago*), 635
Gallinula chloropus, 499
gallopavo (*Meleagris*), 481
gambensis (*Plectropterus*), 203
garrulus (*Bombycilla*), 1113
garrulus (*Coracias*), 970
Garrulus glandarius, 1450
garzetta (*Egretta*), 122
Gavia adamsii, 10
    arctica, 6
    immer, 8
    pacifica, 6, 8
    stellata, 3
Gaviidae, 3
Gaviiformes, 3
Gelochelidon nilotica, 764
genei (*Larus*), 729
gentilis (*Accipiter*), 342
georgiana (*Zonotrichia*), 1637
Geothlypis trichas, 1627
Geronticus eremita, 146
githagineus (*Bucanetes*), 1594
glacialis (*Fulmarus*), 34
glandarius (*Clamator*), 872
glandarius (*Garrulus*), 1450
Glareola maldivarum, 557
    nordmanni, 557
    pratincola, 555
glareola (*Tringa*), 676
Glareolidae, 550
Glareolinae, 555
glaucescens (*Larus*), 752
Glaucidium passerinum, 901
glaucoides (*Larus*), 749
glaucurus (*Eurystomus*), 974
godlewskii (*Anthus*), 1072
goliath (*Ardea*), 134
gracilis (*Prinia*), 1239
graculus (*Pyrrhocorax*), 1464
graeca (*Alectoris*), 453
grallaria (*Fregetta*), 62
grammacus (*Chondestes*), 1635
gravis (*Puffinus*), 47
grayii (*Ardeola*), 115
gregarius (*Vanellus*), 590
grisegena (*Podiceps*), 20
griseldis (*Acrocephalus*), 1271
griseogularis (*Ammoperdix*), 459
griseus (*Limnodromus*), 642
griseus (*Puffinus*), 49
Gruidae, 511
Gruiformes, 482
Grus antigone, 511
    canadensis, 514
    grus, 511
    japonensis, 515
    leucogeranus, 515
grus (*Grus*), 511
grylle (*Cepphus*), 815
Guira guira, 871
guira (*Guira*), 871

Pages 1–1008 appear in Volume 1, pages 1009–1694 in Volume 2

Guiraca caerulea, 1688
gularis (*Cuculus*), 877
gularis (*Egretta*), 121
gulgula (*Alauda*), 1046
gustavi (*Anthus*), 1081
guttatus (*Catharus*), 1210
gutturalis (*Irania*), 1154
Gypaetus barbatus, 308
Gyps africanus, 318
    bengalensis, 314
    fulvus, 315
    rueppellii, 318

haemastica (*Limosa*), 650
Haematopodidae, 535
haematopus (*Himantornis*), 484
Haematopus meadewaldoi, 538
    moquini, 538
    ostralegus, 535
Halcyon smyrnensis, 953
    leucocephala, 954
Haliaeetus albicilla, 303
    leucoryphus, 302
    vocifer, 302
haliaetus (*Pandion*), 386
hasitata (*Pterodroma*), 41
heliaca (*Aquila*), 371
hemilasius (*Buteo*), 360
hemprichii (*Larus*), 709
herodias (*Ardea*), 131
Hesperiphona vespertina, 1613
Heteroscelus brevipes, 684
    incanus, 627
heyi (*Ammoperdix*), 460
hiaticula (*Charadrius*), 564
Hieraaetus fasciatus, 383
    pennatus, 381
Himantopus himantopus, 539
himantopus (*Himantopus*), 539
himantopus (*Micropalama*), 627
Himantornis haematopus, 484
Hippolais caligata, 1277
    icterina, 1282
    languida, 1278
    olivetorum, 1280
    pallida, 1273
    polyglotta, 1284
Hirundapus caudacutus, 939
Hirundinidae, 1053
Hirundo aethiopica, 1064
    daurica, 1064
    fuligula, 1058
    fulva, 1066
    pyrrhonota, 1066
    rupestris, 1059
    rustica, 1061
hirundo (*Sterna*), 779
hispanica (*Oenanthe*), 1183
hispaniolensis (*Passer*), 1506
Histrionicus histrionicus, 259
histrionicus (*Histrionicus*), 259
hodgsoni (*Anthus*), 1077

hornemanni (*Carduelis*), 1578
hortensis (*Sylvia*), 1305
hortulana (*Emberiza*), 1659
humei (*Phylloscopus*), 1327
hybridus (*Chlidonias*), 794
Hydrobates pelagicus, 62
Hydrobatidae, 58
hyemalis (*Clangula*), 261
hyemalis (*Junco*), 1639
Hylocichla mustelina, 1209
hyperboreus (*Larus*), 752
Hypocoliinae, 1116
Hypocolius ampelinus, 1116
hypoleuca (*Ficedula*), 1358
hypoleucos (*Actitis*), 680

iagoensis (*Passer*), 1511
Ibidorhyncha struthersii, 532
ibis (*Bubulcus*), 116
ibis (*Mycteria*), 137
ichthyaetus (*Larus*), 712
Icteridae, 1690
icterina (*Hippolais*), 1282
Icterus bullockii, 1694
    galbula, 1694
    spurius, 1694
    wagleri, 1693
ignicapillus (*Regulus*), 1346
iliaca (*Passerella*), 1636
iliacus (*Turdus*), 1228
immer (*Gavia*), 8
impennis (*Pinguinus*), 815
incanus (*Heteroscelus*), 627
incerta (*Pterodroma*), 40
indicus (*Anser*), 182
indicus (*Vanellus*), 589
infaustus (*Perisoreus*), 1454
inornatus (*Phylloscopus*), 1325
inquieta (*Scotocerca*), 1241
intermedia (*Egretta*), 125
Irania gutturalis, 1154
isabellina (*Oenanthe*), 1175
isabellinus (*Lanius*), 1431
islandica (*Bucephala*), 269
Ixobrychus eurhythmus, 107
    exilis, 104
    minutus, 105

jacobinus (*Clamator*), 872
jamaicensis (*Oxyura*), 282
japonensis (*Grus*), 515
javanica (*Dendrocygna*), 161
juncidis (*Cisticola*), 1237
Junco hyemalis, 1639
junoniae (*Columba*), 850
Jynginae, 980
Jynx ruficollis, 980
    torquilla, 980

Ketupa zeylonensis, 896
krueperi (*Sitta*), 1398

Lagonosticta senegala, 1529

Lagopus lagopus, 430
    mutus, 434
lagopus (*Buteo*), 360
lagopus (*Lagopus*), 430
lanceolata (*Locustella*), 1244
languida (*Hippolais*), 1278
Laniidae, 1428
Laniinae, 1430
Lanius collurio, 1433
    cristatus, 1430
    excubitor, 1440
    excubitorius, 1444
    isabellinus, 1431
    meridionalis, 1442
    minor, 1436
    nubicus, 1447
    schach, 1436
    senator, 1445
lapponica (*Limosa*), 650
lapponicus (*Calcarius*), 1640
Laridae, 709
Larus argentatus, 741
    armenicus, 748
    atricilla, 716
    audouinii, 732
    brunnicephalus, 727
    cachinnans, 746
    canus, 735
    cirrocephalus, 727
    delawarensis, 734
    fuscus, 737
    genei, 729
    glaucescens, 752
    glaucoides, 749
    hemprichii, 709
    hyperboreus, 752
    ichthyaetus, 712
    leucophthalmus, 711
    marinus, 754
    melanocephalus, 714
    minutus, 719
    philadelphia, 723
    pipixcan, 718
    ridibundus, 724
    sabini, 721
ledanti (*Sitta*), 1400
lentiginosus (*Botaurus*), 104
Leptoptilos crumeniferus, 143
lepturus (*Phaethon*), 70
leschenaultii (*Charadrius*), 574
leucocephala (*Halcyon*), 954
leucocephala (*Oxyura*), 285
leucocephalus (*Emberiza*), 1646
leucocephalus (*Cladorhynchus*), 539
leucogaster (*Sula*), 74
leucogenys (*Pycnonotus*), 1109
leucogeranus (*Grus*), 515
leucomelaena (*Sylvia*), 1303
leucomelas (*Calonectris*), 46
leuconotus (*Thalassornis*), 159
leucophrys (*Zonotrichia*), 1637
leucophthalmus (*Larus*), 711

leucopsis (*Branta*), 187
leucoptera (*Loxia*), 1580
leucoptera (*Melanocorypha*), 1027
leucoptera (*Pterodroma*), 37
leucopterus (*Chlidonias*), 799
leucopyga (*Oenanthe*), 1196
leucorhoa (*Oceanodroma*), 64
leucorodia (*Platalea*), 149
leucoryphus (*Haliaeetus*), 302
leucotos (*Dendrocopos*), 1000
leucura (*Oenanthe*), 1198
leucurus (*Vanellus*), 592
lherminieri (*Puffinus*), 56
lichtensteinii (*Pterocles*), 826
Limicola falcinellus, 624
Limnocorax flavirostra, 496
Limnodromus griseus, 642
    scolopaceus, 643
Limosa haemastica, 650
    lapponica, 650
    limosa, 647
limosa (*Limosa*), 647
lineatus (*Buteo*), 353
livia (*Columba*), 839
lobata (*Biziura*), 159
lobatus (*Phalaropus*), 691
Locustella certhiola, 1243
    fasciolata, 1251
    fluviatilis, 1247
    lanceolata, 1244
    luscinioides, 1249
    naevia, 1245
lomvia (*Uria*), 810
longicauda (*Bartramia*), 662
longicaudus (*Stercorarius*), 701
Lophodytes cucullatus, 274
lorenzii (*Phylloscopus*), 1336
Loxia curvirostra, 1582
    leucoptera, 1580
    pytyopsittacus, 1587
    scotica, 1586
ludovicianus (*Pheucticus*), 1686
lugens (*Oenanthe*), 1193
lugubris (*Parus*), 1378
Lullula arborea, 1041
Lunda cirrhata, 825
Luscinia calliope, 1147
    cyane, 1152
    luscinia, 1143
    megarhynchos, 1145
    svecica, 1149
luscinia (*Luscinia*), 1143
luscinioides (*Locustella*), 1249
luteus (*Passer*), 1515
Lymnocryptes (*minimus*), 633

maccormicki (*Catharacta*), 708
Machaerhamphus alcinus, 289
Macronectes giganteus, 32
    halli, 32
macroura (*Zenaida*), 864
macrourus (*Circus*), 335

macrourus (*Urocolius*), 869
macularia (*Actitis*), 683
maculatus (*Pipilo*), 1634
madeira (*Pterodroma*), 37
media (*Gallinago*), 638
magna (*Sturnella*), 1692
magnificens (*Fregata*), 99
magnolia (*Dendroica*), 1622
major (*Dendrocopos*), 993
major (*Parus*), 1393
major (*Podiceps*), 17
major (*Quiscalus*), 1692
malabarica (*Euodice*), 1534
Malaconotinae, 1428
maldivarum (*Glareola*), 557
manyar (*Ploceus*), 1526
marginalis (*Porzana*), 496
marila (*Aythya*), 247
marina (*Pelagodroma*), 60
marinus (*Larus*), 754
maritima (*Calidris*), 618
Marmaronetta angustirostris, 232
marmoratus (*Brachyramphus*), 820
martinica (*Porphyrula*), 502
martius (*Dryocopus*), 989
mauretanicus (*Puffinus*), 53
mauri (*Calidris*), 606
maxima (*Sterna*), 769
meadewaldoi (*Haematopus*), 538
medius (*Dendrocopos*), 998
megala (*Gallinago*), 642
megarhynchos (*Luscinia*), 1145
Melanitta fusca, 266
  nigra, 263
  perspicillata, 266
melanocephala (*Ardea*), 127
melanocephala (*Arenaria*), 686
melanocephala (*Emberiza*), 1679
melanocephala (*Sylvia*), 1296
melanocephalus (*Larus*), 714
Melanocorypha bimaculata, 1026
  calandra, 1024
  leucoptera, 1027
  yeltoniensis, 1029
melanoleuca (*Tringa*), 672
melanophris (*Diomedea*), 28
melanopogon (*Acrocephalus*), 1251
melanothorax (*Sylvia*), 1299
melanotos (*Calidris*), 615
melanura (*Cercomela*), 1166
melba (*Tachymarptis*), 940
Meleagrididae, 481
Meleagris gallopavo, 481
Melierax metabates, 340
melodia (*Melospiza*), 1637
Melospiza melodia, 1637
melpoda (*Estrilda*), 1530
merganser (*Mergus*), 279
Mergellus albellus, 274
Mergini, 259
Mergus merganser, 279
  serrator, 277

meridionalis (*Lanius*), 1442
Meropidae, 963
Merops apiaster, 966
  orientalis, 963
  persicus, 964
  pusillus, 963
merula (*Turdus*), 1215
metabates (*Melierax*), 340
metallicus (*Anthreptes*), 1421
mexicanus (*Carpodacus*), 1596
mexicanus (*Quiscalus*), 1692
Micronisus gabar, 342
Micropalama himantopus, 627
migrans (*Milvus*), 295
migratoria (*Eophona*), 1610
migratorius (*Ectopistes*), 865
migratorius (*Turdus*), 1234
Miliaria calandra, 1681
Milvus migrans, 295
  milvus, 298
milvus (*Milvus*), 298
Mimidae, 1125
Mimus polyglottos, 1125
minimus (*Catharus*), 1211
minimus (*Lymnocryptes*), 633
minor (*Chordeiles*), 936
minor (*Dendrocopos*), 1002
minor (*Fregata*), 99
minor (*Lanius*), 1436
minor (*Phoenicopterus*), 156
minor (*Scolopax*), 644
minuta (*Calidris*), 607
minutilla (*Calidris*), 612
minutus (*Anous*), 803
minutus (*Ixobrychus*), 105
minutus (*Larus*), 719
minutus (*Numenius*), 653
mira (*Scolopax*), 644
Mirafra cordofanica, 1011
mlokosiewiczi (*Tetrao*), 440
Mniotilta varia, 1615
moabiticus (*Passer*), 1509
modularis (*Prunella*), 1128
moesta (*Oenanthe*), 1188
mollis (*Pterodroma*), 38
mollissima (*Somateria*), 251
Molothrus ater, 1691
monacha (*Oenanthe*), 1194
monachus (*Aegypius*), 321
monachus (*Myiopsitta*), 867
monachus (*Necrosyrtes*), 314
monedula (*Corvus*), 1468
mongolicus (*Bucanetes*), 1592
mongolus (*Charadrius*), 572
monorhis (*Oceanodroma*), 66
montanella (*Prunella*), 1131
montanus (*Parus*), 1379
montanus (*Passer*), 1513
Monticola saxatilis, 1201
  solitarius, 1204
montifringilla (*Fringilla*), 1545
Montifringilla nivalis, 1522

moquini (*Haematopus*), 538
morinellus (*Charadrius*), 577
Morus bassanus, 77
  capensis, 79
Motacilla aguimp, 1106
  alba, 1103
  cinerea, 1100
  citreola, 1098
  flava, 1094
motacilla (*Seiurus*), 1626
Motacillidae, 1070
moussieri (*Phoenicurus*), 1163
muraria (*Tichodroma*), 1408
murina (*Pyrrhula*), 1609
Muscicapa dauurica, 1349
  striata, 1349
Muscicapidae, 1349
mustelina (*Hylocichla*), 1209
mutus (*Lagopus*), 434
Mycteria ibis, 137
Myiopsitta monachus, 867
Myrmecocichla aethiops, 1174
mystacea (*Sylvia*), 1295

naevia (*Locustella*), 1245
naevia (*Zoothera*), 1208
naevosa (*Stictonetta*), 159
nana (*Sylvia*), 1301
naumanni (*Falco*), 391
naumanni (*Turdus*), 1219
nebularia (*Tringa*), 670
nebulosa (*Strix*), 913
Necrosyrtes monachus, 314
Nectarinia osea, 1422
Nectariniidae, 1419
neglecta (*Pterodroma*), 37
neglectus (*Phylloscopus*), 1335
Neophron percnopterus, 311
Neotis denhami, 522
  nuba, 522
Netta rufina, 236
Nettapus coromandelianus, 203
neumayer (*Sitta*), 1406
niger (*Chlidonias*), 796
nigra (*Ciconia*), 138
nigra (*Melanitta*), 263
nigriceps (*Diomedea*), 30
nigriceps (*Eremopterix*), 1013
nigricollis (*Podiceps*), 24
nigrogularis (*Phalacrocorax*), 86
nilotica (*Gelochelidon*), 764
nipalensis (*Aquila*), 370
nisoria (*Sylvia*), 1306
nisus (*Accipiter*), 345
nivalis (*Montifringilla*), 1522
nivalis (*Plectrophenax*), 1642
noctua (*Athene*), 903
nordmanni (*Glareola*), 557
novaeseelandiae (*Anthus*), 1070
noveboracensis (*Seiurus*), 1626
nuba (*Neotis*), 522
nubicus (*Caprimulgus*), 928

nubicus (*Lanius*), 1447
Nucifraga caryocatactes, 1460
Numenius arquata, 658
  borealis, 654
  minutus, 653
  phaeopus, 655
  tenuirostris, 657
Numida meleagris, 479
Numididae, 479
Nyctea scandiaca, 897
Nycticorax nycticorax, 109
nycticorax (*Nycticorax*), 109
nyroca (*Aythya*), 242

obscurus (*Turdus*), 1218
obsoleta (*Rhodospiza*), 1591
occidentalis (*Pelecanus*), 93
oceanicus (*Oceanites*), 58
Oceanites oceanicus, 58
Oceanodroma castro, 67
  leucorhoa, 64
  monorhis, 66
ochropus (*Tringa*), 674
ochruros (*Phoenicurus*), 1157
ocularis (*Prunella*), 1132
oedicnemus (*Burhinus*), 546
Oena capensis, 861
Oenanthe alboniger, 1195
  cypriaca, 1183
  deserti, 1185
  finschii, 1187
  hispanica, 1183
  isabellina, 1175
  leucopyga, 1196
  leucura, 1198
  lugens, 1193
  moesta, 1188
  monacha, 1194
  oenanthe, 1178
  picata, 1191
  pleschanka, 1180
  xanthoprymna, 1190
oenanthe (*Oenanthe*), 1178
oenas (*Columba*), 842
olivacea (*Piranga*), 1632
olivaceus (*Vireo*), 1537
olivetorum (*Hippolais*), 1280
olor (*Cygnus*), 162
onocrotalus (*Pelecanus*), 93
Onychognathus tristramii, 1489
orientalis (*Acrocephalus*), 1272
orientalis (*Merops*), 963
orientalis (*Phylloscopus*), 1331
orientalis (*Pterocles*), 832
orientalis (*Streptopelia*), 859
Oriolidae, 1424
Oriolus auratus, 1425
  oriolus, 1424
oriolus (*Oriolus*), 1424
oryzivorus (*Dolichonyx*), 1690
osea (*Nectarinia*), 1422
ostralegus (*Haematopus*), 535

Scientific Names [7]

Otididae, 519
Otis tarda, 529
Otus brucei, 890
  scops, 891
otus (Asio), 915
Oxyura jamaicensis, 282
  leucocephala, 285
Oxyurini, 283

pacifica (Gavia), 6, 8
pacificus (Apus), 947
pacificus (Puffinus), 46
Pagophila eburnea, 761
pallasi (Emberiza), 1676
pallida (Hippolais), 1273
pallidus (Apus), 945
palmarum (Dendroica), 1623
palpebrata (Phoebetria), 32
paludicola (Acrocephalus), 1254
paludicola (Riparia), 1053
palumbus (Columba), 846
palustris (Acrocephalus), 1263
palustris (Parus), 1375
Pandion haliaetus, 386
Pandionidae, 386
Panurus biarmicus, 1362
paradisaea (Sterna), 782
paradisea (Anthropoides), 516
paradoxus (Syrrhaptes), 836
parasiticus (Stercorarius), 699
Paridae, 1375
Parula americana, 1617
  pitiayumi, 1617
Parulidae, 1615
Parus ater, 1385
  caeruleus, 1388
  cinctus, 1382
  cristatus, 1383
  cyanus, 1392
  lugubris, 1378
  major, 1393
  montanus, 1379
  palustris, 1375
parva (Ficedula), 1353
parva (Porzana), 490
parvus (Cypsiurus), 950
Passer domesticus, 1503
  hispaniolensis, 1506
  iagoensis, 1511
  luteus, 1515
  moabiticus, 1509
  montanus, 1513
  simplex, 1511
Passerella iliaca, 1636
Passeridae, 1503
Passeriformes, 1009
Passerina amoena, 1689
  ciris, 1689
  cyanea, 1689
passerinum (Glaucidium), 901
pecuarius (Charadrius), 567
pelagica (Chaetura), 939

pelagicus (Hydrobates), 62
Pelagodroma marina, 60
Pelecanidae, 93
Pelecaniformes, 70
Pelecanus crispus, 96
  occidentalis, 93
  onocrotalus, 93
  rufescens, 98
pelegrinoides (Falco), 424
pendulinus (Remiz), 1416
penelope (Anas), 207
pennatus (Hieraaetus), 381
pensylvanica (Dendroica), 1619
percnopterus (Neophron), 311
Perdix dauurica, 467
  perdix, 464
perdix (Perdix), 464
peregrina (Vermivora), 1616
peregrinus (Falco), 419
Perisoreus infaustus, 1454
Pernis apivorus, 290
  ptilorhyncus, 293
persicus (Merops), 964
personata (Eophona), 1610
perspicillata (Melanitta), 266
petechia (Dendroica), 1618
Petronia petronia, 1519
  xanthocollis, 1518
petronia (Petronia), 1519
petrosus (Anthus), 1089
Phaenicophaeinae, 882
phaeopus (Numenius), 655
Phaethon aethereus, 70
  lepturus, 70
  rubricauda, 70
Phaethontidae, 70
Phalacrocoracidae, 80
Phalacrocorax africanus, 89
  aristotelis, 84
  auritus, 83
  carbo, 80
  nigrogularis, 86
  pygmeus, 87
Phalaropodinae, 689
Phalaropus fulicarius, 692
  lobatus, 691
Phasianidae, 447
Phasianus colchicus, 472
  versicolor, 472
Pheucticus ludovicianus, 1686
philadelphia (Larus), 723
philadelphicus (Vireo), 1537
Philomachus pugnax, 628
philomelos (Turdus), 1225
phoebe (Sayornis), 1010
Phoebetria fusca, 32
  palpebrata, 32
phoeniceus (Agelaius), 1693
Phoenicopteridae, 153
Phoenicopteriformes, 153
Phoenicopterus chilensis, 156
  minor, 156

ruber, 153
Phoenicurus erythrogaster, 1164
  erythronotus, 1155
  ochruros, 1157
  moussieri, 1163
  phoenicurus, 1161
phoenicurus (Phoenicurus), 1161
Phylloscopus bonelli, 1329
  borealis, 1322
  collybita, 1337
  coronatus, 1319
  fuscatus, 1328
  humei, 1327
  inornatus, 1325
  lorenzii, 1336
  neglectus, 1335
  orientalis, 1331
  proregulus, 1324
  schwarzi, 1327
  sibilatrix, 1333
  subviridis, 1325
  trochiloides, 1320
  trochilus, 1340
Pica pica, 1457
pica (Pica), 1457
picata (Oenanthe), 1191
Picidae, 979
Piciformes, 979
Picinae, 983
Picoides tridactylus, 1005
Picus canus, 983
  vaillantii, 988
  viridis, 986
pilaris (Turdus), 1222
Pinguinus impennis, 815
Pinicola enucleator, 1602
pinus (Carduelis), 1568
pinus (Dendroica), 1624
Pipilo erythrophthalmus, 1633
  maculatus, 1634
pipixcan (Larus), 718
Piranga olivacea, 1632
  rubra, 1631
pitiayumi (Parula), 1617
Platalea ajaja, 144
  alba, 152
  leucorodia, 149
platurus (Anthreptes), 1419
platyrhynchos (Anas), 218
Plectropterus gambensis, 203
Plectrophenax nivalis, 1642
Plegadis falcinellus, 144
pleschanka (Oenanthe), 1180
Ploceidae, 1526
Ploceus cucullatus, 1526
  manyar, 1526
Pluvialis apricaria, 581
  dominica, 579
  fulva, 580
  squatarola, 584
Pluvianellus socialis, 532
Pluvianus aegyptius, 550

Podiceps auritus, 22
  cristatus, 17
  grisegena, 20
  major, 17
  nigricollis, 24
podiceps (Podilymbus), 13
Podicipedidae, 13
Podicipediformes, 13
Podilymbus podiceps, 13
podobe (Cercotrichas), 1139
polyglotta (Hippolais), 1284
polyglottos (Mimus), 1125
Polysticta stelleri, 256
pomarina (Aquila), 363
pomarinus (Stercorarius), 696
Porphyrio porphyrio, 504
porphyrio (Porphyrio), 504
Porphyrula alleni, 502
  martinica, 502
Porzana carolina, 490
  marginalis, 496
  parva, 490
  porzana, 488
  pusilla, 494
porzana (Porzana), 488
pratensis (Anthus), 1082
pratincola (Glareola), 555
Prinia gracilis, 1239
Procellariidae, 32
Procellariiformes, 27
proregulus (Phylloscopus), 1324
Prunella atrogularis, 1133
  collaris, 1134
  modularis, 1128
  montanella, 1131
  ocularis, 1132
Prunellidae, 1128
Psittacidae, 866
Psittaciformes, 866
psittacula (Cyclorrhynchus), 820
Psittacula krameri, 866
Pterocles alchata, 834
  coronatus, 828
  exustus, 831
  lichtensteinii, 826
  orientalis, 832
  senegallus, 829
Pteroclididae, 826
Pteroclidiformes, 826
Pterodroma arminjoniana, 37, 40
  feae, 39
  hasitata, 41
  incerta, 40
  leucoptera, 37
  madeira, 37
  mollis, 38
  neglecta, 37
ptilorhyncus (Pernis), 293
Puffinus assimilis, 55
  carneipes, 47
  gravis, 47
  griseus, 49

Pages 1–1008 appear in Volume 1, pages 1009–1694 in Volume 2

lherminieri, 56
mauretanicus, 53
pacificus, 46
puffinus, 51
yelkouan, 54
puffinus (Puffinus), 51
pugnax (Philomachus), 628
purpurea (Ardea), 132
purpureus (Carpodacus), 1687
pusilla (Calidris), 605
pusilla (Emberiza), 1667
pusilla (Porzana), 494
pusilla (Spizella), 1635
pusilla (Wilsonia), 1628
pusillus (Merops), 963
pusillus (Serinus), 1548
Pycnonotidae, 1109
Pycnonotus barbatus, 1111
  cafer, 1112
  leucogenys, 1109
  xanthopygos, 1110
pygargus (Circus), 337
pygmeus (Phalacrocorax), 87
Pyrrhocorax graculus, 1464
  pyrrhocorax, 1466
pyrrhocorax (Pyrrhocorax), 1466
pyrrhonota (Hirundo), 1066
Pyrrhula murina, 1609
  pyrrhula, 1606
pyrrhula (Pyrrhula), 1606
pytyopsittacus (Loxia), 1587

Quelea quelea, 1528
quelea (Quelea), 1528
querquedula (Anas), 226
Quiscalus major, 1692
  mexicanus, 1692
  quiscula, 1692
quiscula (Quiscalus), 1692

Rallidae, 484
ralloides (Ardeola), 113
Rallus aquaticus, 484
rapax (Aquila), 368
razae (Alauda), 1047
Recurvirostra avosetta, 542
Recurvirostridae, 539
reevesii (Syrmaticus), 471
Regulus calendula, 1342
  ignicapillus, 1346
  regulus, 1342
  satrapa, 1348
  teneriffae, 1345
regulus (Regulus), 1342
Remiz pendulinus, 1416
Remizidae, 1416
repressa (Sterna), 785
Rhamphocoris clotbey, 1022
rhipidurus (Corvus), 1486
Rhodopechys sanguinea, 1589
rhodopyga (Estrilda), 1531
Rhodospiza obsoleta, 1591

Rhodostethia rosea, 757
ridibundus (Larus), 724
Riparia cincta, 1058
  paludicola, 1053
  riparia, 1055
riparia (Riparia), 1055
'*risoria*' *(Streptopelia)*, 852
Rissa brevirostris, 709
  tridactyla, 758
rosea (Rhodostethia), 757
roseogrisea (Streptopelia), 852
roseus (Carpodacus), 1600
roseus (Sturnus), 1498
rossii (Anser), 184
Rostratula benghalensis, 532
Rostratulidae, 532
rubecula (Erithacus), 1140
ruber (Eudocimus), 144
ruber (Phoenicopterus), 153
rubetra (Saxicola), 1167
rubicilla (Carpodacus), 1600
rubra (Piranga), 1631
rubricauda (Phaethon), 70
rubripes (Anas), 222
rudis (Ceryle), 959
rueppelli (Gyps), 318
rueppelli (Sylvia), 1300
rufa (Alectoris), 455
rufa (Anhinga), 91
rufescens (Calandrella), 1034
rufescens (Pelecanus), 98
ruficollis (Branta), 192
ruficollis (Calidris), 607
ruficollis (Caprimulgus), 932
ruficollis (Corvus), 1481
ruficollis (Jynx), 980
ruficollis (Stelgidopteryx), 1053
ruficollis (Tachybaptus), 14
ruficollis (Turdus), 1220
rufina (Netta), 236
rufinus (Buteo), 359
rufum (Toxostoma), 1126
rupestris (Hirundo), 1059
rustica (Emberiza), 1665
rustica (Hirundo), 1061
rusticola (Scolopax), 644
rusticolus (Falco), 417
ruticilla (Setophaga), 1625
rutila (Emberiza), 1669
Rynchopidae, 804
Rynchops flavirostris, 804

sabini (Larus), 721
Sagittarius serpentarius, 288
sandvicensis (Sterna), 773
sandwichensis (Ammodramus), 1635
sanguinea (Rhodopechys), 1589
sarda (Sylvia), 1286
satrapa (Regulus), 1348
saturatus (Cuculus), 878
saundersi (Sterna), 793
saxatilis (Monticola), 1201

Saxicola caprata, 1173
  dacotiae, 1169
  rubetra, 1167
  torquata, 1170
Sayornis phoebe, 1010
scandiaca (Nyctea), 897
schach (Lanius), 1436
schoeniclus (Emberiza), 1672
schoenobaenus (Acrocephalus), 1255
schwarzi (Phylloscopus), 1327
scirpaceus (Acrocephalus), 1265
scolopaceus (Limnodromus), 643
Scolopacidae, 599
Scolopacinae, 644
Scolopax minor, 644
  mira, 644
  rusticola, 644
scops (Otus), 891
scotica (Loxia), 1586
Scotocerca inquieta, 1241
Seiurus aurocapillus, 1625
  motacilla, 1626
  noveboracensis, 1626
semipalmata (Anseranas), 159
semipalmatus (Catoptrophorus), 685
semipalmatus (Charadrius), 566
semitorquata (Ficedula), 1355
senator (Lanius), 1445
senegala (Lagonosticta), 1529
senegala (Tchagra), 1428
senegalensis (Burhinus), 548
senegalensis (Centropus), 884
senegalensis (Streptopelia), 860
senegallus (Pterocles), 829
Serinus canaria, 1554
  citrinella, 1556
  pusillus, 1548
  serinus, 1550
  syriacus, 1552
serinus (Serinus), 1550
serpentarius (Sagittarius), 288
serrator (Mergus), 278
Setophaga ruticilla, 1625
sibilatrix (Anas), 210
sibilatrix (Phylloscopus), 1333
sibirica (Zoothera), 1207
sibiricus (Uragus), 1606
signata (Eremopterix), 1012
similis (Anthus), 1076
simplex (Passer), 1511
sinensis (Sturnus), 1491
Sitta canadensis, 1401
  europaea, 1402
  krueperi, 1398
  ledanti, 1400
  neumayer, 1406
  tephronota, 1404
  whiteheadi, 1399
Sittidae, 1398
skua (Catharacta), 704
smithii (Anas), 229
smyrnensis (Halcyon), 953

socialis (Pluvianellus), 532
solitarius (Monticola), 1204
Somateria fischeri, 255
  mollissima, 251
  spectabilis, 253
Somateriini, 251
sparverius (Falco), 396
spectabilis (Somateria), 253
Sphyrapicus varius, 992
Spiloptila clamans, 1241
spinoletta (Anthus), 1086
spinosus (Vanellus), 587
spinus (Carduelis), 1564
Spiza americana, 1686
Spizella pusilla, 1635
splendens (Corvus), 1472
spodocephala (Emberiza), 1645
sponsa (Aix), 204
spurius (Icterus), 1694
squamiceps (Turdoides), 1368
squatarola (Pluvialis), 584
stagnatilis (Tringa), 668
Steganopus tricolor, 689
Stelgidopteryx ruficollis, 1053
stellaris (Botaurus), 101
stellata (Gavia), 3
stelleri (Polysticta), 256
stentoreus (Acrocephalus), 1267
stenura (Gallinago), 640
Stercorariidae, 696
Stercorarius longicaudus, 701
  parasiticus, 699
  pomarinus, 696
Sterna albifrons, 790
  aleutica, 784
  anaethetus, 787
  bengalensis, 772
  bergii, 770
  caspia, 766
  dougallii, 777
  elegans, 776
  forsteri, 785
  fuscata, 789
  hirundo, 779
  maxima, 769
  paradisaea, 782
  repressa, 785
  sandvicensis, 773
  saundersi, 793
Sternidae, 764
stewarti (Emberiza), 1653
Stictonetta naevosa, 159
stolidus (Anous), 802
strepera (Anas), 212
Streptopelia decaocto, 853
  orientalis, 859
  '*risoria*', 852
  roseogrisea, 852
  senegalensis, 860
  turtur, 856
striata (Dendroica), 1623
striata (Muscicapa), 1349

striatus (Butorides), 111
Strigidae, 889
Strigiformes, 886
Striginae, 907
striolata (Emberiza), 1655
Strix aluco, 907
    butleri, 910
    nebulosa, 913
    uralensis, 911
struthersii (Ibidorhyncha), 532
Struthio camelus, 1
Struthionidae, 1
Struthioniformes, 1
sturmii (Ardeirallus), 108
Sturnella magna, 1692
Sturnidae, 1489
sturninus (Sturnus), 1491
Sturnus roseus, 1498
    sinensis, 1491
    sturninus, 1491
    unicolor, 1496
    vulgaris, 1492
subbuteo (Falco), 404
subminuta (Calidris), 612
subruficollis (Tryngites), 628
subviridis (Phylloscopus), 1325
Sula dactylatra, 73
    leucogaster, 74
    sula, 73
sula (Sula), 73
Sulidae, 73
Surnia ulula, 899
svecica (Luscinia), 1149
swainsoni (Buteo), 353
sylvatica (Turnix), 482
Sylvia atricapilla, 1316
    borin, 1314
    cantillans, 1293
    communis, 1310
    conspicillata, 1291
    curruca, 1308
    deserticola, 1290
    hortensis, 1305
    leucomelaena, 1303
    melanocephala, 1296
    melanothorax, 1299
    mystacea, 1295
    nana, 1301
    nisoria, 1306
    rueppelli, 1300
    sarda, 1286
    undata, 1288
Sylviidae, 1235
synoicus (Carpodacus), 1599
Synthliboramphus antiquus, 820
    wumizusume, 820
syriacus (Dendrocopos), 996
syriacus (Serinus), 1552
Syrmaticus reevesii, 471
Syrrhaptes paradoxus, 836

Tachybaptus ruficollis, 14
Tachycineta bicolor, 1053
Tachymarptis aequatorialis, 940
    melba, 940
Tadorna cana, 197
    ferruginea, 197
    tadorna, 198
tadorna (Tadorna), 198
Tadornini, 195
tahapisi (Emberiza), 1659
tarda (Otis), 529
Tarsiger cyanurus, 1153
Tchagra senegala, 1428
tectus (Vanellus), 589
temminckii (Calidris), 610
temminckii (Cursorius), 554
teneriffae (Regulus), 1345
tenuirostris (Numenius), 657
tephronota (Sitta), 1404
Terathopius ecaudatus, 326
Tetrao mlokosiewiczi, 440
    tetrix, 437
    urogallus, 442
Tetraogallus caspius, 451
    caucasicus, 449
Tetraonidae, 427
Tetrax tetrax, 519
tetrax (Tetrax), 519
tetrix (Tetrao), 437
teydea (Fringilla), 1544
Thalassornis leuconotus, 159
theklae (Galerida), 1040
Thraupidae, 1631
Threskiornis aethiopicus, 148
Threskiornithidae, 144
thula (Egretta), 120
Tichodroma muraria, 1408
Tichodromadidae, 1408
tigrina (Dendroica), 1621
Timaliidae, 1362
tinnunculus (Falco), 393
torda (Alca), 812
Torgos tracheliotus, 318
torquata (Saxicola), 1170
torquatus (Turdus), 1212
torquilla (Jynx), 980
totanus (Tringa), 665
Toxostoma rufum, 1126
tracheliotus (Torgos), 318
trichas (Geothlypis), 1627
tricollaris (Charadrius), 569
tricolor (Egretta), 120
tricolor (Steganopus), 689
tridactyla (Rissa), 758
tridactylus (Picoides), 1005
Tringa erythropus, 662
    flavipes, 672
    glareola, 676
    melanoleuca, 672
    nebularia, 670
    ochropus, 674
    solitaria, 673

stagnatilis, 668
totanus, 665
Tringinae, 647
tristis (Acridotheres), 1500
tristramii (Onychognathus), 1489
trivialis (Anthus), 1079
trocaz (Columba), 848
trochiloides (Phylloscopus), 1320
trochilus (Phylloscopus), 1340
troglodytes (Estrilda), 1531
Troglodytes troglodytes, 1122
troglodytes (Troglodytes), 1122
Troglodytidae, 1122
tropica (Fregetta), 62
Tryngites subruficollis, 628
Turdidae, 1137
Turdoides altirostris, 1365
    caudatus, 1367
    fulvus, 1370
    squamiceps, 1368
Turdus iliacus, 1228
    merula, 1215
    migratorius, 1234
    naumanni, 1219
    obscurus, 1218
    philomelos, 1225
    pilaris, 1222
    ruficollis, 1220
    torquatus, 1212
    unicolor, 1212
    viscivorus, 1230
Turnicidae, 482
Turnix sylvatica, 482
turtur (Streptopelia), 856
Tympanuchus cupido, 427
Tyrannidae, 1009
Tyto alba, 886
Tytonidae, 886

ulula (Surnia), 899
undata (Sylvia), 1288
undulata (Chlamydotis), 525
unicolor (Apus), 942
unicolor (Sturnus), 1496
unicolor (Turdus), 1212
Upupa epops, 976
Upupidae, 976
Uraeginthus bengalus, 1530
Uragus sibiricus, 1606
uralensis (Strix), 911
urbica (Delichon), 1066
Uria aalge, 806
    lomvia, 810
Urocolius macrourus, 869
urogallus (Tetrao), 442
urophasianus (Centrocercus), 427
ustulatus (Catharus), 1210

vaillantii (Picus), 988
valisineria (Aythya), 240

Vanellinae, 587
Vanellus gregarius, 590
    indicus, 589
    leucurus, 592
    spinosus, 587
    tectus, 589
    vanellus, 593
vanellus (Vanellus), 593
varia (Mniotilta), 1615
varius (Sphyrapicus), 992
Vermivora chrysoptera, 1615
    peregrina, 1616
verreauxii (Aquila), 380
versicolor (Phasianus), 472
vespertina (Hesperiphona), 1613
vespertinus (Falco), 397
viduata (Dendrocygna), 160
virens (Dendroica), 1620
Vireo flavifrons, 1536
    olivaceus, 1537
    philadelphicus, 1537
Vireonidae, 1536
virescens (Butorides), 113
virescens (Empidonax), 1009
virginianus (Colinus), 448
virgo (Anthropoides), 516
viridis (Picus), 986
viscivorus (Turdus), 1230
vocifer (Haliaeetus), 302
vociferus (Charadrius), 567
vulgaris (Sturnus), 1492

wagleri (Icterus), 1693
whiteheadi (Sitta), 1399
Wilsonia canadensis, 1629
    citrina, 1627
    pusilla, 1628
wumizusume (Synthliboramphus), 820

Xanthocephalus xanthocephalus, 1693
xanthocephalus (Xanthocephalus), 1693
xanthocollis (Petronia), 1518
xanthoprymna (Oenanthe), 1190
xanthopygos (Pycnonotus), 1110
Xenus cinereus, 678

yelkouan (Puffinus), 54
yeltoniensis (Melanocorypha), 1029

Zenaida macroura, 864
zeylonensis (Ketupa), 896
Zonotrichia albicollis, 1639
    georgiana, 1637
    leucophrys, 1637
Zoothera dauma, 1206
    naevia, 1208
    sibirica, 1207

ENGLISH NAMES

Accentor, Alpine, 1134
   Black-throated, 1133
   Radde's, 1132
   Siberian, 1131
Albatross, Black-browed, 27
   Black-footed, 30
   Grey-headed, 30
   Light-mantled Sooty, 32
   Royal, 31
   Shy, 29
   Sooty, 32
   Wandering, 30
   Yellow-nosed, 29
Auk, Great, 815
   Little, 817
Auklet, Crested, 819
   Parakeet, 820
Avadavat, Red, 1533
Avocet, 542

Babbler, Arabian, 1368
   Common, 1367
   Fulvous, 1370
   Iraq, 1365
Bateleur, 326
Bee-eater, 966
   Blue-cheeked, 964
   Little, 963
   Little Green, 963
Bird, Secretary, 288
Bittern, 101
   American, 104
   Dwarf, 108
   Least, 104
   Little, 105
   Schrenck's Little, 107
Blackbird, 1215
   Red-winged, 1693
   Rusty, 1692
   Yellow-headed, 1693
Blackcap, 1316
Blackstart, 1166
Bluetail, Red-flanked, 1153
Bluethroat, 1149
Bobolink, 1690
Bobwhite, 448
Booby, Brown, 74
   Masked, 73
   Red-footed, 73
Brambling, 1545
Brant, 189
Bufflehead, 269
Bulbul, Common, 1111
   Red-vented, 1112
   White-cheeked, 1109
   Yellow-vented, 1110
Bullfinch, 1606
   Azores, 1609
Bunting, Black-faced, 1645

   Black-headed, 1679
   Chestnut, 1669
   Cinereous, 1658
   Cinnamon-breasted Rock, 1659
   Cirl, 1651
   Corn, 1681
   Cretzschmar's, 1663
   Grey-necked, 1662
   House, 1655
   Indigo, 1689
   Lapland, 1640
   Lazuli, 1689
   Little, 1667
   Meadow, 1655
   Ortolan, 1659
   Painted, 1689
   Pallas's Reed, 1676
   Pine, 1646
   Red-headed, 1677
   Reed, 1672
   Rock, 1653
   Rustic, 1665
   Snow, 1642
   White-capped, 1653
   Yellow-breasted, 1670
   Yellow-browed, 1664
Bush-lark, Kordofan, 1011
Bustard, Arabian, 527
   Denham's, 522
   Great, 529
   Little, 519
   Nubian, 522
   see also Houbara
Buzzard, 353
   Crested Honey, 293
   Honey, 290
   Long-legged, 359
   Rough-legged, 360
   Upland, 360

Canary, 1554
Canvasback, 240
Capercaillie, 442
Catbird, Gray, 1127
Chaffinch, 1539
   Blue, 1544
Chat, Ant, 1174
   see also Stonechat, Whinchat
Chicken, Prairie, 427
Chiffchaff, 1337
   Caucasian, 1336
Chough, 1466
   Alpine, 1464
Chukar, 452
Coot, 506
   American, 509
   Crested, 509
Cordon-bleu, Red-cheeked, 1530

Cormorant, 80
   Double-crested, 83
   Great, 80
   Long-tailed, 89
   Pygmy, 87
   Socotra, 86
Corncrake, 496
Coucal, Senegal, 884
Courser, Cream-coloured, 552
   Temminck's, 554
Cowbird, Brown-headed, 1691
Crake, Baillon's, 494
   Black, 496
   Little, 490
   Spotted, 488
   Striped, 496
   see also Corncrake
Crane, 511
   Demoiselle, 516
   Manchurian, 515
   Sandhill, 514
   Sarus, 511
   Siberian White, 515
   Stanley, 516
Creeper, Brown, 1412
Crossbill, 1582
   Parrot, 1587
   Red, 1582
   Scottish, 1586
   Two-barred, 1580
   White-winged, 1580
Crow, Carrion, 1478
   Hooded, 1478
   House, 1472
   Pied, 1481
Cuckoo, 875
   African, 877
   Black-billed, 882
   Didric, 874
   Great Spotted, 872
   Guira, 871
   Jacobin, 872
   Oriental, 878
   Yellow-billed, 882
Curlew, 658
   Eskimo, 654
   Slender-billed, 657
   Stone, 546

Darter, 91
Dickcissel, 1686
Dipper, 1118
Diver, Black-throated, 6
   Great Northern, 8
   Pacific, 6, 8
   Red-throated, 3
   White-billed, 10
Dotterel, 577
Dove, African Collared, 852

   Collared, 853
   Laughing, 860
   Mourning, 864
   Namaqua, 861
   Rock, 839
   Rufous Turtle, 859
   Stock, 842
   Turtle, 856
   Yellow-eyed Stock, 845
Dovekie, 817
Dowitcher, Long-billed, 643
   Short-billed, 642
Duck, Black, 222
   Black-headed, 282
   Falcated, 210
   Ferruginous, 242
   Freckled, 159
   Fulvous Whistling, 160
   Lesser Whistling, 161
   Long-tailed, 261
   Mandarin, 204
   Musk, 159, 282
   Red-billed, 225
   Ring-necked, 241
   Ruddy, 282
   Tufted, 244
   White-backed, 159
   White-faced Whistling, 160
   White-headed, 285
   Wood, 204
Dunlin, 620
Dunnock, 1128

Eagle, African Fish, 302
   Bonelli's, 383
   Booted, 381
   Golden, 376
   Imperial, 371
   Lesser Spotted, 363
   Pallas's Fish, 302
   Short-toed, 324
   Spanish Imperial, 375
   Spotted, 365
   Steppe, 370
   Tawny, 368
   Verreaux's, 380
   White-tailed, 303
Egret, Cattle, 116
   Great, 126
   Great White, 126
   Intermediate, 125
   Little, 122
   Snowy, 120
Eider, 251
   King, 253
   Spectacled, 255
   Steller's, 256

Falcon, Amur, 397 (pl.)

Pages 1–1008 appear in Volume 1, pages 1009–1694 in Volume 2

English names

Barbary, 424
Eleonora's, 407
Red-footed, 397
Sooty, 409
see also Gyrfalcon, Lanner, Peregrine, Saker
Fieldfare, 1222
Finch, Citril, 1556
    Crimson-winged, 1589
    Desert, 1591
    House, 1596
    Mongolian Trumpeter, 1592
    Purple, 1596, 1687
    Snow, 1522
    Trumpeter, 1594
    see also Bullfinch, Chaffinch, Firefinch, Goldfinch, Greenfinch, Hawfinch, Rosefinch
Firecrest, 1346
Firefinch, Red-billed, 1529
Flamingo, Chilean, 156
    Greater, 153
    Lesser, 156
Flicker, Northern, 983
Flycatcher, Acadian, 1009
    Brown, 1349
    Collared, 1357
    Grey-streaked, 1351
    Pied, 1358
    Red-breasted, 1353
    Semi-collared, 1355
    Sooty, 1351
    Spotted, 1349
    Yellow-bellied, 1009
Francolin, Black, 461
    Double-spurred, 463
    Erckel's, 464
Freira, 37
Frigatebird, Ascension, 99
    Great, 99
    Lesser, 99
    Magnificent, 99
Fulmar, 34
    Northern, 34

Gadwall, 212
Gallinule, Allen's, 502
    American Purple, 502
    Common, 499
    Purple, 504
Gannet, 77
    Australasian, 79
    Cape, 79
Garganey, 226
Godwit, Bar-tailed, 650
    Black-tailed, 647
    Hudsonian, 650
Goldcrest, 1342
    Canary Islands, 1345
Goldeneye, 272
    Barrow's, 269
Goldfinch, 1561

European, 1561
Gon-gon, 39
Goosander, 279
Goose, Bar-headed, 182
    Barnacle, 187
    Bean, 170
    Brent, 189
    Canada, 185
    Egyptian, 195
    Greylag, 179
    Lesser White-fronted, 178
    Magpie, 159
    Pink-footed, 173
    Red-breasted, 192
    Ross's, 184
    Snow, 182
    Spur-winged, 203
    White-fronted, 175
    see also Pygmy-goose
Goshawk, 342
    Dark Chanting, 340
    Gabar, 342
Grackle, Boat-tailed, 1692
    Common, 1692
    Great-tailed, 1692
    Tristram's, 1489
Grebe, Black-necked, 24
    Eared, 24
    Great, 17
    Great Crested, 17
    Horned, 22
    Little, 14
    Pied-billed, 13
    Red-necked, 20
    Slavonian, 22
Greenfinch, 1557
Greenshank, 670
Grosbeak, Blue, 1688
    Evening, 1613
    Japanese, 1610
    Pine, 1602
    Rose-breasted, 1686
    Yellow-billed, 1610
Grouse, Black, 437
    Caucasian Black, 440
    Hazel, 428
    Pinnated, 427
    Red, 430
    Sage, 427
    Willow, 430
Guillemot, 806
    Black, 815
    Brünnich's, 810
Guineafowl, Helmeted, 479
Gull, Armenian, 748
    Audouin's, 732
    Black-headed, 724
    Bonaparte's, 723
    Brown-headed, 727
    Common, 735
    Franklin's, 718
    Glaucous, 752
    Glaucous-winged, 752

Great Black-backed, 754
Great Black-headed, 712
Grey-headed, 727
Herring, 741
Iceland, 749
Ivory, 761
Laughing, 716
Lesser Black-backed, 737
Little, 719
Mediterranean, 714
Mew, 735
Ring-billed, 734
Ross's, 757
Sabine's, 721
Slender-billed, 729
Sooty, 709
White-eyed, 711
Yellow-legged, 746
Gyrfalcon, 417

Harlequin, 259
Harrier, Hen, 331
    Marsh, 328
    Montagu's, 337
    Northern, 331
    Pallid, 335
Hawfinch, 1610
Hawk, Bat, 289
    Red-shouldered, 353
    Swainson's, 353
Hemipode, Andalusian, 482
Heron, Black, 119
    Black-crowned Night, 109
    Black-headed, 127
    Chinese Pond, 115
    Goliath, 134
    Great Blue, 131
    Green, 113
    Green-backed, 111
    Grey, 128
    Indian Pond, 115
    Little Blue, 118
    Night, 109
    Purple, 132
    Squacco, 113
    Tricolored, 120
    Western Reef, 121
Hobby, 404
Hoopoe, 976
Houbara, 525
Hypocolius, Grey, 1116

Ibis, Bald, 146
    Glossy, 144
    Sacred, 148
    Scarlet, 144
Ibis-bill, 532

Jackdaw, 1468
    Daurian, 1471
Jaeger, Long-tailed, 701
    Parasitic, 699
    Pomarine, 696

Jay, 1450
    Siberian, 1454
Junco, Dark-eyed, 1639

Kestrel, 393
    American, 396
    Grey, 409
    Lesser, 391
Killdeer, 567
Kingfisher, 956
    Belted, 960
    Grey-headed, 954
    Pied, 959
    White-breasted, 953
Kinglet, Golden-crowned, 1348
    Ruby-crowned, 1342
Kite, Black, 295
    Black-winged, 294
    Red, 298
Kittiwake, 758
    Red-legged, 709
Knot, 599
    Great, 599
    Red, 599

Lammergeier, 308
Lanner, 411
Lapwing, 593
Lark, Asian Short-toed, 1036
    Bar-tailed Desert, 1015
    Bimaculated, 1026
    Black, 1029
    Black-crowned Finch, 1013
    Calandra, 1024
    Chestnut-headed Finch, 1012
    Crested, 1037
    Desert, 1017
    Dunn's, 1014
    Dupont's, 1020
    Hoopoe, 1019
    Horned, 1047
    Hume's, 1034
    Lesser Short-toed, 1034
    Razo, 1047
    Shore, 1047
    Short-toed, 1032
    Temminck's Horned, 1052
    Thekla, 1040
    Thick-billed, 1022
    White-winged, 1027
    see also Bush-lark, Skylark, Woodlark
Linnet, 1568
Longspur, Lapland, 1640
Loon, Arctic, 6
    Common, 8
    Red-throated, 3
    Yellow-billed, 10

Magpie, 1457
    Azure-winged, 1456
    Black-billed, 1457
Mallard, 218

Pages 1–1008 appear in Volume 1, pages 1009–1694 in Volume 2

English names

Marabou, 143
Martin, African Rock, 1058
   Banded, 1058
   Brown-throated Sand, 1053
   Crag, 1059
   House, 1066
   Pale Crag, 1058
   Sand, 1055
Meadowlark, Eastern, 1692
Merganser, Common, 279
   Hooded, 274
   Red-breasted, 277
Merlin, 400
Mockingbird, Northern, 1125
Moorhen, 499
Mousebird, Blue-naped, 869
Murre, Common, 806
   Thick-billed, 810
Murrelet, Ancient, 820
   Japanese, 820
   Marbled, 820
Myna, Common, 1500

Nighthawk, Common, 936
Nightingale, 1145
   Thrush, 1143
Nightjar, 929
   Egyptian, 933
   Golden, 933
   Nubian, 928
   Red-necked, 932
Noddy, Black, 803
   Brown, 802
Nutcracker, 1460
Nuthatch, 1402
   Algerian, 1400
   Corsican, 1399
   Eastern Rock, 1404
   Krüper's, 1398
   Red-breasted, 1401
   Rock, 1406

Oldsquaw, 261
Oriole, African Golden, 1425
   Baltimore, 1694
   Black-vented, 1693
   Bullock's, 1694
   Golden, 1424
   Orchard, 1694
Osprey, 386
Ostrich, 1
Ouzel, Ring, 1212
Ovenbird, 1625
Owl, Barn, 886
   Boreal, 923
   Brown Fish, 896
   Eagle, 893
   Great Grey, 913
   Hawk, 899
   Hume's Tawny, 910
   Little, 903
   Long-eared, 915
   Marsh, 921

Pygmy, 901
   Scops, 891
   Short-eared, 918
   Snowy, 897
   Striated Scops, 890
   Tawny, 907
   Tengmalm's, 923
   Ural, 911
Oystercatcher, 535
   African Black, 538
   Canary Islands, 538

Parakeet, Monk, 867
   Ring-necked, 866
Partridge, 464
   Barbary, 457
   Daurian, 467
   Gray, 464
   Red-legged, 455
   Rock, 453
   Sand, 460
Parula, Northern, 1617
   Tropical, 1617
Pelican, Brown, 93
   Dalmatian, 96
   Pink-backed, 98
   White, 93
Peregrine, 419
Petrel, Atlantic, 40
   Black-capped, 41
   Bulwer's, 41
   Cape, 36
   Collared, 37
   Fea's, 39
   Herald, 37, 40
   Jouanin's, 42
   Kermadec, 37
   Northern Giant, 32
   Soft-plumaged, 38
   Southern Giant, 32
   Storm, 62
   Zino's, 37
   see also Storm-Petrel
Phalarope, Grey, 692
   Northern, 691
   Red, 692
   Red-necked, 691
   Wilson's, 689
Pheasant, 472
   Golden, 475
   Green, 472
   Lady Amherst's, 476
   Reeves's, 471
   Ring-necked, 472
Phoebe, Eastern, 1010
Pigeon, Bolle's Laurel, 848
   Laurel, 850
   Long-toed, 848
   Passenger, 865
   see also Woodpigeon
Pintail, 223
   Bahama, 225
Pipit, Berthelot's, 1075

Blyth's, 1072
Buff-bellied, 1092
Long-billed, 1076
Meadow, 1082
Olive-backed, 1077
Pechora, 1081
Red-throated, 1084
Richard's, 1070
Rock, 1089
Tawny, 1072
Tree, 1079
Water, 1086
Plover, American Golden, 579
   Black-bellied, 584
   Blackhead, 589
   Caspian, 575
   Crab, 544
   Egyptian, 550
   Golden, 581
   Greater Sand, 574
   Grey, 584
   Kentish, 569
   Kittlitz's, 567
   Lesser Sand, 572
   Little Ringed, 561
   Magellanic, 532
   Pacific Golden, 580
   Red-wattled, 589
   Ringed, 564
   Semipalmated, 566
   Snowy, 569
   Sociable, 590
   Spur-winged, 587
   Three-banded, 569
   White-tailed, 592
Pochard, 238
   Red-crested, 236
Pratincole, Black-winged, 557
   Collared, 555
   Oriental, 557
Ptarmigan, 434
   Rock, 434
   Willow, 430
Puffin, 821
   Atlantic, 821
   Tufted, 825
Pygmy-goose, Cotton, 203

Quail, 467
   California, 447
Quelea, Red-billed, 1528

Rail, Nkulenga, 484
   Water, 484
   see also Sora
Raven, 1483
   Brown-necked, 1481
   Fan-tailed, 1486
   White-necked, 1488
Razorbill, 812
Redhead, 238
Redpoll, 1574
   Arctic, 1578

Common, 1574
   Hoary, 1578
Redshank, 665
   Spotted, 662
Redstart, 1161
   American, 1625
   Black, 1157
   Eversmann's, 1155
   Güldenstädt's, 1164
   Moussier's, 1163
Redwing, 1228
Robin, 1140
   American, 1234
   Black Bush, 1139
   Rufous Bush, 1137
   Siberian Blue, 1152
   White-throated, 1154
Roller, 970
   Abyssinian, 972
   Broad-billed, 974
   Indian, 973
Rook, 1475
Rosefinch, Common, 1596
   Great, 1600
   Long-tailed, 1606
   Pallas's, 1600
   Scarlet, 1596
   Sinai, 1599
Rubythroat, Siberian, 1147
Ruff, 628

Saker, 414
Sanderling, 602
Sandgrouse, Black-bellied, 832
   Chestnut-bellied, 831
   Crowned, 828
   Lichtenstein's, 826
   Pallas's, 836
   Pin-tailed, 834
   Spotted, 829
Sandpiper, Baird's, 614
   Broad-billed, 624
   Buff-breasted, 628
   Common, 680
   Curlew, 617
   Green, 674
   Least, 612
   Marsh, 668
   Pectoral, 615
   Purple, 618
   Semipalmated, 605
   Sharp-tailed, 615
   Solitary, 673
   Spotted, 683
   Stilt, 627
   Terek, 678
   Upland, 662
   Western, 606
   White-rumped, 613
   Wood, 676
Sapsucker, Yellow-bellied, 992
Scaup, 247
   Greater, 247

Pages 1–1008 appear in Volume 1, pages 1009–1694 in Volume 2

Lesser, 249
Scoter, Black, 263
   Common, 263
   Surf, 266
   Velvet, 266
   White-winged, 266
See-see, 459
Serin, 1550
   Red-fronted, 1548
   Syrian, 1552
Shag, 84
Shearwater, Audubon's, 56
   Balearic, 53
   Cape Verde, 45
   Cory's, 42
   Flesh-footed, 47
   Great, 47
   Little, 55
   Manx, 51
   Sooty, 49
   Streaked, 46
   Wedge-tailed, 46
   Yelkouan, 54
Shelduck, 198
   Cape, 197
   Ruddy, 197
Shikra, 348
Shoveler, 230
   Cape, 229
   Northern, 230
Shrike, Brown, 1430
   Great Grey, 1440
   Grey-backed Fiscal, 1444
   Isabelline, 1431
   Lesser Grey, 1436
   Long-tailed, 1436
   Masked, 1447
   Red-backed, 1433
   Southern Grey, 1442
   Woodchat, 1445
Silverbill, African, 1534
   Indian, 1534
Siskin, 1564
   Pine, 1568
Skimmer, African, 804
Skua, 704
   Arctic, 699
   Great, 704
   Long-tailed, 701
   Pomarine, 696
   South Polar, 708
Skylark, 1043
   Small, 1046
Smew, 274
Snipe, 635
   Great, 638
   Jack, 633
   Painted, 532
   Pintail, 640
   Swinhoe's, 642
Snowcock, Caspian, 451
   Caucasian, 449
Sora, 490

Sparrow, Dead Sea, 1509
   Desert, 1511
   Field, 1635
   Fox, 1636
   House, 1503
   Iago, 1511
   Lark, 1635
   Pale Rock, 1517
   Rock, 1519
   Savannah, 1635
   Song, 1637
   Spanish, 1506
   Sudan Golden, 1515
   Swamp, 1637
   Tree, 1513
   White-crowned, 1637
   White-throated, 1639
   Yellow-throated, 1518
Sparrowhawk, 345
   Levant, 349
Spoonbill, 149
   African, 152
   Roseate, 144
Starling, 1492
   Daurian, 1491
   Grey-backed, 1491
   Rose-coloured, 1498
   Spotless, 1496
Stilt, Banded, 539
   Black-necked, 539
   Black-winged, 539
Stint, Little, 607
   Long-toed, 612
   Red-necked, 607
   Temminck's, 610
Stonechat, 1170
   Canary Islands, 1169
   Pied, 1173
Stork, Black, 138
   White, 140
   Yellow-billed, 137
Storm-Petrel, Black-bellied, 62
   Leach's, 64
   Madeiran, 67
   Swinhoe's, 66
   White-bellied, 62
   White-faced, 60
   Wilson's, 58
Sunbird, Nile Valley, 1421
   Palestine, 1422
   Pygmy, 1419
Swallow, 1061
   Bank, 1055
   Barn, 1061
   Cave, 1066
   Cliff, 1066
   Ethiopian, 1064
   Red-rumped, 1064
   Rough-winged, 1053
   Tree, 1053
Swan, Bewick's, 166
   Black, 165
   Mute, 162

   Tundra, 166
   Whooper, 168
Swift, 943
   Alpine, 940
   Cape Verde, 941
   Chimney, 939
   Little, 949
   Mottled, 940
   Needle-tailed, 939
   Pacific, 947
   Pallid, 945
   Palm, 950
   Plain, 942
   White-rumped, 947

Tanager, Scarlet, 1632
   Summer, 1631
Tattler, Grey-tailed, 684
   Wandering, 627
Tchagra, Black-crowned, 1428
Teal, 215
   Baikal, 214
   Blue-winged, 228
   Cape, 217
   Cinnamon, 228
   Green-winged, 215
   Marbled, 232
Tern, Aleutian, 784
   Arctic, 782
   Black, 796
   Bridled, 787
   Caspian, 766
   Common, 779
   Elegant, 776
   Forster's, 785
   Gull-billed, 764
   Least, 790
   Lesser Crested, 772
   Little, 790
   Roseate, 777
   Royal, 769
   Sandwich, 773
   Saunders', 793
   Sooty, 789
   Swift, 770
   Whiskered, 794
   White-cheeked, 785
   White-winged Black, 799
Thick-knee, Senegal, 548
Thrasher, Brown, 1126
Thrush, Bicknell's, 1211
   Black-throated, 1220
   Blue Rock, 1204
   Dusky, 1219
   Eye-browed, 1218
   Gray-cheeked, 1211
   Hermit, 1210
   Mistle, 1230
   Naumann's, 1219
   Red-throated, 1220
   Rock, 1201
   Siberian, 1207
   Song, 1225

   Swainson's, 1210
   Tickell's, 1212
   Varied, 1208
   White's, 1206
   Wood, 1209
   see also Waterthrush
Tit, Azure, 1392
   Bearded, 1362
   Blue, 1388
   Coal, 1385
   Crested, 1383
   Great, 1393
   Long-tailed, 1372
   Marsh, 1375
   Penduline, 1416
   Siberian, 1382
   Sombre, 1378
   Willow, 1379
Towhee, Rufous-sided, 1633
   Spotted, 1634
Treecreeper, 1411
   Short-toed, 1414
Tree-Pipit, Olive, 1077
Tropicbird, Red-billed, 70
   Red-tailed, 70
   White-tailed, 70
   Yellow-billed, 70
Turkey, 481
Turnstone, 686
   Black, 686
   Ruddy, 686
Twite, 1571

Veery, 1211
Vireo, Philadelphia, 1537
   Red-eyed, 1537
   Yellow-throated, 1536
Vulture, African White-backed, 318
   Black, 321
   Egyptian, 311
   Griffon, 315
   Hooded, 314
   Lappet-faced, 318
   Rüppell's, 318
   White-backed, 314
   see also Lammergeier

Wagtail, African Pied, 1106
   Citrine, 1098
   Grey, 1100
   Pied, 1103
   White, 1103
   Yellow, 1094
Wallcreeper, 1408
Warbler, Aquatic, 1254
   Arabian, 1303
   Arctic, 1322
   Barred, 1306
   Basra Reed, 1271
   Bay-breasted, 1624
   Black-and-white, 1615
   Blackburnian, 1620

[14] English names

Blackpoll, 1623	Magnolia, 1622	Wood, 1333	Whinchat, 1167
Black-throated Blue, 1620	Marmora's, 1286	Yellow, 1618	Whitethroat, 1310
Black-throated Green, 1620	Marsh, 1263	Yellow-browed, 1325	Lesser, 1308
Blyth's Reed, 1260	Melodious, 1284	Yellow-rumped, 1623	Wigeon, 207
Bonelli's, 1329	Ménétries's, 1295	Waterthrush, Louisiana, 1626	American, 210
Booted, 1277	Moustached, 1251	Northern, 1626	Chiloë, 210
Brooks's Leaf, 1325	Olivaceous, 1273	Waxbill, Black-rumped, 1531	European, 207
Canada, 1629	Olive-tree, 1280	Common, 1531	Willet, 685
Cape May, 1621	Oriental Reed, 1272	Crimson-rumped, 1531	Woodcock, 644
Cape Verde Cane, 1262	Orphean, 1305	Orange-cheeked, 1530	Amami, 644
Cetti's, 1235	Paddyfield, 1258	Waxwing, 1113	American, 644
Chestnut-sided, 1619	Pallas's, 1324	Bohemian, 1113	Woodlark, 1041
Clamorous Reed, 1267	Pallas's Grasshopper, 1243	Cedar, 1115	Woodpecker, Black, 989
Cricket, 1241	Palm, 1623	Weaver, Streaked, 1526	Great Spotted, 993
Cyprus, 1299	Pine, 1624	Village, 1526	Green, 986
Dartford, 1288	Plain Willow, 1335	Wheatear, 1178	Grey-headed, 983
Desert, 1301	Radde's, 1327	Black, 1198	Lesser Spotted, 1002
Dusky, 1328	Reed, 1265	Black-eared, 1183	Levaillant's Green, 988
Eastern Bonelli's, 1331	River, 1247	Cyprus, 1183	Middle Spotted, 998
Eastern Crowned Leaf, 1319	Rüppell's, 1300	Desert, 1185	Syrian, 996
Fan-tailed, 1237	Sardinian, 1296	Eastern Pied, 1191	Three-toed, 1005
Garden, 1314	Savi's, 1249	Finsch's, 1187	White-backed, 1000
Golden-winged, 1615	Scrub, 1241	Hooded, 1194	Woodpigeon, 846
Graceful, 1239	Sedge, 1255	Hume's, 1195	Wren, 1122
Grasshopper, 1245	Spectacled, 1291	Isabelline, 1175	Winter, 1122
Gray's Grasshopper, 1251	Subalpine, 1293	Mourning, 1193	Wryneck, 980
Great Reed, 1269	Tennessee, 1616	Pied, 1180	Red-breasted, 980
Greenish, 1320	Thick-billed, 1273	Red-rumped, 1188	
Hooded, 1627	Tristram's, 1290	Red-tailed, 1190	Yellowhammer, 1648
Hume's Leaf, 1327	Upcher's, 1278	White-crowned Black, 1196	Yellowlegs, Greater, 672
Icterine, 1282	Willow, 1340	Whimbrel, 655	Lesser, 672
Lanceolated, 1244	Wilson's, 1628	Little, 653	Yellowthroat, Common, 1627

Pages 1–1008 appear in Volume 1, pages 1009–1694 in Volume 2

DEUTSCHE NAMEN

Aaskrähe, 1478
Abendkernbeißer, 1613
Adlerbussard, 359
Akaziendrossling, 1370
Akaziengrasmücke, 1303
Aleutenseeschwalbe, 784
Alexandersegler, 941
Alpenbraunelle, 1134
Alpendohle, 1464
Alpenkrähe, 1466
Alpenschneehuhn, 434
Alpensegler, 940
Alpenstrandläufer, 620
Ameisenschmätzer, 1174
Amsel, 1215
Antarktikskua, 708
Arabertrappe, 527
Armenienmöwe, 748
Atlasgrasmücke, 1290
Atlasgrünspecht, 988
Audubonsturmtaucher, 56
Auerhuhn, 442
Austernfischer, 535
   Kanarischer, 538
Australspornpieper, 1070
Azorengimpel, 1609
Aztekenmöwe, 716
Azurbischof, 1688

Bacchusreiher, 115
Bachstelze, 1103
Bairdstrandläufer, 614
Balearensturmtaucher, 53
Balkanlaubsänger, 1331
Baltimoretrupial, 1694
Bartgeier, 308
Bartkauz, 913
Bartlaubsänger, 1327
Bartmeise, 1362
Bartrebhuhn, 467
Basrarohrsänger, 1271
Baßtölpel, 77
Baumfalke, 404
Baumpieper, 1079
Bechsteindrossel, 1220
Bekassine, 635
Bengalengeier, 314
Bergammer, 1659
Bergbraunelle, 1131
Bergente, 247
   Kleine, 249
Bergfink, 1545
Berggimpel, 1600
Berghänfling, 1571
Bergkalanderlerche, 1026
Berglaubsänger, 1329
Bergpieper, 1086
Bergstrandläufer, 606
Beringmöwe, 752

Beutelmeise, 1416
Bienenfresser, 966
Bindenkreuzschnabel, 1580
Bindenseeadler, 302
Bindenstrandläufer, 627
Bindentaucher, 13
Birkenzeisig, 1574
Birkhuhn, 437
Blaßfuß-Sturmtaucher, 47
Bläßgans, 175
Bläßhuhn, 506
   Amerikanisches, 509
Blaßspötter, 1273
Blauelster, 1456
Blauflügelente, 228
Blaukehlchen, 1149
Blaumeise, 1388
Blaumerle, 1204
Blaunachtigall, 1152
Blaunacken-Mausvogel, 869
Blauracke, 970
Blaureiher, 118
Blaurücken-Waldsänger, 1620
Blauschwanz, 1153
Blauwangenspint, 964
Bluthänfling, 1568
Blutschnabelweber, 1528
Blutspecht, 996
Bobolink, 1690
Bonapartemöwe, 723
Borstenrabe, 1486
Brachpieper, 1072
Brachvogel, Großer, 658
Brandgans, 198
Brandseeschwalbe, 773
Brauenwaldsänger, 1615
Braunbauch-Flughuhn, 831
Braunkehlchen, 1167
Braunkehl-Uferschwalbe, 1053
Braunkopfammer, 1677
Braunkopf-Kuhstärling, 1691
Braunkopfmöwe, 727
Braunliest, 953
Braunrücken-Goldsperling, 1515
Braunschnäpper, 1349
Braunwürger, 1430
Brautente, 204
Brillenente, 266
Brillengrasmücke, 1291
Brillenstärling, 1693
Bronzesultanshuhn, 502
Brookslaubsänger, 1325
Bruchwasserläufer, 676
Buchentyrann, 1009
Buchfink, 1539
Büffelkopfente, 269
Bulwersturmvogel, 41
Buntfalke, 396
Buntfuß-Sturmschwalbe, 58
Buntspecht, 993

Buschrohrsänger, 1260
Buschspötter, 1277

Carolinasumpfhuhn, 490
Carolinataube, 864
Chileflamingo, 156
Chinarohrsänger, 1272
Chukarhuhn, 452
Cistensänger, 1237

Dachsammer, 1637
Diademrotschwanz, 1163
Diamantfasan, 476
Dickschnabellumme, 810
Dickschnabel-Rohrsänger, 1273
Dickzissel, 1686
Dohle, 1468
Doppelschnepfe, 638
Doppelspornfrankolin, 463
Dorfweber, 1526
Dorngrasmücke, 1310
Dornspötter, 1278
Dreiband-Regenpfeifer, 569
Dreifarbenreiher, 120
Dreizehenmöwe, 758
Dreizehenspecht, 1005
Drosselrohrsänger, 1269
Drosseluferläufer, 683
Drosselwaldsänger, 1626
Dunkelente, 222
Dunkellaubsänger, 1328
Dünnschnabel-Brachvogel, 657
Dünnschnabelmöwe, 729
Dupontlerche, 1020

Eichelhäher, 1450
Eichenlaubsänger, 1335
Eiderente, 251
Eilseeschwalbe, 770
Einfarbdrossel, 1212
Einfarbsegler, 942
Einfarbstar, 1496
Einödgimpel, 1599
Einödlerche, 1014
Einsiedlerdrossel, 1210
Eisente, 261
Eismöwe, 752
Eissturmvogel, 34
Eistaucher, 8
Eisvogel, 956
Eleonorenfalke, 407
Elfenbeinmöwe, 761
Elster, 1457
Elsterdohle, 1471
Elstersteinschmätzer, 1191
Erckelfrankolin, 464
Erddrossel, 1206
Erlenzeisig, 1564
Erznektarvogel, 1421
Eskimobrachvogel, 654

Fahlbürzel-Steinschmätzer, 1188
Fahlente, 217
Fahlkauz, 910
Fahlkehlschwalbe, 1064
Fahlsegler, 945
Fahlsperling, 1517
Fahlstirnschwalbe, 1066
Falkenraubmöwe, 701
Fasan, 472
Feldlerche, 1043
   Kleine, 1046
Feldrohrsänger, 1258
Feldschwirl, 1245
Feldsperling, 1513
Felsenhuhn, 457
Felsenkleiber, 1406
Felsenschwalbe, 1059
Felsensteinschmätzer, 1187
Felsentaube, 839
Fichtenammer, 1646
Fichtenkreuzschnabel, 1582
Fichtenwaldsänger, 1620
Fichtenzeisig, 1568
Fischadler, 386
Fischmöwe, 712
Fischuhu, 896
Fitis, 1340
Flußregenpfeifer, 561
Flußseeschwalbe, 779
Flußuferläufer, 680
Forsterseeschwalbe, 785
Fuchsammer, 1636

Gabarhabicht, 342
Gänsegeier, 315
Gänsesäger, 279
Gartenbaumläufer, 1414
Gartengrasmücke, 1314
Gartenrotschwanz, 1161
Gaukler, 326
Gebirgsstelze, 1100
Gelbaugentaube, 845
Gelbbauch-Saftlecker, 992
Gelbbrauenammer, 1664
Gelbbrauen-Laubsänger, 1325
Gelbbrust-Pfeifgans, 160
Gelbkehlsperling, 1518
Gelbkehlvireo, 1536
Gelbnasenalbatros, 29
Gelbscheitel-Waldsänger, 1619
Gelbschenkel, Großer, 672
   Kleiner, 669
Gelbschnabelkuckuck, 882
Gelbschnabel-Sturmtaucher, 42
Gelbschnabeltaucher, 10
Gelbschopflund, 825
Gelbspötter, 1282
Gelbsteißbülbül, 1110
Gerfalke, 417
Gimpel, 1606

Pages 1–1008 appear in Volume 1, pages 1009–1694 in Volume 2

Deutsche Namen

Girlitz, 1550
Glanzkrähe, 1472
Gleitaar, 294
Glockenreiher, 119
Gluckente, 214
Goldammer, 1648
Goldfasan, 475
Goldflügel-Waldsänger, 1616
Goldhähnchen-Laubsänger, 1324
Goldkuckuck, 874
Goldregenpfeifer, 581
   Amerikanischer, 579
   Pazifischer, 580
Goldschnepfe, 532
Goldspecht, 983
Goldwaldsänger, 1618
Goliathreiher, 134
Grasammer, 1635
Grasläufer, 628
Grauammer, 1681
Graubrust-Strandläufer, 615
Graubülbül, 1111
Graubürzel-Singhabicht, 340
Graudrossling, 1368
Graufischer, 959
Graugans, 179
Graukehl-Sumpfhuhn, 496
Graukopfalbatros, 30
Graukopfliest, 954
Graukopfmöwe, 727
Graumantelwürger, 1444
Grauortolan, 1663
Graureiher, 128
Graurückendommel, 108
Grauschnäpper, 1349
Grauschwanz-Wasserläufer, 684
Grauspecht, 983
Grauwangendrossel, 1211
Großtrappe, 529
Grünbrust-Nektarvogel, 1419
Grünlaubsänger, 1320
Grünling, 1557
Grünreiher, 113
Grünschenkel, 670
Grünspecht, 986
Grünwaldsänger, 1620
Gryllteiste, 815
Gürtelfischer, 960

Habicht, 342
Habichtsadler, 383
Habichtskauz, 911
Häherkuckuck, 872
Hakengimpel, 1602
Halbringschnäpper, 1355
Hallsturmvogel, 32
Halsbanddrossel, 1208
Halsbandfrankolin, 461
Halsbandschnäpper, 1357
Halsbandsittich, 866
Harlekinlerche, 1012
Haselhuhn, 428
Haubenlerche, 1037

Haubenmeise, 1383
Haubentaucher, 17
Hausammer, 1655
Hausrotschwanz, 1157
Haussegler, 949
Haussperling, 1503
Heckenbraunelle, 1128
Heckensänger, 1137
Heidelerche, 1041
Helmperlhuhn, 479
Hemprichmöwe, 709
Heringsmöwe, 737
Hinduracke, 973
Hirtenmaina, 1500
Hirtenregenpfeifer, 567
Höckerschwan, 162
Hohltaube, 842
Hopfkuckuck, 878
Hudsonschnepfe, 650

Ibis, Heiliger, 148
Indianergoldhähnchen, 1348
Indigofink, 1689
Isabellsteinschmätzer, 1175
Isabellwürger, 1431

Jakobinerkuckuck, 872
Javapfeifgans, 161
Jerichonektarvogel, 1422
Jouaninsturmvogel, 42
Jungfernkranich, 516
Junko, 1639

Kabylenkleiber, 1400
Kaffernadler, 380
Kaffernsegler, 947
Kafferntrappe, 522
Kaiseradler, 371
   Spanischer, 375
Kalanderlerche, 1024
Kammbläßhuhn, 509
Kampfläufer, 628
Kanadagans, 185
Kanadakleiber, 1401
Kanadakranich, 514
Kanadareiher, 131
Kanadawaldsänger, 1629
Kanarengirlitz, 1554
Kanarenpieper, 1075
Kanarenschmätzer, 1169
Kaplöffelente, 229
Kapohreule, 921
Kappenammer, 1679
Kappengeier, 314
Kappensäger, 274
Kappensteinschmätzer, 1194
Kapsturmvogel, 36
Kaptäubchen, 861
Kaptölpel, 79
Kapuzenwaldsänger, 1627
Kapverden-Rohrsänger, 1262
Kapverdensturmtaucher, 45
Kapverdensturmvogel, 39

Karmingimpel, 1596
Kaspikönigshuhn, 451
Katzenvogel, 1127
Kaukasusbirkhuhn, 440
Kaukasuskönigshuhn, 449
Kaukasuslaubsänger, 1336
Keilschwanz-Regenpfeifer, 567
Keilschwanz-Sturmtaucher, 46
Kermadecsturmvogel, 37
Kernbeißer, 1610
Kiebitz, 593
Kiebitzregenpfeifer, 584
Kiefernkreuzschnabel, 1587
Klapperammer, 1635
Klappergrasmücke, 1308
Kleiber, 1402
Kleinspecht, 1002
Kletterwaldsänger, 1615
Klippenkleiber, 1404
Knackerlerche, 1022
Knäkente, 226
Knutt, 599
   Großer, 599
Kohlmeise, 1393
Kolbenente, 236
Kolkrabe, 1483
Königsalbatros, 31
Königsfasan, 471
Königsseeschwalbe, 769
Korallenmöwe, 732
Kordofanlerche, 1011
Kormoran, 80
Kornweihe, 331
Koromandelzwergente, 203
Korsenkleiber, 1399
Krabbentaucher, 817
Kragenente, 259
Kragentrappe, 525
Krähenscharbe, 84
Kranich, 511
Krauskopfpelikan, 96
Kreuzschnabel, Schottischer, 1586
Krickente, 215
Krokodilwächter, 550
Kronenflughuhn, 828
Kronenlaubsänger, 1319
Kronwaldsänger, 1623
Kuckuck, 875
Kuhreiher, 116
Kurzfangsperber, 349
Kurzschnabelgans, 173
Kurzzehenlerche, 1032
Küstenreiher, 121
Küstenseeschwalbe, 782

Lachmöwe, 724
Lachseeschwalbe, 764
Lachtaube, 852
Langschnabelpieper, 1076
Langschwanzdrossling, 1367
Langzehen-Strandläufer, 612
Lannerfalke, 411
Lapplandmeise, 1382

Lasurmeise, 1392
Laufhühnchen, 482
Lazulifink, 1689
Lerchenstärling, 1692
Löffelente, 230
Löffler, 149
   Afrikanischer, 152
Lorbeertaube, 850
   Bolles, 848

Madeirasturmvogel, 37
Madeirawellenläufer, 67
Magellantaucher, 17
Magnolienwaldsänger, 1622
Mandarinente, 204
Mandarinstar, 1491
Mandschurendommel, 107
Mangrovereiher, 111
Mantelmöwe, 754
Manyarweber, 1526
Marabu, 143
Mariskensänger, 1251
Marmelente, 232
Maskenammer, 1645
Maskengrasmücke, 1300
Maskenkernbeißer, 1610
Maskentölpel, 73
Maskenwürger, 1447
Mauerläufer, 1408
Mauersegler, 943
Mäusebussard, 353
Meerstrandläufer, 618
Mehlschwalbe, 1066
Meisengimpel, 1606
Meisenwaldsänger, 1617
Merlin, 400
Misteldrossel, 1230
Mittelmeer-Steinschmätzer, 1183
Mittelmeer-Sturmtaucher, 54
Mittelreiher, 125
Mittelsäger, 277
Mittelspecht, 998
Moabsperling, 1509
Mohrenlerche, 1029
Mohrenschwarzkehlchen, 1173
Mohrensumpfhuhn, 496
Mönchsgeier, 321
Mönchsgrasmücke, 1316
Mönchsittich, 867
Mönchswaldsänger, 1628
Mongolenbussard, 360
Mongolengimpel, 1592
Mongolenregenpfeifer, 572
Mongolenstar, 1491
Moorente, 242
Moorschneehuhn, 430
Mornellregenpfeifer, 577

Nachtfalke, 936
Nachtigall, 1145
Nachtreiher, 109
Naumanndrossel, 1219
Neuntöter, 1433

Pages 1–1008 appear in Volume 1, pages 1009–1694 in Volume 2

Deutsche Namen

Nilgans, 195
Nimmersatt, 137
Noddi, 802
Nonnensteinschmätzer, 1180
Nubiertrappe, 522

Odinshühnchen, 691
Ohrengeier, 318
Ohrenlerche, 1047
Ohrenscharbe, 83
Ohrentaucher, 22
Olivenspötter, 1280
Orangebäckchen, 1530
Orientbrachschwalbe, 557
Orientseeschwalbe, 793
Orientturteltaube, 859
Orpheusgrasmücke, 1305
Orpheusspötter, 1284
Ortolan, 1659

Paddyreiher, 115
Pallasammer, 1676
Palmenwaldsänger, 1623
Palmsegler, 950
Palmtaube, 860
Papageitaucher, 821
Papstfink, 1689
Pazifikpieper, 1092
Pazifiksegler, 947
Petschorapieper, 1081
Pfeifente, 207
    Nordamerikanische, 210
Pfuhlschnepfe, 650
Pharaonenziegenmelker, 933
Philadelphiavireo, 1537
Phoebe, 1010
Pieperwaldsänger, 1625
Pirol, 1424
Plüschkopfente, 255
Polarbirkenzeisig, 1578
Polarmöwe, 749
Prachteiderente, 253
Prachtfregattvogel, 99
Prachtnachtschwalbe, 933
Prachttaucher, 6
Präriebussard, 353
Prärieläufer, 662
Präriemöwe, 718
Provencegrasmücke, 1288
Purpurgrackel, 1692
Purpurhuhn, 504
Purpurreiher, 132

Rainammer, 1635
Rallenreiher, 113
Raubseeschwalbe, 766
Raubwürger, 1440
    Südlicher, 1442
Rauchschwalbe, 1061
Rauhfußbussard, 360
Rauhfußkauz, 923
Razolerche, 1047
Rebhuhn, 464

Regenbrachvogel, 655
Reiherente, 244
Reiherläufer, 544
Rennvogel, 552
Rieddrossling, 1365
Riedscharbe, 89
Riesenalk, 815
Riesenrotschwanz, 1164
Riesenschwirl, 1251
Riesensturmvogel, 32
Riesentafelente, 240
Ringdrossel, 1212
Ringelgans, 189
Ringeltaube, 846
Ringschnabelente, 241
Ringschnabelmöwe, 734
Rohrammer, 1672
Rohrdommel, 101
    Nordamerikanische, 104
Rohrschwirl, 1249
Rohrweihe, 328
Rosaflamingo, 153
Rosapelikan, 93
Rosenbrust-Kernknacker, 1686
Rosengimpel, 1600
Rosenmöwe, 757
Rosenseeschwalbe, 777
Rosenstar, 1498
Rostbürzel-Steinschmätzer, 1190
Rostgans, 197
Rostsperling, 1511
Roststärling, 1692
Rotaugenvireo, 1537
Rotdrossel, 1228
Rötelammer, 1669
Rötelfalke, 391
Rötelgrundammer, 1633
Rötelpelikan, 98
Rötelschwalbe, 1064
Rotflügel-Brachschwalbe, 555
Rotflügelgimpel, 1589
Rotfußfalke, 397
Rotfußtölpel, 73
Rothalsgans, 192
Rothalstaucher, 20
Rothals-Ziegenmelker, 932
Rothuhn, 455
Rotkehlchen, 1140
Rotkehlpieper, 1084
Rotkehl-Strandläufer, 607
Rotkopfwürger, 1445
Rotlappenkiebitz, 589
Rotmilan, 298
Rotschenkel, 665
Rotschnabelalk, 820
Rotschnabelente, 225
Rotschnabel-Tropikvogel, 70
Rotschulterbussard, 353
Rotschulterstärling, 1693
Rotstirngirlitz, 1548
Rubingoldhähnchen, 1342
Rubinkehlchen, 1147
Rüppellseeschwalbe, 772

Rußbülbül, 1112
Rußheckensänger, 1139
Rußseeschwalbe, 789

Saatgans, 170
Saatkrähe, 1475
Säbelschnäbler, 542
Saharaohrenlerche, 1052
Saharasteinschmätzer, 1196
Salzlerche, 1036
Samtente, 266
Samtkopf-Grasmücke, 1296
Sanderling, 602
Sandflughuhn, 832
Sandlerche, 1015
Sandregenpfeifer, 564
    Amerikanischer, 566
Sandstrandläufer, 605
Sardengrasmücke, 1286
Savannenadler, 368
Schachwürger, 1436
Schafstelze, 1094
Scharlachtangare, 1632
Scheckente, 256
Schelladler, 365
Schellente, 272
Scherenschnabel, Afrikanischer, 804
Schieferdrossel, 1207
Schieferfalke, 409
Schikrasperber, 348
Schildrabe, 1481
Schilfrohrsänger, 1255
Schlagschwirl, 1247
Schlammläufer, Großer, 643
    Kleiner, 642
Schlammtreter, 685
Schlangenadler, 324
Schlangenhalsvogel, 91
Schlegelsturmvogel, 40
Schleiereule, 886
Schmarotzerraubmöwe, 699
Schmetterlingsastrild, 1530
Schmuckreiher, 120
Schmuckseeschwalbe, 776
Schmutzgeier, 311
Schnäpperwaldsänger, 1625
Schnatterente, 212
Schneeammer, 1642
Schnee-Eule, 897
Schneefink, 1522
Schneegans, 182
Schneekranich, 515
Schopffalk, 819
Schopfwachtel, 447
Schopfwespenbussard, 293
Schornsteinsegler, 939
Schreiadler, 363
Schreiseeadler, 302
Schuppengrasmücke, 1299
Schuppenkopfprinie, 1241
Schwalbenmöwe, 721
Schwanzmeise, 1372

Schwarzbrauenalbatros, 27
Schwarzflügel-Brachschwalbe, 557
Schwarzfußalbatros, 30
Schwarzhalsreiher, 127
Schwarzhalstaucher, 24
Schwarzkehlbraunelle, 1133
Schwarzkehlchen, 1170
Schwarzkopfmöwe, 714
Schwarzkopf-Ruderente, 282
Schwarzkopf-Steinschmätzer, 1195
Schwarzmilan, 295
Schwarzrücken-Steinschmätzer, 1193
Schwarzschnabelkuckuck, 882
Schwarzschnabel-Sturmtaucher, 51
Schwarzschopfkiebitz, 589
Schwarzschwan, 165
Schwarzschwanz, 1166
Schwarzspecht, 989
Schwarzstirnwürger, 1436
Schwarzstorch, 138
Seeadler, 303
Seeregenpfeifer, 569
Seggenrohrsänger, 1254
Seidenreiher, 122
Seidensänger, 1235
Seidenschwanz, 1113
Seidenwürger, 1116
Senegalamarant, 1529
Senegalracke, 972
Senegaltriel, 548
Senegaltschagra, 1428
Sichelente, 210
Sichelstrandläufer, 617
Sichler, 144
Silberalk, 820
Silberhalstaube, 848
Silberkopfammer, 1653
Silbermöwe, 741
Silberreiher, 126
Silberschnabel, Afrikanischer, 1534
    Indischer, 1534
Singammer, 1637
Singdrossel, 1225
Singschwan, 168
Skua, 704
Smaragdspint, 963
Sokotrakormoran, 86
Sommergoldhähnchen, 1346
Sommertangare, 1631
Spatelente, 269
Spatelraubmöwe, 696
Sperber, 345
Sperbereule, 899
Sperbergeier, 318
Sperbergrasmücke, 1306
Sperlingskauz, 901
Spießbekassine, 640
Spießente, 223
Spießflughuhn, 834

Pages 1–1008 appear in Volume 1, pages 1009–1694 in Volume 2

[18] Deutsche Namen

Spitzschwanz-Strandläufer, 615
Spornammer, 1640
Sporngans, 203
Spornkiebitz, 587
Spornkuckuck, 884
Spottdrossel, 1125
    Rote, 1126
Sprosser, 1143
Sprosserrotschwanz, 1155
Stachelschwanzsegler, 939
Star, 1492
Steinadler, 376
Steinbraunelle, 1132
Steinhuhn, 453
Steinkauz, 903
Steinlerche, 1017
Steinortolan, 1662
Steinrötel, 1201
Steinschmätzer, 1178
Steinschwalbe, 1058
Steinsperling, 1519
Steinwälzer, 686
Stelzenläufer, 539
Stentorrohrsänger, 1267
Steppenadler, 370
Steppenflughuhn, 836
Steppenkiebitz, 590
Steppenpieper, 1072
Steppenweihe, 335
Sterntaucher, 3
Stieglitz, 1561
Stockente, 218
Strandpieper, 1089
Strauß, 1
Streifengans, 182
Streifenohreule, 890
Streifenprinie, 1239
Streifenschwirl, 1243
Streifenwaldsänger, 1623
Strichelschwirl, 1244
Stummellerche, 1034
Sturmmöwe, 735
Sturmschwalbe, 62
Sturmtaucher, Dunkler, 49
    Großer, 47
Sturmvogel, Kleiner, 55
Sumpfammer, 1637
Sumpfhuhn, Kleines, 490
Sumpfläufer, 624
Sumpfmeise, 1375
Sumpfohreule, 918
Sumpfrohrsänger, 1263
Sumpfschwalbe, 1053
Swinhoewellenläufer, 66

Tafelente, 238
Tamariskengrasmücke, 1295
Tannenhäher, 1460
Tannenmeise, 1385
Teichhuhn, 499
Teichrohrsänger, 1265
Teichwasserläufer, 668
Temminckrennvogel, 554
Temminckstrandläufer, 610
Teneriffagoldhähnchen, 1345
Terekwasserläufer, 678
Teufelssturmvogel, 41
Teydefink, 1544
Theklalerche, 1040
Thorshühnchen, 692
Tibetlerche, 1034
Tienschan-Laubsänger, 1327
Tigerfink, 1533
Tigerwaldsänger, 1621
Tordalk, 812
Trauerente, 263
Trauermeise, 1378
Trauerschnäpper, 1358
Trauerseeschwalbe, 796
Trauersteinschmätzer, 1198
Triel, 546
Tristramstar, 1489
Tropfenflughuhn, 829
Trottellumme, 806
Truthuhn, 481
Tüpfelsumpfhuhn, 488
Türkenammer, 1658
Türkenkleiber, 1398
Türkentaube, 853
Turmfalke, 393
Turteltaube, 856

Uferschnepfe, 647
Uferschwalbe, 1055
Uhu, 893
Unglückshäher, 1454

Virginiawachtel, 448

Wacholderdrossel, 1222
Wachtel, 467
Wachtelkönig, 496
Waglertrupial, 1693
Waldammer, 1665
Waldbaumläufer, 1411
Waldbekassine, 642
Walddrossel, 1209
Waldkauz, 907
Waldlaubsänger, 1333
Waldohreule, 915

Waldpieper, 1077
Waldrapp, 146
Waldschnepfe, 644
Waldwasserläufer, 674
Wanderalbatros, 30
Wanderdrossel, 1234
Wanderfalke, 419
Wanderlaubsänger, 1322
Wandertaube, 865
Wasseramsel, 1118
Wasserläufer, Dunkler, 662
    Einsamer, 673
Wasserralle, 484
Weichfeder-Sturmvogel, 38
Weidenammer, 1670
Weidengelbkehlchen, 1627
Weidenmeise, 1379
Weidensperling, 1506
Weißaugenmöwe, 711
Weißbart-Grasmücke, 1293
Weißbart-Seeschwalbe, 794
Weißbauch-Sturmschwalbe, 62
Weißbauchtölpel, 74
Weißbrauendrossel, 1218
Weißbrauen-Uferschwalbe, 1058
Weißbürzel-Strandläufer, 613
Weißflügelgimpel, 1591
Weißflügellerche, 1027
Weißflügel-Seeschwalbe, 799
Weißflügel-Sturmvogel, 37
Weißgesicht-Sturmschwalbe, 60
Weißgesicht-Sturmtaucher, 46
Weißhand-Kernbeißer, 1610
Weißkappenalbatros, 29
Weißkehlammer, 1639
Weißkehlsänger, 1154
Weißkopfmöwe, 746
Weißkopfnoddi, 803
Weißkopf-Ruderente, 285
Weißohrbülbül, 1109
Weißrückenspecht, 1000
Weißschwanzkiebitz, 592
Weißstirnlerche, 1013
Weißstorch, 140
Weißwangengans, 187
Weißwangen-Seeschwalbe, 785
Wellenastrild, 1531
Wellenflughuhn, 826
Wellenläufer, 64
Wendehals, 980
Wermutregenpfeifer, 575
Wespenbussard, 290
Wiedehopf, 976
Wiesenammer, 1655
Wiesenpieper, 1082
Wiesenstrandläufer, 612

Wiesenweihe, 337
Wintergoldhähnchen, 1342
Wilsondrossel, 1211
Wilsonwassertreter, 689
Witwenpfeifgans, 160
Witwenstelze, 1106
Würgfalke, 414
Wüstenfalke, 424
Wüstengimpel, 1594
Wüstengrasmücke, 1301
Wüstenhuhn, Arabisches, 460
    Persisches, 459
Wüstenläuferlerche, 1019
Wüstenprinie, 1241
Wüstenrabe, 1481
Wüstenregenpfeifer, 574
Wüstensperling, 1511
Wüstensteinschmätzer, 1185

Zaunammer, 1651
Zaunkönig, 1122
Zederngirlitz, 1552
Zedernseidenschwanz, 1115
Ziegenmelker, 929
    Nubischer, 928
Zilpzalp, 1337
Zimtracke, 974
Zippammer, 1653
Zitronengirlitz, 1556
Zitronenstelze, 1098
Zügelseeschwalbe, 787
Zwergadler, 381
Zwergammer, 1667
Zwergbrachvogel, 653
Zwergdommel, 105
    Amerikanische, 104
Zwergdrossel, 1210
Zwergflamingo, 156
Zwerggans, 178
Zwergmöwe, 719
Zwergohreule, 891
Zwergsäger, 274
Zwergscharbe, 87
Zwergschnäpper, 1353
Zwergschneegans, 184
Zwergschnepfe, 633
Zwergschwan, 166
Zwergseeschwalbe, 790
Zwergstrandläufer, 607
Zwergsultanshuhn, 502
Zwergsumpfhuhn, 494
Zwergtaucher, 14
Zwergtrappe, 519
Zypernsteinschmätzer, 1183

Pages 1–1008 appear in Volume 1, pages 1009–1694 in Volume 2

NOMBRES ESPAÑOLES

Abanto-marino antártico, 32
   subantártico, 32
Abejaruco común, 966
   oriental, 963
   papirrojo, 964
Abejero europeo, 290
   orientale, 293
Abubilla, 976
Acentor alpino, 1134
   alpino persa, 1132
   común, 1128
   gorginegro, 1133
   siberiano, 1131
Agachadiza común, 635
   chica, 633
   china, 642
   real, 638
   uralense, 640
Agateador común, 1414
   norteño, 1411
Aguatero, 532
Aguila-azor perdicera, 383
Aguila cafre, 380
   esteparia, 370
   imperial ibérica, 375
   imperial oriental, 371
   moteada, 365
   pescadora, 386
   pomerana, 363
   rapaz, 368
   real, 376
   volatinera, 326
Aguililla calzada, 381
Aguilucho cenizo, 337
   lagunero occidental, 328
   pálido, 331
   papialbo, 335
Aguja colinegra, 647
   colipinta, 650
   hudsonica, 650
Agujeta escolopácea, 643
   gris, 642
Albatros cabecigrís, 30
   clororrinco, 29
   frentiblanco, 29
   ojeroso, 27
   patinegro, 30
   real, 31
   viajero, 30
Alca común, 812
   gigante, 815
Alcaraván común, 546
   senegalés, 548
Alcatraz atlántico, 77
   de El Cabo, 79
Alcaudón cabecinegro, 1436
   colirrojo, 1430
   común, 1445
   chico, 1436

dorsirrojo, 1433
enmascarado, 1447
isabel, 1431
real, 1440
real meridional, 1442
Alción cabeciblanco, 954
   común, 960
   de Esmirna, 953
Alcotán europeo, 404
Alimoche común, 311
   sombrío, 314
Alondra cabecinegra, 1013
   cariblanca, 1052
   común, 1043
   cornuda, 1047
   de Dunn, 1014
   de Dupont, 1019
   ibis, 1019
   india, 1046
   piquigruesa, 1022
Alzacola común, 1137
   negro, 1139
Ameritordo común, 1693
Ampelis europeo, 1113
   gris, 1116
Ánade azulón, 218
   friso, 212
   piquirrojo, 225
   rabudo, 223
   sombrío, 222
Andarríos bastardo, 676
   chico, 680
   grande, 674
   maculado, 683
   solitario, 673
   de Terek, 678
Aninga común, 91
Ánsar campestre, 170
   careto, 175
   común, 179
   chico, 178
   indio, 182
   nival, 182
   piquicorto, 173
   de Ross, 184
Arao aliblanco, 815
   de Brünnich, 810
   común, 806
Araocillo antiguo, 820
Archibebe claro, 670
   común, 665
   fino, 668
   gris, 684
   oscuro, 662
   patigualdo chico, 672
   patigualdo grande, 672
   semipalmeado, 685
Arrendajo común, 1450
   funesto, 1454
Autillo común, 891

pálido, 890
Avefría cabecinegra, 589
   de carúncula, 589
   común, 593
   espolada, 587
Avestruz, 1
Avetorillo común, 105
   manchú, 107
   panamericano, 104
   plomizo, 108
Avetoro común, 101
   lentiginoso, 104
Avión cinchado, 1058
   común, 1066
   roquero, 1059
   roquero africano, 1058
   roquero americano, 1066
   zapador, 1055
   zapador africano, 1054
Avoceta, 542
Avutarda arabe, 527
   común, 529
   núbica, 522
Azor común, 342
Azor-lagartijero oscuro, 340

Barnacla canadiense, 185
   cariblanca, 187
   carinegra, 189
   cuellirroja, 192
Becada, 644
Bengalí rojo, 1533
   senegalés, 1529
Bigotudo, 1362
Bijirita sombría, 1620
Bisbita acuático americano, 1092
   arbóreo, 1079
   de Blyth, 1072
   caminero, 1075
   campestre, 1072
   común, 1082
   costero, 1089
   gorgirrojo, 1084
   de Hodgson, 1077
   del Pechora, 1081
   piquilargo, 1076
   ribereño, 1086
   de Richard, 1070
Bolsero de Baltimore, 1694
Búho chico, 915
   nival, 897
   pescador, 896
   real, 893
Buitre dorsiblanco bengalí, 314
   leonado, 315
   moteado, 318
   negro, 321
   orejudo, 318
Buitrón común, 1237

desertícola, 1241
elegante, 1239
Bulbul capirotado, 1110
   cariblanco, 1109
   naranjero, 1111
   ventrirrojo, 1112
Busardo calzado, 360
   chapulinero, 353
   hombrorrojo, 353
   mongol, 360
   moro, 359
   ratonero, 353
Buscarla fluvial, 1247
   de Gray, 1251
   lanceolada, 1244
   de Pallas, 1243
   pintoja, 1245
   unicolor, 1249

Calamón de Allen, 502
   común, 504
Calamoncillo de la Martinica, 502
Calandria aliblanca, 1027
   bimaculada, 1026
   común, 1024
   negra, 1029
Camachuelo carminoso, 1596
   común, 1606
   desertícola, 1591
   ensangrentado, 1589
   grande, 1600
   de Pallas, 1600
   picogrueso, 1602
   del Sinaí, 1599
   trompetero, 1594
Canario, 1554
Canastera alinegra, 557
   común, 555
   de las Maldivas, 557
Candelita, 1625
Candelo escarlata, 1632
   tricolor, 1686
   unicolor, 1631
Cárabo común, 907
   lapón, 913
   oriental, 910
   uralense, 911
Carbonero común, 1393
   garrapinos, 1385
   lapón, 1382
   lúgubre, 1378
   palustre, 1375
   sibilino, 1379
Cardenal azul, 1688
Carpintero dorado, 983
Carraca abisínica, 972
   común, 970
   india, 973
   piquigualda, 974
Carricerín cejudo, 1254

Pages 1–1008 appear in Volume 1, pages 1009–1694 in Volume 2

común, 1255
real, 1251
Carricero agrícola, 1258
de Blyth, 1260
de Cabo Verde, 1262
común, 1265
picogordo, 1273
políglota, 1263
ruidoso, 1267
tordal, 1269
Cascanueces, 1460
Cerceta de alfanjes, 210
aliazul, 228
del Baikal, 214
carretona, 226
común, 215
de El Cabo, 217
pardilla, 232
Cernícalo americano, 396
patirrojo, 397
primilla, 391
vulgar, 393
Cigüeña blanca, 140
negra, 138
Cigüeñuela, 539
Cisne cantor, 168
chico, 166
negro, 165
vulgar, 162
Codorniz común, 467
Cogujada común, 1037
montesina, 1040
Coliazul cejiblanco, 1153
oidorrojo, 1530
Colibrí pigmeo, 1419
Colimbo de Adams, 10
ártico, 6
chico, 3
grande, 8
Colín de California, 447
de Virginia, 448
Colinegro real, 1166
Colirrojo coronado, 1164
diademado, 1163
de Eversmann, 1155
real, 1161
tizón, 1157
Collalba de Chipre, 1183
desértica, 1185
isabel, 1175
fúnebre, 1193
gris, 1178
de Hume, 1195
negra, 1198
negra de Brehm, 1196
oriental, 1187
pechinegra, 1194
persa, 1190
pía, 1180
pía oriental, 1191
rubia, 1183
de Tristram, 1188
Combatiente, 628

Cormorán africano, 89
grande, 80
moñudo, 84
orejudo, 83
pigmeo, 87
de Socotora, 86
Corneja común, 1478
india, 1472
pía, 1481
Corredor común, 552
Correlimos acuminado, 615
de Baird, 614
de Bartram, 662
de Bering, 599
de Bonaparte, 613
canelo, 628
común, 620
cuellirrojo, 607
falcinelo, 624
gordo, 599
de Maur, 606
menudo, 607
oscuro, 618
pectoral, 615
semipalmeado, 605
siberiano, 612
de Temminck, 610
tridáctilo, 602
zancolín, 627
zarapatín, 617
Cotorra cata, 867
Cotorrita de Kramer, 866
Críalo común, 872
etíope, 872
Cucal senegalés, 884
Cuco cobrizo, 874
común, 875
oriental, 878
piquigualdo, 882
piquinegro, 882
Cuchara común, 230
Cuervo colicorto, 1486
común, 1483
desertícola, 1481
Culebrera europea, 324
Curruca arabe, 1303
cabecinegra, 1296
capirotada, 1316
carrasqueña, 1293
gavilana, 1306
de Menetries, 1295
mirlona, 1305
mosquitera, 1314
rabilarga, 1288
de Rüppell, 1300
sahariana, 1301
sarda, 1286
tomillera, 1291
de Tristram, 1290
ustulada, 1299
zarcera, 1310
zarcerilla, 1308

Chagra, 1428
Charlatán común, 1690
Charrán de las Aleutianas, 784
ártico, 782
bengalés, 772
de Berg, 770
cariblanco, 785
común, 779
elegante, 776
embridado, 787
de Forster, 785
pardelo, 802
patinegro, 773
real, 769
rosado, 777
sombrío, 789
Charrancito común, 790
de Saunders, 793
Chingolito llanero, 1635
Chingolo arlequín, 1635
gorgiblanco, 1639
melodioso, 1637
pantanero, 1637
piquiblanco, 1637
pizarroso, 1639
punteado, 1633
sabanero, 1635
zorruno, 1636
Chivi ojirrojo, 1537
Chochín, 1122
Chorlitejo asiático, 575
culirrojo, 567
chico, 561
grande, 564
mongol chico, 572
mongol grande, 574
pastor, 567
patinegro, 569
semipalmeado, 566
tricollar, 569
Chorlito carambolo, 577
coliblanco, 592
dorado común, 581
dorado chico, 579
dorado siberiano, 580
gris, 584
social, 590
Chotacabras americano, 936
dorado, 933
egipcio, 933
gris, 929
núbico, 928
pardo, 932
Chova piquigualda, 456
piquirroja, 1466

Droma cangrejero, 544

Eider de anteojos, 255
común, 251
chico, 256
real, 253
Elanio común, 294

Escribano aureolado, 1670
cabeciblanco, 1646
cabecigrís, 1662
cabecinegro, 1679
carirrojo, 1677
cejigualdo, 1664
ceniciento, 1663
cerillo, 1648
cinéreo, 1658
cioide, 1655
enmascarado, 1645
herrumbroso, 1669
hortelano, 1659
lapón, 1640
montesino, 1653
nival, 1642
de Pallas, 1676
palustre, 1672
pigmeo, 1667
rústico, 1665
sahariano, 1655
soteño, 1651
de Stewart, 1653
tahapis, 1659
Esmerejón, 400
Espátula africana, 152
común, 149
Estornino chino, 1491
dáurico, 1491
irisado, 1489
negro, 1496
pinto, 1492
rosado, 1498
Estrilda común, 1531
Esturnela pechigualda, 1692

Faisán dorado, 475
de Lady Amherst, 476
venerado, 471
vulgar, 472
Falaropo picofino, 691
picogrueso, 692
de Wilson, 689
Fibi primordial, 1010
Flamenco común, 153
enano, 156
Focha americana, 509
común, 506
cornuda, 509
Frailecillo común, 821
de tirabuzones, 825
Francolín biespolado, 463
ventrinegro, 461
Fulmar boreal, 34
Fumarel aliblanco, 799
cariblanco, 794
común, 796

Gallo-lira caucasiano, 440
común, 437
Ganga común, 834
coronada, 828
de Lichtenstein, 826

moruna, 831
moteada, 829
de Pallas, 836
Gansito asiático, 203
Ganso espolonado, 203
del Nilo, 195
Garceta azabache, 119
azul, 118
común, 122
dimorfa, 121
grande, 126
intermedia, 125
nívea, 120
tricolor, 120
Garcilla bueyera, 116
cangrejera, 113
china, 115
india, 115
Garcita azulada, 111
verde, 113
Garza azulada, 131
cabecinegra, 127
goliat, 134
imperial, 132
real, 128
Gavilán común, 345
chikra, 348
gabar, 342
griego, 349
Gavión cabecinegro, 712
común, 754
Gaviota de alas glaucas, 752
argéntea, 741
armenia, 748
de Audouin, 732
de Bonaparte, 723
cabecigrís, 727
cabecinegra, 714
cabecinegra indostánica, 727
cana, 735
de Delaware, 734
enana, 719
de Franklin, 718
fuliginosa, 709
hiperbórea, 752
marfil, 761
ojiblanca, 711
patiamarilla, 746
picofina, 729
polar, 749
reidora, 724
reidora americana, 716
de Ross, 757
de Sabine, 721
sombría, 737
tridáctila, 758
Golondrina canadiense, 1053
común, 1061
dáurica, 1064
etiópica, 1064
Gorrialondra tuticastaña, 1012
Gorrión alpino, 1522
aureo, 1515

cuadrillero, 1686
chillón, 1519
doméstico, 1503
grande, 1511
del Mar Muerto, 1509
molinero, 1513
moruno, 1506
pálido, 1517
pintado, 1518
sahariano, 1511
Graja, 1475
Grajilla común, 1468
dáurica, 1471
Grévol común, 428
Grulla blanca siberiana, 515
canadiense, 514
común, 511
damisela, 516
Guión de codornices, 496
negro, 496

Halcón borní, 411
de Eleonora, 407
gerifalte, 417
peregrino, 419
pizarroso, 409
sacre, 414
tagarote, 424
Herrerillo capuchino, 1383
cíaneo, 1392
común, 1388
Hubara, 525

Ibis eremita, 146
sagrado, 148

Jilguero, 1561

Lagópodo alpino, 434
común, 430
Lavandera blanca, 1103
boyera, 1094
cascadeña, 1100
cetrina, 1098
pía, 1106
Lechuza campestre, 918
común, 886
gavilana, 899
mora, 921
de Tengmalm, 923
Lúgano común, 1564
pinariego, 1568

Malvasía cabeciblanca, 285
canela, 282
Marabú africano, 143
Martinete común, 109
Martín pescador, 956
pescador pío, 959
Mérgulo crestado, 819
lorito, 820
marino, 817
Milano negro, 295

real, 298
Miná común, 1500
Mirlo acuático, 1118
de collar, 1212
común, 1215
Mito, 1372
Mochuelo común, 903
chico, 901
Monjita pico-de-plata, 1534
Morito común, 144
Mosquitero bilistado, 1325
boreal, 1322
de Brooks, 1325
común, 1337
coronado, 1319
iraní, 1335
musical, 1340
de Pallas, 1324
papialbo, 1329
de Schwartz, 1327
silbador, 1333
sombrío, 1328
troquiloide, 1320

Negrón careto, 266
común, 263
especulado, 266
Nodi negro, 803

Oropéndola, 1424
Ortega, 832
Ostrero canario, 538
común, 535

Págalo antártico, 708
grande, 704
parásito, 699
pomarino, 696
rabero, 701
Pagaza piconegra, 764
piquirroja, 766
Paíño boreal, 64
europeo, 62
de Madeira, 67
pechialbo, 60
de Swinhoe, 66
ventriblanco, 62
de Wilson, 58
Pájaro gato, 1127
moscón, 1416
Paloma bravía, 839
migratoria, 865
rabiche, 850
torcaz, 846
torqueza, 848
turqué, 848
zurita, 842
zurita oriental, 845
Papamoscas cerrojillo, 1358
collarino, 1357
gris, 1349
papirrojo, 1353
pardo, 1349

semicollarino, 1355
verde, 1009
Pape arcoiris, 1689
azulejo, 1689
Pardela de Audubon, 56
canosa, 46
capirotada, 47
cenicienta, 42
cenicienta de Edwards, 45
chica, 55
Mediterránea, 54
Mediterránea occidentál, 53
del Pacífico, 46
paticlara, 47
pichoneta, 51
sombría, 49
Pardillo ártico, 1578
común, 1568
piquigualdo, 1571
sizerín, 1574
Paserina lázuli, 1689
Pato arlequín, 259
colorado, 236
havelda, 261
joyuyo, 204
mandarín, 204
Pavo, 481
Pechiazul común, 1149
Pelícano ceñudo, 96
común, 93
rosado, 98
Perdigallo del Caspio, 451
caucasiano, 449
Perdiz chucar, 452
dáurica, 467
desértica, 460
gorgigrís, 459
griega, 453
moruna, 457
pardilla, 464
roja, 455
Petirrojo común, 1140
turco, 1154
Petrel antillano, 41
de Bulwer, 41
damero, 36
freira, 37
gon-gon, 39
de Gould, 37
de Jouanin, 42
de las Kermadec, 37
de Schlegel, 40
Pico chupador, 992
dorsiblanco, 1000
mediano, 998
menor, 1002
picapinos, 993
sirio, 996
tridáctilo, 1005
Picogordo común, 1610
nipón, 1610
piquigualdo, 1610
vespertino, 1613

Pages 1–1008 appear in Volume 1, pages 1009–1694 in Volume 2

[22] Nombres españoles

Picoplata de Malabar, 1534
Pigargo europeo, 303
   de Pallas, 302
   vocinglero, 302
Pintada común, 479
Pinzón azul, 1544
   común, 1539
   real, 1545
Piquero enmascarado, 73
   pardo, 74
   patirrojo, 73
Piquituerto común, 1582
   escocés, 1586
   franjeado, 1580
   lorito, 1587
Pirrula europea de las Azores, 1609
   mongólica, 1592
   rabilargo, 1606
Pito cano, 983
   negro, 989
   real, 986
   real bereber, 988
Playerito menudo, 612
Pluvial egipcio, 550
Polla de agua, 499
Polluela bastarda, 490
   carolina, 490
   culirroja, 496
   chica, 494
   pintoja, 488
Porrón acollarado, 241
   albeola, 269
   bastardo, 247
   bola, 249
   coacoxtle, 240
   europeo, 238
   islándico, 269
   moñudo, 244
   osculado, 272
   pardo, 242

Quebrantahuesos, 308
Quelea común, 1528

Rabihorcado magnífico, 99
Rabijunco etéreo, 70
Rabilargo, 1456
Rabitojo mongol, 939
Rascón, 484

Rayador africano, 804
Reinita alidorada, 1616
   amarilla, 1618
   atigrada, 1621
   cejiblanca, 1622
   coronada, 1623
   de costillas castañas, 1619
   charquera, 1626
   encapuchada, 1627
   gorgigualda, 1627
   listada, 1623
   montana, 1625
   noteña, 1617
   palmera, 1623
   papinegra, 1620
   peregrina, 1615
   trepadora, 1615
Reyezuelo crestado, 1342
   listado, 1346
   de oro, 1348
   sencillo, 1342
   tinerfeño, 1345
Robín americano, 1234
Roquero rojo, 1201
   solitario, 1204
Ruiseñor bastardo, 1235
   calíope, 1147
   coliazul, 1152
   común, 1145
   ruso, 1143

Serreta capuchona, 274
   chica, 274
   grande, 279
   mediana, 277
Silbón americano, 210
   europeo, 207
Silvia de birrete del norte, 1628
   de Blackburn, 1620
   canadiense, 1629
Sinsonte castaño, 1126
   norteño, 1125
Sisón común, 519
   de Denham, 522
Somormujo cuellirrojo, 20
   lavanco, 17
   macachón, 17
Suimanga palestina, 1422
   rabilarga, 1421
Suirirí bicolor, 160
   cariblanco, 160
   de Java, 161

Tántalo africano, 137
Tarabilla canaria, 1169
   común, 1170
   norteña, 1167
   pía, 1173
   termitera, 1174
Tarro blanco, 198
   canelo, 197
Tejedor cabecinegro, 1526
   listado, 1526
Terrera de Cabo Verde, 1047
   colinegra, 1015
   común, 1032
   de Hume, 1034
   marismeña, 1034
   mongólica, 1036
   sahariana, 1017
Torcecuello, 980
Tordalino arábigo, 1368
   colilargo, 1367
   iraquí, 1365
   rojizo, 1370
Tordo cabecicafé, 1691
   cabecidorado, 1693
   canadiense, 1692
Torillo, 482
Tórtola común, 856
   de El Cabo, 861
   oriental, 859
   roseogrís, 852
   senegalesa, 860
   turca, 853
Totovía, 1041
Trepador argelino, 1400
   armenio, 1404
   azul, 1402
   canadiense, 1401
   corso, 1399
   de medalla, 1398
   rupestre, 1406
Trepariscos, 1408
Triguero, 1681
Turpial de Wagler, 1693

Urogallo común, 442
Urraca, 1457

Vencejo asiático, 947
   de Cabo Verde, 941
   común, 943

culiblanco cafre, 947
culiblanco chico, 949
   espinoso, 939
   pálido, 945
   palmero, 950
   real, 940
   unicolor, 942
Verdecillo carinegro, 1548
   común, 1550
   sirio, 1552
Verderón común, 1557
   serrano, 1556
Vireo de Filadelfia, 1537
   de garganta amarilla, 1536
   ojirrojo, 1537
Vuelvepiedras, 686

Zampullín cuellinegro, 24
   cuellirrojo, 22
   chico, 14
   picogrueso, 13
Zánate común, 1692
Zarapito chico, 653
   esquimal, 654
   fino, 657
   real, 658
   trinador, 655
Zarcero común, 1284
   escita, 1277
   grande, 1280
   icterino, 1282
   pálido, 1273
   de Upcher, 1278
Zenaida huilota, 864
Zootera abigarrada, 1208
Zorzal alirrojo, 1228
   carigrís, 1211
   común, 1225
   común americano, 1210
   charlo, 1230
   charlo americano, 1209
   chico, 1210
   dorado, 1206
   de Naumann, 1219
   papinegro, 1220
   real, 1222
   rojigrís, 1218
   siberiano, 1207
   solitario, 1211
   unicolor, 1212

Pages 1–1008 appear in Volume 1, pages 1009–1694 in Volume 2

NOMS FRANÇAIS

Accenteur alpin, 1134
   à gorge noire, 1133
   montanelle, 1131
   mouchet, 1128
   de Radde, 1132
Agrobate podobé, 1139
   roux, 1137
Aigle de Bonelli, 383
   botté, 381
   criard, 365
   ibérique, 375
   impérial, 371
   pomarin, 363
   ravisseur, 368
   royal, 376
   des steppes, 370
   de Verreaux, 380
Aigrette ardoisée, 119
   bleue, 118
   garzette, 122
   Grande, 126
   intermédiaire, 125
   mélanocéphale, 127
   neigeuse, 120
   des récifs, 121
   tricolore, 120
Albatros à cape blanche, 29
   hurleur, 30
   à nez jaune, 29
   à pieds noirs, 30
   royal, 31
   à sourcils noirs, 27
   à tête grise, 30
Alcyon pie, 959
Alouette bilophe, 1052
   calandre, 1024
   calandrelle, 1032
   des champs, 1043
   de Clotbey, 1022
   de Dunn, 1014
   gulgule, 1046
   haussecol, 1047
   de Hume, 1034
   du Kordofan, 1011
   leucoptère, 1027
   lulu, 1041
   monticole, 1026
   nègre, 1029
   pispolette, 1034
   de Razo, 1047
   de Swinhoe, 1036
Amarante du Sénégal, 1529
Ammomane élégante, 1015
   isabelline, 1017
Anhinga roux, 91
Anserelle de Coromandel, 203
Arlequin plongeur, 259
Astrild à joues orange, 1530

   ondulé, 1531
Autour gabar, 342
   des palombes, 342
   sombre, 340
Autruche d'Afrique, 1
Avocette élégante, 542

Balbuzard pêcheur, 386
Barge hudsonienne, 650
   à queue noire, 647
   rousse, 650
Bartramie des champs, 662
Bateleur des savanes, 326
Bécasse des bois, 644
Bécasseau d'Alaska, 606
   de l'Anadyr, 599
   de Baird, 614
   de Bonaparte, 613
   cocorli, 617
   à cou roux, 607
   échasse, 627
   falcinelle, 624
   à longs doigts, 612
   maubèche, 599
   minuscule, 612
   minute, 607
   à queue pointue, 615
   rousset, 628
   sanderling, 602
   semipalmé, 605
   tacheté, 615
   de Temminck, 610
   variable, 620
   violet, 618
Bécassin à bec court, 642
   à long bec, 643
Bécassine double, 638
   des marais, 635
   à queue pointue, 640
   sourde, 633
   de Swinhoe, 642
Bec-croisé bifascié, 1580
   d'Ecosse, 1586
   perroquet, 1587
   des sapins, 1582
Bec-en-ciseaux d'Afrique, 804
Bengali rouge, 1533
Bergeronnette citrine, 1098
   grise, 1103
   pie, 1106
   printanière, 1094
   des ruisseaux, 1100
Bernache du Canada, 185
   à cou roux, 192
   cravant, 189
   nonnette, 187
Bihoreau gris, 109
Blongios mandchou, 107
   nain, 105

   Petit, 104
   de Sturm, 108
Bondrée apivore, 290
   orientale, 293
Bouscarle de Cetti, 1235
Bouvreuil des Açores, 1609
   pivoine, 1606
Bruant auréole, 1670
   à calotte blanche, 1646
   cannelle, 1659
   cendré, 1658
   cendrillard, 1663
   des champs, 1635
   chanteur, 1637
   à cou gris, 1662
   à couronne blanche, 1637
   fauve, 1636
   fou, 1653
   à gorge blanche, 1639
   jaune, 1648
   à joues marron, 1635
   lapon, 1640
   à longue queue, 1655
   des marais, 1637
   masqué, 1645
   mélanocéphale, 1679
   nain, 1667
   des neiges, 1642
   ortolan, 1659
   de Pallas, 1676
   des prés, 1635
   proyer, 1681
   des roseaux, 1672
   roux, 1669
   rustique, 1665
   à sourcils jaunes, 1664
   de Stewart, 1653
   striolé, 1655
   à tête rousse, 1677
   zizi, 1651
Bulbul d'Arabie, 1110
   à ventre rouge, 1112
   des jardins, 1111
   à joues blanches, 1109
Busard cendré, 337
   pâle, 335
   des roseaux, 328
   Saint-Martin, 331
Buse de Chine, 360
   à épaulettes, 353
   féroce, 359
   pattue, 360
   de Swainson, 353
   variable, 353
Butor d'Amérique, 104
   étoilé, 101

Caille des blés, 467

Calliope sibérienne, 1147
Canard à bec rouge, 225
   du Cap, 217
   carolin, 204
   chipeau, 212
   colvert, 218
   à faucilles, 210
   à front blanc, 210
   huppé, 204
   mandarin, 204
   noir, 222
   pilet, 223
   siffleur, 207
   de Smith, 229
   souchet, 230
Capucin bec-d'argent, 1534
   bec-de-plomb, 1534
Cardinal à poitrine rose, 1686
Carouge à épaulettes, 1693
   à tête jaune, 1693
Cassenoix moucheté, 1460
Chardonneret élégant, 1561
Chevalier aboyeur, 670
   arlequin, 662
   bargette, 678
   criard, 672
   culblanc, 674
   gambette, 665
   grivelé, 683
   guignette, 680
   à pattes jaunes, 672
   semipalmé, 685
   de Sibérie, 684
   solitaire, 673
   stagnatile, 668
   sylvain, 676
Chevêche d'Athéna, 903
Chevêchette d'Europe, 901
Chocard à bec jaune, 1464
Choucas de Daourie, 1471
   des tours, 1468
Chouette de Butler, 910
   épervière, 899
   hulotte, 907
   lapone, 913
   de l'Oural, 911
   de Tengmalm, 923
Cigogne blanche, 140
   noire, 138
Cincle plongeur, 1118
Circaète Jean-le-Blanc, 324
Cisticole des joncs, 1237
Cochevis huppé, 1037
   de Thékla, 1040
Colin de Californie, 447
   de Virginie, 448
Coliou huppé, 869
Combattant varié, 628
Conure veuve, 867

Pages 1–1008 appear in Volume 1, pages 1009–1694 in Volume 2

Noms français

Corbeau brun, 1481
    familier, 1472
    freux, 1475
    Grand, 1483
    pie, 1481
    à queue courte, 1486
Cordonbleu à joues rouges, 1530
Cormoran africain, 89
    à aigrettes, 83
    Grand, 80
    huppé, 84
    pygmée, 87
    de Socotra, 86
Corneille noire, 1478
Coucal du Sénégal, 884
Coucou didric, 874
    geai, 872
    gris, 875
    jacobin, 872
    oriental, 878
Coulicou à bec jaune, 882
    à bec noir, 882
Courlis à bec grêle, 657
    cendré, 658
    corlieu, 655
    esquimau, 654
    nain, 653
Courvite isabelle, 552
    de Temminck, 554
Crabier chevelu, 113
    chinois, 115
    de Gray, 115
Cratérope écaillé, 1368
    fauve, 1370
    d'Inde, 1367
    d'Irak, 1365
Crave à bec rouge, 1466
Crécerelle d'Amérique, 396
Cygne chanteur, 168
    noir, 165
    de Bewick, 166
    tuberculé, 162

Damier du Cap, 36
Dendrocygne fauve, 160
    siffleur, 161
    veuf, 160
Dickcissel d'Amérique, 1686
Dindon sauvage, 481
Drome ardéole, 544
Dromoïque du désert, 1241
Durbec des sapins, 1602

Echasse blanche, 539
Effraie des clochers, 886
Eider à duvet, 251
    à lunettes, 255
    de Steller, 256
    à tête grise, 253
Elanion blanc, 294
Engoulevent d'Amérique, 936
    à collier roux, 932
    du désert, 933

doré, 933
d'Europe, 929
de Nubie, 928
Epervier d'Europe, 345
    à pieds courts, 349
    shikra, 348
Erismature rousse, 282
    à tête blanche, 285
Etourneau de Daourie, 1491
    mandarin, 1491
    roselin, 1498
    sansonnet, 1492
    unicolore, 1496

Faisan de Colchide, 472
    doré, 475
    de Lady Amherst, 476
    vénéré, 471
Faucon de Barbarie, 424
    concolore, 409
    crécerelle, 393
    crécerellette, 391
    d'Eléonore, 407
    émerillon, 400
    gerfaut, 417
    hobereau, 404
    kobez, 397
    lanier, 411
    pèlerin, 419
    sacre, 414
Fauvette d'Arabie, 1303
    de l'Atlas, 1290
    babillarde, 1308
    de Chypre, 1299
    épervière, 1306
    grisette, 1310
    des jardins, 1314
    à lunettes, 1291
    mélanocéphale, 1296
    de Ménétries, 1295
    naine, 1301
    orphée, 1305
    passerinette, 1293
    pitchou, 1288
    de Rüppell, 1300
    sarde, 1285
    à tête noire, 1316
Flamant du Chili, 156
    nain, 156
    rose, 153
Fou de Bassan, 77
    brun, 74
    du Cap, 79
    masqué, 73
    à pieds rouges, 73
Foulque d'Amérique, 509
    caronculée, 509
    macroule, 506
Francolin d'Erckel, 464
    à double éperon, 463
    noir, 461
Frégate superbe, 99
Fuligule à bec cerclé, 241

à dos blanc, 240
milouin, 238
milouinan, 247
morillon, 244
nyroca, 242
à tête noire, 249
Fulmar boréal, 34
    géant, 32
    de Hall, 32

Gallinule poule-d'eau, 499
Ganga cata, 834
    couronné, 828
    de Lichtenstein, 826
    tacheté, 829
    unibande, 832
    à ventre brun, 831
Garrot albéole, 269
    d'Islande, 269
    à oeil d'or, 272
Geai des chênes, 1450
Gélinotte des bois, 428
Glaréole à ailes noires, 557
    à collier, 555
    orientale, 557
Gobemouche brun, 1349
    à collier, 1357
    à demi-collier, 1355
    gris, 1349
    nain, 1353
    noir, 1358
Goéland à ailes blanches, 749
    à ailes grises, 752
    argenté, 741
    d'Arménie, 748
    d'Audouin, 732
    à bec cerclé, 734
    de Bering, 752
    bourgmestre, 752
    brun, 737
    cendré, 735
    de Hemprich, 709
    ichthyaète, 712
    à iris blanc, 711
    leucophée, 746
    marin, 754
    railleur, 729
Goglu des prés, 1690
Gorgebleue à miroir, 1149
Grand-duc d'Europe, 893
Gravelot à collier interrompu, 569
    Grand, 564
    kildir, 567
    de Leschenault, 574
    mongol, 572
    pâtre, 567
    Petit, 561
    semipalmé, 566
    à triple collier, 569
Grèbe à bec bigarré, 13
    castagneux, 14
    à cou noir, 24
    esclavon, 22

Grand, 17
huppé, 17
jougris, 20
Grimpereau des bois, 1411
    des jardins, 1414
Grive à ailes rousses, 1219
    des bois, 1209
    à collier, 1208
    dorée, 1206
    à dos olive, 1210
    draine, 1230
    fauve, 1211
    à gorge noire, 1220
    à gorge rousse, 1220
    à joues grises, 1211
    litorne, 1222
    mauvis, 1228
    musicienne, 1225
    de Naumann, 1219
    obscure, 1218
    de Sibérie, 1207
    solitaire, 1210
Grosbec casse-noyaux, 1610
    errant, 1613
    masqué, 1610
    migrateur, 1610
Grue du Canada, 514
    cendrée, 511
    demoiselle, 516
    de Sibérie, 515
Guêpier d'Europe, 966
    d'Orient, 963
    de Perse, 964
Guifette leucoptère, 799
    moustac, 794
    noire, 796
Guillemot de Brünnich, 810
    à cou blanc, 820
    à miroir, 815
    de Troïl, 806
Guiraca bleu, 1688
Gypaète barbu, 308

Harelde boréale, 261
Harfang des neiges, 897
Harle bièvre, 279
    couronné, 274
    huppé, 277
    piette, 274
Héron cendré, 128
    garde-boeufs, 116
    goliath, 134
    Grand, 131
    mélanocéphale, 127
    pourpré, 132
    strié, 111
    vert, 113
Hibou du Cap, 921
    des marais, 918
    moyen-duc, 915
Hirondelle bicolore, 1053
    à collier, 1058
    d'Ethiopie, 1064

Noms français [25]

de fenêtre, 1066
à front blanc, 1066
isabelline, 1058
paludicole, 1054
de rivage, 1055
de rochers, 1059
rousseline, 1064
rustique, 1061
Huîtrier pie, 535
des Canaries, 538
Huppe fasciée, 976
Hypocolius gris, 1116
Hypolaïs bottée, 1277
ictérine, 1282
des oliviers, 1280
pâle, 1273
polyglotte, 1284
d'Upcher, 1278

Ibis chauve, 146
falcinelle, 144
sacré, 148
Iranie à gorge blanche, 1154

Jaseur d'Amérique, 1115
boréal, 1113
Junco ardoisé, 1639

Kétoupa brun, 896

Labbe, Grand, 704
à longue queue, 701
de McCormick, 708
parasite, 699
pomarin, 696
Lagopède alpin, 434
des saules, 430
Linotte à bec jaune, 1571
mélodieuse, 1568
Locustelle fasciée, 1251
fluviatile, 1247
lancéolée, 1244
luscinioïde, 1249
de Pallas, 1243
tachetée, 1245
Loriot d'Europe, 1424
Lusciniole à moustaches, 1251

Macareux huppé, 825
moine, 821
Macreuse brune, 266
à front blanc, 266
noire, 263
Marabout d'Afrique, 143
Marouette de Baillon, 494
de Caroline, 490
ponctuée, 488
poussin, 490
rayée, 496
Martin-chasseur de Smyrne, 953
à tête grise, 954
Martinet cafre, 947
du Cap-Vert, 941

épineux, 939
des maisons, 949
noir, 943
pâle, 945
des palmes, 950
ramoneur, 939
de Sibérie, 947
unicolore, 942
à ventre blanc, 940
Martin-pêcheur d'Europe, 956
d'Amérique, 960
Martin triste, 1500
Mergule nain, 817
Merle d'Amérique, 1234
noir, 1215
à plastron, 1212
unicolore, 1212
Mésange azurée, 1392
bleue, 1388
boréale, 1379
charbonnière, 1393
huppée, 1383
lapone, 1382
à longue queue, 1372
lugubre, 1378
noire, 1385
nonnette, 1375
Mésangeai imitateur, 1454
Milan noir, 295
royal, 298
Moineau blanc, 1511
domestique, 1503
doré, 1515
à dos roux, 1511
espagnol, 1506
friquet, 1513
à gorge jaune, 1518
de la Mer Morte, 1509
pâle, 1517
soulcie, 1519
Moinelette à front blanc, 1013
d'Oustalet, 1012
Monticole bleu, 1202
de roche, 1201
Moqueur chat, 1127
polyglotte, 1125
roux, 1126
Moucherolle phébi, 1010
vert, 1009
Mouette atricille, 716
blanche, 761
de Bonaparte, 723
de Franklin, 718
mélanocéphale, 714
pygmée, 719
rieuse, 724
de Ross, 757
de Sabine, 721
à tête grise, 727
du Tibet, 727
tridactyle, 758

Nette rousse, 236

Niverolle alpine, 1522
Noddi brun, 802
noir, 803

Océanite de Castro, 67
culblanc, 64
frégate, 60
de Swinhoe, 66
tempête, 62
à ventre blanc, 62
de Wilson, 58
Oedicnème criard, 546
du Sénégal, 548
Oie à bec court, 173
cendrée, 179
des moissons, 170
naine, 178
des neiges, 182
rieuse, 175
de Ross, 184
à tête barrée, 182
Oie-armée de Gambie, 203
Oriole de Baltimore, 1694
cul-noir, 1693
Ouette d'Egypte, 195
Outarde arabe, 527
barbue, 529
canepetière, 519
de Denham, 522
houbara, 525
nubienne, 522

Panure à moustaches, 1362
Paruline à ailes dorées, 1616
bleue, 1620
à calotte noire, 1628
du Canada, 1629
à capuchon, 1627
à collier, 1617
couronnée, 1625
à couronne rousse, 1623
à croupion jaune, 1623
flamboyante, 1625
à flancs marron, 1619
à gorge noire, 1620
à gorge orangée, 1620
jaune, 1618
masquée, 1627
noir et blanc, 1615
obscure, 1615
rayée, 1623
des ruisseaux, 1626
à tête cendrée, 1622
tigrée, 1621
Passerin azuré, 1689
indigo, 1689
nonpareil, 1689
Pélican blanc, 93
frisé, 96
gris, 98
Perdrix bartavelle, 453
choukar, 452
de Daourie, 467

gambra, 457
grise, 464
de Hey, 460
rouge, 455
si-si, 459
Perruche à collier, 866
Petit-duc de Bruce, 890
scops, 891
Pétrel de Bulwer, 41
diablotin, 41
gongon, 39
de Gould, 37
de Jouanin, 42
des Kermadec, 37
de Madère, 37
de Schlegel, 40
soyeux, 38
Phaéton à bec rouge, 70
Phalarope à bec étroit, 691
à bec large, 692
de Wilson, 689
Phragmite aquatique, 1254
des joncs, 1255
Pic cendré, 983
à dos blanc, 1000
épeiche, 993
épeichette, 1002
flamboyant, 983
de Levaillant, 988
maculé, 992
mar, 998
noir, 989
syriaque, 996
tridactyle, 1005
vert, 986
Pie bavarde, 1457
bleue, 1456
Pie-grièche brune, 1430
à dos gris, 1444
écorcheur, 1433
grise, 1440
isabelle, 1431
masquée, 1447
méridionale, 1442
à poitrine rose, 1436
schach, 1436
à tête rousse, 1445
Pigeon biset, 839
de Bolle, 848
colombin, 842
d'Eversmann, 845
des lauriers, 850
migrateur, 865
ramier, 846
trocaz, 848
Pingouin, Grand, 815
torda, 812
Pinson des arbres, 1539
bleu, 1544
du Nord, 1545
Pintade de Numidie, 479
Pipit des arbres, 1079
de Berthelot, 1075

Pages 1–1008 appear in Volume 1, pages 1009–1694 in Volume 2

Noms français

à dos olive, 1077
farlousane, 1092
farlouse, 1082
de Godlewski, 1072
à gorge rousse, 1084
à long bec, 1076
maritime, 1089
de la Petchora, 1081
de Richard, 1070
rousseline, 1072
spioncelle, 1086
Plongeon arctique, 6
 à bec blanc, 10
 catmarin, 3
 imbrin, 8
Pluvian fluviatile, 550
Pluvier argenté, 584
 asiatique, 575
 bronzé, 579
 doré, 581
 fauve, 580
 guignard, 577
Pouillot de Bonelli, 1329
 boréal, 1322
 de Brooks, 1325
 brun, 1328
 fitis, 1340
 à grands sourcils, 1325
 de Hume, 1327
 de Lorenz, 1336
 modeste, 1335
 oriental, 1331
 de Pallas, 1324
 de Schwarz, 1327
 siffleur, 1333
 de Temminck, 1319
 véloce, 1337
 verdâtre, 1320
Prinia à front écailleux, 1241
 gracile, 1239
Puffin des Anglais, 51
 d'Audubon, 56
 des Baléares, 53
 du Cap-Vert, 45
 cendré, 42
 fouquet, 46
 fuligineux, 49
 leucomèle, 46
 majeur, 47
 à pieds pâles, 47
 semblable, 55
 yelkouan, 54
Pygargue de Pallas, 302
 à queue blanche, 303
 vocifère, 302

Quiscale bronzé, 1692
 rouilleux, 1692
Râle à bec jaune, 496
 d'eau, 484
 des genêts, 496
Rémiz penduline, 1416
Rhynchée peinte, 532
Robin à flancs roux, 1153
Roitelet à couronne dorée, 1348
 à couronne rubis, 1342
 huppé, 1342
 de Ténérife, 1345
 à triple bandeau, 1346
Rolle violet, 974
Rollier d'Abyssinie, 972
 d'Europe, 970
 indien, 973
Roselin à ailes roses, 1589
 cramoisi, 1596
 githagine, 1594
 de Lichtenstein, 1591
 à longue queue, 1606
 de Mongolie, 1592
 rose, 1600
 du Sinaï, 1599
 tacheté, 1600
Rossignol bleu, 1152
 philomèle, 1145
 progné, 1143
Rougegorge familier, 1140
Rougequeue d'Eversmann, 1155
 à front blanc, 1161
 de Güldenstädt, 1164
 de Moussier, 1163
 noir, 1157
Rousserolle des buissons, 1260
 du Cap-Vert, 1262
 effarvatte, 1265
 à gros bec, 1273
 d'Irak, 1271
 isabelle, 1258
 d'Orient, 1272
 stentor, 1267
 turdoïde, 1269
 verderolle, 1263
Rufipenne de Tristram, 1489

Sarcelle à ailes bleues, 228
 élégante, 214
 d'été, 226
 d'hiver, 215
 marbrée, 232
Serin des Canaries, 1554
 cini, 1550
 à front rouge, 1548
 syriaque, 1552
Sirli du désert, 1019

de Dupont, 1020
Sittelle corse, 1399
 kabyle, 1400
 de Krüper, 1398
 de Neumayer, 1406
 à poitrine rousse, 1401
 des rochers, 1404
 torchepot, 1402
Sizerin blanchâtre, 1578
 flammé, 1574
Souimanga du Nil, 1421
 de Palestine, 1422
 pygmée, 1419
Spatule d'Afrique, 152
 blanche, 149
Starique cristatelle, 819
 perroquet, 820
Sterne des Aléoutiennes, 784
 arctique, 782
 bridée, 787
 caspienne, 766
 caugek, 773
 de Dougall, 777
 élégante, 776
 de Forster, 785
 fuligineuse, 789
 hansel, 764
 huppée, 770
 à joues blanches, 785
 naine, 790
 pierregarin, 779
 royale, 769
 de Saunders, 793
 voyageuse, 772
Sturnelle des prés, 1692
Syrrhapte paradoxal, 836

Tadorne de Belon, 198
 casarca, 197
Talève d'Allen, 502
 sultane, 504
 violacée, 502
Tangara écarlate, 1632
 vermillon, 1631
Tantale ibis, 137
Tarier des Canaries, 1169
 pâtre, 1170
 pie, 1173
 des prés, 1167
Tarin des aulnes, 1564
 des pins, 1568
Tchagra à tête noire, 1428
Tétraogalle du Caucase, 449
 de Perse, 451
Tétras du Caucase, 440
 Grand, 442

lyre, 437
Tichodrome échelette, 1408
Tisserin gendarme, 1526
 manyar, 1526
Tohi à flancs roux, 1633
Torcol fourmilier, 980
Tournepierre à collier, 686
Tourterelle des bois, 856
 maillée, 860
 masquée, 861
 orientale, 859
 rieuse, 852
 triste, 864
 turque, 853
Traquet brun, 1174
 à capuchon, 1194
 de Chypre, 1183
 du désert, 1185
 deuil, 1193
 de Finsch, 1187
 de Hume, 1195
 isabelle, 1175
 motteux, 1178
 oreillard, 1183
 pie, 1180
 à queue noire, 1166
 à queue rousse, 1190
 rieur, 1198
 à tête blanche, 1196
 à tête grise, 1188
 variable, 1191
Travailleur à bec rouge, 1528
Troglodyte mignon, 1122
Turnix mugissant, 482

Vacher à tête brune, 1691
Vanneau éperonné, 587
 huppé, 593
 indien, 589
 à queue blanche, 592
 sociable, 590
 à tête noire, 589
Vautour charognard, 314
 chaugoun, 314
 fauve, 315
 moine, 321
 oricou, 318
 percnoptère, 311
 de Rüppell, 318
Venturon montagnard, 1556
Verdier d'Europe, 1557
Viréo à gorge jaune, 1536
 à oeil rouge, 1537
 de Philadelphie, 1537

Pages 1–1008 appear in Volume 1, pages 1009–1694 in Volume 2

NOMI IN ITALIANO

Airone azzurro americano, 131
  azzurro minore, 118
  bianco maggiore, 126
  bianco mezzano, 125
  cenerino, 128
  golia, 134
  guardabuoi, 116
  della Luisiana, 120
  rosso, 132
  schienaverde, 111
  schistaceo, 121
  testanera, 127
  verde, 113
Albanella minore, 337
  pallida, 335
  reale, 331
Albastrello, 668
Albatros beccogiallo, 29
  cauto, 29
  dai piedi neri, 30
  dal sopracciglio nero, 27
  reale, 31
  testagrigia, 30
  urlatore, 30
Alca impenne, 815
  minore crestata, 819
  minore pappagallo, 820
Allocco, 907
  di Hume, 910
  di Lapponia, 913
  degli Urali, 911
Allodola, 1043
  beccocurvo, 1019
  beccoforte, 1022
  capinera, 1013
  cornuta africana, 1052
  del deserto, 1017
  del deserto codafasciata, 1015
  di Dunn, 1014
  del Dupont, 1020
  golagialla, 1047
  orientale, 1046
  di Razo, 1047
  testacastana, 1012
Alzavola, 215
  asiatica, 214
  del Capo, 217
Amaranto beccorosso, 1529
Anatra beccorosso, 225
  falcata, 210
  mandarina, 204
  marmorizzata, 232
  nera americana, 222
  sposa, 204
Anhinga africana, 91
Aquila anatraia maggiore, 365
  anatraia minore, 363
  del Bonelli, 383
  del Bonelli, 383

imperiale, 371
  imperiale spagnola, 375
  di mare, 303
  di mare del Pallas, 302
  minore, 381
  rapace, 368
  reale, 376
  delle steppe, 370
  urlatrice africana, 302
  di Verreaux, 380
Assiolo, 891
  di Bruce, 890
Astore, 342
  cantante scuro, 340
  gabar, 342
Astrilde, 1531
Averla bruna, 1430
  capirossa, 1445
  cenerina, 1436
  dal dorso rossiccio, 1436
  isabellina, 1431
  maggiore, 1440
  maggiore meridionale, 1442
  mascherata, 1447
  piccola, 1433
Avocetta, 542
Avvoltoio monaco, 321
  orecchiuto, 318

Balestruccio, 1066
Balia caucasica, 1355
  dal collare, 1357
  nera, 1358
Ballerina bianca, 1103
  gialla, 1100
  nera africana, 1106
Barbagianni, 886
Basettino, 1362
Beccaccia, 644
  dorata, 532
  di mare, 535
  di mare delle Canarie, 538
Beccaccino, 635
  codastretta, 640
  di Swinhoe, 642
Beccafico, 1314
Beccamoschino, 1237
Beccapesci, 773
  di Berg, 770
Becco d'argento africano, 1534
  d'argento indiano, 1534
  a cesoie africano, 804
Beccofrusone, 1113
Beccogrosso azzurro, 1688
  pettorosa, 1686
Bengalino, 1533
Berta dell'Atlantico, 47
  di Audubon, 56
  di Bulwer, 41
  grigia, 49

di Jouanin, 42
  maggiore, 42
  maggiore di Edwards, 45
  minore, 51
  minore delle Baleari, 53
  minore fosca, 55
  minore mediterranea, 54
  del Pacifico, 46
  piedicarnicini, 47
  striata, 46
Biancone, 324
Bigia grossa, 1305
  del mar Rosso, 1303
  padovana, 1306
Bigiarella, 1308
Bobolink, 1690
Bulbul capinero, 1110
  golanera, 1111
  guancebianche, 1109
  dal sottocoda rosso, 1112

Calandra, 1024
  asiatica, 1026
  nera, 1029
  siberiana, 1027
Calandrella, 1032
  di Hume, 1034
  della Mongolia, 1036
Calandro, 1072
  maggiore, 1070
Calliope, 1147
Canapiglia, 212
Canapino, 1284
  asiatico, 1277
  levantino, 1280
  maggiore, 1282
  pallido, 1273
  di Upcher, 1278
Canarino, 1554
Cannaiola, 1265
  di Blyth, 1260
  di Capo Verde, 1262
  di Jerdon, 1258
  verdognola, 1263
Cannareccione, 1269
  beccoforte, 1273
  stentoreo, 1267
Capinera, 1316
Capovaccaio, 311
  pileato, 314
Cappellaccia, 1037
  di Thekla, 1040
Cardellino, 1561
Casarca, 219
Cavaliere d'Italia, 539
Cesena, 1222
  di Naumann, 1219
Chiurlo boreale, 654
  maggiore, 658
  minore, 653

piccolo, 655
Chiurlottello, 657
Ciagra del Senegal, 1428
Cicogna bianca, 140
  nera, 138
Cigno minore, 166
  nero, 165
  reale, 162
  selvatico, 168
Cincia bigia, 1375
  bigia alpestre, 1379
  dal ciuffo, 1383
  dalmatina, 1378
  mora, 1385
  siberiana, 1382
Cinciallegra, 1393
Cinciarella, 1388
  azzurra, 1392
Ciuffolotto, 1606
  delle Azzorre, 1609
  del Pallas, 1600
  delle pinete, 1602
  scarlatto, 1596
  scarlatto maggiore, 1600
  siberiano, 1606
  del Sinai, 1599
Civetta, 903
  capogrosso, 923
  nana, 901
Codanera, 1241
Codazzurro, 1153
Codibugnolo, 1372
Codirosso, 1161
  algerino, 1163
  americano, 1625
  di Eversmann, 1155
  di Güldenstädt, 1164
  spazzacamino, 1157
Codirossone, 1201
Codone, 223
Colino della California, 447
  della Virginia, 448
Colomba di Bolle, 848
  dei lauri, 850
  migratrice, 865
  trocaz, 848
Colombaccio, 846
Colombella, 842
  occhigialli, 845
Combattente, 628
Cordirossone, 1201
Cordon blu guancerosse, 1530
Cormorano, 80
  africano, 89
  di Socotra, 86
Cornacchia, 1478
  delle case, 1472
Corriere asiatico, 575
  grosso, 564

Pages 1–1008 appear in Volume 1, pages 1009–1694 in Volume 2

di Kittlitz, 567
di Leschenault, 574
della Mongolia, 572
piccolo, 561
semipalmato, 566
dai tre collari, 569
vocifero, 567
Corrione biondo, 552
Corvo, 1475
    bianco e nero, 1481
    coda a ventaglio, 1486
    del collobruno, 1481
    imperiale, 1483
Coturnice, 453
    orientale, 452
Crociere, 1582
    fasciato, 1580
    delle pinete, 1587
    di Scozia, 1586
Croccolone, 638
Cuculo, 875
    americano, 882
    americano occhirossi, 882
    bianco e nero, 872
    dal ciuffo, 872
    dorato di Levaillant, 874
    orientale, 878
    del Senegal, 884
Culbianco, 1178
    isabellino, 1175
Cutrettola, 1094
    testagialla orientale, 1098

Damigella di Numidia, 516
Dendrocigna facciabianca, 160
    fulva, 160
    indiana, 161
Dendroica di Blackburn, 1620
    di Blackpoll, 1623
    blu gola nera, 1620
    di Capo May, 1621
    coronata, 1623
    fianchicastani, 1619
    gialla, 1618
    delle magnolie, 1622
    delle palme, 1623
    verdastra, 1620
Droma, 544

Edredone, 251
    dagli occhiali, 255
    di Steller, 256
Euristomo africano, 974

Fagiano comune, 472
    dorato, 475
    di Lady Amherst, 476
    di monte, 437
    di monte del Caucaso, 440
    venereto, 471
Falaropo beccolargo, 692
    beccosottile, 691
    di Wilson, 689

Falco cuculo, 397
    giocoliere, 326
    di palude, 328
    pellegrino, 419
    pecchiaiolo, 290
    pecchiaiolo orientale, 293
    pescatore, 386
    della regina, 407
    unicolore, 409
Falcone di Barberia, 424
Fanello, 1568
    nordico, 1571
Faraona, 479
Febe orientale, 1010
Fenicottero, 153
    minore, 156
Fetonte beccorosso, 70
Fiorrancino, 1346
    americano, 1348
Fischione, 207
    americano, 210
Fistione turco, 236
Folaga, 506
    americana, 509
    cornuta, 509
Forapaglie, 1255
    castagnolo, 1251
    macchiettato, 1245
Francolino, 461
    dal doppio sperone, 463
    di Erckel, 464
    di monte, 428
Fraticello, 790
    di Saunders, 793
Fratino, 569
Fregata magnifica, 99
Fringuello, 1539
    alpino, 1522
    delle Canarie, 1544
Frosone, 1610
    codanera, 1610
    mascherato, 1610
    vespertino, 1613
Frullino, 633
Fulmaro, 34

Gabbiano di Armenia, 748
    di Bonaparte, 723
    comune, 724
    corallino, 714
    corso, 732
    eburneo, 761
    di Franklin, 718
    glauco, 752
    di Hemprich, 709
    d'Islanda, 749
    occhibianchi, 711
    del Pallas, 712
    reale, 746
    reale nordico, 741
    roseo, 729
    di Ross, 757
    di Sabine, 721

sghignazzante, 716
testabruna, 727
testagrigia, 727
tridattilo, 758
Gabbianello, 719
Gallina prataiola, 519
Gallinella d'acqua, 499
    nera, 496
Gallo cedrone, 442
Gambecchio, 607
    americano, 612
    di Baird, 614
    collorosso, 607
    frullino, 624
    minore, 612
    nano, 610
Ganga, 832
    ventrecastano, 831
Garrulo arabo, 1368
    comune, 1367
    fulvo, 1370
    iracheno, 1365
Garzetta, 122
    ardesia, 119
    nivea, 120
Gavina, 735
    del Delaware, 734
Gazza, 1457
    azzurra, 1456
    marina, 812
    marina minore, 817
Gendarme, 1526
Germano reale, 218
Gheppio, 393
    americano, 396
Ghiandaia, 1450
    marina, 970
    marina abissina, 972
    marina indiana, 973
    siberiana, 1454
Gipeto, 308
Girfalco d'Islanda, 417
Gobbo della Giamaica, 282
    rugginoso, 285
Gracchio alpino, 1464
    americano, 1692
    corallino, 1466
Grandule, 834
    coronata, 828
    di Lichtenstein, 826
    del Senegal, 829
Grifone, 315
    del Bengala, 314
    di Rüppell, 318
Grillaio, 391
Gru, 511
    canadese, 514
    siberiana, 515
Gruccione, 966
    egiziano, 964
    verde, 963
Guardiano dei Coccodrilli, 550

Gufo comune, 915
    delle nevi, 897
    di palude, 918
    di palude africano, 921
    pescatore bruno, 896
    reale, 893

Ibis eremita, 146
    sacro, 148
Ipocolio, 1116
Ittero alirosse, 1693
    di Baltimora, 1694
    testagialla, 1693
    di Wagler, 1693

Labbo, 699
    codalunga, 701
Lanario, 411
Locustella fluviatile, 1247
    di Gray, 1251
    lanceolata, 1244
    di Pallas, 1243
Lodolaio, 404
Lucherino, 1564
    dei pini, 1568
Luì bianco, 1329
    boreale, 1322
    di Brooks, 1325
    coronato di Temminck, 1319
    forestiero, 1325
    grosso, 1340
    grosso orientale, 1335
    del Pallas, 1324
    piccolo, 1337
    di Radde, 1327
    scuro, 1328
    verdastro, 1320
    verde, 1333

Magnanina, 1288
    sarda, 1286
Maina comune, 1500
Marabú, 143
Marangone africano, 89
    dal ciuffo, 84
    dalla doppia cresta, 83
    minore, 87
Martin pescatore, 956
    pescatore bianco e nero, 959
    pescatore fasciato, 960
    pescatore di Smirne, 953
    pescatore testagrigia, 954
Marzaiola, 226
    americana, 228
Merlo, 1215
    acquaiolo, 1118
    americano, 1692
    dal collare, 1212
Mestolone, 230
Migliarino del Pallas, 1676
    di palude, 1672
Mignattaio, 144
Mignattino, 796

alibianche, 799
piombato, 794
Mimo poliglotta, 1125
rossiccio, 1126
Ministro, 1689
Molotro testabruna, 1691
Monachella, 1183
dal cappuccio, 1194
di Cipro, 1183
codarossa, 1190
del deserto, 1185
dorsonero, 1180
di Finsch, 1187
di Hume, 1195
lamentosa, 1193
nera, 1198
nera testabianca, 1196
testagrigia, 1188
variabile, 1191
Moretta, 244
arlecchino, 259
codona, 261
dal collare, 241
grigia, 247
grigia minore, 249
tabaccata, 242
Moriglione, 238
dorsobianco, 240
Mugnaiaccio, 754

Nettarina metallica, 1421
della Palestina, 1422
pigmea, 1419
Nibbio bianco, 294
bruno, 295
reale, 298
Nitticora, 109
Nocciolaia, 1460

Oca del Canada, 185
collorosso, 192
colombaccio, 189
egiziana, 195
facciabianca, 187
granaiola, 170
indiana, 182
lombardella, 175
lombardella minore, 178
delle nevi, 182
pigmea indiana, 203
di Ross, 184
selvatica, 179
dallo sperone, 203
zamperosee, 173
Occhiocotto, 1296
di Cipro, 1299
di Ménétries, 1295
Occhione, 546
del Senegal, 548
Orcheto marino, 263
Orco marino, 266
dagli occhiali, 266
Organetto, 1574

artico, 1578
Ortolano, 1659
grigio, 1663
Ossifraga, 32
Otarda, 529
araba, 527
di Denham, 522
nubiana, 522

Pagliarolo, 1254
Pantana, 670
Papa lazuli, 1689
della Luisiana, 1689
Parrocchetto dal collare, 866
monaco, 867
Parula americana, 1617
bianca e nera, 1615
del Canada, 1629
dal cappuccio, 1627
golagialla, 1627
di Wilson, 1628
Passera, 1503
di Capo Verde, 1511
del deserto, 1511
dorato, 1515
lagia, 1519
lagia indiana, 1518
lagia pallida, 1517
del Mar Morto, 1509
mattugia, 1513
sarda, 1506
scopaiola, 1128
scopaiola asiatica, 1131
Passerella variabile, 1636
Passero calandra, 1635
dei campi, 1635
cantore, 1637
corona bianca, 1637
golabianca, 1639
delle paludi, 1637
delle praterie, 1635
solitario, 1204
Pavoncella, 593
armata, 587
codabianca, 592
gregaria, 590
indiana, 589
testanera, 589
Pellicano, 93
riccio, 96
rossiccio, 98
Pendolino, 1416
Peppola, 1545
Pernice bianca, 434
bianca nordica, 430
del deserto, 459
di mare, 555
di mare asiatica, 557
di mare orientale, 557
rossa, 455
delle sabbie, 460
sarda, 457
Pesciaiola, 274

Pettazzurro, 1149
Pettegola, 665
Pettirosso, 1140
golabianca, 1154
Picchio cenerino, 983
dorato, 983
dorsobianco, 1000
di Levaillant, 988
muraiolo, 1408
muratore, 1402
muratore algerino, 1400
muratore corso, 1399
muratore di Krüper, 1398
muratore pettofulvo, 1401
muratore di roccia, 1406
muratore di roccia orientale, 1404
nero, 989
rosso maggiore, 993
rosso mezzano, 998
rosso minore, 1002
rosso di Siria, 996
tridattilo, 1005
ventregiallo, 992
verde, 986
Piccione selvatico, 839
Pigliamosche, 1349
bruno asiatico, 1349
pettirosso, 1353
Piovanello, 617
beccosottile, 599
maggiore, 599
pancianera, 620
tridattilo, 602
violetto, 618
Pipilo fianchirossi, 1633
Piro piro asiatico, 684
piro boschereccio, 676
piro codalunga, 662
piro culbianco, 674
piro dorsobianco, 613
piro fulvo, 628
piro macchiato, 683
piro occidentale, 606
piro pettorale, 615
piro pettorossiccio, 643
piro pettorossiccio minore, 642
piro piccolo, 680
piro semipalmato, 605
piro siberiano, 615
piro solitario, 673
piro Terek, 678
piro zampelunghe, 627
Pispola, 1082
beccolungo, 1076
di Berthelot, 1075
golarossa, 1084
della Pechora, 1081
Pispoletta, 1034
Pittima di Hudson, 650
minore, 650
reale, 647
Piviere dorato, 581

dorato del Pacifico, 580
orientale, 579
tortolino, 577
Pivieressa, 584
Podilimbo, 13
Poiana, 353
degli altipiani, 360
calzata, 360
codabianca, 359
spallerosse, 353
di Swainson, 353
Porciglione, 484
Pollo sultano, 504
sultano di Allen, 502
sultano americano, 502
Prinia gracile, 1239
Prispolone, 1079
di Blyth, 1072
indiano, 1077
Procellaria del Capo, 36
Pterodroma alibianche, 37
dal cappuccio, 41
di Kermadec, 37
di Madeira, 37
di Schlegel, 40
Pulcinella dai ciuffi, 825
di mare, 821

Quaglia, 467
tridattila, 482
Quattrocchi, 272
d'Islanda, 269
minore, 269
Quelea, 1528

Rampichino, 1414
alpestre, 1411
Re degli edredoni, 253
Re di quaglie, 496
Regolo, 1342
americano, 1342
di Tenerife, 1345
Rigogolo, 1424
Rondine, 1061
arboricola, 1053
etiopica, 1064
montana, 1059
rossiccia, 1064
rupestre africana, 1058
rupestre americana, 1066
Rondone, 943
di Alexander, 941
cafro, 947
codaforcuta, 947
indiano, 949
maggiore, 940
pallido, 945
delle palme africano, 950
spinoso, 939
spinoso dei camini, 939
unicolore, 942

Sacro, 414

Salciaiola, 1249
Saltimpalo, 1170
    delle Canarie, 1169
    nero e bianco, 1173
Sassicola codanera, 1166
    mangiaformiche, 1174
Schiribilla, 490
    grigiata, 494
    striata, 496
Scricciolo, 1122
Seiuro coronato, 1625
    del nord, 1626
Sgarza cinese, 115
    ciuffetto, 113
    indiana, 115
Shikra, 348
Silvia del Rüppel, 1300
Sirratte, 836
Smergo dal ciuffo, 274
    maggiore, 279
    minore, 277
Smeriglio, 400
Sordone, 1134
    golanera, 1133
    di Radde, 1132
Sparviere, 345
    levantino, 349
Spatola, 149
    africana, 152
Spioncello, 1086
    marino, 1089
    del Pacifico, 1092
Spiza americana, 1686
Starna, 464
    asiatica, 467
Stercorario maggiore, 704
    di McCormick, 708
    mezzano, 696
Sterna aleutina, 784
    codalunga, 782
    comune, 779
    di Dougall, 777
    elegante, 776
    di Forster, 785
    guancebianche, 785
    maggiore, 766
    reale, 769
    dalle redini, 787
    di Rüppel, 772
    scura, 789

stolida bruna, 802
stolida nera, 803
zampenere, 764
Sterpazzola, 1310
    nana, 1301
    di Sardegna, 1291
    di Tristram, 1290
Sterpazzolina, 1293
Stiaccino, 1167
Stillozzo, 1681
Storno, 1492
    della Cina, 1491
    di Dauria, 1491
    nero, 1496
    roseo, 1498
    di Tristram, 1489
Strolaga beccogiallo, 10
    maggiore, 8
    mezzana, 6
    minore, 3
Struzzo, 1
Sturnella orientale, 1692
Succiacapre, 929
    collorosso, 932
    dorato, 933
    isabellino, 933
    della Nubia, 928
    sparviero, 936
Sula, 77
    del Capo, 79
    fosca, 74
    mascherata, 73
    piedirossi, 73
Svasso collorosso, 20
    cornuto, 22
    maggiore, 17
    maggiore beccolungo, 17
    piccolo, 24

Tacchino, 481
Taccola, 1468
    di Dauria, 1471
Tanagra estiva, 1631
    scarlatta, 1632
Tantalo africano, 137
Tarabusino, 105
    americano, 104
    nano, 108
    orientale, 107
Tarabuso, 101
    americano, 104
Tessitore striato, 1526

Tetraogallo del Caspio, 451
    del Caucaso, 449
Tiranno acadico, 1009
Topino, 1055
    africano, 1054
    dai sopraccigli bianchi, 1058
Torcicollo, 980
Tordela, 1230
Tordo di Baird, 1211
    boschereccio, 1209
    bottaccio, 1225
    dorato, 1206
    eremita, 1210
    golanera, 1220
    migratore, 1234
    oscuro, 1218
    sassello, 1228
    siberiano, 1207
    di Swainson, 1210
    di Tickell, 1212
    usignolo bruno, 1211
    vario, 1208
Tortora, 856
    dal collare africana, 852
    dal collare orientale, 853
    lamentosa americana, 864
    maschera di ferro, 861
    orientale, 859
    delle palme, 860
Totano moro, 662
    semipalmato, 685
    zampegialle maggiore, 672
    zampegialle minore, 672
Tottavilla, 1041
Trombettiere, 1594
    alirosse, 1589
    di Lichtenstein, 1591
    mongolo, 1592
Tuffetto, 14

Ubara, 525
Uccello gatto, 1127
    delle tempeste, 62
    delle tempeste di Castro, 67
    delle tempeste codaforcuta, 64
    delle tempeste facciabianca, 60
    delle tempeste di Swinhoe, 66
    delle tempeste ventrebianco, 62
    delle tempeste di Wilson, 58
Ulula, 899

Upupa, 976
Uria, 806
    di Brünnich, 810
    minore, 820
    nera, 815
Usignolo, 1145
    d'Africa, 1137
    azzurro siberiano, 1152
    di fiume, 1235
    maggiore, 1143
    podobè, 1139

Venturone, 1556
Verdone, 1557
Vermivora alidorate, 1615
    del Tennessee, 1616
Verzellino, 1550
    fronterossa, 1548
    di Siria, 1552
Vireo di Filadelfia, 1537
    fronte gialla, 1536
    occhirossi, 1537
Volpoca, 198
Voltapietre, 686
Voltolino, 488
    americano, 490

Zafferano, 737
Zigolo ardesia, 1639
    boschereccio, 1665
    capinero, 1679
    capobianco, 1653
    delle case, 1655
    cenerino, 1658
    dal collare, 1670
    collogrigio, 1662
    giallo, 1648
    golarossa, 1646
    di Lapponia, 1640
    mascherato, 1645
    minore, 1667
    muciatto, 1653
    muciatto orientale, 1655
    nero, 1651
    delle nevi, 1642
    nocciola, 1669
    pettocannella, 1659
    dal sopracciglio giallo, 1664
    testa aranciata, 1677

NEDERLANDSE NAMEN

Aalscholver, 80
   Arabische, 86
   Geoorde, 83
Aasgier, 311
Alaska-strandloper, 606
Albatros, Grote, 30
Aleoetenstern, 784
Alk, 812
   Kleine, 817
Alpengierzwaluw, 940
Alpenheggemus, 1134
Alpenkauw, 1464
Alpenkraai, 1466
Alpensneeuwhoen, 434
Appelvink, 1610
Arend, Zwarte, 380
Arendbuizerd, 359
Atlasgrasmus, 1290
Auerhoen, 442
Avonddikbek, 1613
Azoren-goudvink, 1609
Azuurmees, 1392

Baardgrasmus, 1293
Baardman, 1362
Baardpatrijs, 467
Babbelaar, Arabische, 1368
   Bruingele, 1370
   Irakese, 1365
   Gewone, 1367
Balkanbergfluiter, 1331
Balkansperwer, 349
Balkanvliegenvanger, 1355
Baltimore-troepiaal, 1694
Bandijsvogel, 960
Barmsijs, 1574
Basra-karekiet, 1271
Bastaardarend, 365
Bateleur, 326
Beflijster, 1212
Bergeend, 198
Bergfluiter, 1329
Bergheggemus, 1131
Berghoen, Kaspisch, 451
   Kaukasisch, 449
Bergleeuwerik, 1026
Beringmeeuw, 752
Bijeneter, 966
   Groene, 964
   Kleine Groene, 963
Bisschop, Blauwe, 1688
Bladkoning, 1325
   Brooks', 1325
   Hume's, 1327
Blauwborst, 1149
Blauwfazantje, 1530
Blauwnekmuisvogel, 869
Blauwstaart, 1153
Blauwvleugeltaling, 228

Bobolink, 1690
Bobwhite, 448
Boerenzwaluw, 1061
Bokje, 633
Bontbek, Amerikaanse, 566
Bontbekplevier, 564
Boomklever, 1402
   Algerijnse, 1400
   Canadese, 1401
   Corsicaanse, 1399
   Turkse, 1398
Boomkruiper, 1414
Boomleeuwerik, 1041
Boompieper, 1079
   Groene, 1077
Boomvalk, 404
Boomzwaluw, 1053
Bosgors, 1665
Boskoekoek, 878
Boslijster, Amerikaanse, 1209
Bosrietzanger, 1263
Bosruiter, 676
   Amerikaanse, 673
Bosuil, 907
   Palestijnse, 910
Boszanger, Bruine, 1328
   Noordse, 1322
   Pallas', 1324
   Radde's, 1327
Braamsluiper, 1308
Brandgans, 187
Breedbekstrandloper, 624
Brilduiker, 272
   IJslandse, 269
Brileider, 255
Brilgrasmus, 1291
Brilstern, 786
Brilzee-eend, 266
Bronskopeend, 210
Bruinkeelortolaan, 1663
Bruinkopgors, 1677
Bruinkopkoevogel, 1691
Bruinkopmees, 1382
Bruinkopmeeuw, 727
Bruinnekraaf, 1481
Bruinruggoudmus, 1515
Buffelkopeend, 269
Buidelmees, 1416
Buizerd, 353
   Mongoolse, 360
Burgemeester, Grote, 752
   Kleine, 749
Buulbuul, Arabische, 1111
   Grauwe, 1111

Carolina-eend, 204
Casarca, 197
Cirlgors, 1651
Citroenkanarie, 1556

Citroenkwikstaart, 1098
Coromandeleend, 203
Cyprusgrasmus, 1299
Cyprus-tapuit, 1183

Dennensijs, 1568
Diadeemroodstaart, 1163
Dickcissel, 1686
Diederikkoekoek, 874
Dikbekfuut, 13
Diksnavelleeuwerik, 1022
Diksnavelrietzanger, 1273
Dodaars, 14
Donsstormvogel, 38
Draaihals, 980
Driebandplevier, 569
Drieteenmeeuw, 758
Drieteenspecht, 1005
Drieteenstrandloper, 602
Duinpieper, 1072
Dunbekmeeuw, 729
Dunbekwulp, 657
Dwergaalscholver, 87
   Afrikaanse, 89
Dwergarend, 381
Dwergflamingo, 156
Dwerggans, 178
Dwerggors, 1667
Dwerglijster, 1210
Dwergmeeuw, 719
Dwergooruil, 891
   Gestreepte, 890
Dwergstern, 790
   Saunders, 793
Dwergtjiftjaf, 1335
Dwerguil, 901
Dwergwulp, 653

Eend, Wilde, 218
   Zwarte, 222
Eider, Stellers, 256
Eidereend, 251
Ekster, 1457
   Blauwe, 1456
Elftiran, Groene, 1009
Epauletspreeuw, 1693
Eskimowulp, 654

Fazant, 472
Fitis, 1340
   Grauwe, 1320
Flamingo, 153
Fluiteend, Indische, 161
   Rosse, 160
Fluiter, 1333
Franjepoot, Grauwe, 691
   Grote, 689
   Rosse, 692
Frankolijn, Barbarijse, 463
   Zwarte, 461

Frater, 1571
Fregatvogel, Amerikaanse, 99
Fuut, 17
   Geoorde, 24
   Grote, 17

Gaai, 1450
Gabarhavik, 342
Gans, Canadese, 185
   Grauwe, 179
   Indische, 182
   Ross', 184
Geelbekscharrelaar, 974
Geelborstvireo, 1536
Geelbrauwgors, 1664
Geelbuiksapspecht, 992
Geelgors, 1648
Geelkoptroepiaal, 1693
Geelneusalbatros, 29
Geelpootmeeuw, 746
Geelpootruiter, Grote, 672
   Kleine, 672
Geelsnavelduiker, 10
Geelsnavelkoekoek, 882
Geelstuitzanger, 1623
Geelvleugelzanger, 1616
Gent, Bruine, 74
Gier, Bengaalse, 314
   Rüppells, 318
   Vale, 315
Giervalk, 417
Gierzwaluw, 943
   Kaapverdische, 941
   Siberische, 947
   Vale, 945
Glanskop, 1375
Glanstroepiaal, 1692
Gors, Grauwe, 1681
   Grijze, 1653
   Rosse, 1669
   Stewarts, 1653
Goudfazant, 475
Goudhaantje, 1342
   Amerikaans, 1348
Goudlijster, 1206
Goudplevier, 581
   Amerikaanse, 579
   Kleine, 580
Goudsnip, 532
Goudvink, 1606
Graspieper, 1082
Grasmus, 1310
   Rüppells, 1300
   Sardijnse, 1286
Graszanger, 1237
Griel, 546
   Senegalese, 548
Grijskopalbatros, 30
Grijskopijsvogel, 954

Pages 1–1008 appear in Volume 1, pages 1009–1694 in Volume 2

[32] Nederlandse namen

Grijskopmeeuw, 727
Grijskopspecht, 983
Grijsrugklapekster, 1444
Grijswangdwerglijster, 1211
Groenling, 1557
Groenpootruiter, 670
Grondspecht, Gele, 983
Grutto, 647
  Rode, 650
  Rosse, 650

Haakbek, 1602
Halsbandparkiet, 866
Harlekijneend, 259
Havik, 342
Havikarend, 383
Hazelhoen, 428
Heggemus, 1128
Herdersplevier, 567
Heremietibis, 146
Heremietlijster, 1210
Holenduif, 842
  Oosterse, 845
Honingzuiger, Kleine, 1419
  Palestijnse, 1422
Hop, 976
Houtduif, 846
Houtsnip, 644
Huisgierzwaluw, 949
Huisgors, 1655
Huiskraai, 1472
Huismus, 1503
Huiszwaluw, 1066

Ibis, Heilige, 148
  Zwarte, 144
IJsduiker, 8
IJseend, 261
IJsgors, 1640
IJsvogel, 956
  Bonte, 959
Indigogors, 1689
Isabelklauwier, 1431
Isabeltapuit, 1175
Isabeltortel, 852
Ivoormeeuw, 761

Jager, Grote, 704
  Kleine, 699
  Kleinste, 701
  Middelste, 696
Jakobijnkoekoek, 872
Jan-van-Gent, 77
  Kaapse, 79
Jufferkraan, 516
Junco, Grijze, 1639

Kaffergierzwaluw, 947
Kalanderleeuwerik, 1024
Kalkoen, 481
Kanarie, 1554
  Europese, 1550
  Syrische, 1552

Kanoet, Grote, 599
Kanoetstrandloper, 599
Kapgier, 314
Kapzanger, 1627
Karekiet, Chinese, 1272
  Grote, 1269
  Indische, 1267
  Kleine, 1265
Katvogel, 1127
Kaukasus-tjiftjaf, 1336
Kauw, 1468
  Daurische, 1471
Keep, 1545
Keizerarend, 371
  Spaanse, 375
Kemphaan, 628
Kerkuil, 886
Kermadec-stormvogel, 37
Kiekendief, Blauwe, 331
  Bruine, 328
  Grauwe, 337
Kievit, 593
  Indische, 589
Killdeerplevier, 567
Klapekster, 1440
  Kleine, 1436
Klauwier, Bruine, 1430
  Grauwe, 1433
Klifzwaluw, Amerikaanse, 1066
Kluut, 542
Kneu, 1568
Knobbelmeerkoet, 509
Knobbelzwaan, 162
Koekoek, 875
Koereiger, 116
Kokardezaagbek, 274
Kokmeeuw, 724
  Kleine, 723
Kolgans, 175
Koningsalbatros, 31
Koningseider, 253
Koningsfazant, 471
Koningsstern, 769
Koolmees, 1393
Koperwiek, 1228
Kordofanleeuwerik, 1011
Korhoen, 437
  Kaukasisch, 440
Kortbekzeekoet, 810
Kortteenleeuwerik, 1032
  Kleine, 1034
  Mongoolse, 1036
Kraagtrap, 525
Kraai, 1478
Kraanvogel, 511
  Canadese, 514
  Siberische Witte, 515
Krabplevier, 544
Krakeend, 212
Kramsvogel, 1222
Krekelprinia, 1241
Krekelzanger, 1247
  Grote, 1251

Kroeskoppelikaan, 96
Krokodilwachter, 550
Krombekstrandloper, 617
Kroonboszanger, 1319
Krooneend, 236
Kroonzandhoen, 828
Kruisbek, 1582
  Grote, 1587
  Schotse, 1586
Kuifaalscholver, 84
Kuifalk, 819
Kuifduiker, 22
Kuifeend, 244
Kuifkoekoek, 872
Kuifkwartel, Californische, 447
Kuifleeuwerik, 1037
Kuifmees, 1383
Kuifpapegaaiduiker, 825
Kuifstern, Californische, 776
  Grote, 770
Kwak, 109
Kwartel, 467
Kwartelkoning, 496
Kwikstaart, Afrikaanse Witte, 1106
  Gele, 1094
  Grote Gele, 1100
  Witte, 1103

Lachmeeuw, 716
Lachstern, 764
Lady Amherstfazant, 476
Lammergier, 308
Langsnavelpieper, 1076
Langstaartklauwier, 1436
Langstaartroodmus, 1606
Lannervalk, 411
Laplanduil, 913
Laurierduif, 850
  Bolle's, 848
Lazuligors, 1689
Leeuwerik, Dunns, 1014
  Duponts, 1019
  Thekla, 1040
  Tibetaanse, 1034
  Zwarte, 1029
Lepelaar, 149
  Afrikaanse, 152
Lijster, Bonte, 1208
  Bruine, 1219
  Grote, 1230
  Naumanns, 1219
  Siberische, 1207
  Tickells, 1212
  Vale, 1218
Loodbekje, 1534

Madeira-gierzwaluw, 942
Madeira-stormvogel, 37
Madeira-stormvogeltje, 67
Magnolia-zanger, 1622
Mandarijneend, 204
Mandarijnspreeuw, 1491
Mangrovereiger, 111

Mantelmeeuw, Grote, 754
  Kleine, 737
Manyarwever, 1526
Maquiszanger, 1241
Maraboe, Afrikaanse, 143
Marmereend, 232
Maskerdikbek, 1610
Maskerduif, 861
Maskergent, 73
Maskergors, 1645
Maskerklauwier, 1447
Maskerzanger, Gewone, 1627
Matkop, 1379
Meerkoet, 506
  Amerikaanse, 509
Mees, Zwarte, 1385
Meeuw, Armeense, 748
  Audouins, 732
  Franklins, 718
  Hemprichs, 709
  Ross', 757
Merel, 1215
Miertapuit, Bruine, 1174
Moabmus, 1509
Moerasgors, 1637
Moerassneeuwhoen, 430
Monniksgier, 321
Monniksparkiet, 867
Monnikstapuit, 1194
Morinelplevier, 577
Mus, Kaapverdische, 1511
  Spaanse, 1506

Nachtegaal, 1145
  Blauwe, 1152
  Noordse, 1143
Nachtzwaluw, 929
  Amerikaanse, 936
  Egyptische, 933
  Goudgele, 933
  Moorse, 932
  Nubische, 928
Nijlgans, 195
Nijlhoningzuiger, 1421
Nimmerzat, Afrikaanse, 137
Noddy, 802
Nonnetje, 274
Notenkraker, 1460

Oehoe, 893
Oeraluil, 911
Oeverloper, 680
  Amerikaanse, 683
Oeverpieper, 1090
Oeverzwaluw, 1055
  Vale, 1054
Ooievaar, 140
  Zwarte, 138
Oorgier, 318
Orpheusgrasmus, 1305
Orpheusspotvogel, 1284
Ortolaan, 1659
Ovenvogel, 1625

Pages 1–1008 appear in Volume 1, pages 1009–1694 in Volume 2

Nederlandse namen

Paapje, 1167
Palmgierzwaluw, Afrikaanse, 950
Palmtortel, 860
Palmzanger, 1623
Papegaai-alk, 820
Papegaaiduiker, 821
Parelduiker, 6
Parelhoen, 479
Patrijs, 464
    Barbarijse, 457
    Rode, 455
Pelikaan, Kleine, 98
    Roze, 93
Pestvogel, 1113
Petsjorapieper, 1081
Philadelphia-vireo, 1537
Phoebetiran, 1010
Picata-tapuit, 1191
Pieper, Berthelots, 1075
    Grote, 1070
    Mongoolse, 1072
Pijlstaart, 223
Pijlstormvogel, Audubons, 56
    Australische Grote, 47
    Gestreepte, 46
    Grauwe, 49
    Grote, 47
    Kaapverdische, 45
    Kleine, 55
    Kuhls, 42
    Noordse, 51
    Turkse, 54
    Vale, 53
Pimpelmees, 1388
Plevier, Kaspische, 575
    Kleine, 561
    Mongoolse, 572
Poelruiter, 668
Poelsnip, 638
Porseleinhoen, 488
    Afrikaans, 496
    Zwart, 496
Prairie-buizerd, 353
Prinia, Gestreepte, 1239
Provence-grasmus, 1288
Purpergors, 1689
Purperhoen, Afrikaans, 502
    Amerikaans, 502
Purperkoet, 504
Purpurreiger, 132
Putter, 1561

Raaf, 1483
Ralreiger, 113
    Chinees, 115
    Indische, 115
Ransuil, 915
Razo-leeuwerik, 1047
Regenwulp, 655
Reiger, Amerikaanse Blauwe, 131
    Blauwe, 128
    Groene, 113
    Kleine Blauwe, 118

Zwarte, 119
Renvogel, 552
    Temmincks, 554
Reuzenalk, 815
Reuzenreiger, 134
Reuzenstern, 766
Reuzenstormvogel, Noordelijke, 32
    Zuidelijke, 32
Reuzenzwartkopmeeuw, 712
Rietgans, 170
    Kleine, 173
Rietgors, 1672
    Pallas', 1676
Rietzanger, 1255
    Kaapverdische, 1262
Rifreiger, Westelijke, 121
Ringmus, 1513
Ringsnaveleend, 241
Ringsnavelmeeuw, 734
Roek, 1475
Roerdomp, 101
    Noordamerikaanse, 104
Roestflankzanger, 1619
Roodbekwever, 1528
Roodborst, 1140
    Perzische, 1154
Roodborstkardinaal, 1686
Roodborstlijster, 1234
Roodborsttapuit, 1170
    Canarische, 1169
    Zwarte, 1173
Roodbuikbuulbuul, 1112
Roodbuikzandhoen, 831
Roodflanktowie, 1633
Roodhalsfuut, 20
Roodhalsgans, 192
Roodkeelduiker, 3
Roodkeellijster, 1220
Roodkeelnachtegaal, 1147
Roodkeelpieper, 1084
Roodkeelstrandloper, 607
Roodkopklauwier, 1445
Roodkroonhaantje, 1342
Roodmus, 1596
    Grote, 1600
    Pallas', 1600
Roodoogvireo, 1537
Roodoorgors, 1635
Roodpootgent, 73
Roodpootvalk, 397
Roodschouderbuizerd, 353
Roodsnavelkeerkringvogel, 70
Roodsnavelpijlstaart, 225
Roodstaart, Amerikaanse, 1625
    Eversmanns, 1155
    Gekraagde, 1161
    Zwarte, 1157
Roodstaartgors, 1636
Roodstaarttapuit, 1190
Roodstuittapuit, 1188
Roodstuitzwaluw, 1064
Roodvoorhoofdkanarie, 1548

Rotgans, 189
Rotsduif, 839
Rotsklever, 1406
    Grote, 1402
Rotskruiper, 1408
Rotslijster, Blauwe, 1204
    Rode, 1201
Rotsmus, 1519
    Bleke, 1517
    Indische, 1518
Rotszwaluw, 1059
    Vale, 1058
Rouwmees, 1378
Rouwtapuit, 1193
Ruigpootbuizerd, 360
Ruigpootuil, 923
Ruiter, Bartrams, 662
    Blonde, 628
    Siberische Grijze, 684
    Zwarte, 662

Sahelscharrelaar, 972
Sakervalk, 414
Savannah-gors, 1635
Savanne-arend, 368
Schaarbek, Afrikaanse, 804
Scharrelaar, 970
    Indische, 973
Schildraaf, 1481
Scholekster, 535
    Canarische Zwarte, 538
Schoorsteengierzwaluw, 939
Schreeuwarend, 363
Shikra-sperwer, 348
Sijs, 1564
Sinaï-roodmus, 1599
Sint Helenafazantje, 1531
Slangehalsvogel, Afrikaanse, 91
Slangenarend, 324
Slechtvalk, 419
    Barbarijse, 424
Slobeend, 230
Smelleken, 400
Smient, 207
    Amerikaanse, 210
Smyrnagors, 1658
Smyrna-ijsvogel, 953
Sneeuwgans, 182
Sneeuwgors, 1642
Sneeuwuil, 897
Sneeuwvink, 1522
Snip, Grote Grijze, 643
    Kleine Grijze, 642
    Siberische, 642
Snor, 1249
Sora-ral, 490
Sparrezanger, 1620
Specht, Groene, 986
    Grote Bonte, 993
    Kleine Bonte, 1002
    Levaillants, 988
    Middelste Bonte, 998

Syrische Bonte, 996
    Zwarte, 989
Sperwer, 345
Sperwergrasmus, 1306
Sperweruil, 899
Spoorkoekoek, Senegalese, 884
Spoorwiekgans, 203
Sporenkievit, 587
Spotlijster, 1125
    Rosse, 1126
Spotvogel, 1282
    Griekse, 1280
    Grote Vale, 1278
    Kleine, 1277
    Vale, 1273
Spreeuw, 1492
    Daurische, 1491
    Roze, 1498
    Tristrams, 1489
    Zwarte, 1496
Sprinkhaanzanger, 1245
    Kleine, 1244
    Siberische, 1243
Staartmees, 1372
Steenarend, 376
Steenheggemus, 1132
Steenloper, 686
Steenortolaan, 1662
Steenpatrijs, 453
    Aziatische, 452
Steenuil, 903
Stekelstaart, Rosse, 282
Stekelstaartgierzwaluw, 939
Stekelstaartsnip, 640
Steltkluut, 539
Steltstrandloper, 627
Steppe-arend, 370
Steppeklapekster, 1442
Steppehoen, 836
Steppekiekendief, 335
Steppekievit, 590
Steppevorkstaartplevier, 557
Stern, Arabische, 785
    Bengaalse, 772
    Bonte, 788
    Dougalls, 777
    Forsters, 785
    Grote, 773
    Noordse, 782
    Zwarte, 796
Stormmeeuw, 735
Stormvogel, Bulwers, 41
    Goulds, 37
    Jouanins, 42
    Kaapse, 36
    Kaapverdische, 39
    Noordse, 34
    Schlegels, 40
Stormvogeltje, 62
    Bont, 60
    Chinees, 66
    Vaal, 64
    Wilsons, 58

Nederlandse namen

Strandleeuwerik, 1047
    Temmincks, 1052
Strandloper, Amerikaanse Kleine, 612
    Bairds, 614
    Bonapartes, 613
    Bonte, 620
    Gestreepte, 615
    Grijze, 605
    Kleine, 607
    Paarse, 618
    Siberische, 615
    Temmincks, 610
Strandplevier, 569
Struikrietzanger, 1260
Struisvogel, 1

Tafeleend, 238
    Grote, 240
Taigaboomkruiper, 1411
Taigagaai, 1454
Taiga-strandloper, 612
Taling, Kaapse, 217
    Siberische, 214
Tapuit, 1178
    Blonde, 1183
    Bonte, 1180
    Finsch', 1187
    Hume's, 1195
    Zwarte, 1198
Tenerife-goudhaantje, 1345
Tennessee-zanger, 1615
Terekruiter, 678
Textorwever, Grote, 1526
Tijgervinkje, 1533
Tijgerzanger, 1621
Tjiftjaf, 1337
Toppereend, 247
    Kleine, 249
Torenvalk, 393
    Amerikaanse, 396
    Kleine, 391
Tortel, Oosterse, 859
    Turkse, 853
Trap, Arabische, 527
    Denhams, 522
    Grote, 529
    Kleine, 519
    Nubische, 522
Trekduif, 865
Treurduif, 864
Treurmaina, 1500
Trocazduif, 848
Troepiaal, Waglers, 1693

Zwarte, 1692
Tuinfluiter, 1314
Tureluur, 665

Valk, Eleonora's, 407
Vechtkwartel, Gestreepte, 482
Veery, 1211
Veldgors, 1635
Veldleeuwerik, 1043
    Kleine, 1046
Veldrietzanger, 1258
Velduil, 918
    Afrikaanse, 921
Vink, 1539
    Blauwe, 1544
Vinkleeuwerik, Somalische, 1012
Visarend, 386
Visdief, 779
Visuil, Bruine, 896
Vliegenvanger, Bonte, 1358
    Bruine, 1349
    Grauwe, 1349
    Kleine, 1353
Vorkstaartmeeuw, 721
Vorkstaartplevier, 555
    Oosterse, 557
Vuurgoudhaantje, 1346
Vuurvinkje, 1529

Waaierstaart, Rosse, 1137
    Zwarte, 1139
Waaierstaartraaf, 1486
Waterhoen, 499
    Klein, 490
    Kleinst, 494
Waterlijster, Noordse, 1626
Waterpieper, 1086
    Pacifische, 1092
Waterral, 484
Waterrietzanger, 1254
Watersnip, 635
Waterspreeuw, 1118
Weidegors, 1655
Wenkbrauwalbatros, 27
Wespendief, 290
    Afrikaans, 108
    Asiatische, 293
Wielewaal, 1424
Wigstaartpijlstormvogel, 46
Wilgengors, 1670
Willet, 685
Winterkoning, 1122
Wintertaling, 215
Witbandkruisbek, 1580
Witbandleeuwerik, 1019
Witbandzeearend, 302
Witbrauwzwaluw, 1058

Witbuikreiger, 120
Witbuikstormvogeltje, 62
Witbuikzandhoen, 834
Witgatje, 674
Withalsvliegenvanger, 1357
Witkaakweidespreeuw, 1692
Witkapalbatros, 29
Witkapnoddy, 803
Witkeelgors, 1639
Witkopeend, 285
Witkopgors, 1646
Witkruingors, 1637
Witkruinroodstaart, 1164
Witkruintapuit, 1196
Witoogeend, 242
Witoogmeeuw, 711
Witoorbuulbuul, 1109
Witrugspecht, 1000
Witstaartkievit, 592
Witstuitbarmsijs, 1578
Witvleugeldikbek, 1610
Witvleugelleeuwerik, 1027
Witvleugelstern, 799
Witwangfluiteend, 160
Witwangstern, 794
Woestijngrasmus, 1301
Woestijnleeuwerik, 1017
    Rosse, 1015
Woestijnmus, 1511
Woestijnpatrijs, Arabische, 460
    Perzische, 459
Wocstijnplevier, 574
Woestijntapuit, 1185
Woestijnvalk, 409
Woestijnvink, 1594
    Mongoolse, 1592
    Rode, 1589
    Vale, 1591
Woestijnzandhoen, 829
Wouw, Grijze, 294
    Rode, 298
    Zwarte, 295
Wouwaapje, 105
    Afrikaans, 108
    Amerikaanse, 104
    Mandsjoerijs, 107
Wulp, 658

Zaagbek, Grote, 279
    Middelste, 277
Zandhoen, Lichtensteins, 826
Zanger, Blauwe, 1617
    Bonte, 1615
    Canadese, 1629

Cetti's, 1235
    Gekraagde Groene, 1620
    Gele, 1618
    Wilsons, 1628
Zanggors, 1637
Zanghavik, Donkere, 340
Zanglijster, 1225
Zeearend, 303
    Afrikaanse, 302
Zee-eend, Grote, 266
    Zwarte, 263
Zeekoet, 806
    Zwarte, 815
Zevenstrepengors, 1659
Zijdestaart, 1116
Zilveralk, 820
Zilverbekje, 1534
Zilvermeeuw, 741
Zilverplevier, 584
Zilverreiger, Amerikaanse Kleine, 120
    Grote, 126
    Kleine, 122
    Middelste, 125
Zomertaling, 226
Zomertangare, 1631
Zomertortel, 856
Zuidpooljager, 708
Zwaan, Kleine, 166
    Wilde, 168
    Zwarte, 165
Zwaluw, Ethiopische, 1064
Zwartbuikzandhoen, 832
Zwartkapstormvogel, 41
Zwartkeelheggemus, 1133
Zwartkeellijster, 1220
Zwartkeelzanger, Blauwe, 1620
Zwartkop, 1316
    Arabische, 1303
    Kleine, 1296
    Ménétries', 1295
Zwartkopgors, 1679
Zwartkopkievit, 589
Zwartkopmeeuw, 714
Zwartkopreiger, 127
Zwartkoprietzanger, 1251
Zwartkopzanger, 1623
Zwartkruintsjagra, 1428
Zwartkruinvinkleeuwerik, 1013
Zwartsnavelkoekoek, 882
Zwartstaart, 1166
Zwartevleugeltangare, 1632
Zwartvoetalbatros, 30

SVENSKA NAMN

Aftonfalk, 397
Aftonstenknäck, 1613
Akadtyrann, 1009
albatross, Gråhuvad, 30
   Gråkindad, 29
   Mindre, 29
   Svartbrynad, 27
   Svartfotad, 30
Alfågel, 261
Alförrädare, 256
Alkekung, 817
Alpjärnsparv, 1134
Alpkaja, 1464
Alpkråka, 1466
Alpseglare, 940
Altairödstjärt, 1155
Amarant, 1529
Amurstare, 1491
and, Rödnäbbad, 225
Arabpetrell, 42
Arabskriktrast, 1368
Arabsotmås, 709
Arabtrapp, 527
Aspsavsugare, 992
Atlantpetrell, 39
Atlassångare, 1290
Audubonlira, 56
Aztektärna, 776
Azordomherre, 1609
Azurmes, 1392

Bacchushäger, 115
Backsvala, 1055
   Brunstrupig, 1053
Balkanflugsnappare, 1355
Balkanhök, 349
Balkanmes, 1378
Balkanspett, 996
Baltimoretrupial, 1694
Bandhavsörn, 302
beckasinsnäppa, Mindre, 642
   Större, 643
Bengalgam, 314
Berberfalk, 424
Bergand, 247
   Mindre, 249
Bergfink, 1545
Berghöna, 452
Berglärka, 1047
Bergortolan, 1662
Bergrödstjärt, 1164
Bergsångare, 1329
Bergsökenfink, 1589
Berguv, 893
Beringtärna, 784
Bivråk, 290
Biätare, 966
   Grön, 964
Björktrast, 1222

Blodnäbbsvävare, 1528
Blåhake, 1149
Blåhäger, 118
Blåkråka, 970
   Brednäbbad, 974
   Indisk, 973
Blåmes, 1388
Blånäktergal, 1152
Blåskata, 1456
Blåstjärt, 1153
Blåtjocknäbb, 1688
Blåtrast, 1204
Bläsand, 207
   Amerikansk, 210
Bläsgås, 175
Bobolink, 1690
Bofink, 1539
Brokpetrell, 36
Bronsibis, 144
Brudand, 204
Brunand, 238
   Svartnäbbad, 240
Brunglada, 295
Brunsula, 74
Brunsångare, 1328
Bruntrast, 1219
Bruntörnskata, 1430
Brushane, 628
Buffelhuvud, 269
bulbyl, Rödgumpad, 1112
   Vitkindad, 1109
Buskskvätta, 1167
   Svart, 1173
   Svarthakad, 1170
busksparv, Brunsidad, 1633
Busksångare, 1260
busktörnskata, Svartkronad, 1428
Byvävare, 1526
båtstjärt, Mindre, 1692
Bälteskungsfiskare, 960
Bändelkorsnäbb, 1580

Canadavireo, 1537
Cettisångare, 1235
Chileflamingo, 156
Citronsiska, 1556
citronsångare, Svarthakad, 1627
Citronärla, 1098
Cypernstenskvätta, 1183
Cypernsångare, 1299

Dalripa, 430
Dammsnäppa, 668
Denhamtrapp, 522
Diademrödstjärt, 1163
Diamantfasan, 476
Dickcisselsparv, 1686
Domherre, 1606

dopping, Svarthalsad, 24
   Tjocknäbbad, 13
Drillsnäppa, 680
Dubbelbeckasin, 638
Dubbeltrast, 1230
Dupontlärka, 1021
duva, Spetsstjärtad, 864
Duvhök, 342
Dvärgbeckasin, 633
dvärgbiätare, Grön, 963
Dvärggransångare, 1335
Dvärglira, 55
Dvärglärka, 1034
Dvärgmås, 719
Dvärgrördrom, 105
   Afrikansk, 108
   Amerikansk, 104
   Kinesisk, 107
Dvärgskarv, 87
Dvärgsnäppa, 612
Dvärgsnögås, 184
Dvärgsolfågel, 1419
Dvärgsparv, 1667
Dvärgspov, 653
Dvärgsumphöna, 494
Dvärgsävsparv, 1676
Dvärguv, 891
   Blek, 890
Dvärgörn, 381
dykand, Rödhuvad, 236
   Vitögd, 242

Ejder, 251
Eksångare, 1273
Eleonorafalk, 407
Enkelbeckasin, 635
Entita, 1375
Eremitibis, 146
Eremitskogstrast, 1210
Eskimåspov, 654

Falknattskärra, 936
Fasan, 472
fibi, Grå, 1010
finklärka, Brunhuvad, 1012
   Svartkronad, 1013
Finschstenskvätta, 1187
Fiskgjuse, 386
Fiskmås, 735
Fisktärna, 779
fiskuv, Brun, 896
fiskörn, Afrikansk, 302
Fjällabb, 701
Fjällgås, 178
Fjällpipare, 577
Fjällripa, 434
Fjälluggla, 897
Fjällvråk, 360

fjärilsfink, Rödkindad, 1530
Flamingo, 153
   Mindre, 156
Flikstrandpipare, 566
Flodsångare, 1247
   Större, 1251
Flodtjockfot, 548
flugsnappare, Grå, 1350
   Mindre, 1353
   Svartvit, 1358
flyghöna, Brunbukig, 831
   Svartbukig, 832
   Vitbukig, 834
Fläckdrillsnäppa, 683
Fläckskogstrast, 1209
Forsärla, 1100
frankolin, Marockansk, 463
   Svart, 461
Fregattfågel, 99
Fregattstormfågel, 60
Fältpiplärka, 1072
Fältsångare, 1258

Gabarhök, 342
Garfågel, 815
Glada, 298
   Svartvingad, 294
Glasögonejder, 255
Glasögonflugsnappare, 1349
Glasögonsångare, 1291
Gluttsnäppa, 670
Goliathäger, 134
Gouldpetrell, 37
Gransångare, 1337
   Kaukasisk, 1336
Gravand, 198
Gråfiskare, 959
Grågam, 321
Grågås, 179
gråhäger, Amerikansk, 131
Gråhakedopping, 20
Grålira, 49
Gråsiska, 1574
gråsnäppa, Sibirisk, 684
Gråsparv, 1503
Gråspett, 983
Gråsångare, 1277
Grårast, 1212
Gråtrut, 741
Gräsand, 218
Gräshoppsångare, 1245
grässparv, Gulbrynad, 1635
Grässångare, 1237
Grönbena, 676
Grönfink, 1557
Gröngöling, 986
   Afrikansk, 988
Grönhäger, 111
Grönhämpling, 1552

Pages 1–1008 appear in Volume 1, pages 1009–1694 in Volume 2

[36] Svenska namn

Grönsiska, 1564
Grönsångare, 1333
gulbena, Mindre, 672
　　Större, 672
Guldfasan, 475
Guldgök, 874
Guldnattskärra, 933
Guldsparv, 1515
Guldspett, 983
Guldtrast, 1206
Gulhake, 1627
Gulhämpling, 1550
　　Rödpannad, 1548
Gulsparv, 1648
Gulärla, 1094
Gycklarörn, 326
Gyllensparv, 1670
gås, Rödhalsad, 192
　　Sporrvingad, 203
　　Vitkindad, 187
Gåsgam, 315
Gärdsmyg, 1122
Gök, 875
　　Svartvit, 872
Göktyta, 980

hackspett, Mindre, 1002
　　Större, 993
　　Tretåig, 1005
　　Vitryggig, 1000
Halsbandsflugsnappare, 1357
Halsbandsparakit, 866
Halsbandspetrell, 38
Havslöpare, 58
Havssula, 77
Havstrut, 754
Havsörn, 303
Helenaastrild, 1531
Hjälmpärlhöna, 479
Hornuggla, 915
Hudsonspov, 650
Huskråka, 1472
Hussparv, 1655
Hussvala, 1066
Häcksparv, 1651
Häger, 128
　　Svarthuvad, 127
　　Trefärgad, 120
Hägerpipare, 544
Hämpling, 1568
Härfågel, 976
Härmsångare, 1282
härmtrast, Nordlig, 1125
　　Rödbrun, 1126
Höglandslärka, 1034
hök, Bandvingad, 353
Höksångare, 1306
Hökuggla, 899
Hökörn, 383

ibis, Helig, 148

ibisstork, Afrikansk, 137
iltärna, Mindre, 772
　　Större, 770
Indigofink, 1689
Irakskriktrast, 1365
Isabellastenskvätta, 1175
Isabellatörnskata, 1431
Islandsknipa, 269
islom, Svartnäbbad, 8
　　Vitnäbbad, 10
Ismås, 761

Jaktfalk, 417
Jorduggla, 918
Jungfrutrana, 516
Järnsparv, 1128
　　Kaukasisk, 1132
　　Sibirisk, 1131
　　Svartstrupig, 1133
Järpe, 428
Jättedopping, 17
jättestormfågel, Sydlig, 32

Kabylnötväcka, 1400
Kafferseglare, 947
Kaja, 1468
Kalanderlärka, 1024
　　Asiatisk, 1026
Kalkon, 481
Kamskrake, 274
Kamsothöna, 509
Kanadagås, 185
Kanadaskogssångare, 1629
Kanariebofink, 1544
Kanariebuskskvätta, 1169
Kanarieduva, 848
Kanariefågel, 1554
Kanariepiplärka, 1075
Kanariestrandskata, 538
Kapand, 217
Kapduva, 861
Kapskedand, 229
Kapsula, 79
Kapuggla, 921
Kap Verde-seglare, 941
Kap Verde-sångare, 1262
kardinal, Brokig, 1686
Karibpetrell, 41
Karolinasumphöna, 490
Kattfågel, 1127
Kattuggla, 907
Kaveldunsångare, 1251
Kejsarörn, 371
　　Spansk, 375
Kermadecpetrell, 37
Kilstjärtslira, 46
Kittlitzpipare, 567
Klippduva, 839
Klipphöna, 457
Klippkaja, 1471
Klippkattugla, 910

Klippnötväcka, 1406
　　Östlig, 1404
Klippsparv, 1653
　　Afrikansk, 1659
Klippsvala, 1059
　　Blek, 1058
Klippörn, 380
Knipa, 272
Knölsvan, 162
Kohäger, 116
Koltrast, 1215
Kopparand, 285
　　Amerikansk, 282
Kornknarr, 496
Kornsparv, 1681
Korp, 1483
　　Kortstjärtad, 1486
Korprall, 496
korsnäbb, Mindre, 1582
　　Skotsk, 1586
　　Större, 1587
Korttålärka, 1032
　　Asiatisk, 1036
kostare, Brunhuvad, 1691
Kragalka, 820
Kragtrapp, 525
Kricka, 215
　　Gulkindad, 214
Krokodilväktare, 550
Kronflyghöna, 828
Kronsångare, 1319
Kråka, 1478
　　Svartvit, 1481
Kungsalbatross, 31
Kungsfiskare, 956
　　Gråhuvad, 954
Kungsfågel, 1342
　　Brandkronad, 1346
　　Guldkronad, 1348
　　Rödhuvad, 1342
Kungsfågelsångare, 1324
Kungstärna, 769
Kungsörn, 376
Kusthäger, 121
Kustpipare, 584
Kustsnäppa, 599
　　Större, 599
kärrhök, Blå, 331
　　Brun, 328
Kärrsångare, 1263
Kärrsnäppa, 620
Kärrtärna, 785

Labb, 699
　　Bredstjärtad, 696
Ladusvala, 1061
Lagerduva, 850
Lammgam, 308
Lappmes, 1382
Lappsparv, 1640
Lappuggla, 913
Lavskrika, 1454

Lazulifink, 1689
Levantbulbyl, 1110
lira, Gulnäbbad, 42
　　Ljusfotad, 47
　　Mindre, 51
　　Större, 47
　　Vithuvad, 46
　　Vitmagad, 62
Ljungpipare, 581
Lotushäger, 113
Lundsångare, 1320
Lunnefågel, 821
Långtåsnäppa, 612
lärka, Tjocknäbbad, 1022
　　Vitvingad, 1027
Lärkfalk, 404
Lärksparv, 1635
Lövsångare, 1340

Madeiraduva, 848
Madeirakungsfågel, 1345
Madeirapetrell, 37
Magnoliaskogssångare, 1622
Mandarinand, 204
Mandarinstare, 1491
Marabustork, 143
Marmorand, 232
maskstenknäck, Mindre, 1610
　　Större, 1610
Masksula, 73
Masktörnskata, 1447
Medelhavslira, 54
Medelhavsstenskvätta, 1183
Mellanskarv, 84
Mellanspett, 998
Messångare, 1617
Minervauggla, 903
Mongolfink, 1592
Mongolpipare, 572
Mongolpiplärka, 1072
Mongolvråk, 360
Morkulla, 644
Mosnäppa, 610
Munkgam, 314
Munkparakit, 867
Munkstenskvätta, 1194
Murkrypare, 1408
musfågel, Blånackig, 869
myna, Brun, 1500
Myrsnäppa, 624
Myrspov, 650
Myrtrupial, 1692
mås, Brunhuvad, 727
　　Gråhuvad, 727
　　Långnäbbad, 729
　　Ringnäbbad, 734
　　Sotvingad, 716
　　Svarthuvad, 714
　　Tretåig, 758
Mästersångare, 1305

Natthäger, 109

Pages 1–1008 appear in Volume 1, pages 1009–1694 in Volume 2

Nattskärra, 929
 Nubisk, 928
 Rödhalsad, 932
Nilgås, 195
Nilsolfågel, 1421
noddy, Svart, 803
 Vitpannad, 802
Nordsångare, 1322
Nunnestenskvätta, 1180
Näktergal, 1143
 Vitstrupig, 1154
Nötkråka, 1460
Nötskrika, 1450
Nötväcka, 1402
 Korsikansk, 1399
 Krüpers, 1398
 Rödbröstad, 1401

Oceanlöpare, 67
Olivsångare, 1280
Orientseglare, 947
Orientkriktrast, 1367
Orientstenskvätta, 1191
Orientsångare, 1278
Orientvisseland, 161
ormhalsfågel, Afrikansk, 91
Ormvråk, 353
Ormörn, 324
Orre, 437
 Kaukasisk, 440
Ortolansparv, 1659

Palestinasolfågel, 1422
Palmduva, 860
palmfågel, Grå, 1116
Palmseglare, 950
Papegojalka, 820
Papyrussångare, 1267
Pelikan, 93
 Krushuvad, 96
 Rosaryggad, 98
petrell, Spetsstjärtad, 41
 Vitbukig, 40
Pilfink, 1513
Pilgrimsfalk, 419
pipare, Kaspisk, 575
Piparsnäppa, 662
piplärka, Långnäbbad, 1076
 Rödstrupig, 1084
 Sibirisk, 1077
 Större, 1070
piplärksångare, Nordlig, 1626
 Rödkronad, 1625
Plymhäger, 125
Polyglottsångare, 1284
Praktand, 210
Praktejder, 253
Prinia, streckad, 1239
Provencesångare, 1288
Prutgås, 189
Prärielöpare, 628

Präriemås, 718
Prärietrana, 514
Prärievråk, 353
Pungmes, 1416
Purpurhäger, 132
Purpurhöna, 504
Pavefink, 1689
Pärluggla, 923

Rallbeckasin, 532
Rallhäger, 113
Rapphöna, 464
Razolärka, 1047
regngök, Gulnäbbad, 882
 Svartnäbbad, 882
Ringand, 241
Ringduva, 846
Ringtrast, 1212
Rishäger, 115
Rosenfink, 1596
 Långstjärtad, 1606
 Sibirisk, 1600
 Större, 1600
Rosenmås, 757
Rosenstare, 1498
Rosentärna, 777
Roskarl, 686
Rostand, 197
Rostgumpsvala, 1064
Rostskogstrast, 1211
Rostsparv, 1663
Rubinnäktergal, 1147
Rüppelgam, 318
Råka, 1475
Rävsparv, 1636
Rödahavssotmås, 711
Rödahavssångare, 1303
Rödbena, 665
Rödfalk, 391
Rödflikvipa, 589
Rödhake, 1140
Rödhöna, 455
Rödspov, 647
Rödstjärt, 1161
 Amerikansk, 1625
 Svart, 1157
Rödtangara, 1632
Rödvingetrast, 1228
Rödvingetrupial, 1693
Rördrom, 101
 Amerikansk, 104
Rörhöna, 499
Rörsångare, 1265
rörvävare, Guldkronad, 1526

Saharaskriktrast, 1370
Sahelsångare, 1241
Salskrake, 274
Salskrake, 274
Sammetshätta, 1296
 Östlig, 1295

Sandlöpare, 602
Sandsnäppa, 605
Sandtärna, 764
Sandökenlärka, 1015
Savannblåkråka, 972
Savannörn, 368
saxnäbb, Afrikansk, 804
seglare, Enfärgad, 942
Senegalmarkgök, 884
Shikra, 348
Sibirbeckasin, 640
Sidensvans, 1113
Silkeshäger, 122
Sillgrissla, 806
Silltrut, 737
silvernäbb, Afrikansk, 1534
 Indisk, 1534
Silvertärna, 782
simsnäppa, Brednäbbad, 692
 Smalnäbbad, 691
Sinaiglansstare, 1489
Sinairosenfink, 1599
Sjöorre, 263
skarv, Långstjärtad, 89
Skata, 1457
Skatgök, 872
Skedand, 230
Skedstork, 149
 Afrikansk, 152
Skogsduva, 842
Skogssnäppa, 674
 Amerikansk, 673
skogssångare, Blåryggad, 1620
 Brunhättad, 1623
 Brunkindad, 1621
 Brunsidig, 1619
 Grön, 1620
 Gul, 1618
 Guldvingad, 1616
 Gulgumpad, 1623
 Orangestrupig, 1620
 Svartkronad, 1628
 Svartvit, 1615
 Vitkindad, 1623
skogstrast, Beigekindad, 1210
 Gråkindad, 1211
Skorstensseglare, 939
Skrattduva, 852
Skrattmås, 724
Skrikstrandpipare, 567
skrikörn, Mindre, 363
 Större, 365
Skräntärna, 766
Skäggdopping, 17
Skäggmes, 1362
Skäggtärna, 794
Skärfläcka, 542
Skärpiplärka, 1089
Skärsnäppa, 618
Slagfalk, 411
Slaguggla, 911
Smutsgam, 311
Smyrnakungsfiskare, 953

Smådopping, 14
Smålom, 3
Småskrake, 277
Småsnäppa, 607
Småspov, 655
Småtrapp, 519
Småtärna, 790
 Persisk, 793
Snatterand, 212
snäppa, Gulbröstad, 614
 Rödhalsad, 607
 Spetsstjärtad, 615
Snöfink, 1522
Snöfågel, 1639
Snögås, 182
Snöhäger, 120
snöhöna, Kaspisk, 451
 Kaukasisk, 449
Snösiska, 1578
Snösparv, 1642
Snötrana, 515
Socotraskarv, 86
Sommargylling, 1424
Sommartangara, 1631
Sorgstenskvätta, 1193
Sotfalk, 409
Sothöna, 506
 Amerikansk, 509
Sottärna, 788
 Gråryggad, 787
sparv, Brunryggad, 1511
 Gråhuvad, 1645
 Gulbrynad, 1664
 Gulgrå, 1658
 Rödbrun, 1669
 Spansk, 1506
 Svarthakad, 1653
 Svarthuvad, 1679
 Vitkronad, 1637
 Vitstrupig, 1639
Sparvhök, 345
Sparvuggla, 901
Spetsbergsgrissla, 810
Spetsbergsgås, 173
Spillkråka, 989
Sporrvipa, 587
spov, Smalnäbbad, 657
Spovsnäppa, 617
Springhöna, 482
Stare, 1492
Starrsångare, 1243
Steglits, 1561
Stenfalk, 400
Stenhöna, 453
Stenknäck, 1610
Stenskvätta, 1178
 Rödgumpad, 1190
 Rödstjartad, 1188
 Svart, 1198
 Svartvit, 1195
 Vitkronad, 1196
Stensparv, 1519
 Blek, 1517

Pages 1–1008 appear in Volume 1, pages 1009–1694 in Volume 2

Svenska namn

Gulstrupig, 1518
stensvala, Rödkindad, 1066
Stentrast, 1201
Stenökenlärka, 1017
Stjärtand, 223
Stjärtmes, 1372
stork, Svart, 138
  Vit, 140
Storlabb, 704
Storlom, 6
Stormfågel, 34
Stormsvala, 62
  Klykstjärtad, 64
  Swinhoes, 66
Storskarv, 80
Storskrake, 279
Storspov, 658
Stortrapp, 529
strandpipare, Mindre, 561
  Större, 564
  Svartbent, 569
Strandskata, 535
Strimflyghöna, 826
Stripgås, 182
Struts, 1
Strömand, 259
Strömstare, 1118
Stubbstjärtseglare, 949
Styltlöpare, 539
Styltsnäppa, 627
Stäpphök, 335
Stäpphöna, 836
Stäppsparv, 1677
Stäppvipa, 590
Stäppörn, 370
sula, Rödfotad, 73
sultanhöna, Amerikansk, 502
  Mindre, 502
sumphöna, Mindre, 490
  Rostgumpad, 496
  Småfläckig, 488
Sumpvipa, 592
svan, Svart, 165
Svartand, 222
Svarthakedopping, 22
Svarthäger, 119
Svarthätta, 1316
Svartlärka, 1029
Svartmes, 1385
Svartsnäppa, 662
Svartstare, 1496
Svartstjärt, 1166
Svarttärna, 796

Svärta, 266
  Vitnackad, 266
Sydnäktergal, 1145
Sydpolslabb, 708
sångare, Rödstrupig, 1293
  Sardinsk, 1286
  Svarthakad, 1300
  Tjocknäbbad, 1273
sånghök, Mindre, 340
Sånglärka, 1043
  Mindre, 1046
Sångsparv, 1637
Sångsvan, 168
  Mindre, 166
Sädesärla, 1103
  Afrikansk, 1106
Sädgås, 170
Sävsparv, 1672
Sävsångare, 1255

Taggstjärtseglare, 939
Taigabeckasin, 642
Taigagök, 878
Taigasångare, 1325
Taigatrast, 1220
Talgoxe, 1393
Tallbit, 1602
Tallsiska, 1568
Tallsparv, 1646
Talltita, 1379
Taltrast, 1225
Tamarisksparv, 1509
Tatarfalk, 414
Teklalärka, 1040
Tennesseesångare, 1615
Tereksnäppa, 678
termitbussksvätta, Svart, 1174
Tigerfink, 1533
Tjockfot, 546
Tjäder, 442
Tobisgrissla, 815
tofsalka, Större, 819
Tofslunne, 825
Tofslärka, 1037
Tofsmes, 1383
Tofsvipa, 593
  Svarthuvad, 589
Tordmule, 812
Tornfalk, 393
  Amerikansk, 396
Tornseglare, 943
  Blek, 945
Tornuggla, 886
Trana, 511

trapp, Nubisk, 522
trast, Blåryggad, 1208
  Gråhalsad, 1219
  Sibirisk, 1207
Trastsångare, 1269
tropikfågel, Rödnäbbad, 70
trupial, Gulhuvad, 1693
trut, Armenisk, 748
  Gulfotad, 746
  Rödnäbbad, 732
  Svarthuvad, 712
  Vitvingad, 749
Trädgårdsbulbyl, 1111
Trädgårdssångare, 1314
Trädgårdsträdkrypare, 1414
Trädkrypare, 1411
Trädlärka, 1041
Trädmås, 723
Trädnäktergal, 1137
  Svart, 1139
Trädpiplärka, 1079
Trädsvala, 1053
Träsksparv, 1637
Träsksångare, 1244
tundrapipare, Amerikansk, 579
  Sibirisk, 580
Tundrapiplärka, 1081
Tundrasnäppa, 606
Turkduva, 853
Turkistanduva, 845
Turturduva, 856
  Större, 859
Tuvsnäppa, 615
tärna, Kentsk, 773
  Vitkindad, 785
  Vitvingad, 799
Tärnmås, 721
Törnskata, 1433
  Rostgumpad, 1436
  Rödhuvad, 1445
  Svartpannad, 1436
Törnsångare, 1310

Vadarsvala, 555
  Orientalisk, 557
  Svartvingad, 557
Vaktel, 467
  Vitstrupig, 448
Vandringsalbatross, 30
Vandringsduva, 865
Vandringstrast, 1234
Varfågel, 1440

Vassångare, 1249
Vattenpiplärka, 1086
Vattenrall, 484
Vattensångare, 1254
Videsparv, 1665
Videsångare, 1327
Vigg, 244
Vinterhämpling, 1571
vireo, Gulstrupig, 1536
  Rödögd, 1537
visseland, Brun, 160
  Vitmaskad, 160
Visselhöna, 459
Vitgumpsnäppa, 613
Vittrut, 752

Willetsnäppa, 685
Wilsonsimsnäppa, 689

Åkersparv, 1635
Årta, 226
  Blåvingad, 228

Ägretthäger, 126
Ängshök, 337
Ängspiplärka, 1082
Ängssparv, 1655
ängsstare, Östlig, 1692
Ärtsångare, 1308

Ökenberglärka, 1052
Ökenfink, 1591
Ökenflyghöna, 829
Ökenhöna, 460
Ökenkorp, 1481
ökenlärka, Streckad, 1014
Ökenlöpare, 552
Ökenlöplärka, 1019
Ökennattskärra, 933
Ökenpipare, 574
Ökensnårsångare, 1241
Ökensparv, 1511
Ökenstenskvätta, 1185
Ökensångare, 1301
Ökentrumpetare, 1594
Örnvråk, 359
Örongam, 318
Öronskarv, 83

Pages 1–1008 appear in Volume 1, pages 1009–1694 in Volume 2

РУССКИЕ НАЗВАНИЯ

Авадават, красный, 1533
Авдотка, 546
  африканская, 548
Аист, белый, 140
  черный, 138
Альбатрос, белошапочный, 29
  желтоклювый, 29
  королевский, 31
  сероголовый, 30
  странствующий, 30
  чернобровый, 27
  черноногий, 30
Амарант, обыкновенный, 1529
Амадина, малабарская, 1534
  серебряноклювая, 1534
Астрильд, волнистый, 1531
  красноухий, 1530
  оранжевощекий, 1530
  тигровый, 1533

Баклан, большой, 80
  длиннохвостый, 89
  малый, 87
  ушастый, 83
  хохлатый, 84
  черногорлый, 86
Балобан, 414
  Бартрамия, 662
Бегунок, египетский, 550
  обыкновенный, 552
  Темминка, 554
Беркут, 376
Бекас, 635
  азиатский, 640
  цветной, 532
Белобровик, 1228
Белобрюшка, 820
Береговушка, 1055
  белобровая, 1058
  малая, 1053
Бормотушка, 1277
Бородач, 308
Бургомистр, 752
Буревестник, бледноногий, 47
  большой пестробрюхий, 47
  желтоклювый, 42
  Зеленого мыса, 45
  клинохвостый, 46
  -крошка, 55
  мавританский, 53
  малый, 51
  Одюбона, 56
  пестроголовый, 46
  северный гигантский, 32
  серый, 49
  средиземноморский, 54
  южный гигантский, 32

Бюльбюль, белощекий, 1109
  желтопоясничный, 1110
  обыкновенный, 1111
  розовобрюхий, 1112

Вальдшнеп, 644
Варакушка, 1149
Веретенник, большой, 647
  длинноклювый бекасовидный, 643
  канадский, 650
  короткоклювый бекасовидный, 642
  малый, 650
Вертишейка, 980
Вильсония, канадская, 1629
  капюшонная, 1628
  малая, 1628
Виреон, желтозобый, 1536
  красноглазый, 1537
  тонкоклювый, 1537
Водорез, африканский, 804
Волчок, 105
  американский, 104
  амурский, 107
  сероспинный, 108
Воробей, домовый, 1503
  желтогорлый, 1518
  желтый, 1515
  каменный, 1519
  короткопалый, 1517
  месопотамский, 1509
  полевой, 1513
  пустынный, 1511
  черногрудый, 1506
  Яго, 1511
Ворон, 1483
  блестящий, 1472
  пегий, 1481
  пустынный, 1481
  трубастый, 1486
Ворона, черная, 1478
  щетинистая, 1486
Выпь, 101
  американская, 104
Вьюрок, буланый, 1591
  канареечный, 1550
  королевский, 1548
  лимонный, 1556
  сирийский, 1552
  снежный, 1522
Вяхирь, 846

Гага-гребенушка, 253
  обыкновенная, 251
  очковая, 255
  сибирская, 256
Гагара, белоклювая, 10
  краснозобая, 3

  полярная, 8
  чернозобая, 6
Гагарка, 812
  бескрылая, 815
Гаичка, буроголовая, 1379
  сероголовая, 1382
  средиземноморская, 1378
  черноголовая, 1375
Галка, 1468
  альпийская, 1464
  даурская, 1471
Галстучник, 564
  перепончатопалый, 566
Гаршнеп, 633
Гирья, 203
Глупыш, 34
Глухарь, 442
Гоголь, 272
  исландский, 269
  малый, 269
Голубок, капский, 36
  морской, 729
Голубь, бурый, 845
  канарский, 848
  лавровый, 850
  серебристый, 848
  сизый, 839
  странствующий, 865
Горихвостка, американская, 1625
  белобровая, 1163
  краснобрюхая, 1164
  красноспинная, 1155
  обыкновенная, 1161
  -чернушка, 1157
Горлица, большая, 859
  дикая смеющаяся, 852
  капская, 861
  кольчатая, 853
  малая, 860
  обыкновенная, 856
  плачущая, 864
Гракл, обыкновенный, 1692
  Тристрамов, 1489
Грач, 1475
Гренадерка, 1383
Гриф, бенгальский, 314
  ушастый, 318
  черный, 321
Грязовик, 624
Гуирака, голубая, 1688
Гуменник, 170
  короткоклювый, 173
Гусь, белолобый, 175
  белый, 182
  горный, 182
  нильский, 195
  обыкновенный шпорцевый, 203

Росса, 184
  серый, 179

Дербник, 400
Деряба, 1230
Джек, 525
Древесница, желтая, 1618
Дрозд, американский лесной, 1209
  белозобый, 1212
  бурый, 1219
  вертлявый, 1211
  краснозобый, 1220
  малый, 1211
  Науманна, 1219
  одноцветный, 1212
  оливковый, 1218
  -отшельник, 1210
  ошейниковый, 1208
  певчий, 1225
  пестрый, 1206
  пестрый каменный, 1201
  Свенсонов, 1210
  сибирский, 1207
  синий каменный, 1204
  странствующий, 1234
  чернозобый, 1220
  черный, 1215
Дрофа, 529
  арабская, 527
  кафрская африканская, 522
  нубийская, 522
Дубонос, 1610
  большой черноголовый, 1610
  вечерний американский, 1613
  малый черноголовый, 1610
Дубровник, 1670
Дупель, 638
  лесной, 642
Дутыш, 615
Дятел, африканский зеленый, 988
  белоспинный, 1000
  большой пестрый, 993
  зеленый, 986
  золотой шилоклювый, 983
  малый пестрый, 1002
  седой, 983
  сирийский, 996
  -сосун, желтобрюхий, 992
  средний, 998
  трехпалый, 1005

Жаворонок, белокрылый, 1027
  белолобый воробьиный, 1013

[40] Русские названия

двупятнистый, 1026
Дюпона, 1020
кордофанский
    кустарниковый, 1011
коротколювый хохлатый,
    1040
малый, 1032
малый вьюрковый, 1014
малый полевой, 1046
малый рогатый, 1052
острова Разо, 1047
пестрокрылый пустынный,
    1019
полевой, 1043
пустынный, 1017
пятнистый воробьиный,
    1012
рогатый, 1047
серый, 1034
солончаковый, 1036
степной, 1024
толстоклювый, 1022
тонкоклювый, 1034
хохлатый, 1037
чернохвостый вьюрковый,
    1015
черный, 1028
Желна, 989
Желтозобик, 628
Жулан, рыжехвостый, 1431
    сибирский, 1430
Журавль, канадский, 514
    -красавка, 516
    серый, 511

Завирушка, альпийская, 1134
    лесная, 1128
    пестрая, 1132
    сибирская, 1131
    черногорлая, 1133
Зарничка, тусклая, 1327
Зарянка, 1140
Зеленушка, 1557
Зимняк, 360
Зимородок, 956
    красноносый, 953
    малый пегий, 959
    ошейниковый, 960
    сероголовый, 954
Змеешейка, африканская, 91
Змееяд, 324
Зуек, большеклювый, 574
    двугалстучный, 567
    каспийский, 575
    малый, 561
    малый морской, 567
    монгольский, 572
    морской, 569
    трехполосый, 569
    ходуличниковый, 627
Зяблик, 1539
    голубой, 1544

Ибис, краснощекий, 146
    священный, 148
Иволга, 1424
Иглохвост, дымчатый, 939
Индюк, 481

Казарка, белощекая, 187
    канадская, 185
    краснозобая, 192
    черная, 189
Кайра, толстоклювая, 810
    тонкоклювая, 806
Каменка, белогузая, 1196
    белохвостая, 1198
    белочерная, 1195
    златогузая, 1190
    кипрская, 1183
    краснопоясничная, 1188
    -монашка, 1194
    обыкновенная, 1178
    -плешанка, 1180
    -плясунья, 1175
    пустынная, 1185
    траурная, 1193
    черная, 1191
    чернопегая, 1183
    черношейная, 1187
Каменушка, 259
Камнешарка, 686
Камышница, 499
    пупурная, 502
Камышовка-барсучок, 1255
    болотная, 1263
    веерхвостая, 1237
    вертлявая, 1254
    восточная, 1272
    дроздовидная, 1269
    индийская, 1258
    иракская, 1271
    короткоперая, 1262
    садовая, 1260
    толстоклювая, 1273
    тонкоклювая, 1251
    тростниковая, 1265
    туркестанская, 1267
    широкохвостая, 1235
Канарейка, 1554
Канюк, красноплечий, 353
    -курганник, 359
    обыкновенный, 353
    Свенсонов, 353
Каравайка, 144
Кардинал, индиговый
    овсянковый, 1689
    красногрудый
    дубоносовый, 1686
    лазурный овсянковый, 1689
    расписной овсянковый,
    1689
Качурка, белобрюхая, 62
    белолицая, 60
    вилохвостая, 66
    мадейрская, 67

    прямохвостая, 62
    северная, 64
Кваква, 109
    американская зеленая, 113
    зеленая, 111
Кедровка, 1460
Кеклик, азиатский, 452
    европейский, 453
Клест, белокрылый, 1580
    -еловик, 1582
    -сосновик, 1587
Клинтух, 842
Клуша, 737
Клушица, 1466
Клювач, африканский, 137
Князек, 1392
Кобчик, 397
Козодой, буланый, 933
    золотой, 933
    красношейный, 932
    малый, 936
    нубийский, 928
    обыкновенный, 929
Колпица, 149
    африканская, 152
Колючехвост, 939
Конек, американский, 1092
    Бертелота, 1075
    горный, 1086
    длинноклювый, 1076
    забайкальский, 1072
    краснозобый, 1084
    лесной, 1079
    луговой, 1082
    полевой, 1072
    пятнистый, 1077
    сибирский, 1081
    скальный, 1089
    степной, 1070
Коноплянка, 1568
Конюга, большая, 819
Коростель, 496
Королек, 1342
    золотоголовый, 1348
    канарский, 1345
    красноголовый, 1346
    рубиноголовый, 1342
Коршун, дымчатый, 294
    красный, 298
    черный, 295
Косатка, 210
Крапивник, 1122
Краснозобик, 617
Крачка, алеутская, 784
    американская речная, 785
    аравийская, 785
    белокрылая, 799
    белощекая, 794
    большая хохлатая, 770
    бурокрылая, 787
    королевская, 769
    малая, 790
    малая хохлатая, 772

    мекранская, 793
    обыкновенная, 779
    пестроносая, 773
    полярная, 782
    розовая, 777
    темная, 789
    чайконосая, 764
    черная, 796
    элегантная, 776
Кречет, 417
Кречетка, 590
Кроншнеп, большой, 658
    -малютка, 653
    средний, 655
    тонкоклювый, 657
    эскимосский, 654
Крохаль, большой, 279
    средний, 277
    хохлатый, 274
Кряква, 218
Кукша, 1454
Кулик-воробей, 607
    -красношейка, 607
    -сорока, 535
    -сорока, канарский, 538
Кукушка, белобрюхая, 874
    глухая, 878
    двухцветная, 872
    желтоклювая
    американская, 882
    обыкновенная, 875
    хохлатая, 872
    черноклювая
    американская, 882
    шпорцевая, 884
Курганник, мохноногий, 360
Куропатка, белая, 430
    берберийская каменная,
    457
    бородатая, 467
    калифорнийская, 447
    песчаная, 460
    пустынная, 459
    рыжая горная, 455
    серая, 464
    тундряная, 434

Лазоревка, 1388
Ланнер, 411
Ласточка, американская
    древесная, 1053
    африканская скалистая,
    1058
    белолобая, 1066
    городская, 1066
    деревенская, 1061
    рыжепоясничная, 1064
    скалистая, 1059
    эфиопская, 1064
Лебедь-кликун, 168
    малый, 166
    тундровый, 166
    черный, 165

Pages 1–1008 appear in Volume 1, pages 1009–1694 in Volume 2

-шипун, 162
Лунь, болотный, 328
  луговой, 337
  полевой, 331
  степной, 335
Луток, 274
Лысуха, 506
  американская, 509
  хохлатая, 509
Люрик, 817

Майна, обыкновенная, 1500
Мандаринка, 204
Марабу, африканский, 143
Могильник, 371
  испанский, 375
Моевка, 758
Мородунка, 678
Морянка, 261
Московка, 1385
Мухоловка-белошейка, 1357
  восточная белоглазая, 1009
  малая, 1353
  -пеструшка, 1358
  полуошейниковая, 1355
  серая, 1349
  ширококлювая, 1349

Нектарница, карликовая, 1419
  металлическая короткохвостая, 1421
  палестинская, 1422
Неясыть, африканская, 910
  бородатая, 913
  длиннохвостая, 911
  обыкновенная, 907
Нодди, 802
  черный, 803
Нырок, белоглазый, 242
  длинноносый красноголовый, 240
  красноголовый, 238
  красноносый, 236

Овсянка, белобровая, 1637
  белошапочная, 1646
  белошейная воробьиная, 1639
  болотная, 1637
  горная, 1653
  длиннохвостая, 1655
  домовая, 1655
  желтобровая, 1664
  желчная, 1677
  камышовая, 1672
  каштановая, 1659
  красноклювая, 1663
  -крошка, 1667
  малая воробьиная, 1635
  мелодичная, 1637
  обыкновенная, 1648

  огородная, 1651
  пестрогрудая, 1636
  поляная, 1676
  -ремез, 1665
  рыжая, 1669
  саванная, 1635
  садовая, 1659
  седоголовая, 1645
  серая, 1658
  скалистая, 1662
  Стюарта, 1653
  черноголовая, 1679
Огарь, 197
Океанник, темнобрюхий, 58
Олуша, бурая, 74
  голуболицая, 73
  капская, 79
  красноногая, 73
  северная, 77
Оляпка, 1118
Орел, африканский степной, 368
  -карлик, 381
  кафрский, 380
  -скоморох, 326
  степной, 370
  ястребиный, 383
Орлан-белохвост, 303
  -долгохвост, 302
  -крикун, 302
Осоед, 290
  хохлатый, 293

Парула, белоглазая, 1617
Пастушок, африканский черный, 496
  водяной, 484
Певун, еловый лесной, 1620
  желтогорлый, 1627
  желтошапочный лесной, 1619
  зеленый лесной, 1620
  зеленый пеночковый, 1616
  златокрылый пеночковый, 1615
  золотоголовый дроздовый, 1625
  магнолиевый лесной, 1622
  миртовый, 1623
  пальмовый лесной, 1623
  пегий, 1615
  пестрогрудый лесной, 1623
  речной, 1626
  синеспинный лесной, 1620
  тигровый лесной, 1621
Пеганка, 198
Пеликан, красноспинный, 98
  кудрявый, 96
  розовый, 93
Пеночка, бурая, 1328
  -весничка, 1340

  восточная светлобрюхая, 1331
  гималайская, 1325
  -зарничка, 1325
  зеленая, 1320
  иранская, 1335
  кавказская, 1336
  корольковая, 1324
  светлобрюхая, 1329
  светлоголовая, 1319
  -таловка, 1322
  -теньковка, 1337
  толстоклювая, 1327
  -трещотка, 1333
Перевозчик, 680
  американский, 683
Перепел, 467
  виргинский, 448
Перепелятник, 345
Пересмешка, 1282
  бледная, 1273
  многоголосая, 1284
  пустынная, 1278
  средиземноморская, 1280
Пересмешник, коричневый, 1126
  североамериканский певчий, 1125
Песочник, белохвостый, 610
  большой, 599
  Бонапартов, 613
  Бэрдов, 614
  длиннопалый, 612
  исландский, 599
  -крошка, 612
  малый, 605
  морской, 618
  острохвостый, 615
  перепончатопалый, 606
Песчанка, 602
Пигалица, белохвостая, 592
Пиранга, алая, 1631
  красно-черная, 1632
Пискулька, 178
Пищуха, 1411
  короткопалая, 1414
Плавунчик, большой, 689
  круглоносый, 691
  плосконосый, 692
Поганка, большая, 17
  каролинская, 13
  красношейная, 22
  малая, 14
  серощекая, 20
  черношейная, 24
Погоныш, 488
  каролинский, 490
  -крошка, 494
  малый, 490
  полосатый, 496
Подорлик, большой, 365
  малый, 363

Подорожник, лапландский, 1640
Поморник, большой, 704
  длиннохвостый, 701
  короткохвостый, 699
  средний, 696
  южнополярный, 708
Поползень, 1402
  алжирский, 1400
  большой скалистый, 1404
  канадский, 1401
  корсиканский, 1399
  малый скалистый, 1406
  черноголовый, 1398
Попугай Крамера, ожерелоголовый, 866
  -монах, 867
Поручейник, 668
Просянка, 1681
Птица, кошачья, 1127
  -мышь, синешапочная, 869
Пуночка, 1642
Пустельга, американская, 396
  обыкновенная, 393
  степная, 391

Ремез, 1416
Ржанка, американская, 579
  бурокрылая, 580
  золотистая, 581
  рачья, 544
Рябинник, 1222
Рябок, белобрюхий, 834
  краснобрюхий, 831
  пустынный, 829
  рыжешапочный, 828
  чернобрюхий, 832
  чернолобый, 826
Рябчик, 428

Савка, 285
  американская, 282
Саджа, 836
Сапсан, 419
Сверчок, обыкновенный, 1245
  певчий, 1243
  пятнистый, 1244
  речной, 1247
  соловьиный, 1249
  таежный, 1251
Свиристель, 1113
  американский, 1115
Свиязь, 207
  американская, 210
Сизоворонка, 970
  абиссинская, 972
  бенгальская, 973
Синехвостка, 1153
Синица, большая, 1393
  длиннохвостая, 1367
  усатая, 1362

Синьга, 263
Сип, африканский, 318
  белоголовый, 315
Сипуха, 886
Скворец, 1492
  малый, 1491
  розовый, 1498
  серый, 1491
  черный, 1496
Скопа, 386
Скотоцерка, 1241
Славка, атласская, 1290
  белоусая, 1295
  -завирушка, 1308
  изящная, 1239
  кипрская, 1299
  масличная, 1296
  очковая, 1291
  певчая, 1305
  пегая, 1303
  провансальская, 1288
  пустынная, 1301
  рыжегрудая, 1293
  Рюппеля, 1300
  садовая, 1314
  сардинская, 1286
  серая, 1310
  -черноголовка, 1316
  чешуйчатая, 1241
  ястребиная, 1306
Снегирь, 1606
  азорский, 1609
  длиннохвостый, 1606
  монгольский, 1592
  пустынный, 1594
Сова, африканская ушастая, 921
  белая, 897
  болотная, 918
  ушастая, 915
  ястребиная, 899
Совка, буланая, 890
Сойка, 1450
Сокол Элеоноры, 407
Соловей-белошейка, 1154
  -красношейка, 1147
  обыкновенный, 1143
  синий, 1152
  тугайный, 1137
  черный тугайный, 1139
  южный, 1145
Сорока, 1457
  голубая, 1456
Сорокопут, длиннохвостый, 1436
  -жулан, 1433
  иберийский серый, 1442
  красноголовый, 1445
  маскированный, 1447
  свиристелевый, 1116
  ròòòñåò сероплечий, 1444
  серый, 1440

чернолобый, 1436
Сосновик, шотландский, 1586
Спиза, американская, 1686
Сплюшка, 891
Старик, 820
Стенолаз, 1408
Стервятник, 311
  бурый, 314
Стерх, 515
Страус, 1
Стрепет, 519
Стриж, белобрюхий, 940
  белогузый, 947
  белопоясничный, 947
  бледный, 945
  Зеленого мыса, 941
  малый, 949
  пальмовый, 950
  тусклый, 942
  черный, 943
Султанка, 504
  бронзовая, 502
Сыч, воробьиный, 901
  домовый, 903
  мохноногий, 923

Тайфунник, белокрылый, 37
  Гон-Гон, 39
  длиннохвостый, 41
  Жуанэна, 42
  кермадекский, 37
  мадейрский, 37
  мягкоперый, 38
  черношапочный, 41
  Шлегеля, 40
Тауи, восточный, 1633
Тетерев, 437
  кавказский, 440
Тетеревятник, 342
Тимелия, арабская дроздовая, 1368
  длиннохвостая дроздовая, 1367
  камышевая дроздовая, 1365
  сахарная дроздовая, 1370
Тиркушка, восточная, 557
  луговая, 555
  степная, 557
Ткач, большой масковый, 1526
  красноклювый, 1528
  маньярский, 1526
Топорик, 825
Травник, 665
Трехперстка, африканская, 482
Трупиал, балтиморский цветной, 1694
  буроголовый коровий, 1691
  восточный луговой, 1692
  желтоголовый, 1693
  красноплечий черный, 1693

ржавчатый малый, 1692
чернобрюхий цветной, 1693
Трясогузка, африканская, 1106
  белая, 1103
  горная, 1100
  желтая, 1094
  желтоголовая, 1098
Тулес, 584
Тупик, 821
Турач, 461
  суданский, 464
Турпан, 266
  пестроносый, 266
Турухтан, 628
Тювик, европейский, 349
  туркестанский, 348

Удод, 976
Улар, кавказский, 449
  каспийский, 451
Улит, большой, 670
  желтоногий, 672
  -отшельник, 673
  перепончатопалый, 685
  пестрый, 672
  сибирский пепельный, 684
Утка, вдовушка, 160
  индийская свистящая, 161
  каролинская, 204
  рыжая свистящая, 160
  серая, 212
  черная, 222

Фазан, 472
  алмазный, 476
  золотой, 475
  пестрый китайский, 471
Фаэтон, красноклювый, 70
Феб, восточный, 1010
Филин, 893
  бурый рыбный, 896
Фифи, 676
Фламинго, красный, 153
  малый, 156
  чилийский, 156
Франколин, двушпорцевый, 463
Фрегат, великолепный, 99

Ходулочник, 539
Хондеста, 1635
Хохотун, черноголовый, 712
Хохотунья, 746
Хрустан, 577

Цапля, белая американская, 120
  белокрылая, 115
  береговая, 121
  большая белая, 126
  большая голубая, 131
  восточная желтая, 115

египетская, 116
желтая, 113
исполинская, 134
малая белая, 122
малая голубая, 118
рыжая, 132
серая, 128
средняя белая, 125
трехцветная, 120
черная, 119
чернойшейная, 127
Цесарка, 479

Чагра, черноголовая, 1428
Чайка, армянская, 748
  белая, 761
  белоглазая, 711
  большая морская, 754
  Бонапарта, 723
  буроголовая, 727
  вилохвостая, 721
  кольцеклювая, 734
  малая, 719
  Одуэна, 732
  озерная, 724
  полярная, 749
  розовая, 757
  серебристая, 741
  сероголовая, 727
  серокрылая, 752
  сизая, 735
  смеющаяся, 716
  -трубочист, 709
  черноголовая, 714
  Франклина, 718
Чеглок, 404
  серебристый, 409
Чеграва, 766
Чекан, африканский, 1174
  канарский, 1169
  луговой, 1167
  черноголовый, 1170
  черный, 1173
Чернеть, американская морская, 249
  кольчатая, 241
  морская, 247
  хохлатая, 244
Чернозобик, 620
Чернохвостка, 1166
Черныш, 674
Чечевичник, краснокрылый, 1589
Чечевица, бледная, 1599
  большая, 1600
  обыкновенная, 1596
  сибирская, 1600
Чечетка, горная, 1571
  обыкновенная, 1574
  тундряная, 1578
Чибис, 593
  индийский, 589
  чернохохлый, 589

шпорцевый, 587
Чиж, 1564
  сосновый, 1568
Чирок, капский, 217
  -клоктун, 214
  мраморный, 232
  -свистунок, 215
  синекрылый, 228
  -трескунок, 226

Чистик, 815
Чомга, 17

Шахин, 424
Шилоклювка, 542
Шилохвость, 223
  красноклювая, 225
Широконоска, 230
  африканская, 229

Широкорот, желтоклювый, 974

Щегол, 1561
Щеголь, 662
Щур, 1602
Щурка, зеленая, 964
  золотистая, 966

малая, 963

Юла, 1041
Юнко, темноглазый, 1639
Юрок, 1545

Ястреб, певчий, 340
  -габар, певчий, 342

Legend

— Boundary of the Western Palearctic

▢ Land over 450 metres (1500 feet)

Scale: 0–1000 miles / 0–1600 km

Labels

North Atlantic

Newfoundland, Greenland, Republic of Ireland, Azores, Portugal, Spain (Duero, Tagus, Guadiana), Madeira, Canary Is., Morocco, Western Sahara, Algeria, C Blanc, Banc d'Arguin, Cape Verde Is., Mauritania, The Gambia, Senegal, Guinea Bissau, Mali